CHILTON®

FORD

MECHANICAL SERVICE

2005 Edition

THOMSON
DELMAR LEARNING™

Australia • Canada • Mexico • Singapore • Spain • United Kingdom • United States

THOMSON

DELMAR LEARNING

Chilton®

Ford Mechanical Service

2005 Edition

Vice President, Technology and Trades SBU:
Alar Elken

Executive Director, Professional Business Unit:
Gregory L. Clayton

Publisher, Professional Business Unit:
David Koontz

Marketing Director:
Beth A. Lutz

Marketing Specialist:
Brian McGrath

Production Director:
Mary Ellen Black

Production Manager:
Larry Main

Production Editor:
Elizabeth Hough

Chilton Publishing Assistant:
Paula Baillie

Editorial Assistant:
Kristen Shenfield

Editors:
Terry Blomquist
Timothy A. Crain
Matthew Frederick
Thomas A. Mellon
Richard J. Rivele
Christine L. Sheeky

Cover Design:
Melinda Possinger

Printed in the United States of America
1 2 3 4 5 XX 06 05 04

For more information contact
Thomson Delmar Learning
Executive Woods
5 Maxwell Drive, PO Box 8007,
Clifton Park, NY 12065-8007
Or find us on the World Wide Web at
www.delmarlearning.com

ISBN: 1-4018-6719-7

ISSN: 1548-0887

NOTICE TO THE READER

Publisher does not warrant or guarantee any of the products described herein or perform any independent analysis in connection with any of the product information contained herein. Publisher does not assume, and expressly disclaims, any obligation to obtain and include information other than that provided to it by the manufacturer.

The reader is expressly warned to consider and adopt all safety precautions that might be indicated by the activities herein and to avoid all potential hazards. By following the instructions contained herein, the reader willingly assumes all risks in connection with such instructions.

The publisher makes no representation or warranties of any kind, including but not limited to, the warranties of fitness for particular purpose or merchantability, nor are any such representations implied with respect to the material set forth herein, and the publisher takes no responsibility with respect to such material. The publisher shall not be liable for any special, consequential, or exemplary damages resulting, in whole or part, from the readers' use of, or reliance upon, this material.

Table of Contents

Sections

Model Index

EDITORIAL POLICY

Manufacturer and Model Coverage

This manual does not cover every Ford Motor Company model that is currently available on the market. Rather, the Chilton editorial staff makes judicious decisions as to which models warrant coverage, based on which vehicles are serviced by most technicians.

Model Year Information

This manual is published toward the end of the year prior to the edition year. Every effort is made to gather current data from the Original Vehicle Manufacturers (OEMs) when they publish it. Different OEMs choose to release their new model information at different times of the year. Indeed, the same OEM can publish information early one season and late the next season. As a result, not all models are equally current when each edition of this manual is published.

Although information in this manual is based on industry sources and is as complete as possible at the time of publication, some vehicle manufacturers may make changes which cannot be included here. Information on very late models may not be available in some circumstances. While striving for total accuracy, the publisher cannot assume responsibility for any errors, changes, or omissions that may occur in the compilation of this data.

Safety Notice

Proper service and repair procedures are vital to the safe, reliable operation of all motor vehicles, as well as the personal safety of those performing the repairs. This manual outlines procedures for servicing and repairing vehicles using safe, effective methods. The procedures contain many NOTES, WARNINGS and CAUTIONS which should be followed along with standard safety procedures to reduce the possibility of personal injury or improper service which could damage the vehicle or compromise its safety.

Repair procedures, tools, parts, and technician skill and experience vary widely. It is not possible to anticipate all conceivable ways or conditions under which vehicles may be serviced, or to provide cautions for all possible hazards that may result. Standard and accepted safety precautions and equipment should be used when handling toxic or flammable fluids, and safety goggles or other protection should be used during cutting, grinding, chiseling, prying, or any other process that can cause material removal or projectiles.

Some procedures require the use of tools specially designed for a specific purpose. Before substituting another tool or procedure, you must be completely satisfied that neither your personal safety, nor the performance of the vehicle will be endangered.

LOCATING AND USING THE INFORMATION

Organization

To find where a particular model section or procedure is located, look in the Table of Contents. Main topics are listed with the page number on which they may be found. Following the main topics is an alphabetical listing of all of the procedures within the section and their page numbers.

Part Numbers

Part numbers listed in this book are not recommendations by the publisher for any product by brand name. They are references that can be used with interchanges manuals and aftermarket supplier catalogs to locate each brand supplier's discrete part number.

Special Tools

Special tools are recommended by the vehicle manufacturer to perform specific jobs. When necessary, special tools are referred to in the text by the part number of the tool manufacturer. These tools may be purchased, under the appropriate part number, from your local dealer or regional distributor, or an equivalent tool can be purchased locally from a tool supplier or parts outlet. Before substituting any tool for the one recommended, read the previous Safety Notice.

ACKNOWLEDGEMENT

This publication contains material that is reproduced and distributed under a license from Ford Motor Company. No further reproduction or distribution of the Ford Motor Company material is allowed without the express written permission of Ford Motor Company.

The publisher would like to express appreciation to Ford Motor Company for its assistance in producing this publication.

FORD, LINCOLN AND MERCURY

1

Aviator • Explorer • Explorer Sport • Explorer Sport-Trac • Mountaineer

SPECIFICATION CHARTS

ENGINE AND VEHICLE IDENTIFICATION

Engine							Model Year	
Code ①	Liters (cc)	Cu. In.	Cyl.	Fuel Sys.	Type	Eng. Mfg.	Code ②	Year
E	4.0 (4000)	244	6	MFI	SOHC	Ford	1	2001
H	4.6 (4601)	281	8	SFI	DOHC	Ford	2	2002
P	5.0 (4949)	302	8	MFI	OHV	Ford	3	2003
W	4.6 (4601)	281	8	MFI	SOHC	Ford	4	2004
							5	2005

MFI: Multi-port Fuel Injection

SFI: Sequential Multi-port Fuel Injection

DOHC: Dual Overhead Camshafts

OHV: Overhead Valve

SOHC: Single Overhead Camshaft

① 8th digit of the Vehicle Identification Number (VIN)

② 10th digit of the Vehicle Identification Number (VIN)

67197-EXPL-C01

GENERAL ENGINE SPECIFICATIONS

Year	Model	Engine Displ. Liters	Engine VIN	Net Horsepower @ rpm	Net Torque @ rpm (ft. lbs.)	Bore x Stroke (in.)	Com-pression Ratio	Oil Pressure @ rpm
2001	Explorer	4.0	E	160@4000	225@2500	3.81x3.39	9.0:1	40-60@2000
	Explorer	5.0	P	210@4500	280@3500	4.00x3.00	9.0:1	40-60@2500
	Mountaineer	4.0	E	160@4000	225@2500	3.81x3.39	9.0:1	40-60@2000
	Mountaineer	5.0	P	210@4500	280@3500	4.00x3.00	9.0:1	40-60@2500
	Explorer Sport	4.0	E	160@4000	225@2500	3.81x3.39	9.0:1	40-60@2000
	Explorer Sport-Trac	4.0	E	160@4000	225@2500	3.81x3.39	9.0:1	40-60@2000
2002	Explorer	4.0	E	210@5250	240@3000	3.81x3.39	9.0:1	40-60@2000
	Explorer	4.6	W	239@4750	282@4000	3.55x3.54	9.3:1	20-45@1500
	Mountaineer	4.0	E	210@5250	240@3000	3.81x3.39	9.0:1	40-60@2000
	Mountaineer	4.6	W	239@4750	282@4000	3.55x3.54	9.3:1	20-45@1500
	Explorer Sport	4.0	E	160@4000	225@2500	3.81x3.39	9.0:1	40-60@2000
	Explorer Sport-Trac	4.0	E	160@4000	225@2500	3.81x3.39	9.0:1	40-60@2000
2003	Explorer	4.0	E	210@5250	240@3000	3.81x3.39	9.0:1	40-60@2000
	Explorer	4.6	W	239@4750	282@4000	3.55x3.54	9.3:1	20-45@1500
	Explorer Sport	4.0	E	160@4000	225@2500	3.81x3.39	9.0:1	40-60@2000
	Explorer Sport-Trac	4.0	E	160@4000	225@2500	3.81x3.39	9.0:1	40-60@2000
	Mountaineer	4.0	E	210@5250	240@3000	3.81x3.39	9.0:1	40-60@2000
	Mountaineer	4.6	W	239@4750	282@4000	3.55x3.54	9.3:1	20-45@1500
	Aviator	4.6	H	302@5750	300@3250	3.55x3.54	10.1:1	20-45@1500
2004	Explorer	4.0	E	210@5250	240@3000	3.81x3.39	9.0:1	40-60@2000
	Explorer	4.6	W	239@4750	282@4000	3.55x3.54	9.3:1	20-45@1500
	Explorer Sport-Trac	4.0	E	160@4000	225@2500	3.81x3.39	9.0:1	40-60@2000
	Mountaineer	4.0	E	210@5250	240@3000	3.81x3.39	9.0:1	40-60@2000
	Mountaineer	4.6	W	239@4750	282@4000	3.55x3.54	9.3:1	20-45@1500
	Aviator	4.6	H	302@5750	300@3250	3.55x3.54	10.1:1	20-45@1500

67197-EXPL-C02

GASOLINE ENGINE TUNE-UP SPECIFICATIONS

Year	Engine Displacement Liters	Engine VIN	Spark Plug Gap (in.)	Ignition Timing (deg.) ①		Fuel Pump (psi) ②	Idle Speed (rpm)		Valve Clearance	
				MT	AT		MT	AT	In.	Ex.
2001	4.0	E	0.052-0.056	10B	10B	57-73	①	①	HYD	HYD
	5.0	P	0.052-0.056	—	10B	57-73	①	①	HYD	HYD
2002	4.0	E	0.052-0.056	10B	10B	57-73	①	①	HYD	HYD
	4.6	W	0.052-0.056	—	10B	57-73	①	①	HYD	HYD
2003	4.0	E	0.052-0.056	10B	10B	57-73	①	①	HYD	HYD
	4.6	H	0.052-0.056	—	10B	25-65	①	①	HYD	HYD
	4.6	W	0.052-0.056	—	10B	57-73	①	①	HYD	HYD
2004	4.0	E	0.052-0.056	10B	10B	60-65	①	①	HYD	HYD
	4.6	W	0.052-0.056	—	10B	60-65	①	①	HYD	HYD
	4.6	H	0.052-0.056	—	10B	25-65	①	①	HYD	HYD

NOTE: The Vehicle Emission Control Information label often reflects specification changes changes made during production.

The label figures must be used if they differ from those in this chart.

B: Before top dead center

HYD: Hydraulic

① Idle speed and ignition timing are electronically controlled and cannot be adjusted

② Key on; engine off

67197-EXPL-C03

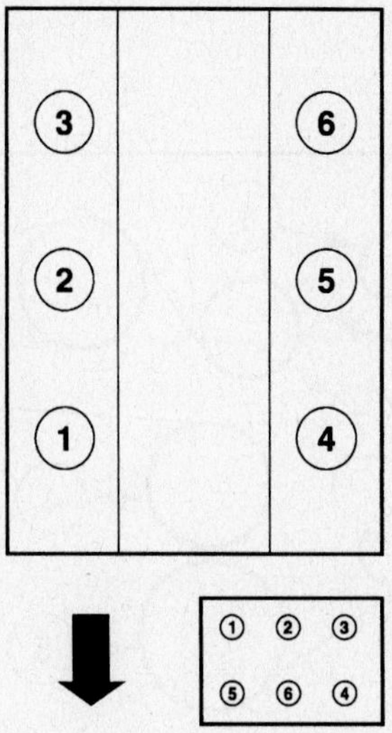

67197-EXPL-GDD

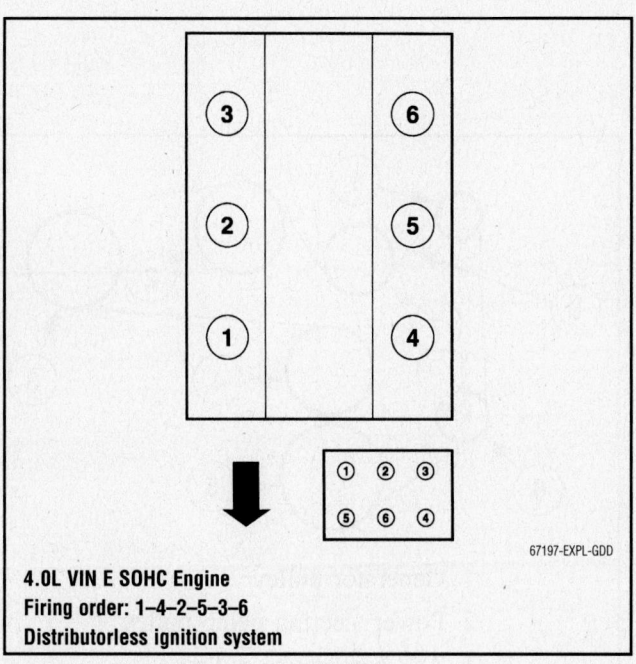

4.0L VIN E SOHC Engine
Firing order: 1–4–2–5–3–6
Distributorless ignition system

79243G53

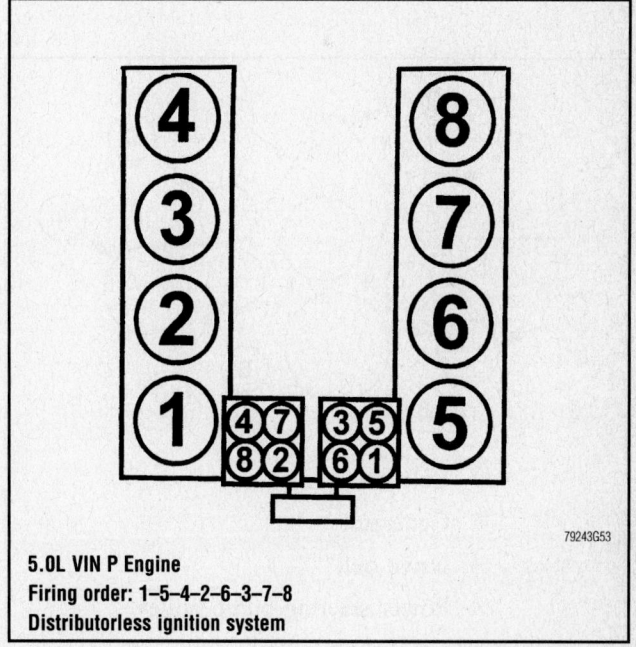

5.0L VIN P Engine
Firing order: 1–5–4–2–6–3–7–8
Distributorless ignition system

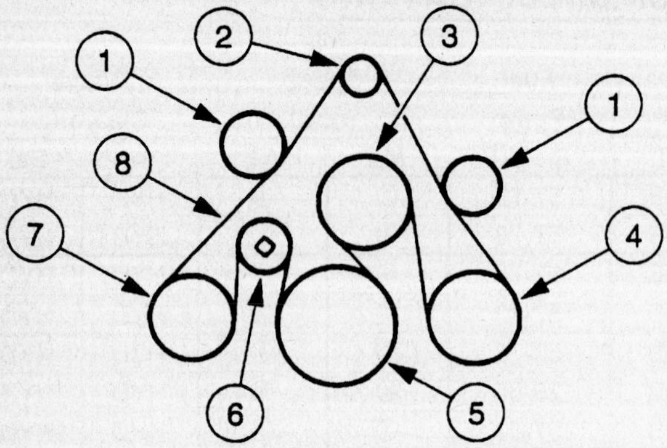

1. Idler
2. Alternator
3. Water Pump
4. Power Steering
5. Crankshaft
6. Drive Belt Tensioner
7. A/C Pulley
8. Drive Belt

79244G86

Accessory drive belt routing —2002 4.6L VIN W SOHC engine with A/C

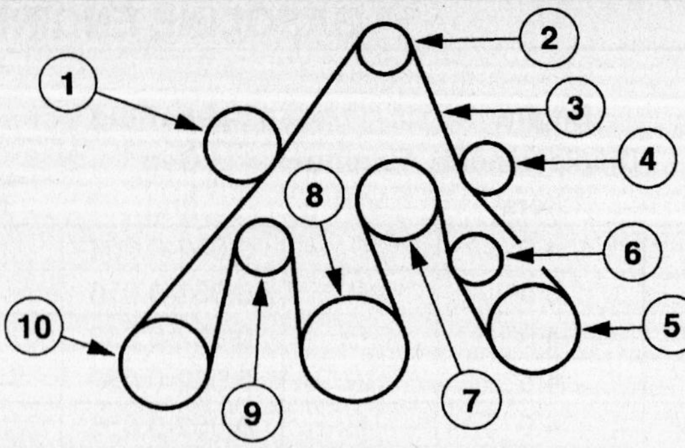

1 Belt idler pulley
2 Generator pulley
3 Drive belt
4 Belt idler pulley
5 Power steering pump pulley
6 Belt idler pulley
7 Coolant pump pulley
8 Crankshaft pulley
9 Drive belt tensioner pulley
10 A/C compressor pulley

67197-EXPL-GBB

Accessory drive belt routing —2003— 4.6L VIN E SOHC engine

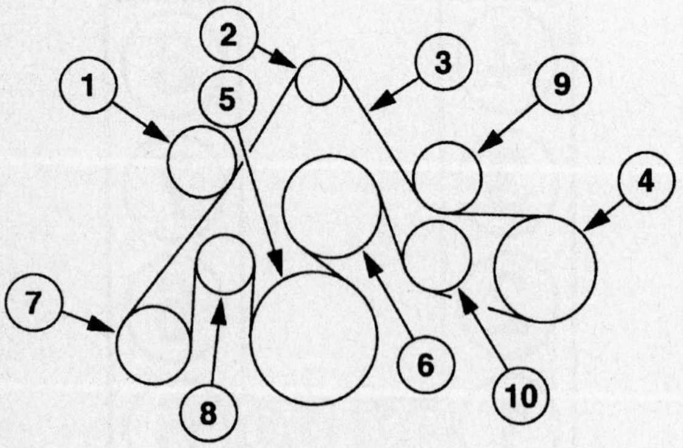

1 Belt idler pulley
2 Generator
3 Drive belt
4 Power steering pump pulley
5 Crankshaft pulley
6 Water pump pulley
7 A/C compressor
8 Drive belt tensioner pulley
9 Belt idler pulley
10 Belt idler pulley

67197-EXPL-GAA

Accessory drive belt routing —4.6L VIN H DOHC engine

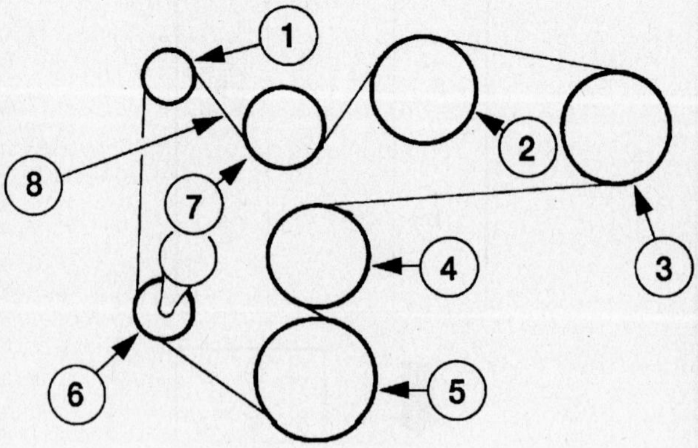

1 Generator pulley
2 Power steering pump pulley
3 A/C compressor pulley
4 Coolant pump pulley
5 Crankshaft damper
6 Drive belt tensioner pulley
7 Belt idler pulley
8 Drive belt

67197-EXPL-GCC

Accessory drive belt routing —4.0L VIN E SOHC engine

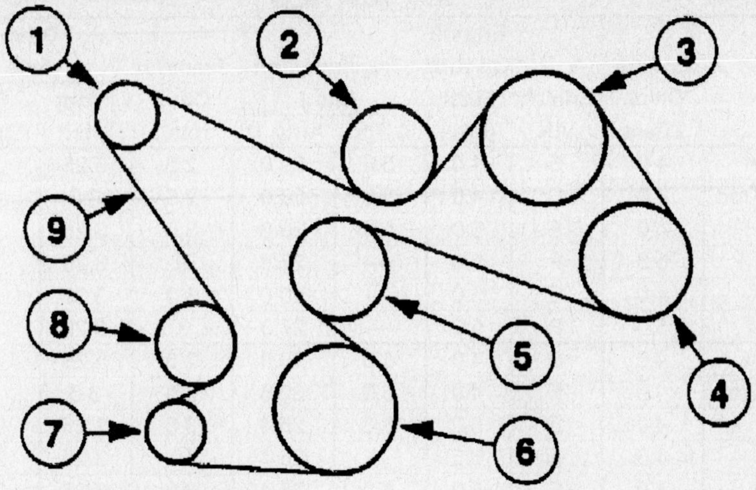

1. Alternator pulley
2. Belt idler pulley
3. Power steering pulley
4. A/C compressor pulley
5. Water pump
6. Crankshaft pulley
7. Belt idler pulley
8. Drive belt tensioner
9. Drive belt

79244G19

Accessory drive belt routing —5.0L VIN P engine

CAPACITIES

Year	Model	Engine Displ. Liters	Engine VIN	Engine Oil with Filter (qts.)	Transmission (pts.) 5-Spd	Transmission (pts.) Auto. ①	Transfer Case (pts.)	Drive Axle Front (pts.)	Drive Axle Rear (pts.)	Fuel Tank (gal.)	Cooling System (qts.)
2001	Explorer Sport	4.0	E	4.0	5.6	20.0	2.5	3.25	②	23.0	14.0
	Explorer Sport-Trac	4.0	E	4.0	5.6	20.0	2.5	3.6	②	23.0	14.0
	Explorer	4.0	E	5.0	5.6	20.0	3.0	3.25	5.5	③	14.0
	Explorer	5.0	P	5.0	—	27.8	④	3.25	5.5	21.0	15.7
	Mountaineer	4.0	E	5.0	—	20.0	3.0	3.25	5.5	21.0	14.0
	Mountaineer	5.0	P	5.0	—	27.8	④	3.25	5.5	21.0	15.7
2002	Explorer Sport	4.0	E	4.0	5.6	20.0	2.5	3.25	②	23.0	14.0
	Explorer Sport-Trac	4.0	E	4.0	5.6	20.0	2.5	3.6	②	23.0	14.0
	Explorer	4.0	E	5.0	5.0	25.4	3.0	3.25	2.75	⑤	⑥
	Explorer	4.6	W	6.5	—	25.4	④	3.25	2.75	22.5	⑦
	Mountaineer	4.0	E	5.0	5.0	25.4	3.0	3.25	2.75	22.5	⑥
	Mountaineer	4.6	W	6.5	—	25.4	④	3.25	2.75	22.5	⑦
2003	Explorer	4.0	E	5.0	5.0	25.4	3.0	2.70	3.50	⑤	⑥
	Explorer Sport	4.0	E	4.0	NA	⑧	2.5	3.25	②	23.0	14.0
	Explorer Sport-Trac	4.0	E	4.0	NA	⑧	2.5	3.25	②	23.0	14.0
	Explorer	4.6	W	6.5	—	25.4	④	2.70	3.50	22.5	⑦
	Mountaineer	4.0	E	5.0	5.0	25.4	3.0	2.70	3.50	22.5	⑥
	Mountaineer	4.6	W	6.5	—	25.4	④	2.70	3.50	22.5	⑦
	Aviator	4.6	H	6.0	—	25.4	2.6	2.70	3.50	22.5	19.6
2004	Explorer	4.0	E	5.0	5.0	25.4	3.0	2.70	3.50	22.5	⑥
	Explorer	4.6	W	6.5	—	25.4	④	2.70	3.50	22.5	⑦
	Explorer Sport-Trac	4.0	E	4.0	NA	⑧	2.5	3.25	②	23.0	14.0
	Mountaineer	4.0	E	5.0	5.0	25.4	3.0	2.70	3.50	22.5	⑥
	Mountaineer	4.6	W	6.5	—	25.4	④	2.70	3.50	22.5	⑦
	Aviator	4.6	H	6.0	—	25.4	2.6	2.70	3.50	22.5	19.6

NA: Information not available

NOTE: All capacities are approximate. Add fluid gradually and check to be sure a proper fluid level is obtained.

① Dry fill

② With Traction-Lok: 5.25 pts.
without Traction-Lok: 5.5 pts.

③ 4-door Explorer: 21.0
Explorer sport Trac: 20.5

④ Part time: 3.0
Full time: 2.6

⑤ 4-door Explorer: 22.5
Explorer Sport: 17.5
Explorer Sport Trac: 20.5

⑥ w/o auxiliary heater: 16.3
w/auxiliary heater: 18.3

⑦ w/o auxiliary heater: 18.6
w/auxiliary heater: 20.1

⑧ 4x2: 20.0 pts.
4x4: 20.6 pts.

67197-EXPL-C04

VALVE SPECIFICATIONS

Year	Engine Displ. Liters	Engine VIN	Seat Angle (deg.)	Face Angle (deg.)	Spring Test Pressure (lbs. @ in.)	Spring Installed Height (in.)	Stem-to-Guide Clearance (in.)		Stem Diameter (in.)	
							Intake	Exhaust	Intake	Exhaust
2001	4.0	E	45	45	202-225@ 1.413-1.445	1.569- 1.601	0.0010- 0.0020	0.0010- 0.0020	0.2740- 0.2748	0.2730- 0.2740
	5.0	P	45	44	200@1.20	①	0.0010- 0.0027	0.0015- 0.0032	0.3415- 0.3423	0.3410- 0.3418
2002	4.0	E	45	45	202-225@ 1.413-1.445	1.569- 1.601	0.0010- 0.0020	0.0010- 0.0020	0.2740- 0.2748	0.2730- 0.2740
	4.6	W	44.5- 45.1	45.25- 45.75	162-180@ 1.134	1.675	0.0008- 0.0027	0.0018- 0.0037	0.2746- 0.2750	0.2736- 0.2740
2003	4.0	E	45	45	202-225@ 1.413-1.445	1.569- 1.601	0.0010- 0.0020	0.0010- 0.0020	0.2740- 0.2748	0.2730- 0.2740
	4.6	H	45	45.5	160@1.03	1.4228	0.0008- 0.0027	0.0018- 0.0037	0.2746- 0.2754	0.2736- 0.2744
	4.6	W	44.5- 45.1	45.25- 45.75	162-180@ 1.134	1.675	0.0008- 0.0027	0.0018- 0.0037	0.2746- 0.2750	0.2736- 0.2740
2004	4.0	E	45	45	202-225@ 1.413-1.445	1.569- 1.601	0.0010- 0.0020	0.0010- 0.0020	0.2740- 0.2748	0.2730- 0.2740
	4.6	H	45	45.5	160@1.03	1.4228	0.0008- 0.0027	0.0018- 0.0037	0.2746- 0.2754	0.2736- 0.2744
	4.6	W	44.5- 45.1	45.25- 45.75	162-180@ 1.134	1.675	0.0008- 0.0027	0.0018- 0.0037	0.2746- 0.2750	0.2736- 0.2740

① Intake: 1.75-1.81 in.
Exhaust: 1.59 in.

67197-EXPL-C05

CRANKSHAFT AND CONNECTING ROD SPECIFICATIONS

All measurements are given in inches.

			Crankshaft				Connecting Rod		
Year	Engine Displ. Liters	Engine VIN	Main Brg. Journal Dia.	Main Brg. Oil Clearance	Shaft End-play	Thrust on No.	Journal Diameter	Oil Clearance	Side Clearance
2001	4.0	X	2.2433-2.2441	0.0008-0.0015	0.0020-0.0120	3	2.1252-2.1260	0.0003-0.0024	0.0002-0.0025
	4.0	E	2.2430-2.2440	0.0008-0.0015	0.0020-0.0125	3	2.7252-2.7260	0.0003-0.0024	0.0036-0.0106
	5.0	P	2.2482-2.2490	0.0008-0.0015	0.0040-0.0080	3	2.1228-2.1236	0.0008-0.0015	0.0010-0.0020
2002	4.0	E	2.2430-2.2440	0.0008-0.0015	0.0020-0.0125	3	2.7252-2.7260	0.0003-0.0024	0.0036-0.0106
	4.6	W	2.6500-2.6570	0.0009-0.0018	0.0051-0.0118	3	2.0867-2.0870	0.0010-0.0027	0.0059-0.0177
2003	4.0	E	2.2430-2.2440	0.0008-0.0015	0.0020-0.0125	3	2.7252-2.7260	0.0003-0.0024	0.0036-0.0106
	4.6	H	2.6567-2.6576	NA	0.0051-0.0118	3	2.0859-2.0867	0.0011-0.0027	0.0059-0.0177
	4.6	W	2.6500-2.6570	0.0009-0.0018	0.0051-0.0118	3	2.0867-2.0870	0.0010-0.0027	0.0059-0.0177
2004	4.0	E	2.2430-2.2440	0.0008-0.0015	0.0020-0.0125	3	2.7252-2.7260	0.0003-0.0024	0.0036-0.0106
	4.6	H	2.6567-2.6576	NA	0.0051-0.0118	3	2.0859-2.0867	0.0011-0.0027	0.0059-0.0177
	4.6	W	2.6500-2.6570	0.0009-0.0018	0.0051-0.0118	3	2.0867-2.0870	0.0010-0.0027	0.0059-0.0177

NA: Information not available

67197-EXPL-C06

PISTON AND RING SPECIFICATIONS

All measurements are given in inches.

Year	Engine Displ. Liters	Engine VIN	Piston Clearance	Ring Gap			Ring Side Clearance		
				Top Comp.	Bottom Comp.	Oil Control	Top Comp.	Bottom Comp.	Oil Control
2001	4.0	E	0.0008-0.0019	0.008-0.018	0.015-0.024	0.015-0.055	0.0160-0.0300	0.0120-0.0260	SNUG
	5.0	P	0.0012-0.0020	0.010-0.020	0.018-0.028	0.010-0.040	0.0013-0.0033	0.0013-0.0033	SNUG
2002	4.0	E	0.0008-0.0019	0.015-0.023	0.015-0.023	0.015-0.055	0.0010-0.0030	0.0010-0.0030	SNUG
	4.6	W	0.0005-0.0010	0.0394	0.0394	0.0500	0.0012-0.0031	0.0012-0.0031	SNUG
2003	4.0	E	0.0008-0.0019	0.015-0.023	0.015-0.023	0.015-0.055	0.0010-0.0030	0.0010-0.0030	SNUG
	4.6	H	0.0004	0.006-0.0120	0.012-0.0220	0.006-0.0260	0.0004-0.0009	0.0012-0.0032	SNUG
	4.6	W	0.0005-0.0010	0.0394	0.0394	0.0500	0.0012-0.0031	0.0012-0.0031	SNUG
2004	4.0	E	0.0008-0.0019	0.015-0.023	0.015-0.023	0.015-0.055	0.0010-0.0030	0.0010-0.0030	SNUG
	4.6	H	0.0004	0.006-0.0120	0.012-0.0220	0.006-0.0260	0.0004-0.0009	0.0012-0.0032	SNUG
	4.6	W	0.0005-0.0010	0.0394	0.0394	0.0500	0.0012-0.0031	0.0012-0.0031	SNUG

67197-EXPL-C07

TORQUE SPECIFICATIONS

All readings in ft. lbs.

Year	Engine Displ. Liters	Engine VIN	Cylinder Head Bolts	Main Bearing Bolts	Rod Bearing Bolts	Crankshaft Damper Bolts	Flywheel Bolts	Manifold Intake *	Manifold Exhaust	Spark Plugs	Oil Pan Drain Plug
2001	4.0	E	①	67-74	19-24	②	54-64	13	15-18	15	19
	5.0	P	③	61-68	19-24	110-130	75-85	④	26-32	15	NA
2002	4.0	E	⑤	72	⑥	⑦	75-85	7	16	15	19
	4.6	W	⑧	⑨	⑩	⑪	59	18	15	12	10
2003	4.0	E	⑤	72	⑥	⑦	75-85	7	16	15	19
	4.6	H	⑧	⑨	⑩	⑪	59	7	15	15	NA
	4.6	W	⑧	⑨	⑩	⑪	59	18	15	12	10
2004	4.0	E	⑤	72	⑥	⑦	75-85	7	16	15	19
	4.6	H	⑧	⑨	⑩	⑪	59	7	15	15	NA
	4.6	W	⑧	⑨	⑩	⑪	59	18	15	12	10

NA: Information not available

* NOTE: Applies to Lower Manifold only.

① Step 1: 28 ft. lbs.
Step 2: Plus 90 degrees
Step 3: Plus 90 degrees

② Step 1: 20-28 ft. lbs.
Step 2: Loosen two turns
Step 3: 20-25 ft. lbs.

③ Step 1: 25-35 ft. lbs.
Step 2: 45-55 ft. lbs.
Step 3: Plus 90 degrees

④ Step 1: 89 inch lbs. (10 Nm)
Step 2: 24 ft. lbs. (32 Nm)

⑤ Step 1: 24 ft. lbs.
Step 2: plus 90 degrees

⑥ Step 1: 15 ft. lbs.
Step 2: +90 degrees

⑦ Step 1: 37 ft. lbs.
Step 2: plus 90 degrees

⑧ Step 1: 30 ft. lbs.
Step 2: +90 degrees
Step 3: back off 1 full turn
Step 4: 30 ft. lbs.
Step 5: +90 degrees
Step 6: +90 additional degrees

⑨ See the procedure in the text.

⑩ Step 1: 17 ft. lbs.
Step 2: 32 ft. lbs.
Step 3: +90-120 degrees

⑪ Step 1: 66 ft. lbs.
Step 2: back off 1 full turn
Step 3: 37 ft. lbs.
Step 4: +90 degrees
Do not exceed 148 ft. lbs.

67197-EXPL-C08

WHEEL ALIGNMENT

			Caster		Camber		
Year	Model		Range (+/-Deg.)	Preferred Setting (Deg.)	Range (+/-Deg.)	Preferred Setting (Deg.)	Toe-in (in.)
2001	Explorer		1.00	+4.20	0.50	-0.50	0.12+/-0.25
	Explorer Sport		1.00	①	0.70	-0.50	0.12+/-0.25
	Explorer Sport Trac		1.00	①	0.70	-0.50	0.12+/-0.25
	Mountaineer		1.00	+4.20	0.50	-0.50	0.12+/-0.25
2002	Explorer	F	1.00	②	0.80	-0.50	0.10+/-0.25
		R	—	—	0.80	-0.50	0.10+/-0.25
	Explorer Sport		1.00	①	0.70	-0.50	0.12+/-0.25
	Explorer Sport Trac		1.00	①	0.70	-0.50	0.12+/-0.25
	Mountaineer	F	1.00	②	0.80	-0.50	0.10+/-0.25
		R	—	—	0.80	-0.50	0.10+/-0.25
2003	Aviator	F	1.00	+5.3	0.08	-0.50	0.24+/-0.25
		R	—	—	0.80	-0.95	0.16+/-0.25
	Explorer	F	1.00	②	0.80	-0.50	0.10+/-0.25
		R	—	—	0.80	-0.50	0.10+/-0.25
	Explorer Sport		1.00	①	0.70	-0.50	0.12+/-0.25
	Explorer Sport Trac		1.00	①	0.70	-0.50	0.12+/-0.25
	Mountaineer	F	1.00	②	0.80	-0.50	0.10+/-0.25
		R	—	—	0.80	-0.50	0.10+/-0.25
2004	Aviator	F	1.00	+5.3	0.08	-0.50	0.24+/-0.25
		R	—	—	0.25	+0.80	0.16+/-0.25
	Explorer	F	1.00	L +5.1; R+5.3	0.80	-0.50	0.10+/-0.25
		R	—	—	0.80	-0.90	0.10+/-0.25
	Explorer Sport-Trac		1.00	L +3.15; R +4.05	0.70	-0.50	0.12+/-0.25
	Mountaineer	F	1.00	L +5.1; R+5.3	0.80	-0.50	0.10+/-0.25
		R	—	—	0.80	-0.90	0.10+/-0.25

① Left side: +3.95
Right side: +4.45

② Left side: +5.1
Right side: +5.3

67197-EXPL-C09

TIRE, WHEEL AND BALL JOINT SPECIFICATIONS

Year	Model	OEM Tires Standard	OEM Tires Optional	Tire Pressures (psi) Front	Tire Pressures (psi) Rear	Wheel Size	Ball Joint Inspection	Lugnut Toque ft. lbs.
2001	Explorer	P225/70R15	P235/70R15	②	②	7-JJ	0.030 in. ①	100
			P255/70R16	②	②	7-JJ		
	Explorer Sport	P235/70R15	P255/70R16	②	②	7-JJ	0.030 in. ①	100
	Explorer Sport Trac	P235/70R15	P255/70R16	②	②	7-JJ	0.030 in. ①	100
	Mountaineer	P225/70R15	P235/70R15	②	②	7-JJ	0.030 in. ①	100
			P255/70R16	②	②	7-JJ		
2002	Explorer	P225/70R16	P245/70R16	②	②	7-JJ	0.030 in. ①	100
			P255/70R16	②	②	7-JJ		
			P255/70HR16	②	②	7-JJ		
	Explorer Sport	P235/70R15	P255/70R16	②	②	7-JJ	0.030 in. ①	100
	Explorer Sport Trac	P235/70R15	P255/70R16	②	②	7-JJ	0.030 in. ①	100
	Mountaineer	P225/70R16	P245/70R16	②	②	7-JJ	0.030 in. ①	100
			P255/70R16	②	②	7-JJ		
			P255/70HR16	②	②	7-JJ		
2003	Aviator	P245/65HR17	none	②	②	7.5	0.030 in. ①	100
	Explorer	P225/70R16	P245/70R16	②	②	7-JJ	0.030 in. ①	100
			P255/70R16	②	②	7-JJ		
			P255/70HR16	②	②	7-JJ		
	Explorer Sport	P235/70R15	P255/70R16	②	②	7-JJ	0.030 in. ①	100
	Explorer Sport Trac	P235/70R15	P255/70R16	②	②	7-JJ	0.030 in. ①	100
	Mountaineer	P225/70R16	P245/70R16	②	②	7-JJ	0.030 in. ①	100
			P255/70R16	②	②	7-JJ		
			P255/70HR16	②	②	7-JJ		
2004	Aviator	P245/65HR17	none	②	②	7.5	0.030 in. ①	100
	Explorer						0.030 in. ①	100
	Eddie Bauer	P245/65R17	none	②	②	NA		
	Limited	P245/65R17	none	②	②	NA		
	NBX	P245/65R17	none	②	②	NA		
	XLS	P235/70R16	none	②	②	7		
	XLS Sport	P235/70R16	none	②	②	7		
	XLT	P235/70R16	none	②	②	7		
	XLT Sport	P245/65R17	none	②	②	NA		
	Explorer Sport-Trac						0.030 in. ①	100
	Adrenalin	P255/70R16	none	②	②	7		
	XLS	P235/70R16	none	②	②	7		
	XLT	P235/70R16	none	②	②	7		
	XLT Premium	P255/70R16	none	②	②	7		
	Mountaineer						0.030 in. ①	100
	Convenience	P235/70R16	none	②	②	NA		
	Luxury	P245/65R17	none	②	②	NA		
	Premier	P245/65R17	none	②	②	NA		

NA: Information not available

OEM: Original Equipment Manufacturer

STD: Standard

OPT: Optional

① Both upper and lower

② See placard on vehicle

67197-EXPL-C10

BRAKE SPECIFICATIONS

All measurements in inches unless noted

			Brake Disc			Brake Drum Diameter				Brake Caliper	
Year	Model		Original Thickness	Min. Thickness	Max. Runout	Original Inside Dia.	Max. Wear Limit	Max. Machine Dia.	Minimum Lining Thickness	Bracket Bolts (ft. lbs.)	Mounting Bolts (ft. lbs.)
2001	Explorer	F	1.020	0.980	0.0005	—	—	—	0.100	72-97	21-26
		R	0.480	0.440	0.0024	—	—	—	0.039	80	20
	Explorer	F	1.020	0.980	0.0016	—	—	—	0.100	72-97	21-26
	Sport	R	0.480	0.440	0.0024	10.00	10.09	10.06	0.039	80	20
	Explorer	F	1.020	0.980	0.0005	—	—	—	0.100	72-97	21-26
	Sport-Trac	R	0.480	0.440	0.0024	—	—	—	0.039	80	20
	Mountaineer	F	1.020	0.980	0.0005	—	—	—	0.100	72-97	21-26
		R	0.480	0.440	0.0024	—	—	—	0.039	80	20
2002	Explorer	F	1.020	0.980	0.0005	—	—	—	0.100	83	24
		R	0.480	0.440	0.0024	—	—	—	0.039	—	24
	Explorer	F	1.020	0.980	0.0016	—	—	—	0.100	72-97	21-26
	Sport	R	0.480	0.440	0.0024	10.00	10.09	10.06	0.039	80	20
	Explorer	F	1.020	0.980	0.0005	—	—	—	0.100	72-97	21-26
	Sport-Trac	R	0.480	0.440	0.0024	—	—	—	0.039	80	20
	Mountaineer	F	1.020	0.980	0.0005	—	—	—	0.100	72-97	21-26
		R	0.480	0.440	0.0024	—	—	—	0.039	80	24
2003	Explorer	F	1.020	0.980	0.0005	—	—	—	0.100	155	24
		R	0.480	0.440	0.0024	—	—	—	0.039	—	24
	Explorer	F	1.020	0.980	0.0016	—	—	—	0.100	72-97	21-26
	Sport	R	0.480	0.440	0.0024	—	—	—	0.039	80	20
	Explorer	F	1.020	0.980	0.0005	—	—	—	0.100	72-97	21-26
	Sport-Trac	R	0.480	0.440	0.0024	—	—	—	0.039	80	20
	Mountaineer	F	1.020	0.980	0.0005	—	—	—	0.100	155	24
		R	0.480	0.440	0.0024	—	—	—	0.039	—	24
	Aviator	F	NA	1.040	NA	—	—	—	0.118	155	32
		R	NA	0.430	NA	—	—	—	0.118	—	24
2004	Aviator	F	NA	1.040	NA	—	—	—	0.118	155	32
		R	NA	0.430	NA	—	—	—	0.118	—	24
	Explorer	F	NA	0.960	NA	—	—	—	0.118	100	24
		R	NA	0.430	NA	—	—	—	0.118	—	24
	Explorer	F	NA	0.980	NA	—	—	—	0.100	72-97	21-26
	Sport-Trac	R	NA	0.440	0.0024	—	—	—	0.039	80	20
	Mountaineer	F	NA	0.960	NA	—	—	—	0.118	100	24
		R	NA	0.430	NA	—	—	—	0.118	—	24

NA: Information not available

NOTE: Due to changes made during production, refer to manufacturer's specifications if they differ from those in this chart

67197-EXPL-C11

SCHEDULED MAINTENANCE INTERVALS

2001-03 Ford Explorer, Explorer Sport, Explorer Sport-Trac, Mercury Mountaineer
2003 Lincoln Aviator

TO BE SERVICED	TYPE OF SERVICE	VEHICLE MILEAGE INTERVAL (x1000)												
		5	10	15	20	25	30	35	40	45	50	55	60	65
Engine oil & filter	R	✓	✓	✓	✓	✓	✓	✓	✓	✓	✓	✓	✓	✓
Tires	Rotate	✓	✓	✓	✓	✓	✓	✓	✓	✓	✓	✓	✓	✓
Auto trans. fluid	I			✓			✓			✓			✓	
Brake pads/shoes	I			✓			✓			✓			✓	
Coolant hoses	S/I			✓			✓			✓			✓	
Steering linkage	I			✓			✓			✓			✓	
Suspension and driveshaft	I			✓			✓			✓			✓	
Ball joints (2wd)	I			✓			✓			✓			✓	
Wheels ①	I						✓						✓	
Cabin air filter	R			✓			✓			✓			✓	
Climate controlled seat filter	R						✓						✓	
Ball joints (2wd)	L			✓			✓			✓			✓	
Exhaust system	I						✓						✓	
Engine air filter	R						✓						✓	
Fuel filter	R						✓						✓	
Green coolant ②	R									✓				
Wheel bearings (2wd)	L												✓	
Manual trans. fluid	R												✓	
Accessory drive belts	I	every 100,000 miles												
Spark plugs	R	every 100,000 miles												
PCV valve	R	every 100,000 miles												
Yellow coolant	R	5 years or 100,000 miles, then evey 3 years or 50,000 miles												
Orange coolant	R	every 150,000 miles												
Auto trans fluid	R	every 150,000 miles												
Differential fluid	R	every 150,000 miles												
Accessory drive belts	R	every 150,000 miles if not previously done so												
Transfer case fluid	R	every 150,000 miles												

R: Replace S: Service I: Inspect L: Lubricate

① Inspect wheel ends for end play and noise

① Recommended, but not required in Calif.

② Change every 30,000 miles or 36 months thereafter

Special Operating Condition Requirements

When towing a trailer or using a camper or car-top carrier:

Change engine oil and install a new oil filter every 4,800 km (3,000 miles) or 3 months.

Change transfer case fluid every 96,000 km (60,000 miles).

Change manual transmission fluid as required.

Inspect and lubricate U-joints as required.

During extensive idling and/or low speed driving for long distances, as in heavy commercial use such as delivery, taxi, patrol car or livery:

Change engine oil and install a new oil filter, lube front lower control arm and steering linkage ball joints with zerk fittings (if equipped) every 4,800 km (3,000 miles) or 3 months.
Inspect brake system and check battery electrolyte level (Patrol cars) every 8,000 km (5,000 miles).
Install a new fuel filter every 24,000 km (15,000 miles).
Change automatic transmission fluid, lubricate 4x2 wheel bearings, install new grease seals and adjust bearings every 48,000 km (30,000 miles).
Install new spark plugs and change transfer case fluid every 96,000 km (60,000 miles).
Install a new cabin air filter as required.

67197-EXPL-C12

SCHEDULED MAINTENANCE INTERVALS

2001-03 Ford Explorer, Explorer Sport, Explorer Sport-Trac, Mercury Mountaineer
2003 Lincoln Aviator
Footnotes Continued

When operating in dusty conditions such as unpaved or dusty roads:

Change engine oil and install a new oil filter every 4,800 km (3,000 miles) or 3 months.

Install a new fuel filter every 24,000 km (15,000 miles).

Change automatic transmission fluid every 48,000 km (30,000 miles).

Change transfer case fluid every 96,000 km (60,000 miles).

Install a new engine air filter as required.

Install a new cabin air filter as required.

When operating in off-road conditions:

Change automatic transmission fluid every 48,000 km (30,000 miles).

Change transfer case fluid every 96,000 km (60,000 miles).

Install a new cabin air filter as required.

Inspect and lubricate U-joints.

Inspect and lubricate steering linkage ball joints with zerk fittings.

67197-EXPL-C13

SCHEDULED MAINTENANCE INTERVALS

2004 Lincoln Aviator, Ford Explorer, Explorer Sport-Trac, Mercury Mountaineer

TO BE SERVICED	TYPE OF SERVICE	VEHICLE MILEAGE INTERVAL (x1000)												
		5	10	15	20	25	30	35	40	45	50	55	60	65
Engine oil & filter	R	✓	✓	✓	✓	✓	✓	✓	✓	✓	✓	✓	✓	✓
Tires	Rotate	✓	✓	✓	✓	✓	✓	✓	✓	✓	✓	✓	✓	✓
Auto trans. fluid	I			✓			✓			✓			✓	
Brake pads/shoes	I			✓			✓			✓			✓	
Coolant hoses	S/I			✓			✓			✓			✓	
Steering linkage	I			✓			✓			✓			✓	
Suspension, driveshaft	I			✓			✓			✓			✓	
Cabin air filter	R			✓			✓			✓			✓	
Ball joints (2wd)	L			✓			✓			✓			✓	
Wheels ①	I						✓						✓	
Exhaust system	I						✓						✓	
Engine air filter	R						✓						✓	
Fuel filter	R						✓						✓	
Climate controlled seat filter	R						✓						✓	
Accessory drive belts	I	every 100,000 miles												
Premium Gold coolant	R	5 years or 100,000 miles, then every 3 years or 30,000 miles												
Spark plugs	R	every 100,000 miles												
PCV valve	R	every 100,000 miles												
Coolant, exc. Premium Gold	R	every 105,000 miles												
Front wheel bearings and seals (2wd)	R	every 150,000 miles if not already done so												
Manual trans. fluid	R	every 120,000 miles												
Auto trans fluid	R	every 150,000 miles												
Differential fluid	R	every 150,000 miles												
Accessory drive belts	R	every 150,000 miles if not already done so												

R: Replace S: Service I: Inspect L: Lubricate

① Inspect wheel ends for end play and noise

Special Operating Condition Requirements

When towing a trailer or using a camper or car-top carrier:

Change engine oil and install a new oil filter every 4,800 km (3,000 miles) or 3 months.

Change transfer case fluid every 96,000 km (60,000 miles).

Change manual transmission fluid as required.

Inspect and lubricate U-joints as required.

During extensive idling and/or low speed driving for long distances, as in heavy commercial use such as delivery, taxi, patrol car or livery:

Change engine oil and install a new oil filter, lube front lower control arm and steering linkage ball joints with zerk fittings (if equipped) every 4,800 km (3,000 miles) or 3 months.

Inspect brake system and check battery electrolyte level (Patrol cars) every 8,000 km (5,000 miles).

Install a new fuel filter every 24,000 km (15,000 miles).

Change automatic transmission fluid, lubricate 4x2 wheel bearings, install new grease seals and adjust bearings every 48,000 km (30,000 miles).

Install new spark plugs and change transfer case fluid every 96,000 km (60,000 miles).

Install a new cabin air filter as required.

When operating in dusty conditions such as unpaved or dusty roads:

Change engine oil and install a new oil filter every 4,800 km (3,000 miles) or 3 months.

Install a new fuel filter every 24,000 km (15,000 miles).

Change automatic transmission fluid every 48,000 km (30,000 miles).

Change transfer case fluid every 96,000 km (60,000 miles).

Install a new engine air filter as required.

Install a new cabin air filter as required.

67197-EXPL-C14

SCHEDULED MAINTENANCE INTERVALS

2004 Lincoln Aviator, Ford Explorer, Explorer Sport-Trac, Mercury Mountaineer

Footnotes Continued

When operating in off-road conditions:

Change automatic transmission fluid every 48,000 km (30,000 miles).

Change transfer case fluid every 96,000 km (60,000 miles).

Install a new cabin air filter as required.

Inspect and lubricate U-joints.

Inspect and lubricate steering linkage ball joints with zerk fittings.

67197-EXPL-C15

PRECAUTIONS

Before servicing any vehicle, please be sure to read all of the following precautions, which deal with personal safety, prevention of component damage, and important points to take into consideration when servicing a motor vehicle:

- Never open, service or drain the radiator or cooling system when the engine is hot; serious burns can occur from the steam and hot coolant.
- Observe all applicable safety precautions when working around fuel. Whenever servicing the fuel system, always work in a well-ventilated area. Do not allow fuel spray or vapors to come in contact with a spark, open flame, or excessive heat (a hot drop light, for example). Keep a dry chemical fire extinguisher near the work area. Always keep fuel in a container specifically designed for fuel storage; also, always properly seal fuel containers to avoid the possibility of fire or explosion. Refer to the additional fuel system precautions later in this section.
- Fuel injection systems often remain pressurized, even after the engine has been turned **OFF**. The fuel system pressure must be relieved before disconnecting any fuel lines. Failure to do so may result in fire and/or personal injury.
- Brake fluid often contains polyglycol ethers and polyglycols. Avoid contact with the eyes and wash your hands thoroughly after handling brake fluid. If you do get brake fluid in your eyes, flush your eyes with clean, running water for 15 minutes. If eye irritation persists, or if you have taken brake fluid internally, IMMEDIATELY seek medical assistance.
- The EPA warns that prolonged contact with used engine oil may cause a number of skin disorders, including cancer! You should make every effort to minimize your exposure to used engine oil. Protective gloves should be worn when changing oil. Wash your hands and any other exposed skin areas as soon as possible after exposure to used engine oil. Soap and water, or waterless hand cleaner should be used.
- All new vehicles are now equipped with an air bag system, often referred to as a Supplemental Restraint System (SRS) or Supplemental Inflatable Restraint (SIR) system. The system must be disabled before performing service on or around system components, steering column, instrument panel components, wiring and sensors. Failure to follow safety and disabling procedures could result in accidental air bag deployment, possible personal injury and unnecessary system repairs.
- Always wear safety goggles when working with, or around, the air bag system. When carrying a non-deployed air bag, be sure the bag and trim cover are pointed away from your body. When placing a non-deployed air bag on a work surface, always face the bag and trim cover upward, away from the surface. This will reduce the motion of the module if it is accidentally deployed. Refer to the additional air bag system precautions later in this section.
- Clean, high quality brake fluid from a sealed container is essential to the safe and proper operation of the brake system. You should always buy the correct type of brake fluid for your vehicle. If the brake fluid becomes contaminated, completely flush the system with new fluid. Never reuse any brake fluid. Any brake fluid that is removed from the system should be discarded. Also, do not allow any brake fluid to come in contact with a painted surface; it will damage the paint.
- Never operate the engine without the proper amount and type of engine oil; doing so WILL result in severe engine damage.
- Timing belt maintenance is extremely important! Many models utilize an interference-type, non-freewheeling engine. If the timing belt breaks, the valves in the cylinder head may strike the pistons, causing potentially serious (also time-consuming and expensive) engine damage. Refer to the maintenance interval charts in the front of this manual for the recommended replacement interval for the timing belt, and to the timing belt section for belt replacement and inspection.
- Disconnecting the negative battery cable on some vehicles may interfere with the functions of the on-board computer system(s) and may require the computer to undergo a relearning process once the negative battery cable is reconnected.
- When servicing drum brakes, only disassemble and assemble one side at a time, leaving the remaining side intact for reference.

ENGINE REPAIR

Alternator

REMOVAL & INSTALLATION

5.0L Engines

1. Before servicing the vehicle, refer to the precautions in the beginning of this section.
2. Remove or disconnect the following:
   - Negative battery cable
   - Air cleaner outlet tube
   - Drive belt
   - Electrical connectors from the alternator
   - A/C manifold and tube bracket aside
   - Wiring harness to alternator push pin
   - Alternator

To install:

3. Install or connect the following:
   - Alternator. Torque the bolts to 40 ft. lbs. (55 Nm).
   - Push pin for the alternator wiring harness
   - A/C manifold and tube bracket. Torque the bolt to 106 inch lbs. (12 Nm).
   - Electrical connectors to the alternator
   - Drive belt
   - Air cleaner outlet tube
   - Negative battery cable

4.0L SOHC Engine

1. Before servicing the vehicle, refer to the precautions in the beginning of this section.
2. Remove or disconnect the following:
   - Negative battery cable
   - Air cleaner outlet tube
   - Accessory drive belt
   - Electrical connectors
   - Wiring harness-to-generator pin-type retainer
   - Stud bolt, the bolts and the alternator
3. To install, reverse the removal procedure. Torque all mounting bolts to 35 ft. lbs. (47Nm).

4.6L DOHC and SOHC Engines

1. Before servicing the vehicle, refer to the precautions in the beginning of this section.
2. Remove or disconnect the following:
   - Negative battery cable
   - Drive belt
   - Electrical connectors

- Stud bolts, the bolts and the generator bracket
- Bolts and remove the alternator

3. To install, reverse the removal procedure. Torque the mounting bolts to 18 ft. lbs. (25Nm).

Ignition Timing

ADJUSTMENT

The ignition timing is preset to 10 degrees Before Top Dead Center (BTDC) and is not adjustable.

Engine Assembly

REMOVAL & INSTALLATION

4.0L SOHC Engines

CAUTION

If the fuel supply manifold is used as a leverage device, damage may occur to the supply manifold. Care must be taken when working around the fuel supply manifold.

- Accelerator cable from engine
- Speed control cable from engine
- Radiator, the fan blade, and the fan shroud
- Accessory bracket bolts and position bracket aside
- Alternator wiring
- Wiring harness retainer and position generator wiring away from engine
- Engine electrical connector
- PCM connector
- PCM ground wire
- Engine ground wire
- Brake booster vacuum hose
- A/C high pressure switch electrical connector
- Bolt and position the A/C lines aside

➡Heater hose will be removed with engine.

- Heater hoses
- Fuel line
- Starter motor
- Engine oil
- Oil drain plug
- Transmission portion of wiring harness
- RH and LH heated oxygen sensor connectors
- Transmission control connector
- Output shaft speed sensor connector
- Digital transmission range sensor connector
- Catalyst monitor sensor electrical connector
- Transmission/transfer case portion of the wiring harness from any routing clips or pushpins. Route transmission/transfer case portion of the wiring harness to top of engine.
- Bolt, and position the transmission cooling line bracket aside
- A/C line bracket nut and position it aside
- Power steering return hose
- Power steering pressure hose
- Vapor management valve hose connector
- Eight bolts and the LH and the RH engine support insulator nuts

➡The lifting eyes should be installed on the exhaust manifold studs for number three and number four cylinders.

1. Install the lifting eyes.
2. Install the spreader bar to the lifting eyes.
3. Attach a floor crane to the spreader bar and remove the engine.
4. Installation is the reverse of removal. Observe the following torques:
    - Left and right engine insulator nuts: 81 ft. lbs. (110 Nm)
    - Engine mount nuts: 59 ft. lbs. (80Nm)
    - Transmission-to-engine bolts: 35 ft. lbs. (47Nm)
    - Torque converter nuts: 35 ft. lbs. (47Nm)

4.6L DOHC and SOHC Engines

1. Before servicing the vehicle, refer to the precautions in the beginning of this section.
2. Relieve the fuel system pressure.
3. Drain the cooling system.
4. Drain the engine oil.
5. Properly discharge the A/C system.
6. Remove or disconnect the following:
    - Battery ground cable
    - Hood
    - Accelerator cable from the throttle body and all clips
    - Speed control cable from the throttle body and all clips
    - Air cleaner and the air cleaner outlet pipe
    - Radiator and transmission cooler
    - A/C condenser core

➡Access the heater control valve bracket through the wheel well.

- Heater control valve bracket bolt
- Coolant hoses
- All vacuum hoses. Position the heater control valve assembly aside.
- A/C tubes from the A/C compressor
- Fuel charging wiring harness connections
- Fuel charging wiring harness from the clips on the bulkhead. Release the fuel charging wiring harness pin-type retainer.
- All connections from the evaporative emissions (EVAP) canister purge solenoid
- Brake booster vacuum hose
- Fuel tube
- Alternator connections and the wiring anchors
- Wiring harness side of the left and right heated exhaust gas oxygen (HEGO) sensor connectors
- Power steering return hose. Plug the power steering reservoir.
- Power steering pressure tube from the power steering pump
- Brackets
- Oil pressure sender electrical connector
- CKP sensor electrical connector and the wiring anchor
- A/C compressor electrical connector and the wiring anchor
- Wiring harness routing clip and wiring anchor
- Starter
- On 4x4 vehicles, the bolt from the right side axle housing bushing bolt
- Right side and the left side stabilizer bar bracket nuts
- Left side lower axle housing bushing bolt
- Left side upper axle housing bushing bolt
- Right lower mount nut and washer
- Left lower mount nut and washer
- Transmission inspection cover, torque converter bolts and the transmission-to-engine block bolts

7. Install the right side Lifting Eye.
8. Install the left side Lifting Eye.
9. Install the Spreader Bar to the Lifting Eyes.
10. Attach a floor crane to the Spreader Bar and remove the engine.

➡If engine disassembly is to be carried out, remove the rear crankshaft oil seal slinger and the crankshaft rear oil seal.

11. Installation is the reverse of removal. Observe the following torques:
   - Lower mount nuts: 59 ft. lbs. (81Nm)
   - Left side upper axle housing bushing bolt: 49 ft. lbs. (66Nm)
   - Left side axle housing bushing bolt: 49 ft. lbs. (66Nm)
   - Stabilizer bar bracket nuts: 30 ft. lbs. (40Nm)
   - Right side axle housing bushing bolt: 49 ft. lbs. (66Nm)

5.0L Engine

1. Before servicing the vehicle, refer to the precautions in the beginning of this section.
2. Relieve the fuel system pressure.
3. Drain the cooling system.
4. Drain the engine oil.
5. Properly discharge the A/C system.
6. Remove or disconnect the following:
   - Battery
   - Drive belt
   - Fan shroud
   - Air cleaner outlet tube
   - A/C condenser core
   - Upper radiator hose
   - Power steering reservoir and move it aside
   - Power steering pump
   - Spark plug wire bracket from the A/C compressor
   - A/C compressor and bracket
   - Alternator
   - Wide-open A/C cut-off switch electrical connector
   - Vapor Management Valve (VMV) hose
   - Lower steering column shaft and move it aside
   - Left and right hand side vacuum connections
   - Accelerator and speed control cables
   - Accelerator cable bracket
   - Powertrain Control Module (PCM) connector
   - PCM ground connector
   - Engine bulkhead connector
   - Heater hoses
   - Transmission fill tube
   - Power steering cooler and move it aside
   - Ground cable from the engine front cover
   - Ground strap from the lower intake manifold
   - Transmission cooler lines from the retainer on the right engine mount
   - Starter
   - Transmission inspection cover
   - Torque converter nuts
   - Transmission
   - Left and right side Heated Oxygen (HO_2S) sensor electrical connectors
   - Transmission bulkhead connector
   - Brake booster vacuum supply line at the left upper intake manifold
   - Low oil level sensor electrical connector
   - Oil bypass filter
   - Lower radiator hose
   - Exhaust manifolds
   - Left and right side motor mounts and install the lifting brackets to the exhaust manifold studbolts on No. 1–No. 8 cylinders
   - Engine from the vehicle
   - Flywheel, if equipped

To install:

7. Install or connect the following:
   - Flywheel. Torque the new bolts to 85 ft. lbs. (115 Nm).
   - Engine. Torque the motor mount bolts to 109 ft. lbs. (148 Nm).
   - Exhaust manifolds
   - Lower radiator hose
   - Oil bypass filter
   - Brake booster vacuum supply line to the left side upper intake connection
   - Transmission bulkhead connector
   - Left and right HO_2S sensor connectors
   - Transmission
   - Torque converter bolts. Torque the bolts to 38 ft. lbs. (51 Nm).
   - Transmission inspection cover
   - Starter
   - Transmission cooler lines to the retainer by the right side motor mount
   - Battery-to-starter relay cable
   - Ground strap to the engine front cover
   - Ground strap to the rear of the lower intake manifold
   - Ground cable to the engine front cover
   - Power steering cooler
   - Transmission fill tube
   - Heater hoses
   - Engine bulkhead connector. Torque the bolt to 89 inch lbs. (10 Nm).
   - PCM and body ground connectors. Torque the bolt to 89 inch lbs. (10 Nm).
   - Accelerator and speed control cables to the clip
   - Accelerator cable bracket. Torque the upper bolts to 15 ft. lbs. (20 Nm) and lower the bolt to 80 inch lbs. (9 Nm).
   - Accelerator and sped control cables to the throttle linkage
   - Accelerator control shield. Torque the bolt to 106 inch lbs. (12 Nm).
   - Left and right side vacuum connections
   - Lower power steering column shaft. Torque the bolt to 40 ft. lbs. (55 Nm).
   - VMV hose
   - Fuel lines
   - Wide-open A/C cutoff switch electrical connector
   - Alternator and electrical connectors
   - A/C compressor
   - Spark plug wire bracket. Torque the nut to 89 inch lbs. (10 Nm).
   - Power steering pump. Torque the bolts to 21 ft. lbs. (29 Nm).
   - Power steering reservoir. Torque the bolts to 106 inch lbs. (12 Nm).
   - Upper radiator hose
   - Fan blade
   - Drive belt
   - A/C condenser core
   - Air cleaner outlet tube
   - Battery and cables
8. Recharge the A/C system.
9. Fill the cooling system.
10. Fill the engine with clean oil.
11. Run the engine and check for leaks and proper operation.
12. Check and adjust the front end alignment.

Water Pump

REMOVAL & INSTALLATION

4.0L SOHC Engine

1. Before servicing the vehicle, refer to the precautions in the beginning of this section.
2. Drain the cooling system.
3. Remove or disconnect the following:
   - Fan shroud
   - Accessory drive belt
   - Idler pulley
   - Water bypass hose
   - Heater hose
   - Lower radiator hose
   - Water pump pulley
   - Water pump

**** WARNING**

Use care when scraping the water pump-to-engine block mating surfaces. Gouges in the aluminum could form leak paths.

4. Clean all the sealing surfaces.

5. To install, reverse the removal procedure. Torque the water pump bolts to 89 inch lbs. (10Nm). Torque the pulley bolts to 18 ft. lbs. (25Nm).

4.6L DOHC and SOHC Engines

1. Before servicing the vehicle, refer to the precautions in the beginning of this section.
2. Drain the cooling system.
3. Remove or disconnect the following:
   - Engine cooling fan
   - Drive belt
   - Water pump pulley
   - Water pump
4. Discard the O-ring seal.
5. To install, reverse the removal procedure. Install a new O-ring seal and lubricate with engine coolant.
6. Torque the pump and pulley bolts to 18 ft. lbs. (25Nm).

5.0L Engines

1. Before servicing the vehicle, refer to the precautions in the beginning of this section.
2. Drain the cooling system.
3. Remove or disconnect the following:
   - Negative battery cable
   - Fan shroud
   - Drive belt
   - Water bypass hose
   - Heater hose from the water pump
   - Engine control sensor wiring and move it aside
   - Water pump pulley
   - Water pump from the engine front cover
   - Water pump inlet hose
   - Water pump

To install:

4. Clean the mounting surfaces of the pump and front cover thoroughly. Remove all traces of gasket material.
5. Install or connect the following:
   - Apply adhesive gasket sealer to both sides of a new gasket and place the gasket on the pump.
   - Inlet hose to the water pump. Torque the clamps to 11 ft. lbs. (15 Nm).
   - Water pump. Torque the bolts to 20 ft. lbs. (28 Nm).
   - Water pump pulley. Torque the bolts to 20 ft. lbs. (28 Nm).
   - Engine control sensor wiring
   - Heater hose
   - Water bypass tube. Torque the clamps to 11 ft. lbs. (15 Nm).
   - Dive belt
   - Fan shroud
   - Negative battery cable
6. Fill the cooling system.
7. Start the vehicle and check for leaks, repair if necessary.

Heater Core

REMOVAL & INSTALLATION

Aviator, Explorer, Explorer Sport Trac, Mountaineer

2001

1. Drain the radiator.
2. Remove the heater water hoses.

⁂ WARNING

Electronic modules are sensitive to static electrical charges. If exposed to these charges, damage may result.

3. Remove the steering column.
4. Disconnect the brake pedal position (BPP) switch electrical connector.
5. If equipped, disconnect the clutch pedal position (CPP) switch electrical connector.
6. Remove the LH and RH cowl side trim panels.
7. Disconnect the electrical connectors and the ground wires on the RH cowl panel.
8. Disconnect the power distribution box from the bracket and position aside.
9. Disconnect the bulkhead wiring harness connectors from inside the engine compartment.
10. Remove the bulkhead connector insulator.
11. Unclip the bulkhead electrical connectors from the dash panel.
12. Remove the passenger side air bag module.
13. Disconnect the blend door actuator electrical connector.
14. Disconnect the climate control vacuum harness connector.
15. Disconnect the radio antenna cable in-line connector.
16. Raise the glove compartment.
17. Remove the instrument panel defroster opening grille.
18. Remove the instrument panel cowl top bolts.
19. If equipped, remove the upper series floor console.
20. Remove the instrument panel brace bolt from under the steering column opening.
21. Remove the windshield side garnish mouldings.
22. Remove the RH instrument panel cowl side bolt.
23. Remove the instrument panel fuse panel door.
24. Remove the LH instrument panel cowl side bolts.
25. Position the instrument panel away from the dash panel.
26. Disconnect the instrument panel to body harness.

➡Two technicians are required to carry out this step.

27. Remove the instrument panel.
28. Recover the refrigerant.
29. Disconnect the A/C cycling switch.
30. Disconnect the blower electrical connectors.
31. Position the speed control servo aside.
32. Remove the nuts and the screws from the windshield washer reservoir/coolant recovery reservoir. Set the reservoir aside.
33. Disconnect the A/C manifold and tube from the suction accumulator/drier.
34. Disconnect the condenser to evaporator tube.
35. Disconnect the A/C system vacuum harness and the A/C evaporator housing mounting nut inside the vehicle.
36. Remove the A/C evaporator housing.
37. Remove the powertrain control module (PCM).
38. Remove the PCM heat sink.
39. Remove the heater air plenum nuts from the engine side of the dash panel.
40. Remove the heater core cover to air plenum screws.
41. Lift off the cover.
42. Remove the heater core.
43. To install, reverse the removal procedure. Be sure to install a new oval foam seal around the heater core inlet and outlet tubes.

2002

1. Deactivate the supplemental restraints system (SRS).
2. Release the two floor console front clips (one each side).
3. On vehicles with manual transmission, remove the gearshift lever handle.
4. Remove the console finish panel mat.
5. On vehicles with automatic transmission:
   a. If equipped, remove the ashtray assembly.
   b. Disconnect the electrical connector.
   c. Remove the console finish panel screw.

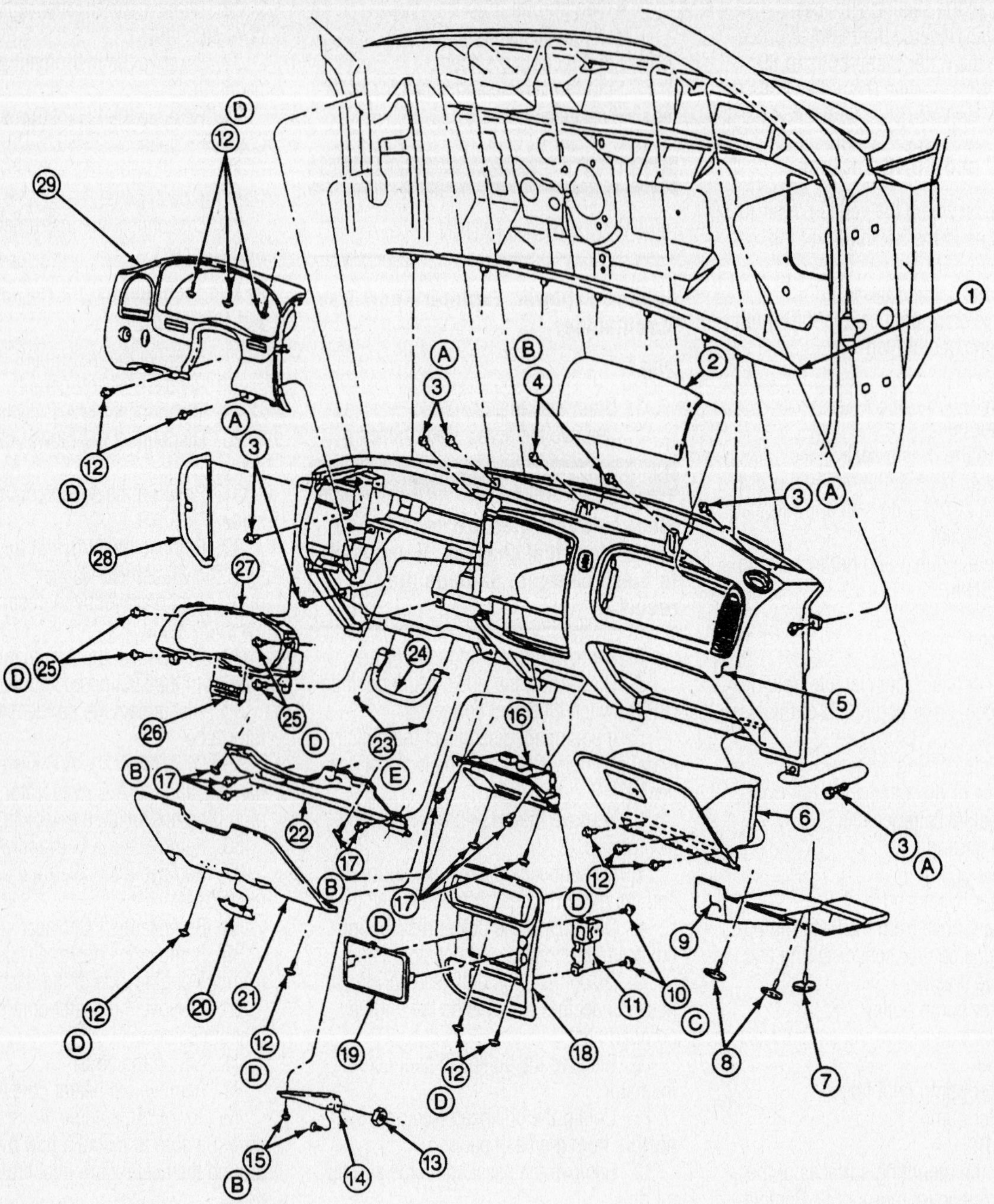

Exploded view of the instrument panel—Explorer

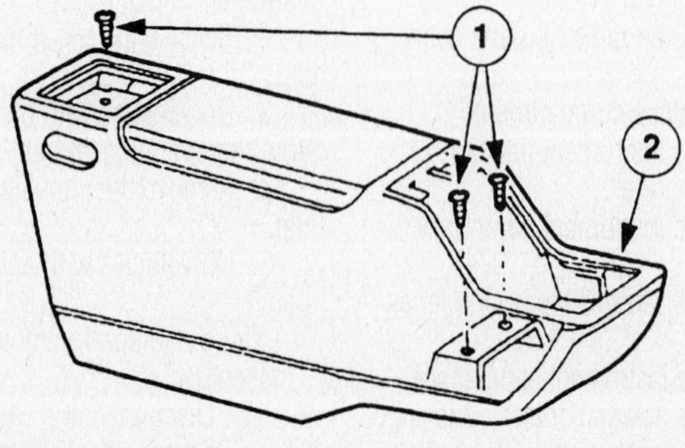

View of the upper series floor console—Explorer

6. Remove the console finish panel by lifting up at the rear and sliding it rearward. Disconnect the electrical connector(s).
7. Remove the floor console front screws.
8. If equipped, disconnect the electrical connector and release the wiring harness locators.
9. Remove the floor console center bolts.
10. Remove the two floor console access covers (one each side).
11. Remove the two floor console rear bolts (one each side).

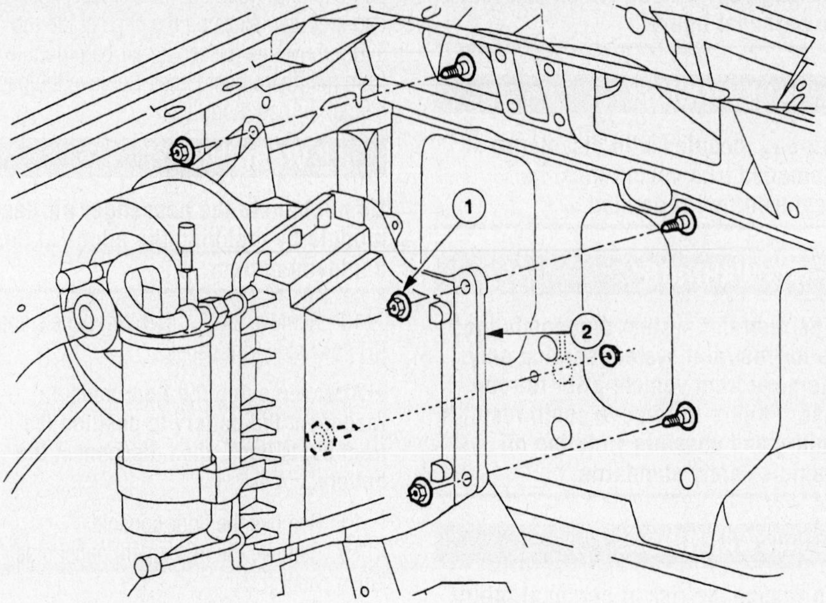

Evaporator housing—Explorer

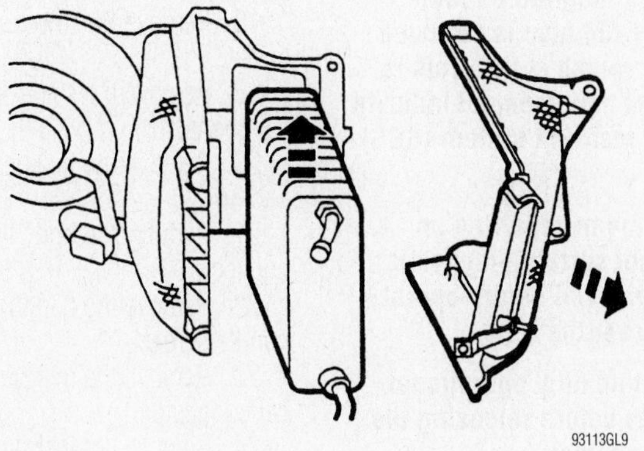

Evaporator core—Explorer

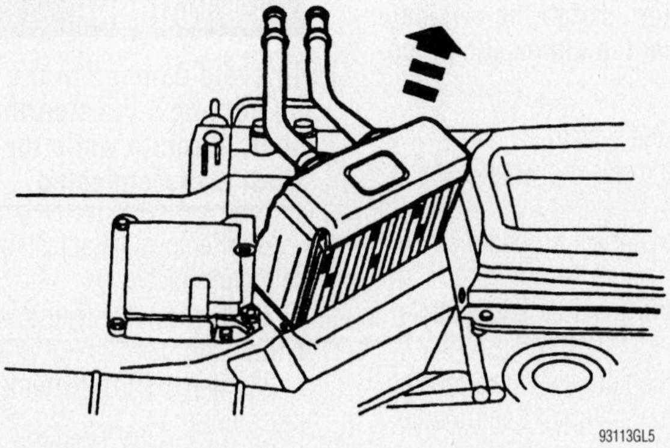

Heater core—Explorer

12. Remove the floor console by lifting up at the rear of the console and sliding it rearward.
13. Remove the screws and remove the snow shield.
14. Crimp off the coolant hoses and disconnect from the heater core.
15. Remove the instrument panel center support brackets.
16. Remove the assist handle bolt covers.
17. Remove the bolts and the passenger assist handle.
18. Remove the windshield side garnish moldings.
19. Remove the defroster opening grill.
20. Remove the upper instrument panel support bolts.
21. Remove the exterior cowl grill.
22. Remove the instrument panel support bolt.
23. Remove the two instrument panel side finish panels.
24. Lower the glove compartment.
25. Loosen the LH instrument panel side support bolts until half the threads are exposed.

⁂ WARNING

Be sure the instrument panel is properly supported to avoid possible damage to the instrument panel wiring or components. To avoid damage to the instrument panel wiring or components, do not use excessive force when moving the instrument panel away from the dash panel. The instrument panel may be supported by installing threaded rods or equivalent, with the same diameter and thread pitch, in place of the instrument panel support bolts.

26. Properly support the instrument panel and remove the RH instrument panel support bolts to allow the instrument panel to be pulled away from the dash panel.
27. Remove the center console floor duct.
28. Remove the screws and remove the RH floor duct.
29. Remove the screws and remove the LH floor duct.
30. Disconnect the electrical connectors and detach the pin-type retainers.
31. Remove the screws and remove the heater core tube cover.
32. Remove the RH heater core cover screws.
33. Remove the LH heater core cover screws and remove the heater core cover.

➡The instrument panel will need to be moved to facilitate removal of the heater core.

34. Remove the heater core.
35. To install, reverse the removal procedure. Lubricate the coolant hoses with coolant hose lubricant or plain water only if needed. Top-off the engine coolant level. For additional information. Reactivate the supplemental restraints system (SRS).

2003–04

FRONT

➡If an evaporator core leak is suspected, the evaporator core must be vacuum leak tested before it is removed from the vehicle.

➡Installation of a new accumulator is not required when repairing the air conditioning system except when there is physical evidence of contamination from a failed A/C compressor or damage to the accumulator.

➡New O-ring seals, lubricated in clean mineral oil, must be installed before reconnecting any A/C fitting which has been disconnected.

➡Lubricate the coolant hoses with plain water only if needed.

1. Recover the refrigerant.

⁂ WARNING

Always wear safety glasses when repairing an air bag supplemental restraint system (SRS) vehicle and when handling an air bag module. This will reduce the risk of injury in the event of an accidental deployment.

⁂ WARNING

Carry a live air bag module with the air bag and trim cover pointed away from your body. This will reduce the risk of injury in the event of an accidental deployment.

⁂ WARNING

Do not set a live air bag module down with the trim cover face down. This will reduce the risk of injury in the event of an accidental deployment.

⁂ WARNING

After deployment, the air bag surface can contain deposits of sodium hydroxide, a product of the gas generant combustion that is irritating to the skin. Wash your hands with soap and water afterwards.

⁂ WARNING

Never probe the connectors on the air bag module. Doing so can result in air bag deployment, which can result in personal injury.

⁂ WARNING

Air bag modules with discolored or damaged trim covers must be replaced, not repainted.

⁂ WARNING

The restraint system diagnostic tool is for restraint system service only. Remove from vehicle prior to road use. Failure to remove could result in injury and possible violation of vehicle safety standards.

⁂ WARNING

To reduce the risk of personal injury, do not use any memory saver devices.

➡The air bag warning lamp illuminates when the RCM fuse is removed and the ignition switch is ON. This is normal operation and does not indicate a supplemental restraint system (SRS) fault.

➡After diagnosing or repairing an SRS, the restraint system diagnostic tools must be removed before operating the vehicle over the road.

➡The SRS must be fully operational and free of faults before releasing the vehicle to the customer.

➡Repair is made by installing a new part only. If the new part does not correct the condition, install the original part and perform the diagnostic procedure again.

2. Depower the system.
3. Remove the steering wheel back cover plugs.
4. Remove the driver air bag module bolts.
5. Remove the driver air bag module connectors.
6. Remove the horn switch connector.
7. Remove the driver air bag module.
8. Open the glove compartment.
9. Through the glove compartment opening, release the tab and disconnect the passenger air bag module electrical connector.
10. Remove the parts in the order indicated in the following illustration and table.
11. Remove the passenger air bag module nuts.
12. Reaching one hand into the glove box opening, push out on the passenger air bag module, releasing the clips at the top and remove the passenger air bag module from the instrument panel. Remove the passenger air bag module.

⁂ WARNING

Do not handle the passenger air bag module by grabbing the edges of the deployment door.

13. Remove the A-pillar trim panels and RH cowl side trim panel.

➡After removing the floor console, reconnect the battery to position the front seats back, then disconnect the battery.

14. Remove the floor console.
15. Remove the instrument panel side finish panel (LH).
16. Remove the steering column opening cover screws.
17. Remove the steering column opening cover
18. Remove the lower steering column cover.
19. Remove the steering column electrical connector.
20. Remove the bulkhead electrical connector.
21. Remove the body harness electrical connector.
22. Remove the parking brake release handle screws.
23. Remove the instrument panel electrical connector.
24. Remove the instrument panel electrical connector.
25. Remove the pinch bolt.

⁂ WARNING

To avoid damage to the clockspring, do not allow the steering column shaft to rotate while the intermediate shaft is disconnected.

26. Remove the adjustable pedal electrical connector
27. Remove the instrument panel center brace bolts.
28. Remove the instrument panel center brace nuts.
29. Remove the instrument panel center brace.
30. Remove the transmission range selector lever cable.
31. Remove the bulkhead electrical connector.
32. Remove the body harness electrical connector.

33. Remove the ground strap bolts.
34. Remove the ground straps.
35. Remove the inertia switch electrical connector.
36. Remove the instrument panel electrical connectors.
37. Remove the instrument panel center brace bolts (RH).
38. Remove the instrument panel center brace nuts (RH).
39. Remove the instrument panel center brace (RH).
40. Remove the instrument panel bolt.
41. Remove the instrument panel defroster grille.
42. Remove the sunload sensor electrical connector.

✻✻ WARNING

To avoid damaging the sun load sensor electrical connector, remove the grille just enough to remove the connector.

43. Remove the instrument panel bolts.
44. Remove the antenna lead-in cable.
45. Remove the RCM electrical connector.
46. Remove the ground strap bolt.
47. Remove the instrument panel bolts.

✻✻ WARNING

To avoid damage to the instrument panel, an assistant is required to support the panel before carrying out this step.

✻✻ WARNING

Before removing the instrument panel make sure all electrical connector wiring is free and not hindered.

48. Remove the instrument panel.
49. Drain the engine coolant.
50. Remove the engine appearance cover.
51. Remove the EGR vacuum regulator solenoid electrical connector.
52. Remove the vacuum hoses.
53. Remove the EGR vacuum regulator solenoid mounting nuts.
54. Remove the EGR vacuum regulator solenoid.
55. Remove the wire harness.
56. Remove the heater tube bracket nut.
57. Compress the clamps and disconnect the heater hoses.
58. Remove the A/C line bracket nut at the inner fender well.
59. Remove the A/C line bracket nut at the dash panel.
60. Remove the disconnect the evaporator outlet line fitting.
61. Remove the vacuum connector.
62. Remove the heater core and evaporator core housing nut.
63. Remove the heater core and evaporator core housing.
64. Remove the floor ducts.
65. Remove the housing brace.
66. Remove the heater tube cover.
67. Remove the heater tube seal.
68. Remove the heater core cover.
69. Remove the heater core.
70. To install, reverse the removal procedure.

✻✻ WARNING

The clockspring electrical connectors are unique and cannot be reversed when connected to the driver air bag module. Match the electrical connector key to the keyway in the driver air bag module. Do not force the electrical connectors into the driver air bag module.

71. Match the electrical connector key to the keyway in the driver air bag module and connect the electrical connectors.

✻✻ WARNING

The passenger air bag module nuts must be torqued in the sequence shown.

➡Make sure the battery negative cable is still disconnected before continuing with the installation portion of this procedure.

72. Install the components in the order indicated in the following illustration and table.
73. Install the passenger air bag module.
74. Connect the passenger air bag module electrical connector.
75. Close the glove compartment.
76. Repower the system.
77. Make sure the retaining clips are in place before positioning the passenger air bag module to the instrument panel.
78. Lubricate the refrigerant system with the correct amount of clean PAG oil.
79. Fill and bleed the engine cooling system.
80. Evacuate, leak test and charge the refrigerant system.

REAR

➡Lubricate the coolant hoses with plain water only if needed.

1. Position the vehicle on a hoist with the gear selector in NEUTRAL.
2. Using suitable tools, clamp-off the underbody heater hoses at the floorpan bracket.
3. Remove the clamps
4. Remove the auxiliary line floorpan bracket nuts.
5. Remove the line bracket screws.
6. Remove the line bracket.

➡The screw and line bracket are located inside the vehicle above the floorpan line bracket.

7. Remove the auxiliary harness electrical connector.
8. Remove the auxiliary housing bolts.
9. Remove the auxiliary housing nut.
10. Remove the blend door actuator electrical connector.
11. Remove the blend door actuator screws.
12. Remove the auxiliary blend door actuator.
13. Remove the temperature blend door actuator screws.
14. Remove the auxiliary temperature blend door actuator.
15. Remove the heater core cover.
16. Remove the clamps.
17. Remove the auxiliary heater core.
18. To install, reverse the removal procedure.
19. Fill the engine cooling system.

Cylinder Head

REMOVAL & INSTALLATION

4.0L SOHC Engine

➡If only one cylinder head is to be removed, only follow the procedures that apply. The following tools, or their equivalents are absolutely necessary to properly perform this procedure:

- Cam Chain Tensioner tool T97T-6K254-A
- Cam Gear Removal tool T97T-6256-F
- Cam Gear Torque adapter T97T-6256-G
- Camshaft Gear Positioning/Holding tool T97T-6256-B
- Camshaft Gear Positioning/Holding tool adapter T97T-6256-A
- Camshaft holding tool T97T-6256-C

- Crankshaft holding tool T97T-6303-A
- Camshaft holding tool adapter T97T-6256-D

1. Before servicing the vehicle, refer to the precautions in the beginning of this section.
2. Properly relieve the fuel system pressure.
3. Drain the cooling system.
4. Remove or disconnect the following:
   - Negative battery cable
   - Lower intake manifold
   - Fan blade and shroud
   - Valve cover
   - Roller followers, if equipped
   - Drive belt
   - Upper radiator hose and tube
   - Alternator electrical connectors
   - Alternator mounting bracket
   - Engine accessory bracket and move it aside
   - Camshaft Position (CMP) electrical connector
   - Crankshaft Position (CKP) sensor electrical connector
   - Engine Coolant Temperature (ECT) sensor electrical connector
   - Coil pack electrical connector
   - Exhaust Gas Recirculation (EGR) valve electrical connector
   - EGR valve bracket and move it aside
   - Heater hoses
   - Fuel injector electrical connectors
   - Water bypass hose
   - Thermostat housing
   - Spark plug wires
   - Fuel injection supply manifold
   - Fuel injectors
   - Crankcase vent separator spring
   - Oil dipstick housing
   - Exhaust manifold
   - Hydraulic chain tensioner
   - Cassette retaining bolt
   - Camshaft sprocket
   - Cylinder head and discard the gasket

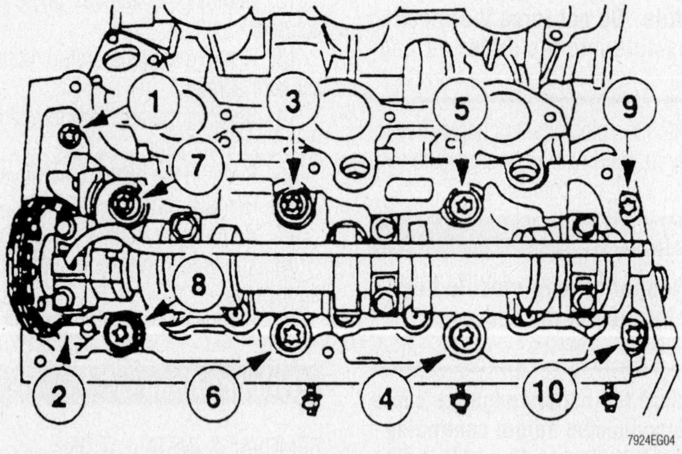

The correct cylinder head bolt loosening sequence must be used to prevent warpage—4.0L SOHC engine

Cylinder head bolt torque sequence—4.0L SOHC engine

To install:

5. Thoroughly clean all gasket mating surfaces. Remove all traces of old gasket material, oil, grease or dirt.
6. Insure that the rubber band is holding the right-hand chain to the cassette.
7. Install a new head gasket and the cylinder head.
8. Torque the new cylinder head bolts in sequence as follows:

2001
- Step 1: 28 ft. lbs. (38 Nm).
- Step 2: Plus 90 degrees.
- Step 3: Plus an additional 90 degrees.

2002
- Step 1: 24 ft. lbs. (32 Nm).
- Step 2: Plus 90 degrees.

9. Install or connect the following:
   - Camshaft sprocket in the cassette and make certain that the camshaft sprocket turns freely on the camshaft
   - Cassette retaining bolt. Torque the bolt to 89 inch lbs. (10 Nm).
   - Exhaust manifold
   - Oil level indicator tube. Torque the bolt to 18 ft. lbs. (25 Nm).
   - Crankcase vent separator and spring
   - Thermostat housing. Torque the bolts to 8 ft. lbs. (11 Nm).
   - Water bypass hose
   - Heater hoses
   - EGR bracket. Torque the bolt to 89 inch lbs. (10 Nm).
   - EGR tube. Torque the nut to 30 ft. lbs. (40 Nm).
   - ECT sensor electrical connector
   - Electrical harness retainer. Torque the bolt to 89 inch lbs. (10 Nm).
   - CKP and CMP electrical connectors
   - Accessory bracket. Torque the bolts to 31 ft. lbs. (42 Nm).
   - Alternator mounting bracket. Torque the bolts to 31 ft. lbs. (42 Nm).
   - Alternator and electrical connectors
   - Drive belt
   - Fan shroud
   - Roller followers
   - Valve cover
   - Lower intake manifold
   - Negative battery cable
10. Change the engine oil and filter.
11. Refill the cooling system.
12. Start the engine and check for leaks, repair if necessary.

4.6L SOHC Engine

1. Before servicing the vehicle, refer to the precautions in the beginning of this section.

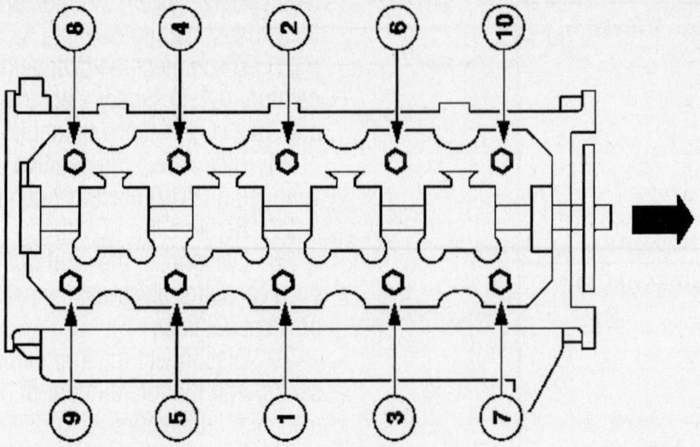

Left cylinder head bolt torque sequence—4.6L SOHC Engine

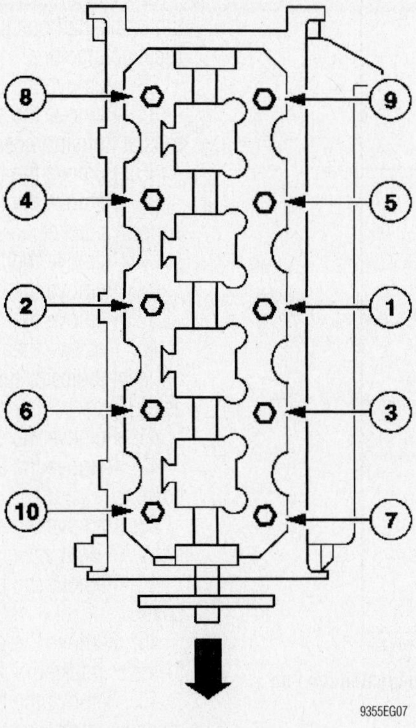

Right cylinder head bolt torque sequence—4.6L SOHC Engine

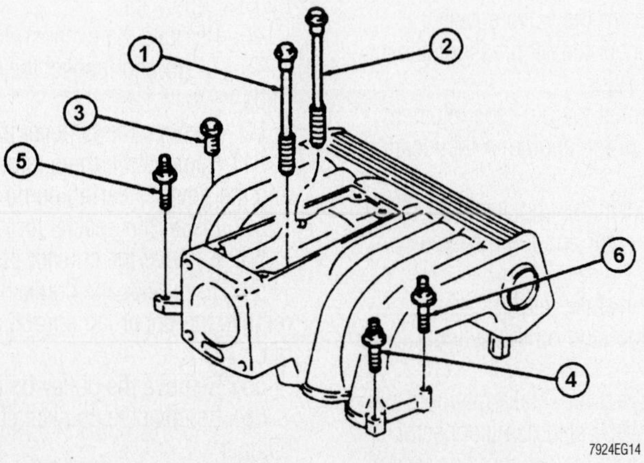

Tighten valve cover bolts in the sequence shown—4.6L SOHC Engine

2. Drain the cooling system.
3. Remove or disconnect the following:
   - Negative battery cable
   - Intake manifold
   - Timing chains
   - Exhaust manifolds
   - Water heater tube back and remove. Discard the O-ring seal.

➡The cylinder head bolts must be discarded with new bolts installed. They are tighten-to-yield designed and cannot be reused. Do not use metal scrapers, wire brushes, power abrasive discs or other abrasive means to clean the sealing surfaces. These tools cause scratches and gouges that make leak paths. Use a plastic scraping tool to remove all traces of the head gasket.

   - Bolts and the RH cylinder head
   - Bolts and the LH cylinder head

4. Installation is the reverse of removal. Observe the following notes and torques:
   a. Install the cylinder head on the head gasket and loosely install the bolts.
   b. Make sure to tighten the bolts in sequence in six stages.
      - Step 1: Tighten to 40 Nm (30 lb-ft).
      - Step 2: Tighten an additional 85-95 degrees.
      - Step 3: Back out all bolts one full turn (360 degrees).
      - Step 4: Tighten to 40 Nm (30 lb-ft).
      - Step 5: Tighten an additional 85-95 degrees.
      - Step 6: Tighten an additional 85-95 degrees.
   c. Install a new O-ring seal and position the water heater tube forward. Lubricate the O-ring seal with premium engine coolant.

4.6L DOHC Engine

1. Remove the engine.
2. Remove the flexplate and the spacer plate.
3. Mount the engine on a suitable engine stand.
4. Remove the two bolts and the pin-type retainer from the power steering reservoir bracket.
5. Remove the power steering reservoir and bracket.
6. Disconnect the battery cable harness ground.
7. Disconnect the electrical connector and the battery cable from the alternator.
8. Disconnect the two retainers from the engine front cover studs.
9. Disconnect the two retainers from the A/C compressor.

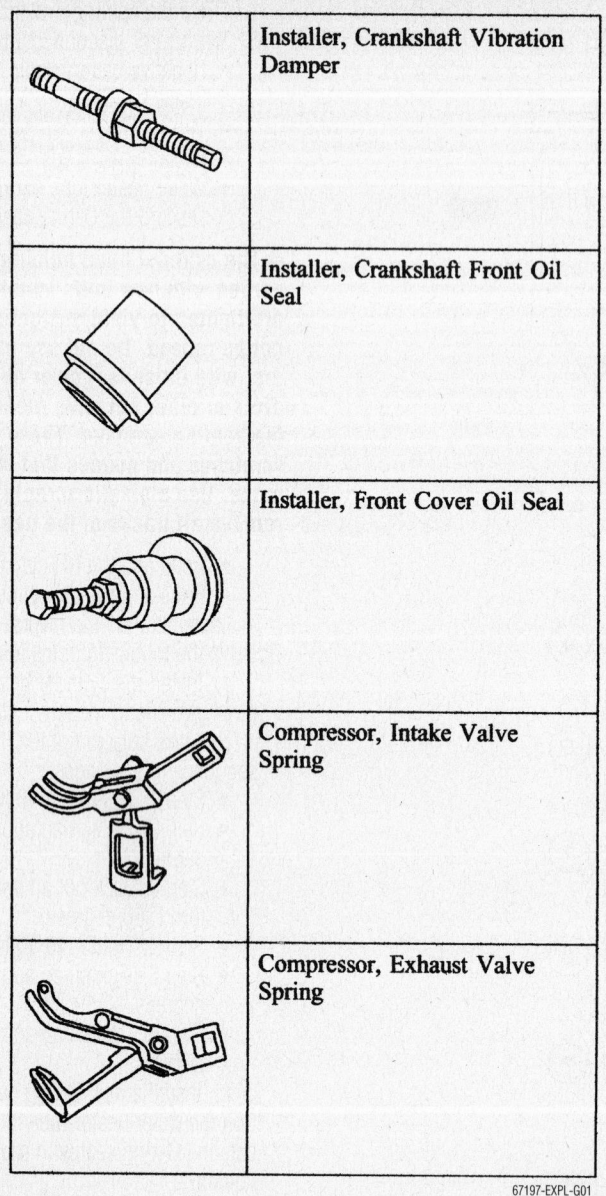

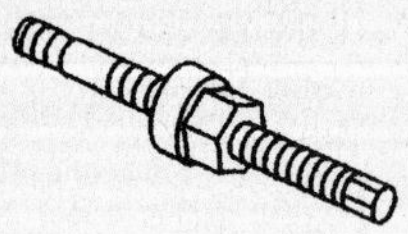	Installer, Crankshaft Vibration Damper
	Installer, Crankshaft Front Oil Seal
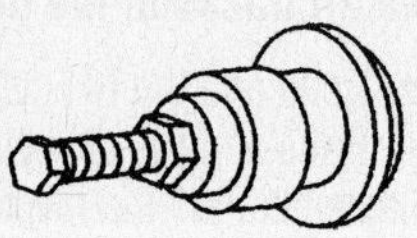	Installer, Front Cover Oil Seal
	Compressor, Intake Valve Spring
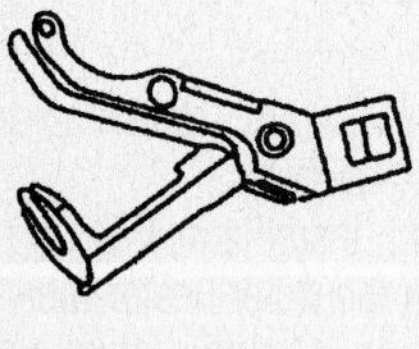	Compressor, Exhaust Valve Spring

67197-EXPL-G01

These tools, or their equivalents, are necessary for cylinder head removal and installation on the 4.6L DOHC engine

10. Remove the bolt and the battery cable harness from the right engine mount.
11. Disconnect the electrical connector and the vacuum hose from the fuel pressure sensor.
12. Disconnect the vacuum harness from the intake manifold and remove the harness.
13. Remove the left and right coil covers.
14. Disconnect the right and left ignition coil electrical connectors.
15. Disconnect the radio interference capacitor electrical connector.
16. Disconnect the eight fuel injector electrical connectors.
17. Separate the fuel charging wiring harness retainers from the fuel injection supply manifold.
18. Disconnect the camshaft position (CMP) sensor electrical connector and the two retainers from the valve cover.
19. Disconnect the oil pressure sensor electrical connector.
20. Disconnect the pin-type retainers from the back of the engine in the locations shown.
21. Disconnect the knock sensor (KS) electrical connector and the pin-type retainer.
22. Disconnect the wiring harness retainer from the stud on the cylinder head.
23. Disconnect the intake manifold runner control (IMRC) electrical connector.
24. Disconnect the exhaust gas recirculation (EGR) system module tube nut from the exhaust manifold.
25. Disconnect the cylinder head temperature (CHT) sensor electrical connector and the two pin-type retainers.
26. Disconnect the positive crankcase ventilation (PCV) electrical connector.
27. Remove the PCV tube.
28. Disconnect the right radio interference capacitor electrical connector and the pin-type retainer.
29. Disconnect the throttle position (TP) sensor and the idle air control (IAC) sensor electrical connectors.
30. Disconnect the two retainers from the right valve cover studs.
31. Disconnect the engine coolant temperature (ECT) sensor electrical connector.
32. Disconnect the A/C compressor and the crankshaft position (CKP) sensor electrical connectors.
33. Remove the upper radiator hose.
34. Remove the coolant bypass-to-thermostat housing hose.
35. Remove the accessory drive belt.
36. Remove the bolts and the generator support bracket.
37. Remove the bolts and the generator.
38. Remove the crossover tube.
39. Remove the eight ignition coils.
40. Remove the nut and the bolt from the oil level indicator tube and position the tube aside.
41. Remove the engine lifting eyes.
42. Remove the bolts and the left valve cover.
43. Remove the coolant hose retainers from the right valve cover.
44. Remove the bolts and the right valve cover.
45. Remove the retainers and the upper-to-lower intake bracket.
46. Remove the bolts in the sequence shown and remove the upper and lower intake.
47. Remove the bolt and the coolant bypass tube.
48. Remove the exhaust manifolds.
49. Clean and inspect the exhaust manifolds.
50. Remove the oil level indicator tube.
51. Remove the three remaining bolts and the power steering pump.
52. Install the special tool.
53. Remove the coolant pump pulley.
54. To remove the crankshaft pulley bolt, remove the upper fan shroud and electric fan.
55. Remove the pulley bolt.
56. Remove the crankshaft vibration damper.
57. Remove the crankshaft front seal.

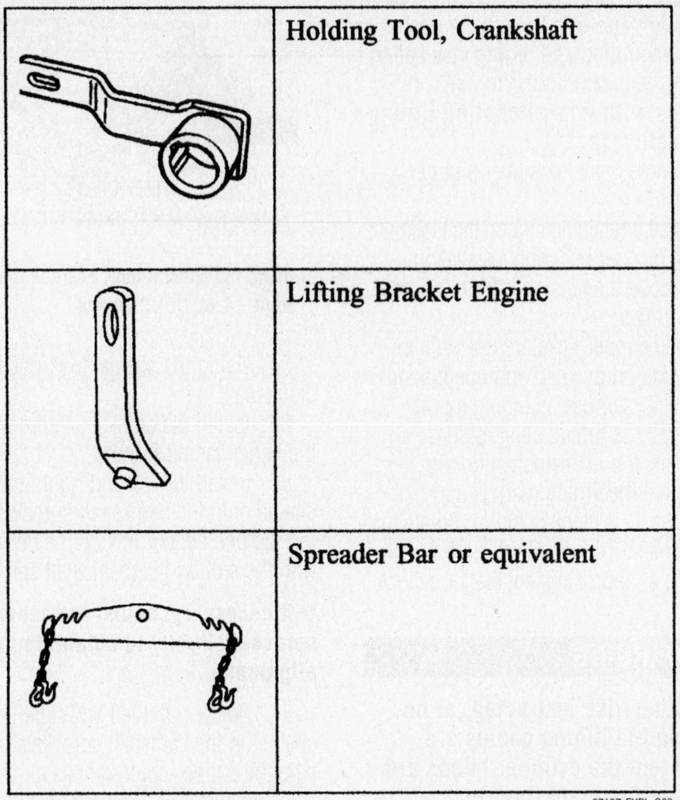

67197-EXPL-G02

These tools, or their equivalents, are necessary for cylinder head removal and installation on the 4.6L DOHC engine

58. Remove the bolt and the belt idler pulley.
59. Remove the four front oil pan-to-timing cover bolts.
60. Remove the engine front cover fasteners in the sequence shown.
61. Remove the crankshaft position sensor pulse wheel.

➡Mark the roller follower locations. Reused roller followers must be installed in their original locations.

62. Position the piston of the cylinder in which the roller followers are being removed at the bottom of the stroke and camshaft lobe at the base circle.
63. Compress the intake valve spring and remove the roller follower.
64. Compress the exhaust valve spring and remove the roller follower.
65. Repeat the above steps to remove all the roller followers.
66. Remove the bolts and the right and left timing chain tensioners.
67. Remove the right and left timing chain tensioner arms.
68. Remove the right and the left timing chains and the crankshaft sprocket.
69. Remove the right and left timing chain guides.
70. Remove the right and left cylinder heads. Discard the cylinder head bolts. Discard the cylinder head gaskets.
71. Clean and inspect the cylinder heads.

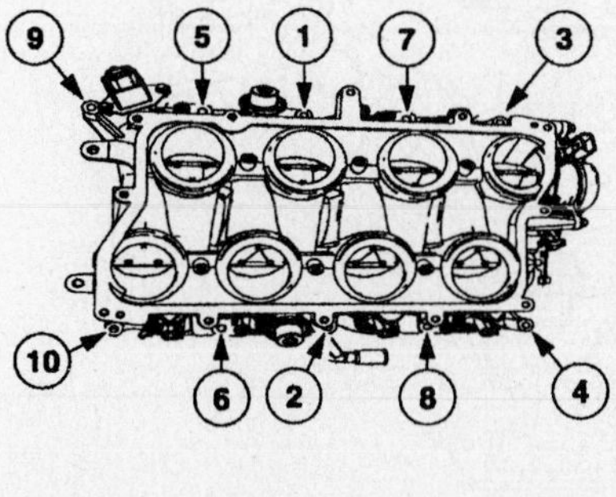

67197-EXPL-G03

Intake manifold bolt removal sequence—4.6L DOHC engine

To install:

➡Before cylinder head installation, use metal surface cleaner and a suitable plastic or wooden scraper to clean the sealing surfaces. All sealing surfaces must be clean. Make sure coolant and oil passages are clear.

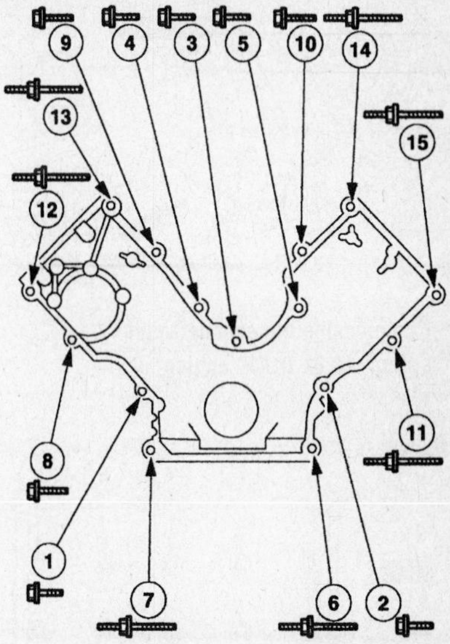

1 Bolt, Hex Flange Head Pilot, M8 x 1.25 x 53
2 Bolt, Hex Flange Head Pilot, M8 x 1.25 x 53
3 Bolt, Hex Flange Head Pilot, M8 x 1.25 x 53
4 Bolt, Hex Flange Head Pilot, M8 x 1.25 x 53
5 Bolt, Hex Flange Head Pilot, M8 x 1.25 x 53
6 Stud, Hex Shldr Pilot, M8 x 1.25 x 50 — M6 x 1 x 10
7 Stud and Washer, Hex Head Pilot, M8 x 1.25 x 60 — M6 x 1 x 26
8 Bolt, Hex Flange Head Pilot, M8 x 1.25 x 53
9 Bolt, Hex Flange Head Pilot, M8 x 1.25 x 53
10 Bolt, Hex Flange Head Pilot, M8 x 1.25 x 53
11 Stud, Hex Shldr Pilot, M8 x 1.25 x 65 — M8 x 1.25 x 26
12 Stud, Hex Head Pilot, M8 x 1.25 x 65 — M8 x 1.25 x 16
13 Stud, Hex Shldr Pilot, M8 x 1.25 x 65 — M8 x 1.25 x 26
14 Stud, Hex Shldr Pilot, M8 x 1.25 x 65 — M8 x 1.25 x 26
15 Stud, Hex Shldr Pilot, M8 x 1.25 x 65 — M8 x 1.25 x 26

67197-EXPL-G04

Front cover bolt removal sequence—4.6L DOHC engine

All vehicles

72. Install new cylinder head gaskets.

➡Lubricate the new bolt heads and threads. Use clean engine oil.

73. Install the left and right cylinder

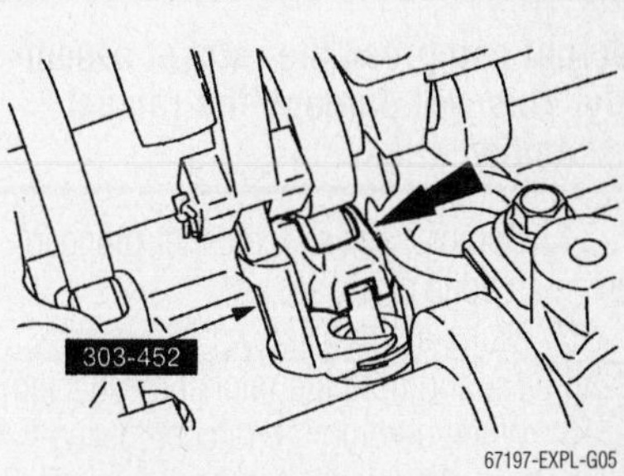

67197-EXPL-G05

Compressing the intake valve spring—4.6L DOHC engine

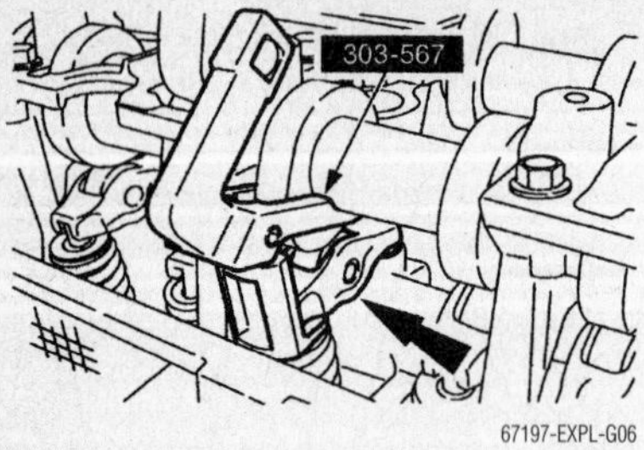

Compressing the exhaust valve spring—4.6L DOHC engine

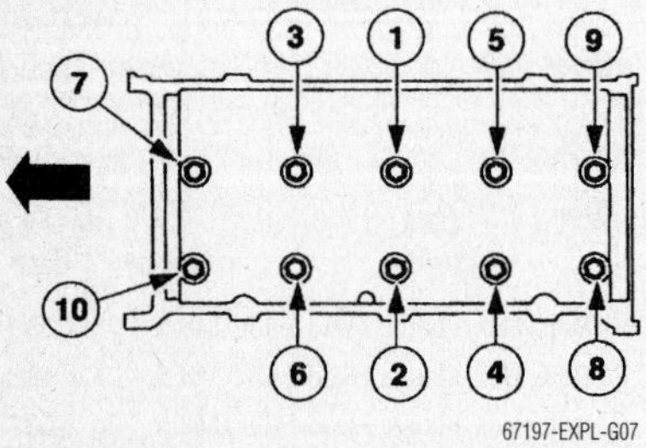

Cylinder head bolt torque sequence—4.6L DOHC engine

heads and tighten the bolts in six stages in the sequence shown.

- Stage 1: Tighten to 40 Nm (30 ft. lbs.).
- Stage 2: Tighten an additional 90 degrees.
- Stage 3: Loosen the bolts a minimum of one full turn.
- Stage 4: Tighten to 40 Nm (30 ft. lbs.).
- Stage 5: Tighten an additional 90 degrees.
- Stage 6: Tighten an additional 90 degrees.

Engines with ratcheting timing chain tensioners

⁂ WARNING

Timing chain procedures must be followed exactly or damage to valves and pistons will result.

⁂ WARNING

Do not compress the ratchet assembly. This will damage the ratchet assembly.

74. Compress each tensioner plunger, using an edge of a vise.
75. Using a small screwdriver or pick, push back and hold the ratchet mechanism.
76. While holding the ratchet mechanism, push the ratchet arm back into the tensioner housing.
77. Install a paper clip into the hole of each tensioner housing to hold the ratchet assembly and plunger in during installation. Remove the tensioner from the vise.

Engines with non-ratcheting timing chain tensioners

78. Compress the tensioner plunger, using a vise.
79. Install a retaining clip on the tensioner to hold the plunger in during installation.
80. Remove the tensioner from the vise.

All engines

81. If the copper links are not visible, mark one link on one end and one link on the other end, and use as timing marks.
82. Install the timing chain guides.
83. Rotate the left camshaft timing sprocket until the timing mark is approximately at the 12 o'clock position. Rotate the right camshaft timing sprocket until the timing mark is at approximately the 11 o'clock position.

⁂ WARNING

Unless otherwise instructed, at no time when the timing chains are removed and the cylinder heads are installed is the crankshaft or camshaft to be rotated. Severe piston and valve damage will occur.

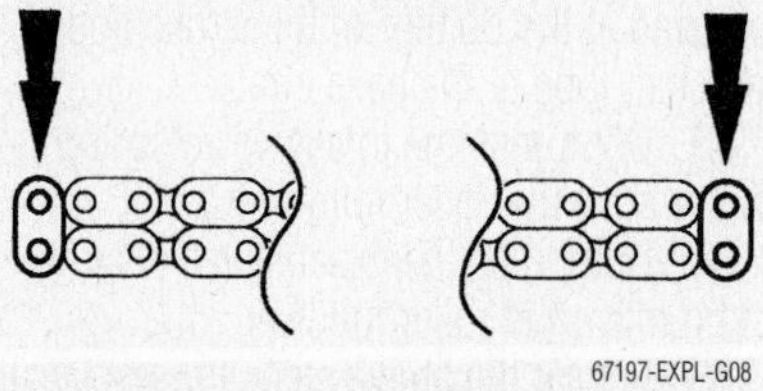

If the copper links are not visible, mark one link on one end and one link on the other end, and use as timing marks—4.6L DOHC engine

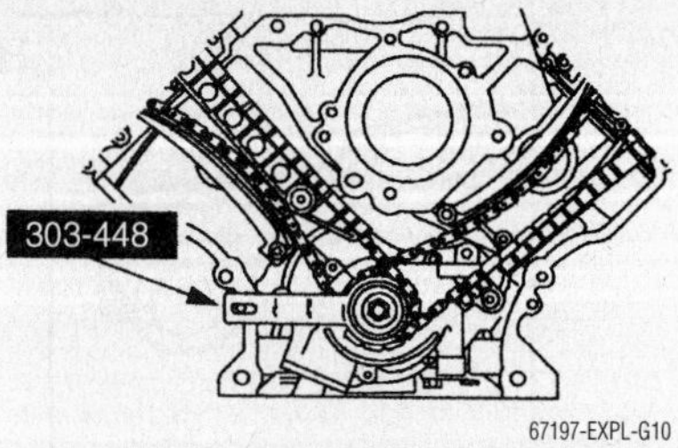

Using the special tool, position the crankshaft—4.6L DOHC engine

84. Using the special tool, position the crankshaft.
85. Install the crankshaft sprocket with the flange facing forward.
86. Install the left and right timing chains onto the crankshaft sprocket, aligning the one copper link on the timing chain with the slot on the crankshaft sprocket.

➡If necessary, adjust the camshaft sprocket slightly to obtain timing mark alignment.

87. Position the left and right timing chains on the camshaft sprockets. Make sure the copper-colored links align with the camshaft sprocket timing marks.
88. Position the left and right timing chain tensioner arms on the dowel pins. Position the timing chain tensioner assemblies, and install the bolts.

Engines with ratcheting timing chain tensioners

89. Remove the retaining clip from the timing chain tensioners.

Engines with non-ratcheting timing chain tensioners

90. Remove the retaining clips from the timing chain tensioners.

All engines

91. As a post-check, verify correct alignment of all timing marks.

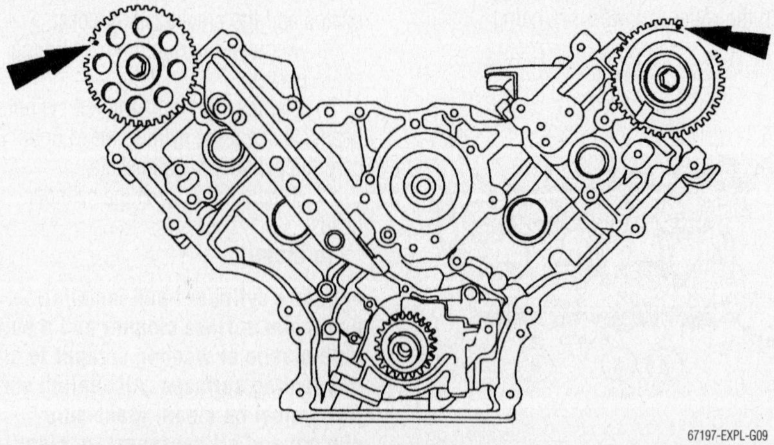

Rotate the left camshaft sprocket so that the timing mark is at the 12:00 o'clock position. Rotate the right camshaft sprocket so that the timing mark is at the 11:00 o'clock position.

92. Position the crankshaft sensor ring on the crankshaft.

➡If the engine front cover is not secured within four minutes, the sealant must be removed and the sealing area cleaned with metal surface cleaner. Allow to dry until there is no sign of wetness, or four minutes, whichever is longer. Failure to follow this procedure can result in future oil leakage.

93. Apply silicone gasket and sealant in the locations shown.

94. Install the engine front cover and tighten the bolts in the sequence shown.

95. Install the four oil pan bolts and tighten in the sequence shown.

- Stage 1: Tighten to 2 Nm (18 inch lbs.).
- Stage 2: Tighten to 20 Nm (15 ft. lbs.).
- Stage 3: Tighten an additional 60 degrees.

⁂ WARNING

If reusing the roller followers, install them in their original locations.

96. Position the piston of the cylinder in which the roller followers are being installed at the bottom of the stroke and the camshaft lobe at the base circle.

97. Compress the intake valve spring and install the roller follower.

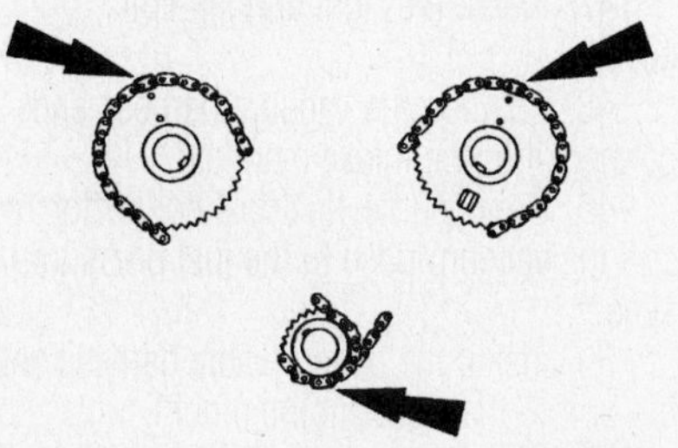

Verify timing mark alignment—4.6L DOHC engine

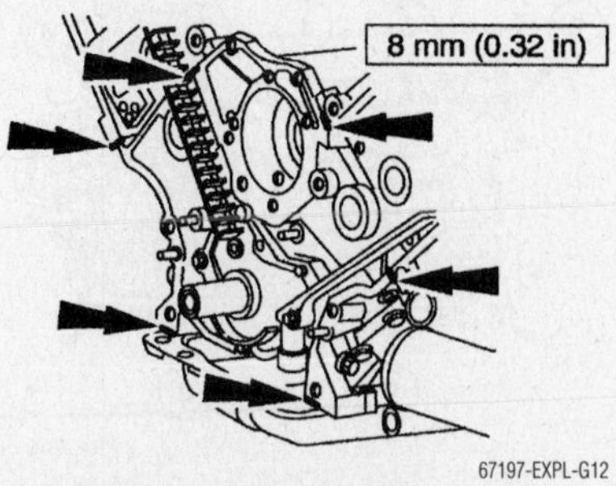

Apply silicone gasket and sealant in the locations shown—4.6L DOHC engine

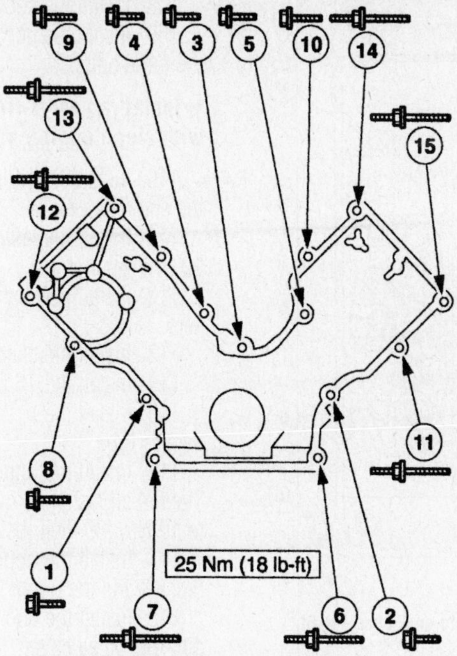

1 Bolt, Hex Flange Head Pilot, M8 x 1.25 x 53
2 Bolt, Hex Flange Head Pilot, M8 x 1.25 x 53
3 Bolt, Hex Flange Head Pilot, M8 x 1.25 x 53
4 Bolt, Hex Flange Head Pilot, M8 x 1.25 x 53
5 Bolt, Hex Flange Head Pilot, M8 x 1.25 x 53
6 Stud, Hex Shldr Pilot, M8 x 1.25 x 50 — M6 x 1 x 10
7 Stud and Washer, Hex Head Pilot, M8 x 1.25 x 60 — M6 x 1 x 26
8 Bolt, Hex Flange Head Pilot, M8 x 1.25 x 53
9 Bolt, Hex Flange Head Pilot, M8 x 1.25 x 53
10 Bolt, Hex Flange Head Pilot, M8 x 1.25 x 53
11 Stud, Hex Shldr Pilot, M8 x 1.25 x 65 — M8 x 1.25 x 26
12 Stud, Hex Head Pilot, M8 x 1.25 x 65 — M8 x 1.25 x 16
13 Stud, Hex Shldr Pilot, M8 x 1.25 x 65 — M8 x 1.25 x 26
14 Stud, Hex Shldr Pilot, M8 x 1.25 x 65 — M8 x 1.25 x 26
15 Stud, Hex Shldr Pilot, M8 x 1.25 x 65 — M8 x 1.25 x 26

67197-EXPL-G13

Front cover bolt torque sequence—4.6L DOHC engine

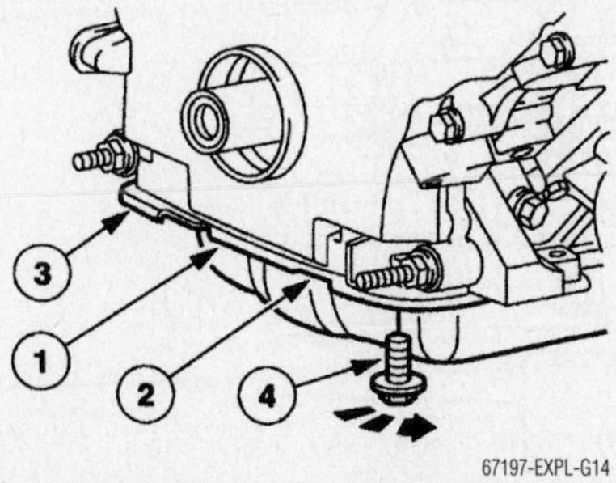

Install the four oil pan bolts and tighten in the sequence shown—4.6L DOHC engine

98. Compress the exhaust valve spring and install the roller follower.

99. Repeat the above steps to install all of the roller followers.

➡If the valve cover is not secured within four minutes, the sealant must be removed and the sealing area cleaned with metal surface cleaner. Allow to dry until there is no sign of wetness, or four minutes, whichever is longer. Failure to follow this procedure can result in future oil leakage.

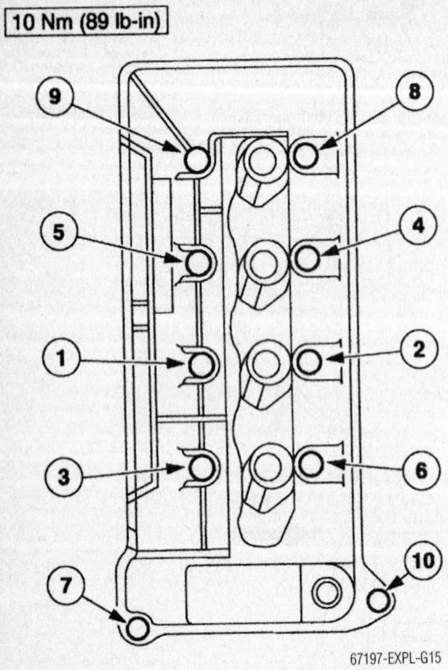

Valve cover bolt torque sequence—4.6L DOHC engine

100.Apply silicone gasket and sealant in the locations shown.

101. Install the valve covers and tighten the bolts in the sequence shown.

102. Install the belt idler pulley.

103. Lubricate the front oil seal and the engine front cover with clean engine oil.

104. Install the front oil seal.

105. Apply silicone gasket and sealant to the Woodruff key on the crankshaft pulley.

106. Install the crankshaft pulley.

107. Install the crankshaft pulley bolt and tighten the bolt in four stages.

- Stage 1: Tighten to 90 Nm (66 ft. lbs.).
- Stage 2: Loosen the bolt one full turn.
- Stage 3: Tighten to 50 Nm (37 ft. lbs.).
- Stage 4: Tighten an additional 90 degrees.

108. Install the coolant pump pulley.

109. Install the power steering pump and the three bolts.

➡Install a new O-ring and lubricate with clean engine oil.

110. Install the oil level indicator tube in the engine block.

111. Install the fasteners and tighten to 10 Nm (89 inch lbs.).

112. Install new exhaust manifold gaskets.

113. Install the exhaust manifolds.

114. Install the 16 nuts.

115. Install the bolt and the coolant bypass tube.

116. Install the upper and lower intake. Tighten all fasteners in the sequence shown to 10 Nm (89 inch lbs.).

117. Install the upper-to-lower intake bracket and the nuts.

118. Install the coolant hose retainers on the right valve cover.

119. Install the engine lifting eyes.

120. Install the eight ignition coils.

121. Install the crossover tube and the nuts. Connect the coolant hose.

122. Install the generator and the bolts.

123. Install the generator support bracket and the bolts.

124. Install the accessory drive belt.

125. Install the coolant bypass-to-thermostat housing hose.

126. Install the upper radiator hose.

127. Position the engine wiring harness on the engine.

128. Connect the A/C compressor and the crankshaft position (CKP) sensor electrical connectors.

129. Connect the engine coolant temperature (ECT) sensor electrical connector.

130. Connect the two harness retainers to the right valve cover studs.

131. Connect the throttle position (TP) sensor and the idle air control (IAC) sensor electrical connectors.

132. Connect the right radio interference capacitor electrical connector and the pin-type retainer.

133. Install the positive crankcase ventilation (PCV) tube.

134. Connect the PCV electrical connector.

135. Connect the cylinder head temperature (CHT) sensor electrical connector and the two pin-type retainers.

136. Connect the exhaust gas recirculation (EGR) system module tube to the exhaust manifold.

137. Connect the intake manifold runner control (IMRC) electrical connector.

138. Connect the wiring harness retainer to the stud on the cylinder head.

139. Connect the KS electrical connector and the pin-type retainer.

140. Connect the pin-type retainers to the back of the engine in the locations shown.

141. Connect the oil pressure sensor electrical connector.

142. Connect the camshaft position (CMP) sensor electrical connector and the two retainers to the valve cover.

143. Install the fuel charging wiring harness retainers to the fuel injection supply manifold.

144. Connect the eight fuel injector electrical connectors.

145. Connect the radio interference capacitor electrical connector.

146. Connect the eight ignition coil electrical connectors.

147. Install the right and left coil covers.

148. Position the vacuum harness and connect it to the intake manifold.

149. Connect the electrical connector and the vacuum hose to the fuel pressure sensor.

150. Install the battery cable harness and the bolt to the right engine mount.

151. Connect the two retainers to the A/C compressor.

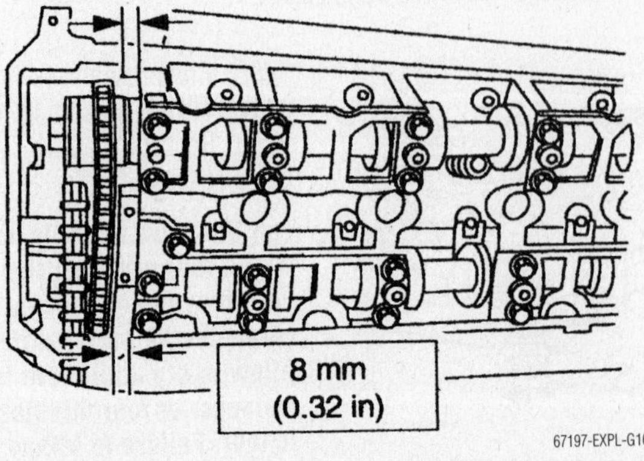

Apply silicone gasket and sealant in the locations shown—4.6L DOHC engine

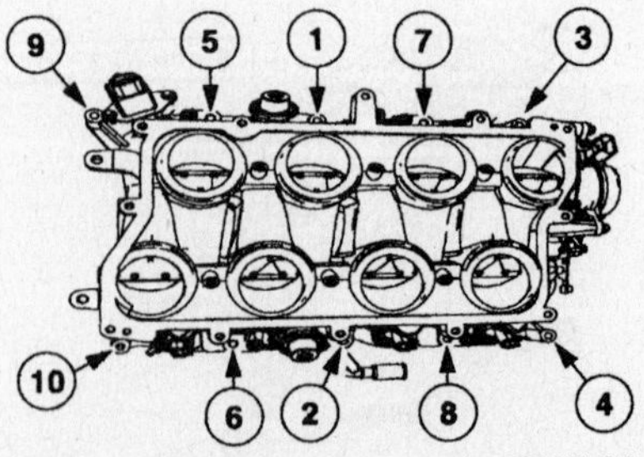

Intake manifold bolt torque sequence—4.6L DOHC engine

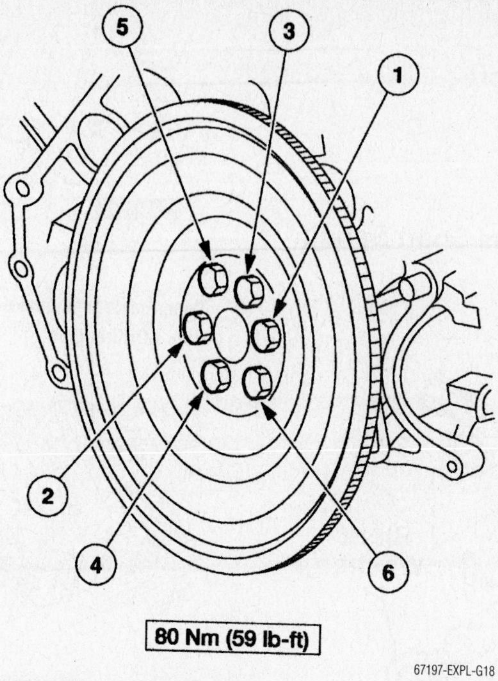

Flexplate bolt torque sequence—4.6L DOHC engine

152. Connect the two retainers to the engine front cover studs.

153. Connect the electrical connector and the battery cable to the alternator.

154. Connect the battery cable harness ground.

155. Position the power steering reservoir and install the two bolts and the pin-type retainer.

156. Install the remaining bolt and the hose to the power steering reservoir.

157. Remove the engine from the stand.

158. Install the spacer plate and the flexplate.

159. Install the engine.

5.0L Engine

1. Before servicing the vehicle, refer to the precautions in the beginning of this section.
2. Drain the cooling system.
3. Remove or disconnect the following:
   - Negative battery cable
   - Lower intake manifold
   - Valve cover
   - Matchmark the rocker arms and push rods to ease installation and remove the rocker arm fulcrums
   - Rocker arms and fulcrum guides
   - Exhaust manifold
   - A/C compressor and power steering bracket, left side cylinder head only
   - Alternator electrical connectors, right side cylinder head only
   - Transmission oil fill tube from the rear of the right side cylinder head
   - Push rods and make certain to matchmark them
   - Cylinder head and discard the gasket

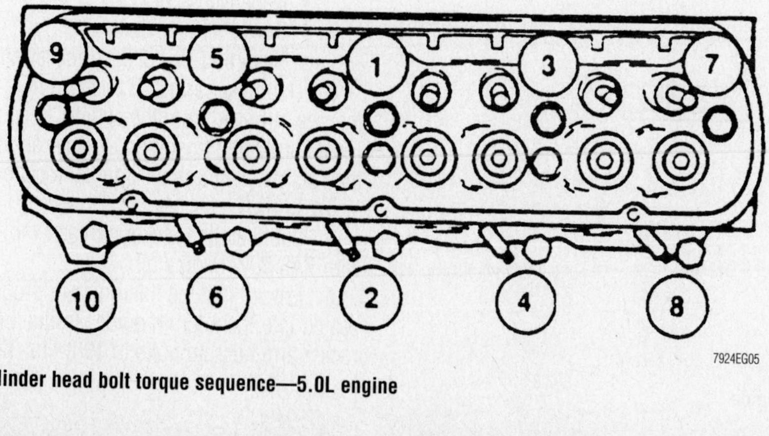

Cylinder head bolt torque sequence—5.0L engine

To install:

4. Thoroughly clean all the gasket mating surfaces.
5. Position a new cylinder head gasket to the engine block, then install the cylinder head.
6. Install new cylinder head bolts, and torque in 3 steps:
   a. Step 1: 30 ft. lbs. (40 Nm).
   b. Step 2: 50 ft. lbs. (68 Nm).
   c. Step 3: Plus an additional 90 degrees.
7. Install or connect the following:
   - Lubricate and install the push rods in their original positions
   - Rocker arms and fulcrum guides
   - Rocker arm fulcrums. Torque the bolts to 25 ft. lbs. (34 Nm).
   - Valve cover
   - Transmission oil fill tube. Torque the bolt to 25 ft. lbs. (34 Nm).
   - Alternator bracket, if removed. Torque the bolts to 48 ft. lbs. (65 Nm).
   - Alternator electrical connectors, if removed
   - Exhaust manifold
   - Lower intake manifold
   - Negative battery cable
8. Fill the cooling system
9. Fill the engine with clean oil and replace the filter.
10. Start the engine and check for leaks, repair if necessary.

Rocker Arms/Roller Followers

REMOVAL & INSTALLATION

4.0L SOHC Engines

➡A special tool is required to compress the valve spring.

1. Before servicing the vehicle, refer to the precautions in the beginning of this section.
2. Disconnect the negative battery cable.
3. Remove the valve cover.
4. Rotate the camshaft so that the base circle of the cam is against the cam follower you intend to remove.

➡If removing more than one cam follower, label them so they can be returned to their original position.

5. Using special tool T97T-6565-A depress the valve spring. Slide the cam follower over the lash adjuster and out from under the camshaft.

To install:

6. Compress the valve spring and slide the roller follower into position.

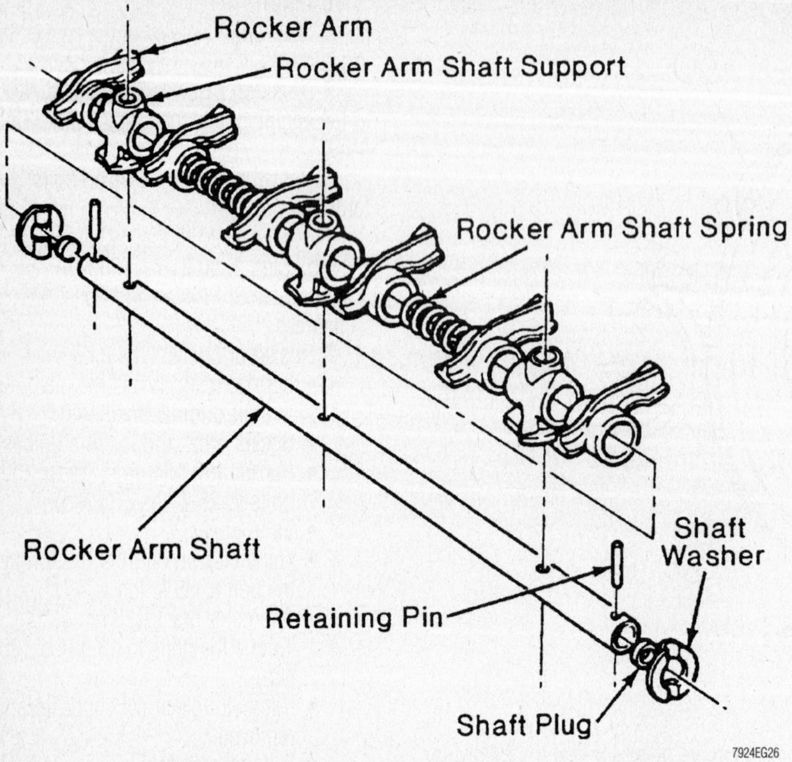

Rocker arm and shaft assembly—4.0L SOHC engine

7. Release the tension from the spring.
8. Install the valve cover and connect the negative battery cable.

4.6L SOHC Engine

1. Before servicing the vehicle, refer to the precautions in the beginning of this section.
2. Remove the valve cover.
3. Position the piston of the cylinder being repaired at the bottom of the stroke.
4. Install the special tool between the valve spring coils to prevent valve stem seal damage.

➡The roller followers are positional. Mark the followers for installation in their original locations.

- Compress the valve springs and remove the camshaft roller followers.

5. Installation is the reverse of removal.

➡There are several kinds of bolts used in fastening the valve cover. The bolts must be installed in their original locations.

4.6L DOHC Engine

➡This step is necessary for camshaft removal and installation only.

1. Remove the engine front cover, timing gears, chain and tensioners.
2. Remove the valve covers.

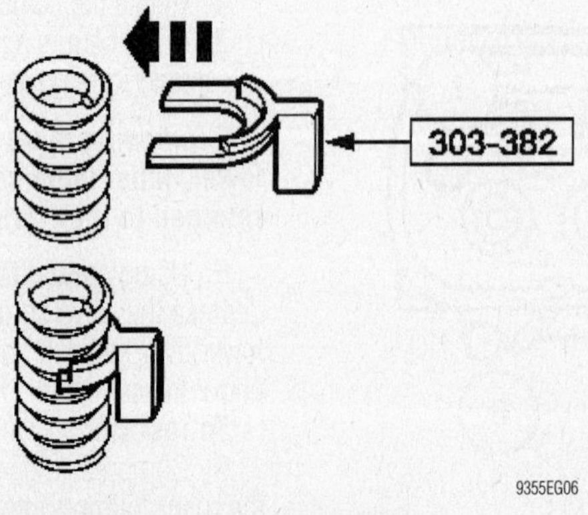

Install this tool to prevent seal damage—4.6L SOHC Engine

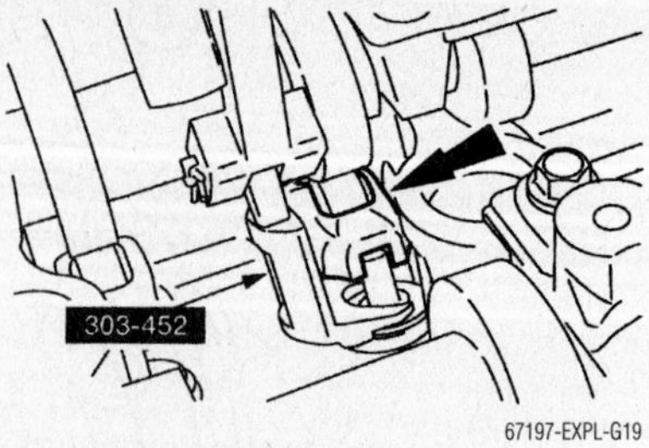

Compressing the intake valve spring—4.6L DOHC engine

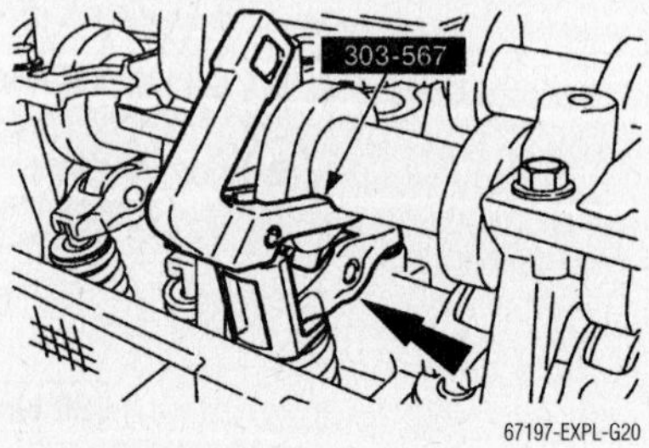

Compressing the exhaust valve spring—4.6L DOHC engine

➡Position the piston of the cylinder to be serviced at the bottom of the stroke and the camshaft at base circle.

3. Using the special tool, partially compress the intake valve spring and remove the camshaft roller follower.
4. Using the special tool, partially compress the exhaust valve spring and remove the camshaft roller follower.

➡Mark the location of the hydraulic lash adjusters to make sure they are assembled in their original position.

5. Remove the hydraulic lash adjusters.
6. Installation is the reverse of removal.

5.0L Engines

1. Before servicing the vehicle, refer to the precautions in the beginning of this section.
2. Remove or disconnect the following:
   - Negative battery cable
   - Rocker arm covers
   - Retaining bolt at each rocker arm
3. The rocker arm and pushrod may then be removed from the engine. Keep all rocker arms and pushrods in order so they may be installed in their original locations.

To install:

4. Lubricate the rocker arm assemblies with SAE 50W engine oil.
5. Ensure that the fulcrums are properly seated into the fulcrum guide. Torque the rocker arm fulcrum bolts to 19 ft. lbs. (26 Nm).

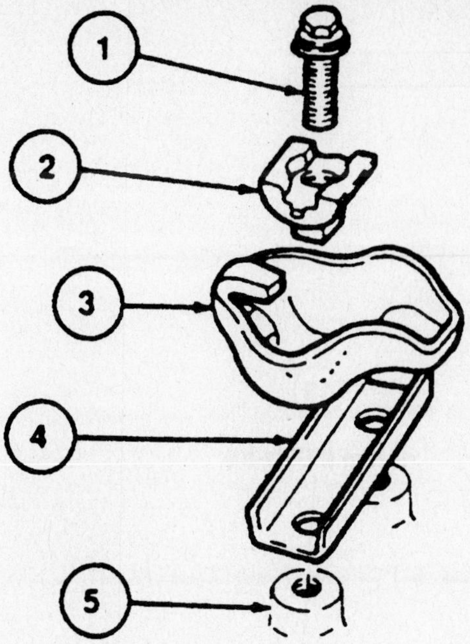

Exploded view of the rocker arm assembly. Notice the fulcrum guide between the pedestals used for extra stability—5.0L engine

6. Install the rocker arm covers and connect the negative battery cable.

Intake Manifold

REMOVAL & INSTALLATION

4.0L SOHC Engine

2001

1. Before servicing the vehicle, refer to the precautions in the beginning of this section.
2. Remove or disconnect the following:
   - Negative battery cable
   - Air cleaner-to-intake tube
   - Accelerator splash shield
   - Accelerator and, if equipped with cruise control, speed control cables from the throttle control cam
   - Accelerator cable retaining bracket from the upper intake manifold
   - Label and disengage all vacuum and electrical connections on the intake manifold.
   - Upper intake manifold attaching bolts
   - Lift up on the manifold and remove both fuel Vapor Management Valve (VMV) hoses
   - Intake manifold and discard the gasket

To install:

➡Ford does not specify a sequence for either upper or lower intake manifolds, but it is recommended that you start tightening in the middle and work your way out to the ends. Repeat the tightening sequence several times until the bolts will no longer turn at the specified torque.

3. Install the lower intake manifold. Torque the bolts to 13 ft. lbs. (18 Nm) for 2001.
4. Position the upper manifold on the lower manifold.
5. Install or connect the following:
   - Attach both VMV hoses to the manifold
   - Upper manifold attaching bolts. Torque the bolts to 62 inch lbs. (7 Nm).
   - Attach any vacuum and electrical connections that were removed
   - Accelerator cable bracket to the intake and the cable (or cables if equipped with cruise control) to the throttle cam
   - Accelerator splash shield
   - Air cleaner-to-intake supply tube
   - Negative battery cable
6. Start the vehicle and check for leaks, repair if necessary.

2002

1. Before servicing the vehicle, refer to the precautions in the beginning of this section.
2. Remove or disconnect the following:
   - Negative battery cable
   - Splash shield
   - Air cleaner outlet pipe
   - Accelerator and speed control cables from the throttle body
   - Accelerator cable and the speed control cable from the bracket. Position the cables away from the intake manifold.
   - All the vacuum hoses
   - KS electrical connector
   - IAC valve electrical connector
   - TP sensor electrical connector
   - EGR transducer electrical connector
   - Wiring anchor
   - EGR tube from the EGR valve
   - Wiring harness pin-type retainer
   - Spark plug wires
   - Intake manifold retainers, then slightly reposition the manifold to access the EGR vacuum regulator (EVR) solenoid.
   - EVR solenoid vacuum and electrical connections
   - Intake manifold
   - To install, reverse the removal procedure. Torque the bolts to 89 inch lbs. (10 Nm). There is no particular sequence.

2003–04

1. Disconnect the battery ground cable.
2. Remove the air cleaner outlet pipe.
3. Remove the positive crankcase ventilation (PCV) tube
4. Remove the brake booster vacuum supply hose
5. Remove the engine main vacuum harness-to-intake manifold fitting
6. Remove the evaporative emissions (EVAP) return tube
7. Remove the EVAP tube pin-type retainer
8. Remove the exhaust gas recirculation (EGR) system module electrical connector
9. Remove the wiring harness pin-type retainer
10. Remove the engine main vacuum harness-to-EGR system module fitting
11. Remove the EGR tube fitting
12. Remove the throttle position (TP) sensor electrical connector
13. Remove the electronic throttle body electrical connector
14. Remove the electronic throttle body-to-intake manifold bolt
15. Remove the intake manifold mounting bolts
16. Remove the intake manifold
17. Remove the intake manifold gaskets
18. To install, reverse the removal procedure.

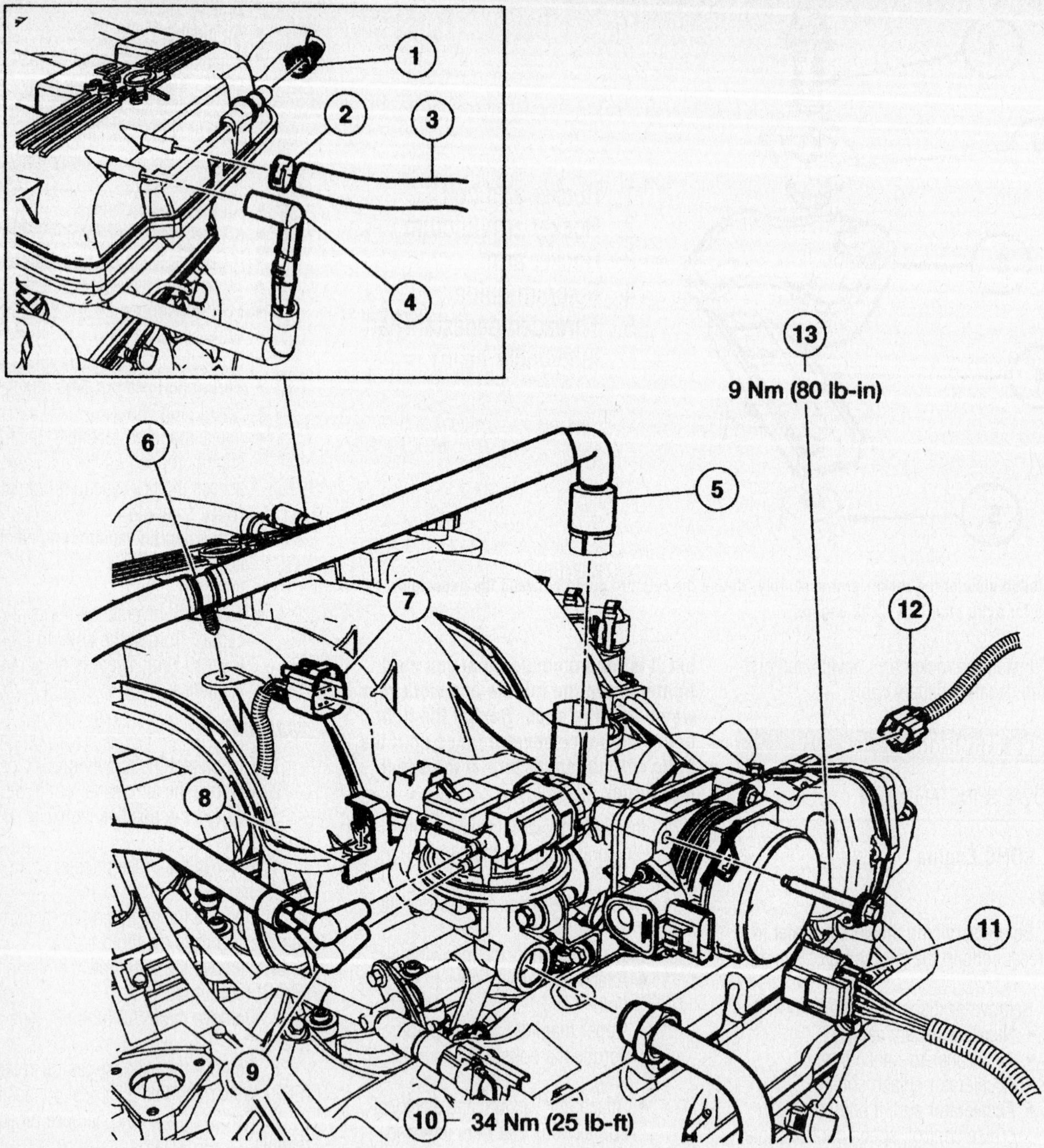

1 Positive crankcase ventilation (PCV) tube
2 Brake booster vacuum supply hose clamp
3 Brake booster vacuum supply hose
4 Engine main vacuum harness-to-intake manifold fitting
5 Evaporative emissions (EVAP) return tube
6 EVAP tube pin-type retainer
7 Exhaust gas recirculation (EGR) system module electrical connector
8 Wiring harness pin-type retainer
9 Engine main vacuum harness-to-EGR system module fitting
10 EGR tube fitting
11 Throttle position (TP) sensor electrical connector
12 Electronic throttle body electrical connector
13 Electronic throttle body-to-intake manifold bolt
14 Intake manifold mounting bolts
15 Intake manifold
16 Intake manifold gaskets

67197-EXPL-G21

Parts to be removed prior to intake manifold removal—2003-04—4.0L SOHC engine

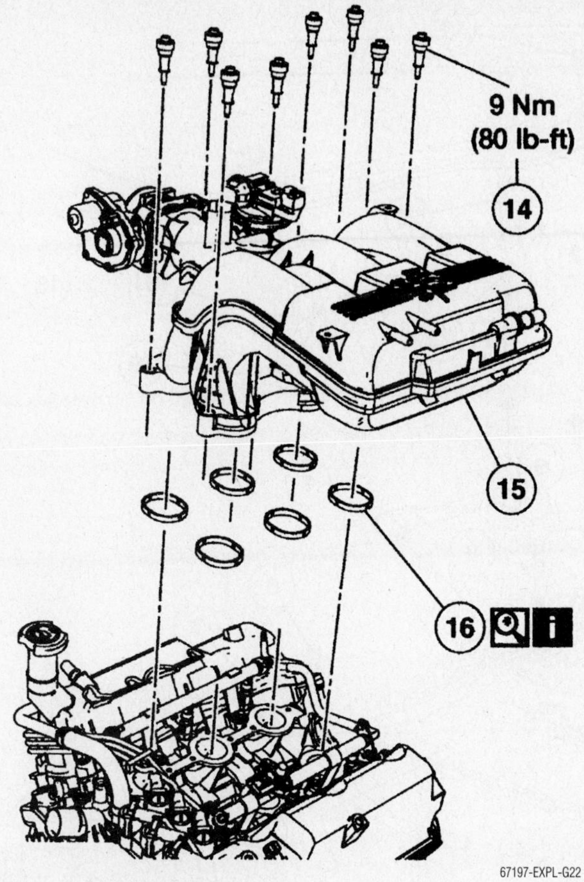

Intake manifold removal—2003–04—4.0L SOHC engine

CAUTION

Do not use metal scrapers, wire brushes, power abrasive discs or other abrasive means to clean the sealing surfaces. These tools can cause scratches and gouges which can make leak paths.

19. Clean the sealing surfaces, inspect the gaskets and install new gaskets if necessary. There is no special torque sequence.

4.6L SOHC Engine

1. Before servicing the vehicle, refer to the precautions in the beginning of this section.
2. Remove or disconnect the following:
   - Negative battery cable
   - Throttle body
   - Ignition coils
   - Fuel injection supply manifold
   - Upper radiator hose and the heater hose
   - Alternator
   - Accelerator cable bracket
   - Throttle body adapter
   - Thermostat housing
   - Intake manifold bolts
   - Intake manifold. Tilt the manifold to slide it out from underneath the exhaust gas recirculation (EGR) tube.
   - To install, reverse the removal procedure. Tighten retaining bolts is sequence shown.

4.6L DOHC Engine

UPPER

1. Disconnect the battery ground cable.
2. Drain the cooling system.
3. Remove the air cleaner outlet pipe.
4. Remove the exhaust gas recirculation (EGR) system module.
5. Remove the positive crankcase ventilation (PCV) valve hose.
6. Remove the brake booster vacuum hose.
7. Remove the evaporative emissions (EVAP) hose.
8. Remove the coolant hose clamp.
9. Remove the PCV coolant hose.
10. Remove the coolant hose clamp.
11. Remove the PCV coolant hose.
12. Remove the accelerator cable.
13. Remove the speed control cable.
14. Remove the accelerator cable return spring.
15. Remove the accelerator cable pin-type retainer.
16. Remove the accelerator cable bracket.
17. Remove the idle air control (IAC) valve electrical connector.
18. Remove the throttle position (TP) sensor electrical connector.
19. Remove the fuel rail shield.
20. Remove the radio capacitor.

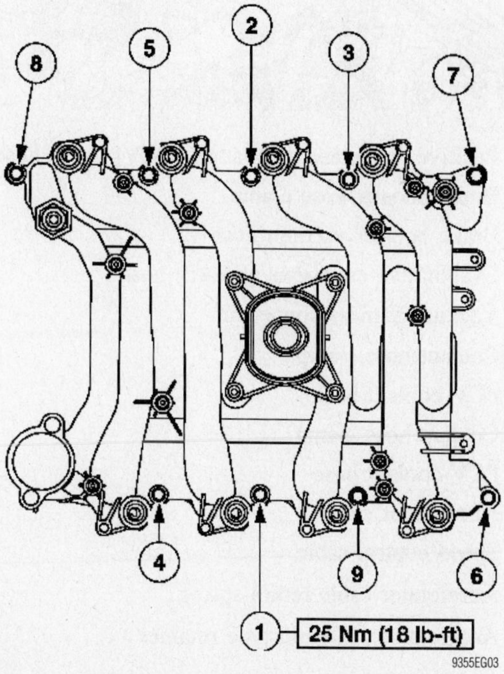

Tighten the manifold bolts in the sequence shown—4.6L SOHC Engine

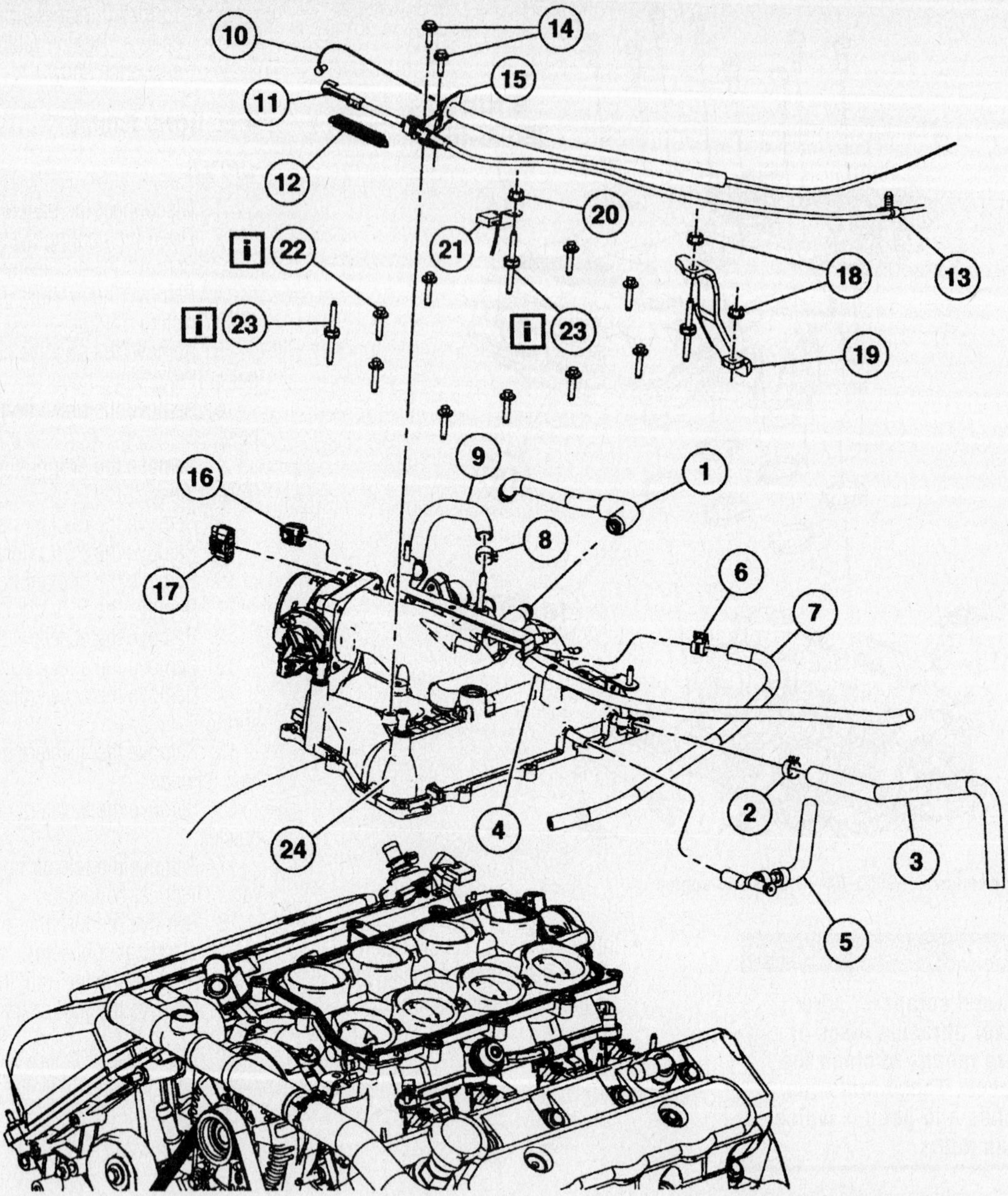

1 Positive crankcase ventilation (PCV) valve hose
2 Brake booster hose clamp
3 Brake booster vacuum hose
4 Evaporative emissions (EVAP) hose
5 Vacuum harness connector
6 Coolant hose clamp
7 PCV coolant hose
8 Coolant hose clamp
9 PCV coolant hose
10 Accelerator cable
11 Speed control cable
12 Accelerator cable return spring
13 Accelerator cable pin-type retainer
14 Accelerator cable bracket bolts
15 Accelerator cable bracket (position aside)
16 Idle air control (IAC) valve electrical connector
17 Throttle position (TP) sensor electrical connector
18 Fuel rail shield nuts
19 Fuel rail shield
20 Radio capacitor nut
21 Radio capacitor (position aside)
22 Upper intake manifold bolts
23 Upper intake manifold studs
24 Upper intake manifold

67197-EXPL-G23

Upper intake manifold and related parts—4.6L DOHC engine

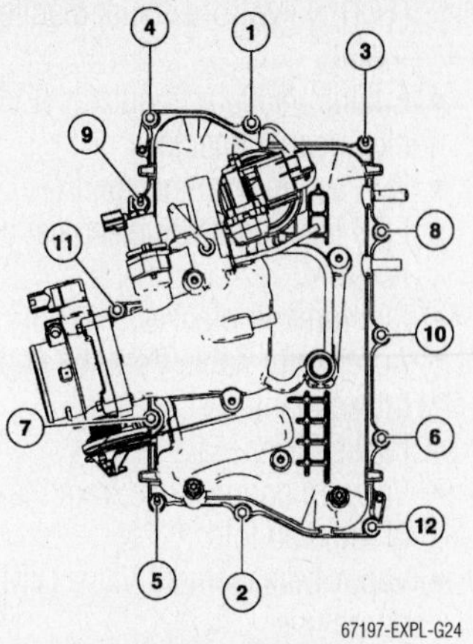

67197-EXPL-G24

Upper intake manifold bolt torque sequence—4.6L DOHC engine

21. Remove the upper intake manifold bolts.
22. Remove the upper intake manifold studs.
23. Remove the upper intake manifold.
24. To install, reverse the removal procedure.

➡Locations 5 and 9 use studs.

25. Tighten the bolts and studs in the sequence shown to 10 Nm (89 inch lbs.).

LOWER

1. Remove the upper intake manifold.
2. Disconnect the spring lock couplings.
3. Remove the cooling fan.
4. Remove the accessory drive belt tensioner.
5. Remove the accessory drive belt.
6. Remove the wire harness retainer.
7. Remove the generator.
8. Remove the generator bracket studs.
9. Remove the upper coolant hose.
10. Remove the coolant bypass hose.
11. Remove the thermostat housing coolant hose.
12. Remove the engine coolant temperature (ECT) sensor electrical connector.
13. Remove the coolant crossover tube.
14. Remove the vacuum hose.
15. Remove the engine harness retainers.
16. Remove the fuel rail pressure (FRP) sensor electrical connector.

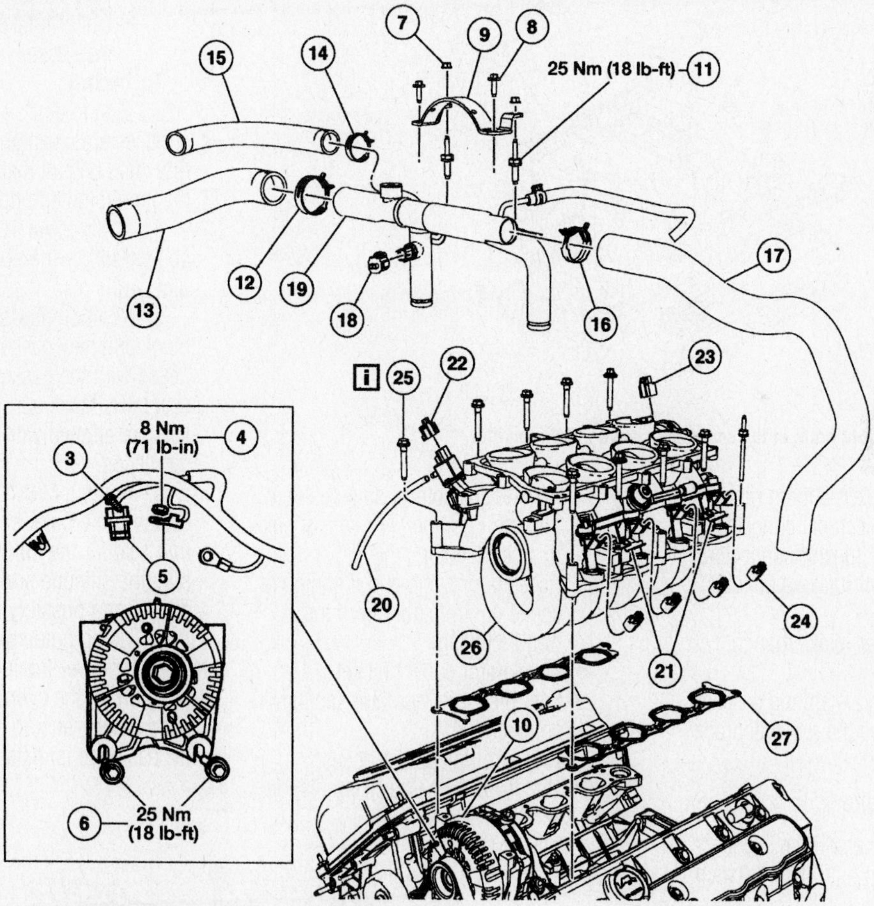

3 Wire harness retainer
4 Generator bracket nuts
5 Electrical connectors (two)
6 Generator bracket bolts
7 Generator bracket nuts
8 Generator bracket bolts
9 Generator bracket
10 Generator
11 Studs
12 Hose clamp
13 Upper coolant hose
14 Hose clamp
15 Coolant hose
16 Hose clamp
17 Thermostat housing coolant hose
18 Engine coolant temperature (ECT) sensor electrical connector
19 Coolant crossover tube
20 Vacuum hose
21 Engine harness retainers
22 Fuel rail pressure (FRP) sensor electrical connector
23 Intake manifold runner control (IMRC) motor electrical connector
24 Fuel injector connectors
25 Lower intake manifold bolts
26 Lower intake manifold assembly
27 Intake manifold gaskets

67197-EXPL-G25

Lower intake manifold and related parts

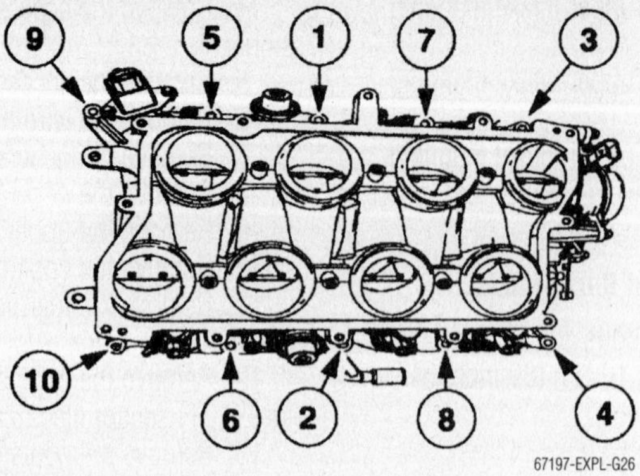

Lower intake manifold torque sequence—4.6L DOHC engine

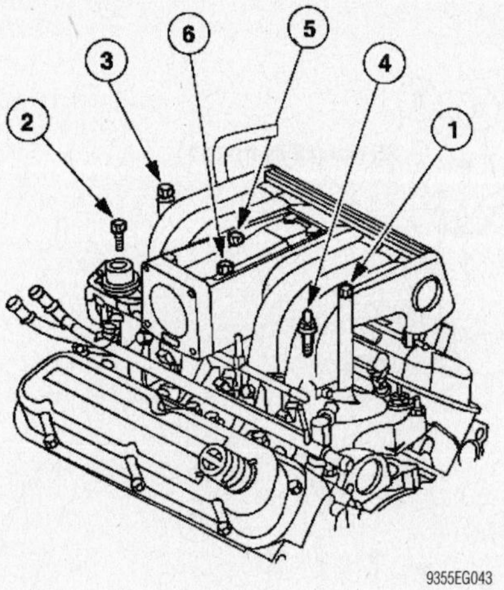

Loosen the upper manifold bolts in the sequence shown—5.0L engine

17. Remove the intake manifold runner control (IMRC) motor electrical connector.
18. Remove the fuel injector connectors.
19. Remove the lower intake manifold bolts.
20. Remove the lower intake manifold assembly.
21. Remove the intake manifold gaskets.
22. To install, reverse the removal procedure.

➡Hand-start all fasteners.

23. Tighten all fasteners in the sequence shown in the illustration to 10 Nm (89 inch lbs.).

5.0L Engine

UPPER

1. Before servicing the vehicle, refer to the precautions in the beginning of this section.
2. Remove or disconnect the following:
   - Negative battery cable
   - Air cleaner outlet tube
   - Idle Air Control (IAC) valve electrical connector
   - Accelerator control splash shield
   - Throttle Position (TP) sensor electrical connector
   - Accelerator cable and if equipped, speed control cable from the throttle linkage
   - Accelerator cable bracket
   - Fuel pressure regulator vacuum connection
   - Pressure transducer hoses
   - Upper Exhaust Gas Recirculation (EGR) valve-to-exhaust manifold tubing
   - Engine Vacuum Regulator (EVR) electrical connector
   - EVR vacuum connector
   - EGR back pressure electrical connector
   - Ignition coil bracket
   - Accelerator cable from the upper intake manifold clips
   - Intake cover plate
   - Vacuum connections from the front of the manifold
   - Vapor Management Valve (VMV) purge line
   - Brake booster vacuum supply line
   - Positive Crankcase Ventilation (PCV) hose
   - PCV heater hoses
   - Upper intake manifold and discard the gasket

To install:

3. Ensure that all of the gasket mating surfaces are clean and free of grease, oil or dirt. Also ensure that the EGR passages in the manifolds and heads are clear.
4. Apply a 1⁄16 in. (1.6mm) bead of silicone sealer to the points where the cylinder block rails meet the cylinder heads.
5. Position new seals on the cylinder block and new gaskets on the cylinder heads with the gaskets interlocked with the seal tabs. Make sure the holes in the gaskets are aligned with the holes in the cylinder heads.
6. Apply a 1⁄16 in. (1.6mm) bead of sealer to the outer end of each intake manifold seal for the full width of the seal. Make sure the silicone sealer will not fall into the engine and possibly block oil passages.
7. Using guide pins to ease installation, carefully lower the intake manifold into position on the cylinder block and cylinder heads. Also, ensure that the water pump bypass hose is installed at the same time.

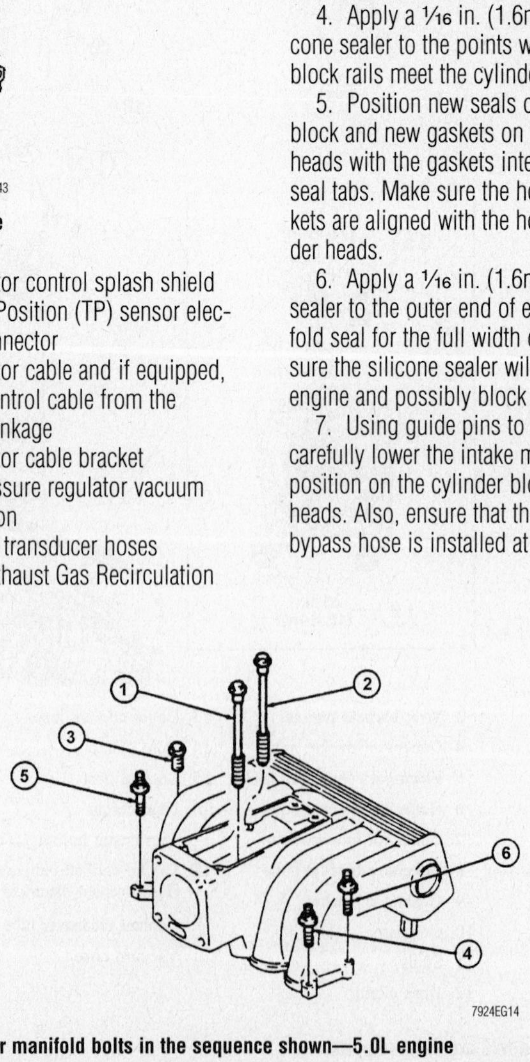

Tighten the upper manifold bolts in the sequence shown—5.0L engine

8. Install or connect the following:
   - Intake manifold and new gasket and hand tighten the bolts
   - Upper intake vacuum connections
   - PCV tube and heater hoses
   - Brake booster vacuum supply line
   - Torque the intake manifold bolts to 18 ft. lbs. (25 Nm).
   - Intake manifold cover plate. Torque the bolts to 15 ft. lbs. (20 Nm).
   - Accelerator cable bracket. Torque the upper bolt to 15 ft. lbs. (20 Nm) and lower bolt to 80 inch lbs. (9 Nm).
   - Accelerator cable and speed control cable to the throttle linkage, if equipped
   - Ignition coils and bracket
   - EVR electrical connector
   - EVR vacuum connector
   - EGR valve vacuum connector
   - Fuel pressure regulator vacuum connector
   - Upper EGR valve-to-exhaust manifold connector. Torque the fastener to 25 ft. lbs. (34 Nm).
   - EGR back pressure transducer electrical connector
   - TP sensor electrical connector
   - IAC valve electrical connector
   - Accelerator control splash shield. Torque the bolt to 89 inch lbs. (10 Nm).
   - Air cleaner outlet tube
   - Negative battery cable

LOWER

1. Before servicing the vehicle, refer to the precautions in the beginning of this section.
2. Drain the cooling system.
3. Remove or disconnect the following:
   - Negative battery cable
   - Radiator overflow hose and set it aside
   - Upper intake manifold
   - Water bypass hose
   - Engine Coolant Temperature (ECT) electrical connector
   - Wire harness retainer nut
   - Heater hoses
   - Positive Crankcase Ventilation (PCV) hoses
   - Upper radiator hose
   - Fuel injector electrical connectors
   - Ground strap at the rear of the lower intake manifold
   - Water temperature indicator sender electrical connector
   - Camshaft Position (CMP) sensor and the camshaft synchronizer
   - Bolts from the lower intake manifold
   - Lower intake manifold and discard the gaskets

To install:

4. Ensure that all of the gasket mating surfaces are clean and free of grease, oil or dirt. Also ensure that the EGR passages in the manifolds and heads are clear.
5. Apply a 1⁄16 in. (1.6mm) bead of silicone sealer to the points where the cylinder block rails meet the cylinder heads.
6. Position new seals on the cylinder block and new gaskets on the cylinder heads with the gaskets interlocked with the seal tabs. Make sure the holes in the gaskets are aligned with the holes in the cylinder heads.
7. Apply a 1⁄16 in. (1.6mm) bead of sealer to the outer end of each intake manifold seal for the full width of the seal. Make sure the silicone sealer will not fall into the engine and possibly block oil passages.
8. Install the lower intake manifold and tighten the bolts in 2 steps as follows:
   a. 89 inch lbs. (10 Nm).
   b. 24 ft. lbs. (32 Nm).
9. Install or connect the following:
   - CMP sensor and camshaft synchronizer
   - Water temperature indicator sender electrical connector
   - Fuel injector electrical connectors
   - Fuel line
   - Water bypass hose
   - Upper radiator hose
   - Heater hoses
   - Wire harness retainer nut. Torque to the bolt to 18 ft. lbs. (25 Nm).
   - ECT electrical connector
   - Water heater bypass hose to the heater tube
   - PCV hoses to the heater tube
   - Upper intake manifold
   - Radiator overflow hose
   - Negative battery cable
10. Fill the cooling system.
11. Start the vehicle and check for leaks, repair if necessary.

Exhaust Manifold

REMOVAL & INSTALLATION

4.0L SOHC Engine

1. Before servicing the vehicle, refer to the precautions in the beginning of this section.
2. Remove or disconnect the following:
   - Negative battery cable
   - Exhaust inlet pipe-to-manifold attaching bolts
   - Differential Pressure Feedback EGR (DPFE) transducer hoses, left side manifold only
   - Exhaust Gas Recirculation (EGR) tube from the manifold and valve, left side manifold only
   - Exhaust manifold and discard the gasket

To install:

3. Clean the gasket mating surfaces.
4. Install or connect the following:
   - New gasket and the exhaust manifold. Torque the bolts to 16 ft. lbs. (22 Nm).
   - EGR tube to the manifold. Torque the fastener to 30 ft. lbs. (40 Nm) left side manifold only
   - DPFE transducer hoses, left side manifold only
   - Exhaust inlet pipe-to-manifold

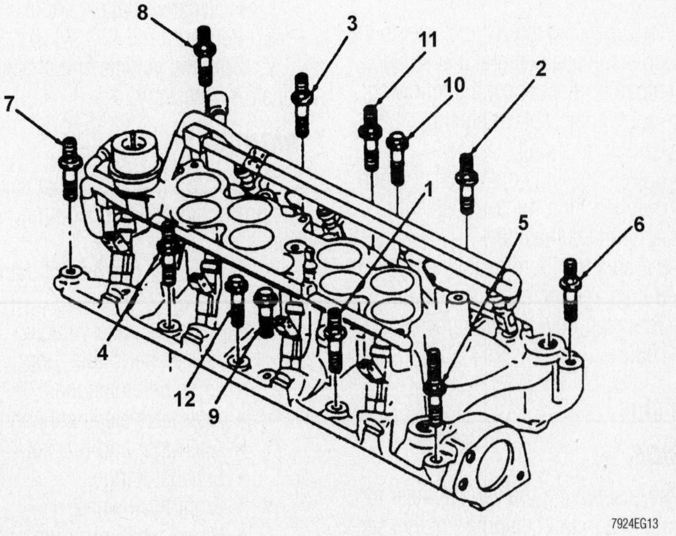

Tighten the lower manifold bolts in the sequence shown—5.0L engine

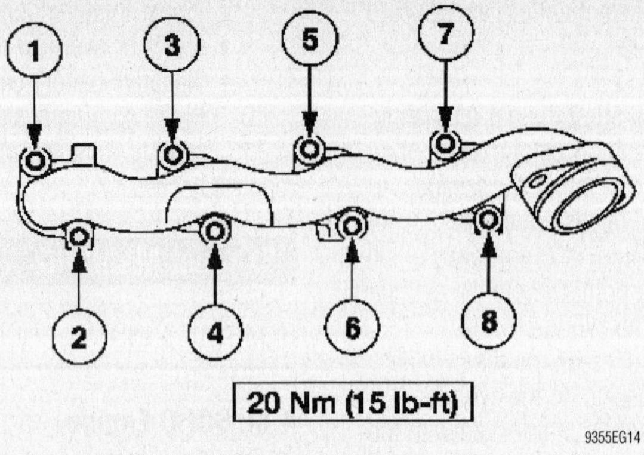

Exhaust manifold torque sequence—4.6L SOHC Engine

attaching bolts. Torque the bolts to 30 ft. lbs. (40 Nm).
- Negative battery cable

5. Start the vehicle and check for leaks, repair if necessary.

4.6L SOHC and DOHC Engines

RIGHT SIDE

1. Before servicing the vehicle, refer to the precautions in the beginning of this section.
2. Remove or disconnect the following:
   - Front fender splash shield
   - Heater control valve bracket bolt
   - Vacuum hose and heater hose. Position the heater control valve assembly aside.
   - Exhaust manifold heat shield
   - Exhaust gas recirculation tube from the exhaust manifold
   - Eight nuts and the exhaust manifold
3. Installation is the reverse of removal. Tighten the exhaust manifold nuts in the sequence shown.

LEFT SIDE

1. Before servicing the vehicle, refer to the precautions in the beginning of this section.
2. Remove or disconnect the following:
   - Front fender splash shield
   - Engine oil runoff
   - Lower steering column shaft pinch bolt. Position the lower steering column shaft aside.
   - Exhaust manifold heat shield
   - 8 nuts and the exhaust manifold
   - To install, reverse the removal procedure.

5.0L Engine

LEFT SIDE

1. Before servicing the vehicle, refer to the precautions in the beginning of this section.
2. Discharge and recover the A/C system.
3. Remove or disconnect the following:
   - Negative battery cable
   - Spark plug wires and retaining brackets
   - A/C manifold and tube
   - Condenser to evaporator tube from the A/C condenser core
   - Oil level indicator tube
   - Left wheel
   - Wheel well apron pin retainers
   - Exhaust manifold and discard the gasket

To install:

4. Clean the gasket mating surfaces.
5. Install or connect the following:
   - New gasket and the exhaust manifold. Torque the bolts to 30 ft. lbs. (40 Nm).
   - Wheel well pin retainers
   - Left wheel
   - Oil level indicator tube. Torque the bolt to 18 ft. lbs. (25 Nm).
   - Condenser to evaporator tube
   - A/C manifold and tube
   - Spark plug wires and brackets
   - Negative battery cable
6. Recharge the A/C system.
7. Start the vehicle and check for leaks, repair if necessary.

RIGHT SIDE

1. Before servicing the vehicle, refer to the precautions in the beginning of this section.
2. Remove or disconnect the following:
   - Negative battery cable
   - Air cleaner outlet tube
   - Drive belt tensioner
   - Alternator electrical connectors
   - Alternator and bracket
   - Exhaust flange
   - Right front wheel
   - Wheel well apron pin retainers
   - Exhaust Gas Recirculation (EGR) valve from the manifold tube
   - Spark plug wires and retaining brackets
   - Exhaust manifold heat shield
   - Exhaust manifold and discard the gasket

To install:

3. Clean the gasket mating surfaces.
4. Install or connect the following:
   - New gasket and the exhaust manifold. Torque the bolts to 30 ft. lbs. (40 Nm).
   - Exhaust manifold heat shield. Torque the bolts to 61 inch lbs. (7 Nm).
   - Spark plug wires and retaining brackets
   - EGR valve to the manifold tube
   - Wheel well apron pin retainers
   - Right front wheel
   - Exhaust flange. Torque the nuts to 26 ft. lbs. (34 Nm).
5. Install the alternator and bracket. Torque the bolts in sequence as follows:
   a. Bolts N0. 1–2 to 26 ft. lbs. (34 Nm).
   b. Bolt N0. 3 to 48 ft. lbs. (65 Nm).
   c. Torque the remaining bolt to 26 ft. lbs. (34 Nm).
6. Install or connect the following:
   - Drive belt tensioner
   - Air cleaner outlet tube
   - Negative battery cable
7. Start the vehicle and check for leaks, repair if necessary.

Camshaft and Valve Lifters

➡Although Ford suggests that this component is removable while the engine is installed in the vehicle, depending on the particular options with which your truck is equipped, working clearance may be extremely tight and this procedure may be much easier to perform with the engine removed. Before commencing, read through this procedure and make certain enough clearance, or working room, exists with the engine in the vehicle; if there is not enough space, the engine should be removed.

REMOVAL & INSTALLATION

4.0L SOHC Engine

1. Before servicing the vehicle, refer to the precautions in the beginning of this section.
2. Remove or disconnect the following:

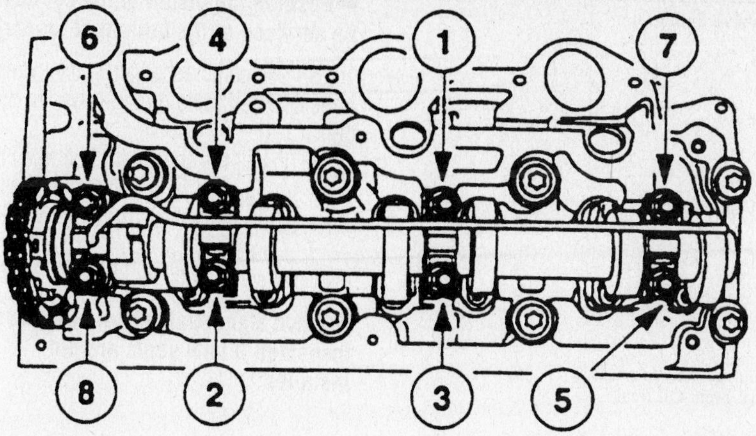

Use the proper sequence to prevent damage to the camshaft both when installing and removing the bearing caps—4.0L SOHC engine

- Negative battery cable for safety
- Valve cover
- Hydraulic camshaft tensioner

➡**The right-hand camshaft sprocket bolt uses left-hand threads.**

3. For the right-hand camshaft use the Cam Gear Torque Adapter tool T97T-6256-F, to remove the camshaft sprocket bolt.
4. For the left-hand camshaft, remove the sprocket bolt.

➡**When removing the followers, label them so that they may be returned to their original positions.**

5. Using the Valve Spring Compressor tool ST1330-A, remove the camshaft roller followers.
6. Install or connect the following:
   - Camshaft bearing cap bolts and the oil rail
   - Camshaft

To install:

7. Lubricate all of the moving parts with SAE 50W engine oil.
8. Install camshaft onto the cylinder head.
9. Position the oil rail and install the bearing caps and bolts. Torque the bolts in 2 steps:
   a. Step 1—53.5 inch lbs. (6 Nm).
   b. Step 2—11–12.5 ft. lbs. (15–17 Nm).
10. Install or connect the following:
   - Camshaft followers
   - Camshaft sprocket bolt and hand tighten the bolt
   - Camshaft Chain Tensioner T97T-6K254-A in the hole that the hydraulic chain tensioner was in
11. Turn the crankshaft one revolution clockwise until No. 1 piston is Top Dead Center (TDC).
12. Install or connect the following:
   - Crankshaft Holding tool T97T-6303-A on the crankshaft to keep it from turning
   - Position the timing slot on the rear of the camshaft to fit Camshaft Holding tool T97T-6256-C and install the holding tool on the rear of the head
   - Camshaft Gear Holding tool T97T-6256-B and Camshaft Gear Holding tool T97T-6256-A on the front of the cylinder head to securely hold the camshaft gear
   - Tighten the camshaft sprocket bolt to 63 ft. lbs. (85 Nm).
13. Remove the Camshaft Chain Tensioner tool and install the hydraulic chain tensioner, tighten the tensioner to 35–39 ft. lbs. (47–53 Nm).
14. Remove the special tools from the engine.
15. Install or connect the following:
   - Valve cover
   - Negative battery cable
16. Start the engine check for leaks and repair if necessary.

4.6L SOHC Engine

1. Before servicing the vehicle, refer to the precautions in the beginning of this section.
2. Remove or disconnect the following:

WARNING

At no time, when the timing chains are removed and the cylinder heads are installed may the crankshaft or camshaft be rotated. Severe piston and valve damage will occur.

- Timing chains

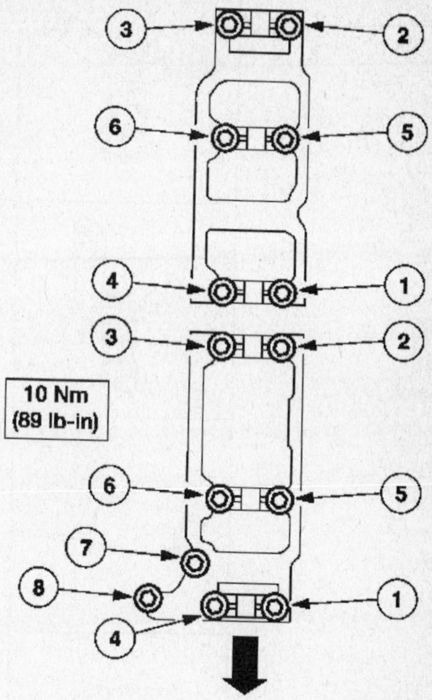

Camshaft bearing bolt torque sequence—4.6L SOHC Engine

- Camshaft roller followers
- Camshaft sprocket
- The 13 camshaft bearing cap bolts
- Camshaft bearing caps ladders
- Camshaft from the cylinder head

➡**The valve tappets are positional. Mark each valve tappet for installation in its original location.**

- Valve tappets

To install:

3. Lubricate the camshaft journals with clean engine oil.
4. Install the camshaft onto the cylinder head.
5. Lubricate the camshaft bearing caps with clean engine oil.
6. Install the camshaft bearing caps and loosely install the bolts.
7. Tighten the bolts in the sequence shown.
8. Install the camshaft sprocket. Tighten the sprocket bolt in two stages.
   - Step 1: Tighten to 40 Nm (30 ft. lbs.)
   - Step 2: Tighten an additional 90 degrees.
9. Install the roller followers
10. Install the timing chains

4.6L DOHC Engine

1. Remove the engine front cover, timing gears, chain and tensioners.
2. Remove the valve covers.
3. Remove the roller followers.
4. Remove the spark plugs.

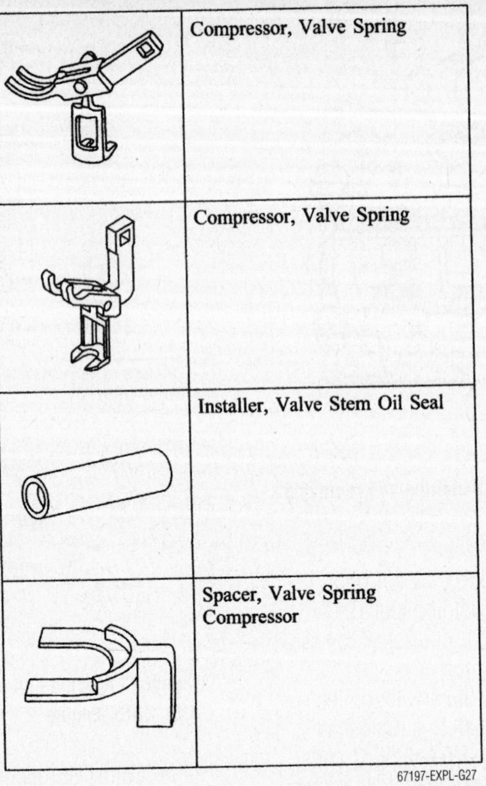

Compressor, Valve Spring
Compressor, Valve Spring
Installer, Valve Stem Oil Seal
Spacer, Valve Spring Compressor
67197-EXPL-G27

These tools, or their equivalents, are necessary for camshaft and lifter removal—4.6L DOHC engine

➡**Position the piston of the cylinder to be serviced at the bottom of the stroke.**

5. Use compressed air in the cylinder to be serviced to hold both valves in position.

6. Install the special tool between the valve spring coils to protect the exhaust valve stem seal from damage.

7. Compress the valve spring. Remove the valve spring retainer keys.

➡**Valve stem seals should be visually inspected if new seals are not installed.**

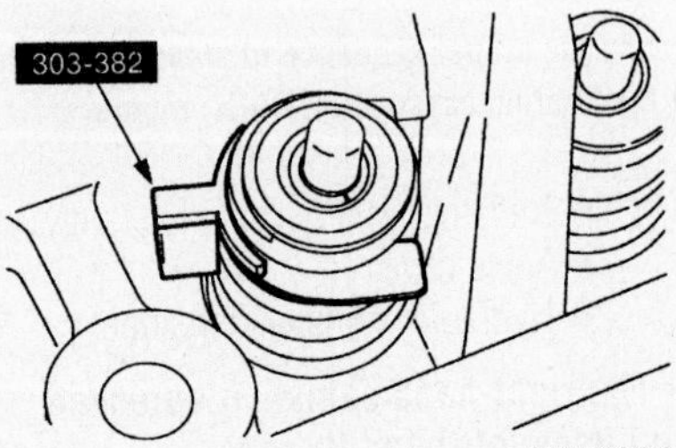

Install the special tool between the valve spring coils to protect the exhaust valve stem seal from damage

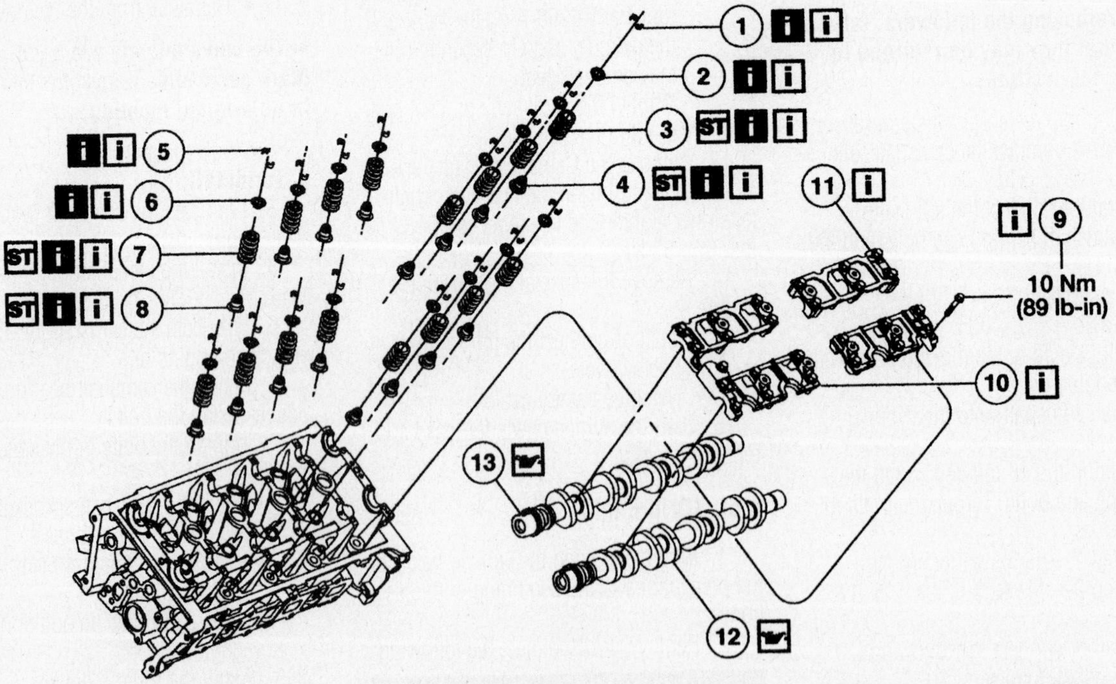

1 Exhaust valve spring retainer keys
2 Exhaust valve spring retainers
3 Exhaust valve springs
4 Exhaust valve stem seals
5 Intake valve spring retainer keys
6 Intake valve spring retainers
7 Intake valve springs
8 Intake valve stem seals
9 Camshaft bearing cap bolts (outboard row)
10 Camshaft bearing caps (exhaust)
11 Camshaft bearing caps (intake)
12 Camshaft (exhaust)
13 Camshaft (intake)

67197-EXPL-G28

Valve train exploded view—4.6L DOHC engine

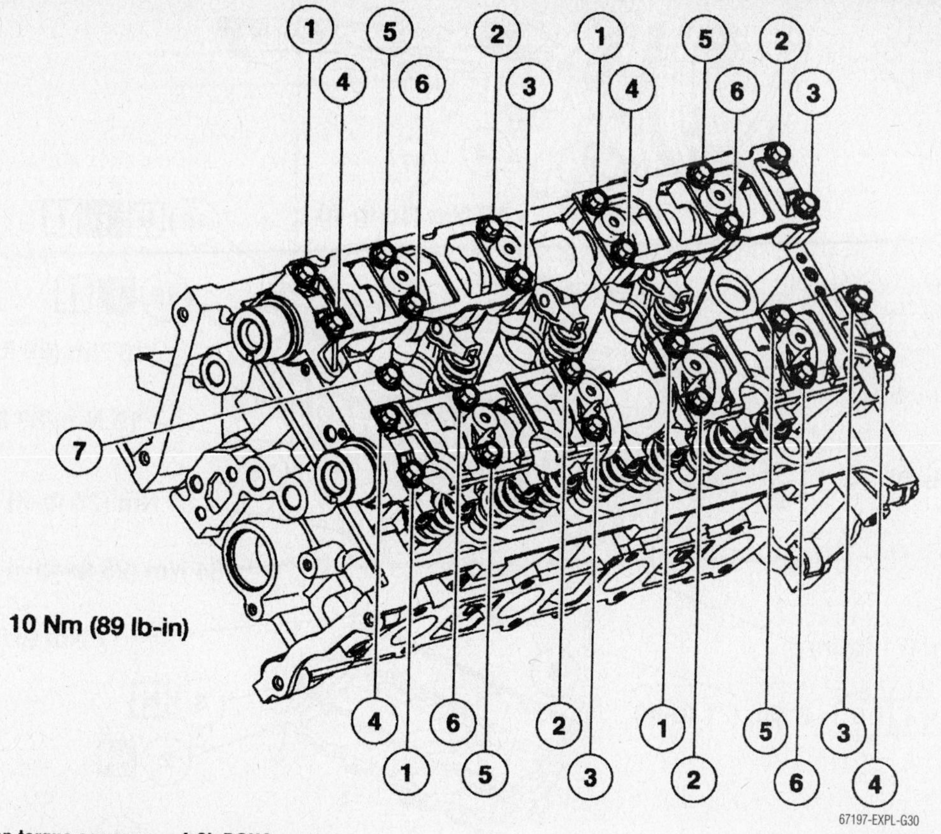

Camshaft bearing cap torque sequence—4.6L DOHC engine

8. Remove the special tool, the valve spring retainer, the valve spring and the valve stem seal.
9. Remove the camshaft bearing cap bolts.
10. Remove the camshaft bearing caps.
11. Remove the camshafts.
12. To install, reverse the removal procedure.
13. Install the camshaft bearing caps and tighten the fasteners in the sequence shown.

5.0L Engine

1. Before servicing the vehicle, refer to the precautions in the beginning of this section.
2. Remove or disconnect the following:
   - Negative battery cable
   - Timing chain cover
   - Camshaft sprocket and chain assembly
   - Upper and lower intake manifolds
   - Both valve covers
   - Loosen the rocker arm bolts and rotate the rocker arms to the side
   - Pushrods in sequence so that they may be installed to their original positions
   - Lifters
   - Camshaft thrust plate bolts and the plate
   - Camshaft from the engine, taking care not to damage the bearings, lobes or journals

To install:

3. Apply SAE 50W engine oil to the camshaft lobes and journals.
4. Install or connect the following:
   - Camshaft
5. Apply SAE 50W engine oil to the camshaft thrust plate. Position the thrust plate with the groove toward the block and install the retaining bolts. Torque to 10–12 ft. lbs. (13–16 Nm).
6. Apply SAE 50W engine oil to the valve tappets and install them. If reusing the old lifters, place them in their original positions.
7. Install or connect the following:
   - Pushrods to their original positions
   - Rocker arms
   - Valve covers
   - Lower and upper intake manifolds
   - Camshaft sprocket and chain assembly. Ensure that the timing marks on the cam and crankshaft sprockets are aligned.
   - Timing chain cover
   - Negative battery cable
8. Start the engine, check for leaks and repair if necessary.

Oil Pan

REMOVAL & INSTALLATION

4.0L SOHC Engine

➡The 4.0L SOHC engine does not use an oil pan in the conventional sense. There is a separate access panel that unbolts from what would be considered the oil pan (which is now known as the ladder frame).

1. Before servicing the vehicle, refer to the precautions in the beginning of this section.
2. Drain the engine oil.
3. Remove or disconnect the following:
   - Negative battery cable
   - Oil pan and discard the gasket

To install:

4. Install or connect the following:
   - New gasket and oil pan. Torque the bolts to 80 inch lbs. (9 Nm).
   - Negative battery cable
5. Fill the engine with clean oil.
6. Start the vehicle and check for leaks, repair if necessary.

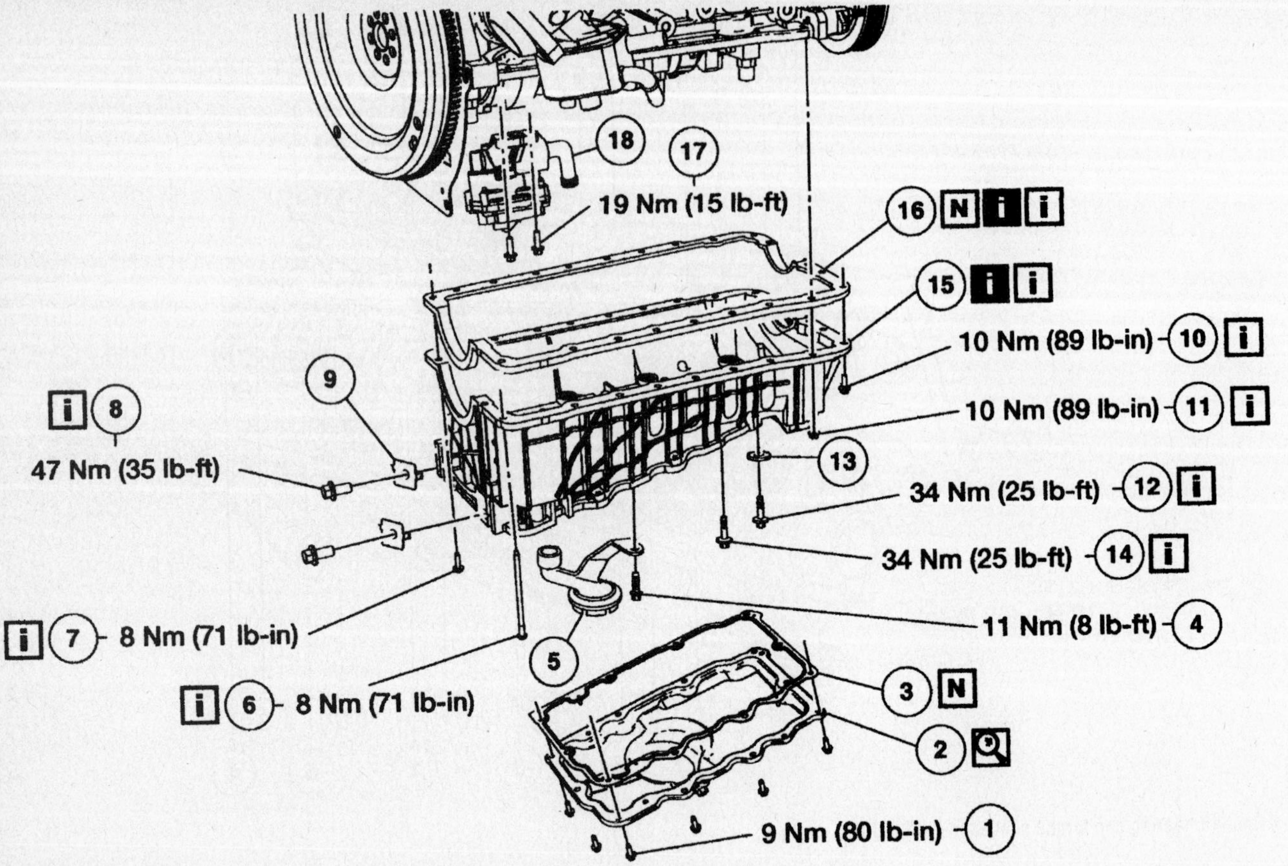

1 Oil pan bolt
2 Oil pan
3 Oil pan gasket
4 Oil pump pickup tube bolt
5 Oil pump pickup tube
6 Lower block cradle bolt
7 Lower block cradle bolt
8 Bell housing-to-lower block cradle bolt
9 Bell housing-to-lower block cradle spacer
10 Lower block cradle bolt
11 Lower block cradle nut
12 Lower block cradle bolt
13 Washer
14 Lower block cradle bolt
15 Lower block cradle (4x2 vehicles)
15 Lower block cradle (4x4 vehicles)
16 Lower block cradle gasket
17 Oil pump bolt
18 Oil pump

67197-EXPL-G35

Oil pan, pump and related parts—4.0L SOHC engine

4.6L SOHC Engine

1. Before servicing the vehicle, refer to the precautions in the beginning of this section.
2. Drain the engine oil.
3. Remove or disconnect the following:
   - Front axle
   - Oil pan

➡Do not use metal scrapers, wire brushes, power abrasive discs, or other abrasive means to clean the sealing surfaces. These may cause scratches and gouges resulting in leak paths. Use a plastic scraper to clean the sealing surfaces.

4. Remove and discard the oil pan gasket. Clean the sealing surfaces with metal surface cleaner. Allow the surfaces to dry for four minutes or until there is no sign of wetness, whichever is longer. Failure to do so can cause future oil leaks.

To install:

➡If the oil pan and gasket are not secured within four minutes of sealer application, the sealant must be removed and the sealing surfaces cleaned with metal surface cleaner.

5. Apply silicone gasket and sealant in the two places shown.
6. Position the new oil pan gasket and the oil pan and loosely install the bolts.
7. Tighten the bolts in three stages in the sequence shown.

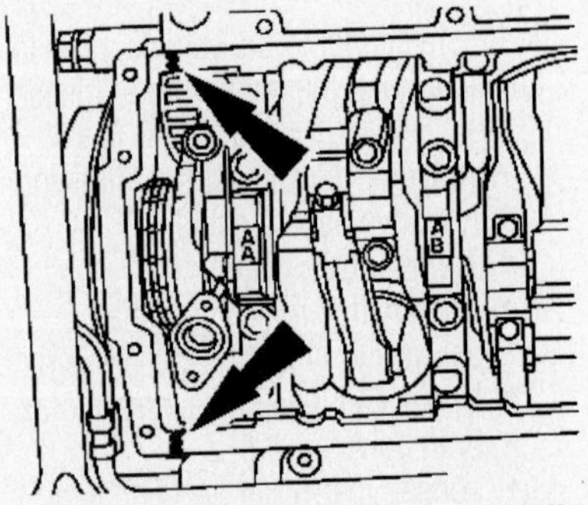

Sealant application—4.6L SOHC Engine

- Step 1: Tighten to 2 Nm (18 inch lbs.)
- Step 2: Tighten to 20 Nm (15 ft. lbs.)

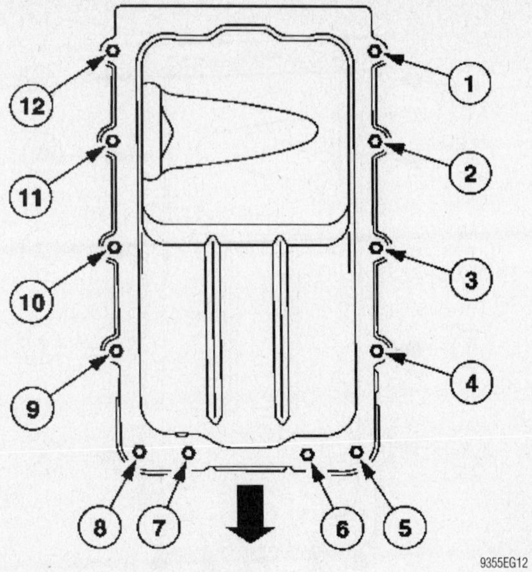

Oil pan bolt torque sequence—4.6L SOHC Engine

- Step 3: Tighten an additional 60 degrees.

8. Install the front axle.

4.6L DOHC Engine

1. With the vehicle in NEUTRAL position on a hoist.
2. Drain the engine oil.
3. Remove the front stabilizer bar.

➡Carry out this step only if the oil pump is being removed.

4. Remove the engine front cover.
5. Remove the front axle bolts.

⁂ WARNING

Do not allow the inner halfshaft boots to contact the lower control arms. Damage to the boots can occur.

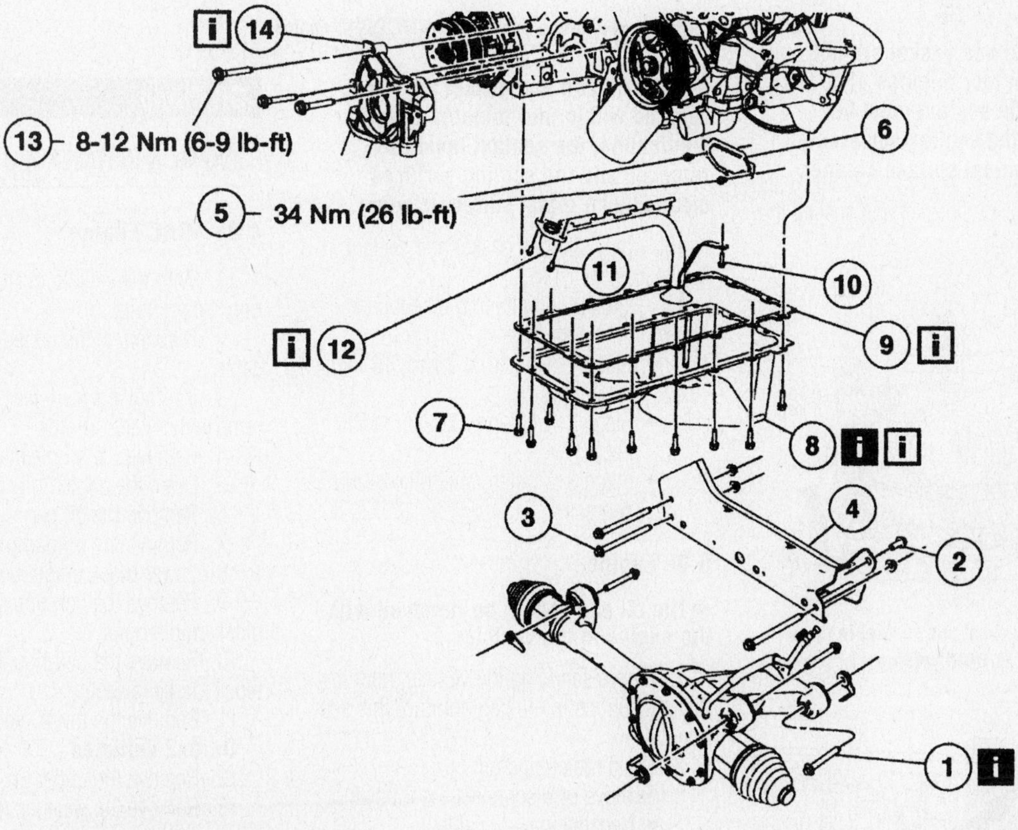

1	Front axle bolts (if equipped)	8	Oil pan
2	Crossmember bolt	9	Oil pan gasket
3	Crossmember bolts	10	Oil pump screen and pickup tube bolt
4	Crossmember	11	Oil pump screen and pickup tube bolts
5	Inspection cover bolts	12	Oil pump screen and pickup tube
6	Inspection cover	13	Oil pump bolts
7	Oil pan bolts	14	Oil pump

67197-EXPL-G31

Oil pan, pump and related parts—4.6L DOHC engine

6. Support the front axle using an appropriate jack. Lower the front axle approximately 6 inches.
7. Remove the crossmember bolts.
8. Remove the crossmember.
9. Remove the inspection cover.
10. Remove the oil pan bolts
11. Oil pan.
12. Remove the oil pan gasket.
13. To install, reverse the removal procedure.

⁂ WARNING

Do not use metal scrapers, wire brushes, power abrasive discs, or other abrasive means to clean the sealing surfaces. These can cause scratches and gouges resulting in leak paths. Use a plastic scraper to clean the sealing surfaces.

14. Clean the sealing surfaces with metal surface cleaner.

➡If the oil pan and gasket are not secured within four minutes of sealer application, the sealant must be removed and the sealing surfaces cleaned with metal surface cleaner.

15. Apply silicone gasket and sealant in two places.

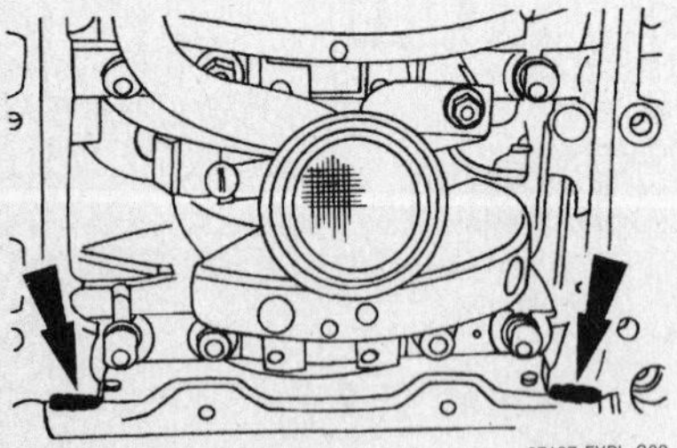

Apply silicone gasket and sealant in these two places—4.6L DOHC engine

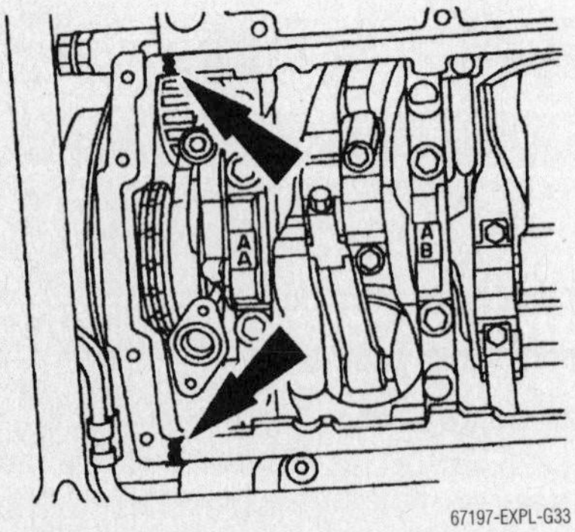

Apply silicone gasket and sealant in these two places—4.6L DOHC engine

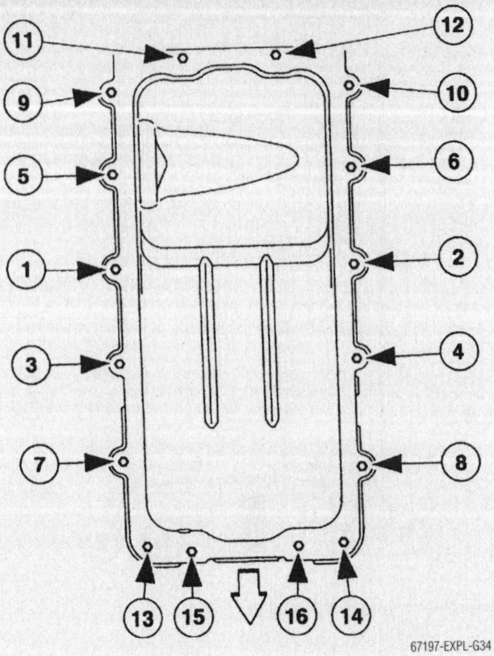

Oil pan bolt torque sequence—4.6L DOHC engine

➡If the oil pan and gasket are not secured within four minutes of sealer application, the sealant must be removed and the sealing surfaces cleaned with metal surface cleaner.

16. Apply silicone gasket and sealant in two places.
17. Tighten the bolts in the sequence shown in three stages:
    - Stage 1: Tighten to 2 Nm (18 inch lbs).
    - Stage 2: Tighten to 20 Nm (15 ft. lbs.).
    - Stage 3: Tighten an additional 60 degrees.

5.0L Engine

➡The oil pan cannot be removed with the engine in the vehicle.

1. Before servicing the vehicle, refer to the precautions in the beginning of this section.
2. Drain the engine oil.
3. Remove or disconnect the following:
    - Negative battery cable
    - Engine from the vehicle
    - Oil pan and discard the gasket

To install:

4. Install or connect the following:
    - Oil pan with a new gasket. Torque the bolts to 18 ft. lbs. (25 Nm).
    - Engine to the vehicle
    - Negative battery cable
5. Fill the engine with clean oil.
6. Start the vehicle and check for leaks, repair if necessary.

Oil Pump

REMOVAL & INSTALLATION

4.0L SOHC Engine

1. With the vehicle in NEUTRAL, position it on a hoist.
2. Disconnect the negative battery cable.
3. If removing the lower block cradle, remove the starter motor.
4. Remove the air cleaner outlet tube.
5. Drain the engine oil.
6. Remove the oil pan
7. Remove the oil pump pickup tube
8. Lower block cradle bolts
9. Remove the bell housing-to-lower block cradle bolts.
10. Remove the bell housing-to-lower block cradle spacer.
11. Remove the lower block cradle nuts

On 4x2 vehicles

12. Remove the upper air deflector.
13. Remove the weatherstrip.

➡Not all vehicles will have screws installed in the radiator fan shroud.

14. Remove and discard the screws. Remove the bolts and the fan shroud.
15. Remove the bolt.
16. Install the RH lifting eye using the previously removed bolt.
17. Install lifting tools.
18. Remove the motor mount nuts.
19. Raise the engine.

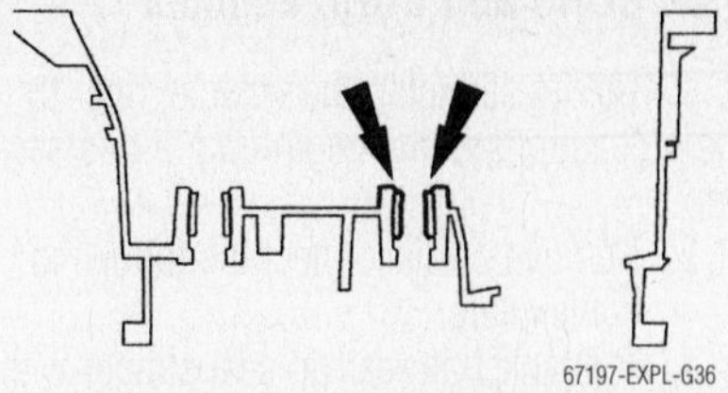

Back the set screws off until they are below the lower block cradle boss—4.0L SOHC engine

WARNING

Failure to back off the set screws can result in damage to the lower block cradle.

20. Back the set screws off until they are below the lower block cradle boss.

On 4x4 vehicles

21. Remove the upper air deflector.
22. Remove the weatherstrip.

➡Not all vehicles will have screws installed in the radiator fan shroud.

23. Remove and discard the screws. Remove the bolts and the fan shroud.
24. Remove the bolt.
25. Install the RH lifting eye using the previously removed bolt.

➡This is not a typical setup. Only the right side of the motor will be raised.

26. Install the special tools.
27. Remove the four bolts and the cross-member.
28. Remove the RH motor mount nut.
29. Remove the LH side through bolt and nut.
30. Raise the engine.

WARNING

Failure to back off the set screws can result in damage to the lower block cradle.

31. Back the set screws off until they are below the lower block cradle boss.

➡Use a floor jack to support the front axle assembly.

32. Remove the bolt from the right side of the axle housing. Discard the nut and bolt.
33. Remove the LH side lower axle housing bolt and nut. Discard the bolt and nut.
34. Remove the LH side upper axle housing axle bolt and nut. Discard the bolt and nut. Lower the axle.
35. Remove the lower block cradle.
36. Lower block cradle gasket.
37. Oil pump bolt
38. Oil pump.

➡Gasket material as well as silicone sealant may be present in the cavities in the main bearing cap. This material must be removed completely prior to assembly.

39. Clean the sealant from the cavities on the rear main bearing cap.

To install:

➡If the block cradle gasket is not secured within four minutes of the sealant application, the sealant must be removed and the sealing area cleaned with metal surface cleaner. Allow the silicone to dry until there is no sign of wetness, or four minutes, whichever is longer. Failure to follow this procedure can cause future oil leakage.

40. Apply silicone in the six places shown.
41. Position the lower block cradle and gasket assembly.
42. Install and hand-tighten the outer bolts and nuts.
43. Install and hand-tighten the rear lower block cradle bolts.
44. Install the two rear lower block cradle-to-bell housing bolts.

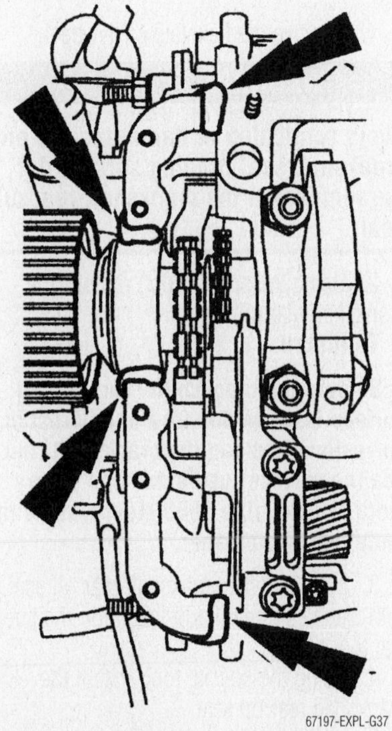

Apply silicone in the six places shown—4.0L SOHC engine

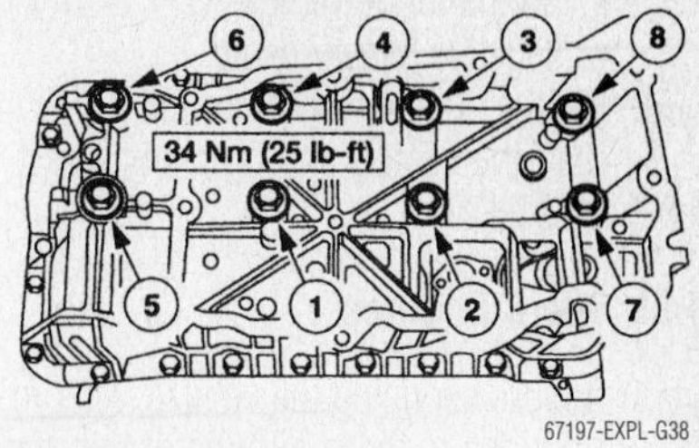

Lower block cradle bolt torque sequence—4.0L SOHC engine

➡The lower block cradle to the cylinder block alignment must be within a maximum mismatch of 0.25 mm (0.01 in) lower block cradle underflush or 0.05 mm (0.00196 in) lower block cradle protrusion.

45. Using a straightedge, align the transmission face of the lower block cradle with the rear face of the cylinder block.
46. Tighten the outer bolts.
47. Tighten the rear lower block cradle bolts.
48. Tighten the eight inserts.
49. Install the two silver covered bolts and new washer seals. Hand-tighten them at this time.
50. Install and hand-tighten the six remaining bolts.
51. Tighten the lower block cradle bolts in two stages:
    - Stage 1: Tighten to 15 Nm (11 ft. lbs.).
    - Stage 2: Tighten to 34 Nm (25 ft. lbs.).
52. To install, reverse the removal procedure.
53. Fill the engine with clean engine oil.

5.0L Engines

➡The oil pump cannot be removed with the engine in the vehicle.

1. Before servicing the vehicle, refer to the precautions in the beginning of this section.
2. Drain the engine oil.
3. Remove or disconnect the following:
    - Engine from the vehicle
    - Oil pan
4. Remove the 2 oil pump attaching bolts and the pump.

To install:

5. Submerge the pump in clean engine oil to prime it.
6. Install the oil pump. Torque the bolts to 23–31 ft. lbs. (30–43 Nm).
7. Install or connect the following:
    - Oil pick-up tube
    - Oil pan

- Engine to the vehicle
- Negative battery cable

8. Fill the engine with clean oil.
9. Start the vehicle and check for leaks, repair if necessary.

4.6L SOHC Engines

1. Before servicing the vehicle, refer to the precautions in the beginning of this section.
2. Drain the engine oil.
3. Remove or disconnect the following:
   - Negative battery cable
   - Timing chains
   - Oil pan
   - Three bolts and the oil pump screen cover and tube
   - Oil pump

To install:

➡Lubricate the new O-ring seal with clean engine oil.

4. Clean and inspect the mating surfaces. Install a new O-ring seal.
5. Position the oil pump.
6. Loosely install the bolts.
7. Tighten the bolts to 89 inch lbs. (10Nm).
8. Install the three oil pump screen and cover bolts. Torque the bolts to 18 ft. lbs. (25Nm).
9. Install the timing chains.
10. Install the oil pan.

4.6L DOHC Engine

1. With the vehicle in NEUTRAL position on a hoist.
2. Drain the engine oil.
3. Remove the front stabilizer bar.

➡Carry out this step only if the oil pump is being removed.

4. Remove the engine front cover.
5. Remove the oil pan.
6. Remove the oil pump screen and pickup tube.
7. Remove the oil pump bolts.
8. Remove the oil pump.
9. To install, reverse the removal procedure.

➡Oil pump must be held against block until bolts are tightened.

10. Align the inner rotor of the oil pump assembly to align with the flats on the crankshaft, and slide the oil pump toward the block until it is seated against the block.
11. Rotate the oil pump assembly to align with the bolt holes.

➡Lubricate the new O-ring seal with clean engine oil before installation.

12. Install O-ring seal.

Rear Main Seal

REMOVAL & INSTALLATION

5.0L Engines

1. Remove the flexplate or flywheel.

⁂ WARNING

Use care to avoid scratching or damaging the oil seal surface or leakage may occur.

2. Using a sharp awl, punch one hole into the crankshaft rear oil seal metal surface between the seal lip and the cylinder block.
3. Screw the threaded end of the special tool into the oil seal. Use the special tool to remove the crankshaft rear oil seal.

To install:

4. Lubricate the outer lips and the inner seal on the crankshaft rear oil seal with clean engine oil.
5. Using the special tool, install the crankshaft rear oil seal. Alternate bolt tightening to correctly seat the crankshaft rear oil seal.
6. Install the flexplate or flywheel.

4.0L SOHC Engine

1. Remove the flexplate or flywheel.

⁂ WARNING

Avoid scratching or damaging the oil crankshaft seal running surface during removal of the crankshaft rear oil seal.

2. Using the special tool, remove the crankshaft rear oil seal.

To install:

➡Be sure the crankshaft rear sealing surface is clean and free of any rust or corrosion. To clean the crankshaft rear sealing surface, use extra-fine emery cloth or extra-fine 0000 steel wool with metal surface cleaner.

3. Lubricate the crankshaft rear oil seal with clean engine oil and install on the special tool.
4. Using the special tool, install the crankshaft rear oil seal.
5. Install the flexplate or flywheel.

4.6L SOHC and DOHC Engines

1. Before servicing the vehicle, refer to the precautions in the beginning of this section.
2. Remove or disconnect the following:
   - Flexplate
   - Crankshaft rear oil seal slinger with a slide hammer
   - Rear oil seal with a slide hammer
3. Installation is the reverse of removal.

Note the following:

- Lubricate the inner lip of the rear crankshaft seal with clean engine oil.
- Use the two Crankshaft Rear Oil Seal Installers to install the rear oil seal.
- Using the two Crankshaft Rear Oil Seal Installers and the Crankshaft Rear Oil Slinger Installer, install the crankshaft rear oil slinger.

Timing Chain, Sprockets, Front Cover and Seal

REMOVAL & INSTALLATION

4.0L SOHC Engine

1. Before servicing the vehicle, refer to the precautions in the beginning of this section.
2. Drain the engine oil.
3. Remove or disconnect the following:
   - Negative battery cable
   - Engine from the vehicle
   - Oil pan
   - Engine front cover
   - Cylinder heads
4. Lock the jackshaft tensioner by installing a pin.
   - Jackshaft sprocket and chain assembly
   - Left front cassette retaining bolt
   - Cassette chain and tensioner assembly
   - Rear jackshaft plug from the engine
   - Right rear cassette retaining bolt and spacer
   - Right rear cassette chain and tensioner
   - Timing chain (s)

To install:

5. Install or connect the following:
   - Timing chain(s)
   - Right rear cassette chain, tensioner and sprocket
   - Jackshaft sprocket and chain on the engine and remove the tensioner pin

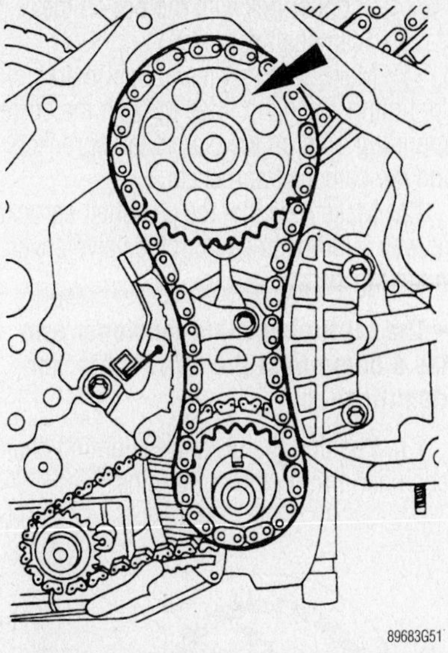

Remove the jackshaft sprocket—4.0L SOHC Engine

6. Torque the jackshaft sprocket bolt in 2 stages:
   a. 32–35 ft. lbs. (43–47 Nm).
   b. Turn an additional 65 degrees.
7. Install or connect the following:
   - Cylinder heads
   - Front cover
   - Oil pan
   - Engine to the vehicle
   - Negative battery cable
8. Fill the engine with clean oil.
9. Start the vehicle, check for leaks and repair if necessary.

4.6L SOHC Engine

SEAL ONLY

1. Before servicing the vehicle, refer to the precautions in the beginning of this section.
2. Remove or disconnect the following:
   - Crankshaft pulley
   - Front oil seal with a puller

To install:

3. Lubricate the front cover and the front oil seal inner lip with clean engine oil.
4. Use a seal driver to install the front oil seal into the engine front cover.
5. Install the crankshaft pulley.

FRONT COVER

1. Before servicing the vehicle, refer to the precautions in the beginning of this section.
2. Drain the engine oil and coolant.
3. Remove or disconnect the following:
   - Battery ground cable
   - Both valve covers
   - Radiator and the transmission cooler
   - Drive belt
   - Water pump
   - CMP sensor
   - Upper radiator hose bracket
   - EGR vacuum regulator solenoid
   - Drive belt tensioner
   - Idler pulleys
   - Crankshaft pulley
   - Coolant hose clamps
   - CKP sensor
   - Oil pan to front cover bolts
   - Crankshaft front oil seal
   - Timing cover bolts and studs
   - Front cover from the cylinder block

To install:

➡If the engine front cover is not secured within four minutes, the sealant must be removed and the sealing area cleaned with metal surface cleaner.

4. Apply silicone gasket and sealant along the cylinder head to block surface and the oil pan to cylinder block surface.
5. Install the engine front cover on the front cover to cylinder block dowel and loosely install the bolts.
6. Tighten the front cover bolts and stud bolts in the sequence shown.
7. Install a new crankshaft front oil seal.
8. Loosely install the pan-to-cover bolts, then, tighten the bolts in two stages, inner bolts first.

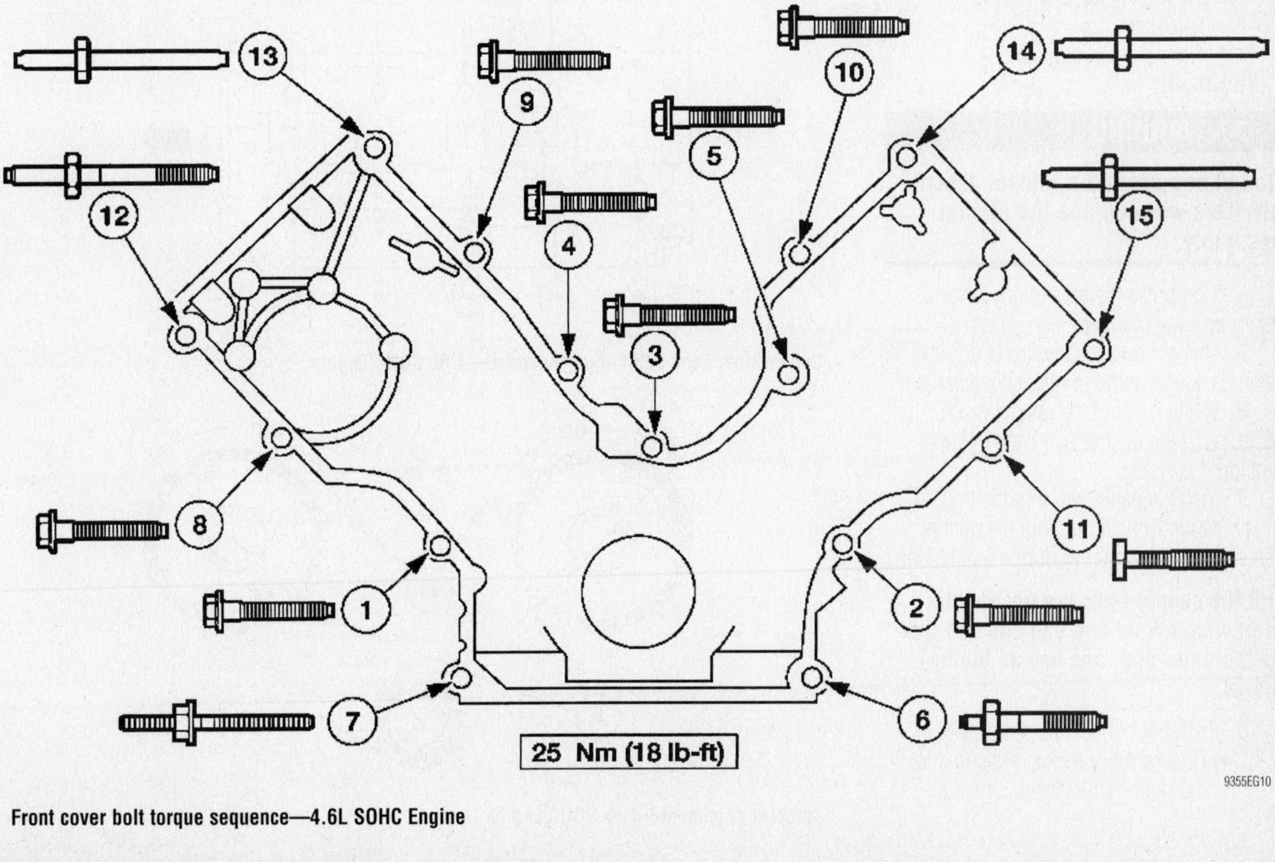

Front cover bolt torque sequence—4.6L SOHC Engine

- Step 1: Tighten to 20 Nm (15 ft. lbs.)
- Step 2: Tighten an additional 90 degrees

9. The remainder of installation is the reverse of removal.

TIMING CHAINS

1. Before servicing the vehicle, refer to the precautions in the beginning of this section.
2. Remove or disconnect the following:

⁂ WARNING

Since the engine is not free-wheeling, timing procedures must be followed exactly or piston and valve damage may occur.

- Front cover
- Crankshaft sensor ring from the crankshaft

3. Rotate the crankshaft until both camshaft key ways are 90 degrees from the valve cover surface. Make sure the copper links line up with the dots on the camshaft sprocket.
4. Install the special tools 303-380 and 303-413 on the camshaft.
5. Remove or disconnect the following:
   - LH timing chain tensioner
   - RH timing chain tensioner
   - LH and RH timing chain tensioner arm from the dowel pins
   - Timing chains and crankshaft sprocket
   - Timing chain guides

To install:

⁂ WARNING

Do not compress the ratchet assembly. This will damage the ratchet assembly.

6. Compress the tensioner plunger, using an edge of a vise.
7. Using a small screwdriver or pick, push back and hold the ratchet mechanism.
8. While holding the ratchet mechanism, push the ratchet arm back into the tensioner housing.
9. Install a paper clip into the hole in the tensioner housing to hold the ratchet assembly and plunger in during installation.

➡If the copper links are not visible, mark one link on one end and one link on the other end, and use as timing marks.

10. Install or connect the following:
    - Crankshaft sprocket, making sure the timing mark faces forward
    - Timing chain guides
    - LH (inner) timing chain on the crankshaft sprocket, aligning the copper link with the dot on the crankshaft sprocket
    - LH (inner) timing chain on the camshaft sprocket, aligning the copper link with the dot on the camshaft sprocket
    - RH (outer) timing chain on the crankshaft sprocket, aligning the copper link with the dot on the crankshaft sprocket
    - RH (outer) timing chain on the camshaft sprocket, aligning the copper link with the dot on the camshaft sprocket
11. Make sure that the copper marks on the timing chain are lined up with the corresponding dots on the crankshaft sprockets and the camshaft sprockets.
12. Make sure that the camshaft sprocket keyway is 90 degrees from the valve cover mounting surface.

➡The LH timing chain tensioner arm has a bump near the dowel hole, for identification.

13. Position the LH and RH timing chain tensioner arm on the dowel pins.

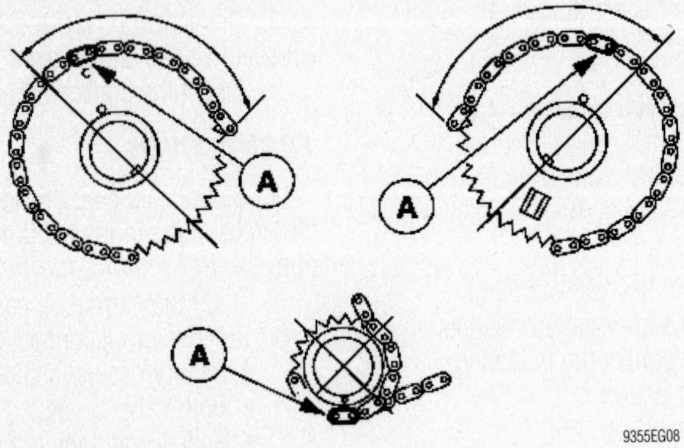

Copper link alignment—4.6L SOHC Engine

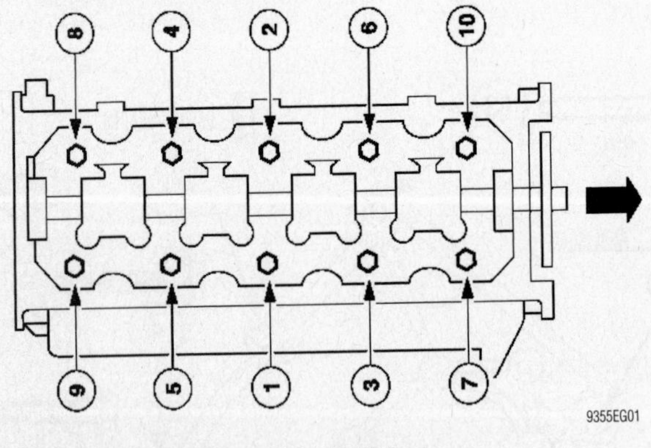

Left cylinder head bolt torque sequence—4.6L SOHC Engine

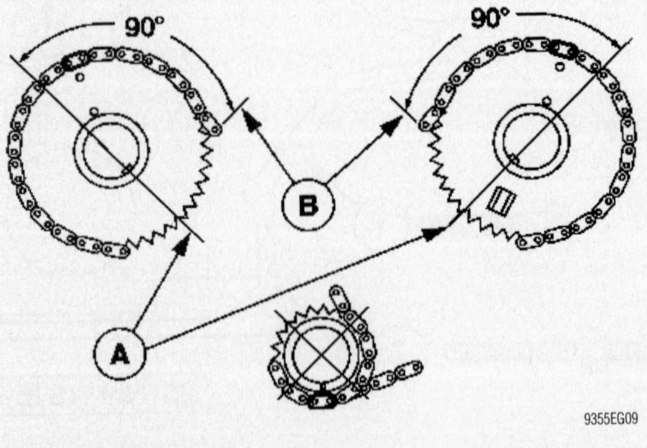

Sprocket alignment—4.6L SOHC Engine

14. Position the RH timing chain tensioner and install the bolts.
15. Position the LH timing chain tensioner and install the bolts.
16. Remove both the RH and LH retaining pins from the timing chain tensioner.
17. Remove the special tools from the camshaft.
18. Install the crankshaft sensor ring on the crankshaft.
19. Install the engine front cover.

4.6L DOHC Engine

1. Disconnect the battery ground cable.
2. Drain the engine oil.
3. Remove the valve covers.
4. If removing the timing chains, gears or other valve train components, remove the camshaft roller followers.
5. Remove the cooling fan.
6. Remove the accessory drive belt.
7. Remove the accessory drive belt tensioner.
8. Remove the accessory drive belt idler pulleys.
9. Remove the power steering reservoir bracket.
10. Remove the CMP sensor.
11. Remove the oil filter.
12. Remove the power steering hose bracket.
13. Remove the power steering pump.
14. Remove the A/C compressor electrical connector.
15. Remove the battery wiring harness.
16. Remove the A/C compressor.
17. Remove the transmission tube clip.
18. Remove the crankshaft position (CKP) sensor.
19. Remove the water pump pulley.
20. Remove the crankshaft pulley.

➡To remove the crankshaft pulley bolt, remove the upper fan shroud and electric fan.

21. Remove the crankshaft seal.
22. Remove the engine front cover.

➡Remove the four oil pan-to-engine front cover bolts.

23. Remove the engine front cover gaskets.
24. Remove the crankshaft sensor ignition pulse ring.

⁂ WARNING

Since the engine is not free-wheeling, timing procedures must be followed exactly or piston and valve damage can occur.

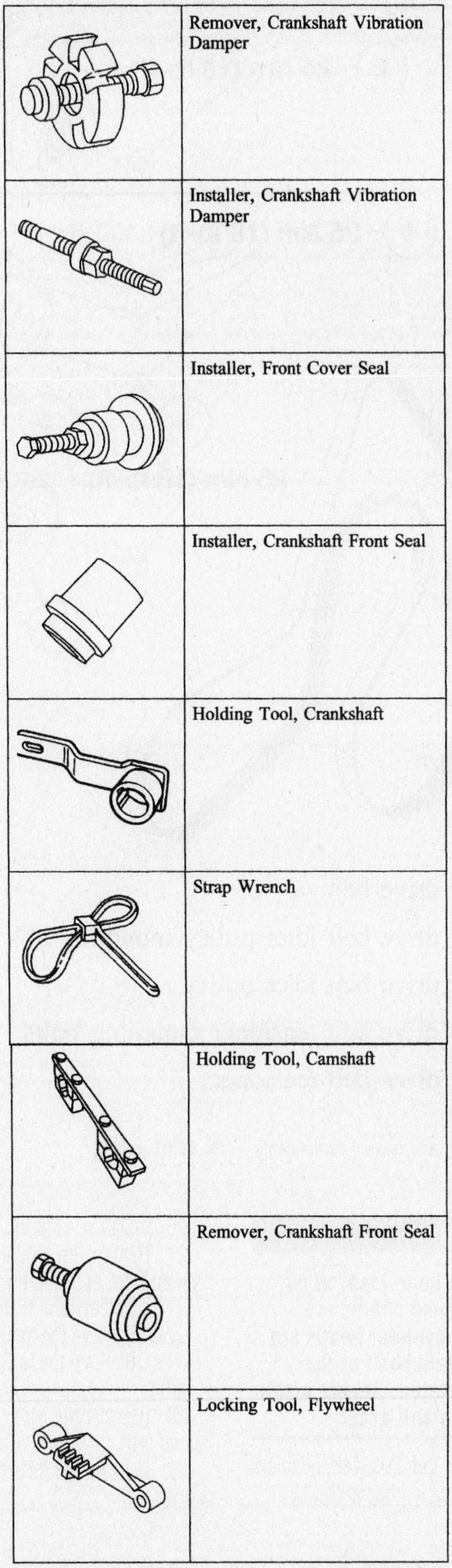
Remover, Crankshaft Vibration Damper
Installer, Crankshaft Vibration Damper
Installer, Front Cover Seal
Installer, Crankshaft Front Seal
Holding Tool, Crankshaft
Strap Wrench
Holding Tool, Camshaft
Remover, Crankshaft Front Seal
Locking Tool, Flywheel

67197-EXPL-G40

These tools, or their equivalents, are necessary for timing chain and front cover replacement—4.6L DOHC engine

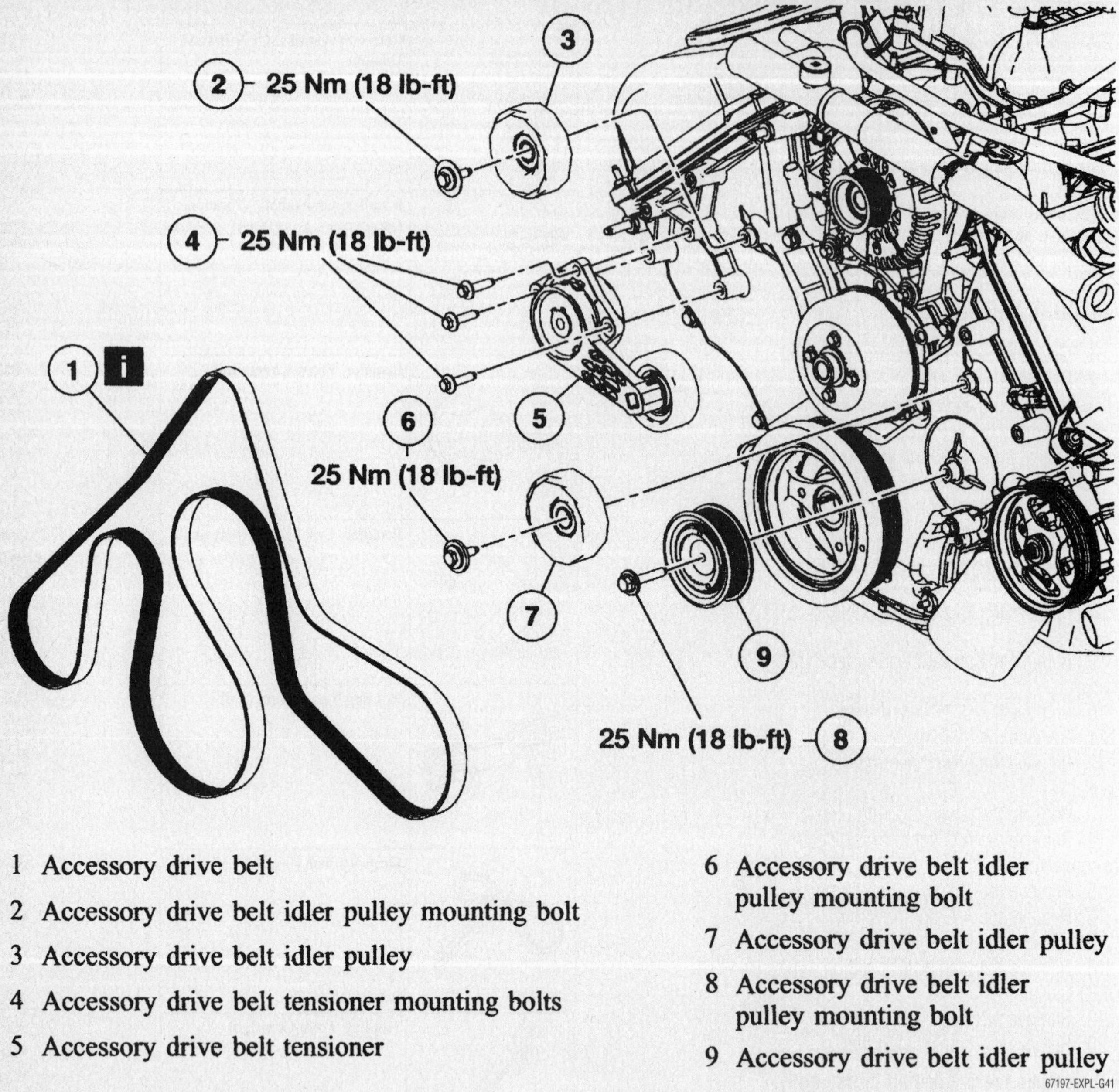

Accessory drive belt and related components—4.6L DOHC engine

WARNING

Unless otherwise instructed, at no time when the timing chains are removed and the cylinder heads are installed is the crankshaft or the camshaft to be rotated. Severe piston and valve damage will occur.

25. Using special tool 303-448, position the crankshaft with the keyway at the 12 o'clock position.
26. Install special tool 303-446.
27. Remove the timing chain tensioning system from both timing chains.
    a. Remove the bolts.
    b. Remove the timing chain tensioners.
    c. Remove the timing chain tensioner arms.
28. Remove the right timing chain tensioner arm.
29. Remove the left timing chain tensioner arm.
30. Remove the right camshaft sprocket mounting bolt.
31. Remove the right camshaft sprocket washer.
32. Remove the right camshaft sprocket.
33. Remove the left camshaft sprocket mounting bolt.
34. Remove the left camshaft sprocket washer.
35. Remove the crankshaft gear.
36. Remove the left timing chain.

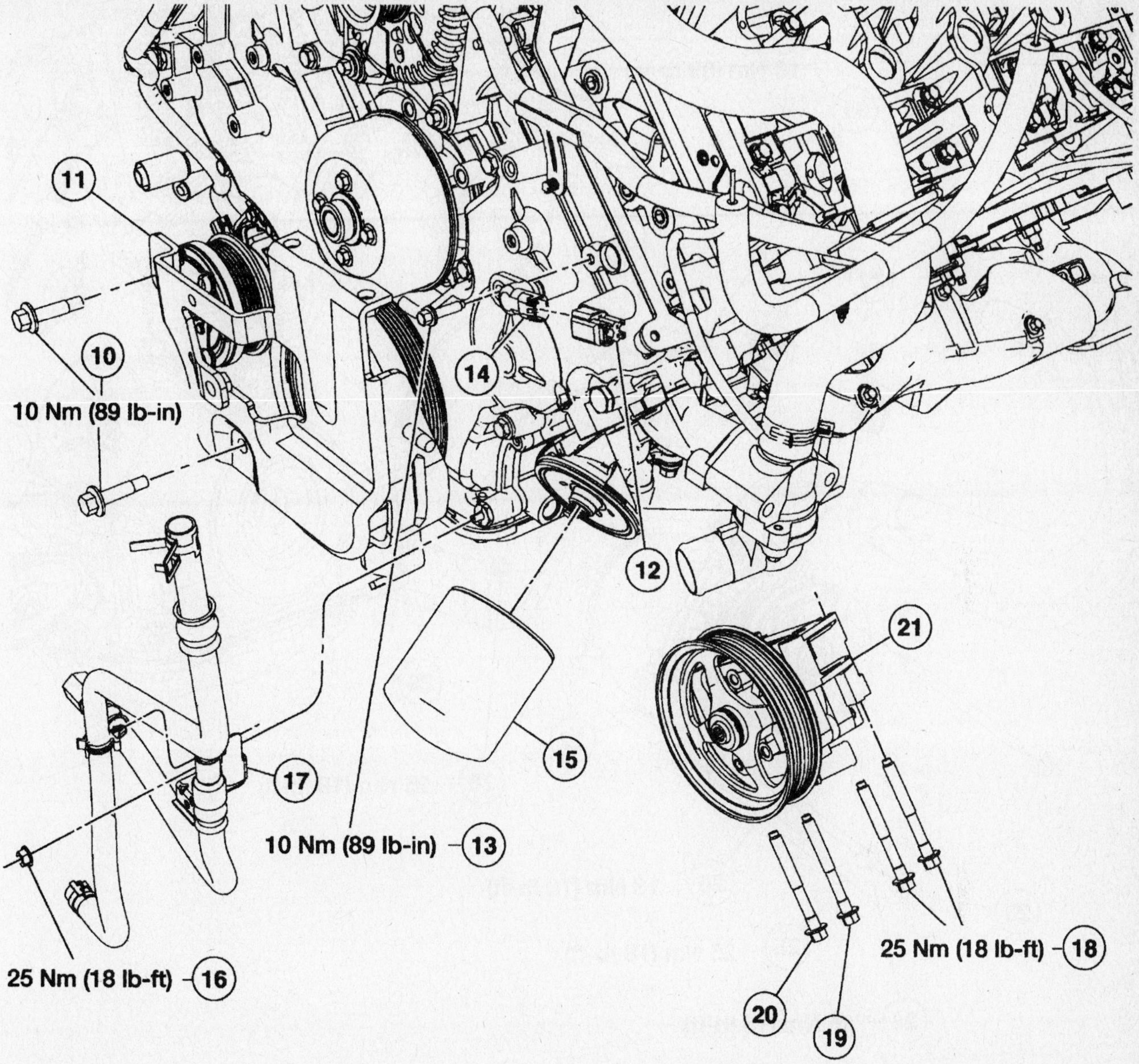

10 Power steering reservoir bracket bolts
11 Power steering reservoir bracket
12 Camshaft position (CMP) sensor electrical connector
13 CMP sensor mounting bolt
14 CMP sensor
15 Oil filter
16 Power steering hose bracket mounting nut
17 Power steering hose bracket
18 Power steering pump mounting bolts
19 Power steering pump mounting bolt
20 Power steering pump mounting bolt
21 Power steering pump

67197-EXPL-G42

Power steering pump and related components—4.6L DOHC engine

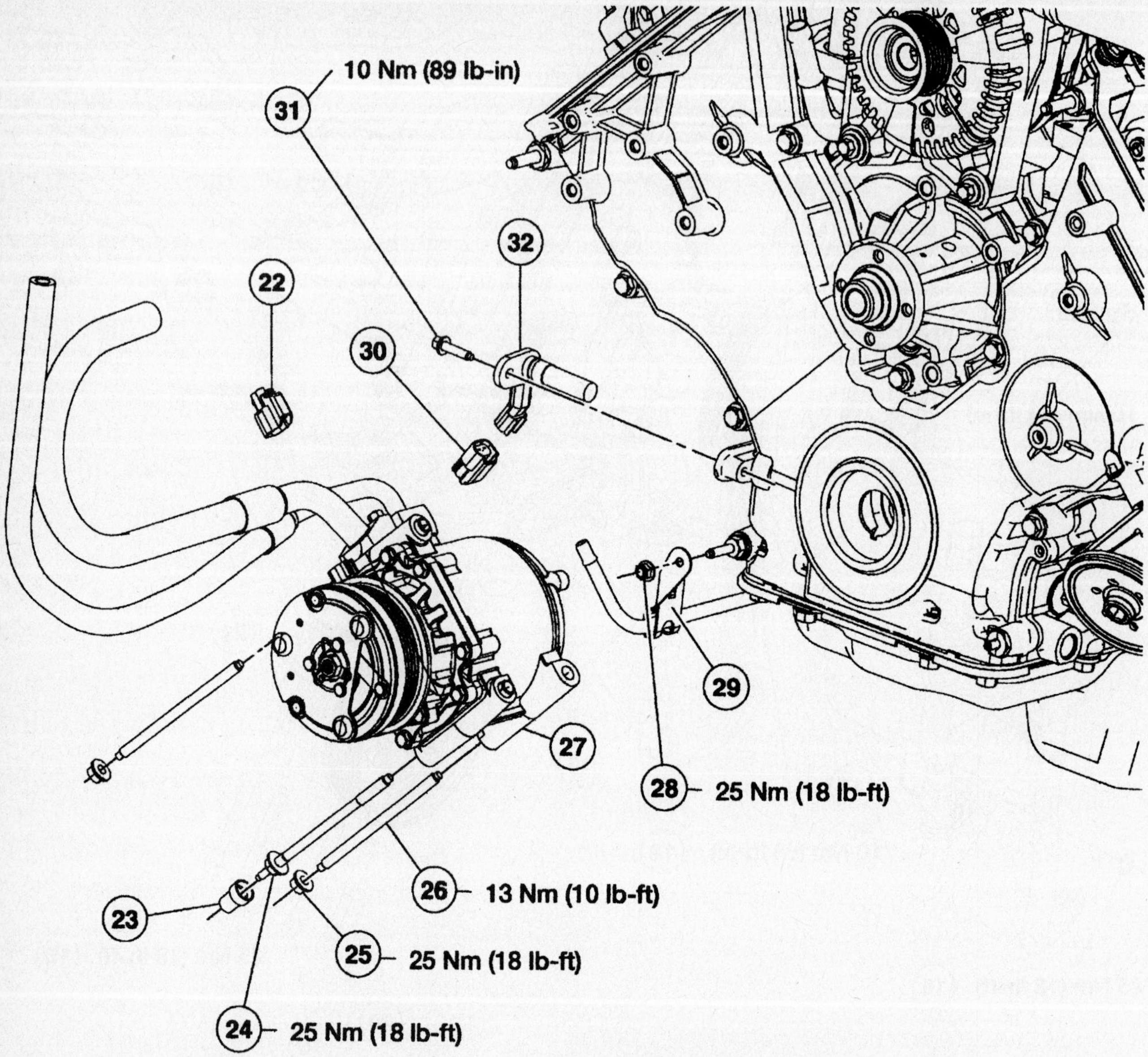

22 A/C compressor electrical connector
23 Battery wiring harness
24 A/C compressor mounting stud
25 A/C compressor mounting nuts
26 A/C compressor mounting studs
27 A/C compressor (position aside)
28 Transmission tube clip mounting nut
29 Transmission tube clip
30 Crankshaft position sensor electrical connector
31 Crankshaft position sensor mounting bolt
32 Crankshaft position (CKP) sensor

67197-EXPL-G43

A/C compressor and related components—4.6L DOHC engine

37. Remove the left camshaft sprocket.
38. Remove the left timing chain guide.
39. Remove the right timing chain guide.
40. Remove the right intake camshaft gear mounting bolt.
41. Remove the right intake camshaft gear washer.
42. Remove the right intake camshaft gear spacer.
43. Remove the left intake camshaft gear mounting bolt.
44. Remove the left intake camshaft gear washer.
45. Remove the left intake camshaft gear spacer.
46. Remove the left secondary timing chain tensioner.

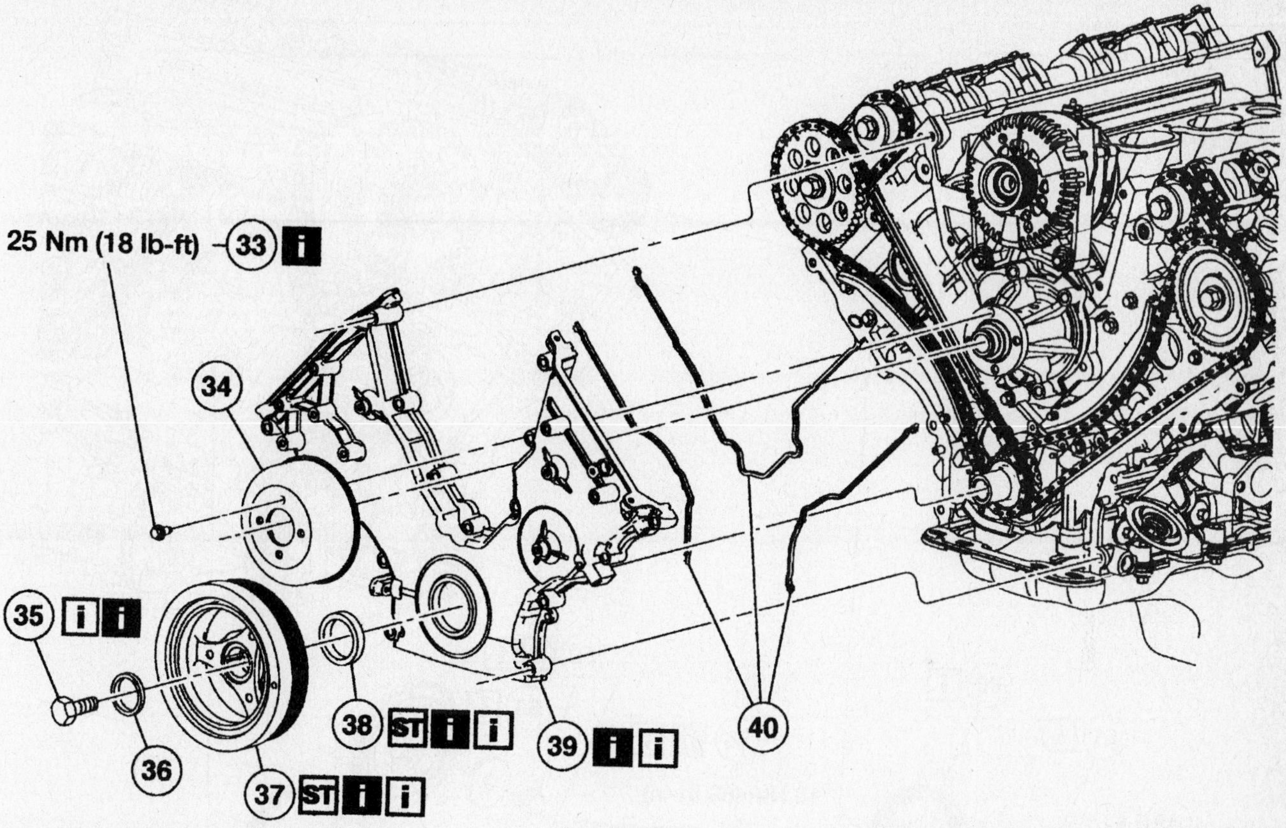

33 Water pump pulley bolts
34 Water pump pulley
35 Crankshaft pulley bolt
36 Crankshaft pulley washer
37 Crankshaft pulley
38 Crankshaft seal
39 Engine front cover
40 Engine front cover gaskets

67197-EXPL-G44

Front cover removal—4.6L DOHC engine

47. Remove the right secondary timing chain tensioner.
48. Remove the right exhaust secondary camshaft sprocket.
49. Remove the right secondary camshaft timing chain.
50. Remove the right intake secondary camshaft sprocket.
51. Remove the left intake secondary camshaft sprocket.
52. Remove the left exhaust secondary camshaft sprocket.
53. Remove the left secondary camshaft timing chain.

To install:

54. Installation is the reverse of removal. Observe the following notes:

Timing Drive Components Installation

55. Compress the tensioner and install the retaining pin. Install the tensioners.

**** CAUTION**

Timing marks must be at the 12 o'clock position and indexed at the 6 o'clock position.

56. Install the camshaft sprockets and chain as an assembly.
57. Install special tool 303-446.
58. Install the camshaft spacer, washer and bolt, and hand-tighten the bolt.
59. Install the camshaft gear, washer and bolt, and hand-tighten the bolt.
60. Tighten the bolts in two stages:
- Stage 1: Tighten to 40 Nm (30 ft. lbs.).
- Stage 2: Tighten an additional 90 degrees.

**** CAUTION**

Timing chain procedures must be followed exactly or damage to valves and pistons will result.

**** CAUTION**

Do not compress the ratchet assembly. This will damage the ratchet assembly.

61. On engines with ratcheting timing chain tensioners only.
  a. Compress each tensioner plunger, using an edge of a vise.
  b. Using a small screwdriver or pick, push back and hold the ratchet mechanism.
  c. While holding the ratchet mechanism, push the ratchet arm back into the tensioner housing.
  d. Install a paper clip into the hole of each tensioner housing to hold the

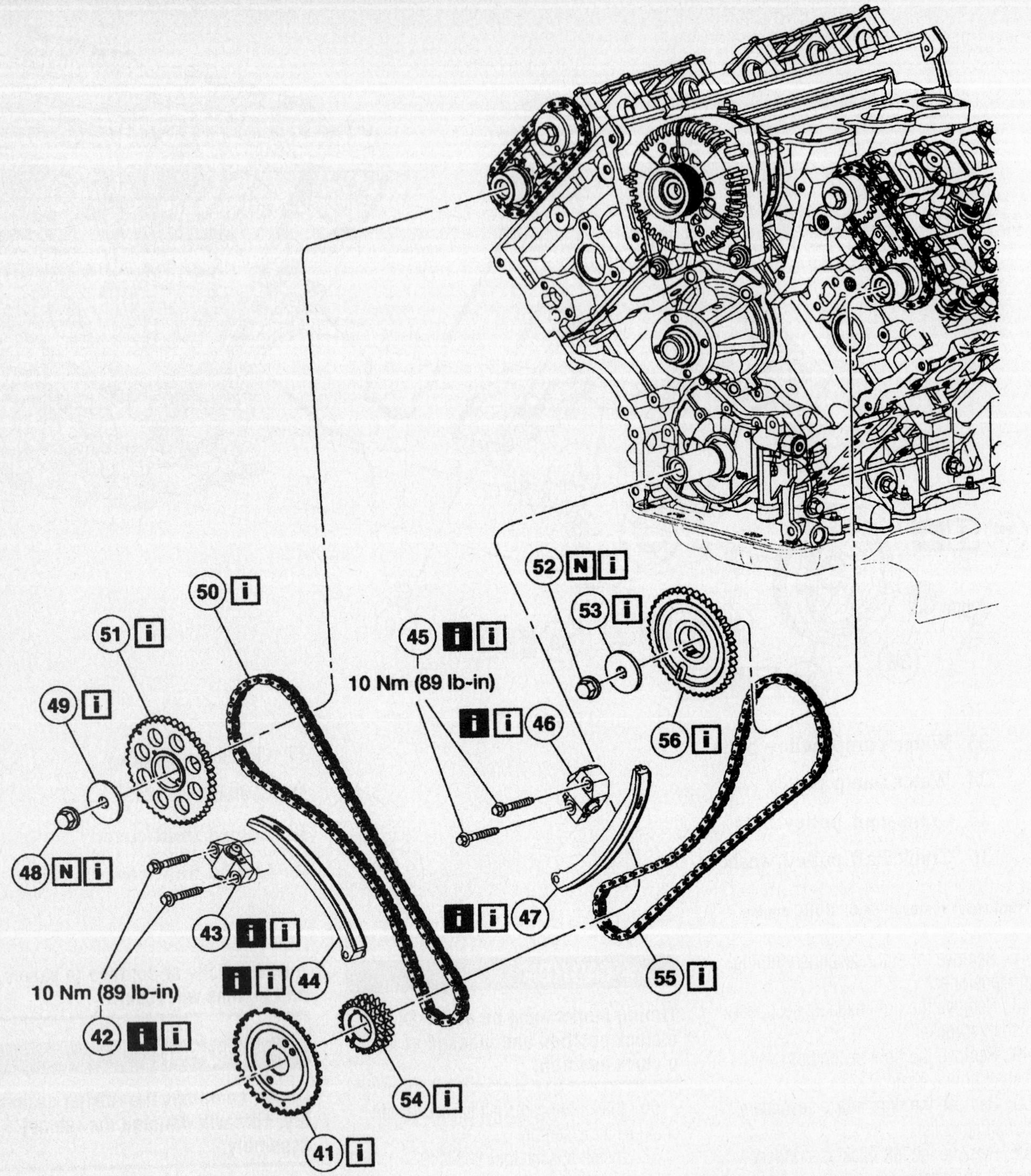

41 Crankshaft sensor ignition pulse ring
42 RH timing chain tensioner mounting bolts
43 RH timing chain tensioner
44 RH timing chain tensioner arm
45 LH timing chain tensioner mounting bolts
46 LH timing chain tensioner
47 LH timing chain tensioner arm
48 RH camshaft sprocket mounting bolt
49 RH camshaft sprocket washer
50 RH timing chain
51 RH camshaft sprocket
52 LH camshaft sprocket mounting bolt
53 LH camshaft sprocket washer
54 Crankshaft gear
55 LH timing chain
56 LH camshaft sprocket

Timing chains and related parts—4.6L DOHC engine

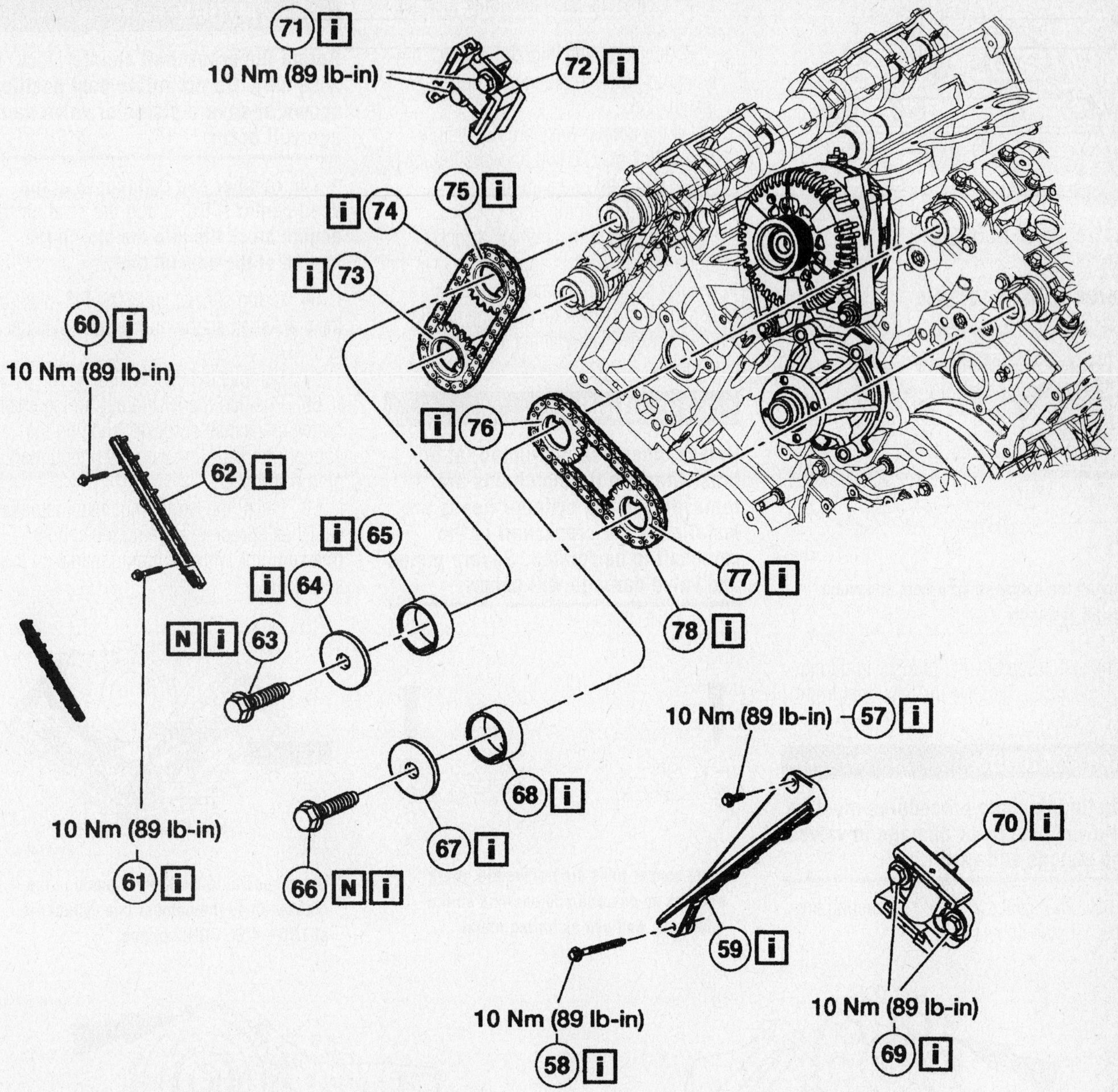

57 LH timing chain guide upper mounting bolt
58 LH timing chain guide lower mounting bolt
59 LH timing chain guide
60 RH timing chain guide upper mounting bolt
61 RH timing chain guide lower mounting bolt
62 RH timing chain guide
63 RH intake camshaft gear mounting bolt
64 RH intake camshaft gear washer
65 RH intake camshaft gear spacer
66 LH intake camshaft gear mounting bolt
67 LH intake camshaft gear washer
68 LH intake camshaft gear spacer
69 LH secondary timing chain tensioner mounting bolts
70 LH secondary timing chain tensioner
71 RH secondary timing chain tensioner mounting bolts
72 RH secondary timing chain tensioner
73 RH exhaust secondary camshaft sprocket
74 RH secondary camshaft timing chain
75 RH intake secondary camshaft sprocket
76 LH intake secondary camshaft sprocket
77 LH exhaust secondary camshaft sprocket
78 LH secondary camshaft timing chain

67197-EXPL-G46

Secondary timing chains and related parts—4.6L DOHC engine

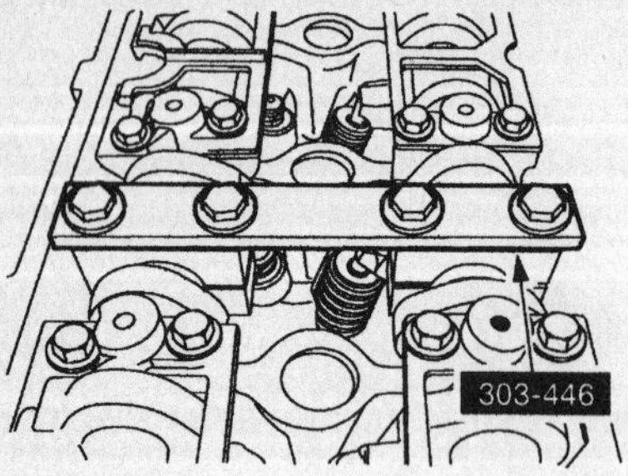

67197-EXPL-G47

Install special tool 303-446

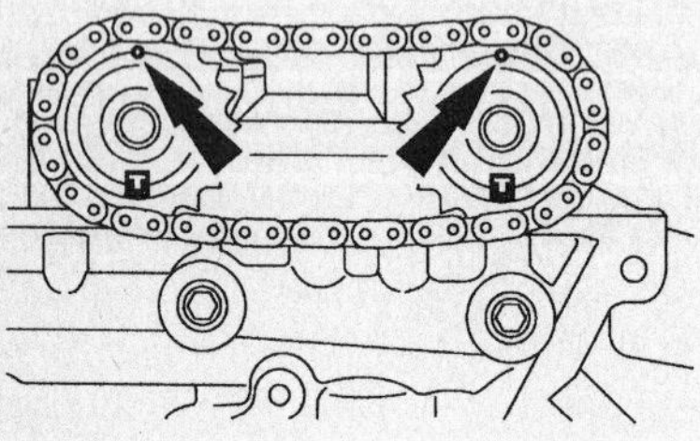

67197-EXPL-G48

Install the camshaft sprockets and chain as an assembly

ratchet assembly and plunger in during installation. Remove the tensioner from the vise.

⁂ CAUTION

The timing chain procedures must be followed exactly or damage to valves and pistons will result.

62. On engines with non-ratcheting timing chain tensioners only.

a. Compress each tensioner plunger, using a vise.

b. Install a retaining clip on each tensioner to hold the plunger in during installation.

63. If the copper links are not visible, mark one link on one end and one link on the other end, and use as timing marks.

64. Install the timing chain guides.

65. Rotate the left camshaft sprocket until the timing mark is approximately at the 12 o'clock position. Rotate the right camshaft timing sprocket until the timing mark is approximately at the 11 o'clock position.

⁂ CAUTION

Unless otherwise instructed, at no time when the timing chains are removed and the cylinder heads are installed is the crankshaft or the camshaft to be rotated. Severe piston and valve damage will occur.

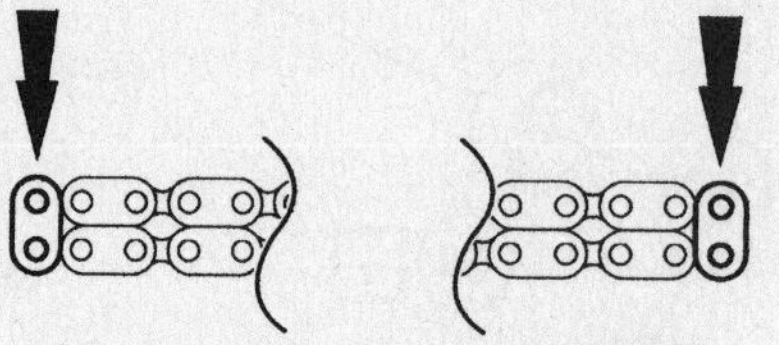

67197-EXPL-G49

If the copper links are not visible, mark one link on one end and one link on the other end, and use as timing marks

⁂ CAUTION

Rotate the crankshaft counterclockwise only. Do not rotate past position shown or severe piston or valve damage will occur.

➡The number one cylinder is at top dead center (TDC) when the stud on the engine block fits into the slot in the handle of the special tool.

66. Using special tool 303-448, position the crankshaft so the number one cylinder is at TDC.

67. Remove the special tool.

68. Position the left (inner) timing chain on the crankshaft sprocket, aligning the copper (marked) link with the timing mark on the sprocket.

69. Install the left timing chain onto the camshaft sprocket, aligning the copper (marked) link with the timing marks on the sprocket.

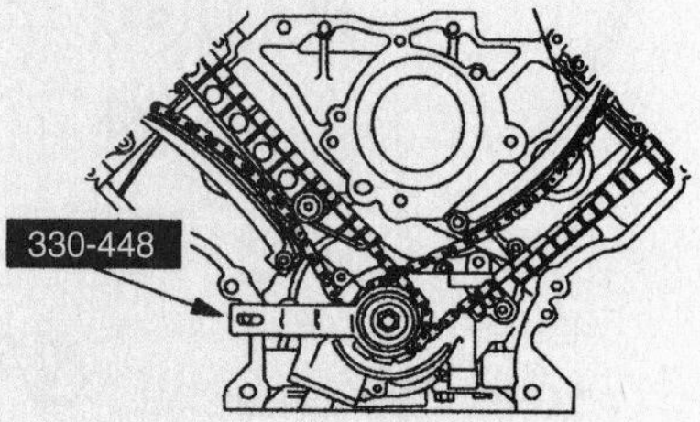

67197-EXPL-G51

Using special tool 303-448, position the crankshaft so the number one cylinder is at TDC—4.6L DOHC engine

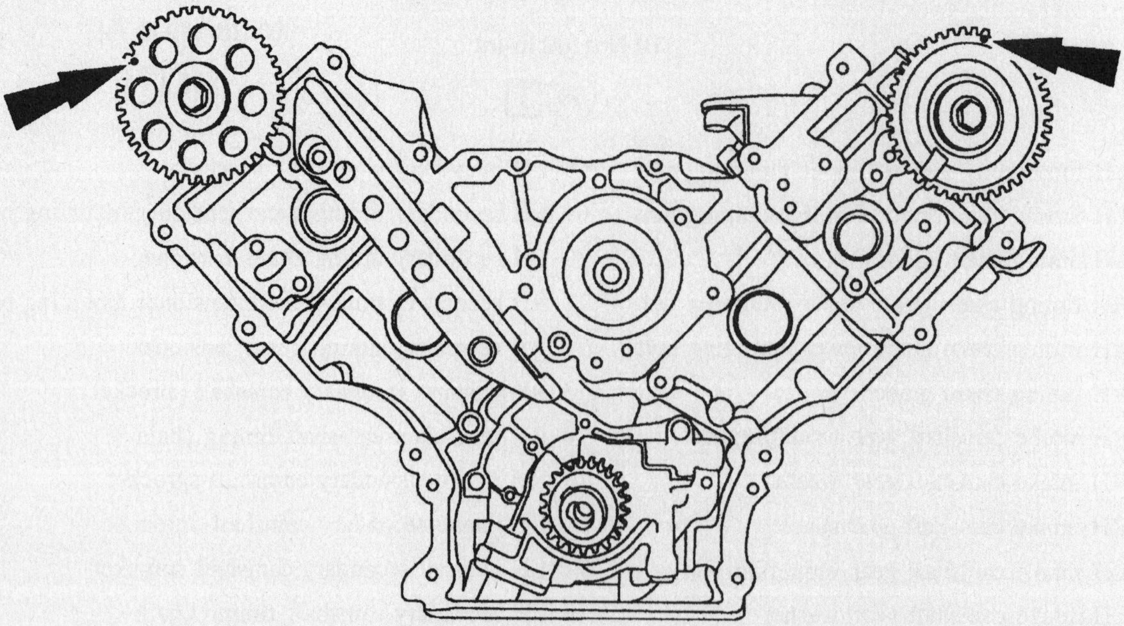

67197-EXPL-G50

Rotate the left camshaft sprocket until the timing mark is approximately at the 12 o'clock position. Rotate the right camshaft timing sprocket until the timing mark is approximately at the 11 o'clock position—4.6L DOHC engine

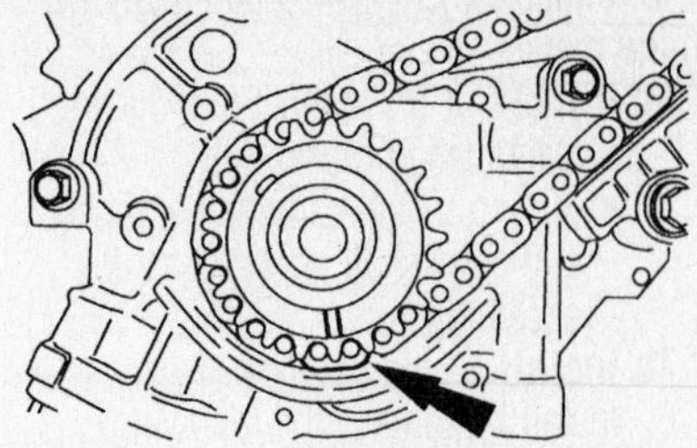

67197-EXPL-G52

Position the left (inner) timing chain on the crankshaft sprocket, aligning the copper (marked) link with the timing mark on the sprocket—4.6L DOHC engine

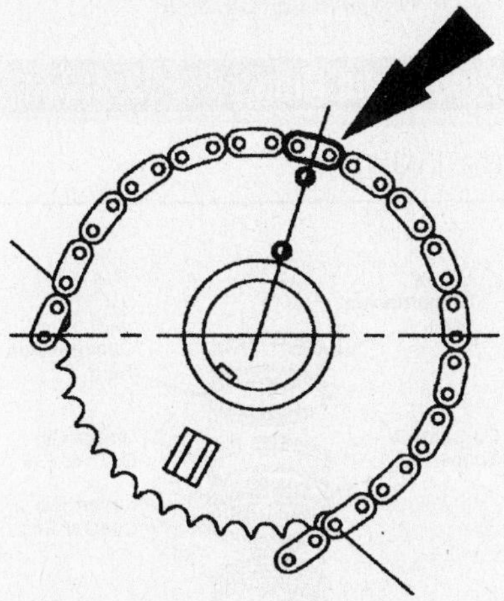

67197-EXPL-G53

Install the left timing chain onto the camshaft sprocket, aligning the copper (marked) link with the timing marks on the sprocket—4.6L DOHC engine

➡The left timing chain tensioner arm has a bump near the dowel hole for identification.

70. Position the left timing chain tensioner arms on the dowel pin and install the left timing chain tensioner.

71. On engines with ratcheting timing chain tensioners only, remove the retaining clip from the left timing chain tensioner.

72. On engines with non-ratcheting timing chain tensioners only, remove the retaining clip from the left timing chain tensioner.

73. Position the right (outer) timing chain on the crankshaft sprocket, aligning the copper (marked) link with the timing mark on the sprocket.

74. Install the right timing chain onto the camshaft sprocket, aligning the copper (marked) link with the timing marks on the sprocket.

75. Position the right timing chain tensioner arms on the dowel pin and install the right timing chain tensioner.

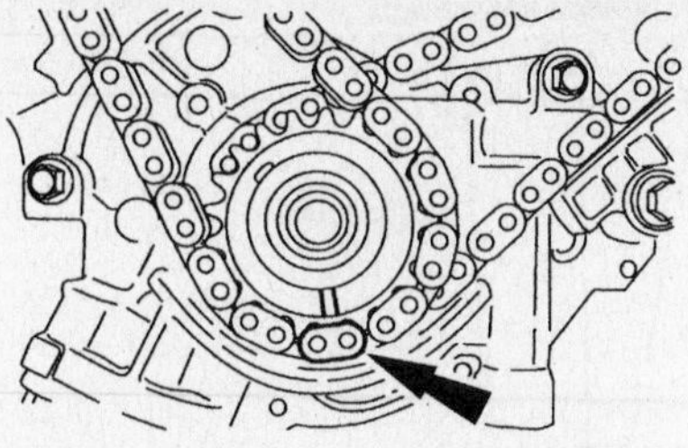

67197-EXPL-G54

Position the right (outer) timing chain on the crankshaft sprocket, aligning the copper (marked) link with the timing mark on the sprocket—4.6L DOHC engine

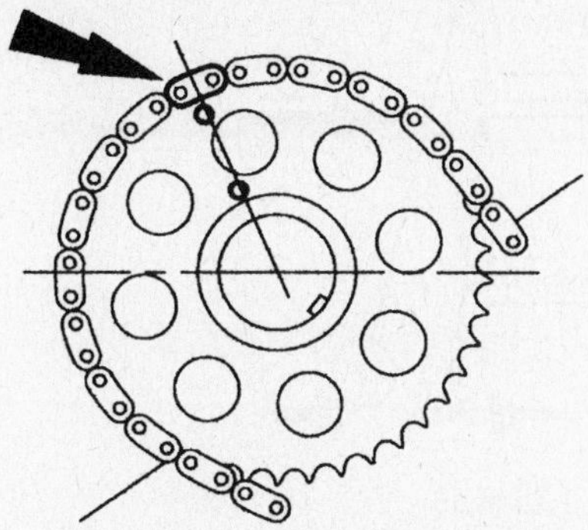

67197-EXPL-G55

Install the right timing chain onto the camshaft sprocket, aligning the copper (marked) link with the timing marks on the sprocket—4.6L DOHC engine

76. On engines with ratcheting timing chain tensioners only, remove the retaining clip from the right timing chain tensioner.

77. On engines with non-ratcheting timing chain tensioners only, remove the retaining clips from the right timing chain tensioner.

➡Applies to all engines.

78. Make sure that the copper (marked) chain links are lined up with the dots on the crankshaft sprockets and the camshaft sprocket.

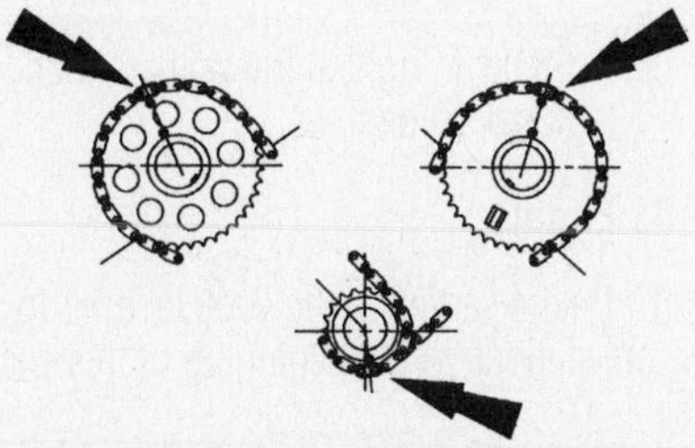

67197-EXPL-G56

Make sure that the copper (marked) chain links are lined up with the dots on the crankshaft sprockets and the camshaft sprocket—4.6L engine

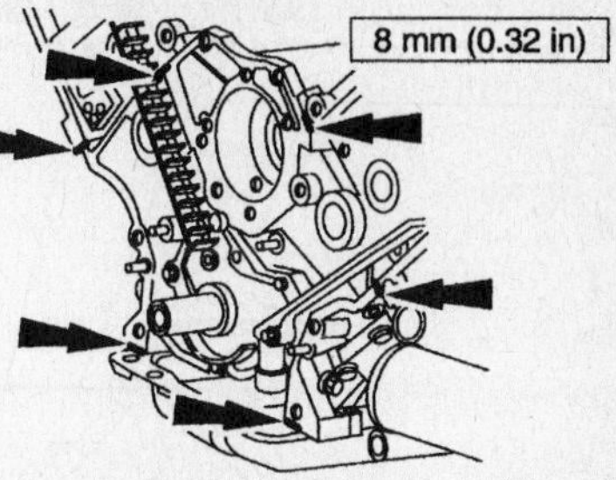

67197-EXPL-G58

Apply sealant in the locations shown—4.6L engine

Engine Front Cover Installation

➡If the engine front cover is not secured within four minutes, the sealant must be removed and the sealing area cleaned with metal surface cleaner. Allow to dry until there is no sign of wetness, or four minutes, whichever is longer. Failure to follow these instructions can result in future oil leakage.

79. Apply sealant in the locations shown.

80. Install the engine front cover seals and the engine front cover, and tighten the fasteners in the sequence shown.

81. Tighten the oil pan bolts in the sequence shown in three stages:

- Stage 1: Tighten to 2 Nm (18 inch lbs.).
- Stage 2: Tighten to 20 Nm (15 ft. lbs.).
- Stage 3: Tighten an additional 60 degrees.

Crankshaft Seal Installation

82. Lubricate the engine front cover and the crankshaft seal inner lip with clean engine oil.

83. Using the special tools, install the crankshaft seal.

Crankshaft Pulley Bolt Installation

➡The crankshaft pulley must be installed within four minutes after applying the sealant.

84. Apply sealant to the Woodruff key slot on the crankshaft pulley.

85. Using the special tool, install the crankshaft pulley.

86. Install the bolt and the washer. Tighten the bolt in four stages:

- Stage 1: Tighten the bolt to 90 Nm (66 ft. lbs.).
- Stage 2: Loosen the bolt one full turn.
- Stage 3: Tighten the bolt to 50 Nm (37 ft. lbs.).

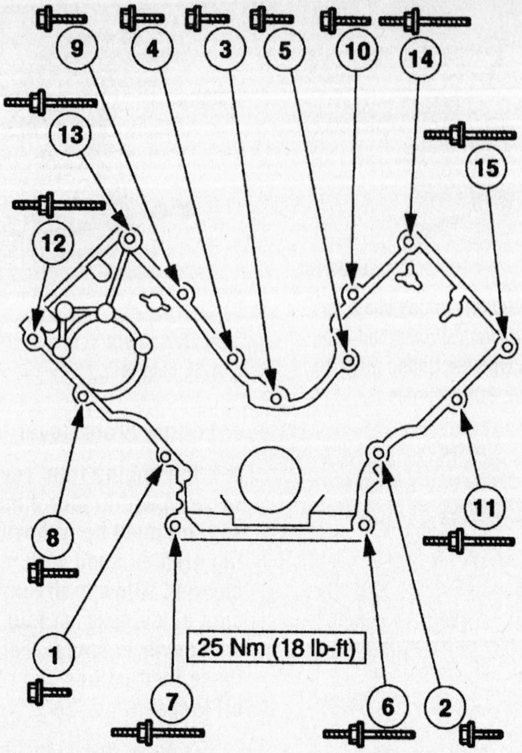

1 Bolt, Hex Flange Head Pilot, M8 x 1.25 x 53
2 Bolt, Hex Flange Head Pilot, M8 x 1.25 x 53
3 Bolt, Hex Flange Head Pilot, M8 x 1.25 x 53
4 Bolt, Hex Flange Head Pilot, M8 x 1.25 x 53
5 Bolt, Hex Flange Head Pilot, M8 x 1.25 x 53
6 Stud, Hex Shldr Pilot, M8 x 1.25 x 50 — M6 x 1 x 10
7 Stud and Washer, Hex Head Pilot, M8 x 1.25 x 60 — M6 x 1 x 26
8 Bolt, Hex Flange Head Pilot, M8 x 1.25 x 53
9 Bolt, Hex Flange Head Pilot, M8 x 1.25 x 53
10 Bolt, Hex Flange Head Pilot, M8 x 1.25 x 53
11 Stud, Hex Shldr Pilot, M8 x 1.25 x 65 — M8 x 1.25 x 26
12 Stud, Hex Head Pilot, M8 x 1.25 x 65 — M8 x 1.25 x 16
13 Stud, Hex Shldr Pilot, M8 x 1.25 x 65 — M8 x 1.25 x 26
14 Stud, Hex Shldr Pilot, M8 x 1.25 x 65 — M8 x 1.25 x 26
15 Stud, Hex Shldr Pilot, M8 x 1.25 x 65 — M8 x 1.25 x 26

67197-EXPL-G57

Front cover bolt torque sequence—4.6L engine

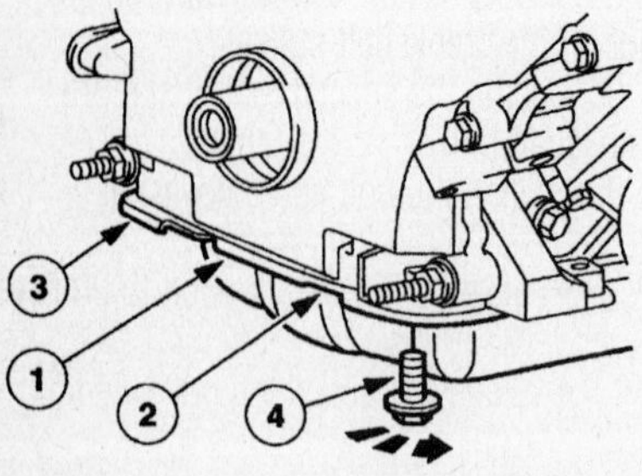

Oil pan bolt torque sequence—4.6L engine

- Stage 4: Tighten the bolt an additional 90 degrees.

5.0L Engine

1. Before servicing the vehicle, refer to the precautions in the beginning of this section.
2. Drain the engine oil.
3. Remove or disconnect the following:
   - Negative battery cable
   - Engine front cover
4. Rotate the crankshaft and align the timing marks.
5. Remove or disconnect the following:
   - Camshaft sprocket bolt
   - Timing chain, camshaft sprocket and crankshaft sprocket as an assembly

To install:

6. Install or connect the following:
   - Timing chain, camshaft and crankshaft sprockets as an assembly
7. Align the timing marks.
   - Camshaft sprocket bolt. Torque the bolt to 45 ft. lbs. (61 Nm).
   - Engine front cover
   - Negative battery cable

Piston and Ring

POSITIONING

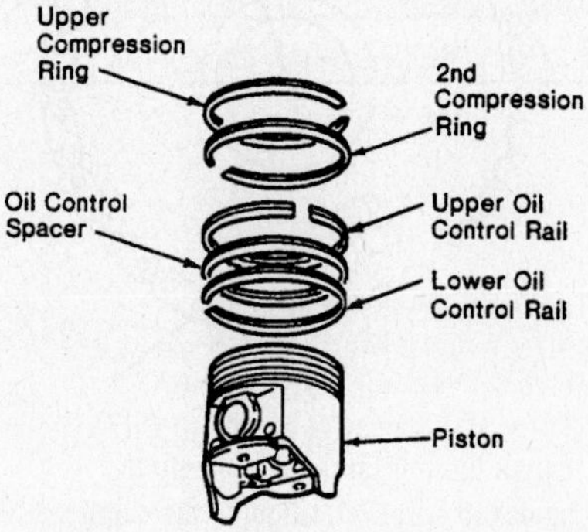

Piston ring positioning

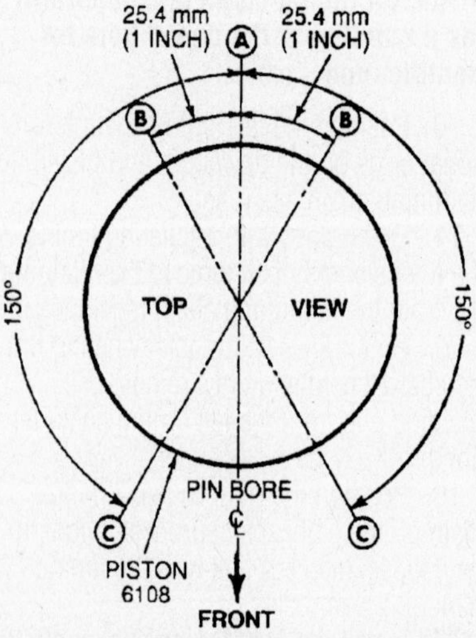

Piston ring end gap spacing

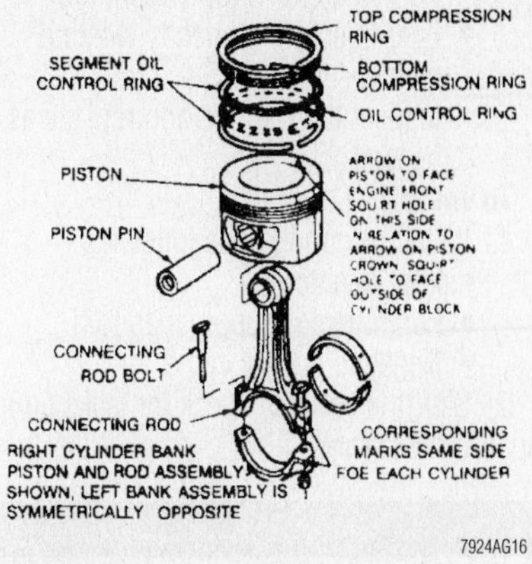

Piston and connecting rod positioning on 4.0L

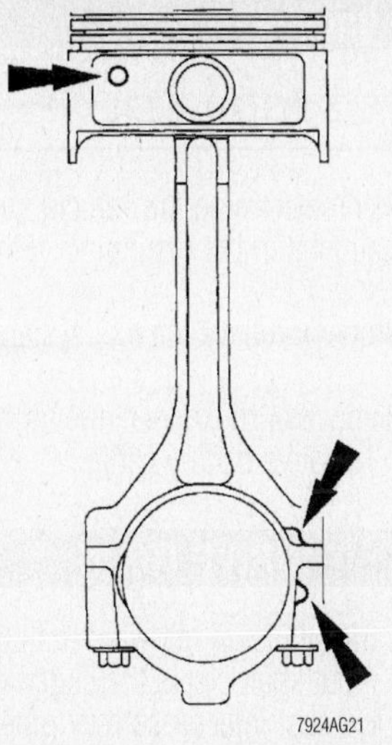

Piston and connecting rod positioning on 4.6L

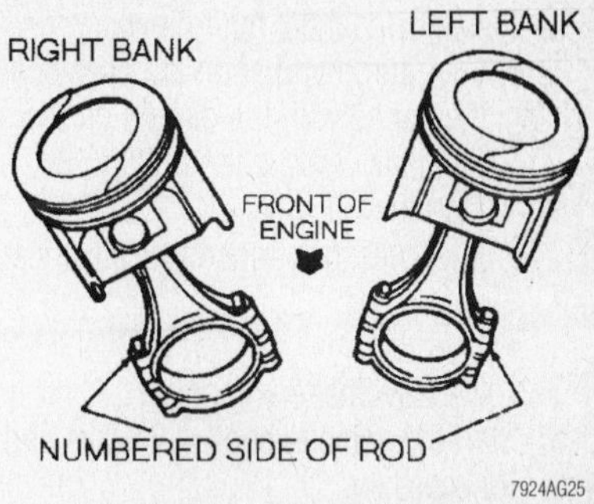

Piston and connecting rod positioning on 5.0L

FUEL SYSTEM

Fuel System Service Precautions

Safety is the most important factor when performing not only fuel system maintenance, but any type of maintenance. Failure to conduct maintenance and repairs in a safe manner may result in serious personal injury or death. Work on a vehicle's fuel system components can be accomplished safely and effectively by adhering to the following rules and guidelines.

- To avoid the possibility of fire and personal injury, always disconnect the negative battery cable unless the repair or test procedure requires that battery voltage be applied.
- Always relieve the fuel system pressure prior to disconnecting any fuel system component (injector, fuel rail, pressure regulator, etc.) fitting or fuel line connection. Exercise extreme caution whenever relieving fuel system pressure, to avoid exposing your skin, face and eyes to fuel spray. Please be advised that fuel under pressure may penetrate the skin or any part of the body that it contacts.
- Always place a shop towel or cloth around the fitting or connection prior to loosening to absorb any excess fuel due to spillage. Ensure that all fuel spillage is quickly remove from engine surfaces. Ensure that all fuel-soaked cloths or towels are deposited into a flame-proof waste container with a lid.
- Always keep a dry chemical (Class B) fire extinguisher near the work area.
- Do not allow fuel spray or fuel vapors to come into contact with a light bulb, spark or open flame.
- Always use a second wrench when loosening or tightening fuel line connection fittings. This will prevent unnecessary stress and torsion to fuel piping. Always follow the proper torque specifications.
- Always replace worn fuel fitting O-rings with new ones. Do not substitute fuel hose where rigid pipe is installed.

Relieving Fuel System Pressure

All engines are equipped with a pressure relief valve located on the fuel supply manifold. Remove the fuel tank cap and attach fuel pressure gauge T80L-9974-B, to the valve to release the fuel pressure. Be sure to drain the fuel into a suitable container and to avoid gasoline spillage. If a pressure gauge is not available, disconnect the vacuum hose from the fuel pressure regulator and attach a hand-held vacuum pump. Apply about 25 in. Hg (84 kPa) of vacuum to the regulator to vent the fuel system pressure into the fuel tank through the fuel return hose. Note that this procedure will remove the fuel pressure from the lines, but not the fuel. Take precautions to avoid the risk of fire and use clean rags to soak up any spilled fuel when the lines are disconnected.

An alternate method of relieving the fuel system pressure involves disconnecting the inertia switch.

Fuel Filter

REMOVAL & INSTALLATION

1. Before servicing the vehicle, refer to the precautions in the beginning of this section.
2. Properly relieve the fuel system pressure.
3. Remove or disconnect the following:
   - Negative battery cable
   - Fuel lines
   - Fuel filter from the support

To install:

4. Install or connect the following:
   - Fuel filter to the support
   - Fuel lines
   - Negative battery cable
5. Start the vehicle, check for leaks and repair if necessary.

Fuel Pump

REMOVAL & INSTALLATION

1. Before servicing the vehicle, refer to the precautions in the beginning of this section.
2. Properly relieve the fuel system pressure.
3. Remove or disconnect the following:
   - Negative battery cable
   - Fuel tank
   - Fuel pressure transducer electrical connector
   - Fuel pump assembly

To install:

4. Install or connect the following:
   - Fuel pump and align the arrow on the flange with the dimple on the fuel tank. Torque the bolts to 80 inch lbs. (9 Nm).
   - Fuel pressure transducer electrical connector
   - Fuel tank
   - Negative battery cable
5. Start the vehicle, check for leaks and repair if necessary.

Fuel Injectors

REMOVAL & INSTALLATION

1. Before servicing the vehicle, refer to the precautions in the beginning of this section.
2. Properly relieve the fuel system pressure.
3. Remove or disconnect the following:
   - Negative battery cable
   - Fuel injection supply manifold
   - Fuel injectors by gently twisting them
   - Inspect the O-rings and replace as needed

To install:

4. Install or connect the following:
   - Fuel injectors
   - Fuel injector supply manifold
   - Negative battery cable
5. Start the vehicle, check for leaks and repair if necessary.

DRIVE TRAIN

Transmission Assembly

REMOVAL & INSTALLATION

Manual Transmission

2001

1. Before servicing the vehicle, refer to the precautions in the beginning of this section.
2. Drain the transmission fluid.
3. Place the transmission in **Neutral**.
4. Remove or disconnect the following:
   - Negative battery cable
   - Gearshift lever assembly from the control housing
5. On 2WD vehicles, matchmark the driveshaft to the rear axle flange. Position a drain pan under the rear of the transmission. Remove the driveshaft-to-rear axle flange fasteners and pull the driveshaft rearward to disengage it from the transmission.
   - Heated Oxygen (HO_2S) sensor
   - Back-up lamp switch electrical connector
   - Fan shroud
   - Clutch hydraulic line at the clutch housing
   - Speedometer from the transfer case/extension housing and place a wood block on a service jack and position the jack under the engine oil pan
   - Transfer case on 4WD vehicles
   - Starter
6. Position a transmission jack under the transmission.
   - Transmission-to-engine retaining bolts and washers
   - Transmission mount and damper to the crossmember
   - Crossmember and lower the engine jack slightly to angle the transmission assembly. Work the clutch housing off the locating dowels and slide the clutch housing and the transmission rearward until the input shaft clears the clutch disc
   - Exhaust inlet cross over pipe
   - Transmission from the vehicle

To install:

7. Check that the mating surfaces of the clutch housing, engine rear and dowel holes are free of burrs, dirt and paint.
8. Place the transmission on the transmission jack. Position the transmission under the vehicle, then raise it into position. Align the input shaft splines with the clutch disc splines and work the transmission forward onto the locating dowels.
9. Install or connect the following:
   - Transmission-to-engine retaining bolts and washers. Tighten the retaining bolts to 30–41 ft. lbs. (40–55 Nm).
   - Exhaust inlet cross over pipe and remove the transmission jack
   - Right side transmission mount. Torque the bolt to 81 ft. lbs. (110 Nm).
   - Rear crossmember. Torque the bolts to 53 ft. lbs. (72 Nm).
   - Starter motor and tighten the attaching nuts
   - Transfer case on 4WD vehicles. Torque the bolts to 87 ft. lbs. (119 Nm).
   - Rear driveshaft
   - Starter motor, back-up lamp switch connectors
   - Hydraulic clutch line and bleed the system
   - Speedometer cable
   - Gearshift lever assembly
   - Negative battery cable
10. Fill the transmission fluid to the proper level.
11. Check for proper shifting and operation of the transmission.

2002–04

1. Before servicing the vehicle, refer to the precautions in the beginning of this section.
2. Disconnect the negative battery cable

➡Do not remove the gearshift lever knob.

3. Remove the screw and lift the console finish panel to access the gearshift lever nut. Remove the gearshift lever nut, then remove the upper gearshift lever and console finish panel as an assembly.
4. Install the nut on the front of the lever, then tighten the nut to remove the eccentric stud out of the gearshift lever.
5. place the transmission in NEUTRAL.
6. If the transmission is being disassembled, drain the transmission fluid.
7. Remove or disconnect the following:
   - Front support
   - Starter
   - Rear driveshaft and position it aside
   - Transfer case
8. Using a suitable high lift jack, support the transmission. Securely strap the jack to the transmission.
9. Remove or disconnect the following:
   - RH crossmember cover, then the four bolts
   - LH crossmember bolts
   - Heat shields from the crossmember
   - Nuts, then lower the crossmember from the vehicle
   - Hydraulic line from the clutch slave cylinder
   - Output shaft speed (OSS) sensor electrical connector
   - Reverse lamp switch electrical connector
   - Wiring harness brackets from the transmission, then disconnect the heated oxygen sensor (HO2S) electrical connectors
   - The two rear three way catalytic converter (TWC) bolts
   - Front TWC bolts and the TWC
   - The eight transmission-to-engine bolts
   - Transmission.

To install:

10. Raise and position the transmission to the engine and clutch.
11. Install the eight transmission-to-engine bolts. Torque the bolts to 44 ft. lbs. (50Nm).
12. Position the front three way catalytic converter (TWC) in the vehicle and install the bolts. Do not tighten the bolts at this time.
13. Using a new gasket and nuts, install the two rear three way catalytic converter (TWC) bolts. Tighten the rear bolts, then the front bolts at this time. Tighten the bolts to 30 ft. lbs. (40Nm).
14. Install or connect the following:
    - Heated oxygen sensor (HO2S) electrical connectors
    - RH and LH crossmember bolts and the RH crossmember cover. Torque the crossmember nuts to 46 ft. lbs. (63Nm).
    - Transmission mount nuts. Torque the nuts to 72 ft. lbs. (98Nm).
    - Heat shields to the crossmember
    - Wiring harness brackets and attach the wiring harness
    - Front support
    - Hydraulic line to the clutch slave cylinder
    - Starter
    - Transfer case
    - Driveshaft
15. Position the upper gearshift lever, the outer gearshift lever boot and console finish panel into the console, then install the nut.
16. Connect the battery ground cable.
17. Bleed the clutch hydraulic system.

Automatic Transmission

2001

1. Before servicing the vehicle, refer to the precautions in the beginning of this section.
2. Drain the transmission fluid.
3. Place the transmission in **Neutral**.
4. Remove or disconnect the following:
    - Negative battery cable
    - Fluid level indicator
    - Fan shroud
    - Transfer case on 4WD vehicles
    - Rear driveshaft after matchmarking the yoke and axle flange
    - Starter
    - Access cover and adapter plate bolts from the lower left side of the converter housing
    - Flywheel-to-converter attaching nuts. Use a socket and breaker bar on the crankshaft pulley attaching bolt. Rotate the pulley clockwise as viewed from the front to gain access to each of the nuts.
    - Shifter cable
    - Transmission wire harness
    - Heated Oxygen (HO_2S) sensor connector
    - Transmission connector
    - Digital Transmission Range (TR) sensor connector
    - Catalytic converter
    - Speedometer cable and/or vehicle speed sensor from the transfer case (4WD) or extension housing (2WD)
    - Transmission cooler lines
    - Engine rear support-to-crossmember bolts and the crossmember-to-frame side support attaching nuts and bolts
    - Crossmember
    - Transmission mount
    - Transmission upper fill tube
    - Vent tube on 4WD vehicles
    - Converter housing-to-engine bolts
5. Move the transmission to the rear so it disengages from the dowel pins and the converter is disengaged from the flywheel. Lower the transmission from the vehicle.
6. Remove the torque converter from the transmission, if necessary.

To install:

7. Install the converter on the transmission.

✲✲ WARNING

Before installing an automatic transmission, always check that the torque converter is fully seated into the transmission. Typically, the converter has notches or tangs on the hub that must engage the transmission fluid pump. If they are not engaged in the pump, the transmission will not mate to the engine properly, as the converter will be holding it away. Severe damage to the pump, converter or transmission casting can occur if the transmission-to-engine bolts are tightened to force the transmission to mate to the engine.

Proper installation of the converter requires full engagement of the converter hub in the pump gear. To accomplish this, the converter must be pushed and at the same time rotated through what feels like 2 notches or bumps. When fully installed, rotation of the converter will usually result in a clicking noise heard, caused by the converter surface touching the housing to case bolts.

For reference, a properly installed converter will have a distance from the converter pilot nose from face-to-converter housing outer face of 13⁄32–9⁄16 in. (10.5–14.5mm).

8. Rotate the converter so that the drive studs are in alignment with the holes in the flywheel.
9. Move the converter and transmission assembly forward into position, being careful not to damage the flywheel and converter pilot. The converter housing is piloted into position by the dowels in the rear of the engine block.

➡During this move, to avoid damage, do not allow the transmission to get into a nose down position as this will cause the converter to move forward and disengage from the pump gear.

10. Install or connect the following:
    - Converter housing to engine. Torque the bolts to 30–41 ft. lbs. (40–55 Nm). The 2 longer bolts are located at the dowel holes.
    - Upper fluid filler tube and bracket. Torque the bolt to 41 ft. lbs. (55 Nm).
    - Transfer case, 4WD vehicles
    - Exhaust bracket. Torque the bolts to 64 ft. lbs. (87 Nm).
    - Crossmember. Torque the bolts to 87 ft. lbs. (118 Nm).
    - Transmission mount. Torque the bolts to 81 ft. lbs. (110 Nm).
    - Transmission cooler lines. Torque the bolts to 23 ft. lbs. (31 Nm).
    - Starter. Torque the bolts to 30 ft. lbs. (40 Nm).
    - Transmission wire harness
    - HO_2S sensor
    - Transmission connector
    - TR sensor connector
    - Shift cable and bracket
    - Driveshaft. Torque the bolts to 95 ft. lbs. (129 Nm).
    - Shroud. Torque the bolts to 71 inch lbs. (8 Nm).
    - Fluid level indicator
    - Negative battery cable
11. Fill the transmission to the proper level.
12. Start the vehicle and check for leaks, repair if necessary.

2002–04

1. Before servicing the vehicle, refer to the precautions in the beginning of this section.
2. Place the transmission in **Neutral**.
3. Remove or disconnect the following:
    - Negative battery cable

- Transfer case
- Three-way catalytic converter

4. If transmission disassembly is necessary, drain the transmission fluid.
5. Support the transmission with a transmission jack.
6. Remove or disconnect the following:
   - RH heat shield
   - LH heat shield bolt
   - Plastic cover from the right side of the frame rail
   - Plastic shield on the right side of the crossmember near the fuel tank
   - The two upper crossmember bolts (two on each side)
   - RH side lower crossmember bolts
   - LH side lower crossmember bolts
   - Transmission support insulator nuts and remove the crossmember
   - The two bolts and remove the transmission support insulator
   - Shift cable and bracket
   - Turbine shaft speed (TSS) sensor, output shaft speed (OSS) sensor and intermediate shaft speed (ISS) sensor electrical connectors

➡Clean the area around connector to prevent contamination of the solenoid body connector.

7. Remove or disconnect the following:
   - Screw from the solenoid body connector and disconnect the connector
   - Harness retainers
   - Left side heated oxygen sensor (HO2S) connector from the bracket
   - Digital transmission range (TR) sensor connector
   - Starter

✲✲ WARNING

Do not damage the cooler tubes

- Transmission cooler tubes
- Access cover

➡Make an identifying mark on the nut, stud, and adapter plate to allow for correct installation.

- The 4 torque converter nuts
- The 7 engine-to-transmission retaining bolts

8. Lower the transmission from the vehicle.

To install:

9. Raise and position the transmission.
10. Align the flexplate to converter marks made at removal.
11. Install seven engine to transmission retaining bolts. Torque to 35 ft. lbs. (48 Nm).
12. Install the four torque converter nuts. Torque to 28 ft. lbs. (48Nm).
13. Install or connect the following:
    - Access cover
    - Transmission cooler tubes
    - Starter
    - Digital transmission range (TR) sensor connector
    - Left side heated oxygen sensor (HO2S) to the bracket
    - Wire harness, and install the retainers

✲✲ WARNING

Damage will occur to the solenoid body assembly if the screw is tightened above 44 inch lbs. (5Nm). Always install new O-ring seals on vehicle harness connector. Clean the area around connector to prevent contamination of the solenoid body connector. Use petroleum jelly to lubricate the O-ring seals to aid in the installation process.

14. Install and lubricate new O-ring seals on the transmission connector and connect the connector.
15. Install or connect the following:
    - Turbine shaft speed (TSS) sensor, output shaft speed (OSS) sensor, and intermediate shaft speed (ISS) sensor electrical connectors.
    - Shift cable and bracket
    - Three-way catalytic converter
    - Transmission support insulator. Torque to 68 ft. lbs. (90Nm).
    - Crossmember in place and loosely install the two nuts to hold up the crossmember.
    - Crossmember bolts. Torque to 52 ft. lbs. (70Nm).
    - Rear transmission support nuts. Torque to 52 ft. lbs. (70Nm).
    - Heat shields
    - Plastic cover to the right side of the frame rail
    - Plastic shield on the right side of the crossmember near the fuel tank
    - Transfer case

➡Ford recommends, if the vehicle was not equipped with a fluid filter, install a Fluid Filter Service Kit (XC3Z-7B155-AA). Follow the instructions supplied in the kit. If the vehicle was equipped with a fluid filter, install a new filter (XC3Z-7B155-AB).

16. Connect the battery ground cable.
17. Verify that the shift cable is correctly adjusted.

Clutch

REMOVAL & INSTALLATION

1. Before servicing the vehicle, refer to the precautions in the beginning of this section.
2. Remove or disconnect the following:
   - Negative battery cable
   - Transmission

➡If the parts are to be reused, index-mark the clutch pressure plate to the flywheel.

- Bolts, clutch pressure plate and the clutch disc

To install:

3. Lubricate the transmission input shaft pilot bearing with grease.
4. Adjust the clutch pressure plate:
   a. Using a suitable press, press

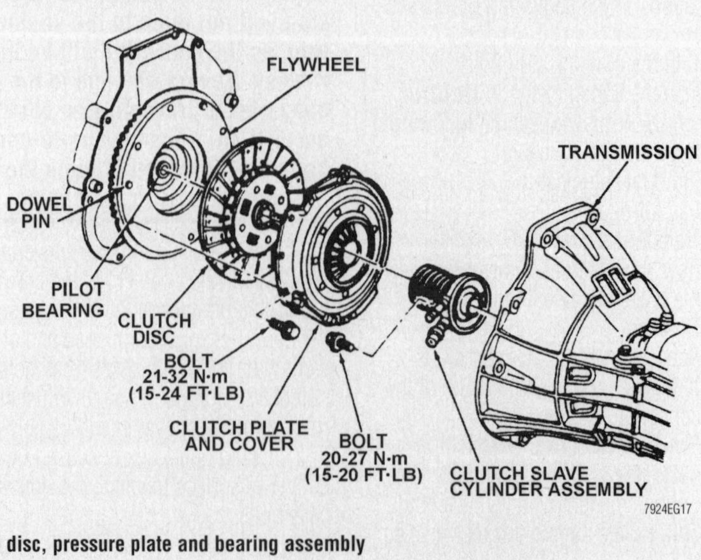

Clutch disc, pressure plate and bearing assembly

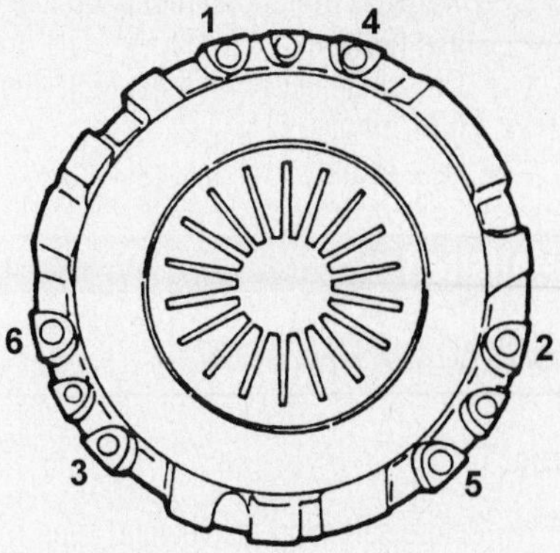

Tighten the bolts gradually in the correct sequence to avoid warping the pressure plate

downward on the fingers until the adjusting ring moves freely.

b. Rotate the adjusting ring counterclockwise to compress the tension springs. Hold the adjusting ring in this position.

c. Release the pressure on the fingers. The adjusting ring will stay in the reset position.

5. Position the clutch disc on the flywheel.

6. Align the clutch disc and the clutch pressure plate. Install the bolts and tighten in a star pattern sequence. Torque to 21 ft. lbs. (28Nm).

➡An "L" is stamped by three bolt holes on the clutch pressure plate. Tighten these to specification first to ensure correct clutch alignment.

7. Install the transmission.

ADJUSTMENT

Because the clutch is hydraulically driven, there is no adjustment required.

In the event the clutch pedal develops a squeak or uneven feel when depressing, spray the pedal bushing assembly with penetrating oil and work the pedal back-and-forth.

Hydraulic Clutch System

BLEEDING

The following procedure is recommended for bleeding the clutch hydraulic system installed on the vehicle. It is recommended that the original clutch tube, with quick-connect fitting be replaced when servicing the hydraulic system, because air can be trapped in the quick-connect fitting and prevent complete bleeding of the system. The replacement tube does not include a quick-connect fitting.

1. Before servicing the vehicle, refer to the precautions in the beginning of this section.
2. Clean the dirt and grease from the dust cap.
3. Remove the cap and diaphragm and fill the reservoir to the top with approved brake fluid C6AZ-19542-AA or BA, (ESA-M6C25-A).

➡To keep brake fluid from entering the clutch housing, route a suitable rubber tube of appropriate inside diameter from the bleed screw to a container.

4. Loosen the bleed screw, located in the slave cylinder body, next to the inlet connection. Fluid will now begin to move from the master cylinder down the tube to the slave cylinder.

➡The reservoir must be kept full at all times during the bleeding operation, to ensure no additional air enters the system.

5. Observe the bleed screw outlet. When the slave cylinder is full, a steady stream of fluid will flow from the outlet port. Tighten the bleed screw.
6. Depress the clutch pedal to the floor and hold for 1–2 seconds. Release the pedal as rapidly as possible. The pedal must be released completely. Pause for 1–2 seconds. Repeat 10 times.
7. Check the fluid level in the reservoir. The fluid should be level with the step when the diaphragm is removed.
8. Hold the pedal to the floor, slightly open the bleed screw to allow any additional air to escape. Close the bleed screw, then release the pedal.
9. Check the fluid in the reservoir. The hydraulic system should now be fully bled, and should actuate the clutch.
10. Check the vehicle by starting, pushing the clutch pedal to the floor and selecting reverse gear. There should be no grating of gears. If there is, and the hydraulic system still contains air; repeat the bleeding procedure.

Transfer Case Assembly

REMOVAL & INSTALLATION

Automatic Shift

1. Before servicing the vehicle, refer to the precautions in the beginning of this section.
2. Place the transmission in **Neutral**.
3. Remove or disconnect the following:
   - Negative battery cable
   - Skid plate

➡Drain the transfer case if disassembly is necessary.

➡Match-mark the front and rear driveshaft yokes and pinion flange and the rear driveshaft yoke and rear output flange.

   - Rear driveshaft
   - Front driveshaft
   - Vent tube
   - Shift motor electrical connector
4. Using a suitable high lift jack, support the transfer case.
5. Remove or disconnect the following:
   - RH crossmember cover, then the four bolts
   - The four LH crossmember bolts
   - Heat shields from the crossmember
   - Transmission mount nuts
   - The seven bolts and separate the transfer case from the extension housing
6. Lower the transfer case from the vehicle.
7. Remove and discard the transfer case-to-extension housing gasket. Clean the gasket surfaces.
8. To install, reverse the removal procedure. Use a new transfer case gasket.
9. Torque the transfer case-to-extension housing bolts to 30 ft. lbs. (40Nm).
10. Torque the crossmember bolts to 46 ft. lbs. (63Nm).
11. Torque the mount nuts to 72 ft. lbs. (98Nm).

All Wheel Drive

1. Before servicing the vehicle, refer to the precautions in the beginning of this section.
2. Place the transmission in **Neutral**.
3. Disconnect the negative battery cable.
4. Matchmark the front and rear driveshafts.
5. If transfer case disassembly is necessary, remove the drain plug and drain the fluid.
6. Remove or disconnect the following:

⁂ WARNING

Always disconnect the front driveshaft from the transfer case first. Otherwise, the weight of the driveshaft can pinch the boot between the shaft and the boot can and cause the boot to tear.

- Front driveshaft
- Rear driveshaft
- Skid plate
- Hose from the vent
- The two bolts retaining the heat shield to the crossmember. Loosen, but do not remove, the two nuts retaining the transmission mount to the crossmember.
- The three bolts and the heat shield

7. Position a high lift jack under the transfer case.
8. Remove or disconnect the following:
   - The four upper nuts and bolts retaining the crossmember to the frame
   - The two lower bolts retaining the crossmember to the frame
   - The two nuts retaining the transmission mount to the crossmember
   - The crossmember
   - The two bolts, the transmission mount, and the exhaust hanger

➡**Support the transmission.**

   - The seven bolts retaining the transfer case to the transmission

9. Separate the transfer case from the extension housing and the output shaft.
10. Lower the transfer case from the vehicle.
11. Remove and discard the gasket, and clean the mating surfaces.
12. To install, reverse the removal procedure. Note the following:
    - Install a new gasket on the transfer case.
    - Watch for obstructions when installing the transfer case in the vehicle.
    - Install the original transfer case to transmission bolts. If new bolts are required, be sure to install only the specific Ford transfer case to transmission bolts.
    - Transfer case-to-extension: 30 ft. lbs. (40Nm)
    - Transmission mount bolts: 66 ft. lbs. (90Nm)
    - Crossmember bolts/nuts: 52 ft. lbs. (70Nm)
    - Heat shield: 18 ft. lbs. (25Nm)

Halfshaft

REMOVAL & INSTALLATION

Front

1. Before servicing the vehicle, refer to the precautions in the beginning of this section.
2. Remove or disconnect the following:
3. Loosen the front axle wheel hub retainer.
   - Wheel and tire assembly
   - Hub retainer and the washer. Discard the front axle wheel hub retainer.
   - The two bolts and position the disc brake caliper aside

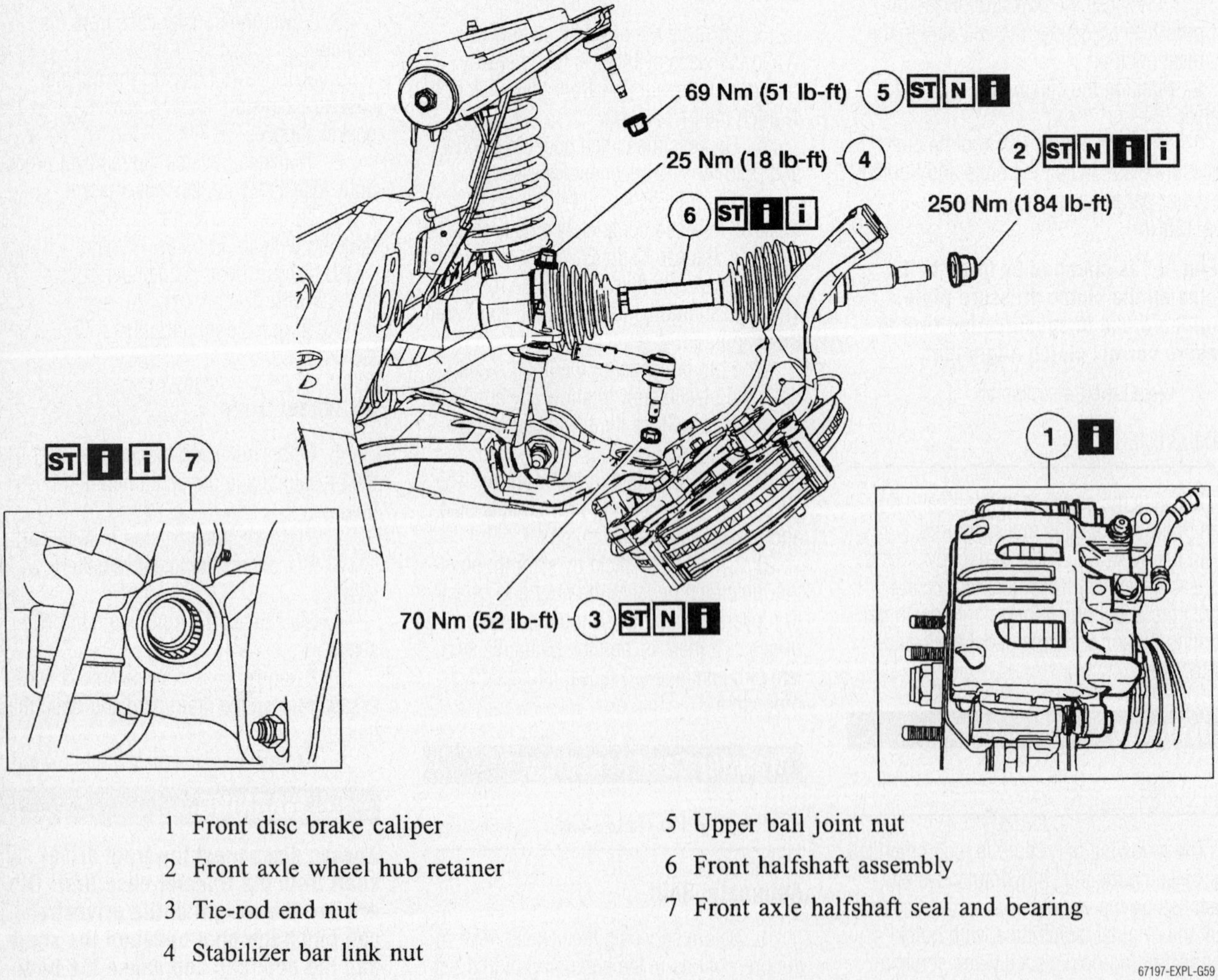

Front axle halfshaft

- Tie-rod end from the knuckle. Discard the nut.
- Stabilizer bar link. Discard the nut.

⁂ WARNING

Do not allow the knuckle to hang freely. It is possible to overextend and internally separate each inner CV joint from its housing.

- Upper ball joint from the knuckle

⁂ WARNING

Do not use a hammer to separate the outboard CV joint from the hub. Damage to the threads and internal CV joint components may result.

4. Press the outboard CV joint until it is loose in the hub.
5. Remove the outboard CV joint from the hub.

⁂ WARNING

Do not damage the axle shaft oil seal or the machined sealing surface on the inboard CV joint housing.

➡A circlip retains the inboard CV joint housing to the differential side gear in the axle.

6. On the left side, pry the LH inboard CV joint housing from the differential side gear.
7. On the right side, disengage the RH inboard CV joint housing from the axle tube.
8. Pull the halfshaft and the axle shaft away from the axle tube, and separate the inboard CV joint housing from the axle shaft.
9. Remove the halfshaft assembly from the vehicle.

⁂ WARNING

Do not damage the axle shaft oil seal, the machined sealing surface on the inboard CV joint housing, or the axle shaft splines.

10. To install, reverse the removal procedure.
11. Always install the halfshaft with a new retainer circlip and a new front axle wheel hub retainer.
12. On the right side, check the retainer circlip engagement after reseating the axle shaft and after installing the halfshaft in the axle. On the LH side, check the retainer circlip engagement after installing the halfshaft in the axle. When seated, the retainer circlip will lock the axle shaft and the inboard CV joint housing to the axle.

⁂ WARNING

Never use power tools to tighten the front axle wheel hub retainer. Torque the retainer to 184 ft. lbs. (250 Nm).

➡It may be necessary to support the front suspension lower arm to be able to connect the upper ball joint to the knuckle.

Rear

1. Before servicing the vehicle, refer to the precautions in the beginning of this section.
2. Remove or disconnect the following:

⁂ WARNING

Do not loosen the rear axle wheel hub retainer until after the wheel and tire assembly are removed from the vehicle. Wheel bearing damage will occur if the wheel bearing is unloaded with the weight of the vehicle applied.

- Wheel and tire assembly

➡Have an assistant press the brake pedal to keep the axle from rotating.

- Hub retainer and the washer
- Caliper. Position the disc brake caliper out of the way.
- Brake disc
- Bolt retaining the parking brake cable bracket to the frame

⁂ WARNING

Using a rubber hose approximately 37.5 mm (1.5 in) long, cover the stabilizer link bolt threads and nut to prevent boot damage when removing the halfshaft assembly from the vehicle.

⁂ WARNING

The bolt that retains the upper ball joint to the knuckle is longer and it has fewer threads than the bolt that retains the toe link to the knuckle. Switching these bolts during installation will prevent the pinch arms on the knuckle from correctly retaining the toe link to the knuckle. This may cause the toe link to separate from the knuckle during vehicle operation. Failure to follow these instructions may result in personal injury.

- Pinch bolts and disconnect the toe link and upper ball joint from the knuckle

⁂ WARNING

Using a wood stick, approximately 450 mm (18 in) long and 25 mm (1 in) wide, support the rear suspension upper arm to prevent boot damage when removing the halfshaft assembly from the vehicle

➡Do not use a hammer to separate the outboard CV joint from the hub. Damage to the threads and internal CV joint components may result. Once the outboard CV joint separates from the hub the knuckle will continue to pivot until the brake backing plate presses against the suspension lower arm. To prevent damage to the brake backing plate, immediately after separating the outboard CV joint from the hub, rest the knuckle on a cushioned support that is tall enough to keep the backing plate from pressing against the suspension lower arm.

3. Press the outboard CV joint until it is loose in the hub.
4. Separate the outboard CV joint from the hub.
5. Rest the knuckle on a cushioned support.

⁂ WARNING

Do not damage the axle shaft oil seal or the machined sealing surface on the inboard CV joint housing.

➡A circlip retains the inboard CV joint housing to the differential side gear in the axle.

6. Using the special tool, disengage the inboard CV joint housing from the differential side gear.

⁂ WARNING

To prevent damage to the axle shaft oil seal, install the special tool 205-461 (seal protector) before removing the inboard CV joint housing from the axle.

7. Remove the halfshaft assembly from the vehicle.
8. To install, reverse the removal procedure.
9. To prevent damage to the axle shaft oil seal, install the seal protector before positioning the inboard CV joint housing in the axle.

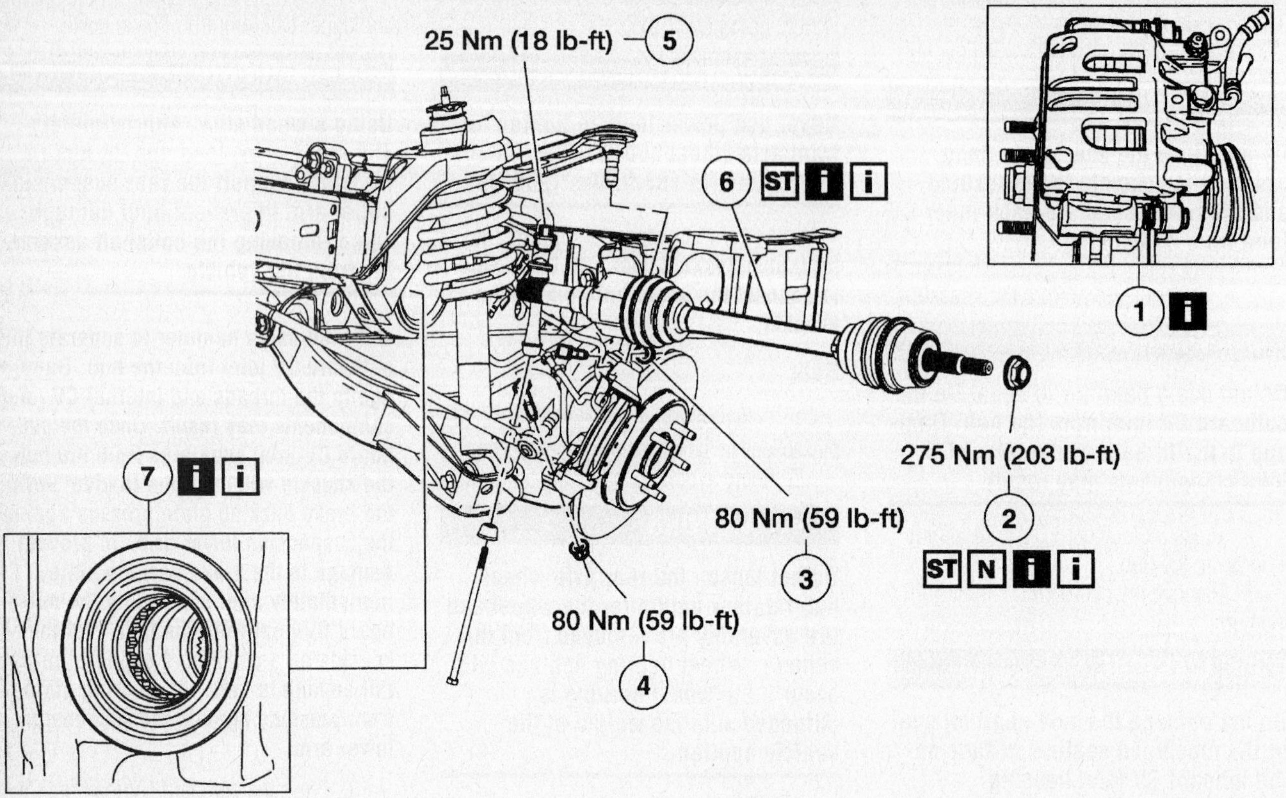

1 Rear disc brake caliper
2 Rear axle wheel hub retainer and washer
3 Upper ball joint pinch bolt nut
4 Toe link pinch bolt nut
5 Sway bar stud nut
6 Rear halfshaft assembly
7 Rear stub shaft housing bearing and seal

67197-EXPL-G60

Rear axle halfshaft

10. Always install the halfshaft with a new retainer circlip and a new rear axle wheel hub retainer.

➡Never use power tools to tighten the rear axle wheel hub retainer. Torque the retainer to 203 ft. lbs. (275Nm).

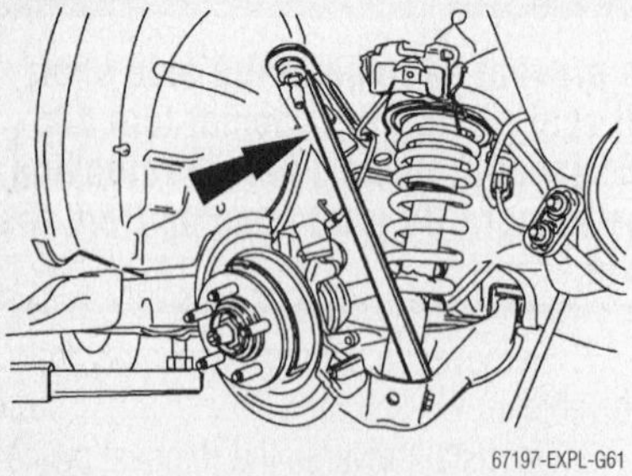

67197-EXPL-G61

Using a wood stick, approximately 450 mm (18 in) long and 25 mm (1 in) wide, support the rear suspension upper arm to prevent boot damage when removing the halfshaft assembly from the vehicle

CV-Joints

OVERHAUL

Front Shaft

1. Before servicing the vehicle, refer to the precautions in the beginning of this section.
2. Remove or disconnect the following:
   - Halfshaft
   - The two inboard boot clamps
   - Inboard halfshaft boot off the inboard CV joint housing
3. Separate the CV joint from the CV joint housing.
4. Mark the shaft and the inboard CV joint to ease alignment during assembly.
5. Remove the snap ring.
6. Remove the CV joint.
7. Inspect the stop ring for wear or damage. Install a new stop ring as necessary.
8. Remove the inboard halfshaft boot from the shaft assembly.
9. Remove the two outboard boot clamps.
10. Remove the outboard halfshaft boot.

➡If grease is contaminated, clean and inspect the joint for wear. If worn or damaged, install a new outboard CV joint and shaft assembly.

To assemble

11. Pack the outboard CV joint with grease.

➡Clean the halfshaft boot mounting surfaces of access grease before positioning the halfshaft boot into place.

12. Position the outboard halfshaft boot and outboard boot clamps.
13. Tighten the through-bolt until the installer is in the closed position.
14. Use the CV Boot Clamp Installer to install the outboard CV joint boot clamps.

15. Position the boot clamp on the halfshaft.
16. Position the inboard halfshaft boot.
17. Install a new stop ring, if necessary.
18. Install the CV joint on the halfshaft.
19. Install the snap-ring.
20. Lubricate the three CV joint needle bearings with grease.
21. Fill the inboard CV joint housing with 235 grams of grease.
22. Position the CV joint housing onto the CV joint.

➡Remove any excess grease from the inboard halfshaft boot mating surface before positioning into place.

23. Position the inboard halfshaft boot and boot clamp.
24. Insert a wood wand to relieve built-up air pressure in the halfshaft boot.
25. Use the CV Boot Clamp Installer to install the inboard boot clamps.
26. Install the front wheel halfshaft.

Rear Shaft

1. Before servicing the vehicle, refer to the precautions in the beginning of this section.
2. Remove or disconnect the following:
3. Remove the halfshaft.
4. For the inboard CV joint:
   a. Remove and discard the boot clamps.
   b. Remove the inboard CV joint housing.
   c. Remove and discard the retainer circlip.
   d. Slide the boot away from the CV joint.
   e. Using a suitable 3-jaw puller, remove the CV joint.
   f. Remove and discard the tri-lobe insert and the boot.
5. For the outboard CV joint:
   a. Remove and discard the boot clamps.
   b. Remove and discard the boot.

➡Do not disassemble the side shaft assembly. Install a new halfshaft assembly, if the components are worn/damaged.

6. Inspect the grease packed in the inboard CV joint and the outboard CV joint for contamination. Rub some of the grease from each joint between two fingers. Any gritty feeling indicates contamination. Wash all of the grease from the inboard CV joint, the inboard CV joint housing, the outboard CV joint, and the interconnecting shaft. Thoroughly dry all of the components, and inspect them for wear and damage. Discard the assembly, if necessary. Proceed as follows only if not discarding the assembly.
7. If necessary, remove and discard the outboard dust seal. Tap uniformly around the seal to separate it from the joint.
8. On the inboard end:
   a. Remove and discard the retainer circlip.
   b. If necessary, remove and discard the inboard dust seal. Tap uniformly around the seal to separate it from the housing.

To assemble:

9. For the outboard CV joint:
   a. Slide the boot on the interconnecting shaft.
   b. Pack the outboard CV joint with 225 grams (5.29 ounces) of grease.
   c. Spread any remaining grease evenly inside the boot.
   d. Clean any excess grease from the boot mounting surfaces before installing the boot.
   e. Install the boot by seating it in the groove in the CV joint housing.
   f. Tighten the through-bolt until the special tool is in the closed position.
   g. Using the special tool, install both boot clamps.
   h. If removed, use the special tools to install the dust seal.

⁂ WARNING

Do not overexpand or twist the circlip during installation.

   i. Install the retainer circlip.
10. Install the halfshaft in a soft jaw vise.
11. For the inboard CV joint:
   a. Position the clamp on the interconnecting shaft.
   b. Position the boot on the interconnecting shaft.

➡The lip on the end of the tri-lobe insert must seat against the end of the boot.

   c. Install the tri-lobe insert.

➡One side of the inboard CV joint has a champher cut in the edge of joint at the inner diameter near the splines. Install the inboard CV joint so that the champher faces the outboard end of the halfshaft.

   d. Install the CV joint.
   e. Install the retainer circlip.
   f. Pack the inboard CV joint housing with 250 grams (5.88 ounces) of grease.
   g. Spread any remaining grease evenly inside the boot and on the CV joint.
   h. Clean any excess grease from the boot mounting surfaces before installing the boot.
   i. Install the inboard CV joint housing, seating the boot in the groove in the housing.
12. Set the halfshaft assembled length to 28.44 inches (722.4mm).
13. Hold the inner joint to prevent the assembled length from changing, and insert a wood wand between the boot and the joint to equalize the pressure.
14. Tighten the through-bolt until the special tool is in the closed position.
15. Install both boot clamps.
16. If removed, install the dust seal.
17. Install the halfshaft.

Front Axle Tube Bearing

REMOVAL & INSTALLATION

1. Before servicing the vehicle, refer to the precautions in the beginning of this section.
2. Remove or disconnect the following:
   - Right-hand halfshaft
   - Right-hand axle shaft
   - Axle seal, with a slide hammer
   - Axle tube bearing, with a slide hammer

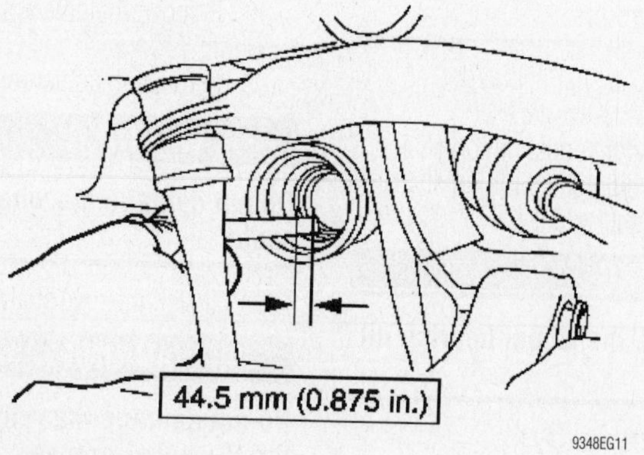

Front axle tube bearing depth

3. Clean the bearing and seal surfaces of any foreign debris.

To install:

4. Use an axle bearing replacer and the handle to replace the RH axle tube bearing.
5. Check the bearing depth as shown.
6. Use an axle seal replacer and the handle to replace the axle tube seal.

➡Care should be taken not to damage the axle seal surface.

7. Install the axle shaft.
8. Refill the front drive axle to proper level using SAE 80W90.
9. Install the RH halfshaft.

Rear Axle Shaft, Bearing and Seal

REMOVAL & INSTALLATION

Ford 8.8 inch Ring Gear Solid Axle

1. Raise and support the vehicle.
2. Remove the wheel and tire assembly.
3. Loosen the bolts and drain the lubricant from the rear axle housing.
4. Remove the cover bolts.
5. Remove the differential housing cover.
6. Remove the rear brake disc or brake drum and shoes.

**** WARNING**

Turning the differential case or an axle shaft with the differential pinion shaft removed will cause the differential pinion gears to fall out of the assembly and damage the components.

7. Remove the differential pinion shaft lock bolt.
8. Remove the differential pinion shaft.

**** WARNING**

Do not damage the rubber O-ring in the U-washer groove.

9. Push the axle shaft inboard.
10. Remove the U-washer.
11. Install the differential pinion shaft.
12. Install the differential pinion shaft lock bolt finger-tight.

**** WARNING**

Do not damage the wheel bearing oil seal.

13. Remove the axle shaft.

➡If the wheel bearing oil seal is leaking, the axle housing vent may be plugged with foreign material.

➡If only a new seal needs to be installed, use care to avoid damaging the seal bore.

14. Using a suitable seal remover, remove the axle shaft oil seal. Discard the oil seal.
15. Inspect the rear wheel bearing and axle shaft for wear or damage.
16. If necessary, using the special tools, remove the rear wheel bearing.

To install:

17. Lubricate the new rear wheel bearing.
    a. For 4.0L SOHC and 4.0L EI with limited slip rear axles and 5.0L vehicles, use SAE 75W-140 High Performance Rear Axle Lubricant F1TZ-19580-B or equivalent meeting Ford specification WSL-M2C192-A.
    b. For 4.0L SOHC and 4.0L EI with conventional rear axles, use SAE 80W-90 Premium Rear Axle Lubricant XY-80W90-QL or equivalent meeting Ford specification WSP-M2C197-A.
18. Using a bearing driver, install the rear wheel bearing.
19. Lubricate the lip of the new wheel bearing oil seal. Use Premium Long-Life Grease XG-1-C or equivalent meeting Ford specification ESA-M1C75-B.
20. Using a seal driver, install the wheel bearing oil seal.
21. Lubricate the lip of the wheel bearing oil seal. Use Premium Long-Life Grease XG-1-C or equivalent meeting Ford specification ESA-M1C75-B.

**** WARNING**

Turning the differential case or an axle shaft with the differential pinion shaft removed will cause the differential pinion gears to fall out of the assembly and damage the components.

22. Remove the differential pinion shaft lock bolt.
23. Remove the differential pinion shaft.

**** WARNING**

Do not damage the wheel bearing oil seal.

24. Install the axle shaft.

**** WARNING**

Do not damage the rubber O-ring in the U-washer groove.

25. Position the U-washer on the button end of the axle shaft.
26. Pull the axle shaft outward.

**** WARNING**

If a new pinion shaft lock bolt is unavailable, coat the threads with Threadlock and Sealer EOAZ-19554-AA or equivalent meeting Ford specification WSK-M2G351-A5.

27. Align the hole in the differential pinion shaft with the differential pinion shaft lock bolt hole.
28. Install a new differential pinion shaft lock bolt. Torque to 22 ft. lbs. (30 Nm).
29. Install the rear brake disc or shoes and drum.

➡The machined surfaces on the differential housing and the differential housing cover must be clean and free of oil before applying the silicone sealant. Cover the inside of the rear axle prior to cleaning the machined surface to prevent contamination.

30. Clean the cover gasket mating surfaces.

➡Install the differential housing cover within 15 minutes of applying the silicone, or it will be necessary to apply new sealant.

31. Apply a continuous bead of sealant to the differential housing cover. Use Clear Silicone Rubber D6AZ-19562-AA or equivalent meeting Ford specification ESB-M4G92-A.

➡If possible, allow one hour before filling the axle with lubricant to allow the silicone sealant to cure.

32. Position the differential housing cover. Install the bolts. Torque to 33 ft. lbs. (45 Nm).
33. Fill the rear axle housing with the specified lubricant type and quantity.

➡For Traction-Lok® axles, first fill the rear axle with 118 ml (4 oz) of Additive Friction Modifier C8AZ-19B546-A or equivalent meeting Ford specification EST-M2C118-A.

➡Service refill capacities are determined by filling the rear axle to the level shown.

34. Fill the rear axle with 2.4 liters (5.0 pints) of lubricant and install the fill plug.
    a. For conventional axles, use SAE 80W-90 Premium Rear Axle Lubricant XY-80W90-QL or equivalent meeting Ford specification WSP-M2C197-A.

b. For Traction-Lok® axles, use SAE 75W-140 Synthetic Rear Axle Lubricant F1TZ-19580-B or equivalent meeting Ford specification WSL-M2C192-A.

35. Install the wheel and tire assembly.

Pinion Seal

REMOVAL & INSTALLATION

Solid Rear Axle

1. Before servicing the vehicle, refer to the precautions in the beginning of this section.
2. Drain the axle housing fluid.
3. Remove or disconnect the following:
   - Negative battery cable
   - Rear wheels
   - Driveshaft
   - Brake calipers and pads or brake drum

➡**The brake calipers and pads or brake drum must be removed so that there is no additional drag when measuring pinion bearing preload.**

4. Use an inch lb. torque wrench and measure and record the amount of torque required to maintain pinion rotation through several revolutions.
5. Remove or disconnect the following:
   - Pinion flange
   - Pinion seal
   - Pinion bearing
   - Collapsible spacer

To install:

➡**Use a new collapsible spacer and flange nut for assembly.**

6. Install or connect the following:
   - Collapsible spacer
   - Pinion bearing
   - Pinion seal
   - Pinion flange
7. Rotate the pinion flange occasionally while tightening the flange nut to make sure the pinion bearings seat correctly.
8. Take frequent bearing preload torque readings. Tighten the flange nut to achieve the preload torque readings originally recorded.

**** CAUTION**

Never loosen the pinion nut to reduce bearing preload. If it is necessary to reduce bearing preload, install a new collapsible spacer and pinion nut.

9. Install or connect the following:
   - Driveshaft
   - Brake calipers and pads or brake drum
   - Wheels
   - Negative battery cable
10. Fill the differential with gear lubricant and check for leaks.

Halfshaft Rear Axle

1. Before servicing the vehicle, refer to the precautions in the beginning of this section.
2. Drain the axle housing fluid.
3. Remove or disconnect the following:
   - Rear wheel and tire assemblies
   - Brake caliper and support bracket from the knuckle as an assembly. Wire the caliper and support bracket assembly out of the way.

➡**Matchmark the driveshaft flange and rear axle pinion flange to maintain initial balance during installation.**

4. Disconnect and position the driveshaft out of the way.
5. Install an inch/pound torque wrench on the nut and record the torque necessary to maintain rotation of the drive pinion gear through several revolutions.
6. Remove and discard the pinion flange nut.

➡**Matchmark the rear axle pinion flange and drive pinion gear stem to maintain initial balance during installation.**

7. Remove the rear axle pinion flange.
8. Force up on the metal flange of the rear axle drive pinion seal. Install gripping pliers and strike with a hammer until the rear axle drive pinion seal is removed.

To install:

9. Lubricate the new rear drive pinion seal with grease.

➡**If the rear axle drive pinion seal becomes misaligned during installation, remove the rear axle drive pinion seal and install a new seal.**

10. Drive in the rear axle drive pinion seal.
11. Inspect the rear axle pinion flange seal journal for rust, nicks and scratches prior to installing the flange. Polish the seal journal with fine crocus cloth, if necessary.
12. Lubricate the rear axle pinion flange splines.
13. Install the rear axle pinion flange.

**** WARNING**

Do not under any circumstance loosen the nut to reduce preload. If it is necessary to reduce preload, install a new differential drive pinion collapsible spacer and nut.

14. Rotate the pinion occasionally to make sure the pinion bearings seat correctly. Take frequent pinion bearing torque preload readings by rotating the drive pinion gear with an inch/pound torque wrench.
15. If the preload recorded prior to disassembly is lower than the specification for used bearings, then tighten the nut to specification. If the preload recorded prior to disassembly is higher than the specification for used bearings, then tighten the nut to the original reading as recorded.

Pinion bearing preload used bearings: 8–14 inch lbs. (0.9–1.16 Nm); new bearings: 16–29 inch lbs. (1.8–3.2 Nm).

16. Connect the driveshaft. Torque the bolts to 83 ft. lbs. (112 Nm).
17. Install the rear brake calipers.
18. Install the rear wheel and tire assemblies.

Front Axle

➡**This operation disturbs the differential pinion bearing preload. Carefully reset the preload during assembly.**

**** CAUTION**

The electrical power to the air suspension system must be shut off prior to hoisting, jacking or towing an air suspension vehicle. This can be accomplished by turning off the air suspension switch located in the rear jack storage area. Failure to do so can result in unexpected inflation or deflation of the air springs, which can result in shifting of the vehicle during these operations.

1. Before servicing the vehicle, refer to the precautions in the beginning of this section.
2. Index-mark the front driveshaft and pinion flange.
3. Remove or disconnect the following:
   - Front driveshaft from the pinion flange, and position it aside

➡**Do not allow the driveshaft to hang unsupported.**

4. Using a Nm (inch-pound) torque wrench, measure the torque required to maintain pinion rotation. Record the measurement.
5. Index-mark the pinion flange and the pinion stem.

6. Hold the pinion flange while removing the nut.
7. Place a drain pan under the differential housing.
8. Using a puller, remove the pinion flange.
9. Inspect the pinion flange for burrs and damage. Inspect the end of the pinion flange that contacts the bearing cone, the nut counterbore, and the seal surface for nicks. Discard the pinion flange as necessary.
10. Using a seal remover and impact slide hammer, remove the pinion seal.
11. Remove the front axle drive pinion shaft oil slinger and the differential pinion bearing.
12. Remove and discard the collapsible spacer.

To install:

13. Verify that the splines on the pinion stem are free of burrs. If burrs are evident, remove them with a fine crocus cloth. Work in a rotating motion to wipe the pinion clean.
14. Clean the pinion seal bore.
15. Install a new collapsible spacer.
16. Install the original differential pinion bearing and the front axle drive pinion shaft oil slinger.
17. Lubricate the pinion seal. Use Motorcraft SAE 80W90 Thermally Stable 4x4 Axle Lubricant meeting Ford specification WSP-M2C197-A.
18. Install the pinion seal.
19. Lubricate the pinion flange splines. Use Motorcraft SAE 80W90 Thermally Stable 4x4 Axle Lubricant meeting Ford specification WSP-M2C197-A.

➡Never use a metal hammer on the pinion flange or install the flange with power tools. If necessary, use a plastic hammer to tap on a tight fitting flange.

- Align the index marks and install the pinion flange.
- Install the new nut hand-tight.

➡Do not loosen the nut to reduce preload. Install a new collapsible spacer and nut if preload reduction is necessary.

20. Use the special tool to hold the pinion flange while tightening the nut to set the preload.
21. Tighten the nut, rotating the pinion occasionally to ensure the differential pinion bearings are seating correctly. Take frequent differential pinion bearing preload readings by rotating the pinion with a Nm (inch-pound) torque wrench. The final reading must be 0.56 Nm (5 inch lbs.) more than the initial reading taken during removal.
22. Align the index marks and position the front driveshaft.
23. Install the universal joint spider retainers and bolts.
24. Check the fluid level and, if necessary, fill the axle to specification. Use Motorcraft SAE 80W90 Thermally Stable 4x4 Axle Lubricant meeting Ford specification WSP-M2C197-A.
25. Lower the vehicle.
26. If so equipped, reactivate the air suspension.

STEERING AND SUSPENSION

Air Bag

PRECAUTIONS

- Always wear safety glasses when servicing an air bag vehicle, and when handling an air bag.
- Never attempt to service the steering wheel or steering column on an air bag-equipped vehicle without first properly disarming the air bag system. The air bag system should be properly disarmed whenever ANY service procedure in this manual indicates that you should do so.
- When carrying a live air bag module, always make sure the bag and trim cover are pointed away from your body. In the unlikely event of an accidental deployment, the bag will then deploy with minimal chance of injury.
- When placing a live air bag on a bench or other surface, always face the bag and trim cover up, away from the surface. This will reduce the motion of the air bag if it is accidentally deployed.
- If you should come in contact with a deployed air bag, be advised that the air bag surface may contain deposits of sodium hydroxide, which is a product of the gas combustion and is irritating to the skin. Always wear gloves and safety glasses when handling a deployed air bag, and wash your hands with mild soap and water afterwards.

DISARMING THE SYSTEM

1. Before servicing the vehicle, refer to the precautions in the beginning of this section.
2. Disconnect the negative battery cable from the battery.
3. Disconnect the positive battery cable from the battery.
4. Wait 1 minute. This time is required for the back-up power supply in the air bag diagnostic monitor to completely drain. The system is now disarmed.

ARMING THE SYSTEM

1. Before servicing the vehicle, refer to the precautions in the beginning of this section.
2. Connect the positive battery cable.
3. Connect the negative battery cable.
4. Stand outside the vehicle and carefully turn the ignition to the **RUN** position. Be sure that no part of your body is in front of the air bag module on the steering wheel, to prevent injury in case of an accidental air bag deployment.
5. Ensure the air bag indicator light turns off after approximately 6 seconds. If the light does not illuminate at all, does not turn off, or starts to flash, test the system.

Power Rack and Pinion Steering Gear

REMOVAL & INSTALLATION

2001 Explorer and Mountaineer; 2001–03 Explorer Sport; 2001–04 Explorer Sport-Trac

WITH 4.0L ENGINE

1. Place front wheels in the straight ahead position. Do not lock the steering column.
2. Remove the front wheel and tire assemblies.
3. Disconnect the power steering return line hose at the power steering fluid cooler. Drain the fluid into a suitable container.
4. Remove the power steering fluid cooler to crossmember nuts.
5. Disconnect the power steering return hose. Remove the power steering fluid cooler.
6. Loosen the tie-rod end jam nuts.
7. Remove and discard the cotter pins and nuts.

⁂ WARNING

Do not damage the tie-rod boot when installing the special tool.

➡Remove the adapter from the ball end of the special tool. Apply a small amount of grease to the tie-rod end stud and the ball of the special tool.

8. Using the special tool, separate the tie-rod ends from the wheel knuckles.
9. Remove the tie-rod ends. Count and record the number of turns required to remove the tie-rod end.
10. Remove the front stabilizer bar link nuts from the front suspension lower arms.
11. Remove the front stabilizer bar link bolts and the front stabilizer bar links.
12. Remove the front stabilizer bar bolts and brackets.
13. Remove the front stabilizer bar.
14. Remove the stabilizer bar insulator.
15. Disconnect the power steering pressure hose. Discard the O-ring seal.
16. Disconnect the power steering return hose. Discard the O-ring seal.
17. Plug ends of fluid lines and ports in steering gear to prevent damage and entry of dirt.

⁂ WARNING

Do not allow the intermediate shaft to rotate while it is disconnected from the steering gear or damage to the clockspring can result. If there is evidence that the intermediate shaft has rotated, the clockspring must be removed and recentered.

18. Remove the pinch bolt and detach the intermediate shaft from the gear.

⁂ WARNING

Hold the tops of the steering gear to crossmember stud bolts to avoid damaging the steering gear fluid transfer tubes.

19. Remove the steering gear nuts.
20. Remove the stud bolts and washers.
21. Rotate the steering gear control valve housing toward the front of the vehicle.
22. Turn the steering gear input shaft to the right until the stop is reached.
23. Move the steering gear as far to the RH side of the vehicle as possible.
24. Move the LH front wheel spindle tie-rod forward to clear the frame crossmember.
25. Remove the steering gear from the vehicle.

To install:

➡New Teflon® seals must be installed any time the lines are disconnected from the steering gear.

26. Using Teflon seal special installation tool 211-D027 (D90P-3517-A) or equivalent, install new O-ring seals on the power steering return hose and power steering pressure hose.

➡Make sure the steering gear input shaft is turned to the left until the stop is reached.

➡Handle the steering gear with caution to avoid damage to the fluid transfer tubes and to avoid dimples in the tie-rod boots.

27. Turn the steering gear input shaft to the right until the stop is reached. Note the number of turns.

➡Make sure the steering gear control valve housing is turned toward the front of the vehicle.

28. Install the steering gear into the RH opening of the crossmember.
29. Move the steering gear as far to the RH side of the vehicle as possible.
30. Move the LH front wheel spindle tie-rod into the opening in the crossmember and move the steering gear into position.
31. To place the steering gear in the straight ahead position, turn the steering gear input shaft to the left by half the number of turns recorded previously.
32. Rotate the steering gear control valve housing toward the rear of the vehicle.
33. Install the steering gear to crossmember washers and stud bolts. The dished side of the washers face down.

⁂ WARNING

Hold the tops of the steering gear to crossmember stud bolts to avoid damaging the steering gear fluid transfer tubes.

34. Install the nuts. Torque to 111 ft. lbs. (150 Nm).

⁂ WARNING

Do not allow the intermediate shaft to rotate while it is disconnected from the steering gear or damage to the clockspring can result. If there is evidence that the intermediate shaft has rotated, the clockspring must be removed and recentered.

35. Connect the intermediate shaft to the steering gear input shaft. Install a new lower steering column pinch bolt. Torque to 36 ft. lbs. (49 Nm).
36. Install the power steering return hose and then tighten tube nut to 22 ft. lbs. (30 Nm).
37. Install the power steering pressure hose and tighten the tube nut to 22 ft. lbs. (30 Nm).
38. Install the power steering fluid cooler. Torque to 59 ft. lbs. (80 Nm).

➡In the event the self-tapping bolts cannot be installed in the frame, there is a kit available with flag nuts. Consult your parts department for further information.

39. Install the front stabilizer bar.
40. Install the front stabilizer bar link bolts and the front stabilizer bar links.
41. Install the front stabilizer bar link nuts to the front suspension lower arms. Torque to 18 ft. lbs. (25 Nm).
42. Install the tie-rod ends on the front wheel spindle tie-rod. Rotate the tie-rod end the number of turns recorded during removal.
43. Position the tie-rod ends on the steering knuckles.
44. Install the castellated nuts. Torque to 66 ft. lbs. (90 Nm).
45. Install the new cotter pins. Check that the brake dust shields are not bent and are not in contact with the outer tie-rod boot seals.
46. Tighten the tie-rod jam nuts. Torque to 59 ft. lbs. (80 Nm).
47. Install the front wheel and tire assemblies.
48. Fill and leak check the power steering system.
49. Check the wheel alignment.

WITH 5.0L ENGINE

1. Place the front wheels in the straight ahead position. Do not lock the steering column.

⁂ CAUTION

The electrical power to the air suspension system must be shut off prior to hoisting, jacking or towing an air suspension vehicle. This can be accomplished by turning off the air suspension switch located in the rear jack storage area. Failure to do so can result in unexpected inflation or deflation of the air springs, which can result in shifting of the vehicle during these operations.

2. Raise the vehicle.
3. Remove the engine oil cooler.

4. Remove the front wheel and tire assemblies.
5. Remove the inner fender splash shields.
6. Remove the top motor mount nuts.
7. Remove the bolts and the radiator air deflector.
8. Loosen the LH tie-rod end jam nuts.
9. Disconnect the tie-rod ends.
10. Remove the LH tie-rod end. Count and record the number of turns required to remove the LH tie-rod end.
11. Remove the front stabilizer bar. Mark the driver side end of the stabilizer bar for correct installation.
12. Remove the power steering fluid cooler to crossmember nuts.
13. Disconnect the power steering return line hose at the power steering fluid cooler. Allow the system to drain.
14. Disconnect the power steering return hose and remove the power steering fluid cooler.
15. Disconnect the steering gear lines. Discard the O-ring seals.
16. Plug the ends of all fluid lines removed and ports in steering gear to prevent damage and entry of dirt.
17. Rotate the steering column shaft to access the Intermediate shaft pinch bolt. Remove the pinch bolt.
18. Lower the vehicle.
19. Turn the steering wheel back to the straight ahead position. Turn the ignition key to the locked position.
20. Raise the vehicle.

✻✻ CAUTION

Do not rotate the steering wheel when the lower steering column shaft is disconnected, or damage to the air bag sliding contact will result.

21. Disconnect the intermediate shaft from the steering gear input shaft.
22. Raise the engine. Position a block of wood between a screw jack and the bottom of the oil pan to avoid damaging the oil pan while raising the engine.

✻✻ CAUTION

Hold the tops of the steering gear to crossmember stud bolts to avoid damaging the steering gear fluid transfer tubes.

23. Remove the steering gear nuts.
24. Remove the steering gear stud bolts and washers.
25. Remove the steering gear to crossmember insulator bushings.
26. Rotate the steering gear control valve housing toward the front of the vehicle.
27. Turn the steering gear input shaft to the right until the stop is reached.
28. Move the steering gear as far to the RH side of the vehicle as possible.
29. Move the LH front wheel spindle tie-rod forward to clear the crossmember. Turn the steering gear input shaft to the left until the stop is reached.
30. Remove the steering gear from the vehicle.

To install:

➡New Teflon® seals must be installed any time the lines are disconnected from the steering gear.

31. Using Teflon seal special installation tool 211-D027 (D90P-3517-A) or equivalent, install new O-ring seals on the power steering return hose and power steering pressure hose.

➡Make sure the steering gear input shaft is turned to the left until the stop is reached.

➡Handle the steering gear with caution to avoid damage to fluid transfer tubes and to avoid dimples in the tie-rod boots.

32. Turn the steering gear input shaft to the right until the stop is reached. Note the number of turns required.

➡Make sure the steering gear control valve housing is turned toward the front of the vehicle.

33. Install the steering gear into the RH opening of the crossmember.
34. Move the steering gear as far to the RH side of the vehicle as possible.
35. Move the LH front wheel spindle tie-rod into the opening in the crossmember and move the steering gear into position.
36. To place the steering gear in the straight ahead position, turn the steering gear input shaft to the left by half the number of turns recorded previously.

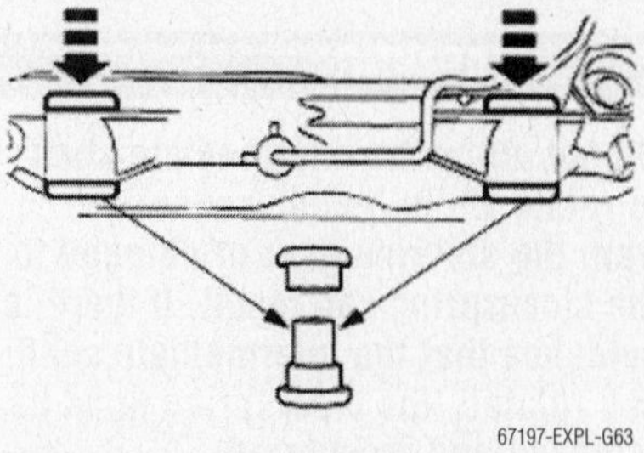

5.0L steering gear insulator bushing installation

37. Rotate the steering gear control valve housing toward the rear of the vehicle.
38. Install the steering gear to crossmember insulator bushings as shown. The large end of metal sleeve must be positioned downward. Check that the mounting surfaces on the crossmember are clean and free of debris.
39. Install the steering gear to crossmember washers and stud bolts.

✻✻ CAUTION

Hold the tops of the steering gear to crossmember stud bolts to avoid damaging the steering gear fluid transfer tubes.

40. Install the nuts. Torque to 111 ft. lbs. (150 Nm).
41. Lower the engine.

✻✻ CAUTION

Do not rotate the steering wheel when the lower steering column shaft is disconnected, or damage to the air bag sliding contact will result.

42. Connect the intermediate shaft to the steering gear input shaft. Install the pinch bolt. Torque to 36 ft. lbs. (49 Nm).
43. Connect the steering gear fluid lines. Torque to 22 ft. lbs. (30 Nm).
44. Connect the power steering return hose to the power steering fluid cooler.
45. Position the power steering fluid cooler and install the power steering fluid cooler to crossmember nuts. Torque to 59 ft. lbs. (80 Nm).
46. Connect the power steering return line hose to the power steering fluid cooler.
47. Install the engine oil cooler. Torque to 40–60 ft. lbs. (54–88 Nm).
48. Install the front stabilizer bar. Orientate the front stabilizer bar as noted during removal.
49. Install the tie-rod end. Rotate the tie-rod end the number of turns recorded during removal.
50. Position the tie-rod ends to the steering knuckles.
51. Install the castellated nuts. Torque to 66 ft. lbs. (90 Nm).
52. Install the new cotter pins. Check that brake dust shields are not bent and are not in contact with the outer tie-rod boot seals.
53. Tighten the LH tie-rod end jam nut.
54. Install the radiator air deflector. Torque to 22 ft. lbs. (30 Nm).
55. Install the top engine mount nuts. Torque to 111 ft. lbs. (150 Nm).

56. Install the inner fender splash shields.

57. Install the front wheel and tire assemblies.

➡If equipped with air suspension, reactivate the system by turning on the air suspension switch.

58. Lower the vehicle.

59. Refill and bleed the engine cooling system.

60. Fill and leak check the power steering system.

61. Check the wheel alignment.

2002–04 Explorer and Mountaineer

➡New O-ring seals must be installed any time the lines are disconnected from the steering gear.

1. With the vehicle in NEUTRAL, position the vehicle on a hoist.

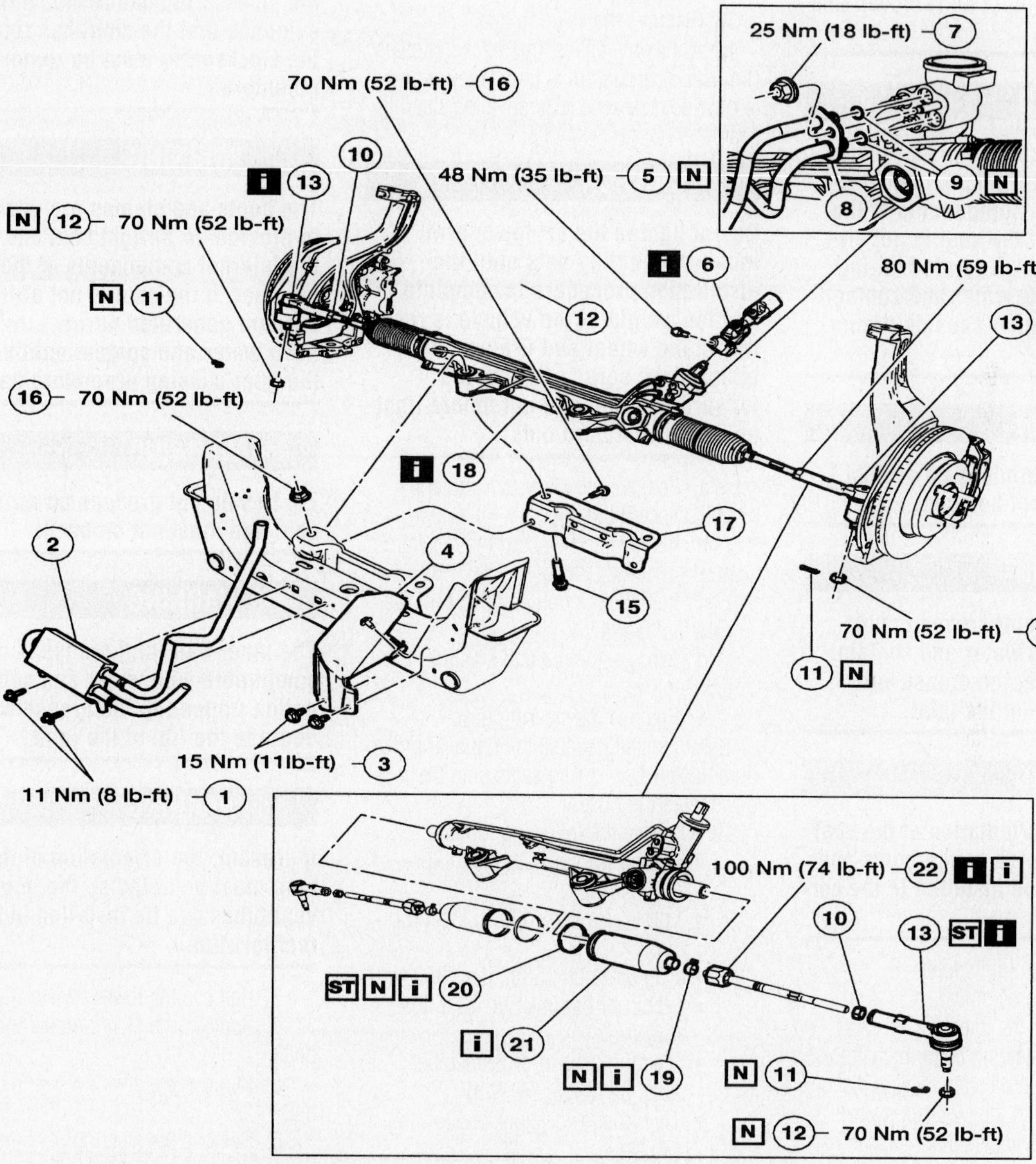

1 Fluid cooler-to-crossmember bolts	9 O-ring	17 Bracket
2 Fluid cooler	10 Tie-rod jam nuts (loosen only)	18 Steering gear
3 Oil shield-to-crossmember bolts	11 Cotter pins	19 Outer bellows clamp
4 Oil shield	12 Tie-rod nuts	20 Inner bellows clamp
5 Intermediate shaft-to-steering gear bolt	13 Outer tie-rods	21 Steering gear bellows
6 Intermediate shaft (detach only)	14 Steering gear-to-crossmember nuts	22 Spindle tie-rod
7 Clamp plate-to-steering gear nut	15 Steering gear-to-crossmember bolts	
8 Steering line clamp plate	16 Bracket-to-crossmember bolts	

67197-EXPL-G64

Steering gear and related parts—2002–04 Explorer and Mountaineer

2. Remove the wheel and tire assembly.

CAUTION

Do not allow the steering column shaft to rotate while the intermediate shaft is disconnected or damage to the clockspring can result. If there is evidence that the shaft has rotated, the clockspring must be removed and recentered.

CAUTION

The boots and clamps are designed to provide an airtight seal and protect the internal components of the steering gear. If the seal is not airtight the vacuum generated during turning will draw water and contamination into the gear causing premature damage.

CAUTION

Zip ties do not produce an airtight seal and must not be used.

CAUTION

The inner ball joint grease is not compatible with water and contamination trapped in the grease will degrade the life of the joint.

CAUTION

If present, the orientation of the vent tube must be noted so the boots and vent tubes can be installed in the correct location.

3. Remove the fluid cooler.
4. Remove the oil shield.
5. Disconnect the intermediate shaft. Do not allow the steering column shaft to rotate while the intermediate shaft is disconnected or damage to the clockspring can result. If there is evidence that the shaft has rotated, the clockspring must be removed and recentered.
6. Remove the steering line clamp plate. Discard the O-rings.
7. Loosen the tie-rod jam nuts.
8. Remove the tie-rod nuts.
9. Remove the outer tie-rod end from the knuckles. If repairing the RH side it will be necessary to pull back the LH inner tie rod boot to hold the steering gear.
10. Note the number of times the tie-rod ends turn for assembly reference.
11. After removing the tie-rod end, remove the tie-rod end jam nut from the front wheel spindle tie-rod.
12. Remove the steering gear-to-crossmember nuts.
13. Remove the steering gear-to-crossmember bolts.
14. Remove the bracket-to-crossmember bolts.
15. Remove the bracket.
16. Remove the steering gear.
17. On 4x4 vehicles, the following steps () must be carried out at the LH lower arm to provide clearance to remove the steering gear.

CAUTION

Do not tighten the LH lower arm inboard mounting nuts until the installation procedure is complete and the weight of the vehicle is resting on the wheel and tire assemblies. Make sure to tighten the forward flag bolt and nut before tightening the rearward nuts.

a. Remove the lower arm nut and bolt. Discard the nut.
b. Remove and discard the rearward nuts.
c. Remove the bolt and flag nut. Discard the flag nut.
d. Remove the nut and stabilizer bar link.
e. Remove the steering gear.

18. To install, reverse the removal procedure. Install new O-ring seals onto the power steering hoses.
19. Observe the following torques:
- Lower arm nut/bolt: 295 ft. lbs. (400 Nm)
- Lower arm nuts: 111 ft. lbs. (150 Nm)
- Flag bolt/nut: 258 ft. lbs. (350 Nm)
- Stabilizer bar link nut: 18 ft. lbs. (25 Nm)
- Steering gear-to-crossmember nuts: 52 ft. lbs. (70 Nm)
- Tie rod end-to-knuckle: 52 ft. lbs. (70 Nm)
- Tie rod end jam nut: 59 ft. lbs. (80 Nm)
- Bracket-to-crossmember bolts: 52 ft. lbs. (70 Nm)

20. Fill and leak test the system.
21. Check and, if necessary, align the front end.

Aviator

➡New O-ring seals must be installed any time the lines are disconnected from the steering gear.

1. With the transmission in NEUTRAL, position the vehicle on a hoist.
2. Remove the wheel and tire assembly.

WARNING

Do not allow the steering column shaft to rotate while the intermediate shaft is disconnected or damage to the clockspring can result. If there is evidence that the shaft has rotated, the clockspring must be removed and recentered.

WARNING

The boots and clamps are designed to provide an airtight seal and protect the internal components of the steering gear. If the seal is not airtight the vacuum generated during turning will draw water and contamination into the gear causing premature damage.

WARNING

Zip ties do not produce an airtight seal and must not be used.

WARNING

The inner ball joint grease is not compatible with water and contamination trapped in the grease will degrade the life of the joint.

WARNING

If present, the orientation of the vent tube must be noted so the boots and vent tubes can be installed in the correct location.

3. Remove the lower arm-to-frame nuts.
4. Remove the steering gear mounting bolt
5. Remove the steering gear mounting nut.
6. Remove the steering gear mounting bracket bolt.
7. Remove the steering gear mounting bracket.
8. Remove the steering column pinch bolt.
9. Remove the variable assist power steering (VAPS) switch electrical connector.
10. Remove the power steering hose mounting plate nuts.
11. Disconnect the power steering hoses.
12. Discard the O-rings.
13. Remove the tie-rod end-to-knuckle nut. If repairing the RH side it will be neces-

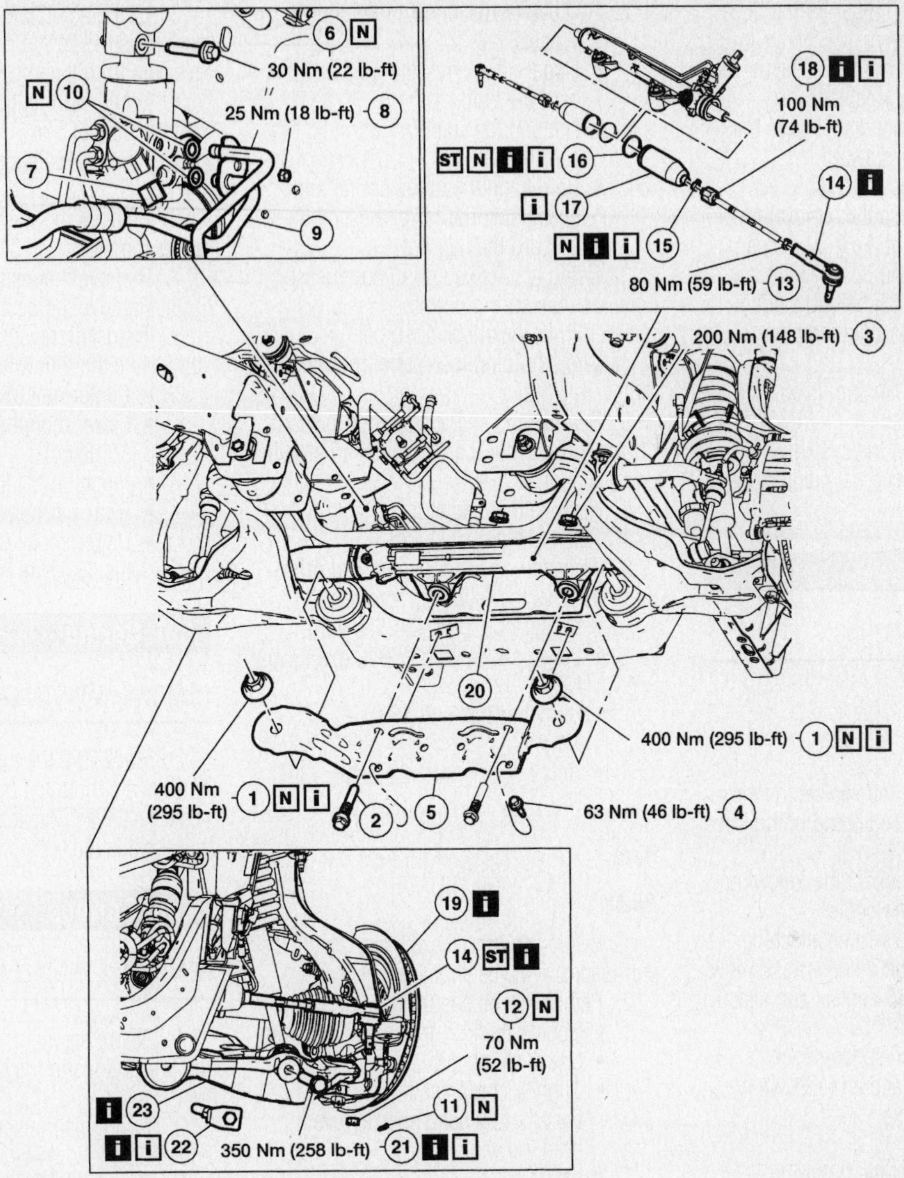

67197-EXPL-G62

1 Lower arm-to-frame nut
2 Steering gear mounting bolt
3 Steering gear mounting nut
4 Steering gear mounting bracket bolt
5 Steering gear mounting bracket
6 Steering column pinch bolt
7 Variable assist power steering (VAPS) switch electrical connector
8 Power steering hose mounting plate nut
9 Power steering hoses
10 O-ring
11 Cotter pin
12 Tie-rod end-to-wheel knuckle nut
13 Tie-rod end jam nut
14 Tie-rod end
15 Outer bellows clamp
16 Inner bellows clamp
17 Steering gear bellows
18 Inner tie-rod
19 Brake disc
20 Steering gear
21 Shock absorber bolt
22 Lower arm-to-frame bolt
23 Lower arm

67197-EXPL-G62A

Steering gear and related parts—Aviator

sary to pull back the LH inner tie rod boot to hold the steering gear.

14. Note the number of times the tie-rod ends turn for assembly reference.
15. Remove the brake disc. Rotate the front of the brake rotor outward.
16. Remove the steering gear.
17. To install, reverse the removal procedure. Final tightening of the front suspension components should be carried out at or near the ride height (curb height) setting. Install new O-ring seals onto the power steering hoses.
18. See the accompanying illustration for relevant fastener torques.
19. Fill and leak test the system.
20. Check and, if necessary, adjust the front toe.

Shock Absorber

REMOVAL & INSTALLATION

Front

2001

1. Before servicing the vehicle, refer to the precautions in the beginning of this section.
2. Remove or disconnect the following:
   - Negative battery cable
   - Upper shock-to-frame attaching nut, washer and insulator assembly
   - Lower shock-to-control arm attaching nuts
   - Slightly compress the shock absorber by hand and remove it from the vehicle

To install:

3. Install or connect the following:
   - Position the lower washer and insulator on the shock absorber rod and position the shock absorber to the upper frame bracket mount
   - Position the upper insulator and washer on the shock absorber rod and install the attaching nut loosely.
   - Position the lower shock absorber mounting studs into the control arm and install the attaching nuts loosely.
   - Torque the lower shock attaching nuts to 15–21 ft. lbs. (21–29 Nm), and the upper shock attaching bolts to 30–40 ft. lbs. (40–55 Nm).
   - Negative battery cable

2002–04

1. Before servicing the vehicle, refer to the precautions in the beginning of this section.
2. Remove or disconnect the following:
   - Wheel
   - Upper shock mounting nuts. Discard the nuts.
   - Nut and the stabilizer bar link. Discard the nut.
   - Bolt, flagnut and the spring and shock absorber as an assembly. Discard the flag nut.
3. Using a suitable spring compressor, compress the spring until the tension is released from the shock absorber.
4. While holding the flats of the washer, remove the nut.
5. Remove the shock absorber. Discard the nut, remove the washer, bushing and the upper mount.
6. Remove the insulator.
7. Remove the dust shield.
8. To install, reverse the removal procedure. Observe the following torques:
   - Center nut: 37 ft. lbs. (50 Nm)
   - Lower shock bolt: 258 ft. lbs. (350 Nm).
   - Sway bar link nut: 18 ft. lbs. (25 Nm)
   - Upper shock nuts: 22 ft. lbs. (30 Nm)

Rear

2001

1. Before servicing the vehicle, refer to the precautions in the beginning of this section.
2. Remove or disconnect the following:
   - Upper shock-to-frame attaching nut
   - Lower shock nut
   - Slightly compress the shock absorber by hand and remove it from the vehicle

To install:

3. Install or connect the following:
   - Shock absorber upper end and nut
   - Shock absorber lower end and nut
   - Torque the upper and lower shock attaching nuts to 53 ft. lbs. (72Nm)

2002–04

1. Before servicing the vehicle, refer to the precautions in the beginning of this section.
2. Remove or disconnect the following:
   - Wheels
   - Upper shock mounting nuts. Discard the nuts.
   - Nut and the stabilizer bar link. Discard the nut.
   - Ball joint pinch bolt. Discard the nut.
   - Bolt, flag nut and the shock absorber and spring as an assembly. Discard the flag nut.
3. Using a suitable spring compressor, compress the spring until the tension is released from the shock absorber.
4. While holding the flats of the washer, remove the nut.
5. Remove the shock absorber. Discard the nut.
6. Remove the washer, bushing and the upper mount.
7. Remove the insulator.
8. Remove the dust shield.
9. To install, reverse the removal procedure. Note the following torques:
   - Center nut: 37 ft. lbs. (50 Nm)
   - Lower mounting bolt: 184 ft. lbs. (250 Nm)
   - Pinch bolt: 111 ft. lbs. (150 Nm)
   - Sway bar nut: 18 ft. lbs. (25 Nm)
   - Upper shock mounting nuts: 22 ft. lbs. (30 Nm)

Coil Spring

REMOVAL & INSTALLATION

For 2002–04, the front and rear coil springs are mounted on the shock absorbers. See Shock absorber Removal and Installation.

Leaf Springs

REMOVAL & INSTALLATION

1. Before servicing the vehicle, refer to the precautions in the beginning of this section.
2. Turn the air suspension switch off, if equipped.
3. Remove or disconnect the following:
   - Negative battery cable
   - Rear wheels and support the rear axle
   - Separate the rear spring from the axle and position the spring plate aside
   - Rear spring

To install:

4. Install or connect the following:
   - On Explorer and Mountaineer, torque the rear spring mounting bolts to 85 ft. lbs. (115 Nm) and the single bolt to 66 ft. lbs. (90 Nm).
   - On 2001 Explorer Sport and Sport-Trac, The leaf spring forward mounting bolt will have two different tightening specifications based upon the type of frame coating that is used. The E-coated frame is black and very smooth. The wax-

coated frame is gray and very sticky. With an E-coated frame, the bolt torque is 129 ft. lbs. (175 Nm); with the wax-coated frame, the torque is 85 ft. lbs. (115 Nm).
- On the 2002 Explorer Sport and Sport-Trac, the forward bolt is torque to 129 ft. lbs. (175 Nm); the rear bolt to 85 ft. lbs. (115 Nm).
- On the 2003–04 Explorer Sport and Sport-Trac, the forward bolt is torque to 166 ft. lbs. (225 Nm); the rear bolt to 85 ft. lbs. (115 Nm).
- Properly position the spring plate and install the U-bolts. Torque the bolts to 76 ft. lbs. (103 Nm).
- Rear wheels and remove the rear axle support
- Negative battery cable

5. Turn the air suspension switch on, if equipped.

Torsion Bar

REMOVAL & INSTALLATION

1. Before servicing the vehicle, refer to the precautions in the beginning of this section.
2. Remove or disconnect the following:
   - Negative battery cable
   - Torsion bar cover plate and measure the length of the torsion bar adjustment bolt
   - Relieve torsion bar tension
   - Torsion bar adjustment bolt
   - Torsion bar and insulator

To install:

3. Install or connect the following:
   - Torsion bar in the front suspension lower arm
   - Torsion bar adjuster and position the insulator
4. Preload the torsion bar and install a **NEW**adjuster bolt. Turn the bolt until it reaches the measurement made during the removal procedure.
   - Torsion bar cover plate. Torque the bolts to 46 ft. lbs. (63 Nm).
   - Negative battery cable

Upper Ball Joint

REMOVAL & INSTALLATION

The upper ball joints are integral with the control arm. If the ball joint is defective, the entire control arm must be replaced.

Lower Ball Joint

REMOVAL & INSTALLATION

2-Wheel Drive

The ball joints on the Aviator are not replaceable.

1. With the vehicle in NEUTRAL, position it on a hoist.
2. Remove the wheel and tire assembly.

⁂ WARNING

Do not allow the disc brake caliper to hang suspended from the brake hose. Provide a suitable support.

3. Remove the caliper support bracket bolts, then position the caliper and support bracket aside.
4. Disconnect the ABS electrical connector.
5. Unclip the front ABS wire from the vehicle frame.
6. Using a suitable jack, support the front suspension lower arm.
7. Remove the tie-rod end castellated nut. Remove and discard the cotter pin and the castellated nut.

⁂ WARNING

Do not use a hammer to separate the tie-rod from the wheel knuckle or damage to the wheel knuckle will result.

⁂ WARNING

Do not damage the tie-rod boot when installing the special tool.

8. Separate the tie-rod end from the front wheel knuckle.
9. Remove the lower ball joint castellated nut.
10. Separate the front wheel knuckle from the front suspension lower arm. Then, loosely install the lower ball joint castellated nut.
11. Remove the pinch bolt and nut.
12. Remove the hand-tightened lower ball joint castellated nut, then remove the front wheel knuckle.
13. Remove the snap ring from the ball joint. Discard the snap ring.
14. Using a suitable ball joint remover tool, remove the ball joint.

⁂ WARNING

Do not damage the ball joint boot when installing the special tool.

➡Clean and inspect the control arm ball joint bore for damage before installing a new ball joint.

➡Make sure the new ball joint snap ring is fully seated.

15. To install, reverse the removal procedure. Always install new castellated nuts and cotter pins. Torque the ball joint nut to 98 ft. lbs. (133 Nm) for 2001 Explorer and Mountaineer models and 2001–04 Explorer Sport and Sport-Trac models; 129 ft. lbs. (175 Nm) for 2002–04 Explorer and Mountaineer models. Torque the tie rod end ball stud nut to 52 ft. lbs. (70 Nm). Torque the pinch bolt to 41 ft. lbs. (55 Nm).

4-Wheel Drive

1. With the vehicle in NEUTRAL, position it on a hoist.
2. Remove the wheel and tire assembly.

⁂ WARNING

Do not reuse the torque prevailing design hub nut and washer assembly.

3. Remove and discard the hub nut and washer assembly.

⁂ WARNING

Do not allow the disc brake caliper to hang suspended from the brake hose. Provide a suitable support.

4. Remove the caliper support bracket bolts, then position the caliper and support bracket aside.
5. Remove the brake disc.

⁂ WARNING

Do not use a hammer to separate the outboard front wheel halfshaft joint from the wheel hub. Damage to the outboard CV joint stub shaft threads and internal CV joint components may result.

6. Separate the outboard front wheel halfshaft joint from the wheel hub.
7. Disconnect the ABS electrical connector.
8. Unclip the front ABS wire from the vehicle frame.
9. Using a suitable jack, support the front suspension lower arm.

⁂ WARNING

Secure the front axle shaft to prevent it from overextending. Failure to do so can cause damage to the front axle shaft.

10. Support the front axle shaft with wire.
11. Remove the tie-rod end castellated nut. Discard the cotter pin. Discard the castellated nut.

**** WARNING**

Do not use a hammer to separate the tie-rod from the wheel knuckle or damage to the wheel knuckle will result.

**** WARNING**

Do not damage the tie-rod boot when installing the special tool.

12. Separate the tie-rod end from the front wheel knuckle.
13. Remove the lower ball joint castellated nut. Discard the cotter pin. Discard the castellated nut.
14. Separate the front wheel knuckle from the front suspension lower arm then, loosely install the lower ball joint castellated nut.
15. Remove the pinch bolt and nut.
16. Remove the hand-tightened lower ball joint castellated nut, then remove the front wheel knuckle.
17. Remove the snap ring from the ball joint. Discard the snap ring.
18. Using a suitable ball joint remover tool, remove the ball joint.

**** WARNING**

Do not damage the ball joint boot when installing the special tool.

➡Clean and inspect the control arm ball joint bore for damage before installing a new ball joint.

➡Make sure the new ball joint snap ring is fully seated.

19. To install, reverse the removal procedure. Always install new castellated nuts and cotter pins. Observe the following torques:
    - Hub nut to 122 ft. lbs. (220 Nm) for 2001 Explorer and Mountaineer models; 162 ft. lbs. (220 Nm) for 2001–04 Explorer Sport and Sport-Trac models; 184 ft. lbs. (250 Nm) for 2002–04 Explorer and Mountaineer models.
    - Tie rod end nut to 52 ft. lbs. (70 Nm)
    - Ball joint nut to 98 ft. lbs. (133 Nm) for 2001 Explorer and Mountaineer models; 2001–04 Explorer Sport and Sport-Trac models; 129 ft. lbs. (175 Nm) for 2002–04 Explorer and Mountaineer models
    - Pinch bolt to 41 ft. lbs. (55 Nm).

Upper Control Arm

REMOVAL & INSTALLATION

2001

1. Before servicing the vehicle, refer to the precautions in the beginning of this section.
2. Turn off the air suspension switch, if equipped.
3. Remove or disconnect the following:
    - Negative battery cable
    - Front wheel
    - Pinch bolt and nut from the spindle
    - Upper control arm

To install:

4. Install or connect the following:
    - Upper control arm. Torque the bolts to 112 ft. lbs. (153 Nm).
    - Pinch bolt and nut to the front spindle. Torque the bolt to 46 ft. lbs. (63 Nm).
    - Front wheel
    - Negative battery cable
    - Turn on the air suspension switch, if equipped.

2002–02 Explorer and Mountaineer; 2002–04 Explorer Sport and Sport-Trac

1. Before servicing the vehicle, refer to the precautions in the beginning of this section.
2. Remove or disconnect the following:
    - Wheel
    - Upper ball joint from the wheel knuckle. Remove and discard the nut.
    - Arm-to-frame nuts and shims. Discard the nuts.
    - Upper arm
3. To install, reverse the removal procedure. Torque the upper arm nuts to 111 ft. lbs. (150 Nm). Torque the ball joint nut to 41 ft. lbs. (55 Nm).
4. Check and, if necessary, align the front end.

2003–04 Aviator, Explorer and Mountaineer

1. On 4x4 vehicles, loosen the axle retainer nut.

➡The wheel speed sensor electrical connectors are located in the engine compartment secured to the fender aprons.

2. Disconnect the wheel speed sensor.
3. Remove the wheel and tire assembly.
4. Remove the upper ball joint-to-wheel knuckle nut. Separate the ball joint from the knuckle with a ball joint driver.

**** WARNING**

Do not use a hammer to separate the ball joint from the wheel knuckle or damage to the wheel knuckle can result.

5. Remove the upper arm-to-frame nuts.
6. Remove the set shims.
7. Remove the upper arm.
8. To install, reverse the removal procedure. Check and, if necessary, align the front end. Torque the ball joint nut to 38 ft. lbs. (52 Nm); the upper arm nuts to 111 ft. lbs. (150 Nm).

UPPER CONTROL ARM BUSHING REPLACEMENT

The control arm bushings are not serviceable. If they require service, the upper or lower arm must be replaced.

Lower Control Arm

REMOVAL AND & INSTALLATION

2001

1. Before servicing the vehicle, refer to the precautions in the beginning of this section.
2. Turn off the air suspension switch, if equipped.
3. Remove or disconnect the following:
    - Negative battery cable
    - Front wheel
    - Brake rotor shield
    - Shock absorber
    - Torsion bar
    - Lower ball joint from the spindle
    - Lower control arm

To install:

4. Install or connect the following:
    - Lower control arm assembly to the crossmember and hand tighten the bolts at this time
    - Lower ball joint to the spindle. Torque the new castle nut to 21 ft. lbs. (29 Nm).
    - Torsion bar. Torque the bolts to 21 ft. lbs. (29 Nm). Check and adjust the ride height.
    - Shock absorber. Torque the bolts to 21 ft. lbs. (29 Nm).
    - Brake rotor shield

- Torque the lower control arm bolts to 148 ft. lbs. (200 Nm)
- Front wheel
- Negative battery cable

5. Turn on the air suspension switch, if equipped.
6. Check and adjust the front end alignment as needed.

2002–04

1. Before servicing the vehicle, refer to the precautions in the beginning of this section.
2. Remove or disconnect the following:

➡For reference during the installation procedure, measure the distance from the lip of the fender to the center of the wheel hub with the vehicle in a level static ground position.

- Wheel
- Upper shock absorber mounting nuts. Discard the nuts.
- Stabilizer bar connecting link. Discard the nut.
- Bolt, flag nut and the spring and shock assembly. Discard the flag nut.
- Separate the ball joint from the wheel knuckle. Discard the nut.
- Arm-to-frame nuts. Discard the nuts.
- Arm-to-knuckle bolt. Discard the nut.

➡On 4x4 vehicles, make sure that the crimped area of the outer CV boot clamp is not positioned downward or it will interfere with the removal of the arm.

- Lower arm

➡Using a suitable jack stand, raise the suspension until the distance between the lip of the fender and the center of the wheel hub is equal to the measurement taken in the removal procedure before tightening the inboard lower arm mountings. Make sure that the forward mounting is tightened first.

3. To install, reverse the removal procedure. Observe the following torques:
   - Control arm-to-frame flag bolt: 295 ft. lbs. (400 Nm)

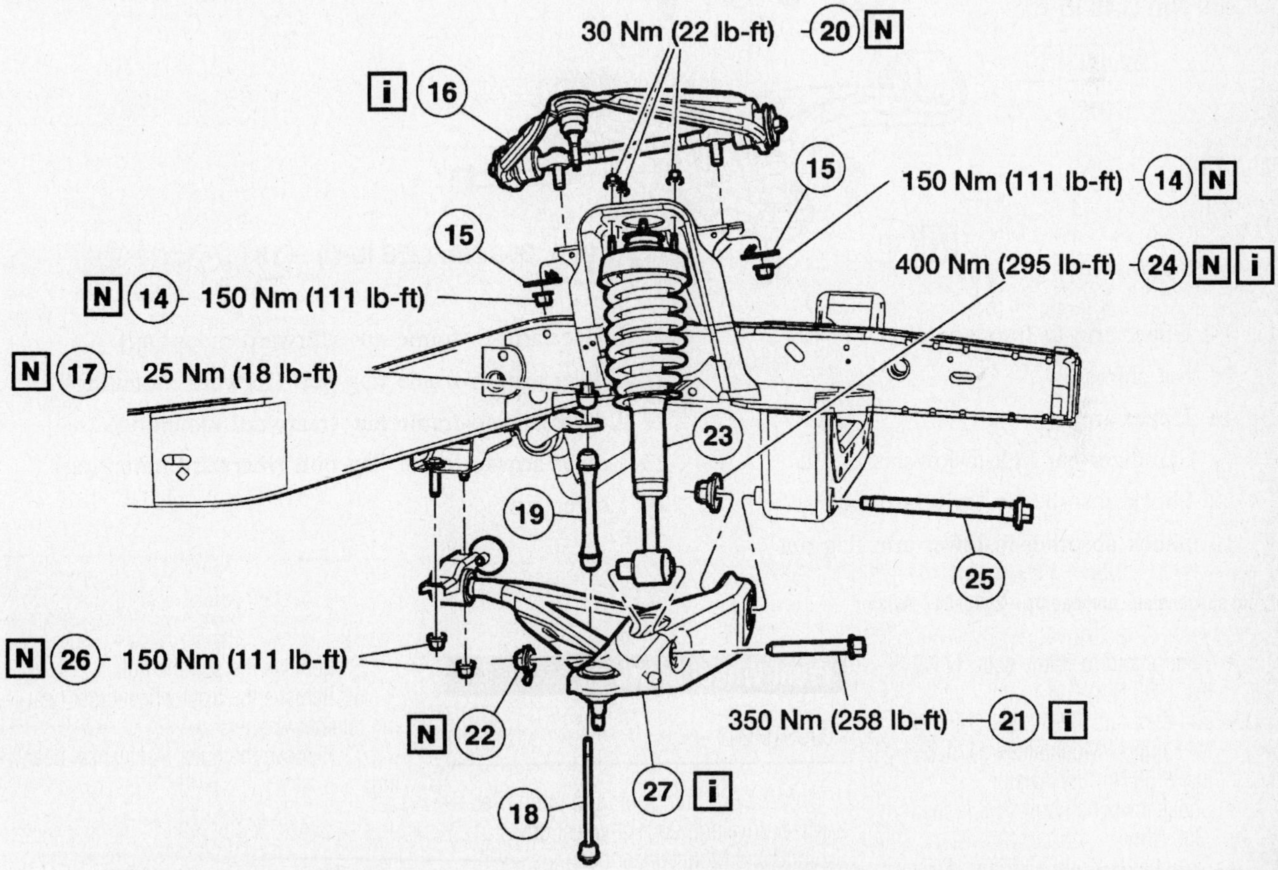

14 Upper arm-to-frame nuts
15 Set shims
16 Upper arm RH/LH
17 Nut and grommet
18 Stud
19 Stabilizer bar link
20 Shock absorber upper mount-to-frame nuts
21 Shock absorber-to-lower arm bolt
22 Shock absorber-to-lower arm flag nut
23 Shock absorber and spring assembly
24 Lower arm-to-frame nut (forward mounting)
25 Lower arm-to-frame flag bolt (forward mounting)
26 Lower arm-to-frame nuts (rearward mounting)
27 Lower Arm

67197-EXPL-G68

Front suspension components—2003–04 Explorer and Mountaineer

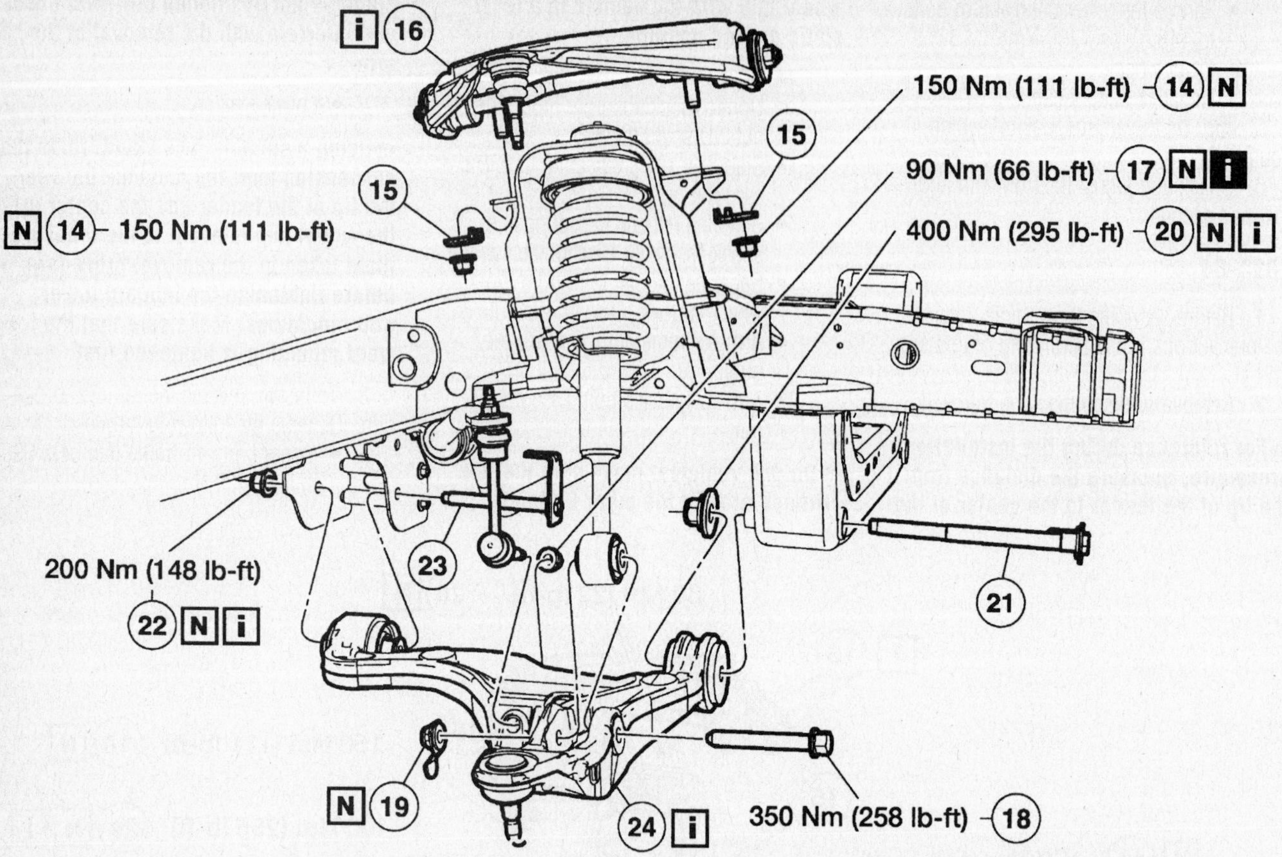

14 Upper arm-to-frame nuts
15 Set shims
16 Upper arm
17 Stabilizer bar link-to-lower arm nut
18 Shock absorber-to-lower arm bolt
19 Shock absorber-to-lower arm flag nut
20 Lower arm-to-frame nut (forward mounting)
21 Lower arm-to-frame flag bolt (forward mounting)
22 Lower arm-to-frame nut (rearward mounting)
23 Lower arm-to-frame flag bolt (rearward mounting)
24 Lower arm

67197-EXPL-G69

Front suspension components—2003–04—Aviator

- Control arm-to-frame nuts: 111 ft. lbs. (150 Nm)
- Ball joint nut: 129 ft. lbs. (175 Nm) for Explorer/Mountaineer; 111 ft. lbs. (150 Nm) for Aviator
- Shock lower flag bolt: 258 ft. lbs. (350 Nm)
- Sway bar link nut: 18 ft. lbs. (25 Nm)
- Upper shock mounting nuts: 22 ft. lbs. (30 Nm)

4. Check and, if necessary, align the front end.

LOWER CONTROL ARM BUSHING REPLACEMENT

The control arm bushings are not serviceable. If they require service, the upper or lower arm must be replaced.

Front Wheel Bearings

ADJUSTMENT

Only the Explorer Sport and Sport-Trac 2-wheel drive and 2001 2-wheel drive Explorer and Mountaineer bearings are adjustable.

2001–04 2-Wheel Drive Explorer Sport and Sport-Trac, 2001 2-Wheel Drive Explorer and Mountaineer

1. Remove the front disc brake caliper anchor plate.

➡Match-mark the brake disc and hub.

2. Remove the hub grease cap.
3. Remove the cotter pin.
4. Remove the nut retainer.
5. Remove the spindle nut.
6. Remove the front wheel outer bearing retainer washer.
7. Remove the outer front wheel bearing.

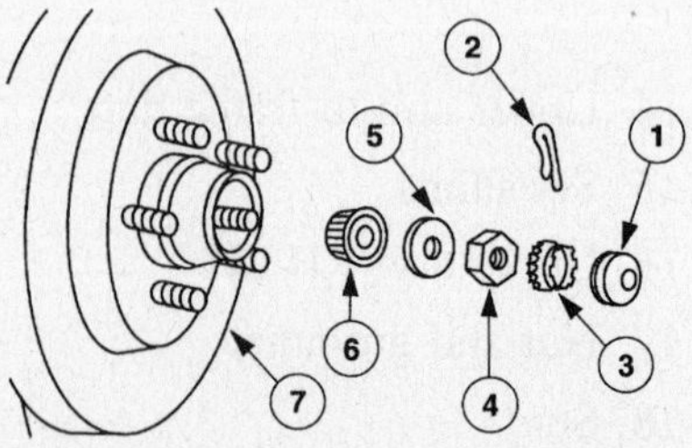

67197-EXPL-G65

Front wheel bearings—2001–04 2-wheel drive Explorer Sport and Sport-Trac and 2001 2-Wheel Drive Explorer and Mountaineer

8. Remove the brake disc and hub.
9. Remove the wheel hub grease seal and wheel bearing.

To install:

10. Clean and inspect the front wheel bearings and the brake disc and hub. Use Metal Brake Parts Cleaner F6AZ-2C410-AB, or equivalent.
11. Lubricate the front wheel bearings. Use Premium Long-Life Grease XG-1-C or -K or equivalent meeting Ford specification ESA-M1C75-B.
12. Install the inner front wheel bearing.
13. Install a new wheel hub grease seal.
14. Position the brake disc and hub.
15. Install the outer front wheel bearing.
16. Install the front wheel outer bearing retainer washer.
17. Install the spindle nut.
18. Tighten the spindle nut to 21 ft. lbs. (29 Nm) while rotating the brake disc and hub.
19. Loosen the spindle nut one-half turn.
20. Tighten the spindle nut to 17 inch lbs. (2 Nm) while rotating the brake disc and hub.
21. Install the nut retainer.
22. Install the cotter pin.
23. Install the hub grease cap.
24. Install the front disc brake caliper anchor plate.

REMOVAL & INSTALLATION

2001–04 2-wheel drive Explorer Sport and Sport-Trac; 2001 2-Wheel Drive Explorer and Mountaineer

1. Remove the front disc brake caliper anchor plate.

➡Match-mark the brake disc and hub.

2. Remove the hub grease cap.
3. Remove the cotter pin.
4. Remove the nut retainer.
5. Remove the spindle nut.
6. Remove the front wheel outer bearing retainer washer.
7. Remove the outer front wheel bearing.
8. Remove the brake disc and hub.
9. Remove the wheel hub grease seal and inner wheel bearing.
10. Place the hub on a solid wood work surface.
11. Using a suitable drift, drive out the inner bearing race; then, in a similar fashion, drive out the outer bearing race.

To install:

12. Clean and inspect the front wheel bearings and the brake disc and hub. Use Metal Brake Parts Cleaner F6AZ-2C410-AB, or equivalent.
13. Pack the front wheel bearings. Use Premium Long-Life Grease XG-1-C or -K or equivalent meeting Ford specification ESA-M1C75-B.
14. Lubricate the bearing race bore. Using a suitable driver, drive in a new outer bearing race until it fully seats; then, in a similar fashion, drive a new inner race into position.
15. Pack the hub bore with the same grease used to pack the bearings.
16. Install the inner front wheel bearing.
17. Install a new wheel hub grease seal.
18. Position the hub on the spindle.
19. Install the outer front wheel bearing.
20. Install the front wheel outer bearing retainer washer.
21. Install the spindle nut.
22. Tighten the spindle nut to 21 ft. lbs. (29 Nm) while rotating the brake disc and hub.
23. Loosen the spindle nut one-half turn.
24. Tighten the spindle nut to 17 inch lbs. (2 Nm) while rotating the brake disc and hub.
25. Install the nut retainer.
26. Install the cotter pin.
27. Install the hub grease cap.
28. Install the front disc brake caliper anchor plate.

2001 4-Wheel Drive Explorer, Explorer Sport, Explorer Sport-Trac and Mountaineer

⁂ WARNING

If equipped, always turn off the Automatic Ride Control (ARC) service switch before lifting the vehicle off of the ground. Failure to do so could damage the ARC system components.

1. Before servicing the vehicle, refer to the precautions in the beginning of this section.
2. Remove or disconnect the following:
   - Negative battery cable
   - Front wheels
   - Front disc brake caliper, bracket and rotor
   - Brake rotor splash shield
   - Antilock Brake System (ABS) sensor and wire harness from the steering knuckle, if equipped
   - Front wheel hub nut and washer
   - Wheel hub/bearing to steering knuckle retaining bolts
   - Hub and bearing assembly

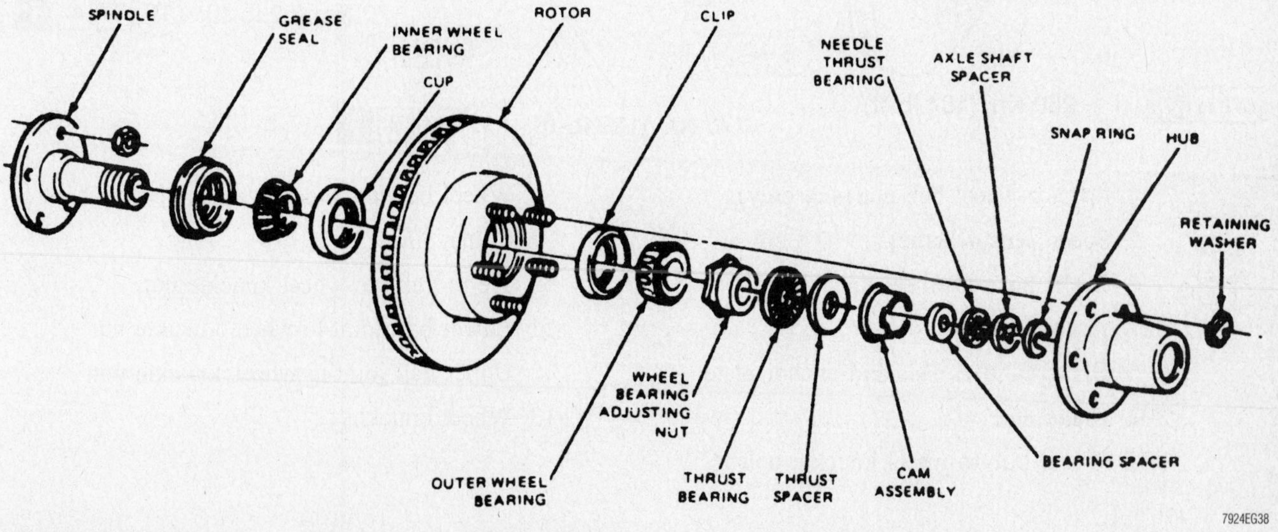

Exploded view of the wheel bearing and automatic locking hub assembly

⁂ WARNING

Never reuse the wheel hub nut and washer. This nut is a torque prevailing design and cannot be reused.

➡The hub shaft is a slip fit into the wheel hub and bearing; a press is not normally required.

3. Ensure that the wheel hub shaft can be pushed inwards. If not, assemble a press to the front wheel studs and press the wheel hub shaft inwards slightly to break it loose.

To install:

4. Install or connect the following:
   - ABS sensor to the wheel hub then position the hub to the front axle shaft and steering knuckle
   - Retaining bolts. Torque the bolts to 74–96 ft. lbs. (100–130 Nm).
   - Position the hub on the driveshaft and into the steering knuckle. Torque the bolts to 89 inch lbs. (10 Nm).
   - Brake rotor splash shield
   - ABS sensor. Torque the bolt to 89 inch lbs. (10 Nm).
   - Hub washer and nut and tighten to 157–213 ft. lbs. (212–288 Nm)
   - Front brake rotor, bracket and caliper
   - Front wheels
   - Negative battery cable

2002–04

2-WHEEL DRIVE

⁂ WARNING

If equipped, always turn off the Automatic Ride Control (ARC) service switch before lifting the vehicle off of the ground. Failure to do so could damage the ARC system components.

1. Before servicing the vehicle, refer to the precautions in the beginning of this section.
2. Remove or disconnect the following:

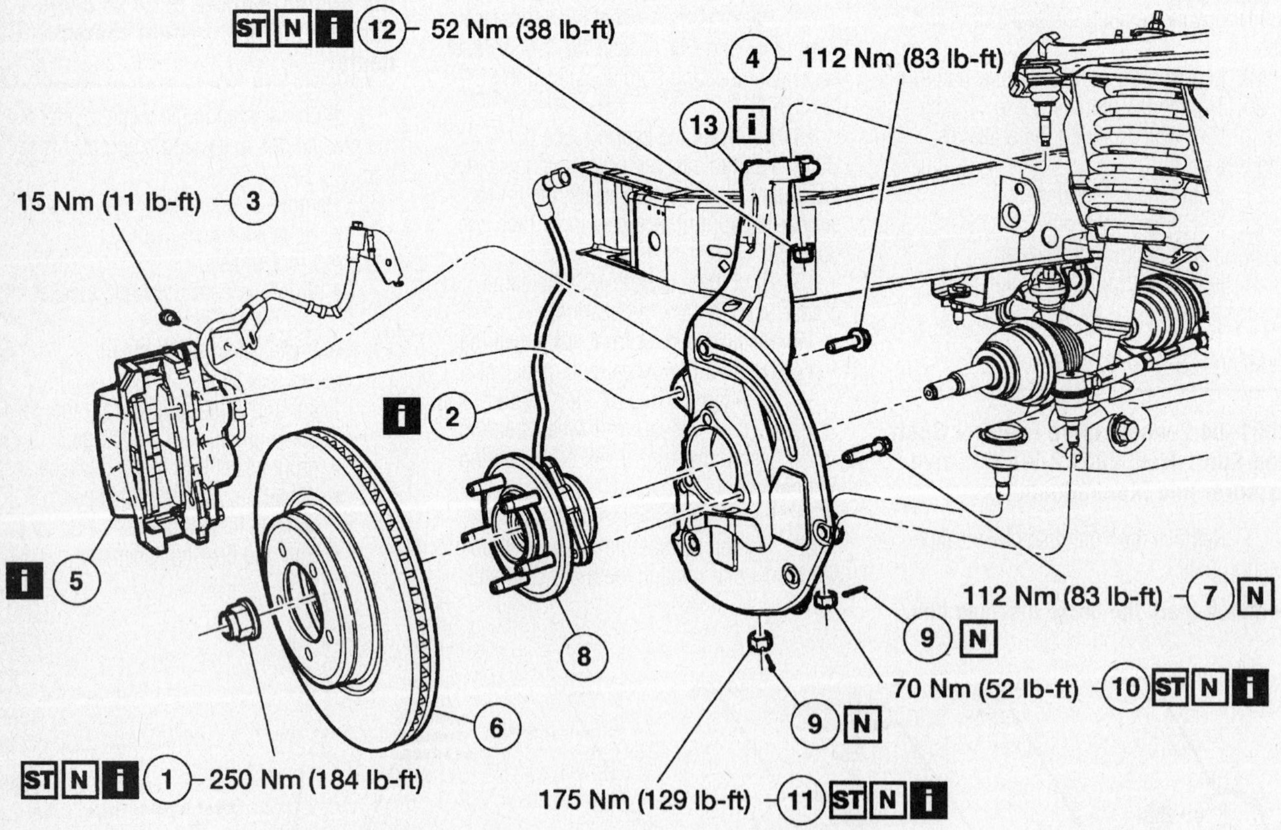

1	Axle-to-wheel hub nut (4x4 only)	8	Wheel bearing and hub assembly
2	Speed sensor harness	9	Cotter pins
3	Brake hose-to-wheel knuckle bolt	10	Tie-rod end-to-wheel knuckle nut
4	Anchor plate bolt	11	Lower ball joint-to-wheel knuckle nut
5	Brake caliper, pads and anchor plate	12	Upper ball joint-to-wheel knuckle nut
6	Brake disc	13	Wheel knuckle
7	Wheel hub-to-wheel knuckle bolt		

67197-EXPL-G66

Front hub and related parts—2003–04 Explorer and Mountaineer

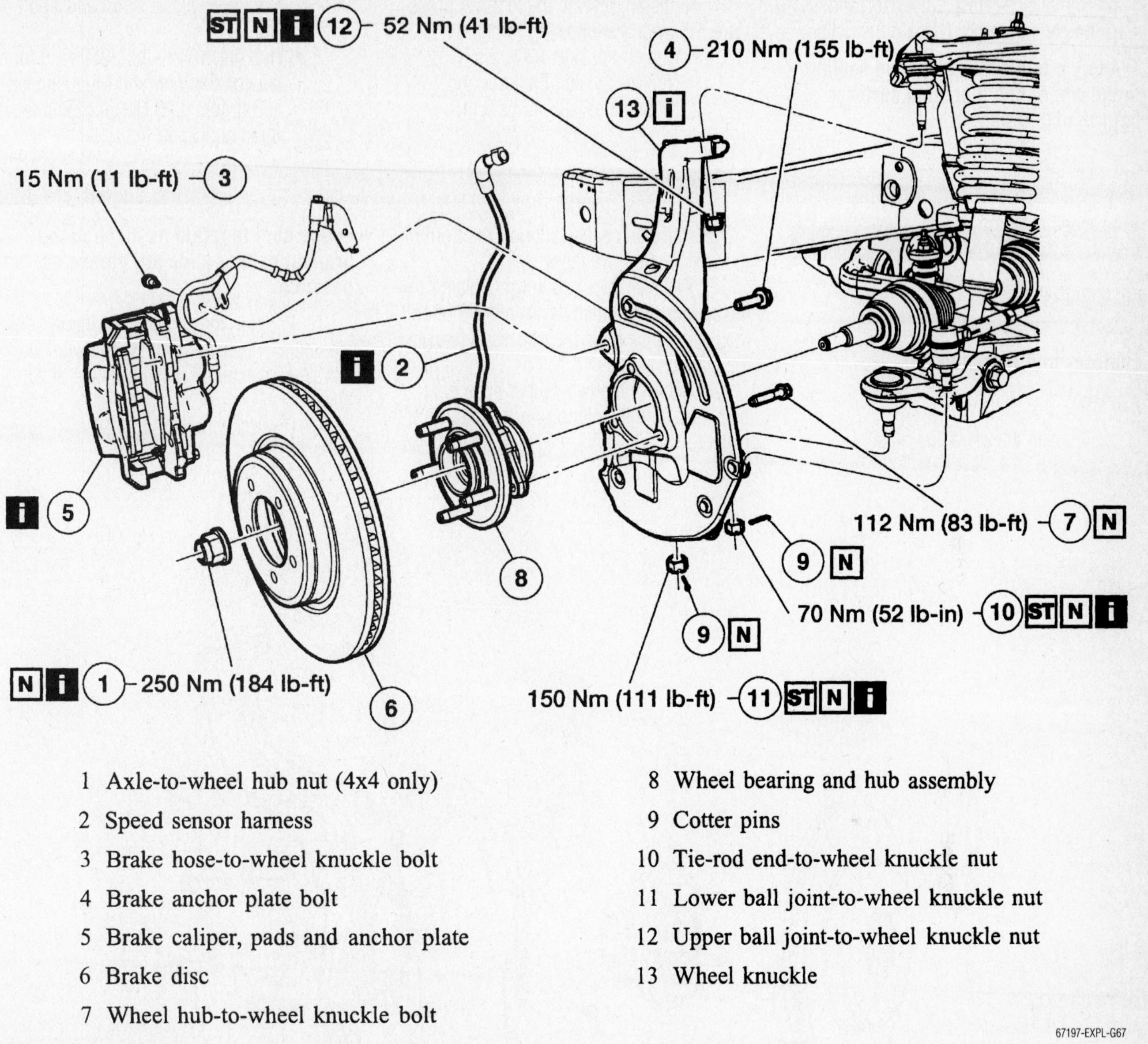

Front hub and related parts—2003–04—Aviator

➡The wheel speed sensor connectors are located in the engine compartment and are secured to the fender aprons.

- Wheel speed sensor connector
- Brake disc
- Wiring harness from the retainers
- Bolts, wheel hub and sensor as an assembly. Discard the bolts.

3. To install, reverse the removal procedure.

➡Apply a thin coat of silicone sealant to the wheel hub mounting surfaces before installation.

4. Torque the hub-to-knuckle bolts to 83 ft. lbs. (112 Nm)

4-WHEEL DRIVE

✲✲ WARNING

If equipped, always turn off the Automatic Ride Control (ARC) service switch before lifting the vehicle off of the ground. Failure to do so could damage the ARC system components.

1. Before servicing the vehicle, refer to the precautions in the beginning of this section.
2. Remove or disconnect the following:
   - Hub nut. Discard the nut.

➡The wheel speed sensor connectors are located in the engine compartment and are secured to the fender aprons.

- Wheel speed sensor connector
- Brake disc

✲✲ WARNING

Do not use a hammer to separate the outboard CV joint from the hub. Damage to the threads and internal CV joint components may result.

3. Press the outboard CV joint until it is loose in the hub.
4. Detach the harness from the retainers.

✲✲ WARNING

Do not overextend the CV joint and boots when removing the wheel hub.

5. Remove the bolts, wheel hub and sensor as an assembly. Discard the bolts.

➡Apply a thin coat of silicone sealant to the wheel hub mounting surfaces before installation.

6. To install, reverse the removal procedure. Observe the following torques:
- Hub-to-knuckle: 83 ft. lbs. (112 Nm) for Aviator, Explorer and Mountaineer; 85 ft. lbs. (115 Nm) for Explorer Sport and Sport-Trac.
- Hub nut: 184 ft. lbs. (250 Nm) for Aviator, Explorer and Mountaineer; 162 ft. lbs. (220 Nm) for Explorer Sport and Sport-Trac.

BRAKES

Brake Caliper

REMOVAL & INSTALLATION

Explorer and Mountaineer

FRONT

1. Loosen the wheel lug nuts.
2. Raise and safely support the front of the vehicle. Remove the wheel.
3. Place an 8 in. (203mm) C-clamp on the caliper and tighten the clamp to bottom the caliper pistons in their bores. Remove the clamp.
4. Remove the two caliper slide pin bolts and lift the caliper from the anchor plate.

➡Use care to retain as much of the original caliper slide pin grease as possible.

5. Position the caliper on a frame member or suspend it with some wire. Do not allow the caliper to hang by the brake hose.
6. Disconnect and plug the brake hose

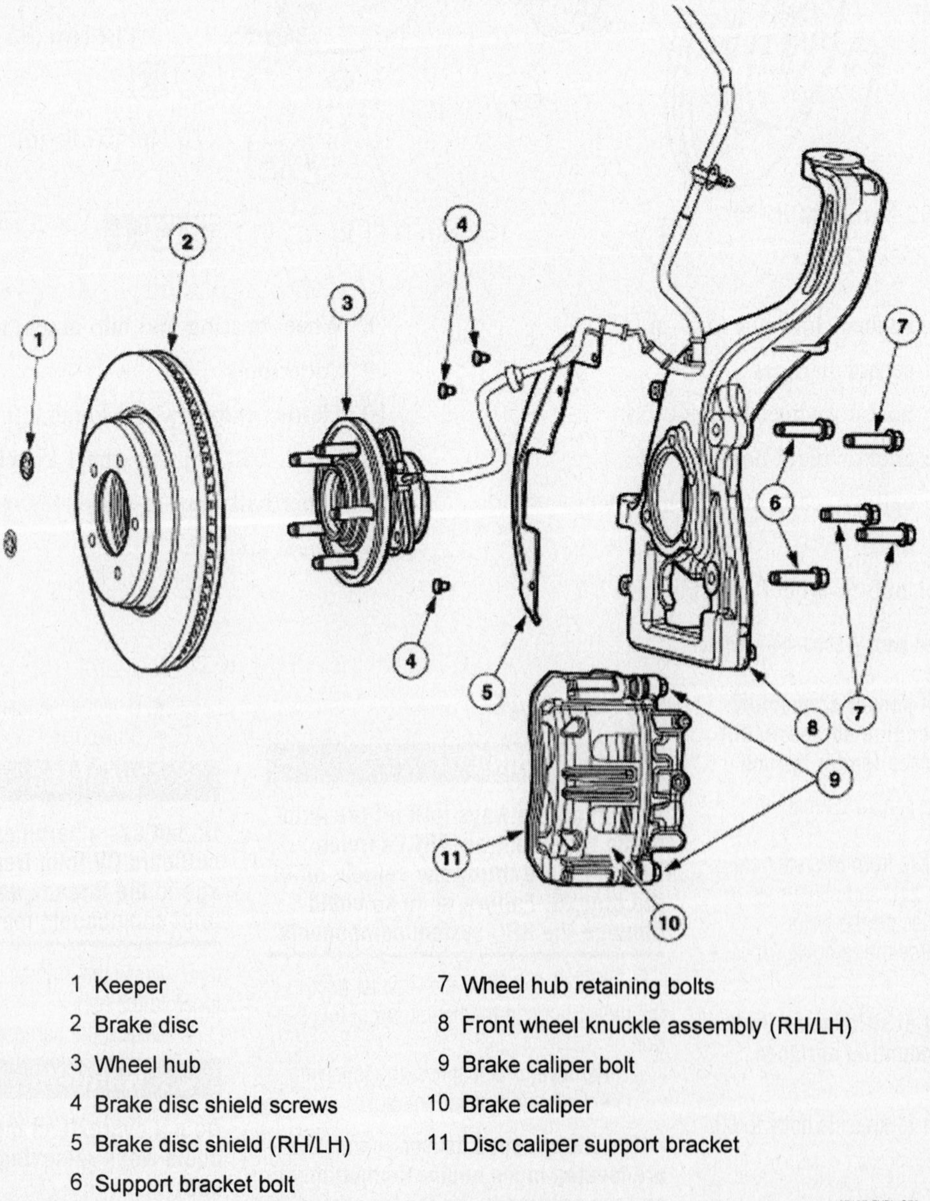

Front brake components—2002 Explorer and Mountaineer

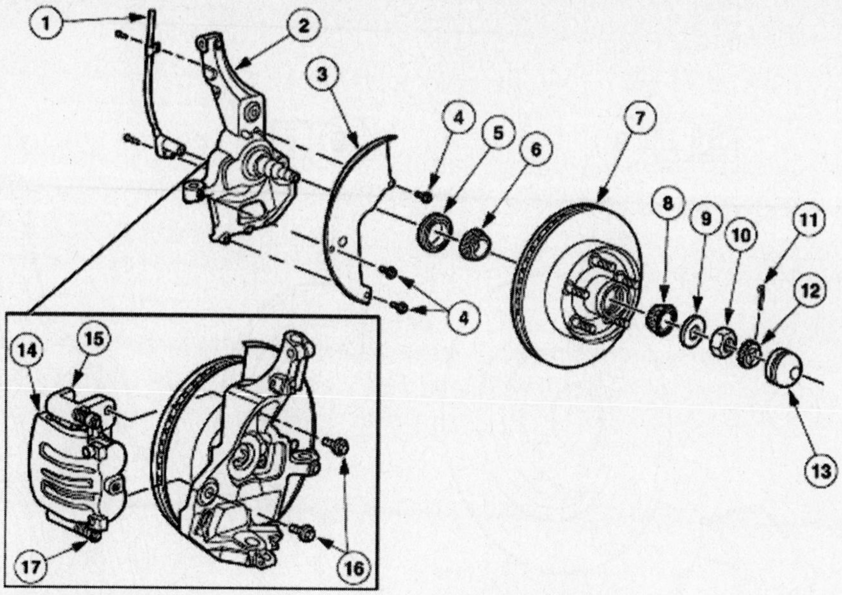

1 Front brake anti-lock sensor
2 Front wheel spindle
3 Brake disc shield
4 Bolt
5 Grease seal
6 Front wheel bearing
7 Brake disc and hub
8 Front wheel bearing
9 Front wheel outer bearing retainer washer
10 Hub spindle nut
11 Cotter pin
12 Nut retainer
13 Hub grease cap
14 Disc brake caliper
15 Front disc brake caliper anchor plate
16 Caliper anchor plate bolts
17 Disc brake caliper bolt

67197-EXPL-G71

Front brake components—2001 2-wheel drive Explorer and Mountaineer; 2001–04 2-wheel drive Explorer Sport and Sport-Trac

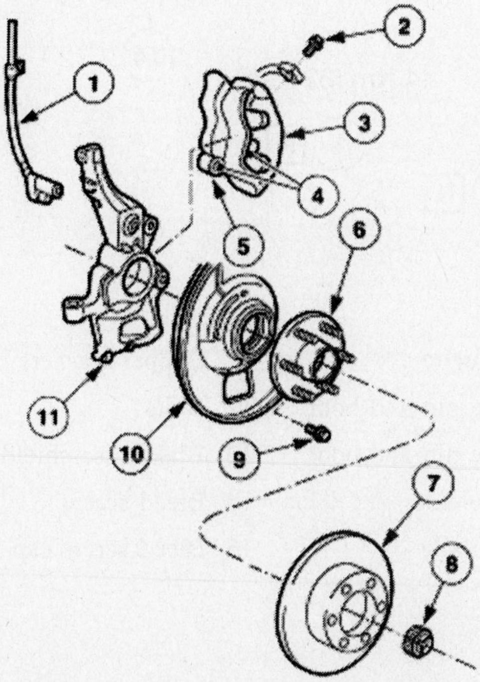

1 Front brake anti-lock sensor
2 Front brake hose bolt
3 Disc brake caliper
4 Pads
5 Front disc brake caliper anchor plate
6 Wheel hub
7 Brake disc
8 Front axle wheel hub retainer
9 Bolt
10 Brake disc shield
11 Front wheel knuckle

67197-EXPL-G72

Front brake components—2001 4-wheel drive Explorer and Mountaineer; 2002–04 Explorer Sport and Sport-Trac

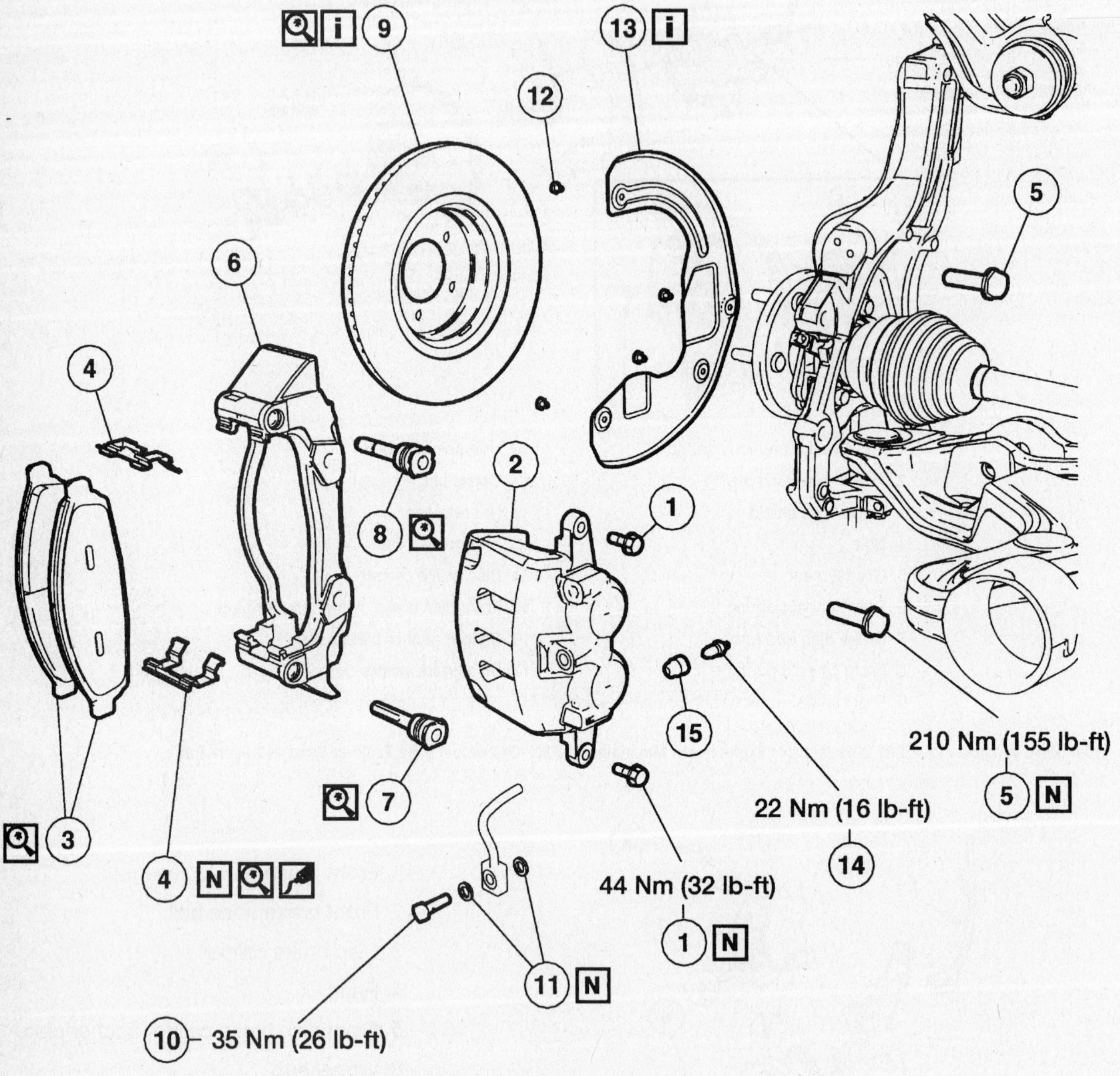

1 Bolts	6 Anchor	11 Copper washers
2 Brake caliper	7 Guide pin and boot	12 Bolts
3 Brake disc pads	8 Guide pin and boot	13 Brake disc shield (LH/RH)
4 Slippers	9 Brake disc	14 Bleed screw
5 Brake caliper anchor bracket bolt	10 Flow bolt	15 Bleed screw cap

67197-EXPL-G74

Front brake components—Aviator

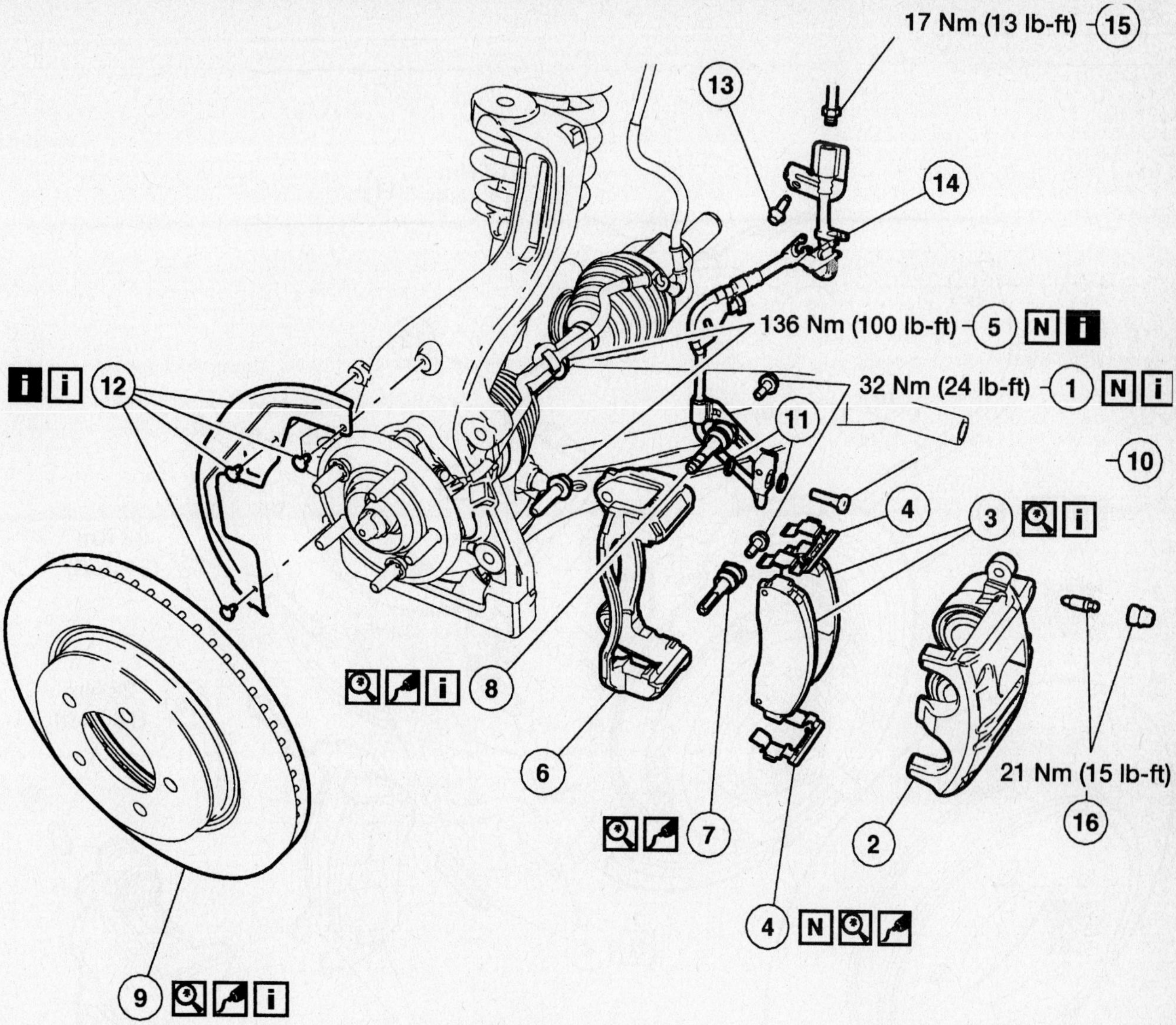

1 Caliper bolts	6 Anchor bracket	11 Copper washers
2 Brake caliper (RH/LH)	7 Guide pin and boot	12 Brake disc shield
3 Brake disc pads	8 Locating pin and boot	13 Brake hose bracket bolt
4 Slippers	9 Brake disc	14 Brake hose (RH/LH)
5 Anchor bracket bolt kit	10 Flow bolt	15 Brake tube

67197-EXPL-G73

Front brake components—2003–04 Explorer and Mountaineer

at the caliper. Remove the caliper from the rotor.

To install:

7. Position the caliper over the brake pads and align the slide pin mounting holes.

8. Install the slide pin bolts and tighten them to 21–26 ft. lbs. (30–36 Nm) for all except Aviator; for Aviator, torque the bolts to 32 ft. lbs. (44 Nm).

9. Install the caliper brake hose using new washers. Tighten the bolt to 23–29 ft. lbs. (40 Nm).

10. Install the wheel and snug the lug nuts.

11. Lower the vehicle and tighten the lug nuts to 100 ft. lbs. (135 Nm).

➡**The first couple of times you apply the brakes, the pedal may go to the floor. Continue to pump the brake pedal until it feels firm.**

12. Start the engine and apply the brakes several times to readjust the caliper pistons. Ensure that the pedal feels firm before operating the vehicle.

REAR

1. Siphon part of the brake fluid out of the master cylinder to avoid overflow when the caliper piston is pressed into the caliper bore.

2. Raise the vehicle and support it safely. Remove the wheel and tire assembly.

3. Position an 8 in. (20cm) C-clamp on the caliper and tighten the clamp to move the caliper piston into the bore approximately ⅛ in. (3mm). Remove the clamp.

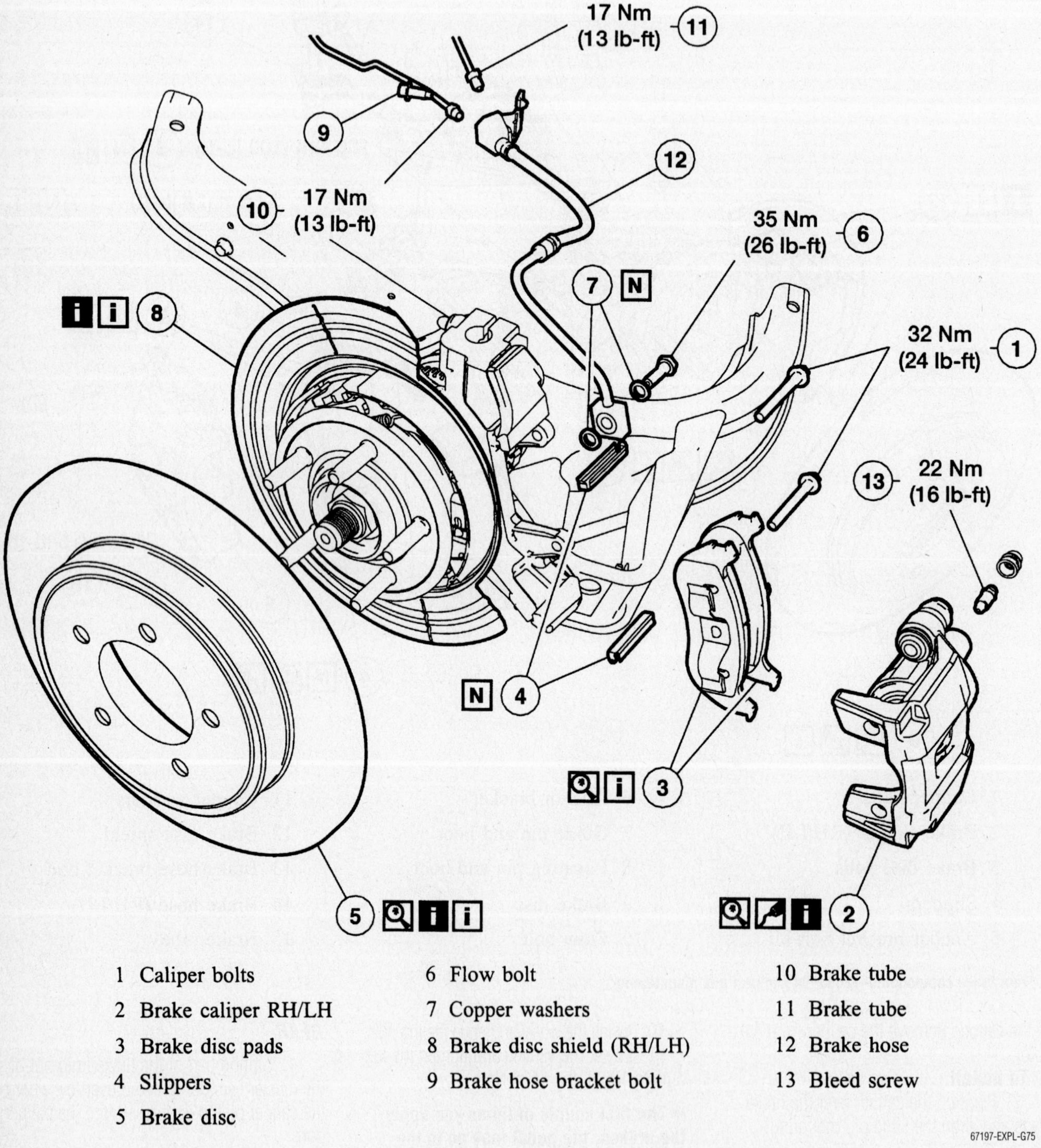

Rear disc brake components—2003–04 Explorer and Mountaineer

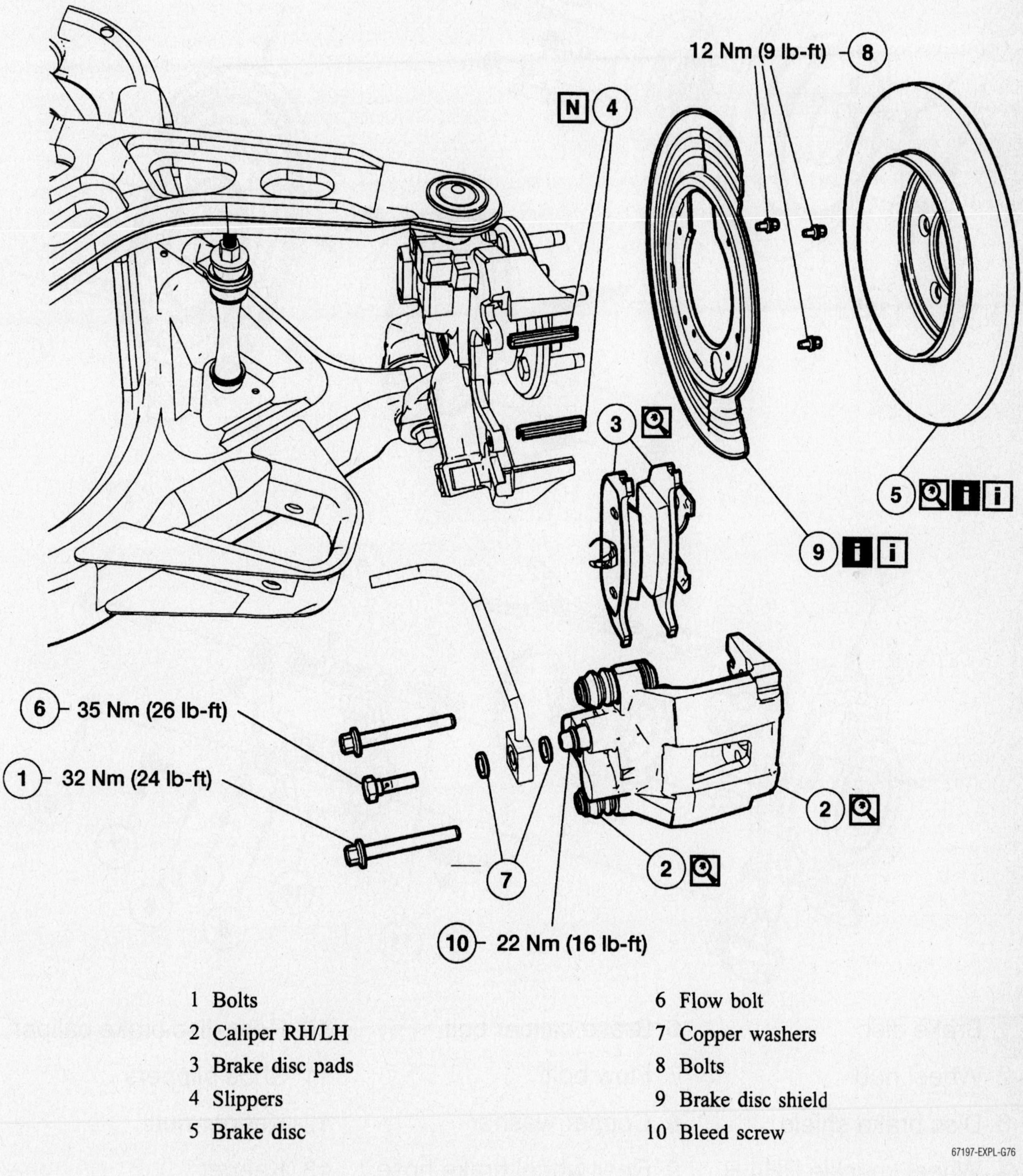

1 Bolts
2 Caliper RH/LH
3 Brake disc pads
4 Slippers
5 Brake disc
6 Flow bolt
7 Copper washers
8 Bolts
9 Brake disc shield
10 Bleed screw

67197-EXPL-G76

Rear disc brake components—Aviator

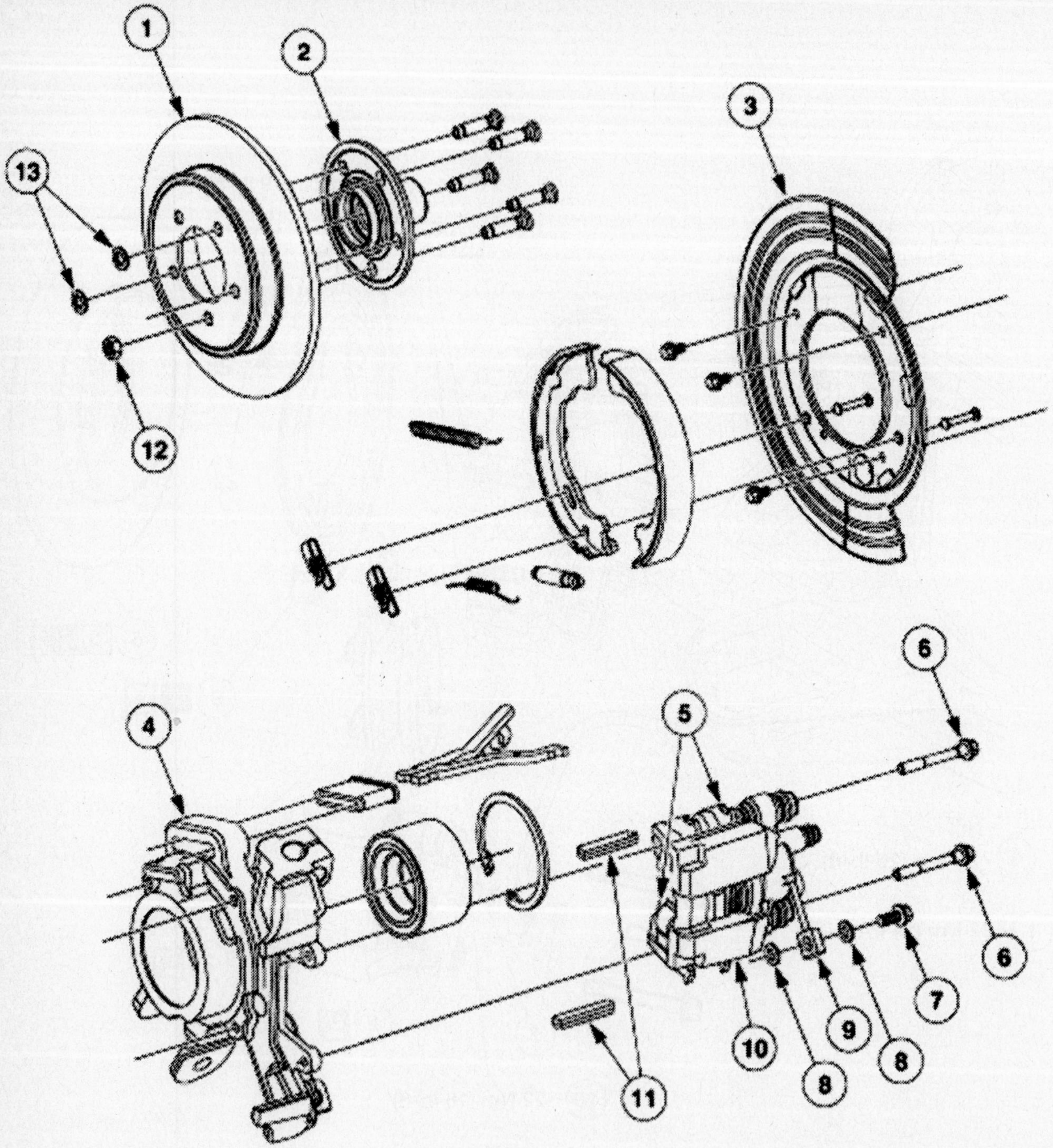

1 Brake disc
2 Wheel hub
3 Disc brake shield
4 Wheel knuckle RH/LH
5 Brake pads
6 Brake caliper bolt
7 Flow bolt
8 Copper washer
9 Rear wheel brake hose
10 Rear disc brake caliper
11 Shoe slippers
12 Keeper nut
13 Keeper

67197-EXPL-G77

Rear disc brake components—2002 Explorer and Mountaineer

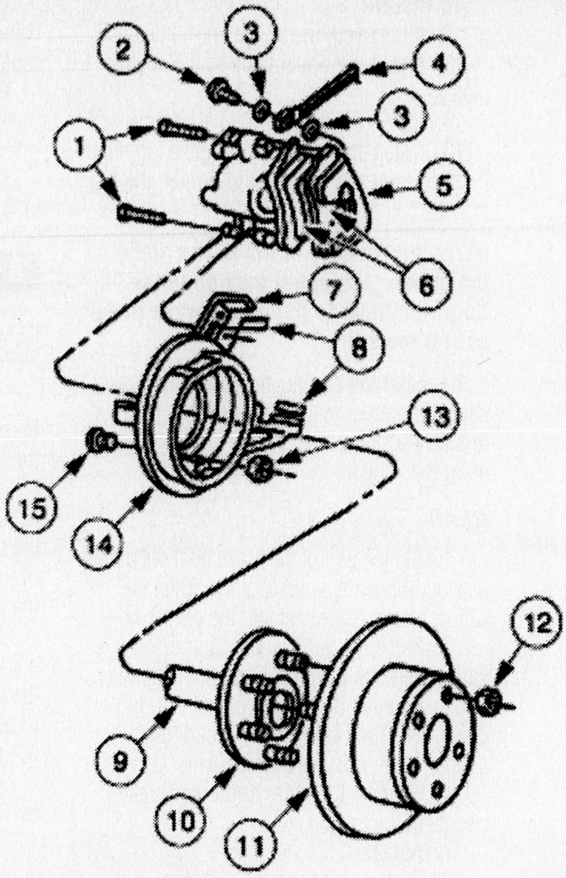

1 Brake caliper bolt
2 Flow bolt
3 Copper washer
4 Rear wheel brake hose
5 Rear disc brake caliper
6 Brake pads
7 Adapter assy
8 Shoe slippers
9 Axle shaft
10 Wheel stud
11 Brake disc
12 Keeper nut
13 Rear wheel disc brake adapter nut
14 Rear wheel disc brake adapter
15 Rear wheel disc brake adapter bolt

67197-EXPL-G78

Rear disc brake components—2001 Explorer and Mountaineer; 2001–04 Explorer Sport and Sport-Trac

➡Do not pry the piston away from the rotor.

4. Clean excess dirt from the retainer bolt area.
5. Using a Torx® socket, remove the 2 retainer bolts securing the caliper to the bracket and adapter plate.
6. Disconnect and plug the brake hose at the caliper. Remove the caliper from the rotor.

To install:

7. Make sure the caliper mounting surfaces are free of dirt. Lubricate the caliper grooves with disc brake caliper grease and install the caliper.
8. Position the caliper to the bracket and secure in place with the retainer bolts. Tighten the bolts to:
    - 2001 Explorer and Mountaineer: 20 ft. lbs. (27 Nm)
    - 2001–04 Explorer Sport and Sport-Trac: 20 ft. lbs. (27 Nm)
    - 2002–04 Explorer and Mountaineer: 24 ft. lbs. (32 Nm)
    - 2003–04 Aviator: 24 ft. lbs. (32 Nm)
9. Install the caliper brake hose using new washers. Tighten the bolt to 23
10. Fill and bleed the brake system.
11. Install the wheel and tire assembly and lower the vehicle. Check the brake fluid level and check the brakes for proper operation.

Disc Brake Pads

REMOVAL & INSTALLATION

Explorer and Mountaineer

FRONT

1. Raise and safely support the front of the vehicle. Remove the wheel.
2. Place an 8 in. (203mm) C-clamp on the caliper and tighten the clamp to bottom the caliper pistons in their bores. Remove the clamp.
3. Remove the two caliper slide pin bolts and lift the caliper from the anchor plate.

➡Use care to retain as much of the original caliper slide pin grease as possible.

4. Position the caliper on a frame member or suspend it with some wire. Do not allow the caliper to hang by the brake hose.
5. Remove the brake pads and, if necessary, the anti-rattle clips from the anchor plate.
6. Remove the shims, if any, from the brake pads for re-use.

To install:

7. If removed, install the anti-rattle clips.
8. Install the brake pads to the anchor plate.
9. Install the caliper.
10. Install the wheel and snug the lug nuts.
11. Lower the vehicle and tighten the lug nuts to 100 ft. lbs. (135 Nm).

➡The first couple of times you apply the brakes, the pedal may go to the floor. Continue to pump the brake pedal until it feels firm.

12. Start the engine and apply the brakes several times to readjust the caliper pistons. Ensure that the pedal feels firm before operating the vehicle.

REAR

1. Siphon part of the brake fluid out of the master cylinder to avoid overflow when the caliper piston is pressed into the caliper bore.
2. Raise the vehicle and support it safely. Remove the wheel and tire assembly.
3. Remove the brake caliper, but do not disconnect the brake hose. Secure the caliper aside with mechanic's wire.
4. Remove the inner and outer brake pad from the caliper.

To install:

5. Bottom out the caliper piston in the caliper bore using an 8 in. (20cm) C-clamp or equivalent and a worn out inner brake pad or block of wood to push against the piston. Do not attempt to bottom out the piston with the outer brake pad installed.
6. Position the inboard brake pad in the caliper and press the retainer spring fully into the caliper piston.
7. Start one end of the outboard brake shoe and lining on the caliper and rotate it down until the locating lugs and the retainer spring are fully seated.
8. Install new shoe slippers on the rear wheel disc brake adapter.
9. Install the caliper.
10. Install the wheel and tire assembly and lower the vehicle. Apply the brakes several times before moving the vehicle to seat the pads.
11. Check the brake fluid level. Check the brakes for proper operation.

Brake Drums

REMOVAL & INSTALLATION

1. Raise and safely support the vehicle. Remove the wheel and tire assembly.
2. Remove the retaining nuts, if equipped, and remove the brake drum.
3. Inspect the brake drum surface for wear, scoring and runout. Machine or replace, as necessary.

To install:

4. Install the brake drum and secure in place with the retainer nuts, if equipped.
5. Adjust the rear brakes.
6. Install the wheel. Lower the vehicle.

Brake Shoes

REMOVAL & INSTALLATION

1. Raise and safely support the vehicle. Remove the wheel and tire assembly and the brake drum.
2. Pull backward on the adjusting lever cable to disengage the adjusting lever from the adjusting screw. Move the outboard side of the adjusting screw upward and back off the pivot nut as far as it will go.
3. Pull the adjusting lever, cable and automatic adjuster spring down and toward the rear to unhook the pivot hook from the large hole in the secondary shoe web. Do not pry the pivot hook from the hole.
4. Remove the automatic adjuster spring and adjusting lever.
5. Remove the secondary shoe-to-anchor spring using a suitable brake spring removal/installation tool. Using the tool, remove the primary shoe-to-anchor spring and unhook the cable anchor. Remove the anchor pin plate, if equipped.
6. Remove the cable guide from the secondary shoe.
7. Remove the shoe hold-down springs, shoes, adjusting screw, pivot nut and socket. Note the color and position of each hold-down spring so they can be reassembled in the same position.
8. Remove the parking brake link and spring. Disconnect the parking brake cable from the parking brake lever.
9. Remove the secondary brake shoe. On 9 in. (22.8cm) rear brakes, remove the parking brake lever from the shoe. On 10 in. (25.4cm) rear brakes, remove the retainer clip and spring washer and remove the parking brake lever.

To install:

10. Clean the backing plate ledge pads and sand lightly. Apply a light coating of high temperature lithium grease to the points where the brake shoes touch the backing plate. Lubricate the adjusting cable eye and the anchor pin area.
11. Install the parking brake lever on the secondary shoe. On 10 in. (25.4cm) brakes, secure with the spring washer and retaining clip.

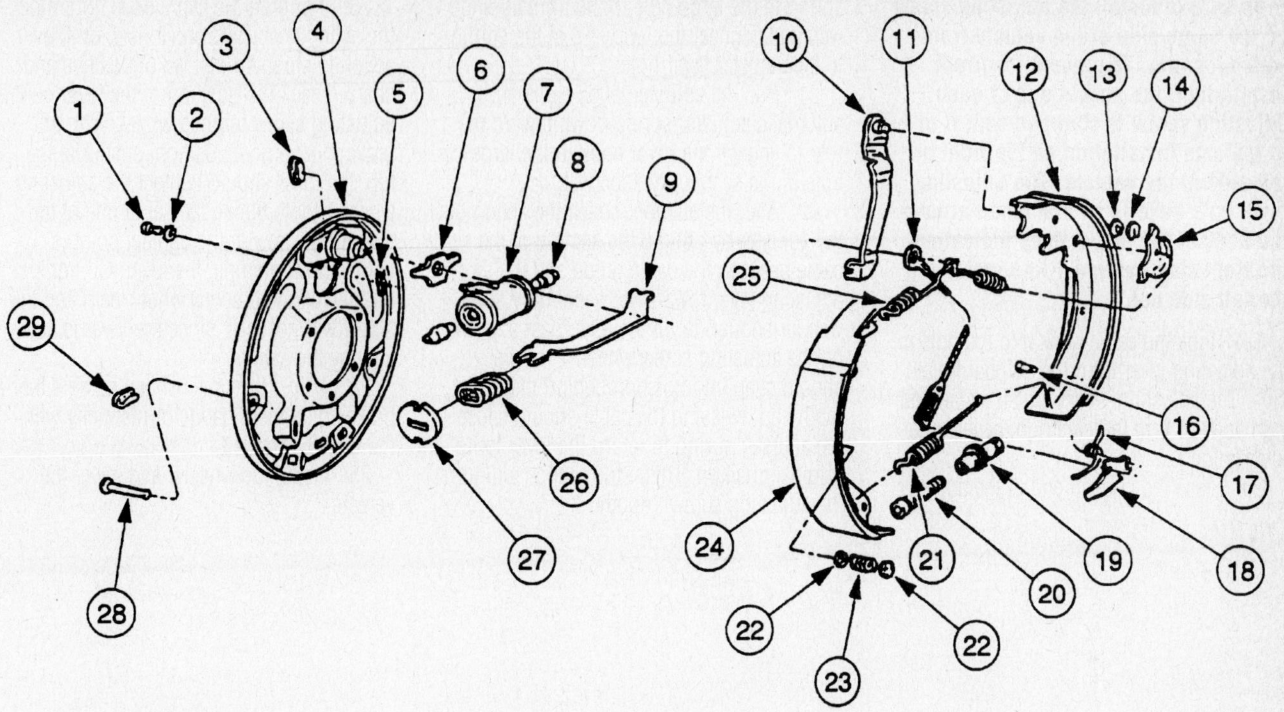

1	Wheel Cylinder-to-Backing Plate Bolt (2 Req'd)	12	Rear Brake Shoe and Lining, Secondary	22	Brake Shoe Hold-Down Spring Cup
2	Washer	13	Washer	23	Brake Shoe Hold-Down Spring
3	Inspection Hole Cover	14	Parking Brake Lever Pin Retainer	24	Rear Brake Shoe and Lining, Primary
4	Brake Backing Plate	15	Cable Guide	25	Brake Shoe Retracting Spring, Short
5	Lining Inspection Hole	16	Adjusting Lever Pin	26	Parking Brake Link Spring
6	Anchor Pin Guide Plate	17	Adjusting Lever Return Spring	27	Parking Brake Spring Retainer
7	Rear Wheel Cylinder	18	Brake Shoe Adjusting Lever	28	Brake Shoe Hold-Down Spring Pin
8	Wheel Cylinder Brake Shoe Link	19	Brake Shoe Adjusting Screw Nut	29	Brake Adjusting Hole Cover
9	Parking Brake Strut	20	Brake Adjuster Screw		
10	Parking Brake Lever	21	Brake Shoe Adjusting Screw Spring		
11	Brake Shoe Adjusting Lever Cable				

93026G21

Exploded view of the rear brake shoes and components—Explorer Sport and Sport-Trac

12. Position the brake shoes on the backing plate and install the hold-down spring pins, springs and cups. Install the parking brake link, spring and washer. Connect the parking brake cable to the parking brake lever.

13. Install the anchor pin plate, if equipped, and place the cable anchor over the anchor pin with the crimped side toward the backing plate.

14. Install the primary shoe-to-anchor spring using the brake spring removal/installation tool.

15. Install the cable guide on the secondary shoe with the flanged hole fitted into the hole in the secondary shoe. Thread the cable around the cable guide groove.

➡Make sure the cable is positioned in the groove and not between the guide and shoe web.

16. Install the secondary shoe-to-anchor (long) spring.

➡Make sure the cable end is not cocked or binding on the anchor pin when installed. All parts should be flat on the anchor pin.

17. Apply high temperature lithium grease to the threads and the socket end of the adjusting screw. Turn the adjusting screw into the adjusting pivot nut to the end of the threads and then loosen, ½ turn.

18. Place the adjusting socket on the screw and install the assembly between the shoe ends with the adjusting screw nearest the secondary shoe.

➡Be sure to install the adjusting screw on the same side of the vehicle from which it came. To prevent incorrect installation, the socket end of each adjusting screw is stamped with R or L, to indicate installation on the right or left side of the vehicle. The adjusting pivot nuts have lines machined around the body of the nut, 2 lines indicating the right side nut and 1 line indicating the left side nut.

19. Hook the cable hook into the hole in the adjusting lever from the outboard plate side. The adjusting levers are also stamped with an **R** or **L** to indicate right or left side installation.
20. Place the hooked end of the adjuster spring in the large hole in the primary shoe web and connect the loop end of the spring to the adjuster lever hole.
21. Pull the adjuster lever, cable and automatic adjuster spring down toward the rear to engage the pivot hook in the large hole in the secondary shoe web.
22. After installation, check the action of the adjuster by pulling the section of the cable between the cable guide and the adjusting lever toward the secondary shoe web far enough to lift the lever past a tooth on the adjusting screw wheel. The lever should snap into position behind the next tooth and releasing the cable should cause the adjuster spring to return the lever to its original position. This return action will turn the adjusting screw 1 tooth.
23. If pulling the cable does not produce the action described previously, or if lever action is sluggish instead of positive and sharp, check the position of the lever on the adjusting screw toothed wheel. With the brake in a vertical position, anchor at the top, the lever should contact the adjusting wheel 1 tooth above the centerline of the adjusting screw. If the contact point is below the centerline, the lever will not lock on the adjusting screw wheel teeth and the screw will not turn, since the lever is actuated by the cable.
24. Adjust the brake shoes using either a brake adjustment gauge or manually with the drums installed.
25. Install the wheels, and lower the vehicle.

FORD AND LINCOLN

Blackwood • E-Series Vans • F-Series Pickups

2

SPECIFICATION CHARTS

ENGINE AND VEHICLE IDENTIFICATION

Engine							Model Year	
Code ①	Liters (cc)	Cu. In.	Cyl.	Fuel Sys.	Type	Eng. Mfg.	Code ②	Year
2	4.2 (4195)	256	6	MFI	OHV	Ford	1	2001
3	5.4 (5409)	330	8	SFI	SOHC	Ford	2	2002
5	5.4 (5409)	330	8	EFI	SOHC	Ford	3	2003
6	4.6 (4588)	280	8	MFI	SOHC	Ford	4	2004
A	5.4 (5409)	330	8	EFI	DOHC	Ford	5	2005
F	7.3 (7292)	445	8	TDI	OHV	Navistar		
D	6.8 (6802)	415	10	EFI	SOHC	Ford		
L	5.4 (5409)	330	8	EFI	SOHC	Ford		
M	5.4 (5409)	330	8	EFI ③	SOHC	Ford		
P	6.0 (5921)	365	8	TDI	OHV	Navistar		
S	6.8 (6802)	415	10	MFI	SOHC	Ford		
W	4.6 (4588)	280	8	MFI	SOHC	Ford		
Z	5.4 (5409)	330	8	EFI ④	SOHC	Ford		

MFI: Multi-port Fuel Injection

DI: Direct Injection Turbo-Diesel

EFI: Electronic Fuel Injection

SFI: Sequential Fuel Injection

OHV: Overhead Valve

SOHC: Single Overhead Camshaft

① 8th digit of the Vehicle Identification Number (VIN)

② 10th digit of the Vehicle Identification Number (VIN)

③ Natural Gas Vehicle (NGV)

④ Bi-fuel Vehicle (Natural Gas/Propane)

67197-EFSE-C01

GENERAL ENGINE SPECIFICATIONS

Year	Model	Engine Displ. Liters	Engine VIN	Net Horsepower @ rpm	Net Torque @ rpm (ft. lbs.)	Bore x Stroke (in.)	Com-pression Ratio	Oil Pressure @ rpm
2001	E-150	4.2	2	205@4400	255@3000	3.81x3.74	9.3:1	40@2500
		4.6	6	210@4400	290@3250	3.55x3.54	9.0:1	20-45@1500
		4.6	W	210@4400	290@3250	3.55x3.54	9.0:1	20-45@1500
		5.4	L	235@4250	330@3000	3.55x4.17	9.0:1	40-70@1500
	E-250	4.2	2	205@4400	255@3000	3.81x3.74	9.3:1	40@2500
		5.4	L	235@4250	330@3000	3.55x4.17	9.0:1	40-70@1500
	E-350	5.4	L	235@4250	330@3000	3.55x4.17	9.0:1	40-70@1500
		6.8	S	265@4250	410@2750	4.09x4.17	9.0:1	1000@4000
		7.3	F	210@3000	425@2000	4.11x4.18	17.5:1	40-70@3000
	F-150	4.2	2	205@4400	255@3000	3.81x3.74	9.3:1	40@2500
		4.6	6	210@4400	290@3250	3.55x3.54	9.0:1	20-45@1500
		4.6	W	210@4400	290@3250	3.55x3.54	9.0:1	20-45@1500
		5.4	3	235@4250	330@3000	3.55x4.17	9.0:1	40-70@1500
		5.4	L	235@4250	330@3000	3.55x4.17	9.0:1	40-70@1500
	F-250	4.6	6	210@4400	290@3250	3.55x3.54	9.0:1	20-45@1500
		4.6	W	210@4400	290@3250	3.55x3.54	9.0:1	20-45@1500
		5.4	L	235@4250	330@3000	3.55x4.17	9.0:1	40-70@1500
	F-350	5.4	L	235@4250	330@3000	3.55x4.17	9.0:1	40-70@1500
		6.8	S	265@4250	410@2750	4.09x4.17	9.0:1	40-75@2000
		7.3	F	210@3000	425@2000	4.11x4.18	17.5:1	40-70@3000
2002	E-150	4.2	2	205@4400	255@3000	3.81x3.74	9.3:1	40@2500
		4.6	6	210@4400	290@3250	3.55x3.54	9.0:1	20-45@1500
		4.6	W	210@4400	290@3250	3.55x3.54	9.0:1	20-45@1500
		5.4	L	235@4250	330@3000	3.55x4.17	9.0:1	40-70@1500
	E-250	4.2	2	205@4400	255@3000	3.81x3.74	9.3:1	40@2500
		5.4	L	235@4250	330@3000	3.55x4.17	9.0:1	40-70@1500
	E-350	5.4	L	235@4250	330@3000	3.55x4.17	9.0:1	40-70@1500
		6.8	S	265@4250	410@2750	4.09x4.17	9.0:1	40-75@2000
		7.3	F	210@3000	425@2000	4.11x4.18	17.5:1	40-70@3000
	F-150	4.2	2	205@4400	255@3000	3.81x3.74	9.3:1	40@2500
		4.6	W	210@4400	290@3250	3.55x3.54	9.0:1	20-45@1500
		5.4	3	235@4250	330@3000	3.55x4.17	9.0:1	40-70@1500
		5.4	L	235@4250	330@3000	3.55x4.17	9.0:1	40-70@1500
	F-250	4.6	6	210@4400	290@3250	3.55x3.54	9.0:1	20-45@1500
		4.6	W	210@4400	290@3250	3.55x3.54	9.0:1	20-45@1500
		5.4	L	235@4250	330@3000	3.55x4.17	9.0:1	40-70@1500
	F-350	5.4	L	235@4250	330@3000	3.55x4.17	9.0:1	40-70@1500
		6.8	S	265@4250	410@2750	4.09x4.17	9.0:1	40-75@2000
		7.3	F	210@3000	425@2000	4.11x4.18	17.5:1	40-70@3000
2003	E-150	4.2	2	205@4400	255@3000	3.81x3.74	9.3:1	40@2500
		5.4	L	235@4250	330@3000	3.55x4.17	9.0:1	40-70@1500
	E-250	4.2	2	205@4400	255@3000	3.81x3.74	9.3:1	40@2500
		5.4	L	235@4250	330@3000	3.55x4.17	9.0:1	40-70@1500

67197-EFSE-C02

GENERAL ENGINE SPECIFICATIONS

Year	Model	Engine Displ. Liters	Engine VIN	Net Horsepower @ rpm	Net Torque @ rpm (ft. lbs.)	Bore x Stroke (in.)	Com-pression Ratio	Oil Pressure @ rpm
2003	E-350	5.4	L	235@4250	330@3000	3.55x4.17	9.0:1	40-70@1500
(cont.)		6.8	S	265@4250	410@2750	4.09x4.17	9.0:1	40-75@2000
		7.3	F	210@3000	425@2000	4.11x4.18	17.5:1	40-70@3000
	Blackwood	5.4	A	300@5000	355@2750	3.55x4.17	9.0:1	40-70@1500
	F-150	4.2	2	205@4400	255@3000	3.81x3.74	9.3:1	40@2500
		4.6	W	210@4400	290@3250	3.55x3.54	9.0:1	20-45@1500
		5.4	3	235@4250	330@3000	3.55x4.17	9.0:1	40-70@1500
		5.4	L	235@4250	330@3000	3.55x4.17	9.0:1	40-70@1500
	F-250	4.6	6	210@4400	290@3250	3.55x3.54	9.0:1	20-45@1500
		4.6	W	210@4400	290@3250	3.55x3.54	9.0:1	20-45@1500
		5.4	L	235@4250	330@3000	3.55x4.17	9.0:1	40-70@1500
	F-350	5.4	L	235@4250	330@3000	3.55x4.17	9.0:1	40-70@1500
		6.8	S	265@4250	410@2750	4.09x4.17	9.0:1	40-75@2000
		7.3	F	210@3000	425@2000	4.11x4.18	17.5:1	40-70@3000
2004	F-150	4.2	2	202@4800	255@3400	3.81x3.74	9.3:1	40@2500
		4.6	W	210@4400	290@3250	3.55x3.54	9.0:1	20-45@1500
		5.4	5	300@5000	365@3750	3.55x4.17	9.8:1	40-60@2000
	Lightning	5.4	3	380@4750	450@3250	3.55x4.17	8.4:1	40-60@2000
	F-Super Duty	5.4	L	235@4250	330@3000	3.55x4.17	9.0:1	40-70@1500
		6.0	P	325@3300	560@2000	3.74x4.13	18.0:1	24@1200
		6.8	D	265@4250	410@2750	4.09x4.17	9.0:1	40-75@2000
		6.8	S	265@4250	410@2750	4.09x4.17	9.0:1	40-75@2000
	E-150 Wagon	4.6	W	210@4400	290@3250	3.55x3.54	9.0:1	20-45@1500
		5.4	L	235@4250	330@3000	3.55x4.17	9.0:1	40-70@1500
	E-350 Wagon	5.4	L	235@4250	330@3000	3.55x4.17	9.0:1	40-70@1500
		6.0	P	325@3300	560@2000	3.74x4.13	18.0:1	24@1200
		6.8	S	265@4250	410@2750	4.09x4.17	9.0:1	40-75@2000
	E-150 Cargo	4.6	W	210@4400	290@3250	3.55x3.54	9.0:1	20-45@1500
		5.4	L	235@4250	330@3000	3.55x4.17	9.0:1	40-70@1500
		6.0	P	325@3300	560@2000	3.74x4.13	18.0:1	24@1200
	E-250 Cargo	4.6	W	210@4400	290@3250	3.55x3.54	9.0:1	20-45@1500
		5.4	L	235@4250	330@3000	3.55x4.17	9.0:1	40-70@1500
		6.0	P	325@3300	560@2000	3.74x4.13	18.0:1	24@1200
	E-350 Cargo	5.4	L	235@4250	330@3000	3.55x4.17	9.0:1	40-70@1500
		6.0	P	325@3300	560@2000	3.74x4.13	18.0:1	24@1200
		6.8	S	265@4250	410@2750	4.09x4.17	9.0:1	40-75@2000

67197-EFSE-C03

GASOLINE ENGINE TUNE-UP SPECIFICATIONS

Year	Engine Displacement Liters	Engine VIN	Spark Plug Gap (in.)	Ignition Timing (deg.) ①		Fuel Pump (psi) ②	Idle Speed (rpm) ③		Valve Clearance	
				MT	AT		MT	AT	In.	Ex.
2001	4.2	2	0.052-0.056	10B	10B	30-45	NA	NA	HYD	HYD
	4.6	6	0.052-0.056	10B	10B	30-45	NA	NA	HYD	HYD
	4.6	W	0.052-0.056	10B	10B	30-45	NA	NA	HYD	HYD
	5.4	L	0.052-0.056	10B	10B	28-45	NA	NA	HYD	HYD
	5.4	3	0.052-0.056	10B	10B	28-45	NA	NA	HYD	HYD
	6.8	S	0.052-0.055	10B	10B	28-45	NA	NA	HYD	HYD
2002	4.2	2	0.052-0.056	10B	10B	30-45	NA	NA	HYD	HYD
	4.6	6	0.052-0.056	10B	10B	30-45	NA	NA	HYD	HYD
	4.6	W	0.052-0.056	10B	10B	30-45	NA	NA	HYD	HYD
	5.4	L	0.052-0.056	10B	10B	28-45	NA	NA	HYD	HYD
	5.4	3	0.052-0.056	10B	10B	28-45	NA	NA	HYD	HYD
	6.8	S	0.052-0.055	10B	10B	28-45	NA	NA	HYD	HYD
2003	4.2	2	0.052-0.056	10B	10B	30-45	NA	NA	HYD	HYD
	4.6	6	0.052-0.056	10B	10B	30-45	NA	NA	HYD	HYD
	4.6	W	0.052-0.056	10B	10B	30-45	NA	NA	HYD	HYD
	5.4	A	0.052-0.056	10B	10B	28-45	NA	NA	HYD	HYD
	5.4	L	0.052-0.056	10B	10B	28-45	NA	NA	HYD	HYD
	5.4	3	0.052-0.056	10B	10B	28-45	NA	NA	HYD	HYD
	6.8	S	0.052-0.055	10B	10B	28-45	NA	NA	HYD	HYD
2004	4.2	2	0.052-0.056	10B	10B	30-45	NA	NA	HYD	HYD
	4.6	W	0.040-0.050	10B	10B	30-45	NA	NA	HYD	HYD
	5.4	5	0.052-0.056	10B	10B	28-55	NA	NA	HYD	HYD
	5.4	3	0.052-0.056	—	10B	28-50	NA	NA	HYD	HYD
	5.4	L	0.052-0.056	10B	10B	28-45	NA	NA	HYD	HYD
	6.8	S	0.052-0.056	10B	10B	28-45	NA	NA	HYD	HYD

NOTE: The Vehicle Emission Control Information label often reflects specification changes changes made during production. The label figures must be used if they differ from this chart.

B: Before top dead center

HYD: Hydraulic

NA: Information not Available

① Ignition timing is preset and cannot be adjusted

② With engine running

③ Idle speed is electronically controlled and cannot be adjusted

67197-EFSE-C04

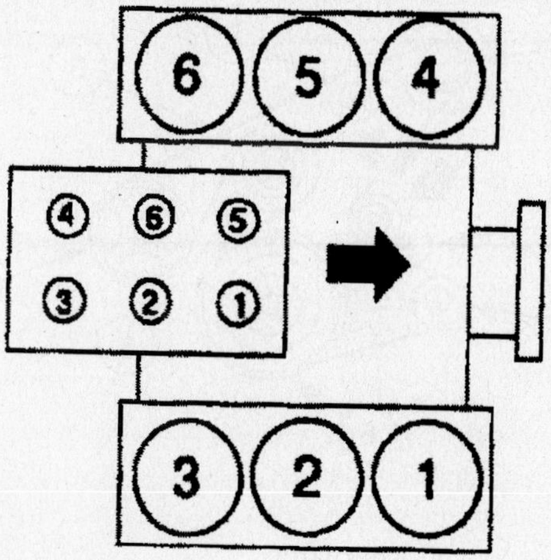

4.2L Engine
Firing Order: 1–4–2–5–3–6
Distributorless ignition system

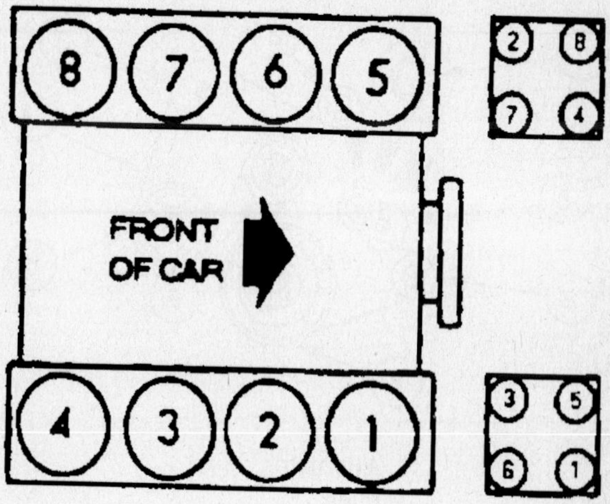

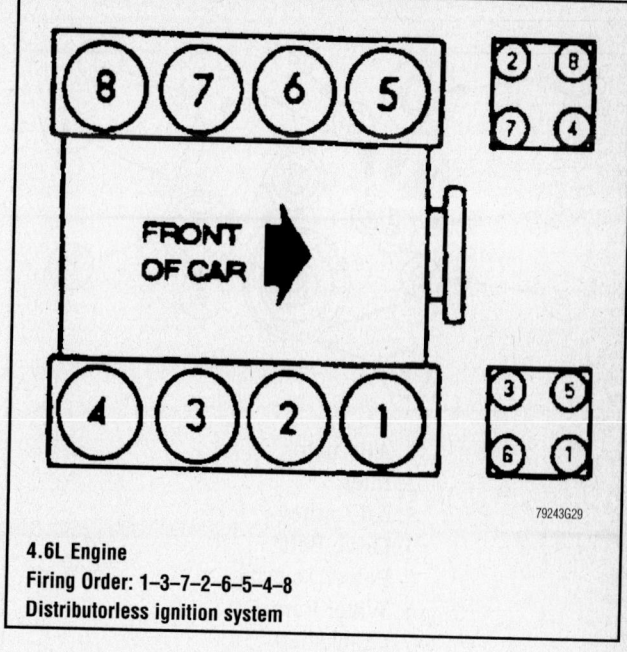

4.6L Engine
Firing Order: 1–3–7–2–6–5–4–8
Distributorless ignition system

5.4L Engine
Firing Order: 1–3–7–2–6–5–4–8
Distributorless ignition system; one coil per cylinder

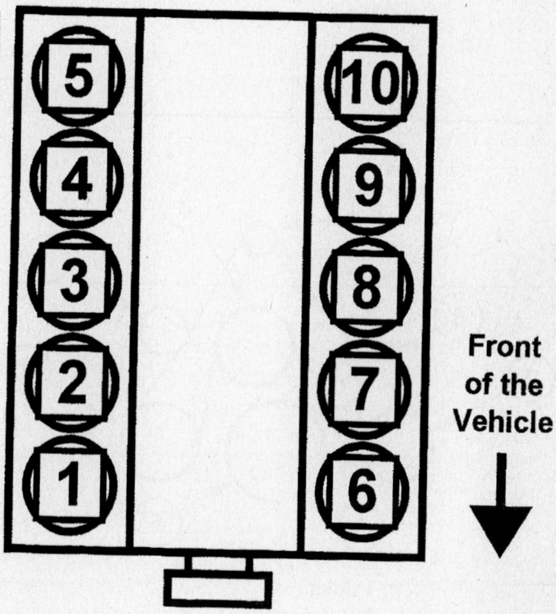

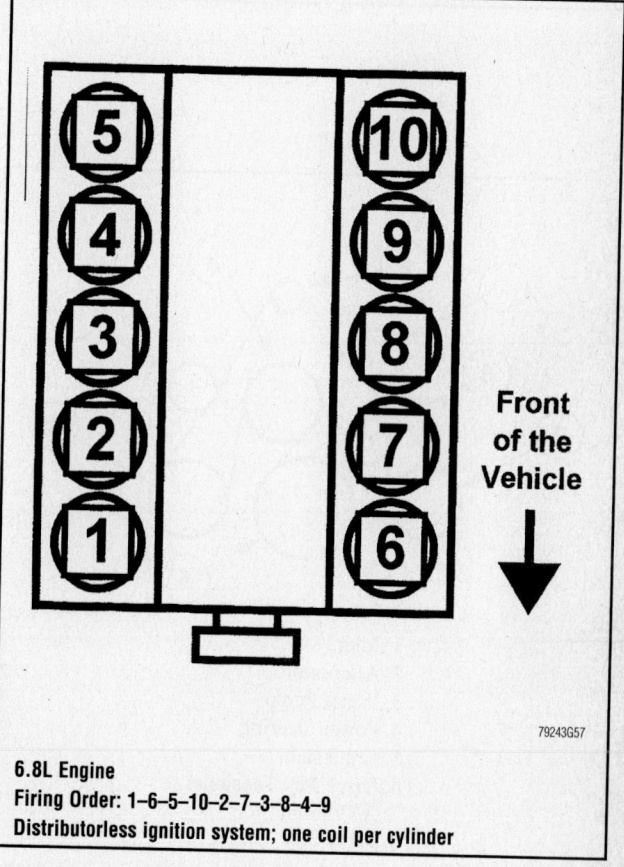

6.8L Engine
Firing Order: 1–6–5–10–2–7–3–8–4–9
Distributorless ignition system; one coil per cylinder

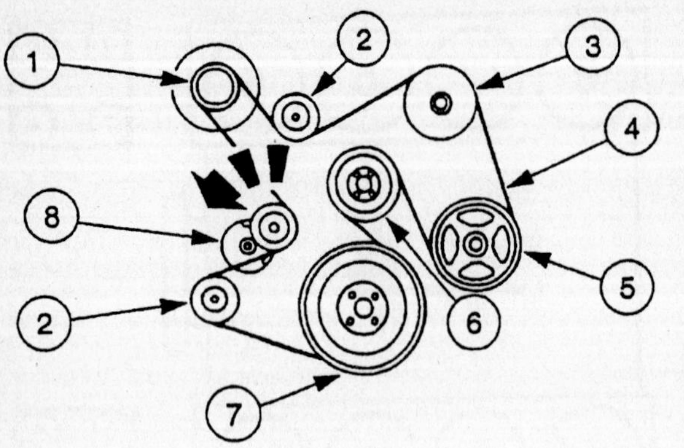

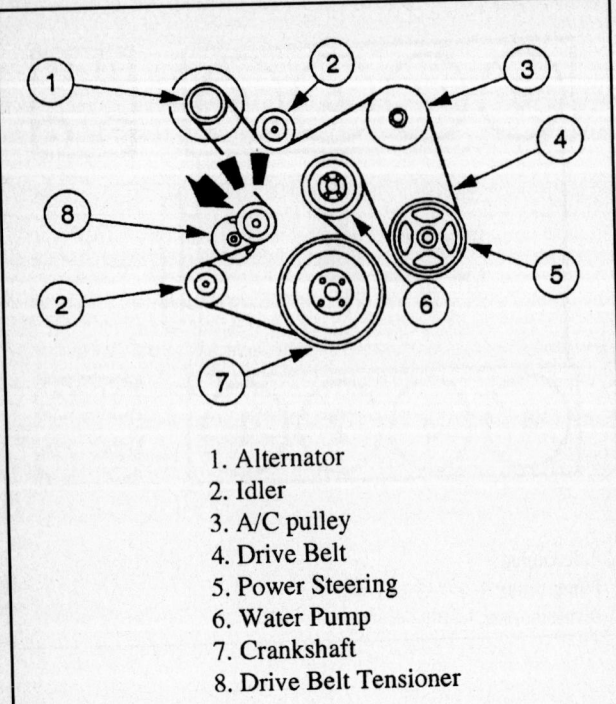

1. Alternator
2. Idler
3. A/C pulley
4. Drive Belt
5. Power Steering
6. Water Pump
7. Crankshaft
8. Drive Belt Tensioner

79244G88

Accessory drive belt routing—4.2L engine with A/C

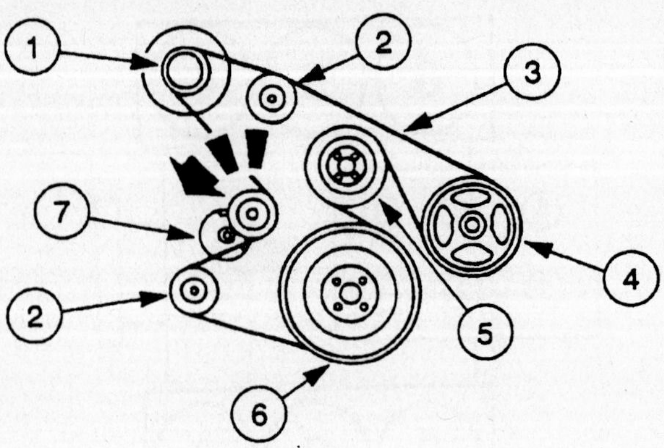

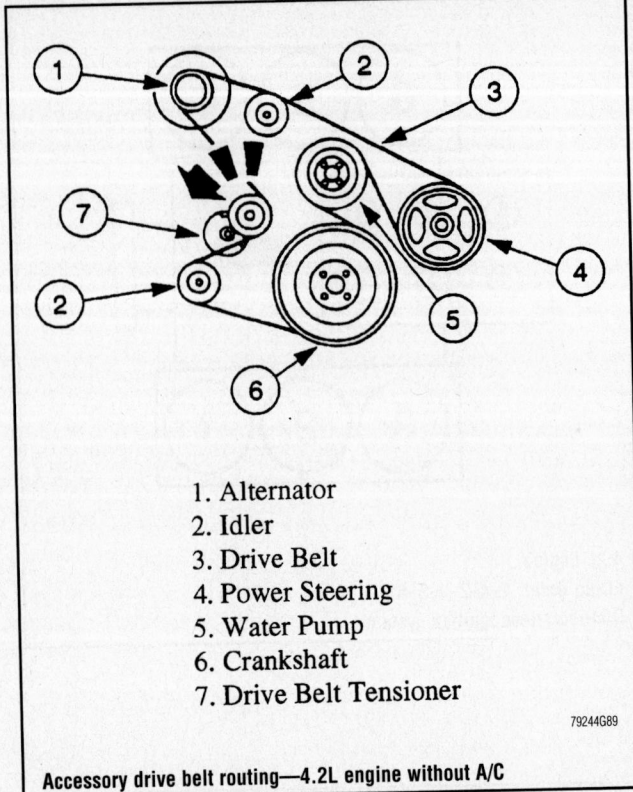

1. Alternator
2. Idler
3. Drive Belt
4. Power Steering
5. Water Pump
6. Crankshaft
7. Drive Belt Tensioner

79244G89

Accessory drive belt routing—4.2L engine without A/C

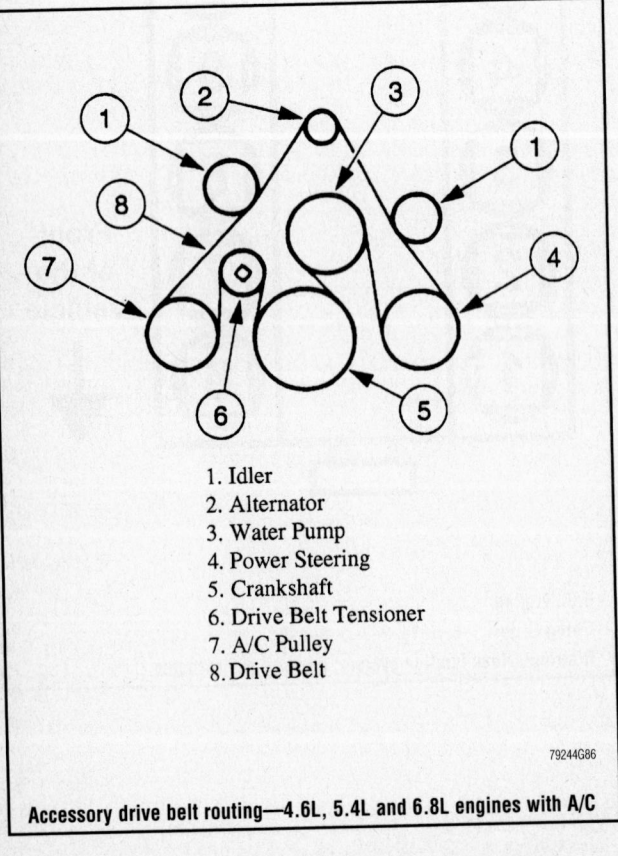

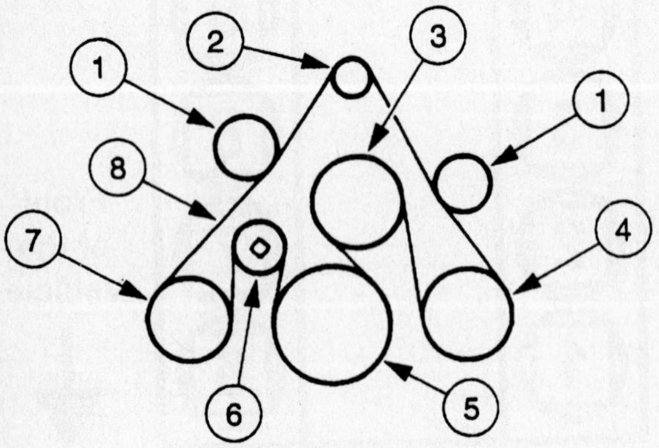

1. Idler
2. Alternator
3. Water Pump
4. Power Steering
5. Crankshaft
6. Drive Belt Tensioner
7. A/C Pulley
8. Drive Belt

79244G86

Accessory drive belt routing—4.6L, 5.4L and 6.8L engines with A/C

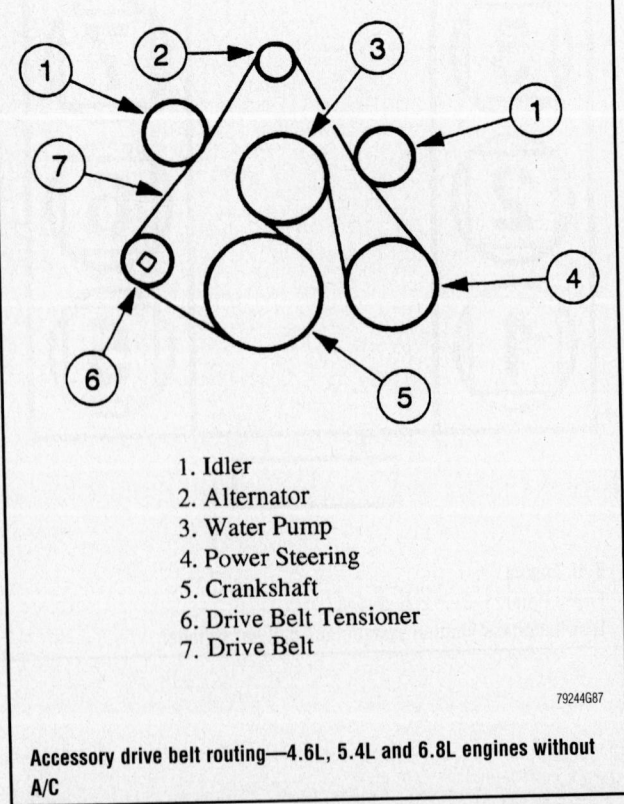

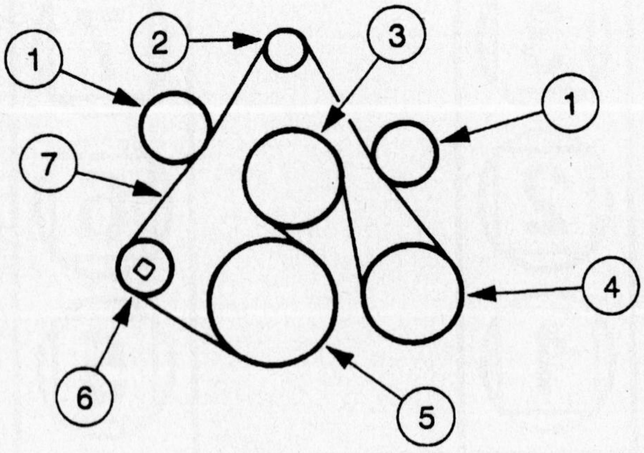

1. Idler
2. Alternator
3. Water Pump
4. Power Steering
5. Crankshaft
6. Drive Belt Tensioner
7. Drive Belt

79244G87

Accessory drive belt routing—4.6L, 5.4L and 6.8L engines without A/C

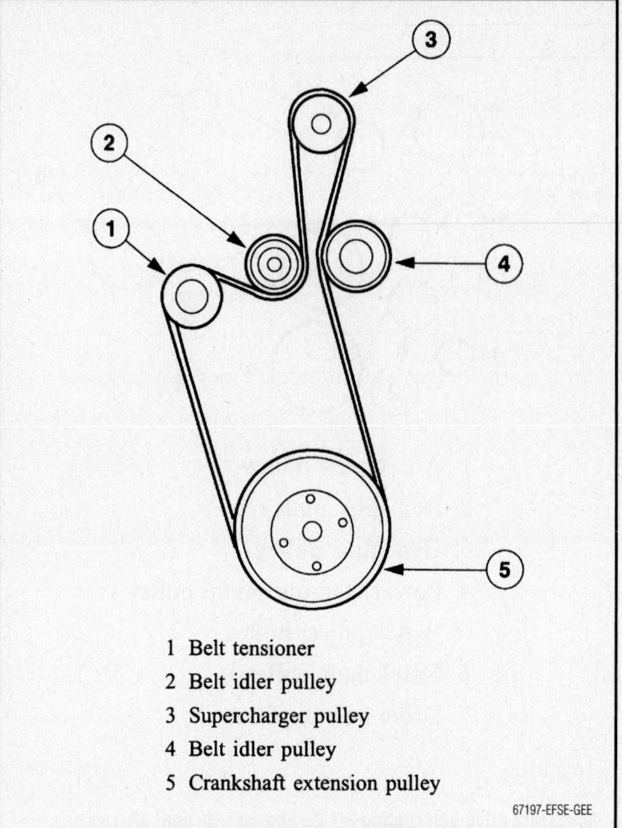

Supercharger drive belt routing—5.4L engines

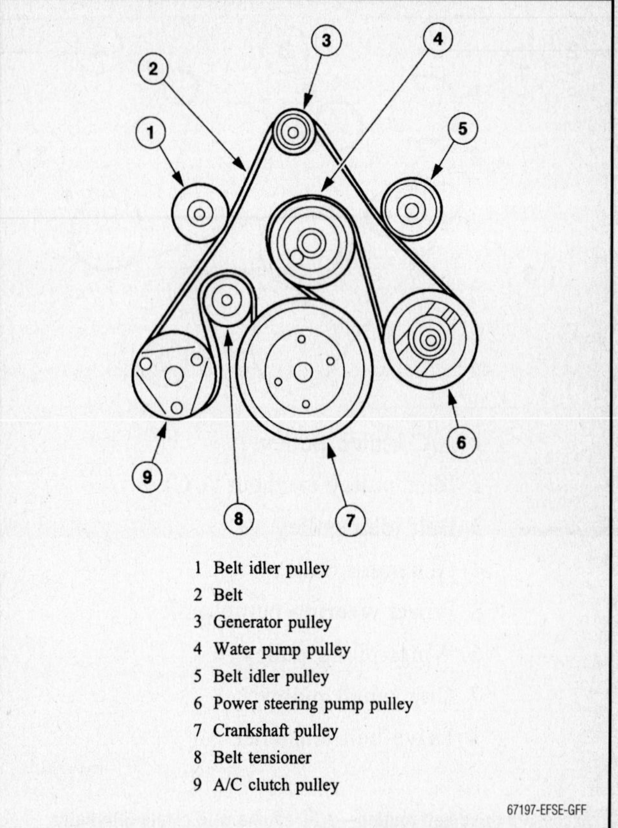

Accessory drive belt routing—5.4L supercharged engines

DIESEL ENGINE TUNE-UP SPECIFICATIONS

Year	Engine Displ. Liters	Engine VIN	Valve Clearance Intake (in.)	Valve Clearance Exhaust (in.)	Injection Pump Setting (deg.)	Injection Nozzle Pressure (psi) New	Injection Nozzle Pressure (psi) Used	Idle Speed (rpm)	Cranking Compression Pressure (psi)
2001	7.3	F	HYD	HYD	①	1875	1425	②	③
2002	7.3	F	HYD	HYD	①	1875	1425	②	③
2003	7.3	F	HYD	HYD	①	1875	1425	②	③
2004	6.0	P	HYD	HYD	①	④	④	②	NA

NOTE: The Vehicle Emission Control Information label often reflects specification changes made during production. The label figures must be used if they differ from those in this chart

HYD: Hydraulic

B: Before top dead center

NA: Not Available

① PCM controlled

② See underhood emission label

③ Compression pressure in the lowest cylinder must be at least 75% of the highest cylinder

Minimum pressure: 195 psi

Maximum pressure: 440 psi

④ Pump output pressure: 450-4,000 psi

67197-EFSE-C05

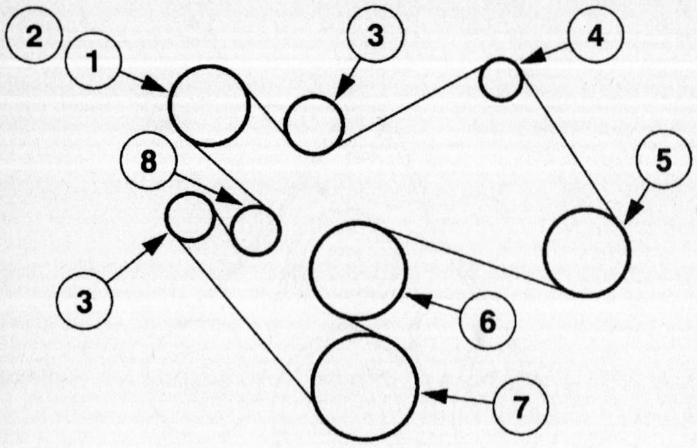

1 A/C clutch pulley
2 Idler pulley (without A/C)
3 Belt idler pulley
4 Generator pulley
5 Power steering pump pulley
6 Water pump pulley
7 Crankshaft pulley
8 Drive belt tensioner

67197-EFSE-GAA

Accessory drive belt routing—7.3L engine with single alternator

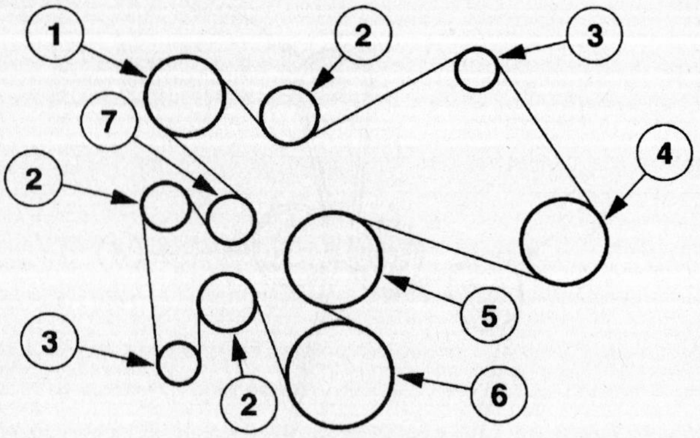

1 A/C clutch pulley
2 Belt idler pulley
3 Generator pulley
4 Power steering pump pulley
5 Water pump pulley
6 Crankshaft pulley
7 Drive belt tensioner

67197-EFSE-GBB

Accessory drive belt routing—7.3L engine with dual alternator

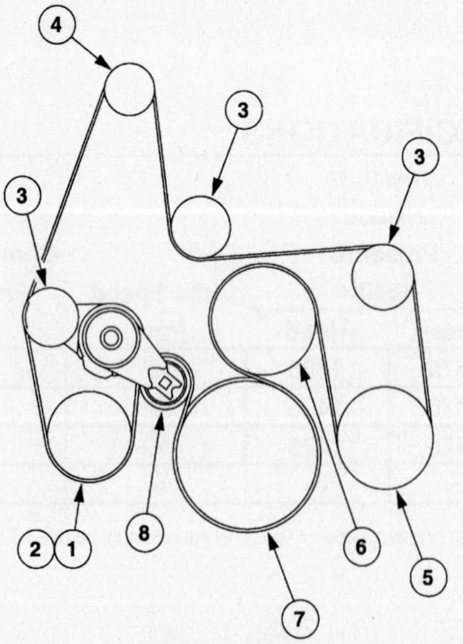

1 A/C clutch pulley
2 Idler pulley (without A/C)
3 Belt idler pulley
4 Generator pulley
5 Power steering pump pulley
6 Water pump pulley
7 Crankshaft pulley
8 Drive belt tensioner

67197-EFSE-GCC

Accessory drive belt routing—6.0L engine with single alternator

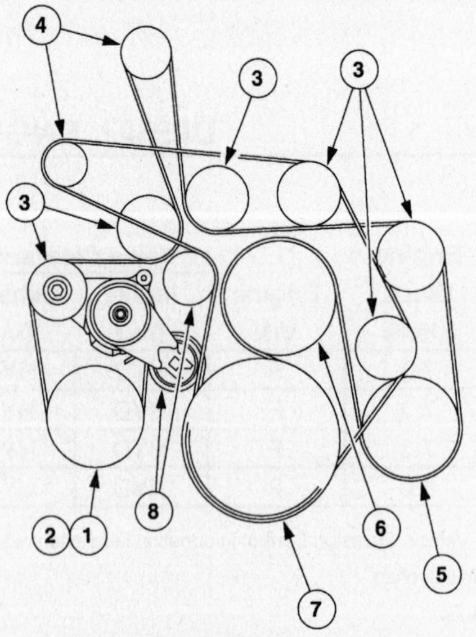

1 A/C clutch pulley
2 Idler pulley (without A/C)
3 Belt idler pulley
4 Generator pulley
5 Power steering pump pulley
6 Water pump pulley
7 Crankshaft pulley
8 Drive belt tensioner

67197-EFSE-GDD

Accessory drive belt routing—6.0L engine with dual alternator

CAPACITIES

Year	Model	Engine Displ. Liters	Engine VIN	Engine Oil with Filter (qts.)	Transmission (pts.) 5-Spd	Transmission (pts.) Auto.	Transfer Case (pts.)*	Drive Axle Front (pts.)	Drive Axle Rear (pts.)	Fuel Tank (gal.)	Cooling System (qts.)
2001	E-150	4.2	2	6.0	—	①	—	—	6.0	35.0	②
		4.6	6	6.0	—	①	—	—	6.0	35.0	②
		4.6	W	6.0	—	①	—	—	6.0	35.0	②
		5.4	L	7.0	—	①	—	—	6.0	35.0	②
	E-250	4.2	2	6.0	—	①	—	—	6.0 ③	35.0	②
		5.4	L	7.0	—	①	—	—	6.0 ③	35.0	②
	E-350	5.4	L	7.0	—	①	—	—	6.0 ③	35.0	②
		6.8	S	7.0	—	①	—	—	6.0 ③	35.0	②
		7.3	F	14.0	—	①	—	—	6.0 ③	35.0	②
	F-150	4.2	2	6.0	7.6	26.0	4.0	3.7	5.5	24.5 ④	②
		4.6	6	6.0	7.6	26.0	4.0	3.7	5.5	24.5 ④	②
		4.6	W	6.0	7.6	26.0	4.0	3.7	5.5	24.5 ④	②
		5.4	3	7.0	7.6	①	4.2	6.0	6.0	30.0	②
		5.4	L	7.0	7.6	①	4.2	6.0	6.0	30.0	②
	F-250	4.6	6	6.0	7.6	26.0	4.0	3.7	5.5	24.5 ④	②
		4.6	W	6.0	7.6	26.0	4.0	3.7	5.5	24.5 ④	②
		5.4	L	7.0	7.6	①	4.2	6.0	6.0	30.0	②
	F-350	5.4	L	7.0	7.6	①	4.2	6.0	6.0	30.0	②
		6.8	S	6.0	7.6	①	4.2	6.0 ⑤	6.0 ③	35.0	②
		7.3	F	14.0	7.6	①	4.2	6.0 ⑤	6.0 ③	35.0	23.0
2002	E-150	4.2	2	6.0	—	①	—	—	6.0	35.0	②
		4.6	6	6.0	—	①	—	—	6.0	35.0	②
		4.6	W	6.0	—	①	—	—	6.0	35.0	②
		5.4	L	7.0	—	①	—	—	6.0	35.0	②
	E-250	4.2	2	6.0	—	①	—	—	6.0 ③	35.0	②
		5.4	L	7.0	—	①	—	—	6.0 ③	35.0	②
	E-350	5.4	L	7.0	—	①	—	—	6.0 ③	35.0	②
		6.8	S	6.0	—	①	—	—	6.0 ③	35.0	②
		7.3	F	14.0	—	①	—	—	6.0 ③	35.0	23.0
	F-150	4.2	2	6.0	7.6	26.0	4.0	3.7	5.5	24.5 ④	②
		4.6	W	6.0	7.6	26.0	4.0	3.7	5.5	24.5 ④	②
		5.4	3	7.0	7.6	①	4.2	6.0	6.0	30.0	②
		5.4	L	7.0	7.6	①	4.2	6.0	6.0	30.0	②
	F-250	4.6	W	6.0	7.6	26.0	4.0	3.7	5.5	24.5 ④	②
		5.4	L	7.0	7.6	①	4.2	6.0	6.0	30.0	②
	F-350	5.4	L	7.0	7.6	①	4.2	6.0	6.0	30.0	②
		6.8	S	6.0	7.6	①	4.2	6.0 ⑤	6.0 ③	35.0	②
		7.3	F	14.0	7.6	①	4.2	6.0 ⑤	6.0 ③	35.0	23.0
2003	E-150	4.2	2	6.0	—	①	—	—	6.0	35.0	②
		4.6	6	6.0	—	①	—	—	6.0	35.0	②
		4.6	W	6.0	—	①	—	—	6.0	35.0	②
		5.4	L	7.0	—	①	—	—	6.0	35.0	②
	E-250	4.2	2	6.0	—	①	—	—	6.0 ③	35.0	②
		5.4	L	7.0	—	①	—	—	6.0 ③	35.0	②

67197-EFSE-C06

CAPACITIES

Year	Model	Engine Displ. Liters	Engine VIN	Engine Oil with Filter (qts.)	Transmission (pts.) 5-Spd	Transmission (pts.) Auto.	Transfer Case (pts.)*	Drive Axle Front (pts.)	Drive Axle Rear (pts.)	Fuel Tank (gal.)	Cooling System (qts.)
2003 (cont.)	E-350	5.4	L	7.0	—	①	—	—	6.0 ③	35.0	②
		6.8	S	6.0	—	①	—	—	6.0 ③	35.0	②
		7.3	F	14.0	—	①	—	—	6.0 ③	35.0	23.0
	Blackwood	5.4	A	6.0	—	15.9	—	—	6.0	25.0	19.2
	F-150	4.2	2	6.0	7.6	26.0	4.0	3.7	5.5	24.5 ④	②
		4.6	W	6.0	7.6	26.0	4.0	3.7	5.5	24.5 ④	②
		5.4	3	7.0	7.6	①	4.2	6.0	6.0	30.0	②
		5.4	L	7.0	7.6	①	4.2	6.0	6.0	30.0	②
	F-250	4.6	W	6.0	7.6	26.0	4.0	3.7	5.5	24.5 ④	②
		5.4	L	7.0	7.6	①	4.2	6.0	6.0	30.0	②
	F-350	5.4	L	7.0	7.6	①	4.2	6.0	6.0	30.0	②
		6.8	S	6.0	7.6	①	4.2	6.0 ⑤	6.0 ③	35.0	②
		7.3	F	14.0	7.6	①	4.2	6.0 ⑤	6.0 ③	35.0	23.0
2004	F-150	4.2	2	6.0	7.6	⑦	4.0	3.7	5.5	⑥	②
		4.6	W	6.0	—	27.8	4.0	3.6	5.5	⑥	19.8
		5.4	5	6.0	—	27.8	4.0	3.6	5.5	⑥	20.6
		5.4	3	6.0	—	34.2	—	—	NA	24.5	19.4 ⑧
	F-Super Duty	5.4	L	6.0	11.6	⑨	NA	5.9	⑩	26.4	⑪
		6.0	P	15.0	11.6	⑨	NA	5.9	⑩	27.5	⑪
		6.8	S	6.0	11.6	⑨	NA	5.9	⑩	28.5	⑪
	E-150 Wagon	4.6	W	6.0	—	⑫	—	—	⑬	35.0	⑮
		5.4	L	6.0	—	⑫	—	—	⑬	35.0	⑯
	E-350 Wagon	5.4	L	6.0	—	⑫	—	—	⑬	35.0	⑯
		6.0	P	15.0	—	⑫	—	—	⑬	35.0	⑰
		6.8	S	6.0	—	⑫	—	—	⑬	35.0	⑱
	E-150 Cargo	4.6	W	6.0	—	⑫	—	—	⑬	35.0	⑮
		5.4	L	6.0	—	⑫	—	—	⑬	35.0	⑯
		6.0	P	15.0	—	⑫	—	—	⑬	35.0	⑰
	E-250 Cargo	4.6	W	6.0	—	⑫	—	—	⑬	35.0	⑮
		5.4	L	6.0	—	⑫	—	—	⑬	35.0	⑯
		6.0	P	15.0	—	⑫	—	—	⑬	35.0	⑰
	E-350 Cargo	5.4	L	6.0	—	⑫	—	—	⑬	⑭	⑯
		6.0	P	15.0	—	⑫	—	—	⑬	⑭	⑰
		6.8	S	6.0	—	⑫	—	—	⑬	⑭	⑱

67197-EFSE-C07

CAPACITIES

Footnotes

NA: Information not available

NOTE: All capacities are approximate. Add fluid gradually and check to be sure a proper fluid level is obtained.

* Overhaul

① With 4R70W: 28 pts.
With E40D: 32.0 pts.
With4R100: 35.4 pts.

② 4.2L. with standard cooling: 23.3 qts.
4.2L. with aux cooling: 25.4 qts.
4.6L. with standard cooling: 25.0 qts.
4.6L. with aux cooling: 27.2 qts.
5.4L. with standard cooling: 29.0 qts.
5.4L. with aux cooling: 31.0 qts.
6.8L. with standard cooling: 30.6 qts.
6.8L. with aux cooling: 32.8 qts.

③ Heavy duty: 7.5 pts.

④ Also available with a 30 gallon
tank with 8 ft. box

⑤ Without PTO: 4.2 pts.
With PTO: 12.0 pts.

⑥ Regular cab w/126 in. wheel base and 6.5 ft. bed: 26.0
Super cab w/132 inch wheel base and 5.5 ft. bed: 26.0
Crew cab w/144 in. wheel base and 6.5 ft. bed: 27.0
Regular cab w/144 in. wheel base and 8 ft. bed: 27.0
SuperCab w/163 in. wheel base and 8 ft. bed: 27.0
Optional for SuperCab w/144 in. wheel base and 6.5 ft. bed: 35.7
Optional for Regular Cab w/144 in. wheel base and 8 ft. bed: 35.7
Optional for SuperCab w/163 in. wheel base and 8 ft. bed: 35.7
Heritage edition:
4x4 short wheelbase Regular Cab: 24.5
4x2 Regular Cab w/short wheelbase: 25.0
4x2 SuperCab w/short wheelbase: 25.0
4x4 SuperCab w/short wheelbase: 25.0
4x2 Regular Cab w/long wheelbase: 30.0
4x4 Regular Cab w/long wheelbase: 30.0
4x2 SuperCab w/long wheelbase: 30.0
4x4 SuperCab w/long wheelbase: 30.0
Lightning: 24.5

⑦ 4R100: 34.2 pts.
4R70W: 27.8 pts.

⑧ Plus 4.2 qts. For supercharger

⑨ 4R100: 34.2 pts.
TorqShift: 38.4 pts.

⑩ Dana 80
Conventional and Trutrac: 8.5 pts.
Trac-Lok: 8.0 pts.; first fill with 0.5 pt. of friction modifier
Dana S110: 16 pts.
Dana S135: 24.5 pts.
Ford 10.5 inch: NA

⑪ Wide frame vehicles with standard bed midship tank: 38
Wide frame vehicles with short bed midship tank: 29
Narrow frame chassis cab with aft of axle standard equipment: 40
Narrow frame chassis cab with optional midship tank: 19

⑫ 4R100: 34.2 pts.
4R70W/4R75W: 27.8 pts.
TorqShift: 38.4 pts.

⑬ Ford 8.8 inch: 5.7 pts.
Ford 9.5 inch: 5.7 pts.
Traction-Lok includes 4 oz. of friction modifier
Dana 60-1U: 6.3 pts.
Dana 70-2U: 6.6 pts.
Dana 70-HD: 7.4
Limited slip includes 7 oz. of friction modifier

⑭ E-350 Cargo: 35.0
Commercial chassis-cab: 37.0 standard; 55.0 optional

⑮ Without rear heat: 23.8 qts.
With rear heat: 26.0 qts.

⑯ Without rear heat: 27.8 quarts
With rear heat: 29.8 quarts

⑰ Without rear heat: 22.2 qts.
With rear heat: 24.9 qts.

⑱ Without rear heat: 29.4 qts.
With rear heat: 31.6 qts.

67197-EFSE-C08

VALVE SPECIFICATIONS

Year	Engine Displ. Liters	Engine VIN	Seat Angle (deg.)	Face Angle (deg.)	Spring Test Pressure (lbs. @ in.)	Spring Installed Height (in.)	Stem-to-Guide Clearance (in.) Intake	Stem-to-Guide Clearance (in.) Exhaust	Stem Diameter (in.) Intake	Stem Diameter (in.) Exhaust
2001	4.2	2	44.75	45.67	224@1.16	1.566-1.637	0.0450-0.0900	0.0015-0.0033	0.2738-0.2728	0.2751-0.2741
	4.6	6	45	45.5	132@1.100	1.570	0.0008-0.0027	0.0018-0.0037	0.2750-0.2746	0.2740-0.2736
	4.6	W	45	45.5	132@1.100	1.570	0.0008-0.0027	0.0018-0.0037	0.2750-0.2746	0.2740-0.2736
	5.4	3	45	45.5	150@1.10	1.570	0.0008-0.0027	0.0018-0.0037	0.2754-0.2746	0.2744-0.2736
	5.4	L	45	45.5	150@1.10	1.570	0.0008-0.0027	0.0018-0.0037	0.2754-0.2746	0.274-0.2736
	6.8	S	44.50-45.25	45.25-45.75	150@1.10	1.570	0.0008-0.0027	0.0018-0.0037	0.2754-0.2746	0.2744-0.2735
	7.3	F	①	①	200@1.38	②	0.0055	0.0055	0.3119-0.3126	0.3119-0.3126
2002	4.2	2	44.75	45.67	224@1.16	1.566-1.637	0.0450-0.0900	0.0015-0.0033	0.2738-0.2728	0.2751-0.2741
	4.6	6	45	45.5	132@1.100	1.570	0.0008-0.0027	0.0018-0.0037	0.2750-0.2746	0.2740-0.2736
	4.6	W	45	45.5	132@1.100	1.570	0.0008-0.0027	0.0018-0.0037	0.2750-0.2746	0.2740-0.2736
	5.4	3	45	45.5	150@1.10	1.570	0.0008-0.0027	0.0018-0.0037	0.2754-0.2746	0.2744-0.2736
	5.4	L	45	45.5	150@1.10	1.570	0.0008-0.0027	0.0018-0.0037	0.275-0.2746	0.274-0.2736
	6.8	S	44.50-45.25	45.25-45.75	150@1.10	1.570	0.0008-0.0027	0.0018-0.0037	0.2754-0.2746	0.2744-0.2735
	7.3	F	①	①	200@1.38	②	0.0055	0.0055	0.3119-0.3126	0.3119-0.3126
2003	4.2	2	44.75	45.67	224@1.16	1.566-1.637	0.0450-0.0900	0.0015-0.0033	0.2738-0.2728	0.2751-0.2741
	4.6	6	45	45.5	132@1.100	1.570	0.0008-0.0027	0.0018-0.0037	0.2750-0.2746	0.2740-0.2736
	4.6	W	45	45.5	132@1.100	1.570	0.0008-0.0027	0.0018-0.0037	0.2750-0.2746	0.2740-0.2736
	5.4	A	44.5-45	45.25-45.75	150@1.10	1.575	0.0008-0.0027	0.0018-0.0037	0.2754-0.2746	0.2744-0.2736
	5.4	3	45	45.5	150@1.10	1.570	0.0008-0.0027	0.0018-0.0037	0.2754-0.2746	0.2744-0.2736
	5.4	L	45	45.5	150@1.10	1.570	0.0008-0.0027	0.0018-0.0037	0.2754-0.2746	0.2744-0.2736
	6.8	S	44.50-45.25	45.25-45.75	150@1.10	1.570	0.0008-0.0027	0.0018-0.0037	0.2754-0.2746	0.2744-0.2735
	7.3	F	①	①	200@1.38	②	0.0055	0.0055	0.3119-0.3126	0.3119-0.3126

67197-EFSE-C09

VALVE SPECIFICATIONS

Year	Engine Displ. Liters	Engine VIN	Seat Angle (deg.)	Face Angle (deg.)	Spring Test Pressure (lbs. @ in.)	Spring Installed Height (in.)	Stem-to-Guide Clearance (in.) Intake	Stem-to-Guide Clearance (in.) Exhaust	Stem Diameter (in.) Intake	Stem Diameter (in.) Exhaust
2004	4.2	2	44.75	45.67	224@1.16	1.566-1.637	0.0450-0.0900	0.0015-0.0033	0.2738-0.2728	0.2751-0.2741
	4.6	W	45.5	45.25-45.75	132@1.103	1.563-1.586	0.0008-0.0027	0.0018-0.0037	0.2754-0.2746	0.2744-0.2736
	5.4	3	45.5	45.25-45.75	171@1.14	1.6654-1.689	0.0008-0.0027	0.0018-0.0037	0.2754-0.2746	0.2744-0.2736
	5.4	5	44.5-45	45.5	171@1.22	1.660	0.0010-0.0020	0.0030-0.0040	0.2350-0.2360	0.2340-0.2350
	5.4	L	45.5	45.25-45.75	171@1.34	1.6654-1.689	0.0008-0.0027	0.0018-0.0037	0.2754-0.2746	0.2744-0.2736
	6.0	P	①	①	191@1.51	1.820	0.0055 max.	0.0055 max.	0.2720-0.2735	0.2720-0.2735
	6.8	S	44.5-45	45.25-45.75	150@1.10	1.575	0.0008-0.0027	0.0018-0.0037	0.2754-0.2746	0.2744-0.2736

① Intake: 30
Exhaust: 37.5
② Intake: 1.767 in.
Exhaust: 1.833 in.

67197-EFSE-C10

PISTON AND RING SPECIFICATIONS

All measurements are given in inches.

	Engine			Ring Gap			Ring Side Clearance		
Year	Displ. Liters	Engine VIN	Piston Clearance	Top Compression	Bottom Compression	Oil Control	Top Compression	Bottom Compression	Oil Control
2001	4.2	2	0.0007-0.0018	0.001-0.002	0.001-0.002	0.006-0.026	0.0012-0.0031	0.0012-0.0031	SNUG
	4.6	6	0.0005-0.0010	0.010-0.020	0.010-0.020	0.006-0.026	0.0016-0.0031	0.0012-0.0031	SNUG
	4.6	W	0.0005-0.0010	0.010-0.020	0.010-0.020	0.006-0.026	0.0016-0.0031	0.0012-0.0031	SNUG
	5.4	3	0.0000-0.0010	0.005-0.011	0.010-0.016	0.006-0.026	0.0012-0.0037	0.0012-0.0037	SNUG
	5.4	L	0.0000-0.0010	0.005-0.011	0.010-0.016	0.006-0.026	0.0012-0.0037	0.0012-0.0037	SNUG
	6.8	S	0.0000-0.0010	0.005-0.011	0.010-0.016	0.006-0.026	0.0012-0.0037	0.0012-0.0037	SNUG
	7.3	F	0.0044-0.0057	0.014-0.024	0.062-0.072	0.012-0.024	0.0013-0.0033	0.0013-0.0033	SNUG
2002	4.2	2	0.0007-0.0018	0.001-0.002	0.001-0.002	0.006-0.026	0.0012-0.0031	0.0012-0.0031	SNUG
	4.6	6	0.0005-0.0010	0.010-0.020	0.010-0.020	0.006-0.026	0.0016-0.0031	0.0012-0.0031	SNUG
	4.6	W	0.0005-0.0010	0.010-0.020	0.010-0.020	0.006-0.026	0.0016-0.0031	0.0012-0.0031	SNUG
	5.4	3	0.0000-0.0010	0.005-0.011	0.010-0.016	0.006-0.026	0.0012-0.0037	0.0012-0.0037	SNUG
	5.4	L	0.0000-0.0010	0.005-0.011	0.010-0.016	0.006-0.026	0.0012-0.0037	0.0012-0.0037	SNUG
	6.8	S	0.0000-0.0010	0.005-0.011	0.010-0.016	0.006-0.026	0.0012-0.0037	0.0012-0.0037	SNUG
	7.3	F	0.0044-0.0057	0.014-0.024	0.062-0.072	0.012-0.024	0.0013-0.0033	0.0013-0.0033	SNUG
2003	4.2	2	0.0007-0.0018	0.001-0.002	0.001-0.002	0.006-0.026	0.0012-0.0031	0.0012-0.0031	SNUG
	4.6	6	0.0005-0.0010	0.010-0.020	0.010-0.020	0.006-0.026	0.0016-0.0031	0.0012-0.0031	SNUG
	4.6	W	0.0005-0.0010	0.010-0.020	0.010-0.020	0.006-0.026	0.0016-0.0031	0.0012-0.0031	SNUG
	5.4	A	0.0000-0.0010	0.005-0.011	0.010-0.016	0.006-0.026	0.0012-0.0037	0.0012-0.0037	SNUG
	5.4	3	0.0000-0.0010	0.005-0.011	0.010-0.016	0.006-0.026	0.0012-0.0037	0.0012-0.0037	SNUG
	5.4	L	0.0000-0.0010	0.005-0.011	0.010-0.016	0.006-0.026	0.0012-0.0037	0.0012-0.0037	SNUG
	6.8	S	0.0000-0.0010	0.005-0.011	0.010-0.016	0.006-0.026	0.0012-0.0037	0.0012-0.0037	SNUG
	7.3	F	0.0044-0.0057	0.014-0.024	0.062-0.072	0.012-0.024	0.0013-0.0033	0.0013-0.0033	SNUG

67197-EFSE-C11

PISTON AND RING SPECIFICATIONS

All measurements are given in inches.

Year	Engine Displ. Liters	Engine VIN	Piston Clearance	Ring Gap Top Compression	Ring Gap Bottom Compression	Ring Gap Oil Control	Ring Side Clearance Top Compression	Ring Side Clearance Bottom Compression	Ring Side Clearance Oil Control
2004	4.2	2	0.0007-0.0018	0.001-0.002	0.001-0.002	0.006-0.026	0.0012-0.0031	0.0012-0.0031	SNUG
	4.6	W	0.0005-0.0010	0.010-0.020	0.010-0.020	0.006-0.026	0.0012-0.0028	0.0012-0.0028	0.0018-0.0077
	5.4	5	0.0010-0.0018	0.005-0.011	0.010-0.016	0.006-0.026	0.0012-0.0020	0.0012-0.0031	SNUG
	5.4	3	0.0010-0.0018	0.005-0.011	0.010-0.016	0.006-0.026	0.0012-0.0020	0.0012-0.0031	SNUG
	5.4	L	0.0000-0.0010	0.006-0.0150	0.010-0.020	0.006-0.026	0.0002-0.0013	0.0012-0.0031	SNUG
	6.0	P	0.0017-0.0036	0.011-0.0210	0.055-0.0650	0.009-0.0190	NA	NA	NA
	6.8	S	0.0000-0.0010	0.005-0.011	0.010-0.016	0.006-0.026	0.0012-0.0037	0.0012-0.0037	SNUG

NA: Information not available

67197-EFSE-C12

CRANKSHAFT AND CONNECTING ROD SPECIFICATIONS

All measurements are given in inches.

Year	Engine Displ. Liters	Engine VIN	Crankshaft Main Brg. Journal Dia.	Crankshaft Main Brg. Oil Clearance	Crankshaft Shaft End-play	Crankshaft Thrust on No.	Connecting Rod Journal Dia.	Connecting Rod Oil Clearance	Connecting Rod Side Clearance
2001	4.2	2	2.5190-2.5198	0.0008-0.0015	0.0080-0.0078	3	2.3103-2.3111	0.0003-0.0024	0.0047-0.0193
	4.6	6	2.6500-2.6570	0.0011-0.0026	0.0051-0.0120	5	2.0870-2.8670	0.0011-0.0026	0.0006-0.0177
	4.6	W	2.6500-2.6570	0.0011-0.0026	0.0051-0.0120	5	2.0870-2.8670	0.0011-0.0026	0.0006-0.0177
	5.4	3	2.6568-2.6576	0.0009-0.0019	0.0015-0.0030	5	2.0859-2.0867	0.0010-0.0025	0.0006-0.0177
	5.4	L	2.6568-2.6576	0.0009-0.0019	0.0015-0.0030	5	2.0859-2.0867	0.0010-0.0025	0.0006-0.0177
	6.8	S	2.6568-2.6576	0.0009-0.0019	0.0015-0.0030	5	2.0859-2.0867	0.0010-0.0025	0.0006-0.0177
	7.3	F	3.1228-3.1236	0.0018-0.0036	0.0025-0.0085	4	2.4980-2.4990	0.0015-0.0045	0.0120-0.0240
2002	4.2	2	2.5190-2.5198	0.0008-0.0015	0.0080-0.0078	3	2.3103-2.3111	0.0003-0.0024	0.0047-0.0193
	4.6	6	2.6500-2.6570	0.0011-0.0026	0.0051-0.0120	5	2.0870-2.8670	0.0011-0.0026	0.0006-0.0177
	4.6	W	2.6500-2.6570	0.0011-0.0026	0.0051-0.0120	5	2.0870-2.8670	0.0011-0.0026	0.0006-0.0177
	5.4	3	2.6568-2.6576	0.0009-0.0019	0.0015-0.0030	5	2.0859-2.0867	0.0010-0.0025	0.0006-0.0177
	5.4	L	2.6568-2.6576	0.0009-0.0019	0.0015-0.0030	5	2.0859-2.0867	0.0010-0.0025	0.0006-0.0177
	6.8	S	2.6568-2.6576	0.0009-0.0019	0.0015-0.0030	5	2.0859-2.0867	0.0010-0.0025	0.0006-0.0177
	7.3	F	3.1228-3.1236	0.0018-0.0036	0.0025-0.0085	4	2.4980-2.4990	0.0015-0.0045	0.0120-0.0240
2003	4.2	2	2.5190-2.5198	0.0008-0.0015	0.0080-0.0078	3	2.3103-2.3111	0.0003-0.0024	0.0047-0.0193
	4.6	6	2.6500-2.6570	0.0011-0.0026	0.0051-0.0120	5	2.0870-2.8670	0.0011-0.0026	0.0006-0.0177
	4.6	W	2.6500-2.6570	0.0011-0.0026	0.0051-0.0120	5	2.0870-2.8670	0.0011-0.0026	0.0006-0.0177
	5.4	A	2.6568-2.6576	0.0009-0.0019	0.0015-0.0030	5	2.0859-2.0867	0.0010-0.0025	0.0006-0.0177
	5.4	3	2.6568-2.6576	0.0009-0.0019	0.0015-0.0030	5	2.0859-2.0867	0.0010-0.0025	0.0006-0.0177
	5.4	L	2.6568-2.6576	0.0009-0.0019	0.0015-0.0030	5	2.0859-2.0867	0.0010-0.0025	0.0006-0.0177
	6.8	S	2.6568-2.6576	0.0009-0.0019	0.0015-0.0030	5	2.0859-2.0867	0.0010-0.0025	0.0006-0.0177
	7.3	F	3.1228-3.1236	0.0018-0.0036	0.0025-0.0085	4	2.4980-2.4990	0.0015-0.0045	0.0120-0.0240

67197-EFSE-C13

CRANKSHAFT AND CONNECTING ROD SPECIFICATIONS

All measurements are given in inches.

Year	Engine Displ. Liters	Engine VIN	Crankshaft Main Brg. Journal Dia.	Crankshaft Main Brg. Oil Clearance	Crankshaft Shaft End-play	Crankshaft Thrust on No.	Connecting Rod Journal Dia.	Connecting Rod Oil Clearance	Connecting Rod Side Clearance
2004	4.2	2	2.5190-2.5198	0.0008-0.0015	0.0080-0.0078	3	2.3103-2.3111	0.0003-0.0024	0.0047-0.0193
	4.6	W	2.6500-2.6570	0.0011-0.0026	0.0051-0.0120	5	2.0859-2.0867	0.0010-0.0027	0.0006-0.0177
	5.4	3	2.6568-2.6576	0.0009-0.0019	0.0015-0.0030	5	2.0859-2.0867	0.0010-0.0025	0.0049-0.0187
	5.4	5	2.6568-2.6576	0.0009-0.0019	0.0030-0.0148	5	2.0859-2.0867	0.0010-0.0025	0.0049-0.0187
	5.4	L	2.6568-2.6576	0.0009-0.0019	0.0030-0.0148	5	2.0859-2.0867	0.0010-0.0025	0.0049-0.0187
	6.0	P	3.1500-3.1880	NA	0.0087	NA	2.7160-2.7170	NA	0.012-0.0240
	6.8	S	2.6568-2.6576	0.0009-0.0019	0.0015-0.0030	5	2.0859-2.0867	0.0010-0.0025	0.0006-0.0177

NA: Information not available

① Journal 1: 0.0004 - 0.0022 inches

Journals 2, 3, 4 and 5: 0.0009 - 0.0027 inches

67197-EFSE-C14

TORQUE SPECIFICATIONS

All readings in ft. lbs.

Year	Engine Displ. Liters	Engine VIN	Cylinder Head Bolts	Main Bearing Bolts	Rod Bearing Bolts	Crankshaft Damper Bolts	Flywheel Bolts	Manifold Intake *	Manifold Exhaust	Spark Plugs	Oil Pan Drain Plug
2001	4.2	2	①	81-88	②	103-117	54-63	③	15-22	8-14	19
	4.6	6	④	⑤	29-33	⑥	54-64	⑦	18	7-14	10
	4.6	W	④	⑤	29-33	⑥	54-64	⑦	18	7-14	10
	5.4	3	④	⑧	⑨	⑥	54-64	⑩	18	9-20	10
	5.4	L	④	⑧	⑨	⑥	54-64	⑩	18	9-20	10
	6.8	S	④	⑧	⑨	⑥	54-64	⑩	17-20	7-14	10
	7.3	F	⑪	95	70	90	89	18	45	—	NA
2002	4.2	2	①	81-88	②	103-117	54-63	③	15-22	8-14	19
	4.6	6	④	⑤	29-33	⑥	54-64	⑦	18	7-14	10
	4.6	W	④	⑤	29-33	⑥	54-64	⑦	18	7-14	10
	5.4	3	④	⑧	⑨	⑥	54-64	⑩	18	9-20	10
	5.4	L	④	⑧	⑨	⑥	54-64	⑩	18	9-20	10
	6.8	S	④	⑧	⑨	⑥	54-64	⑩	17-20	7-14	10
	7.3	F	⑪	95	70	90	89	18	45	—	NA
2003	4.2	2	①	⑫	⑬	103-117	54-63	⑭	30	8-14	19
	4.6	6	④	⑮	29-33	⑥	54-64	⑦	18	7-14	10
	4.6	W	④	⑮	29-33	⑥	54-64	⑦	18	7-14	10
	5.4	A	④	⑧	⑬	⑯	54-64	⑩	18	9-20	10
	5.4	3	④	⑧	⑨	⑰	54-64	⑩	18	9-20	10
	5.4	L	④	⑧	⑨	⑰	54-64	⑩	18	9-20	10
	6.8	S	④	⑧	⑨	⑰	54-64	⑩	17-20	7-14	10
	7.3	F	⑪	95	70	90	89	18	45	—	NA
2004	4.2	2	①	⑫	⑬	118	59	⑭	24	11	19
	4.6	W	④	⑧	⑱	⑰	59	⑲	18	11	10
	5.4	3	④	⑧	⑨	⑥	54-64	⑩	18	9-20	10
	5.4	5	④	⑧	⑬	⑰	59	⑦	18	25	10
	5.4	L	④	⑧	⑱	⑰	59	⑳	18	13	10
	6.0	P	㉒	㉒	㉓	㉒	69	8	28 ㉑	—	18
	6.8	S	④	⑧	⑱	⑰	59	⑦	18	7.5	10

NA: Information not available

* NOTE: Applies to Lower Manifold only.

① Step 1: 14 ft. lbs.
Step 2: 29 ft. lbs.
Step 3: 36 ft. lbs.

② Step 1: 29 ft. lbs.
Step 2: Plus 90 degrees

③ Tighten bolts to 71-101 inch lbs.

④ Step 1: 30 ft. lbs.
Step 2: Plus 85-95 degrees
Step 3: Plus 85-95 degrees

Footnotes Continued on next page

⑪ Step 1: 65 ft. lbs.
Step 2: 85 ft. lbs.
Step 3: 105 ft. lbs.

⑫ Step 1: 37 ft. lbs.
Step 2: plus 120 degrees

⑬ Step 1: 18 ft. lbs.
Step 2: 33 ft. lbs.
Step 3: plus 90-120 degrees

⑭ Step 1: 44 inch lbs.
Step 2: 89 inch lbs.

67197-EFSE-C15

Torque Chart

Footnotes Continued

⑤ Jack screws:
Step 1: 45 inch lbs.
Step 2: 98 inch lbs.
Cross-mounted cap bolts:
Step 1: 24 ft. lbs.
Step 2: Plus 90 degrees

⑥ Step 1: 88 ft. lbs.
Step 2: Loosen bolt
Step 3: 39 ft. lbs.
Step 4: Plus 90 degrees

⑦ Step 1: 18 inch lbs.
Step 2: 89 inch lbs. lbs.

⑧ Vertical bolts:
Step 1: 30 ft. lbs.
Step 2: plus 90 degrees
Side bolts:
Step 1: 22 ft. lbs.
Step 2: plus 90 degrees

⑨ Step 1: 27-32 ft. lbs.
Step 2: Plus 90 degrees

⑩ Step 1: 18 inch lbs.
Step 2: 71-106 inch lbs.

⑮ Windsor
Step 1 Vertical and side bolts: 22 ft. lbs.
Step 2 Vertical and side bolts: plus 90 degrees
Romeo
Vertical bolts
Step 1: 30 ft. lbs.
Step 2: plus 90 degrees
Jackscrews
Step 1: 44 inch lbs.
Step 2: 89 inch lbs.
Side bolts: 15 ft. lbs.

⑯ Step 1: 66 ft. lbs.
Step 2: Loosen 1 turn
Step 3: 37 ft. lbs.
Step 4: plus 90 degrees

⑰ Step 1: 66 ft. lbs.
Step 2: loosen 1 full turn
Step 3: 37 ft. lbs.
Step 4: plus 90 deg. Without exceeding 148 ft. lbs.

⑱ Step 1: 32 ft. lbs.
Step 2: 105 degrees

⑲ 89 inch lbs.

⑳ Step 1: 18 inch lbs.
Step 2: 18 ft. lbs.

㉑ Apply high temp nickel anti-seize to the threads

㉒ See the procedure in the text

㉓ Step 1: 33 ft. lbs.
Step 2: 50 ft. lbs.

67197-EFSE-C16

BRAKE SPECIFICATIONS

All measurements in inches unless noted

			Brake Disc			Brake Drum Diameter			Brake Caliper	
Year	Model		Original Thickness	Minimum Thickness	Maximum Runout	Original Inside Diameter	Max. Wear Limit	Maximum Machine Diameter	Bracket Bolts (ft. lbs.)	Mounting Bolts (ft. lbs.)
2001	E-150		1.160	0.960	0.0025	11.03	11.09	11.06	141-191	22-26
	E-250		1.300	1.100	0.0003	12.00	12.09	12.06	141-191	16-30
	E-350	F	1.300	1.100	0.0003	12.00	12.09	12.06	141-191	16-30
		R	NA	1.120	0.0015	—	—	—	128	27
	F-150	F	①	②	0.0025	11.03	11.09	11.06	136	21-26
		R	NA	0.470	NA	—	—	—	80	20
	F-250	F	①	②	0.0025	12.00	12.09	12.06	166	42
		R	NA	1.100	NA	—	—	—	128	27
	F-350	F	①	②	0.0025	12.00	12.09	12.06	166	42
		R	NA	1.100	NA	—	—	—	128	27
2002	E-150		1.160	0.960	0.0025	11.03	11.09	11.06	141-191	22-26
	E-250	F	1.300	1.100	0.0003	12.00	12.09	12.06	141-191	16-30
		R	NA	1.120	0.0015	—	—	—	128	27
	E-350	F	1.300	1.100	0.0003	12.00	12.09	12.06	141-191	16-30
		R	NA	1.120	0.0015	—	—	—	128	27
	F-150	F	①	②	0.0025	11.03	11.09	11.06	136	21-26
		R	NA	0.470	NA	—	—	—	80	20
	F-250	F	①	②	0.0025	12.00	12.09	12.06	166	42
		R	NA	1.000	NA	—	—	—	128	27
	F-350	F	①	②	0.0025	12.00	12.09	12.06	166	42
		R	NA	1.000	NA	—	—	—	128	27
2003	E-150		1.160	0.960	0.0025	11.03	11.09	11.06	141-191	22-26
	E-250	F	1.300	1.100	0.0003	12.00	12.09	12.06	141-191	16-30
		R	NA	1.120	0.0015	—	—	—	128	27
	E-350	F	1.300	1.100	0.0003	12.00	12.09	12.06	141-191	16-30
		R	NA	1.120	0.0015	—	—	—	128	27
	Blackwood	F	NA	③	0.0025	—	—	—	148	20
		R	NA	0.480	0.0002	—	—	—	NA	20
	F-150	F	NA	③	0.0025	—	—	—	136	21-26
		R	NA	0.480	0.0002	NA	NA	11.29	80	20
	F-250	F	①	②	0.0025	12.00	12.09	12.06	166	42
		R	NA	1.000	NA	—	—	—	128	27
	F-350	F	①	②	0.0025	12.00	12.09	12.06	166	42
		R	NA	1.000	NA	—	—	—	128	27
2004	F-150	F	NA	1.12	NA	—	—	—	148	47
		R	NA	0.72	NA	—	—	—	NA	22
	Heritage & Lightning	F	NA	HD 1.09 LD 0.965	NA	—	—	—	136	21-26
		R	NA	0.48	NA	—	11.29	—	80	20
	F-250	F	NA	1.41	NA	—	—	—	166	42
		R	NA	1.10	NA	—	—	—	④	27
	F-350	F	NA	1.41	NA	—	—	—	166	42
		R	NA	1.10	NA	—	—	—	④	27

67197-EFSE-C17

BRAKE SPECIFICATIONS

All measurements in inches unless noted

			Brake Disc			Brake Drum Diameter			Brake Caliper	
Year	Model		Original Thickness	Minimum Thickness	Maximum Runout	Original Inside Diameter	Max. Wear Limit	Maximum Machine Diameter	Bracket Bolts (ft. lbs.)	Mounting Bolts (ft. lbs.)
2004 (cont.)	E-150	F	NA	1.15	NA	—	—	—	129	24
		R	NA	0.80	NA	—	—	—	129	24
	E-250	F	NA	1.10	NA	—	—	—	166	16-30
		R	NA	1.10	NA	—	—	—	⑤	27
	E-350	F	NA	1.10	NA	—	—	—	166	23
		R	NA	1.10	NA	—	—	—	⑤	27

NOTE: Due to changes made during production, refer to manufacturer's specifications if they differ from those in this chart

NA: Information not available

① 1.020 inches for 4x2
1.220 inches for 4x4

② 0.972 inch for 4x2
1.090 inches for 4x4

③ LD: 0.965 inch
HD: 1.090 inch

④ Adapter plate:
With Dana axle: 100
With Ford axle: 150
Anchor plate: 128

⑤ Support bracket support bolts: 100 ft. lbs.
Anchor plate: 129 ft. lbs.

67197-EFSE-C18

WHEEL ALIGNMENT

			Caster		Camber		
Year	Model		Range (+/-Deg.)	Preferred Setting (Deg.)	Range (+/-Deg.)	Preferred Setting (Deg.)	Toe-in (Deg.)
2001	E-Series	150	2.75	+4.00	0.05	0.25	0.30+/-0.25
		250-350	2.75	+4.00	0.05	0.05	0.06+/-0.25
	F-150	4x2	1.00	+6.2	0.07	0.30	0.06+/-0.25
		4x4	1.00	+4.06	0.07	0.10	0.20+/-0.25
		Lightning	—	+.007	—	0.05	0.10+/-0.25
	F-250/350	4x2	2.00	+4.00	1.00	+0.62	0.03+/-0.25
		4x4	2.00	+3.50	1.00	+0.25	0.03+/-0.25
2002	E-Series	150	2.75	+4.00	0.05	0.25	0.30+/-0.25
		250-350	2.75	+4.00	0.05	0.05	0.06+/-0.25
	F-150	4x2	1.00	+6.2	0.07	0.30	0.06+/-0.25
		4x4	1.00	+4.06	0.07	0.10	0.20+/-0.25
		Lightning	—	+.007	—	0.05	0.10+/-0.25
	F-250/350	4x2	2.00	+4.00	1.00	+0.62	0.03+/-0.25
		4x4	2.00	+3.50	1.00	+0.25	0.03+/-0.25
2003	E-Series	150	2.75	+4.00	0.05	0.25	0.30+/-0.25
		250-350	2.75	+4.00	0.05	0.05	0.06+/-0.25
	Blackwood		1.00	①	0.07	-0.30	0.03+/-0.25
	F-150	4x2	1.00	①	0.07	-0.30	0.03+/-0.15
		4x4	1.00	②	0.07	-0.10	0.20+/-0.25
		Lightning	—	③	—	-0.50	-0.05+/-0.15
	F-250/350	4x2	2.00	+4.00	1.00	+0.62	0.03+/-0.25
		4x4	2.00	+3.50	1.00	+0.25	0.03+/-0.25
2004	F-150 Reg. Cab	4x2	1.00	+4.4	0.75	-0.20	0.15+/-0.25
	F-150 Super Cab	4x2	1.00	+4.6	0.75	-0.20	0.15+/-0.25
	F-150 Crew Cab	4x2	1.00	+4.6	0.75	-0.20	0.15+/-0.25
	F-150 Reg. Cab	4x4	1.00	+4.3	0.75	-0.20	0.15+/-0.25
	F-150 Super Cab	4x4	1.00	+4.6	0.75	-0.20	0.15+/-0.25
	F-150 Crew Cab	4x4	1.00	+4.6	0.75	-0.20	0.15+/-0.25
	F-150 Heritage	4x2	1.00	L +6.2 R +6.7	0.70	-0.30	0.03+/-0.15
		4x4	1.00	L +4.6 R +5.3	0.70	-0.10	0.10+/-0.15
	F-150 Lightning		NA	L +6.7 R +7.2	NA	-0.50	-0.05+/-0.15
	F-250/350	4x2	2.00	+4.00	1.00	0.62	0.03+/-0.25 ④
		4x4	2.00	+3.50	1.00	+0.25	0.03+/-0.25 ④
	E-150	Left	2.75	+4.00	0.50	+0.25	0.06+/-0.25
		Right	2.75	+4.50	—	—	—
	E-250	Left	2.75	+4.00	0.50	+0.50	0.06+/-0.25
		Right	2.75	+4.50	—	—	—
	E-350	Left	2.75	+4.00	0.50	+0.50	0.06+/-0.25
		Right	2.75	+4.50	—	—	—
	E-Super Duty	Left	2.75	+4.00	0.50	+0.50	0.06+/-0.25
		Right	2.75	+4.50	—	—	—

NA: Information not available

① Left: +6.2 Right: +6.7

② Left: +4.6 Right: +5.3

③ Left: +6.7 Right: +7.2

④ At curb ride height

67197-EFSE-C19

TIRE, WHEEL AND BALL JOINT SPECIFICATIONS

				Tire Pressures				Wheel Lug
		OEM Tires		(psi.)		Wheel	Ball Joint	Torque
Year	Model	Standard	Optional	Front	Rear	Size	Inspection	(ft. lbs.)
2001	E-150 Club Wagon	P235/75R15XL	none	①	①	6-JJ	0.030 in. ②	③
	E-150 Van	P225/75R15SL	none	①	①	7-J	0.030 in. ②	③
	E-250	LT225/75RX16	none	①	①	7-K	0.030 in. ②	③
	E-350	LT225/75RX16	none	①	①	7-K	0.030 in. ②	③
	E-350 Super Duty	LT245/75RX16E	none	①	①	7-K	0.030 in. ②	③
	F-150	P235/70R16	P255/70R16	①	①	7-J	0.030 in. ②	③
			P265/70R17	①	①		0.030 in. ②	③
	F-250	LT245/75R16	P255/70R16	①	①	7-J	0.030 in. ②	③
	F-350	P235/85R16	none	①	①	7-J	0.030 in. ②	③
2002	E-150 Club Wagon	P235/75R15XL	none	①	①	6-JJ	0.030 in. ②	③
	E-150 Van	P225/75R15SL	none	①	①	7-J	0.030 in. ②	③
	E-250	LT225/75RX16	none	①	①	7-K	0.030 in. ②	③
	E-350	LT225/75RX16	none	①	①	7-K	0.030 in. ②	③
	E-350 Super Duty	LT245/75RX16E	none	①	①	7-K	0.030 in. ②	③
	F-150	P235/70R16	P255/70R16	①	①	7-J	0.030 in. ②	③
			P265/70R17	①	①		0.030 in. ②	③
	F-250	LT245/75R16	P255/70R16	①	①	7-J	0.030 in. ②	③
	F-350	P235/85R16	none	①	①	7-J	0.030 in. ②	③
2003	E-150 Club Wagon	P235/75R15	none	①	①	6-JJ	0.030 in. ②	④
	E-350 Club Wagon	LT225/75R16E	LT245/75R16E	①	①	7JJ	0.030 in. ②	④
	E-150 cargo	P235/75R15E	none	①	①	7JJ	0.030 in. ②	④
	E-250 cargo	LT225/75R16E	none	①	①	7-K	0.030 in. ②	④
	E-350 cargo	LT245/75R16E	none	①	①	7-K	0.030 in. ②	④
	E-350 Super Duty	LT245/75RX16E	none	①	①	7-K	0.030 in. ②	④
	F-150 XL, XLT	P235/70R16	P255/70R16	①	①	7-J	0.030 in. ②	④
			P265/70R17					④
	F-150 Harley	P275/45VR20	none	①	①	9-J	0.030 in. ②	④
	F-150 King Ranch							④
	2wd	P275/60R17	none	①	①	7.5J	0.030 in. ②	④
	4wd	P265/70R17	none	①	①	7.5J		④
	F-150Lariat							④
	2wd	P275/60R17	none	①	①	7.5J	0.030 in. ②	④
	4wd	P265/70R17	none	①	①	7.5J		④
	F-150 Lightning	P295/45ZR18	none	①	①	9.5J	0.030 in. ②	④
	F-250 Lariat	LT265/75R16E	none	①	①	7-J	0.030 in. ②	④
	F-250 XL, XLT	LT235/85R16E	LT265/75R16E	①	①	7-J	0.030 in. ②	④
	F-350	LT265/75R16E	none	①	①	7-J	0.030 in. ②	④
2004	F-150 Reg. Cab							
	XL 2wd	P235/70R17	P255/65R17	①	①	NA	0.060 in. ②	150
	XL 4wd	P235/75R17	P255/70R17	①	①	NA	0.060 in. ②	150
	XLT 2wd	P235/70R17	P255/65R17	①	①	NA	0.060 in. ②	150
	XLT 4wd	P235/75R17	P255/70R17	①	①	NA	0.060 in. ②	150
	STX 2wd	P255/65R17	none	①	①	NA	0.060 in. ②	150
	STX 4wd	P255/70R17	none	①	①	NA	0.060 in. ②	150

67197-EFSE-C20

TIRE, WHEEL AND BALL JOINT SPECIFICATIONS

		OEM Tires		Tire Pressures (psi.)				Wheel Lug
Year	Model	Standard	Optional	Front	Rear	Wheel Size	Ball Joint Inspection	Torque (ft. lbs.)
2004 (cont.)	FX4	P255/70R17	LT245/65R18	①	①	NA	0.060 in. ②	150
			P275/65R18	①	①	NA	0.060 in. ②	150
	F-150 SuperCab							
	XL 2wd	P235/70R17	P255/65R17	①	①	NA	0.060 in. ②	150
	XL 4wd	P235/75R17	P255/70R17	①	①	NA	0.060 in. ②	150
	XLT 2wd	P235/70R17	P255/65R17	①	①	NA	0.060 in. ②	150
	XLT 4wd	P235/75R17	P255/70R17	①	①	NA	0.060 in. ②	150
	STX 2wd	P255/65R17	none	①	①	NA	0.060 in. ②	150
	STX 4wd	P255/70R17	none	①	①	NA	0.060 in. ②	150
	Lariat 2wd	P265/60R18	LT245/65R18	①	①	NA	0.060 in. ②	150
	Lariat 4wd	P275/65R18	LT245/65R18	①	①	NA	0.060 in. ②	150
	FX4	P255/70R17	LT245/65R18	①	①	NA	0.060 in. ②	150
			P275/65R18	①	①	NA	0.060 in. ②	150
	F-150 Crew Cab							
	XLT 2wd	P255/65R17	none	①	①	NA	0.060 in. ②	150
	XLT 4wd	P255/70R17	none	①	①	NA	0.060 in. ②	150
	Lariat 2wd	P265/60R18	LT245/65R18	①	①	NA	0.060 in. ②	150
	Lariat 4wd	P275/65R18	LT245/65R18	①	①	NA	0.060 in. ②	150
	FX4	P255/70R17	LT245/65R18	①	①	NA	0.060 in. ②	150
			P275/65R18	①	①	NA	0.060 in. ②	150
	F-150 Heritage Reg. Cab							
	XL 2wd	P255/70R16	P235/70R16	①	①	NA	0.060 in. ②	④
	XL 4wd	P255/70R16	P235/70R16	①	①	NA	0.060 in. ②	④
			LT245/70R16	①	①	NA	0.060 in. ②	④
	XLT 2wd	P255/70R16	P235/70R16	①	①	NA	0.060 in. ②	④
	XLT 4wd	P255/70R16	P235/70R16	①	①	NA	0.060 in. ②	④
			LT245/70R16	①	①	NA	0.060 in. ②	④
	SuperCab							
	XL 2wd	P235/70R16	none	①	①	NA	0.060 in. ②	④
	XL 4wd	P235/70R16	LT245/75R16	①	①	NA	0.060 in. ②	④
			P235/70R16	①	①	NA	0.060 in. ②	④
	XLT 2wd	P235/70R16	none	①	①	NA	0.060 in. ②	④
	XL 4wd	P235/70R16	LT245/75R16	①	①	NA	0.060 in. ②	④
			P235/70R16	①	①	NA	0.060 in. ②	④
	F-150 SVT Lightning	P295/45ZR18	none	①	①	NA	0.060 in. ②	④
	F-250 Lariat	LT265/75R16E	none	①	①	NA	0.030 in. ②	165
	F-250 XL, XLT	LT235/85R16E	LT265/75R16E	①	①	NA	0.030 in. ②	165
	F-350 SRW	LT265/75R16E	none	①	①	NA	0.030 in. ②	165
	F-350 DRW	LT235/85R16E	none	①	①	NA	0.030 in. ②	165
	E-150 Wagon	P235/70R16	none	①	①	NA	0.030 in. ②	③
	E-350 Wagon	LT225/75R16E	none	①	①	NA	0.030 in. ②	③
	E-350 Ext. Wagon	LT245/75R16E	none	①	①	NA	0.030 in. ②	③
	E-150 Cargo	P235/70R16	none	①	①	NA	0.030 in. ②	③

67197-EFSE-C21

TIRE, WHEEL AND BALL JOINT SPECIFICATIONS

		OEM Tires		Tire Pressures (psi.)				Wheel Lug
Year	Model	Standard	Optional	Front	Rear	Wheel Size	Ball Joint Inspection	Torque (ft. lbs.)
2004	E-250 Cargo	LT245/75R16E	none	①	①	NA	0.030 in. ②	③
(cont.)	E-350 cargo	LT245/75R16E	none	①	①	NA	0.030 in. ②	③

NA: Information not available
OEM: Original Equipment Manufacturer
PSI: Pounds Per Square Inch
STD: Standard
OPT: Optional
SRW: Single rear wheels
DRW: Dual rear wheels
① See placard on vehicle
② Both upper and lower
③ 5 lug: 74-133 ft. lbs.
8 lug: 126-170 ft. lbs.
④ 12mm: 100 ft. lbs.
14mm: 150 ft. lbs.

67197-EFSE-C22

SCHEDULED MAINTENANCE INTERVALS

2001-02 E-150 & F-150

TO BE SERVICED	TYPE OF SERVICE	VEHICLE MILEAGE INTERVAL (x1000)												
		5	10	15	20	25	30	35	40	45	50	55	60	65
Engine oil & filter	R	✓	✓	✓	✓	✓	✓	✓	✓	✓	✓	✓	✓	✓
Tires	Rotate	✓	✓	✓	✓	✓	✓	✓	✓	✓	✓	✓	✓	✓
Auto trans. fluid	I			✓			✓			✓			✓	
Brake pads/shoes	I			✓			✓			✓			✓	
Coolant hoses	S/I			✓			✓			✓			✓	
Steering linkage	I			✓			✓			✓			✓	
Cabin air filter	R			✓			✓			✓			✓	
Ball joints (2wd)	L			✓			✓			✓			✓	
Exhaust system	I						✓						✓	
Engine air filter	R						✓						✓	
Fuel filter ①	R						✓						✓	
Auto trans fluid (4-speed)	R						✓						✓	
Green coolant ②	R									✓				
Wheel bearings (2wd)	L												✓	
Manual trans. fluid	R												✓	
Spark plugs	R	every 100,000 miles												
PCV valve	R	every 100,000 miles												
Orange coolant	R	every 150,000 miles												
Auto trans fluid (5-speed)	R	every 150,000 miles												
Differential fluid	R	every 150,000 miles												
Accessory drive belts	R	every 150,000 miles												
Transfer case fluid	R	every 150,000 miles												

R: Replace S: Service I: Inspect L: Lubricate

① Recommended, but not required in Calif.

② Change every 30,000 miles or 36 months thereafter

FREQUENT OPERATION MAINTENANCE (SEVERE SERVICE)

If a vehicle is operated under any of the following conditions it is considered severe service:

- Towing a trailer or using a camper or car-top carrier.

- Repeated short trips of less than 5 miles in temperatures below freezing, or trips of less than 10 miles in any temperature.

- Extensive idling or low-speed driving for long distance as in heavy commercial use, such as delivery, taxi or police cars.

- Operating on rough, muddy or salt-covered roads.

- Operating on unpaved or dusty roads.

- Driving in extremely hot (over 90°) conditions.

Engine oil & filter: replace every 3000 miles.
Air cleaner filter: service or inspect every 6000 miles.
Exhaust system: check every 6000 miles.
Automatic transmission fluid & filter: change every 30,000 miles.
Transfer case fluid: change every 60,000 miles
Fule filter: change every 15,000 miles
Spark plugs: change every 60,000 miles
2wd front wheel bearings: lubricate every 30,000 miles

67197-EFSE-C23

SCHEDULED MAINTENANCE INTERVALS

2001-02 E-250/350 & F-250/350—Gasoline Engines

		VEHICLE MILEAGE INTERVAL (x1000)																			
TO BE SERVICED	TYPE OF SERVICE	5	10	15	20	25	30	35	40	45	50	55	60	65	70	75	80	85	90	95	100
Accessory drive belt	S/I																				✓
Air cleaner filter ①②	R						✓						✓						✓		
Automatic transmission fluid ③	R						✓						✓						✓		
Engine coolant ④⑤	R										✓						✓				
Engine cooling system hoses, clamps & coolant ⑥	S/I			✓			✓			✓			✓			✓			✓		
Engine oil & filter	R	✓	✓	✓	✓	✓	✓	✓	✓	✓	✓	✓	✓	✓	✓	✓	✓	✓	✓	✓	✓
Exhaust system	S/I			✓			✓			✓			✓			✓			✓		
Front wheel bearings	S/I & L																		✓		
Front/rear axle lubricant ⑦	R																				✓
Fuel filter ⑧	R						✓						✓						✓		
PCV valve	R	Every 120,000 miles																			
Rotate tires	S/I	✓	✓	✓	✓	✓	✓	✓	✓	✓	✓	✓	✓	✓	✓	✓	✓	✓	✓	✓	✓
Spark plugs	R																				✓
Steering linkage, suspension, driveshaft U joints	S/I & L	✓	✓	✓	✓	✓	✓	✓	✓	✓	✓	✓	✓	✓	✓	✓	✓	✓	✓	✓	✓

R: Replace S/I: Service or Inspect

① Perform this at the mileage shown or every 30 months, whichever occurs first.

② 7.3L DIT Diesel engine: the air filter should be replaced when the restriction gauge is in the red zone.

③ Except the E40D transmission.

④ Drain, flush and refill the cooling system initially at 50,000 miles or 48 months, whichever occurs first, then every 30,000 miles or 30 months thereafter.

⑤ 7.3L DIT Diesel engine: add 4 pints of FW-15 each time the coolant is replaced.

⑥ 7.3L DIT Diesel engine: add 8-10 oz. of FW-15 to the engine coolant every 15,000 miles.

⑦ The axle lubricant must be replaced every 100,000 miles of if the axle has been submerged under water.

⑧ 7.3L DIT Diesel engine: the fuel filter should be replaced when the restriction lamp is illuminated.

FREQUENT OPERATION MAINTENANCE (SEVERE SERVICE)

If a vehicle is operated under any of the following conditions it is considered severe service:

- Towing a trailer or using a camper or car-top carrier.

- Repeated short trips of less than 5 miles in temperatures below freezing, or trips of less than 10 miles in any temperature.

- Extensive idling or low-speed driving for long distances as in heavy commercial use, such as delivery, taxi or police cars.

- Operating on rough, muddy or salt-covered roads.

- Operating on unpaved or dusty roads.

Engine oil & filter: replace every 3000 miles.

Tires: rotate and inspect every 6000 miles.

Steering linkage, suspension, U-joints: lubricate every 6000 miles.

Exhaust system: inspect for leaks or damage every 12,000 miles.

Fuel filter: replace every 15,000 miles.

Automatic transmission fluid: change ever 21,000 miles.

Front wheel bearings (2WD): inspect and repack every 30,000 miles.

Rear axle lubricant (E-Super Duty only): replace every 30,000 miles.

Spark plugs (except 4.2L engine): replace every 60,000 miles.

PCV valve: replace every 60,000 miles.

Accessory drive belt: inspect every 60,000 miles.

Spark plugs: replace every 99,000 miles.

67197-EFSE-C24

SCHEDULED MAINTENANCE INTERVALS

2003 Blackwood, E-150 & F-150

		VEHICLE MILEAGE INTERVAL (x1000)												
TO BE SERVICED	TYPE OF SERVICE	5	10	15	20	25	30	35	40	45	50	55	60	65
Engine oil & filter	R	✓	✓	✓	✓	✓	✓	✓	✓	✓	✓	✓	✓	✓
Tires	Rotate	✓	✓	✓	✓	✓	✓	✓	✓	✓	✓	✓	✓	✓
Auto trans. fluid	I			✓			✓			✓			✓	
Brake pads/shoes	I			✓			✓			✓			✓	
Wheels	I ①			✓			✓			✓			✓	
Coolant hoses	S/I			✓			✓			✓			✓	
Steering linkage	I			✓			✓			✓			✓	
Suspension	I			✓			✓			✓			✓	
Cabin air filter	R			✓			✓			✓			✓	
Ball joints (2wd)	I			✓			✓			✓			✓	
Exhaust system	I						✓						✓	
Engine air filter	R						✓						✓	
Fuel filter ②	R						✓						✓	
Auto trans fluid (4R100)	R						✓						✓	
Green coolant ③	R									✓				
Accessory drive belts	I	every 100,000 miles												
Spark plugs	R	every 100,000 miles												
Yellow coolant ⑥	R	every 5 years or 100,000 miles												
PCV valve ④	R	every 100,000 miles												
PCV valve ⑤	R	every 120,000 miles												
Auto trans fluid (exc. 4R100)	R	every 150,000 miles												
Differential fluid	R	every 150,000 miles												
Accessory drive belts	R	every 150,000 miles if not previously changed												

R: Replace S: Service I: Inspect L: Lubricate

① Inspect for end play and noise

② Recommended, but not required in Calif.

③ Change every 30,000 miles or 36 months thereafter

④ Vehicle under 6,000 lbs GVW

⑤ Vehicles over 6,000 lbs GVW

⑥ Then every 3 years or 50,000 thereafter

Special Operating Condition Requirements

When towing a trailer or using a camper or car-top carrier:

Change engine oil and install a new oil filter every 4,800 km (3,000 miles), 3 months or 200 hours of engine operation (whichever occurs first).

Change transfer case fluid every 96,000 km (60,000 miles).

Change manual transmission fluid as required.

Inspect and lubricate U-joints as required.

During extensive idling and/or low speed driving for long distances, as in heavy commercial use such as delivery, taxi, patrol car or livery:

Change engine oil and install a new oil filter every 4,800 km (3,000 miles), 3 months or 200 hours of engine operation (whichever occurs first).

Lube front lower control arm and steering linkage ball joints with zerk fittings (if equipped) every 4,800 km (3,000 miles) or 3 months.

Inspect brake system and check battery electrolyte level (Patrol cars) every 8,000 km (5,000 miles).

Install a new fuel filter every 24,000 km (15,000 miles).

Change automatic transmission fluid, lubricate 4x2 wheel bearings, install new grease seals and adjust bearings every 48,000 km (30,000 miles). If equipped, change the in-line service installed transmission fluid filter.

Install new spark plugs and change transfer case fluid every 96,000 km (60,000 miles).

Install a new cabin air filter as required.

When operating in dusty conditions such as unpaved or dusty roads:

Change engine oil and install a new oil filter every 4,800 km (3,000 miles) or 3 months.

67197-EFSE-C25

SCHEDULED MAINTENANCE INTERVALS
2003 Blackwood, E-150 & F-150
Footnotes Continued

Install a new fuel filter every 24,000 km (15,000 miles).

Change automatic transmission fluid every 48,000 km (30,000 miles). If equipped, change the in-line service installed transmission fluid filter.

Change transfer case fluid every 96,000 km (60,000 miles).

Install a new engine air filter as required.

Install a new cabin air filter as required.

When operating in off-road conditions:

Change automatic transmission fluid every 48,000 km (30,000 miles). If equipped, change the in-line service installed transmission fluid filter.

Change transfer case fluid every 96,000 km (60,000 miles).

Install a new cabin air filter as required.

Inspect and lubricate U-joints.

Inspect and lubricate steering linkage ball joints with zerk fittings.

67197-EFSE-C26

SCHEDULED MAINTENANCE INTERVALS

2003 E-250/350 & F-250/350—Gasoline Engines

TO BE SERVICED	TYPE OF SERVICE	VEHICLE MILEAGE INTERVAL (x1000)																			
		5	10	15	20	25	30	35	40	45	50	55	60	65	70	75	80	85	90	95	100
Accessory drive belt	S/I																				✓
Air cleaner filter ①	R						✓						✓						✓		
Automatic transmission fluid ②	R						✓						✓						✓		
Engine coolant ③	R										✓						✓				
Engine cooling system hoses, clamps & coolant	S/I			✓			✓			✓			✓			✓			✓		
Engine oil & filter	R	✓	✓	✓	✓	✓	✓	✓	✓	✓	✓	✓	✓	✓	✓	✓	✓	✓	✓	✓	✓
Exhaust system	S/I			✓			✓			✓			✓			✓			✓		
Front wheel bearings	S/I & L																		✓		
Front/rear axle lubricant ④	R																				✓
Fuel filter	R						✓						✓						✓		
PCV valve	R	Every 120,000 miles																			
Rotate tires	S/I	✓	✓	✓	✓	✓	✓	✓	✓	✓	✓	✓	✓	✓	✓	✓	✓	✓	✓	✓	✓
Spark plugs	R																				✓
Steering linkage, suspension, driveshaft U joints	S/I & L	✓	✓	✓	✓	✓	✓	✓	✓	✓	✓	✓	✓	✓	✓	✓	✓	✓	✓	✓	✓

R: Replace S/I: Service or Inspect

① Perform this at the mileage shown or every 30 months, whichever occurs first.

② Except the E40D transmission.

④ Drain, flush and refill the cooling system initially at 50,000 miles or 48 months, whichever occurs first, then every 30,000 miles or 30 months thereafter.

Special Operating Condition Requirements

When towing a trailer or using a camper or car-top carrier:

Change engine oil and install a new oil filter every 4,800 km (3,000 miles), 3 months or 200 hours of engine operation (whichever occurs first).

Change transfer case fluid every 96,000 km (60,000 miles).

Change manual transmission fluid as required.

Inspect and lubricate U-joints as required.

During extensive idling and/or low speed driving for long distances, as in heavy commercial use such as delivery, taxi, patrol car or livery:

Change engine oil and install a new oil filter every 4,800 km (3,000 miles), 3 months or 200 hours of engine operation (whichever occurs first).

Lube front lower control arm and steering linkage ball joints with zerk fittings (if equipped) every 4,800 km (3,000 miles) or 3 months.

Inspect brake system and check battery electrolyte level (Patrol cars) every 8,000 km (5,000 miles).

Install a new fuel filter every 24,000 km (15,000 miles).

Change automatic transmission fluid, lubricate 4x2 wheel bearings, install new grease seals and adjust bearings every 48,000 km (30,000 miles). If equipped, change the in-line service installed transmission fluid filter.

Install new spark plugs and change transfer case fluid every 96,000 km (60,000 miles).

Install a new cabin air filter as required.

When operating in dusty conditions such as unpaved or dusty roads:

Change engine oil and install a new oil filter every 4,800 km (3,000 miles) or 3 months.

Install a new fuel filter every 24,000 km (15,000 miles).

Change automatic transmission fluid every 48,000 km (30,000 miles). If equipped, change the in-line service installed transmission fluid filter.

Change transfer case fluid every 96,000 km (60,000 miles).

Install a new engine air filter as required.

Install a new cabin air filter as required.

When operating in off-road conditions:

Change automatic transmission fluid every 48,000 km (30,000 miles). If equipped, change the in-line service installed transmission fluid filter.

Change transfer case fluid every 96,000 km (60,000 miles).

Install a new cabin air filter as required.

Inspect and lubricate U-joints.

Inspect and lubricate steering linkage ball joints with zerk fittings.

67197-EFSE-C27

SCHEDULED MAINTENANCE INTERVALS

All 2003 Vehicles with the 7.3L Diesel

		VEHICLE MILEAGE INTERVAL (x1000)																			
TO BE SERVICED	TYPE OF SERVICE	5	10	15	20	25	30	35	40	45	50	55	60	65	70	75	80	85	90	95	100
Accessory drive belt	S/I																				✓
Air cleaner filter	R	When the restriction gauge is in the red zone																			
Automatic transmission fluid ①	R						✓						✓						✓		
Engine coolant ②③	R										✓						✓				
Engine cooling system hoses, clamps & coolant ④	S/I			✓			✓			✓			✓			✓			✓		
Engine oil & filter	R	✓	✓	✓	✓	✓	✓	✓	✓	✓	✓	✓	✓	✓	✓	✓	✓	✓	✓	✓	✓
Exhaust system	S/I			✓			✓			✓			✓			✓			✓		
Front wheel bearings	S/I & L																		✓		
Front/rear axle lubricant ⑤	R																				✓
Fuel filter	R	When the retsriction lamp is illuminated																			
PCV valve	R	Every 120,000 miles																			
Rotate tires	S/I	✓	✓	✓	✓	✓	✓	✓	✓	✓	✓	✓	✓	✓	✓	✓	✓	✓	✓	✓	✓
Steering linkage, suspension, driveshaft U joints	S/I & L	✓	✓	✓	✓	✓	✓	✓	✓	✓	✓	✓	✓	✓	✓	✓	✓	✓	✓	✓	✓

R: Replace S/I: Service or Inspect

① Except the E40D transmission.

② Drain, flush and refill the cooling system initially at 50,000 miles or 48 months, whichever occurs first, then every 30,000 miles or 30 months thereafter.

③ Add 4 pints of FW-15 each time the coolant is replaced.

④ Add 8-10 oz. of FW-15 to the engine coolant every 15,000 miles.

⑤ The axle lubricant must be replaced every 100,000 miles of if the axle has been submerged under water.

Special Operating Condition Requirements

When towing a trailer or using a camper or car-top carrier:

Change engine oil and install a new oil filter every 4,800 km (3,000 miles), 3 months or 200 hours of engine operation (whichever occurs first).

Change transfer case fluid every 96,000 km (60,000 miles).

Change manual transmission fluid as required.

Inspect and lubricate U-joints as required.

During extensive idling and/or low speed driving for long distances, as in heavy commercial use such as delivery, taxi, patrol car or livery:

Change engine oil and install a new oil filter every 4,800 km (3,000 miles), 3 months or 200 hours of engine operation (whichever occurs first).

Lube front lower control arm and steering linkage ball joints with zerk fittings (if equipped) every 4,800 km (3,000 miles) or 3 months.

Inspect brake system and check battery electrolyte level (Patrol cars) every 8,000 km (5,000 miles).

Install a new fuel filter every 24,000 km (15,000 miles).

Change automatic transmission fluid, lubricate 4x2 wheel bearings, install new grease seals and adjust bearings every 48,000 km (30,000 miles). If equipped, change the in-line service installed transmission fluid filter.

Install new spark plugs and change transfer case fluid every 96,000 km (60,000 miles).

Install a new cabin air filter as required.

When operating in dusty conditions such as unpaved or dusty roads:

Change engine oil and install a new oil filter every 4,800 km (3,000 miles) or 3 months.

Install a new fuel filter every 24,000 km (15,000 miles).

Change automatic transmission fluid every 48,000 km (30,000 miles). If equipped, change the in-line service installed transmission fluid filter.

Change transfer case fluid every 96,000 km (60,000 miles).

Install a new engine air filter as required.

Install a new cabin air filter as required.

When operating in off-road conditions:

Change automatic transmission fluid every 48,000 km (30,000 miles). If equipped, change the in-line service installed transmission fluid filter.

Change transfer case fluid every 96,000 km (60,000 miles).

Install a new cabin air filter as required.

Inspect and lubricate U-joints.

Inspect and lubricate steering linkage ball joints with zerk fittings.

67197-EFSE-C30

SCHEDULED MAINTENANCE INTERVALS

2004 F-150

TO BE SERVICED	TYPE OF SERVICE	VEHICLE MILEAGE INTERVAL (x1000)												
		5	10	15	20	25	30	35	40	45	50	55	60	65
Engine oil & filter	R	✓	✓	✓	✓	✓	✓	✓	✓	✓	✓	✓	✓	✓
Tires	Rotate	✓	✓	✓	✓	✓	✓	✓	✓	✓	✓	✓	✓	✓
Wheels	I ①			✓			✓			✓			✓	
Auto trans. fluid	I			✓			✓			✓			✓	
Brake pads/shoes	I			✓			✓			✓			✓	
Coolant hoses	S/I			✓			✓			✓			✓	
Steering linkage	I			✓			✓			✓			✓	
Suspension	I			✓			✓			✓			✓	
Driveshaft	I			✓			✓			✓			✓	
Cabin air filter	R			✓			✓			✓			✓	
Ball joints (2wd)	L			✓			✓			✓			✓	
Exhaust system	I						✓						✓	
Engine air filter	R						✓						✓	
Fuel filter	R						✓						✓	
Auto trans fluid (4R100)	R						✓						✓	
Front wheel bearings (2wd)	L/Adj												✓	
Front wheel bearings grease seal (2wd)	R												✓	
Accessory drive belts	I	every 100,000 miles												
Front wheel bearings (2wd)	R	at 150,000 miles, if not previously done so												
Spark plugs	R	every 100,000 miles												
PCV valve	R	every 120,000 miles												
Premium Gold coolant	R	every 5 years or 100,000 miles												
Auto trans fluid (all exc. 4R100)	R	every 150,000 miles												
Differential fluid	R	every 150,000 miles												
Accessory drive belts	R	every 150,000 miles, if not previously done so												

R: Replace S: Service I: Inspect L: Lubricate

① Inspect for end play and noise

Special Operating Condition Requirements

When towing a trailer or using a camper or car-top carrier:

Change engine oil and install a new oil filter every 4,800 km (3,000 miles), 3 months or 200 hours of engine operation (whichever occurs first).

Change transfer case fluid every 96,000 km (60,000 miles).

Change manual transmission fluid as required.

Inspect and lubricate U-joints as required.

During extensive idling and/or low speed driving for long distances, as in heavy commercial use such as delivery, taxi, patrol car or livery:

Change engine oil and install a new oil filter every 4,800 km (3,000 miles), 3 months or 200 hours of engine operation (whichever occurs first).

Lube front lower control arm and steering linkage ball joints with zerk fittings (if equipped) every 4,800 km (3,000 miles) or 3 months.

Inspect brake system and check battery electrolyte level (Patrol cars) every 8,000 km (5,000 miles).

Install a new fuel filter every 24,000 km (15,000 miles).

Change automatic transmission fluid, lubricate 4x2 wheel bearings, install new grease seals and adjust bearings every 48,000 km (30,000 miles). If equipped, change the in-line service installed transmission fluid filter.

Install new spark plugs and change transfer case fluid every 96,000 km (60,000 miles).

Install a new cabin air filter as required.

67197-EFSE-C31

SCHEDULED MAINTENANCE INTERVALS

2004 F-150

Footnotes Continued

When operating in dusty conditions such as unpaved or dusty roads:

Change engine oil and install a new oil filter every 4,800 km (3,000 miles) or 3 months.

Install a new fuel filter every 24,000 km (15,000 miles).

Change automatic transmission fluid every 48,000 km (30,000 miles). If equipped, change the in-line service installed transmission fluid filter.

Change transfer case fluid every 96,000 km (60,000 miles).

Install a new engine air filter as required.

Install a new cabin air filter as required.

When operating in off-road conditions:

Change automatic transmission fluid every 48,000 km (30,000 miles). If equipped, change the in-line service installed transmission fluid filter.

Change transfer case fluid every 96,000 km (60,000 miles).

Install a new cabin air filter as required.

Inspect and lubricate U-joints.

Inspect and lubricate steering linkage ball joints with zerk fittings.

67197-EFSE-C32

SCHEDULED MAINTENANCE INTERVALS

2004 Gasoline F-250/350 Super Duty

TO BE SERVICED	TYPE OF SERVICE	VEHICLE MILEAGE INTERVAL (x1000)												
		5	10	15	20	25	30	35	40	45	50	55	60	65
Engine oil & filter	R	✓	✓	✓	✓	✓	✓	✓	✓	✓	✓	✓	✓	✓
Tires	Rotate	✓	✓	✓	✓	✓	✓	✓	✓	✓	✓	✓	✓	✓
Wheels	I ①	✓	✓	✓	✓	✓	✓	✓	✓	✓	✓	✓	✓	✓
Auto trans. fluid	I			✓			✓			✓			✓	
Brake pads, hoses, etc.	I			✓			✓			✓			✓	
Coolant hoses	I			✓			✓			✓			✓	
Steering linkage	I/L			✓			✓			✓			✓	
Cabin air filter	R			✓			✓			✓			✓	
Ball joints (2wd)	L			✓			✓			✓			✓	
Driveshaft	I			✓			✓			✓			✓	
4x4 front axle U-joints	L			✓			✓			✓			✓	
Exhaust system	I						✓						✓	
Engine air filter	R						✓						✓	
Fuel filter	R						✓						✓	
Auto trans fluid (4R100)	R						✓						✓	
4x2 front wheel bearings	Adj.												✓	
4x2 front wheel bearing grease seals	R												✓	
Accessory drive belts	I	Every 100,000 miles												
Front wheel bearings (2wd)	R	at 150,000 miles, if not previously done so												
4x4 front hub needle bearings	L	every 120,000 miles												
Spark plugs	R	every 100,000 miles												
PCV valve	R	every 120,000 miles												
Manual trans. Fluid	R	every 120,000 miles												
Premium Gold coolant	R	every 3 years or 100,000 miles and every 50,000 miles thereafter												
Coolant (exc. Premium Gold)	R	every 135,000 miles												
Auto trans fluid (exc. 4R100)	R	every 150,000 miles												
Differential fluid	R	at 150,000 miles then every 50,000 thereafter												
Transfer case fluid	R	every 150,000 miles												
Accessory drive belts	R	every 150,000 miles, if not previously done so												

R: Replace S: Service I: Inspect L: Lubricate Adj: adjust

① Inspect for end play and noise

Special Operating Condition Requirements

When towing a trailer or using a camper or car-top carrier:

Change engine oil and install a new oil filter every 4,800 km (3,000 miles), 3 months or 200 hours of engine operation (whichever occurs first).

Change transfer case fluid every 96,000 km (60,000 miles).

Change manual transmission fluid as required.

Inspect and lubricate U-joints as required.

During extensive idling and/or low speed driving for long distances, as in heavy commercial use such as delivery, taxi, patrol car or livery:

Change engine oil and install a new oil filter every 4,800 km (3,000 miles), 3 months or 200 hours of engine operation (whichever occurs first).

Lube front lower control arm and steering linkage ball joints with zerk fittings (if equipped) every 4,800 km (3,000 miles) or 3 months.

Inspect brake system and check battery electrolyte level (Patrol cars) every 8,000 km (5,000 miles).

Install a new fuel filter every 24,000 km (15,000 miles).

Change automatic transmission fluid, lubricate 4x2 wheel bearings, install new grease seals and adjust bearings every 48,000 km (30,000 miles). If equipped, change the in-line service installed transmission fluid filter.

67197-EFSE-C33

SCHEDULED MAINTENANCE INTERVALS
2004 Gasoline F-250/350 Super Duty
Footnotes Continued

Install new spark plugs and change transfer case fluid every 96,000 km (60,000 miles).

Install a new cabin air filter as required.

When operating in dusty conditions such as unpaved or dusty roads:

Change engine oil and install a new oil filter every 4,800 km (3,000 miles) or 3 months.

Install a new fuel filter every 24,000 km (15,000 miles).

Change automatic transmission fluid every 48,000 km (30,000 miles). If equipped, change the in-line service installed transmission fluid filter.

Change transfer case fluid every 96,000 km (60,000 miles).

Install a new engine air filter as required.

Install a new cabin air filter as required.

When operating in off-road conditions:

Change automatic transmission fluid every 48,000 km (30,000 miles). If equipped, change the in-line service installed transmission fluid filter.

Change transfer case fluid every 96,000 km (60,000 miles).

Install a new cabin air filter as required.

Inspect and lubricate U-joints.

Inspect and lubricate steering linkage ball joints with zerk fittings.

67197-EFSE-C34

SCHEDULED MAINTENANCE INTERVALS

2004 E-150

TO BE SERVICED	TYPE OF SERVICE	VEHICLE MILEAGE INTERVAL (x1000)												
		5	10	15	20	25	30	35	40	45	50	55	60	65
Engine oil & filter	R	✓	✓	✓	✓	✓	✓	✓	✓	✓	✓	✓	✓	✓
Tires	Rotate	✓	✓	✓	✓	✓	✓	✓	✓	✓	✓	✓	✓	✓
Wheels	I ①	✓	✓	✓	✓	✓	✓	✓	✓	✓	✓	✓	✓	✓
Auto trans. fluid	I			✓			✓			✓			✓	
Brake pads/shoes	I			✓			✓			✓			✓	
Coolant hoses	S/I			✓			✓			✓			✓	
Steering linkage	I			✓			✓			✓			✓	
Suspension	I			✓			✓			✓			✓	
Driveshaft	I			✓			✓			✓			✓	
Cabin air filter	R			✓			✓			✓			✓	
Ball joints (2wd)	I/L			✓			✓			✓			✓	
Exhaust system	I						✓						✓	
Engine air filter	R						✓						✓	
Fuel filter	R						✓						✓	
Auto trans fluid (4R100) ②	R						✓						✓	
Front wheel bearings (2wd) ③	L/Adj												✓	
Front wheel bearings grease seal (2wd)	R												✓	
Accessory drive belts	I	every 100,000 miles												
Spark plugs	R	every 100,000 miles												
Premium Gold coolant ⑤	R	at 3 years or 100,000 miles												
Coolant (exc. Premium Gold)	R	every 105,000 miles												
PCV valve ④	R	every 120,000 miles												
Front wheel bearings (2wd)	R	at 150,000 miles, if not previously done so												
Auto trans fluid (exc. 4R100) ②	R	every 150,000 miles												
Differential fluid	R	every 150,000 miles												
Accessory drive belts	R	every 150,000 miles, if not previously done so												

R: Replace S: Service I: Inspect L: Lubricate Adj: Adjust

① Inspect for end play and noise

② Change both the in-line and installed filters

③ And every 15,000 miles thereafter

④ Under 6,000 lbs. GVW, replace every 100,000 miles

⑤ And every 50,000 miles thereafter

Special Operating Condition Requirements

When towing a trailer or using a camper or car-top carrier:

Change engine oil and install a new oil filter every 4,800 km (3,000 miles), 3 months or 200 hours of engine operation (whichever occurs first).

Change transfer case fluid every 96,000 km (60,000 miles).

Change manual transmission fluid as required.

Inspect and lubricate U-joints as required.

During extensive idling and/or low speed driving for long distances, as in heavy commercial use such as delivery, taxi, patrol car or livery:

Change engine oil and install a new oil filter every 4,800 km (3,000 miles), 3 months or 200 hours of engine operation (whichever occurs first).

Lube front lower control arm and steering linkage ball joints with zerk fittings (if equipped) every 4,800 km (3,000 miles) or 3 months.

Inspect brake system and check battery electrolyte level (Patrol cars) every 8,000 km (5,000 miles).

Install a new fuel filter every 24,000 km (15,000 miles).

67197-EFSE-C35

SCHEDULED MAINTENANCE INTERVALS

2004 E-150

Footnotes Continued

Change automatic transmission fluid, lubricate 4x2 wheel bearings, install new grease seals and adjust bearings every 48,000 km (30,000 miles). If equipped, change the in-line service installed transmission fluid filter.

Install new spark plugs and change transfer case fluid every 96,000 km (60,000 miles).

Install a new cabin air filter as required.

When operating in dusty conditions such as unpaved or dusty roads:

Change engine oil and install a new oil filter every 4,800 km (3,000 miles) or 3 months.

Install a new fuel filter every 24,000 km (15,000 miles).

Change automatic transmission fluid every 48,000 km (30,000 miles). If equipped, change the in-line service installed transmission fluid filter.

Change transfer case fluid every 96,000 km (60,000 miles).

Install a new engine air filter as required.

Install a new cabin air filter as required.

When operating in off-road conditions:

Change automatic transmission fluid every 48,000 km (30,000 miles). If equipped, change the in-line service installed transmission fluid filter.

Change transfer case fluid every 96,000 km (60,000 miles).

Install a new cabin air filter as required.

Inspect and lubricate U-joints.

Inspect and lubricate steering linkage ball joints with zerk fittings.

67197-EFSE-C36

SCHEDULED MAINTENANCE INTERVALS

2004 Gasoline E-250 and 350

TO BE SERVICED	TYPE OF SERVICE	VEHICLE MILEAGE INTERVAL (x1000)												
		5	10	15	20	25	30	35	40	45	50	55	60	65
Engine oil & filter	R	✓	✓	✓	✓	✓	✓	✓	✓	✓	✓	✓	✓	✓
Tires	Rotate	✓	✓	✓	✓	✓	✓	✓	✓	✓	✓	✓	✓	✓
Wheels	I ①	✓	✓	✓	✓	✓	✓	✓	✓	✓	✓	✓	✓	✓
Auto trans. fluid	I			✓			✓			✓			✓	
Brake pads, hoses, etc.	I			✓			✓			✓			✓	
Coolant hoses	I			✓			✓			✓			✓	
Steering linkage	I/L			✓			✓			✓			✓	
Cabin air filter	R			✓			✓			✓			✓	
Ball joints (2wd)	L			✓			✓			✓			✓	
Driveshaft	I			✓			✓			✓			✓	
4x4 front axle U-joints	L			✓			✓			✓			✓	
Exhaust system	I						✓						✓	
Engine air filter	R						✓						✓	
Fuel filter	R						✓						✓	
Auto trans fluid (4R100)	R						✓						✓	
4x2 front wheel bearings	Adj.												✓	
4x2 front wheel bearing grease seals	R												✓	
Accessory drive belts	I	Every 100,000 miles												
Spark plugs	R	every 100,000 miles												
Premium Gold coolant	R	every 3 years or 100,000 miles and every 50,000 miles thereafter												
4x4 front hub needle bearings	L	every 120,000 miles												
PCV valve	R	every 120,000 miles												
Manual trans. Fluid	R	every 120,000 miles												
Coolant (exc. Premium Gold)	R	every 135,000 miles												
Front wheel bearings (2wd)	R	at 150,000 miles, if not previously done so												
Auto trans fluid (exc. 4R100)	R	every 150,000 miles												
Differential fluid	R	at 150,000 miles then every 50,000 thereafter												
Transfer case fluid	R	every 150,000 miles												
Accessory drive belts	R	every 150,000 miles, if not previously done so												

R: Replace S: Service I: Inspect L: Lubricate Adj: adjust

① Inspect for end play and noise

Special Operating Condition Requirements

When towing a trailer or using a camper or car-top carrier:

Change engine oil and install a new oil filter every 4,800 km (3,000 miles), 3 months or 200 hours of engine operation (whichever occurs first).

Change transfer case fluid every 96,000 km (60,000 miles).

Change manual transmission fluid as required.

Inspect and lubricate U-joints as required.

During extensive idling and/or low speed driving for long distances, as in heavy commercial use such as delivery, taxi, patrol car or livery:

Change engine oil and install a new oil filter every 4,800 km (3,000 miles), 3 months or 200 hours of engine operation (whichever occurs first).

Lube front lower control arm and steering linkage ball joints with zerk fittings (if equipped) every 4,800 km (3,000 miles) or 3 months.

Inspect brake system and check battery electrolyte level (Patrol cars) every 8,000 km (5,000 miles).

Install a new fuel filter every 24,000 km (15,000 miles).

Change automatic transmission fluid, lubricate 4x2 wheel bearings, install new grease seals and adjust bearings every 48,000 km (30,000 miles). If equipped, change the in-line service installed transmission fluid filter.

67197-EFSE-C37

SCHEDULED MAINTENANCE INTERVALS

2004 Gasoline E-250 and 350

Footnotes Continued

Install new spark plugs and change transfer case fluid every 96,000 km (60,000 miles).

Install a new cabin air filter as required.

When operating in dusty conditions such as unpaved or dusty roads:

Change engine oil and install a new oil filter every 4,800 km (3,000 miles) or 3 months.

Install a new fuel filter every 24,000 km (15,000 miles).

Change automatic transmission fluid every 48,000 km (30,000 miles). If equipped, change the in-line service installed transmission fluid filter.

Change transfer case fluid every 96,000 km (60,000 miles).

Install a new engine air filter as required.

Install a new cabin air filter as required.

When operating in off-road conditions:

Change automatic transmission fluid every 48,000 km (30,000 miles). If equipped, change the in-line service installed transmission fluid filter.

Change transfer case fluid every 96,000 km (60,000 miles).

Install a new cabin air filter as required.

Inspect and lubricate U-joints.

Inspect and lubricate steering linkage ball joints with zerk fittings.

67197-EFSE-C38

SCHEDULED MAINTENANCE INTERVALS

All 2004 Vehicles with the 6.0L Deisel Engine

TO BE SERVICED	TYPE OF SERVICE	VEHICLE MILEAGE INTERVAL (x1000)												
		7.5	15	22.5	30	37.5	45	52.5	60	67.5	75	82.5	90	97.5
Engine oil & filter	R	✓	✓	✓	✓	✓	✓	✓	✓	✓	✓	✓	✓	✓
Tires	Rotate	✓	✓	✓	✓	✓	✓	✓	✓	✓	✓	✓	✓	✓
Air filter minder	I ①	✓	✓	✓	✓	✓	✓	✓	✓	✓	✓	✓	✓	✓
Wheels	I ②	✓	✓	✓	✓	✓	✓	✓	✓	✓	✓	✓	✓	✓
Brake pads, hoses, etc.	I		✓		✓		✓		✓		✓		✓	
Coolant hoses	I		✓		✓		✓		✓		✓		✓	
Steering linkage and suspension	I/L		✓		✓		✓		✓		✓		✓	
Cabin air filter	R		✓		✓		✓		✓		✓		✓	
Ball joints (2wd)	L		✓		✓		✓		✓		✓		✓	
Driveshaft	I/L		✓		✓		✓		✓		✓		✓	
4x4 front axle U-joints	L		✓		✓		✓		✓		✓		✓	
Exhaust system and heat shields	I		✓		✓		✓		✓		✓		✓	
Engine air filter	R				✓				✓				✓	
Fuel filters ③	R		✓		✓		✓		✓		✓		✓	
Auto trans fluid ④	R				✓				✓				✓	
4x2 front wheel bearings	L								✓					
4x2 front wheel bearing grease seals	R								✓					
Accessory drive belts	I													✓
4x4 front hub needle bearings	L								✓					
Manual trans. Fluid	R								✓					
Coolant	R										✓			✓
Rear differential fluid ⑤	R													✓
Transfer case fluid	R	every 150,000 miles												
4x2 front wheel bearings	R	every 150,000 miles												
Accessory drive belts	R	every 150,000 miles, if not previously done so												

R: Replace S: Service I: Inspect L: Lubricate Adj: adjust

① Reset after new filter is installed

② Inspect for end play and noise

③ Frame-mounted and engine

④ Including external and in-line filters

⑤ Dana axles using non-synthetic fluid only

Special Operating Condition Requirements

When towing a trailer or using a camper or car-top carrier:

Change engine oil and install a new oil filter every 4,800 km (3,000 miles), 3 months or 200 hours of engine operation (whichever occurs first).

Change transfer case fluid every 96,000 km (60,000 miles).

Change manual transmission fluid as required.

Inspect and lubricate U-joints as required.

During extensive idling and/or low speed driving for long distances, as in heavy commercial use such as delivery, taxi, patrol car or livery:

Change engine oil and install a new oil filter every 4,800 km (3,000 miles), 3 months or 200 hours of engine operation (whichever occurs first).

Lube front lower control arm and steering linkage ball joints with zerk fittings (if equipped) every 4,800 km (3,000 miles) or 3 months.

Inspect brake system and check battery electrolyte level (Patrol cars) every 8,000 km (5,000 miles).

Install a new fuel filter every 24,000 km (15,000 miles).

67197-EFSE-C39

SCHEDULED MAINTENANCE INTERVALS
All 2004 Vehicles with the 6.0L Deisel Engine
Footnotes Continued

Change automatic transmission fluid, lubricate 4x2 wheel bearings, install new grease seals and adjust bearings every 48,000 km (30,000 miles). If equipped, change the in-line service installed transmission fluid filter.

Change transfer case fluid every 96,000 km (60,000 miles).

Install a new cabin air filter as required.

When operating in dusty conditions such as unpaved or dusty roads:

Change engine oil and install a new oil filter every 4,800 km (3,000 miles) or 3 months.

Install a new fuel filter every 24,000 km (15,000 miles).

Change automatic transmission fluid every 48,000 km (30,000 miles). If equipped, change the in-line service installed transmission fluid filter.

Change transfer case fluid every 96,000 km (60,000 miles).

Install a new engine air filter as required.

Install a new cabin air filter as required.

When operating in off-road conditions:

Change automatic transmission fluid every 48,000 km (30,000 miles). If equipped, change the in-line service installed transmission fluid filter.

Change transfer case fluid every 96,000 km (60,000 miles).

Install a new cabin air filter as required.

Inspect and lubricate U-joints.

Inspect and lubricate steering linkage ball joints with zerk fittings.

67197-EFSE-C40

PRECAUTIONS

Before servicing any vehicle, please be sure to read all of the following precautions, which deal with personal safety, prevention of component damage and important points to take into consideration when servicing a motor vehicle:

- Never open, service or drain the radiator or cooling system when the engine is hot; serious burns can occur from the steam and hot coolant.
- Observe all applicable safety precautions when working around fuel. Whenever servicing the fuel system, always work in a well-ventilated area. Do not allow fuel spray or vapors to come in contact with a spark, open flame, or excessive heat (a hot drop light, for example). Keep a dry chemical fire extinguisher near the work area. Always keep fuel in a container specifically designed for fuel storage; also, always properly seal fuel containers to avoid the possibility of fire or explosion. Refer to the additional fuel system precautions later in this section.
- Fuel injection systems often remain pressurized, even after the engine has been turned **OFF**. The fuel system pressure must be relieved before disconnecting any fuel lines. Failure to do so may result in fire and/or personal injury.
- Brake fluid often contains polyglycol ethers and polyglycols. Avoid contact with the eyes and wash your hands thoroughly after handling brake fluid. If you do get brake fluid in your eyes, flush your eyes with clean, running water for 15 minutes. If eye irritation persists, or if you have taken brake fluid internally, seek medical assistance IMMEDIATELY.
- The EPA warns that prolonged contact with used engine oil may cause a number of skin disorders, including cancer! You should make every effort to minimize your exposure to used engine oil. Protective gloves should be worn when changing oil. Wash your hands and any other exposed skin areas as soon as possible after exposure to used engine oil. Soap and water, or waterless hand cleaner should be used.
- All new vehicles are now equipped with an air bag system, often referred to as a Supplemental Restraint System (SRS) or Supplemental Inflatable Restraint (SIR) system. The system must be disabled before performing service on or around system components, steering column, instrument panel components, wiring and sensors. Failure to follow safety and disabling procedures could result in accidental air bag deployment, possible personal injury and unnecessary system repairs.
- Always wear safety goggles when working with, or around, the air bag system. When carrying a non-deployed air bag, be sure the bag and trim cover are pointed away from your body. When placing a non-deployed air bag on a work surface, always face the bag and trim cover upward, away from the surface. This will reduce the motion of the module if it is accidentally deployed. Refer to the additional air bag system precautions later in this section.
- Clean, high quality brake fluid from a sealed container is essential to the safe and proper operation of the brake system. You should always buy the correct type of brake fluid for your vehicle. If the brake fluid becomes contaminated, completely flush the system with new fluid. Never reuse any brake fluid. Any brake fluid that is removed from the system should be discarded. Also, do not allow any brake fluid to come in contact with a painted surface; it will damage the paint.
- Never operate the engine without the proper amount and type of engine oil; doing so WILL result in severe engine damage.
- Timing belt maintenance is extremely important! Many models utilize an interference-type, non-freewheeling engine. If the timing belt breaks, the valves in the cylinder head may strike the pistons, causing potentially serious (also time-consuming and expensive) engine damage. Refer to the maintenance interval charts in the front of this manual for the recommended replacement interval for the timing belt and to the timing belt section for belt replacement and inspection.
- Suspension fasteners are critical parts because they affect performance of vital components and systems and their failure can result in major service expense. Install new parts with the same part number or an equivalent part if installation is necessary. Do not use an installation part of lesser quality or substitute design. Torque values must be used as specified during reassembly to make sure of correct retention of these parts.
- Disconnecting the negative battery cable on some vehicles may interfere with the functions of the on-board computer system(s) and may require the computer to undergo a relearning process once the negative battery cable is reconnected.
- When servicing drum brakes, only disassemble and assemble one side at a time, leaving the remaining side intact for reference.
- Only an MVAC-trained, EPA-certified automotive technician should service the air conditioning system or its components.

GASOLINE ENGINE REPAIR

➡Disconnecting the negative battery cable on some vehicles may interfere with the functions of the on board computer systems and may require the computer to undergo a relearning process, once the negative battery cable is reconnected.

Alternator

REMOVAL

4.2L Engine

1. Before servicing the vehicle, refer to the precautions in the beginning of this section.
2. Before servicing the vehicle, refer to the precautions in the beginning of this section.
3. Remove or disconnect the following:
   - Negative battery cable
   - Drive belt
   - Alternator electrical connectors
   - Alternator stator and voltage regulator connectors
   - Alternator battery cable nut and the cable
   - Alternator bolts and the alternator

To install:

4. Install or connect the following:
   - Alternator and tighten the bolts to 35 ft. lbs. (47 Nm)
   - Alternator battery cable and the nut
   - Alternator stator and voltage regulator connectors
   - Alternator electrical connectors
   - Drive belt
   - Negative battery cable

4.6L Engines

1. Before servicing the vehicle, refer to the precautions in the beginning of this section.
2. Remove or disconnect the following:
   - Negative battery cable
   - Drive belt
   - Alternator bracket bolts
   - Ignition wire from the alternator
   - Alternator bolts and the alternator

- Alternator electrical connectors
- Alternator stator and voltage regulator connectors
- Alternator battery cable nut and the cable

To install:

3. Install or connect the following:
   - Alternator battery cable and the nut
   - Alternator stator and voltage regulator connectors
   - Alternator electrical connectors
   - Alternator and the bolts, tighten to 18 ft. lbs. (25 Nm)
   - Ignition wire from the alternator
   - Alternator bracket bolts and tighten to 89 inch lbs. (10 Nm)
   - Drive belt
   - Negative battery cable

Lightning Models

1. Before servicing the vehicle, refer to the precautions in the beginning of this section.
2. Remove or disconnect the following:
   - Negative battery cable
   - Bolt retaining the degas bottle and position the bottle aside
   - Supercharger drive belt
   - Accessory drive belt
   - Upper alternator bracket
   - Lower alternator brackets
   - Upper bracket bolts
   - Electrical connector from the alternator bracket
   - Right hand lower bolt
   - Alternator bracket
   - Alternator electrical connectors
   - Alternator

To install:

3. Before servicing the vehicle, refer to the precautions in the beginning of this section.
4. Installation is the reverse of removal, please note the following torques:
   a. All alternator bracket bolts to 89 inch lbs. (10 Nm).
   b. Right hand lower bolt to 18 ft. lbs. (25 Nm).

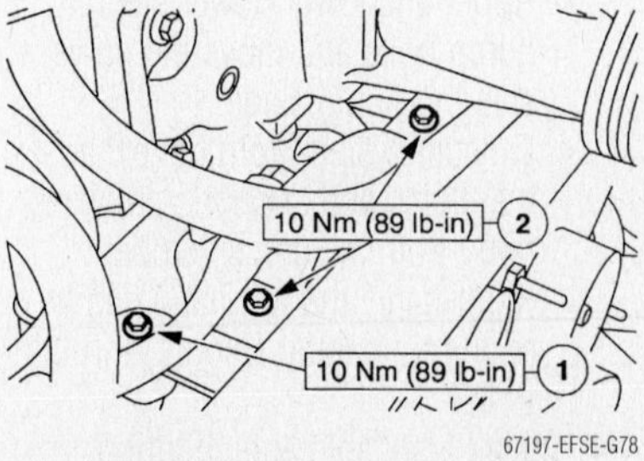

Alternator mounting bracket installation—supercharged engine

5.4L and 6.8L Engines

1. Before servicing the vehicle, refer to the precautions in the beginning of this section.
2. Remove or disconnect the following:
   - Negative battery cable
   - Air cleaner assembly
   - Mass Air Flow (MAF) sensor electrical connector
   - Accessory drive belt
   - Alternator electrical connectors
   - Bolts retaining the upper intake plenum to the alternator bracket
   - Alternator bolts and the alternator

To install:

3. Before servicing the vehicle, refer to the precautions in the beginning of this section.
4. Installation is the reverse of removal, please note the following torques:
   a. Bolts retaining the upper intake plenum to the alternator bracket. Tighten to 80–106 inch lbs. (7–9 Nm).
   b. Alternator bolts and the alternator. Tighten to 16–21 ft. lbs. (22–28 Nm).

Ignition Timing

ADJUSTMENT

Distributorless Ignition System

Base timing for distributorless ignition engines is set at the factory at 10 degrees BTDC and is not adjustable.

Engine Assembly

REMOVAL & INSTALLATION

F-150 4.2L Engine

✲✲ CAUTION

Fuel injection systems remain under pressure, even after the engine has been turned OFF. The fuel system pressure must be relieved before disconnecting any fuel lines. Failure to do so may result in fire and/or personal injury.

1. Before servicing the vehicle, refer to the precautions in the beginning of this section.
2. Remove or disconnect the following:
   - Both battery cables, negative cable first
   - Hood
   - Coolant
   - Refrigerant, using approved equipment
3. Relieve the fuel system pressure as follows:
   a. Remove the fuel tank fill cap to relieve the pressure in the fuel tank.
   b. Remove the cap from the fuel pressure relief valve located on the fuel injection supply manifold.
   c. Attach a fuel pressure gauge to the relief valve and drain the fuel through the drain tube into a suitable container.
   d. After the fuel system pressure is relieved, remove the fuel pressure gauge and install the cap on the relief valve. Secure the fuel tank fill cap.
4. Remove or disconnect the following:
   - Engine cooling fan, shroud and radiator
   - Engine air cleaner outlet tube
   - Accelerator and cruise control cables at the throttle body
   - VMV hose
   - Manifold vacuum connection
   - Intake Manifold Runner Control (IMRC) vacuum connectors, fuel pressure regulator vacuum connector, IMRC solenoid vacuum connector and vacuum reservoir connector
   - Exhaust Gas Recirculation (EGR) valve vacuum connector.
   - The 3 power steering reservoir retaining bolts and position aside
   - Air conditioning compressor manifold bolt and disconnect, then position the air conditioning lines aside.
   - The 4 power steering pump retaining bolts and position the pump aside.
   - Alternator electrical harness connectors. Remove the positive battery cable nut and disconnect the battery cable.
   - Electrical harness connectors to the fuel injectors
   - Wires at the spark plugs
   - Both heater hoses
   - Brake booster vacuum hose
   - EGR Differential Pressure Feedback (DPFE) transducer hose
   - Breather tube from the cylinder head cover
   - Upper intake manifold
   - Fuel supply and return lines, and remove the fuel injection supply manifold.
   - Block heater cable
   - Exhaust system from the exhaust manifolds and support with wire hung from the crossmember.

- Starter motor
- Transmission from the vehicle. If equipped with a manual transmission, remove the clutch assembly.
- Right-hand and left-hand engine support insulator through-bolts

5. Install a suitable engine lifting bracket and connect suitable engine lifting equipment to the lifting brackets.

6. Carefully raise the engine out of the engine compartment and position on a work stand. Remove the engine lifting equipment.

To install:

7. Install the engine lifting brackets. Support the engine using a suitable floor crane installed to the lifting equipment and remove the engine from the work stand.

8. Carefully lower the engine into the engine compartment aligning the engine support insulators.

9. Remove the engine lifting equipment and brackets.

10. Raise and safely support the vehicle.

11. Install or connect the following:
- Left-hand and right-hand engine support insulator through-bolts and tighten them to 51–67 ft. lbs. (68–92 Nm).
- Transmission, and the clutch assembly
- Starter motor
- Exhaust pipes to the exhaust manifolds and tighten to 30 ft. lbs. (41 Nm).
- Block heater
- Heater cable
- Alternator electrical harness connectors and install the positive battery cable to the retaining stud. Tighten the retaining nut to 96 inch lbs. (11 Nm).
- Fuel injectors and the fuel injection supply manifold
- New upper intake gasket and install the upper intake manifold.
- Electrical harness connectors to the fuel injectors
- Power steering pump in position and install 4 retaining bolts. Tighten the bolts to 17–20 ft. lbs. (22–28 Nm).

➡**Ensure that the air conditioning manifold O-rings are in place.**

- Air conditioning manifold to the compressor and install the retaining bolt. Tighten the bolt to 14–18 ft. lbs. (18–24 Nm).
- Power steering reservoir and 3 retaining bolts. Tighten the bolts to 107 inch lbs. (12 Nm).
- VMV hose
- IMRC vacuum connectors, fuel pressure regulator vacuum connector, IMRC solenoid vacuum connector and vacuum reservoir connector.
- EGR valve vacuum connector
- Brake booster hose
- The 2 EGR DPFE transducer hoses
- Manifold vacuum connection
- Accelerator cable and speed control cable in position, if equipped. Tighten the speed control cable retaining bolt to 72 inch lbs. (8 Nm) and accelerator cable retaining bolt to 25 inch lbs. (3 Nm).
- Radiator, cooling fan and shroud
- Engine air cleaner outlet tube
- Engine oil
- Coolant; bleed the system
- Battery cables, negative cable last

12. Start the engine and allow to reach normal operating temperature while checking for leaks.

13. Check all fluid levels.

14. Properly evacuate and recharge the air conditioning system using approved equipment.

15. Install the hood, aligning the marks that were made during removal.

16. Road test the vehicle and check the engine and transmission for proper operation.

2001–03 F-Series 4.6L, 5.4L, 6.8L Engines and 2004 Heritage Edition

✻✻ CAUTION

Fuel injection systems remain under pressure, even after the engine has been turned OFF. The fuel system pressure must be relieved before disconnecting any fuel lines. Failure to do so may result in fire and/or personal injury.

1. Before servicing the vehicle, refer to the precautions in the beginning of this section.

2. Remove or disconnect the following:
- Both battery cables, negative cable first
- Hood
- Coolant
- Refrigerant, using approved equipment

3. Relieve the fuel system pressure as follows:

a. Remove the fuel tank fill cap to relieve the pressure in the fuel tank.

b. Remove the cap from the fuel pressure relief valve located on the fuel injection supply manifold.

c. Attach a fuel pressure gauge to the relief valve and drain the fuel through the drain tube into a suitable container.

d. After the fuel system pressure is relieved, remove the fuel pressure gauge and install the cap on the relief valve. Secure the fuel tank fill cap.

- Engine cooling fan, shroud and radiator
- Accessory drive belt
- Engine air cleaner outlet tube
- Intake manifold assembly
- Bulkhead connector cover and disconnect the bulkhead connector.
- The 3 power steering reservoir bracket retaining bolts and move the reservoir aside.
- The 2 Differential Pressure Feedback (DPFE) transducer hoses
- Upper and lower EGR valve to exhaust manifold tube fittings and remove the tube.
- Heater water hose
- Ignition coil, radio capacitor and CMP sensor electrical harness connectors
- Both ignition coils and mounting bracket bolts and remove the coil and bracket assemblies.
- Starter motor
- The 3 lower radiator air deflector screws. Remove the 5 clips and remove the air deflector.
- Air conditioning compressor electrical harness connector
- Air conditioning manifold-to-compressor bolt and remove the manifold and tube assembly.
- The 3 air conditioning compressor retaining bolts and remove the air conditioning compressor.
- Fluid cooler hoses from the block mounted clip
- On vehicles with automatic transmissions: the inspection cover, torque converter bolts and transmission-to-engine retaining bolts.
- On vehicles with manual transmission: the transmission and the clutch assembly.
- Upper and lower power steering pump bolts and move the power steering pump aside.
- Exhaust system from the exhaust manifolds and support with wire hung from the crossmember.
- Right-hand and left-hand engine support insulator (mount) through-bolts

4. Install a suitable engine lifting bracket and connect suitable engine lifting equipment to the lifting brackets.

5. Carefully raise the engine out of the engine compartment and place on a work stand. Remove the engine lifting equipment.

To install:

6. Install the engine lifting brackets. Support the engine using a suitable floor crane installed to the lifting equipment and remove the engine from the work stand.
7. Carefully lower the engine into the engine compartment. Start the converter pilot into the flywheel and align the paint marks on the flywheel and torque converter. Be sure the studs on the torque converter align with the holes in the flywheel.
8. Fully engage the engine to the transmission and lower onto the engine support insulators.
9. Remove the engine lifting equipment and brackets.
10. Raise and safely support the vehicle.
11. Install or connect the following:
   - If equipped with a manual transmission: the clutch and transmission assemblies.
   - The 6 engine-to-transmission retaining bolts and tighten to 30–44 ft. lbs. (40–60 Nm).
   - The engine support insulator through-bolts and tighten to 15–22 ft. lbs. (20–30 Nm).
   - The 4 torque converter retaining nuts and tighten to 22–25 ft. lbs. (20–30 Nm).
   - Transmission housing cover to the cylinder block
   - Exhaust pipes to the exhaust manifolds and tighten to 30 ft. lbs. (41 Nm)
   - Power steering pump in position on the cylinder block and install 4 retaining nuts. Tighten to 15–20 ft. lbs. (20–30 Nm).
   - Starter motor
   - Transmission fluid cooler hoses into the cylinder block mounted clip
   - Air conditioning compressor in position and install the 3 retaining bolts. Tighten the bolts to 15–22 ft. lbs. (20–30 Nm).
   - Air conditioning manifold and tube assembly on the compressor and install the retaining bolt. Tighten the bolt to 14–18 ft. lbs. (18–24 Nm).
   - Air conditioning compressor clutch electrical harness connector
   - Lower radiator air deflector
   - Ignition coil and bracket assemblies. Tighten the retaining nuts to 15–23 ft. lbs. (20–30 Nm).
   - Ignition coil, radio capacitor and CMP sensor electrical harness connectors
   - Rear heater water hose and compress and slide the clamp in position.
   - EGR valve to exhaust manifold tube and tighten the upper and lower fittings to 26–33 ft. lbs. (35–45 Nm).
   - DPFE transducer hoses.
   - Power steering pump reservoir in position and install the 3 retaining bolts. Tighten the bolts to 71–107 inch lbs. (8–12 Nm).
   - Engine bulkhead connector and install the retaining bolt. Tighten the bolt to 36–50 inch lbs. (4–6 Nm). Install the cover.
   - Intake manifold assembly
   - Accessory drive belt
   - Radiator, cooling fan and shroud
   - Engine air cleaner outlet tube
   - Engine oil
12. Fill and bleed the engine cooling system.
13. Connect both battery cables, negative cable last.
14. Start the engine and allow to reach normal operating temperature while checking for leaks.
15. Check all fluid levels.
16. Properly evacuate and recharge the air conditioning system using approved equipment.
17. Install the hood, aligning the marks that were made during removal.
18. Road test the vehicle and check the engine and transmission for proper operation.

2004 F-150 4.6L Engine, Exc. Heritage Edition

1. Before servicing the vehicle, refer to the precautions in the beginning of this section.
2. With the vehicle in **Neutral**, position it on a hoist.
3. Remove the hood.
4. Remove the cowl.
5. Discharge and recover the A/C system.
6. Remove the cooling module.
7. Remove the intake manifold.
8. Remove the powertrain control module (PCM) and the support bracket.
9. Remove the starter.
10. Disconnect the electrical connector and remove the bolt and the ground strap.
11. Disconnect the electrical connectors.
12. Remove the nut and disconnect the A/C manifold and tube assembly and position aside.
13. Disconnect the heater hose.
14. If equipped, disconnect the block heater electrical connector.
15. Remove the nut and the A/C manifold and tube assembly support bracket.
16. Disconnect the degas bottle coolant hose.
17. Remove the power steering fluid reservoir and support bracket.
18. Remove the bolts.
19. Position the power steering fluid reservoir and bracket aside.
20. Remove the bolts and position the power steering pump aside.

Vehicles with early build 4.6L engines, 2WD

21. Disconnect the power steering pressure switch electrical connector.

All vehicles

22. Remove the bolts and position the A/C compressor aside.
23. Remove the nut and the transmission cooler tube support bracket.
24. Remove the bolt and the starter electrical harness support bracket.
25. Remove the bolts and the flexplate inspection cover.
26. Remove the cylinder block opening cover.
27. Remove the torque converter-to-flexplate nuts.
28. Discard the nuts.

➡**The upper two transmission-to-engine bolts will be removed later.**

29. Remove the lower five transmission-to-engine bolts.
30. Remove the nut and the transmission fluid filler tube.
31. Disconnect the heated exhaust gas oxygen sensor (HO2S) electrical connectors.
32. Remove the four exhaust manifold flange nuts.
33. Remove the RIGHT motor mount nut.
34. Remove the LEFT motor mount bolt.
35. Install the special tool.
36. Using a suitable floor crane, remove the engine assembly from the vehicle.

To install:

37. Using a suitable floor crane, position the engine assembly into the vehicle.
38. Position and install and the left engine mount bolt. Torque to 148 ft. lbs. (200 Nm).
39. Install the right motor mount nut. Torque to 148 ft. lbs. (200 Nm).

➡**Align the engine-to-transmission dowels before installing the engine-to-transmission bolts.**

40. Install the lower five transmission-to-engine bolts. Torque to 35 ft. lbs. (48 Nm).

41. Install the four exhaust manifold-to-catalytic converter nuts. Torque to 30 ft. lbs. (40 Nm).

42. Connect the heated exhaust gas oxygen sensor (HO_2S) electrical connectors.

➡Lubricate the O-ring seals with clean transmission fluid.

43. Install the transmission fluid filler tube and nut.

44. Install the torque converter-to-flexplate nuts. Torque to 26 ft. lbs. (35 Nm).

45. Install the cylinder block opening cover.

46. Install the starter wiring harness support bracket and bolt.

47. Install the flexplate inspection cover. Torque to 25 ft. lbs. (34 Nm).

48. Install the upper two transmission-to-engine bolts. Torque to 35 ft. lbs. (48 Nm).

49. Install the transmission cooler tube support bracket and the nut.

50. Position the power steering pump and install the bolts. Torque to 18 ft. lbs. (25 Nm).

Vehicles with early build 4.6L engines, 2WD

51. Connect the power steering pressure switch electrical connector.

All vehicles

52. Position the power steering reservoir and install the two lower bolts.

53. Install the power steering reservoir support bracket and nut.

54. Connect the degas bottle coolant hose.

55. Position the A/C compressor and install the bolts. Torque to 18 ft. lbs. (25 Nm).

56. Position the A/C manifold and tube assembly support bracket and install the nut.

57. If equipped, connect the block heater electrical connector.

58. Connect the heater coolant hose.

59. Connect the A/C manifold and tube assembly and install the nut. Torque to 18 ft. lbs. (25 Nm).

60. Connect the electrical connectors.

61. Install the ground strap and the bolt and connect the electrical connector.

62. Remove the special tool.

63. Install the intake manifold.

64. Install the cooling module.

65. Install the starter.

66. Install the powertrain control module (PCM) and the support bracket.

67. Install the wiper cowl.

68. Install the hood and the four bolts.

69. Fill the engine with clean engine oil.

70. Evacuate and charge the A/C system.

71. Fill and bleed the engine cooling system.

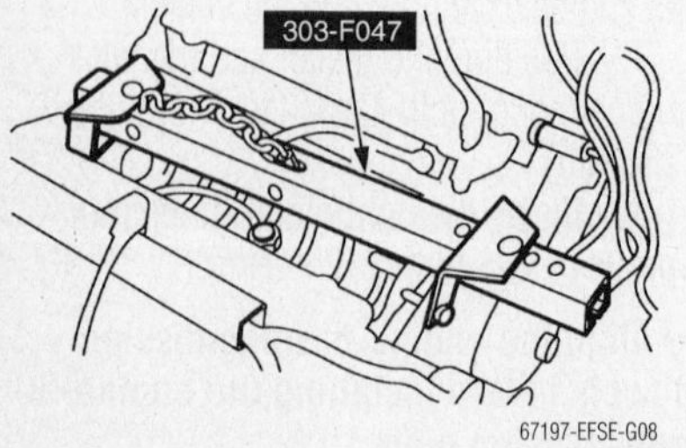

Special tool 303-F047

2004 F-150 5.4L Engine, Exc. Heritage Edition

1. Before servicing the vehicle, refer to the precautions in the beginning of this section.

2. With the vehicle in **Neutral**, position it on a hoist

3. Remove the hood.

4. Remove the cowl.

5. Discharge and recover the A/C system.

6. Remove the radiator.

7. Remove the intake manifold.

8. Remove the powertrain control module (PCM) and the support bracket.

9. Remove the starter.

10. Disconnect the electrical connector and remove the bolt and the ground strap.

11. Disconnect the electrical connectors.

12. Disconnect the heater hose.

13. Remove the nut and disconnect the A/C manifold and tube assembly.

14. Remove the nut and the A/C manifold and tube assembly and support bracket.

15. Remove the nut and the transmission cooler tube support bracket.

16. Remove the bolt and disconnect the A/C manifold tube assembly and the electrical connector. Discard the A/C manifold tube assembly O-ring seals.

17. Remove the bolts and the A/C compressor.

18. If equipped, disconnect the block heater electrical connector.

19. Disconnect the coolant hose.

20. Remove the power steering reservoir support bracket nut.

21. Remove the bolts and position the power steering reservoir assembly aside.

22. Remove the bolts and position the power steering pump aside.

23. Remove the bolt and the starter electrical harness support bracket.

24. Remove the bolts and the flexplate inspection cover.

25. Remove the cylinder block opening cover.

26. Remove the torque converter-to-flexplate nuts. Discard the nuts.

27. Remove the lower five transmission-to-engine bolts. The upper two transmission-to-engine bolts will be removed later.

28. Remove the nut and the transmission fluid filler tube.

29. Disconnect the heated exhaust gas oxygen sensor electrical connectors.

30. Remove the four exhaust manifold flange nuts.

31. Remove the right motor mount nut.

32. Remove the left motor mount bolt.

33. Install a suitable tool.

34. Using a suitable floor crane remove the engine assembly from the vehicle.

To install:

35. Using a suitable floor crane, position the engine assembly into the vehicle.

➡Align the engine-to-transmission dowels before installing the engine mount bolt.

36. Position and install and the left engine mount bolt. Torque to 148 ft. lbs. (200 Nm).

37. Install the right motor mount nut. Torque to 76 ft. lbs. (103 Nm).

38. Install the lower five transmission-to-engine bolts. Torque to 35 ft. lbs. (48 Nm).

39. Install the four exhaust manifold to catalytic converter nuts. Torque to 30 ft. lbs. (48 Nm).

40. Connect the heated exhaust gas oxygen sensor electrical connectors.

➡Lubricate the O-ring seals with transmission fluid.

41. Install the transmission fluid filler tube and nut.

42. Install the torque converter-to-flexplate nuts. Torque to 26 ft. lbs. (35 Nm).

43. Install the cylinder block opening cover.

44. Install the starter electrical harness support bracket and the bolt.

45. Install the flexplate inspection cover. Torque to 25 ft. lbs. (34 Nm).

46. Install the upper two transmission-to-engine bolts. Torque to 35 ft. lbs. (48 Nm).

47. Position the power steering pump and install the bolts. Torque to 18 ft. lbs. (25 Nm).

48. Position the power steering reservoir and install the two lower bolts.

49. Install the power steering reservoir support bracket nut.

50. Position the A/C compressor and install the bolts. Torque to 18 ft. lbs. (25 Nm).

➡Install new A/C manifold tube O-ring seals.

51. Connect the A/C compressor electrical connector and the manifold tube assembly and install the bolt.
52. Install the transmission cooler tube support bracket and the nut.
53. Position the A/C manifold and tube assembly support bracket and install the nut.
54. Connect the A/C manifold and tube assembly and install the nut.
55. If equipped, connect the block heater electrical connector.
56. Connect the heater hose.
57. Connect the coolant hose.
58. Connect the electrical connectors.
59. Install the ground strap and the bolt and connect the electrical connector.
60. Remove the lifting bracket.
61. Install the intake manifold.
62. Install the radiator.
63. Install the starter.
64. Install the powertrain control module (PCM) and the support bracket.
65. Install the cowl.
66. Install the hood. Torque to 22 ft. lbs. (30 Nm).
67. Fill the engine with clean engine oil.
68. Evacuate and charge the A/C system.
69. Fill and bleed the engine cooling system.

2004 F-250/350 5.4L Engine

1. Before servicing the vehicle, refer to the precautions in the beginning of this section.
2. Disconnect the battery ground cable.
3. Discharge and recover the air conditioning system
4. Remove the radiator grille supports.
5. Remove the A/C condenser core.
6. Remove the radiator, fan shroud and engine cooling fan.
7. Disconnect the 42-pin and the 16-pin connectors and remove the nut and the harness mounting bracket.
8. Remove the intake manifold.
9. At the oil cooler water inlet, disconnect and set aside the lower radiator hose.
10. Remove the bolts and position the power steering pump aside.
11. Disconnect the connector and position the harness aside.
12. Disconnect the A/C cycling switch, remove the harness from the studs and position aside.
13. Disconnect the suction hose at the accumulator.
14. Disconnect the heater hose.
15. Disconnect the left and right heated exhaust gas oxygen sensor connectors.
16. Remove the A/C compressor.
17. Raise and support the vehicle.
18. Remove the nuts.

Manual transmission vehicles

19. Remove the transmission and clutch.

All vehicles

20. Drain the engine oil and remove the oil bypass filter.
21. Loosen the threaded shaft and position the oil cooler aside.

Automatic transmission vehicles

22. Remove the starter motor.
23. Remove the nuts and position the transmission oil cooler lines and bracket aside.

All vehicles

24. Remove the stud bolt and position the ground strap aside.
25. Remove the nut and position the wiring bracket aside.

Automatic transmission vehicles

26. Remove the flywheel inspection cover.
27. Remove the access plug.
28. Remove and discard the four torque converter nuts.
29. Remove the four lower transmission to engine bolts.

All vehicles

30. Lower the vehicle.
31. Remove the nut retaining the ground strap, and on automatic transmission equipped vehicles remove the transmission oil filler tube.
32. Remove the two studs and the heater water tube.

Automatic transmission vehicles

33. Remove the three upper transmission to engine bolts.

All vehicles

34. Install the Modular Engine Lift Bracket and support the engine with a floor crane.
35. Remove the engine support insulator nuts.

Automatic transmission vehicles

36. Support the transmission with a jack.

All vehicles

37. Remove the engine from the vehicle.

To install:

All vehicles

38. Position the engine in the vehicle and remove the Modular Engine Lift Bracket and the jack supporting the transmission.
39. Install and tighten the nuts to 66 ft. lbs. (90 Nm).
40. On automatic transmission-equipped vehicles, install the transmission oil level tube, and the ground strap on all vehicles.

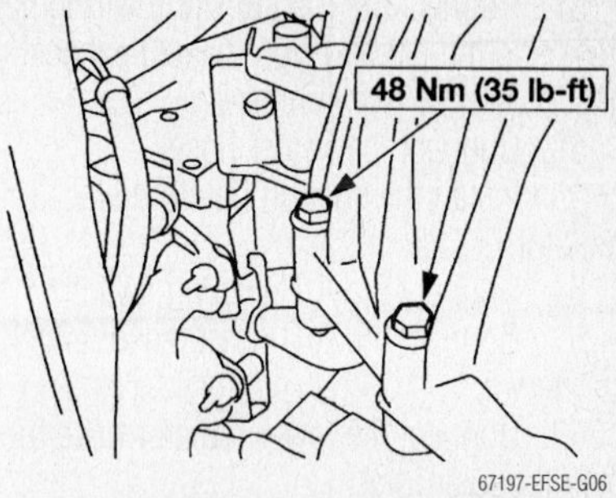

Lower transmission-to-engine bolts—5.4L engine

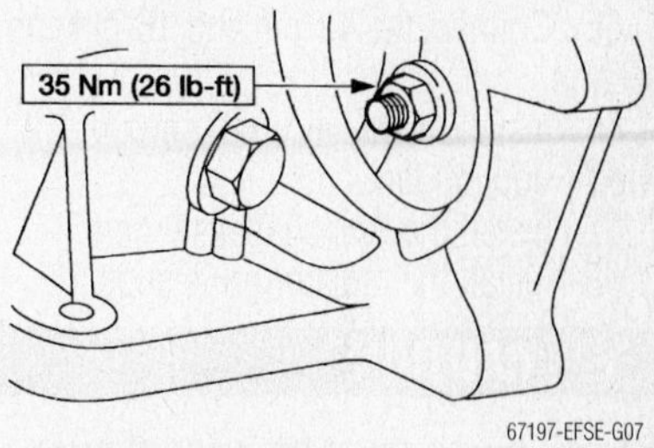

Torque converter nuts

41. Install the heater water tube and the two studs.
42. Raise and support the vehicle.

Automatic transmission vehicles

43. Install the lower transmission to engine bolts. Torque to 35 ft. lbs. (48 Nm).
44. Install and tighten the four new nuts retaining the torque converter. Torque to 26 ft. lbs. (35 Nm).
45. Install the access plug.
46. Install the flywheel inspection cover. Torque to 25 ft. lbs. (34 Nm).

All vehicles

47. Install the wiring bracket.

Automatic transmission vehicles

48. Install the starter motor.

All vehicles

49. Install the oil cooler to the oil filter adapter and install the oil bypass filter. Torque to 43 ft. lbs. (58 Nm).

Manual transmission vehicles

50. Install the clutch and transmission.

All vehicles

51. Install the head pipes and torque the nuts to 30 ft. lbs. (40 Nm).
52. Install the A/C compressor.
53. Install the ground strap.

Automatic transmission vehicles

54. Install the transmission cooler line bracket.
55. Lower the vehicle.
56. Install the remaining transmission-to-engine bolts and fuel line bracket. Torque to 35 ft. lbs. (48 Nm).

57. Connect the transmission wiring harness and the left and right heated exhaust gas oxygen sensor connectors.
58. Connect the heater hose.
59. Install the suction line to the accumulator.
60. Connect the A/C cycling switch.
61. Connect the A/C compressor clutch connector.
62. Connect the lower radiator hose to the oil cooler inlet.
63. Install the power steering pump. Torque the mounting bolts to 18 ft. lbs. (25 Nm).
64. Install the intake manifold.
65. Connect the 42-pin and 16-pin connectors.
66. Install the radiator, fan shroud and engine cooling fan.
67. Install the A/C condenser core.
68. Install the radiator grille support.

CAUTION

The oil pump must be primed prior to starting the engine.

69. Fill all fluids to the correct levels.
70. Connect the battery ground cable.
71. Start the engine and check for leaks. Stop the engine and recheck the fluid levels.
72. Evacuate and recharge the A/C system.

2004 F-250/350 6.8L Engine

1. Before servicing the vehicle, refer to the precautions in the beginning of this section.
2. Disconnect the battery ground cable.
3. Discharge and recover the air conditioning system.
4. Remove the radiator grille support.
5. Remove the A/C condenser core.
6. Remove the radiator, fan shroud and engine cooling fan.
7. Disconnect the 42-pin and the 16-pin connectors and remove the nut and the harness mounting bracket.
8. Remove the intake manifold.
9. At the oil cooler water inlet, disconnect and set aside the lower radiator hose.
10. Remove the bolts and position the power steering pump aside.
11. Disconnect the connector and position the harness aside.
12. Disconnect the A/C cycling switch. Remove the harness from the studs and position aside.
13. Disconnect the suction hose at the accumulator.
14. Disconnect the left and right heated exhaust gas oxygen sensor connectors.
15. Remove the A/C compressor.
16. Remove the nuts.

Manual transmission vehicles

17. Remove the transmission and clutch.

All vehicles

18. Drain the engine oil and remove the oil bypass filter.
19. Loosen the threaded shaft and position the oil cooler aside.

Automatic transmission vehicles

20. Remove the starter motor.
21. Remove the bolts and position the transmission oil cooler lines and bracket aside.

All vehicles

22. Remove the stud bolt and position the ground strap aside.
23. Remove the nut and position the wiring bracket aside.

Automatic transmission vehicles

24. Remove the flywheel inspection cover.
25. Remove the access plug.
26. Remove and discard the four torque converter bolts.
27. Remove the four lower transmission-to-engine bolts.

All vehicles

28. Lower the vehicle.
29. Remove the nut retaining the ground strap, and on automatic transmission-equipped vehicles, remove the transmission oil filler tube.
30. Disconnect the heater hose at the rear of the engine.
31. Remove the two studs and the heater water tube.

Automatic transmission vehicles

32. Remove the three upper transmission-to-engine bolts.

All vehicles

33. Install the Modular Engine Lift Bracket and support the engine with a suitable floor crane.
34. Remove the engine support insulator nuts.

Automatic transmission vehicles

35. Support the transmission with a jack.

All vehicles

36. Remove the engine from the vehicle.
37. Remove the bolts, and remove the flywheel.
38. Using a puller set, remove the rear crankshaft oil slinger.
39. Using a puller set, remove the crankshaft rear oil seal.
40. Remove the bolts and remove the crankshaft rear oil seal retainer plate.
41. Mount the engine on an engine stand.
42. Remove the Modular Engine Lift Bracket.

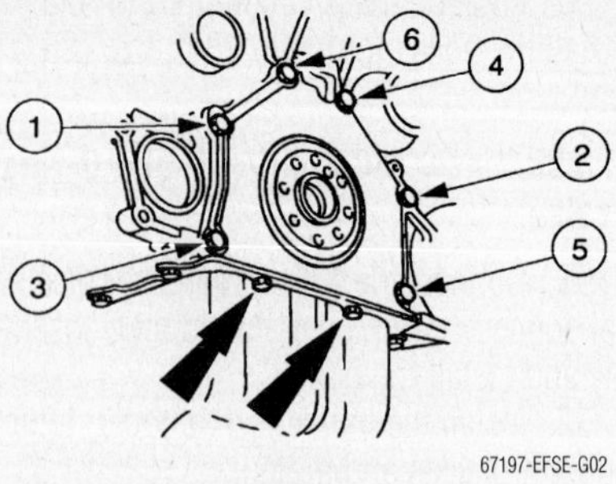

Retainer plate installation—6.8L engine

To install:

43. Attach the Modular Engine Lift Bracket.
44. Remove the engine from the engine stand.

CAUTION

Do not use metal scrapers, wire brushes, power abrasive discs or other abrasive means to clean the aluminum retainer plate. These tools cause scratches and gouges, which make leak paths. Use a plastic scraping tool to remove all traces of old sealant.

45. Clean and inspect the mating surfaces of the cylinder block and the retainer plate.

➡If not secured within four minutes, sealant must be removed and the sealing area cleaned with metal surface cleaner. Allow to dry until there is no sign of wetness, or four minutes, whichever is longer. Failure to follow this procedure can cause future oil leakage.

46. Apply a bead of silicone gasket and sealant around the rear oil seal retainer sealing surface.
47. Install the retainer plate. Tighten the bolts in two steps.
    a. Step 1: Tighten the retainer plate bolts to 10 Nm (89 lb-in), in the sequence shown.
    b. Step 2: Tighten the oil pan bolts to 20 Nm (20 lb-ft), then tighten an additional 90 degrees.
48. Use the two crankshaft rear oil seal installers to install the crankshaft rear oil seal.
49. Using the two crankshaft rear oil seal installers and the crankshaft rear oil slinger installer, install the crankshaft rear oil slinger.
50. Install the flywheel. Torque to 54–64 ft. lbs. (73–87 Nm).
51. Position the engine in the vehicle

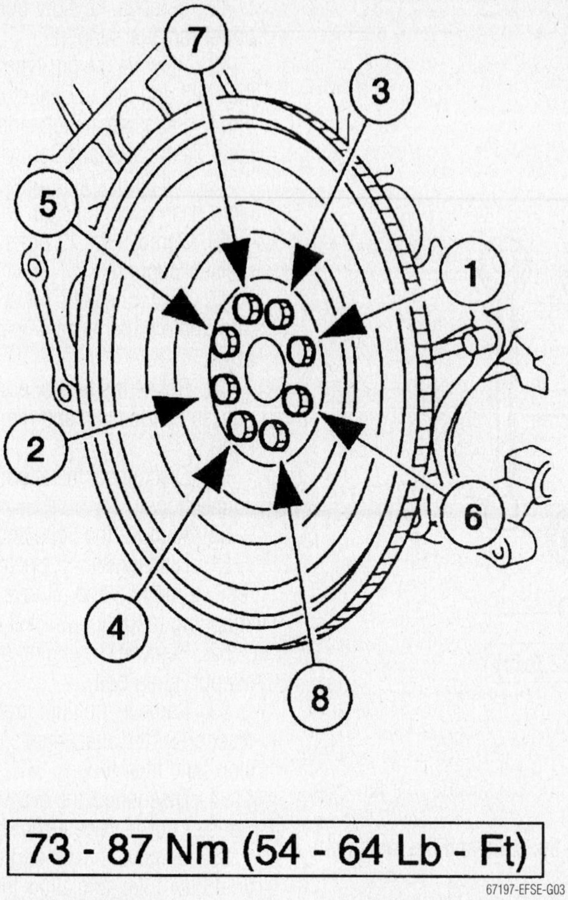

Flywheel installation—6.8L Engine

and remove the Modular Engine Lift Bracket and the jack supporting the transmission.

52. Install and tighten the nuts to 66 ft. lbs. (90 Nm).

53. Install the heater water tube and the two studs. Torque to 20 ft. lbs. (25 Nm).

54. On automatic transmission-equipped vehicles, install the transmission oil level indicator tube and the ground strap on all vehicles.

55. Raise and support the vehicle.

Automatic transmission vehicles

56. Install the lower transmission to engine bolts.

57. Install the access plug.

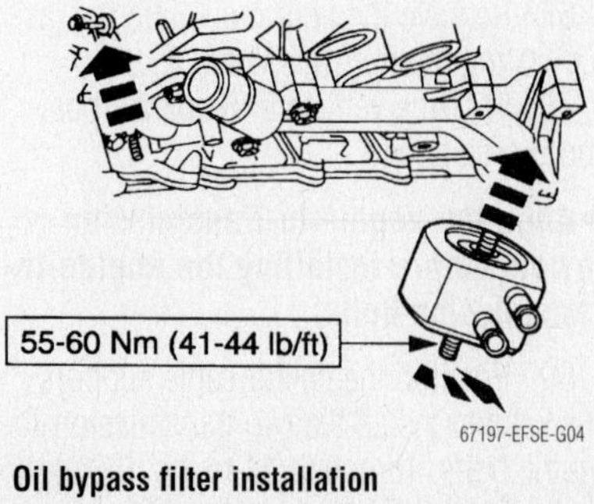

Oil bypass filter installation

58. Install and tighten the four new nuts retaining the torque converter. Torque to 25–34 ft. lbs. (34–46 Nm).

59. Install the flywheel inspection cover.

All vehicles

60. Install the wiring bracket.

Automatic transmission vehicles

61. Install the starter motor.

All vehicles

62. Install the oil cooler to the oil filter adapter and install the oil bypass filter.

Manual transmission vehicles

63. Install the clutch and transmission

All vehicles

64. Install the head pipes and tighten the nuts to 30 ft. lbs. (40 Nm).

65. Install the ground strap.

Automatic transmission vehicles

66. Install the transmission cooler line bracket.

All vehicles

67. Install the A/C compressor.

68. Lower the vehicle.

69. Position back and install the heater hose at the rear of the engine.

Automatic transmission vehicles

70. Install the remaining transmission-to-engine bolts and fuel line bracket.

All vehicles

71. Connect the transmission wiring harness and the left and right heated exhaust gas oxygen sensor connectors.

72. Install the suction line to the accumulator.

73. Connect the A/C cycling switch.

74. Connect the connector and position the harness aside.

75. Install the power steering pump. Torque the mounting bolts to 20 ft. lbs. (25 Nm).

76. Position back and connect the lower radiator hose at the oil cooler water inlet.

77. Install the intake manifold.

78. Install the harness mounting bracket and connect the 42-pin and 16-pin connectors.

79. Install the radiator.

80. Install the A/C condenser core.

81. Install the radiator grille support.

82. Fill all fluids to the correct levels.

83. Connect the battery ground cable.

84. Start the engine and check for leaks. Stop the engine and recheck the fluid levels.

85. Evacuate and recharge the A/C system.

E-Series 4.6L and 5.4L engines

1. Before servicing the vehicle, refer to the precautions in the beginning of this section.
2. Recover the air conditioning system.
3. With the vehicle in **Neutral**, position it on a hoist.
4. Remove the intake manifold.
5. Remove the front bumper.
6. Disconnect the lower radiator hose from the radiator.
7. Remove the pushpin retainers and the splash shield.
8. Disconnect the transmission cooler hoses and drain the fluid into a suitable container.
9. Disconnect the exhaust system from the left and right exhaust manifolds.
10. Drain the engine oil.
11. Disconnect the oil cooler hoses and set aside.
12. Remove the oil filter.
13. Loosen the threaded insert and remove the oil cooler from the oil filter adapter.
14. Remove the starter motor solenoid terminal cover.
15. Disconnect the starter motor electrical connections.
16. Remove the starter motor.
17. Remove the flexplate inspection cover.
18. Remove the cylinder block opening cover.

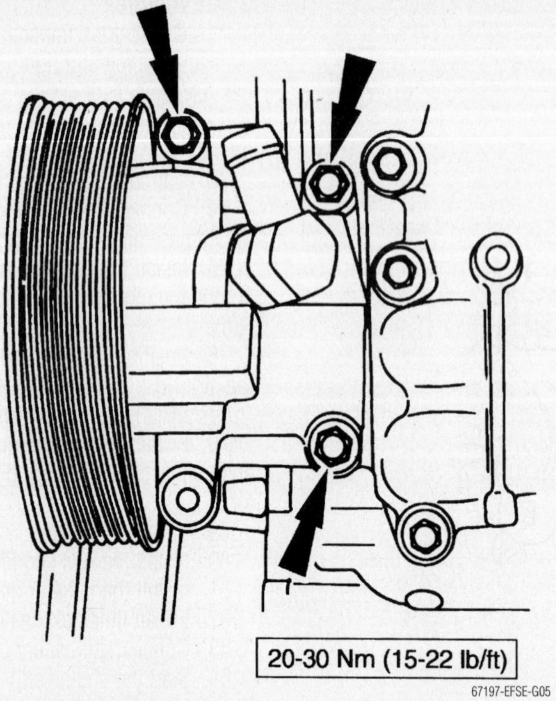

Power steering pump installation—6.8L engine

19. Remove the four nuts retaining the torque converter to the flexplate. Rotate the crankshaft to access all of the torque converter nuts. Discard the torque converter nuts.

Vehicles equipped with 4R70E transmission

20. Disconnect the shift cables and the mounting bracket.

All vehicles

21. Remove the two transmission-to-engine bolts and position the shifter cable support bracket, mounting bracket and cable aside.
22. Remove the right lower transmission-to-engine bolt.
23. Remove the four engine mount nuts.
24. Disconnect the power steering reservoir hose and the power steering pressure tube at the power steering pump.
25. Remove the nut and position the transmission fluid cooler support bracket aside.
26. Remove the valance panel.
27. Disconnect the upper radiator hose and the degas hose from the radiator.

⁂ CAUTION

The large clutch assembly nut has a right-hand thread and must be rotated counterclockwise to remove it.

28. Remove the fan blade assembly and fan clutch.
29. Disconnect the degas hose retainer from the fan shroud.
30. Remove the fan shroud, the fan, and the fan clutch.
31. Remove the bolts and the radiator support brackets.
32. Remove the radiator from the vehicle.
33. Disconnect the A/C pressure cutoff switch electrical connector.

⁂ CAUTION

Use a wrench on each side of the fitting to prevent damage to the A/C fitting.

34. Disconnect the compressor suction tube.

⁂ CAUTION

Use a wrench on each side of the fitting to prevent damage to the A/C fitting.

35. Disconnect the compressor discharge tube.
36. Disconnect the condenser-to-evaporator tube.
37. Remove the A/C condenser core.
38. Set the hood latch aside.
39. Remove the bolts and set the power steering reservoir aside.
40. Remove the nuts and the battery feed cable at the power distribution box.
41. Set aside the battery feed wiring harness to the power distribution box. Disengage the routing clips.
42. Remove the eight bolts and the upper radiator support.
43. Remove the pin-type retainers and the right and left air deflectors.
44. Set aside the left side of the headlamp dash panel junction wiring harness.
45. Set aside the right side of the headlamp dash panel junction wiring harness.
46. Remove the 12 bolts and the lower radiator support.
47. Disconnect the lower radiator hose from the oil filter adapter water inlet and set aside.
48. Rotate the tensioner clockwise to release belt tension and remove the drive belt.
49. Disconnect the alternator electrical connections.
50. Remove the bolts and the alternator.
51. Disconnect the electrical connectors, remove the nut, and remove the ground strap and mounting bracket.
52. Remove the engine oil filler tube support strap bolt.
53. Remove the fluid level indicator and disconnect the attachments at the transmission fluid filler tube.
54. Disconnect the evaporative emission canister purge valve electrical connector.
55. Remove the nut and the transmission fluid filler tube, and allow the fluid to drain into a suitable container.
56. Release the wiring retainers from the heater outlet tube bracket.
57. Disconnect the hose from the heater outlet tube.
58. Remove the heater outlet tube studs.
59. Remove the heater outlet tube.
60. Release the transmission wiring harness connector retainer from the bracket.
61. Disconnect the transmission wiring harness and the left and right heated oxygen sensor (HO_2S) electrical connectors.
62. Remove the four upper transmission-to-engine bolts and the fuel line, and the left HO_2S brackets.
63. Install a lifting tool.
64. Support the transmission with a jack.
65. Remove the engine from the vehicle.

To install:

66. Position the engine in the vehicle.
67. Remove the floor crane and the jack supporting the transmission.
68. With the vehicle in **Neutral**, position it on a hoist.

➡Align the engine-to-transmission dowels before installing the engine-to-transmission bolts.

69. Position the shifter cable support bracket and install the two transmission-to-engine bolts. Torque to 44 ft. lbs. (60 Nm).

70. Install the right lower transmission-to-engine bolt. Torque to 44 ft. lbs. (60 Nm).

Vehicles equipped with 4R70E transmission

71. Install the mounting bracket and connect the shift cable. Torque to 18 ft. lbs. (25 Nm).

All engines

72. Install four new nuts to retain the torque converter to the flexplate. Rotate the crankshaft to access all of the torque converter nuts. Torque to 26 ft. lbs. (35 Nm).

73. Install the cylinder block opening cover.

74. Install the flexplate inspection cover. Torque to 26 ft. lbs. (35 Nm).

75. Install the four engine mount nuts. Torque to 66 ft. lbs. (90 Nm).

76. Install the starter motor.

77. Connect the starter motor electrical connections.

78. Install the starter motor solenoid terminal cover.

79. Mount the exhaust system to the left and right exhaust manifolds. Torque to 30 ft. lbs. (40 Nm).

➡Make sure the tab on the oil filter adapter nests into the notch in the oil cooler.

80. Position the oil cooler on the oil filter adapter and install the threaded insert.

81. Install a new oil filter.

82. Connect the oil cooler hoses.

Romeo engine (4.6L)

83. Position the transmission fluid cooler tube support bracket and install the nut.

Windsor engine (5.4L)

84. Position the fluid cooler tube support bracket and install the nut.

All engines

85. Using a suitable tool, install a new Teflon® seal on the power steering pressure tube.

86. Connect the power steering pressure tube and the power steering reservoir hose to the power steering pump.

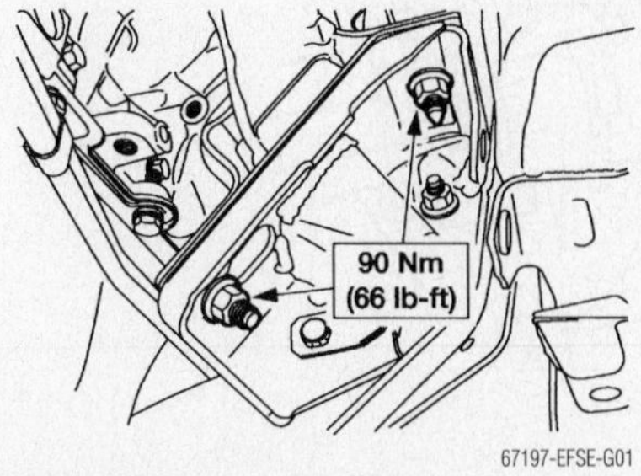

Engine mount nuts— E-Series 4.6L and 5.4L engines

87. Install and tighten the oil drain plug.

88. Position the fuel line and the left heated oxygen sensor (HO_2S) brackets. Install the four upper transmission-to-engine bolts. Torque to 44 ft. lbs. (60 Nm).

89. Connect the transmission wiring harness and the left and right HO_2S electrical connectors. Install the transmission wiring harness connector retainer in the bracket.

➡Do not reuse the O-ring seals. Lubricate the new O-ring seals with clean engine coolant before installing the heater outlet tube.

90. Insert the heater outlet tube over the new seals.

Romeo engines (4.6L)

91. Install the heater outlet tube studs. Torque to 18 ft. lbs. (25 Nm).

➡Lubricate the O-ring seals with clean transmission fluid before installing the transmission fluid filler tube.

92. Install the transmission fluid filler tube.

Windsor engine (5.4L)

93. Install the heater outlet tube studs.

➡Lubricate the O-ring seals with clean transmission fluid before installing the transmission fluid filler tube.

94. Install the transmission fluid filler tube.

All engines

95. Connect the hose to the heater outlet tube.

96. Install the wiring retainers in the heater outlet tube bracket.

97. Connect the evaporative emission canister purge valve electrical connector.

98. Connect the attachments to the transmission fluid filler tube and install the fluid level indicator.

99. Install the engine oil filler tube support strap bolt.

100. Position the mounting bracket, install the ground strap and nut, and connect the electrical connectors.

101. Position the alternator and install the bolts.

102. Connect the alternator electrical connections.

103. Rotate the tensioner clockwise and install the drive belt. Refer to the decal on the upper radiator air deflector for belt routing.

104. Connect the lower radiator hose to the oil filter adapter coolant inlet.

105. Position the lower radiator support and install the twelve bolts.

106. Install the right side of the headlamp dash panel junction wiring harness.

107. Install the left side of the headlamp dash panel junction wiring harness.

108. Position the right and left air deflectors and install the pin-type retainers.

109. Position the upper radiator support and install the eight bolts.

110. Route the battery feed wiring harness to the power distribution box and insert the routing clips.

111. Install the battery feed cable to the power distribution box.

112. Install the power steering reservoir.

113. Install the hood latch.

114. Install the A/C condenser core.

**** CAUTION**

Use a wrench on each side of the fitting to prevent damage to the A/C fitting.

115. Connect the condenser core refrigerant tubes. Torque the condenser-to-evaporator fitting to 13 ft. lbs. (18 Nm); the compressor discharge tube to 29 ft. lbs. (39 Nm).

**** CAUTION**

Use a wrench on each side of the fitting to prevent damage to the A/C fitting.

116. Connect the compressor suction tube. Torque to 35 ft. lbs. (47 Nm).

117. Connect the A/C pressure cutoff switch electrical connector.

118. Position the radiator in the vehicle.

119. Position the radiator support brackets and install the bolts.

120. Position the fan shroud and the fan and fan clutch in the vehicle, and install the bolts.

121. Insert the degas hose retainer in the fan shroud.

122. Install the fan blade assembly and fan clutch.

123. Connect the upper radiator hose and the degas hose to the radiator.

124. Install the valance panel.

125. Connect the transmission cooler tubes.

126. Position the splash shield and install the pushpin retainers.

127. Connect the lower radiator hose to the radiator.

128. Install the front bumper.

**** CAUTION**

The oil pump must be primed prior to starting the engine.

129. Fill the engine with clean engine oil.

130. Install the intake manifold.

131. Fill all fluids to the correct levels.
132. Start the engine and check for leaks. Stop the engine and recheck the fluid levels.
133. Evacuate and recharge the air conditioning system.

E-Series 6.8L Engine

1. Before servicing the vehicle, refer to the precautions in the beginning of this section.
2. Discharge and recover the air conditioning system.
3. Disconnect the 42-pin and the two 16-pin connectors and remove the nut and the harness mounting bracket.
4. Remove the intake manifold.
5. Remove the radiator right and left air deflector, the radiator grille support, the valance panel and the upper and lower radiator supports.
6. Remove the radiator, fan shroud and engine cooling fan.
7. Remove the headlamps and side marker lamps.
8. Remove the A/C condenser core.
9. Remove the drive belt.
10. Disconnect the power steering reservoir hose at the power steering pump and let drain into a drain pan.
11. Disconnect the power steering pressure hose.
12. At the oil cooler water inlet, disconnect and set aside the lower radiator hose.
13. Disconnect the suction hose at the accumulator.
14. Disconnect the transmission harness and the left and right heated exhaust gas oxygen sensor connectors.
15. Remove the upper transmission-to-engine bolts and the fuel tube bracket.
16. Raise and support the vehicle.
17. Remove the nuts.
18. Drain the engine oil and remove the oil filter.
19. Disconnect the oil cooler hoses and position aside.
20. Loosen the threaded shaft and remove the oil cooler from the oil filter adapter.
21. Remove the starter motor.
22. Remove the bolt from the A/C compressor manifold and position the manifold aside.
23. Remove the bolts and position the shift cable and bracket aside.
24. Remove and discard the six nuts retaining the torque converter to the flexplate.
25. Remove the engine support insulator nuts.
26. Remove the remaining transmission to engine bolts.
27. Lower the vehicle.
28. Install the Modular Engine Lift Bracket.
29. Support the transmission with a jack.
30. Remove the engine from the vehicle.

To install:

31. Position the engine in the vehicle and remove the Modular Engine Lift Bracket and the jack supporting the transmission.
32. Raise and support the vehicle.
33. Install and tighten the nuts.
34. Install the fuel tube line bracket and the remaining transmission to engine bolts.
35. Install and tighten the six new nuts retaining the torque converter.
36. Install the starter motor.
37. Position the A/C compressor manifold assembly and install the bolt.
38. Install and tighten the nuts.
39. Connect the power steering pressure hose.
40. Install the oil cooler to the oil filter adapter and install the oil filter.
41. Position the oil cooler hoses and install the hose clamps.
42. Connect the lower radiator hose.
43. Lower the vehicle.
44. Connect the transmission wiring harness and the left and right heated exhaust gas oxygen sensor connectors.
45. Install the intake manifold.
46. Install the suction line to the accumulator.
47. Install the power steering reservoir hose at the power steering pump.
48. Connect the 16 and 42-pin connectors.
49. Install the drive belt.
50. Install the upper and lower radiator supports, valance panel, radiator grille support and the radiator right and left air deflectors.
51. Install the A/C condenser core.
52. Install the radiator.
53. Install the side marker lamps and headlamps.
54. Install the engine air cleaner assembly and tubes.
55. Fill all fluids to the correct levels.
56. Connect the battery ground cable.
57. Start the engine and check for leaks. Stop the engine and recheck the fluid levels.
58. Evacuate and recharge the A/C system.

Water Pump

REMOVAL & INSTALLATION

All Except Supercharger Cooling

1. Before servicing the vehicle, refer to the precautions in the beginning of this section.
2. Remove or disconnect the following:
   - Negative battery cable
   - Radiator, fan blade assembly and fan shroud
   - Accessory drive belt
   - Water pump pulley
   - Heater hose from the water pump
   - Water pump bolts and nuts. Note the locations of the bolts if different lengths.
   - Water pump stud bolt, the water pump and the water pump housing gasket. Discard the water pump housing gasket.

To install:

3. Before installing the water pump, be sure to completely clean the water pump mounting surfaces of all dirt, grime and old gasket material.
4. Install or connect the following:
   - Water pump onto the engine with a new gasket. Install the water pump stud bolt temporarily finger-tight.

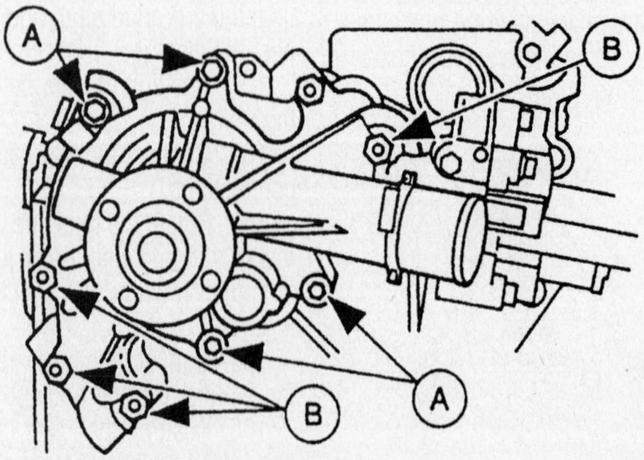

When removing the water pump, note the locations of the mounting bolts (A) and nuts (B)—4.2L engine

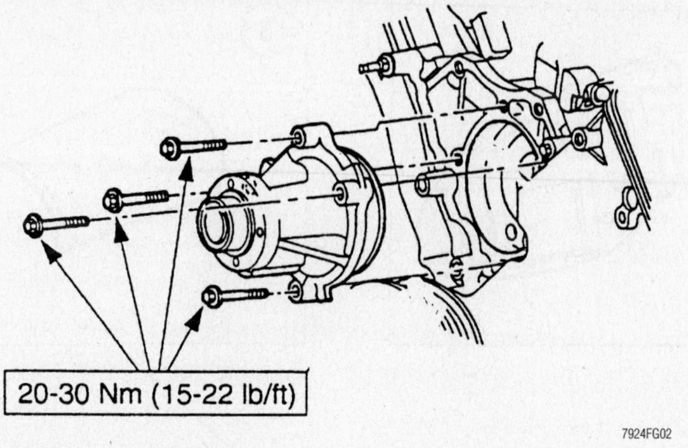

Exploded view of the water pump mounting—4.6L, 5.4L and 6.8L engines, exc. 5.4L Supercharged engine

- Water pump mounting nuts and bolts temporarily finger-tight, then tighten all water pump housing fasteners to 18 ft. lbs. (25 Nm).
- Water outlet tube for the heater, if equipped
- Water pump pulley and accessory drive belt
- Fan shroud, fan blade assembly and the radiator
- Coolant
- Negative battery cable

5. Start the engine and check for any fluid leaks.
6. If necessary, bleed the cooling system.

5.4L Supercharger Cooling

1. Before servicing the vehicle, refer to the precautions in the beginning of this section.
2. Disconnect the battery ground cable.
3. Drain the coolant.
4. Remove the retainer and position the inner fender well aside.
5. Disconnect the water pump electrical connector.
6. Disconnect the coolant hoses.

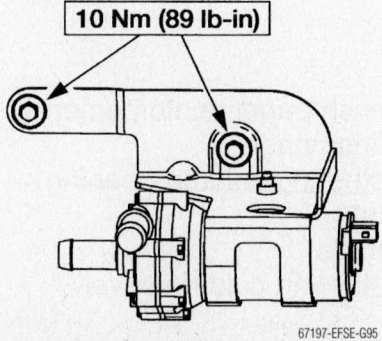

Supercharger cooling pump installation

7. Remove the bolts and the water pump.
8. To install, reverse the removal procedure.

Heater Core

REMOVAL & INSTALLATION

2001–02 E-Series Vans

1. Before servicing the vehicle, refer to the precautions in the beginning of this section.
2. Drain the cooling system into a clean container for reuse.
3. Disconnect the battery ground cable.
4. Disconnect the quick disconnect heater hose couplings at the heater core by performing the following procedure:

WARNING

The engine must be off, fully cool and the cooling system fully depressurized before attempting to disconnect any heater water hoses. Failure to comply with this warning can result in serious injury or burns from hot liquid escaping out of the engine cooling system.

a. Depressurize the engine cooling system.
b. Push the heater water hose toward the tube to fully expose the locking tabs.

➡When compressing the white coupling retainer, the quick disconnect tool must be perpendicular to and on the highest point of the coupling.

c. Push the quick disconnect tool over the coupling retainer windows to compress the retainer locking tabs.
d. A slight twisting motion while pulling on the heater water hose may be necessary to assist in the removal.
e. Pull the heater water hose away from the heater core tube.
f. Plug the heater water hose.
g. Remove the white coupling retainer from the tube.
h. Spread the retainer tabs apart and slide the retainer off the tube.
i. Discard the retainer.

5. Plug the heater water hoses with a suitable ⅝ or ¾ inch plug.
6. Remove the engine cover.
7. Remove the instrument panel finish panel by performing the following procedure:

CAUTION

Electronic modules are sensitive to static electrical charges. If exposed to these charges, damage may result.

a. Remove the driver's side air bag module by removing or disconnecting the following items:

➡To deplete the backup power supply energy, disconnect the battery ground cable and wait at least 1 minute. Be sure to disconnect auxiliary batteries and power supplies (if equipped).

- Make sure the wheels are in the straight-ahead position.
- Driver's air bag module.
- Both back cover plugs.
- Both driver's air bag module screws.
- Horn electrical connector.
- Air bag sliding contact electrical connector.
- Driver's air bag module.

b. Remove the passenger's side air bag modules by removing or disconnecting the following items:

- Instrument panel finish panel and reinforcement.
- Passenger's air bag module retaining nuts.

➡Using a ⅜ inch x 4 inch slotted screwdriver, carefully slide the head of the screwdriver under the right bottom edge of the door and lift upward, separating the door from the clip. Separate the rest of the door from the clips by lifting the door with your hands.

- Passenger's air bag module electrical connector and module.
- The 6 offset fasteners used for

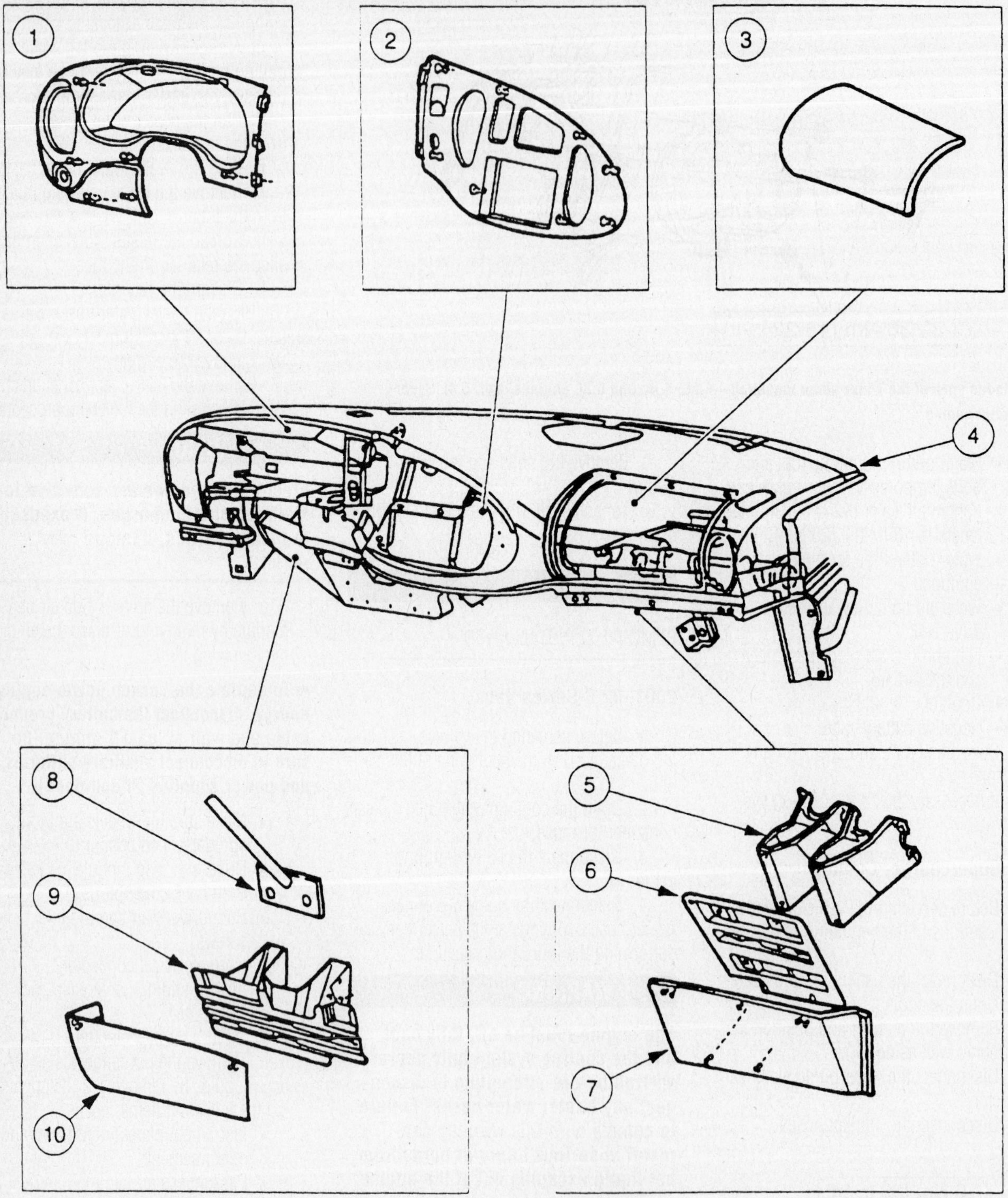

1. Instrument panel cluster finish panel
2. Instrument panel center finsh panel
3. passenger air bag opening cover
4. Instrument panel
5. Instrument panel dash brace
6. Instrument panel finsh panel reinforcement
7. Instrument panel finsh panel
8. Instrument panel steering column opening cover reinforcement
9. Driver side knee brace
10. Instrument panel steering column cover

93113GM4

Exploded view of the instrument panel—Ford Econoline

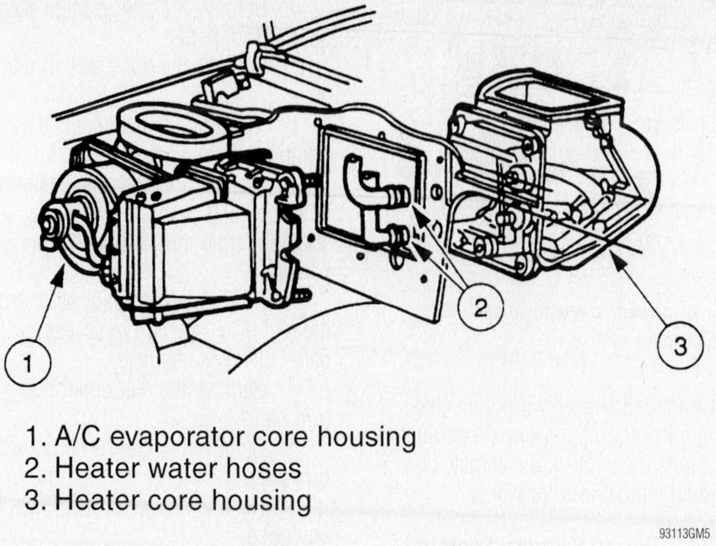

View of the heater housing and evaporator housing—Ford Econoline

retaining the air bag module door from the instrument panel.

c. Remove the steering column

d. Remove the engine cover.

e. Unfasten the engine cover latches and remove the engine cover.

f. Remove the instrument panel cowl top screws.

g. Remove the upper center instrument panel access panel.

h. Disconnect the climate control vacuum harness connector.

i. Remove the left-hand upper access panel.

j. Remove the instrument panel cowl top nut.

k. Remove the alternative fuel control module (AFCM).

l. Remove the instrument panel dash brace.

m. Remove the 6 dash brace bolts.

n. Remove the instrument panel dash brace.

o. From under the hood, position the coolant reservoir aside.

p. Remove the bolts.

q. Position the coolant reservoir aside.

r. Remove the main wiring assembly/lower bulkhead connector by performing the following:

- Loosen the bulkhead connector bolt.
- Disconnect the bulkhead connector clips.
- Remove the bulkhead connector.

s. Remove the right-hand A-pillar lower trim panel.

t. Disconnect the electronic crash sensor (ECS) module electrical connector by performing the following:

- Remove the locking clip.

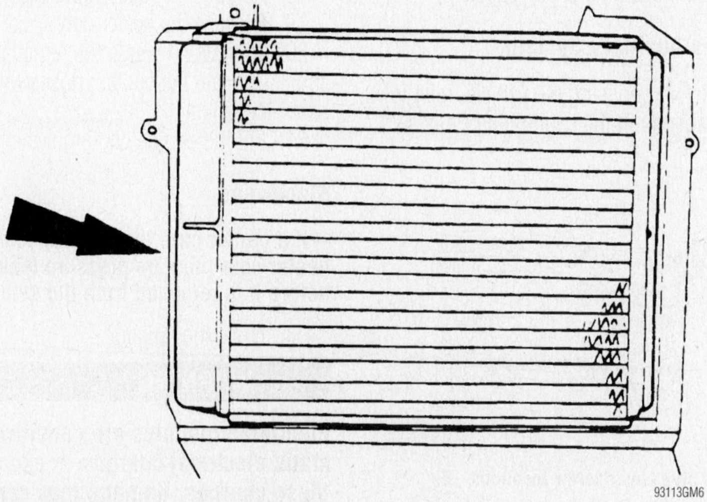

View of the heater core—Ford Econoline

- Disconnect the connector.

u. Disconnect the right-hand instrument panel cowl side wiring electrical connectors.

v. Disconnect the antenna cable from the radio and release the cable locators from the instrument panel.

w. Remove the right-hand instrument panel bolts.

x. Remove the left-hand instrument panel bolts.

➡Removing the instrument panel from the vehicle requires 2 technicians.

y. Remove the instrument panel from the vehicle through the passenger side door.

8. Remove the heater core cover.
9. Remove the 7 screws.
10. Remove the heater core cover.
11. Remove and discard the heater core case seal.

➡Use care not to spill the coolant remaining in the heater core during removal.

12. Remove the heater core.

To install:

13. Install the instrument panel finish panel by performing the following procedure:

⁂ CAUTION

Electronic modules are sensitive to static electrical charges. If exposed to these charges, damage may result.

➡Verify that the 4-way locator on the left-hand side is properly seated.

a. Position the instrument panel on the center dash panel.

b. Install the left-hand instrument panel bolts.

c. Install the right-hand instrument panel bolts.

d. Engage the antenna cable locators into the instrument panel and connect it to the radio.

e. Connect the right-hand instrument panel cowl side wiring electrical connectors.

f. Connect the ECS module electrical connector by performing the following:

- Connect the connector.
- Install the locking clip.

g. Install the right-hand A-pillar lower trim panel.

h. From inside the engine compartment, install the main wiring assembly/lower bulkhead connector by performing the following:

- Position the bulkhead connector.
- Secure the clips.
- Tighten the bolt.

i. Install the coolant reservoir by performing the following:

- Position the reservoir.
- Install the bolts.

j. Install the instrument panel dash brace by performing the following:

- Position the instrument panel dash brace.
- Install the 6 bolts.

k. Install the AFCM.

l. Install the instrument panel cowl top nut.

m. Install the left-hand upper access panel.

n. Connect the climate control vacuum harness connector.

o. Install the upper center instrument panel access panel.

p. Install the instrument panel cowl top screws.

q. Install the engine cover.

r. Position the engine cover in the vehicle and fasten the cover latches.

s. Install the steering column.

t. Install the driver's side air bag module by installing or connecting the following items:

- Driver's air bag module.
- Air bag sliding contact electrical connector.
- Horn electrical connector.
- Both driver's air bag module screws.
- Both back cover plugs.

u. Install the passenger's side air bag modules by installing or connecting the following items:

- The 6 offset fasteners used for retaining the air bag module door from the instrument panel.
- Passenger's air bag module electrical connector and module.
- Passenger's air bag module retaining nuts.
- Instrument panel finish panel and reinforcement.

14. Connect the quick disconnect heater hose couplings at the heater core by performing the following procedure:

a. Clean the tubes and lubricate them with rubber insulator lube.

b. Install a new coupling retainer, spacer, and lubricated O-ring seals into the quick disconnect coupling housing.

c. Push the heater water hose with a quick disconnect coupling onto the tube.

d. Make sure the coupling is fully engaged by lightly pulling on the heater water hose.

15. Refill the cooling system.

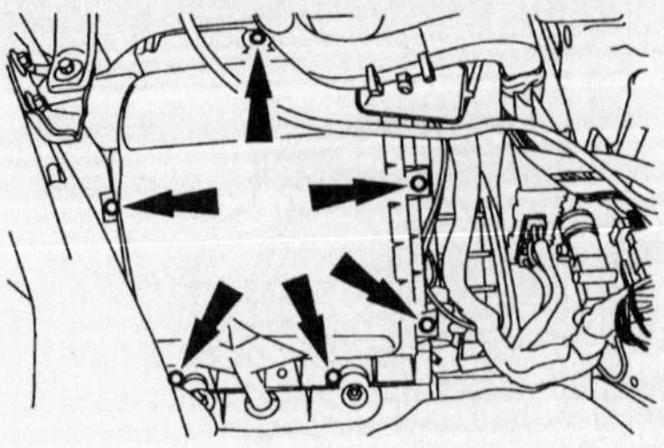
67197-EFSE-G31

Heater core cover screw locations—E-Series primary system

16. Connect the battery ground cable.

17. Run the engine to normal operating temperatures; then, check the climate control operation and check for leaks.

2003–04 E-Series Primary System

1. Before servicing the vehicle, refer to the precautions in the beginning of this section.

2. Clamp off and disconnect the heater hoses at the heater core.

3. Remove the instrument panel finish panel.

4. Remove the bolts and the instrument panel reinforcement.

5. Remove the bolts and the instrument panel reinforcement bracket.

6. Remove the screws and position the wire harness aside.

7. Remove the screws and the heater core cover.

➡Use care not to spill the coolant remaining in the heater core during removal.

8. Remove the heater core.

9. To install, reverse the removal procedure. Clean and lubricate the coolant hoses with plain water only, if needed.

10. Fill the engine cooling system.

E-Series Auxiliary System

1. Before servicing the vehicle, refer to the precautions in the beginning of this section.

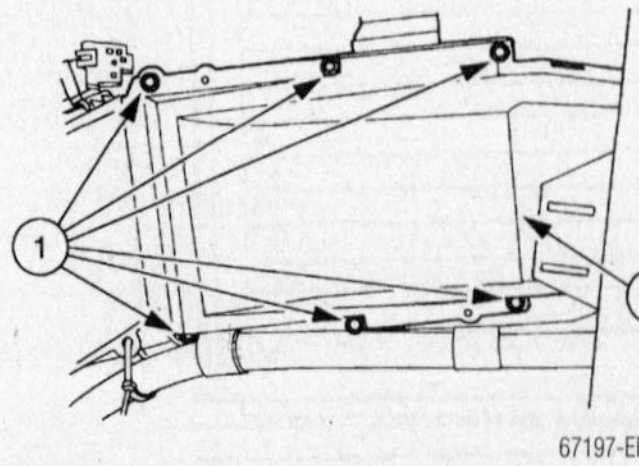

67197-EFSE-G32

Heater core cover screw locations—E-Series auxiliary system

2. Remove the rear seats to access the quarter trim panels.

3. Remove the body side trim finish panel.

4. If equipped, disconnect the power-point electrical connector.

5. Remove the rear seat belt shoulder strap opening covers. Pry the seat belt shoulder strap covers from the upper trim panels.

6. Pull the rear door weatherstrip off at the rear door opening to access the trim panel pin-type retainers.

7. Remove the rear upper body side trim panel.

8. Remove the rear lower body side trim panel.

9. Remove the window latches, if so equipped.

10. Position the driver seat to the full forward position.

11. Remove the front seat belt guide cover.

12. Remove the nut and front safety belt guide.

13. Remove the front upper body side trim panel.

14. Remove the front lower body side trim panel.

15. To install, reverse the removal procedure.

a. To install the front upper body side trim panel, start at the lower pillar, then work to the second lower pillar.

b. To install the rear lower body side trim panel, start with the clip to the heater housing, then snap the trim panel into the panel clips.

c. To install the rear upper body side trim panel, start with the rear lower portion of the panel at the pillar.

16. Remove the heater core cover.

17. Clamp off and disconnect the heater hoses.

18. Remove the heater core case seal.

19. Remove the heater core.

20. To install, reverse the removal procedure. Lubricate the coolant hoses with plain water, if needed.

21. Fill the engine cooling system.

Blackwood

➡If a heater core leak is suspected, the heater core must be pressure leak tested before it is removed from the vehicle.

1. Drain the radiator.

⁂ WARNING

Electronic modules are sensitive to static electrical charges. If exposed to these charges, damage may result.

2. Move the adjustable pedals to the full forward position.

3. Position the seats as needed to remove the console.

4. Remove the console finish panel, disconnect the wiring and remove the 4 bolts.

**** CAUTION**

Always wear safety glasses when repairing an air bag supplemental restraint system (SRS) vehicle and when handling an air bag module. This will reduce the risk of injury in the event of an accidental deployment.

**** CAUTION**

Carry a live air bag module with the air bag and deployment door pointed away from your body. This will reduce the risk of injury in the event of an accidental deployment.

**** CAUTION**

Do not set a live air bag module down with the deployment door face down. This will reduce the risk of injury in the event of an accidental deployment.

**** CAUTION**

After deployment, the air bag surface can contain deposits of sodium hydroxide, a product of the gas generant combustion that is irritating to the skin. Wash your hands with soap and water afterwards.

**** CAUTION**

Never probe the connectors on the air bag module. Doing so can result in air bag deployment, which can result in personal injury.

**** CAUTION**

Air bag modules with discolored or damaged deployment doors must be installed new, not repainted.

**** CAUTION**

To avoid accidental deployment and possible personal injury, the backup power supply must be depleted before repairing or replacing any front or side air bag supplemental restraint system (SRS) components and before servicing, replacing, adjusting or striking components near the front or side air bag sensors, such as doors, instrument panel, console, door latches, strikers, seats and hood latches.

The side air bag sensors are located at or near the base of the B-pillar.

To deplete the backup power supply energy, disconnect the battery ground cable and wait at least one minute. Be sure to disconnect auxiliary batteries and power supplies (if equipped).

5. Disconnect the battery ground cable and wait at least one minute.

**** CAUTION**

To reduce the risk of serious personal injury, read and follow all warnings, cautions, notes, and instructions in the supplemental restraint system (SRS) deactivation/reactivation procedure.

6. Deactivate the supplemental restraint system (SRS). For additional information, refer to the Suspension and Steering part of this chapter.

7. Release the passenger air bag module electrical connector pin-type retainer.

8. Remove the bolt.

9. Push in the two glove compartment door tabs and position downward.

10. Remove the two lower passenger air bag module retaining bolts.

11. Remove the passenger air bag module.

12. Remove the steering column.

a. Remove the two back cover plugs from the steering wheel.

b. Remove the two driver air bag module retaining bolts.

c. Disconnect the driver air bag module electrical connector.

d. Disconnect the horn switch electrical connector.

e. Remove the driver air bag module.

f. Turn the steering wheel to the straight ahead position and turn the ignition switch to the OFF position.

g. Remove the steering wheel bolt.

**** WARNING**

Removing the steering wheel without using a puller can damage the column bearings.

h. Use a puller to remove the steering wheel.

i. Remove the clockspring.

**** WARNING**

Do not allow the steering column shaft to rotate while the intermediate shaft is disconnected or damage to the clockspring can result.

j. If there is evidence that the shaft has rotated, the clockspring must be removed and recentered.

k. Remove the reinforcement.

l. Remove the multifunction switch.

m. Remove the ignition switch connector.

n. Remove the selector lever indicator cable.

o. Remove the intermediate shaft bolt.

p. Remove all electrical connectors.

q. Remove the column mounting nuts (4).

r. Remove the shift cable.

s. Remove the column.

13. Remove the junction block splash shield.

14. Remove the bolts and disconnect the cable ends from the starter relay.

15. Remove the bolts and position aside the junction block bracket.

16. Disconnect the heater core hose couplings.

17. Remove the A/C plenum demister adapter.

18. Remove the screws and position aside the relay bracket.

19. Disconnect the vacuum line.

20. Remove the plenum chamber top.

21. Remove the blend door assembly from the case.

22. Remove the heater core.

23. To install, reverse the removal procedure. Lubricate the coolant hoses with coolant hose lubricant or plain water only, if needed. Observe the following torques:

- Steering wheel bolt: 28 ft. lbs. (38 Nm)
- Intermediate shaft bolt: 35 ft. lbs. (48 Nm)

2001–03 F-150; 2004 Heritage Edition

1. Before servicing the vehicle, refer to the precautions in the beginning of this section.

2. If equipped with power seats, move the seats fully rearward.

3. Disconnect the negative battery cable.

CAUTION

After disconnecting the negative battery cable, wait for 1 minute for the SRS module to deplete its energy.

4. Drain the cooling system into a clean container for reuse.
5. Remove the instrument panel by performing the following procedure:
   a. If equipped with automatic transmission, move the shift lever to the **1** position to ease removal.
   b. At the lower driver's side instrument panel, release the electrical connector push button clip and move the connector aside.
   c. Remove the fuse panel door.
   d. Remove the hood latch release handle screws and move the hood release handle aside.
   e. Remove the parking brake release handle screws and move the parking brake release handle aside.
   f. If equipped, remove the 2 instrument panel floor duct panel push clips and release the expander clip.
   g. Remove the lower steering column cover bolts and the cover.
   h. Remove both front door scuff plates.
   i. Remove both side cowl trim panels.
   j. Disconnect the electrical connector from the Brake Pedal Position (BPP) switch.
   k. Remove the radio ground and the GEM/CTM ground bolts.
   l. Disconnect the left side instrument panel main wiring harness connector.
   m. In the engine compartment, remove the bulkhead wiring harness connector bolts and disconnect the wiring connectors.
   n. In the engine compartment, release the 6 locking tabs and remove the bulkhead electrical connector from the instrument panel.
   o. Disconnect the air bag diagnostic monitor electrical connector.
   p. Disconnect the inertia fuel shutoff switch electrical connector.
   q. Remove the right side ground bolts.
   r. Disconnect the right side instrument panel wiring harness connectors.
   s. Disconnect the electronic blend door actuator electrical connector.
   t. Disconnect the climate control head vacuum harness connector.
   u. Remove the steering column opening cover reinforcement nuts and the cover reinforcement.
   v. At the base of the steering column, disconnect the air bag sliding contact and the anti-theft sensor electrical connectors.
   w. At the steering column, disconnect the remaining electrical connectors.
   x. If equipped with a transmission range indicator, remove the bolt and disconnect the cable.
   y. Remove the steering column-to-instrument panel nuts and lower the steering column.
   z. Remove and disconnect the radio.
   aa. Remove the instrument panel relay cover and disconnect the autolamp sensor electrical connector.
   bb. Remove the glove box.
   cc. At the passenger's air bag module, remove the screws, disconnect the electrical connector and remove the air bag module.

Place the air bag module in a safe place with the front facing upward.

   a. Remove the right side assist handle screw covers, the screws and the handle.
   b. At both doors, pull back the weatherstrip seals and remove the windshield garnish moldings.
   c. Remove the instrument panel reinforcement bolt below the left side corner of the glove box.
   d. Through the air bag module opening, remove the instrument panel bolts.
   e. Remove the upper instrument panel cowl covers and bolts.
   f. At the relay bracket, remove the instrument panel bolt.
   g. At the lower left side of the cigar lighter, remove the instrument panel bolt.
   h. At the both sides, remove the instrument panel-to-cowl side nuts.
   i. At the steering column opening, remove the instrument panel bolts.
   j. Remove the upper instrument panel floor brace bolt.
   k. Loosen the instrument panel brace bolts and nut.
   l. Using an assistant, remove the instrument panel.
6. Compress the holding tabs and disconnect the heater hoses from the heater core.
7. Remove the air conditioning plenum screw and the air conditioning plenum demister adapter.
8. Disconnect the vacuum line.

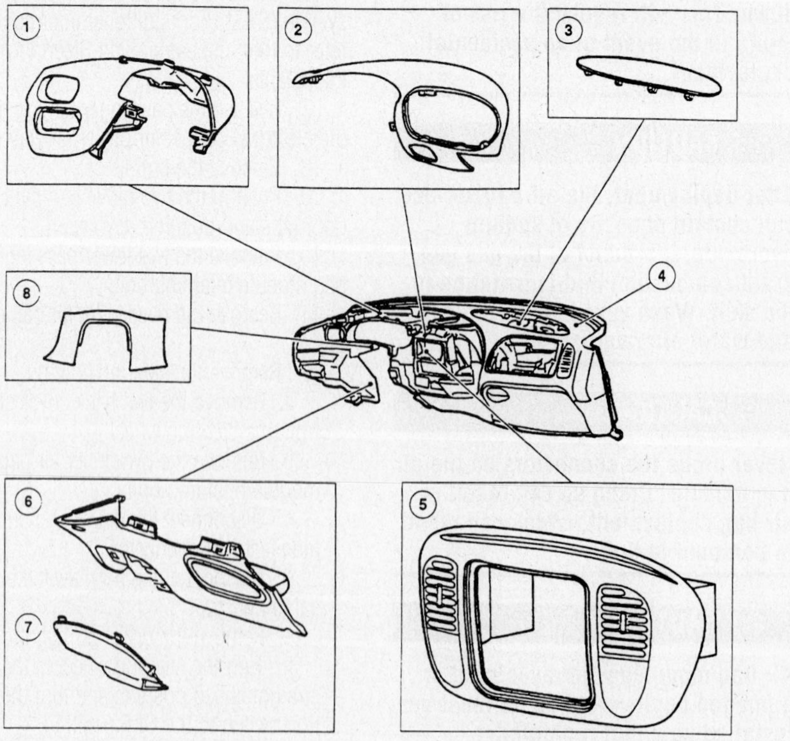

1. Instrument panel finish panel
2. Instrument cluster panel
3. Instrument panel relay cover
4. Instrument panel
5. Center instrument panel finish panel
6. Instrument panel steering column cover, lower
7. Instrument panel fuse door
8. Steering column opening cover

93113GL0

Exploded view of the instrument panel components—2001–03 Ford F-150 and Blackwood; 2004 Heritage Edition

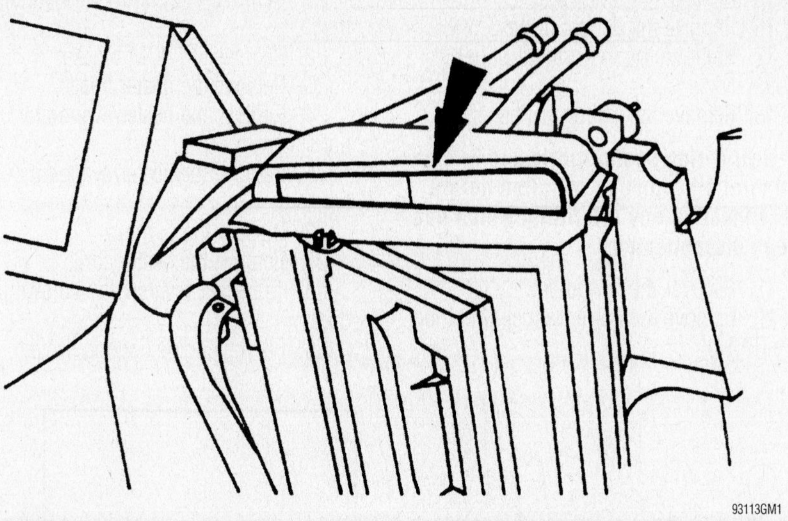

View of the heater core—2001–03 Ford F-150 and Blackwood; 2004 Heritage Edition

9. Remove the heater core bracket screws and the bracket.
10. Remove the 13 heater housing plenum camber cover screws and the heater housing plenum chamber cover.
11. Remove the blend door assembly from the heater housing.
12. Remove the heater core.

To install:

13. Install the heater core.
14. Install the blend door assembly to the heater housing.
15. Install the 13 heater housing plenum camber cover and the heater housing plenum chamber cover screws.
16. Install the heater core bracket and the bracket screws.
17. Connect the vacuum line.
18. Install the air conditioning plenum demister adapter and the air conditioning plenum screw.
19. Connect the heater hoses to the heater core.
20. Install the instrument panel by performing the following procedure:
    a. Using an assistant, install the instrument panel.
    b. Tighten the instrument panel brace bolts and nut.
    c. Install the upper instrument panel floor brace bolt.
    d. At the steering column opening, install the instrument panel bolts.
    e. At the both sides, install the instrument panel-to-cowl side nuts.
    f. At the lower left side of the cigar lighter, install the instrument panel bolt.
    g. At the relay bracket, install the instrument panel bolt.
    h. Install the upper instrument panel cowl bolts and covers.
    i. Through the air bag module opening, install the instrument panel bolts.
    j. Install the instrument panel reinforcement bolt below the left side corner of the glove box.
    k. At both doors, install the windshield garnish moldings and the weatherstrip seals.
    l. Install the right side assist handle, the screws and the handle screw covers.
    m. At the passenger's air bag module, connect the electrical connector and install the air bag module.
    n. Install the glove box.
    o. Install the instrument panel relay cover and connect the autolamp sensor electrical connector.
    p. Install and connect the radio.
    q. Install the steering column-to-instrument panel and the steering column nuts.
    r. If equipped with a transmission range indicator, connect the cable and install the bolt.
    s. At the steering column, connect the remaining electrical connectors.
    t. At the base of the steering column, connect the air bag sliding contact and the anti-theft sensor electrical connectors.
    u. Install the steering column opening cover reinforcement and the cover reinforcement nuts.
    v. Connect the climate control head vacuum harness connector.
    w. Connect the electronic blend door actuator electrical connector.
    x. Connect the right side instrument panel wiring harness connectors.
    y. Install the right side ground bolts.
    z. Connect the inertia fuel shutoff switch electrical connector.
    aa. Connect the air bag diagnostic monitor electrical connector.
    bb. In the driver's compartment, install the bulkhead electrical connector to the instrument panel.
    cc. In the engine compartment, install the bulkhead wiring harness connector bolts and connect the wiring connectors.
    dd. Connect the left side instrument panel main wiring harness connector.
    ee. Install the radio ground and the GEM/CTM ground bolts.
    ff. Connect the electrical connector to the Brake Pedal Position (BPP) switch.
    gg. Install both side cowl trim panels.
    hh. Install both front door scuff plates.
    ii. Install the lower steering column cover and the cover bolts.
    jj. If equipped, install the 2 instrument panel floor duct panel push clips and the expander clip.
    kk. Install the parking brake release handle and the parking brake release handle aside screws.
    ll. Install the hood latch release handle screws.
    mm. Install the fuse panel door.
    nn. At the lower driver's side instrument panel, install the connectors.
21. Refill the cooling system.
22. Connect the negative battery cable.
23. Run the engine to normal operating temperatures; then, check the climate control operation and check for leaks.

2004 F-150

1. Before servicing the vehicle, refer to the precautions in the beginning of this section.

➡Lubricate the coolant hoses with plain water only if needed.

⁂ WARNING

To reduce the risk of serious personal injury, read and follow all warnings, cautions, notes and instructions in the supplemental restraint system (SRS) deactivation/reactivation procedure.

2. Remove the driver and passenger air bag modules.
3. Remove the A-pillar trim panels and right cowl side trim panel.
4. If equipped with a floor console, remove the floor console.
5. Remove the steering column opening cover.

6. Remove the steering column lower cover.
7. Remove the steering column upper cover.
8. Remove the instrument panel center finish panel
9. Remove the instrument panel finish panel.
10. Parking brake release handle.
11. Remove the hood release handle.
12. Remove the electrical connectors.
13. Remove the center brace.
14. Remove the transmission range cable.
15. Remove the pinch bolt.
16. Remove the defrost grilles.
17. Remove the instrument panel bolt.
18. Remove the instrument panel.

➡New O-ring seals, lubricated in clean mineral oil, must be installed before reconnecting any A/C fitting which has been disconnected.

19. Recover the refrigerant.
20. Remove the powertrain control module (PCM).
21. Remove the battery and battery tray.
22. Remove the PCM bracket.
23. Remove the heater hoses
24. Remove the heater/evaporator core housing nuts.
25. Remove the heater/evaporator core housing.
26. Discard the O-rings.
27. Remove the heater core.
28. To install, reverse the removal procedure.
29. Fill the engine cooling system.

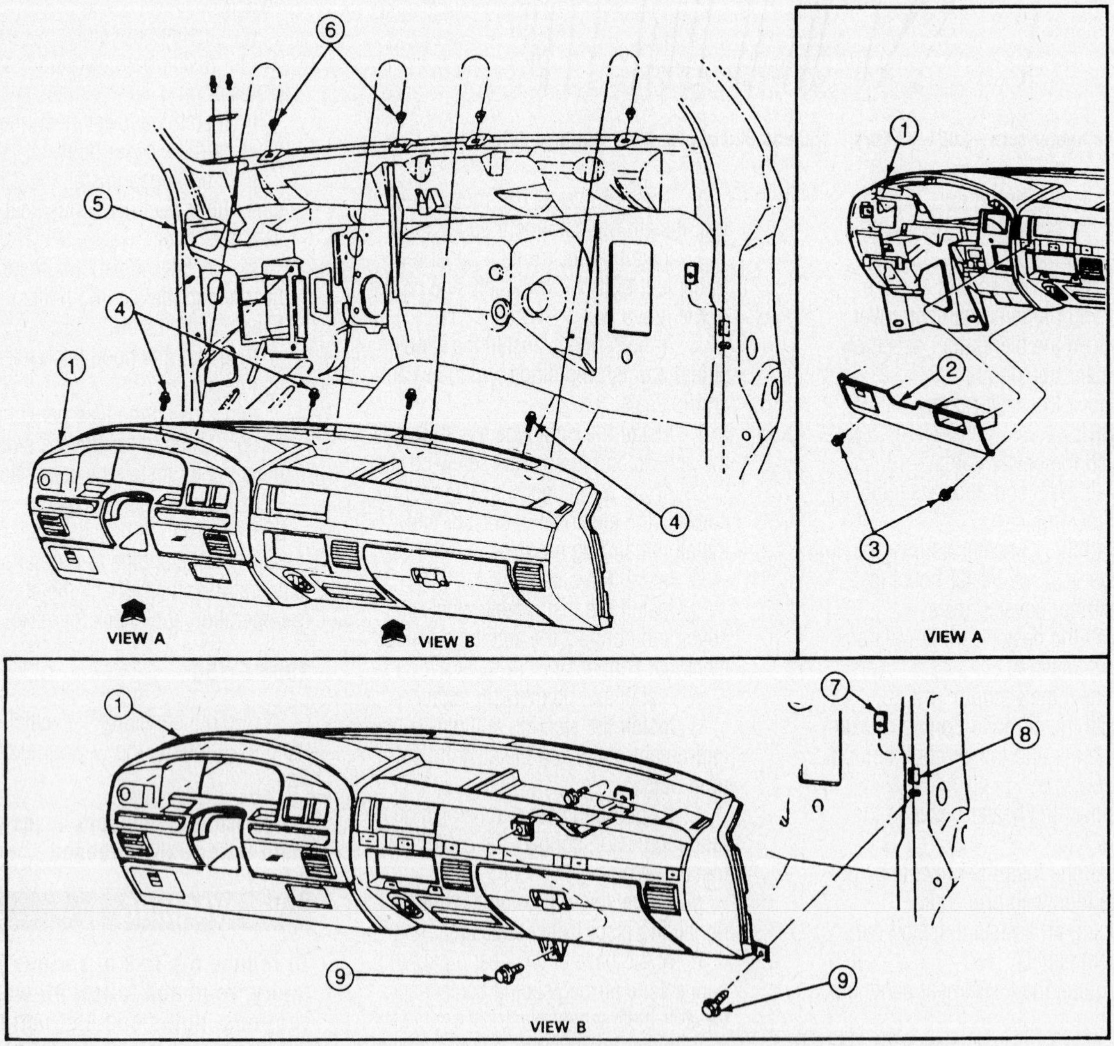

Item	Description
1	Instrument Panel
2	Steering Column Opening Cover
3	Screw(s) 2.0-3.0 N•m (19.0-26.0 In-Lb)
4	Screw(s) 2.0-2.4 N•m (18.0-21.0 In-Lb)

Item	Description
5	Cowl Side Panel (LH)
6	Nut Insert (4 Req'd)
7	U-Nut
8	Cowl Side Panel (RH)
9	Screw(s) 2.0-2.4 N•m (18.0-21.0 In-Lb)

88280G13

Instrument panel installation—2001–03 F-250HD

2001–03 F-250

1. Before servicing the vehicle, refer to the precautions in the beginning of this section.
2. If equipped with power seats, move the seats fully rearward.
3. Disconnect the negative battery cable.

CAUTION

After disconnecting the negative battery cable, wait for 1 minute for the SRS module to deplete its energy.

4. Drain the cooling system into a clean container for reuse.
5. Remove the instrument panel by performing the following procedure:
    a. If equipped with automatic transmission, move the shift lever to the **1** position to ease removal.
    b. At the lower driver's side instrument panel, release the electrical connector push button clip and move the connector aside.
    c. Remove the fuse panel door.
    d. Remove the hood latch release handle screws and move the hood release handle aside.
    e. Remove the parking brake release handle screws and move the parking brake release handle aside.
    f. If equipped, remove the 2 instrument panel floor duct panel push clips and release the expander clip.
    g. Remove the lower steering column cover bolts and the cover.
    h. Remove both front door scuff plates.
    i. Remove both side cowl trim panels.
    j. Disconnect the electrical connector from the Brake Pedal Position (BPP) switch.
    k. Remove the radio ground and the GEM/CTM ground bolts.
    l. Disconnect the left side instrument panel main wiring harness connector.
    m. In the engine compartment, remove the bulkhead wiring harness connector bolts and disconnect the wiring connectors.
    n. In the engine compartment, release the 6 locking tabs and remove the bulkhead electrical connector from the instrument panel.
    o. Disconnect the air bag diagnostic monitor electrical connector.
    p. Disconnect the inertia fuel shutoff switch electrical connector.
    q. Remove the right side ground bolts.
    r. Disconnect the right side instrument panel wiring harness connectors.
    s. Disconnect the electronic blend door actuator electrical connector.
    t. Disconnect the climate control head vacuum harness connector.
    u. Remove the steering column opening cover reinforcement nuts and the cover reinforcement.
    v. At the base of the steering column, disconnect the air bag sliding contact and the anti-theft sensor electrical connectors.
    w. At the steering column, disconnect the remaining electrical connectors.
    x. If equipped with a transmission range indicator, remove the bolt and disconnect the cable.
    y. Remove the steering column-to-instrument panel nuts and lower the steering column.
    z. Remove and disconnect the radio.
    aa. Remove the instrument panel relay cover and disconnect the autolamp sensor electrical connector.
    bb. Remove the glove box.
    cc. At the passenger's air bag module, remove the screws, disconnect the electrical connector and remove the air bag module.

Place the air bag module in a safe place with the front facing upward.

a. Remove the right side assist handle screw covers, the screws and the handle.
b. At both doors, pull back the weatherstrip seals and remove the windshield garnish moldings.
c. Remove the instrument panel reinforcement bolt below the left side corner of the glove box.
d. Through the air bag module opening, remove the instrument panel bolts.
e. Remove the upper instrument panel cowl covers and bolts.
f. At the relay bracket, remove the instrument panel bolt.
g. At the lower left side of the cigar lighter, remove the instrument panel bolt.
h. At the both sides, remove the instrument panel-to-cowl side nuts.
i. At the steering column opening, remove the instrument panel bolts.

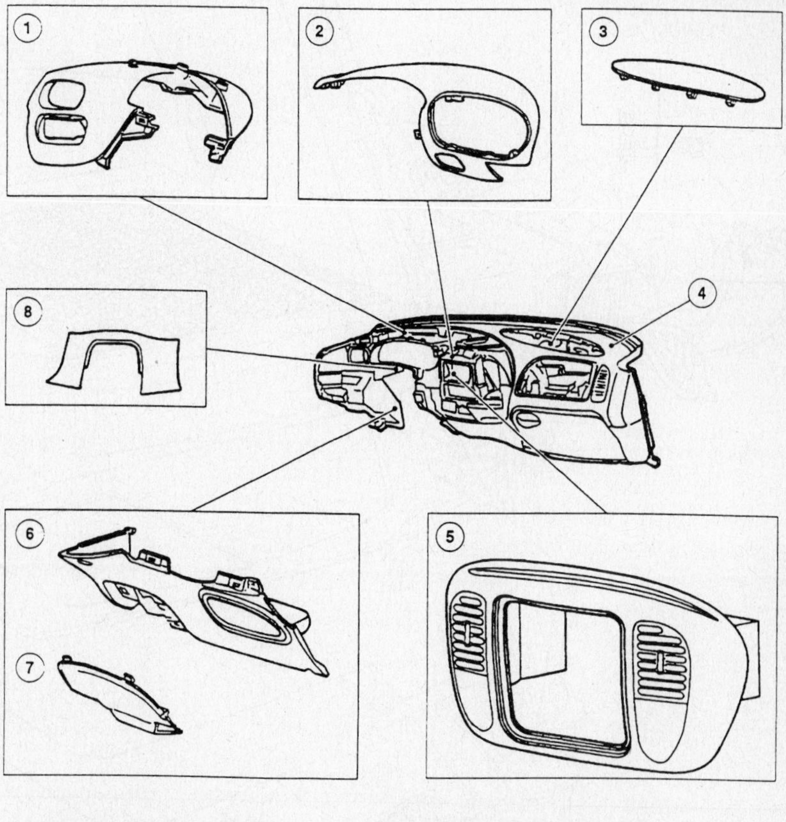

1. Instrument panel finish panel
2. Instrument cluster panel
3. Instrument panel relay cover
4. Instrument panel
5. Center instrument panel finish panel
6. Instrument panel steering column cover, lower
7. Instrument panel fuse door
8. Steering column opening cover

93113GL0

Exploded view of the instrument panel components—2001–03 F-250

j. Remove the upper instrument panel floor brace bolt.

k. Loosen the instrument panel brace bolts and nut.

l. Using an assistant, remove the instrument panel.

6. Compress the holding tabs and disconnect the heater hoses from the heater core.

7. Remove the air conditioning plenum screw and the air conditioning plenum demister adapter.

8. Disconnect the vacuum line.

9. Remove the heater core bracket screws and the bracket.

10. Remove the 13 heater housing plenum camber cover screws and the heater housing plenum chamber cover.

11. Remove the blend door assembly from the heater housing.

12. Remove the heater core.

To install:

13. Install the heater core.

14. Install the blend door assembly to the heater housing.

15. Install the 13 heater housing plenum camber cover and the heater housing plenum chamber cover screws.

16. Install the heater core bracket and the bracket screws.

17. Connect the vacuum line.

18. Install the air conditioning plenum demister adapter and the air conditioning plenum screw.

19. Connect the heater hoses to the heater core.

20. Install the instrument panel by performing the following procedure:

a. Using an assistant, install the instrument panel.

b. Tighten the instrument panel brace bolts and nut.

c. Install the upper instrument panel floor brace bolt.

d. At the steering column opening, install the instrument panel bolts.

e. At the both sides, install the instrument panel-to-cowl side nuts.

f. At the lower left side of the cigar lighter, install the instrument panel bolt.

g. At the relay bracket, install the instrument panel bolt.

h. Install the upper instrument panel cowl bolts and covers.

i. Through the air bag module opening, install the instrument panel bolts.

j. Install the instrument panel reinforcement bolt below the left side corner of the glove box.

k. At both doors, install the wind-

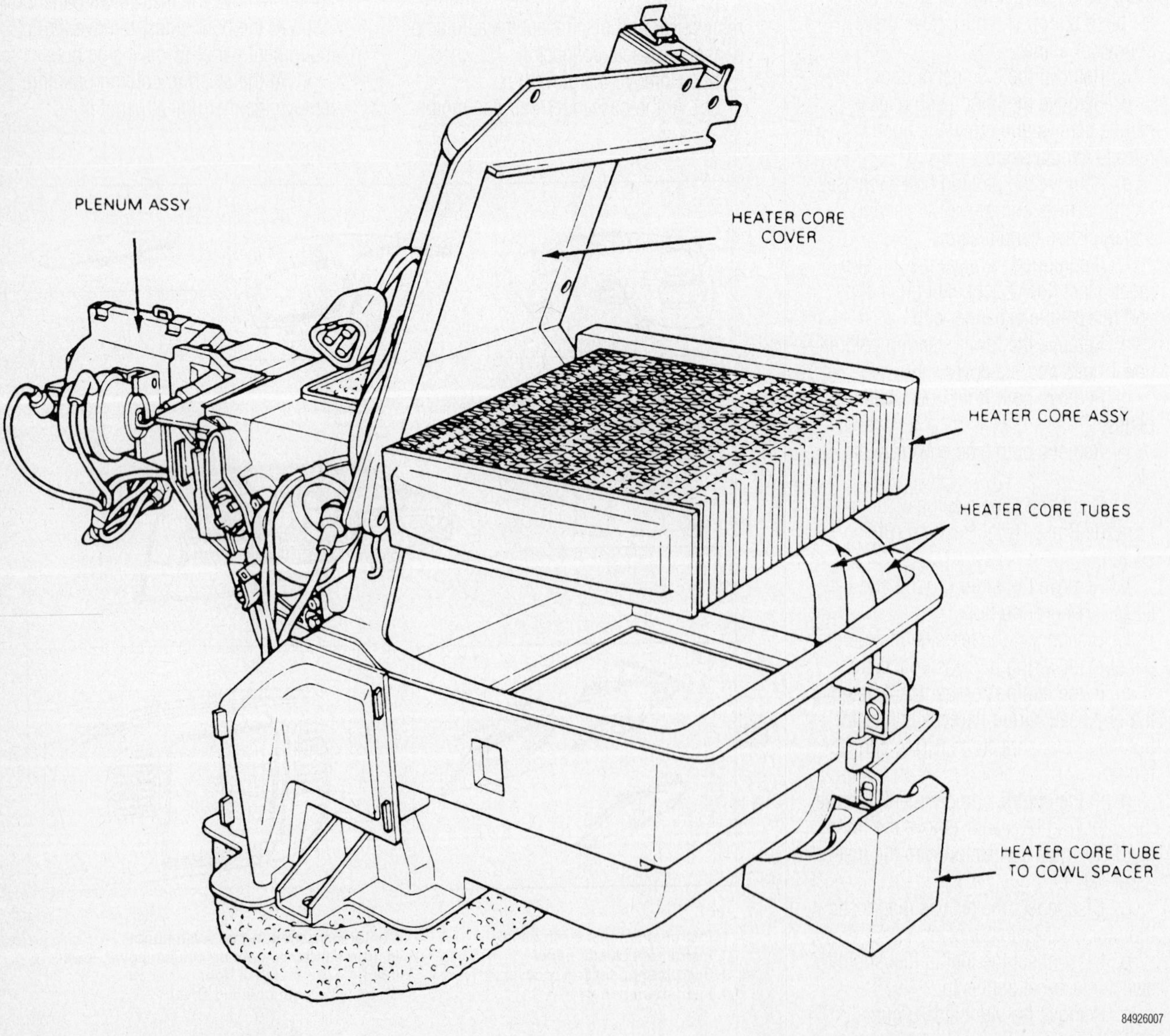

Heater core removal—2001–03 F-250

shield garnish moldings and the weatherstrip seals.

l. Install the right side assist handle, the screws and the handle screw covers.

m. At the passenger's air bag module, connect the electrical connector and install the air bag module.

n. Install the glove box.

o. Install the instrument panel relay cover and connect the autolamp sensor electrical connector.

p. Install and connect the radio.

q. Install the steering column-to-instrument panel and the steering column nuts.

r. If equipped with a transmission range indicator, connect the cable and install the bolt.

s. At the steering column, connect the remaining electrical connectors.

t. At the base of the steering column, connect the air bag sliding contact and the anti-theft sensor electrical connectors.

u. Install the steering column opening cover reinforcement and the cover reinforcement nuts.

v. Connect the climate control head vacuum harness connector.

w. Connect the electronic blend door actuator electrical connector.

x. Connect the right side instrument panel wiring harness connectors.

y. Install the right side ground bolts.

z. Connect the inertia fuel shutoff switch electrical connector.

aa. Connect the air bag diagnostic monitor electrical connector.

bb. In the driver's compartment, install the bulkhead electrical connector to the instrument panel.

cc. In the engine compartment, install the bulkhead wiring harness connector bolts and connect the wiring connectors.

dd. Connect the left side instrument panel main wiring harness connector.

ee. Install the radio ground and the GEM/CTM ground bolts.

ff. Connect the electrical connector to the Brake Pedal Position (BPP) switch.

gg. Install both side cowl trim panels.

hh. Install both front door scuff plates.

ii. Install the lower steering column cover and the cover bolts.

jj. If equipped, install the 2 instrument panel floor duct panel push clips and the expander clip.

kk. Install the parking brake release handle and the parking brake release handle aside screws.

ll. Install the hood latch release handle screws.

mm. Install the fuse panel door.

nn. At the lower driver's side instrument panel, install the connectors.

21. Refill the cooling system.

22. Connect the negative battery cable.

23. Run the engine to normal operating temperatures; then, check the climate control operation and check for leaks.

2001–03 F-350 and Super Duty

1. Before servicing the vehicle, refer to the precautions in the beginning of this section.

2. If equipped with power seats, move the seats fully rearward.

3. Disconnect the negative battery cable.

**** CAUTION**

After disconnecting the negative battery cable, wait for 1 minute for the SRS module to deplete its energy.

4. Drain the cooling system into a clean container for reuse.

5. Remove the instrument panel by performing the following procedure:

a. If equipped with automatic transmission, move the shift lever to the **1** position to ease removal.

b. At the lower driver's side instrument panel, release the electrical connector push button clip and move the connector aside.

c. Remove the fuse panel door.

d. Remove the hood latch release handle screws and move the hood release handle aside.

e. Remove the parking brake release handle screws and move the parking brake release handle aside.

f. If equipped, remove the 2 instrument panel floor duct panel push clips and release the expander clip.

g. Remove the lower steering column cover bolts and the cover.

h. Remove both front door scuff plates.

i. Remove both side cowl trim panels.

j. Disconnect the electrical connector from the Brake Pedal Position (BPP) switch.

k. Remove the radio ground and the GEM/CTM ground bolts.

l. Disconnect the left side instrument panel main wiring harness connector.

m. In the engine compartment, remove the bulkhead wiring harness connector bolts and disconnect the wiring connectors.

n. In the engine compartment, release the 6 locking tabs and remove the bulkhead electrical connector from the instrument panel.

o. Disconnect the air bag diagnostic monitor electrical connector.

p. Disconnect the inertia fuel shutoff switch electrical connector.

q. Remove the right side ground bolts.

r. Disconnect the right side instrument panel wiring harness connectors.

s. Disconnect the electronic blend door actuator electrical connector.

t. Disconnect the climate control head vacuum harness connector.

u. Remove the steering column opening cover reinforcement nuts and the cover reinforcement.

v. At the base of the steering column, disconnect the air bag sliding contact and the anti-theft sensor electrical connectors.

w. At the steering column, disconnect the remaining electrical connectors.

x. If equipped with a transmission range indicator, remove the bolt and disconnect the cable.

y. Remove the steering column-to-instrument panel nuts and lower the steering column.

z. Remove and disconnect the radio.

aa. Remove the instrument panel relay cover and disconnect the autolamp sensor electrical connector.

bb. Remove the glove box.

cc. At the passenger's air bag module, remove the screws, disconnect the electrical connector and remove the air bag module.

Place the air bag module in a safe place with the front facing upward.

a. Remove the right side assist handle screw covers, the screws and the handle.

b. At both doors, pull back the weatherstrip seals and remove the windshield garnish moldings.

c. Remove the instrument panel reinforcement bolt below the left side corner of the glove box.

d. Through the air bag module opening, remove the instrument panel bolts.

e. Remove the upper instrument panel cowl covers and bolts.

f. At the relay bracket, remove the instrument panel bolt.

g. At the lower left side of the cigar lighter, remove the instrument panel bolt.

h. At the both sides, remove the instrument panel-to-cowl side nuts.

i. At the steering column opening, remove the instrument panel bolts.

j. Remove the upper instrument panel floor brace bolt.

k. Loosen the instrument panel brace bolts and nut.

l. Using an assistant, remove the instrument panel.

6. Compress the holding tabs and disconnect the heater hoses from the heater core.

7. Remove the air conditioning plenum screw and the air conditioning plenum demister adapter.

8. Disconnect the vacuum line.

9. Remove the heater core bracket screws and the bracket.

10. Remove the 13 heater housing plenum camber cover screws and the heater housing plenum chamber cover.

11. Remove the blend door assembly from the heater housing.

12. Remove the heater core.

To install:

13. Install the heater core.

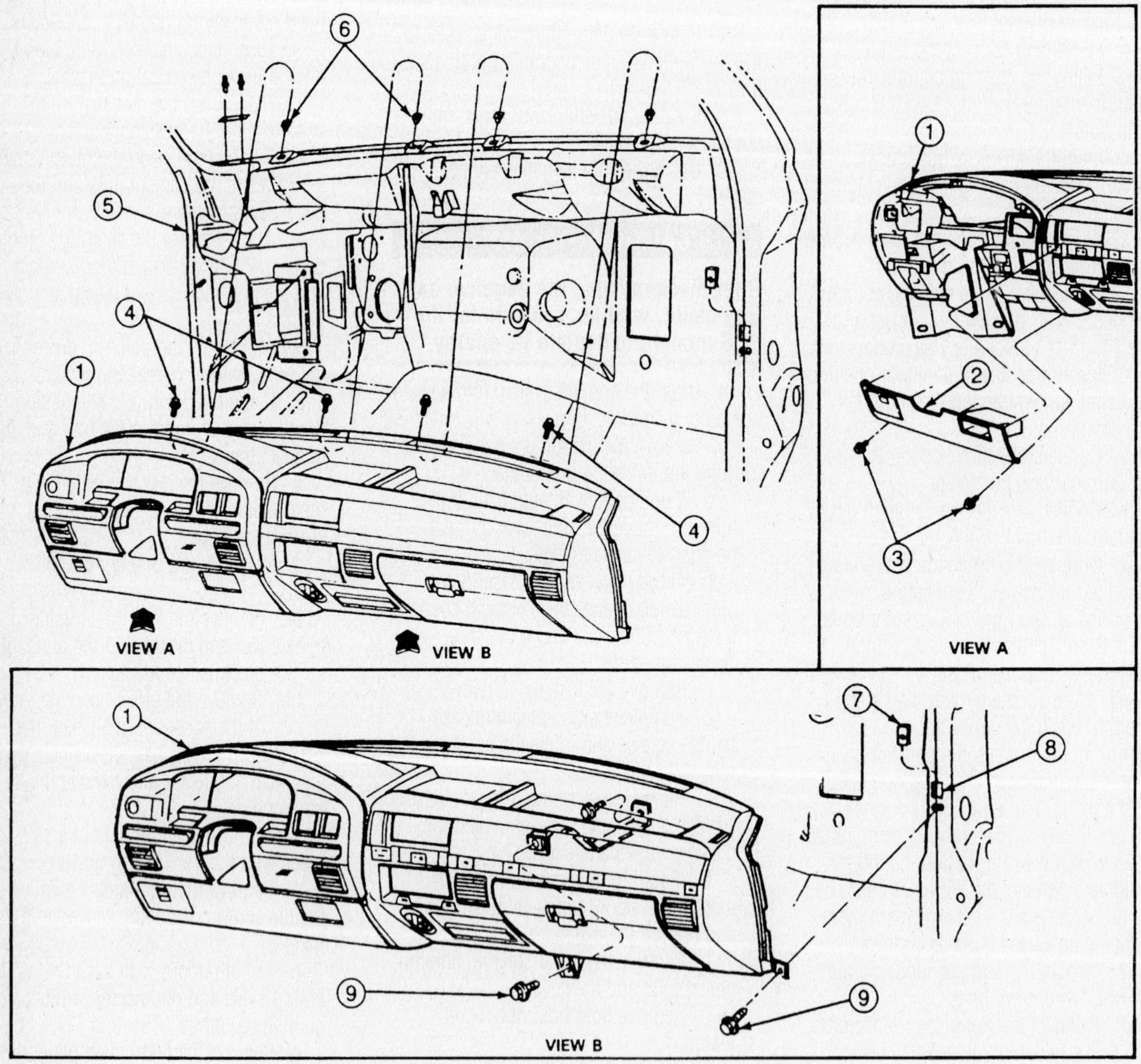

Item	Description
1	Instrument Panel
2	Steering Column Opening Cover
3	Screw(s) 2.0-3.0 N•m (19.0-26.0 In-Lb)
4	Screw(s) 2.0-2.4 N•m (18.0-21.0 In-Lb)

Item	Description
5	Cowl Side Panel (LH)
6	Nut Insert (4 Req'd)
7	U-Nut
8	Cowl Side Panel (RH)
9	Screw(s) 2.0-2.4 N•m (18.0-21.0 In-Lb)

88280G13

Instrument panel installation—2001–03 F-350 and F-Super Duty

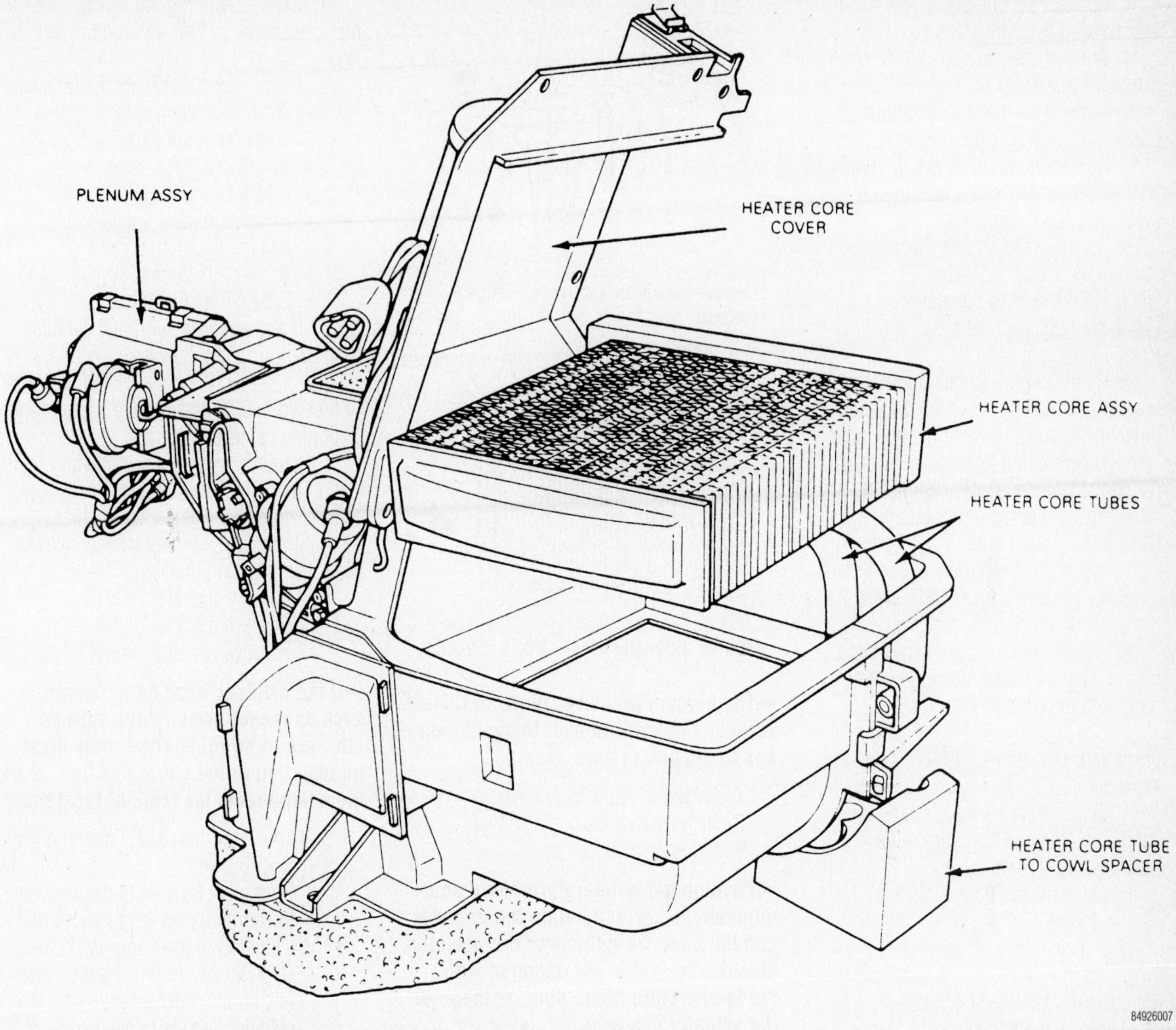

Heater core removal—2001–03 F-350 and F-Super Duty models

14. Install the blend door assembly to the heater housing.
15. Install the 13 heater housing plenum camber cover and the heater housing plenum chamber cover screws.
16. Install the heater core bracket and the bracket screws.
17. Connect the vacuum line.
18. Install the air conditioning plenum demister adapter and the air conditioning plenum screw.
19. Connect the heater hoses to the heater core.
20. Install the instrument panel by performing the following procedure:
   a. Using an assistant, install the instrument panel.
   b. Tighten the instrument panel brace bolts and nut.
   c. Install the upper instrument panel floor brace bolt.
   d. At the steering column opening, install the instrument panel bolts.
   e. At the both sides, install the instrument panel-to-cowl side nuts.
   f. At the lower left side of the cigar lighter, install the instrument panel bolt.
   g. At the relay bracket, install the instrument panel bolt.
   h. Install the upper instrument panel cowl bolts and covers.
   i. Through the air bag module opening, install the instrument panel bolts.
   j. Install the instrument panel reinforcement bolt below the left side corner of the glove box.
   k. At both doors, install the windshield garnish moldings and the weatherstrip seals.
   l. Install the right side assist handle, the screws and the handle screw covers.
   m. At the passenger's air bag module, connect the electrical connector and install the air bag module.
   n. Install the glove box.
   o. Install the instrument panel relay cover and connect the autolamp sensor electrical connector.
   p. Install and connect the radio.
   q. Install the steering column-to-instrument panel and the steering column nuts.
   r. If equipped with a transmission range indicator, connect the cable and install the bolt.
   s. At the steering column, connect the remaining electrical connectors.
   t. At the base of the steering column, connect the air bag sliding contact and the anti-theft sensor electrical connectors.
   u. Install the steering column opening cover reinforcement and the cover reinforcement nuts.

v. Connect the climate control head vacuum harness connector.

w. Connect the electronic blend door actuator electrical connector.

x. Connect the right side instrument panel wiring harness connectors.

y. Install the right side ground bolts.

z. Connect the inertia fuel shutoff switch electrical connector.

aa. Connect the air bag diagnostic monitor electrical connector.

bb. In the driver's compartment, install the bulkhead electrical connector to the instrument panel.

cc. In the engine compartment, install the bulkhead wiring harness connector bolts and connect the wiring connectors.

dd. Connect the left side instrument panel main wiring harness connector.

ee. Install the radio ground and the GEM/CTM ground bolts.

ff. Connect the electrical connector to the Brake Pedal Position (BPP) switch.

gg. Install both side cowl trim panels.

hh. Install both front door scuff plates.

ii. Install the lower steering column cover and the cover bolts.

jj. If equipped, install the 2 instrument panel floor duct panel push clips and the expander clip.

kk. Install the parking brake release handle and the parking brake release handle aside screws.

ll. Install the hood latch release handle screws.

mm. Install the fuse panel door.

nn. At the lower driver's side instrument panel, install the connectors.

21. Refill the cooling system.
22. Connect the negative battery cable.
23. Run the engine to normal operating temperatures; then, check the climate control operation and check for leaks.

2004 F-250/350

1. Before servicing the vehicle, refer to the precautions in the beginning of this section.
2. Clamp off and disconnect the heater water hoses from the heater core.
3. Disengage the stops and lower the glove compartment door.

➡Do not attempt to bend any part of the panel/floor door lever. It is brittle and will break.

4. Remove the panel/floor door vacuum control motor.
5. Remove the electronic blend door actuator and bracket assembly.
6. Remove the core cover screws.

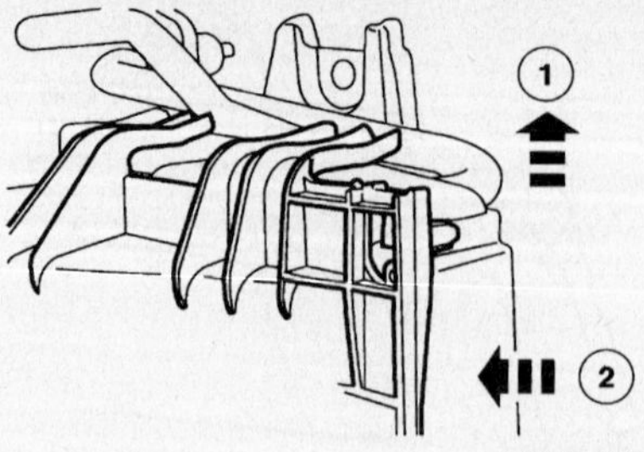

Remove the heater core cover—2004 F-250/350

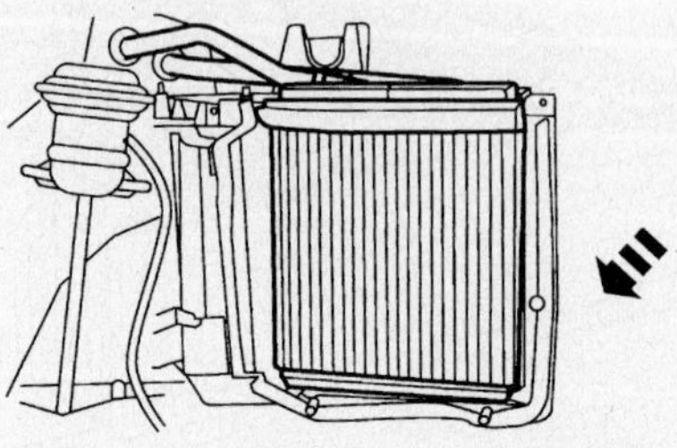

Remove the heater core—2004 F-250/350

➡The heater core cover must be raised vertically before removal to avoid damage to the heater core housing.

7. Raise the heater core cover.
8. Remove the heater core cover.
9. Remove the heater core.

➡Position the temperature blend door manually to correctly align the actuator and the door. Do not power the actuator electrically. If it is not engaged with the temperature blend door, damage to the actuator can occur.

➡Add gasket between housing and cover before installing cover.

10. To install, reverse the removal procedure.
11. Fill the engine cooling system.

Cylinder Head

REMOVAL & INSTALLATION

4.2L Engine

⁂ CAUTION

Fuel injection systems remain under pressure, even after the engine has been turned OFF. The fuel system pressure must be relieved before disconnecting any fuel lines. Failure to do so may result in fire and/or personal injury.

1. Before servicing the vehicle, refer to the precautions in the beginning of this section.
2. Remove or disconnect the following:
   - Air conditioning system, using approved equipment
   - Negative battery cable
   - Coolant
   - Upper and lower intake manifolds and related components
   - Rocker arm covers
   - Exhaust manifold
3. If removing the left-hand cylinder head, perform the following:
   a. Position the power steering pump reservoir aside and remove the air conditioning compressor.
   b. Remove and support the air conditioning compressor bracket and power steering pump aside.
4. If removing the right-hand cylinder head, perform the following:
   a. Remove the alternator.
   b. Remove the idler pulley.
   c. Remove the alternator bracket.

➡If the cylinder head components, such as rocker arms, valve springs, etc., are to be reinstalled, they must be installed in the same position. Mark the components for original location.

5. Remove or disconnect the following:
   - Rocker arms
   - Pushrods. Be sure to label or mark the components removed for reinstallation in their original location.
   - Cylinder head bolts. New cylinder head bolts must be used when the cylinder head is reinstalled.
   - Cylinder head. Remove the cylinder head gasket and discard.

To install:

6. Clean and inspect the cylinder head for flatness.
7. Install a new cylinder head gasket on the cylinder block with the small hole to the front of the engine.
8. Install the cylinder head.

⁂ WARNING

Always use new cylinder head bolts for installation. Lubricate the cylinder head bolts with clean engine oil prior to installation. Be sure to tighten the cylinder head bolts in three (3) steps.

9. Install new cylinder head bolts and torque the bolts following the proper tightening sequence in 3 steps to the following values:

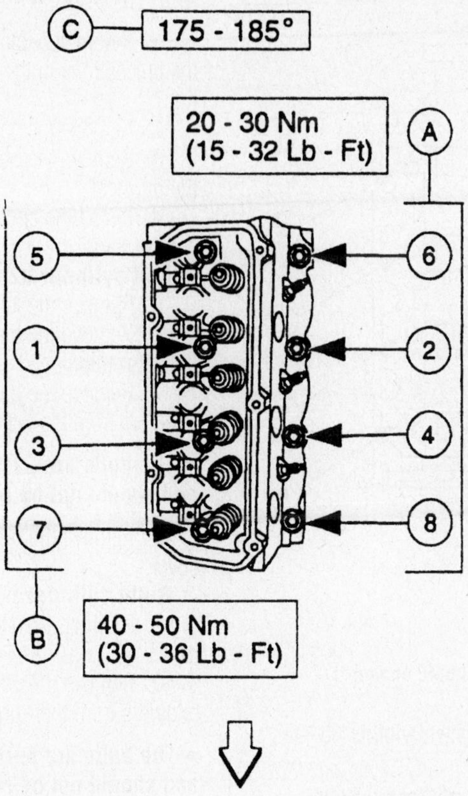

Tighten the cylinder head bolts in the following sequence—4.2L engine

a. Step 1: 14 ft. lbs. (20 Nm).
b. Step 2: 29 ft. lbs. (40 Nm).
c. Step 3: 36 ft. lbs. (50 Nm).

⁂ WARNING

Do not loosen all of the cylinder head bolts at one time. Each cylinder head bolt must be loosened and the final tightening performed prior to loosening the next bolt in the sequence.

10. In the same sequence as used previously, loosen the cylinder head bolt 3 turns, then tighten the cylinder head bolt to the specific value according to its length. The short bolts (A) should be tightened to 15–22 ft. lbs. (20–30 Nm) and the long bolts (B) to 30–36 ft. lbs. (40–50 Nm). Finally, tighten each cylinder head bolt, in sequence, an additional 175–185 degrees.

11. Lubricate the pushrods with clean engine oil prior to installation, then install them into their original positions.

12. Install the rocker arms. Tighten the rocker arm mounting bolts to 23–29 ft. lbs. (30–40 Nm).

13. If the valvetrain components were replaced with new components, inspect the valve clearance.

14. If installing the right-hand cylinder head, perform the following:
a. Position the alternator bracket in place, then install the 2 long bolts to 31–39 ft. lbs. (41–54 Nm). Install the short bolt and tighten to 18–22 ft. lbs. (24–31 Nm).
b. Install the idler pulley. Tighten the center retaining bolt to 35–46 ft. lbs. (47–63 Nm).
c. Install the alternator.

15. If installing the left-hand cylinder head, complete the following steps:
a. Position the air conditioning compressor bracket and power steering pump in place, then start the air conditioning compressor bracket bolt. Install the 3 compressor bracket bolts to 30–40 ft. lbs. (40–55 Nm). Then, install the 2 compressor bracket nuts to 16–21 ft. lbs. (21–29 Nm).
b. Install the air conditioning compressor.
c. Install the power steering pump reservoir. Tighten the hold-down bolts to 80–107 inch lbs. (9–12 Nm).

16. Install or connect the following:
- Exhaust manifold
- The 2 rocker arm covers. Inspect the rocker arm cover gaskets for damage prior to installation; replace them if necessary.
- Lower intake manifold and related components
- Upper intake manifold and related components

17. Drain the engine oil into a suitable container and replace the oil filter.

18. Fill the engine with the proper amount of engine oil.

19. Fill the cooling system.

20. Evacuate and recharge the air conditioning system using approved equipment.

21. Connect the negative battery cable.

22. Start the engine and check for any fluid or vacuum leaks.

2001–03 E- and F-Series; 2004 Heritage Editions 4.6L, 5.4L and 6.8L Engines

⁂ CAUTION

Fuel injection systems remain under pressure, even after the engine has been turned OFF. The fuel system pressure must be relieved before disconnecting any fuel lines. Failure to do so may result in fire and/or personal injury.

➡To correctly tighten the cylinder head bolts an angle torque wrench is needed.

1. Before servicing the vehicle, refer to the precautions in the beginning of this section.
2. Remove or disconnect the following:
- Air conditioning system, using approved equipment
- Negative battery cable
- Cylinder head covers
- Intake manifold
- Timing chains from the engine
- Exhaust manifolds
- The 2 heater hose retaining bolts, then compress and slide the hose clamp back to remove the heater water hose.
- Any remaining hoses, electrical connections or cables
- The cylinder head bolts, then lift the cylinder head from the engine block. Discard the cylinder head gasket and clean the engine block surface.

To install:

⁂ WARNING

Cylinder head bolts must be replaced with new ones. They are torque-to-yield designed and cannot be reused.

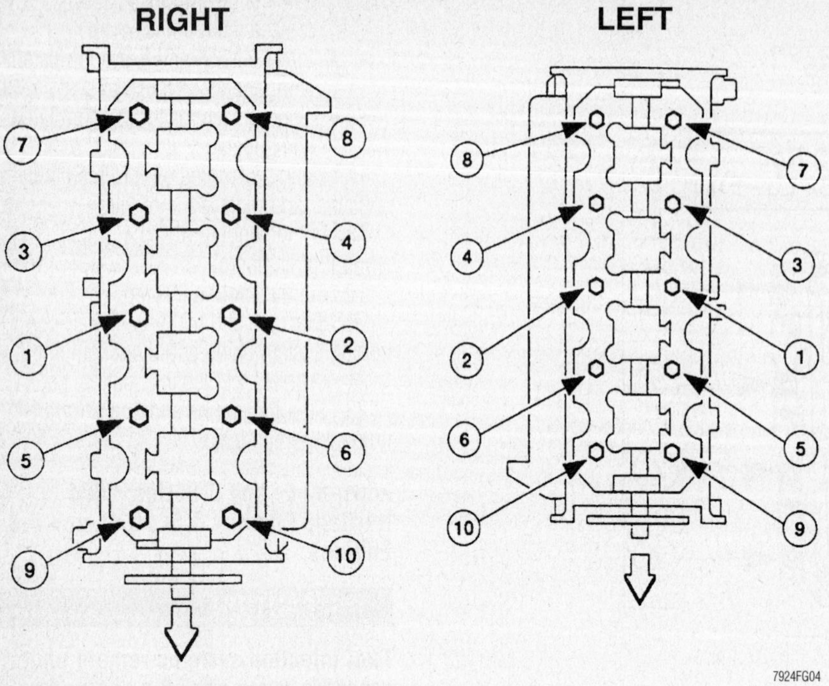

Tighten the cylinder head bolts using 3 steps in this sequence—4.6L and 5.4L engine

3. Turn the crankshaft to position the keyway at the 12 o'clock position.

4. Clean and inspect the cylinder head for damage or warpage. Install the cylinder head gasket over the dowel pins. Then, install the cylinder head onto the engine block. Loosely install NEW cylinder head bolts.

➡**Be sure to tighten the head bolts in 3 steps.**

5. Tighten the cylinder head bolts in the correct sequence using 3 steps, as follows:
   a. Step 1—30 ft. lbs. (40 Nm).
   b. Step 2—tighten an additional 85–95 degrees.
   c. Step 3—tighten another 85–95 degrees.

6. Install or connect the following:
   - Heater hose
   - Exhaust manifolds
   - Timing chains
   - Intake manifold
   - Cylinder head covers
   - Electrical connections and cables
   - Refrigerant
   - Engine oil
   - Coolant
   - Negative battery cable

7. Start the engine and check for any fluid or vacuum leaks.

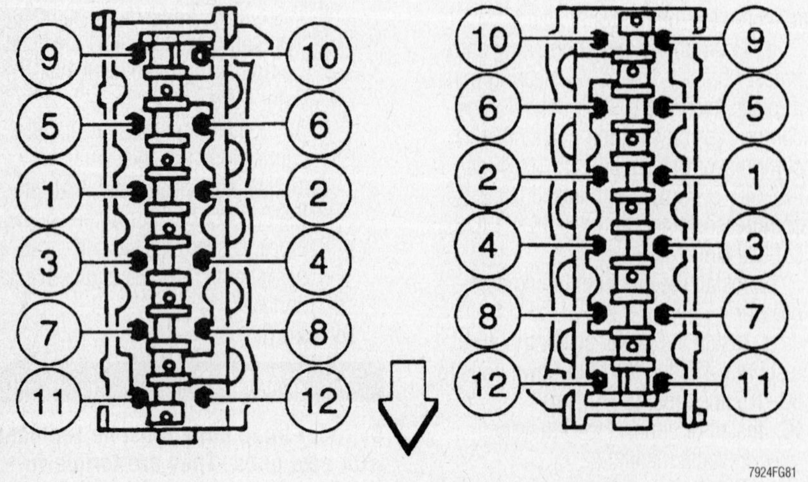

Tighten the cylinder head bolts using 3 steps in this sequence—6.8L engine

2004 E- and F-150 4.6L Engine, Exc. 2004 Heritage Edition

1. Before servicing the vehicle, refer to the precautions in the beginning of this section.
2. Remove the engine.
3. Remove the bolts and the flexplate or the flywheel.
4. Install the engine onto a suitable engine stand.

Left cylinder head

5. Remove the left exhaust manifold.
6. Remove the bolt and the oil level indicator tube.
7. Remove the engine wiring harness retainers from the left valve cover studs.

➡**The bolts are part of the valve cover and should not be removed.**

8. Remove the bolts and the left valve cover.

Right cylinder head

9. Remove the right exhaust manifold.
10. Remove the engine wiring harness retainers from the right valve cover studs.

➡**The bolts are part of the valve cover and should not be removed.**

11. Remove the bolts and the right valve cover.
12. Disconnect the coolant hoses from the heater outlet tube.
13. Remove the heater outlet tube studs.
14. Remove the heater outlet tube and discard the O-ring seal.

All cylinder heads

15. Remove the bolts, the coolant pump pulley and the accessory drive belt idler pulleys.
16. Remove the bolts and the accessory drive belt tensioner.
17. Disconnect the left and right radio ignition interference capacitor electrical connectors.
18. Remove the nuts and the two radio interference capacitors.
19. Remove the bolt and the left CMP sensor.
20. Connect the crankshaft position (CKP) sensor electrical connector.
21. Remove the bolt and washer and using a suitable tool, remove the crankshaft pulley. Discard the crankshaft bolt.
22. Using a suitable tool, remove the crankshaft seal.
23. Remove the four oil pan front bolts.
24. Remove the bolts.

⁂ WARNING

Do not use metal scrapers, wire brushes, power abrasive discs or other abrasive means to clean the sealing surfaces. These tools cause scratches and gouges which make leak paths. Use a plastic scraping tool to remove all traces of old sealant.

25. Remove the engine front cover from the front cover to cylinder block dowel.
26. Remove the engine front cover gaskets.
27. Clean the mating surfaces with silicone gasket remover and metal surface prep. Follow the directions on the packaging.
28. Inspect the mating surfaces.

⁂ WARNING

Use care when removing the spark plugs.

➡Use compressed air to remove any foreign material from the spark plug well before removing the spark plugs.

29. Remove the eight spark plugs.
30. Install a suitable tool between the valve spring coils to prevent valve stem seal damage.

➡The camshaft roller followers must be reinstalled in their original locations. Record the camshaft roller follower locations.

➡Position the cam lobe away from the camshaft roller follower prior to removing each camshaft roller follower.

31. Use a suitable tool to compress the valve springs, and remove the camshaft roller followers.
32. Remove the special tool.

➡The camshaft roller followers must be reinstalled in their original locations.

33. Repeat the previous three steps for each of the roller followers.

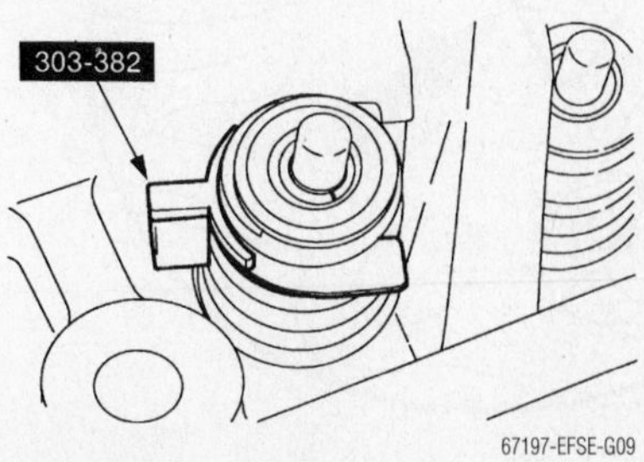

Special tool installation

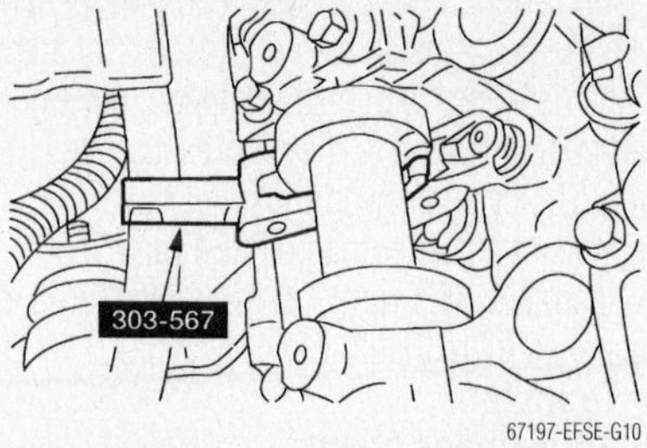

Special tool installation to compress the valve spring

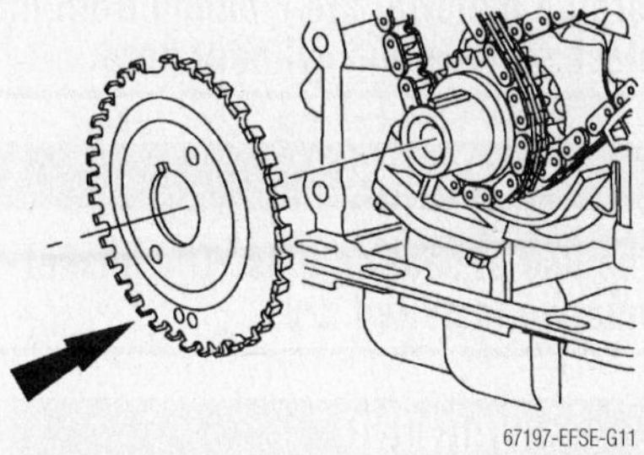

Senor ring

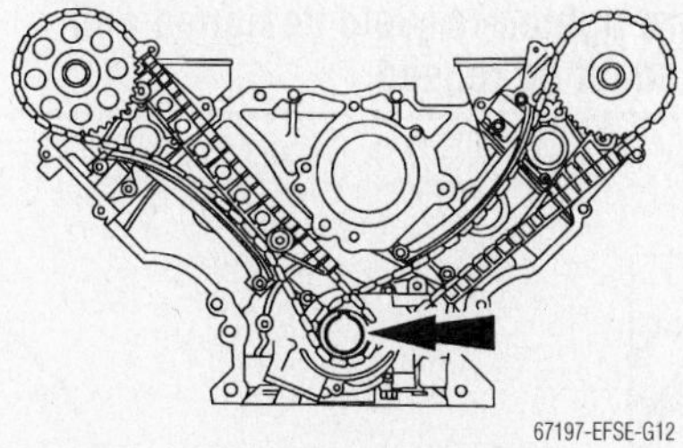

Keyway at the 12 o'clock position

34. Remove the crankshaft sensor ring from the crankshaft.
35. Position the crankshaft with the keyway at the 12 o'clock position.
36. Remove the timing chain tensioning system from both timing chains.
37. Remove the right timing chain from the camshaft sprocket.
38. Remove the right timing chain from the crankshaft sprocket.
39. Remove the left timing chain from the camshaft sprocket.
40. Remove the left timing chain and crankshaft sprocket(s).
41. Remove both timing chain guides.
42. Install the lifting tools on both ends of the cylinder head being serviced.

⁂ WARNING

The cylinder head must be cool before removing it from the engine. Cylinder head warpage can result if a warm or hot cylinder head is removed.

⁂ WARNING

Place clean shop towels over exposed engine cavities. Carefully remove the towels so foreign material is not dropped into the engine.

⁂ WARNING

The cylinder head bolts must be discarded and new bolts must be installed. They are tighten-to-yield designed and cannot be reused.

⁂ WARNING

Do not use metal scrapers, wire brushes, power abrasive discs or other abrasive means to clean the sealing surfaces. These tools cause scratches and gouges that make leak paths. Use a plastic scraping tool to remove all traces of the head gasket.

⁂ WARNING

Aluminum surfaces are soft and can be scratched easily. Never place the cylinder head gasket surface, unprotected, on a bench surface.

43. Remove the bolts and the cylinder head. Discard the cylinder head gasket. Discard the cylinder head bolts.

⁂ WARNING

Do not use metal scrapers, wire brushes, power abrasive discs or other abrasive means to clean the sealing surfaces. These tools cause scratches and gouges that make leak paths. Use a plastic scraping tool to remove all traces of the head gasket.

⁂ WARNING

Observe all warnings or cautions and follow all application directions contained on the packaging of the silicone gasket remover and the metal surface prep.

➡If there is no residual gasket material present, metal surface prep can be used to clean and prepare the surfaces.

44. Clean the cylinder head-to-cylinder block mating surfaces of both the cylinder head and the cylinder block.
45. Remove any large deposits of silicone or gasket material with a plastic scraper.

46. Apply silicone gasket remover, following package directions, and allow to set for several minutes.

47. Remove the silicone gasket remover with a plastic scraper. A second application of silicone gasket remover may be required if residual traces of silicone or gasket material remain.

48. Apply metal surface prep, following package directions, to remove any remaining traces of oil or coolant, and to prepare the surfaces to bond with the new gasket. Do not attempt to make the metal shiny. Some staining of the metal surfaces is normal.

➡Make sure all cylinder head surfaces are clear of any gasket material, RTV, oil and coolant. The cylinder head surface must be clean and dry before running a flatness check.

➡Use a straightedge that is calibrated by the manufacturer to be flat with 0.005 mm (0.0002 in.) per running foot length. For example, if the straightedge is 61 cm (24 in.) long, the machine edge must be flat with 0.010 mm (0.0004 in.) from end to end.

49. Support the cylinder head on a bench with the head gasket side up. Inspect all areas of the deck face with a straightedge, paying particular attention to the oil pressure feed area. The cylinder head must not have depressions deeper than 0.0254 mm (0.001 in.) across a 38.1 mm (1.5 in.) square area, or scratches more than 0.0254 mm (0.001 in.).

To install:

All cylinder heads

**** WARNING**

Make sure all coolant residue and foreign material are cleaned from the block surface and cylinder bore.

**** WARNING**

The use of sealing aids. The gasket must be installed dry.

**** WARNING**

The cylinder head bolts must be discarded and new bolts installed. They are tighten-to-yield designed and cannot be reused.

➡Do not turn the crankshaft until instructed to do so.

50. Using the lifting tools, position the cylinder head gaskets and cylinder heads over the dowels and install the cylinder head bolts loosely.

51. Tighten the bolts in the sequence shown.

a. Stage 1: Tighten to 40 Nm (30 ft. lbs.).

b. Stage 2: Tighten an additional 90 degrees.

c. Stage 3: Tighten an additional 90 degrees.

All cylinder heads

**** WARNING**

Timing chain procedures must be followed exactly or damage to valves and pistons will result.

52. Compress the tensioner plunger, using a vise.

53. Install a retaining clip on the tensioner to hold the plunger in during installation.

54. Remove the tensioner from the vise.

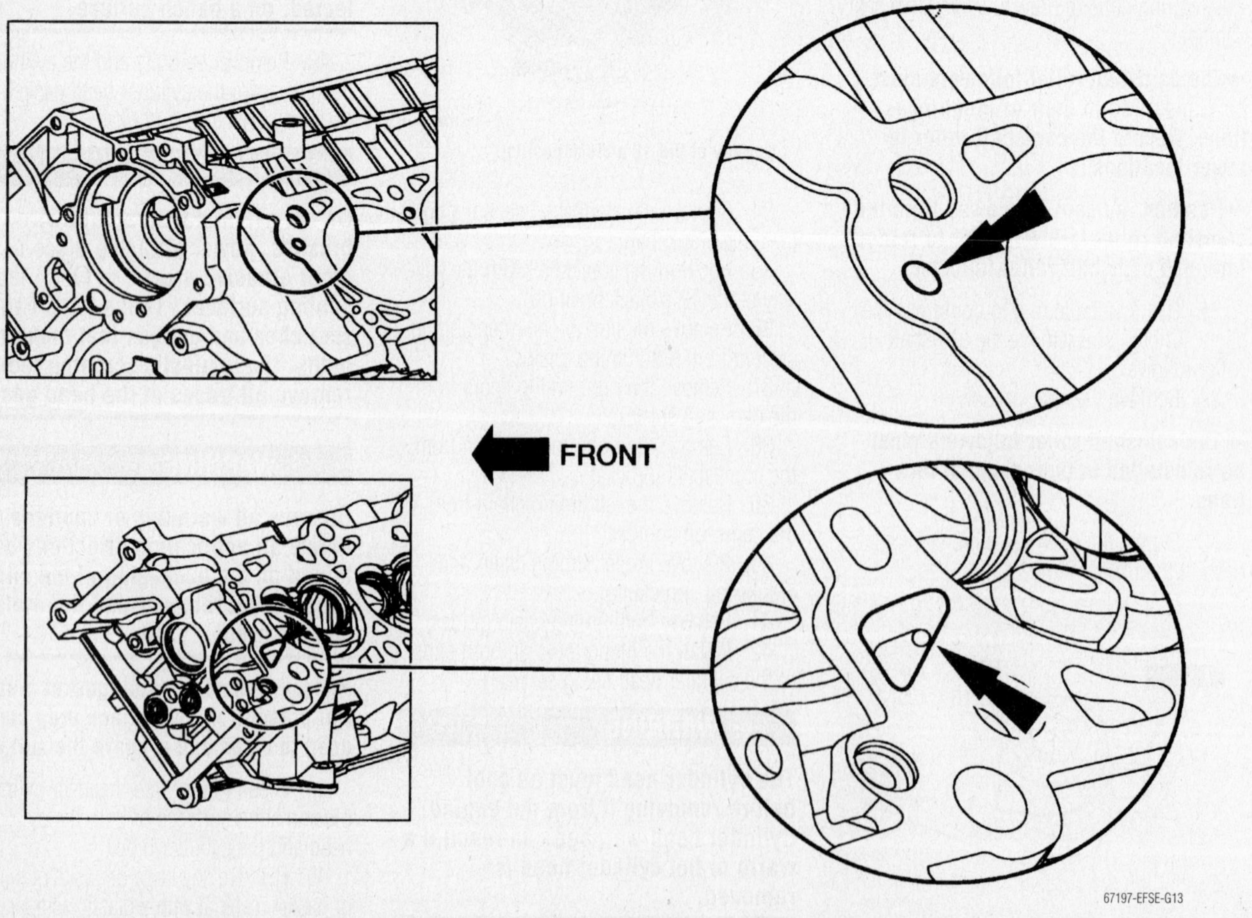

When checking flatness, pay particular attention to these areas

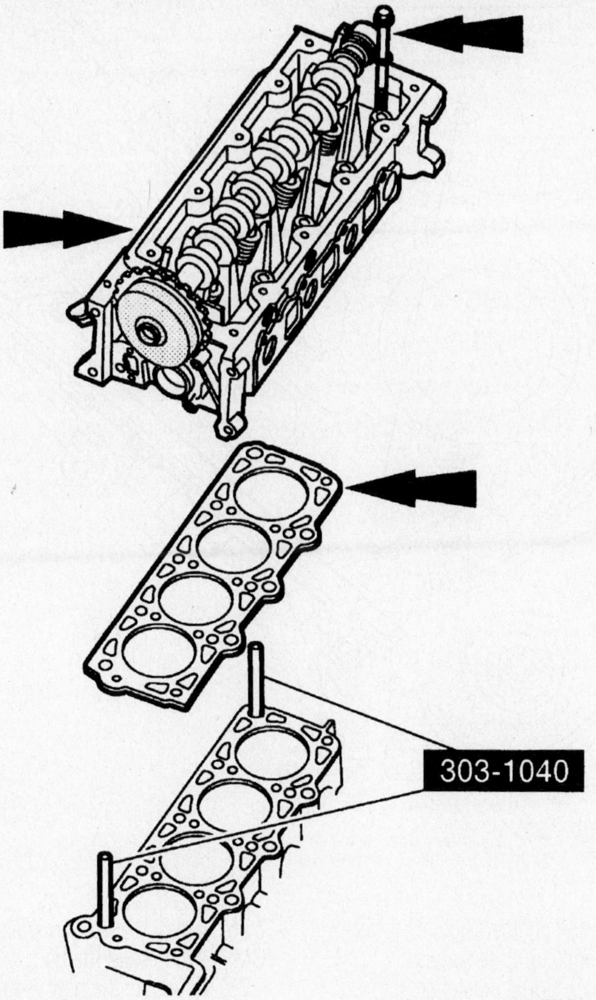

Dowel installation

If the copper links are not visible, mark one link on one end and one link on the other end, and use as timing marks.

55. Install the crankshaft sprocket, making sure the flange faces forward.
56. Position the left timing chain guide.
57. Install and tighten the left bolts.
58. Position the right timing chain guide.
59. Install and tighten the right bolts.
60. Rotate the right camshaft sprocket until the timing mark is approximately at the 11 o'clock position. Rotate the left camshaft sprocket until the timing mark is approximately at the 12 o'clock position.

➡The number one cylinder is at top dead center (TDC) when the stud on the engine block fits into the slot in the handle of the special tool.

61. Position the crankshaft so the number one cylinder is at TDC with the special tool.
62. Remove the Crankshaft Holding Tool.
63. Position the left (inner) timing chain on the crankshaft sprocket, aligning the copper (marked) link with the timing mark on the sprocket.
64. Install the left timing chain on the camshaft sprocket, aligning the copper (marked) link with the timing marks on the sprocket.

➡The left timing chain tensioner arm has a bump near the dowel hole for identification.

65. Position the left timing chain tensioner arm on the dowel pin and install the left timing chain tensioner.
66. Remove the retaining clip from the left timing chain tensioner. Torque to 18 ft. lbs. (25 Nm).
67. Position the right (outer) timing chain on the crankshaft sprocket, aligning the copper (marked) link with the timing mark on the sprocket.
68. Install the right timing chain on the camshaft sprocket, aligning the copper (marked) link with the timing marks on the sprocket.
69. Position the right timing chain tensioner arm on the dowel pin and install the right timing chain tensioner. Torque to 18 ft. lbs. (25 Nm).
70. Remove the retaining clip from the right timing chain tensioner. Make sure that the copper (marked) chain links are lined up with the dots on the crankshaft sprockets and the camshaft sprocket.
71. Install the crankshaft sensor ring on the crankshaft.

⁂ WARNING

Do not use metal scrapers, wire brushes, power abrasive discs or other abrasive means to clean the sealing surfaces. These tools cause scratches and gouges which make leak paths. Use a plastic scraping tool to remove all traces of old sealant.

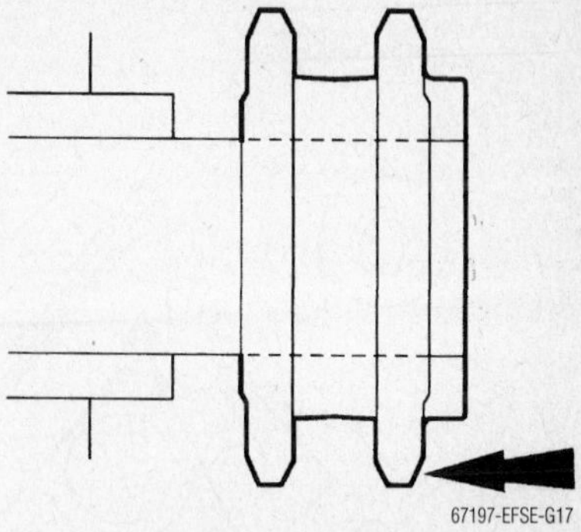

Install the crankshaft sprocket, making sure the flange faces forward

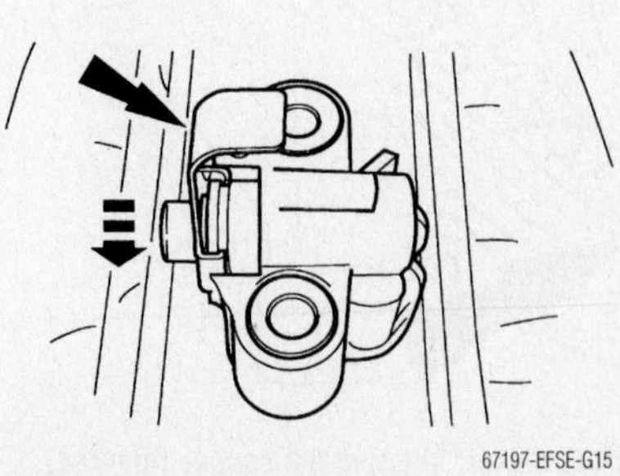

Plunger held by a retaining clip

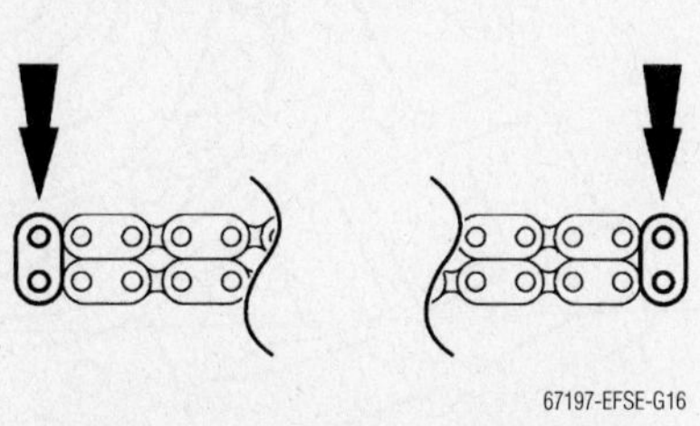

If the copper links are not visible, mark one link on one end and one link on the other end, and use as timing marks

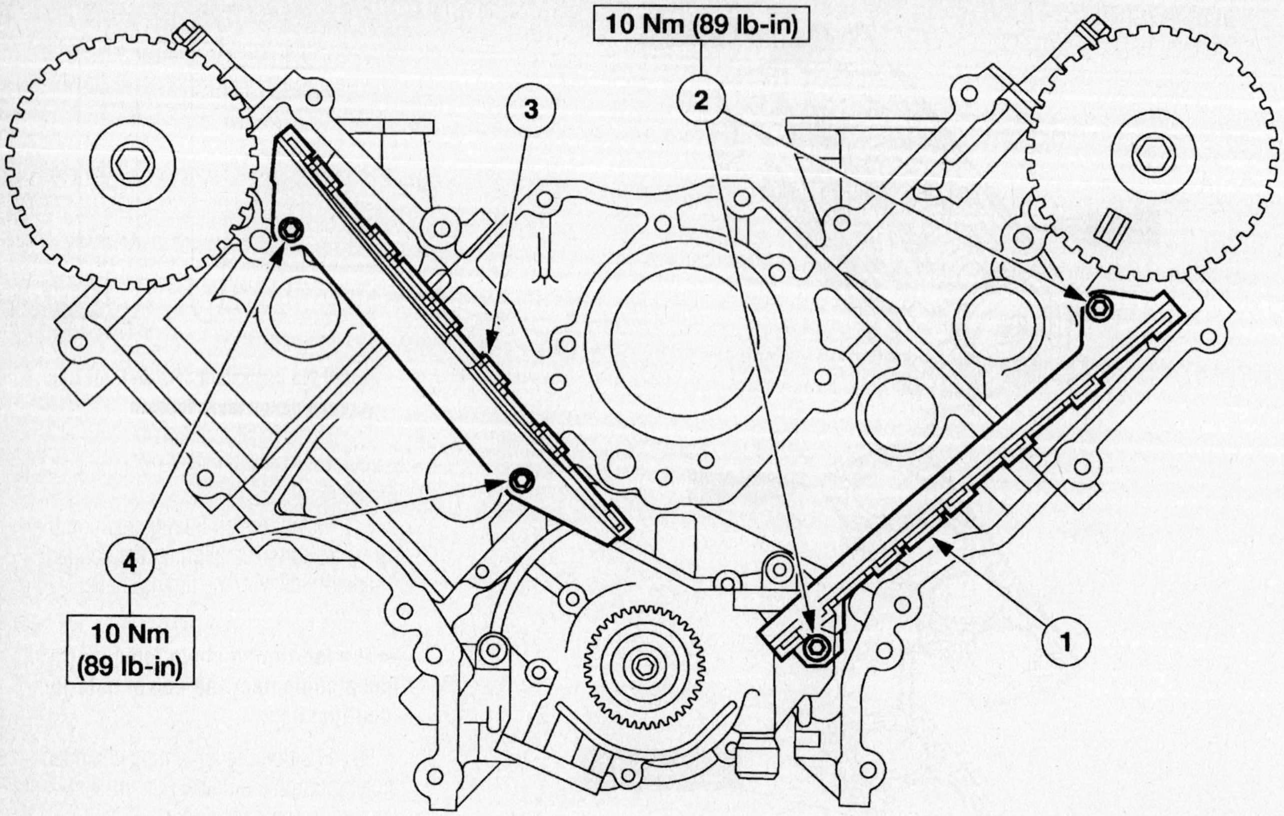

Timing chain guide installation—4.6L engine

➡If the engine front cover is not secured within four minutes, the sealant must be removed and the sealing area cleaned. To clean the sealing area, use silicone gasket remover and metal surface prep. Follow the directions on the packaging. Failure to follow this procedure can cause future oil leakage.

➡Make sure that the engine front cover gasket is in place on the engine front cover before installation.

72. Apply a bead of silicone gasket and sealant along the cylinder head-to-cylinder block surface and the oil pan-to-cylinder block surface, at the locations shown.

73. Install a new engine front cover gasket on the engine front cover. Position the engine front cover. Install the fasteners finger-tight.

74. Tighten the engine front cover fasteners in sequence in three stages.

a. Stage 1: Tighten fasteners 1 through 5 to 25 Nm (18 ft. lbs.).

b. Stage 2: Tighten fasteners 6 and 7 to 25 Nm (18 ft. lbs.).

c. Stage 3: Tighten fasteners 8 through 15 to 25 Nm (18 ft. lbs.).

75. Install the left camshaft position (CMP) sensor and the bolt.

76. Lubricate the new O-ring seal with clean engine oil prior to installation.

77. Lubricate the engine front cover and the crankshaft seal inner lip with clean engine oil. Use a driver to install the crankshaft seal into the engine front cover.

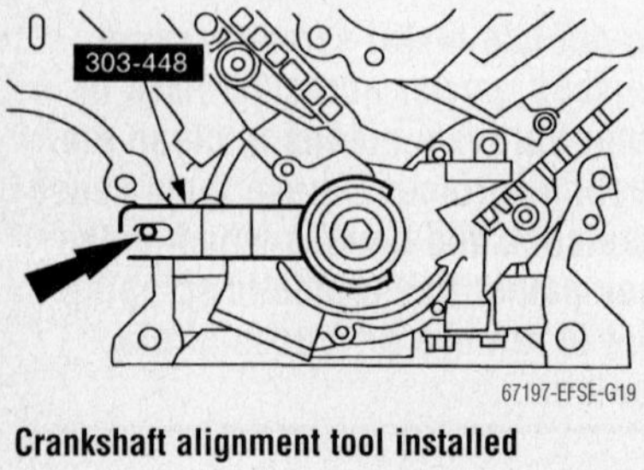

Crankshaft alignment tool installed

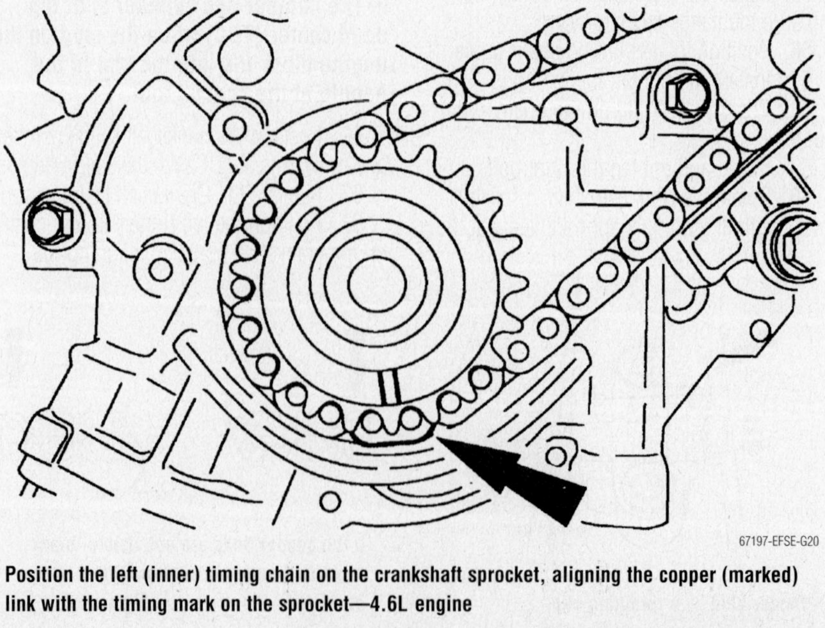

Position the left (inner) timing chain on the crankshaft sprocket, aligning the copper (marked) link with the timing mark on the sprocket—4.6L engine

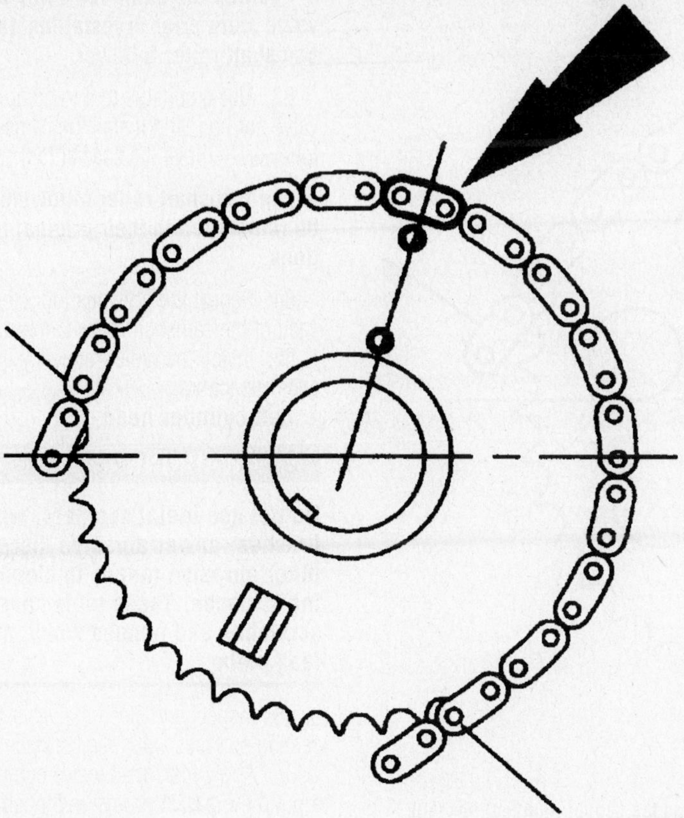

Install the left timing chain on the camshaft sprocket, aligning the copper (marked) link with the timing marks on the sprocket—4.6L engine

➡If not secured within four minutes, the sealant must be removed and the sealing area cleaned. To clean the sealing area, use silicone gasket remover and metal surface prep. Follow the directions on the packaging. Failure to follow this procedure can cause future oil leakage.

78. Apply silicone gasket and sealant to the Woodruff key slot on the crankshaft pulley. Use the special tool to install the crankshaft pulley.

79. Tighten the new crankshaft pulley bolt in four stages.
   a. Stage 1: Tighten to 90 Nm (66 ft. lbs.).
   b. Stage 2: Loosen 360 degrees.
   c. Stage 3: Tighten to 50 Nm (37 ft. lbs.).
   d. Stage 4: Tighten an additional 90 degrees.

80. Install the three accessory drive belt idler pulleys, the coolant pump pulley and the bolts.

81. Position the accessory drive belt tensioner and install the bolts.

82. Install a suitable tool between the valve spring coils to prevent valve stem seal damage.

➡The camshaft roller followers must be reinstalled in their original locations.

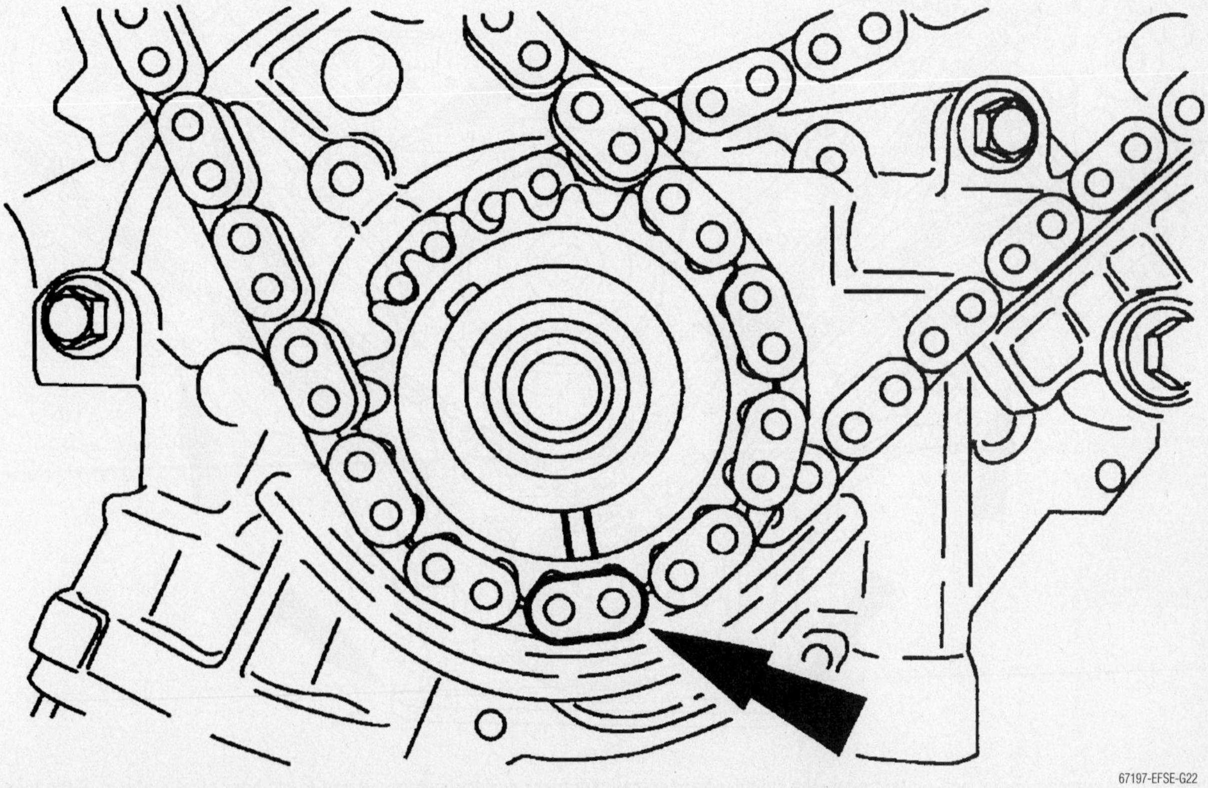

Position the right (outer) timing chain on the crankshaft sprocket, aligning the copper (marked) link with the timing mark on the sprocket—4.6L engine

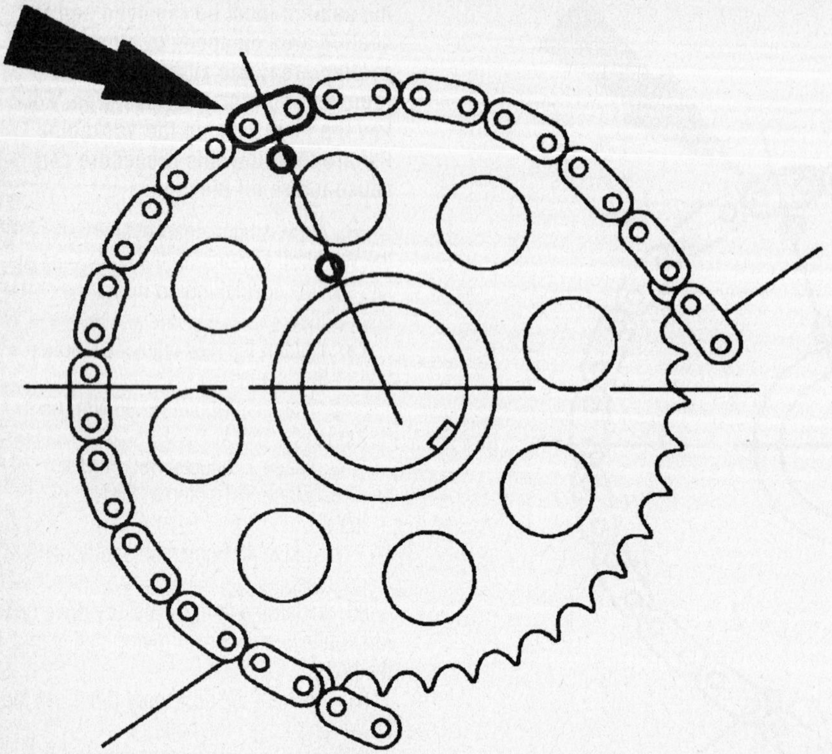

Install the right timing chain on the camshaft sprocket, aligning the copper (marked) link with the timing marks on the sprocket—4.6L engine

➡Position the cam lobe away from the valve stem prior to installing each camshaft roller follower.

83. Use a suitable tool to compress the valve springs, and install the camshaft roller follower. Remove the special tool.

➡The camshaft roller followers must be reinstalled in their original locations.

84. Repeat the previous four steps for each of the camshaft roller followers.
85. Install the radio frequency interference capacitors.

Left cylinder head

⁂ WARNING

Do not use metal scrapers, wire brushes, power abrasive discs or other abrasive means to clean sealing surfaces. These tools cause scratches and gouges which make leak paths.

86. Inspect and clean the valve cover sealing surfaces with metal surface cleaner.
87. Apply instant adhesive completely around the gasket groove in the left valve cover.

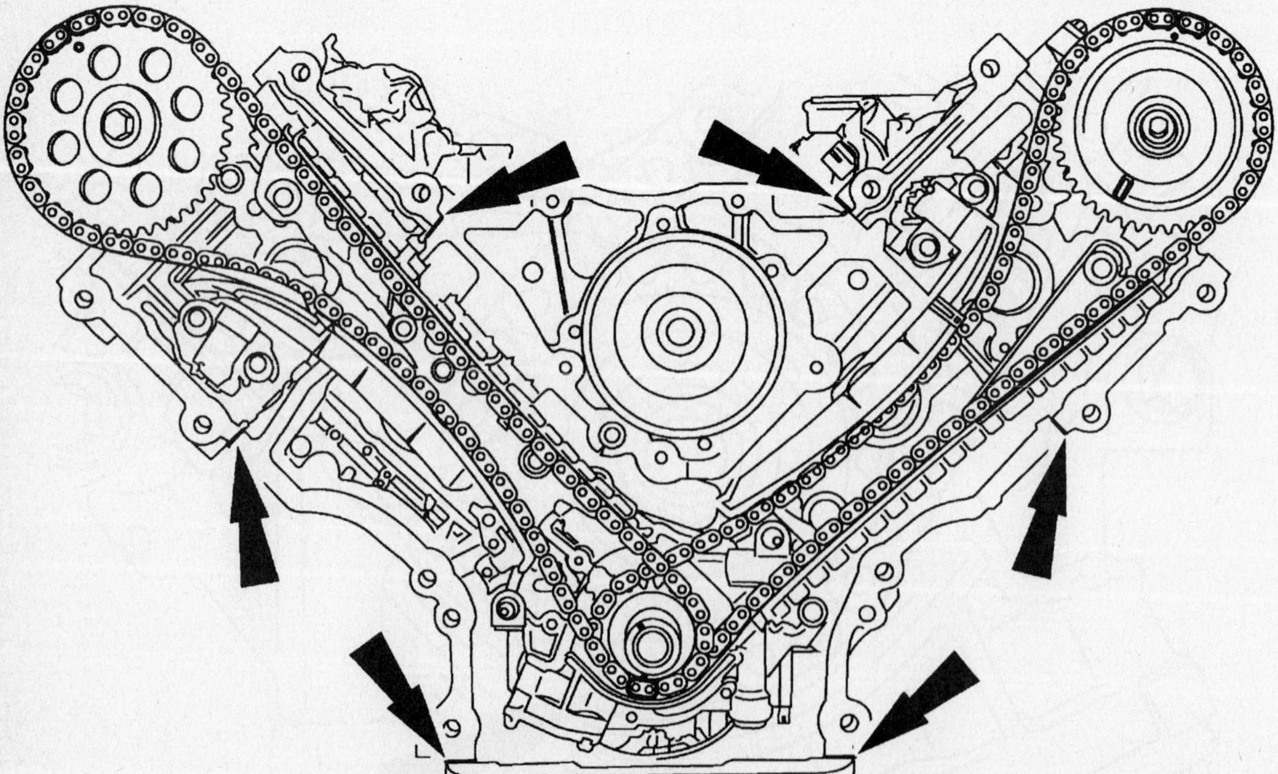

Apply a bead of silicone gasket and sealant along the cylinder head-to-cylinder block surface and the oil pan-to-cylinder block surface, at the locations shown—4.6L engine

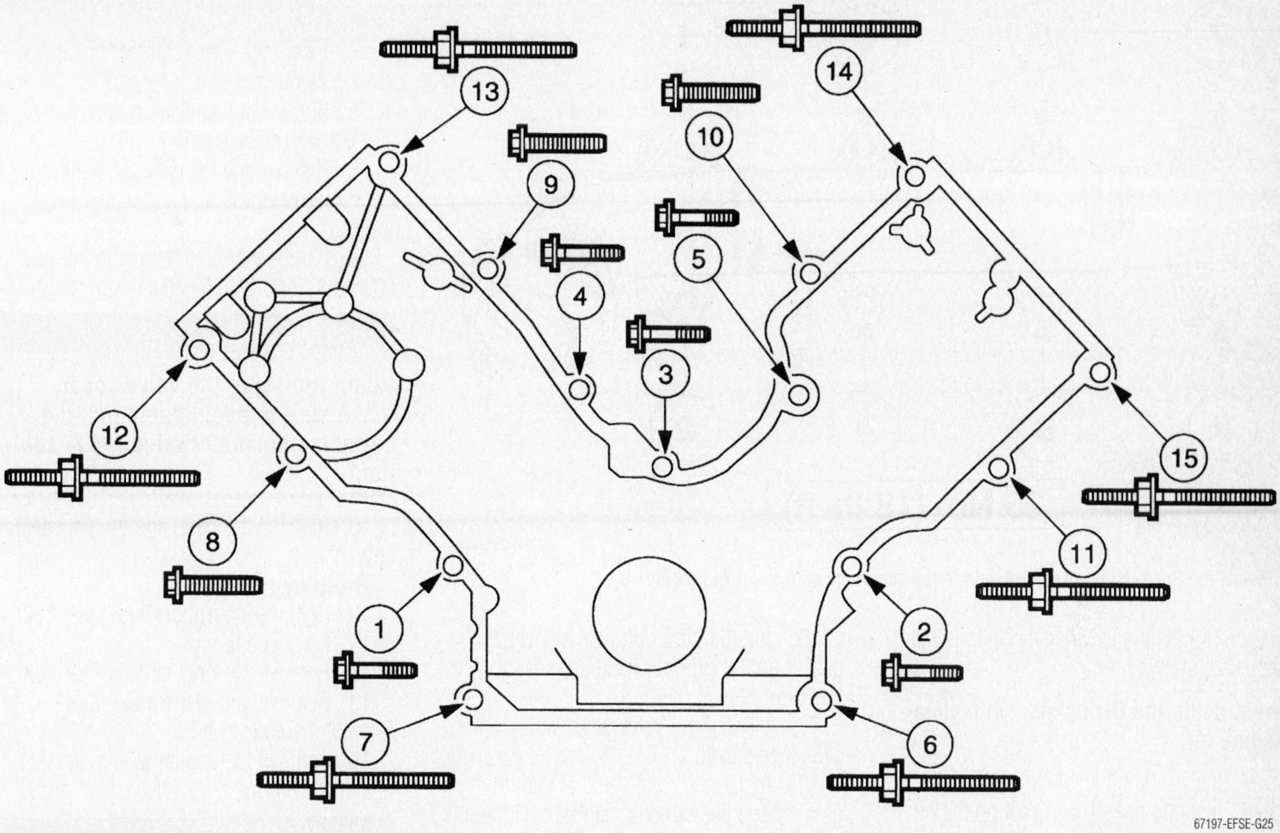

Front cover fastener tightening sequence—4.6L engine

88. Install the new valve cover gasket.

➡**If not secured within four minutes, the sealant must be removed and the sealing area cleaned. To clean the sealing area, use silicone gasket remover and metal surface prep.**

89. Follow the directions on the packaging. Failure to follow this procedure can cause future oil leakage.

90. Apply silicone gasket and sealant in two places where the engine front cover meets the cylinder head.

91. Position the left valve cover and gasket on the cylinder head and install the bolts loosely.

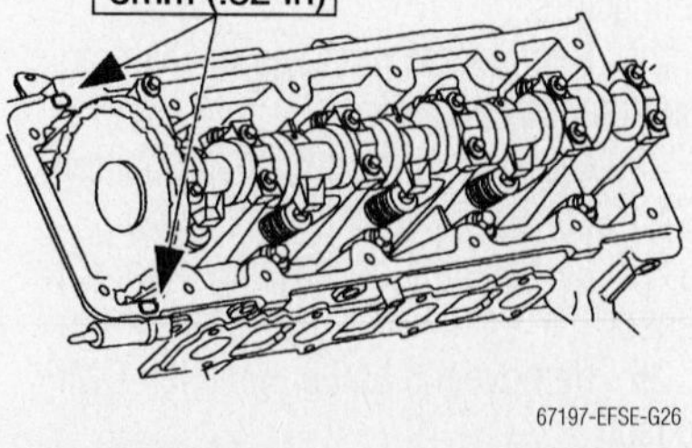

Apply silicone gasket and sealant in two places where the engine front cover meets the cylinder head

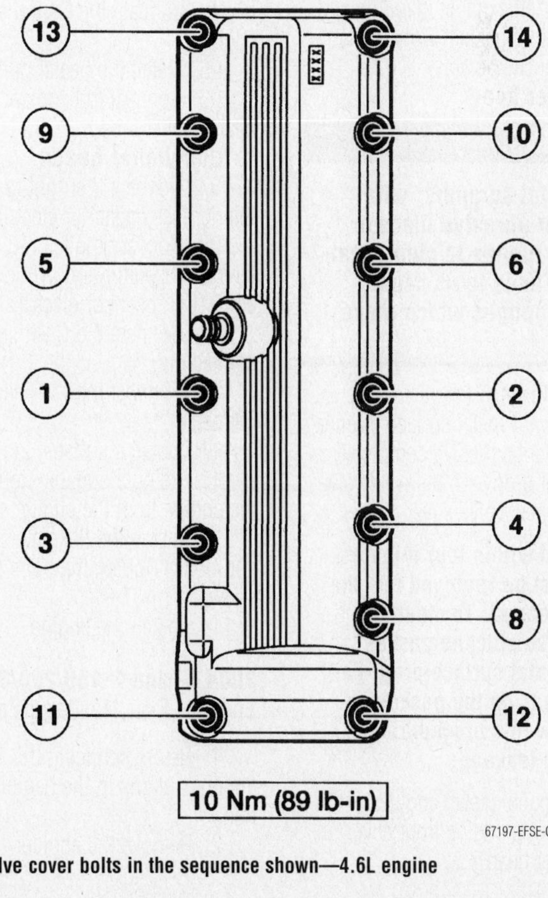

Tighten the valve cover bolts in the sequence shown—4.6L engine

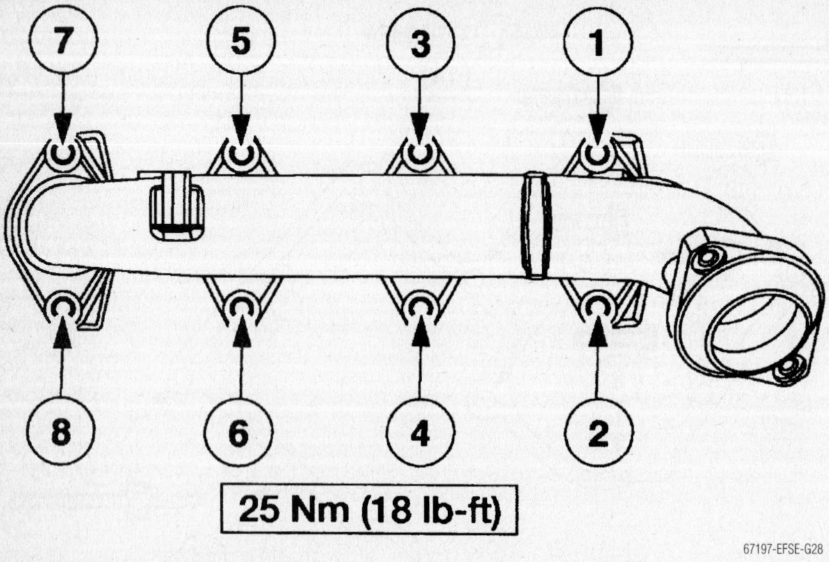

Install the left exhaust manifold nuts in the sequence shown—4.6L engine

92. Tighten the valve cover bolts in the sequence shown.

➡**Lubricate the O-ring seal with clean engine oil.**

93. Install the oil level indicator tube.
94. Install a new O-ring seal on the oil level indicator tube.
95. Install the oil level indicator tube.
96. Install the bolt.
97. Install the left exhaust manifold and the exhaust manifold gasket. Tighten the nuts in the sequence shown.

Right cylinder head

⁂ WARNING

Do not use metal scrapers, wire brushes, power abrasive discs or other abrasive means to clean sealing surfaces. These tools cause scratches and gouges which make leak paths.

98. Inspect and clean the valve cover sealing surfaces with metal surface cleaner.
99. Apply instant adhesive completely around the gasket groove in the right valve cover. Install the new valve cover gasket.

➡**If not secured within four minutes, the sealant must be removed and the sealing area cleaned. To clean the sealing area, use silicone gasket remover and metal surface prep. Follow the directions on the packaging. Failure to follow this procedure can cause future oil leakage.**

100. Apply silicone gasket and sealant in two places where the engine front cover meets the cylinder head.
101. Position the right valve cover and gasket on the cylinder head and install the bolts loosely.
102. Tighten the valve cover bolts in the sequence shown.
103. Install the right exhaust manifold gaskets and the exhaust manifold. Tighten the nuts in the sequence shown.
104. Slide the heater outlet tube forward with a new O-ring seal into the cylinder block. Lubricate the O-ring seal with engine coolant.
105. Install the heater outlet tube studs.
106. Connect the coolant hoses to the heater outlet tube.

All cylinder heads

107. Connect the right radio ignition interference capacitor electrical connector.
108. Connect the left radio ignition interference capacitor and cylinder head temperature (CHT) sensor electrical connectors.
109. Connect the CMP sensor electrical connectors.
110. Connect the CKP sensor electrical connector.
111. Install a suitable tool.
112. Using a suitable floor crane, remove the engine from the engine stand.
113. Install the flexplate or the flywheel and bolts. Tighten the bolts in the sequence shown.
114. Install the engine.

2004 E- and F-150/250/350 5.4L Engine, Exc. Heritage Edition

1. Before servicing the vehicle, refer to the precautions in the beginning of this section.
2. Remove the engine.
3. Remove the bolts and the flexplate or the flywheel.
4. Install the engine onto a suitable engine stand.
5. Remove the special tool.

Left cylinder head

6. Remove the left exhaust manifold.
7. Remove the bolt and the oil level indicator tube.
8. Remove the engine wiring harness retainers from the left valve cover studs.

⁂ WARNING

When removing the valve cover, make sure to avoid damaging the variable camshaft timing (VCT) solenoid.

9. Remove the bolts and the left valve cover.

Right cylinder head

10. Remove the right exhaust manifold.
11. Remove the nuts.
12. Remove the right exhaust manifold.
13. Remove and discard the right exhaust manifold gasket.
14. Remove the engine wiring harness retainers from the right valve cover studs.

⁂ WARNING

When removing the valve cover, make sure to avoid damaging the variable camshaft timing (VCT) solenoid.

15. Remove the bolts and the right valve cover.
16. Remove the stud.

All cylinder heads

17. Remove the bolts, the coolant pump pulley and the accessory drive belt idler pulleys.
18. Remove the bolt and washer and using a puller set, remove the crankshaft pulley. Discard the crankshaft bolt.
19. Using a suitable tool, remove the crankshaft seal.
20. Remove the bolts and the accessory drive belt tensioner.
21. Disconnect the left and right radio ignition interference capacitor electrical connectors.
22. Remove the nuts and the two radio interference capacitors.
23. Disconnect the camshaft position (CMP) sensor electrical connectors.
24. Remove the bolt and the right CMP sensor.
25. Remove the bolt and the left CMP sensor.
26. Disconnect the crankshaft position (CKP) sensor electrical connector.

	Alignment Pins, Cylinder Head 303-1040 (SR-015486)
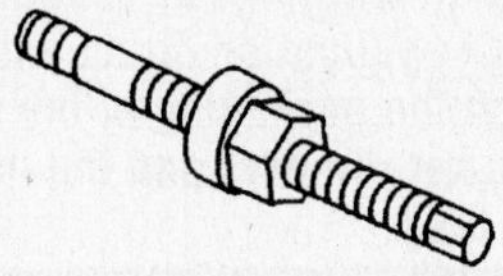	Installer, Crankshaft Vibration Damper 303-102 (T74P-6316-B)
	Installer, Crankshaft Front Oil Seal 303-635
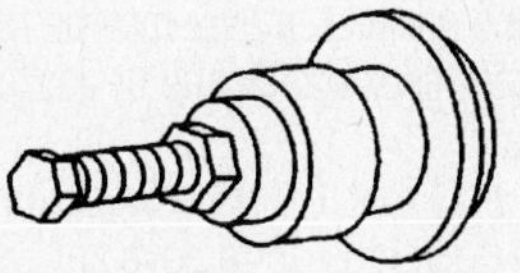	Installer, Front Cover Oil Seal 303-335 (T88T-6701-A)
	Modular Engine Lift Bracket 303-F047 (014-00073) or equivalent
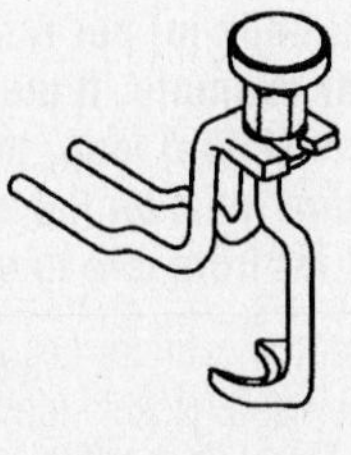	Compressor, Valve Spring 303-1039

67197-EFSE-G33

Special tools needed for this procedure. These tools are referred to in the following procedure

27. Remove the oil pan front bolts.
28. Remove the bolts.

⁂ WARNING

Do not use metal scrapers, wire brushes, power abrasive discs or other abrasive means to clean the sealing surfaces. These tools cause scratches and gouges which make leak paths. Use a plastic scraping tool to remove all traces of old sealant.

29. Remove the engine front cover from the front cover-to-cylinder block dowels.
30. Remove the engine front cover gaskets.
31. Clean the mating surfaces with silicone gasket remover and metal surface prep. Follow the directions on the packaging.
32. Inspect the mating surfaces.

⁂ WARNING

Do not allow the valve keepers to fall off of the valve or the valve can drop into the cylinder.

➡It may be necessary to push the valve down while compressing the valve spring.

➡The roller followers must be installed in their original positions.

33. Using a suitable tool, remove all of the roller followers. Record the roller follower positions.
34. Position the crankshaft keyway at the 12 o'clock position.
35. Remove the bolts, the left timing chain tensioner and tensioner arm.
36. Remove the bolts, the right timing chain tensioner and tensioner arm.
37. Remove the ignition pulse wheel from the crankshaft.
38. Remove the right timing chain from the camshaft sprocket.
39. Remove the right timing chain from the crankshaft sprocket.
40. Remove the left timing chain from the camshaft sprocket.
41. Remove the left timing chain and crankshaft sprocket.
42. Remove both timing chain guides.

⁂ WARNING

Use only hand tools to remove the camshaft phaser sprocket assembly or damage can occur to the camshaft or camshaft phaser sprocket.

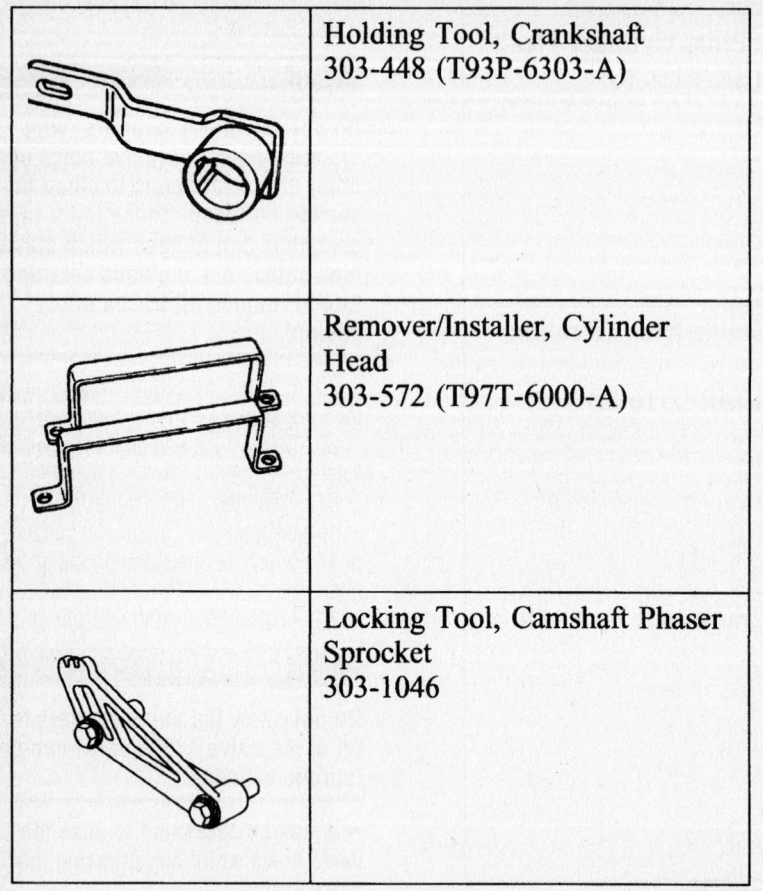

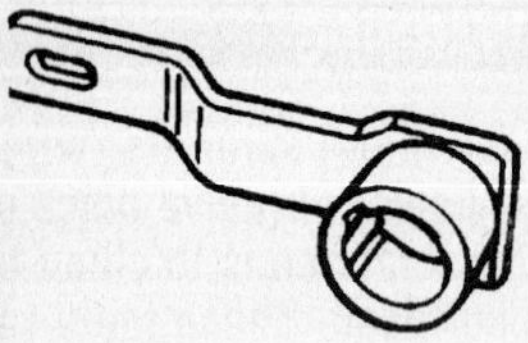	Holding Tool, Crankshaft 303-448 (T93P-6303-A)
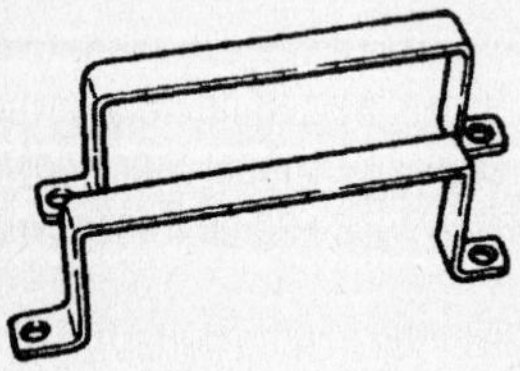	Remover/Installer, Cylinder Head 303-572 (T97T-6000-A)
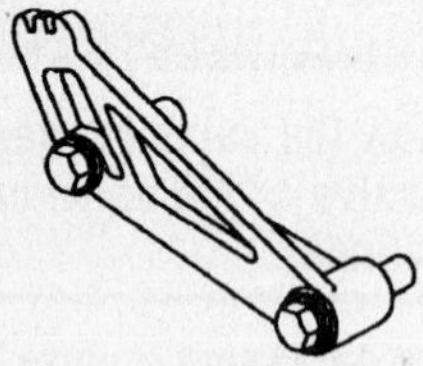	Locking Tool, Camshaft Phaser Sprocket 303-1046

67197-EFSE-G34

Special tools needed for this procedure. These tools are referred to in the following procedure

43. If disassembly of the cylinder head is required, using a suitable tool, loosen the camshaft phaser sprocket bolt.

44. Install a suitable tool onto the left cylinder head. Install a suitable tool onto the right cylinder head.

**** WARNING**

The cylinder head must be cool before removing it from the engine. Cylinder head warpage can result if a warm or hot cylinder head is removed.

**** WARNING**

Do not use the variable camshaft timing (VCT) phaser sprocket as a lifting point or leveraging device when removing the cylinder head or damage to the VCT phaser sprocket can occur.

**** WARNING**

Place clean shop towels over exposed engine cavities. Carefully remove the towels so foreign material is not dropped into the engine.

**** WARNING**

The cylinder head bolts must be discarded and new bolts must be installed. They are tighten-to-yield designed and cannot be reused.

**** WARNING**

Do not use metal scrapers, wire brushes, power abrasive discs or other abrasive means to clean the sealing surfaces. These tools cause scratches and gouges that make leak paths. Use a plastic scraping tool to remove all traces of the head gasket.

**** WARNING**

Aluminum surfaces are soft and can be scratched easily. Never place the cylinder head gasket surface, unprotected, on a bench surface.

45. Remove the bolts and the cylinder head. Discard the cylinder head gasket. Discard the cylinder head bolts.

**** WARNING**

Do not use metal scrapers, wire brushes, power abrasive discs or other abrasive means to clean the sealing surfaces. These tools cause scratches and gouges that make leak paths. Use a plastic scraping tool to remove all traces of the head gasket.

**** WARNING**

Observe all warnings or cautions and follow all application directions contained on the packaging of the silicone gasket remover and the metal surface prep.

➡If there is no residual gasket material present, metal surface prep can be used to clean and prepare the surfaces.

46. Clean the cylinder head-to-cylinder block mating surfaces of both the cylinder head and the cylinder block.

47. Remove any large deposits of silicone or gasket material with a plastic scraper.

48. Apply silicone gasket remover, following package directions, and allow to set for several minutes.

49. Remove the silicone gasket remover with a plastic scraper. A second application of silicone gasket remover may be required if residual traces of silicone or gasket material remain.

50. Apply metal surface prep, following package directions, to remove any remaining traces of oil or coolant, and to prepare the surfaces to bond with the new gasket. Do not attempt to make the metal shiny. Some staining of the metal surfaces is normal.

➡Make sure all cylinder head surfaces and engine block surfaces are clear of any gasket material, RTV, oil and coolant. The cylinder head and engine block surfaces must be clean and dry before running a flatness check.

➡Use a straightedge that is calibrated by the manufacturer to be flat within 0.005 mm (0.0002 in) per running foot of length. For example, if the straightedge is 61 cm (24 in) long, the machined edge must be flat with 0.010 mm (0.0004 in) from end to end.

51. Support the cylinder head on a bench with the head gasket side up. Inspect all areas of the deck face with a straightedge, paying particular attention to the oil pressure feed area. The cylinder head must

not have depressions deeper than 0.0254 mm (0.001 in) across a 38.1 mm (1.5 in) square area, or scratches more than 0.0254 mm (0.001 in).

To install:

All cylinder heads

**** WARNING**

Make sure all coolant residue and foreign material are cleaned from the block surface and cylinder bore.

**** WARNING**

The use of sealing aids (aviation cement, copper spray, and glue) is not permitted. The gasket must be installed dry.

**** WARNING**

The cylinder head bolts must be discarded and new bolts installed. They are tighten-to-yield designed and cannot be reused.

**** WARNING**

Do not allow the cylinder head alignment pins to contact the cylinder head gasket or cylinder head sealing surfaces or damage can occur to the cylinder head or cylinder head gasket.

➡Do not turn the crankshaft until instructed to do so.

52. Using the cylinder head alignment pins, position the cylinder head gasket and cylinder head onto the dowels and install the cylinder head bolts loosely.
53. Tighten the bolts in the sequence shown.
    a. Stage 1: Tighten to 40 Nm (30 ft. lbs.).
    b. Stage 2: Tighten an additional 90 degrees.
    c. Stage 3: Tighten an additional 90 degrees.
54. Remove the special tool.

**** WARNING**

Timing chain procedures must be followed exactly or damage to valves and pistons will result.

55. Compress the tensioner plunger, using a vice.
56. Install a retaining clip on the tensioner to hold the plunger in during installation.

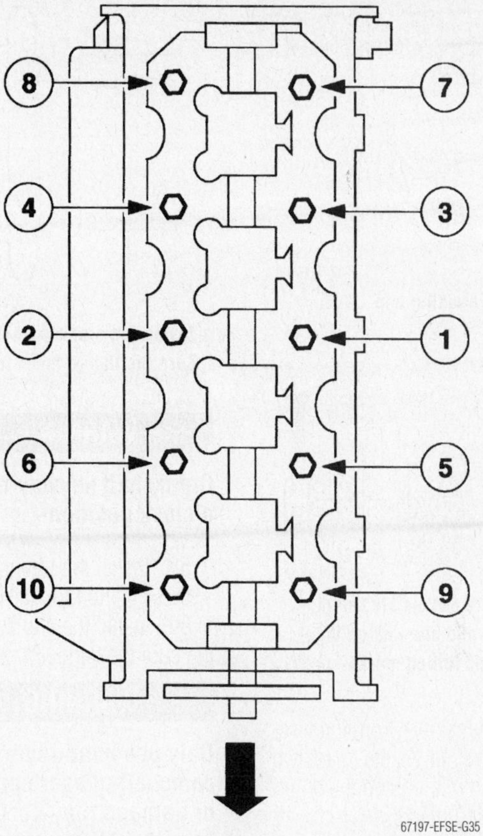

Left cylinder head torque sequence—5.4L engine

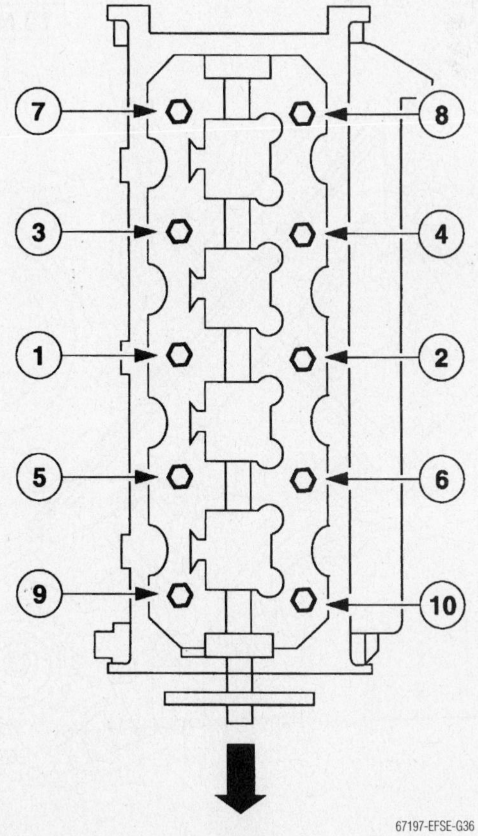

Right cylinder head torque sequence—5.4L engine

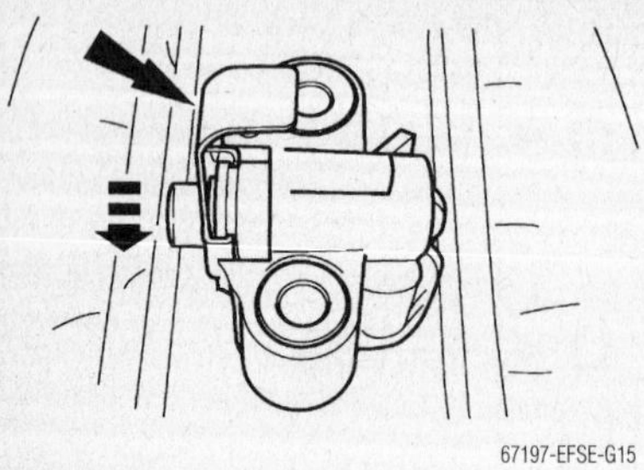

Plunger held by a retaining clip

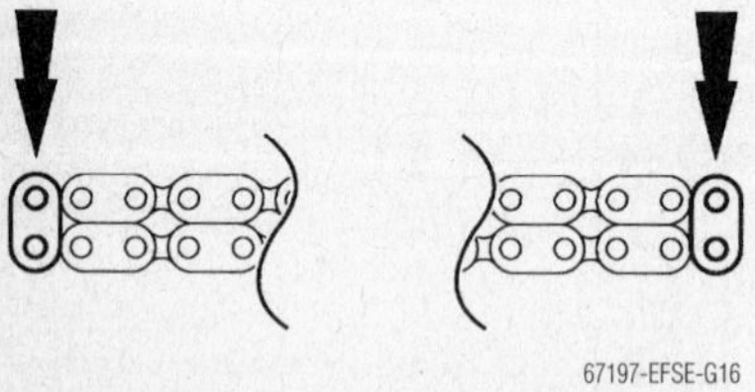

If the copper links are not visible, mark one link on one end and one link on the other end, and use as timing marks

57. Remove the tensioner from the vise. If the copper links are not visible, mark two links on one end and one link on the other end, and use as timing marks.

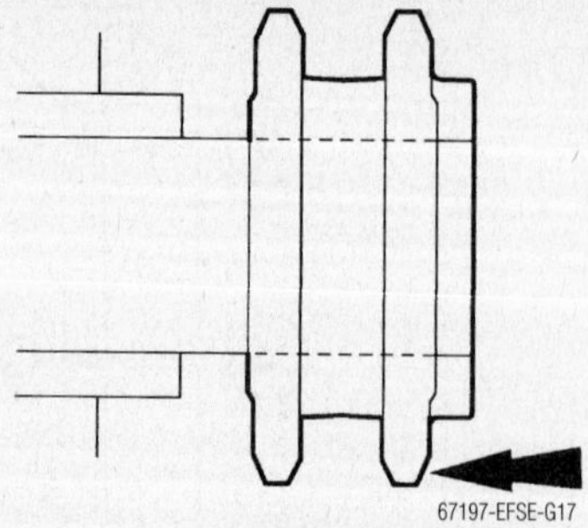

Install the crankshaft sprocket, making sure the flange faces forward

WARNING

Crankshaft keyway must be in the 12 o'clock position.

58. Install the timing chain guides.
59. Position the left timing chain guide.
60. Install the crankshaft sprocket, making sure the flange faces forward.

WARNING

Only use hand tools to install the camshaft phaser sprocket assembly or damage may occur to the camshaft or camshaft phaser unit.

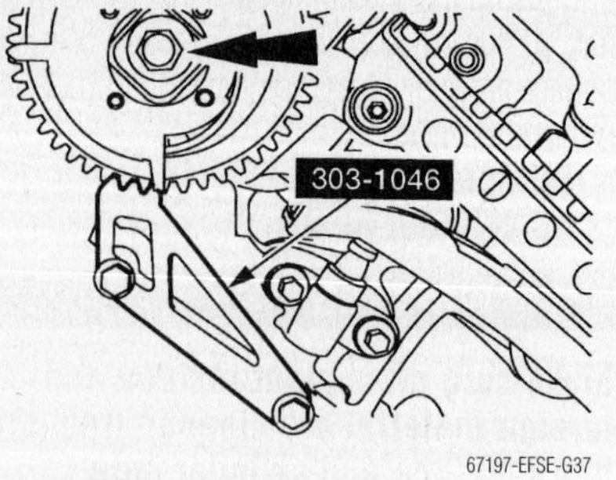

Camshaft phaser sprocket holding tool—5.4L engine

➡This step is only required if cylinder head was disassembled.

61. Using a suitable tool, tighten the bolt in two stages:
 a. Stage 1: Tighten to 40 Nm (30 ft. lbs.).
 b. Stage 2: Tighten an additional 90 degrees.
62. Remove the special tool.
63. Position the lower end of the left (inner) timing chain on the crankshaft sprocket, aligning the timing mark on the outer flange of the crankshaft sprocket with the single copper (marked) link on the chain.

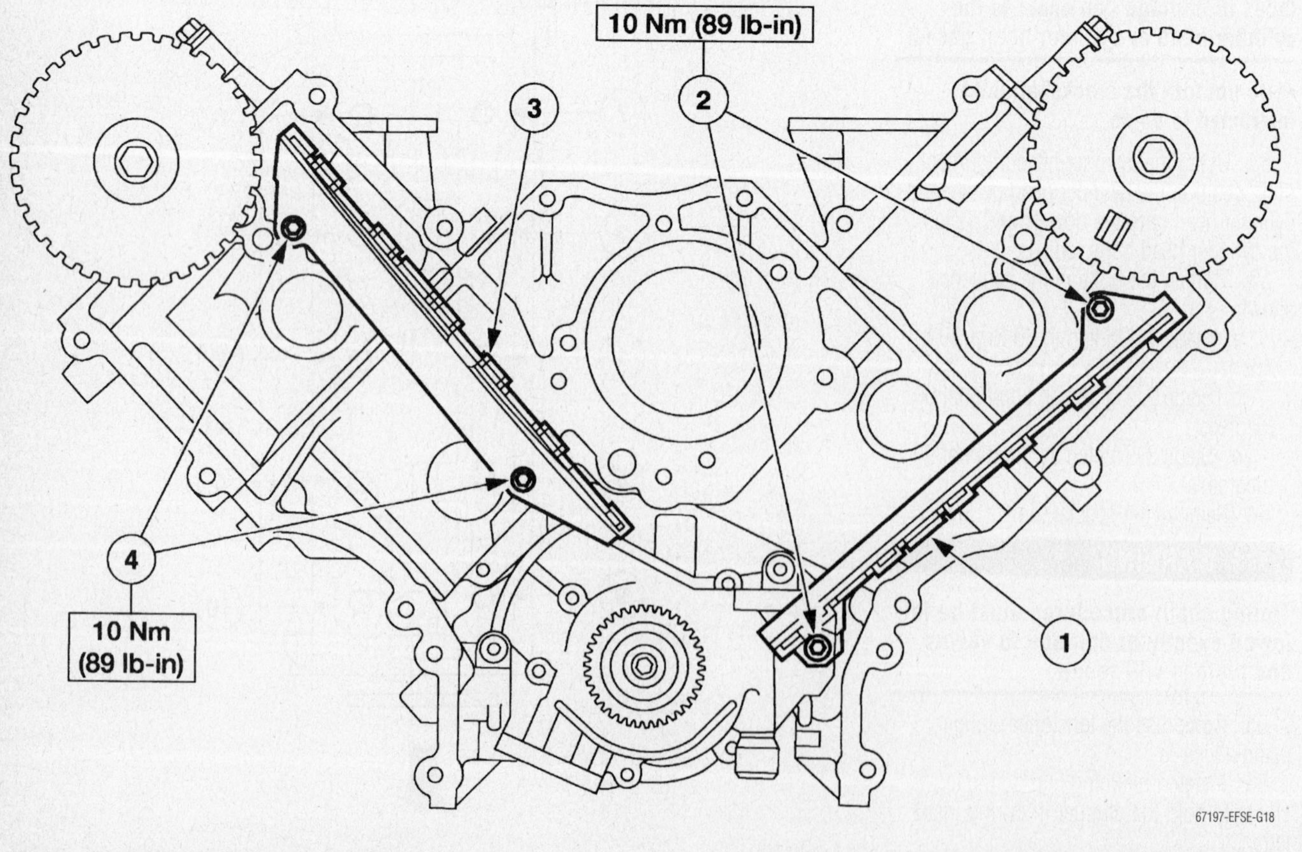

Timing chain guide installation—5.4L engine

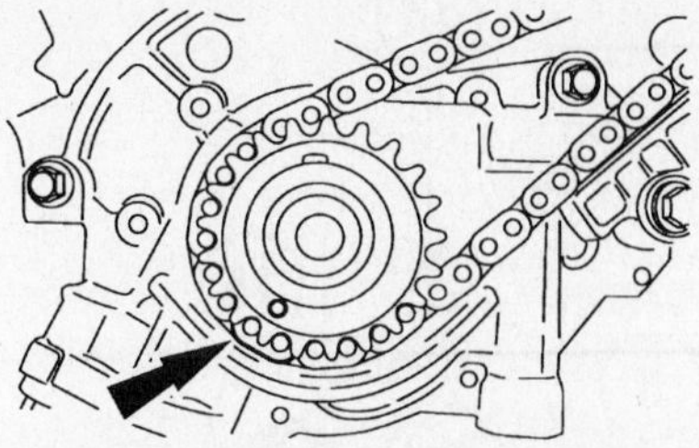

67197-EFSE-G38

Position the lower end of the left (inner) timing chain on the crankshaft sprocket, aligning the timing mark on the outer flange of the crankshaft sprocket with the single copper (marked) link on the chain—5.4L engine

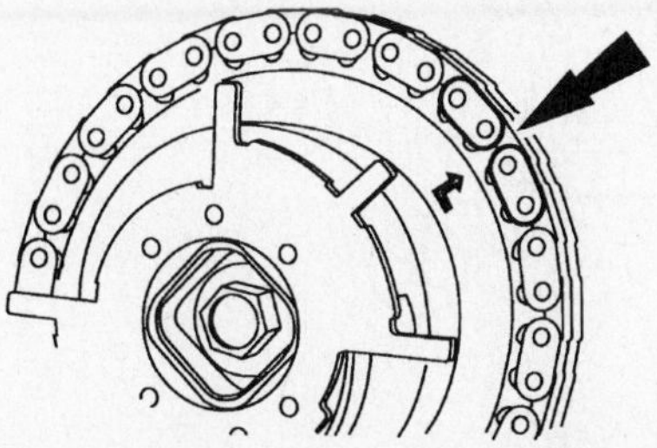

67197-EFSE-G39

Position the timing chain on the camshaft sprocket with the camshaft sprocket timing mark positioned between the two copper (marked) chain links—5.4L engine

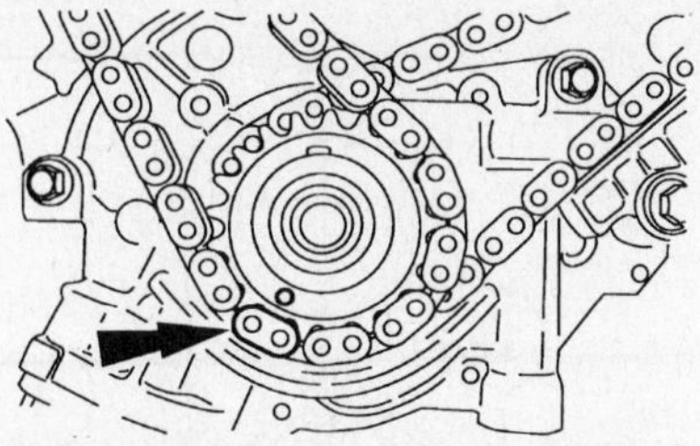

67197-EFSE-G40

Position the lower end of the right (outer) timing chain on the crankshaft sprocket, aligning the timing mark on the sprocket with the single copper (marked) chain link—5.4L engine

➡Make sure the upper half of the timing chain is below the tensioner arm dowel.

Position the timing chain on the camshaft sprocket with the camshaft sprocket timing mark positioned between the two copper (marked) chain links.

➡The left timing chain tensioner arm has a bump near the dowel hole for identification.

64. Position the left timing chain tensioner arm on the dowel pin and install the left timing chain tensioner.

65. Remove the retaining clip from the left timing chain tensioner.

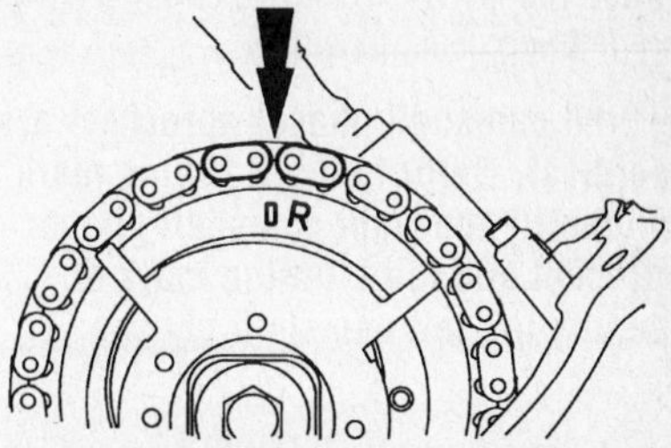

67197-EFSE-G41

Position the right timing chain on the camshaft sprocket. Make sure the camshaft sprocket timing mark is positioned between the two copper (marked) chain links—5.4L engine

66. Position the lower end of the right (outer) timing chain on the crankshaft sprocket, aligning the timing mark on the sprocket with the single copper (marked) chain link.

➡The lower half of the timing chain must be positioned above the tensioner arm dowel.

67. Position the right timing chain on the camshaft sprocket. Make sure the camshaft sprocket timing mark is positioned between the two copper (marked) chain links.

68. Position the right timing chain tensioner arm on the dowel pin and install the right timing chain tensioner.

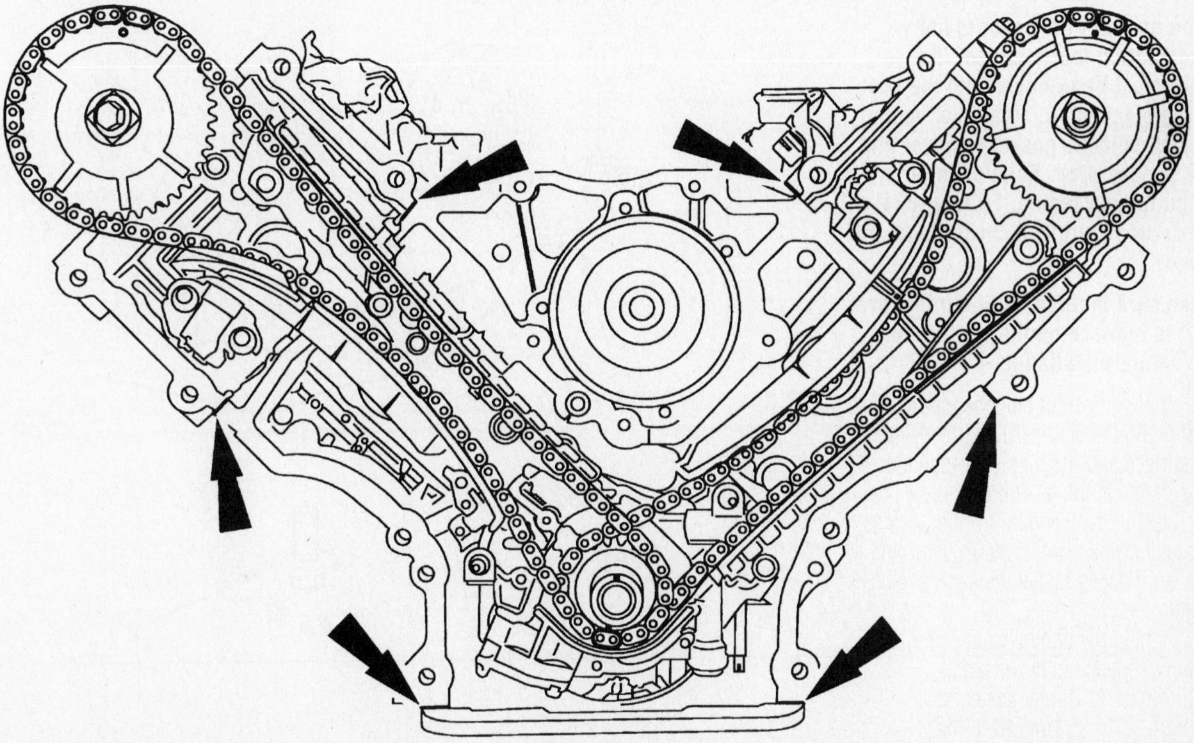

67197-EFSE-G42

Apply a bead of silicone gasket and sealant along the cylinder head-to-cylinder block surface and the oil pan-to-cylinder block surface, at the locations shown—5.4L engine

69. Remove the retaining clip from the right timing chain tensioner.

➡Both camshaft phaser sprockets are identical. Refer to the R timing mark to identify the right camshaft phaser sprocket and the L timing mark to identify the left camshaft phaser sprocket.

70. As a post-check, verify correct alignment of all timing marks. Make sure the R and L timing marks on the sprockets correspond to the above note.

71. Install the crankshaft sensor ring on the crankshaft.

➡Lubricate the camshaft roller followers using clean engine oil.

➡Using the mark on each camshaft roller follower, make sure it is returned to its original position.

72. Using a suitable tool, install all of the camshaft roller followers.

✲✲ WARNING

Do not use metal scrapers, wire brushes, power abrasive discs or other abrasive means to clean the sealing surfaces. These tools cause scratches and gouges which make leak paths. Use a plastic scraping tool to remove all traces of old sealant.

➡If the engine front cover is not secured within four minutes, the sealant must be removed and the sealing area cleaned. To clean the sealing area, use silicone gasket remover and metal surface prep. Follow the directions on the packaging. Failure to follow this procedure can cause future oil leakage.

➡Make sure that the engine front cover gasket is in place on the engine front cover before installation.

73. Apply a bead of silicone gasket and sealant along the cylinder head-to-cylinder block surface and the oil pan-to-cylinder block surface, at the locations shown.

74. Install a new engine front cover gasket on the engine front cover. Position the engine front cover. Install the fasteners finger-tight.

75. Tighten the engine front cover fasteners in sequence in two stages.

a. Stage 1: Tighten fasteners 1 through 15 to 25 Nm (18 ft. lbs.).

b. Stage 2: Tighten fasteners 6 and 7 to 48 Nm (35 ft. lbs.).

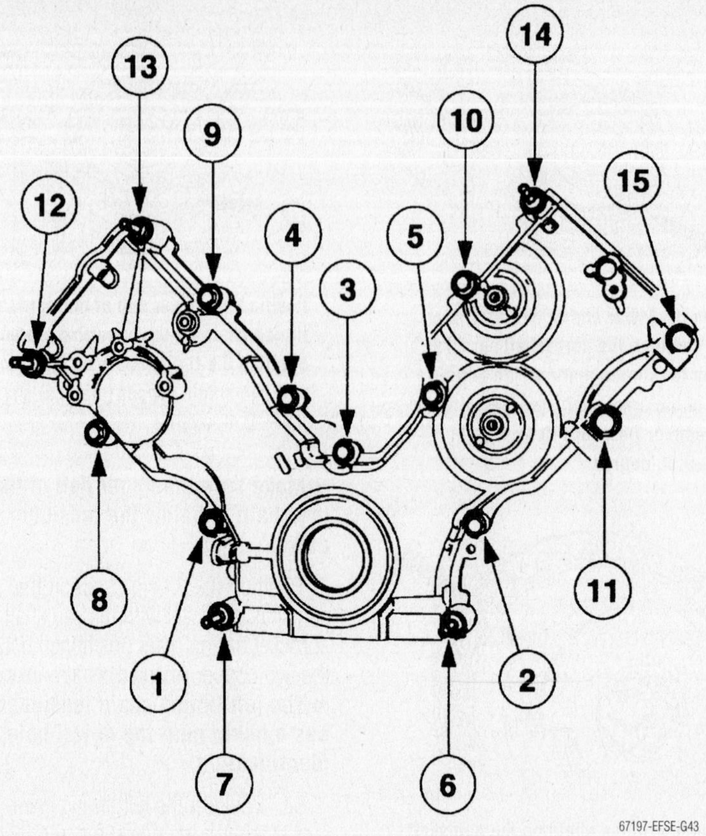

Front cover fastener tightening sequence—5.4L engine

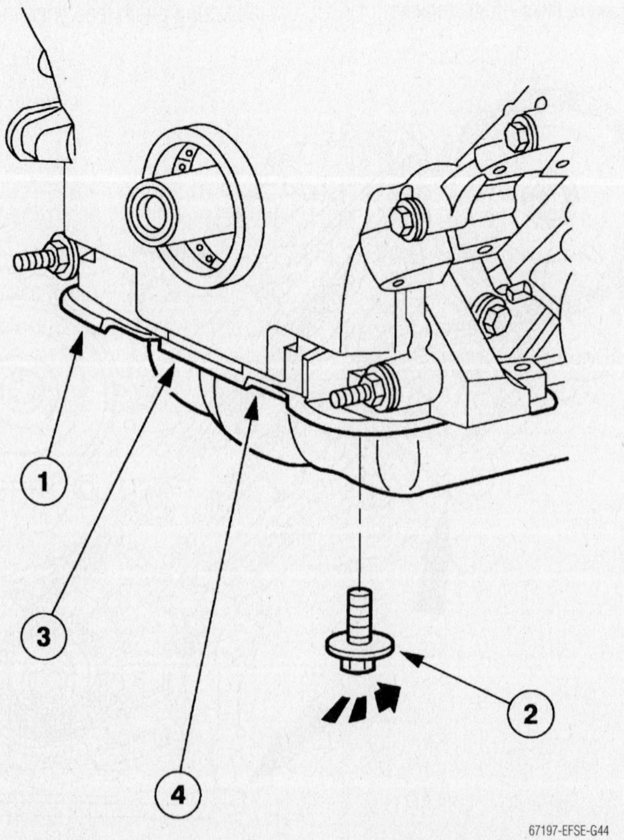

Loosely install the pan-to-case bolts, then tighten the bolts in two stages, in the sequence shown—5.4L engine

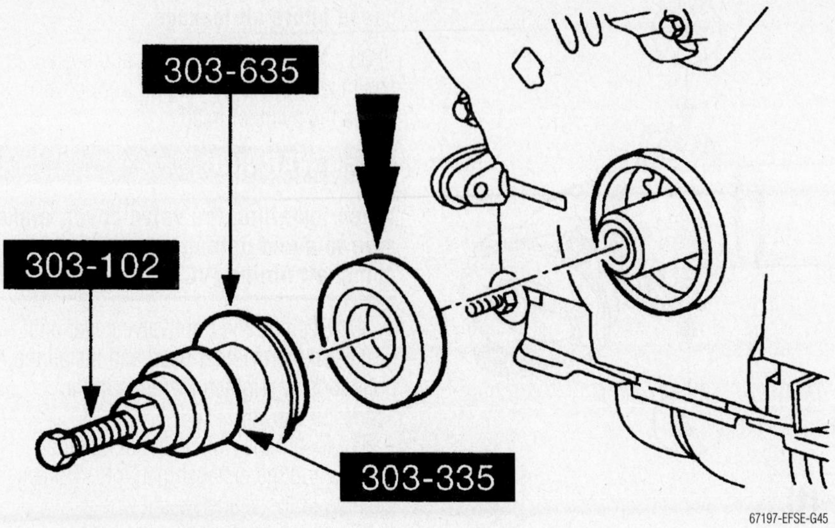

Crankshaft seal installation tools

c. Loosely install the pan-to-case bolts, then tighten the bolts in two stages, in the sequence shown.

d. Stage 1: Tighten to 20 Nm (15 ft. lbs.).

e. Stage 2: Tighten an additional 60 degrees.

76. Install the left camshaft position (CMP) sensor and the bolt.

77. Lubricate the new O-ring seal with clean engine oil prior to installation.

78. Install the right CMP sensor and the bolt.

79. Lubricate the new O-ring seal with clean engine oil prior to installation.

80. Lubricate the engine front cover and the crankshaft seal inner lip with clean engine oil.

81. Use the special tools to install the crankshaft seal into the engine front cover.

82. If not secured within four minutes, the sealant must be removed and the sealing area cleaned. To clean the sealing area, use silicone gasket remover and metal surface prep. Follow the directions on the packaging. Failure to follow this procedure can cause future oil leakage.

83. Apply silicone gasket and sealant to the Woodruff key slot on the crankshaft pulley. Use a suitable tool to install the crankshaft pulley.

84. Tighten the new crankshaft pulley bolt in four stages.

a. Stage 1: Tighten to 90 Nm (66 ft. lbs.).

b. Stage 2: Loosen 360 degrees.

c. Stage 3: Tighten to 50 Nm (37 ft. lbs.).

d. Stage 4: Tighten an additional 90 degrees.

85. Install the three accessory drive belt idler pulleys, the coolant pump pulley and the bolts. Torque all bolts to 18 ft. lbs. (25 Nm).

86. Position the accessory drive belt tensioner and install the bolts. Torque all bolts to 18 ft. lbs. (25 Nm).

87. Install the radio frequency interference capacitors.

Left cylinder head

WARNING

Do not use metal scrapers, wire brushes, power abrasive discs or other abrasive means to clean sealing surfaces. These tools cause scratches and gouges which make leak paths. Use a plastic scraping tool to remove all traces of old sealant.

WARNING

Inspect and clean the valve cover sealing surfaces with silicone gasket remover and metal surface prep. Follow the directions on the packaging.

➡If not secured within four minutes, the sealant must be removed and the sealing area cleaned. To clean the sealing area, use silicone gasket remover and metal surface prep. Follow the directions on the packaging. Failure to follow this procedure can cause future oil leakage.

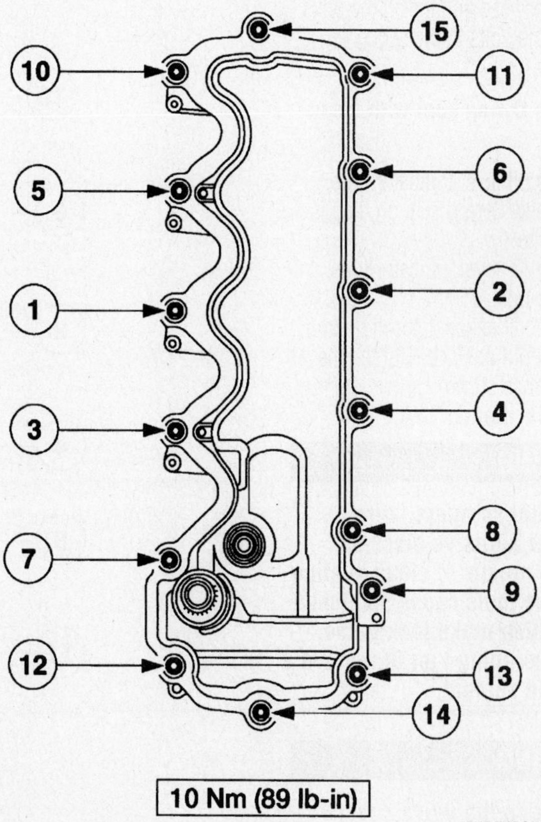

Left valve cover torque sequence—5.4L engine

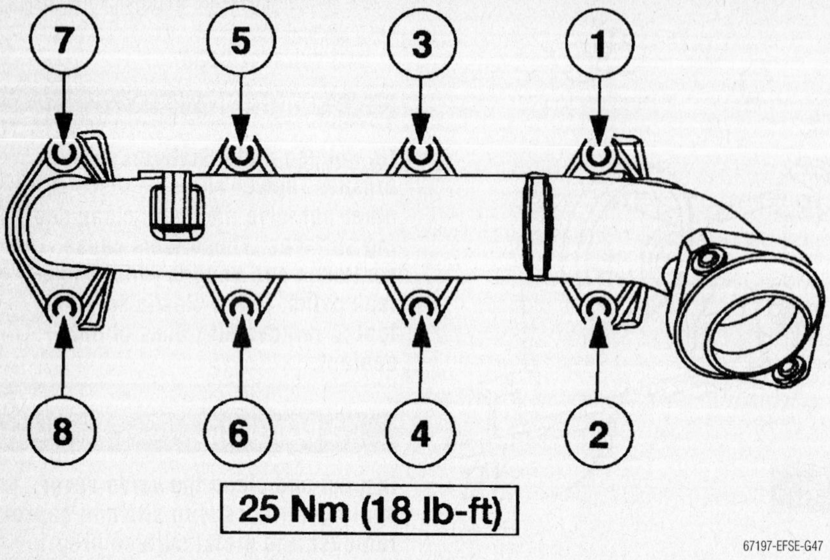

Left exhaust manifold torque sequence—5.4L engine

88. Apply silicone gasket and sealant in two places where the engine front cover meets the cylinder head.

WARNING

When installing the valve cover, make sure to avoid damaging the variable camshaft timing (VCT) solenoid.

89. Position the left valve cover and gasket on the cylinder head and install the bolts loosely. Tighten the bolts in the sequence shown.

➡Lubricate the O-ring seal with clean engine oil.

90. Install the oil level indicator tube.
91. Install a new O-ring seal on the oil level indicator tube.
92. Install the oil level indicator tube.
93. Install the bolt.
94. Install the left exhaust manifold and the exhaust manifold gaskets. Tighten the nuts in the sequence shown.

Right cylinder head

WARNING

Do not use metal scrapers, wire brushes, power abrasive discs or other abrasive means to clean sealing surfaces. These tools cause scratches and gouges which make leak paths. Use a plastic scraping tool to remove all traces of old sealant.

WARNING

Inspect and clean the valve cover sealing surfaces with silicone gasket remover and metal surface prep. Follow the directions on the packaging.

➡If not secured within four minutes, the sealant must be removed and the sealing area cleaned. To clean the sealing area, use silicone gasket remover and metal surface prep. Follow the directions on the packaging. Failure to follow this procedure can cause future oil leakage.

95. Apply silicone gasket and sealant in two places where the engine front cover meets the cylinder head.

WARNING

When installing the valve cover, make sure to avoid damaging the variable camshaft timing (VCT) solenoid.

96. Position the right valve cover and gasket on the cylinder head and install the bolts loosely. Tighten the bolts in the sequence shown.
97. Install the right exhaust manifold gaskets and the exhaust manifold. Tighten the nuts in the sequence shown.
98. Install the heater outlet tube stud.

All cylinder heads

99. Position the electrical harness on the valve cover and connect the engine wiring harness retainers to the valve cover studs.
100. Connect the right radio ignition interference capacitor electrical connector.
101. Connect the left radio ignition interference capacitor and cylinder head temperature (CHT) sensor electrical connectors.
102. Connect the CMP sensor electrical connectors.

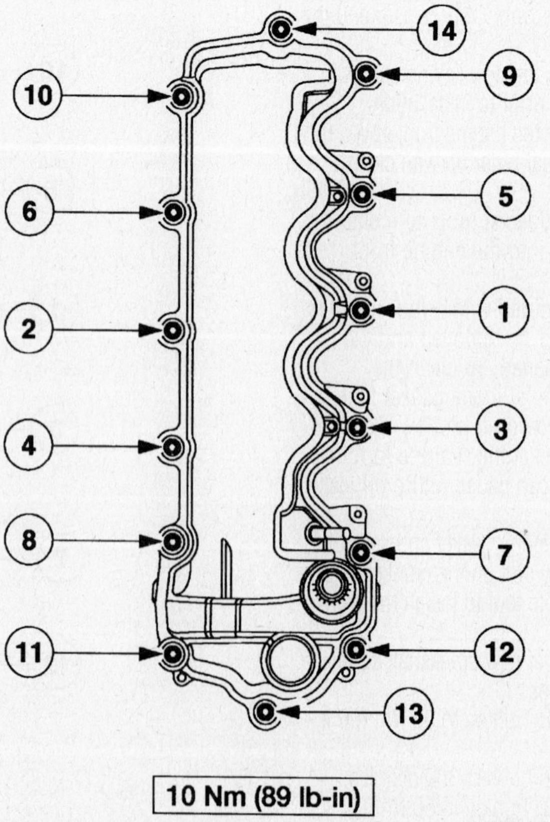

Right valve cover torque sequence—5.4L engine

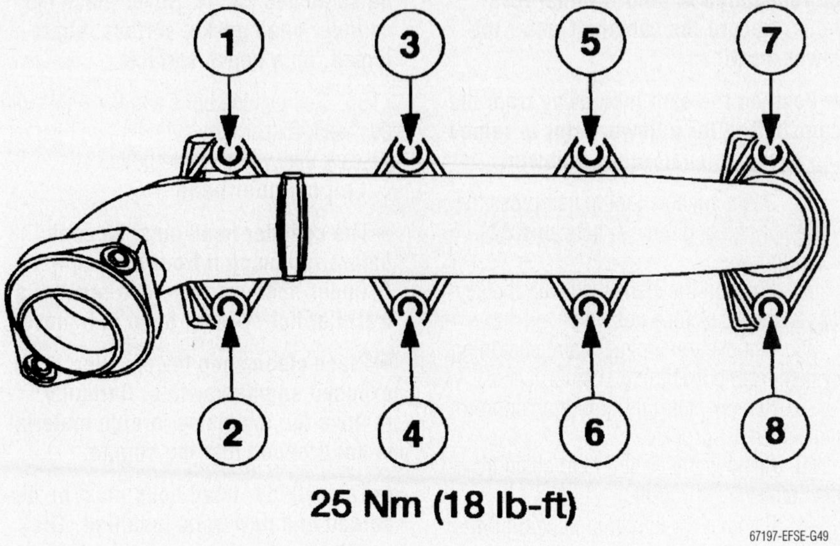

Right exhaust manifold torque sequence—5.4L engine

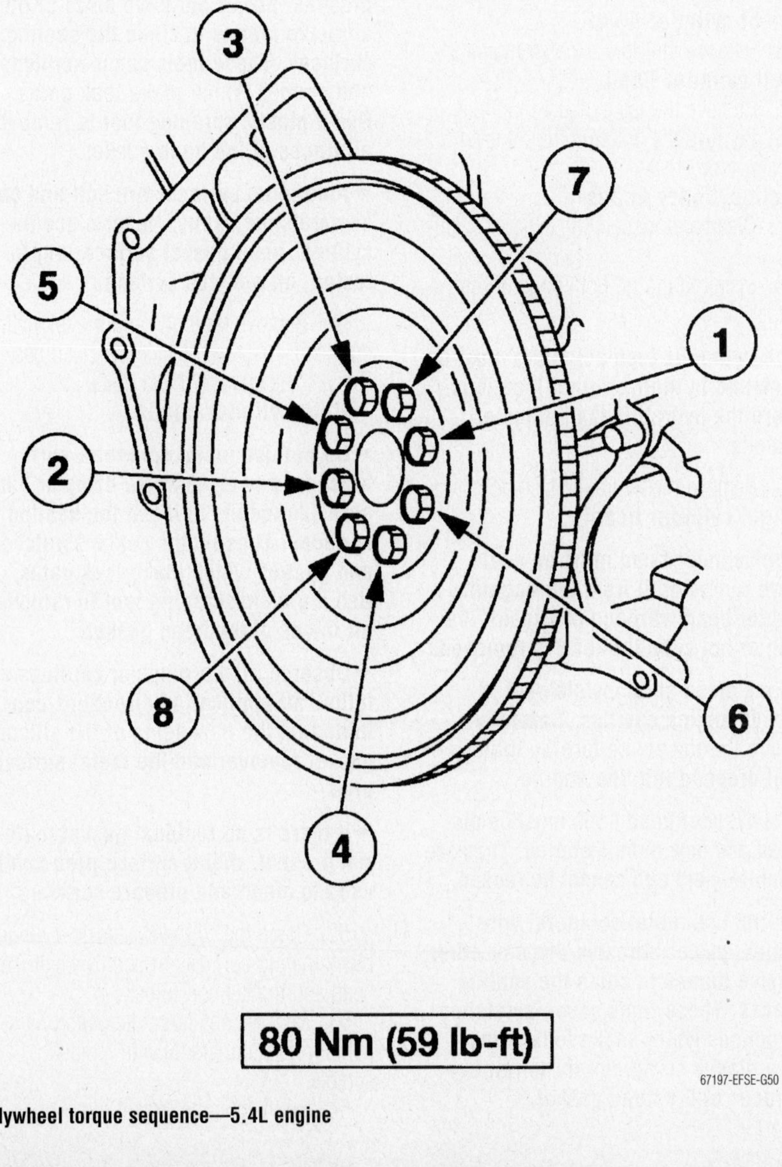

Flywheel torque sequence—5.4L engine

103. Connect the CKP sensor electrical connector.

104. Install a suitable tool.

105. Using a suitable floor crane remove the engine from the engine stand.

106. Install the flexplate or the flywheel and bolts. Tighten the bolts in the sequence shown.

107. Install the engine.

2004 E- and F-250/350 6.8L Engine

1. Before servicing the vehicle, refer to the precautions in the beginning of this section.
2. Remove the engine.
3. Remove the bolts and the flexplate or the flywheel.
4. Mount the engine on a suitable work stand.
5. Remove the right engine mount.
6. Remove the cylinder block drain plugs and drain the coolant into a suitable container.
7. Disconnect the left radio frequency interference capacitor and cylinder head temperature (CHT) sensor electrical connectors.
8. Disconnect the camshaft position (CMP) electrical connector.
9. Disconnect the right radio frequency interference capacitor electrical connector.
10. Disconnect the knock sensor electrical connector.
11. Disconnect the crankshaft position (CKP) sensor electrical connector.
12. Disconnect the oil pressure switch electrical connector.
13. Remove the nuts and position the wiring harness aside.
14. Disconnect all of the harness routing clips and connector retainers. Remove the engine control sensor wiring harness.
15. Remove the nuts and the two radio interference capacitors.
16. Remove the crankcase ventilation tube from the left valve cover.
17. Remove the positive crankcase ventilation (PCV) valve and hose from the right valve cover.

➡Do not use metal scrapers, wire brushes, power abrasive discs or other abrasive means to clean the sealing surfaces. These tools cause scratches and gouges which make leak paths. Use a plastic scraping tool to remove all traces of old sealant.

➡The bolts are part of the valve cover and should not be removed.

18. Fully loosen the bolts and remove the left valve cover.

19. Clean the valve cover mating surface of the cylinder head with silicone gasket remover and metal surface prep. Follow the directions on the packaging.

20. Inspect the valve cover gasket. If the gasket is damaged, remove and discard the gasket. Clean the valve cover gasket groove with soap and water or a suitable solvent.

➡Do not use metal scrapers, wire brushes, power abrasive discs or other abrasive means to clean the sealing surfaces. These tools cause scratches and gouges which make leak paths. Use a plastic scraping tool to remove all traces of old sealant.

➡The bolts are part of the valve cover and should not be removed.

21. Fully loosen the bolts and remove the right valve cover.

22. Clean the valve cover mating surfaces of the cylinder head with silicone gasket remover and metal surface prep. Follow the directions on the packaging.

23. Inspect the valve cover gasket. If the gasket is damaged, remove and discard the gasket. Clean the valve cover gasket groove with soap and water or a suitable solvent.

24. Remove the bolt and the idler pulley.

25. Remove the coolant pump pulley.

26. Using a suitable tool remove and discard the crankshaft pulley bolt.

27. Using a suitable tool, remove the crankshaft front seal.

28. Remove the four oil pan bolts.

➡Correct fastener location is essential for assembly procedure. Record fastener location.

29. Remove the fasteners.

30. Remove the engine front cover from the cylinder block.

31. Remove the crankshaft sensor ring from the crankshaft.

➡The caps must be marked for installation in their original locations or damage to the engine may occur.

32. Remove the six bolts and remove the balance shaft bearing caps.

33. Remove the balance shaft.

➡Use care when removing the spark plugs.

➡Use compressed air to remove any foreign material from the spark plug well before removing the spark plugs.

34. Remove the 10 spark plugs.

35. Install a suitable tool between the valve spring coils to prevent valve stem seal damage.

➡The camshaft roller followers must be reinstalled in their original locations. Record the camshaft roller follower locations.

➡Position the cam lobe away from the camshaft roller follower prior to removing each camshaft roller follower.

36. Use a suitable tool to compress the valve springs and remove the camshaft roller followers.

37. Position the crankshaft with the keyway at the 12 o'clock position.

38. Remove the timing chain tensioning system from both timing chains.

39. Remove the right timing chain from the camshaft sprocket.

40. Remove the right timing chain from the crankshaft sprocket.

41. Remove the left timing chain from the camshaft sprocket.

42. Remove the left timing chain from the crankshaft sprocket.

43. Remove both timing chain guides.

Right cylinder head

44. Remove the right exhaust manifold.

Left cylinder head

45. Remove the left exhaust manifold.

46. Remove the bolt and the oil level indicator tube.

Both cylinder heads

47. Clean and inspect the exhaust manifolds.

48. Install lifting on both ends of the cylinder head.

➡The hydraulic lash adjusters must be reinstalled in their original locations. Record the hydraulic lash adjuster locations.

49. Remove the hydraulic lash adjusters.

Right cylinder head

➡The cylinder head must be cool before removing it from the engine. Cylinder head warpage can result if a warm or hot cylinder head is removed.

➡Place clean shop towels over exposed engine cavities. Carefully remove the towels so foreign material is not dropped into the engine.

➡The cylinder head bolts must be discarded and new bolts installed. They are tighten-to-yield and cannot be reused.

➡Do not use metal scrapers, wire brushes, power abrasive discs or other abrasive means to clean the sealing surfaces. These tools cause scratches and gouges which make leak paths. Use a plastic scraping tool to remove all traces of the head gasket.

➡Aluminum surfaces are soft and can be scratched easily. Never place the cylinder head gasket surface, unprotected, on a bench surface.

50. Remove the bolts and the right cylinder head. Discard the cylinder head gasket. Discard the cylinder head bolts.

Left cylinder head

➡The cylinder head must be cool before removing it from the engine. Cylinder head warpage can result if a warm or hot cylinder head is removed.

➡Place clean shop towels over exposed engine cavities. Carefully remove the towels so foreign material is not dropped into the engine.

➡The cylinder head bolts must be discarded and new bolts installed. They are tighten-to-yield and cannot be reused.

➡Do not use metal scrapers, wire brushes, power abrasive discs or other abrasive means to clean the sealing surfaces. These tools cause scratches and gouges which make leak paths. Use a plastic scraping tool to remove all traces of the head gasket.

➡Aluminum surfaces are soft and can be scratched easily. Never place the cylinder head gasket surface, unprotected, on a bench surface.

51. Remove the bolts and the left cylinder head. Discard the cylinder head gasket. Discard the cylinder head bolts.

Both cylinder heads

➡Do not use metal scrapers, wire brushes, power abrasive discs or other abrasive means to clean the sealing surfaces. These tools cause scratches and gouges which make leak paths. Use a plastic scraping tool to remove all traces of the head gasket.

➡Observe all warnings or cautions and follow all application directions contained on the packaging of the silicone gasket remover and the metal surface prep.

➡If there is no residual gasket material present, metal surface prep can be used to clean and prepare surfaces.

52. Clean the cylinder head-to-cylinder block mating surfaces of both the cylinder head and the cylinder block.

53. Remove any large deposits of silicone or gasket material with a plastic scraper.

54. Apply silicone gasket remover, fol-

lowing package directions, and allow to set for several minutes.

55. Remove the silicone gasket remover with a plastic scraper. A second application of silicone gasket remover may be required if residual traces of silicone or gasket material remain.

56. Apply metal surface prep, following package directions, to remove any remaining traces of oil or coolant, and to prepare the surfaces to bond with the new gasket. Do not attempt to make the metal shiny. Some staining of the metal surface is normal.

➡Make sure all cylinder head surfaces are clear of any gasket material, RTV, oil and coolant. The cylinder head surface must be clean and dry before running a flatness check.

➡Use a straightedge that is calibrated by the manufacturer to be flat with 0.005 mm (0.0002 in.) per running foot length. For example, if the straightedge is 61 cm (24 in.) long, the machined edge must be flat with 0.010 mm (0.0004 in.) from end to end.

57. Support the cylinder head on a bench with the head gasket side up. Inspect all areas of the deck face with a straightedge, paying particular attention to the oil pressure feed area. The cylinder head must not have depressions deeper than 0.0254 mm (0.001 in.) across a 38.1 mm (1.5 in.) square area, or scratches more than 0.0254 mm (0.001 in.).

To install:

Both cylinder heads

➡Make sure all coolant residue and foreign material is cleaned from the block surface and the cylinder bore.

➡The use of sealing aids is not permitted. The gasket must be installed dry.

➡The new gasket has a film coating which is crucial to the gasket's ability to seal properly. Do not scratch the gasket.

58. Install the head gaskets over the dowel pins. Position the heads.

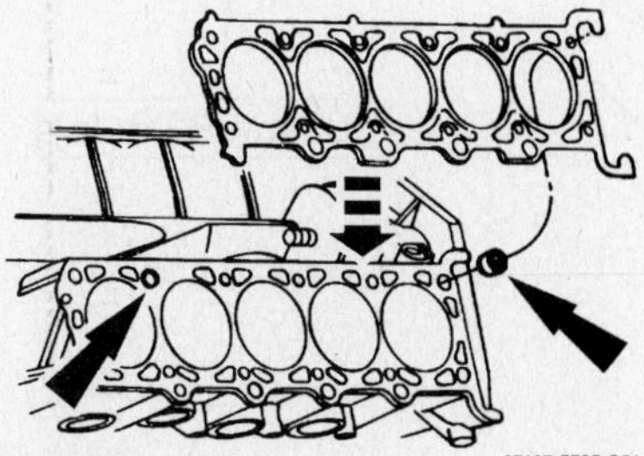

Dowel locations—6.8L engine

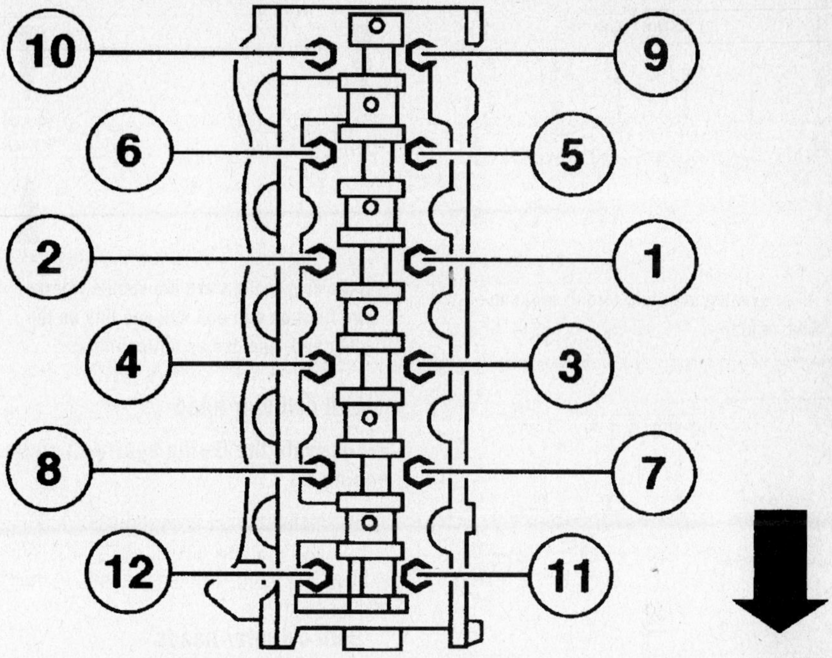

Left cylinder head torque sequence—6.8L engine

➡The new cylinder head bolts must be lightly oiled with a rag and allowed to drain for a few minutes prior to installation.

59. Loosely install cylinder head bolts.

60. Tighten the cylinder head bolts in three steps in the sequence shown.

a. Step 1: Tighten the bolts to 40 Nm (30 ft. lbs.).

b. Step 2: Tighten the bolts an additional 90 degrees.

c. Step 3: Tighten the bolts an additional 90 degrees.

61. Remove the Lifting Handles from both ends of the cylinder head.

➡Lubricate the hydraulic lash adjusters with clean engine oil.

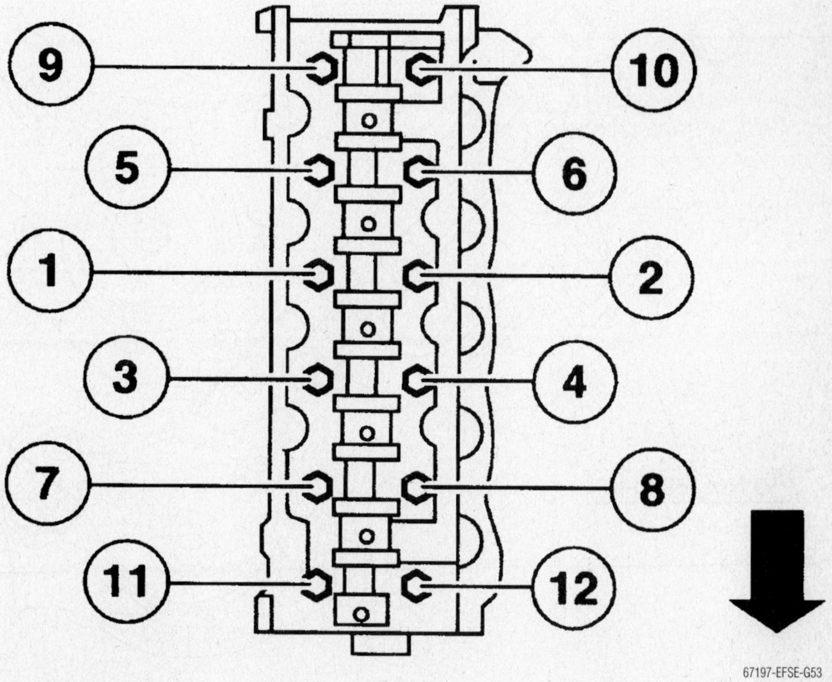

Right cylinder head torque sequence—6.8L engine

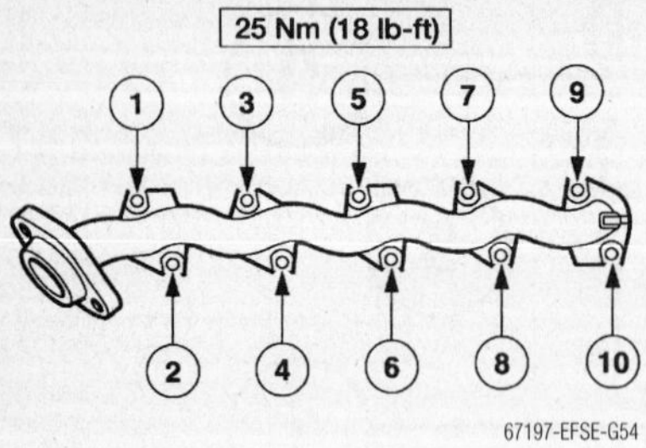

Right exhaust manifold torque sequence—6.8L engine

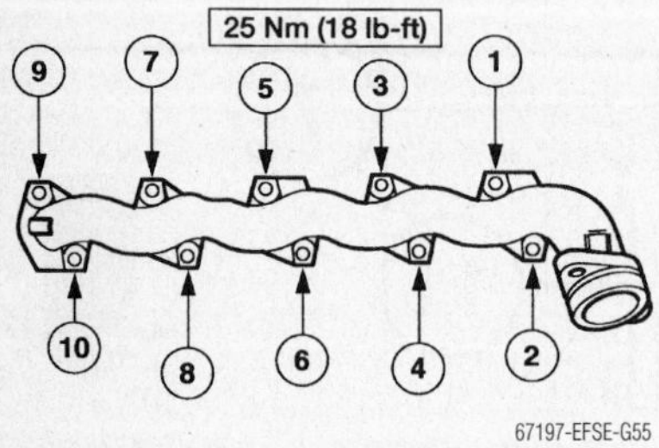

Left exhaust manifold torque sequence—6.8L engine

62. Install the hydraulic lash adjusters in their original locations.

Right cylinder head

63. Using a new gasket, install the right exhaust manifold. Tighten the nuts in the sequence shown.

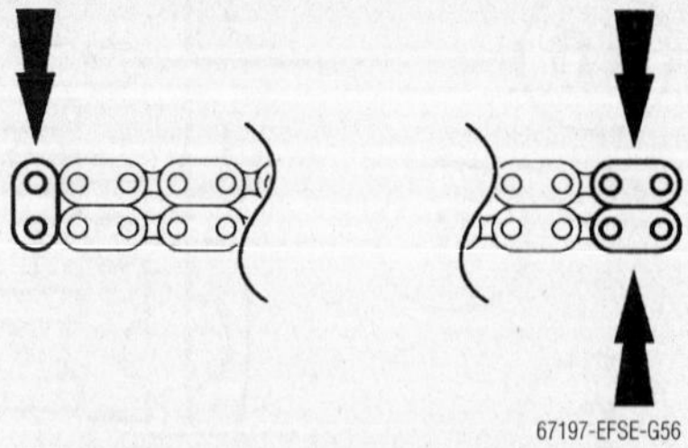

If the copper links are not visible, mark two links on one end and one link on the other end, and use as timing marks

Left cylinder head

➡Lubricate the O-ring seal with clean engine oil.

64. Install the oil level indicator tube.
65. Using a new gasket, install the left exhaust manifold. Tighten the nuts in the sequence shown.

Both cylinder heads

➡The timing chain procedures must be followed exactly or damage to the valves and pistons will result.

66. Compress the tensioner plunger, using an edge of a vise.
67. Install a retaining clip on the tensioner to hold the plunger in during installation.
68. Remove the tensioner from the vise.
69. If the copper links are not visible, mark two links on one end and one link on the other end, and use as timing marks.
70. Install the timing chain guides.
71. Pre-position the camshafts. Rotate the left camshaft until the timing mark is approximately at 12 o'clock. Rotate the right camshaft until the timing mark is approximately at 11 o'clock.

➡Rotate the crankshaft counterclockwise only. Do not rotate past position shown or severe piston and valve damage can occur.

➡The number one piston is at top dead center (TDC) when the stud on the engine block fits into the slot in the handle of the crankshaft alignment tool 303-448, or equivalent.

72. Position the crankshaft so the number one cylinder is at TDC with a suitable tool.
73. Remove the crankshaft holding tool.
74. Install the crankshaft sprocket, making sure the flange faces forward.
75. Position the lower end of the left timing chain on the crankshaft sprocket, aligning the timing mark on the outer flange of the crankshaft sprocket with the single copper (marked) link on the chain.

➡Make sure the upper half of the timing chain is below the tensioner arm dowel.

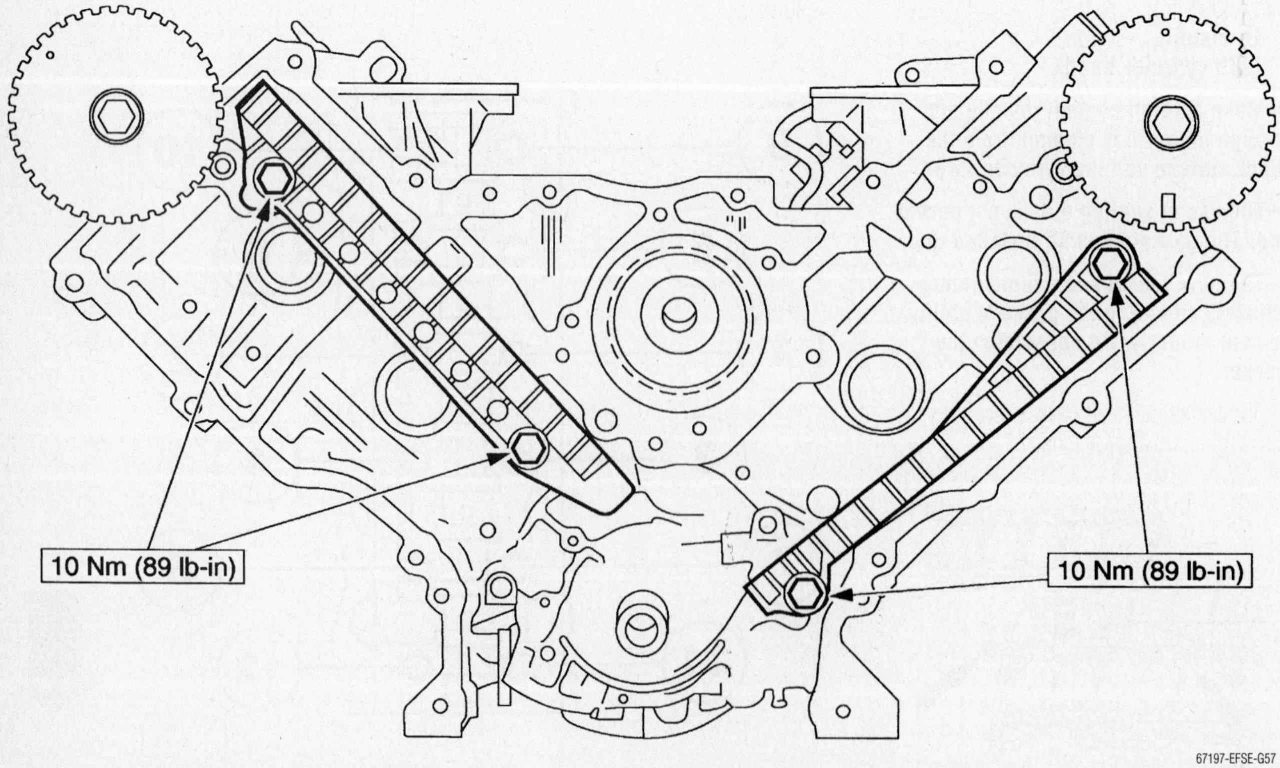

Timing chain guides installed—6.8L engine

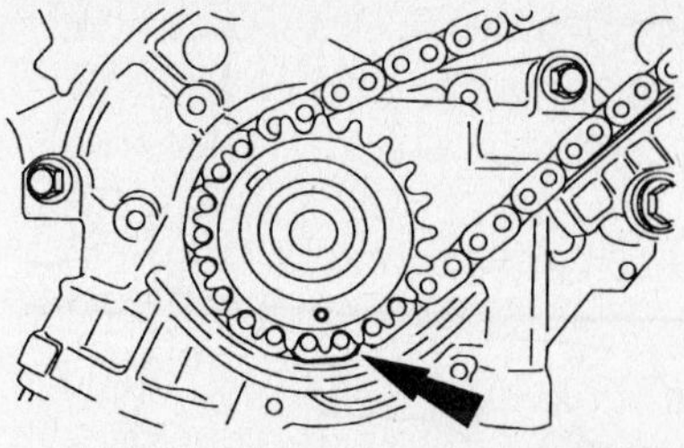

67197-EFSE-G58

Position the lower end of the left timing chain on the crankshaft sprocket, aligning the timing mark on the outer flange of the crankshaft sprocket with the single copper (marked) link on the chain—6.8L engine

76. Position the timing chain on the camshaft sprocket with the camshaft sprocket timing mark positioned between the two copper (marked) chain links.

➡The left timing chain tensioner arm has a bump near the dowel hole for identification.

77. Position the left timing chain tensioner arm on the dowel pin and install the left timing chain tensioner.
78. Remove the retaining clip from the left timing chain tensioner.

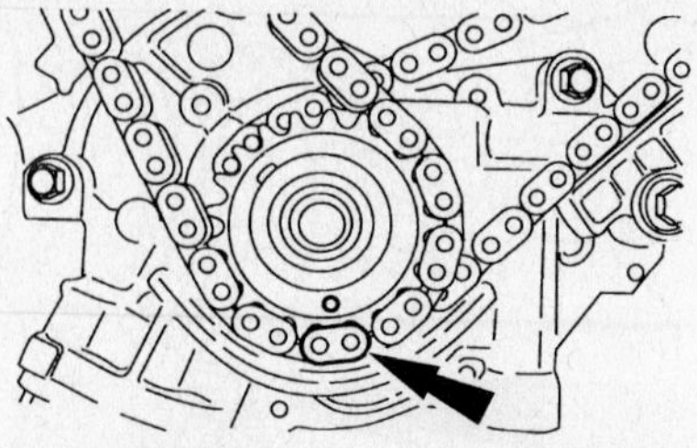

67197-EFSE-G60

Position the lower end of the right timing chain on the crankshaft sprocket, aligning the timing mark on the outer flange of the crankshaft sprocket with the single copper (marked) link on the chain—6.8L engine

79. Position the lower end of the right timing chain on the crankshaft sprocket, aligning the timing mark on the outer flange of the crankshaft sprocket with the single copper (marked) link on the chain.

➡The lower half of the timing chain must be positioned above the tensioner arm dowel.

80. Install the right timing chain on the camshaft sprocket. Make sure the camshaft sprocket timing mark is positioned between the two copper (marked) chain links.

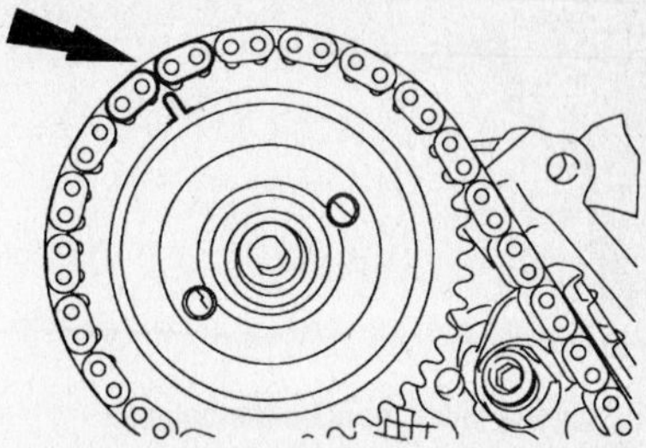

67197-EFSE-G61

Install the right timing chain on the camshaft sprocket. Make sure the camshaft sprocket timing mark is positioned between the two copper (marked) chain links—6.8L engine

81. Position the right timing chain tensioner arm on the dowel pin and install the right timing chain tensioner.
82. Remove the retaining clip from the right timing chain tensioner.
83. As a post-check, verify correct alignment of all timing marks.
84. Install a suitable tool between the valve spring coils to prevent valve stem seal damage.

➡Lubricate the camshaft roller followers, using clean engine oil.

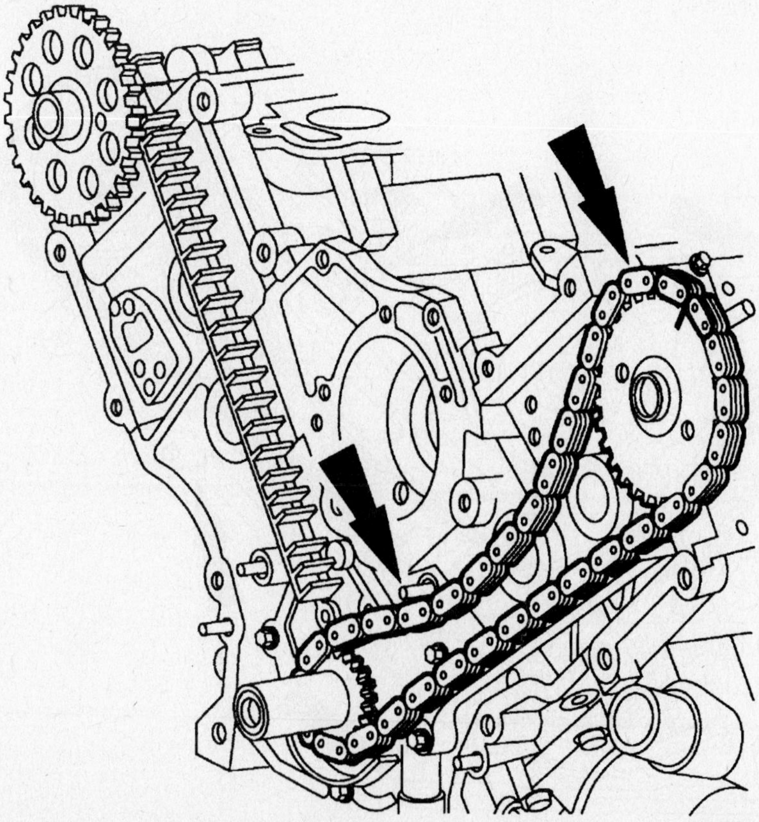

67197-EFSE-G59

Position the timing chain on the camshaft sprocket with the camshaft sprocket timing mark positioned between the two copper (marked) chain links—6.8L engine

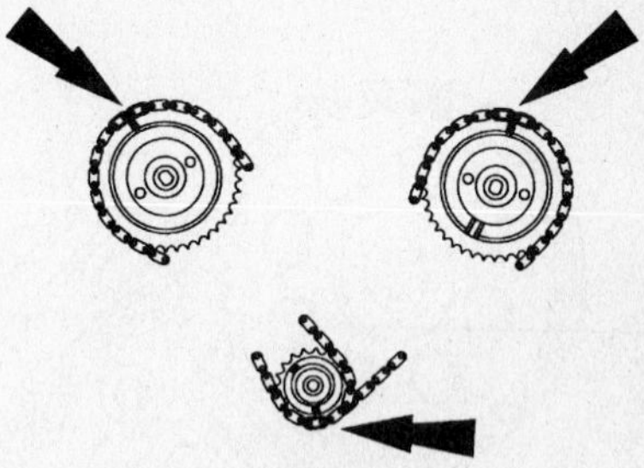

67197-EFSE-G62

As a post-check, verify correct alignment of all timing marks—6.8L engine

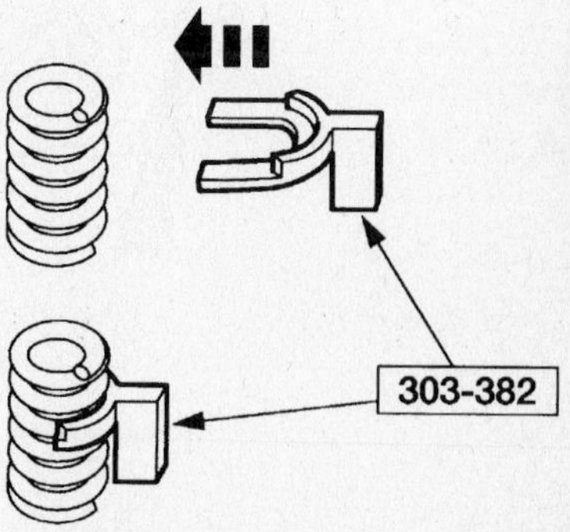

67197-EFSE-G63

Install this tool, or equivalent, between the valve spring coils to prevent valve stem seal damage

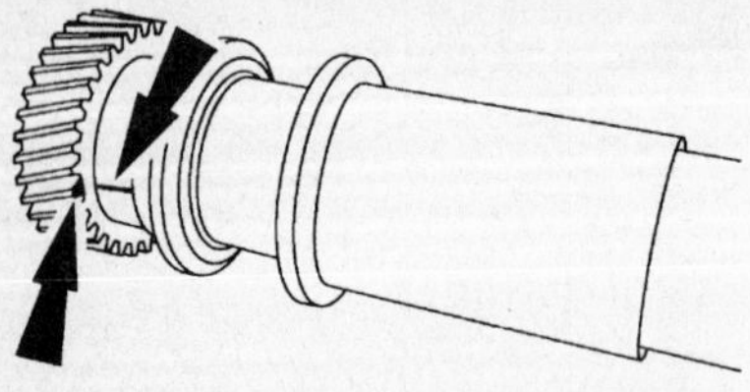

Using the index mark on the balance shaft, mark the corresponding gear tooth with chalk—6.8L engine

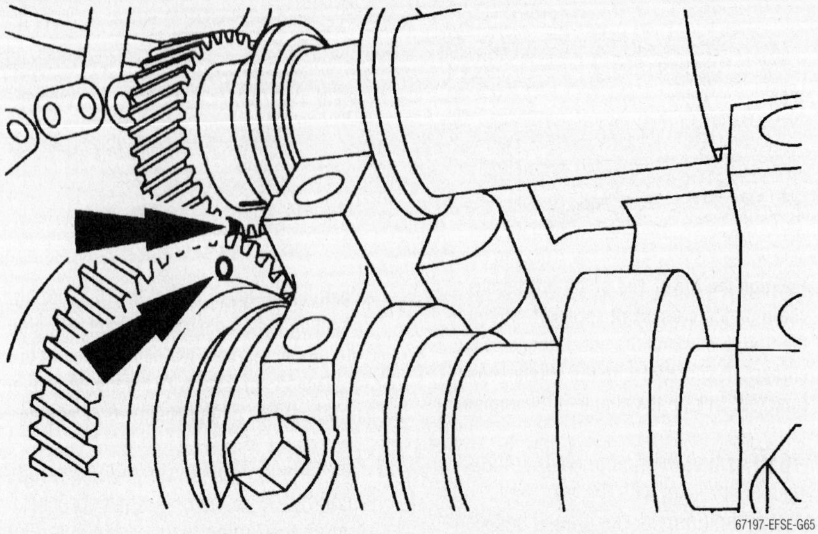

It may be necessary to use an inspection mirror to see the marks. Align the chalk mark on the balance shaft with the camshaft timing mark as shown—6.8L engine

➡Position the cam lobe away from the camshaft roller follower location prior to installing each camshaft roller follower.

85. Install a suitable tool. Compress the valve spring. Install the camshaft roller followers in their original locations. Remove the tool.

➡When installing the spark plugs, use care not to exceed the recommended torque.

86. Install the 10 spark plugs. Tighten the spark plugs to 18 Nm (13 ft. lbs.).
87. Using the index mark on the balance shaft, mark the corresponding gear tooth with chalk.
88. Position the balance shaft on the journals.

➡It may be necessary to use an inspection mirror to see the marks. Align the chalk mark on the balance shaft with the camshaft timing mark as shown.

89. Install the bearing caps in their original locations. Install the bolts and tighten the bolts in the sequence shown.
90. Install the crankshaft sensor ring on the crankshaft.

➡If the front cover is not secured within four minutes, the sealant must be removed and the sealing area cleaned. To clean the sealing area, use silicone gasket remover and metal surface prep. Follow the directions on the packaging. Failure to follow this procedure can cause future oil leakage.

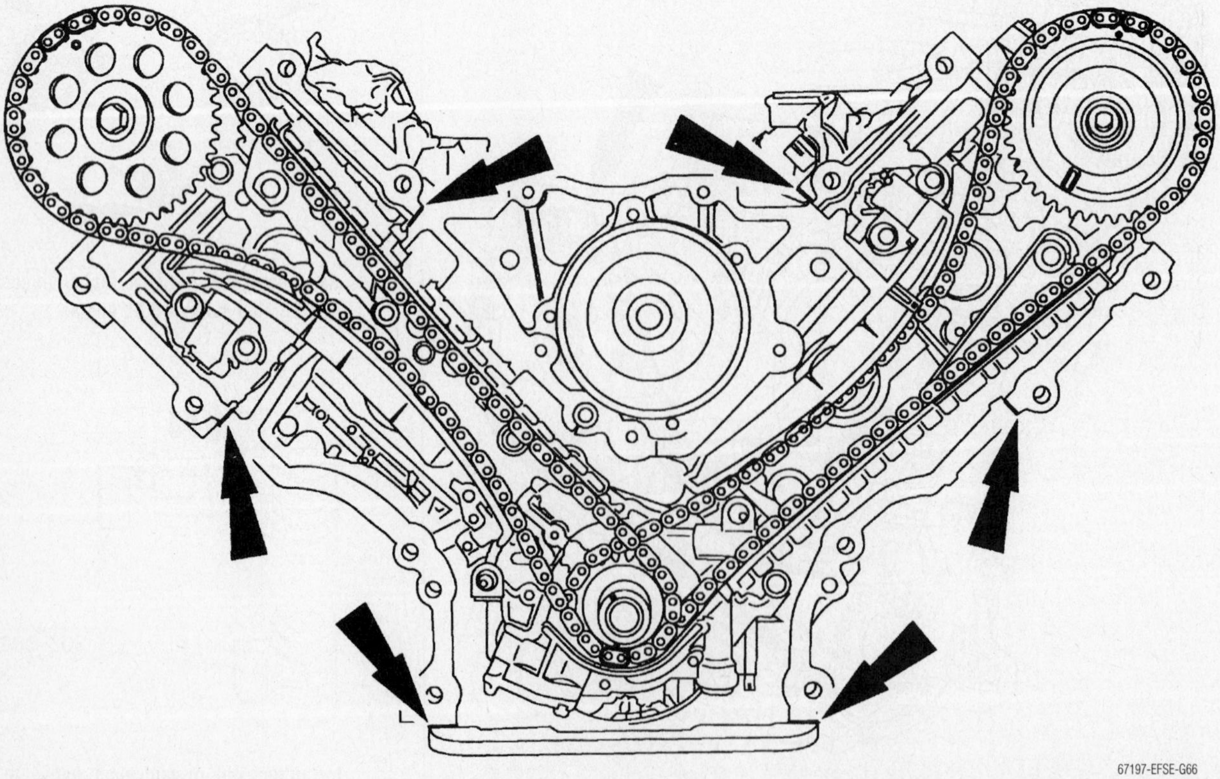

Apply a bead of silicone gasket and sealant along the cylinder head-to-cylinder block surface and the oil pan-to-cylinder block surface, at the locations shown—6.8L engine

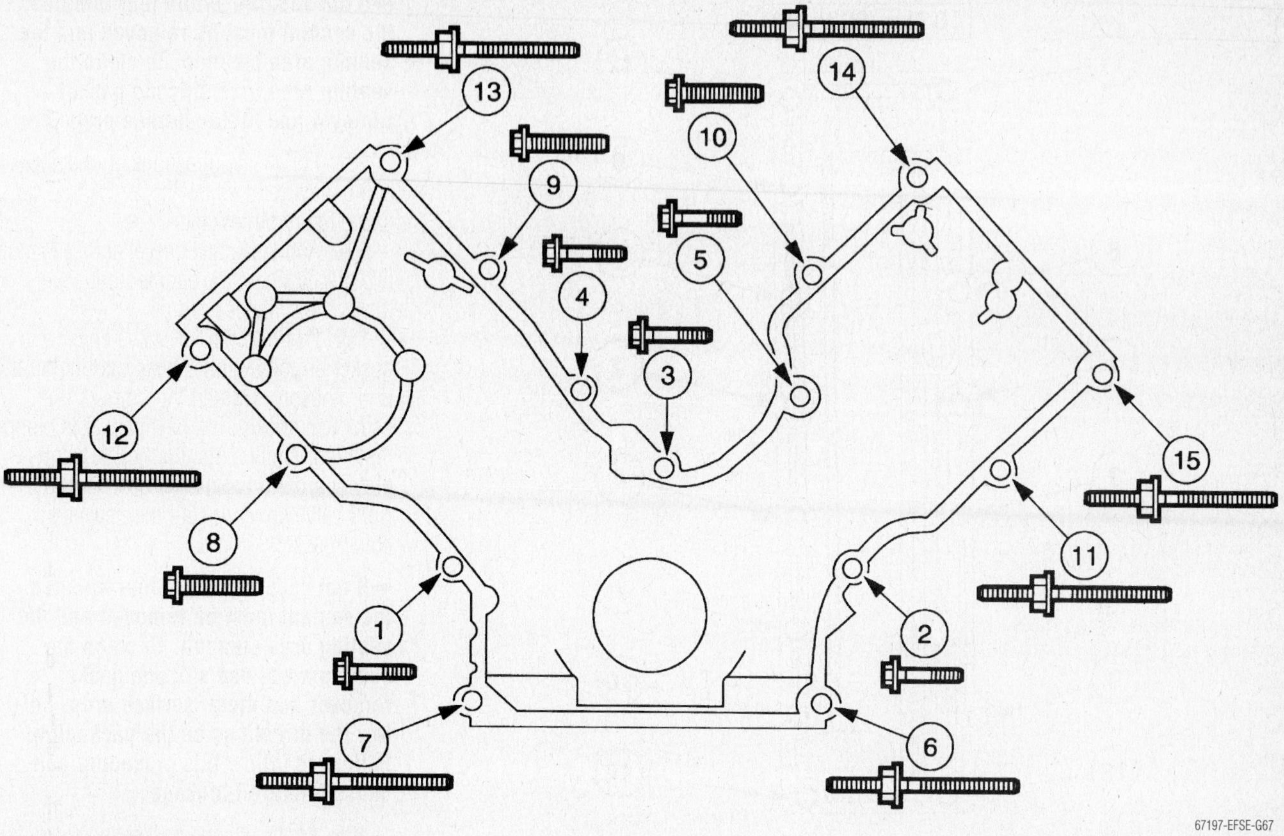

Front cover tightening sequence—6.8L engine

91. Apply a bead of silicone gasket and sealant along the cylinder head-to-cylinder block surface and the oil pan-to-cylinder block surface, at the locations shown.

92. Install a new front cover gasket on the engine front cover. Position the engine front cover. Install the fasteners.

93. Tighten the engine front cover fasteners in sequence in two steps.

a. Step 1: Tighten fasteners 1 through 7 to 25 Nm (18 ft. lbs.).

b. Step 2: Tighten fasteners 6 through 15 to 48 Nm (35 ft. lbs.).

94. Loosely install the bolts, then tighten in two steps in the sequence shown.

a. Step 1: Tighten to 20 Nm (15 ft. lbs.).

b. Step 2: Tighten an additional 60 degrees.

95. Position the belt idler pulley and install the bolt. Torque to 18 ft. lbs. (25 Nm).

96. Lubricate the engine front cover and the crankshaft front seal inner lip with clean engine oil.

97. Using the special tools, install the crankshaft front seal.

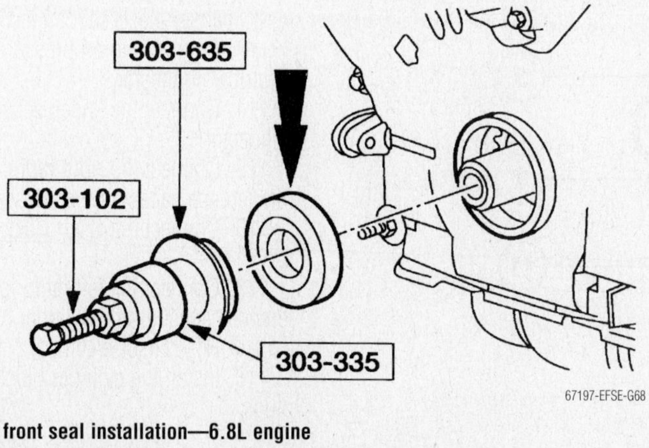

Crankshaft front seal installation—6.8L engine

➡If not secured within four minutes, the sealant must be removed and the sealing area cleaned. To clean the sealing area, use silicone gasket remover and metal surface prep. Follow the directions on the packaging. Failure to follow this procedure can cause future oil leakage.

98. Apply silicone gasket and sealant to the Woodruff key slot on the crankshaft pulley. Use a suitable tool to install the crankshaft pulley.

➡Use a suitable strap wrench to hold the pulley while tightening the bolt.

99. Tighten the new crankshaft bolt in four steps.

a. Step 1: Tighten to 90 Nm (66 ft. lbs.).

b. Step 2: Loosen 360 degrees.

c. Step 3: Tighten to 50 Nm (37 ft. lbs.).

d. Step 4: Tighten an additional 90 degrees.

100. Position the water pump pulley on the water pump and install the bolts. Torque to 18 ft. lbs. (25 Nm).

101. If a new gasket is being installed, apply instant adhesive completely around the gasket groove in the left valve cover. Install the new valve cover gasket.

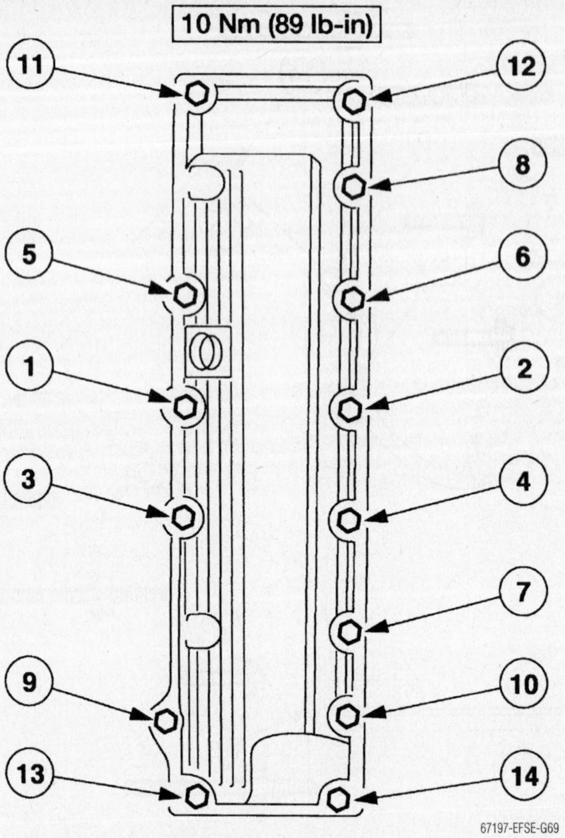

Left valve cover torque sequence—6.8L engine

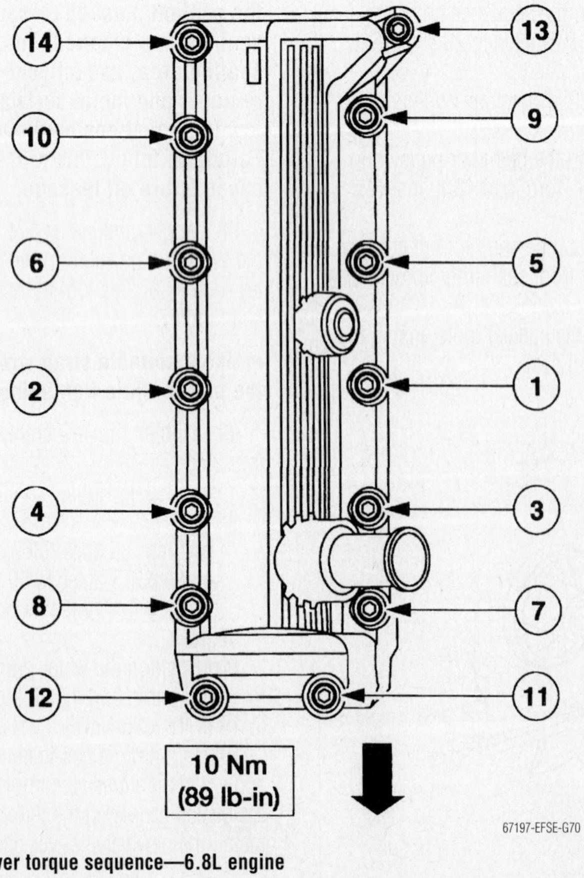

Left valve cover torque sequence—6.8L engine

➡If not secured within four minutes, the sealant must be removed and the sealing area cleaned. To clean the sealing area, use silicone gasket remover and metal surface prep.

102. Follow the directions on the packaging. Failure to follow this procedure can cause future oil leakage.

103. Apply silicone gasket and sealant in two places where the engine front cover meets the cylinder head.

104. Position the left valve cover and gasket on the cylinder head and install the bolts loosely. Tighten the bolts in the sequence shown. If a new gasket is being installed, apply instant adhesive completely around the gasket groove in the right valve cover. Install the new valve cover gasket.

➡If not secured within four minutes, the sealant must be removed and the sealing area cleaned. To clean the sealing area, use silicone gasket remover and metal surface prep. Follow the directions on the packaging. Failure to follow this procedure can cause future oil leakage.

105. Apply silicone gasket and sealant in two places where the engine front cover meets the cylinder head.

106. Position the right valve cover and gasket on the cylinder head and install the bolts loosely. Tighten the bolts in the sequence shown.

107. Install the crankcase ventilation tube on the left valve cover.

108. Install the positive crankcase ventilation (PCV) valve and hose in the right valve cover.

109. Install the radio frequency interference capacitors.

110. Position the engine control sensor wiring harness and connect the wiring harness retainers onto the valve cover studs.

111. Install the lifting bracket.

112. Connect the crankshaft position (CKP) sensor electrical connector.

113. Connect the oil pressure switch electrical connector.

114. Connect the knock sensor electrical connector.

115. Connect the right radio ignition interference capacitor electrical connector.

116. Connect the CMP electrical connector.

117. Connect the left radio ignition interference capacitor and cylinder head temperature (CHT) sensor electrical connectors.

118. Install the cylinder block drain plugs.

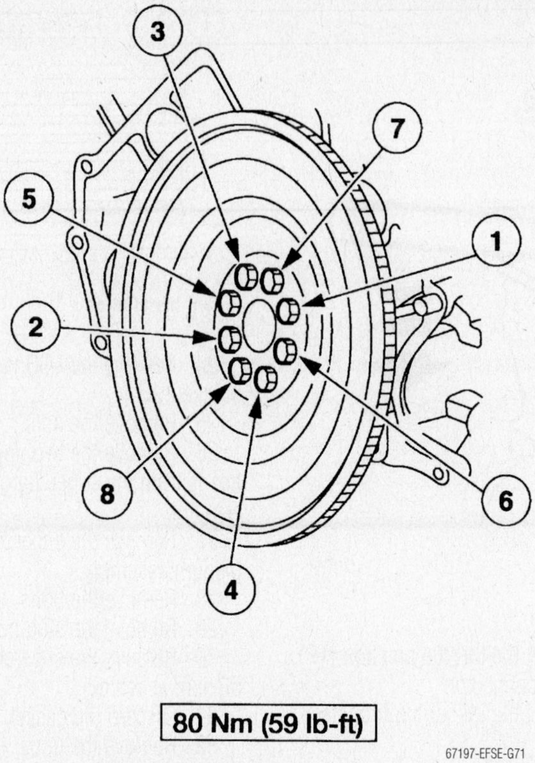

Flexplate/flywheel torque sequence—6.9L engine

119. Install the right engine mount. Tighten the bolts to 63 Nm (46 ft. lbs.).

120. Install a suitable tool and remove the engine from the work stand.

121. Install the flexplate or the flywheel and the bolts in the sequence shown.

122. Install the engine.

Rocker Arms

REMOVAL & INSTALLATION

4.2L Engine

1. Before servicing the vehicle, refer to the precautions in the beginning of this section.

➡If removing more than 1 rocker arm, mark the components for proper location.

2. Before servicing the vehicle, refer to the precautions in the beginning of this section.
3. Disconnect the negative battery cable.
4. Remove the lower intake manifold.
5. Remove the rocker arm cover.
6. Remove the rocker arm hold-down bolt, then remove the rocker arm from the cylinder head.

To install:

7. Position the rocker arms in place, then install the hold-down bolts. Tighten the bolts to 23–29 ft. lbs. (30–40 Nm).
8. Install the rocker arm cover and the lower intake manifold.
9. Connect the negative battery cable.

Camshaft Roller Follower

REMOVAL & INSTALLATION

4.6L and 5.4L Engines

1. Before servicing the vehicle, refer to the precautions in the beginning of this section.
2. Disconnect the negative battery cable.
3. Remove the camshaft covers.
4. Position the piston of the cylinder being serviced at the bottom of its travel.

➡Two different valve spring compressor tools are used for this procedure. Valve Spring Compressor (T91P-6565-A) is used on the exhaust camshaft and Valve Spring Compressor (T93P-6565-A) is used on the intake camshaft.

5. Compress the valve spring and remove the follower.

To install:

6. Position the piston of the cylinder being serviced at the bottom of its travel.
7. Apply clean engine oil to the rocker arm, valve stem tip and tappet bore.

➡Valve tappet should have no more than 1/16 inch (1.5mm) of travel before installing the rocker arm.

8. Compress the valve spring using the correct tool and install the follower.

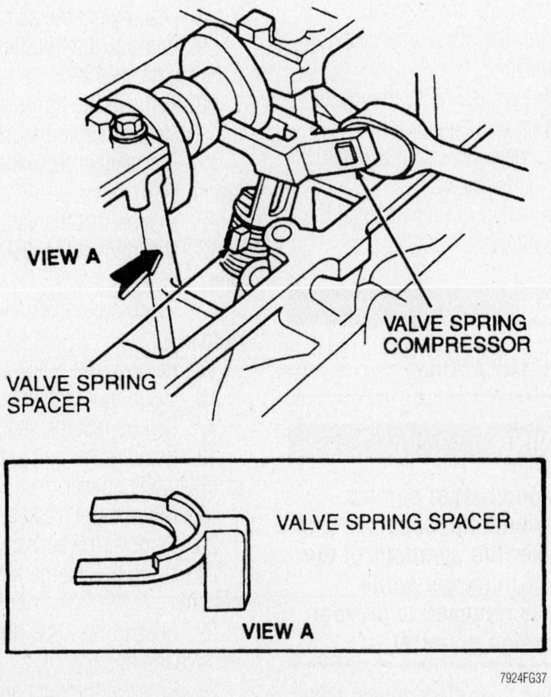

Using the proper tool, compress the valve spring and remove the follower—4.6L and 5.4L engine

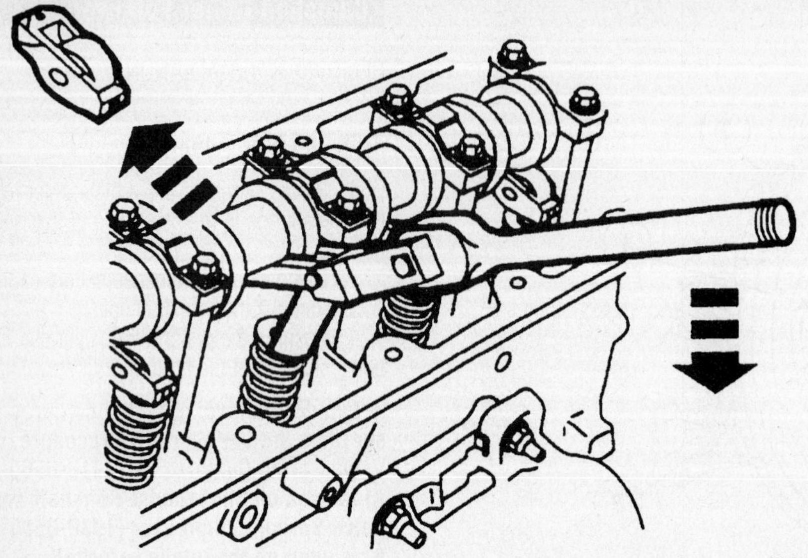

7924FG82

Compress the valve spring and remove the rocker arm—6.8L engine

6.8L Engine

1. Before servicing the vehicle, refer to the precautions in the beginning of this section.
2. Disconnect the negative battery cable.
3. Remove the camshaft covers.
4. Position the base circle of the camshaft lobe on the rocker arm to be serviced. Also, be sure the piston is not at the top of its travel near the valve.
5. Compress the valve spring and remove the follower.

To install:

6. Position the base circle of the camshaft lobe over the place where the follower is to be installed.
7. Apply clean engine oil to the rocker arm, valve stem tip and tappet bore.
8. Compress the valve using a suitable tool and install the follower.
9. Install the rocker arm covers and the remaining components.

Supercharger

REMOVAL & INSTALLATION

⁂ WARNING

Before any supercharger service, clean areas around the supercharger assembly. Cover the openings of the engine and supercharger while supercharger is removed to prevent damage by foreign material.

1. Disconnect the battery ground cable.
2. Remove the air cleaner outlet tube.
3. Disconnect the throttle position (TP) sensor electrical connector.
4. Disconnect the idle air control electrical connector.
5. Disconnect the vacuum lines and the vent hose.

➡Pay attention to the throttle return spring orientation. It must be reinstalled in the correct orientation.

6. Disconnect the throttle control linkages.
7. Disconnect the speed control actuator cable.
8. Disconnect the accelerator cable.
9. Remove the throttle return spring from the throttle lever.
10. Remove the three nuts and position the accelerator cable bracket aside.
11. Disconnect the exhaust gas recirculation (EGR) valve.
12. Disconnect the vacuum line.
13. Disconnect the EGR valve tube fitting.
14. Disconnect the brake booster vacuum line.
15. Remove the bracket bolt.
16. Disconnect the vacuum line.
17. Disconnect the vacuum line.
18. Remove the bolts and remove the throttle body spacer.
19. Discard the gasket.
20. Remove the accessory drive belt.
21. Drain the charge air cooling (CAC) system.
22. Loosen the CAC hose clamps and remove the hoses.
23. Remove the nuts and position the CAC water supply and return tubes aside.

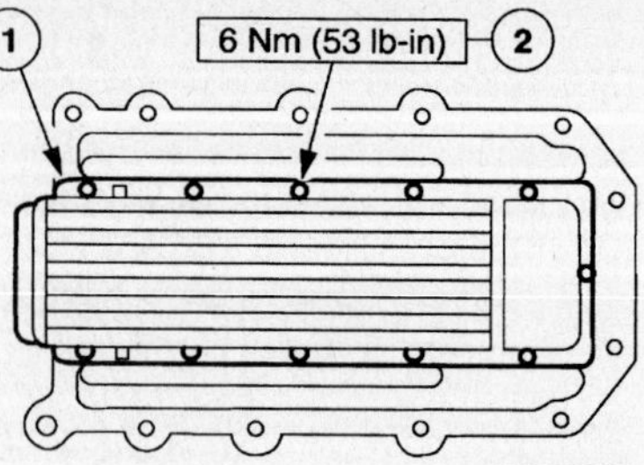

67197-EFSE-G93

CAC installation

24. Remove the EGR vacuum regulator bracket.
25. Remove the nuts.
26. Remove the two upper vacuum hoses from the supercharger bypass valve actuator.
27. Remove the supercharger and plenum assembly.
28. Remove the bolts.
29. Remove the stud bolts.
30. Remove the supercharger and plenum assembly.
31. Discard the gasket.
32. Remove the upper intake manifold.
33. Remove the vacuum hose.
34. Remove the CAC hoses.
35. Remove the bolts and the upper intake manifold.
36. Discard the gasket.
37. Remove the CAC.
38. Remove and discard the bolts.
39. Remove the CAC.
40. Discard the gasket.

To install:

➡New bolts must be used during assembly of the charge air cooler (CAC).

41. Install the CAC.
42. Install a new gasket.
43. Install the new bolts.
44. Install the upper intake manifold.
45. Install the upper intake manifold and the four rear bolts loosely.
46. Install the CAC hoses.
47. Install a new gasket.
48. Install the vacuum hose.
49. Install a new gasket and install the supercharger and plenum assembly, bolts, and the stud bolts. Tighten the retainers in two steps, in the sequence shown
    a. Step 1: Tighten to 2 Nm (18 inch lbs.).
    b. Step 2: Tighten 45 Nm (33 ft. lbs.).
50. Install the two upper vacuum hoses to the supercharger bypass valve actuator.
51. Install the EGR vacuum regulator bracket and install the nuts.

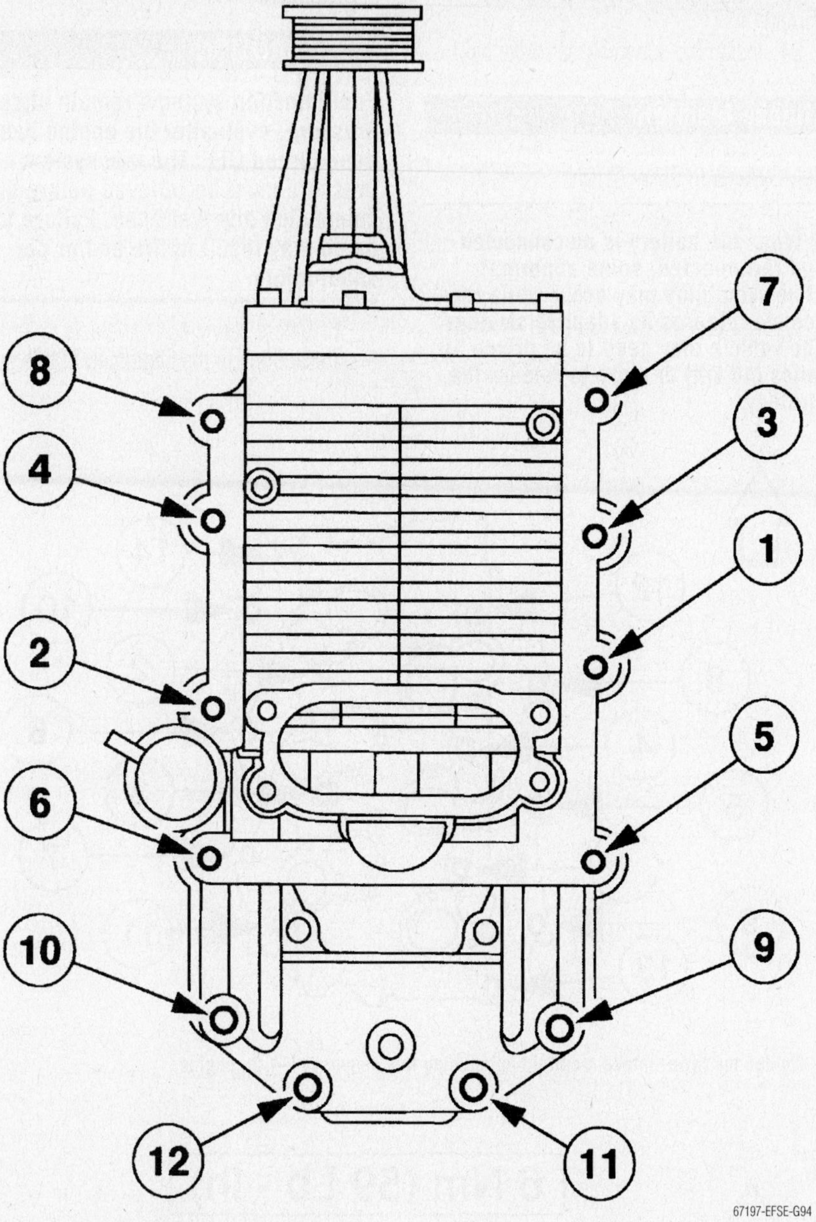

Supercharger fastener tightening sequence

52. Position the bracket and install the nuts.
53. Position the CAC water supply and return tubes and install the nuts.
54. Install the CAC hoses and the clamps.
55. Fill the CAC cooling system.
56. Install the accessory drive belt.

⁂ WARNING

The throttle body spacer sealing surfaces are soft metals. Carefully clean both sealing surfaces.

➡Inspect the throttle body spacer gasket and install a new gasket if necessary.

57. Install a new gasket and install the throttle body spacer.
58. Tighten the bolts in two steps.
  a. Step 1: Tighten to 10 Nm (89 inch lbs.).
  b. Step 2: Tighten an additional 90 degrees.
59. Connect the vacuum lines.
60. Connect the brake booster vacuum line.
61. Connect the vacuum line.
62. Install the bracket bolt.
63. Connect the exhaust gas recirculation (EGR) valve.
64. Connect the vacuum line.
65. Connect the EGR vacuum tube fitting.
66. Install the accelerator cable bracket and install the three bolts.

➡Pay attention to the throttle return spring orientation. It must be reinstalled in the correct orientation.

67. Connect the throttle control linkages.
68. Connect the speed control actuator cable.
69. Connect the accelerator cable.
70. Install the throttle return spring to the throttle lever.
71. Connect the vacuum lines and the vent hose.
72. Connect the idle air control electrical connector.
73. Connect the throttle position (TP) sensor electrical connector.
74. Install the air cleaner outlet tube.
75. Connect the battery ground cable.

Supercharger Drive Pulleys

REMOVAL & INSTALLATION

1. With the vehicle in **Neutral**, position it on a hoist.
2. Remove the supercharger drive belt.
3. Remove the fan blade, clutch and shroud.
4. Remove the nut and position the shield aside.
5. Remove the transmission cooler line clip.
6. Remove the two nuts and one bolt.
7. Remove the starter motor.
8. Remove the accessory drive belt.

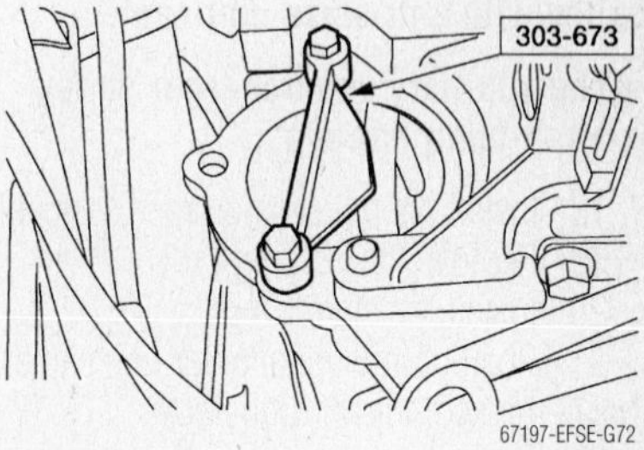

Flywheel holding tool

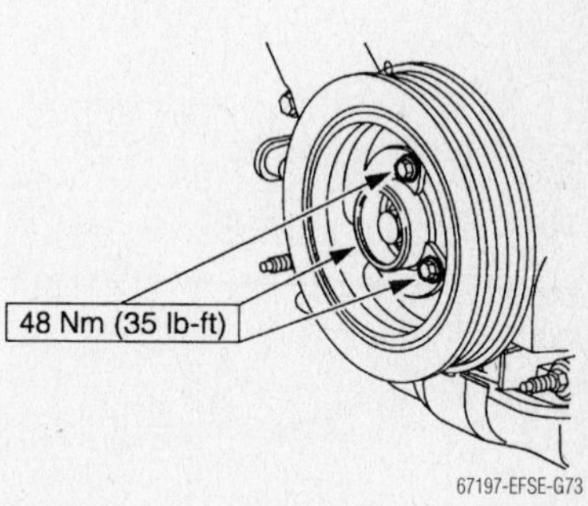

Supercharger pulley adapter

➡The auxiliary supercharger pulley has left-hand threads.

9. Remove the pulley and brace assembly.
10. Remove the supercharger pulley adapter.
11. Remove and discard the bolt.
12. Using a suitable tool, remove the crankshaft pulley.

To install:

➡If the crankshaft pulley is not secured within four minutes, the sealant must be removed and the sealing area cleaned. To clean the sealing area, use silicone gasket remover and metal surface prep. Follow the directions on the packaging. Failure to follow this procedure can cause future oil leakage.

13. Apply silicone gasket and sealant to the Woodruff key slot on the crankshaft pulley.
14. Using a suitable tool, install the crankshaft pulley.
15. Install a new bolt and washer. Tighten the bolt in four steps.
    a. Step 1: Tighten to 90 Nm (66 ft. lbs.).
    b. Step 2: Loosen the bolt.
    c. Step 3: Tighten to 50 Nm (37 ft. lbs.).
    d. Step 4: Tighten an additional 90 degrees.
16. Install the supercharger pulley adapter.
17. Install the accessory drive belt.

➡Coat the threads of the supercharger pulley with anti-seize lubricant.

➡The auxiliary supercharger pulley has left-hand threads.

18. Install the pulley and brace assembly.
19. Remove the holding tool.
20. Install the starter motor.
21. Position the transcooler line clip and install the two nuts and one bolt.
22. Position the shield and install the nut.

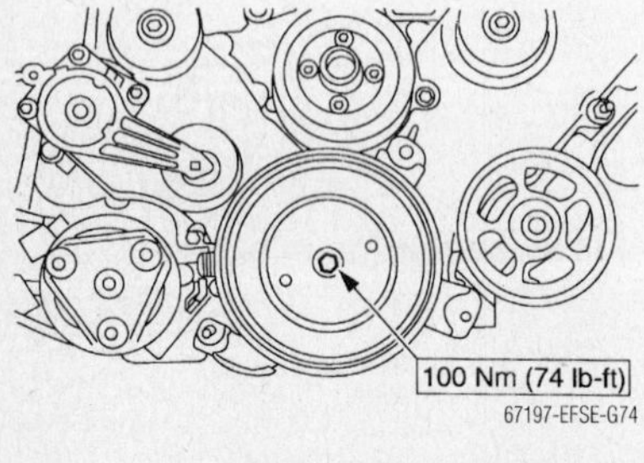

Pulley and brace assembly installation

23. Install the fan blade, clutch and shroud.
24. Install the supercharger drive belt.

Intake Manifold

REMOVAL & INSTALLATION

➡When the battery is disconnected and reconnected, some abnormal drive symptoms may occur while the vehicle relearns its adaptive strategy. The vehicle may need to be driven 10 miles (16 km) or more to relearn the strategy.

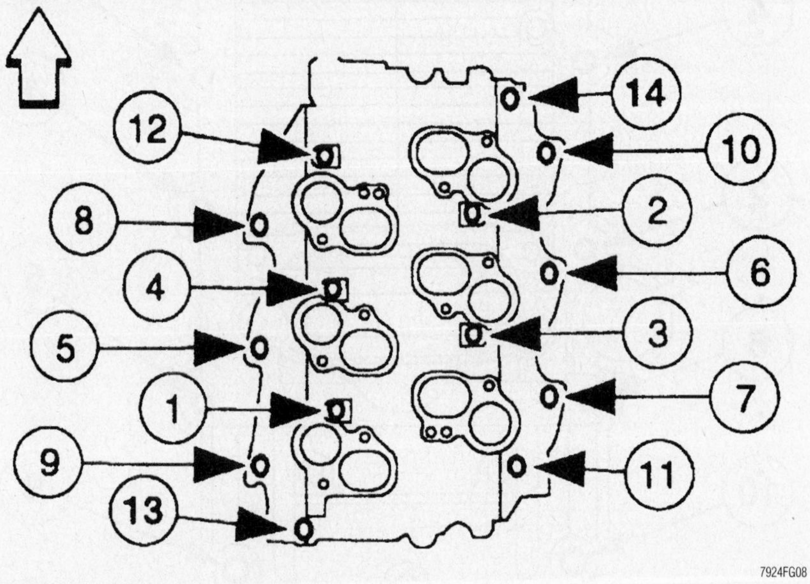

Tighten the lower intake manifold bolts using this sequence—4.2L engine

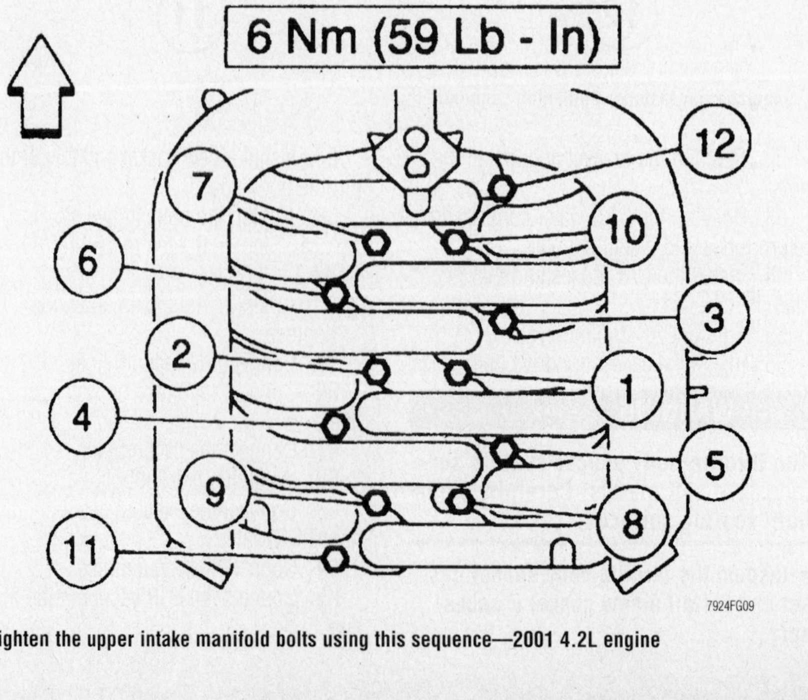

Tighten the upper intake manifold bolts using this sequence—2001 4.2L engine

4.2L Engine

✻✻ CAUTION

Fuel injection systems remain under pressure, even after the engine has been turned OFF. The fuel system pressure must be relieved before disconnecting any fuel lines. Failure to do so may result in fire and/or personal injury.

1. Before servicing the vehicle, refer to the precautions in the beginning of this section.

2. Remove or disconnect the following:
   - Engine air cleaner outlet tube
   - Ignition coil electrical connector
   - Radio ignition interference capacitor electrical connector
   - Spark plug wires
   - Accelerator control splash shield
   - Accelerator cable end, if equipped, the speed control actuator cable end
   - Accelerator cable and actuator cable aside, after removing the hold-down bolts
   - Vapor Management Valve (VMV) hose
   - Brake booster vacuum hose
   - Manifold vacuum connection
   - Positive Crankcase Ventilation (PCV) valve from the rocker arm cover
   - Engine Vacuum Regulator (EVR) bracket aside
   - Throttle Position (TP) sensor and Idle Air Control (IAC) valve electrical connectors
   - Breather from the rocker arm cover
   - The 12 upper intake manifold retaining bolts, then lift the manifold off of the engine and discard the intake manifold upper gasket.
   - The 6 fuel injector electrical connectors
   - The engine coolant temperature sensor and the water temperature indicator sending unit electrical connectors.
   - Exhaust Gas Recirculation (EGR) valve vacuum hose
   - EGR valve tube upper fitting
   - Radiator hose from the lower intake manifold
   - Intake Manifold Runner Control (IMRC) actuator brackets aside
   - Fuel pressure regulator vacuum line
   - Fuel system pressure
   - Fuel lines
   - Water pump bypass hose

➡Remove the lower intake manifold with the fuel injection supply manifold and fuel injectors as one unit.

   - The 6 long bolts and the 8 short bolts, then lift the lower intake manifold off of the engine.

3. Remove and discard the lower intake manifold sealing components.

To install:

4. Clean all components of dirt, grease and old gasket material.
5. Install the lower intake manifold front and rear end seals as follows:

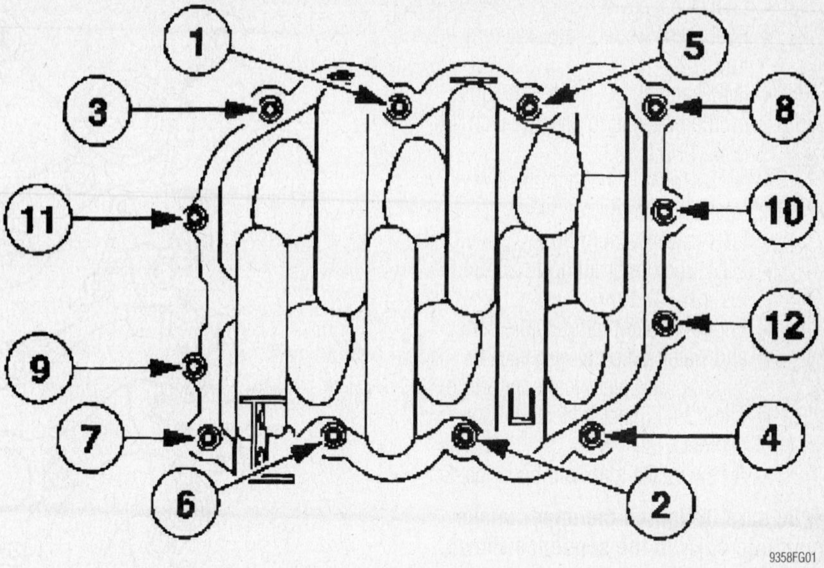

Tighten the upper intake manifold bolts using this sequence—2002–04 4.2L engine

a. Apply a bead of RTV sealant to the intake manifold front and rear end seal mounting points.
b. Install the lower intake manifold front and rear end seals.

6. Install new lower intake manifold gaskets onto the cylinder heads.

➡The lower intake manifold must be installed within 15 minutes of applying sealant.

7. Apply a bead of RTV silicone gasket sealant to the end of the lower intake manifold end seals, where they stop on the cylinder head surface. Position the intake manifold onto the engine block and cylinder heads.
8. Install the lower intake manifold mounting bolts in the correct positions. Refer to the illustration for the correct placement of the long (A) and the short (B) mounting bolts.

➡Be sure to tighten the intake manifold bolts in 2 steps.

9. Tighten the lower intake manifold mounting bolts following the tightening sequence:
   a. Step 1: 44 inch lbs. (5 Nm).
   b. Step 2: 71–101 inch lbs. (8.0–11.5 Nm).
10. Install or connect the following:
   - Water bypass hose

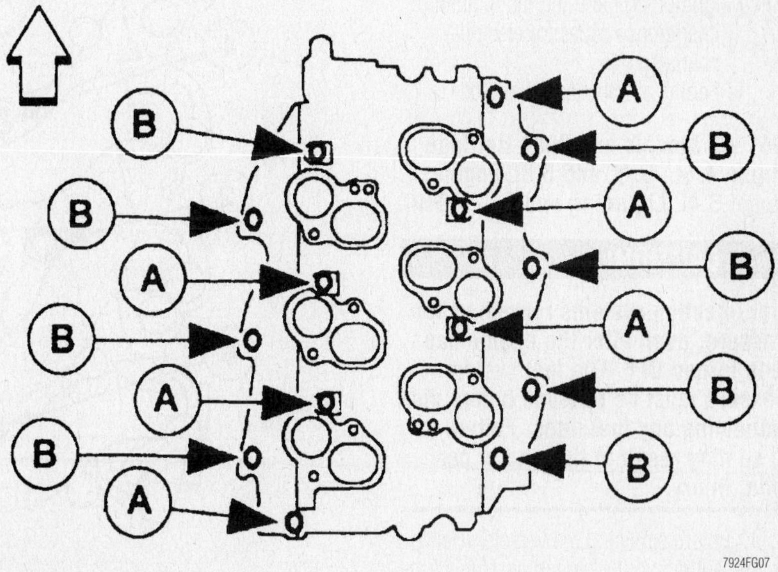

Make certain to install the long bolts (A) and the short bolts (B) in the correct lower intake manifold holes—4.2L engine

- Fuel lines
- Fuel pressure regulator vacuum line
- IMRC actuators. Tighten the bolts on the brackets to 71–102 inch lbs. (8.0–11.5 Nm).
- Upper radiator hose to the lower intake manifold
- EGR valve vacuum hose
- EGR tube upper fitting to 25–34 ft. lbs. (37–47 Nm).
- Engine coolant temperature sensor and water temperature indicator sending unit electrical connectors
- The 6 fuel injector electrical connectors
- A new intake manifold upper gasket

➡Be sure to tighten the upper intake manifold bolts in the sequence shown.

11. Position the upper intake manifold onto the lower intake manifold, then tighten the upper intake manifold bolts following the tightening sequence as follows:
    - 59 inch lbs. (6 Nm).
    - 72–96 inch lbs. (8.0–11.5 Nm).
12. Install or connect the following:
    - Breather into the rocker arm cover
    - TP sensor and IAC valve electrical connectors
    - EVR bracket
    - Manifold vacuum connection
    - PCV valve
    - Brake booster vacuum hose
    - Vapor Management Valve (VMV) hose
    - Accelerator and speed actuator cables
    - Accelerator control splash shield
    - Spark plug wires
    - Ignition coil and the radio ignition interference capacitor electrical connectors
    - Engine air cleaner outlet tube

2001–03 Models and 2004 Heritage Edition 4.6L, 5.4L and 6.8L Engine, Except 5.4L Lightning and 5.4L DOHC

✻✻ CAUTION

Fuel injection systems remain under pressure, even after the engine has been turned OFF. The fuel system pressure must be relieved before disconnecting any fuel lines. Failure to do so may result in fire and/or personal injury.

1. Before servicing the vehicle, refer to the precautions in the beginning of this section.

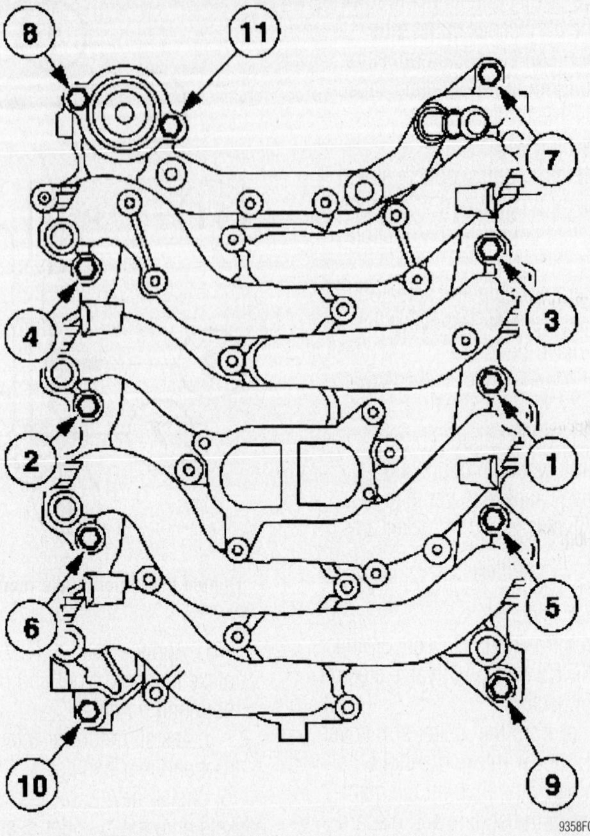

Tighten the bolts in 2 steps using this sequence—2001–03 4.6L shown, 5.4L engine similar

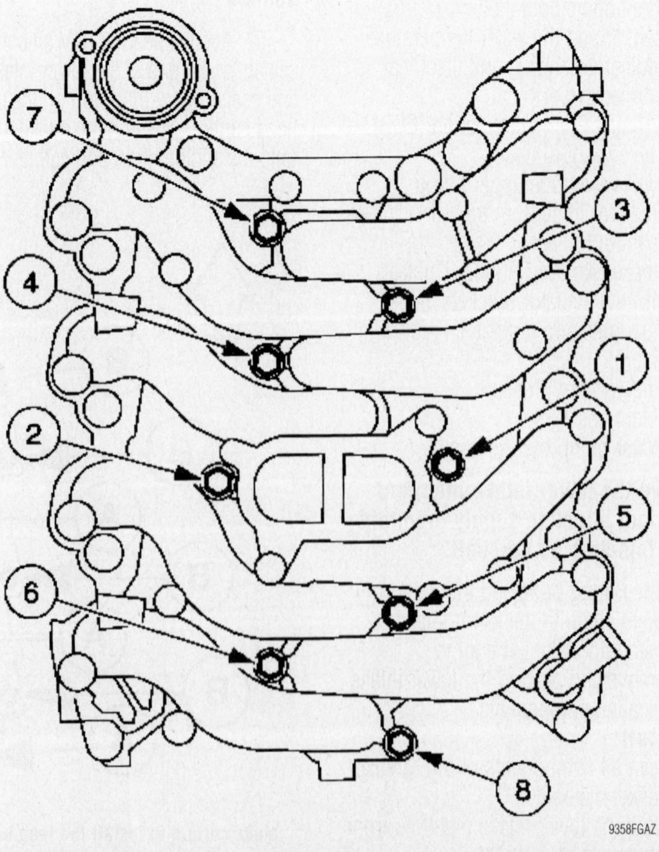

Tighten the bolts in 2 steps using this sequence—2001 4.6L shown, 5.4L engine similar

2. Remove or disconnect the following:
- Negative battery cable
- Engine cover, if equipped
- Fuel system pressure
- Coolant
- Upper radiator hose from the intake manifold
- Accelerator cable from the bracket and the throttle body cam
- Speed control actuator cable from the throttle body
- Throttle cable return spring
- All vacuum hoses, fuel lines and electrical wires from the throttle body and intake manifold
- Accelerator bracket from the throttle body
- Exhaust Gas Recirculation (EGR) valve-to-exhaust manifold tube
- Brake booster vacuum hose bracket
- Two evaporative emission canister purge valve nuts and position the valve aside
- Four bolts, and the throttle body and adapter as an assembly
- Fuel injector electrical connectors
- Bolts and the eight ignition coils
- On the 6.8L engine: the radio interference capacitors from the left side of the intake manifold
- Alternator upper support bracket
- Heater hose from the intake manifold
- Water thermostat housing and thermostat. Discard the O-ring seal.
- Intake manifold bolts
- Intake manifold from the engine, then detach the Intake Manifold Tuning Valve (IMTV) electrical connector. Remove and discard the upper intake manifold gaskets.
- Upper-to-lower intake manifold bolts, then separate the upper intake manifold from the lower intake manifold. Discard the old gasket.

To install:

3. Position the lower intake manifold gasket and the upper intake manifold onto the lower intake manifold, then loosely install the upper-to-lower intake manifold bolts.

➡Be sure to tighten the lower-to-upper manifold bolts in 2 steps.

4. Tighten the 8 lower-to-upper intake manifold bolts in 2 steps following the tightening sequence as follows:
- 18 inch lbs. (2 Nm).
- 18 ft. lbs. (25 Nm).

5. Position the 2 upper intake manifold gaskets on the cylinder heads. Set the upper

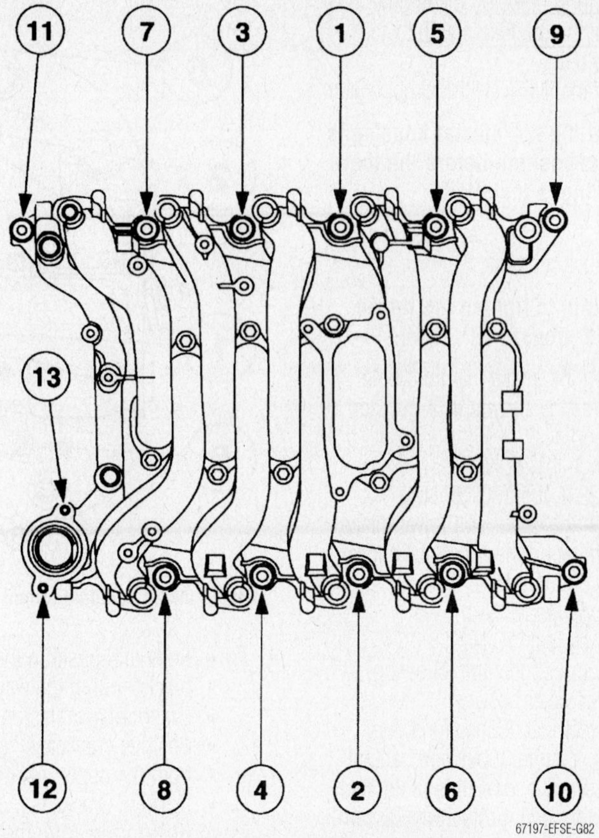

Tighten the bolts in 2 steps using this sequence—2001–03 6.8L engine

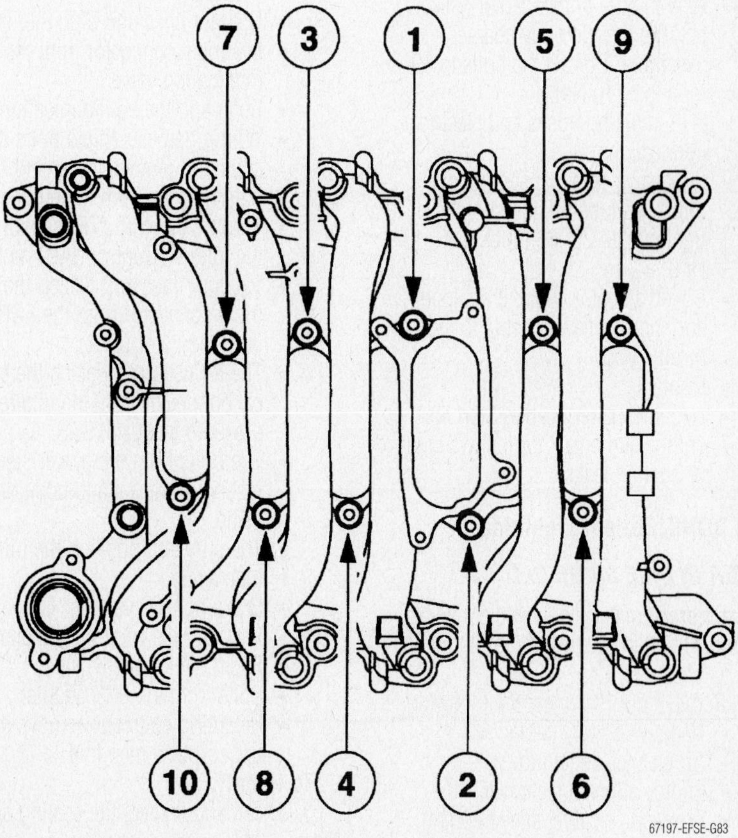

Tighten the bolts in 2 steps using this sequence—2002–03 6.8L engine

intake manifold in place on the engine, then loosely install the 9 intake manifold-to-cylinder head bolts.

6. Attach the IMTV electrical connector.

➡Check that the thermostat housing is in the correct position before the thermostat housing is installed.

7. Install the thermostat housing and start the 2 housing bolts.

➡Make certain to tighten the intake manifold in 2 steps.

8. Tighten the intake manifold bolts using the sequence shown, in 2 steps on all except 6.8L engines.
 - 18 inch lbs. (2 Nm).
 - 15–22 ft. lbs. (20–30 Nm)

9. Tighten the upper and lower intake manifold bolts using the sequence shown, in 2 steps on 6.8L engines.
 - 18 inch lbs. (2 Nm).
 - 89 inch lbs. (10 Nm).

10. Install or connect the following:
 - Heater water hose
 - All electrical connections, fuel lines, vacuum tubes and coolant hoses to the intake manifold, fuel injectors and throttle body assembly.
 - Alternator support bracket
 - Ignition coils
 - New throttle body adapter gasket and the throttle body adapter assembly. Tighten the bolts to 89 inch lbs. (10 Nm).
 - All remaining hoses and electrical connections
 - EGR valve-to-exhaust manifold tube. The tube fittings should be tightened to 26–33 ft. lbs. (35–45 Nm).
 - Speed actuator cable, if equipped, and the accelerator cable to the throttle body.
 - Coolant
 - Negative battery cable. Start the engine and check for fuel, vacuum or coolant leaks.

5.4L DOHC, Except Lightning

UPPER INTAKE MANIFOLD

1. Before servicing the vehicle, refer to the precautions in the beginning of this section.

2. Remove or disconnect the following:
 - Negative battery cable
 - Air cleaner outlet tube
 - Engine appearance cover
 - Accelerator cable, speed control cable, and the return spring
 - Bolts and position the cables and bracket out of the way
 - Evaporative emission return line
 - Positive crankcase ventilation tube from the upper intake manifold
 - Positive Crankcase Ventilation (PCV) valve from the valve cover
 - PCV valve tube from the water heated fitting and remove the tube assembly
 - Coolant lines and plug the lines.
 - Electrical connector from the communication valve
 - Bolts and the communication valve
 - Wiring harness shield bolts and 5 clips and remove the shield
 - Balance tube from the engine
 - Throttle Position (TP) sensor and the Idle Air Control (IAC) motor
 - Vacuum lines and detach the electrical connector from the EGR vacuum regulator (EVR).
 - The 2 hoses and detach the electrical connector from the differential pressure feedback EGR.
 - The bolts from the power steering reservoir bracket and position it aside.
 - The stud and position the oil fill tube aside.
 - Brake booster vacuum line
 - Vacuum line from the Exhaust Gas Recirculation (EGR) valve
 - Bolts from the EGR adapter
 - Retaining bolts and remove the upper intake manifold.

To install:

3. Clean and inspect the sealing surfaces.

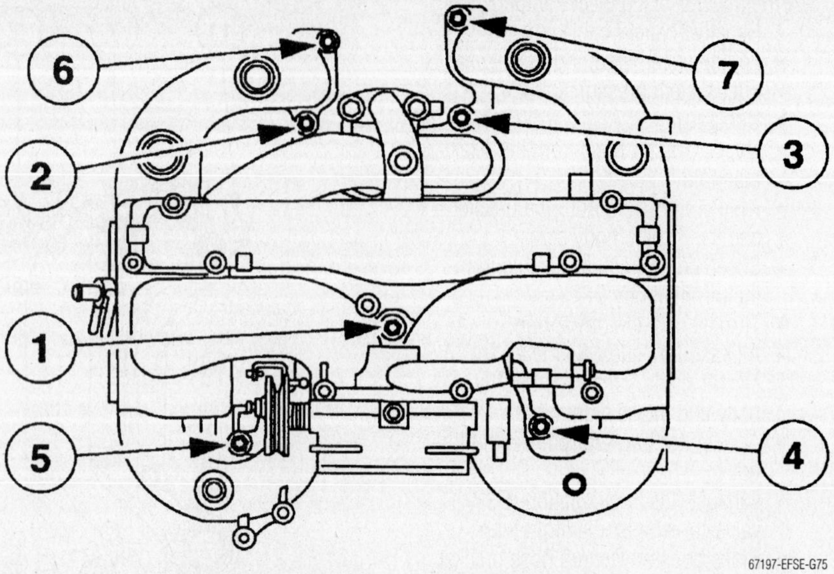

Upper intake manifold torque sequence—5.4L DOHC, exc. Lightning

4. Install or connect the following:
 - Upper intake manifold: Tighten the bolts to 89 inch lbs. (10 Nm) and then an additional 90 degrees in the sequence shown
 - EGR adapter with a new gasket
 - Vacuum line to the EGR valve
 - Brake booster vacuum line
 - Oil fill tube and install the stud
 - Power steering reservoir bracket and install the bolts.
 - EGR valve at the exhaust manifold
 - The 2 hoses and the electrical connector to the differential pressure feedback EGR
 - Vacuum lines and the electrical connector to the EVR
 - IAC valve and the TP sensor
 - Balance tube on the engine
 - Communication valve and the bolts
 - Wiring harness shield, the bolts and 5 clips
 - Electrical connector to the communication valve
 - Tube assembly and connect the PCV valve tube to the water-heated fitting.
 - Coolant hoses
 - PCV valve in the valve cover
 - EVAP return line
 - Cables and bracket and install the bolts
 - Accelerator cable, speed control cable, and the return spring
 - Engine appearance cover and the 3 bolts
 - Engine air cleaner and outlet tube

LOWER INTAKE MANIFOLD

⁂ CAUTION

Fuel injection systems remain under pressure, even after the engine has been turned OFF. The fuel system pressure must be relieved before disconnecting any fuel lines. Failure to do so may result in fire and/or personal injury.

1. Before servicing the vehicle, refer to the precautions in the beginning of this section.
2. Remove or disconnect the following:
   - Negative battery cable
   - Coolant
   - Upper intake manifold
   - Fuel system pressure
   - Fuel lines
   - Engine water bypass hose
   - Electrical connector from the water temperature indicator sender
   - Upper radiator hose, the heater water inlet hose and the heated PCV water fitting inlet hose.
   - The 4 bolts and the upper alternator support bracket
   - The 8 fuel injectors
   - Vacuum line from the fuel injector pressure regulator and position out of the way.
   - Lower intake manifold
   - Radio ignition interference capacitors

To install:

3. Remove and inspect the gaskets, install new gaskets if necessary.
4. Clean the sealing surfaces.
5. Install or connect the following:
   - Radio ignition interference capacitors

➡Bolts should be hand-started, positions 7–12 first then 1–6.

   - Lower intake manifold: Tighten the bolts to 89 inch lbs. (10 Nm) and then an additional 90 degrees in the sequence shown.
   - Water temperature indicator sensor
   - Vacuum harness and connect the vacuum line to the fuel injector pressure regulator.
   - Fuel injectors
   - Upper alternator support bracket and the 4 bolts
   - Heater water inlet hose, the upper radiator hose and the water heated fitting inlet hose
   - Engine water bypass return hose
   - Fuel lines
   - Upper intake manifold
   - Coolant

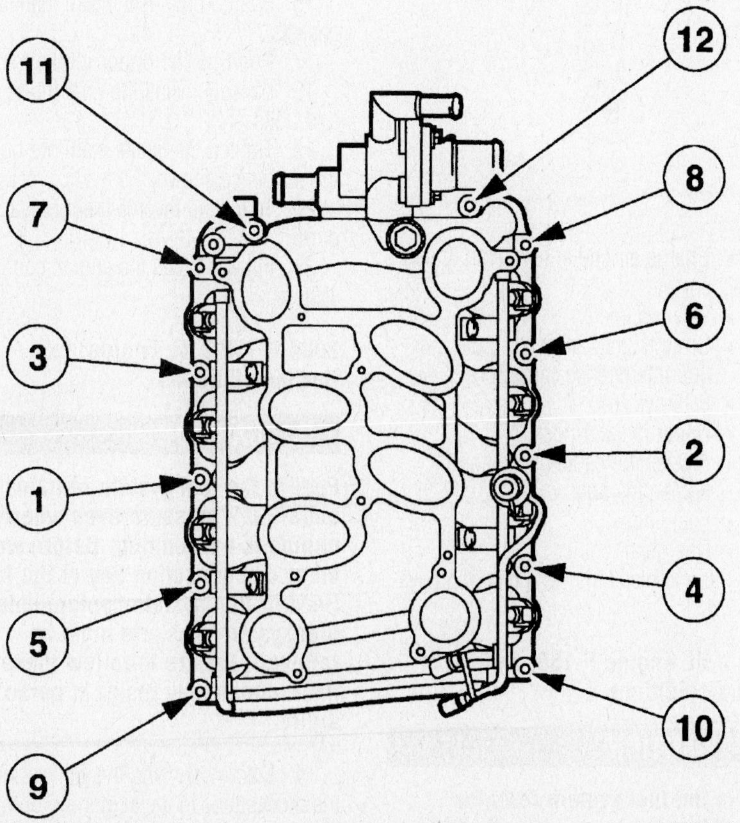

Lower intake manifold torque sequence—5.4L DOHC, exc. Lightning

5.4L Lightning

⁂ CAUTION

Fuel injection systems remain under pressure, even after the engine has been turned OFF. The fuel system pressure must be relieved before disconnecting any fuel lines. Failure to do so may result in fire and/or personal injury.

1. Before servicing the vehicle, refer to the precautions in the beginning of this section.
2. Remove or disconnect the following:
   - Negative battery cable
   - Charge air cooler
   - Fuel system pressure
   - Fuel lines
   - Coolant
   - Accelerator cable bracket retaining bolts and remove the bracket.
   - Upper radiator hose from the thermostat housing

⁂ WARNING

Do not disconnect the PCV hose system from the intake. Installation can not be carried out with the intake in place.

   - Positive Crankcase Ventilation (PCV) system
   - Ground strap and both radio ignition interference capacitors
   - Vacuum line
   - Charge air cooler temperature sensor connector
   - Heater hose
   - Fuel injector electrical connectors
   - Vapor management valve vacuum line
   - Vacuum line near the brake vacuum booster
   - Fuel injection supply manifold
   - Ignition coil connectors
   - Ignition coils
   - Accessory drive belt
   - Alternator bracket
   - Intake manifold

To install:

3. Position the gaskets.

⁂ CAUTION

If the PCV system hose becomes disconnected, the intake manifold will have to be removed to reattach the hose.

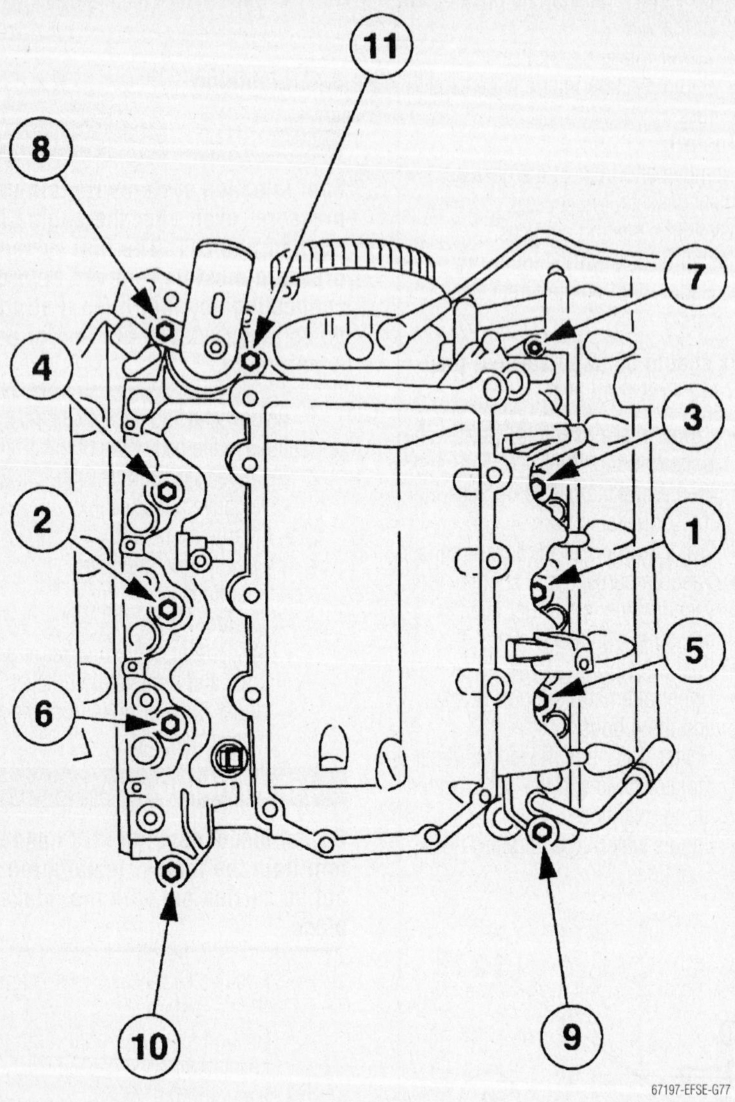

Intake manifold bolt torque—5.4L supercharged engine

4. Make sure that the PCV system hose is securely connected.
5. Install or connect the following:
   - Intake manifold
   - Thermostat outlet
   - Intake manifold bolts. Tighten the bolts in 2 stages. Stage 1: tighten to 18 inch lbs. (2 Nm). Stage 2: tighten to 19 ft. lbs. (25 Nm).
   - Alternator bracket
   - Accessory drive belt
   - Ignition coils
   - Ignition coil connectors
   - Fuel injection supply manifold and install the bolts.
   - Vacuum line near the brake vacuum booster
   - Vapor management valve vacuum line
   - Fuel injector electrical connectors
   - Heater hose
   - Charge air cooler temperature sensor
   - Vacuum line
   - Ground strap and both radio ignition interference capacitors
   - PCV system
   - Upper radiator hose
   - Accelerator cable bracket and tighten the bolts.
   - Fuel lines
   - Charge air cooler
6. Fill and bleed the engine cooling system.

2004 4.6L Engine F-150, Exc. Heritage Editions

⁂ CAUTION

Fuel in the fuel system remains under high pressure even when the engine is not running. Before servicing or disconnecting any of the fuel lines or fuel system components, the fuel system pressure must be relieved to prevent accidental spraying of fuel, causing personal injury or a fire hazard.

1. Before servicing the vehicle, refer to the precautions in the beginning of this section.
2. Drain the cooling system.
3. Release the fuel system pressure.
4. Remove the air cleaner outlet tube.
5. Remove the upper radiator hose.
6. Remove the heater hoses.
7. Remove the vapor management valve.
8. Remove the throttle position sensor.
9. Remove the power steering reservoir and bracket.
10. Remove all related hoses and electrical connectors.
11. Remove the alternator and brackets.
12. Remove the ignition coils.
13. Remove the thermostat.
14. Remove the manifold and gaskets.
15. Clean the mating surfaces of the cylinder head and the intake manifold with metal surface prep and silicone gasket remover. Follow the directions on the packaging.
16. Position the new intake manifold gaskets.
17. Position the upper intake manifold.
18. Loosely install the nine intake manifold bolts.
19. Tighten the intake manifold bolts in the sequence shown.
20. To install, reverse the removal procedure.
21. Fill and bleed the engine cooling system.

2004 F-150 5.4L Engine, Exc. Heritage Editions

⁂ CAUTION

Fuel in the fuel system remains under high pressure even when the engine is not running. Before working on or disconnecting any of the fuel lines or fuel system components, the fuel system pressure must be relieved. Failure to follow these instructions may result in personal injury.

1. Before servicing the vehicle, refer to the precautions in the beginning of this section.

WARNING

If the supply manifold is used as a leverage device, damage may occur to the supply manifold. Care must be taken when working around the supply manifold.

2. With the vehicle in **Neutral**, position it on a hoist.
3. Relieve the fuel system pressure.
4. Drain the engine cooling system.
5. Remove the air cleaner and the air cleaner inlet pipe and resonator.
6. Remove the alternator.
7. Remove the radiator hoses.
8. Remove the heater hoses.
9. Remove the PCV tube.
10. Remove or disconnect all electrical wires and hoses related to intake manifold removal.
11. Remove the coolant crossover assembly.
12. Remove the intake manifold bolts. Lift off the manifold and discard the gaskets.
13. Clean and inspect the sealing surfaces with silicone gasket remover and metal surface prep.

CAUTION

Do not use metal scrapers, wire brushes, power abrasive discs or other abrasive means to clean the sealing surfaces. These tools cause scratches and gouges which make leak paths. Use a plastic scraping tool to remove all traces of old sealant.

14. Install new intake manifold gaskets onto the intake manifold.
15. Position the intake manifold and loosely install the bolts.
16. Tighten the bolts in two steps in the sequence shown.
    a. Step 1: Tighten to 2 Nm (18 lb-in).
    b. Step 2: Tighten to 10 Nm (89 lb-in).
17. The remainder of installation is the reverse of the removal procedure.

2004 F-250/350 5.4L Engine

WARNING

Fuel in the fuel system remains under high pressure even when the engine is not running.

1. Before servicing the vehicle, refer to the precautions in the beginning of this section.

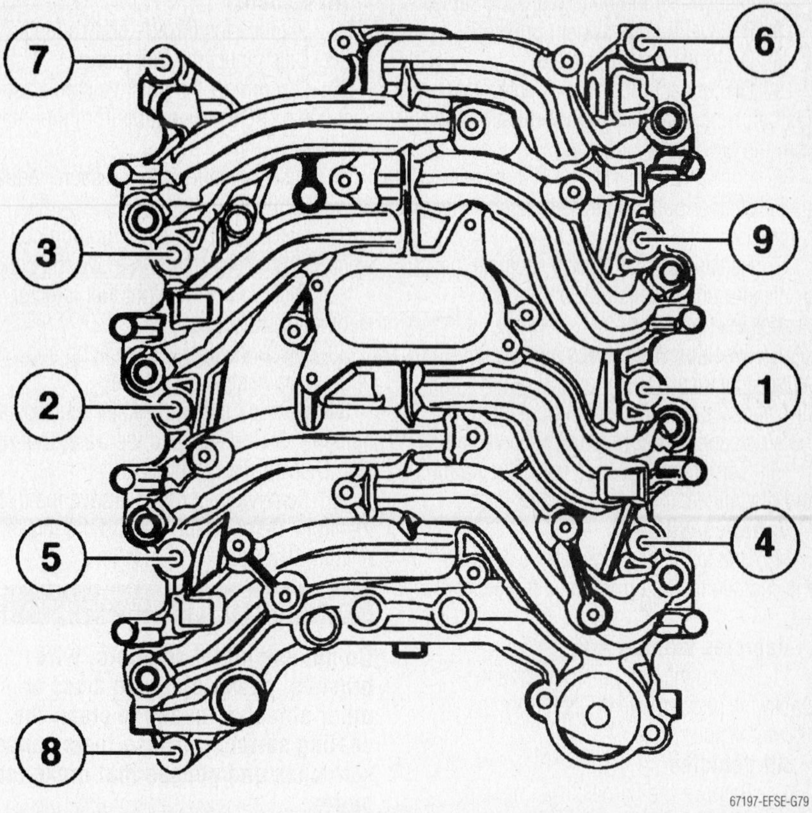

4.6L engine intake manifold torque sequence—2004 models, except Heritage editions

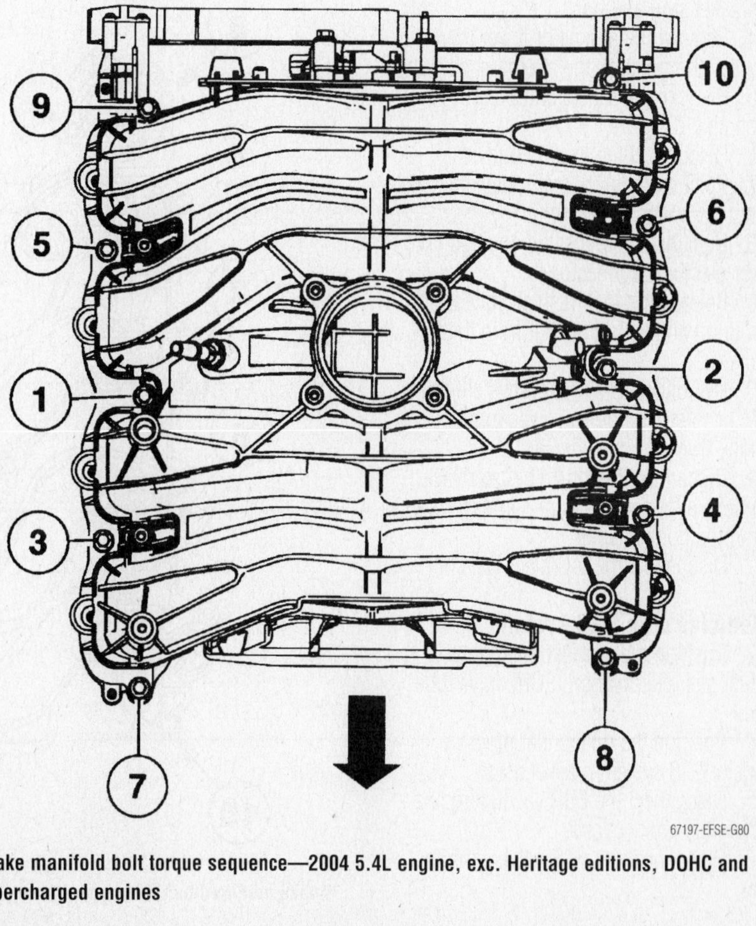

Intake manifold bolt torque sequence—2004 5.4L engine, exc. Heritage editions, DOHC and supercharged engines

2. Disconnect the battery ground cable.
3. Relieve the fuel system pressure.
4. Drain the cooling system.
5. Disconnect the idle air control (IAC) fresh air tube and crankcase ventilation tube from the air cleaner outlet pipe.
6. Loosen the clamp and disconnect the air cleaner outlet pipe from the throttle body.
7. Release the clamp and separate the air cleaner outlet housing from the air cleaner inlet housing.
8. Disconnect the mass air flow (MAF) sensor electrical connector and remove the air cleaner, air cleaner outlet housing and the air cleaner outlet pipe from the vehicle.
9. Compress and slide the hose clamp and disconnect the upper radiator hose.

Vehicles with A/C

10. Rotate the belt tensioner clockwise and remove the belt from the alternator pulley.

Vehicles without A/C

11. Rotate the belt tensioner counterclockwise and remove the belt from the alternator pulley.

All vehicles

12. Remove the harness routing clips from the alternator stud and the engine front cover.
13. Remove the alternator upper mounting bracket stud and bolt.
14. Remove the alternator lower mounting bolts and position the alternator aside.
15. Remove the accelerator control splash shield.
16. Disconnect the throttle body cam.
17. Position the accelerator cable bracket aside.
18. Disconnect the throttle position (TP) sensor electrical connector.
19. Release the IAC fresh air tube from the routing clip and remove it from the vehicle.
20. Disconnect the evaporative emission (EVAP) canister purge valve vacuum hose from the throttle body adapter.
21. Disconnect the brake booster vacuum hose and main vacuum harness from the throttle body adapter.
22. Disconnect the IAC electrical connector.

Vehicles with EGR system

23. Position the exhaust manifold to exhaust gas recirculation (EGR) valve tube aside.
24. Position the differential pressure feedback EGR system bracket aside.
25. Disconnect the EGR vacuum regulator solenoid connections.
26. Disconnect the EGR valve vacuum hose.

All vehicles

27. Remove the throttle body adapter.
28. Disconnect the fuel tubes.
29. Disconnect the positive crankcase ventilation (PCV) valve tube from the intake manifold.
30. Disconnect the fuel pressure regulator vacuum hose.
31. Disconnect the eight ignition coil electrical connectors and remove the coils.
32. Disconnect the eight fuel injector electrical connectors.
33. Disconnect the hose clamp and remove the heater water hose.
34. Remove the bolts, water thermostat housing, O-ring seal and the water thermostat. Discard the O-ring seal.
35. Remove the bolts. Remove the intake manifold. Remove and discard the intake manifold gaskets.

⁂ CAUTION

Do not use metal scrapers, wire brushes, power abrasive discs or other abrasive means to clean the sealing surfaces. These tools cause scratches and gouges that make leak paths.

36. Clean all mating surfaces.

To install:

All vehicles

37. Position the intake manifold gaskets.
38. Position the intake manifold. Loosely install the nine bolts.
39. Install the thermostat. Install a new O-ring seal.

➡The thermostat housing bolts are tightened in sequence with the intake manifold bolts. Do not tighten the thermostat housing bolts during thermostat installation.

40. Install the water thermostat housing and loosely install the bolts.
41. Tighten the bolts in two steps, in the sequence shown.
    a. Step 1: Tighten to 2 Nm (18 inch lbs.).
    b. Step 2: Tighten to 25 Nm (18 ft. lbs.).
42. Install the heater water hose and position the clamp.
43. Connect the eight fuel injector electrical connectors.
44. Install the eight ignition coils.
45. Connect the eight ignition coil electrical connectors.

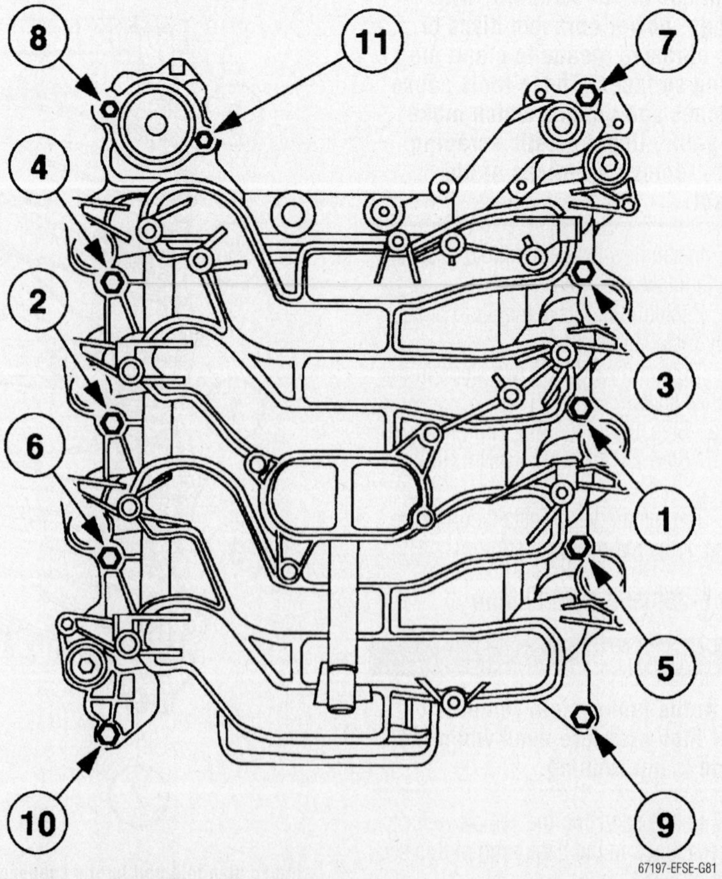

Intake manifold bolt torque sequence—2004 5.4L engine

46. Position the vacuum harness and connect the fuel pressure regulator vacuum hose.

47. Connect the PCV valve tube to the intake manifold.

48. Connect the fuel tubes and install the safety clips.

49. Install a new throttle body adapter gasket.

50. Install the throttle body adapter.

Vehicles with EGR system

51. Connect the EGR valve vacuum hose.

52. Connect the EGR vacuum regulator solenoid connections.

53. Install the differential pressure feedback EGR system bracket.

54. Connect the exhaust manifold to EGR valve tube upper fitting.

55. Tighten both fittings starting at the top in three steps. Hand-tighten.

56. Tighten the upper fitting to 40–60 Nm (30–44 ft. lbs.).

57. Tighten the lower fitting to 40–60 Nm (30–44 ft. lbs.).

All vehicles

58. Connect the IAC electrical connector.

59. Connect the brake booster vacuum hose and main vacuum harness to the throttle body adapter.

60. Connect the evaporative emission canister purge valve vacuum hose to the throttle body adapter.

61. Install the IAC fresh air tube and insert it in the routing clip.

62. Connect the TP sensor electrical connector.

63. Install the accelerator cable bracket.

64. Connect the throttle body cam. .

65. Install the accelerator control splash shield.

66. Position the alternator and install the lower mounting bolts.

67. Install the alternator upper mounting bracket stud and bolt.

68. Install the harness routing clips on the alternator stud and the engine front cover.

Vehicles without A/C

69. Rotate the belt tensioner counterclockwise and install the belt on the alternator pulley.

Vehicles with A/C

70. Rotate the belt tensioner clockwise and install the belt on the alternator pulley.

All vehicles

71. Install the upper radiator hose and reposition the clamp.

72. Position the air cleaner, air cleaner outlet housing and the air cleaner outlet pipe in the vehicle and connect the MAF sensor electrical connector.

73. Join the air cleaner outlet housing to the air cleaner inlet housing and secure the clamp.

74. Connect the air cleaner outlet pipe to the throttle body and tighten the clamp.

75. Connect the IAC fresh air tube and crankcase ventilation tube to the air cleaner outlet pipe.

76. Fill the cooling system.

77. Connect the battery ground cable.

2004 F-250/350 6.8L Engine

⁂ WARNING

Fuel in the fuel system remains under high pressure, even when the engine is not running.

1. Before servicing the vehicle, refer to the precautions in the beginning of this section.
2. Relieve the fuel system pressure.
3. Disconnect the battery ground cable.
4. Drain the engine cooling system.
5. Remove the air cleaner outlet tube
6. Remove the alternator.
7. Disconnect the fuel lines.
8. Compress and slide the hose clamp and disconnect the water outlet hose.
9. Remove the accelerator cable splash shield.
10. Disconnect the accelerator and speed control cables and remove the return spring.
11. Position the accelerator cable bracket aside.
12. Disconnect the positive crankcase valve (PCV) tube and the vacuum lines.
13. Disconnect the heated PCV coolant hoses.
14. Remove the bolt and position the right radio ignition interference capacitor aside.
15. Disconnect the vacuum line.
16. Disconnect the five right fuel injector electrical connectors and five ignition coil electrical connectors.
17. Disconnect the idle air control (IAC) motor electrical connector and the bypass hose.
18. Disconnect the heater hose.
19. Remove the throttle body (TB) adapter.
20. Disconnect the five left fuel injector electrical connectors and ignition coil connectors.
21. Disconnect the engine coolant temperature (ECT) sensor electrical connectors.
22. Remove the ignition coils.

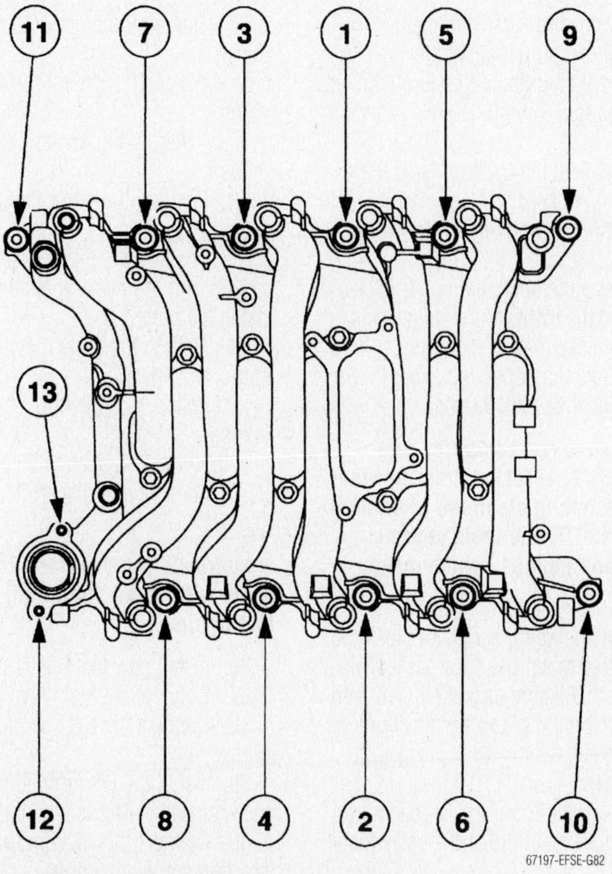

Lower intake manifold bolt torque sequence—2004 6.8L engine

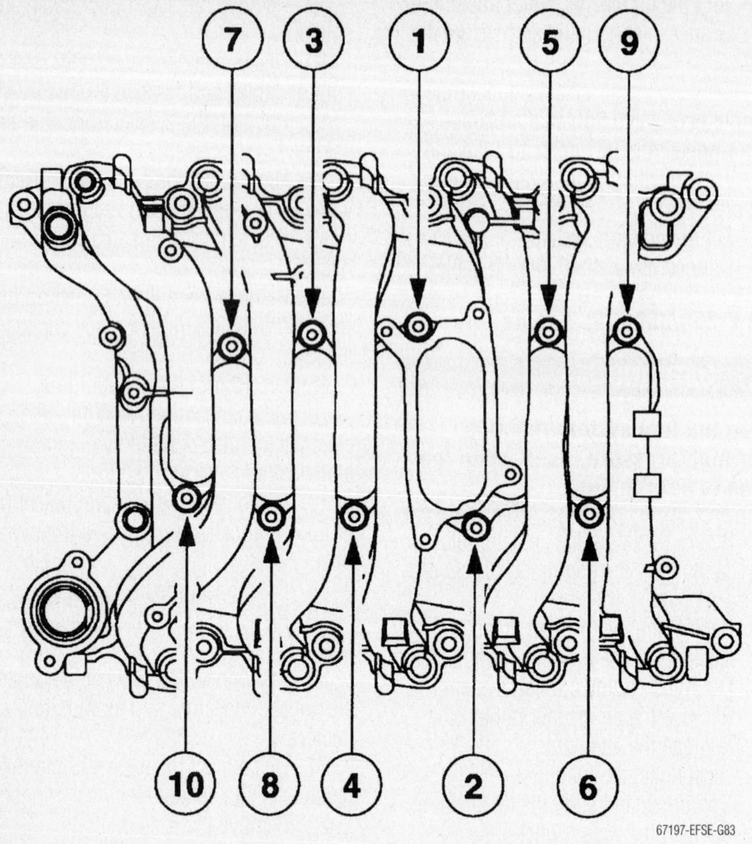

Upper intake manifold bolt torque sequence—2004 6.8L engine

23. Disconnect the vacuum connector from the fuel pressure regulator.
24. Remove the bolt and position the left ignition radio ignition interference capacitor aside.
25. Disconnect the cylinder head temperature (CHT) sensor.
26. Remove the water thermostat and housing.
27. Remove the bolts, the upper intake manifold and the intake manifold gaskets. Discard the intake manifold gaskets.
28. Separate the upper and lower intake manifolds. Remove the 10 bolts.

➡Do not use metal scrapers, wire brushes, power abrasive discs or other abrasive means to clean the aluminum retainer plate. These tools cause scratches and gouges, which make leak paths.

29. Clean all mating surfaces. Clean the intake manifold mating surface of the cylinder heads with silicone gasket remover and metal surface prep. Follow the directions on the packaging.

To install:

30. Position the lower intake manifold to the upper intake manifold and loosely install the 10 bolts.
31. Tighten the bolts in the sequence shown.
   a. Step 1: Tighten to 2 Nm (18 inch lbs.).
   b. Step 2: Tighten to 10 Nm (89 inch lbs.).
32. Position the water thermostat and housing.
33. Position the upper intake manifold gaskets and the intake manifold, and loosely install the bolts.
34. Tighten the bolts in two steps, in the sequence shown.
   a. Step 1: Tighten to 2 Nm (18 inch lbs.).
   b. Step 2: Tighten to 25 Nm (18 ft. lbs.).
35. Remove and discard the fuel line O-ring seals.

➡Lubricate the new fuel line O-ring seals with clean engine oil prior to installation.

36. Install new fuel line O-ring seals.
37. Connect the fuel lines.
38. Connect the CHT sensor electrical connector.
39. Install the left radio ignition interference capacitor electrical connector.
40. Connect the vacuum line to the fuel injection supply manifold.
41. Connect the ECT sensor electrical connector.
42. Install the ignition coils.
43. Connect the five left fuel injector electrical connectors and five left ignition coil connectors.
44. Position a new TB adapter press-in-place gasket.
45. Position the TB adapter.
46. Install the four bolts in two steps:
   a. Step 1: Tighten to 9 Nm (80 inch lbs.).
   b. Step 2: Tighten an additional 90 degrees.
47. Connect the coolant hose.
48. Connect the IAC motor bypass hose and connector.
49. Connect the five right fuel injector electrical connectors and five right ignition coil connectors.
50. Connect the vacuum line.
51. Install the right radio ignition interference capacitor.
52. Connect the PCV tube and the vacuum lines.
53. Connect the heated PCV coolant hoses.
54. Install the accelerator bracket.
55. Connect the accelerator and speed control cables and return spring.
56. Install the accelerator cable splash shield.
57. Connect the engine water outlet hose and position the hose clamp.
58. Install the alternator.
59. Install the air cleaner outlet tube.
60. Connect the battery ground cable.
61. Fill and bleed the cooling system.

E-Series 4.6L Engine

1. Before servicing the vehicle, refer to the precautions in the beginning of this section.
2. Disconnect the battery ground cable.
3. Remove the interior engine cover.
4. Disconnect the fuel hose spring lock couplings.
5. Drain the cooling system.
6. Position the engine air cleaner (ACL).
7. Disconnect the MAF sensor electrical connector and remove the ACL.
8. Disconnect the two hoses.
9. Loosen the clamp and remove the air cleaner outlet pipe.
10. Remove the accelerator cable snow shield.
11. Disconnect the accelerator cable.
12. Disconnect the speed control actuator cable.
13. Remove the throttle return spring.

14. Compress and slide the hose clamp and disconnect the upper radiator hose.
15. Disconnect the coolant hose.
16. Disconnect the coolant hose.
17. Disconnect the throttle position (TP) sensor.
18. Disconnect the differential pressure feedback exhaust gas recirculation (EGR) system.
19. Remove the differential pressure feedback EGR system bracket.
20. Remove the bolt and the bracket.
21. Remove the bolts and the alternator upper support bracket.
22. Remove the bolt retaining the transmission fluid filler tube support bracket.
23. Remove the bolts and position the accelerator cable bracket and cables aside.
24. If equipped, disconnect the auxiliary heater hoses and position aside.
25. Disconnect and remove the idle air control (IAC) valve fresh air hose.
26. Disconnect the evaporative emission canister purge valve hose and vacuum hose.
27. Remove the two evaporative emission canister purge valve nuts and position the valve aside.
28. Disconnect the fuel pressure sensor vacuum hose and the electrical connector.
29. Disconnect the EGR vacuum regulator solenoid connections.
30. Disconnect the EGR valve vacuum hose.
31. Disconnect the IAC valve electrical connector.
32. Disconnect the brake booster and main engine vacuum harness and position aside.
33. Disconnect the differential pressure feedback EGR sensor hoses from the EGR valve.
34. Remove the exhaust manifold-to-EGR valve tube.
35. Remove the four bolts, and the throttle body and adapter as an assembly. Discard the throttle body adapter gasket.
36. Disconnect and remove the positive crankcase ventilation hose.
37. Disconnect the eight ignition coil electrical connectors.
38. Disconnect the eight fuel injector electrical connectors.
39. Remove the eight bolts and the eight ignition coils.
40. Remove the bolts, the thermostat housing and the thermostat. Disconnect the O-ring seal.
41. Remove the nine bolts retaining the intake manifold.
42. Remove the intake manifold. Discard the intake manifold gaskets.
43. Inspect the throttle body, the intake manifold, and their sealing surfaces for damage.

To install:

➡Do not use metal scrapers, wire brushes, power abrasive discs or any other abrasive means to clean the sealing surfaces. These tools cause scratches and gouges which can cause leak paths. Use a plastic scraping tool to clean these surfaces.

44. Clean and inspect all sealing surfaces.
45. Position the intake manifold gaskets and the intake manifold.
46. Loosely install the nine bolts.
47. Install the thermostat.
48. Install a new O-ring seal.

➡The thermostat housing bolts are tightened in sequence with the intake manifold bolts.

49. Loosely install the thermostat housing and bolts.
50. Tighten the bolts in two steps, in the sequence shown.
    a. Step 1: Tighten to 2 Nm (18 inch lbs.).
    b. Step 2: Tighten to 25 Nm (18 ft. lbs.).
51. Install the eight ignition coils and bolts.
52. Connect the eight fuel injector electrical connectors.
53. Connect the eight ignition coil electrical connectors.
54. Install the PCV hose and valve.
55. Install the new throttle body adapter gasket.
56. Install the throttle body and adapter assembly.
57. Install the exhaust manifold-to-EGR valve tube and tighten the fitting in two steps.
    a. Hand-tighten the fittings.
    b. Tighten the fittings to 50 Nm (37 ft. lbs.).
58. Connect the differential pressure EGR sensor hoses to the EGR valve tube.
59. Connect the brake booster and the main engine vacuum harnesses.
60. Connect the IAC electrical connector.
61. Connect the EGR vacuum hose.
62. Connect the EGR vacuum regulator solenoid connections.
63. Connect the fuel pressure sensor vacuum hose and electrical connector.
64. Install the evaporative emission canister purge valve and bracket.
65. Connect the evaporative emission canister purge valve hoses.
66. Connect the IAC fresh air hose.
67. If equipped, connect the auxiliary heater hoses.
68. Connect the fuel hose spring lock couplings.
69. Install the accelerator cable bracket.
70. Install the transmission fluid filler tube bracket and bolt.
71. Install the alternator upper support bracket.
72. Install the accelerator cable routing bracket.
73. Install the differential pressure feedback EGR system bracket.
74. Connect the differential pressure feedback EGR sensor.
75. Connect the TP sensor.
76. Connect the coolant hose.
77. Connect the coolant hose.
78. Connect the upper radiator hose.
79. Connect the accelerator cable.
80. Connect the speed control actuator cable.
81. Install the throttle return spring.
82. Install the accelerator cable snow shield.
83. Install the air cleaner outlet pipe and tighten the clamp.
84. Connect the two hoses to the air cleaner outlet pipe.
85. Connect the MAF sensor and position the ACL.
86. Install the ACL.
87. Install the interior engine cover.
88. Connect the battery ground cable.
89. Fill and bleed the engine cooling system.

E-Series 5.4L Engine

1. Before servicing the vehicle, refer to the precautions in the beginning of this section.
2. Disconnect the battery ground cable.
3. Remove the interior engine cover.
4. Disconnect the fuel hose spring lock couplings.
5. Drain the cooling system.
6. Position the engine air cleaner (ACL).
7. Disconnect the MAF sensor electrical connector and remove the ACL.
8. Disconnect the two hoses.
9. Loosen the clamp and remove the air cleaner outlet pipe.
10. Remove the accelerator cable snow shield.
11. Disconnect the accelerator cable.
12. Disconnect the speed control actuator cable.
13. Remove the throttle return spring.
14. Compress and slide the hose clamp and disconnect the upper radiator hose.

15. Disconnect the coolant hose.
16. Disconnect the coolant hose.
17. Disconnect the throttle position (TP) sensor.
18. Disconnect the differential pressure feedback exhaust gas recirculation (EGR) system.
19. Remove the differential pressure feedback EGR system bracket.
20. Remove the bolt and the bracket.
21. Remove the bolts and the alternator upper support bracket.
22. Remove the bolt retaining the transmission fluid filler tube support bracket.
23. Remove the bolts and position the accelerator cable bracket and cables aside.
24. If equipped, disconnect the auxiliary heater hoses and position aside.
25. Disconnect and remove the idle air control (IAC) valve fresh air hose.
26. Disconnect the evaporative emission canister purge valve hose and vacuum hose.
27. Remove the two evaporative emission canister purge valve nuts and position the valve aside.
28. Disconnect the fuel pressure sensor vacuum hose and the electrical connector.
29. Disconnect the EGR vacuum regulator solenoid connections.
30. Disconnect the EGR valve vacuum hose.
31. Disconnect the IAC valve electrical connector.
32. Disconnect the brake booster and main engine vacuum harness and position aside.
33. Disconnect the differential pressure feedback EGR sensor hoses from the EGR valve.
34. Remove the exhaust manifold-to-EGR valve tube.
35. Remove the four bolts, and the throttle body and adapter as an assembly. Discard the throttle body adapter gasket.
36. Disconnect and remove the positive crankcase ventilation hose.
37. Disconnect the eight ignition coil electrical connectors.
38. Disconnect the eight fuel injector electrical connectors.
39. Remove the eight bolts and the eight ignition coils.
40. Remove the bolts, the thermostat housing and the thermostat. Disconnect the O-ring seal.
41. Remove the nine bolts retaining the intake manifold.
42. Remove the intake manifold. Discard the intake manifold gaskets.
43. Inspect the throttle body, the intake manifold, and their sealing surfaces for damage.

To install:

➡Do not use metal scrapers, wire brushes, power abrasive discs or any other abrasive means to clean the sealing surfaces. These tools cause scratches and gouges which can cause leak paths. Use a plastic scraping tool to clean these surfaces.

44. Clean and inspect all sealing surfaces.
45. Position the intake manifold gaskets and the intake manifold.
46. Loosely install the nine bolts.
47. Install the thermostat.
48. Install a new O-ring seal.

➡The thermostat housing bolts are tightened in sequence with the intake manifold bolts.

49. Loosely install the thermostat housing and bolts.
50. Tighten the bolts in two steps, in the sequence shown.
    a. Step 1: Tighten to 2 Nm (18 inch lbs.).
    b. Step 2: Tighten to 25 Nm (18 ft. lbs.).
51. Install the eight ignition coils and bolts.
52. Connect the eight fuel injector electrical connectors.
53. Connect the eight ignition coil electrical connectors.
54. Install the PCV hose and valve.
55. Install the new throttle body adapter gasket.
56. Install the throttle body and adapter assembly.
57. Install the exhaust manifold-to-EGR valve tube and tighten the fitting in two steps.
    a. Hand-tighten the fittings.
    b. Tighten the fittings to 50 Nm (37 ft. lbs.).
58. Connect the differential pressure EGR sensor hoses to the EGR valve tube.
59. Connect the brake booster and the main engine vacuum harnesses.
60. Connect the IAC electrical connector.
61. Connect the EGR vacuum hose.
62. Connect the EGR vacuum regulator solenoid connections.
63. Connect the fuel pressure sensor vacuum hose and electrical connector.
64. Install the evaporative emission canister purge valve and bracket.
65. Connect the evaporative emission canister purge valve hoses.
66. Connect the IAC fresh air hose.
67. If equipped, connect the auxiliary heater hoses.
68. Connect the fuel hose spring lock couplings.
69. Install the accelerator cable bracket.
70. Install the transmission fluid filler tube bracket and bolt.
71. Install the alternator upper support bracket.
72. Install the accelerator cable routing bracket.
73. Install the differential pressure feedback EGR system bracket.
74. Connect the differential pressure feedback EGR sensor.
75. Connect the TP sensor.
76. Connect the coolant hose.
77. Connect the coolant hose.
78. Connect the upper radiator hose.
79. Connect the accelerator cable.
80. Connect the speed control actuator cable.
81. Install the throttle return spring.
82. Install the accelerator cable snow shield.
83. Install the air cleaner outlet pipe and tighten the clamp.
84. Connect the two hoses to the air cleaner outlet pipe.
85. Connect the MAF sensor and position the ACL.
86. Install the ACL.
87. Install the interior engine cover.
88. Connect the battery ground cable.
89. Fill and bleed the engine cooling system.

E-Series 6.8L Engine

⁂ WARNING

Fuel in the fuel system remains under high pressure, even when the engine is not running. Do not work on any fuel system component until fuel system pressure has been relieved. Failure to follow these instructions may result in personal injury.

1. Before servicing the vehicle, refer to the precautions in the beginning of this section.
2. Relieve the fuel system pressure.
3. Disconnect the battery ground cable.
4. Drain the engine cooling system.
5. Remove the engine cover.
6. Remove the air cleaner and the air cleaner outlet tube.
7. Remove the alternator.
8. Disconnect the fuel tube spring lock coupling.
9. Compress and slide the hose clamp and disconnect the coolant hose.

10. Remove the accelerator cable splash shield.
11. Disconnect the accelerator and speed control cables and remove the return spring.
12. Position the accelerator cable bracket aside.
13. Disconnect the heated positive crankcase valve (PCV) tube and the vacuum hoses.
14. Disconnect the PCV electrical connector and tube.
15. Disconnect the PCV coolant hoses.
16. Disconnect the vacuum tube.
17. Disconnect the idle air control (IAC) motor electrical connector and remove the bypass hose.
18. Disconnect the throttle position (TP) sensor electrical connector.
19. Disconnect the heater hose.
20. Remove the throttle body (TB) adapter.
21. Disconnect the vacuum connector.
22. Disconnect the fuel rail pressure (FRP) sensor electrical connector.
23. Disconnect the ten fuel injector electrical connectors.
24. Disconnect the ten ignition coil electrical connectors.
25. Remove the nuts and position the right fuel charge wiring harness aside.
26. Remove the nuts and position the left fuel charge wiring harness aside.
27. Remove the ten bolts and the ten ignition coils.
28. Disconnect the engine coolant temperature (ECT) sensor electrical connector.
29. Disconnect the heater hose.
30. Remove the water thermostat and housing.
31. Remove the bolts, the upper intake manifold and the intake manifold gaskets. Discard the intake manifold gaskets.

➡Do not use metal scrapers, wire brushes, power abrasive discs or other abrasive means to clean the aluminum retainer plate. These tools cause scratches and gouges, which make leak paths.

32. Clean the intake manifold mating surface of the cylinder heads with silicone gasket remover and metal surface prep. Follow the directions on the packaging.

To install:

33. Tighten the bolts in the sequence shown.
    a. Step 1: Tighten to 2 Nm (18 inch lbs.).
    b. Step 2: Tighten to 10 Nm (89 inch lbs.).
34. Position the water thermostat and housing. Install a new O-ring seal.
35. Install the upper intake manifold. Position the new upper intake manifold gaskets and the intake manifold, and loosely install the bolts.
36. Tighten the bolts in two steps, in the sequence shown.
    a. Step 1: Tighten to 2 Nm (18 inch lbs.).
    b. Step 2: Tighten to 25 Nm (18 ft. lbs.).
37. Connect the heater hose.
38. Connect the ECT sensor electrical connector.
39. Install the ten bolts and the ten ignition coils.
40. Connect the FRP sensor electrical connector.
41. Connect the vacuum connector.
42. Install the TB adapter.
43. Install a new TB adapter gasket.
44. Position the TB adapter.
45. Install the four bolts and tighten in two steps.
    a. Step 1: Tighten the bolts to 9 Nm (80 inch lbs.).
    b. Step 2: Tighten an additional 90 degrees.
46. Connect the TP sensor electrical connector.
47. Connect the PCV coolant hoses.
48. Connect the heated PCV tube and the vacuum hoses.
49. Connect the IAC motor electrical connector and install the bypass hose.
50. Connect the vacuum tube.
51. Position the left fuel charge wiring harness and install the nuts.
52. Position the right fuel charge wiring harness and install the nuts.
53. Connect the ten ignition coil electrical connectors.
54. Connect the ten fuel injector electrical connectors.
55. Install the PVC tube and connect the electrical connector.
56. Install the accelerator bracket.
57. Connect the accelerator and speed control cables and return spring.
58. Install the accelerator cable splash shield.
59. Connect the coolant hose and position the hose clamp.
60. Connect the fuel tube spring lock coupling.
61. Install the alternator.
62. Install the air cleaner and the air cleaner outlet tube.
63. Install the engine cover.
64. Connect the battery ground cable.
65. Fill and bleed the engine cooling system.

Exhaust Manifold

REMOVAL & INSTALLATION

4.2L Engine

1. Before servicing the vehicle, refer to the precautions in the beginning of this section.
2. Remove or disconnect the following:
    - Negative battery cable
    - For the right-hand manifold: the EGR valve-to-exhaust manifold tube
    - For the left-hand manifold: the oil level indicator tube bracket nut, then remove the oil level indicator tube. Remove and discard the oil level indicator tube O-ring.

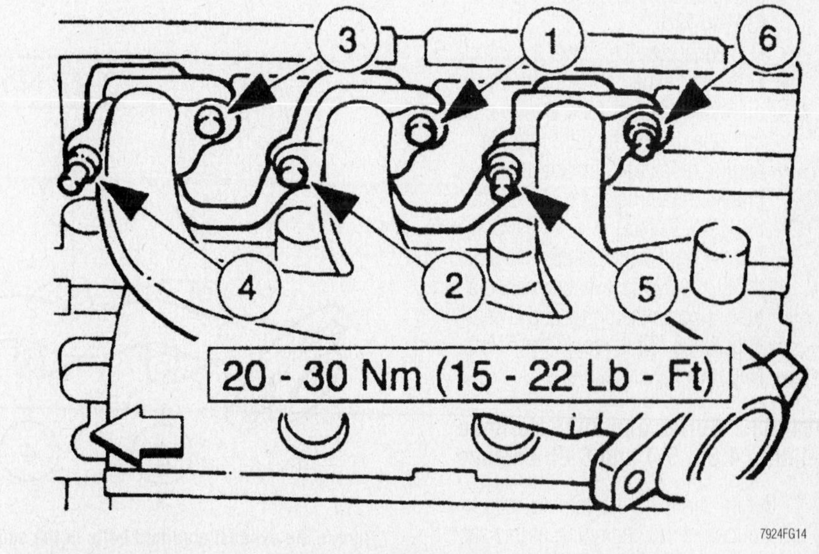

Tighten the left-hand exhaust manifold bolts in the order shown—4.2L engine

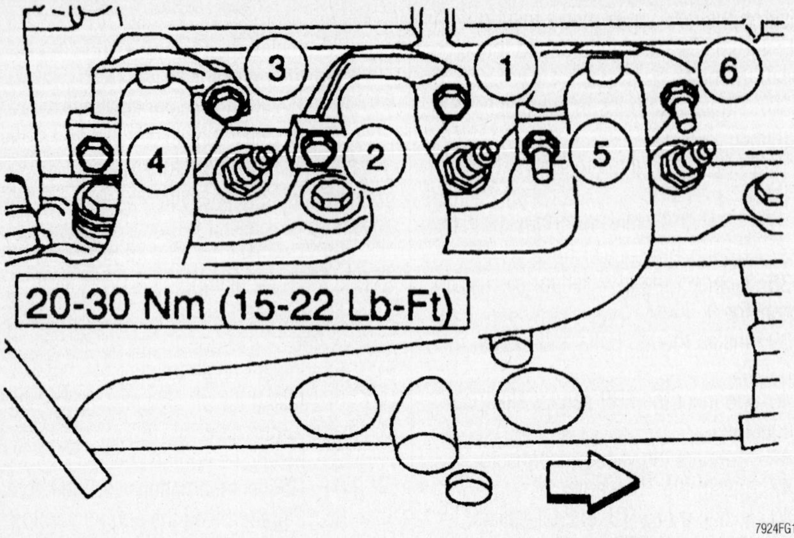

Tighten the right-hand exhaust manifold bolts in the order shown—4.2L engine

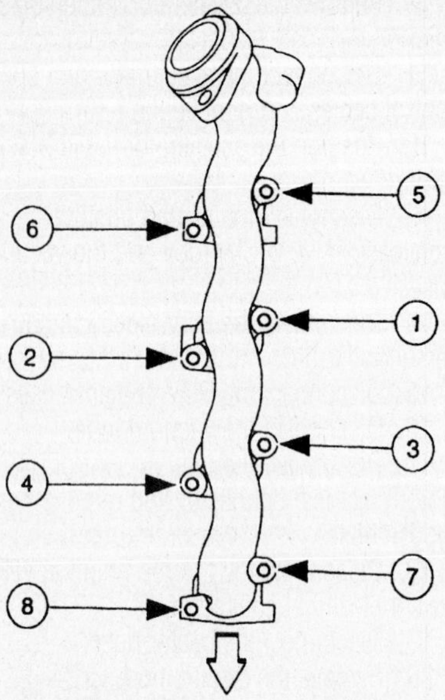

Tighten the exhaust manifold bolts in the sequence shown—2001–03 4.6L engine shown, 5.4L engine similar

- Oxygen Sensor (O_2S) electrical connector
- The 2 catalytic converter-to-exhaust manifold nuts, then disconnect the Y-pipe from the left-hand exhaust manifold.
- Exhaust manifold stud bolts, then remove the manifold mounting bolts
- Exhaust manifold. Remove and discard the exhaust manifold gasket.

To install:

3. Install or connect the following:
   - New exhaust manifold gasket onto the engine, then install the exhaust manifold. Tighten the bolts and stud bolts in the sequence shown to 15–22 ft. lbs. (20–30 Nm).
   - Y-pipe to the exhaust manifold, then install and tighten the catalytic converter nuts to 25–34 ft. lbs. (34–46 Nm).
   - O_2S connector, then lower the vehicle.
   - Left-hand exhaust manifold: a new oil level indicator tube O-ring onto the tube. Insert the tube into the engine block and tighten the bracket retaining nut to 15–22 ft. lbs. (20–30 Nm).
   - For the right-hand exhaust manifold: the EGR valve-to-exhaust manifold tube. Tighten the upper and lower fittings to 25–34 ft. lbs. (34–47 Nm).
   - Negative battery cable

2001–03 F-Series and 2004 Heritage Editions, 4.6L, 5.4 and 6.8L Engines

1. Before servicing the vehicle, refer to the precautions in the beginning of this section.

2. Remove or disconnect the following:
   - Front fender splash shield
   - For the left-hand exhaust manifold: the EGR valve-to-exhaust manifold tube and if equipped, the DPFE gas recirculation transducer hoses.
   - On the 4.6L and 5.4L engines: the catalytic converter-to-exhaust manifold bolts
   - On the 6.8L engine: the front exhaust pipe from the manifold
   - The exhaust manifold mounting nuts, then remove the exhaust manifold itself. Remove and discard the old gasket.

3. Clean and inspect the exhaust manifold for damage.

To install:

4. Position a new gasket and the exhaust manifold onto the engine block.

5. Install the mounting nuts and tighten following the sequence shown.
   - 4.6L and 5.4L engines: 18 ft. lbs. (25 Nm).
   - 6.8L engines: 17–20 ft. lbs. (23–27 Nm).

6. On the 6.8L engine, tighten the exhaust manifold-to-front pipe fasteners to 27–34 ft. lbs. (34–46 Nm).

7. On the 4.6L and 5.4L engines, attach the catalytic converter to the exhaust manifold, install the catalytic converter-to-exhaust manifold bolts and tighten to 25–34 ft. lbs. (34–46 Nm).

8. For the left-hand exhaust manifold, install the DPFE transducer hoses if equipped, and the EGR valve-to-exhaust

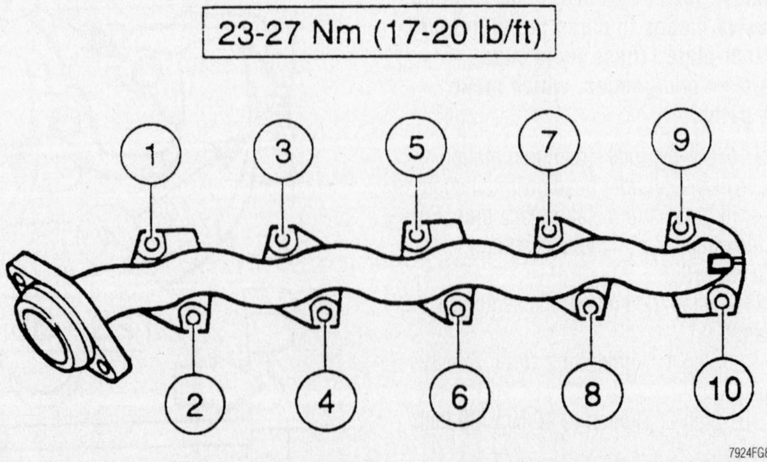

Tighten the exhaust manifold bolts in the sequence shown—right side of 2001–03 6.8L engine shown

manifold tube. Tighten the upper and lower fittings to 26–33 ft. lbs. (35–45 Nm).

9. Install the front fender splash shield.
10. Lower the vehicle to the ground.

2004 F-150 4.6L Engine, Exc. Heritage Editions

RIGHT SIDE

1. Before servicing the vehicle, refer to the precautions in the beginning of this section.
2. With the vehicle in **Neutral**, position it on a hoist.
3. Remove the starter.
4. Remove the right inner fenderwell.
5. Disconnect the Y-pipe at the manifold.
6. Remove the manifold. Discard the gasket.

➡Do not use metal scrapers, wire brushes, power abrasive discs, or other abrasive means to clean the sealing surfaces. These may cause scratches and gouges resulting in leak paths. Use a plastic scraper to clean the sealing surfaces.

7. Clean the sealing surfaces with metal surface prep.

➡Install a new exhaust manifold gasket.

8. Position the right exhaust manifold and tighten the nuts in the sequence shown.
9. To install, reverse the removal procedure.

LEFT SIDE

1. Before servicing the vehicle, refer to the precautions in the beginning of this section.

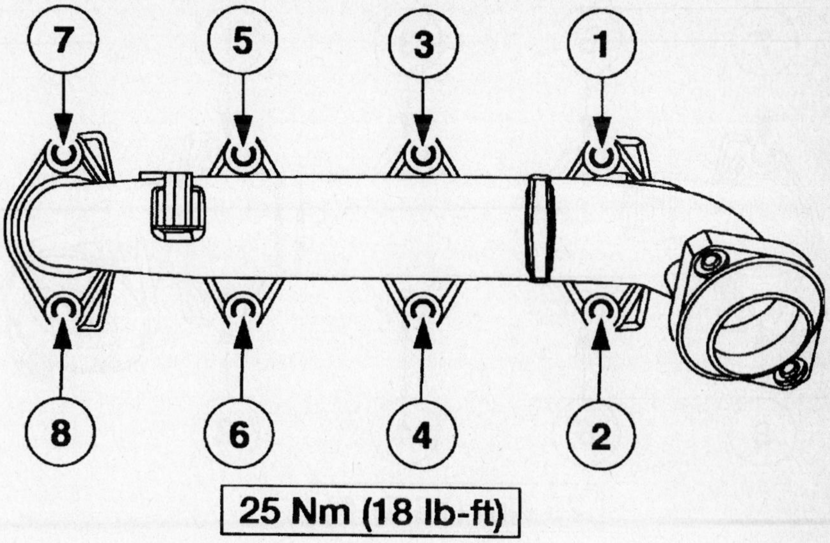

Left exhaust manifold torque sequence—2004 4.6L engines, exc. Heritage edition

2. With the vehicle in **Neutral**, position it on a hoist.
3. Remove the left inner fenderwell.
4. Disconnect the Y-pipe at the manifold.
5. Disconnect the EGR tube.
6. Remove the manifold. Discard the gasket.

➡Do not use metal scrapers, wire brushes, power abrasive discs, or other abrasive means to clean the sealing surfaces. These may cause scratches and gouges resulting in leak paths. Use a plastic scraper to clean the sealing surfaces.

7. Clean the sealing surfaces with metal surface prep.

➡Install a new exhaust manifold gasket.

8. Position the left exhaust manifold and tighten the exhaust manifold nuts in the sequence shown.
9. To install, reverse the removal procedure.

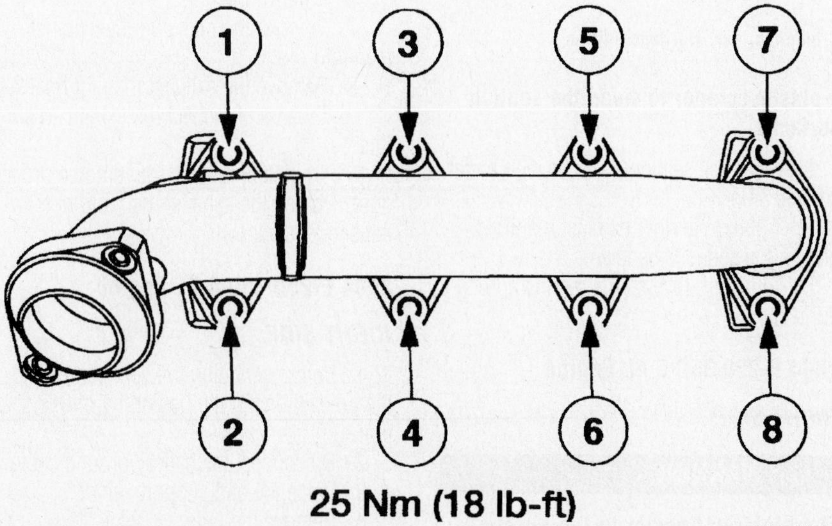

Right exhaust manifold torque sequence—2004 4.6L engines, exc. Heritage edition

2004 F-150 5.4L Engine, Exc. Heritage Editions

LEFT SIDE

1. Before servicing the vehicle, refer to the precautions in the beginning of this section.
2. With the vehicle in **Neutral**, position it on a hoist.
3. Remove the front and rear oxygen sensors.
4. Disconnect the Y-pipes from the manifolds.
5. Remove the manifold. Discard the gasket.

➡Do not use metal scrapers, wire brushes, power abrasive discs, or other abrasive means to clean the sealing surfaces. These may cause scratches and gouges resulting in leak paths. Use a plastic scraper to clean the sealing surfaces.

6. Clean the sealing surfaces with metal surface prep.
7. Install the manifold with a new gasket.
8. Tighten the exhaust manifold nuts in the sequence shown.
9. To install, reverse the removal procedure.

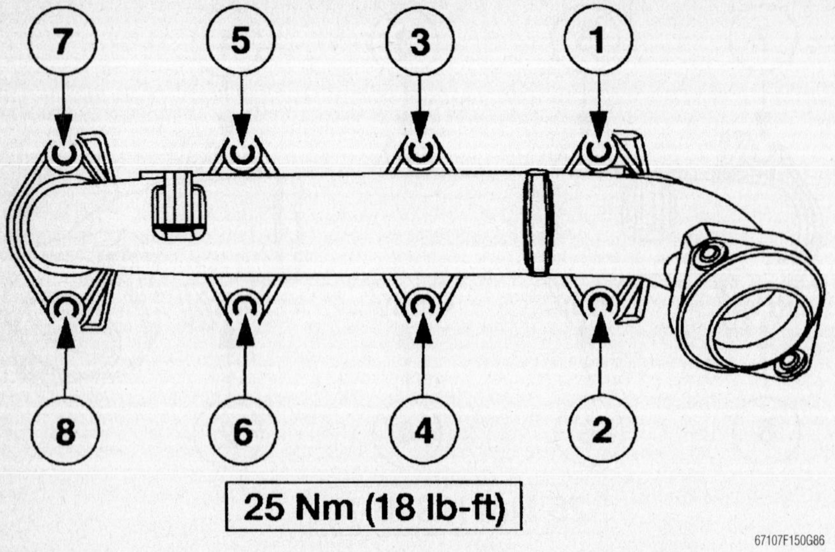

Left exhaust manifold torque sequence—2004 5.4L engines, exc. Heritage edition

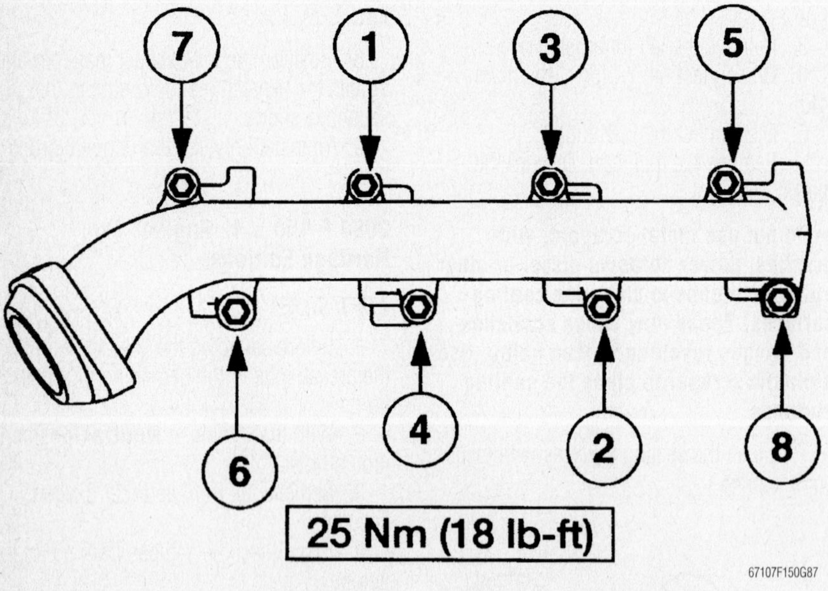

Right exhaust manifold torque sequence—2004 5.4L engines, exc. Heritage edition

RIGHT SIDE

1. Before servicing the vehicle, refer to the precautions in the beginning of this section.
2. With the vehicle in **Neutral**, position it on a hoist.
3. Remove the starter.
4. Disconnect the Y-pipes at the manifolds.
5. Remove the manifold. Discard the gasket.

➡Do not use metal scrapers, wire brushes, power abrasive discs, or other abrasive means to clean the sealing surfaces. These may cause scratches and gouges resulting in leak paths. Use a plastic scraper to clean the sealing surfaces.

6. Clean the sealing surfaces with metal surface prep.
7. Tighten the right exhaust manifold nuts in the sequence shown.
8. To install, reverse the removal procedure.

2004 F-250/350 5.4L Engine

RIGHT SIDE

WARNING

The electrical power to the air suspension system must be shut off prior to hoisting, jacking or towing an air suspension vehicle. This can be accomplished by turning off the air suspension switch located in the right kick panel area. Failure to do so may result in unexpected inflation or deflation of the air springs which may result in shifting of the vehicle during these operations.

1. Before servicing the vehicle, refer to the precautions in the beginning of this section.
2. Raise and support the vehicle.
3. On 4x4 vehicles, remove the front wheel opening molding.
4. Remove the front fender splash shield.
5. Remove the three way catalytic converter to exhaust manifold nuts.
6. Remove the eight nuts and the exhaust manifold.
7. Remove and discard the exhaust manifold gasket.
8. Clean and inspect the exhaust manifold.
9. To install, reverse the removal procedure. Tighten the exhaust manifold nuts in the sequence shown.

LEFT SIDE

1. Before servicing the vehicle, refer to the precautions in the beginning of this section.
2. If equipped, remove the exhaust manifold to exhaust gas recirculation (EGR) valve tube.
3. Raise and support the vehicle.
4. On 4x4 vehicles, remove the front wheel opening molding.
5. Remove the front fender splash shield.
6. Remove the nuts.
7. Remove the nuts and the exhaust manifold.
8. Remove the exhaust manifold gasket.
9. Clean and inspect the exhaust manifold.
10. To install, reverse the removal procedure. Tighten the exhaust manifold nuts in the sequence shown.

2004 F-250/350 6.8L Engine

RIGHT SIDE

1. Before servicing the vehicle, refer to the precautions in the beginning of this section.
2. Disconnect the battery ground cable.
3. Raise the and support vehicle.
4. Remove the nuts.
5. Remove the right front wheel and tire.

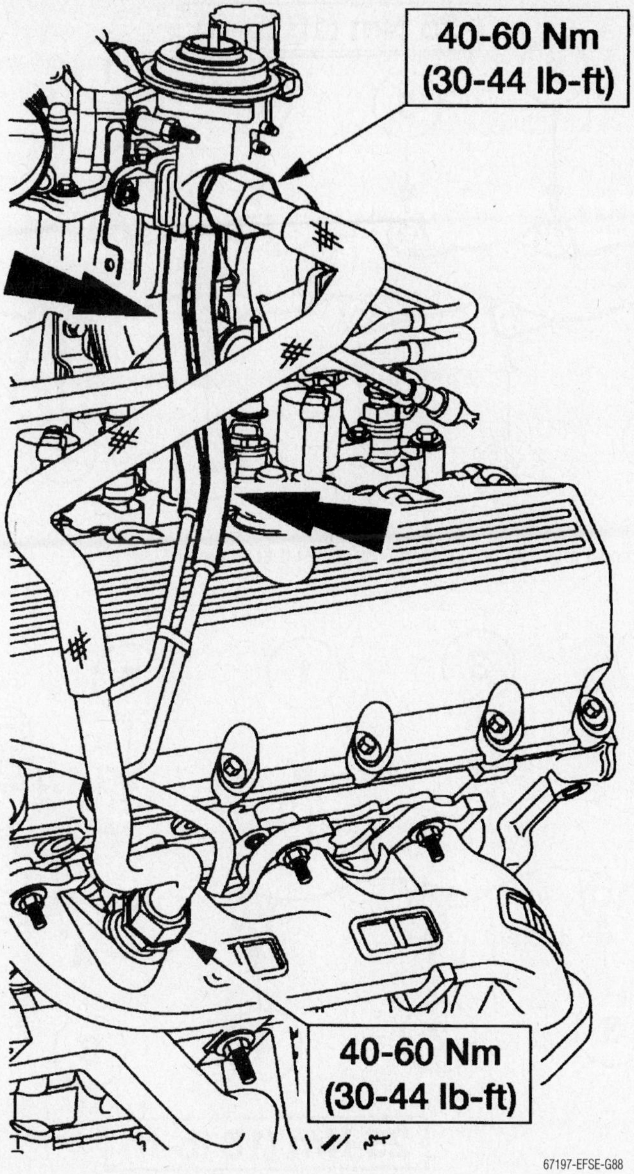

EGR valve tube installation—2004 F-250/350 w/5.4L engine

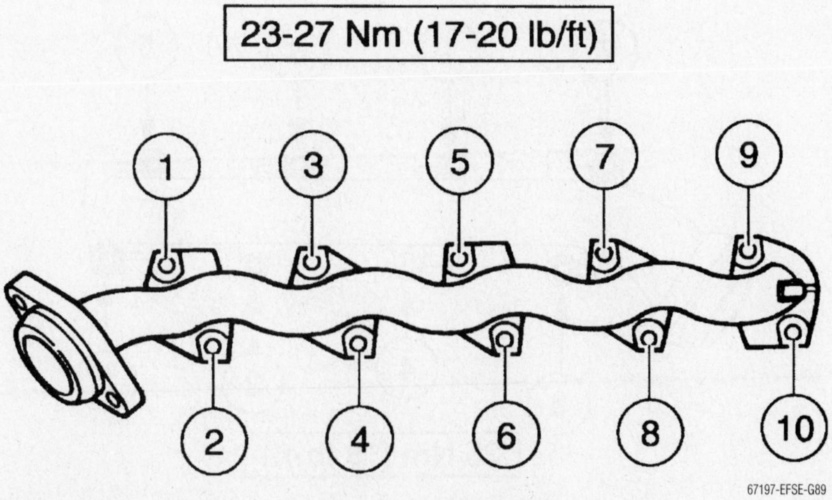

Right side exhaust manifold torque sequence—2004 6.8L engine

6. Remove the right front inner fender well.

7. Remove the nuts, the exhaust manifold and the gasket. Discard the gasket.

8. Clean and inspect the exhaust manifold.

9. To install, reverse the removal procedure.

LEFT SIDE

1. Before servicing the vehicle, refer to the precautions in the beginning of this section.

2. With the vehicle in **Neutral**, position it on a hoist.

3. Remove the air cleaner outlet tube.

4. Remove the nuts.

5. Remove the left front wheel and tire.

6. Remove the left front inner fender well.

7. Remove the nuts, the exhaust manifold and the gasket. Discard the exhaust manifold gasket.

8. Clean and inspect the exhaust manifold.

9. To install, reverse the removal procedure.

E-Series 4.6L and 5.4L Engines

LEFT SIDE

1. Before servicing the vehicle, refer to the precautions in the beginning of this section.

All engines

2. With the vehicle in **Neutral**, position it on a hoist.

3. Remove the engine cover.

Romeo engine (4.6L)

4. Disconnect the differential pressure feedback exhaust gas recirculation (EGR) sensor hoses from the EGR valve tube.

5. Remove the EGR valve tube.

Windsor engine (5.4L)

6. Disconnect the differential pressure feedback EGR sensor hoses from the EGR valve tube.

7. Disconnect the upper and lower fittings and remove the EGR valve tube.

8. Remove the front fender splash shield.

9. Remove the nuts.

10. Remove the nuts and the exhaust manifold.

11. Remove and discard the exhaust manifold gasket.

12. Clean and inspect the exhaust manifold.

To install:

All engines

13. Position the exhaust manifold gasket.

14. Position the exhaust manifold and loosely install the nuts.
15. Tighten the nuts in the sequence shown.
16. Install the three-way catalytic converter to exhaust manifold nuts.

Windsor engine (5.4L)

17. Install the EGR valve fittings in two steps.
   a. Step 1: Hand-tighten the fittings.
   b. Step 2: Tighten the lower fitting to 50 Nm (37 ft. lbs.).
18. Connect the differential pressure feedback EGR sensor hoses to the EGR valve tube.

Romeo engine (4.6L)

19. Install the EGR valve tube in two steps.
   a. Step 1: Connect the upper and lower fittings and hand-tighten.
   b. Step 2: Tighten the fittings to 40 Nm (30 ft. lbs.).
20. Install the differential feedback EGR sensor hoses to the EGR valve tube.

All engines

21. Install the front fender splash shield.
22. Install the engine cover.

RIGHT SIDE

1. Before servicing the vehicle, refer to the precautions in the beginning of this section.
2. With the vehicle in **Neutral**, position it on a hoist.
3. Remove the front fender splash shield.
4. Remove the three-way catalytic converter to exhaust manifold nuts.
5. Remove the eight nuts and the exhaust manifold.
6. Remove and discard the exhaust manifold gasket.
7. Clean and inspect the exhaust manifold.
8. To install, reverse the removal procedure. Tighten the exhaust manifold nuts in the sequence shown.

E-Series 6.8L Engine

RIGHT SIDE

1. Before servicing the vehicle, refer to the precautions in the beginning of this section.
2. With the vehicle in **Neutral**, position it on a hoist.
3. Remove the engine cover.
4. Remove the nuts.
5. Remove the 10 nuts, the exhaust manifold, and the exhaust manifold gasket.
6. Discard the exhaust manifold gasket.

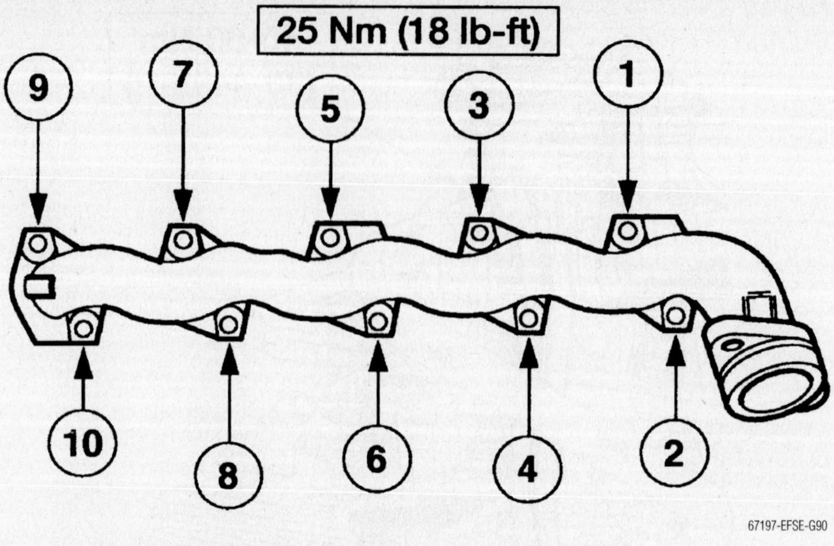

Left side exhaust manifold torque sequence—2004 6.8L engine

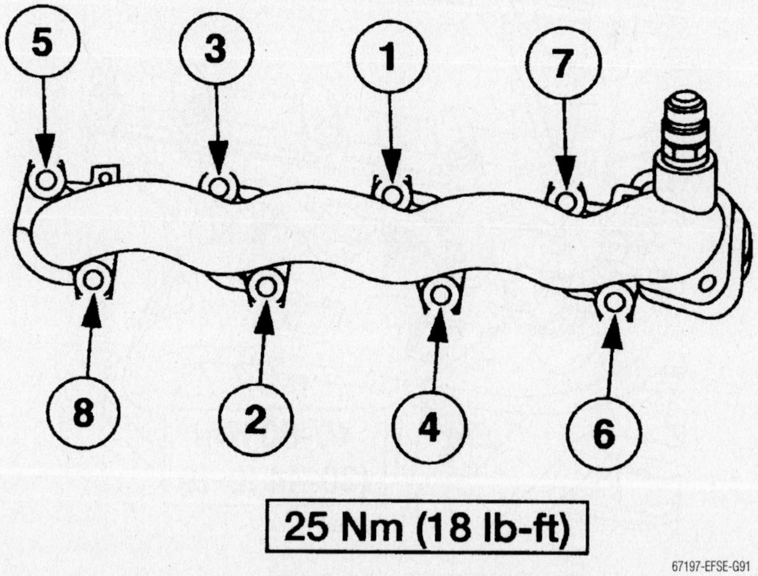

Left exhaust manifold torque sequence—E-Series with 4.6L or 5.4L engine

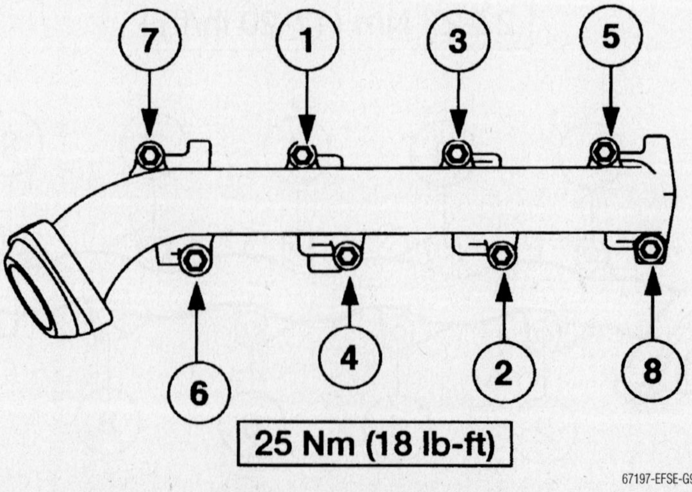

Right exhaust manifold torque sequence—E-Series with 4.6L or 5.4L engine

7. Clean and inspect the exhaust manifold.

8. To install, reverse the removal procedure.

9. Install a new exhaust manifold gasket. Tighten the exhaust manifold nuts in the sequence shown.

LEFT SIDE

1. Before servicing the vehicle, refer to the precautions in the beginning of this section.

2. With the vehicle in **Neutral**, position it on a hoist.

3. Remove the engine cover.

4. Remove the nuts.

5. Remove the exhaust manifold nuts and the exhaust manifold. Discard the exhaust manifold gaskets.

6. Clean and inspect the exhaust manifold.

7. To install, reverse the removal procedure. Tighten the exhaust manifold nuts in the sequence shown.

Camshaft and Valve Lifters

REMOVAL & INSTALLATION

4.2L Engine

1. Before servicing the vehicle, refer to the precautions in the beginning of this section.

2. Remove or disconnect the following:
   - Negative battery cable
   - Lower intake manifold
   - Rocker arm cover
   - Rocker arm hold-down bolt, then remove the rocker arm from the cylinder head.
   - Pushrods
   - Valve lifters by pulling them up out of their bores
   - Timing chain and sprockets
   - Camshaft key from the end of the camshaft, then slide the engine dynamic balance shaft drive gear off the camshaft.
   - The 2 camshaft thrust plate retaining bolts (1), then remove the thrust plate (2). Remove the camshaft spacer (3), then slide the camshaft (4) out of the front of the engine block. Be cautious not to gouge or scratch the camshaft bearing journals.

To install:

3. Lubricate the camshaft with engine oil prior to installation.

4. Carefully slide the camshaft into the

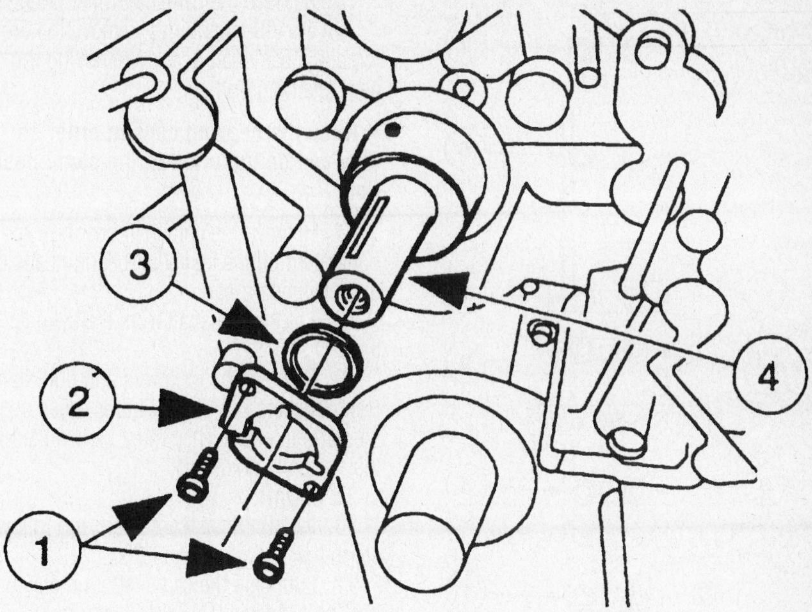

Exploded view of the camshaft retaining hardware—4.2L engine

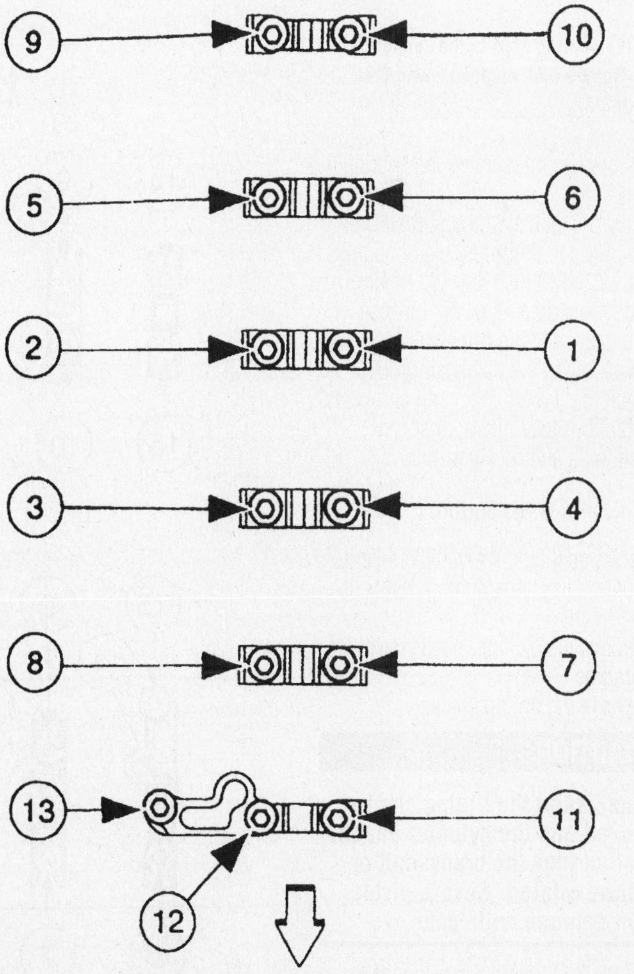

Tighten the bearing caps in the sequence shown—Windsor 4.6L engine shown, 5.4L engine similar

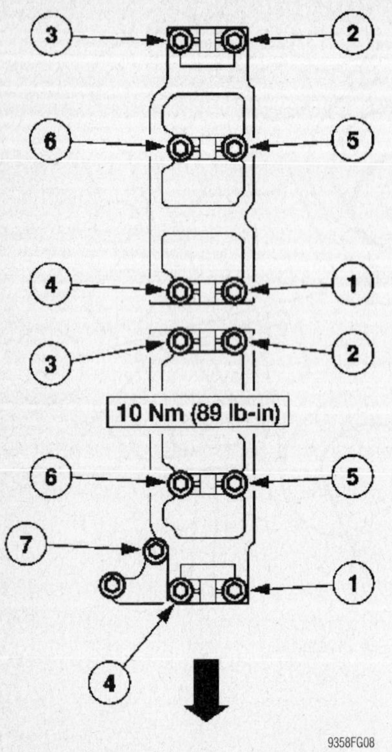

Tighten the bearing caps in the sequence shown—Romeo 4.6L engine shown, 5.4L engine similar

camshaft bore. Do not scratch the bearing surfaces.

5. Install the camshaft thrust plate with the spacer. Tighten the thrust plate mounting bolts to 72–120 inch lbs. (8–14 Nm).

6. Slide the engine dynamic balance shaft drive gear onto the camshaft. Install the camshaft key to the camshaft groove.

7. Install the timing chain and sprockets.

8. Install the valve lifters, pushrods, intake manifolds and rocker arm covers.

4.6L, 5.4L and 6.8L Engines

1. Before servicing the vehicle, refer to the precautions in the beginning of this section.

2. Remove the cylinder head covers from the engine.

3. Remove the timing chain.

⁂ CAUTION

At no time, when the timing chains are removed and the cylinder heads are installed may the crankshaft or camshaft be rotated. Severe piston and valve damage will occur.

4. On the 6.8L engine, remove the 6 bolts securing the balance shaft to the cylinder head and remove the shaft.

5. Remove the camshaft roller lifters.

6. On VIN W engines, remove the timing chain camshaft gear by removing the gear retaining bolt.

➡Keep the bearing caps in order so they can be installed in the same position.

7. Remove the camshaft bearing cap bolts, then lift the camshaft bearing caps off of the cylinder head.

8. Lift the camshaft from the cylinder head.

9. Remove the followers and pull the lash adjusters out of their bores. Keep all the parts in order. They must be installed in their original positions.

To install:

10. Install the lash adjusters and followers in their original positions.

11. Lubricate the camshaft journals and bearing caps with SAE 5W30 engine oil. On the 6.8L engine, lubricate the balance shaft journals and bearing caps with the same lubricant.

12. Lower the camshaft onto the camshaft bearing journals.

13. Install the camshaft bearing caps, then loosely install the bearing cap bolts.

14. Tighten the camshaft bearing cap mounting bolts, in the sequence shown for the particular engine, to 71–107 inch lbs. (8–12 Nm).

15. On the 6.8L engine, align the timing marks and position the balance shaft on the journals, then install the bearing caps. Tighten the bolts in sequence to 89 inch lbs. (10 Nm).

16. On engines equipped with bolt on sprockets, install the camshaft timing chain gear by tightening the retaining bolts as follows:

   a. M10 Step 1: 30 ft. lbs. (40 Nm).
   b. M10 Step 2: Tighten 90 degrees.
   c. M12 Step 1: 90 ft. lbs. (120 Nm).

17. Install the valve lifters.

18. Install the timing chain and sprockets, if applicable.

19. Install the cylinder head covers.

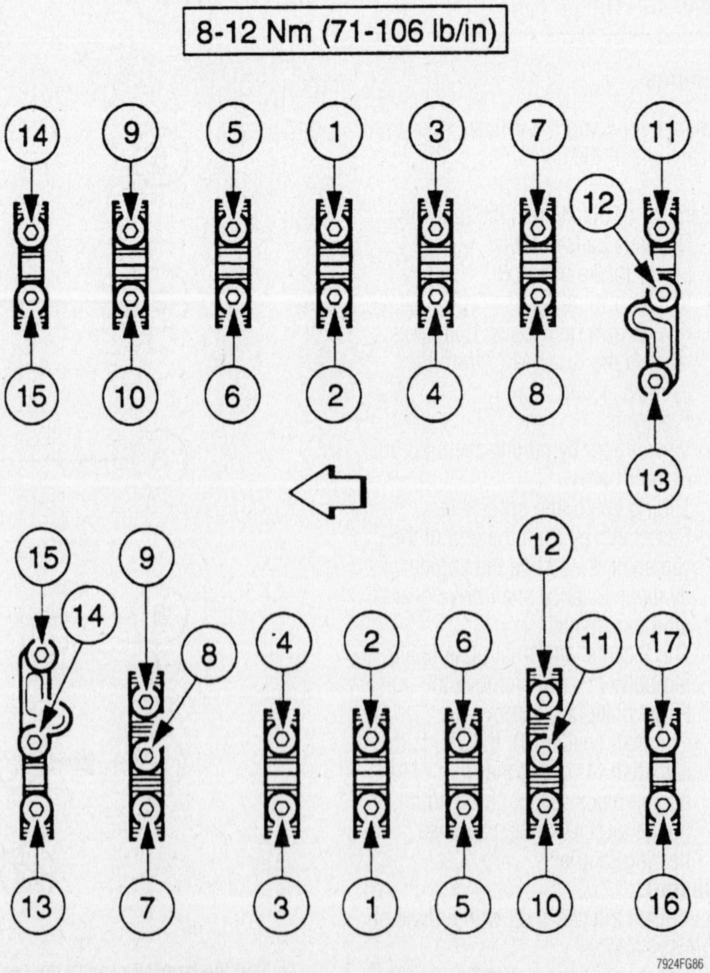

Camshaft bearing cap bolt tightening sequence—6.8L engine

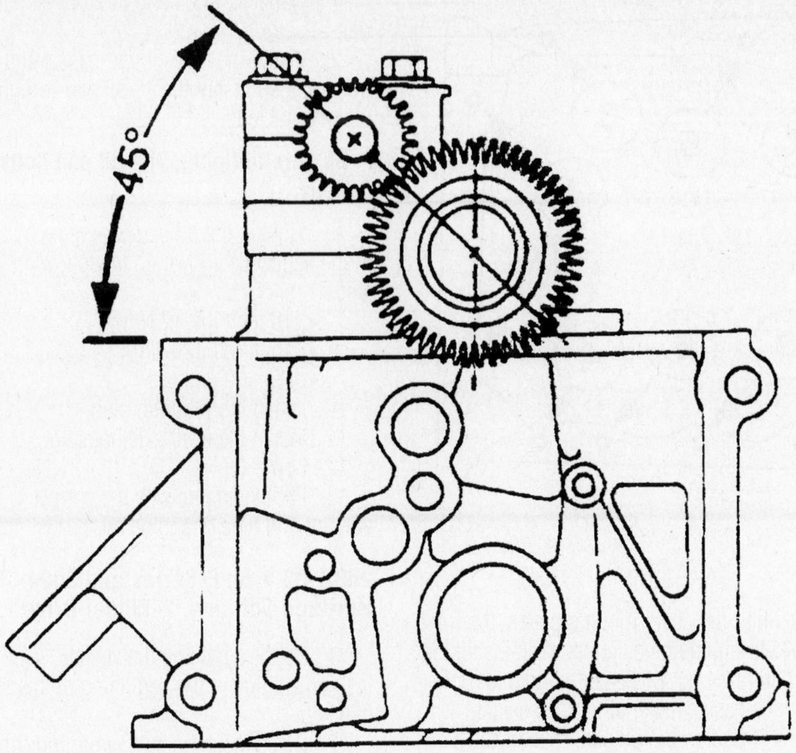

Be sure to align the balance shaft timing mark with the mark on the camshaft gear—6.8L engine

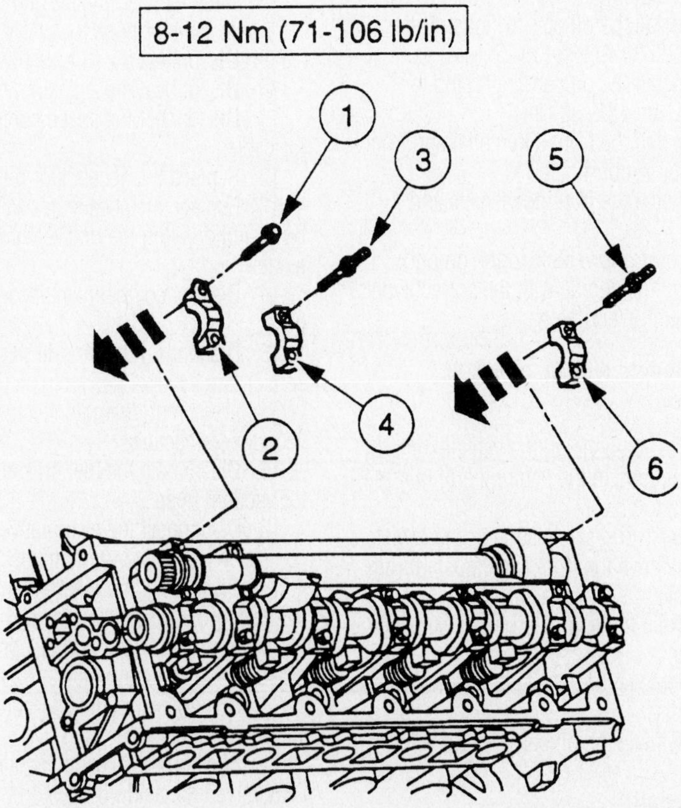

Tighten the balance shaft bearing cap bolts in the sequence shown—6.8L engine

Valve Lash

ADJUSTMENT

4.2L, 4.6L, 5.4L and 6.8L Engines

The 4.2L, 4.6L, 5.4L and 6.8L engines do not require valve lash adjusting, because they utilize hydraulic lash components in their valve actuation systems. The 4.2L engine uses hydraulic valve lifters, whereas the 4.6L, 5.4L and 6.8L engines utilize hydraulic lash adjusters, all of which automatically adjust the valve lash. No valve lash adjustment is necessary.

Starter Motor

REMOVAL & INSTALLATION

1. Before servicing the vehicle, refer to the precautions in the beginning of this section.
2. Remove or disconnect the following:
   - Negative battery cable
   - Starter motor electrical connections
   - Starter motor bolts and the motor

To install:

3. Installation is the reverse of removal, tighten the starter motor bolts to 16–20 ft. lbs. (22–28 Nm).

Oil Pan

REMOVAL & INSTALLATION

4.2L Engine

4X2 MODELS

1. Before servicing the vehicle, refer to the precautions in the beginning of this section.
2. Remove or disconnect the following:
   - Engine
   - Dipstick tube
   - Oil filter
   - The 15 oil pan-to-cylinder block mounting bolts and the oil pan

To install:

➡If the oil pan is not installed within 15 minutes, remove the sealer and reapply.

3. Temporarily install 2 locator dowels in 2 of the oil pan-to-engine block corner mounting bolt holes.
4. Clean and apply sealant to the rear main bearing cap, the oil pan mating surface, the front cover mounting area, the front cover-to-engine block joints, then install the

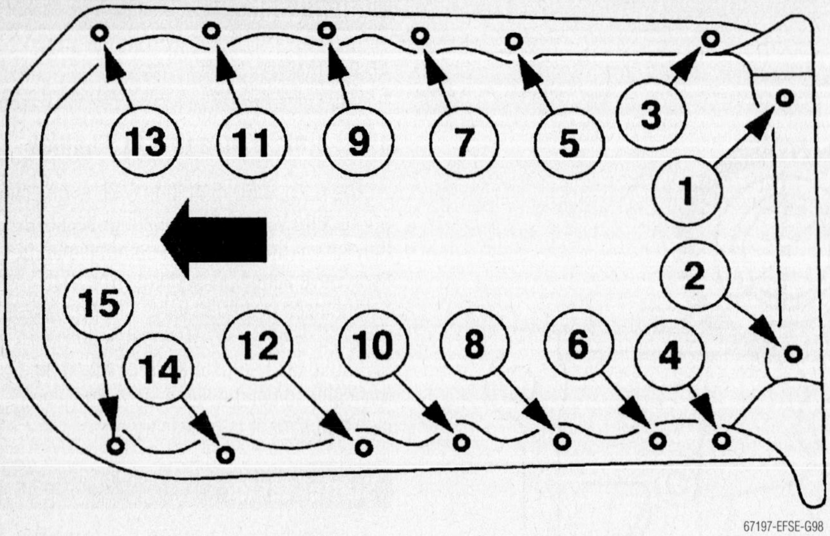

Oil pan torque sequence—4.2L engine

oil pan rear seal. Make certain to use RTV silicone gasket material for the sealant.

5. Position the oil pan, then install 13 of the oil pan mounting bolts loosely.

6. Remove the 2 locator dowels and install the remaining 2 oil pan-to-engine block bolts.

7. In sequence, tighten the 15 oil pan mounting bolts to 44 inch lbs. (5 Nm), then retighten the bolts to 89 inch lbs. (10 Nm).

- Oil filter
- Dipstick tube
- Engine

8. Fill the engine with the correct type and amount of engine oil.

4X4 MODELS

1. Before servicing the vehicle, refer to the precautions in the beginning of this section.

2. Remove or disconnect the following:

- Engine oil
- The 2 front wheel driveshafts and joints, if so equipped.
- Front differential from the vehicle
- Front differential support
- The 3 oil pan-to-transmission bolts
- The 15 oil pan-to-cylinder block mounting bolts, then lower the oil pan

To install:

➡If the oil pan is not installed within 15 minutes, remove the sealer and reapply.

3. Temporarily install 2 locator dowels in 2 of the oil pan-to-engine block corner mounting bolt holes.

4. Clean and apply sealant to the rear main bearing cap, the oil pan mating surface, the front cover mounting area, the front cover-to-engine block joints, then install the oil pan rear seal. Make certain to use RTV silicone gasket material for the sealant.

5. Position the oil pan, then install 13 of the oil pan mounting bolts loosely.

6. Remove the 2 locator dowels and install the remaining 2 oil pan-to-engine block bolts.

7. In sequence, tighten the 15 oil pan mounting bolts to 44 inch lbs. (5 Nm), then retighten the bolts to 89 inch lbs. (10 Nm).

8. Install the oil pan-to-transmission bolts to 28–38 ft. lbs. (38–51 Nm).

9. Install the oil pan drain plug to 16–22 ft. lbs. (22–30 Nm).

10. Install the front differential and front differential support.

11. Install the front driveshafts and joints.

12. Lower the vehicle to the ground.

13. Fill the engine with the correct type and amount of engine oil.

2001 Models w/4.6L and 5.4L Engines

1. Before servicing the vehicle, refer to the precautions in the beginning of this section.

2. Raise and safely support the vehicle.

3. Remove the front axle housing from the vehicle.

4. Drain the engine oil into a suitable container.

5. Remove the 16 oil pan-to-engine block bolts.

6. Remove the oil pan and old oil pan gasket.

To install:

7. Clean the oil pan and engine block mating surfaces of oil and old gasket material.

8. Install the new oil pan gasket and the oil pan, then install the 16 oil pan-to-engine block bolts loosely.

➡Be sure to tighten the oil pan bolts in 3 steps.

9. Tighten the oil pan-to-engine bolts in the sequence shown, in the following 3 steps:

- 18 inch lbs. (2 Nm)
- 15 ft. lbs. (20 Nm)
- Plus 60 degrees

10. Install the oil drain plug.

11. Install the front axle housing.

12. Lower the vehicle.

13. Fill the engine with the correct amount and type of engine oil.

2002–03 4.6L F-Series and 2004 Heritage Editions, 2-Wheel Drive

1. Before servicing the vehicle, refer to the precautions in the beginning of this section.

2. Disconnect the battery ground cable.

3. Drain the engine cooling system.

4. Remove the radiator air deflector.

5. Remove the screws.

6. Remove the radiator air deflector.

7. Remove the engine air cleaner and the air cleaner outlet pipe.

8. Slide the clamp back and disconnect the upper radiator hose.

9. Disconnect the throttle body cam.

10. Disconnect the accelerator cable.

11. Disconnect the speed control actuator cable.

12. Replace the accelerator return spring.

13. Replace the accelerator cable bracket bolts and position the bracket and cables aside.

14. Disconnect the main vacuum harness.

15. Disconnect the throttle position (TP) sensor and exhaust gas recirculation (EGR) vacuum regulator solenoid electrical and vacuum connectors.

16. Disconnect the fuel pressure regulator vacuum hose.

17. Disconnect the evaporative emission canister purge valve vacuum hose.

18. Disconnect the differential pressure feedback EGR system electrical connector.

19. Replace the nut and disconnect the brake booster vacuum tube and bracket.

20. Disconnect the idle air control (IAC) electrical connector.

21. Position the exhaust manifold to EGR valve tube aside.

22. Disconnect the upper fitting.

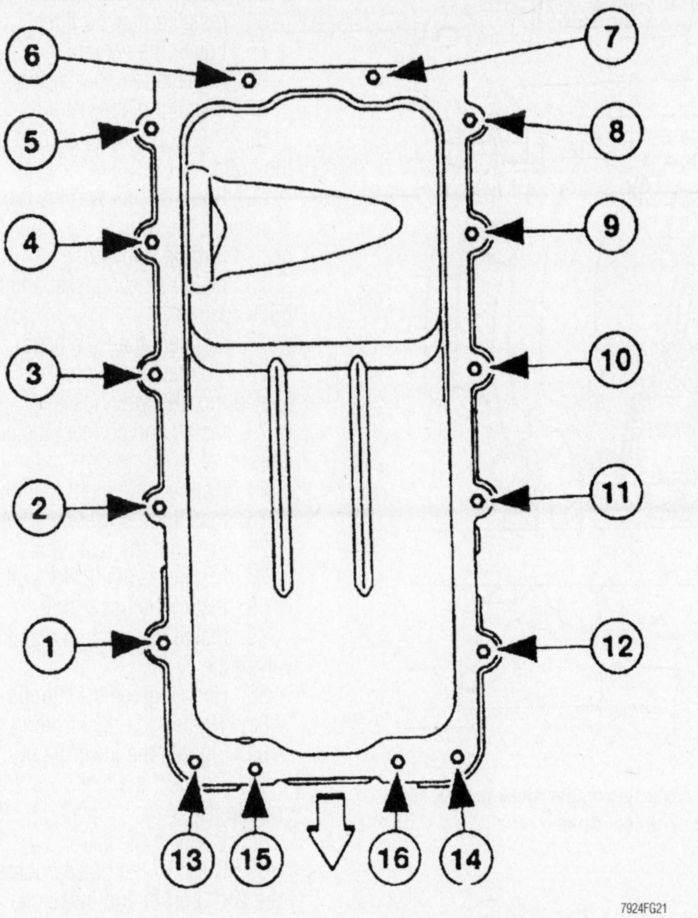

Tighten the oil pan-to-engine block bolts in 3 steps following the sequence shown—2001 models w/4.6L and 5.4L engines

23. Loosen the lower fitting. Position the tube aside.
24. Disconnect the fuel hose spring lock couplings.
25. Disconnect the heater hose and the heated throttle body hose.
26. Disconnect the positive crankcase ventilation (PCV) hose.
27. Replace the four bolts and move the throttle body adapter forward slightly.
28. Disconnect the heated throttle body hose and remove the throttle body adapter.
29. Discard the throttle body adapter-to-intake manifold gasket.
30. Inspect the throttle body and adapter for damage.
31. Position the power steering reservoir aside.
32. Replace the upper bolt.
33. Replace the lower bolts.
34. Position the reservoir aside.
35. Disconnect the eight fuel injectors.
36. Disconnect and remove the eight ignition coils.
37. Replace the alternator.
38. Position the fan shroud aside.
39. Replace the bolts.
40. Position the fan shroud aside.
41. Install a suitable lifting tool on the engine using the alternator mounting bolts.
42. Raise and support the vehicle.
43. Replace the drain plug and drain the engine oil.
44. Replace the right motor mount bolt.
45. Replace the left motor mount bolt.
46. Lower the vehicle.
47. Raise the engine using a suitable tool.
48. Raise the vehicle on the hoist.

➡Be careful when removing the oil pan gasket. The oil pan gasket is reusable if it is not damaged.

49. Replace the oil pan bolts.
50. Position the oil pan aside to gain access to the oil pump screen and pickup tube.
51. Replace the bolts and the oil pump screen and pickup tube.

To install:

➡Do not use metal scrapers, wire brushes, power abrasive discs or other abrasive means to clean the sealing surfaces. These tools cause scratches and gouges which make leak paths. Use a plastic scraping tool to remove all traces of old sealant.

52. Clean the sealing surfaces with silicone gasket remover and metal surface prep. Follow the directions on the packaging. Inspect the mating surfaces.
53. Position the oil pan gasket, the oil pan and the oil pump screen and pickup tube together to the cylinder block.

➡Make sure the O-ring is in place and not damaged. A missing or damaged O-ring can cause foam in the lubrication system, low oil pressure, and severe engine damage.

➡Install a new O-ring and lubricate with clean engine oil.

Position the oil pump screen and pickup tube and install the bolts.

➡If the oil pan is not secured within four minutes, the sealant must be removed and the sealing area cleaned. To clean the sealing area, use silicone gasket remover and metal surface prep. Follow the directions on the packaging. Failure to follow this procedure can cause future oil leakage.

54. Apply the silicone gasket and sealant at the engine front cover-to-cylinder block mating surface.

➡If the oil pan is not secured within four minutes, the sealant must be removed and the sealing area cleaned. To clean the sealing area, use silicone gasket remover and metal surface prep. Follow the directions on the packaging. Failure to follow this procedure can cause future oil leakage.

55. Apply the silicone gasket and sealant at the rear oil seal retainer-to-cylinder block sealing surface.
56. Position the oil pan gasket.
57. Install the bolts and tighten in three steps, in the sequence shown.
    a. Step 1: Tighten to 2 Nm (18 inch lbs.).
    b. Step 2: Tighten to 20 Nm (15 ft. lbs.).
    c. Step 3: Tighten an additional 60 degrees.
58. Lower the vehicle.
59. Lower the engine and remove the special tool.
60. Raise the vehicle on the hoist.
61. Install the right motor mount bolt.
62. Install the left motor mount bolt.

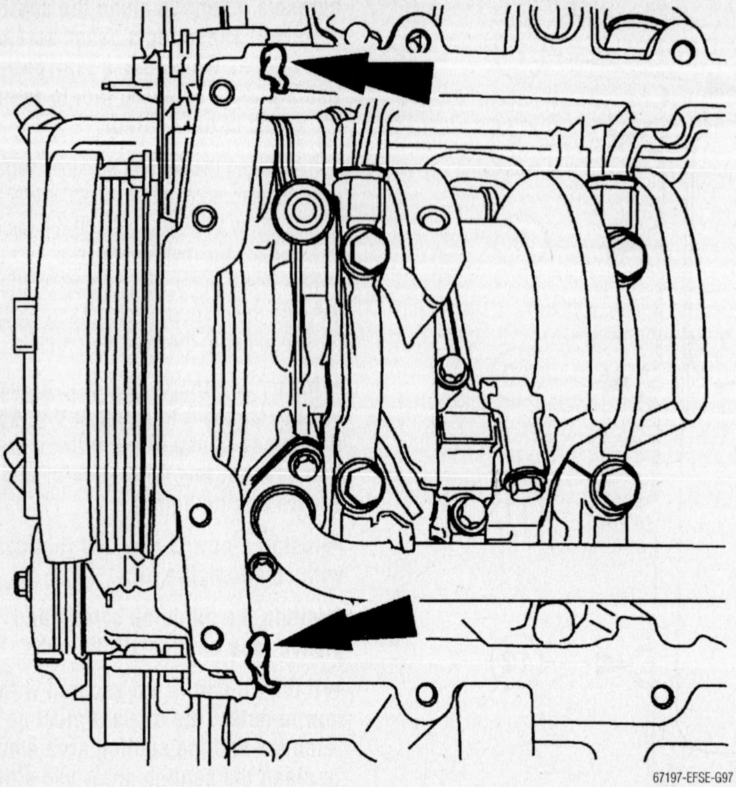

Apply the silicone gasket and sealant at the engine front cover-to-cylinder block mating surface—2002–03 4.6L F-Series and 2004 Heritage Editions, 2-Wheel Drive

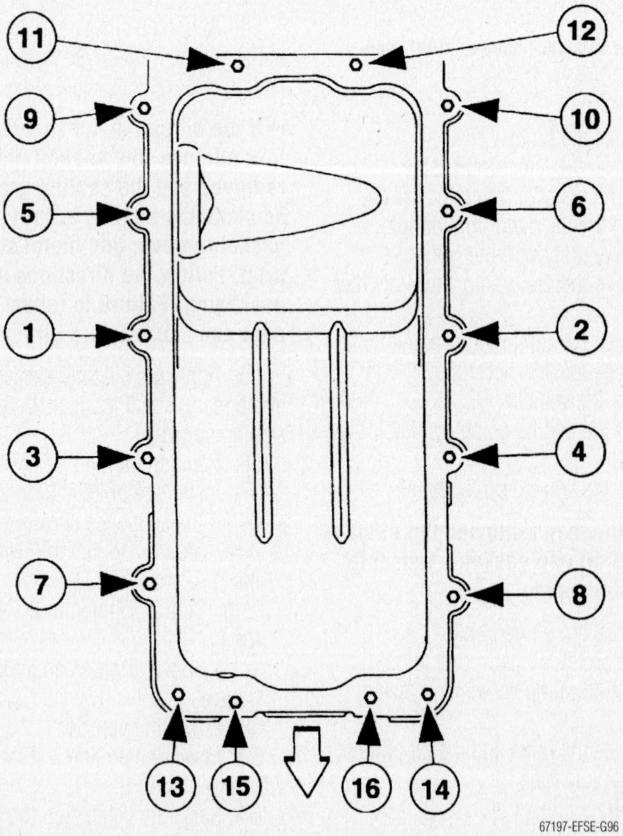

Oil pan bolt torque sequence—2002–03 4.6L F-Series and 2004 Heritage Editions, 2-Wheel Drive

63. Install the drain plug.
64. Install a new oil filter.
65. Lower the vehicle.
66. Replace the special tool.
67. Install the alternator.
68. Position the fan shroud and install the bolts.
69. Roughly position the throttle body adapter in the vehicle and connect the heated throttle body hose.
70. Install a new gasket and the throttle body adapter.
71. Connect the PCV hose.
72. Connect the heater hose and the heated throttle body hose.
73. Install and connect the eight ignition coils.
74. Connect the eight fuel injectors.
75. Install the power steering reservoir.
76. Position the reservoir.
77. Install the two lower bolts.
78. Install the upper bolt.
79. Install the exhaust manifold to EGR valve tube.
80. Hand-tighten the fittings.
81. Tighten the upper fitting.
82. Tighten the lower fitting.
83. Connect the fuel hose spring lock couplings.
84. Connect the IAC.
85. Connect the brake booster vacuum tube and bracket and install the nut.
86. Connect the differential pressure feedback EGR system electrical connectors.
87. Connect the evaporative emission canister purge valve vacuum hose.
88. Connect the fuel pressure regulator vacuum hose.
89. Connect the TP sensor electrical connector and the EGR vacuum regulator solenoid electrical and vacuum connectors.
90. Connect the main vacuum harness.
91. Install the accelerator cable bracket and the bolts.
92. Connect the throttle body cam.
93. Connect the accelerator cable.
94. Connect the speed control actuator cable.
95. Install the throttle return spring.
96. Connect the upper radiator hose and position the clamp.
97. Install the engine air cleaner and the air cleaner outlet pipe.
98. Install the radiator air deflector.
99. Connect the battery ground cable.

➡The oil pump must be primed prior to starting the engine.

100. Fill the engine with clean engine oil.
101. Fill and bleed the engine cooling system.

2002–03 5.4L F-Series and 2004 Heritage Editions, 2-Wheel Drive

1. Before servicing the vehicle, refer to the precautions in the beginning of this section.
2. With the vehicle in **Neutral**, position it on a hoist.
3. Remove the radiator air deflector.
4. Remove the screws.
5. Remove the radiator air deflector.
6. Remove the air cleaner outlet pipe.
7. Remove the alternator.
8. Disconnect the acceleration controls from the throttle body cam.
9. Disconnect the accelerator cable.
10. Disconnect the speed control actuator cable.
11. Remove the accelerator return spring.
12. Remove the accelerator cable bracket bolts and position the bracket and cables aside.
13. Disconnect the positive crankcase ventilation (PCV) hose from the throttle body adapter.
14. Disconnect the evaporative emission (EVAP) canister purge valve hose.
15. Remove the crankcase ventilation tube and the idle air control (IAC) valve fresh air tube.
16. Disconnect the exhaust gas recirculation (EGR) valve vacuum hose and the main vacuum harness.
17. Disconnect the throttle position sensor electrical connector and the EGR vacuum regulator solenoid connections.
18. Disconnect the fuel pressure regulator vacuum hose.
19. Disconnect the EVAP canister purge valve vacuum hose.
20. Remove the bolt and disconnect the brake booster vacuum tube and bracket.
21. Remove the nut.
22. Disconnect the IAC electrical connector.
23. Disconnect the upper EGR valve tube fitting.
24. Remove the four bolts and set aside the throttle body adapter.
25. Discard the throttle body adapter gasket.
26. Position the fan shroud aside.
27. Remove the bolts.
28. Position the fan shroud aside.
29. Install a suitable tool on the engine using the alternator mounting bolts.
30. Install a suitable tool on the engine.
31. Remove the starter solenoid terminal cover.
32. Disconnect the starter motor electrical connections.
33. Remove the starter motor.
34. Remove the stud bolt.
35. Remove the two bolts.
36. Remove the starter.
37. Remove the drain plug and drain the engine oil.
38. Remove the right motor mount bolt.
39. Remove the left motor mount bolt.
40. Raise the engine using a cane.

➡**Be careful when removing the oil pan gasket. The oil pan gasket is reusable if it is not damaged.**

41. Remove the oil pan bolts.
42. Position the oil pan aside to gain access to the oil pump screen and pickup tube.
43. Remove the bolt securing the rear of the oil pump screen and pickup tube.
44. Remove the bolts securing the front of the oil pump screen and pickup tube.
45. Remove the oil pan and the oil pickup tube together.

To install:

⁂ WARNING

Do not use metal scrapers, wire brushes, power abrasive discs or other abrasive means to clean the sealing surfaces. These tools cause scratches and gouges which make leak paths. Use a plastic scraping tool to remove all traces of old sealant.

46. Clean the sealing surfaces with silicone gasket remover and metal surface prep. Follow the directions on the packaging. Inspect the mating surfaces.
47. Position the oil pan gasket, the oil pan and the oil pump screen and pickup tube together to the cylinder block.

⁂ WARNING

Make sure the O-ring is in place and not damaged. A missing or damaged O-ring can cause foam in the lubrication system, low oil pressure, and severe engine damage.

48. Install a new O-ring and lubricate with clean engine oil.
49. Position the oil pump screen and pickup tube and install the bolts.

➡**If the oil pan is not secured within four minutes, the sealant must be removed and the sealing area cleaned. To clean the sealing area, use silicone gasket remover and metal surface prep. Follow the directions on the packaging. Failure to follow this procedure can cause future oil leakage.**

50. Apply silicone gasket and sealant at the engine front cover-to-cylinder block mating surface.

➡**If the oil pan is not secured within four minutes, the sealant must be removed and the sealing area cleaned. To clean the sealing area, use silicone gasket remover and metal surface prep. Follow the directions on the packaging. Failure to follow this procedure can cause future oil leakage.**

51. Apply silicone gasket and sealant at the rear oil seal retainer-to-cylinder block sealing surface.
52. Position the oil pan gasket.
53. Install the bolts and tighten in three steps, in the sequence shown.
    a. Step 1: Tighten to 2 Nm (18 inch lbs.).
    b. Step 2: Tighten to 20 Nm (15 ft. lbs.).
    c. Step 3: Tighten an additional 60 degrees.
54. Lower the engine and remove the crane.
55. Install the right motor mount bolt.
56. Install the left motor mount bolt.
57. Install the drain plug.
58. Install the starter motor.
59. Position the starter.
60. Install the two bolts.
61. Install the stud bolt.
62. Connect the starter motor electrical connections.
63. Install the starter motor solenoid terminal cover.
64. Install a new oil filter.
65. Position the fan shroud and install the bolts.
66. Install a new throttle body adapter gasket.
67. Position the throttle body adapter and install the four bolts.
68. Connect the upper EGR valve tube fitting.
69. Connect the IAC electrical connector.
70. Install the nut.
71. Connect the brake booster vacuum tube and bracket and install the bolt.
72. Connect the EVAP canister purge valve vacuum hose.
73. Connect the fuel pressure regulator vacuum hose.
74. Connect the TP sensor electrical connector and the EGR vacuum regulator solenoid electrical and vacuum connectors.
75. Connect the EGR valve vacuum hose and the main vacuum harness vacuum connections.
76. Install the crankcase ventilation tube and the IAC fresh air tube.

77. Connect the EVAP canister purge valve hose.
78. Connect the PCV hose to the throttle body adapter.
79. Install the accelerator cable bracket and the bolts.
80. Connect the accelerator controls to the throttle body cam.
81. Connect the accelerator cable.
82. Connect the speed control actuator cable.
83. Install the throttle return spring.
84. Install the alternator.
85. Install the air cleaner outlet pipe.
86. Install the radiator air deflector.

2002–03 F-Series and 2004 Heritage Editions, 4.6L and 5.4L Engines w/4-Wheel Drive

1. Before servicing the vehicle, refer to the precautions in the beginning of this section.
2. With the vehicle in **Neutral**, position it on a hoist.
3. Drain the engine oil.
4. Support the front axle housing with a jack stand.
5. Remove front axle housing support.
6. Remove the front axle mount bolt.
7. Remove the axle support bolts.
8. Remove the axle support.

WARNING

Use care when lowering the front axle housing, or the vacuum hoses to the axle solenoid may become disconnected or damaged.

9. Remove the right and front axle housing mount bolts and lower the axle to allow clearance for the oil pan to be removed.
10. Remove the oil pan.

WARNING

Do not use metal scrapers, wire brushes, power abrasive discs, or other abrasive means to clean the sealing surfaces. These may cause scratches and gouges resulting in leak paths. Use a plastic scraper to clean the sealing surfaces.

11. Remove and discard the oil pan gasket. Clean the sealing surfaces with silicone gasket remover and metal surface prep. Follow the directions on the packaging. Inspect the mating surfaces.

To install:

➡If the oil pan is not secured within four minutes, the sealant must be removed and the sealing area cleaned. To clean the sealing area, use silicone gasket remover and metal surface prep. Follow the directions on the packaging. Failure to follow this procedure can cause future oil leakage.

12. Apply silicone gasket and sealant in the locations shown.

➡If the oil pan is not secured within four minutes, the sealant must be removed and the sealing area cleaned. To clean the sealing area, use silicone gasket remover and metal surface prep. Follow the directions on the packaging. Failure to follow this procedure can cause future oil leakage.

13. Apply silicone gasket and sealant in the locations shown.
14. Position the new oil pan gasket and the oil pan and loosely install the bolts.
15. Tighten the bolts in three steps in the sequence shown.
    a. Step 1: Tighten to 2 Nm (18 inch lbs.).
    b. Step 2: Tighten to 20 Nm (15 ft. lbs.).
    c. Step 3: Tighten an additional 60 degrees.
16. Position the front axle housing and loosely install the bolts.
17. Position the axle support and install the bolts.
18. Tighten the bolts to 66 ft. lbs. (90 Nm).
19. Remove the jack stand.
20. Install a new oil filter.

WARNING

The oil pump must be primed prior to starting the engine.

21. Fill the engine with clean engine oil.
22. Start the engine and inspect for leaks.

2001–03 6.8L Engine

CAUTION

Fuel injection systems remain under pressure, even after the engine has been turned OFF. The fuel system pressure must be relieved before disconnecting any fuel lines. Failure to do so may result in fire and/or personal injury.

1. Before servicing the vehicle, refer to the precautions in the beginning of this section.
2. Remove or disconnect the following:

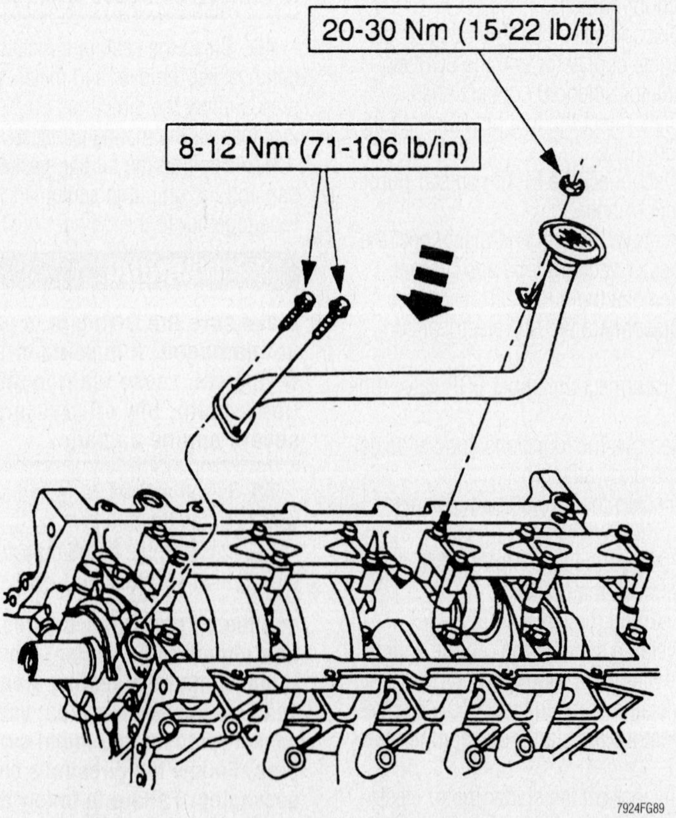

Oil pump pick-up tube and screen assembly—2001–03 6.8L engine

- Negative battery cable
- Fuel system pressure
- Coolant
- Upper radiator hose from the intake manifold
- Air cleaner outlet tube
- Accelerator cable from the bracket and the throttle body cam
- Speed control actuator cable from the throttle body
- All vacuum hoses, fuel lines and electrical wires from the throttle body and intake manifold
- Brake booster vacuum hose bracket
- EGR valve-to-exhaust manifold tube and disconnect the vacuum line
- Connector and vacuum line from the Engine Vacuum Regulator (EVR) solenoid
- The 4 bolts and the throttle body adapter
- Upper fan shroud mounting screws and position the shroud toward the engine.
- The alternator and install the Modular Engine Support Bracket on the engine using the mounting holes.
- Lower engine mount-to-frame nuts
- Turbine Shaft Speed (TSS) and Output Shaft Speed (OSS) sensors from the transmission. Plug the openings.

3. Lower the vehicle to the floor and raise the engine using a hoist attached to the support bracket.
4. Install an engine support fixture with a J hook to keep the engine raised, then remove the hoist.
5. Remove or disconnect the following:
   - Engine oil and filter
   - Dual converter Y-pipe and the flywheel inspection cover
   - Driveshaft and the 2 transmission mounting nuts
   - Transmission. Be sure to support the transmission along the rails of the pan to avoid damage.
   - Oil pan mounting bolts and partially lower the pan.
   - Oil pump pick-up tube and screen assembly and allow it to drop into the pan.
   - Oil pan towards the rear of the vehicle

To install:

⁂ WARNING

To prevent possible oil leaks, use only a plastic scraper to clean the oil pan mounting surface.

6. Clean the oil pan-to-engine mounting surface.
7. Place the oil pump pick-up tube and screen assembly in the oil pan, then position the pan and gasket near the engine.
8. Install the oil pump pick-up tube and screen assembly.
   a. Tighten the nut to 15–22 ft. lbs. (20–30 Nm).
   b. Tighten the 2 bolts to 71–106 inch lbs. (8–12 Nm).
9. Apply a bead of silicone sealant to the areas where the front cover and rear bearing cap meet the engine block.
10. Install the oil pan. Tighten the bolts in sequence using 3 steps as follows:
    - 18 inch lbs. (2 Nm)
    - 15 ft. lbs. (20 Nm)
    - Plus an additional 90°
11. Lower the transmission and install the 2 mounting nuts. Tighten the nuts to 60–80 ft. lbs. (81–108 Nm).
12. Install the driveshaft and the TSS and OSS sensors.
13. Install the flywheel cover and dual converter Y-pipe.
14. Install the oil bypass filter.
15. Lower the vehicle and remove the engine support fixture.
16. Install the engine mounting nuts. Tighten the nuts to 66 ft. lbs. (90 Nm).
17. Remove the modular engine support bracket and install the alternator.
18. Install the fan shroud.
19. Use a new gasket and install the throttle body adapter.
20. Tighten the bolts in 2 steps:
    - 71–88 inch lbs. (8–10 Nm).
    - 85–95 degrees.
21. Connect the EVR solenoid harness and vacuum line.
22. Attach the vacuum line to the EGR valve.
23. Install the EGR valve-to-exhaust manifold tube. Tighten the fittings to 55 ft. lbs. (41 Nm).
24. Install the EGR transducer.
25. Install all remaining components in the reverse of the removal.

⁂ WARNING

Operating the engine without the proper amount and type of engine oil will result in severe engine damage.

26. Fill the engine with SAE 5W30 oil.
27. Fill and bleed the cooling system.

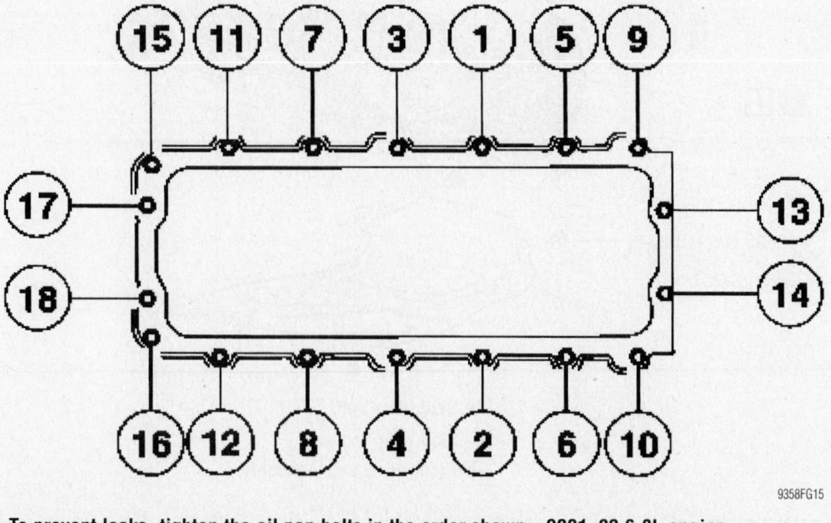

To prevent leaks, tighten the oil pan bolts in the order shown—2001–03 6.8L engine

2004 F-150 w/4.6L or 5.4L Engine, Exc. Heritage Editions

2-WHEEL DRIVE

1. Before servicing the vehicle, refer to the precautions in the beginning of this section.
2. With the vehicle in **Neutral**, position it on a hoist.
3. Drain the engine oil.
4. Remove the oil level sensor electrical connector.
5. Remove the oil pan.

⁂ CAUTION

Do not use metal scrapers, wire brushes, power abrasive discs, or other abrasive means to clean the sealing surfaces. These may cause scratches and gouges resulting in leak paths. Use a plastic scraper to clean the sealing surfaces.

6. Remove and discard the oil pan gasket. Clean the sealing surfaces with silicone gasket remover and metal surface prep. Follow the directions on the packaging. Inspect the mating surfaces.

7. To install, reverse the removal procedure.

➡If the oil pan is not secured within four minutes, the sealant must be removed and the sealing area cleaned. To clean the sealing area, use silicone gasket remover and metal surface prep. Follow the directions on the packaging. Failure to follow this procedure can cause future oil leakage.

8. Apply silicone gasket and sealant in the locations shown.

9. Position the new oil pan gasket and the oil pan and loosely install the bolts.

10. Tighten the bolts in three steps in the sequence shown.

  a. Step 1: Tighten to 2 Nm (18 inch lbs.).

  b. Step 2: Tighten to 20 Nm (15 ft. lbs.).

  c. Step 3: Tighten an additional 60 degrees.

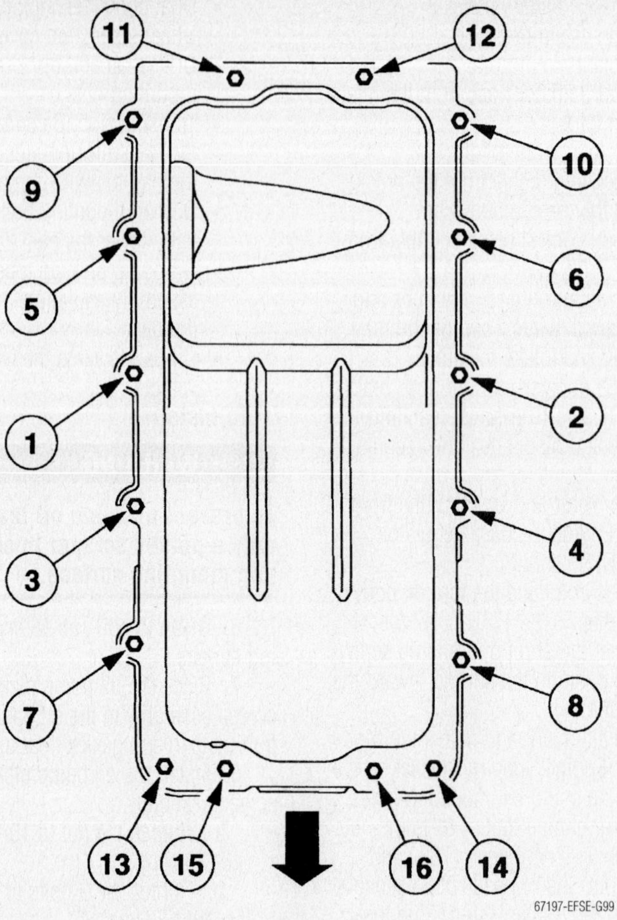

Oil pan torque sequence—2004 4.6L or 5.4L engine

4-WHEEL DRIVE

1. With the vehicle in **Neutral**, position it on a hoist.

2. Drain the engine oil.

3. Remove the skid plate.

4. Index mark the front driveshaft flange to the front axle pinion flange.

⁂ CAUTION

Do not let the driveshaft hang unsupported.

5. Disconnect and support the front driveshaft.

6. Remove the four bolts.

7. Support the front driveshaft.

8. Remove the bolts and the crossmember.

9. Position a suitable hydraulic jack under the front axle. Securely strap the jack to the axle.

➡Rotate the steering column so the pinch bolt for the steering column coupling allows clearance for the isolator bolt.

10. Remove the upper front axle carrier mounting bushing bolt.

11. Remove the axle shaft housing carrier bushing bolt.

12. Remove the lower front axle carrier mounting bushing bolt.

13. Lower the front axle assembly.

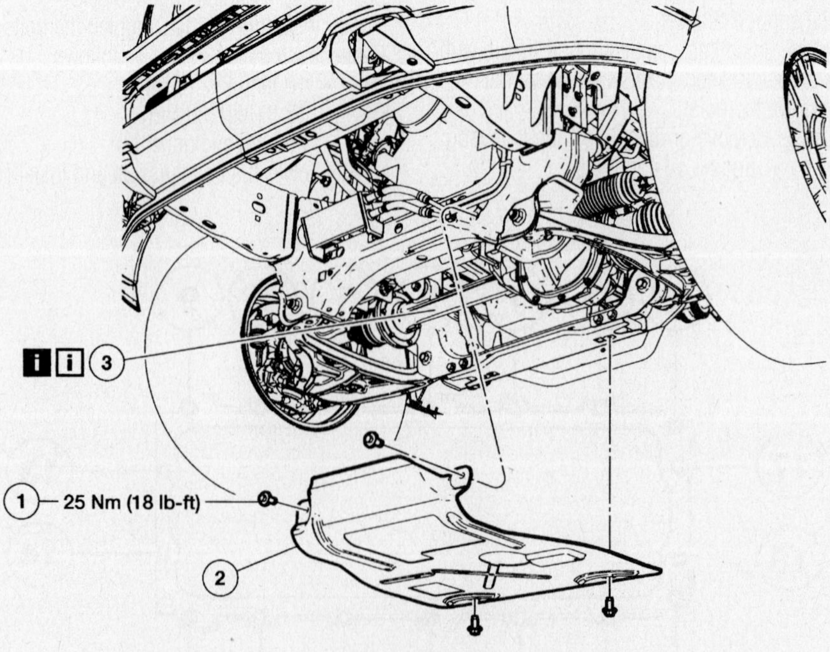

Skid plate removal

14. Disconnect the oil level sensor.
15. Remove the oil pan bolts and lower the pan. Discard the gasket.

CAUTION

Do not use metal scrapers, wire brushes, power abrasive discs, or other abrasive means to clean the sealing surfaces. These may cause scratches and gouges resulting in leak paths.

16. Use a plastic scraper to clean the sealing surfaces.
17. Remove and discard the oil pan gasket. Clean the sealing surfaces with silicone gasket remover and metal surface prep. Follow the directions on the packaging. Inspect the mating surfaces.

To install

➡If the oil pan is not secured within four minutes, the sealant must be removed and the sealing area cleaned. To clean the sealing area, use silicone gasket remover and metal surface prep. Follow the directions on the packaging. Failure to follow this procedure can cause future oil leakage.

18. Apply silicone gasket and sealant in the locations shown.
19. Position the new oil pan gasket and the oil pan and loosely install the bolts.
20. Tighten the bolts in three steps in the sequence shown.
   a. Step 1: Tighten to 2 Nm (18 inch lbs.).
   b. Step 2: Tighten to 20 Nm (15 ft. lbs.).
   c. Step 3: Tighten an additional 60 degrees.
21. Position the front axle.

➡Rotate the steering column so the pinch bolt for the steering column coupling allows clearance for the isolator bolt.

22. Install the upper front axle carrier mounting bushing bolt. Torque to 66 ft. lbs. (90 Nm).
23. Install the axle shaft housing carrier bushing nut and bolt. Torque to 66 ft. lbs. (90 Nm).
24. Install the lower front axle carrier mounting bushing bolt. Torque to 66 ft. lbs. (90 Nm).
25. Align the index marks, connect the front driveshaft and install the bolts. Torque to 82 ft. lbs. (111 Nm).
26. Install the crossmember and the bolts. Torque to 66 ft. lbs. (90 Nm).
27. Install the skid plate.
28. Add the correct amount of engine oil. Install a new filter.

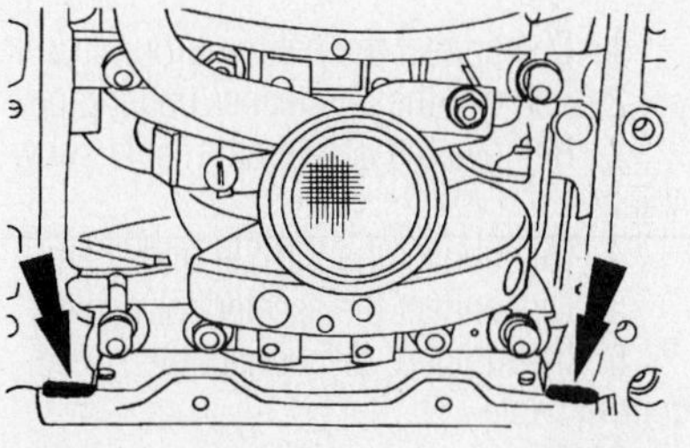

67197-EFSE-G102

Silicone sealant application

2004 F-250/350

5.4L ENGINE W/MANUAL TRANSMISSION

1. Before servicing the vehicle, refer to the precautions in the beginning of this section.
2. With the vehicle in **Neutral**, position it on a hoist.
3. Remove the air cleaner outlet tube.
4. Remove the accelerator cable splash shield.
5. Disconnect the throttle body cable.
6. Disconnect the accelerator cable.
7. Disconnect the speed control cable.
8. Remove the throttle return spring.
9. Position the cable mounting bracket aside.
10. Remove the bolts.
11. Position the cable mounting bracket aside.
12. Disconnect the hose and electrical connector from the idle air control (IAC) valve.
13. Remove the routing clip for the IAC hose.
14. Disconnect the throttle position (TP) sensor connector.
15. Remove the two bolts retaining the vacuum transducer bracket to the intake manifold. Position the bracket and transducer out of the way.
16. If equipped, disconnect the exhaust gas recirculation (EGR) valve.
17. Loosen the EGR tube nut.
18. Disconnect the EGR tube.
19. Disconnect the vacuum line from the EGR valve.
20. If equipped, disconnect the EGR vacuum regulator (EVR) connections.
21. Disconnect electrical connector.
22. Disconnect vacuum connector.
23. Remove the four bolts. Position aside the throttle body, throttle body spacer, EGR valve, EVR and IAC valve as an assembly.
24. Hold the accelerator pedal in place and disconnect the accelerator cable from the accelerator pedal and shaft.
25. Disconnect the accelerator cable routing clip from the dash panel stud.
26. Remove the bolts and pull the accelerator cable through the bulkhead.
27. Remove the alternator.

CAUTION

The large clutch assembly nut has a right-hand thread and must be rotated counterclockwise to remove it.

28. Remove the fan blade assembly and the fan clutch from the coolant pump pulley.
29. Use a suitable tool to hold the coolant pump pulley.
30. Using a suitable tool, remove the fan blade and the fan clutch from the coolant pump pulley.
31. Install a lifting eye on the engine using the alternator mounting holes.
32. Install a lifting bar.
33. Disconnect the heated oxygen sensor electrical connectors.
34. Remove the exhaust Y-pipe flange nuts.
35. Remove the nuts and the exhaust Y-pipe.
36. Remove the engine mount nuts.
37. Remove the stud bolt and the ground strap from the engine.
38. Remove the starter motor.
39. Remove the nut and bracket from the engine front cover stud and position the wire harness aside.
40. Remove the transmission mount nuts.
41. Drain the engine oil and remove the oil filter.
42. Remove the oil cooler threaded insert and position the oil cooler aside.

CAUTION

Use care when raising the engine, so that the engine does not touch the body.

43. Using a crane, raise the engine.
44. Remove the bolts and remove the right engine mount from the engine.
45. Remove the bolts and partially lower the oil pan.
46. Remove the bolts from the oil pump screen and pickup tube.
47. Remove the rear bolt and the oil pump screen and pickup tube and let them drop into the oil pan.

➡To remove the oil pan, position the rear of the pan to the left side of the vehicle and rotate slightly to the right while pulling outward.

48. Remove the oil pan and the oil pan gasket.
49. Clean the inside of the oil pan thoroughly and inspect the oil pan gasket.

CAUTION

Do not use metal scrapers, wire brushes, power abrasive discs or other abrasive means to clean the sealing surfaces. These tools cause scratches and gouges, which make leak paths. Use a plastic scraping tool to remove all traces of sealant.

50. Clean the mating surfaces for the oil pan and the engine with silicone gasket remover and metal surface prep. Follow the directions on the packaging.

To install:

➡The oil pump screen and pickup tube must be in the oil pan when the oil pan is positioned in the vehicle.

➡To install the oil pan, position the oil pan to the left rear side of the engine and rotate slightly while sliding the front of the pan toward the right side of the engine.

51. Position the oil pan gasket and the oil pan in the vehicle from the rear of the engine.

CAUTION

Make sure the O-ring is in place and not damaged. A missing or damaged O-ring can cause foam in the lubrication system, low oil pressure and severe engine damage.

➡Clean and inspect the mating surfaces and install a new O-ring. Lubricate with clean engine oil.

52. Position the oil pump screen and pickup tube on the engine. Install the bolts. Torque the base bolts to 89 inch lbs. (10 Nm); the support bolt to 18 ft. lbs. (25 Nm).

➡If not secure within four minutes, the sealant must be removed and the sealing area cleaned. To clean the sealing area, follow the directions provided on the packaging of the silicone gasket remover and the metal surface prep. Failure to follow this procedure can cause future oil leakage.

53. Apply a bead of silicone gasket and sealant at the crankshaft rear seal retainer-to-cylinder block surface.

➡If not secure within four minutes, the sealant must be removed and the sealing area cleaned. To clean the sealing area, follow the directions provided on the packaging of the silicone gasket remover and the metal surface prep. Failure to follow this procedure can cause future oil leakage.

54. Apply a bead of silicone gasket and sealant at the engine front cover-to-cylinder block surface.

➡Be sure to tighten the bolts in two steps.

55. Tighten the bolts in the sequence shown.
  a. Step 1: Tighten to 20 Nm (15 ft. lbs.).
  b. Step 2: Tighten an additional 90 degrees.
56. Position the right engine mount on the engine and install the bolts. Tighten the bolts to 80 Nm (59 ft. lbs.).
57. Using a crane, lower the engine and remove the crane.
58. Install the starter motor.
59. Position the oil cooler and install the threaded insert.
60. Install a new oil filter.

➡Make sure that both the engine and transmission mounts are correctly seated before installing the mounting nuts.

61. Install the engine mount nuts.
62. Install the transmission mount nuts.
63. Position the wire harness back and install the bracket and nut on the engine front cover stud.
64. Install the ground strap to the engine.
65. Position the exhaust Y-pipe and install the nuts on the exhaust manifold studs.
66. Install the exhaust Y-pipe flange nuts.
67. Connect the heated oxygen sensor electrical connectors.
68. Remove the special tool.
69. Install the fan blade assembly and the fan clutch to the coolant pump pulley.
70. Install the fan blade and the fan clutch.
71. Use a suitable tool to hold the coolant pump pulley.
72. Using a suitable tool, tighten the fan clutch nut to the coolant pump pulley. Torque to 41 ft. lbs. (55 Nm).
73. Install the alternator.
74. Install the accelerator cable bracket.
75. Position the accelerator cable through the bulkhead. Install the bolts.
76. Connect the accelerator cable routing clip to the dash panel stud.
77. Connect the accelerator cable to the accelerator pedal and shaft.
78. Install a new throttle body adapter gasket.
79. Position the throttle body adapter back.
80. Position the throttle body adapter.
81. Install the four bolts. Torque to 89 inch lbs. (10 Nm).
82. Connect the EGR valve vacuum hose.
83. Connect the EGR vacuum regulator solenoid connections.
84. Install the differential pressure feedback EGR system bracket.
85. Connect the exhaust manifold to EGR valve tube upper fitting.
86. Tighten both fittings starting at the top in three steps.
  a. Hand-tighten.
  b. Tighten the upper fitting to 40–60 Nm (30–44 ft. lbs.).
  c. Tighten the lower fitting to 40–60 Nm (30–44 ft. lbs.).
87. Connect the IAC electrical connector.
88. Install the IAC fresh air tube and insert it in the routing clip.
89. Connect the TP sensor electrical connector.
90. Install the accelerator cable bracket.
91. Position the accelerator cable bracket.
92. Install the bolts.
93. Connect the throttle body cam.
94. Connect the accelerator cable.
95. If equipped, connect speed control cable.
96. Install the return spring.
97. Install the accelerator control splash shield.
98. Install the air cleaner outlet tub.

CAUTION

The oil pump must be primed prior to starting the engine.

99. Fill the engine with clean engine oil.

5.4L ENGINE W/AUTOMATIC TRANSMISSION

1. Disconnect the battery ground cable.
2. Remove the air cleaner outlet pipe.
3. Remove the accelerator cable snow shield.
4. Disconnect the throttle body cam.
5. Disconnect the accelerator cable.
6. If equipped, disconnect the speed control cable.
7. Remove the throttle return spring.

8. Position the accelerator cable bracket aside.
9. Remove the bolts.
10. Position the bracket and cables aside.
11. Release the idle air control (IAC) fresh air tube from the routing clip and remove it from the vehicle.
12. Disconnect the IAC electrical connector.
13. If equipped, disconnect the differential pressure feedback exhaust gas recirculation (EGR) system electrical connector.
14. Disconnect the throttle position (TP) sensor electrical connector.
15. If equipped, disconnect the differential pressure feedback EGR system vacuum hoses.
16. If equipped, disconnect the exhaust manifold to EGR valve tube fittings and remove the tube.
17. If equipped, remove the differential pressure feedback EGR system bracket.
18. If equipped, disconnect the EGR valve vacuum hose.
19. If equipped, disconnect the EGR vacuum regulator solenoid electrical and vacuum connections.
20. Position aside the throttle body adapter.
21. Remove the four bolts.
22. Position aside the throttle body adapter.
23. Remove and discard the throttle body adapter gasket.
24. Remove the alternator.

✱✱ WARNING

The large clutch assembly nut has a right-handed thread and must be rotated counterclockwise to remove it.

25. Using a suitable tool, remove the fan and fan clutch from the coolant pump pulley.
26. Install the modular engine support bracket on the engine using the alternator mounting holes.
27. Raise and support the vehicle.
28. Remove the engine mount nuts.

✱✱ WARNING

Damage to the turbine shaft speed/output shaft speed (TSS/OSS) sensors may occur and cause the transmission or torque converter operational concerns if the transmission is raised prior to removing TSS/OSS.

✱✱ WARNING

Sensor bosses must be cleaned prior to removal and then plugged to prevent contamination from damaging internal components.

29. If the vehicle is equipped with TSS and OSS, remove the sensors and install plugs in the transmission.
30. Loosely install the nuts and the crossmember.
31. Loosely install the left crossmember nuts.
32. Loosely install the right crossmember nuts.
33. Loosely install the crossmember bolts.
34. Remove the jack from under the transmission.
35. Lower the vehicle.
36. Raise the engine using the Heavy Duty Engine Support.
37. Raise and support the vehicle.
38. Drain the engine oil and remove the oil bypass filter.
39. Disconnect the heated oxygen sensor (HO_2S) electrical connectors.
40. Remove the exhaust Y-pipe flange nuts.
41. Remove the nuts and the exhaust Y-pipe.
42. Remove the flywheel inspection plate.

✱✱ WARNING

Support the transmission on the oil pan rails only or internal transmission damage can occur.

43. Support the transmission with a transmission jack.
44. Remove the transmission mount nuts.
45. Raise the transmission.
46. Remove the bolts and partially lower the oil pan.
47. Remove the bolts from the oil pump screen and pickup tube and let them drop into the oil pan.
48. Remove the rear bolt and the oil pump screen and pickup tube and let them drop into the oil pan
49. Remove the oil pan and the oil pan gasket from the rear of the engine.
Clean the oil pan thoroughly and inspect the oil pan gasket.

✱✱ WARNING

Do not use metal scrapers, wire brushes, power abrasive discs or other abrasive means to clean the sealing surface. These tools cause scratches and gouges, which make leak paths. Use a plastic scraping tool to remove all traces of sealant.

50. Clean the mating surfaces for the oil pan with silicone gasket remover and metal surface prep. Follow the directions on the packaging.
51. Installation

➡The oil pump screen and pickup tube must be in the oil pan when the oil pan is positioned in the vehicle.

52. Position the oil pan gasket and the oil pan in the vehicle from the rear of the engine.

✱✱ WARNING

Make sure the O-ring is in place and not damaged. A missing or damaged O-ring can cause foam in the lubrication system, low oil pressure and severe engine damage.

➡Clean and inspect the mating surfaces and install a new O-ring. Lubricate with clean engine oil.

53. Position the oil pump screen and pickup tube on the engine.
54. Install the bolts.

➡If not secure within four minutes, the sealant must be removed and the sealing area cleaned. To clean the sealing area, follow the directions provided on the packaging of the silicone gasket remover and the metal surface prep. Failure to follow this procedure can cause future oil leakage.

55. Apply a bead of silicone gasket and sealant at the crankshaft rear seal retainer-to-cylinder block surface.

➡If not secure within four minutes, the sealant must be removed and the sealing area cleaned. To clean the sealing area, follow the directions provided on the packaging of the silicone gasket remover and the metal surface prep. Failure to follow this procedure can cause future oil leakage.

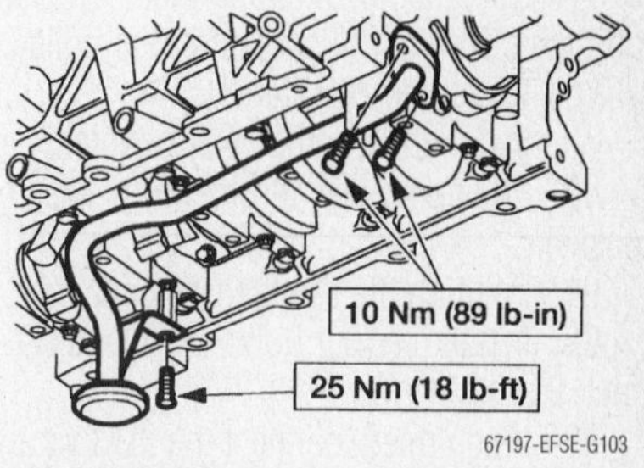

Oil pump installation—4.6L or 5.4L

56. Apply a bead of silicone gasket and sealant at the engine front cover-to-cylinder block surface.

➡**Be sure to tighten the bolts in two steps.**

57. Tighten the bolts in the sequence shown.
  a. Stage 1: Tighten to 20 Nm (15 ft. lbs.).
  b. Stage 2: Tighten an additional 90 degrees.
58. Lower the transmission.
59. Remove the High Lift Transmission Jack.
60. Install the flywheel inspection plate.
61. Install the oil bypass filter.
62. Lower the vehicle.
63. Align the engine mount studs and lower the engine, using the Heavy Duty Engine Support.
64. Remove the Heavy Duty Engine Support.
65. Raise and support the vehicle.
66. Install the engine mount nuts.
67. Install a suitable transmission jack and support the transmission.
68. Remove the right crossmember nuts.
69. Remove the left crossmember nuts.
70. Remove the nuts and the crossmember.
71. If the vehicle is equipped with TSS and turbine/OSS, remove the plugs and install the sensors in the transmission.
72. Position the exhaust Y-pipe and install the nuts on the exhaust manifold studs.
73. Install the exhaust Y-pipe flange nuts.
74. Connect the heated oxygen sensor electrical connectors.
75. Lower the vehicle.
76. Remove the modular engine support bracket from the engine.

✲✲ WARNING

The large clutch assembly nut has a right-handed thread and must be rotated counterclockwise to remove it.

77. Using a suitable tool, install the fan and fan clutch to the coolant pump pulley.
78. Install the alternator.
79. Install the throttle body adapter.
80. Position a new throttle body adapter gasket.
81. Position the throttle body adapter.
82. Install the four bolts. Torque to 89 inch lbs. (10 Nm).
83. If equipped, connect the EGR vacuum regulator solenoid electrical and vacuum connections.
84. If equipped, connect the EGR valve vacuum hose.
85. If equipped, install the differential pressure feedback EGR system bracket. Position the bracket. Install the bolts.
86. If equipped, install the exhaust manifold to EGR valve tube.
  a. Hand-tighten the fittings.
  b. Tighten the upper fitting to 40–60 Nm (30–44 ft. lbs.).
  c. Tighten the lower fitting to 40–60 Nm (30–44 ft. lbs.).
87. If equipped, connect the differential pressure feedback EGR system vacuum lines.

If equipped, connect the differential pressure feedback EGR system electrical connector.

88. Connect the TP sensor electrical connector.
89. Connect the IAC electrical connector.
90. Install the IAC fresh air tube and insert it in the routing clip.
91. Install the accelerator cable bracket.
92. Position the bracket and cables.
93. Install the bolts.
94. Connect the throttle body cam.
95. Connect the accelerator cable.
96. If equipped, connect the speed control cable.
97. Install the throttle return spring.
98. Install the accelerator cable snow shield.
99. Install the air cleaner outlet pipe.
100. Connect the battery cable.

✲✲ WARNING

The oil pump must be primed prior to starting the engine.

101. Fill the engine with clean engine oil.

6.8L ENGINE W/MANUAL TRANSMISSION

1. With the vehicle in **Neutral**, position it on a hoist.
2. Remove the air cleaner outlet tube.
3. Remove the accelerator cable splash shield.
4. Disconnect the accelerator and speed control cables and remove the return spring.
5. Position the accelerator cable bracket aside.
6. Remove the bolts.
7. Position the bracket and cables aside.
8. Disconnect the positive crankcase ventilation (PCV) tube and the vacuum hoses.
9. Disconnect the vacuum hose.
10. Disconnect the idle air control (IAC) motor electrical connector and the bypass hose.
11. Remove the routing clip for the IAC hose.
12. Disconnect the throttle position (TP) sensor electrical connector.
13. Position aside the throttle body adapter.
14. Remove the four adapter bolts.
15. Position aside the throttle body adapter.
16. Remove and discard the throttle body adapter gasket.
17. Hold the accelerator pedal in place and disconnect the accelerator cable from the accelerator pedal and shaft.
18. Disconnect the accelerator cable routing clip from the dash panel stud.
19. Remove the bolts and pull the accelerator cable through the bulkhead.
20. Remove the alternator.

✲✲ WARNING

The large clutch assembly nut has a right-hand thread and must be rotated counterclockwise to remove it.

21. Using the special tools, remove the cooling fan and the fan clutch from the coolant pump pulley.
22. Install a suitable tool on the engine using the alternator mounting holes.
23. Install a suitable tool.
24. Disconnect the heated oxygen sensor (HO_2S) electrical connectors.
25. Remove the exhaust Y-pipe flange nuts.
26. Remove the nuts and the exhaust Y-pipe.
27. Remove the engine mount nuts.
28. Remove the starter motor.
29. Remove the stud bolt and the ground strap from the engine.
30. Remove the nut and bracket from the engine front cover stud and position aside the wire harness.
31. Remove the transmission mount nuts.
32. Drain the engine oil and remove the oil filter.
33. Remove the oil cooler threaded insert and position the oil cooler aside.

✲✲ WARNING

Use care when raising the engine, so that the engine does not touch the body.

34. Using a suitable tool, raise the engine.
35. Remove the bolts and remove the right engine mount from the engine.
36. Remove the bolts and partially lower the oil pan.
37. Remove the oil pump screen and pickup tube and spacer and let them drop into the oil pan.

➡To remove the oil pan, position the rear of the pan to the left side of the vehicle and rotate slightly to the right while pulling outward.

38. Remove the oil pan and the oil pan gasket.
39. Clean the inside of the oil pan thoroughly and inspect the oil pan gasket.

⁂ WARNING

Do not use metal scrapers, wire brushes, power abrasive discs or other abrasive means to clean the sealing surfaces. These tools cause scratches and gouges, which make leak paths. Use a plastic scraping tool to remove all traces of sealant.

40. Clean the mating surfaces for the oil pan and the engine with silicone gasket remover and metal surface prep. Follow the directions on the packaging.

To install:

➡The oil pump screen and pickup tube must be in the oil pan when the oil pan is positioned in the vehicle.

➡To install the oil pan, position the oil pan to the left rear side of the engine and rotate slightly while sliding the front of the pan toward the right side of the engine.

41. Position the oil pan gasket and the oil pan in the vehicle from the rear of the engine.

⁂ WARNING

Make sure the O-ring is in place and not damaged. A missing or damaged O-ring can cause foam in the lubrication system, low oil pressure and severe engine damage.

➡Clean and inspect the mating surfaces and install a new O-ring. Lubricate with clean engine oil.

42. Place the oil pump screen and pickup tube in the pan.
43. Install and tighten the spacer. Torque to 18 ft. lbs. (25 Nm).

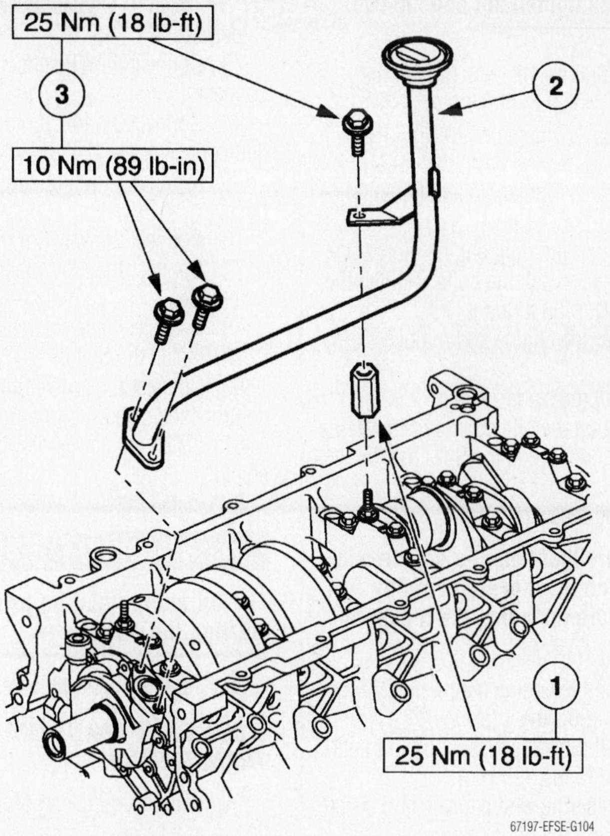

Oil pump pick-up tube installation—6.8L engine

44. Position the oil pump screen and pickup tube on the engine.
45. Install the bolts. Torque the tube base bolts to 89 inch lbs. (10 Nm); the bracket bolt to 18 ft. lbs. (25 Nm).
46. Apply a bead of silicone gasket and sealant at the crankshaft rear seal retainer-to-cylinder block surface.

➡If not secure within four minutes, the sealant must be removed and the sealing area cleaned. To clean the sealing area, follow the directions provided on the packaging of the silicone gasket remover and the metal surface prep. Failure to follow this procedure can cause future oil leakage.

47. Apply a bead of silicone gasket and sealant at the engine front cover-to-cylinder block surface.

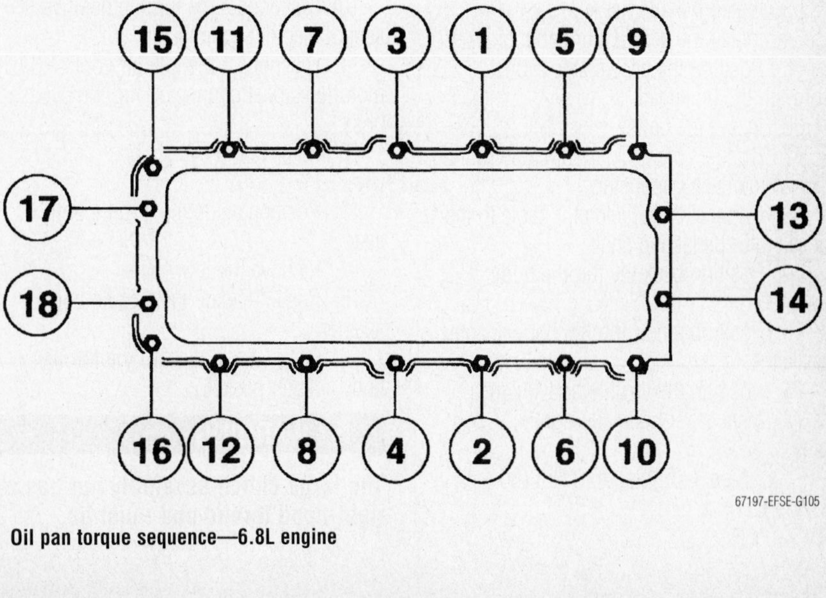

Oil pan torque sequence—6.8L engine

➡**Be sure to tighten the bolts in two steps.**

48. Position the oil pan. Tighten the bolts in the sequence shown.
   a. Step 1: Tighten to 20 Nm (15 ft. lbs.).
   b. Step 2: Tighten an additional 90 degrees.
49. Position the right engine mount on the engine and install the bolts. Tighten the bolts to 80 Nm (59 ft. lbs.).
50. Lower the engine and remove the lifting tool.
51. Install the starter motor.
52. Position the oil cooler and install the threaded insert. Torque to 43 ft. lbs. (58 Nm).
53. Install a new oil filter.

➡**Make sure that both the engine and transmission mounts are correctly seated before installing the mounting nuts.**

54. Install the engine mount nuts. Torque to 66 ft. lbs. (90 Nm).
55. Install the transmission mount nuts. Torque to 70 ft. lbs. (95 Nm).
56. Position the wire harness and install the bracket and nut on the engine front cover stud.
57. Install the ground strap to the engine.
58. Position the exhaust Y-pipe and install the nuts on the exhaust manifold studs.
59. Install the exhaust Y-pipe flange nuts.
60. Connect the heated oxygen sensor electrical connectors.
61. Remove the special tool.
62. Using the special tools, install the cooling fan and the fan clutch to the coolant pump pulley.
63. Install the alternator.
64. Install the accelerator cable bracket.
65. Position the accelerator cable through the bulkhead.
66. Install the bolts.
67. Connect the accelerator cable routing clip to the dash panel stud.
68. Connect the accelerator cable to the accelerator pedal and shaft.
69. Position back the throttle body adapter.
70. Position a new throttle body adapter gasket.
71. Position the throttle body adapter.
72. Install four bolts and tighten in two steps.
   a. Step 1: Tighten to 9 Nm (80 inch lbs.).
   b. Step 2: Tighten an additional 90 degrees.
73. Connect the TP sensor electrical connector.
74. Connect the IAC motor bypass hose and connector.
75. Install the routing clip for the IAC hose.
76. Connect the vacuum hose.
77. Connect the PCV tube and the vacuum hoses.
78. Position the accelerator bracket.
79. Install the bolts.
80. Connect the accelerator and speed control cables and return spring.
81. Install the accelerator cable splash shield.
82. Install the air cleaner outlet tube.

⁂ WARNING

The oil pump must be primed prior to starting the engine.

83. Fill the engine with clean engine oil.

6.8L ENGINE W/AUTOMATIC TRANSMISSION

1. With the vehicle in **Neutral**, position it on a hoist.
2. Disconnect the battery ground cable.
3. Remove the air cleaner outlet tube.
4. Remove the accelerator cable splash shield.
5. Disconnect the accelerator and speed control cables and return spring.
6. Position the accelerator cable bracket aside.
7. Remove the bolts. Position the bracket and cables aside.
8. Disconnect the positive crankcase ventilation (PCV) tube and the vacuum hoses.
9. Disconnect the vacuum hose.
10. Disconnect the throttle positive (TP) sensor electrical connector.
11. Disconnect the idle air control (IAC) motor electrical connector and the bypass hose.
12. Remove the bypass hose routing clip.
13. Position aside the throttle body adapter.
14. Remove the four bolts.
15. Position aside the throttle body adapter.
16. Remove and discard the throttle body adapter gasket.

⁂ WARNING

The large clutch assembly nut has a right-hand thread and must be rotated counterclockwise to remove it.

17. Using a suitable tool, remove the fan and fan clutch from the coolant pump pulley.
18. Remove the alternator.
19. Install a suitable tool on the engine, using the alternator mounting holes.
20. Raise and support the vehicle.
21. Remove the engine mount nuts.

⁂ WARNING

Damage to the TSS/OSS may occur and cause the transmission or torque converter operational concerns if the transmission is raised prior to removing TSS/OSS.

⁂ WARNING

Sensor bosses must be cleaned prior to removal and then plugged to prevent contamination from damaging internal components.

22. If the vehicle is equipped with turbine shaft speed (TSS) and output shaft sensors (OSS), remove the sensors and install plugs in the transmission.
23. Loosely install the nuts and the crossmember.
24. Loosely install the left crossmember nuts.
25. Loosely install the right crossmember nuts.
26. Loosely install the crossmember bolts.
27. Remove the jack assembly supporting the transmission.
28. Raise the engine, using a suitable tool.
29. Drain the engine oil and remove the oil bypass filter.
30. Remove the bolt and position the oil cooler aside.
31. Disconnect the heated oxygen sensor electrical connectors.
32. Remove the exhaust Y-pipe flange nuts.
33. Remove the nuts and the exhaust Y-pipe.
34. Remove the flywheel inspection plate.

⁂ WARNING

Support the transmission on the oil pan rails only or internal transmission damage can occur.

35. Support the transmission with a transmission jack.

36. Remove the transmission mount nuts.
37. Using the transmission jack, raise the transmission.
38. Remove the nut and detach the transmission cooling tube bracket.
39. Remove the bolts and partially lower the oil pan.
40. Remove the bolts retaining the oil pump screen cover and pickup tube.
41. Remove the pickup tube and spacer and let them drop into the oil pan
42. Remove the oil pan and the oil pan gasket from the rear of the engine.

⁂ WARNING

Do not use metal scrapers, wire brushes, power abrasive discs or other abrasive means to clean the sealing surface. These tools cause scratches and gouges, which make leak paths. Use a plastic scraping tool to remove all traces of sealant.

43. Clean the mating surfaces and thoroughly clean the oil pan.

To install:

➡Clean and inspect the oil pump screen cover and tube mating surfaces and install a new O-ring. Lubricate the O-ring with clean engine oil.

➡The oil pump screen and pickup tube must be in the oil pan when the oil pan is positioned in the vehicle.

44. Position the oil pan gasket and the oil pan in the vehicle from the rear of the engine.
45. Position the oil pump screen and pickup tube.
46. Install and tighten the spacer. Torque to 18 ft. lbs. (25 Nm).
47. Insert the oil pump screen pickup tube into the oil pump and position the support bracket over the spacer.
48. Install and tighten the bolts. Torque the pump base bolts to 89 inch lbs. (10 Nm); the bracket bolt to 18 ft. lbs. (25 Nm).

➡If not secure within four minutes, the sealant must be removed and the sealing area cleaned. To clean the sealing area, follow the directions provided on the packaging of the silicone gasket remover and the metal surface prep. Failure to follow this procedure can cause future oil leakage.

49. Apply a bead of silicone gasket and sealant where the front cover and rear crankshaft seal retainer fit to the engine block.

➡Be sure to tighten the bolts in two steps.

50. Position the oil pan. Tighten the bolts in the sequence shown.
    a. Step 1: Tighten to 20 Nm (15 ft. lbs.).
    b. Step 2: Tighten an additional 90 degrees.
51. Attach the transmission cooling tube bracket and install the nut.
52. Lower the transmission.
53. Loosely install the transmission mount nuts.
54. Install the flywheel inspection plate. Torque to 25 ft. lbs. (34 Nm).
55. Position the engine oil cooler adapter and install the bolt. Torque to 43 ft. lbs. (58 Nm).
56. Install the oil bypass filter.
57. Align the engine mount studs and lower the engine.
58. Remove the lifting tool.
59. Install the engine mount nuts. Torque to 66 ft. lbs. (90 Nm).
60. Install a suitable transmission jack and support the transmission.
61. Install the right crossmember nuts.
62. Install the left crossmember nuts.
63. Install the nuts and the crossmember.
64. If the vehicle is equipped with TSS and turbine/OSS, remove the plugs and install the sensors in the transmission.
65. Position the exhaust Y-pipe and install the nuts on the exhaust manifold studs.
66. Install the exhaust Y-pipe flange nuts. Torque to 39 ft. lbs. (40 Nm).
67. Connect the heated oxygen sensor electrical connectors.

⁂ WARNING

The large clutch assembly nut has a right-hand thread and must be rotated counterclockwise to remove it.

68. Using a suitable tool, install the fan and fan clutch to the coolant pump pulley. Torque to 84 –112 ft. lbs. (113–153 Nm).
69. Install the alternator.
70. Position the throttle body adapter.
71. Position a new throttle body adapter gasket.
72. Position the throttle body adapter.
73. Install the four bolts and tighten in two steps.
    a. Step 1: Tighten to 9 Nm (80 inch lbs.).
    b. Step 2: Tighten an additional 90 degrees.
74. Connect the IAC electrical connector and the bypass hose.
75. Install the bypass hose routing clip.
76. Connect the vacuum hose.
77. Connect the PCV tube and the vacuum hoses.
78. Connect the TP sensor electrical connector.
79. Install the accelerator cable bracket.
80. Position the bracket and cables. Install the bolts.
81. Connect the accelerator and speed control cables and return spring.
82. Install the accelerator cable splash shield.
83. Install the air cleaner outlet tube.
84. Connect the battery cable.

⁂ WARNING

The oil pump must be primed prior to starting the engine.

85. Fill the engine with clean engine oil.

E-series

4.6L AND 5.4L ENGINES

1. Before servicing the vehicle, refer to the precautions in the beginning of this section.
2. Remove the intake manifold.
3. Remove the fan shroud and the engine cooling fan.
4. Remove the retainers and the shield.
5. Remove the nut and position the transmission fluid filler tube aside.
6. Release the wiring retainers from the heater outlet tube bracket.
7. Release the knock sensor electrical connector retainer.

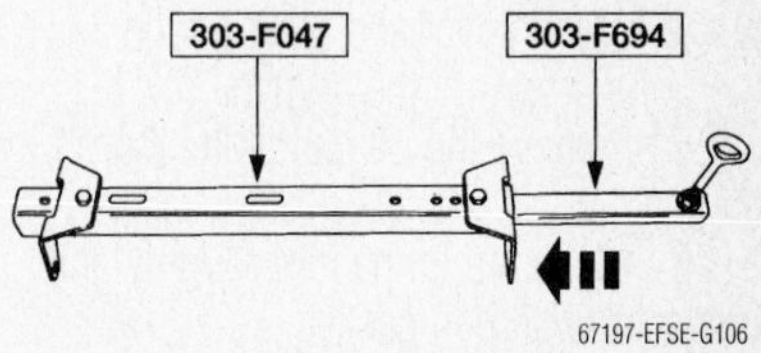

Assemble the lifting tools, 303-F047 and 303-F694, or equivalent

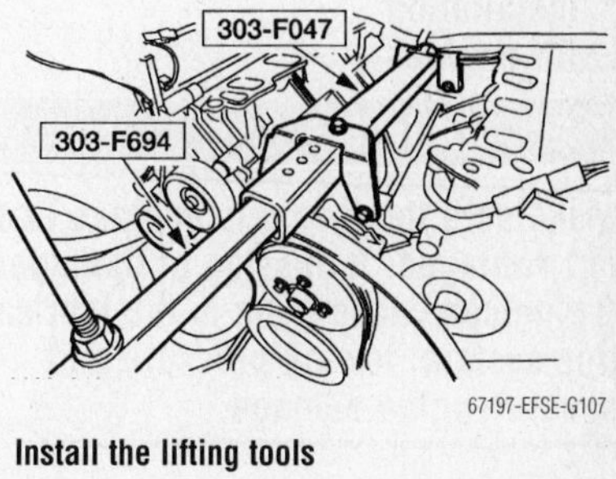

Install the lifting tools

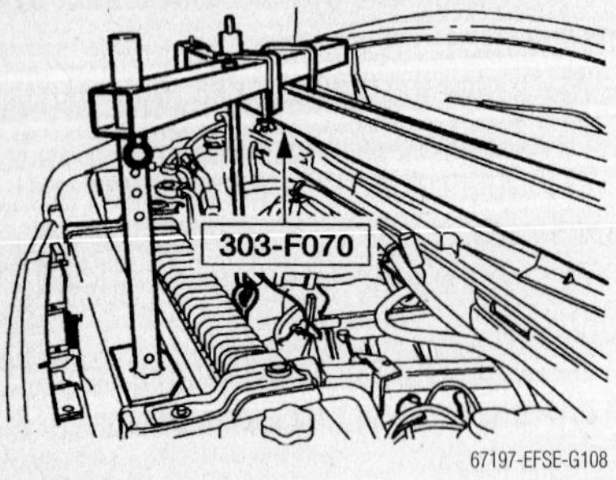

Install the support tools

8. Release the engine harness routing clip retainer.
9. Disconnect the hose from the heater outlet tube.
10. Remove the heater outlet tube studs.
11. Remove the heater outlet tube.
12. Remove the two transmission-to-engine bolts.
13. Assemble the lifting tools, 303-F047 and 303-F694, or equivalent.
14. Install the lifting tools.
15. Install the special tool and support the engine.
16. Raise and support the vehicle.
17. Drain the engine oil, and remove the oil filter.
18. Remove the four engine mount nuts.
19. Remove the flexplate inspection plate.
20. Lower the vehicle.
21. Using the Heavy Duty Engine Support, raise the engine 10¼ inches from the crankshaft pulley to the lower edge of the No. 1 crossmember.
22. Raise the vehicle.
23. Remove the bolts and partially lower the oil pan.
24. Remove the bolts retaining the oil pump screen cover and pickup tube and let the bolts drop into the oil pan.
25. Remove the oil pan at the rear of the engine.
26. Clean the oil pan thoroughly and inspect the oil pan gasket.
27. Clean the mating surfaces for the oil pan with silicone gasket remover and metal surface prep. Follow the directions on the packaging.

Installation

All engines

⁂ WARNING

Make sure the O-ring is in place and not damaged. A missing or damaged O-ring can cause foam in the lubrication system, low oil pressure and severe engine damage.

➡Clean and inspect the mating surfaces and install a new O-ring. Lubricate the O-ring with clean engine oil.

28. Install the oil pump screen and pickup tube. Torque the tube base bolts to 89 inch lbs. (10 Nm); the bracket bolt to 18 ft. lbs. (25 Nm).

➡If not secured within four minutes, the sealant must be removed and the sealing area cleaned. To clean the sealing area, use silicone gasket remover and metal surface prep. Follow the directions on the packaging. Failure to follow this procedure can cause future oil leakage.

29. Apply a bead of silicone gasket and sealant at the crankshaft rear seal retainer-to-cylinder block surface.
30. Apply a bead of silicone gasket and sealant at the engine front cover-to-cylinder block surface.
31. Position the pan. Tighten the bolts in three steps, in the sequence shown.
  a. Stage 1: Tighten to 2 Nm (18 inch lbs.).
  b. Stage 2: Tighten to 20 Nm (15 ft. lbs.).
  c. Stage 3: Tighten an additional 90 degrees.

32. Install a new oil filter.
33. Lower the vehicle.
34. Lower the engine and remove the special tools.
35. Raise the vehicle.
36. Install the four engine mount nuts. Torque to 66 ft. lbs. (90 Nm).
37. Install the flexplate inspection plate. Torque to 25 ft. lbs. (34 Nm).
38. Lower the vehicle.
39. Install the transmission-to-engine bolts. Torque to 44 ft. lbs. (60 Nm).

➡Do not reuse the O-ring seals. Lubricate the new O-ring seals with clean engine coolant before installing the heater outlet tube.

40. Insert the heater outlet tube over the new seals.

Romeo engines (4.6L)

41. Install the heater outlet tube studs.
42. Position the transmission fluid filler tube and install the nut.

Windsor engine (5.4L)

43. Install the heater outlet tube studs.
44. Position the transmission fluid filler tube and install the nut.

All engines

45. Connect the hose to the heater outlet tube.

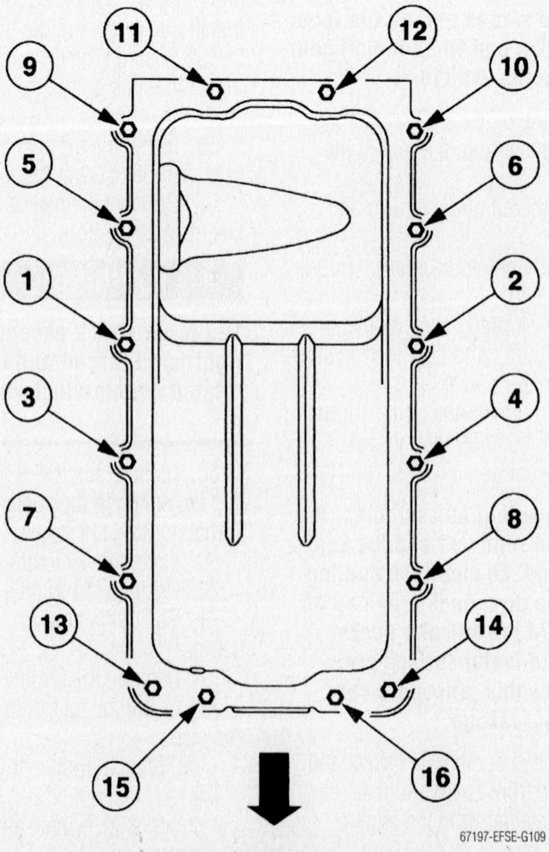

Oil pan torque sequence—E-Series w/4.6L or 5.4L engine

46. Install the wiring retainers in the heater outlet tube bracket.
47. Install the knock sensor electrical connector retainer.
48. Install the engine harness routing clip retainer.
49. Install the retainers and the shield.
50. Install the fan shroud and the engine cooling fan.
51. Install the intake manifold.

WARNING

The oil pump must be primed prior to starting the engine.

52. Fill the engine with clean engine oil.
53. Start the engine and check for leaks.

6.8L ENGINES

1. With the vehicle in **Neutral**, position it on a hoist.
2. Remove the intake manifold.
3. Remove the engine cooling fan and fan shroud.
4. Remove the retainers and the shield.
5. Remove the nut attaching the transmission oil level tube.
6. Disconnect the wiring harness retainer.
7. Remove the stud bolts and position the heater outlet tube aside.
8. Remove the two transmission-to-engine bolts.
9. Disconnect the oil fill tube from the right valve cover.
10. Remove the bolt and the oil fill tube.
11. Assemble the special tools.
12. Install the special tools.
13. Install the special tool and support the engine.
14. Drain the engine oil and remove the oil filter.
15. Remove the right and left nuts retaining the front engine support insulators to the front engine support bracket.
16. Remove the flywheel inspection plate.
17. Using the special tool, raise the engine.
18. Remove the bolts and partially lower the oil pan.
19. Remove the bolts retaining the oil pump screen cover and pickup tube and allow it to drop into the oil pan. Also remove the tube and spacer and allow it to drop into the oil pan.
20. Remove the oil pan.

WARNING

Do not use metal scrapers, wire brushes, power abrasive discs or other abrasive means to clean the sealing surfaces. These tools cause scratches and gouges, which make leak paths. Use a plastic scraping tool to remove all traces of sealant.

21. Clean the mating surfaces for the oil pan and the engine with silicone gasket remover and metal surface prep. Follow the directions on the packaging.

To install:

➡The oil pump screen and pickup tube must be in the oil pan when the oil pan is positioned in the vehicle.

22. Position the oil pan gasket and the oil pan in the vehicle from the rear of the engine.

WARNING

Make sure the O-ring is in place and not damaged. A missing or damaged O-ring can cause foam in the lubrication system, low oil pressure and severe engine damage.

➡Clean and inspect the mating surfaces and install a new O-ring. Lubricate the O-ring with clean engine oil.

23. Position the oil pump screen cover and pickup tube.
24. Install and tighten the spacer. Torque to 18 ft. lbs. (25 Nm).
25. Insert the oil pump screen pickup tube into the oil pump and position the support bracket over the spacer.
26. Install and tighten the bolts. Torque the tube base bolts to 89 inch lbs. (10 Nm); the bracket bolt to 18 ft. lbs. (25 Nm).
27. Apply a bead of silicone gasket and sealant at the crankshaft rear seal retainer-to-cylinder block surface.

➡If not secure within four minutes, the sealant must be removed and the sealing area cleaned. To clean the sealing area, follow the directions provided on the packaging of the silicone gasket remover and the metal surface prep. Failure to follow this procedure can cause future oil leakage.

28. Apply a bead of silicone gasket and sealant at the engine front cover-to-cylinder block surface.

➡Position the oil pan. Tighten the bolts in three steps.

29. Tighten the bolts in the sequence shown.
    a. Step 1: Tighten to 2 Nm (18 inch lbs.).
    b. Step 2: Tighten to 20 Nm (15 ft. lbs.).
    c. Step 3: Tighten an additional 90 degrees.
30. Lower the engine and remove the special tools.
31. Install the flywheel inspection plate.
32. Tighten the engine support insulators.

➡Do not reuse the O-ring seal.

33. Install the new O-ring seal and the heater outlet tube.
34. Lubricate the O-ring seal with engine coolant.
35. Hand-tighten the heater outlet tube upper stud and install the lower stud. Tighten the studs to 40 Nm (30 ft. lbs.).
36. Remove the nut attaching the transmission oil level tube.
37. Connect the wiring harness retainer.
38. Connect the oil fill tube to the right valve cover.
39. Install the oil fill tube support bracket bolt.
40. Install the transmission-to-engine bolts.
41. Install the shield and the retainers.
42. Install the intake manifold.
43. Fill the engine with clean engine oil.
44. Install the engine cooling fan and fan shroud.
45. Fill and bleed the engine cooling system.

Oil Pump

REMOVAL & INSTALLATION

4.2L Engine

1. Before servicing the vehicle, refer to the precautions in the beginning of this section.
2. Disconnect the negative battery cable.
3. Raise and support the front of the vehicle.
4. Drain the engine oil into a suitable container and dispose.
5. Remove the oil filter.
6. Remove the 6 oil pump bolts, then remove the oil pump drive gear, the oil pump driven gear, the oil pump O-ring and the oil pump itself. Discard the used oil pump O-ring.
7. Inspect the oil pump components for damage or excessive wear.
8. Check the oil pump face for warpage with a flat edge ruler. The face cannot

exhibit more than 0.00157 inch (0.04mm) of distortion.

9. Remove the plug over the oil pressure relief; valve.

10. Remove the oil pressure relief valve ball and spring, then clean the parts.

To install:

➡Lubricate the parts with clean engine oil before assembly.

11. Assemble the oil pressure relief valve ball and spring with a new plug.

12. Install the oil pump, along with a new O-ring, the oil pump driven gear and the drive gear. Install and tighten the 6 oil pump mounting bolts to the torque value specifications indicated in the illustration.

13. Apply a film of clean engine oil to the rubber O-ring on the new filter, then install the filter onto the filter mount.

14. Install the oil pan drain plug.

15. Lower the vehicle.

16. Fill the engine with the correct amount and type of new engine oil.

17. Connect the negative battery cable.

18. Start the engine and make certain that the oil light on the instrument panel extinguishes within 6–8 seconds after the engine starts.

4.6L and 5.4L Engines

1. Before servicing the vehicle, refer to the precautions in the beginning of this section.
2. Disconnect the negative battery cable.
3. Remove or disconnect the following:
   - Crankshaft sprocket
   - Oil pan
   - Bolts and the oil pump

To install:

4. Clean and inspect the mating surfaces.
5. Install or connect the following:
   - Oil pump, and the bolts loosely. Tighten the bolts in the sequence illustrated to 89 inch lbs. (10 Nm).
   - Oil pan
   - Crankshaft sprocket

6.8L Engine

1. Before servicing the vehicle, refer to the precautions in the beginning of this section.
2. Disconnect the negative battery cable.
3. Remove the engine front cover and crankshaft sprocket.
4. Drain the engine oil into a suitable container.
5. Remove the oil pan.
6. Remove the 3 oil pump mounting

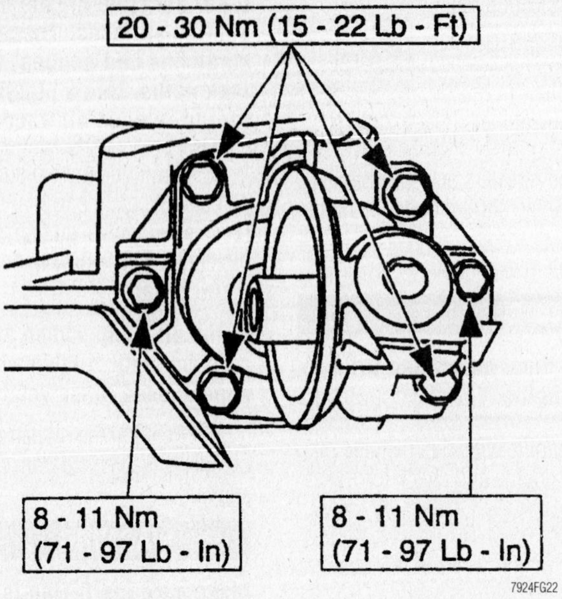

Tighten the oil pump mounting bolts to the specifications shown—4.2L engines

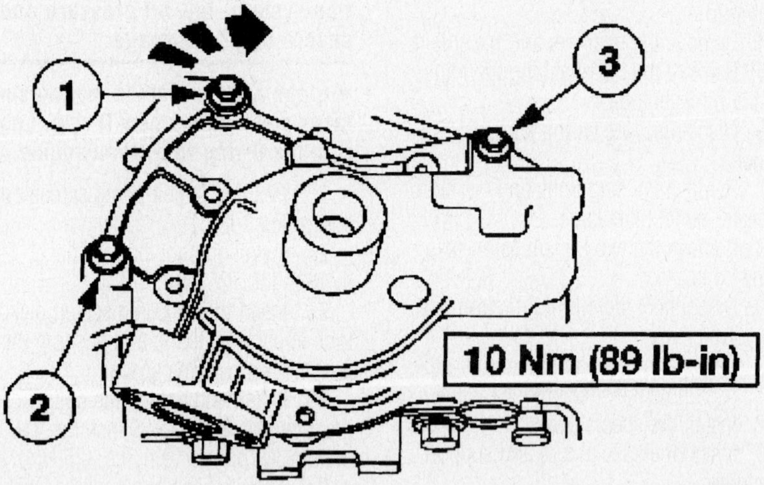

Tighten the oil pump mounting bolts in the sequence shown—4.6L and 5.4L engines

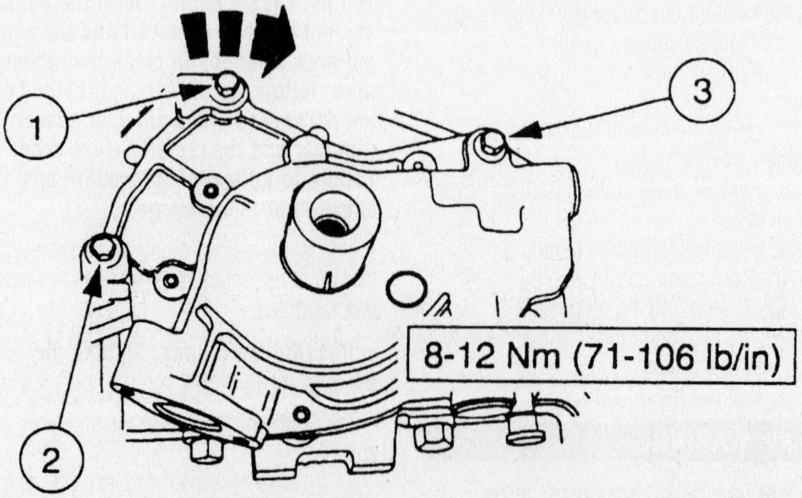

Be sure to tighten the oil pump mounting bolts in the sequence shown—6.8L engine

bolts, then remove the oil pump from the engine.

To install:

7. Clean and inspect the mating surfaces.
8. Install the oil pump and loosely install the oil pump mounting bolts. Tighten the bolts in the sequence shown to 71–106 inch lbs. (8–12 Nm).
9. Install the oil pan.
10. Install the crankshaft sprocket and timing chains.
11. Install the front cover.
12. Refill the engine oil with the recommended engine oil and amount.
13. Install the negative battery cable.

Rear Main Seal

REMOVAL & INSTALLATION

2001–02 Engines and 2003 Engines without Retainer Plate

If the crankshaft rear oil seal replacement is the only operation being performed, it can be done in the vehicle as detailed in the following procedure. If the oil seal is being replaced in conjunction with a rear main bearing replacement, the engine must be removed from the vehicle and installed on a work stand.

1. Before servicing the vehicle, refer to the precautions in the beginning of this section.
2. Disconnect the negative battery cable.
3. Remove the transmission from the vehicle.
4. Remove the flywheel/flexplate. If equipped, remove the crankshaft oil slinger from the crankshaft.
5. Use an awl to punch 2 holes in the crankshaft rear oil seal. Punch the holes on opposite sides of the crankshaft and just above the bearing cap-to-cylinder block split line.
6. Install a sheet metal screw in each hole. Use 2 small prybars to pry against both screws at the same time to remove the crankshaft rear oil seal. It may be necessary to place small blocks of wood against the cylinder block to provide a fulcrum point for the prybars. Use caution throughout this procedure to avoid scratching or otherwise damaging the crankshaft oil seal surface.
7. Clean the oil seal recess in the cylinder block and main bearing cap.

To install:

8. Clean, inspect and polish the rear oil seal rubbing surface on the crankshaft.
9. Coat the new oil seal and the crankshaft with a light film of engine oil.
10. Start the seal in the recess with the seal lip facing forward and install it with a seal driver. Keep the tool straight with the centerline of the crankshaft and install the seal until the tool contacts the cylinder block surface. Remove the tool and inspect the seal to be sure it was not damaged during installation.
11. If equipped, install the crankshaft oil slinger.
12. Position the flywheel on the crankshaft flange. Coat the threads of the flywheel attaching bolts with Loctite® and install the bolts. Tighten the bolts in sequence across from each other to 75–85 ft. lbs. (102–115 Nm).
13. Install the transmission, following the recommended procedure.
14. Install the negative battery cable.

2003 Engines with Retainer Plate and All 2004 Engines

1. Before servicing the vehicle, refer to the precautions in the beginning of this section.
2. Remove the oil pan.
3. Remove the flexplate.
4. Using a puller set, remove the crankshaft rear oil seal slinger.
5. Using a puller set, remove the crankshaft rear seal.
6. Remove the six bolts and the crankcase rear seal retainer.

To install:

***** CAUTION**

Do not use metal scrapers, wire brushes, power abrasive discs or other abrasive means to clean the sealing surfaces. These tools cause scratches and gouges which make leak paths. Use a plastic scraping tool to remove all traces of old sealant.

➡Clean the sealing surfaces with silicone gasket remover and metal surface prep.

7. Follow the directions on the packaging. Failure to follow this procedure can cause future oil leakage.
8. Clean and inspect the mating surface.

***** CAUTION**

Do not use metal scrapers, wire brushes, power abrasive discs or other abrasive means to clean the sealing surfaces. These tools cause scratches and gouges which make leak paths. Use a plastic scraping tool to remove all traces of old sealant.

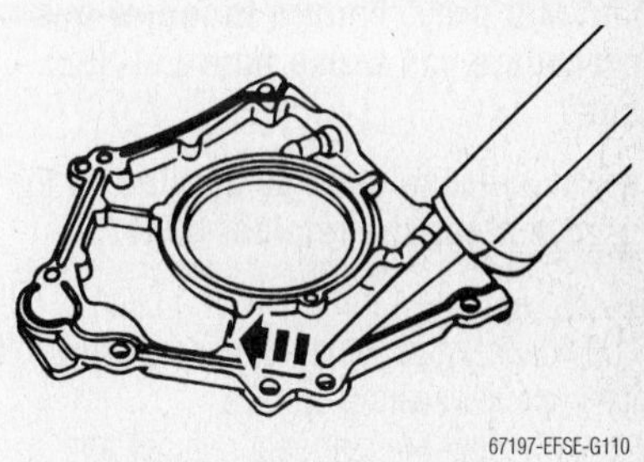
67197-EFSE-G110

The silicone must be applied on the groove along the retainer plate

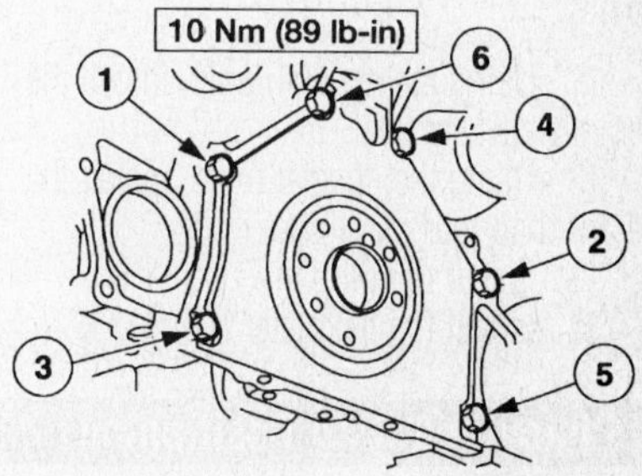

67197-EFSE-G111

Tighten the bolts in the sequence shown

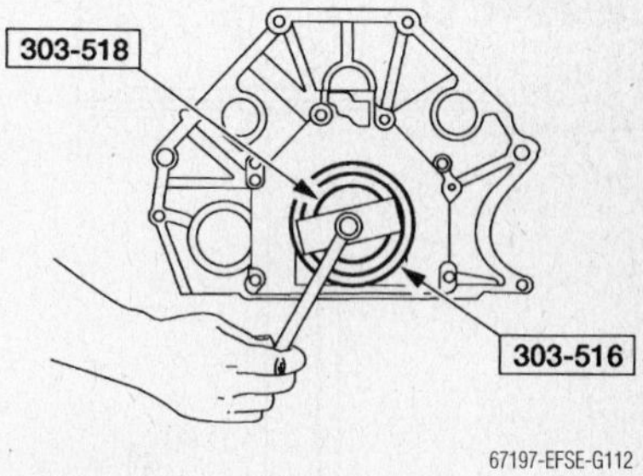

67197-EFSE-G112

Using the special tools, install the crankshaft rear seal

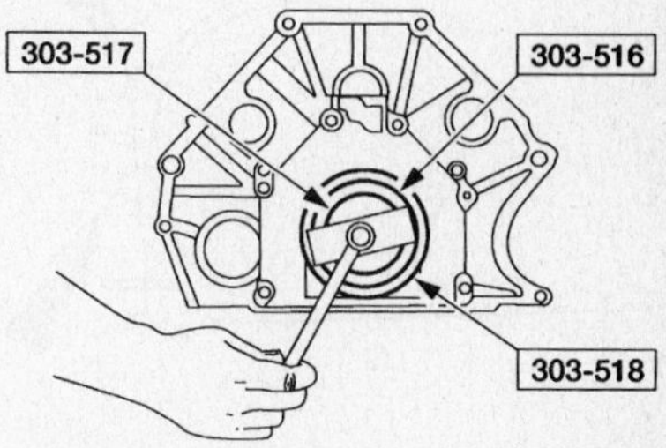

67197-EFSE-G113

Using the special tools, install the crankshaft rear oil slinger

➡**If the rear crankshaft seal retaining plate is not secured within four minutes, the sealant must be removed and the sealing area cleaned. To clean the sealing area, follow the directions provided on the packaging of the silicone gasket remover and the metal surfacer prep. Failure to follow this procedure can cause future oil leakage.**

➡**The silicone must be applied on the groove along the retainer plate.**

9. Apply a 4 mm (0.16 in.) bead of silicone gasket and sealant around the rear oil seal retainer sealing surface.
10. Install the rear seal retainer and loosely install the six bolts. Tighten the bolts in the sequence shown.

➡**Lubricate the inner lip of the rear crankshaft seal with clean engine oil.**

11. Using the special tools, install the crankshaft rear seal.
12. Using the special tools, install the crankshaft rear oil slinger.
13. Install the oil pan.
14. Install the flexplate.

Piston and Ring Positioning

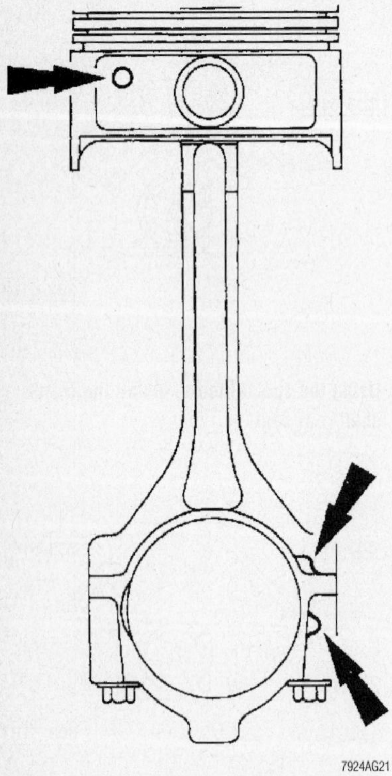

4.2L and 4.6L engines—piston and connecting rod front mark locations

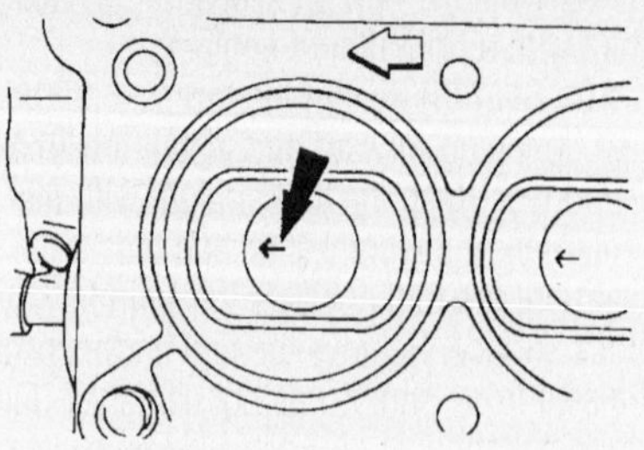

4.2L and 4.6L engines—piston-to-engine orientation

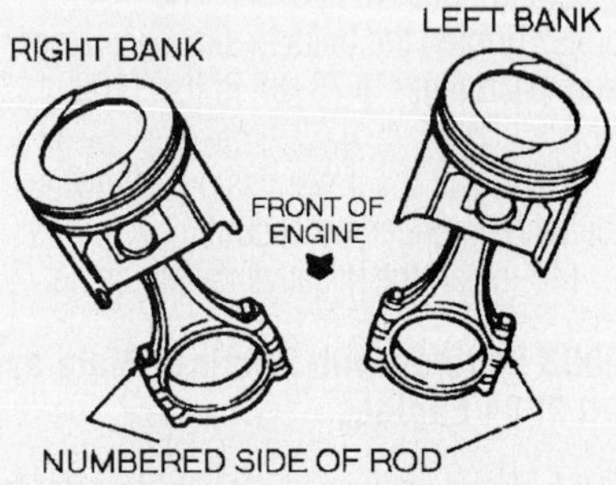

5.4L and 6.8L engines—piston and connecting rod assembly positioning

Timing Chain, Sprockets and Front Cover

REMOVAL & INSTALLATION

4.2L Engine

1. Before servicing the vehicle, refer to the precautions in the beginning of this section.
2. Remove or disconnect the following:
   - Negative battery cable
   - Accessory drive belt
   - Coolant
   - Radiator, fan blade assembly and fan shroud
   - Water pump
   - Exhaust Gas Recirculation (EGR) valve vacuum hose
   - EGR tube upper fitting
   - EGR valve and adapter assembly
   - Wiring harness from the heater water outlet tube
   - Heater water outlet bolt and position the outlet tube aside
   - Camshaft Position (CMP) sensor electrical harness connector and mark the position of the connector for proper installation.
3. Rotate the crankshaft until the Top Dead Center (TDC) timing mark lines up with the timing mark.

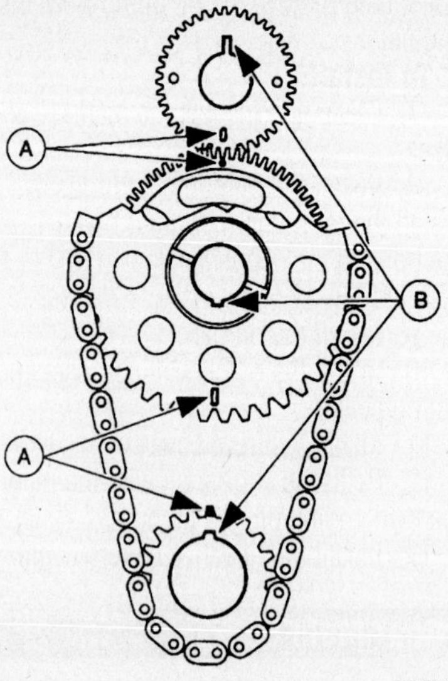

After removing the CMP drive gear, be sure the timing marks (A) and keyways (B) are aligned—4.2L engine

- The 2 bolts retaining the CMP and remove the CMP from the camshaft synchronizer.
- Camshaft synchronizer adjustment bolt and remove the camshaft synchronizer.

➡**The oil pump drive shaft may come out with the camshaft synchronizer.**

- Engine oil
- Crankshaft pulley and damper
- Engine oil pan
- The 2 engine front cover stud bolts, front cover bolt and cap screw
- Engine front cover and gasket off the dowels and discard the gasket.
- CMP sensor drive gear bolt and drive gear

4. Be sure the timing marks and keyways align.
5. Compress and install a retaining pin to hold the timing chain tensioner.
6. Slide both sprockets and timing chain forward and remove as an assembly.
7. Remove the 3 bolts retaining the timing chain tensioner and remove the tensioner.
8. Check the timing chain and sprockets for excessive wear. Replace if necessary.

To install:

9. Before installation, clean and inspect all parts. Clean the gasket material from the engine oil pan, cylinder block and engine front cover.
10. Place the timing chain tensioner in position and install the 3 retaining bolts. Tighten the bolts to 72–120 inch lbs. (8–14 Nm).

11. Verify that the balance shaft timing gears are in correct alignment.

12. Slide both sprockets and the timing chain onto the camshaft and crankshaft with the timing marks aligned. Install the CMP sensor drive gear and bolt. Tighten the bolt to 30–36 ft. lbs. (40–50 Nm).

13. Remove the retaining pin.

14. Inspect the engine front cover seal for wear or damage and replace if necessary.

15. Install or connect the following:

- Engine front cover gasket and front cover onto the guide studs
- The 2 engine front cover stud bolts, front cover bolt and cap screw. Tighten the bolts to 15–22 ft. lbs. (20–30 Nm).
- Engine oil pan
- Crankshaft damper and pulley

⁂ WARNING

A Synchro Positioning Tool must be used prior to installation. Failure to using this procedure will result in the fuel system being out of time, possibly causing engine damage.

16. Install Synchro Positioning Tool T89P-12200-A, or equivalent, on the camshaft synchronizer by rotating the tool until it engages the notch in the housing.

17. Install the camshaft synchronizer housing assembly so that the arrow on the tool is 54 degrees from the centerline of the engine.

18. Install the adjustment bolt and tighten to 15–22 ft. lbs. (21–30 Nm).

19. Remove the tool and position the CMP sensor. Install the 2 bolts and 40–70 inch lbs. (5–8 Nm) and install the CMP electrical harness connector.

20. Install or connect the following:

- Water pump
- Fan shroud, fan blade assembly and radiator
- Accessory drive belt

21. Fill the crankcase with the correct amount and type of engine oil.

22. Fill and bleed the engine cooling system.

23. Connect the negative battery cable.

24. Start the engine and check for coolant and oil leaks.

25. Road test the vehicle and check for proper engine operation.

DYNAMIC BALANCE SHAFT

1. Remove the timing chain.

2. Remove the engine dynamic balance shaft.

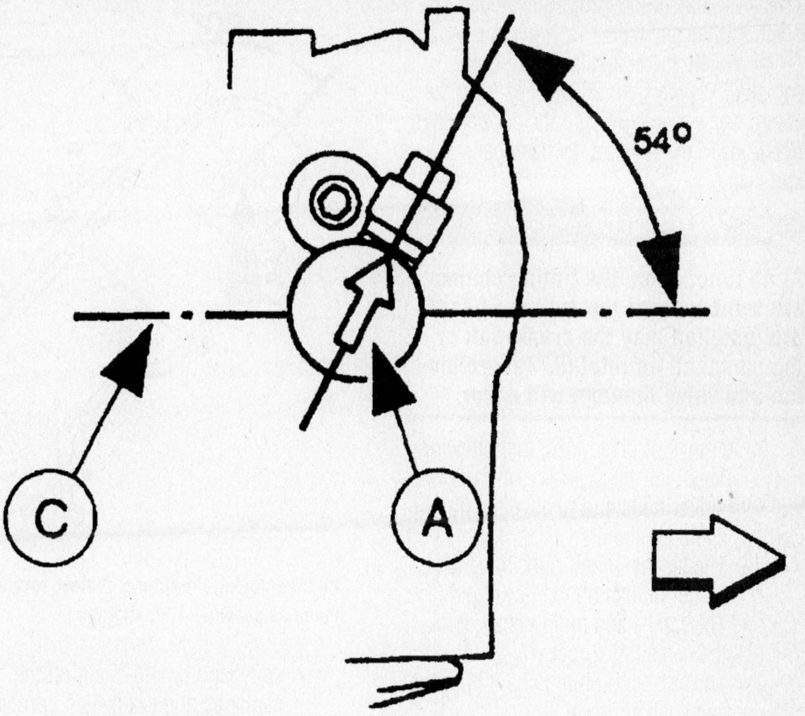

Install the camshaft synchronizer (A) in the orientation shown (white arrow points to front of engine)—4.2L engines

3. Remove the bolts.

4. Remove the balance shaft drive gear, thrust plate and engine dynamic balance shaft as an assembly.

To install:

➡If correctly aligned, the engine dynamic balance shaft keyway will be at 12 o'clock and the camshaft keyway will be at 6 o'clock on the camshaft.

5. Turn the camshaft so that the timing mark is at 12 o'clock and install the engine dynamic balance shaft assembly into the cylinder block. Turn the engine balance shaft driven gear so that the timing mark on the camshaft gear lines up with the timing mark on the engine balance shaft drive gear.

6. Install the bolts.

7. Install the timing chain.

4.6L Engine

1. Before servicing the vehicle, refer to the precautions in the beginning of this section.

2. Remove or disconnect the following:

- Negative battery cable
- Radiator, fan blade and fan shroud assembly
- Accessory drive belt
- Water pump pulley
- Electrical harness connectors from both ignition coils
- Both ignition coils with their brackets attached
- Left-hand and right-hand cylinder head covers
- The 2 upper power steering pump retaining bolts
- The 2 lower power steering pump retaining bolts and move the pump aside.
- Crankshaft Position (CKP) sensor electrical harness connector. Remove the retaining bolt and remove the CKP sensor.
- Engine oil
- The 4 oil pan-to-engine front cover retaining bolts
- Crankshaft damper retaining bolt and washer from the crankshaft
- Damper from the crankshaft
- Camshaft position (CMP) sensor retaining bolt and remove the CMP sensor
- Idler pulley bolt and remove the pulley
- The 3 belt tensioner retaining bolts and remove the tensioner
- The 8 engine front cover retaining bolts and the 7 nuts. Swing the top of the cover out off the dowel pins and remove the cover.
- Sensor ring from the crankshaft

3. Use Camshaft Positioning Tool T91P-6256-A and Camshaft Positioning

Adapters T92P-6256-A or equivalents, to position the camshaft.

4. Rotate the crankshaft until both camshaft keyways are 90 degrees from the cam cover surface. Be sure the copper links line up with the dots on the camshaft sprockets.

**** WARNING**

At no time, when the timing chains are removed and the cylinder heads are installed may the crankshaft or the camshaft be rotated. Severe piston and valve damage will occur.

5. Remove or disconnect the following:
- The 2 left-hand and right-hand tensioner bolts and remove the timing chain tensioners
- The left-hand and right-hand tensioner guides off the dowel pins
- The right-hand timing chain from the camshaft sprocket
- The left-hand timing chain from the camshaft sprocket
- The left-hand and right-hand timing chain guide bolts and remove the timing chain guides.

6. If necessary, remove the camshaft gear bolt and remove the camshaft gear.

To install:

7. Examine the timing chains, looking for the copper links. If the copper links are not visible, lay the chain on a flat surface and pull the chain taught until the opposite sides of the chain contact one another. Mark the links at each end of the chain and use these marks in place of the copper links.

➡If the engine jumped time, damage has been done to valves and possibly pistons and/or connecting rods. Any damage must be corrected before installing the timing chains.

8. Install or connect the following:
- Camshaft gears and tighten the retaining bolt to 81–95 ft. lbs. (110–130 Nm)
- Left-hand and right-hand timing chain guides and retaining bolts. Tighten the retaining bolts to 71–106 inch lbs. (8–12 Nm).
- Left-hand crankshaft sprocket with the tapered part of the sprocket facing away from the engine block

➡The crankshaft sprockets are identical. They may only be installed one way, with the tapered part of the sprockets facing each other. Ensure that the keyway and timing marks on the crankshaft sprockets are aligned.

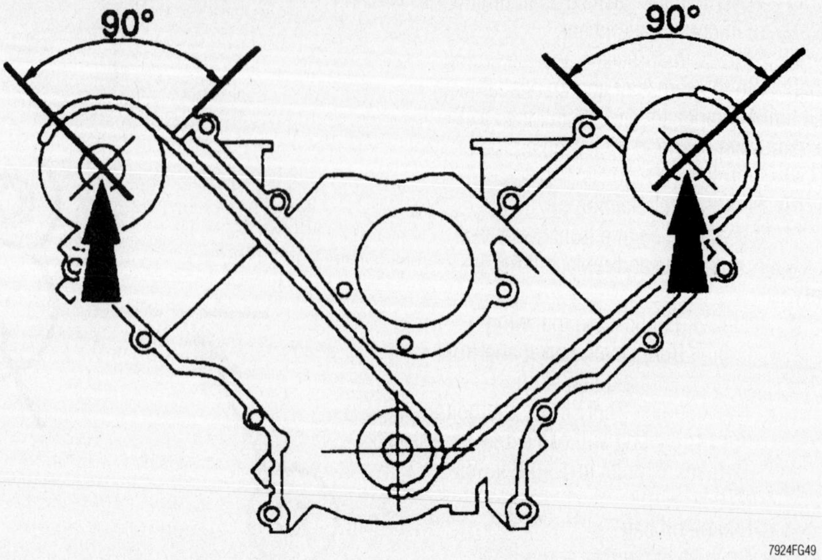

When removing the timing chains, rotate the crankshaft so that the camshaft keyways are positioned as shown—4.6L engines

- Left-hand timing chain on the camshaft and crankshaft sprockets. Be sure the copper links of the timing chain line up with the timing marks on both sprockets.
- Right-hand crankshaft sprocket with the tapered part of the sprocket facing the left-hand crankshaft sprocket
- Right-hand timing chain on the camshaft and crankshaft sprockets. Be sure the copper links of the timing chain line up with the timing marks on both sprockets.

9. It is necessary to bleed the timing chain tensioners before installation. Proceed as follows:

a. Place the timing chain tensioner in a soft-jawed vise.

b. Using a small pick or similar tool, hold the ratchet lock mechanism away from the ratchet stem and slowly compress the tensioner plunger by rotating the vise handle.

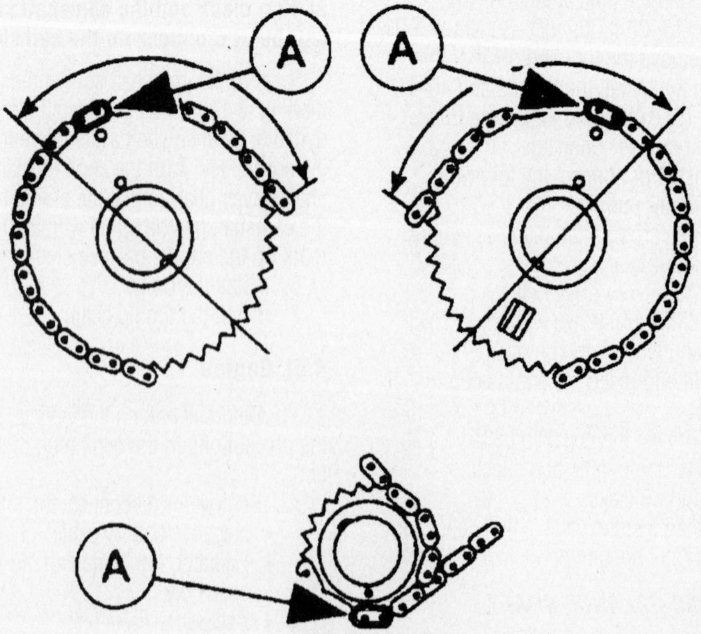

When installing the timing chains, make certain that the copper colored links (A) are aligned with the timing marks—4.6L engine

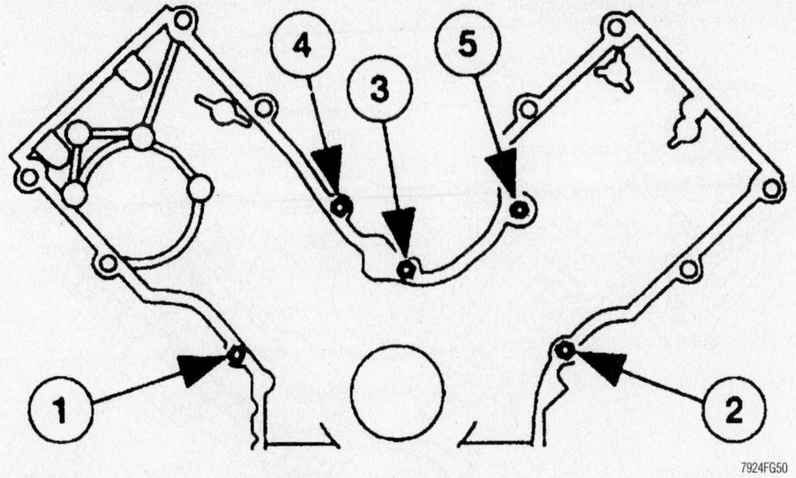

Tighten the first 5 front cover fasteners in the sequence shown—4.6L engine

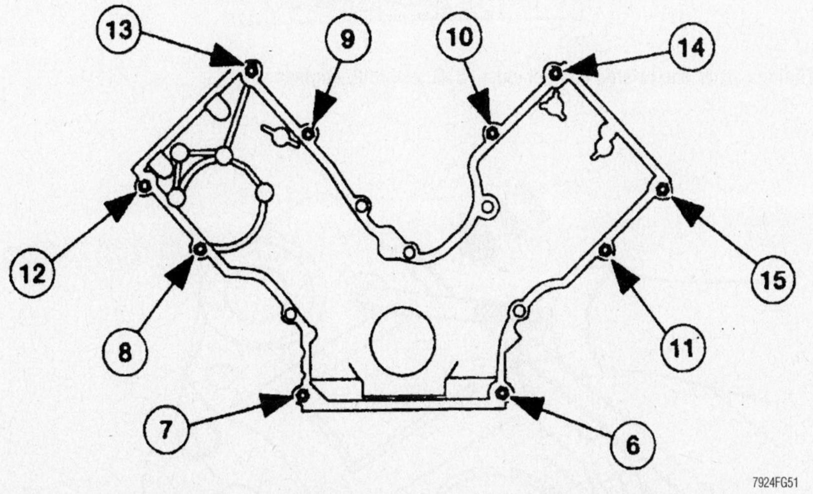

Continue tightening the remaining fasteners in the sequence shown here—4.6L engine

⁂ WARNING

The tensioner must be compressed slowly or damage to the internal seals will result.

c. Once the tensioner plunger bottoms in the tensioner bore, continue to hold the ratchet lock mechanism and push down on the ratchet stem until flush with the tensioner face.

d. While holding the ratchet stem flush to the tensioner face, release the ratchet lock mechanism and install a paper clip or similar tool in the tensioner body to lock the tensioner in the collapsed position.

e. The paper clip must not be removed until the timing chain, tensioner, tensioner arm and timing chain guide are completely installed on the engine.

10. Install or connect the following:
- Left-hand and right-hand timing chain tensioner guides on the dowel pins
- Left-hand and right-hand timing chain tensioners in position and install the retaining bolts. Tighten the bolts to 15–22 ft. lbs. (20–30 Nm).

11. Remove the retaining pins from the timing chain tensioners.

12. Remove Camshaft Positioning Tool T91P-6256-A and Camshaft Positioning Adapters T92P-6256-A or equivalents from the camshaft.

13. Install or connect the following:
- Crankshaft sensor ring on the crankshaft
- Apply a bead of silicone sealer along the cylinder head-to-cylinder block and the oil pan-to-cylinder block sealing surfaces
- Engine front cover carefully onto the dowel pins. Tighten the engine front cover bolts, in sequence, to 15–22 ft. lbs. (20–30 Nm).
- Idler pulley in position and install the retaining bolt. Tighten the bolt to 15–22 ft. lbs. (20–30 Nm).
- Drive belt tensioner and the 3 retaining bolts. Tighten the bolts to 15–22 ft. lbs. (20–30 Nm).
- CMP sensor and the retaining bolt. Tighten the bolt to 106 inch lbs. (12 Nm).
- The 4 oil pan-to-engine front cover bolts and tighten, in sequence, in 2 steps: Tighten the bolts to 15 ft. lbs. (20 Nm). Rotate the bolts an additional 60 degrees.
- CKP sensor and the retaining bolt. Tighten the bolt to 106 inch lbs. (12 Nm). Connect the CKP sensor electrical harness connector.
- Power steering pump and the 2 upper and 2 lower retaining bolts. Tighten the bolts to 15–20 ft. lbs. (20–30 Nm).
- Ignition coil and brackets on the engine front cover and install the bracket bolts. Tighten the bolts to 15–22 ft. lbs. (20–30 Nm).
- Ignition coil and capacitor electrical harness connectors
- CMP electrical harness connector
- Water pump pulley and tighten the bolts to 15–22 ft. lbs. (20–30 Nm)
- Radiator, fan blade and fan shroud assembly
- Accessory drive belt

14. Refill the engine oil with the recommended type of engine oil and the correct amount.

15. Install the negative battery cable.

16. Start the engine and check for leaks.

17. Road test the vehicle and check for proper engine operation.

5.4L and 6.8L Engines, Except Lightning

1. Before servicing the vehicle, refer to the precautions in the beginning of this section.

2. Remove or disconnect the following:
- Negative battery cable
- Radiator, fan blade and fan shroud assembly
- Accessory drive belt
- Water pump pulley
- Electrical harness connectors from both ignition coils
- Both ignition coils with their brackets attached
- The 2 upper power steering pump retaining bolts

- The 2 lower power steering pump retaining bolts and move the pump aside.
- Crankshaft Position (CKP) sensor electrical harness connector
- Retaining bolt and the CKP sensor.

3. Drain the engine oil.
- The 4 oil pan-to-engine front cover retaining bolts
- The crankshaft damper retaining bolt and washer from the crankshaft
- Crankshaft damper from the crankshaft
- Camshaft Position (CMP) sensor retaining bolt and remove the CMP sensor
- Idler pulley bolt and remove the pulley
- The 3 belt tensioner bolts and remove the tensioner
- The 8 engine front cover retaining bolts and the 7 nuts. Swing the top of the cover out off the dowel pins and remove the cover.
- Sensor ring from the crankshaft

4. Use Camshaft Positioning Tool T96T-6256-A or equivalent, to position the camshaft.

5. Rotate the crankshaft until both camshaft keyways are 90 degrees from the cam cover surface. Be sure the copper links line up with the dots on the camshaft sprockets.

⁂ WARNING

At no time, when the timing chains are removed and the cylinder heads are installed may the crankshaft or the camshaft be rotated. Severe piston and valve damage will occur.

6. Install Camshaft Holding Tool T96T-6256-B or equivalent, on the camshaft.

7. Remove or disconnect the following:
- 2 left-hand and right-hand tensioner bolts and the timing chain tensioners
- Left-hand and right-hand tensioner guides off the dowel pins
- Right-hand timing chain from the camshaft sprocket
- Left-hand timing chain from the camshaft sprocket
- Left-hand and right-hand timing chain guide bolts, and the timing chain guides

To install:

8. Examine the timing chains, looking for the copper links. If the copper links are not visible, lay the chain on a flat surface and pull the chain taught until the opposite sides of the chain contact one another. Mark the links at each end of the chain and use these marks in place of the copper links.

➡If the engine jumped time, damage has been done to valves and possibly pistons and/or connecting rods. Any damage must be corrected before installing the timing chains.

9. Install or connect the following:
- Left-hand and right-hand timing chain guides and retaining bolts. Tighten the retaining bolts to 89 inch lbs. (10 Nm).
- Left-hand crankshaft sprocket with the tapered part of the sprocket facing away from the engine block.

➡The crankshaft sprockets are identical. They may only be installed one way, with the tapered part of the sprockets facing each other. Ensure that the keyway and timing marks on the crankshaft sprockets are aligned.

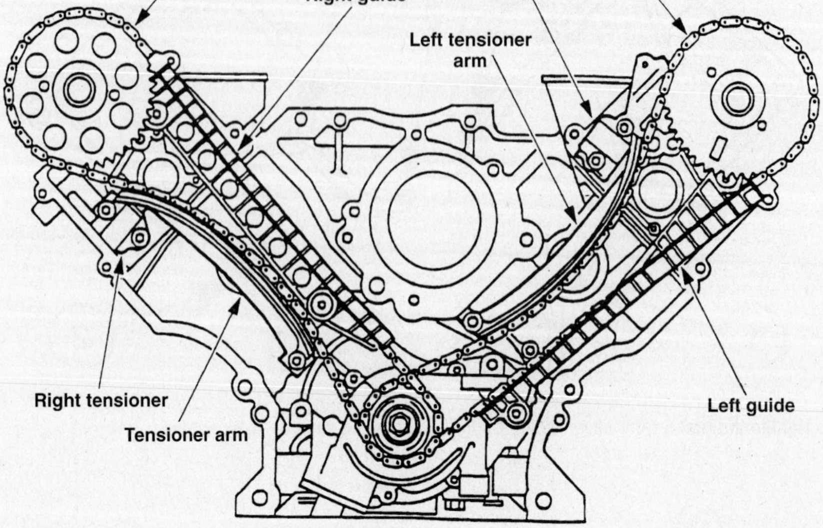

Timing chains and related components—5.4L and 6.8L engines

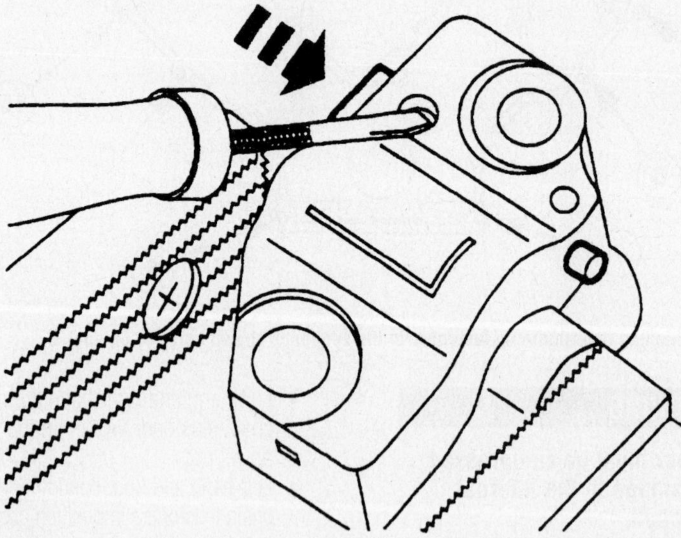

Compress the tensioner while holding the ratchet mechanism with a suitable tool—5.4L and 6.8L engines

7924FG92

If the copper links are not visible, mark one link on one end of the chain and 2 links on the opposite end of the chain—5.4L and 6.8L engines

- Left-hand timing chain on the camshaft and crankshaft sprockets. Be sure the 1 copper link aligns with the mark on the crankshaft sprocket and the 2 copper links align with the mark on the camshaft sprocket.
- Right-hand crankshaft sprocket with the tapered part of the sprocket facing the left-hand crankshaft sprocket
- Right-hand timing chain on the camshaft and crankshaft sprockets. Be sure the copper links of the timing chain line up with the timing marks on both sprockets.

10. It is necessary to bleed the timing chain tensioners before installation. Proceed as follows:

a. Place the timing chain tensioner in a soft-jawed vise.

b. Using a small pick or similar tool, hold the ratchet lock mechanism away from the ratchet stem and slowly compress the tensioner plunger by rotating the vise handle.

**** WARNING**

The tensioner must be compressed slowly or damage to the internal seals will result.

c. Once the tensioner plunger bottoms in the tensioner bore, continue to hold the ratchet lock mechanism and push down on the ratchet stem until flush with the tensioner face.

d. While holding the ratchet stem flush to the tensioner face, release the ratchet lock mechanism and install a paper clip or similar tool in the tensioner body to lock the tensioner in the collapsed position.

e. The paper clip must not be removed until the timing chain, tensioner, tensioner arm and timing chain guide are completely installed on the engine.

11. Install the left-hand and right-hand timing chain tensioner guides on the dowel pins.

12. Place the left-hand and right-hand timing chain tensioners in position and install the retaining bolts. Tighten the bolts to 15–22 ft. lbs. (20–30 Nm).

13. Remove the retaining pins from the timing chain tensioners.

14. Remove Cam Holding Tool T96T-6256-B or equivalent, from the camshaft.

15. Install the crankshaft sensor ring on the crankshaft.

16. Apply silicone gasket along the cylinder head-to-cylinder block and engine oil pan-to-cylinder block sealing surfaces.

17. Install or connect the following:

- Engine front cover carefully onto the dowel pins
- Engine front cover bolts, in sequence, in the following manner: bolts 1 through 7: 15–22 ft. lbs. (20–30 Nm); bolts 6 through 15: 29–40 ft. lbs. (40–55 Nm)
- Idler pulley in position and install the retaining bolt. Tighten the bolt to 15–22 ft. lbs. (20–30 Nm).
- Drive belt tensioner in position and install 3 retaining bolts. Tighten the bolts to 15–22 ft. lbs. (20–30 Nm).
- CMP sensor in position and install the retaining bolt. Tighten the bolt to 106 inch lbs. (12 Nm).
- Damper on the crankshaft. Ensure the crankshaft key and keyway are aligned.
- 4 front oil pan-to-engine front cover retaining bolts and tighten in sequence in 2 steps: Tighten the bolts to 15 ft. lbs. (20 Nm); Rotate the bolts an additional 60 degrees.
- CKP sensor in position and install the retaining bolt. Tighten the bolt to 106 inch lbs. (12 Nm).
- CKP sensor electrical harness connector
- Power steering pump with 2 upper and 2 lower retaining bolts. Tighten the bolts to 15–20 ft. lbs. (20–30 Nm).
- Ignition coil and brackets on the engine front cover and install the bracket bolts. Tighten the bolts to 15–22 ft. lbs. (20–30 Nm).
- Ignition coil and capacitor electrical harness connectors
- CMP electrical harness connector

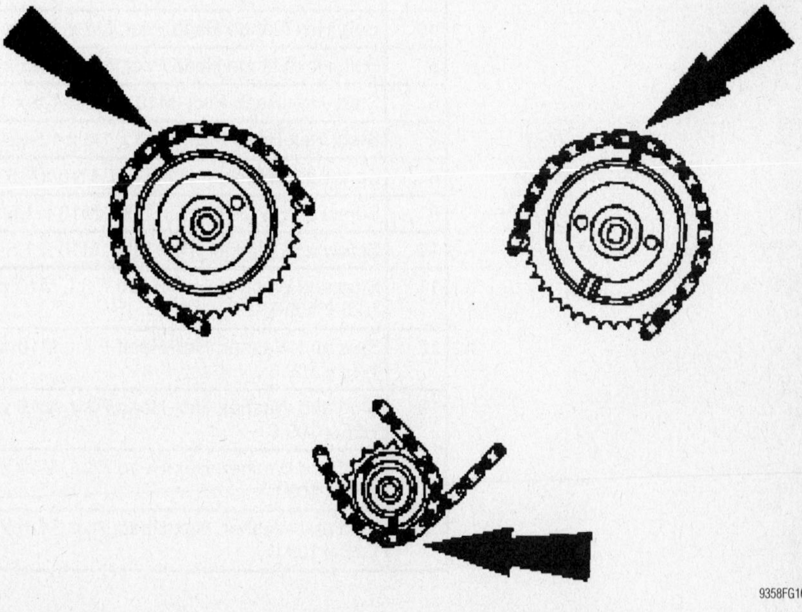
9358FG16

Check the alignment of the timing marks as shown——5.4L and 6.8L engines

- Water pump pulley and tighten the bolts to 15–22 ft. lbs. (20–30 Nm)
- Radiator
- Negative battery cable

18. Refill the engine oil with the recommended type of engine oil and the correct amount.
19. Start the engine and check for leaks.
20. Road test the vehicle and check for proper engine operation.

5.4L Supercharged Engine

1. Before servicing the vehicle, refer to the precautions in the beginning of this section.
2. Remove or disconnect the following:
   - Negative battery cable
   - Drive belt
   - Nuts and position the under vehicle shield aside
   - Starter motor

➡The auxiliary supercharger pulley has left-hand threads.

   - Pulley and brace assembly
   - Supercharger pulley adapter
   - Crankshaft pulley
   - Front cover oil seal

To install:

3. Clean the engine front cover seal bore, then lubricate the seal bore and the seal lip with clean engine oil.
4. Install the new front cover oil seal. Make sure the seal is installed evenly and straight.

➡The crankshaft pulley must be installed within 4 minutes after applying the silicone.

5. Apply silicone sealant to the woodruff key slot on the crankshaft pulley.

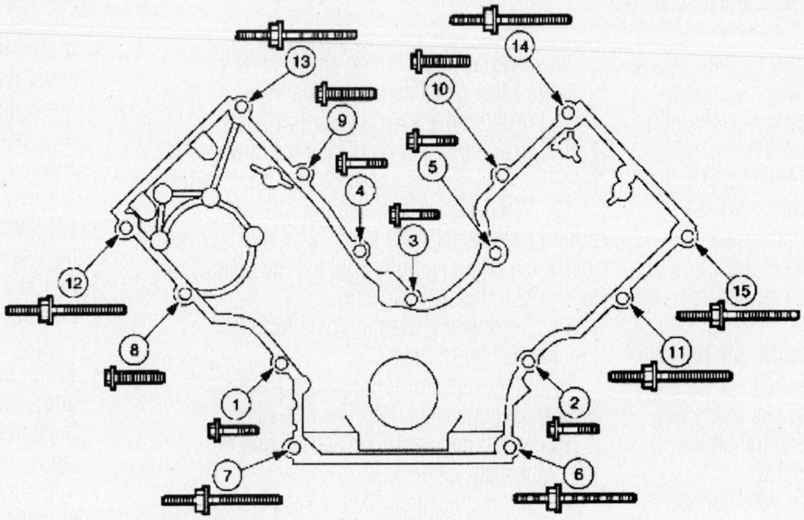

Item	Description
1	Bolt, Hex Flange Head Pilot, M8 x 1.25 x 53
2	Bolt, Hex Flange Head Pilot, M8 x 1.25 x 53
3	Bolt, Hex Flange Head Pilot, M8 x 1.25 x 53
4	Bolt, Hex Flange Head Pilot, M8 x 1.25 x 53
5	Bolt, Hex Flange Head Pilot, M8 x 1.25 x 53
6	Stud, Hex-Head Pilot, M10 x 1.5 x 1.5 x 103.1
7	Stud, Hex-Head Pilot, M10 x 1.5 x 1.5 x 103.1
8	Screw and Washer, Hex Pilot, M10 x 1.5 x 57.5
9	Screw and Washer, Hex Pilot, M10 x 1.5 x 57.5
10	Screw and Washer, Hex Pilot, M10 x 1.5 x 57.5
11	Stud and Washer, Hex-Head Pilot, M10 x 1.5 x M8 x 1.25 x 109.6
12	Stud and Washer, Hex-Head Pilot, M10 x 1.5 x M8 x 1.25 x 109.6
13	Stud and Washer, Hex-Head Pilot, M10 x 1.5 x M8 x 1.25 x 109.6
14	Stud and Washer, Hex-Head Pilot, M10 x 1.5 x M8 x 1.25 x 109.6
15	Stud and Washer, Hex-Head Pilot, M10 x 1.5 x M8 x 1.25 x 109.6

9358FG17

Install the timing cover and tighten the bolts to specification——5.4L and 6.8L engines

6. Using the special tool, install the crankshaft pulley.

7. Install the bolt and washer. Tighten the bolt in 4 stages.

- 66 ft. lbs. (90 Nm)
- Loosen the bolt
- 34–39 ft. lbs. (47–53 Nm)
- + 85–90 degrees.

8. Install the supercharger pulley adapter.

➡Coat the threads of the supercharger pulley with High Temperature Nickel Anti-Seize Lubricant F6AZ-9L494-AA.

➡The auxiliary supercharger pulley has left-hand threads.

9. Install the pulley and brace assembly.

10. Remove the special tool.

11. Install the starter motor.

12. Position the transmission cooler line clip and install the 2 nuts and 1 bolt.

13. Position the shield and install the nut.

14. Lower the vehicle.

15. Install the drive belt.

16. Connect the negative battery cable.

DIESEL ENGINE REPAIR

Alternator

REMOVAL & INSTALLATION

F-Series w/6.0L Engine

SINGLE ALTERNATOR

1. Before servicing the vehicle, refer to the precautions in the beginning of this section.

2. Disconnect the batteries.

3. Remove the cooling fan.

4. Rotate the accessory drive belt tensioner clockwise and remove the drive belt from the alternator pulley.

5. Disconnect the alternator electrical connectors.

6. Remove the bolts and the alternator.

7. To install, reverse the removal procedure. Torque the mounting bolts to 35 ft. lbs. (47 Nm).

DUAL ALTERNATORS

1. Before servicing the vehicle, refer to the precautions in the beginning of this section.

➡The following applies to the secondary alternator. For primary alternator, see the Single Alternator procedure.

2. Disconnect the batteries.

3. Remove the coolant fan.

4. Rotate the accessory drive belt tensioner clockwise and remove the drive belt from the alternator pulley.

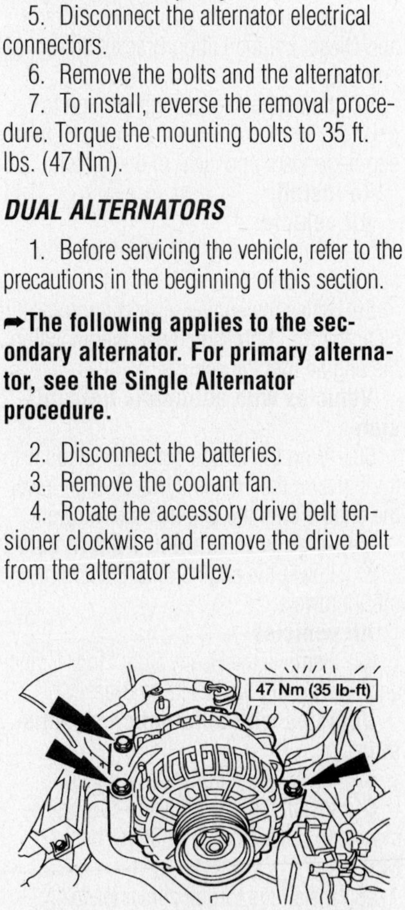

Single alternator mounting—6.0L w/single alternator

5. Remove the nut and position the alternator B+ cable aside.

6. Loosen the 2 clamps and remove the turbo-to-air cooler pipe.

7. Disconnect the alternator electrical connectors.

8. Remove the bolts and the alternator.

9. To install, reverse the removal procedure. Torque the mounting bolts to 35 ft. lbs. (47 Nm).

E-Series w/6.0L Engine

SINGLE ALTERNATOR

1. Before servicing the vehicle, refer to the precautions in the beginning of this section.

➡This procedure also applies to the primary alternator of the dual alternator system.

2. Disconnect the dual batteries.

3. Position the cowl wire harness aside.

4. Remove the bolt and position the manifold absolute pressure (MAP) sensor aside.

5. Remove the bolt and position the ground strap aside.

6. Remove the 2 bolts and position the power steering fluid reservoir aside.

7. Remove the 3 bolts and position the cowl wire harness aside.

8. Remove the 3 bolts, 2 nuts and the intercooler tube bracket.

9. Loosen the 2 clamps and remove the intercooler tube.

10. Rotate the accessory drive belt tensioner clockwise and remove the drive belt from the alternator pulley.

11. Remove the transmission dipstick tube bracket nut from the alternator stud bolt and position the bracket aside.

12. Remove the alternator.

➡Make sure to install the engine-to-body ground strap under the front alternator stud bolt.

13. To install, reverse the removal procedure. Torque the mounting bolts to 35 ft. lbs. (47 Nm).

DUAL ALTERNATORS

➡This procedure applies to the secondary alternator of the dual alternator system.

1. Disconnect the dual batteries.

2. Position the cowl wire harness aside.

3. Remove the bolt and position the manifold absolute pressure (MAP) sensor aside.

4. Remove the bolt and position the ground strap aside.

5. Remove the 2 bolts and position the power steering fluid reservoir aside.

6. Remove the 3 bolts and position the cowl wire harness aside.

7. Remove the 3 bolts, 2 nuts and the intercooler tube bracket.

8. Loosen the 2 clamps and remove the intercooler tube.

9. Rotate the accessory drive belt tensioner clockwise and remove the drive belt from the alternator pulley.

10. Disconnect the alternator electrical connectors.

11. Remove the bolts and the alternator.

12. To install, reverse the removal procedure. Torque the mounting bolts to 35 ft. lbs. (47 Nm).

7.3L Engine

SINGLE ALTERNATOR

1. Before servicing the vehicle, refer to the precautions in the beginning of this section.

2. Remove or disconnect the following:

- Negative battery cable
- Air cleaner, if necessary
- Drive belt from the alternator pulley
- Electrical harness connectors at the alternator assembly
- Alternator bolts and the alternator from the vehicle

To install:

3. Install or connect the following:

- Alternator in position
- Alternator retaining bolts and tighten to 35 ft. lbs. (47 Nm).

4. Install all remaining components in the reverse order of removal.

DUAL ALTERNATOR

1. Before servicing the vehicle, refer to the precautions in the beginning of this section.
2. Remove or disconnect the following:
   - Single (upper) alternator
   - Electrical harness connectors at the alternator assembly
   - Alternator bolts and the alternator from the vehicle

To install:

3. Install or connect the following:
   - Alternator in position
   - Alternator retaining bolts and tighten to 35 ft. lbs. (47 Nm).
   - Single (upper) alternator

Engine Assembly

REMOVAL & INSTALLATION

F-Series w/6.0L Engine

1. Before servicing the vehicle, refer to the precautions in the beginning of this section.
2. With the vehicle in **Neutral**, position it on a hoist.
3. Disconnect the left and right battery ground cables.

Vehicles with manual transmission

➡On vehicles equipped with manual transmissions, the transmission must be removed before the engine can be removed.

4. Remove the transmission.
5. Remove the clutch.

All vehicles

6. Remove the air cleaner assembly.
7. Remove the radiator.
8. Remove the charge air cooler.

Vehicles with A/C

9. Remove the A/C condenser assembly.

All vehicles

10. Remove the parking lamp and the headlamp assemblies.
11. Remove the radiator grille, the radiator grille opening panel, and the upper radiator core supports.
12. Remove the front bumper.
13. Disconnect the transmission cooler hoses.
14. Remove the bolts and the transmission oil cooler.
15. Remove the bolts and position the power steering cooler out of the way.
16. Remove the intake manifold.
17. Remove the accessory drive idler pulley and bolt.
18. Disconnect the heater hose at the coolant pump.
19. Remove the battery cable cover.
20. Remove the nut and position the cable out of the way.
21. Disconnect the left glow plug electrical connector and wire retainer.
22. Disconnect the camshaft position (CMP) sensor electrical connector.
23. Remove the clips and disconnect the fuel lines.
24. Remove the lower radiator hose clamp and the hose.
25. Remove the power steering upper mounting bolts.

➡Bolts need to be removed evenly.

26. Remove the bolts and position the power steering pump out of the way.
27. Remove the nut and the battery cable bracket.
28. Remove the ground stud and the ground cable.
29. If equipped, disconnect the injection control pressure (ICP) sensor electrical connector
30. Disconnect the glow plug electrical connectors.
31. Disconnect the right glow plug electrical connector and wire retainer.

Vehicles with A/C

32. Disconnect the A/C high pressure switch and the A/C clutch electrical connectors.
33. Disconnect the air conditioning manifold lines from the A/C compressor.
34. Remove the A/C compressor.

All vehicles

35. Disconnect the crankshaft position (CKP) sensor electrical connector and position the wiring aside.

Vehicles with automatic transmission

36. Remove the automatic transmission fluid indicator and tube.

All vehicles

37. Remove the ground strap at the back of the right head.
38. Remove the solenoid cap and disconnect the starter wiring.
39. Disconnect the block heater electrical connector.
40. Position the block heater and starter wiring harness out of the way.

Vehicles with automatic transmission

41. Remove the torque converter cover.
42. Remove the torque converter nuts.

All vehicles

43. Remove the bolts for the turbocharger adapter pipe.
44. Remove the motor mount nuts.
45. Loosen the nuts at the turbocharger adapter pipe flange.

Vehicles with automatic transmission

46. Remove the nine bell housing bolts.

All vehicles

47. Remove the turbocharger adapter pipe.
48. Remove the nuts and the fuel injector control module bracket.
49. Remove the left rear valve cover stud.
50. Remove the transmission cooler line bracket.
51. Secure the turbocharger outlet pipe.
52. Remove the manufacturer's lifting eye.
53. Install the engine lifting eye on the right cylinder head.
54. Install the front lifting brackets.

Vehicles with automatic transmission

55. Position a suitable jack under the transmission.

All vehicles

56. Install the Heavy Duty Floor Crane and Diesel Engine Lifting Bracket on the engine.
57. Raise the engine high enough to clear the No. 1 crossmember and pull the engine forward and clear of the vehicle.

To install:

All vehicles

58. With the vehicle in **Neutral**, position it on a hoist.
59. Raise the engine high enough to clear the No. 1 crossmember, then position the engine into the vehicle.

Vehicles with automatic transmission

60. Align the torque converter studs with the holes in the engine flywheel, then lower the engine onto the engine mount towers.

Vehicles with manual transmission

61. Lower the engine onto the engine mount towers.

All vehicles

62. Remove the Heavy Duty Floor Crane and the Diesel Engine Lifting Bracket.

Vehicles with automatic transmission

63. Remove the transmission jack.
64. Install the transmission-to-engine mounting bolts. Torque to 35 ft. lbs. (47 Nm).
65. Install the torque converter-to-flywheel retaining nuts. Torque to 26 ft. lbs. (35 Nm).
66. Install the flywheel housing cover.

All vehicles

67. Install the left and right side engine mount retaining nuts. Torque to 76 ft. lbs. (103 Nm).
68. Remove the engine lifting eye from the right side cylinder head.
69. Install the manufacturer's lifting bracket.
70. Remove the two engine lift adapters.
71. Position the turbocharger outlet pipe.
72. Install the transmission cooler tube bracket and nut.
73. Install the left rear valve cover stud.
74. Install the fuel injector control module bracket and nuts.
75. Position the turbocharger adapter pipe.

➡Do not tighten bolts at this time.

➡Apply anti-seize lubricant to the threads prior to installing the bolts.

76. Install the bolts for the turbocharger adapter pipe.
77. Position back the block heater and starter wiring.
78. Connect the block heater electrical connector.
79. Connect the starter wiring and install the solenoid cap.
80. Connect the ground strap on the right head and install the bolt.

Vehicles with automatic transmission

81. Install the automatic transmission fluid indicator and tube.

All vehicles

82. Position back the wiring and connect the crankshaft position (CKP) sensor electrical connector.

Vehicles with A/C

83. Install the A/C compressor. Torque to 18 ft. lbs. (25 Nm).
84. Position the A/C compressor manifold and install the bolt. Torque to 15 ft. lbs. (21 Nm).
85. Connect the A/C high pressure switch and the clutch electrical connectors.

All vehicles

86. Connect the right glow plug electrical connector and wire retainer.
87. Connect the glow plug module electrical connectors.
88. If equipped, connect the injection control pressure (ICP) sensor electrical connector.
89. Connect the ground cable and the ground stud.
90. Install the battery cable bracket and the nut.

➡The lower bolts need to be installed evenly.

91. Position the power steering pump and install the lower bolts. Torque to 18 ft. lbs. (25 Nm).
92. Install the power steering pump upper bolts. Torque to 18 ft. lbs. (25 Nm).
93. Install the lower radiator hose and clamp.
94. Connect the fuel lines and install the clips.
95. Connect the camshaft position (CMP) sensor electrical connector.
96. Connect the left glow plug electrical connector and wire retainer.
97. Connect the battery crossover cable.
98. Install the battery cable cover.
99. Connect the heater hose at the coolant pump.
100. Install the accessory drive idler pulley and bolt. Torque to 35 ft. lbs. (47 Nm).
101. Install the intake manifold.
102. Tighten the turbocharger adapter pipe at the flanges. Torque to 20 ft. lbs. (27 Nm).
103. Tighten the nuts at the turbocharger adapter pipe flange. Torque to 35 ft. lbs. (47 Nm).
104. Position and install the power steering cooler and bolts.
105. Install the transmission oil cooler.
106. Connect the transmission cooler hoses.
107. Install the front bumper.
108. Install the upper radiator core support, radiator grill opening panel and the radiator grill.
109. Install the headlamp and the parking lamp assemblies.

Vehicles with A/C

110. Install the A/C condenser assembly.

All vehicles

111. Install the charge air cooler.
112. Install the radiator.
113. Install the air cleaner assembly.

Vehicles with manual transmission

114. Install the clutch assembly.
115. Install the transmission.

All vehicles

116. Fill the motor with clean engine oil.
117. Connect the left and right battery cables.
118. Fill the cooling system.

Vehicles with automatic transmission

119. Check and fill the automatic transmission.

E-Series w/6.0L Engine

1. Before servicing the vehicle, refer to the precautions in the beginning of this section.
2. With the vehicle in **Neutral**, position it on a hoist.
3. Disconnect the battery ground cable.
4. Remove the A/C condenser assembly.
5. Remove the front bumper.
6. Remove the radiator grille support.
7. Remove the bolts, disconnect and remove the air bag sensor. Position the headlamp wiring harness aside.

➡Mark the latch before removal.

8. Remove the bolts and hood latch.
9. Remove the bolts and the radiator core supports.
10. Remove the intake manifold.
11. Remove the right fan stator stand-off. Remove the bolt and position the cable aside.
12. Remove the bolt and disconnect the A/C manifold. Disconnect the A/C electrical connector.
13. Remove and discard the O-ring seals.
14. Cap or plug the A/C openings as needed.
15. Disconnect the A/C pressure switch and remove the A/C manifold from the tee block.
16. Remove and discard the O-ring seals.
17. Cap or plug the A/C openings as needed.
18. Remove the bolts and A/C compressor.
19. Disconnect the heater hose. Position the heater hose and tube aside.
20. Remove the bolt and idler pulley.
21. Disconnect the engine coolant fill hose from the retaining clip. Loosen the clamp and remove the engine coolant fill hose.
22. Disconnect the vacuum pump hose and remove the engine coolant fill hose clip.

➡The bolt behind the pulley is accessed thorough the pulley with an extension and a short socket.

23. Remove the bolts and the vacuum pump.
24. Remove the power steering upper mounting bolt.
25. Remove the left fan stator stand-off.

➡The front bolt will remain in the power steering pump.

26. Remove the bolts and position the power steering pump aside.
27. Remove the lower radiator hose.
28. Disconnect the fuel lines.
29. Disconnect the crankshaft position

(CKP) sensor electrical connector and retaining clips.

30. Disconnect the injection control pressure (ICP) sensor electrical connector.

31. Disconnect the glow plug electrical connector and wire retainers.

32. Disconnect the camshaft position sensor electrical connector and retaining clips.

33. Disconnect the two electrical connectors and wire retainers at the transmission.

34. Disconnect the glow plug module electrical connectors.

⁂ WARNING

Do not use power tools when removing the nut.

35. Position back the boot and disconnect the B+ wire.

36. Disconnect the exhaust back pressure sensor electrical connector and push-pin retainer.

37. Disconnect the left glow plug electrical connector.

38. Disconnect the two electrical connectors.

39. Disconnect the two powertrain control module (PCM) electrical connectors and position the engine wiring harness aside.

40. Remove the bolts and the oil filter assembly.

41. Remove the bolt at the transmission filter housing.

42. Remove the retaining nut and remove the transmission cooling tubes.

43. Remove the torque converter cover.

44. Remove and discard the torque converter nuts.

45. Disconnect the block heater electrical connector.

46. Remove the retaining nut and position the cable bracket aside.

47. Remove the starter solenoid protective cap.

48. Disconnect the starter motor electrical connections.

49. Remove the bolts and the starter motor.

50. Remove the cylinder block drain plugs and drain the coolant from the block.

➡Apply clean engine oil to the O-ring seal prior to installing.

51. Install the cylinder block drain plugs.

52. Remove the motor mount nuts.

53. Remove the lower engine-to-transmission bolts.

54. Remove the mounting bolts from the lifting bracket and remove the tubes.

55. Disconnect the glow plug wire retainer. Remove the retaining nut and the oil indicator and tube.

56. Remove and discard the O-ring seal.

57. Disconnect the exhaust back pressure tube at the exhaust manifold.

58. Remove the retaining nuts and glow plug module.

➡Power stud removed for clarity.

59. Remove the retaining nuts and the glow plug module mounting bracket.

➡Mark the location of the studs prior to removal.

60. Remove the bolts and the left valve cover.

61. Remove the valve cover gasket.

62. Cover the cylinder head after the valve cover is removed.

63. Remove the ICP sensor.

64. Plug or cap the opening as needed.

➡Do not tighten the bolts at this time.

65. Install the special tools.

66. Remove the turbocharger exhaust pipe.

67. Remove the factory installed engine lifting bracket.

68. Install the special tool.

69. Install the engine lifting attachment and lifting crane.

70. Remove the oil pan drain plug and drain the engine oil.

➡It may be necessary to raise the engine for oil pan removal. Disconnect the linkage from the engine lifting attachment to raise the engine.

71. Remove the bolts and position back the oil pan until the oil pick up tube bolts are accessible.

72. Remove the bolts and let the oil pick-up tube go into the oil pan. Remove the oil pan.

73. Remove the press-in-place gasket and discard.

74. Remove and discard the oil pick-up tube O-ring seal.

75. Position a suitable transmission jack under the transmission.

76. Remove the upper engine-to-transmission bolts.

77. Raise the engine. Remove the four bolts and left motor mount.

78. Remove the bolts and the right motor mount.

➡The engine must be moved to the driver's side for removal.

79. Remove the engine from the vehicle.

To install:

80. With the vehicle in **Neutral**, position it on a hoist.

➡The engine will need to be to the left of the center line to install.

81. Raise the engine high enough to clear the No.1 crossmember, and position the engine into the vehicle.

82. Position the motor mounts in the vehicle.

83. Align the torque converter studs with the holes in the flywheel and push the engine in. Install the top engine-to-transmission bolts. Torque to 35 ft. lbs. (47 Nm).

84. Lower the engine until it is just above the motor mounts. Install the retaining bolts in the right motor mount. Torque to 59 ft. lbs. (80 Nm).

85. Install the four retaining bolts in the left motor mount. Torque to 59 ft. lbs. (80 Nm).

86. Remove the transmission jack.

87. Install a new O-ring seal on the oil pick-up tube and position the oil pick-up tube in the oil pan.

88. Install a new press-in-place gasket into the upper oil pan.

89. Position the oil pan in the vehicle.

90. Install the oil pick-up tube and bolts.

91. Position the oil pan and install the bolts.

92. Clean and inspect the oil pan drain plug and gasket. Install new, if necessary.

93. Install the oil pan drain plug.

94. Lower the engine. Remove the lifting crane and the engine lifting attachment.

95. Remove the special tool.

96. Install the factory engine lifting bracket.

97. Loosely install the turbocharger exhaust pipe.

98. Remove the special tools.

➡Remove the plug or cap as needed.

99. Install the injection control pressure (ICP) sensor.

➡Remove the protective covering from the cylinder head.

100. Install the left valve cover gasket, valve cover and bolts.

101. Install the glow plug module mounting bracket and retaining nuts.

102. Install the glow plug module and retaining nuts.

103. Connect the exhaust back pressure tube at the exhaust manifold.

➡Install a new O-ring seal and coat with clean engine oil.

104. Install the oil indicator and tube. Install the retaining nut. Connect the glow plug wire retainer.

105. Position the fuel tubes and install the mounting bolts for the fuel tube bracket.

106. Install the lower engine-to-transmission bolts. Torque to 35 ft. lbs. (47 Nm).

➡The top retaining nut must be tightened first.

107. Install the motor mount nuts. Torque to 66 ft. lbs. (90 Nm).

108. Install the starter motor and bolts.

109. Connect the starter motor electrical connections.

110. Install the starter motor solenoid protective cap.

111. Position the cable bracket and install the retaining nut.

112. Connect the block heater electrical connector.

113. Install the new torque converter nuts. Torque to 26 ft. lbs. (35 Nm).

114. Install the torque converter cover.

➡Inspect the O-ring seals. Install new seals, if necessary.

115. Position the transmission cooler tubes and install the retaining nut.

116. Position the tube lock and install the retaining bolt.

117. Install the oil filter assembly and bolts. Torque to 18 ft. lbs. (25 Nm).

118. Position the engine wiring harness and connect the powertrain control module (PCM) electrical connectors.

119. Connect the two electrical connectors.

120. Connect the left glow plug electrical connector.

121. Connect the push pin retainer and exhaust back pressure sensor electrical connector.

WARNING

Do not use power tools when installing the nut. Use care not to overtighten the nut during installation.

122. Connect the B+ wire and reposition the boot.

123. Connect the two electrical connectors and wire retainers at the transmission.

124. Connect the glow plug module electrical connectors.

125. Connect the camshaft position sensor electrical connector and retaining clips.

126. Position the wiring and connect the glow plug electrical connector. Connect the ICP sensor electrical connector.

127. Connect the crankshaft position (CKP) sensor electrical connector and retaining clips.

128. Connect the fuel lines and install the retaining clips.

129. Install the lower radiator hose.

130. Position the power steering pump and install the bolts. Torque to 18 ft. lbs. (25 Nm).

131. Install the left fan stator stand-off.

132. Install the power steering upper mounting bolt. Torque to 18 ft. lbs. (25 Nm).

➡The bolt behind the pulley is accessed through the pulley with an extension and a short socket.

133. Install the vacuum pump and bolts. Torque to 35 ft. lbs. (47 Nm).

134. Connect the vacuum pump hose and install the engine coolant fill hose clip.

135. Install the engine coolant fill hose and tighten the clamp. Connect the engine fill hose into the retaining clip

136. Install the idler pulley and bolt. Torque to 35 ft. lbs. (47 Nm).

137. Position the heater hose back and connect.

138. Install the A/C compressor and bolts. Torque to 18 ft. lbs. (25 Nm).

➡Remove the caps or plugs as needed. Install new O-ring seals and lubricate with PAG oil.

139. Install the A/C manifold at the tee block. Torque to 35 ft. lbs. (47 Nm).Connect the high pressure switch electrical connector.

➡Remove the caps or plugs as needed. Install new O-ring seals and lubricate with PAG oil.

140. Position the A/C manifold and install the bolt. Connect the A/C clutch electrical connector.

141. Position back the cable and install the bolt. Install the right fan stator stand-off.

142. Install the intake manifold.

143. Install the radiator core supports and bolts.

144. Position the hood latch and install the bolts.

145. Position the headlamp wiring harness and connect the air bag sensor. Install the bolts for the air bag sensor.

146. Install the valance panel, radiator grille support and grille opening panel reinforcement.

147. Install the front bumper.

148. Install the A/C condenser assembly.

149. Fill the engine with clean engine oil.

150. Connect the battery ground cable.

151. Check and top off the transmission fluid level.

7.3L Engine

CAUTION

The fuel system remains under pressure, even after the engine has been turned OFF. The fuel system pressure must be relieved before disconnecting any fuel lines. Failure to do so may result in fire and/or personal injury.

1. Before servicing the vehicle, refer to the precautions in the beginning of this section.
2. Remove or disconnect the following:
   - Hood
   - Coolant
   - Engine oil
   - Negative battery cable
   - Air cleaner and intake duct assembly
   - Upper grille support bracket
   - Upper air conditioning condenser mounting bracket
   - Refrigerant, using approved equipment to remove the condenser
   - Radiator fan shroud halves
   - Fan and clutch assembly
   - Radiator hoses and the transmission cooler lines, if equipped
   - Condenser. Cap all openings immediately!
   - Radiator
   - Power steering pump and position it out of the way
   - Fuel supply line heater and alternator wires at the alternator
   - Oil pressure sending unit wire at the sending unit
   - Sender from the firewall and lay it on the engine
   - Accelerator cable and the speed control cable, if equipped, from the injection pump
   - Cable bracket with the cables attached, from the intake manifold and position it out of the way
   - Transmission kickdown rod from the injection pump, if equipped
   - Main wiring harness connector from the right side of the engine and the ground strap from the rear of the engine
   - Fuel system pressure
   - Fuel return hose from the left rear of the engine
   - The 2 upper transmission-to-engine attaching bolts
   - Heater hoses
   - Water temperature sender wire
   - Overheat light switch wire and position the wire out of the way

- Battery ground cables from the front of the engine and the cables from the starter
- Fuel inlet line, and plug the fuel line at the fuel pump.
- Exhaust pipe at the exhaust manifold
- Engine insulators from the no. 1 crossmember
- Flywheel inspection plate and the 4 converter-to-flywheel attaching nuts, if equipped with automatic transmission

3. Support the transmission on a jack.
4. Remove the 4 lower transmission attaching bolts.
5. Attach an engine lifting sling and remove the engine from the vehicle.

To install:

6. Lower the engine into vehicle.
7. Align the converter to the flexplate and the engine dowels to the transmission.
8. Install the engine mount bolts and tighten them to 80 ft. lbs. (109 Nm).
9. Remove the engine lifting sling.
10. Install the 4 lower transmission attaching bolts. Tighten the bolts to 65 ft. lbs. (88 Nm).
11. Remove transmission jack.
12. Raise and support the front end.
13. Install or connect the following:
    - 4 converter-to-flywheel attaching nuts, if equipped with an automatic transmission. Tighten the nuts to 34 ft. lbs. (47 Nm).
    - Flywheel inspection plate. Tighten the bolts to 60–90 inch lbs. (6.7–10 Nm).
    - Exhaust pipe at the exhaust manifold
    - Fuel inlet line
    - Battery ground cables to the front of the engine
    - Starter cables at the starter
    - Overheat light switch wire
    - Water temperature sender wire
    - Heater hoses
    - The 2 upper transmission-to-engine attaching bolts. Tighten the bolts to 65 ft. lbs. (88 Nm).
    - Fuel return hose at the left rear of the engine
    - Main wiring harness connector at the right side of the engine and the ground strap from the rear of the engine
    - Transmission kickdown rod at the injection pump, if equipped
    - Accelerator cable and the speed control cable, if equipped, at the injection pump
    - Cable bracket with the cables attached, to the intake manifold
    - Oil pressure sending unit
    - Oil pressure sending unit wire at the sending unit
    - Fuel supply line heater and alternator wires at the alternator
    - Power steering pump
    - Radiator
    - Condenser
    - Radiator hoses and the transmission cooler lines, if equipped
    - Fan and clutch assembly
    - Radiator fan shroud halves
    - Upper grille support bracket and upper air conditioning condenser mounting bracket
    - Air cleaner and intake duct assembly
14. Refill the engine oil with the recommended type of engine oil and the correct amount.
15. Connect the negative battery cable.
16. If equipped with air conditioning, charge the system.
17. Fill the cooling system.
18. Install the hood.

Water Pump

REMOVAL & INSTALLATION

F-Series w/6.0L Engine

1. Before servicing the vehicle, refer to the precautions in the beginning of this section.
2. With the vehicle in **Neutral**, position it on a hoist.
3. Remove the radiator and shroud assembly.
4. Disconnect the cooling fan electrical connector. Unclip and position the fan wiring aside.

➡Use a hole in the fan hub to prevent the fan from turning.

5. Using the hub tool, loosen the cooling fan and clutch.
6. Remove the cooling fan assembly.
7. Remove the stator bolts and the stator.

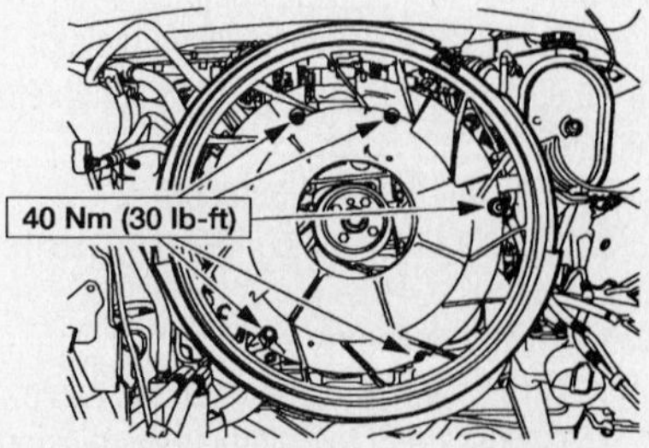

Remove the stator bolts and the stator

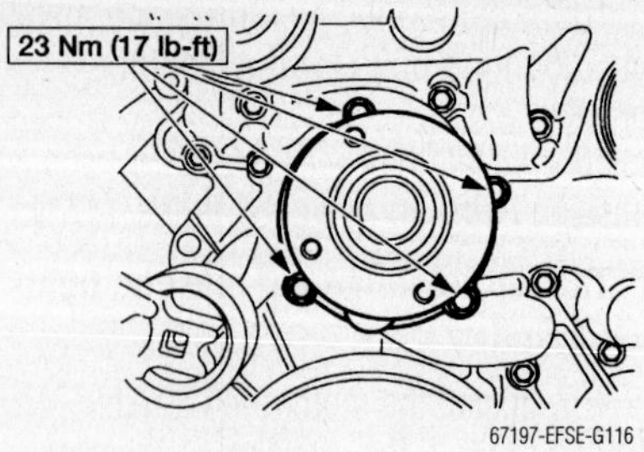

Water pump mounting

8. Rotate the accessory drive belt tensioner clockwise and position the drive belt aside.
9. Remove the bolts and the coolant pump pulley.
10. Remove the bolts and the coolant pump.
11. Remove and discard the O-ring seal.
12. Clean and inspect the coolant pump mounting.
13. To install, reverse the removal procedure.
14. Observe the following torques:
    - Stator bolts: 30 ft. lbs. (40 Nm)
    - Pulley bolts: 23 ft. lbs. (31 Nm)
    - Water pump bolts: 17 ft. lbs. (23 Nm)

E-Series w/6.0L Engine

1. Before servicing the vehicle, refer to the precautions in the beginning of this section.
2. Remove the radiator.
3. Disconnect the electrical connector. Release the wiring from the stator.

➡Use a hole in the fan hub to prevent the fan from turning.

4. Using the special tool, loosen the fan clutch by turning the wrench counterclockwise. Remove the cooling fan and clutch.
5. Remove the bolts and cooling fan stator assembly.
6. Remove the accessory drive belt.
7. Remove the bolts and the coolant pump pulley.
8. Remove the bolts and the coolant pump.
9. Remove and discard the O-ring seal.
10. Clean and inspect the coolant pump mounting.
11. To install, reverse the removal procedure.

7.3L Engine

1. Before servicing the vehicle, refer to the precautions in the beginning of this section.

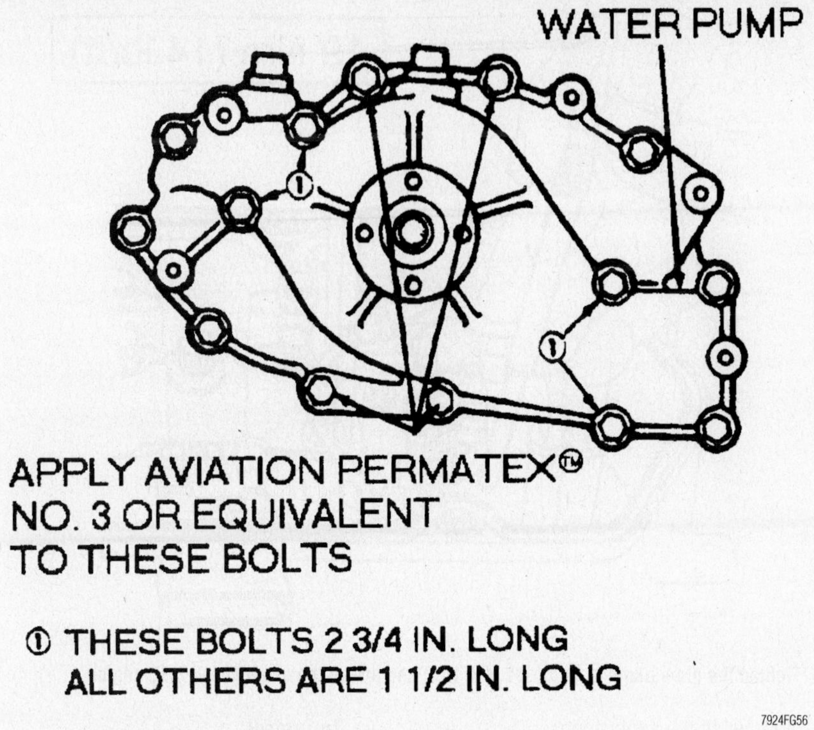
WATER PUMP
APPLY AVIATION PERMATEX®
NO. 3 OR EQUIVALENT
TO THESE BOLTS
① THESE BOLTS 2 3/4 IN. LONG
ALL OTHERS ARE 1 1/2 IN. LONG
7924FG56

Apply RTV sealant to the bolts indicated—7.3L diesel engine

2. Disconnect both battery ground cables.
3. Drain the cooling system.
4. Remove the radiator shroud and the fan clutch and fan.

➡The fan clutch bolts are right-hand thread. Remove them by turning counter-clockwise.

5. Loosen, but do not remove, the water pump pulley bolts.
6. Remove the drive belt.
7. Remove the water pump pulley bolts and remove the pulley from the pump.
8. Detach the Engine Coolant Temperature (ECT) sensor connector.
9. Remove the heater hose from the water pump.
10. Remove the bolts attaching the water pump to the front cover and lift off the pump.

To install:

11. Thoroughly clean the mating surfaces of the pump and front cover.
12. Using a new gasket, position the water pump over the dowel pins and into place on the front cover.
13. Install the attaching bolts. Tighten the bolts to 15 ft. lbs. (20 Nm).
14. Connect the heater hose to the pump.
15. Attach the ECT sensor connector.
16. Install the water pump pulley and start the water pump pulley bolts.
17. Install the drive belt.
18. Install and tighten the pulley retaining bolts to 12–18 ft. lbs. (16–24 Nm).
19. Install the fan and fan shroud assembly.
20. Fill and bleed the cooling system.
21. Connect the battery ground cables.
22. Start the engine and check for leaks.

Glow Plugs

REMOVAL & INSTALLATION

6.0L Engine

F-SERIES EARLY BUILD

1. Before servicing the vehicle, refer to the precautions in the beginning of this section.
2. If servicing the right glow plugs, remove the evaporator core housing.
3. Disconnect the glow plug electrical connector.
4. Remove the glow plug buss bar.

➡If coolant residue is found on the glow plug, a new glow plug sleeve may have to be installed.

5. Remove the glow plug.

To install:

6. Install the glow plug. Torque to 14 ft. lbs. (19 Nm).

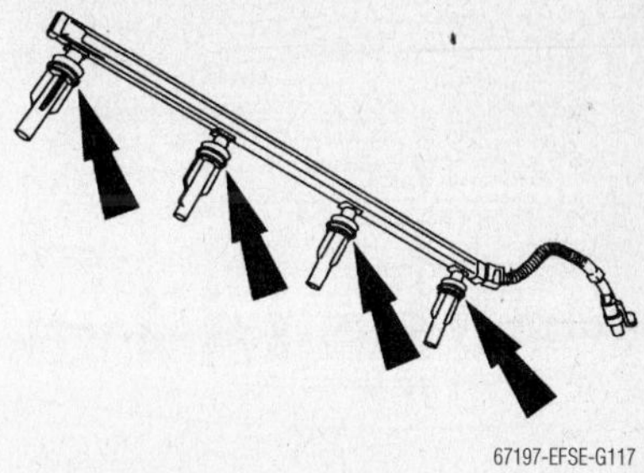
67197-EFSE-G117

Apply clean engine oil to the O-rings—early build 6.0L engines

7. Clean and inspect the O-rings and install new if necessary.
8. Apply clean engine oil to the O-rings.
9. Install the glow plug buss bar.
10. Connect the glow plug electrical connector.
11. If removed, install the evaporator core housing.

F-SERIES LATE BUILD

1. Before servicing the vehicle, refer to the precautions in the beginning of this section.
2. Disconnect the glow plug electrical connector.

⁂ WARNING

Do not pull on the wiring to remove the glow plug connector or damage may occur.

3. Remove the glow plug harness.

➡If coolant residue is found on the glow plug, a new glow plug sleeve may have to be installed.

4. Remove the glow plug.
5. Installation
6. Install the glow plug. Torque to 14 ft. lbs. (19 Nm).
7. Clean and inspect the O-rings and install new if necessary.
8. Apply clean engine oil to the O-rings.

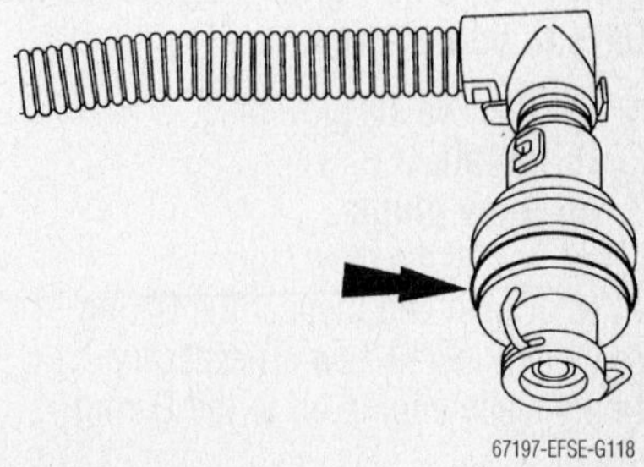
67197-EFSE-G118

Apply clean engine oil to the O-rings—late build 6.0L engines

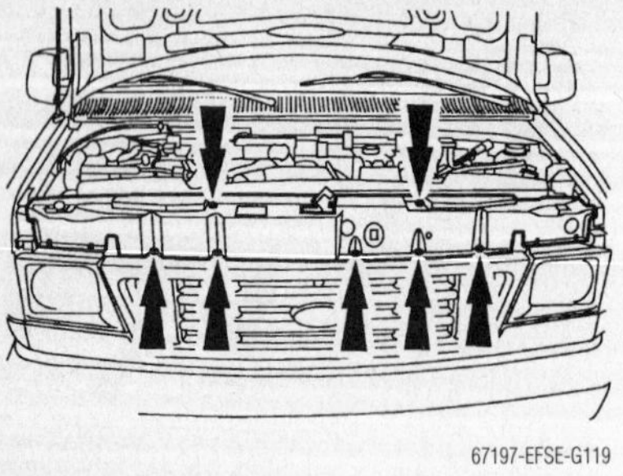

Air deflector removal

9. Install the glow plug harness.
10. Connect the glow plug electrical connector.

E-SERIES

Number 2 glow plug

1. Remove the retainers and the air deflector.
2. Remove the power steering reservoir bracket retainers.
3. Remove the power steering fluid indicator and retainers. Remove the power steering reservoir mounting bracket. Install the power steering fluid indicator and position the power steering reservoir aside.
4. Disconnect the coolant hoses from the air cleaner outlet pipe.
5. Loosen the clamps and remove the air cleaner outlet pipe.

Number 1, 3-8 glow plugs

6. Remove the engine cover.

All glow plugs

**** WARNING**

Do not pull on the wiring to remove the glow plug connector or damage may occur.

➡Only the number 2 glow plug is accessed from under the hood, all others are accessed from inside the cab.

7. Remove the glow plug harness as needed.

➡If coolant residue is found on the glow plug, a new glow plug sleeve may have to be installed.

8. Remove the glow plug.

To install:

All glow plugs

9. Install the glow plug.
10. Clean and inspect the O-ring seals and install new if necessary. Apply clean engine oil to the O-ring seals.
11. Install the glow plug harness.

Number 1, 3-8 glow plug

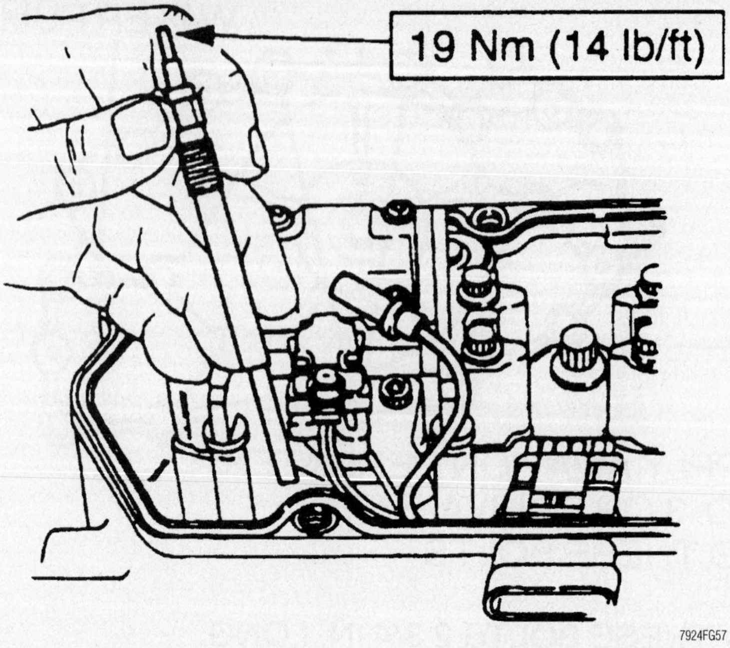

Tighten the glow plugs to 14 ft. lbs. (19 Nm) and attach the connector—7.3L engine

12. Install the engine cover.

Number 2 glow plug

13. Install the air cleaner outlet pipe and tighten the clamps.
14. Connect the coolant hoses to the air cleaner outlet pipe.
15. Remove the power steering fluid indicator. Position the power steering reservoir mounting bracket. Install the retainers and power steering fluid indicator.
16. Install the power steering reservoir bracket retainers.
17. Position the air deflector and install the retainers.

7.3L Engine

**** CAUTION**

The red-striped wiring harness carries 115v direct current. Severe electrical shock may be received. DO NOT pierce.

1. Before servicing the vehicle, refer to the precautions in the beginning of this section.
2. Disconnect the negative battery cable.
3. Remove the rocker arm cover.
4. Disconnect the glow plug electrical leads using a pair of pliers.
5. Remove the glow plugs by unscrewing them from the cylinder head with a 10mm socket and wrench.
6. Inspect the tips of the plugs for any evidence of distortion or missing tip ends; replace them if necessary.

To install:

7. Install the glow plug into the cylinder head. Tighten the glow plugs to 14 ft. lbs. (19 Nm).
8. Attach the glow plug electrical connector. Be sure that the glow plug wiring is routed to avoid moving components in the engine bay.
9. Install the rocker arm cover.

➡When the battery is disengaged and reconnected, some abnormal drive symptoms may occur while the vehicle relearns its adaptive strategy. The vehicle may need to be driven 10 miles (16 km) or more to relearn this strategy.

10. Connect the negative battery cable.

Cylinder Head

REMOVAL & INSTALLATION

E-Series w/6.0L Engine

1. Before servicing the vehicle, refer to the precautions in the beginning of this section.
2. Remove the engine.
3. Remove the bolts and flexplate.
4. Mount the engine on an engine stand.
5. Remove the rear lifting eye.
6. Remove the lifting tools.

**** WARNING**

Do not pull on the wiring to remove the glow plug connector or damage may occur.

7. Remove the glow plug harness.
8. Remove the eight glow plugs.
9. Prior to removing the exhaust manifolds, inspect the exhaust manifold for warpage with a feeler gauge between the manifold and cylinder head. Record the measurement and compare with the specifications.
10. Remove the bolts and the exhaust manifolds.

Left cylinder head

11. Remove the left cylinder head banjo fitting and fuel line. Discard the sealing washers.
12. Remove the bolts and turbocharger heat shield.
13. Remove the protective covering from the left cylinder head.

Right cylinder head

➡Mark the location of the stud bolts.

14. Remove the right valve cover.
15. Remove the stud bolts, bolts and valve cover.
16. Clean and inspect the gaskets.
17. Install a new gasket if necessary.
18. Clean and inspect the sealing surfaces.
19. Remove the right high pressure oil rail-to-valve cover gasket.

Both cylinder heads

20. Remove the bolts and the high pressure oil rail.

⁂ WARNING

Do not attempt to apply battery voltage to the fuel injector or damage to the fuel injector will occur.

21. Using a 19 mm (0.74 in.) socket, push the fuel injector electrical connector out of the rocker arm carrier.

➡If engine oil is found in the engine coolant or engine coolant is found in the combustion chambers, new injector sleeves may need to be installed.

22. Remove the bolt, the fuel injector hold down and the fuel injector.
23. Remove the crankcase-to-head tube assembly.
24. Remove and discard the O-ring seals.
25. Remove and discard the 20 inner cylinder head bolts.
26. Remove the 16 bolts and the rocker arm assemblies.
27. Remove the bolts and rocker arm carrier.
28. Clean and inspect the gasket.
29. Install new gaskets if necessary.
30. Clean and inspect the sealing surfaces.
31. Mark the 16 valve bridges with a permanent marker and remove.
32. Mark and remove the 16 push rods.
33. Remove the outer cylinder head bolts.
34. Install the lifting tool.
35. Remove the cylinder head.
36. Remove and discard the cylinder head gasket and dowels.
37. Clean and inspect the gasket sealing surfaces.

To install:

Both cylinder heads

➡Use care to avoid scratching the blue compound on the cylinder head gasket.

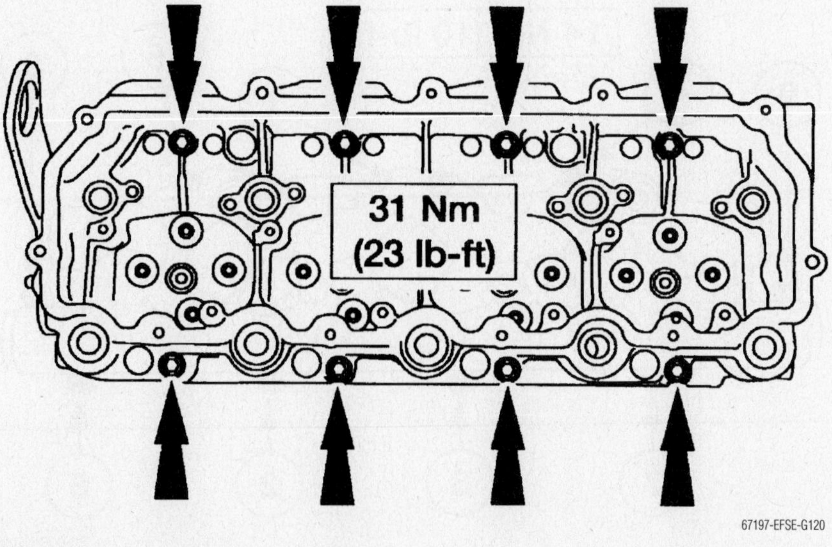

Rocker arm carrier installation—6.0L engine

38. Install a new cylinder head gasket with the part number facing upward and verify the top five holes and the head gasket push rod holes line up.
39. Install new dowels and the cylinder head gasket.
40. Using the special tools, install the cylinder head on the engine.
41. Install the lifting bracket.
42. Install the bolts.
43. Install the cylinder head.
44. Install the outer cylinder head bolts finger-tight.

➡Higher mileage engines require push rods to be cleaned so the copper colored end of the push rod can be identified.

45. Apply clean engine oil to each end of the push rod. Insert them into their respective positions with the copper colored end up.

➡Coat the end of each valve stem with clean engine oil.

46. Install the 16 valve bridges.
47. Install the rocker arm carrier.
48. Install the press-in-place gasket.
49. Install the rocker arm carrier and bolts.

⁂ WARNING

Rotate the crankshaft until the damper locating dowel notch is in the six o'clock position or engine damage can occur.

➡Apply clean engine oil to the top center of each valve bridge.

50. Install the rocker arm assemblies and 16 bolts. Torque to 23 ft. lbs. (31 Nm).

⁂ WARNING

Using too much engine oil on the threads of the cylinder head bolts can cause damage to the threads and poor sealing. Using anti-seize compounds, grease or any other lubricants other than engine oil on the cylinder head bolt threads can affect the true torque value of the bolts.

➡Lightly lubricate the new cylinder head bolt threads and flanges with clean engine oil.

51. Install the 20 inner cylinder head retaining bolts finger-tight.
52. Tighten the head bolts in the following sequence.
  a. Tighten bolts 1 through 10 to 88 Nm (65 ft. lbs.).

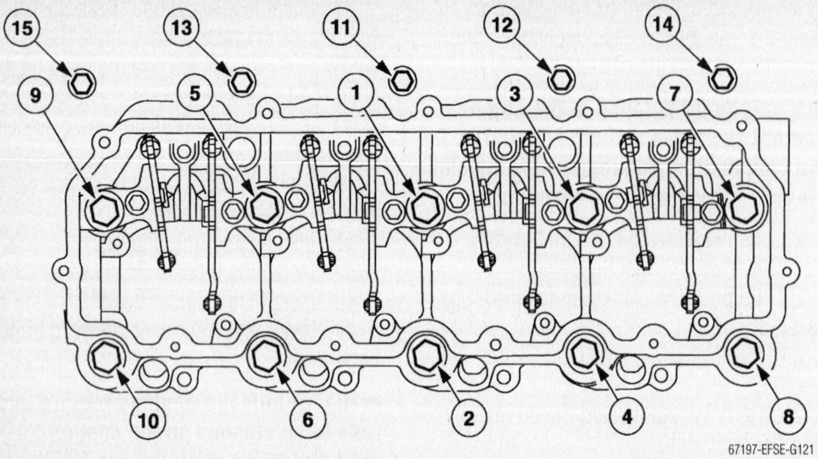

Cylinder head bolt torque sequence—6.0L engine

b. Tighten bolts 1, 3, 5, 7 and 9 to 115 Nm (85 ft. lbs.).

c. Tighten bolts in sequence 1 through 10, clockwise 90 degrees.

d. Tighten bolts in sequence 1 through 10, a second time, clockwise 90 degrees.

e. Tighten bolts in sequence 1 through 10, a third time, clockwise 90 degrees.

f. Tighten bolts 11 through 15 to 24 Nm (18 ft. lbs.).

g. Tighten bolts 11 through 15 to 31 Nm (23 ft. lbs.).

➡Install new O-ring seals and apply clean engine oil.

53. Install a crankcase-to-head tube assembly.

⁂ WARNING

If the fuel injector oil inlet D-ring is damaged, a new fuel injector must be installed.

➡Lubricate the fuel injector and O-ring seals liberally with clean engine oil.

54. Install new O-ring seals and a copper washer on the fuel injector.

55. Install the fuel injector, the fuel injector hold-down and bolt.

⁂ WARNING

Be sure the injector wiring is clear of all moving parts or engine damage can occur.

56. Install the fuel injector electrical connector into the rocker carrier.

➡Apply clean engine oil to the top fuel injector O-ring seals before installing the high pressure oil rail.

57. Position the high pressure oil rail on the injectors.

58. Place the high pressure oil rail on top of the carrier so that the four single ball tubes are engaging the fuel injector lead angle.

59. Insert three guide bolts, two on the ends of the straight side of the high pressure oil rail and one in the middle of the wavy side of the high pressure rail. Install the guide studs six to seven turns.

60. Press the high pressure oil rail into the fuel injectors.

61. Inspect that the high pressure oil rail mounting feet are flat against the mounting surface.

62. Loosely install the six bolts.

63. Install the remaining bolts and tighten.

64. Remove the three guide bolts.

65. Loosely install the three remaining bolts.

66. Tighten the nine bolts in the sequence shown.

Right cylinder head

67. Install the right high pressure oil rail-to-valve cover gasket.

⁂ WARNING

To prevent engine damage, do not use air powered tools when installing the valve cover.

68. Install the right valve cover. Torque to 71 inch lbs. (8 Nm).

69. Install the right exhaust manifold and the bolts.

➡Apply anti-seize lubricant to the bolt threads prior to installing. Torque to 28 ft. lbs. (39 Nm).

➡Start installing the bolts with the second bolt from the rear on the top. The hole diameter is smaller, therefore allowing alignment of the remaining bolts.

Left cylinder head

70. Cover the left cylinder head with an appropriate covering.

71. Install the turbocharger heat shield and bolts.

➡Install new sealing washers.

72. Install the left cylinder head fuel line. Torque to 28 ft. lbs. (39 Nm).

➡Start installing the bolts with the second bolt from the rear on the top. The hole diameter is smaller, therefore allowing alignment of the remaining bolts.

➡Apply anti-seize lubricant to the bolt threads prior to installing.

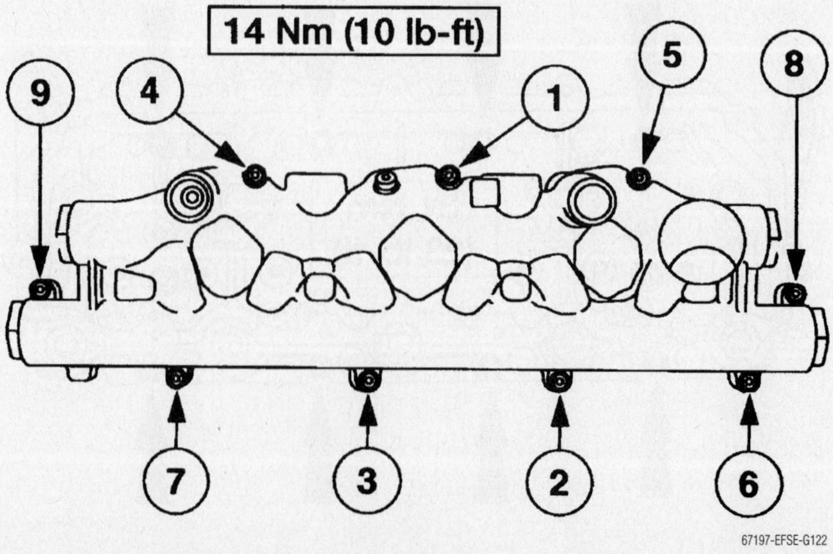

High pressure oil rail torque sequence—6.0L engine

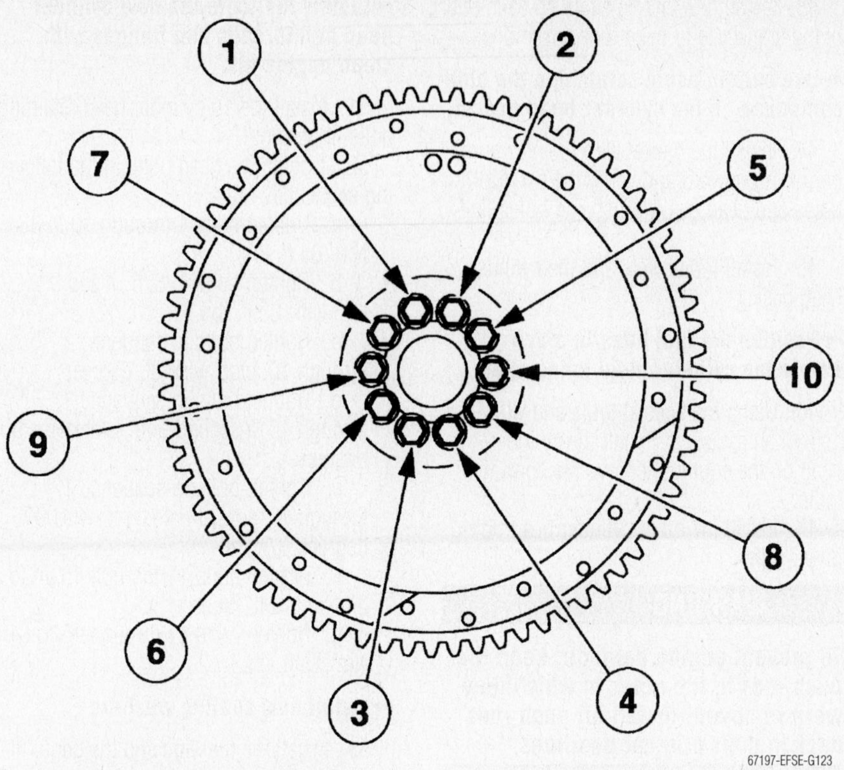

Flexplate torque sequence—6.0L engine

73. Install the left exhaust manifold and the bolts. Torque to 28 ft. lbs. (39 Nm).

Both cylinder heads

74. Install the eight glow plugs.

➡Clean and inspect the glow plug connector O-ring seals and install new as necessary.

➡Apply clean engine oil to the O-ring seals.

75. Install the glow plug harness.
76. Install the special tools.
77. Install the special tool.
78. Install the lifting crane and remove the engine from the engine stand.
79. Install the flexplate and the bolts.
80. Tighten the bolts in the following sequence.
    a. Tighten all bolts to 5 Nm (44 inch lbs.).
    b. Tighten all bolts to 94 Nm (69 ft. lbs.) in the sequence shown.
81. Install the engine.

F-Series w/6.0L Engine

LEFT CYLINDER HEAD

1. Before servicing the vehicle, refer to the precautions in the beginning of this section.
2. With the vehicle in **Neutral**, position it on a hoist.
3. Drain the engine oil.
4. Remove the intake manifold.
5. Remove the left cylinder block drain plug.
6. Remove the fuel injector control module mounting bracket.
7. Disconnect the glow plug connector and position aside.
8. Remove the nut and position the oil level indicator and tube aside.

➡Mark the position of the valve cover bolts for valve cover bolt installation.

9. Remove the eleven bolts and the valve cover. Remove the valve cover gasket.

Early build

10. Disconnect the high-pressure oil rail supply line at the high-pressure oil rail.
11. Remove the bolts and the high-pressure oil rail.
12. Disconnect and remove the high-pressure oil supply line.
13. Remove and discard the crankcase-to-head tube assembly.
14. Remove the glow plug buss bar.

Late build

➡Do not remove the oil rail end plugs or acoustic wave attenuator port fitting. Service parts are not available to support the components.

15. Remove the bolts and the high-pressure oil rail.

➡The rings on the crankcase-to-head tube are to be used to pry the tube assembly from the branch tube assembly or the oil rail assembly.

16. Remove the crankcase-to-head tube assembly.
17. Remove and discard the O-rings.

⁂ WARNING

Do not pull on the glow plug wire or damage may occur.

18. Remove the glow plug harness.

All vehicles

19. Remove the four glow plugs.

⁂ WARNING

Do not attempt to apply battery voltage to the fuel injector or damage to the fuel injector will occur.

20. Using a 19 mm socket, push the fuel injector electrical connector out of the rocker arm carrier.
21. Prior to removing the injector assembly, insert clean shop towels in the oil drain holes adjacent to each glow plug.

⁂ WARNING

Failure to account for all snaprings or pieces prior to placing the vehicle back in service can cause engine damage. A missing snapring can be ingested into the lube oil system causing severe engine damage.

⁂ WARNING

To prevent engine damage, do not use air tools to remove the fuel injectors. The snapring that extracts the injector can dislodge and fall into the oil drain hole.

➡There is no need to drain the fuel rail.

➡If engine oil is found in the engine coolant or engine coolant is found in the combustion chambers, a new injector sleeve may need to be installed.

22. Remove the bolt, fuel injector hold-down and fuel injector.

➡If a snapring or piece of a snapring is missing from the injector hold-down assembly, it must be located prior to removing the shop towels.

23. Remove the shop towels.
24. Remove the nuts and bolts from the turbocharger adapter pipe.

25. Remove the turbocharger adapter pipe.

➡Remove and discard the O-ring.

26. Remove the oil level indicator and tube.
27. Remove the bolts and turbocharger heat shield.
28. Disconnect the exhaust backpressure tube at the exhaust manifold.
29. Remove the retaining nuts. Remove the exhaust backpressure bracket assembly.
30. Remove the retaining bolt from the fuel line bracket and position the fuel lines aside.
31. Remove the bolt and the idler pulley.
32. Disconnect the heater hose from the front cover and position aside.
33. Remove the fan shroud mounting stud.

➡Remove and discard the sealing washers.

34. Remove the banjo fitting and the fuel line.
35. Remove and discard the inner cylinder head bolts.
36. Remove the eight bolts and the rocker arm assemblies.
37. Mark the eight valve bridges with a permanent marker and remove.

⁂ WARNING

To prevent engine damage, keep the push rods in the order in which they were removed. Install all push rods back in their original positions.

38. Mark the location and remove the eight push rods.
39. Remove the outer cylinder head bolts.
40. Install the special tool and the lifting crane. With the help of an assistant, remove the cylinder from the vehicle.
41. Remove and discard the cylinder head gasket and dowels.
42. Clean and inspect the gasket sealing surfaces.

To install:

All vehicles, late build

⁂ WARNING

Install new D-ring seals on the crankcase-to-head tube. It requires several hours after installation for the D-ring seals to relax back to their original size. If the tube assembly is installed before the D-ring seals have relaxed, damage to the D-ring seals can occur.

43. Install new D-ring seals on each end of the crankcase to head tube assemblies.

➡Use care to avoid scratching the blue compound on the cylinder head gasket.

44. Install the gasket with the part number facing upward and verify the five top holes and head gasket push rod holes line up.
45. Install the dowels and the cylinder head gasket.

➡Position the fuel lines in place before the cylinder head is installed.

46. Using the special tools and with the help of an assistant, install the cylinder head on the engine. Remove the special tools.
47. Install the outer cylinder head bolts finger-tight.

⁂ WARNING

To prevent engine damage, keep the push rods in the order in which they were removed. Install all push rods back in their original positions.

➡Higher mileage engines require push rods to be cleaned so the copper colored end of the push rod can be identified.

48. Apply clean engine oil to each end of the push rod. Insert them into their respective positions with the copper-colored end up.

➡Coat the end of each valve stem with clean engine oil.

49. Install the eight valve bridges.

⁂ WARNING

Rotate the crankshaft until the damper locating dowel notch is in the six o'clock position or engine damage can occur.

➡Apply clean engine oil to the top center of each valve bridge.

50. Install the rocker arm assemblies and eight bolts.

⁂ WARNING

Using too much engine oil on the threads of the cylinder head bolts can cause damage to the threads and poor sealing. Using anti-seize compounds, grease or any other lubricants other than engine oil on the cylinder head bolt threads can affect the true torque value of the bolts.

➡Lightly lubricate the new cylinder head bolt threads and flanges with clean engine oil.

51. Install the 10 cylinder head retaining bolts finger-tight.
52. Tighten the head bolts in the following sequence.
  a. Tighten bolts 1 through 10 to 88 Nm (65 ft. lbs.).
  b. Tighten bolts 1, 3, 5, 7 and 9 to 115 Nm (85 ft. lbs.).
  c. Tighten bolts in sequence 1 through 10, clockwise 90 degrees.
  d. Tighten bolts in sequence 1 through 10, a second time, clockwise 90 degrees.
  e. Tighten bolts in sequence 1 through 10, a third time, clockwise 90 degrees.
  f. Tighten bolts 11 through 15 an to 24 Nm (18 ft. lbs.).
  g. Tighten bolts 11 through 15 to 31 Nm (23 ft. lbs.).

➡Install new sealing washers.

53. Install the fuel line and the banjo fitting.
54. Install the fan shroud mounting stud.
55. Position back and connect the heater hose.
56. Install the idler pulley and bolt.
57. Position the exhaust backpressure bracket assembly and install the retaining nuts.
58. Tighten the exhaust backpressure tube fitting at the exhaust manifold.
59. Position the fuel lines and install the fuel line bracket retaining bolt.

➡Install a new O-ring and apply clean engine oil.

60. Install the oil level indicator and tube.
61. Install the turbocharger heat shield and bolts.
62. Position the turbocharger adapter pipe.

➡Apply anti-seize lubricants to the bolt threads prior to installing the bolts.

➡Do not tighten until after the turbocharger is installed.

63. Install the nuts and bolts in the adapter pipe.
64. Install the four glow plugs.

⁂ WARNING

If the fuel injector oil inlet D-ring is damaged, a new fuel injector must be installed.

➡**Lubricate the fuel injector and O-rings liberally with clean engine oil.**

65. Install new O-rings and copper washer on the fuel injector.

✲✲ WARNING

To prevent engine damage, do not use air tools to install the fuel injectors. The snapring that extracts the injector can dislodge and fall into the oil drain hole.

66. Install the fuel injector, fuel injector hold-down and bolt. Torque to 24 ft. lbs. (33 Nm).

✲✲ WARNING

To prevent engine damage, be sure the injector wiring is clear of all moving parts.

67. Install the fuel injector electrical connector into the rocker carrier.
68. Apply engine oil to the top fuel injector O-rings.

Early build

➡**Install a new lower O-ring.**

69. Install a new crankcase to head tube assembly. Torque to 33 ft. lbs. (45 Nm).
70. Install the high-pressure oil rail and bolts.
71. Install the high-pressure oil rail.
72. Install the bolts finger tight.
73. Tighten the bolts in the sequence shown.
74. Install the high-pressure oil line.

➡**Clean and inspect the O-rings and install new as necessary.**

➡**Apply clean engine oil to the O-rings before installing.**

75. Install the glow plug buss bar.

Late build

✲✲ WARNING

To prevent engine damage, check that the crankcase-to-head tube assemblies bottom out in the branch tube assembly. The oil rail, crankcase-to-head tube and the fuel injectors will not function correctly if the tube is not bottomed out.

➡**Install new O-rings and .**

76. Apply clean engine oil and install the crankcase-to-head tube assembly.

➡**Apply clean engine oil to the top fuel injector D-ring before installing the high pressure oil rail.**

77. Position the oil rail on the fuel injectors.
78. Place the oil rail on top of the carrier so that the four single ball tubes are engaging the injector lead angle.
79. Insert three guide bolts, two on the ends of the straight side of the oil rail and one in the middle of the wavy side of the oil rail. Install the guide studs six to seven turns.
80. Press the oil rail into the fuel injectors.
81. Inspect that the oil rail mounting feet are flat against the mounting surface.
82. Loosely install the six bolts.
83. Install the oil rail retaining bolts.
84. Remove the three guide bolts.
85. Loosely install the three remaining bolts.
86. Tighten the bolts in the sequence shown.

➡**Clean and inspect the O-rings and install new as necessary.**

➡**Apply clean engine oil to the O-rings prior to installing.**

87. Install the glow plug harness.

All vehicles

✲✲ WARNING

To prevent engine damage, do not use air powered tools when installing the valve cover.

➡**Clean and inspect the valve cover gasket. Install a new gasket if necessary.**

88. Position the valve cover gasket. Install the valve cover and 11 bolts.
89. Position back the oil level indicator and install the nut.
90. Connect the glow plug connector.
91. Install the fuel injector control module mounting bracket.

➡**Install a new oil filter.**

92. Fill the crankcase with clean engine oil.

➡**Lightly lubricate the O-ring with clean engine oil before installing.**

93. Install the cylinder block drain plug.
94. Install the intake manifold.

RIGHT CYLINDER HEAD

All vehicles

1. With the vehicle in **Neutral**, position it on a hoist.
2. Drain the engine oil.
3. Remove the intake manifold.
4. Remove the fuel injectors.
5. Remove the starter.
6. Remove the cylinder drain block.
7. Remove the nuts and bolts from the turbocharger adapter pipe.
8. Remove the turbocharger adapter pipe.

Early build

9. Remove and discard the crankcase to head tube assembly.
10. Remove the glow plug buss bar.

Late build

➡**The rings on the crankcase-to-head tube are to be used to pry the tube assembly from the branch tube assembly or the oil rail assembly.**

11. Remove the crankcase-to-head tube assembly.
12. Remove and discard the O-rings.

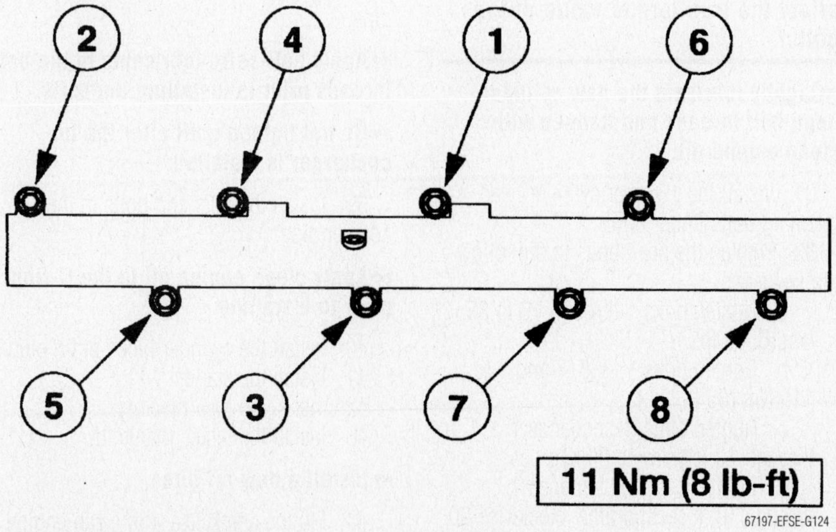

High pressure oil rail torque installation—F-Series w/6.0L engine

WARNING

Do not pull on the wiring to remove the glow plug connector or damage may occur.

13. Remove the glow plug harness.

All vehicles

14. Remove the four glow plugs.
15. Remove and discard the 10 inner cylinder head bolts.
16. Remove the eight bolts and the rocker arm assemblies.
17. Mark the eight valve bridges with a permanent marker and remove.

WARNING

To prevent engine damage, keep the push rods in the order in which they were removed. Install all push rods back in their original positions.

18. Mark the location and remove the eight push rods.
19. Remove the outer cylinder head bolts.
20. Install the special tool and the lifting crane. With the help of an assistant, remove the cylinder head from the vehicle.
21. Remove and discard the cylinder head gasket and dowels.
22. Clean and inspect the gasket sealing surfaces.

To install:

All vehicles, late build

WARNING

Install new D-ring seals on the crankcase-to-head tube. It requires several hours after installation for the D-ring seals to relax back to their original size. If the tube assembly is installed before the D-ring seals have relaxed, damage to the D-ring seals can occur.

23. Install new D-ring seals on each end of the crankcase to head tube assemblies.

➡Use care to avoid scratching the blue compound on the cylinder head gasket.

24. Install a new cylinder head gasket with the part number facing up and verify the top five holes and the head gasket push rod holes line up.
25. Install the dowels and the cylinder head gasket.
26. Using the special tools and with the help of an assistant, install the cylinder head on the engine. Remove the special tools.
27. Install the outer cylinder head bolts finger-tight.

WARNING

To prevent engine damage, keep the push rods in the order in which they were removed. Install all push rods back in their original positions.

➡Higher mileage engines require push rods to be cleaned so the copper colored end of the push rod can be identified.

28. Apply clean engine oil to each end of the push rod. Insert them into their respective positions with the copper colored end up.

➡Coat the end of each valve stem with clean engine oil.

29. Install the eight valve bridges.

WARNING

Rotate the crankshaft until the damper locating dowel notch is in the six o'clock position or engine damage can occur.

➡Apply clean engine oil to the top center of each valve bridge.

30. Install the rocker arm assemblies and eight bolts.

WARNING

Using too much engine oil on the threads of the cylinder head bolts can cause damage to the threads and poor sealing. Using anti-seize compounds, grease or any other lubricants other than engine oil on the cylinder head bolt threads can affect the true torque value of the bolts.

➡Lightly lubricate the new cylinder head bolt threads and flanges with clean engine oil.

31. Install the 10 inner cylinder head retaining bolts finger-tight.
32. Tighten the head bolts in the following sequence.
    a. Tighten bolts 1 through 10 to 88 Nm (65 ft. lbs.).
    b. Tighten bolts 1, 3, 5, 7 and 9 to 115 Nm (85 ft. lbs.).
    c. Tighten bolts in sequence 1 through 10, clockwise 90 degrees.
    d. Tighten bolts in sequence 1 through 10, a second time, clockwise 90 degrees.
    e. Tighten bolts in sequence 1 through 10, a third time, clockwise 90 degrees.
    f. Tighten bolts 11 through 15 to 24 Nm (18 ft. lbs.).
    g. Tighten bolts 11 through 15 to 31 Nm (23 ft. lbs.).
33. Install the four glow plugs.

Late build

➡Clean and inspect the glow plug connector O-rings and install new as necessary.

➡Apply clean engine oil to the O-rings.

34. Install the glow plug harness.

WARNING

To prevent engine damage, check that the crankcase-to-head tube assemblies bottom out in the branch tube assembly. The oil rail, crankcase-to-head tube and the fuel injectors will not function correctly if the tube is not bottomed out.

35. Apply clean engine oil and install the crankcase-to-head tube assembly.

Early build

➡Clean and inspect the glow plug buss bar O-rings and install new as necessary.

➡Apply clean engine oil to the O-rings before installing.

36. Install the glow plug buss bar.

➡Install a new lower O-ring.

37. Install a new crankcase to head tube assembly.

All vehicles

38. Position the turbocharger adapter pipe.

➡Apply anti-seize lubricants to the bolt threads prior to installing the bolts.

➡Do not tighten until after the turbocharger is installed.

39. Install the nuts and bolts in the adapter pipe.

➡Apply clean engine oil to the O-ring prior to installing.

40. Install the cylinder block drain plug.
41. Install the starter.
42. Install the fuel injectors.
43. Install the intake manifold.

➡Install a new oil filter.

44. Fill the crankcase with clean engine oil.

F-Series w/7.3L Engine

CAUTION

The fuel system remains under pressure, even after the engine has been turned OFF. The fuel system pressure must be relieved before disconnecting any fuel lines. Failure to do so may result in fire and/or personal injury.

1. Before servicing the vehicle, refer to the precautions in the beginning of this section.
2. Remove or disconnect the following:
   - Negative battery cables
   - Bumper
   - Hood
3. Recover the refrigerant using approved equipment.
4. Drain the coolant.
5. Position a suitable container at the end of the water drain tube and move the fuel filter drain lever to drain.
6. Remove or disconnect the following:
   - Cooling module assembly
   - Cooling fan
   - Drive belt
   - Two nuts and bolts from the exhaust manifold and the exhaust adapter pipe
   - Intake Air Temperature (IAT) sensor electrical connector
   - Turbocharger inlet pipe
   - Engine air cleaner and the air cleaner intake pipe as an assembly
   - Four nuts and the engine air cleaner support brackets
   - Charge air cooler outlet pipe
   - Turbocharger outlet pipe
   - A/C pressure cutoff switch electrical connector, A/C cycling switch electrical connector, and the A/C manifold and tube
   - O-ring seals
7. Plug or cap the A/C manifold and tube, and the A/C compressor ports.
   - Wiring harness
   - Generator electrical connector and electrical lead
   - Belt tensioner
   - Bolt and the idler pulley
   - Heater water hose
   - Fuel tubes
   - Water-In-Fuel (WIF) sensor and fuel heater electrical connectors
   - Water drain tube
   - Two bolts and fuel filter/water separator
   - High-pressure oil hose
   - Fuel injector/glow plug nine-pin electrical connector
   - Compressor manifold and turbocharger outlet pipe
   - Clamp and the left half of the compressor manifold
   - Right half of the compressor manifold
   - Fuel tube
   - Valve cover and gasket
   - Electrical connectors from the fuel injectors
   - Electrical leads from the glow plugs
   - Rocker arms and the push rods
   - Oil drain plugs
   - Oil deflector from the fuel injector hold-down plate
   - Outboard fuel injector hold-down bolts
   - Hold-down plates from the inner hold-down bolts
   - Fuel injectors
   - The 4 outboard fuel injector hold-down bolts, retaining screws and 4 oil deflectors

WARNING

Remove the oil drain plugs prior to removing the injectors or oil could enter the combustion chamber, which could result in hydrostatic lock and severe engine damage.

   - Oil rail drain plugs
   - Fuel injectors using Injector Remover No. T94T-9000-aH1, or equivalent. Position the tool's fulcrum beneath the fuel injector hold-down plate and over the edge of the cylinder head. Install the remover screw in the threaded hole of the fuel injector plate (see illustration). Tighten the screw to lift out the injector from its bore. Place the injector in a suitable protective sleeve such as Rotunda Injector Protective Sleeve, No. 014-00933-2 or equivalent, and set the injector in a suitable holding rack.

➡During removal the injector tab (located above the fuel injector) must be bent completely flat and flush with the cowl and heat shield.

8. Remove or disconnect the following:
   - The 4 glow plugs
9. Loosen the cylinder head bolts.
10. Carefully lift the cylinder head out of the engine compartment and remove the head gaskets.

➡To prepare a good seat for the fuel injector O-rings, use a suitable injector sleeve brush to clean any debris from the bore.

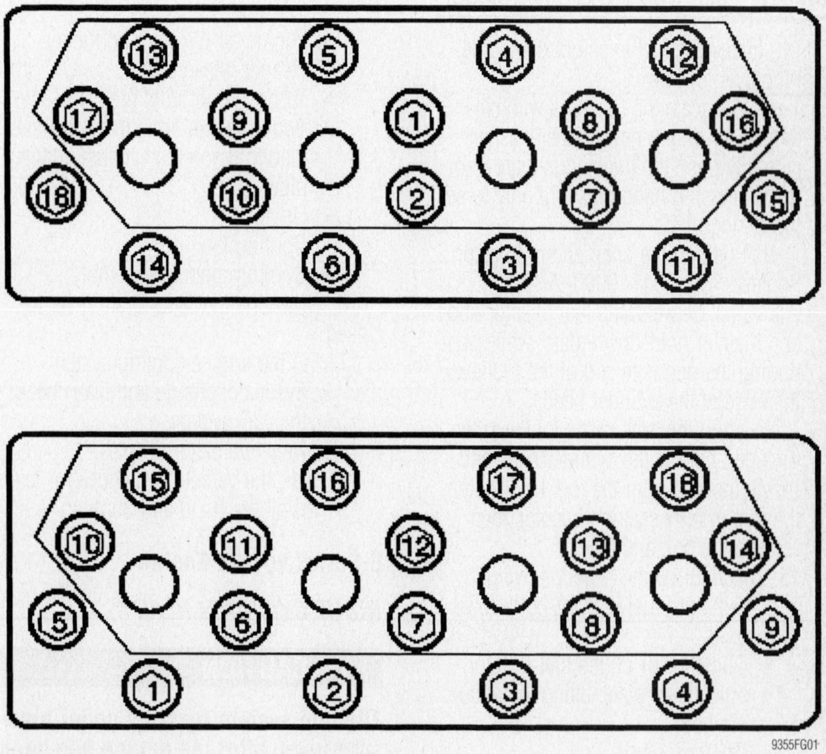

Cylinder head torque sequence—7.3L diesel engines

11. Carefully clean the cylinder block and head mating surfaces.

12. Install or connect the following:
- Cylinder head gasket on the engine block and carefully lower the cylinder head in place
- Cylinder head bolt and torque in 3 steps using the sequence shown in the illustration as follows:

a. Step 1: 65 ft. lbs. (88 Nm).
b. Step 2: 85 ft. lbs. (115 Nm).
c. Step 3: 95 ft. lbs. (129 Nm).

➡Lubricate the threads and the mating surfaces of the bolt heads and washers with engine oil.

13. Install or connect the following:
- Glow plugs

⁂ CAUTION

Remove all oil and fuel from the cylinders before installing the fuel injectors. Failure to do so can cause hydro-static lock, resulting in severe engine damage.

⁂ CAUTION

Do not strike the top of the fuel injector to seat the injector in the cylinder head fuel injector bore. Damage to the fuel injector will occur. Use hand pressure on the top of the fuel injector until the fuel injector hold-down plate is flush with the cylinder head.

14. Install the fuel injectors using special tools as follows:

a. Lubricate the injectors with clean engine oil. Using new copper washers, carefully push the injectors square into the bore using hand pressure only to seat the O-rings.

b. Position the open end of Injector Replacer, No. T94T–9000–AH2, or equivalent between the fuel injector body and injector hold-down plate, while positioning the opposite end of the tool over the edge of the cylinder head.

c. Align the hole in the tool with the threaded hole in the cylinder head and install the bolt from the tool kit. Tighten the bolt to fully seat the injector, then remove the bolt and tool.

15. Install or connect the following:
- Outboard fuel injector hold-down bolts
- Oil deflector on the fuel injector hold-down plate, using the shoulder bolt
- Oil drain plugs
- Rocker arms and push rods
- Valve cover gasket
- Glow plug electrical leads
- Fuel injector electrical connectors
- Valve cover
- Fuel tube
- Compressor manifold and turbocharger outlet pipe
- Right half of the compressor manifold and tighten the clamps
- Left half of the compressor manifold and the clamp
- IAT sensor electrical connector
- Fuel injector/glow plug nine-pin electrical connector and high-pressure oil hose
- Fuel filter/water separator and the two bolts
- Fuel tubes
- Water drain tube and the fuel supply tube
- Fuel tubes, the WIF sensor electrical connector, and the fuel heater electrical connector
- Heater water hose
- Idler pulley and bolt
- Belt tensioner and bolt
- Generator electrical connector and electrical lead
- Wring harness
- A/C manifold and tube and the A/C cycling switch electrical connector
- Turbocharger outlet pipe
- Charge air cooler outlet pipe
- Engine air cleaner brackets and four nuts
- Engine air cleaner and the air cleaner intake pipe
- Turbocharger inlet pipe
- Two nuts and bolts in the exhaust manifold and the exhaust adapter pipe
- Drive belt
- Cooling fan
- Cooling module assembly

16. Move the fuel filter drain lever to closed.

17. Fill the engine cooling system.
- Evacuate, charge and leak check the A/C system.
- Negative battery cables.

18. Start the vehicle and check for leaks.

19. Install the hood and bumper.

E-Series w/7.3L Engine

RIGHT CYLINDER HEAD

⁂ CAUTION

The fuel system remains under pressure, even after the engine has been turned OFF. The fuel system pressure must be relieved before disconnecting any fuel lines. Failure to do so may result in fire and/or personal injury.

1. Before servicing the vehicle, refer to the precautions in the beginning of this section.

2. Remove or disconnect the following:
- Negative battery cable
- Coolant
- Engine oil
- Vacuum hose from the right intake manifold
- Both electrical harness connectors from the valve cover
- Valve cover and gasket
- Intake manifold covers
- Fuel injector electrical injectors
- Fuel lines from the heads
- Fuel supply assembly
- Heater hose
- Dipstick tube bracket retainer at the right exhaust manifold
- Rocker arms and pushrods

3. Loosen the 4 inboard fuel injector hold-down bolts.

4. Remove or disconnect the following:
- The 4 outboard fuel injector hold-down bolts, retaining screws and 4 oil deflectors

⁂ WARNING

Remove the oil drain plugs prior to removing the injectors or oil could enter the combustion chamber, which could result in hydrostatic lock and severe engine damage.

- Oil rail drain plugs
- Fuel injectors using Injector Remover No. T94T-9000-aH1, or equivalent. Position the tool's fulcrum beneath the fuel injector hold-down plate and over the edge of the cylinder head. Install the remover screw in the threaded hole of the fuel injector plate (see illustration). Tighten the screw to lift out the injector from its bore. Place the injector in a suitable protective sleeve such as Rotunda Injector Protective Sleeve, No. 014-00933-2 or equivalent, and set the injector in a suitable holding rack.

➡During removal the injector tab (located above the fuel injector) must be bent completely flat and flush with the cowl and heat shield.

5. Use a vacuum pump, remove the oil and fuel left over in the injector bores.

6. Remove or disconnect the following:
   - The 4 glow plugs
   - High pressure oil pump supply line from the cylinder head
   - Grille opening reinforcement, headlamp assembly, radiator, oil reservoir and fuel filter assembly
   - Exhaust back pressure line
7. Loosen the glow plug relay bracket retainers and disengage the ground wire.
8. Disconnect the fuel return line at the front of the cylinder head.
9. Loosen the cylinder head bolts.
10. Carefully lift the cylinder head out of the engine compartment and remove the head gaskets.

To install:

➡To prepare a good seat for the fuel injector O-rings, use a suitable injector sleeve brush to clean any debris from the bore.

11. Carefully clean the cylinder block and head mating surfaces.
12. Position the cylinder head gasket on the engine block and carefully lower the cylinder head in place.
13. Install the cylinder head bolts and torque in 3 steps using the sequence shown in the illustration as follows:
    a. Step 1: 65 ft. lbs. (88 Nm).
    b. Step 2: 85 ft. lbs. (115 Nm).
    c. Step 3: 95 ft. lbs. (129 Nm).

➡Lubricate the threads and the mating surfaces of the bolt heads and washers with engine oil.

14. Install or connect the following:
    - Fuel return line to the cylinder head
    - Glow plug relay bracket and ground wire, then tighten the retainers
    - Exhaust back pressure line
    - High pressure fuel supply line and tighten the fitting to 19 ft. lbs. (26 Nm)
    - Heater hose to the cylinder head
    - Manifold hoses
    - Fuel supply lines to the rear of the cylinder head
    - Banjo bolt through the fuel line and into the pump. Tighten the pump to 40 ft. lbs. (55 Nm).
    - Glow plugs, coated with anti-seize compound. Tighten the glow plugs to 14 ft. lbs. (19 Nm).
15. Install the fuel injectors using special tools as follows:
    a. Lubricate the injector O-rings with clean engine oil. Using new copper washers, carefully push the injectors square into the bore using hand pressure only to seat the O-rings.
    b. Position the open end of Injector Replacer, No. T94T-9000-aH2, or equivalent, between the fuel injector body and injector hold-down plate, while positioning the opposite end of the tool over the edge of the cylinder head.
    c. Align the hole in the tool with the threaded hole in the cylinder head and install the bolt from the tool kit. Tighten the bolt to fully seat the injector, then remove the bolt and tool.
16. Install or connect the following:
    - Oil rail drain plugs and tighten them to 53 inch lbs. (6 Nm)
    - The 4 outboard fuel injector hold-down bolts, the 4 oil deflectors and retaining screws. Tighten them to 120 inch lbs. (12 Nm).
    - The 4 inboard fuel injector hold-down bolts and tighten them to 120 inch lbs. (12 Nm)
    - Oil deflectors
17. Turn the engine by hand until the timing mark is at the 11 o'clock position as viewed from the front.
18. Dip the pushrod ends in clean engine oil and install the pushrods with the copper colored ends toward the rocker arms, making sure the pushrods are fully seated in the tappet pushrod seats.
19. Install or connect the following:
    - Rocker arms and posts in their original positions. Apply multi-purpose grease to the valve stem tips. Install the rocker arm posts, bolts and tighten to 27 ft. lbs. (37 Nm).
    - Valve cover gasket
    - Wiring to the fuel injectors and glow plugs
    - Valve cover, tightening the bolts to 97 inch lbs. (11 Nm)
    - Both electrical harness connectors to the valve cover
    - Vacuum hose to the right intake valve manifold cover
20. Refill the engine coolant with the recommended type and the correct amount.
21. Refill the engine oil with the recommended type of engine oil and the correct amount.
22. Install the engine in the van.
23. Connect the negative battery cable.

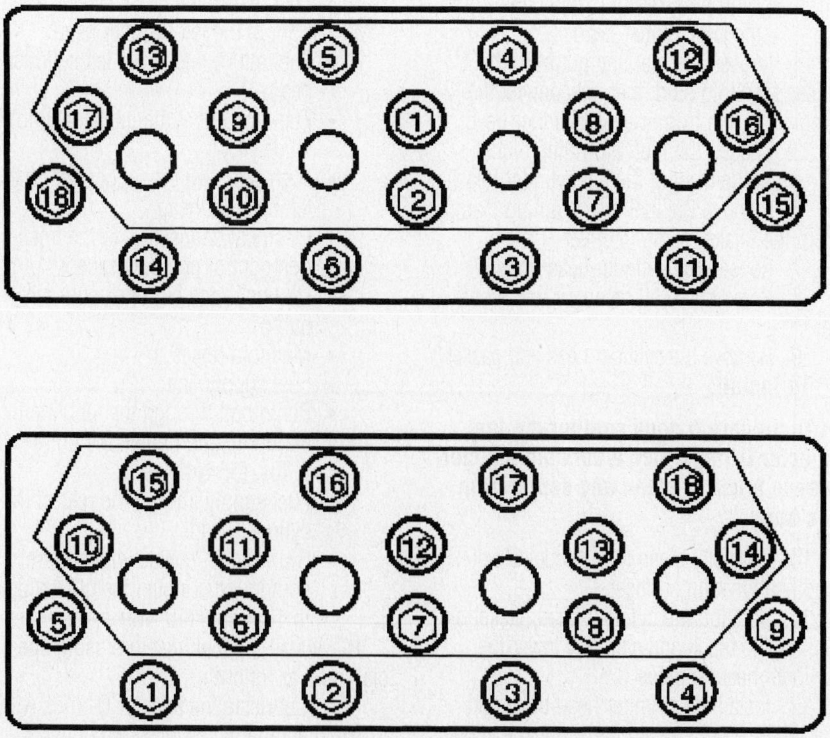

9355FG01

Cylinder head torque sequence—7.3L diesel engines

LEFT CYLINDER HEAD

CAUTION

Fuel system remains under pressure, even after the engine has been turned OFF. The fuel system pressure must be relieved before disconnecting any fuel lines. Failure to do so may result in fire and/or personal injury.

1. Before servicing the vehicle, refer to the precautions in the beginning of this section.

2. Remove or disconnect the following:
- Negative battery cable
- Coolant
- Engine oil
- Wiring harness bracket
- Electrical connections from the valve cover gasket, then remove the valve cover
- Electrical connections from the fuel injectors and glow plugs
- Valve cover gasket
- Rocker arms and pushrods

➡Be sure to note the location of each part prior to removal, as each reinstalled part must be returned to their original location.

- The 4 inboard fuel injector hold-down bolts

⁂ WARNING

Remove the oil drain plugs prior to removing the injectors or oil could enter the combustion chamber, which could result in hydrostatic lock and severe engine damage.

- The oil rail drain plugs

⁂ CAUTION

Be sure to retrieve the fuel injector copper washer, located at the tip of the injector during removal.

- The 4 outboard fuel injector hold-down bolts, retaining screws and 4 oil deflectors
- Fuel injectors using Injector Remover No. T94T-9000-aH1, or equivalent. Position the tool's fulcrum beneath the fuel injector hold-down plate and over the edge of the cylinder head. Install the remover screw in the threaded hole of the fuel injector plate (see illustration). Tighten the screw to lift out the injector from its bore. Place the injector in a suitable protective sleeve such as Rotunda Injector Protective Sleeve, No. 014-00933-2 or equivalent, and set the injector in a suitable holding rack.

3. Use a vacuum pump to remove the oil and fuel left over in the injector bores.
- The 4 glow plugs
- Fuel system pressure
- Fuel supply lines from the rear of the cylinder head

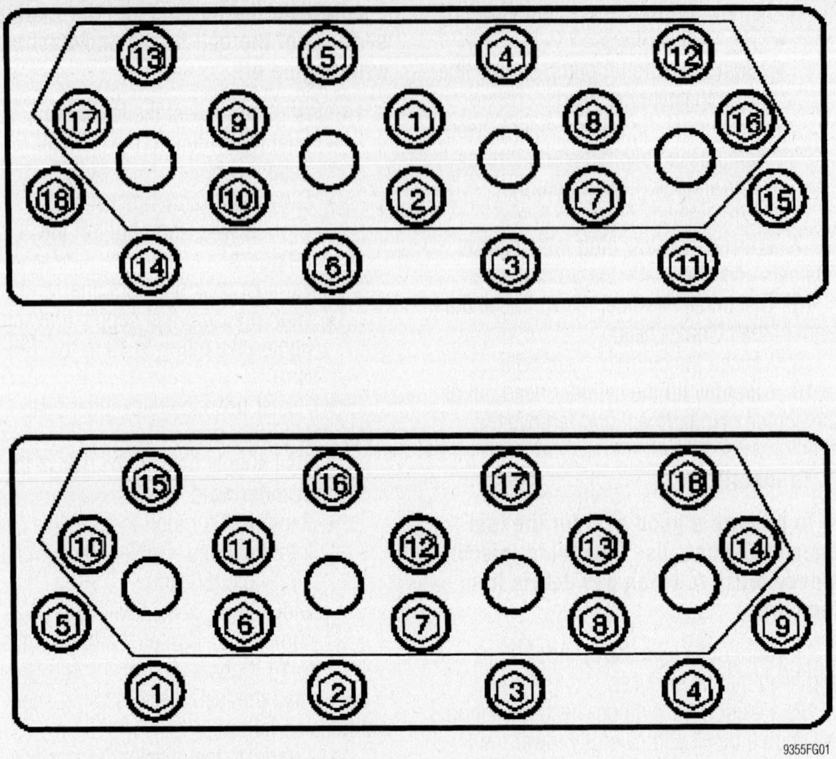

Cylinder head torque sequence—7.3L diesel engines

- Banjo bolt from the fuel line at the pump
- Oil line from the high pressure oil pump
- Electrical connection from the injection control pressure sensor
- High pressure oil supply line from the left cylinder head

4. Loosen the fuel line nut from the intake manifold stud, then disconnect the fuel return line from the left cylinder head.

5. Loosen the fuel return line block screws at the front of the left cylinder head.

6. Remove the fuel line retaining clamp from the intake manifold cover.

7. Loosen the cylinder head bolts.

8. Remove the oil reservoir and fuel filter.

9. Remove the cylinder head and gasket.

To install:

➡To prepare a good seat for the fuel injector O-rings, use a suitable injector sleeve brush to clean any debris from the bore.

10. Carefully clean the cylinder block and head mating surfaces.

11. Position the cylinder head gasket on the engine block and carefully lower the cylinder head in place.

12. Install the cylinder head bolts and torque in 3 steps using the sequence shown in the illustration as follows:

a. Step 1: 65 ft. lbs. (88 Nm).
b. Step 2: 85 ft. lbs. (115 Nm).
c. Step 3: 95 ft. lbs. (129 Nm).

13. Install or connect the following:
- Fuel line retaining clamp to the intake manifold cover
- Fuel return line block screws at the front of the left cylinder head
- Fuel return line to the left cylinder head
- Fuel line nut at the intake manifold stud
- High pressure oil supply line to the left cylinder head
- Electrical connection to the injection control pressure sensor
- Oil line to the high pressure oil pump
- Manifold hoses
- Fuel supply line
- Banjo bolt through the fuel line into the pump and tighten the bolt to 40 ft. lbs. (55 Nm)
- Fuel supply lines at the rear of the cylinder heads
- Glow plugs, coated with anti-seize compound. Tighten the glow plugs to 14 ft. lbs. (19 Nm).

14. Install the fuel injectors using special tools as follows:

a. Lubricate the injector O-rings with clean engine oil. Using new copper washers, carefully push the injectors square into the bore using hand pressure only to seat the O-rings.

b. Position the open end of Injector Replacer, No. T94T-9000-aH2 or equivalent between the fuel injector body and injector hold-down plate, while positioning the opposite end of the tool over the edge of the cylinder head.

c. Align the hole in the tool with the threaded hole in the cylinder head and install the bolt from the tool kit. Tighten the bolt to fully seat the injector, then remove the bolt and tool.

15. Install or connect the following:
 - The 4 outboard fuel injector hold-down bolts, the 4 oil deflectors and retaining screws. Tighten them to 120 inch lbs. (12 Nm).
 - Oil deflectors and tighten the bolts to 120 inch lbs. (12 Nm)
 - Oil rail drain plugs and tighten them to 53 inch lbs. (6 Nm)
 - The 4 inboard fuel injector hold-down bolts and tighten them to 120 inch lbs. (12 Nm)

16. Turn the engine over by hand until the timing mark is at the 11 o'clock position as viewed from the front.

17. Dip the pushrod ends in clean engine oil and install the pushrods with the copper colored ends toward the rocker arms, making sure the pushrods are fully seated in the tappet pushrod seats.

18. Install the rocker arms and posts in their original positions. Apply multipurpose grease to the valve stem tips. Install the rocker arm posts, bolts and tighten to 27 ft. lbs. (37 Nm).

19. Install the valve cover gasket.

20. Connect the wiring to the fuel injectors and glow plugs.

21. Install the valve cover, tightening the bolts to 97 inch lbs. (11 Nm).

22. Connect both electrical harness connectors to the valve cover.

23. Install the engine in the van.

24. Connect the negative battery cable.

Rocker Arms

REMOVAL & INSTALLATION

6.0L Engines

⁂ WARNING

To prevent engine damage, always install new head bolts when installing the cylinder head, even if removing only one bolt. It will be necessary to install a new cylinder head gasket as well as all cylinder head bolts.

1. Before servicing the vehicle, refer to the precautions in the beginning of this section.

2. Remove the cylinder head.

➡Be careful when removing the rocker arm. Do not lose the rocker arm steel ball.

3. Push down on the rocker arm. Pull out on the bottom releasing the ball.

4. Remove the steel ball from the rocker arm.

5. Remove the rocker arm clip from the fulcrum plate.

6. Clean and inspect all parts for pitting or scuffing. Install new rocker arm and valve bridges, as needed.

To install:

7. Install the rocker arm clip on the fulcrum plate.

8. Place the steel ball in the rocker arm cup and lubricate with clean engine oil.

9. Push down on the rocker arm to compress the rocker arm clip. Insert the ball and release the pressure. Verify free movement of the rocker arm on the fulcrum.

10. Install the cylinder head.

7.3L Engines

1. Before servicing the vehicle, refer to the precautions in the beginning of this section.

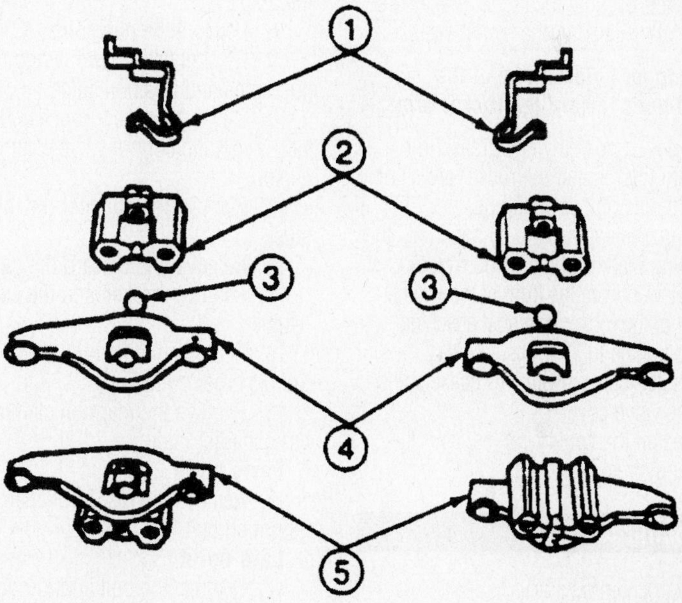

Item	Description
1	Snap Retaining Clip
2	Rocker Arm Pedestal
3	Rocker Arm Ball
4	Rocker Arm
5	Rocker Arm Assembly

7924FG58

Exploded view of the rocker arm assembly—7.3L diesel engines

2. Disconnect the ground cables from both batteries.

3. Remove the valve cover attaching screws and remove both valve covers.

4. Remove the valve rocker arm post mounting bolts. Remove the rocker arms and posts in order and mark them with tape so they can be installed in their original positions.

5. If the cylinder heads are to be removed, then the pushrods can now be removed. Make a holder for the pushrods out of a piece of wood or cardboard and remove the pushrods in order. It is very important that the pushrods be reinstalled in their original order. The pushrods can remain in position if no further disassembly is required.

To install:

6. If the pushrods were removed, install them in their original locations. Make sure they are fully seated in the tappet seats.

➡The copper colored end of the pushrod goes toward the rocker arm.

7. Apply a polyethylene grease to the valve stem tips. Install the rocker arms and posts in their original positions.

8. Turn the engine over by hand until the valve timing mark is at the 11:00 o'clock position, as viewed from the front of the engine. Install all of the rocker arm post attaching bolts and tighten to 20 ft. lbs. (27 Nm).

9. Install new valve cover gaskets and install the valve cover.

10. Install the battery cables, start the engine and check for leaks.

Turbocharger

REMOVAL & INSTALLATION

6.0L Engines

F-SERIES

1. Before servicing the vehicle, refer to the precautions in the beginning of this section.

2. Remove the air cleaner assembly.

✲✲ WARNING

To maintain the integrity of the coolant and the cooling system, add Motorcraft Premium Gold Engine Coolant VC-7-A (in Oregon VC-7-B) or equivalent meeting Ford specification WSS-M97B51-A1 (yellow color). Always fill the cooling system with the same type of coolant that was drained from the system. Do not mix coolant types.

Do not add orange-colored Motorcraft Specialty Orange Engine Coolant VC-2, or equivalent meeting WSS-M97B44-D. Mixing coolants may degrade the coolant's corrosion protection.

Do not add alcohol, methanol, brine, or any engine coolants mixed with alcohol or methanol antifreeze. These can cause engine damage from overheating or freezing.

3. Disconnect the engine vent and radiator vent hoses.

4. Remove the bolts and position the degas bottle aside.

5. Loosen the clamp at the turbocharger.

6. Remove the nuts and the turbocharger intake tube.

7. Disconnect the charge air cooler inlet pipe.

8. Remove the push pins.

9. Disconnect the two wiring harness push pins and position aside.

10. Disconnect the turbocharger variable vane hydraulic control valve electrical connector.

11. Remove the bolts for the oil supply tube.

12. Remove and discard the gasket.

13. Remove the bolt and the wire retainer.

14. Remove the marmon clamp from the turbocharger outlet.

15. Remove the marmon clamp from the turbocharger inlet.

Early build

16. Remove the rear turbocharger mounting bolt.

Late build

17. Remove the bolt and turbocharger oil supply tube.

18. Remove and discard the O-ring.

19. Remove the rear turbocharger mounting bolt.

All vehicles

20. Remove the front mounting bolts.

21. Position the turbocharger and remove the turbocharger drain tube.

22. Remove and discard the drain tube O-rings.

23. Remove the turbocharger. Remove the bolts and the turbocharger pedestal.

To install:

All vehicles

24. Position the turbocharger on the turbocharger pedestal.

➡Install new O-rings and apply clean engine oil.

25. Position the turbocharger and install the turbocharger drain tube.

Late build

26. Install the turbocharger and the rear mounting bolt. Torque to 28 ft. lbs. (38 Nm).

➡Install a new O-ring and apply clean engine oil.

27. Install the oil supply tube. Torque to 10 ft. lbs. (13 Nm).

28. Position the locking bracket and install the bolt.

Early build

29. Install the turbocharger and the rear mounting bolt. Torque to 28 ft. lbs. (38 Nm).

30. Install the oil feed tube.

All vehicles

31. Install the turbocharger front mounting bolts. Torque to 28 ft. lbs. (38 Nm).

32. Install the turbocharger inlet marmon clamp.

33. Install the turbocharger exhaust marmon clamp.

34. Position the wire retainer and install the bolt.

35. Prelubricate the oil inlet hole of the turbocharger assembly with clean engine oil and spin the compressor wheel several times to coat the bearing with oil.

36. Position the oil feed tube and install the bolts.

37. Install a new gasket.

38. Connect the turbocharger variable vane hydraulic control valve electrical connector.

39. Position the wiring harness and connect the push pins.

40. Install the push pins.

41. Connect the charge air cooler inlet pipe.

42. Install the turbocharger intake tube.

E-SERIES

1. Remove the engine cover.

2. Remove the retainers and the air deflector.

3. Remove the power steering reservoir bracket retainers.

4. Remove the power steering fluid indicator and retainers. Remove the power steering mounting bracket. Install the power steering fluid indicator and position the power steering reservoir aside.

5. Disconnect the coolant hoses from the air cleaner outlet pipe.

6. Loosen the clamps and remove the air cleaner outlet pipe.

7. Disconnect the turbocharger intake tube breather hose.

8. Loosen the clamp at the turbocharger.

9. Remove the retaining nuts and position the heater hose bracket aside.

10. Remove the stud bolts and the turbocharger intake tube.
11. Loosen the turbocharger outlet clamp.
12. Remove the turbocharger outlet clamp.
13. Loosen the pipe at the exhaust clamp and position aside.
14. Remove the left bolts for the turbocharger adapter pipe.
15. Remove the right bolts for the turbocharger adapter pipe.
16. Remove the exhaust gas recirculation (EGR) cooler clamp.
17. Remove the turbocharger inlet clamp and turbocharger adapter pipe.
18. Disconnect the VGT actuator electrical connector and retaining clip.
19. Remove the bolts from the turbocharger oil feed tube.
20. Remove and discard the gasket.
21. Remove the bolt and turbocharger oil feed tube.
22. Remove and discard the O-ring seals.
23. Remove the right turbocharger mounting bolts.
24. Remove the left turbocharger mounting bolts.

⁂ WARNING

Use care not to damage the turbocharger outlet hose when removing the turbocharger.

25. Position the turbocharger to remove the turbocharger drain tube. Remove the turbocharger from the vehicle.
26. Remove and discard the O-ring seals.
27. Remove the plug and drain the oil filter assembly and tubes.
28. Disconnect the wiring push-pin retainer.
29. Remove the bolt and retaining clamp at the oil filter adapter.
30. Remove the bolt and retaining clamp at the oil filter assembly.

➡Use care not to bend the tubes, when removing the oil filter tubes.

31. Remove the bolt and retaining clamp. Remove the oil filter tubes.
32. Remove the bolts and turbocharger pedestal.

To install:

➡Install new drain tube O-ring seals and apply clean engine oil.

➡Position the turbocharger outlet flange into the outlet hose as the turbocharger is positioned on the engine.

33. Position the turbocharger in the vehicle. Install the turbocharger drain tube and turbocharger.
34. Install the left turbocharger mounting bolts. Torque to 28 ft. lbs. (38 Nm).
35. Install the right turbocharger mounting bolts. Torque to 28 ft. lbs. (38 Nm).
36. Prelubricate the oil inlet hole of the turbocharger assembly with clean engine oil and spin the compressor wheel several times to coat the bearings with oil.

➡Use a new gasket when installing the oil feed tube.

37. Install the turbocharger oil feed tube, gasket and bolts. Torque to 18 ft. lbs. (25 Nm).

➡Install a new O-ring seal and apply clean engine oil.

38. Install the turbocharger oil feed tube. Position the clamp and install the bolt.
39. Connect the VGT actuator electrical connector and retaining clip.
40. Position the turbocharger adapter pipe and install the turbocharger inlet clamp.
41. Install the EGR cooler clamp.
42. Install the right bolts for the turbocharger adapter pipe. Torque to 20 ft. lbs. (27 Nm).
43. Install the left bolts for the turbocharger adapter pipe.
44. Position back the turbocharger exhaust pipe. Install the turbocharger exhaust clamp.
45. Tighten the retaining nuts at the exhaust clamp.
46. Tighten the turbocharger outlet clamp.
47. Install the turbocharger intake tube.
48. Inspect the O-ring seals for the oil filter tube ends, replace as necessary.
49. Check and top off the engine oil after the engine is running

7.3L Engines

1. Before servicing the vehicle, refer to the precautions in the beginning of this section.
2. Remove or disconnect the following:
   - Negative battery cable
   - The 2 air intake tube assembly bolts, clamps at the turbocharger, crankcase breather assembly, engine air cleaner and air intake tube and hoses.
   - Exhaust outlet clamp from the turbocharger
   - Engine charge exhaust pipe bolt from the transmission, if so equipped
   - Bolts and nuts from the catalytic converter-to-engine charge exhaust pipe, if so equipped
3. Loosen 2 bolts retaining the turbocharger exhaust inlet pipe to the left exhaust manifold.
4. For automatic transmissions, remove the bolts retaining the left turbocharger exhaust inlet pipe to the turbocharger exhaust inlet adapter.
5. Loosen the 2 bolts retaining the turbocharger exhaust inlet pipe to the right exhaust manifold.
6. Remove or disconnect the following:
   - Lower bolt retaining the right turbocharger exhaust inlet pipe to the turbocharger exhaust inlet adapter
7. For automatic transmissions, the upper bolts retaining the right and left turbocharger exhaust inlet pipes to the turbocharger exhaust inlet adapter.
   - Right engine lift hook and bolt
   - Air inlet hose clamp and hose at the turbocharger and lay the hose aside
   - Compressor manifold
   - The 4 bolts retaining the turbocharger pedestal assembly to the cylinder block
   - Turbocharger assembly and detach all electrical connectors from it

➡If the turbocharger is not being removed for service, install the Fuel/Oil turbo Protector Cap Set T94T-9395-AH or equivalent.

   - Oil gallery O-rings

To install:

8. Install or connect the following:
   - Oil gallery O-rings
   - Turbocharger electrical connectors
   - Turbocharger assembly
   - The 4 bolts retaining the turbocharger pedestal assembly to the engine block. Tighten the bolts to 18 ft. lbs. (25 Nm).
   - 4 bolts retaining the right and left turbocharger exhaust inlet pipes to the turbocharger exhaust inlet adapter, loosely.
   - Compressor manifold, intake manifold hoses and clamps. Be sure the compressor outlet seal is in position.
   - Right engine lift hook and bolt
   - The 2 right and left lower bolts (retaining the turbocharger exhaust inlet pipes to the turbocharger exhaust inlet adapter) to 36 ft. lbs. (49 Nm)

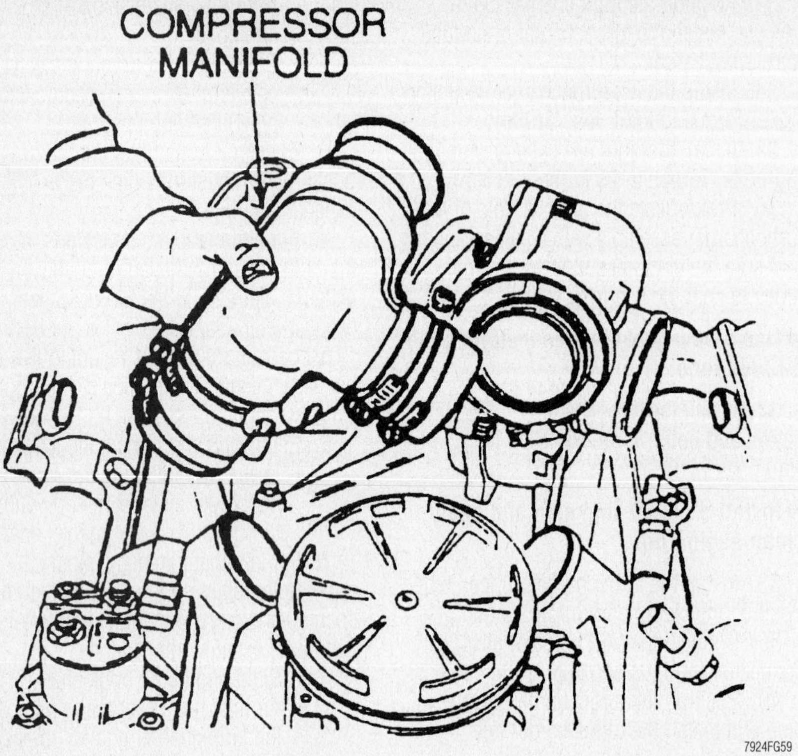

After loosening the clamps, the compressor manifold can be removed—7.3L diesel engines

- The 4 right and left bolts and nuts (retaining the turbocharger exhaust inlet pipes to the exhaust manifolds) to 36 ft. lbs. (49 Nm)
- Catalytic converter to the engine charge exhaust pipe bolts and nuts
- The 2 right and left upper bolts (retaining the turbocharger exhaust inlet pipes to the turbocharger exhaust inlet adapter) to 36 ft. lbs. (49 Nm)
- Exhaust outlet clamp to the turbocharger
- Air intake tube and hose assembly
- Negative battery cable

Intake Manifold

REMOVAL & INSTALLATION

6.0L Engine

E-SERIES

1. Before servicing the vehicle, refer to the precautions in the beginning of this section.

All vehicles

2. Disconnect the battery ground cable.
3. Remove the cooling fan stator.
4. Remove the turbocharger pedestal.

Vehicles with dual alternator

5. Remove the accessory drive belt.
6. Remove the bolt and the accessory drive belt tensioner.
7. Remove the accessory drive belt.
8. Remove the bolts, bracket and accessory drive belt idler pulley.
9. Remove the bolts and the accessory drive belt tensioner.
10. Disconnect the alternator electrical connector and the B+ wire.
11. Remove the bolts and the alternator with mounting bracket.

Vehicles with single alternator

12. Remove the accessory drive belt.

All vehicles

13. Remove the bolt and position the ground wire aside. Disconnect the electrical connector push pin.
14. Remove the conduit and position the wiring aside.
15. Disconnect the locking tab.
16. Remove the bolts.
17. Disconnect the push pin and remove the wiring harness from the conduit.
18. Remove the bolts for the charge air cooler tube and oil fill tube.
19. Disconnect the oil fill tube at the valve cover.
20. Remove the retaining nuts, charge air cooler tube, oil fill tube and bracket.
21. Disconnect the alternator electrical connector and the B+ wire. Disconnect the wiring push pin.
22. Remove the retaining nut for the transmission fluid indicator and tube.
23. Disconnect the push pin retainer. Remove the retaining nut and position the transmission fluid indicator and tube aside.
24. Remove the three bolts, ground wire and the alternator.
25. Remove the bolts and position the heater hose tube aside.
26. Remove and discard the O-ring.

➡It will be necessary to position back or remove the heat insulating wrap.

27. Remove the retaining nut and disconnect the wiring retainer and the injection pressure regulator valve electrical connector.
28. Disconnect the exhaust gas recirculation (EGR) valve electrical connector.
29. Disconnect the engine oil pressure (EOP) sensor electrical connector.
30. Disconnect the engine oil temperature (EOT) sensor electrical connector.
31. Disconnect the EGR throttle position control module electrical connector.
32. Disconnect the EGR throttle position sensor electrical connector.
33. Disconnect the pin-type retainer and engine coolant temperature (ECT) sensor.
34. Remove the ECT sensor.
35. Plug or cap the opening as needed.
36. Disconnect the intake air temperature (IAT2) sensor electrical connector.
37. Remove the IAT2 sensor.
38. Plug or cap the opening as needed.
39. Disconnect the eight fuel injector electrical connectors and wiring connectors.
40. Disconnect the harness retainers. Position the engine wiring harness as needed for intake manifold removal.
41. Disconnect the manifold absolute pressure (MAP) sensor hose.
42. Disconnect the engine coolant vent hose.
43. Remove the secondary fuel filter and remove all fuel from the filter housing.
44. Disconnect the fuel tube fittings at the secondary fuel filter.
45. Remove and discard the copper sealing washers.
46. Disconnect the fuel tube at the fuel filter housing. Remove the retaining nut.
47. Remove the banjo bolt and fuel tube.
48. Discard the copper sealing washers.
49. Remove the bolts and the secondary fuel filter assembly.
50. Remove the nuts and turbocharger heat shield.

➡Align the flat edge with the index feature located on the coolant supply port.

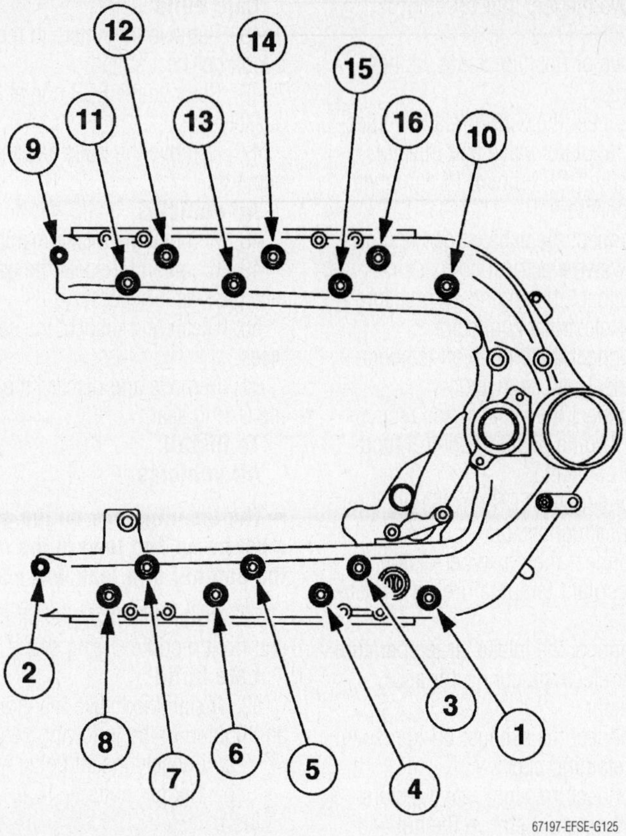

Intake manifold torque sequence—6.0L engine

51. Pull the EGR cooler clamp forward, twist and then slide the EGR cooler hose rearward to remove.
52. Remove the bolts and the intake manifold.
53. Remove the intake manifold gaskets.
54. Clean and inspect the gaskets. Install new gaskets if necessary.
55. Clean and inspect the sealing surfaces.
56. Remove and discard the front module O-ring seal.

To install:

➡The locating tabs on the gaskets must be positioned upward and toward the center of the engine, or a leak will occur.

57. Install the intake manifold gaskets. Install a new front module O-ring seal.
58. Install the intake manifold and bolts and tighten in the following sequence.
   a. Loosely install bolts 1–8.
   b. Tighten bolts 9–16 to 11 Nm (8 ft. lbs.).
   c. Tighten all bolts to 11 Nm (8 ft. lbs.).
59. Slide the EGR cooler hose forward and rotate the flat to lock.
60. Install the turbocharger heat shield and nuts.
61. Install the secondary fuel filter assembly and bolts. Toque to 18 ft. lbs. (25 Nm).

➡Install new copper sealing washers.

62. Install the fuel line and banjo bolt. Torque to 28 ft. lbs. (38 Nm).
63. Install the retaining nut and connect the fuel tube at the fuel filter assembly. Torque to 19 ft. lbs. (26 Nm).

➡Install new copper sealing washers on the banjo fitting.

64. Reposition and connect the fuel tubes at the fuel filter housing.
65. Install the secondary fuel filter and cover. Torque to 10 ft. lbs. (14 Nm).

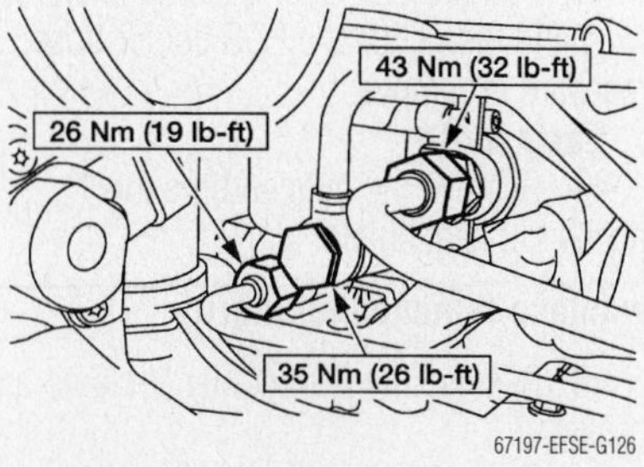

Reposition and connect the fuel tubes at the fuel filter housing

66. Connect the engine coolant vent hose.
67. Connect the MAP sensor hose.
68. Position back the engine wiring harness as needed. Connect the harness retainers.
69. Connect the eight fuel injector electrical connectors.
70. Remove the plug or cap. Install the IAT2 sensor.
71. Connect the IAT2 sensor electrical connector.
72. Remove the plug or cap. Install the ECT sensor.
73. Connect the ECT sensor electrical connector and the pin-type retainer.
74. Connect the EGR throttle position sensor electrical connector.
75. Connect the EGR throttle position control module electrical connector.
76. Connect the EOT sensor electrical connector.
77. Connect the EOP sensor electrical connector.
78. Connect the EGR valve electrical connector.
79. Connect the injector pressure regulator valve electrical connector and position back the heat insulating wrap. Install the wiring retainer and retaining nut.

➡Install a new O-ring seal and apply clean engine coolant

80. Install the heater tube and bolts.
81. Install the alternator, ground wire and the three bolts.
82. Position back the transmission fluid indicator and tube and install the retaining nut.
83. Connect the push pin retainer.
84. Install the retaining nut for the transmission fluid indicator and tube.
85. Connect the alternator B+ wire and electrical connector. Position back the boot. Connect the wiring push pin.
86. Position the charge air cooler tube, oil fill tube and bracket. Install the retaining nuts.
87. Connect the oil fill tube at the valve cover.
88. Install the bolts for the charge air cooler tube and oil fill tube.
89. Install the conduit and bolts.
90. Install the wiring harness into the conduit and connect the push pin.
91. Install the bolts.
92. Connect the locking tab.
93. Connect the electrical connector push pin. Position back the ground wire and install the bolt.

Vehicles with single alternator

94. Install the accessory drive belt.

Vehicles with dual alternator

95. Install the alternator with mounting bracket and bolts.
96. Connect the alternator B+ wire and electrical connector. Position the boot.
97. Install the accessory drive belt tensioner and bolts.
98. Position the accessory drive belt idler pulley. Install the bracket and bolts.
99. Install the accessory drive belt.
100. Install the accessory drive belt tensioner and bolt.
101. Install the accessory drive belt.

All vehicles

102. Install the cooling fan stator.
103. Install the turbocharger pedestal and turbocharger.
104. Connect the battery ground cable.

F-SERIES

1. Remove the auxiliary battery.
2. Remove the cooling fan stator.
3. Remove the degas bottle.
4. Remove the upper radiator hose.
5. Remove the turbocharger-to-charge air cooler duct.

Vehicles with dual alternators

6. Remove the accessory drive belt.
7. Remove the bolt and the accessory drive belt tensioner.
8. Remove the accessory drive belt.
9. Remove the bolts, bracket and accessory drive belt idler pulley.
10. Remove the bolts and the accessory drive belt tensioner.
11. Disconnect the wire retainer, alternator electrical connector and B+ wire.
12. Remove the bolts and the alternator with mounting bracket.

Vehicles with single alternator

13. Remove the accessory drive belt.

All vehicles

14. Position the boot back and disconnect the alternator B+ wire and electrical connector.
15. Remove the three bolts and the alternator.
16. Remove the turbocharger and the turbocharger pedestal.

Early build

17. Loosen the clamps and remove the charge air cooler duct.

Late build

18. Loosen the clamps and remove the charge air cooler duct.
19. Remove the bolts and position the heater hose tube aside.
20. Remove and discard the O-ring.
21. Disconnect the manifold absolute pressure (MAP) sensor hose.
22. Disconnect the engine coolant vent hose.
23. Remove the fuel injector control module.
24. Remove or position aside the heat insulating wrap.
25. Disconnect the wiring retainer, injector pressure regulator valve and injector control pressure (ICP) sensor (if equipped) electrical connectors.
26. Disconnect the exhaust gas recirculation (EGR) valve electrical connector.
27. Disconnect the engine oil pressure (EOP) sensor electrical connector.
28. Disconnect the engine oil temperature sensor electrical connector.
29. Disconnect the exhaust gas recirculation (EGR) throttle position control module electrical connector.
30. Disconnect the EGR throttle position sensor electrical connector.
31. Disconnect the pin-type retainer and engine coolant temperature (ECT) sensor.
32. Disconnect the intake air temperature (IAT2) sensor electrical connector and wiring connector.
33. Disconnect the exhaust backpressure sensor and retaining clip.
34. Disconnect the eight fuel injectors electrical connectors. Remove the nut and the fuel injector wiring harness.

➡It is necessary to remove the fuel filter and drain the housing.

35. Disconnect the fuel line fittings.

➡It is necessary to remove the oil filter and drain the housing.

36. Remove the four bolts and the oil filter housing.
37. Remove the bolt and the oil filter return tube.
38. Remove the fuel line.
39. Remove the bolt, and the banjo fitting from the fuel line.
40. Discard the sealing washers and remove the fuel line.

➡Align the flat edge with the index feature located on the coolant supply port.

41. Pull the EGR cooler clamp forward, twist and then slide the EGR cooler hose rearward to remove.

Early build

42. Remove the nuts and the turbocharger heat shield.

➡Intake removed for clarity.

43. Remove the EGR cooler V-clamp and gasket.
44. Remove the bolts and the intake manifold.

Late build

45. Remove the nuts and the turbocharger heat shield.
46. Remove the EGR cooler V-clamp and gasket.
47. Remove the bolts and the intake manifold.

All vehicles

48. Remove the intake manifold gaskets.
49. Clean and inspect the gaskets. Install new gaskets if necessary.
50. Clean and inspect the sealing surfaces.
51. Remove and discard the front module O-ring seal.

To install:

All vehicles

➡The locating tabs on the gaskets must be up and toward the center of the engine, or a leak will occur.

52. Install the intake manifold gaskets and front module O-ring seal.

Late build

53. Install the intake manifold and bolts and tighten in the following sequence.
    a. Loosely install bolts 1–8.
    b. Tighten bolts 9–16 to 11 Nm (8 ft. lbs.).
    c. Tighten all bolts to 11 Nm (8 ft. lbs.) in the sequence shown.
54. Install the gasket and the EGR cooler V-clamp.
55. Install the turbocharger heat shield and the nuts.

Early build

56. Install the intake manifold and bolts and tighten in the following sequence.
    a. Loosely install bolts 1–8.
    b. Tighten bolts 9–16 to 11 Nm (8 ft. lbs.).
    c. Tighten all bolts to 11 Nm (8 ft. lbs.) in the sequence shown.
57. Install the gasket and the EGR cooler V-clamp.
58. Install the turbocharger heat shield and the nuts.

All vehicles

59. Slide the EGR cooler hose forward and rotate flat to lock.
60. Install the fuel line.
61. Install new sealing washers.
62. Install the fuel line.
63. Install the banjo fitting and the bolt.
64. Install the oil filter return tube and bolt.
65. On new oil filter return tubes, tighten to 6 Nm (53 inch lbs.).
66. On used oil filter return tubes, tighten to 3 Nm (27 inch lbs.).
67. Install the oil filter housing and the four bolts.

68. Clean and inspect the housing O-rings. Install new O-rings if necessary.
69. Connect the fuel line fittings.
70. Position the fuel injector harness and install the retaining nut. Connect the eight fuel injectors electrical connectors.
71. Connect the retaining clip and exhaust backpressure sensor electrical connector.
72. Connect the wiring retainer and the IAT2 sensor electrical connector.
73. Connect the engine coolant temperature (ECT) sensor and the pin-type retainer.
74. Connect the EGR throttle position sensor electrical connector.
75. Connect the EGR throttle position control module electrical connector.
76. Connect the engine oil temperature (EOT) sensor electrical connector.
77. Connect the engine oil pressure (EOP) sensor electrical connector.
78. Connect the EGR valve electrical connector.
79. Connect the wiring retainer, injector pressure regulator and ICP sensor (if equipped) electrical connectors.
80. Position back or install the heat insulating wrap.
81. Install the fuel injector control module.
82. Connect the engine coolant vent hose and clamp.
83. Connect the MAP sensor hose.

➡Install a new O-ring on the heater hose tube.

84. Install the heater hose tube and bolts.

Early build

85. Install the charge air cooler duct and tighten the clamps.

Late build

86. Install the charge air cooler duct and tighten the clamp.
87. Install the turbocharger pedestal and turbocharger.
88. Install the alternator and the three bolts.
89. Connect the alternator B+ wire land electrical connector and position the boot back.

Vehicles with single alternator

90. Install the accessory drive belt.

Vehicles with dual alternators

91. Install the alternator with mounting bracket and bolts.
92. Connect the B+ wire, alternator electrical connector and wire retainer.
93. Install the accessory drive belt tensioner and bolts.
94. Position the accessory drive belt idler pulley. Install the bracket and bolts.
95. Install the accessory drive belt.
96. Install the accessory drive belt tensioner and bolt.
97. Install the accessory drive belt.

All vehicles

98. Install the turbocharger-to-charge air cooler duct.
99. Install the upper radiator hose.
100. Install the degas bottle.
101. Install the auxiliary battery.
102. Install the cooling fan stator

7.3L Engine

⁑ CAUTION

The fuel system remains under pressure, even after the engine has been turned OFF. The fuel system pressure must be relieved before disconnecting any fuel lines. Failure to do so may result in fire and/or personal injury.

1. Before servicing the vehicle, refer to the precautions in the beginning of this section.
2. Relieve the fuel system pressure.
3. Remove or disconnect the following:
   - Both battery ground cables
   - Air cleaner and install clean rags into the air intake of the intake manifold. It is important that no dirt or foreign objects get into the intake.
   - Injection pump
   - Fuel return hose from No. 7 and No. 8 rear nozzles and remove the return hose to the fuel tank.
   - Engine wiring harness from the engine

➡The engine harness ground cables must be removed from the back of the left cylinder head.

   - Bolts attaching the intake manifold to the cylinder heads and remove the manifold.
   - Crankcase Depression Regulator (CDR) valve tube grommet from the valley pan
   - Bolts attaching the valley pan strap to the front of the engine block and remove the strap.
   - Valley pan drain plug and remove the valley pan.

To install:

4. Apply a ⅛ inch (3mm) bead of RTV sealer to each end of the cylinder block.

➡The RTV sealer should be applied immediately prior to the valley pan installation.

5. Install or connect the following:
   - Valley pan drain plug, CDR valve tube and new grommet into the valley pan
   - New O-ring and new back-up ring on the CDR valve
   - Valley pan strap on the front of the valley pan
   - Intake manifold and tighten the bolts to 24 ft. lbs. (33 Nm) using the sequence illustrated
   - Engine wiring harness and the engine ground wire located to the rear of the left cylinder head
   - Injection pump
   - No. 7 and No. 8 fuel return hoses and the fuel tank return hose
   - Air cleaner
   - Battery ground cables to both batteries

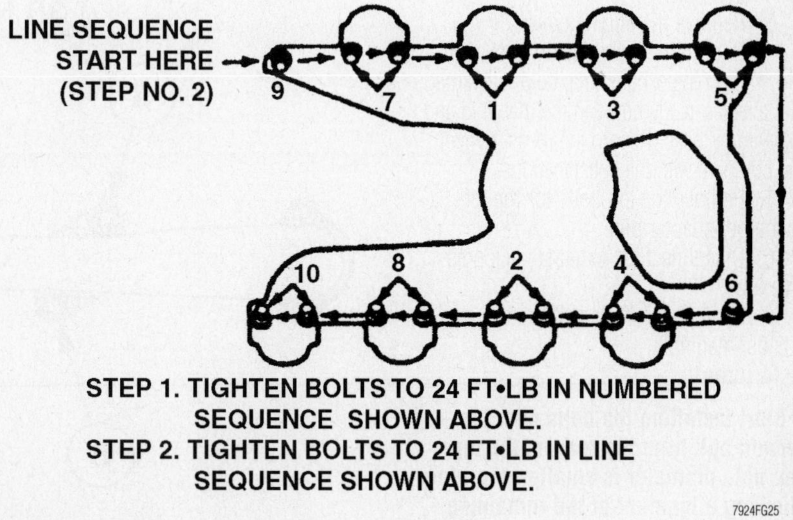
LINE SEQUENCE
START HERE
(STEP NO. 2)
9
7
1
3
5
10
8
2
4
6
STEP 1. TIGHTEN BOLTS TO 24 FT•LB IN NUMBERED SEQUENCE SHOWN ABOVE.
STEP 2. TIGHTEN BOLTS TO 24 FT•LB IN LINE SEQUENCE SHOWN ABOVE.
7924FG25

Intake manifold bolt torque sequence—7.3L diesel engines

➡If necessary, purge the nozzle high pressure lines of air by loosening the connector ½ to 1 turn and cranking the engine until solid stream of fuel, devoid of any bubbles, flows from the connection.

6. Run the engine and check for oil and fuel leaks.

⁂ CAUTION

Keep eyes and hands away from the nozzle spray. Fuel spraying from the nozzle under high pressure can penetrate the skin.

7. Check and adjust the injection pump timing.

Exhaust Manifold

REMOVAL & INSTALLATION

6.0L Engine

F-SERIES

1. Before servicing the vehicle, refer to the precautions in the beginning of this section.
2. Remove the engine.
3. Mount the engine on a work stand.
4. Remove the manifold(s).
5. Installation is the reverse of removal.
6. Inspect the manifold and head surfaces of warpage.
7. Installation is the reverse of removal.

E-SERIES—LEFT SIDE

1. Before servicing the vehicle, refer to the precautions in the beginning of this section.
2. With the vehicle in **Neutral**, position it on a hoist.
3. Remove the engine cover.
4. Prior to removing the exhaust manifold, inspect the exhaust manifold for warpage with a feeler gauge between the manifold and the cylinder head. Record the measurement and compare with the specifications.
5. Remove the left bolts for the turbocharger adapter pipe.
6. Disconnect the exhaust back pressure tube.
7. Remove the bolts and the left exhaust manifold.

To install:

➡Start installing the bolts with the second bolt from the rear on the top. The hole diameter is smaller, therefore allowing alignment of the remaining bolts.

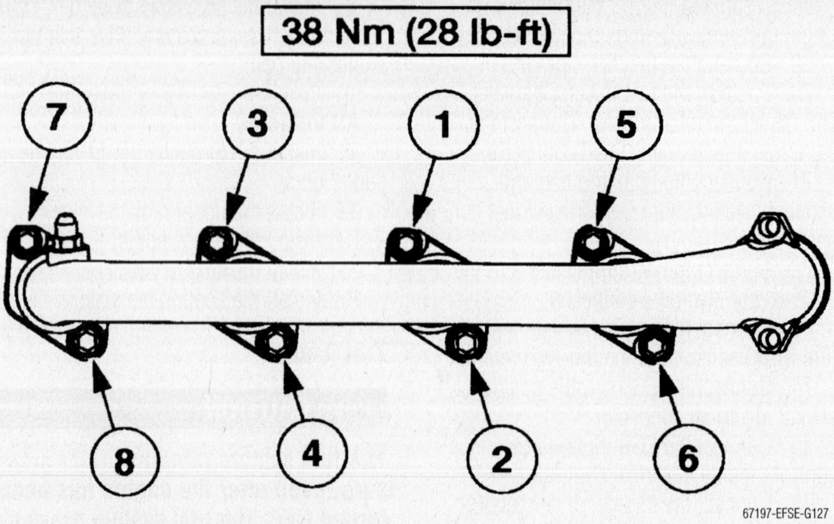

Left exhaust manifold torque sequence—6.0L engine

➡Apply anti-seize lubricant to the bolt threads prior to installing the bolts.

8. Install the left exhaust manifold and bolts. Tighten the bolts in the sequence shown.
9. Connect the exhaust back pressure tube at the exhaust manifold.
10. Install the left bolts for the turbocharger adapter pipe.
11. Install the engine cover.

E-SERIES—RIGHT SIDE

1. With the vehicle in **Neutral**, position it on a hoist.
2. Disconnect the battery ground cable.
3. Remove the engine cover.
4. Prior to removing the exhaust manifold, inspect the exhaust manifold for warpage with a feeler gauge between the manifold and the cylinder head. Record the measurement and compare with the specifications.
5. Remove the right bolts for the turbocharger adapter pipe.
6. Remove the bolts and the right exhaust manifold.

To install:

➡Start installing the bolts with the second bolt from the rear on the top. The hole diameter is smaller, therefore allowing alignment of the remaining bolts.

➡Apply anti-seize lubricant to the bolt threads prior to installing the bolts.

7. Install the right exhaust manifold and bolts. Tighten the bolts in the sequence shown.
8. Install the right bolts for the turbocharger adapter pipe.

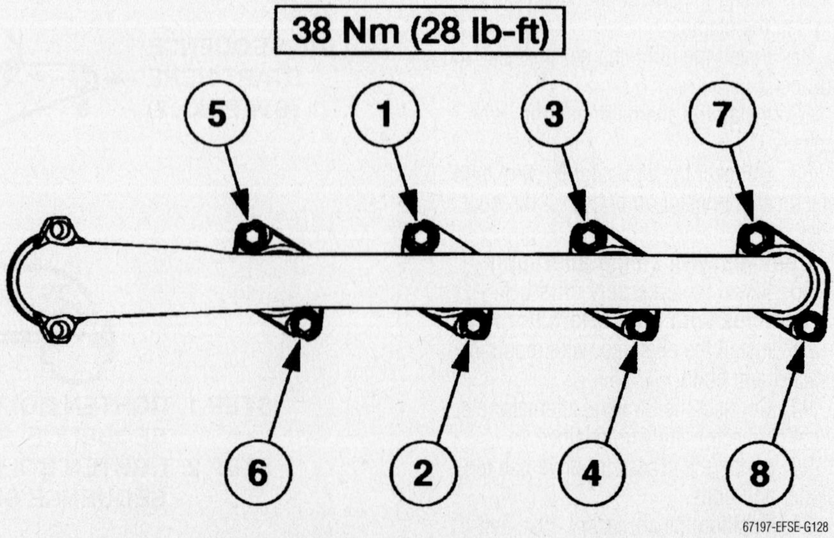

Right exhaust manifold torque sequence—6.0L engine

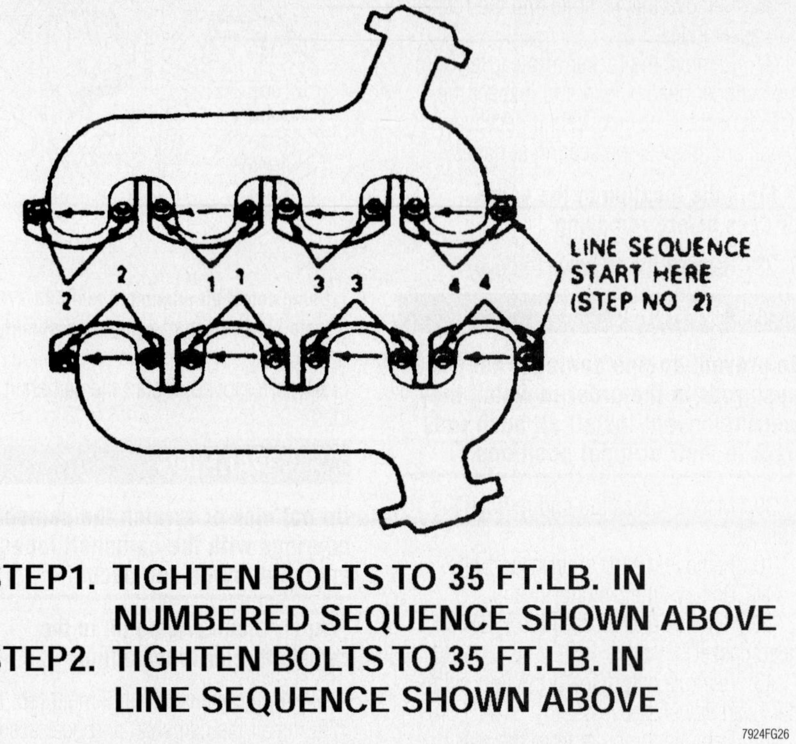
LINE SEQUENCE START HERE (STEP NO 2)
2
2
1
1
3
3
4
4
STEP1. TIGHTEN BOLTS TO 35 FT.LB. IN NUMBERED SEQUENCE SHOWN ABOVE
STEP2. TIGHTEN BOLTS TO 35 FT.LB. IN LINE SEQUENCE SHOWN ABOVE
7924FG26

Tighten the manifold bolts in the proper sequence—7.3L diesel engines

9. Install the engine cover.
10. Connect the battery ground cable.

7.3L Engine

1. Before servicing the vehicle, refer to the precautions in the beginning of this section.
2. Disconnect the ground cables from both batteries.
3. Raise the vehicle and safely support it.
4. Disconnect the muffler inlet pipe from the exhaust manifolds.
5. If removing the right manifold, lower the vehicle. When removing the left manifold, raise the vehicle and remove the manifold from underneath. Bend the tabs on the manifold attaching bolts, then remove the bolts and manifold.

To install:

6. Before installing, clean all mounting surfaces on the cylinder heads and the manifold. Apply an anti-seize compound on the manifold both threads and install the left manifold, using a new gasket and new locking tabs.
7. Tighten the bolts to 45 ft. lbs. (61 Nm) and bend the tabs over the flats on the bolt heads to prevent the bolts from loosening.
8. Raise the vehicle to install the right manifold. Install the right manifold steps 5 and 6.
9. Connect the inlet pipes to the manifold and tighten. Lower the vehicle, connect the batteries and run the engine to check for exhaust leaks.

Camshaft and Valve Lifters

REMOVAL & INSTALLATION

6.0L Engine

1. Before servicing the vehicle, refer to the precautions in the beginning of this section.
2. Mount the engine on an engine stand.
3. Remove the serpentine belt idler.
4. Remove the serpentine belt tensioner.
5. Remove and discard the O-rings from the oil filter base.
6. Remove the exhaust gas recirculation (EGR) cooler coolant supply port cover. Clean and inspect the gaskets. Install new gaskets if necessary. Clean and inspect the sealing surfaces.

⁂ WARNING

In the event of a catastrophic engine failure, always install a new oil cooler cover assembly (with oil cooler). Foreign material cannot be removed from the oil cooler.

➡The oil cooler is replaced as an assembly.

7. Remove the oil cooler assembly. Clean and inspect the gaskets. Install a new gasket if necessary. Clean and inspect the sealing surfaces.
8. Remove the oil pump inlet strainer. Clean and inspect for tears and other damage.
9. Remove bolts and the turbocharger heat shield.
10. Remove the high-pressure oil pump cover. Use a thin gasket scraper to separate the cover from the crankcase. Clean and inspect the gaskets. Install a new gasket if necessary. Clean and inspect the sealing surfaces.
11. Remove the bolts from the high-pressure oil pump discharge pipe.
12. Disconnect and remove the high-pressure oil pump discharge pipe.
13. Remove and discard the D-ring seal.
14. Remove and discard the high-pressure pump O-ring seal.
15. Remove the bolts and the high-pressure oil pump.
16. Remove and discard the lower O-ring seal.
17. Remove the glow plug buss bar.
18. Remove the eight glow plugs.

➡Mark the location of the stud bolts.

19. Remove the valve covers. Clean and inspect the gaskets. Install a new gasket if necessary. Clean and inspect the sealing surfaces.
20. Remove the bolts and the coolant pump pulley.
21. Remove the coolant pump.
22. If equipped, remove the bolts and the dual alternator pulley.
23. Prior to removing the crankshaft damper, check the crankshaft vibration damper runout.

➡Pry the crankshaft forward at the same point to eliminate possible error caused by crankshaft end play.

24. Rotate the crankshaft 90 degrees. Pry the crankshaft forward. Record the measurement. Repeat every 90 degrees. If the runout exceeds 0.002 inch, install a new crankshaft vibration damper.

⁂ WARNING

To prevent engine damage, you must always replace all four bolts when installing the vibration damper.

✲✲ CAUTION

To avoid personal injury, support the vibration damper during mounting bolt removal. The damper can slide off the nose of the crankshaft very easily.

25. Remove the bolts and the crankcase vibration damper. Discard the bolts.
26. Punch two holes in the seal. Remove the crankshaft seal with a slide-hammer.

➡Production engine will not have a wear sleeve. If equipped, remove the crankshaft damper wear sleeve.

27. Remove the oil pump body. Remove and discard the O-ring seal.

➡Mark the front of each drive rotor for correct reassembly orientation.

28. Remove the inner and outer oil pump drive rotors.
29. Remove the engine front cover. Clean and inspect the gaskets. Install new gaskets if necessary. Clean and inspect the sealing surfaces.
30. Using a quick-disconnect tool, disconnect the high-pressure oil rail supply line at the high-pressure oil rail.
31. Remove the bolts and the high pressure oil rail. Disconnect and remove the high-pressure oil supply line.

✲✲ WARNING

Do not attempt to put battery voltage to the fuel injector or damage to the fuel injector will occur.

32. Using a 19 mm socket, push the fuel injector electrical connector out of the rocker arm carrier.

✲✲ WARNING

To prevent engine damage, do not use air tools to remove the fuel injectors. The clip that extracts the injector can dislodge and fall into the oil drain hole.

➡If engine oil is found in the engine coolant or engine coolant is found in the combustion chambers, new injector sleeve may need to be installed.

33. Remove the bolt, the fuel injector hold down and the fuel injector.
34. Remove and discard the crankcase-to-head tube assembly.
35. Remove the inner head bolts from both cylinder heads.
36. Remove the 16 bolts and the rocker arm assemblies.
37. Remove the rocker arm carrier from the cylinder head. Clean and inspect the gaskets. Install new gaskets if necessary. Clean and inspect the sealing surfaces.

➡Mark the location of the valve bridges before removing.

38. Remove the 16 valve bridges.

✲✲ WARNING

To prevent engine damage, keep the push rods in the order in which they were removed. Install all push rods back in their original positions.

39. Mark the location and remove the 16 push rods.
40. Remove the 10 outer head bolts.
41. Remove the cylinder heads.
42. Remove and discard the cylinder head gasket.
43. Remove and discard the four cylinder head dowel sleeves.
44. Remove the bolts from the rear engine tube assembly.
45. Remove the bolt and the rear engine tube assembly.

✲✲ WARNING

To prevent engine damage, keep the cam followers in the order in which they were removed. Install all cam followers back in their original positions.

46. Remove the bolts and the roller follower guides. Remove the hydraulic cam followers.
47. Install the special tool and measure the camshaft gear backlash. Install a new camshaft gear if backlash is not within specification.
48. Install a dial indicator and measure the camshaft end play. Install a new camshaft thrust plate if end play is not within 0.002–0.008 inch.
49. Remove the bolt and the camshaft position (CMP) sensor.

✲✲ WARNING

Do not knick or scratch the camshaft bearings with the camshaft lobes or engine damage will occur.

50. Remove the thrust plate mounting bolts and remove the camshaft and gear.

To install:

➡Check alignment of the oil holes after installing the bearings.

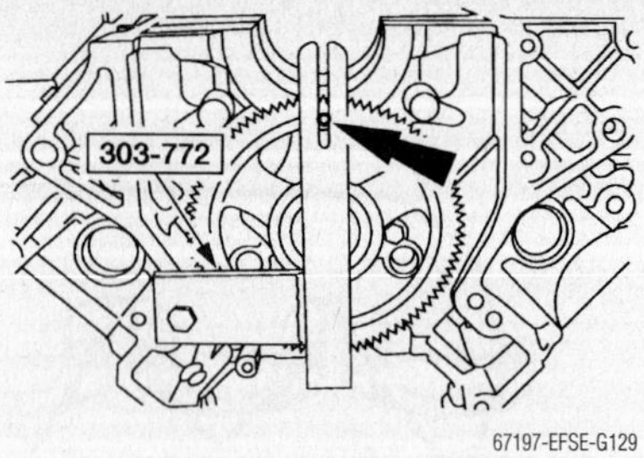

Using camshaft alignment tool 303-772, install the camshaft and gear assembly

51. If removed, install the camshaft bearings.

✲✲ WARNING

Do not nick or scratch the camshaft bearings with the camshaft lobes or engine damage can occur.

➡Apply clean engine oil to the camshaft prior to installing.

52. Using camshaft alignment tool 303-772, install the camshaft and gear assembly. Aligning it with the crankshaft. Install the thrust plate mounting bolts. Torque to 23 ft. lbs. (31 Nm).
53. The remainder of installation is the reverse of removal.

7.3L Engine

Ford recommends removing the diesel engine from the vehicle for camshaft removal.

➡A camshaft removal tool, Ford part no. T65L-6250-a and adapter 14-0314 is needed to remove the diesel camshaft.

1. Before servicing the vehicle, refer to the precautions in the beginning of this section.
2. Remove the intake manifold and valley pan, if equipped.
3. Remove the rocker covers, and either remove the rocker arm shafts or loosen the rockers on their pivots and remove the pushrods. The pushrods must be reinstalled in their original positions.
4. Remove the valve lifters in sequence with a magnet. They must be replaced in their original positions.
5. Remove the timing gear cover, timing gear and sprockets.

To install:

6. liberally coat the camshaft with oil before installing it. Slide the camshaft into the engine very carefully so as not to scratch the bearing bores with the camshaft lobes. Install the camshaft thrust plate and

tighten the attaching screws to 10–12 ft. lbs. (13–16 Nm). Measure the camshaft end-play. If the end-play is more than 0.009 inch (0.228mm), replace the thrust plate. Assemble the remaining components in the reverse order of removal.

7. Install the timing gear and front cover.
8. Install the valve lifters. They must be replaced in their original positions.
9. Install the pushrods, the rocker arms and the rocker arm covers.
10. Install the intake manifold and valley pan, if equipped.

Valve Lash

ADJUSTMENT

Valve lash on the diesel engine is not adjustable.

Starter Motor

REMOVAL & INSTALLATION

6.0L Engine

1. Before servicing the vehicle, refer to the precautions in the beginning of this section.

**** WARNING**

When performing maintenance on the starting system, be aware heavy gauge leads are connected to the battery. Make sure protective caps are in place when maintenance is complete.

2. Disconnect the battery ground cable.
3. Raise and support the vehicle.
4. Remove starter solenoid protective cap.
5. Disconnect the starter motor electrical connections.
6. Remove the bolts and the starter.
7. To install, reverse the removal procedure. Torque the bolts to 18 ft. lbs. (25Nm).

7.3L Engine

1. Before servicing the vehicle, refer to the precautions in the beginning of this section.
2. Remove or disconnect the following:
   - Negative battery cable
   - Starter motor electrical connections
   - Starter motor bolts and the motor

To install:

3. Installation is the reverse of removal, tighten the starter motor bolts to 16–20 ft. lbs. (22–28 Nm).

Oil Pan

REMOVAL & INSTALLATION

F-Series w/6.0L Engine

1. Before servicing the vehicle, refer to the precautions in the beginning of this section.
2. With the vehicle in **Neutral**, position it on a hoist.
3. Disconnect the negative battery cable.

**** WARNING**

Never remove the pressure relief cap while the engine is operating or when the cooling system is hot. Failure to follow these instructions can result in damage to the cooling system or engine or result in personal injury. To avoid having scalding hot coolant or steam blow out of the degas bottle when removing the pressure relief cap, wait until the engine has cooled then wrap a thick cloth around the pressure relief cap and turn it slowly. Step back while the pressure is released from the cooling system. When certain all the pressure has been released, (still with a cloth) turn and remove the pressure relief cap. Failure to follow these instructions can result in personal injury.

**** CAUTION**

The coolant must be removed in a suitable, clean container for reuse. If the coolant is contaminated, it must be recycled or disposed of correctly and the system filled with new coolant.

➡Less than 80% of coolant capacity can be recovered with the engine in the vehicle. Dirty, rusty, or contaminated coolant requires replacement.

4. Place a suitable container below the radiator draincock. If equipped, disconnect the coolant return hose at the fluid cooler.
5. Remove the fill cap from the degas bottle.
6. Remove the oil pan drain plug and drain the engine oil.
7. Loosen the exhaust pipe retaining nuts.
8. Remove the motor mount retaining nuts.
9. Open the radiator draincock.
10. Disconnect the lower radiator hose.
11. Disconnect the transmission cooler tubes.
12. Remove the air cleaner assembly.
13. On vehicles with metal ducts, remove the charge air cooler duct.
14. On vehicles with blow molded ducts, loosen the clamps and remove the charge air cooler duct.
15. Remove the radiator support brackets.
16. Remove the 4 pin-type retainers and pull back the sight shield.
17. With the sight shield pulled back, remove the 3 pin-type retainers and the 2 wiring retainers and position the harness rearward out of the way.
18. Disconnect the upper radiator hose and the radiator overflow hose.
19. Remove the radiator and shroud as an assembly.
20. Disconnect the cooling fan electrical connector. Unclip and position the fan and wiring aside.

➡Use a hole in the fan hub to prevent the fan from turning.

21. Using the special tool 303-591, or equivalent, loosen the fan clutch. Remove the cooling fan.
22. Remove the bolts from the stator, remove the stator.
23. On vehicles with dual alternator perform the following:
    a. Remove the secondary accessory drive belt.
    b. Remove the bolt and the secondary accessory drive belt tensioner.
    c. Remove the primary accessory drive belt.
    d. Remove the bolts, bracket and secondary accessory drive belt idler pulley.
    e. Remove the bolts and the primary accessory drive belt tensioner.
    f. Disconnect the wire retainer, secondary alternator electrical connector and B+ wire.
    g. Remove the bolts and the secondary alternator with mounting bracket.
24. On vehicles with single alternator, remove the primary accessory drive belt.
25. Remove the bolts and position the alternator back.
26. Support the hood and disconnect the hood lift assemblies.
27. Install two lifting eyes 303-D030, or equivalent, on the right hand cylinder head.
28. Remove the retaining bolt for the fuel lines.

29. Disconnect the heater hose at the coolant pump.
30. Remove the fan shroud mounting stud.
31. Install one lifting eye 303-D030 on the left hand cylinder head.

⁂ CAUTION

Do not use the special tool to raise the engine. Damage to the special tool or vehicle may occur.

➡The ball studs may have to be removed.

➡This procedure requires a second bolt hook assembly. The tools are available through Rotunda tools with the following numbers: bolt hook 303-F070-6, handle 303-F070-8, bracket 303-F070-7 and washer 303-F070-12004.

32. Install the special tool.

➡The engine must be raised evenly.

33. Using a lifting crane, raise the engine until the turbo charger is about to touch the cowl. Secure the engine with the special tool.
34. Remove the transmission cooler line bracket.
35. Remove the bolts and position back the oil pan until the oil pick-up tube bolts are accessible.
36. Remove the bolts and let the oil pick-up tube go into the oil pan. Remove the oil pan.
37. Remove the press-in-place gasket and discard.
38. Clean and inspect the sealing surfaces.
39. Remove and discard the oil pick-up tube O-ring.

To install:

40. Install a new O-ring on the oil pick-up tube and position the oil pick-up tube in the oil pan.
41. Install a new press-in-place gasket into the upper oil pan.
42. Position the oil pan in the vehicle.
43. Install the oil pick-up tube and bolts. Tighten to 10 ft. lbs. (13 Nm).
44. Position the oil pan and install the bolts. Tighten to 10 ft. lbs. (13 Nm).
45. Install the transmission cooler tube bracket and nut. Tighten to 18 inch lbs. (10 Nm).
46. Using the special tool 303-F070, lower the engine.
47. Remove the lifting eyes 303-D030 and the special tool.
48. Install the fan shroud mounting stud. Tighten to 30 ft. lbs. (40 Nm).
49. Connect the heater hose at the coolant pump.
50. Install the retaining bolt for the fuel lines. Tighten to 10 ft. lbs. (10 Nm).
51. If removed, install the ball studs.
52. Connect the hood lifts and remove the hood support.
53. Position the primary alternator and install the bolts.
54. On a vehicle with a single alternator, install the primary accessory drive belt.
55. On vehicles with dual alternators, perform the following:
    a. Install the secondary alternator with mounting bracket and bolts. Tighten to 35 ft. lbs. (47 Nm).
    b. Connect the B+ wire, secondary alternator electrical connector and wire retainer.
    c. Install the primary accessory drive belt tensioner and bolts. Tighten to 18 ft. lbs. (25 Nm).
    d. Position the secondary accessory drive belt idler pulley. Install the bracket and bolts. Tighten to 18 ft. lbs. (25 Nm).
    e. Install the primary accessory drive belt.
    f. Install the secondary accessory drive belt tensioner and bolt. Tighten to 18 ft. lbs. (25 Nm).
    g. Install the secondary accessory drive belt.
56. Install the stator and stator bolts. Tighten to 30 ft. lbs. (40 Nm).
57. Install the cooling fan clutch. Use the special tool 303-591 to tighten the cooling fan clutch. Tighten to 98 ft. lbs. (133 Nm).
58. Position and clip the cooling fan wiring. Connect the cooling fan electrical connector.
59. Install the radiator and shroud as an assembly.
60. Connect the upper radiator hose and the radiator overflow hose.
61. Position the harness wiring, install the 2 wiring retainers and 3 pin-type retainers.
62. Position the sight shield and install the 4 pin-type retainers.
63. Install the radiator support brackets. Tighten to 22 ft. lbs. (30 Nm).
64. On vehicles with metal ducts, install the charge air cooler duct and tighten the clamps.
65. On vehicles with blow molded ducts, install the charge air cooler duct and tighten the clamps.
66. Install the air cleaner assembly.
67. Connect the transmission cooler tubes. Tighten to the specifications illustrated.
68. Close the radiator draincock. Connect the lower radiator hose.
69. Install the motor mount retaining nuts. Tighten to 76 ft. lbs. (103 Nm).
70. Tighten the exhaust pipe retaining nuts to 35 ft. lbs. (47 Nm).
71. Clean and inspect the oil pan drain plug and gasket, install new if necessary.
72. Install the oil pan drain plug and tighten to 18 ft. lbs. (25 Nm).
73. Connect the negative battery cables.
74. Fill the engine with clean engine oil.
75. Fill the coolant system.
76. Run the engine and check for leaks

E-Series w/6.0L Engine

1. Before servicing the vehicle, refer to the precautions in the beginning of this section.
2. With the vehicle in **Neutral**, position it on a hoist.
3. Disconnect the battery ground cable.
4. Remove the engine cover.
5. Remove the cooling fan stator.
6. Remove the A/C compressor.
7. Remove the power steering pump upper mounting bolt.
8. Remove the left fan stator stand-off.

➡The front bolt will remain in the power steering pump.

9. Remove the bolts and position the power steering pump aside.
10. Remove the oil pan drain plug.
11. Loosen the exhaust pipe retaining nuts.
12. Remove the motor mount retaining nuts.
13. Remove the right fan stator stand-off. Remove the bolt and position the cable aside.
14. Install the special tools.
15. Install the engine lifting attachment and floor crane.

⁂ CAUTION

Use care when raising the engine to avoid engine or body damage

16. Raise the engine.
17. Remove the bolts and position back the oil pan until the oil pick up tube bolts are accessible.
18. Remove the bolts and let the oil pick-up tube go into the oil pan. Remove the oil pan.
19. Remove and discard the oil pick-up tube O-ring seal.

20. Remove the press-in-place gasket and discard.
21. Clean and inspect the sealing surfaces.

Installation

22. Install a new press-in-place gasket into the upper oil pan.
23. Install a new O-ring seal on the oil pick-up tube and position the oil pick-up tube in the oil pan.
24. Position the oil pan in the vehicle.
25. Install the oil pan pick-up tube and bolts.
26. Install the oil pan and bolts.
27. Lower the engine.
28. Remove the floor crane and engine lifting attachment.
29. Remove the special tools.
30. Position back the cable and install the bolt. Install the right fan stator stand-off.

➡Tighten the top retaining nut first.

31. Install the motor mount retaining nuts.
32. Tighten the exhaust pipe retaining nuts.
33. Clean and inspect the oil pan drain plug and gasket. Install new, if necessary.
34. Install the oil drain plug.
35. Reposition the power steering pump and install the power steering pump bolts.
36. Install the left fan stator stand-off.
37. Install the upper power steering mounting bolt.
38. Fill the engine with clean engine oil.
39. Install the engine cover.
40. Connect the battery ground cable.
41. Install the A/C compressor.
42. Install the cooling fan stator.
43. Run the engine and check for leaks.

7.3L Engine

1. Before servicing the vehicle, refer to the precautions in the beginning of this section.
2. Remove the engine.
3. Remove the oil pan bolts.
4. Lower the oil pan.

➡The oil pan is sealed to the crankcase with RTV silicone sealant in place of a gasket. It may be necessary to separate the pan from the crankcase with a utility knife. Also, the crankshaft may have to be turned to allow the pan to clear the crankshaft throws.

5. Clean the pan and crankcase mating surfaces thoroughly.

To install:

6. Apply a ⅛ in. (3mm) bead of RTV silicone sealant to the pan mating surfaces, and a ¼ in. (6mm) bead on the front and rear covers and in the corners; you have 15 minutes within which to install the pan!
7. Install the locating dowels into position.
8. Position the pan on the engine and install the pan bolts loosely.
9. Remove the dowels.
10. Tighten the pan bolts to:
    a. ¼ in.-20 bolts: 84 inch lbs. (10 Nm).
    b. 5⁄16 in.-18 bolts: 14 ft. lbs. (19 Nm).
    c. ⅜ in.-16 bolts: 24 ft. lbs. (33 Nm).
11. Install the engine.

Oil Pump

REMOVAL & INSTALLATION

6.0L Engine

1. Before servicing the vehicle, refer to the precautions in the beginning of this section.
2. With the vehicle in **Neutral**, position it on a hoist.
3. Disconnect the battery ground cable(s).
4. Remove the cooling fan
5. On vehicles with dual alternator, perform the following:
    a. Remove the dual alternator accessory drive belt.
    b. Remove the bolts and the dual alternator pulley.
6. Remove the accessory drive belt.
7. Check the crankshaft vibration damper runout.
8. Remove the paint from the face of the crankshaft vibration damper at four points 90 degrees apart.
9. Attach dial indicator 100-002 to the cylinder block. Position the tool on one of the unpainted surfaces.
10. Using a suitable tool, pry the crankshaft forward. Zero the dial indicator.

➡Pry the crankshaft forward only to eliminate possible error caused by crankshaft end play.

11. Rotate the crankshaft 90 degrees. Pry the crankshaft forward. Record the measurement. Repeat at each unpainted surface.
12. If the runout exceeds specification, install a new crankshaft vibration damper.

⁂ CAUTION

To avoid personal injury, support the vibration damper during mounting bolt removal. The damper can slide off the nose of the crankshaft very easily.

13. Remove the bolts and the crankshaft vibration damper.
14. Discard the bolts.
15. Punch two holes in the seal.
16. Using the special tool 303-D060, remove the crankshaft seal.

➡Production engine will not have a wear sleeve.

17. If equipped, remove the crankshaft damper wear sleeve using tool 303-762.
18. Remove the bolts and the gerotor cover. Remove and discard the O-ring seal.

➡Mark the front of the inner and outer gerotor for correct reassembly.

19. Remove the inner and outer gerotors.
20. Inspect the oil pump components and replace as necessary.
21. Inspect the oil pump for excessive metal particles.
22. Inspect the oil pump for gouging, cracks or deep scratches.
23. Inspect the oil pump inner and outer gear rotors for damage or excessive wear.

To install:

➡Install the gears with marks pointing outward.

24. Lubricate the inner gear with lithium assembly grease and install onto the crankshaft. Lubricate the outer gear with lithium assembly grease and mesh with the inner gear rotor in the oil pump housing. Wipe off the excess assembly grease.
25. Install a new O-ring seal.
26. Install the gerotor cover and bolts. Tighten to 71 inch lbs. (8 Nm).
27. Thoroughly clean the crankshaft front seal mounting surface.
28. Apply Threadlock 262® to the outer circumference of the leading edge of the crankshaft.

➡New seal and wear sleeve must not be separated.

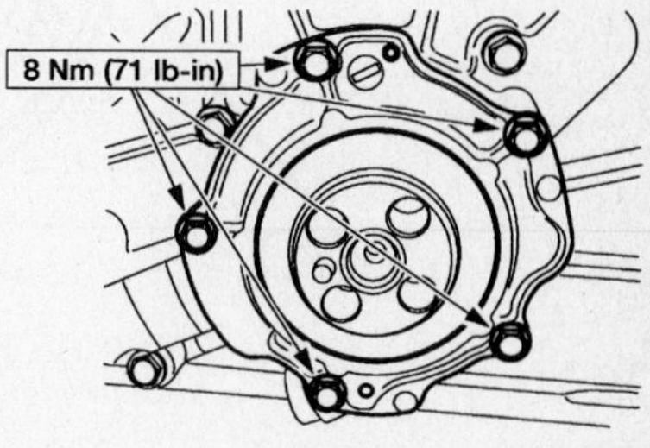

Oil pump mounting—6.0L engine

29. Using the special tool 303-361, install the oil seal and wear sleeve assembly.

⁂ CAUTION

To prevent engine damage, you must always install four new bolts when installing the vibration damper.

➡Do not use anti-seize compounds, grease or any lubricants. Lubricants have an adverse effect on the torque results.

30. Install the crankshaft vibration damper and bolts.
31. Tighten the bolts in a criss-cross sequence as follows:
    a. Tighten the bolts to 50 ft. lbs. (68 Nm).
    b. Tighten the bolts an additional 90 degrees.
32. Install the accessory drive belt.
33. On vehicles with dual alternator, perform the following:
    a. Install the dual alternator pulley and bolts. Tighten to 35 ft. lbs. (47 Nm).
    b. Install the dual alternator accessory drive belt.
34. Install the cooling fan stator
35. Connect the battery ground cables.

7.3L Engine

1. Before servicing the vehicle, refer to the precautions in the beginning of this section.
2. Disconnect the negative battery cable.
3. Remove the oil pan.
4. Remove the oil pick-up tube from the pump.
5. Unbolt and remove the oil pump.

To install:

6. Assemble the pick-up tube and pump. Use a new gasket.
7. Install the oil pump and tighten the bolts to 14 ft. lbs. (19 Nm).
8. Install the oil pick up tube.
9. Install the oil pan.
10. Connect the negative battery cable.

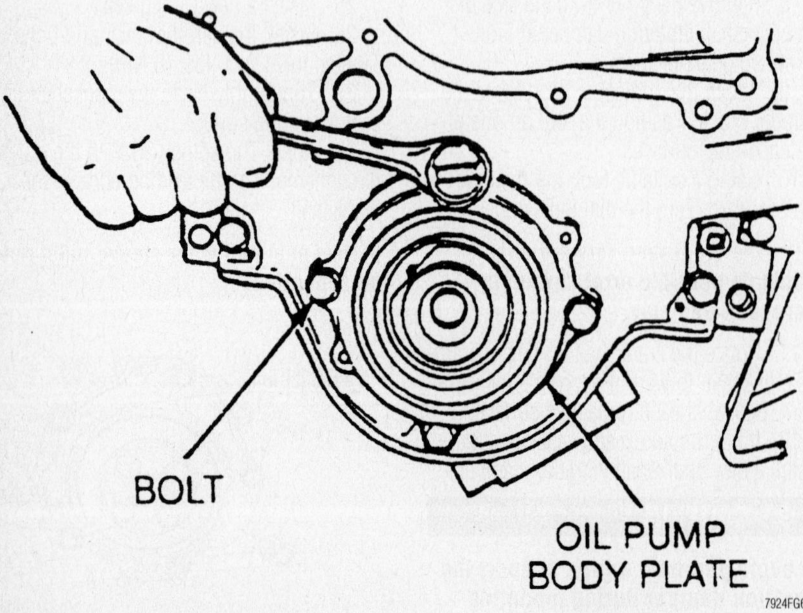

The oil pump is mounted on the cylinder block with 4 bolts—7.3L diesel engine

Rear Main Seal

REMOVAL & INSTALLATION

6.0L Engine

1. Before servicing the vehicle, refer to the precautions in the beginning of this section.
2. Remove the transmission.
3. Remove the bolts.
4. Remove the flexplate or flywheel.

➡Use extreme care when removing the flywheel front adapter to prevent damage to the alignment dowel pin.

5. Remove the flywheel front adapter.

⁂ CAUTION

To prevent engine damage, do not remove the rear primary crankshaft flange bolts under any circumstances. If the flange is removed and reinstalled, it will result in engine vibration and premature transmission component wear.

6. Punch two holes in the rear main seal, across from each other.
7. Using the puller tool 100-001, remove the rear main seal.

➡Production engines will not have a wear sleeve.

8. If equipped with a crankshaft wear sleeve, use the tool 303-771 to remove the crankshaft rear wear sleeve.
9. Clean and inspect the crankshaft sealing surface.

To install:

➡The crankshaft rear oil seal and wear sleeve are installed as an assembly.

➡Lubricate the outer diameter of the rubber seal with a solution of dish soap and water (approximately 50/50 mix) prior to assembly. Do not use any other type of lubricant.

10. Apply a bead of Threadlock 262® around the circumference of the outer rear edge of the secondary crankshaft flange.
11. Using tool 303-770, install the crankshaft rear oil seal.
12. Install the flywheel front adapter.
13. Install the flexplate or flywheel.
14. Install the bolts. Snug all bolts to 44 inch lbs. (5 Nm), then tighten all bolts to 69 ft. lbs. (94 Nm) in the sequence illustrated.
15. Install the transmission.

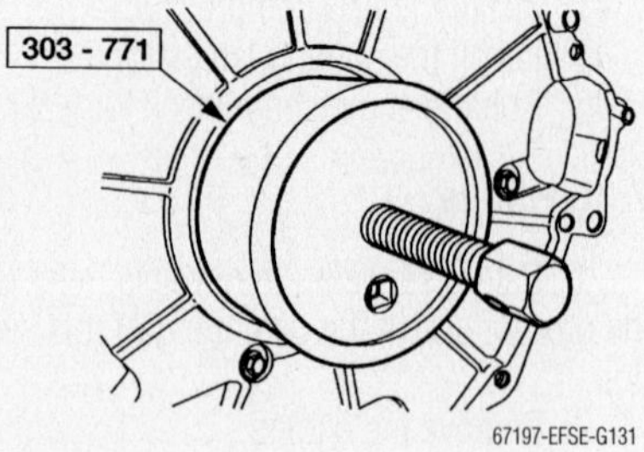

Rear main seal wear sleeve removal—6.0L engine

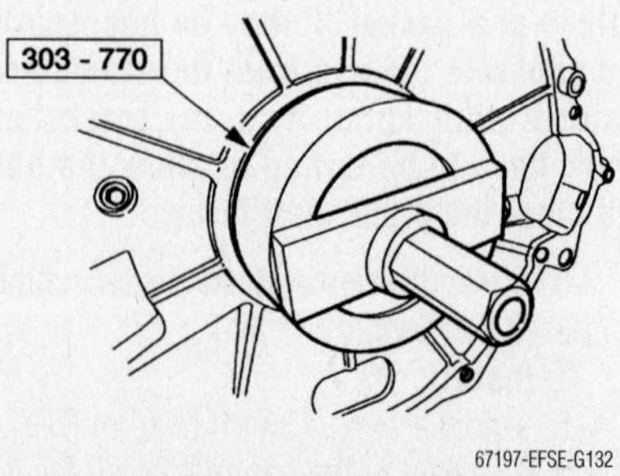

Rear main seal installation—6.0L engine

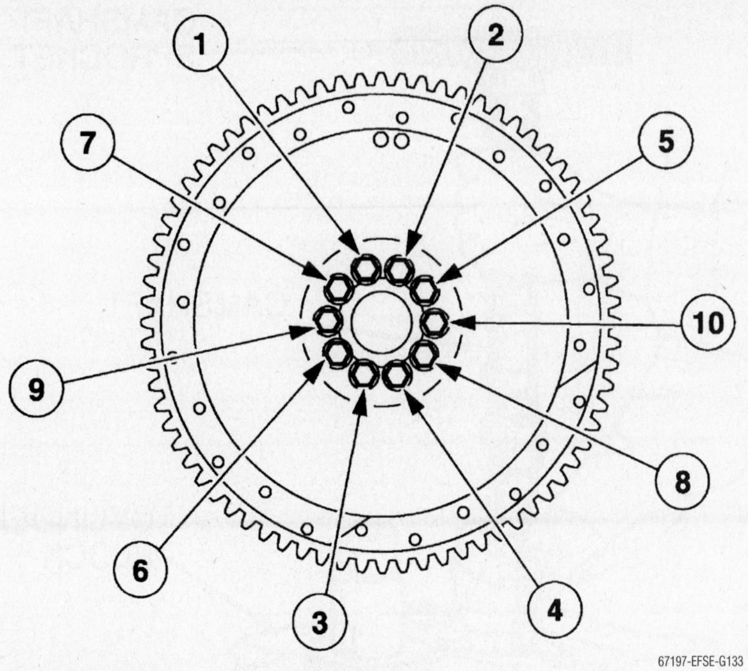

Flywheel installation torque sequence—6.0L engine

7.3L Engine

1. Before servicing the vehicle, refer to the precautions in the beginning of this section.
2. Remove the transmission.
3. Remove the flywheel.
4. Loosen the crankshaft rear oil seal bolts and remove the seal.
5. Clean the seal mating surfaces.

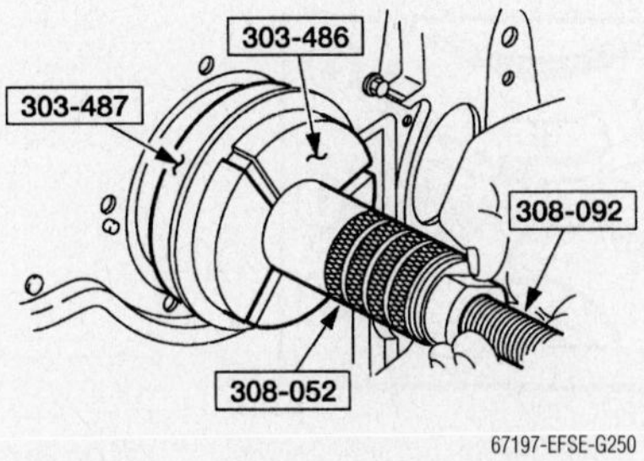

Assemble the seal and wear ring removal tools, then remove the wear ring—7.3L diesel engine

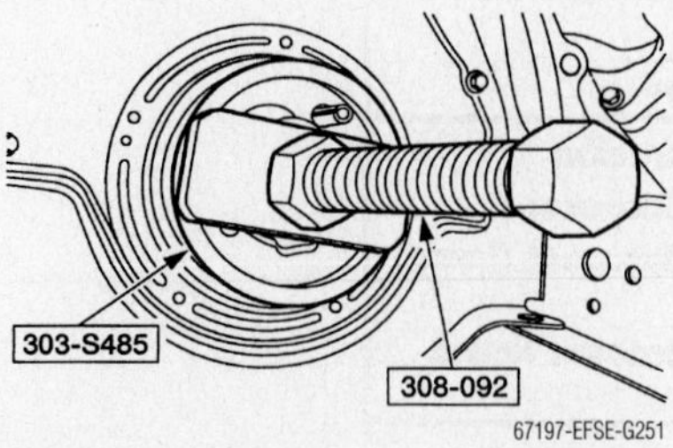

Rear main seal installation—7.3L diesel engine

6. If installing the old seal, inspect it for damage.
7. Using crankshaft wear ring removal tool T94T-6701-AH1, forcing screw T84T-7025-B, remover tube T77J-7025-B and wear ring remover sleeve T94T-6701-AH2 (refer to the illustration), or their equivalents, remove the wear ring.

To install:

8. Apply RTV silicone sealant to the seal retaining ring and the seal retaining bolts.
9. Using seal replacers T94T-6701-AH3 and T94T-AH4, driver sleeve T79T-6316-A4 (part of T79T-6316-A) and guide pins T94P-7000-P or their equivalents, install the wear ring and oil seal.
10. Install and tighten the seal retaining bolts.
11. Remove the installation tools and install the flywheel.
12. Install the transmission.

Piston and Ring Positioning

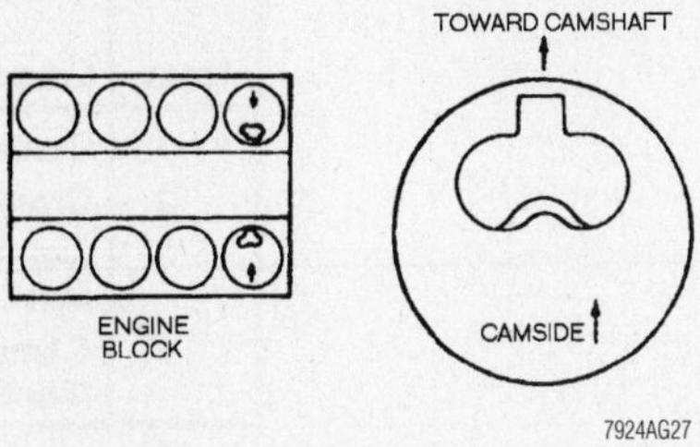

7.3L Diesel engines—piston-to-engine orientation

Timing Gears, Front Cover and Seal

REMOVAL & INSTALLATION

6.0L Engine

1. Before servicing the vehicle, refer to the precautions in the beginning of this section.
2. Remove the intake manifold.
3. Disconnect the exhaust pressure tube at the exhaust.
4. Remove the nuts and the exhaust pressure bracket assembly.

➡Remove the thermostat housing only if a new front cover is being installed.

5. Remove the stud bolts and the thermostat housing.
6. Remove and discard the O-ring.
7. Disconnect the engine coolant fill hose.
8. Disconnect the lower radiator hose.
9. Remove the stator stand-off bolt.

➡Upper bolts shown, lower bolts similar.

10. Remove the four bolts and position aside the power steering pump.
11. Remove the bolts and the accessory drive belt tensioner.
12. Remove the bolts and the accessory drive idler pulleys.

➡Remove the coolant pump pulley only if a new front cover is being installed.

13. Remove the coolant pump pulley.

➡Remove the coolant pump only if a new front cover is being installed.

14. Remove the bolts and the coolant pump.
15. Remove and discard the O-ring.
16. Remove the nut and the battery cable bracket.
17. Remove the oil pump.
18. Check the crankshaft vibration damper runout.
19. Remove the paint from the face of the crankshaft vibration damper at four points 90 degrees apart.
20. Attach the special tool to the cylinder block. Position the special tool on one of the unpainted surfaces.
21. Using a suitable tool, pry the crankshaft forward. Zero the dial indicator.

➡Pry the crankshaft forward at the same point to eliminate possible error caused by crankshaft end play.

22. Rotate the crankshaft 90 degrees. Pry the crankshaft forward. Record the measurement. Repeat at each unpainted surface. If the runout exceeds specification, install a new crankshaft vibration damper.

✻✻ WARNING

To prevent engine damage, you must always install four new bolts when installing the vibration damper.

✻✻ WARNING

To avoid personal injury, support the vibration damper during mounting bolt removal. The damper can slide off the nose of the crankshaft very easily.

23. Remove the bolts and the crankcase vibration damper.
24. Discard the bolts.
25. Remove the bolts and the front cover.

✻✻ WARNING

Sealant is used where the crankcase and lower crankcase meet. Failure to cut the sealant could result in pulling the lower crankcase seal out while removing the front cover gasket.

26. Use a thin blade scraper to cut the sealant where the crankcase and the lower crankcase meet. Remove and discard the front cover gasket.
27. Clean and inspect the sealing surfaces.
28. Punch two holes in the seal.
29. Using the special tool, remove the crankshaft seal.

➡Production engine will not have a wear sleeve.

30. If equipped, remove the crankshaft damper wear sleeve.
31. Remove the thrust plate mounting bolts and remove the camshaft and gear.

To install:

32. Install the camshaft and gear assembly. Using the special tool, align the camshaft timing mark as shown. Install the thrust plate mounting bolts.
33. Thoroughly clean the crankshaft front seal mounting surface.
34. Apply Threadlock 262® to the outer circumference of the leading edge of the crankshaft.

➡New seal and wear sleeve must not be separated.

35. Using the special tool, install the oil seal and wear sleeve assembly.

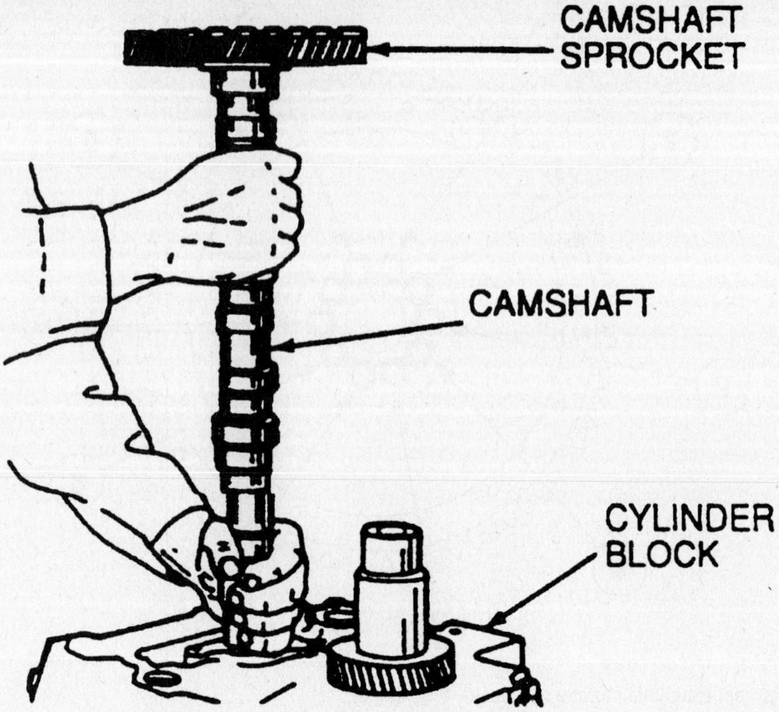

The camshaft gear is removed with the camshaft, then pressed off-7.3L diesel engines

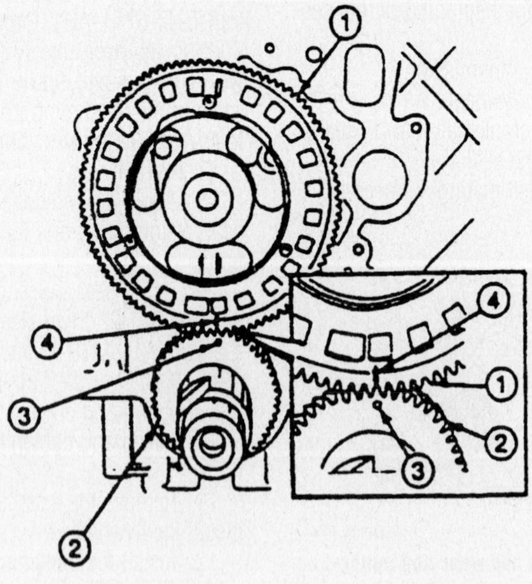

Item	Description
1	Camshaft Sprocket
2	Crankshaft Sprocket
3	Crankshaft Sprocket Timing Mark
4	Camshaft Sprocket Timing Mark

7924FG64

Be sure the timing marks are aligned as illustrated—7.3L diesel engines

36. If removed, install the front cover crankcase dowels into the cylinder block.

➡**Use guide studs to aid in installation. Studs must be fabricated locally.**

37. Install the guide studs.
38. Apply a bead of sealant at the seam where the crankcase and the lower crankcase meet.
39. Install a new engine front cover gasket.
40. Install the engine front cover and bolts.

⁂ WARNING

To prevent engine damage, you must always install four new bolts when installing the vibration damper.

➡**Do not use anti-seize compounds, grease or any lubricants. Lubricants have an adverse effect on the torque results.**

41. Install the crankshaft vibration damper and bolts.
42. Tighten the bolts in the sequence shown.
  a. Tighten the bolts to 68 Nm (50 ft. lbs.).
  b. Tighten the bolts an additional 90 degrees.
43. Install the oil pump.
44. Install the battery cable bracket and nut.

➡**Install a new O-ring on the coolant pump pulley.**

45. If removed, install the coolant pump and bolts.
46. If removed, install the coolant pump pulley and bolts.
47. Install the accessory drive idler pulleys and bolts.
48. Install the accessory drive belt tensioner and bolts.
49. Position back the power steering pump and install the bolts.
50. Install the stator stand-off bolt.
51. Connect the lower radiator coolant hose.
52. Connect the engine coolant hose.

➡**Install a new O-ring.**

53. If removed, install the thermostat housing and stud bolts.
54. Install the exhaust pressure bracket assembly and retaining nuts.
55. Connect the exhaust pressure tube fitting at the exhaust manifold.
56. Install the intake manifold.

7.3L Engine

➡**The crankshaft gear sprocket is not serviced separately from the crankshaft. Do not try to remove the sprocket or you will damage the crankshaft.**

Remove the camshaft sprocket as follows:

1. Before servicing the vehicle, refer to the precautions in the beginning of this section.
2. Disconnect the negative battery cable.
3. Remove the camshaft.
4. Use a press to remove the sprocket from the camshaft.
5. Remove the thrust plate and sprocket key.
6. Inspect the camshaft and related parts for wear and damage.

To install:

7. Clean the nose of the camshaft and install the thrust plate.
8. Place the key in the keyway on the camshaft.
9. Heat the sprocket in an oven to 500°F (260°C).
10. Remove the sprocket from the oven, align the sprocket keyway with the camshaft key and install the sprocket on the camshaft until it is fully seated. Allow the camshaft assembly to cool before installation
11. Install the camshaft in the engine and align the timing marks on the gears.
12. Connect the negative battery cable.

GASOLINE FUEL SYSTEM

Fuel System Service Precautions

Safety is the most important factor when performing not only fuel system maintenance, but any type of maintenance. Failure to conduct maintenance and repairs in a safe manner may result in serious personal injury or death. Maintenance and testing of the vehicle's fuel system components can be accomplished safely and effectively by adhering to the following rules and guidelines.

- To avoid the possibility of fire and personal injury, always disconnect the negative battery cable unless the repair or test procedure requires that battery voltage be applied.
- Always relieve the fuel system pressure prior to disconnecting any fuel system component (injector, fuel rail, pressure regulator, etc.), fitting or fuel line connection. Exercise extreme caution whenever relieving fuel system pressure, to avoid exposing skin, face and eyes to fuel spray. Please be advised that fuel under pressure may penetrate the skin or any part of the body that it contacts.
- Always place a shop towel or cloth around the fitting or connection prior to loosening to absorb any excess fuel due to spillage. Ensure that all fuel spillage (should it occur) is quickly removed from engine surfaces. Ensure that all fuel soaked cloths or towels are deposited into a suitable waste container.
- Always keep a dry chemical (Class B) fire extinguisher near the work area.
- Do not allow fuel spray or fuel vapors to come into contact with a spark or open flame.
- Always use a back-up wrench when loosening and tightening fuel line connection fittings. This will prevent unnecessary stress and torsion to fuel line piping. Always follow the proper torque specifications.
- Always replace worn fuel fitting O-rings with new. Do not substitute fuel hose or equivalent where fuel pipe is installed.

Fuel System Pressure

RELIEVING

➡**A fuel pressure gauge is needed to correctly perform this procedure.**

⁂ CAUTION

Fuel injection systems remain under pressure, even after the engine has been turned OFF. The fuel system pressure must be relieved before disconnecting any fuel lines. Failure to do so may result in fire and/or personal injury.

1. Before servicing the vehicle, refer to the precautions in the beginning of this section.
2. Disconnect the negative battery cable and remove the fuel filler cap.
3. Remove the cap from the pressure relief valve on the fuel supply manifold. Install a fuel pressure gauge to the pressure relief valve.

4. Direct the gauge drain hose into a suitable container and depress the pressure relief button.
5. Remove the gauge and replace the cap on the pressure relief valve.

➡**As an alternate method on models except F-150 and Blackwood, disconnect the inertia switch and crank the engine for 15–20 seconds until the pressure is relieved.**

Fuel Filter

REMOVAL & INSTALLATION

2001–03

⁂ CAUTION

Fuel injection systems remain under pressure, even after the engine has been turned OFF. The fuel system pressure must be relieved before disconnecting any fuel lines. Failure to do so may result in fire and/or personal injury.

1. Before servicing the vehicle, refer to the precautions in the beginning of this section.
2. Disconnect the negative battery cable and relieve the fuel system pressure.
3. Disconnect the fuel lines from the fuel filter. Have a drain pan handy to catch any residual fuel once the lines are separated. On newer models, disconnect the fuel lines from the filter as follows:
   a. Disconnect the safety clip from the male hose.
   b. Install and push the fuel line disconnect tool into the female fitting.
   c. Separate the male and female fittings.
   d. Inspect the fuel lines for any damage after the fuel is finished draining.
4. Remove the fuel filter from the bracket and the retainer, if equipped. Note the direction of the flow arrow so the replacement filter can be installed correctly.

To install:

5. Position the fuel filter into the mounting bracket with the flow arrow pointing in the correct direction.
6. Install the fuel lines to the fuel filter. On newer models, align and push the male tube into the female fitting until a click is heard. Pull on the fitting to ensure that it is fully engaged, then install the safety clip.
7. Lower the vehicle to the ground.

➡**When the battery has been disconnected and reconnected, some abnormal drive symptoms may occur while the PCM relearns its adaptive strategy. The vehicle may need to be driven 10 miles (16 km) or more to relearn the strategy.**

8. Connect the negative battery cable.

2004

1. Before servicing the vehicle, refer to the precautions in the beginning of this section.
2. Relieve the fuel system pressure.
3. Disconnect the fuel lines from the fuel filter.
4. Remove the fuel filter.
5. Loosen the fuel filter clamp screw.

➡**Make sure that an audible click is heard when installing the fuel lines. Pull back on the fuel lines to confirm engagement.**

6. To install, reverse the removal procedure.

Fuel Pump

REMOVAL & INSTALLATION

F-Series

⁂ CAUTION

Fuel injection systems remain under pressure, even after the engine has been turned OFF. The fuel system pressure must be relieved before disconnecting any fuel lines. Failure to do so may result in fire and/or personal injury.

1. Before servicing the vehicle, refer to the precautions in the beginning of this section.
2. Remove or disconnect the following:
   - Negative battery cable
   - Fuel pressure
   - Fuel tank skid plate bolts and lower the skid plate.
   - Fuel
   - Fuel tank filler pipe hose from the tank
   - Fuel tank filler pipe vent hose from the tank
   - Fuel lines from the fuel pump
   - Front fuel tank connections
   - Rear Evaporative Emissions (EVAP) hose clamp and the hose

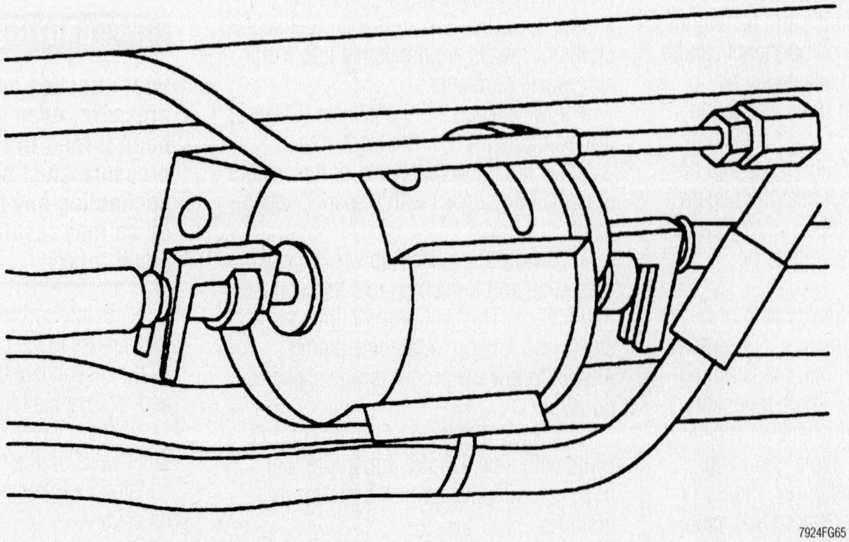

Typical fuel filter mounting along an under vehicle frame rail

- Electrical connector from the fuel pump

3. Support the fuel tank with a jack.
   - Fuel tank support strap bolts and the fuel tank straps
   - Fuel tank
   - Fuel pump bolts, if equipped with a metal tank
   - Fuel pump retaining ring, if equipped with a plastic tank
4. Squeeze the locking tabs and remove the fuel pump.

To install:

5. Install the fuel pump.
6. Tighten the fuel pump bolts to 80–107 inch lbs. (9–12 Nm), if equipped.
7. Install the fuel pump retaining ring if equipped with a plastic tank
8. Install the fuel tank.
   a. Tighten the fuel tank strap bolts to 25–33 ft. lbs. (34–46 Nm).
   b. Tighten the skid plate bolts to 10–13 ft. lbs. (13–17 Nm).
9. Connect the negative battery.

E-Series

CAUTION

Fuel injection systems remain under pressure, even after the engine has been turned OFF. The fuel system pressure must be relieved before disconnecting any fuel lines. Failure to do so may result in fire and/or personal injury.

1. Before servicing the vehicle, refer to the precautions in the beginning of this section.
2. Remove or disconnect the following:
   - Negative battery cable
   - Fuel system pressure
   - Fuel tank
   - Fuel tank filler pipe vent hose and fuel tank filler pipe from the tank
3. Support the fuel tank with a jack.
   - The 2 fuel tank support strap nuts
   - 2 fuel tank support straps
4. Lower the fuel tank to allow access to the electrical connections
   - Fuel tank connections
   - Fuel and electrical connections from the fuel pump
   - Fuel tank
   - Fuel tank screws/nuts, fuel pump and sender

To install:

5. Install or connect the following:
   - Fuel sender and fuel pump into the fuel tank. Tighten the screws/nuts.
6. Raise the fuel tank.
   - Fuel and electrical connections to the fuel pump
   - Fuel tank connections
   - Fuel tank support straps and tighten the nuts to 13–17 ft. lbs. (17–23 Nm)
   - Fuel tank filler pipe vent hose and the fuel tank filler pipe to the tank
   - Negative battery cable

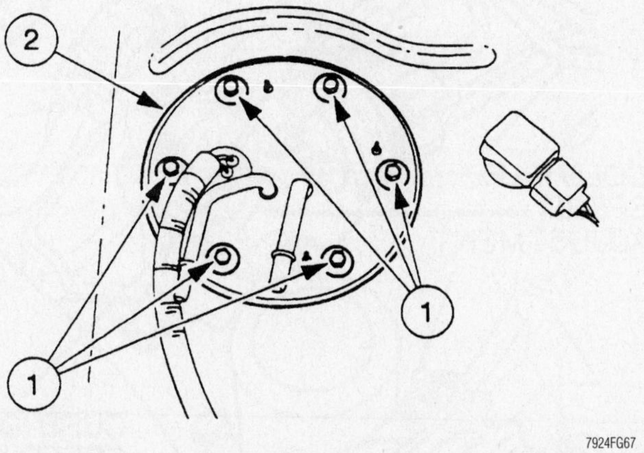

Remove the mounting bolts (1), then lift the fuel pump assembly (2) out of the tank—E-Series

Fuel Injector

REMOVAL & INSTALLATION

CAUTION

Fuel injection systems remain under pressure, even after the engine has been turned OFF. The fuel system pressure must be relieved before disconnecting any fuel lines. Failure to do so may result in fire and/or personal injury.

1. Before servicing the vehicle, refer to the precautions in the beginning of this section.
2. Remove or disconnect the following:
   - Negative battery cable
   - Engine cover, if equipped
   - Air cleaner assembly
3. Relieve the fuel pressure.
   - Upper intake manifold
   - Fuel pressure regulator valve vacuum hose
   - Fuel lines
   - Fuel injector electrical connectors
   - Fuel injector electrical harness from the fuel injector manifold
   - Fuel injection supply manifold
   - Fuel rail bolts and the rail

CAUTION

The fuel injectors are deposit-resistant. Do not clean. Use O-rings made of special fuel resistant material. Use of ordinary O-rings can cause the fuel system to leak.

- Fuel injector retaining clips and the fuel injectors. Inspect the fuel injector O-rings and if necessary.

To install:

➡Make sure the injector clips must engage the upper groove on the injectors or fuel leaks may occur.

4. Lubricate the new O-rings with clean engine oil to aid installation.
5. Install or connect the following:
   - Fuel injectors and the rail. Tighten the rail bolts to 89 inch lbs. (10 Nm).
   - Fuel injection supply manifold
   - Fuel injector electrical harness from the fuel injector manifold
   - Fuel injector electrical connectors
   - Fuel lines
   - Fuel pressure regulator valve vacuum hose
   - Upper intake manifold
   - Air cleaner assembly
   - Engine cover, if equipped
   - Negative battery cable

DIESEL FUEL SYSTEM

Fuel System Service Precautions

Safety is the most important factor when performing not only fuel system maintenance but any type of maintenance. Failure to conduct maintenance and repairs in a safe manner may result in serious personal injury or death. Maintenance and testing of the vehicle's fuel system components can be accomplished safely and effectively by adhering to the following rules and guidelines.

- To avoid the possibility of fire and personal injury, always disconnect the negative battery cable unless the repair or test procedure requires that battery voltage be applied.
- Always relieve the fuel system pressure prior to disconnecting any fuel system component (injector, fuel rail, pressure regulator, etc.), fitting or fuel line connection. Exercise extreme caution whenever relieving fuel system pressure, to avoid exposing skin, face and eyes to fuel spray. Please be advised that fuel under pressure may penetrate the skin or any part of the body that it contacts.
- Always place a shop towel or cloth around the fitting or connection prior to loosening to absorb any excess fuel due to spillage. Ensure that all fuel spillage (should it occur) is quickly removed from engine surfaces. Ensure that all fuel soaked cloths or towels are deposited into a suitable waste container.
- Always keep a dry chemical (Class B) fire extinguisher near the work area.
- Do not allow fuel spray or fuel vapors to come into contact with a spark or open flame.
- Always use a back-up wrench when loosening and tightening fuel line connection fittings. This will prevent unnecessary stress and torsion to fuel line piping. Always follow the proper torque specifications.
- Always replace worn fuel fitting O-rings with new. Do not substitute fuel hose or equivalent where fuel pipe is installed.

Fuel System Pressure

RELIEVING

6.0L Engine

1. Before servicing the vehicle, refer to the precautions in the beginning of this section.

**** CAUTION**

Before removing the fuel tank filler cap, turn the fuel tank filler cap ¼ to

2. Remove the Schrader valve cap and install the fuel pressure gauge 310-012.
3. Open the manual valve slowly on the fuel pressure gauge 310-012 and relieve the fuel pressure. This will drain some fuel out of the system; place the fuel in a suitable container.

7.3L Engine

**** CAUTION**

Before removing the fuel tank filler cap, turn the fuel tank filler cap ¼ to ¾ turn counterclockwise and wait for the tank pressure to be relieved. Personal injury may result if the fuel tank filler cap is removed without the pressure fully relieved.

1. Before servicing the vehicle, refer to the precautions in the beginning of this section.
2. Remove the fuel tank filler cap to relieve any pressure in the fuel tank.
3. When servicing the fuel lines, loosen the fuel fitting to allow any residual fuel line pressure to be relieved.

Idle Speed

ADJUSTMENT

6.0L Engine

The fuel system is controlled by the PCM.

7.3L Engine

1. Before servicing the vehicle, refer to the precautions in the beginning of this section.
2. Place the transmission in Neutral (manual transmissions) or **P** (automatic transmissions).
3. Bring the engine up to normal operating temperature.

➡Idle speed is measured with the manual transmission in Neutral or the automatic transmission in D.

4. Ensure that the curb idle adjusting screw is against the stop. If not, correct the vehicle linkage.
5. Check curb idle speed. Curb idle speed is specified on the Vehicle Emissions Control Information (VECI) decal on the underside of the vehicle's hood. Adjust the idle speed to specification using the idle speed adjusting screw.

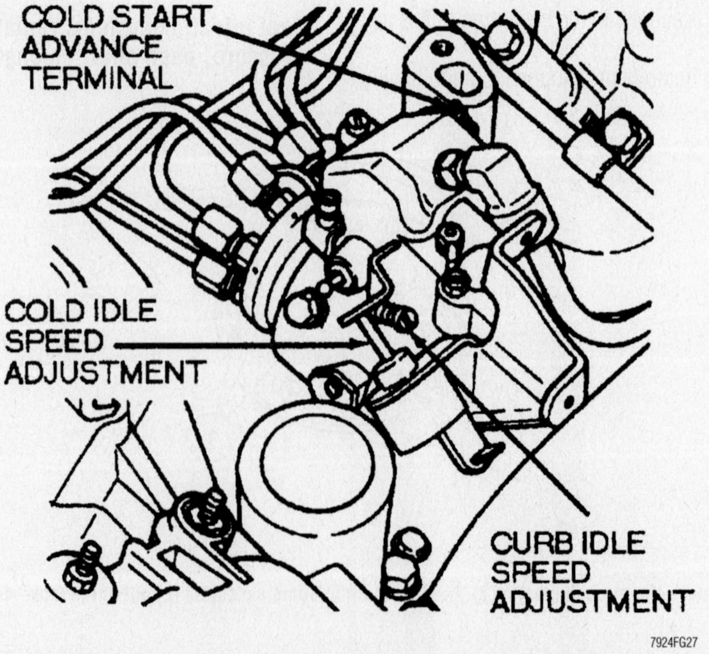

Raise or lower the curb idle speed by turning the curb idle speed adjusting screw—7.3L diesel engines

6. Place the transmission in Neutral (manual) or **P**. Rev the engine momentarily, then place the transmission in the specified gear and recheck the idle speed. Adjust again if necessary.

7. Remove the tachometer and close the hood.

Fuel Filter/Water Separator

DRAINING WATER

7.3L Engine

1. Before servicing the vehicle, refer to the precautions in the beginning of this section.
2. Properly relieve the fuel system pressure.
3. Shut the engine **OFF**. Failure to shut the engine **OFF** before draining the separator will cause air to enter the system.
4. Unscrew the vent on the top center of the separator unit 2 ½ –3 turns.
5. Unscrew the drain screw on the bottom of the separator 1 ½ –2 turns and drain the water into an appropriate container.
6. After the water is completely drained, close the water drain finger-tight.
7. Tighten the vent until snug, then turn it an additional ¼ turn.
8. Start the engine and check the "Water in Fuel" indicator light; it should not be lit. If it is lit and continues to stay so, there is a problem somewhere else in the fuel system.

REMOVAL & INSTALLATION

7.3L Engine

✲✲ CAUTION

The fuel system remains under pressure, even after the engine has been turned OFF. The fuel system pressure must be relieved before disconnecting any fuel lines. Failure to do so may result in fire and/or personal injury.

1. Before servicing the vehicle, refer to the precautions in the beginning of this section.
2. Remove or disconnect the following:
   - Negative battery cable
   - Turbocharger assembly
   - Baffle and the air inlet crossover manifold
3. Place a suitable container under the drain hose and open the filter drain.
   - The 2 capscrews securing the fuel filter base to the crankcase
   - Water drain hose from the filter
   - Fuel system pressure
   - Fuel outlet hose, located between the fuel and filter housing, and the fuel return hose from the fuel pressure regulator valve.
   - The 2 fuel supply hoses that connect the regulator block to the cylinder head fuel rails
4. Loosen the clamp at the fuel pump end of the hose, which connects the fuel filter to the inlet of high pressure stage at the fuel pump.
5. Disengage the wiring harness from the right side of the filter housing.
6. Disengage the electrical connections from the Water In Fuel (WIF) sensor and the fuel heater.
7. Remove the fuel filter.
8. Use a prybar to remove the fuel filter cap and the filter element will come out with the cap.
9. Depress the element locking tabs and remove the element from the cap.

To install:

10. Clean the mating surfaces and install the filter element onto the cap, making sure the tabs engage.
11. Install the filter gap and press down firmly, but gently, to engage it.
12. Install or connect the following:
   - Wiring connections to the fuel heater and WIF sensor
   - Wiring harness to the filter housing
13. Tighten the clamp at the fuel pump end of the hose, which connects the fuel filter to the inlet of high pressure stage at the fuel pump.
14. Engage the 2 fuel supply hoses that connect the regulator block to the cylinder head fuel rails.
15. Install or connect the following:
   - Fuel outlet hose, located between the fuel and filter housing, and the fuel return hose from the fuel pressure regulator valve.
   - Water drain hose to the filter
   - The 2 capscrews securing the fuel filter base to the crankcase
   - Air inlet crossover manifold and baffle
   - Turbocharger assembly
   - Negative battery cable

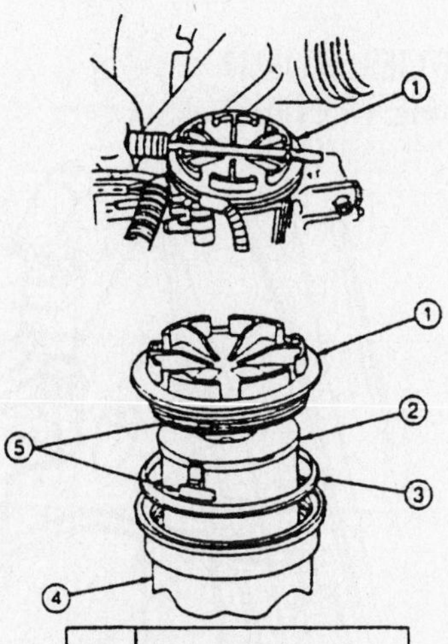

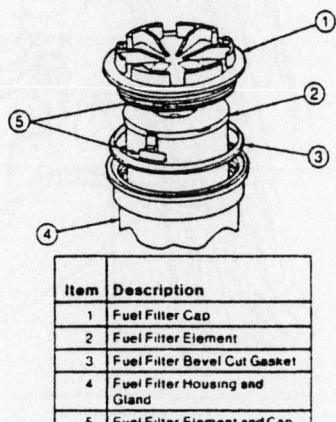

Item	Description
1	Fuel Filter Cap
2	Fuel Filter Element
3	Fuel Filter Bevel Cut Gasket
4	Fuel Filter Housing and Gland
5	Fuel Filter Element and Cap Locking Tabs

7924FG68

Use a prytool to remove the fuel filter cap to gain access to the filter element-7.3L diesel fuel systems

Fuel Conditioning Module—6.0L Engine

The Fuel Condition Module contains the fuel pump and water separator.

DRAINING

✲✲ CAUTION

Smoking or open flame of any type must not be present when working near fuel or fuel vapor.

1. Disconnect both battery ground cables.
2. Raise and support the vehicle.
3. Open the fuel/water separator drain valve to release the fuel pressure.

REMOVAL & INSTALLATION

✲✲ CAUTION

Smoking or open flame of any type must not be present when working near fuel or fuel vapor.

67197-EFSE-G134

Fuel/water separator drain valve—6.0L engine

1. Disconnect both battery ground cables.
2. Raise and support the vehicle.
3. Open the fuel/water separator drain valve to release the fuel pressure.
4. Disconnect the electrical connectors.
5. Disconnect the fuel pump electrical connector.
6. Disconnect the fuel warmer electrical connector.
7. Disconnect the water-in-fuel electrical connector.
8. Disconnect the fuel hoses.
9. Remove the fuel hose retaining clips and discard. Disconnect the fuel hoses from the fuel pump.
10. Press in the retaining clips and release the fuel hoses.
11. Remove the mounting nuts and the fuel conditioning module.
12. To install, reverse the removal procedure. Torque the nuts to 11 ft. lbs. (15 Nm).

Fuel Filter—6.0L Engine

REMOVAL & INSTALLATION

**** CAUTION**

Do not smoke or carry lighted tobacco or open flame of any type when working on or near any fuel related component. Highly flammable mixtures are always present and can be ignited, resulting in possible personal injury.

1. Relieve the fuel system pressure.
2. Disconnect the fuel lines from the fuel filter.
3. Remove the fuel filter.
4. Loosen the fuel filter clamp screw.

➡Make sure that an audible click is heard when installing the fuel lines. Pull back on the fuel lines to confirm engagement.

5. To install, reverse the removal procedure.

Fuel filter—6.0L engine

Injection Pump

REMOVAL & INSTALLATION

7.3L Engine

**** CAUTION**

The fuel system remains under pressure, even after the engine has been turned OFF. The fuel system pressure must be relieved before disconnecting any fuel lines. Failure to do so may result in fire and/or personal injury.

1. Before servicing the vehicle, refer to the precautions in the beginning of this section.
2. Remove or disconnect the following:
   - Negative battery cable
   - Turbocharger assembly
   - Fuel system pressure
   - Fuel line banjo bolt at the pump
   - Fuel line fittings at the rear of the cylinder heads
   - Fuel lines
   - The 3 hose clamps at the injection pump fittings
   - Water drain hose at the fuel filter
   - Filter and position it forward
   - Injection pump retaining bolts, then lift the pump out of the crankcase bore
   - Injection pump tappet from the crankcase bore

To install:

3. Rotate the engine so the injection pump eccentric is on the base circle.
4. Install or connect the following:
   - Injection pump tappet in the base of the injection pump
   - O-ring on the injection pump base
   - Injection pump and tighten the bolts to 19–27 ft. lbs. (26–37 Nm)
   - Fuel filter and connect the water drain hose
   - The 3 fuel hoses at the front of the injection pump
   - The fuel line clamps
   - Fuel filter retaining bolts
   - Fuel line assembly and new seal rings at the rear of the pump
   - Fuel line fittings at the rear of the cylinder heads
   - Fuel line banjo fitting at the pump. Tighten the bolt to 18 ft. lbs. (24 Nm).
   - Turbocharger assembly
   - Negative battery cable

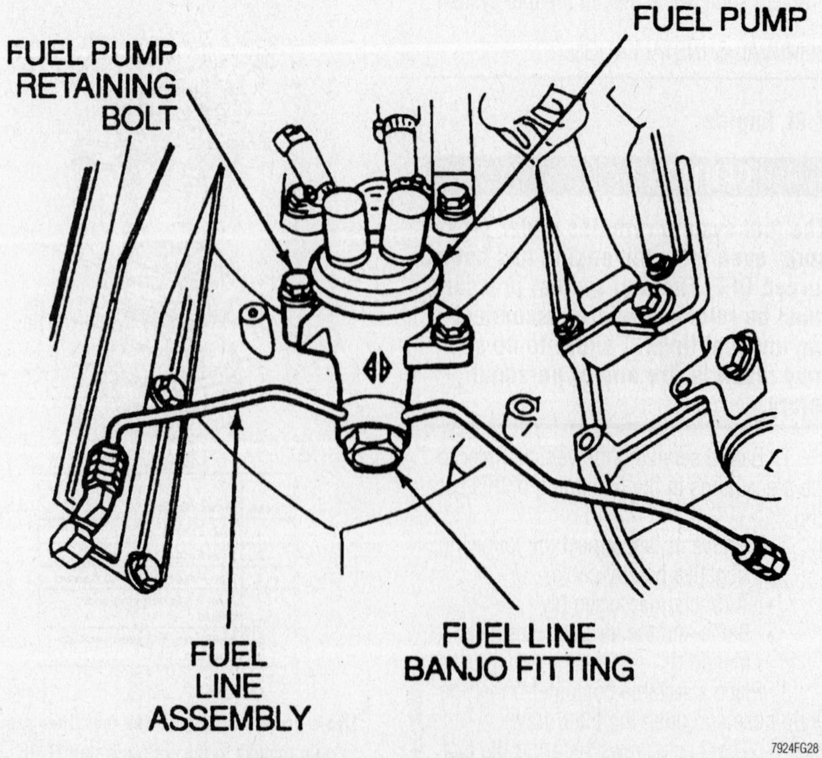

Injection pump assembly components—7.3L diesel engines

Oil Pump

REMOVAL & INSTALLATION

6.0L Engine

EARLY BUILD

1. Before servicing the vehicle, refer to the precautions in the beginning of this section.
2. Remove the intake manifold.
3. Remove the turbocharger heat shield.
4. Remove the bolts and the high-pressure oil pump cover. Use a thin gasket scraper to separate the cover from the crankcase.
5. Remove and discard the press-in-place gasket.
6. Position the dial indicator with bracketry onto the oil pump drive and check the oil pump drive gear backlash.
7. Remove the bolts from the high-pressure oil pump discharge pipe.
8. Using the special tool, disconnect

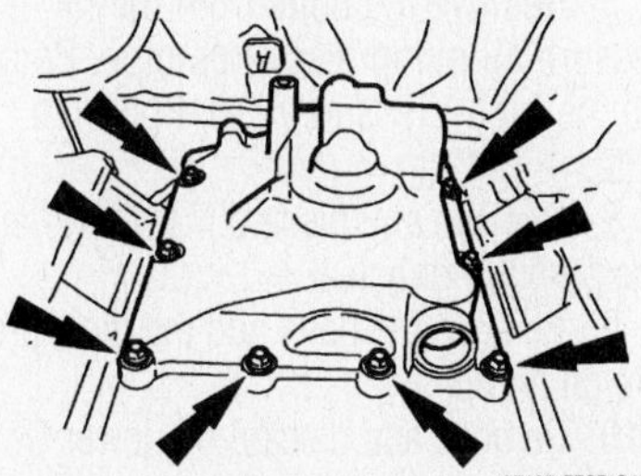

Remove the high-pressure oil pump cover

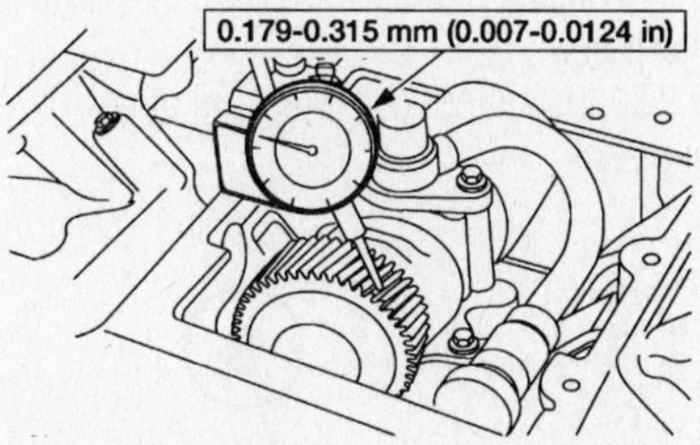

Check the oil pump drive gear backlash

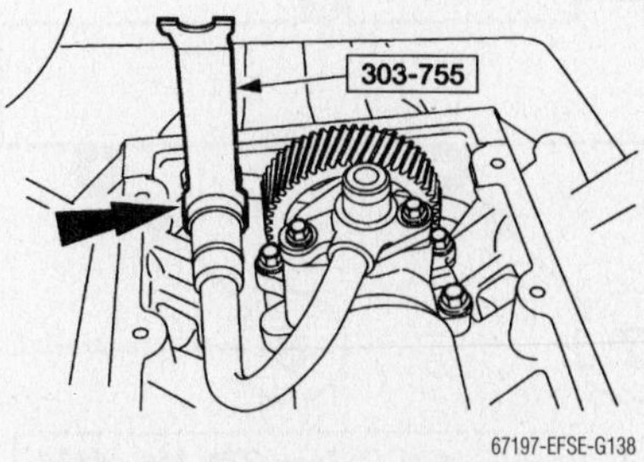

Using the special tool, disconnect and remove the high-pressure oil pump discharge pipe

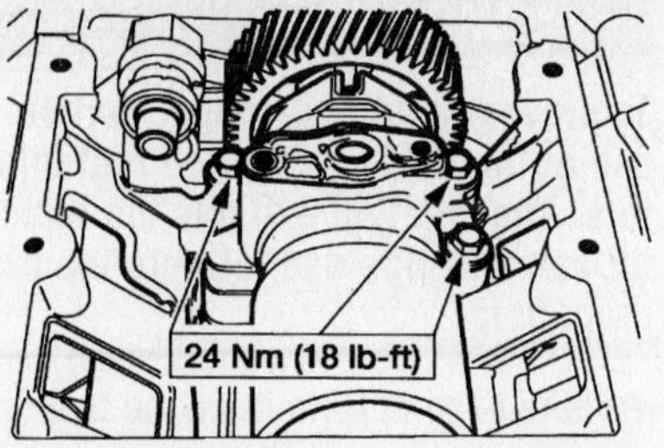

Install the high pressure oil pump

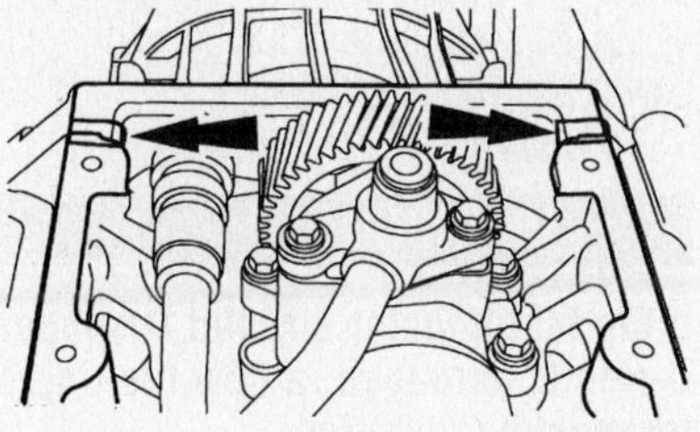

Clean the cover mounting surface and apply sealer at the seams

and remove the high-pressure oil pump discharge pipe.

9. Remove and discard the D-shaped O-ring seal.
10. Remove and discard the high-pressure pump O-ring seal.
11. Remove the bolts and the high-pressure oil pump.
12. Remove and discard the lower O-ring seal.

To install:

13. Install a new lower O-ring seal.
14. Install the high-pressure oil pump and bolts. Torque to 18 ft. lbs. (24 Nm).
15. Install the high-pressure pump O-ring seal.
16. Install the oil pump discharge pipe.
17. Install the bolts for the oil discharge pipe.
18. Position the dial indicator with bracketry onto the oil pump drive and check the oil pump drive gear backlash.
19. Install a new D-ring seal on the high-pressure discharge pipe. Torque to 71 inch lbs. (8 Nm).
20. Install a new press-in-place gasket in the high-pressure pump cover.
21. Clean the cover mounting surface and apply sealer at the seams.
22. Install the high-pressure pump cover and bolts. Torque to 8 ft. lbs. (11 Nm).
23. Install the turbocharger heat shield.
24. Install the intake manifold.

LATE BUILD

1. Remove the intake manifold.
2. Remove the turbocharger heat shield.
3. Remove the bolts and the high-pressure oil pump cover.
4. Use a thin gasket scraper to separate the cover from the crankcase.
5. Remove and discard the press-in-place gasket.
6. Position the dial indicator with bracketry onto the oil pump drive and check the oil pump drive gear backlash.
7. Remove the bolts from the high-pressure oil pump discharge pipe. Position the high-pressure discharge pipe aside.
8. Remove and discard the high-pressure pump O-ring seal.
9. Remove the bolts and the high-pressure oil pump.
10. Remove and discard the lower O-ring seal.

To install:

11. Install a new lower O-ring seal.
12. Install the high-pressure oil pump and bolts.
13. Install the high-pressure pump O-ring seal.
14. Position back the high-pressure discharge tube and install the bolts.
15. Position the dial indicator with bracketry onto the oil pump drive and check the oil pump drive gear backlash.
16. Install a new D-ring seal on the high-pressure discharge pipe.
17. Install a new press-in-place gasket in the high-pressure pump cover. Clean the cover mounting surface and apply sealer at the seams.
18. Install the high-pressure pump cover and bolts.
19. Install the turbocharger heat shield.
20. Install the intake manifold.

Fuel Injectors

REMOVAL & INSTALLATION

CAUTION

Observe all applicable safety precautions when working around fuel. Whenever servicing the fuel system, always work in a well ventilated area. Do not allow fuel spray or vapors to come in contact with a spark or open flame. Keep a dry chemical fire extinguisher near the work area. Always keep fuel in a container specifically designed for fuel storage; also, always properly seal fuel containers to avoid the possibility of fire or explosion.

6.0L Engine

EARLY BUILD

1. Before servicing the vehicle, refer to the precautions in the beginning of this section.
2. If removing the right fuel injectors, remove the evaporator case.
3. Remove the valve cover.
4. Disconnect the fuel injector electrical connector.
5. Disconnect the high-pressure oil rail supply line at the high-pressure oil rail.
6. Remove the bolts and the high-pressure oil rail.
7. Disconnect and remove the high-pressure oil supply line.

⁂ WARNING

Do not attempt to apply battery voltage to the fuel injector or damage to the fuel injector will occur.

8. Using a 19 mm socket, push the fuel injector electrical connector out of the rocker arm carrier.

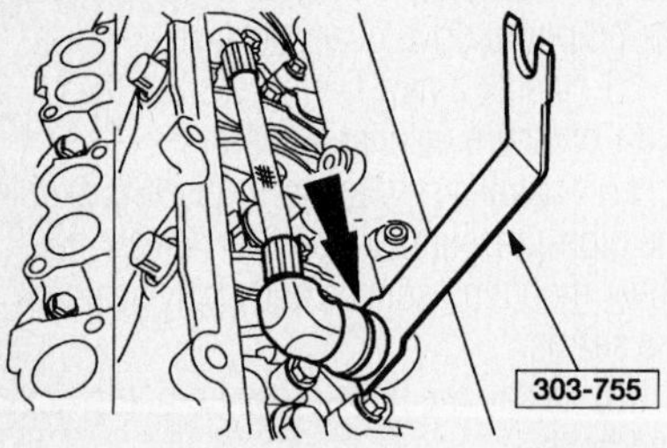

Disconnect the high-pressure oil rail supply line at the high-pressure oil rail—6.0L engine

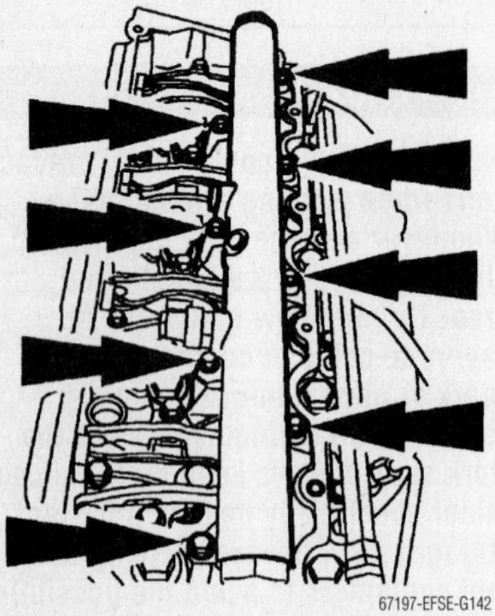

Remove the bolts and the high-pressure oil rail

⁂ WARNING

To prevent engine damage, do not use air tools to remove the fuel injectors. The clip that extracts the injector can dislodge and fall into the oil drain hole.

➡There is no need to drain the fuel rail.

➡If engine coolant is found in the combustion chambers, It may be necessary to install a new injector sleeve.

9. Remove the bolt, fuel injector hold-down clamp and fuel injector.

To install:

⁂ WARNING

If the fuel injector oil inlet D-shaped O-ring is damaged, a new fuel injector must be installed.

10. Install new O-ring seals and copper washer on the fuel injector. Lubricate the fuel injector and O-ring seals liberally with clean engine oil.

⁂ WARNING

To prevent engine damage, do not use air tools to install the fuel injectors. The clip that extracts the injector can dislodge and fall into the oil drain hole.

11. Install the fuel injector, fuel injector hold-down clamp and bolt. Torque to 24 ft. lbs. (33 Nm).
12. Install the fuel injector electrical connector into the rocker carrier.
13. Apply engine oil to the top fuel injector O-ring seals.
14. Install the high-pressure oil rail and bolts.
15. Install the high-pressure oil rail.
16. Install the bolts finger tight.
17. Tighten the bolts in the sequence shown.
18. Install the high-pressure oil line.
19. Connect the fuel injector electrical connector.
20. Install the valve covers.
21. If removed, install the evaporator case.

LATE BUILD

1. Remove the intake manifold.
2. Remove the turbocharger heat shield.
3. Remove the bolts and the high-pressure oil pump cover.
4. Use a thin gasket scraper to separate the cover from the crankcase.
5. Remove and discard the press-in-place gasket.
6. Position the dial indicator with bracketry onto the oil pump drive and check the oil pump drive gear backlash.
7. Remove the bolts from the high-pressure oil pump discharge pipe. Position the high-pressure discharge pipe aside.
8. Remove and discard the high-pressure pump O-ring seal.
9. Remove the bolts and the high-pressure oil pump.
10. Remove and discard the lower O-ring seal.

To install:

11. Install a new lower O-ring seal.
12. Install the high-pressure oil pump and bolts.
13. Install the high-pressure pump O-ring seal.

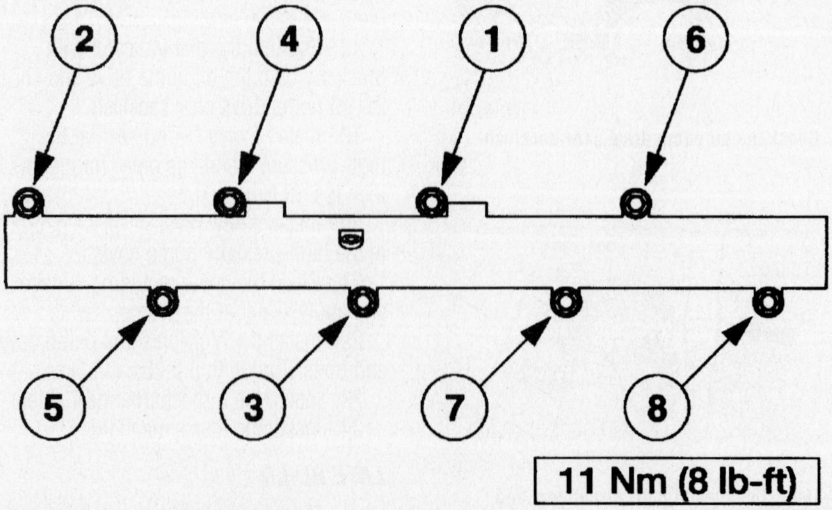

High pressure oil rail torque sequence—6.0L engine

14. Position back the high-pressure discharge tube and install the bolts.
15. Position the dial indicator with bracketry onto the oil pump drive and check the oil pump drive gear backlash.
16. Install a new D-ring seal on the high-pressure discharge pipe.
17. Install a new press-in-place gasket in the high-pressure pump cover.
18. Clean the cover mounting surface and apply sealer at the seams.
19. Install the high-pressure pump cover and bolts.
20. Install the turbocharger heat shield.
21. Install the intake manifold.

7.3L Engine

1. Before servicing the vehicle, refer to the precautions in the beginning of this section.

⁂ CAUTION

The red-striped wires on the DI Turbo carry 115 volts DC. A severe electrical shock may be given. Do not pierce the wires.

⁂ WARNING

Do not pierce the wires or damage to the harness could occur.

Special tools required:
- Slide Hammer, No. T50T–100–A, or equivalent
- Injector Remover, No. T94T–9000–AH1, or equivalent
- Injector Replacer, No. T94T–9000–AH2, or equivalent

2. Remove or disconnect the following:
- Valve cover
- Fuel injector electrical connector

⁂ WARNING

Remove the oil drain plugs prior to removing the injectors or oil could enter the combustion chamber, which could result in hydrostatic lock and severe engine damage.

- Oil rail drain plugs
- Retaining screw and oil deflector. The shoulder bolt on the inboard side of the fuel injector does not require removal.
- Outboard fuel injector retaining bolt
- Heater distribution box screws, nuts and clip
- Outer half of the case (to service No. 4 fuel injector only)

3. Remove the fuel injector using Injector Remover No. T94T–9000–AH1, or equivalent. Position the tool's fulcrum beneath the fuel injector hold-down plate and over the edge of the cylinder head. Install the remover screw in the threaded hole of the fuel injector plate (see illustration). Tighten the screw to lift out the injector from its bore. Place the injector in a suitable protective sleeve such as Rotunda Injector Protective Sleeve, No. 014–00933–2, and set the injector in a suitable holding rack.
4. Remove the fuel injector sleeves, if required. Insert the Injector Sleeve Tap Plug, 014–00934–3 into the injector sleeve to prevent debris from entering the combustion chamber. Insert Injector Sleeve Tap Pilot into the fuel injector sleeve and tighten 1–1 ½ turns. Attach Slide Hammer T50T–100–A to the Injector Sleeve Tap 014–00934–1 and 014–00934–2 Injector Sleeve Tap Pilot and remove the fuel injector sleeve from the bore.
5. Use Rotunda Injector Sleeve Brush 104–00934–A, or equivalent to clean the injector bore of any sealant residue. Make sure to remove any debris.

To install:

6. If removed, install the fuel injector sleeves using Rotunda Sleeve Replacer, No. 014–00934–4, or equivalent. Apply Threadlock, No. 262–E2FZ–19554–B, or equivalent to the fuel injector sleeves. Using a rubber mallet, tap on the tool to seat the injector bore. Remove the tool and remove any residue sealant.
7. Clean the fuel injector sleeve using a suitable sleeve brush set. Clean any debris from the sleeve.
8. Clean the injector bore with a lint-free shop towel.
9. Install the fuel injectors using special tools as follows:
   a. Lubricate the injectors with clean engine oil. Using new copper washers, carefully push the injectors square into the bore using hand pressure only to seat the O-rings.
   b. Position the open end of Injector Replacer, No. T94T–9000–AH2, or equivalent between the fuel injector body and injector hold-down plate, while positioning the opposite end of the tool over the edge of the cylinder head.
   c. Align the hole in the tool with the threaded hole in the cylinder head and install the bolt from the tool kit. Tighten the bolt to fully seat the injector, then remove the bolt and tool.
10. Install or connect the following:
- Outer half of the heater distribution box and retaining hardware (for No. 4 injector only)
- Oil deflector and bolt. Tighten the bolt to 108 inch lbs. (12 Nm).
- Fuel rail drain plug, tightening it to 96 inch lbs. (11 Nm)
- Oil rail drain plug, tightening it to 53 inch lbs. (6 Nm)
- Heater distribution box
- Fuel injector wiring harness
- Valve cover

DRIVE TRAIN

Automatic Transmission Assembly

REMOVAL & INSTALLATION

E-Series

4R100 TRANSMISSION

1. Before servicing the vehicle, refer to the precautions in the beginning of this section.
2. Disconnecting the negative battery cable on some vehicles may interfere with the functions of the on-board computer system(s) and may require the computer to undergo a relearning process once the negative battery cable is reconnected.
3. Remove or disconnect the following:
   - Air cleaner air intake duct assembly
   - Transmission fluid level indicator
   - Engine cover
   - Vapor management valve assembly and position it aside, gas engines only
   - Fluid filler tube from the short fluid inlet tube, and secure it aside
   - Fluid level indicator bracket nut, on 7.3L diesel engines
   - Fluid level indicator tube from the short fluid inlet tube and position it aside, on 7.3L diesel engines
   - Electrical harness connectors

➡**Note the length of each bolt for correct location during installation.**

   - Transmission-to-engine retaining bolts
   - Driveshaft
   - Transmission-mounted parking brake assembly, if equipped
   - Vehicle Speed Sensor (VSS)
   - Transmission Range (TR) sensor connector
   - Heat shield, if equipped
   - Solenoid body connector
   - Shift cable at the transmission and position it aside
   - Starter motor
   - Cylinder block opening cover
   - Torque converter-to-flexplate retaining nuts and discard

⁂ CAUTION

Be sure not to raise the back of the transmission too high. If it makes contact with the underbody, damage to the TSS sensor can occur.

4. Install a high lift transmission jack with the special jack adapter 014-0763 to the transmission.
   - Rear transmission mount nuts
   - Transmission support crossmember
   - Fluid cooler tubes at the Cooler Bypass Valve (CBV), plug all fittings and position the tubes aside
   - Right and left hand transmission-to-engine bolts
   - Flexplate inspection cover
   - Transmission-to-engine bolts
   - Transmission from the vehicle

To install:

5. Installation is the reverse of removal, please note the following specs:
6. Transmission-to-engine bolts to 45 ft. lbs. (61 Nm).
7. Torque converter-to-flexplate nuts to 26 ft. lbs. (35 Nm).
8. Flexplate inspection cover nuts to 25 ft. lbs. (34 Nm) on gasoline engines and 15 ft. lbs. (20 Nm) on 7.3L diesel engines.
9. Fluid filler tube into the short fluid inlet tube. Tighten the nut to 26 ft. lbs. (35 Nm).
10. Refill all transmissions with the correct amount of Motorcraft MERCON® automatic transmission fluid.
11. Connect the negative battery cable.

4R70W TRANSMISSION

1. Before servicing the vehicle, refer to the precautions in the beginning of this section.
2. Disconnecting the negative battery cable on some vehicles may interfere with the functions of the on-board computer system(s) and may require the computer to undergo a relearning process once the negative battery cable is reconnected.
3. Remove or disconnect the following:
   - Air cleaner air intake duct assembly
   - Transmission fluid level indicator
   - Engine cover
   - Fluid filler tube, on 4.6L and 5.4L engines
   - Fluid filler tube from the transmission and position aside, on 4.2L engines
   - Electrical harness connectors
   - Torque converter housing-to-engine retaining bolts. Position the fuel and electrical harness bracket aside.
   - Fluid cooler tubes and plug all fittings and position the tubes aside
4. Mark the driveshaft flange and the rear companion flange for correct alignment during assembly.
   - Driveshaft
   - Shift cable and bracket
   - Access plugs
   - Starter motor
   - Fasteners and the front A/C deflector assembly, if equipped
   - Four torque converter-to-flexplate retaining nuts and discard the bolts
   - Catalyst monitoring sensor harness connectors, if equipped
   - Right hand heat shield
   - Exhaust Gas Recirculation (EGR) vacuum reservoir from the frame and position aside, if equipped
5. Attach a high lift transmission jack to the transmission.
   - Crossmember
6. Support the transmission support crossmember.
   - Transmission insulator and retainer-to-crossmember retaining nuts and the transmission support crossmember
   - Clamp and hanger assembly, if equipped
   - Transmission range sensor electrical connector
   - Output Shaft Speed (OSS) sensor connector
   - Solenoid body connector
   - Flexplate inspection plate
   - Transmission bolts

⁂ CAUTION

Watch for clearance between the transmission and the right side heat shield retaining bracket as well as other obstructions.

7. Separate and back the transmission away from the engine, positioning the extension housing above the exhaust crossover pipe.
8. Hold the torque converter firmly in its installed position.

⁂ CAUTION

Watch for clearance between the transmission and the engine.

9. Carefully tilt the front of the transmission downward.

⁂ CAUTION

Watch for clearance between the transmission and the exhaust crossover pipe.

10. Lower the transmission enough to clear the engine. Then, carefully pull the jack with the transmission toward the front of the vehicle, lowering the transmission as required until it is clear of the exhaust crossover pipe.

11. Tilt the front of the transmission upward until level.

12. Remove the transmission with the jack from under the vehicle.

To install:

13. Installation is the reverse of removal, please note the following specs:

14. Transmission-to-engine bolts to 45 ft. lbs. (61 Nm).

15. Torque converter-to-flexplate nuts to 26 ft. lbs. (35 Nm).

16. Flexplate inspection cover nuts to 15 ft. lbs. (20 Nm).

17. Refill all transmissions with the correct amount of Motorcraft MERCON® automatic transmission fluid.

18. Connect the negative battery cable.

TORQSHIFT

1. With the vehicle in **Neutral**, position it on a hoist.
2. Disconnect both battery ground cables.
3. Remove the fluid level indicator.
4. Remove the driveshaft.
5. If transmission disassembly is required, drain the transmission fluid. Remove the drain plug and allow the fluid to drain
6. Install the drain plug.
7. Install a suitable high-lift transmission jack.
8. If equipped, remove the two wire harness retainers from the rear crossmember and position the wire harness out of the way.
9. If equipped, disconnect the transmission-mounted parking brake linkage.
10. Disconnect the return spring from the pin.
11. Remove the hair pin retaining clip from the pin.
12. Remove the pin from the actuating lever and move the actuating cable away from the actuating lever.
13. Compress the spring on the actuating cable and remove the cable from the cable housing.
14. Disconnect the transmission fluid cooler tubes form the filter assembly.
15. Remove the shift cable.
16. Remove the transmission fluid filter bracket nut.
17. Remove the transmission fluid filter bracket bolt, position the bracket aside.
18. Disconnect the shift cable from the manual lever.
19. Disconnect the wire harness from the shift cable bracket.
20. Remove the bolts and position the shift cable and bracket aside.
21. Disconnect the oil cooler tubes from the filter housing.
22. Remove the bolts and remove the filter housing.
23. Loosen the bolt and disconnect the solenoid body electrical connector.
24. Disconnect and position the wire harness out of the way.
25. Disconnect the output shaft speed (OSS) sensor electrical connector.
26. Disconnect the turbine shaft and intermediate shaft combination speed sensor electrical connector.
27. Disconnect the wire harness from the side of the transmission.
28. Remove the cylinder block opening cover in order to gain access to the torque converter nuts.
29. Use a suitable tool to rotate the crankshaft in order to gain access to the flexplate nuts.
30. Remove and discard the six torque nuts.
31. Remove the left crossmember nuts.
32. Remove the left crossmember bracket.
33. Remove the right crossmember nuts.
34. Remove the right crossmember bracket.
35. Remove the rear transmission mount nuts and remove the crossmember.
36. Remove the rear transmission mount from the extension housing.
37. Disconnect the wire harness from the bracket.
38. Remove the nine transmission-to-engine mounting bolts and position the wire harness bracket out of the way.
39. Slide the transmission back enough to install the special tool.
40. If equipped, remove the transmission-mounted parking brake.
41. Remove the transmission-mounted parking retaining bolts.
42. Keeping the vent in an upward position, remove the transmission-mounted parking brake assembly.
43. Remove and discard the transmission-mounted parking brake gasket.
44. Disconnect the transmission fluid cooler tubes and filter housing from the case.
45. If overhauling the transmission or installing a new transmission, carry out the transmission fluid cooler backflushing and cleaning.

To install:

46. Loosely position the fluid cooler tubes in place.
47. Loosely install the fluid filter bracket and bolt.
48. Tighten the transmission fluid cooler tube nuts and remove the bracket bolt. Torque to 30 ft. lbs. (40 Nm).
49. If equipped, install the transmission-mounted parking brake.
50. Install a new transmission-mounted parking brake gasket.
51. Keeping the vent in an upward position, install the transmission-mounted parking brake assembly.
52. Install the transmission-mounted parking retaining bolts. Torque to 41 ft. lbs. (55 Nm).

⁂ WARNING

Prior to the installation of the assembly, the torque converter pilot hub must be correctly lubricated or damage to the torque converter or the engine crankshaft can occur.

53. Lubricate the torque converter pilot hub with multi-purpose grease.
54. If the special tool has not been installed during the assembly of the transmission, install the special tool to hold the torque converter in place while moving and positioning the transmission in place. Once the transmission is in place, prior to bolting it to the engine, remove the special tool.
55. Position the transmission in place. While raising the transmission up into the engine compartment, align the fluid filler tube with the stub tube on the transmission using the fluid level indicator as a guide.
56. While installing the transmission to the engine, align the torque converter studs with the mounting holes in the flexplate.
57. Position the electrical wiring harness bracket in place and install the eight transmission-to-engine bolts. Torque to 35 ft. lbs. (47 Nm).
58. Connect the wire harness to the bracket.
59. Using the special tool, rotate the crankshaft to gain access to the torque converter studs.
60. Install the new torque converter-to-flexplate nuts. Torque to 26 ft. lbs. (35 Nm).
61. Install the cylinder block opening cover.
62. Install the wire harness retaining clip into the transmission on the left side.
63. Connect the wire harness on the right side of the transmission.
64. Install the oil filter housing.

65. Connect the oil cooler lines.
66. Install the transmission mount. Torque to 69 ft. lbs. (94 Nm).
67. Position the crossmember to the transmission mount and loosely install the nut.
68. Install the left crossmember bracket.
69. Install the left crossmember nuts. Torque to 60 ft. lbs. (81 Nm).
70. Install the right crossmember bracket.
71. Install the right crossmember nuts. Torque to 60 ft. lbs. (81 Nm).
72. Tighten the rear insulator nuts. Torque to 69 ft. lbs. (94 Nm).
73. Connect the output shaft speed (OSS) sensor electrical connector.
74. Connect the intermediate shaft and turbine shaft combination speed sensor electrical connector.
75. Connect the solenoid body electrical connector.
76. Connect the shift cable.
77. Install the cable housing bracket with one stud bolt and one bolt.
78. Install the bolt and nut for the filter housing and bracket.
79. Connect the wire harness to the bracket.
80. Install the shift cable to the manual lever.
81. Install the fluid cooler lines to the filter housing.
82. If equipped, install the wire harness and install the two retaining clips to the rear of the crossmember.
83. If equipped, connect the transmission-mounted parking brake.
84. Install the parking cable into the bracket.
85. Position the cable end onto the lever.
86. Install the retaining pin and clip.
87. Install the return spring.
88. Install the rear driveshaft.

✲✲ WARNING

If installing a newly overhauled or a new transmission, a new transmission fluid cooler remote filter must be installed. Otherwise transmission failure can occur.

89. Install a new transmission fluid cooler remote filter.
90. Use the following guidelines for the in-line transmission fluid filter:
    a. If the transmission was overhauled and the vehicle was equipped with an in-line fluid filter, install a new in-line fluid filter.
    b. If the transmission was overhauled and the vehicle was not equipped with an in-line fluid filter, install a new in-line fluid filter kit.
    c. If the transmission is being installed for a non-internal repair, do not install an in-line filter or filter kit.
    d. If installing a new or a Ford authorized remanufactured transmission, install the in-line transmission fluid filter that is supplied.

Prior to lowering the vehicle, install a new in-line transmission filter or a filter kit.

91. Connect the battery ground cable.
92. Adjust the shift linkage. Verify that the vehicle starts in **Park** and **Neutral** and the **Reverse** lamps illuminate in **Reverse**.
93. Install the fluid level indicator.
94. With the engine running and the transmission at normal operating temperature 66–77øC (150–170øF), check and adjust the transmission fluid level, and check for any leaks. If fluid is needed, add fluid in increments of 0.24 liter (0.5 pint) until the correct level is achieved (fluid should be in the cross-hatched area of the fluid level indicator).

2001–03 F-Series and 2004 Heritage Edition

1. Before servicing the vehicle, refer to the precautions in the beginning of this section.
2. Remove or disconnect the following:
    - Negative battery cable
    - Front driveshaft and transfer case, on 4-wheel drive vehicles
    - All cables, connectors and fluid lines that may interfere with transmission removal. Tag them if helpful for installation.
    - Torque converter from the flexplate and disconnect the shift linkage
    - Transmission fluid
3. Position a transmission jack under the transmission and safety-chain the case to the jack.
    - Driveshaft
    - Transmission rear mount
    - Crossmember
    - The transmission-to-engine block bolts

✲✲ CAUTION

The torque converter will fall out of the transmission if it is tilted forward. Keep a hand on it while lowering the transmission out of the vehicle.

4. Roll the transmission rearward until the input shaft clears, lower the jack and remove the transmission.

To install:

5. Carefully raise the transmission to the engine or bell housing.
6. Roll the transmission forward and into position.
7. Install or connect the following:
    - Tighten the bolts to 65 ft. lbs. (87 Nm) for the diesel or to 50 ft. lbs. (67 Nm) for 2001 gasoline engines or 35 ft. lbs. (47 Nm) on 2002–03 gasoline engines.
    - Crossmember and tighten the bolts to 59 ft. lbs. (80 Nm).
    - Transmission rear insulator and lower retainer. Tighten the bolts to 60 ft. lbs. (81 Nm).
8. The rest of the installation is the reverse of removal.
9. Refill all transmissions with the correct amount of Motorcraft MERCON® automatic transmission fluid.
10. Connect the negative battery cable.

2004 F-150

2-WHEEL DRIVE

1. Before servicing the vehicle, refer to the precautions in the beginning of this section.
2. Disconnect the battery ground cable.
3. With the vehicle in **Neutral**, position it on a hoist
4. Remove the rear driveshaft.
5. Drain the transmission fluid.
6. Loosen the transmission fluid pan bolts and allow the fluid to drain. After the fluid has drained, remove the bolts.
7. Remove the transmission fluid pan and transmission fluid pan gasket. Drain the rest of the fluid from the pan.

➡The transmission fluid pan gasket is reusable. Clean and inspect for damage. If not damaged, the gasket should be reused.

8. Install the transmission fluid pan and gasket.

➡If removing the transmission for a transmission related failure, it is not necessary to torque the transmission fluid pan back onto the transmission case. If removing the transmission for a non-related transmission failure it is necessary to re-torque the transmission fluid pan.

9. Install the bolts.
10. Remove the starter motor solenoid terminal cover.

11. Disconnect the starter motor electrical connectors and the ground wire.
12. Remove the starter motor.
13. Remove the rubber plug to access the nuts.
14. Remove the four torque converter nuts. Rotate either the front of the crankshaft or the flexplate to access all of the torque converter retaining nuts.
15. Remove the transmission inspection cover.
16. Remove the shift cable and bracket.
17. Disconnect the transmission electrical connectors.
18. Disconnect the solenoid body assembly electrical connector.
19. Using a suitable transmission jack, support the transmission and secure the transmission to it with a safety strap.
20. Remove the left bolt retaining the heat shield to the transmission support crossmember.
21. Remove the right bolt retaining the heat shield to the transmission support crossmember.
22. Loosen, but do not remove, the transmission mount-to-crossmember nuts.
23. Remove the two bolts for the exhaust hanger.
24. Remove the four crossmember-to-frame nuts and bolts (two on each side).
25. Remove the transmission mount-to-crossmember nuts, and remove the crossmember.
26. Remove the two bolts and the rear transmission mount.
27. Remove the exhaust hanger.
28. Disconnect the transmission fluid cooler tubes.
29. Remove the seven transmission-to-engine bolts.
30. Position the fuel lines and bracket aside.

**** CAUTION**

The torque converter is heavy and can result in injury if it falls out of the transmission. Secure the torque converter in the transmission. If the torque converter is dropped, a new one must be installed.

31. Slide the transmission rearward enough to install the holding tool.

**** CAUTION**

The transmission must be secured with a safety chain or strap.

➡The front of the transmission must be lowered for the transmission to clear the exhaust system during removal.

32. Continue moving the transmission rearward, while gradually lowering the front of the transmission. As the transmission jack is being lowered, move the transmission forward and remove it from the vehicle.
33. If the transmission is to be overhauled or if installing a new transmission, carry out transmission backflushing and cleaning.

To install:

**** WARNING**

The torque converter cover is piloted into position to the engine crankshaft by dowels in the rear of the engine block. The torque converter must rest squarely and loosely against the flexplate. This indicates that the torque converter pilot is not binding in the engine crankshaft.

**** CAUTION**

The transmission must be secured with a safety chain or strap.

**** CAUTION**

The torque converter is heavy and can result in injury if it falls out of the transmission. Secure the torque converter in the transmission. If the torque converter is dropped, a new one must be installed.

➡The front of the transmission must be lowered for the transmission to clear the exhaust system during installation.

34. Continue moving the transmission rearward, while gradually raising the front of the transmission. As the transmission jack is being raised, move the transmission back and install it in the vehicle.
35. Align the orange balancing marks between the torque converter studs and the flexplate bolt holes.
36. Remove the special tool.

➡While positioning the transmission to the back of the engine, use the transmission filler tube as a guide and install the transmission filler tube into the case.

37. Install the seven transmission-to-engine bolts.
38. Position the fuel line and bracket.
39. Install the bolts.
40. Connect the transmission fluid cooler tubes.
41. Position the rear transmission insulator and install the bolts.
42. Install the exhaust hanger onto the exhaust crossover pipe.
43. Position the crossmember and loosely install the mount-to-crossmember nuts.
44. Install the four crossmember-to-frame nuts and bolts (two on each side).
45. Tighten the rear transmission mount nuts.
46. Install the bolt retaining the left heat shield to the transmission support crossmember.
47. Install the bolt retaining the right heat shield to the transmission support crossmember.
48. Install the bolts retaining the exhaust hanger to the crossmember.
49. Connect the solenoid body assembly electrical connector.
50. Connect the transmission electrical connectors.
51. Install the shift cable bracket and connect the transmission shift linkage.
52. Install the transmission inspection cover.
53. Install the torque converter nuts.
54. Install the rubber access plug.
55. Install the starter motor.
56. Connect the starter motor electrical connectors and the ground wire.
57. Install the starter motor solenoid terminal cover.
58. Install the rear driveshaft.
59. Use the following guidelines for the in-line transmission fluid filter:
    a. If the transmission was overhauled and the vehicle was equipped with an in-line fluid filter, install a new in-line fluid filter.
    b. If the transmission was overhauled and the vehicle was not equipped with an in-line fluid filter, install a new in-line fluid filter kit.
    c. If the transmission is being installed for a non-internal repair, do not install an in-line filter or filter kit.
    d. If installing a new or a Ford authorized remanufactured transmission, install the in-line transmission fluid filter that is supplied.
60. Prior to lowering the vehicle, install a new in-line transmission filter or a filter kit.
61. Connect the battery ground cable.
62. Fill the transmission with clean automatic transmission fluid and inspect for correct operation.
63. Check the fluid filter for any leaks.

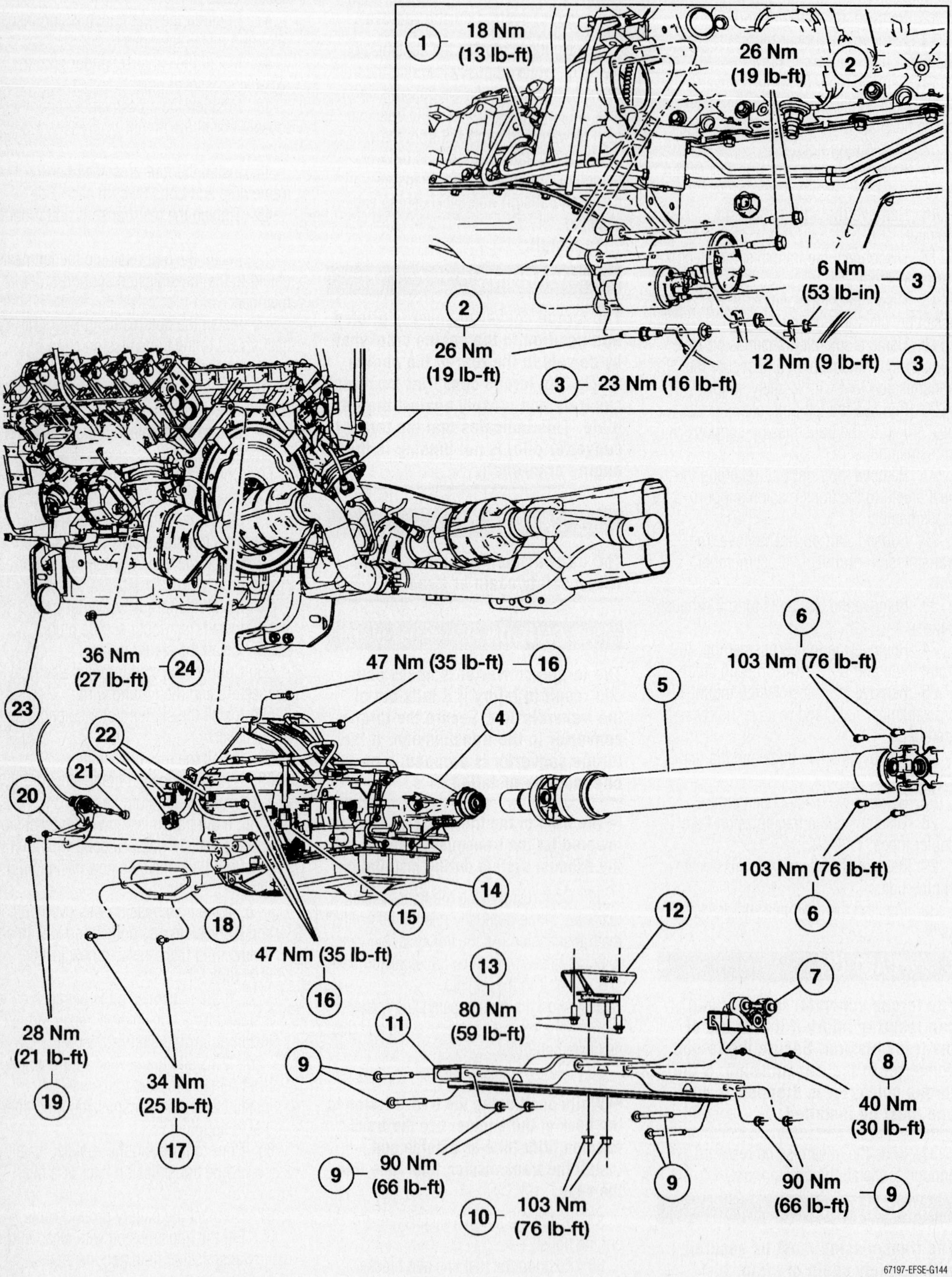

2004 4R70E/4R75E mounting—2-wheel drive

1 Transmission fluid cooler tube nuts
2 Starter motor bolts
3 Electrical connector nuts
4 Transmission assembly
5 Rear driveshaft
6 Rear driveshaft retaining bolts
7 Exhaust hanger
8 Exhaust hanger bolts
9 Rear crossmember bolts and nuts
10 Rear transmission mount nuts
11 Rear crossmember
12 Transmission rear support
13 Transmission rear support bolts
14 Transmission fluid pan
15 Transmission solenoid body connector
16 Transmission retaining bolts
17 Inspection cover shield bolts
18 Inspection cover
19 Transmission range selector lever cable bolts
20 Transmission range selector lever cable and bracket
21 Transmission range selector lever cable end
22 Transmission electrical connectors
23 Torque converter nut rubber access plug
24 Torque converter nuts

4-WHEEL DRIVE

1. Disconnect the battery ground cable.
2. With the vehicle in **Neutral**, position it on a hoist.
3. Drain the transmission fluid.
4. Loosen the transmission fluid pan bolts and allow the fluid to drain. After the fluid has drained, remove the bolts.
5. Remove the transmission fluid pan and transmission fluid pan gasket. Drain the rest of the transmission fluid from the fluid pan.

➡The transmission fluid pan gasket is reusable. Clean and inspect for damage. If not damaged, the gasket should be reused.

6. Install the transmission fluid pan and gasket.
7. Position the transmission fluid pan gasket.
8. Position the transmission fluid pan.

➡If removing the transmission for a transmission related failure, it is not necessary to torque the transmission fluid pan back onto the transmission case. If removing the transmission for a non-related transmission failure it is necessary to re-torque the transmission fluid pan.

9. Install the bolts.
10. Using a suitable transmission jack, support the transmission and secure the transmission to it with a safety strap.
11. Index-mark the front flange of the front driveshaft.
12. Index-mark the rear flange of the front driveshaft at the transfer case.
13. Remove the front driveshaft shield.
14. Remove the transfer case.
15. Remove the front driveshaft.
16. Remove the left bolt retaining the heat shield to the transmission support crossmember.
17. Remove the right bolt retaining the heat shield to the transmission support crossmember.
18. Loosen, but do not remove, the transmission mount-to-crossmember nuts.
19. Remove the two bolts for the exhaust hanger.
20. Remove the four crossmember-to-frame nuts and bolts (two on each side).
21. Remove the transmission mount-to-crossmember nuts, and remove the crossmember.
22. Remove the two bolts and the rear transmission mount.
23. Remove the exhaust hanger.
24. Remove the starter motor solenoid terminal cover.
25. Disconnect the starter motor electrical connectors and the ground wire.
26. Remove the starter motor.
27. Remove the rubber plug to access the nuts.
28. Remove the four nuts. Rotate the crankshaft/flexplate assembly to access all the nuts.
29. Remove the transmission inspection cover.
30. Remove the shift cable and bracket.
31. Disconnect the solenoid body assembly electrical connector.
32. Disconnect the transmission electrical connectors.
33. Disconnect the transmission fluid cooler tubes.
34. Remove the seven transmission-to-engine bolts.
35. Position the fuel lines and bracket aside.

⁂ CAUTION

The torque converter is heavy and can result in injury if it falls out of the transmission. Secure the torque converter in the transmission. If the torque converter is dropped, a new one must be installed.

36. Slide the transmission rearward enough to install a torque converter holding tool.

⁂ CAUTION

The transmission must be secured with a safety chain or strap.

➡The front of the transmission must be lowered for the transmission to clear the exhaust system during removal.

37. Continue moving the transmission rearward, while gradually lowering the front of the transmission. As the transmission jack is being lowered, move the transmission forward and remove it from the vehicle.
38. If the transmission is to be overhauled or if installing a new transmission, carry out transmission backflushing and cleaning.

To install:

⁂ CAUTION

The torque converter cover is piloted into position to the engine crankshaft by dowels in the rear of the engine block. The torque converter must rest squarely and loosely against the flexplate. This indicates that the torque converter pilot is not binding in the engine crankshaft.

39. When installing a new transmission, install a torque converter holding tool.

⁂ CAUTION

The transmission must be secured with a safety chain or strap.

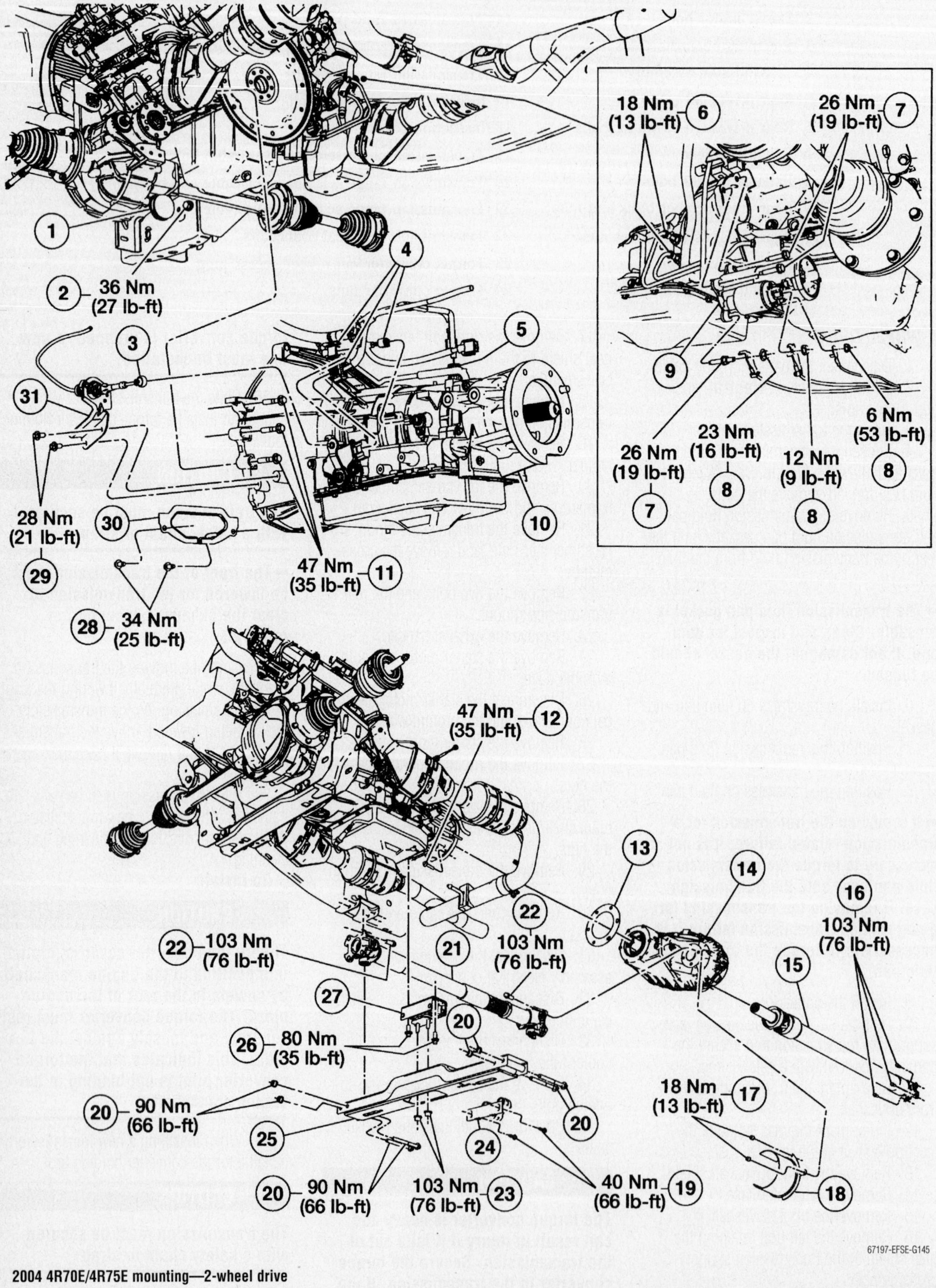

2004 4R70E/4R75E mounting—2-wheel drive

1 Torque converter nut rubber access plug
2 Torque converter nuts
3 Transmission range selector lever cable end
4 Transmission electrical connectors
5 Transmission solenoid body connector
6 Transmission fluid cooler tube nuts
7 Starter motor bolts
8 Starter motor electrical connector nuts
9 Starter motor
10 Transmission assembly
11 Transmission retaining bolts
12 Transfer case retaining bolts
13 Transfer case gasket
14 Transfer case
15 Rear driveshaft
16 Rear driveshaft retaining bolts
17 Front driveshaft shield nuts
18 Front driveshaft shield
19 Exhaust hanger bolts
20 Rear crossmember bolts and nuts
21 Front driveshaft
22 Front driveshaft retaining bolts
23 Rear transmission mount nuts
24 Exhaust hanger
25 Rear crossmember
26 Transmission rear support bolts
27 Transmission rear support
28 Inspection cover shield bolts
29 Transmission range selector lever cable bracket bolts
30 Inspection cover shield
31 Transmission range selector lever cable bracket

✱✱ CAUTION

The torque converter is heavy and can result in injury if it falls out of the transmission. Secure the torque converter in the transmission. If the torque converter is dropped, a new one must be installed.

➡The front of the transmission must be lowered for the transmission to clear the exhaust system during installation.

40. Continue moving the transmission rearward, while gradually raising the front of the transmission. As the transmission jack is being raised, move the transmission back and install it in the vehicle.
41. Align the orange balancing marks between the torque converter studs and the flexplate bolt holes.
42. Remove the special tool.

➡While positioning the transmission to the back of the engine use the transmission filler tube as a guide and install the transmission filler tube into the case.

43. Install the seven transmission-to-engine bolts.
44. Position the fuel line and bracket.
45. Install the bolts.
46. Connect the transmission fluid cooler tubes.
47. Install the exhaust hanger onto the exhaust crossover pipe.
48. Position the rear transmission insulator and install the bolts.
49. Position the crossmember and loosely install the mount-to-crossmember nuts.
50. Install the four crossmember-to-frame nuts and bolts (two on each side).
51. Tighten the rear transmission mount nuts.
52. Install the bolt retaining the left heat shield to the transmission support crossmember.
53. Install the bolt retaining the right heat shield to the transmission support crossmember.
54. Install the bolts retaining the exhaust hanger to the crossmember.
55. Connect the solenoid body assembly electrical connector.
56. Connect the transmission electrical connectors.
57. Install the shift cable bracket and connect the transmission shift linkage.
58. Install the transmission inspection cover.
59. Install the torque converter nuts.
60. Install the rubber access plug.
61. Install the starter motor.
62. Connect the starter motor electrical connectors and the ground wire.
63. Install the starter motor solenoid terminal cover.
64. Position the front driveshaft in place.
65. Install the transfer case.
66. Align the marks made during removal, install the front driveshaft flange and the four bolts.
67. Align the marks made during removal, install the rear driveshaft flange and the four bolts.
68. Install the front driveshaft shield.
69. Install the rear driveshaft.
70. Use the following guidelines for the in-line transmission fluid filter:
    a. If the transmission was overhauled and the vehicle was equipped with an in-line fluid filter, install a new in-line fluid filter.
    b. If the transmission was overhauled and the vehicle was not equipped with an in-line fluid filter, install a new in-line fluid filter kit.
    c. If the transmission is being installed for a non-internal repair, do not install an in-line filter or filter kit.
    d. If installing a new or a Ford authorized remanufactured transmission, install the in-line transmission fluid filter that is supplied.
71. Prior to lowering the vehicle, install a new in-line transmission filter or a filter kit.
72. Connect the battery ground cable.
73. Fill the transmission with clean automatic transmission fluid and inspect for correct operation.
74. Check the fluid filter for any leaks.

2004 F-250 and 350

4R100

1. Before servicing the vehicle, refer to the precautions in the beginning of this section.
2. With the vehicle in **Neutral** position it on a hoist.

➡When the battery has been disconnected and reconnected, some abnormal drive symptoms can occur while the vehicle relearns its adaptive strat-

egy. The customer needs to be notified that they may experience slightly different upshifts (either soft or firm) and that this is a temporary condition that will eventually return to normal operating condition.

3. Disconnect the battery ground cable.

4x4 vehicles

4. Remove the transfer case assembly.

4x2 vehicles

5. Remove the rear driveshaft.

All vehicles

6. If transmission disassembly is required, drain the transmission fluid.

Vehicles equipped with a fluid pan drain plug

7. Remove the drain plug and allow the fluid to drain.
8. Install the drain plug.

Vehicles not equipped with a fluid pan drain plug

➡**Do not discard the gasket unless damaged. This is a reusable gasket.**

9. Remove the transmission fluid pan and gasket.

➡**The transmission pan gasket is reusable unless damaged.**

➡**Apply a light coat of petroleum jelly to hold the gasket to the fluid pan.**

10. Position the gasket onto the cleaned fluid pan. Make sure the magnet is positioned over the dimple in the fluid pan.

✲✲ CAUTION

Mixing the 4x2 and the 4x4 style transmission fluid filters and transmission pan assembly components can cause transmission damage.

11. Install the correct pan with gasket for this application. Alternately tighten the bolts.

✲✲ CAUTION

Make sure securing straps or the transmission jack adapter do not touch the cooler bypass valve (CBV). Do not use the CBV as a handle. Damage to the CBV can cause a leak.

12. Install a suitable transmission jack and support the transmission.

4x4 vehicles

13. Remove the jack stand after installing a suitable transmission jack.

4x2 vehicles

14. Remove the right crossmember nuts.
15. Remove the left crossmember nuts.
16. Remove the nuts and the crossmember.
17. Remove the transmission mount.

Vehicles equipped with a transmission-mounted parking brake

18. Disconnect and lubricate.
19. Disconnect the parking brake lever return spring from the parking brake lever.
20. Apply penetrating oil to the adjusting clevis, jam nut and the threads on the front parking brake cable and conduit.
21. Disconnect the front parking brake cable and conduit.
22. Loosen the jam nut.
23. Remove the clevis locking pin.
24. Remove the clevis pin.
25. Remove the adjusting clevis from the parking brake lever.
26. Compress the retainer, and remove the front parking brake cable and conduit from the cable bracket.

All vehicles

➡**If the vehicle is equipped with a power take-off unit, all or part of the PTO unit will need to be removed.**

27. Disconnect the shift cable from the transmission.
28. Disconnect the shift cable from the manual lever.
29. Remove the shift cable bracket from the transmission and position aside.
30. Disconnect the digital transmission range (TR) sensor connector and the wire loom from the shift cable bracket.
31. Disconnect the solenoid body connector.
32. Disconnect the turbine shaft speed (TSS) sensor and the output shaft speed (OSS) sensor.
33. Remove the wiring harness from the transmission and position aside.
34. Remove the flexplate inspection cover.
35. Remove the starter motor.
36. Remove the cylinder block opening cover.
37. Remove and discard the torque converter-to-flexplate nuts. 5.4L engine has four torque converter-to-flexplate nuts. 6.8L engine has six torque converter-to-flexplate nuts.
38. Position a suitable drain pan and disconnect the transmission fluid cooler tubes from the cooler bypass valve.
39. Remove the seven transmission-to-engine mounting bolts.
40. Gently rock the transmission side-to-side to disengage it from the locator dowels.
41. Move the transmission and the transmission jack rearward to clear the engine flexplate.

✲✲ WARNING

The torque converter is heavy and can result in injury if it falls out of the transmission. Secure the torque converter in the transmission. If the torque converter is dropped, a new one must be installed.

42. Hold the torque converter in place.

✲✲ CAUTION

Use care while removing the transmission to avoid obstructions.

✲✲ CAUTION

Do not use the cooler bypass valve as a handle. Damage to the cooler bypass valve assembly can occur or damage to the case can result.

✲✲ CAUTION

If a safety strap is being used to hold the transmission to the high-lift transmission jack, place the strap behind the cooler bypass valve (CBV) to prevent damage to the cooler bypass valve.

43. Lower the transmission out of the vehicle.

Vehicles equipped with a transmission-mounted parking brake

44. Remove the transmission-mounted parking brake.
45. Keep the parking brake vent in the upward position to prevent contamination of the brake shoes and linings.
46. Remove the six bolts, parking brake assembly and the gasket from the extension housing. Discard the bolts and the gasket. Clean the mating surfaces.

All vehicles

47. If the transmission is being overhauled or if installing a new or remanufactured transmission, carry out the transmission fluid cooler—backflushing and cleaning.

To install:

✲✲ CAUTION

Prior to the installation of the transmission, the fluid, cooler lines and the cooler bypass valve must be cleaned. Transmission failure can occur if this procedure is not followed.

CAUTION

A new transmission oil-to-air (OTA) cooler must be installed if the transmission was overhauled or exchanged due to a failure of the transmission. Transmission failure can occur if this procedure is not followed.

48. Inspect the wiring harness and the connectors for damage, terminal condition, corrosion and seal integrity. Repair or install new as required.

Vehicles equipped with a transmission-mounted parking brake

49. If removed, install the parking brake. Position the parking brake assembly with a new gasket on the transmission extension housing.
50. Install six new bolts. Torque to 41 ft. lbs. (55 Nm).

All vehicles

CAUTION

Prior to the installation of the transmission, the torque converter pilot hub must be lubricated or damage to the torque converter or the engine crankshaft can occur.

51. Lubricate the torque converter pilot hub with multi-purpose grease.
52. Raise the transmission into place.

CAUTION

Do not use the cooler bypass valve as a handle. Damage to the cooler bypass valve assembly can occur or damage to the case can result.

CAUTION

Be careful not to raise the transmission up too far. The sensors can make contact with the underbody of the vehicle and cause damage to the sensors. Sensor failure or leakage can occur.

➡While raising the transmission up into the engine compartment, make sure to align the fluid filler tube with the stub tube on the transmission, using the dipstick as a guide.

53. Position the transmission.
54. Remove the holding tool.

CAUTION

Do not allow the torque converter drive flats to disengage from the pump gear. Use care not to damage the flexplate and the converter pilot. The torque converter must rest squarely against the flexplate, indicating the converter pilot is not binding in the crankshaft.

55. While installing the transmission to the engine, align the torque converter studs with the mounting holes in the flexplate.
56. Install seven transmission-to-engine bolts. Torque to 35 ft. lbs. (47 Nm).
57. Install the new torque converter-to-flexplate nuts. 5.4L engines have four torque converter nuts. 6.8L engines have six torque converter nuts. Torque to 26 ft. lbs. (35 Nm).
58. Install the cylinder block opening cover.
59. Install the flexplate inspection cover and bolts. Torque to 25 ft. lbs. (34 Nm).
60. Install the starter motor.
61. Connect the solenoid pack electrical connector.
62. Connect the digital transmission (TR) sensor connector.
63. Connect the turbine shaft speed (TSS) sensor and the output shaft speed (OSS) sensor.

➡If the vehicle is equipped with a power take-off (PTO) unit, all or part of the PTO unit will need to be installed.

64. Install the cable housing bracket. Torque to 18 ft. lbs. (25 Nm).
65. Install the shift cable to the manual lever.
66. Install the transmission fluid cooler tubes to the cooler bypass valve.

Vehicle equipped with a transmission-mounted parking brake

67. Position the front parking brake cable and conduit, and press the retainer into the cable bracket until it snaps into place.
68. Set the adjusting clevis. Loosen the jam nut several turns. Position the parking brake lever in the applied position. Tighten or loosen the adjusting clevis until the adjusting clevis hole lines up with the parking brake lever hole, then loosen the adjusting clevis to the specification.
69. Install the pins, and tighten the jam nut.
70. Install the clevis pin through the adjusting clevis and the parking brake lever.
71. Install the locking pin in the clevis pin. Tighten the jam nut.
72. Install the parking brake lever return spring.

4x4 vehicles

73. Support the extension housing with a jack stand and remove the transmission jack.
74. Install the transfer case.

4x2 vehicles

75. Install the transmission mount. Torque to 70 ft. lbs. (95 Nm).
76. Position the crossmember to the transmission mount and loosely install the nuts.

All vehicles

77. Install the crossmember bolts. Torque to 52 ft. lbs. (70 Nm).
78. Install the nuts. Torque to 52 ft. lbs. (70 Nm).

4x2 vehicles

79. Remove the transmission jack. Tighten the nuts. Torque to 69 ft. lbs. (94 Nm).
80. Install the rear driveshaft.
81. Use the following guidelines for the in-line transmission fluid filter:
   a. If the transmission was overhauled and the vehicle was equipped with an in-line fluid filter, install a new in-line fluid filter.
   b. If the transmission was overhauled and the vehicle was not equipped with an in-line fluid filter, install a new in-line fluid filter kit.
   c. If the transmission is being installed for a non-internal repair, do not install an in-line filter or filter kit.
   d. If installing a new or a Ford-authorized remanufactured transmission, install the in-line transmission fluid filter that is supplied.
82. Prior to lowering the vehicle, install a new in-line transmission filter or a filter kit.

➡When the battery has been disconnected and reconnected, some abnormal drive symptoms can occur while the vehicle relearns its adaptive strategy. The customer needs to be notified that they may experience slightly different upshifts (either soft or firm) and this is a temporary condition that will eventually return to normal operating condition.

83. Connect the battery ground cable.
84. Fill the transmission to the correct level with clean automatic transmission fluid.

TORQSHIFT

1. With the vehicle in **Neutral**, position it on a hoist.
2. Disconnect both battery ground cables.

4x4 vehicles

3. Remove the transfer case assembly.

4x2 vehicles

4. Remove the driveshaft.

All vehicles

5. If transmission disassembly is required, drain the transmission fluid. Remove the drain plug and allow the fluid to drain

6. Install the drain plug.

7. Install a suitable high-lift transmission jack.

4x4 vehicles

8. Remove the jack stand from under the extension housing after installing the transmission jack.

4x2 vehicles

9. Remove the wire harness from the rear crossmember.

10. Remove the rear transmission mount nuts.

All vehicles

11. Remove the shift cable.

12. Disconnect the shift cable from the manual lever.

13. Remove the bolts and position the shift cable and bracket out of the way.

14. Loosen the bolt and disconnect the solenoid body electrical connector.

15. Disconnect the output shaft speed (OSS) sensor electrical connector.

16. Disconnect the turbine shaft and intermediate shaft combination speed sensor electrical connector.

17. Disconnect the right and left wire harness from the side of the transmission.

4x2 vehicles

18. Remove the left crossmember bolts and nuts.

19. Loosen, but do not remove, the bolt and nut indicated.

20. Remove the right crossmember bracket bolts and nuts.

21. Remove the bracket bolts and nuts.

22. Remove the bracket.

23. Remove the right crossmember bolts and nuts.

24. Remove the left crossmember bolt, nut and the crossmember.

25. Remove the bolt and nut.

26. Remove the crossmember.

27. Remove the rear transmission mount from the extension housing.

28. If equipped with dual alternators, rotate the tensioner and remove the outer accessory drive belt from the crankshaft pulley.

All vehicles

29. Remove the cylinder block opening cover in order to gain access to the torque converter nuts.

➡Using a suitable strap wrench, rotate the crankshaft pulley to gain access to the torque converter nuts.

30. Remove and discard the torque converter nuts.

31. While holding the case fitting, disconnect the rear fluid cooler tube nut and tube.

32. While holding the case fitting, disconnect the front fluid cooler tube nut and tube.

33. Remove the nine transmission-to-engine mounting bolts.

34. Slide the transmission back enough to install the special tool.

35. If the transmission is being overhauled or if installing a new or remanufactured transmission, carry out the transmission fluid cooler backflushing and cleaning.

To install:

✻✻ CAUTION

Prior to the installation of a new or overhauled transmission, the fluid cooler lines must be cleaned. Otherwise transmission failure can occur.

✻✻ CAUTION

Prior to the installation of a new or overhauled transmission, a new transmission fluid cooler remote filter must be installed. Otherwise transmission failure can occur.

36. If necessary, install a new transmission fluid cooler remote filter.

✻✻ CAUTION

Prior to the installation of the assembly, the torque converter pilot hub must be correctly lubricated or damage to the torque converter or the engine crankshaft can occur.

37. Lubricate the torque converter pilot hub with multi-purpose grease.

38. If the holding special tool has not been installed during the assembly of the transmission, install the tool to hold the torque converter in place while moving and positioning the transmission in place. Once the transmission is in place, prior to bolting it to the engine, remove the special tool.

39. Position the transmission in place. While raising the transmission up into the engine compartment, align the fluid filler tube with the stub tube on the transmission using the fluid level indicator as a guide.

40. While installing the transmission to the engine, align the torque converter studs with the mounting holes in the flexplate.

41. Install nine transmission-to-engine bolts. Torque to 35 ft. lbs. (47 Nm).

42. Using a suitable strap wrench to rotate the crankshaft and pulley, install the new torque converter-to-flexplate nuts. Torque to 26 ft. lbs. (35 Nm).

43. Install the cylinder block opening cover.

44. If equipped with dual alternators, rotate the tensioner and install the outer accessory drive belt onto the crankshaft pulley.

All applications

45. Install the front transmission fluid cooler tube.

46. Install the rear transmission fluid cooler tube.

4x4 applications

47. Support the extension housing with a jack stand and remove the transmission jack.

4x2 applications

48. Install the transmission mount. Torque to 69 ft. lbs. (94 Nm).

49. Position the crossmember to the transmission mount and loosely install the nut.

50. Install the left crossmember bolts. Torque to 60 ft. lbs. (81 Nm).

51. Install the right crossmember bolts. Torque to 60 ft. lbs. (81 Nm).

52. Install the bracket and loosely install the bolts.

53. Install the bracket.

54. Loosely install the bolts.

55. Tighten the right crossmember bolts. Torque to 60 ft. lbs. (81 Nm).

56. Tighten the rear insulator nuts. Torque to 69 ft. lbs. (94 Nm).

All applications

57. Connect the right and left wiring harness to the side of the transmission.

58. Connect the output shaft speed (OSS) sensor electrical connector.

59. Connect the intermediate shaft and turbine shaft combination speed sensor electrical connector.

60. Connect the solenoid body electrical connector.

➡If the vehicle is equipped with a power take-off unit, all or part of the PTO unit will need to be installed.

61. Connect the shift cable.

62. Install the cable housing bracket.

63. Install the shift cable to the manual lever.

4x2 applications

64. Reconnect the wire harness to the frame.

4x4 applications

65. If equipped, install the transfer case.

4x2 applications

66. Install the rear driveshaft.

All applications

➡Use the following guidelines for the in-line transmission filter:

a. If the transmission was overhauled and the vehicle was equipped with an in-line fluid filter, install a new in-line fluid filter.

b. If the transmission was overhauled and the vehicle was not equipped with the in-line fluid filter, install a new in-line fluid filter.

c. If the transmission is being installed for a non-internal repair, do not install an in-line filter or filter kit.

d. If installing a new or a Ford-authorized remanufactured transmission, install the in-line transmission fluid filter that is supplied.

67. Prior to lowering the vehicle, install a new in-line transmission filter or a filter kit.

68. Connect the battery ground cable.

69. Adjust the shift linkage. Verify that the vehicle starts in **Park** and **Neutral** and the **Reverse** lamps illuminate in **Reverse**.

70. With the engine running and the transmission at normal operating temperature 66n dash>77°C (150–170°F), check and adjust the transmission fluid level, and check for any leaks. If fluid is needed, add fluid in increments of 0.24L (0.5 pint) until the correct level is achieved (fluid should be in the cross-hatched area of the fluid level indicator).

Manual Transmission

REMOVAL & INSTALLATION

2001–02 F-Series

1. Before servicing the vehicle, refer to the precautions in the beginning of this section.

2. Remove or disconnect the following:
   - Negative battery cable
   - Shifter boot and lever from inside the vehicle
   - Front driveshaft and transfer case, on 4-wheel drive vehicles
   - All cables, connectors and fluid lines that may interfere with transmission removal. Tag them if helpful for installation.
   - Transmission fluid
   - Driveshaft
   - Transmission rear mount
   - Crossmember
   - The transmission-to-engine block bolts
   - Bolts securing the transmission to the bell housing

3. Position a transmission jack under the transmission and safety-chain the case to the jack.

⁂ CAUTION

The torque converter will fall out of the transmission if it is tilted forward. Keep a hand on it while lowering the transmission out of the vehicle.

4. Roll the transmission rearward until the input shaft clears, lower the jack and remove the transmission.

To install:

5. Carefully raise the transmission to the engine or bell housing.

6. Roll the transmission forward and into position.

7. Install or connect the following:
   - Tighten the bolts to 50 ft. lbs. (64 Nm) for 2001 models or 39 ft. lbs. (54 Nm) for 2002–03 models.
   - Crossmember and tighten the bolts to 59 ft. lbs. (80 Nm).
   - Transmission rear insulator and lower retainer. Tighten the bolts to 60 ft. lbs. (81 Nm).

8. The rest of the installation is the reverse of removal.

9. Refill all transmissions with the correct amount of Motorcraft MERCON® automatic transmission fluid.

10. Connect the negative battery cable.

2003 F-Series, 2004 F-150

1. Before servicing the vehicle, refer to the precautions in the beginning of this section.

2. Disconnect the battery ground cable.

3. Remove the four screws and pull the gearshift lever upward.

4. Remove the upper gearshift lever.

5. Remove the four bolts and the lower gearshift lever boot.

6. Remove the bolts and the lower gearshift lever.

7. Raise and support the vehicle.

8. If the transmission is being disassembled, drain the transmission fluid.

9. Remove the fuel line bracket from the rear of the transmission.

10. Disconnect the wiring harness from the transmission, the transfer case (if equipped) and the heated oxygen sensors (HO_2S).

11. Remove the starter.

➡Index-mark the rear driveshaft yoke and the pinion flange to maintain initial driveshaft balance during installation.

12. Remove the rear driveshaft.

Four wheel drive vehicles

13. Remove the skid plate.

➡Index-mark the front driveshaft yoke and pinion flange and at the transfer case.

14. Remove the front driveshaft.

15. Remove the torsion bars and the torsion bar crossmember.

16. Remove the transfer case.

17. Remove the transfer case heat shield.

All vehicles

18. Disconnect the clutch hydraulic line.

19. Support the transmission.

20. Securely strap the jack to the transmission.

21. Remove the two heat shield bolts from the crossmember.

22. Remove the crossmember nuts.

23. Remove the transmission mount nuts, then remove the crossmember.

24. Remove the nut.

25. Remove the bolts and the transmission mount and exhaust bracket.

Four wheel drive vehicles

26. Remove the bolt and disconnect the transfer case shift lever.

All vehicles

27. Remove the two engine oil pan-to-transmission bolts.

28. Remove the six engine-to-transmission bolts.

29. The upper bolt secures the fuel line bracket.

30. Remove the muffler from the exhaust hanger brackets.

31. Remove the tail pipe from the exhaust hanger insulator.

32. Tilt the exhaust downward and position it on top of the rear axle.

33. Remove the transmission.

34. Move the transmission rearward until the input shaft is clear of the clutch. Tilt the front of the transmission downward while lowering from the vehicle.

To install

35. Using the special tool, raise and position the transmission.

36. Install the six engine-to-transmission bolts. Tighten to 44 ft. lbs. (59 Nm).

37. Install the two engine oil pan-to-transmission bolts. Tighten to 32 ft. lbs. (44 Nm).

38. Raise the transmission, then install the muffler and tail pipe into the exhaust hanger insulators.

39. Install the transmission mount and exhaust bracket.
40. Position the transmission mount.
41. Install the bolts. Tighten to 60–80 ft. lbs. (87–110 Nm).
42. Install the nut. Tighten to 32 ft. lbs. (44 Nm).
43. Install the crossmember.
44. Position the crossmember.
45. Install the bolts. Tighten to 58–75 ft. lbs. (78–103 Nm).
46. Install the transmission mount bolts. Tighten to 60–80 ft. lbs. (87–110 Nm).
47. Install the two heat shield bolts to the crossmember. Tighten to 10–12 ft. lbs. (13–17 Nm).
48. Remove the transmission support from the transmission.
49. Connect the clutch hydraulic line.

Four wheel drive vehicles

50. Connect the transfer case shift lever and install the bolt.
51. Install the transfer case.
52. Install the torsion bars and the torsion bar crossmember.

➡Align the index marks made during removal.

53. Install the front driveshaft.
54. Install the transfer case heat shield.
55. Install the skid plate.

All vehicles

➡Align the index marks made during removal.

56. Install the rear driveshaft.
57. Install the starter.
58. Connect the wiring harness to the transmission, the transfer case (if equipped) and the heated oxygen sensors (HO_2S).
59. Install the fuel line bracket.
60. Install the lower gearshift lever.
61. Install the lower gearshift lever boot.
62. Install the upper gearshift lever.
63. Install the upper gearshift lever boot.
64. Connect the battery ground cable.
65. If the transmission was disassembled, refill with transmission fluid.
66. Check the vehicle ride height. Adjust as necessary.

2004 F-250 and 350

1. Before servicing the vehicle, refer to the precautions in the beginning of this section.
2. Remove the four screws and the outer shift lever boot.
3. Remove the upper gearshift lever.
4. Remove the lower shift lever boot.
5. Remove the lower gearshift and shift housing.

Vehicles with a manual shift lever

6. Shift the transfer case into 4H.
7. Remove the screws that attach the bezel and boot assembly to the floor.
8. Remove the bolt that attaches the shift lever to the transfer case control lever assembly, and remove the shift lever, and the bezel and boot assembly.
9. Slide the bezel and boot assembly upward on the shift lever.
10. Remove the bolt, the shift lever, and the bezel and boot assembly.

All vehicles

11. Raise and support the vehicle.
12. If the transmission is being disassembled, drain the transmission fluid.
13. Remove the starter.
14. Disconnect the rear driveshaft and position it aside.
15. Remove the transfer case, if equipped.
16. Remove any power take-off (PTO) equipment, if equipped.
17. Support the transmission.
18. Securely strap the jack to the transmission.
19. Remove the transmission mount nuts.
20. Remove the right crossmember nuts.
21. Remove the left crossmember bolts.
22. Disconnect the reverse lamp switch electrical connector.
23. Disconnect wiring harness from the transmission.
24. Remove the clutch slave cylinder and position it aside.
25. Push the clutch slave cylinder inward, then rotate counterclockwise 45 degrees to remove.
26. Disconnect the transmission cooling tubes.
27. Remove the dust cover bolts.
28. Remove the transmission-to-engine bolts. For vehicles equipped with 7.3L diesel engines, remove six bolts. For vehicles equipped with gasoline engines, remove seven bolts.
29. Remove the transmission. Move the transmission rearward until the input shaft is clear of the clutch, then lower from the vehicle.

To install:

30. Using the transmission jack, raise and position the transmission to the engine and clutch.
31. Install the transmission-to-engine bolts. For vehicles equipped with 7.3L diesel engines, install six bolts. For vehicles equipped with gasoline engines, install seven bolts. Torque to 46 ft. lbs. (63 Nm).
32. Install the left crossmember bolts. Torque to 52 ft. lbs. (70 Nm).
33. Install the right crossmember nuts. Torque to 52 ft. lbs. (70 Nm).
34. Install the transmission mount nuts. Torque to 60 ft. lbs. (81 Nm).
35. Install the engine plate bolts. Torque to 21 ft. lbs. (28 Nm).
36. Remove the transmission jack.
37. Connect the transmission cooler tubes.
38. Install the clutch slave cylinder. Rotate the clutch slave cylinder clockwise 45 degrees to lock in position.
39. Install the starter.
40. Install the transfer case, if equipped. If the transfer case control lever assembly was removed from the transmission, it must be correctly aligned.
41. Connect the driveshaft.
42. Install any power take-off (PTO) equipment, if equipped.
43. Connect the reverse lamp switch electrical connector.
44. Connect the wiring harness to the transmission.
45. Refill the transmission to specification.
46. Lower the vehicle.

➡Do not use a silicone sealing compound.

➡Do not wait longer than ten minutes to tighten the six bolts due to the rapid cure time of the sealant.

47. Install the lower gearshift lever and shift housing assembly. Apply gasket maker to the shift housing and the main case.
48. Install the lower shift lever boot.
49. Apply Threadlock and sealer to the gearshift lever bolts. Install the upper gearshift lever.
50. Install the screws.

Vehicles with a manual shift lever

51. Position the shift lever with the bezel and boot assembly and install the bolt.
52. Position the bezel and boot assembly and install the screws. Verify the shift sequence from 2H to 4L to 2H.

Clutch

REMOVAL & INSTALLATION

2001–03 F-Series, Exc. Diesel, and 2004 F-150

1. Before servicing the vehicle, refer to the precautions in the beginning of this section.
2. Disconnect the negative battery cable.
3. Raise and safely support the vehicle.

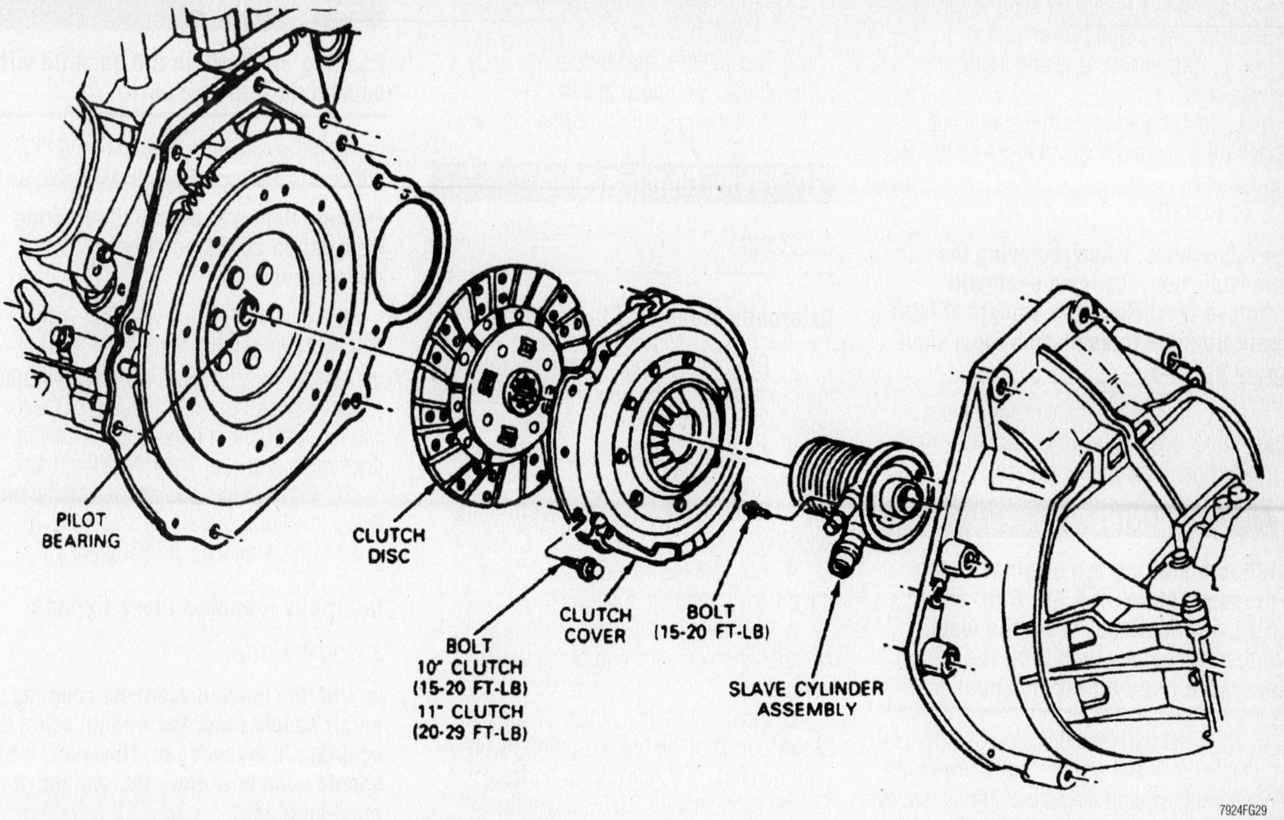

Typical clutch assembly with internal slave cylinder—2001–03

4. On vehicles with the externally mounted slave cylinder, remove the clutch slave cylinder. On vehicles with an internally mounted slave cylinder, disengage the quick-disconnect coupling with a spring coupling tool.

5. Remove the transmission.

6. On gasoline engine models, remove the starter. Remove the flywheel housing attaching bolts and remove the housing. On 7.3L diesel engine models, remove the cover, then remove the release lever and bearing from the clutch housing. To remove the release lever:

a. Remove the dust boot.

b. Push the release lever forward to compress the slave cylinder.

c. Remove the slave cylinder by prying on the steel clip to free the tangs while pulling the cylinder clear.

d. Remove the release lever by pulling it outward.

7. Mark the pressure plate and cover assembly and the flywheel so that they can be reinstalled in the same relative position.

8. Loosen the pressure plate and cover attaching bolts evenly in a staggered sequence a turn at time until the pressure plate springs are relieved of their tension. Remove the attaching bolts.

9. Remove the pressure plate and cover assembly and the clutch disc from the flywheel.

10. Inspect the flywheel for wear, damage and flatness.

To install:

11. Position the clutch disc on the flywheel so that an aligning tool or spare transmission mainshaft can enter the clutch pilot bearing and align the disc.

12. When reinstalling the original pressure plate and cover assembly, align the assembly and flywheel according to the marks made during removal. Position the pressure plate and cover assembly on the flywheel, align the pressure plate and disc, and install the retaining bolts. Tighten the bolts in an alternating sequence a few turns at a time until the proper torque is reached:

a. On 2001 models with a 10 inch clutch: 15–20 ft. lbs. (20–27 Nm).

b. On 2001 models with a 11 inch clutch: 20–29 ft. lbs. (27–39 Nm).

c. On 2002–03 models with a 11 inch clutch: 35–46 ft. lbs. (47–63 Nm).

13. Remove the tool used to align the clutch disc.

14. With the clutch fully released, apply a light coat of grease on the sides of the driving lugs.

15. Position the clutch release bearing and the bearing hub on the release lever. Install the release lever on the fulcrum in the flywheel housing. Apply a light coating of grease to the release lever fingers and the fulcrum. Fill the groove of the release bearing hub with grease.

16. If the flywheel housing has been removed, position it against the rear engine cover plate and install the attaching bolts and tighten them to 40–50 ft. lbs. (54–68 Nm).

17. Install the starter motor, if removed.

18. Install the transmission.

19. Install the slave cylinder and bleed the system.

20. Connect the negative battery cable.

2004 F-250 and 350 and 2001–03 w/Diesel Engines

1. Before servicing the vehicle, refer to the precautions in the beginning of this section.

2. Remove the transmission.

3. Index-mark the clutch pressure plate and the flywheel, if reinstalling these parts.

4. Remove the bolts, the clutch pressure plate, and the clutch disc.

5. Inspect the transmission input shaft pilot bearing for:

a. Misalignment and looseness in the crankshaft (gasoline engine) or flywheel (7.3L diesel engine).

b. Needle rollers for scoring, discoloration, wear, and broken rollers.

c. Seal for damage and lubricant leakage.

6. Install a new transmission input shaft pilot bearing if any of these conditions are present.

To install:

➡Sometimes, when removing the transmission, the input shaft will remove a considerable amount of lubricant from the transmission input shaft pilot bearing.

7. Lubricate the transmission input shaft pilot bearing, as necessary. Use Krytox® High-Temperature Grease.

⁂ WARNING

When installing the original clutch pressure plate on 5.4L, 6.0L and 6.8L applications, reset the wear indicator before installing the clutch pressure plate on the flywheel.

8. Reset the wear indicator. Using a suitable press and adapter, press downward on the fingers until the adjusting ring moves freely.

9. Rotate the adjusting ring counterclockwise to compress the tension springs. Hold the adjusting ring in this position.

10. Release the pressure on the fingers. The adjusting ring will now stay in the reset position.

11. Position the clutch disc on the flywheel and the clutch alignment tool in the pilot bearing to align the clutch disc. The 5.4L/6.8L engines accept a 1¼ inch input shaft. The 6.0L engines accept a 1⅜ inch input shaft with 0.98 inch pilot bearing inner diameter. The 7.3L engines accept a 1⅜ inch input shaft with 0.67 inch pilot bearing inner diameter.

➡Align the index marks if installing the original clutch pressure plate and flywheel.

12. Install the clutch pressure plate. Position the clutch pressure plate on the dowels.

➡The 7.3L diesel engine flywheel has two dowels. The gasoline engine flywheel has three dowels.

13. Using a clutch alignment tool, align the clutch disc and the pressure plate.

14. Install the bolts and tighten in a star pattern sequence to:

- 5.4L and 6.8L engines: 33 ft. lbs. (45 Nm)
- 6.0L and 7.3L engines: 21 ft. lbs. (28 Nm)

15. Remove the special tool.
16. Install the transmission.
17. Test the system for normal operation.

Hydraulic Clutch System

BLEEDING

Externally Mounted Slave Cylinder

1. Before servicing the vehicle, refer to the precautions in the beginning of this section.
2. Clean the reservoir cap and the slave cylinder connection.
3. Remove the slave cylinder from the housing.
4. Using a 3⁄32 inch punch, drive out the pin that holds the tube in place.
5. Remove the tube from the slave cylinder and place the end of the tube in a container.
6. Hold the slave cylinder so the connector port is at the highest point, by tipping it about 30 degrees from horizontal. Fill the cylinder with DOT 3 brake fluid through the port. It may be necessary to rock the cylinder or slightly depress the pushrod to expel all the air.

⁂ CAUTION

Pushing too hard on the pushrod will spurt fluid from the port!

7. When all air is expelled—no more bubble are seen—install the slave cylinder.

➡Some fluid will be expelled during installation as the pushrod is depressed.

8. Remove the reservoir cap. Some fluid will run out of the tube end into the container. Pour fluid into the reservoir until a steady stream of fluid runs out of the tube and the reservoir is filled. Quickly install the diaphragm and cap. The flow should stop.
9. Connect the tube and install the pin. Check the fluid level.
10. Check the clutch operation.

Internally Mounted Slave Cylinder

EXCEPT F-150

➡With the quick-disconnect coupling, no air should enter the system when the coupling is disengaged. However, if air should somehow enter the system, it must be bled.

1. Before servicing the vehicle, refer to the precautions in the beginning of this section.

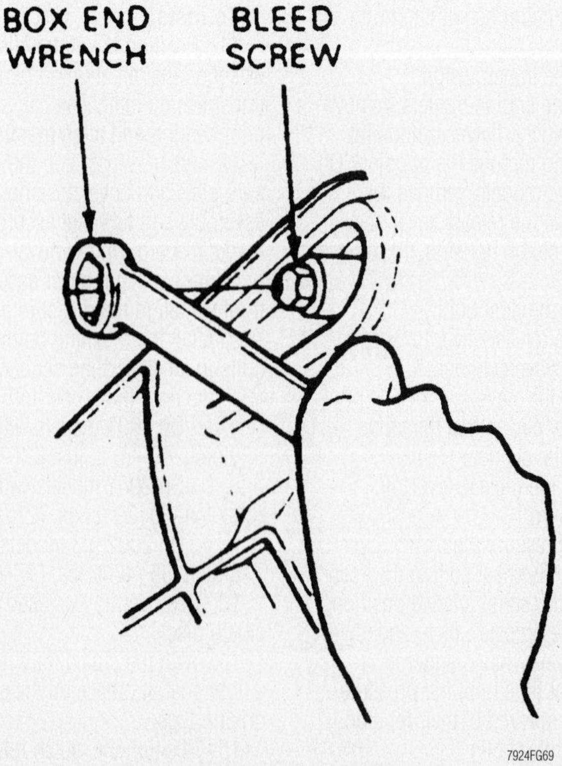

Bleed screw location for internally mounted slave cylinders

2. Remove the reservoir cap and diaphragm. Fill the reservoir with DOT 3 brake fluid.

3. Connect a piece of rubber tubing to the slave cylinder bleed screw. Place the other end in a container.

4. Loosen the bleed screw. Gravity will force fluid from the master cylinder to flow down to the slave cylinder, forcing air out of the bleed screw. When a steady stream—no bubbles—flows out, the system is bled. Close the bleed screw.

➡Check periodically to be sure the master cylinder reservoir doesn't run dry.

5. Add fluid to fill the master cylinder reservoir.

6. Fully depress the clutch pedal. Release it as quickly as possible. Pause for 2 seconds. Repeat this procedure 10 times.

7. Check the fluid level. Refill it if necessary. It should be kept full.

8. Repeat Steps 5 and 6, 5 more times.

9. Install the diaphragm and cap.

10. Have an assistant hold the pedal to the floor while you crack the bleed screw—not too far—just far enough to expel any trapped air. Close the bleed screw, then release the pedal.

11. Check, and if necessary, fill the reservoir.

F-150

➡Be sure to keep the clutch master cylinder reservoir full of brake fluid during the bleeding process to prevent air from entering the clutch master cylinder.

1. Before servicing the vehicle, refer to the precautions in the beginning of this section.

2. Raise and safely support the front of the vehicle.

➡It is necessary to have the assistance of a helper to bleed this system.

3. Fill the clutch system reservoir with DOT 3 brake fluid.

4. Have your assistant depress the clutch pedal rapidly for 5–10 strokes.

5. Wait 1–3 minutes.

6. Repeat Steps 3 and 4 3 more times.

7. Loosen the bleeder screw on the transmission for the slave cylinder.

8. Have the helper fully depress the clutch pedal and hold it down.

9. Tighten the bleeder screw.

10. The helper should now release the clutch pedal.

11. Apply pressure to the clutch pedal. If the clutch pedal travels more than 6–7 inches (15.3–17.7 cm), repeat the bleeding process.

Transfer Case Assembly

REMOVAL & INSTALLATION

2001–02

⁂ CAUTION

The catalytic converter is located beside the transfer case. Due to the extreme high temperatures generated by the converter, be careful when removing the transfer case, or personal injury may result.

1. Before servicing the vehicle, refer to the precautions in the beginning of this section.

2. Remove or disconnect the following:
- Negative battery cable
- Fluid from the transfer case
- 4WD indicator switch wire connector at the transfer case
- Skid plate from the frame, if equipped
- Front driveshaft from the front output yoke
- Rear driveshaft from the rear output shaft yoke
- Speedometer driven gear from the transfer case rear bearing retainer
- Retaining rings and shift rod from the transfer case shift lever
- Vent hose from the transfer case
- Heat shield from the frame

3. Support the transfer case with a transmission jack.

4. Remove the bolts retaining the transfer case to the transmission adapter.

5. Lower the transfer case from the vehicle.

To install:

6. When installing place a new gasket between the transfer case and the adapter.

7. Raise the transfer case with the transmission jack so the transmission output shaft aligns with the splined transfer case input shaft. Install the bolts retaining the transfer case to the adapter.

8. Remove the transmission jack from the transfer case.

9. Install or connect the following:
- Rear driveshaft to the rear output shaft yoke. Tighten the bolts to 15 ft. lbs. (20 Nm).
- Shift lever to the transfer case and install the retaining nut
- Speedometer driven gear to the transfer case
- 4WD indicator switch wire connector at the transfer case
- Front driveshaft to the front output yoke. Tighten the bolts to 15 ft. lbs. (20 Nm).
- Heat shield to the frame crossmember and the mounting lug on the transfer case. Install and tighten the retaining bolts.
- Skid plate to the frame

10. Install the drain plug. Remove the filler plug and install 6 pts. (2.8L) of Dexron®II transmission.

11. Connect the negative battery cable.

2003 and 2004 Heritage Editions

WITH 4R70W TRANSMISSION

1. Before servicing the vehicle, refer to the precautions in the beginning of this section.

2. With the vehicle in **Neutral**, raise and support the vehicle.

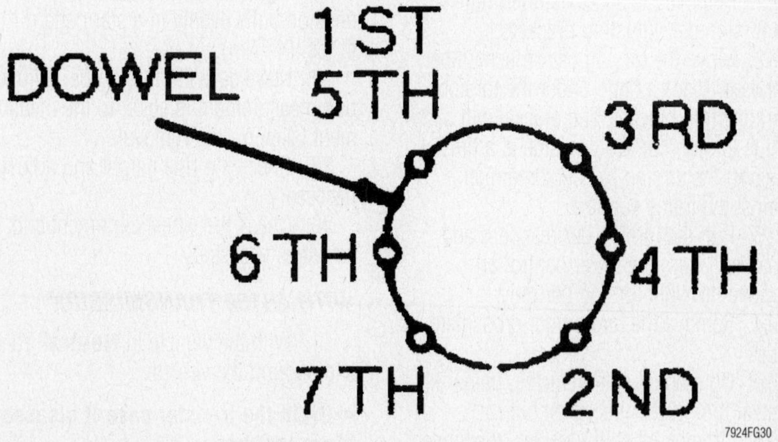

Transfer case-to-adapter bolt torque sequence—Borg-Warner model 13–45

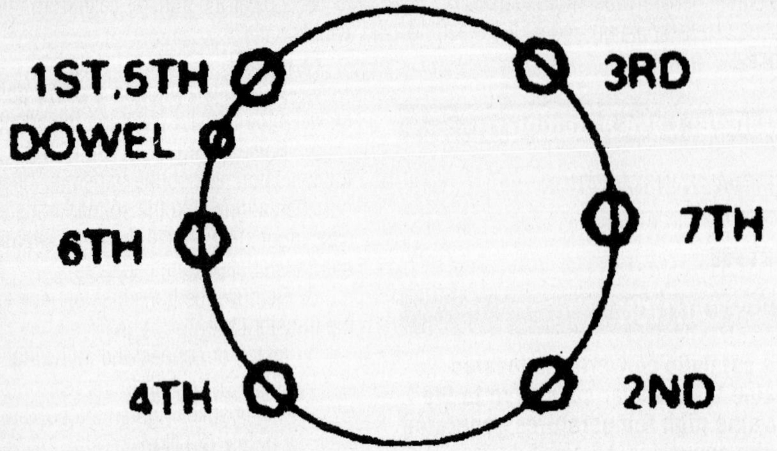

Transfer case-to-adapter bolt torque sequence—Borg-Warner 13-56 electronic and manual shift transfer case used in F-Series

➡Drain the transfer case if disassembly is required.

3. Remove the drain plug and drain the transfer case.
4. Remove the skid plate, if equipped.

➡To maintain initial driveshaft balance, index-mark the rear driveshaft.

5. Remove the rear driveshaft.
6. Remove the two nuts and the front driveshaft shield.

➡To maintain initial driveshaft balance, index-mark the front driveshaft.

7. Remove the front driveshaft.
8. Make preliminary adjustment references. Index-mark the torsion bars to the torsion bar crossmember. Measure and record the torsion bar adjuster bolt length.
9. Remove the bolts.
10. Using the special tools, remove the torsion bar adjustment nuts. Tighten the special tool until the torsion bar adjuster lifts off the adjuster nut.
11. Remove the six bolts, then remove the torsion bar adjusting brackets.
12. Move the torsion bar crossmember rearward. Place a hand under the torsion bar adjusting bracket. The bracket will slide off the torsion bar. Do one side at a time. Position the torsion bar crossmember against the frame support.
13. Pull the torsion bar rearward and separate it from the lower control arm. Remove the right torsion bar only.
14. Remove the torsion bar crossmember.
15. On manual shift vehicles, using the special tool, disconnect the shift rod.
16. On manual shift vehicles, disconnect the 4WD indicator switch electrical connector, if equipped.
17. Disconnect the electric shift motor electrical connector, if equipped.
18. Disconnect the vent hose.
19. Using a suitable jack, support the transfer case. Secure the transfer case to the jack with a safety strap.
20. Remove the six transfer case-to-transmission bolts and separate the transfer case from the transmission.
21. Lower the transfer case from the vehicle.

➡Carefully clean the gasket surfaces. Nicks and gouges cause fluid leaks.

22. Remove the transfer case-to-transmission gasket, and clean the mating surfaces.

To install:

23. To install, reverse the removal procedure.
24. Install a new transfer case-to-transmission mounting gasket.
25. Fill the transfer case.
26. Tighten the transfer case-to-transmission bolts evenly in a star pattern to 35 ft. lbs. (47 Nm).
27. Make sure the torsion bar adjuster bolt measurement is equal to the measurement taken during removal.
28. Check the ride height and adjust as necessary.
29. Check the wheel alignment and adjust as necessary.

WITH 4R100 TRANSMISSION

1. With the vehicle in **Neutral**, raise and support the vehicle.

➡Drain the transfer case if disassembly is required.

2. Remove the drain plug and drain the transfer case.
3. Remove the skid plate, if equipped.

➡To maintain initial driveshaft balance, index-mark the rear driveshaft.

4. Remove the rear driveshaft.
5. Remove the two nuts and the front driveshaft shield.

➡To maintain initial driveshaft balance, index-mark the front driveshaft.

6. Remove the front driveshaft.
7. Make preliminary adjustment references. Index-mark the torsion bars to the torsion bar crossmember. Measure and record the torsion bar adjuster bolt length.
8. Remove the torsion bar adjustment nuts. Tighten the tool until the torsion bar adjuster lifts off the adjuster nut.
9. Remove the six bolts, then remove the torsion bar adjusting brackets.
10. Move the torsion bar crossmember rearward. Place a hand under the torsion bar adjusting bracket. The bracket will slide off the torsion bar. Do one side at a time. Position the torsion bar crossmember against the frame support.
11. Pull the torsion bar rearward and separate it from the lower control arm. Remove the right torsion bar only.
12. Remove the torsion bar crossmember.
13. On manual shift vehicles, using the special tool, disconnect the shift rod.
14. On manual shift vehicles, disconnect the 4WD indicator switch electrical connector, if equipped.
15. Disconnect the electric shift motor electrical connector, if equipped.
16. Disconnect the vent hose.
17. Using a suitable jack, support the transfer case. Secure the transfer case to the jack with a safety strap.
18. Remove the two exhaust heat shield-to-crossmember bolts.
19. Loosen, but do not remove, the transmission mount-to-crossmember nuts.
20. Remove the six crossmember-to-frame nuts and bolts.
21. Remove the transmission mount-to-crossmember nuts and the crossmember.
22. Remove the transmission mount and the exhaust hanger.
23. Position a suitable jack stand under the extension housing.
24. Remove the six transfer case-to-transmission bolts and separate the transfer case from the transmission.
25. Lower the transfer case from the vehicle.

26. Remove the transfer case-to-transmission gasket and clean the mounting surfaces.

To install:

27. To install, reverse the removal procedure.
28. Install a new transfer case-to-transmission mounting gasket.
29. Fill the transfer case.
30. Tighten the transfer case-to-transmission bolts evenly in a star pattern to 35 ft. lbs. (47 Nm).
31. Make sure the torsion bar adjuster bolt measurement is equal to the measurement taken during removal.
32. Check the ride height and adjust as necessary.
33. Check the wheel alignment and adjust as necessary.

2004 F-150, Exc. Heritage Editions

MECHANICAL SHIFT

1. Before servicing the vehicle, refer to the precautions in the beginning of this section.

➡To maintain initial driveshaft balance, index-mark the rear driveshaft.

2. Remove the rear driveshaft.
3. Remove the skid plate, if equipped.
4. Drain the fluid if the transfer case is to be disassembled.
5. Disconnect the transfer case wire harness
6. Remove the front driveshaft shield.
7. Remove the front driveshaft.
8. Remove the transfer case shift linkage.
9. Remove the vent tube.
10. Position a suitable transmission jack to the transfer case. Securely strap the transfer case to the jack.
11. Remove the transfer case.
12. Clean the mating surfaces of the transfer case and the transmission.
13. To install, reverse the removal procedure.
14. Tighten the transmission-to-transfer case bolts evenly in a star pattern.
15. Fill the transfer case.

ELECTRONIC SHIFT

➡To maintain initial driveshaft balance, index-mark the rear driveshaft.

1. Remove the rear driveshaft.
2. Remove the skid plate, if equipped.

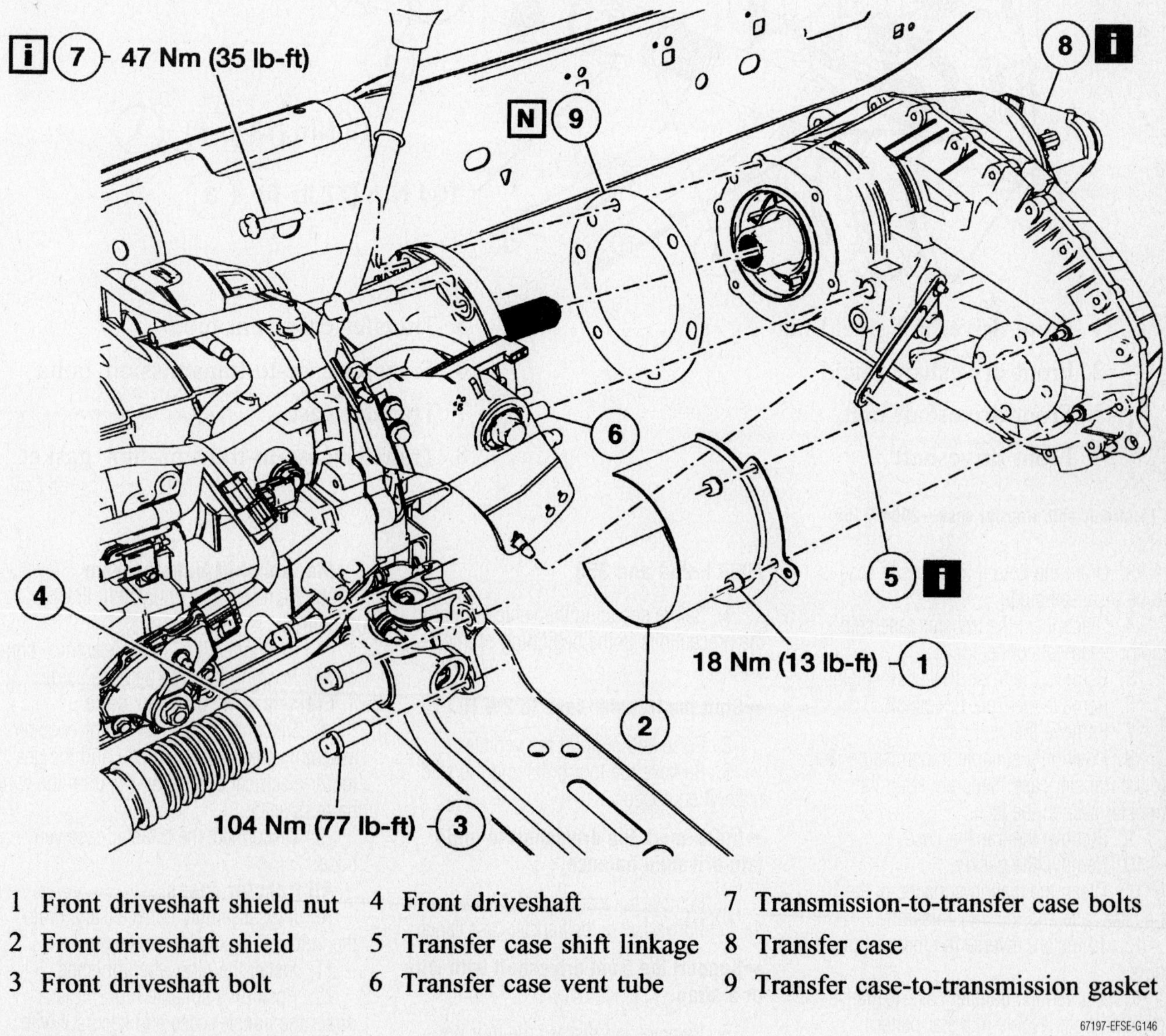

Mechanical shift transfer case—2004 F-150

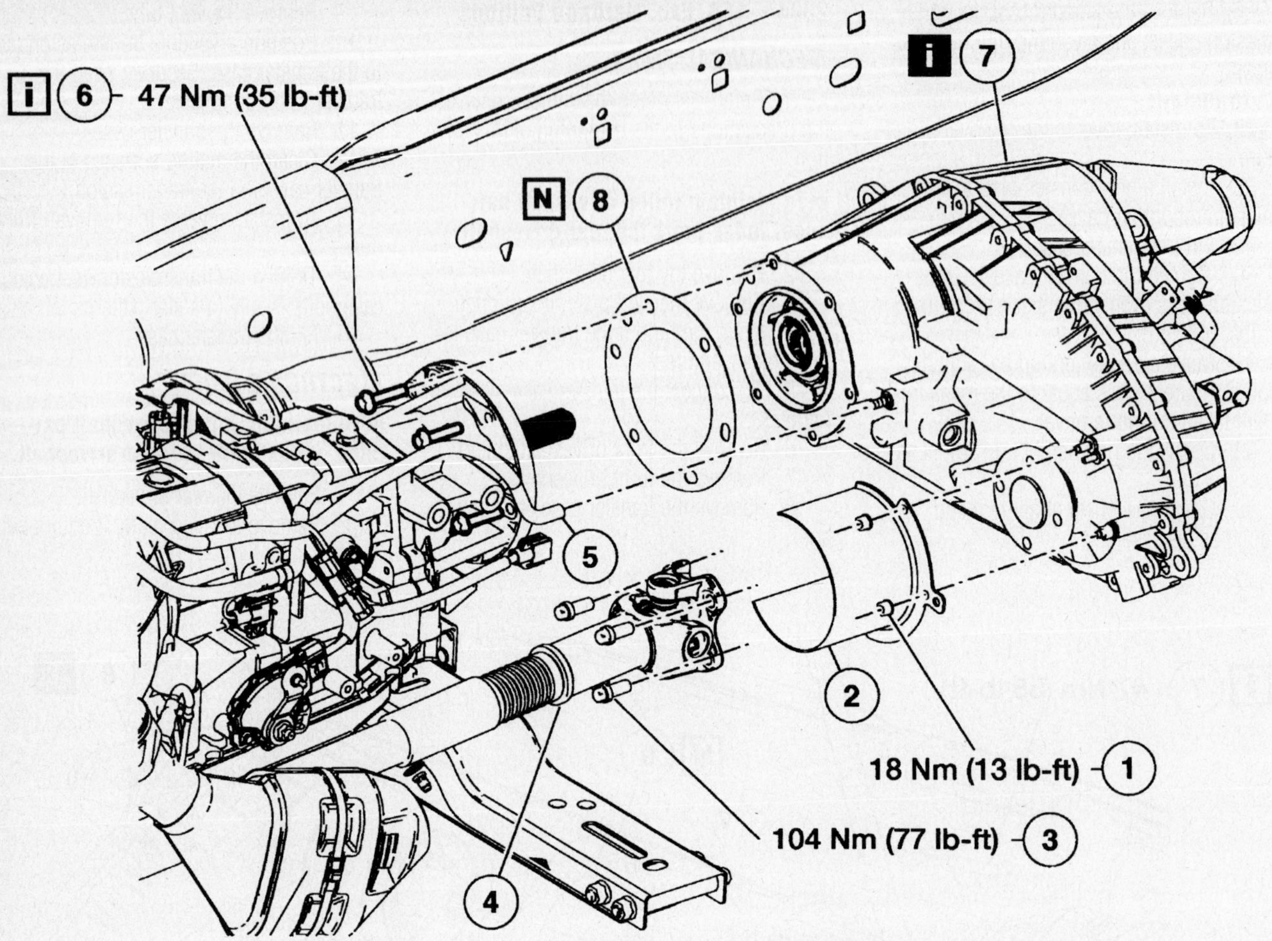

1 Front driveshaft shield nut
2 Front driveshaft shield
3 Front driveshaft bolt
4 Front driveshaft
5 Transfer case vent tube
6 Transfer case-to-transmission bolts
7 Transfer case
8 Transfer case-to-transmission gasket

67197-EFSE-G151

Electronic shift transfer case—2004 F-150

3. Drain the fluid if the transfer case is to be disassembled.
4. Disconnect the transfer case shift motor electrical connector.
5. Remove the front driveshaft shield.
6. Remove the front driveshaft.
7. Remove the vent tube.
8. Position a suitable transmission jack to the transfer case. Securely strap the transfer case to the jack.
9. Remove the transfer case.
10. Remove the gasket.
11. Clean the mating surfaces of the transfer case and the transmission.
12. To install, reverse the removal procedure.
13. Tighten the transfer case-to-transmission bolts evenly in a star pattern.
14. Fill the transfer case.

2004 F-250 and 350

1. Before servicing the vehicle, refer to the precautions in the beginning of this section.

➡Shift the transfer case to 2W HI.

2. Raise and support the vehicle.
3. Remove the four bolts and the skid plate, if equipped.

➡Index-mark the driveshaft to maintain driveline balance.

4. Remove the rear driveshaft. Index-mark the front driveshaft to the transfer case flange.

➡Support the front driveshaft with wire or a strap.

5. Remove and discard the four bolts and position the front driveshaft aside.

Manual shift transfer case

6. Remove the manual shift linkage, if equipped.
7. Disconnect the switch electrical connector. Position the wire harness aside.

Electric shift transfer case

8. Disconnect the gear motor encoder assembly electrical connector and the gear motor electrical connector. Position the wire harness aside.
9. Disconnect the transfer case vent hose.

All transfer cases

10. If disassembly is necessary, drain the fluid into a suitable container.
11. Install the plug when finished.
12. Position a suitable high-lift jack under the transfer case and secure it with safety straps.

With automatic transmission

13. Remove the four right crossmember bolts.
14. Remove the three left crossmember bolts.
15. Remove the nuts and the crossmember.
16. Remove the transmission mount.
17. Position a suitable jack stand under the extension housing.

All vehicles

18. Remove the six transfer case-to-transmission bolts.
19. Separate the transfer case from the extension housing. Pull the transfer case rearward, then lower the transfer case from the vehicle.

➡Carefully clean the gasket surfaces. Nicks and gouges cause fluid leaks.

20. Remove the transfer case-to-transmission gasket. Clean the mating surfaces, using metal surface cleaner.

To install:

21. Install a new mounting gasket.

⁂ CAUTION

Secure the transfer case to the high-lift jack with a safety strap.

22. Raise the transfer case into position.
23. Install the six bolts retaining the transfer case to the extension housing. Torque to 37 ft. lbs. (50 Nm).

Automatic transmission vehicles

24. Remove the jack stand from the extension housing.
25. Install the transmission mount. Torque to 70 ft. lbs. (95 Nm).
26. Position the crossmember and loosely install the two nuts.
27. Install the three left crossmember bolts. Torque to 52 ft. lbs. (70 Nm).
28. Install the right crossmember bolts. Torque to 52 ft. lbs. (70 Nm).

All vehicles

29. Remove the high-lift jack.

Automatic transmission vehicles

30. Tighten the transmission mount-to-crossmember nuts. Torque to 69 ft. lbs. (92 Nm).

All electric shift transfer cases

31. Connect the vent hose.
32. Connect the two gear motor encoder assembly electrical connectors.

All manual shift transfer cases

33. Connect the 3-position mode switch harness connector.
34. Connect the manual shift linkage.

All vehicles

➡Align the index-marks when installing the driveshaft.

35. Connect the front driveshaft to the transfer case and install the four new bolts. Torque to 82 ft. lbs. (111 Nm).
36. Install the rear driveshaft.
37. If equipped, install the skid plate and the four bolts.
38. If drained, fill the transfer case.

Halfshaft

REMOVAL & INSTALLATION

2001–03 F-150, Blackwood and F-250

The F-250 Heavy Duty and Super Duty models do not have halfshafts.

1. Before servicing the vehicle, refer to the precautions in the beginning of this section.
2. Remove or disconnect the following:
   - Front wheels
   - Front hub cotter pin, retainer and nut
3. Using a floor hydraulic jack, support the lower suspension arm.
   - Upper ball joint cotter pin and castle nut
   - Knuckle from the front suspension upper arm
4. Lower the lower suspension arm and steering knuckle slightly to facilitate easier halfshaft removal.
5. Remove the 2 disc caliper mounting bolts, then lift the front disc caliper off of the front disc brake caliper anchor plate and position aside. Do not allow the caliper to hang by the brake hose; suspend it from the vehicle's frame with strong cord or wire.
6. Remove the 6 front halfshaft-to-differential bolts.

⁂ WARNING

Use care to avoid damaging the hub seal when removing the front halfshaft.

7. Remove the inboard end of the halfshaft from the differential case or extension axle case. Separate the front halfshaft and joints from the hub, then remove the halfshaft and joints from the vehicle.

To install:

8. Slide the halfshaft outboard end into the hub, making sure that the splines engage.
9. Situate the inboard end of the halfshaft against the front differential flange and install the 6 halfshaft-to-differential bolts. Tighten the halfshaft bolts to 60 ft. lbs. (82 Nm).
10. Install the front disc brake caliper onto the rotor and anchor plate, then install and tighten the 2 caliper mounting bolts to 21–26 ft. lbs. (28–36 Nm).
11. Lift the lower suspension arm and steering knuckle up until the upper ball joint stud is inserted into the steering knuckle. Install the upper ball joint castle nut and tighten to 57–76 ft. lbs. (77–104 Nm). Install a new cotter pin.
12. Install the hub nut onto the halfshaft and tighten the hub nut to 221 ft. lbs. (300 Nm).

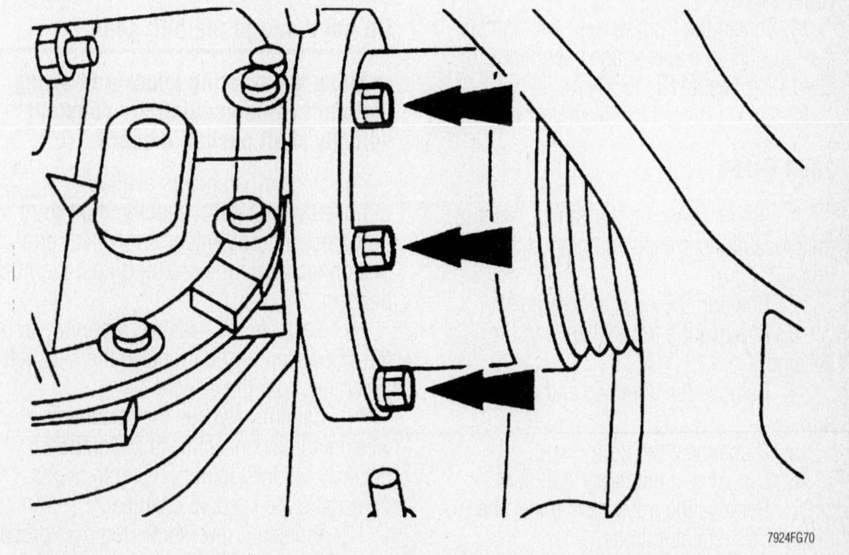

Halfshaft-to-differential mounting bolts (3 of the 6 bolts shown)—2001–03 F-150 and Blackwood

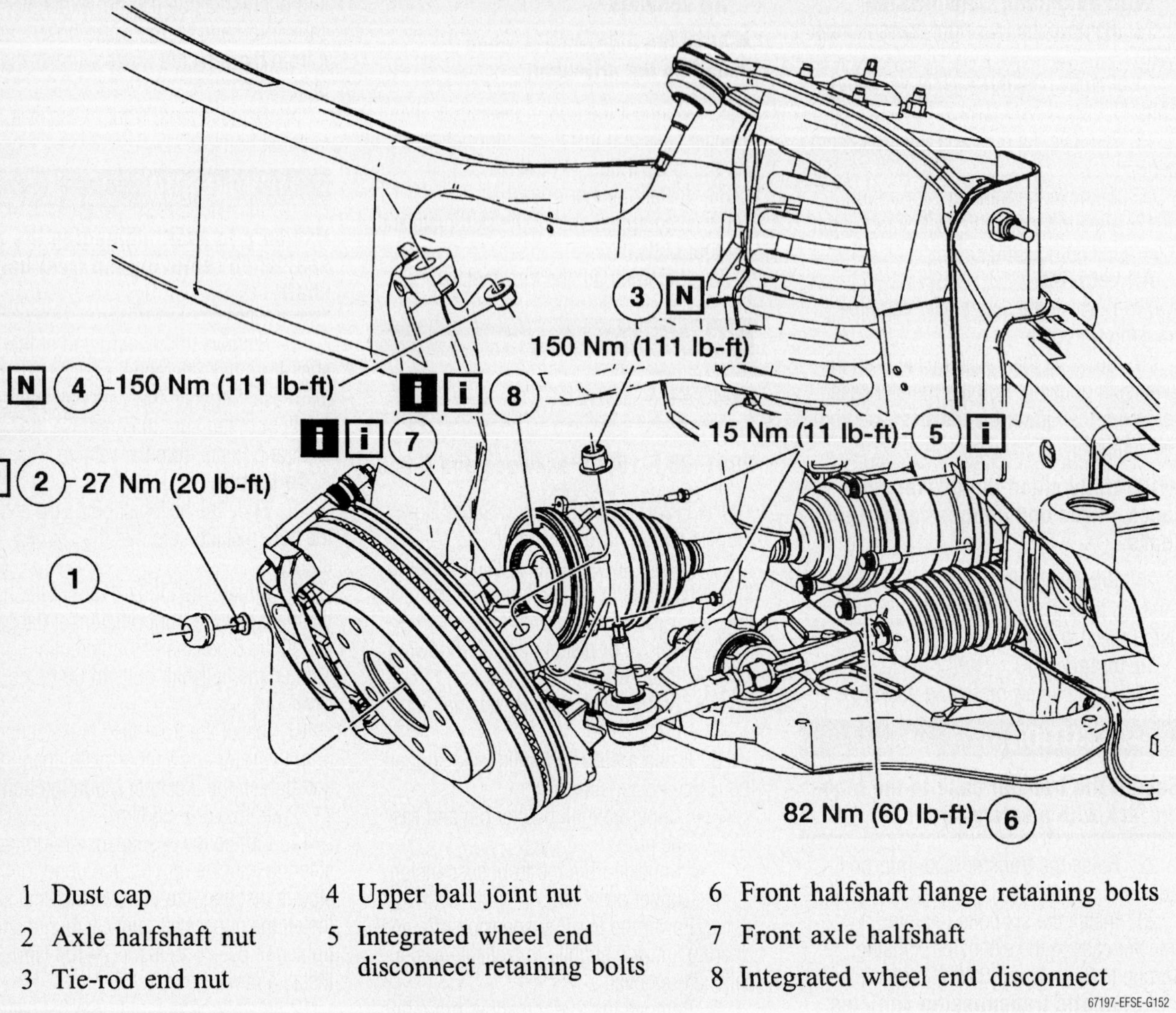

Front axle halfshaft assembly—2004 F-150, exc. Heritage Editions

13. Install the hub nut retainer and a new cotter pin.
14. Install the front wheels and tighten the lug nuts in a star-shaped sequence to 83–112 ft. lbs. (113–153 Nm).
15. Lower the vehicle to the ground.

2004 F-150

1. Before servicing the vehicle, refer to the precautions in the beginning of this section.
2. Position the vehicle on a hoist.
3. Remove the front wheel and tire assembly.
4. Remove the dust cap and axle shaft nut.
5. Disconnect the tie rod end.
6. Disconnect the upper ball joint.
7. Remove the integrated wheel end disconnect retaining bolts.
8. Remove the halfshaft flange retaining bolts.

WARNING

Do not damage the hub seal.

➡Allow the steering knuckle to swing outboard while keeping the constant velocity shaft pushed inboard.

9. Once clearance is available, remove the constant velocity shaft joint outboard end and integrated wheel end disconnect from the steering knuckle hub bearing.
10. Separate the halfshaft assembly from the axle assembly and remove the halfshaft assembly from the vehicle.
11. Carefully remove the integrated wheel end disconnect from the outboard constant velocity joint housing to avoid damage to the vacuum chamber.
12. To install, reverse the removal procedure. Take note of the following:

➡Maintain a clean work surface.

13. Compress the integrated wheel end disconnect on the bench to collapse the vacuum chamber.
14. While the integrated wheel end disconnect is collapsed, install a vacuum cap on the vacuum port.

WARNING

Do not install the integrated wheel end disconnect in the knuckle. It must be installed on the outer constant velocity joint housing.

WARNING

Do not dislodge the integrated wheel end disconnect seal spring when installing the integrated wheel end on the outer constant velocity joint housing.

15. Install the integrated wheel end disconnect on the outer constant velocity joint housing

16. Install the front axle halfshaft in the vehicle. Install the halfshaft flange retaining bolts.

⁂ WARNING

Verify the spline engagement by checking for spline lash before tightening the retainers of the integrated wheel end disconnect and the front axle halfshaft retaining nut.

17. Install and tighten the integrated wheel end disconnect retaining bolts.

18. Verify the front axle lubricant level is to specifications

CV-Joint

OVERHAUL

➡Before continuing with this procedure, make sure to have available a new CV-joint boot kit, for each CV-joint being serviced. The outer CV-joint cannot be disassembled, only the boot can be replaced.

Inner CV-Joint And Boot

1. Before servicing the vehicle, refer to the precautions in the beginning of this section.
2. Remove the halfshaft assembly from the vehicle.
3. Clamp the halfshaft in a vise equipped with jaw caps to prevent damage to machined surfaces. Do not allow the vise jaws to contact the boot or its clamp.
4. Slide 2 inboard clamp protectors off the boot clamps.
5. Carefully remove 2 boot clamps, and slide the boot off the inner CV-joint and housing.
6. Remove the CV-joint retaining ring and remove the housing.
7. Mark the inner race and the ball cage for assembly.
8. Remove 6 cage balls.
9. Remove the snapring.
10. Remove the inner race and ball cage.
11. Clean all parts in suitable parts cleaning solvent and inspect for wear.

To install:

12. Place the boot and the small boot clamp and protector on the shaft.
13. Place the ball cage on the shaft with the tapered end toward the outer CV-joint.

➡Line up the marks made at disassembly.

14. Position the inner race on the driveshaft in the position marked on disassembly.
15. Install the snapring.
16. Lubricate and position 6 balls with suitable CV-joint grease.
17. Place the boot protector and boot clamp on the CV-joint housing. Fill the housing with 8.29 ounces of suitable CV-joint grease.
18. Place the housing to the cage and bearings and install the retaining ring.
19. Remove any excess grease from the mating surface and position the boot and clamp.
20. Adjust the CV-joint to boot spacing to 16.43 in. (417.25mm) for 2001–03; 16.93 in. (429.92mm) for 2004.
21. After adjusting the CV-joint to boot spacing, insert a dull bladed screwdriver blade to relieve built up air pressure in the boot.
22. Use CV Boot Clamp Installer T95P-3514-A or equivalent, to install the boot clamps.
23. Place the clamp protectors over the boot clamps.
24. Install the halfshaft into the vehicle.
25. Road test the vehicle and check for proper operation.

Outer Boot

1. Before servicing the vehicle, refer to the precautions in the beginning of this section.
2. Remove the inner CV-joint and housing from the halfshaft.
3. Remove the inner boot from the halfshaft.
4. Remove the outer boot clamp protectors and carefully remove the boot clamps.
5. Remove the outer boot from the halfshaft and inspect the grease for contamination.
6. If the grease is contaminated, clean and inspect the joint for wear. Replace the joint and shaft if worn or damaged.

To install:

7. Place the new CV-joint boot on the shaft.
8. Using 5.82 ounces (165 grams) of suitable CV-joint grease, pack the outer CV-joint with grease, then spread the remaining grease inside the boot.
9. Clean the boot mounting surface and position the boot in the joint grooves.
10. Place the clamps in position and use CV Boot Clamp Installer T95P-3514-A or equivalent, to install the boot clamps.
11. Place the clamp protectors over the boot clamps.
12. Install the inner boot on the halfshaft.
13. Install the inner CV-joint and housing on the halfshaft.
14. Install the halfshaft in the vehicle.
15. Road test the vehicle and check for proper operation.

Front Drive Axle Shaft and Seal

REMOVAL & INSTALLATION

See Front Hub and Bearing, in this chapter.

Rear Axle Drive Shaft and Seal

REMOVAL & INSTALLATION

See Rear Hub and Bearing, in this chapter.

Front Differential Pinion Seal

REMOVAL & INSTALLATION

2001–04 F-150

⁂ WARNING

This operation disturbs the pinion bearing preload. Carefully reset the preload during assembly.

➡The front drive axle must be in Neutral before beginning this procedure.

1. Raise and support the vehicle.
2. Remove the front differential support.
3. Match-mark the front driveshaft to the axle universal joint flange.

⁂ WARNING

Do not allow the driveshaft to hang unsupported.

4. Disconnect and support the front driveshaft.

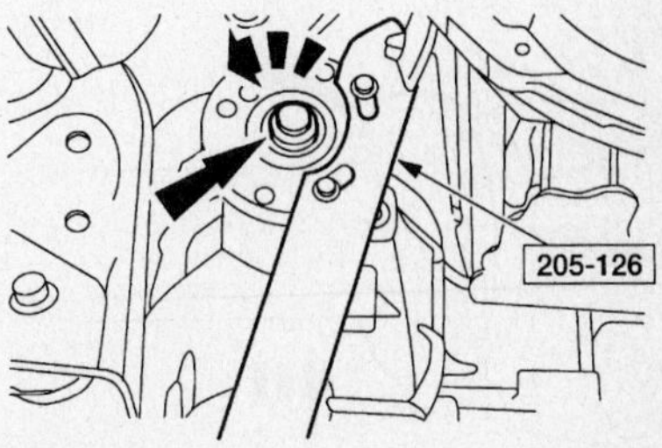

67197-EFSE-G234

Flange holding tool—Ford 8.8 inch, 9.75 inch and 10.25 inch Ring Gear Axle

5. Using an Nm (inch-pound) torque wrench, measure the torque necessary to maintain pinion rotation. Record the measurement for reference during installation.

6. Match-mark the axle universal joint flange to the pinion stem.

7. Install a holding tool, and loosen, but do not remove the pinion nut.

✲✲ WARNING

Before proceeding, place a drain pan under the differential carrier.

8. With the pinion nut still engaged by a few threads, use a 2-jawed pull to separate the axle universal joint flange from the pinion gear.

9. Remove the nut and the flange.

10. Inspect the axle universal joint flange for burrs, the nut counterbore and the seal contact surface for nicks, and the bearing cone contact area for damage. Install a new flange if necessary.

11. Check the pinion stem splines for burrs. If burrs are evident, remove them with a fine crocus cloth. Working in a rotating motion, wipe the pinion clean.

12. Remove the pinion seal.

To install:

13. Clean the pinion seal bore, and use the Pinion Seal Replacer to install the pinion seal.

14. Install the front axle universal joint flange.

15. Lubricate the axle universal joint flange splines and the pinion seal. Use Motorcraft SAE 75W-90 Premium 4x4 Front Axle Lubricant XY-75W90-TQL or equivalent meeting Ford specification WSP-M2C201-A.

✲✲ WARNING

Never install the axle universal joint flange with a hammer or power tools.

➡Disregard the scribe marks if installing a new flange.

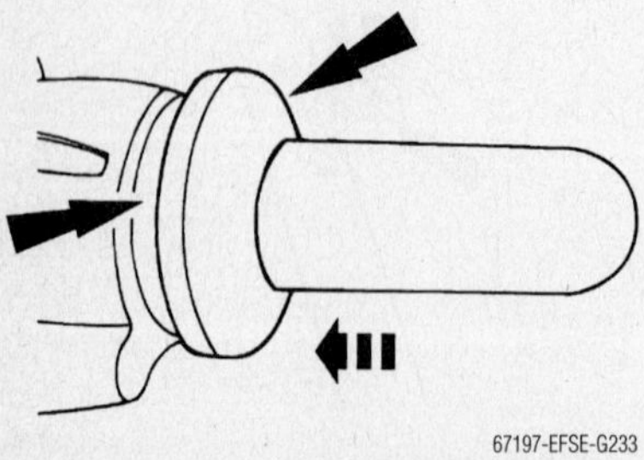

Installing the pinion seal—Ford 8.8 inch, 9.75 inch and 10.25 inch Ring Gear Axle

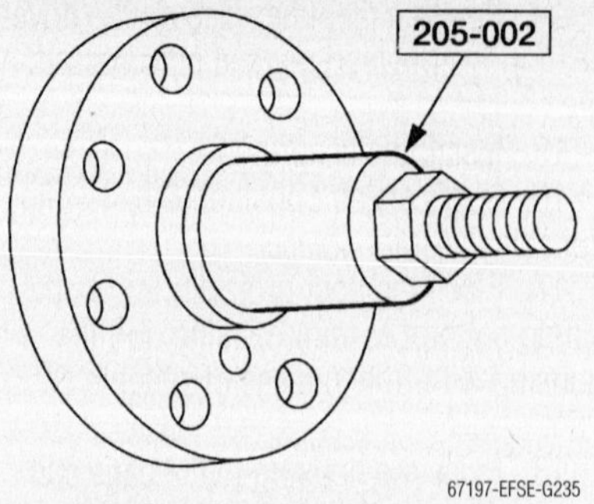

Install the axle universal joint flange—Ford 8.8 inch, 9.75 inch and 10.25 inch Ring Gear Axle

16. Align the index-marks and position the axle universal joint flange on the pinion shaft.

➡Rotate the pinion gear occasionally to make sure the pinion bearings seat correctly.

17. Using the special tool, install the axle universal joint flange.

✲✲ WARNING

Do not loosen the pinion nut to reduce preload under any circumstance. If it is necessary to reduce preload, install a new drive pinion collapsible spacer and pinion nut.

18. Install the special tool, and tighten the pinion nut as follows:

a. Rotate the pinion gear occasionally to make sure the pinion bearings are seating correctly. Take frequent pinion bearing torque preload readings by rotating the pinion gear with a Nm (inch/pound) torque wrench.

b. If the preload recorded prior to disassembly is lower than the specification for used bearings, then tighten the pinion nut to the specification. If the preload recorded prior to disassembly is higher than the specification for used bearings, then tighten the pinion nut to the original reading as recorded.

- Pinion bearing torque preload (used pinion bearing): 0.9–1.5 Nm (8–14 inch lbs.)
- Pinion bearing torque preload (new pinion bearing): 1.8–3.3 Nm (16–29 inch lbs.)

19. Align the index-marks then attach the front driveshaft. Torque to 83 ft. lbs. (112 Nm).

20. Install the front differential support.

21. Check and, if necessary, fill the differential. Use Motorcraft SAE 75W-90 Premium 4x4 Front Axle Lubricant XY-75W90-TQL, or equivalent, meeting Ford specification WSP-M2C201-A.

22. Lower the vehicle.

2001–04 F-250 and 350 Dana 50 and 60 Axles

1. With the vehicle in **Neutral**, raise and support the vehicle.

2. Match-mark the front driveshaft and the front axle flange to maintain driveline balance.

3. Disconnect the front driveshaft from the front axle flange, and position it aside.

4. Rotate the pinion with a Nm (inch lb.) torque wrench. Record the torque necessary to maintain rotation of the pinion through several revolutions.

5. Remove and discard the nut and washer. Use the special tool to prevent the flange from turning while removing the nut.

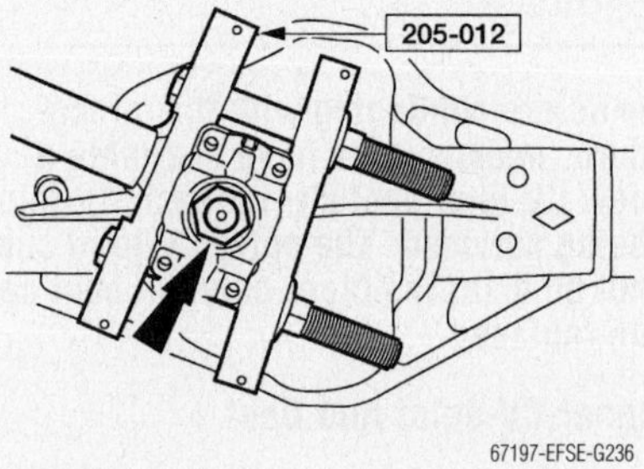

Flange holding tool—F-250/350 Dana 50 and 60 Axles

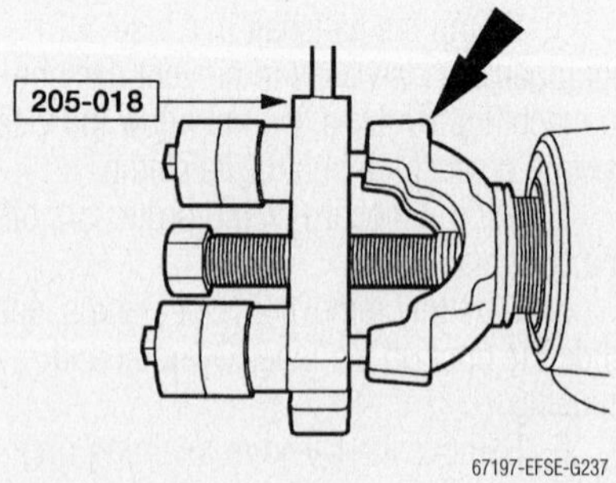

Flange removal—F-250/350 Dana 50 and 60 Axles

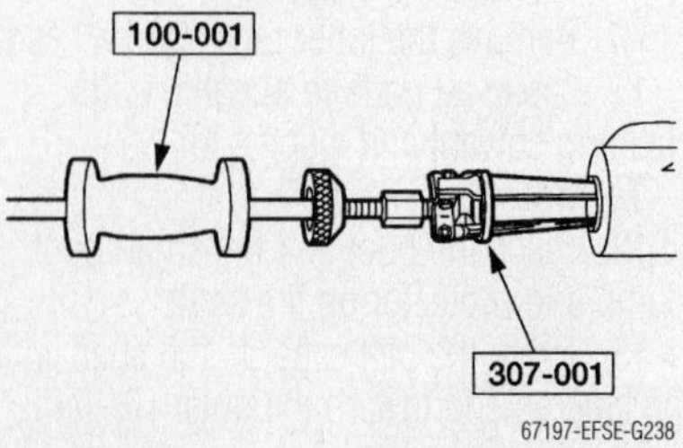

Pinion seal removal—F-250/350 Dana 50 and 60 Axles

➡**Match-mark the flange and the pinion shaft.**

6. Using the special tool, remove the flange.
7. Using the special tools, remove the pinion seal. Discard the seal.
8. Clean and inspect the following:
a. The seal mounting surface.
b. The flange lugs and the flange end that contacts the bearing cone.
c. Verify that the flange nut counterbore and the seal contact surfaces are smooth and free of nicks.

To install:

9. Using a suitable driver, install the pinion seal. Lightly coat the pinion seal lip with lubricant.

**** WARNING**

Never use a metal hammer on the pinion flange or install the flange with power tools. If necessary, use a plastic hammer to tap on a tight fitting flange.

➡**Align the index marks.**

10. Lightly coat the flange splines and seal mating area with lubricant, then install the flange with a new washer and nut.

**** WARNING**

Never back off the pinion nut to reduce preload. If preload reduction is necessary, install a new collapsible spacer and pinion nut.

11. Tighten the pinion nut as follows:
a. Use the special tool to prevent the flange from turning while tightening the nut. Remove the special tool when taking pinion bearing torque preload readings.
b. Take frequent pinion bearing torque preload readings.

➡**Never back off the pinion nut to reduce preload. If preload reduction is necessary, install a new collapsible spacer and pinion nut.**

c. For new pinion bearing installation, tighten the pinion nut to a rotating torque of 1.7–3.4 Nm (15–30 inch lbs.) Pinion nut torque range is 217–678 Nm (160–500 ft. lbs.).
d. For original pinion bearing installation, the reading must be 0.56 Nm (5 lb-in) more than the initial reading taken during the disassembly procedure.

12. Connect the front driveshaft to the front axle flange. Torque to 26 ft. lbs. (35 Nm).
13. Check and, if necessary, fill the axle with the specified lubricant.
14. Lower the vehicle.

Rear Differential Pinion Seal

REMOVAL & INSTALLATION

Ford 8.8 inch, 9.75 inch and 10.25 inch Ring Gear Axle

1. Raise and support the vehicle.
2. Remove the rear wheel and tire assemblies.

**** WARNING**

Remove the brake drums or discs to prevent brake drag during drive pinion bearing preload adjustment.

3. Remove the brake discs or drums.
4. Mark the driveshaft flange and pinion flange for correct alignment during installation.
5. Remove the four bolts.

**** WARNING**

The driveshaft centering socket yoke fits tightly on the rear axle pinion flange pilot. Never hammer on the driveshaft or any of its components to disconnect the yoke from the flange.

6. Disconnect the driveshaft centering socket yoke from the rear axle pinion flange. Position the driveshaft out of the way.
7. Install a Nm (inch-pound) torque wrench on the pinion nut and record the torque required to maintain rotation of the pinion through several revolutions.

**** WARNING**

After removal of the pinion nut, discard it. A new nut must be used for installation.

8. Use the special tool to hold the pinion flange while removing the pinion nut.
9. Mark the driveshaft pinion flange in relation to the drive pinion stem to make sure of correct alignment during installation.
10. Using the special tool, remove the pinion flange.

To install:

11. Lubricate the pinion flange splines. Use SAE 75W-140 High Performance Rear Axle Lubricant F1TZ-19580-B or equivalent meeting Ford specification WSL-M2C192-A.

➡**Disregard the scribe marks if a new pinion flange is being installed.**

12. Align the pinion flange with the drive pinion shaft.
13. With the drive pinion in place in the rear axle housing, install the pinion flange using the special tool.
14. Position the new pinion nut.

**** WARNING**

Under no circumstances is the pinion nut to be backed off to reduce preload. If reduced preload is required, a new collapsible spacer and pinion nut must be installed.

15. Using the special tool to hold the pinion flange, tighten the pinion nut as follows:
a. Rotate the pinion occasionally to make sure the cone and roller bearings are seating correctly.
b. Install a Nm (inch-pound) torque wrench on the pinion nut.
c. Rotating the pinion through several revolutions, take frequent cone and roller bearing torque preload readings until the original recorded preload reading is obtained.
d. If the original recorded preload is lower than specifications, tighten to the appropriate specifications for used bearings. If the preload is higher than specification, tighten the nut to the original reading as recorded.
- Pinion bearing torque preload (used pinion bearing): 0.9–1.5 Nm (8–14 inch lbs.)
- Pinion bearing torque preload (new pinion bearing): 1.8–3.3 Nm (16–29 inch lbs.)

16. Position the rear driveshaft and align the marks on the pinion flange.

**** WARNING**

The driveshaft centering socket yoke fits tightly on the rear axle pinion flange pilot. To make sure that the yoke seats squarely on the flange, tighten the bolts evenly in a cross pattern as shown.

17. Install the bolts and tighten to 83 ft. lbs. (112 Nm)
18. Install the brake disc or drum.
19. Install the rear wheel and tire assemblies.

Ford 10.5 inch Ring Gear Axle

➡**The rear wheels and brake calipers must be removed to prevent brake drag during drive pinion bearing preload adjustment.**

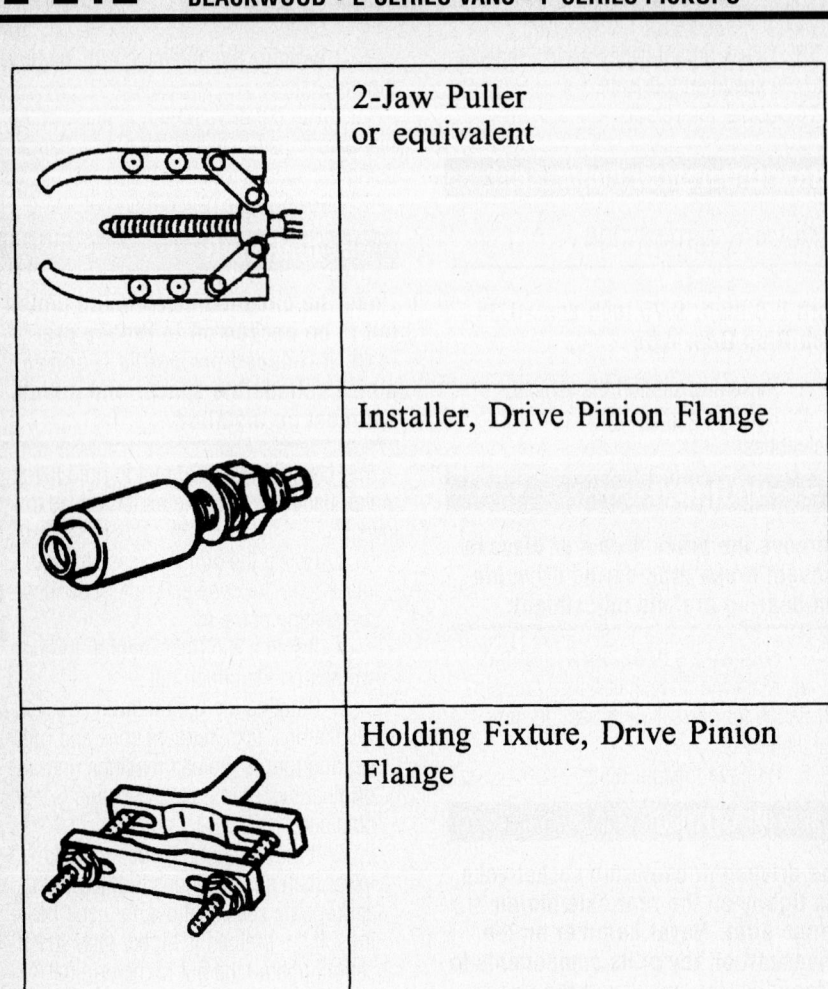

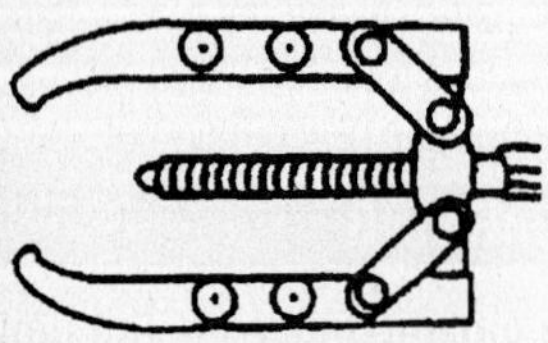	2-Jaw Puller or equivalent
	Installer, Drive Pinion Flange
	Holding Fixture, Drive Pinion Flange
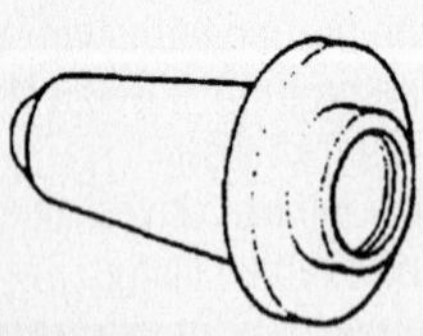	Installer, Drive Pinion Oil Seal

67197-EFSE-G240

Tools necessary for this job—Ford 10.5 inch ring gear axle pinion seal replacement

1. Remove the rear brake calipers.
2. Remove the driveshaft.
3. Install an Nm (inch-pound) torque wrench on the pinion nut, and record the rotational torque required to maintain rotation of the pinion through several revolutions.

➡**After removal of the pinion nut, discard it. A new nut must be used for installation.**

4. Use a flange holder to hold the pinion flange while removing the pinion nut.
5. Mark the pinion flange in relation to the drive pinion stem to ensure proper alignment during installation.
6. Use a 2-jaw puller to remove the pinion flange.
7. Force up on the metal flange of the rear axle drive pinion seal. Install locking pliers to the seal flange and strike with a hammer until the rear axle drive pinion seal is removed.

To install:

8. Lubricate the new pinion seal. Use Premium Long-Life Grease XG-1-C or equivalent meeting Ford specification ESA-M1C75-B.

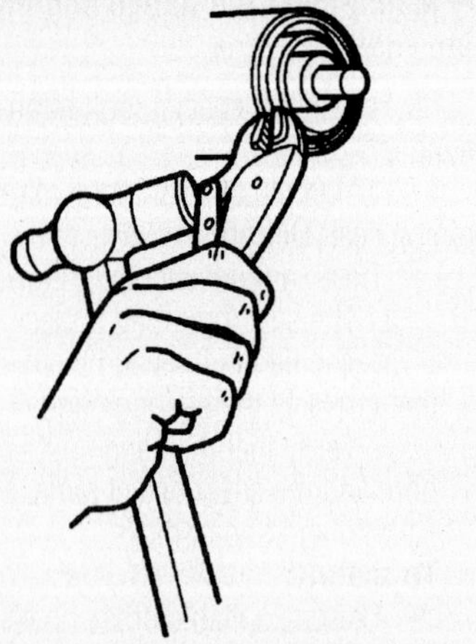

67197-EFSE-G241

Pinion seal removal—Ford 10.5 inch ring gear axle

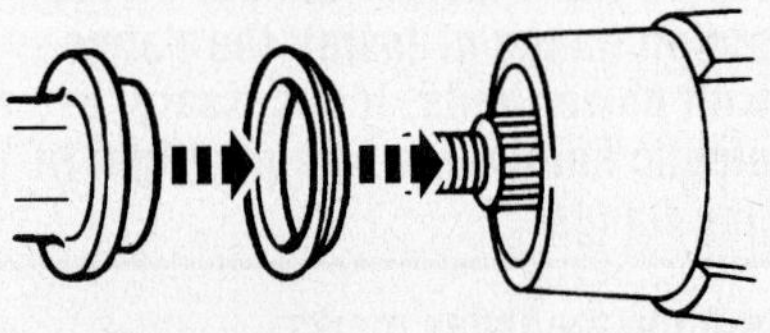

67197-EFSE-G242

Pinion seal installation—Ford 10.5 inch ring gear axle

➡**If the rear axle drive pinion seal becomes misaligned during installation, remove the rear axle drive pinion seal and replace it with a new seal.**

9. Use the Pinion Seal Replacer to install the rear axle drive pinion seal.
10. Lubricate the pinion flange splines. Use SAE 75W-140 Synthetic Rear Axle Lubricant F1TZ-19580-B or equivalent meeting Ford specification WSL-M2C192-A.

➡**Disregard the scribe marks if a new pinion flange is being installed.**

11. Align the pinion flange with the drive pinion shaft.
12. With the pinion flange in place in the rear axle housing, install the pinion flange using the Companion Flange Replacer.
13. Position the new pinion nut.

➡**Under no circumstances is the pinion nut to be backed off to reduce preload. If reduced preload is required, a new collapsible spacer and pinion nut must be installed.**

14. Use the Flange Holder to hold the pinion flange while tightening the pinion nut.
    a. Tighten the pinion nut, rotating the pinion occasionally to make sure the

cone and roller bearings are seating properly. Take frequent cone and roller bearing torque preload readings until the originally recorded preload reading is obtained by rotating the pinion with an Nm (inch-pound) torque wrench.

b. If the original recorded preload is lower than specifications, tighten to the appropriate specification for used bearings. If the preload is higher than specification, tighten the nut to the original reading as recorded.

c. Pinion bearing preload (used pinion bearing): 0.9–1.5Nm (8–14 inch lbs)

d. Pinion bearing preload (new pinion bearing): 1.8–3.3Nm (16–29 inch lbs.)

e. Initial minimum breakaway torque (Traction-Lok®): 27Nm (20 ft. lbs)

15. Install the driveshaft.
16. Install the brake calipers.

Dana 80 Rear Axle

1. Raise the vehicle on a hoist or raise the rear end of the vehicle with a jack. Install safety stands under the frame rails and lower the jack or hoist far enough to allow the rear axle to drop into the rebound position for working clearance.

➡To maintain driveline balance, mark the driveshaft components so they can be reinstalled in their original positions.

2. Disconnect the driveshaft at the rear axle, and position it aside.

➡Index-mark the flange to the pinion shaft.

3. While using a flange holding tool to prevent the flange or yoke from turning, remove the pinion nut.
4. Using a 2-jaw puller, remove the flange or yoke.
5. Using a bushing remover and slide hammer, remove the pinion seal.
6. Clean the rear axle pinion seal seat.

To install:

➡If the pinion seal becomes cocked during installation, remove the seal and install a new one. Make sure the garter spring remains in place during assembly. If the spring is dislodged, a new pinion seal must be installed.

7. Install the seal using a suitable driver. Coat the pinion seal rubber lips with lubricant.
8. Using the special tool, 205-285, or equivalent, install the pinion flange.

⁂ WARNING

Always install a new washer and locknut.

9. Install the new washer and locknut. Torque to 470 ft. lbs. (637 Nm).
10. Install the driveshaft at the rear axle. Observe the following torques:
    - Split pin yoke: 26 ft. lbs. (35 Nm)
    - Circular flange: 82 ft. lbs. (111 Nm)

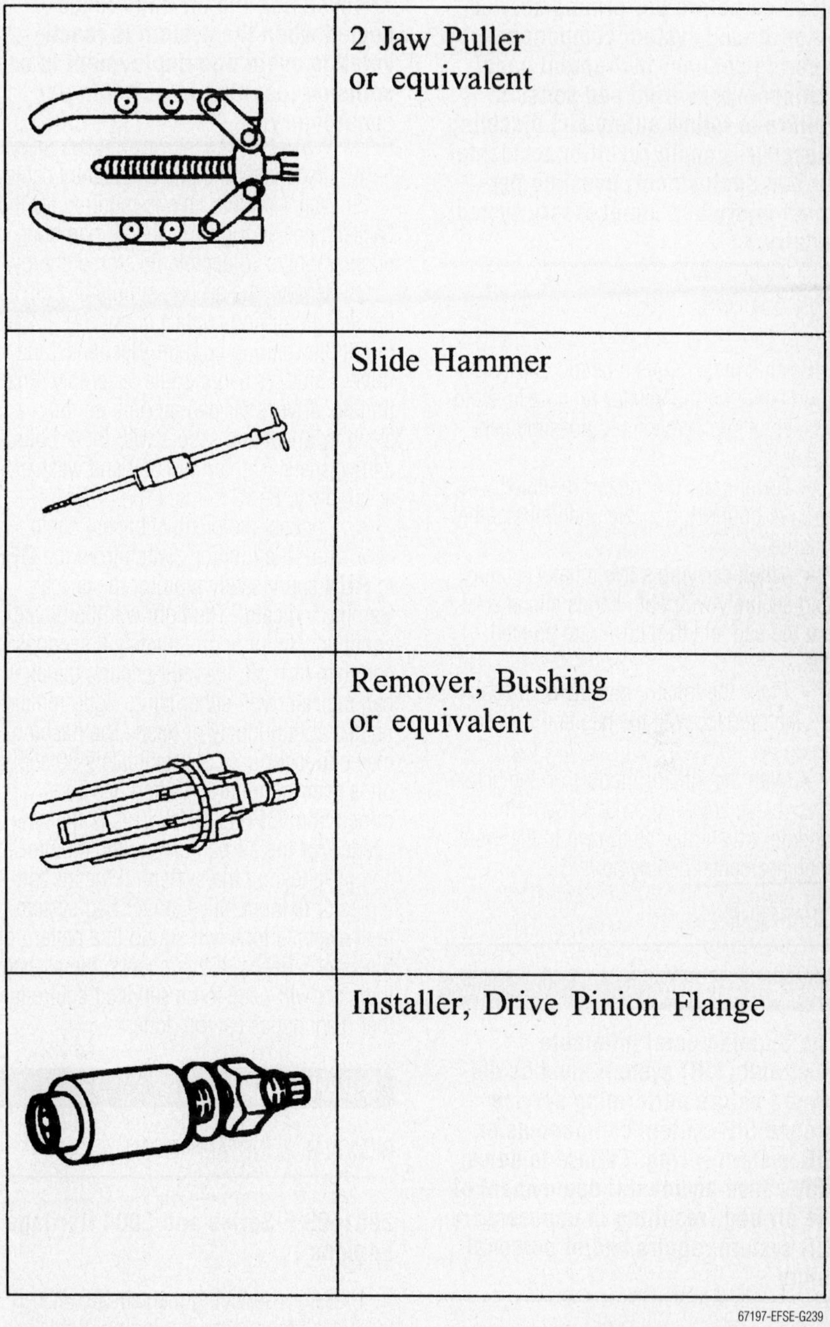

Tools necessary for this job—Dana 80 axle

STEERING AND SUSPENSION

Air Bag

✻✻ CAUTION

Some vehicles are equipped with an air bag system. The system must be disabled before performing service on or around system components, steering column, instrument panel components, wiring and sensors. Failure to follow safety and disabling procedures could result in accidental air bag deployment, possible personal injury and unnecessary system repairs.

PRECAUTIONS

Several precautions must be observed when handling the inflator module to avoid accidental deployment and possible personal injury:

- Never carry the inflator module by the wires or connector on the underside of the module.
- When carrying a live inflator module, hold securely with both hands and ensure that the bag and trim cover are pointed away.
- Place the inflator module on a bench or other surface with the bag and trim cover facing up.
- With the inflator module on the bench, never place anything on or close to the module, which may be thrown in the event of an accidental deployment.

DISARMING

✻✻ CAUTION

The Supplemental Inflatable Restraint (SIR) system must be disarmed before performing service around SIR system components or SIR system wiring. Failure to do so may cause accidental deployment of the air bag, resulting in unnecessary SIR system repairs and/or personal injury.

The positive battery cable must be disconnected for a minimum of 1 minute before beginning any air bag work to de-energize the back-up power supply. It is a good idea to disengage both the positive and negative battery cables to ensure that the Air Bag system is definitely discharged.

ARMING THE SYSTEM

✻✻ WARNING

If the air bag simulators have been used, the air bag simulators must be removed and the air bags reconnected when the system is reactivated to avoid non-deployment in a collision resulting in possible personal injury.

1. Disconnect the positive battery cable.
2. Wait 1 minute, this is required for the back-up power supply in the air bag diagnostic monitor to deplete its stored energy.
3. Remove the air bag simulator from the air bag sliding contact connector at the top of the steering column. Reconnect the driver's side air bag module assembly. Position the driver's air bag module on the steering wheel and secure with the 2 bolts and washers. Tighten the bolt and washer assembly to 8–10 ft. lbs. (10–14 Nm).
4. Connect the positive battery cable.
5. Turn the ignition switch from the **OFF** to **RUN** and visually monitor the air bag warning indicator. The light will illuminate continuously for approximately 6 seconds and then turn off. If a fault occurs, the air bag indicator will either fail to light, remain lighted continuously or flash. The flashing may not occur until approximately 30 seconds after the ignition switch has been turned from **OFF** to **RUN**. This is the time needed for the air bag diagnostic monitor to complete testing the system. If the air bag indicator is inoperative, an air bag system fault exists, a tone will sound in a pattern of 5 sets of 5 beeps. If this occurs, the air bag indicator will need to be serviced before further diagnostics can be done.

Steering Gear

REMOVAL & INSTALLATION

2001–03 F-Series and 2004 Heritage Editions

1. Before servicing the vehicle, refer to the precautions in the beginning of this section.
2. Remove or disconnect the following:
   - Skid plate
   - Lower radiator air deflector
   - Pitman arm cotter pin and castellated nut from the drag link
   - Pitman arm from the drag link
   - Dust cover from the steering shaft valve housing
   - Intermediate shaft pinch bolt and slide the shaft off the steering gear input shaft.
   - Power steering pressure hoses at the steering gear
   - Steering gear-to-frame rail retaining bolts
   - Steering gear
3. If replacing or servicing the steering gear, match mark the sector shaft arm to the sector shaft and remove the steering gear sector shaft arm retaining nut and lockwasher.
4. Remove the sector shaft arm.

To install:

5. If removed, install the steering gear sector shaft arm to the sector shaft aligning the match marks made during removal. Install the retaining nut and lockwasher and tighten to 170–228 ft. lbs. (234–316 Nm) on 2001 models or 173–233 ft. lbs. (234–316 Nm) on 2002–03 models and 2004 Heritage Editions.
6. Install or connect the following:
   - Steering gear into position
   - 3 retaining bolts and tighten them to 50–68 ft. lbs. (68–92 Nm
   - Power steering pressure hoses to the steering gear using new seals, if necessary
   - Intermediate shaft on the steering gear input shaft
   - Shaft pinch bolt. Tighten the pinch bolt to 30–42 ft. lbs. (41–57 Nm).
   - Dust cover over the steering shaft valve housing
   - Pitman arm to the drag link
   - Castellated nut and tighten to 57–76 ft. lbs. (77–104 Nm). Install a new cotter pin.
   - Radiator air deflector and secure with the retaining screws and push clips
   - Skid plate and secure it with the retaining bolts
7. Lower the vehicle.
8. Fill and bleed the power steering system.
9. Road test the vehicle and check the steering system for proper operation.

Blackwood

1. Before servicing the vehicle, refer to the precautions in the beginning of this section.

✲✲ CAUTION

The electrical power to the air suspension system must be shut off prior to hoisting or jacking an air suspension vehicle. This can accomplished by turning off the air suspension switch located below the glove box in the lower right of the passenger foot well. Failure to do so can result in unexpected inflation or deflation of the air springs, which can result in the shifting of the vehicle during these operations.

2. Turn the air suspension switch to the OFF position.

✲✲ WARNING

Do not allow the steering column shaft to rotate while the intermediate shaft is disconnected or damage to the clockspring can result. If there is evidence that the shaft has rotated, the clockspring must be removed and re-centered.

3. Remove or disconnect the following:
   - Pitman arm cotter pin and castellated nut from the drag link
   - Pitman arm from the drag link
   - Dust cover from the steering shaft valve housing
   - Intermediate shaft pinch bolt and slide the shaft off the steering gear input shaft.
   - Power steering pressure hoses at the steering gear. Remove and discard the original seals.
   - Steering gear-to-frame rail retaining bolts
   - Steering gear and spacer.
4. If replacing or servicing the steering gear, match mark the sector shaft arm to the sector shaft and remove the steering gear sector shaft arm retaining nut and lockwasher.
5. Remove the sector shaft arm.
6. Installation is the reverse of removal. Install a new seal on the hose fittings at the power steering gear. When installing the power steering lines, use special installer tool, Installer Set, Teflon® Seal 214-D027 (D90P-3517-A), or equivalent. Stretch the seal over the special tool until it is large enough to slip over the tube nut.
7. Observe the following torques:
   - Steering gear mounting bolts: 199 ft. lbs. (270 Nm)
   - Intermediate shaft pinch bolt: 35 ft. lbs. (48 Nm)
   - Sector shaft-to-gear: 203 ft. lbs. (275 Nm)
   - Sector shaft-to-drag link: 66 ft. lbs. (90 Nm)

2004 F-150, Exc. Heritage Editions

1. Before servicing the vehicle, refer to the precautions in the beginning of this section.

➡New O-ring seals must be installed any time the lines are disconnected from the steering gear.

2. With the transmission in **Neutral**, raise and support the vehicle.
3. Remove the skid plate.

✲✲ WARNING

The boots and clamps are designed to produce an airtight seal and protect the internal components of the steering gear. If the seal is not airtight, the vacuum generated during turning will draw water and contamination into the gear, causing premature damage.

✲✲ WARNING

Zip ties do not provide an airtight seal and must not be used.

✲✲ WARNING

The inner ball joint grease is not compatible with water, and contamination trapped in the grease will degrade the life of the joint.

✲✲ WARNING

If present, the orientation of the vent tube must be noted so the boots and vent tubes can be installed in the correct location.

4. Remove the tire and wheel assembly.
5. Remove the lower shaft-to-steering gear bolt.
6. Disconnect the lower shaft.

➡Do not allow the steering column shaft to rotate while the lower shaft is disconnected or damage to the clockspring can result. If there is evidence that the shaft has rotated, the clockspring must be removed and re-centered. Follow these steps to center the clockspring.

a. Hold the clockspring outer housing stationary.

✲✲ WARNING

Overturning will destroy the clockspring. The internal ribbon wire acts as the stop and can be broken from its internal connection.

b. While turning the rotor counterclockwise, carefully feel for the ribbon wire to run out of length and for a slight resistance. Stop turning at this point.

c. Turn the clockspring clockwise approximately three turns. This is the center point of the clockspring.

d. Do not allow the rotor to turn from this position. Two pieces of tape may be applied to the clockspring to prevent accidental rotation, until the clockspring is installed.

7. Remove the clamp plate-to-steering gear bolt
8. Remove the power steering high pressure hose.
9. Remove the O-ring seal.
10. Remove the power steering return hose
11. Remove the O-ring seal.
12. Remove the tie-rod nuts.
13. Remove the outer tie-rods. When repairing the right side, it is necessary to pull back the left inner tie-rod boot to hold the steering gear. Note the number of times the tie-rod ends turn for assembly reference. After removing the tie-rod end, remove the tie-rod end jam nut from the front wheel spindle tie-rod
14. Remove the bracket-to-crossmember nuts.
15. Remove the bracket-to-crossmember bolts.
16. Remove the steering gear-to-crossmember nuts.
17. Remove the steering gear-to-crossmember bolts.
18. Remove the bracket.
19. Remove the steering gear. Remove the steering gear through the left side wheel opening.
20. To install, reverse the removal procedure.
21. Fill and leak test the system.
22. Check and, if necessary, align the front end.
23. Install new O-ring seals onto the power steering hoses.

2004 F-250 and 350

1. Before servicing the vehicle, refer to the precautions in the beginning of this section.

➡New O-ring seals must be installed any time the lines are disconnected from the steering gear.

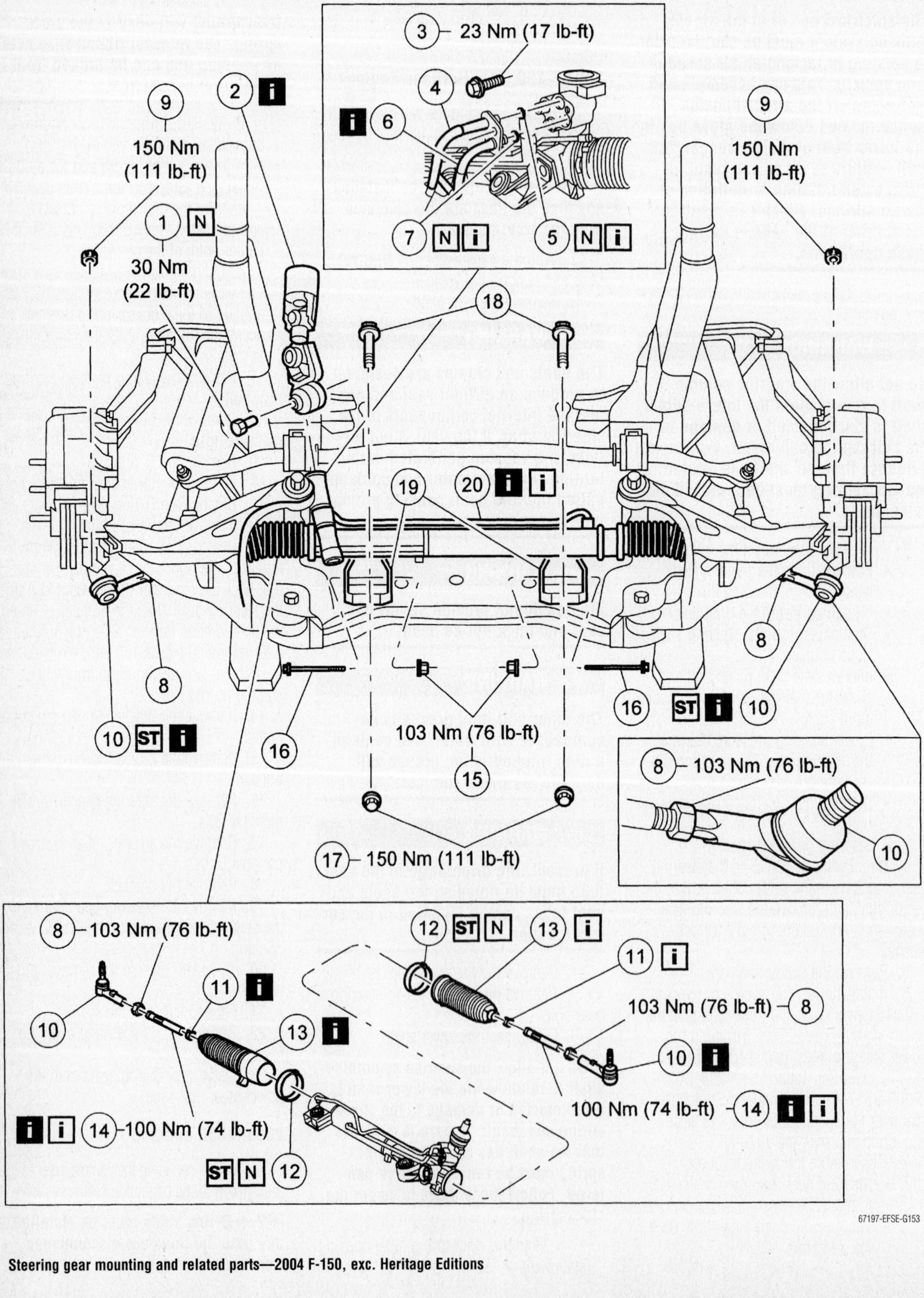

Steering gear mounting and related parts—2004 F-150, exc. Heritage Editions

1 Lower shaft-to-steering gear bolt
2 Lower shaft (disconnect only)
3 Clamp plate-to-steering gear bolt
4 Power steering high pressure hose
5 O-ring seal
6 Power steering return hose
7 O-ring seal
8 Tie-rod lock nuts (loosen only)
9 Tie-rod nuts
10 Outer tie-rods
11 Outer bellows clamp
12 Inner bellows clamp
13 Steering gear bellows
14 Inner tie-rod
15 Bracket-to-crossmember nuts
16 Bracket-to-crossmember bolts
17 Steering gear-to-crossmember nuts
18 Steering gear-to-crossmember bolts
19 Bracket
20 Steering gear

2. Remove the air cleaner assembly.
3. Disengage the steering coupling shield from the line fitting and slide upward on the steering shaft.
4. Remove the pinch bolt. Turn the steering wheel as necessary to access the bolt. Make sure the steering column is locked.

WARNING

During installation, do not over-tighten the fittings.

5. Disconnect the lines. Discard the O-ring seals.
6. Raise the vehicle.
7. Remove the cotter pin and nut. Discard the cotter pin.
8. Using the special tool, disconnect the drag link.
9. Remove the bolts, washers and the steering gear.
10. Remove the nut.
11. Using a puller, remove the steering gear sector shaft arm.
12. To install, reverse the removal procedure.
13. Note the following torques:
- Pinch bolt: 36 ft. lbs. (48 Nm)
- Power steering lines: 26 ft. lbs. (35 Nm)
- Pitman arm-to-drag link: 66 ft. lbs. (90 Nm)
- Gear mounting bolts: 59 ft. lbs. (80 Nm)
- Pitman arm-to-gear: 199 ft. lbs. (270 Nm)

14. Install a new high pressure hose O-ring seal and a new return hose O-ring seal.
15. Fill and leak check the system.

2001–03 E-Series

1. Before servicing the vehicle, refer to the precautions in the beginning of this section.
2. Place the wheels in the straight-ahead position.
3. Remove or disconnect the following:
- Pressure and return lines. Cap the openings.
- Splash shield from the flex coupling
- Flex coupling at the gear
- Pitman arm from the sector shaft
- Steering gear mounting bolts
- Steering gear. It may be necessary to work it free of the flex coupling.

To install:

4. Install or connect the following:
- Splash shield on the steering gear lugs
- Flex coupling into place on the steering shaft. Be sure the steering wheel spokes are still horizontal.

5. Center the steering gear input shaft with the indexing flat facing downward.
- Steering gear input shaft into the flex coupling and into place on the frame side rail
- Flex coupling bolt and tighten it to 30 ft. lbs. (41 Nm)
- Gear mounting bolts and tighten them to 65 ft. lbs. (88 Nm)
- Pitman arm. Tighten the nut to 230 ft. lbs. (312 Nm).
- Pressure and return lines. Tighten the lines to 25 ft. lbs. (34 Nm).
- Flex coupling shield

6. Fill the steering reservoir.
7. Run the engine and turn the steering wheel lock-to-lock several times to expel air. Check for leaks.

2004 E-Series

1. Before servicing the vehicle, refer to the precautions in the beginning of this section.

➡**New O-ring seals must be installed any time the lines are disconnected from the steering gear.**

2. Place the front wheels in the straight-ahead position and the ignition switch in the OFF position.

WARNING

Do not allow the steering column shaft to rotate while the intermediate shaft is disconnected or damage to the clockspring can result. If there is evidence that the shaft has rotated, the clockspring must be removed and recentered.

3. Remove the bolt and detach the intermediate shaft from the gear.
4. Disconnect the lines. Discard the O-ring seals.
5. With vehicle in **Neutral**, place on a hoist.
6. Remove and discard the cotter pin and nut.
7. Using a puller, separate the drag link.
8. Remove the bolts and the steering gear.
9. Secure the steering gear in a vise and remove the nut and lock-washer.
10. Using the special tool, remove the steering gear sector shaft arm.
11. To install, reverse the removal procedure.
12. Note the following torques:
- Pinch bolt: 18 ft. lbs. (25 Nm)
- Power steering lines: 15 ft. lbs. (20 Nm)
- Drag link-to-Pitman arm: 68 ft. lbs. (92 Nm)
- Pitman arm-to-gear: 199 ft. lbs. (270 Nm)

13. Install a new high pressure hose O-ring seal and a new return hose O-ring seal.
14. Fill and leak check the system.

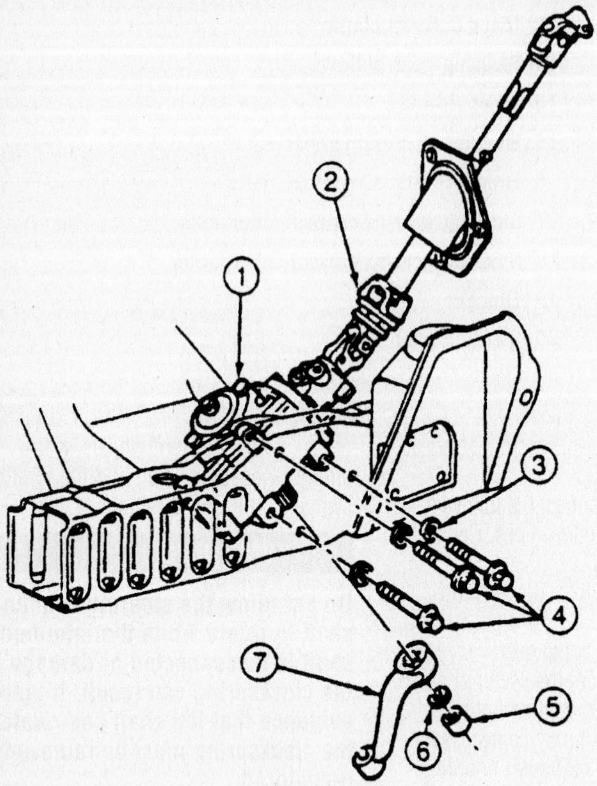

Item	Description
1	Steering Gear
2	Lower Steering Column Shaft
3	Washer
4	Bolt 73-90 N-m (54-66 Ft-Lb)
5	Nut 230-310 N-m (170-228 Ft-Lb)
6	Washer
7	Steering Gear Sector Shaft Arm

The steering gear is mounted on the left frame rail as show—E-Series

Front Shock Absorber

REMOVAL & INSTALLATION

2001–03 F-Series and 2004 Heritage Editions

1. Before servicing the vehicle, refer to the precautions in the beginning of this section.
2. If equipped with 2-wheel drive, perform the following:
   a. Hold the shock absorber stem and remove the nut from the top of the shock.
   b. Raise and safely support the vehicle.
   c. Remove the 2 lower retaining nuts and remove the shock absorber.
3. If equipped with 4-wheel drive, perform the following:
   a. Hold the shock absorber stem and remove the nut, washer and bushing from the top of the shock absorber stud.
   b. Raise and support the vehicle.
   c. Remove the lower retaining nut and bolt.
   d. Remove the shock absorber from the vehicle.

To install:

4. On 4-wheel drive models, perform the following:
   a. Install the washer and bushing to the top stem of the shock absorber.
   b. Place the shock absorber up through the coil spring.
   c. Install the 2 lower retaining nuts. Tighten the nuts to 19–25 ft. lbs. (26–34 Nm).
   d. Lower the vehicle.
   e. Install the bushing, washer and retaining nut to the top of the shock absorber stud. Tighten the nut to 34–46 ft. lbs. (47–63 Nm).
5. On 4-wheel drive models, perform the following:
   a. Install the washer and bushing to the top stem of the shock absorber.
   b. Place the shock absorber up through the coil spring.
   c. Install the lower retaining nut and bolt. Tighten to 57–76 ft. lbs. (77–104 Nm).
   d. Lower the vehicle.
   e. Install the shock absorber upper bushing, washer and retaining nut. Tighten the nut to 22–29 ft. lbs. (30–40 Nm).
6. Road test the vehicle and check for proper operation.

2004 F-250 and 350

2-WHEEL DRIVE

1. Before servicing the vehicle, refer to the precautions in the beginning of this section.

**** CAUTION**

All vehicles are equipped with gas-pressurized shock absorbers which will extend unassisted. Do not apply heat or flame to the shock absorbers during removal or component servicing.

**** CAUTION**

Suspension fasteners are critical parts because they affect performance of vital components and systems and their failure can result in major service expense. They must be replaced with the same part number or an equivalent part if replacement is necessary. Do not use a replacement part of lesser quality or substitute design. Torque values must be used as specified during reassembly to ensure proper retention of these parts.

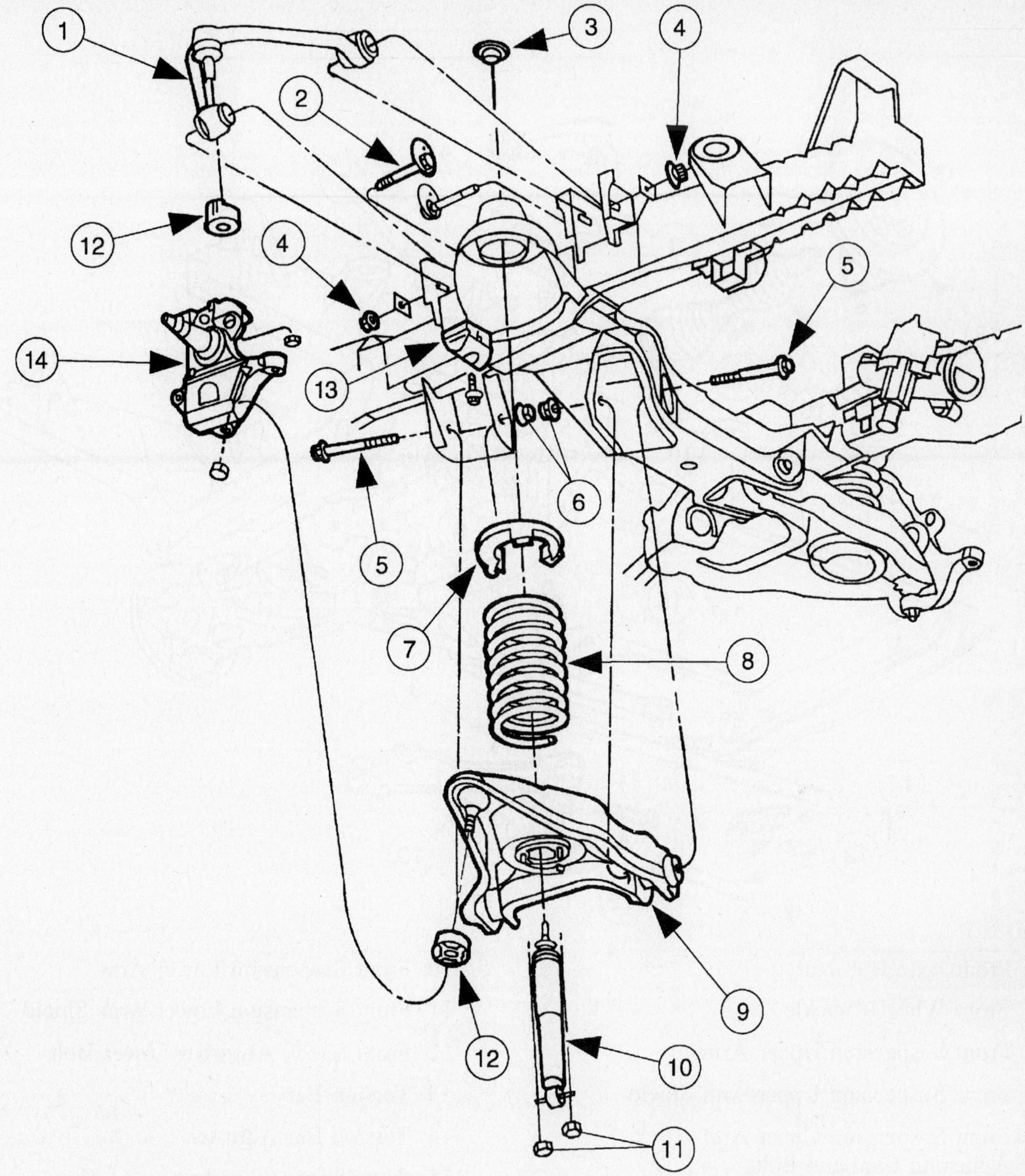

1 Front Suspension Upper Arm
2 Front Suspension Upper Arm Cams
3 Shock Absorber Insulator Assy
4 Front Suspension Cam Nut
5 Front Suspension Lower Arm Bolt
6 Front Suspension Lower Arm Nut
7 Front Spring Insulator
8 Front Coil Spring
9 Front Suspension Lower Arm
10 Front Shock Absorber
11 Shock Absorber Nut
12 Ball Joint Shield
13 Front Suspension Bumper
14 Front Wheel Spindle

67197-EFSE-G154

2-wheel drive front suspension—2001–03 F-Series and 2004 Heritage Editions

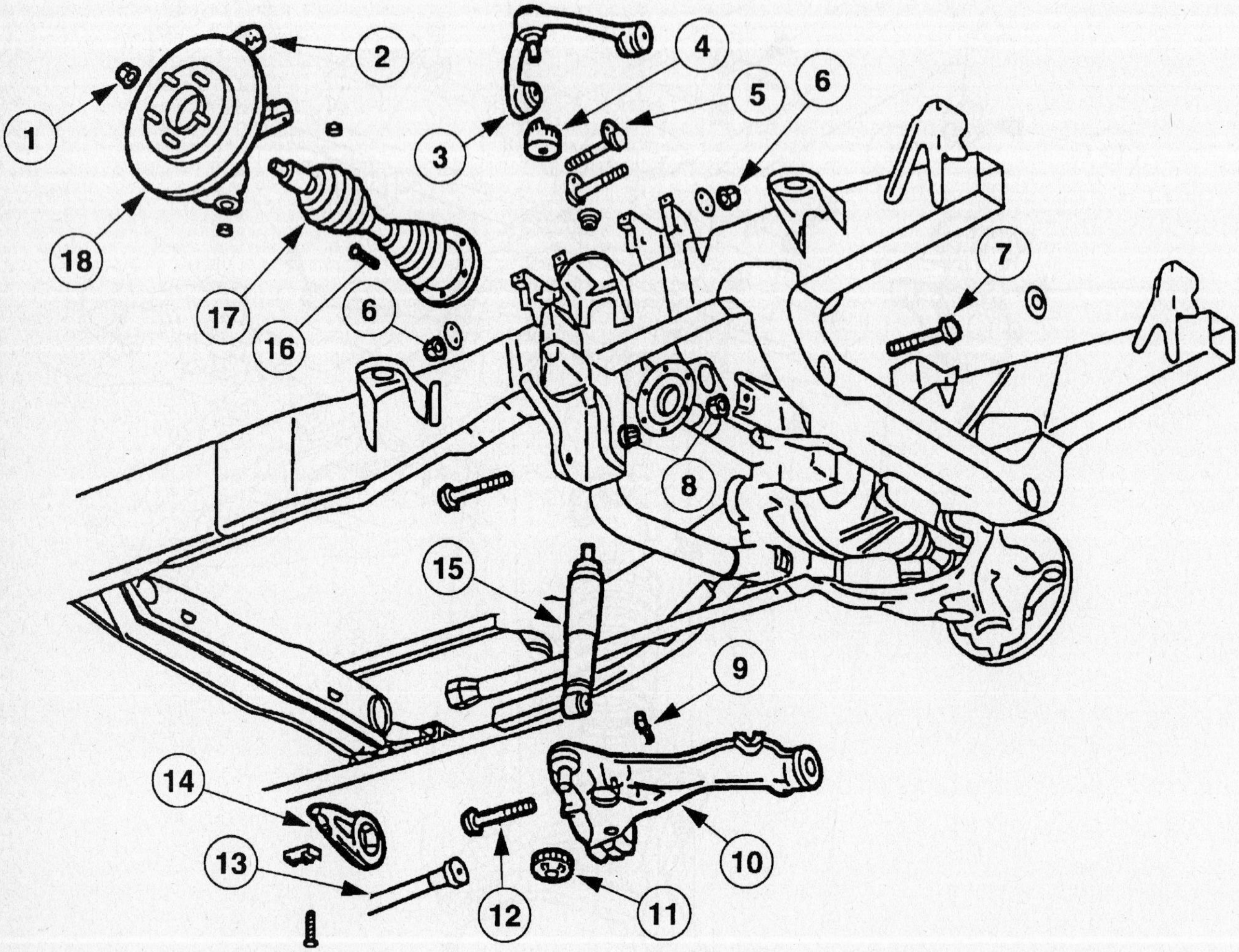

1 Front Axle Hub Nut
2 Front Wheel Knuckle
3 Front Suspension Upper Arm
4 Front Suspension Upper Arm Shield
5 Front Suspension Upper Arm Adjusting Cam and Bolts
6 Front Suspension Upper Arm Nuts
7 Front Suspension Lower Arm Bolts
8 Front Suspension Lower Arm Nuts
9 Front Shock Absorber Lower Nut
10 Front Suspension Lower Arm
11 Front Suspension Lower Arm Shield
12 Front Shock Absorber Lower Bolt
13 Torsion Bar
14 Torsion Bar Adjuster
15 Front Shock Absorber
16 Front Wheel Driveshaft and Joint Bolt
17 Front Wheel Driveshaft and Joint
18 Brake Disc

67197-EFSE-G155

4-wheel drive front suspension—2001–03 F-150 and 2004 Heritage Editions

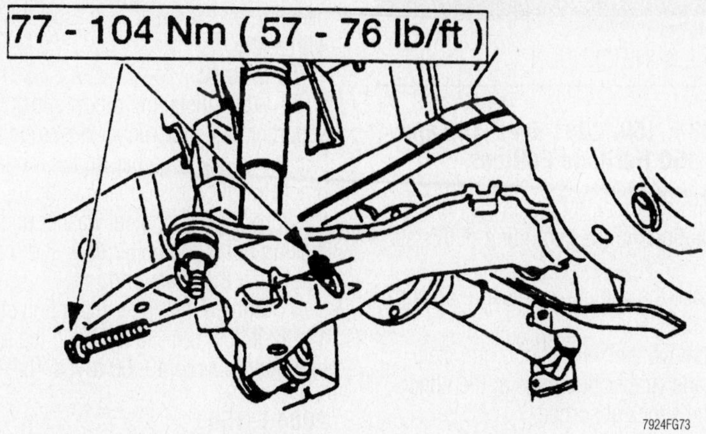

Lower shock absorber mounting—models with torsion bar suspension

2. Raise the hood and remove the upper shock absorber retaining nut and upper shock absorber insulator.
3. Raise and support the vehicle.
4. Remove the lower shock absorber retaining nut and remove the shock absorber.
5. Using new fasteners, follow the removal procedure in reverse order. Observe the following torques:

- Upper nut: 30 ft. lbs. (40 Nm)
- Lower nut: 60 ft. lbs. (80 Nm)

4-WHEEL DRIVE

1. Raise and support the vehicle.
2. Disconnect the shock absorber lower mount.
3. Disconnect the shock absorber upper mount.
4. To install, reverse the removal procedure. Tighten the upper and lower nuts to 76 ft. lbs. (103 Nm).

E-Series

1. Before servicing the vehicle, refer to the precautions in the beginning of this section.
2. Raise the vehicle and secure on support stands.
3. Remove the self-locking nut, steel washer, and rubber bushings at the upper end of the shock absorber.
4. Remove the bolt and nut at the lower end and remove the shock absorber.

To install:

5. When installing a new shock absorber, use new rubber bushings. Position the shock absorber on the mounting brackets with the stud end at the top. Install the upper bushing, steel washer and self-locking nut at the upper end, and the bolt and nut at the lower end.
6. Tighten the upper mounting studs to 18–22 ft. lbs. (24–30 Nm) and the lower mounting nuts to 40–60 ft. lbs. (54–82 Nm).

Front Strut

REMOVAL & INSTALLATION

2004 F-150, exc. Heritage Editions

1. Before servicing the vehicle, refer to the precautions in the beginning of this section.
2. Remove the wheel and tire assembly.

➡**Use the hex holding feature to prevent the ball studs (upper control arm, lower control arm tie-rod end, and stabilizer bar links) from turning while removing and installing the nuts.**

⁂ WARNING

Do not hammer the ball studs to separate them from the wheel knuckle or stabilizer bar. Doing so can cause damage. Lightly tap the wheel knuckle or stabilizer bar to loosen the joints.

3. Remove the strut upper mount nuts.
4. Remove the lower mount nut/bolt.
5. Remove the nut and detach the tie-rod from the wheel knuckle. Discard the nut.
6. Remove the strut.
7. To install, reverse the removal procedure. Tighten the upper strut nuts to 35 ft. lbs. (48 Nm); the lower strut-to-arm bolts to 351 ft. lbs. (475 Nm).
8. Do not tighten the fasteners until the installation procedure is complete and the weight of the vehicle is resting on the wheel and tire assemblies.

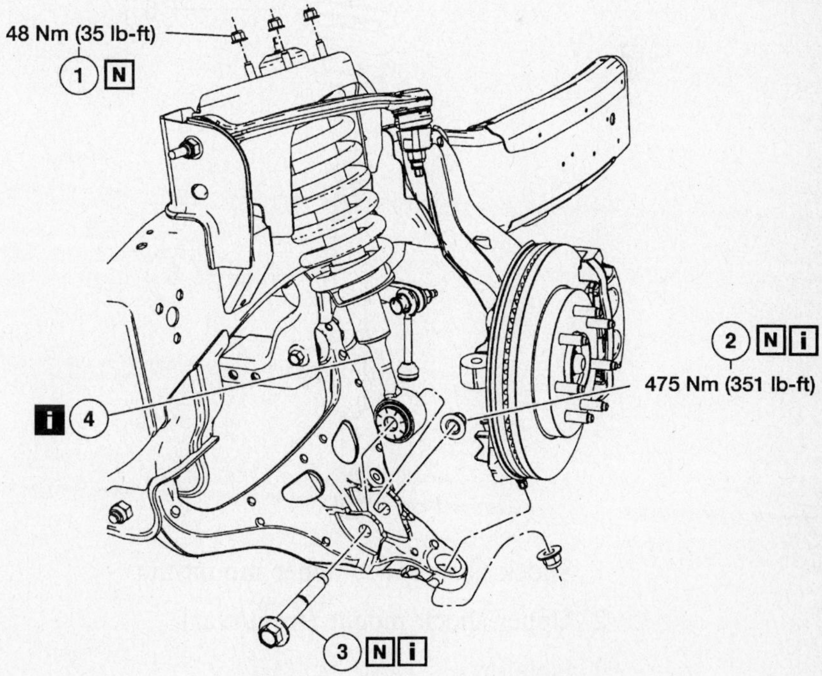

1 Shock absorber upper mount-to-frame nuts
2 Shock absorber-to-lower arm nut
3 Shock absorber-to-lower arm bolt
4 Shock absorber and spring assembly

67197-EFSE-G156

Front strut mounting— exc. Heritage Editions

DISASSEMBLY

1. Remove the strut and spring assembly.
2. Using a suitable spring compressor, compress the spring until the tension is released from the shock absorber.
3. Remove the shock absorber-to-upper mount nut
4. Remove the upper shock mount
5. Remove the isolator
6. Remove the dust tube
7. Remove the coil spring
8. Remove the shock absorber
9. To assemble, reverse the disassembly procedure.

Rear Shock Absorber

REMOVAL & INSTALATION

2001–03 F-150, 2001–04 E-150 and 2004 F-150 Heritage Editions

1. Before servicing the vehicle, refer to the precautions in the beginning of this section.
2. Raise the vehicle and secure on support stands.
3. Remove the self-locking nut, steel washer, and rubber bushings at the upper end of the shock absorber.
4. Remove the bolt and nut at the lower end and remove the shock absorber. If needed, raise the rear axle assembly slightly with a jack.

To install:

5. When installing a new shock absorber, use new rubber bushings. Position the shock absorber on the mounting brackets with the stud end at the top. Install the upper bushing, steel washer and self-locking nut at the upper end, and the bolt and nut at the lower end.
6. Tighten the upper mounting studs to 18–22 ft. lbs. (24–30 Nm) and the lower mounting nuts to 40–60 ft. lbs. (54–82 Nm).

2004 F-150

1. Before servicing the vehicle, refer to the precautions in the beginning of this section.
2. Raise the vehicle and secure on support stands.
3. Remove the self-locking nut, steel washer, and rubber bushings at the upper end of the shock absorber.
4. Remove the bolt and nut at the lower end and remove the shock absorber. If needed, raise the rear axle assembly slightly with a jack.

To install:

5. When installing a new shock absorber, use new rubber bushings. Position the shock absorber on the mounting brackets with the stud end at the top. Install the upper bushing, steel washer and self-locking nut at the upper end, and the bolt and nut at the lower end.
6. Tighten the upper mounting studs to 18–22 ft. lbs. (24–30 Nm) and the lower mounting nuts to 40–60 ft. lbs. (54–82 Nm).

2001–04 E- and F-250 and 350

WARNING

Suspension fasteners are critical parts because they affect performance of vital components and systems and their failure can result in major service expense. Install new parts with the same part number or an equivalent part if installation is necessary. Do not use an installation part of lesser quality or substitute design. Torque values must be used as specified during reassembly to make sure of correct retention of these parts.

1. Before servicing the vehicle, refer to the precautions in the beginning of this section.

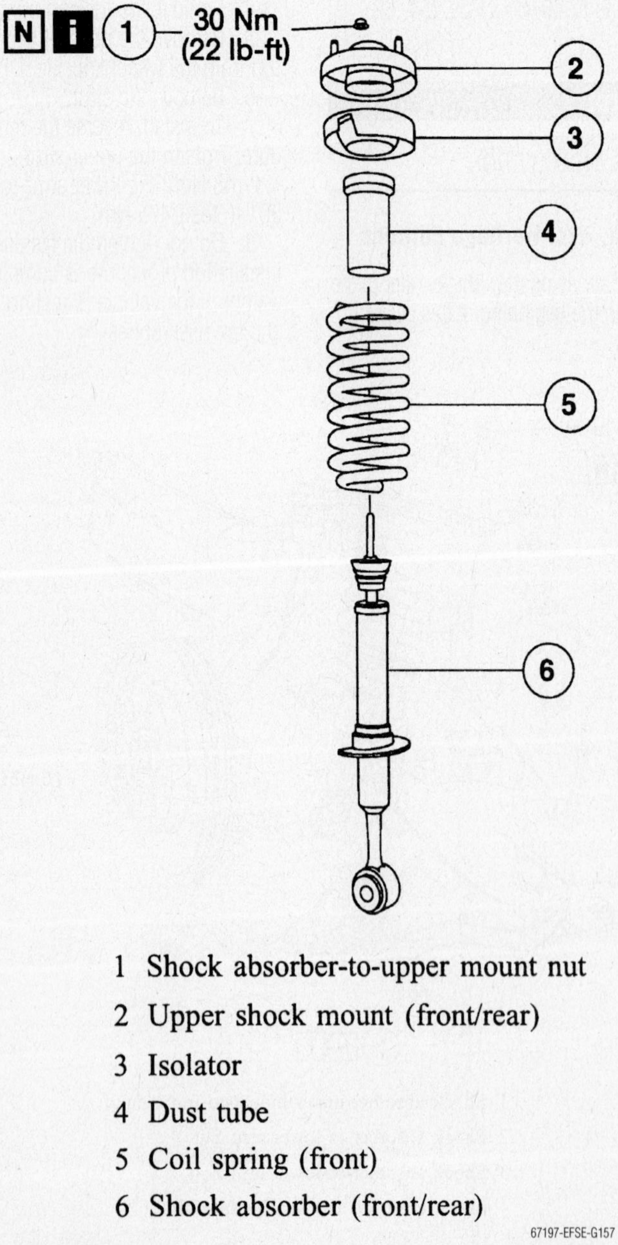

Front strut disassembled

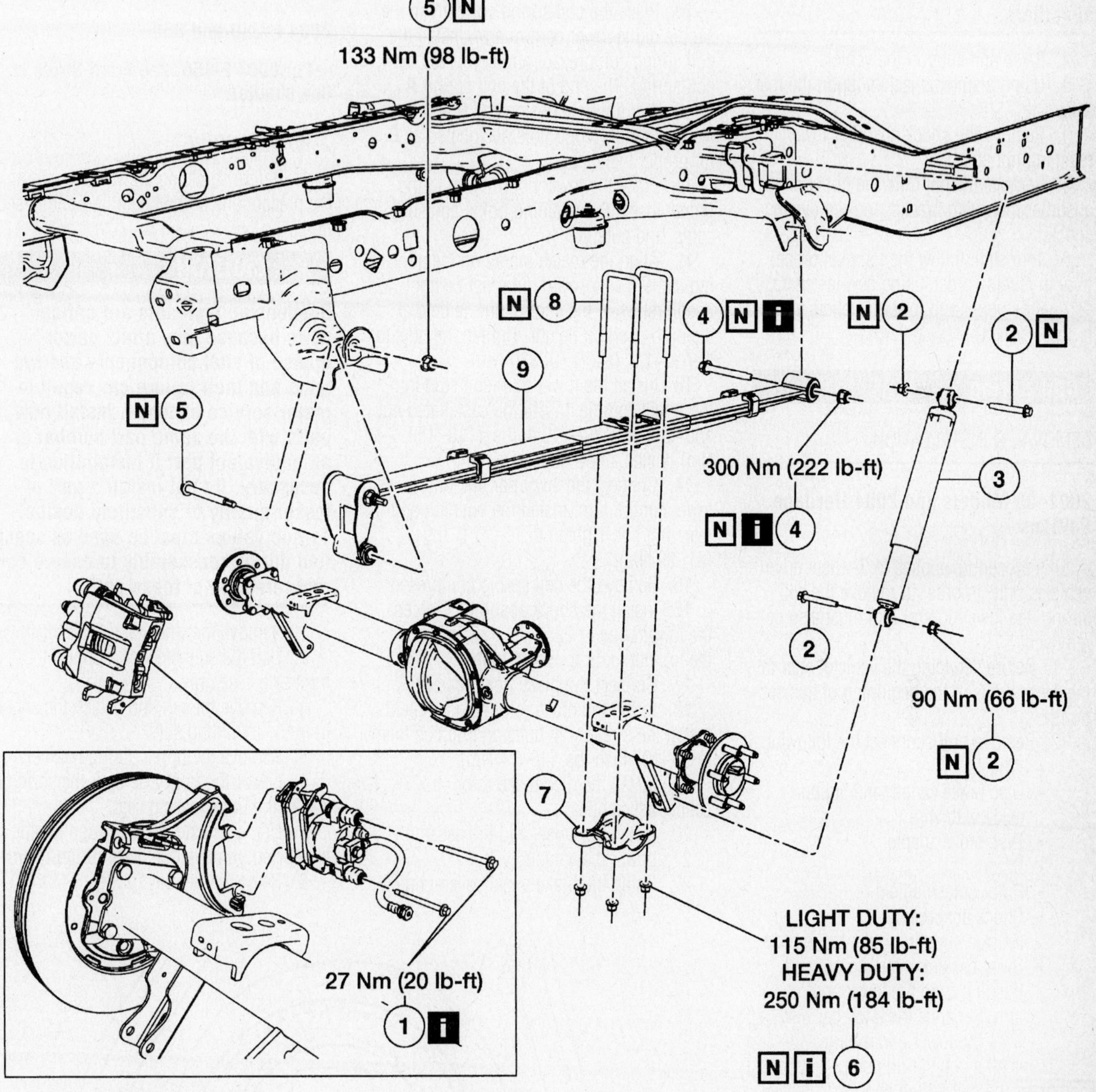

1 Brake caliper and bolts
2 Shock absorber bolts and nuts
3 Shock absorber
4 Spring front eye bolt and nut
5 Spring shackle bolt and nut (rear)
6 U-bolt nuts (light duty/heavy duty)
7 U-bolt plate
8 U-bolts
9 Rear spring assembly

67197-EFSE-G158

Rear suspension components—2004 F-150, exc. Heritage Editions

WARNING

The low pressure gas shock absorbers are charged with nitrogen gas. Do not attempt to open, puncture or apply heat to shock absorbers.

2. Raise and support the vehicle.
3. Using a suitable jack, support the rear axle.
4. Remove the shock absorber lower retaining nut and bolt.
5. Remove the nut from the upper shock absorber mounting bracket and remove the shock.
6. To install, follow the removal procedure in reverse order, using new fasteners. Torque the upper and lower fasteners to 46 ft. lbs. (62 Nm).

Coil Spring

REMOVAL & INSTALLATION

2001–03 Models and 2004 Heritage Editions

This procedure applies to 2-wheel drive vehicles only. In order to remove the coil spring, the lower control arm must also be removed.

1. Before servicing the vehicle, refer to the precautions in the beginning of this section.
2. Remove or disconnect the following:
   - Wheels
   - Disc brake caliper and support aside with wire
   - Disc brake adapter
   - Rotor
   - Rotor splash shield
   - Shock absorber
   - Bracket supporting the brake hose
   - Sway bar link retaining nut and bushing from the lower control arm. Separate the sway bar link from the lower control arm.
3. Install a coil spring compressor and compress the coil spring enough to relieve the tension of the spring between the upper and lower control arms.
4. Remove the cotter pin and castellated nut from the lower ball joint. Separate the lower ball joint from the wheel spindle.
5. Matchmark the lower control arm alignment cams for installation reference.
6. Remove the lower control arm retaining nuts and bolts.
7. Remove the lower control arm and the compressed coil spring as an assembly.
8. Loosen the coil spring compressor and remove the coil spring from the lower control arm.
9. Inspect the coil spring and replace as needed.

To install:

10. Place the coil spring correctly in the saddle of the lower control arm. Install the coil spring compressor and compress the coil spring. The end of the coil spring **A**, must cover the hole designated **B** and be visible in the second hole designated as **C**, for proper installation.
11. Place the lower control arm to the frame. Install the retaining bolts, adjusting cams, and nuts.
12. Align the match marks on the adjusting cams. The forward nut must be tightened first while the control arm is held at the curb position height. Tighten the nuts to 197–241 ft. lbs. (270–330 Nm).
13. Install the lower ball joint stud into the wheel spindle. Install the castellated nut and tighten to 83–113 ft. lbs. (113–153 Nm). Install a new cotter pin.
14. Connect the sway bar link to the lower control arm. Install the bushing and retaining nut. Tighten to 15–21 ft. lbs. (21–29 Nm).
15. Remove the coil spring compressor.
16. Install the shock absorber. Tighten the lower bolts to 22 ft. lbs. (32 Nm) and the top nut to 45 ft. lbs. (61 Nm).
17. Connect the brake hose bracket.
18. Install the brake rotor splash shield. Install the 3 retaining bolts and tighten them to 90–107 inch lbs. (10–14 Nm).
19. Install the disc brake rotor and caliper assemblies.
20. Install the wheel and tire assembly.
21. Lower the vehicle.
22. Pump the brake pedal several times to position the brake pads prior to moving the vehicle.
23. Check the alignment and adjust if out of specification.
24. Road test the vehicle and check for proper operation.

2004 F-250 and 350

➡For 2004 F-150, see Front Struts in this chapter.

2-WHEEL DRIVE

1. Before servicing the vehicle, refer to the precautions in the beginning of this section.

CAUTION

Suspension fasteners are critical parts because they affect performance of vital components and systems and their failure can result in major service expense. Install new parts with the same part number or an equivalent part if installation is necessary. Do not install a part of lesser quality or substitute design. Torque values must be used as specified during reassembly to ensure correct retention of these parts.

2. Remove the wheel and tire assembly.
3. Using a suitable jack, support the front axle assembly.
4. Remove the nut and detach the shock from the mounting stud.
5. Remove the upper spring retainer.
6. Lower the front axle until the spring is free of the upper spring seat.
7. Using an extension through the top of the spring, remove the lower spring retainer.
8. Remove the front spring.

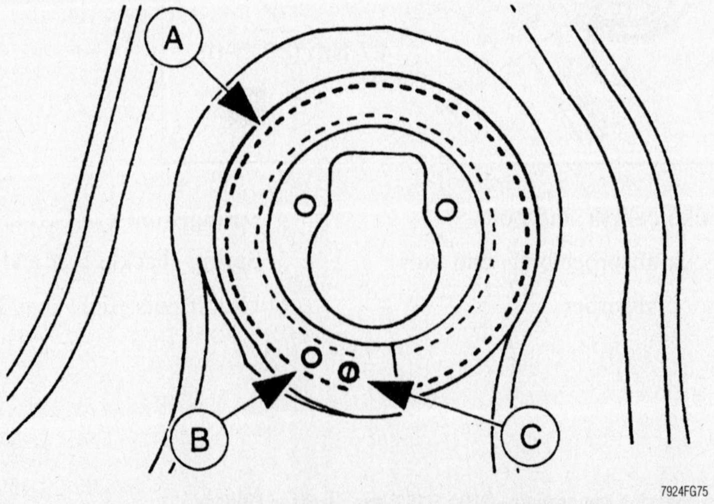

Be sure the coil spring is mounted correctly in the lower control arm

9. Using new fasteners, reverse the removal procedure. Tighten the lower spring retainer to 99 ft. lbs. (133 Nm); the upper retainer to 26 ft. lbs. (35 Nm).

Torsion Bar

REMOVAL & INSTALLATION

2001–03 F-150 4-Wheel Drive and 2004 Heritage Editions

1. Raise the vehicle on a hoist.
2. Apply a liberal amount of penetrating oil to the torsion bar adjuster and to the torsion bar socket at the lower control arm.
3. Make preliminary adjustment references.
 a. Mark the relation of the torsion bars to crossmember support.
 b. Measure and record the torsion bar adjuster bolt lengths.
4. Relieve the torsion bar tension.
 a. Remove the torsion bar adjuster bolt, 1
 b. Install the special tools T96T-5310-A and T95T-5310-A on the crossmember support and to the torsion bar adjuster, 2
 c. Tighten the special tool until the torsion bar adjuster lifts off the adjuster nut, 3
 d. Remove the torsion bar adjuster nut, 4
 e. Remove the special tool, 5. Repeat on the opposite side.

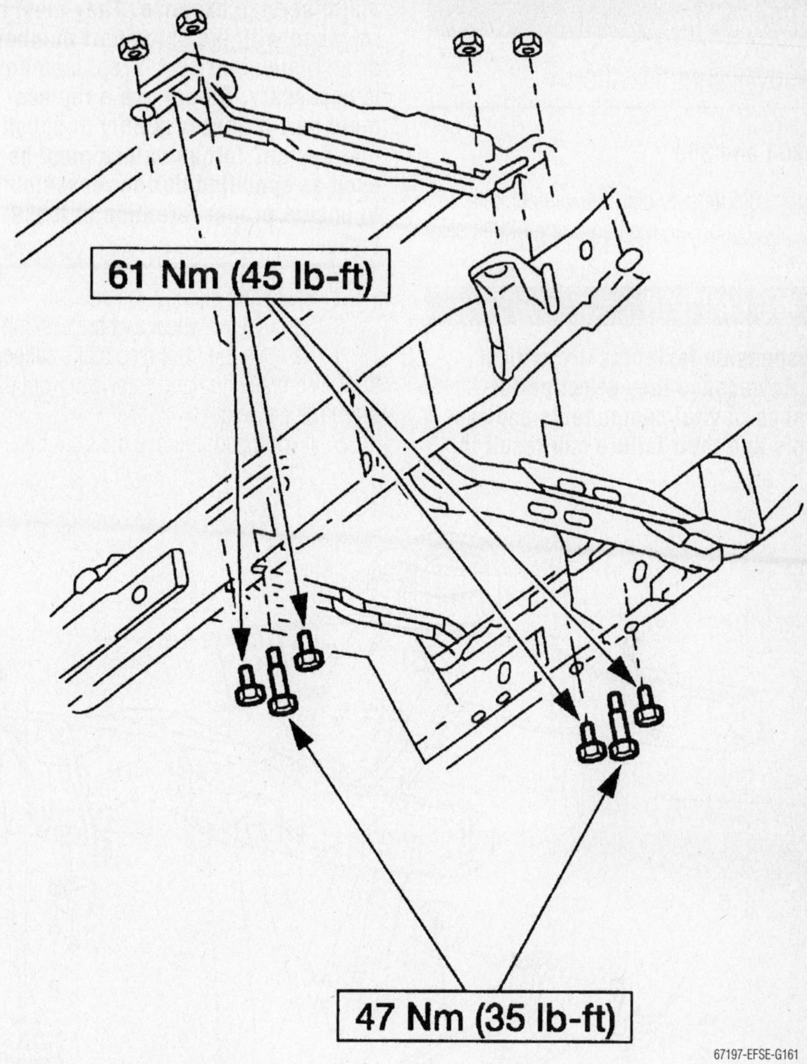

Install the crossmember bolts

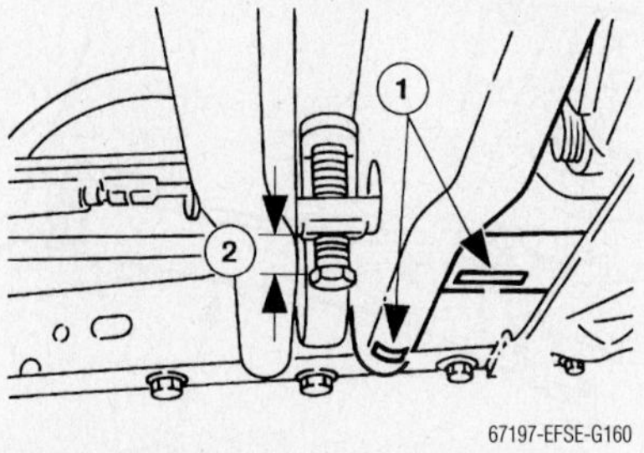

Mark the relation of the torsion bars to crossmember support

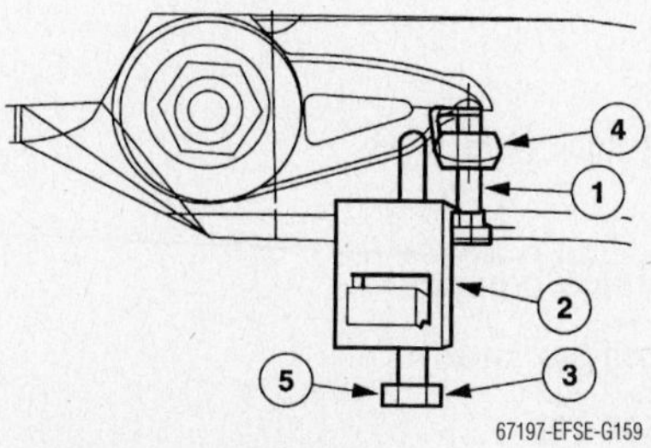

Special tools assembled

5. Remove the retainers from the torsion bar crossmember support.
6. Position the torsion bar crossmember support rearward and remove the torsion bar adjusting bracket.
7. Position the torsion bar rearward and separate it from the lower control arm. Position the torsion bar forward and remove the bar.

To install:

8. Position the torsion bar to the vehicle and install the bar to the lower control arm.
9. Reposition the torsion bar crossmember support. Install the torsion bar adjusting brackets to the torsion bars.
10. Install the crossmember support retaining bolts.
11. Load the torsion bar.
 a. Install special tools T95T-5310-A and T96T-5310-A on the crossmember support and to the torsion bar adjuster.
 b. Tighten the special tool to load the torsion bar.
 c. Install the torsion bar adjuster nuts.
 d. Loosely install the torsion bar adjuster bolt.
 e. Remove the special tools. Repeat on the opposite side.
12. Make an initial adjustment.
 a. Turn the torsion bar adjusters until the reference marks on the torsion bars and crossmember support align.
 b. Measure the torsion bar adjuster bolt lengths. The measurement must equal the measurement taken during removal.
13. Lower the vehicle.
14. Check the ride height. Adjust as necessary.
15. Check the wheel alignment. Adjust as necessary.

Radius Arm

REMOVAL & INSTALLATION

F-250 and 350

1. Before servicing the vehicle, refer to the precautions in the beginning of this section.

CAUTION

Suspension fasteners are critical parts because they affect performance of vital components and systems and their failure can result in major service expense. They must be replaced with the same part number or an equivalent part if replacement is necessary. Do not use a replacement part of lesser quality or substitute design. Torque values must be used as specified during reassembly to ensure proper retention of these parts.

2. Raise and support the vehicle.
3. Remove the wheel and tire assembly.
4. Remove the front disc brake caliper, front disc brake hub and rotor and front disc brake rotor shield.
5. If equipped, remove the front disc brake ABS sensor retainer bolt and the ABS sensor harness bracket bolt from the front wheel spindle. Position the ABS sensor out of the way.
6. Remove the cotter pin and the castellated nut from the tie rod end. Discard the cotter pin.
7. Using the Pitman Arm Puller, remove the tie rod end.
8. Remove the cotter pin and the lower ball joint castellated nut. Discard the cotter pin.
9. Remove the pinch bolt and the camber adjuster from the upper ball joint.

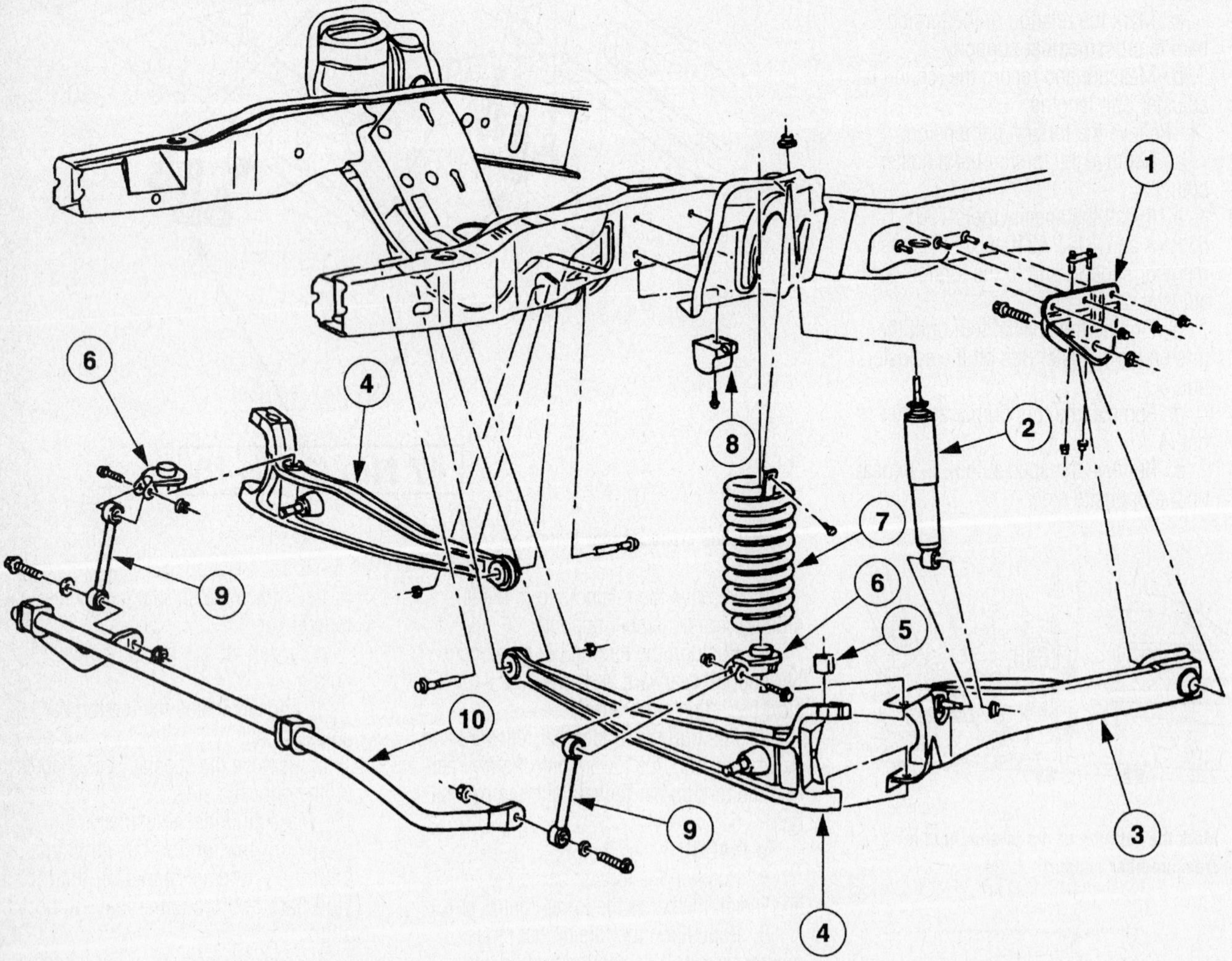

1 Radius arm bracket (LH/RH)
2 Shock absorber
3 Radius arm (LH/RH)
4 Front axle I-beam (LH/RH)
5 Upper ball joint adjuster
6 Lower spring seat (LH/RH)
7 Coil spring
8 Front axle jounce bumper
9 Front stabilizer bar link
10 Front stabilizer bar

67197-EFSE-G162

2-wheel drive F-250 and 350 front suspension

✲✲ WARNING

To prevent damage to the ball joint seal and the ball joint socket, do not use a pickle fork-type remover to loosen the ball joints.

10. Remove the front wheel spindle.
  a. Strike the lower end of the axle to loosen the ball joint.
  b. Remove the front wheel spindle.
11. Using a suitable jack, support the front axle.
12. Remove the front coil spring.
13. Remove the lower spring insulator.
14. Remove the lower spring seat.
15. Remove the radius arm-to-axle bolt and nut.
16. Remove the radius arm-to-frame bracket pivot bolt and nut. Remove the radius arm.

➡In order to obtain clearance for the pivot bolt removal when servicing the right front axle, it may be necessary to raise the side of the vehicle to relieve the weight on the suspension.

17. Remove the pivot nut and bolt.

To install:

18. Inspect the pivot bushing for wear or damage. Replace as necessary.
19. Position the front axle into the pivot bracket and install a new pivot bolt and nut hand-tight.
20. Inspect the radius arm pivot bushing for wear or damage. Replace as necessary.

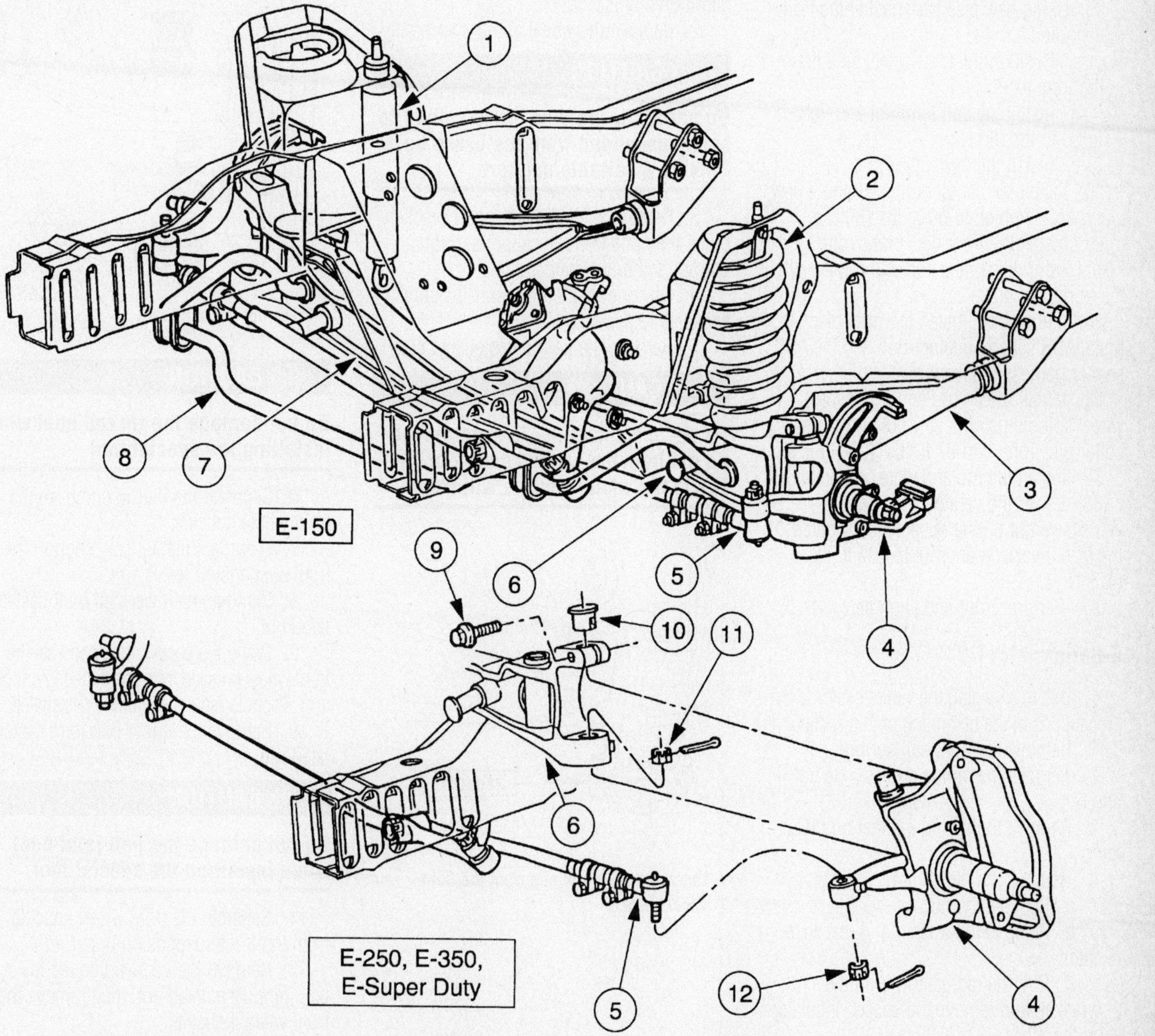

1 Shock absorber
2 Front coil spring
3 Radius arm
4 Front wheel spindle
5 Tie-rod end
6 Front axle (LH)
7 Front axle (RH)
8 Front stabilizer bar
9 Upper ball joint pinch bolt
10 Caster/camber insert
11 Lower ball joint nut
12 Tie-rod end-to-spindle nut

67197-EFSE-G167

E-Series front suspension

21. Install a new radius arm-to-axle bolt and nut hand-tight.
22. Install a new radius arm-to-frame bracket pivot bolt and nut. Tight to 185 ft. lbs. (250 Nm).
23. Tighten the radius arm-to-axle bolt and nut to 295 ft. lbs. (400 Nm).
24. Install the lower spring seat.
25. Install the lower spring insulator.
26. Install the front coil spring.

➡Tighten the ball joint nut further, if necessary, in order to insert the new cotter pin.

27. Using new fasteners, install the front wheel spindle in the front axle.
   a. Position the front wheel spindle in the front axle.
   b. Install the ball joint nut and tighten to 60 ft. lbs (80 Nm).
   c. Install the new cotter pin.
28. Install the camber adjuster and a new pinch bolt. Tighten to 60 ft. lbs (80 Nm).
29. Install the front disc brake rotor shield, front disc brake hub and rotor and the front disc brake caliper.
30. If equipped, install the front disc brake ABS sensor, retainer bolt and the ABS sensor harness bracket bolt.
31. Install the tie rod end in the front wheel spindle using a new nut and a new cotter pin. Torque to 67 ft. lbs. (90 Nm).
32. Install the tire and wheel assembly.
33. Lower the vehicle and with the vehicle weight on the suspension, tighten the axle pivot bolt and nut to 130 ft. lbs. (175 Nm).
34. Perform front end alignment.

E-Series

1. Before servicing the vehicle, refer to the precautions in the beginning of this section.
2. Remove the front coil spring.
3. Remove the radius arm-to-axle nut and bolt.
4. Remove the spring retainer and insulator.
5. Remove the radius arm.
   a. Remove the nut.
   b. Remove the washer and rear insulator.
   c. Pull the radius arm from the bracket, and remove the spacer, insulator and washer.

➡Inspect the bushings and install new as necessary.

6. To install, reverse the removal procedure. Torque the pivot nut to 221 ft. lbs. (300 Nm); the bracket end nut to 98 ft. lbs. (133 Nm).

Ball Joints

REMOVAL & INSTALLATION

2001–03 2-Wheel Drive F-150 and 2004 Heritage Edition

➡The upper ball joints are not replaceable and are serviced with the control arm.

LOWER

1. With the vehicle in **Neutral**, position it on a hoist.
2. Remove the wheel and tire assembly.

CAUTION

Do not allow the disc brake caliper to hang suspended from the brake hose. Provide a suitable support.

3. Remove the caliper support bracket bolts, then position the caliper support bracket and brake caliper aside.
4. Disconnect the front anti-lock brake sensor wire assembly.
5. Remove the tie rod castellated nut.

CAUTION

Do not use a hammer to separate the tie rod from the wheel knuckle or damage to the wheel knuckle will result.

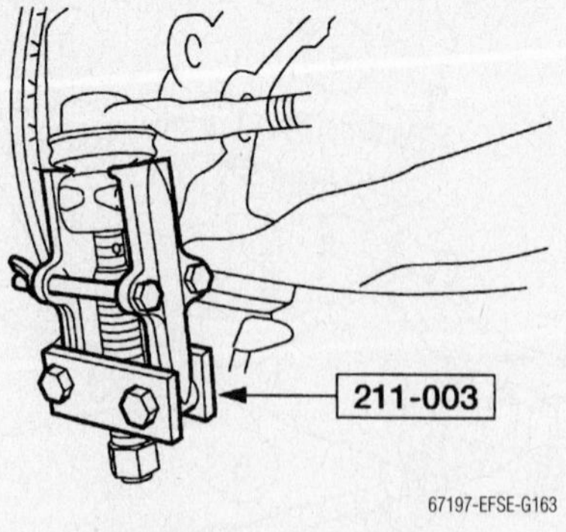

67197-EFSE-G163

Separate the tie rod end from the front wheel knuckle

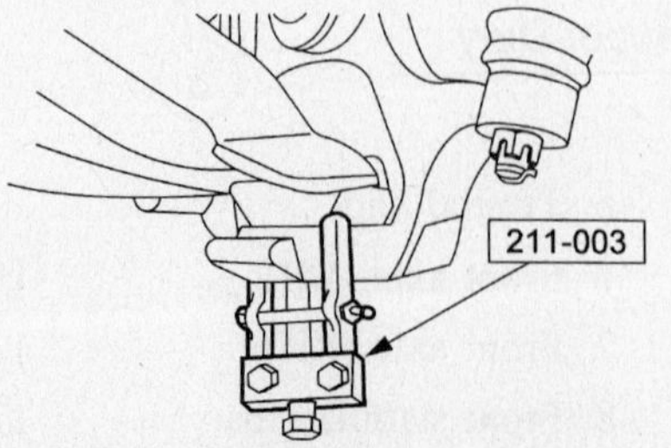

67197-EFSE-G164

Separate the knuckle from the lower arm

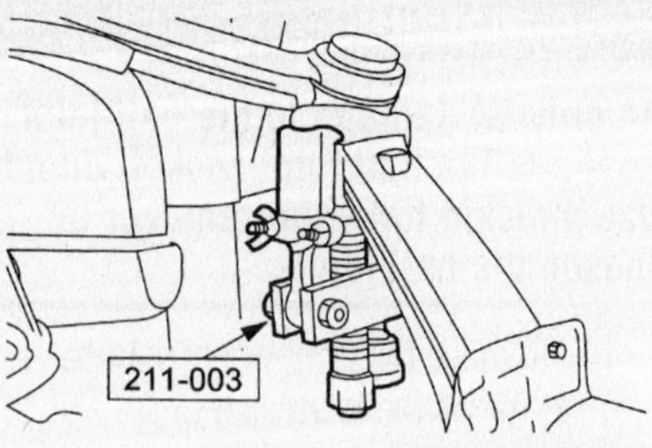

67197-EFSE-G165

Separate the knuckle from the upper arm

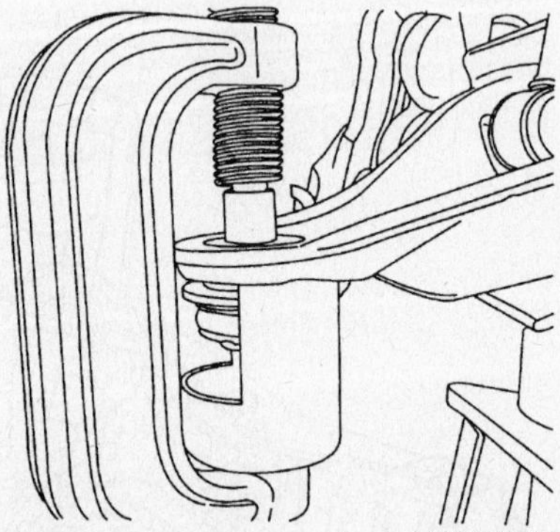

67197-EFSE-G166

Press out the ball joint

CAUTION

Do not damage the tie rod boot when installing the special tool.

6. Separate the tie rod end from the front wheel knuckle.
7. Using a suitable jack, support the front suspension lower arm.
8. Remove the lower ball joint castellated nut.
9. Using the special tool, separate the front wheel knuckle from the suspension lower arm. Then, loosely reinstall the castellated nut.
10. Remove the upper ball joint castellated nut.

CAUTION

Do not damage the ball joint boot when installing the special tool.

11. Separate the front wheel knuckle from the front suspension upper arm.
12. Remove the hand-tightened lower ball joint castellated nut, then remove the front wheel knuckle.
13. Remove the snap ring from the ball joint. Discard the snapring.
14. Press out the ball joint.

CAUTION

Do not damage the ball joint boot when installing the special tool.

➡Clean and inspect the control arm ball joint bore for damage before installing a new ball joint.

➡Make sure the new ball joint snap ring is fully seated.

15. To install, reverse the removal procedure. Always install new castellated nuts and cotter pins. Observe the following torques:

- Caliper support bracket: 83 ft. lbs. (112 Nm)
- Tie rod end ball stud: 70 ft. lbs. (95 Nm)
- Lower ball joint stud nut: 98 ft. lbs. (133 Nm)
- Upper ball joint stud nut: 66 ft. lbs. (90 Nm)

2001–03 4-Wheel Drive F-150 and 2004 Heritage Editions

➡The upper ball joints are not replaceable and are serviced with the control arm.

LOWER

1. Before servicing the vehicle, refer to the precautions in the beginning of this section.
2. With the vehicle in **Neutral**, position it on a hoist.
3. Remove the wheel and tire assembly.
4. Remove and discard the cotter pin. Remove the retainer. Remove the hub nut.

⁂ WARNING

Do not allow the disc brake caliper to hang suspended from the brake hose. Provide a suitable support.

5. Remove the caliper support bracket bolts, then position the caliper support bracket and caliper aside.
6. Remove the brake disc.
7. Disconnect the front anti-lock brake sensor (ABS) wire assembly.
8. Remove the tie-rod castellated nut. Discard the cotter pin and nut.

⁂ WARNING

Do not use a hammer to separate the tie-rod from the wheel knuckle or damage to the wheel knuckle will result. Do not damage the tie-rod boot when installing the special tool.

9. Using the special tool, separate the tie-rod end from the front wheel knuckle.
10. Using a suitable jack, support the front suspension lower arm.
11. Remove the lower ball joint castellated nut. Discard the cotter pin. Discard the castellated nut.
12. Separate the front wheel knuckle from the suspension lower arm. Then, loosely install the castellated nut.
13. Remove the upper ball joint castellated nut. Discard the cotter pin. Discard the castellated nut.
14. Separate the front wheel knuckle from the front suspension upper arm.
15. Separate the halfshaft from the hub, then remove the hand-tightened lower ball joint castellated nut. Remove the front wheel knuckle. Position the halfshaft aside and support with wire.
16. Remove the snapring from the ball joint. Discard the snapring.
17. Using a suitable ball joint remover tool and receiver cup, remove the ball joint.

➡Do not damage the ball joint boot when installing the special tool.

➡Clean and inspect the control arm ball joint bore for damage before installing a new ball joint.

➡Make sure the new ball joint snapring is fully seated.

18. To install, reverse the removal procedure. Always install new castellated nuts and cotter pins. Observe the following torques:

- Upper ball joint nut: 66 ft. lbs. (90 Nm).
- Lower ball joint nut: 98 ft. lbs. (133 Nm).
- Tie rod end nut: 70 ft. lbs. (95 Nm).
- Brake caliper support bracket: 136 ft. lbs. (185 Nm).
- Hub nut: 221 ft. lbs. (300 Nm).

2001–04 2-Wheel Drive F-250 and 350

UPPER AND LOWER

1. Before servicing the vehicle, refer to the precautions in the beginning of this section.
2. Raise and support the vehicle.
3. Remove the wheel and tire assembly.
4. Remove the disc brake caliper and the front disc brake hub and rotor.
5. Remove the front disc brake rotor shield.
6. If equipped, remove the ABS sensor retaining bolt, ABS sensor harness retaining bolt and the ABS sensor. Position out of the way.
7. Disconnect the tie rod end.
   a. Remove and discard the cotter pin.
   b. Remove the castellated nut.
   c. Using the Pitman Arm Puller, remove the tie rod end.
8. Remove the pinch bolt.
9. Remove the camber adjuster.

⁂ WARNING

To prevent damage to the ball joint seal and the ball joint socket, do not use a pickle fork-type remover to loosen the ball joints.

10. Remove the front wheel spindle.
    a. Remove and discard the cotter pin.
    b. Loosen, but do not remove, the castellated nut.
    c. Strike the lower end of the front axle to loosen the ball joint.
    d. Remove the castellated nut and the front wheel spindle.
11. Position the front wheel spindle in a vise, and remove the snapring from the lower ball joint.

⁂ WARNING

To avoid damage to the components, do not use heat to aid ball joint removal.

12. Using the ball joint press tool and suitable receiver cup, remove the lower ball joint from the front wheel spindle.
13. Using the ball joint press tool and suitable receiver cup, remove the upper ball joint.

To install:

⁂ WARNING

To avoid damage to components, do not use heat to aid installation.

➡Clean the wheel knuckle ball joint bores.

➡The lower ball joint must be installed first.

14. Using the ball joint press with suitable receiver cups, install the lower ball joint.
15. Using the ball joint press with suitable receiver cups, install the upper ball joint.
16. Install the snapring in the groove at the bottom of the ball joint.

➡Tighten the ball joint nut further, if necessary, in order to insert a new cotter pin.

17. Using new fasteners, follow the removal procedure in reverse order.
18. Check the front end alignment. Observe the following torques:

- Ball joint stud nut: 99 ft. lbs. (133 Nm)
- Pinch bolt: 60 ft. lbs. (80 Nm)
- Tie rod end stud nut: 67 ft. lbs. (90 Nm)

2001–04 4-Wheel Drive F-250 and 350

UPPER

1. Raise and support the vehicle.
2. Remove the wheel and tire assembly.
3. Remove the front brake disc.
4. Remove the wheel hub and bearing.
5. Using a drift, drive the axle shaft main seal out of the wheel knuckle.
6. Remove the axle shaft and main seal.
7. Remove the tie-rod end castellated nut.
8. Disconnect the tie-rod end from the wheel knuckle.
9. Remove the upper ball joint castellated nut and the insert.
10. Remove the lower ball joint nut.
11. Remove the knuckle.
12. Clean and inspect the wheel knuckle ball joint bores.
13. Place the wheel knuckle into a suitable vise.

➡Always remove the lower ball joint first.

14. Remove the lower ball joint.

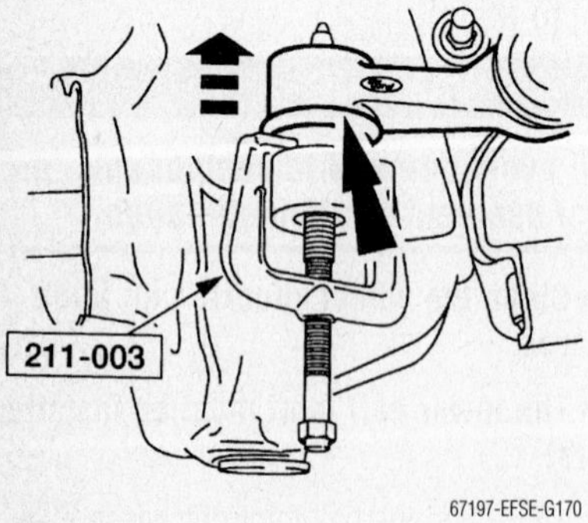

Disconnect the tie-rod end from the wheel knuckle—2001–04 F-250 and 350

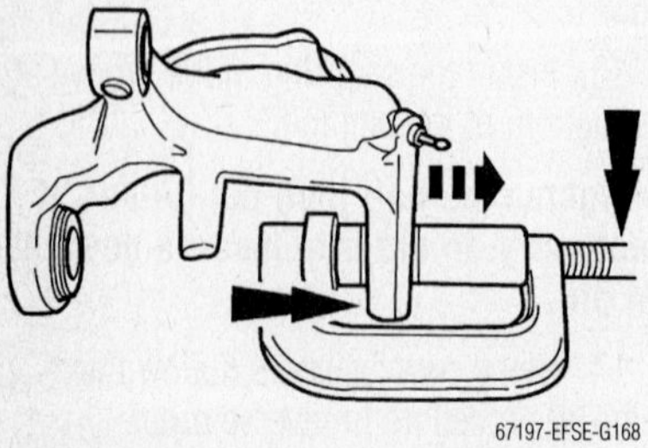

Removing the upper ball joint—2001–04 F-250 and 350

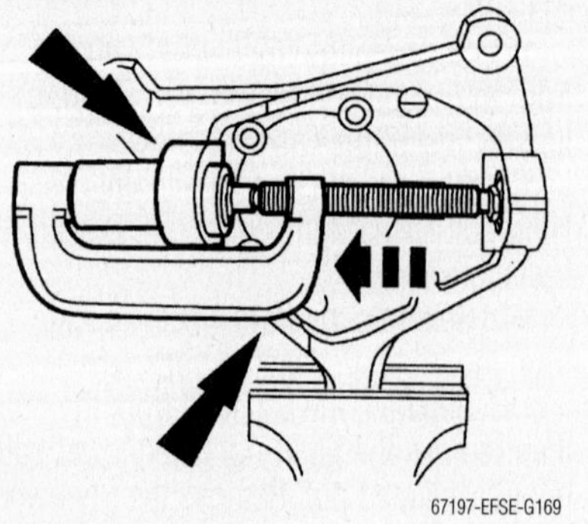

Installing the upper ball joint—2001–04 F-250 and 350

15. Using a suitable ball joint press, remove the upper ball joint.

To install:

16. Clean the wheel knuckle ball joint bores.
17. Using a suitable ball joint press, install the upper ball joint.
18. Install the lower ball joint.
19. Install the wheel knuckle.
20. Position the wheel knuckle onto the axle housing.
21. Install the nut onto the lower ball joint. Do not tighten the nut at this time.
22. Install the insert and the castellated nut onto the upper ball joint. Do not tighten the nut at this time.
23. Tighten the lower ball joint retaining nut. Pre-tighten the nut to 47 Nm (35 ft. lbs.).

➡Do not loosen the castellated nut to install the cotter pin.

24. Tighten the upper ball joint castellated nut. Torque to 69 ft. lbs. (94 Nm).
25. Install the cotter pin. If necessary, tighten the castellated nut until the cotter pin can be installed.
26. Tighten the lower ball joint nut to 204 Nm (150 ft. lbs.).
27. Position the tie-rod end into the wheel knuckle.

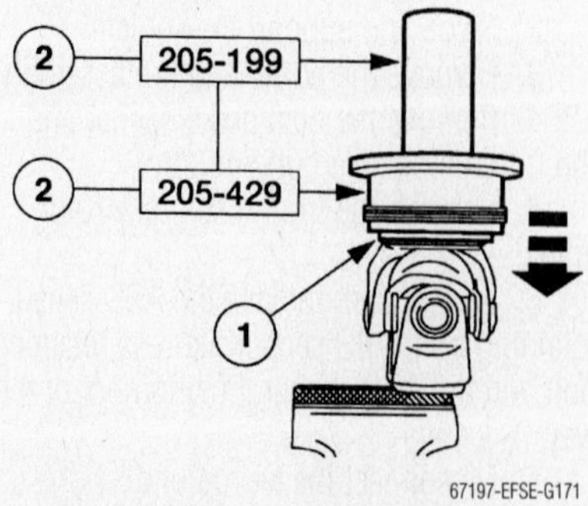

Seat the main seal onto the axle shaft—2001–04 F-250 and 350

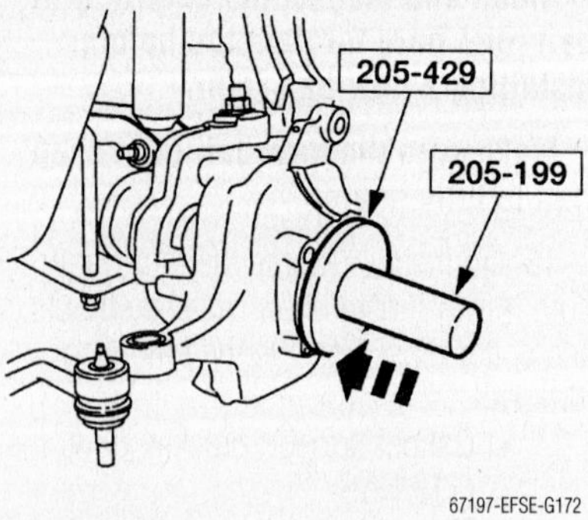

Install the main seal into the wheel knuckle—2001–04 F-250 and 350

28. Install and tighten the castellated nut. Torque to 52 ft. lbs. (70 Nm).
29. Install the cotter pin.
30. Position the main seal onto the axle shaft.
31. Using the special tools and a hammer, seat the main seal onto the axle shaft.
32. Position the axle shaft into the axle housing.
33. Using the special tools and a hammer, install the main seal into the wheel knuckle.
34. Install the wheel hub and bearing.
35. Install the front brake disc.
36. Install the wheel and tire assembly.

LOWER

1. Raise and support the vehicle.
2. Remove the wheel and tire assembly.
3. Remove the front brake disc.
4. Remove the wheel hub and bearing.
5. Using a drift, drive the axle shaft main seal out of the wheel knuckle.
6. Remove the axle shaft and main seal.
7. Remove the tie-rod end castellated nut.
8. Disconnect the tie-rod end from the wheel knuckle.
9. Remove the upper ball joint castellated nut and the insert.
10. Remove the lower ball joint nut.

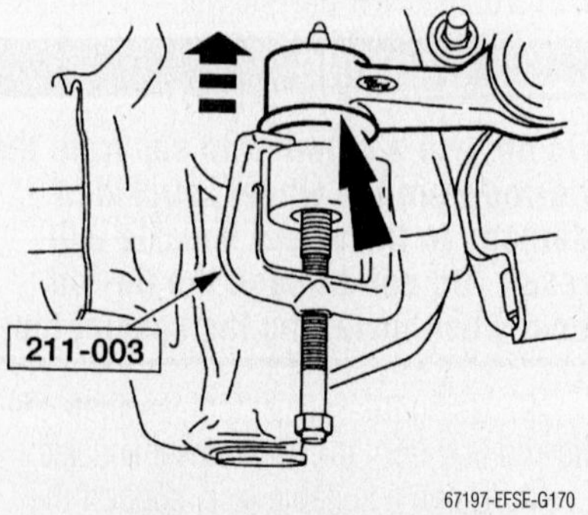

Disconnect the tie-rod end from the wheel knuckle—2001–04 F-250 and 350

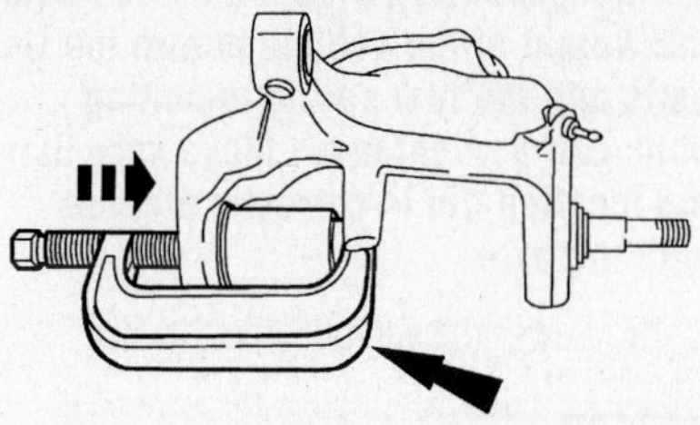

Lower ball joint removal—2001–04 F-250 and 350

11. Remove the knuckle.
12. Clean and inspect the wheel knuckle ball joint bores.
13. Place the wheel knuckle into a suitable vise.
14. Remove the lower ball joint.

To install:

15. Clean the wheel knuckle ball joint bores.
16. Install the lower ball joint.
17. Install the wheel knuckle.
18. Position the wheel knuckle onto the axle housing.
19. Install the nut onto the lower ball joint. Do not tighten the nut at this time.
20. Install the insert and the castellated nut onto the upper ball joint. Do not tighten the nut at this time.
21. Tighten the lower ball joint retaining nut. Pre-tighten the nut to 47 Nm (35 ft. lbs.).

➡Do not loosen the castellated nut to install the cotter pin.

22. Tighten the upper ball joint castellated nut. Torque to 69 ft. lbs. (94 Nm).
23. Install the cotter pin. If necessary, tighten the castellated nut until the cotter pin can be installed.
24. Tighten the lower ball joint nut to 204 Nm (150 ft. lbs.).
25. Position the tie-rod end into the wheel knuckle.
26. Install and tighten the castellated nut. Torque to 52 ft. lbs. (70 Nm).
27. Install the cotter pin.

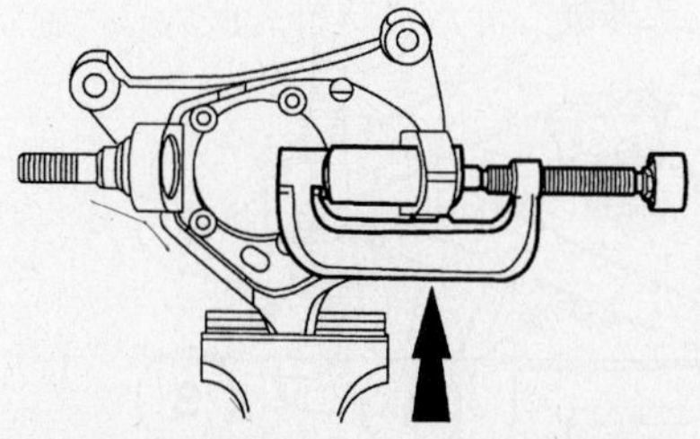

Lower ball joint installation—2001–03 F-250 and 350

28. Position the main seal onto the axle shaft.
29. Using the special tools and a hammer, seat the main seal onto the axle shaft.
30. Position the axle shaft into the axle housing.
31. Using the special tools and a hammer, install the main seal into the wheel knuckle.
32. Install the wheel hub and bearing.
33. Install the front brake disc.
34. Install the wheel and tire assembly.

2001 E-Series

UPPER AND LOWER

1. Before servicing the vehicle, refer to the precautions in the beginning of this section.
2. Raise and support the vehicle.
3. Remove the wheel and tire assembly.
4. Remove the brake disc shield.
5. Disconnect the tie rod end using the Pitman Arm Puller.
6. Remove the cotter pin and the lower ball joint nut. Discard the cotter pin.
7. Remove the pinch bolt and the camber adjuster.

✲✲ WARNING

To prevent damage to the ball joint seal and the ball joint socket, do not use a pickle fork-type remover to loosen the ball joints.

8. Remove the front wheel spindle.
   a. Strike the lower end of the front axle to loosen the ball joint.
   b. Remove the front wheel spindle.
9. Position the front wheel spindle in a vise, and remove the snapring from the lower ball joint.

✲✲ WARNING

To avoid damage to the components, do not use heat to aid ball joint removal.

10. Using the special tool and suitable receiver cup, remove the lower ball joint from the front wheel spindle.
11. Using the special tool and suitable receiver cup, remove the upper ball joint.

To install:

✲✲ WARNING

To avoid damage to components, do not use heat to aid installation.

➡Clean the wheel knuckle ball joint bores.

➡The lower ball joint must be installed first.

12. Using the special tool with suitable receiver cups, install the lower ball joint.
13. Using the special tool with suitable receiver cups, install the upper ball joint.
14. Install the snapring in the groove at the bottom of the ball joint.

➡Tighten the ball joint nut further, if necessary, in order to insert the new cotter pin.

15. Position the front wheel spindle in the front axle.
16. Install the ball joint nut. Torque to 110–150 ft. lbs. (149–203 Nm).
17. Install a new cotter pin.
18. Install the camber adjuster and pinch bolt. Torque to 65–85 ft. lbs. (87–119 Nm).
19. Install the brake disc shield, brake disc and hub and the disc brake caliper.
20. Using a new cotter pin, install the tie rod end in the front wheel spindle. Torque to 57–75 ft. lbs. (77–103 Nm).
21. Install the tire and wheel assembly.
22. Check the toe setting.

2002–04 E-Series

UPPER AND LOWER

1. Before servicing the vehicle, refer to the precautions in the beginning of this section.
2. Remove the brake disc.
3. Remove the wheel speed sensor.
4. Remove and discard the cotter pin and nut.
5. Using the special tool, disconnect the tie-rod end.
6. Remove and discard the cotter pin and the nut.
7. Remove the pinch bolt adjuster.

✲✲ WARNING

To prevent damage to the ball joint seal and the ball joint socket, do not use a pickle fork-type remover to loosen the ball joints.

8. Remove the front wheel spindle.
   a. Strike the lower end of the front axle to loosen the ball joint.
   b. Remove the front wheel spindle.
9. Position the front wheel spindle in a vise, and remove the snapring from the lower ball joint.

✲✲ WARNING

To avoid damage to the components, do not use heat to aid ball joint removal.

10. Using the special tool and suitable receiver cup, remove the lower ball joint from the front wheel spindle.

11. Using the special tool and suitable receiver cup, remove the upper ball joint.

To install:

❊❊ WARNING

To avoid damage to components, do not use heat to aid installation.

➡Clean the wheel knuckle ball joint bores.

➡The lower ball joint must be installed first.

12. Using the special tool with suitable receiver cups, install the lower ball joint.

13. Using the special tool with suitable receiver cups, install the upper ball joint.

14. Install the snapring in the groove at the bottom of the ball joint.

15. To install, reverse the removal procedure.

16. Observe the following torques:

- Pinch bolt: 77 ft. lbs. (104 Nm)
- Lower ball stud nut: 130 ft. lbs. (176 Nm)
- Toe rod end nut: 66 ft. lbs. (90 Nm)

17. Check and, if necessary, align the front end.

2004 F-150, Exc. Heritage Editions

The upper and lower ball joints are serviced as part of the control arms and can't be replaced separately.

Front Leaf Spring

REMOVAL & INSTALLATION

2001–04 F-250 and 350 4-Wheel Drive

1. Remove the wheel and tire assembly.

2. Position jack stands under the axle housing.

3. Remove the left shock absorber.

❊❊ CAUTION

Never reuse U-bolts. The U-bolts are a torque-to-yield design and cannot be retightened. Failure to use a new U-bolt can result in loose or broken springs and suspension components.

4. Remove the two U-bolts and the front spring spacer. Discard the U-bolts.

5. Disconnect the front driveshaft.

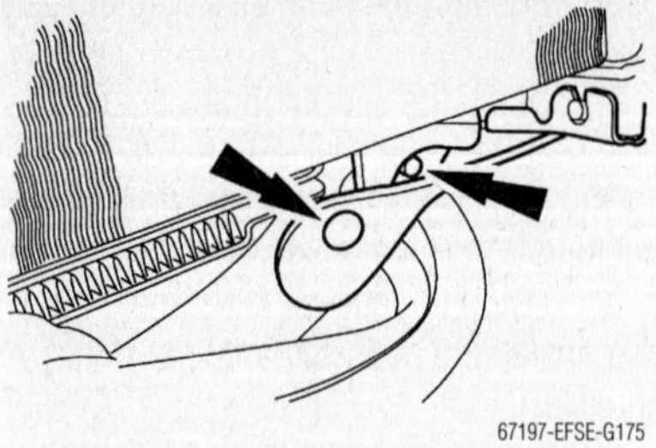

Detach the pushpin and remove the bolt and condenser bracket

6. Remove the nut and bolt. Position the spring shackle away from the spring.

7. Detach the pushpin and remove the bolt and condenser bracket.

8. Remove the nut and bolt. Using a hammer, tap the spring until the spring is free from the hanger bracket.

9. Remove the spring from the vehicle.

To install:

10. Position the spring in the vehicle.

11. Position the spring into the hanger.

12. Install the bolt and nut. Do not tighten the nut and bolt at this time.

13. Install the bracket and bolt and the pushpin. Torque to 80 inch lbs. (9 Nm).

14. Position the shackle.

15. Install the bolt and nut. Do not tighten the nut and bolt at this time.

➡The U-bolts insert through the front spring cap on the left-hand side and through the axle assembly bracket on the right-hand side.

16. Position the front spring spacer and the new U-bolts.

17. Install the four nuts. Do not tighten the nuts at this time.

18. Install the shock, bolt and nut onto the spring cap. Torque to 76 ft. lbs. (103 Nm).

➡The suspension must be loaded with the weight of the vehicle before the U-bolts and the leaf spring mounting bolts can be tightened. Make sure that the locating pin is correctly aligned with the axle.

19. Lower the vehicle onto the jack stands until the front suspension is supporting the weight of the vehicle.

➡The U-bolts must be tightened in sequence.

20. Tighten the U-bolts following the sequence shown. Torque to 99 ft. lbs. (133 Nm).

21. Tighten the leaf spring rear retaining bolt and nut to 185 ft. lbs. (250 Nm).

22. Tighten the leaf spring front retaining bolt and nut to 203 ft. lbs. (275 Nm).

23. Raise the vehicle and remove the jack stands.

24. Position the driveshaft onto the pinion flange.

25. Install the two straps and the four bolts. Torque to 76 ft. lbs. (103 Nm).

26. Install the wheel and tire assembly.

Rear Leaf Spring

REMOVAL & INSTALLATION

2001 E-Series

1. Before servicing the vehicle, refer to the precautions in the beginning of this section.

2. Remove the wheel and tire assembly.

3. Support the rear axle with a suitable jack.

4. Disconnect the lower end of the rear shock absorber.

5. Remove the nuts. Remove the U-bolts. Remove the rear spring plate.

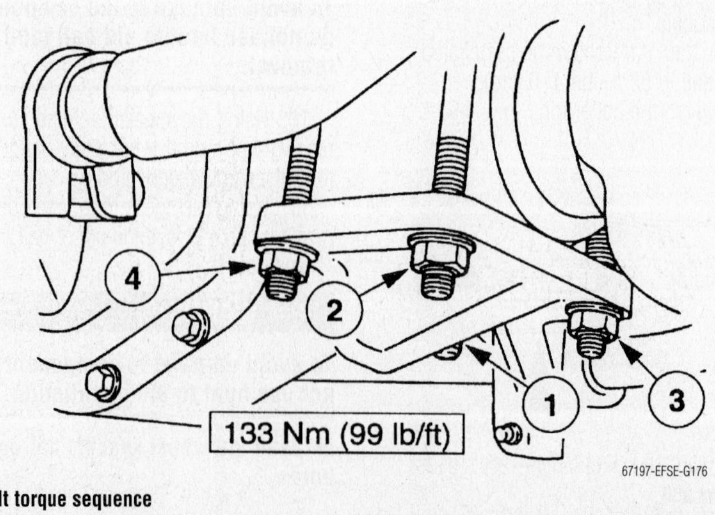

U-bolt torque sequence

6. Remove the front shackle bolt and nut.
7. Remove the rear shackle bolt and nut.
8. Remove the rear spring.
9. Installation is the reverse of removal. Observe the following torques:

- E-150 U-bolt nuts: 73–97 ft. lbs. (98–132 Nm)
- E-250/350/Super Duty U-bolt nuts: 110–150 ft. lbs. (149–203 Nm)
- E-150 front shackle bolt/nut: 111–136 ft. lbs. (150–189 Nm)
- E-250/350/Super Duty front shackle bolt/nut: 240–289 ft. lbs. (325–393 Nm)
- Rear shackle bolt/nut, all: 57–76 ft. lbs. (77–103 Nm)
- Shock absorber lower end: 52 ft. lbs. (70 Nm)

2002–03 E-Series

1. Before servicing the vehicle, refer to the precautions in the beginning of this section.
2. Remove the wheel and tire assembly.
3. Support the rear axle with a suitable jack.
4. Disconnect the lower end of the rear shock absorber.
5. Remove the nuts. Remove the U-bolts. Remove the rear spring plate.
6. Remove the front shackle bolt and nut.
7. Remove the rear shackle bolt and nut.
8. Remove the rear spring.
9. Installation is the reverse of removal. Observe the following torques:

- E-150 U-bolt nuts: 85 ft. lbs. (115 Nm)
- E-250/350/Super Duty U-bolt nuts: 130 ft. lbs. (176 Nm)
- E-150 front shackle bolt/nut: 124 ft. lbs. (168 Nm)
- E-250/350/Super Duty front shackle bolt/nut: 265 ft. lbs. (359 Nm)
- Rear shackle bolt/nut, all: 66 ft. lbs. (90 Nm)
- Shock absorber lower end: 52 ft. lbs. (70 Nm)

2004 E-Series

1. Before servicing the vehicle, refer to the precautions in the beginning of this section.
2. Remove the wheel and tire assembly.
3. Support the rear axle with a suitable jack.
4. Disconnect the lower end of the rear shock absorber.
5. Remove the nuts. Remove the U-bolts. Remove the rear spring plate.
6. Remove the front shackle bolt and nut.
7. Remove the rear shackle bolt and nut.
8. Remove the rear spring.
9. Installation is the reverse of removal. Observe the following torques:

- E-150 U-bolt nuts: 85 ft. lbs. (115 Nm)
- E-250/350/Super Duty U-bolt nuts: 130 ft. lbs. (176 Nm)
- E-150 front shackle bolt/nut: 148 ft. lbs. (200 Nm)
- E-250/350/Super Duty front shackle bolt/nut: 295 ft. lbs. (400 Nm)
- Rear shackle bolt/nut, all: 85 ft. lbs. (115 Nm)
- Shock absorber lower end: 52 ft. lbs. (70 Nm)

2001–03 F-150 and 2004 Heritage Editions, exc. Blackwood

✲✲ WARNING

Do not use heat to loosen a seized wheel nut. Heat can damage the wheel and the wheel bearings.

1. Before servicing the vehicle, refer to the precautions in the beginning of this section.
2. Loosen the rear wheel nuts.
3. Remove the rear wheel center caps.
4. With the weight of the vehicle on the wheels, loosen but do not remove the wheel nuts.
5. Raise and support the vehicle.
6. Remove the wheel and tire assembly.
7. Use a jack to support the rear axle.

✲✲ WARNING

Lower the rear axle only enough to gain access to the rear spring.

8. Remove the nuts. Remove the U-bolts. Remove the rear spring plates.
9. Carefully lower the rear axle. On 4x4 vehicles, remove the rear spring spacer.
10. Remove the front and rear shackle bolts and nuts. Remove the rear spring.

To install:

11. Position the rear spring. Install the shackle nuts and bolts. Torque the front shackle bolt to 157–212 ft. lbs. (213–288 Nm); the rear shackle bolt to 73–97 ft. lbs. (98–132 Nm).
12. On 4-wheel drive vehicles, position the rear spring spacer.
13. Position the U-bolts.
14. Install the rear spring plate.
15. Install the nuts. Torque to 73–97 ft. lbs. (98–132 Nm).
16. Remove the jack.
17. Install the wheel and tire assembly.

Blackwood

✲✲ CAUTION

Do not remove an air spring or solenoid valve when there is pressure in the system. Do not remove any components supporting an air spring without either venting the air or providing support for the air spring to prevent vehicle damage or personal injury.

✲✲ CAUTION

The electrical power to the air suspension system must be shut off prior to hoisting or jacking an air suspension vehicle. This can be accomplished by turning off the air suspension switch located below the glove box in the lower right of the passenger foot well. Failure to do so can result in unexpected inflation or deflation of the air springs, which can result in shifting of the vehicle during these operations.

1. Using a scan tool, vent the rear air springs. For additional information, see the Air Spring procedure in this chapter.

✲✲ CAUTION

The venting procedure must be followed before proceeding!

2. Turn the air suspension switch to the OFF position.
3. Raise and support the rear of the vehicle.
4. Support the suspension with a floor jack and lower the axle to relieve tension from the springs.
5. Remove the wheel and tire assemblies.
6. Remove the U-bolt nuts.
7. Remove the U-bolt plate.
8. Remove the U-bolts.
9. Remove the shock absorber lower nut.
10. Disconnect the height sensor linkage.

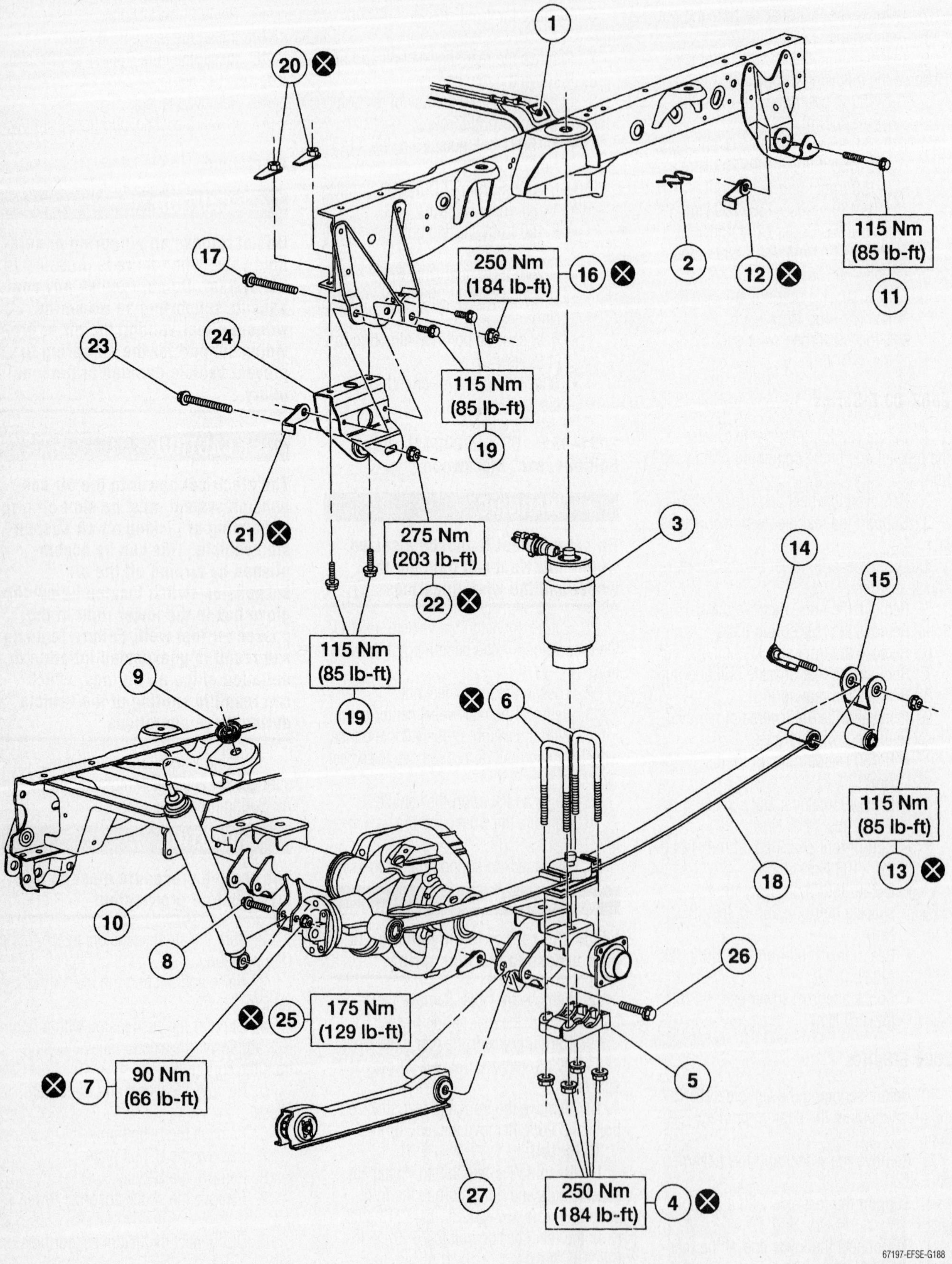

Blackwood rear suspension with air springs

1 Air spring supply line
2 Air spring retainer clip
3 Air spring
4 U-bolt nuts (4)
5 U-bolt plate
6 U-bolts (6)
7 Shock absorber lower nut
8 Shock absorber lower bolt
9 Shock absorber upper nut and insulator
10 Shock abosorber
11 Shackle-to-frame bracket bolt
12 Shackle-to-frame bracket flag nut
13 Leaf spring-to-shackle nut
14 Leaf spring-to-shackle flag bolt
15 Shackle
16 Leaf spring-to-frame bracket nut
17 Leaf spring-to-frame bracket bolt
18 Leaf spring
19 Control arm bracket bolts (4)
20 Control arm bracket flag nuts (2)
21 Control arm bracket flag nuts (2)
22 Control arm-to-bracket nut
23 Control arm-to-bracket bolt
24 Control arm bracket
25 Control arm-to-axle nut
26 Control arm-to-axle bolt
27 Control arm

11. Remove the shock absorber lower bolt.
12. Remove the lower the axle until it is no longer making contact with the leaf spring.
13. Remove the leaf spring-to-shackle flag bolt.
14. Remove the shackle.
15. Remove the leaf spring-to-frame bracket nut.
16. Remove the leaf spring-to-frame bracket.
17. Remove the leaf spring bolt.
18. To install, reverse the removal procedure. Tighten the leaf spring and shackle nuts and bolts with the vehicle at curb height. Tighten the U-bolt nuts in a criss-cross pattern.
19. Observe the following torques:
- U-bolt nuts: 184 ft. lbs. (250 Nm)
- Shock absorber lower nut: 66 ft. lbs. (90 Nm)
- Leaf spring-to-shackle flag bolt: 88 ft. lbs. (115 Nm)
- Leaf spring-to-frame bracket nut: 184 ft. lbs. (250 Nm)

2004 F-150, Exc. Heritage Editions

1. Before servicing the vehicle, refer to the precautions in the beginning of this section.
2. Remove the wheel and tire assembly.

⁂ WARNING

Lower the rear axle only enough to gain access to the rear spring.

3. Use the jack to support and lower the rear axle housing assembly.
4. Remove the brake caliper and bolts

⁂ WARNING

Do not allow the brake caliper to hang from the brake hose.

5. Position the brake caliper aside using mechanic's wire.
6. Disconnect the shock absorber at the lower end.
7. Remove the spring front eye bolt and nut.

➡Lower the fuel tank to gain access to the spring shackle bolt.

8. Remove the spring shackle bolt and nut (rear)
9. Remove the U-bolt nuts
10. Remove the U-bolt plate
11. Remove the U-bolts
12. Remove the rear spring assembly
13. To install, reverse the removal procedure.

➡Tighten the U-bolt nuts in a cross pattern in 3 even steps.

14. Observe the following torques:
- U-bolt nuts. Light duty: 85 ft. lbs. (115 Nm)
- U-bolt nuts. Heavy duty: 184 ft. lbs. (250 Nm)
- Spring front eye bolt: 222 ft. lbs. (300 Nm)
- Spring rear shackle bolt: 98 ft. lbs. (133 Nm)
- Shock absorber nut: (66 ft. lbs. (90 Nm)

2001–04 F-250, 350, Super Duty

⁂ CAUTION

Suspension fasteners are critical parts because they affect performance of vital components and systems and their failure can result in major service expense. Install new parts with the same part number or an equivalent part if installation is necessary. Do not use an installation part of lesser quality or substitute design. Torque values must be used as specified during reassembly to make sure of correct retention of these parts.

1. Before servicing the vehicle, refer to the precautions in the beginning of this section.
2. Raise and support the vehicle.
3. Remove the wheel and tire assembly.
4. Support the rear axle with a suitable jack.
5. Remove the U-bolt retaining nuts and remove the U-bolts.
6. Remove the rear spring upper plate.
7. Remove the nut and bolt from the rear spring front hanger bracket.
8. Remove the lower nut and bolt from the rear spring shackle bracket. Remove the rear spring assembly.
9. If equipped, remove the auxiliary spring and auxiliary spring spacer.
10. Using new fasteners, follow the removal procedure in reverse order. Observe the following torques:
- U-bolt nuts: 185 ft. lbs. (250 Nm)
- Front and rear spring hanger/shackle bolts: 185 ft. lbs. (250 Nm)

Air Spring

REMOVAL & INSTALLATION

Blackwood

⁂ CAUTION

Do not remove an air spring or solenoid valve when there is pressure in the system. Do not remove any components supporting an air spring without either venting the air or providing support for the air spring to prevent vehicle damage or personal injury.

⁂ CAUTION

The electrical power to the air suspension system must be shut off prior to hoisting or jacking an air suspension vehicle. This can be accomplished by turning off the air suspension switch located below the glove box in the lower right of the passenger foot well. Failure to do so can result in unexpected inflation or deflation of the air springs, which can result in shifting of the vehicle during these operations.

⁂ CAUTION

Disconnecting an air line that is connected to the air compressor can cause personal injury or damage to components as high pressure air is vented uncontrolled.

DEPRESSURIZING THE SYSTEM

Using the scan tool, Worldwide Diagnostic System (WDS) 418-F224, New Generation STAR (NGS) Tester 418-F052 or equivalent scan tool, vent the rear air springs as follows:

1. Turn the air suspension switch on.
2. Connect the scan tool to the data link connector (DLC) and follow the directions.
3. Select Air Suspension Module under Active Command Mode:
   - VENT REAR to deflate the rear down.
   - LIFT REAR to inflate the rear up.

Calibration

Initialize System—Clear B2140 DTC

4. Turn the ignition key OFF and turn it back to RUN; exit the vehicle, close all doors and allow the system to vent the vehicle down to kneel height (approximately 30 seconds).
5. Connect the scan tool to the data link connector (DLC) and follow the directions.
6. Select Air Suspension Module.

➡Do not save rear ride heights. The air suspension module has pre-calibrated values already stored. Refer to Rear Ride Height Mechanical Resetting.

7. Select the "Save Calibration Values (Store Ride Height)" scan tool command to calibrate the air suspension module. Trigger through the warning message(s) and save "initialization" (trigger from OFF to ON).

REMOVAL & INSTALLATION

1. Turn the air suspension switch to the OFF position.
2. Remove the wheel and tire assemblies.
3. Compress the quick connect lock ring while pulling outward on the air supply line. Remove the air spring supply line
4. Remove the air spring retainer clip
5. Remove the air spring
6. To install, reverse the removal procedure.

Ride Height Sensor

REMOVAL & INSTALLATION

Blackwood

⁂ CAUTION

The electrical power to the air suspension system must be shut off prior to hoisting, jacking or towing an air suspension vehicle. This can be accomplished by turning off the air suspension switch located in the RH kick panel area. Failure to do so may result in unexpected inflation or deflation of the air springs which may result in shifting of the vehicle during these operations.

1. Raise and support the vehicle.
2. Disconnect the wiring.
3. Disconnect the sensor arm.

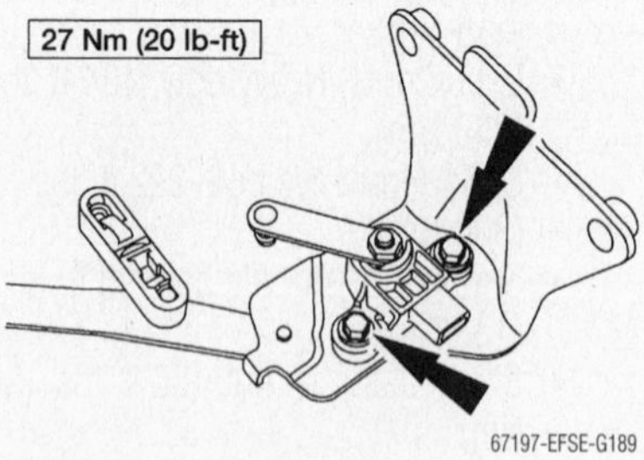

Height sensor installation

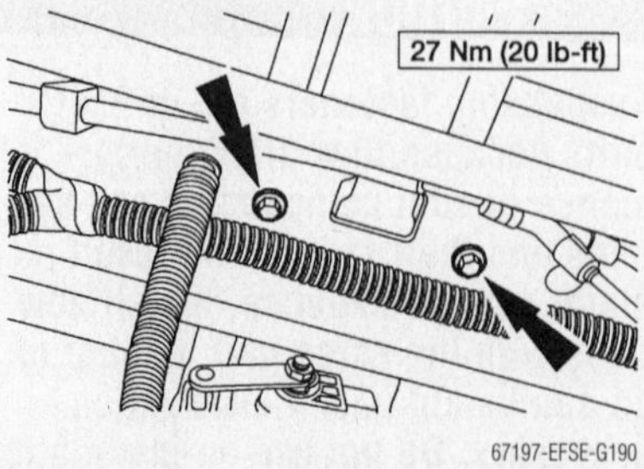

Height sensor mounting bracket installation

4. Remove the bolts and the mounting bracket.
5. Disconnect the sensor arm.
6. Remove the bolts and the sensor.
7. To install, reverse the removal procedure. Torque the bolts to 20 ft. lbs. (27 Nm).

Spindle

REMOVAL & INSTALLATION

E-Series

1. Remove the brake disc.
2. Remove the wheel speed sensor.
3. Remove and discard the cotter pin and nut.
4. Disconnect the tie-rod end.
5. Remove and discard the cotter pin and the nut.
6. Remove the pinch bolt adjuster.

⁂ WARNING

To prevent damage to the ball joint seal and the ball joint socket, do not use a pickle fork-type remover to loosen the ball joints.

7. Strike the lower end of the front axle to loosen the ball joint.
8. Remove the front wheel spindle.
9. To install, reverse the removal procedure. Observe the following torques:
   - Tie rod ball stud nut: 66 ft. lbs. (90 Nm)
   - Lower ball joint nut: 130 ft. lbs. (176 Nm)
   - Pinch bolt adjuster bolt: 77 ft. lbs. (104 Nm)
10. Check and, if necessary, align the front end.

2001–03 F-150 2-Wheel Drive and 2004 Heritage Editions

1. Raise the vehicle on a hoist.
2. Remove the wheel and tire assembly.
3. Remove the brake disc shield.
4. Use a suitable jack to support the front suspension lower arm.

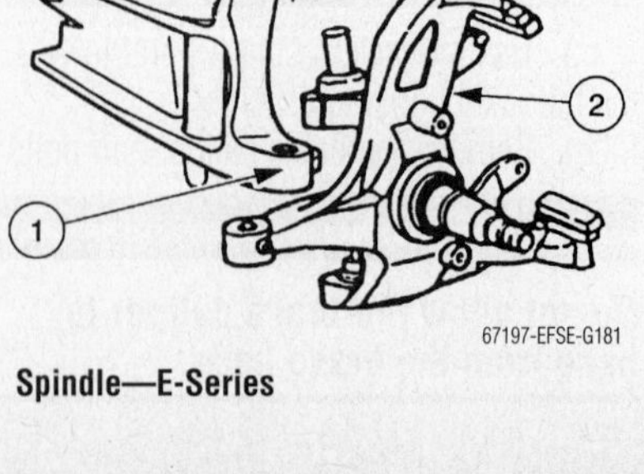

Spindle—E-Series

5. Remove the upper ball joint castellated nut.

6. Use a pitman arm puller to separate the ball joint from the front wheel spindle.

➡Do not allow the shaft to turn while removing the nut. Hold the flats using a suitable wrench or damage to the shock can result.

7. Remove the upper shock absorber nut.

8. Remove two shock absorber lower nuts.

9. Remove the front shock absorber.

10. Remove the tie rod castellated nut.

11. Use a pitman arm puller to separate the tie rod end from the front wheel spindle.

12. Use the coil spring compressor to compress the coil spring.

13. Remove lower ball joint castellated nut.

14. Use a pitman arm puller to separate the lower ball joint from the front wheel spindle.

15. Remove the front wheel spindle.

To install:

16. Install the lower ball joint castellated nut. Torque to 83–112 ft. lbs. (113–153 Nm). Install a new cotter pin.

17. Install the upper ball joint castellated nut. Torque to 57–76 ft. lbs. (77–103 Nm). Install a new cotter pin.

18. Remove the coil spring compressor.

19. Install the tie rod castellated nut. Torque to 57–76 ft. lbs. (77–103 Nm).Install a new cotter pin.

20. Install the front shock absorber and the two shock absorber lower nuts. Torque to 22–29 ft. lbs. (30–40 Nm).

21. Install the upper shock absorber nut. Torque to 35–46 ft. lbs. (47–63 Nm).

22. Install the brake disc shield.

23. Install the tire and wheel assembly

24. Inspect and adjust the front end alignment.

2001–04 F-250 and 350 2-Wheel Drive

1. Raise and support the vehicle.

2. Remove the wheel and tire assembly.

3. Remove the disc brake caliper and the front disc brake hub and rotor.

4. Remove the front disc brake rotor shield.

5. If equipped, remove the ABS sensor retaining bolt, ABS sensor harness retaining bolt and the ABS sensor. Position out of the way.

6. Remove and discard the cotter pin. Remove the castellated nut. Using a pitman arm puller, remove the tie rod end.

7. Remove the pinch bolt.

8. Remove the camber adjuster.

⁂ WARNING

To prevent damage to the ball joint seal and the ball joint socket, do not use a pickle fork-type remover to loosen the ball joints.

9. Remove and discard the cotter pin. Loosen, but do not remove, the castellated nut. Strike the lower end of the front axle to loosen the ball joint.

10. Remove the castellated nut and the front wheel spindle.

➡Tighten the ball joint nut further, if necessary, in order to insert a new cotter pin.

11. Using new fasteners, follow the removal procedure in reverse order. Observe the following torques:

- Lower ball joint nut: 99 ft. lbs. (133 Nm)
- Pinch bolt: 60 ft. lbs. (80 Nm)
- Tie rod ball stud nut: 67 ft. lbs. (90 Nm)

12. Check the front end alignment.

Knuckle

REMOVAL & INSTALLATION

2001–03 F-150 4-Wheel Drive and 2004 Heritage Edition

1. Remove the wheel hub.

2. Remove the retaining bolts. Remove the torsion bar cover plate.

3. Make an alignment mark on the torsion bar and the torsion bar crossmember support. Measure and record the length.

4. Position the torsion bar tool and the adapters.

5. Tighten the torsion bar tool until the torsion bar adjuster lifts off of the adjustment bolt.

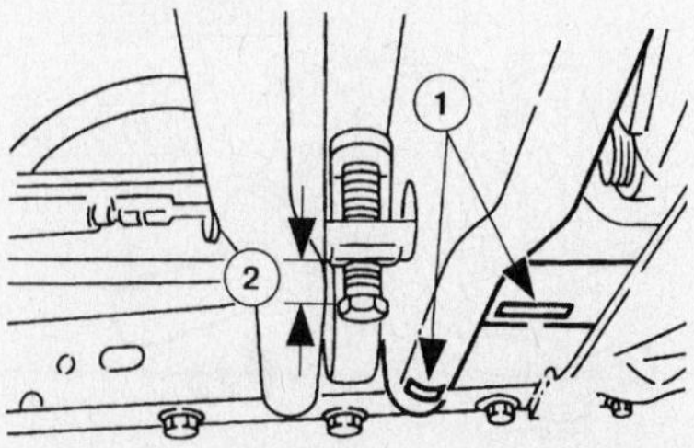

67197-EFSE-G182

Make an alignment mark on the torsion bar and the torsion bar crossmember support

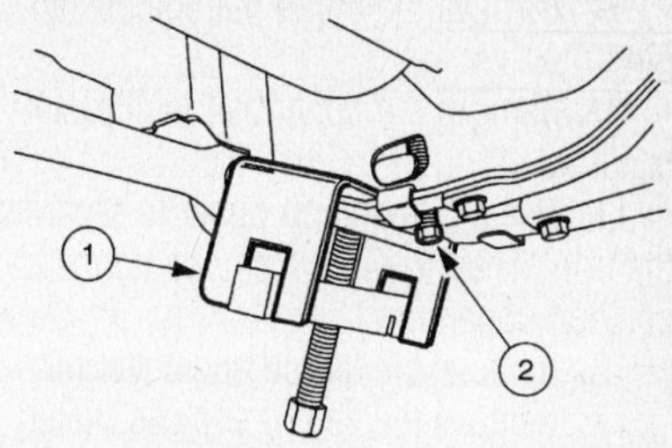

67197-EFSE-G183

Tighten the torsion bar tool until the torsion bar adjuster lifts off of the adjustment bolt

⁂ CAUTION

The torsion bar adjustment bolt is coated with dry Loctite® and must be replaced if it is backed off or removed. Failure to do so can cause the adjustment bolt to loosen up during operation and a loss of vehicle alignment.

6. Remove the torsion bar adjustment bolt and nut.

7. Loosen the torsion bar adjustment tool until the tension is off the torsion bar.

8. If equipped with 4 wheel anti-lock brake system, safely reposition the anti-lock sensor wire.

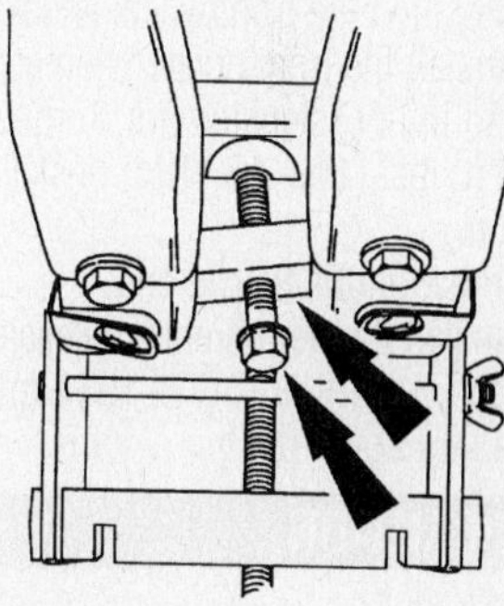

67197-EFSE-G184

Remove the torsion bar adjustment bolt and nut

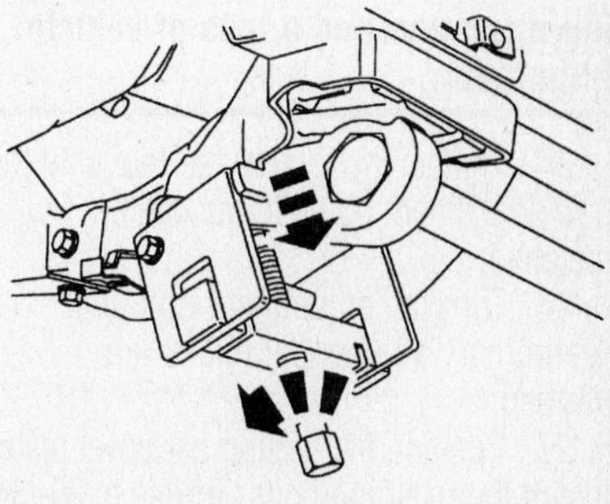

67197-EFSE-G185

Loosen the torsion bar adjustment tool until the tension is off the torsion bar

9. Remove the upper ball joint cotter pin.
10. Remove the upper ball joint castellated nut.
11. Use a pitman arm puller to separate the front wheel knuckle from the front suspension upper arm.
12. Remove the tie rod end cotter pin.
13. Remove the tie rod end castellated nut.
14. Use a pitman arm puller to separate the tie rod end. Separate the tie rod end from the front wheel knuckle.
15. Remove the lower ball joint cotter pin.
16. Remove the lower ball joint castellated nut.
17. Use a pitman arm puller to separate the front wheel knuckle from the front suspension lower arm.

**** WARNING**

Suspend the front wheel driveshaft and joint with wire.

18. Remove the front wheel knuckle.

To install:

19. Position the front wheel knuckle.
20. Install the front suspension arm upper ball joint castellated nut. Torque to 57–75 ft. lbs. (77–103 Nm). Install a new cotter pin.
21. Position the front wheel knuckle.
22. Install the front suspension arm lower ball joint castellated nut. Torque to 83–112 ft. lbs. (113–153 Nm). Install a new cotter pin.
23. Position the tie rod end.
24. Install the tie rod end castellated nut. Torque to 57–75 ft. lbs. (77–103 Nm). Install a new cotter pin.

**** CAUTION**

The torsion bar adjustment bolt is coated with dry Loctite® and must be replaced if it is backed off or removed. Failure to do so can cause the adjustment bolt to loosen up during operation and a loss of vehicle alignment.

25. Tighten the torsion bar tool until the new adjustment bolt and nut can be installed.
26. Turn the adjustment bolt until preliminary adjustment marks are aligned.
27. Position the torsion bar cover plate. Install the retaining bolts. Torque to 25–34 ft. lbs. (34–46 Nm).
28. Install the wheel hub.
29. If equipped with 4WABS, install the anti-lock sensor wire bracket bolt.
30. Adjust the ride height.

2001–04 F-250 and 350 4-Wheel Drive

1. Raise and support the vehicle.
2. Remove the wheel and tire assembly.
3. Remove the front brake disc.
4. Remove the wheel hub and bearing.
5. Using a drift, drive the axle shaft main seal out of the wheel knuckle.
6. Remove the axle shaft and main seal.
7. Remove the tie-rod end castellated nut.
8. Using the special tool, disconnect the tie-rod end from the wheel knuckle.
9. Remove the upper ball joint castellated nut and the insert.
10. Remove the lower ball joint nut.
11. Remove the knuckle.
12. Clean and inspect the wheel knuckle ball joint bores.

To install:

13. Position the wheel knuckle onto the axle housing.
14. Install the nut onto the lower ball joint. Do not tighten the nut at this time.
15. Install the insert and the castellated nut onto the upper ball joint. Do not tighten the nut at this time.
16. Tighten the lower ball joint retaining nut. Pre-tighten the nut to 47 Nm (35 ft. lbs.).

➡Do not loosen the castellated nut to install the cotter pin.

17. Tighten the upper ball joint castellated nut. Torque to 69 ft. lbs. (94 Nm).
18. Install the cotter pin. If necessary, tighten the castellated nut until the cotter pin can be installed.
19. Tighten the lower ball joint nut to 204 Nm (150 ft. lbs.).
20. Position the tie-rod end into the wheel knuckle.
21. Install and tighten the castellated nut. Torque to 52 ft. lbs. (70 Nm).
22. Install the cotter pin.
23. Position the main seal onto the axle shaft.
24. Using the special tools and a hammer, seat the main seal onto the axle shaft.
25. Position the axle shaft into the axle housing.
26. Using the special tools and a hammer, install the main seal into the wheel knuckle.
27. Install the wheel hub and bearing.
28. Install the front brake disc.
29. Install the wheel and tire assembly.

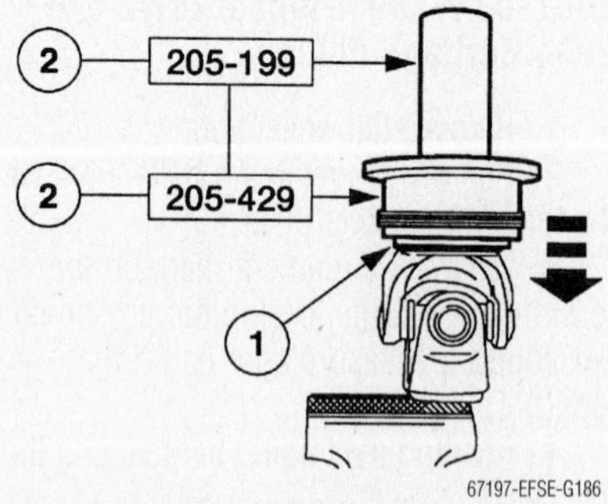

Seat the main seal onto the axle shaft

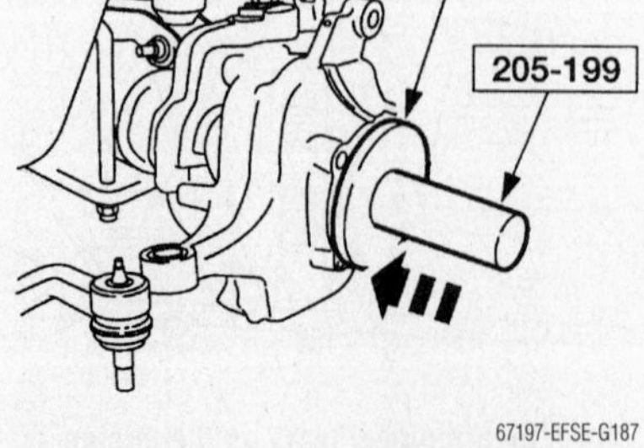

Install the main seal into the wheel knuckle

2004 F-150 2-Wheel Drive, exc. Heritage Editions

1. Disconnect the wheel speed sensor.
2. Remove the wheel and tire assembly.

➡Use the hex holding feature to prevent the ball studs (upper control arm, lower control arm tie-rod end, and stabilizer bar links) from turning while removing and installing the nuts.

**** WARNING**

Do not hammer the ball studs to separate them from the wheel knuckle or stabilizer bar. Doing so can cause damage. Lightly tap the wheel knuckle or stabilizer bar to loosen the joints.

3. Remove the anchor plate bolts.
4. Remove the brake caliper, pads and anchor plate.

**** WARNING**

Do not allow the brake caliper to hang by the flexible brake hose.

5. Support the caliper to the vehicle.
6. Remove the axle-to-wheel hub nut, retainer and cotter pin.
7. Remove the brake disc.

8. Remove the tie-rod end-to-wheel knuckle nut.

➡**The upper ball joint, the lower ball joint, and the tie-rod end ball studs do not require special tools to be removed from the wheel knuckle.**

9. Remove the lower ball joint-to-wheel knuckle nut.
10. Remove the upper ball joint-to-wheel knuckle nut.
11. To install, reverse the removal procedure. Observe the following torques:
   - Anchor plate bolts: 148 ft. lbs. (200 Nm)
   - Hub nut: 296 ft. lbs. (400 Nm)
   - Lower ball joint nut: 111 ft. lbs. (148 Nm)
   - Upper ball joint nut: 85 ft. lbs. (115 Nm)
   - Tie rod end stud nut: 111 ft. lbs. (148 Nm)
12. Clean the axle (including the threaded end) of any dirt or grease using the specified solvent before installing the brake disc.

⁂ WARNING

Do not apply lubricant to the threaded part of the axle.

13. To ease installation, lubricate the inner race of the bearing assembly using the specified lubricant.

2004 F-150 4-Wheel Drive, exc. Heritage Edition

1. Loosen the axle retainer nut.

➡**The wheel speed sensor electrical connectors are located in the engine compartment secured to the fender aprons.**

2. Disconnect the wheel speed sensor.
3. Remove the wheel and tire assembly.

⁂ WARNING

Use the hex holding feature to prevent the ball studs (upper control arm, lower control arm tie-rod end, and stabilizer bar links) from turning while removing and installing the nuts.

⁂ WARNING

Do not hammer the ball studs to separate them from the wheel knuckle or stabilizer bar. Doing so can cause damage. Lightly tap the wheel knuckle or stabilizer bar to loosen the joints.

4. Remove the axle-to-wheel hub nut. Separate the outboard CV joint from the wheel hub.
5. Remove the wheel speed sensor harness connector.
6. Remove the anchor plate bolt.
7. Remove the brake caliper, pads and anchor plate.

⁂ WARNING

Do not allow the brake caliper to hang from the hose or damage to the hose can occur.

8. Position the caliper, pads and anchor plate aside.
9. Remove the brake disc.
10. Remove the wheel hub-to-wheel knuckle bolts.
11. Remove the wheel bearing and hub assembly.
12. Remove the tie-rod end-to-wheel knuckle nut.
13. Remove the lower ball joint-to-wheel knuckle nut.
14. Remove the upper ball joint-to-wheel knuckle nut.
15. Remove the knuckle.
16. To install, reverse the removal procedure. Observe the following torques:
   - Axle hub nut: 20 ft. lbs. (27 Nm)
   - Anchor plate bolt: 148 ft. lbs. (200 Nm)
   - Hub-to-knuckle bolts: 148 ft. lbs. (200 Nm)
   - Lower ball joint nut: 111 ft. lbs. (150 Nm)
   - Upper ball joint nut: 85 ft. lbs. (115 Nm)
17. Check and, if necessary, align the front end.

Upper Control Arm

REMOVAL & INSTALLATION

2001–03 F-150 and 2004 Heritage Editions

2-WHEEL DRIVE

1. Before servicing the vehicle, refer to the precautions in the beginning of this section.
2. Raise the vehicle on a hoist.
3. Remove the wheel and tire assembly
4. Use a suitable jack to support the front suspension lower arm.
5. Mark the position of the front suspension upper arm cam.
6. Remove the upper ball joint castellated nut.
7. Use a Pitman Arm Puller to separate the ball joint from the front wheel spindle.
8. Remove the two nuts and the two bolts. Remove the front suspension upper arm.

To install:

9. Align the marks made during removal on the front suspension upper arm cams.

➡**The forward front suspension upper arm nut must be tightened first while the control arms are held at the curb position ride height.**

10. Position the front suspension upper arm. Install the two bolts and nuts. Torque to 84–112 ft. lbs. (113–153 Nm).
11. Install the upper ball joint castellated nut. Torque to 57–76 ft. lbs. (77–103 Nm). Install a new cotter pin.
12. Install the tire and wheel assembly.
13. Inspect and adjust the front end alignment.

4-WHEEL DRIVE

1. Raise and support the vehicle.
2. Remove the wheel and tire assembly.
3. Use a suitable jack to support the lower arm.
4. If equipped with 4-wheel anti-lock brake system (4WABS), safely reposition the anti-lock sensor wire.
5. Remove the upper ball joint cotter pin.
6. Remove the upper ball joint castellated nut.
7. Use a Pitman Arm Puller to separate the front suspension upper arm from the front wheel knuckle.
8. Mark the position of the camber adjustment cam.
9. Remove the nuts. Remove the bolts. Remove the front suspension upper arm.

To install:

➡**Align the marks made during removal on the camber adjustment cam before tightening the nuts.**

10. Position the front suspension upper arm. Install the bolts. Install the nuts. Torque to 83–112 ft. lbs. (113–153 Nm).
11. Position the front wheel knuckle.
12. Install the upper ball joint castellated nut. Torque to 57–75 ft. lbs. (77–103 Nm). Install a new cotter pin.
13. Position the anti-lock sensor wire. Install the bolt. Torque to 60–80 inch lbs. (7–9 Nm).

✱✱ WARNING

When the wheel is installed, always remove any corrosion, dirt or foreign material present on the mounting surfaces of the wheels or the surfaces of the wheel hub, brake drum or brake disc that contact the wheel. Installing wheels without proper metal-to-metal contact at the wheel mounting surfaces can cause the wheel nut to loosen and the wheel to come off while the vehicle is in motion, causing loss of control.

14. Clean the wheel hub and mounting surfaces.
15. Install the tire and wheel assembly.
16. Lower the vehicle.
17. Check the front end alignment.

2004 F-150, Exc. Heritage Editions

2-WHEEL DRIVE

1. Before servicing the vehicle, refer to the precautions in the beginning of this section.
2. Disconnect the wheel speed sensor.
3. Remove the wheel and tire assembly.

➡Use the hex holding feature to prevent the ball studs (upper control arm, lower control arm tie-rod end, and stabilizer bar links) from turning while removing and installing the nuts.

✱✱ WARNING

Do not hammer the ball studs to separate them from the wheel knuckle or stabilizer bar. Doing so can cause damage. Lightly tap the wheel knuckle or stabilizer bar to loosen the joints.

4. Remove the anchor plate bolt.
5. Remove the brake caliper, pads and anchor plate.

✱✱ WARNING

Do not allow the brake caliper to hang by the flexible brake hose.

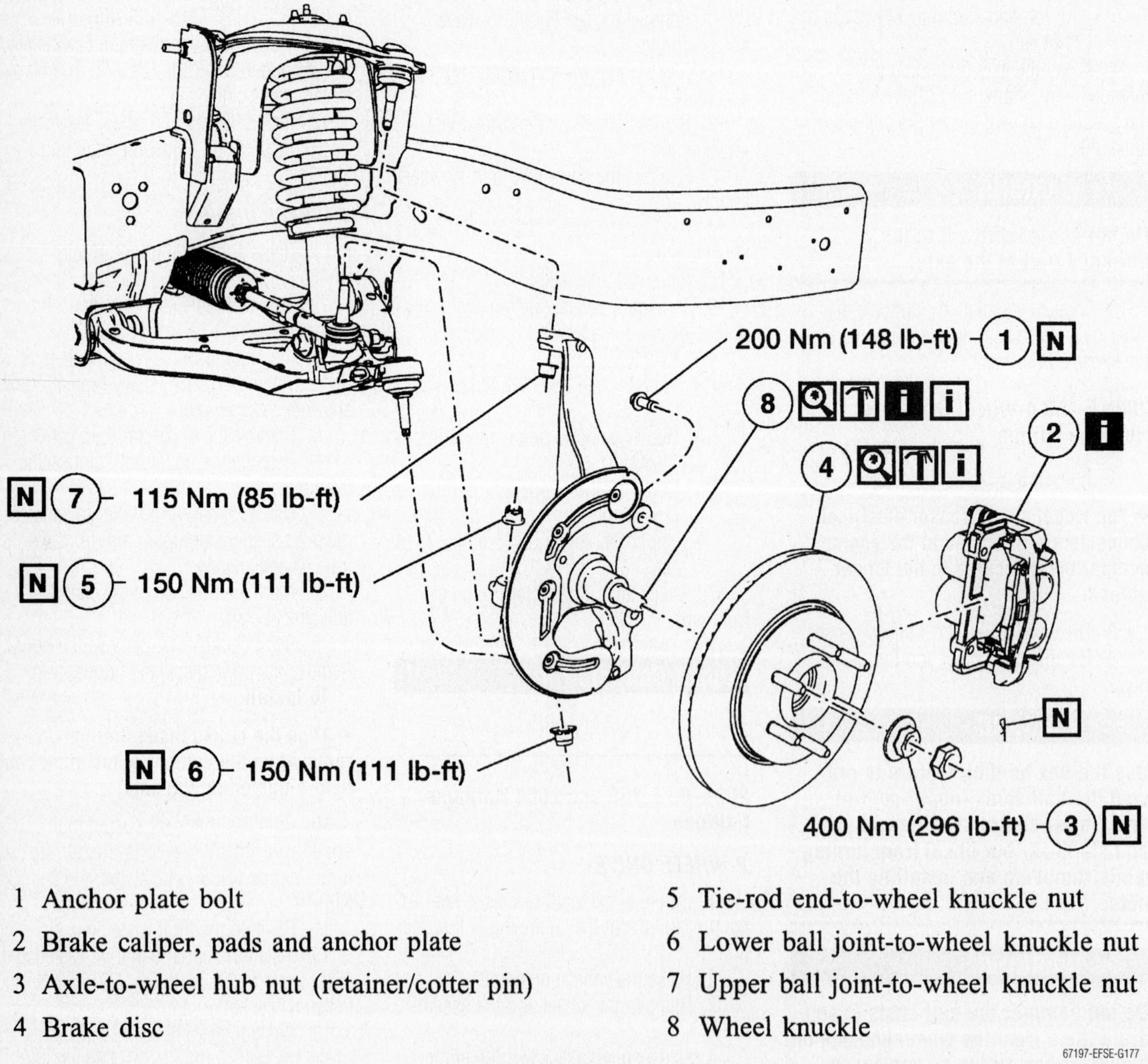

Spindle and hub components—2004 F-150 2-Wheel Drive, Exc. Heritage Editions

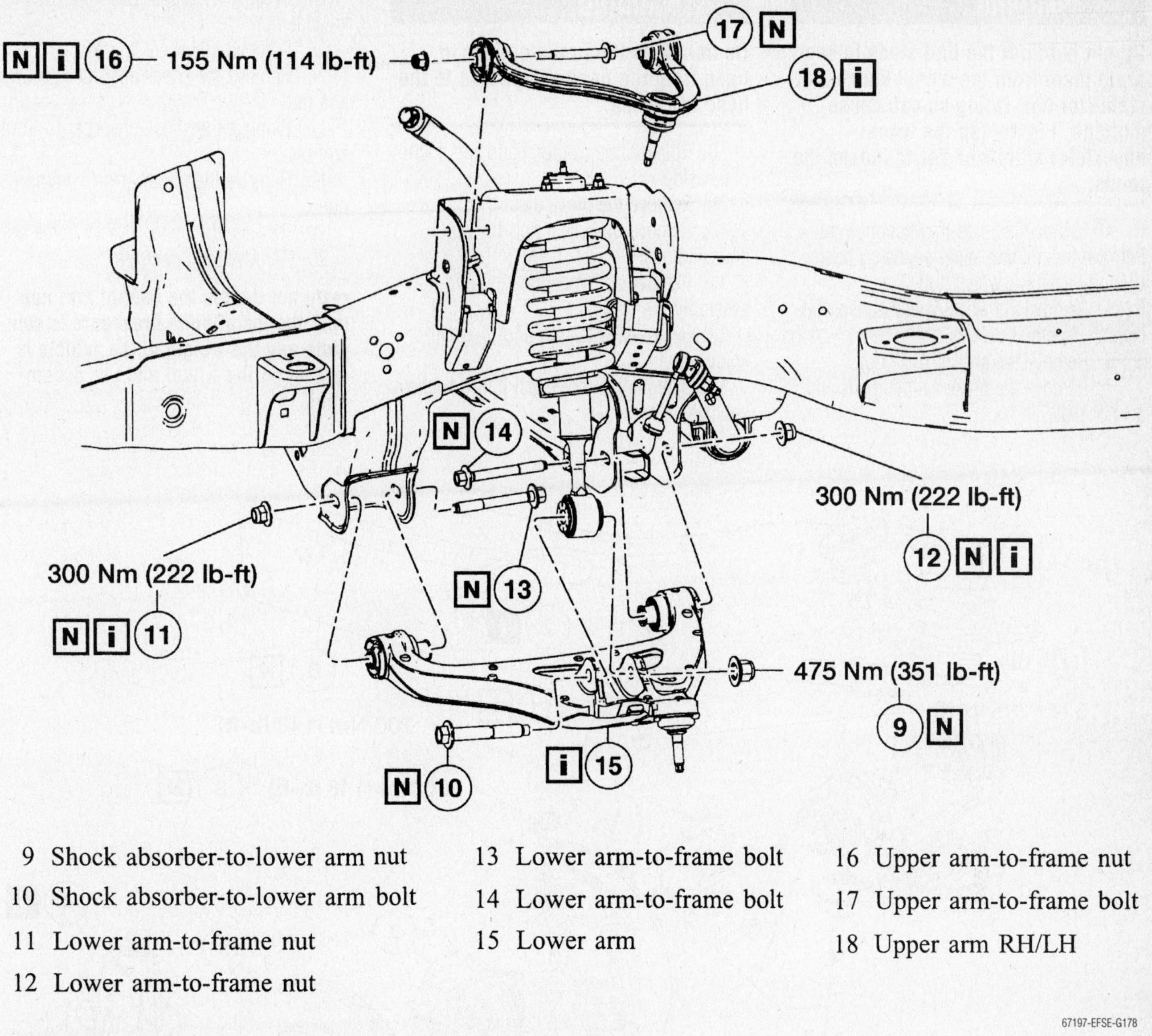

Front suspension components—2004 F-150 2-Wheel drive, Exc. Heritage Editions

6. Support the caliper to the vehicle.
7. Remove the axle-to-wheel hub nut, retainer and cotter pin.
8. Remove the brake disc.
9. Remove the tie-rod end-to-wheel knuckle nut.
10. Remove the lower ball joint-to-wheel knuckle nut.
11. Remove the upper ball joint-to-wheel knuckle nut.
12. Remove the wheel knuckle.
13. Remove the strut-to-lower arm nut.
14. Remove the strut-to-lower arm bolt.
15. Remove the upper arm-to-frame nut.
16. Remove the upper arm-to-frame bolt.
17. Remove the upper arm.
18. To install, reverse the removal procedure. Observe the following torques:

- Control arm-to-frame: 114 ft. lbs. (155 Nm)
- Upper control arm-to-knuckle: 85 ft. lbs. (115 Nm)
- Lower control arm-to-knuckle: 111 ft. lbs. (150 Nm)
- Tie rod end-to-knuckle: 111 ft. lbs. (150 Nm)
- Hub nut: 296 ft. lbs. (400 Nm)
- Anchor plate bolts: 148 ft. lbs. (200 Nm)

**** CAUTION**

Do not apply lubricant to the threaded part of the axle.

19. To ease installation, lubricate the inner race of the bearing assembly using the specified lubricant.

4-WHEEL DRIVE

1. Loosen the axle retainer nut.

➡The wheel speed sensor electrical connectors are located in the engine compartment secured to the fender aprons.

2. Disconnect the wheel speed sensor.
3. Remove the wheel and tire assembly.

**** WARNING**

Use the hex holding feature to prevent the ball studs (upper control arm, lower control arm tie-rod end, and stabilizer bar links) from turning while removing and installing the nuts.

⁂ WARNING

Do not hammer the ball studs to separate them from the wheel knuckle or stabilizer bar. Doing so can cause damage. Lightly tap the wheel knuckle or stabilizer bar to loosen the joints.

4. Remove the axle-to-wheel hub nut. Remove the nut and separate the outboard CV-joint from the wheel hub.
5. Remove the wheel speed sensor harness connector.
6. Remove the anchor plate bolt.
7. Remove the brake caliper, pads and anchor plate .

⁂ WARNING

Do not allow the brake caliper to hang from the hose or damage to the hose can occur.

8. Position the caliper, pads and anchor plate aside.
9. Remove the brake disc.
10. Remove the wheel hub-to-wheel knuckle bolts.
11. Remove the wheel bearing and hub assembly.
12. Remove the tie-rod end-to-wheel knuckle nut.
13. Remove the lower ball joint-to-wheel knuckle nut.
14. Remove the upper ball joint-to-wheel knuckle nut.
15. Remove the wheel knuckle.
16. Remove the shock absorber-to-lower arm nut.
17. Remove the shock absorber-to-lower arm bolt.
18. Remove the upper arm-to-frame nut.
19. Remove the upper arm-to-frame bolt.
20. Remove the upper arm.

➡Do not tighten the control arm nuts until the installation procedure is complete and the weight of the vehicle is resting on the wheel and tire assemblies.

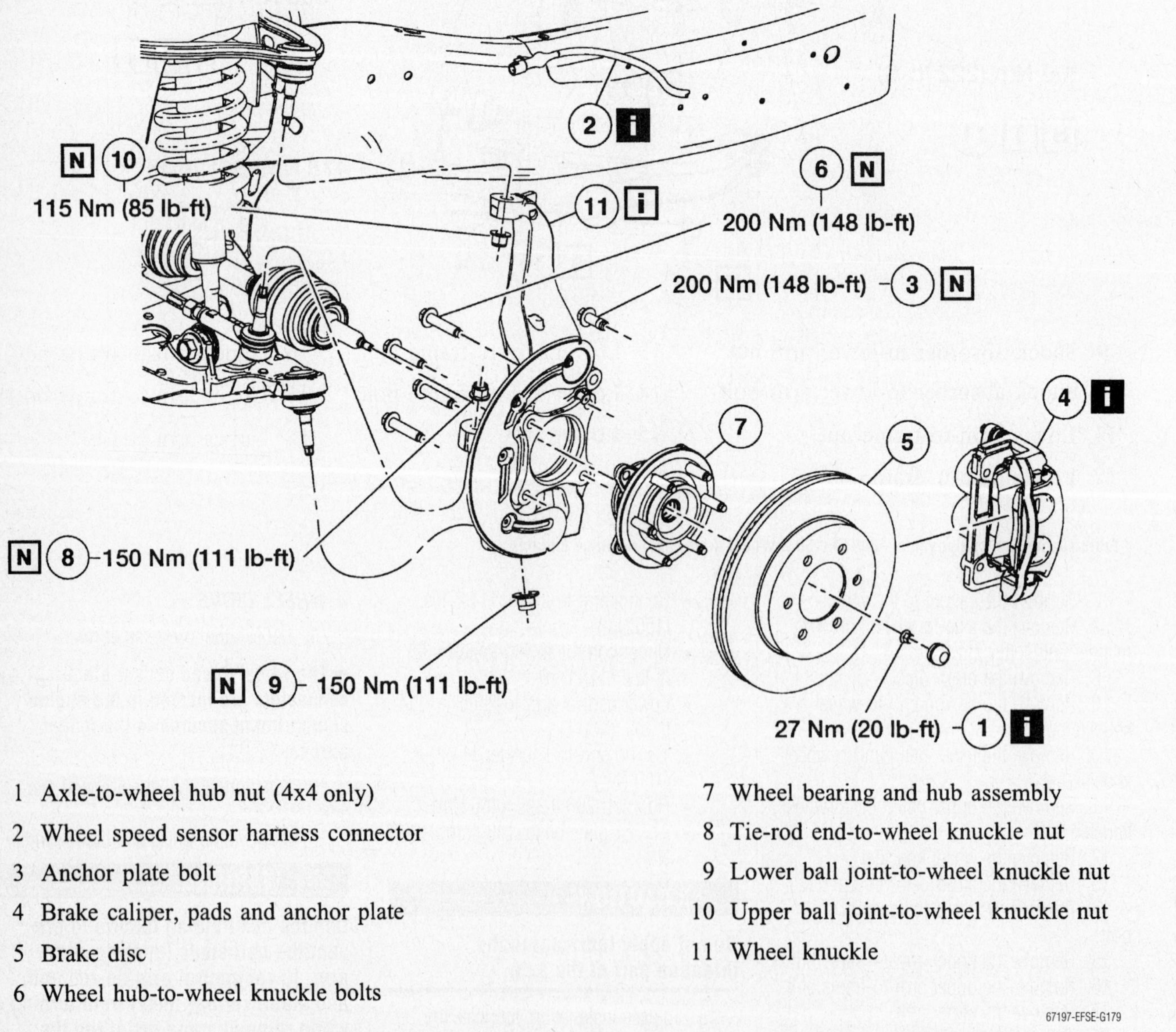

Spindle and hub components—2004 F-150 4-Wheel Drive, Exc. Heritage Editions

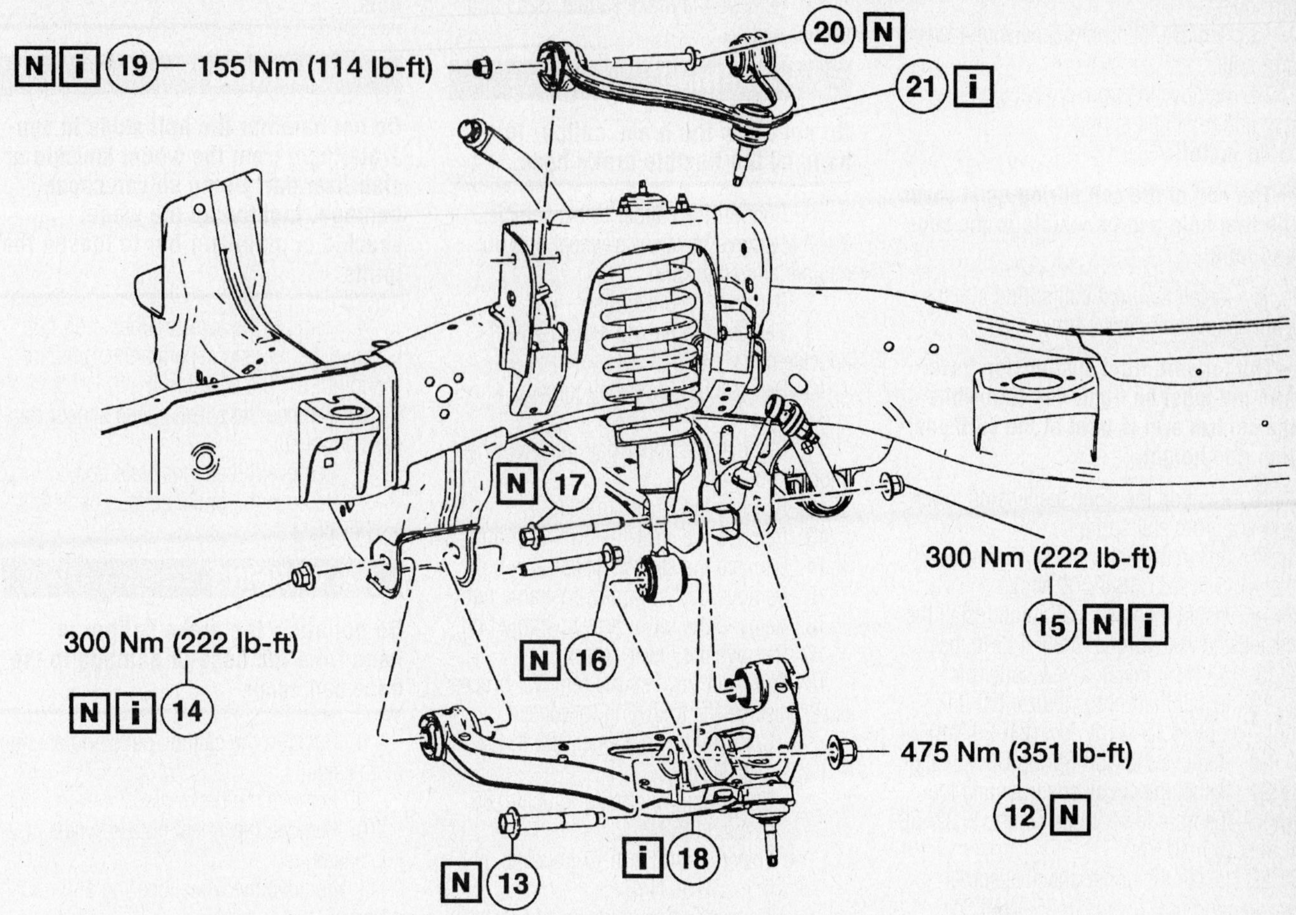

12 Shock absorber-to-lower arm nut
13 Shock absorber-to-lower arm bolt
14 Lower arm-to-frame nut
15 Lower arm-to-frame nut
16 Lower arm-to-frame bolt
17 Lower arm-to-frame bolt
18 Lower arm
19 Upper arm-to-frame nut
20 Upper arm-to-frame bolt
21 Upper arm RH/LH

67197-EFSE-G180

Front suspension components—2004 F-150 4-Wheel drive, Exc. Heritage Editions

21. To install, reverse the removal procedure.
22. Observe the following torques:
    - Upper arm-to-frame: 114ft. lbs. (155 Nm)
    - Strut-to-lower arm: 351 ft. lbs. (475 Nm)
    - Upper arm-to-knuckle: 85 ft. lbs. (115 Nm)
    - Lower arm-to-knuckle: 111 ft. lbs. (150 Nm)
    - Tie rod end-to-knuckle: 111 ft. lbs. (150 Nm)
    - Anchor plate: 148 ft. lbs. (200 Nm)
23. Check and, if necessary, align the front end.

Lower Control Arm

REMOVAL & INSTALLATION

2001–03 F-150 and 2004 Heritage Editions

1. Before servicing the vehicle, refer to the precautions in the beginning of this section.
2. Raise the vehicle on a hoist.
3. Remove the wheel and tire assembly
4. Remove the brake disc shield.

⁂ WARNING

Do not allow the shaft to turn while removing the nut. Hold the flats using a suitable wrench or damage to the shock can result.

5. Remove the upper shock absorber nut and washer.
6. Remove the front shock absorber.
7. Remove the brake hose bracket screw and bracket from the front suspension lower arm.
8. Remove the front stabilizer bar link nut.
9. Use the Coil Spring Compressor to compress the coil spring.
10. Remove the lower ball joint castellated nut. Discard the cotter pin.
11. Use a Pitman Arm Puller to separate the lower ball joint from the front wheel spindle.

12. Remove the front suspension lower arm nuts.
13. Remove the front suspension lower arm bolts.
14. Remove the front suspension lower arm and front coil spring.

To install:

➡The end of the coil spring must cover the first hole and be visible in the second hole.

15. Install the front coil spring into the front suspension lower arm.

➡The forward front suspension lower arm nut must be tightened first while the control arm is held at the curb position ride height.

16. Position the front suspension lower arm and front coil spring.
17. Install the bolts and nuts. Torque to 121–148 ft. lbs. (164–200 Nm).
18. Install the lower ball joint. Install the castellated nut. Torque to 83–112 ft. lbs. (113–153 Nm). Install a new cotter pin.
19. Install the front stabilizer bar link nut. Torque to 15–21 ft. lbs. (21–29 Nm).
20. Remove the Coil Spring Compressor.
21. Install the shock absorber and the shock absorber lower nuts. Torque to 22–29 ft. lbs. (30–40 Nm).
22. Install the upper shock absorber washer and nut. Torque to 35–46 ft. lbs. (47–63 Nm).
23. Install the brake hose bracket screw.
24. Install the brake disc shield.
25. Install the tire and wheel assembly.
26. Inspect and adjust the front end alignment.

2004 F-150, Exc. Heritage Editions

2-WHEEL DRIVE

1. Before servicing the vehicle, refer to the precautions in the beginning of this section.
2. Disconnect the wheel speed sensor.
3. Remove the wheel and tire assembly.

➡Use the hex holding feature to prevent the ball studs (upper control arm, lower control arm tie-rod end, and stabilizer bar links) from turning while removing and installing the nuts.

⁂ WARNING

Do not hammer the ball studs to separate them from the wheel knuckle or stabilizer bar. Doing so can cause damage. Lightly tap the wheel knuckle or stabilizer bar to loosen the joints.

4. Remove the anchor plate bolt.
5. Remove the brake caliper, pads and anchor plate.

⁂ WARNING

Do not allow the brake caliper to hang by the flexible brake hose.

6. Support the caliper to the vehicle.
7. Remove the axle-to-wheel hub nut, retainer and cotter pin.
8. Remove the brake disc.
9. Remove the tie-rod end-to-wheel knuckle nut.
10. Remove the lower ball joint-to-wheel knuckle nut.
11. Remove the upper ball joint-to-wheel knuckle nut.
12. Remove the wheel knuckle.
13. Remove the strut-to-lower arm nut.
14. Remove the strut-to-lower arm bolt.
15. Remove the lower arm-to-frame nut.
16. Remove the lower arm-to-frame bolt.
17. Remove the lower arm.
18. To install, reverse the removal procedure. Observe the following torques:
    - Control arm-to-frame: 222 ft. lbs. (300 Nm)
    - Upper control arm-to-knuckle: 85 ft. lbs. (115 Nm)
    - Lower control arm-to-knuckle: 111 ft. lbs. (150 Nm)
    - Tie rod end-to-knuckle: 111 ft. lbs. (150 Nm)
    - Hub nut: 296 ft. lbs. (400 Nm)
    - Anchor plate bolts: 148 ft. lbs. (200 Nm)

⁂ CAUTION

Do not apply lubricant to the threaded part of the axle.

19. To ease installation, lubricate the inner race of the bearing assembly using the specified lubricant.

4-WHEEL DRIVE

1. Loosen the axle retainer nut.

➡The wheel speed sensor electrical connectors are located in the engine compartment secured to the fender aprons.

2. Disconnect the wheel speed sensor.
3. Remove the wheel and tire assembly.

⁂ WARNING

Use the hex holding feature to prevent the ball studs (upper control arm, lower control arm tie-rod end, and stabilizer bar links) from turning while removing and installing the nuts.

⁂ WARNING

Do not hammer the ball studs to separate them from the wheel knuckle or stabilizer bar. Doing so can cause damage. Lightly tap the wheel knuckle or stabilizer bar to loosen the joints.

4. Remove the axle-to-wheel hub nut. Remove the nut and separate the outboard CV-joint from the wheel hub.
5. Remove the wheel speed sensor harness connector.
6. Remove the anchor plate bolt.
7. Remove the brake caliper, pads and anchor plate .

⁂ WARNING

Do not allow the brake caliper to hang from the hose or damage to the hose can occur.

8. Position the caliper, pads and anchor plate aside.
9. Remove the brake disc.
10. Remove the wheel hub-to-wheel knuckle bolts.
11. Remove the wheel bearing and hub assembly.
12. Remove the tie-rod end-to-wheel knuckle nut.
13. Remove the lower ball joint-to-wheel knuckle nut.
14. Remove the upper ball joint-to-wheel knuckle nut.
15. Remove the wheel knuckle.
16. Remove the shock absorber-to-lower arm nut.
17. Remove the shock absorber-to-lower arm bolt.
18. Remove the lower arm-to-frame nut.
19. Remove the lower arm-to-frame bolt.
20. Remove the lower arm.

➡Do not tighten the control arm nuts until the installation procedure is complete and the weight of the vehicle is resting on the wheel and tire assemblies.

21. To install, reverse the removal procedure.
22. Observe the following torques:
    - Lower arm-to-frame: 222 ft. lbs. (300 Nm)
    - Strut-to-lower arm: 351 ft. lbs. (475 Nm)

- Upper arm-to-knuckle: 85 ft. lbs. (115 Nm)
- Lower arm-to-knuckle: 111 ft. lbs. (150 Nm)
- Tie rod end-to-knuckle: 111 ft. lbs. (150 Nm)
- Anchor plate: 148 ft. lbs. (200 Nm)

23. Check and, if necessary, align the front end.

Front Wheel Bearings

ADJUSTMENT

E-Series, 2-Wheel Drive 2001–03 F-150 and 2-Wheel Drive 2004 Heritage Editions

CAUTION

If equipped with the automatic air suspension system, the service switch near the right kick panel must be turned OFF before raising the vehicle for service.

1. Before servicing the vehicle, refer to the precautions in the beginning of this section.
2. Raise and safely support the vehicle.
3. Support the front end.
4. Remove the wheel cover, if equipped.
5. Remove the grease cap.

➡Check the wheel bearings for sufficient grease.

6. Remove the cotter pin and retaining washer. Back off the spindle nut. Discard the cotter pin.
7. Adjust the wheel bearings as follows:
   a. Tighten the spindle nut to 30 ft. lbs. (40 Nm) while rotating the brake disc clockwise to seat the wheel bearings.
   b. Back off the nut 2 full turns.
   c. While rotating the disc counterclockwise, tighten the nut to 17–24 ft. lbs. (23–34 Nm).
   d. Back off the spindle nut about ½ turn.
   e. Tighten the spindle nut to 17 inch lbs. (2 Nm).
8. Install the retaining washer so the castellations are aligned with the cotter pin hole. Install a new cotter pin.
9. Check the wheel and tire assembly for proper rotation, then install the grease cap. If the wheel still does not rotate properly, inspect and clean or replace the wheel bearings and cups.
10. Install the wheel cover, if equipped.
11. Lower the vehicle.
12. Road test the vehicle and check for proper operation.

2001–04 2-Wheel Drive F-250, 350 and Super Duty

To check the wheel bearing adjustment, raise the front of the vehicle. Grasp the tire at the sides, and alternately push inward and pull outward on the tire. If any looseness is felt, adjust the front wheel bearings as follows.

1. Before servicing the vehicle, refer to the precautions in the beginning of this section.
2. Remove the hub cap from the hub.
3. Remove the cotter pin and the castellated nut.
4. While rotating the wheel, tighten the adjusting nut to 21 ft. lbs. (28 Nm) to seat the bearings.
5. Back off the adjusting nut until loose (120–180 degrees).
6. While rotating the wheel, tighten the adjusting nut to 18 inch lbs. (2Nm). Torque required to rotate the hub should be 18 inch lbs. (2 Nm).
7. Install the castellated nut and insert a new cotter pin.
8. Install the hub cap.

2004 F-150

The hub and bearing assembly is not adjustable.

REMOVAL & INSTALLATION

E-Series, 2-Wheel Drive 2001–03 F-150 and 2-Wheel Drive 2004 Heritage Editions

The hub is part of the disc brake rotor and cannot be serviced separately. The inner and outer wheel bearing and races are serviced individually. Be sure to have a new hub grease seal when servicing the wheel bearings.

1. Before servicing the vehicle, refer to the precautions in the beginning of this section.
2. Remove or disconnect the following:
   - Wheels
   - Caliper
   - Brake pads and anti-rattle clips
   - Hub grease cap, cotter pin, retainer washer and the spindle nut
   - Wheel bearing retainer washer and the outer wheel bearing
   - Brake hub and rotor assembly
   - Grease seal
   - Inner wheel bearing
3. Clean and inspect the wheel bearings and races for unusual wear or damage. Replace parts as necessary.
4. Inspect the hub and brake rotor assembly. If required, the hub and brake rotor assembly must be replaced as a unit.

To install:

5. If needed, pack the wheel bearing with a suitable high temperature wheel bearing grease before assembly.
6. Install or connect the following:
   - Inner wheel bearing in the hub and brake rotor assembly
   - New grease seal
   - Hub and rotor assembly on the wheel spindle and install the outer wheel bearing.
   - Retainer washer and the spindle nut
7. Adjust the wheel bearings as follows as described above.
8. Install or connect the following:
   - Retaining washer, so the castellations are aligned with the cotter pin hole. Install a new cotter pin.
   - Anti-rattle clips
   - Disc brake pads
   - Caliper
   - Wheels. Tighten the lug nuts to 83–112 ft. lbs. (113–153 Nm).
9. Check the wheel and tire assembly for proper rotation, then install the grease cap.
10. Lower the vehicle.
11. Road test the vehicle and check for proper operation.

2001–03 F-150 4-Wheel Drive

The wheel bearings are of the cartridge design and are an integral part of the hub assembly. The bearings are permanently lubricated and require no maintenance or adjustments. If required, a new hub assembly must be installed.

1. Before servicing the vehicle, refer to the precautions in the beginning of this section.
2. Remove the wheel and tire assembly.
3. Remove the brake disc.
4. Remove the cotter pin. Remove the retainer. Remove the front wheel hub nut.
5. If equipped with 4-wheel anti-lock brake system (4WABS), remove the brake disc shield.
6. If equipped with 4WABS, remove the front brake anti-lock sensor.

WARNING

Do not overextend CV-joint and boots when removing the hub and bearing assembly.

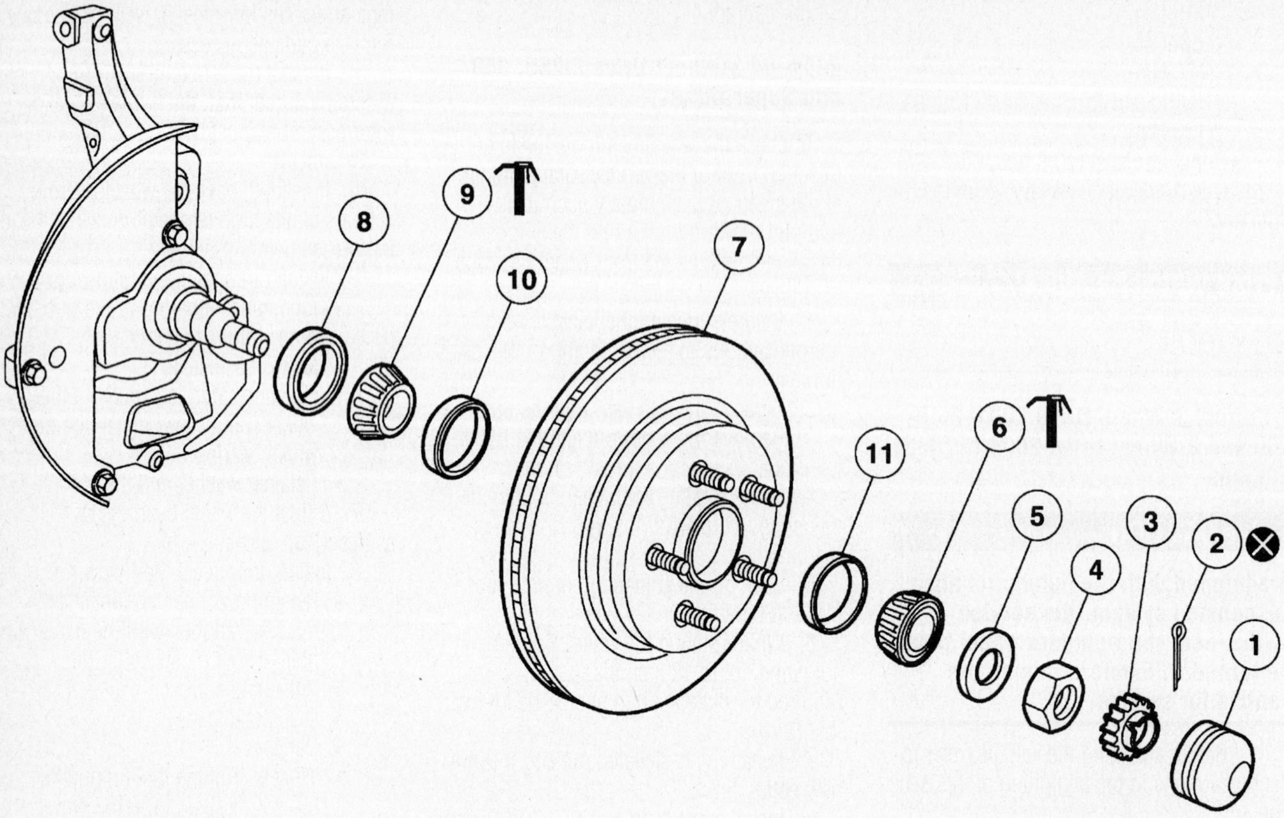

1 Grease cap	5 Washer	9 Inner wheel bearing
2 Cotter pin	6 Outer wheel bearing	10 Inner bearing cup
3 Retainer	7 Brake disc and hub assembly	11 Outer bearing cup
4 Spindle nut	8 Grease seal	

67197-EFSE-G191

Front hub and bearings— E-Series, 2-wheel drive 2001–03 F-150 and 2-wheel drive 2004 Heritage Editions; 2-wheel drive F-250/350 similar

➡The CV-joint is a slip fit into the wheel hub and bearing. A puller will not normally be required.

7. Remove the bolts. Push the CV-joint inboard. Remove the wheel hub.
8. Remove the seal.

Install:

9. Install the seal.
10. Position the wheel hub on the front wheel driveshaft and joint and into the front wheel knuckle.
11. Install the bolts. Torque to 110–148 ft. lbs. (149–201 Nm).
12. If equipped with 4WABS, install the front brake anti-lock sensor. Torque to 62–80 inch lbs. (7–9 Nm).
13. Position the brake disc shield.
14. Install the bolts. Torque to 80–107 inch lbs. (9–12 Nm).
15. Install the front axle hub nut. Torque to 188–254 ft. lbs. (255–345 Nm).
16. Install the retainer.
17. Install the cotter pin.
18. Install the brake disc.

⁂ WARNING

When the wheel is installed, always remove any corrosion, dirt or foreign material present on the mounting surfaces of the wheels or the surfaces of the wheel hub, brake drum or brake disc that contact the wheel. Installing wheels without proper metal-to-metal contact at the wheel mounting surfaces can cause the wheel nut to loosen and the wheel to come off while the vehicle is in motion, causing loss of control.

19. Clean the wheel hub and mounting surfaces.
20. Install the tire and wheel assembly.
21. Lower the vehicle.

2001–04 2-Wheel Drive F-250, 350 and Super Duty

1. Raise and support the vehicle.
2. Remove the front wheel and tire assemblies.
3. Remove the front disc brake caliper and rotor, and position the caliper out of the way.
4. Remove the hub cap from the hub assembly.
5. Remove the cotter pin, adjusting nut and flat washer.

➡Inspect the condition of the spindle and nut threads to ensure a free turning nut when reassembling.

6. Remove the outer bearing cone and roller assembly, and pull the hub assembly from the spindle.
7. Using care not to damage the bear-

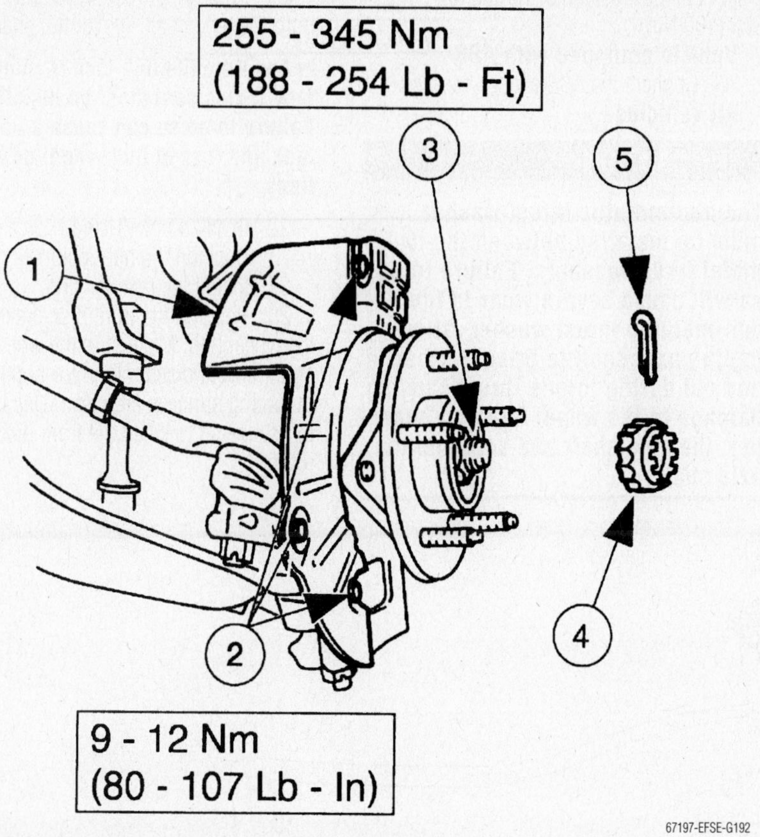

Front hub/bearing installation—2001–03 F-150 4-Wheel Drive

ing cage, use a suitable slide hammer and bearing seal remover to remove the inner bearing cone and bearing seal.

To install:

➡Do not spin the bearing dry with compressed air.

➡Remove all traces of lubricant from the bearings, hub and axle spindle. Inspect bearings and bearing cups for pitting, spalling or unusual wear. If either bearings or bearing cups are worn or damaged, replace both bearings and bearing cups.

➡It is recommended that bearings and bearing cups be replaced in sets. If cups are worn or damaged, install the inner and outer bearing cups in the hub with an appropriate bearing cup driver tool. Check for proper seating of new bearing cups by trying to insert a 0.38-mm (0.0015-inch) feeler gauge between the bottom face of the cup and wheel hub seat. You should not be able to insert the feeler gauge.

8. Remove all burrs, nicks or scratches from the shoulder of the spindle and seal bore in the hub with emery cloth.
9. Pack the inside of the hub with lithium-base wheel bearing grease such as Motorcraft Premium Long-Life Grease XG-1-C or -K or equivalent meeting Ford specification ESA-M1C75-B. Fill the hub until the grease is flush with the inside diameters of both bearing cups.
10. Pack the bearing cone and roller assemblies with wheel bearing grease. Use a bearing packer for this operation. If a packer is not available, work as much lubricant as possible between the rollers and cages.

⁂ WARNING

Keep the hub centered on the spindle to prevent damage to the grease seal or spindle threads.

11. Place the inner bearing cone and roller assembly in the inner cup and install the wheel bearing hub seal, using a suitable seal replacer. Make sure seal is fully seated and lubricated.
12. Install the hub assembly.
13. Install the outer bearing cone and roller assembly and the flat washer on the spindle and install the adjusting nut. Adjust the bearings. Install a new cotter pin.
14. Install the hub cap.
15. Install the front disc brake caliper and rotor.
16. Install the front wheel and tire assemblies.
17. Lower the vehicle.

2001–04 4-Wheel Drive F-250, 350 and Super Duty

1. Before servicing the vehicle, refer to the precautions in the beginning of this section.

All vehicles

2. Remove the wheel and tire assembly.
3. Remove the two front disc brake caliper anchor plate bolts.
4. Remove the front disc brake caliper anchor plate and position aside.

➡If excessive force must be used during brake rotor removal, the brake rotors should be checked for lateral runout prior to installation.

5. On F-250 and F-350 4x4 SRW vehicles, remove the rotor.
6. On DRW vehicles, remove the eight hub plate nuts. Remove the hub plate. Remove the rotor.
7. Remove the retainer ring. Pull outward and remove the hub lock.
8. Remove the snapring. Remove the three thrust washers.

Vehicles equipped with ABS

➡Do not remove the ABS sensor from the bearing.

9. Disconnect the ABS wheel sensor harness.

All vehicles

➡The wheel hub and bearing is a slip fit design and should not require a puller to remove it.

10. Remove the four lock nuts. Remove the wheel hub and bearing.
11. If necessary, remove the brake disc shield.

Vehicles with ABS

12. If necessary, remove the bolt and the ABS sensor.

All vehicles

➡If necessary, position the hub in a soft-jawed vise.

13. Install two nuts on the studs and use the inner nut to remove the studs.
14. Remove and discard the O-ring.

To install:

All vehicles

➡Any time the wheel hub is removed for any reason, a new O-ring seal must be installed. Failure to do so can cause a vacuum leak and loss of four wheel drive operations.

15. Install a new O-ring.

➡**Position the hub in a soft-jawed vise.**

16. Install two nuts on the studs and use the outer nut to install the studs.

Vehicles equipped with ABS

17. Position the ABS sensor and install the bolt. Torque to 13 ft. lbs. (18 Nm).

All vehicles

18. Position the brake disc shield.

➡**Apply a coat of Ford High Temperature 4x4 Front Axle and Wheel Bearing Grease E8TZ-19590-A meeting Ford specification ESA-M1C198-A to the O-ring area of the wheel hub and bearing before installing the hub and bearing.**

19. Position the wheel hub and bearing. Install the four lock nuts. Torque to 133 ft. lbs. (180 Nm).

Vehicle equipped with ABS

20. Connect the ABS sensor harness.

All vehicles

CAUTION

The non-metallic thrust washer must be installed between the two metal thrust washers. Failure to do so will cause severe wear to the non-metallic thrust washer, allowing the axle shaft to travel further in and out during torque thrust causing damage to the wheel hub and bearing, the axle shaft end seal and the axle shaft.

21. Position the three thrust washers onto the axle shaft. Install the snapring.

➡**Any time the hub lock is removed, a new O-ring seal must be installed. Failure to do so can cause a vacuum leak and loss of four wheel drive functions.**

22. Install a new O-ring.
23. Position the hub lock.
24. Install the retainer ring.
25. Position the front disc brake rotor to the wheel hub. Make sure the wheel hub and the front disc brake rotor braking and mounting surfaces are clean. Use brake parts cleaner to clean the front disc brake rotor.

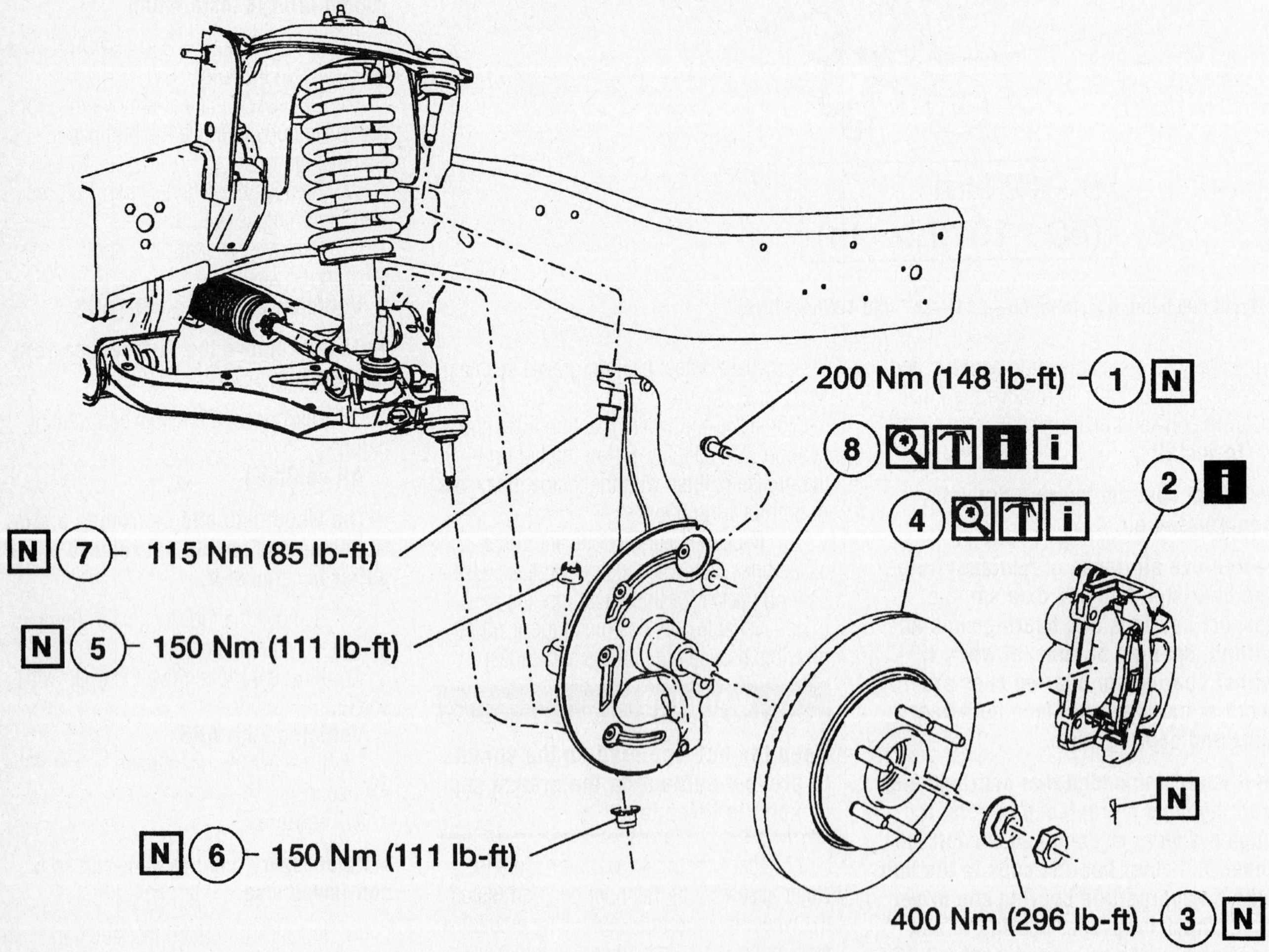

1 Anchor plate bolt
2 Brake caliper, pads and anchor plate
3 Axle-to-wheel hub nut (retainer/cotter pin)
4 Brake disc
5 Tie-rod end-to-wheel knuckle nut
6 Lower ball joint-to-wheel knuckle nut
7 Upper ball joint-to-wheel knuckle nut
8 Wheel knuckle

67197-EFSE-G193

Front hub/bearing installation—2004 2-wheel drive F-150, exc. Heritage Editions

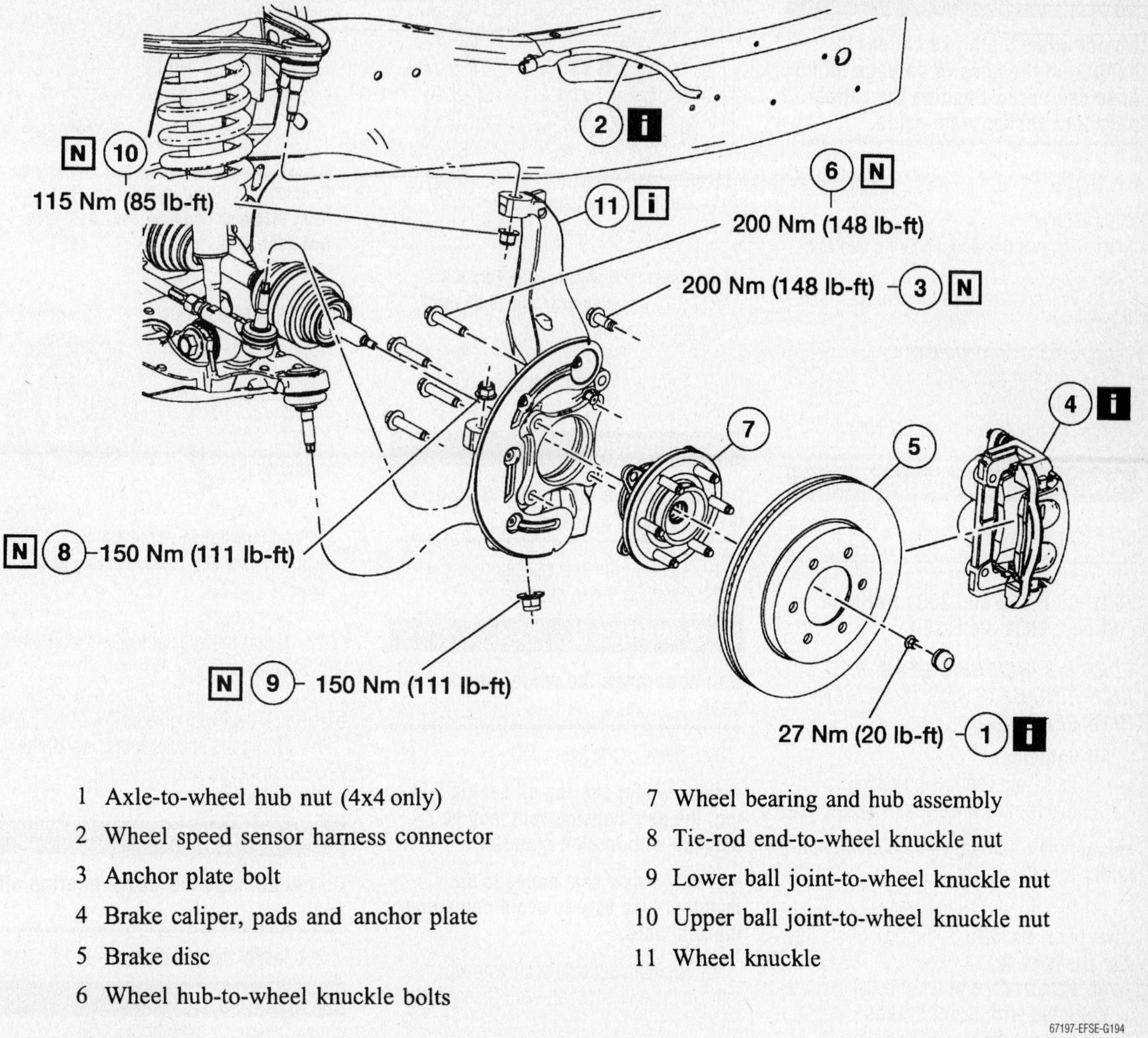

Front hub/bearing assembly—2004 F-150 4-wheel drive, exc. Heritage Edition

26. For DRW vehicles, install the front wheel hub extender and nuts. Torque to 130 ft. lbs. (176 Nm).
27. Position back the front disc brake caliper and anchor plate. Install the front disc brake caliper anchor plate bolts. Torque to 166 ft. lbs. (225 Nm).
28. Install the wheel and tire assembly.
29. Test the system for normal operation.

2004 2-Wheel Drive F-150, exc. Heritage Editions

1. Disconnect the wheel speed sensor.
2. Remove the wheel and tire assembly.
3. Remove the anchor plate bolts.
4. Remove the brake caliper, pads and anchor plate. Support the caliper to the vehicle.
5. Remove the axle-to-wheel hub nut, retainer and cotter pin.
6. Remove the brake disc/hub assembly.
7. To install, reverse the removal procedure.

CAUTION

Do not apply lubricant to the threaded part of the axle.

8. To ease installation, lubricate the inner race of the bearing assembly using the specified lubricant.
9. Observe the following torques:
   - Hub nut: 296 ft. lbs. (400 Nm)
   - Anchor plate bolts: 148 ft. lbs. (200 Nm)

2004 F-150 4-Wheel Drive, exc. Heritage Edition

1. Loosen the axle retainer nut.

➡The wheel speed sensor electrical connectors are located in the engine compartment secured to the fender aprons.

2. Disconnect the wheel speed sensor.
3. Remove the wheel and tire.
4. Remove the axle-to-wheel hub nut and, using a puller, separate the outboard CV joint from the wheel hub. assembly.
5. Remove the wheel speed sensor harness connector.
6. Remove the anchor plate bolts.
7. Remove the brake caliper, pads and anchor plate.

CAUTION

Do not allow the brake caliper to hang from the hose or damage to the hose can occur. Position the caliper, pads and anchor plate aside.

8. Remove the brake disc.
9. Remove the wheel hub-to-wheel knuckle bolts.
10. Remove the wheel bearing and hub assembly.
11. To install, reverse the removal procedure.
12. Observe the following torques:
    - Anchor plate bolts: 148 ft. lbs. (200 Nm)
    - Hub nut: 20 ft. lbs. (27 Nm)

Rear Wheel Bearings

REMOVAL & INSTALLATION

2001–03 F-150 and 2004 Heritage Editions; 2001–04 E-150

FORD 8.8 INCH RING GEAR, 9.75 INCH RING GEAR AND 10.25 INCH RING GEAR AXLES

All Vehicles

1. Raise and support the vehicle.
2. Remove the wheel and tire assembly.

➡Empty the lubricant into a clean container for reuse.

3. Remove the 10 differential housing cover bolts and drain the lubricant from the rear axle housing.
4. Remove the differential housing cover.

Vehicles with drum brakes

5. Remove the rear brake drums.

Vehicles with disc brakes

6. Remove the rear disc brake caliper. Wire the rear disc brake caliper aside.
7. Remove the rear brake disc.

All vehicles

8. Remove and discard the differential pinion shaft lock bolt.

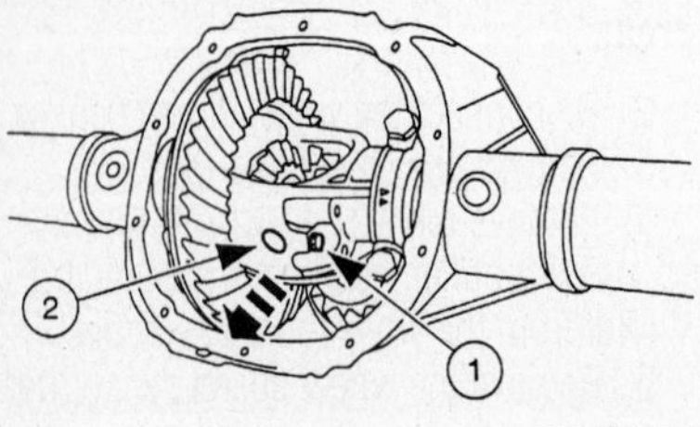

Lock bolt (1); pinion shaft (2)—Ford 8.8 inch, 9.75 inch ring gear, and 10.25 inch ring gear axles

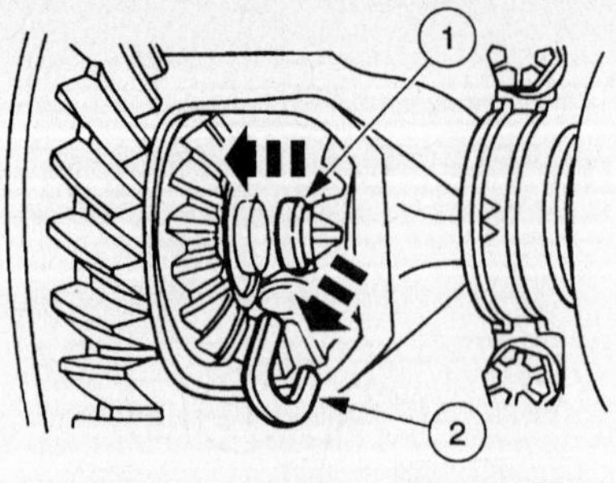

Axle shaft (1); U-washer (2)—Ford 8.8 inch, 9.75 inch ring gear and 10.25 inch ring gear axles

9. Remove the differential pinion shaft.

WARNING

Do not damage the rubber O-rings in the axle shaft grooves.

10. Push in the axle shafts.
11. Remove the U-washers.

WARNING

Do not damage the wheel bearing oil seal.

12. Remove the axle shaft.

➡If the wheel bearing oil seal is leaking, the axle housing vent may be plugged with foreign material.

➡If only a new seal needs to be installed, use care to avoid damaging the seal bore.

13. Using a suitable seal remover, remove the axle shaft oil seal. Discard the oil seal.
14. Inspect the rear wheel bearing and axle shaft for wear or damage.
15. If necessary, using the special tools, remove the rear wheel bearing.

To install:

16. Lubricate the new rear wheel bearing with rear axle lubricant.

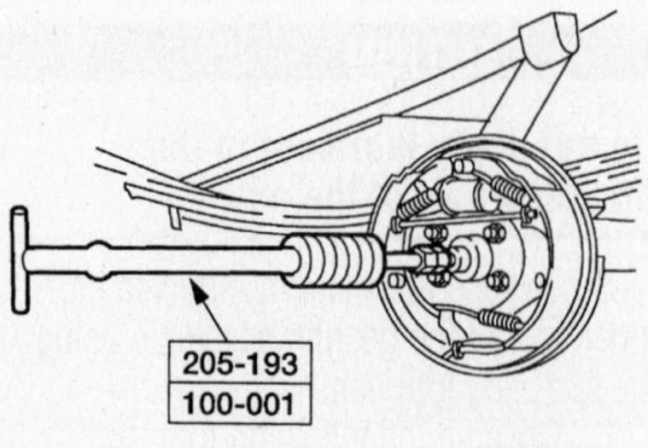

Rear wheel bearing removal—Ford 8.8 inch, 9.75 inch ring gear and 10.25 inch ring gear axles

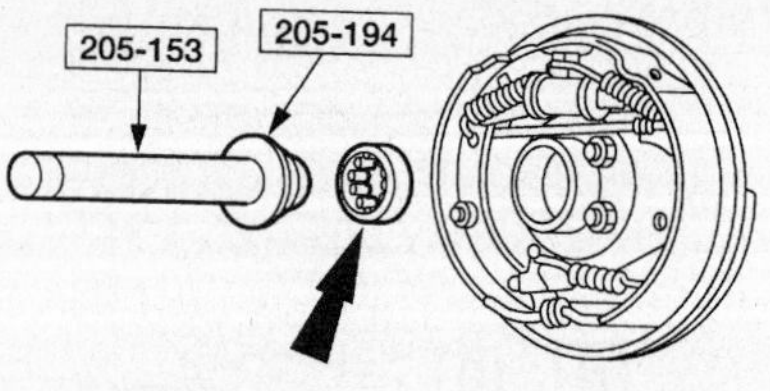

Rear wheel bearing installation—Ford 8.8 inch, 9.75 inch ring gear and 10.25 inch ring gear axles

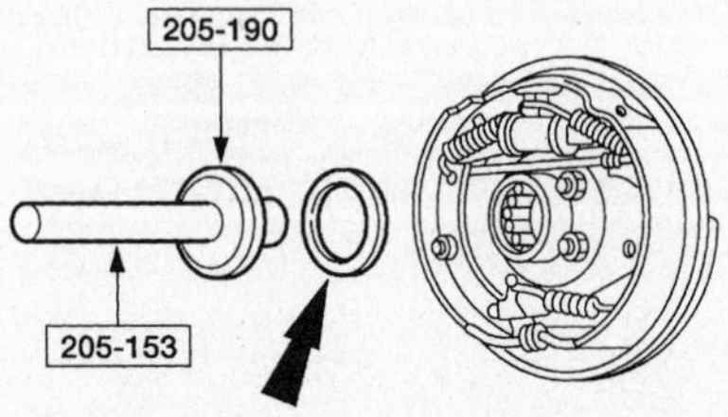

Oil seal installation—Ford 8.8 inch, 9.75 inch ring gear and 10.25 inch ring gear axles

17. Using the special tools, install the rear wheel bearing.
18. Lubricate the lip of the new wheel bearing oil seal with grease.
19. Using the special tools, install the wheel bearing oil seal.

All vehicles

WARNING

Do not damage the wheel bearing oil seal.

20. Install the axle shaft.

WARNING

Do not damage the rubber O-rings in the U-washer grooves.

21. Position the two U-washers on the button end of the axle shafts.
22. Pull the axle shafts outward.

➡If a new pinion shaft lock bolt is unavailable coat the threads with Threadlock prior to installation.

23. Align the hole in the differential pinion shaft with the case lock bolt hole.
24. Install a new differential pinion shaft lock bolt. Torque to 15–30 ft. lbs. (20–40 Nm).

Vehicles with drum brakes

25. Install the rear brake drums.

Vehicles with disc brakes

26. Install the rear brake disc.
27. Install the rear disc brake caliper.

➡Clean the gasket mating surface of the rear axle and the differential housing cover.

28. Apply a new continuous bead of sealant to the differential housing cover.

➡The differential housing cover must be installed within 15 minutes of application of the silicone, or new sealant must be applied. If possible, allow one hour before filling with lubricant to make sure the silicone sealant has correctly cured.

29. Install the differential housing cover.
30. Install the 10 differential housing cover bolts. Torque to 33 ft. lbs. (45 Nm).
31. Fill the rear axle housing with 2.37 liters (5 pints) with the specified lubricant.

✲✲ CAUTION

Always remove any corrosion, dirt or foreign material present on the mounting surfaces of the wheel or the surface of the wheel hub or brake drum or disc that contacts the wheel. Installing wheels without correct metal-to-metal contact at the wheel mounting surfaces can cause the lug nut to loosen and the wheel to come off while the vehicle is in motion, causing loss of control.

32. Clean the wheel hub and mounting surfaces.
33. Install the tire and wheel assembly.

2001–04 E-250/350 and F-250/350

DANA 60 AND 70 SEMI-FLOATING AXLES

1. Raise the vehicle on a hoist or raise the rear end of the vehicle with a jack. Install safety stands under the frame rails and lower the jack or hoist enough to allow the rear axle to drop into the rebound position for working clearance.
2. Remove the rear wheel and tire assembly.
3. Remove the brake disc.
4. Remove the differential housing cover and drain the lubricant. Clean the gasket material from the differential housing cover and the differential housing.

✲✲ WARNING

The differential assembly is equipped with either a Loctite® coated differential pinion shaft lock screw or a differential pinion shaft lock screw with torque prevailing threads. The Loctite® treated differential pinion shaft lock screw has a 5⁄32 inch hexagram socket head. Never, under any circumstance, reuse a Loctite® coated screw after removing it. Always discard the screw and install a new one. The torque prevailing differential pinion shaft lock screw has a 12-point drive head. This type of screw is reusable for no more than four installations. When in doubt about the number of installations of a torque prevailing differential pinion shaft lock screw, discard it and install a new screw.

5. Remove the lock screw.
6. Remove the differential pinion shaft. The pinion shaft is a slip-fit design and is removable by hand.
7. Push the flanged end of the axle shaft toward the center of the axle and remove the U-washer.

✲✲ WARNING

Do not damage the wheel bearing oil seal when removing the axle shaft.

✲✲ WARNING

Do not rotate the differential side gears when removing the axle shaft. Rotating the differential side gears causes the differential pinion gears and differential pinion thrust washers to turn to the differential case opening and fall out of the differential case.

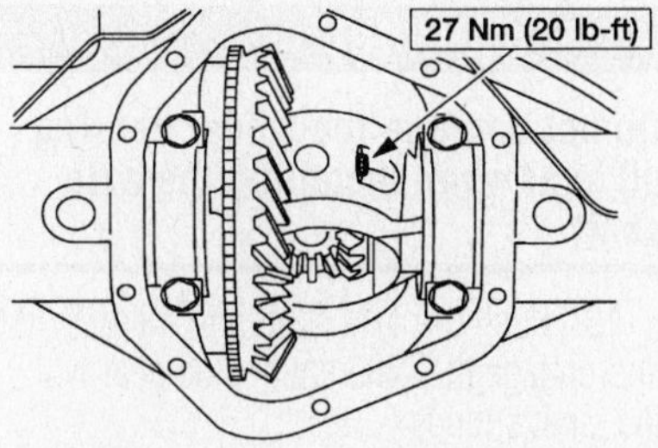

67197-EFSE-G219

Lock screw—Dana 60 and 70 semi-floating axle

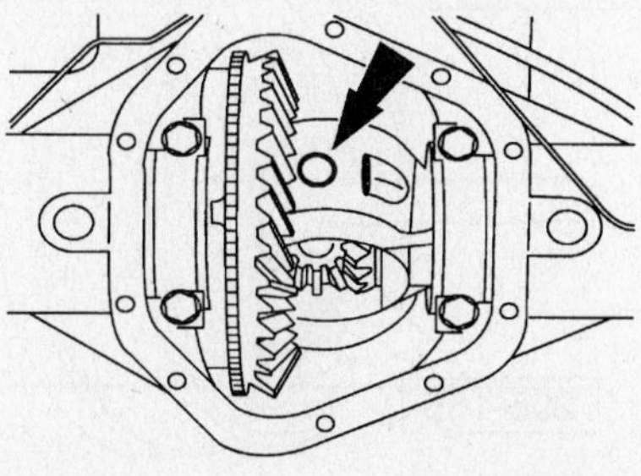

67197-EFSE-G220

Pinion shaft—Dana 60 and 70 semi-floating axle

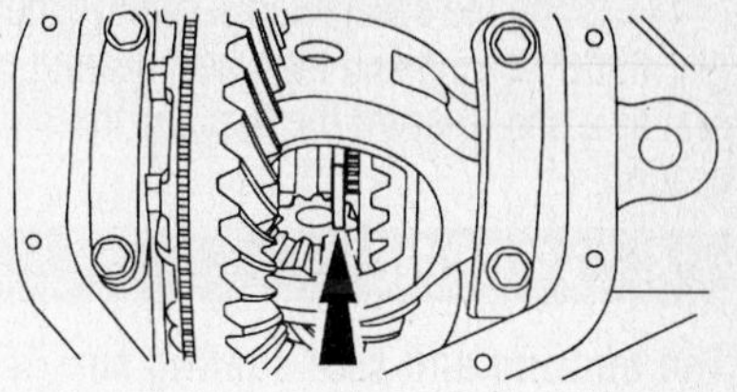

67197-EFSE-G221

U-washer—Dana 60 and 70 semi-floating axle

8. Remove the axle shaft.

✲✲ WARNING

Install the differential pinion shaft and the old lock screw (finger-tight) to prevent the differential side gears and differential pinion gears from rotating and falling out of the differential case.

9. Install the differential pinion shaft and the old lock screw in the differential case.

✲✲ WARNING

It is not necessary to remove the rear wheel bearing if only installing a new wheel bearing oil seal. To remove only the wheel bearing oil seal, pry it from the axle tube. Do not damage the seal seating surface. If removing the rear wheel bearing and the oil seal, proceed to the following step.

10. Remove the wheel bearing oil seal from the axle tube. Discard the seal.

✲✲ CAUTION

Make sure protective eyewear is in place. Failure to follow these instructions may result in personal injury.

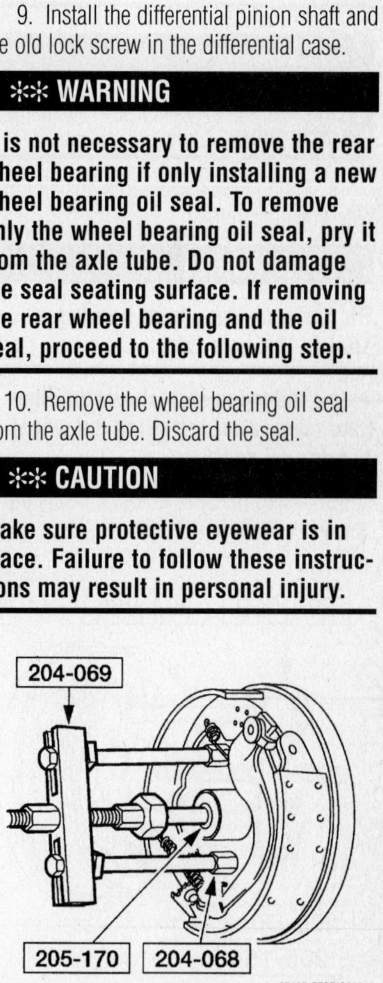

67197-EFSE-G222

Rear wheel bearing removal—Dana 60 and 70 semi-floating axle

11. Using the special tools, remove the rear wheel bearing and the wheel bearing oil seal as a unit. Discard the seal and the bearing.

WARNING

The bearing and seal seating surfaces must be free from burrs, nicks, and spalling.

12. Clean and inspect the bore in the axle tube. Wipe the bore in the axle tube with emery cloth to smooth the surface. Clean the bore with a standard metal-cleaning solvent. Wipe the bore with a soft, lint free cloth to remove any foreign material.

To install:

13. If removed, lubricate the new rear wheel bearing with the specified axle lubricant.

WARNING

Install the rear wheel bearing with the identification numbers facing outward.

WARNING

Do not cock the rear wheel bearing in the axle tube.

14. Using the special tools, install the rear wheel bearing.

WARNING

Do not cock the wheel bearing oil seal in the axle tube.

➡The following step shows an alternate method for installing the wheel bearing oil seal.

15. Using the special tools, install the new wheel bearing oil seal.

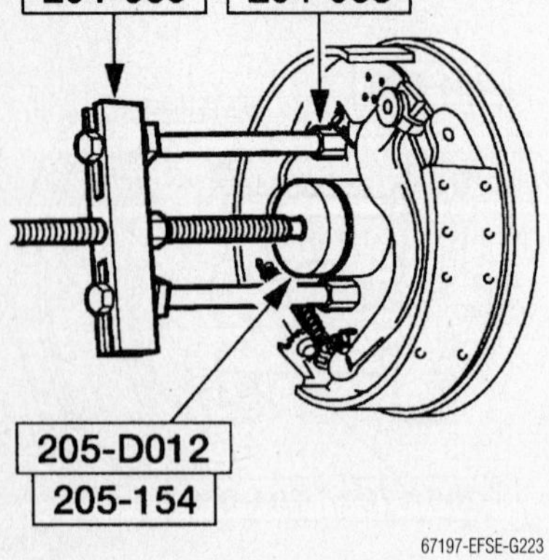

Rear wheel bearing installation—Dana 60 and 70 semi-floating axle

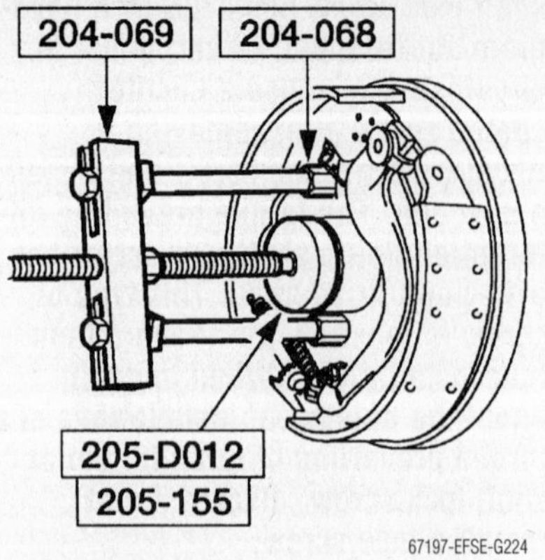

Oil seal installation—Dana 60 and 70 semi-floating axle

WARNING

Do not cock the wheel bearing oil seal in the axle tube bore.

➡This step is an alternate method for installing the wheel bearing oil seal. Carry out this step if wheel bearing oil seal installation was not done in the previous step.

16. Using the special tools, install the new wheel bearing oil seal.

17. Lubricate the cavity between the wheel bearing oil seal lips and the rear wheel bearing with grease.

18. Remove the lock screw and the differential pinion shaft.

WARNING

Do not damage the wheel bearing oil seal when installing the axle shaft.

19. Push the axle shaft into the axle tube and engage the differential side gear with the shaft splines.

20. Push the axle shaft toward the center of the axle and install the U-washer. Pull the axle shaft outward until the U-washer locks into the differential side gear.

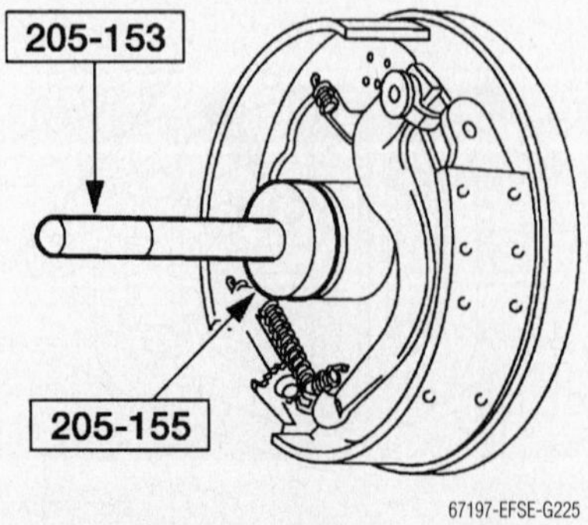

Alternate oil seal installation—Dana 60 and 70 semi-floating axle

21. Align the differential pinion shaft lock screw hole with the hole in the differential case. Correctly position the differential pinion thrust washers. Install the differential pinion shaft.

➡The threads in the differential case and on the lock screw must be free of dirt and oil.

22. Install the new lock screw. Torque to 20 ft. lbs. (27 Nm).

WARNING

Clean the mounting surface on the differential housing cover and the differential housing with a suitable solvent to remove all traces of oil film.®

WARNING

The differential housing cover uses silicone rubber sealant material as a gasket. Install the differential housing cover within 15 minutes of applying the silicone or it will be necessary to apply new sealant.

23. Apply a continuous bead of silicone sealant to the differential housing cover mounting surface.

24. Assemble two bolts into the differential housing cover at the eight o'clock and two o'clock positions. Using the two bolts as a guide, position the differential housing cover on the differential housing.

25. Install the remaining bolts. Tighten the bolts alternately and evenly. Tighten Grade 5 bolts to 47 Nm (35 ft. lbs. Tighten Grade 8 bolts to 61 Nm (45 ft. lbs.

WARNING

Allow 1 hour cure time before filling the axle with the correct amount of specified lubricant and operating the vehicle.®

WARNING

For limited-slip axles, first fill the axle with 7 ounces of friction modifier.

26. Fill the differential housing with 3.0L (6.3 pints) of the specified rear axle lubricant. For additional information, refer to Specifications in this section.

27. Install the brake disc.

28. Install the wheel and tire assembly.

29. Lower the vehicle.

2001–04 F-250 and 350 Ford Full-Floating Axle

1. Set the parking brake.
2. Loosen the retaining bolts.
3. Raise the vehicle to the desired working height, keeping the axle parallel with the floor.
4. Release the parking brake.
5. Remove the wheel(s).
6. Remove the brake caliper and rotor on the single rear wheel axle.
7. Remove the retaining bolts and axle shaft.

➡The hub nuts are right-hand thread (right hub) and left-hand thread (left hub). Each hub nut is stamped RH or LH.

8. Install the Ford Axle Locknut Socket so that the drive tangs of the tool engage the four slots in the hub nut.

**** WARNING**

Discard the hub nut if the hub nut comes apart during removal.

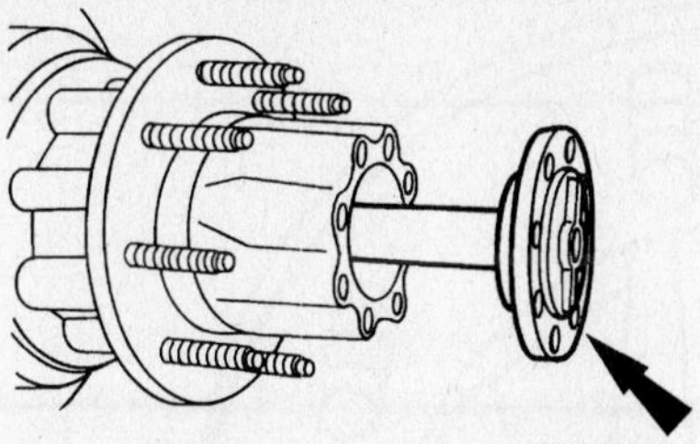

67197-EFSE-G211

Axle shaft removal—Ford full-floating axle

**** WARNING**

Under no circumstances are power tools to be used when performing these operations.

➡The hub nut will ratchet during this operation.

9. Remove the hub nut (counterclockwise for right-hand thread; clockwise for left-hand thread).
10. Install the Step Plate.
11. Install the 2-Jaw Puller and loosen the rear hub to the point of removal.

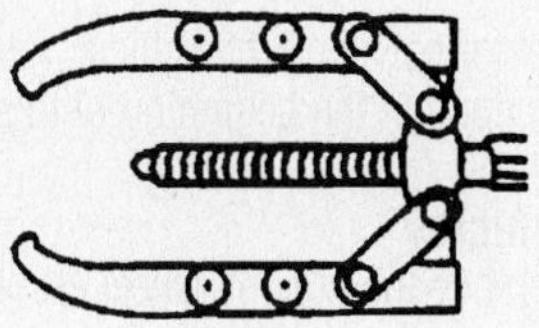	2-Jaw Puller or Equivalent
	Socket, Ford Axle Locknut
	Step Plate or Equivalent

67197-EFSE-G210

Tools needed for the following hub removal on the Ford Full-Floating axle

**** WARNING**

Do not drop the outer hub bearing when removing the hub.

12. Remove the rear hub assembly.

**** WARNING**

Install a new hub seal each time the hub assembly is removed.

➡The inner bearing is located behind the hub seal.

13. Pack each bearing and replace the hub seals.

**** WARNING**

Use extreme care not to scratch or gouge the seal or bearing surfaces.

14. If after hub removal, the hub seal or seal inner sleeve remains on the spindle, remove using the Step Plate and the 2-Jaw Puller.
15. Inspect the seal surface and inner shoulder for scratches and damage. Remove all scratches, gouges or galling damage with No. 600 or finer crocus cloth.

To install:

➡Clean the spindle thoroughly after removing the rear hub.

16. Coat the spindle with axle lubricant.

**** WARNING**

The hub bearings must be prelubed prior to installation.

17. Fill the hub cavity with 1 oz. of axle lubricant.

**** WARNING**

Use extreme care not to damage the hub seal by allowing it to contact the spindle during installation.

➡Coat the spindle and hub seal inside diameter with axle lubricant.

➡Installing the rear hub in this manner causes the outer bearing to act as a pilot making the installation easier.

18. Push the rear hub and outer bearing onto the spindle as an assembly. Hold the outer bearing seated and use the bearing as a pilot.

**** WARNING**

Install a new hub nut if the hub nut comes apart during installation.

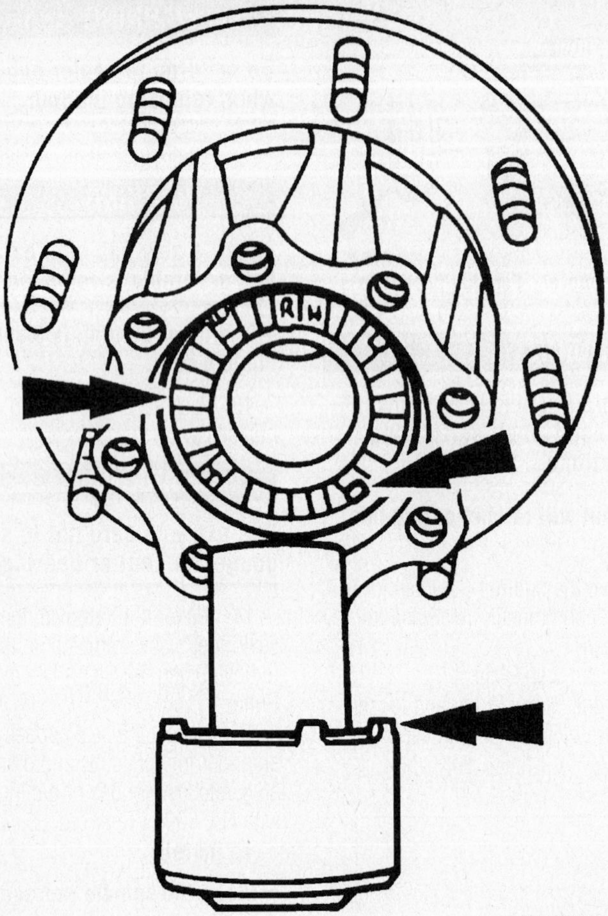

67197-EFSE-G212

Install the Ford Axle Locknut Socket so that the drive tangs of the tool engage the four slots in the hub nut

➡Make sure the hub nut tab is located in the keyway prior to thread engagement.

19. Install the hub nut on the spindle. Turn the hub nut clockwise for right-hand thread or counterclockwise for left-hand thread.
20. Position the Ford Axle Locknut Socket on the hub nut.

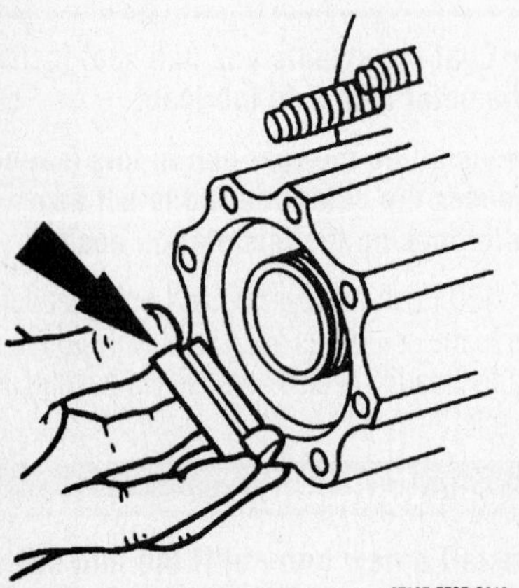

67197-EFSE-G213

Install the Step Plate

⁂ WARNING

Under no circumstances are power tools to be used when performing these operations.

➡The hub nut will ratchet as torque is applied.

21. Tighten the hub nut, rotating the rear hub occasionally while tightening. Torque to 60 ft. lbs. (81 Nm).
22. Adjust hub nuts as follows:
    a. For new bearings, ratchet back five teeth or notches (1/8 turn) on the hub nut. Five notches must be felt during this operation in order to have performed it correctly.
    b. For used bearings, ratchet back seven teeth or notches (1/6 turn) on the hub nut. Seven notches must be felt during this operation to have performed it correctly.
23. Inspect the axle shaft O-ring seal for cracks, nicks or wear and replace it if required.

➡Lubricate the O-ring seal with lubricant prior to installation of axle shaft.

24. Install the axle shaft.

➡Coat the threads of the retaining bolts with Stud and Bearing Mount E0AZ-19554-BA or equivalent meeting Ford specification WSK-M2G349-A1.

25. Install and tighten the retaining bolts until they seat.

⁂ WARNING

Remember, the last step of this procedure is to tighten the axle shaft bolts to specification, after the wheel lug nuts have been tightened.

26. Install the brake rotor and caliper on the single rear wheel axles.
27. Install the wheels and tires but do not tighten the lug nuts to specification at this time.
28. Check the axle lubricant level.
29. Lower the vehicle.
30. Tighten the wheel lug nuts.
31. Tighten the axle shaft retaining bolts. Torque to 80 ft. lbs. (109 Nm).

2001–04 E-350 and F-350 w/Dual Rear Wheels and Dana Axle

1. Before servicing the vehicle, refer to the precautions in the beginning of this section.

All vehicles

2. Remove the tire and wheel assembly.
3. Remove the anchor plate.
4. Remove the axle shaft.

Dana 70

➡Make sure that the drive tangs on the special tool engage the four slots of the hub nut.

5. Using special tool 205-282, or equivalent, remove the hub nut.

All vehicles

6. Remove the outer rear wheel bearing.

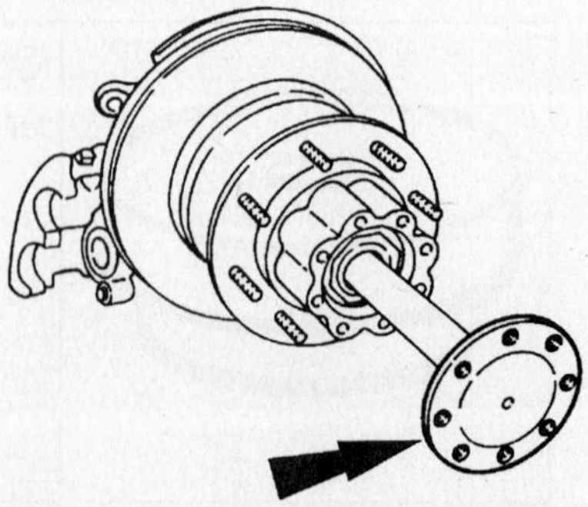

67197-EFSE-G243

Remove the axle shaft

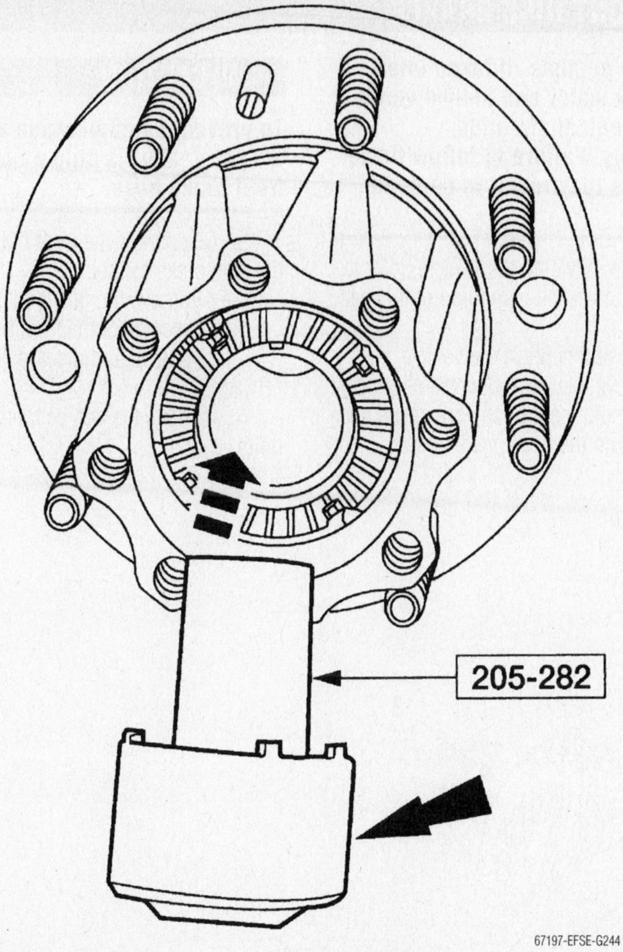

Using special tool 205-282, or equivalent, remove the hub nut

7. Remove the rear hub and brake disc assembly.
8. Remove the bolts and separate the rear hub from the rear brake disc.
9. Inspect the rear hub for the following:
   - Cracks and damage around the bolt holes.
   - Oversized holes.

To install:

⁂ WARNING

Install a new rear hub seal after removing the rear hub from the axle. A damaged or worn seal can permit bearing lubricant to reach the brake linings, resulting in ineffective brake operation. Failure to follow these instructions may result in personal injury.

⁂ WARNING

Clean and remove any dirt or foreign material in the rear hub bolt holes.

10. Install a new rear hub seal.
11. Position the rear brake disc on the rear hub and install the bolts. Torque to 66–88 ft. lbs. (89–119 Nm).

⁂ WARNING

Thoroughly clean the spindle. Wrap the spindle threads with electrician's tape to prevent damage while installing the rear hub and brake disc assembly.

⁂ WARNING

Lightly coat the spindle and pack each rear wheel bearing with Premium Long-Life Grease XG-1-C or equivalent meeting Ford specification ESA-M1C75-B.

12. Prepare the spindle for rear hub installation.
13. Slide the rear hub and brake disc assembly over the axle housing spindle. Remove the electrician's tape.
14. Install the outer rear wheel bearing.
15. Start the hub nut making sure that the tab aligns correctly in the keyway prior to thread engagement.

➡Apply inward pressure to the socket to separate the ratcheting components of the hub nut.

16. To adjust the bearings, tighten the nut to 70 ft. lbs. (95 Nm).
17. Back off the nut 90 degrees.
18. Tighten the nut to 18 ft. lbs. (24 Nm). To verify that there is no side-to-side end play, attach a magnetically mounted dial indicator to the spindle end and place the dial indicator tip on the outboard surface of the hub. Check for side-to-side end play. Final bearing adjustment has zero end play. The maximum torque to rotate the hub is 2.3 Nm (20 inch lbs.) when end play is zero.
19. Install the axle shaft.
20. Install the anchor plate.
21. Install the tire and wheel assembly.

FRONT DISC BRAKES

Brake Caliper

REMOVAL & INSTALLATION

2001–03 E-150

** CAUTION

Brake fluid contains polyglycol ethers and polyglycols. Avoid contact with eyes. Wash hands thoroughly after handling. If brake fluid contacts eyes, flush eyes with running water for 15 minutes. Get medical attention if irritation persists. If taken internally, drink water and induce vomiting. Get medical attention immediately. Failure to follow these instructions may result in personal injury.

1. Before servicing the vehicle, refer to the precautions in the beginning of this section.
2. Raise and support the vehicle.
3. Remove the tire and wheel assembly.
4. If so equipped, unclip the speed sensor wiring from the front brake hose.

** CAUTION

To prevent contamination and reduce air entry, always plug a disconnected front brake hose.

5. Disconnect the front brake hose from the disc brake caliper.
   a. Remove the flow bolt.
   b. Disconnect the front brake hose.
   c. Remove and discard the copper washers.
6. Remove the two disc brake caliper bolts.

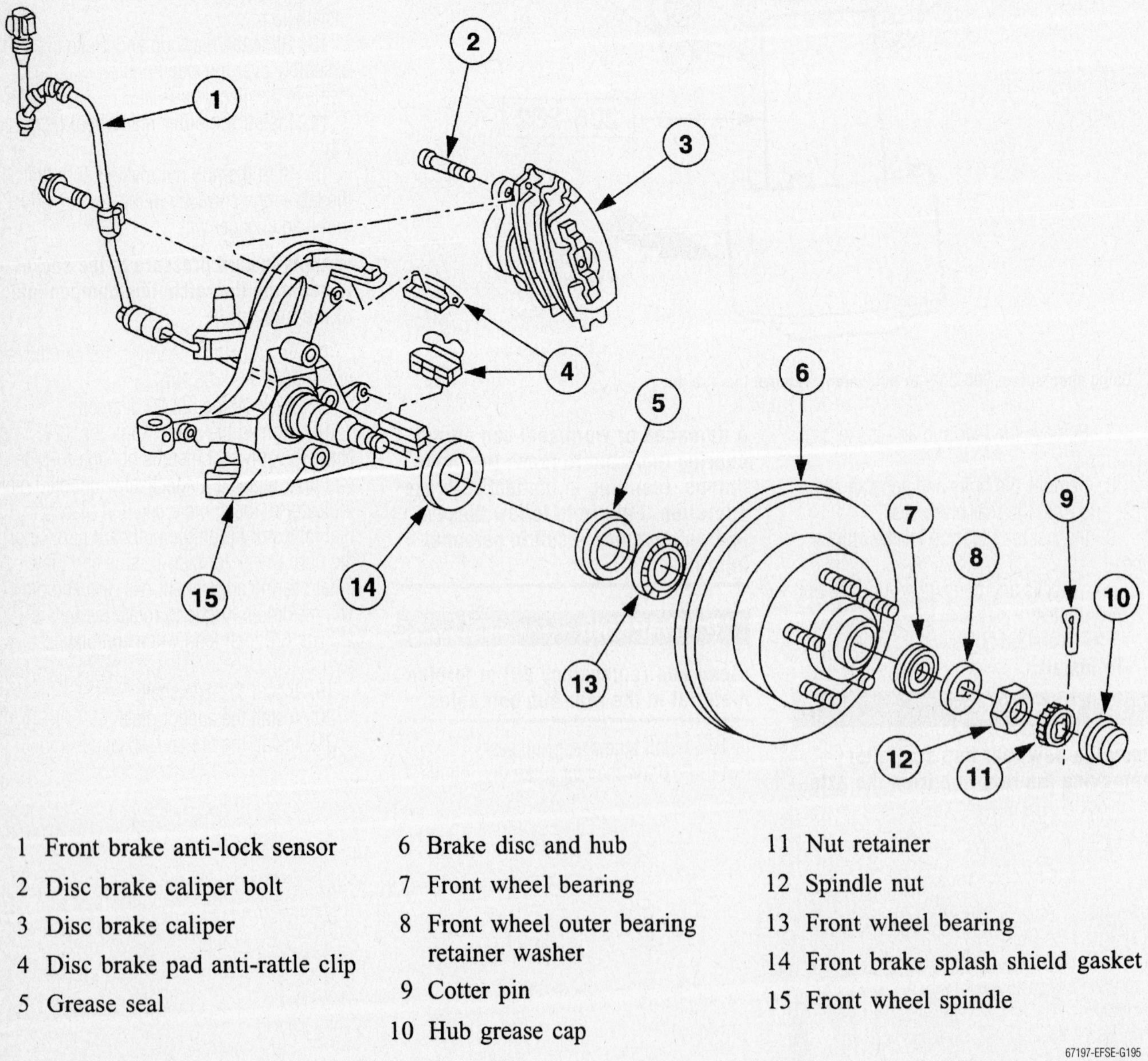

Front disc brake components—2001–03 E-150

7. Remove the disc brake caliper from the front wheel spindle.

8. Inspect the disc brake caliper for brake fluid leakage. If the disc brake caliper is leaking, it must be rebuilt or replaced.

To install:

CAUTION

To prevent deterioration of the caliper sleeve boots, do not use petroleum-based lubricant.

9. Fill the rubber caliper sleeve boots with silicone brake caliper grease.

10. Install the disc brake caliper and the front brake hose.

a. Position the disc brake caliper in the front wheel spindle.

b. Install and tighten the disc brake caliper bolts. Torque to 22–26 ft. lbs. (30–36 Nm).

c. Using new copper washers, install the front brake hose and the bolt. Torque to 22–30 ft. lbs. (30–40 Nm).

11. If so equipped, attach the speed sensor wiring to the front brake hose.

12. Bleed the brake system.

13. Install the wheel and tire assembly.

14. Inspect the brake system operation.

2004 E-150

1. Before servicing the vehicle, refer to the precautions in the beginning of this section.

2. Remove the tire and wheel assembly.

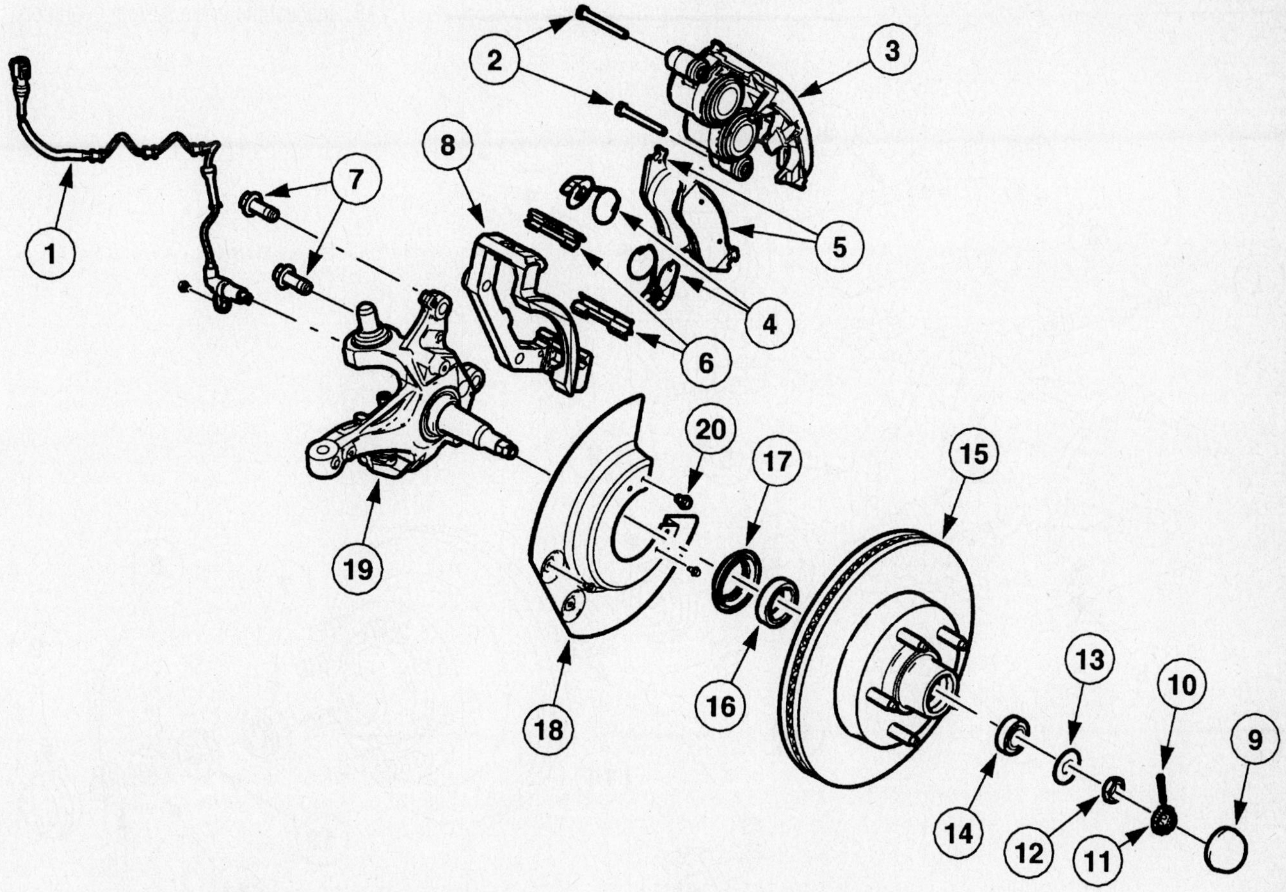

1 Front brake anti-lock sensor	11 Nut retainer
2 Disc brake caliper bolt	12 Spindle nut
3 Disc brake caliper	13 Front wheel outer bearing retainer washer
4 Disc brake pad retaining springs	14 Front wheel bearing/cup
5 Front disc pads	15 Brake disc and hub
6 Disc brake pad anti-rattle clip	16 Front wheel bearing/cup
7 Disc brake caliper anchor bracket bolts	17 Grease seal
8 Disc brake caliper anchor bracket	18 Backing plate
9 Hub grease cap	19 Front wheel spindle
10 Cotter pin	20 Backing plate screw

67197-EFSE-G198

Front disc brake components—2004 E-150

3. Release the speed sensor wiring from the front brake hose.

CAUTION

To prevent contamination and reduce air entry, always plug a disconnected front brake hose.

4. Disconnect the front brake hose from the disc brake caliper.
   a. Remove the flow bolt.
   b. Disconnect the front brake hose.
   c. Remove and discard the copper washers.

5. Remove the two disc brake caliper bolts.
6. Remove the disc brake caliper from the front disc brake caliper anchor plate.
7. Inspect the disc brake caliper for brake fluid leakage. If the disc brake caliper is leaking, it must be rebuilt or installed new.

To install:

CAUTION

To prevent deterioration of the caliper sleeve boots, do not use petroleum-based lubricant.

8. Fill the rubber caliper sleeve boots with caliper grease.
9. Position the disc brake caliper in the front disc brake caliper anchor plate.
10. Install and tighten the disc brake caliper bolts. Torque to 24 ft. lbs. (32 Nm).
11. Using new copper washers, install the front brake hose and the bolt. Torque to 26 ft. lbs. (35 Nm).
12. Attach the speed sensor wiring to the front brake hose.
13. Bleed the brake system.
14. Install the wheel and tire assembly.
15. Inspect the brake system operation.

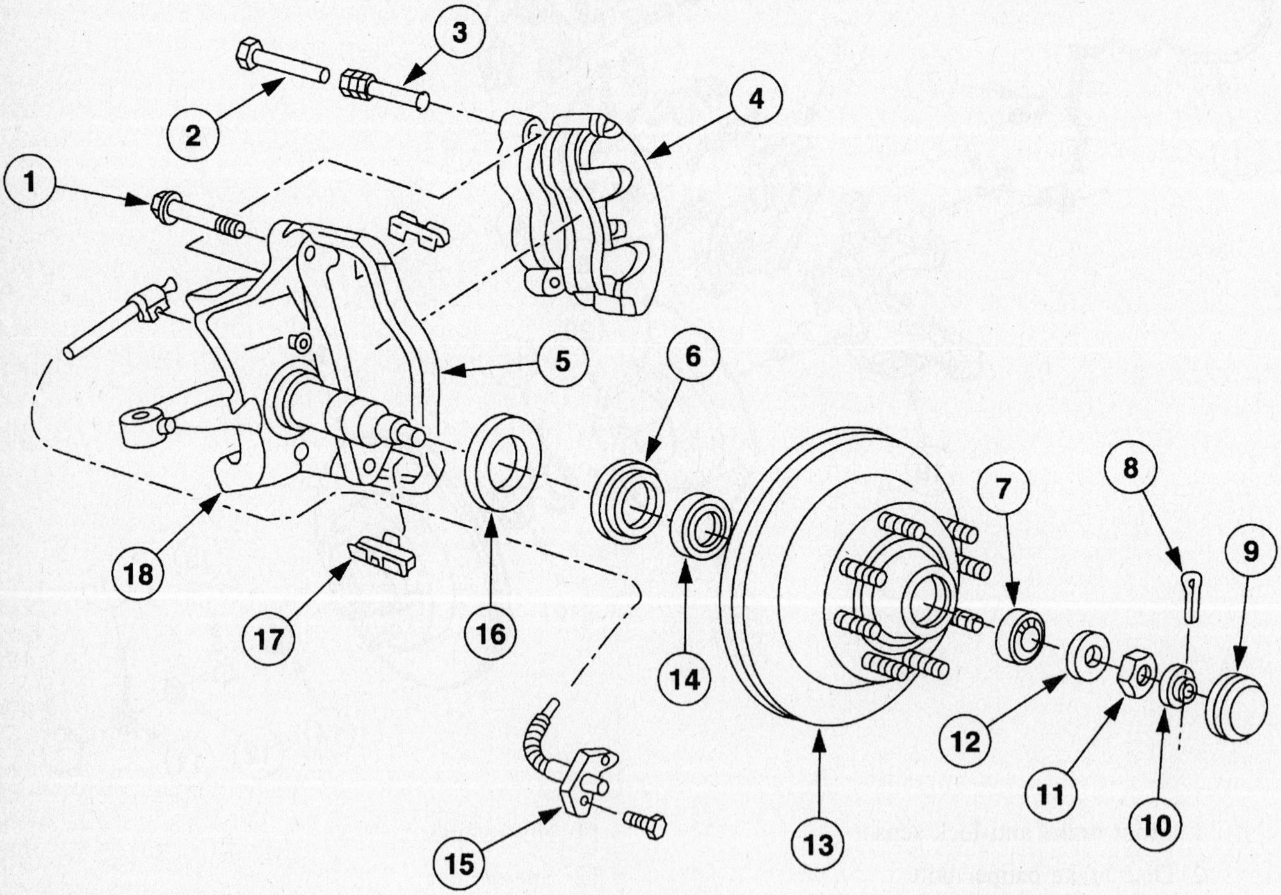

1 Anchor bracket-to-spindle bolt
2 Disc brake caliper bolt
3 Caliper bolt sleeve
4 Disc brake caliper
5 Front disc brake caliper anchor plate
6 Wheel hub grease seal
7 Front wheel bearing
8 Cotter pin
9 Hub grease cap
10 Nut retainer
11 Spindle nut
12 Front wheel outer bearing retainer washer
13 Brake disc and hub
14 Bearing cone and roller
15 Front brake anti-lock sensor
16 Front brake splash shield gasket
17 Disc brake pad anti-rattle clip
18 Front wheel spindle

67197-EFSE-G196

Front disc brake components—E-250, 350 single rear wheel

2001–04 E-250, 350

CAUTION

Brake fluid contains polyglycol ethers and polyglycols. Avoid contact with eyes. Wash hands thoroughly after handling. If brake fluid contacts eyes, flush eyes with running water for 15 minutes. Get medical attention if irritation persists. If taken internally, drink water and induce vomiting. Get medical attention immediately. Failure to follow these instructions may result in personal injury.

1. Before servicing the vehicle, refer to the precautions in the beginning of this section.
2. Raise and support the vehicle.
3. Remove the tire and wheel assembly.
4. If so equipped, unclip the speed sensor wiring from the front brake hose.

CAUTION

To prevent contamination and reduce air entry, always plug a disconnected front brake hose.

5. Disconnect the front brake hose from the disc brake caliper.
   a. Remove the flow bolt.
   b. Disconnect the front brake hose.
   c. Remove and discard the copper washers.
6. Remove the two disc brake caliper bolts.
7. Remove the disc brake caliper from the anchor plate.
8. Inspect the disc brake caliper for brake fluid leakage. If the disc brake caliper is leaking, it must be rebuilt or replaced.

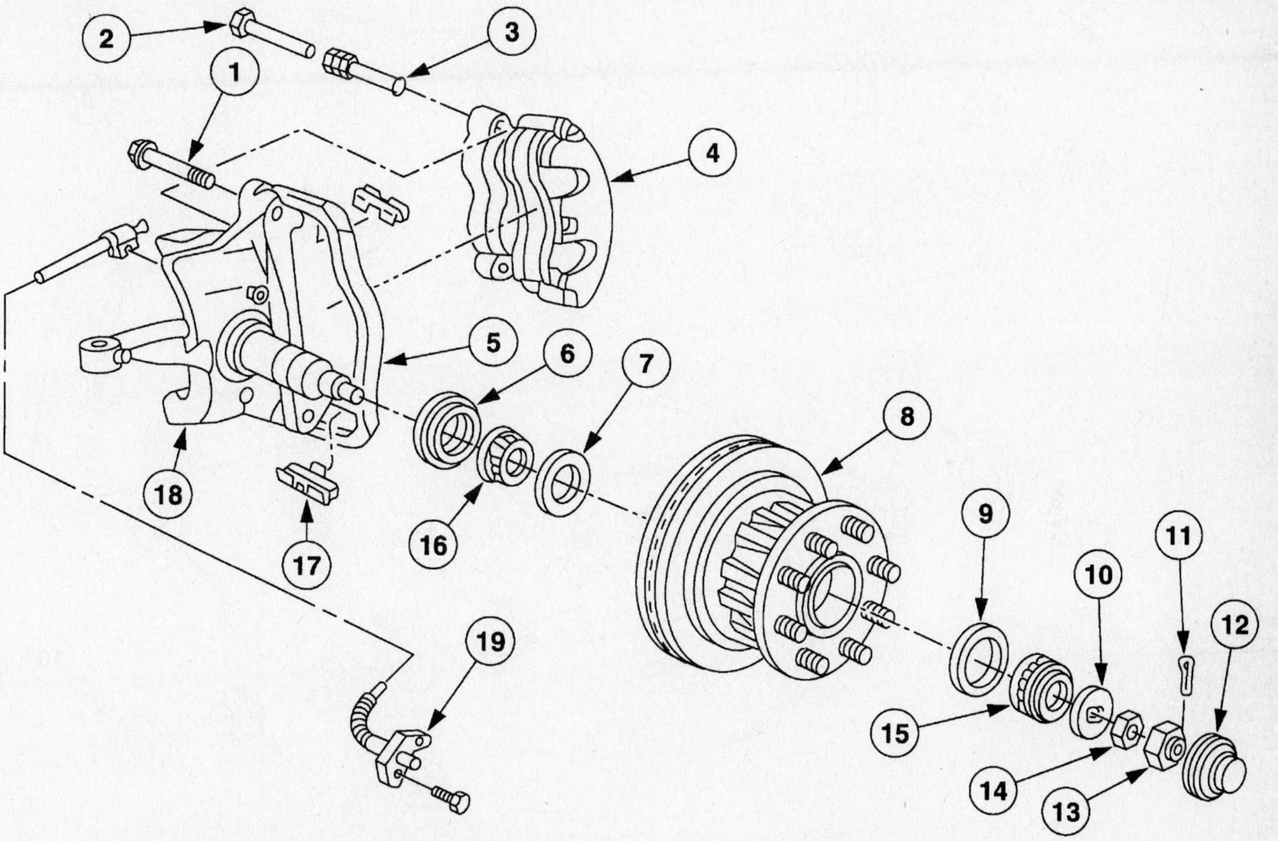

1 Anchor bracket-to-spindle bolt
2 Disc brake caliper bolt
3 Caliper bolt sleeve
4 Disc brake caliper
5 Front disc brake caliper anchor plate
6 Wheel hub grease seal
7 Bearing cup, inner
8 Brake disc and hub (dual rear wheel)
9 Bearing cup, outer
10 Front wheel outer bearing retainer washer
11 Cotter pin
12 Hub grease cap
13 Nut retainer
14 Spindle nut
15 Front wheel bearing (outer)
16 Front wheel bearing (inner)
17 Disc brake pad anti-rattle clip
18 Front wheel spindle
19 Front brake anti-lock sensor

67197-EFSE-G197

Front disc brake components—E-250, 350 dual rear wheel

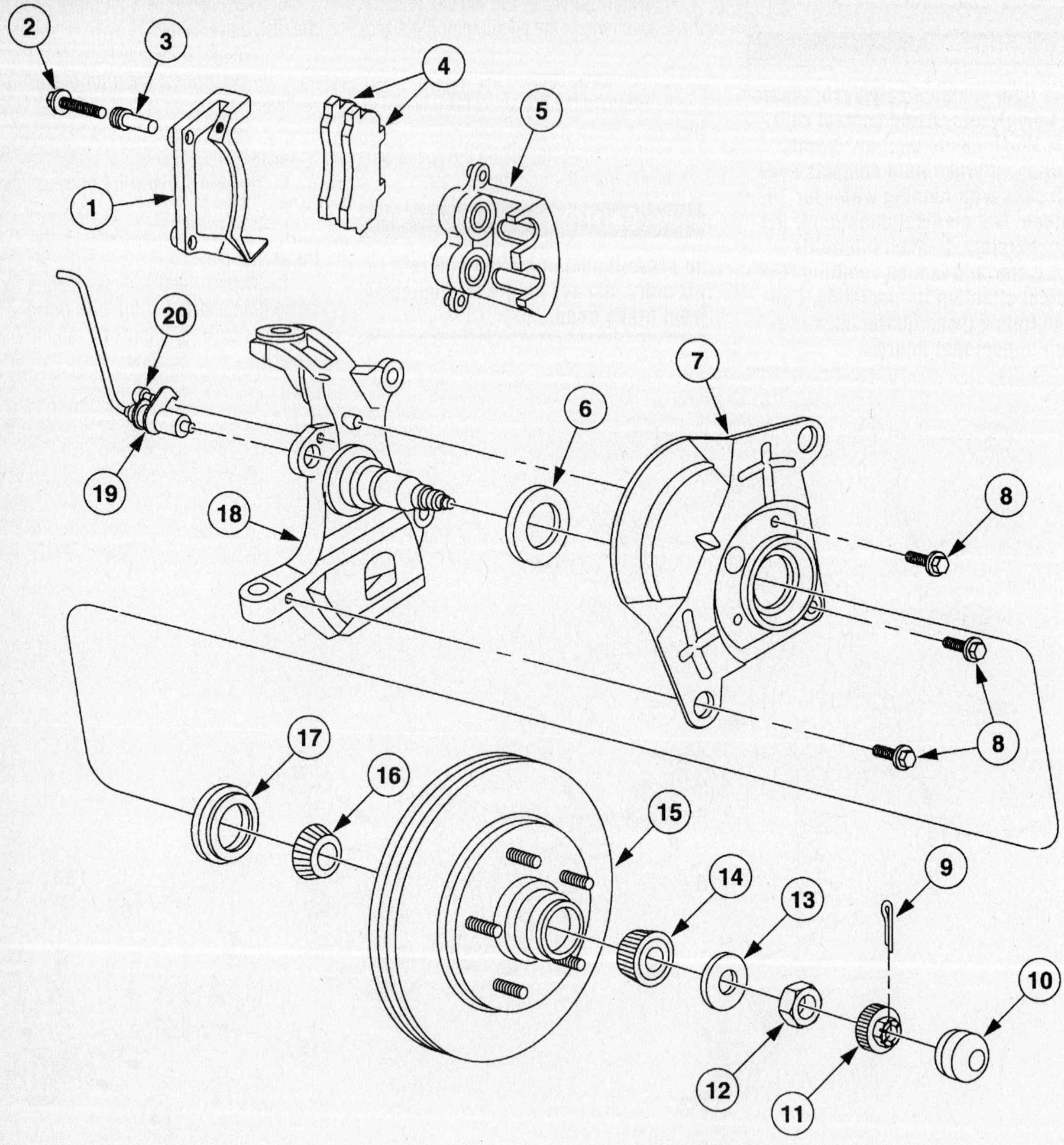

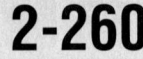

1 Front disc brake caliper anchor plate
2 Disc brake caliper bolt
3 Pin assembly
4 Pads
5 Disc brake caliper
6 Front brake splash shield gasket
7 Brake disc shield
8 Brake disc shield bolts
9 Cotter pin
10 Hub grease cap
11 Nut retainer
12 Spindle nut
13 Front wheel outer bearing retainer washer
14 Front wheel bearing (outer)
15 Brake disc and hub
16 Front wheel bearing (inner)
17 Wheel hub grease seal
18 Front wheel spindle
19 Front brake anti-lock sensor
20 Anti-lock brake sensor bolt

67197-EFSE-G199

Front disc brake components—2001–03 2-Wheel Drive F-150 and 2004 Heritage Editions

To install:

CAUTION

To prevent deterioration of the caliper sleeve boots, do not use petroleum-based lubricant.

9. Fill the rubber caliper sleeve boots with silicone brake caliper grease.
10. Install the disc brake caliper and the front brake hose.
   a. Position the disc brake caliper on the anchor plate.
   b. Install and tighten the disc brake caliper bolts. Torque to 16–30 ft. lbs. (22–40 Nm) for 2001–03 models; 23 ft. lbs. (31 Nm) for 2004 models.
   c. Using new copper washers, install the front brake hose and the bolt. Torque to 22–30 ft. lbs. (30–40 Nm) for 2001–03 models; 28 ft. lbs. (35 Nm) for 2004 models.
11. If so equipped, attach the speed sensor wiring to the front brake hose.
12. Bleed the brake system.
13. Install the wheel and tire assembly.
14. Inspect the brake system operation.

2001–03 F-150 and 2004 Heritage Editions

1. Before servicing the vehicle, refer to the precautions in the beginning of this section.

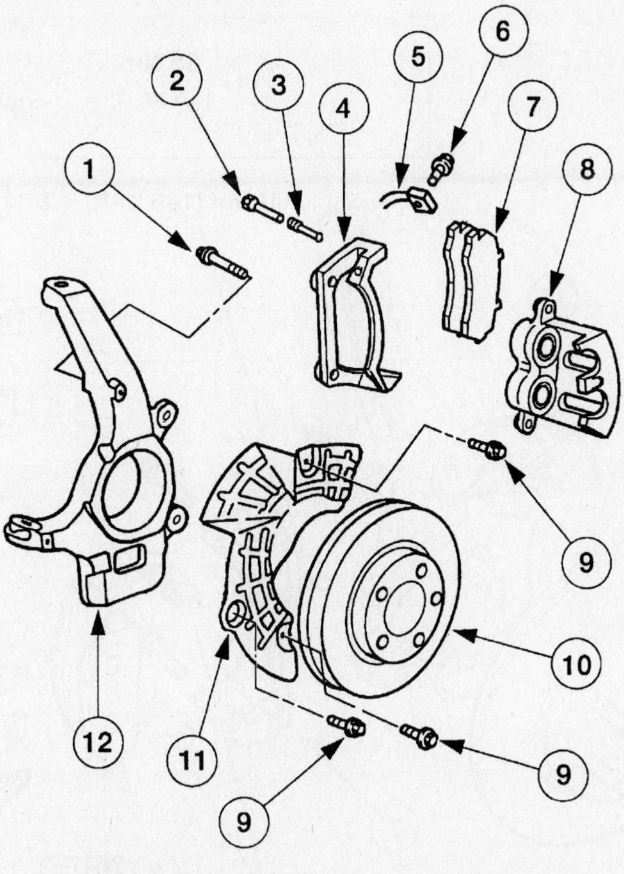

1 Front disc brake caliper anchor plate bolt
2 Disc brake caliper bolt
3 Pin assembly
4 Front disc brake caliper anchor plate
5 Front brake hose
6 Front brake hose bolt
7 Pads
8 Disc brake caliper
9 Bolt
10 Brake disc
11 Brake disc shield
12 Front wheel knuckle

67197-EFSE-G200

Front disc brake components—2001–03 4-Wheel Drive F-150 and 2004 Heritage Editions

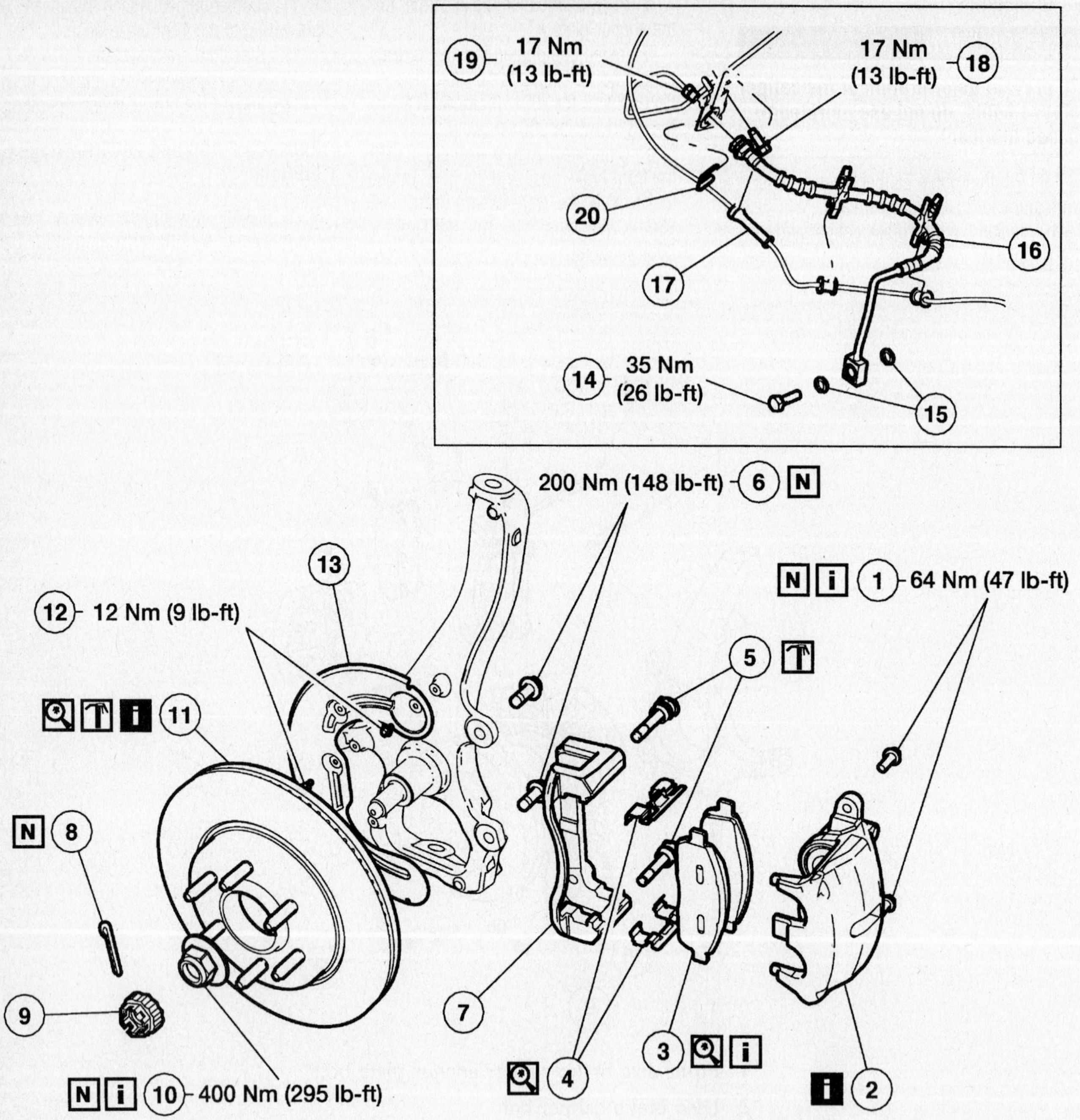

1 Caliper bolt (2 required for each side)	11 Brake disc (heavy duty and light duty)
2 Brake caliper (RH/LH)	12 Dust shield bolts (3 required each side)
3 Brake disc pads (1 kit LH and RH)	13 Dust shield (RH/LH)
4 Slippers	14 Flow bolt
5 Guide pin and boot (2 required for each side)	15 Copper washers (2 required each side)
6 Anchor bracket bolts (2 required for each side)	16 Brake line (RH/LH)
7 Anchor bracket	17 Anti-lock brake sensor cable
8 Cotter pin	18 Brake line bracket bolt
9 Retainer	19 Brake line fitting
10 Spindle nut (1 each side)	20 Retainer clip

67197-EFSE-G201

Front disc brake components—2004 F-150 2-Wheel Drive, exc. Heritage Edition

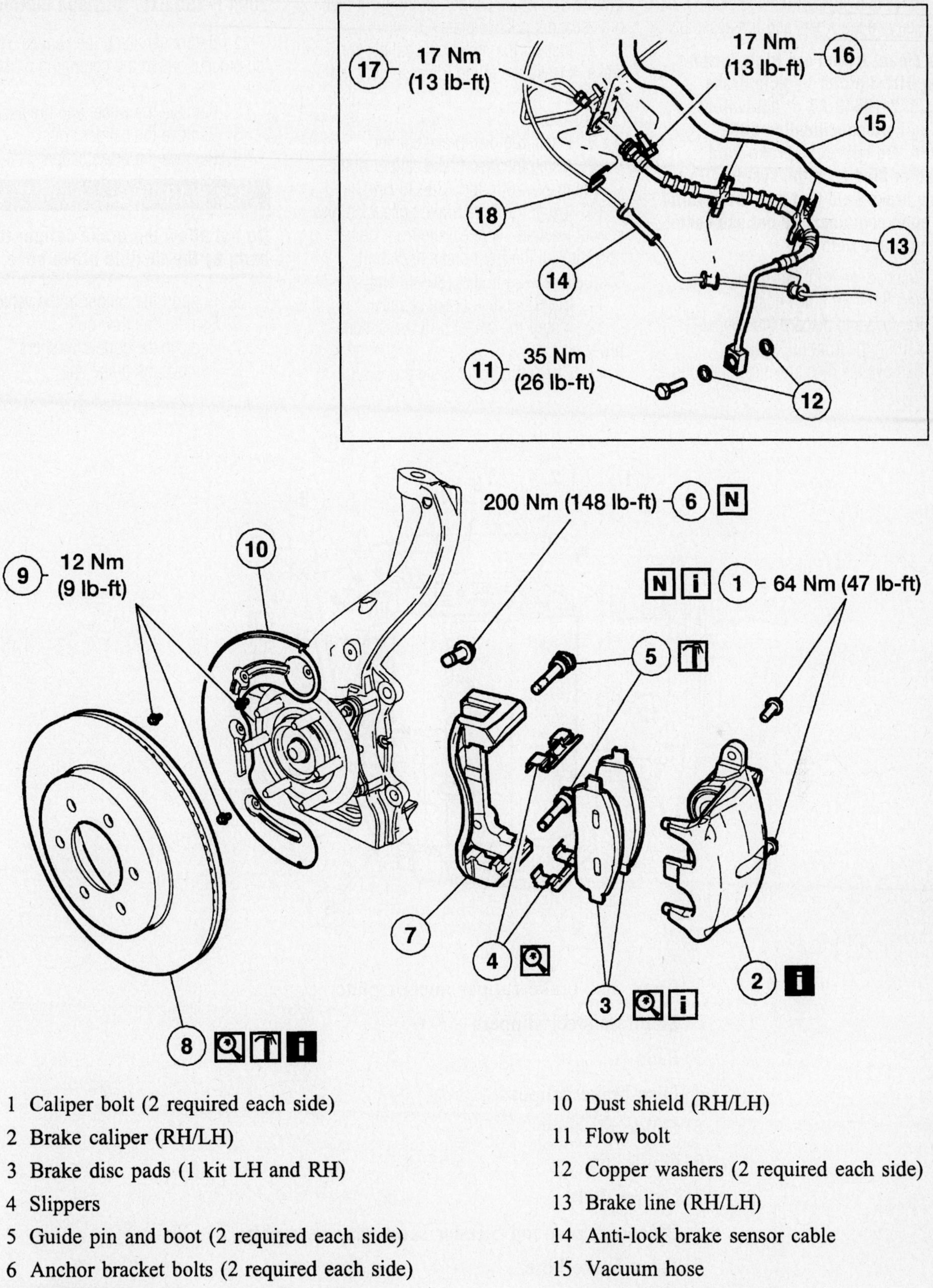

1 Caliper bolt (2 required each side)
2 Brake caliper (RH/LH)
3 Brake disc pads (1 kit LH and RH)
4 Slippers
5 Guide pin and boot (2 required each side)
6 Anchor bracket bolts (2 required each side)
7 Anchor bracket
8 Brake disc (heavy duty and light duty)
9 Dust shield bolts (3 required each side)
10 Dust shield (RH/LH)
11 Flow bolt
12 Copper washers (2 required each side)
13 Brake line (RH/LH)
14 Anti-lock brake sensor cable
15 Vacuum hose
16 Brake line bracket bolt
17 Brake line fitting
18 Retainer clip

67197-EFSE-G202

Front disc brake components—2004 F-150 4-Wheel Drive, exc. Heritage Edition

2. Raise and support the vehicle.
3. Remove the wheel and tire assembly.

➡Use clean, fresh Ford High Performance DOT 3 Motor Vehicle Brake Fluid C6AZ-19542-AB or equivalent meeting Ford specification ESA-M6C25A. Do not reuse brake fluid drained or bled from the system. Do not use brake fluid that has been stored in an open container. Do not use contaminated brake fluid.

4. Remove the front brake hose bolt. Disconnect the front brake hose.
5. Remove and discard the copper washers. Plug the front brake hose.
6. Remove the disc brake caliper bolts. Lift the disc brake caliper off the front disc brake caliper anchor plate.
7. Inspect the disc brake caliper for leaks. If leaks are found, disassembly is required.

To install:

8. Install the disc brake caliper.
9. Install the disc brake caliper bolts. Torque to 21–26 ft. lbs. (28–36 Nm).
10. Install the front brake hose. Use new copper washers; connect the front brake hose. Install the front brake hose bolt. Torque to 23–29 ft. lbs. (30–40 Nm).
11. Bleed the disc brake caliper.
12. Install the wheel and tire assembly.
13. Check the brake system operation.

2004 F-150 Exc. Heritage Editions

1. Before servicing the vehicle, refer to the precautions in the beginning of this section.
2. Remove the wheel and tire assembly.
3. Remove the caliper bolts
4. Remove the brake caliper

⁂ WARNING

Do not allow the brake caliper to hang by the flexible brake hose.

5. Support the caliper to the vehicle.
6. Remove the flow bolt
7. Remove the copper washers
8. Remove the brake line

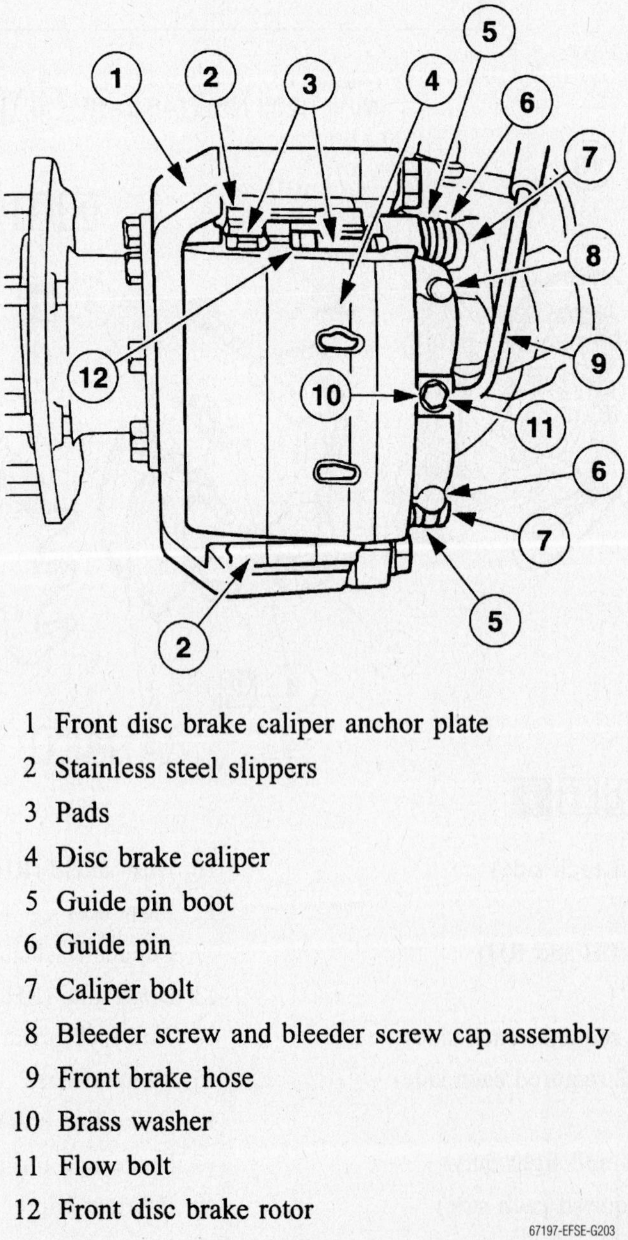

Front disc brake—F-250/350

9. To install, reverse the removal procedure. Tighten the bottom caliper bolt and then the top caliper bolt. Always use new copper washers.
10. Observe the following torques:
- Caliper mounting bolts: 47 ft. lbs. (64 Nm)
- Brake line flow bolt: 26 ft. lbs. (35 Nm)

11. Bleed the brake system.

2001–04 F-250, 350

1. Before servicing the vehicle, refer to the precautions in the beginning of this section.
2. Raise and support the vehicle.
3. Remove the wheel and tire assembly.
4. Remove the brake hose bolt.
5. Disconnect the brake hose.
6. Remove and discard the copper washers. Plug the brake hose.
7. Remove the disc brake caliper pin bolts.
8. Lift the disc brake caliper from the disc brake caliper anchor plate.
9. Inspect the disc brake caliper for leaks. If leaks are found, disassembly is required.

To install:

10. Install the disc brake caliper.
11. Install the disc brake caliper pin bolts. Torque to 42 ft. lbs. (56 Nm).
12. Using new copper washers, connect the brake hose. Install the brake hose bolt. Torque to 26 ft. lbs. (35 Nm).
13. Bleed the brake system.
14. Install the wheel and tire assembly.
15. Fill the brake master cylinder reservoir with clean DOT 3 motor vehicle brake fluid. Install the brake master cylinder filler cap.
16. Inspect the brake system operation.

Brake Pads

REMOVAL & INSTALLATION

2001–03 E-Series

1. Before servicing the vehicle, refer to the precautions in the beginning of this section.
2. Remove and discard enough brake fluid from the brake master cylinder to allow room for the brake fluid displaced when the caliper piston is pressed to the bottom of the caliper bore.
3. Raise and support the vehicle.
4. Remove the tire and wheel assembly.
5. If so equipped, unclip the speed sensor wiring from the front brake hose.
6. Remove the two disc brake caliper bolts.

⁂ WARNING

Do not allow the disc brake caliper to hang from the front brake hose. Use wire to support the disc brake caliper from a convenient underbody component.

7. Remove the disc brake caliper.
8. Remove the pads.

To install:

9. If installing new pads, use a C-clamp and a pad to press the caliper piston to the bottom of the caliper bore. Install the pads.

⁂ CAUTION

To prevent deterioration of the caliper sleeve boots, do not use petroleum-based lubricant.

10. Fill the rubber caliper sleeve boots with silicone brake caliper grease.
11. Install the disc brake caliper.
12. If so equipped, attach the speed sensor wiring to the front brake hose.
13. Install the tire and wheel assembly. Before driving, pump the brake pedal a few times to seat the pads.

2004 E-Series

1. Before servicing the vehicle, refer to the precautions in the beginning of this section.
2. Remove and discard enough brake fluid from the brake master cylinder to allow room for the brake fluid displaced when the caliper piston is pressed to the bottom of the caliper bore.
3. Remove the tire and wheel assembly.
4. Release the speed sensor wiring from the front brake hose.
5. Remove the two disc brake caliper bolts.

⁂ WARNING

Do not allow the disc brake caliper to hang from the front brake hose. Use wire to support the disc brake caliper from a convenient underbody component.

6. Remove and support the disc brake caliper.
7. Remove the brake pad retaining springs.
8. Remove the brake pads from the brake caliper anchor bracket.

To install:

➡Bottoming of the caliper piston is not necessary if reusing the original pads.

9. If installing new pads, use a suitable tool and a pad to press the caliper piston to the bottom of the caliper bore.

⁂ CAUTION

To prevent deterioration of the caliper sleeve boots, do not use petroleum-based lubricant.

10. Fill the rubber caliper sleeve boots with caliper grease.
11. Install the brake pads.
12. Install the brake pad retaining springs.
13. Install the disc brake caliper. Install and tighten the disc brake caliper bolts.
14. Attach the speed sensor wiring to the front brake hose.
15. Install the tire and wheel assembly.

2001–03 F-150 and 2004 Heritage Editions

1. Before servicing the vehicle, refer to the precautions in the beginning of this section.
2. Raise and support the vehicle.
3. Remove the wheel and tire assembly.
4. Inspect the pads for wear or contamination.

⁂ CAUTION

When removing the front disc brake caliper, never allow it to hang from the brake hose. Provide a suitable support.

5. Remove the two disc brake caliper bolts.
6. Lift the disc brake caliper off the front disc brake caliper anchor plate.
7. Inspect the disc brake caliper for leaks. If leaks are found, disassembly is required.
8. Remove the pads.
9. Remove the disc brake anti-rattle clips.

To install:

⁂ CAUTION

Do not allow grease, oil, brake fluid or other contaminants to contact the pad lining material. Do not install contaminated pads.

10. Install the disc brake anti-rattle clips.
11. Install the pads.
12. Install the disc brake caliper.

13. Install the disc brake caliper bolts. Torque to 21–26 ft. lbs. (28–36 Nm).
14. Install the tire and wheel assembly.
15. Check brake operation.

2004 F-150, Exc. Heritage Editions

1. Before servicing the vehicle, refer to the precautions in the beginning of this section.
2. Remove the wheel and tire assembly.
3. Remove the caliper bolt
4. Remove the brake caliper

CAUTION

Do not allow the brake caliper to hang by the flexible brake hose.

5. Support the caliper to the vehicle.
6. Remove the brake disc pads
7. Remove the slippers
8. To install, reverse the removal procedure. If replacing the brake pads, remove the brake fluid in the master cylinder reservoir until it is half filled.
9. If replacing the brake pads, use a suitable tool to push the brake caliper pistons into the caliper bore.

➡Tighten the bottom caliper bolt first and then the top caliper bolt. Torque to 47 ft. lbs. (64 Nm).

10. If the hydraulic system was opened, bleed the brake system.

2001–04 F-250, 350

1. Before servicing the vehicle, refer to the precautions in the beginning of this section.
2. Raise and support the vehicle.
3. Remove the wheel and tire assembly.
4. ®Inspect the pads for wear or contamination.

➡When removing the front disc brake caliper, never allow it to hang from the brake hose. Provide a suitable support.

5. Remove the two disc brake caliper bolts.
6. Lift the disc brake caliper off the front disc brake caliper anchor plate.
7. Measure the brake disc and hub thickness. Install a new brake disc and hub if not within specification.
8. If necessary, resurface the brake disc. Ford recommends on-vehicle brake disc machining. Follow lathe manufacturer's instructions.
9. Inspect the disc brake caliper for leaks. If leaks are found, disassembly is required.
10. If necessary, remove the V-springs.
11. Remove the pads.
12. Remove the disc brake anti-rattle clips.

To install:

CAUTION

Do not allow grease, oil, brake fluid or other contaminants to contact the pad lining material. Do not install contaminated pads.

13. Install the disc brake anti-rattle clips.
14. Install the pads.
15. Install the V-springs.

CAUTION

Use care not to damage the bleeder screw or front wheel disc brake shield.

16. Install the disc brake caliper.
17. Install the disc brake caliper bolts.
18. Install the tire and wheel assembly.
19. Check brake operation.

Brake Rotor

REMOVAL & INSTALLATION

2001–04 E-150

1. Before servicing the vehicle, refer to the precautions in the beginning of this section.
2. Remove the pads.
3. Remove the hub grease cap.
4. Remove the cotter pin.
5. Remove the nut retainer.
6. Remove the spindle nut.
7. Remove the front wheel outer bearing retainer washer.
8. Remove the outer front wheel bearing.
9. Remove the brake disc and hub.
10. Remove the wheel hub grease seal.
11. Remove the inner front wheel bearing.

B>To install:

12. Thoroughly clean and inspect the front wheel bearings and the brake disc and hub.
13. Lubricate the front wheel bearings.
14. Install the inner front wheel bearing.
15. Install a new wheel hub grease seal.
16. Position the brake disc and hub.
17. Install the outer front wheel bearing.
18. Install the front wheel outer bearing retainer washer.
19. Install the spindle nut.
20. Adjust the wheel bearings as described in this chapter.
21. Install the nut retainer.
22. Install the cotter pin.
23. Install the hub grease cap.
24. Install the pads.

2001–04 E-250, 350; 2-Wheel Drive F-250, 350

1. Before servicing the vehicle, refer to the precautions in the beginning of this section.
2. Remove the front disc brake caliper anchor plate.
3. Remove the hub grease cap.
4. Remove the cotter pin.
5. Remove the nut retainer.
6. Remove the spindle nut.
7. Remove the front wheel outer bearing retainer washer.
8. Remove the outer front wheel bearing.
9. Remove the brake disc and hub.
10. Remove the wheel hub grease seal.
11. Remove the inner front wheel bearing.

To install:

12. Thoroughly clean and inspect the front wheel bearings and the brake disc and hub.
13. Lubricate the front wheel bearings.
14. Install the inner front wheel bearing.
15. Install a new wheel hub grease seal.
16. Position the brake disc and hub.
17. Install the outer front wheel bearing.
18. Install the front wheel outer bearing retainer washer.
19. Install the spindle nut.
20. Adjust the wheel bearings as described in this chapter.
21. Install the nut retainer.
22. Install the cotter pin.
23. Install the hub grease cap.
24. Install the front disc brake caliper anchor plate.

2001–03 F-150 and 2004 Heritage Editions

1. Before servicing the vehicle, refer to the precautions in the beginning of this section.
2. Remove the front disc brake caliper anchor plate.
3. On 4x4 vehicles, remove the brake disc.
4. On 4x2 vehicles, remove the brake disc and hub as follows:
   a. Remove the hub grease cap.
   b. Remove the cotter pin.
   c. Remove the nut retainer.
   d. Remove the spindle nut.

e. Remove the front wheel outer bearing retainer washer.
f. Remove the outer front wheel bearing).
g. Remove the brake disc and hub.
5. If necessary on 4x2 vehicles, remove the front wheel bearing.
a. Remove the wheel hub grease seal.
b. Remove the front wheel bearing.

To install:

6. On 4x4 vehicles, position the brake disc to the wheel hub. Use Metal Brake Parts Cleaner F3AZ-19579-SA or equivalent to clean the brake disc and hub.
7. On 4x2 vehicles, thoroughly clean and inspect the front wheel bearings and the brake disc and hub. Use Metal Brake Parts Cleaner F3AZ-19579-SA or equivalent.
8. On 4x2 vehicles, lubricate the front wheel bearings. Use Premium Long-Life Grease XG-1-C or -K or equivalent meeting Ford specification ESA-M1C75-B.
9. On 4x2 vehicles, install a new wheel hub grease seal.
a. Install the inner front wheel bearing.
b. Install a new wheel hub grease seal.
10. On 4x2 vehicles, install the brake disc and hub.
a. Position the brake disc and hub.
b. Install the outer front wheel bearing.
c. Install the front wheel outer bearing retainer washer.
d. Install the spindle nut.
11. While rotating the brake disc and hub, tighten the spindle nut to 30 ft. lbs. (40 Nm).
12. Loosen the spindle nut two turns.
13. Tighten the spindle nut 17–24 ft. lbs. (23–34 Nm) while rotating the brake disc and hub.
14. Loosen the spindle nut ½ turn.
15. Tighten the spindle nut to 17 inch lbs. (2 Nm) while rotating the brake disc and hub.
16. On 4x2 vehicles, install the following components:
a. Install the nut retainer.
b. Install the cotter pin.
c. Install the hub grease cap.
17. Install the front disc brake caliper anchor plate.

2004 F-150 Exc. Heritage Editions

1. Before servicing the vehicle, refer to the precautions in the beginning of this section.
2. Remove the wheel and tire assembly.
3. Remove the anchor bracket bolts and position the caliper and bracket aside.
4. Remove the cotter pin
5. Remove the retainer
6. Remove the spindle nut
7. Remove the brake disc
8. To install, reverse the removal procedure.
9. Tighten the spindle nut to 295 ft. lbs. (400 Nm). Rotate the hub counterclockwise 5 rotations and then recheck the torque.
10. If the hydraulic system was opened, bleed the brake system.

2001–04 4-Wheel Drive F-250, 350

1. Before servicing the vehicle, refer to the precautions in the beginning of this section.
2. Remove the disc brake caliper anchor plate.

➡Perform this step for F-250 and F-350 4x4 SRW vehicles.

3. Remove the rotor.

➡Perform this step for DRW vehicles.

➡If excessive force must be used during brake rotor removal, the brake rotors should be checked for lateral runout prior to installation.

4. Remove the eight hub extender nuts, the hub plate and the rotor.
a. Remove the eight hub plate nuts.
b. Remove the hub plate.
c. Remove the rotor.

To install:

5. Position the front disc brake rotor to the wheel hub. Make sure the wheel hub and the front disc brake rotor braking and mounting surfaces are clean. Use brake parts cleaner to clean the front disc brake rotor.

➡Perform this step for DRW vehicles only.

6. Install the front wheel hub extender and nuts.
7. Install the front disc brake caliper anchor plate.

Anchor Plate

REMOVAL & INSTALLATION

2001–03 F-150 and 2004 Heritage Editions; 2001–04 F-250, 350; 2002–04 E-150; 2001–04 E-250, 350

1. Before servicing the vehicle, refer to the precautions in the beginning of this section.
2. Remove the pads.
3. Remove the two front disc brake caliper anchor plate bolts.
4. Remove the front disc brake caliper anchor plate.

To install:

5. Position the front disc brake caliper anchor plate.
6. Install the front disc brake caliper anchor plate bolts. Observe the following torques:
   - F-150: 136 ft. lbs. (185 Nm)
   - F-250 and 350: 166 ft. lbs. (225 Nm)
   - 2001–03 E-Series: 141–191 ft. lbs. (191–259 Nm)
   - 2004 E-150: 129 ft. lbs. (175 Nm)
   - 2004 E-250 and 350: 166 ft. lbs. (225 Nm)
7. Install the pads.

2004 F-150 Exc. Heritage Editions

1. Before servicing the vehicle, refer to the precautions in the beginning of this section.
2. Remove the wheel and tire assembly.
3. Remove the brake caliper.

⁂ WARNING

Do not allow the brake caliper to hang by the flexible brake hose.

4. Support the caliper to the vehicle.
5. Remove the brake disc pads.
6. Remove the slippers.
7. Remove the guide pin and boot.
8. Remove the anchor bracket bolts.
9. Remove the anchor bracket.
10. To install, reverse the removal procedure. Torque the anchor bracket bolts to 148 ft. lbs. (200 Nm).

➡Tighten the bottom caliper bolt first and then the top caliper bolt.

REAR DISC BRAKES

Brake Caliper

REMOVAL & INSTALLATION

2001 E-350, 2002–04 E-250, 350

1. Before servicing the vehicle, refer to the precautions in the beginning of this section.
2. Raise and support the vehicle.
3. Remove the wheel and tire assembly.
4. Remove the caliper pin bolts.
5. Remove the flow bolt and discard the copper washers.
6. Remove the rear disc brake caliper.

To install:

➡Use new copper washers on the flow bolt.

7. Follow the removal procedure in reverse order.
8. Torque the flow bolt to 26 ft. lbs. (35 Nm).
9. Torque the caliper pin bolts to 27 ft. lbs. (36 Nm).
10. Bleed the brake system.

2004 E-150

1. Before servicing the vehicle, refer to the precautions in the beginning of this section.
2. Remove the wheel and tire assembly.
3. Remove the brake hose flow bolt and

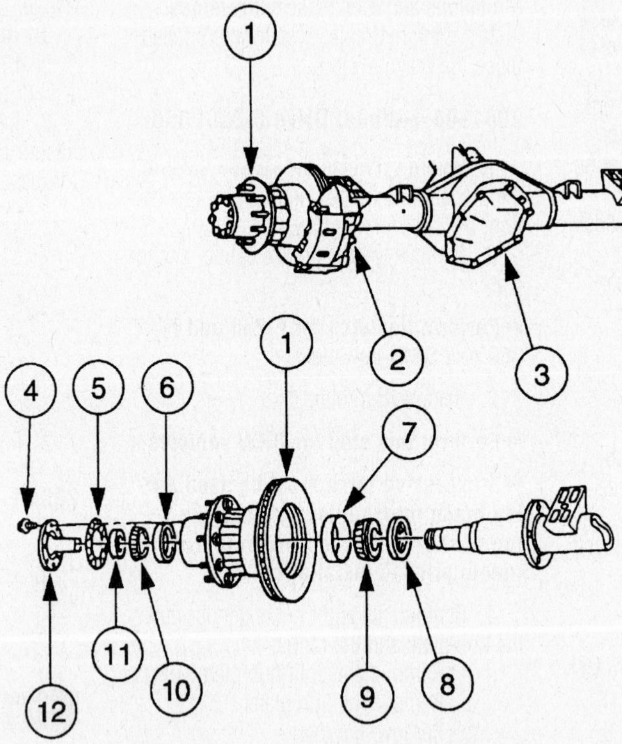

1 Rear brake disc and hub assembly
2 Rear disc brake caliper
3 Dana full-floating axle — Model 80
4 Axle shaft-to-rear hub bolt
5 Rear wheel gasket
6 Outer bearing cup
7 Inner bearing cup
8 Inner hub seal
9 Rear wheel bearing inner cone and roller
10 Rear wheel bearing outer cone and roller
11 Wheel bearing lock nut
12 Axle shaft

67197-EFSE-G204

Rear disc brake exploded view—2001 E-350, 2002–04 E-250, 350

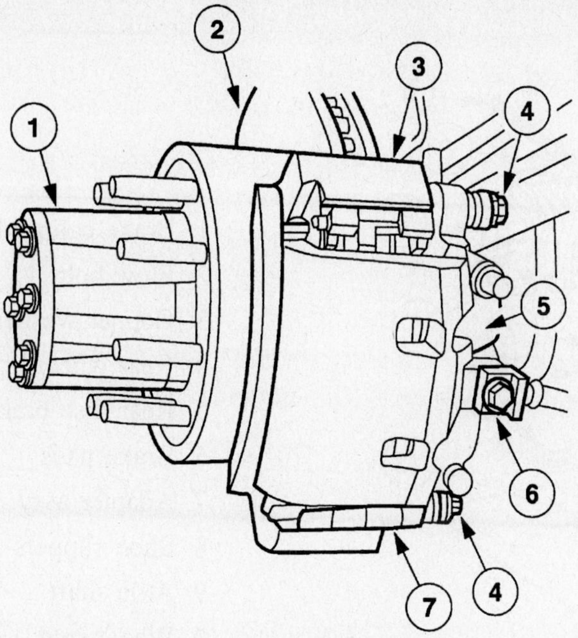

1 Hub
2 Brake disc
3 Caliper anchor plate
4 Caliper pin bolts
5 Caliper
6 Brake hose
7 Stainless steel slippers

67197-EFSE-G205

Rear disc brake caliper—2001 E-350, 2002–04 E-250, 350

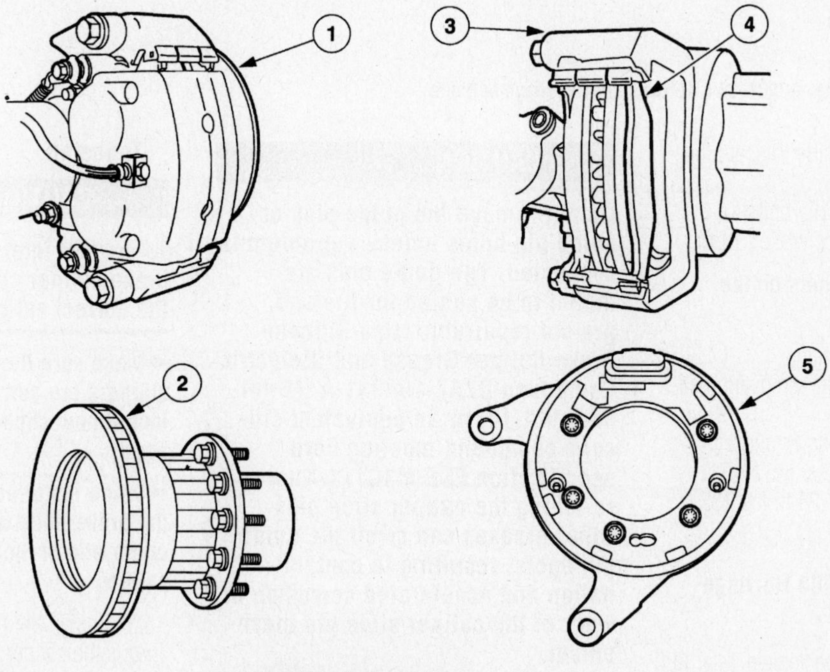

1 Rear disc brake caliper
2 Rear brake disc and hub assembly
3 Rear disc brake caliper anchor plate
4 Brake pads
5 Rear wheel disc brake adapter

67197-EFSE-G206

Rear disc brake and hub—2001 E-350, 2002–04 E-250, 350

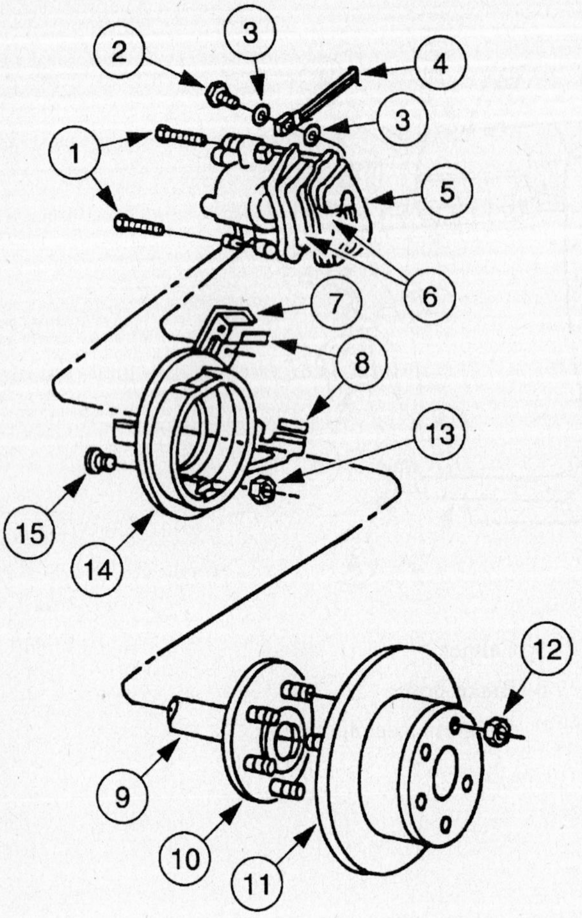

1 Brake caliper bolt
2 Flow bolt
3 Copper washer
4 Rear wheel brake hose
5 Rear disc brake caliper
6 Brake pads
7 Adapter assy
8 Shoe slippers
9 Axle shaft
10 Wheel stud
11 Brake disc
12 Keeper nut
13 Rear wheel disc brake adapter nut
14 Rear wheel disc brake adapter
15 Rear wheel disc brake adapter bolt

67197-EFSE-G207

Rear disc brake components—2001–03 F-150 and 2004 Heritage Editions

position the brake hose aside. Discard the copper washers.

4. Remove the caliper pin bolts.
5. Remove the rear disc brake caliper.

➡**Use new copper washers on the brake hose flow bolt.**

6. Bleed the brake system.
7. To install, reverse the removal procedure.
8. Observe the following torques:
   - Flow bolt: 26 ft. lbs. (35 Nm)
   - Caliper pin bolts: 24 ft. lbs. (32 Nm)

2001–03 F-150 and 2004 Heritage Editions

1. Before servicing the vehicle, refer to the precautions in the beginning of this section.
2. Raise and support the vehicle.
3. Remove the wheel and tire assembly.
4. Remove bolt. Disconnect the rear wheel brake hose. Remove the copper washers and plug the brake hose.

✲✲ CAUTION

Do not remove the guide pins or guide pin boots unless a problem is suspected. The guide pins are meant to be sealed for life and are not repairable. Use Silicone Brake Caliper Grease and Dielectric Compound D7AZ-19A331-A (Motorcraft WA-10) or an equivalent silicone compound meeting Ford specification ESE-M1C171-A for re-lubing the caliper slide pins. Other greases can swell the guide pin boots, resulting in contamination and accelerated corrosion or wear of the caliper slide pin mechanism.

5. Remove the brake caliper bolts.
6. Lift the rear disc brake caliper off the rear disc brake caliper anchor plate.
7. Inspect the rear disc brake caliper for leaks. If leaks are found, disassembly is required.

To install:

✲✲ CAUTION

To prevent interference with rear disc brake caliper operation, install only the correct caliper bolt.

➡**Make sure the stainless steel shoe slippers are correctly positioned. Install new slippers if worn or damaged.**

➡**When installed, the locator notch on the brake pads will be located at the upper end of the rear disc brake caliper.**

8. Install the rear disc brake caliper. Torque the caliper bolts to 20 ft. lbs. (27 Nm).
9. Connect the brake hose and install the caliper flow bolt. Use new copper washers. Torque to 23–29 ft. lbs. (30–40 Nm).
10. Bleed the disc brake caliper.
11. Install the wheel and tire assembly.
12. Verify correct brake operation.

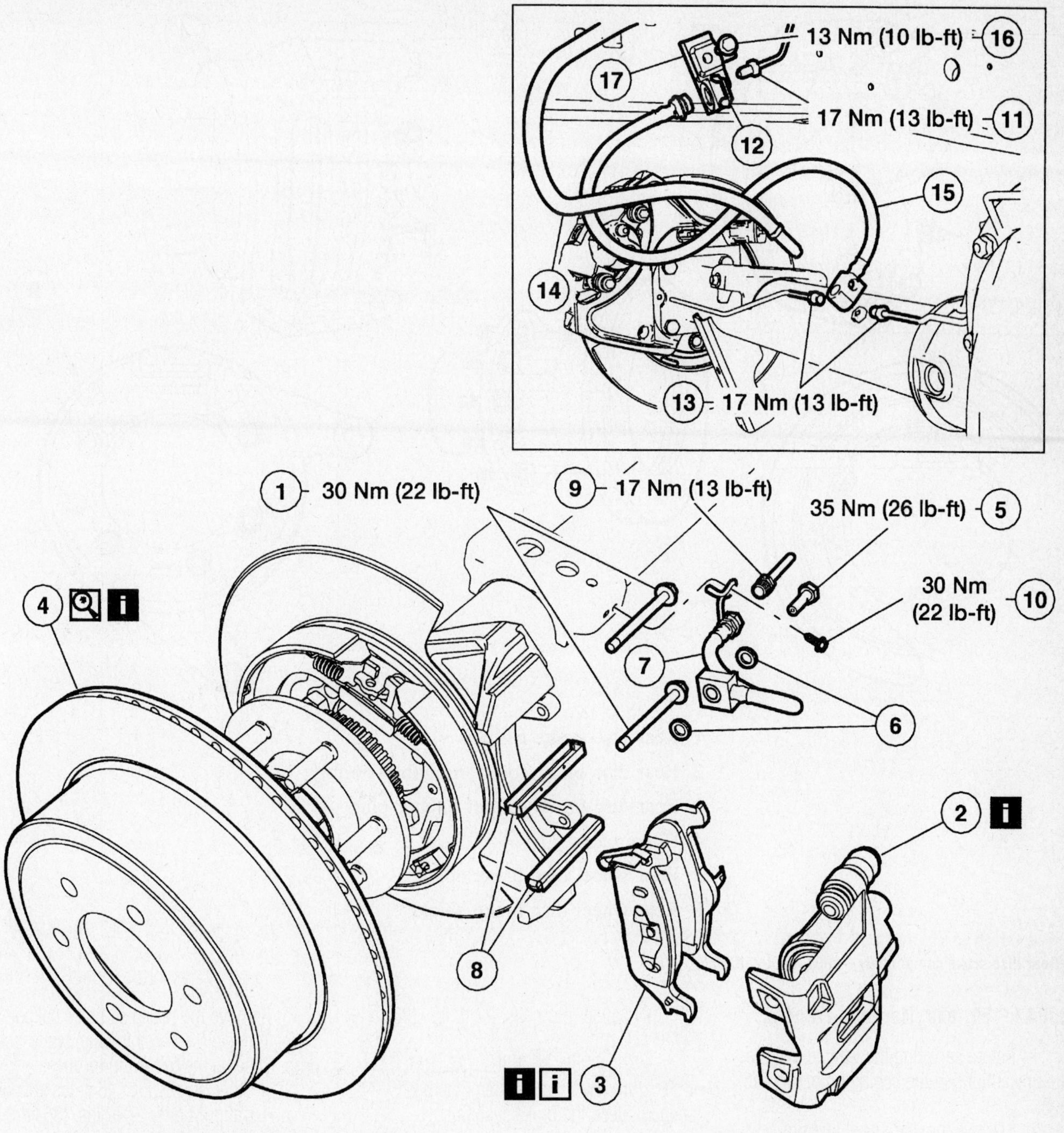

1 Caliper bolts (2 required each side)
2 Brake caliper
3 Disc brake pads (1 kit)
4 Brake disc
5 Flow bolt
6 Brass washers (2 required each side)
7 Brake hose
8 Slippers (2 required each side)
9 Brake line fitting
10 Rear brake hose bracket bolt
11 Brake line fitting
12 Retainer clip
13 Brake line fittings
14 Rear axle vent tube
15 Rear brake jounce hose
16 Rear brake jounce hose bracket bolt
17 Rear brake jounce hose bracket

67197-EFSE-G208

Rear disc brake components—2004 F-150, exc. Heritage Editions

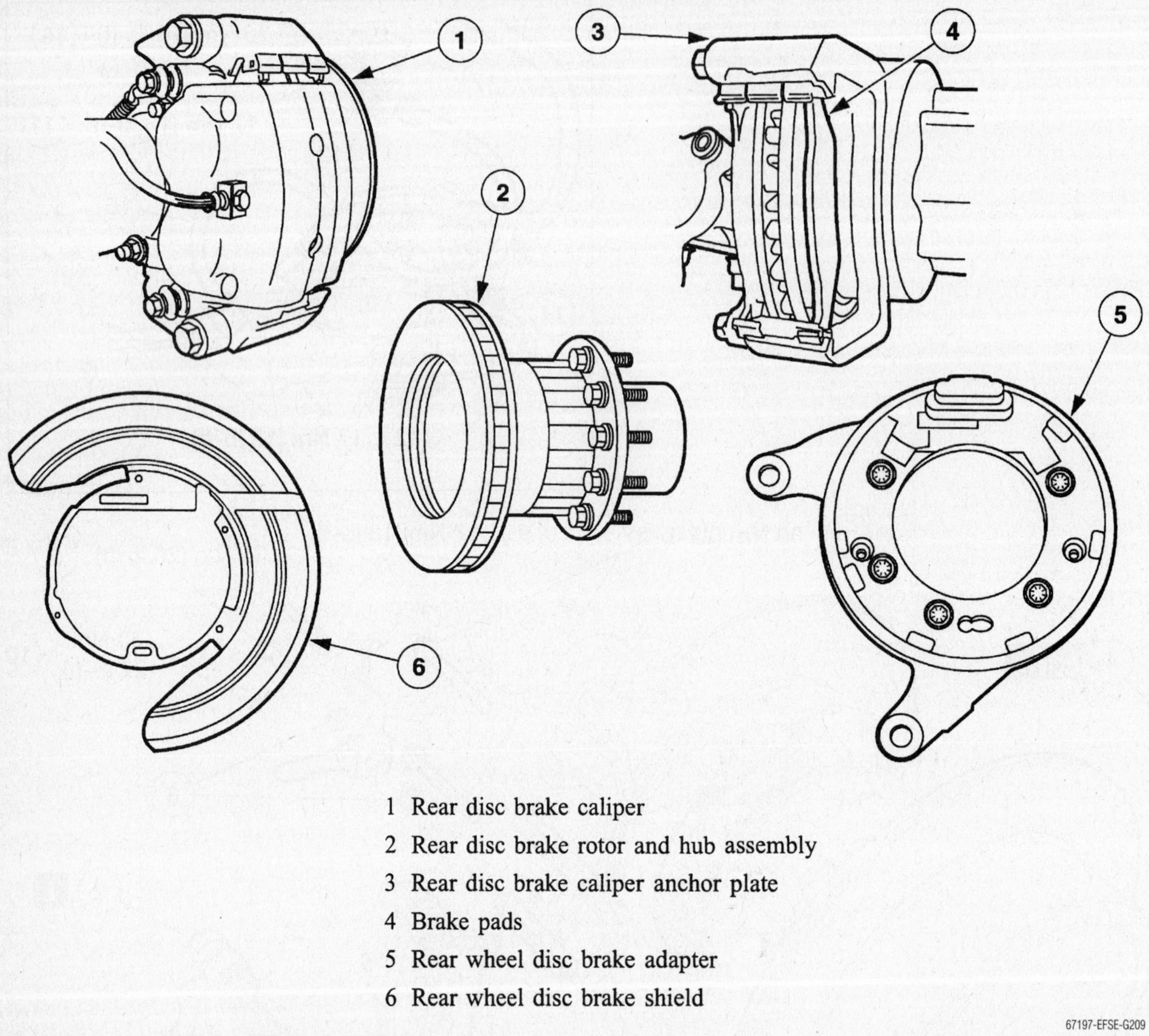

Rear disc brake components—2001–04 F-250, 350

2004 F-150, exc. Heritage Editions

1. Before servicing the vehicle, refer to the precautions in the beginning of this section.
2. Remove the rear wheel and tire assembly.
3. Remove the caliper bolts
4. Remove the brake caliper

⁂ CAUTION

Do not allow the brake caliper to hang by the flexible brake hose.

5. Support the caliper to the vehicle.
6. Remove the flow bolt
7. Remove the brass washers
8. Remove the brake hose
9. To install, reverse the removal procedure. Observe the following torques:
   - Caliper pin bolts: 22 ft. lbs. (30 Nm)
   - Flow bolt: 26 ft. lbs. (35 Nm)
10. If the hydraulic system has been opened, bleed the brake system.

2001–04 F-250, 350

1. Before servicing the vehicle, refer to the precautions in the beginning of this section.

All vehicles

2. Raise and support the vehicle.
3. Remove the wheel and tire assembly.

Forward-of-axle brake calipers

4. Remove the nuts and the stone shield.

All vehicles

5. Remove the caliper pin bolts.
6. Remove the banjo bolt. Remove and discard the copper washers.
7. Remove the rear disc brake caliper.
8. To install, reverse the removal procedure. Observe the following torques:
   - Stone shield nuts: 46 ft. lbs. (62 Nm)
   - Caliper bolts: 27 ft. lbs. (36 Nm)
   - Banjo bolt: 37 ft. lbs. (50 Nm)

➡Prior to installing the banjo bolt, install new copper washers.

9. Bleed the brake system.

Brake Pads

REMOVAL & INSTALLATION

2001 E-350, 2002–04 E-250, 350

1. Before servicing the vehicle, refer to the precautions in the beginning of this section.

2. Remove the brake master cylinder filler cap. Check brake fluid level in brake master cylinder reservoir. Remove fluid until brake master cylinder reservoir is half full.

3. Raise and support the vehicle.

4. Remove the wheel and tire assembly.

5. Inspect the brake pads for wear or contamination. If worn, damaged or past specification, install new components.

6. Remove the caliper pin bolts.

⁂ CAUTION

Never allow the rear disc brake caliper to hang from the brake hose. Provide suitable support.

7. Remove the rear disc brake caliper.

8. Remove the brake pads and rail clips.

9. Inspect the disc brake caliper for leaks. If leaks are found, disassembly is required.

10. Installation is the reverse of removal. Observe the following torques:

- Caliper bolts: 27 ft. lbs. (36 Nm)

2004 E-150

1. Before servicing the vehicle, refer to the precautions in the beginning of this section.

2. Clean the area and remove the brake cylinder filler cap. Check the brake fluid level in the brake master cylinder reservoir. Remove the fluid until the brake master cylinder reservoir is half full.

3. Remove the wheel and tire assembly.

4. Inspect the brake pads for wear or contamination. If worn, damaged or past specification, install new brake pads.

5. Remove the caliper pin bolts.

⁂ CAUTION

Never allow the rear disc brake caliper to hang from the brake hose. Provide suitable support.

6. Position aside the rear disc brake caliper aside.

7. Remove the brake pad clips and the brake pads.

8. If installing new brake pads, use a suitable tool to push the brake caliper pistons into the caliper bore.

9. Inspect the rear disc brake caliper for leaks. If leaks are found, disassembly is required.

10. To install, reverse the removal procedure. Torque the caliper pin bolts to 24 ft. lbs. (32 Nm).

2001–03 F-150, 2004 Heritage Edition

1. Before servicing the vehicle, refer to the precautions in the beginning of this section.

2. Remove brake fluid in the master cylinder reservoir until the reservoir is half full.

⁂ CAUTION

Install new pads if worn to or past the specified thickness above the metal backing plate or rivets. Install new pads in complete axle sets.

3. Inspect the brake pads for wear or contamination. Install new pads if worn to or past specification.

➡It is not necessary to remove the rear wheel brake hose when performing this procedure.

4. Remove the rear disc brake caliper.

⁂ CAUTION

Do not allow grease, oil, brake fluid or other contaminants to contact the brake pads.

5. Remove the brake pads.

6. Retract the caliper piston into the rear disc brake caliper.

7. Remove and discard the slippers.

To install:

⁂ CAUTION

Stainless steel slipper replacement is mandatory with the service brake pad installation, even if the slippers appear undamaged. Make sure the slippers are correctly positioned with the slipper ends snug against the outboard end of the anchor plate rail.

8. Clean the slipper mating surface, and install the new slippers.

⁂ WARNING

Install new brake pads in full axle sets. Do not install new brake pads on only one side of vehicle.

9. Install the brake pads.

10. Install the rear disc brake caliper.

11. Verify correct brake operation.

2004 F-150, Exc. Heritage Editions

1. Before servicing the vehicle, refer to the precautions in the beginning of this section.

2. Remove the rear wheel and tire assembly.

3. Remove the caliper bolts

4. Remove the brake caliper

⁂ WARNING

Do not allow the brake caliper to hang by the flexible brake hose.

5. Support the caliper to the vehicle.

6. Remove the disc brake pads

⁂ WARNING

Do not us any tools to separate the inner brake pad away from the caliper piston. Damage to the caliper piston may occur.

7. Remove the rear brake pads from the caliper.

8. To install, reverse the removal procedure.

⁂ CAUTION

Install a new lining if worn to or past the specified thickness above the metal backing plate or if the pad is damaged. Install new pads in complete axle sets.

9. If replacing the brake pads, use a suitable suction device, remove the brake fluid in the master cylinder reservoir until it is half filled.

10. If replacing the brake pads, use a suitable tool to push the brake caliper pistons into the caliper bore.

11. If the hydraulic system has been opened, bleed the brake system.

2001–04 F-250, 350

1. Before servicing the vehicle, refer to the precautions in the beginning of this section.

All vehicles

2. Remove the brake master cylinder filler cap. Check brake fluid level in brake master cylinder reservoir. Remove fluid until brake master cylinder reservoir is half full.

3. Remove the wheel and tire assembly.

4. Inspect the brake pads for wear or contamination. For additional information, refer to Specifications in this section.

Forward-of-axle brake calipers

5. Remove the nuts and the stone shield.

All vehicles

6. Remove the caliper pin bolts.

⁂ CAUTION

Never allow the rear disc brake caliper to hang from the brake hose. Provide suitable support.

7. Remove the rear disc brake caliper.
8. Remove the brake pads and rail clips.
9. Inspect the disc brake caliper for leaks. If leaks are found, disassembly is required.
10. To install, reverse the removal procedure. Torque the stone shield nuts to 46 ft. lbs. (62 Nm). Torque the caliper pin bolts to 27 ft. lbs. (36 Nm).

Brake Rotor

REMOVAL & INSTALLATION

2001–03 F-150 and 2004 Heritage Editions; 2004 E-150

1. Remove the rear disc brake caliper.

➡If the rear brake disc binds on the rear parking brake shoe and linings, remove the adjustment hole access plug and contract the parking brake shoe and lining.

2. Remove the rear brake disc.
3. Measure the rear brake disc, and resurface if the disc surface has been scored. Install a new rear brake disc if below minimum thickness specification.
4. Installation is the reverse of removal.

2004 F-150, exc. Heritage Editions

1. Remove the rear wheel and tire assembly.
2. Remove the caliper.
3. Remove the disc brake pads.
4. Remove the brake disc.
5. To install, reverse the removal procedure.

⁂ CAUTION

Make sure to apply a thin coat of anti-seize lubricant to the hub flange only. Do not allow the lubricant to make contact with the wheel studs, brake pads or brake disc.

6. Apply a thin coat of anti-seize lubricant to the wheel hub flange.

2001–04 E-350
2002–04 E-250 Single Rear Wheel
2001–F-250 and 350 with Ford 10.5 inch axle and Single Rear Wheel

1. Before servicing the vehicle, refer to the precautions in the beginning of this section.
2. Remove the anchor plate.
3. Match-mark the brake disc and one stud.
4. Remove the brake disc.
5. Installation is the reverse of removal.

2001–04 F-250 and 350 Ford Full-Floating Axle

See the procedure under Rear Wheel Bearings, above.

2001–04 E-350 and F-350 w/Dual Rear Wheels and Dana Axle

See the procedure under Rear Wheel Bearings, above.

Anchor Plate

REMOVAL & INSTALLATION

2001–04 E-350
2002–04 E-250
2004 E-150

1. Before servicing the vehicle, refer to the precautions in the beginning of this section.
2. Remove the brake pads.
3. Remove the anchor plate bolts and remove the rear disc brake caliper anchor plate.
4. To install, reverse the removal procedure. Torque the anchor plate bolts to 128 ft. lbs. (173 Nm); 129 ft. lbs. for 2004 E-150.

2001–04 F-250 and 350

1. Before servicing the vehicle, refer to the precautions in the beginning of this section.
2. Remove the brake pads.

➡Forward-of-axle rear disc brake caliper anchor plates are mounted using stud bolts.

3. Remove the bolts and the rear disc brake caliper anchor plate.

To install:

4. Position the rear disc brake caliper anchor plate. Install the bolts. Torque to 128 ft. lbs. (173 Nm).
5. Install the brake pads.

DRUM BRAKES

Drum

REMOVAL & INSTALLATION

2001–03 E-150 and 250

1. Before servicing the vehicle, refer to the precautions in the beginning of this section.

⁂ CAUTION

To reduce the possibility of uneven braking, always replace rear brake shoes and linings at both ends of an axle.

2. Remove the wheel and tire assembly.

⁂ WARNING

Use of a brake drum puller or a torch is not recommended. Brake drum distortion can result.

➡If the brake drum is rusted to the axle shaft pilot diameter, tap the center of the brake drum between the wheel studs.

3. Remove and discard the spring nut.
4. Pull the brake drum off the axle shaft.

➡If brake drums will not come off, follow these steps.

a. Using a screwdriver, move the brake shoe adjusting lever off the brake adjuster screw.
b. Move the adjustment tool handle downward to loosen the brake shoe adjusting screw nut.
c. Use a Brake Adjustment Tool to loosen the brake shoe adjusting screw nut.

5. Installation is the reverse of removal. Adjust the brake shoes if necessary.

2001–03 F-150 and 2004 Heritage Edition

1. Before servicing the vehicle, refer to the precautions in the beginning of this section.

➡Make sure the parking brake control is fully released.

2. Relieve tension on parking brake system.
a. Pull the front parking brake cable and conduit.

b. Insert a suitable retainer pin.

3. Raise and support vehicle.

4. Remove the wheel and tire assembly.

⁂ WARNING

Use of a brake drum puller or a torch is not recommended. Brake drum distortion may result.

➡If the brake drum is rusted to the axle shaft flange, tap the center of the brake drum.

5. Remove the brake drum.

➡If the brake drum will not come off, follow these steps.

a. Disengage the brake shoe adjusting lever (2A176) from the brake adjuster screw.

b. Use a brake adjustment tool to loosen the brake adjuster screw.

To install:

6. Install the brake drum.

7. If necessary, adjust the rear brakes.

⁂ WARNING

Always remove any corrosion, dirt or foreign material present on the mounting surfaces of the wheel or the surface of the wheel hub or brake drum that contacts the wheel. Installing wheels without correct metal to metal contact at the wheel mounting surfaces can cause the wheel nuts to loosen and could allow the wheel to come off while the vehicle is in motion, causing loss of control.

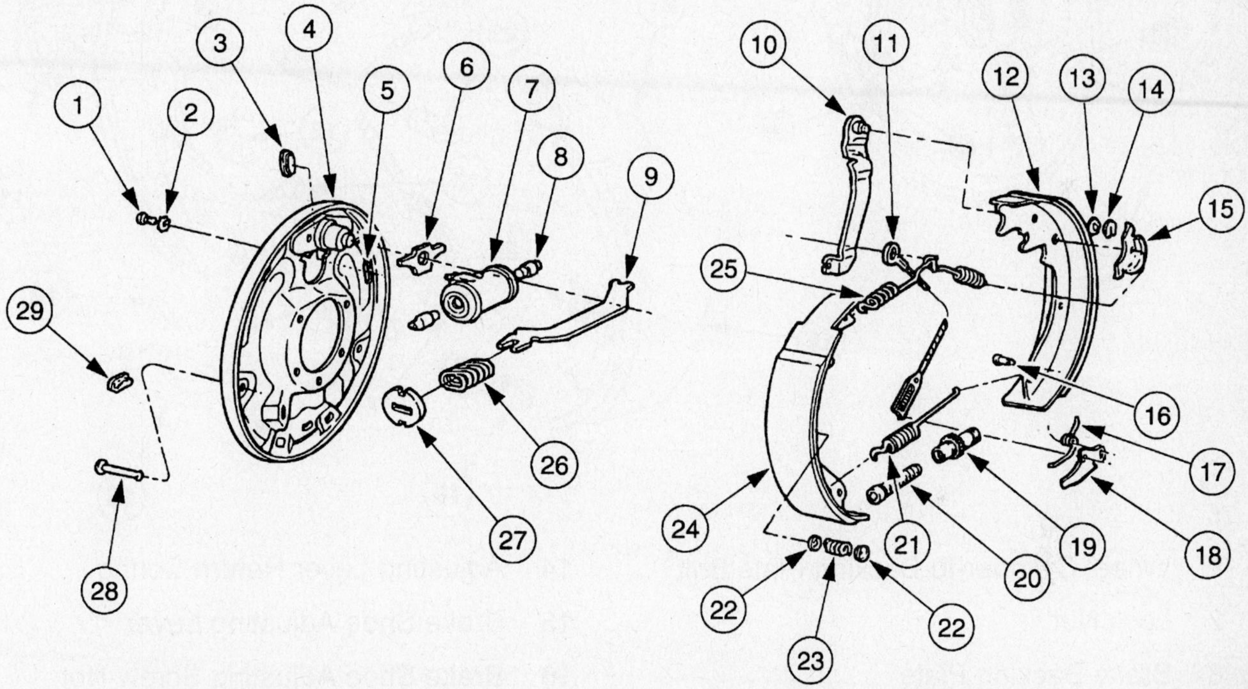

1 Wheel Cylinder-to-Backing Plate Bolt
2 Washer
3 Inspection Hole Cover
4 Brake Backing Plate
5 Lining Inspection Hole
6 Anchor Pin Guide Plate
7 Rear Wheel Cylinder
8 Wheel Cylinder Brake Shoe Link
9 Parking Brake Strut
10 Parking Brake Lever
11 Brake Shoe Adjusting Lever Cable
12 Rear Brake Shoe and Lining, Secondary
13 Washer
14 Parking Brake Lever Pin Retainer
15 Cable Guide
16 Adjusting Lever Pin
17 Adjusting Lever Return Spring
18 Brake Shoe Adjusting Lever
19 Brake Shoe Adjusting Screw Nut
20 Brake Adjuster Screw
21 Brake Shoe Adjusting Screw Spring
22 Brake Shoe Hold-Down Spring Cup
23 Brake Shoe Hold-Down Spring
24 Rear Brake Shoe and Lining, Primary
25 Brake Shoe Retracting Spring, Short
26 Parking Brake Link Spring
27 Parking Brake Spring Retainer
28 Brake Shoe Hold-Down Spring Pin
29 Brake Adjusting Hole Cover

67197-EFSE-G226

Drum brake components—2001–03 F-150 and 2004 Heritage Editions and 2001–03 E-150

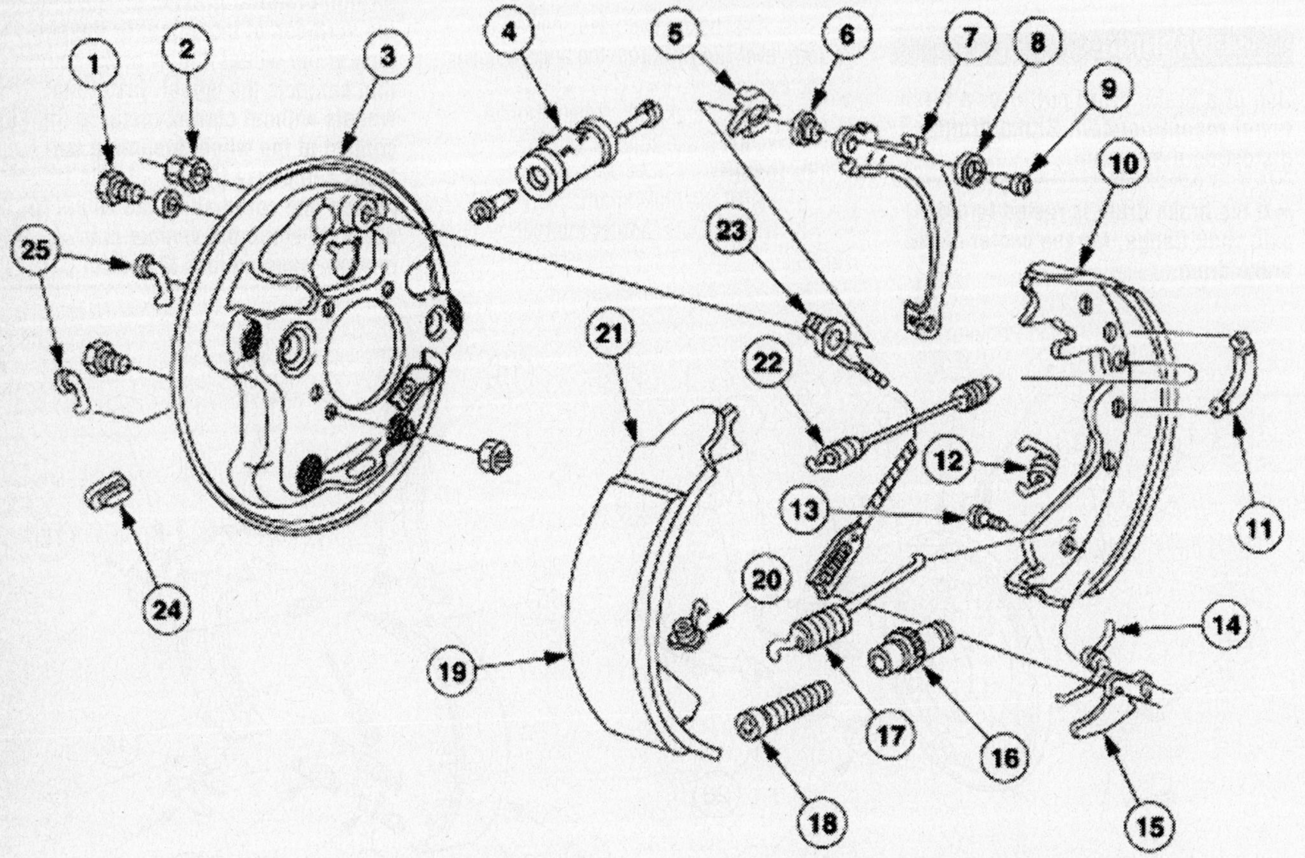

1	Wheel Cylinder-to-Backing Plate Bolt	14	Adjusting Lever Return Spring
2	Lock Nut	15	Brake Shoe Adjusting Lever
3	Brake Backing Plate	16	Brake Shoe Adjusting Screw Nut
4	Rear Wheel Cylinder	17	Brake Shoe Adjusting Screw Spring
5	Brake Shoe Anchor Pin Guide Plate	18	Brake Adjuster Screw
6	Parking Brake Link Spring	19	Shoe Lining
7	Parking Brake Lever	20	Brake Shoe Hold-Down Spring
8	Parking Brake Lever Pin Retainer	21	Rear Brake Shoe and Lining, Primary
9	Parking Brake Lever Bolt	22	Brake Shoe Retracting Spring
10	Rear Brake Shoe and Lining, Secondary	23	Brake Shoe Adjusting Lever Cable
11	Cable Guide	24	Brake Adjusting Hole Cover
12	Brake Shoe Hold-Down Spring	25	Brake Shoe Hold-Down Spring Pin
13	Adjusting Lever Pin		

67197-EFSE-G227

Drum brake components—2001 E-250

8. Clean the wheel hub mounting surface.
9. Install the tire and wheel assembly.
10. Lower the vehicle.
11. Remove the pin in the parking brake control and verify correct parking brake operation.
12. Check brake operation before driving.

Brake Shoes

REMOVAL & INSTALLATION

2001 E-250

2001–03 E-150
2001–03 F-150
2004 Heritage Edition

1. Before servicing the vehicle, refer to the precautions in the beginning of this section.
2. Remove the brake drum.
3. Inspect the rear brake assembly for the following:
   - the rear wheel cylinder for excessive leakage that can contaminate brake system parts, and rebuild as necessary.
   - the rear brake shoes and linings for contamination, and replace as necessary.
   - the rear brake shoes and linings for minimum thickness, above the backing plate or rivets, and replace as necessary.
   - the springs for heat discoloration, and replace as necessary.
   - the adjusting lever contact with the brake adjuster screw.

➡To aid installation, note the locations of the short and long brake shoe retracting springs.

4. Remove the long brake shoe retracting spring.
5. Remove the short brake shoe retracting spring.
6. Disconnect the brake shoe adjusting lever cable from the brake shoe adjusting lever.
7. Remove the brake shoe adjusting lever cable.
8. Remove the cable guide.
9. Remove the adjusting lever return spring and the brake shoe adjusting lever.
10. Remove the brake shoe adjusting screw spring.
11. Remove the brake adjuster screw assembly.

➡The parking brake link spring and the brake parking spring retainer will come off with the parking brake strut.

12. Remove the brake shoe anchor pin guide plate, parking brake strut and the parking brake lever pin retainer.
13. Remove the brake shoe hold-down springs.
14. Remove the rear brake shoes and linings.
15. Remove the brake shoe hold-down spring pins.
16. Pull back the parking brake cable spring and disconnect the parking brake lever.

To install

17. Compress the parking brake cable spring and attach the parking brake lever.
18. Clean and lubricate the brake backing plate. Use Silicone Brake Caliper Grease and Dielectric Compound D7AZ-19A331-A (Motorcraft WA-10) or an equivalent silicone compound meeting Ford specification ESE-M1C171-A.
19. Attach the parking brake lever to the rear brake shoe and lining and secure the parking brake lever pin retainer.
20. Position the rear brake shoes and linings.
21. Install the brake shoe hold-down spring pins and the brake shoe hold-down springs.
22. Install the parking brake link spring and the parking brake spring retainer on the parking brake strut.
23. Install the parking brake strut.
24. Install the brake shoe anchor pin guide plate.
25. Place the brake shoe adjusting lever cable over the anchor pin, with crimped side in.
26. Install the short brake shoe retracting spring.

➡Make sure the brake shoe adjusting lever cable is positioned in the cable guide groove.

27. Install the cable guide.
28. Position the brake shoe adjusting lever cable in the cable guide groove.
29. Install the long brake shoe retracting spring.

➡To prevent incorrect installation, the socket end of each brake adjuster screw is stamped R or L.

30. Apply Silicone Brake Caliper Grease and Dielectric Compound D7AZ-19A331-A (Motorcraft WA-10) or an equivalent silicone compound meeting Ford specification ESE-M1C171-A.
31. Install the brake adjuster screw into the brake shoe adjusting screw nut to the end of the threads, and then loosen one-half turn.
32. Install the brake shoe adjusting screw socket on the brake shoe adjusting screw nut.
33. Position the brake adjuster screw assembly.
34. Install the brake shoe adjusting screw spring.
35. Position the brake shoe adjusting lever cable.
36. Position the adjusting lever return spring and the brake shoe adjusting lever.
37. Hook the brake shoe adjusting lever cable to the brake shoe adjusting lever.
38. Test the operation of the automatic self-adjuster.
    a. Pull the brake shoe adjusting lever cable and check that the brake shoe adjusting lever rotates the brake shoe adjuster assembly.
    b. Release the brake shoe adjusting lever cable and check that the brake shoe adjusting lever advances to the next notch on the brake shoe adjusting screw nut.
39. Adjust the brakes.
40. Install the brake drum.

Wheel Cylinder

REMOVAL & INSTALLATION

1. Remove the brake shoes and linings.
2. Disconnect the brake line fitting.
3. Remove the two rear wheel cylinder bolts.
4. Remove the rear wheel cylinder.
5. Follow the removal procedure in reverse order.
6. Torque the wheel cylinder bolts to 9–13 ft. lbs. (12–18 Nm); torque the brake line fitting to 11–14 ft. lbs. (15–20 Nm).
7. Bleed the brake system.

BRAKE SYSTEM SERVICE

Drum Brakes

BRAKE SHOE ADJUSTMENT

Brake Drums Removed

➡After adjusting the rear brake shoes and linings, check the parking brake for proper operation. Make sure the parking brake cable equalizer operates freely.

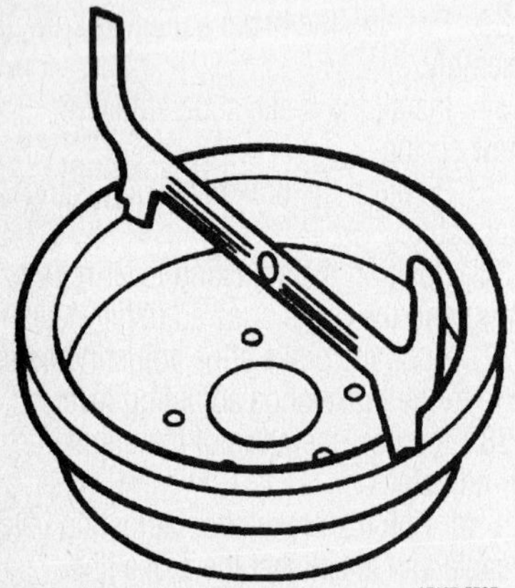

67197-EFSE-

Measuring the brake drum inside diameter

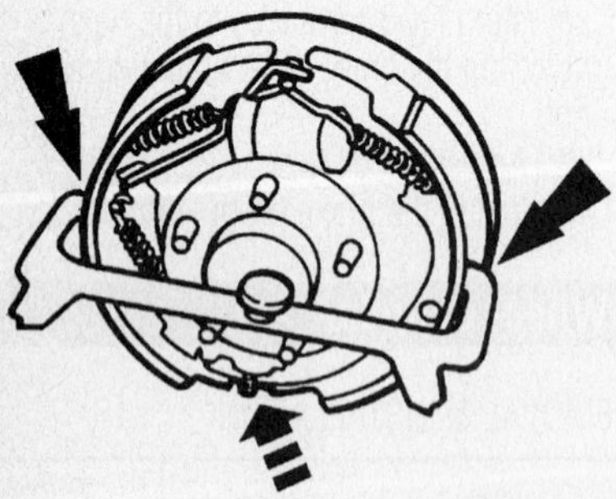

67197-EFSE-G229

Measuring overall brake shoe width

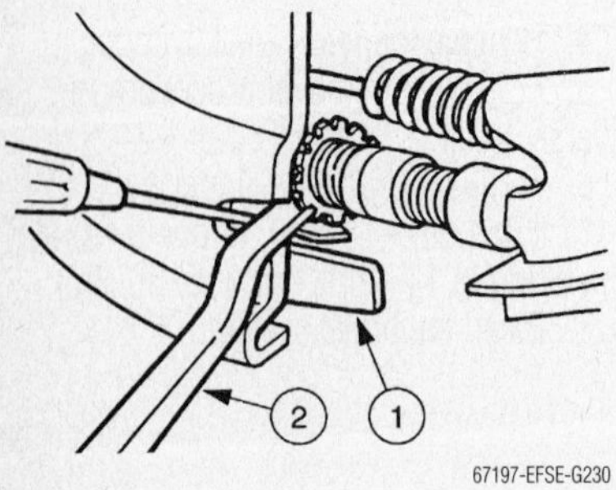

67197-EFSE-G230

Adjusting the brakes with the drum installed

1. Measure the brake drum inside diameter. Use a brake adjustment gauge.
2. Rotate the brake adjuster screw until the rear brake shoes and linings touch the tool.

Brake Drums Installed

3. Raise and support the vehicle.
4. At the back of the brake backing plate, remove the brake adjusting hole cover.
5. Rotate the brake shoe adjusting screw nut.
6. Using a screwdriver, move the brake shoe adjusting lever off the brake shoe adjusting screw nut.

➡Move the adjustment tool handle upward to tighten the brake adjuster screw.

7. Turn the brake shoe adjusting screw nut until the brake drum begins to drag. Then loosen the brake adjuster screw until the brake drum rotates freely.
8. Replace the brake adjusting hole cover.
9. Lower the vehicle.

Disc and Drum Brakes

BRAKE BLEEDING

2001–03

1. Before servicing the vehicle, refer to the precautions in the beginning of this section.

Master Cylinder Priming—In-Vehicle or Bench

➡When any part of the hydraulic system has been disconnected for repair or replacement, air can enter the system and cause spongy brake pedal action. This requires bleeding of the hydraulic system after it has been properly connected. The hydraulic system can be bled manually or with pressure bleeding equipment.

➡When the brake master cylinder has been replaced or the system has been emptied, or partially emptied, it should be primed to prevent air from entering the system.

2. For in-vehicle priming, disconnect the brake lines.
3. For bench priming, mount the brake master cylinder in a vise.
4. Install short brake tubes with the ends submerged in the brake master cylinder reservoir, and fill the brake master cylinder reservoir with High Performance DOT 3 Motor Vehicle Brake Fluid C6AZ-19542-AB or equivalent DOT 3 fluid meeting Ford specification ESA-M6C25-A.

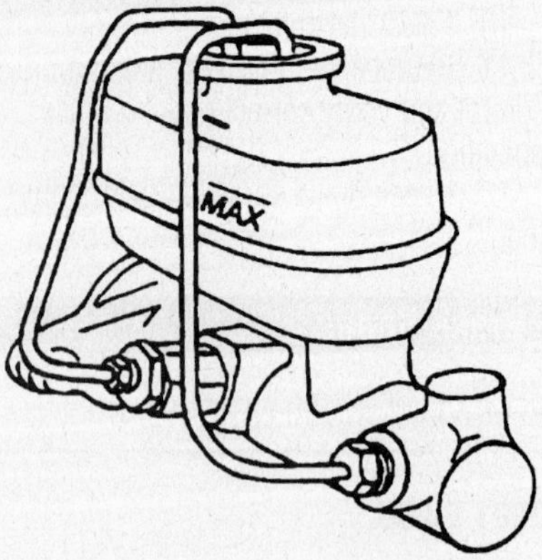

67197-EFSE-G231

In-vehicle bleeding the master cylinder

5. Have an assistant pump the brake pedal, or slowly depress the primary piston until clear fluid flows from both brake tubes, without air bubbles.
6. If the brake master cylinder is being primed at the bench, install it in the vehicle.
7. Remove the short brake tubes, and install the brake outlet tubes.
8. Bleed each brake tube at the brake master cylinder as follows:
   a. Have an assistant pump the brake pedal, and then hold firm pressure on the brake pedal.
   b. Loosen the rearmost brake tube fittings until a stream of brake fluid comes out. While the assistant maintains pressure on the brake pedal, tighten the brake tube fitting.
   c. Repeat this operation until clear, bubble-free fluid comes out.
   d. Refill the brake master cylinder reservoir as necessary. Repeat the bleeding operation at the front brake tube.
9. Tighten the brake lines at the master cylinder to 18 ft. lbs. (25 Nm) for 2001–03 models; 13 ft. lbs. (17 Nm) for 2004 models.

Four Wheel Anti-Lock Brake System (4WABS) Hydraulic Control Unit (HCU)

➡This procedure needs to be performed only if the 4-wheel anti-lock brake (4WABS) hydraulic control unit (HCU) has been replaced or if air is suspected in the HCU.

10. Clean all dirt from and remove the brake master cylinder filler cap, and fill the

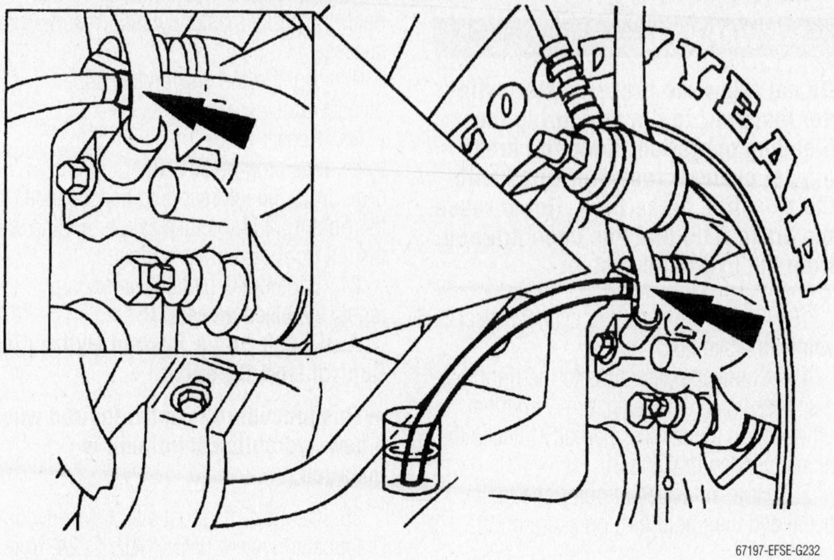

Bleeding a caliper

brake master cylinder reservoir with the specified brake fluid.

11. Connect a clear waste line to the right rear bleeder screw and the other end in a container partially filled with recommended brake fluid.

12. With the right rear bleeder screw open, cycle the brake pedal until no more air is seen in the waste line.

13. Tighten the right rear bleeder screw, and disconnect the waste line.

14. Repeat Steps 2, 3 and 4 for the left rear bleeder screw, the right front disc brake caliper bleeder screw, and the left front disc brake caliper bleeder screw, in that order.

15. Connect Worldwide Diagnostic System (WDS) 418-F224, New Generation STAR (NGS) Tester 418-F052, or equivalent diagnostic tool DCL cable adapter into the vehicle data link connector (DLC) under the dash, and follow the scan tool instructions.

16. Repeat the system bleed procedure as outlined in Steps 1 through 5.

Caliper or Wheel Cylinder

➡It is not necessary to do a complete brake system bleed if only the disc brake caliper or drum brake wheel cylinder was disconnected.

17. Place a box end wrench on the disc brake caliper or wheel cylinder bleeder screw. Attach a rubber drain tube to the bleeder screw, and submerge the free end of the tube in a container partially filled with clean brake fluid.

18. Have an assistant pump the brake pedal and then hold firm pressure on the brake pedal.

19. Loosen the bleeder screw until a stream of brake fluid comes out. While the assistant maintains pressure on the brake pedal, tighten the bleeder screw. Repeat until clear, bubble-free fluid comes out. Refill the brake master cylinder reservoir as necessary.

20. Tighten the bleeder screw.

2004

1. Before servicing the vehicle, refer to the precautions in the beginning of this section.

➡When any part of the hydraulic system has been disconnected for repair or new installation, air may get into the system and cause spongy brake pedal action. This requires bleeding of the hydraulic system after it has been correctly connected. The hydraulic system can be gravity bled, manually bled or with pressure bleeding equipment.

Master Cylinder, Bench

2. Support the brake master cylinder body in a vise and fill the brake master cylinder reservoir with specified brake fluid.

➡Original equipment lines are not intended to be used during this procedure.

3. Install short brake tubes with the ends submerged in the brake master cylinder reservoir.

4. Slowly press the primary piston until clear fluid flows from both brake tubes, without air bubbles.

5. Remove the short brake tubes and plug the brake tube ports.

Master Cylinder—In Vehicle

⁂ WARNING

Do not allow the brake master cylinder reservoir to run dry during the bleeding operation. Keep the brake master cylinder reservoir filled with the specified brake fluid. Never reuse the brake fluid that has been drained from the hydraulic system.

➡When a new brake master cylinder has been installed or the system has been emptied, or partially emptied, it should be primed to prevent air from getting into the system.

6. Disconnect the brake master cylinder outlet tubes.

➡Original equipment lines are not intended to be used during this procedure.

7. Install short brake tubes with ends submerged in the brake master cylinder reservoir and fill the brake master cylinder reservoir with brake fluid.

8. Have an assistant pump the brake pedal until clear fluid flows from both brake tubes without air bubbles.

9. Remove the short brake tubes and install the brake outlet tubes.

10. Bleed each brake tube at the brake master cylinder as follows:

a. Have an assistant pump the brake pedal and then hold firm pressure on the brake pedal.

b. Loosen the rear brake tube fittings until a stream of brake fluid comes out. Have an assistant maintain pressure on the brake pedal while tightening the brake tube fitting.

c. Repeat this operation until clear, bubble-free fluid comes out.

d. Refill the brake master cylinder reservoir as necessary. Repeat the bleeding operation at the front brake tube.

11. While the assistant maintains pressure on the brake pedal, tighten the brake tubes.

Gravity Bleeding

⁂ WARNING

Do not allow the brake master cylinder reservoir to run dry during the bleeding operation. Keep the brake master cylinder reservoir filled with the specified brake fluid. Never reuse the brake fluid that has been drained from the hydraulic system.

➡When a new brake master cylinder has been installed or the system has been emptied, or partially emptied, it should be primed to prevent air from getting into the system.

12. Fill the brake master cylinder reservoir with brake fluid.
13. Connect a clear tube to the right rear disc brake caliper bleeder screw and the other end in a container partially filled with recommended brake fluid.
14. Open the bleeder screw and leave open until clear bubble-free brake fluid flows. Refill the brake master cylinder reservoir as necessary.
15. Tighten the disc brake caliper bleeder screw.
16. Repeat Steps 1 through 4 for the three remaining brake calipers, going in order from the left rear disc brake caliper to the right front disc brake caliper ending with the left front disc brake caliper.
17. If the brake pedal feels spongy, repeat the bleed procedure.

Manual Bleeding

⁂ WARNING

Do not allow the brake master cylinder reservoir to run dry during the bleeding operation. Keep the brake master cylinder reservoir filled with the specified brake fluid. Never reuse the brake fluid that has been drained from the hydraulic system.

18. Fill the brake master cylinder reservoir with brake fluid.
19. Connect a clear tube to the right rear disc brake caliper bleeder screw and the other end in a container partially filled with recommended brake fluid.
20. Have an assistant pump the brake pedal and then hold firm pressure on the brake pedal.
21. Loosen the disc brake caliper bleeder screw until a stream of brake fluid comes out. Have an assistant maintain pressure on the brake pedal while tightening the disc brake caliper bleeder screw. Repeat until clear, bubble-free fluid comes out. Refill the brake master cylinder reservoir as necessary.
22. Tighten the disc brake caliper bleeder screw.
23. Repeat Steps 1 through 5 for the three remaining brake calipers, going in order from the left rear disc brake caliper to the right front disc brake caliper ending with the left front disc brake caliper.
24. If the brake pedal feels spongy, repeat the bleed procedure.

Anti-Lock Brake System Hydraulic Control Unit Bleeding

➡This procedure is only required when a new hydraulic control unit is installed.

25. Connect diagnostic tool Worldwide Diagnostic System (WDS) 418-F224, New Generation STAR (NGS) Tester 418-F052, or equivalent diagnostic tool and follow the ABS system bleed instructions.
26. Use the gravity bleed or manual bleed procedure(s) to bleed the system. Begin at the right rear caliper.

LINCOLN

Continental

3

SPECIFICATION CHARTS

ENGINE AND VEHICLE IDENTIFICATION

Engine						
Code ①	Liters (cc)	Cu. In.	Cyl.	Fuel Sys.	Engine Type	Eng. Mfg.
V	4.6 (4593)	281	8	SFI	DOHC	Ford

Model Year	
Code ②	Year
1	2001
2	2002
3	2003
4	2004
5	2005

OHV: Overhead Valves

DOHC: Double Overhead Camshafts

SFI: Sequential Fuel Injection

① 8th digit of the Vehicle Identification Number (VIN)

② 10th digit of the Vehicle Identification Number (VIN)

67197-CONT-C01

GENERAL ENGINE SPECIFICATIONS

Year	Model	Engine Displacement Liters	Engine ID/VIN	Net Horsepower @ rpm	Net Torque @ rpm (ft. lbs.)	Bore x Stroke (in.)	Com-pression Ratio	Oil Pressure @ rpm
2001	Continental	4.6	V	275@5750	275@4750	3.55x3.54	9.8:1	33@1500
2002	Continental	4.6	V	275@5750	275@4750	3.55x3.54	9.8:1	33@1500

SFI: Sequential Fuel Injection

67197-CONT-C02

ENGINE TUNE-UP SPECIFICATIONS

Year	Engine Displacement Liters	Engine ID/VIN	Spark Plug Gap (in.)	Ignition Timing (deg.)	Fuel Pump (psi) A	Idle Speed (rpm)	Valve Clearance	
							Intake	Exhaust
2001	4.6	V	0.052-0.056	10B	30-45	B	HYD	HYD
2002	4.6	V	0.052-0.056	10B	30-45	B	HYD	HYD

NOTE: The Vehicle Emission Control Information label often reflects specification changes made during production. The label figures must be used if they differ from those in this chart.

B: Before Top Dead Center

HYD: Hydraulic

NA: Not Adjustable

A Fuel pressure with engine running, pressure regulator vacuum hose connected

B Refer to Vehicle Emission Control Information label

67197-CONT-C03

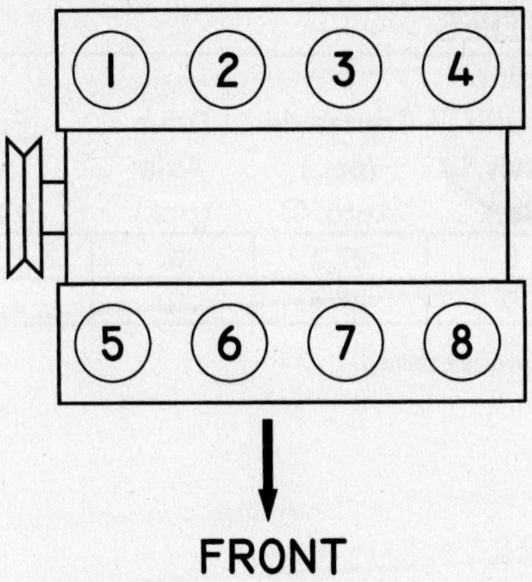

93003G03

4.6L Engine
Firing order: 1–3–7–2–6–5–4–8
Distributorless ignition system (One coil on each cylinder)

Drive Belt Routing

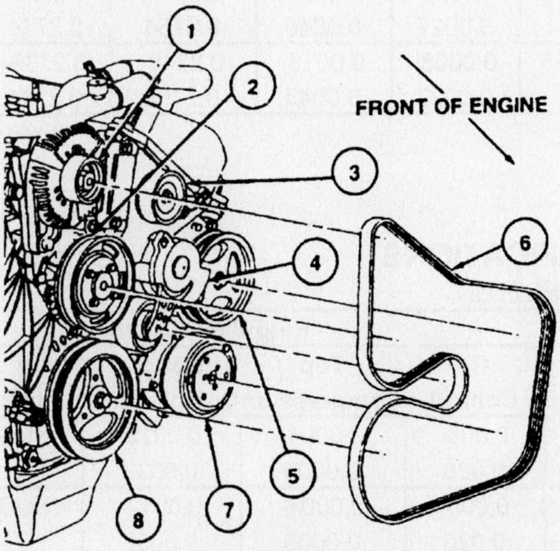

1. Generator
2. Water pump pulley
3. Belt idler pulley
4. Drive belt tensioner
5. Power steering pump
6. Drive belt
7. A/C compressor
8. Crankshaft pulley

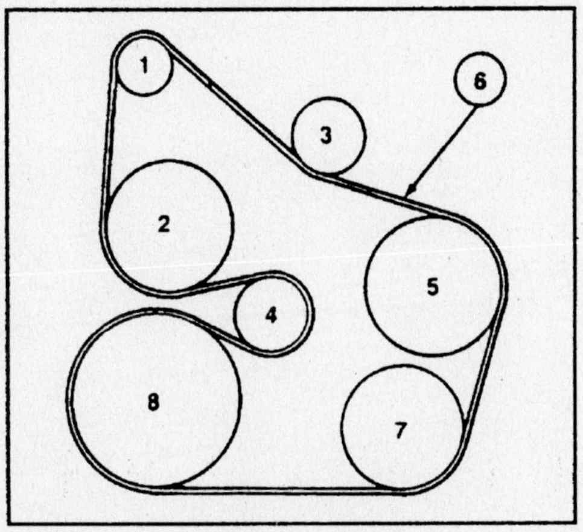

79224G26

Serpentine accessory drive belt routing—4.6L (VIN V) engine

CAPACITIES

Year	Model	Engine Displacement Liters	Engine ID/VIN	Engine Oil with Filter (qts.)	Transaxle (pts.) Auto. ①	Drive Axle (pts.)	Fuel Tank (gal.)	Cooling System (qts.)
2001	Continental	4.6	V	6.0	27.4	②	20.0	14.3
2002	Continental	4.6	V	6.0	27.4	②	20.0	14.3

NOTE: All capacities are approximate. Add fluid gradually and ensure a proper fluid level is obtained.

① Includes torque converter

② Included in transaxle capacity

67197-CONT-C04

VALVE SPECIFICATIONS

Year	Engine Displacement Liters	Engine ID/VIN	Seat Angle (deg.)	Face Angle (deg.)	Spring Test Pressure (lbs. @ in.)	Spring Installed Height (in.)	Stem-to-Guide Clearance (in.)		Stem Diameter (in.)	
							Intake	Exhaust	Intake	Exhaust
2001	4.6	V	45	45.5	160@1.103	1.425	0.0008-0.0027	0.0018-0.0045	0.2746-0.2754	0.2736-0.2744
2002	4.6	V	45	45.5	160@1.103	1.425	0.0008-0.0027	0.0018-0.0045	0.2746-0.2754	0.2736-0.2744

67197-CONT-C05

PISTON AND RING SPECIFICATIONS

All measurements are given in inches.

Year	Engine Displacement Liters	Engine ID/VIN	Piston Clearance	Ring Gap			Ring Side Clearance		
				Top Compression	Bottom Compression	Oil Control	Top Compression	Bottom Compression	Oil Control
2001	4.6	V	0.0007-0.0018	0.010-0.020	0.010-0.020	0.006-0.026	0.0004-0.0009	0.0012-0.0032	SNUG
2002	4.6	V	0.0007-0.0018	0.010-0.020	0.010-0.020	0.006-0.026	0.0004-0.0009	0.0012-0.0032	SNUG

67197-CONT-C06

CRANKSHAFT AND CONNECTING ROD SPECIFICATIONS

All measurements are given in inches.

Year	Engine Displacement Liters	Engine ID/VIN	Crankshaft Main Brg. Journal Dia.	Crankshaft Main Brg. Oil Clearance	Crankshaft Shaft End-play	Crankshaft Thrust on No.	Connecting Rod Journal Diameter	Connecting Rod Oil Clearance	Connecting Rod Side Clearance
2001	4.6	V	2.6580-2.6576	0.0010-0.0018	0.0051-0.0119	5	2.0859-2.0867	0.0011-0.0027	0.0059 0.0177
2002	4.6	V	2.6580-2.6576	0.0010-0.0018	0.0051-0.0119	5	2.0859-2.0867	0.0011-0.0027	0.0059 0.0177

67197-CONT-C07

TORQUE SPECIFICATIONS

All readings in ft. lbs.

Year	Engine Displacement Liters	Engine ID/VIN	Cylinder Head Bolts	Main Bearing Bolts	Rod Bearing Auto.	Crankshaft Damper Bolts	Flywheel Bolts	Manifold Intake	Manifold Exhaust	Spark Plugs	Oil Pan Drain Plug
2001	4.6	V	①	②	③	④	54-64	⑤	⑥	7-15	10
2002	4.6	V	①	②	③	④	54-64	⑤	⑥	7-15	10

① Step 1: 28-31 ft. lbs.
Step 2: Plus 85-95 degrees
Step 3: Loosen all bolts 360 degrees
Step 4: 28-31 ft. lbs.
Step 5: Plus 85-95 degrees
Step 6: Plus 85-95 degrees

② Step 1: Main bearing cap bolts: 6-9 ft. lbs.
Step 2: Main bearing cap bolts, outer: 16-21 ft. lbs.
Step 3: Main bearing cap bolts, inner: 27-32 ft. lbs.
Step 4: Rotate main bearing cap bolts 85-95 degrees
Step 5: Main cap adjusting screws 4 ft. lbs. then 7.5 ft. lbs.
Step 6: Main cap side bolts: 7 ft. lbs. then 14-17 ft. lbs.

③ Step 1: 5 ft. lbs.
Step 2: 10 ft. lbs.
Step 3: 18-25 ft. lbs.
Step 4: Plus 85-95 degrees

④ Step 1: 77-99 ft. lbs.
Step 2: Loosen 360 degrees
Step 3: 35-39 ft. lbs.
Step 4: Plus 85-95 degrees

⑤ Step 1: Four inside short bolts: 9-11 ft. lbs.
Step 2: All other bolts: 13-16 ft. lbs.
Step 3: All bolts plus 85-95 degrees

⑥ Studs: 8-9 ft. lbs.
Nuts: 14-16 ft .lbs.

67197-CONT-C08

WHEEL ALIGNMENT

Year	Model		Caster Range (+/-Deg.)	Caster Preferred Setting (Deg.)	Camber Range (+/-Deg.)	Camber Preferred Setting (Deg.)	Toe-in (in.)
2001	Continental	F	0.75	0	0.70	0	-0.20 +/- 0.25
		R	—	—	0.70	0	0.20 +/- 0.25
2002	Continental	F	0.75	0	0.70	0	-0.20 +/- 0.25
		R	—	—	0.70	0	0.20 +/- 0.25

67197-CONT-C09

TIRE, WHEEL AND BALL JOINT SPECIFICATIONS

		OEM Tires		Tire Pressures (psi)		Wheel	Ball Joint	Lug
Year	Model	Standard	Optional	Front	Rear	Size	Inspection	Nuts
2001	Continental	P225/60HR16	P225/60VR16	30	30	7-JJ	①	95
2002	Continental	P225/60HR16	P225/60VR16	30	30	7-JJ	①	95

OEM: Original Equipment Manufacturer

PSI: Pounds Per Square Inch

① Replace if any measurable movement is found.

67197-CONT-C10

BRAKE SPECIFICATIONS

All measurements in inches unless noted

						Brake Drum				Brake Caliper	
			Brake Disc			Original	Max.	Maximum	Minimum	Bracket	Mounting
			Original	Minimum	Maximum	Inside	Wear	Machine	Lining	Bolts	Bolts
Year	Model		Thickness	Thickness	Run-out	Diameter	Limit	Diameter	Thickness	(ft. lbs.)	(ft. lbs.)
2001	Continental	F	1.020	0.974	0.003	—	—	—	0.039	65-87	25
		R	0.550	0.502	0.001	—	—	—	0.039	—	25
2002	Continental	F	1.020	0.974	0.003	—	—	—	0.039	65-87	25
		R	0.550	0.502	0.001	—	—	—	0.039	—	25

NOTE: Follow specifications stamped on the rotor if figures differ from those in this chart.

F: Front

R: Rear

67197-CONT-C11

SCHEDULED MAINTENANCE INTERVALS

Lincoln Continental

TO BE SERVICED	TYPE OF SERVICE	VEHICLE MILEAGE INTERVAL (X1000)																			
		5	10	15	20	25	30	35	40	45	50	55	60	65	70	75	80	85	90	95	100
Engine oil & filter	R	✓	✓	✓	✓	✓	✓	✓	✓	✓	✓	✓	✓	✓	✓	✓	✓	✓	✓	✓	✓
Rotate tires	S/I	✓		✓		✓		✓		✓		✓		✓		✓		✓		✓	
Engine coolant protection, hoses & clamps	S/I			✓			✓			✓			✓			✓			✓		
Passenger compartment air filter	R			✓			✓			✓			✓			✓			✓		
Air cleaner filter	R						✓						✓						✓		
Automatic transaxle fluid & filter	R						✓						✓						✓		
Brake lines & connections	S/I						✓						✓						✓		
Exhaust heat shields	S/I						✓						✓						✓		
Front and rear disc brake pads & rotors	S/I						✓						✓						✓		
Accessory drive belt(s)	S/I												✓								
Engine coolant ①	R										✓						✓				
Spark plugs	R																				✓
PCV valve	R												✓								

① Engine coolant - change initially at 50,000 miles & thereafter every 30,000 miles.

R - Replace S/I - Service and Inspect

FREQUENT OPERATION MAINTENANCE (SEVERE SERVICE)

If a vehicle is operated under any of the following conditions it is considered severe service:

- Extremely dusty areas.

- 50% or more of the vehicle operation is in 32°C (90°F) or higher temperatures, or constant operation in temperatures below 0°C (32°F).

- Prolonged idling (vehicle operation in stop and go traffic)..

- Frequent short running periods (engine does not warm to normal operating temperatures).

- Police, taxi, delivery usage or trailer towing usage.

Oil & oil filter - change every 3000 miles.

Rotate tires at 6000 miles & every 9000 miles thereafter.

Air cleaner element service or inspect every 15,000 miles.

Automatic transaxle fluid & filter - change every 21,000 miles.

67197-CONT-C12

PRECAUTIONS

Before servicing any vehicle, please be sure to read all of the following precautions, which deal with personal safety, prevention of component damage, and important points to take into consideration when servicing a motor vehicle:

- Never open, service or drain the radiator or cooling system when the engine is hot; serious burns can occur from the steam and hot coolant.
- Observe all applicable safety precautions when working around fuel. Whenever servicing the fuel system, always work in a well-ventilated area. Do not allow fuel spray or vapors to come in contact with a spark, open flame, or excessive heat (a hot drop light, for example). Keep a dry chemical fire extinguisher near the work area. Always keep fuel in a container specifically designed for fuel storage; also, always properly seal fuel containers to avoid the possibility of fire or explosion. Refer to the additional fuel system precautions later in this section.
- Fuel injection systems often remain pressurized, even after the engine has been turned **OFF**. The fuel system pressure must be relieved before disconnecting any fuel lines. Failure to do so may result in fire and/or personal injury.
- Brake fluid often contains polyglycol ethers and polyglycols. Avoid contact with the eyes and wash your hands thoroughly after handling brake fluid. If you do get brake fluid in your eyes, flush your eyes with clean, running water for 15 minutes. If eye irritation persists, or if you have taken brake fluid internally, IMMEDIATELY seek medical assistance.
- The EPA warns that prolonged contact with used engine oil may cause a number of skin disorders, including cancer. You should make every effort to minimize your exposure to used engine oil. Protective gloves should be worn when changing oil. Wash your hands and any other exposed skin areas as soon as possible after exposure to used engine oil. Soap and water, or waterless hand cleaner should be used.
- All new vehicles are now equipped with an air bag system, often referred to as a Supplemental Restraint System (SRS) or Supplemental Inflatable Restraint (SIR) system. The system must be disabled before performing service on or around system components, steering column, instrument panel components, wiring and sensors. Failure to follow safety and disabling procedures could result in accidental air bag deployment, possible personal injury and unnecessary system repairs.
- Always wear safety goggles when working with, or around, the air bag system. When carrying a non-deployed air bag, be sure the bag and trim cover are pointed away from your body. When placing a non-deployed air bag on a work surface, always face the bag and trim cover upward, away from the surface. This will reduce the motion of the module if it is accidentally deployed. Refer to the additional air bag system precautions later in this section.
- Clean, high quality brake fluid from a sealed container is essential to the safe and proper operation of the brake system. You should always buy the correct type of brake fluid for your vehicle. If the brake fluid becomes contaminated, completely flush the system with new fluid. Never reuse any brake fluid. Any brake fluid that is removed from the system should be discarded. Also, do not allow any brake fluid to come in contact with a painted surface; it will damage the paint.
- Never operate the engine without the proper amount and type of engine oil; doing so WILL result in severe engine damage.
- Timing belt maintenance is extremely important. Many models utilize an interference-type, non-freewheeling engine. If the timing belt breaks, the valves in the cylinder head may strike the pistons, causing potentially serious (also time-consuming and expensive) engine damage. Refer to the maintenance interval charts in the front of this manual for the recommended replacement interval for the timing belt, and to the timing belt section for belt replacement and inspection.
- Disconnecting the negative battery cable on some vehicles may interfere with the functions of the on-board computer system(s) and may require the computer to undergo a relearning process once the negative battery cable is reconnected.
- When servicing drum brakes, only disassemble and assemble one side at a time, leaving the remaining side intact for reference.

ENGINE REPAIR

➡Disconnecting the negative battery cable on some vehicles may interfere with the functions of the on board computer system. The computer may undergo a relearning process once the negative battery cable is reconnected.

Alternator

REMOVAL

1. Before servicing the vehicle, refer to the precautions in the beginning of this section.
2. Remove or disconnect the following:
   - Negative battery cable
   - Coolant recovery reservoir
   - Accessory drive belt
   - Alternator bracket
   - Alternator

INSTALLATION

1. Install or connect the following:
   - Alternator. Tighten the bolts to 15–22 ft. lbs. (20–30 Nm).
   - Alternator bracket. Tighten the nuts to 71–106 inch lbs. (8–12 Nm).
   - Accessory drive belt
   - Coolant recovery reservoir
   - Negative battery cable

Ignition Timing

ADJUSTMENT

The ignition timing is controlled by the Powertrain Control Module (PCM) and is not adjustable.

Engine Assembly

REMOVAL & INSTALLATION

➡Disable the air suspension before raising the vehicle. The switch is located on the left side of the luggage compartment.

1. Before servicing the vehicle, refer to the precautions in the beginning of this section.
2. Turn the air suspension switch to the **OFF** position.
3. Recover the A/C refrigerant.
4. Drain the cooling system.
5. Remove or disconnect the following:
   - Negative battery cable
   - Hood
   - Steering column intermediate shaft

- Engine appearance cover
- Intake Air Temperature (IAT) sensor connector
- Air cleaner outlet tube
- Upper motor mount
- Fuel line
- Dash panel ground straps
- Powertrain Control Module (PCM) connector
- Mass Air Flow (MAF) sensor connector
- Traction control motor connector
- Throttle Position (TP) sensor connector
- Cruise control cable
- Accelerator cable and bracket
- Shift selector cable
- Engine control wiring harness (3 connectors) and bracket
- Intake manifold vacuum lines
- Transaxle cooler lines
- Power steering return hose at the fluid reservoir
- Alternator wiring harness
- Radiator splash shield
- Heater hoses
- Radiator hoses
- Front wheels
- Suspension height sensor links
- Lower ball joints
- Stabilizer bar links
- Outer tie rod ends
- Halfshafts
- Heated Oxygen (HO_2S) sensor connectors
- Dual converter Y-pipe
- Block heater connector
- Power steering cooler lines
- A/C compressor lines
- Starter motor
- Subframe brackets
- Torque converter shield and torque converter

6. Support the powertrain from below and remove the subframe bolts.
7. Raise the vehicle away from the powertrain.
8. Attach an engine hoist to the powertrain.
9. Remove or disconnect the following:
   - Power steering sensor connector
   - Power steering return line from the steering gear
   - Turbine shaft speed sensor connector
   - Transaxle range sensor connector
   - Transaxle control harness connector
   - Left and right engine support insulators
   - Right engine support insulator bracket
10. Remove the transaxle flange bolts and separate the engine from the transaxle.

To install:

➡When installing suspension components and halfshafts, use new nuts, bolts, circlips and split pins.

11. Install or connect the following:
    - Transaxle. Tighten the flange bolts to 25–34 ft. lbs. (34–46 Nm).
    - Right engine support insulator bracket. Tighten the bolts to 16–21 ft. lbs. (22–29 Nm).
    - Left and right engine support insulators. Tighten the through-bolts to 64–88 ft. lbs. (87–119 Nm).
    - Transaxle control harness connector
    - Transaxle range sensor connector
    - Turbine shaft speed sensor connector
    - Power steering return line from the steering gear
    - Power steering sensor connector
12. Lower the vehicle on to the powertrain assembly. Use 2 pieces of ¾ inch outside diameter pipe in the alignment holes behind the front subframe mounts to align the subframe to the body. Tighten the subframe mounting bolts to 57–76 ft. lbs. (77–103 Nm).
13. Install or connect the following:
    - Torque converter shield and torque converter. Tighten the torque converter nuts to 20–34 ft. lbs. (27–46 Nm).
    - Subframe brackets
    - Starter motor
    - A/C compressor lines
    - Power steering cooler lines
    - Block heater connector
    - Dual converter Y-pipe
    - HO_2S sensor connectors
    - Halfshafts. Tighten the hub retainer nuts to 170–202 ft. lbs. (230–275 Nm).
    - Outer tie rod ends. Tighten the nuts to 35–46 ft. lbs. (47–63 Nm).
    - Stabilizer bar links. Tighten the nuts to 30–40 ft. lbs. (40–55 Nm).
    - Lower ball joints. Tighten the nuts to 50–68 ft. lbs. (68–92 Nm).
    - Suspension height sensor links
    - Front wheels
    - Radiator hoses
    - Heater hoses
    - Radiator splash shield
    - Alternator wiring harness
    - Power steering return hose at the fluid reservoir
    - Transaxle cooler lines
    - Intake manifold vacuum lines
    - Engine control wiring harness (3 connectors) and bracket
    - Shift selector cable
    - Accelerator cable and bracket
    - Cruise control cable
    - TP sensor connector
    - Traction control motor connector
    - MAF sensor connector
    - PCM connector
    - Dash panel ground straps
    - Fuel line
    - Upper motor mount
    - Air cleaner outlet tube
    - IAT sensor connector
    - Engine appearance cover
    - Steering column intermediate shaft
    - Hood
    - Negative battery cable
14. Fill the cooling system.
15. Check and adjust fluid levels as necessary.
16. Recharge the A/C system.
17. Turn the air suspension switch to the **ON** position.
18. Run the engine and check for leaks.

➡Whenever the subframe is removed or lowered, the wheel alignment should be checked.

Water Pump

REMOVAL & INSTALLATION

1. Before servicing the vehicle, refer to the precautions in the beginning of this section.
2. Drain the cooling system.
3. Remove or disconnect the following:
   - Negative battery cable
   - Coolant reservoir

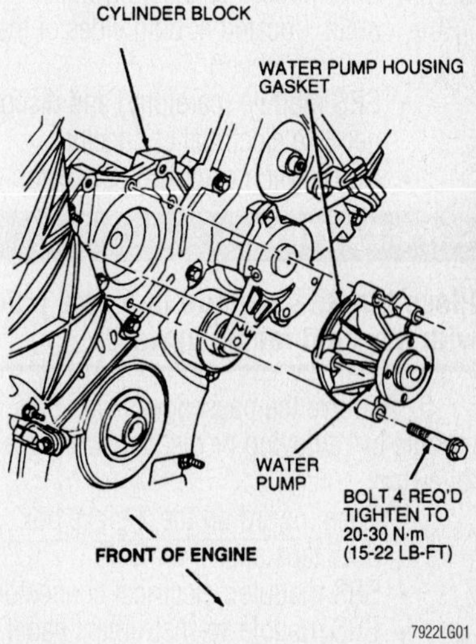

Exploded view of the water pump mounting

- Accessory drive belt
- Alternator
- Water pump pulley
- Water pump

To install:

4. Install or connect the following:
   - Water pump. Use a new O-ring seal and tighten the bolts to 15–22 ft. lbs. (20–30 Nm).
   - Water pump pulley. Tighten the bolts to 15–22 ft. lbs. (20–30 Nm).
   - Alternator
   - Accessory drive belt
   - Coolant reservoir
   - Negative battery cable
5. Fill the cooling system.
6. Start the engine and check for leaks.

Heater Core

REMOVAL & INSTALLATION

1. Before servicing the vehicle, refer to the precautions in the beginning of this section.
2. Disconnect the negative battery cable.

⁂ CAUTION

After disconnecting the negative battery cable, wait for at least 1 minute for the SRS or air bag module to deplete its energy.

3. Drain the cooling system into a clean container for reuse.
4. Remove the driver's side air bag module by removing or disconnecting the following:
   - SRS module-to-steering wheel bolts, (located at both sides of the steering wheel)
   - SRS module (carefully) and disconnect the electrical connector
   - Horn switch electrical connector

⁂ CAUTION

Place the SRS module in a safe place with the front facing upward.

5. Remove the passenger's side SRS module by removing or disconnecting the following:
   - Push inward on the 2 glove box door tabs and lower it
   - SRS module's electrical connector
   - SRS module-to-instrument panel bolts and the module

⁂ CAUTION

Place the SRS module in a safe place with the front facing upward

6. Remove the instrument panel by removing or disconnecting the following:
   - Floor or mini console (if equipped)
   - Rear seat climate control air duct sleeve (if equipped)
   - Left side instrument panel insulator pushpins and the insulator; then, disconnect the courtesy lamp
   - Instrument panel steering column cover screws and the cover
   - Pull the hood release handle, remove the screws and move it aside
   - Pull the parking brake handle, remove the bolts, the parking brake release handle and the cable
   - Steering column opening cover reinforcement-to-instrument panel bolts and the cover reinforcement
   - Release the cable and the conduit from the parking brake actuator
   - Steering column
   - Loosen the bolt and disconnect the left side outboard bulkhead electrical connector
   - Loosen the bolt and disconnect the left side inboard bulkhead electrical connector
   - Heated seat switch electrical connectors (if equipped)
   - 2 steering column mounting support-to-instrument cowl brace bolts
   - Instrument panel dash brace-to-instrument panel bolt
   - Instrument panel dash brace nut and move the brace aside
   - Right side instrument panel insulator pushpins and the insulator; then, disconnect the courtesy lamp
   - Vacuum harness connector
   - In-line electrical harness connector
   - Scuff plate and the cowl trim panel (located on the right side)
   - Antenna connector
   - Brake shift interlock actuator cable
   - Pry out the instrument panel defroster opening grille assembly, disconnect the 2 electrical connectors and remove the assembly
   - Upper instrument panel-to-cowl screws
   - Instrument panel support-to-cowl side nut (located on the right side)
   - Instrument panel support-to-cowl side bolts (located on the left side); then, loosen the instrument panel support-to-cowl side captive bolt
   - Pull the instrument panel rearward; then, disconnect the Electronic Air Temperature Control (EATC) hose from the heater plenum and make sure all electrical connectors are disconnected
   - Instrument panel with the help of an assistant

➡Check the upper cowl clips for damage; if necessary, replace them.

7. Loosen the screw and disconnect the PCM electrical connector.
8. Disconnect the heater hoses from the heater core.
9. Remove or disconnect the following:
   - 2 metal cover-to-heater/air conditioning housing screws and the metal cover (located below the heater core cover)
   - Electrical harness connector (at the heater core cover); then, remove the air conditioning electronic blend door actuator-to-heater/air conditioning housing screws and the actuator
   - Air conditioning air intake flue damper assist spring

⁂ WARNING

Do not bend any part of the air conditioning damper door shaft for it is brittle and will break.

10. Depress the air conditioning damper door shaft locking ramp, disconnect it from the air temperature control door shaft, swing the locking ramp counterclockwise and remove it from the air conditioning damper door shaft.
11. Move the air conditioning damper door shaft counterclockwise and remove it.
12. Remove or disconnect the following:
    - Heater core cover-to-heater/air conditioning housing screws and the cover
    - Heater core cover seal
    - Heater core from the heater/air conditioning housing

To install:

13. Install or connect the following:
    - Heater core to the heater/air conditioning housing
    - Heater core cover seal
    - Heater core cover and the cover-to-heater/air conditioning housing screws
    - Air conditioning damper door shaft and move it clockwise and install it
    - Air conditioning damper door shaft locking ramp, swing the locking

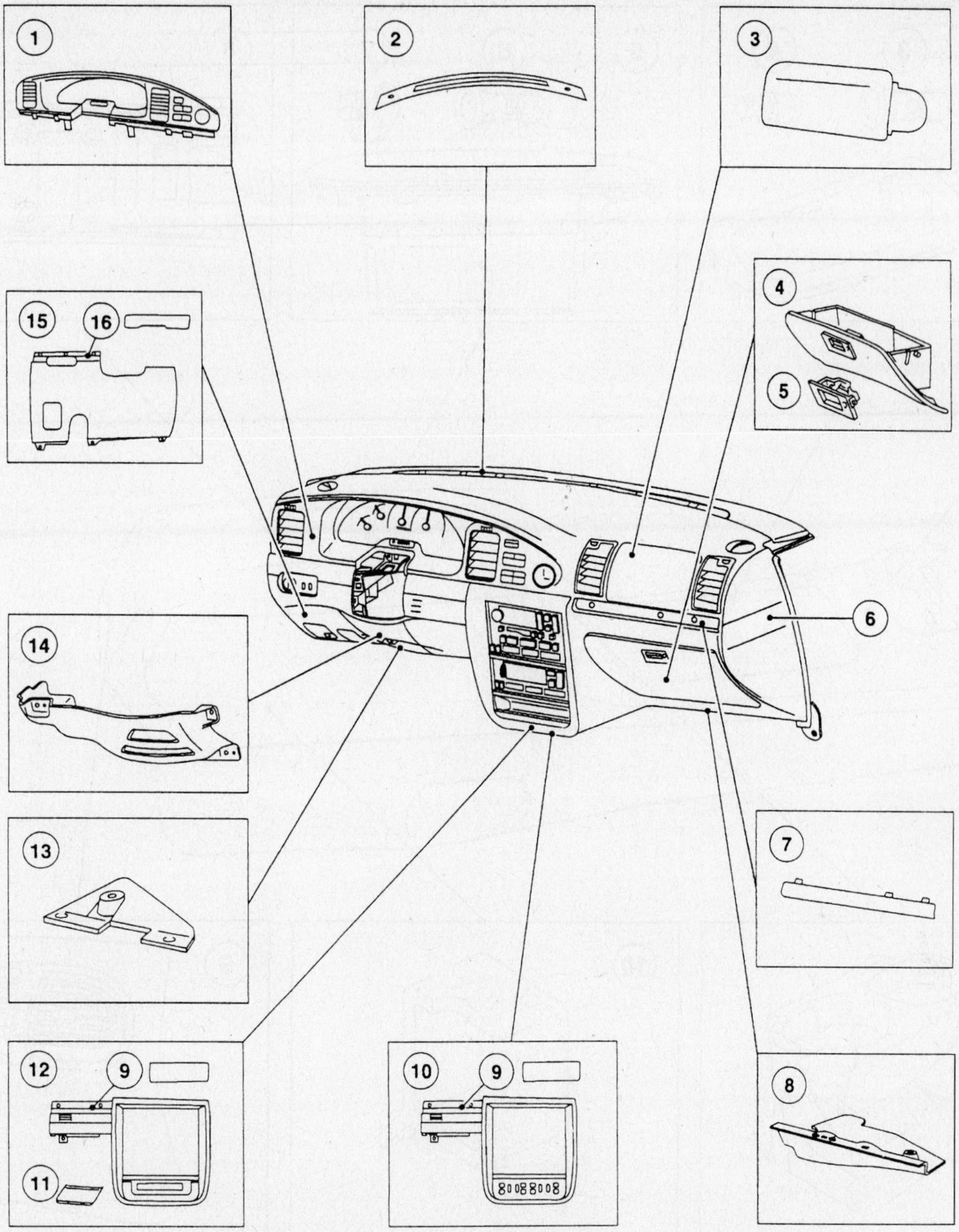

1 Instrument Panel Cluster Finish Panel
2 Instrument Panel Defroster Opening Grille
3 Passenger Side Air Bag Module
4 Glove Compartment
5 Glove Compartment Door Latch
6 Instrument Panel
7 Instrument Panel Finish Panel (RH) (Wood)
8 Instrument Panel Insulator (RH)
9 Instrument Panel Finish Panel (Wood)
10 Instrument Panel Center Finish Panel (Service, Heated Seat)
11 Utility Compartment Lower Mat
12 Instrument Panel Center Finish Panel (Base)
13 Instrument Panel Insulator (LH)
14 Instrument Panel Steering Column Opening Cover Reinforcement
15 Instrument Panel Steering Column Cover
16 Instrument Panel Finish Panel (LH) (Wood)

93111G99

Exploded view of the instrument panel and related components—Continental

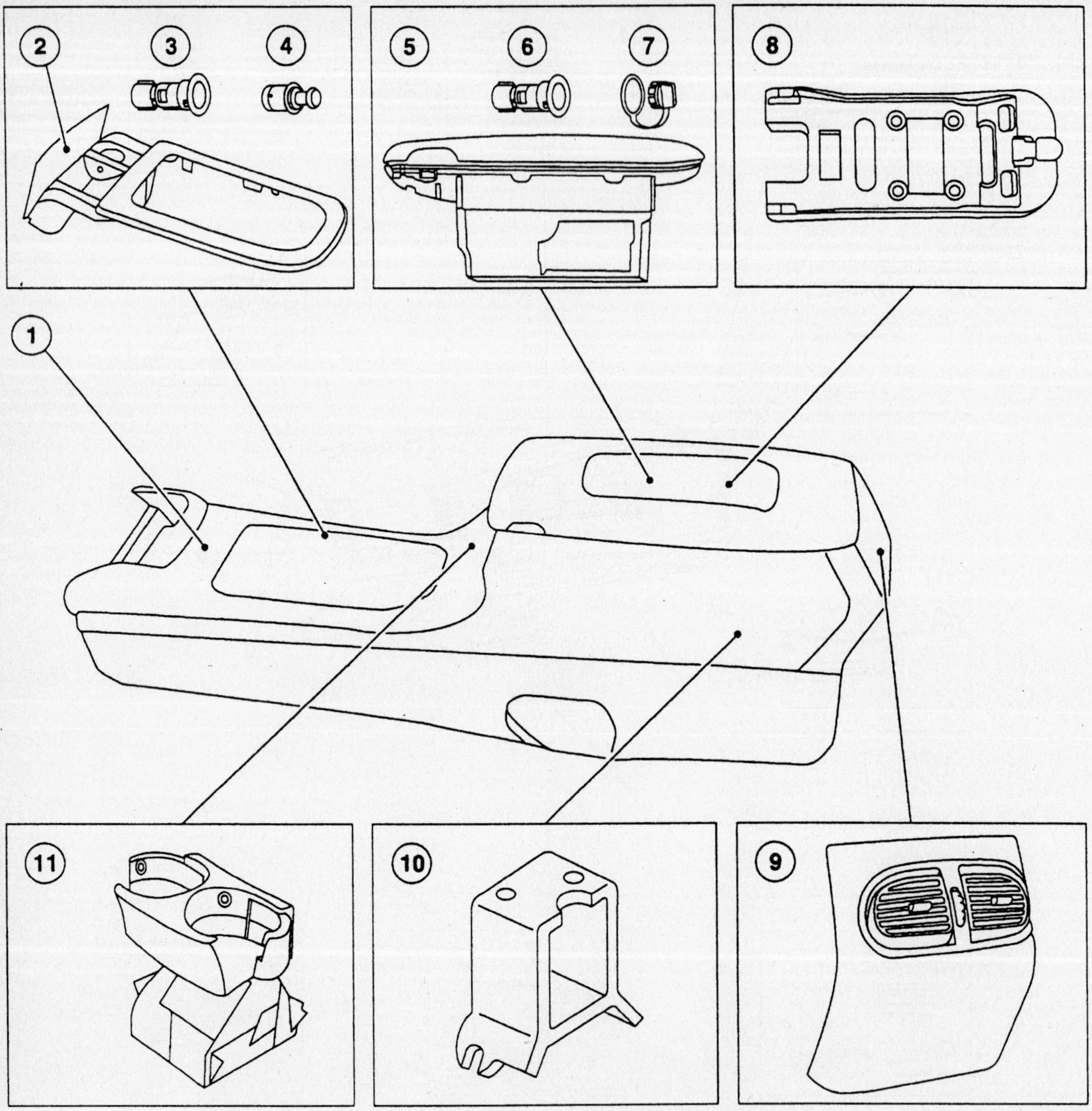

1 Console Panel
2 Console Top Panel
3 Cigar Lighter Socket and Retainer
4 Cigar Lighter Knob and Element
5 Glove Compartment
6 Auxiliary Electric Power Socket
7 Auxiliary Electrical Power Socket Cap
8 Handset Cradle
9 A/C Register
10 Instrument Panel Console Bracket
11 Utility Tray Beverage Holder

93111G00

Exploded view of the floor console and related components—Continental

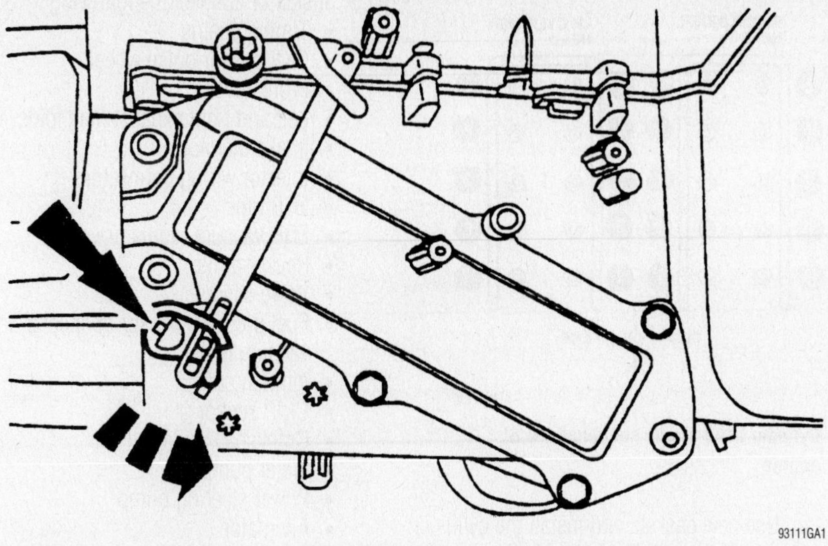

View of the air temperature control door shaft and heater core cover—Continental

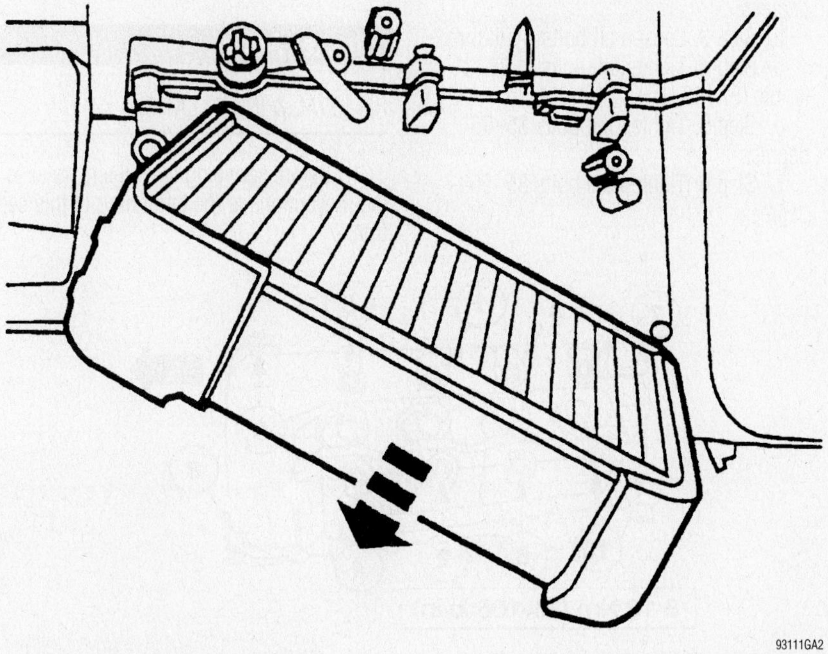

View of the heater core—Continental

ramp clockwise and connect it to the air temperature control door shaft
- Air conditioning air intake flue damper assist spring

⁂ WARNING

Do not bend any part of the air conditioning damper door shaft for it is brittle and will break.

14. At the heater core cover, install the air conditioning electronic blend door actuator and the actuator-to-heater/air conditioning housing screws; then, connect the electrical harness connector.
15. Install or connect the following:
    - Metal cover and the 2 metal cover-to-heater/air conditioning housing screws (located below the heater core cover)
    - Heater hoses to the heater core
    - PCM electrical connector and tighten the screw
16. Install the instrument panel by installing or connecting the following:
    - Instrument panel with the help of an assistant
    - Pull the instrument panel rearward; then, connect the Electronic Air Temperature Control (EATC) hose to the heater plenum and make sure all electrical connectors are connected
    - Instrument panel support-to-cowl side bolts and the instrument panel support-to-cowl side captive bolt (located on the left side)
    - Instrument panel support-to-cowl side nut (located on the right side)
    - Upper instrument panel-to-cowl screws
    - 2 electrical connectors and the instrument panel defroster opening grille assembly
    - Brake shift interlock actuator cable
    - Antenna connector
    - Scuff plate and the cowl trim panel (located on the right side)
    - In-line electrical harness connector
    - Vacuum harness connector
    - Right side instrument panel insulator and secure the insulator with the pushpins; then, connect the courtesy lamp
    - Instrument panel dash brace and the brace nut
    - Instrument panel dash brace-to-instrument panel bolt
    - 2 steering column mounting support-to-instrument cowl brace bolts
    - Heated seat switch electrical connectors (if equipped)
    - Left side inboard bulkhead electrical connector and tighten the bolt
    - Left side outboard bulkhead electrical connector and tighten the bolt
    - Steering column
    - Cable and the conduit to the parking brake actuator
    - Steering column opening cover reinforcement and the cover reinforcement-to-instrument panel bolts
    - Parking brake handle, the bolts and the cable
    - Hood release handle and the screws
    - Instrument panel steering column cover and the cover screws
    - Left side instrument panel insulator and secure the insulator with the pushpins; then, connect the courtesy lamp
    - Rear seat climate control air duct sleeve, if equipped
    - Floor or mini console, if equipped
17. Install the passenger's side SRS module by installing or connecting the following:

- SRS module and torque the module-to-instrument panel bolts to 62–97 inch lbs. (7–11 Nm)
- SRS module's electrical connector
- Glove box door

18. Install the driver's side air bag module by installing or connecting the following:
    - Horn switch electrical connector
    - Electrical connector and install the SRS module
    - SRS module-to-steering wheel bolts (located at both sides of the steering wheel), and torque the bolts to 90–122 inch lbs. (10–14 Nm)
19. Refill the cooling system.
20. Connect the negative battery cable.
21. Operate the engine to normal operating temperatures; then, check the climate control operation and check for leaks.

Cylinder Head

REMOVAL & INSTALLATION

1. Before servicing the vehicle, refer to the precautions in the beginning of this section.
2. Remove the engine from the vehicle and mount it on a suitable workstand.
3. Remove or disconnect the following:
    - Accessory drive belt and tensioner
    - Idler pulley
    - Engine Coolant Temperature (ECT) sensor connector
    - Water bypass tube
    - Alternator
    - Water pump
    - Power steering pump
    - Crankshaft pulley
    - Camshaft Position (CMP) sensor
    - Ignition coils
    - Fuel pressure sensor connector and vacuum line
    - Valve covers
    - Rocker arms
    - Exhaust Gas Recirculation (EGR) vacuum regulator valve
    - EGR tube
    - Injector wiring connectors
    - Intake manifold
    - Left and right exhaust manifolds
    - Front cover
    - Crankshaft Position (CKP) sensor pulse wheel
    - Timing chains
    - Cylinder heads

To install:

➡The cylinder head bolts are a torque-to-yield design and cannot be reused.

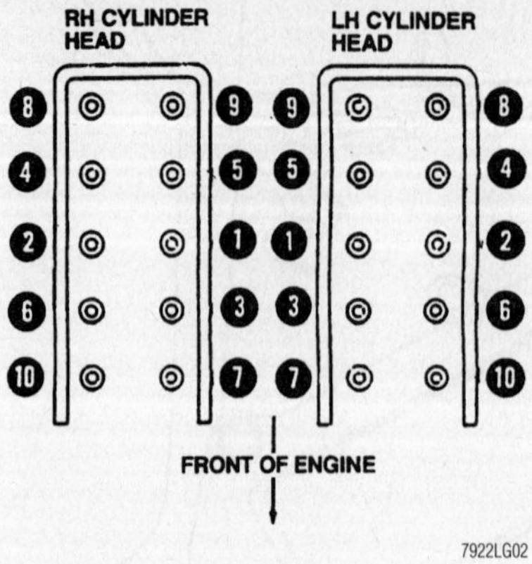

Cylinder head torque sequence—4.6L engine

4. Use new gaskets and install the cylinder heads. Tighten the cylinder head bolts in sequence as follows:
    a. Step 1: 28–31 ft. lbs. (37–43 Nm).
    b. Step 2: Tighten the bolts 85–95 degrees.
    c. Step 3: Loosen all bolts 1 full turn.
    d. Step 4: Tighten all bolts to 28–31 ft. lbs. (37–43 Nm).
    e. Step 5: Tighten the bolts 85–95 degrees.
    f. Step 6: Tighten the bolts 85–95 degrees.
5. Install or connect the following:
    - Timing chains
    - CKP sensor pulse wheel
    - Front cover
    - Left and right exhaust manifolds
    - Intake manifold
    - Injector wiring connectors
    - EGR tube
    - EGR vacuum regulator valve
    - Rocker arms
    - Valve covers
    - Fuel pressure sensor connector and vacuum line
    - Ignition coils
    - CMP sensor
    - Crankshaft pulley
    - Water pump
    - Power steering pump
    - Alternator
    - Water bypass tube
    - ECT sensor connector
    - Idler pulley
    - Accessory drive belt and tensioner

Rocker Arms

REMOVAL & INSTALLATION

1. Before servicing the vehicle, refer to the precautions at the beginning of this section.

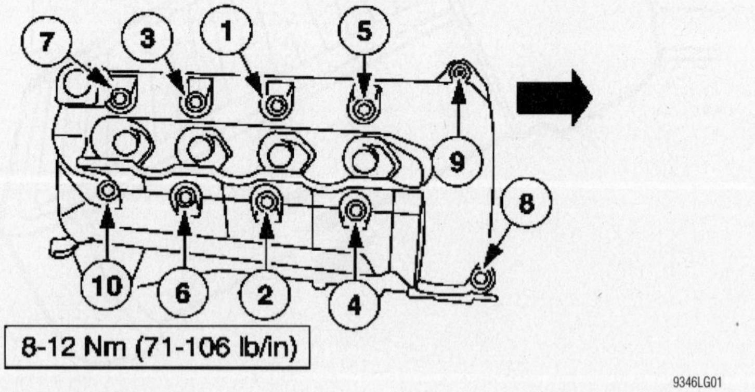

Right valve cover torque sequence

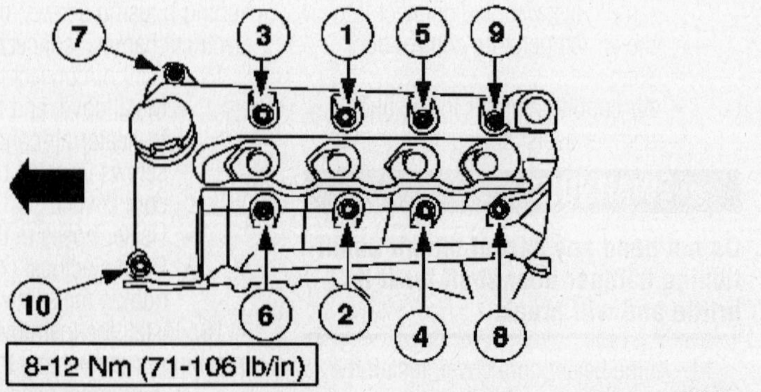

Left valve cover torque sequence

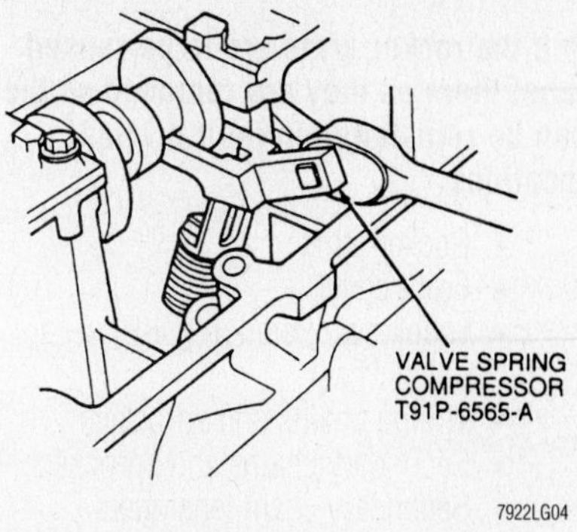

Rocker arm service tool—exhaust valve tool shown

2. Remove or disconnect the following:
   - Negative battery cable
   - Ignition coils
   - Valve covers
3. Rotate the crankshaft so that the piston on the cylinder to be serviced is at bottom dead center with the valves closed.
4. Install special tool Valve Spring Compressor T91P-6565-A for exhaust valves, and Valve Spring Compressor T93P-6565-A for intake valves.
5. Compress the valve spring and remove the rocker arm. Repeat for each arm to be removed.

➡If the rocker arms are to be reused, ensure that they are installed in the same position that they were removed from.

To install:

6. Compress the valve spring and install the rocker arm. Repeat for each arm to be installed.
7. Install or connect the following:
   - Valve covers
   - Ignition coils
   - Negative battery cable
8. Start the engine and check for proper operation.

Intake Manifold

REMOVAL & INSTALLATION

1. Before servicing the vehicle, refer to the precautions in the beginning of this section.
2. Drain the cooling system.
3. Remove or disconnect the following:
   - Negative battery cable
   - Engine appearance cover
   - Air cleaner outlet tube
   - Fuel lines
   - Fuel pressure sensor vacuum and wiring connectors
   - Upper radiator hose
   - Heater hose
   - Engine Coolant Temperature (ECT) sensor connector
   - Water bypass tube
   - Cruise control cable
   - Accelerator cable and bracket
   - Crankcase vent hose
   - Chassis vacuum supply line
   - Secondary air injection vacuum lines
   - Exhaust Gas Recirculation (EGR) vacuum regulator
   - EGR valve
   - Intake Manifold Runner Control (IMRC) cables and actuator
   - Fuel injector connectors
   - Throttle Position (TP) sensor connector
   - Idle Air Control (IAC) connector
   - Fuel temperature sensor connector
   - Fuel injection supply manifold
   - Intake manifold. Loosen the bolts in three steps in the sequence shown.
   - IMRC housings

To install:

4. Install or connect the following:
   - IMRC housings to the intake manifold
   - Fuel supply manifold and tighten the bolts to 71–106 inch lbs. (8–12 Nm)

➡Install the long bolts and studs in the outer holes. Install the short bolts and studs in the inner holes.

5. Install the intake manifold assembly. Tighten the bolts in sequence as follows:

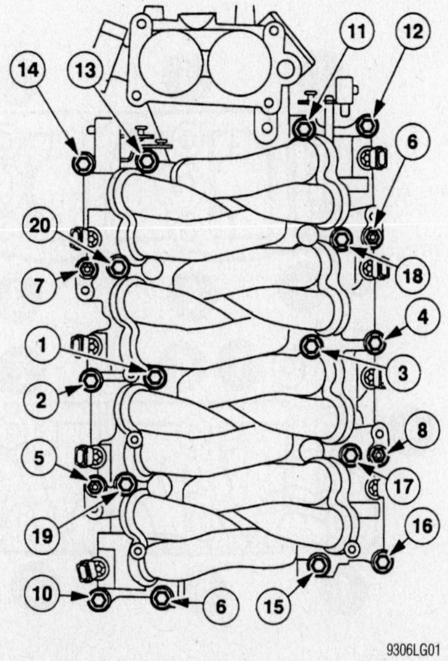

Intake manifold bolt removal sequence

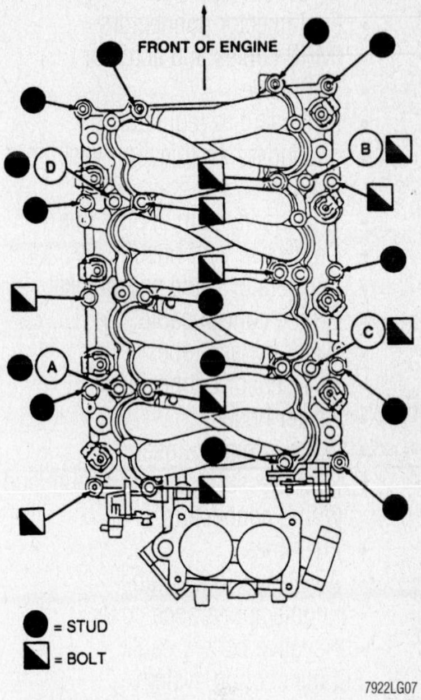

IMRC torque sequence—4.6L engine

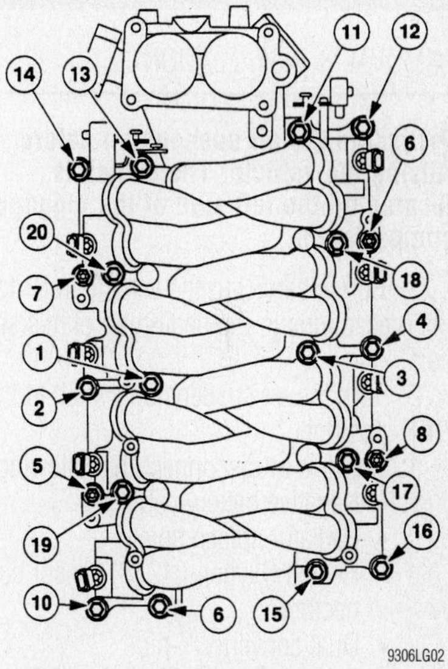

Intake manifold torque sequence—4.6L engine

a. Step 1: Tighten bolts 5, 7, 9, and 11 to 9–11 ft. lbs. (12–15 Nm).

b. Step 2: Tighten all other bolts to 13–16 ft. lbs. (18–22 Nm).

c. Step 3: Tighten all bolts in sequence 85–95 degrees.

6. Tighten the IMRC housing bolts as follows:

a. Step 1: 71–89 inch lbs. (8–10 Nm).

b. Step 2: Plus 85–95 degrees.

7. Install or connect the following:
   - Fuel temperature sensor connector
   - IAC connector

- TP sensor connector
- Fuel injector connectors
- IMRC cables and actuator
- EGR valve
- EGR vacuum regulator
- Secondary air injection vacuum lines
- Chassis vacuum supply line
- Crankcase vent hose
- Accelerator cable and bracket
- Cruise control cable
- Water bypass tube
- ECT sensor connector
- Heater hose
- Upper radiator hose
- Fuel pressure sensor vacuum and wiring connectors
- Fuel lines
- Air cleaner outlet tube
- Engine appearance cover
- Negative battery cable

8. Fill the cooling system.
9. Start the engine and check for leaks.

Exhaust Manifold

REMOVAL & INSTALLATION

➡Disable the air suspension before raising the vehicle. The switch is located on the left side of the luggage compartment.

1. Before servicing the vehicle, refer to the precautions in the beginning of this section.
2. Turn the air suspension switch to the **OFF** position.
3. Remove or disconnect the following:
   - Negative battery cable
   - Radiator splash shield
   - Heated Oxygen (HO_2S) sensor connectors
   - Dual converter Y-pipe
   - Exhaust Gas Recirculation (EGR) tube
   - Secondary air injection tubes
   - Exhaust manifolds

To install:

4. Install or connect the following:
   - Exhaust manifolds. Tighten the nuts in sequence to 13–16 ft. lbs. (18–22 Nm).
   - Secondary air injection tubes. Tighten the nuts to 25–34 ft. lbs. (34–46 Nm).
   - EGR tube. Tighten the nut to 30–33 ft. lbs. (40–45 Nm).
   - Dual converter Y-pipe
   - HO_2S sensor connectors
   - Radiator splash shield
   - Negative battery cable

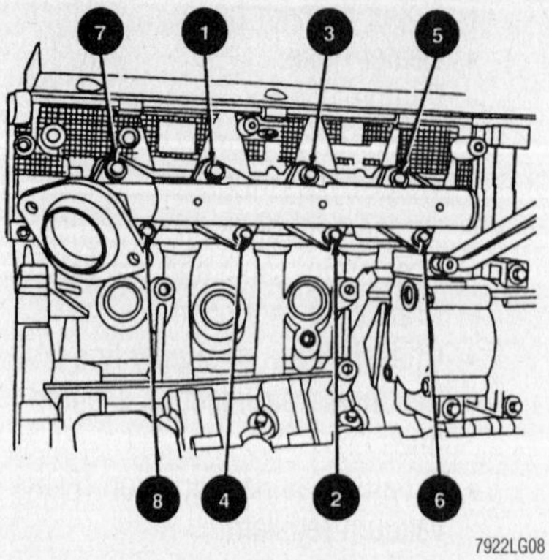

Exhaust manifold torque sequence—right side shown, left side similar

5. Turn the air suspension switch to the **ON** position.
6. Start the engine and check for leaks.

Camshaft and Valve Lifters

REMOVAL & INSTALLATION

1. Before servicing the vehicle, refer to the precautions in the beginning of this section.
2. Remove the engine from the vehicle and mount it on a suitable workstand.
3. Remove or disconnect the following:
   - Accessory drive belt and tensioner
   - Idler pulley
   - Alternator
   - Power steering pump
   - Water pump
   - Crankshaft pulley
   - Crankshaft Position (CKP) sensor
   - Camshaft Position (CMP) sensor
   - Ignition coils
   - Valve covers

➡If the rocker arms are to be reused, label them as they are removed so they can be reinstalled in their original locations.

   - Rocker arms
   - Front cover
   - Crankshaft Position (CKP) sensor pulse wheel
   - Timing chains and sprockets
   - Secondary chains and sprockets
   - Secondary chain tensioners
   - Camshaft cap assemblies
   - Camshafts

To install:

➡The exhaust camshaft cap assembly outboard bolts are shorter than the other cap assembly bolts.

4. Install the camshafts and camshaft cap assemblies in their original positions.
5. Tighten the camshaft cap assembly bolts in sequence as follows:
   a. Step 1: Tighten all bolts to 71–106 inch lbs. (8–12 Nm)
   b. Step 2: Loosen all bolts 2 turns
   c. Step 3: Tighten all bolts to 71–106 inch lbs. (8–12 Nm)
6. Install or connect the following:
   - Secondary chain tensioners
   - Secondary chains and sprockets
   - Timing chains and sprockets
   - CKP sensor pulse wheel
   - Front cover
   - Rocker arms in the original locations
   - Valve covers
   - Ignition coils
   - CMP sensor

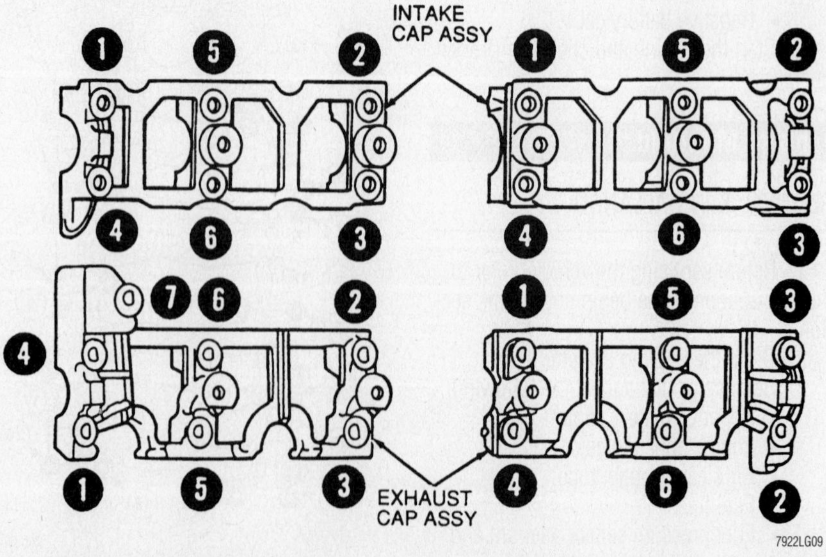

Camshaft cap assembly torque sequence

- CKP sensor
- Crankshaft pulley
- Water pump
- Power steering pump
- Alternator
- Idler pulley
- Accessory drive belt and tensioner

7. Remove the engine from the workstand and install in the vehicle.
8. Start the engine and check for leaks.

Valve Lash

ADJUSTMENT

The 4.6L DOHC engine uses hydraulic valve lash adjusters that do not require any adjustment.

Starter Motor

REMOVAL & INSTALLATION

➡Disable the air suspension before raising the vehicle. The switch is located on the left side of the luggage compartment.

1. Before servicing the vehicle, refer to the precautions in the beginning of this section.
2. Turn the air suspension switch to the **OFF** position.
3. Remove or disconnect the following:
   - Negative battery cable
   - Radiator splash shield
   - Starter solenoid connections
   - Starter

To install:

4. Install or connect the following:
   - Starter. Tighten the bolt and stud to 15–20 ft. lbs. (20–27 Nm).
   - Starter solenoid connections
   - Radiator splash shield
   - Negative battery cable
5. Turn the air suspension switch to the **ON** position.

Oil Pan

REMOVAL & INSTALLATION

➡Disable the air suspension before raising the vehicle. The switch is located on the left side of the luggage compartment.

1. Before servicing the vehicle, refer to the precautions in the beginning of this section.
2. Turn the air suspension switch to the **OFF** position.

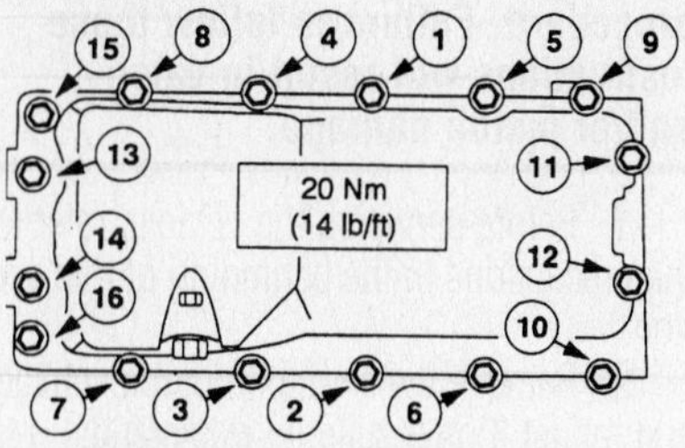

Oil pan bolt torque sequence

3. Remove or disconnect the following:
   - Negative battery cable
   - Oil level dipstick
   - Dual converter Y-pipe
   - Oil level sensor connector
   - Power steering pressure hose bracket
   - Oil pan

To install:

4. Apply a bead of silicone sealer to the oil pan flange. Also apply a bead of sealer to the front cover/cylinder block joint and fill the grooves on both sides of the rear main seal cap.
5. Use a new gasket and install the oil pan. Tighten the bolts in sequence as follows:
   a. Step 1: 14 ft. lbs. (20 Nm)
   b. Step 2: Plus 60 degrees
6. Install or connect the following:
   - Power steering pressure hose bracket
   - Oil level sensor connector
   - Dual converter Y-pipe
   - Oil level dipstick
   - Negative battery cable
7. Fill the crankcase with the correct type and quantity of engine oil.
8. Turn the air suspension switch to the **ON** position.
9. Start the engine and check for leaks.

Oil Pump

REMOVAL & INSTALLATION

1. Before servicing the vehicle, refer to the precautions in the beginning of this section.
2. Remove the engine from the vehicle and mount it on a suitable workstand.
3. Remove or disconnect the following:
   - Accessory drive belt and tensioner
   - Idler pulley
   - Alternator
   - Power steering pump
   - Water pump
   - Crankshaft pulley

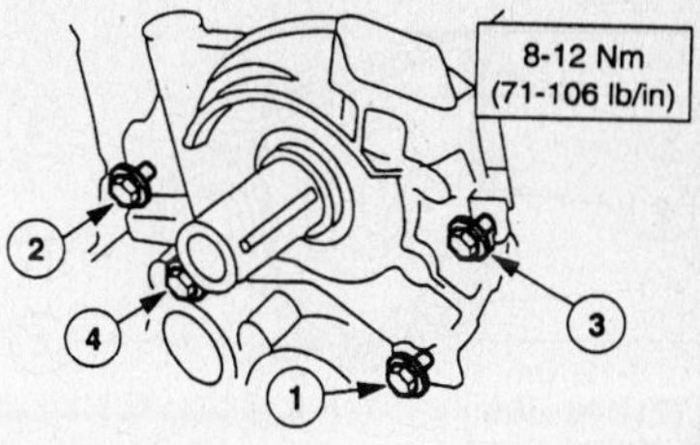

Oil pump mounting bolt sequence

   - Camshaft Position (CMP) sensor
   - Ignition coils
   - Valve covers
   - Rocker arms
   - Front cover
   - Crankshaft Position (CKP) sensor pulse wheel
   - Timing chains and sprockets
   - Oil pan
   - Oil pump pickup screen and tube
   - Oil pump

To install:

4. Install or connect the following:
   - Oil pump. Tighten the bolts in sequence to 71–106 inch lbs. (8–12 Nm).
   - Oil pump pickup screen and tube. Use a new O-ring seal and tighten the mounting bolts to 71–106 inch lbs. (8–12 Nm). Tighten the bracket bolt to 15–22 ft. lbs. (20–30 Nm).
   - Oil pan
   - Timing chains and sprockets
   - CKP sensor pulse wheel
   - Front cover
   - Rocker arms
   - Valve covers
   - Ignition coils
   - CMP sensor
   - Crankshaft pulley
   - Water pump
   - Power steering pump
   - Alternator
   - Idler pulley
   - Accessory drive belt and tensioner
5. Remove the engine from the workstand and install in the vehicle.
6. Restore all fluid levels.
7. Run the engine and check for leaks and proper operation.

Rear Main Seal

REMOVAL & INSTALLATION

1. Before servicing the vehicle, refer to the precautions in the beginning of this section.
2. Remove or disconnect the following:

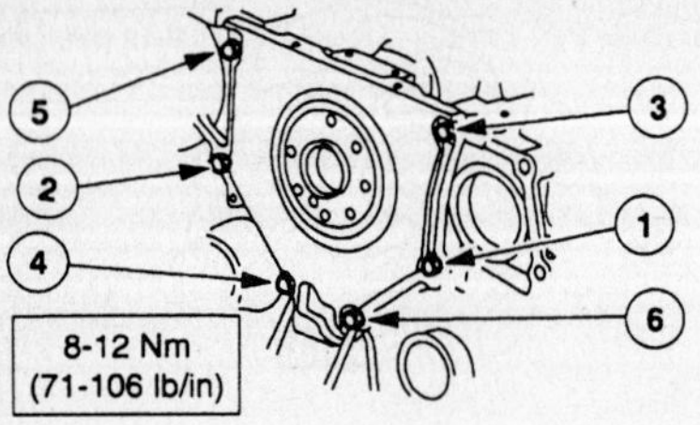

Oil seal retainer torque sequence

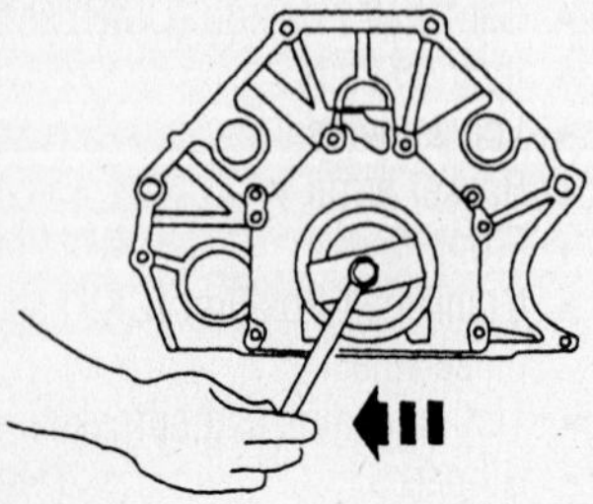

Use Seal Installer T82L-6701-A and adapter T91P-6701-A to press the seal into the retainer

- Transaxle
- Flywheel
- Oil seal retainer
- Oil seal

To install:

3. Apply a 0.060 inch (1.5mm) continuous bead of silicone gasket sealer to the cylinder block.
4. Install or connect the following:
   - Oil seal retainer. Tighten the bolts in sequence to 71–106 inch lbs. (8–12 Nm).
   - Oil seal. Use special tool Seal Installer T82L-6701-A and adapter T91P-6701-A to press the seal into place.
   - Flywheel. Tighten the bolts to 54–64 ft. lbs. (73–87 Nm).
   - Transaxle
5. Check and adjust all fluid levels as necessary.
6. Start the engine and check for leaks.

Timing Chain, Sprockets, Front Cover and Seal

REMOVAL & INSTALLATION

⁂ WARNING

This is an interference engine. When the timing chains are removed and the cylinder heads are installed, the crankshaft and/or camshafts must not be rotated unless as directed in this procedure. Failure to follow these instructions will result in valve and/or piston damage.

1. Before servicing the vehicle, refer to the precautions in the beginning of this section.
2. Remove the engine from the vehicle and mount it on a suitable workstand.
3. Remove or disconnect the following:
   - Accessory drive belt and tensioner
   - Idler pulley
   - Alternator
   - Power steering pump
   - Water pump
   - Crankshaft pulley
   - Front crankshaft seal
   - Crankshaft Position (CKP) sensor
   - Camshaft Position (CMP) sensor
   - Engine control sensor wiring harness
   - Ignition coils
   - Valve covers

➡If the rocker arms are to be reused, label them as they are removed so they can be reinstalled in their original locations.

   - Rocker arms
   - Front cover
4. Rotate the crankshaft to place the piston for No. 1 cylinder at Top Dead Center (TDC) on its compression stroke.
5. Remove or disconnect the following:
   - Right primary timing chain tensioner
   - Right primary timing chain tensioner arm and chain guide
   - Right primary timing chain and sprockets
   - Left primary timing chain tensioner
   - Left primary timing chain tensioner arm and chain guide
   - Left primary timing chain and sprockets

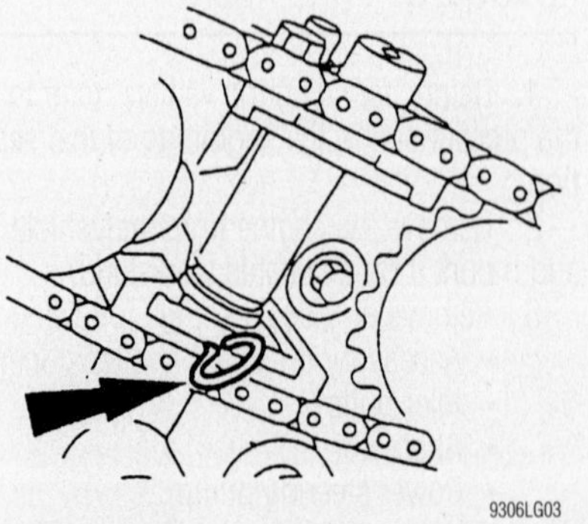

Secondary timing chain tensioner and locking pin

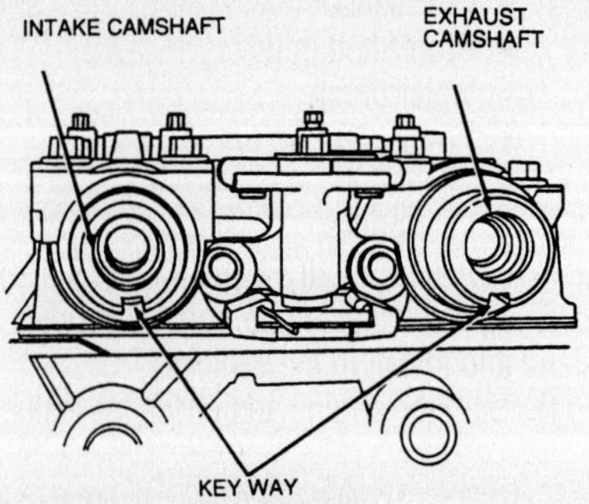

Position the camshafts for timing chain installation

6. Compress the secondary chain tensioners and install locking pins.
7. Remove or disconnect the following:
   - Left and right secondary timing chains and sprockets
   - Camshaft Holding and Camshaft Positioning tools

To install:

8. Position the camshafts as shown for timing chain installation.
9. Install the secondary timing chains and sprockets. Do not tighten the sprocket bolts at this time.

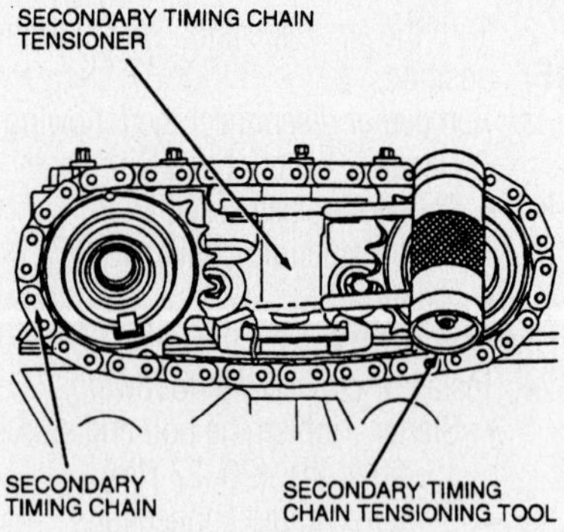

Secondary timing chain tensioning tool

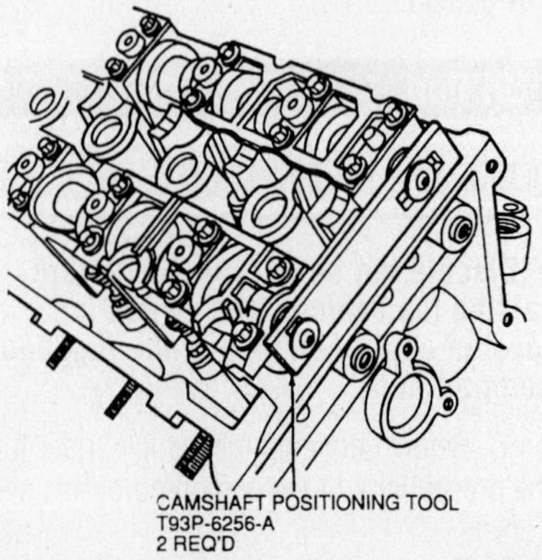

Install the Camshaft Positioning tool

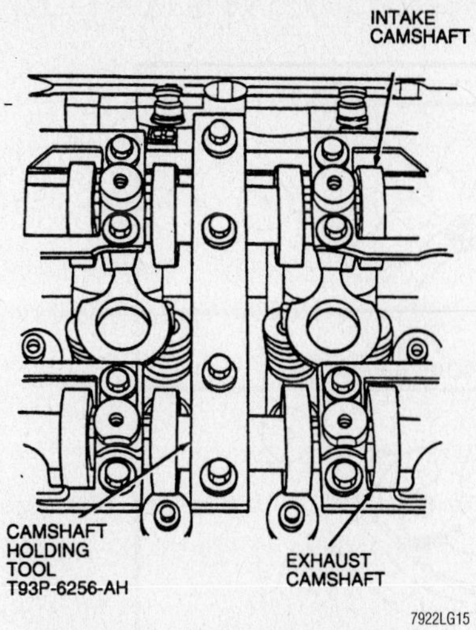

Install the Camshaft Holding tool to secure the camshafts and prevent damage to the Camshaft Positioning Tool

10. Tension the secondary timing chains with the special tool Secondary Timing 11.0. Chain Tensioning Tool T93P-6256-BH.

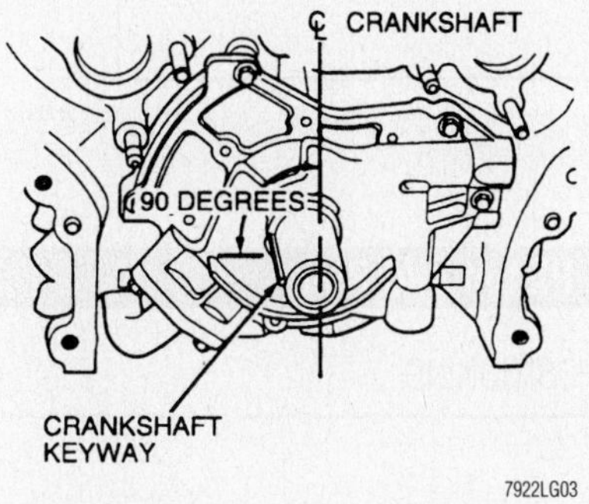

Turn the crankshaft to the 9 o'clock position for timing chain installation

11. Install Camshaft Positioning tool T93P-6256-A in the rear D-slots of the camshaft.

12. Install Camshaft Holding tool T93P-6256-AH onto the camshafts to keep the camshafts from rotating and to prevent damaging the camshaft positioning tool.

13. Tighten the secondary timing chain sprocket bolts to 81–95 ft. lbs. (110–130 Nm).

14. Remove the secondary timing chain tensioner locking pins and remove the tensioner tool.

15. Position the crankshaft so that the keyway is at 9 o'clock as shown.

16. Compress the primary timing chain tensioners and install locking pins.

17. Install the crankshaft timing sprockets arranged as shown.

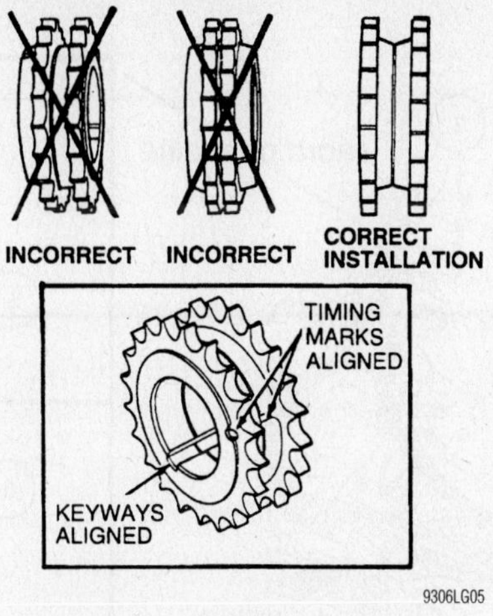

Crankshaft timing sprocket installation

18. Align the copper colored primary timing chain links with the sprocket timing marks as shown.

19. Install or connect the following:

- Left primary timing chain and sprockets
- Left primary timing chain tensioner arm and chain guide
- Left primary timing chain tensioner
- Right primary timing chain and sprockets
- Right primary timing chain tensioner arm and chain guide
- Right primary timing chain tensioner

20. Tighten the camshaft sprocket bolts to 81–95 ft. lbs. (110–130 Nm).

21. Remove the primary timing chain tensioner locking pins.

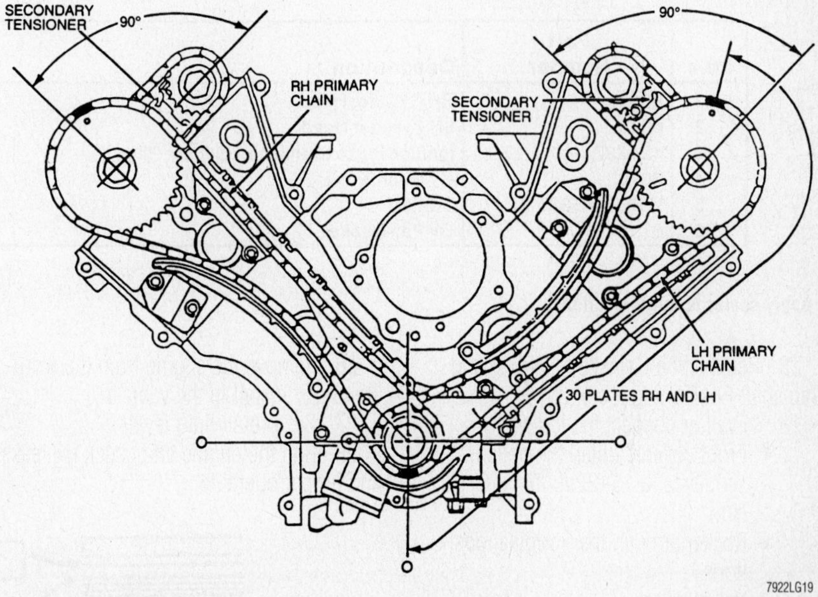

Correct alignment of the primary timing chains and sprockets

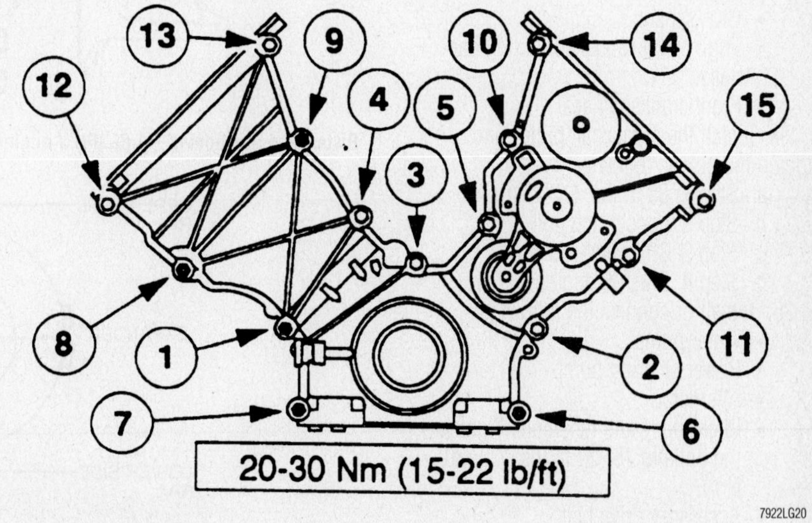

Front cover torque sequence

Sealer Location

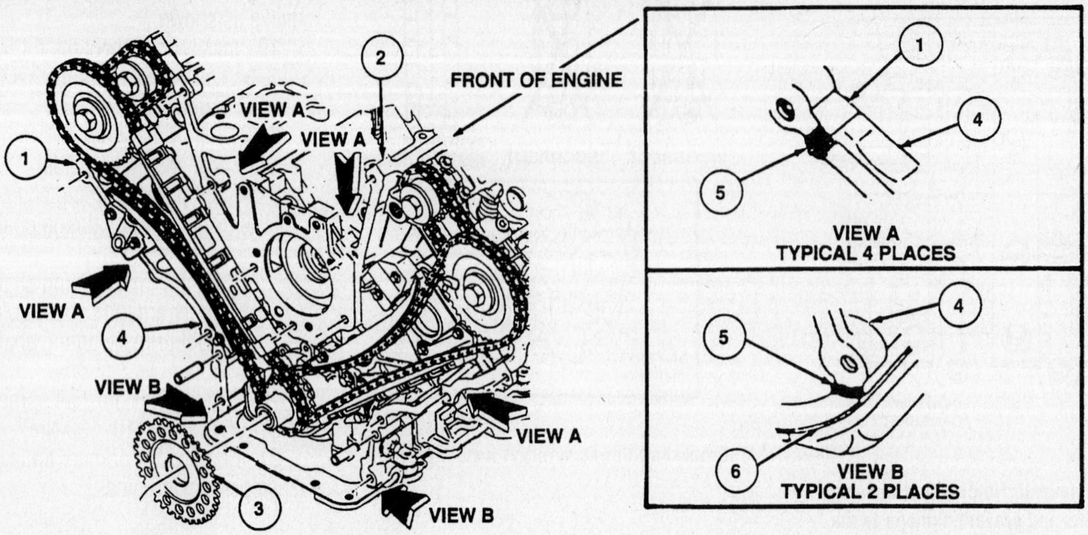

Item	Part Number	Description
1	6049	RH Cylinder Head
2	6049	LH Cylinder Head
3	12A227	Ignition Pulse Crankshaft Sensor Ring
4	6010	Cylinder Block
5	WSE-M4G320-A2	Sealer
6	6710	Oil Pan Gasket

9306LG06

Apply sealer to these locations

22. Remove the Camshaft Holding and Camshaft Positioning tools.
23. Install or connect the following:
    - Front cover. Tighten the bolts in sequence to 15–22 ft. lbs. (20–30 Nm).
    - Rocker arms in their original positions
    - Valve covers
    - Ignition coils
    - CMP sensor
    - CKP sensor
    - Engine control sensor wiring harness
    - Front crankshaft seal
24. Install the crankshaft pulley and tighten the bolt as follows:
    a. Step 1: 89 ft. lbs. (120 Nm)
    b. Step 2: Loosen the bolt
    c. Step 3: 39 ft. lbs. (53 Nm)
    d. Step 4: Plus 90 degrees
25. Install or connect the following:
    - Water pump
    - Power steering pump
    - Alternator
    - Idler pulley and tensioner. Tighten the bolts to 15–22 ft. lbs. (20–30 Nm).
    - Accessory drive belt
26. Remove the engine from the workstand and install in the vehicle.
27. Restore all fluid levels.
28. Run the engine and check for leaks and proper operation.

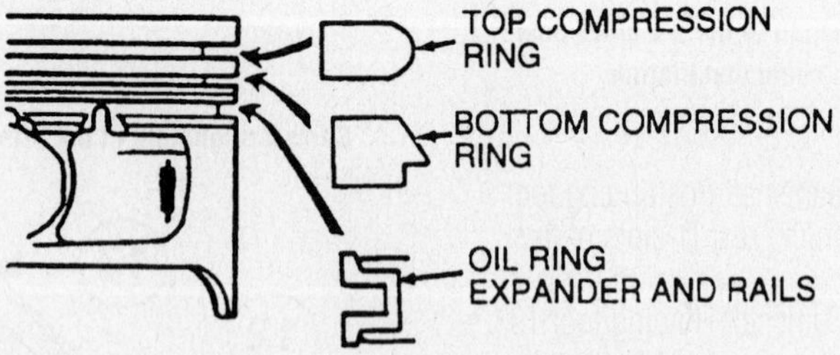

7922AG34

Piston ring positioning—4.6L VIN V engine

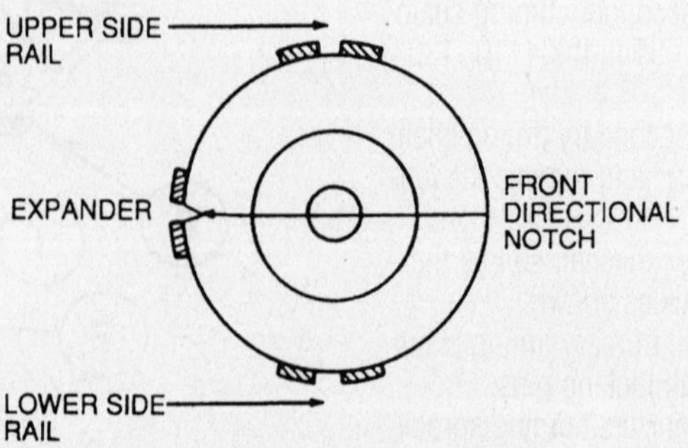

7922AG33

Piston positioning and ring end-gap spacing—4.6L VIN V engine

Piston and Ring

POSITIONING

FUEL SYSTEM

Fuel System Service Precautions

Safety is the most important factor when performing not only fuel system maintenance but any type of maintenance. Failure to conduct maintenance and repairs in a safe manner may result in serious personal injury or death. Maintenance and testing of the vehicle's fuel system components can be accomplished safely and effectively by adhering to the following rules and guidelines.

- To avoid the possibility of fire and personal injury, always disconnect the negative battery cable unless the repair or test procedure requires that battery voltage be applied.
- Always relieve the fuel system pressure prior to disconnecting any fuel system component (injector, fuel rail, pressure regulator, etc.), fitting or fuel line connection. Exercise extreme caution whenever relieving fuel system pressure, to avoid exposing skin, face and eyes to fuel spray. Please be advised that fuel under pressure may penetrate the skin or any part of the body that it contacts.
- Always place a shop towel or cloth around the fitting or connection prior to loosening to absorb any excess fuel due to spillage. Ensure that all fuel spillage (should it occur) is quickly removed from engine surfaces. Ensure that all fuel soaked cloths or towels are deposited into a suitable waste container.
- Always keep a dry chemical (Class B) fire extinguisher near the work area.
- Do not allow fuel spray or fuel vapors to come into contact with a spark or open flame.
- Always use a back-up wrench when loosening and tightening fuel line connection fittings. This will prevent unnecessary stress and torsion to fuel line piping. Always follow the proper torque specifications.
- Always replace worn fuel fitting O-rings with new. Do not substitute fuel hose or equivalent, where fuel pipe is installed.

Fuel System Pressure

RELIEVING

1. Before servicing the vehicle, refer to the precautions in the beginning of this section.
2. Disconnect the negative battery cable.
3. Remove the fuel tank fill cap to relieve the pressure in the fuel tank.
4. Remove the cap from the Schrader valve located on the fuel supply manifold.
5. Attach Fuel Pressure Gauge T80L-9974-A to the valve and drain the fuel through the drain tube into a suitable container.
6. After the fuel system pressure is relieved, remove the fuel pressure gauge and install the cap on the Schrader valve.
7. Install the fuel tank fill cap.
8. Connect the negative battery cable only after system repairs are completed.

Fuel Filter

REMOVAL & INSTALLATION

➡Disable the air suspension before raising the vehicle. The switch is located on the left side of the luggage compartment.

1. Before servicing the vehicle, refer to the precautions in the beginning of this section.
2. Turn the air suspension service switch **OFF**.
3. Disconnect the negative battery cable.
4. Relieve the fuel system pressure.
5. Disconnect the fuel lines.
6. Loosen the filter retaining clamp and remove the fuel filter.

To install:

7. Install the fuel filter with the flow arrow facing the proper direction and tighten the filter retaining clamp.
8. Push the fuel lines on to the filter fittings until an audible click is heard.
9. Connect the negative battery cable.
10. Start the engine and check for fuel leaks and proper operation.
11. Turn the air suspension switch to the **ON** position.

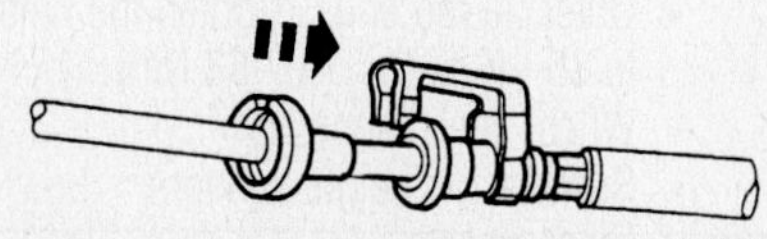

7922LG21

Slide the special tool into the fitting to disengage it

Fuel Pump

REMOVAL & INSTALLATION

➡Disable the air suspension before raising the vehicle. The switch is located on the left side of the luggage compartment.

1. Before servicing the vehicle, refer to the precautions in the beginning of this section.
2. Turn the air suspension switch to the **OFF** position.
3. Relieve the fuel system pressure.
4. Drain the fuel tank.
5. Remove or disconnect the following:
   - Fuel fill and vent hoses
   - Fuel supply line
   - Fuel pump module wiring connector
   - Fuel tank
   - Fuel pump module locking ring
6. Pull the fuel pump module up and out of the fuel tank until the locking tabs for the fuel pump module are accessible. Squeeze both locking tabs together and remove the fuel pump module from the fuel tank.

To install:

7. Install or connect the following:
   - Fuel pump module. Use a new O-ring seal and push the module into the tank so that the locking tabs engage.
   - Fuel pump module lock ring
   - Fuel tank. Tighten the tank strap bolts to 22–30 ft. lbs. (30–40 Nm).
   - Fuel pump module wiring connector
   - Fuel supply line
   - Fuel fill and vent hoses

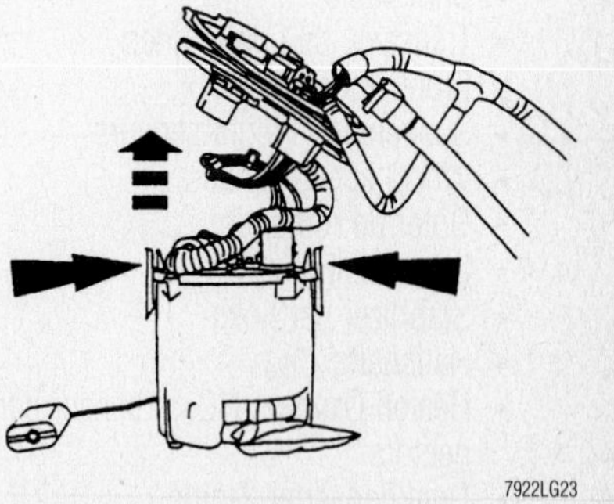

7922LG23

To remove the fuel pump, reach into the tank and press the locking tabs inward

8. Add 10 gallons of clean fuel to the tank.
9. Turn the air suspension switch to the **ON** position.
10. Start the engine and check for leaks.

Fuel Injector

REMOVAL & INSTALLATION

1. Before servicing the vehicle, refer to the precautions in the beginning of this section.
2. Relieve the fuel system pressure.
3. Remove or disconnect the following:
   - Negative battery cable
   - Fuel pressure sensor vacuum line and wiring connector
   - Fuel injector wiring connectors
   - Fuel supply line
   - Fuel supply manifold with injectors attached
4. Remove the clips and the injectors from the fuel supply manifold.

To install:

5. Install the injectors with new O-ring seals.
6. Install the fuel supply manifold and tighten the bolts as follows:
   a. Step 1: 15 ft. lbs. (20 Nm)
   b. Step 2: Plus 85–95 degrees
7. Install or connect the following:
   - Fuel supply line
   - Fuel injector wiring connectors
   - Fuel pressure sensor vacuum line and wiring connector
   - Negative battery cable
8. Start the engine and check for leaks.

DRIVE TRAIN

Transaxle Assembly

REMOVAL & INSTALLATION

➡Disable the air suspension before raising the vehicle. The switch is located on the left side of the luggage compartment.

1. Before servicing the vehicle, refer to the precautions in the beginning of this section.
2. Turn the air suspension switch to the **OFF** position.
3. Attach a powertrain support fixture to the engine lifting eyes.
4. Drain the transaxle fluid.
5. Remove or disconnect the following:
   - Battery and battery tray
   - Intake Air Temperature (IAT) sensor connector
   - Mass Air Flow (MAF) sensor connector
   - Air cleaner assembly
   - Transaxle control harness connector
   - Transaxle range sensor
   - Turbine shaft speed sensor connector
   - Shift cable
   - Transaxle oil cooler lines
   - Front wheels
   - Suspension height sensors
   - Wheel speed sensors
   - Outer tie rod ends
   - Lower ball joints
   - Stabilizer bar links
   - Halfshafts
   - Heated Oxygen (HO_2S) sensor connectors
   - Dual converter Y-pipe
   - Starter
   - Left, right and rear engine support insulators
   - Subframe
   - Transaxle housing cover
   - Torque converter
   - Transaxle flange bolts
6. Separate the transaxle from the engine and lower it from the vehicle.

To install:

➡When installing suspension components and halfshafts, use new nuts, bolts, circlips and split pins.

7. Install or connect the following:
   - Transaxle. Tighten the flange bolts to 25–34 ft. lbs. (34–46 Nm).
   - Torque converter. Tighten the nuts to 20–34 ft. lbs. (27–46 Nm).
   - Transaxle housing cover
8. Install the subframe. Use 2 pieces of ¾ inch outside diameter pipe in the alignment holes behind the front subframe mounts to align the subframe to the body. Tighten the subframe mounting bolts to 57–76 ft. lbs. (77–103 Nm).
9. Install or connect the following:
   - Left, right and rear engine support insulators. Tighten the through-bolts to 64–88 ft. lbs. (87–119 Nm).
   - Starter. Tighten the bolt and stud to 15–21 ft. lbs. (21–29 Nm).
   - Dual converter Y-pipe
   - HO_2S sensor connectors
   - Halfshafts
   - Stabilizer bar links. Tighten the nuts to 30–40 ft. lbs. (40–55 Nm).
   - Lower ball joints. Tighten the pinch bolts to 50–68 ft. lbs. (68–92 Nm).
   - Outer tie rod ends. Tighten the nuts to 35–46 ft. lbs. (47–63 Nm).
   - Wheel speed sensors
   - Suspension height sensors
   - Front wheels
   - Transaxle oil cooler lines
   - Shift cable
   - Turbine shaft speed sensor connector
   - Transaxle range sensor
   - Transaxle control harness connector
   - Air cleaner assembly
   - MAF sensor connector
   - IAT sensor connector
   - Battery and battery tray
10. Fill the transaxle.
11. Check for leaks and proper operation.

➡Whenever the subframe is removed or lowered, the wheel alignment should be checked.

Halfshaft

REMOVAL & INSTALLATION

➡Disable the air suspension before raising the vehicle. The switch is located on the left side of the luggage compartment.

1. Before servicing the vehicle, refer to the precautions in the beginning of this section.
2. Turn the air suspension switch to the **OFF** position.
3. Remove or disconnect the following:
   - Negative battery cable
   - Front wheels
   - Suspension height sensors
   - Wheel speed sensors
   - Outer tie rod ends
   - Lower ball joints
   - Stabilizer bar links
4. Separate the halfshafts from the wheel hubs and the transaxle and remove the halfshafts.

To install:

➡When installing suspension components and halfshafts, use new nuts, bolts, circlips and split pins.

5. Install or connect the following:
   - Halfshafts. Tighten the hub retainer nuts to 180–200 ft. lbs. (245–270 Nm).
   - Stabilizer bar links. Tighten the nuts to 30–40 ft. lbs. (40–55 Nm).
   - Lower ball joints. Tighten the pinch bolts to 50–68 ft. lbs. (68–92 Nm).
   - Outer tie rod ends. Tighten the nuts to 35–46 ft. lbs. (47–63 Nm).
   - Wheel speed sensors
   - Suspension height sensors
   - Front wheels
6. Turn the air suspension switch to the **ON** position.
7. Road test the vehicle and check for proper operation.

CV-Joint

REMOVAL AND REPLACEMENT

Inner Tripod Joint

The inner CV-joint is serviced with the halfshaft as an assembly. The inner CV-joint boot can be serviced by removing the outer CV-joint.

Outer CV-Joint

1. Before servicing the vehicle, refer to the precautions in the beginning of this section.
2. Place the halfshaft in a vise.
3. Remove the CV-joint boot clamps and slide the boot away from the joint.
4. Drive the CV-joint off the halfshaft with a brass drift and a hammer.

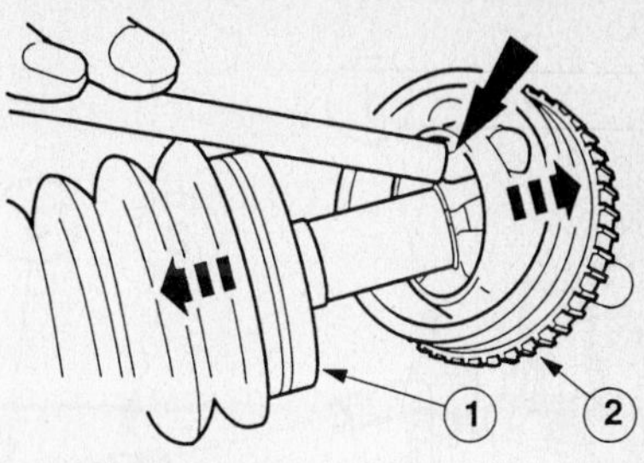

Removing the outer CV-joint

To install:

5. Replace the snapring.
6. Fill the CV-joint with fresh grease and slide the joint onto the halfshaft.
7. Use new clamps and install the CV-joint boot.

STEERING AND SUSPENSION

Air Bag

CAUTION

Some vehicles are equipped with an air bag system. The system must be disarmed before performing service on, or around, system components, the steering column, instrument panel components, wiring and sensors. Failure to follow the safety precautions and the disarming procedure could result in accidental air bag deployment, possible injury and unnecessary system repairs.

PRECAUTIONS

Several precautions must be observed when handling the inflator module to avoid accidental deployment and possible personal injury.

- Never carry the inflator module by the wires or connector on the underside of the module.
- When carrying a live inflator module, hold securely with both hands, and ensure that the bag and trim cover are pointed away.
- Place the inflator module on a bench or other surface with the bag and trim cover facing up.
- With the inflator module on the bench, never place anything on or close to the module that may be thrown in the event of an accidental deployment.

DISARMING

1. Before servicing the vehicle, refer to the precautions in the beginning of this section.
2. Disconnect both battery cables from the battery, negative cable first.
3. Wait 1 minute before proceeding with the service procedure. This is the time required for the back-up power supply in the air bag diagnostic monitor to deplete its stored energy.
4. After service is completed, reconnect the battery cables, negative cable last.
5. Turn the ignition switch to the **RUN** position. The air bag indicator should light continuously for approximately 6 seconds, then turn **OFF**. If the indicator fails to light, flashes or remains lit continuously, there is a fault in the air bag system.

Power Rack and Pinion Steering Gear

REMOVAL & INSTALLATION

CAUTION

Do not rotate the steering wheel when the intermediate shaft is disconnected. Damage to the air bag sliding contact will result.

➡Disable the air suspension before raising the vehicle. The switch is located on the left side of the luggage compartment.

1. Before servicing the vehicle, refer to the precautions in the beginning of this section.
2. Turn the air suspension switch to the **OFF** position.
3. Remove or disconnect the following:
   - Negative battery cable
   - Steering column intermediate shaft
   - Powertrain Control Module (PCM) harness connector and ground straps
   - Upper motor mount
   - Front wheels
   - Outer tie rod ends
   - Suspension height sensor
   - Heated Oxygen (HO_2S) sensor connectors
   - Dual converter Y-pipe
   - Subframe brackets
   - Steering gear attachment nuts
   - Rear subframe bolts. Lower the rear of the subframe about 4 inches for access.
   - Steering gear heat shield and bracket
   - Steering gear harness connectors
4. Rotate the steering gear to clear the mounting bolts from the subframe, and pull the gear to the left to access the fluid lines.
   - Power steering pressure and return lines
   - Steering gear through the left fender opening

To install:

5. Install or connect the following:
   - Steering gear
   - Power steering pressure and return

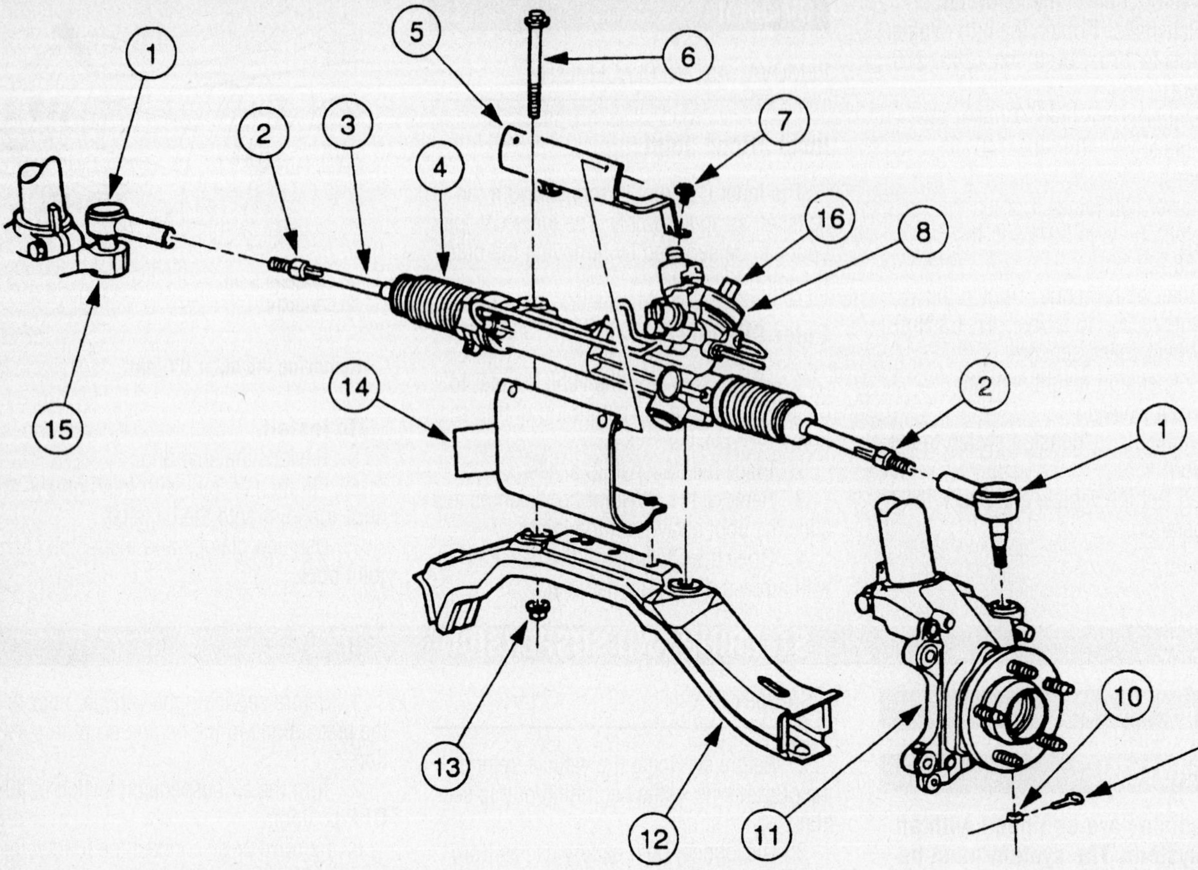

1 Tie Rod End
2 Nut
3 Front Wheel Spindle Tie Rod (2 Req'd)
4 Front Suspension Steering Ball Stud Dust Seal
5 Power Steering Hose Bracket
6 Bolt (2 Req'd)
7 Screw (2 Req'd)
8 Steering Gear
9 Screw (2 Req'd)
10 Nut (2 Req'd)
11 Front Wheel Knuckle (LH)
12 Crossmember
13 Nut (2 Req'd)
14 Steering Shaft U-Joint Shield
15 Front Wheel Knuckle (RH)
16 Power Steering Gear Input Shaft and Control

7922LG26

Exploded view of the power rack and pinion steering gear mounting

lines. Use new O-ring seals and tighten the fittings to 25–30 ft. lbs. (33–41 Nm).

- Steering gear harness connectors
- Steering gear heat shield and bracket
- Rear subframe bolts. Tighten the bolts to 57–76 ft. lbs. (77–103 Nm).
- Steering gear attachment nuts. Tighten the nuts to 84–112 ft. lbs. (113–153 Nm).
- Subframe brackets
- Dual converter Y-pipe
- HO_2S sensor connectors
- Suspension height sensor
- Outer tie rod ends. Use new nuts and tighten to 35–46 ft. lbs. (47–63 Nm).
- Front wheels
- Upper motor mount
- PCM harness connector and ground straps
- Steering column intermediate shaft. Tighten the lower bolt to 30–38 ft. lbs. (41–51 Nm), and the upper bolt to 24–30 ft. lbs. (33–41 Nm).
- Negative battery cable

6. Fill the power steering system.

7. Turn the air suspension switch to the **ON** position.

8. Check the system for leaks and proper operation. Adjust the toe setting as necessary.

Strut

REMOVAL & INSTALLATION

Front

➡Disable the air suspension before raising the vehicle. The switch is located on the left side of the luggage compartment.

1. Before servicing the vehicle, refer to the precautions in the beginning of this section.
2. Turn the air suspension switch to the **OFF** position.
3. Remove or disconnect the following:
   - Negative battery cable
   - Front wheels
   - Left and right splash shields
   - Wheel speed sensors
   - Strut actuator connectors, if equipped with semi-active suspension
   - Suspension height sensors
   - Brake line brackets
   - Stabilizer bar links
   - Lower ball joints
   - Steering knuckle pinch bolts
4. Separate the steering knuckles from the struts and remove the struts from the vehicle.

To install:

➡When installing suspension components, use new nuts, bolts, circlips and split pins.

5. Install or connect the following:
   - Struts. Tighten the shock tower nuts to 23–29 ft. lbs. (30–40 Nm), and the steering knuckle pinch bolts to 73–97 ft. lbs. (98–132 Nm).
   - Stabilizer bar links. Tighten the nuts to 30–40 ft. lbs. (40–55 Nm).
   - Lower ball joints. Tighten the pinch bolts to 50–68 ft. lbs. (68–92 Nm).
   - Brake line brackets. Tighten the screws to 11 ft. lbs. (15 Nm).
   - Suspension height sensors
   - Strut actuator connectors, if equipped with semi-active suspension
   - Wheel speed sensors
   - Left and right splash shields
   - Front wheels
   - Negative battery cable
6. Turn the air suspension switch to the **ON** position.
7. Road test the vehicle and check for proper operation.

Shock Absorber

REMOVAL & INSTALLATION

Rear

➡Disable the air suspension before raising the vehicle. The switch is located on the left side of the luggage compartment.

1. Before servicing the vehicle, refer to the precautions in the beginning of this section.
2. Turn the air suspension switch to the **OFF** position.
3. Remove or disconnect the following:
   - Negative battery cable
   - Luggage compartment side trim panels
   - Rear wheels
   - Shock absorber actuator harness connectors, if equipped with semi-active suspension
   - Shock absorber assemblies

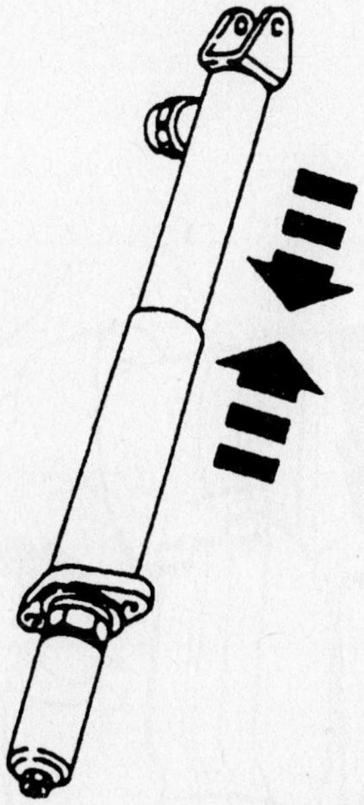

7922LG28

To bleed the air from the shock, turn it upside down, then compress and expand it several times

To install:

4. If replacing the shock absorbers, transfer the mass dampers and mounting brackets.
5. Turn the shock absorber upside down. Compress and expand the piston several times to remove the air from the working cylinder.
6. Install or connect the following:
   - Shock absorber assemblies. Tighten the upper bolts to 25–34 ft. lbs. (34–46 Nm), and the lower bolts to 58–68 ft. lbs. (68–92 Nm).
   - Shock absorber actuator harness connectors, if equipped with semi-active suspension
   - Rear wheels
   - Luggage compartment side trim panels
   - Negative battery cable
7. Turn the air suspension system **ON**.

Coil Spring

REMOVAL & INSTALLATION

Front

1. Before servicing the vehicle, refer to the precautions in the beginning of this section.
2. Remove the strut assembly from the vehicle.
3. Install a spring compressor and retract the spring until the upper mount rotates freely.
4. Remove or disconnect the following:
   - Flange nut
   - Strut mounting bracket
   - Strut bearing plate
   - Coil spring

To install:

5. Install or connect the following:
   - Coil spring
   - Strut bearing plate
   - Strut mounting bracket
   - Flange nut. Tighten the nut to 63–70 ft. lbs. (85–95 Nm).
6. Remove the spring compressor and ensure that the spring seats correctly in the insulators.
7. Install the strut assembly to the vehicle.

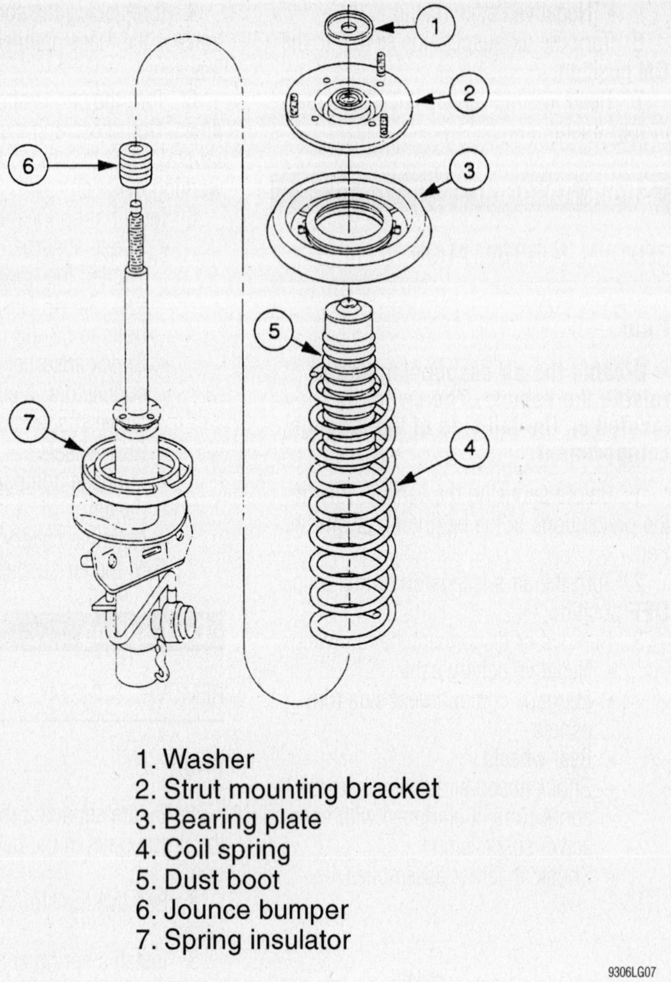

Exploded view of the front strut and coil spring assembly

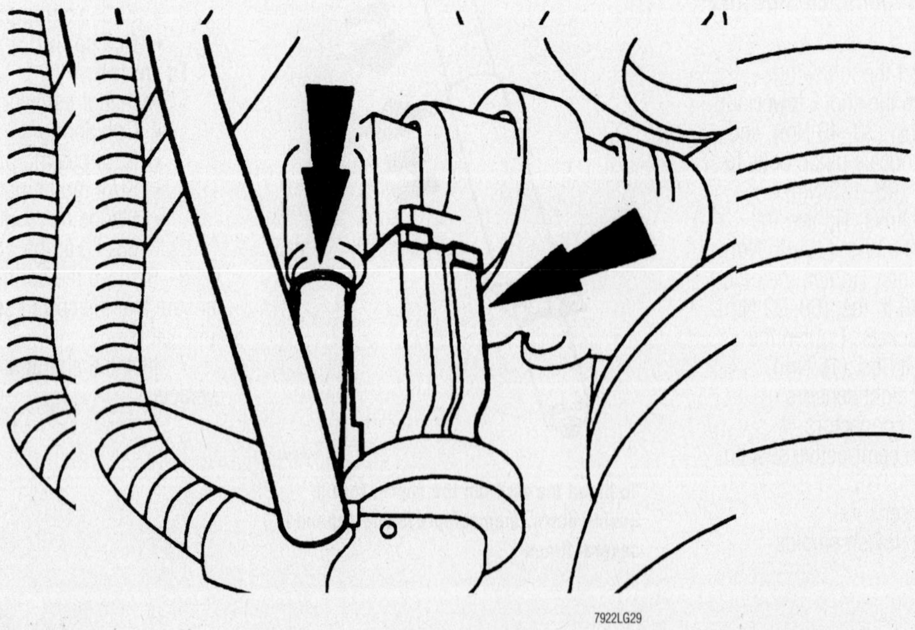

Air spring solenoid connectors

Air Spring

REMOVAL & INSTALLATION

Rear

**** CAUTION**

Do not attempt to service the rear suspension without first deflating the air springs.

➡Servicing the air suspension requires a scan tool with bi-directional capabilities to deflate the air springs.

1. Properly deflate the air springs.
2. Turn the air suspension switch to the **OFF** position. The switch located on the right kick panel.
3. Remove the rear wheels.
4. Disconnect the air spring solenoid wiring connector and air line.
5. Depress the locking tabs and remove the air spring from the vehicle.

To install:

6. Install or connect the following:
   - Air spring. Press the plastic tab through the suspension until it clicks into place.
   - Solenoid air line and electrical connectors
   - Rear wheels
7. Turn the air suspension switch to the **ON** position and check for proper operation.

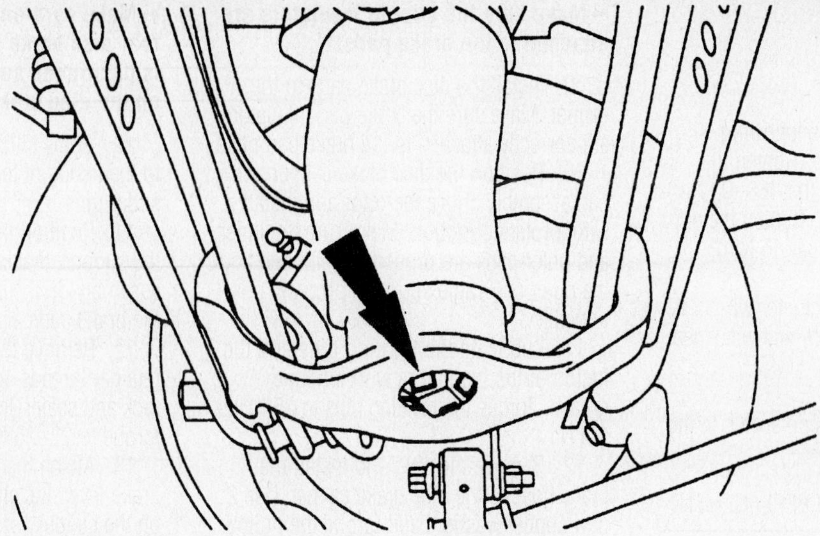

Depress the plastic locking tabs to disengage the air spring from the suspension

Lower Ball Joints

REMOVAL & INSTALLATION

The lower ball joint is an integral part of the steering knuckle. If the lower ball joint is found to be defective, the entire steering knuckle must be replaced.

Upper Control Arm

REMOVAL & INSTALLATION

Rear

**** CAUTION**

Do not attempt to service the rear suspension without first deflating the air springs.

➡Servicing the air suspension requires a scan tool with bi-directional capabilities to deflate the air springs.

1. Properly deflate the air springs.
2. Turn the air suspension switch to the **OFF** position. The switch located on the right kick panel.
3. Support the lower control arm on a jackstand.
4. Remove or disconnect the following:
   - Rear wheel
   - Brake hose bracket
   - Upper ball joint
   - Upper control arm

To install:

➡Use new mounting nuts and bolts.

5. Install the upper control arm. Tighten the ball joint nut to 50–67 ft. lbs. (68–92 Nm), then tighten the control arm mounting bolts to 73–97 ft. lbs. (98–132 Nm).
6. Install or connect the following:
   - Brake hose bracket
   - Rear wheel
7. Turn the air suspension switch to the **ON** position and check for proper operation.
8. Check the wheel alignment and adjust as necessary.

CONTROL ARM BUSHING REPLACEMENT

The control arm bushings are serviced with the control arm as an assembly.

Lower Control Arm

REMOVAL & INSTALLATION

Front

➡Disable the air suspension before raising the vehicle. The switch is located on the left side of the luggage compartment.

1. Before servicing the vehicle, refer to the precautions in the beginning of this section.
2. Turn the air suspension switch to the **OFF** position.
3. Remove or disconnect the following:
   - Front wheel
   - Wheel speed sensor wiring harness
   - Height sensor
   - Lower ball joint
   - Lower control arm strut
   - Lower control arm

To install:

4. Install or connect the following:
   - Lower control arm. Tighten the pivot bolt to 73–98 ft. lbs. (98–132 Nm).
   - Lower control arm strut. Tighten the nut to 73–98 ft. lbs. (98–132 Nm).
   - Lower ball joint. Use a new nut and tighten to 50–67 ft. lbs. (68–92 Nm).
   - Wheel speed sensor wiring harness
   - Height sensor
   - Front wheel
5. Turn the air suspension switch to the **ON** position.
6. Check the wheel alignment and adjust as necessary.

Rear

**** CAUTION**

Do not attempt to service the rear suspension without first deflating the air springs.

➡Disable the air suspension before raising the vehicle. The switch is located on the left side of the luggage compartment.

1. Before servicing the vehicle, refer to the precautions in the beginning of this section.
2. Turn the air suspension switch to the **OFF** position.
3. Remove or disconnect the following:
   - Rear wheel
   - Shock absorber actuator wiring, if equipped with semi-active suspension

- Air spring
- Stabilizer bar link
- Lower control arm

To install:

4. Install or connect the following:
   - Lower control arm. Tighten the bolts to 50–68 ft. lbs. (68–92 Nm).
   - Stabilizer bar link. Tighten the nut to 25–34 ft. lbs. (34–46 Nm).
   - Air spring
   - Shock absorber actuator wiring, if equipped with semi-active suspension
   - Rear wheel
5. Turn the air suspension switch to the **ON** position.

CONTROL ARM BUSHING REPLACEMENT

The control arm bushings are serviced with the control arm as an assembly.

Brake Caliper

REMOVAL & INSTALLATION

CAUTION

The air suspension switch, located in the left side of the luggage compartment, must be turned OFF before raising the vehicle. Failure to do so may result in unexpected inflation or deflation of the air springs that may result in shifting of the vehicle during service.

Front

1. Before servicing the vehicle, refer to the precautions in the beginning of this section.
2. Remove brake fluid from the brake master cylinder reservoir until the reservoir is ½ full.
3. Turn the air suspension switch, located in the left side of the luggage compartment, to the **OFF** position.
4. Remove the wheel and tire assembly.
5. Mark the disc brake caliper to ensure that it is reinstalled in the correct location.
6. Remove the hollow bolt connecting the brake hose to the disc brake caliper and plug the brake hose. Discard the 2 copper sealing washers.
7. Remove the caliper locating pins and lift the caliper off the rotor using a rotating motion.

To install:

8. Retract the disc brake caliper piston fully in the piston bore, using an old brake pad or block of wood and a C-clamp.

➡Make sure the clip-on insulators are attached to the brake pads.

9. Install the disc brake pads to the caliper. Make sure the brake pad insulators are correctly attached to the brake pad plate.
10. Position the disc brake caliper and pad assembly above the rotor and install it with a rotating motion. Make sure the inner and outer pads are properly positioned and the outer anti-rattle spring is properly positioned.
11. Lubricate the locating pins and the inside of the insulators with silicone grease. Torque the locating pins to 25 ft. lbs. (34 Nm).
12. Remove the plug and install the brake hose to the disc brake caliper. Use 2 new copper washers and torque the hollow bolt to 30–40 ft. lbs. (41–54 Nm).
13. Bleed the brake system, filling the master cylinder as required.
14. Install the wheel and tire assembly; torque the nuts to 85–104 ft. lbs. (115–142 Nm).
15. Pump the brake pedal several times to position the brake pads prior to moving the vehicle.
16. Turn the air suspension service switch to the **ON** position.
17. Road test the vehicle and check for proper brake system operation.

Rear

1. Before servicing the vehicle, refer to the precautions in the beginning of this section.
2. Remove brake fluid from the brake master cylinder reservoir until the reservoir is ½ full.
3. Turn the air suspension switch, located in the left side of the luggage compartment, to the **OFF** position.
4. Remove the wheel and tire assembly.
5. Remove the retaining bolt and disconnect the brake hose from the caliper assembly. Discard the copper sealing washers.
6. Remove the retaining clip from the parking brake at the caliper. Disengage the parking brake cable end from the lever arm.
7. Lift the rear disc brake caliper away from the rear disc support bracket.
8. Remove the disc brake caliper locating pins and boots from the rear disc support bracket.

To install:

9. Using rear caliper piston adjuster tool T87P-2588-A, rotate the rear disc brake piston and adjuster clockwise until fully seated.

➡Make sure one of the 2 slots in the rear disc brake piston and adjuster face is positioned so it will engage the nib on the disc brake pad.

10. Apply silicone dielectric compound to the inside of the slider pin boots and the slider pins.
11. Position the slider pins and boots in the support bracket. Position the caliper assembly on the support bracket. Make sure the brake pads are installed correctly.
12. Remove the residue from the pin retainer threads and apply 1 drop of threadlock and sealer. Install the pin retainers and torque to 23–26 ft. lbs. (31–35 Nm).
13. Attach the cable end to the parking brake lever. Install the cable retaining clip on the caliper assembly.
14. Using new washers, connect the brake flex hose to the caliper. Torque the retaining bolt to 40 ft. lbs. (54 Nm).
15. Bleed the brake system, filling the master cylinder as required.
16. Install the wheel and tire assembly; torque the nuts to 85–104 ft. lbs. (115–142 Nm).
17. Pump the brake pedal several times to position the brake pads prior to moving the vehicle.
18. Turn the air suspension service switch to the **ON** position.
19. Road test the vehicle and check for proper brake system operation.

Disc Brake Pads

REMOVAL & INSTALLATION

CAUTION

The air suspension switch, located in the left side of the luggage compartment, must be turned OFF before raising the vehicle. Failure to do so may result in unexpected inflation or deflation of the air springs that may result in shifting of the vehicle during service.

Front

1. Before servicing the vehicle, refer to the precautions in the beginning of this section.
2. Remove the master cylinder reservoir cap and check the fluid level in the reservoir. Remove brake fluid until the reservoir is ½ full. Discard the removed fluid.
3. Turn the air suspension switch, located in the left side of the luggage compartment, to the **OFF** position.

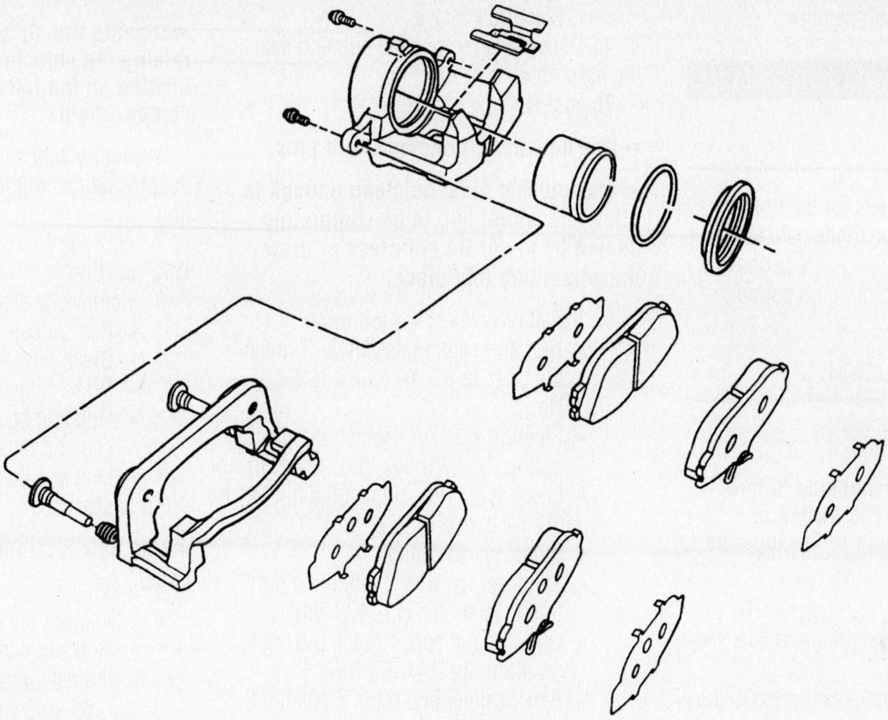

Front disc brake caliper, pads and related components

4. Remove the wheel and tire assembly.

5. Remove the disc brake caliper locating pins. Lift the caliper assembly from the anchor plate and rotor using a rotating motion.

6. Suspend the caliper inside the fender housing with wire. Do not allow the caliper to hang from the brake hose.

7. Remove the inner and outer brake pads. Inspect the rotor braking surfaces for scoring and machine as necessary.

To install:

8. Use a C-clamp and an old brake pad or block of wood to seat the caliper piston in its bore.

9. Remove any rust buildup from the inside of the caliper in the brake pad contact area.

10. Install the inner pad in the caliper piston.

11. Install the outer pad onto the anchor plate. Make sure the clips are properly seated.

➡Make sure the insulators are installed on the brake pads.

12. Install the disc brake caliper onto the anchor plate.

13. Install caliper locating pins and torque to 23–28 ft. lbs. (31–38 Nm).

14. Install wheel and tire assembly and torque lugs nuts to 85–104 ft. lb. (115-142 Nm)

15. Pump the brake pedal several times prior to moving the vehicle to position the brake pads to the rotor.

16. Refill the master cylinder reservoir as necessary, using only clean DOT 3 brake fluid from a closed container.

17. Turn the air suspension service switch to the **ON** position.

18. Road test the vehicle and check the brake system for proper operation.

Rear

1. Remove the master cylinder reservoir cap and check the fluid level in the reservoir. Remove brake fluid until the reservoir is ½ full. Discard the removed fluid.

2. Turn the air suspension switch, located in the left side of the luggage compartment, to the **OFF** position.

3. Remove the wheel and tire assembly.

4. Remove the screw retaining the brake hose bracket to the frame side rail.

5. Remove the retaining clip from the parking brake cable at the disc brake caliper. Remove the cable end from the parking brake lever.

6. Remove the upper disc brake caliper locating pin at the support bracket. Rotate the caliper away from the rotor.

7. Remove the disc brake pads.

8. Inspect the rotor braking surfaces for scoring and machine as necessary.

To install:

9. Using Rear Caliper Piston Adjuster T87P-2588-A, rotate the piston clockwise until it is fully seated. Make sure one of the slots in the piston face is positioned so it will engage the nib on the brake pad.

10. Install the brake pads in the support bracket. Rotate the caliper assembly over the rotor into position on the support bracket. Make sure the brake pads are installed correctly.

11. Remove the residue from the rear brake pin retainer bolt threads and apply 1 drop of a suitable threadlock sealer. Install and torque the disc brake caliper locating pin to 23–26 ft. lbs. (31–35 Nm).

12. Attach the cable end to the parking brake lever. Install the cable retaining clip on the caliper assembly. Position the brake flex hose and bracket assembly to the side rail, and install the retaining screw. Torque to 11 ft. lbs. (16 Nm).

13. Install the wheel and tire assembly and torque lug nuts to 85–104 ft. lbs. (115–142 Nm).

14. Pump the brake pedal several times prior to moving the vehicle, to position the brake pads to the rotor.

15. Refill the master cylinder reservoir if necessary, using only clean DOT 3 brake fluid from a closed container.

16. Turn the air suspension service switch to the **ON** position.

17. Road test the vehicle and check the brake system for proper operation.

Wheel Bearings

ADJUSTMENT

There is no adjustment for the front or rear wheel bearings due to the nature of their design. These bearings are permanently lubricated and require no periodic maintenance.

REMOVAL & INSTALLATION

Front

➡Disable the air suspension before raising the vehicle. The switch is located on the left side of the luggage compartment.

1. Before servicing the vehicle, refer to the precautions in the beginning of this section.
2. Turn the air suspension switch to the **OFF** position.
3. Remove or disconnect the following:
   - Front wheel
   - Hub retainer nut
   - Brake caliper and rotor
   - Outer tie rod end
   - Stabilizer bar link
   - Wheel speed sensor
   - Lower ball joint
   - Steering knuckle
4. Unbolt and remove the hub and bearing assembly.

To install:

➡Use new nuts, bolts and split pins.

➡The knuckle must be clean enough to allow the wheel hub to be completely seated by hand. Do not press or draw the wheel hub into place.

5. Install or connect the following:
   - Hub and bearing assembly. Tighten the bolts to 61–78 ft. lbs. (83–107 Nm).
   - Steering knuckle. Tighten the pinch bolt to 72–97 ft. lbs. (98–132 Nm).
   - Lower ball joint. Tighten the nut to 50–67 ft. lbs. (68–92 Nm).
   - Wheel speed sensor
   - Stabilizer bar link. Tighten the nut to 57–75 ft. lbs. (77–103 Nm).
   - Outer tie rod end. Tighten the nut to 35–46 ft. lbs. (47–63 Nm).
   - Brake caliper and rotor. Tighten the caliper anchor bracket bolts to 65–87 ft. lbs. (88–118 Nm).
   - Hub retainer nut. Tighten the nut to 170–202 ft. lbs. (230–275 Nm).
   - Front wheel
6. Turn the air suspension switch to the **ON** position.

Rear

➡Disable the air suspension before raising the vehicle. The switch is located on the left side of the luggage compartment.

1. Before servicing the vehicle, refer to the precautions in the beginning of this section.
2. Turn the air suspension switch to the **OFF** position.
3. Remove or disconnect the following:
   - Rear wheel
   - Brake hose bracket
   - Brake caliper and rotor
   - Hub and bearing assembly grease cap and retaining nut
   - Hub and bearing assembly

To install:

➡Use new retaining nuts and grease caps

4. Install or connect the following:
   - Hub and bearing assembly. Tighten the retaining nut to 188–254 ft. lbs. (255–345 Nm).
   - Grease cap
   - Brake caliper and rotor
   - Brake hose bracket
   - Rear wheel
5. Turn the air suspension switch to the **ON** position.

MERCURY

4

Cougar

SPECIFICATION CHARTS

ENGINE AND VEHICLE IDENTIFICATION

Engine						
Code ①	Liters (cc)	Cu. In.	Cyl.	Fuel Sys.	Engine Type	Eng. Mfg.
3	2.0 (1999)	122	4	SFI	DOHC	Ford
L	2.5 (2507)	153	6	SFI	DOHC	Ford
G	2.5 (2507)	153	6	SFI	DOHC	Ford

Model Year	
Code ②	Year
1	2001
2	2002

SFI: Sequential Fuel Injection

DOHC: Double Overhead Camshaft

① 8th digit of VIN

② 10th digit of VIN

7197-COUG-C01

GENERAL ENGINE SPECIFICATIONS

Year	Model	Engine Displacement Liters	Engine ID/VIN	Net Horsepower @ rpm	Net Torque @ rpm (ft. lbs.)	Bore x Stroke (in.)	Com-pression Ratio	Oil Pressure @ rpm
2001	Cougar	2.0	3	125@6000	130@4500	3.39x3.46	9.6:1	20-45@1500
		2.5	L	170@6200	165@4200	3.25x3.13	9.7:1	25-45@1500
	Cougar S	2.5	G	196@6750	168@5500	3.25x3.13	10.25:1	20-45@1500
2002	Cougar	2.0	3	125@6000	130@4500	3.39x3.46	9.6:1	20-45@1500
		2.5	L	170@6200	165@4200	3.25x3.13	9.7:1	25-45@1500
	Cougar S	2.5	G	196@6750	168@5500	3.25x3.13	10.25:1	20-45@1500

SFI: Sequential Fuel Injection

7197-COUG-C02

ENGINE TUNE-UP SPECIFICATIONS

Year	Engine Displacement Liters	Engine ID/VIN	Spark Plug Gap (in.)	Ignition Timing (deg.)		Fuel Pump (psi) ①	Idle Speed (rpm)		Valve Clearance	
				MT	AT		MT	AT	Intake	Exhaust
2001	2.0	3	0.050	10B	10B	37-41	②	②	HYD	HYD
	2.5	G	0.054	10B	—	37-41	②	—	HYD	HYD
	2.5	L	0.054	10B	10B	37-41	②	②	HYD	HYD
2002	2.0	3	0.050	10B	10B	37-41	②	②	HYD	HYD
	2.5	G	0.054	10B	—	37-41	②	—	HYD	HYD
	2.5	L	0.054	10B	10B	37-41	②	②	HYD	HYD

NOTE: The Vehicle Emission Control Information label often reflects specification changes made during production. The label figures must be used if they differ from those in this chart.

B: Before Top Dead Center

HYD: Hydraulic

① Fuel pressure with engine running, pressure regulator vacuum hose connected

② Refer to Vehicle Emission Control Information (VECI) label

7197-COUG-C03

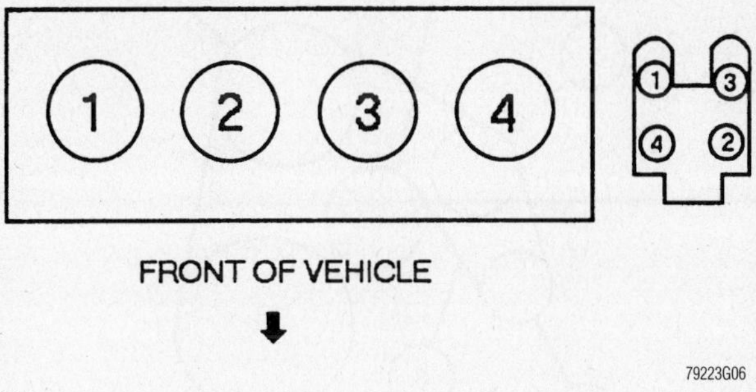

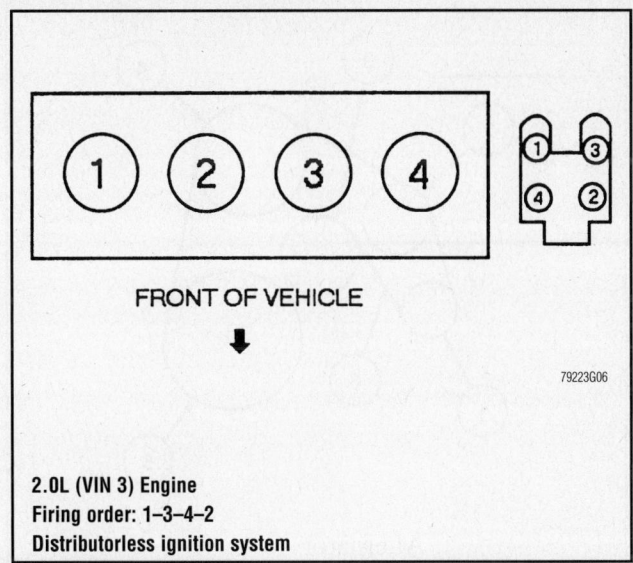

2.0L (VIN 3) Engine
Firing order: 1–3–4–2
Distributorless ignition system

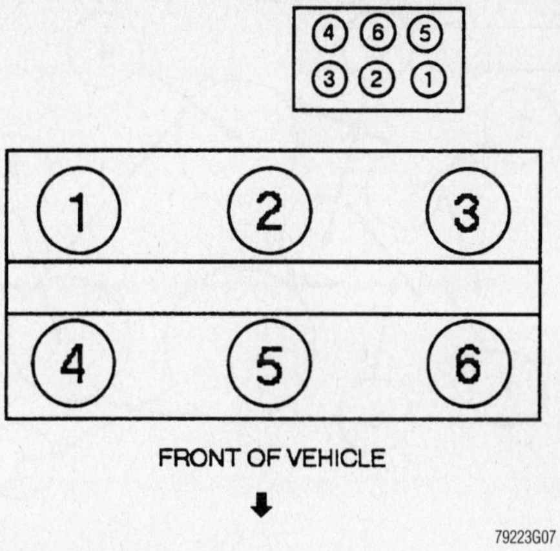

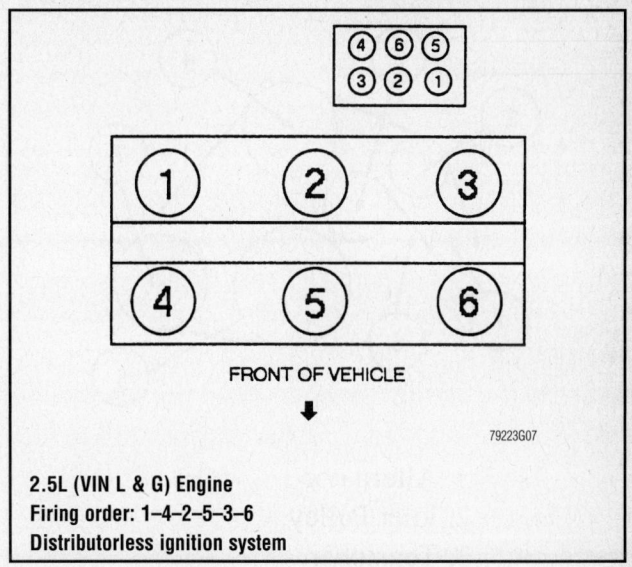

2.5L (VIN L & G) Engine
Firing order: 1–4–2–5–3–6
Distributorless ignition system

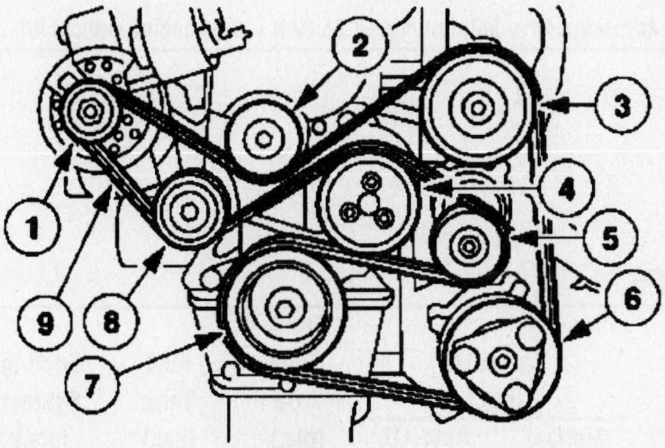

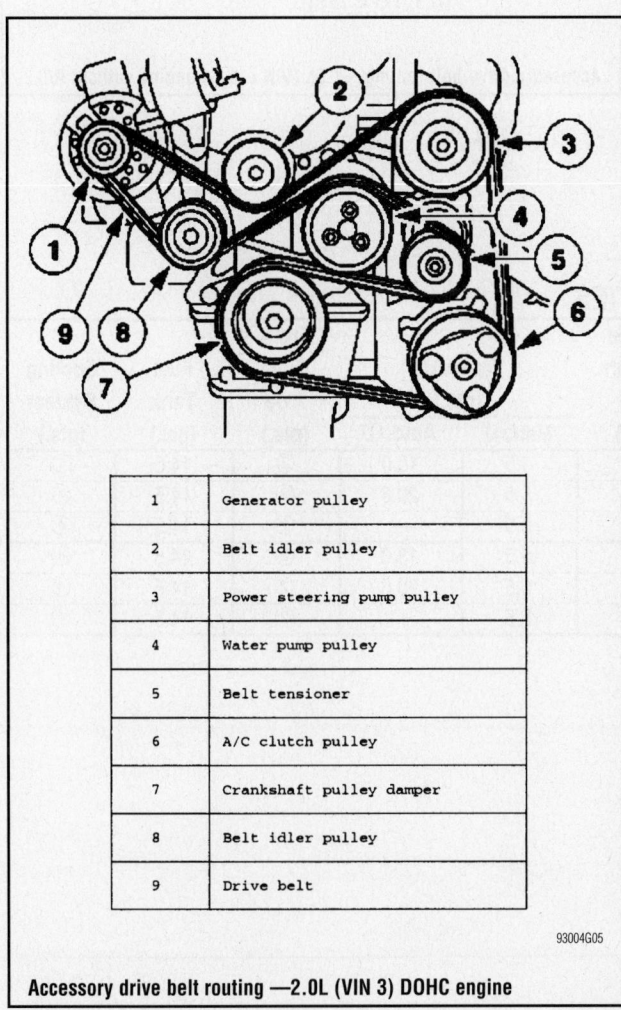

1	Generator pulley
2	Belt idler pulley
3	Power steering pump pulley
4	Water pump pulley
5	Belt tensioner
6	A/C clutch pulley
7	Crankshaft pulley damper
8	Belt idler pulley
9	Drive belt

93004G05

Accessory drive belt routing —2.0L (VIN 3) DOHC engine

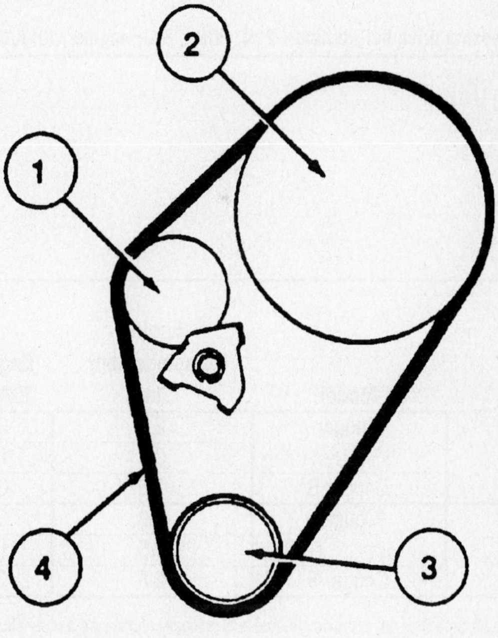

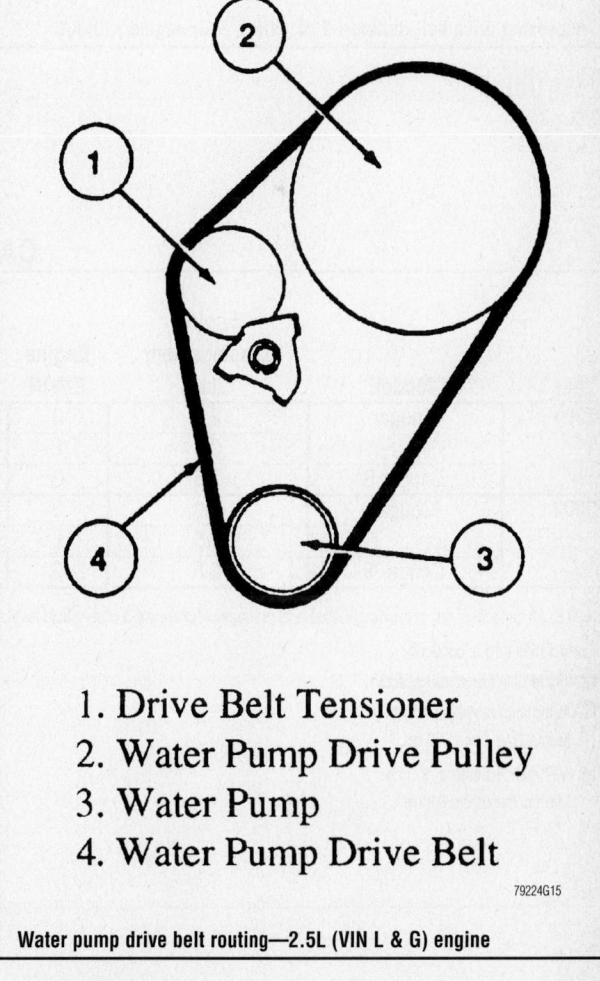

1. Drive Belt Tensioner
2. Water Pump Drive Pulley
3. Water Pump
4. Water Pump Drive Belt

79224G15

Water pump drive belt routing—2.5L (VIN L & G) engine

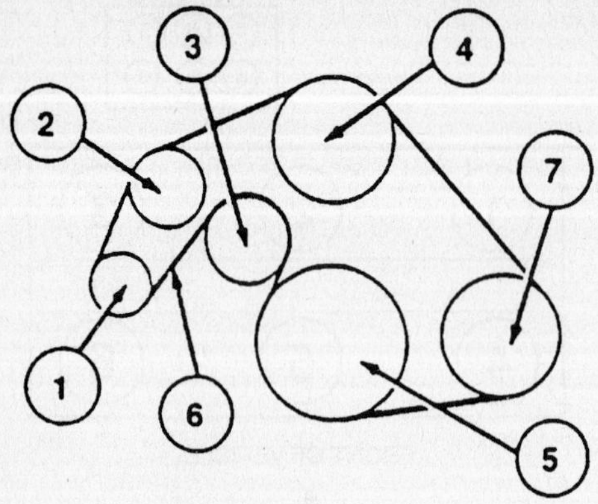

1. Alternator
2. Idler Pulley
3. Tensioner
4. Power Steering Pulley
5. Crankshaft Pulley
6. Drive Belt
7. A/C Compressor

79224G16

Accessory drive belt routing—2.5L (VIN L & G) engine with A/C

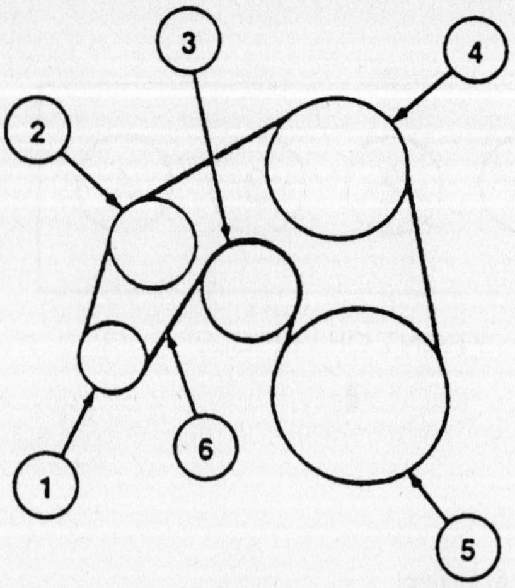

1. Alternator
2. Idler Pulley
3. Tensioner
4. Power Steering Pulley
5. Crankshaft Pulley
6. Drive Belt

79224G17

Accessory drive belt routing—2.5L (VIN L & G) engine without A/C

CAPACITIES

Year	Model	Engine Displacement Liters	Engine ID/VIN	Engine Oil with Filter (qts.)	Transaxle (pts.) Manual	Transaxle (pts.) Auto. ①	Front Drive Axle (pts.)	Fuel Tank (gal.)	Cooling System (qts.)
2001	Cougar	2.0	3	4.5	5.5	18.0	②	14.5	④
		2.5	L	5.8	5.5	20.6	②	14.5	④
	Cougar S	2.5	G	5.8	5.5	—	②	14.5	④
2002	Cougar	2.0	3	4.5	5.5	18.0	②	14.5	④
		2.5	L	5.8	5.5	20.6	②	14.5	④
	Cougar S	2.5	G	5.8	5.5	—	②	14.5	④

NOTE: All capacities are approximate. Add fluid gradually and ensure a proper fluid level is obtained.

① Includes torque converter

② Included in transaxle capacity

③ Automatic transaxle: 7.5 qts.
Manual transaxle: 7.0 qts.

④ Automatic transaxle: 9.1 qts.
Manual transaxle: 8.9 qts.

7197-COUG-C04

VALVE SPECIFICATIONS

Year	Engine Displacement Liters	Engine ID/VIN	Seat Angle (deg.)	Face Angle (deg.)	Spring Test Pressure (lbs. @ in.)	Spring Installed Height (in.)	Stem-to-Guide Clearance (in.) Intake	Stem-to-Guide Clearance (in.) Exhaust	Stem Diameter (in.) Intake	Stem Diameter (in.) Exhaust
2001	2.0	3	45	45	NA	1.346	0.0007-0.0025	0.0014-0.0032	0.2373-0.2379	0.2366-0.2372
	2.5	G	44.75	45.5	153@1.18	1.570	0.0007-0.0027	0.0017-0.0037	0.2350-0.2358	0.2343-0.2350
	2.5	L	44.75	45.5	153@1.18	1.570	0.0007-0.0027	0.0017-0.0037	0.2350-0.2358	0.2343-0.2350
2002	2.0	3	45	45	NA	1.346	0.0007-0.0025	0.0014-0.0032	0.2373-0.2379	0.2366-0.2372
	2.5	G	44.75	45.5	153@1.18	1.570	0.0007-0.0027	0.0017-0.0037	0.2350-0.2358	0.2343-0.2350
	2.5	L	44.75	45.5	153@1.18	1.570	0.0007-0.0027	0.0017-0.0037	0.2350-0.2358	0.2343-0.2350

NA: Not Available

7197-COUG-C05

CRANKSHAFT AND CONNECTING ROD SPECIFICATIONS

All measurements are given in inches.

Year	Engine Displacement Liters	Engine ID/VIN	Crankshaft Main Brg. Journal Dia.	Crankshaft Main Brg. Oil Clearance	Crankshaft Shaft End-play	Crankshaft Thrust on No.	Connecting Rod Journal Diameter	Connecting Rod Oil Clearance	Connecting Rod Side Clearance
2001	2.0	3	2.2827-2.2835	0.0008-0.0017	0.0035-0.0100	3	1.8425-1.8504	0.0006-0.0028	0.0035-0.0126
	2.5	G	2.4670-2.4790	0.0009-0.0019	0.0040-0.0090	4	1.9670-1.9680	0.0010-0.0025	0.0039-0.0118
	2.5	L	2.4670-2.4790	0.0009-0.0019	0.0040-0.0090	4	1.9670-1.9680	0.0010-0.0025	0.0039-0.0118
2002	2.0	3	2.2827-2.2835	0.0008-0.0017	0.0035-0.0100	3	1.8425-1.8504	0.0006-0.0028	0.0035-0.0126
	2.5	G	2.4670-2.4790	0.0009-0.0019	0.0040-0.0090	4	1.9670-1.9680	0.0010-0.0025	0.0039-0.0118
	2.5	L	2.4670-2.4790	0.0009-0.0019	0.0040-0.0090	4	1.9670-1.9680	0.0010-0.0025	0.0039-0.0118

7197-COUG-C07

PISTON AND RING SPECIFICATIONS

All measurements are given in inches.

Year	Engine Displacement Liters	Engine ID/VIN	Piston Clearance	Ring Gap Top Compression	Ring Gap Bottom Compression	Ring Gap Oil Control	Ring Side Clearance Top Compression	Ring Side Clearance Bottom Compression	Ring Side Clearance Oil Control
2001	2.0	3	0.0008-0.0016	0.008-0.010	0.012-0.020	0.016-0.055	0.0016-0.0028	0.0008-0.0021	Snug
	2.5	G	0.0005-0.0009	0.004-0.010	0.011-0.017	0.006-0.026	0.0015-0.0029	0.0015-0.0033	Snug
	2.5	L	0.0005-0.0009	0.004-0.010	0.011-0.017	0.006-0.026	0.0015-0.0029	0.0015-0.0033	Snug
2002	2.0	3	0.0008-0.0016	0.008-0.010	0.012-0.020	0.016-0.055	0.0016-0.0028	0.0008-0.0021	Snug
	2.5	G	0.0005-0.0009	0.004-0.010	0.011-0.017	0.006-0.026	0.0015-0.0029	0.0015-0.0033	Snug
	2.5	L	0.0005-0.0009	0.004-0.010	0.011-0.017	0.006-0.026	0.0015-0.0029	0.0015-0.0033	Snug

7197-COUG-C08

TORQUE SPECIFICATIONS

All readings in ft. lbs.

Year	Engine Displacement Liters	Engine ID/VIN	Cylinder Head Bolts	Main Bearing Bolts	Rod Bearing Bolts	Crankshaft Damper Bolts	Flywheel Bolts	Manifold Intake	Manifold Exhaust	Spark Plugs	Oil Pan Drain Plug
2001	2.0	3	①	55-66	②	81-89	80-87	12-15	13-16	9-13	9-11
	2.5	G	③	④	⑤	⑥	54-64	6-9	13-16	7-15	9-11
	2.5	L	③	④	⑤	⑥	54-64	6-9	13-16	7-15	9-11
2002	2.0	3	①	55-66	②	81-89	80-87	12-15	13-16	9-13	9-11
	2.5	G	③	④	⑤	⑥	54-64	6-9	13-16	7-15	9-11
	2.5	L	③	④	⑤	⑥	54-64	6-9	13-16	7-15	9-11

NOTE: Always follow proper torque patterns. Stretch bolts are used in all procedures that require rotating the fastener a certain number of degrees.
The bolts stretch and cannot be reused. For reassembly, replace with new fastners.

① Step 1: 15-22 ft. lbs.
Step 2: 30-37 ft. lbs.
Step 3: Tighten 90-120 degrees

② Step 1: 22-25 ft. lbs.
Step 2: Tighten each bolt 85-95 degrees

③ Step 1: 27-32 ft. lbs.
Step 2: Tighten 85-95 degrees
Step 3: Loosen bolts then repeat Step 1
Step 4: Tighten 85-95 degrees
Step 5: Repeat Step 4

④ Step 1: 2.0-3.6 ft. lbs.
Step 2: Push crankshaft rearward.
Lightly seat crankshaft washer forward
Step 3: Outer cap bolts: 16-21 ft. lbs.
Step 4: Inner cap bolts: 27-32 ft. lbs.
Step 5: Tighten inner and outer cap bolts 85-95 degrees
Step 6: Remaining bolts: 15-22 ft. lbs.

⑤ 26-33 ft. lbs. plus 90-120 degrees

⑥ Step 1: 89 ft. lbs.
Step 2: Loosen bolt
Step 3: 35-39 ft. lbs.
Step 4: Tighten 85-95 degrees

7197-COUG-C06

WHEEL ALIGNMENT

			Caster		Camber		
Year	Model		Range (+/-Deg.)	Preferred Setting (Deg.)	Range (+/-Deg.)	Preferred Setting (Deg.)	Toe-in (in.)
2001	Cougar	F	1.10	+2.52	1.38	-0.61	0 +/- 0.05
		R	—	—	1.28	-0.70	0.8 +/- 0.05
2002	Cougar	F	1.10	+2.52	1.38	-0.61	0 +/- 0.05
		R	—	—	1.28	-0.70	0.8 +/- 0.05

7197-COUG-C09

TIRE, WHEEL AND BALL JOINT SPECIFICATIONS

		OEM Tires		Tire Pressures (psi)				
Year	Model	Standard	Optional	Front	Rear	Wheel Size	Ball Joint Inspection	Lug Nut (ft. lbs.)
2001	Cougar	P205/60R15	P215/50R16	35	35	6-JJ	①	63
2002	Cougar	P205/60R15	P215/50R16	35	35	6-JJ	①	63

OEM: Original Equipment Manufacturer

PSI: Pounds Per Square Inch

STD: Standard

OPT: Optional

① Replace if any measurable movement is found

7197-COUG-C10

BRAKE SPECIFICATIONS
MERCURY COUGAR
All measurements in inches unless noted

			Brake Disc			Brake Drum				
Year	Model		Original Thickness	Minimum Thickness	Maximum Run-out	Original Inside Diameter	Max. Wear Limit	Maximum Machine Diameter	Minimum Lining Thickness	Brake Caliper Mounting Bolts (ft. lbs.)
2001	Cougar	F	0.950	0.870	0.006	—	—	—	0.125	20
		R	0.790	0.710	0.006	8.00	NA	8.04	0.125	30
2002	Cougar	F	0.950	0.870	0.006	—	—	—	0.125	20
		R	0.790	0.710	0.006	8.00	NA	8.04	0.125	30

NOTE: Follow specifications stamped on rotor or drum if figures differ from those in this chart.

NA: Not Available

F: Front

R: Rear

7197-COUG-C11

SCHEDULED MAINTENANCE INTERVALS

Mercury—Cougar

TO BE SERVICED	TYPE OF SERVICE	VEHICLE MILEAGE INTERVAL (x1000)												
		5	10	15	20	25	30	35	40	45	50	55	60	65
Engine oil & filter	R	✓	✓	✓	✓	✓	✓	✓	✓	✓	✓	✓	✓	✓
Rotate tires	S/I	✓		✓		✓		✓		✓		✓		✓
Front & rear brakes	S/I		✓		✓		✓		✓		✓		✓	
Cooling system, hoses, clamps & coolant strength	S/I			✓			✓			✓			✓	
Passenger compartment air filter	R				✓				✓				✓	
Air cleaner element	R						✓						✓	
Automatic transaxle fluid & filter	S/I												✓	
Exhaust heat shields	S/I						✓						✓	
Accessory drive belt(s)	S/I						✓						✓	
Fuel lines & hoses	S/I						✓						✓	
Crankcase emission filter (2.0L)	R						✓							
Engine coolant ①	R										✓			
Spark plugs ②③	R												✓	
PCV valve	R												✓	

R: Replace S/I: Service or Inspect

① Change initially at 50,000 miles & every 30,000 miles thereafter.

② 2.0L: replace every 60,000 miles.

③ 2.5L: replace every 100,000 miles.

FREQUENT OPERATION MAINTENANCE (SEVERE SERVICE)

If a vehicle is operated under any of the following conditions it is considered severe service:

- Extremely dusty areas.

- 50% or more of the vehicle operation is in 32°C (90°F) or higher temperatures, or constant operation in temperatures below 0°C (32°F).

- Prolonged idling (vehicle operation in stop and go traffic).

- Frequent short running periods (engine does not warm to normal operating temperatures).

- Police, taxi, delivery usage or trailer towing usage.

Oil & filter change: change every 3000 miles.

Front & rear brakes: check every 9000 miles.

Rotate tires at 6000 miles & every 9000 miles thereafter.

Air cleaner element: check every 15,000 miles.

Passenger compartment air filter: change every 18,000 miles.

Automatic transaxle fluid & filter: change every 30,000 miles.

Spark plugs: replace every 60,000 miles.

7197-COUG-C12

PRECAUTIONS

Before servicing any vehicle, please be sure to read all of the following precautions, which deal with personal safety, prevention of component damage, and important points to take into consideration when servicing a motor vehicle:

- Never open, service or drain the radiator or cooling system when the engine is hot; serious burns can occur from the steam and hot coolant.
- Observe all applicable safety precautions when working around fuel. Whenever servicing the fuel system, always work in a well-ventilated area. Do not allow fuel spray or vapors to come in contact with a spark, open flame, or excessive heat (a hot drop light, for example). Keep a dry chemical fire extinguisher near the work area. Always keep fuel in a container specifically designed for fuel storage; also, always properly seal fuel containers to avoid the possibility of fire or explosion. Refer to the additional fuel system precautions later in this section.
- Fuel injection systems often remain pressurized, even after the engine has been turned **OFF**. The fuel system pressure must be relieved before disconnecting any fuel lines. Failure to do so may result in fire and/or personal injury.
- Brake fluid often contains polyglycol ethers and polyglycols. Avoid contact with the eyes and wash your hands thoroughly after handling brake fluid. If you do get brake fluid in your eyes, flush your eyes with clean, running water for 15 minutes. If eye irritation persists, or if you have taken brake fluid internally, IMMEDIATELY seek medical assistance.
- The EPA warns that prolonged contact with used engine oil may cause a number of skin disorders, including cancer. You should make every effort to minimize your exposure to used engine oil. Protective gloves should be worn when changing oil. Wash your hands and any other exposed skin areas as soon as possible after exposure to used engine oil. Soap and water, or waterless hand cleaner should be used.
- All new vehicles are now equipped with an air bag system, often referred to as a Supplemental Restraint System (SRS) or Supplemental Inflatable Restraint (SIR) system. The system must be disabled before performing service on or around system components, steering column, instrument panel components, wiring and sensors. Failure to follow safety and disabling procedures could result in accidental air bag deployment, possible personal injury and unnecessary system repairs.
- Always wear safety goggles when working with, or around, the air bag system. When carrying a non-deployed air bag, be sure the bag and trim cover are pointed away from your body. When placing a non-deployed air bag on a work surface, always face the bag and trim cover upward, away from the surface. This will reduce the motion of the module if it is accidentally deployed. Refer to the additional air bag system precautions later in this section.
- Clean, high quality brake fluid from a sealed container is essential to the safe and proper operation of the brake system. You should always buy the correct type of brake fluid for your vehicle. If the brake fluid becomes contaminated, completely flush the system with new fluid. Never reuse any brake fluid. Any brake fluid that is removed from the system should be discarded. Also, do not allow any brake fluid to come in contact with a painted surface; it will damage the paint.
- Never operate the engine without the proper amount and type of engine oil; doing so WILL result in severe engine damage.
- Timing belt maintenance is extremely important. Many models utilize an interference-type, non-freewheeling engine. If the timing belt breaks, the valves in the cylinder head may strike the pistons, causing potentially serious (also time-consuming and expensive) engine damage. Refer to the maintenance interval charts in the front of this section for the recommended replacement interval for the timing belt, and to the timing belt procedure in this section for belt replacement and inspection.
- Disconnecting the negative battery cable on some vehicles may interfere with the functions of the on-board computer system(s) and may require the computer to undergo a relearning process once the negative battery cable is reconnected.
- When servicing drum brakes, only disassemble and assemble one side at a time, leaving the remaining side intact for reference.
- Only an MVAC-trained, EPA-certified automotive technician should service the air conditioning system or its components.

ENGINE REPAIR

➡**Disconnecting the negative battery cable on some vehicles may interfere with the operation of the on-board computer system. The computer may undergo a relearning process once the negative battery cable is reconnected.**

Alternator

REMOVAL

2.0L Engine

1. Before servicing the vehicle, refer to the precautions in the beginning of this section.
2. Remove or disconnect the following:
    - Negative battery cable
    - Air intake resonator
    - Alternator electrical connections
    - Ground cable
    - Right front wheel
    - Right splash shield
    - Radiator splash shield
    - Accessory drive belt
    - Alternator. Remove the upper mounting bolt last.

➡**There is a limited amount of space in which to remove the alternator. Do not damage any hoses or cables.**

2.5L Engine

1. Before servicing the vehicle, refer to the precautions in the beginning of this section.
2. Remove or disconnect the following:
    - Negative battery cable
    - Right front wheel
    - Right splash shield
    - Accessory drive belt
    - Right outer tie rod end
    - Alternator wiring connectors
    - Alternator rear support bracket
    - Alternator

INSTALLATION

2.0L Engine

1. Install or connect the following:
    - Alternator and tighten the mounting bolts to 33 ft. lbs. (45 Nm)
    - Accessory drive belt

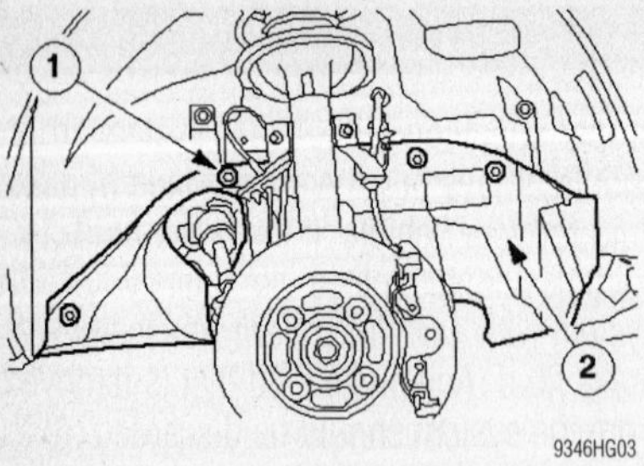

Remove the splash shield retainers (1) and the splash shield (2)—2.5L engines

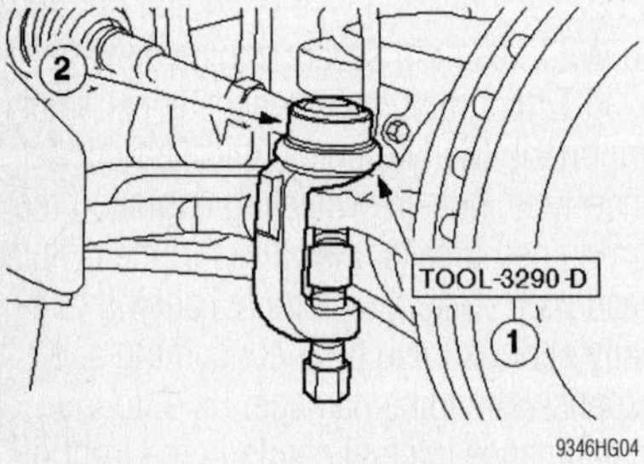

Separate the outer tie rod end from the steering knuckle—2.5L engines

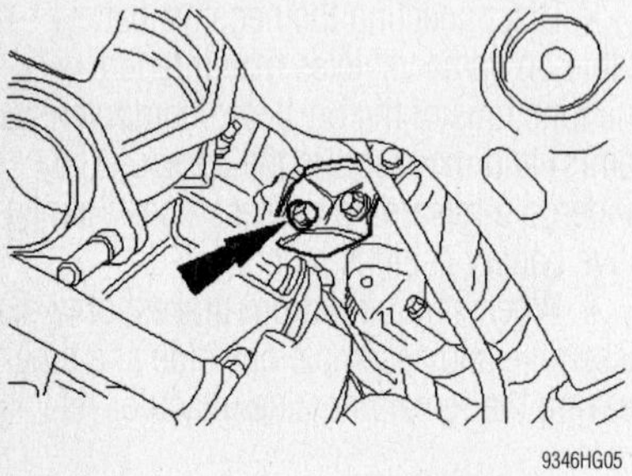

Alternator rear support bracket—2.5L engines

- Radiator splash shield
- Right splash shield
- Right front wheel and tighten the lug nuts to 63 ft. lbs. (86 Nm)
- Ground cable
- Alternator electrical connections
- Air intake resonator
- Negative battery cable

2.5L Engine

Install or connect the following:

- Alternator and tighten the mounting bolts to 33 ft. lbs. (45 Nm)
- Alternator rear support bracket and tighten the bracket bolts to 18 ft. lbs. (25 Nm)
- Alternator wiring connectors
- Right outer tie rod end and tighten the nut to 21 ft. lbs. (28 Nm)
- Accessory drive belt
- Right splash shield
- Right front wheel and tighten the lug nuts to 63 ft. lbs. (86 Nm)
- Negative battery cable

Ignition Timing

ADJUSTMENT

The ignition timing is set at 10 degrees Before Top Dead Center (BTDC) and is not adjustable.

Engine Assembly

REMOVAL & INSTALLATION

2.0L Engine

➡The engine and transaxle are removed as an assembly.

1. Before servicing the vehicle, refer to the precautions in the beginning of this section.
2. Drain the engine oil and coolant.
3. Drain the transaxle fluid.
4. Attach an engine support fixture to the engine lifting eyes.
5. Secure the A/C condenser to the radiator support with safety wire.
6. Loosen the suspension strut lock nuts five turns.
7. Remove or disconnect the following:
    - Battery
    - Mass Air Flow (MAF) sensor connector
    - Intake Air Temperature (IAT) sensor connector
    - Air cleaner outlet tube
    - Air cleaner
    - Exhaust Gas Recirculation (EGR) vacuum supply hose, if equipped
    - EGR pressure sensor wiring, if equipped
    - EGR exhaust tube, if equipped
    - Engine wiring harness connector
    - Accelerator cable
    - Upper drive belt cover
    - Front wheels
    - Left, right and center splash shields
    - Accessory drive belt
    - Stabilizer bar links
    - Outer tie rod ends
    - Lower ball joints
    - Lower radiator retainers
    - Both Heated Oxygen (HO_2S) sensor connectors
    - Power steering hoses
    - Brake booster vacuum line at intake manifold
    - Steering gear and heat shield
    - Right support insulator and bracket
    - Front exhaust pipe
    - Left support insulator
    - A/C accumulator attachment bolts
    - 4 sub-frame bolts and the sub-frame
8. If equipped with a manual transaxle, place the shift lever in neutral and lock it in place with special tool Gear Lever Aligner T97P-7025-A.
    - Clutch hydraulic line
    - Shift cables
    - Reverse lamp switch connector
    - Vehicle Speed (VSS) sensor connector
9. If equipped with an automatic transaxle, remove or disconnect the following:
    - Torque converter
    - Transaxle control wiring connector
    - Range switch connector
    - Ground strap
    - Shift cable
    - VSS and bracket
    - Transaxle cooler lines and fluid tube
    - Turbine Shaft Speed (TSS) sensor connector
10. For all vehicles, remove or disconnect the following:
    - Axle halfshafts
    - Upper and lower radiator hoses
    - 2 ground cables from radiator support
    - Coolant hose bracket and power steering hose
    - Coolant hoses at the recovery tank
    - Radiator and cooling fan
    - A/C compressor
    - Power steering pump
    - Heater hoses and bracket
    - Powertrain Control Module (PCM) connector
    - Vacuum hoses at the intake manifold
    - Battery cables at the battery tray
    - Fuel supply and return lines
11. Support the engine and transaxle assembly from below.
12. Remove or disconnect the following:
    - Top support fixture
    - Coolant recovery tank
    - Front and rear engine mount brackets
13. Raise the vehicle away from the engine and transaxle assembly
14. Support the engine with a shop hoist.
15. Support the transaxle with a transmission jack.
16. Remove or disconnect the following:

- Starter
- Ground cable
- Transaxle flange bolts
- Engine from the transaxle

To install:

➡Replace all snaprings, split pins and self-locking nuts.

17. Install or connect the following:
    - Engine to the transaxle and tighten the flange bolts to 35 ft. lbs. (48 Nm)
    - Ground cable
    - Starter motor and tighten the bolts to 35 ft. lbs. (48 Nm)

18. Lower the vehicle onto the engine and transaxle assembly.

19. Install or connect the following:
    - Front and rear engine mount brackets and tighten the nuts and bolts finger tight
    - Top support fixture
    - Fuel supply and return lines
    - Battery cables to the battery tray. Secure with cable ties.
    - Vacuum hoses to the intake manifold
    - PCM connector
    - Heater hoses and bracket
    - Power steering pump and tighten the bolts to 35 ft. lbs. (47 Nm)
    - A/C compressor and tighten the bolts to 18 ft. lbs. (25 Nm)
    - Radiator and cooling fan
    - Power steering hose and coolant hose bracket
    - 2 ground cables to the radiator support
    - Upper and lower radiator hoses
    - Axle halfshafts

20. If equipped with a manual transaxle, install or connect the following:
    - Clutch hydraulic line
    - Shift cables
    - VSS connector
    - Reverse lamp switch connector

21. Remove the Gear Lever Aligner.

22. If equipped with an automatic transaxle, install or connect the following:
    - TSS connector
    - Torque converter and tighten the nuts to 27 ft. lbs. (36 Nm)
    - Transaxle cooler lines and fluid tube
    - VSS and bracket
    - Shift cable
    - Ground strap
    - Range switch connector
    - Transaxle control wiring connector
    - Powertrain Alignment Gauge T94P-6000-AH to the sub-frame
    - Lower radiator retainers to the sub-frame
    - Sub-frame and attach the front bolts. Do not tighten the bolts at this time.
    - Steering gear to the sub-frame and tighten the bolts to 96 ft. lbs. (130 Nm)
    - Steering gear heat shield
    - Power steering line and coolant line brackets
    - Sub-frame Alignment Pin Set T94P-2100-AH
    - Sub-frame bolts and tighten the bolts in a diagonal pattern to 96 ft. lbs. (130 Nm)
    - Center bolt to the Powertrain Alignment Tool and tighten it to 22 ft. lbs. (30 Nm)
    - Right engine support insulator

23. Tighten the support insulator center bolt to 88 ft. lbs. (120 Nm) and the restrictor mounting bolts to 35 ft. lbs. (48 Nm).
    - Rear engine mounting bracket nut and tighten it to 61 ft. lbs. (83 Nm)

⁂ CAUTION

Do not twist or strain the front engine mounting bracket.

- Front engine mounting bracket and tighten the fasteners to 61 ft. lbs. (83 Nm)

24. Remove the Powertrain Alignment Tool.

25. Install the left engine support insulator. Tighten the center bolt to 88 ft. lbs. (120 Nm) and the restrictor mounting bolts to 35 ft. lbs. (48 Nm).

26. Install or connect the following:
    - Brake booster vacuum hose
    - A/C accumulator attachment bolts
    - Front exhaust pipe. Use a new gasket and tighten the fasteners to 33 ft. lbs. (45 Nm)
    - Attach the power steering lines to the sub-frame
    - HO_2S connectors. Secure the wiring harness with cable ties
    - Lower ball joints and tighten the nuts to 61 ft. lbs. (83 Nm)
    - Outer tie rod ends and tighten the nuts to 19 ft. lbs. (26 Nm)
    - Stabilizer bar links and tighten the fasteners to 35 ft. lbs. (47 Nm)
    - Left, right and center splash shields
    - Front wheels and tighten the lug nuts to 63 ft. lbs. (86 Nm)

27. Remove the top engine support.

28. Install or connect the following:
    - Coolant recovery tank and hoses
    - Ground cables at the radiator support
    - Accelerator cable
    - Accessory drive belt
    - Upper drive belt cover and tighten the bolts to 18 ft. lbs. (24 Nm)
    - EGR exhaust tube, if equipped
    - EGR pressure sensor connector, if equipped
    - EGR vacuum supply, if equipped
    - CKP sensor connector
    - ECT sensor connector
    - Main wiring harness connector
    - Air cleaner and air outlet tube
    - IAT sensor connector
    - MAF sensor connector
    - Battery

29. Tighten the suspension strut locknuts to 34 ft. lbs. (46 Nm).

30. Fill the engine crankcase and cooling system.

31. Fill the transaxle.

32. Start the engine and check for leaks and proper operation.

➡Whenever the vehicle sub-frame is removed or lowered, the wheel alignment should be checked.

2.5L Engine

➡The engine and transaxle are removed as an assembly.

1. Before servicing the vehicle, refer to the precautions in the beginning of this section.

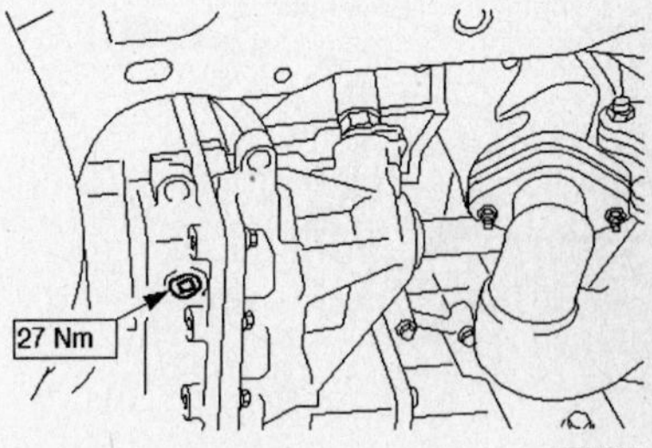

9346HG07

Automatic transaxle drain plug—all models

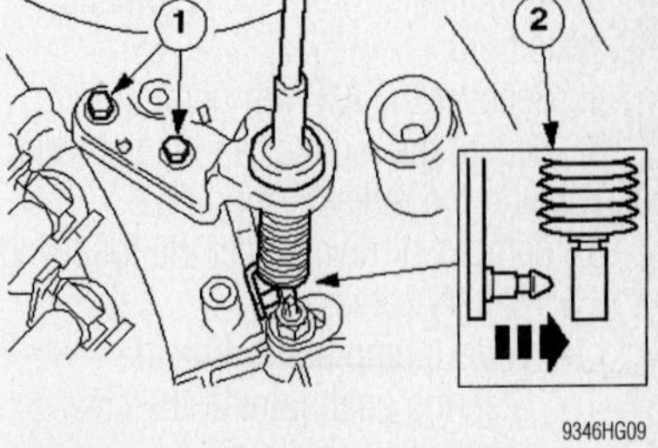

9346HG09

Shift cable bracket (1) and cable (2) removal—2.5L engine

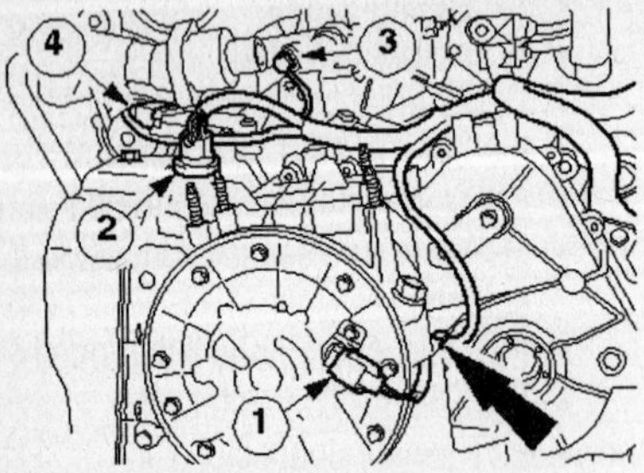

1 Turbine shaft speed (TSS) sensor
2 Transmission control
3 Temperature gauge sender unit
4 Transmission range (TR) sensor

9346HG10

Automatic transaxle harness connections—all models

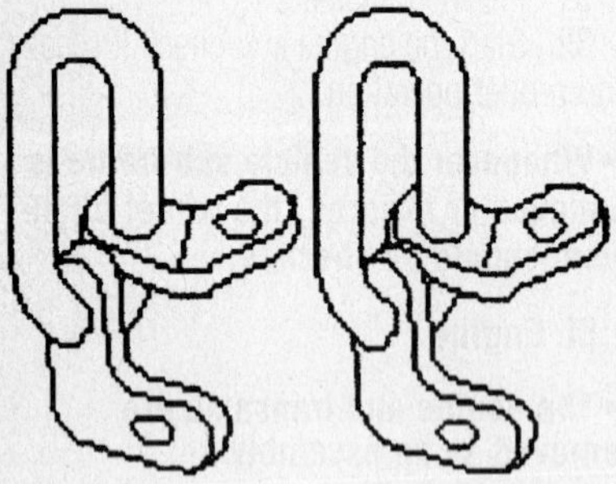

9346HG06

Engine lifting brackets 303-050—2.5L engines

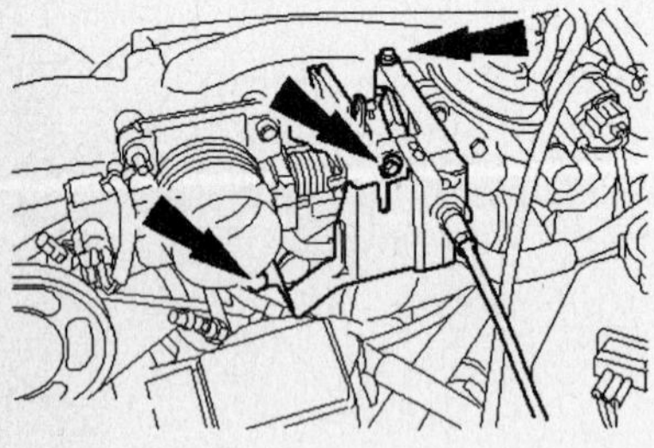

9346HG08

Throttle cable bracket fasteners—2.5L engines

2. Drain the engine cooling system and engine oil.
3. If equipped, drain the automatic transaxle fluid.
4. Recover the A/C refrigerant.
5. Secure the radiator and fan shroud assembly to the radiator support.
6. Remove or disconnect the following:
   - Battery cables
   - Water pump pulley shield
   - Steering shaft joint at the cowl inside the vehicle
   - Air cleaner
   - Catalytic converter
   - Exhaust crossover pipe
   - Front wheels
   - Stabilizer bar links
   - Outer tie rod ends
   - Lower ball joints
   - Axle halfshafts from the hubs
   - A/C accumulator mounting screws
   - Speedometer cable
   - Vehicle Speed (VSS) sensor connector
   - Accelerator cable and bracket
   - Cruise control cable, if equipped
   - Engine control wiring harness connectors
   - Power steering pump reservoir
   - Power steering return hose
   - Power steering pressure hose
   - Powertrain Control Module (PCM) connector and ground strap
   - Coolant recovery tank
7. If equipped with an automatic transaxle, remove or disconnect the following:
   - Shift cable and bracket
   - Transaxle range switch connector
   - Transaxle oil cooler lines
   - Torque converter
8. If equipped with a manual transaxle, remove or disconnect the following:
   - Shift rod and stabilizer bar
   - Clutch slave cylinder fluid pipe
9. For all models, remove or disconnect the following:
   - Power brake booster vacuum pipe
   - Radiator hoses
   - Heater hoses
   - Block heater wiring, if equipped
   - A/C suction and discharge hoses
   - Lower radiator supports
   - A/C compressor harness connector
   - Bumper cover braces
   - Front engine support insulator and bracket
   - Transaxle support insulator
10. Support the powertrain from below.
    - Sub-frame bolts
    - Powertrain from the vehicle
11. Attach an engine hoist to the engine lifting eyes.
12. Remove the left and right engine support insulators and lift the powertrain away from the sub-frame.
13. Support the transaxle from below.
14. Remove or disconnect the following:
    - Axle halfshafts
    - Starter, wiring harness and ground cable
    - Engine from the transaxle

To install:

➡Replace all snaprings, split pins and self-locking nuts.

15. Install or connect the following:
    - Engine to the transaxle and tighten the transaxle flange bolts to 25–34 ft. lbs. (34–46 Nm)
    - Axle halfshafts to the transaxle
    - Starter and wiring harness connectors
    - Ground cable
    - Powertrain Alignment Gauge T94P-6000-AH to the sub-frame and position the powertrain and tighten the center bolt to 20 ft. lbs. (27 Nm)
    - Right engine support insulator. Leave the bolts finger-tight.
    - Sub-frame Alignment Pin Set 94P-2100-AH and raise the sub-frame to the vehicle and tighten the sub-frame bolts to 92–100 ft. lbs. (125–135 Nm)
16. Remove the alignment pins.
    - Front engine support insulator and bracket and tighten the nuts to 84 inch lbs. (10 Nm)

➡The left and right support insulators must be aligned in the middle of the support insulator brackets.

    - Right engine support insulator-to-sub-frame bolts to 30–41 ft. lbs. (41–55 Nm)
    - Front engine support bracket nuts to 52–70 ft. lbs. (70–95 Nm)
    - Engine and transmission support insulator nuts to 30–41 ft. lbs. (41–55 Nm) for automatic transaxles or 52–70 ft. lbs. (70–95 Nm) for manual transaxles
    - Right front engine support insulator through-bolt to 75–102 ft. lbs. (103–137 Nm)
17. Remove the powertrain alignment gauge.
    - Left front engine support insulator to the sub-frame. Tighten the retaining bolts to 30–41 ft. lbs. (41–55 Nm) and the through-bolt to 75–102 ft. lbs. (103–137 Nm).
    - A/C compressor wiring harness
    - Lower radiator supports and tighten the retaining bolts to 71–97 inch lbs. (8–11 Nm)
    - A/C suction and discharge hoses
    - Block heater wiring, if equipped
    - Radiator hoses
    - Heater hoses
    - Power brake booster vacuum pipe
18. If equipped with an automatic transaxle, install or connect the following:
    - Transaxle oil cooler lines
    - Transaxle range switch connector

- Shift cable and bracket and tighten the retaining bolts to 15–19 ft. lbs. (20–25 Nm)
- Torque converter and tighten the retaining nuts in an alternating sequence to 54–64 ft. lbs. (73–87 Nm)

19. If equipped with a manual transaxle, install or connect the following:
    - Clutch slave cylinder fluid pipe
    - Shift rod and stabilizer bar. Tighten the shift rod bolt to 17 ft. lbs. (23 Nm) and the stabilizer nut to 41 ft. lbs. (55 Nm).
20. For all models, install or connect the following:
    - Coolant recovery tank
    - PCM connector and ground strap
    - Power steering pressure hose
    - Power steering return hose
    - Power steering pump reservoir
    - Engine control wiring harness connectors
    - Cruise control cable, if equipped
    - Accelerator cable and bracket and tighten the bolts to 71–106 inch lbs. (8–12 Nm)
    - VSS sensor connector
    - Speedometer cable
    - A/C accumulator mounting screws
    - Axle halfshafts to the front hubs. Use new nuts and tighten to 246 ft. lbs. (340 Nm).
    - Outer tie rod ends and tighten the nuts to 21 ft. lbs. (28 Nm)
    - Lower ball joints and tighten the bolts to 37–43 ft. lbs. (50–58 Nm)
    - Stabilizer bar links and tighten the nuts to 35 ft. lbs. (48 Nm)
    - Front wheels and tighten the lug nuts to 63 ft. lbs. (86 Nm)
    - Exhaust crossover pipe
    - Catalytic converter
    - Air cleaner
    - Steering shaft joint and tighten the bolt to 18 ft. lbs. (24 Nm)
    - Water pump pulley shield
    - Battery cables
21. Fill the crankcase and cooling system.
22. If equipped, fill the automatic transaxle.
23. Check all fluid levels.
24. Evacuate and recharge the air conditioning system.
25. Start the engine and check for leaks and proper operation.

➡Whenever the vehicle sub-frame is removed or lowered, the wheel alignment should be checked.

Water Pump

REMOVAL & INSTALLATION

2.0L Engine

1. Before servicing the vehicle, refer to the precautions in the beginning of this section.
2. Drain the cooling system.
3. Remove or disconnect the following:
    - Negative battery cable
    - Lower radiator hose
    - Accessory drive belt
    - Timing belt
    - Water pump

To install:

4. Install or connect the following:
    - Water pump and tighten the bolts to 12–15 ft. lbs. (16–20 Nm)
    - Timing belt
    - Accessory drive belt
    - Lower radiator hose
    - Negative battery cable
5. Fill the cooling system and inspect for leaks

2.5L Engine

➡The 3 water pump bolts are a torque-to-yield design and must be replaced.

1. Before servicing the vehicle, refer to the precautions in the beginning of this section.
2. Drain the engine cooling system.
3. Remove or disconnect the following:
    - Negative battery cable
    - Water pump pulley shield
    - Water pump drive belt
    - Coolant hoses
    - Water pump and pump housing
4. Separate the water pump from the pump housing.

To install:

5. Install or connect the following:
    - Water pump to the pump housing and tighten the fasteners to 16–18 ft. lbs. (22–25 Nm)
    - Water pump and pump housing to the engine. Use new bolts and tighten them to 11–13 ft. lbs. (15–18 Nm), plus 85–95 degrees.
    - Coolant hoses
    - Water pump drive belt

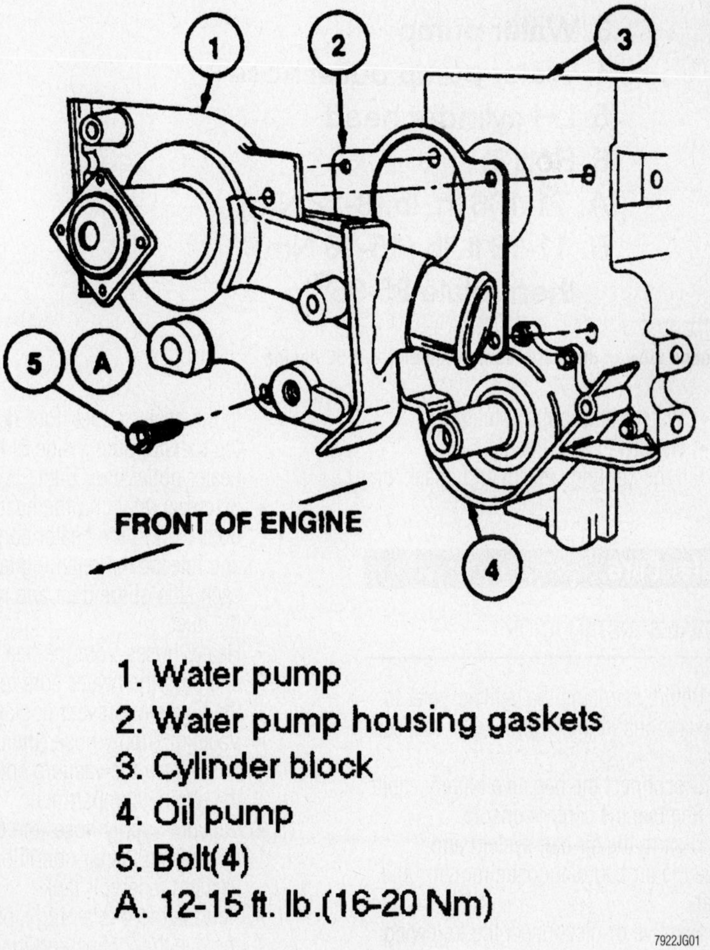

Exploded view of the water pump mounting—2.0L engine

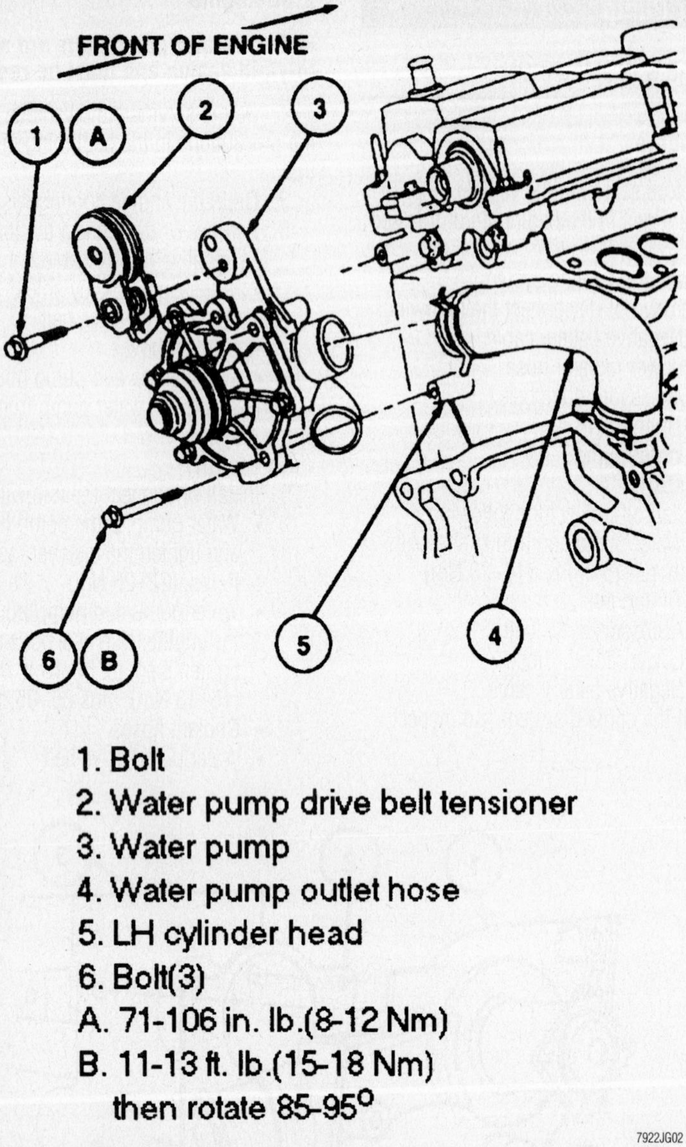

Exploded view of the water pump mounting—2.5L engine

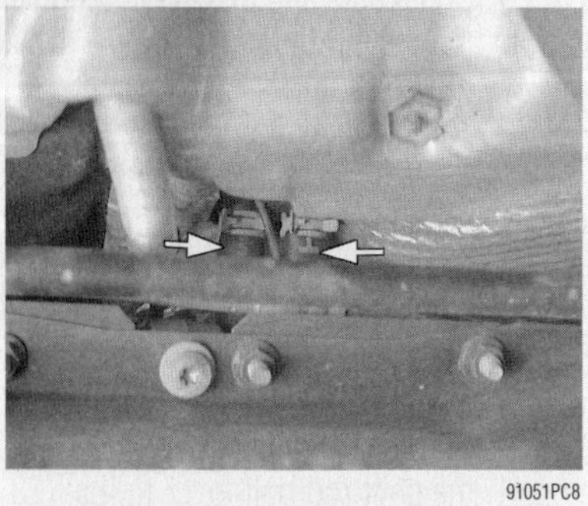

The heater hoses are best accessed from underneath the vehicle

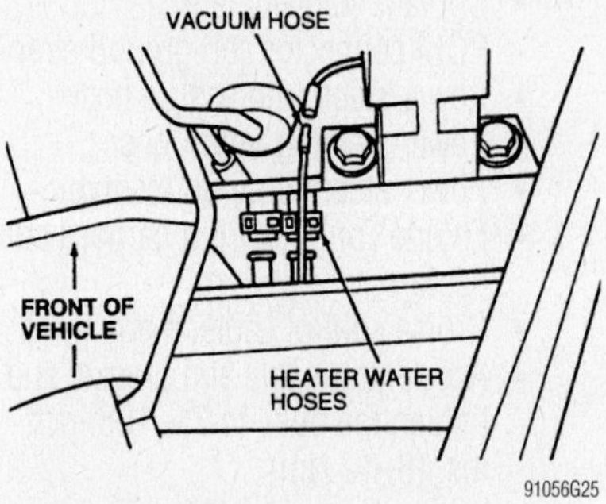

Disconnect the heater hose and vacuum line from underneath the vehicle

- Water pump pulley shield
- Negative battery cable

6. Fill the cooling system and check for leaks.

Heater Core

REMOVAL & INSTALLATION

1. Before servicing the vehicle, refer to the precautions in the beginning of this section
2. Disconnect the negative battery cable.
3. Remove the center console.
4. Disarm the air bag system and remove the air bag diagnostic monitor and bracket.
5. Remove or disconnect the following:
   - Screw retaining the air transfer duct to the heater outlet floor duct. Push the transfer duct inside of the heater outlet floor duct
   - 3 screws retaining the heater outlet floor duct to the heater core cover and release the retaining tabs on each side of the duct and remove the duct
   - Heater hoses from the heater core, and plug the heater core tubes and the hoses to prevent coolant loss.
   - Vacuum supply hose (the black hose) from the vacuum source in the engine compartment
   - Vacuum supply hose (the black hose) from the air conditioning vacuum reservoir tank
   - Release the 4 retaining tabs and remove the 2 clips and the heater core cover containing the heater core
   - Heater dash panel seal and the vacuum hose from the heater core cover
   - Retaining screw and the heater core bracket from the cover
   - Heater core from the cover
   - Heater core case seal from the heater core

To install:

6. Install or connect the following:
   - Heater core case seal on the heater core
   - Heater core into the cover
   - Retaining screw and the heater core bracket onto the cover
   - Heater dash panel seal and the vacuum hose onto the heater core cover
   - 4 retaining tabs, 2 clips and the heater core cover containing the heater core
   - Vacuum supply hose (the black hose) to the air conditioning vacuum reservoir tank
   - Vacuum supply hose (the black hose) to the vacuum source in the engine compartment
   - Unplug the hoses and the core tubes (if installing old core)

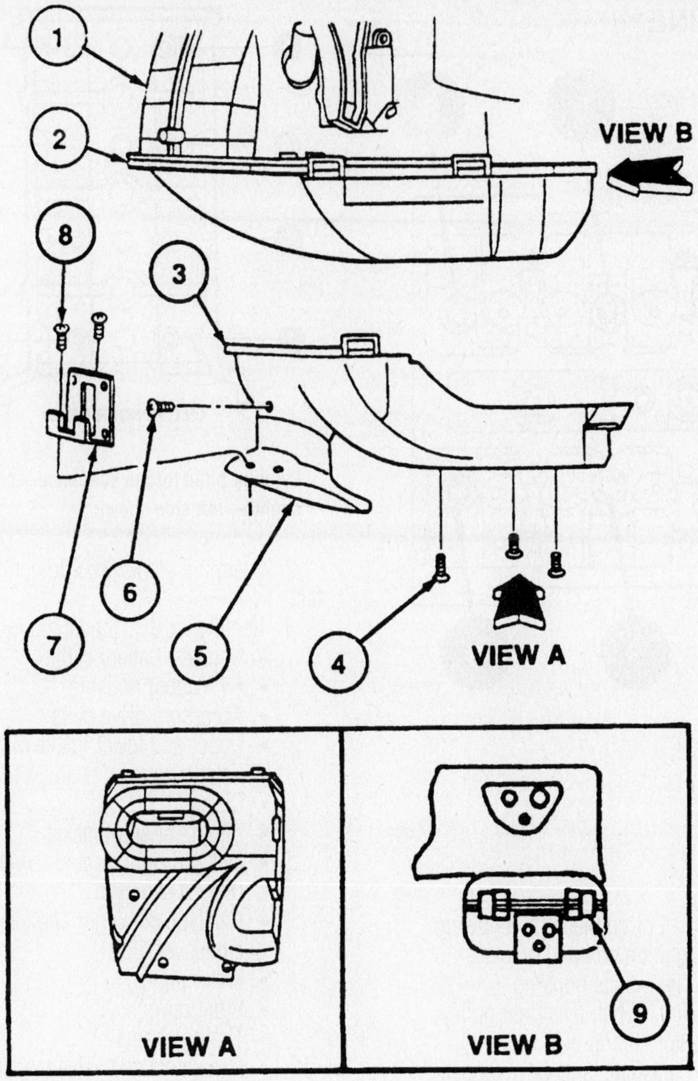

Item	Description
1	A/C Evaporator Housing
2	Heater Core Cover (Part of 19850)
3	Heater Outlet Floor Duct
4	Screw (3 Req'd)
5	Air Transfer Duct (Part of 18C433)
6	Screw (1 Req'd)
7	Air Bag Diagnostic Monitor Bracket
8	Screw (2 Req'd)
9	Clip (2 Req'd)

91056G26

Exploded view of the heater outlet floor duct and related components

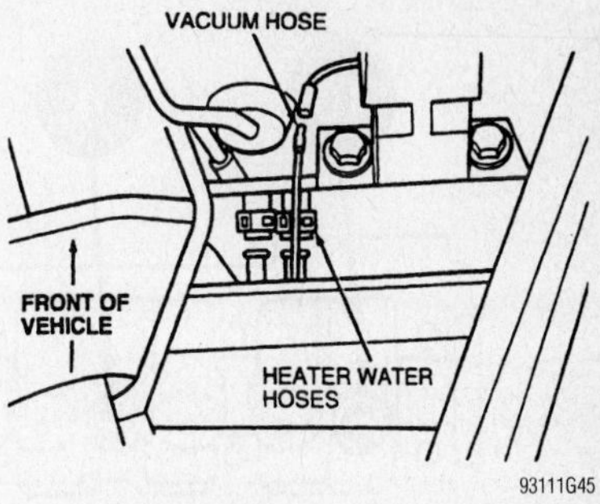

View of the heater and vacuum hoses

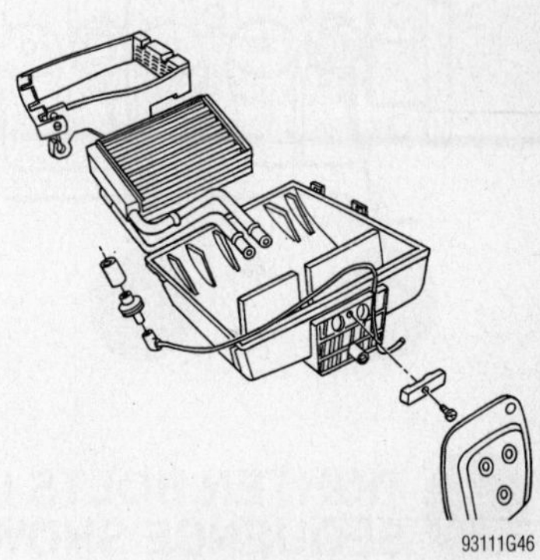

Exploded view of the heater core and related components

- Heater hoses onto the heater core
- Heater outlet floor duct, engage the retaining tabs and tighten the 3 screws retaining the heater outlet floor duct to the heater core cover
- Push the transfer duct out of the heater outlet floor duct
- Tighten the screw retaining the air transfer duct to the heater outlet floor duct
- Air bag diagnostic monitor and bracket
- Center console

7. Connect the negative battery cable.

Cylinder Head

REMOVAL & INSTALLATION

2.0L Engine

➡The cylinder head bolts are a torque-to-yield design and must be replaced.

1. Before servicing the vehicle, refer to the precautions in the beginning of this section.
2. Drain the cooling system.
3. Remove or disconnect the following:
   - Negative battery cable

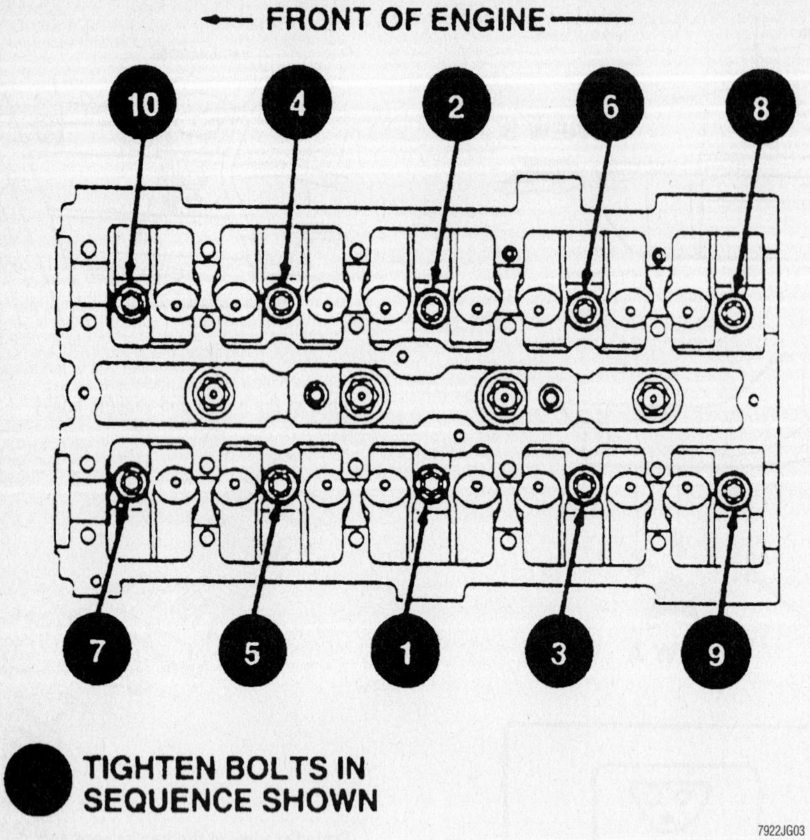

Cylinder head torque sequence—2.0L engine

- Accessory drive belt
- Power steering pump pulley cover
- Power steering pump
- Power steering pump support bracket
- Alternator bracket
- Valve cover
- Intake manifold
- Exhaust manifold
- Front cover
- Timing belt
- Camshafts and valve tappets
- Right lifting eye
- Timing belt tensioner pulley
- Thermostat housing
- Ignition coil and bracket
- Spark plugs
- Cylinder head bolts in the reverse of the installation order
- Cylinder head

To install:

4. Use a new gasket and install the cylinder head.

5. Use new cylinder head bolts and oil the threads.

6. Tighten the cylinder head bolts in sequence as follows:
   a. Step 1: 15–22 ft. lbs. (20–30 Nm)
   b. Step 2: 30–37 ft. lbs. (40–50 Nm)
   c. Step 3: Plus 90–120 degrees

7. Install or connect the following:
   - Ignition coil and bracket
   - Thermostat housing
   - Timing belt tensioner pulley
   - Right lifting eye
   - Camshafts and valve tappets
   - Timing belt
   - Front cover
   - Exhaust manifold
   - Intake manifold
   - Valve cover
   - Alternator bracket
   - Power steering pump support bracket
   - Power steering pump
   - Accessory drive belt
   - Power steering pump pulley cover
   - Spark plugs
   - Negative battery cable

8. Fill the cooling system and check for leaks.

9. Start the engine and check for proper operation.

2.5L Engine

➡The cylinder head bolts are a torque-to-yield design and must be replaced.

1. Before servicing the vehicle, refer to the precautions in the beginning of this section.

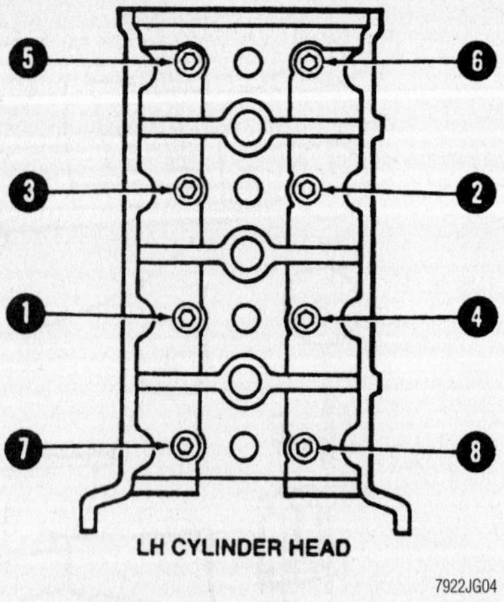

Cylinder head torque sequence—2.5L engine—left side shown

2. Drain the crankcase and cooling system.

3. Remove or disconnect the following:
   - Negative battery cable
   - Air cleaner housing
   - Accessory drive belts
   - Upper and lower intake manifolds
   - Valve covers
   - Oil pan
   - Alternator and bracket
   - Left and right Heated Oxygen (HO_2S) sensors
   - Catalytic converter and exhaust crossover tube
   - Water pump
   - Front cover
   - Timing chains
   - Camshafts and valve tappets
   - Exhaust Gas Recirculation (EGR) pressure sensor
   - EGR transducer
   - EGR tube
   - Coolant bypass
   - Oil dipstick tube
   - Cylinder head bolts in the reverse of the installation order
   - Cylinder heads

To install:

4. Use new gaskets and install the cylinder heads.

5. Use new cylinder head bolts and oil the threads.

6. Tighten the cylinder head bolts in sequence as follows:
   a. Step 1: 27–32 ft. lbs. (37–43 Nm)
   b. Step 2: Plus 85–95 degrees
   c. Step 3: Loosen the bolts, in sequence, 1 full turn
   d. Step 4: 27–32 ft. lbs. (37–43 Nm)
   e. Step 5: Plus 85–95 degrees
   f. Step 6: Plus 85–95 degrees

7. Install or connect the following:

- Oil dipstick tube
- Coolant bypass
- EGR tube
- EGR transducer
- EGR pressure sensor
- Camshafts and valve tappets
- Timing chains
- Front cover
- Water pump
- Catalytic converter and exhaust crossover tube
- Left and right HO_2 sensors
- Alternator and bracket
- Oil pan
- Valve covers
- Upper and lower intake manifolds
- Accessory drive belts
- Air cleaner housing
- Negative battery cable

8. Fill the cooling system and the engine crankcase.
9. Start the engine and check for leaks and proper operation.

Rocker Arms

REMOVAL & INSTALLATION

The 2.0L engine is not equipped with rocker arms, the camshaft directly actuates the valves through a hydraulic lifter.

2.5L Engine

1. Before servicing the vehicle, refer to the precautions in the beginning of this section.
2. Remove or disconnect the following:
   - Battery
   - Upper manifold
   - Intake Manifold Runner Control (IMRC) actuator
   - Spark plug wires and bracket
   - Coolant reservoir breather hose bracket and wiring
   - Fuel injector wiring harness
   - Ignition coil
   - Wiring harness brackets

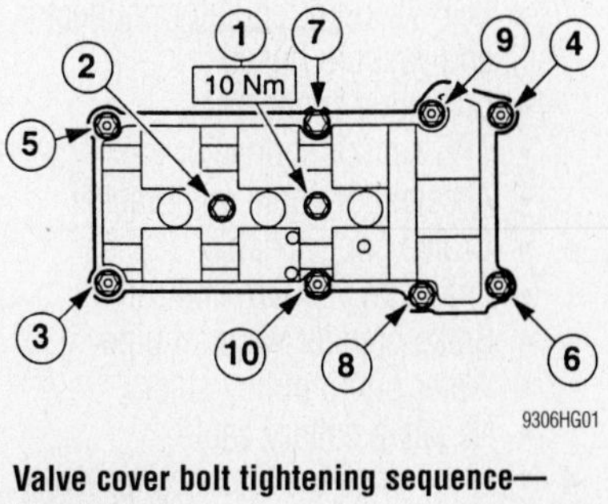

Valve cover bolt tightening sequence—2.5L engine

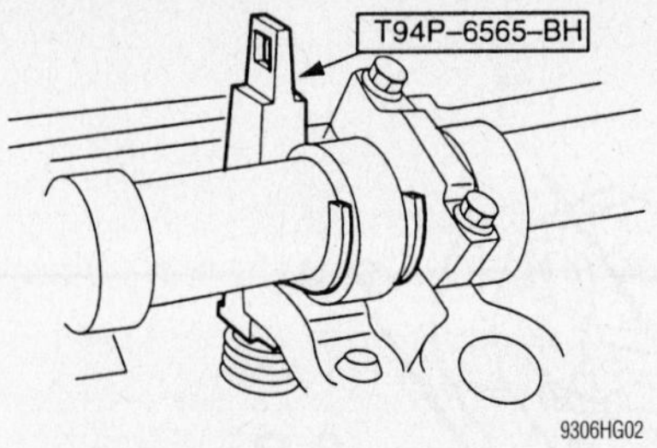

Valve spring compressor—2.5L engine

   - Valve covers. Remove the bolts in reverse of the installation sequence.
   - Spark plugs.
3. Rotate the crankshaft so that the piston on the cylinder to be serviced is at bottom dead center with the valves closed.
4. Install an adapter in the spark plug hole and connect a compressed air supply at 102–144 psi.
5. Install special tool Valve Spring Compressor T94P-6565-BH.
6. Compress the valve spring and remove the roller finger follower. Repeat for each follower to be removed.

➡If the followers are to be reused, ensure that they are installed in the same position that they were removed from.

To install:

7. Compress the valve spring and install the roller finger follower. Repeat for each follower to be installed.
8. Install or connect the following:
   - Spark plugs
   - Valve covers and tighten the bolts in sequence to 88 inch lbs. (10 Nm)
   - Wiring harness brackets
   - Ignition coil
   - Fuel injector wiring harness
   - Coolant reservoir breather hose bracket and wiring
   - Spark plug wires and bracket
   - IMRC actuator
   - Upper manifold
   - Battery
9. Run the engine and check for leaks and proper operation.

Intake Manifold

REMOVAL & INSTALLATION

2.0L Engine

1. Before servicing the vehicle, refer to the precautions in the beginning of this section.
2. Remove or disconnect the following:
   - Negative battery cable
   - Air cleaner outlet pipe
   - Cruise control cable
   - Accelerator cable and bracket
   - Throttle Position (TPS) sensor connector
   - Idle Air Control (IAC) connector
   - Engine Coolant Temperature (ECT) sensor connector
   - Fuel lines
   - Injector wiring harness
   - Fuel supply manifold with injectors attached
   - Engine wiring harness
   - Vacuum hoses
   - Brake booster vacuum pipe
   - Accessory drive belt
   - Alternator
   - Intake manifold

To install:

3. Install or connect the following:
   - Intake manifold with a new gasket and tighten the fasteners to 13 ft. lbs. (18 Nm)
   - Alternator
   - Accessory drive belt
   - Brake booster vacuum pipe
   - Vacuum hoses
   - Engine wiring harness
   - Fuel supply manifold with injectors attached
   - Injector wiring harness
   - Fuel lines
   - Accelerator cable and bracket
   - Cruise control cable
   - ECT sensor connector
   - IAC connector
   - TPS sensor connector
   - Air cleaner outlet pipe
   - Negative battery cable

2.5L Engine

1. Before servicing the vehicle, refer to the precautions in the beginning of this section.
2. Remove or disconnect the following:
   - Negative battery cable
   - Water pump pulley shield
   - Brake booster vacuum pipe
   - Emission vacuum control
   - Cruise control cable
   - Accelerator cable and bracket
   - Throttle Position (TPS) sensor connector
   - Idle Air Control (IAC) valve connector
   - Exhaust Gas Recirculation (EGR) vacuum regulator connector and vacuum hoses

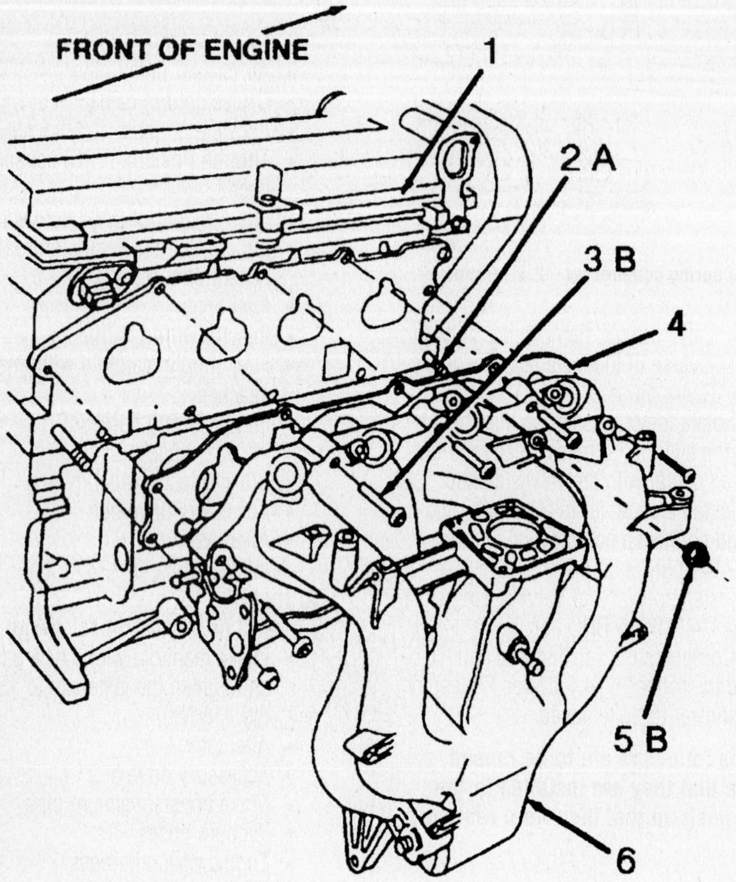

1. Cylinder head
2. Stud (2 req'd)
3. Bolt (8 req'd)
4. Intake manifold gasket
5. Nut (2 req'd)
6. Intake manifold

A. Tighten to 0-10 Nm (0-89 lb. in.)
B. Tighten to 16-20 Nm (12-15 lb. ft.)

7922JG06

Exploded view of the intake manifold mounting—2.0L engine

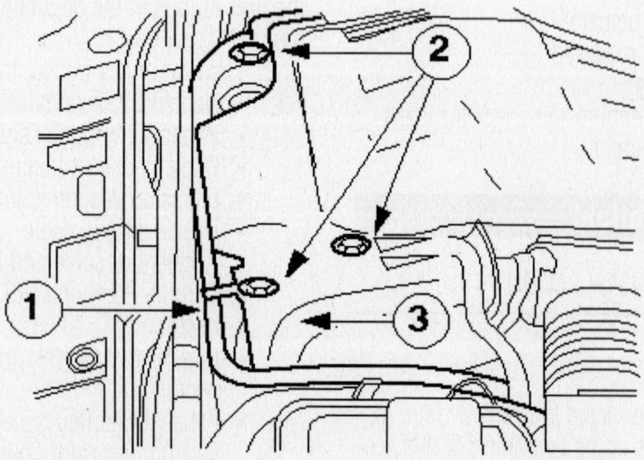

9346HG13

Coolant line (1), Cover bolts (2) and Water Pump Cover (3)—2.5L engine

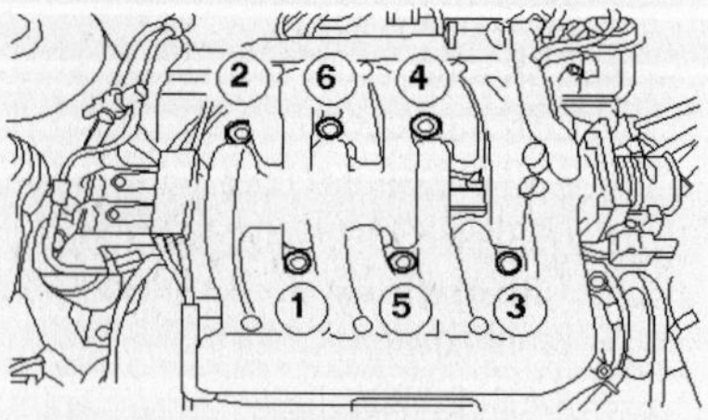

9346HG14

Upper intake manifold loosening sequence—2.5L engine

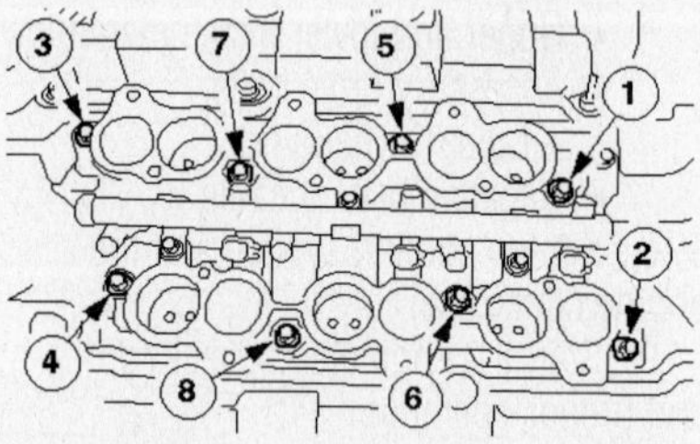

9346HG15

Lower intake manifold loosening sequence—2.5L engine

- PCV valve vacuum hose
- EGR tube
- Intake Manifold Runner Control (IMRC)
- Upper intake manifold. Remove the bolts in the sequence shown.
- Fuel injector wiring harness
- Fuel lines
- Lower intake manifold. Remove the bolts in the sequence shown.

To install:

3. Install or connect the following:
   - Lower intake manifold. Use new gaskets and tighten the bolts to 71–106 inch lbs. (8–12 Nm).
   - Fuel lines
   - Fuel injector wiring harness
   - Upper intake manifold. Use new gaskets and tighten the bolts to 71–106 inch lbs. (8–12 Nm).
   - IMRC
   - EGR tube
   - PCV valve vacuum hose
   - EGR vacuum regulator connector and vacuum hoses
   - IAC valve connector
   - TPS sensor connector
   - Accelerator cable and bracket
   - Cruise control cable
   - Emission vacuum control
   - Brake booster vacuum pipe
   - Water pump pulley shield
   - Negative battery cable
4. Run the engine and check for leaks

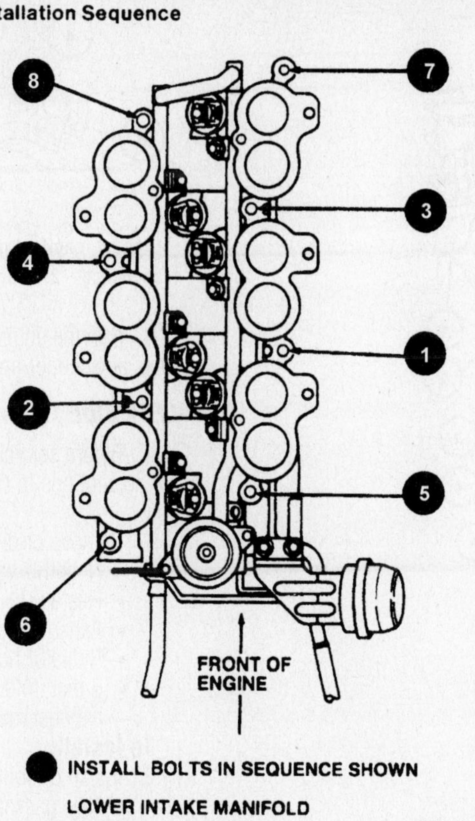

Lower intake manifold torque sequence—2.5L engine

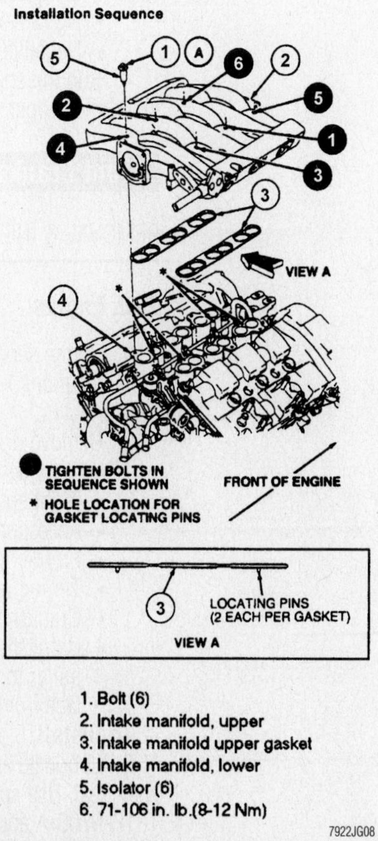

Upper intake manifold torque sequence—2.5L engine

Exhaust Manifold

REMOVAL & INSTALLATION

2.0L Engine

1. Before servicing the vehicle, refer to the precautions in the beginning of this section.
2. Remove or disconnect the following:
   - Negative battery cable
   - Air intake resonators
   - Oil dipstick tube
   - Exhaust manifold heat shield
   - Heated Oxygen (HO_2S) sensor
   - Exhaust Gas Recirculation (EGR) tube and bracket, if equipped
   - Catalytic converter
   - 9 manifold retaining nuts and the exhaust manifold

To install:

3. Install or connect the following:
   - New gasket
   - Exhaust manifold and tighten the nuts to 12 ft. lbs. (16 Nm)
   - Catalytic converter
   - EGR tube and bracket, if equipped
   - HO_2 sensor
   - Exhaust manifold heat shield
   - Oil dipstick tube
   - Air intake resonators
   - Negative battery cable
4. Run the engine and check for leaks and proper operation

2.5L Engine

RIGHT SIDE

1. Before servicing the vehicle, refer to the precautions in the beginning of this section.
2. Remove or disconnect the following:
   - Negative battery cable
   - Alternator and bracket
   - Heated Oxygen (HO_2S) sensor
   - Catalytic converter
   - Halfshaft bearing support bracket
   - Exhaust Gas Recirculation (EGR) tube
   - Exhaust manifold

To install:

3. Install or connect the following:
   - New gasket
   - Exhaust manifold and tighten the fasteners to 15 ft. lbs. (20 Nm)
   - EGR tube
   - Halfshaft bearing support bracket
   - Catalytic converter
   - HO_2S sensor
   - Alternator and bracket
   - Negative battery cable

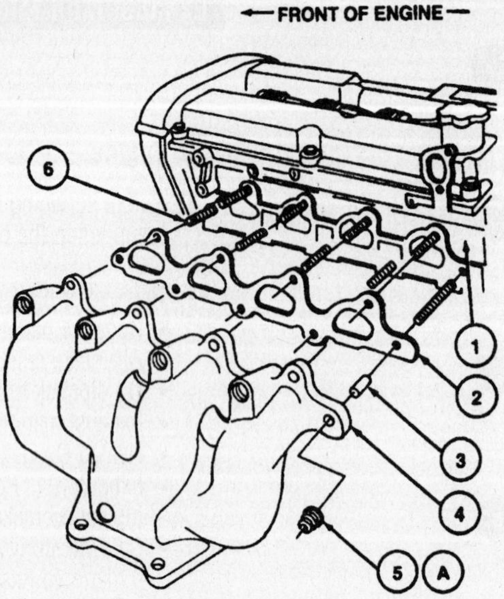

1 Cylinder Head
2 Exhaust Manifold Gasket
3 Spacer
4 Exhaust Manifold
5 Nut (9 Req'd)
6 Stud (9 Req'd)
A Tighten to 14-17 N·m (13-16 Lb-Ft)

7922JG09

Exhaust manifold and related components—2.0L engine

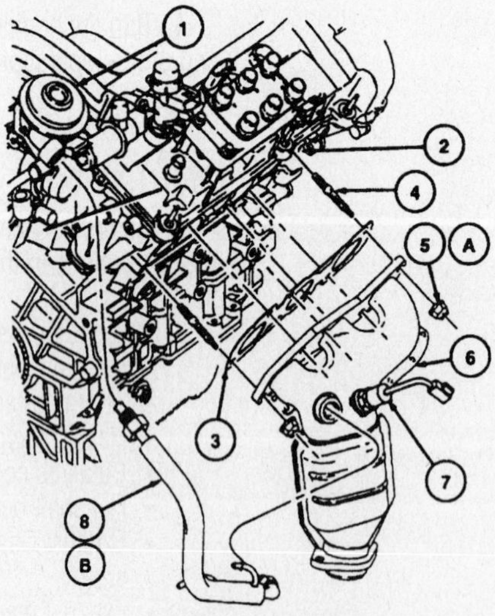

1 EGR Valve
2 RH Cylinder Head
3 Exhaust Manifold Gasket
4 Stud Bolt (6 Req'd)
5 Nut (6 Req'd)
6 RH Exhaust Manifold
7 Heated Oxygen Sensor
8 EGR Valve to Exhaust Manifold Tube
A Tighten to 18-22 N·m (13-16 Lb-Ft)
B Tighten to 35-45 N·m (26-33 Lb-Ft)

7922JG10

Exploded view of the right-side exhaust manifold mounting—2.5L engine

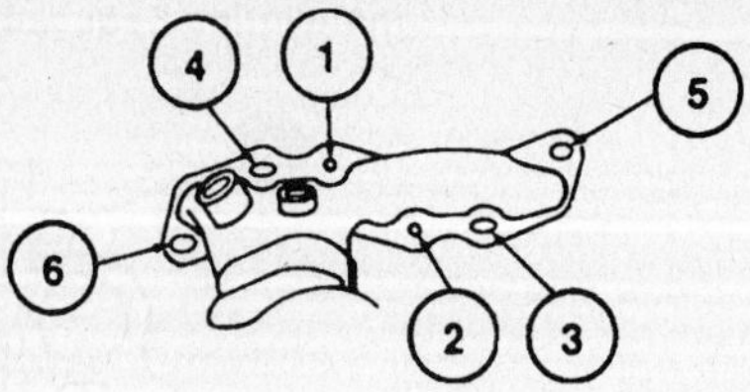

7922JG11

Exhaust manifold mounting bolt tightening sequence—2.5L engine

4. Run the engine and check for exhaust leaks and proper operation.

LEFT SIDE

1. Before servicing the vehicle, refer to the precautions in the beginning of this section.
2. Remove or disconnect the following:
   - Negative battery cable
   - Heated Oxygen (HO_2S) sensor
   - Exhaust crossover tube
   - Radiator hose bracket
   - 6 manifold retaining nuts and the exhaust manifold

To install:

3. Install or connect the following:
   - New gasket
   - Exhaust manifold and tighten the nuts to 15 ft. lbs. (20 Nm)
   - Radiator hose bracket
   - Exhaust crossover tube
   - HO_2S sensor
   - Negative battery cable
4. Run the engine and check for exhaust leaks and proper operation.

Front Crankshaft Seal

REMOVAL & INSTALLATION

2.0L Engine

1. Before servicing the vehicle, refer to the precautions in the beginning of this section.
2. Remove or disconnect the following:
   - Negative battery cable
   - Right splash shield
   - Accessory drive belt
   - Front cover
   - Timing belt
   - Crankshaft sprocket
   - Crankshaft front seal. Use a seal puller to protect the crankshaft surface from damage.

To install:

3. Lubricate and install the front crankshaft seal. Use special tool Seal Replacer T81P-6700-A and the crankshaft pulley bolt to press the seal into the oil pump housing.

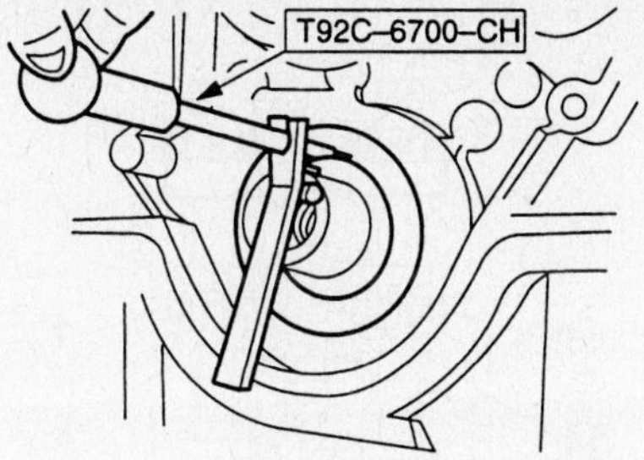

7922JG12

Removing the crankshaft front seal with Ford seal remover T92C-6700-CH—2.0L engine

4. Install or connect the following:
   - Crankshaft sprocket
   - Timing belt
   - Front cover
   - Accessory drive belt
   - Right splash shield
   - Negative battery cable
5. Run the engine and check for leaks and proper operation.

Camshaft and Valve Lifters

REMOVAL & INSTALLATION

2.0L Engine

1. Before servicing the vehicle, refer to the precautions in the beginning of this section.

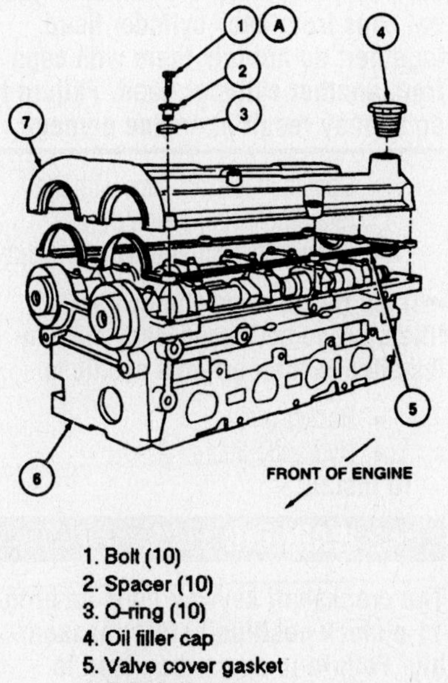

1. Bolt (10)
2. Spacer (10)
3. O-ring (10)
4. Oil filler cap
5. Valve cover gasket
6. Cylinder head
7. Valve cover
A. 53-71 in. lb.(6-8 Nm)

7922JG13

Valve cover and related components—2.0L engine

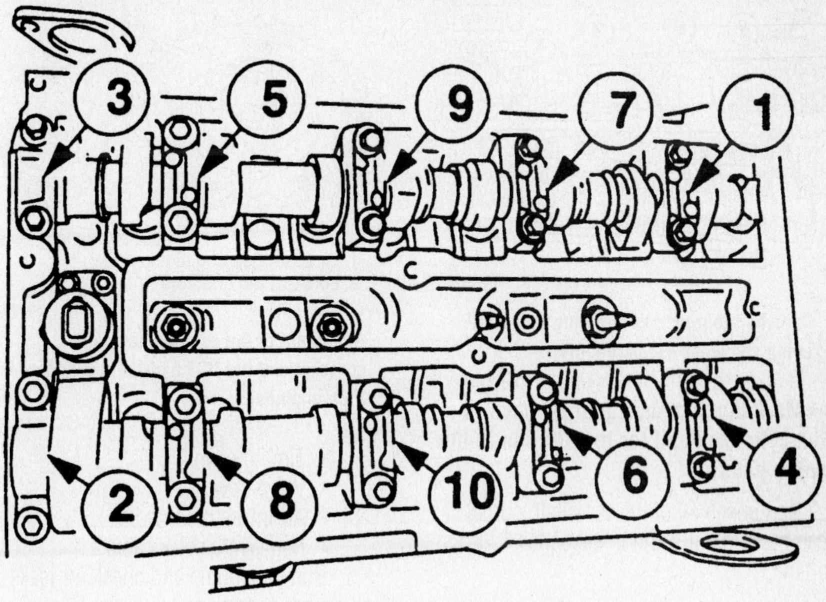

7922JG15

Remove the bearing cap bolts in pairs using the sequence shown—2.0L engine

2. Remove or disconnect the following:
   - Negative battery cable
   - Air intake resonators
   - Valve cover
   - Timing cover
   - Timing belt
   - Intake camshaft sprocket
   - Exhaust gear oil plug
   - Exhaust camshaft sprocket and Variable Camshaft Timing (VCT) hydraulic cylinder
   - Oil feed flange bolts

1. Bolt(20)
2. Camshaft journal cap(8)
3. Cylinder head
4. Camshaft
5. Camshaft journal thrust cap(2)
A. 13-15 ft. lb.(17-21 Nm)

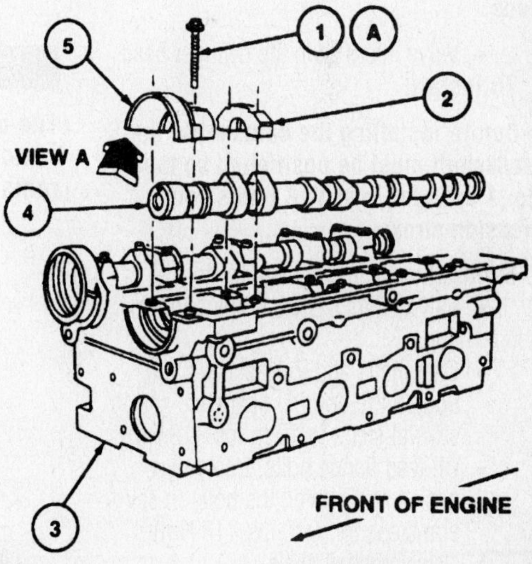

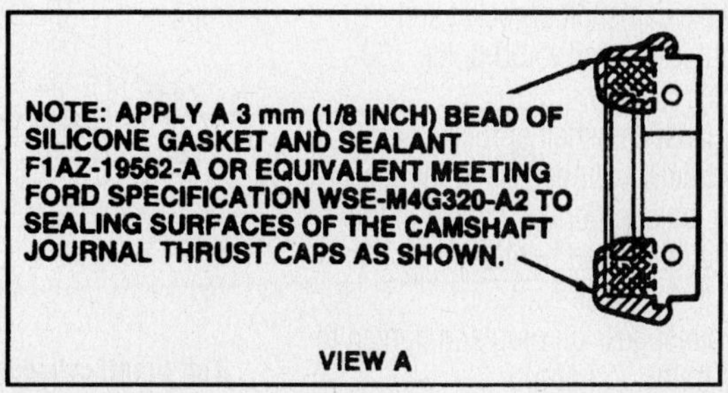

7922JG14

During assembly, apply sealant to the camshaft journal thrust cap as shown—2.0L engine

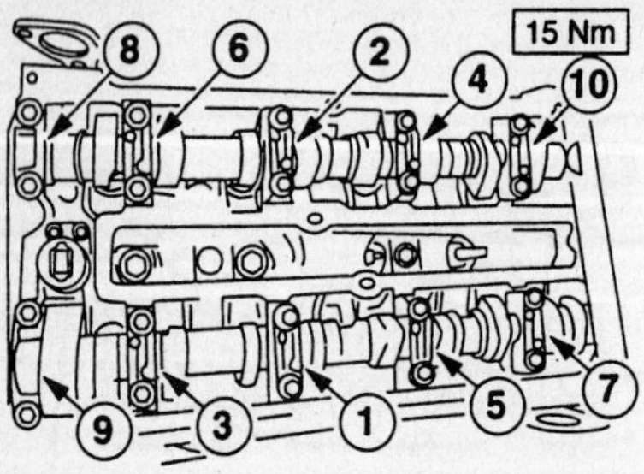

7922JG16

Camshaft journal cap retaining bolt tightening sequence—2.0L engine

➡Mark the camshaft journal caps to the cylinder head for installation in the same position.

3. Loosen all of the camshaft journal cap bolts in pairs and in sequence 1 turn at a time.

➡Remove the camshaft journal thrust caps last.

4. Remove or disconnect the following:
   - All of the camshaft journal caps
   - Intake and exhaust camshafts and the camshaft front seals from the cylinder head

➡If the valve lifters are to be reused, mark their locations to ensure that they will be installed into their correct positions.

   - Valve lifters from the cylinder head

To install:

➡Before installing the camshafts, the crankshaft must be positioned so that No. 1 cylinder is at TDC on its compression stroke.

5. Install or connect the following:
   - Valve lifters in their original positions
   - Camshafts and journal caps and tighten the bolts in sequence and in several steps to 10 ft. lbs. (13 Nm)
   - Oil feed flange bolts. Use a new gasket and tighten the bolts in several steps to 10 ft. lbs. (13 Nm).
   - New camshaft seals
   - Intake camshaft sprocket and tighten the bolt to 50 ft. lbs. (68 Nm)
   - Exhaust camshaft sprocket and VCT hydraulic cylinder. Align the locating tab and bore as shown in the illustration and tighten the bolt to 88 ft. lbs. (120 Nm).
   - Exhaust gear oil plug and tighten to 27 ft. lbs. (37 Nm)
   - Timing belt

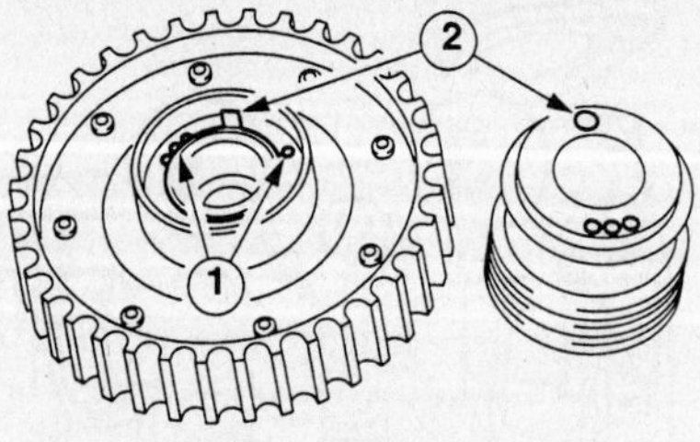

1. Oil passages and match mark
2. Locating tab and bore

9306HG03

Alignment of the exhaust camshaft sprocket and the VCT hydraulic cylinder—2.0L engines

   - Timing cover
   - Valve cover
   - Air intake resonators
   - Negative battery cable

6. Run the engine and check for leaks and proper operation.

2.5L Engine

1. Before servicing the vehicle, refer to the precautions in the beginning of this section.
2. Remove or disconnect the following:
   - Negative battery cable
   - Upper intake manifold
   - Valve covers
   - Accessory drive belts
   - Front cover and timing chains

⁂ WARNING

The camshaft thrust caps must be removed before loosening the remaining camshaft journal cap bolts

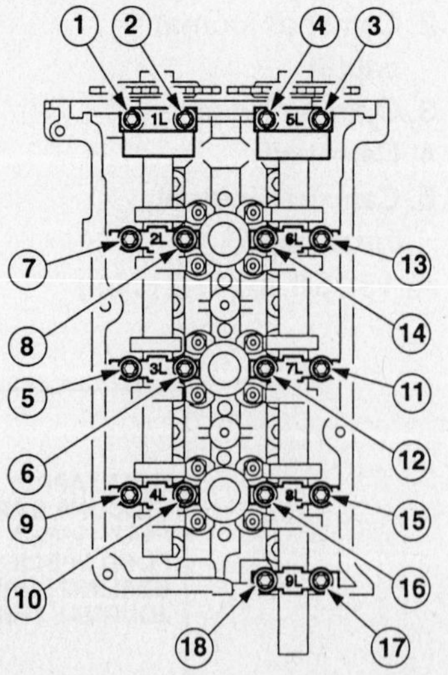

7922JG18

Left (front) cylinder head camshaft journal cap retaining bolt loosening sequence—2.5L engine

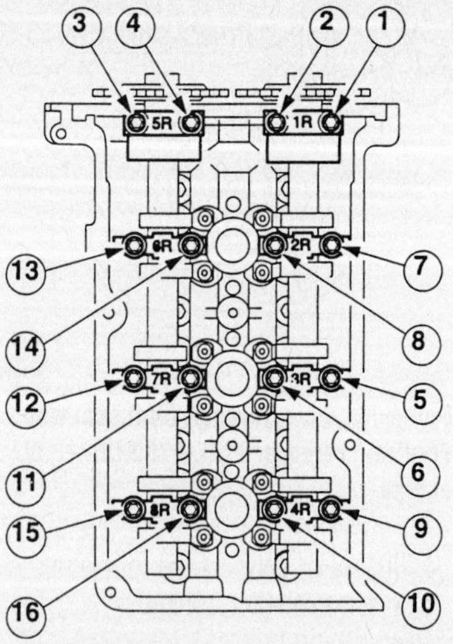

7922JG19

Right (rear) cylinder head camshaft journal cap retaining bolt loosening sequence—2.5L engine

to ensure that the thrust caps are not damaged.

   - Camshaft thrust caps
   - Camshaft journal caps. Loosen the bolts in sequence and in several passes to allow the camshaft to raise off the cylinder head evenly.

⁂ WARNING

The camshaft journal caps and cylinder heads are numbered to ensure that they are assembled in their original positions. Keep the camshaft journal caps from each cylinder head together; do not mix them with caps from another cylinder head. Failure to do so may result in engine damage.

   - Camshaft journal caps with the retaining bolts installed
   - Camshafts from the cylinder head.

➡If the rocker arms and hydraulic lifters are to be reused, they must be installed in their original positions.

   - Rocker arms.
   - Hydraulic lifters

To install:

⁂ WARNING

The crankshaft keyway must be at the 11 o'clock position before reassembly. Failure to do so may lead to engine damage.

3. Rotate the crankshaft so that the keyway is at the 11 o'clock position for installation of the camshafts.

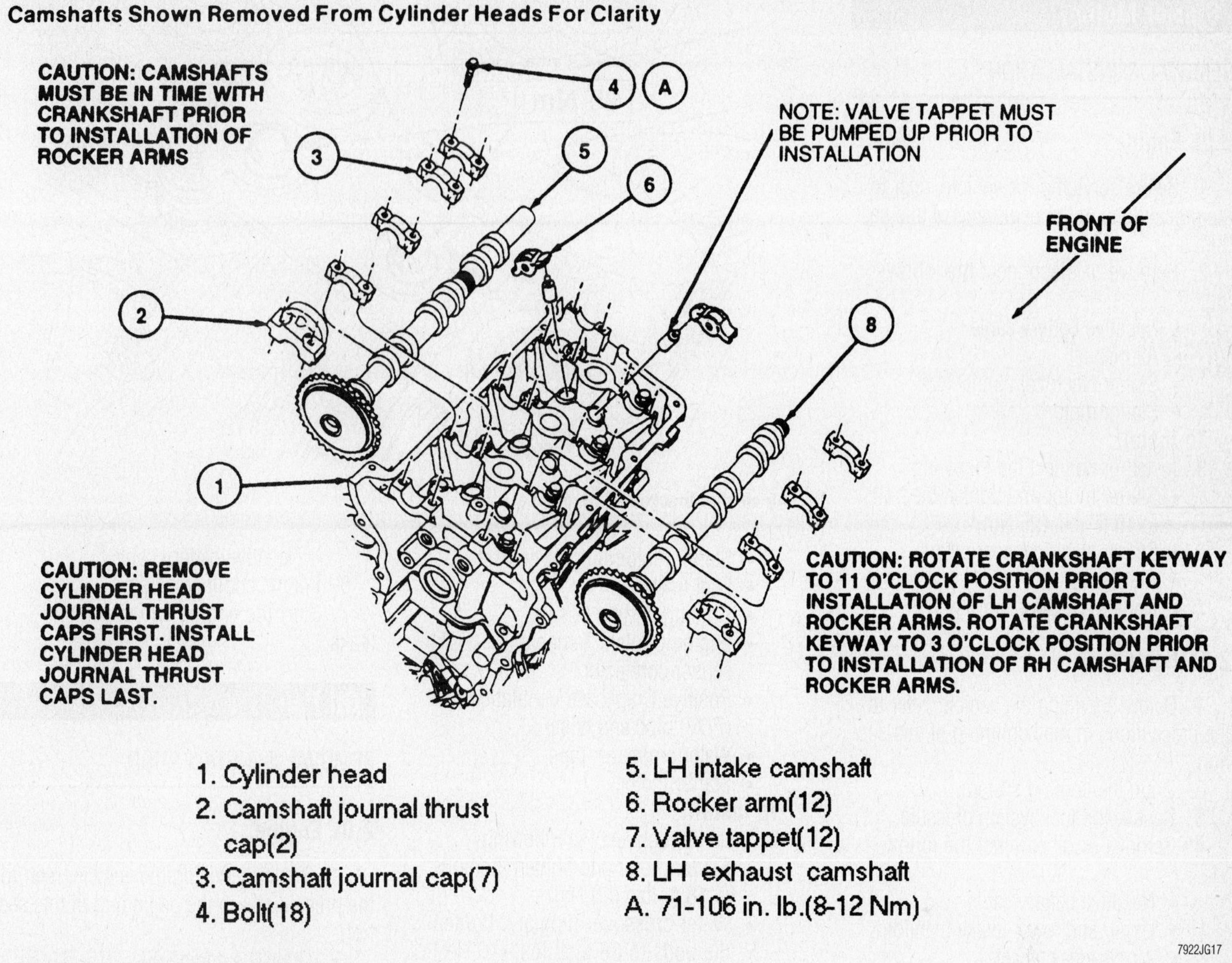
Camshafts Shown Removed From Cylinder Heads For Clarity
CAUTION: CAMSHAFTS MUST BE IN TIME WITH CRANKSHAFT PRIOR TO INSTALLATION OF ROCKER ARMS
NOTE: VALVE TAPPET MUST BE PUMPED UP PRIOR TO INSTALLATION
FRONT OF ENGINE
CAUTION: REMOVE CYLINDER HEAD JOURNAL THRUST CAPS FIRST. INSTALL CYLINDER HEAD JOURNAL THRUST CAPS LAST.
CAUTION: ROTATE CRANKSHAFT KEYWAY TO 11 O'CLOCK POSITION PRIOR TO INSTALLATION OF LH CAMSHAFT AND ROCKER ARMS. ROTATE CRANKSHAFT KEYWAY TO 3 O'CLOCK POSITION PRIOR TO INSTALLATION OF RH CAMSHAFT AND ROCKER ARMS.
1. Cylinder head
2. Camshaft journal thrust cap(2)
3. Camshaft journal cap(7)
4. Bolt(18)
5. LH intake camshaft
6. Rocker arm(12)
7. Valve tappet(12)
8. LH exhaust camshaft
A. 71-106 in. lb.(8-12 Nm)
7922JG17

Exploded view of camshaft mounting—2.5L engine

4. Install or connect the following:
   - Hydraulic lifters
   - Rocker arms.
   - Camshafts. Align the sprocket timing marks.

➡Do not install the camshaft journal thrust caps until the rocker arms and timing chains have been installed and the camshaft journal caps are secured into position.

   - All camshaft journal caps except the thrust caps.
   - Timing chains. Tighten the camshaft journal cap bolts in reverse of the loosening order and in several steps to 71–106 inch lbs. (8–12 Nm).
   - Thrust caps and tighten the bolts to 71–106 inch lbs. (8–12 Nm)
   - Front cover
   - Accessory drive belts
   - Valve covers
   - Upper intake manifold. Use new gaskets and tighten the bolts to 71–106 inch lbs. (8–12 Nm).
   - Negative battery cable
5. Run the engine and check for leaks and proper engine operation.

Valve Lash

ADJUSTMENT

2.0L Engine

1. Before servicing the vehicle, refer to the precautions in the beginning of this section.
2. Remove the valve cover.
3. Rotate the crankshaft so that the valve is closed and measure the clearance between the lifter shim and the camshaft base circle. Intake valve clearance should be 0.004–0.007 inches. Exhaust valve clearance should be 0.010–0.013 inches.
4. Measure each valve and note the clearance.
5. If adjustment is necessary, remove the camshafts and replace the shims. To obtain the correct shim size, measure the current shim and add the measured clearance. Subtract 0.006 inches for intake valves, and 0.012 inches for exhaust valves.
6. Install the camshafts and valve cover.

2.5L Engine

The lash adjusters are hydraulic and are not adjustable.

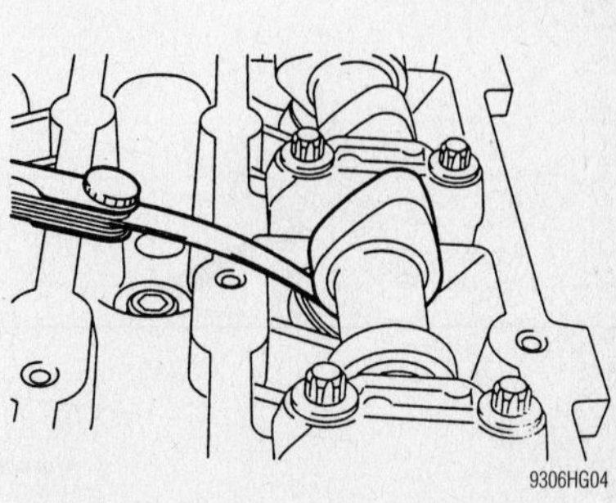
9306HG04

Checking valve clearance—2.0L engine

Starter Motor

REMOVAL & INSTALLATION

2.0L Engine

1. Before servicing the vehicle, refer to the precautions in the beginning of this section.
2. Remove or disconnect the following:
   - Negative battery cable
   - Air cleaner
   - Starter electrical connectors
   - Starter motor

To install:

3. Install or connect the following:
   - Starter motor and tighten the bolts to 18 ft. lbs. (25 Nm)
   - Starter electrical connectors
   - Air cleaner
   - Negative battery cable

2.5L Engine

1. Before servicing the vehicle, refer to the precautions in the beginning of this section.
2. Drain the cooling system.
3. Relieve the fuel system pressure.
4. Remove or disconnect the following:
   - Negative battery cable
   - Upper and lower intake manifold
   - Air cleaner bracket
   - Starter motor harness connectors
   - Shift cable and bracket

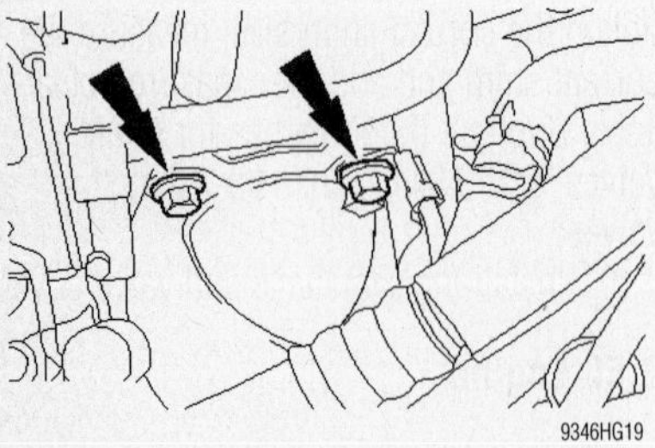

Starter motor upper bolts—2.0L engine

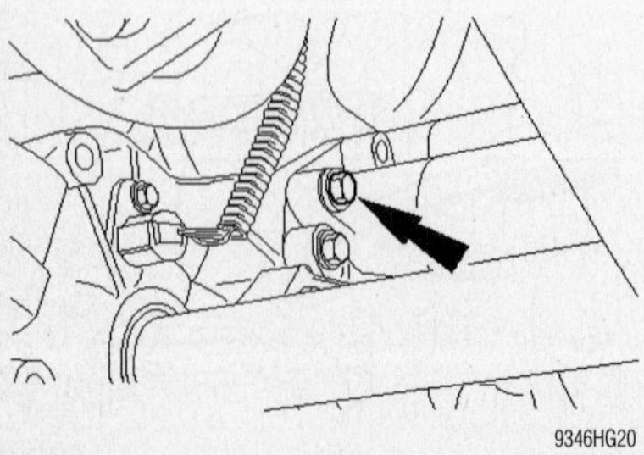

Starter motor lower bolt—2.0L engine

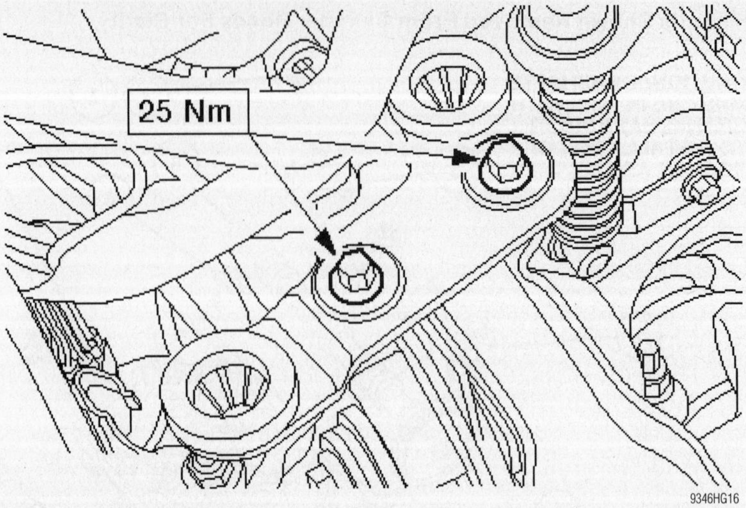

Air cleaner bracket—2.5L engine

   - Starter motor support bracket
   - Fuel lines
   - Cooling system hoses
   - Engine Coolant Temperature (ECT) sensor connector
   - Positive Crankcase Ventilation (PCV) pipe and hose
   - Water crossover pipe
   - Starter motor

To install:

5. Install or connect the following:
   - Starter motor and tighten the bolts to 18 ft. lbs. (25 Nm)
   - Water crossover pipe and tighten the bolts to 88 inch lbs. (10 Nm)
   - PCV pipe and hose
   - ECT sensor connector
   - Cooling system hoses
   - Fuel lines
   - Starter motor support bracket. Tighten the nuts to 13 ft. lbs. (17 Nm) and the bolt to 18 ft. lbs. (25 Nm).
   - Shift cable and bracket
   - Starter motor harness connectors
   - Air cleaner bracket and tighten the bolts to 18 ft. lbs. (25 Nm)
   - Upper and lower intake manifold

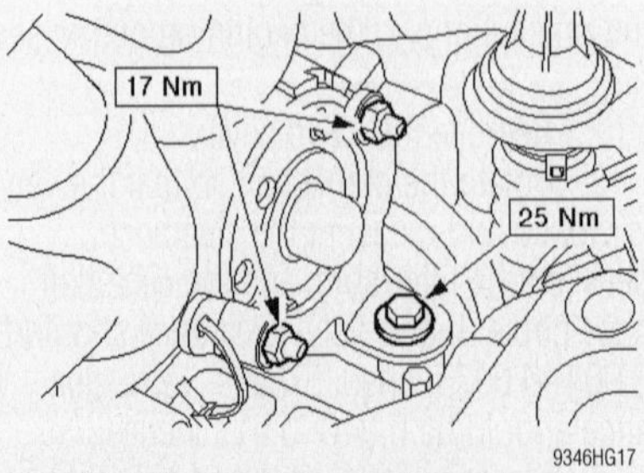

Starter motor support bracket—2.5L engine

   - Negative battery cable
6. Fill the cooling system.
7. Start the engine and check for leaks.

Oil Pan

REMOVAL & INSTALLATION

2.0L Engine

1. Before servicing the vehicle, refer to the precautions in the beginning of this section.
2. Drain the engine oil.
3. Attach an engine support fixture to the engine lifting eyes.
4. Remove or disconnect the following:
   - Negative battery cable
   - Catalytic converter
   - Oil lever sensor connector
   - Heater coolant pipe
   - Lower engine rear plate
   - Left and right engine support insulator center bolts
   - Front engine support bracket

➡Mark the location of the upper front engine support before removing it from the front engine support bracket.

5. Raise the engine to allow room for removal of the engine oil pan.
6. Remove the oil pan.

To install:

➡Apply a bead of silicone gasket sealer to the oil pump parting lines and at the crankshaft rear main seal retainer on the cylinder block.

7. Install or connect the following:
   - Oil pan. Use a new gasket and

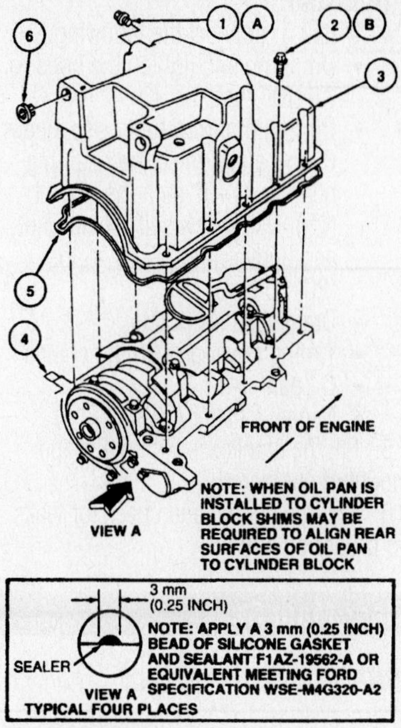
FRONT OF ENGINE
VIEW A
NOTE: WHEN OIL PAN IS INSTALLED TO CYLINDER BLOCK SHIMS MAY BE REQUIRED TO ALIGN REAR SURFACES OF OIL PAN TO CYLINDER BLOCK
3 mm (0.25 INCH)
SEALER
NOTE: APPLY A 3 mm (0.25 INCH) BEAD OF SILICONE GASKET AND SEALANT F1AZ-19562-A OR EQUIVALENT MEETING FORD SPECIFICATION WSE-M4G320-A2
VIEW A
TYPICAL FOUR PLACES

1 Oil pan drain plug
2 Bolt (10 req'd)
3 Oil pan
4 Cylinder block
5 Oil pan gasket
6 Oil pan spacer (as req'd)
A Tighten to 21-28 Nm (15-21 lb-ft)
B Tighten to 20-24 Nm (15-18 lb-ft)

7922JG20

Exploded view of the oil pan mounting—2.0L engine

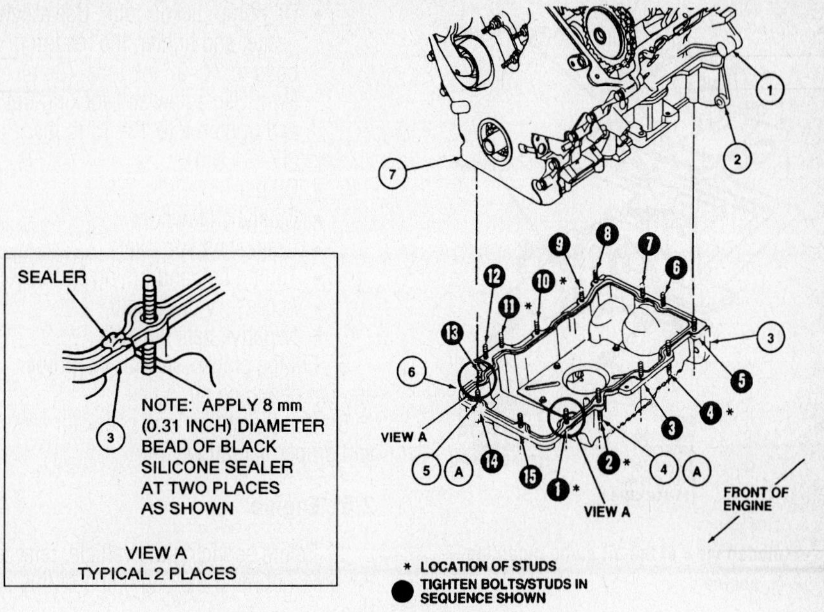
SEALER
NOTE: APPLY 8 mm (0.31 INCH) DIAMETER BEAD OF BLACK SILICONE SEALER AT TWO PLACES AS SHOWN
VIEW A
TYPICAL 2 PLACES
VIEW A
FRONT OF ENGINE
* LOCATION OF STUDS
TIGHTEN BOLTS/STUDS IN SEQUENCE SHOWN

1 Upper cylinder block
2 Lower cylinder block
3 Oil pan
4 Stud bolt (5 req'd)
5 Bolt (10 req'd)
6 Oil pan gasket
7 Engine front cover
A Tighten to 20-30 Nm (15-22 lb-ft)

7922JG21

Exploded view of the oil pan mounting, showing the mounting bolt and stud tightening sequence—2.5L engine

tighten the bolts in several passes to 15–18 ft. lbs. (20–24 Nm), working from the center of the block towards the ends. Tighten the transaxle case bolts to 25–34 ft. lbs. (34–46 Nm).
- Lower engine rear plate

8. Lower the engine and install or connect:
   - Front engine support insulator
   - Left and right engine support insulator center bolts
   - Heater coolant pipe
   - Oil lever sensor connector
   - Catalytic converter
   - Negative battery cable
9. Fill the crankcase with the proper amount of engine oil.
10. Run the engine and check for leaks and proper operation.

2.5L Engine

1. Before servicing the vehicle, refer to the precautions in the beginning of this section.
2. Drain the engine oil.
3. Attach an engine support fixture to the engine lifting eyes.
4. Remove or disconnect the following:
   - Negative battery cable
   - Water pump pulley shield
   - Exhaust crossover pipe and bracket
   - Exhaust heat shields
   - Lower engine rear plate
   - Left and right engine support insulator center bolts
   - Front engine support bracket

➡Mark the location of the upper front engine support before removing it from the front engine support bracket.

5. Raise the engine to allow room for removal of the engine oil pan.
6. Remove the oil pan.

To install:

7. Apply a bead of silicone sealer to the gasket area where the pan meets the parting lines of the lower cylinder block and the front engine cover.
8. Install or connect the following:
   - Oil pan. Use a new gasket and tighten the pan bolts in several passes to 15–22 ft. lbs. (20–30 Nm). Tighten the transaxle case bolts to 25–34 ft. lbs. (34–46 Nm).
   - Lower engine rear plate
9. Lower the engine and install or connect:
   - Front engine support insulator
   - Left and right engine support insulator center bolts
   - Exhaust heat shields
   - Exhaust crossover pipe and bracket
   - Water pump pulley shield
   - Negative battery cable
10. Fill the crankcase with the proper amount of engine oil.
11. Run the engine and check for leaks and proper operation.

Oil Pump

REMOVAL & INSTALLATION

2.0L Engine

1. Before servicing the vehicle, refer to the precautions in the beginning of this section.
2. Drain the engine oil.
3. Remove or disconnect the following:
   - Negative battery cable
   - Accessory drive belt
   - Timing belt cover and timing belt
   - Crankshaft sprocket
   - Catalytic converter
   - Oil pan
   - Oil pump pickup tube
   - Oil filter
   - Oil pump

To install:

➡Clearance between the cylinder block oil pan sealing surface to the oil pump oil pan sealing surface should not exceed 0.012–0.031 in. (0.3–0.8mm).

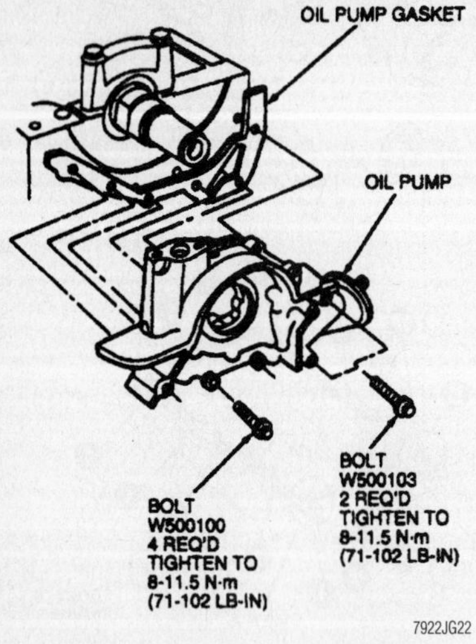

Exploded view of the oil pump mounting—2.0L engine

4. Install or connect the following:
   - Oil pump. Use a straight-edge to align the oil pump oil pan sealing surface with the cylinder block oil pan sealing surface and tighten the oil pump retaining bolts to 71–102 inch lbs. (8–11.5 Nm).
   - Oil filter
   - Oil pump pickup tube. Use a new gasket and tighten the retaining bolts to 71–97 inch lbs. (8–11 Nm). Use a new self-locking nut and tighten it to 13–15 ft. lbs. (17–21 Nm).
   - Oil pan
   - Catalytic converter
   - Crankshaft sprocket
   - Timing belt and timing belt cover
   - Accessory drive belt
   - Negative battery cable
5. Fill the crankcase with the proper amount of engine oil.
6. Run the engine and check for leaks and proper operation.

2.5L Engine

1. Before servicing the vehicle, refer to the precautions in the beginning of this section.
2. Drain the engine oil.
3. Remove or disconnect the following:
   - Negative battery cable
   - Oil pan
   - Timing cover and timing chains
   - Crankshaft sprockets
   - Oil pump pickup tube
   - Oil pump

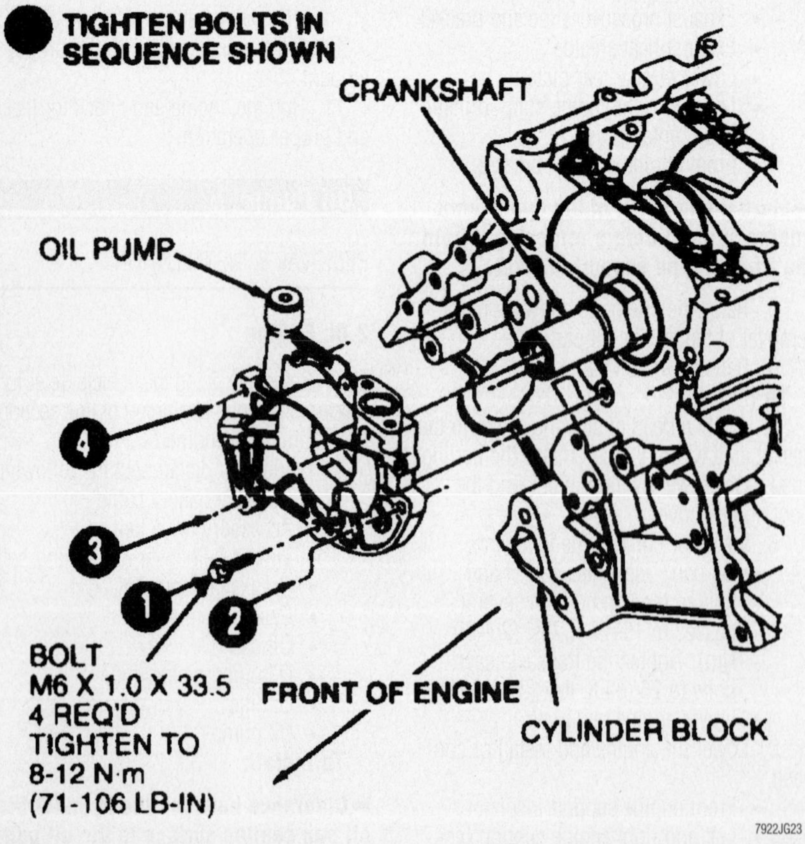

Exploded view of the oil pump mounting, showing the retaining bolt tightening sequence—2.5L engine

To install:

4. Install or connect the following:
   - Oil pump and tighten the bolts to 71–106 inch lbs. (8–12 Nm)
   - Oil pump pickup tube. Use a new O-ring and tighten the retaining bolts to 71–97 inch lbs. (8–11 Nm). Use a new self-locking nut and tighten it to 15–22 ft. lbs. (20–30 Nm).
   - Crankshaft sprockets
   - Timing cover and timing chains
   - Oil pan
   - Negative battery cable
5. Fill the crankcase with the proper amount of engine oil.
6. Run the engine and check for leaks and proper operation.

Rear Main Seal

REMOVAL & INSTALLATION

2.0L Engine

1. Before servicing the vehicle, refer to the precautions in the beginning of this section.
2. Remove or disconnect the following:
   - Negative battery cable
   - Transaxle
   - Clutch, if equipped
   - Flywheel
   - Oil seal

To install:

3. Install or connect the following:
   - Oil seal. Seat the seal flush with the rear of the crankshaft oil seal retainer.
   - Flywheel and tighten the bolts to 82 ft. lbs. (112 Nm)
   - Clutch, if equipped
   - Transaxle
   - Negative battery cable
4. Start the engine and check for leaks.

2.5L Engine

1. Before servicing the vehicle, refer to the precautions in the beginning of this section.
2. Remove or disconnect the following:
   - Negative battery cable
   - Transaxle
   - Clutch, if equipped
   - Flywheel
   - Oil seal. Use special tool Seal Remover T95P-6701-EH and a slide hammer to remove the seal.

To install:

3. Install or connect the following:
   - Oil seal. Press the seal in evenly

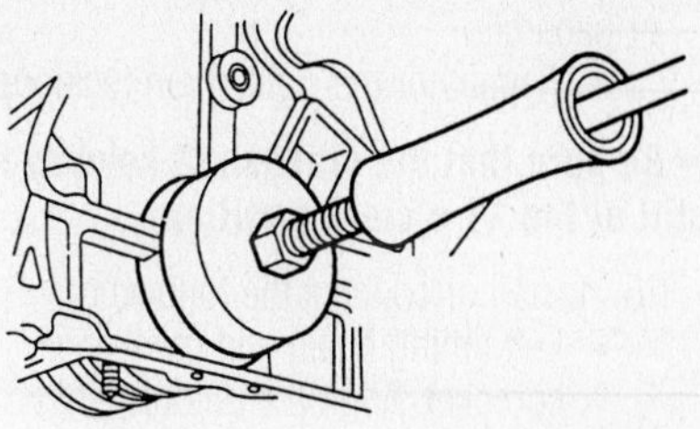

7922JG24

Thread the oil seal remover into the seal, then withdraw the seal using a slide hammer—2.5L engine

until it is flush with the cylinder block.
- Flywheel and tighten the bolts to 59 ft. lbs. (80 Nm)
- Clutch, if equipped
- Transaxle
- Negative battery cable

4. Start the engine and check for leaks.

Timing Chain, Sprockets, Front Cover and Seal

REMOVAL & INSTALLATION

2.5L Engine

1. Before servicing the vehicle, refer to the precautions in the beginning of this section.
2. Attach an engine support fixture to the engine lifting eyes.

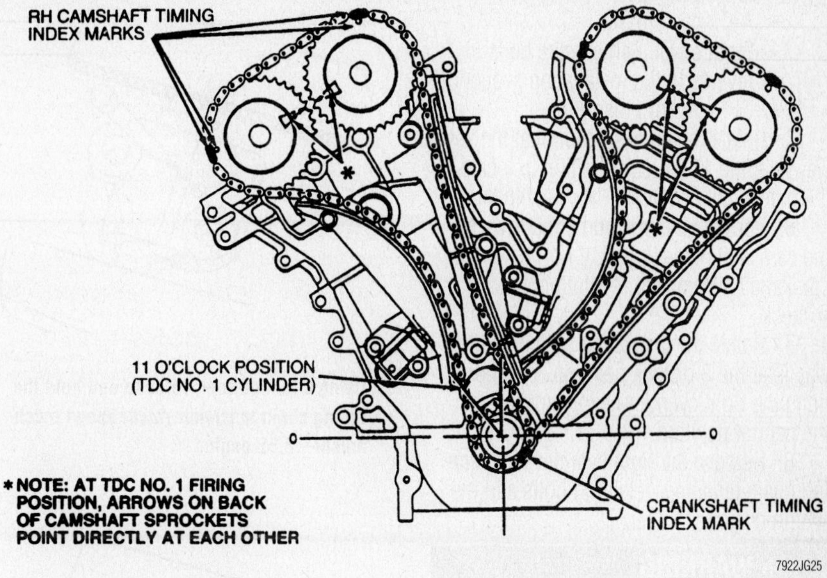
RH CAMSHAFT TIMING INDEX MARKS
11 O'CLOCK POSITION (TDC NO. 1 CYLINDER)
*NOTE: AT TDC NO. 1 FIRING POSITION, ARROWS ON BACK OF CAMSHAFT SPROCKETS POINT DIRECTLY AT EACH OTHER
CRANKSHAFT TIMING INDEX MARK
7922JG25

View of the timing chains and gears, showing the right timing chain alignment marks properly positioned—2.5L engine

3. Drain the engine oil.
4. Remove or disconnect the following:
   - Negative battery cable
   - Upper intake manifold
   - Valve covers
   - Low coolant level sensor connector
   - Coolant recovery reservoir

➡**Mark the position of the upper front engine support bracket before removing.**

- Front engine support insulator and bracket
- Engine wiring harness connectors
- Accessory drive belt
- Power steering pump and bracket
- Right splash shield
- Alternator
- Crankshaft pulley
- Crankshaft Position (CKP) sensor connector
- Oil pan

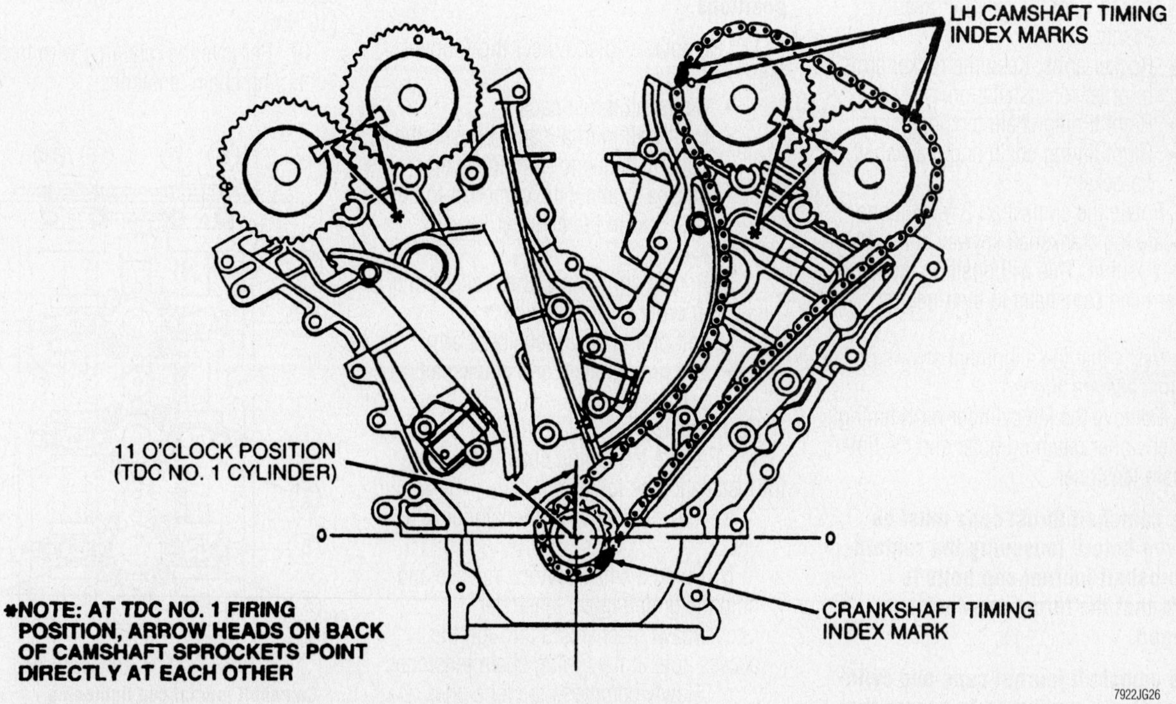
LH CAMSHAFT TIMING INDEX MARKS
11 O'CLOCK POSITION (TDC NO. 1 CYLINDER)
*NOTE: AT TDC NO. 1 FIRING POSITION, ARROW HEADS ON BACK OF CAMSHAFT SPROCKETS POINT DIRECTLY AT EACH OTHER
CRANKSHAFT TIMING INDEX MARK
7922JG26

Left timing chain alignment mark positioning for servicing the chain—2.5L engine

- A/C compressor
- A/C hose bracket
- Front cover. Remove the bolts in reverse of the installation sequence.
- CKP sensor pulse ring

5. Rotate the crankshaft so that the keyway is at the 11 o'clock position to locate the crankshaft at TDC for No. 1 cylinder.

6. Verify that the alignment arrows on the camshafts are aligned. If not, rotate the crankshaft 1 complete revolution and recheck.

7. Rotate the crankshaft so that the keyway is at the 3 o'clock position. This positions the right cylinder head camshafts to the neutral position.

8. Remove the right cylinder head timing chain tensioner retaining bolts and the timing chain tensioner.

⁂ WARNING

The camshaft thrust caps must be removed before loosening the remaining camshaft journal cap bolts to ensure that the thrust caps are not damaged.

➡The camshaft journal caps and cylinder heads are numbered to ensure that they are assembled in their original positions.

9. Remove or disconnect the following:
   - Camshaft thrust caps
   - Camshaft journal caps. Loosen the bolts in sequence and in several passes to allow the camshaft to be raised from the cylinder head evenly.
   - Rocker arms. Keep the rocker arms in order for installation.
   - Right timing chain tensioner arm
   - Right timing chain and crankshaft sprocket

10. Rotate the crankshaft 2 revolutions and locate the crankshaft keyway at the 11 o'clock position. This will position the left cylinder head camshafts to their neutral position.

11. Verify that the alignment arrows on the camshafts are aligned.

12. Remove the left cylinder head timing chain tensioner retaining bolts and the timing chain tensioner.

➡The camshaft thrust caps must be removed before loosening the remaining camshaft journal cap bolts to ensure that the thrust caps are not damaged.

➡The camshaft journal caps and cylinder heads are numbered to ensure that

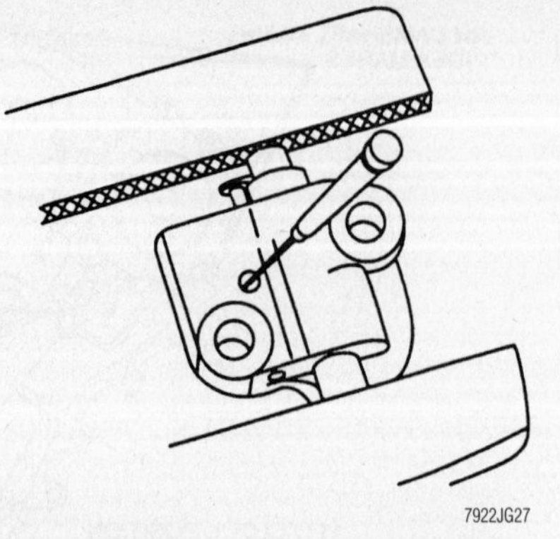

Using a thin prytool, release and hold the timing chain tensioner ratchet/pawl mechanism—2.5L engine

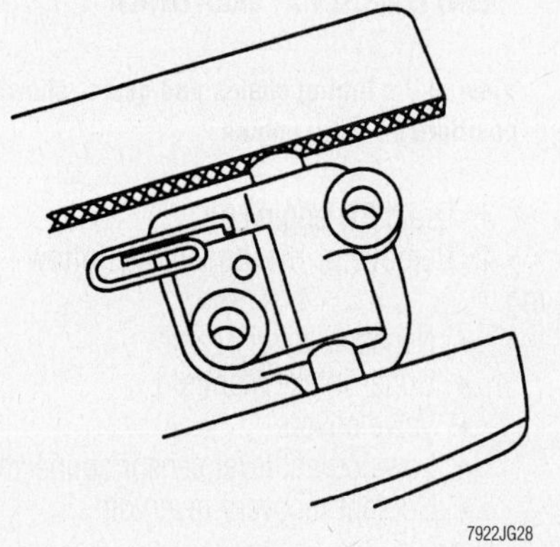

Retain the piston with a 1.5mm wire or paperclip—2.5L engine

they are assembled in their original positions.

13. Remove or disconnect the following:
   - Camshaft thrust caps
   - Camshaft journal caps. Loosen the bolts in sequence and in several passes to allow the camshaft to be raised from the cylinder head evenly.
   - Rocker arms. Keep the rocker arms in order for installation.
   - Left timing chain tensioner arm
   - Left timing chain and crankshaft sprocket

To install:

14. Prepare the timing chain tensioners for installation as follows:

a. Place the left chain tensioner in a vise.

b. Using a small prytool, release and hold the timing chain tensioner ratchet/pawl mechanism through the access hole in the timing chain tensioner.

c. Slowly compress the tensioner.

d. Lock the piston with a 1.5mm wire or paperclip.

e. Repeat for the right chain tensioner.

➡Be sure that the crankshaft keyway is still at the 11 o'clock position.

15. Install or connect the following:
   - Left timing chain and crankshaft sprocket. Align the colored links with the index marks on the camshaft and crankshaft sprockets.
   - Left timing chain tensioner arm
   - Left timing chain tensioner and tighten the retaining bolts to 15–22 ft. lbs. (20–30 Nm)
   - Right timing chain and crankshaft sprocket. Align the colored links with the index marks on the camshaft and crankshaft sprockets.
   - Right timing chain tensioner arm
   - Right timing chain tensioner and tighten the retaining bolts to 15–22 ft. lbs. (20–30 Nm)

➡The crankshaft keyway must be in the 11 o'clock position to install the left cylinder head rocker arms.

➡Do not install the camshaft journal thrust caps until the other journal caps have been installed and tightened.

16. Install the left cylinder head rocker arms in their original positions.

17. Tighten the left camshaft journal caps in the order shown and in several passes to 88 inch lbs. (10 Nm).

18. Install the left camshaft journal thrust caps and tighten the bolts to 88 inch lbs. (10 Nm).

19. Remove the retaining wire from the left timing chain tensioner.

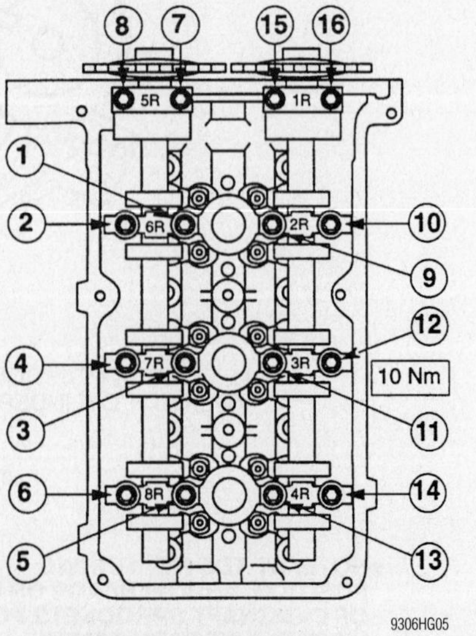

Camshaft journal cap tightening sequence—2.5L engine

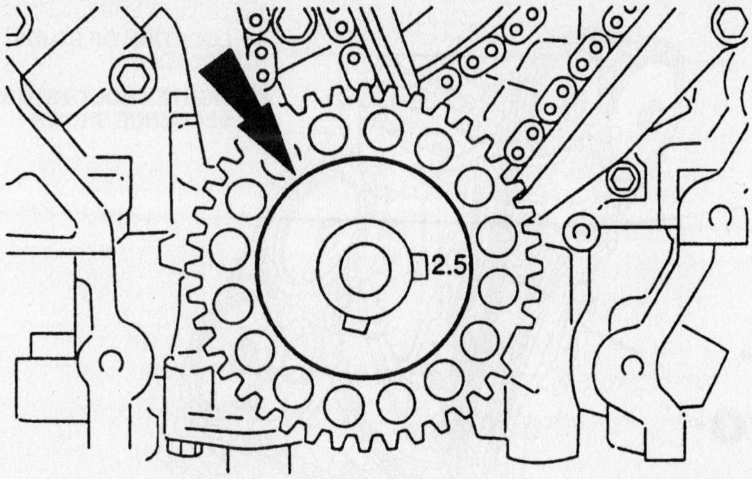

When installing the CKP sensor pulse ring, be sure to use the correct keyway—2.5L engine

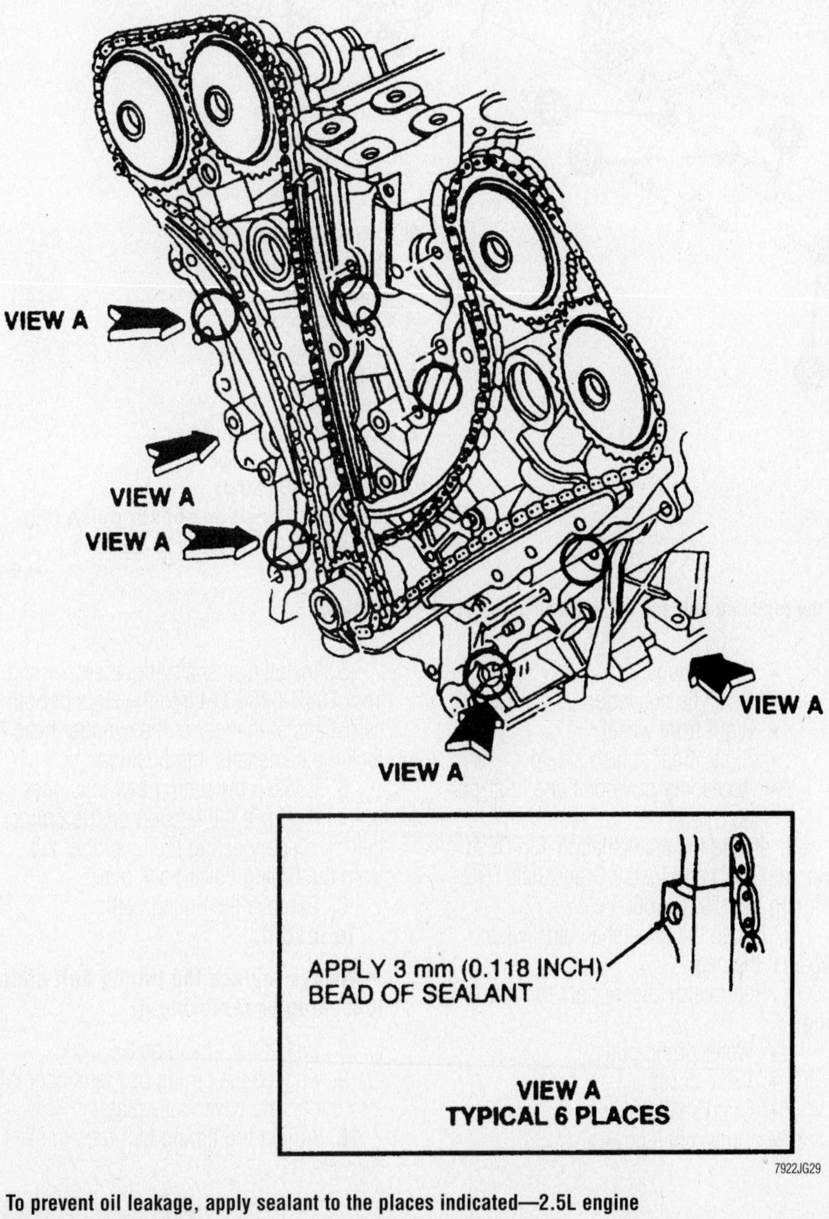

To prevent oil leakage, apply sealant to the places indicated—2.5L engine

➡The crankshaft keyway must be in the 3 o'clock position to install the right cylinder head rocker arms.

20. Rotate the crankshaft so that the keyway is in the 3 o'clock position.
21. Install the right cylinder head rocker arms in their original positions.

➡Do not install the camshaft journal thrust caps until the other journal caps have been installed and tightened.

22. Tighten the right camshaft journal caps in the order shown and in several passes to 88 inch lbs. (10 Nm).
23. Install the right camshaft journal thrust caps and tighten the bolts to 88 inch lbs. (10 Nm).
24. Remove the retaining wire from the right timing chain tensioner.
25. Install the CKP sensor pulse ring. Use the keyway for the 2.5L engine as shown.
26. Replace the crankshaft seal in the front cover with a new one. Apply clean engine oil to the seal lip.
27. Apply silicone sealer to the 6 critical areas shown in View **A**.
28. Place new front cover gaskets onto the dowel pins on the cylinder block and heads.
29. Place the front cover into position.
30. Install the 6 front cover retaining bolts and stud bolts where the silicone sealer was applied.
31. Tighten the bolts and stud bolts until the front cover contacts the cylinder block and heads, then turn the bolts and stud bolts an additional 1/4 turn.
32. Install the remaining front cover retaining bolts and stud bolts.
33. Tighten all of the front cover retaining bolts and stud bolts in sequence to 15–22 ft. lbs. (20–30 Nm).
34. Install or connect the following:
    - A/C hose bracket
    - A/C compressor
    - Oil pan
    - CKP sensor connector
    - Crankshaft pulley
    - Alternator
    - Right splash shield
    - Power steering pump and bracket
    - Accessory drive belt
    - Engine wiring harness connectors
    - Front engine support insulator and bracket
    - Coolant recovery reservoir
    - Low coolant level sensor connector
    - Valve covers
    - Upper intake manifold. Use new

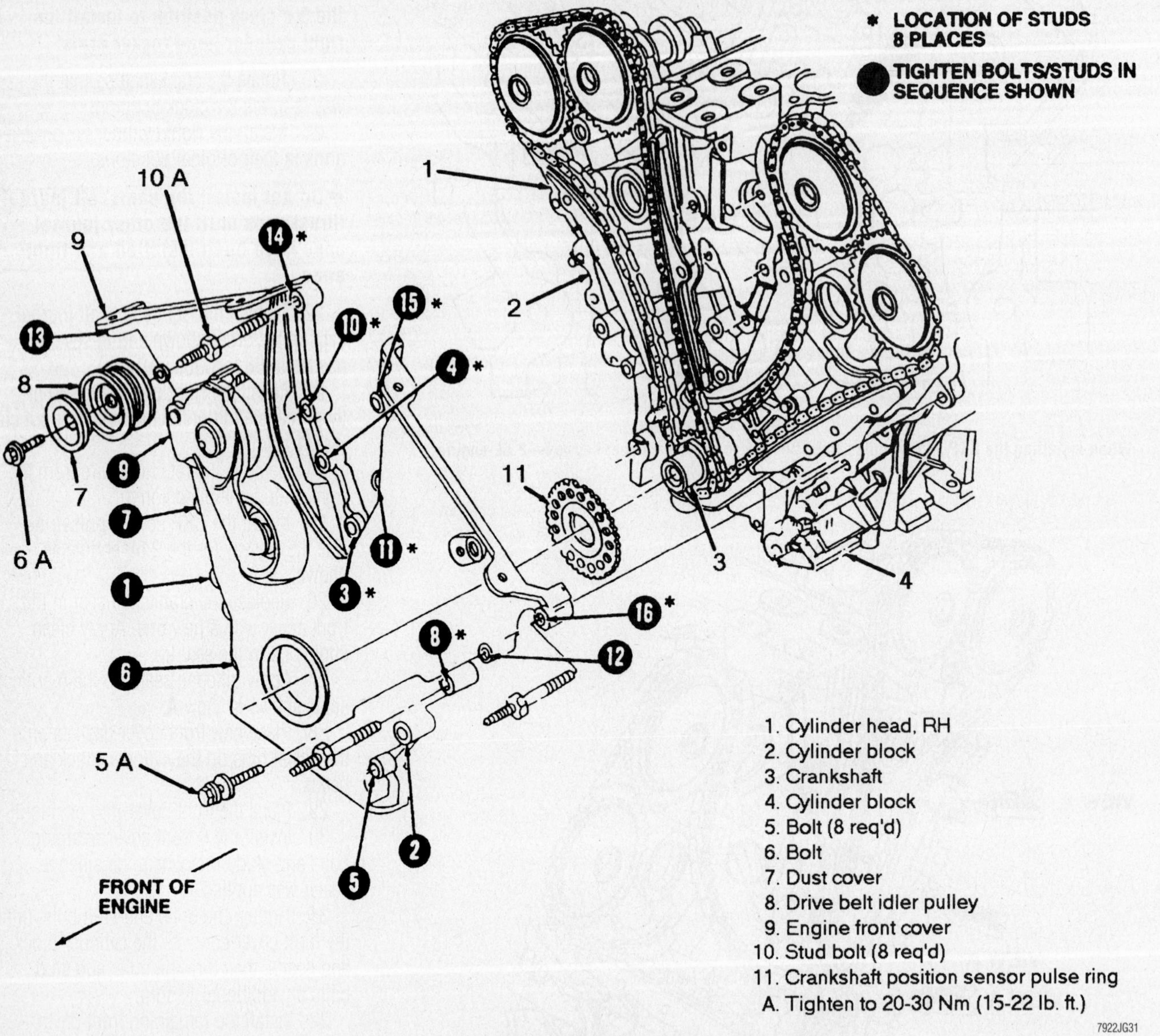

Exploded view of the front cover mounting, showing the retaining bolt and nut tightening sequence—2.5L engine

gaskets and tighten the bolts to 71–106 inch lbs. (8–12 Nm).

- Negative battery cable

35. Fill the engine with the proper amount and grade of oil.

36. Run the engine and check for leaks and proper operation.

Timing Belt

REMOVAL & INSTALLATION

2.0L Engine

1. Remove or disconnect the following:

- Negative battery cable
- Spark plugs
- Catalytic converter
- Right front wheel
- Right inner splash shield
- Accessory drive belt and idler pulley

2. Rotate the crankshaft to Top Dead Center (TDC) and install Crankshaft TDC Timing Peg T97P-6000-A.

3. Rotate the crankshaft clockwise against the peg.

4. Remove or disconnect the following:

- Water pump pulley
- Valve cover
- Crankshaft pulley
- Timing belt covers

5. Install Camshaft Alignment Timing Tool T94P-6256-CH into the slots of both camshafts at the rear of the cylinder head to lock the camshafts into position.

6. Loosen the timing belt tensioner bolt and relieve the tension on the timing belt by disconnecting the tensioner tab from the timing cover back plate.

7. Remove the timing belt.

To install:

➡Always replace the timing belt after loosening or removing it.

8. Install the timing belt.

9. Engage the timing belt tensioner tab into the timing cover back plate.

10. Adjust the timing belt tensioner

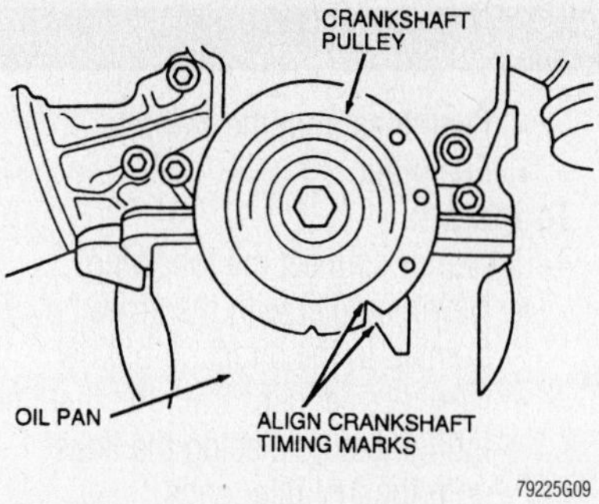

Crankshaft alignment position—2.0L engines

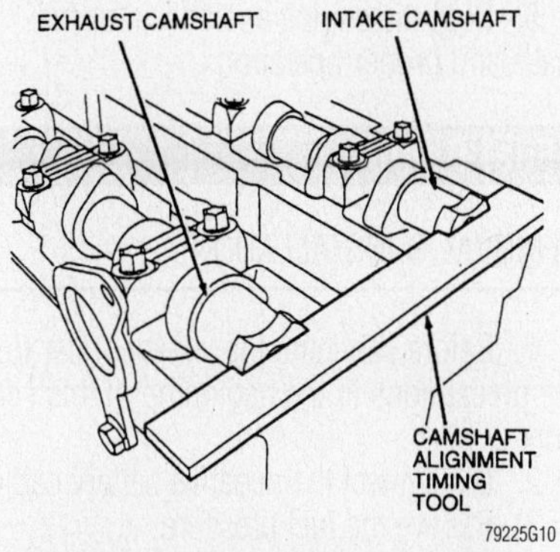

Placement of camshaft alignment timing tool T94P-6256-CH—2.0L engines

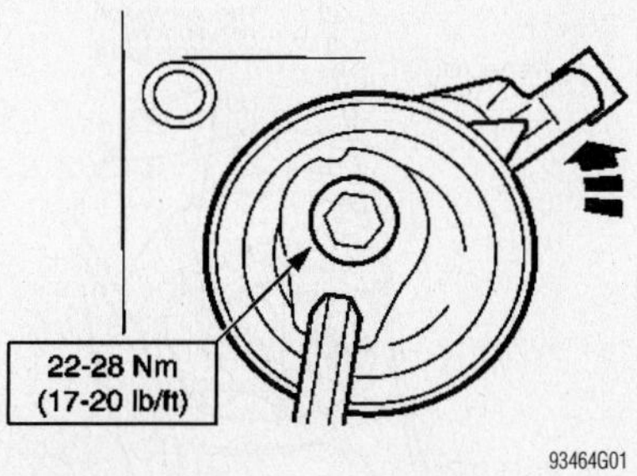

Timing belt tensioner index marks— 2.0L engines

until the index marks are aligned. Tighten the tensioner bolt to 17–20 ft. lbs. (22–28 Nm).

11. Remove the camshaft alignment tool and the crankshaft TDC peg.

12. Install or connect the following:

- Timing belt covers
- Crankshaft pulley. Tighten the bolt to 81–89 ft. lbs. (110–120 Nm).
- Valve cover
- Water pump pulley
- Accessory drive belt and idler pulley
- Right inner splash shield
- Right front wheel
- Catalytic converter
- Spark plugs
- Negative battery cable

13. Start the engine and check for proper operation.

Piston and Ring

POSITIONING

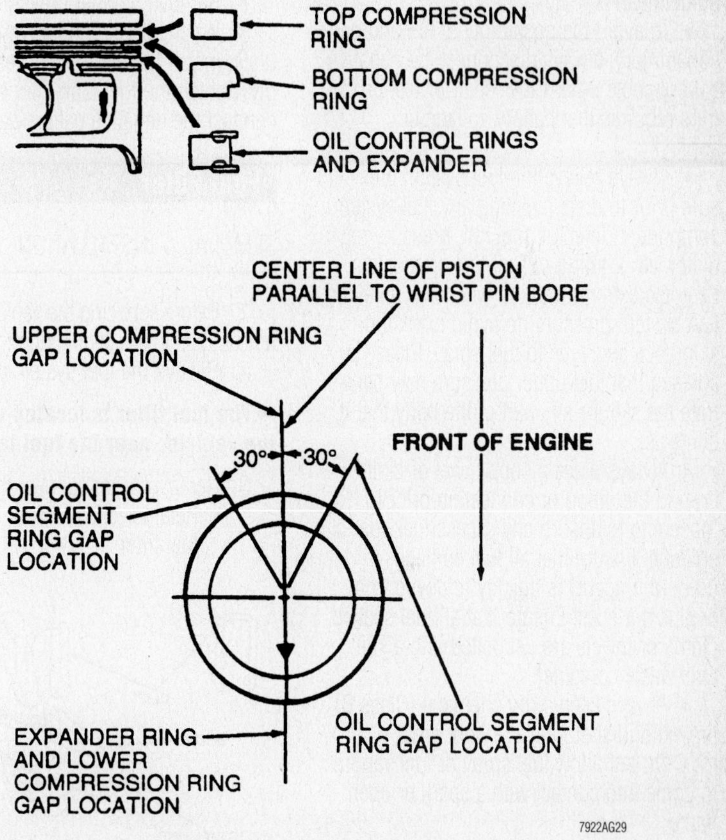

2.0L and 2.5L engines—piston ring positioning, end-gap spacing and piston positioning. The small directional arrow must face the front of the engine.

FUEL SYSTEM

Fuel System Service Precautions

Safety is the most important factor when performing not only fuel system maintenance but also any type of maintenance. Failure to conduct maintenance and repairs in a safe manner may result in serious personal injury or death. Maintenance and testing of the vehicle's fuel system components can be accomplished safely and effectively by adhering to the following rules and guidelines.

- To avoid the possibility of fire and personal injury, always disconnect the negative battery cable unless the repair or test procedure requires that battery voltage be applied.
- Always relieve the fuel system pressure prior to disconnecting any fuel system component (injector, fuel rail, pressure regulator, etc.), fitting or fuel line connection. Exercise extreme caution whenever relieving fuel system pressure, to avoid exposing skin, face and eyes to fuel spray. Please be advised that fuel under pressure may penetrate the skin or any part of the body that it contacts.
- Always place a shop towel or cloth around the fitting or connection prior to loosening to absorb any excess fuel due to spillage. Ensure that all fuel spillage (should it occur) is quickly removed from engine surfaces. Ensure that all fuel soaked cloths or towels are deposited into a suitable waste container.
- Always keep a dry chemical (Class B) fire extinguisher near the work area.
- Do not allow fuel spray or fuel vapors to come into contact with a spark or open flame.
- Always use a back-up wrench when loosening and tightening fuel line connection fittings. This will prevent unnecessary stress and torsion to fuel line piping.
- Always replace worn fuel fitting O-rings with new. Do not substitute fuel hose or equivalent, where fuel pipe is installed.

Before servicing the vehicle, make sure to refer to the precautions in the beginning of this section as well.

Fuel System Pressure

RELIEVING

1. Before servicing the vehicle, refer to the precautions in the beginning of this section.
2. Disconnect the negative battery cable.
3. Remove the engine air cleaner assembly.
4. Loosen the fuel tank filler cap to relieve pressure in the fuel tank.
5. Connect a fuel pressure gauge to the fuel pressure relief valve located on the fuel rail.
6. Open the manual valve on the fuel pressure gauge and drain the fuel through the drain tube into a suitable container.
7. Remove the fuel pressure gauge.
8. When service on the vehicle is complete, install the engine air cleaner assembly, tighten the fuel tank filler cap and connect the negative battery cable.

Fuel Filter

REMOVAL & INSTALLATION

1. Before servicing the vehicle, refer to the precautions in the beginning of this section.
2. Relieve the fuel system pressure.

➡The fuel filter is located underneath the vehicle, near the fuel tank.

3. Remove or disconnect the following:
   - Retainer clips at both ends of the fuel filter
   - Fuel lines from the fuel filter
   - Fuel filter

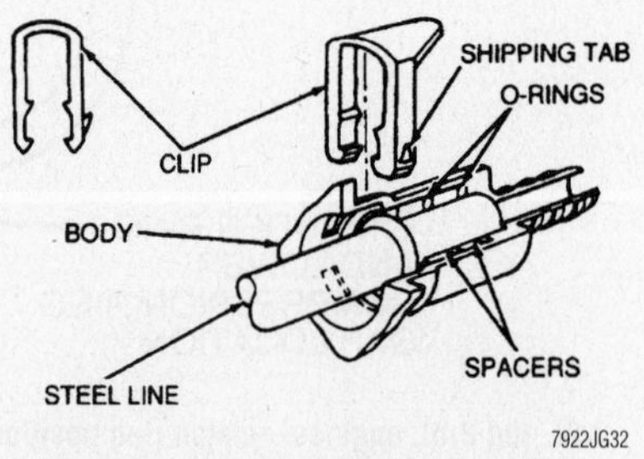

Once the retainer clip has been removed from the connector, the fuel line can be removed from the filter

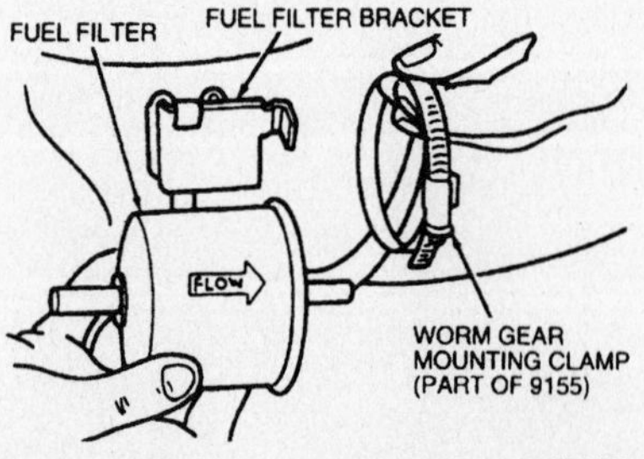

Be sure to install the fuel filter with the arrow pointing in the direction of the fuel flow

To install:

4. Install or connect the following:
   - New fuel filter with the arrow pointed in the direction of flow
   - New retainer clips onto the fuel line fittings before placing the lines onto the fuel filter ends
   - Fuel lines onto the fuel filter until an audible click is heard. Pull on the fitting to verify a good connection.
5. Start the engine and check for fuel leaks and proper operation.

Fuel Pump

REMOVAL & INSTALLATION

1. Before servicing the vehicle, refer to the precautions in the beginning of this section.
2. Disconnect the negative battery cable.
3. Relieve the fuel pressure.
4. Remove or disconnect the following:
   - Rear seat cushion
   - Plastic access panel

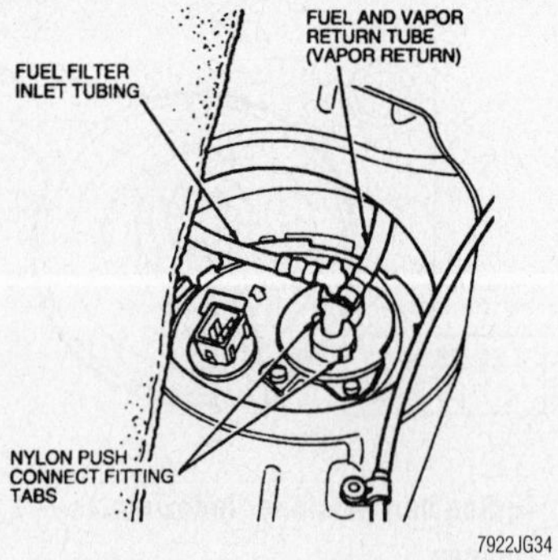

View of fuel pump fittings through the floor pan

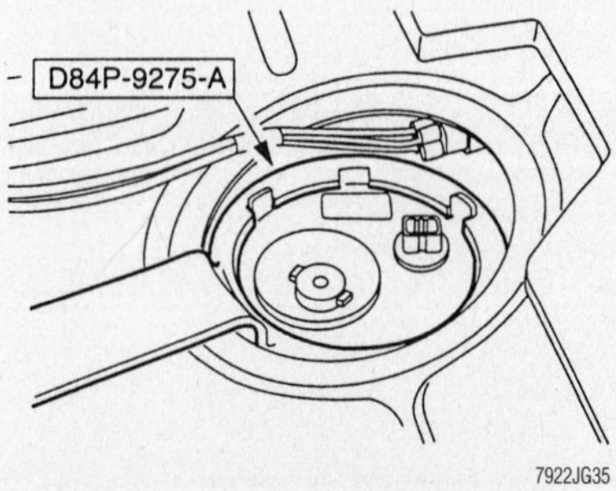

Remove the locking ring from the fuel pump sender with special tool Fuel Tank Sender Wrench D84P-9275-A

- Fuel pump wiring harness connector
- Fuel lines
- Fuel pump module lock ring. Use special tool Fuel Tank Sender Wrench D84P-9275-A.
- Fuel pump module

To install:

5. Install or connect the following:
   - Fuel pump module with a new O-ring seal
   - Fuel pump module lock ring using special tool Fuel Tank Sender Wrench D84P-9275-A
   - Fuel lines
   - Fuel pump wiring harness connector
   - Plastic access panel
   - Rear seat cushion
   - Negative battery cable
6. Start the engine and check for leaks and proper operation.

Fuel Injector

REMOVAL & INSTALLATION

2.0L Engine

1. Before servicing the vehicle, refer to the precautions in the beginning of this section.
2. Relieve the fuel system pressure.
3. Remove or disconnect the following:
   - Negative battery cable
   - Fuel lines and retaining clip
   - Pressure regulator vacuum hose
   - Accelerator cable
   - Fuel injector electrical connectors
   - Fuel supply manifold with the injectors attached
   - Injectors from the supply manifold

To install:

4. Install or connect the following:
   - Fuel injectors using new O-ring seals
   - Fuel supply manifold with the injectors attached and tighten the bolts to 88 inch lbs. (10 Nm)
   - Fuel injector electrical connectors
   - Accelerator cable
   - Pressure regulator vacuum hose
   - Fuel lines and retaining clip
   - Negative battery cable

2.5L Engine

EXCEPT RETURNLESS FUEL DELIVERY SYSTEM

1. Before servicing the vehicle, refer to the precautions in the beginning of this section.
2. Relieve fuel system pressure.
3. Remove or disconnect the following:
   - Negative battery cable
   - Air cleaner outlet tube
   - Fuel lines
   - Upper intake manifold
   - Fuel injector electrical connectors
   - Fuel pressure regulator vacuum line
   - Intake Manifold Runner Control (IMRC) rod
   - Fuel supply manifold with the injectors attached
   - Injectors from the supply manifold

To install:

4. Install or connect the following:
   - Fuel injectors using new O-ring seals
   - Fuel supply manifold with the injectors attached and tighten the bolts to 88 inch lbs. (10 Nm)
   - IMRC rod
   - Fuel pressure regulator vacuum line
   - Fuel injector electrical connectors
   - Upper intake manifold. Use new gaskets and tighten the bolts to 71–106 inch lbs. (8–12 Nm).
   - Fuel lines
   - Air cleaner outlet tube
   - Negative battery cable
5. Start the engine and check for leaks.

RETURNLESS FUEL DELIVERY SYSTEM

1. Before servicing the vehicle, refer to the precautions in the beginning of this section.
2. Relieve fuel system pressure.
3. Remove or disconnect the following:
   - Negative battery cable
   - Air cleaner outlet tube
   - Fuel line
   - Upper intake manifold
   - Fuel injector electrical connectors
   - Fuel pressure sensor electrical connector and vacuum line
   - Intake Manifold Runner Control (IMRC) rod
   - Fuel supply manifold with the injectors attached
   - Injectors from the supply manifold

To install:

4. Install or connect the following:
   - Fuel injectors using new O-ring seals
   - Fuel supply manifold with the injectors attached and tighten the bolts to 88 inch lbs. (10 Nm)
   - IMRC rod
   - Fuel pressure sensor electrical connector and vacuum line
   - Fuel injector electrical connectors
   - Upper intake manifold. Use new gaskets and tighten the bolts to 71–106 inch lbs. (8–12 Nm).
   - Fuel line
   - Air cleaner outlet tube
   - Negative battery cable
5. Start the engine and check for leaks.

DRIVE TRAIN

Transaxle Assembly

REMOVAL & INSTALLATION

Automatic

1. Before servicing the vehicle, refer to the precautions in the beginning of this section.
2. Secure the radiator and fan shroud to the radiator support.
3. Loosen the left and right upper strut mounting nuts 5 turns.
4. Attach an engine support fixture to the engine lifting eyes.
5. Drain the transaxle fluid.
6. Remove or disconnect the following:
   - Battery
   - Air cleaner assembly and mounting bracket
   - Shift cable and bracket
   - Transaxle control wiring harness connector
   - Transaxle range sensor
   - Rear transaxle support insulator
   - Oil dipstick tube and exhaust manifold heat shield (2.0L engine)
   - Water pump pulley shield (2.5L engine)

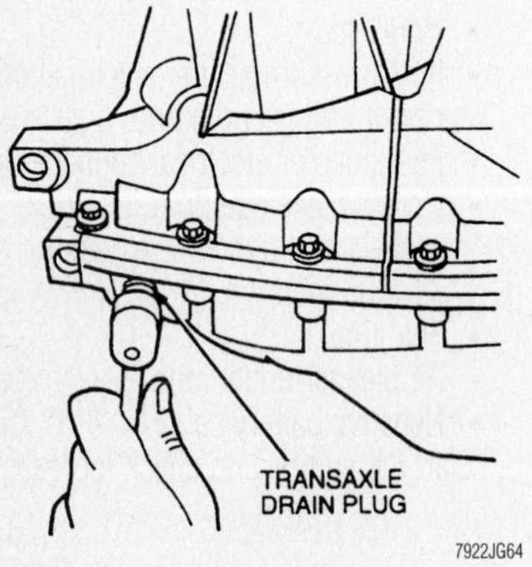

Transaxle drain plug location

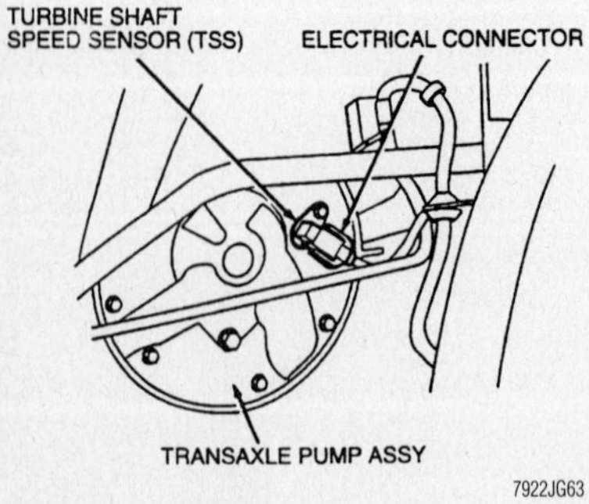

TSS sensor location

   - Steering column pinch bolt
   - Front wheels
   - Left, right and center splash shields
   - Lower ball joints
   - Outer tie rod ends
   - Stabilizer bar links
   - Left and right engine support insulators
   - Power steering oil cooler hoses
   - Bumper cover braces
   - Radiator air deflector
   - Lower radiator supports
   - Exhaust system
   - A/C accumulator mounting bolts
   - Transaxle cooler lines and bracket
   - Power steering hoses. Lower the sub-frame for access.
   - Sub-frame
   - Turbine Speed (TSS) sensor connector
   - Halfshafts and intermediate shaft
   - Vehicle Speed (VSS) sensor connector
   - Torque converter nuts
   - Starter
   - Front engine support insulator bracket
7. Lower the powertrain until the transaxle is level with the left frame member.
8. Support the transaxle on a transmission jack.
9. Remove or disconnect the following:
   - Transaxle flange bolts
   - Transaxle from the engine
   - Transaxle

➡Use care when removing the transaxle to prevent the torque converter from falling out.

To install:

➡Replace all snaprings, split pins and self-locking nuts.

10. Install or connect the following:
    - Transaxle and tighten the flange bolts to 41–50 ft. lbs. (55–68 Nm)
    - Torque converter and tighten the nuts to 23–39 ft. lbs. (31–53 Nm)
    - TSS connector
    - Sub-frame. Attach the power steering hoses before installing the sub-frame bolts.
    - Special tool Sub-frame Alignment Pin Set T95P-2100-AH and the sub-frame bolts. Tighten the sub-frame bolts to 81–110 ft. lbs. (110–150 Nm). Remove the alignment pins.
    - A/C accumulator bolts

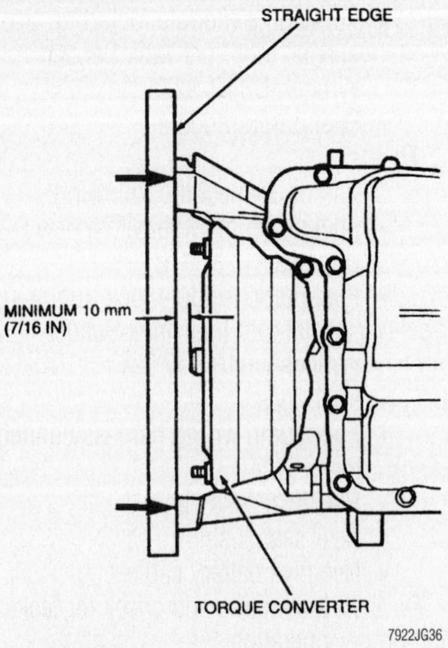

Be sure that the torque converter is properly seated in the transaxle

    - Power steering oil cooler lines
    - Special tool Powertrain Alignment Gauge T94P-6000-AH in place of the left engine support insulator. Tighten the 2 retaining bolts to 20 ft. lbs. (27 Nm) and snug the through-bolt.

➡The left and right support insulators must be aligned in the middle of the support insulator brackets.

    - Right engine support insulator. Tighten the 2 sub-frame retaining bolts to 30–41 ft. lbs. (41–55 Nm) and the through-bolt to 75–102 ft. lbs. (103–137 Nm).
    - Rear engine support insulator and tighten the nuts to 61 ft. lbs. (83 Nm)
    - Front engine support insulator and tighten the nuts to 61 ft. lbs. (83 Nm)
11. Remove the Powertrain Alignment Tool and install the left engine support insulator. Tighten the center bolt to 88 ft. lbs. (120 Nm), and the mounting bolts to 35 ft. lbs. (48 Nm).
    - Transaxle oil cooler lines
    - Starter and tighten the bolts to 43–58 ft. lbs. (59–79 Nm) for the 2.5L engine or 15–20 ft. lbs. (20–27 Nm) for the 2.0L engine
    - VSS sensor connector
    - Halfshafts and intermediate shaft
    - Exhaust system
    - Lower radiator supports
    - Radiator air deflector

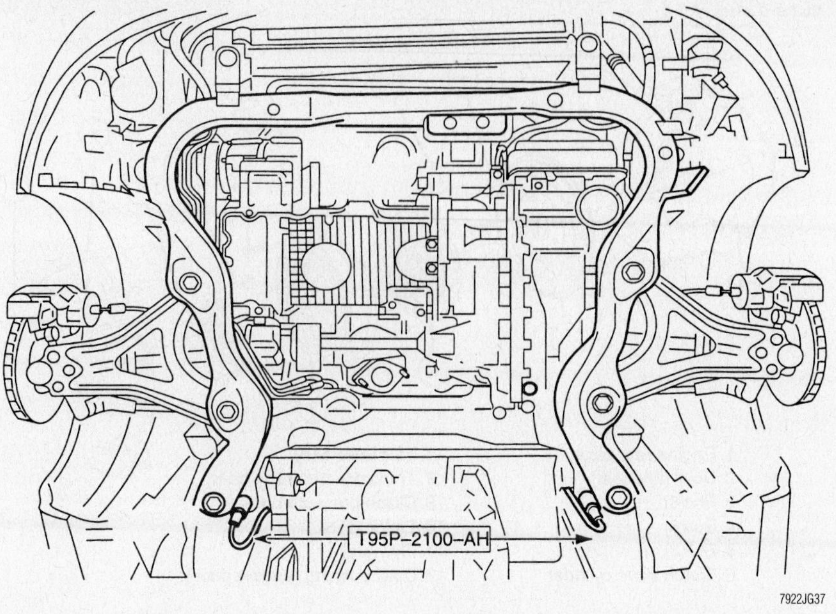

To properly align the sub-frame, install Sub-frame Alignment Pin Set T95P-2100-AH—manual and automatic transaxles

- Bumper cover braces
- Power steering oil cooler hoses
- Left and right engine support insulators
- Stabilizer bar links and tighten the nuts to 35–48 ft. lbs. (47–65 Nm)
- Outer tie rod ends and tighten the nuts to 23–35 ft. lbs. (31–47 Nm)
- Lower ball joints and tighten the pinch bolt to 61 ft. lbs. (83 Nm)
- Left, right and center splash shields
- Front wheels and tighten the lug nuts to 63 ft. lbs. (86 Nm)
- Steering column pinch bolt and tighten the bolt to 15–20 ft. lbs. (20–27 Nm)
- Water pump pulley shield (2.5L engine)
- Oil dipstick tube and exhaust manifold heat shield (2.0L engine)
- Transaxle range sensor
- Transaxle control wiring harness connector
- Shift cable and bracket
- Air cleaner assembly and mounting bracket
- Battery

12. Tighten the upper strut mount nuts to 34 ft. lbs. (46 Nm).
13. Fill the transaxle.
14. Check for leaks and proper operation.

➡Whenever the vehicle sub-frame is removed or lowered, the wheel alignment should be checked.

Manual

1. Before servicing the vehicle, refer to the precautions in the beginning of this section.
2. Attach an engine support fixture to the engine lifting eyes.
3. Secure the radiator and fan shroud to the radiator support.
4. Loosen the front strut upper mount nuts 5 turns.
5. Remove or disconnect the following:
   - Battery
   - Steering column pinch bolt
   - Air cleaner assembly and bracket
   - Front and rear engine support insulators
   - Reverse lamp switch connector
   - Ground strap
   - Left, right and center splash shields
   - Clutch slave cylinder hydraulic line
   - Accessory drive belt cover
   - Power steering pulley shield (2.5L engine)
   - Exhaust crossover pipe (2.5L engine)
   - Catalytic converter
   - Front wheels
   - Vehicle Speed Sensor (VSS)
   - Radiator air deflector
   - Shift rod and stabilizer bar
   - A/C accumulator mounting bolts.
   - Halfshafts and intermediate shaft
   - Left and right engine support insulators
   - Lower ball joints
   - Outer tie rod ends
   - Stabilizer bar links
   - Power steering oil cooler lines
   - Lower radiator supports
   - Bumper cover braces
   - Power steering hoses. Lower the sub-frame for access.
   - Sub-frame
   - Starter
6. Lower the powertrain until the transaxle is level with the left frame member.
7. Support the transaxle on a transmission jack.
8. Remove or disconnect the following:
   - Transaxle flange bolts
   - Transaxle from the engine

To install:

➡Replace all snaprings, split pins and self-locking nuts.

9. Install or connect the following:
   - Transaxle to the engine and tighten the flange bolts to 28–38 ft. lbs. (38–51 Nm)
   - Starter and tighten the bolts to 35 ft. lbs. (48 Nm)
   - Sub-frame. Attach the power steering hoses before installing the sub-frame bolts.
   - Special tool Sub-frame Alignment Pin Set T95P-2100-AH and the sub-frame bolts. Tighten the sub-frame bolts to 81–110 ft. lbs. (110–150 Nm). Remove the alignment pins.
   - A/C accumulator bolts
   - Power steering oil cooler lines
   - Special tool Powertrain Alignment Gauge T94P-6000-AH in place of the left engine support insulator. Tighten the 2 retaining bolts to 20 ft. lbs. (27 Nm) and snug the through-bolt.

➡The left and right support insulators must be aligned in the middle of the support insulator brackets.

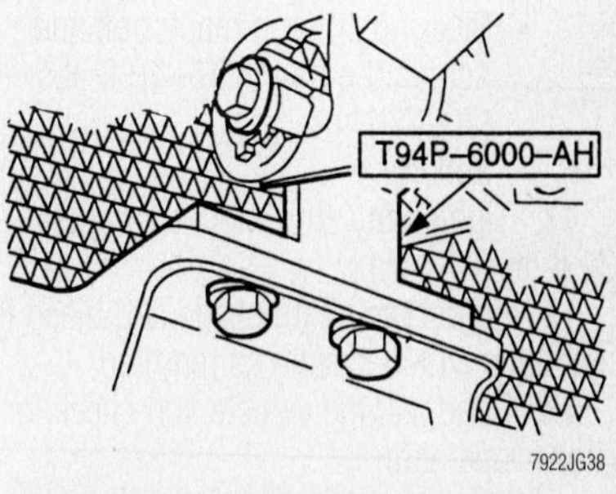

The Powertrain Alignment Gauge (T94P-6000-AH) tool must be installed in the correct position to ensure proper engine/transaxle orientation

- Right engine support insulator. Tighten the 2 sub-frame retaining bolts to 30–41 ft. lbs. (41–55 Nm) and the through-bolt to 75–102 ft. lbs. (103–137 Nm).
- Rear engine support insulator and tighten the nuts to 61 ft. lbs. (83 Nm)
- Front engine support insulator and tighten the nuts to 61 ft. lbs. (83 Nm)

10. Remove the Powertrain Alignment Tool and install the left engine support insulator. Tighten the center bolt to 88 ft. lbs. (120 Nm), and the mounting bolts to 35 ft. lbs. (48 Nm).

- Bumper cover braces
- Lower radiator supports and tighten the bolts to 71–97 inch lbs. (8–11 Nm)
- Power steering oil cooler lines
- Stabilizer bar links and tighten the fasteners to 35–48 ft. lbs. (47–65 Nm)
- Outer tie rod ends and tighten the nuts to 23–35 ft. lbs. (31–47 Nm)
- Lower ball joints and tighten the pinch bolt to 61 ft. lbs. (83 Nm)
- Halfshafts and intermediate shaft
- A/C accumulator mounting bolts
- Shift rod and stabilizer bar. Tighten the shift rod bolt to 14–18 ft. lbs. (19–25 Nm) and the stabilizer bar nut to 28–38 ft. lbs. (38–51 Nm).
- Radiator air deflector
- VSS sensor
- Front wheels and tighten the lug nuts to 63 ft. lbs. (86 Nm)
- Catalytic converter
- Exhaust crossover pipe (2.5L engine)
- Power steering pulley shield (2.5L engine)
- Accessory drive belt cover
- Clutch slave cylinder hydraulic line
- Left, right and center splash shields
- Ground strap
- Reverse lamp switch connector
- Air cleaner assembly and bracket
- Steering column pinch bolt and tighten the bolt to 15–20 ft. lbs. (20–27 Nm)
- Battery

11. Tighten the strut mounting nuts to 34 ft. lbs. (46 Nm).
12. Adjust the shift linkage and bleed the hydraulic clutch system as required.
13. Road test the vehicle and check for proper operation.

➡Whenever the vehicle sub-frame is removed or lowered, the wheel alignment should be checked.

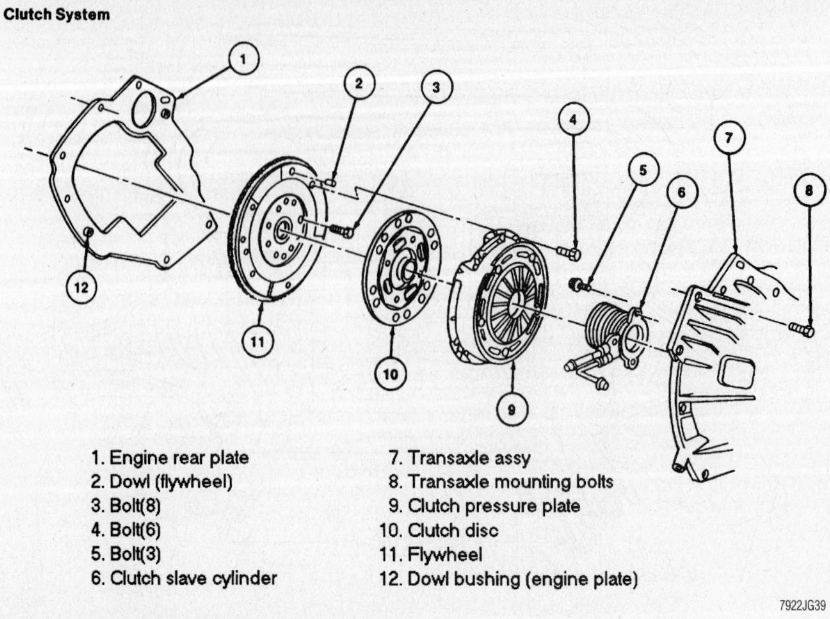

Exploded view of the clutch disc, pressure plate and related component mounting

Clutch

ADJUSTMENTS

These vehicles are equipped with a hydraulic clutch system. No adjustment is necessary.

REMOVAL & INSTALLATION

1. Before servicing the vehicle, refer to the precautions in the beginning of this section.
2. Remove or disconnect the following:
   - Transaxle
   - Clutch pressure plate
   - Clutch disk
   - Flywheel

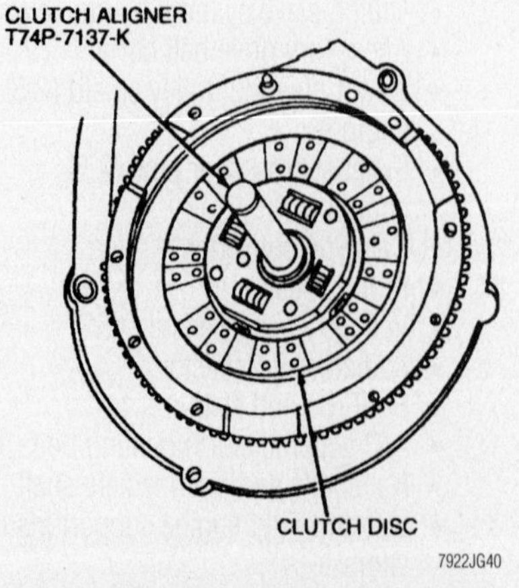

Insert an alignment tool through the clutch disc to ensure that it is centered after the pressure plate is installed

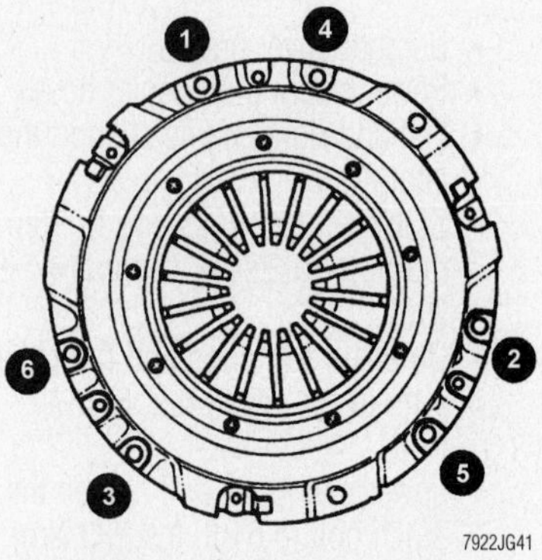

Pressure plate tightening sequence

To install:

3. Install or connect the following:
   - Flywheel and tighten the bolts to 82 ft. lbs. (112 Nm) for 2.0L engines or to 59 ft. lbs. (80 Nm) for 2.5L engines
   - Clutch disk and pressure plate and tighten the pressure plate bolts evenly in several passes to 21 ft. lbs. (29 Nm)
   - Transaxle
4. Bleed the hydraulic clutch system, if required.
5. Check the clutch system for proper operation.

Hydraulic Clutch System

BLEEDING

1. Before servicing the vehicle, refer to the precautions in the beginning of this section.

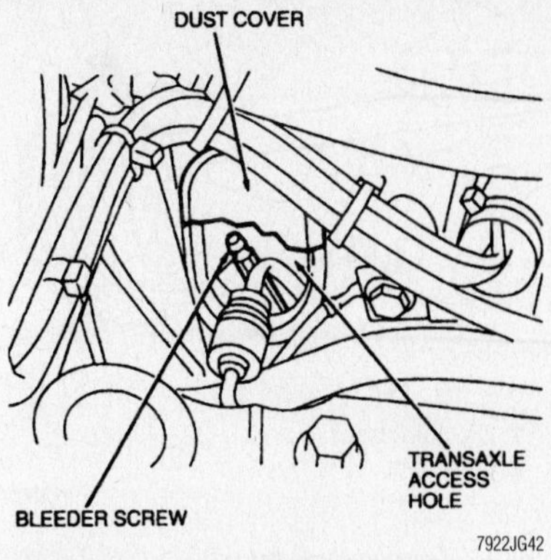

Clutch slave cylinder bleeder valve

2. Remove the negative battery cable, air cleaner outlet tube and the Mass Airflow (MAF) sensor.

➡The brake master cylinder fluid reservoir is also the reservoir for the hydraulic clutch master cylinder.

3. Remove the rubber inspection cover from the bell housing.
4. Connect a hose to the bleeder valve fitting on the clutch slave cylinder. Submerge the other end of the hose into a container of clean brake fluid.
5. Open the bleeder valve and have an assistant depress the clutch pedal.
6. Close the bleeder before releasing the clutch pedal.
7. Repeat the procedure until no more air bubbles are seen.
8. Install the rubber inspection cover to the bell housing.
9. Top off the brake master cylinder fluid reservoir and install the diaphragm and cap securely.
10. Install the MAF sensor and air cleaner outlet tube.
11. Connect the negative battery cable.
12. Check the clutch for proper operation.

Halfshaft

REMOVAL & INSTALLATION

Left Axle Halfshaft

➡Replace all snaprings, split pins and self-locking nuts.

1. Before servicing the vehicle, refer to the precautions in the beginning of this section.
2. Loosen the strut upper mount nut 5 turns.
3. Remove or disconnect the following:
   - Wheel

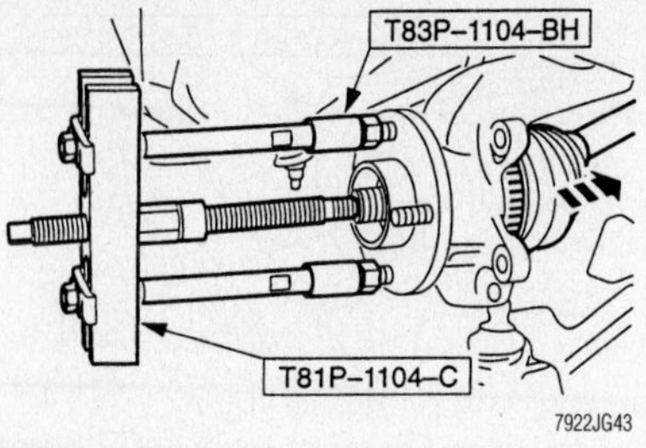

Front Hub Remover/Replacer T81P-1104-C

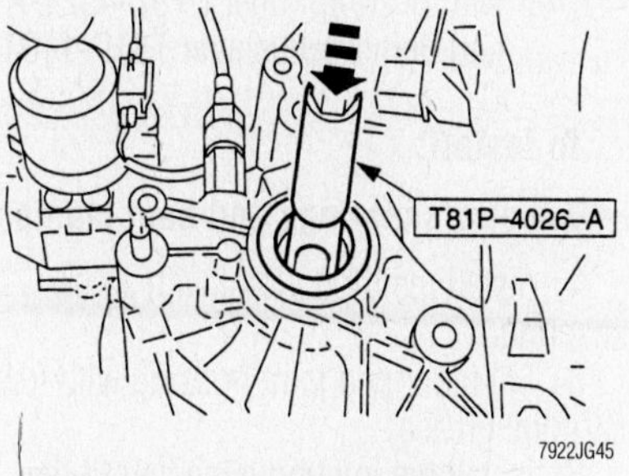

If the intermediate shaft has been removed, install Differential Rotator T81P-4026-A before removing the left halfshaft

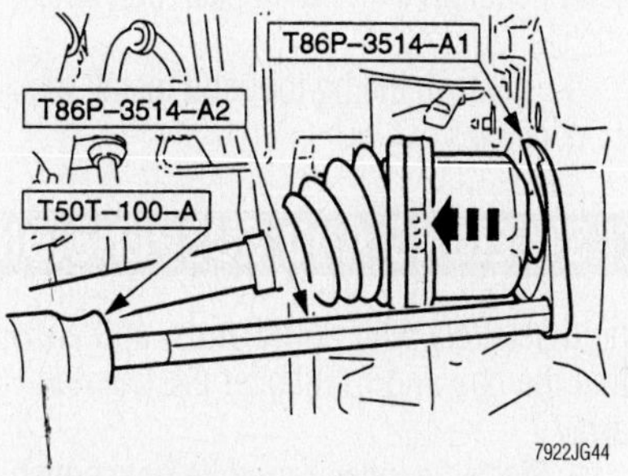

Use a slide hammer to pull the inner CV-joint from the transaxle

   - Splash shield
   - Lower ball joint
   - Outer tie rod end
   - Stabilizer bar link
4. Separate the outer CV-joint and halfshaft from the wheel hub.
5. Install CV-joint puller T86P-3514-A1 between the inner CV-joint and the transaxle case.

➡If the intermediate shaft has already been removed, install Differential Rotator T81P-4026-A into the right side of the differential before removing the left shaft to maintain alignment within the differential.

6. Remove the halfshaft.

To install:

7. Install the halfshaft. A non-metallic mallet may be used to aid in seating the inner CV-joint into the differential side gear. Insert the stub shaft into the wheel hub and tighten the hub nut to 246 ft. lbs. (340 Nm).
8. Install or connect the following:
   - Lower ball joint and tighten the pinch bolt to 61 ft. lbs. (83 Nm)
   - Outer tie rod end and tighten the nut to 20 ft. lbs. (28 Nm)
   - Stabilizer bar link and tighten the nut to 14–23 ft. lbs. (20–32 Nm)
   - Splash shield
   - Wheel and tighten the lug nuts to 63 ft. lbs. (86 Nm)
9. Tighten the strut upper mount nut to 34 ft. lbs. (46 Nm).
10. Road test the vehicle and check for proper operation.

Right Axle Halfshaft

➡Replace all snaprings, split pins and self-locking nuts.

1. Before servicing the vehicle, refer to the precautions in the beginning of this section.
2. Loosen the strut upper mount nut 5 turns.
3. Remove or disconnect the following:
   - Wheel
   - Splash shield
   - Lower ball joint
   - Outer tie rod end
   - Stabilizer bar link
4. Separate the outer CV-joint and halfshaft from the wheel hub.
5. Install CV-joint puller T86P-3514-A1 between the inner CV-joint and the intermediate shaft.

To install:

6. Install the halfshaft. A non-metallic mallet may be used to aid in seating the inner CV-joint into the differential side gear. Insert the stub shaft into the wheel hub and tighten the hub nut to 246 ft. lbs. (340 Nm).
7. Install or connect the following:
   - Lower ball joint and tighten the pinch bolt to 61 ft. lbs. (83 Nm)
   - Outer tie rod end and tighten the nut to 20 ft. lbs. (28 Nm)
   - Stabilizer bar link and tighten the nut to 14–23 ft. lbs. (20–32 Nm)
   - Splash shield
   - Wheel and tighten the lug nuts to 63 ft. lbs. (86 Nm)
8. Tighten the strut upper mount nut to 34 ft. lbs. (46 Nm).
9. Road test the vehicle and check for proper operation.

Intermediate Shaft

1. Before servicing the vehicle, refer to the precautions in the beginning of this section.

2. Remove the right halfshaft.

➡If the left halfshaft has already been removed, install Differential Rotator T81P-4026-A into the left side of the differential before removing the intermediate shaft to maintain alignment within the differential.

3. Remove the intermediate shaft and support bearing.

To install:

4. Install the intermediate shaft and tighten the support bearing nuts to 17–22 ft. lbs. (24–30 Nm).
5. Install the right halfshaft.

CV-Joints

REMOVAL AND REPLACEMENT

Inner Tripod Joint

1. Before servicing the vehicle, refer to the precautions in the beginning of this section.
2. Place the halfshaft in a vise.
3. Remove or disconnect the following:
   - Tripod joint boot
   - Tripod joint housing
   - Tripod assembly snapring
   - Tripod assembly from the halfshaft

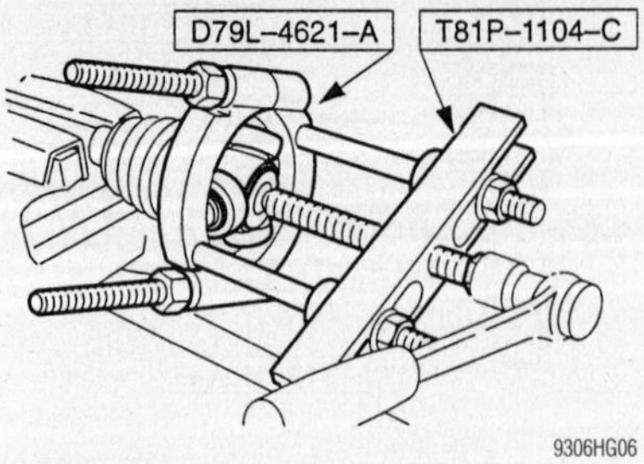

Removing the tripod assembly

with Bearing Puller D79L-4621-A and Remover/Installer T81P-1104-C

To install:

➡Use new snaprings and boot clamps.

4. Install the tripod joint with a new snapring.
5. Fill the tripod joint housing with fresh CV-joint grease.
6. Install the joint housing and fit the boot to the seat groove.
7. Insert a small prytool under the boot seat to permit air to escape.
8. Slide the tripod joint fully into the joint housing, then pull it back out ¾ inches (20mm).
9. Remove the prytool and install the boot clamps.

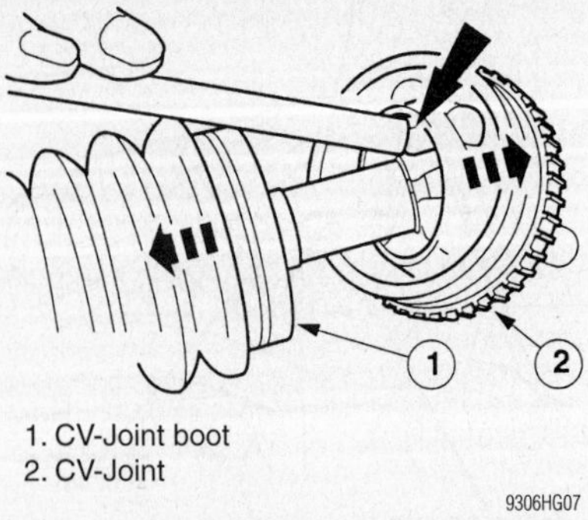

1. CV-Joint boot
2. CV-Joint

Removing the outer CV-joint

Outer CV-Joint

1. Before servicing the vehicle, refer to the precautions in the beginning of this section.
2. Place the halfshaft in a vise.
3. Remove the CV-joint boot clamps and slide the boot away from the joint.
4. Drive the CV-joint off the halfshaft with a brass drift and a hammer.

To install:

5. Replace the snapring.
6. Fill the CV-joint with fresh grease and slide the joint on to the halfshaft.
7. Fit the CV-joint boot to the seat grooves and install new boot clamps.

STEERING AND SUSPENSION

Air Bag

⁂ CAUTION

Some vehicles are equipped with an air bag system. The system must be disarmed before performing service on, or around, system components, the steering column, instrument panel components, wiring and sensors. Failure to follow the safety precautions and the disarming procedure could result in accidental air bag deployment, possible injury and unnecessary system repairs.

PRECAUTIONS

Several precautions must be observed when handling the inflator module to avoid accidental deployment and possible personal injury.

- Never carry the inflator module by the wires or connector on the underside of the module
- When carrying a live inflator module, hold securely with both hands, and ensure that the bag and trim cover are pointed away
- Place the inflator module on a bench or other surface with the bag and trim cover facing up
- With the inflator module on the bench, never place anything on or close to the module that may be thrown in the event of an accidental deployment

Before servicing the vehicle, also be sure to refer to the precautions in the beginning of this section as well

DISARMING

1. Before servicing the vehicle, refer to the precautions in the beginning of this section.
2. Position the vehicle with the front wheels in a straight-ahead position.
3. Disconnect the battery cables, negative cable first.
4. Wait at least 1 minute for the air bag back-up power supply to drain before continuing.
5. Proceed with the repair.

ARMING

1. After repairs are completed, connect the battery cables.
2. Check the functioning of the air bag system by turning the ignition key to the **RUN** position and visually monitoring the air bag indicator lamp in the instrument cluster. The indicator lamp should illuminate for approximately 6 seconds, then turn **OFF**. If the indicator lamp does not illuminate, stays on, or flashes at any time, a fault has been detected by the air bag diagnostic monitor.

Power Rack and Pinion Steering Gear

REMOVAL & INSTALLATION

1. Before servicing the vehicle, refer to the precautions in the beginning of this section.
2. Attach an engine support fixture to the engine lifting eyes.
3. Secure the radiator and fan shroud to the radiator support.

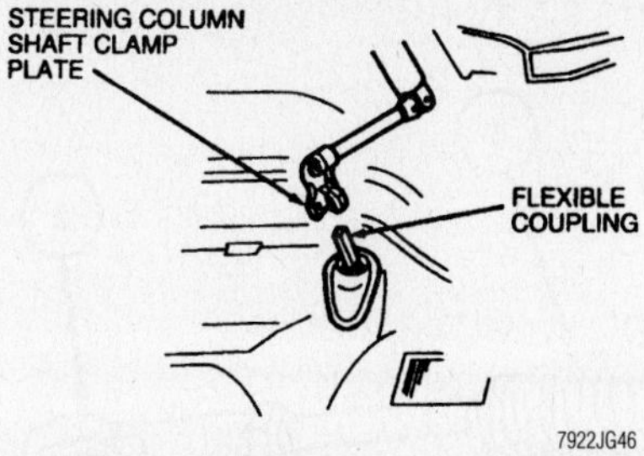

Disconnect the steering column shaft from the flexible coupling

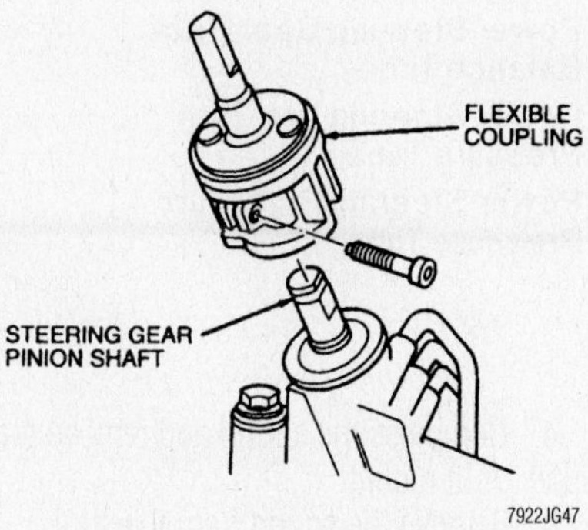

Disengage the flexible coupling from the steering gear pinion shaft

4. Remove or disconnect the following:
   - Negative battery cable
   - Steering gear flexible coupling
   - Left, right and center splash shields
   - Catalytic converter
   - Front wheels
   - Lower ball joints
   - Outer tie rod ends
   - Stabilizer bar links
   - Left and right engine support insulators
   - Power steering oil cooler lines
   - A/C accumulator bolts
   - Lower radiator supports
   - Bumper cover braces
   - Power steering hoses. Lower the sub-frame for access.
   - Sub-frame with steering gear attached
   - Steering gear cover plate
   - Steering gear

To install:

5. Install or connect the following:
   - Steering gear and tighten the bolts to 101 ft. lbs. (137 Nm)
   - Steering gear cover plate and tighten the bolts to 37 ft. lbs. (50 Nm)
   - Sub-frame. Use new seals and attach the power steering hoses before bolting the sub-frame in place.
   - Special tool Sub-frame Alignment Pin Set T95P-2100-AH and the sub-frame bolts. Tighten the sub-frame bolts to 81–110 ft. lbs. (110–150 Nm). Remove the alignment pins.
   - Special tool Powertrain Alignment Gauge T94P-6000-AH in place of the left engine support insulator. Tighten the 2 retaining bolts to 20 ft. lbs. (27 Nm) and snug the through-bolt.

➡The left and right support insulators must be aligned in the middle of the support insulator brackets.

   - Right engine support insulator. Tighten the 2 sub-frame retaining bolts to 30–41 ft. lbs. (41–55 Nm) and the through-bolt to 75–102 ft. lbs. (103–137 Nm).

6. Remove the Powertrain Alignment Tool and install the left engine support insulator and tighten the center bolt to 88 ft. lbs.

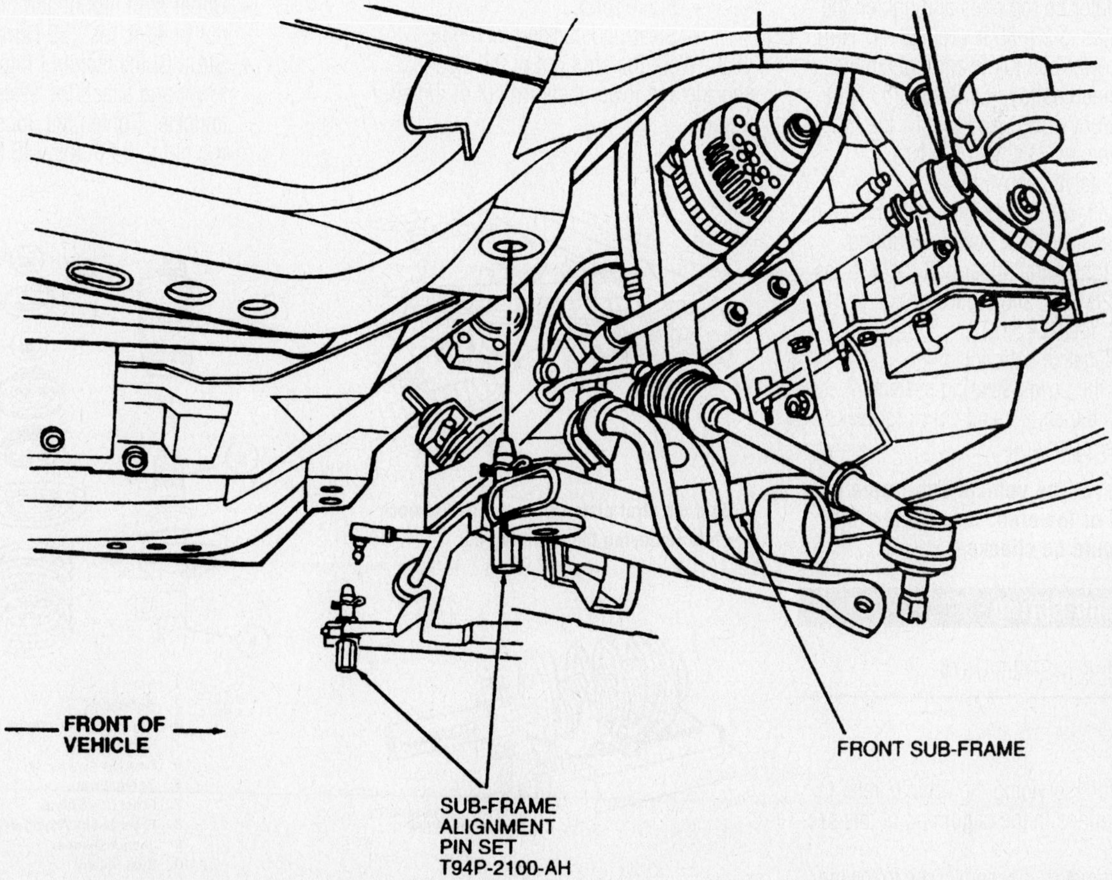

Install Sub-frame Alignment Pin Set T94P-2100-AH into the sub-frame to ensure correct positioning

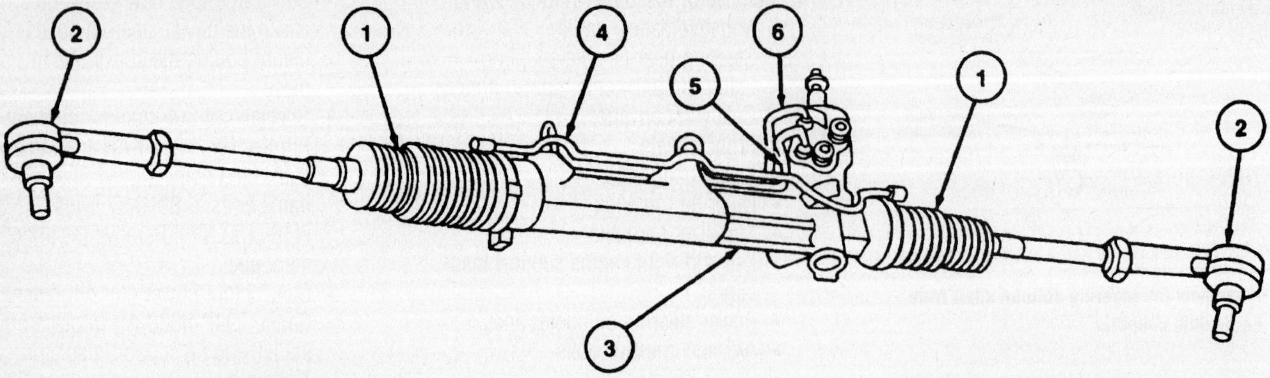

1 Front Suspension Steering Ball Dust Seal
2 Front Wheel Spindle Connecting End
3 Steering Gear
4 Power Steering Gear Rack Balance Tube
5 Power Steering Left Turn Pressure Tube
6 Power Steering Right Turn Pressure Tube

7922JG49

Power rack and pinion steering gear assembly component identification

(120 Nm), and the mounting bolts to 35 ft. lbs. (48 Nm).

- Bumper cover braces
- Lower radiator supports
- A/C accumulator bolts
- Power steering oil cooler lines
- Stabilizer bar links
- Outer tie rod ends and tighten the nuts to 23–35 ft. lbs. (31–47 Nm)
- Lower ball joints and tighten the pinch bolts to 61 ft. lbs. (83 Nm)
- Front wheels and tighten the lug nuts to 63 ft. lbs. (86 Nm)
- Catalytic converter
- Left, right and center splash shields
- Steering gear flexible coupling. Tighten the upper bolt to 21 ft. lbs. (28 Nm), and the lower bolt to 18 ft. lbs. (24 Nm).
- Negative battery cable

7. Fill the power steering system.
8. Run the engine and check for leaks and proper operation.

➡**Whenever the vehicle sub-frame is removed or lowered, the wheel alignment should be checked.**

Strut and Spring

REMOVAL & INSTALLATION

Front

1. Before servicing the vehicle, refer to the precautions in the beginning of this section.
2. Remove or disconnect the following:
   - Negative battery cable
   - Front wheels
   - Upper strut mount nut
   - Stabilizer bar link
   - Brake hose bracket
   - Lower ball joint
   - Wheel speed sensor
   - Brake caliper and caliper support
   - Brake rotor
   - Steering knuckle pinch bolt
3. Work the strut out of the steering knuckle and lower the strut out of the strut tower.

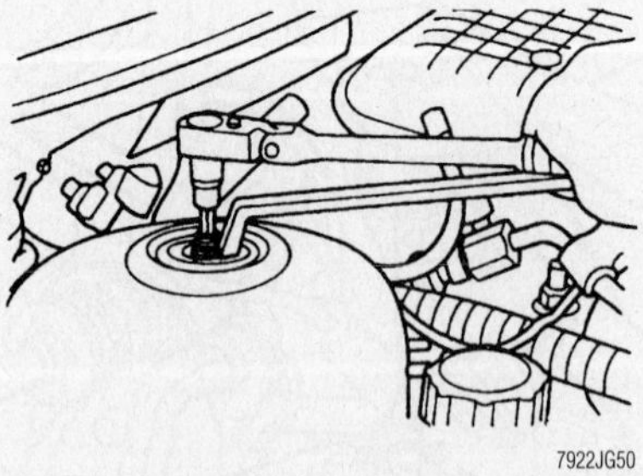

7922JG50

Hold the strut piston with an Allen wrench while removing the retaining nut

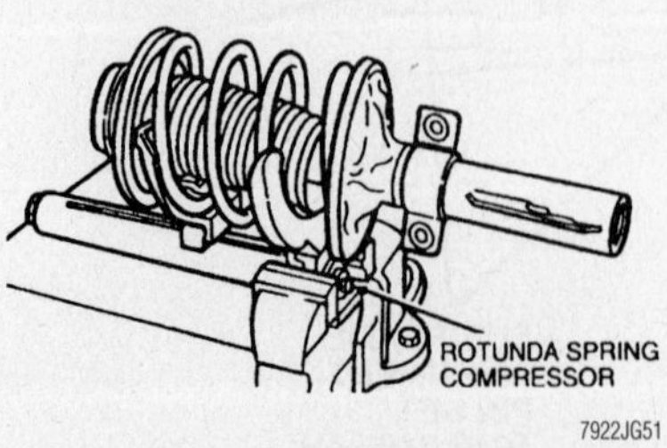

7922JG51

Compress the coil spring assembly before removing the retaining nut

4. Compress the spring and remove the upper strut mount.
5. Remove the spring from the strut.

To install:

6. Transfer parts as necessary and install the spring onto the strut.
7. Install or connect the following:
   - Upper strut mount and tighten the nut to 44 ft. lbs. (59 Nm)
   - Strut. Guide the strut into the strut tower and attach the steering knuckle. Tighten the upper mounting nut to 34 ft. lbs. (46 Nm), and

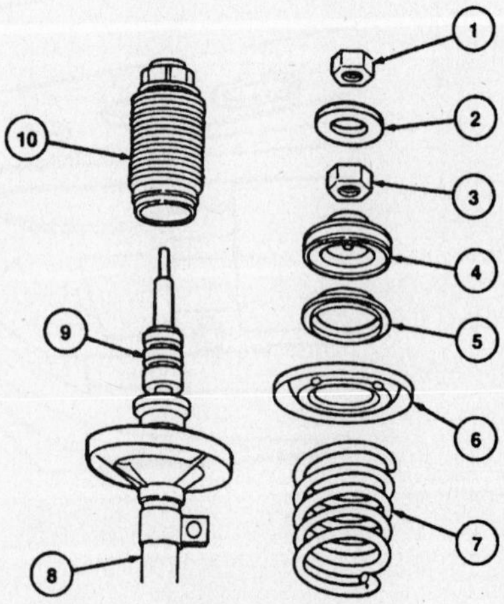

1 Nut
2 Retainer
3 Upper Mount Retainer Nut
4 Upper Mount
5 Bearing
6 Spring Seat
7 Front Coil Spring
8 Front Shock Absorber
9 Jounce Bumper
10 Dust Shield

7922JG52

Exploded view of the strut assembly

the knuckle pinch bolt to 40 ft. lbs. (54 Nm).

- Brake rotor
- Brake caliper support and tighten the bolts to 88 ft. lbs. (120 Nm)
- Wheel speed sensor
- Lower ball joint and tighten the pinch bolt to 61 ft. lbs. (83 Nm)
- Brake hose bracket
- Stabilizer bar link and tighten the nuts to 35–48 ft. lbs. (47–65 Nm)
- Front wheels and tighten the lug nuts to 63 ft. lbs. (86 Nm)
- Negative battery cable

Rear

1. Before servicing the vehicle, refer to the precautions in the beginning of this section.
2. Remove or disconnect the following:
    - Negative battery cable
    - Rear wheels
    - Wheel speed sensor
    - Rear brake hose
    - Stabilizer bar link
    - Tie rod

➡The front and rear control arms must be supported before the removal of the strut attachments.

- Spindle pinch bolt

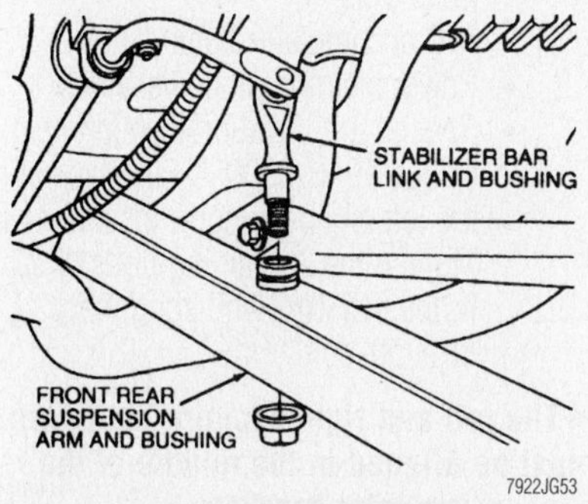

For rear strut removal, disconnect the sway bar link from the control arm . . .

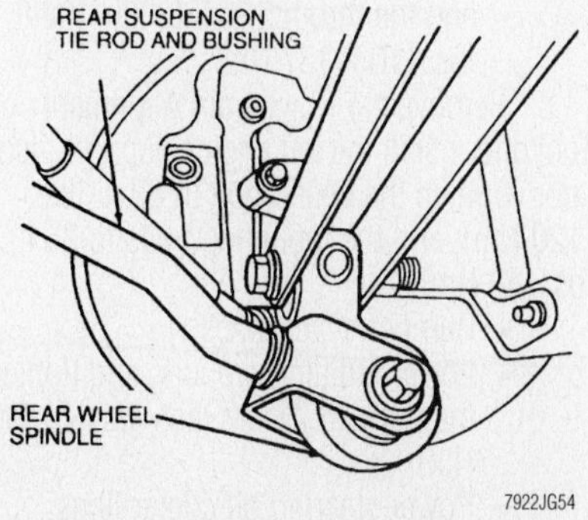

. . . and separate the tie rod from the rear wheel spindle

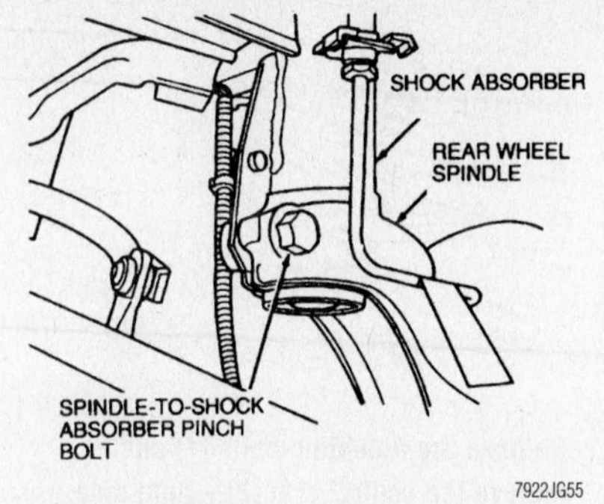

Remove the rear strut pinch bolt to release the strut assembly from the spindle

3. Separate the spindle from the strut by tapping down on the wheel spindle.
4. Compress the coil spring.
5. Remove the 2 top retaining bolts and remove the strut assembly.
6. Compress the coil spring enough to relieve the tension on the spring seat.
7. Remove the top mount nut, rear strut bracket, bushing and spring seat.

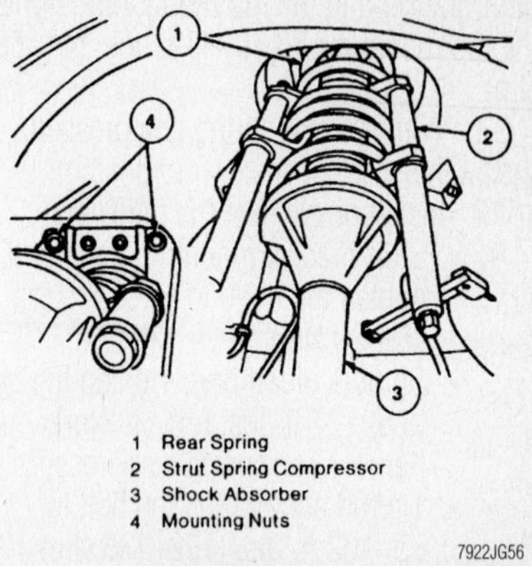

Compress the coil spring, then unthread the 2 top retaining bolts and remove the strut from the vehicle

8. Remove the coil spring from the strut.

To install:

9. Transfer parts as necessary and install the spring to the strut.
10. Install the spring seat, bushing, rear

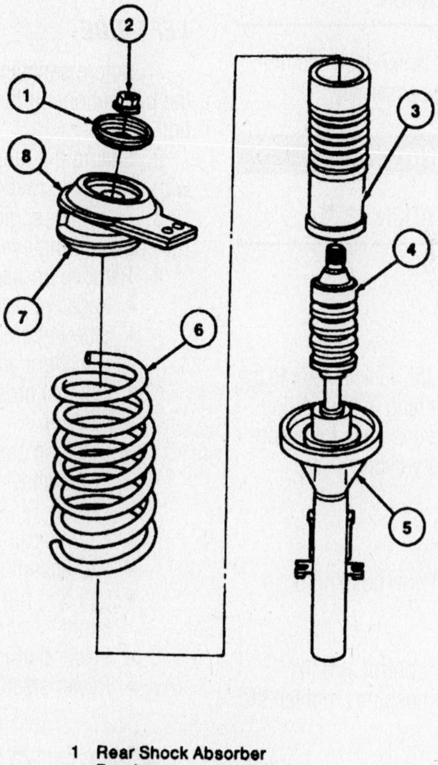

Exploded view of the rear strut assembly

strut bracket and the top mount nut. Tighten the top mount nut to 30–43 ft. lbs. (41–58 Nm).

11. With the coil spring compressed, install the strut assembly into position.

12. Install or connect the following:
- Strut bracket mounting bolts and tighten the bolts to 17–22 ft. lbs. (23–30 Nm)
- Spindle pinch bolt. Tighten the bolt to 52–72 ft. lbs. (70–98 Nm). Remove the spring compressor.
- Tie rod and tighten the bolt to 75–102 ft. lbs. (102–138 Nm)
- Stabilizer bar link
- Rear brake hose
- Wheel speed sensor
- Rear wheels and tighten the lug nuts to 63 ft. lbs. (86 Nm)
- Negative battery cable

13. Bleed the brake system.

14. Check the rear wheel alignment.

Lower Ball Joint

REMOVAL & INSTALLATION

The lower ball joint is serviced with the lower control arm as an assembly.

Lower Control Arm

REMOVAL & INSTALLATION

Front

RIGHT SIDE

1. Before servicing the vehicle, refer to the precautions in the beginning of this section.

2. Remove or disconnect the following:
- Negative battery cable
- Wheel
- Splash shield
- Lower ball joint
- Lower control arm bushing bolts
- Lower control arm

To install:

3. Install the lower control arm. If equipped with vertical bushings, tighten the

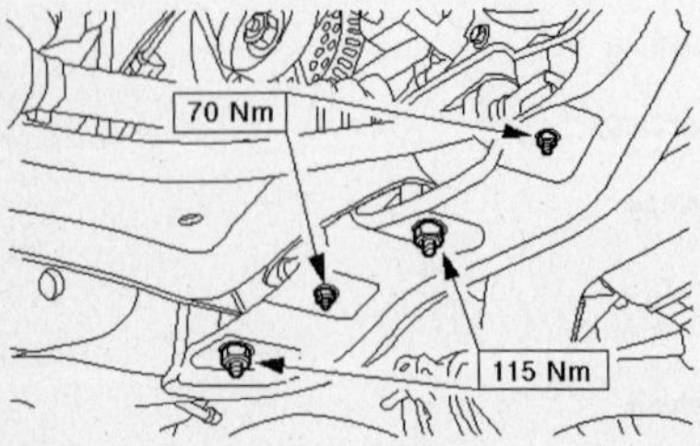

9346HG25

Control arm with horizontal bushings

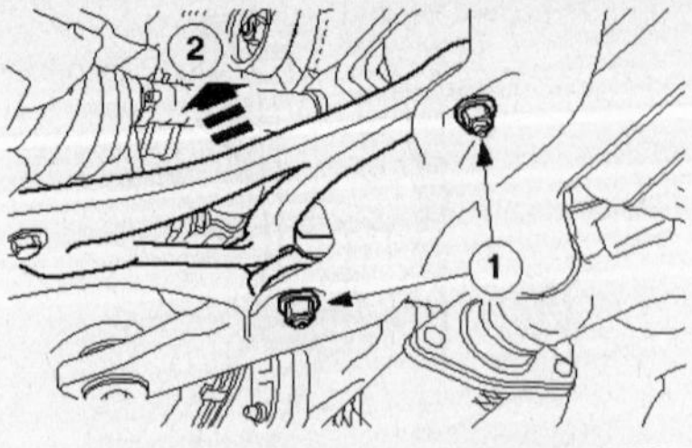

9346HG24

Remove the mounting bolts (1) and remove the control arm (2)—right side with vertical bushings shown

mounting bolts to 85 ft. lbs. (115 Nm). If equipped with horizontal bushings, tighten the large bolts to 85 ft. lbs. (115 Nm) and the small bolts to 52 ft. lbs. 970 Nm).

4. Install or connect the following:
- Lower ball joint and tighten the pinch bolt to 61 ft. lbs. (83 Nm)
- Splash shield
- Wheel and tighten the lug nuts to 63 ft. lbs. (86 Nm)
- Negative battery cable

5. Check the wheel alignment.

LEFT SIDE

1. Before servicing the vehicle, refer to the precautions in the beginning of this section.

2. Secure the radiator and fan shroud with safety wire to the radiator support.

3. Attach an engine support fixture to the engine lifting eyes.

4. Remove or disconnect the following:
- Negative battery cable
- Steering column pinch bolt
- Left, right and center splash shields
- Exhaust crossover pipe (2.5L engine)
- Catalytic converter
- Front wheel
- Vehicle Speed Sensor (VSS)
- Radiator air deflector
- A/C accumulator mounting bolts
- Left and right engine support insulators
- Lower ball joint
- Power steering oil cooler lines

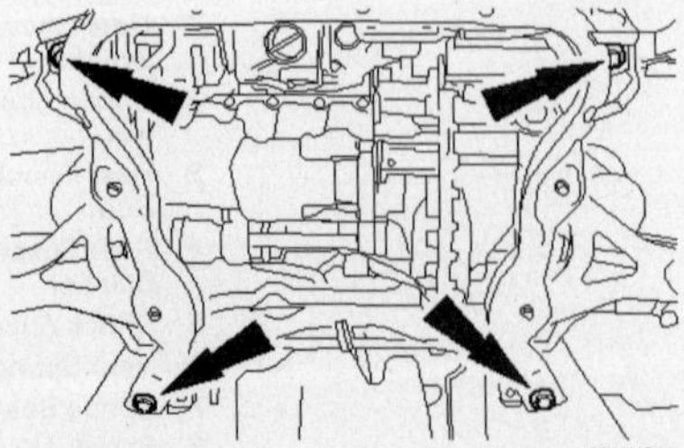
9346HG26

Remove these bolts to lower the sub-frame

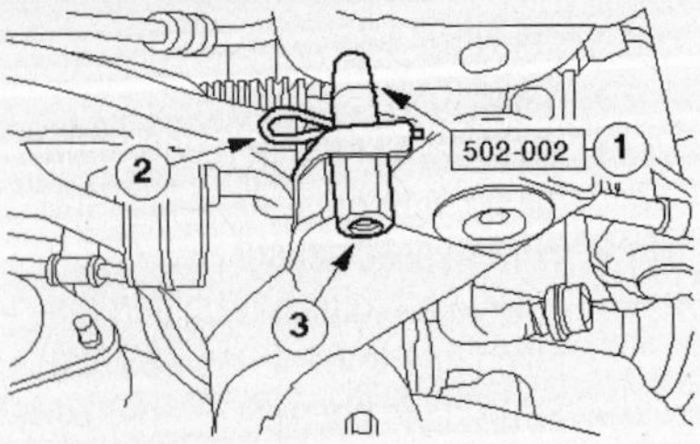

9346HG27

Sub-frame alignment guide pin (1), locking pin (2) and guide pin sleeve (3)

- Lower radiator supports
- Bumper cover braces

5. Unbolt and lower the sub-frame to access the left control arm bushing bolts.

6. Remove the control arm.

To install:

➡Replace all snaprings, split pins and self-locking nuts.

7. Install or connect the following:
- Left control arm and tighten the mounting bolts and nuts to 96 ft. lbs. (130 Nm)
- Sub-frame
- Special tool Sub-frame Alignment Pin Set T95P-2100-AH and the sub-frame bolts. Tighten the sub-frame bolts to 81–110 ft. lbs. (110–150 Nm). Remove the alignment pins.
- A/C accumulator bolts
- Power steering oil cooler lines
- Special tool Powertrain Alignment Gauge T94P-6000-AH in place of the left engine support insulator. Tighten the 2 retaining bolts to 20 ft. lbs. (27 Nm) and snug the through-bolt.

➡The left and right support insulators must be aligned in the middle of the support insulator brackets.

- Right engine support insulator. Tighten the 2 sub-frame retaining bolts to 30–41 ft. lbs. (41–55 Nm) and the through-bolt to 75–102 ft. lbs. (103–137 Nm).

8. Remove the Powertrain Alignment Tool and install the left engine support insulator. Tighten the center bolt to 88 ft. lbs. (120 Nm), and the mounting bolts to 35 ft. lbs. (48 Nm).
- Bumper cover braces
- Lower radiator supports and tighten the bolts to 71–97 inch lbs. (8–11 Nm)
- Power steering oil cooler lines
- Lower ball joint and tighten the pinch bolt to 61 ft. lbs. (83 Nm)

- A/C accumulator mounting bolts
- Radiator air deflector
- VSS sensor
- Front wheel and tighten the lug nuts to 63 ft. lbs. (86 Nm)
- Catalytic converter
- Exhaust crossover pipe (2.5L engine)
- Left, right and center splash shields
- Steering column pinch bolt and tighten the bolt to 15–20 ft. lbs. (20–27 Nm)
- Battery

9. Road test the vehicle and check for proper operation.

➡**Whenever the vehicle sub-frame is removed or lowered, the wheel alignment should be checked.**

Rear

1. Before servicing the vehicle, refer to the precautions in the beginning of this section.
2. Remove or disconnect the following:
   - Rear wheels
   - Rear control arms
   - Stabilizer bar link
   - Exhaust system hangers
   - Wheel spindle tie bar
3. Support the rear crossmember and remove the 4 attaching bolts.
4. Lower the rear crossmember to access the front control arm bolts.
5. Remove the front control arm.

To install:

➡**The final tightening of the rear suspension components is carried out with the full weight of the vehicle on its wheels.**

6. Install or connect the following:
   - Front control arm and tighten the bolts to 62 ft. lbs. (84 Nm)
   - Rear crossmember. Use special tools Sub-frame Alignment Pins T94P-2100-AH, and tighten the crossmember bolts to 88 ft. lbs. (120 Nm).
   - Wheel spindle tie bar and tighten the bolt to 88 ft. lbs. (120 Nm)
   - Exhaust system hangers
   - Stabilizer bar link and tighten the nuts to 26 ft. lbs. (35 Nm)
   - Rear control arms and tighten the bolts to 62 ft. lbs. (84 Nm)
   - Rear wheels and tighten the lug nuts to 63 ft. lbs. (86 Nm)
7. Complete the final tightening of the rear suspension components with the full weight of the vehicle resting on its wheels.

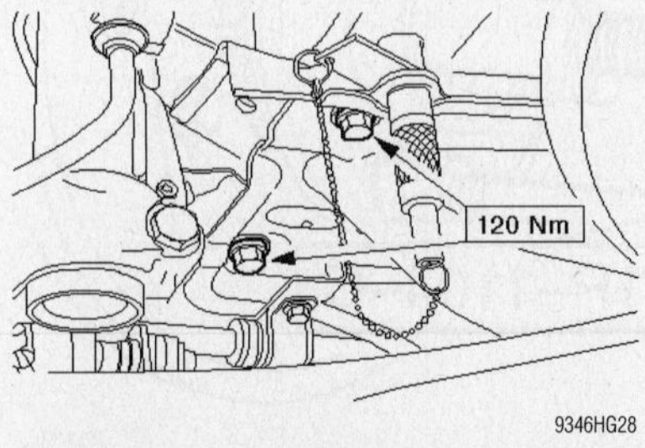

Tighten the rear crossmember bolts with the alignment pins installed

CONTROL ARM BUSHING REPLACEMENT

The control arm bushings are serviced with the control arm as an assembly

Wheel Bearings

ADJUSTMENT

Front and Rear

The wheel bearings are not adjustable. If the bearings make noise or become loose, they must be replaced.

REMOVAL & INSTALLATION

Front

1. Before servicing the vehicle, refer to the precautions in the beginning of this section.

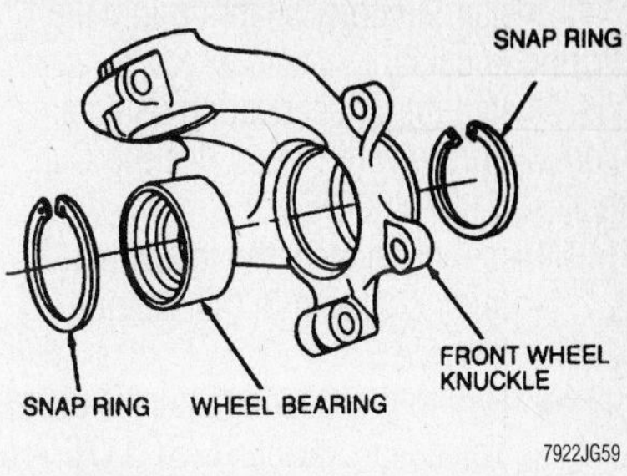

The wheel bearing is retained in the knuckle by 2 snaprings

2. Remove or disconnect the following:
   - Negative battery cable
   - Front wheel
   - Brake caliper support and rotor
   - Wheel speed sensor
   - Outer tie rod end
   - Wheel hub retaining nut
   - Lower ball joint
   - Steering knuckle pinch bolt
   - Halfshaft from the wheel hub
   - Steering knuckle from the vehicle
3. Install Front Hub Remover/Replacer T81P-1104-C or equivalent with the appropriate adapters and separate the hub from the steering knuckle.
   - Inner and outer snaprings securing the wheel bearing
   - Wheel bearing from the steering knuckle. Drive or press the old wheel bearing out as required.

To install:

➡**Replace all snaprings, split pins and self-locking nuts.**

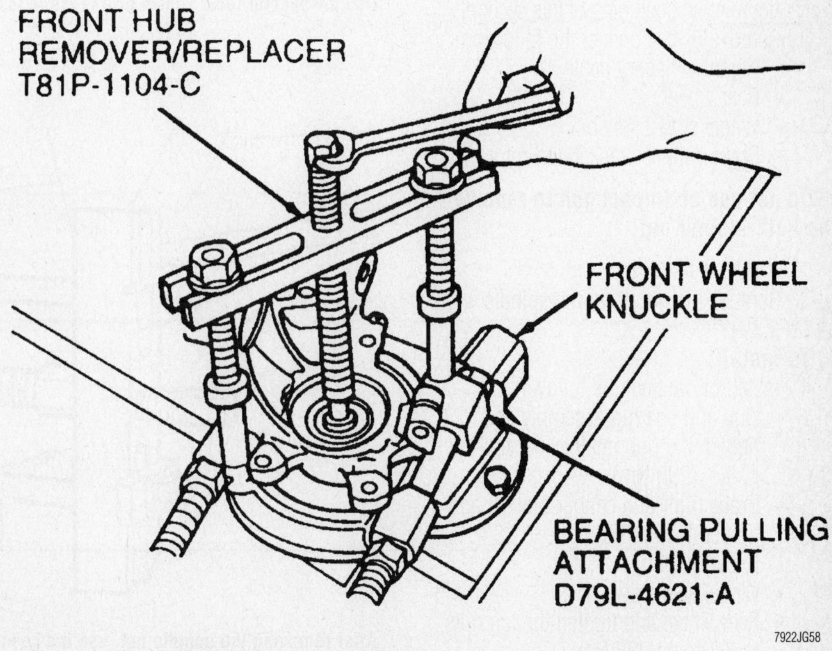

Use gear or bearing pulling tools to remove the hub from the steering knuckle

4. Install the outer snapring into the steering knuckle.

5. Install the wheel bearing using a hydraulic press with Pinion Bearing Cup Replacer T80T-4000-E.

6. Install or connect the following:
- Inner snapring in the steering knuckle
- Hub to the steering knuckle using Threaded Drawbar T75T-1176-A or equivalent
- Steering knuckle. Tighten the strut pinch bolt to 40 ft. lbs. (54 Nm), and the ball joint pinch bolt to 61 ft. lbs. (83 Nm).
- Wheel speed sensor
- Wheel hub retaining nut and tighten the hub nut to 246 ft. lbs. (340 Nm)
- Outer tie rod end and tighten the nut to 18–22 ft. lbs. (24–30 Nm)
- Wheel speed sensor
- Brake caliper support and rotor and tighten the bolts to 88 ft. lbs. (120 Nm)
- Front wheel and tighten the lug nuts to 63 ft. lbs. (86 Nm)
- Negative battery cable

7. Road test the vehicle and check for proper operation.

Rear

➡The wheel bearings are contained within the wheel hub and must be replaced as an assembly.

WITH REAR DISC BRAKES

1. Before servicing the vehicle, refer to the precautions in the beginning of this section.

2. Remove or disconnect the following:
- Negative battery cable
- Rear wheel
- Wheel speed sensor
- Brake caliper bracket and rotor

➡Do not use an impact gun to remove the hub retainer nut.

- Hub retainer nut

3. Remove the hub from the spindle with a hub puller.

To install:

4. Install or connect the following:
- Hub and bearing assembly and tighten the hub retainer nut to 213 ft. lbs. (290 Nm)
- Brake rotor and caliper bracket and tighten the bracket bolts to 88 ft. lbs. (120 Nm)
- Wheel speed sensor
- Rear wheel and tighten the lug nuts to 63 ft. lbs. (86 Nm)

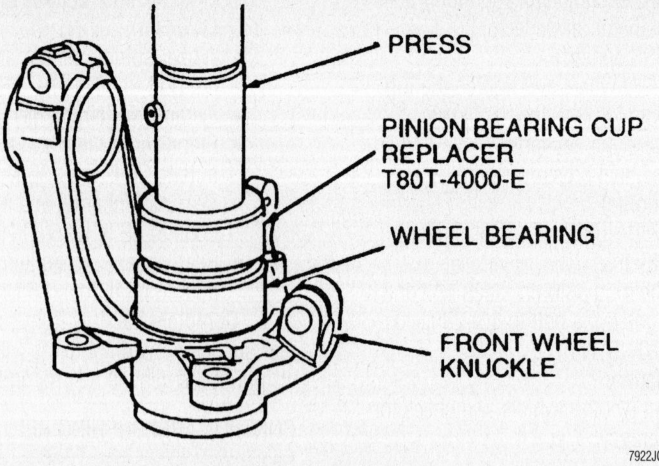

Using a press, install the new wheel bearing in the knuckle

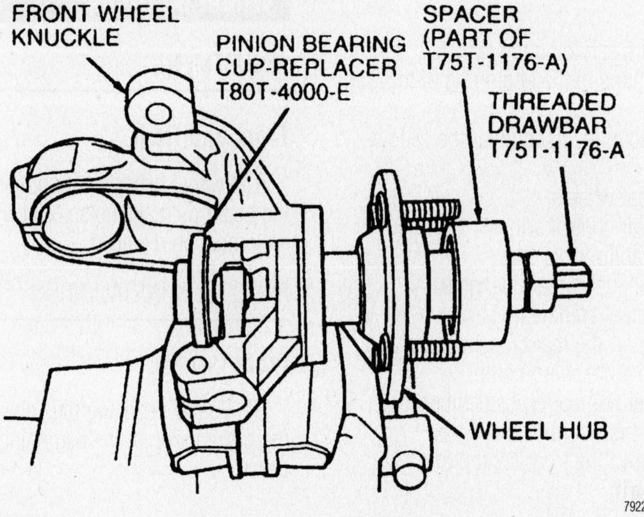

Use the special tools shown or a press to install the hub in the knuckle assembly

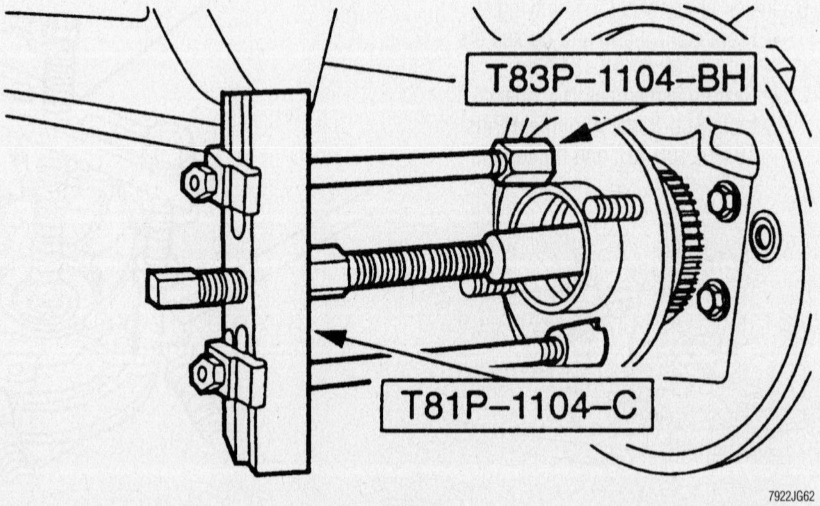

After removing the spindle nut, use the special tools or a puller to remove the rear hub/bearing assembly—all models

• Negative battery cable

5. Road test the vehicle and check for proper operation.

WITH REAR DRUM BRAKES

➡**The wheel bearings are contained within the wheel hub and must be replaced as an assembly.**

1. Before servicing the vehicle, refer to the precautions in the beginning of this section.

2. Remove or disconnect the following:
   - Negative battery cable
   - Rear wheel
   - Brake drum
   - Wheel speed sensor

➡**Do not use an impact gun to remove the hub retainer nut.**

   - Wheel hub retainer nut

3. Remove the hub from the spindle with a hub puller.

To install:

4. Install or connect the following:
   - Hub and bearing assembly and tighten the hub retainer nut to 213 ft. lbs. (290 Nm)
   - Brake drum
   - Wheel speed sensor
   - Rear wheel and tighten the lug nuts to 63 ft. lbs. (86 Nm)
   - Negative battery cable

5. Road test the vehicle and check for proper operation.

BRAKES

Brake Caliper

REMOVAL & INSTALLATION

Front

1. Before servicing the vehicle, refer to the precautions in the beginning of this section.

2. Remove or disconnect the following:
   - Wheel and tire assembly
   - Outer disc brake pad spring clip (anti-rattle clip)
   - 2 locator pin covers and remove the locator pins
   - Hose from its mounting on the strut
   - Caliper off of the brake rotor
   - Inboard disc brake pad from the caliper

3. If the brake caliper is to be removed from the vehicle, place a pan under the caliper to catch the brake fluid for proper disposal. Disconnect the brake hose at the caliper and allow to drain.
   - Brake caliper

4. If the caliper is not to be serviced, tie off the caliper to prevent strain on the brake hose.

To install:

5. If removed, install the brake hose onto the caliper. Torque the fitting to 10 ft. lbs. (14 Nm).

6. Position the inboard disc brake pad into the caliper.

7. Make sure that the outboard disc brake pad is positioned properly.

8. Install or connect the following:
   - Caliper over the brake rotor and position onto the caliper anchor plate
   - 2 caliper locator pins and torque to 20 ft. lbs. (28 Nm)
   - Caliper locator pin covers
   - Outer disc brake pad spring clip
   - Brake hose to the front strut

9. Bleed the brake system of air. Top off the master cylinder when complete.

10. If the brake pedal feels spongy, repeat the brake bleeding procedure.

11. Reinstall the wheel and tire assembly. Torque the lug nuts to 62 ft. lbs. (85 Nm).

12. Pump the brake pedal several times to position the brake pads before attempting to move the vehicle.

13. Road test the vehicle and check for proper brake system operation.

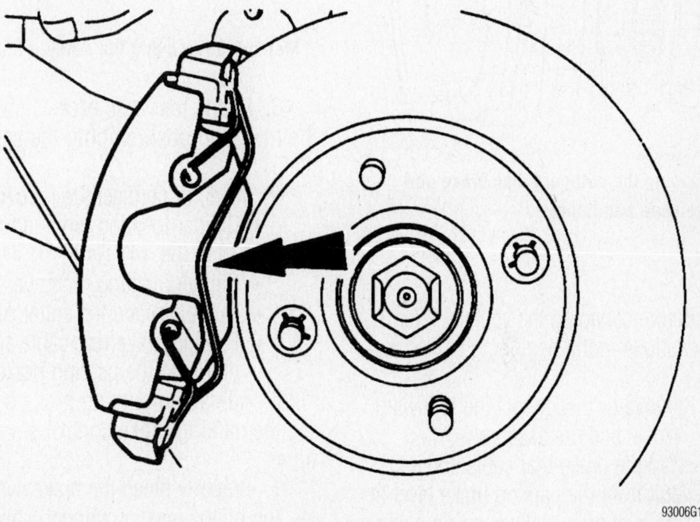

Position of disc brake pad (brake shoe and lining) spring clip

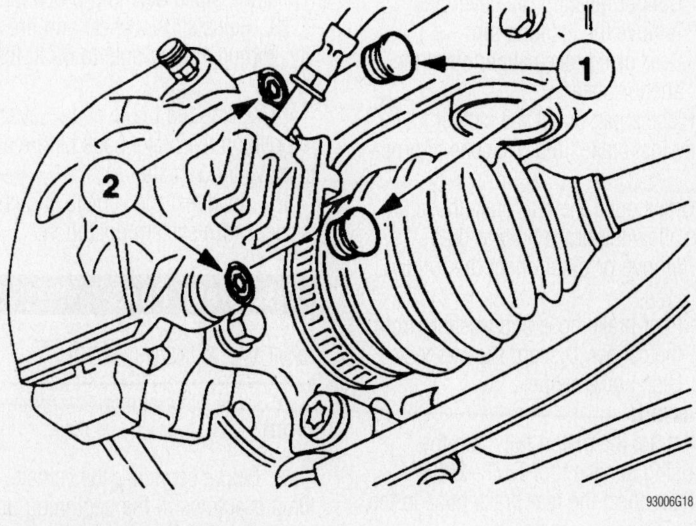

View of disc brake caliper locating pins

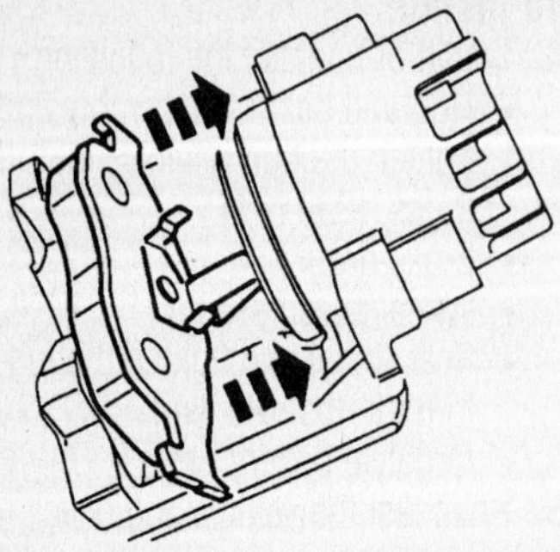

View of inboard disc brake pad (brake shoe and lining)

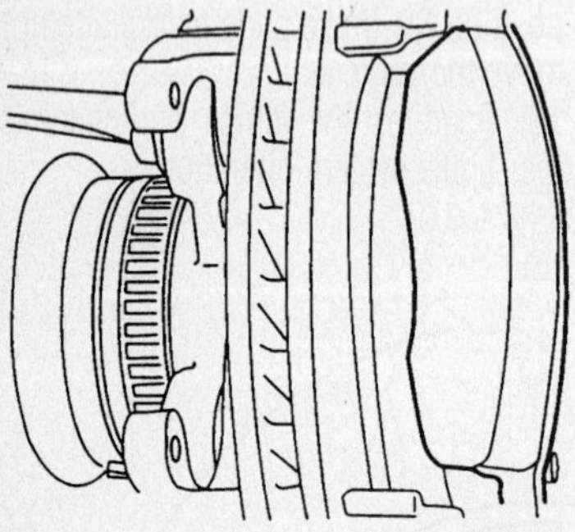

Positioning the outboard disc brake pad (brake shoe and lining)

Rear

1. Before servicing the vehicle, refer to the precautions in the beginning of this section.
2. Remove or disconnect the following:
   - Wheel and tire assembly
   - Parking brake rear cable and conduit from the parking brake lever at the disc brake caliper, using a pair of pliers
   - Cotter pin and guide pin
   - Caliper locating pin cover and remove the locating pin
   - Rear disc brake caliper off of the anchor plate
3. Place a pan under the caliper to catch any lost brake fluid. Dispose of properly.
4. Crack open the rear brake hose fitting and allow the brake fluid to drain.
5. Remove or disconnect the following:
   - Rear brake hose and washers from the caliper. Discard the washers.
   - Disc brake caliper

To install:

6. Retract the piston fully into the caliper using service tool T87P–2588–A.
7. Reconnect the rear brake hose to the caliper using new washers.

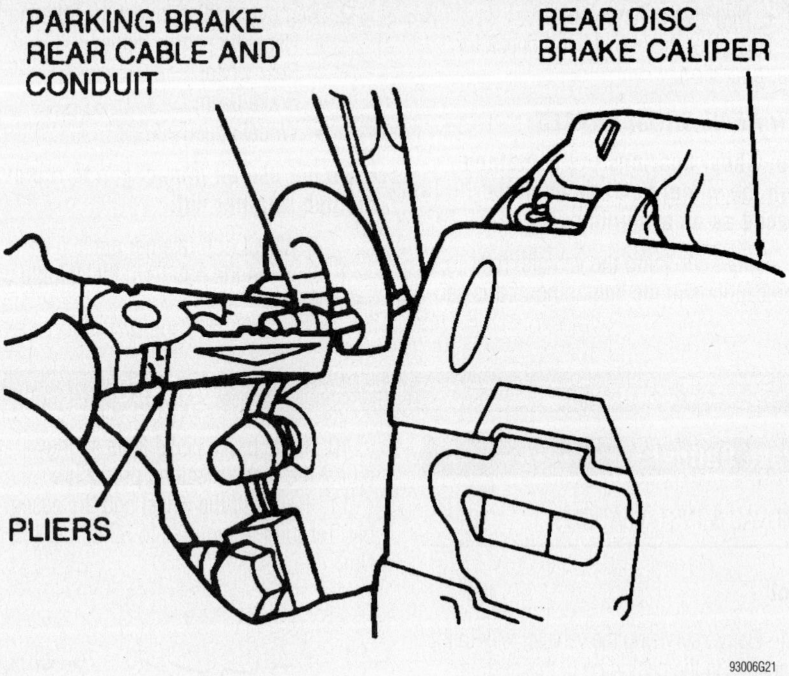

Method of removing the parking brake cable

8. Fit the rear disc brake caliper over the rotor and position onto the anchor plate.
9. Install or connect the following:
   - Caliper locating pin and torque to 30 ft. lbs. (41 Nm)
   - Caliper locating pin cover
   - Guide pin and the cotter pin
   - Parking brake rear cable and conduit onto the parking brake lever
10. Adjust the parking brake by operating the parking brake control several times.
11. Properly bleed the brake system of air. Top off the master cylinder when complete.
12. If the brake pedal feels spongy, repeat the brake bleeding procedure.
13. Reinstall the wheel and tire assembly. Torque the lug nuts to 62 ft. lbs. (85 Nm).
14. Pump the brake pedal several times to position the brake pads before attempting to move the vehicle.
15. Road test the vehicle and check for proper brake system operation.

Disc Brake Pads

REMOVAL & INSTALLATION

Front

1. Before servicing the vehicle, refer to the precautions in the beginning of this section.
2. Remove or disconnect the following:
   - ½ of the brake fluid from the master cylinder reservoir
   - Wheel and tire assembly
   - Outer disc brake pad spring clip (anti-rattle clip)
   - 2 locating pin covers and remove the locating pins
   - Hose from its mounting on the strut
   - Caliper off of the brake rotor and tie it off to prevent damage to the brake hose
   - Outboard disc brake pad from the anchor plate
   - Inboard disc brake pad from the brake caliper

To install:

3. If installing new disc brake pads, use a C-clamp or similar tool to push the caliper piston into the caliper bore. This will allow room for the new pads.
4. Install or connect the following:
   - Inboard disc brake pad into the caliper
   - Outboard disc brake pad into position in the anchor plate
   - Disc brake caliper onto the rotor
   - 2 locating pins and torque to 20 ft. lbs. (28 Nm)
   - Caliper locating pin covers
   - Brake hose to its support on the strut
   - Disc brake pad spring clip
   - Wheel and tire assembly. Torque the lug nuts to 62 ft. lbs. (85 Nm).

5. Pump the brake pedal several times to achieve a good pedal before attempting to move the vehicle.
6. Check the brake fluid level in the master cylinder fluid reservoir and add fluid as necessary.
7. Road test the vehicle and check for proper brake system operation.

Rear

1. Before servicing the vehicle, refer to the precautions in the beginning of this section.
2. Remove or disconnect the following:
   - ½ of the brake fluid from the master cylinder reservoir
   - Wheels
   - Cotter pin and guide pin
   - Caliper locating pin cover and remove the locating pin
3. Swing the rear disc brake caliper away from the brake rotor and anchor plate. There is no need to remove the parking brake cable or brake hose.
   - Inner and outer disc brake pads

To install:

4. If installing new disc brake pads, use rear caliper piston adjuster T87P-2588-A or similar tool, to rotate the rear disc brake piston clockwise, retracting the caliper piston. This will allow room for the new brake pads.
5. Install the inner and outer disc brake pads.
6. Swing the rear disc brake caliper back into position over the disc brake pads.
7. Clean the locating pin threads and apply 1 drop of a thread locking agent or similar sealer.
8. Apply a small amount of disc brake caliper slide grease to the shaft of the locating pin.
9. Install or connect the following:
   - Locating pin and torque to 30 ft. lbs. (41 Nm)
   - Guide pin and the cotter pin
10. Adjust the parking brake by operating the parking brake control several times.
    - Wheel and tire assembly. Torque the lug nuts to 62 ft. lbs. (85 Nm).
11. Adjust the parking brake by operating the parking brake control several times.
12. Pump the brake pedal several times to achieve a good pedal before attempting to move the vehicle.
13. Check the brake fluid level in the master cylinder fluid reservoir and add fluid as necessary.
14. Road test the vehicle and check for proper brake system operation.

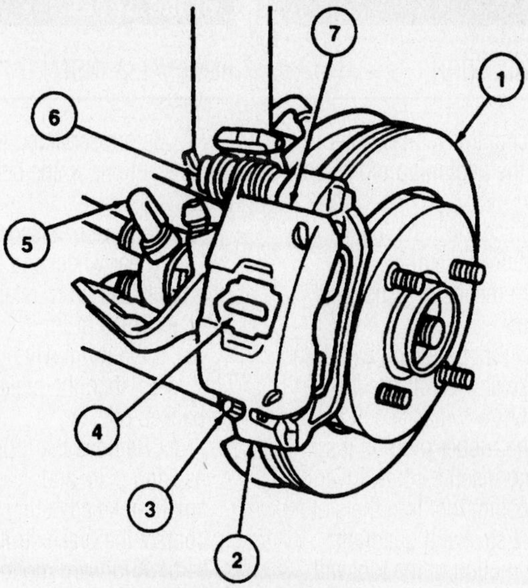

1	Rear Disc Brake Rotor
2	Rear Disc Brake Caliper Anchor Plate
3	Guide Pin
4	Anti-Rattle Clip
5	Parking Brake Lever

93006G26

Rear disc brake components

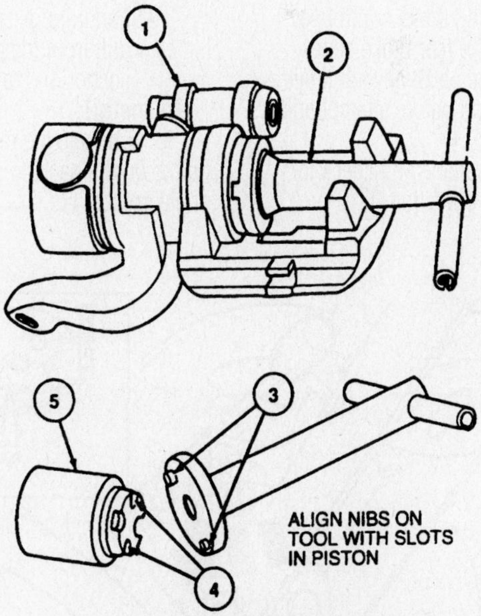

1	Rear Disc Brake Caliper
2	Rear Caliper Piston Adjuster
3	Nibs (Part of T87P-2588-A)
4	Piston Adjuster Slots (Part of 2B588)
5	Rear Disc Brake Caliper Piston and Adjuster

93006G27

Diagram of rear caliper piston and the adjuster tool

Brake Drums

REMOVAL & INSTALLATION

1. Before servicing the vehicle, refer to the precautions in the beginning of this section.
2. Remove or disconnect the following:
   - Wheel and tire assembly
   - Brake drum retainers, if installed
   - Brake drum
3. If the drum will not slide off with light force, then the brake shoes will need to be backed off as follows:
   a. Remove the rubber plug on the backing plate and insert a screwdriver or small brake adjusting tool into the slot to contact the brake strut and quadrant.
   b. A forward motion of the tool will separate the quadrant from the knurled wheel and allow the brake shoes to retract.
   c. Remove the brake drum.

To install:

4. Make sure the brake drum and shoes are clean of any oils or protective coatings.
5. Install or connect the following:
   - Brake drum onto the wheel hub
   - New brake drum retainers if available
   - Wheel and tire assembly. Torque the lug nuts to 62 ft. lbs. (85 Nm).
6. Work the parking brake control several times to adjust the rear brake shoes.
7. Pump the brake pedal several times to assure a good pedal before attempting to move the vehicle.
8. Road test the vehicle and check for proper brake system operation.

Brake Shoes

REMOVAL & INSTALLATION

1. Before servicing the vehicle, refer to the precautions in the beginning of this section.
2. Remove or disconnect the following:
   - Rear wheel and tire assembly
   - Brake drum retainers, if equipped
   - Brake drum
3. If the drum will not slide off with light force, then the brake shoes will need to be backed off:
   a. Remove the rubber plug on the backing plate and insert a screwdriver or small brake adjusting tool into the slot to contact the brake strut and quadrant.
   b. A forward motion of the screwdriver will separate the quadrant from the knurled wheel and allow the brake shoes to retract.
   c. Remove the brake drum.
4. Remove or disconnect the following:
   - Brake shoe hold down springs and the brake shoe hold down pins
   - Brake shoe retracting springs
   - Parking brake cable and conduit from the parking brake lever
   - Brake shoes
   - Rear brake strut and quadrant from the rear brake shoe
   - Parking brake rear cable and conduit from the parking brake cable anchor on the trailing brake shoe.

To install:

5. Lubricate the rear brake shoe contact points on the backing plate with an appropriate grease.

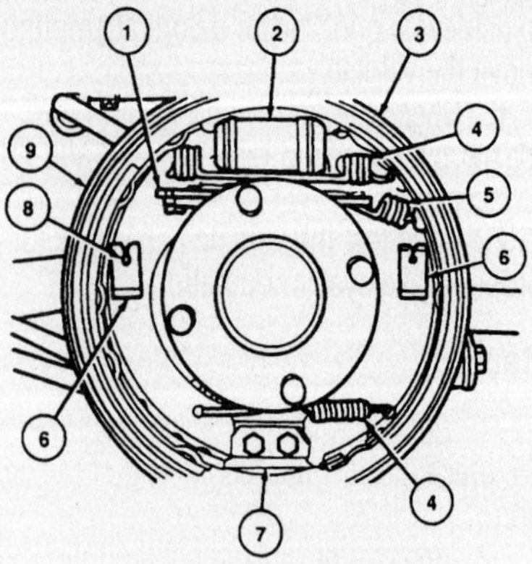

1	Rear Brake Strut and Quadrant
2	Rear Wheel Cylinder
3	Rear Brake Shoe and Lining
4	Brake Shoe Retracting Spring
5	Parking Brake Return Spring
6	Brake Shoe Hold Down Spring
7	Anchor Block
8	Brake Shoe Hold Down Spring Pin
9	Rear Brake Backing Plate

93006G50

Location of brake components

6. Engage the parking brake rear cable and conduit into the parking brake cable anchor on the trailing brake shoe.
7. Position the trailing brake shoe on the backing plate.
8. Engage the rear brake strut and quadrant.
9. Install or connect the following:
   - Parking brake rear spring
   - Brake shoe hold down spring pin and the hold down spring
   - Leading brake shoe into the slot on the rear brake strut and quadrant
   - Brake shoe hold down pin and hold down spring
   - Brake shoe retracting springs
10. Adjust the brake shoes by first measuring the inside drum diameter with an appropriate brake adjustment gauge.
11. Insert a small screwdriver or similar tool into the knurled quadrant of the brake strut and quadrant to adjust the brake shoes to the same measurement of the brake drum by expanding the brake strut and quadrant.
12. Trial fit the brake drum. The shoes should just contact the drum surface when properly adjusted.
13. Make sure that the brake drum and brake shoes are clean of any oils or protective coatings.
14. Install or connect the following:
    - Brake drum
    - New drum retainers, if available

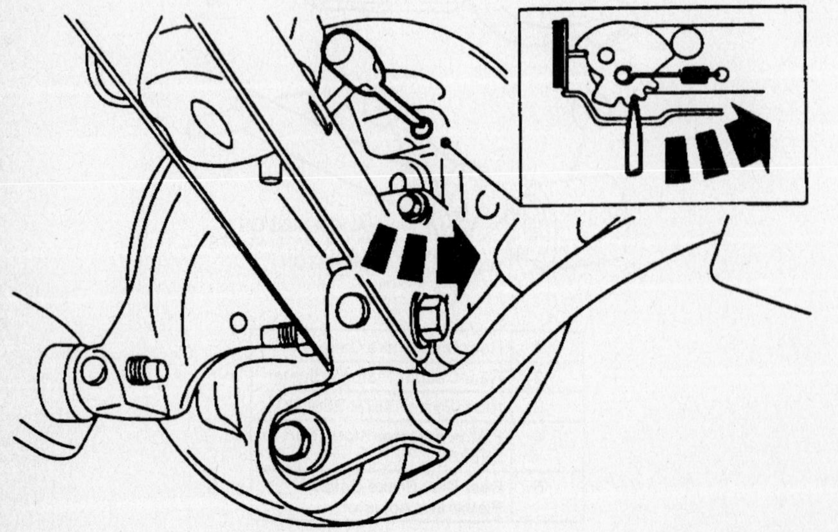

93006G38

Positioning a screwdriver to release the brake shoes

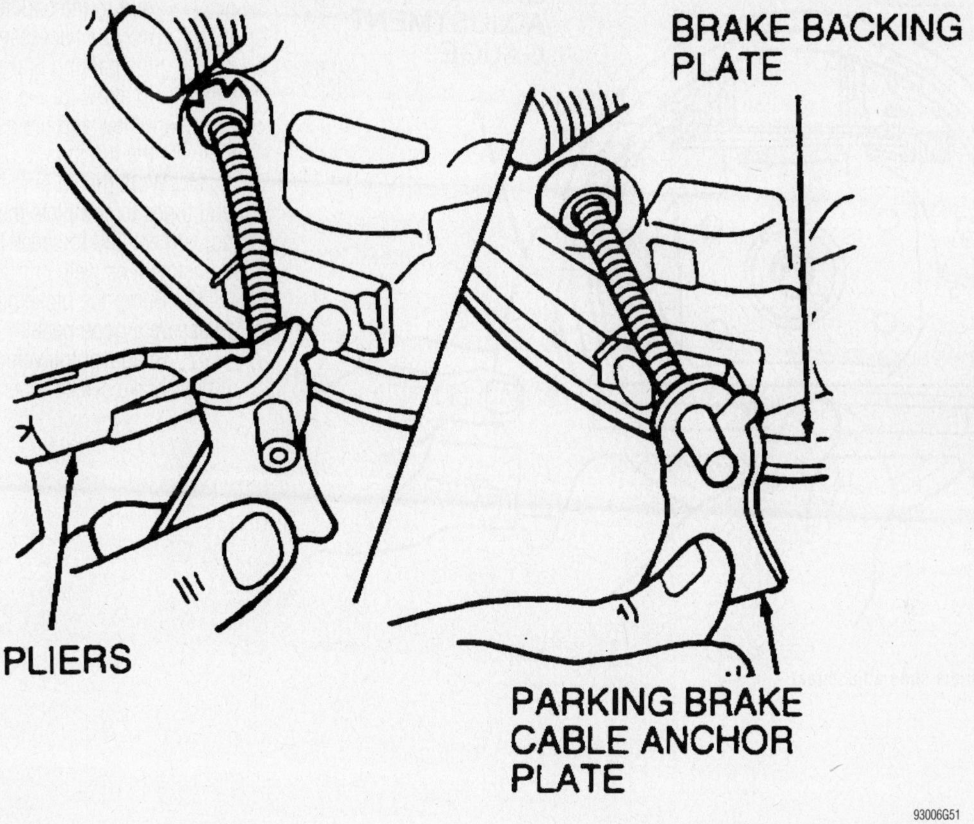

Removal of parking brake cable

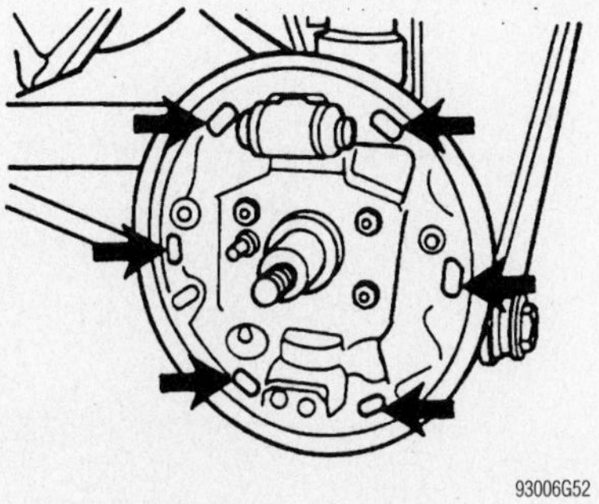

Lubricating points on backing plate

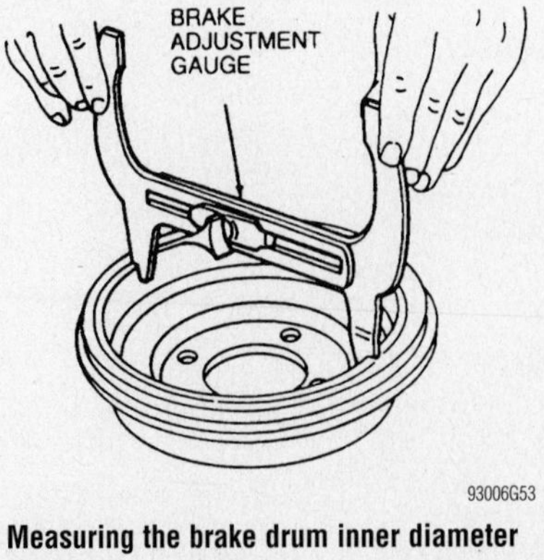

Measuring the brake drum inner diameter

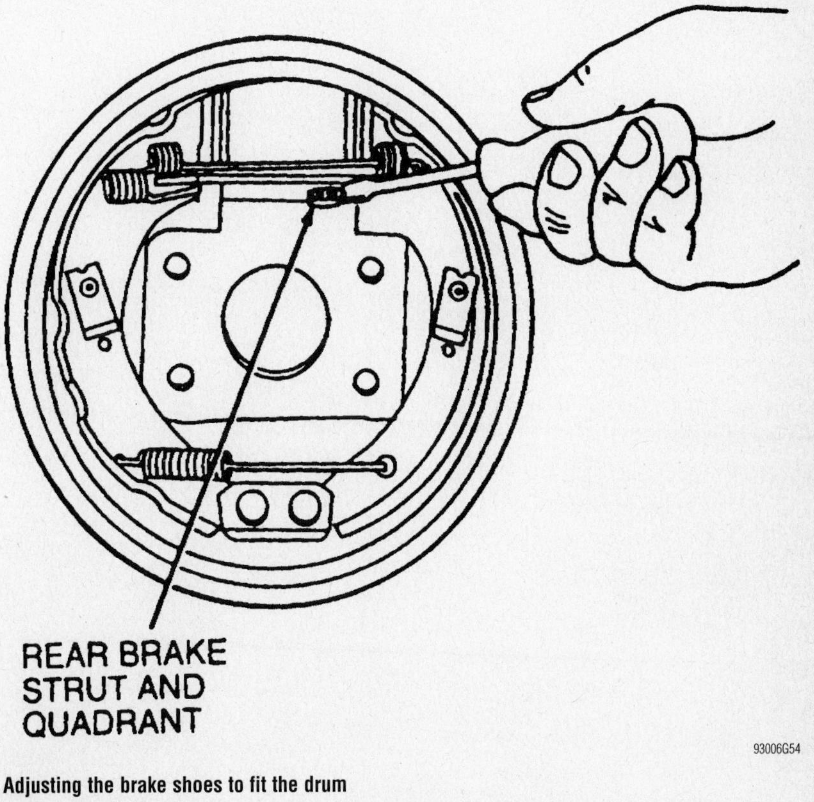

Adjusting the brake shoes to fit the drum

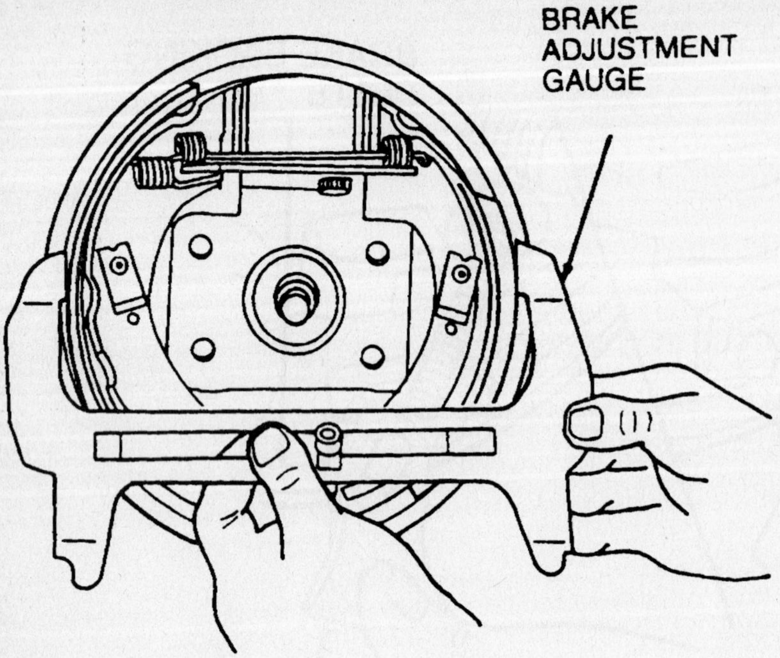

Measuring the brake shoe adjustment

- Parking brake rear cable and conduit to the parking brake lever. It may be necessary to back off on the parking brake cable adjustment to allow for the new brake shoes.
- Wheel and tire assembly. Torque the lug nuts to 62 ft. lbs. (85 Nm).

15. Work the parking brake control several times to complete the brake shoe adjustment and to check the parking brake adjustment as well.

16. Pump the brake pedal several times to assure a good pedal.

17. Road test the vehicle and check for proper brake system operation.

FORD, LINCOLN AND MERCURY

5

Crown Victoria • Grand Marquis • Marauder • Town Car

SPECIFICATION CHARTS

ENGINE AND VEHICLE IDENTIFICATION

Engine						
Code ①	Liters (cc)	Cu. In.	Cyl.	Fuel Sys.	Type	Eng. Mfg.
W	4.6 (4593)	281	8	SFI	SOHC	Ford
V	4.6 (4593)	281	8	SFI	DOHC	Ford

Model Year	
Code ②	Year
1	2001
2	2002
3	2003
4	2004
5	2005

SFI: Sequential Fuel Injection

SOHC: Single Overhead Camshaft

① 8th digit of the Vehicle Identification Number (VIN)

② 10th digit of the Vehicle Identification Number (VIN)

67197-CROW-C01

GENERAL ENGINE SPECIFICATIONS

Year	Model	Engine Displacement Liters	Engine ID/VIN	Net Horsepower @ rpm	Net Torque @ rpm (ft. lbs.)	Bore x Stroke (in.)	Com-pression Ratio	Oil Pressure @ rpm
2001	Crown Victoria	4.6	W	①	②	3.55x3.54	③	20-45@1500
	Grand Marquis	4.6	W	①	②	3.55x3.54	9.4:0	20-45@1500
	Town Car	4.6	W	①	②	3.55x3.54	9.4:0	20-45@1500
2002	Crown Victoria	4.6	W	①	②	3.55x3.54	③	20-45@1500
	Grand Marquis	4.6	W	①	②	3.55x3.54	9.4:0	20-45@1500
	Town Car	4.6	W	①	②	3.55x3.54	9.4:0	20-45@1500
2003	Crown Victoria	4.6	W	①	②	3.55x3.54	③	20-45@1500
	Grand Marquis	4.6	W	①	②	3.55x3.54	9.4:0	20-45@1500
	Town Car	4.6	W	①	②	3.55x3.54	9.4:0	20-45@1500
	Marauder	4.6	V	302@5750	318@4300	3.55x3.54	9.85:0	20-45@1500
2004	Crown Victoria	4.6	W	①	②	3.60x3.60	③	20-45@1500
	Grand Marquis	4.6	W	①	②	3.60x3.60	9.4:0	20-45@1500
	Town Car	4.6	W	①	②	3.60x3.60	9.4:0	20-45@1500
	Marauder	4.6	V	302@5750	318@4300	3.60x3.60	9.85:0	20-45@1500

① Single exhaust: 220@4750
Dual exhaust: 235@4000
Crown Victoria with natural gas: 178@4500

② Single exhaust: 265@3250
Dual exhaust: 275@3250
Crown Victoria with natural gas: 237@3500

③ Gasoline Engine: 9.4:1
Natural Gas Engine: 10.0:1

67197-CROW-C02

ENGINE TUNE-UP SPECIFICATIONS

Year	Engine Displacement Liters	Engine ID/VIN	Spark Plug Gap (in.)	Ignition Timing (deg.)	Fuel Pump (psi) ①	Idle Speed (rpm)	Valve Clearance Intake	Valve Clearance Exhaust
2001	4.6	W	0.054	10B	35-45	②	HYD	HYD
2002	4.6	W	0.054	10B	35-45	②	HYD	HYD
2003	4.6	W	0.054	10B	35-45	②	HYD	HYD
	4.6	V	0.054	10B	35-45	②	HYD	HYD
2004	4.6	W	0.054	10B	35-45	②	HYD	HYD
	4.6	V	0.054	10B	35-45	②	HYD	HYD

NOTE: The Vehicle Emission Control Information label often reflects specification changes made during production. The label figures must be used if they differ from those in this chart.

B: Before Top Dead Center

HYD: Hydraulic

① Fuel pressure with engine running, pressure regulator vacuum hose connected

② Refer to Vehicle Emission Control Information label

67197-CROW-C03

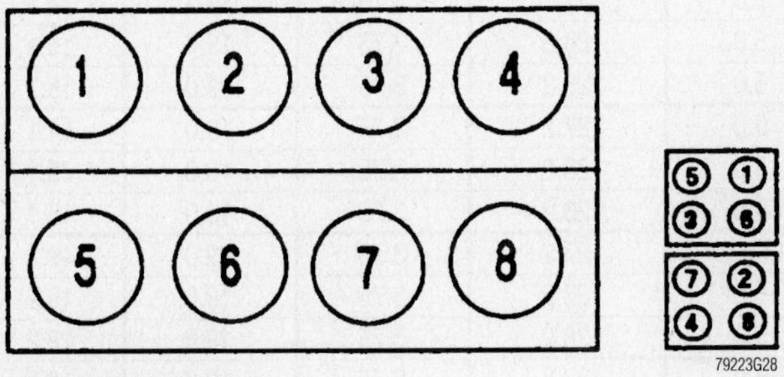

4.6L (VIN W) Engine
Firing order: 1–3–7–2–6–5–4–8
Distributorless ignition system

1. Alternator
2. Water pump
3. Power steering pump
4. Crankshaft
5. A/C compressor
6. Drive belt
7. Tensioner
8. Idler pulley

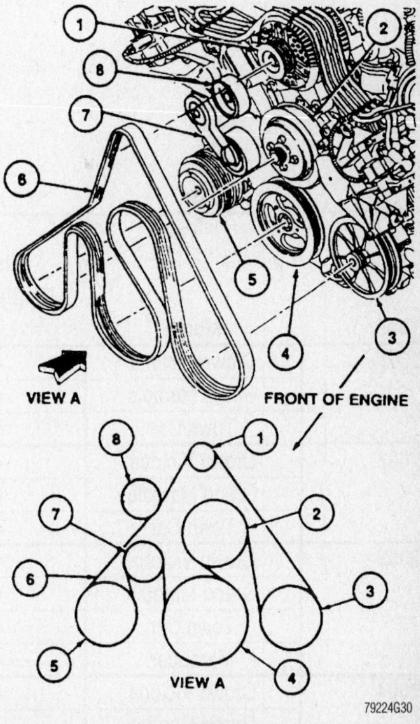

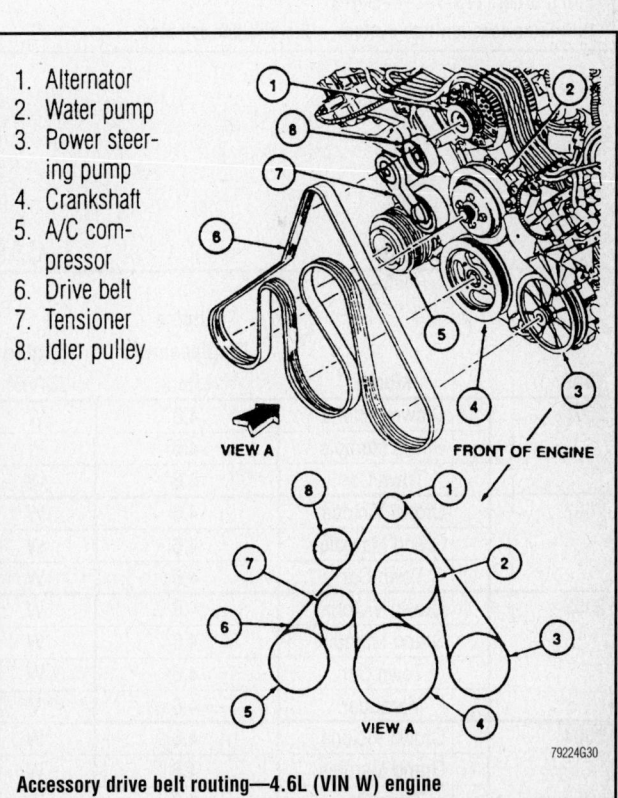

Accessory drive belt routing—4.6L (VIN W) engine

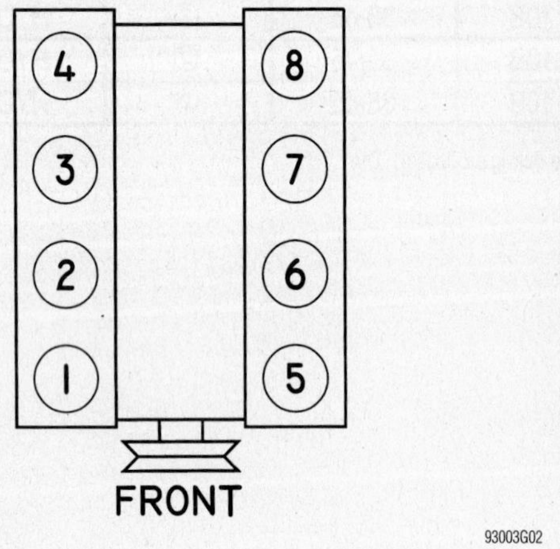

4.6L (VIN V) Engine
Firing order: 1–3–7–2–6–5–4–8
Distributorless ignition system—One coil per cylinder

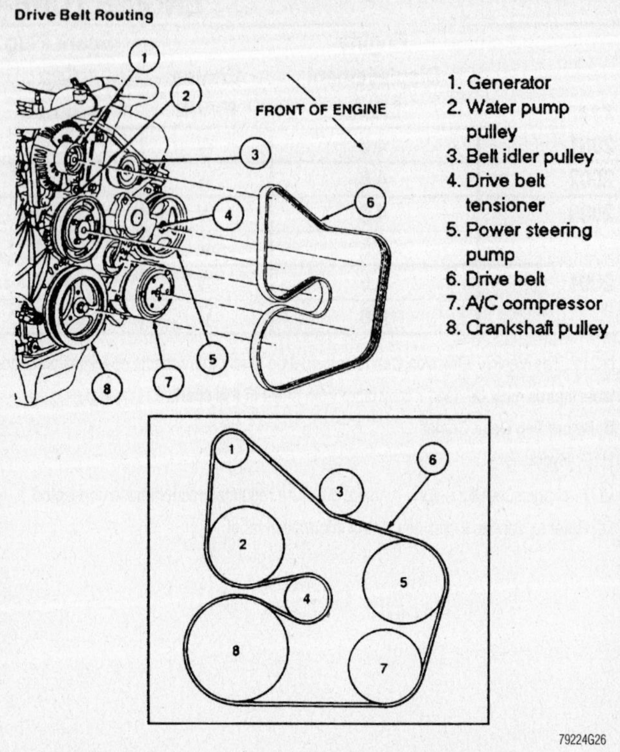

Serpentine accessory drive belt routing— 4.6L (VIN V) engine

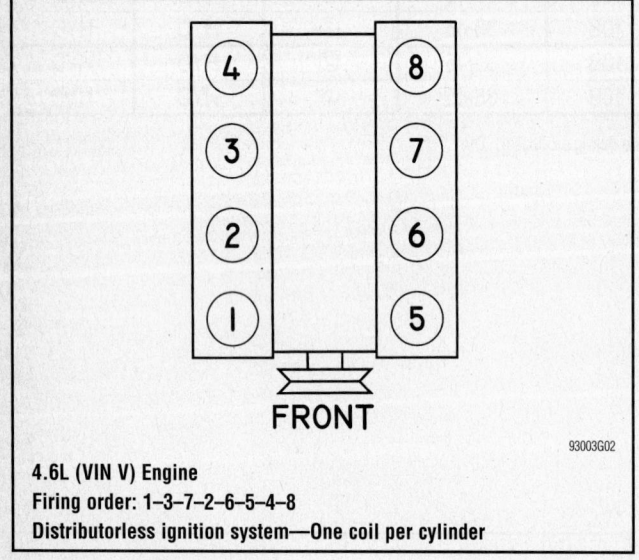

CAPACITIES

Year	Model	Engine Displacement Liters	Engine ID/VIN	Engine Oil with Filter (qts.)	Automatic Transmission (pts.) ①	Rear Drive Axle (pts.)	Fuel Tank (gal.)	Cooling System (qts.)
2001	Crown Victoria	4.6	W	5.0	27.2	3.75	19.0	14.1
	Grand Marquis	4.6	W	5.0	28.2	3.75	19.0	15.1
	Town Car	4.6	W	5.0	28.2	3.75	19.0	15.1
2002	Crown Victoria	4.6	W	5.0	27.2	3.75	19.0	14.1
	Grand Marquis	4.6	W	5.0	28.2	3.75	19.0	15.1
	Town Car	4.6	W	5.0	28.2	3.75	19.0	15.1
2003	Crown Victoria	4.6	W	5.0	27.2	3.75	19.0	19.2
	Grand Marquis	4.6	W	5.0	28.2	3.75	19.0	19.2
	Town Car	4.6	W	5.0	28.2	3.75	19.0	19.2
	Marauder	4.6	V	6.0	27.8	3.75	19.0	18.8
2004	Crown Victoria	4.6	W	5.0	27.2	3.75	19.0	19.2
	Grand Marquis	4.6	W	5.0	28.2	3.75	19.0	19.2
	Town Car	4.6	W	5.0	28.2	3.75	19.0	19.2
	Marauder	4.6	V	6.0	27.8	3.75	19.0	18.8

NOTE: All capacities are approximate. Add fluid gradually and ensure a proper fluid level is obtained.

① Includes torque converter

67197-CROW-C04

VALVE SPECIFICATIONS

Year	Engine Displacement Liters	Engine ID/VIN	Seat Angle (deg.)	Face Angle (deg.)	Spring Test Pressure (lbs. @ in.)	Spring Installed Height (in.)	Stem-to-Guide Clearance (in.) Intake	Stem-to-Guide Clearance (in.) Exhaust	Stem Diameter (in.) Intake	Stem Diameter (in.) Exhaust
2001	4.6	W	45	45.5	132@1.10	1.570	0.0008-0.0027	0.0018-0.0037	0.2746-0.2754	0.2736-0.2744
2002	4.6	W	45	45.5	132@1.10	1.570	0.0008-0.0027	0.0018-0.0037	0.2746-0.2754	0.2736-0.2744
2003	4.6	W	45	45.5	132@1.10	1.570	0.0008-0.0027	0.0018-0.0037	0.2746-0.2754	0.2736-0.2744
	4.6	V	45	45.5	160@1.03	1.660	0.0008-0.0027	0.0018-0.0037	0.2746-0.275	0.2736-0.2744
2004	4.6	W	45	45.5	132@1.10	1.570	0.0008-0.0027	0.0018-0.0037	0.2746-0.2754	0.2736-0.2744
	4.6	V	45	45.5	160@1.03	1.660	0.0008-0.0027	0.0018-0.0037	0.2746-0.275	0.2736-0.2744

67197-CROW-C05

PISTON AND RING SPECIFICATIONS

All measurements are given in inches.

Year	Engine Displacement Liters	Engine ID/VIN	Piston Clearance ①	Ring Gap Top Compression	Ring Gap Bottom Compression	Ring Gap Oil Control	Ring Side Clearance Top Compression	Ring Side Clearance Bottom Compression	Ring Side Clearance Oil Control
2001	4.6	W	0.0002-0.0010	0.005-0.012	0.012-0.022	0.006-0.026	②	0.008-0.0024	0.0010-0.0077
2002	4.6	W	0.0002-0.0010	0.005-0.012	0.012-0.022	0.006-0.026	②	0.008-0.0024	0.0010-0.0077
2003	4.6	W	0.0002-0.0010	0.005-0.012	0.012-0.022	0.006-0.026	②	0.008-0.0024	0.0010-0.0077
	4.6	V	0.0000-0.0010	0.010-0.020	0.009-0.019	0.006-0.026	0.0003-0.0009	0.0012-0.0031	SNUG
2004	4.6	W	0.0002-0.0010	0.005-0.012	0.012-0.022	0.006-0.026	②	0.008-0.0024	0.0010-0.0077
	4.6	V	0.0000-0.0010	0.010-0.020	0.009-0.019	0.006-0.026	0.0003-0.0009	0.0012-0.0031	SNUG

① Measured 1.96 in. (43mm) from the top

② On 10:1 engines: 0.0012-0.0028 in.
On 9:1 engines: 0.0008-0.0024 in.

67197-CROW-C06

CRANKSHAFT AND CONNECTING ROD SPECIFICATIONS

All measurements are given in inches.

Year	Engine Displacement Liters	Engine ID/VIN	Crankshaft Main Brg. Journal Dia.	Crankshaft Main Brg. Oil Clearance	Crankshaft Shaft End-play	Crankshaft Thrust on No.	Connecting Rod Journal Diameter	Connecting Rod Oil Clearance	Connecting Rod Side Clearance
2001	4.6	W	2.6500-2.6570	0.0009-0.0026	0.0051-0.0119	5	2.0870-2.8670	0.0009-0.0026	0.0006-0.0177
2002	4.6	W	2.6500-2.6570	0.0009-0.0026	0.0051-0.0119	5	2.0870-2.8670	0.0009-0.0026	0.0006-0.0177
2003	4.6	W	2.6500-2.6570	0.0009-0.0026	0.0051-0.0119	5	2.0870-2.8670	0.0009-0.0026	0.0006-0.0177
	4.6	V	2.6567-2.6577	0.0001-0.0018	0.0051-0.0119	5	2.0859-2.0867	0.0011-0.0027	0.0006-0.0177
2004	4.6	W	2.6500-2.6570	0.0009-0.0026	0.0051-0.0119	5	2.0870-2.8670	0.0009-0.0026	0.0006-0.0177
	4.6	V	2.6567-2.6577	0.0001-0.0018	0.0051-0.0119	5	2.0859-2.0867	0.0011-0.0027	0.0006-0.0177

67197-CROW-C07

TORQUE SPECIFICATIONS

All readings in ft. lbs.

Year	Engine Displacement Liters	Engine ID/VIN	Cylinder Head Bolts	Main Bearing Bolts	Rod Bearing Bolts	Crankshaft Damper Bolts	Flywheel Bolts	Manifold Intake	Manifold Exhaust	Spark Plugs	Oil Pan Drain Plug
2001	4.6	W	①	②	③	114-121	54-64	15-22	13-16	7-15	10
2002	4.6	W	①	②	③	114-121	54-64	15-22	13-16	7-15	10
2003	4.6	W	①	②	③	114-121	54-64	15-22	13-16	7-15	10
	4.6	V	④	⑤	⑥	114-121	54-64	⑦	13-16	7-15	10
2004	4.6	W	①	②	③	114-121	54-64	15-22	13-16	7-15	10
	4.6	V	④	⑤	⑥	114-121	54-64	⑦	13-16	7-15	10

NOTE: Stretch bolts are used in all procedures that require rotating the fastener a certain number of degrees. The bolts stretch and cannot be reused. For reassembly, replace with new fasteners.

① Step 1: 22-30 ft. lbs.
Step 2: Plus 85-95 degrees
Step 3: Plus 85-95 degrees

② Step 1: Main bearing cap bolts: 22-25 ft. lbs.
Step 2: Rotate each bolt 85-95 degrees
Step 3: Main bearing cap adjusting screws: 44 inch lbs. then 80-90 inch lbs.
Step 4: Main bearing cap side bolts: 7 ft. lbs. then 14-17 ft. lbs.

③ Step 1: 8 ft. lbs.
Step 2: 12 ft. lbs.
Step 3: 25-34 ft. lbs.
Step 4: Rotate 85-95 degrees

④ Step 1: 30 ft. lbs.
Step 2: Tighten 90 degrees
Step 3: Loosen all bolts one (1) turn
Step 4: 30 ft. lbs.
Step 5: Tighten 90 degrees
Step 6: Tighten an additional 90 degrees

⑤ Step 1: Main bearing cap bolts: 6-9 ft. lbs.
Step 2: Main bearing cap bolts, outer: 16-21 ft. lbs.
Step 3: Main bearing cap bolts, inner: 27-32 ft. lbs.
Step 4: Tighten main bearing cap bolts 85-95 degrees
Step 5: Main cap adjusting screws: 4 ft. lbs. then 7.5 ft. lbs.
Step 6: Main cap side bolts: 7 ft. lbs. then 14-17 ft. lbs.

⑥ Step 1: 5ft. lbs.
Step 2: 10 ft. lbs.
Step 3: 18-25 ft. lbs.
Step 4: Tighten 85-95 degrees

⑦ Step 1: Four inside short bolts: 9-11 ft. lbs.
Step 2: All other bolts: 13-16 ft. lbs.
Step 3: Tighten 85-95 degrees

67197-CROW-C08

WHEEL ALIGNMENT

			Caster		Camber		
Year	Model		Range (+/-Deg.)	Preferred Setting (Deg.)	Range (+/-Deg.)	Preferred Setting (Deg.)	Toe-in (in.)
2001	Crown Victoria	F	0.75	+0.50	0.75	0	-0.25 +/- 0.25
		R	—	—	—	—	—
	Town Car	F	0.75	0	0.75	0	-0.12 +/- 0.25
		R	—	—	—	—	—
	Grand Marquis	F	0.75	+0.50	0.75	0	-0.13 +/- 0.25
		R	—	—	—	—	—
2002	Crown Victoria	F	0.75	+0.50	0.75	0	-0.25 +/- 0.25
		R	—	—	—	—	—
	Town Car	F	0.75	0	0.75	0	-0.12 +/- 0.25
		R	—	—	—	—	—
	Grand Marquis	F	0.75	+0.50	0.75	0	-0.13 +/- 0.25
		R	—	—	—	—	—
2003	Crown Victoria	F	0.75	+0.50	0.75	0	-0.25 +/- 0.25
		R	—	—	—	—	—
	Town Car	F	0.75	0	0.75	0	-0.12 +/- 0.25
		R	—	—	—	—	—
	Grand Marquis	F	0.75	+0.50	0.75	0	-0.13 +/- 0.25
		R	—	—	—	—	—
	Marauder	F	0.75	+6.10	0.75	-0.70	-0.12 +/- 0.25
		R	—	—	—	—	—
2004	Crown Victoria	F	0.75	+0.50	0.75	0	-0.25 +/- 0.25
		R	—	—	—	—	—
	Town Car	F	0.75	0	0.75	0	-0.12 +/- 0.25
		R	—	—	—	—	—
	Grand Marquis	F	0.75	+0.50	0.75	0	-0.13 +/- 0.25
		R	—	—	—	—	—
	Marauder	F	0.75	+6.10	0.75	-0.70	-0.12 +/- 0.25
		R	—	—	—	—	—

67197-CROW-C09

TIRE, WHEEL AND BALL JOINT SPECIFICATIONS

Year	Model	OEM Tires		Tire Pressure		Wheel	Ball Joint	Lug Nut
		Standard	Optional	Front	Rear	Size	Inspection	(ft. lbs.)
2001	Crown Victoria	P225/60SR16	P225/60TR16	32	32	7-JJ	U ① L ①②	95
	Crown Victoria Police Special	P225/60SR16	P225/60VR16	35	35	7J	U ① L ①②	95
	Grand Marquis	P225/60SR16	None	32	32	6-JJ	U ① L ①②	95
	Grand Marquis w/Handling package	P225/60TR16	P225/60VR16	35	35	6-JJ	U ① L ①②	95
	Town Car	P225/70R15	P235/60R16	30	30	Std: 6.5-JJ Opt: 7-JJ	U ① L ①②	95
2002	Crown Victoria	P225/60SR16	P225/60TR16	32	32	7-JJ	U ① L ①②	95
	Crown Victoria Police Special	P225/60SR16	P225/60VR16	35	35	7J	U ① L ①②	95
	Grand Marquis	P225/60SR16	None	32	32	6-JJ	U ① L ①②	95
	Grand Marquis w/Handling package	P225/60TR16	P225/60VR16	35	35	6-JJ	U ① L ①②	95
	Town Car	P225/70R15	P235/60R16	30	30	Std: 6.5-JJ Opt: 7-JJ	U ① L ①②	95
2003	Crown Victoria	P225/60SR16	P225/60TR16	32	32	7-JJ	U ① L ①②	95
	Crown Victoria Police Special	P225/60SR16	P225/60VR16	35	35	7J	U ① L ①②	95
	Grand Marquis	P225/60SR16	None	32	32	6-JJ	U ① L ①②	95
	Grand Marquis w/Handling package	P225/60TR16	P225/60VR16	35	35	6-JJ	U ① L ①②	95
	Town Car	P225/70R15	P235/60R16	30	30	Std: 6.5-JJ Opt: 7-JJ	U ① L ①②	95
	Marauder	③	None	30	30	8-JJ	U ① L ①②	95
2004	Crown Victoria	P225/60SR16	P225/60TR16	32	32	7-JJ	U ① L ①②	95
	Crown Victoria Police Special	P225/60SR16	P225/60VR16	35	35	7J	U ① L ①②	95
	Grand Marquis	P225/60SR16	None	32	32	6-JJ	U ① L ①②	95
	Grand Marquis w/Handling package	P225/60TR16	P225/60VR16	35	35	6-JJ	U ① L ①②	95
	Town Car	P225/70R15	P235/60R16	30	30	Std: 6.5-JJ Opt: 7-JJ	U ① L ①②	95
	Marauder	③	None	30	30	8-JJ	U ① L ①②	95

OEM: Original Equipment Manufacturer

PSI: Pounds Per Square Inch

STD: Standard

OPT: Optional

① Replace if any measurable movement is found.

② Do not lift car. Inspect the boss into which the grease fitting is threaded. Replace if the boss is flush or receded below the surface of the ball joint.

③ Front: P235/50ZR18
Rear: P245/55ZR18

67197-CROW-C10

BRAKE SPECIFICATIONS

All measurements in inches unless noted

		Front Brake Disc			Rear Brake Disc			Minimum	Brake Caliper	
		Original	Minimum	Maximum	Original	Minimum	Maximum	Lining	Bracket Bolts	Mounting Bolts
Year	Model	Thickness	Thickness	Run-out	Thickness	Thickness	Run-out	Thickness	(ft. lbs.)	(ft. lbs.)
2001	Crown Victoria	1.024	0.974	0.002	0.550	0.510	0.002	0.125	125-169	21-26
	Grand Marquis	1.024	0.974	0.002	0.550	0.510	0.002	0.125	125-169	21-26
	Town Car	1.024	0.974	0.002	0.550	0.510	0.002	0.125	125-169	21-26
2002	Crown Victoria	1.063	1.010	0.002	NA	0.790	0.003	0.039	118	32
	Grand Marquis	1.063	1.010	0.002	NA	0.790	0.003	0.039	118	32
	Town Car	1.063	1.010	0.002	NA	0.790	0.003	0.039	118	32
2003	Crown Victoria	1.063	1.037	0.002	NA	0.790	0.003	0.039	118	32
	Grand Marquis	1.063	1.037	0.002	NA	0.790	0.003	0.039	118	32
	Town Car	1.063	1.037	0.002	NA	0.790	0.003	0.039	118	32
	Marauder	1.063	1.037	0.002	NA	0.790	0.003	0.039	118	32
2004	Crown Victoria	1.063	1.037	0.002	NA	0.790	0.003	0.039	118	32
	Grand Marquis	1.063	1.037	0.002	NA	0.790	0.003	0.039	118	32
	Town Car	1.063	1.037	0.002	NA	0.790	0.003	0.039	118	32
	Marauder	1.063	1.037	0.002	NA	0.790	0.003	0.039	118	32

NOTE: Follow specifications stamped on rotor or drum if figures differ from those in this chart.

NA: Not Available

67197-CROW-C11

SCHEDULED MAINTENANCE INTERVALS

Ford—Crown Victoria, Mercury—Grand Marquis, Marauder & Lincoln—Town Car

TO BE SERVICED	TYPE OF SERVICE	VEHICLE MILEAGE INTERVAL (x1000)												
		5	10	15	20	25	30	35	40	45	50	55	60	65
Engine oil & filter	R	✓	✓	✓	✓	✓	✓	✓	✓	✓	✓	✓	✓	✓
Rotate tires	S/I	✓		✓		✓		✓		✓		✓		✓
Cooling system, hoses, clamps & coolant strength	S/I			✓			✓			✓			✓	
Lubricate steering linkage	S/I			✓			✓			✓			✓	
Air cleaner element	R						✓						✓	
Automatic transaxle fluid & filter	R						✓						✓	
Spark plugs ①	R													
Exhaust heat shields	S/I						✓						✓	
Fuel filter (NGV Crown Victoria) ②	R					✓					✓			
Front & rear brakes	S/I						✓						✓	
Lubricate suspension (Town Car)	S/I						✓						✓	
Engine coolant ③	R										✓			
PCV valve	R												✓	
Accessory drive belt	S/I												✓	

R: Replace S/I: Service or Inspect

① Replace every 100,000 miles.

② Also drain coalescer assembly. Perform every 24,000 miles for severe service.

③ Change initially at 50,000 miles & thereafter every 30,000 miles.

FREQUENT OPERATION MAINTENANCE (SEVERE SERVICE)

If a vehicle is operated under any of the following conditions it is considered severe service:

- **Extremely dusty areas.**
- **50% or more of the vehicle operation is in 32°C (90°F) or higher temperatures, or constant operation in temperatures below 0°C (32°F).**
- **Prolonged idling (vehicle operation in stop and go traffic).**
- **Frequent short running periods (engine does not warm to normal operating temperatures).**
- **Police, taxi, delivery usage or trailer towing usage.**

Oil & filter change: change every 3000 miles.

Rotate tires at 6000 miles & every 9000 miles thereafter.

Automatic transmission fluid & filter: change every 21,000 miles.

67197-CROW-C12

PRECAUTIONS

Before servicing any vehicle, please be sure to read all of the following precautions, which deal with personal safety, prevention of component damage, and important points to take into consideration when servicing a motor vehicle:

- Never open, service or drain the radiator or cooling system when the engine is hot; serious burns can occur from the steam and hot coolant.
- Observe all applicable safety precautions when working around fuel. Whenever servicing the fuel system, always work in a well-ventilated area. Do not allow fuel spray or vapors to come in contact with a spark, open flame, or excessive heat (a hot drop light, for example). Keep a dry chemical fire extinguisher near the work area. Always keep fuel in a container specifically designed for fuel storage; also, always properly seal fuel containers to avoid the possibility of fire or explosion. Refer to the additional fuel system precautions later in this section.
- Fuel injection systems often remain pressurized, even after the engine has been turned **OFF**. The fuel system pressure must be relieved before disconnecting any fuel lines. Failure to do so may result in fire and/or personal injury.
- Brake fluid often contains polyglycol ethers and polyglycols. Avoid contact with the eyes and wash your hands thoroughly after handling brake fluid. If you do get brake fluid in your eyes, flush your eyes with clean, running water for 15 minutes. If eye irritation persists, or if you have taken brake fluid internally, IMMEDIATELY seek medical assistance.
- The EPA warns that prolonged contact with used engine oil may cause a number of skin disorders, including cancer. You should make every effort to minimize your exposure to used engine oil. Protective gloves should be worn when changing oil. Wash your hands and any other exposed skin areas as soon as possible after exposure to used engine oil. Soap and water, or waterless hand cleaner should be used.
- All new vehicles are now equipped with an air bag system, often referred to as a Supplemental Restraint System (SRS) or Supplemental Inflatable Restraint (SIR) system. The system must be disabled before performing service on or around system components, steering column, instrument panel components, wiring and sensors. Failure to follow safety and disabling procedures could result in accidental air bag deployment, possible personal injury and unnecessary system repairs.
- Always wear safety goggles when working with, or around, the air bag system. When carrying a non-deployed air bag, be sure the bag and trim cover are pointed away from your body. When placing a non-deployed air bag on a work surface, always face the bag and trim cover upward, away from the surface. This will reduce the motion of the module if it is accidentally deployed. Refer to the additional air bag system precautions later in this section.
- Clean, high quality brake fluid from a sealed container is essential to the safe and proper operation of the brake system. You should always buy the correct type of brake fluid for your vehicle. If the brake fluid becomes contaminated, completely flush the system with new fluid. Never reuse any brake fluid. Any brake fluid that is removed from the system should be discarded. Also, do not allow any brake fluid to come in contact with a painted surface; it will damage the paint.
- Never operate the engine without the proper amount and type of engine oil; doing so WILL result in severe engine damage.
- Timing belt maintenance is extremely important. Many models utilize an interference-type, non-freewheeling engine. If the timing belt breaks, the valves in the cylinder head may strike the pistons, causing potentially serious (also time-consuming and expensive) engine damage.
- Disconnecting the negative battery cable on some vehicles may interfere with the functions of the on-board computer system(s) and may require the computer to undergo a relearning process once the negative battery cable is reconnected.
- When servicing drum brakes, only disassemble and assemble one side at a time, leaving the remaining side intact for reference.

ENGINE REPAIR

Alternator

REMOVAL

1. Before servicing the vehicle, refer to the precautions in the beginning of this section.
2. Remove or disconnect the following:
   - Negative battery cable
   - Engine cover
   - Pushpins and wiring harness
   - Mounting bolts
   - Alternator bracket
   - Accessory drive belt
   - Alternator

INSTALLATION

Install or connect the following:

- Alternator. Tighten the mounting bolts to 15–22 ft. lbs. (20–30 Nm).
- Alternator bracket. Tighten the bolts to 71–106 inch lbs. (8–12 Nm).
- Accessory drive belt
- Pushpins and wiring harness
- Engine cover
- Negative battery cable

Ignition Timing

ADJUSTMENT

The 4.6L engines are equipped with a Distributorless Ignition System (DIS). No adjustments are necessary.

Engine Assembly

REMOVAL & INSTALLATION

1. Before servicing the vehicle, refer to the precautions in the beginning of this section.
2. Drain the engine cooling system.
3. Recover the refrigerant from the air conditioning system.
4. Relieve the fuel system pressure.
5. Drain the engine oil.
6. Remove or disconnect the following:
   - Battery
   - Hood
   - Engine cooling fan, shroud and radiator
   - Windshield wiper governor and support bracket
   - Engine air cleaner outlet tube
   - Engine/transmission harness connector from the retaining bracket
   - Accelerator and cruise control cables at the throttle body
   - Electrical connector and vacuum hose from the evaporative emission canister purge valve
   - Positive battery cable from the power distribution box and harness

- Vacuum supply hose from the throttle body adapter vacuum port
- Both heater hoses
- Alternator harness from the front fender apron and the power distribution box
- Air conditioning hoses from the air conditioning compressor
- Power steering control valve harness connector
- Body ground strap from the dash panel
- Exhaust system from the exhaust manifolds and support with wire hung from the crossmember
- Retaining nut from the transmission line bracket
- 3 bolts and 1 stud retaining the engine to the transmission knee braces
- Starter motor
- 4 bolts retaining the power steering pump to the cylinder block and position aside

7. Transmission housing cover from the cylinder block to access the torque converter nuts.

Rotate the crankshaft until each of the 4 nuts is accessible and remove the nuts

8. Remove or disconnect the following:
    - 6 transmission-to-engine retaining bolts
    - Engine support insulator (mount) through-bolts
9. Support the transmission with a floor jack and a block of wood.
    - Bolt retaining the right-hand front engine support insulator to the front engine mount insulator support bracket
10. Install engine lifting bracket to the front of the left-hand cylinder head and to the rear of the right-hand cylinder head. Connect engine lifting equipment to the lifting brackets
11. Raise the engine slightly using a floor crane and carefully separate the engine from the transmission. Do not let the torque converter fall out of the transmission.
12. Carefully lift the engine out of the engine compartment and position on a workstand. Remove the engine lifting equipment.

To install:

13. Engine lifting brackets. Support the engine using a floor crane installed to the lifting equipment and remove the engine from the workstand
14. Lower the engine into the engine compartment. Start the converter pilot into the flywheel and align the paint marks on the flywheel and torque converter. Be sure the studs on the torque converter align with the holes in the flywheel.
15. Fully engage the engine to the transmission and lower onto front engine support insulators.
16. Install or connect the following:
    - Engine lifting equipment and brackets
    - Bolt retaining the right-hand front engine support insulator to the front engine mount insulator support bracket
    - 6 engine-to-transmission retaining bolts and tighten to 30–44 ft. lbs. (40–60 Nm)
    - Front engine support insulator through-bolts and tighten to 15–22 ft. lbs. (20–30 Nm)
    - 4 torque converter retaining nuts and tighten to 22–25 ft. lbs. (20–30 Nm)
    - Transmission housing cover to the cylinder block
    - Power steering pump on the cylinder block and the 4 retaining nuts. Tighten to 15–22 ft. lbs. (20–30 Nm).
    - Starter motor
    - Engine-to-transmission brace and the 3 bolts and 1 stud. Tighten the bolts and stud to 18–31 ft. lbs. (25–43 Nm).
    - Transmission line bracket to the brace stud and 1 retaining nut. Tighten to 15–22 ft. lbs. (20–30 Nm).
    - Exhaust system to the exhaust manifolds. Tighten the 4 nuts to 20–30 ft. lbs. (27–41 Nm). Be sure the exhaust system clears the No. 3 crossmember. Adjust as necessary.
    - Power steering valve harness connector
    - Ground strap to the dash panel
    - Air conditioning lines to the air conditioning compressor
    - Alternator harness at the front fender apron and the power distribution box
    - Both heater hoses
    - Vacuum supply hose to the throttle body adapter vacuum port
    - Positive battery cable to the power distribution box and harness
    - Electrical connector and vacuum hose to the evaporative emission canister purge valve
    - Accelerator and cruise control cables at the throttle body
    - Engine/transmission harness connector to the retaining bracket on the power brake booster
    - Windshield wiper governor and support bracket
    - Fuel supply and return lines
    - Radiator, cooling fan and shroud
    - Engine air cleaner outlet tube
    - Hood
    - Battery
17. Fill the crankcase to the correct level.
18. Fill the cooling system.
19. Start the engine and allow it to reach normal operating temperature.
20. Check for leaks and proper fluid levels.
21. Evacuate and recharge the air conditioning system.
22. Road test the vehicle and check the engine and transmission for proper operation.

Water Pump

REMOVAL & INSTALLATION

1. Before servicing the vehicle, refer to the precautions in the beginning of this section.
2. Drain the cooling system.
3. Remove or disconnect the following:
    - Negative battery cable
    - Cooling fan and shroud
    - Accessory drive belt
    - 4 water pump pulley-to-water pump bolts
    - Pulley
    - 4 water pump-to-engine bolts
    - Water pump

To install:

4. Clean the sealing surfaces of the water pump and block.
5. Install or connect the following:
    - New O-ring, lubricate it with clean antifreeze prior to installation
    - Water pump. Tighten the bolts to 15–22 ft. lbs. (20–30 Nm).
    - Water pump pulley. Tighten the bolts to 15–22 ft. lbs. (20–30 Nm).
    - Accessory drive belt
6. Fill the cooling system.

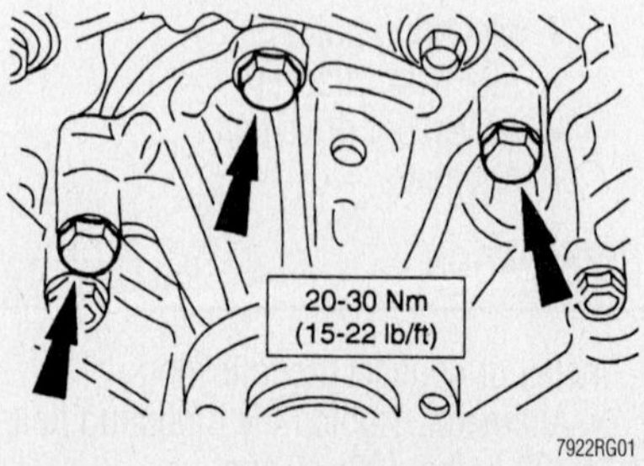

Be sure to tighten the water pump mounting bolts to the specification

7. Operate the engine to normal operating temperatures and check for leaks.

Heater Core

REMOVAL & INSTALLATION

1. Before servicing the vehicle, refer to the precautions in the beginning of this section.
2. If equipped, turn the air suspension service switch to the **OFF** position before raising the vehicle.
3. Drain the cooling system.
4. Remove or disconnect the following:
   - Negative battery cable
   - Front seats
   - Carpeting
   - Rear airflow duct
   - Heater hoses
5. Remove the driver's side air bag module by removing or disconnect the following:
   - SRS module-to-steering wheel bolts, from both sides of the steering wheel.
   - SRS module and disconnect the electrical connector
   - Horn switch electrical connector
6. Remove the passenger's side SRS module by removing or disconnecting the following:
   - Open the glove box and disconnect the glove compartment isolator
   - Push inward on the 2 glove box door tabs and lower it
   - SRS module's electrical connector
   - SRS module-to-instrument panel bolts and the module
7. Remove the instrument panel by removing or disconnecting the following:
   - Speed control servo nuts and move the servo aside
   - Bolt and disconnect the left side bulkhead connector
   - Left side bulkhead connector from the dash panel
   - Windshield washer fluid reservoir screw, and position the reservoir aside
   - Blower motor electrical connector
   - Air conditioning pressure cut-off switch electrical connector
   - In-line electrical harness connector
   - Electronic Automatic Temperature Control (EATC) variable blower motor controller electrical connector
   - Right front wheel
   - Right front fender splash shield bolts and move the shield away from the cowl
   - Wiring harness from the evaporator case
   - Wiring harness from the cowl
   - Right side instrument panel lower insulator pushpins, disconnect the power point electrical connector, remove the courtesy lamp from the socket and remove the insulator
   - Unseat the wiring grommet and feed the wiring harness through the cowl
   - Cowl side trim panels and the windshield side garnish moldings from both sides
   - Locking clip and the Electronic Crash Sensor (ECS) module electrical connector
   - Antenna connector
   - Ground wire bolts from the right side
   - EATC hose at the evaporator housing
   - Bolt and disconnect the bulkhead connector from the right side
   - Pull the door weatherstrip seals away from the instrument panel on both sides
   - Tunnel brace trim cover
   - Climate control head vacuum harness connector
   - Instrument panel tunnel brace nuts and the brace
   - Ground wire bolts from the left side
   - Wiring harness connectors from the left side
   - Parking brake switch electrical connector

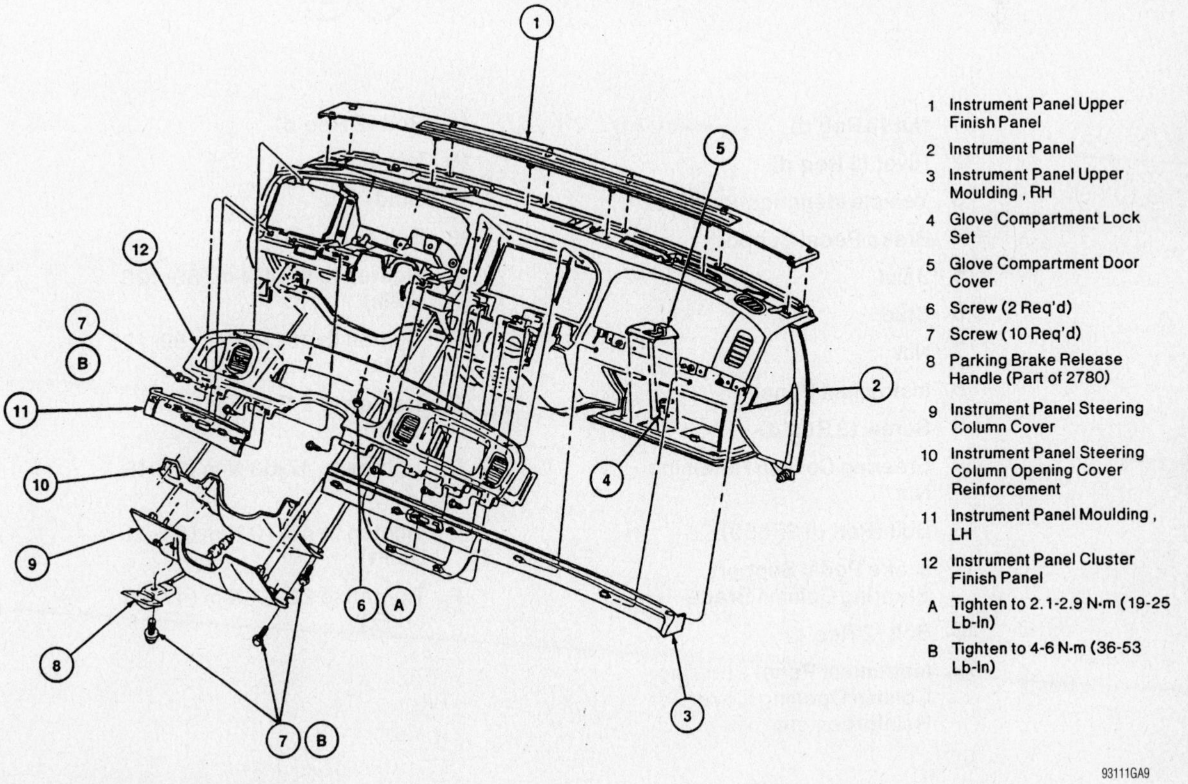

View of the instrument panel finish panels

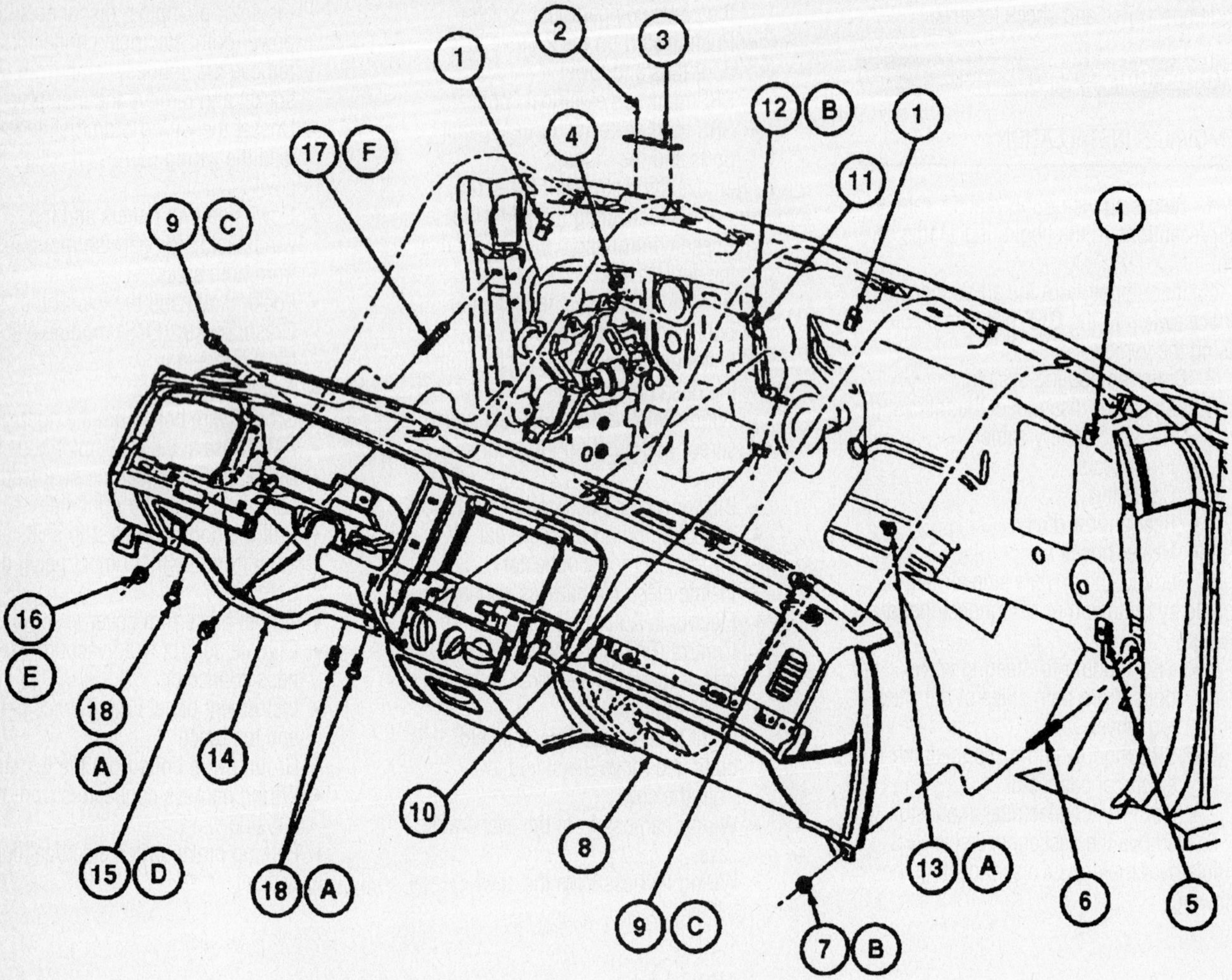

1	Nut (3 Req'd)	15	Bolt (2 Req'd)
2	Rivet (2 Req'd)	16	Nut
3	Vehicle Identification Plate	17	Stud
4	Brake Pedal Support	18	Bolt (3 Req'd)
5	J Nut	A	Tighten to 9-14 N·m (80-123 Lb-In)
6	Stud	B	Tighten to 10-14 N·m (89-123 Lb-In)
7	Nut	C	Tighten to 2-3 N·m (18-26 Lb-In)
8	Instrument Panel	D	Tighten to 47-63 N·m (35-46 Lb-Ft)
9	Screw (3 Req'd)	E	Tighten to 45-70 N·m (34-51 Lb-Ft)
10	Steering Column Retaining Nut	F	Tighten to 22-34 N·m (17-25 Lb-Ft)
11	Bolt (Part of 3F659)		
12	Brake Pedal Support Steering Column Brace		
13	Bolt (2 Req'd)		
14	Instrument Panel Steering Column Opening Cover Reinforcement		

93111GA0

Exploded view of the instrument panel—Crown Victoria and Grand Marquis shown, other models similar

a. Pry out the instrument panel defroster opening grille, disconnect the electrical connectors and the remove the grille.

8. Remove or disconnect the following:
- 3 upper instrument panel screws
- Instrument panel cowl side nut from both sides
- Instrument panel cowl side bolt from the left side
- Instrument panel
- Windshield wiper mounting arm and the pivot shaft
- Electrical connector and remove the upper cowl extension screw and the extension
- Vacuum hoses and the wire harness connector from the evaporative emissions canister purge valve
- Evaporative emissions purge valve nuts and the valve
- Cowl side nut
- Right side rear fender apron screws and reposition the apron
- Evaporator housing nut and screw
- Vacuum hoses and the wiring harness connectors
- Plenum chamber's lower flange nuts
- Plenum chamber's upper flange nut and the plenum chamber
- Heater core cover-to-heater plenum chamber screws and the cover
- Seal and the heater core

To install:

9. Install or connect the following:
- Seal and the heater core
- Heater core cover and the cover-to-heater plenum chamber screws
- Plenum chamber and the plenum chamber's upper flange nut
- Plenum chamber's lower flange nuts
- Vacuum hoses and the wiring harness connectors
- Evaporator housing nut and screw
- Right side rear fender apron and the apron screws
- Cowl side nut

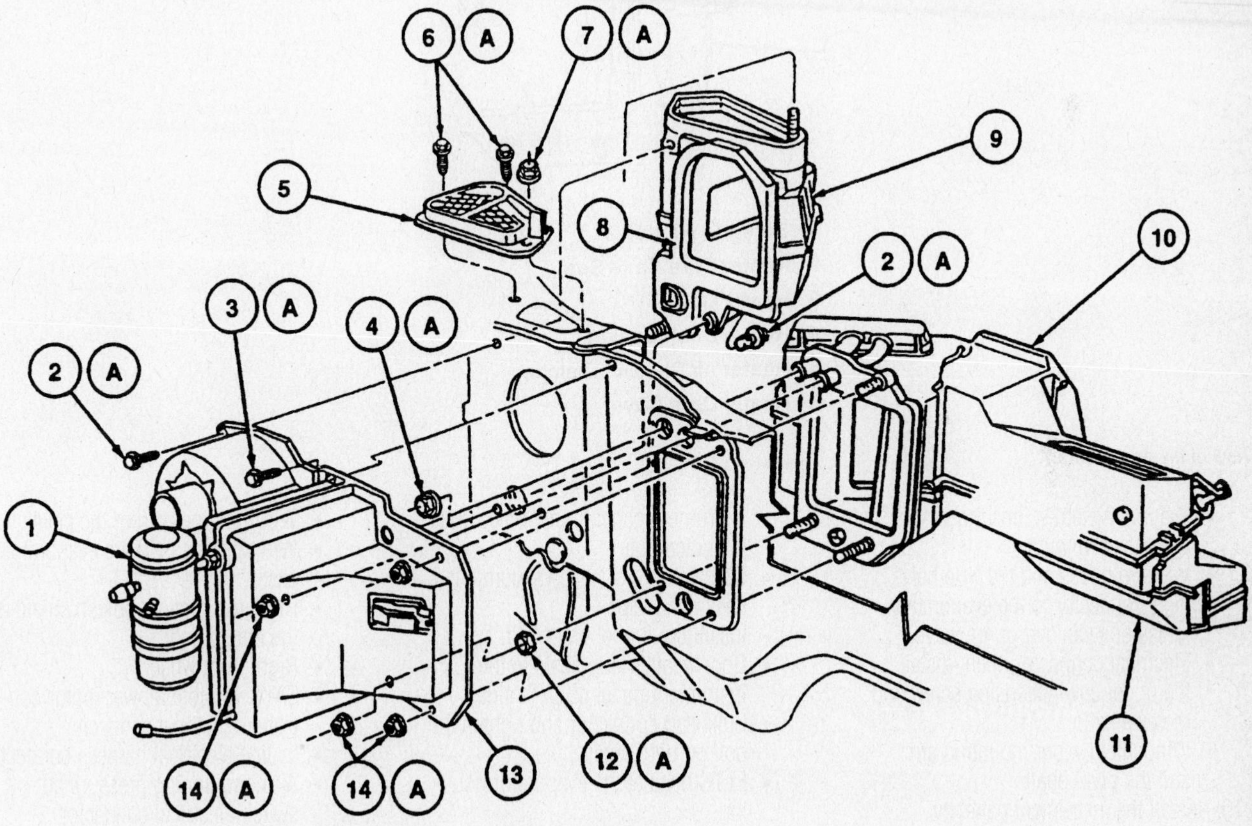

1	Suction Accumulator / Drier
2	Screw (2 Req'd)
3	Screw
4	Nut and Washer Assy
5	A / C Recirculating Air Duct Screen
6	Screw (2 Req'd)
7	Nut and Washer
8	A / C Air Inlet Door Inner Seal
9	A / C Recirculating Air Duct
10	Heater Air Plenum Chamber
11	Heater Outlet Floor Duct
12	Nut
13	A / C Evaporator Core Housing
14	Nut and Washer
A	Tighten to 2.5-3.2 N·m (23-28 Lb-In)

93111GB1

Exploded view of the heater air plenum chamber, the evaporator housing and the air inlet duct

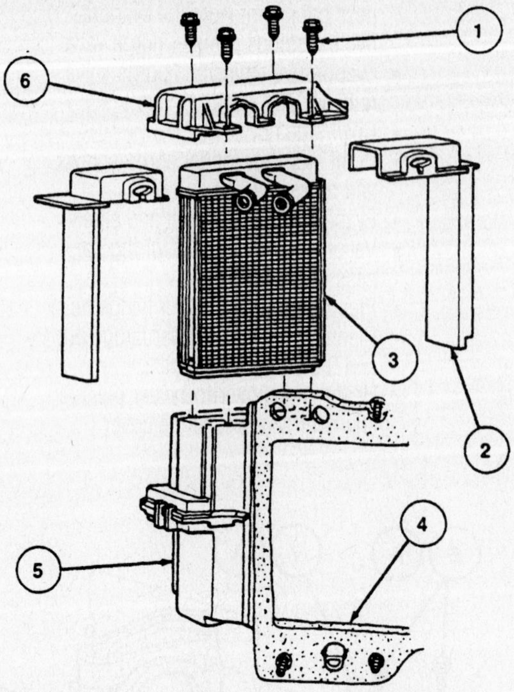

View of the heater core

- Evaporative emissions purge valve and the valve nuts
- Vacuum hoses and the wire harness connector to the evaporative emissions canister purge valve
- Electrical connector; then, install the upper cowl extension screw and the extension
- Windshield wiper mounting arm and the pivot shaft

10. Install the instrument panel by installing or connecting the following:
- Instrument panel
- At the left side, install the instrument panel cowl side bolt
- Instrument panel cowl side nut on both sides
- 3 upper instrument panel screws
- Electrical connectors and the install the instrument panel defroster opening grille
- Parking brake switch electrical connector
- Wiring harness connectors on the left side
- Ground wire bolts on the left side
- Instrument panel tunnel brace and the brace nuts
- Climate control head vacuum harness connector
- Install the tunnel brace trim cover.
- Door weatherstrip seals to the instrument panel on both sides
- Bulkhead connector and tighten the bolt on both sides
- EATC hose at the evaporator housing
- Ground wire bolts on the right side
- Antenna connector
- ECS module electrical connector and the locking clip
- Cowl side trim panels and the windshield side garnish moldings on both sides
- Feed the wiring harness through the cowl and seat the wiring grommet
- Right side instrument panel lower insulator, connect the power point electrical connector, install the courtesy lamp to the socket and secure the insulator with pushpins
- Wiring harness from the cowl
- Wiring harness to the evaporator case
- Right front fender splash shield and the shield bolts
- Right front wheel
- EATC variable blower motor controller electrical connector
- In-line electrical harness connector
- Air conditioning pressure cut-off switch electrical connector
- Blower motor electrical connector
- Windshield washer fluid reservoir and the reservoir screw
- Left side bulkhead connector to the dash panel
- Left side bulkhead connector and tighten the bolt
- Speed control servo and the servo nuts
- Right SRS module and torque the module-to-instrument panel bolts to 62–97 inch lbs. (7–11 Nm)
- Right SRS module electrical connector
- Glove compartment isolator

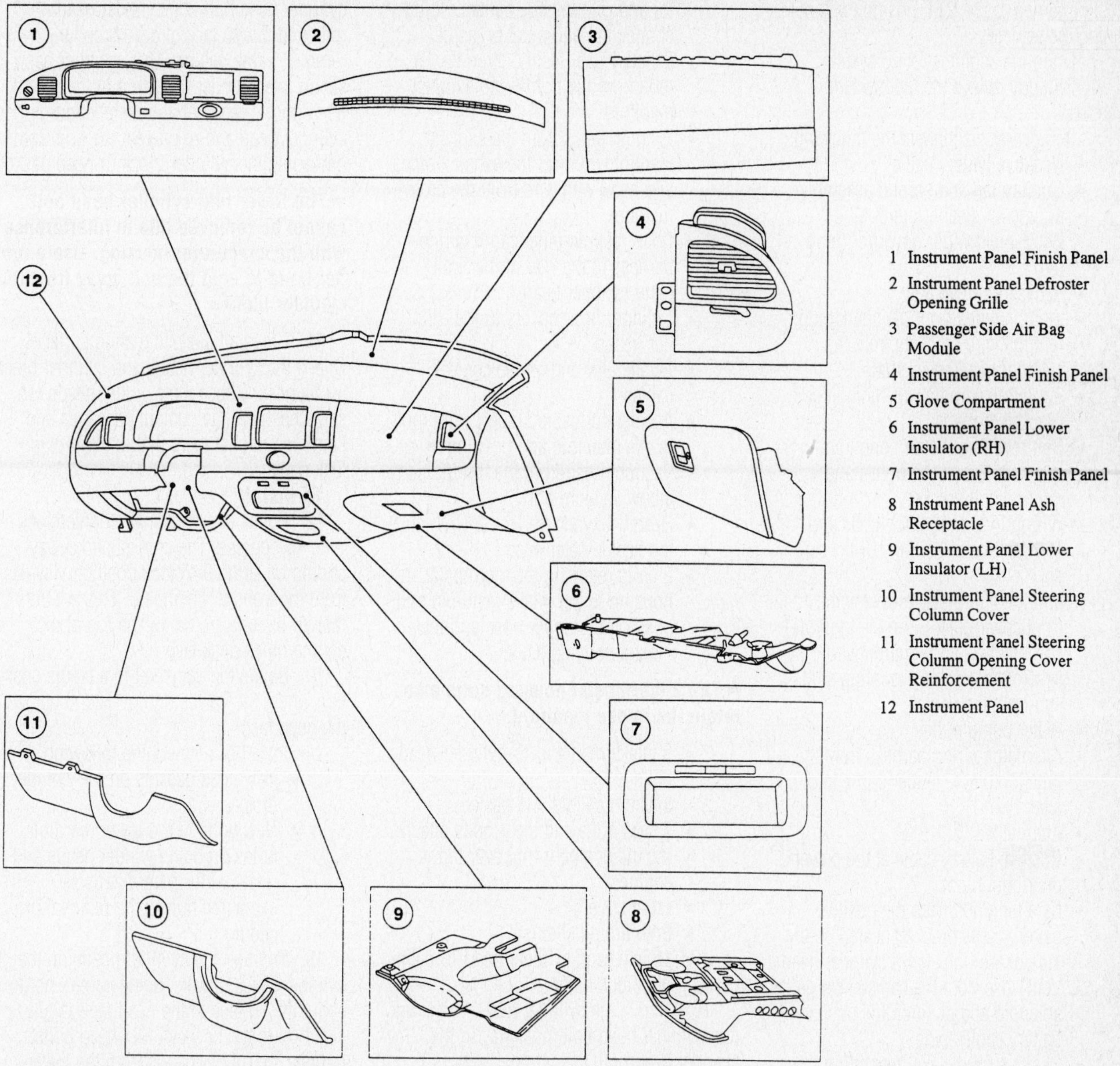

Exploded view of the instrument panel—Town Car

- Glove box door
- Horn switch electrical connector
- Electrical connector and install the left SRS module
- Left SRS module-to-steering wheel bolts on both sides of the wheel, and torque the bolts to 90–122 inch lbs. (10–14 Nm)
- Rear airflow duct
- Carpeting
- Front seats
- Heater hoses

11. Refill the cooling system.
12. Connect the negative battery cable.
13. Operate the engine to normal operating temperature. Check the climate control operation and check for leaks.

Cylinder Head

REMOVAL & INSTALLATION

4.6L (VIN W) SOHC Engine

➡**The cylinder head bolts are a torque-to-yield design and cannot be reused. Always use new cylinder head bolts for assembly.**

1. Before servicing the vehicle, refer to the precautions in the beginning of this section.
2. If equipped with air suspension, the air suspension switch, located on the right-hand side of the luggage compartment,

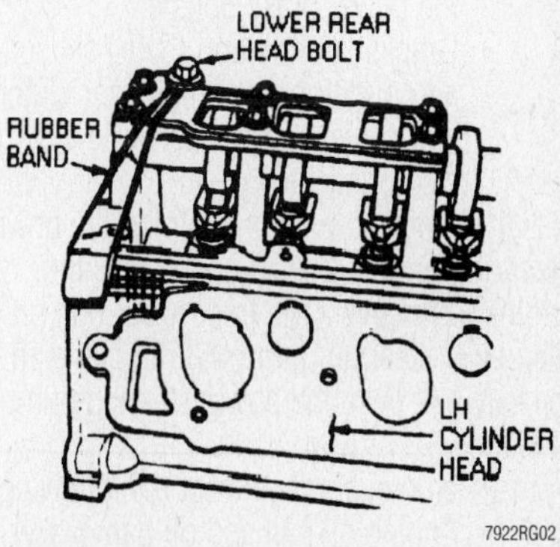

Use a rubber band to support the rear cylinder head bolt to ease removal of the head

must be turned to the **OFF** position before raising the vehicle.

3. Drain the engine cooling system.

4. Properly relieve the fuel system pressure.

5. Remove or disconnect the following:

- Negative battery cable
- Cooling fan and shroud assembly
- Engine air cleaner outlet tube
- Windshield wiper governor (module)
- Accessory drive belt
- Ignition wires from the spark plugs
- Ignition wire brackets from the cylinder head cover studs
- 2 ignition wire tray-to-ignition coil brackets bolts
- Bolt retaining the air conditioning pressure line to the right-hand ignition coil bracket
- Wiring to both ignition coils and the Camshaft Position (CMP) sensor
- Ignition coil brackets-to-engine front cover nuts. Slide the ignition coil brackets and ignition wire assemblies off the mounting studs and from the vehicle
- Water pump pulley
- Alternator wiring harness from the junction block, fender apron and alternator
- Alternator
- Positive battery cable at the power distribution box
- Retaining bolt from the positive battery cable bracket located on the side of the right-hand cylinder head
- Vent hose from the canister purge solenoid and position the positive battery cable aside
- Positive Crankcase Ventilation (PCV) valve from the cylinder head cover
- Engine/transmission harness connector from the retaining bracket on the power brake booster
- Crankshaft Position (CKP) sensor, air conditioning compressor clutch and canister purge solenoid electrical connectors

6. Remove the bolts retaining the power steering pump to the cylinder block and engine front cover. The front lower bolt on the power steering pump will not come all the way out. Wire the power steering pump aside.

7. Remove or disconnect the following:

- Engine oil pan and oil pan gasket
- Crankshaft pulley retaining bolt
- Pulley
- Power steering control valve actuator and oil pressure sensor wiring connectors and position aside.
- Exhaust Gas Recirculation (EGR) tube from the right-hand exhaust manifold
- Exhaust pipes from the exhaust manifolds. Lower the exhaust pipes and hang with wire from the crossmember.
- Bolts retaining the starter wiring harness to the rear of the right-hand cylinder head
- Cylinder head covers to the cylinder heads
- Accelerator and cruise control cables
- Accelerator cable bracket from the intake manifold and position aside
- Vacuum hose from the throttle body elbow vacuum port
- Heated Oxygen Sensor (HO_2S) and the heater water hose
- 2 bolts retaining the thermostat housing to the intake manifold and position the upper hose and thermostat housing aside

➡The 2 thermostat housing bolts also retain the intake manifold.

- 9 intake manifold-to-cylinder heads bolts
- Intake manifold and gaskets
- 7 stud bolts and the 4 bolts attaching the engine front cover to the engine
- Front cover
- Both timing chains
- 10 left-hand cylinder head-to-cylinder block bolts

8. Remove the cylinder head. The lower rear cylinder head bolt must stay in the cylinder head until the cylinder head is removed due to lack of clearance for removal in the vehicle. Use a rubber band to secure the cylinder head bolt in the cylinder head during removal and installation of the cylinder head and to prevent the bolt from damaging the cylinder block or head gasket

➡The lower rear cylinder head bolt cannot be removed due to interference with the power brake booster. Use a rubber band to hold the bolt away from the cylinder block.

9. Remove or disconnect the following:

- Ground strap, 1 stud and 1 bolt retaining the heater return line to the right-hand cylinder head
- 10 right-hand cylinder head-to-cylinder block bolts

10. Remove the cylinder head. The lower rear cylinder head bolt must stay in the cylinder head until the cylinder head is removed due to lack of clearance for removal in the vehicle. Use a rubber band to secure the cylinder head bolt in the cylinder head during removal and installation of the cylinder head and to prevent the bolt from damaging the cylinder block or head gasket.

➡The lower rear cylinder head bolt cannot be removed due to interference with the evaporator housing. Use a rubber band to hold the bolt away from the cylinder block.

11. Clean all gaskets mating surfaces. Check the cylinder heads and cylinder block for flatness. Check the cylinder heads for scratches near the coolant passages and combustion chambers that could provide leak paths.

To install:

12. Rotate the crankshaft counterclockwise 45 degrees. The crankshaft keyway should be at the 9 o'clock position viewed from the front of the engine. This ensures that all pistons are below the top of the engine block deck face.

13. Rotate the camshaft to a stable position where the valves do not extend below the head face.

14. Install or connect the following:

- New head gaskets on the cylinder block
- New bolts in the lower rear bolt holes on both cylinder heads and retain with rubber bands as explained during the removal procedure

15. Position the cylinder heads on the cylinder block dowels, being careful not to score the surface of the head face. Apply clean oil to the new cylinder head bolts, remove the rubber bands from the lower rear bolts and install all bolts hand-tight.

16. Tighten the new cylinder head bolts, in sequence, as follows:

a. Step 1: 28–31 ft. lbs. (37–43 Nm).
b. Step 2: plus 85–95 degrees.
c. Step 3: loosen all bolts at least 1 full turn.
d. Step 4: 27–32 ft. lbs. (37–43 Nm).
e. Step 5: plus 85–95 degrees.
f. Step 6: again, plus 85–95 degrees.

17. Position the heater return hose and install the 2 retaining bolts.

18. Rotate the camshafts using the flats matched at the center of the camshaft until both are in time. Install Camshaft Positioning Tools T91P-6256-A, on the flats of the camshafts to keep them from rotating.

19. Rotate the crankshaft clockwise 45 degrees to position the crankshaft at Top Dead Center (TDC) for the No. 1 cylinder.

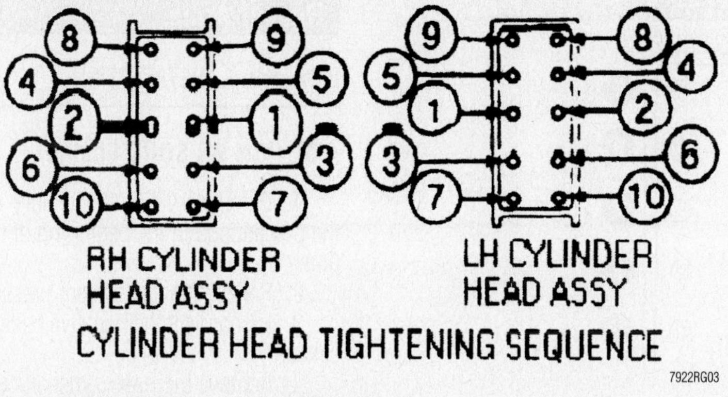

Cylinder head torque sequence—4.6L SOHC engine

➡The crankshaft must only be rotated in the clockwise direction and only as far as TDC.

20. Install or connect the following:
- Both timing chains
- New engine front cover seal and gasket. Apply silicone sealer to the lower corners of the cover where it meets the junction of the engine oil pan and cylinder block and to the points where the cover contacts the junction of the cylinder block and the cylinder heads.
- Engine front cover and the bolts. Tighten to 15–22 ft. lbs. (20–30 Nm).
- New intake manifold gaskets on the cylinder heads. Be sure the alignment tabs on the gaskets are aligned with the holes in the cylinder heads.
- Intake manifold on the cylinder heads and the retaining bolts. Tighten the bolts in sequence, to 15–22 ft. lbs. (20–30 Nm).
- Thermostat, O-ring, thermostat housing and upper hose. Tighten the 2 retaining bolts to 15–22 ft. lbs. (20–30 Nm).
- Heater water hose and both HO_2S sensors
- Vacuum hose to the throttle body adapter vacuum port
- Accelerator cable bracket on the intake manifold
- Accelerator and cruise control cables to the throttle body

21. Apply silicone sealer to both places where the engine front cover meets the cylinder heads.
- Cylinder head covers with new gasket on the cylinder heads. Tighten the bolts and stud bolts to 71–106 inch lbs. (8–12 Nm).
- Starter motor wiring harness to the right-hand cylinder head and tighten the retaining bolt
- Exhaust pipes to the exhaust manifolds. Tighten the 4 nuts to 20–30 ft. lbs. (27–41 Nm).

➡Be sure the exhaust system clears the No. 3 crossmember. Adjust as necessary.

- EGR tube to the right-hand exhaust manifold and tighten the line nut to 26–33 ft. lbs. (35–45 Nm).
- Power steering control valve actuator and oil pressure sensor electrical connectors

22. Apply a small amount of silicone sealer in the rear of the keyway on the crankshaft pulley.
- Pulley on the crankshaft, making sure the crankshaft key and keyway are aligned.

23. Install the crankshaft pulley and tighten the bolt as follows:
   a. Step 1: Tighten to 66 ft. lbs. (90 Nm).
   b. Step 2: Loosen one complete turn.
   c. Step 3: Tighten to 35–39 ft. lbs. (47–53 Nm).
   d. Step 4: Tighten an additional 85–95 degrees.

24. Install or connect the following:
- Engine oil pan and a new gasket
- Power steering pump in position on the cylinder block
- 4 retaining bolts. Tighten the bolts to 15–22 ft. lbs. (20–30 Nm).
- Air conditioning compressor, CKP sensor and canister purge solenoid electrical connectors
- Engine/transmission harness connector on the power brake booster
- PCV valve in the right-hand cylinder head cover and connect the canister purge solenoid vent hose
- Positive battery cable harness on the right-hand cylinder head
- Bolt retaining the cable bracket to the cylinder head
- Positive battery cable at the power distribution box and battery
- Alternator and the 2 retaining bolts. Tighten the bolts to 15–22 ft. lbs. (20–30 Nm).
- 2 bolts retaining the alternator brace to the intake manifold. Tighten to 72–96 inch lbs. (8–12 Nm).
- Water pump pulley. Tighten the bolts to 15–22 ft. lbs. (20–30 Nm).
- Ignition coil brackets and ignition wire assemblies onto the mounting studs
- 7 nuts retaining the ignition coil brackets to the engine front cover and tighten to 15–22 ft. lbs. (20–30 Nm)
- 2 bolts retaining the ignition wire tray to the ignition coil bracket and tighten to 71–106 inch lbs. (8–12 Nm)
- Ignition coil and CMP sensor harness connectors
- Air conditioning pressure line on the right-hand ignition coil bracket and tighten the retaining bolt
- Ignition wires to the spark plugs and the bracket onto the cylinder head cover studs
- Accessory drive belt and the windshield wiper governor
- Fuel supply and return lines
- Cooling fan and shroud
- Engine air cleaner outlet tube
- Negative battery cable

25. Fill the cooling system.

26. If equipped with air suspension, turn the air suspension switch to the **ON** position.

27. Refill the engine with the correct amount of oil and replace the filter.

28. Start the engine and bring to normal operating temperature while checking for leaks.

29. Road test the vehicle and check for proper engine operation.

4.6L (VIN V) DOHC Engine

1. Before servicing the vehicle, refer to the precautions in the beginning of this section.

2. Remove the engine from the vehicle and mount it on a suitable workstand.

3. Remove or disconnect the following:
- Fuel injectors
- Intake manifold
- Front cover
- Timing chains
- Valve covers
- Exhaust manifolds
- Cylinder heads

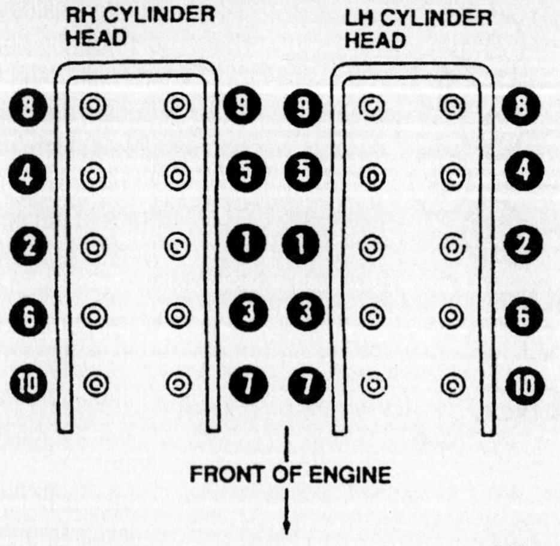

7922NG11

Cylinder head torque sequence—4.6L DOHC engine

To install:

➡Use new cylinder head bolts for assembly.

4. Install the cylinder heads with new gaskets and bolts. Tighten the bolts in sequence as follows:
   a. Step 1: 28–31 ft. lbs. (37–43 Nm)
   b. Step 2: Plus 85–95 degrees
   c. Step 3: Loosen all bolts one full turn
   d. Step 4: 28–31 ft. lbs. (37–43 Nm)
   e. Step 5: Plus 85–95 degrees
   f. Step 6: Plus 85–95 degrees
5. Install or connect the following:
   - Exhaust manifolds
   - Valve covers
   - Timing chains
   - Front cover
   - Intake manifold
   - Fuel injectors
6. Install the engine into the vehicle.

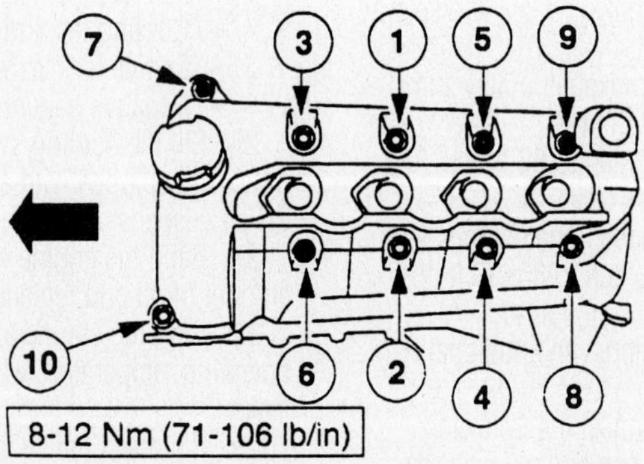

9346GG08

Left valve cover torque sequence—4.6L (VIN V) engine

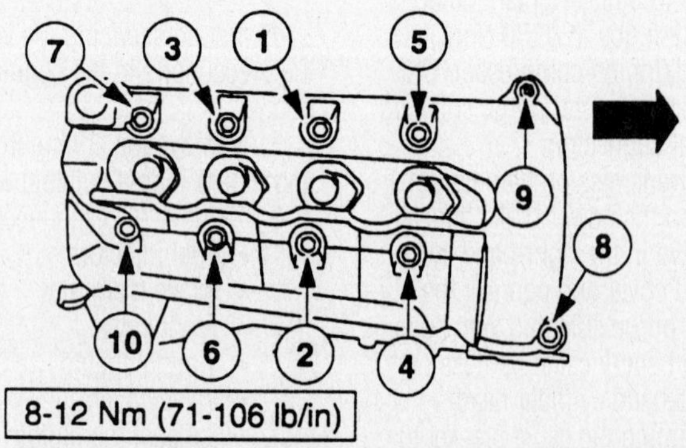

9346GG09

Right valve cover torque sequence—4.6L (VIN V) engine

Rocker Arms

REMOVAL & INSTALLATION

4.6L (VIN W) SOHC Engine

1. Before servicing the vehicle, refer to the precautions in the beginning of this section.
2. Relieve the fuel system pressure.
3. Disconnect the negative battery cable.
4. Remove the right camshaft cover by removing or disconnecting the following:
   - Positive battery cable at the battery and at the power distribution box
   - Retaining bolt from the positive battery cable bracket located on the side of the right cylinder head
   - Crankshaft Position (CKP) sensor, air conditioning compressor clutch and canister purge solenoid connectors. Position the harness aside.
   - Vent hose from the purge solenoid and position the positive battery cable aside
   - Ignition wires from the spark plugs
   - Ignition wire brackets from the camshaft cover studs and position the wires aside
   - PCV valve from the camshaft cover grommet and position aside
   - Bolts and stud bolts and remove the camshaft cover.
5. Remove the left camshaft cover by removing or disconnecting the following:
   - Air inlet tube
   - Fuel lines
   - PSP switch and oil pressure sending unit and position the harness aside
   - 42-pin engine harness connector from the retaining bracket on the brake vacuum booster and position aside
   - Windshield wiper module
   - Ignition wires from the spark plugs
   - Ignition wire brackets from the studs and position the wires aside
   - Camshaft cover
6. Position the piston of the cylinder being serviced at the bottom of its stroke and position the camshaft lobe on the base circle.
7. Install valve spring spacer tool T91P-6565-AH between the spring coils to prevent valve seal damage.
8. Install a valve spring compressor under the camshaft and on top of the valve spring retainer.
9. Compress the valve spring and

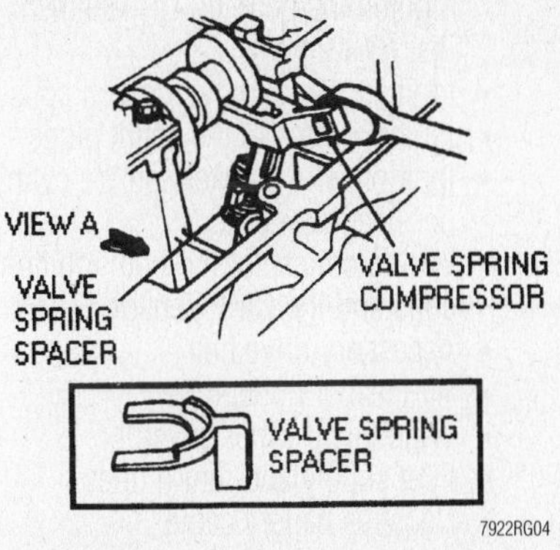

After installing the spring spacer, compress the valve spring and remove the rocker arm

remove the roller follower. Remove the valve spring compressor and spacer.

To install:

10. Apply engine oil to the valve stem and tip and roller follower contact surfaces.
11. Install valve spring spacer tool T91P-6565-AH between the spring coils. Compress the valve spring, and install the roller follower.

➡The piston must be at the bottom of its stroke and the camshaft at the base circle.

12. Remove the valve spring compressor and spacer.
13. Clean the sealing surfaces of the camshaft covers and cylinder heads. Apply silicone sealer to the places where the front cover meets the cylinder head.
14. Position new gaskets onto the camshaft covers and install the covers. Install the bolts and stud bolts and tighten to 72–106 inch lbs. (8–12 Nm).
15. When installing the right camshaft cover, install or connect the following:
    - PCV into the camshaft cover grommet
    - Ignition wire brackets on the studs
    - Wires to the spark plugs
    - Canister purge solenoid, air conditioning compressor clutch and CKP sensor
    - Positive battery cable harness on the right cylinder head
    - Bolt retaining the cable bracket to the cylinder head
    - Positive battery cable at the power distribution box and the battery
16. When installing the left camshaft cover, install or connect the following:
    - Ignition wire brackets on the studs
    - Wires to the spark plugs
    - Windshield wiper module
    - 42-pin and transmission harness connectors
    - Retaining bracket
    - PSP switch and oil pressure sending unit harness.
    - Fuel lines
    - Negative battery cable
17. Start the engine and check for leaks.

4.6L (VIN V) DOHC Engine

1. Before servicing the vehicle, refer to the precautions at the beginning of this section.
2. Remove or disconnect the following:
    - Negative battery cable
    - Ignition coils
    - Valve covers
3. Rotate the crankshaft so that the piston on the cylinder to be serviced is at bottom dead center with the valves closed.
4. Install special tool Valve Spring Compressor T97P-6565-AH for exhaust valves, and Valve Spring Compressor T93P-6565-A for intake valves.
5. Compress the valve spring and remove the rocker arm. Repeat for each arm to be removed.

➡If the rocker arms are to be reused, ensure that they are installed in the same position that they were removed from.

To install:

6. Compress the valve spring and install the rocker arm. Repeat for each arm to be installed.
7. Install or connect the following:
    - Valve covers
    - Ignition coils
    - Negative battery cable
8. Start the engine and check for proper operation.

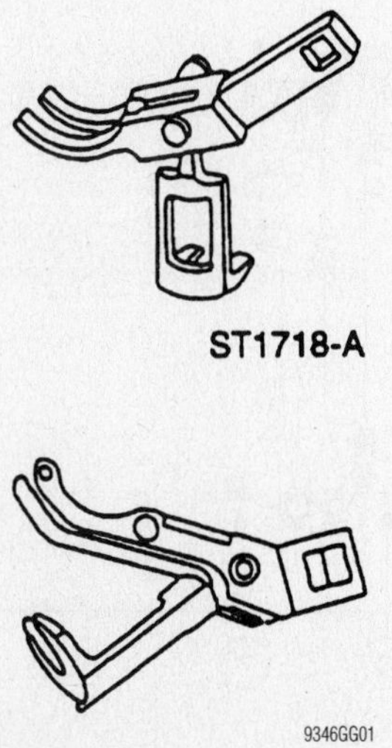

Rocker arm service tools—4.6L (VIN V) engine

Intake Manifold

REMOVAL & INSTALLATION

4.6L (VIN W) SOHC Engine

1. Before servicing the vehicle, refer to the precautions in the beginning of this section.
2. If equipped with air suspension, the air suspension switch, located on the right-hand side of the luggage compartment, must be turned to the **OFF** position before raising the vehicle.
3. Disconnect negative battery cable.
4. Drain the engine cooling system.
5. Properly relieve the fuel system pressure.
6. Remove or disconnect the following:
    - Fuel supply and return lines
    - Windshield wiper governor (module)
    - Engine air cleaner outlet tube
    - Accessory drive belt
    - Ignition wires from the spark plugs
    - Ignition wire brackets from the cylinder head cover studs
    - Ignition coils and the Camshaft Position (CMP) sensor
    - Ignition wires from both ignition coils
    - 2 bolts retaining the ignition wire bracket to the ignition coil brackets
    - Ignition wire assembly
    - Alternator wiring harness from the junction block at the fender apron and alternator
    - Bolts retaining the alternator brace to the intake manifold and the alternator to the cylinder block
    - Alternator
    - Oil pressure sensor and power steering control valve actuator wiring and position the wiring harness aside
    - Exhaust Gas Recirculation (EGR) valve-to-exhaust manifold tube from the right-hand exhaust manifold
    - Engine/transmission harness connector from the retaining bracket on the power brake booster
    - Air conditioning compressor clutch, Crankshaft position (CKP) sensor and the canister purge solenoid wiring connectors
    - Positive Crankcase Ventilation (PCV) valve from the cylinder head cover
    - Canister purge vent hose from the PCV valve

- Accelerator and cruise control cables from the throttle body
- Accelerator cable bracket from the intake manifold and position aside
- Vacuum hose from the throttle body adapter port
- Heated Oxygen Sensor (HO_2S) and the heater water hose
- 2 bolts retaining the thermostat housing to the intake manifold and position the upper hose and thermostat housing aside

➡The 2 thermostat housing bolts are also used to retain the intake manifold.

- 9 bolts retaining the intake manifold to the cylinder heads
- Intake manifold and gaskets

7. If replacing the intake manifold, swap over the necessary parts.

To install:

8. Clean all gaskets mating surfaces.
9. Position new intake manifold gaskets on the cylinder heads. Be sure the alignment tabs on the gaskets are aligned with the holes in the cylinder heads.
10. Install the intake manifold and the 9 retaining bolts. Hand-tighten the right-rear bolt (viewed from the front of the engine) before final tightening, then tighten the bolts, in sequence, to 15–22 ft. lbs. (20–30 Nm).
11. Inspect and if necessary, replace the O-ring seal on the thermostat housing. Position the housing and upper hose and install the 2 retaining bolts. Tighten to 15–22 ft. lbs. (20–30 Nm).
12. Install or connect the following:

- Heater water hose
- HO_2 sensor
- Vacuum hose to the throttle body adapter vacuum port
- Accelerator cable bracket on the intake manifold
- Accelerator and cruise control cables to the throttle body
- PCV valve in the cylinder head cover
- Canister purge solenoid vent hose
- Air conditioning compressor clutch, CKP sensor and canister purge solenoid wiring connectors
- Engine/transmission harness connector the retaining bracket on the power brake booster
- EGR valve-to-exhaust manifold tube to the right-hand exhaust manifold. Tighten the tube nut to 26–33 ft. lbs. (35–45 Nm).
- Power steering control valve actuator
- Oil pressure sensor wiring connectors
- Alternator. Tighten the bolts to 15–22 ft. lbs. (20–30 Nm).
- 2 bolts retaining the alternator brace to the intake manifold and tighten to 71–106 inch lbs. (8–12 Nm)
- Alternator wiring harness to the alternator, right-hand fender apron and junction block
- Ignition wire assembly on the engine
- 2 bolts retaining the ignition wire bracket to the ignition coil brackets. Tighten the bolts to 71–106 inch lbs. (8–12 Nm).
- Ignition wires to the ignition coils
- Ignition wires to the spark plugs
- Ignition wire brackets on the cylinder head cover studs
- Wiring connectors to both ignition coils and the CMP sensor
- Accessory drive belt
- Air cleaner outlet tube
- Windshield wiper governor
- Fuel supply and return lines
- Negative battery cable

13. Fill the engine cooling system.
14. If equipped with air suspension, turn the air suspension switch to the **ON** position.
15. Start the engine and check for leaks.
16. Road test the vehicle and check for proper operation.

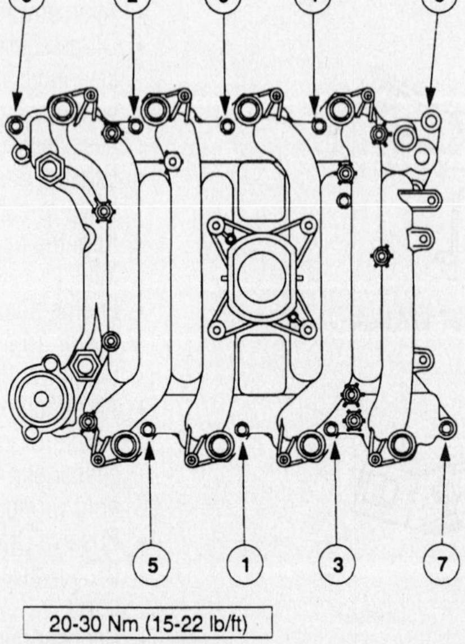

Intake manifold torque sequence—4.6L (VIN W) SOHC engine

4.6L (VIN V) DOHC Engine

1. Before servicing the vehicle, refer to the precautions in the beginning of this section.
2. Drain the cooling system.
3. Relieve the fuel system pressure.
4. Remove or disconnect the following:

- Negative battery cable
- Coolant bypass tube
- Alternator
- Throttle cable
- Cruise control cable
- Evaporative Emissions (EVAP) hose
- Throttle Position (TP) sensor connector
- Positive Crankcase Ventilation (PCV) hose
- Idle Air Control (IAC) valve connector
- Exhaust Gas Recirculation (EGR) tube
- EGR vacuum hose
- Upper intake manifold
- Fuel injection supply manifold
- Lower intake manifold

To install:

5. Install or connect the following:

- Lower intake manifold. Tighten the bolts in sequence to 15–22 ft. lbs. (20–30 Nm).
- Fuel injection supply manifold
- Upper intake manifold. Tighten the bolts to 15–22 ft. lbs. (20–30 Nm).
- EGR vacuum hose
- EGR tube
- IAC valve connector
- PCV hose
- TP sensor connector
- EVAP hose
- Cruise control cable

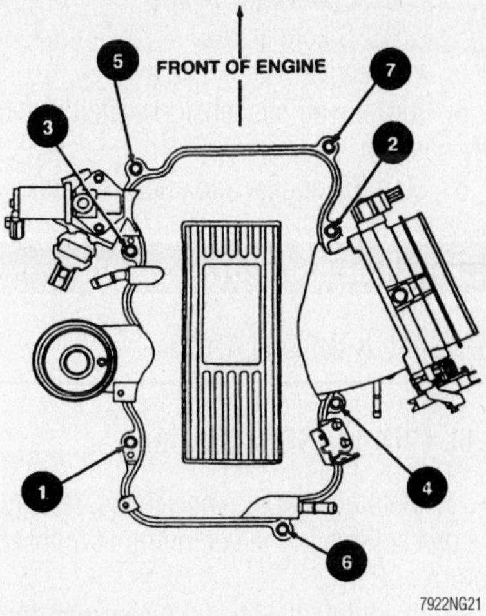

Upper intake manifold torque sequence—4.6L (VIN V) DOHC engine

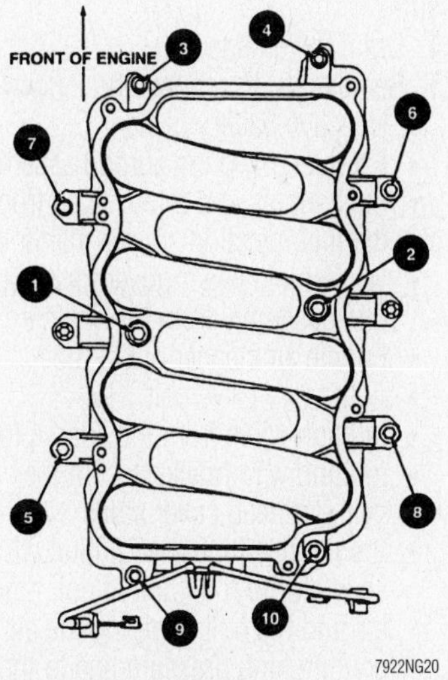

Lower intake manifold torque sequence—4.6L (VIN V) DOHC engine

- Throttle cable
- Alternator
- Coolant bypass tube
- Negative battery cable

6. Fill the cooling system.
7. Start the engine and check for leaks.

Exhaust Manifold

REMOVAL & INSTALLATION

4.6L (VIN W) SOHC Engine

1. Before servicing the vehicle, refer to the precautions in the beginning of this section.
2. Drain the engine cooling system.
3. Relieve the fuel system pressure.
4. Discharge the air conditioning system

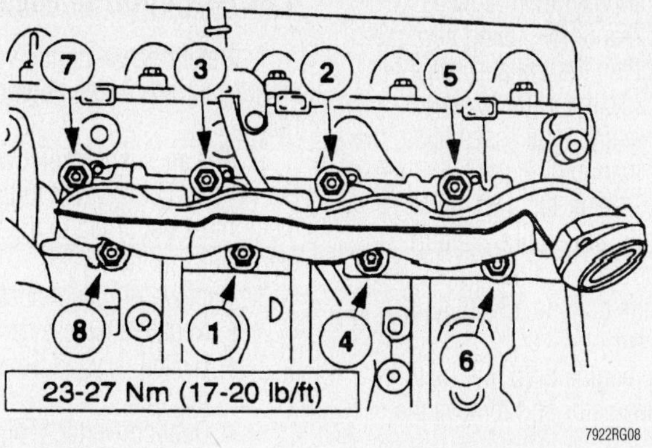

Left exhaust manifold torque sequence—4.6L (VIN W) SOHC engine

5. Remove or disconnect the following:
   - Battery cables
   - Engine air inlet tube
   - Cooling fan and shroud assembly
   - Fuel supply and return lines
   - Upper radiator hose
   - Windshield wiper governor and support bracket
   - Compressor outlet hose at the compressor and the hose assembly-to-right-hand ignition coil bracket bolt. Plug both openings.
   - Engine/transmission harness connector from the retaining bracket on the power brake booster
   - Heater hose
   - Ground strap-to-right-hand cylinder head nut
   - Upper stud and lower bolt retaining the heater hose to the right cylinder head and position aside
   - Heater blower motor switch resistor
   - Bolt retaining the right-hand front engine support insulator to the sub-frame
   - Both Heated Oxygen Sensors (HO_2S)
   - Engine support insulator through-bolts
   - Exhaust Gas Recirculation (EGR) valve-to-exhaust manifold tube nut from the right-hand exhaust manifold.
   - Catalytic converter pipes from both exhaust manifolds. Lower the exhaust system and hang it from the crossmember with wire.
   - Left-hand exhaust manifold
   - Front engine support insulator from the cylinder block and the 8 nuts retaining the exhaust manifold
   - Left-hand exhaust manifold and the 2 manifold gaskets
6. Position an adjustable jackstand and a block of wood under the engine oil pan, rearward of the oil drain hole. Raise the engine approximately 4 inches (100mm).
7. Install or connect the following:
   - 8 exhaust manifold retaining nuts and right-hand exhaust manifold
   - Manifold and gasket

To install:

8. If the exhaust manifolds are being replaced, transfer the heated O_2sensors and tighten to 27–33 ft. lbs. (37–45 Nm). On the right-hand exhaust manifold, transfer the EGR tube connector and tighten to 33–48 ft. lbs. (45–65 Nm).
9. Clean the mating surfaces of the exhaust manifolds and cylinder heads.

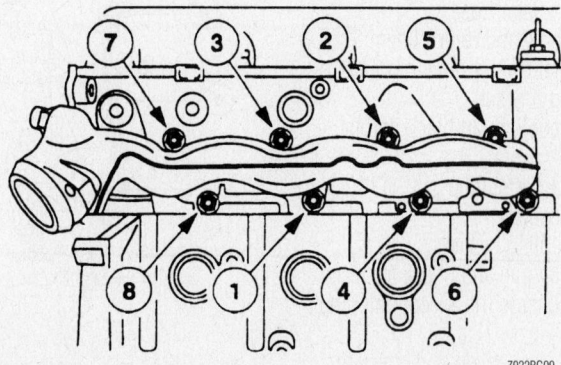

Right exhaust manifold torque sequence—4.6L (VIN W) SOHC engine

10. Install or connect the following:
    - Exhaust manifolds, using new gaskets. Tighten the bolts in sequence to 15–22 ft. lbs. (20–30 Nm).
    - EGR valve and tube assembly to the exhaust manifold. Tighten the line nut to 26–33 ft. lbs. (35–45 Nm).
    - Left-hand front engine support insulator to the cylinder block and tighten the bolts to 15–22 ft. lbs. (20–30 Nm)
11. Lower the engine onto the front engine support insulator and remove the jack.
12. Install or connect the following:
    - Left-hand and right-hand engine support insulator through-bolts and tighten to 15–22 ft. lbs. (20–30 Nm).
    - Catalytic converter pipes to both exhaust manifolds. Tighten the nuts to 20–30 ft. lbs. (27–41 Nm).

➡Be sure the exhaust system clears the No. 3 crossmember. Adjust as necessary.

- Both HO_2S sensors
- Bolt retaining the right-hand front engine support insulator to the sub-frame. Tighten to 15–22 ft. lbs. (20–30 Nm)
- Heater blower motor switch resistor using the 2 retaining screws
- Heater hose in position
- Upper stud and lower bolt and tighten to 15–22 ft. lbs. (20–30 Nm)
- Ground strap onto the stud and tighten the nut to 15–22 ft. lbs. (20–30 Nm)
- Heater hose
- Engine/transmission harness connector
- Retaining bracket on the power brake booster
- Air conditioning compressor outlet hose to the compressor
- Bolt retaining the hose assembly to the right-hand ignition coil bracket
- Upper radiator hose
- Fuel supply and return lines
- Windshield wiper governor and retaining bracket
- Engine cooling fan blade and fan shroud
- Engine air inlet tube
- Battery cables

13. Fill the cooling system.
14. Start the engine and check for leaks.
15. Properly charge the air conditioning system.
16. Road test the vehicle and check for proper operation.

4.6L (VIN V) DOHC Engine

1. Before servicing the vehicle, refer to the precautions in the beginning of this section.
2. The air suspension switch, located on the left-hand side of the luggage compartment, must be turned to the **OFF** position before raising the vehicle.
3. Remove or disconnect the following:
    - Negative battery cable
    - Heated Oxygen Sensor (HO_2S) connectors
    - Dual converter Y-pipe
    - Secondary air injection tubes
    - Exhaust Gas Recirculation (EGR) tube
    - Exhaust manifolds and gaskets

To install:

4. Install or connect the following:
    - Exhaust manifolds with new gaskets. Tighten the nuts in sequence to 14–16 ft. lbs. (18–22 Nm).
    - EGR tube
    - Secondary air injection tubes

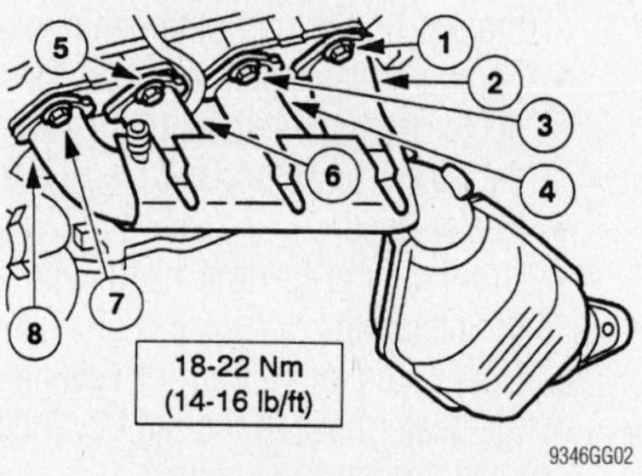

Left exhaust manifold torque sequence—4.6L (VIN V) DOHC engine

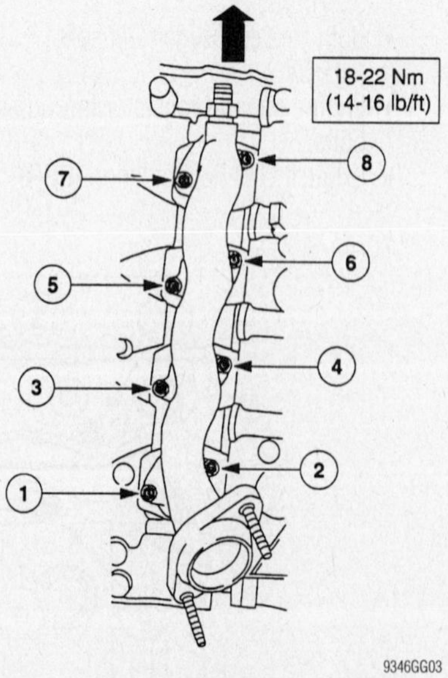

Right exhaust manifold torque sequence—4.6L (VIN V) DOHC engine

- Dual converter Y-pipe
- HO_2S connectors
- Negative battery cable

5. Turn the air suspension switch to the **ON** position.
6. Start the engine and check for leaks.

Camshaft and Valve Lifters

REMOVAL & INSTALLATION

4.6L (VIN W) SOHC Engine

1. Before servicing the vehicle, refer to the precautions in the beginning of this section.
2. If equipped with air suspension, the air suspension switch, must be turned to the **OFF** position before raising the vehicle.
3. Properly relieve the fuel system pressure.
4. Drain the engine oil
5. Remove or disconnect the following:
    - Negative battery cable
    - Fan blade and fan shroud assembly
    - Fuel supply and return lines from the fuel injection supply manifold
    - Windshield wiper governor (module) assembly from the vehicle
    - Engine air cleaner outlet tube
    - Accessory drive belt
    - Ignition wires from the spark plugs
    - Ignition wire brackets from the cylinder head cover studs
    - 2 bolts retaining the ignition wire separator to the ignition coil brackets and the bolt retaining the air conditioning pressure line to the right-hand ignition coil bracket.
    - Connectors from both ignition coils and the Camshaft Position (CMP) sensor
    - Ignition coils with brackets attached
    - Electrical connector from the alternator and at the power distribution box
    - Water pump pulley
    - Positive battery cable at the power distribution box
    - Bolt from the positive battery cable bracket located on the right-hand cylinder head
    - Fuel vapor hose from the EVAP canister purge valve and position the positive battery cable aside
    - Positive Crankcase Ventilation (PCV) valve from the cylinder head cover and position aside
    - Engine/transmission harness connector from the bracket on the power brake booster

- Crankshaft Position (CKP) sensor and air conditioning clutch harness connectors
- Bolts retaining the power steering pump to the cylinder block and wire the pump aside

➡The front lower bolt on the power steering pump will not come all the way out.

- Oil pan
- Crankshaft pulley bolt and washer and crankshaft pulley
- Engine oil filter
- Power steering control valve actuator and oil pressure sensor
- Oil filter adapter
- Cylinder head covers
- Engine front cover
- Timing chains

6. Rotate the crankshaft counterclockwise no more than 45 degrees from Top Dead Center (TDC) to ensure that all pistons are below the top of the engine block deck face.

7. Install a valve spring compressor under the camshaft and on top of one of the valve spring retainers.

8. Install Valve Spring Spacer T91P-6565-AH between the valve spring coils. Be sure that the valve being compressed is on its base circle. Compress the valve spring and remove the rocker arm. Repeat the procedure until all rocker arms are removed.

9. If required, pull the lash adjusters out of their bores in the cylinder head. Note their locations, they must be installed in the same bore they were removed from.

➡Do not mix the camshaft bearing caps. Note the camshaft bearing cap locations for installation.

10. To remove each camshaft, unfasten the 14 bolts retaining the camshaft bearing caps (cluster assemblies) to the cylinder head. Tap upward on the camshaft bearing caps at points near the upper bearing halves and gradually lift the camshaft bearing cap clusters from the cylinder head.

11. Repeat the removal procedure for the opposite cylinder head.

12. Remove the camshaft straight upward to avoid bearing damage.

13. Clean and inspect the camshafts and related components for unusual wear or damage.

To install:

14. Clean and inspect the cylinder head covers, engine front cover and cylinder head sealing surfaces.

15. Apply clean engine oil to the

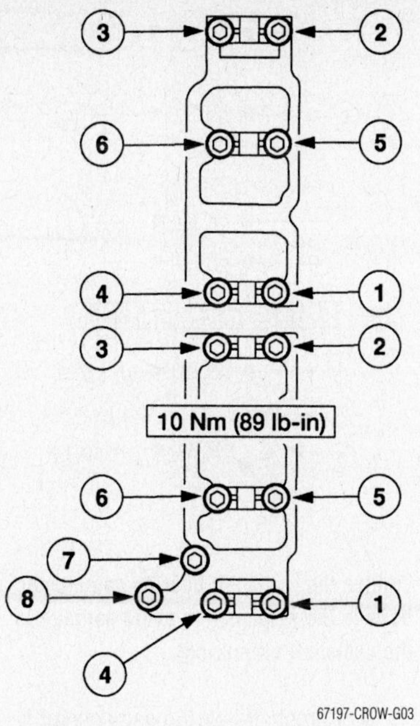

Camshaft bearing cap torque sequence

camshaft journals and lobes. Position the camshafts on the cylinder heads.

16. Install and seat the camshaft bearing cap cluster assemblies. Install and hand start the retaining bolts. Tighten the camshaft cluster retaining bolts in sequence to 71–106 inch lbs. (8–12 Nm). Be sure to tighten each camshaft bearing cap cluster individually.

➡Each camshaft bearing cap cluster assembly is tightened individually.

17. Loosen the camshaft bearing cap cluster retaining bolts approximately 2 turns or until the heads of the bolts are free. Tighten all bolts, again in sequence, to 71–106 inch lbs. (8–12 Nm).

18. Repeat the camshaft bearing cap installation for the opposite cylinder head.

➡The camshafts should turn freely but with a slight drag.

19. Check camshaft end-play as follows:

a. Step 1: Install a dial indicator on the front of the engine. Position it so the indicator foot is resting on the camshaft sprocket bolt or the front of the camshaft.

b. Step 2: Push the camshaft toward the rear of the engine and zero the dial indicator.

c. Step 3: Pull the camshaft forward and release it. Specified end-play is 0.0901–0.006 inch (0.025–0.190mm).

d. Step 4: If end-play is too tight, check for binding or foreign material in the camshaft thrust bearing. If end-play is excessive, check for worn camshaft thrust plate and replace the cylinder head, as required.

e. Step 5: Remove the dial indicator.

20. If removed, install the lash adjusters in their original positions.

21. If necessary, install Camshaft Positioning Tools T92P-6256-A on the flats of the camshafts and install the spacers and camshaft sprockets. Install the bolts and washers and tighten to 81–95 ft. lbs. (110–130 Nm).

22. Install a valve spring compressor under the camshaft and on top of the valve spring retainer.

23. Install Valve Spring Spacer T91P-6565-AH between the valve spring coils. Be sure that the valve being compressed is on its base circle. Compress the valve spring and install the rocker arm. Repeat the procedure until all rocker arms are installed.

24. Rotate the crankshaft clockwise 45 degrees to position the crankshaft at Top Dead Center (TDC).

➡The crankshaft must only be rotated in the clockwise direction and only as far as TDC.

25. Install or connect the following:

- Timing chains
- Engine front cover
- Cylinder head covers. Tighten the cylinder head cover bolts to 71–106 inch lbs. (8–12 Nm).
- Power steering control valve actuator connector
- Oil pressure sensor harness connector

26. Apply silicone sealer to the crankshaft keyway

27. Install the crankshaft pulley and tighten the bolt as follows:

a. Step 1: Tighten to 66 ft. lbs. (90 Nm).

b. Step 2: Loosen one complete turn.

c. Step 3: Tighten to 35–39 ft. lbs. (47–53 Nm).

d. Step 4: Tighten an additional 85–95 degrees.

28. Install or connect the following:

- Engine oil pan
- Power steering pump on the engine and the 4 retaining bolts. Tighten the bolts to 15–22 ft. lbs. (20–30 Nm).
- Air conditioning clutch and CKP sensor
- Evaporative emission canister purge valve harness connector
- Engine/transmission harness connectors on the power brake booster retaining bracket
- PCV valve to the right-hand cylinder head cover

- Positive battery cable harness on the right-hand cylinder head
- Bolt retaining the battery cable bracket to the cylinder head
- Evaporative emission hose to the canister purge valve
- Positive battery cable at the power distribution box
- Water pump pulley and tighten the bolts to 15–22 ft. lbs. (20–30 Nm)
- Ignition coil brackets and ignition wires to the engine front cover. Tighten the retaining nuts to 15–22 ft. lbs. (20–30 Nm).
- Harness connectors to the ignition coils and the CMP sensor
- Air conditioning pressure line on the right-hand ignition coil bracket and the retaining bolt
- Ignition wires to the spark plugs and the brackets onto the cylinder head cover studs
- Accessory drive belt
- Windshield wiper governor
- Fuel supply and return lines
- Fan and shroud assembly
- Negative battery cable

29. Fill the engine cooling system.
30. Fill the crankcase.
31. If equipped with air suspension, turn the air suspension switch to the **ON** position.
32. Start the engine and check for leaks.
33. Road test the vehicle and check for proper engine operation.

4.6L (VIN V) DOHC Engine

1. Before servicing the vehicle, refer to the precautions in the beginning of this section.
2. Drain the cooling system.
3. Remove or disconnect the following:
   - Negative battery cable
   - Windshield wiper module
   - Engine front cover
   - Timing chains and sprockets
   - Valve covers
   - Rocker arms
   - Hydraulic lifters
   - Camshaft bearing cap assembly
   - Camshafts

➡Keep all valvetrain components in order for assembly in their original locations.

To install:

➡The outboard exhaust camshaft bearing cap bolts are shorter than the other bearing cap bolts.

4. Install or connect the following:
   - Camshafts

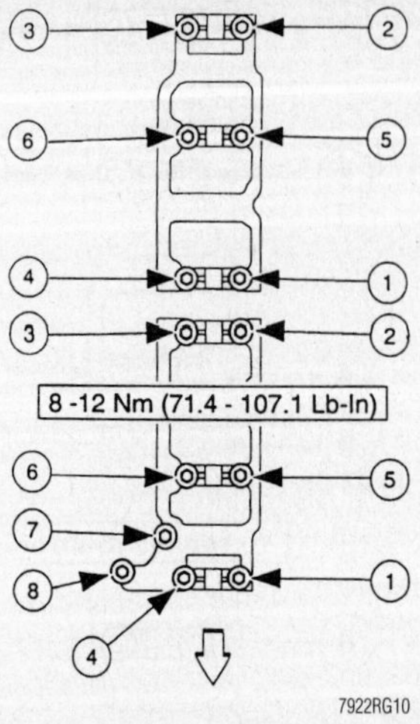

Tighten the camshaft bearing cap cluster bolts in the sequence to avoid damage to the camshaft or bearings

   - Camshaft bearing cap assembly. Tighten the bolts in sequence to 71–106 inch lbs. (8–12 Nm).
   - Hydraulic lifters
   - Rocker arms
   - Valve covers
   - Timing chains and sprockets
   - Engine front cover
   - Windshield wiper module
   - Negative battery cable
5. Fill the cooling system.
6. Start the engine and check for leaks.

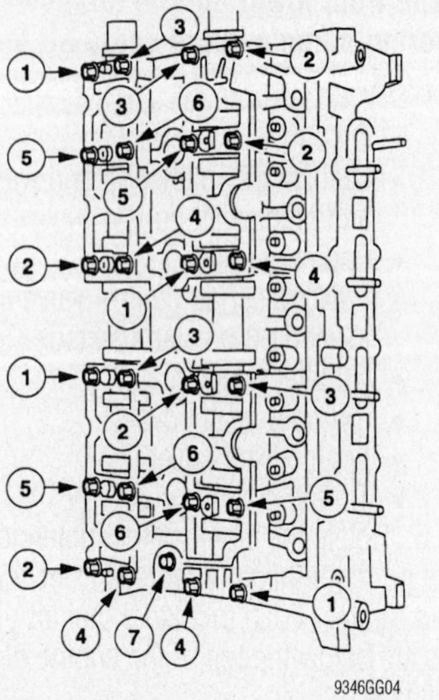

Camshaft bearing cap torque sequence—4.6L (VIN V) engine

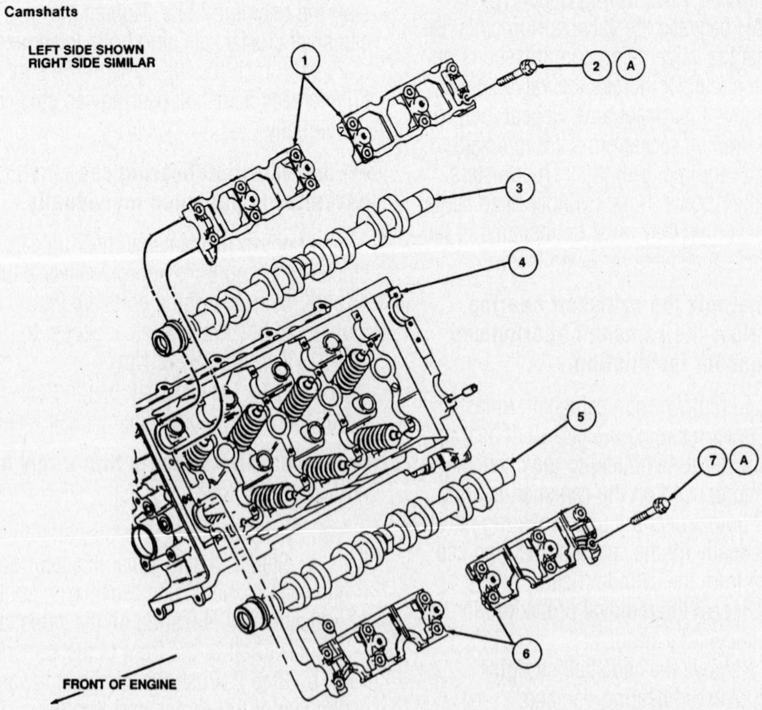

Item	Part Number	Description
1	6B277	Cap, Intake Camshaft
2	N806070	Bolt (19 Req'd Per Cylinder Head)
3	6250	Intake Camshaft
4	6049	LH Cylinder Head

Item	Part Number	Description
5	6250	Exhaust Camshaft
6	6B278	Cap, Exhaust Camshaft
7	N80757	Bolt (6 Req'd Per Cylinder Head)
A	—	Tighten to 8-12 N-m (71-106 Lb-In)

9346GG05

Exploded view of the camshaft mounting—4.6L (VIN V) engine

Valve Lash

ADJUSTMENT

The 4.6L (VIN W and V) engines are equipped with hydraulic lash adjusters. Valve clearance is not adjustable.

Starter Motor

REMOVAL & INSTALLATION

1. Before servicing the vehicle, refer to the precautions in the beginning of this section.
2. Remove or disconnect the following:
   - Negative battery cable
   - Red solenoid safety cap
   - Wires from solenoid
   - 2 upper bolts
   - 1 lower bolt and starter

To install:

3. Install or connect the following:
   - Starter and lower mounting bolt
   - 2 upper mounting bolts. Tighten all 3 bolts to 15–20 ft. lbs. (20–27 Nm).
   - Wires from solenoid and tighten the nut to 40–50 inch lbs.
   - Red solenoid safety cap
   - Negative battery cable

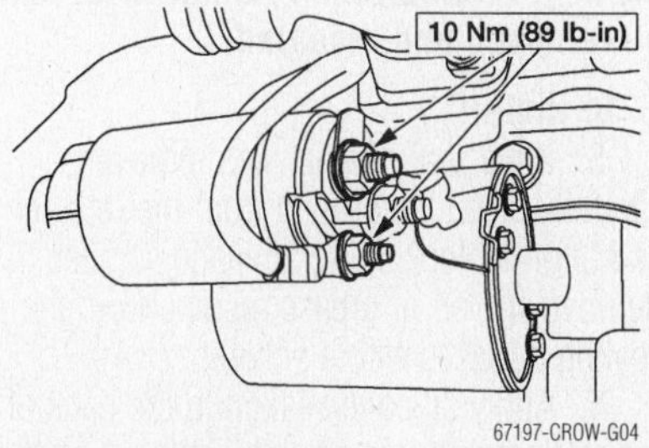

Starter motor mounting

Oil Pan

REMOVAL & INSTALLATION

1. Before servicing the vehicle, refer to the precautions in the beginning of this section.
2. Drain the engine cooling system
3. Properly discharge the air conditioning system.
4. Relieve the fuel system pressure.
5. Remove or disconnect the following:
   - Negative battery cable
   - Engine air cleaner outlet tube
   - Fuel supply and return lines at the fuel injection supply manifold
   - Cooling fan and fan shroud
   - Upper radiator hose
   - Wiper governor and support bracket
   - A/C compressor outlet hose
   - Engine/transmission electrical harness connector from the retaining bracket on the power brake booster
   - Heater water hose
   - Nut retaining the ground strap to the right-hand cylinder head
   - Upper stud and loosen the lower bolt retaining the heater outlet hose to the right-hand cylinder head and position aside
   - Heater blower motor switch resistor
6. Drain the engine oil and reinstall the oil pan drain plug with a new gasket. Tighten the plug to 10–12 ft. lbs. (13–16 Nm).
7. Remove or disconnect the following:
   - Bolt retaining the right-hand engine support insulator to the lower front sub-frame
   - Bolts retaining the left-hand and right-hand front engine support insulators to the engine mount supports
   - Catalytic converter pipes from both exhaust manifolds. Lower the exhaust system and support it with wire from the transmission crossmember.
8. Position a jack and a block of wood under the oil pan, rearward of the oil drain hole. Raise the engine approximately 4 inches (100mm) and insert 2 wood blocks approximately 2½–2¾ inch (60–70mm) thick under each front engine support insulator. Lower the engine onto the wood blocks and remove the jack.
9. Remove the oil pan.
10. If necessary, remove the 2 bolts retaining the oil pick-up tube to the oil pump and remove the bolt retaining the pick-up tube to the main bearing stud spacer. Remove the pick-up tube.

To install:

11. Clean the engine oil pan and inspect for damage. Clean the sealing surfaces of the front cover and cylinder block. Clean and inspect the oil pick-up tube and replace the O-ring.
12. If removed, position the oil pick-up tube on the oil pump and hand start the 2 retaining bolts. Install the bolt retaining the pick-up tube on the main bearing stud spacer, hand tight.
13. Tighten the pick-up tube-to-oil pump bolts to 72–108 inch lbs. (8–12 Nm), then tighten the pick-up tube-to-main bearing stud spacer bolt to 15–22 ft. lbs. (20–30 Nm).
14. Position a new gasket on the oil pan.

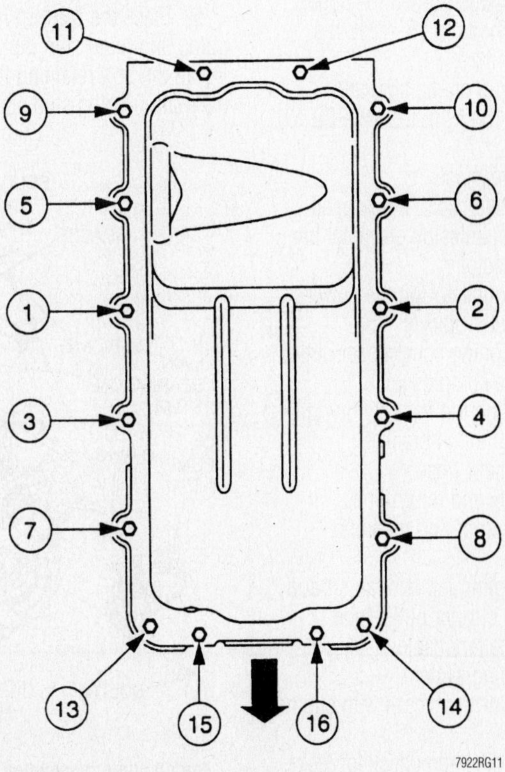

To prevent oil leaks, tighten the oil pan bolts in the sequence shown

Apply silicone sealer to where the front cover meets the cylinder block and the crankshaft rear oil seal and retainer meets the cylinder block. Position the oil pan to the engine and install the retaining bolts. Tighten the bolts in sequence, to 14 ft. lbs. (20 Nm), then rotate the oil pan retaining bolts, in sequence an additional 60 degrees within 4 minutes of applying the silicone sealer.

15. Position the jack and wood block under the engine oil pan, rearward of the oil drain hole, and raise the engine enough to remove the wood blocks. Lower the engine and remove the jack.

16. Install or connect the following:

- Left-hand and right-hand engine support insulator through-bolts and tighten to 15–22 ft. lbs. (20–30 Nm)
- Bolt retaining the right-hand engine support insulator to the lower front sub-frame. Tighten the bolt to 15–22 ft. lbs. (20–30 Nm).
- Exhaust system to the exhaust manifolds and tighten the 4 retaining nuts to 20–30 ft. lbs. (27–41 Nm). Be sure the exhaust system clears the crossmember. Adjust as necessary.
- New engine oil filter
- Heater blower motor switch resistor using the 2 retaining screws
- Heater water hose
- Upper stud and tighten the upper and lower bolts to 15–22 ft. lbs. (20–30 Nm)
- Ground strap on the stud and tighten to 15–22 ft. lbs. (20–30 Nm)
- Heater water hose
- Throttle valve cable, if equipped
- Engine/transmission electrical harness connector
- Harness connector on the power brake booster bracket
- Air conditioning compressor outlet hose to the compressor
- Bolt retaining the hose to the right-hand ignition coil bracket
- Upper radiator hose
- Fuel supply and return lines
- Wiper governor and retaining bracket
- Engine cooling fan and fan shroud
- Engine air cleaner outlet tube
- Negative battery cable

17. Fill the cooling system.

18. Fill the engine crankcase with engine oil.

19. Start the engine and check for leaks.

20. Properly evacuate and recharge the air conditioning system.

21. Road test the vehicle and check for proper engine operation.

Oil Pump

REMOVAL & INSTALLATION

1. Before servicing the vehicle, refer to the precautions in the beginning of this section.

2. Remove or disconnect the following:

- Negative battery cable
- Cylinder head covers
- Engine front cover
- Engine oil pan
- Timing chains
- 2 bolts retaining the oil pick-up tube to the oil pump and the bolt attaching the oil pick-up tube to the main bearing stud spacer
- Pick-up tube
- 4 bolts retaining the oil pump to the cylinder block
- Oil pump

To install:

3. Rotate the inner rotor of the oil pump to align with the flats on the crankshaft and install the oil pump flush with the cylinder block. Install the 4 retaining bolts and tighten to 72–106 inch lbs. (8–12 Nm).

4. Clean the oil pick-up tube and replace the O-ring.

5. Place the pick-up tube on the oil pump and hand start the 2 retaining bolts. Install the bolt retaining the pick-up tube to the main bearing stud spacer hand tight.

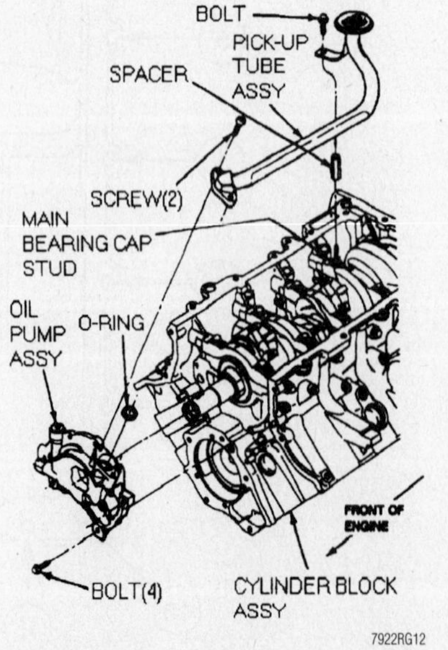

The oil pump is mounted on the crankshaft at the front of the engine

Tighten the pick-up tube-to-oil pump bolts to 72–106 inch lbs. (8–12 Nm). Tighten the pick-up tube to main bearing stud spacer bolt to 15–22 ft. lbs. (20–30 Nm).

6. Install or connect the following:

- New engine oil filter
- Timing chains
- Engine oil pan
- Engine front cover
- Cylinder head covers
- Negative battery cable

7. Fill the crankcase.

8. Start the engine and check for leaks and proper engine oil pressure.

9. Road test the vehicle and check for proper engine operation.

Rear Main Seal

REMOVAL & INSTALLATION

1. Before servicing the vehicle, refer to the precautions in the beginning of this section.

2. Remove or disconnect the following:

- Transmission
- Flexplate or flywheel

3. With a sharp awl, carefully punch a small hole in the metal portion of the seal.

4. Remove the seal using a slide hammer with a sheet metal screw attached.

➡**If the oil leak is coming from around the seal retainer, the retainer must also be removed and resealed.**

To install:

5. If the seal retainer was removed, carefully clean the sealant from the retainer and engine block using a plastic scraper. Remove any oil or grease residue from the sealing surfaces with a solvent.

6. Apply silicone sealant to the back of the retainer and immediately install it on the

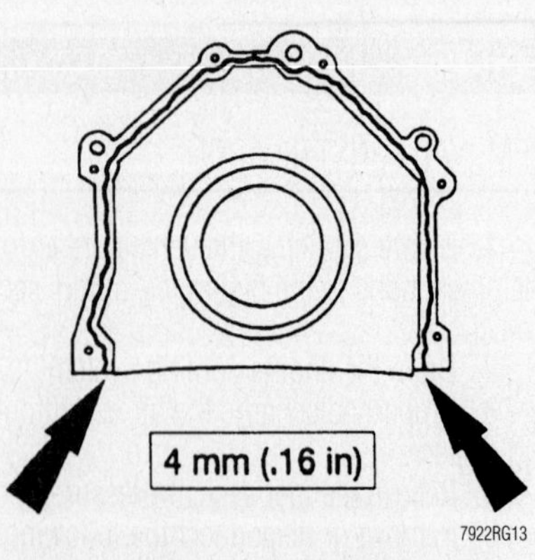

Apply a continuous bead of silicone sealant to the back of the seal retainer before installing it on the engine

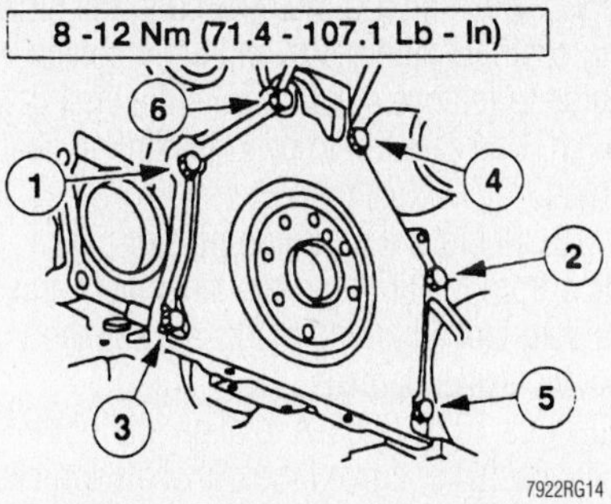

To avoid leakage, be sure to tighten the crankshaft rear oil seal retainer bolts in the correct sequence

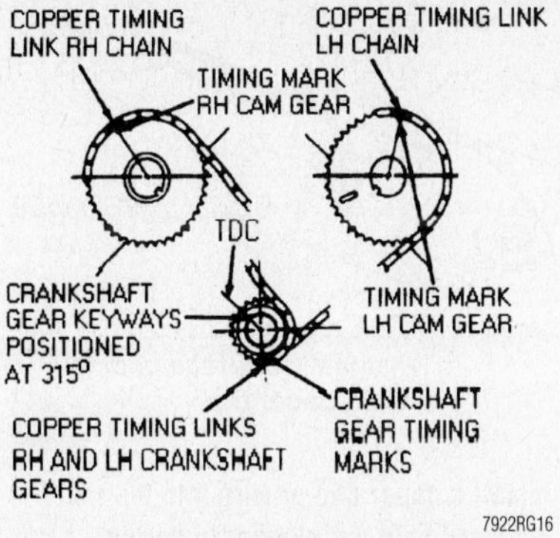

Be sure that the timing marks are aligned when the No. 1 piston is at TDC on compression—4.6L (VIN W) SOHC engine

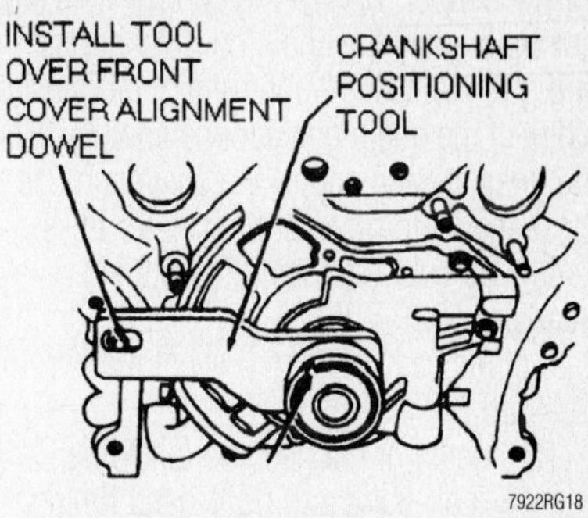

Install the crankshaft positioning tool to be sure the crankshaft does not turn while installing the timing chains—4.6L (VIN W) SOHC engine

engine block. Tighten the bolts in sequence to 71–107 inch lbs. (8–12 Nm).

7. Lubricate the seal and the crankshaft with clean engine oil.
8. Install the seal with the spring side toward the engine.
9. Remove the installation tool.
10. Install or connect the following:
    - Flexplate or flywheel. Tighten the bolts, in a crisscross pattern, to 54–64 ft. lbs. (73–87 Nm).
    - Transmission
    - Negative battery cable
11. Check the engine oil level.
12. Start the engine and check for leaks.

Timing Chain, Sprockets, Front Cover and Seal

REMOVAL & INSTALLATION

4.6L (VIN W) SOHC Engine

1. Before servicing the vehicle, refer to the precautions in the beginning of this section.
2. Drain the engine oil.
3. Remove or disconnect the following:
    - Negative battery cable
    - Cooling fan and shroud
    - Accessory drive belt
    - Water pump pulley
    - Power steering pump
    - Oil pan
    - Crankshaft pulley retaining bolt and washer
    - Crankshaft pulley
    - Bolt retaining the air conditioning pressure line to the right-hand ignition coil bracket
    - Cylinder head covers
    - Wiring at both ignition coils and the Crankshaft Position (CMP) sensor
    - 3 bolts retaining the right-hand ignition coil bracket to the engine front cover. Position the power steering hose aside.
    - 3 nuts retaining the left-hand ignition coil bracket to the engine front cover. Slide both ignition coil brackets and ignition wires off the mounting studs and lay the assembly on top of the engine.
    - Bolts retaining the drive belt idler pulley and the pulley
    - Wiring to the Crankshaft Position (CKP) sensor and the sensor
4. If equipped, remove the retainers for the oil cooler from the engine front cover retaining stud bolts and position the oil cooler aside.
    - 9 stud bolts and the 6 standard engine front cover bolts and the cover
    - Crankshaft oil seal from the cover using a seal driver
    - CKP sensor pulse wheel
5. Rotate the engine to set the piston for No. 1 to Top Dead Center (TDC) on its compression stroke.

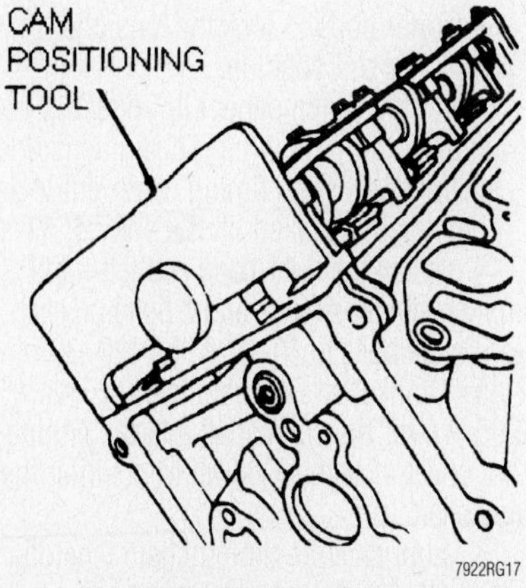

Use the special tool to maintain camshaft position while installing the timing chains—4.6L (VIN W) SOHC engine

6. Install Camshaft Positioning Adapters T92P-6256-A on the flats of both camshafts. This will prevent accidental rotation of the camshafts.
7. Remove or disconnect the following:
    - 2 bolts retaining the tensioner to the right-hand cylinder head and the tensioner
    - Right-hand timing chain tensioner arm
    - 2 bolts retaining the right-hand timing chain guide to the cylinder head and remove the timing chain guide.
    - Right-hand timing chain from the camshaft and crankshaft sprockets
    - Right-hand camshaft sprocket retaining bolt, washer, sprocket and spacer, if necessary
    - 2 bolts retaining the timing chain tensioner to the left-hand cylinder head
    - Timing chain tensioner
    - Left-hand timing chain tensioner arm
    - 2 bolts retaining the timing chain guide to the left-hand cylinder head
    - Timing chain guide
    - Left-hand timing chain from the camshaft and crankshaft sprockets
    - Left-hand camshaft sprocket retaining bolt, washer, sprocket and spacer, if necessary
8. If necessary, note the position of the crankshaft sprockets and remove the crankshaft sprockets by sliding them off the front of the crankshaft.
9. Inspect the plastic running face on the tensioner arms and chain guides. If worn or damaged, inspect the engine oil pan for contamination and thoroughly clean the oil pan. Replace the oil pick-up tube.

To install:

10. Examine the timing chains, looking

for the copper links. If the copper links are not visible, lay the chain on a flat surface and pull the chain taught until the opposite sides of the chain contact one another. Mark the links at each end of the chain and use these marks in place of the copper links.

11. Be sure Camshaft Positioning Adapters T92P-6256-A are installed on the flats of the camshafts to prevent them from rotating.

12. Install or connect the following:
- Left-hand and right-hand timing chain guides and retaining bolts. Tighten the retaining bolts to 71–106 inch lbs. (8–12 Nm).
- Left-hand and right-hand camshaft spacers and sprockets, (if removed) on the camshafts, the washers and retaining bolts but do not tighten at this time.
- Left-hand crankshaft sprocket with the tapered part of the sprocket facing away from the engine block
- Left-hand timing chain on the camshaft and crankshaft sprockets. Be sure the copper links of the timing chain line up with the timing marks on both sprockets.
- Right-hand crankshaft sprocket with the tapered part of the sprocket facing the left-hand crankshaft sprocket, if removed
- Right-hand timing chain on the camshaft and crankshaft sprockets. Be sure the copper links of the timing chain line up with the timing marks on both sprockets.

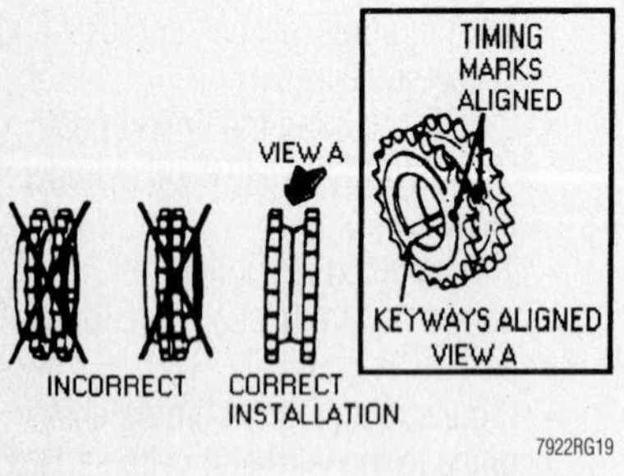

Install the crankshaft sprockets with the tapered sides facing each other

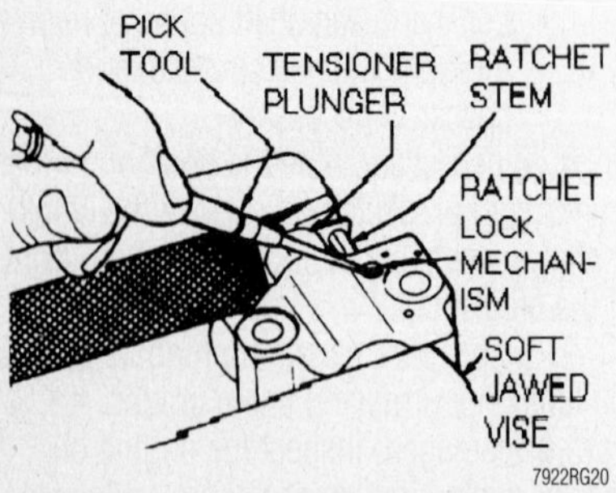

Slowly compress the timing chain tensioner while holding the ratchet lock away from the stem with a suitable tool—4.6L (VIN W) SOHC engine

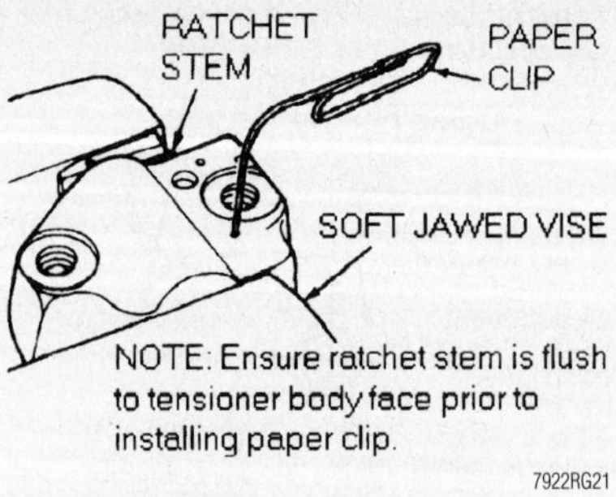

Install a paper clip or wire into the tensioner to hold the plunger in during assembly—4.6L (VIN W) SOHC engine

13. It is necessary to bleed the timing chain tensioners before installation. Proceed as follows:

a. Step 1: position the timing chain tensioner in a soft-jawed vise.

b. Step 2: using a small pick or similar tool, hold the ratchet lock mechanism away from the ratchet stem and slowly compress the tensioner plunger by rotating the vise handle.

⁂ WARNING

The tensioner must be compressed slowly or damage to the internal seals will result.

c. Step 3: once the tensioner plunger bottoms in the tensioner bore, continue to hold the ratchet lock mechanism and push down on the ratchet stem until flush with the tensioner face.

d. Step 4: while holding the ratchet stem flush with the tensioner face, release the ratchet lock mechanism and install a paper clip or similar tool in the tensioner body to lock the tensioner in the collapsed position.

e. Step 5: the paper clip must not be removed until the timing chain, tensioner, tensioner arm and timing chain guide are completely installed on the engine.

14. Install the right-hand and left-hand timing chain tensioners and 2 bolts on each. Tighten the bolts to 15–22 ft. lbs. (20–30 Nm).

15. Crankshaft Positioning Tool T93P-6265-A over the crankshaft and the engine front cover alignment dowel to position the crankshaft.

16. Lubricate the timing chain tensioner arm contact surfaces with clean engine oil and install the right-hand and left-hand tensioner arms on their dowel pins.

17. Position a C-clamp around the timing chain tensioner arm and timing chain guide to remove all slack from the timing chain. Use care not to bend the timing chain guide.

18. Remove the locking pins or paper clips from the timing chain tensioners and be sure that all timing marks are aligned.

19. Using Camshaft Positioning Adapters T92P-6265-A to align and hold the camshafts, tighten the camshaft sprocket retaining bolts to 81–95 ft. lbs. (110–130 Nm).

20. Position a dial indicator in the No. 1 cylinder spark plug hole to measure intake valve lift. The intake valve should be at maximum lift when the crankshaft is at 114 degrees after TDC. If the intake valve lift is not at maximum lift, loosen the camshaft sprocket bolt and repeat the steps detailing the installation of the timing chain tensioners to the tightening of the camshaft sprockets.

21. Remove the camshaft and crankshaft positioning tools.

22. Install a new crankshaft seal in the front cover. Apply engine oil to the lip of the seal.

23. Thoroughly clean the sealing surfaces of the front cover, cylinder block and oil pan. Apply silicone sealer to the points where the cylinder head meets the cylinder block.

24. Install or connect the following:
- Front cover in position using new gaskets
- Retaining bolts and studs in their proper locations. Tighten in sequence to 15–22 ft. lbs. (20–30 Nm) within 4 minutes of applying the silicone sealer.
- Oil cooler to the front cover retaining stud bolts, if equipped
- CKP sensor and attach the harness connector
- Drive belt idler pulley
- Ignition coil brackets and ignition wires as an assembly onto the mounting studs
- Power steering hose and the nuts retaining the coil brackets to the front cover. Tighten the nuts to 15–22 ft. lbs. (20–30 Nm).
- Wiring to both ignition coils and the CMP sensor
- Cylinder head covers
- Air conditioning pressure line on the right-hand ignition coil bracket and tighten the retaining bolt

25. Apply a small amount of silicone sealer in the rear of the keyway in the crankshaft pulley.

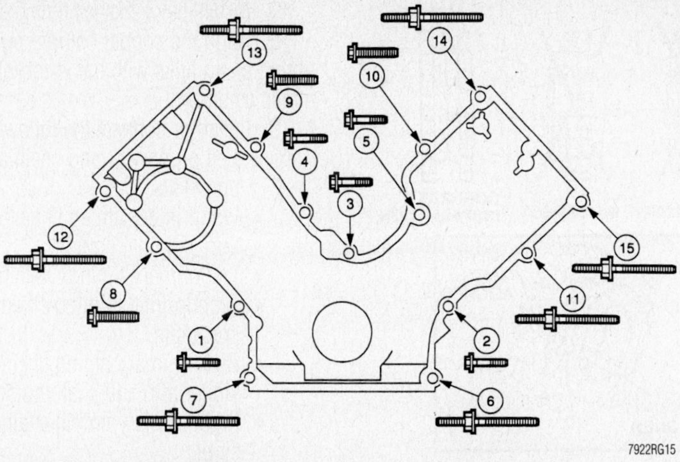

Timing chain front cover bolt tightening sequence—4.6L (VIN W) SOHC engine

26. Install or connect the following:
    - Pulley on the crankshaft
    - Crankshaft pulley bolt and washer and tighten to 114–121 ft. lbs. (155–165 Nm)
    - Oil pan
    - Power steering pump on the engine. Tighten the bolts to 15–22 ft. lbs. (20–30 Nm).
    - Water pump pulley. Tighten the bolts to 15–22 ft. lbs. (20–30 Nm).
    - Accessory drive belt
    - Engine cooling fan and shroud
    - Negative battery cable
27. Fill the engine.
28. Start the engine and check for leaks.
29. Road test the vehicle and check for proper engine operation.

4.6L (VIN V) DOHC Engine

FRONT OIL SEAL

1. Before servicing the vehicle, refer to the precautions in the beginning of this section.
2. Remove or disconnect the following:
    - Negative battery cable
    - Secondary air injection diverter valve
    - Accessory drive belt
    - Crankshaft pulley
    - Front oil seal

To install:

3. Install the front oil seal flush with the front timing cover.
4. Apply silicone sealer to the crankshaft pulley woodruff key slot.
5. Install the crankshaft pulley and tighten the bolt as follows:
    a. Step 1: 66 ft. lbs. (90 Nm)
    b. Step 2: Loosen the bolt one full turn
    c. Step 3: 35–39 ft. lbs. (47–53 Nm)
    d. Step 4: Plus 85–90 degrees
6. Install or connect the following:
    - Accessory drive belt
    - Secondary air injection diverter valve. Tighten the nuts to 71–106 inch lbs. (8–12 Nm).
    - Negative battery cable

FRONT COVER, TIMING CHAINS AND SPROCKETS

1. Before servicing the vehicle, refer to the precautions in the beginning of this section.
2. Drain the cooling system.
3. Drain the engine oil.
4. Remove or disconnect the following:
    - Negative battery cable
    - Water outlet hose
    - Water bypass hose
    - Cooling fan assembly
    - Accessory drive belt
    - Water pump pulley
    - Secondary air injection valve
    - Power steering pump harness connector
    - Starter relay cable and bracket
    - Camshaft Position (CMP) sensor connector
    - Power steering pump and reservoir
    - Crankshaft pulley
    - Valve covers
    - Rocker arms
    - Accessory drive belt tensioner and idler pulley
    - Crankshaft Position (CKP) sensor connector
    - Front cover
    - CKP sensor pulse wheel
5. Rotate the crankshaft so that the No. 1 cylinder is at Top Dead Center (TDC) of the compression stroke.
6. Remove or disconnect the following:
    - Right primary timing chain tensioner
    - Right primary timing chain tensioner arm and chain guide
    - Right primary timing chain and sprockets
    - Left primary timing chain tensioner
    - Left primary timing chain tensioner arm and chain guide
    - Left primary timing chain and sprockets
7. Compress the secondary chain tensioners and install locking pins.
8. Remove the left and right secondary timing chains and sprockets.

To install:

9. Position the camshafts with the keys facing downward.
10. Install the secondary timing chains and sprockets. Do not tighten the sprocket bolts at this time.
11. Tension the secondary timing chains with the Secondary Timing Chain Tensioning tool T93P-6256-BH.
12. Install Camshaft Positioning tool T93P-6256-A in the rear D-slots of the camshaft.
13. Install Camshaft Holding tool T93P-6256-AH onto the camshafts to keep the camshafts from rotating and to prevent damaging the camshaft positioning tool.
14. Tighten the secondary timing chain

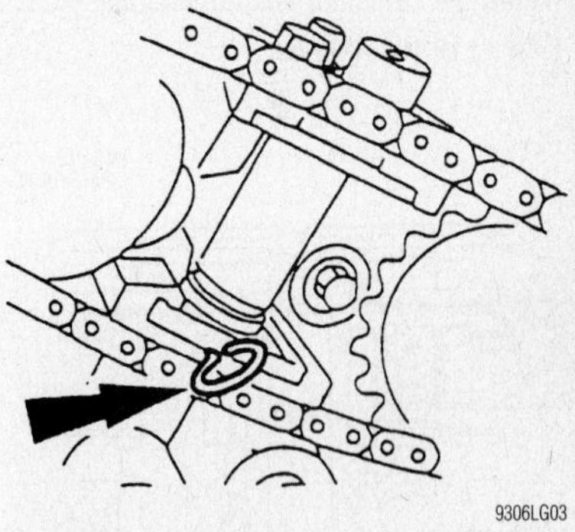

Secondary timing chain tensioner and locking pin—4.6L (VIN V) DOHC engine

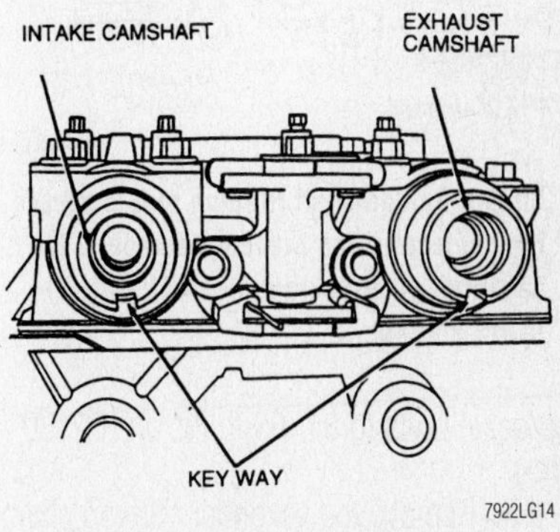

Position the camshafts for timing chain installation—4.6L (VIN V) DOHC engine

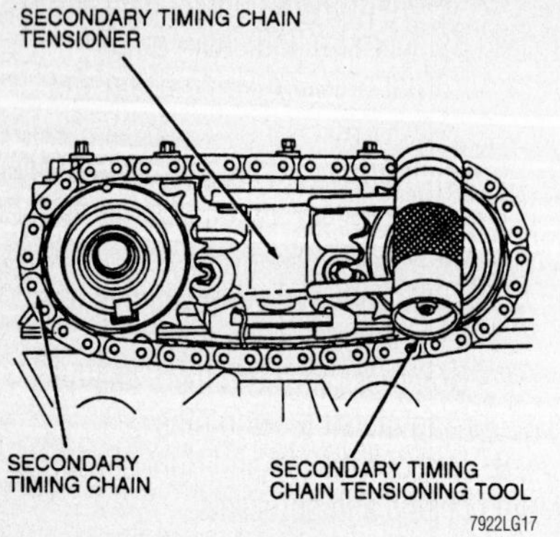

Secondary timing chain tensioning tool—4.6L (VIN V) DOHC engine

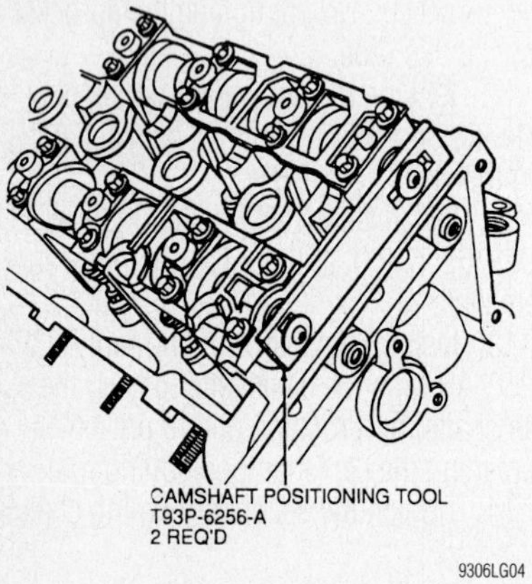

Install the camshaft positioning tool—4.6L (VIN V) DOHC engine

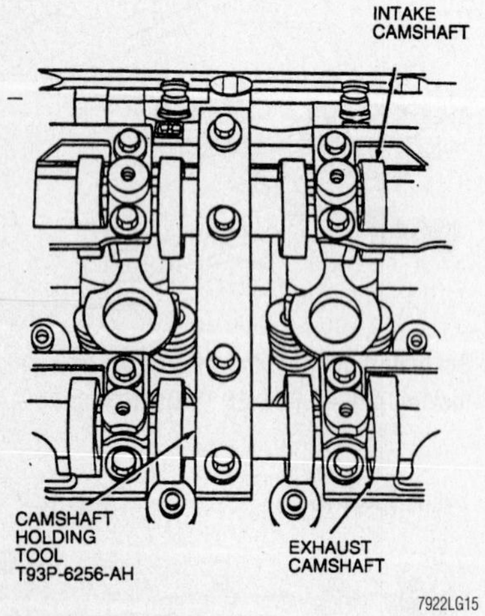

Install the Camshaft Holding tool to secure the camshafts and prevent damage to the camshaft positioning tool—4.6L (VIN V) DOHC engine

sprocket bolts to 81–95 ft. lbs. (110–130 Nm).

15. Remove the secondary timing chain tensioner locking pins and remove the tensioner tool.

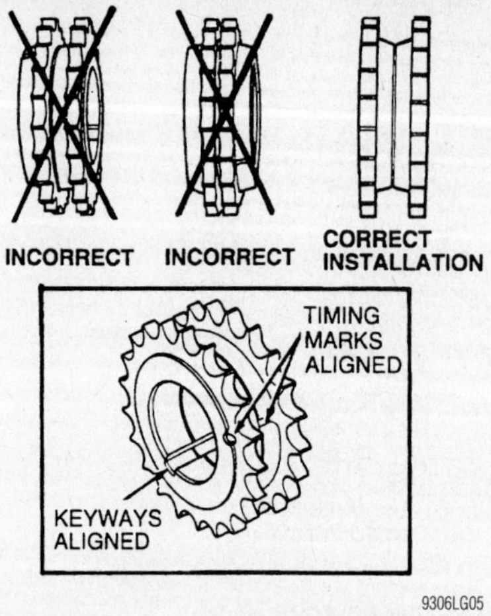

Crankshaft timing sprocket installation—4.6L (VIN V) DOHC engine

16. Compress the primary timing chain tensioners and install locking pins.

17. Install the crankshaft timing sprockets.

18. Align the copper colored primary timing chain links with the sprocket timing marks as shown.

19. Install or connect the following:
- Left primary timing chain and sprockets
- Left primary timing chain tensioner arm and chain guide
- Left primary timing chain tensioner
- Right primary timing chain and sprockets
- Right primary timing chain tensioner arm and chain guide
- Right primary timing chain tensioner

20. Tighten the camshaft sprocket bolts to 81–95 ft. lbs. (110–130 Nm).

21. Remove the primary timing chain tensioner locking pins.

22. Remove the Camshaft Holding and Camshaft Positioning tools.

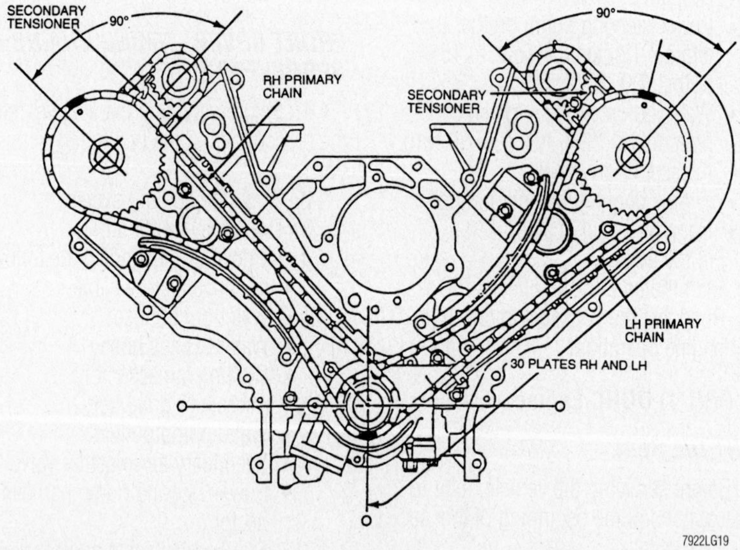

Correct alignment of the primary timing chains and sprockets—4.6L (VIN V) DOHC engine

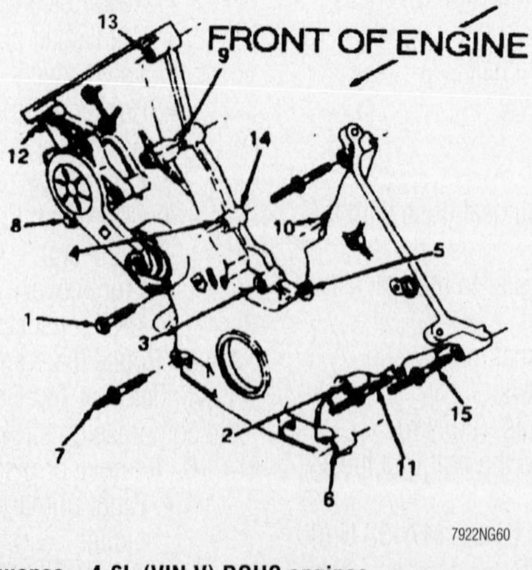

Front cover torque sequence—4.6L (VIN V) DOHC engines

Sealer Location

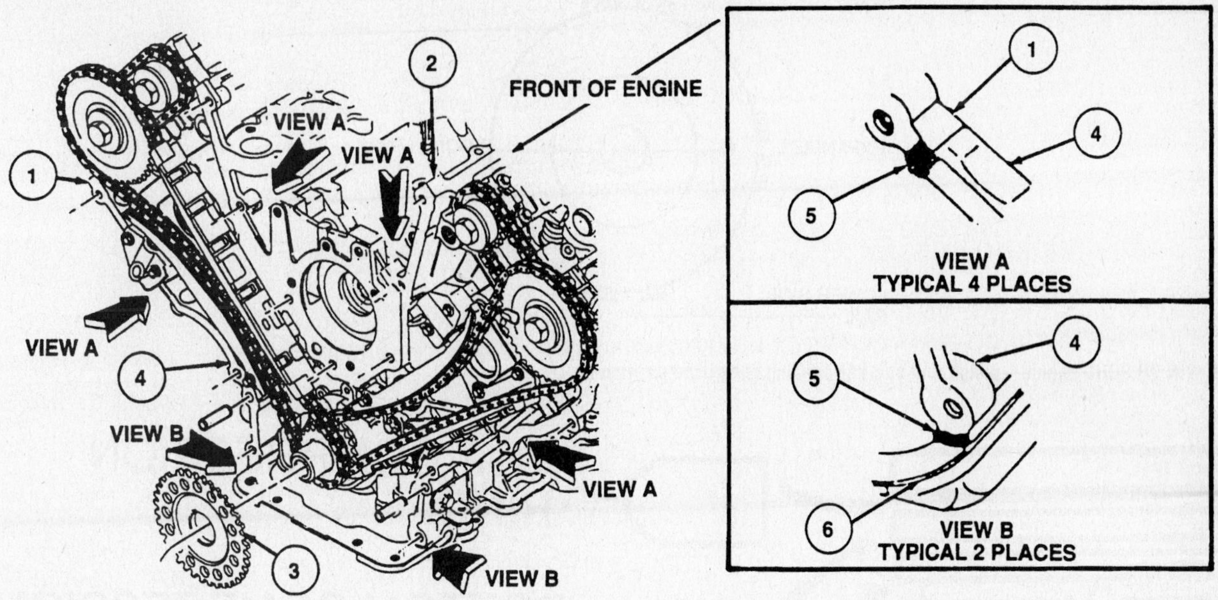

Item	Part Number	Description
1	6049	RH Cylinder Head
2	6049	LH Cylinder Head
3	12A227	Ignition Pulse Crankshaft Sensor Ring
4	6010	Cylinder Block
5	WSE-M4G320-A2	Sealer
6	6710	Oil Pan Gasket

9306LG06

Apply sealer to these locations—4.6L (VIN V) DOHC engines

23. Install or connect the following:
    - CKP sensor pulse wheel
    - Front cover. Apply silicone sealant as shown. Tighten the bolts in sequence to 15–22 ft. lbs. (20–30 Nm).
    - CKP sensor connector
    - Accessory drive belt tensioner and idler pulley. Tighten the bolts to 15–22 ft. lbs. (20–30 Nm).
    - Rocker arms
    - Valve covers
24. Apply silicone sealer to the crankshaft pulley woodruff key slot.
25. Install the crankshaft pulley and tighten the bolt as follows:
    a. Step 1: 66 ft. lbs. (90 Nm)
    b. Step 2: Loosen the bolt one full turn
    c. Step 3: 35–39 ft. lbs. (47–53 Nm)
    d. Step 4: Plus 85–90 degrees
26. Install or connect the following:
    - Power steering pump and reservoir
    - CMP sensor connector
    - Starter relay cable and bracket
    - Power steering pump harness connector
    - Secondary air injection valve
    - Water pump pulley. Tighten the bolts to 15–22 ft. lbs. (20–30 Nm).
    - Accessory drive belt
    - Cooling fan assembly
    - Water bypass hose
    - Water outlet hose
    - Negative battery cable
27. Fill the crankcase to the correct level.
28. Fill the cooling system.
29. Turn the air suspension switch to the **ON** position.
30. Start the engine and check for leaks.

Piston and Ring

POSITIONING

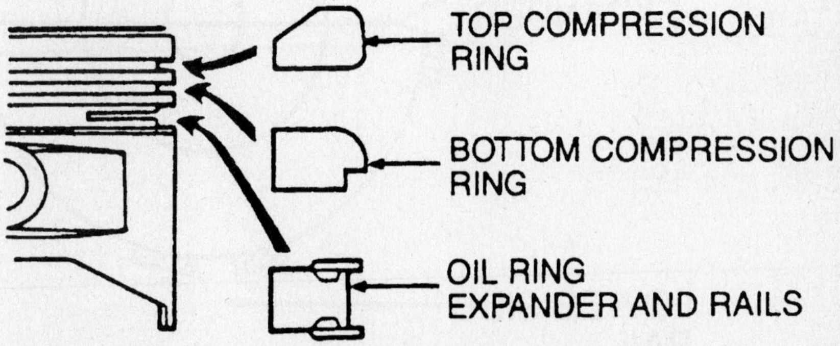

7922AG32

4.6L (VIN W) SOHC engine—piston ring positioning

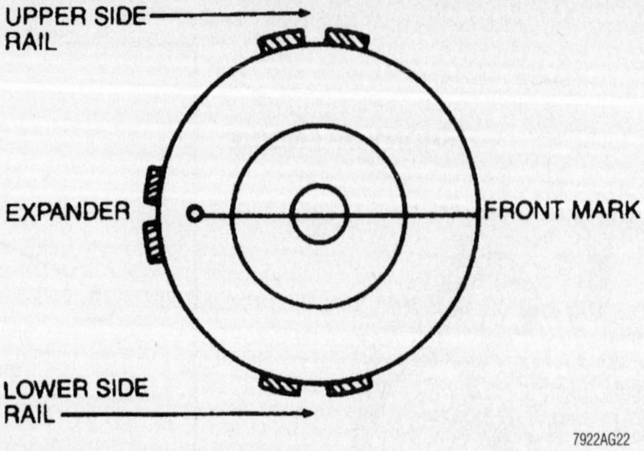

4.6L (VIN W) SOHC engine—piston ring end-gap spacing and piston positioning

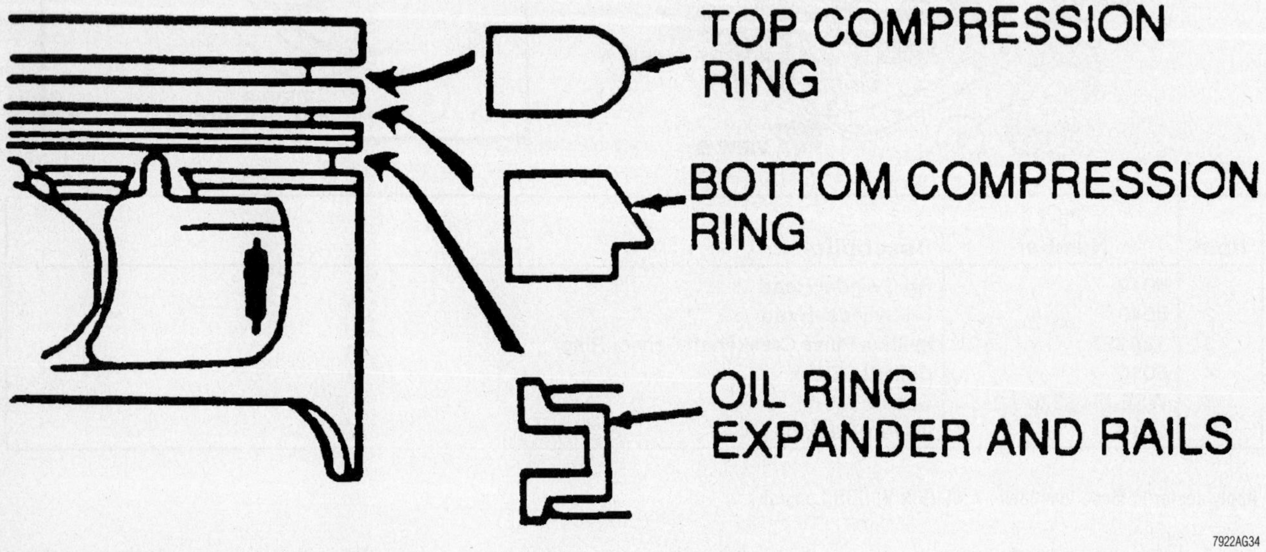

4.6L (VIN V) DOHC engine—piston ring positioning

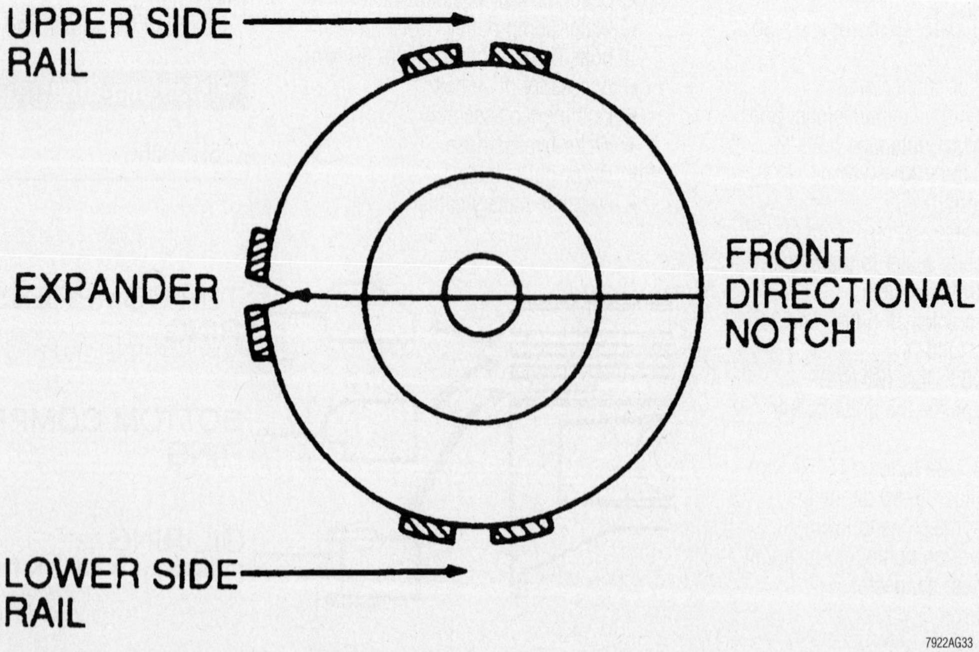

4.6L (VIN V) DOHC engine—piston ring end-gap spacing

FUEL SYSTEM

Fuel System Service Precautions

Safety is the most important factor when performing not only fuel system maintenance but any type of maintenance. Failure to conduct maintenance and repairs in a safe manner may result in serious personal injury or death. Maintenance and testing of the vehicle's fuel system components can be accomplished safely and effectively by adhering to the following rules and guidelines.

- To avoid the possibility of fire and personal injury, always disconnect the negative battery cable unless the repair or test procedure requires that battery voltage be applied.
- Always relieve the fuel system pressure prior to disconnecting any fuel system component (injector, fuel rail, pressure regulator, etc.), fitting or fuel line connection. Exercise extreme caution whenever relieving fuel system pressure, to avoid exposing skin, face and eyes to fuel spray. Please be advised that fuel under pressure may penetrate the skin or any part of the body that it contacts.
- Always place a shop towel or cloth around the fitting or connection prior to loosening to absorb any excess fuel due to spillage. Ensure that all fuel spillage (should it occur) is quickly removed from engine surfaces. Ensure that all fuel soaked cloths or towels are deposited into a waste container.
- Always keep a dry chemical (Class B) fire extinguisher near the work area.
- Do not allow fuel spray or fuel vapors to come into contact with a spark or open flame.
- Always use a back-up wrench when loosening and tightening fuel line connection fittings. This will prevent unnecessary stress and torsion to fuel line piping. Always follow the proper torque specifications.
- Always replace worn fuel fitting O-rings with new. Do not substitute fuel hose or equivalent, where fuel pipe is installed.

Fuel System Pressure

RELIEVING

Fuel supply lines on all fuel injected engines will remain pressurized for some period of time after the engine is shut **OFF**. This pressure must be relieved before servicing the fuel system. Pressure is relieved through the fuel pressure relief valve, located on the fuel rail.

To relieve the fuel system pressure, first remove the fuel tank cap to relieve pressure in the tank, then remove the cap on the fuel pressure relief valve. Attach a fuel pressure gauge and drain the system through the drain tube into a container. Remove the fuel pressure gauge and replace the cap on the relief valve.

Fuel Filter

REMOVAL & INSTALLATION

1. Before servicing the vehicle, refer to the precautions in the beginning of this section.
2. Disconnect the negative battery cable.
3. Relieve the fuel system pressure.
4. If equipped with air suspension, turn the air suspension switch to the **OFF** position.
5. Remove the hairpin clip push connect fittings from both ends of the fuel filter as follows:
   a. Step 1: Inspect the visible internal portion of the fitting for dirt accumulation. If more than a light coating of dust is present, clean the fitting before disassembly.
   b. Step 2: Some adhesion between the seals in the fitting and the filter will occur with time. To separate, twist the fitting on the filter, then push and pull the fitting until it moves freely on the filter.
   c. Step 3: Remove the hairpin clip from the fitting by first bending and breaking the shipping tab. Next, spread the 2 clip legs by hand about ⅛ inch each, to disengage the body and push the legs into the fitting. Lightly pull the triangular end of the clip and work it clear of the filter and fitting.
   d. Step 4: Grasp the fitting and pull in an axial direction to remove the fitting from the filter. Be careful on 90 degree elbow connectors, as excessive side loading could break the connector body.
   e. Step 5: After disassembly, inspect the inside of the fitting for any internal parts such as O-rings and spacers that may have been dislodged from the fitting. Replace any damaged connector.
6. Remove the filter retaining clamp and remove the fuel filter. Note the direction of the flow arrow on the filter, so the replacement filter can be reinstalled in the same position.

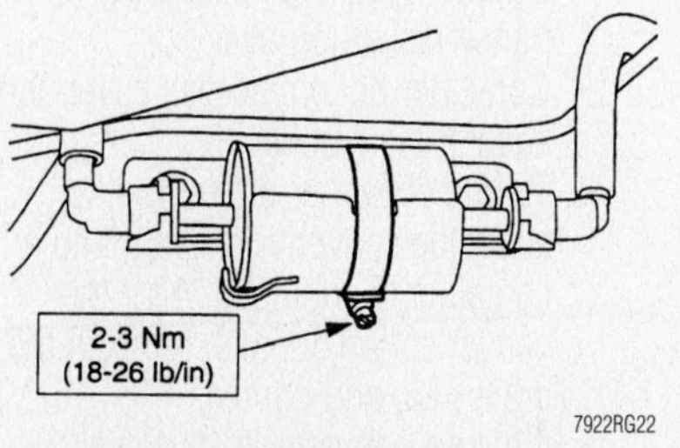

The fuel filter is located near the center of the vehicle on the frame rail

To install:

7. Install or connect the following:
   - Fuel filter with the flow arrow facing the proper direction and tighten the filter retaining clamp
   - Rubber insulator rings on the new filter. Replace the insulator rings if the filter moves freely after the retainer is installed.
   - Filter into the retainer with the flow arrow pointing out the open end of the retainer
   - Retainer on the bracket and tighten the mounting bolts to 27–44 inch lbs. (3–5 Nm)
8. Install the hairpin clip push connect fittings at both ends of the fuel filter as follows:
   a. Step 1: Install a new connector if damage was found. Insert a new clip into any 2 adjacent openings with the triangular portion pointing away from the fitting opening. Install the clip until the legs of the clip are locked on the outside of the body. Piloting with an index finger is necessary.
   b. Step 2: Before installing the fitting on the filter, wipe the filter end with a clean cloth. Inspect the inside of the fitting to be sure it is free of dirt and/or obstructions.
   c. Step 3: Apply a light coating of engine oil to the filter end. Align the fitting and filter axially and push the fitting onto the filter end. When the fitting is engaged, a definite click will be heard. Pull on the fitting to be sure it is fully engaged.
9. If equipped with air suspension, turn the air suspension switch to the **ON** position.
10. Reconnect the negative battery cable.
11. Start the engine and check for fuel leaks and proper operation.

Fuel Pump

REMOVAL & INSTALLATION

1. Before servicing the vehicle, refer to the precautions in the beginning of this section.
2. Disconnect the negative battery cable.
3. Relieve the fuel system pressure.
4. Install a hose into the fuel filler pipe and drain or siphon the fuel into a storage tank designed for fuel storage.
5. Remove any dirt that has accumulated around the fuel pump and fuel lines to prevent the entry of contaminants into the tank during fuel pump removal and installation.
6. Remove or disconnect the following:
   - Fuel supply and return line fittings at the fuel pump using fuel line disconnect tools
   - Fuel pump module electrical connector
   - 6 retaining bolts around the perimeter of the fuel pump module
   - Fuel pump module and seal from the fuel tank

To install:

7. Clean the fuel pump module mounting flange and fuel tank mounting surface.
8. Install or connect the following:
   - New seal and the fuel pump module using care not to damage the inlet filter and fuel sending unit float arm
   - 6 retaining bolts and tighten to 80–107 inch lbs. (9–12 Nm)
   - Fuel pump module electrical connector
   - Fuel supply and return lines to the fuel pump module. Pull on the fuel line fittings to verify engagement.
9. Add a minimum of 10 gallons (38L) of clean fuel to the fuel tank and check for leaks.

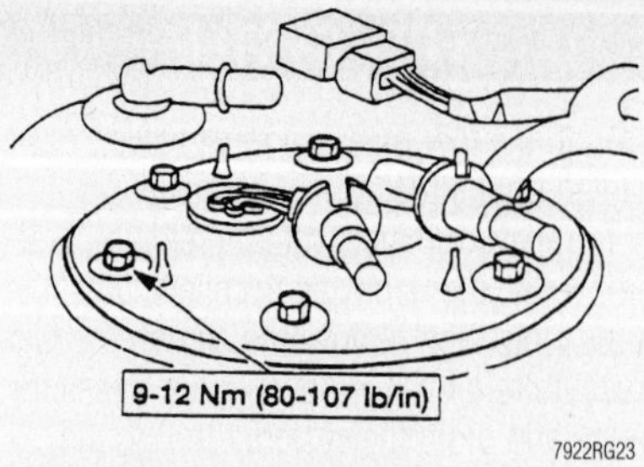

Tighten the fuel pump mounting bolts to 80–107 inch lbs. (9–12 Nm)

10. Install a Fuel pressure gauge to the Schrader valve on the fuel injection supply manifold.
11. Connect the negative battery cable.
12. Cycle the ignition switch from the **OFF** to **ON** position 5–10 times for 3 second intervals or until the fuel pressure gauge shows at least 35 psi (241 kPa).
13. Check for fuel leaks.
14. Remove the fuel pressure gauge.
15. Start the engine and recheck for fuel leaks.
16. Road test the vehicle and check for proper operation.

Fuel Injector

REMOVAL & INSTALLATION

4.6L (VIN W) SOHC Engine

1. Before servicing the vehicle, refer to the precautions in the beginning of this section.
2. Relieve the fuel system pressure.
3. Remove or disconnect the following:
   - Injector supply manifold
   - Wiring
   - Injector by pulling it up and gently rocking it side to side
   - O-rings and discard

To install:

4. Lubricate new O-rings with clean engine oil.
5. Install or connect the following:
   - O-rings
   - Fuel injector using a light, twisting and pushing motion
   - Injector supply manifold
   - Wiring

4.6L (VIN V) DOHC Engine

1. Before servicing the vehicle, refer to the precautions in the beginning of this section.
2. Relieve the fuel system pressure.
3. Remove or disconnect the following:
   - Negative battery cable
   - Air cleaner outlet tube
   - Upper intake manifold
   - Fuel pressure regulator vacuum line
   - Fuel pulse damper vacuum line
   - Brake booster vacuum line
   - Fuel lines
   - Fuel injector harness connectors
   - Injector supply manifold
   - Fuel injectors

To install:

4. Install or connect the following:
   - Fuel injectors with new O-ring seals
   - Injector supply manifold. Tighten the bolts to 71–106 inch lbs. (8–12 Nm).
   - Fuel injector harness connectors
   - Fuel lines
   - Brake booster vacuum line
   - Fuel pulse damper vacuum line
   - Fuel pressure regulator vacuum line
   - Upper intake manifold
   - Air cleaner outlet tube
   - Negative battery cable
5. Start the engine and check for leaks.

DRIVE TRAIN

Transmission Assembly

REMOVAL & INSTALLATION

1. Before servicing the vehicle, refer to the precautions in the beginning of this section.
2. Drain the transmission.
3. If equipped with air suspension, the air suspension switch, located on the right-hand side of the luggage compartment, must be turned to the **OFF** position before raising the vehicle.
4. Remove or disconnect the following:
   - Negative battery cable
   - Exhaust system as necessary for transmission removal
   - Converter bottom access cover and adapter plate bolts
   - Torque converter drain plug, to allow the converter to drain into a container, if equipped. After the converter has drained, reinstall the drain plug and tighten.
   - 4 torque converter-to-flywheel retaining nuts
   - Driveshaft (mark for installation), plug the transmission extension housing to prevent fluid leakage
   - Vehicle Speed Sensor (VSS) or if equipped, the speedometer cable from the transmission extension housing
   - Shift cable from the transmission manual control lever the throttle valve cable from the transmission throttle valve lever, if equipped
   - Transmission wiring harness connectors.

- Starter motor retaining bolts and place the starter motor aside

5. Position a transmission jack under the transmission and raise it enough to allow crossmember removal.
6. Remove or disconnect the following:
   - Engine rear support-to-crossmember bolts and the crossmember-to-frame side support retaining bolts
   - Crossmember and transmission support insulator
7. Lower the transmission jack and allow the transmission to hang.
8. Place a jack to the front of the engine and raise the engine enough to gain access to the 2 upper transmission-to-cylinder block retaining bolts. Do not remove the bolts at this time.
9. Remove or disconnect the following:
   - Transmission cooler lines at the transmission. Plug all openings to keep dirt out.
   - Lower transmission-to-cylinder block retaining bolts
   - Transmission fluid fill tube and plug the opening in the transmission
10. Secure the transmission to the transmission jack with a safety strap or chain.
11. Remove the 2 upper transmission-to-cylinder block bolts.
12. Carefully move the transmission rearward to disengage the bell housing from the dowel pins and the torque converter studs from the flywheel.
13. Remove or disconnect the following:
   - Transmission
   - Torque converter to prevent the converter from dropping out of the transmission causing possible damage or personal injury

To install:

14. Remove the safety stand and block of wood supporting the rear of the engine, if installed.
15. Install or connect the following:
   - Torque converter drain plug to 21–23 ft. lbs. (28–30 Nm), if equipped
   - Torque converter on the transmission and rotate into position to be sure the drive flats are fully engaged in the pump gear. When fully seated, the center of the torque converter should be about 7/16–9/16 inch (10.2–14.4mm) below the transmission mounting surface
16. Mount the transmission on a transmission jack and secure with a safely strap or chain. Raise the transmission and align with the cylinder block dowel pins.
17. Rotate the converter until the studs

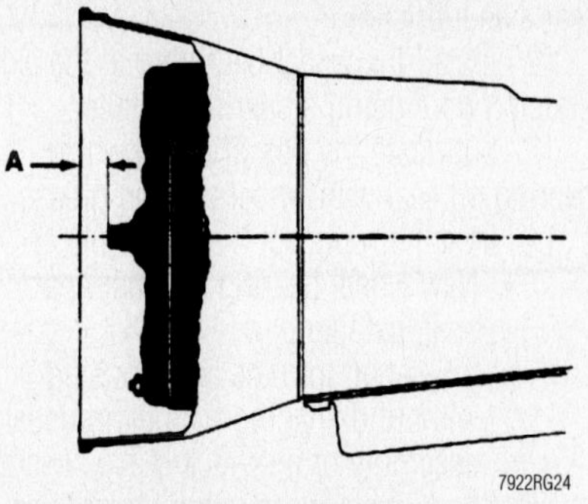

To prevent transmission damage, be sure that the torque converter is fully seated in the front pump of the transmission

and drain plug are in alignment with the holes in the flywheel. Align the orange balancing marks on the converter stud and flywheel bolt hole, if balancing marks are present.

18. Slide the transmission assembly forward into position, being careful not to damage the flywheel and converter pilot.
19. Install or connect the following:
   - 2 transmission housing-to-cylinder block bolts at the engine dowel pin locations. Tighten the bolts to 41–50 ft. lbs. (55–68 Nm).
   - Transmission housing-to-cylinder block bolts. Tighten the bolts to 41–50 ft. lbs. (55–68 Nm).
20. Remove the safety strap or chain from around the transmission.
21. Install or connect the following:
   - Transmission fluid fill tube. Tighten the bolt to 28–38 ft. lbs. (38–51 Nm).
   - Oil cooler lines to the transmission case. Tighten the cooler line fittings to 15–19 ft. lbs. (20–26 Nm).
22. Remove the jack supporting the front of the engine.
23. Install or connect the following:
   - Crossmember using the proper jack
   - Crossmember and transmission support insulators in position
   - Engine rear support-to-crossmember retaining bolts and the cross-member-to-frame side support retaining bolts
24. Remove the transmission jack.
25. Install or connect the following:
   - Transmission wiring harness connectors
   - Starter motor and wiring
   - 4 torque converter-to-flywheel retaining nuts. Tighten to 20–33 ft. lbs. (27–46 Nm).
   - Torque converter access cover and cover plate bolts. Tighten the bolts to 12–16 ft. lbs. (16–22 Nm).
   - Exhaust system
   - VSS and the wiring, or if equipped, the speedometer cable to the transmission extension housing
   - Driveshaft, aligning the marks that were made during removal
   - Shift cable to the transmission manual control lever
   - Throttle valve cable to the transmission throttle valve lever, if equipped
26. If equipped with air suspension, turn the air suspension switch to the **ON** position.
27. Fill the transmission.
28. Start the engine and check the transmission for leakage.
29. Road test the vehicle and check for proper transmission operation.

Axle Shaft, Bearing and Seal

REMOVAL & INSTALLATION

1. Before servicing the vehicle, refer to the precautions in the beginning of this section.
2. If equipped, turn the air suspension service switch to the **OFF** position before raising the vehicle.

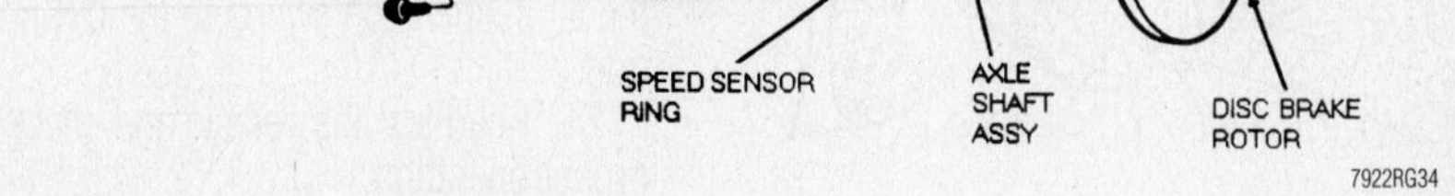

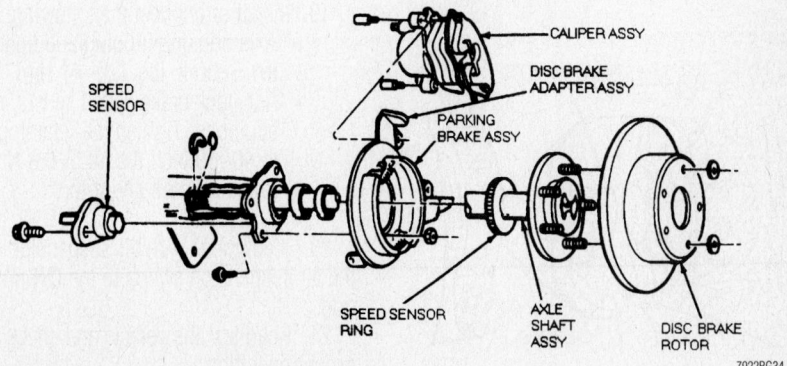

Exploded view of the rear axle shaft assembly

3. Remove or disconnect the following
   - Wheel
   - Brake caliper and rotor
   - Anti-lock brake speed sensor, if equipped
4. Clean all dirt from the area of the axle housing cover.
5. Place a drain pan under the axle housing.
6. Remove or disconnect the following:
   - Axle housing cover retaining bolts and the cover, draining the axle lubricant from the housing
   - Differential pinion shaft lockbolt and the differential pinion shaft
7. Push the flanged end of the axle shaft being removed toward the center of the vehicle
8. Remove or disconnect the following:
   - C-lock from the button end of the axle shaft
   - Axle shaft from the housing, being careful not to damage the oil seal and anti-lock brake sensor ring, if equipped.
9. Insert an axle bearing remover in the axle housing bore and position it behind the wheel bearing so the tangs on the tool engage the bearing outer race.
10. Remove the wheel bearing and seal as an assembly using an impact slide hammer attached to the bearing remover tool

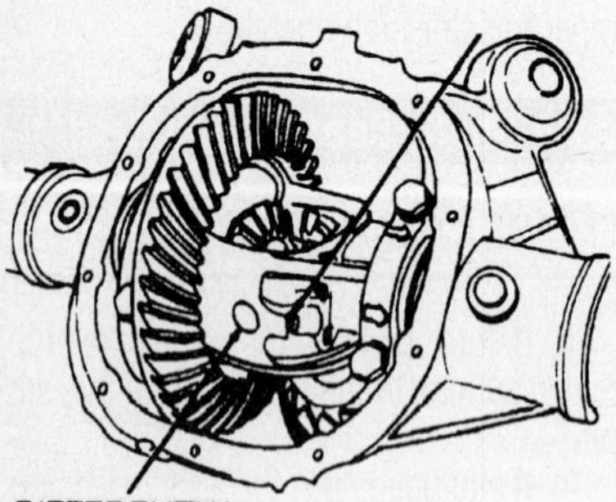

Removal of differential pinion shaft

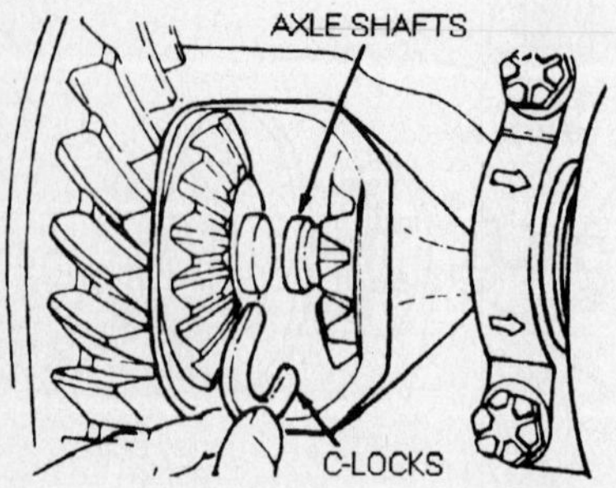

Removing axle shaft C-lock clips

To install:

11. Lubricate the new wheel bearing with rear axle lubricant.
12. Install the wheel bearing into the axle housing bore using a bearing installer.
13. Lubricate the lips of a new wheel bearing oil seal with wheel bearing grease.
14. Install or connect the following:
    - New wheel bearing seal using a seal installer
    - Axle shaft into the axle housing without damaging the bearing/seal assembly or anti-lock brake sensor ring, if equipped. Start the splines into the side gear and push firmly until the button end of the axle shaft can be seen in the differential case.
    - C-lock on the button end of the axle shaft splines, then push the shaft outboard until the shaft splines engage and the C-lock seats in the counterbore of the differential side gear.
    - Differential pinion shaft through the case and pinion gears, aligning the hole in the shaft with the lockbolt hole
    - Apply a thread locking compound to the lockbolt threads and place in the case and pinion shaft. Tighten to 15–30 ft. lbs. (20–41 Nm).
15. Cover the inside of the differential case with a shop rag and clean the sealing surface of the axle housing and the axle housing cover. Remove the shop rag.
16. Apply a ⅛–3⁄16 inch (3.18–4.76mm) wide bead of silicone sealer to the cover.
17. Install the axle housing and bolts and tighten in a crisscross pattern. Final torque the cover retaining bolts to 28–38 ft. lbs. (38–52 Nm).
18. Add the appropriate rear axle lubricant to the axle housing to a level ¼–9⁄16 inch (6–14mm) below the bottom of the fill hole. If equipped with a limited slip differential, add 4 oz. (118.3 ml) of the appropriate friction modifier.
19. Install or connect the following:
    - Axle housing fill plug and tighten to 15–30 ft. lbs. (20–41 Nm)
    - Anti-lock brake speed sensor, if equipped. Tighten the retaining bolt to 40–60 inch lbs. (4.5–6.8 Nm).
    - Brake calipers and rotors
    - Wheel
20. If equipped with air suspension, turn the air suspension switch to the **ON** position.
21. Road test the vehicle and check for proper operation.

Pinion Seal

REMOVAL & INSTALLATION

1. Before servicing the vehicle, refer to the precautions in the beginning of this section.
2. Remove or disconnect the following:
   - Driveshaft
   - Rear wheels
   - Rear brake calipers

➡The rear brake calipers must be removed so that there is no additional drag when measuring pinion bearing preload.

3. Use an inch lb. torque wrench and measure the amount of torque required to maintain pinion rotation through several revolutions.
4. Remove the pinion flange and remove the seal.

To install:

5. Install or connect the following:
   - Pinion seal and flange
   - New pinion flange nut
6. Rotate the pinion flange occasionally while tightening the flange nut to make sure the pinion bearings seat correctly.
7. Take frequent bearing preload torque readings.
8. If the preload recorded prior to disassembly is **lower** than the specification for used bearings, then tighten the pinion flange nut to specification. If the preload recorder prior to disassembly is **higher** than the specification for used bearings, then tighten the pinion flange nut to the original reading as recorded.
9. The pinion bearing preload specifications are as follows:
   a. Used bearings: 8–14 inch lbs. (0.9–1.6 Nm).
   b. New bearings: 16–29 inch lbs. (1.8–3.2 Nm).

CAUTION

Never loosen the pinion nut to reduce bearing preload. If it is necessary to reduce bearing preload, install a new collapsible spacer and pinion nut.

10. Install or connect the following:
    - Driveshaft
    - Brake calipers
    - Rear wheels
11. Fill the differential with gear lubricant and check for leaks.

STEERING AND SUSPENSION

Air Bag

CAUTION

Some vehicles are equipped with an air bag system. The system must be disarmed before performing service on, or around, system components, the steering column, instrument panel components, wiring and sensors. Failure to follow the safety precautions and the disarming procedure could result in accidental air bag deployment, possible injury and unnecessary system repairs.

PRECAUTIONS

Several precautions must be observed when handling the inflator module to avoid accidental deployment and possible personal injury.

- Never carry the inflator module by the wires or connector on the underside of the module.
- When carrying a live inflator module, hold securely with both hands, and ensure that the bag and trim cover are pointed away.
- Place the inflator module on a bench or other surface with the bag and trim cover facing up.
- With the inflator module on the bench, never place anything on or close to the module which may be thrown in the event of an accidental deployment.

DISARMING

1. Before servicing the vehicle, refer to the precautions in the beginning of this section.
2. Position the vehicle with the front wheels in a straight-ahead position.
3. Disconnect both battery cables.
4. Wait at least 1 minute for the air bag back-up power supply to deplete its stored energy before continuing.
5. Proceed with the repair.
6. Once the repair is complete.
7. Reconnect both battery cables.
8. Prove out the air bag system by turning the ignition key to the **RUN** position and visually monitoring the air bag indicator lamp in the instrument cluster. The indicator lamp should illuminate for approximately 6 seconds, then turn **OFF**. If the indicator lamp does not illuminate, stays on, or flashes at any time, a fault has been detected by the air bag diagnostic monitor.

Power Steering Gear

REMOVAL & INSTALLATION

Recirculating Ball Steering Gear

1. Before servicing the vehicle, refer to the precautions in the beginning of this section.
2. If equipped with air suspension, the air suspension switch must be turned to the **OFF** position before raising the vehicle.
3. Center the steering wheel and turn the key to the locked position.
4. Remove or disconnect the following:
   - Negative battery cable
   - Bolt and the intermediate shaft from the steering gear
5. On the Town Car, separate the 2 halves and remove the steering gear cover.
6. Tag the power steering pressure and return lines so they may be reassembled in their original positions.
7. Place a drain pan under the steering gear
8. Remove or disconnect the following:
   - Pressure and return lines. Plug the lines and ports in the gear to prevent the entry of dirt.
   - Pitman arm from the center link using a puller. It is not necessary to remove the Pitman arm from the steering gear.
9. Support the steering gear
10. Remove or disconnect the following:
    - Steering gear-to-frame rail retaining bolts
    - Steering gear

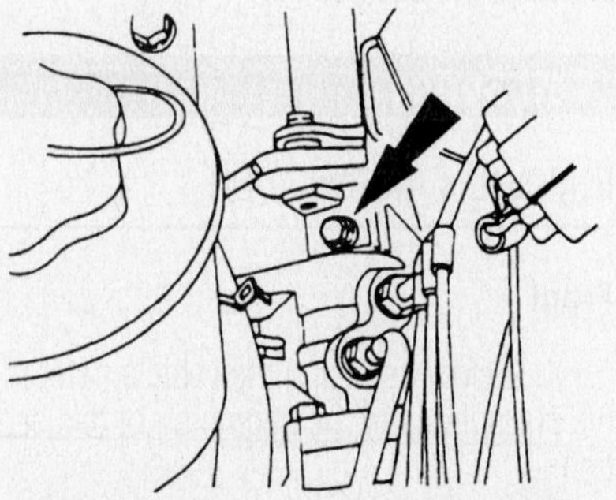

Remove the bolt and separate the intermediate shaft from the steering gear input shaft

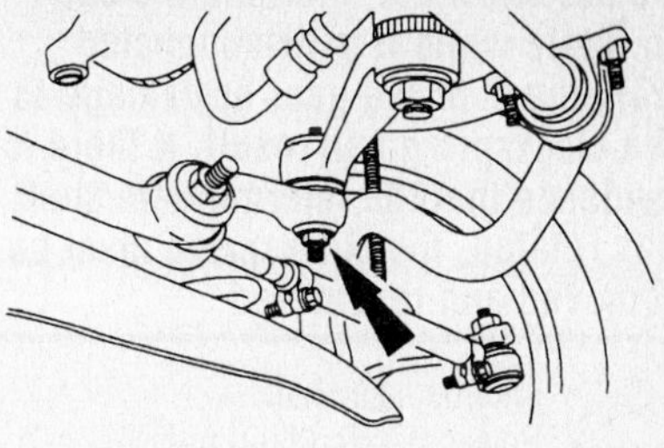

Remove the locknut and separate the Pitman arm from the center link using the appropriate puller

To install:

11. Install or connect the following:
    - Steering gear on the frame rail. Tighten the steering gear-to-frame retaining bolts to 50–67 ft. lbs. (66–90 Nm).
    - Pitman arm to the center link. Tighten the retaining nut to 52–60 ft. lbs. (70–81 Nm).
    - Power steering pressure and return lines to the steering gear and tighten the lines to 12–18 ft. lbs. (16–24 Nm)
    - Intermediate shaft to the steering gear. Tighten the bolt to 31–41 ft. lbs. (41–55 Nm).
    - Negative battery cable
12. If equipped with air suspension, turn the air suspension switch to the **ON** position.
13. Fill the reservoir with the correct power steering fluid and turn the steering wheel from stop-to-stop to distribute the fluid. Check the fluid level and add fluid, if necessary.
14. Start the engine and turn the steering wheel from left to right. Check for leaks.
15. On the Town Car, install the steering gear cover.

Rack and Pinion Steering Gear

1. Before servicing the vehicle, refer to the precautions in the beginning of this section.
2. If equipped with air suspension, the air suspension switch must be turned to the **OFF** position before raising the vehicle.
3. Center the steering wheel and turn the key to the locked position.
4. Remove or disconnect the following:
   - Negative battery cable
   - Front wheels
   - Outer tie rod ends

**** CAUTION**

Do not allow the intermediate shaft to rotate while it is disconnected from the steering gear or damage to the clockspring can result. If there is evidence that the intermediate shaft has rotated, the clockspring must be removed and recentered.

- Intermediate shaft
- Power steering fluid pressure and return lines
- Power steering pressure switch connector
- Steering gear mounting nuts
- Steering gear mounting studs
- Steering gear

To install:

5. Installation is the reverse of the removal procedure, while using the following torque values:
   - Steering gear mounting studs: 15 ft. lbs. (20 Nm)
   - Steering gear mounting nuts: 76 ft. lbs. (103 Nm)
   - Power steering fluid pressure and return lines: 13 ft. lbs. (18 Nm)
   - Intermediate shaft pinch bolt: 20 ft. lbs. (32 Nm)
   - Outer tie rod end nuts: 59 ft. lbs. (80 Nm)

Shock Absorber

REMOVAL & INSTALLATION

Front

1. Before servicing the vehicle, refer to the precautions in the beginning of this section.
2. If equipped with air suspension, the air suspension switch, located on the right-hand side of the luggage compartment, must be turned to the **OFF** position before raising the vehicle.
3. Remove or disconnect the following:
   - Nut, washer and bushing from the upper end of the shock absorber
   - 2 bolts retaining the shock absorber to the lower control arm
   - Shock absorber

To install:

4. Prior to installation, prime the new shock absorber. Fully extend the shock absorber while in the right side up (installed) position. Turn the shock absorber upside down and fully compress it. Repeat the procedure at least 3 times to purge any air trapped in the shock absorber.
5. Install or connect the following:
   - New bushing and washer on the stud on the top of the new shock absorber and position the unit inside the front coil spring
   - 2 lower retaining bolts and tighten them to 10–12 ft lbs. (13–17 Nm).
   - New bushing and washer on the shock absorber top stud
   - New retaining nut. Tighten the retaining nut to 25–34 ft. lbs. (34–46 Nm).
6. If equipped with air suspension, turn the air suspension switch to the **ON** position.

Rear

1. Before servicing the vehicle, refer to the precautions in the beginning of this section.
2. If equipped with air suspension, turn the air suspension service switch **OFF**.
3. Be sure the ignition switch is in the **OFF** position.
4. Support the rear axle assembly with a jack.
5. Remove or disconnect the following:
   - Top retaining nut, washer and bushing
   - Bottom retaining nut and washer
   - Shock absorber

To install:

6. Install or connect the following:
   - Shock absorber so the upper stud enters the hole in the frame
   - Top bushing, washer and retaining nut. Tighten to 26–34 ft. lbs. (34–46 Nm).
7. Extend the shock absorber and place the lower stud through the hole in the bracket
8. Bottom retaining washer and nut. Tighten to 57–75 ft. lbs. (76–103 Nm).
9. Remove the jack from the axle assembly.
10. Turn the air suspension service switch to the **ON** position.

Coil Spring

REMOVAL & INSTALLATION

Front

1. Before servicing the vehicle, refer to the precautions in the beginning of this section.
2. If equipped with air suspension, turn the air suspension service switch to the **OFF** position before raising the vehicle.
3. Remove or disconnect the following:
   - Wheel
   - Shock absorber
   - Center link from the Pitman arm
4. Using a spring compressor perform the following steps:
   a. Step 1: install 1 plate with the pivot ball seat facing downward into the coils of the spring. Rotate the plate so it is flush with the upper surface of the lower arm.
   b. Step 2: Install the other plate with the pivot ball seat facing upward into the coils of the spring. Insert the upper ball nut through the coils of the spring, so the nut rests in the upper plate.
   c. Step 3: Insert the compression rod into the opening in the lower arm, through the upper and lower plate and upper ball nut. Insert the securing pin through the upper ball nut and compression rod.
   d. Step 4: With the upper ball nut secured, turn the upper plate so it walks up the coil until it contacts the upper spring seat. Then, back off ½ turn.
   e. Step 5: Install the lower ball nut and thrust washer on the compression rod and screw on the forcing nut. Tighten the forcing nut until the spring is compressed enough so it is free in its seat.
5. Remove or disconnect the following:
   - 2 lower control arm pivot bolts
   - Lower arm from the frame crossmember
   - Coil spring
6. If a new coil spring is to be installed,

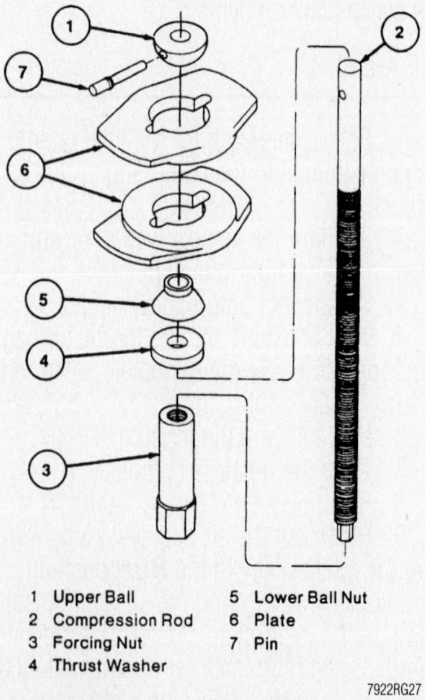

Exploded view of Spring Compressor D78P-5310-A

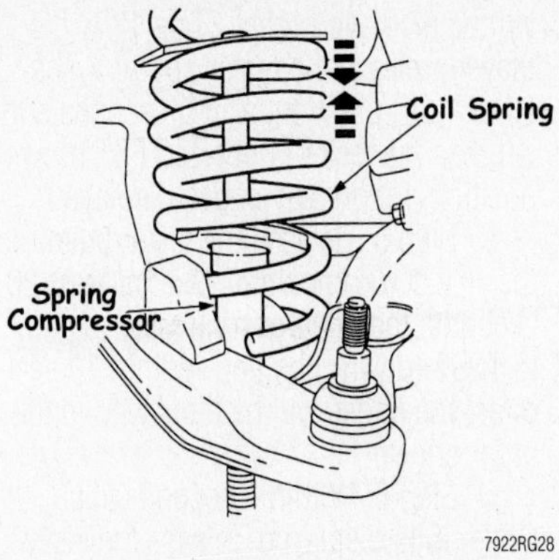

Compress the coil spring until it moves away from its seat

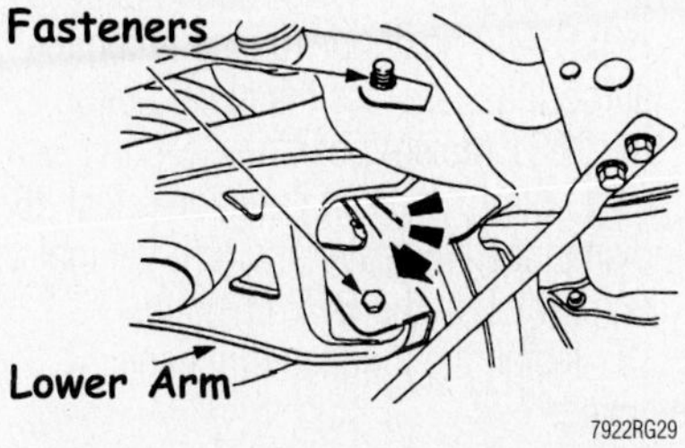

Remove the fasteners attaching the lower arm to the frame, then lower the arm and remove the spring with the compressor

mark the position of the upper and lower plates on the spring with chalk. With an assistant, compress a new spring for installation and measure the compressed length and the amount of curvature of the old spring.

7. Loosen the forcing nut to relieve the spring tension and remove the tools from the spring.

To install:

8. Assemble the spring compressor and locate in the same position as marked during disassembly.
9. Before compressing the coil spring, be sure the upper ball nut securing the pin is inserted properly.
10. Compress the coil spring until the spring height reaches the dimension measured during disassembly.
11. Position the coil spring assembly into the lower arm and position the lower arm into the frame crossmember.
12. Install both the front and rear lower control arm pivot bolts through the frame and lower arm bushings. Tighten the bolts and nuts to 109–148 ft. lbs. (148–201 Nm).
13. Remove the spring compressor from the coil spring.
14. Install or connect the following:
    - Drag link to the Pitman arm. Tighten the retaining nut to 60 ft. lbs. (80 Nm).
    - Shock absorber inside the coil spring
    - Retaining bolts
    - Wheel. Tighten the lug nuts to 85–105 ft. lbs. (115–142 Nm).
15. Place a washer and retaining nut on the shock absorber top stud. Tighten the nut to 25–34 ft. lbs. (34–46 Nm).
16. If equipped with air suspension, turn the air suspension switch to the **ON** position.
17. Check the front end alignment.

Rear

1. Before servicing the vehicle, refer to the precautions in the beginning of this section.
2. Place a hoist under the rear axle housing and raise and safely support the vehicle.
3. Support the frame side rails with 2 jackstands.
4. Remove or disconnect the following:
    - Rear sway bar
    - Lower studs of both rear shock absorbers from the mounting brackets on the axle tube
    - Parking brake cable from the upper arm retainer before lowering the axle housing
5. Lower the axle housing until the coil springs are released. If the axle housing is supported by the hoist, lower the hoist allowing the rear of the vehicle to rest on the jackstands. If the vehicle's axle housing is supported by the jackstands, leave the hoist stationary and lower the jackstands or raise the hoist to release the tension on the coil springs.
6. Remove the coil springs and insulators.

To install:

7. Install or connect the following:
    - Coil spring in the upper and lower seats with an insulator between the upper end of the spring and frame seat
    - Axle housing and connect the lower studs of the shock absorbers to the mounting brackets
    - Parking cable into the upper arm retainer
    - Sway bar
8. Road test the vehicle and check for proper operation.

Air Spring

REMOVAL & INSTALLATION

✲✲ CAUTION

Before servicing any air suspension component, disconnect power to the system by turning the air suspension service switch OFF or by disconnecting the negative battery cable. Do not remove an air spring under any circumstances when there is pressure in the air spring. Do not remove any components supporting an air spring without either exhausting the air or providing support for the air spring.

1. Before servicing the vehicle, refer to the precautions in the beginning of this section.
2. Turn the air suspension switch to the **OFF** position.
3. Raise and safely support the vehicle so the suspension is fully down with no load.
4. Remove or disconnect the following:
    - Heat shield, as required
    - Spring retainer clip
    - Air spring solenoid valve electrical connector
    - Air line
    - Air spring solenoid retainer
5. Rotate the solenoid valve counterclockwise to the first stop.
6. Pull the solenoid valve straight out slowly to the second stop to bleed air from the system.

✲✲ CAUTION

Do not fully release the solenoid until the air is completely bled from the air spring or personal injury may result.

7. After the air is fully bled from the system, rotate the solenoid valve counterclockwise to the third stop and remove the solenoid valve from the solenoid housing. Remove the large O-ring from the solenoid housing.
8. Remove the air spring.

To install:

9. Check the solenoid valve O-rings for cuts or abrasions. Replace the O-rings as required. Lightly grease the O-ring area of the solenoid valve and the larger solenoid housing O-ring with silicone dielectric compound.
10. Insert the solenoid into the air spring end cap and rotate clockwise to the third

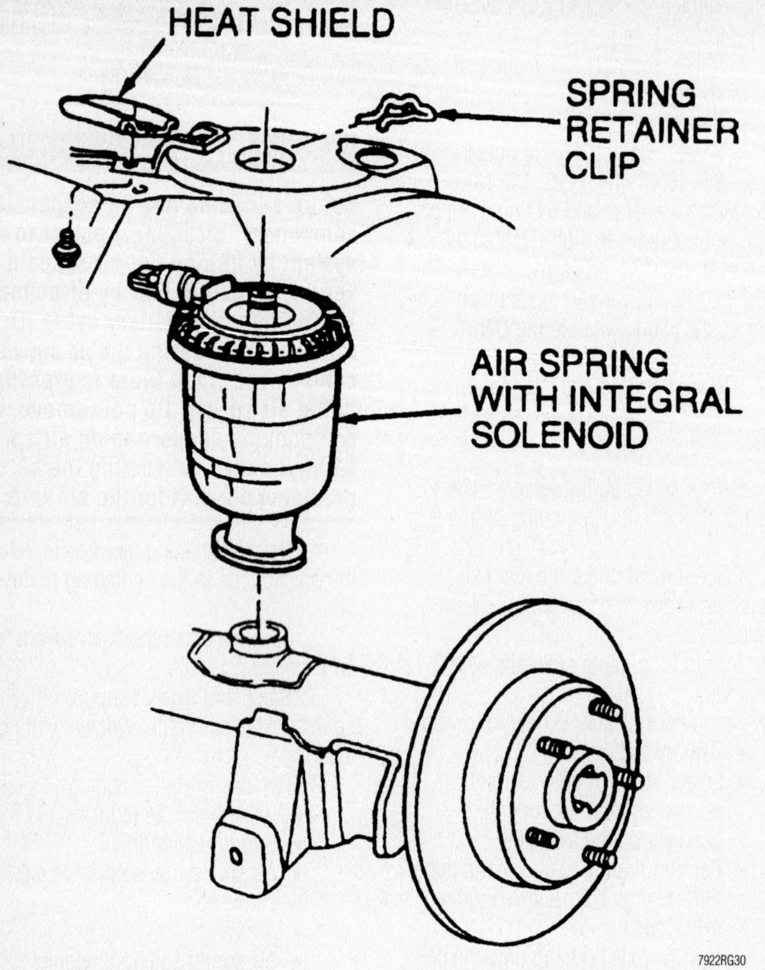

Exploded view of the air spring mounting

stop, push in to the second stop, then rotate clockwise to the first stop.

11. Install or connect the following:
   - Air spring solenoid retainer. Inspect the wiring harness connector and ensure the rubber gasket is in place at the bottom of the connector cavity
   - Air spring into the frame (upper) spring seat, taking care to keep the solenoid air and electrical connections clean and free of damage
   - Push-on ring spring retainer clip to the knob of the spring cap from the top side of the frame spring seat
   - Air line and electrical connector to the solenoid
   - Heat shield to the frame spring seat, if removed
   - Align the air spring piston-to-axle (lower) seat. Squeeze to increase pressure and push downward on the piston, snapping the piston to the axle seat at rebound and supported by the shock absorber.
   - Negative battery cable.

⁂ WARNING

The air springs may be damaged if the suspension is allowed to compress before the spring is inflated.

12. Refill the air spring as follows:

   a. Step 1: Turn the air suspension switch to the **ON** position. The ignition switch must be **ON** and the engine running or a battery charger must be connected to the battery to reduce battery drain.

   b. Step 2: Fold back or remove the right-hand luggage compartment trim panel and connect Super Star II Tester 007–0041-A to the air suspension DLC, which is located near the air suspension switch.

   c. Step 3: Set the tester to EEC-IV/MCU mode. Also set the tester to FAST mode. Release the tester button to the HOLD (up) position and turn the tester **ON**.

   d. Step 4: Depress the tester button to TEST (down) position. A Code 10 will be displayed. Within 2 minutes a Code 13 will be displayed. After Code 13 is displayed, release the tester button to the HOLD (up) position, wait 5 seconds and depress the tester button to TEST (down) position. Ignore any codes displayed.

   e. Step 5: Release the tester button to the HOLD (up) position. Wait at least 20 seconds, then depress the tester button to TEST (down) position. Within 10 seconds, the codes will be displayed in the order shown.

   f. Step 6: Within 4 seconds after Code 26 is displayed, release the tester button to the HOLD (up) position. Waiting longer than 4 seconds may result in Functional Test 31 being entered. The compressor will fill the air springs with air as long as the tester button is in the HOLD (up) position. To stop filling the air springs, depress the tester button to the TEST (down) position.

   g. Step 7: To exit Functional Test 26, disconnect the tester and turn the ignition switch to the **OFF** position.

13. Install the luggage trim panel, if removed.

Upper Ball Joint

REMOVAL & INSTALLATION

1. Before servicing the vehicle, refer to the precautions in the beginning of this section.
2. If equipped with air suspension, the air suspension switch to the **OFF** position before raising the vehicle.
3. Place supports under both sides of the frame just behind the lower control arms.
4. Remove the wheel.
5. Place a floor jack under the lower control arm at the lower ball joint area. The floor jack will support the spring load on the lower control arm.
6. Remove the retaining nut and pinch bolt from the upper ball joint stud.
7. Mark the position of the alignment cams. When replacing the upper ball joint this will approximate the current alignment.
8. Remove or disconnect the following:
   - 2 nuts retaining the upper ball joint to the upper control arm
   - Upper ball joint from the upper control arm and spread the slot in the wheel spindle with a prybar to remove the ball joint stud from the wheel spindle.

To install:

9. Install or connect the following:
   - Upper ball joint to the upper control arm
   - Ball stud into the wheel spindle
   - Upper ball joint pinch bolt and retaining nut. Tighten to 56–77 ft. lbs. (76–104 Nm).
   - Alignment cams to the approximate position at removal. If not marked, install in the neutral positions.
   - 2 nuts retaining the upper ball joint to the upper control arm. Hold the cams and tighten the nuts to 107–129 ft. lbs. (145–175 Nm).
   - Wheel and tire assembly. Tighten the lug nuts in a star pattern to 85–105 ft. lbs. (115–142 Nm).
10. Remove the floor jack from under the lower control arm.
11. If equipped with air suspension, turn the air suspension switch to the **ON** position.
12. Check and adjust the front wheel alignment.

Lower Ball Joint

REMOVAL & INSTALLATION

1. Before servicing the vehicle, refer to the precautions in the beginning of this section.
2. If equipped with air suspension, the air suspension service switch to the **OFF** position before raising the vehicle.
3. Place supports under both sides of the frame behind the lower control arms.
4. Remove or disconnect the following:
   - Wheel
   - Wheel spindle
   - Ball joint boot seal and discard
   - Ball joint

To install:

➡When installing a new ball joint, the protective cover should be left on to protect the ball joint seal during installation. It may be necessary to trim the cover so it can pass through the installation tool.

5. Install the ball joint.
6. Discard the protective cover and be sure the new ball joint is fully seated in the lower control arm. Ensure that the ball joint seal is not damaged.

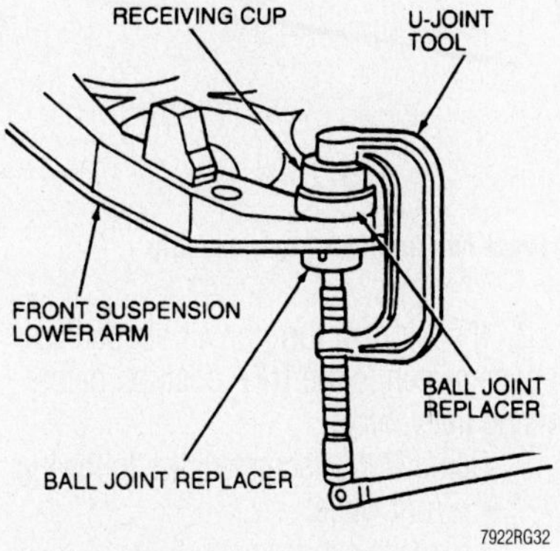

Use a ball joint press to install the new ball joint into the lower control arm

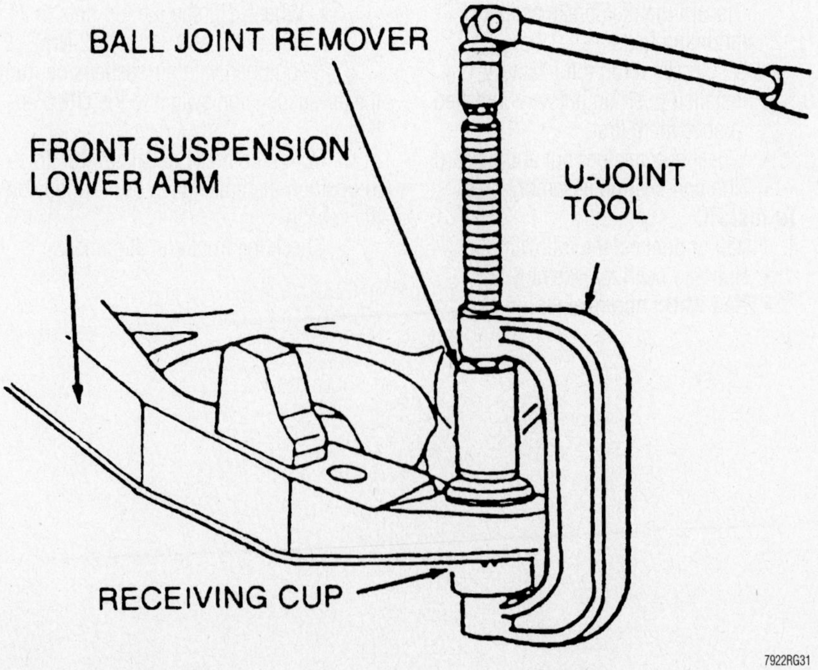

Use a ball joint press to remove the ball joint from the lower control arm

7. Install or connect the following:
   - Wheel spindle
   - Wheel. Tighten the lug nuts to 85–105 ft. lbs. (115–142 Nm).
8. If equipped with air suspension, turn the air suspension service switch to the **ON** position.
9. Check the front end alignment.

Upper Control Arm

REMOVAL & INSTALLATION

1. Before servicing the vehicle, refer to the precautions in the beginning of this section.
2. If equipped with air suspension, the air suspension switch, located on the right-hand side of the luggage compartment, must be turned to the **OFF** position before raising the vehicle.
3. Remove or disconnect the following:
   - Front wheel
   - Upper ball joint. Support the lower control arm.
   - Upper control arm pivot bolts
   - Upper control arm

To install:

4. Install or connect the following:
   - Upper control arm. Tighten the pivot bolts to 110–148 ft. lbs. (148–201 Nm).
   - Upper ball joint. Tighten the pinch bolt to 56–76 ft. lbs. (76–104 Nm).
   - Front wheel
5. If equipped with air suspension, turn the air suspension service switch to the **ON** position.
6. Align the vehicle.

CONTROL ARM BUSHING REPLACEMENT

The control arm bushings are serviced with the control arm as an assembly.

Lower Control Arm

REMOVAL & INSTALLATION

1. Before servicing the vehicle, refer to the precautions in the beginning of this section.
2. If equipped with air suspension, the air suspension switch, located on the right-hand side of the luggage compartment, must be turned to the **OFF** position before raising the vehicle.
3. Remove or disconnect the following:
   - Front wheel
   - Wheel speed sensor
   - Brake caliper and rotor

- Sway bar link
- Shock absorber
- Drag link
- Lower ball joint
- Coil spring
- Lower control arm pivot bolts
- Lower control arm

To install:

4. Install or connect the following:
   - Lower control arm. Tighten the pivot bolts to 110–148 ft. lbs. (148–201 Nm).
   - Coil spring
   - Lower ball joint. Tighten the nut to 110–148 ft. lbs. (148–201 Nm).
   - Drag link. Tighten the nut to 35–46 ft. lbs. (47–63 Nm).
   - Shock absorber
   - Sway bar link
   - Brake caliper and rotor
   - Wheel speed sensor
   - Front wheel
5. If equipped with air suspension, turn the air suspension service switch to the **ON** position.
6. Align the vehicle.

CONTROL ARM BUSHING REPLACEMENT

The control arm bushings are serviced with the control arm as an assembly.

Wheel Bearings

ADJUSTMENT

The front wheel bearings are of a hub unit design and are pre-greased, sealed and require no maintenance. The bearings are preset and cannot be adjusted.

REMOVAL & INSTALLATION

Front

1. Before servicing the vehicle, refer to the precautions in the beginning of this section.

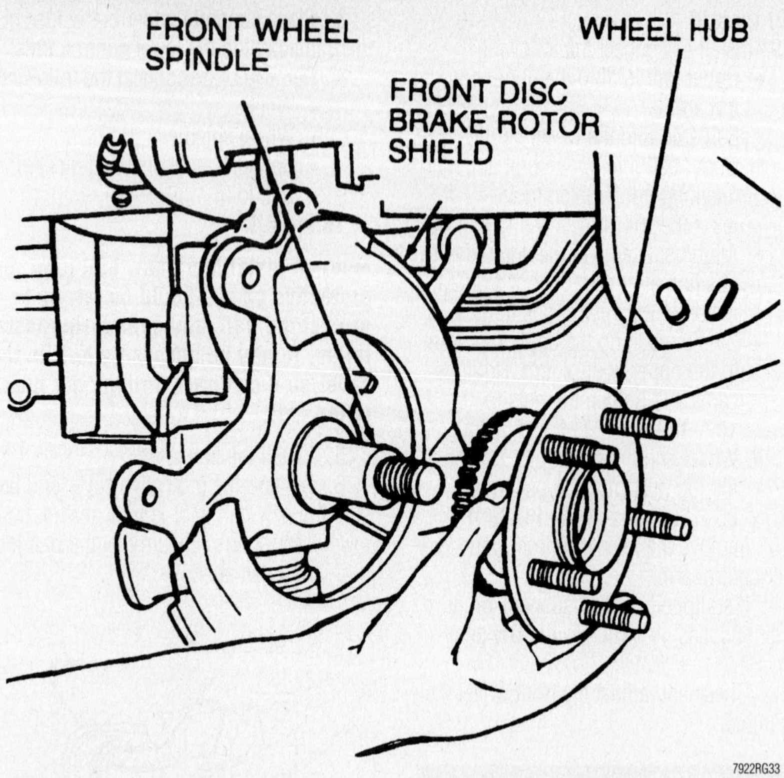

Front hub and bearing assembly

2. If equipped, turn the air suspension service switch to the **OFF** position before raising the vehicle.
3. Remove or disconnect the following
   - Front wheel
   - Grease cap from the hub
   - Disc brake caliper. Suspend the caliper with a length of wire. Do not let it hang from the brake hose. Discard the disc brake caliper mounting bolts.
   - Disc brake rotor. If the factory installed push on nuts are installed, remove them first.
   - Wheel hub retainer nut and discard
   - Hub and bearing assembly

To install:

4. Install or connect the following:
   - Hub and bearing assembly
   - New wheel hub retainer nut and tighten to 189–254 ft. lbs. (255–345 Nm)
   - Disc brake rotor and push on nuts, if equipped
   - New grease cap seal
   - Disc brake caliper using the 2 new disc brake caliper mounting bolts. Tighten the bolts to 125–170 ft. lbs. (170–230 Nm).
   - Wheel. Tighten the lug nuts to 85–104 ft. lbs. (115–142 Nm).
5. If equipped with air suspension, turn the air suspension switch to the **ON** position.
6. Pump the brake pedal several times to position the brake pads prior to moving the vehicle.
7. Check the front end alignment.

BRAKES

Brake Caliper

REMOVAL & INSTALLATION

Front

➡Before continuing with this procedure, make sure to have available, 2 new disc brake caliper guide pin bolts and 2 banjo bolt sealing washers, per caliper. Once removed, these parts loose their torque holding ability or retention capability and must not be reused.

1. Before servicing the vehicle, refer to the precautions in the beginning of this section.
2. If equipped with air suspension, the air suspension switch, located on the right-hand side of the luggage compartment, must be turned to the **OFF** position before raising the vehicle.
3. Remove or disconnect the following:
   - Front wheel and tire assembly
   - Banjo bolt securing the brake hose from the disc brake caliper. Plug the brake hose. Discard the sealing washers.
   - 2 disc brake caliper guide pin bolts and discard. If removing both calipers, mark the right and left sides so they may be reinstalled correctly.
   - Disc brake caliper off of the anchor plate

To install:

4. Retract the disc brake caliper piston fully in the piston bore, using an old brake pad or block of wood and a C-clamp.
5. Install or connect the following:
   - Disc brake pads to the caliper. Make sure that the brake pad insulators are correctly attached to the brake pad plate.
   - Disc brake caliper onto the anchor plate. Make sure the inner and outer pads are properly positioned and the anti-rattle spring is properly positioned. The caliper bleed screw should be positioned on top of the caliper when assembled on the vehicle.
   - 2 new caliper guide pin bolts and torque to 21–26 ft. lbs. (28–26 Nm)
   - Brake hose, after unplugging it, to the disc brake caliper using 2 new copper sealing washers on the banjo bolt. Torque the bolt to 30–40 ft. lbs. (41–54 Nm).
6. Bleed the brake system, filling the master cylinder as required. Only use clean DOT 3 brake fluid from a sealed container.
7. Install the wheel and tire assembly. Torque the lug nuts in a star pattern to 85–104 ft. lbs. (115–142 Nm).
8. If equipped with air suspension, turn the air suspension switch to the **ON** position.
9. Pump the brake pedal several times to position the brake pads prior to moving the vehicle.
10. Road test the vehicle and check for proper brake system operation.

Rear

➡Before continuing with this procedure, make sure to have available, 2 banjo bolt sealing washers, per caliper. Once removed, these parts loose their torque holding ability or retention capability and must not be reused.

1. Before servicing the vehicle, refer to the precautions in the beginning of this section.
2. If equipped with air suspension, the air suspension switch, located on the right-hand side of the luggage compartment, must be turned to the **OFF** position before raising the vehicle.
3. Remove or disconnect the following:
   - Rear wheel and tire assembly
   - Banjo bolt securing the brake hose to the disc brake caliper. Plug the brake hose and discard both sealing washers.
   - 2 disc brake caliper locating bolts. Lift the disc brake caliper off the rotor and anchor plate using a rotating motion.

To install:

4. Retract the disc brake caliper piston fully in the piston bore, using an old brake pad or block of wood and a C-clamp.
5. Install or connect the following:
   - Disc brake pads on the caliper. Make sure that the pads are on the correct side.
   - Caliper assembly above the rotor with the anti-rattle spring located on the lower adapter support arm. Install the caliper over the rotor with a rotating motion.
6. Clean the inner surface of the caliper bushings and locating bolts. Lubricate the caliper locating bolts with a suitable silicone dielectric compound. Install and start the locating bolts by hand only. Torque both bolts to 16–20 ft. lbs. (22–27 Nm).
   - Brake hose, after unplugging it, to the disc brake caliper using 2 new copper sealing washers on the banjo bolt. Torque the bolt to 30–40 ft. lbs. (41–54 Nm).
7. Bleed the brake system, filling the master cylinder as required. Only use clean DOT 3 brake fluid from a sealed container.
8. Replace the rubber rear disc brake bleeder screw cap.
9. Install the wheel and tire assembly. Torque the lug nuts in a star pattern to 85–104 ft. lbs. (115–142 Nm).
10. If equipped with air suspension, turn the air suspension switch to the **ON** position.
11. Pump the brake pedal several times to position the brake pads prior to moving the vehicle.
12. Road test the vehicle and check for proper brake system operation.

Disc Brake Pads

REMOVAL & INSTALLATION

➡Before continuing with this procedure, make sure to have available, 2 new disc brake caliper anchor bracket mounting bolts, per caliper. Once removed, these parts lose their torque holding ability or retention capability and must not be reused.

Front

1. Before servicing the vehicle, refer to the precautions in the beginning of this section.
2. If equipped with air suspension, the air suspension switch, located on the right-hand side of the luggage compartment, must be turned to the **OFF** position before raising the vehicle.
3. Remove or disconnect the following:
   - ½ of the brake fluid from the brake master cylinder reservoir. Properly dispose of the used brake fluid.
   - Front wheel and tire assembly
   - 2 disc brake caliper anchor bracket mounting bolts and discard. Lift the caliper assembly from the disc brake rotor using a rotating motion. Suspend the caliper inside the fender housing with wire. Do not allow the caliper to hang from the brake hose.
   - Inner and outer disc brake pads.

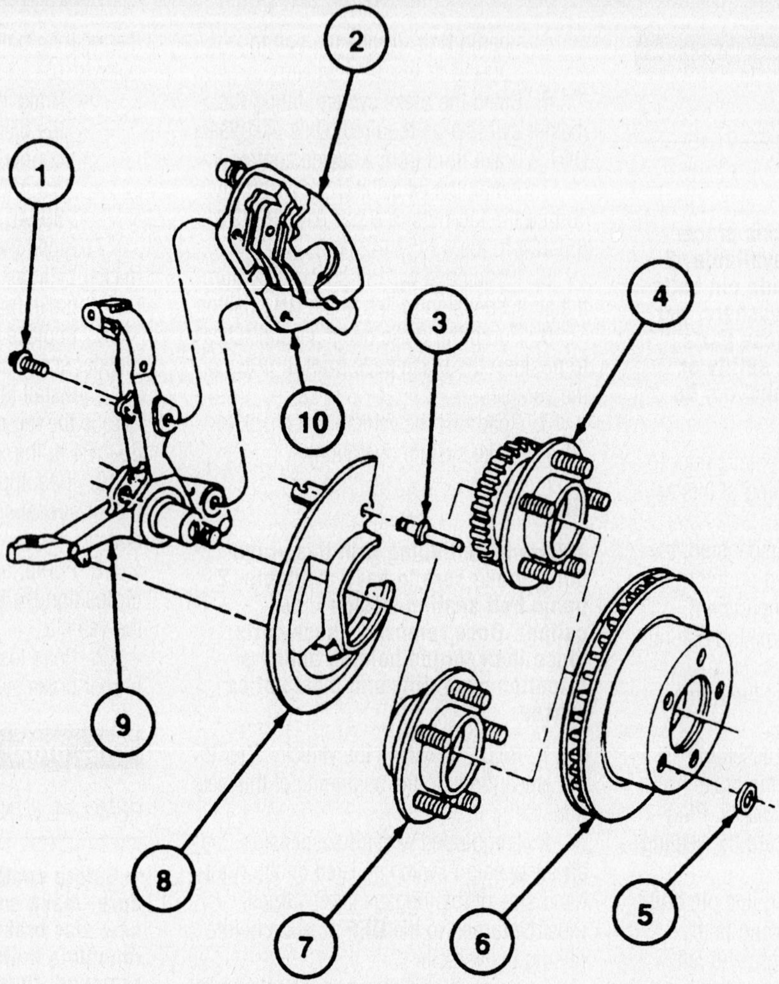

Item	Description
1	Bolt (2 Req'd)
2	Disc Brake Caliper
3	Rivet (3 Req'd)
4	Wheel Hub with ABS
5	Washer (2 Req'd)
6	Front Disc Brake Rotor
7	Wheel Hub Without ABS
8	Front Disc Brake Rotor Shield
9	Front Wheel Spindle
10	Front Disc Brake Caliper Anchor Plate Assembly

93006G83

Exploded view of the front disc brake assembly

Inspect the rotor braking surfaces for scoring and machine as necessary. Refer to the minimum rotor thickness specification when machining. If machining is not necessary, hand-sand the glaze from the braking surfaces with medium grit sandpaper. Make sure to wear an approved respirator.

To install:

4. Use a C-clamp and an old brake pad, wood block to seat the caliper piston in its bore. Do not allow metal or sharp objects to come into direct contact with the plastic caliper piston surface or damage will result.
5. Remove all rust buildup from the inside of the caliper legs.
6. Make sure the anti-rattle spring is seated in the caliper lining inspection opening and that it is installed from the lining side.
7. Install or connect the following:
   - Inner disc brake pad to the caliper piston. Do not bend the pad clips during installation in the piston or distortion and rattles can occur. Install the outer disc brake pad. Make sure the clips are properly seated.
   - Caliper over the rotor and install 2 new anchor bracket mounting bolts. Torque the bolts to 126–169 ft. lbs. (170–230 Nm).
   - Wheel and tire assembly. Torque the lug nuts in a star pattern to 85–104 ft. lbs. (115–142 Nm).
8. If equipped with air suspension, turn the air suspension switch to the **ON** position.
9. Pump the brake pedal prior to moving the vehicle to seat the brake pads.
10. Fill the master cylinder reservoir with clean DOT 3 brake fluid from a closed container.
11. If the disc brake calipers were replaced or repaired be sure to bleed the system.
12. Road test the vehicle and check for proper brake system operation.

Rear

1. Before servicing the vehicle, refer to the precautions in the beginning of this section.
2. If equipped with air suspension, the air suspension switch, located on the right-hand side of the luggage compartment, must be turned to the **OFF** position before raising the vehicle.
3. Remove or disconnect the following:
   - ½ of the brake fluid from the brake master cylinder reservoir. Properly dispose of the used brake fluid.
   - Rear wheel and tire assembly
   - 2 disc brake caliper retaining bolts
   - Caliper off the disc brake rotor and anchor plate using a rotating motion
   - Inner and outer disc brake pads
4. Inspect the disc brake rotor for scoring and wear. Inspect the rotor braking surfaces for scoring and machine as necessary. Refer to the minimum rotor thickness specification when machining. If machining is

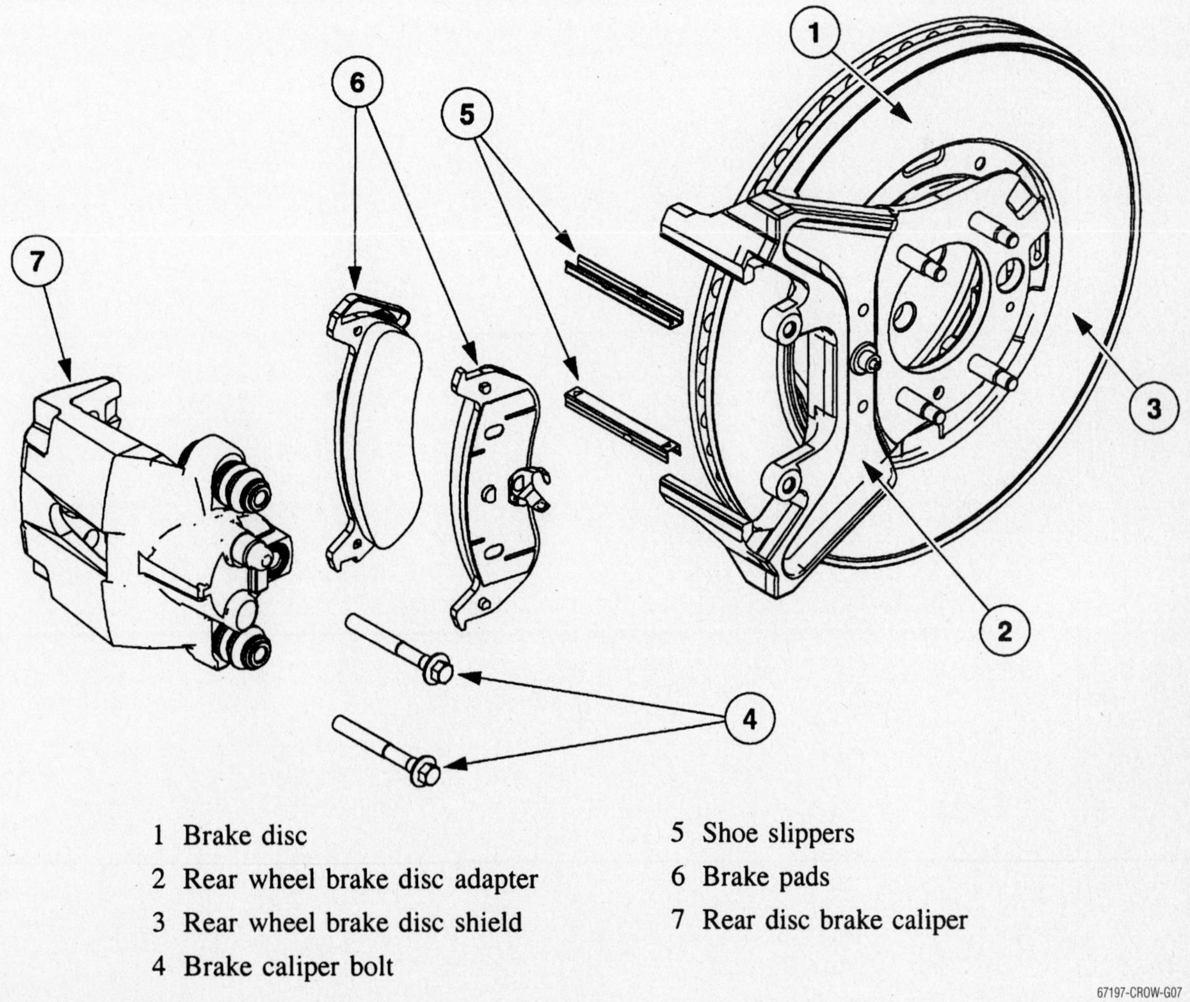

Exploded view of the rear disc brake assembly

not necessary, hand-sand the glaze from the braking surfaces with medium grit sandpaper. Make sure to wear an approved respirator.

To install:

5. Use a C-clamp and an old brake pad, wood block to seat the caliper piston in its bore. Do not allow metal or sharp objects to come into direct contact with the plastic caliper piston surface or damage will result.
6. Remove all rust buildup from the inside of the caliper legs.
7. Install or connect the following:
   - Inner disc brake pad to the caliper piston. Do not bend the pad clips during installation in the piston or distortion and rattles can occur. Install the outer disc brake pad. Make sure the clips are properly seated.

➡Make sure the insulators are installed on the brake pads.

   - Disc brake caliper over the disc brake rotor and install 2 caliper retaining bolts. Torque to 16–20 ft. lbs. (22–27 Nm).
   - Wheel and tire assembly. Torque the lug nuts in a star pattern to 85–104 ft. lbs. (115–142 Nm).

8. If equipped with air suspension, turn the air suspension switch to the **ON** position.
9. Pump the brake pedal prior to moving the vehicle to seat the disc brake pads.
10. Fill the master cylinder reservoir with clean DOT 3 brake fluid from a closed container.
11. If the disc brake calipers were replaced or repaired be sure to bleed the system.
12. Road test the vehicle and check for proper brake system operation.

FORD

Escape

6

SPECIFICATION CHARTS

ENGINE AND VEHICLE IDENTIFICATION

Engine						
Code ①	Liters	Cu. In.	Cyl.	Fuel Sys.	Engine Type	Eng. Mfg.
B	2.0	121	4	SFI	DOHC	Ford
Z	2.3	137	4	SFI	DOHC	Ford
1	3.0	182	6	SFI	DOHC	Ford

Model Year	
Code ②	Year
1	2001
2	2002
3	2003
4	2004
5	2005

SFI: Multi-port Fuel Injection

DOHC: Double Overhead Camshafts

① 8th digit of VIN

② 10th digit of VIN

67197-ESCA-C01

GENERAL ENGINE SPECIFICATIONS

Year	Model	Engine Displacement Liters	Engine VIN	Net Horsepower @ rpm	Net Torque @ rpm (ft. lbs.)	Bore x Stroke (in.)	Com-pression Ratio	Oil Pressure @ rpm
2001	Escape	2.0	B	127@5500	135@4500	3.34x3.46	9.6:1	54-80 ①
	Escape	3.0	1	200@5500	200@4500	3.50x3.13	10.0:1	45 ①
2002	Escape	2.0	B	127@5500	135@4500	3.34x3.46	9.6:1	54-80 ①
	Escape	3.0	1	200@5500	200@4500	3.50x3.13	10.0:1	45 ①
2003	Escape	2.0	B	127@5500	135@4500	3.34x3.46	9.6:1	54-80 ①
	Escape	3.0	1	200@5500	200@4500	3.50x3.13	10.0:1	45 ①
2004	Escape	2.0	B	127@5500	135@4500	3.34x3.46	9.6:1	54-80 ①
	Escape	3.0	1	200@5500	200@4500	3.50x3.13	10.0:1	45 ①
2005	Escape	2.3	Z	153@5800	152@4250	3.44x3.70	NA	29-30@2000
	Escape	3.0	1	200@5500	200@4500	3.50x3.13	10.0:1	11@1500②

SFI: Multi-port Fuel Injection

① The manufacturer does not provide an engine speed specification for oil pump pressure.

② Minimum hot

67197-ESCA-C02

ENGINE TUNE-UP SPECIFICATIONS

Year	Engine Displacement Liters	Engine VIN	Spark Plug Gap (in.)	Ignition Timing (deg.) MT	Ignition Timing (deg.) AT	Fuel Pump (psi)	Idle Speed (rpm) MT	Idle Speed (rpm) AT	Valve Clearance Intake	Valve Clearance Exhaust
2001	2.0	B	0.039-0.043	10 BTDC	—	65②	①	—	HYD.	HYD.
	3.0	1	0.052-0.056	10 BTDC	10 BTDC	65②	①	①	HYD.	HYD.
2002	2.0	B	0.051	10 BTDC	—	65②	①	—	HYD.	HYD.
	3.0	1	0.052-0.056	10 BTDC	10 BTDC	65②	①	①	HYD.	HYD.
2003	2.0	B	0.051	10 BTDC	—	65②	①	—	HYD.	HYD.
	3.0	1	0.052-0.056	10 BTDC	10 BTDC	65②	①	①	HYD.	HYD.
2004	2.0	B	0.051	10 BTDC	—	65②	①	—	HYD.	HYD.
	3.0	1	0.052-0.056	10 BTDC	10 BTDC	65②	①	①	HYD.	HYD.
2005	2.3	Z	0.041-0.045	10 BTDC	10 BTDC	39②	①	①	HYD.	HYD.
	3.0	1	0.052-0.056	10 BTDC	10 BTDC	39②	①	①	HYD.	HYD.

BTDC: Before Top Dead Center

HYD: Hydraulic lash adjusters

NA: Information not available

① Refer to Vehicle Emission Control Information Label

② Key on; engine off

67197-ESCA-C03

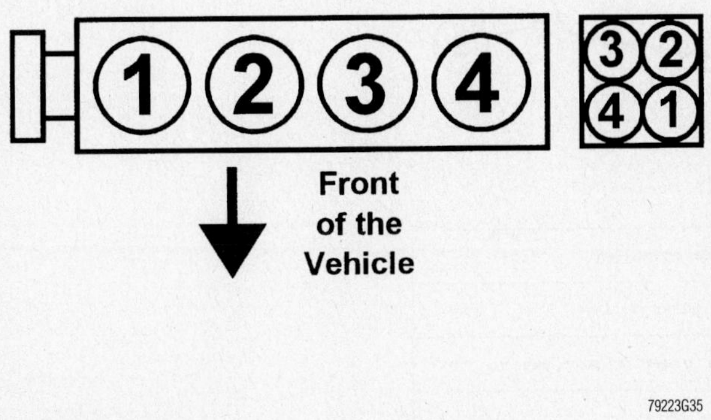

2.0L engine
Firing order: 1-3-4-2
Distributorless ignition

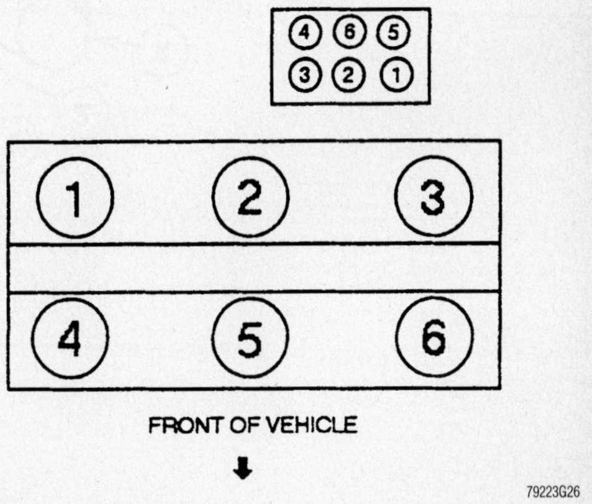

3.0L engine
Firing order: 1-4-2-5-3-6
Distributorless ignition

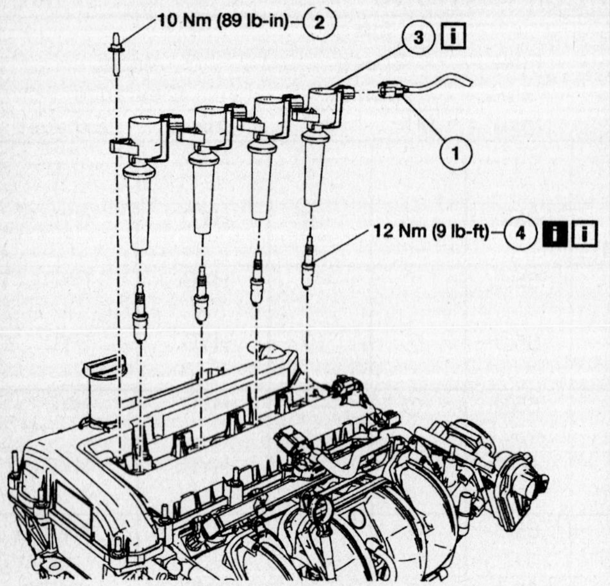

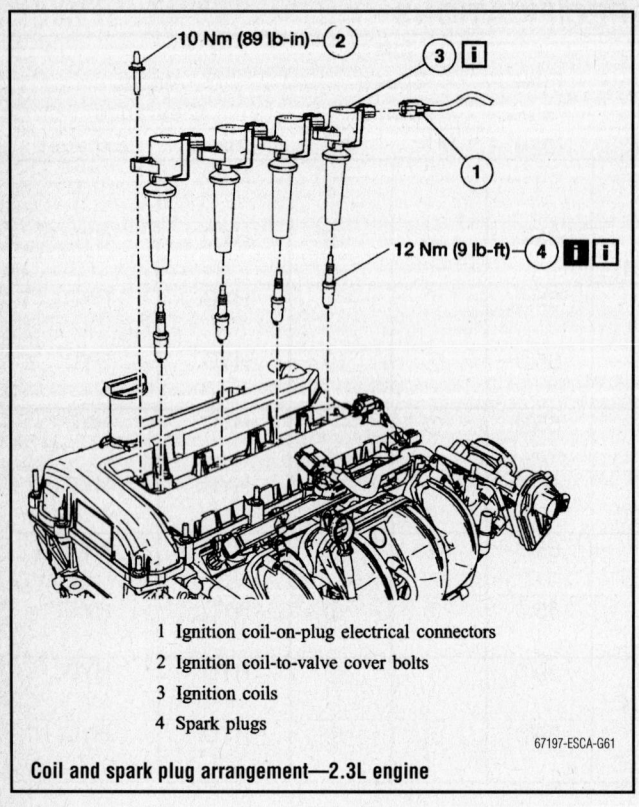

1 Ignition coil-on-plug electrical connectors
2 Ignition coil-to-valve cover bolts
3 Ignition coils
4 Spark plugs

67197-ESCA-G61

Coil and spark plug arrangement—2.3L engine

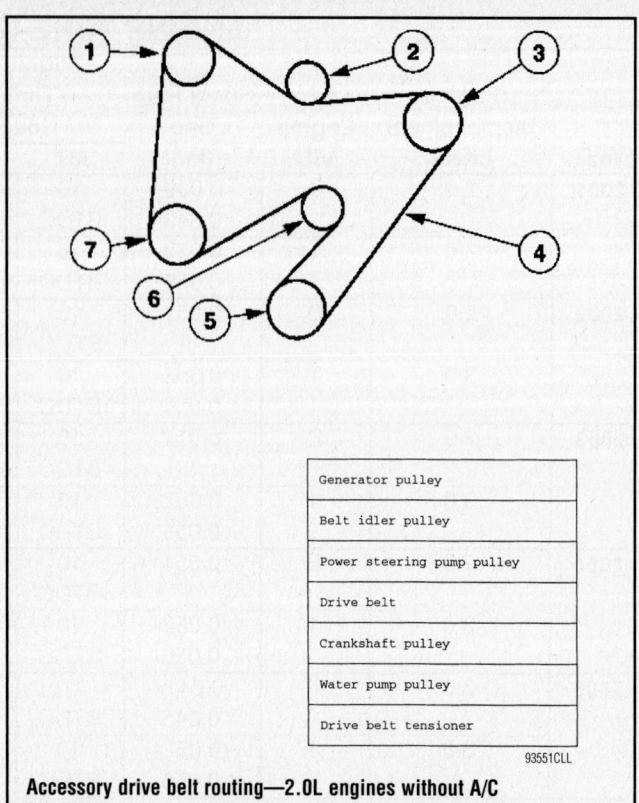

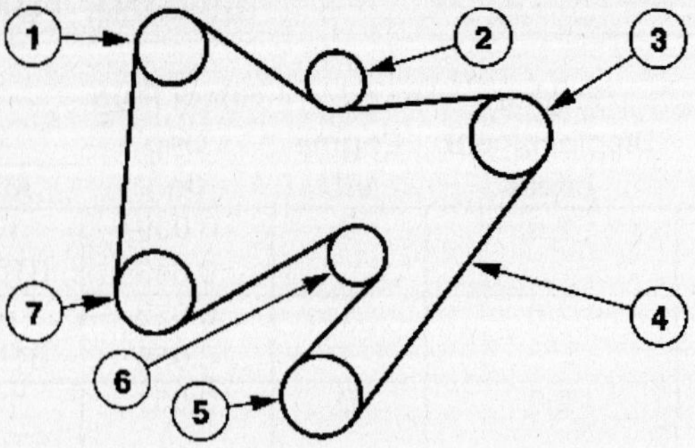

1	Generator pulley
2	Belt idler pulley
3	Power steering pump pulley
4	Drive belt
5	Crankshaft pulley
6	Water pump pulley
7	Drive belt tensioner

93551CLL

Accessory drive belt routing—2.0L engines without A/C

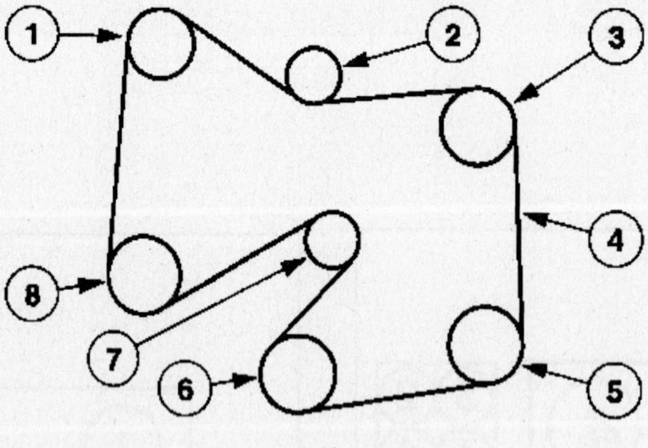

1	Generator pulley
2	Belt idler pulley
3	Power steering pump pulley
4	Drive belt
5	A/C clutch pulley
6	Crankshaft pulley
7	Water pump pulley
8	Drive belt tensioner

93551CLM

Accessory drive belt routing—2.0L engines with A/C

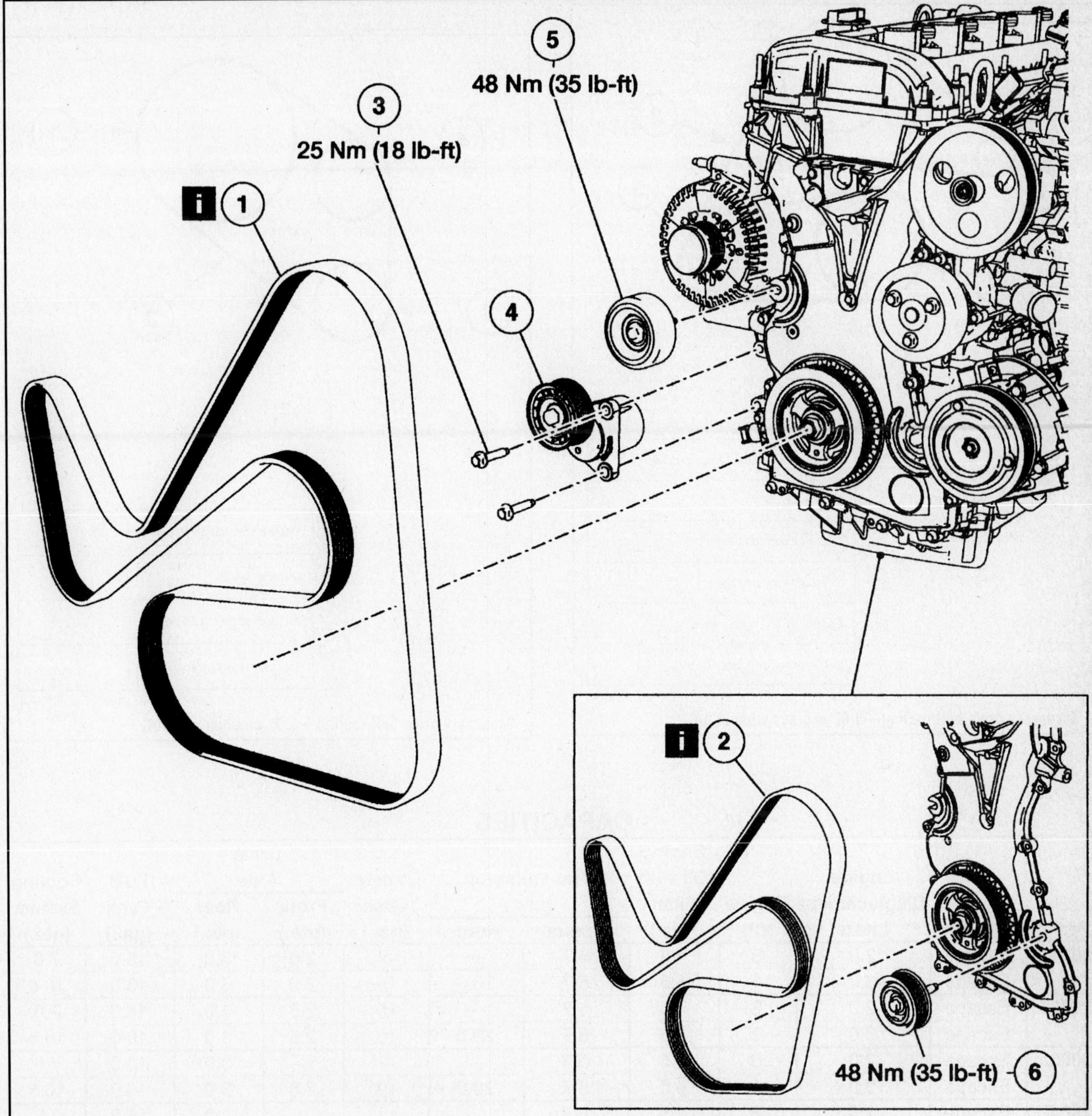

1 Accessory drive belt (with A/C)
2 Accessory drive belt (without A/C)
3 Accessory drive belt tensioner bolts
4 Accessory drive belt tensioner
5 Accessory drive belt idler pulley
6 Accessory drive belt idler pulley (without A/C only)

67197-ESCA-G62

Accessory drive belt routings—2.3L engine

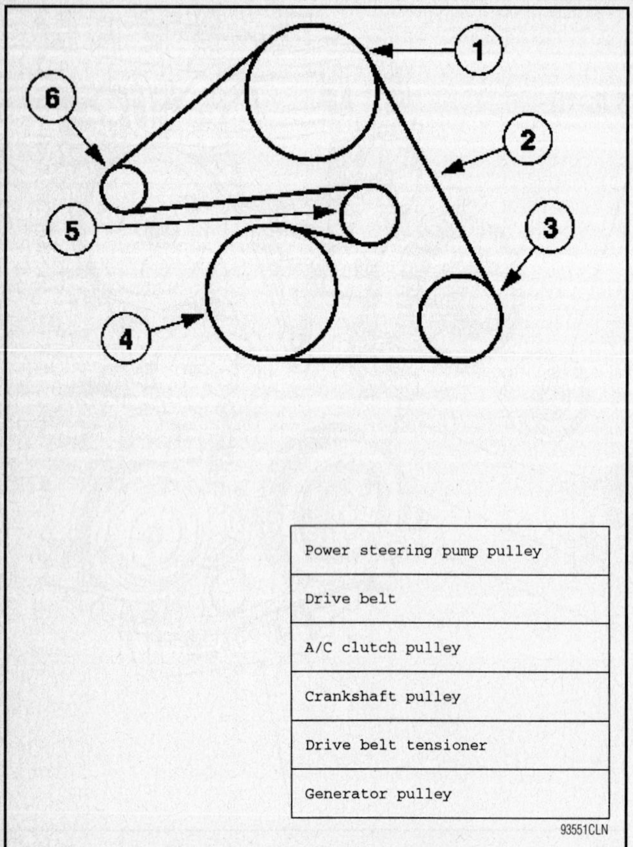

Accessory drive belt routing—3.0L engines without A/C

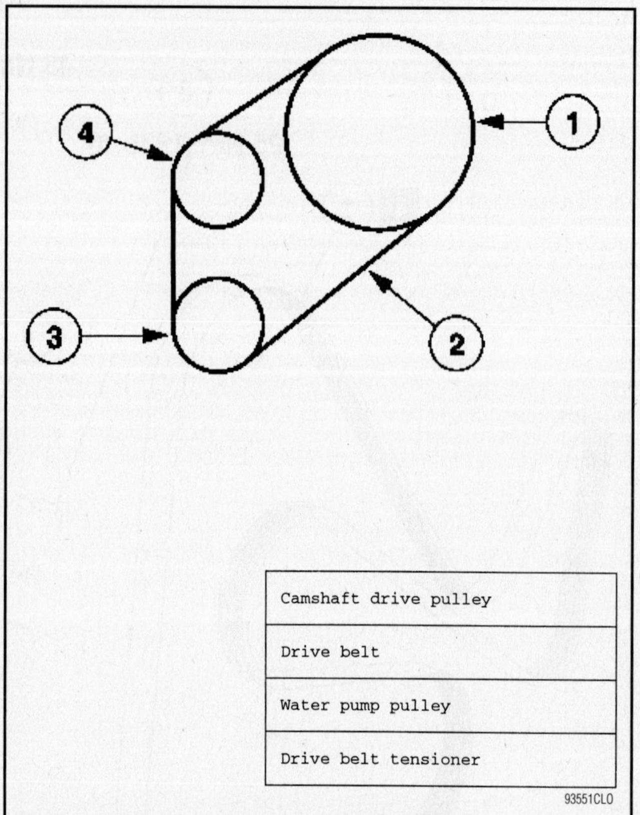

Accessory drive belt routing—3.0L engines with A/C

CAPACITIES

Year	Model	Engine Displacement Liters	Engine VIN	Engine Oil with Filter (qts.)	Transmission (pts.) Manual	Transmission (pts.) Auto.	Transfer Case (pts.)	Drive Axle Front (pts.)	Drive Axle Rear (pts.)	Fuel Tank (gal.)	Cooling System (qts.)
2001	Escape	2.0	B	4.5	5.7	—	①	2.6	3.0	15.0	7.0
	Escape	3.0	1	5.8	5.7	20.0	①	2.6	3.0	16.0	10.5
2002	Escape	2.0	B	4.5	5.7	—	①	2.6	3.0	15.0	7.0
	Escape	3.0	1	5.5	5.7	20.0	①	2.6	3.0	16.0	10.5
2003	Escape	2.0	B	4.5	5.7	—	—	—	3.0	15.0	7.0
	Escape	3.0	1	5.5	5.7	20.0	①	2.6	3.0	16.0	10.5
2004	Escape	2.0	B	4.5	5.7	—	—	—	3.0	15.0	7.0
	Escape	3.0	1	5.5	5.7	20.0	①	2.6	3.0	16.0	10.5
2005	Escape	2.3	B	4.5	NA	—	—	—	3.0	16.5	7.6
	Escape	3.0	1	5.5	NA	20.0	0.75	2.95	3.0	16.5	10.5

NA: Information not available

NOTE: All capacities are approximate. Add fluid gradually and check to be sure a proper fluid level is obtained.

① The transfer case is lubricated for life and is not to be checked unless a leak is suspected or a repair is necessary.

67197-ESCA-C04

VALVE SPECIFICATIONS

Year	Engine Displacement Liters	Engine VIN	Seat Angle (deg.)	Face Angle (deg.)	Spring Test Pressure (lbs. @ in.)	Spring Installed Height (in.)	Stem-to-Guide Clearance (in.) Intake	Stem-to-Guide Clearance (in.) Exhaust	Stem Diameter (in.) Intake	Stem Diameter (in.) Exhaust
2001	2.0	B	45	45	①	1.346	0.0007-0.0025	0.0007-0.0025	0.2374	0.2374
	3.0	1	44.75	45.5	153@ 1.18	1.57	0.0008-0.0027	0.0018-0.0037	0.2352-0.2360	0.2343-0.2350
2002	2.0	B	45	45	①	1.346	0.0007-0.0025	0.0007-0.0025	0.2374	0.2374
	3.0	1	44.75	45.5	153@ 1.18	1.57	0.0008-0.0027	0.0018-0.0037	0.2352-0.2360	0.2343-0.2350
2003	2.0	B	45	45	①	1.346	0.0007-0.0025	0.0007-0.0025	0.2374	0.2374
	3.0	1	44.75	45.5	153@ 1.18	1.57	0.0008-0.0027	0.0018-0.0037	0.2352-0.2360	0.2343-0.2350
2004	2.0	B	45	45	①	1.346	0.0007-0.0025	0.0007-0.0025	0.2374	0.2374
	3.0	1	44.75	45.5	153@ 1.18	1.57	0.0008-0.0027	0.0018-0.0037	0.2352-0.2360	0.2343-0.2350
2005	2.3	Z	45	45	38.6@1.49	1.496	0.0009	0.0011	0.2153-0.2159	0.2151-0.2157
	3.0	1	44.75	45.5	153@ 1.18	1.57	0.0008-0.0027	0.0018-0.0037	0.2352-0.2360	0.2343-0.2350

① Intake: 82.1@ 0.988
Exhaust: 95@ 1.0275

67197-ESCA-C05

CRANKSHAFT AND CONNECTING ROD SPECIFICATIONS

All measurements are given in inches.

	Engine		Crankshaft				Connecting Rod		
Year	Displacement Liters)	Engine VIN	Main Brg. Journal Dia.	Main Brg. Oil Clearance	Shaft End-play	Thrust on No.	Journal Diameter	Oil Clearance	Side Clearance
2001	2.0	B	2.2827-2.2835	①	0.0035-0.0102	3	1.8460-1.8468	0.0006-0.0028	0.0040-0.0110
	3.0	1	2.4791-2.4800	0.0010-0.0018	0.0043-0.0091	3	1.9673-1.9681	0.0010-0.0025	0.0039-0.0118
2002	2.0	B	2.2827-2.2835	①	0.0035-0.0102	3	1.8460-1.8468	0.0006-0.0028	0.0040-0.0110
	3.0	1	2.4791-2.4800	0.0010-0.0018	0.0043-0.0091	3	1.9673-1.9681	0.0011-0.0026	0.0039-0.0118
2003	2.0	B	2.2827-2.2835	①	0.0035-0.0102	3	1.8460-1.8468	0.0006-0.0028	0.0040-0.0110
	3.0	1	2.4791-2.4800	0.0010-0.0018	0.0043-0.0091	3	1.9673-1.9681	0.0011-0.0026	0.0039-0.0118
2004	2.0	B	2.2827-2.2835	①	0.0035-0.0102	3	1.8460-1.8468	0.0006-0.0028	0.0040-0.0110
	3.0	1	2.4791-2.4800	0.0010-0.0018	0.0043-0.0091	3	1.9673-1.9681	0.0011-0.0026	0.0039-0.0118
2005	2.3	Z	2.0460-2.0470	0.0007-0.0013	0.0080-0.0160	NA	1.9673-1.9681	0.0011-0.0026	0.0760-0.1200
	3.0	1	2.4791-2.4800	0.0010-0.0018	0.0043-0.0091	3	1.9673-1.9681	0.0011-0.0026	0.0039-0.0118

NA: Information not available

① Journals 1, 2 and 4: 0.0010 - 0.0017 in.
Journal 3: 0.0012 - 0.0019 in.

67197-ESCA-C06

PISTON AND RING SPECIFICATIONS

All measurements are given in inches.

Year	Engine Displacement Liters	Engine VIN	Piston Clearance	Ring Gap Top Compression	Ring Gap Bottom Compression	Ring Gap Oil Control	Ring Side Clearance Top Compression	Ring Side Clearance Bottom Compression	Ring Side Clearance Oil Control
2001	2.0	B	0.0004-0.0012	0.0100-0.0300	0.0100-0.0300	0.0160-0.0660	0.0015-0.0032	0.0015-0.0035	snug
	3.0	1	0.0005-0.0009	0.0039-0.0098	0.0106-0.0165	0.0059-0.0256	0.0016-0.0030	0.0016-0.0033	snug
2002	2.0	B	0.0004-0.0012	0.0100-0.0300	0.0100-0.0300	0.0160-0.0660	0.0015-0.0032	0.0015-0.0035	snug
	3.0	1	0.0005-0.0009	0.0039-0.0098	0.0106-0.0165	0.0059-0.0256	0.0016-0.0030	0.0016-0.0033	snug
2003	2.0	B	0.0004-0.0012	0.0100-0.0300	0.0100-0.0300	0.0160-0.0660	0.0015-0.0032	0.0015-0.0035	snug
	3.0	1	0.0005-0.0009	0.0039-0.0098	0.0106-0.0165	0.0059-0.0256	0.0016-0.0030	0.0016-0.0033	snug
2004	2.0	B	0.0004-0.0012	0.0100-0.0300	0.0100-0.0300	0.0160-0.0660	0.0015-0.0032	0.0015-0.0035	snug
	3.0	1	0.0005-0.0009	0.0039-0.0098	0.0106-0.0165	0.0059-0.0256	0.0016-0.0030	0.0016-0.0033	snug
2005	2.3	Z	0.0009-0.0017	0.0060-0.0120	0.0120-0.0180	0.0070-0.0270	NA	NA	NA
	3.0	1	0.0005-0.0009	0.0039-0.0098	0.0106-0.0165	0.0059-0.0256	0.0016-0.0030	0.0016-0.0033	snug

NA: Information not available

67197-ESCA-C07

TORQUE SPECIFICATIONS

All readings in ft. lbs.

Year	Engine Displacement Liters	Engine VIN	Cylinder Head Bolts	Main Bearing Bolts	Rod Bearing Bolts	Crankshaft Damper Bolts	Flywheel Bolts	Manifold Intake	Manifold Exhaust	Spark Plugs	Oil Pan Drain Plug
2001	2.0	B	①	②	③	80-87	83	13	12	11	18
	3.0	1	④	⑤	⑥	⑦	59	⑧	15	11	NA
2002	2.0	B	①	②	③	80-87	83	13	12	11	18
	3.0	1	④	⑤	⑥	⑦	59	⑧	15	11	NA
2003	2.0	B	①	②	③	80-87	83	13	12	11	18
	3.0	1	④	⑤	⑥	⑦	59	⑧	15	11	NA
2004	2.0	B	①	②	③	85	83	13	12	11	18
	3.0	1	④	⑤	⑥	⑦	59	⑧	15	11	NA
2005	2.3	Z	⑨	NA	NA	⑩	⑪	13	NA	11	21
	3.0	1	④	⑤	⑥	⑦	59	⑧	15	11	NA

NA: Information not available

① Step 1: 15 ft. lbs. (20 Nm).
Step 2: 30 ft. lbs. (40 Nm).
Step 3: Plus an additional 90 degrees.

② Step 1: 18 ft. lbs.
Step 2: +60 degrees

③ Step 1: 26 ft. lbs.
Step 2: +90 degrees

④ Step 1: 30 ft. lbs. (40 Nm).
Step 2: Tighten the bolts 90 degrees.
Step 3: Loosen the bolts one full turn.
Step 4: 30 ft. lbs. (40 Nm).
Step 5: Tighten the bolts 90 degrees.
Step 6: Tighten the bolts 90 degrees.

⑤ Step 1: Fasteners 1-8: 18 ft. lbs.
Step 2: Fasteners 9-19: 30 ft. lbs.
Step 3: Fasteners 1-16: +90 degrees
Step 4: fasteners 17-22: 18 ft. lbs.

⑥ Step 1: 17 ft. lbs.
Step 2: 32 ft. lbs.

⑦ Step 1: 89 ft. lbs.
Step 2: Loosen 1 full turn
Step 3: 37 ft. lbs.
Step 4: 66 ft. lbs.

⑧ 89 inch lbs.

⑨ Step 1: 44 inch lbs.
Step 2: 11 ft. lbs.
Step 3: 33 ft. lbs.
Step 4: +90 degrees
Step 5: Plus 90 degrees

⑩ Step2: 74 ft. lbs.
Step 2: plus 90 degrees

⑪ Step 1: 37 ft. lbs.
Step 2: 50 ft. lbs
Step 3: 83 ft. lbs.

67197-ESCA-C08

WHEEL ALIGNMENT

Year	Model		Caster Range (+/-Deg.)	Caster Preferred Setting (Deg.)	Camber Range (+/-Deg.)	Camber Preferred Setting (Deg.)	Toe-in (in.)
2001	ALL	F	NA	+1.93	NA	-0.84	0.12+/-0.12
		R	NA	NA	NA	-0.04	0.09+/-0.11
2002	ALL	F	NA	+1.93	NA	-0.84	0.12+/-0.12
		R	NA	NA	NA	-0.04	0.09+/-0.11
2003	ALL	F	NA	+1.93	NA	-0.84	0.12+/-0.12
		R	NA	NA	NA	-0.04	0.09+/-0.11
2004①	4-cyl.	F	1.00	+1.72	1.00	-0.48	-0.08+/-0.32
		R	NA	NA	1.00	+0.13	0.04+/-0.17
	6-cyl.	F	1.00	+1.72	1.00	-0.84	0.23+/-0.32
		R	NA	NA	1.00	+0.13	0.10+/-0.17
2005	4-cyl.	F	1.00	+1.70	1.00	-1.00	-0.23+/-0.23
		R	NA	NA	0.75	+0.10	0.10+/-0.20
	6-cyl.	F	1.00	+1.70	1.00	-1.00	0.23+/-0.32
		R	NA	NA	0.75	+0.10	0.10+/-0.20

NA: Information not available

① Assumes 8 gallons of gas

67197-ESCA-C09

TIRE, WHEEL AND BALL JOINT SPECIFICATIONS

Year	Model	OEM Tires Standard	OEM Tires Optional	Tire Pressures (psi) Front	Tire Pressures (psi) Rear	Wheel Size	Ball Joint Inspection	Lug Nuts (ft. lbs.)
2001	Escape	P225/70SR15	P235/70R16	①	①	NA	0.030 in.	98
2002	Escape	P215/70R16	P225/70SR15 P235/70R16	①	①	NA	0.030 in.	98
2003	Escape	P215/70R16	P225/70SR15 P235/70R16	①	①	NA	0.030 in.	98
2004	Escape XLS	P225/70SR15	NA	①	①	6.5	0.030 in.	98
	Escape XLT	P235/70R16	NA	①	①	7.5	0.030 in.	98
2005	Escape XLS							
	Value	P225/70R15	NA	①	①	6.5	0.030 in.	98
	All others	P225/75R15	NA	①	①	6.5	0.030 in.	98
	XLT	P235/70R16	NA	①	①	7.5	0.030 in.	98
	Limited	P235/70R16	NA	①	①	7.5	0.030 in.	98

OEM: Original Equipment Manufacturer
PSI: Pounds Per Square Inch
STD: Standard
OPT: Optional
NA: Not Available

67197-ESCA-C10

BRAKE SPECIFICATIONS

All measurements in inches unless noted

Year	Model		Brake Disc Original Thickness	Brake Disc Minimum Thickness	Brake Disc Maximum Run-out	Brake Drum Original Inside Diameter	Brake Drum Maximum Machine Diameter	Minimum Lining Thickness	Brake Caliper Bracket Bolts (ft. lbs.)	Brake Caliper Mounting Bolts (ft. lbs.)
2001	Escape		0.940	0.860	0.002	9.00	9.06	0.039	111	26
2002	Escape		0.940	0.860	0.004	9.00	9.06	0.039	111	26
2003	Escape		0.940	0.860	0.002	9.00	9.06	0.039	111	26
2004	Escape		0.940	0.860	0.002	9.00	9.06	0.039	111	26
2005	Escape	F	NA	①	0.004	NA	9.05	0.118	111	②
		R	NA	0.430	0.004	—	—	0.118	—	26

NA: Information not available
① Base brakes: 0.86 in.
With 4-wheel discs: 0.95 in.
② With disc/drum: 26 ft. lbs.
With 4-wheel disc: 33 ft. lbs.

67197-ESCA-C11

SCHEDULED MAINTENANCE INTERVALS

2001-03 Ford Escape

TO BE SERVICED	TYPE OF SERVICE	VEHICLE MILEAGE INTERVAL (x1000)												
		5	10	15	20	25	30	35	40	45	50	55	60	65
Air cleaner filter	R						✓						✓	
Accessory drive belt	S/I	Every 100,000 miles												
Accessory drive belt	R	At 120,000 miles if not previously done so												
Auto. Trans. Fluid level	I			✓			✓			✓			✓	
Auto. Trans. Fluid	②						✓						✓	
Auto. Trans. Fluid	③	Every 150,000 miles												
Brake system ①	S/I			✓			✓			✓			✓	
Cabin air filter	R			✓			✓			✓			✓	
Camshaft belt (2.0L)	R	Every 120,000 miles												
Cooling system hoses and clamps	S/I			✓			✓			✓			✓	
Engine coolant (green coolant)	R	Every 75,000 miles and every 30,000 miles thereafter												
Engine coolant (yellow coolant)	R	At 5 years or 100,000 miles												
Engine oil & filter	R	✓	✓	✓	✓	✓	✓	✓	✓	✓	✓	✓	✓	✓
PCV valve	R	Every 100,000 miles												
Exhaust system & heat shields	S/I						✓						✓	
Fuel filter	R						✓						✓	
Rear axle lubricant	R	Every 150,000 miles												
Rotate tires	S/I	✓	✓	✓	✓	✓	✓	✓	✓	✓	✓	✓	✓	✓
Steering linkage	S/I			✓			✓			✓			✓	
Spark plugs	R	Change at 100,000 miles												
Suspension components	S/I			✓			✓			✓			✓	
Wheels for play and noise	I			✓			✓			✓			✓	

R: Replace S/I: Inspect and service, if necessary L: Lubricate A: Adjust C: Clean

① Inspect the reservoir fluid level, rotor and or drum, brake lines, hoses, calipers and or wheel cylinders

② Change automatic transmission/transaxle fluid on all vehicles equipped with AX4S, 4F50N, 4R100, 4F27E

Special Operating Condition Requirements

When towing a trailer or using a camper or car-top carrier:

Change engine oil and install a new oil filter every 4,800 km (3,000 miles) or 3 months.

Change transfer case fluid every 96,000 km (60,000 miles).

Change manual transmission fluid as required.

Inspect and lubricate U-joints as required.

During extensive idling and/or low speed driving for long distances, as in heavy commercial use such as delivery, taxi, patrol car or livery:

Change engine oil and install a new oil filter, lube front lower control arm and steering linkage ball joints with zerk fittings (if equipped) every 4,800 km (3,000 miles) or 3 months.

Inspect brake system and check battery electrolyte level (Patrol cars) every 8,000 km (5,000 miles).

Install a new fuel filter every 24,000 km (15,000 miles).

Change automatic transmission fluid, lubricate 4x2 wheel bearings, install new grease seals and adjust bearings every 48,000 km (30,000 miles).

Install new spark plugs and change transfer case fluid every 96,000 km (60,000 miles).

Install a new cabin air filter as required.

When operating in dusty conditions such as unpaved or dusty roads:

Change engine oil and install a new oil filter every 4,800 km (3,000 miles) or 3 months.

Install a new fuel filter every 24,000 km (15,000 miles).

Change automatic transmission fluid every 48,000 km (30,000 miles).

Change transfer case fluid every 96,000 km (60,000 miles).

Install a new engine air filter as required

Install a new cabin air filter as required.

67197-ESCA-C12

SCHEDULED MAINTENANCE INTERVALS
2001-03 Ford Escape
Footnotes Continued

When operating in off-road conditions:

Change automatic transmission fluid every 48,000 km (30,000 miles).

Change transfer case fluid every 96,000 km (60,000 miles).

Install a new cabin air filter as required.

Inspect and lubricate U-joints.

Inspect and lubricate steering linkage ball joints with zerk fittings.

67197-ESCA-C13

SCHEDULED MAINTENANCE INTERVALS

2004 Ford Escape

TO BE SERVICED	TYPE OF SERVICE	VEHICLE MILEAGE INTERVAL (x1000)												
		5	10	15	20	25	30	35	40	45	50	55	60	65
Air cleaner filter	R						✓						✓	
Accessory drive belt	S/I	Every 100,000 miles												
Accessory drive belt	R	At 150,000 miles if not previously done so												
Auto. Trans. Fluid level	I			✓			✓			✓			✓	
Auto. Trans. Fluid	②						✓						✓	
Auto. Trans. Fluid	③	Every 150,000 miles												
Ball joints (2wd)	L			✓			✓			✓			✓	
Brake system ①	S/I			✓			✓			✓			✓	
Cabin air filter	R			✓			✓			✓			✓	
Camshaft belt (2.0L)	R	Every 120,000 miles												
Cooling system hoses and clamps	S/I			✓			✓			✓			✓	
Engine coolant (exc. Premium gold)	R	Every 105,000 miles, then every 3 years or 50,000 miles												
Engine coolant (Premium gold)	R	At 5 years or 100,000 miles												
Engine oil & filter	R	✓	✓	✓	✓	✓	✓	✓	✓	✓	✓	✓	✓	✓
PCV valve (external)	R	Every 100,000 miles												
Exhaust system & heat shields	S/I						✓						✓	
Front wheel bearings and seals	R	Every 150,000 miles, if not previously done so												
Fuel filter	R						✓						✓	
Rear axle lubricant	R	Every 150,000 miles												
Rotate tires	S/I	✓	✓	✓	✓	✓	✓	✓	✓	✓	✓	✓	✓	✓
Steering linkage	S/I			✓			✓			✓			✓	
Spark plugs	R	Change at 100,000 miles												
Suspension components	S/I			✓			✓			✓			✓	
Wheels for play and noise	I			✓			✓			✓			✓	

R: Replace S/I: Inspect and service, if necessary L: Lubricate A: Adjust C: Clean

① Inspect the reservoir fluid level, rotor and or drum, brake lines, hoses, calipers and or wheel cylinders

② Change automatic transmission/transaxle fluid and filter on all vehicles equipped with AX4S, 4F50N, 4R100, 4F27E.

③ All transaxles

Special Operating Condition Requirements

When towing a trailer or using a camper or car-top carrier:

Change engine oil and install a new oil filter every 4,800 km (3,000 miles) or 3 months.

Change transfer case fluid every 96,000 km (60,000 miles).

Change manual transmission fluid as required.

Inspect and lubricate U-joints as required.

During extensive idling and/or low speed driving for long distances, as in heavy commercial use such as delivery, taxi, patrol car or livery:

Change engine oil and install a new oil filter, lube front lower control arm and steering linkage ball joints with zerk fittings (if equipped) every 4,800 km (3,000 miles) or 3 months.

Inspect brake system and check battery electrolyte level (Patrol cars) every 8,000 km (5,000 miles).

Install a new fuel filter every 24,000 km (15,000 miles).

Change automatic transmission fluid, lubricate 4x2 wheel bearings, install new grease seals and adjust bearings every 48,000 km (30,000 miles).

Install new spark plugs and change transfer case fluid every 96,000 km (60,000 miles).

Install a new cabin air filter as required.

When operating in dusty conditions such as unpaved or dusty roads:

Change engine oil and install a new oil filter every 4,800 km (3,000 miles) or 3 months.

Install a new fuel filter every 24,000 km (15,000 miles).

Change automatic transmission fluid every 48,000 km (30,000 miles).

Change transfer case fluid every 96,000 km (60,000 miles).

67197-ESCA-C14

SCHEDULED MAINTENANCE INTERVALS

2004 Ford Escape

Footnotes Continued

Install a new engine air filter as required.

Install a new cabin air filter as required.

When operating in off-road conditions:

Change automatic transmission fluid every 48,000 km (30,000 miles).

Change transfer case fluid every 96,000 km (60,000 miles).

Install a new cabin air filter as required.

Inspect and lubricate U-joints.

Inspect and lubricate steering linkage ball joints with zerk fittings.

67197-ESCA-C15

SCHEDULED MAINTENANCE INTERVALS

2005 Ford Escape

TO BE SERVICED	TYPE OF SERVICE	VEHICLE MILEAGE INTERVAL (x1000)												
		5	10	15	20	25	30	35	40	45	50	55	60	65
Air cleaner filter	R						✓						✓	
Accessory drive belt	I ⑤	Every 100,000 miles												
Auto. Trans. fluid level	I			✓			✓			✓			✓	
Auto. Trans. Fluid	③ ④						✓						✓	
Ball joints (2wd)	L			✓			✓			✓			✓	
Brake system ①	S/I			✓			✓			✓			✓	
Cabin air filter	R			✓			✓			✓			✓	
Cooling system hoses and clamps	S/I			✓			✓			✓			✓	
Driveshafts & halfshafts	S/I			✓			✓			✓			✓	
Engine coolant (Premium Gold)	R	Five years or 100,000 miles, then every 3 years or 50,000 miles												
Engine coolant (exc. Premium Gold)	R	Every 105,000 miles												
Engine oil & filter	R	✓	✓	✓	✓	✓	✓	✓	✓	✓	✓	✓	✓	✓
Front wheel bearings and seals (2wd)	R	Every 150,000 miles, if not previously done												
Fuel filter	R						✓						✓	
Man. Trans. Fluid	R	Every 120,000 miles												
PCV valve	S/I	Every 100,000 miles												
Exhaust system & heat shields	S/I						✓						✓	
Rear axle lubricant (4wd)	R	Every 150,000 miles												
Rotate tires	S/I	✓	✓	✓	✓	✓	✓	✓	✓	✓	✓	✓	✓	✓
Steering linkage	S/I			✓			✓			✓			✓	
Spark plugs	R	Change at 100,000 miles												
Suspension components	S/I			✓			✓			✓			✓	
Wheels ②	I			✓			✓			✓			✓	

R: Replace S/I: Inspect and service, if necessary L: Lubricate A: Adjust C: Clean

① Inspect the reservoir fluid level, rotor and or drum, brake lines, hoses, calipers and or wheel cylinders

② Inspect for end play and noise

③ Change automatic transmission/transaxle fluid and filter on all vehicles equipped with 4F50N, 4R100 and 4F27E.

④ Change every 150,000 miles for all transaxles

⑤ Replace at 150,000 miles, if not previously done

Special Operating Condition Requirements

When towing a trailer or using a camper or car-top carrier:

Change engine oil and install a new oil filter every 4,800 km (3,000 miles) or 3 months.

Change transfer case fluid every 96,000 km (60,000 miles).

Change manual transmission fluid as required.

Inspect and lubricate U-joints as required.

During extensive idling and/or low speed driving for long distances, as in heavy commercial use such as delivery, taxi, patrol car or livery:

Change engine oil and install a new oil filter, lube front lower control arm and steering linkage ball joints with zerk fittings (if equipped) every 4,800 km (3,000 miles) or 3 months.

Inspect brake system and check battery electrolyte level (Patrol cars) every 8,000 km (5,000 miles).

Install a new fuel filter every 24,000 km (15,000 miles).

Change automatic transmission fluid, lubricate 4x2 wheel bearings, install new grease seals and adjust bearings every 48,000 km (30,000 miles).

Install new spark plugs and change transfer case fluid every 96,000 km (60,000 miles).

Install a new cabin air filter as required.

67197-ESCA-C16

SCHEDULED MAINTENANCE INTERVALS

2005 Ford Escape

Footnotes Continued

When operating in dusty conditions such as unpaved or dusty roads:

Change engine oil and install a new oil filter every 4,800 km (3,000 miles) or 3 months.

Install a new fuel filter every 24,000 km (15,000 miles).

Change automatic transmission fluid every 48,000 km (30,000 miles).

Change transfer case fluid every 96,000 km (60,000 miles).

Install a new engine air filter as required.

Install a new cabin air filter as required.

When operating in off-road conditions:

Change automatic transmission fluid every 48,000 km (30,000 miles).

Change transfer case fluid every 96,000 km (60,000 miles).

Install a new cabin air filter as required.

Inspect and lubricate U-joints.

Inspect and lubricate steering linkage ball joints with zerk fittings.

67197-ESCA-C17

PRECAUTIONS

Before servicing any vehicle, please be sure to read all of the following precautions, which deal with personal safety, prevention of component damage, and important points to take into consideration when servicing a motor vehicle:

- Never open, service or drain the radiator or cooling system when the engine is hot; serious burns can occur from the steam and hot coolant.
- Observe all applicable safety precautions when working around fuel. Whenever servicing the fuel system, always work in a well-ventilated area. Do not allow fuel spray or vapors to come in contact with a spark, open flame, or excessive heat (a hot drop light, for example). Keep a dry chemical fire extinguisher near the work area. Always keep fuel in a container specifically designed for fuel storage; also, always properly seal fuel containers to avoid the possibility of fire or explosion. Refer to the additional fuel system precautions later in this section.
- Fuel injection systems often remain pressurized, even after the engine has been turned **OFF**. The fuel system pressure must be relieved before disconnecting any fuel lines. Failure to do so may result in fire and/or personal injury.
- Brake fluid often contains polyglycol ethers and polyglycols. Avoid contact with the eyes and wash your hands thoroughly after handling brake fluid. If you do get brake fluid in your eyes, flush your eyes with clean, running water for 15 minutes. If eye irritation persists, or if you have taken brake fluid internally, IMMEDIATELY seek medical assistance.
- The EPA warns that prolonged contact with used engine oil may cause a number of skin disorders, including cancer! You should make every effort to minimize your exposure to used engine oil. Protective gloves should be worn when changing oil. Wash your hands and any other exposed skin areas as soon as possible after exposure to used engine oil. Soap and water, or waterless hand cleaner should be used.
- All new vehicles are now equipped with an air bag system, often referred to as a Supplemental Restraint System (SRS) or Supplemental Inflatable Restraint (SIR) system. The system must be disabled before performing service on or around system components, steering column, instrument panel components, wiring and sensors. Failure to follow safety and disabling procedures could result in accidental air bag deployment, possible personal injury and unnecessary system repairs.
- Always wear safety goggles when working with, or around, the air bag system. When carrying a non-deployed air bag, be sure the bag and trim cover are pointed away from your body. When placing a non-deployed air bag on a work surface, always face the bag and trim cover upward, away from the surface. This will reduce the motion of the module if it is accidentally deployed. Refer to the additional air bag system precautions later in this section.
- Clean, high quality brake fluid from a sealed container is essential to the safe and proper operation of the brake system. You should always buy the correct type of brake fluid for your vehicle. If the brake fluid becomes contaminated, completely flush the system with new fluid. Never reuse any brake fluid. Any brake fluid that is removed from the system should be discarded. Also, do not allow any brake fluid to come in contact with a painted surface; it will damage the paint.
- Never operate the engine without the proper amount and type of engine oil; doing so WILL result in severe engine damage.
- Timing belt maintenance is extremely important! Many models utilize an interference-type, non-freewheeling engine. If the timing belt breaks, the valves in the cylinder head may strike the pistons, causing potentially serious (also time-consuming and expensive) engine damage. Refer to the maintenance interval charts in the front of this manual for the recommended replacement interval for the timing belt, and to the timing belt section for belt replacement and inspection.
- Disconnecting the negative battery cable on some vehicles may interfere with the functions of the on-board computer system(s) and may require the computer to undergo a relearning process once the negative battery cable is reconnected.
- When servicing drum brakes, only disassemble and assemble one side at a time, leaving the remaining side intact for reference.
- Only an MVAC-trained, EPA-certified automotive technician should service the air conditioning system or its components.

ENGINE REPAIR

Distributor

The Escape uses a Direct Ignition System (DIS). No distributor is used.

Alternator

REMOVAL & INSTALLATION

2.0L Engine

1. Remove or disconnect the following:
   - Negative battery cable
   - Drive belt
   - Alternator electrical connectors and loosen the upper alternator bolt while moving the alternator to the rear of the engine
   - Alternator. Torque the lower bolt to 35 ft. lbs. (48Nm). Torque the upper bolt to 18 ft. lbs. (25 Nm).

To install:

2. Install or connect the following:
   - Alternator with the upper bolt in the alternator before installation. Torque the bolts to 18 ft. lbs. (25 Nm).
   - Alternator electrical connectors
   - Drive belt
   - Negative battery cable

2.3L Engine

1. Disconnect the battery.
2. Remove the front end accessory drive belt tensioner. Rotate the front end accessory drive belt tensioner counterclockwise to loosen tension on the front end accessory drive belt.
3. Remove the front end accessory drive belt.
4. Remove the alternator B+ terminal.
5. Remove the alternator electrical connector.
6. Remove the alternator lower air duct bolt.
7. Remove the alternator lower air duct. Press the locking tab to release the lower air duct from the upper air duct.
8. Remove the pin-type retainer.
9. Remove the alternator shield .
10. Remove the alternator stud nut.
11. Remove the alternator stud.
12. Remove the alternator bolts.
13. Remove the alternator.

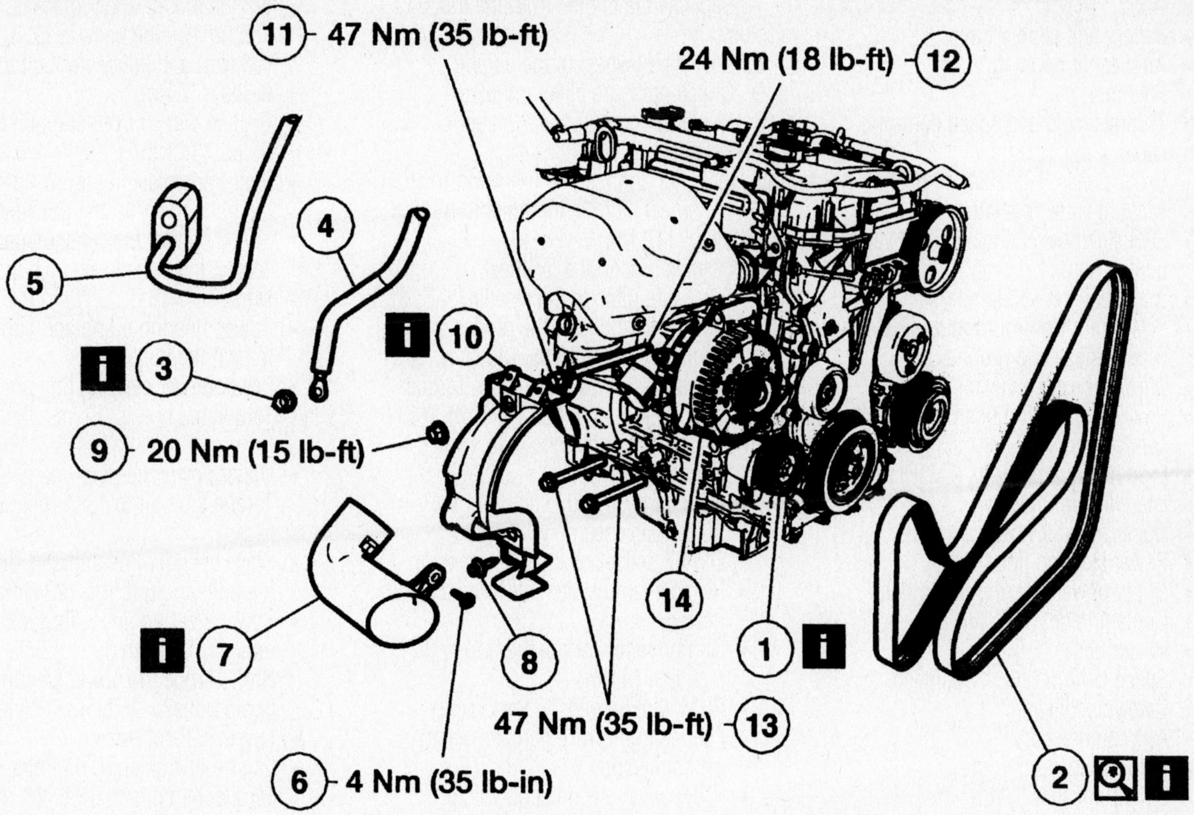

1 Front end accessory drive belt tensioner
2 Front end accessory drive belt
3 Generator B+ terminal nut
4 Generator B+ terminal
5 Generator electrical connector
6 Generator lower air duct bolt
7 Generator lower air duct
8 Pin-type retainer
9 Generator shield nut
10 Generator shield
11 Generator stud nut
12 Generator stud
13 Generator bolts
14 Generator

67197-ESCA-G01

Alternator mounting—2.3L Engine

14. To install, reverse the removal procedure. Observe the following torques:
    - Alternator mounting bolts: 35 ft. lbs. (47 Nm)
    - Alternator stud: 18 ft. lbs. (24 Nm)
    - Stud nut: 35 ft. lbs. (47 Nm)
    - Shield nut: 15 ft. lbs. (20 Nm)
    - Lower air duct bolt: 35 inch lbs. (4 Nm)

3.0L Engine

1. Remove or disconnect the following:
    - Negative battery cable
    - Right side intermediate axle shaft
    - Right side splash shield and retainers
    - Drive belt
    - Alternator electrical connectors
    - Alternator. Torque the mounting and adjusting bolts to 35 ft. lbs. (48Nm).

To install:

2. Install or connect the following:
    - Alternator. Torque the bolts to 35 ft. lbs. (48 Nm).
    - Alternator electrical connectors
    - Drive belt
    - Negative battery cable

Ignition Timing

ADJUSTMENT

Ignition timing is controlled by the Powertrain Control Module (PCM). No adjustment is necessary or possible.

Engine Assembly

REMOVAL & INSTALLATION

2.0L Engine

MANUAL TRANSMISSION

1. Before servicing the vehicle, refer to the precautions in the beginning of this section.
2. Properly recover the air conditioning system refrigerant.
3. Properly relieve the fuel system pressure.
4. Drain the cooling system.
5. Drain the engine oil.
6. Remove or disconnect the following:

- Hood
- Battery and battery tray
- Air cleaner housing
- Fuel lines
- Throttle cable and speed control cable, if equipped
- Exhaust Gas Recirculation (EGR) vacuum valve regulator
- EGR electrical connectors and vacuum hoses
- Brake booster vacuum hose
- Powertrain Control Module (PCM) wire harness and ground
- Wire harness connector
- Power distribution board electrical connectors
- Evaporative emissions (EVAP) canister vacuum lines
- Upper radiator hose
- Power steering line bracket
- Upper power steering pump bolts
- Coolant hose
- Heater hoses
- Speed control unit, if equipped
- Catalytic converter
- A/C compressor
- Both halfshafts
- Shifter linkages
- Block heater electrical connector, if equipped
- Front transmission through bolt
- Engine-to-transmission bolts
- Lower radiator hose
- Power steering pump
- Clutch slave cylinder line from the bracket and move it aside
- Rear transmission mount
- Left side transmission mount
- Lower ground cable
- Engine mount upper bracket
- Engine and transmission as an assembly by using a proper lifting device
- Alternator electrical connectors
- Knock Sensor (KS) electrical connector
- Oil pressure sender electrical connector
- Starter electrical connector
- Vehicle Speed Sensor (VSS) electrical connector
- Park Neutral Position (PNP) electrical connector
- Fuel charging wire harness electrical connector
- PCM wire harness from the bracket
- PCM ground wire
- Back up lamp switch electrical connector
- Wire harness
- Differential Pressure Feedback (DPFEE) EGR sensor

7. Separate the engine from the transmission.
8. Lock the flywheel to the engine.
9. Clutch pressure plate and disc.
10. Flywheel and rear cover plates.

To install:

11. Install or connect the following:
    - Flywheel. Torque the bolts to 83 ft. lbs. (112 Nm).
    - Clutch disc to the flywheel
    - Pressure plate to the flywheel. Torque the bolts in the proper sequence to 18 ft. lbs. (25 Nm).
    - Transmission to the engine. Torque the bolts to 33 ft. lbs. (45 Nm).
    - Starter. Torque the bolts to 18 ft. lbs. (25 Nm).
    - Wire harness and attach it to the powertrain assembly
    - DPFEE sensor electrical connector
    - Reverse lamp switch electrical connector
    - Ground wire. Torque the bolt to 80 inch lbs. (9 Nm).
    - PCM wire harness to the bracket
    - Fuel charging wire harness electrical connector
    - PNP switch electrical connector
    - VSS electrical connector
    - KS, Oil pressure sender and starter electrical connector. Torque the fasteners to 9 ft. lbs. (12 Nm).
    - Alternator electrical connectors. Torque the fasteners to 71 inch lbs. (8 Nm).
    - Powertrain assembly in the vehicle
    - Left side transmission mount. Torque the side bolts to 41 ft. lbs. (55 Nm) and the center bolt to 66 ft. lbs. (90 Nm).
    - Engine mount upper bracket. Torque the side bolts to 72 ft. lbs. (98 Nm) and the center bolt to 57 ft. lbs. (77 Nm).
    - Ground wire. Torque the bolt to 25 ft. lbs. (34 Nm).
    - Rear transmission mount. Torque the bolts to 41 ft. lbs. (55 Nm).
    - Speed control unit, if equipped. Torque the bolts to 89 inch lbs. (10 Nm)..
    - Power steering pump and hand tighten the bolts
    - Lower power steering line bracket. Torque the bolt to 89 inch lbs. (10 Nm).
    - Upper power steering line bolt. Torque the bolt to 15 ft. lbs. (20 Nm).
    - Slave cylinder line and clip. Torque the bolts to 16 ft. lbs. (22 Nm).
    - Power steering lines. Torque the retaining bolts to 89 inch lbs. (10 Nm). Torque the power steering pump bolts to 18 ft. lbs. (25 Nm).
    - Lower radiator hose
    - Engine-to-transmission bolts. Torque the bolts to 33 ft. lbs. (45 Nm).
    - Front transmission through bolt. Torque the bolt to 66 ft. lbs. (90 Nm).
    - Block heater electrical connector
    - Shifter linkages. Torque the upper bolt to 33 ft. lbs. (45 Nm) and the lower bolt to 15 ft. lbs. (20 Nm).
    - Coolant hose
    - Catalytic converter
    - Heater hoses
    - Upper radiator hose

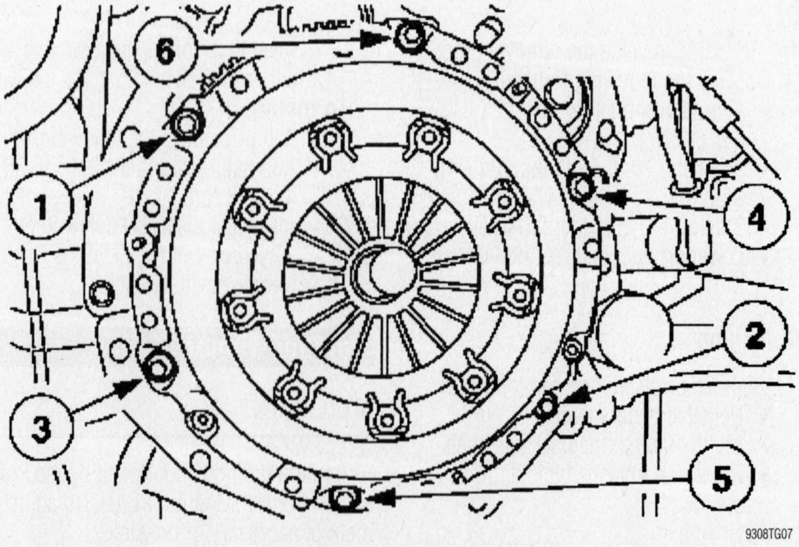

Tighten the pressure plate bolts in the proper sequence–2.0L engine

- EVAP canister vacuum lines
- Power distribution box electrical connector. Torque the fastener to 9 ft. lbs. (12 Nm).
- Wire harness electrical connector
- Ground wires
- PCM wire harness and ground
- Brake booster vacuum supply hose to the intake manifold
- EGR vacuum regulator valve hoses and electrical connector
- Throttle cable and speed control cable, if equipped
- Fuel lines
- Battery tray and battery
- Air cleaner
- Hood

12. Fill the engine with clean oil.
13. Fill and bleed the cooling system.
14. Recharge the A/C system.
15. Start the vehicle, check for leaks and repair if necessary.

2.3L Engine

MANUAL TRANSAXLE

All vehicles

1. With the vehicle in NEUTRAL, position it on a hoist.
2. Release the fuel system pressure.
3. Remove the engine air cleaner and air cleaner outlet pipe.
4. Remove the battery tray.
5. Drain the engine oil.
6. Drain the cooling system.
7. Remove the starter.
8. Remove the catalytic converter.
9. Remove the accessory drive belt.
10. Remove the bolts and the lateral support crossmember.
11. Remove the LH front drive halfshaft.
12. Remove the front drive intermediate halfshaft

4x4 vehicles

13. Remove the six bolts holding the driveshaft to the transfer case.
14. Position the driveshaft aside.

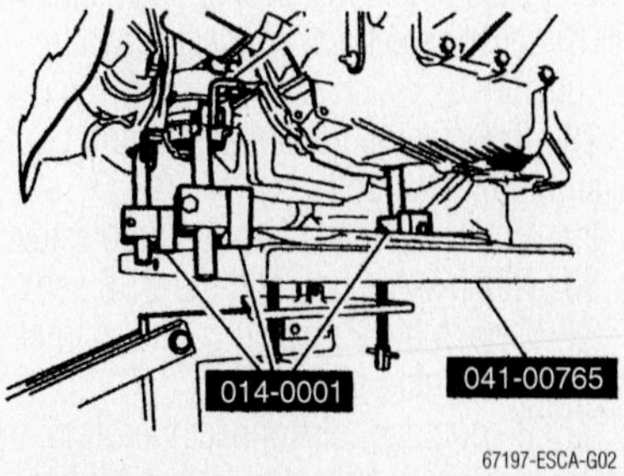

Engine secured to lift table—2.3L engine w/manual transaxle

All vehicles

15. If equipped, remove the bolt and ground eyelet.
16. Remove the power distribution box cover.
17. Remove the nuts and disconnect the cables.
18. Disconnect the electrical connector from the power distribution box.
19. Remove the bolt and disconnect the ground strap. Loosen the bolt and disconnect the 42-pin electrical connector.
20. Detach the wiring harness retainers from the battery tray bracket and position the wiring harness out of the way.
21. Disconnect the clutch hydraulic tube fitting. Detach the tube from the spring clip and position aside.
22. Remove the retaining clips and disconnect the transaxle control cable.
23. Remove the retaining clips and disconnect the transaxle control cable.
24. Disconnect the vehicle speed sensor electrical connector and pin-type retainer.
25. Disconnect the reversing lamp indicator switch and detach the wiring harness retainers.
26. If equipped, disconnect the block heater electrical connector. Detach all the block heater wiring harness retainers and position the wiring harness aside.
27. Disconnect the upper radiator and coolant vent hoses.
28. Remove the nuts and the coolant vent hose brackets. Position the coolant vent hose aside.
29. Detach the heater hose support strap from the stud.
30. Disconnect the heater hoses from the heater core.
31. Remove the retainers and the accelerator cable snow shield.
32. Disconnect the accelerator cable and speed control cable (if equipped).
33. Remove the nut from the accelerator control cable bracket.
34. Remove the nut from the accelerator control cable bracket and position the accelerator control cable and bracket assembly aside.
35. Remove the nut and position the power steering tube and bracket aside.
36. Disconnect the vacuum supply tube and position aside.
37. Disconnect the fuel vapor return tube and position aside.
38. Disconnect the vacuum reservoir tube and position aside.
39. Disconnect the fuel supply tube and retainer and position aside.
40. Detach the electrical connector retainers.
41. Disconnect the powertrain control module (PCM) electrical connectors. Remove the nut and position the harness aside.
42. Remove the bolt and detach the ground wire.
43. Remove the two power steering pump bolts.
44. Disconnect the lower radiator hose from the radiator.
45. Disconnect the A/C compressor electrical connector and remove the four bolts. Position the A/C compressor aside and support the compressor with a length of mechanics wire.
46. Disconnect the power steering pressure (PSP) sensor electrical connector.

➡The bolt under the power steering pressure tube will remain with the power steering pump.

47. Remove the bolts and position the power steering pump aside.
48. Remove the front roll restrictor bolt and the two bolts for the engine support crossmember.
49. Remove the rear nut and the engine support crossmember.

➡The transaxle-to-engine bolts differ in length. Mark the bolts for correct installation.

50. Remove the two transaxle-to-engine bolts.

➡The transaxle-to-engine bolts differ in length. Mark the bolts for correct installation.

51. Remove the two transaxle-to-engine bolts.
52. Using the special tools, secure the engine to the lift table.
53. Remove the engine mount bracket bolt.
54. Remove the nuts and the engine mount bracket.
55. Remove the bolt from the transaxle rear mount.
56. Remove bolt from the LH transaxle mount.
57. Lower the engine and transaxle from the vehicle.
58. Using the engine crane and spreader bar, remove the engine and transaxle from the lift table.

➡The transaxle-to-engine bolts differ in length. Mark the bolts for correct installation.

59. Remove the remaining six engine-to-transaxle bolts and separate the engine and transaxle.

To install:

All vehicles

60. Using the engine crane and spreader bar, position the engine and transaxle together. Install the six upper transaxle-to-engine bolts. Torque to 35 ft. lbs. (48 Nm).

61. Using the engine crane and spreader bar, position the engine and transaxle onto the lift table.

62. Using the special tools, secure the engine to the lift table.

63. Raise the engine and transaxle into the vehicle.

64. Install the bolt in the LH transaxle mount. Torque to 76 ft. lbs. (103 Nm).

65. Install the bolt in the rear transaxle mount. Torque to 85 ft. lbs. (115 Nm).

66. Install the engine mount bracket. Torque to 66 ft. lbs. (90 Nm).

67. Install the engine mount bracket bolt. Torque to 66 ft. lbs. (90 Nm).

68. Install the 4 lower transaxle-to-engine bolts. Torque to 35 ft. lbs. (48 Nm).

69. Install the engine support crossmember and nut. Torque to 66 ft. lbs. (90 Nm).

70. Install the two bolts for the engine support crossmember. Torque to 66 ft. lbs. (90 Nm).

71. Install the front roll restrictor bolt. Torque to 85 ft. lbs. (115 Nm).

➡**The bolt under the power steering pressure tube will remain with the power steering pump.**

72. Position the power steering pump and install the bolts. Torque to 18 ft. lbs. (25 Nm).

73. Connect the power steering pressure (PSP) sensor electrical connector.

74. Install the A/C compressor and connect the A/C compressor electrical connector. Torque to 18 ft. lbs. (25 Nm).

75. Connect the lower radiator hose to the radiator.

76. Install the two lower power steering pump bolts. Torque to 18 ft. lbs. (25 Nm).

77. Install the ground wire and bolt.

78. Connect the powertrain control module (PCM) electrical connectors. Position the harness and install the nut.

79. Attach the electrical connector retainers.

80. Connect the fuel supply tube.

81. Connect the vacuum reservoir tube.

82. Connect the fuel vapor return tube and retainer.

83. Connect the vacuum supply tube.

84. Install the power steering tube and bracket.

85. Position the accelerator control cable and bracket and install the nut.

86. Install the accelerator control cable and bracket and nut.

87. Install the accelerator cable and speed control cable (if equipped).

88. Install the accelerator cable snow shield and retainers.

89. Connect the heater hoses to the heater core.

90. Attach the heater hose support strap to the stud.

91. Position the coolant vent hose and install the coolant vent hose brackets and nuts.

92. Connect the upper radiator and coolant vent hoses.

93. If equipped, route the block heater wiring harness and attach all retainers. Connect the block heater electrical connector.

94. Connect the reversing lamp indicator switch and attach the wiring harness retainers.

95. Connect the vehicle speed sensor (VSS) electrical connector and pin-type retainer.

96. Connect the transaxle control cable and install the retaining clips.

97. Connect the transaxle control cable and install the retaining clips.

98. Connect the clutch hydraulic tube fitting. Attach the tube to the spring clip.

99. Attach the wiring harness retainers to the battery tray bracket.

100. Connect the 42-pin electrical connector and tighten the bolt. Install the ground strap and bolt.

101. Connect the electrical connector to the power distribution box.

102. Connect the cables and install the nuts.

103. Install the power distribution box cover.

104. If equipped, install the ground eyelet and bolt.

4x4 vehicles

105. Install the driveshaft. Torque to 15 ft. lbs. (20 Nm).

All vehicles

106. Install the front drive intermediate halfshaft.

107. Install the LH front drive halfshaft.

108. Install the lateral support crossmember. Torque to 85 ft. lbs. (115 Nm).

109. Install the accessory drive belt.

110. Install the catalytic converter.

111. Install the starter.

112. Install the battery tray and battery.

113. Install the engine air cleaner and air cleaner outlet pipe.

114. Fill the engine with clean engine oil.

115. Fill and bleed the cooling system.

116. Bleed the clutch system.

AUTOMATIC TRANSAXLE

All vehicles

1. With the vehicle in NEUTRAL, position it on a hoist.
2. Release the fuel system pressure.
3. Remove the engine air cleaner and air cleaner outlet pipe
4. Remove the battery tray.
5. Drain the engine oil.
6. Drain the cooling system.
7. Remove the starter.
8. Remove the catalytic converter.
9. Remove the accessory drive belt.
10. Remove the left front drive halfshafts.

4wd vehicles

11. Remove the transfer case.

2wd vehicles

12. Remove the bolts and the lateral support crossmember.
13. Remove the front drive intermediate halfshaft.

All vehicles

14. If equipped, remove the bolt and ground eyelet.
15. Remove the power distribution box cover.
16. Remove the nuts and disconnect the cables.
17. Disconnect the electrical connector from the power distribution box.
18. Remove the bolt and disconnect the ground strap. Loosen the bolt and disconnect the 42-pin electrical connector.
19. Detach the wiring harness retainers from the battery tray bracket and position the wiring harness out of the way.
20. Disconnect the transaxle electrical connector.
21. Disconnect the shift cable from the transaxle manual lever.
22. Position the transaxle control cable and bracket aside.
23. Disconnect the transmission range (TR) sensor electrical connector.
24. Detach the transaxle control harness from the brackets.
25. Disconnect the fluid cooler tube.
26. Disconnect the output shaft speed (OSS) sensor electrical connector (black).
27. Disconnect the turbine shaft speed (TSS) sensor electrical connector (white connector).
28. Remove the transmission fluid cooler retaining bracket bolt.
29. Position the fluid cooler tube aside.
30. Remove the bolt and the OSS sensor.
31. Detach the transaxle control harness from the retaining clip.
32. If equipped, disconnect the block heater electrical connector. Detach all the

block heater wiring harness retainers and position the wiring harness aside.

33. Disconnect the upper radiator and coolant vent hoses.

34. Remove the nuts and the coolant vent hose brackets. Position the coolant vent hose aside.

35. Detach the heater hose support strap from the stud.

36. Disconnect the heater hoses from the heater core.

37. Remove the retainers and the accelerator cable snow shield.

38. Disconnect the accelerator cable and speed control cable (if equipped).

39. Remove the nut from the accelerator control cable bracket.

40. Remove the nut from the accelerator control cable bracket and position the accelerator control cable and bracket assembly aside.

41. Remove the nut and position the power steering tube and bracket aside.

42. Disconnect the vacuum supply tube and position aside.

43. Disconnect the fuel vapor return tube and retainer and position aside.

44. Disconnect the vacuum reservoir tube and position aside.

45. Disconnect the fuel supply tube and position aside.

46. Detach the electrical connector retainers.

47. Disconnect the powertrain control module (PCM) electrical connectors. Remove the nut and position the harness aside.

48. Remove the bolt and detach the ground wire.

49. Remove the two power steering pump bolts.

50. Disconnect the lower radiator hose from the radiator.

51. Disconnect the A/C compressor electrical connector and remove the four bolts. Position the A/C compressor aside and support the compressor with a length of mechanics wire.

52. Disconnect the power steering pressure (PSP) sensor electrical connector.

➡The bolt under the power steering pressure tube will remain with the power steering pump.

53. Remove the bolts and position the power steering pump aside.

54. Remove the front roll restrictor bolt and the two bolts for the engine support crossmember.

55. Remove the rear nut and the engine support crossmember.

➡The transaxle-to-engine bolts differ in length. Mark the bolts for correct installation.

56. Remove the two transaxle-to-engine bolts

➡The transaxle-to-engine bolts differ in length. Mark the bolts for correct installation.

57. Remove the two transaxle-to-engine bolts

58. Using the special tools, secure the engine to the lift table.

59. Remove the engine mount bracket bolt.

60. Remove the nuts and the engine mount bracket.

61. Remove the bolt from the transaxle rear mount.

62. Remove the bolt from the left transaxle mount.

63. Lower the engine and transaxle from the vehicle.

64. Using the engine crane and spreader bar remove the engine and transaxle from the lift table.

65. Remove the starter motor isolator.

66. Remove and discard the four torque converter nuts.

➡The transaxle-to-engine bolts differ in length. Mark the bolts for correct installation.

67. Remove the remaining six engine-to-transaxle bolts and separate the engine and transaxle.

To install:

All vehicles

68. Using the engine crane and spreader bar, position the engine and transaxle together. Install the six upper transaxle-to-engine bolts. Torque to 35 ft. lbs. (48 Nm).

69. Install new torque converter nuts. Torque to 26 ft. lbs. (35 Nm).

70. Install the starter motor isolator.

71. Using the engine crane and spreader bar, position the engine and transaxle onto the lift table.

72. Using the special tools, secure the engine to the lift table.

73. Raise the engine and transaxle into the vehicle.

74. Install the bolt in the left transaxle mount. Torque to 76 ft. lbs. (103 Nm).

75. Install the bolt in the rear transaxle mount. Torque to 85 ft. lbs. (115 Nm).

76. Install the engine mount bracket. Torque to 66 ft. lbs. (90 Nm).

77. Install the engine mount bracket bolt. Torque to 66 ft. lbs. (90 Nm).

78. Install the 4 transaxle-to-engine bolts. Torque to 35 ft. lbs. (48 Nm).

79. Install the engine support crossmember and nut. Torque to 66 ft. lbs. (90 Nm).

80. Install the two bolts for the engine support crossmember. Torque to 66 ft. lbs. (90 Nm).

81. Install the front roll restrictor bolt . Torque to 85 ft. lbs. (115 Nm).

82. Position the power steering pump and install the upper bolts. Torque to 18 ft. lbs. (25 Nm).

83. Connect the power steering pressure (PSP) sensor electrical connector.

84. Install the A/C compressor and connect the A/C compressor electrical connector. Torque to 18 ft. lbs. (25 Nm).

85. Connect the lower radiator hose to the radiator.

86. Install the two lower power steering pump bolts. Torque to 18 ft. lbs. (25 Nm).

87. Install the ground wire and bolt.

88. Connect the powertrain control module (PCM) electrical connectors. Position the harness and install the nut.

89. Attach the electrical connector retainers.

90. Connect the fuel supply tube.

91. Connect the vacuum reservoir tube.

92. Connect the fuel vapor return tube and retainer.

93. Connect the vacuum supply tube.

94. Position the power steering tube and bracket and install the nut.

95. Position the accelerator control cable and bracket and install the nut.

96. Install the accelerator control cable and bracket and nut.

97. Install the accelerator cable and speed control cable (if equipped).

98. Install the accelerator cable snow shield and the retainers.

99. Connect the heater hoses to the heater core.

100. Attach the heater hose support strap to the stud.

101. Position the coolant vent hose and install the coolant vent hose brackets and nuts.

102. Connect the upper radiator and coolant vent hoses.

103. If equipped, route the block heater wiring harness and attach all retainers. Connect the block heater electrical connector.

104. Attach the transaxle control harness to the retaining clip.

105. Install the output shaft speed (OSS) sensor and bolt.

106. Install the fluid cooler tube.

107. Connect the transmission fluid cooler tube.

108. Attach the transaxle control harness to the brackets.

109. Connect the transmission range (TR) sensor electrical connector.

110. Install the transaxle control cable and bracket.

111. Connect the shift cable to the transaxle manual lever.

112. Connect the transaxle electrical connector.

113. Attach the wiring harness retainers to the battery tray bracket.

114. Connect the 42-pin electrical connector and tighten the bolt. Install the ground strap and bolt.

115. Connect the electrical connector to the power distribution box.

116. Connect the cables and install the nuts.

117. Install the power distribution box cover.

118. If equipped, install the ground eyelet and bolt.

2wd vehicles

119. Install the front drive intermediate halfshaft.

120. Install the lateral support crossmember. Torque to 85 ft. lbs. (115 Nm).

4wd vehicles

121. Install the transfer case.

All vehicles

122. Install the left front drive halfshaft.

123. Install the accessory drive belt.

124. Install the catalytic converter.

125. Install the starter.

126. Install the battery tray and battery.

127. Install the engine air cleaner and air cleaner outlet pipe.

128. Fill the engine with clean engine oil.

129. Fill and bleed the cooling system.

3.0L Engine

1. Before servicing the vehicle, refer to the precautions in the beginning of this section.
2. Properly recover the air conditioning system refrigerant.
3. Properly relieve the fuel system pressure.
4. Drain the cooling system.
5. Drain the engine oil.
6. Remove or disconnect the following:
   - Hood
   - Battery and battery tray
   - Air cleaner outlet tube and housing
   - Lower radiator air deflectors
   - Fuel lines
   - Water pump drive belt
   - Accelerator cable and speed control cable, if equipped
   - Vapor Management Valve (VMV)
   - Powertrain Control Module (PCM)
   - PCM ground wire
   - Thermostat housing and hose assembly and move them aside
   - Power distribution box electrical connector
   - Power distribution box cover
   - Nuts and cables from inside the power distribution box
   - Transmission linkage
   - Brake booster vacuum hose
   - Heater hoses
   - Power steering return line
   - Power Steering Pressure (PSP) switch electrical connector
   - Power steering supply line
   - Oil level indicator
   - Catalytic converter
   - A/C compressor
   - Both front wheels
   - Intermediate drive shaft, if equipped
7. Separate both side ball joints.
8. Separate both side tie rod ends from the steering knuckles.
9. Separate both sway bar links from the strut mounts.
10. Separate the struts from the steering knuckles.
11. Remove or disconnect the following:
    - Both wheel speed sensors, if equipped
    - Brake calipers from the steering knuckles and properly support the struts
    - Steering shaft from the rack
    - Transmission line bracket bolt
    - Transmission cooler lines
    - Torque converter inspection cover
    - Torque converter nuts
    - Block heater wiring, if equipped
12. Install a powertrain lifting devise and raise the vehicle.
    - Engine support bracket
    - Transmission support
    - 2 rear subframe bolts
    - 2 subframe side bolts
    - Motor mount support bolts
    - Engine and transmission as an assembly
    - Heated Oxygen (HO_2S) sensor
    - Transmission Range (TR) sensor
    - Transmission harness electronic control switch
    - Transmission control harness from the bracket
    - Starter and wire harness
    - Knock Sensor (KS) electrical connector
    - Output Shaft Speed (OSS) sensor electrical connector
    - HO_2S sensor and Exhaust Gas Recirculation (EGR) tube from the exhaust manifold
    - Alternator and electrical connectors
    - Right side exhaust manifold and gasket
    - Halfshaft support bracket and move it aside
13. Separate the engine from the transmission assembly

To install:

14. Install or connect the following:
    - Powertrain assembly on the subframe
    - Transmission-to-engine bolts. Torque the bolts to 30 ft. lbs. (40 Nm).
    - Halfshaft bracket. Torque the bolts to 18 ft. lbs. (25 Nm).
    - Right side exhaust manifold and new gasket. Torque the bolts to 15 ft. lbs. (25 Nm).
    - Alternator. Torque the larger bolts to 18 ft. lbs. (25 Nm) and smaller bolt to 89 inch lbs. (10 Nm).
    - EGR tube and HO_2S sensor electrical connectors
    - OSS sensor electrical connector
    - KS jumper electrical connector
    - Starter. Torque the bolts to 18 ft. lbs. (25 Nm).
    - Transmission control harness to the bracket. Torque the bolt to 18 ft. lbs. (25 Nm).
    - Transmission harness
    - Transmission range sensor
    - Powertrain assembly
    - Motor mount support. Torque the bolts to 66 ft. lbs. (90 Nm).
    - Subframe side nuts. Torque the nuts to 76 ft. lbs. (103 Nm). Raise the vehicle and support the powertrain assembly with a lifting device.
    - Transmission mount. Torque the bolts to side bolts to 66 ft. lbs. (90 Nm) and the other bolts to 76 ft. lbs. (103 Nm).
    - Motor mount. Torque the bolts to side bolts to 66 ft. lbs. (90 Nm) and the other bolts to 76 ft. lbs. (103 Nm). Remove the powertrain lift.
    - Block heater electrical connector, if equipped
    - Torque converter. Torque the nuts to 27 ft. lbs. (37 Nm).
    - Transmission cover plate and plug
    - Transmission cooler lines
    - Transmission cooler line bracket. Torque the bolt to 15 ft. lbs. (20 Nm).
    - Steering shaft to the rack. Torque the bolt to 18 ft. lbs. (25 Nm).
    - Struts to the steering knuckles. Torque the bolts to 75 ft. lbs. (102 Nm).

- Brake calipers to the steering knuckles
- Wheel speed sensors, if equipped. Torque the bolts to 89 inch lbs. (10 Nm).
- Sway bar links to the strut mount. Torque the bolts to 41 ft. lbs. (55 Nm).
- Tie rods to the steering knuckles. Torque the bolts to 41 ft. lbs. (55 Nm).
- Ball joints. Torque the bolts to 52 ft. lbs. (70 Nm).
- Intermediate drive shaft, if equipped
- Both front wheels
- A/C compressor
- Lower radiator air deflectors
- Catalytic converter
- Oil level indicator dipstick tube
- Power steering line and bracket. Torque the bolt to 13 ft. lbs. (17 Nm).
- PSP switch electrical connector
- Power steering return line
- Heater hoses
- Vacuum lines
- Transmission linkage
- Wire harness cables and nuts to the power distribution box. Torque the nuts to 89 inch lbs. (10 Nm).
- Power distribution box wire harness
- Thermostat housing and connect the hoses
- Ground wire. Torque the bolt to 89 inch lbs. (10 Nm).
- PCM electrical connector
- VMV electrical connector
- Accelerator cable and speed control cable, if equipped
- Air cleaner assembly
- Water pump drive belt
- Battery and tray

15. Fill and bleed the cooling system.
16. Fill the engine with clean oil.
17. Recharge the A/C system.
18. Inspect and top off the power steering fluid.
19. Start the vehicle, check for leaks and repair if necessary.

Water Pump

REMOVAL & INSTALLATION

2.0L Engine

1. Before servicing the vehicle, refer to the precautions in the beginning of this section.
2. Drain the cooling system.

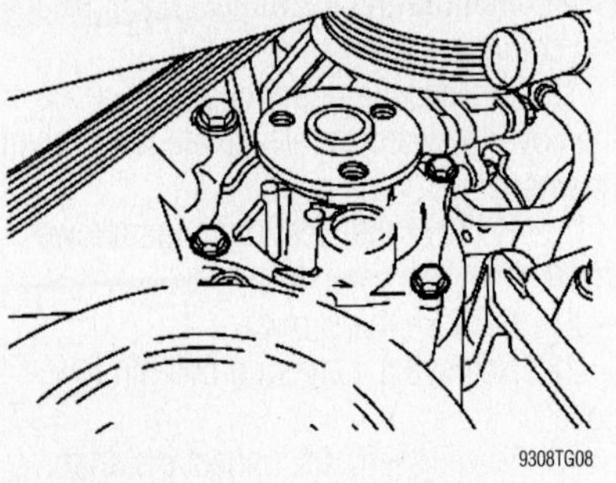

Exploded view of the water pump–2.0L engine

3. Remove or disconnect the following:
   - Negative battery cable
   - Right front wheel
   - Splash shield
   - Drive belt
   - Water pump pulley
   - Water pump

To install:

4. Install or connect the following:
   - Water pump. Torque the bolts to 89 inch lbs. (10 Nm).
   - Water pump pulley. Torque the bolts to 89 inch lbs. (10 Nm).
   - Drive belt
   - Splash shield
   - Right front wheel
   - Negative battery cable

5. Refill the cooling system.
6. Start the vehicle and check for leaks, repair if necessary.

2.3L Engine

1. With the vehicle in NEUTRAL, position it on a hoist.
2. Drain the cooling system.
3. Remove the accessory drive belt.
4. Remove the coolant pump pulley bolts.
5. Remove the coolant pump pulley.
6. Remove the coolant pump bolts.
7. Remove the coolant pump.
8. Remove the coolant pump O-ring seal.
9. To install, reverse the removal procedure. Torque the water pump bolts to 89 inch lbs. (10 Nm). Torque the pulley bolts to 18 ft. lbs. (25 Nm).
10. Fill and bleed the cooling system.

3.0L Engine

1. Before servicing the vehicle, refer to the precautions in the beginning of this section.
2. Drain the cooling system.
3. Remove or disconnect the following:
   - Negative battery cable
   - Air cleaner outlet tube

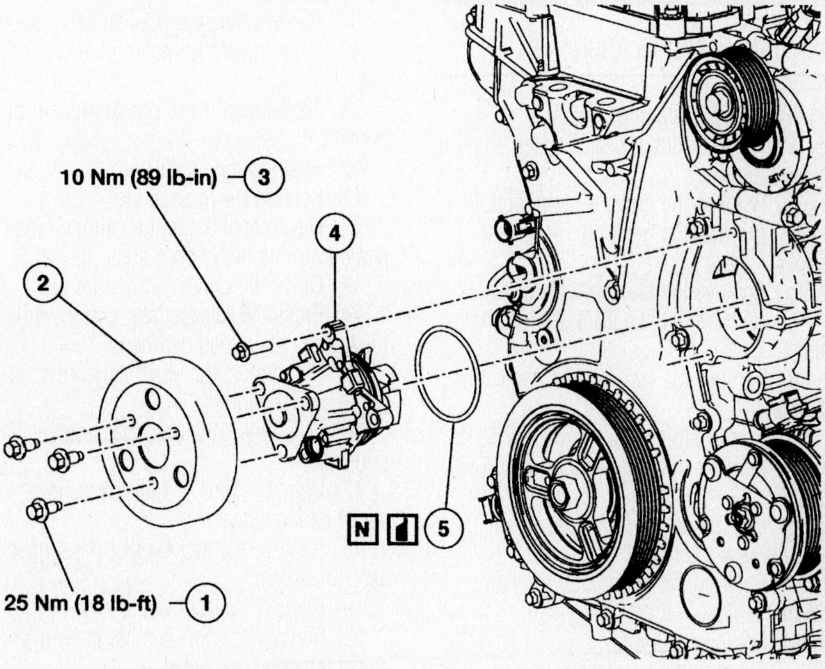

1 Coolant pump pulley bolts
2 Coolant pump pulley
3 Coolant pump bolts
4 Coolant pump
5 Coolant pump O-ring seal

67197-ESCA-G03

Water pump mounting—2.3L engine

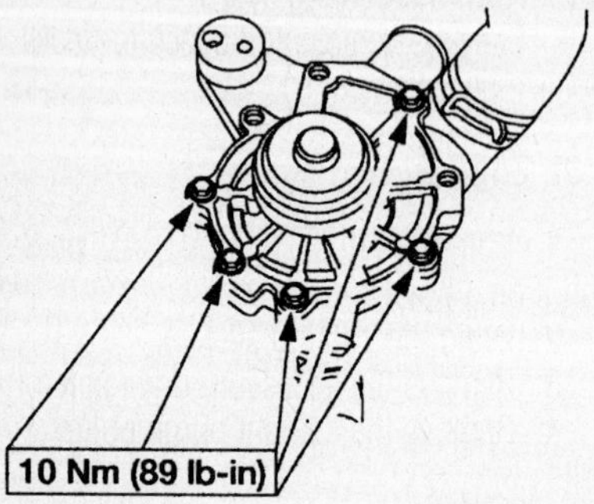

Water pump mounting—3.0L engine

- Water pump belt tensioner
- Coolant hoses
- Water pump
- Water pump from the housing

To install:

- Water pump to the housing. Torque the bolts to 89 inch lbs. (10 Nm).
- Water pump. Torque the bolts to 89 inch lbs. (10 Nm).
- Coolant hoses
- Water pump belt tensioner
- Air cleaner outlet tube
- Negative battery cable

4. Refill the cooling system.
5. Start the vehicle and check for leaks, repair if necessary.

Heater Core

REMOVAL & INSTALLATION

1. Drain the engine coolant.
2. Disconnect the heater hoses from the heater core.
3. Remove the driver air bag module.
4. Remove the two front door scuff plates.
5. Remove the four pin-type retainers.
6. Remove the two front door scuff plates.
7. Remove the two A-pillar lower trim panels.
8. Remove the two pin-type retainers.
9. Remove the two A-pillar lower trim panels.
10. Disconnect the electrical connectors located by the LH cowl.
11. Position the hood latch release handle aside.
12. Remove the bolts.
13. Position the hood latch release handle aside.
14. Remove the utility compartment.
15. Remove the four pin-type retainers.
16. Remove the utility compartment.
17. Disconnect the electrical connector.
18. Remove the instrument panel steering column cover.
19. Release the upper clips and rotate the cover outward to release the lower pivot retainers.
20. Remove the steering column lower cover.
21. Remove the screws.
22. Remove the steering column lower cover.
23. If equipped, disconnect the shift cable.
24. Disconnect the shift cable.
25. Disconnect the shift cable from the retaining bracket.
26. Remove the steering column coupler access cover.
27. Disconnect the steering column coupler.
28. Remove the steering column coupler bolt and nut.
29. Disconnect the steering column coupler.
30. Remove the cover panel.
31. Remove the pin-type retainer.
32. Release the retaining clip.
33. Disconnect the electrical connectors.
34. Disconnect the climate control vacuum harness connector.
35. Disconnect the in-line electrical connector.
36. Remove the four instrument panel center brace bolts.
37. Remove the passenger air bag module.
38. Disconnect the vacuum harness connector.
39. Disconnect the temperature control cable.
40. Position the locator pin.
41. Release the locking tab.
42. Disconnect the temperature control cable from the blend door shaft.
43. Close the glove compartment.
44. Press the release tabs inward while raising the glove compartment.
45. Disconnect the electrical connectors at the blower motor.
46. Disconnect the antenna cable in-line connector.
47. Open the four A-pillar passenger assist handle covers.
48. Remove the two A-pillar passenger assist handles.
49. Remove the four bolts.
50. Remove the two A-pillar passenger assist handles.
51. Remove the two windshield side garnish moldings.
52. Remove the instrument panel cowl top cover.
53. Remove the instrument panel cowl top bolt.
54. Loosen the tilt lever (if equipped) and lower steering column.
55. Position the transmission range selector lever (if equipped) down to provide access to the instrument cluster finish panel and instrument cluster.
56. Remove the screws and the instrument cluster finish panel.
57. Remove the screws.
58. Disconnect the electrical connectors and remove the instrument cluster.
59. Through the instrument cluster opening, remove the instrument panel nut.
60. Remove the two instrument panel finish end panels.
61. Remove the four instrument panel cowl side bolts.

➡This step requires an assistant.

62. Remove the instrument panel.
63. Remove the heater blending door levers.
64. Remove the screw for heater blending door.
65. Remove the levers for the blending door.
66. Remove the heater core.
67. Remove the three screws.
68. Remove the cover for the heater core and pull the heater core out of the housing.

➡Before installing the temperature control cable, make sure the blend door, cable and temperature switch are correctly positioned.

69. To install, reverse the removal procedure.

CAUTION

Electronic modules are sensitive to static electrical charges. If exposed to these charges, damage may result.

CAUTION

Once the new module is installed, it is necessary to download the module configuration information from the scan tool into the new instrument cluster.

Cylinder Head

REMOVAL & INSTALLATION

2.0L Engine

1. Before servicing the vehicle, refer to the precautions in the beginning of this section.

2. Properly relieve the fuel system pressure.
3. Drain the engine oil.
4. Remove or disconnect the following:
- Negative battery cable
- Ignition coil bracket
- Thermostat housing
- Positive Crankcase Ventilation (PCV) tube
- Intake manifold
- Exhaust manifold
- Power steering bracket and move it aside
- Valve tappets
- Engine mount lower bracket
- Engine mount upper bracket
- Cylinder head bolts in the proper sequence and discard the gasket

To install:

5. Install a new head gasket and the cylinder head.
6. Lubricate the cylinder head bolt threads.
7. Torque the cylinder head bolts in the proper sequence as follows:
   a. Step 1: 15 ft. lbs. (20 Nm).
   b. Step 2: 30 ft. lbs. (40 Nm).
   c. Step 3: Plus an additional 90 degrees.
8. Install or connect the following:
- Engine mount upper bracket. Torque the 2 upper bolts to 72 ft. lbs. (98 Nm) and the center bolt to 57 ft. lbs. (77 Nm).
- Engine mount lower bracket. Torque the bolts to 37 ft. lbs. (50 Nm).
- Valve tappets
- Power steering pump bracket. Torque the bolts to 20 ft. lbs. (28 Nm).
- Exhaust manifold
- Intake manifold
- PCV tube
- Thermostat housing
- Ignition coil bracket
- Negative battery cable

9. Fill the engine with clean oil and replace the filter.
10. Start the vehicle and check for leaks, repair if necessary.

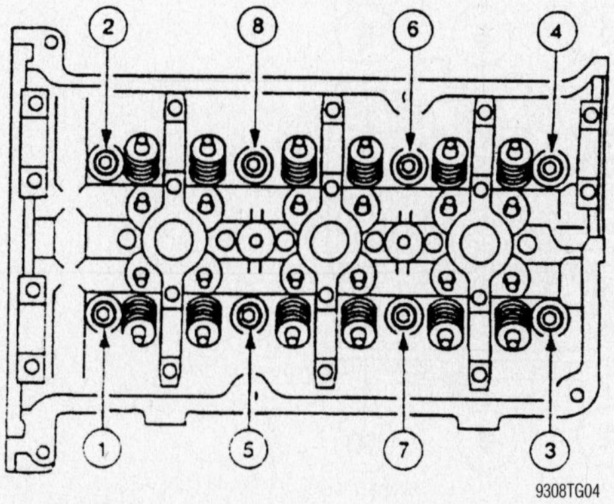

Cylinder head bolt torque sequence 2.0L engine

2.3L Engine

⁂ WARNING

During engine repair procedures, cleanliness is extremely important. Any foreign material, including any material created while cleaning gasket surfaces, that enters the oil passages, coolant passages or the oil pan can cause engine failure.

1. With the vehicle in NEUTRAL, position it on a hoist.
2. Remove the camshafts.
3. Remove the intake manifold.
4. Remove the catalytic converter.
5. Disconnect the radio ignition interference capacitor electrical connector
6. Disconnect the exhaust gas recirculation (EGR) valve electrical connector
7. Remove the upper radiator hose.
8. Remove the EGR coolant tube clamp.
9. Remove the EGR coolant hose.
10. Remove the engine coolant vent hose.
11. Remove the heater hose.
12. Remove the bypass hose.
13. Remove and discard the cylinder head bolts.
14. Remove the cylinder head.
15. Remove the cylinder head gasket.
16. Inspect the cylinder head for distortion.

⁂ WARNING

Do not use metal scrapers, wire brushes, power abrasive discs or other abrasive means to clean the sealing surfaces. These tools cause scratches and gouges that make leak paths. Use a plastic scraping tool to remove all traces of the head gasket.

⁂ WARNING

Observe all warnings or cautions and follow all application directions contained on the packaging of the silicone gasket remover and the metal surface prep.

➡If there is no residual gasket material present, metal surface prep can be used to clean and prepare the surfaces.

17. Clean the cylinder head-to-cylinder block mating surface of both the cylinder head and the cylinder block.
18. Remove any large deposits of silicone or gasket material with a plastic scraper.
19. Apply silicone gasket remover, following package directions, and allow to set for several minutes.
20. Remove the silicone gasket remover with a plastic scraper. A second application of silicone gasket remover may be required if residual traces of silicone or gasket material remain.
21. Apply metal surface prep, following package directions, to remove any traces of oil or coolant, and to prepare the surfaces to bond with the new gasket. Do not attempt to make the metal shiny. Some staining of the metal surfaces is normal.
22. Apply silicone gasket and sealant to the locations shown.
23. Install a new head gasket.

➡The cylinder head bolts are torque-to-yield and must not be reused. New cylinder head bolts must be installed.

➡Lubricate the bolts with clean engine oil prior to installation.

24. Install new cylinder head bolts. Tighten the bolts in the sequence shown in five stages.
   a. Tighten the bolts to 5 Nm (44 inch lbs.).
   b. Tighten the bolts to 15 Nm (11 ft. lbs.).
   c. Tighten the bolts to 45 Nm (33 ft. lbs.).
   d. Turn the bolts 90 degrees.
   e. Turn the bolts an additional 90 degrees.
25. To install, reverse the removal procedure.

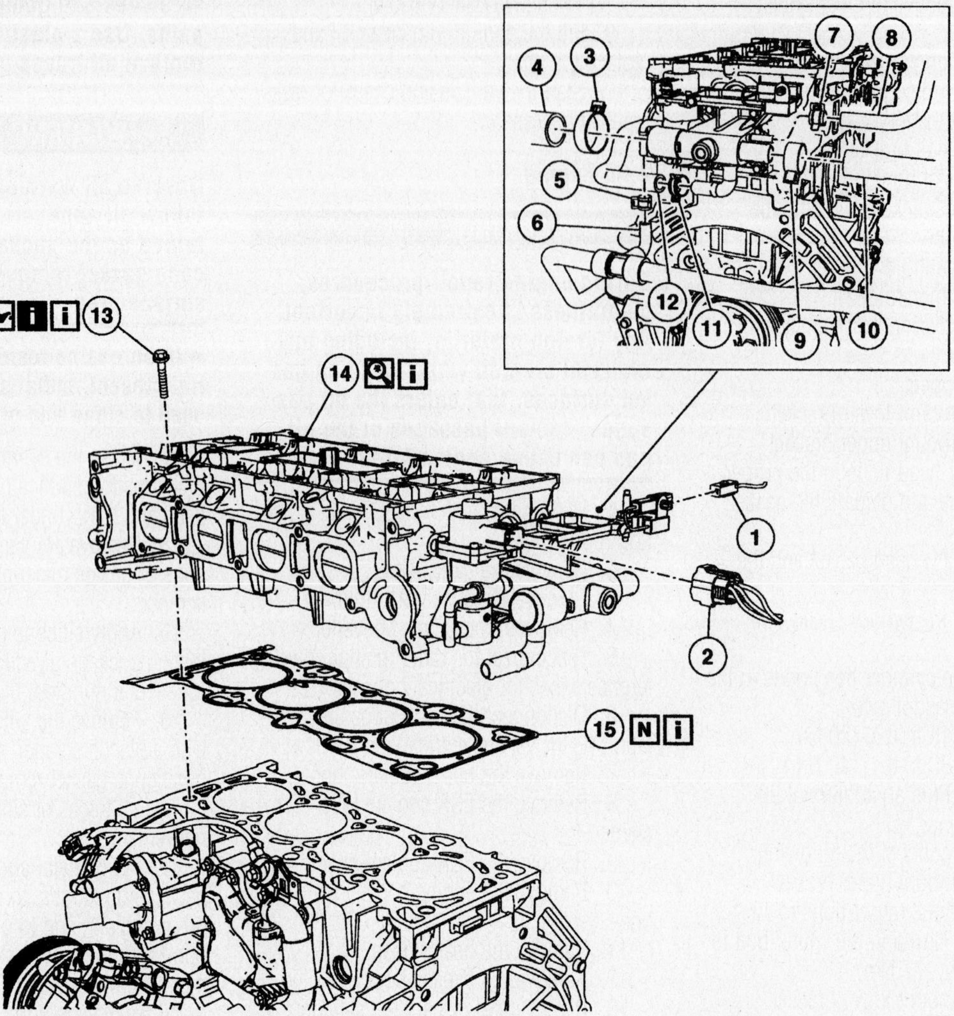

1 Radio ignition interference capacitor electrical connector
2 Exhaust gas recirculation (EGR) valve electrical connector
3 Upper radiator hose clamp
4 Upper radiator hose (position aside)
5 EGR coolant tube clamp
6 EGR coolant hose (part of heater hose) (position aside)
7 Engine coolant vent hose clamp
8 Engine coolant vent hose (position aside)
9 Heater hose clamp
10 Heater hose (position aside)
11 Bypass hose clamp
12 Bypass hose (position aside)
13 Cylinder head bolt
14 Cylinder head
15 Cylinder head gasket

67197-ESCA-G04

Cylinder head removal—2.3L engine

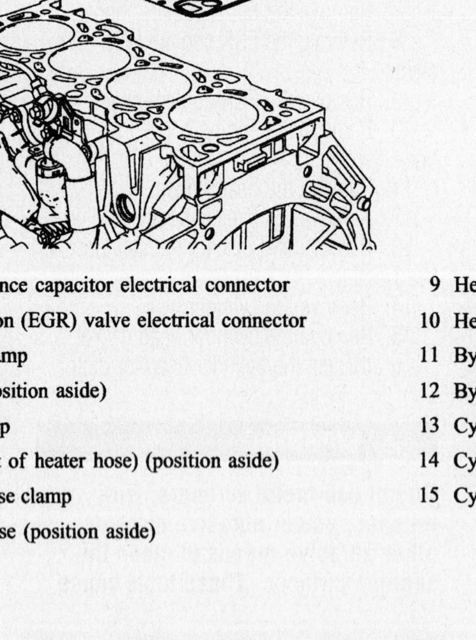

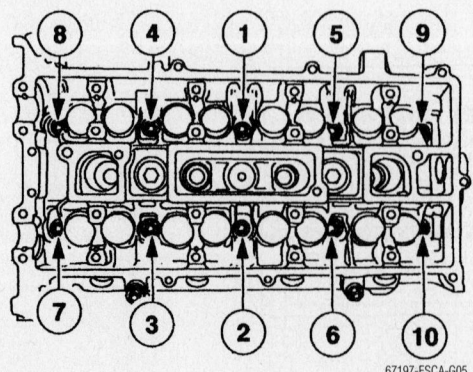

67197-ESCA-G05

Cylinder head bolt torque sequence—2.3L engine

3.0L Engine

The procedure for the left side cylinder head and right side are similar. Changes in the procedure will be noted for either side cylinder head.

1. Before servicing the vehicle, refer to the precautions in the beginning of this section.
2. Properly relieve the fuel system pressure.
   - Drain the cooling system.
3. Remove or disconnect the following:
   - Negative battery cable
   - Camshaft
   - Exhaust Gas Recirculation (EGR) tube, right side only
   - Exhaust manifold
   - Camshaft followers
   - Hydraulic lash adjusters and matchmark them for proper installation
   - Cylinder head bolts in sequence and discard them
   - Cylinder head and discard the gasket

To install:

4. Install a new head gasket and the cylinder head.

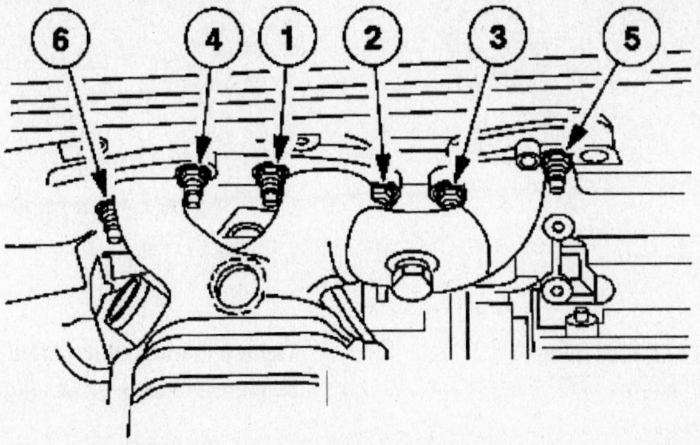

Right side exhaust manifold bolt torque sequence–3.0L engine

5. Lubricate the cylinder head bolt threads.
6. Torque the cylinder head bolts in the proper sequence as follows:
   a. Step 1: 30 ft. lbs. (40 Nm).
   b. Step 2: Additional 90 degrees.
   c. Step 3: Loosen the bolts one full turn.
   d. Step 4: 30 ft. lbs. (40 Nm).
   e. Step 5: Plus an additional 90 degrees.
   f. Step 6: Plus an additional 90 degrees.
7. Install or connect the following:
   - Hydraulic lash adjusters
   - Camshaft followers
   - Camshaft
   - Exhaust manifold. Torque the bolts in sequence to 15 ft. lbs. (20 Nm), right side only
   - EGR tube, right side only
   - Coolant bypass tube
   - Negative battery cable
8. Fill the coolant to the proper level.
9. Start the vehicle and check for leaks, repair if necessary.

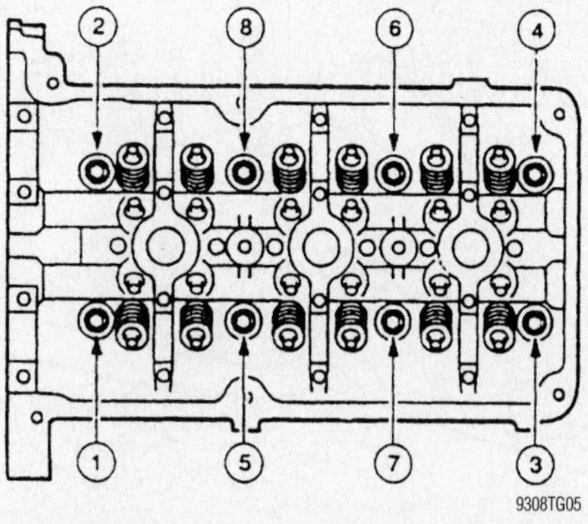

Left side cylinder head bolt torque sequence 3.0L engine

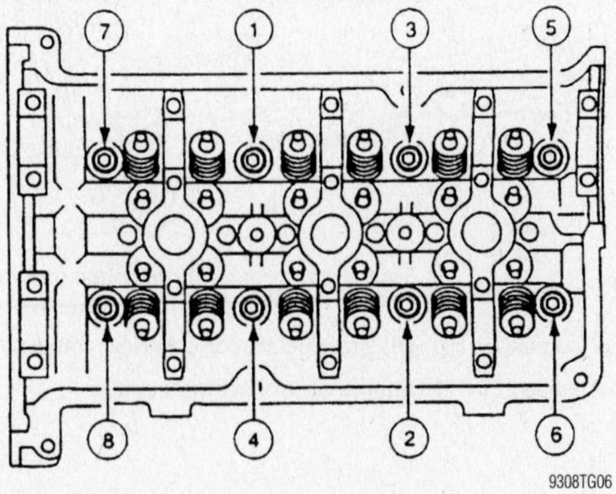

Right side cylinder head bolt torque sequence 3.0L engine

Intake Manifold

REMOVAL & INSTALLATION

2.0L Engine

1. Before servicing the vehicle, refer to the precautions in the beginning of this section.
2. Properly relieve the fuel system pressure.
3. Remove or disconnect the following:
   - Negative battery cable
   - Fuel injection supply manifold
   - Throttle Position (TP) sensor electrical connector
   - Idle Air Control (IAC) electrical connector and unclip the harness from the bracket
   - Main engine control sensor wiring
   - Connector from the bracket
   - Powertrain Control Module (PCM) wire harness from the bracket
   - Brake booster vacuum hose
   - 4 additional vacuum lines
   - Positive Crankcase Ventilation (PCV) hose from the intake manifold
   - Knock Sensor (KS) electrical connector
   - Alternator

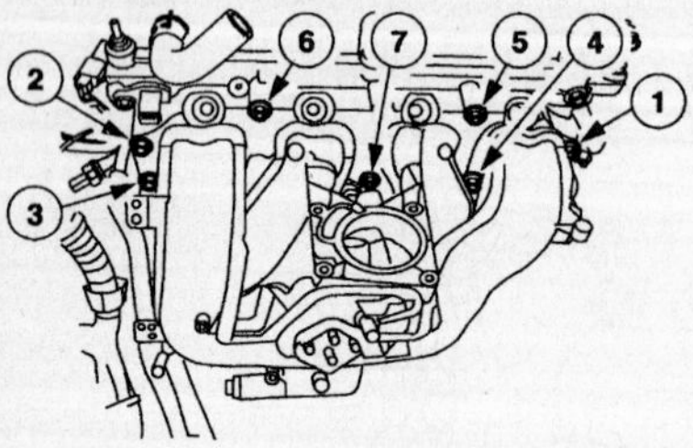

Intake manifold bolt loosening sequence—2.0L engine

- Intake manifold and discard the gasket

4. Clean the mating surfaces.

To install:

5. Install or connect the following:
   - New gasket

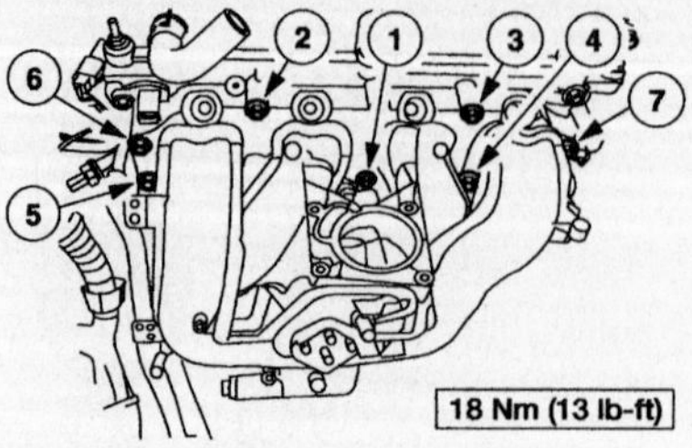

Tighten the intake manifold bolts in the sequence shown—2.0L engine

- Intake manifold. Torque the bolts, in sequence, to 13 ft. lbs. (18 Nm).
- Alternator
- KS electrical connector
- PCV vacuum line
- 4 vacuum lines
- Brake booster vacuum supply hose
- PCM wire harness to the bracket
- Main engine control sensor wiring
- IAC valve electrical connector and attach the harness to the bracket
- TP sensor electrical connector
- Fuel injection supply manifold
- Negative battery cable

6. Start the vehicle and check for leaks, repair if necessary.

2.3L Engine

1. With the vehicle in NEUTRAL, position it on a hoist.
2. Remove the throttle body.
3. Remove the fuel rail.
4. Remove the oil level indicator tube.
5. Remove the vacuum tube.

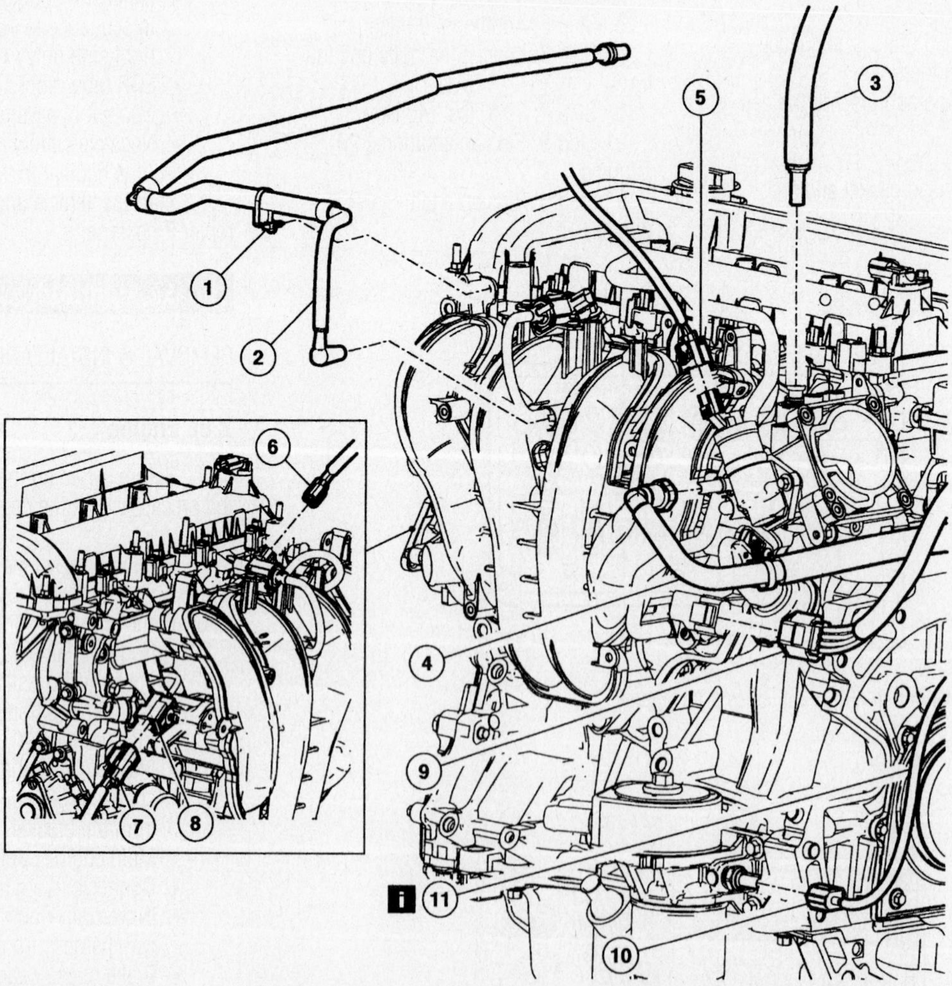

1 Vacuum tube retainer
2 Vacuum tube
3 Vacuum supply hose
4 Fuel vapor return hose
5 Idle air control (IAC) motor electrical connector
6 Swirl control valve electrical connector
7 Knock sensor (KS) electrical connector
8 Pin-type retainer
9 Temperature manifold absolute pressure (TMAP) sensor electrical connector
10 Oil pressure sender electrical connector
11 Engine control wiring harness

67197-ESCA-G06

Intake manifold and related parts—2.3L engine

6. Remove the vacuum supply hose.
7. Remove the fuel vapor return hose.
8. Remove the idle air control (IAC) motor electrical connector.
9. Remove the swirl control valve electrical connector.
10. Remove the knock sensor (KS) electrical connector.
11. Remove the temperature manifold absolute pressure (TMAP) sensor electrical connector.
12. Remove the oil pressure sender electrical connector.
13. Remove the engine control wiring harness.
14. Remove the intake manifold bolts.

➡There are three different size bolts used. Mark the location of the bolts to make sure they are installed in the correct location.

15. Remove the bolts and position the intake manifold aside to access the crankcase vent hose clamp and the EGR tube.
16. Remove the crankcase vent hose.
17. Remove the exhaust gas recirculation (EGR) tube.
18. Remove the intake manifold.
19. Remove the intake manifold gasket.
20. To install, reverse the removal procedure. Torque the intake manifold bolts to 13 ft. lbs. (18 Nm).

3.0L Engine

UPPER

1. Before servicing the vehicle, refer to the precautions in the beginning of this section.
2. Properly relieve the fuel system pressure.
3. Drain the coolant system.
4. Remove or disconnect the following:
   - Negative battery cable
   - Air cleaner outlet tube
   - Engine appearance cover
   - Throttle cable
   - Speed control cable, if equipped
   - Throttle cable bracket
   - Throttle Position (TP) sensor electrical connector
   - Idle Air Control (IAC) valve electrical connector
   - Exhaust Gas Recirculation (EGR) valve vacuum hose and tube
   - EGR vacuum regulator valve electrical connector and hose
   - Chassis vacuum hose
   - Engine vacuum hose
   - Positive Crankcase Ventilation (PCV) hose
   - Vapor Management Valve (VMV) vacuum hose
   - Electrical connectors from the left side of the upper intake manifold
   - Power Steering Pressure (PSP) sensor electrical connector
   - Upper intake manifold and discard the gasket
5. Clean the mating surfaces.

To install:

6. Install or connect the following:
   - New gasket

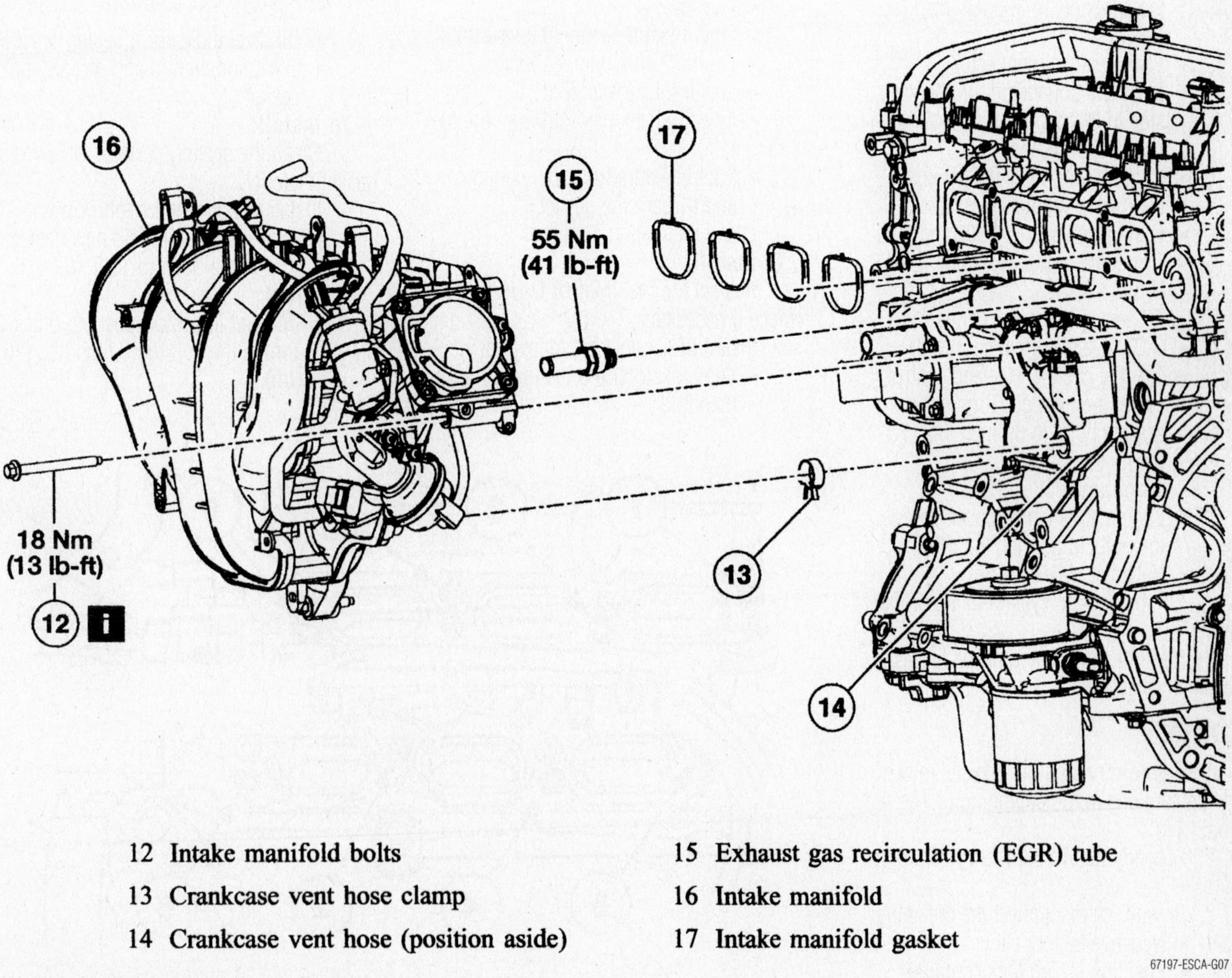

Intake manifold installation—2.3L engine

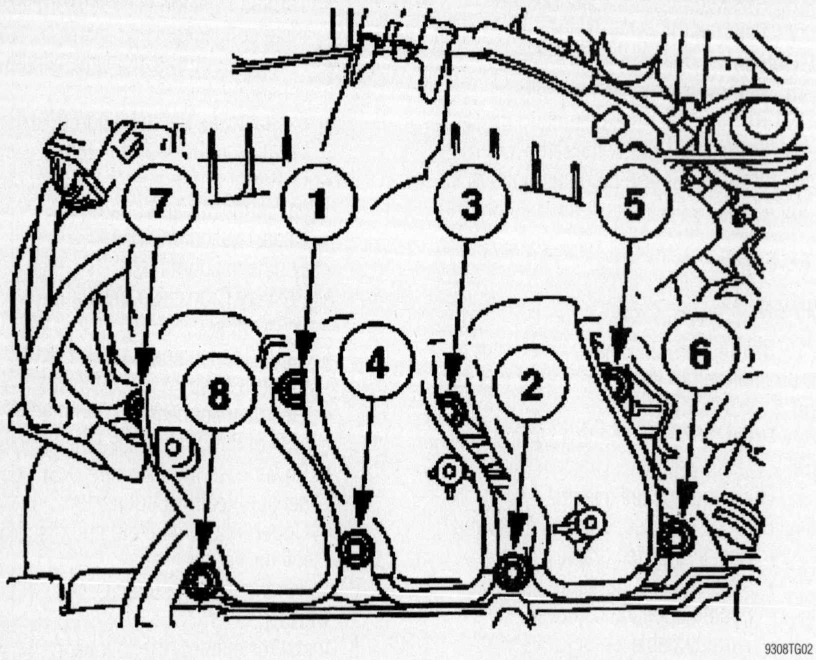

Tighten the upper intake manifold bolts in the sequence shown—3.0L engine

- Intake manifold. Torque the bolts, in sequence, to 89 inch lbs. (10 Nm).
- PSP electrical connector
- Electrical connectors on the left side of the upper intake manifold
- VMV vacuum hose
- Chassis, engine and PCV hoses
- EGR valve vacuum regulator
- EGR valve vacuum hose and tube. Torque the nut to 30 ft. lbs. (40 Nm).
- TP sensor electrical connector
- IAC valve electrical connector
- Throttle cable and speed control cable, if equipped. Torque the bracket bolts to 89 inch lbs. (10 Nm).
- Air cleaner outlet tube
- Engine appearance cover. Torque the bolts to 53 inch lbs. (6 Nm).
- Negative battery cable

7. Fill the coolant system to the proper level.

8. Start the vehicle and check for leaks, repair if necessary.

LOWER

1. Before servicing the vehicle, refer to the precautions in the beginning of this section.

2. Properly relieve the fuel system pressure.

3. Remove or disconnect the following:
   - Negative battery cable
   - Fuel line spring lock coupling
   - Upper intake manifold
   - Fuel rail
   - Fuel injector electrical connectors
   - Fuel pressure damper vacuum line
   - Lower intake manifold
   - Lower intake manifold from the fuel rail
   - Fuel injectors from the manifold and discard the gasket

4. Clean the mating surfaces.

To install:

5. Inspect the fuel injector O-rings and replace if necessary.

6. Install or connect the following:
   - Fuel injectors into the lower intake manifold
   - Fuel rail. Torque the bolts to 89 inch lbs. (10 Nm).
   - New gasket
   - Intake manifold. Torque the bolts, in sequence, to 89 inch lbs. (10 Nm).
   - Fuel rail electrical connectors
   - Fuel injector electrical connectors
   - Fuel pressure damper vacuum line
   - Upper intake manifold
   - Fuel line spring lock coupling
   - Negative battery cable

7. Start the vehicle and check for leaks, repair if necessary.

Exhaust Manifold

REMOVAL & INSTALLATION

2.0L Engine

1. Before servicing the vehicle, refer to the precautions in the beginning of this section.

2. Remove or disconnect the following:
   - Negative battery cable
   - Catalytic converter
   - Oil level indicator tube and bracket
   - Exhaust manifold and discard the gasket

To install:

3. Clean the sealing surfaces of any old gasket material.

4. Install or connect the following:
   - Exhaust manifold and new gasket. Torque the bolts to 12 ft. lbs. (16 Nm).
   - Oil level indicator tube and bracket. Torque the bolt to 89 inch lbs. (10 Nm).

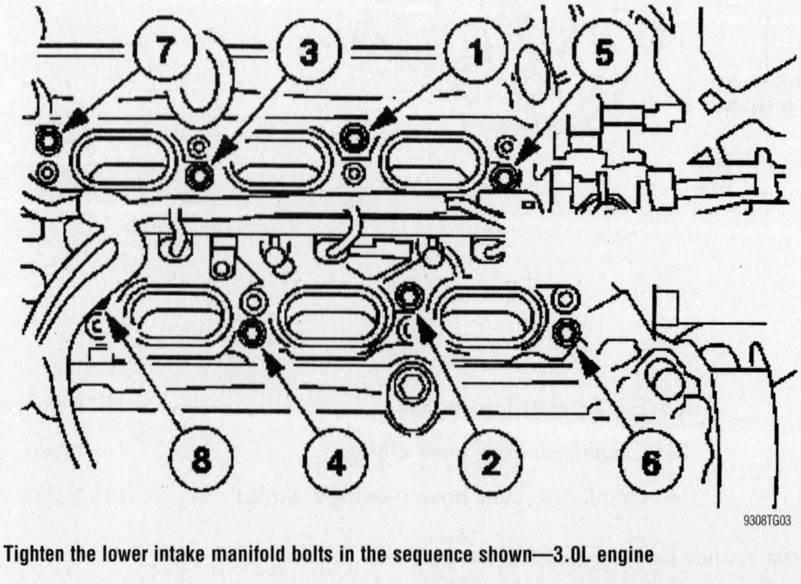

Tighten the lower intake manifold bolts in the sequence shown—3.0L engine

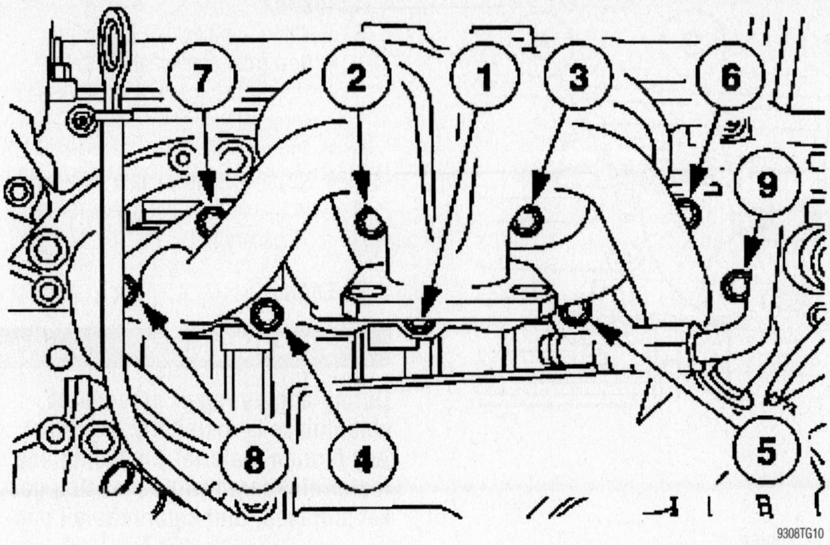

Exhaust manifold bolt torque sequence–2.0L engine

- Catalytic converter
- Negative battery cable

5. Start the vehicle and check for leaks, repair if necessary.

2.3L Engine

⁂ WARNING

Do not use oil or grease-based lubricants on the insulators. They may cause deterioration of the rubber.

⁂ WARNING

Oil or grease-based lubricants on the insulators may cause the exhaust hanger insulator to separate from the exhaust hanger bracket during vehicle operation.

➡Exhaust fasteners are of a torque prevailing design. Use only new fasteners with the same part number as the original. Torque values must be used as specified during reassembly to make sure of correct retention of exhaust components.

1. Remove the flex pipe nuts.
2. Remove the flex pipe gasket.
3. Remove the manifold bracket bolts.
4. Remove the heat shield.
5. Remove the exhaust manifold nuts.
6. Remove the catalyst monitor sensor.
7. Remove the heated oxygen sensor.
8. Remove the exhaust manifold.
9. To install, reverse the removal procedure. Make sure to apply anti-seize lubricant to the threads of the sensors before installation. Failure to tighten the exhaust manifold nuts to specification before installing the manifold bracket bolts will cause the manifold to develop an exhaust gas leak.
10. Observe the following torques:
    - Exhaust manifold-to-head: 35 ft. lbs. (47 Nm)
    - Flex pipe-to-manifold: 18 ft. lbs. (25 Nm)
    - Heated oxygen sensor: 35 ft. lbs. (47 Nm)
    - Catalyst monitor sensor: 30 ft. lbs. (40 Nm)
11. Check the exhaust system for proper alignment.

3.0L Engine

LEFT SIDE

1. Before servicing the vehicle, refer to the precautions in the beginning of this section.
2. Remove or disconnect the following:
    - Negative battery cable
    - Heated Oxygen (HO_2S) sensor and catalyst monitor
    - Splash shield
    - Exhaust crossover pipe
    - Drive belt
    - A/C compressor and move it aside
    - Exhaust manifold and discard the gasket

To install:

3. Clean the sealing surfaces of any old gasket material.
4. Install or connect the following:
    - Exhaust manifold and new gasket. Torque the bolts to 15 ft. lbs. (20 Nm).
    - A/C compressor. Torque the bolts to 18 ft. lbs. (20 Nm).
    - Drive belt
    - Exhaust crossover pipe. Torque the bolts to 30 ft. lbs. (40 Nm).
    - Splash shield. Torque the bolts to 80 inch lbs. (9 Nm).
    - Left side HO_2S sensor and catalyst monitor
    - Negative battery cable
5. Start the vehicle and check for leaks, repair if necessary.

RIGHT SIDE

1. Before servicing the vehicle, refer to the precautions in the beginning of this section.

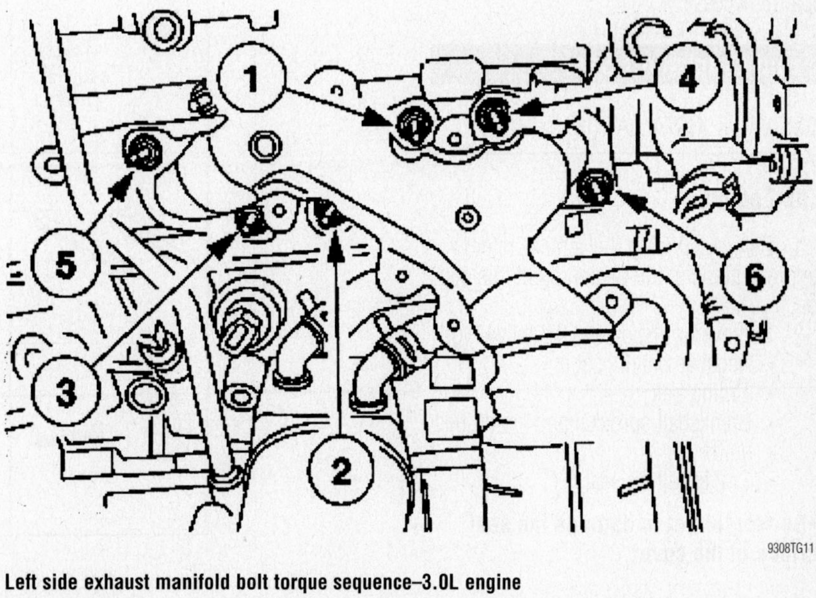

Left side exhaust manifold bolt torque sequence–3.0L engine

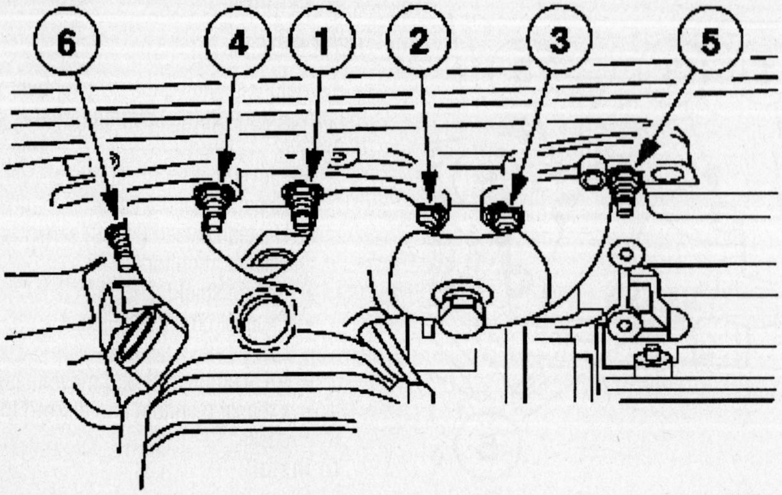

Right side exhaust manifold bolt torque sequence–3.0L engine

2. Remove or disconnect the following:
   - Negative battery cable
   - Exhaust Gas Recirculation (EGR) tube
   - Alternator
   - Right side Heated Oxygen (HO_2S) sensor
   - Right side exhaust manifold and discard the gasket

To install:

3. Clean the sealing surfaces of any old gasket material.
4. Install or connect the following:
   - Exhaust manifold and new gasket. Torque the bolts to 15 ft. lbs. (20 Nm).
   - Right side HO_2S sensor
   - Alternator
   - EGR tube
   - Negative battery cable
5. Start the vehicle and check for leaks, repair if necessary.

Front Crankshaft Seal

REMOVAL & INSTALLATION

2.0L Engine

1. Before servicing the vehicle, refer to the precautions in the beginning of this section.
2. Remove or disconnect the following:
   - Negative battery cable
   - Timing belt
   - Crankshaft sprocket and timing belt guide
   - Crankshaft oil seal

➡**Be careful not to damage the seal surface of the cover.**

To install:

3. Install or connect the following:
   - New front crankshaft oil seal
   - Timing belt guide and crankshaft sprocket
   - Timing belt
   - Negative battery cable
4. Start the vehicle and check for leaks, repair if necessary.

2.3L Engine

⁂ WARNING

During engine repair procedures, cleanliness is extremely important. Any foreign material, including any material created while cleaning gasket surfaces, that enters the oil pas-

	Alignment Plate, Camshaft
	Timing Peg, Crankshaft
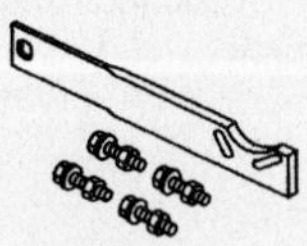	Holding Fixture, Drive Pinion Flange
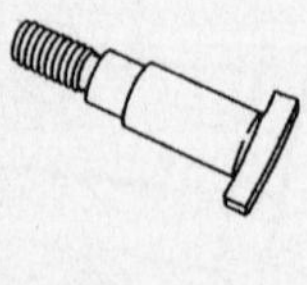	Adapter for 205-126
	Remover, Oil Seal
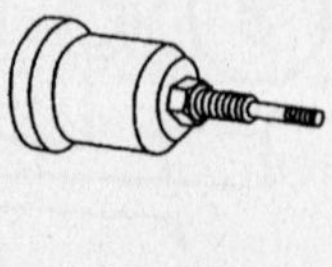	Installer, Front Oil Seal

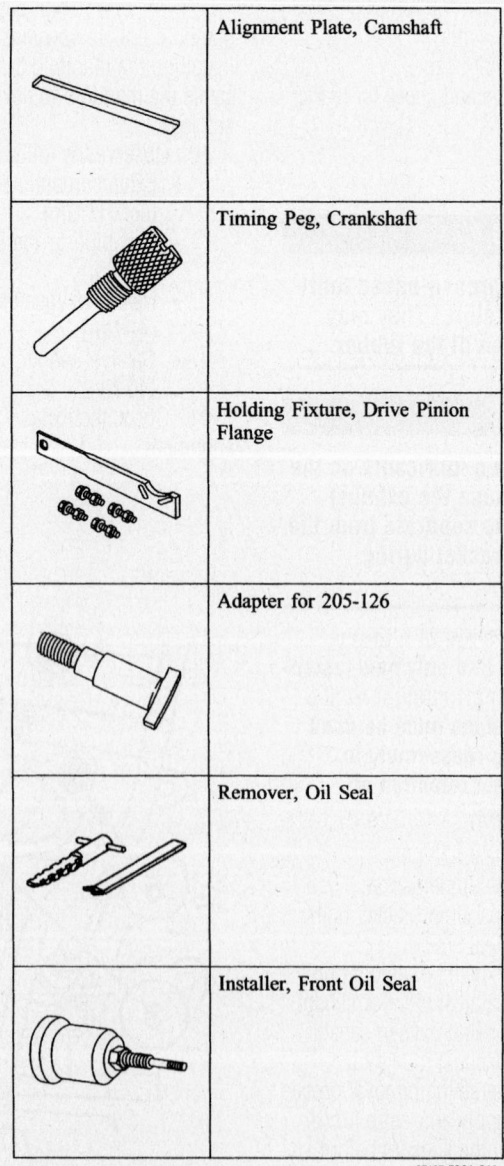

67197-ESCA-G08

Tools necessary for this job—2.3L engine

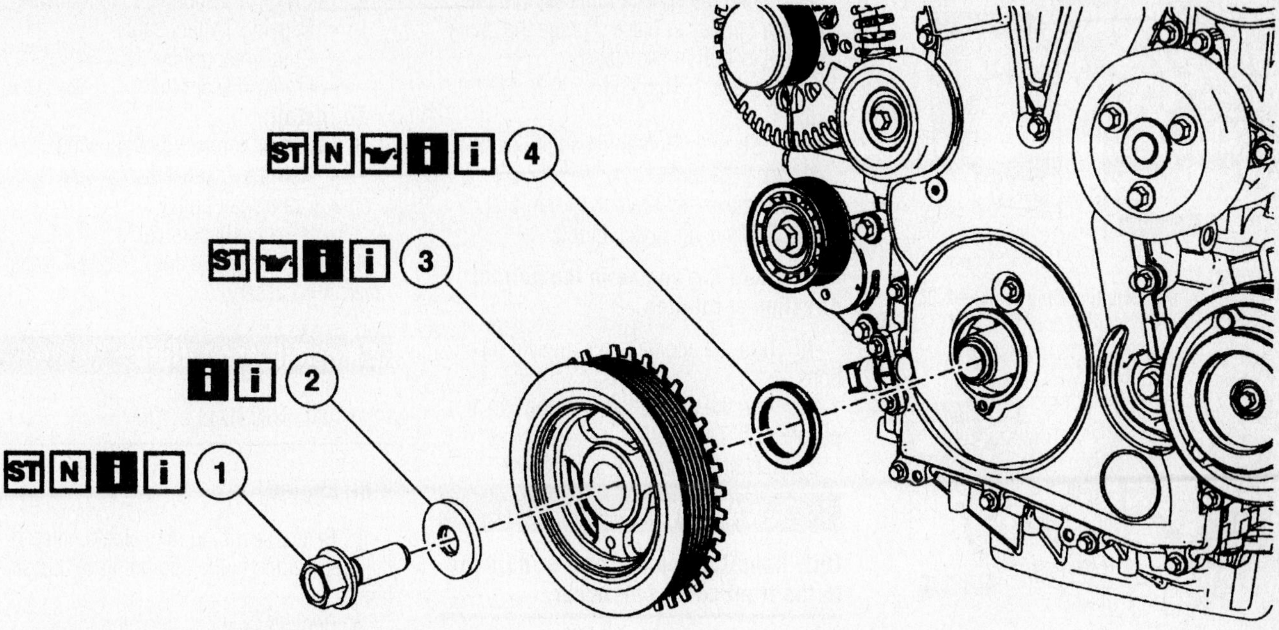

1 Crankshaft pulley bolt
2 Crankshaft pulley washer
3 Crankshaft pulley
4 Crankshaft front seal

67197-ESCA-G09

Crankshaft pulley and seal—2.3L engine

sages, coolant passages or the oil pan can cause engine failure.

WARNING

The crankshaft, the crankshaft sprocket and the pulley are fitted together by friction, using diamond washers between the flange faces on each part. For that reason, the crankshaft sprocket is also unfastened if you loosen the pulley. Therefore, the engine must be retimed each time the damper is removed. Otherwise severe engine damage can occur.

1. With the vehicle in NEUTRAL, position it on a hoist.
2. Remove the accessory drive belt.
3. Remove the valve cover.

WARNING

Failure to position the No. 1 piston at top dead center (TDC) can result in damage to the engine. Turn the engine in the normal direction of rotation only.

4. Using the crankshaft pulley bolt, turn the crankshaft clockwise to position the No. 1 piston at top dead center (TDC).

WARNING

The special tool 303-465 is for camshaft alignment only. Using this tool to prevent engine rotation can result in engine damage.

➡The camshaft timing slots are offset. If the special tool cannot be installed, rotate the crankshaft one complete revolution clockwise to correctly position the camshafts.

5. Install the special tool in the slots on the rear of both camshafts.

➡Installing the special tool in this step will prevent the engine from being rotated in the clockwise direction.

6. Install special tool 303-507.
7. Crankshaft pulley bolt and washer.

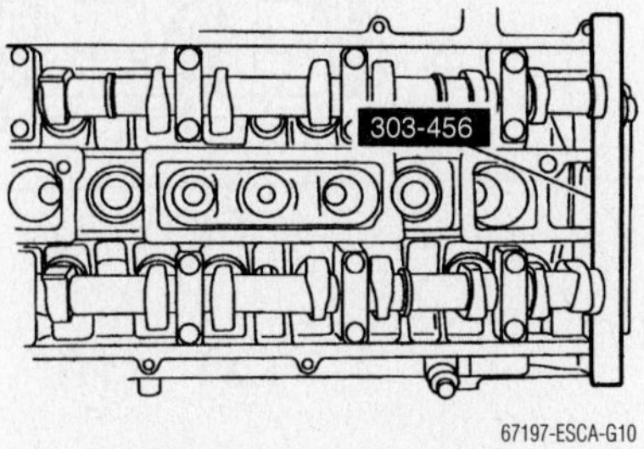

67197-ESCA-G10

Camshaft holding tool—2.3L engine

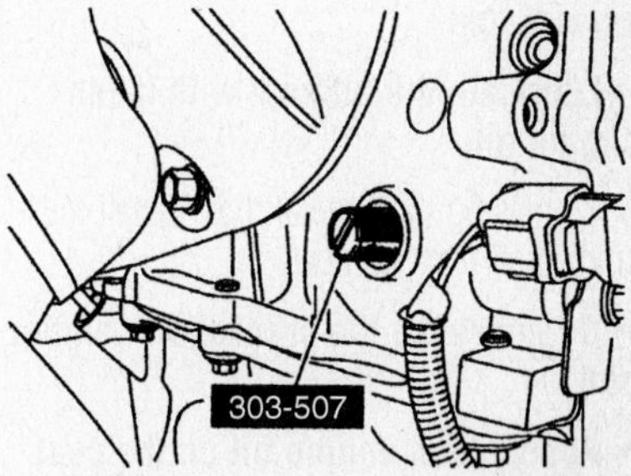

67197-ESCA-G11

Install special tool 303-507—2.3L engine

8. Remove the engine plug bolt.
9. Install the crankshaft holding tools.

WARNING

Failure to hold the crankshaft pulley in place while loosening the bolt can result in damage to the engine.

10. Remove the crankshaft pulley. Remove the crankshaft pulley bolt and washer.
11. Remove the crankshaft pulley.

WARNING

Use care not to damage the engine front cover or the crankshaft when removing the seal.

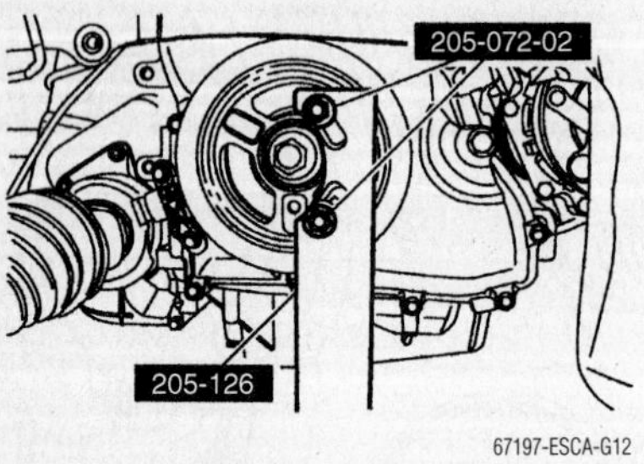

Install the crankshaft holding tools —2.3L engine

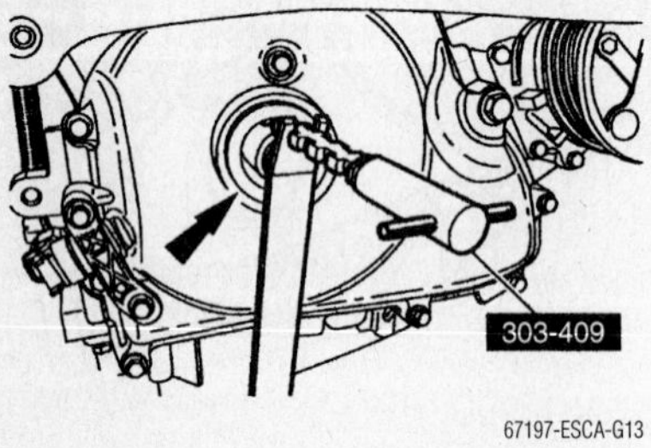

Using the special tool, install the crankshaft front oil seal —2.3L engine

12. Using the special tool, remove the crankshaft front oil seal.

➡**Remove the through-bolt from the special tool.**

➡**Lubricate the oil seal with clean engine oil.**

13. Using the special tool, install the crankshaft front oil seal.

➡**Do not reuse the crankshaft damper bolt.**

➡**Apply clean engine oil on the seal area before installing.**

14. Install the crankshaft pulley and hand-tighten the bolt.

⁂ WARNING

Only hand-tighten the bolt or damage to the front cover can occur.

➡**This step will correctly align the crankshaft pulley to the crankshaft.**

15. Install a standard 6 mm x 18 mm bolt through the crankshaft pulley and thread it into the front cover. Rotate the pulley as necessary to align the bolt holes.

⁂ WARNING

Failure to hold the crankshaft pulley in place while tightening the bolt can cause damage to the engine front cover.

16. Using the special tools to hold the crankshaft pulley in place, tighten the crankshaft pulley bolt in two stages:
 a. Stage 1: Tighten to 100 Nm (74 ft. lbs.).
 b. Stage 2: Tighten an additional 90 degrees (¼ turn).
17. Remove the 6 mm x 18 mm bolt.
18. Remove the special tools.

➡**Only turn the engine in the normal direction of rotation.**

19. Turn the engine two complete revolutions.
20. Turn the crankshaft until the No. 1 piston is at TDC.
21. Install special tool 303-507.

⁂ WARNING

Only hand-tighten the bolt or damage to the front cover can occur.

22. Using the 6 mm x 18 mm bolt, check the position of the crankshaft pulley. If it is not possible to install the bolt, correct the engine timing.
23. Using special tool 303-465, check the position of the camshafts. If it is not possible to install the special tool, correct the engine timing.
24. Remove the 6 mm x 18 mm bolt.
25. Install the engine plug bolt.

3.0L Engine

1. Before servicing the vehicle, refer to the precautions in the beginning of this section.
2. Remove or disconnect the following:
 - Negative battery cable
 - Crankshaft pulley
 - Front oil seal

To install:

3. Install or connect the following:
 - New front crankshaft oil seal
 - Crankshaft pulley
 - Negative battery cable
4. Start the vehicle and check for leaks, repair if necessary.

Camshaft and Lifters

REMOVAL & INSTALLATION

2.0L Engine

1. Before servicing the vehicle, refer to the precautions in the beginning of this section.
2. Remove or disconnect the following:
 - Negative battery cable
 - Camshaft timing sprocket and verify the valve clearance
 - Camshaft journal cap bolts by loosening them in several passes in the proper sequence
 - Camshafts
3. Inspect the camshaft for wear and discard the oil seals

To install:

4. Install or connect the following:
 - Camshaft cam followers, lubricate the bearing journals thoroughly. Torque the caps to 14 ft. lbs. (19 Nm).

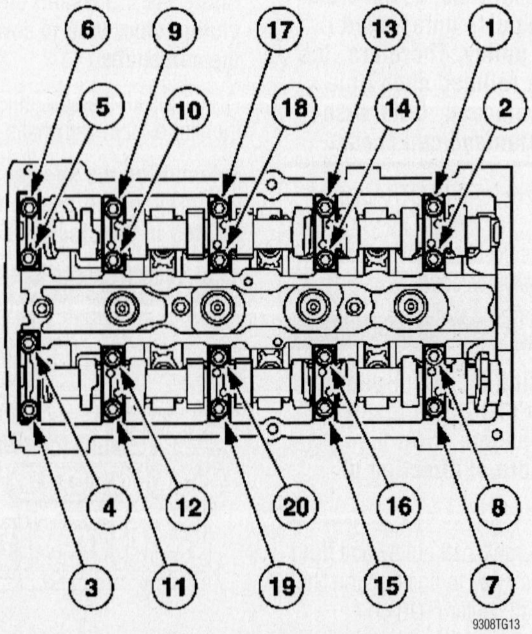

Remove the camshaft bearing caps in sequence–2.0L engine

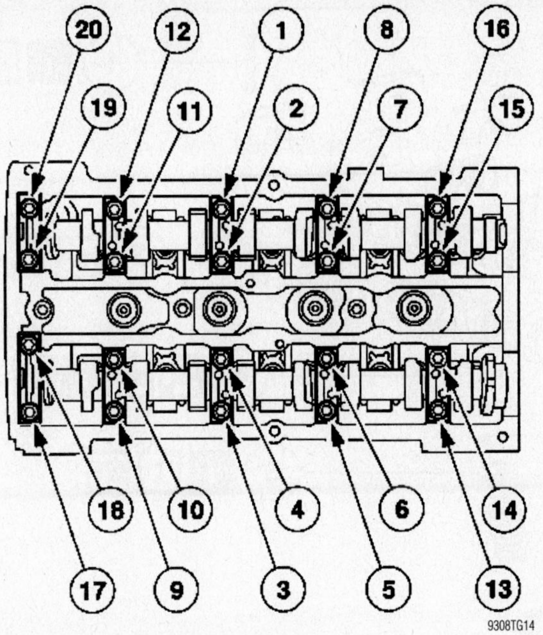

9308TG14

Camshaft bearing cap tightening sequence–2.0L engine

✻✻ WARNING

Turn the engine clockwise only, and only use the crankshaft bolt.

➡Before removing the camshafts, measure the clearance of each valve at base circle, with the lobe pointed away from the tappet. Failure to measure all clearances prior to removing the camshafts will necessitate repeated removal and installation and wasted labor time.

2. Use a feeler gauge to measure the clearance of each valve and record its location.

➡The number on the valve tappet only reflects the digits that follow the decimal. For example, a tappet with the number 0.650 has the thickness of 3.650 mm.

- Exhaust camshaft oil seal
- Camshaft timing sprocket
- Negative battery cable

2.3L Engine

✻✻ WARNING

During engine repair procedures, cleanliness is extremely important. Any foreign material, including any material created while cleaning gasket surfaces, that enters the oil passages, coolant passages or the oil pan can cause engine failure.

✻✻ WARNING

The crankshaft, the crankshaft sprocket and the pulley are fitted together by friction, using diamond washers between the flange faces on each part. For that reason, the crankshaft sprocket is also unfastened if you loosen the pulley. Therefore, the engine must be retimed each time the damper is removed. Otherwise severe engine damage can occur.

1. With the vehicle in NEUTRAL, position it on a hoist.

➡Valve tappets are select fit and the valve clearance must be checked before removing the tappets.

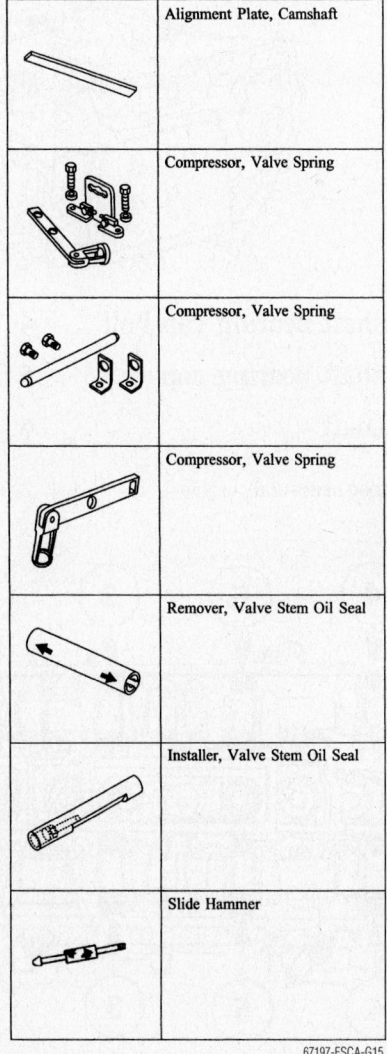

	Alignment Plate, Camshaft
	Compressor, Valve Spring
	Compressor, Valve Spring
	Compressor, Valve Spring
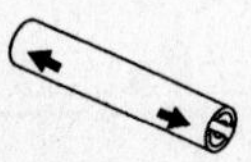	Remover, Valve Stem Oil Seal
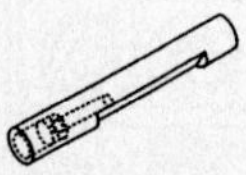	Installer, Valve Stem Oil Seal
	Slide Hammer

67197-ESCA-G15

Tools necessary for camshaft and lifter removal—2.3L engine

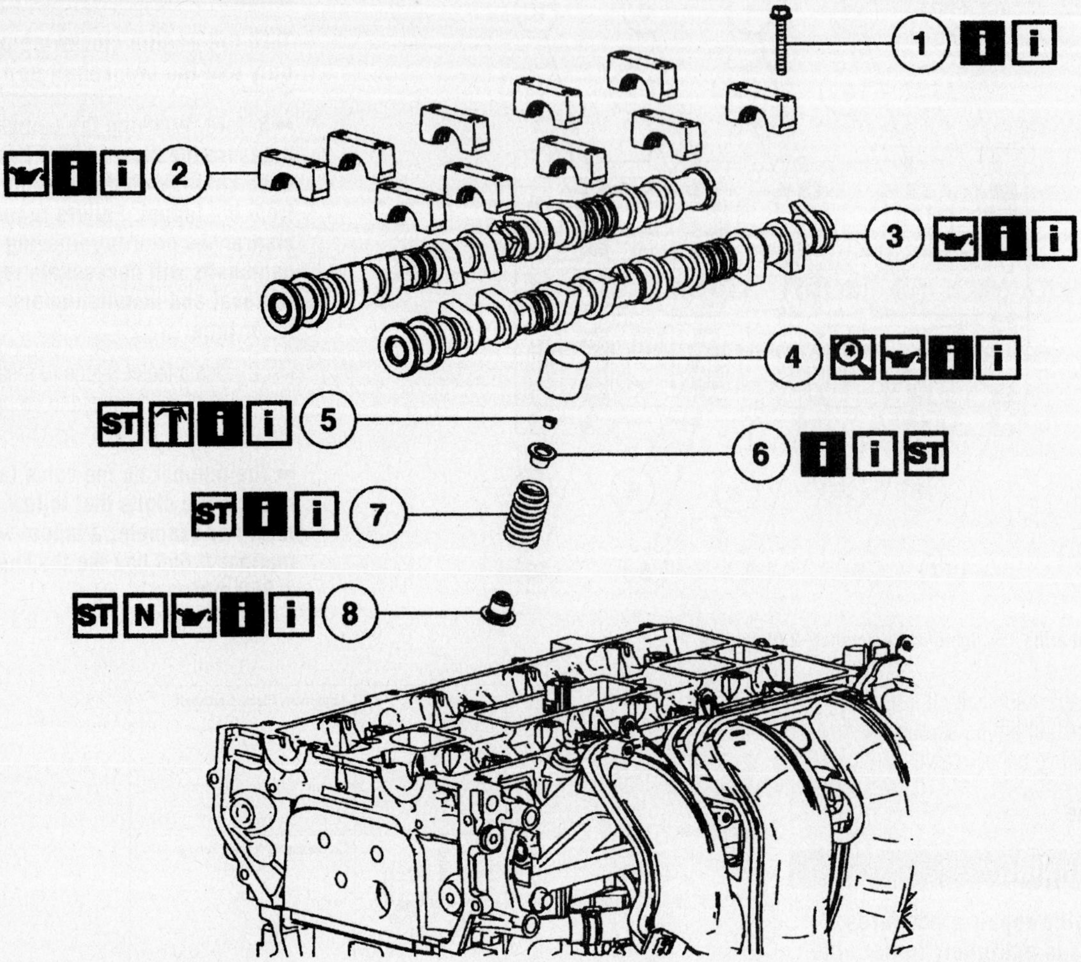

1 Camshaft bearing cap bolt
2 Camshaft bearing cap
3 Camshaft
4 Valve tappet
5 Valve collet
6 Valve spring retainer
7 Valve spring
8 Valve seal

67197-ESCA-G16

Camshafts, lifters and related parts—2.3L engine

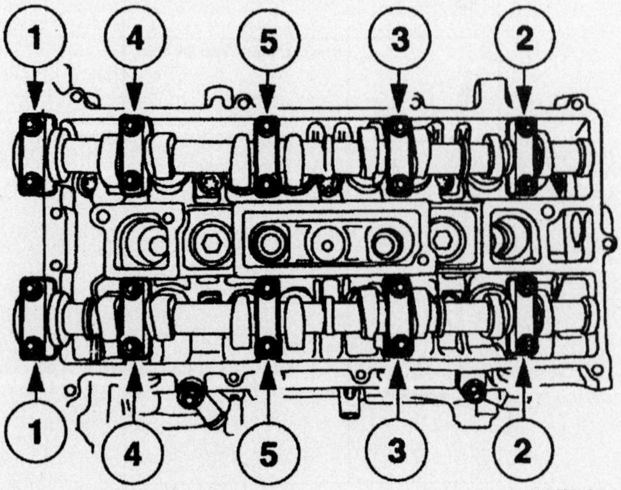

67197-ESCA-G17

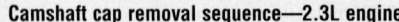

Camshaft cap removal sequence—2.3L engine

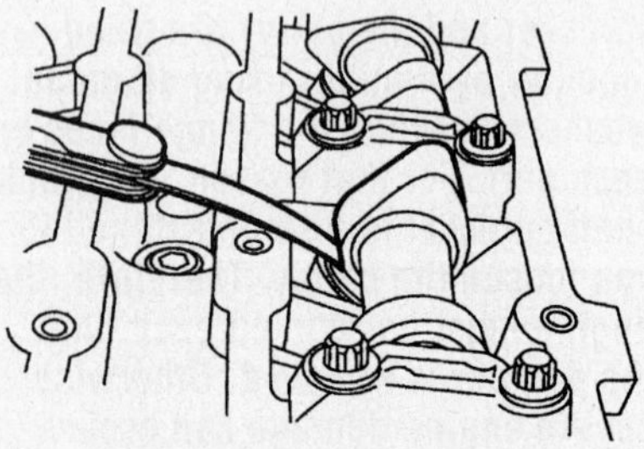

67197-ESCA-G14

Valve clearance check—2.3L engine

➡A midrange clearance is the most desirable:

- Intake: 0.22–0.28 mm (0.008–0.011 inch)
- Exhaust: 0.27–0.33 mm (0.010–0.013 inch)

3. Select tappets using this formula: tappet thickness = measured clearance + the base tappet thickness–most desirable thickness.

4. Select the tappets and mark the installation location.

5. If any tappets do not measure within specifications, install new tappets in these locations.

6. Remove the timing chain and sprockets.

7. Mark the position of the camshaft lobes on the No. 1 cylinder for assembly reference.

⁂ WARNING

Failure to follow the camshaft loosening procedure can result in damage to the camshafts.

8. Loosen the camshaft bearing bolts in the sequence shown, one turn at a time. Repeat until all the tension is released.

9. Remove the camshaft bearing caps.

⁂ WARNING

If the camshafts and valve tappets are to be reused, mark the location of the valve tappets to make sure they are assembled in their original positions.

➡The number on the valve tappets only reflects the digits that follow the decimal. For example, a tappet with the number 0.650 has the thickness of 3.650 mm.

10. Remove the camshafts.

11. Valve tappets.

12. To install, reverse the removal procedure. Coat the valve tappets with clean engine oil and insert them.

⁂ WARNING

Install the camshafts with the alignment slots in the camshafts lined up so the Camshaft Alignment Plate can be installed without rotating the camshafts. Make sure the lobes on the No. 1 cylinder are in the same position as noted in the removal procedure. Rotating the camshafts when the timing chain is removed, or installing the camshafts 180 degrees out of position can cause severe damage to the valves and pistons.

➡Lubricate the camshaft journals and bearing caps with clean engine oil.

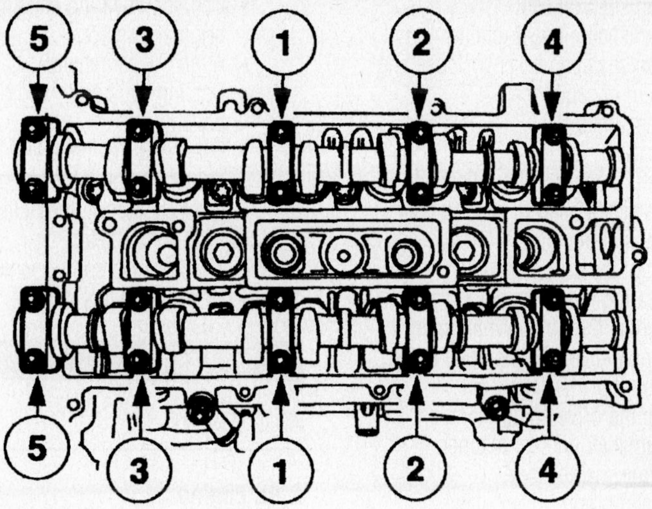

67197-ESCA-G18

Camshaft cap torque sequence—2.3L engine

13. Install the camshafts and bearing caps. Tighten the bolts in the sequence shown in three stages.

a. Stage 1: Tighten the camshaft bearing bolt caps one turn at a time until tight.

b. Stage 2: Tighten the bolts to 7 Nm (62 inch lbs.).

c. Stage 3: Tighten the bolts to 16 Nm (12 ft. lbs.).

3.0L Engine

LEFT SIDE

1. Before servicing the vehicle, refer to the precautions in the beginning of this section.

2. Remove or disconnect the following:

- Negative battery cable
- Water pump belt
- Timing drive components

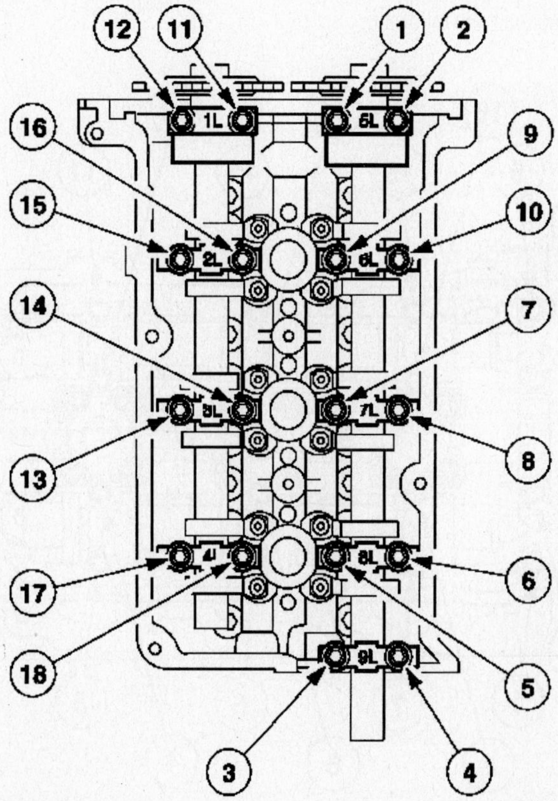

9308TG15

Remove and install the left side camshaft bearing caps in sequence–3.0L engine

- Camshaft oil seal
- Camshaft oil seal retainer
- Camshaft cap bolts by loosening them in sequence
- Camshafts

To install:

3. Install or connect the following:
   - Camshaft bearing caps in their original position
   - Align the camshafts
   - Bearing thrust caps and hand tighten the bolts. When aligned properly, torque the bolts to 89 inch lbs. (10 Nm).
   - Timing drive components
   - Camshaft oil seal retainer
   - Crankshaft oil seal
   - Water pump drive pulley
   - Water pump belt
   - Negative battery cable

RIGHT SIDE

1. Before servicing the vehicle, refer to the precautions in the beginning of this section.
2. Remove or disconnect the following:
   - Negative battery cable
   - Timing drive components
   - Camshaft cap bolts by loosening them in sequence
   - Camshafts caps
   - Camshafts

To install:

3. Install or connect the following:
   - Camshaft bearing caps in their original position
   - Align the camshafts
   - Bearing caps and hand tighten the bolts
   - Bearing thrust caps and hand tighten the bolts. When aligned properly, torque the bolts to 89 inch lbs. (10 Nm).
   - Timing drive components
   - Negative battery cable

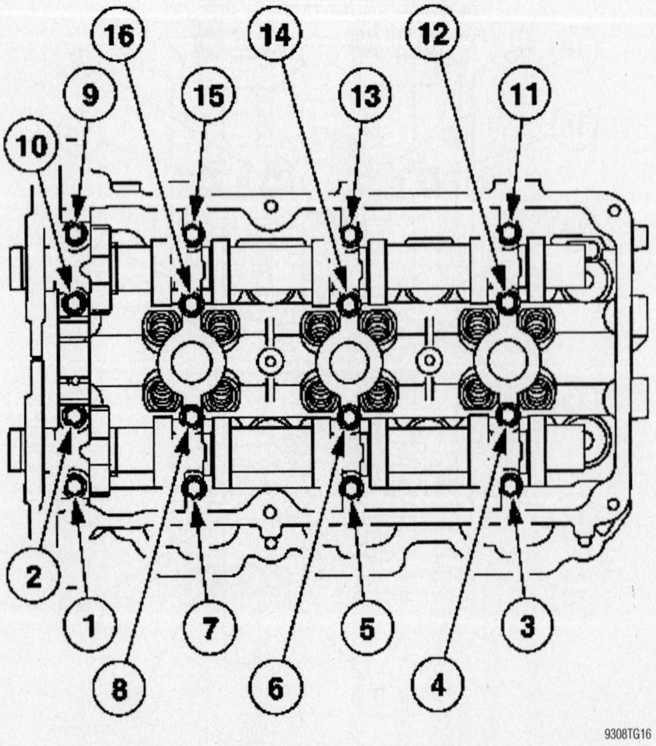

Remove and install the right side camshaft bearing caps in sequence–3.0L engine

Valve Lash

ADJUSTMENT

2.0L Engine

1. Before servicing the vehicle, refer to the precautions in the beginning of this section.
2. Remove or disconnect the following:
   - Negative battery cable
   - Timing belt
3. Measure each valve's clearance at the base circle with the lobe facing away from the tappet.
4. Use a feeler gauge to measure and record each valve's clearance
5. Remove or disconnect the following:
   - Camshafts
   - Valve tappets from the cylinder head
6. A mid range clearance is recommended as follows:
   a. Intake: 0.006 inch (0.15mm).
   b. Exhaust: 0.012 inch (0.3mm).

To install:

7. Install or connect the following:
   - Valve tappets after lubricating them with clean engine oil
   - Camshafts and verify each valve's clearance at the base circle with the lobe facing away from the tappet
   - Timing belt
   - Negative battery cable

2.3L Engine

➡Before removing the camshafts, measure the clearance of each valve at base circle, with the lobe pointed away from the tappet. Failure to measure all clearances prior to removing the camshafts will necessitate repeated removal and installation and wasted labor time.

1. Use a feeler gauge to measure the clearance of each valve and record its location.

➡The number on the valve tappet only reflects the digits that follow the decimal. For example, a tappet with the number 0.650 has the thickness of 3.650 mm.

➡A midrange clearance is the most desirable:

- Intake: 0.22–0.28 mm (0.008–0.011 inch)
- Exhaust: 0.27–0.33 mm (0.010–0.013 inch)

2. Select tappets using this formula: tappet thickness = measured clearance + the base tappet thickness–most desirable thickness.
3. Select the tappets and mark the installation location.
4. If any tappets do not measure within specifications, install new tappets in these locations.

3.0L Engine

1. Before servicing the vehicle, refer to the precautions in the beginning of this section.
2. Remove or disconnect the following:
   - Negative battery cable
   - Camshaft followers
   - Hydraulic lash adjusters

➡Mark the position of the hydraulic lash adjusters to assure they are assembled in their original position

3. Inspect the adjusters for scoring or uneven wear in the bore and replace them as required.

To install:

4. Install or connect the following:
   - Hydraulic lash adjusters after lubricating them with clean engine oil
   - Camshaft followers
   - Negative battery cable

Starter Motor

REMOVAL & INSTALLATION

2.0L Engine

1. Before servicing the vehicle, refer to the precautions in the beginning of this section.
2. Remove or disconnect the following:
   - Negative battery cable
   - Starter bolts
   - Exhaust system, AWD vehicles only
   - Halfshaft support bracket bolts
   - Starter electrical connectors
   - Starter

To install:

3. Install or connect the following:

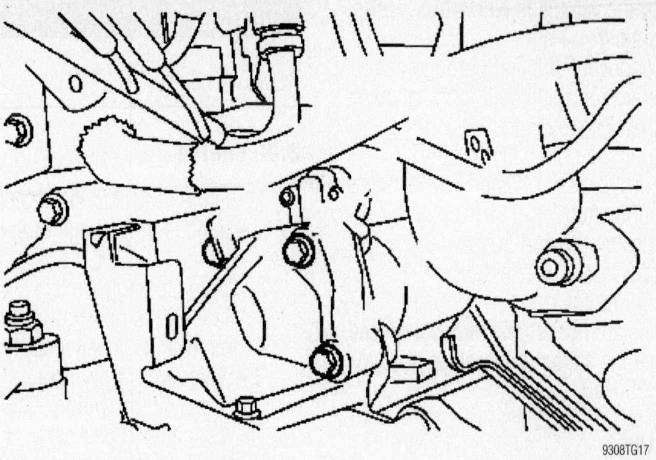

Removal of the starter motor–2.0L engine

   - Starter. Torque bolts to 20 ft. lbs. (27 Nm).
   - Starter electrical connectors
   - Halfshaft support bracket. Torque the bolts to 11 ft. lbs. (15 Nm).
   - Exhaust system on AWD vehicles. Torque the bolts to 18 ft. lbs. (25 Nm).
   - Negative battery cable

2.3L Engine

⁂ WARNING

When performing maintenance on the starting system, be aware that heavy gauge leads are connected directly to the battery. Make sure protective caps are in place when maintenance is completed.

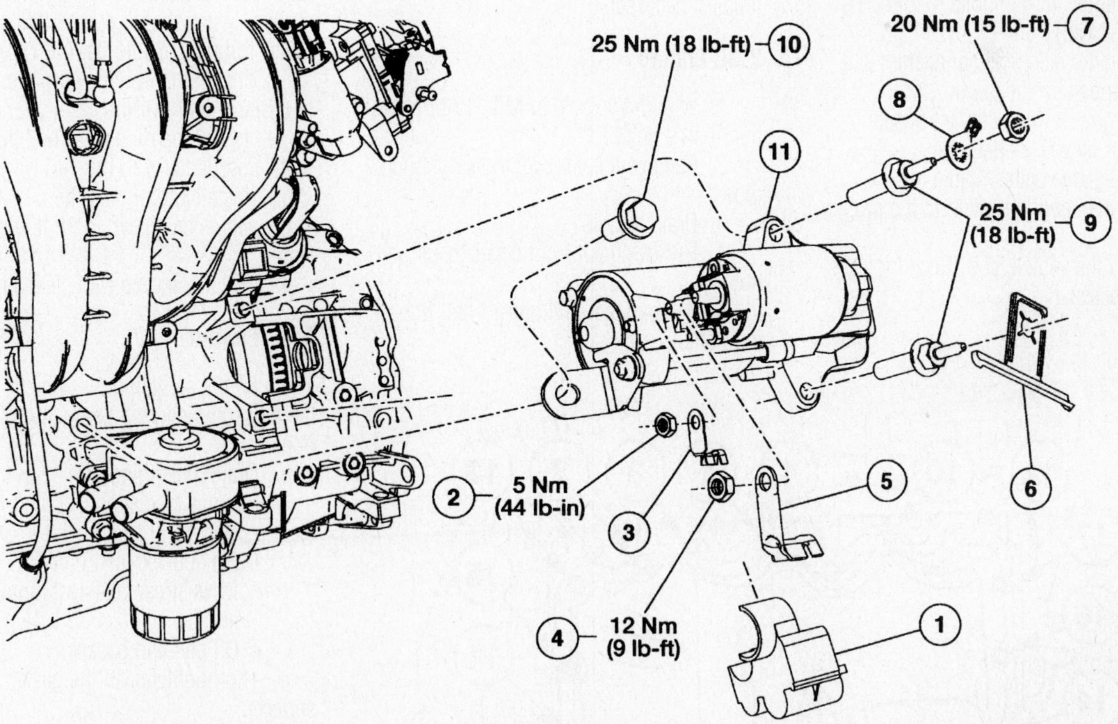

1 Starter motor solenoid terminal cover
2 Starter solenoid wire nut
3 Starter solenoid wire
4 Starter solenoid battery cable nut
5 Starter solenoid battery cable
6 Wiring harness retainer
7 Ground strap nut
8 Ground strap
9 Starter motor stud bolts
10 Starter motor bracket bolt
11 Starter motor

67197-ESCA-G19

Starter mounting—2.3L engine

1. With the vehicle in NEUTRAL, position it on a hoist
2. Disconnect the battery ground cable.
3. Starter motor solenoid terminal cover
4. Starter solenoid wire
5. Starter solenoid battery cable
6. Wiring harness retainer
7. Ground strap
8. Starter motor stud bolts
9. Starter motor bracket bolt
10. Starter motor
11. To install, reverse the removal procedure. Torque the starter and bracket bolts to 18 ft. lbs. (25 Nm).

3.0L Engine

1. Before servicing the vehicle, refer to the precautions in the beginning of this section.
2. Drain the cooling system.
3. Remove or disconnect the following:
   - Negative battery cable
   - Air cleaner outlet tube
   - Coolant hoses and move the thermostat aside
   - Starter electrical connectors
   - Starter

To install:

4. Install or connect the following:
   - Starter. Torque bolts to 20 ft. lbs. (27 Nm).
   - Starter electrical connectors and reposition the thermostat
   - Connect the 4 coolant hoses
   - Air cleaner outlet tube
   - Negative battery cable
5. Fill the cooling system to the proper level.
6. Start the vehicle and check for leaks, repair if necessary.

Oil Pan

REMOVAL & INSTALLATION

2.0L Engine

1. Before servicing the vehicle, refer to the precautions in the beginning of this section.
2. Drain the engine oil.
3. Support the powertrain assembly.
4. Remove or disconnect the following:
   - Negative battery cable
   - Catalytic converter
   - Oil pan and gasket
5. Thoroughly clean the gasket mating surfaces.

To install:

6. Apply silicone sealer to the oil pan.
7. Install a new gasket on the oil pan.
8. Oil pan. Torque the bolts in sequence to:
   a. Step 1: 53 inch lbs. (6 Nm).
   b. Step 2: 106 in lbs. (12 Nm).
9. Install or connect the following:
   - Catalytic converter
   - Negative battery cable
10. Fill the engine with clean oil.
11. Start the engine and check for leaks, repair if necessary.

2.3L Engine

1. With the vehicle in NEUTRAL, position it on a hoist.
2. Remove the oil level indicator and tube.
3. Drain the oil.
4. Remove the 4 front cover-to-oil pan bolts.
5. Remove the 4 oil pan-to-bell housing bolts.
6. Remove the 13 oil pan-to-block bolts.

To install:

⁂ WARNING

CAUTION: Do not use metal scrapers, wire brushes, power abrasive discs or other abrasive means to clean the sealing surfaces. These tools cause scratches and gouges, which make leak paths. Use a plastic scraping tool to remove traces of sealant.

Clean and inspect all mating surfaces.

➡If the oil pan is not secured within four minutes of sealant application the sealant must be removed and the sealing area cleaned with metal surface cleaner. Allow to dry until there is no sign of wetness, or four minutes, whichever is longer. Failure to follow this procedure can cause future oil leakage.

➡The oil pan must be installed and the bolts tightened within four minutes of applying the silicone gasket and sealant.

7. Apply a 2.5 mm bead of silicone gasket and sealant to the oil pan. Install the oil pan. Install the oil pan-to-bell housing bolts. Torque to 35 ft. lbs. (48 Nm).
8. Install the front cover bolts. Torque to 89 inch lbs. (10 Nm).
9. Install the oil pan-to-bell housing bolts. Torque to 35 ft. lbs. (48 Nm).
10. Install and tighten the oil pan bolts in the sequence shown to 18 ft. lbs. (25 Nm).
11. Fill the engine with clean engine oil.

3.0L Engine

1. Before servicing the vehicle, refer to the precautions in the beginning of this section.
2. Drain the engine oil.
3. Remove or disconnect the following:
   - Negative battery cable
   - Flexible exhaust pipe
   - Downstream catalyst monitor sensor
   - Oil pan and gasket
4. Thoroughly clean the gasket mating surfaces.

To install:

5. Apply silicone sealer to the oil pan.
6. Install or connect the following:
   - New gasket on the oil pan
   - Oil pan. Torque the bolts in sequence to 18 ft. lbs. (25 Nm).
   - Flexible exhaust pipe

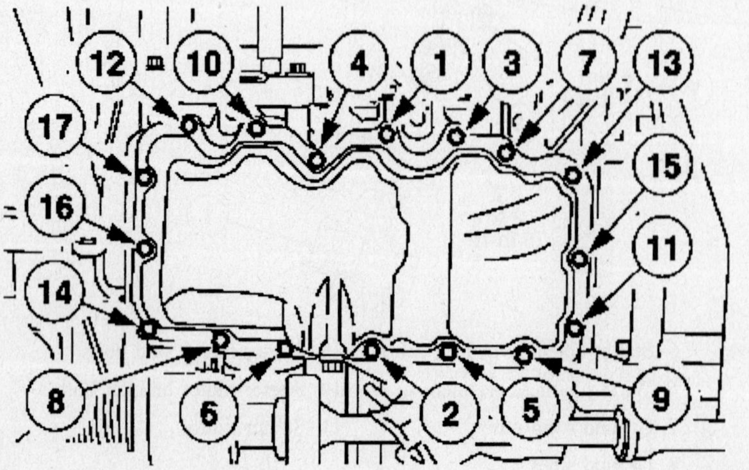

Tighten the oil pan bolts in sequence–2.0L engine

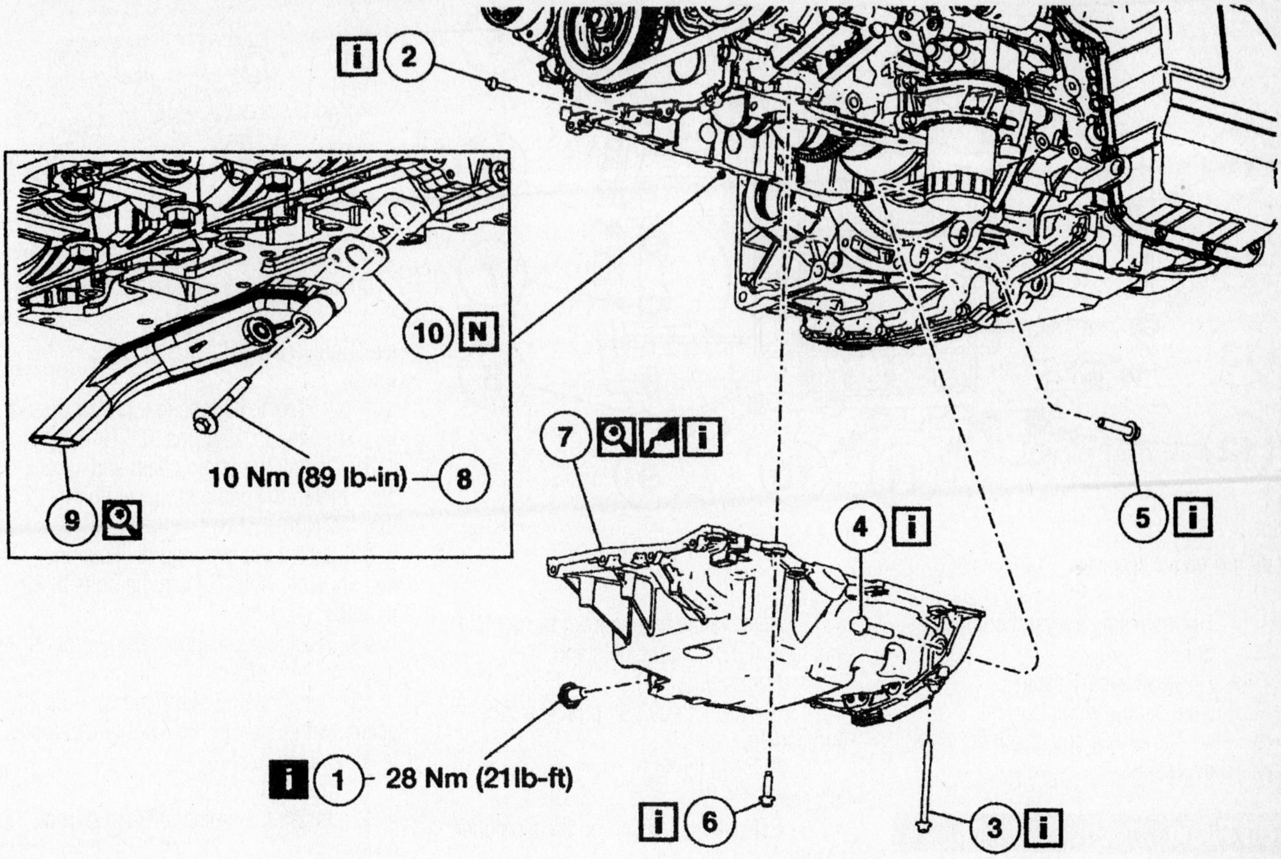

1 Drain plug
2 Engine front cover bolt
3 Oil pan bolt
4 Oil pan-to-bell housing bolt
5 Oil pan-to-bell housing bolt
6 Oil pan bolt
7 Oil pan
8 Oil pump screen and pickup tube bolt
9 Oil pump screen and pickup tube
10 Oil pump screen and pickup tube gasket

67197-ESCA-G20

Oil pan, pump and related parts—2.3L engine

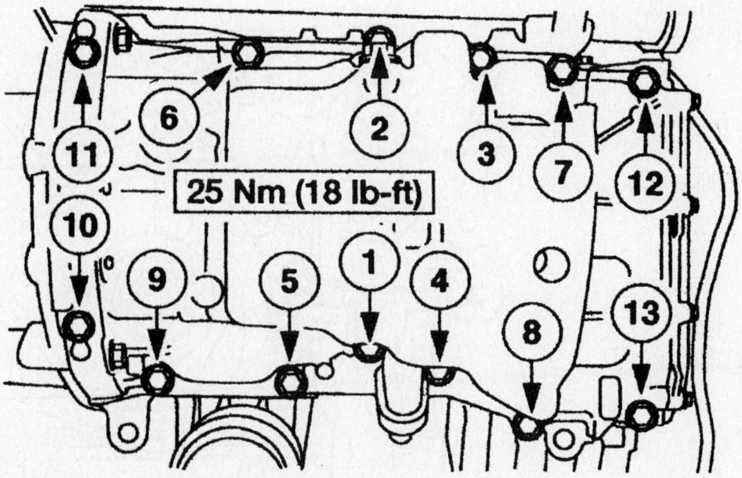

67197-ESCA-G21

Oil pan bolt torque sequence—2.3L engine

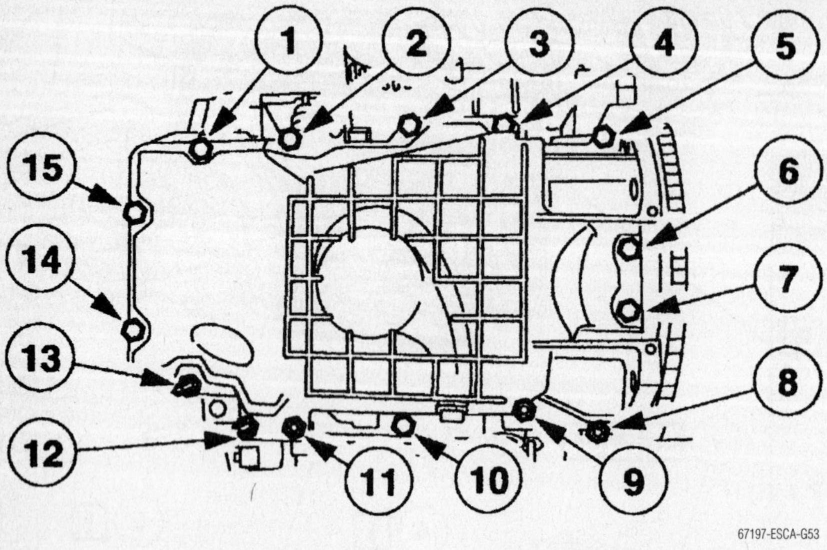

Oil pan torque sequence—3.0L engine

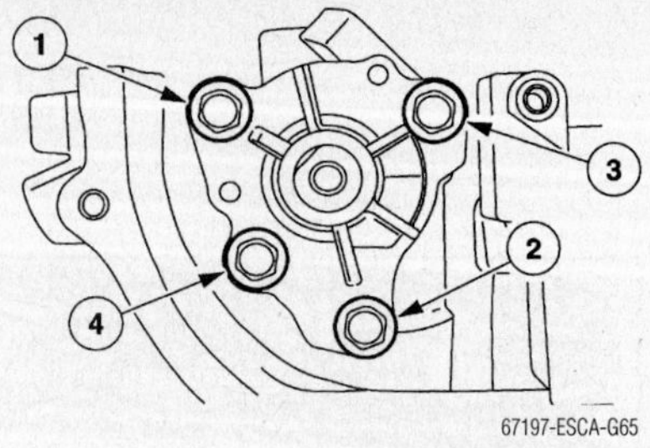

Oil pump torque sequence—2.3L engine

- Downstream catalyst monitor sensor
- Negative battery cable

7. Fill the engine with clean oil.
8. Start the vehicle and check for leaks, repair if necessary.

Oil Pump

REMOVAL & INSTALLATION

2.0L Engine

1. Before servicing the vehicle, refer to the precautions in the beginning of this section.
2. Drain the engine oil.
3. Remove or disconnect the following:
   - Negative battery cable
   - Oil pan
   - Oil pump screen cover and tube
   - Oil pump and discard the gasket
4. Thoroughly clean the gasket mating surfaces.

To install:

5. Install or connect the following:
   - Oil pump screen cover and tube with a new gasket. Torque the bolts to 89 inch lbs. (10 Nm).
   - Oil pump to the oil pan
   - Oil pan
   - Negative battery cable
6. Refill the engine with clean oil.
7. Start the engine and check for leaks; repair if necessary.

2.3L Engine

1. Before servicing the vehicle, refer to the precautions in the beginning of this section.
2. Remove the engine from the vehicle and mount it on an engine stand.
3. Remove the oil pan.
4. Remove the oil pump pickup tube and screen.
5. Remove the front cover and the timing chain.
6. Release the tension on the tensioner spring.
7. Remove the tensioner and the shoulder bolt.
8. Remove the guide.

➡**The oil pump chain sprocket must be held in place.**

9. Remove the oil pump chain and sprockets.
10. Remove the oil pump assembly and gasket.

To install:

11. Install the oil pump with a new gasket. Tighten the bolts in sequence as follows:
    a. Step 1: 89 inch lbs. (10 Nm).
    b. Step 2: 17 ft. lbs. (23 Nm).
12. Install the pump chain and sprockets. Tighten the pump sprocket bolt to 18 ft. lbs. (25 Nm).
13. Install the chain guide, tensioner, and shoulder bolt. Tighten the bolts to 89 inch lbs. (10 Nm).
14. Hook the tensioner spring around the shoulder bolt.
15. Install the oil pump pickup tube and screen with a new gasket. Tighten the bolts to 89 ft. lbs. (10 Nm).
16. Install the oil pan.
17. Install the timing chain and front cover.
18. Install the engine into the vehicle.

3.0L Engine

1. Before servicing the vehicle, refer to the precautions in the beginning of this section.
2. Drain the engine oil.
3. Remove or disconnect the following:
   - Negative battery cable
   - Timing drive components
   - Oil pump screen cover and tube
   - Damper bolt and crankshaft sprockets

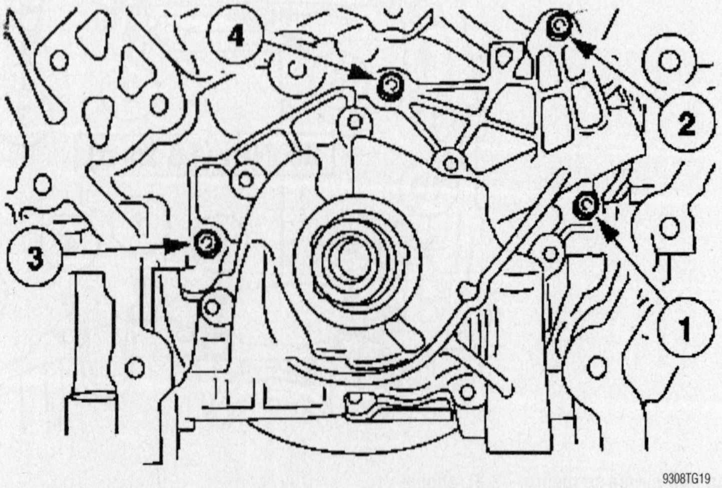

Remove the oil pump bolts in the proper sequence–3.0L engine

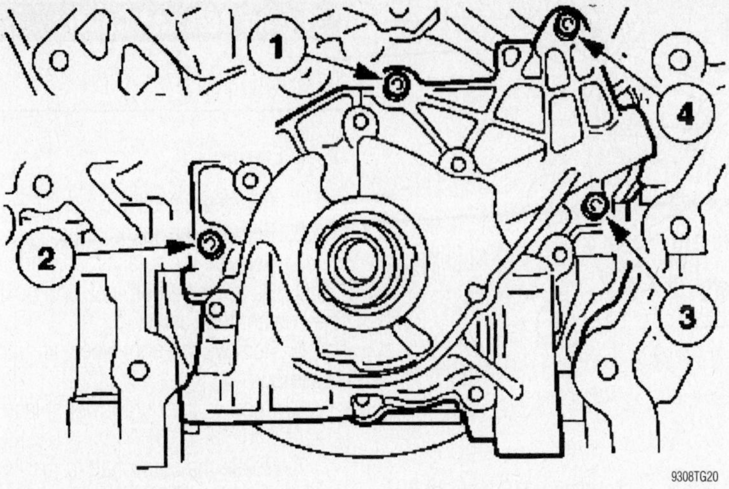

Install the oil pump bolts in the proper sequence–3.0L engine

- Oil pump bolts in the proper sequence

4. Thoroughly clean the gasket mating surfaces.

To install:

5. Install or connect the following:
   - Oil pump and bolts in the proper sequence. Torque the bolts to 89 inch lbs. (10 Nm).
   - Crankshaft sprockets
   - Oil pump screen cover and tube
   - Timing drive components
   - Negative battery cable
6. Refill the engine with clean oil.
7. Start the engine and check for leaks; repair if necessary.

Rear Main Seal

REMOVAL & INSTALLATION

2.0L Engine

1. Before servicing the vehicle, refer to the precautions in the beginning of this section.
2. Remove or disconnect the following:
   - Negative battery cable
   - Flywheel
   - Rear main seal

To install:

3. Coat the oil seal with clean engine oil.
4. Install or connect the following:

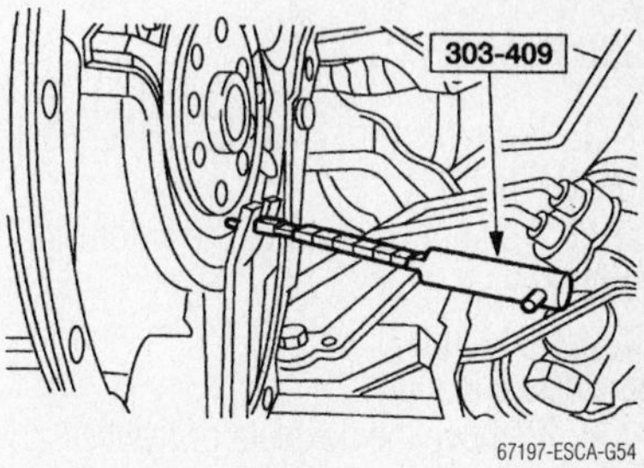

Rear main seal removal—2.0L engine

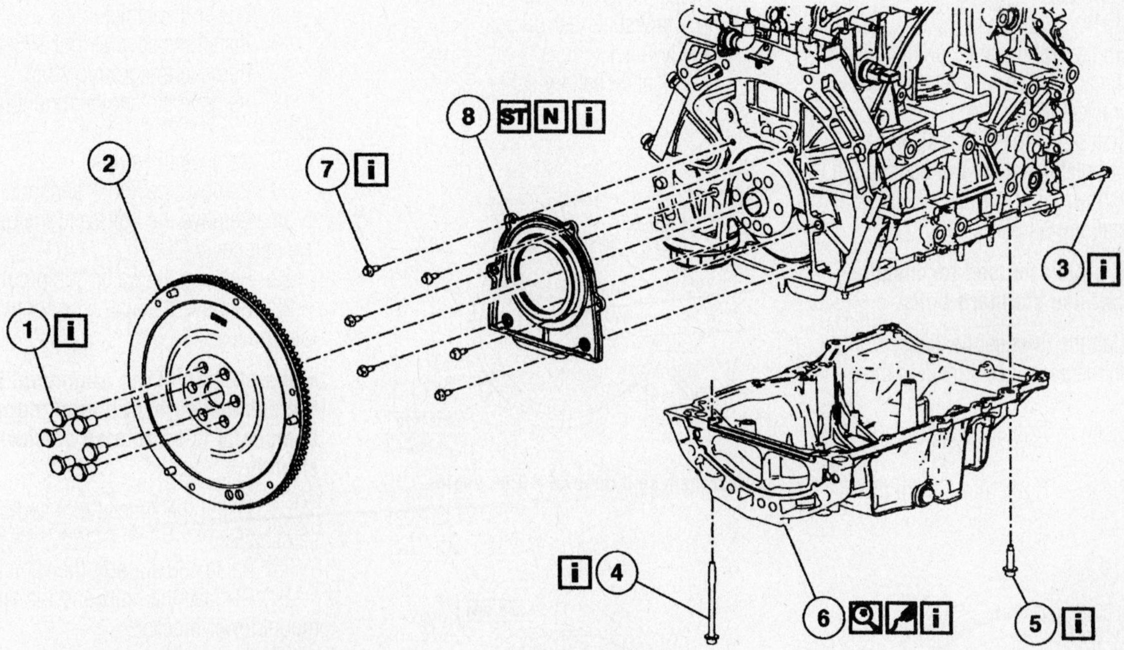

1 Flexplate or flywheel bolt
2 Flexplate or flywheel
3 Engine front cover bolt
4 Oil pan bolt
5 Oil pan bolt
6 Oil pan
7 Crankshaft rear oil seal with retainer plate bolt
8 Crankshaft rear oil seal with retainer plate

67197-ESCA-G22

Rear main seal and related parts—2.3L engine

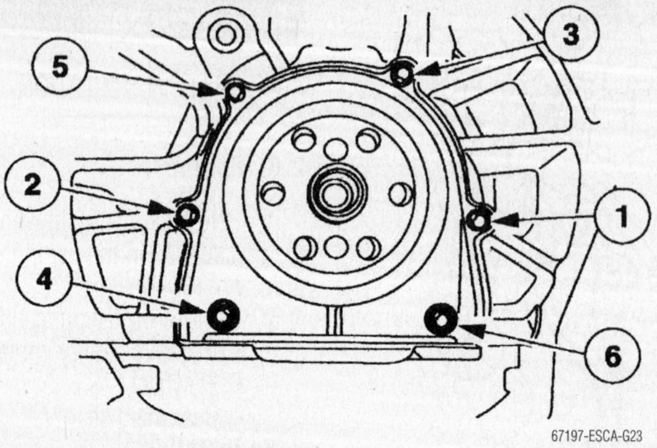

Retainer plate torque sequence—2.3L engine

- Crankshaft rear oil seal
- Flywheel
- Negative battery cable

2.3L Engine

1. With the vehicle in NEUTRAL, position it on a hoist.
2. If equipped, remove the automatic transaxle.
3. If equipped, remove the manual transaxle and clutch.
4. Remove the flexplate or flywheel.
5. Remove the oil pan.
6. Remove the crankshaft rear oil seal with retainer plate

To install:

7. Using a seal installer, position the crankshaft rear oil seal with retainer plate onto the crankshaft.
8. Install the crankshaft rear oil seal with retainer plate. Tighten the bolts in the sequence shown to 10 Nm (89 inch lbs.).
9. Install the oil pan.

➡Special bolts are used for installation. Do not use standard bolts.

10. Install the flywheel/flexplate. Tighten the bolts in the sequence shown in three stages.
    a. Stage 1: Tighten to 50 Nm (37 ft. lbs.).
    b. Stage 2: Tighten to 80 Nm (50 ft. lbs.).
    c. Stage 3: Tighten to 112 Nm (83 ft. lbs.).

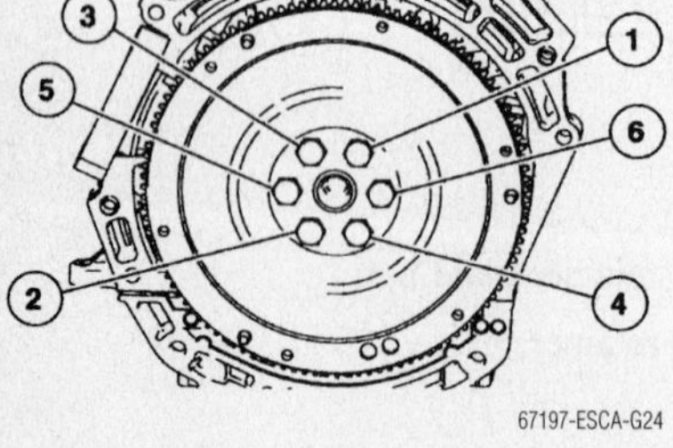

Flywheel torque sequence—2.3L engine

3.0L Engine

1. Before servicing the vehicle, refer to the precautions in the beginning of this section.
2. Remove or disconnect the following:
    - Negative battery cable
    - Flexplate
    - Rear main oil seal

To install:

3. Coat the oil seal with clean engine oil.
4. Install or connect the following:
    - Crankshaft rear oil seal
    - Flywheel
    - Negative battery cable

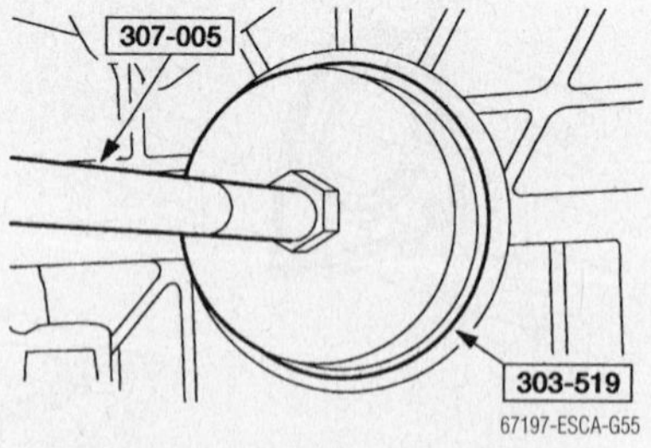

Rear main seal removal—3.0L engine

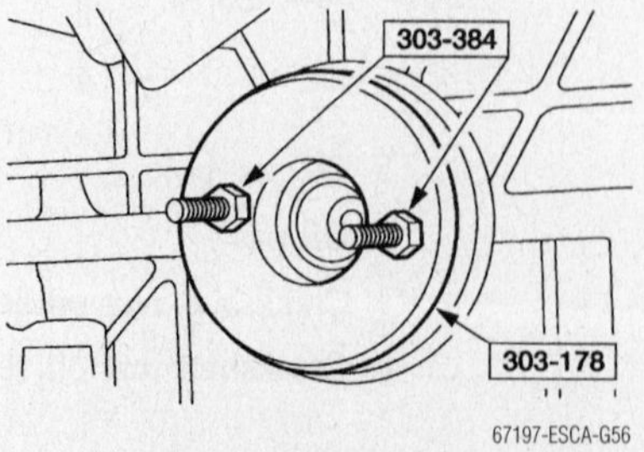

Rear main seal installation—3.0L engine

Timing Belt and Covers

REMOVAL & INSTALLATION

2.0L Engine

1. Remove the valve cover.
2. Remove the spark plugs.
3. Remove the catalytic converter.
4. Remove the bolt, nut, and position the coolant tube aside.
5. Remove the right wheel and tire assembly.
6. Remove the right lower splash shield.
7. Rotate the crankshaft to just before top dead center (TDC) (No. 1 cylinder).
8. Remove the stud.
9. Install the special tool.

➡Make sure the correct (second) notch in the pulley is indexed to the lower cylinder block.

10. Rotate the crankshaft clockwise against the peg to bring it to TDC (No. 1 cylinder).
11. Loosen the water pump pulley bolts.
12. Disconnect the battery ground cable.
13. Remove the crankshaft pulley.
14. Remove the bolts and the lower timing belt cover.
15. Lower the vehicle.
16. Install the special tool 303-F072.
17. Remove the ground strap.
18. Remove the engine mount upper bracket.
19. Remove the studs.
20. Remove the knock sensor connector.
21. Remove the bolts and the upper timing belt cover.
22. Remove the water pump pulley.
23. Remove the accessory drive belt idler pulley.

➡Installation of the alignment tool into the exhaust camshaft may require the camshafts to be rotated clockwise slightly.

24. Install the special tool and align the camshafts.
25. Raise and support the vehicle.
26. Remove the bolts and the engine mount lower bracket.
27. Loosen the timing belt tensioner pulley and allow to slide down to the bottom of its travel.

⁂ CAUTION

If the camshaft timing belt is to be reused, mark the direction of the camshaft timing belt to rotation of

camshaft prior to removal or premature wear or failure may occur.

28. Slide the timing belt off of the camshaft and crankshaft sprockets. the timing belt for wear. Install a new belt if necessary.

To install:

➡Make sure the correct (second) notch in the pulley is indexed to the lower cylinder block.

29. Slide the crankshaft pulley onto the crankshaft and confirm the crankshaft position is at TDC (No. 1 cylinder) by rotating it clockwise against the alignment peg.
30. Remove the crankshaft pulley.
31. Lower the vehicle.
32. Confirm that the timing belt tensioner is installed correctly with the tab positioned in the slot in the inner timing cover.
33. Install the timing belt onto the timing belt sprockets.
34. Adjust the timing belt tensioner.
35. Using a 6 mm Allen wrench, rotate the adjuster counterclockwise and align the marks as shown.
36. Tighten the tensioner pulley bolt.
37. Raise the vehicle.
38. Install the front engine mount lower bracket.
39. Install the accessory drive belt idler pulley.
40. Install the water pump pulley.
41. Hand tighten the bolts.
42. Lower the vehicle.
43. Install the timing belt covers. Torque the cover bolts to 62 inch lbs. (7Nm).
44. Tighten the water pump pulley bolts.
45. Remove the special tool.
46. Install the stud. Torque to 25 ft. lbs. (34Nm).
47. Install the coolant tube.
48. Install the catalytic converter.
49. Remove the special tool.
50. Install the valve cover.
51. Install the spark plugs.

Timing Chain, Gears, Front Cover and Seal

REMOVAL & INSTALLATION

2.3L Engine

⁂ CAUTION

During engine repair procedures, cleanliness is extremely important. Any foreign material, including any material created while cleaning gasket surfaces, that enters the oil passages, coolant passages or the oil pan can cause engine failure.

⁂ CAUTION

The crankshaft, the crankshaft sprocket and the pulley are fitted together by friction, using diamond washers between the flange faces on each part. For that reason, the crankshaft sprocket is also unfastened if you loosen the pulley. Therefore, the engine must be retimed each time the damper is removed. Otherwise severe engine damage can occur.

1. With the vehicle in NEUTRAL, position it on a hoist.
2. Remove the accessory drive belt and idler pulleys.
3. Remove the engine mount.
4. Remove the valve cover.

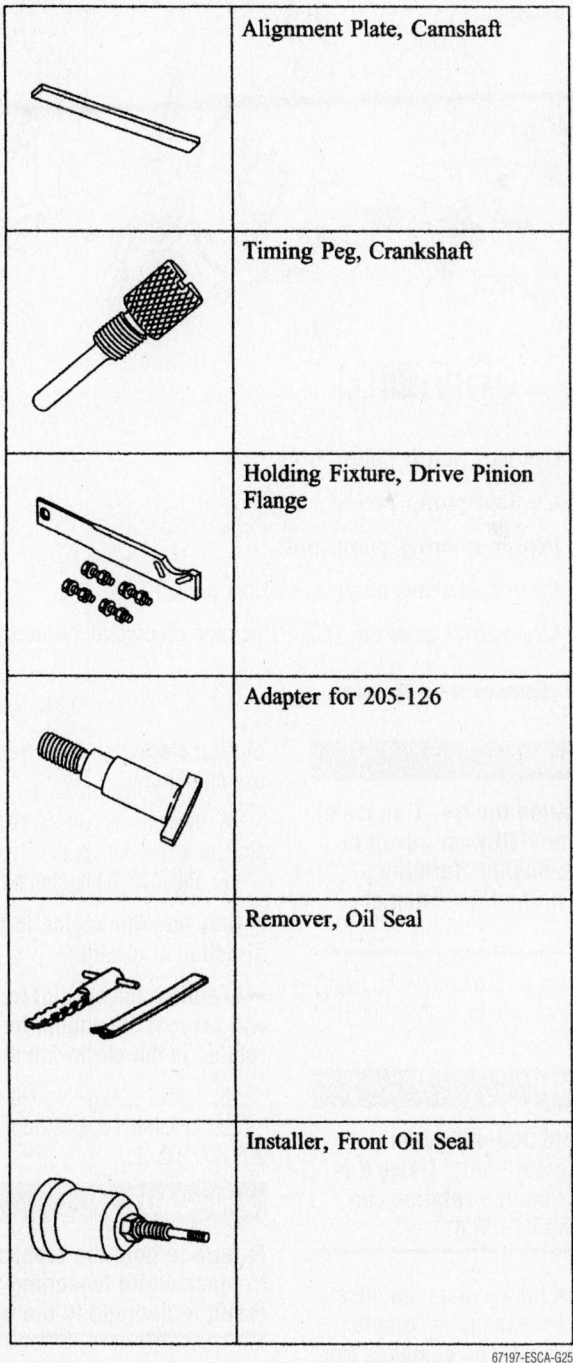

	Alignment Plate, Camshaft
	Timing Peg, Crankshaft
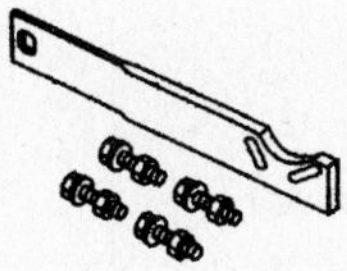	Holding Fixture, Drive Pinion Flange
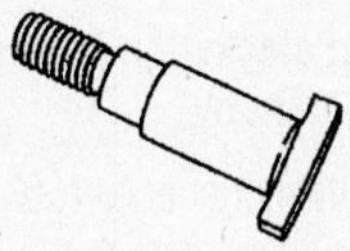	Adapter for 205-126
	Remover, Oil Seal
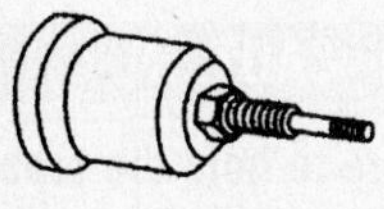	Installer, Front Oil Seal

67197-ESCA-G25

Tools needed for timing chain and gears replacement—2.3L engine

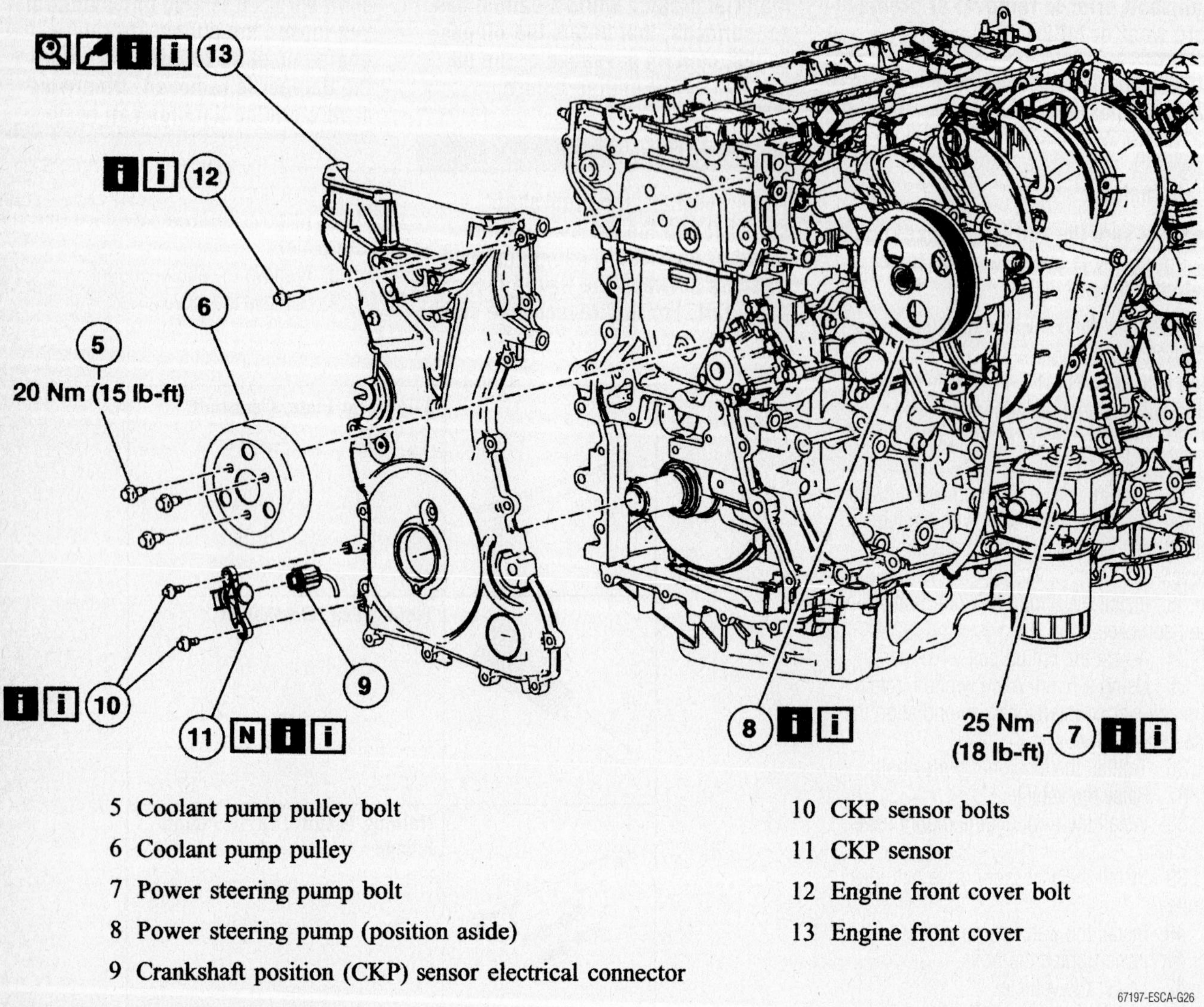

Front cover and related parts—2.3L engine

CAUTION

Failure to position the No. 1 piston at top dead center (TDC) can result in damage to the engine. Turn the engine in the normal direction of rotation only.

5. Using the crankshaft pulley bolt, turn the crankshaft clockwise to position the No. 1 piston at TDC.

CAUTION

The special tool 303-465 is for camshaft alignment only. Using this tool to prevent engine rotation can result in engine damage.

➡The camshaft timing slots are offset. If the special tool cannot be installed, rotate the crankshaft one complete revolution clockwise to correctly position the camshafts.

6. Install special tool 303-465 in the slots on the rear of both camshafts.
7. Remove the engine plug bolt.

➡Only turn the engine in the normal direction of rotation.

➡Installing the special tool in this step will prevent the engine from being rotated in the clockwise direction.

8. Install special tool 303-507.
9. Install the special tools 205-126 and 205-072-02.

CAUTION

Failure to hold the crankshaft pulley in place while loosening the bolt can result in damage to the engine.

10. Remove the crankshaft pulley bolt and washer.
11. Remove the crankshaft pulley.
12. Remove the crankshaft front seal.
13. Remove the coolant pump pulley.
14. Remove the power steering pump and position it aside.

➡The bolt under the power steering pressure tube will remain with the power steering pump.

15. Remove the CKP sensor.

➡Whenever the crankshaft position (CKP) sensor is removed, a new one must be installed, using the alignment jig supplied with the new part.

16. Remove the engine front cover bolts (there are 22).
17. Remove the engine front cover.
18. Remove the timing chain tensioner. Compress the timing chain tensioner, and

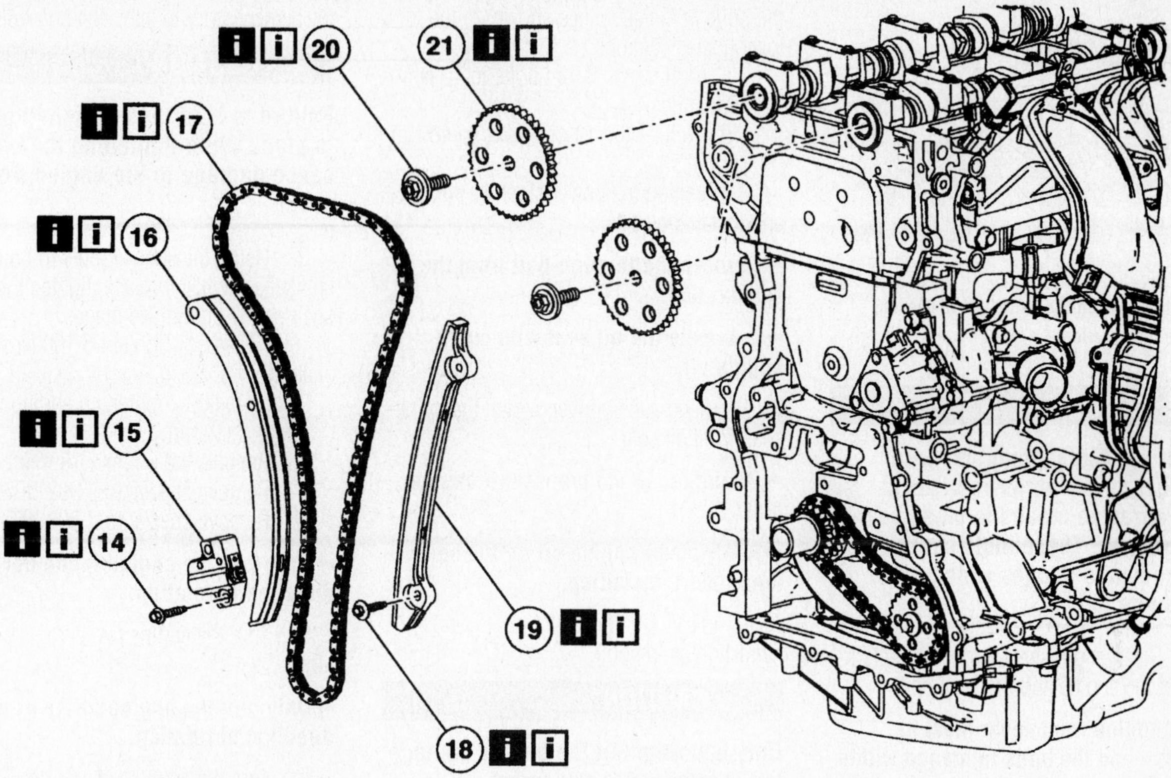

14 Timing chain tensioner bolt
15 Timing chain tensioner
16 RH timing chain guide
17 Timing chain
18 LH timing chain guide bolt
19 LH timing chain guide
20 Camshaft sprocket bolt
21 Camshaft sprocket

67197-ESCA-G27

Timing chain and related parts—2.3L engine

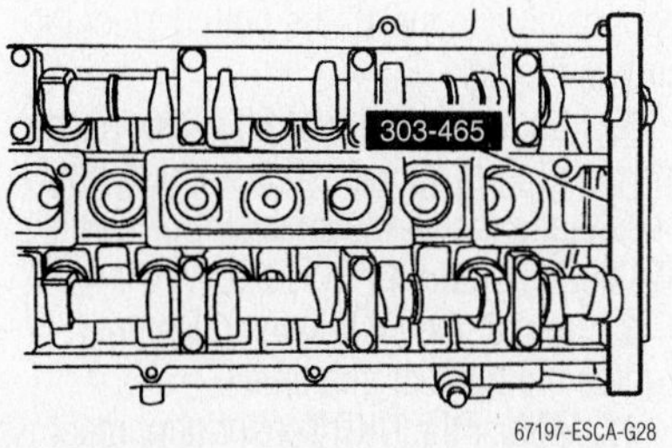

Install special tool 303-465 in the slots on the rear of both camshafts

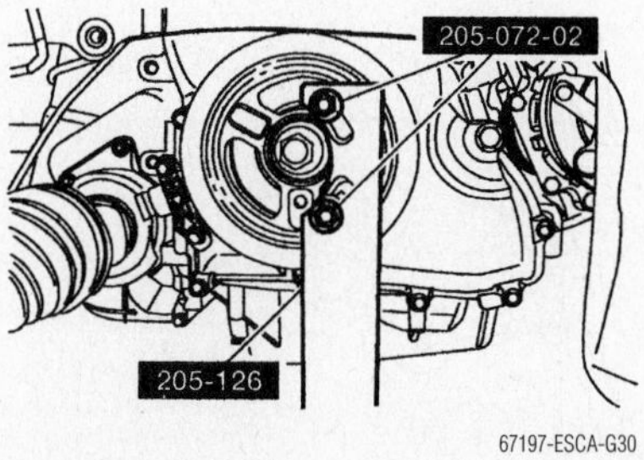

Install the special tools 205-126 and 205-072-02

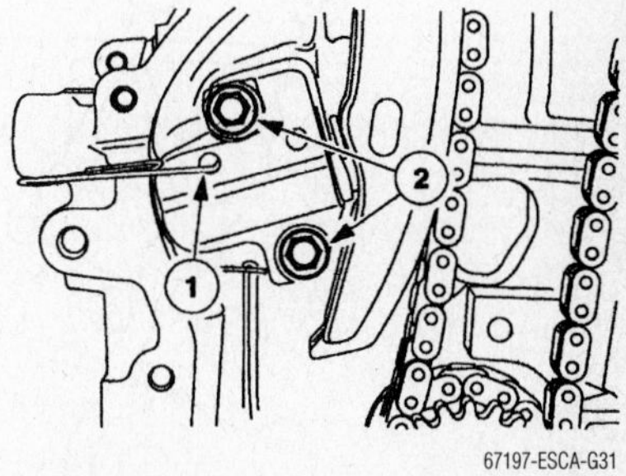

Compress the timing chain tensioner, and insert a paper clip into the hole to retain the tensioner

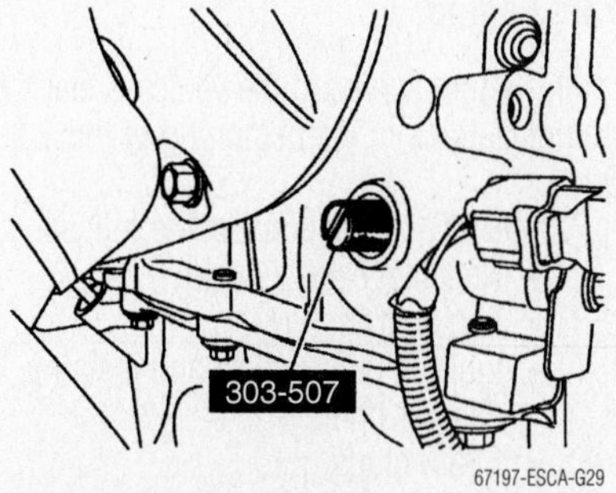

Install special tool 303-507

insert a paper clip into the hole to retain the tensioner.

19. Remove the RH timing chain guide.
20. Remove the timing chain.
21. Remove the LH timing chain guide.
22. Remove the camshaft sprocket bolts.
23. Remove the camshaft sprockets.

⁂ CAUTION

Do not rely on the Camshaft Alignment Plate to prevent camshaft rotation. Damage to the tool or the camshaft can occur. Use the flats on the camshaft to prevent camshaft rotation.

To install:

24. Installation is the reverse of removal. Note the following:

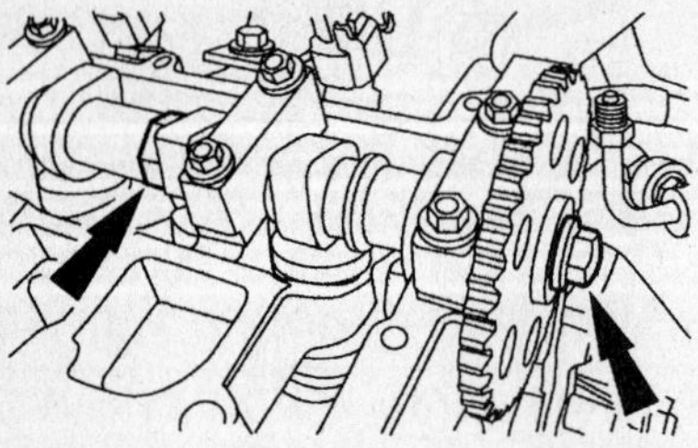

67197-ESCA-G32

Use the flats on the camshaft to prevent camshaft rotation

✻✻ CAUTION

Do not use metal scrapers, wire brushes, power abrasive disks or other abrasive means to clean sealing surfaces. These tools cause scratches and gouges which make leak paths.

25. Clean and inspect the mounting surfaces of the engine and the front cover.

➡The engine front cover must be installed and the bolts tightened within four minutes of applying the silicone gasket and sealant.

26. Apply a 2.5 mm bead of silicone gasket and sealant to the cylinder head and oil pan joint areas. Apply a 2.5 mm bead of silicone gasket and sealant to the front cover.

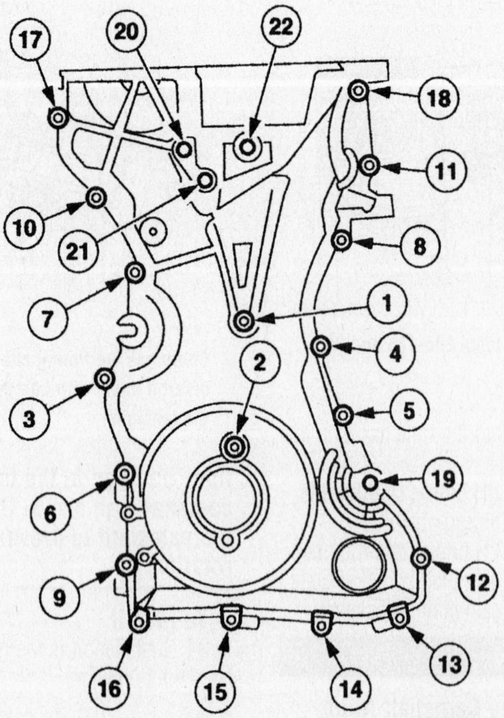

67197-ESCA-G33

Front cover bolt torque sequence—2.3L engine

27. Install the engine front cover. Tighten the bolts in the sequence shown, to the following specifications:
   a. Tighten the 8 mm bolts to 10 Nm (89 inch lbs.).
   b. Tighten the 13 mm bolts to 48 Nm (35 ft. lbs.).
28. Position the power steering pump and install the bolts.

➡Remove the through-bolt from the special tool.

➡Lubricate the oil seal with clean engine oil.

29. Using a seal driver, install the crankshaft front oil seal.

➡Do not reuse the crankshaft damper bolt.

➡Apply clean engine oil on the seal area before installing.

30. Install the crankshaft pulley and hand-tighten the bolt.

✻✻ CAUTION

Only hand-tighten the bolt or damage to the front cover can occur.

➡This step will correctly align the crankshaft pulley to the crankshaft.

31. Install a standard 6 mm x 18 mm bolt through the crankshaft pulley and thread it into the front cover. Rotate the pulley as necessary to align the bolt holes.

✻✻ CAUTION

Failure to hold the crankshaft pulley in place while tightening the bolt can cause damage to the engine front cover.

32. Using the special tools to hold the crankshaft pulley in place, tighten the crankshaft pulley bolt in two stages:
   a. Stage 1: Tighten to 100 Nm (74 ft. lbs.).
   b. Stage 2: Tighten an additional 90 degrees (1/4 turn).
33. Remove the 6 mm x 18 mm bolt.
34. Remove special tool 303-507.
35. Remove special tool 303-465.

➡Only turn the engine in the normal direction of rotation.

36. Turn the engine two complete revolutions.

➡Only turn the engine in the normal direction of rotation.

37. Turn the crankshaft until the No. 1 piston is at TDC.
38. Install special tool 303-507.

✻✻ CAUTION

Only hand-tighten the bolt or damage to the front cover can occur.

39. Using the 6 mm x 18 mm bolt, check the position of the crankshaft pulley. If it is not possible to install the bolt, correct the engine timing.
40. Using special tool 303-465, check the position of the camshafts. If it is not possible to install the special tool, correct the engine timing.
41. Install the CKP sensor. Do not tighten the bolts at this time.
42. Adjust the CKP sensor alignment jig and tighten the bolts.
43. Remove the 6 mm x 18 mm bolt.
44. Install the engine plug bolt.

3.0L Engine

1. Before servicing the vehicle, refer to the precautions in the beginning of this section.
2. Remove or disconnect the following:
   - Negative battery cable
   - Engine front cover
   - Ignition pulse wheel and install a damper bolt
   - Spark plugs
3. Rotate the crankshaft clockwise to

position the keyway at the 11 o'clock position and the camshafts in the correct positions. The No. 1 cylinder will be at Top Dead Center (TDC).

4. Rotate the crankshaft clockwise 120 degrees to the 3 o'clock position to locate the right side camshafts in the neutral position.

5. Remove or disconnect the following:
- Right side timing chain and tensioner
- Tensioner arm and timing chain guide

6. Rotate the crankshaft clockwise 2 times to position the keyway at the 11 o'clock position. This will position the left side camshafts in the neutral position.

7. Verify that the left side crankshafts are in the neutral position and mark the link position on the crankshaft sprocket.

8. Remove or disconnect the following:
- Left side timing chain and tensioner
- Tensioner arm and timing chain guide
- Damper bolt and crankshaft sprockets

To install:

9. Install the crankshaft sprockets.

10. Position the timing chain tensioner in a soft jaw vise. Hold the ratchet lock mechanism away from the ratchet stem and slowly compress the timing chain tensioner

11. If the timing marks on the chain are not visible, use a permanent marker to mark the left and right side timing chains. Mark the timing chains in the following sequence:

a. Mark any link to use as the crankshaft timing mark.

b. Count 29 links from the crankshaft timing mark and mark the link as the exhaust cam sprocket timing mark.

c. Continue counting to 42 and mark the link as the intake sprocket timing mark

12. Install the guide. Torque the bolts to 18 ft. lbs. (25 Nm).

13. Install the left side timing chain and align the chain in the following sequence:

a. Mark any link to use as the crankshaft timing mark.

b. Count 29 links from the crankshaft timing mark and mark the link as the exhaust cam sprocket timing mark.

c. Continue counting to 42 and mark the link as the intake sprocket timing mark

14. Install or connect the following:
- Left side timing chain and tensioner arm. Torque the bolts to 18 ft. lbs. (25 Nm).
- Crankshaft damper bolt and rotate the keyway to the 3 o'clock position

15. Verify that the right side camshafts are properly positioned and install the right side timing chain and guide. Torque the bolts to 18 ft. lbs. (25 Nm).

16. Make certain that the timing chain aligns with the marks on the camshaft and crankshaft sprockets

✱✱ CAUTION

Install the pulse wheel with the keyway in the slot stamped 20–25–34Y–30M (Color Blur).

17. Install or connect the following:
- Right side timing chain tensioner and arm. Torque the bolts to 18 ft. lbs. (25 Nm) and remove the damper bolt
- Ignition pulse wheel
- Spark plugs
- Engine front cover
- Negative battery cable

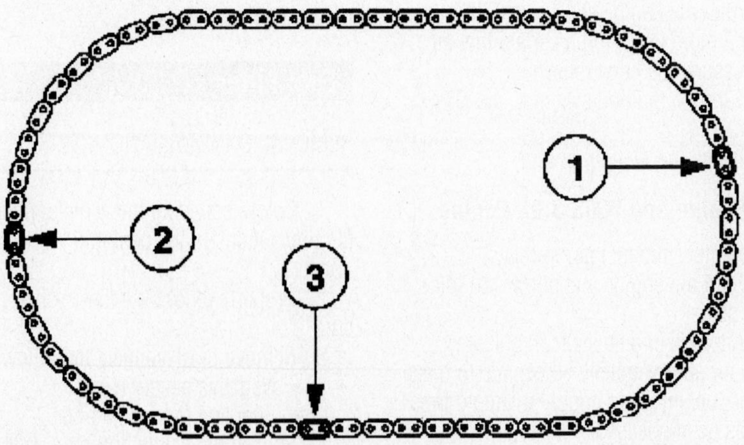

Mark the timing chain in the proper sequence–3.0L engine

Piston and Ring

POSITIONING

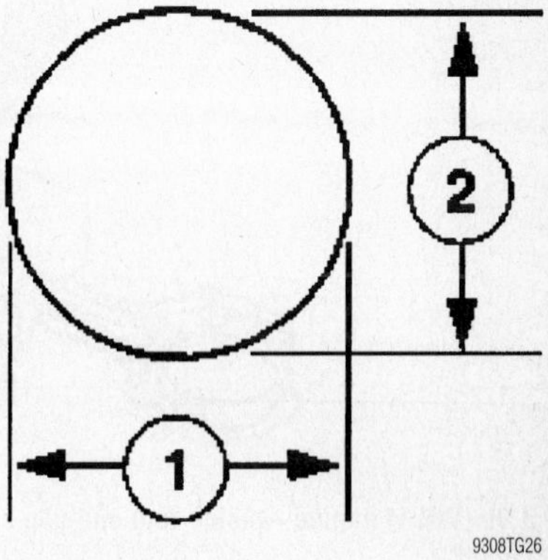

2.0L and 2.3L engine —piston ring end-gap spacing

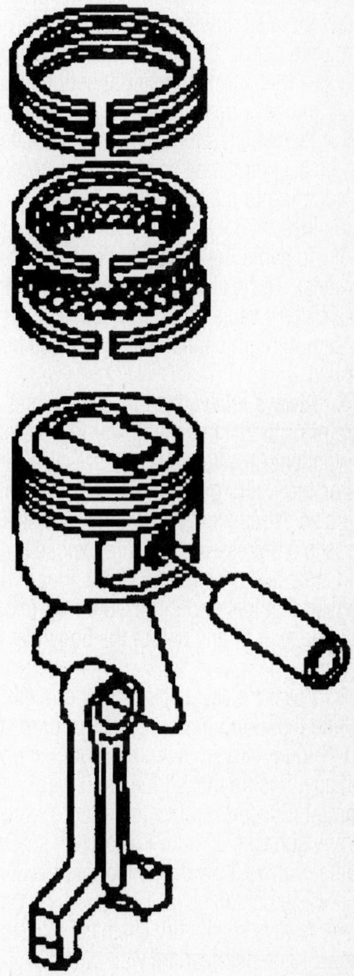

2.0L and 2.3L engine piston and connecting rod positioning ring end-gap spacing

3.0L (VIN 1) engine—piston ring end-gap spacing

FUEL SYSTEM

Fuel System Service Precautions

Safety is the most important factor when performing not only fuel system maintenance but also any type of maintenance. Failure to conduct maintenance and repairs in a safe manner may result in serious personal injury or death. Maintenance and testing of the vehicle's fuel system components can be accomplished safely and effectively by adhering to the following rules and guidelines.

1. To avoid the possibility of fire and personal injury, always disconnect the negative battery cable unless the repair or test procedure requires that battery voltage be applied.
2. Always relieve the fuel system pressure prior to disconnecting any fuel system component (injector, fuel rail, pressure regulator, etc.), fitting or fuel line connection. Exercise extreme caution whenever relieving fuel system pressure, to avoid exposing skin, face and eyes to fuel spray. Please be advised that fuel under pressure may penetrate the skin or any part of the body that it contacts.
3. Always place a shop towel or cloth around the fitting or connection prior to loosening to absorb any excess fuel due to spillage. Ensure that all fuel spillage (should it occur) is quickly removed from engine surfaces. Ensure that all fuel soaked cloths or towels are deposited into a suitable waste container.
4. Always keep a dry chemical (Class B) fire extinguisher near the work area.
5. Do not allow fuel spray or fuel vapors to come into contact with a spark or open flame.
6. Always use a backup wrench when loosening and tightening fuel line connection fittings. This will prevent unnecessary stress and torsion to fuel line piping.
7. Always replace worn fuel fitting O-rings with new. Do not substitute fuel hose or equivalent, where fuel pipe is installed.

Before servicing the vehicle, make sure to refer to the precautions in the beginning of this section as well.

Fuel System Pressure

RELIEVING

2.0L Engine

1. Before servicing the vehicle, refer to the precautions in the beginning of this section.
2. Remove or disconnect the following:
3. Remove the fuel pump relay and start the engine.
4. After the engine stalls, crank the engine 2 more times to be certain that all fuel pressure has been relieved.
5. Turn the ignition switch to the **OFF**-position.
6. Install the fuel pump relay.

2.3L Engine and 2005 3.0L Engine

1. Remove the fuel pump relay.
2. Start the engine and allow it to idle until it stalls.
3. After the engine stalls, crank the engine for approximately 5 seconds to make sure the fuel injection supply manifold pressure has been released.
4. Turn the ignition switch to the OFF position.
5. When fuel system service is complete, install the fuel pump relay.

➡It may take more than one key cycle to pressurize the fuel system.

6. Cycle the ignition key and wait three seconds to pressurize the fuel system. Check for leaks before starting the engine.
7. Start the vehicle and check the fuel system for leaks.

2001–04 3.0L Engine

1. Before servicing the vehicle, refer to the precautions in the beginning of this section.
2. Remove or disconnect the following:
3. Remove the schrader valve cap at the end of the fuel injection supply manifold and attach a fuel pressure gauge.
4. Open the manual valve slowly and drain the fuel into a suitable container.
5. Continue draining the fuel system to relieve fuel pressure.

Fuel Filter

REMOVAL & INSTALLATION

1. Before servicing the vehicle, refer to the precautions in the beginning of this section.
2. Properly relieve the fuel system pressure.
3. Remove or disconnect the following:
   - Negative battery cable
   - Fuel line to the fuel filter
4. Loosen the clamp and remove the filter

To install:

5. Install or connect the following:
   - New clips to the fuel lines
   - Fuel filter and tighten the clamp
   - Fuel lines to the fuel filter
   - Negative battery cable
6. Start the vehicle and check for leaks, repair if necessary.

Fuel Pump

REMOVAL & INSTALLATION

2001–04

1. Before servicing the vehicle, refer to the precautions in the beginning of this section.
2. Properly relieve the fuel system pressure.
3. Remove or disconnect the following:
   - Negative battery cable
   - Gas cap to relieve any additional fuel pressure
   - Left rear seat cushion and lift the access cover on the scuff plate
   - Pin type retainers and move the carpet aside
   - Screws from the fuel pump module access cover
   - Fuel pump module electrical connectors
   - Fuel and vapor lines from the fuel tank
   - Fuel pump module and discard the gasket

To install:

4. Install or connect the following:
   - New fuel pump module gasket
   - Fuel pump module. Torque the module to 60 ft. lbs. (81 Nm).
   - Fuel and vapor lines to the fuel tank
   - Fuel pump module electrical connectors
   - Fuel pump module access cover and tighten the screws securely
   - Pin type retainers and reposition the carpet
   - Left rear seat cushion
   - Gas cap
   - Negative battery cable
5. Start the engine and check for leaks, repair if necessary.

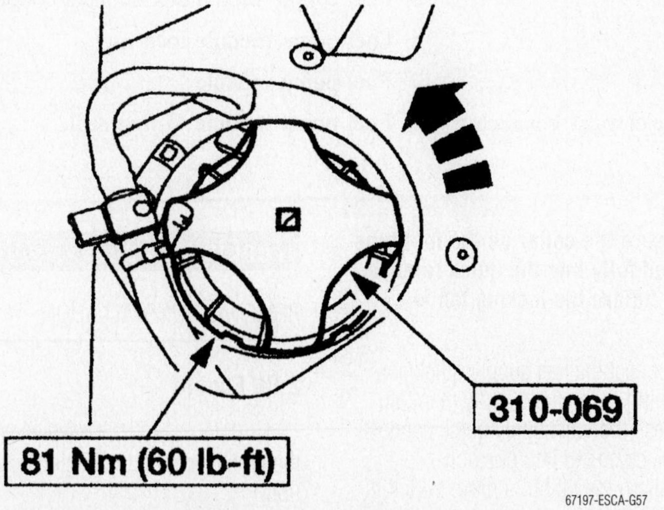

Fuel pump module removal/installation—2001–04 models

2005

1. With the vehicle in NEUTRAL, position it on a hoist.
2. Release the fuel system pressure.
3. Disconnect the battery ground cable.
4. If removing the fuel tank on a four wheel drive vehicle, it is necessary to lower the exhaust system from the catalytic convertor back. Support the exhaust system with a suitable stand, release the three rear exhaust hangers and carefully lower the exhaust system to allow enough clearance to remove the fuel tank.
5. If removing the fuel tank on a four wheel drive vehicle, remove the rear driveshaft.
6. Lift the LH rear seat cushion, position the carpet aside and remove the screws and the fuel pump module access cover.
7. Disconnect the fuel pump module electrical connector
8. Fuel vapor control tube assembly valve electrical connector
9. Using a suitable fuel pump lock ring remover, rotate the lock ring counterclockwise and remove.

**** WARNING**

The fuel pump module must be handled carefully to avoid damage to the float arm and filter.

**** WARNING**

Some fuel will remain in the fuel pump module after draining the fuel tank. Carefully drain the fuel pump module into a suitable container.

10. Prior to completely removing the fuel pump, position it aside and using the special tool and a suitable fuel recovery system, drain the fuel tank.
11. To release the bottom-mounted fuel pump module, reach into the fuel pump module opening and squeeze the retainer tabs on the pump module housing and pull upward.
12. Remove the fuel pump module O-ring seal

To install:

13. Turn the ignition key to the ON position to pressurize the fuel system.
14. Visually inspect the fuel system for leaks.

**** WARNING**

Make sure the fuel tube clicks into place when installing the tube. To make sure the tube is fully seated, pull on the tube.

➡Apply clean engine oil to the end of the tube before inserting the tube into the connector.

15. Install the fuel tube quick release coupling.

**** WARNING**

Inspect the surfaces of the fuel pump module flange and fuel tank O-ring contact surfaces. Do not polish or adjust the O-ring contact area of the fuel pump flange or fuel tank. Install a new fuel pump module or fuel tank if the O-ring contact area is bent, scratched or corroded.

**** WARNING**

Make sure to install a new fuel pump module O-ring and lock ring.

16. Lubricate and install a new fuel pump module O-ring seal upon installing the fuel pump module.

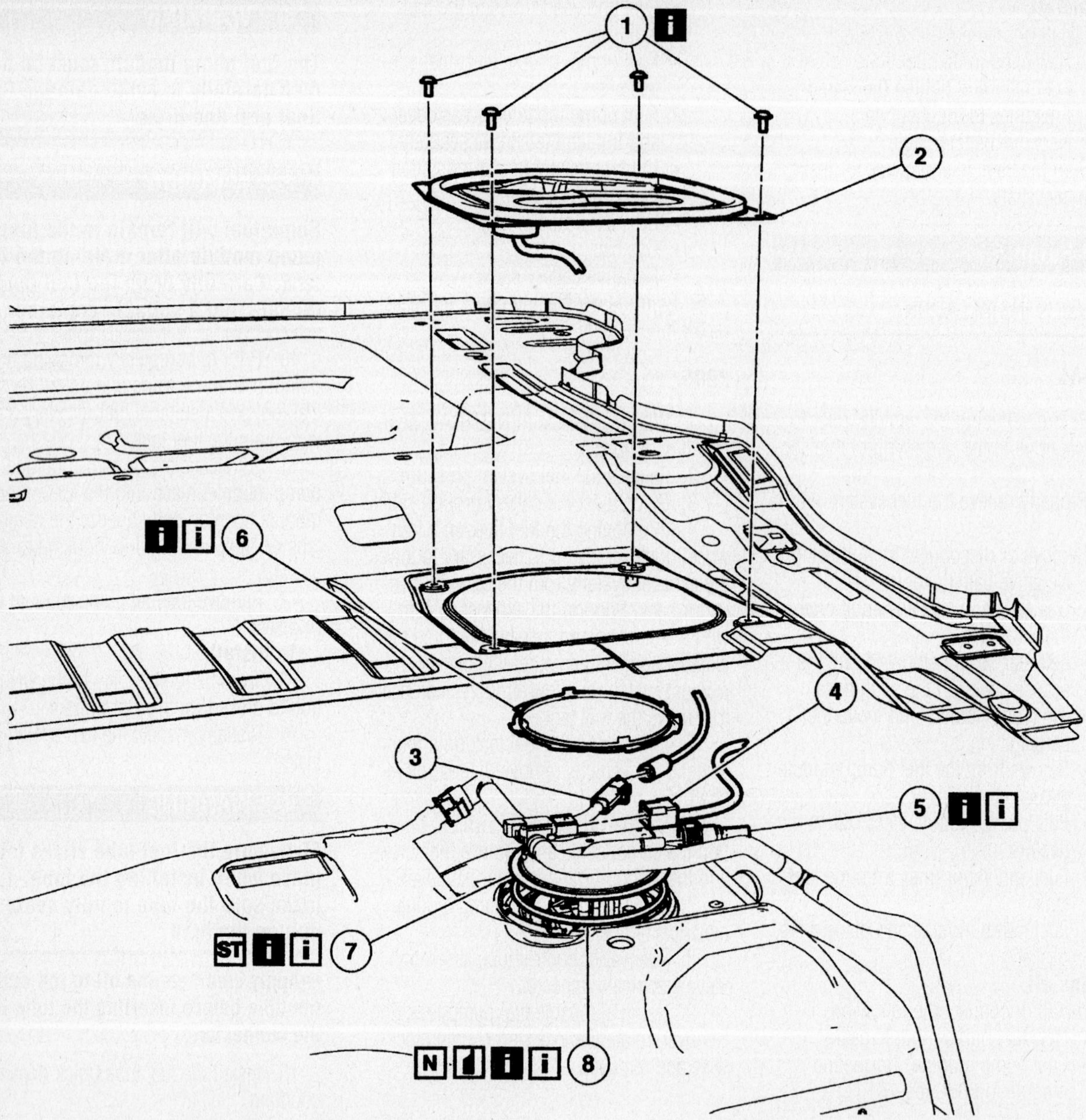

1 Fuel pump module access cover screws
2 Fuel pump module access cover
3 Fuel pump module electrical connector
4 Fuel vapor control tube assembly valve electrical connector
5 Fuel supply tube quick connect coupling
6 Fuel pump module lock ring
7 Fuel pump module
8 Fuel pump module O-ring seal

67197-ESCA-G34

Access to fuel pump module—2005 models

17. When installing the fuel pump module, make sure to align the locator tabs on the fuel tank mounting flange.

➡Be sure the aligning tabs of the fuel pump module unit are positioned in the slot before tightening the lock ring.

18. Holding the fuel pump module O-ring seal in place, rotate the lock ring clockwise until it stops against the retainer tabs.

➡Make sure the collar on the fuel tube is inserted fully into the quick release coupling before the locking tab is locked.

19. Connect the fuel supply quick connect coupling to the fuel supply manifold.
20. Press the fuel supply quick connect coupling locking tab into position.
21. Pull on the fitting to make sure it is fully engaged.

Fuel Injectors

REMOVAL & INSTALLATION

2.0L Engine

1. Before servicing the vehicle, refer to the precautions in the beginning of this section.
2. Release the fuel system pressure.

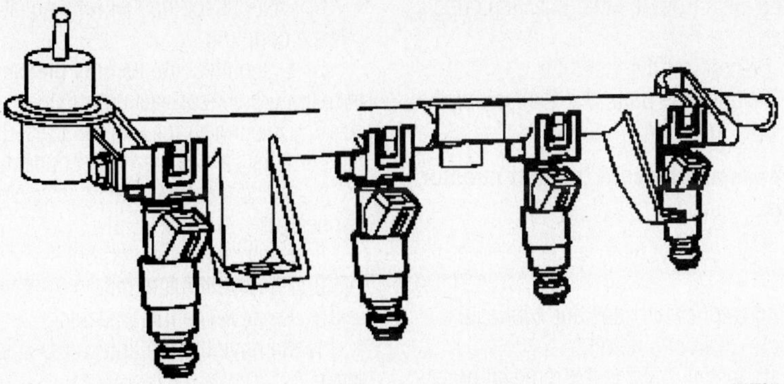

9308TG22

Remove the fuel injectors from the fuel supply manifold–2.0L engine

3. Remove or disconnect the following:
   - Negative battery cable
   - Fuel injection supply manifold
   - Retaining clips and gently twist the fuel injector out of the manifold

4. Check the O-rings and replace if damaged.

To install:

5. Install or connect the following:
   - Fuel injector(s) using new O-rings lubricated with clean engine oil
   - Fuel injector into the supply manifold
   - Retaining clips when the fuel injectors are seated properly
   - Fuel injection supply manifold
   - Negative battery cable

6. Start the vehicle and check for leaks, repair if necessary.

2.3L Engine

⁂ WARNING

Do not smoke or carry lighted tobacco or open flame of any type when working on or near fuel-related components. Highly flammable vapors are always present and can ignite. Failure to follow these instructions can result in personal injury.

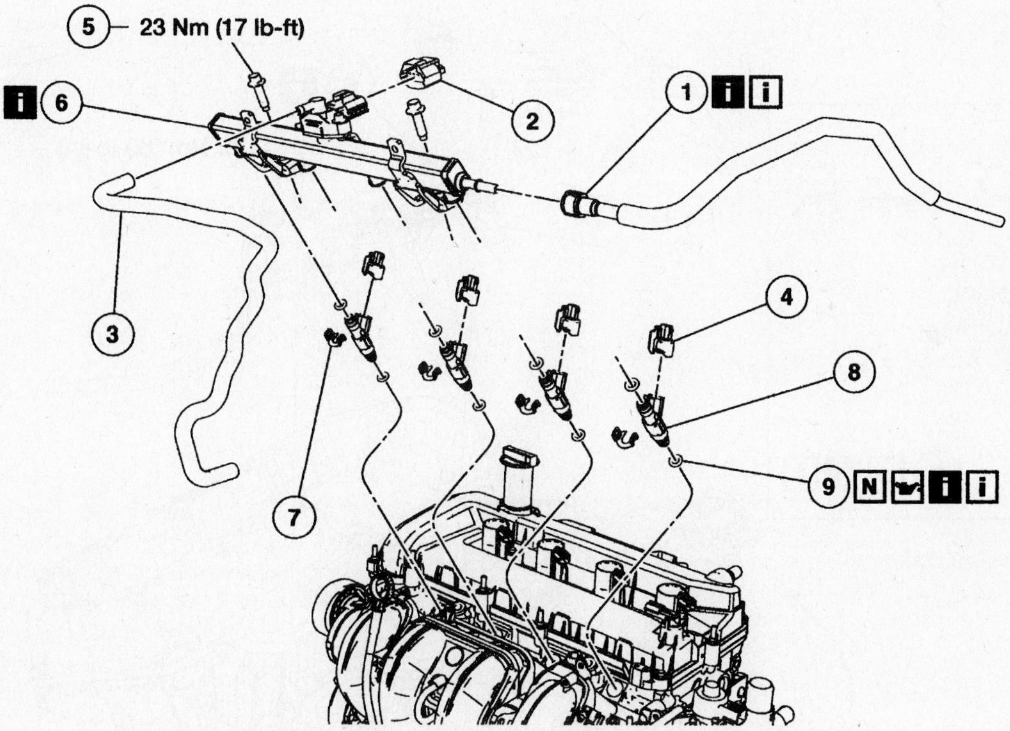

1 Fuel tube quick release coupling (position aside)
2 Fuel rail pressure and temperature sensor electrical connector
3 Fuel rail pressure and temperature sensor vacuum tube (position aside)
4 Fuel injector electrical connectors
5 Fuel rail bolts
6 Fuel rail
7 Fuel injector clips
8 Fuel injectors
9 Fuel injector O-ring seals

67197-ESCA-G35

Fuel rail and injectors—2.3L engine

⁂ WARNING

This procedure involves fuel handling. Be prepared for fuel spillage at all times and always observe fuel handling precautions. Failure to follow these instructions can result in personal injury.

1. Before servicing the vehicle, refer to the precautions in the beginning of this section.
2. Disconnect the battery ground cable.
3. Release the fuel pressure.
4. Disconnect the fuel injector electrical connectors.
5. Disconnect the fuel pressure regulator electrical connector and the vacuum hose.
6. Disconnect the fuel injector harness retaining clips from the fuel injection supply manifold.
7. Disconnect the fuel tube.
8. Remove the bolts and the fuel injection supply manifold.

➡Remove and discard the fuel injector O-rings.

9. If necessary, remove the fuel injectors.
10. Install new O-rings and lubricate them with clean engine oil.
11. To install, reverse the removal procedure. Tighten the fuel supply manifold bolts to 18 ft. lbs. (25 Nm).

3.0L Engine

1. Release the fuel system pressure
2. Disconnect the battery ground cable.
3. Remove the upper intake manifold.
4. Disconnect the fuel tube quick release coupling.
5. Disconnect the fuel rail pressure and temperature sensor vacuum tube
6. Disconnect the fuel rail pressure and temperature sensor electrical connector.
7. Disconnect the fuel injector electrical connectors.
8. Remove the fuel rail bolts.
9. Remove the fuel rail.
10. Remove the fuel injector.
11. Remove the fuel injector O-ring seals.
12. To install, reverse the removal procedure. Install new fuel injector O-ring seals.
13. Lubricate the new O-ring seals with clean engine oil before installing. Torque the fuel rail bolts to 89 inch lbs. (10 Nm).

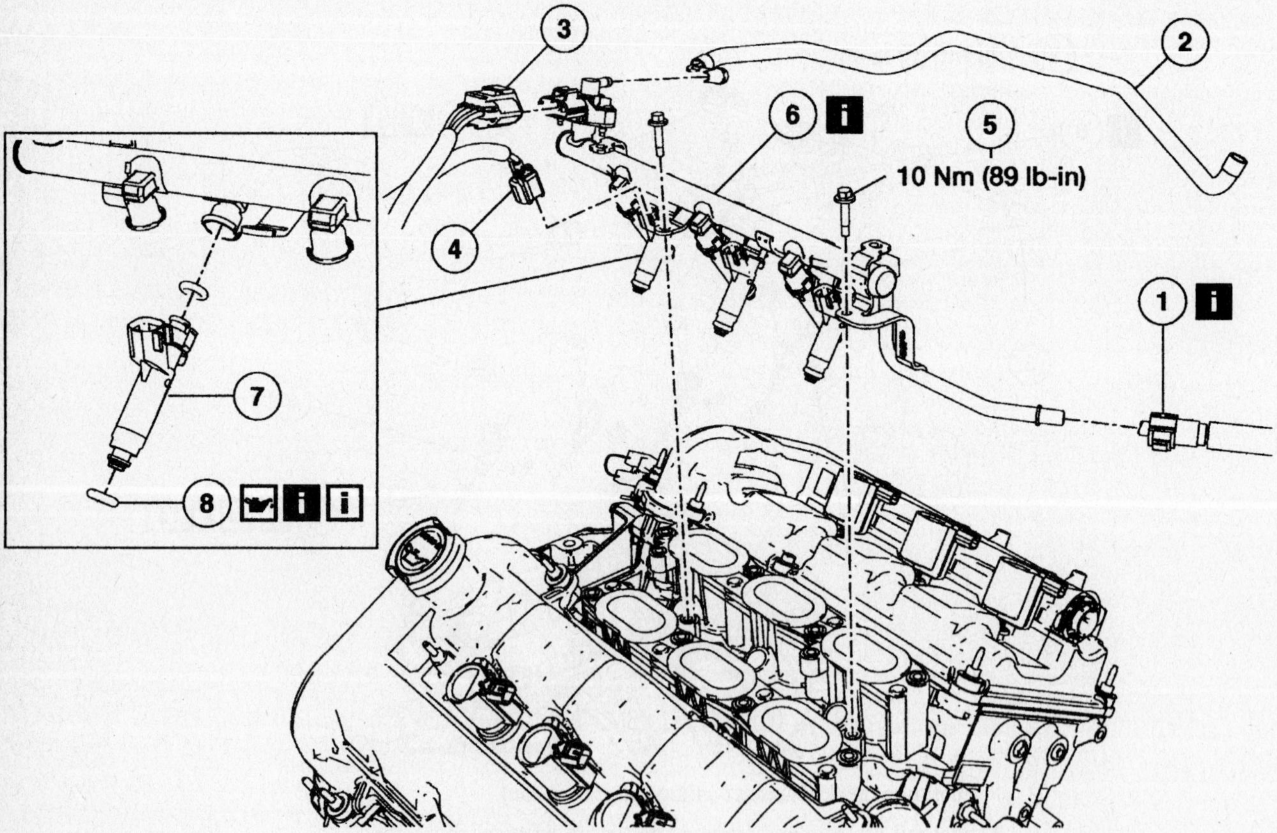

1 Fuel tube quick release coupling (position aside)
2 Fuel rail pressure and temperature sensor vacuum tube (position aside)
3 Fuel rail pressure and temperature sensor electrical connector
4 Fuel injector electrical connector
5 Fuel rail bolt
6 Fuel rail
7 Fuel injector
8 Fuel injector O-ring seals

67197-ESCA-G36

Fuel rail and injectors—3.0L engine

DRIVE TRAIN

Transmission Assembly

REMOVAL & INSTALLATION

Manual Transmission

1. Before servicing the vehicle, refer to the precautions in the beginning of this section.
2. Drain the transmission fluid.
3. Remove or disconnect the following:
    - Battery cables
    - Battery and tray
    - Mass Air Flow (MAF) sensor electrical connector
    - Accelerator cable from the air cleaner outlet tube
    - Emission management tube and hose
    - Crankcase ventilation hose
    - Air cleaner outlet tube
    - Air cleaner housing
    - Back-up lamp switch electrical connector
    - Front wire harness bracket and move it aside
    - Front wire harness bracket spacer
    - Wire harness from the rear harness bracket
    - Park Neutral Position (PNP) electrical connector
    - Rear wire harness bracket and move it aside
    - Vehicle Speed Sensor (VSS) electrical connector
    - Clutch slave cylinder line from the bracket and move it aside while properly supporting the engine
    - Left side transmission support insulator and bracket
    - Rear transmission support insulator
    - Front transmission support insulator and bracket
    - Starter and move it aside
    - Top transmission flywheel housing bolts
    - Front transmission flywheel housing bolts
    - Transfer case, if equipped
    - Left side halfshaft
    - Rear transmission support insulator bracket
    - Shifter linkage and stabilizer bar
    - Transverse crossmember
    - Front to aft crossmember
    - Left side splash shield and properly support the transmission
    - Remaining transmission flywheel housing bolts
    - Transmission and separate the right side halfshaft from the transmission

To install:

4. Align the right side half shaft to the transmission and position the transmission to the engine.
5. Install or connect the following:
    - Transmission flywheel housing bolts. Torque the bolts to 33 ft. lbs. (45 Nm) and remove the transmission support
    - Left side splash shield
    - Front-to-aft crossmember. Torque the bolts to 66 ft. lbs. (90 Nm).
    - Transverse crossmember. Torque the bolts to 85 ft. lbs. (115 Nm).
    - Shifter linkage. Torque the bolt to 15 ft. lbs. (20 Nm).
    - Stabilizer bar. Torque the bolt to 30 ft. lbs. (40 Nm).
    - Rear transmission support bracket. Torque the bolts to 66 ft. lbs. (90 Nm).
    - Left side halfshaft
    - Transfer case, if equipped
    - Front transmission flywheel housing bolts. Torque the bolts to 33 ft. lbs. (45 Nm).
    - Top transmission flywheel housing bolts. Torque the bolts to 33 ft. lbs. (45 Nm).
    - Starter. Torque the bolts to 33 ft. lbs. (45 Nm).
    - Front transmission support insulator and bracket. Torque the lower bolt to 66 ft. lbs. (90 Nm) and the 3 upper bolts to 41 ft. lbs. (55 Nm).
    - Rear transmission support insulator bolt. Torque the bolt to 66 ft. lbs. (90 Nm).
    - Left side transmission support insulator bracket. Torque the bolts to 66 ft. lbs. (90 Nm).
    - Left side transmission support insulator. Torque the large bolt to 66 ft. lbs. (90 Nm) and the 3 remaining bolts to 41 ft. lbs. (55 Nm).
    - Clutch slave cylinder. Torque the bolt to 15 ft. lbs. (20 Nm).
    - Clutch slave cylinder line to the bracket and install the retaining clip
    - VSS electrical connector
    - Rear wire harness bracket. Torque the bolts to 80 inch lbs. (9 Nm).
    - PNP switch electrical connector
    - Wire harness to the rear bracket
    - Front wire harness bracket spacer and bracket. Torque the bolt to 9 ft. lbs. (12 Nm).
    - Back-up lamp switch electrical connector
    - Air cleaner housing
    - MAF sensor electrical connector
    - Air cleaner outlet tube
    - Crankcase ventilation hose
    - Emission management tube and hose
    - Accelerator cable to the air cleaner outlet tube
    - Battery and tray
    - Both battery cables
6. Fill the transmission to the proper level.
7. Start the vehicle and check for leaks, repair if necessary.

Automatic Transmission

1. Before servicing the vehicle, refer to the precautions in the beginning of this section.
2. Remove or disconnect the following:
    - Battery cables
    - Battery and tray
    - Breather tube
    - Mass Air Flow (MAF) sensor
    - Intake tube and air cleaner cover
    - Air cleaner assembly
    - Transmission Range (TR) sensor
    - Heated Oxygen ($H0_2S$) sensors
    - Transmission harness connector and bracket
    - Wire harness bracket spacer and move the bracket aside
    - Shift cable
    - Shift cable bracket and move the bracket aside
    - Starter electrical connectors
    - Starter
    - Electrical connectors from the valve cover and install an engine support bar
    - Upper transmission retaining bolts
    - Left side upper transmission mounting plate
    - Rear transmission mount
    - Right side engine mount bolt and slightly raise the engine
    - Both front wheels and splash shields
    - Right side halfshaft and intermediate shaft assembly after matchmarking them
    - Cross brace
    - Center exhaust pipe and rubber hanger
    - Front exhaust pipe and flange
    - Rear exhaust pipe flange
    - Driveshaft

- PTU vent tube
- Lower transmission bracket
- Access cover
- Flexplate nuts
- Output Shaft Speed (OSS) sensor
- Turbine Shaft Speed (TSS) sensor
- Fluid cooler tube and move it aside
- Fluid cooler line and install a transmission jack
- Bolts from the PTU unit
- Transmission with the PTU unit attached

To install:

3. Install or connect the following:
   - Transmission with the PTU unit. Torque the engine-to-transmission mounting bolts to 30 ft. lbs. (40 Nm).
   - Fluid cooler line. Torque the fastener to 17 ft. lbs. (23 Nm) and remove the transmission jack.
   - Fluid cooler tube. Torque the bolt to 17 ft. lbs. (23 Nm).
   - OSS sensor
   - TSS sensor
   - Flexplate nuts. Torque the nuts to 27 ft. lbs. (36 Nm).
   - Access cover
   - Cross brace. Torque the bolts to 96 ft. lbs. (130 Nm).
   - Transmission bracket. Torque the bolts to 30 ft. lbs. (40 Nm).
   - PTU vent tube
   - Driveshaft. Torque the bolts to 15 ft. lbs. (20 Nm).
   - Exhaust pipe and flange. Torque the bolts to 21 ft. lbs. (29 Nm).
   - Exhaust pipe and rubber hanger. Torque the bolts to 21 ft. lbs. (29 Nm).
   - Left side halfshaft assembly
   - Right side halfshaft and intermediate shaft assembly by aligning the matchmarks
   - Splash shields
   - Both front wheels and lower the engine on to the right side engine mount
   - Right side engine mount bolt. Torque the bolt to 89 ft. lbs. (120 Nm).
   - Rear transmission mount. Torque the upper bolt to 89 ft. lbs. (120 Nm) and the lower bolts to 35 ft. lbs. (45 Nm).
   - Transmission mount assemble. Torque the bolts to 30 ft. lbs. (40 Nm) and remove the engine support bar.
   - Electrical connectors to the valve cover
   - Starter. Torque the bolts to 20 ft. lbs. (27 Nm).
   - Starter electrical connectors
   - Shifter cable and bracket. Torque the bolt to 14 ft. lbs. (19 Nm) and connect the shifter cable.
   - Wire harness and install the harness bracket spacer
   - Wire harness bracket. Torque the bolt to 89 inch lbs. (10 Nm).
   - HO_2S sensor
   - TR sensor and make certain it is properly aligned
   - Air cleaner assembly
   - Intake tube and air cleaner cover
   - Breather tube
   - MAF sensor
   - Battery tray
   - Battery and cables

4. Fill the transmission with clean fluid to the proper level.

5. Start the vehicle and check for leaks, repair if necessary

Clutch

ADJUSTMENTS

The clutch is hydraulically driven and therefore no adjustment is required.

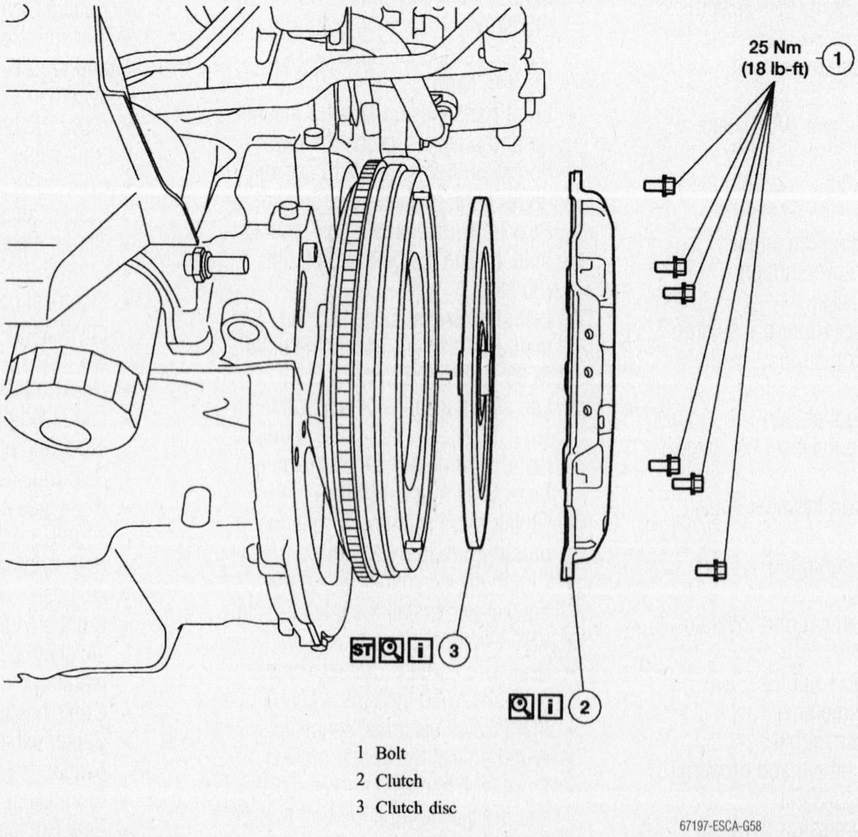

Clutch components

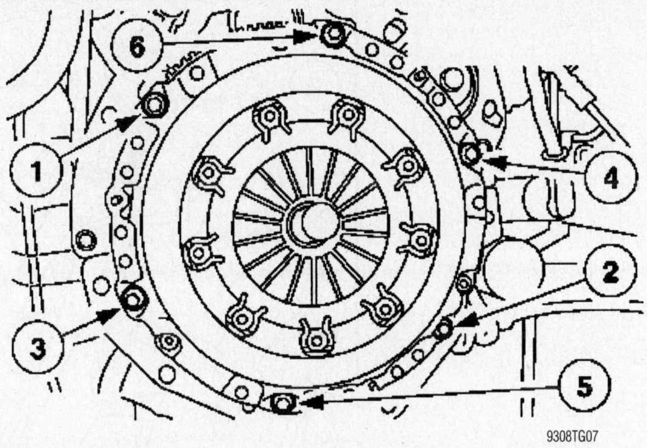

Torque the pressure plate bolts in the proper sequence

REMOVAL & INSTALLATION

1. Before servicing the vehicle, refer to the precautions in the beginning of this section.
2. Remove or disconnect the following:
   - Negative battery cable
   - Transmission and lock the flywheel to the engine with special tool 303–103
   - Pressure plate bolts by loosening them evenly
   - Clutch pressure plate and disc
3. Clean the pressure plate and inspect it for burn marks, scores, flatness or ridges, replace if damaged.
4. Inspect the pressure plate diaphragm finger for wear, replace if damaged.
5. Measure the depth of the rivet heads. Minimum depth is 0.012 inch (0.3mm).
6. Inspect the clutch disc for signs of wear and replace if needed.
7. Check the clutch disc runout. Replace the disc if not with specification: 0.027 inch (0.7mm).

To install:

8. Install or connect the following:
   - Clutch disc to the flywheel
   - Pressure plate to the flywheel. Torque the bolts in sequence to 21 ft. lbs. (29 Nm).
   - Transmission
   - Negative battery cable
9. Check the transmission fluid level and top off if necessary.

Hydraulic Clutch System

BLEEDING

The following procedure is recommended for bleeding the clutch hydraulic system installed on the vehicle. It is recommended that the original clutch tube, with quick-connect fitting be replaced when servicing the hydraulic system, because air can be trapped in the quick-connect fitting and prevent complete bleeding of the system.

1. Before servicing the vehicle, refer to the precautions in the beginning of this section.
2. Clean the dirt and grease from the dust cap.
3. Remove the cap and diaphragm and fill the reservoir ¾ of the way with approved brake fluid C6AZ-19542-AB or DOT 3 equivalent fluid (ESA-M6C25-A).
4. Loosen the bleeder screw cover from the slave cylinder and attach a hose to the screw.
5. Place the hose in a container and slowly pump the clutch pedal several times.
6. With the clutch pedal depressed, loosen the bleeder screw to release the fluid and air.
7. Remove the hose and tighten the bleeder screw.
8. Repeat this procedure until all the air is removed from the hydraulic system.

REMOVAL & INSTALLATION

Master Cylinder

2001–04

CAUTION

Note the location and orientation of the clutch push rod retainer. If any damage to the retainer occurs during removal or installation, replace the retainer clip. The retainer must be correctly installed during assembly.

1. Disconnect the clutch master cylinder rod from the clutch pedal.
2. Remove the clutch pedal bracket nut.

CAUTION

Brake fluid contains polyglycol ethers and polyglycols. Avoid contact with eyes. Wash hands thoroughly after handling. If brake fluid contacts eyes, flush eyes with running water for 15 minutes. Get medical attention if irritation persists. If taken internally, drink water and induce vomiting. Get medical attention immediately. Failure to follow these instructions may result in personal injury.

CAUTION

Brake fluid is harmful to painted and plastic surfaces. If brake fluid is spilled onto a painted or plastic surface, wash it immediately with water.

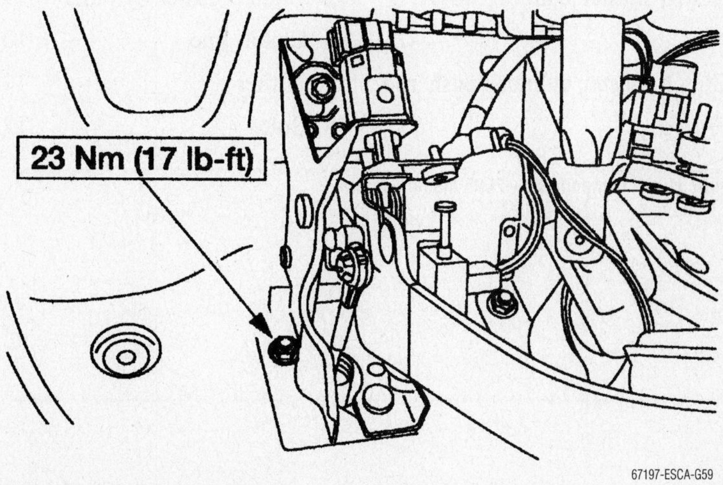

Clutch pedal bracket—2001–04 models

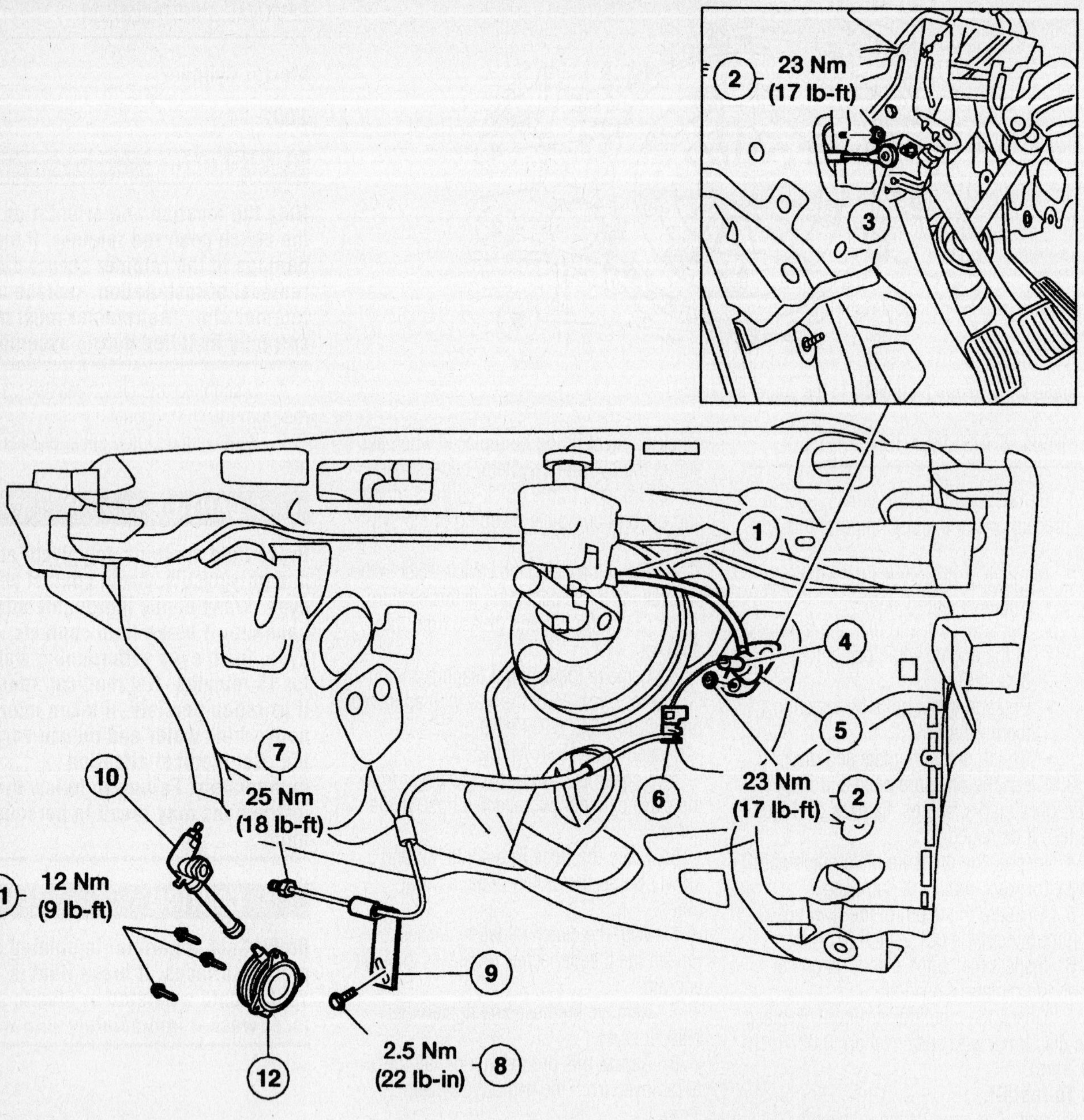

1 Clutch master cylinder hose	5 Clutch master cylinder	9 Clutch line bracket
2 Nut	6 Clutch line	10 Clutch slave cylinder to clutch line adapter
3 Clutch master cylinder push rod	7 Fitting	11 Bolt
4 Fitting	8 Bolt	12 Clutch slave cylinder

67197-ESCA-G60

Hydraulic clutch components—2005 models shown

3. Disconnect the clutch master cylinder hose from the clutch master cylinder and plug it to prevent excess fluid loss.
4. Disconnect the clutch slave cylinder tube from the clutch master cylinder.
5. Remove the clutch master cylinder nut.
6. Remove the clutch master cylinder.
7. To install, reverse the removal procedure. Make sure that the push rod retainer is correctly positioned. Observe the following torques:
   - Clutch pedal bracket nut: 17 ft. lbs. (23 Nm).
   - Slave cylinder tube-to-master cylinder: 22 ft. lbs. (30 Nm).
   - Master cylinder mounting nuts: 17 ft. lbs. (23 Nm).
8. Bleed the air from the system.

2005

⁂ WARNING

Brake fluid is harmful to painted and plastic surfaces. If brake fluid is spilled onto a painted or plastic surface, wash it immediately with water.

➡If removing the clutch line, remove the engine air cleaner.

➡If removing the clutch slave cylinder or the clutch slave cylinder-to-clutch line adapter, remove the transaxle.

1. Remove the clutch master cylinder hose
2. Remove the clutch master cylinder push rod
3. Remove the fitting at the master cylinder.
4. Remove the clutch master cylinder

⁂ WARNING

Make sure the O-rings are properly positioned on the hydraulic line fittings or leaks may occur.

5. To install, reverse the removal procedure. Torque the mounting nuts to 17 ft. lbs. (23 Nm).
6. Fill and bleed the system.

Slave Cylinder

2001–04

⁂ WARNING

Brake fluid is harmful to painted and plastic surfaces. If brake fluid is spilled onto a painted or plastic surface, wash it immediately with water.

1. Disconnect the clutch slave cylinder tube and plug it to prevent excess fluid loss.
2. Remove the clutch slave cylinder bolts.
3. Remove the clutch slave cylinder.
4. To install, reverse the removal procedure. Observe the following torques:
   - Slave cylinder mounting bolts: 15 ft. lbs. (20 Nm)
   - Fluid line-to-slave cylinder: 18 ft. lbs. (25 Nm)
5. Bleed the air from the system.

2005

⁂ WARNING

Brake fluid is harmful to painted and plastic surfaces. If brake fluid is spilled onto a painted or plastic surface, wash it immediately with water.

➡If removing the clutch line, remove the engine air cleaner.

➡If removing the clutch slave cylinder or the clutch slave cylinder to clutch line adapter, remove the transaxle.

1. Remove the clutch slave cylinder to clutch line adapter.
2. Remove the bolts.
3. Remove the clutch slave cylinder.

➡Make sure the O-rings are properly positioned on the hydraulic line fittings or leaks may occur.

4. To install, reverse the removal procedure.
5. Fill and bleed the system.

Transfer Case

REMOVAL & INSTALLATION

2001–02 w/2.0L Engine

1. Remove the driveshaft.
2. Remove the four bolts and remove the crossmember brace.
3. Remove the two bolts and the two nuts.
4. Remove the transfer case.
5. Remove and discard the O-ring seal.

To install:

6. Install a new O-ring seal.
7. Install the transfer case.
8. Install the two bolts and the two nuts. Torque to 33 ft. lbs. (45 Nm).
9. Install the crossmember brace and the four bolts. Torque to 30 ft. lbs. (40 Nm).
10. Install the driveshaft.

2001–04

1. Disconnect the battery.
2. Drain the transfer case.
3. Remove the driveshaft.
4. Remove the 4 bolts and the crossmember brace.
5. Remove the generator.
6. Disconnect the RH catalyst monitor.
7. Remove and discard the 2 nuts and separate the flexible Y-pipe from the manifold.
8. Remove and discard the nuts. Discard the gasket.
9. Remove and discard the nuts. Discard the gasket.
10. Remove the flexible pipe. Disconnect the hanger.
11. Disconnect the EGR valve to exhaust manifold tube at the manifold.
12. Disconnect the RH heated oxygen sensor.
13. Remove the RH exhaust manifold. Discard the gasket.
14. Remove the support bracket.
15. Remove the 3 transfer case bolts.

➡The transfer case driven gear seal must be replaced whenever the link shaft or transfer case is removed from the vehicle.

➡If necessary, replace the RH differential fluid seal.

16. Remove the bolt and the transfer case.

To install:

➡The transfer case driven gear seal must be replaced whenever the link shaft or transfer case is removed from the vehicle. If necessary, replace the RH differential fluid seal.

17. Install the transfer case. Install the bolts. Torque to 33 ft. lbs. (45 Nm).
18. Install the side mount bolt. Torque to 33 ft. lbs. (45 Nm).
19. Install the bracket and the 6 bolts. Torque to 30 ft. lbs. (40 Nm).
20. Install the RH exhaust manifold.
21. Connect the RH heated oxygen sensor.
22. Connect the EGR to exhaust manifold tube. Torque to 30 ft. lbs. (40 Nm).
23. Install the flex pipe and connect the hanger.
24. Install the gasket and the nuts. Torque to 37 ft. lbs. (50 Nm).
25. Install the flex pipe gasket and the nuts. Torque to 37 ft. lbs. (50 Nm).
26. Connect the flexible Y-pipe to the exhaust manifold and install the two nuts. Torque to 30 ft. lbs. (40 Nm).

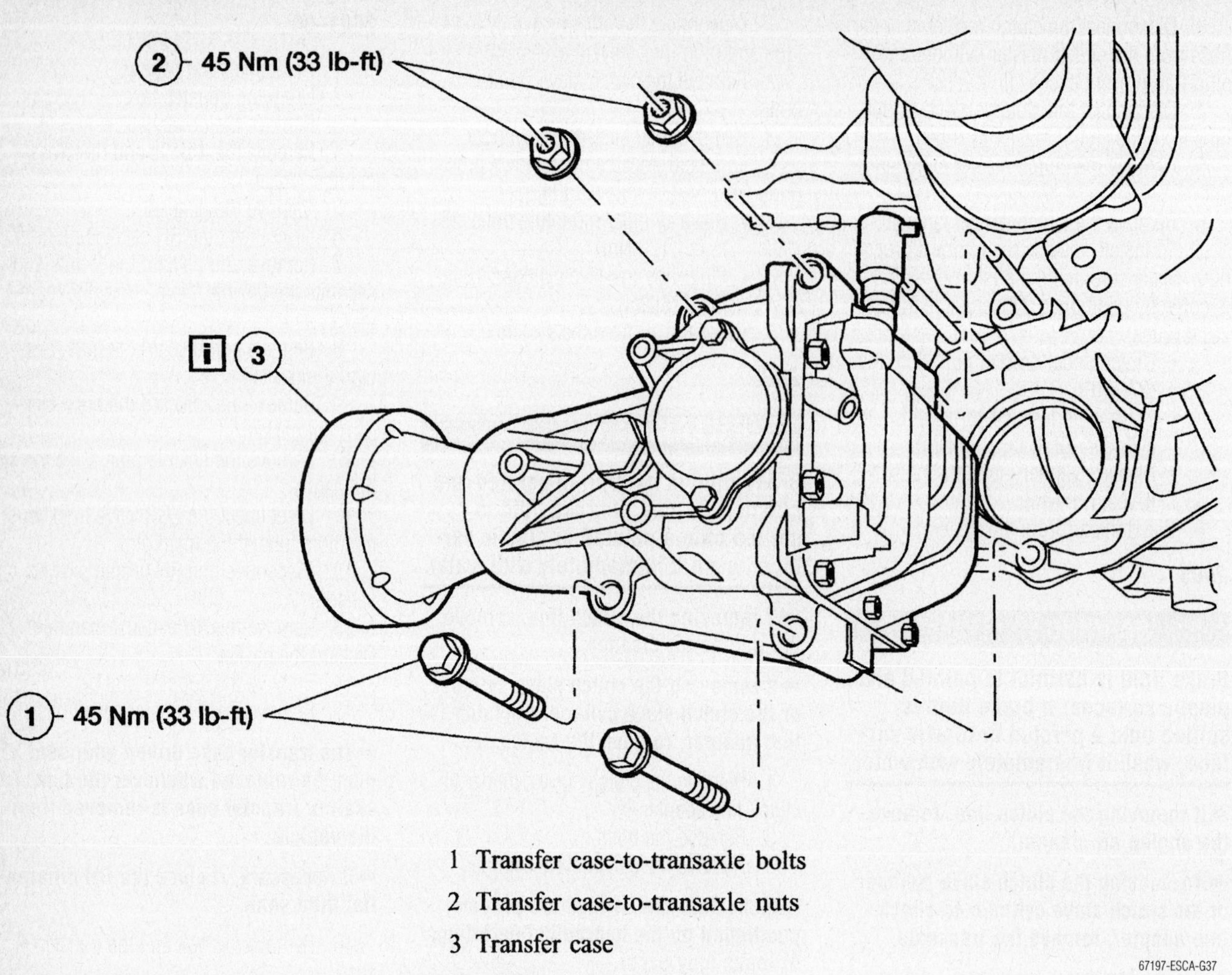

Transfer case mounting with manual transaxle—2005 model shown

27. Connect the RH catalyst monitor sensor.
28. Install the generator.
29. Install the crossmember brace and the four bolts. Torque to 30 ft. lbs. (40 Nm).
30. Check the transfer case fluid level.
31. Install the driveshaft.
32. Connect the battery ground cable.
33. Check the transaxle fluid level.

2005

WITH MANUAL TRANSAXLE

1. Remove the driveshaft.
2. Remove the 4 bolts and the crossmember brace.
3. Remove the transfer case-to-transaxle bolts.
4. Remove the transfer case-to-transaxle nut.
5. Remove the transfer case.
6. To install, reverse the removal procedure.

WARNING

The O-ring must be properly installed before mating the transfer case to the manual transaxle. Failure to properly install the O-ring may cause the O-ring to be damaged resulting in transaxle oil leak.

7. Install a new O-ring seal. Observe the following torques:
   - Crossmember bolts: 30 ft. lbs. (40 Nm).
   - Transfer case mounting nuts: 33 ft. lbs. (45 Nm).
   - Transfer case mounting bolts: 33 ft. lbs. (45 Nm).

WITH AUTOMATIC TRANSAXLE

1. Disconnect the battery.
2. Drain the transfer case.
3. Remove the front RH intermediate shaft.
4. Remove the driveshaft.
5. Remove the 4 bolts and the crossmember brace.
6. Remove the generator.
7. Remove the exhaust as required.
8. Remove the heat shield.
9. Remove the transfer case-to-transaxle bolts.
10. Remove the transfer case.

WARNING

A new transfer case driven gear seal must be installed whenever the intermediate shaft or transfer case is removed from the vehicle.

➡If necessary, replace the RH differential fluid seal.

11. To install, reverse the removal procedure. Observe the following torques:
    - Crossmember bolts: 30 ft. lbs. (40 Nm).
    - Transfer case-to-transaxle bolts: 33 ft. lbs. (45 Nm).

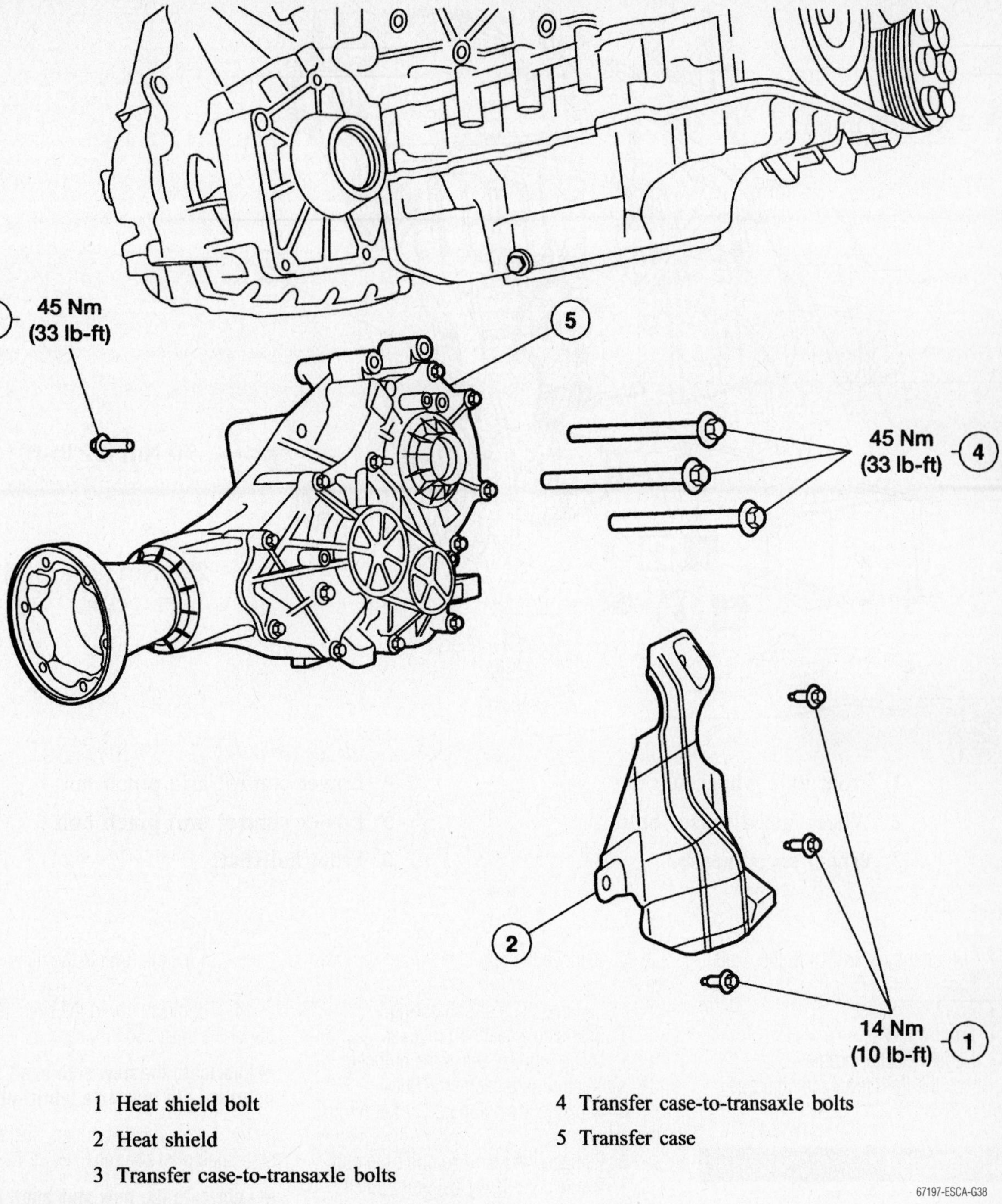

Transfer case mounting with automatic transaxle—2005 model shown

Front Halfshaft

REMOVAL & INSTALLATION

1. Before servicing the vehicle, refer to the precautions in the beginning of this section.
2. Place the transmission in the PARK position.
3. Remove or disconnect the following:
   - Negative battery cable
   - Front wheel
   - Front brake disc
   - Front axle wheel hub nut and discard the nut
   - Tie rod end and separate the lower ball from the steering knuckle
   - Halfshaft from the steering knuckle
   - Halfshaft

To install:

4. When seated properly, the halfshaft bearing retainer circlip will snap into the differential side gear groove.
5. Position the halfshaft and joint so that the splines align with differential side gear splines. Push the halfshaft into side gear.
6. Install or connect the following:
   - Halfshaft into the steering knuckle
   - Lower ball joint to steering knuckle. Torque the pinch bolt to 52 ft. lbs. (70 Nm).
   - Tie rod end. Torque the nut to 41 ft. lbs. (55 Nm).
   - New front axle wheel hub nut.

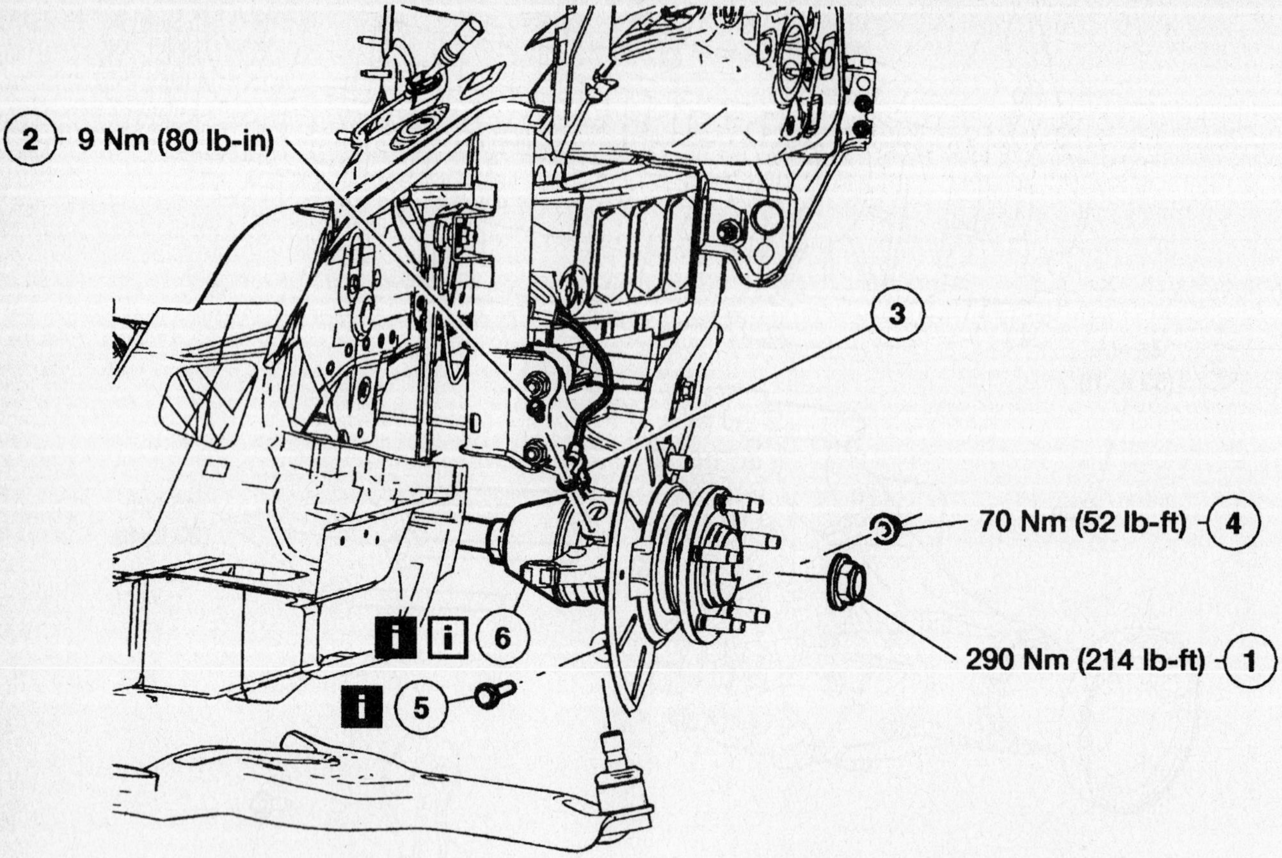

Front halfshaft

Torque the nut to 214 ft. lbs. (290 Nm).

- Front brake disc
- Front wheel
- Negative battery cable

7. Check the fluid level and adjust as needed.

Rear Halfshaft

REMOVAL & INSTALLATION

1. Place the selector lever in NEUTRAL.
2. Raise and support the vehicle.

WARNING

Do not loosen the rear axle wheel hub retainer until after the wheel and tire assembly are removed from the vehicle. Wheel bearing damage will occur if the wheel bearing is unloaded with the weight of the vehicle applied.

3. Remove the rear brake drum or brake disc.
4. Remove the rear coil spring.
5. Rear axle wheel hub retainer.
6. Using a puller, press the outboard CV joint until it is loose in the hub.
7. Separate the outboard CV joint from the hub.
8. Remove the rear axle hub assembly.
9. Remove the lower ball joint pinch bolt nut.

WARNING

Do not use a hammer to separate the rear axle halfshaft assembly from the hub. Damage to the threads and the internal CV joint can result.

10. Using a prybar, remove the halfshaft.
11. Using a puller, remove the stub and shaft seal.
12. Using a slidehammer and adapter, remove the stub shaft pilot bearing and seal.
13. To install, reverse the removal procedure.
14. Fill the axle with the specified quantity of the specified lubricant.

➡Lubricate the new stub shaft pilot bearing with rear axle lubricant.

15. Using suitable drivers, install the stub shaft pilot bearing.

➡Lubricate the new stub shaft pilot bearing housing seal with grease.

16. Using the special tools, install the stub shaft pilot bearing housing seal.
17. Install a new circlip on the inboard CV joint.
18. Install the halfshaft end into the hub assembly.
19. Observe the following torques:
    - Lower ball joint: 85 ft. lbs. (115 Nm).
    - Halfshaft nut: 214 ft. lbs. (290 Nm).

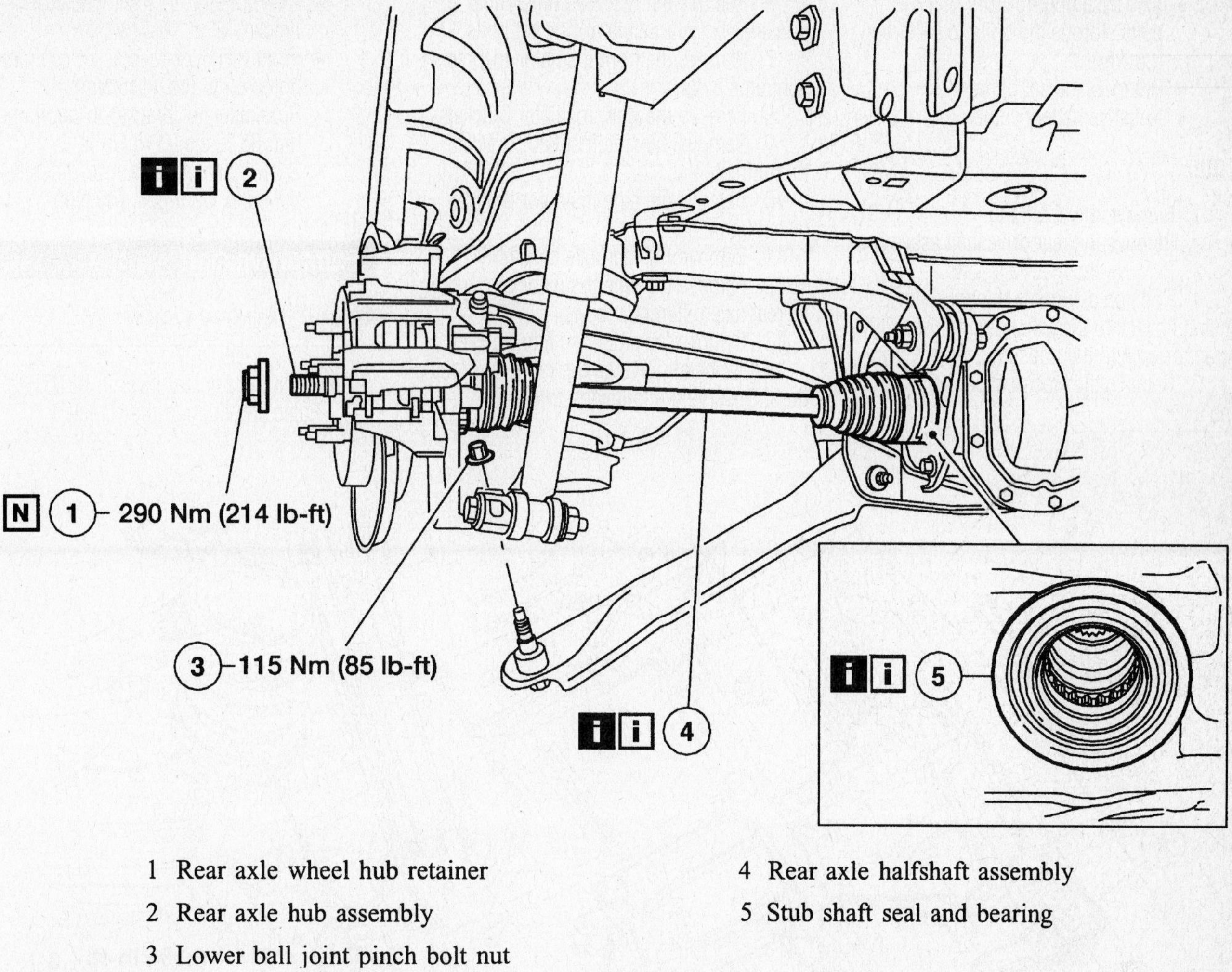

Rear halfshaft assembly

CV-Joints

OVERHAUL

1. Before servicing the vehicle, refer to the precautions in the beginning of this section.
2. Remove or disconnect the following:
   - Negative battery cable
   - Halfshaft and secure it in a soft-jawed vise
   - Inboard halfshaft boot clamp
   - Boot from the inboard CV-joint housing
   - Tripod joint from the CV-joint housing and matchmark the tripod joint to the halfshaft
   - Snapring and boot from the halfshaft
   - Outboard halfshaft boot clamps
   - Outboard boot back to expose the CV-joint and matchmark the joint to the halfshaft
   - Outboard CV-joint from the halfshaft
   - Halfshaft retainer circlip and discard it
   - Boot from the halfshaft

To install:

3. Install or connect the following:
   - Outboard CV-joint and boot
   - New halfshaft bearing circlip
   - Inboard CV-joint to the halfshaft
   - Outboard halfshaft boot forward on to the outboard CV-joint
   - New outboard halfshaft boot clamps
   - Inboard halfshaft boot
   - Tripod joint on the halfshaft by aligning the matchmarks
   - New snapring to the tripod joint and lubricate the needle bearings while filling the housing with CV-joint grease, E43Z–19590–A
   - Inboard halfshaft boot with new clamps
   - Halfshaft
   - Negative battery cable

Rear Drive Axle Housing

REMOVAL & INSTALLATION

2001–04

1. Before servicing the vehicle, refer to the precautions in the beginning of this section.
2. Remove or disconnect the following:
   - Negative battery cable
   - Rotary blade coupling
   - Rear halfshafts
   - Axle assembly-to-front bracket bolts
   - Rear axle-to-side bracket bolts
   - Axle assembly

To install:

3. Install or connect the following:
   - Axle assembly
   - Rear axle-to-side-bearing bolts. Torque the bolts to 59 ft. lbs. (80 Nm).

- Axle assembly-to-front bracket bolts. Torque the bolts to 59 ft. lbs. (80 Nm).
- Rotary blade coupling
- Negative battery cable

2005

1. Remove the spare tire.
2. Remove the rear driveshaft assembly.
3. Remove the rear halfshafts.
4. Position a suitable transmission hydraulic jack to the axle housing. Securely strap the jack to the housing.
5. Remove the electrical connector at the axle.
6. Remove the rear axle differential housing-to-front insulator bracket bolts.
7. Remove the front insulator-to-bracket subframe bolts.
8. Remove the front insulator brackets.
9. Remove the side insulator bracket-to-subframe nut.
10. Remove the side insulator bracket-to-subframe bolt.
11. Remove the rear axle assembly.
12. Remove the side insulator bracket-to-rear axle differential bolts.
13. Remove the side insulator bracket.
14. To install, reverse the removal procedure. Observe the following torques:

- Axle housing-to-front insulator bracket: 59 ft. lbs. (80 Nm)
- Front insulator-to-bracket subframe bolts: 85 ft. lbs. (115 Nm)
- Side insulator bracket-to-subframe nut: 85 ft. lbs. (115 Nm)
- Side insulator bracket-to-differential bolts: 59 ft. lbs. (80 Nm)

Front Driveshaft

REMOVAL & INSTALLATION

1. Place the selector lever in NEUTRAL.
2. Raise and support the vehicle.

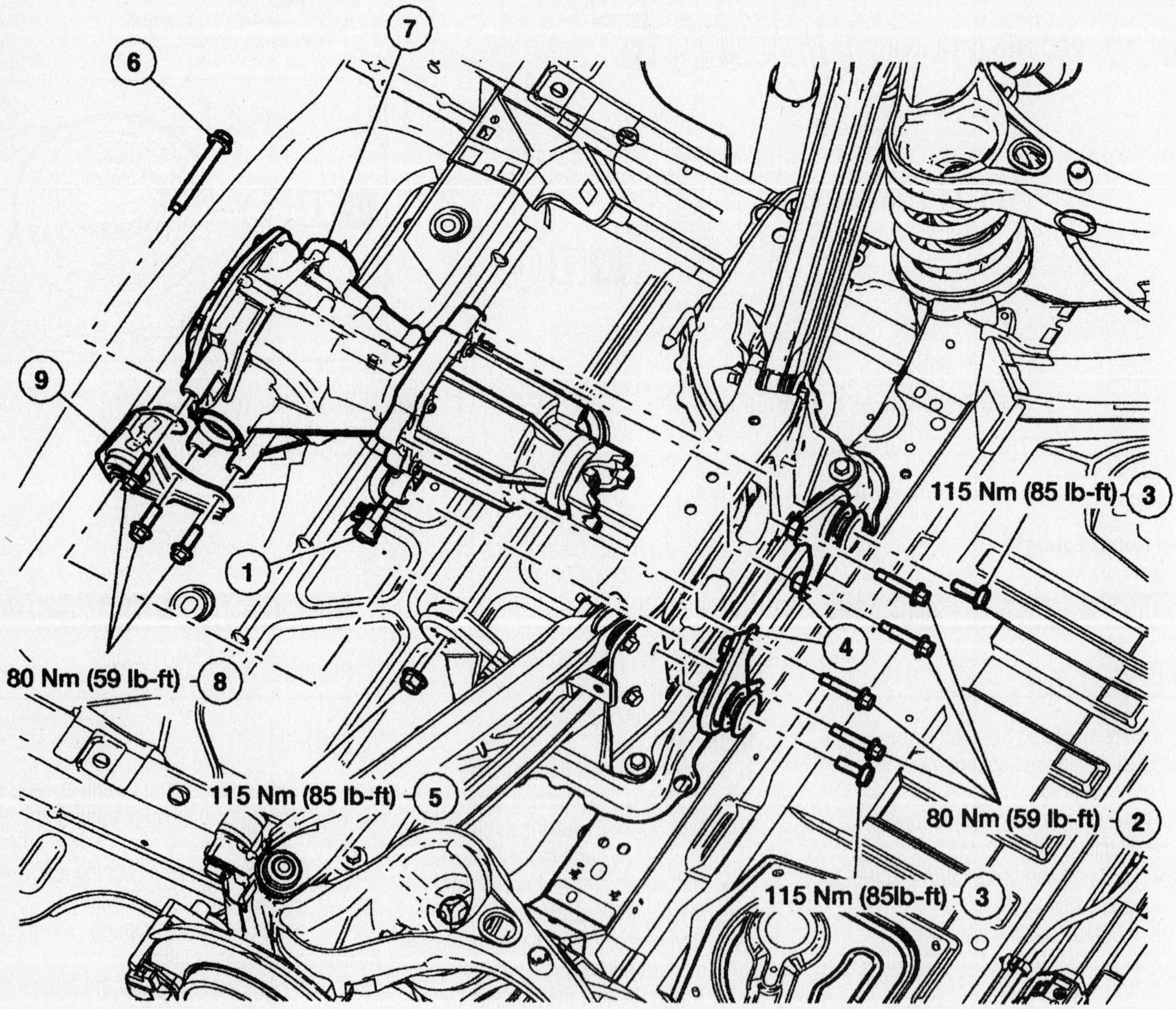

1 Electrical connector
2 Rear axle differential housing-to-front insulator bracket bolts
3 Front insulator-to-bracket subframe bolts
4 Front insulator brackets
5 Side insulator bracket-to-subframe nut
6 Side insulator bracket-to-subframe bolt
7 Rear axle assembly
8 Side insulator bracket-to-rear axle differential bolts
9 Side insulator bracket

67197-ESCA-G41

Rear drive axle removal—2005 models shown

3. Remove the front driveshaft-to-transfer case bolts.
4. Remove the universal joint cap bolts. Index-mark the pinion and yoke to the rear driveshaft before removing the bolts.
5. Remove the universal joint cap straps.
6. Remove the front driveshaft.
7. To install, reverse the removal procedure.

Rear Driveshaft

REMOVAL & INSTALLATION

1. Place the selector lever in NEUTRAL.
2. Raise and support the vehicle.
3. Remove the ground strap bolt
4. Remove the universal joint cap bolts. Index-mark the pinion and yoke to the rear driveshaft before removing the bolts.
5. Remove the universal joint cap straps.
6. Remove the center bearing support nuts.
7. Remove the universal joint cap bolts.
8. Remove the universal joint cap straps.
9. Remove the rear driveshaft.
10. To install, reverse the removal procedure.

Intermediate Shaft

REMOVAL & INSTALLATION

➡If removing the intermediate shaft in order to repair a separate component, it should only be removed as an assembly with the RH front halfshaft.

1. Remove the RH front halfshaft.
2. Remove the inner halfshaft bearing retainer nuts
3. Remove the intermediate shaft
4. To install, reverse the removal procedure. Apply a thin coat of grease to the splines of the intermediate shaft.
5. Verify the front axle lubricant level is to specifications.

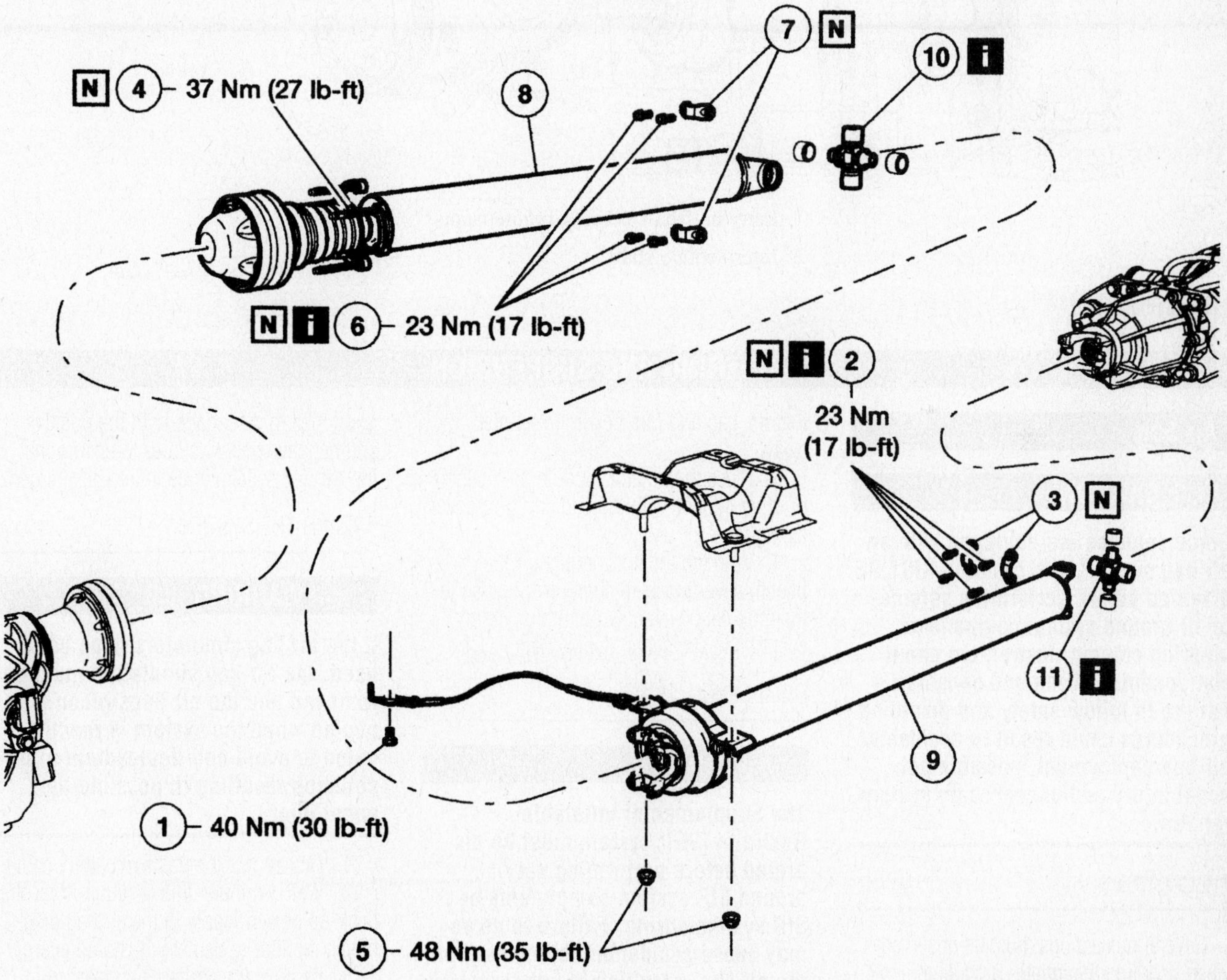

1 Ground strap bolt	5 Center bearing support nuts	9 Rear driveshaft
2 Universal joint cap bolts	6 Universal joint cap bolts	10 Front driveshaft U-joint
3 Universal joint cap straps	7 Universal joint cap straps	11 Rear driveshaft U-joint
4 Front driveshaft-to-transfer case bolts	8 Front driveshaft	

67197-ESCA-G42

Rear driveshaft assembly

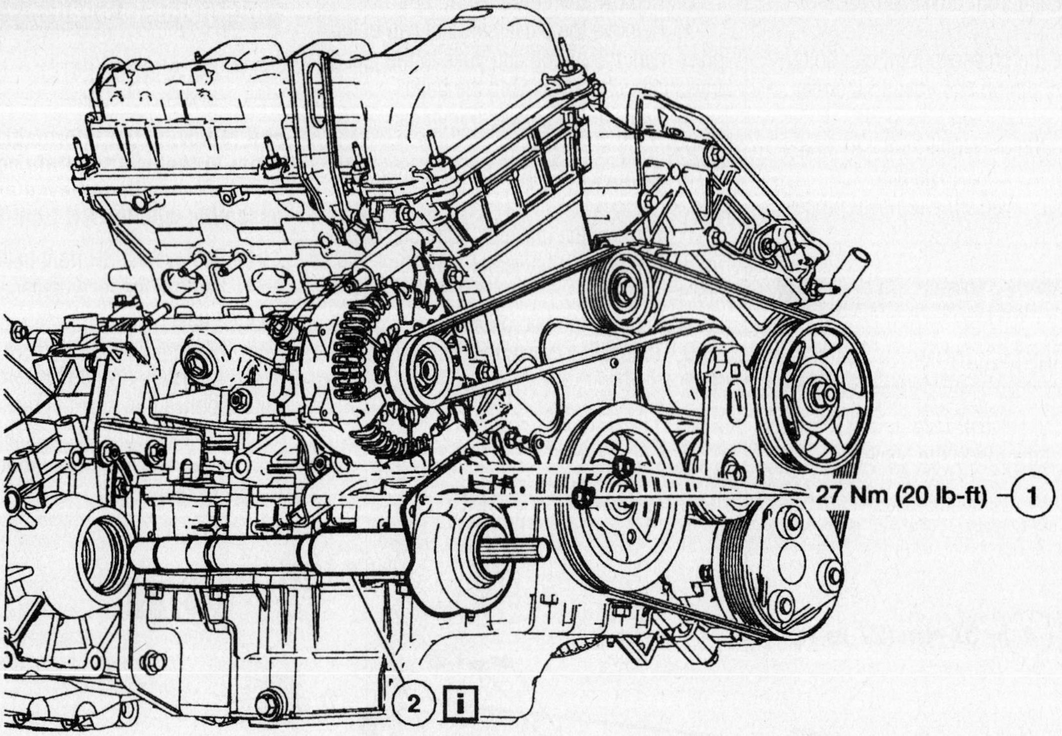

Intermediate shaft

STEERING AND SUSPENSION

Air Bag

** CAUTION

Some vehicles are equipped with an air bag system. The system MUST BE disabled before performing service on or around system components, steering column, instrument panel components, wiring and sensors. Failure to follow safety and disabling procedures could result in accidental air bag deployment, possible personal injury and unnecessary system repairs.

PRECAUTIONS

Several precautions must be observed when handling the inflator module to avoid accidental deployment and possible personal injury:

1. Never carry the inflator module by the wires or connector on the underside of the module.
2. When carrying a live inflator module, hold securely with both hands and ensure that the bag and trim cover are pointed away.
3. Place the inflator module on a bench or other surface with the bag and trim cover facing up.
4. With the inflator module on the bench, never place anything on or close to the module, which may be thrown in the event of an accidental deployment.

DISARMING

** CAUTION

The Supplemental Inflatable Restraint (SIR) system must be disarmed before performing service around SIR system components or SIR system wiring. Failure to do so may cause accidental deployment of the air bag, resulting in unnecessary SIR system repairs and/or personal injury.

The positive battery cable must be disconnected for a minimum of 1 minute before beginning any air bag work to de-energize the back-up power supply. It is a good idea to disengage both the positive and negative battery cables to ensure that the Air Bag system is definitely discharged.

ARMING THE SYSTEM

** WARNING

If the air bag simulators have been used, the air bag simulators must be removed and the air bags reconnected when the system is reactivated to avoid non-deployment in a collision resulting in possible personal injury.

1. Disconnect the positive battery cable.
2. Wait 1 minute, this is required for the back-up power supply in the air bag diagnostic monitor to deplete its stored energy.
3. Remove the air bag simulator from the air bag sliding contact connector at the top of the steering column. Reconnect the driver's side air bag module assembly. Position the driver's air bag module on the steering wheel and secure with the 2 bolts and washers. Tighten the bolt and washer assembly to 8–10 ft. lbs. (10–14 Nm).

4. Connect the positive battery cable.

5. Turn the ignition switch from the **OFF** to **RUN** and visually monitor the air bag warning indicator. The light will illuminate continuously for approximately 6 seconds and then turn off. If a fault occurs, the air bag indicator will either fail to light, remain lighted continuously or flash. The flashing may not occur until approximately 30 seconds after the ignition switch has been turned from **OFF** to **RUN**. This is the time needed for the air bag diagnostic monitor to complete testing the system. If the air bag indicator is inoperative, an air bag system fault exists, a tone will sound in a pattern of 5 sets of 5 beeps. If this occurs, the air bag indicator will need to be serviced before further diagnostics can be done.

Steering Gear

REMOVAL & INSTALLATION

1. Before servicing the vehicle, refer to the precautions in the beginning of this section.

2. Place the steering wheel in the

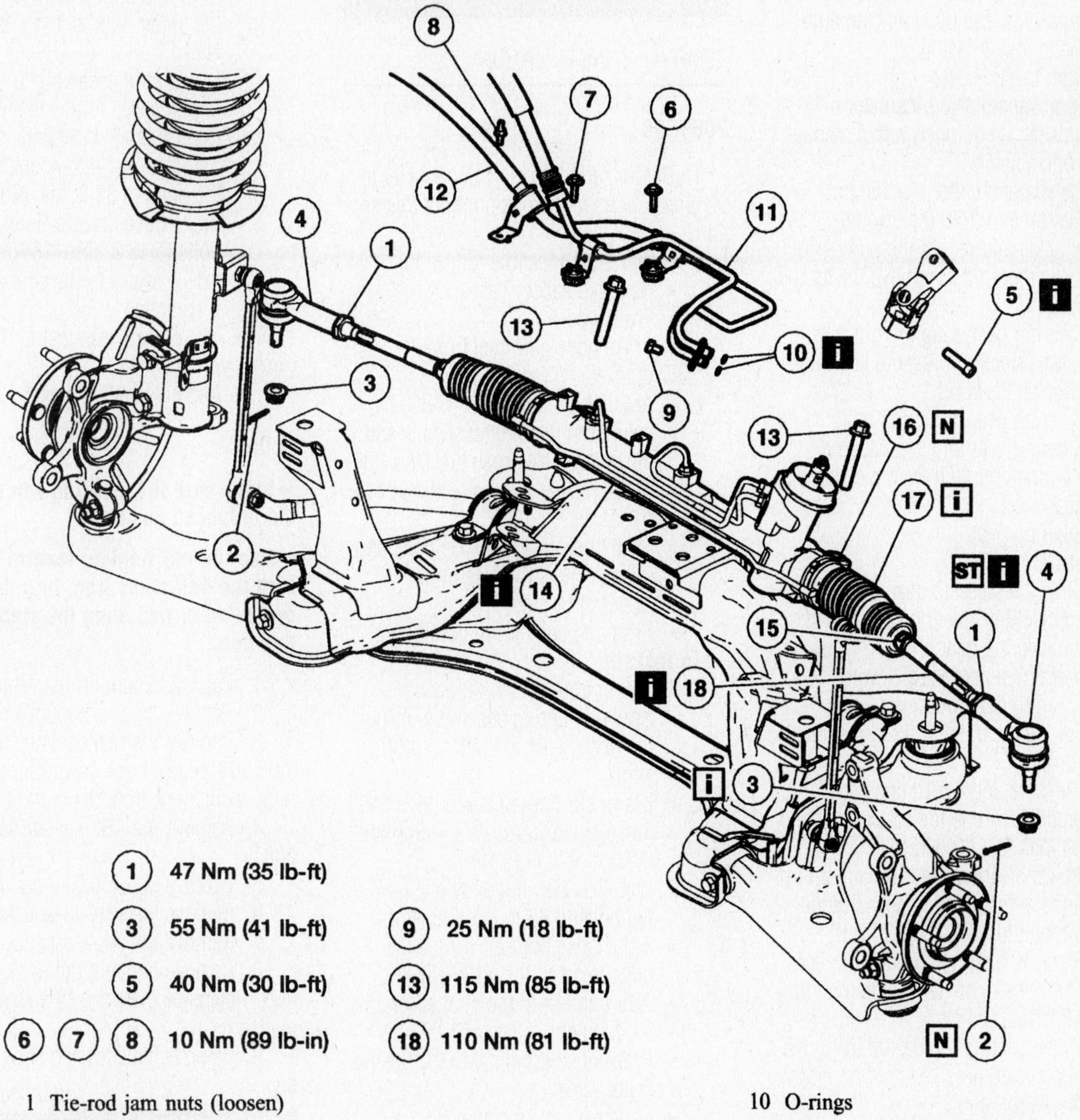

1 Tie-rod jam nuts (loosen)
2 Cotter pins
3 Tie-rod end nuts
4 Tie-rod end, outer
5 Steering column coupling-to-steering gear pinch bolt
6 Power steering return line bracket-to-subframe bolt
7 Power steering pressure line bracket-to-steering gear bolt
8 Power steering return line bracket-to-steering gear stud
9 Power steering line clamp plate bolt
10 O-rings
11 Power steering pressure line
12 Power steering return line
13 Steering gear mounting bolts
14 Steering gear
15 Tie-rod boot clamp, outer
16 Tie-rod boot clamp, inner
17 Tie-rod boot
18 Tie-rod, inner

67197-ESCA-G44

Steering gear mounting

straight-ahead position. Lock the steering wheel in place, using a steering wheel holder.

➡**Locking the steering wheel keeps the clockspring in alignment position.**

3. Drain the power steering fluid.
4. Remove or disconnect the following:
   - Negative battery cable
   - Rear transmission insulator
   - Rear transmission insulator bracket, if equipped with an automatic transmission
   - Both front wheels
   - Rear transmission insulator bracket, if equipped with a manual transmission
   - Tie rod end cotter pin and nut
   - Tie rod end from the steering knuckle and record the number of turns required to remove the tie rod end
   - Steering gear coupling pinch bolt
   - Power steering pressure and return lines and bracket
   - Steering gear mounting bolts
   - Steering gear and separate the steering coupling from the steering gear shaft
   - Steering gear

To install:

- Slide the steering gear rearward to connect the steering coupling to the steering gear shaft

5. Install or connect the following:
   - Steering gear mounting bolts. Torque the bolts to 93 ft. lbs. (126 Nm).
   - Pressure and return lines and bracket. Torque the bracket bolts to 89 inch lbs. (10 Nm).
   - Power steering pressure and return lines to the steering gear. Torque the bolt to 18 ft. lbs. (25 Nm).
   - Steering gear pinch bolt and reposition the boot. Torque the bolt to 18 ft. lbs. (25 Nm).
   - Tie rod end to the tie rod using the number of turns required to remove the tie rod end
   - Jam nuts. Torque the nuts to 35 ft. lbs. (47 Nm).
   - Tie rod end to the steering knuckle. Torque the nut to 41 ft. lbs. (57 Nm) and install a new cotter pin
   - Rear transmission insulator bracket. Torque the bolts to 66 ft. lbs. (90 Nm).
   - Both front wheels
   - Rear transmission insulator bracket. Torque the bolts to 66 ft. lbs. (90 Nm).
   - Rear transmission insulator. Torque the bolts to 66 ft. lbs. (90 Nm).
   - Negative battery cable
6. Fill and bleed the power steering system.
7. Start the vehicle and check for leaks, repair if necessary.
8. Check and adjust the front end alignment.

Strut

REMOVAL & INSTALLATION

2001–03

1. Before servicing the vehicle, refer to the precautions in the beginning of this section.
2. Install or connect the following:
   - Negative battery cable
   - Front wheel
   - Brake hose grommet from the bracket
   - Antilock Brake System (ABS) harness from the strut assembly and move the brake hose bracket aside
   - Stabilizer bar link nut and move the bar aside
   - Strut to steering knuckle bolts and support the strut assembly
   - Upper strut nuts
   - Strut and coil spring assembly

To install:

3. Install or connect the following:
   - Strut and spring assembly. Torque the upper nuts to 59 ft. lbs. (80 Nm).
   - Lower strut assembly to the steering knuckle. Torque the lower bolts to 85 ft. lbs. (115 Nm).
   - Stabilizer bar into position. Torque the bolts to 35 ft. lbs. (48 Nm).
   - Brake hose bracket. Torque the bolts to 14 ft. lbs. (18 Nm).
   - ABS harness to the strut assembly, if equipped
   - Brake hose grommet to the bracket
   - Front wheel
   - Negative battery cable

2004

1. Before servicing the vehicle, refer to the precautions in the beginning of this section.
2. Install or connect the following:
   - Negative battery cable
   - Front wheel
   - Brake hose grommet from the bracket
   - Antilock Brake System (ABS) harness from the strut assembly and move the brake hose bracket aside
   - Stabilizer bar link nut and move the bar aside
   - Strut to steering knuckle bolts and support the strut assembly
   - Upper strut nuts
   - Strut and coil spring assembly

To install:

3. Install or connect the following:
   - Strut and spring assembly. Torque the upper nuts to 41 ft. lbs. (55 Nm).
   - Lower strut assembly to the steering knuckle. Torque the lower bolts to 85 ft. lbs. (115 Nm).
   - Stabilizer bar into position. Torque the bolts to 35 ft. lbs. (48 Nm).
   - Brake hose bracket. Torque the bolts to 13 ft. lbs. (18 Nm).
   - ABS harness to the strut assembly, if equipped
   - Brake hose grommet to the bracket
   - Front wheel
   - Negative battery cable

2005

➡**Make sure the steering wheel is in the unlocked position.**

➡**Use the hex holding feature to prevent the ball studs from turning while removing or installing the stabilizer bar link nuts.**

1. Raise and support the vehicle.
2. Remove the brake jounce hose clip.
3. Remove the brake jounce hose. Pull the brake jounce hose downward slightly to remove the hose from the bracket.
4. Remove the ABS sensor harness bolt.
5. Remove the stabilizer bar link nut.
6. Remove the strut-to-knuckle nuts.
7. Remove the strut-to-knuckle bolts.
8. Remove the strut upper bushing nuts. Reference mark the strut mounting plate nuts.
9. Remove the strut and spring assembly.

⁂ WARNING

Do not allow the axle shaft to move outboard. Over-extension of the tripod CV joint can result in separation of internal parts, causing failure of the axle shaft.

10. To install, reverse the removal procedure.
11. Align the strut mounting plate nuts to the reference marks.

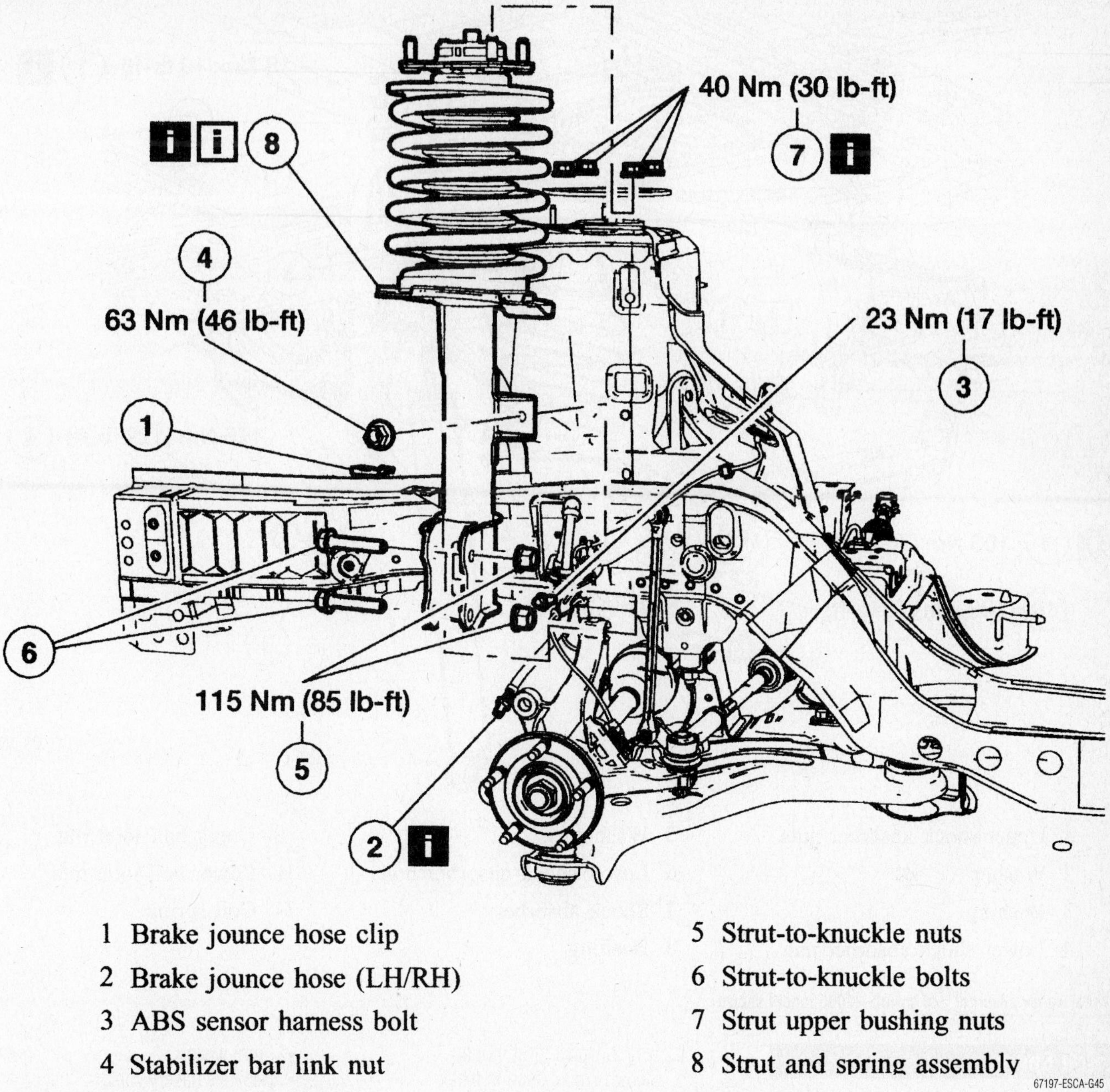

Front strut installation—2005 model shown

12. Check the front end alignment and adjust as necessary.

Shock Absorber

REMOVAL & INSTALLATION

Rear

2001–04

1. Before servicing the vehicle, refer to the precautions in the beginning of this section.
2. Remove or disconnect the following:
   - Negative battery cable
   - Rear quarter trim panel
   - Upper shock absorber nut and raise the vehicle enough to relax the suspension
   - Lower shock absorber nut
   - Shock absorber

To install:

3. Install or connect the following:
   - Shock absorber. Torque the lower nut to 85 ft. lbs. (115 Nm).
   - Upper shock absorber nut. Torque the nut to 13 ft. lbs. (18 Nm).
   - Rear quarter trim panel
   - Negative battery cable

2005

1. Remove the wheel and tire assemblies.
2. Remove the rear quarter trim panel. Remove the upper shock absorber nut, bushing and washer.
3. Remove the lower shock absorber nut, bolt and washer.
4. Remove the shock absorber and bushing.
5. To install, reverse the removal procedure. Torque the upper nut to 13 ft. lbs.; the lower nut to 129 ft. lbs. (175 Nm).

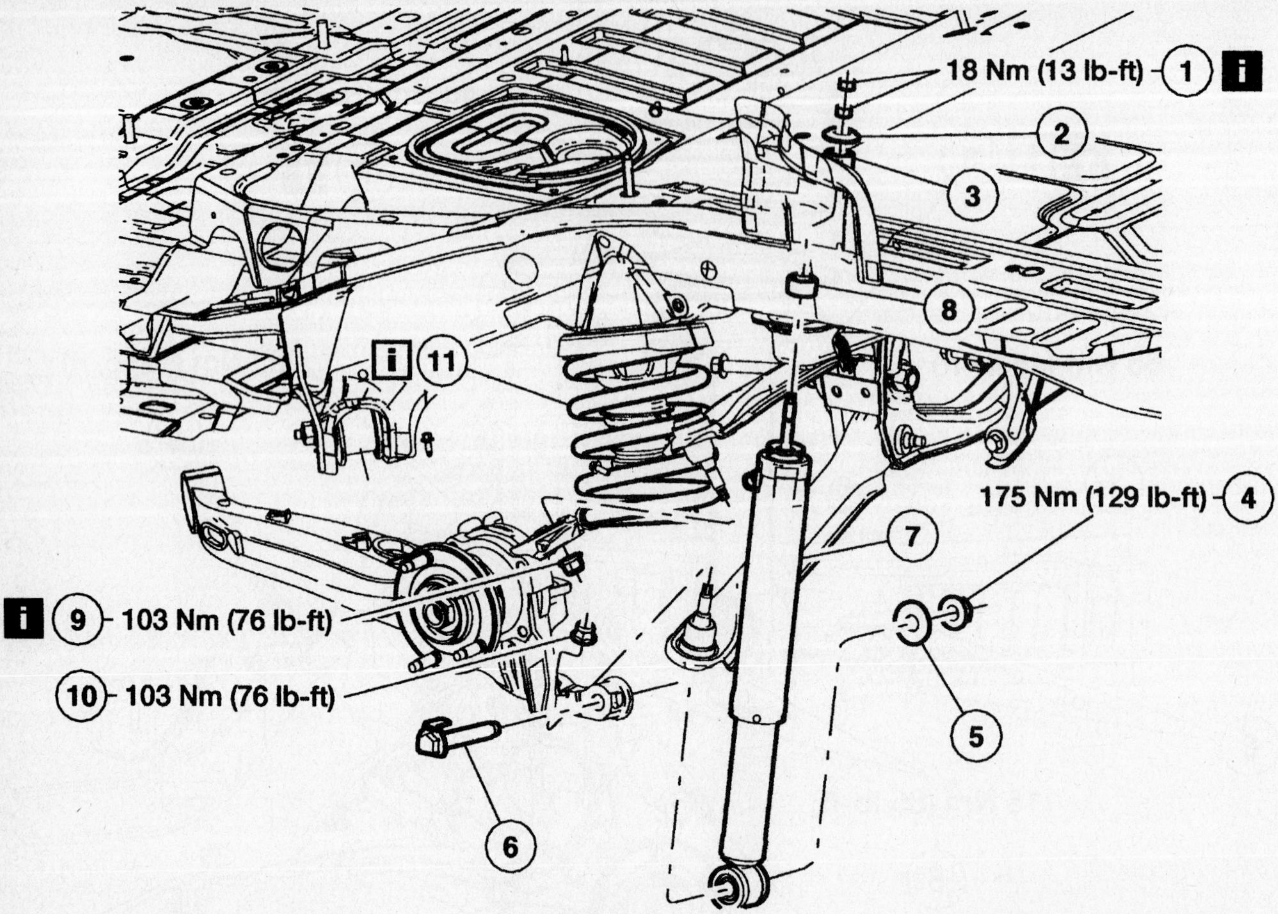

1 Upper shock absorber nuts
2 Washer
3 Bushing
4 Lower shock absorber nut
5 Washer
6 Lower shock absorber bolt
7 Shock absorber
8 Bushing
9 Upper ball joint nut
10 Lower ball joint nut
11 Coil spring

67197-ESCA-G46

Rear shock absorber and spring—2005 model shown

Coil Spring

REMOVAL & INSTALLATION

Front

1. Before servicing the vehicle, refer to the precautions in the beginning of this section.
2. Install or connect the following:
   - Negative battery cable
   - Front wheel
   - Strut and spring assembly and mount the strut assembly in a holding fixture and compress the coil spring using a suitable tool
   - Strut piston rod nut
3. Coil spring by disassembling the strut in the following sequence:
   a. Step 1: Metal sheet plate.
   b. Step 2: Upper strut mount.
   c. Step 3: Thrust bearing plate.
   d. Step 4: Thrust bearing.
   e. Step 5: Upper spring seat.
   f. Step 6: Upper spring seat isolator.
   g. Step 7: Coil spring.
   h. Step 8: Dust boot.
   i. Step 9: Rubber bump stopper.
   j. Step 10: Lower spring seat.

To install:

   - Assemble the strut assembly in the reverse order of the removal procedure
4. Install or connect the following:
   - Strut piston rod nut. Torque the nut to 76 ft. lbs. (103 Nm) and remove the assembly from the holding fixture
   - Strut and spring assembly
   - Front wheel
   - Negative battery cable

Rear

2001–04

1. Before servicing the vehicle, refer to the precautions in the beginning of this section.
2. Remove or disconnect the following:
   - Wheel and install 1 lug nut to retain the brake drum
   - Brake line from the wheel cylinder
   - Brake line bracket
   - Bolts from the Antilock Braking System (ABS) sensor bracket and move the sensor aside, if equipped
   - Rear knuckle and loosen the inside upper and lower arm bolts
   - Shock absorber lower nut

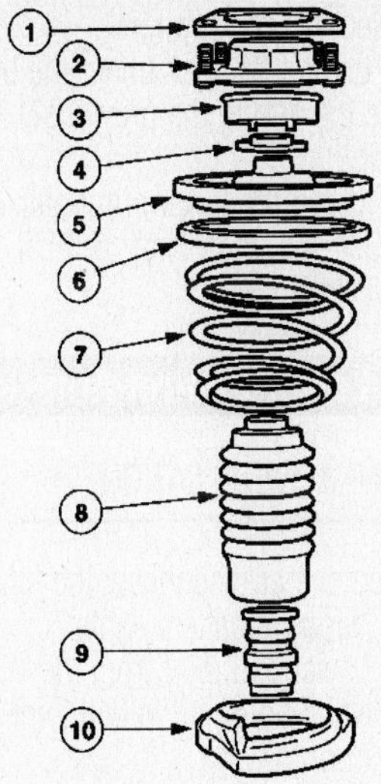

1 Metal sheet plate
2 Upper strut mount
3 Thrust bearing plate
4 Thrust bearing
5 Upper spring seat
6 Upper spring seat isolator
7 Spring
8 Dust boot
9 Rubber bump stopper
10 Lower spring seat

9308TG23

Disassemble the strut assembly in the proper sequence

- Spring

To install:

3. Install or connect the following:
   - Spring to the shock absorber
   - Lower shock absorber nut. Torque the nut to 85 ft. lbs. (115 Nm).
   - Inside upper and lower arm bolts. Torque the bolts to 85 ft. lbs. (115 Nm).
   - ABS sensor bracket, if equipped. Torque the bolts to 80 inch lbs. (9 Nm).
   - Brake line bracket. Torque the bolt to 15 ft. lbs. (20 Nm).
   - Brake line to the wheel cylinder. Torque the fastener to 11 ft. lbs. (15 Nm) and remove the lug nut
   - Wheel
   - Negative battery cable

2005

1. Remove the wheel and tire assemblies.
2. Remove the lower shock absorber nut and washer
3. Remove the lower shock absorber bolt
4. Support the wheel knuckle.
5. Remove the upper ball joint nut.
6. Remove the upper arm inner bolt.
7. Loosen the lower arm inner bolt.

➡Note the position of the coil spring insulator and coil spring for installation.

8. Carefully lower the wheel knuckle support.
9. Remove the coil spring
10. To install, reverse the removal procedure.
11. Align the coil spring and coil spring insulator to the previously noted position.
12. Observe the following torques:
    - Stabilizer bar link nut: 46 ft. lbs. (63 Nm)
    - Strut-to-knuckle nuts: 85 ft. lbs. (115 Nm)
    - Upper ball joint nut: 76 ft. lbs. (103 Nm)
    - Lower shock absorber nut: 129 ft. lbs. (175 Nm)

Front Lower Control Arm

REMOVAL & INSTALLATION

2001–2003

1. Remove or disconnect the following:
   - Negative battery cable

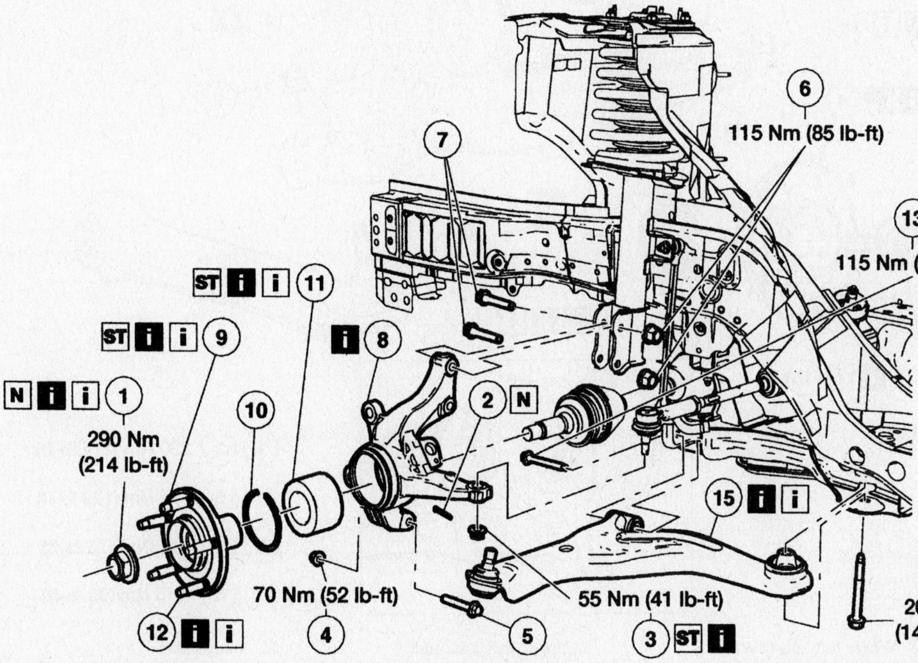

1 Wheel hub nut	6 Strut-to-knuckle nuts	11 Bearing
2 Cotter pin	7 Strut-to-knuckle bolts	12 Wheel stud
3 Tie rod end-to-knuckle nut	8 Wheel knuckle (LH/RH)	13 Lower arm bolt (front)
4 Lower ball joint pinch bolt	9 Wheel hub	14 Lower arm bolt (rear)
5 Lower ball joint pinch bolt	10 Snap ring	15 Lower arm

67197-ESCA-G48

Front lower control arm assembly—2004–05 model shown

- Front wheel
- Lower ball joint from the knuckle and support the subframe
- Lower control arm

To install:

2. Install or connect the following:
   - Lower control arm bolts and hand tighten them
   - Pinch bolt to the wheel knuckle. Torque the nut to 52 ft. lbs. (70 Nm) and remove the subframe support
   - Front wheel and jounce the vehicle
3. Torque the inner lower control arm bolt to 148 ft. lbs. (200 Nm) and outer bolt 85 ft. lbs. (115 Nm).

2004–05

1. Remove the wheel.
2. Lift the lower arm with a floor jack until the vehicle starts to lift.
3. Record the ride height. It's measure from the center of the halfshaft to the fender lip.
4. Remove the floor jack.
5. Disconnect the ball joint from the knuckle.
6. Support the sub-frame and remove the lower arm.

To install:

7. Install the lower arm, with the bolts loose.
8. Connect the ball joint. Torque the bolt to 52 ft. lbs. (70 Nm).
9. Remove the support.
10. Position the jack under the ball joint and raise the arm to the previously recorded ride height.
11. Tighten the lower arm bolts. Horizontal 85 ft. lbs. (115 Nm); vertical 148 ft. lbs. (200 Nm).

Rear Lower Control Arm

REMOVAL & INSTALLATION

1. Remove or disconnect the following:
   - Negative battery cable
   - Lower ball joint from the knuckle while holding the ball joint stud from moving

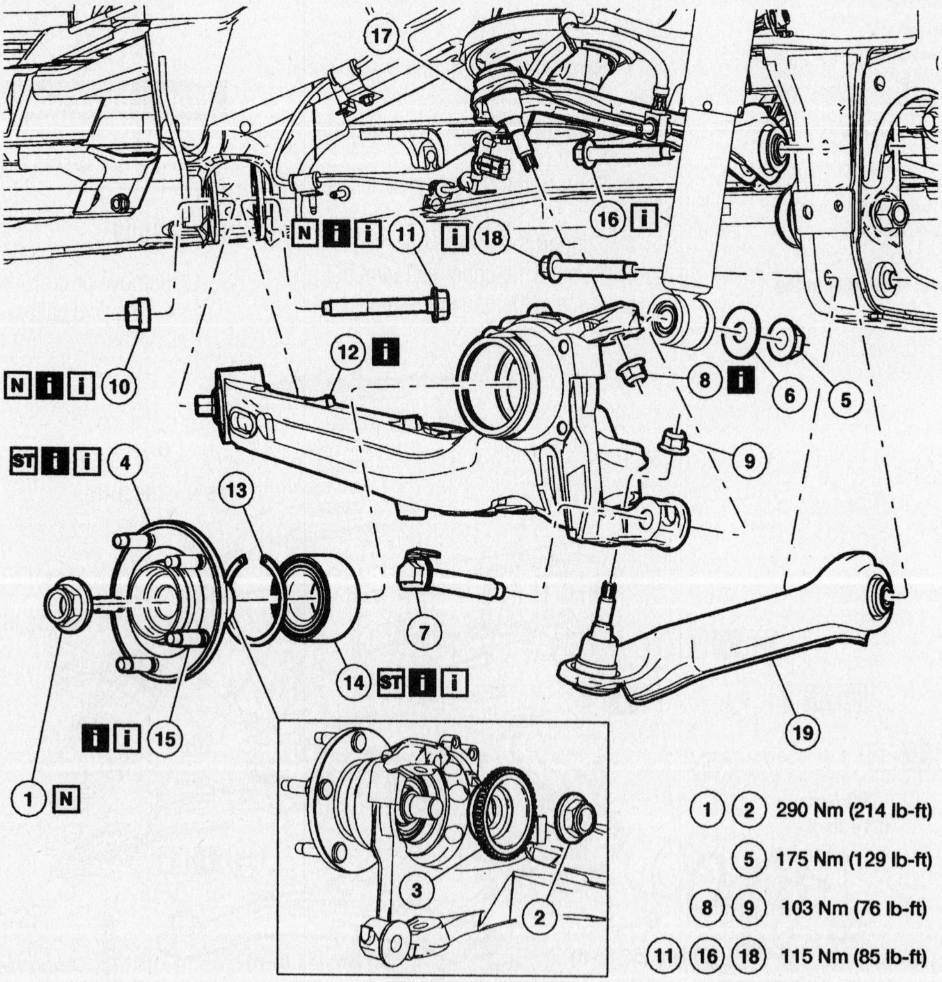

1 Wheel hub nut (4WD)
2 Wheel hub nut (2WD)
3 ABS sensor ring (2WD)
4 Wheel hub
5 Lower shock absorber nut
6 Washer
7 Lower shock absorber bolt
8 Upper ball joint nut
9 Lower ball joint nut
10 Cam nut
11 Wheel knuckle bolt
12 Knuckle assembly (LH/RH)
13 Wheel bearing snap ring
14 Wheel bearing
15 Wheel stud
16 Upper arm inner bolt
17 Upper arm
18 Lower arm inner bolt
19 Lower arm

67197-ESCA-G47

Rear lower control arm and related parts

- Lower ball joint nut
- Lower control arm
- Lower control arm inner bolt

To install:

2. Install or connect the following:
   - Lower control arm inner bolt
   - Lower control arm. Torque the bolts to 85 ft. lbs. (115 Nm).
   - Lower ball joint nut
   - Lower ball joint the knuckle. Torque the ball joint nut to 76 ft. lbs. (103 Nm).
   - Rear wheel
   - Negative battery cable

Rear Upper Control Arm

REMOVAL & INSTALLATION

1. Remove the wheel and tire.

➡It may be necessary to hold the ball joint stud to keep it from turning while removing the nut.

2. Separate the upper arm from the wheel knuckle. Remove the upper ball joint nut.
3. Remove the upper arm inner bolt.
4. Remove the upper arm.
5. To install, reverse the removal procedure. Observe the following torques:
   - Ball joint nut: 76 ft. lbs. (103 Nm)
   - Lower arm bolts: 85 ft. lbs. (115 Nm)

Wheel Bearings

REMOVAL & INSTALLATION

Front

1. Before servicing the vehicle, refer to the precautions in the beginning of this section.
2. Remove or disconnect the following:
   - Negative battery cable
   - Front wheel
   - Brake disc
   - Wheel hub nut
   - Tie rod end cotter pin and nut
   - Tie rod end from the knuckle
   - Antilock Brake System (ABS) sensor bolt and move the sensor aside, if equipped
   - Lower ball joint from the knuckle
   - Halfshaft from the knuckle and properly support the halfshaft
   - Steering knuckle
3. Press the hub from the wheel bearing and knuckle
4. Press the inner wheel bearing race from the knuckle and remove the snapring

➡The above step will not be necessary if the inner race remains with the knuckle.

5. Press the outer wheel bearing race from the knuckle

To install:

6. Install or connect the following:
   - Wheel bearing into the steering knuckle
   - Snapring
   - Wheel hub into the wheel bearing by using a press
   - Steering knuckle. Torque the bolts to 85 ft. lbs. (115 Nm).
   - Halfshaft into the wheel hub
   - Pinch bolt to knuckle. Torque the nut to 52 ft. lbs. (70 Nm).
   - Ball joint stud into the knuckle
   - ABS sensor. Torque the bolt to 80 inch lbs. (9 Nm), if equipped
   - Tie rod end to the knuckle. Torque the nut to 41 ft. lbs. (55 Nm).
   - New cotter pin to the tie rod end nut
   - Wheel hub. Torque the nut to 214 ft. lbs. (290 Nm).
   - Brake disc
   - Front wheel
   - Negative battery cable

Rear

2WD VEHICLES

1. Before servicing the vehicle, refer to the precautions in the beginning of this section.
2. Remove or disconnect the following:
   - Negative battery cable
   - Rear wheel
   - Rear brake drum
   - Wheel hub nut
   - Wheel hub
   - Inner wheel bearing race from the hub
   - Snapring
   - Wheel bearing outer race from the knuckle

To install:

3. Install or connect the following:
   - Wheel bearing in to the knuckle
   - Snapring
   - Wheel hub into the wheel bearing
   - Wheel hub nut. Torque the nut to 214 ft. lbs. (290 Nm).
   - Brake drum
   - Rear wheel
   - Negative battery cable

4WD VEHICLES

1. Before servicing the vehicle, refer to the precautions in the beginning of this section.
2. Remove or disconnect the following:
   - Negative battery cable
   - Rear wheel
   - Rear brake shoes
   - Rear halfshaft nut and loosen the halfshaft from the hub
   - Wheel hub and place it in a vise
   - Inner wheel bearing race from the hub
   - Antilock Brake System (ABS) sensor bracket and move the sensor aside, if equipped
   - Parking brake cable from the steering knuckle
   - Brake line from the wheel cylinder and support the knuckle
   - Lower shock absorber nut
   - Lower ball joint by holding the ball joint stud
   - Upper ball joint
   - Coil spring while noting the location of the insulator
   - Steering knuckle cam
   - Steering knuckle
   - Snapring and press out the outer wheel bearing race from the knuckle

To install:

3. Install or connect the following:
   - New wheel bearing into the steering knuckle
   - Snapring to the knuckle
   - Wheel hub
   - Steering knuckle cam and hand tighten the bolt
   - Coil spring
   - Shock absorber lower nut. Torque the nut to 85 ft. lbs. (115 Nm)for 2001–04 models; 129 ft. lbs. (175 Nm).
   - Upper ball joint. Torque the nut to 76 ft. lbs. (103 Nm).
   - Lower ball joint. Torque the nut to 76 ft. lbs. (103 Nm). Align the steering knuckle cam and torque the bolt to 85 ft. lbs. (115 Nm).
   - Brake line to the wheel cylinder. Torque the brake line bracket bolt to 15 ft. lbs. (20 Nm) and the brake line fastener to 11 ft. lbs. (15 Nm).
   - Parking brake cable to the backing plate. Torque the bolt to 16 ft. lbs. (22 Nm).
   - ABS sensor bracket. Torque the bolt to 80 inch lbs. (9 Nm), if equipped

- Halfshaft nut. Torque the nut to 214 ft. lbs. (290 Nm).
- Brake shoes
- Rear wheel
- Negative battery cable

4. Fill and bleed the brake system.
5. Check and adjust the wheel alignment as needed.

BRAKES

Brake Caliper

REMOVAL & INSTALLATION

Front

2001–04

1. Remove the wheel and tire assembly.
2. Remove the brake caliper clip.
3. Remove the brake caliper bolt caps and bolts.
4. Position the caliper aside.
5. Disconnect and cap the brake line from the caliper and remove the caliper. To install, reverse the removal procedure. Torque the mounting bolts to 26 ft. lbs. (35Nm). Torque the brake line to 15 ft. lbs. (20Nm).
6. Bleed the brake system.

2005

1. Remove the wheel and tire assembly.
2. Remove the brake caliper clip.
3. Remove the brake caliper dust boot caps.
4. Remove the brake caliper guide bolts.

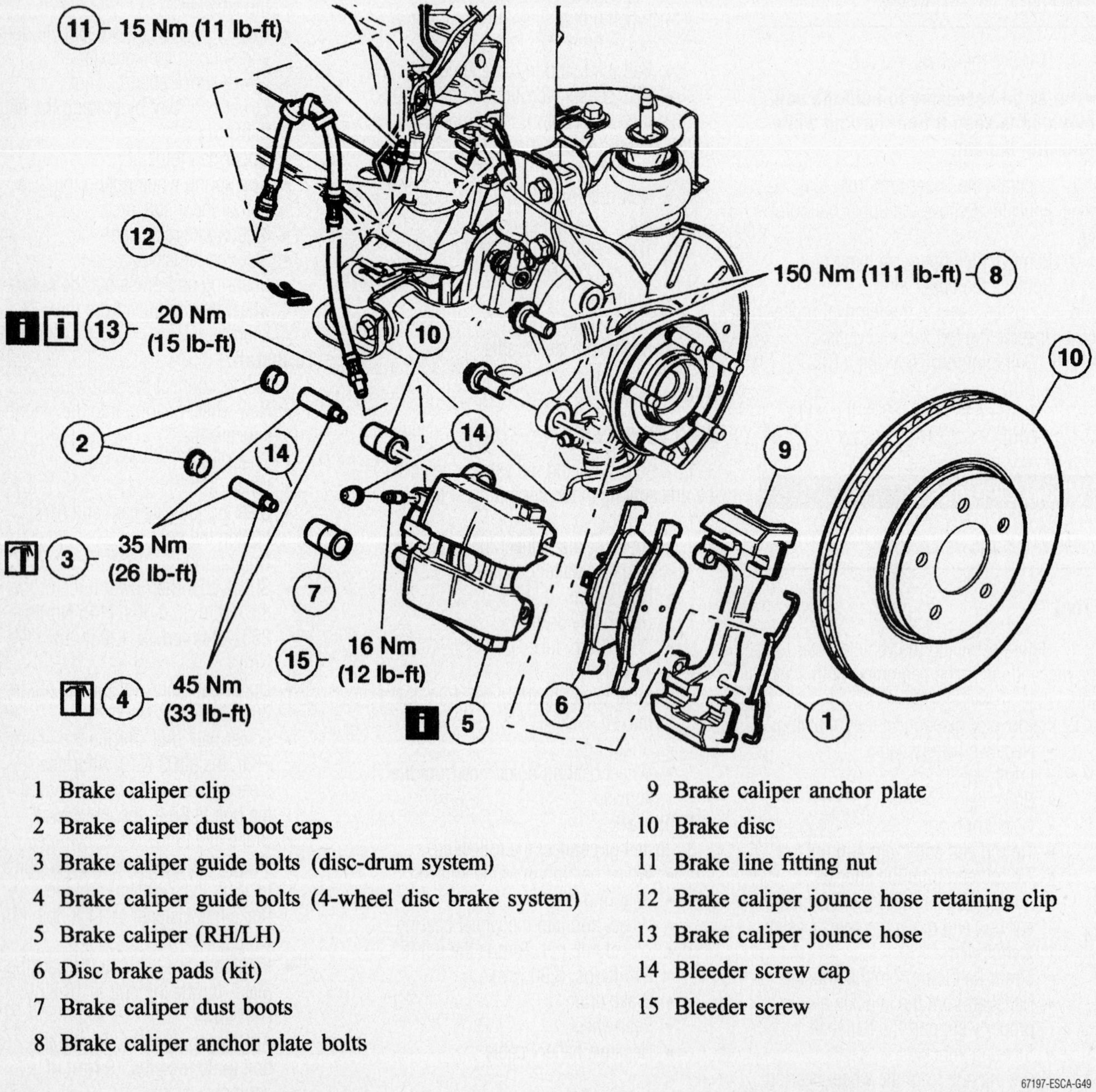

Front caliper installation—2005 model shown

5. Remove the brake caliper.

✲✲ CAUTION

Do not allow the brake caliper to hang by the flexible brake hose.

6. Remove the disc brake pads.
7. Remove the brake caliper jounce hose. Loosen the jounce hose fitting prior to removing the brake caliper.
8. To install, reverse the removal procedure.
9. Torque the caliper pin bolts to 26 ft. lbs. (35 Nm) on disc/drum systems; 33 ft. lbs. (45 Nm) on 4-wheel disc systems.
10. If the hydraulic system was opened, bleed the brake system.

➡Thread the brake caliper jounce hose onto the brake caliper before installing the brake caliper.

➡Make sure that the brake caliper jounce hose is not twisted.

11. Position the brake caliper to the anchor plate and tighten the brake caliper jounce hose.

Rear

1. Remove the wheel and tire assembly.
2. Remove the brake caliper guide bolts.

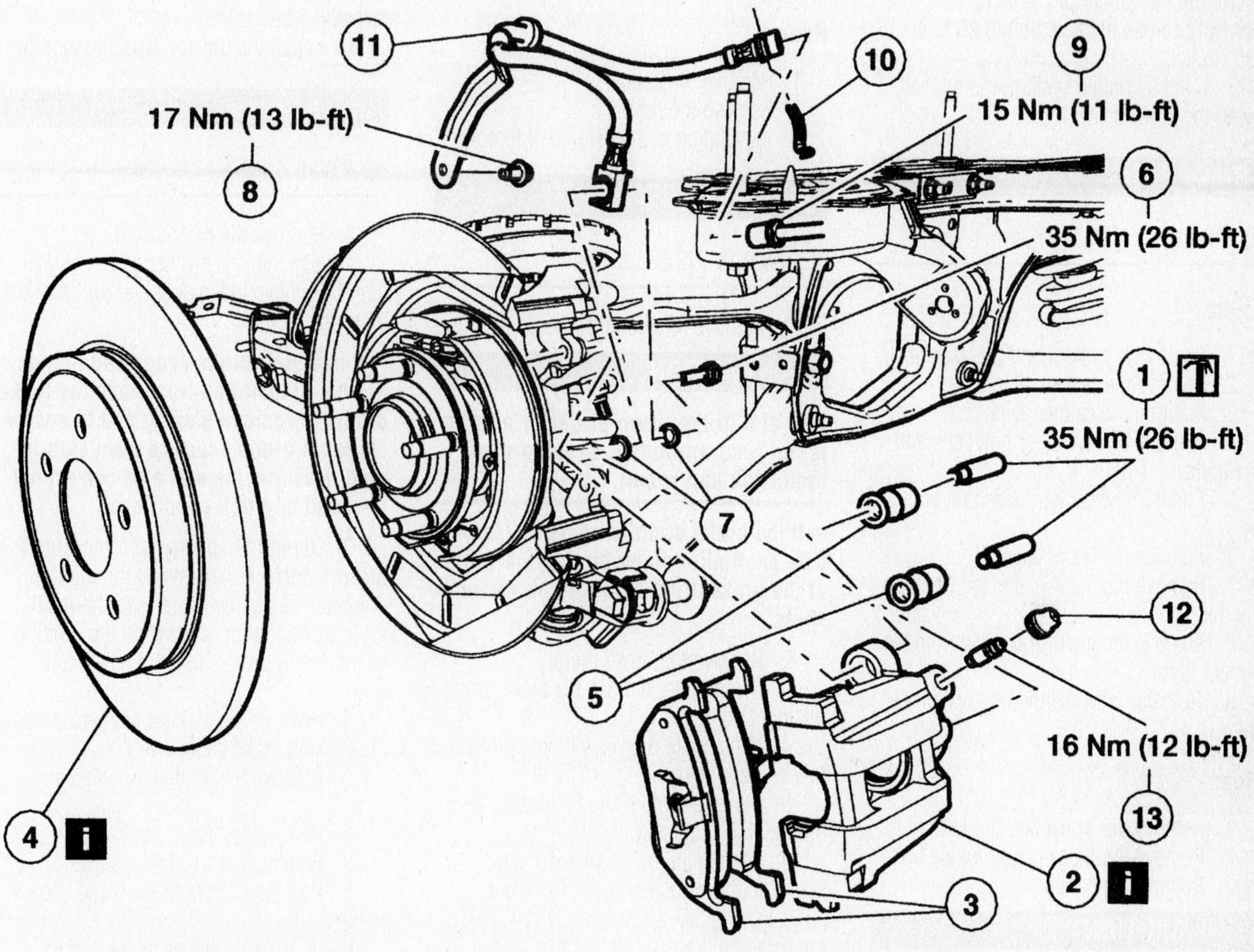

1 Brake caliper guide bolts
2 Caliper (RH/LH)
3 Brake disc pads
4 Brake disc
5 Brake caliper guide bolt
6 Brake caliper hose flow bolt
7 Copper washers
8 Brake caliper jounce hose bracket bolt
9 Brake line fitting nut
10 Brake caliper jounce hose retaining clip
11 Jounce hose (RH/LH)
12 Bleeder screw cap
13 Bleeder screw

67197-ESCA-G50

Rear caliper installation—2005 model shown

3. Remove the caliper.

⁂ CAUTION

Do not allow the brake caliper to hang by the flexible brake hose.

4. Remove the brake disc pads.
5. Remove the brake caliper hose flow bolt.
6. Remove and discard the copper washers.
7. To install, reverse the removal procedure. Use new copper washers.

Torque the caliper pin bolts to 26 ft. lbs. (35 Nm); torque the flow bolt to 26 ft. lbs. (35 Nm).

8. If the hydraulic system was opened, bleed the brake system.

Brake Pads

REMOVAL & INSTALLATION

Front

1. Remove the wheel and tire assembly.
2. Remove the brake caliper clip.
3. Position the caliper aside.
4. Remove brake caliper bolt caps and the bolts.
5. Position the caliper aside and support.
6. Remove the brake pads.
7. Remove the outer brake pad from the anchor.
8. Remove the inner brake pad from the caliper piston.
9. To install, reverse the removal procedure.

Rear

1. Remove the wheel and tire assembly.
2. Remove the brake caliper guide bolts.
3. Remove the caliper.

⁂ CAUTION

Do not allow the brake caliper to hang by the flexible brake hose.

4. Remove the brake disc pads.
5. To install, reverse the removal procedure.
6. If the hydraulic system was opened, bleed the brake system.

Brake Rotor

REMOVAL & INSTALLATION

Front

1. Remove the brake caliper anchor plate.
2. Remove the brake disc retaining clips (if equipped) and the brake disc.
3. To install, reverse the removal procedure. Torque the anchor plate bolts to 111 ft. lbs. (150Nm).

Rear

1. Remove the caliper.
2. Remove the rotor.
3. Installation is the reverse of removal.

Brake Drum

REMOVAL & INSTALLATION

1. Remove the tire and wheel assembly.

⁂ CAUTION

Use of a brake drum puller or a torch is not recommended. Brake drum distortion can result.

➡If the brake drum is rusted to the axle shaft pilot diameter, tap the center of the brake drum between the wheel studs.

2. Remove the brake drum.
3. If equipped, remove the brake drum retaining clips.
4. If the brake drums will not come off, follow these steps.
5. Move the brake shoe adjusting lever off the brake adjuster screw.
6. Loosen the brake shoe adjuster screw nut by adjusting the nut upward.
7. Using the special tool, 134-R0191, measure the brake drum inside diameter.
8. Install a new brake drum if the maximum inside diameter exceeds the specification shown on the outside of the brake drum.

To install:

⁂ WARNING

Whenever a wheel is installed, always remove any corrosion, dirt or foreign material present on the mounting surfaces of the wheel or the surface of the wheel hub, brake drum or brake disc that contacts the wheel. Installing wheels without correct metal-to-metal contact at the wheel mounting surfaces can cause the wheel nuts to loosen and the wheel to come off while the vehicle is in motion, causing loss of control. Failure to follow these instructions may result in personal injury.

9. Clean the wheel hub mounting surface and wheel pilot.
10. Install the tire and wheel assembly.

Brake Shoes

REMOVAL & INSTALLATION

1. Remove the brake drum.
2. Use the Brake/Clutch/Service Vacuum to remove brake dust and dirt from the brake assemblies.

➡If new rear brake shoes and linings are being installed, resurface the brake drums to remove glazing and to ensure an equal friction surface from side-to-side. Resurfacing will also correct out-of-round and bell conditions.

3. Using the special tool, measure the braking surface diameter. If the inside diameter measures more than the maximum specification shown on the outside of the brake drum, install a new brake drum.
4. Remove the parking brake cable from the parking brake cable lever.
5. Remove the hold-down clips and pins.
6. Remove the lower spring.
7. Remove the rear brake shoes.
8. Pull the bottom of the brake shoe forward.
9. Release the upper return spring.
10. Remove both brake shoes together.
11. Remove the self adjuster lever.
12. Remove the self adjuster and spring assembly.
13. Return the self adjuster to the fully seated position.
14. Remove the parking brake lever.
15. Remove the horseshoe clip.
16. Remove the parking brake lever.

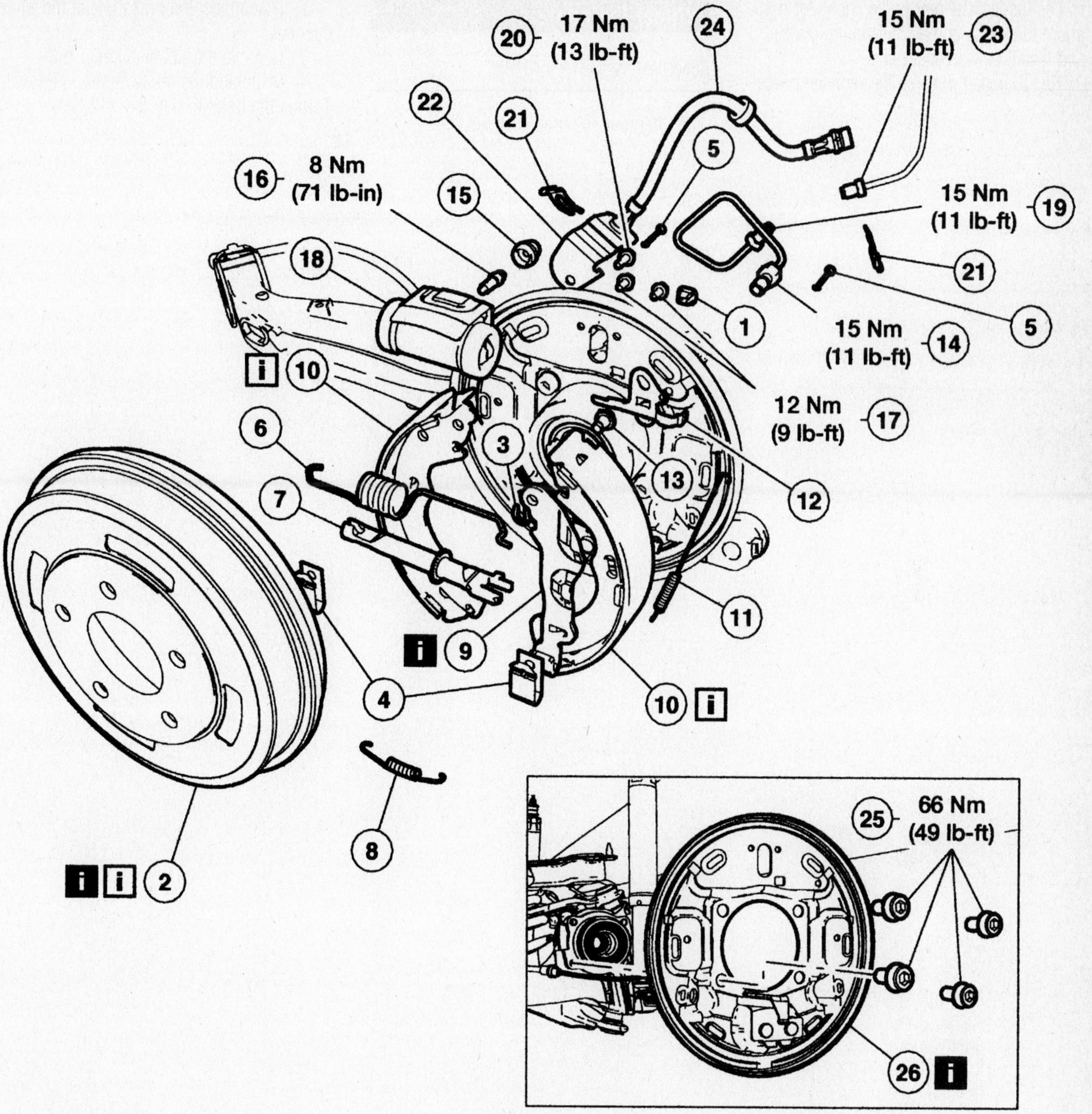

1 Plug	10 Brake shoe (kit)	19 Brake line fitting nut
2 Brake drum	11 Adjuster lever spring	20 Jounce hose bracket bolt
3 Parking brake lever clip	12 Adjuster lever (LH/RH)	21 Jounce hose retaining clips
4 Brake shoe retaining clips	13 Pivot pin (part of 2200)	22 Jounce hose bracket
5 Brake shoe retaining pins	14 Brake line fitting nut	23 Brake line fitting nut
6 Upper return spring	15 Bleeder screw cap	24 Jounce hose (LH/RH)
7 Adjuster assembly (LH/RH)	16 Bleeder screw	25 Backing plate bolts
8 Lower return spring	17 Wheel cylinder bolts	26 Backing plate
9 Parking brake actuator lever (LH/RH)	18 Wheel cylinder	

67197-ESCA-G51

Drum brake exploded view

17. Inspect the rear brake shoes for minimum thickness above the backing plate, and install new as necessary.
18. To install, reverse the removal procedure.

Wheel Cylinder

REMOVAL & INSTALLATION

1. Remove the brakes shoes.
2. Disconnect the brake line at the wheel cylinder.
3. Remove the wheel cylinder bolts.
4. Installation is the reverse of removal. Torque the bolts to 9 ft. lbs. (12 Nm).

FORD

Escort ZX2

7

SPECIFICATION CHARTS

ENGINE AND VEHICLE IDENTIFICATION

Engine						
Code ①	Liters (cc)	Cu. In.	Cyl.	Fuel Sys.	Type	Eng. Mfg.
3	2.0 (1990)	121	4	SFI	DOHC	Ford

SFI: Sequential Fuel Injection

DOHC: Double Overhead Camshafts

SOHC: Single Overhead Camshaft

① 8th digit of the VIN

② 10th digit of the VIN

Model Year	
Code ②	Year
1	2001
2	2002
3	2003
4	2004
5	2005

67197-ESCO-C01

GENERAL ENGINE SPECIFICATIONS

Year	Model	Engine Displacement Liters (VIN)	Net Horsepower @ rpm	Net Torque @ rpm (ft. lbs.)	Bore x Stroke (in.)	Com-pression Ratio	Oil Pressure @ rpm
2001	ZX2	2.0 (3)	125@5500	130@4000	3.34x3.46	10.0:1	35-65@2000
2002	ZX2	2.0 (3)	125@5500	130@4000	3.34x3.46	10.0:1	35-65@2000
2003	ZX2	2.0 (3)	125@5500	130@4000	3.34x3.46	10.0:1	35-65@2000

SFI: Sequential Fuel Injection

67197-ESCO-C02

ENGINE TUNE-UP SPECIFICATIONS

Year	Engine Displacement Liters (VIN)	Spark Plug Gap (in.)	Ignition Timing (deg.)		Fuel Pump (psi)	Idle Speed (rpm)		Valve Clearance	
			MT	AT		MT	AT	In.	Ex.
2001	2.0 (3)	0.052-0.056	10B	10B	31-38 ①	②	②	HYD	HYD
2002	2.0 (3)	0.052-0.056	10B	10B	31-38 ①	②	②	HYD	HYD
2003	2.0 (3)	0.052-0.056	10B	10B	31-38 ①	②	②	HYD	HYD

NOTE: The Vehicle Emission Control Information label often reflects specification changes made during production. The label figures must be used if they differ from those in this chart.

B: Before Top Dead Center

HYD: Hydraulic

① Fuel pressure with engine running, pressure regulator vacuum hose connected

② Refer to Vehicle Emission Control Information label

67197-ESCO-C03

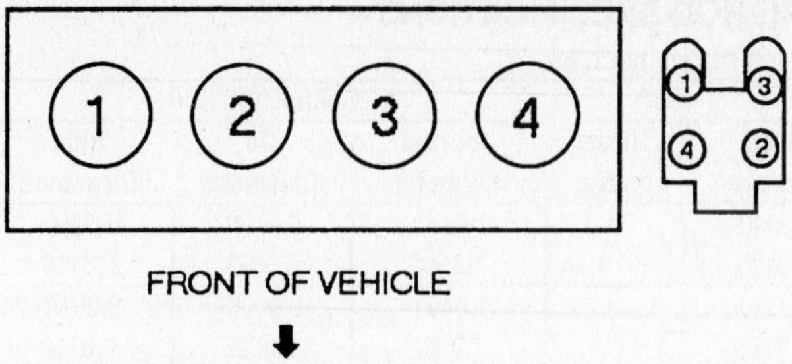

79223G06

2.0L VIN 3 Engine
Firing order: 1–3–4–2
Distributorless ignition system

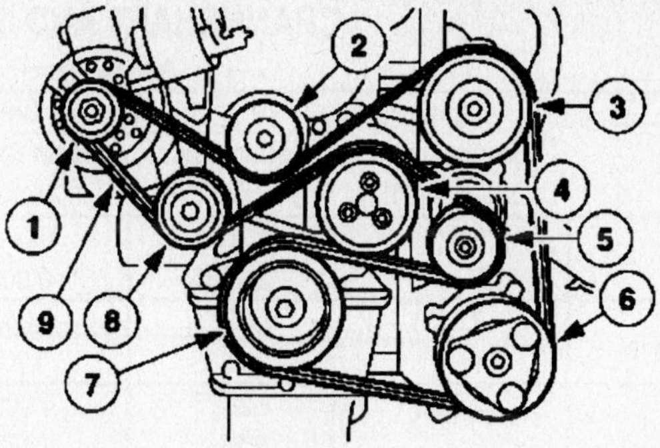

1	Generator pulley
2	Belt idler pulley
3	Power steering pump pulley
4	Water pump pulley
5	Belt tensioner
6	A/C clutch pulley
7	Crankshaft pulley damper
8	Belt idler pulley
9	Drive belt

93004G05

Accessory drive belt routing —Ford/Mercury 2.0L (VIN 3) DOHC engine

CAPACITIES

Year	Model	Engine Displacement Liters (VIN)	Oil with Filter (qts.)	Transmission (pts.) Manual	Transmission (pts.) Auto. ①	Front Axle (pts.)	Fuel Tank (gal.)	Cooling System (qts.)
2001	ZX2	2.0 (3)	4.0	6.7	13.4	②	13.2	④
2002	ZX2	2.0 (3)	4.0	6.7	13.4	②	13.2	④
2003	ZX2	2.0 (3)	4.0	6.7	13.4	②	13.2	④

Note: All capacities are approximates. Add fluid gradually and ensure a proper fluid level is obtained.

① Includes torque converter

② Included in transaxle capacity

③ Manual transaxle 7.9
Automatic transaxle 5.8

④ Manual transaxle 7.0
Automatic transaxle 7.5

67197-ESCO-C04

VALVE SPECIFICATIONS

Year	Engine Displacement Liters (VIN)	Seat Angle (deg.)	Face Angle (deg.)	Spring Test Pressure (lbs. @ in.)	Spring Installed Height (in.)	Stem-to-Guide Clearance (in.) Intake	Stem-to-Guide Clearance (in.) Exhaust	Stem Diameter (in.) Intake	Stem Diameter (in.) Exhaust
2001	2.0 (3)	45	45	NA	1.346	0.0007-0.0025	0.0014-0.0032	0.2373-0.2379	0.2366-0.2372
2002	2.0 (3)	45	45	NA	1.346	0.0007-0.0025	0.0014-0.0032	0.2373-0.2379	0.2366-0.2372
2003	2.0 (3)	45	45	NA	1.346	0.0007-0.0025	0.0014-0.0032	0.2373-0.2379	0.2366-0.2372

67197-ESCO-C05

CRANKSHAFT AND CONNECTING ROD SPECIFICATIONS

All measurements are given in inches.

Year	Engine Displacement Liters (VIN)	Crankshaft Main Brg. Journal Dia.	Crankshaft Main Brg. Oil Clearance	Crankshaft Shaft End-play	Crankshaft Thrust on No.	Connecting Rod Journal Diameter	Connecting Rod Oil Clearance	Connecting Rod Side Clearance
2001	2.0 (3)	2.2827-2.2835	0.0008-0.0026	0.0040-0.0120	3	1.8461-1.8500	0.0006-0.0028	0.0040-0.0110
2002	2.0 (3)	2.2827-2.2835	0.0008-0.0026	0.0040-0.0120	3	1.8461-1.8500	0.0006-0.0028	0.0040-0.0110
2003	2.0 (3)	2.2827-2.2835	0.0008-0.0026	0.0040-0.0120	3	1.8461-1.8500	0.0006-0.0028	0.0040-0.0110

67197-ESCO-C06

PISTON AND RING SPECIFICATIONS

All measurements are given in inches.

Year	Engine Displacement Liters (VIN)	Piston Clearance	Ring Gap Top Compression	Ring Gap Bottom Compression	Ring Gap Oil Control	Ring Side Clearance Top Compression	Ring Side Clearance Bottom Compression	Ring Side Clearance Oil Control
2001	2.0 (3)	0.0010-0.0022	0.012-0.022	0.012-0.022	0.010-0.039	0.0015-0.0032	0.0015-0.0035	—
2002	2.0 (3)	0.0010-0.0022	0.012-0.022	0.012-0.022	0.010-0.039	0.0015-0.0032	0.0015-0.0035	—
2003	2.0 (3)	0.0010-0.0022	0.012-0.022	0.012-0.022	0.010-0.039	0.0015-0.0032	0.0015-0.0035	—

67197-ESCO-C07

TORQUE SPECIFICATIONS

All readings in ft. lbs.

Year	Engine Displacement Liters (VIN)	Cylinder Head Bolts	Main Bearing Bolts	Rod Bearing Bolts	Crankshaft Damper Bolts	Flywheel Bolts	Manifold Intake	Manifold Exhaust	Spark Plugs	Oil Pan Drain Plug
2001	2.0 (3)	①	55-65	②	80-87	71-76	11-12	10-12	10-12	22
2002	2.0 (3)	①	55-65	②	80-87	71-76	11-12	10-12	10-12	22
2003	2.0 (3)	①	55-65	②	80-87	71-76	11-12	10-12	10-12	22

NOTE: Always follow proper torque patterns

NOTE: Stretch bolts are used in all procedures that require rotating the fastener a certain number of degrees. The bolts stretch and cannot be reused. For reassembly, replace with new fastner

① Do not reuse cylinder head bolts.

Step 1: Tighten bolts, in sequence, to 15 ft. lbs.

Step 2: Tighten bolts in sequence to 40 ft. lbs.

Step 3: Turn all bolts, in sequence, plus an additional 105 degrees

② Step 1: 22-25 ft. lbs.

Step 2: Rotate each bolt 85-95 degrees

67197-ESCO-C08

WHEEL ALIGNMENT

			Caster		Camber		
Year	Model		Range (+/-Deg.)	Preferred Setting (Deg.)	Range (+/-Deg.)	Preferred Setting (Deg.)	Toe-in (in.)
2001	ZX2	F	1.00	+2.20	1.00	+0.40	0.08 +/- 0.10
		R	—	—	1.00	+1.18	0.08 +/- 0.10
2002	ZX2	F	1.00	+2.20	1.00	+0.40	0.08 +/- 0.10
		R	—	—	1.00	+1.18	0.08 +/- 0.10
2003	ZX2	F	1.00	+2.20	1.00	+0.40	0.08 +/- 0.10
		R	—	—	1.00	+1.18	0.08 +/- 0.10

67197-ESCO-C09

TIRE, WHEEL AND BALL JOINT SPECIFICATIONS

		OEM Tires		Tire Pressures (psi)		Wheel	Ball Joint	Lug
Year	Model	Standard	Optional	Front	Rear	Size	Inspection	Nut
2001	ZX2	P185/65R14	P185/60R15	32	32	5.5-JJ	0.030 in. ①	76
2002	ZX2	P185/65R14	P185/60R15	32	32	5.5-JJ	0.030 in. ①	76
2003	ZX2	P185/65R14	P185/60R15	32	32	5.5-JJ	0.030 in. ①	76

OEM: Original Equipment Manufacturer

PSI: Pounds Per Square Inch

STD: Standard

OPT: Optional

① Maximum radial tolerance in inches

② Replace if any measurable movement is found.

67197-ESCO-C10

BRAKE SPECIFICATIONS

All measurements in inches unless noted

			Brake Disc			Brake Drum Diameter				Brake Caliper
Year	Model		Original Thickness	Minimum Thickness	Maximum Runout	Original Inside Diameter	Max. Wear Limit	Maximum Machine Diameter	Minimum Lining Thickness	Mounting Bolts (ft. lbs.)
2001	ZX2	F	0.870	0.790	0.004	—	—	—	0.080	36-43
		R	0.350	0.280	0.004	7.87	7.95	7.91	0.040	—
2002	ZX2	F	0.870	0.790	0.004	—	—	—	0.080	36-43
		R	0.350	0.280	0.004	7.87	7.95	7.91	0.040	—
2003	ZX2	F	0.870	0.790	0.004	—	—	—	0.080	36-43
		R	0.350	0.280	0.004	7.87	7.95	7.91	0.040	—

NOTE: Follow specifications stamped on rotor or drum if figures differ from those in this chart.

NA: Not Available

F: Front

R: Rear

67197-ESCO-C11

SCHEDULED MAINTENANCE INTERVALS

Ford—Escort ZX2

TO BE SERVICED	TYPE OF SERVIC	VEHICLE MILEAGE INTERVAL (x1000)																			
		5	10	15	20	25	30	35	40	45	50	55	60	65	70	75	80	85	90	95	100
Engine oil & filter	R	✓	✓	✓	✓	✓	✓	✓	✓	✓	✓	✓	✓	✓	✓	✓	✓	✓	✓	✓	✓
Tires ①	S/I	✓		✓		✓		✓		✓		✓		✓		✓		✓		✓	
Air Cleaner	R						✓						✓						✓		
Spark Plugs	R																				✓
Drive Belts	S/I												✓								
Cooling system	S/I			✓			✓			✓			✓			✓			✓		
Engine coolant	R										✓						✓				
PCV valve	R												✓								
Exhaust heat shields	S/I						✓						✓						✓		
Brake linings & drums	S/I						✓						✓						✓		
Brake line hoses & connections	S/I						✓						✓						✓		
Front ball joints	S/I						✓						✓						✓		
Bolts & nuts on chassis body	S/I						✓						✓						✓		
Steering linkage operation	S/I						✓						✓						✓		
Brake pads & rotor	S/I						✓						✓						✓		
Clutch pedal operation	S/I						✓						✓						✓		
Halfshaft dust boots	S/I						✓						✓						✓		

R: Replace   S/I: Inspect and service, if needed

① Rotate, inspect the tire tread for wear, and adjust air pressure.

FREQUENT OPERATION MAINTENANCE (SEVERE SERVICE)

If a vehicle is operated under any of the following conditions it is considered severe service:

- Extremely dusty areas.

- 50% or more of the vehicle operation is in 32°C (90°F) or higher temperatures, or constant operation in temperatures below 0°C (32°F).

- Prolonged idling (vehicle operation in stop and go traffic).

- Frequent short running periods (engine does not warm to normal operating temperatures).

- Police, taxi, delivery usage or trailer towing usage.

Oil & filter change: change every 3000 miles.

Rotate tires at 6000 miles & every 9000 miles thereafter.

Automatic transmission fluid & filter: change every 21,000 miles.

67197-ESCO-C12

PRECAUTIONS

Before servicing any vehicle, please be sure to read all of the following precautions, which deal with personal safety, prevention of component damage, and important points to take into consideration when servicing a motor vehicle:

- Never open, service or drain the radiator or cooling system when the engine is hot; serious burns can occur from the steam and hot coolant.
- Observe all applicable safety precautions when working around fuel. Whenever servicing the fuel system, always work in a well-ventilated area. Do not allow fuel spray or vapors to come in contact with a spark, open flame or excessive heat (a hot drop light, for example). Keep a dry chemical fire extinguisher near the work area. Always keep fuel in a container specifically designed for fuel storage; also, always properly seal fuel containers to avoid the possibility of fire or explosion. Refer to the additional fuel system precautions later in this section.
- Fuel injection systems often remain pressurized, even after the engine has been turned **OFF**. The fuel system pressure must be relieved before disconnecting any fuel lines. Failure to do so may result in fire and/or personal injury.
- Brake fluid often contains polyglycol ethers and polyglycols. Avoid contact with the eyes and wash your hands thoroughly after handling brake fluid. If you do get brake fluid in your eyes, flush your eyes with clean, running water for 15 minutes. If eye irritation persists, or if you have taken brake fluid internally, IMMEDIATELY seek medical assistance.
- The EPA warns that prolonged contact with used engine oil may cause a number of skin disorders, including cancer! You should make every effort to minimize your exposure to used engine oil. Protective gloves should be worn when changing oil. Wash your hands and any other exposed skin areas as soon as possible after exposure to used engine oil. Soap and water, or waterless hand cleaner should be used.
- All new vehicles are now equipped with an air bag system. The system must be disabled before performing service on or around system components, steering column, instrument panel components, wiring and sensors. Failure to follow safety and disabling procedures could result in accidental air bag deployment, possible personal injury and unnecessary system repairs.
- Always wear safety goggles when working with, or around, the air bag system. When carrying a non-deployed air bag, be sure the bag and trim cover are pointed away from your body. When placing a non-deployed air bag on a work surface, always face the bag and trim cover upward, away from the surface. This will reduce the motion of the module if it is accidentally deployed. Refer to the additional air bag system precautions later in this section.
- Clean, high quality brake fluid from a sealed container is essential to the safe and proper operation of the brake system. You should always buy the correct type of brake fluid for your vehicle. If the brake fluid becomes contaminated, completely flush the system with new fluid. Never reuse any brake fluid. Any brake fluid that is removed from the system should be discarded. Also, do not allow any brake fluid to come in contact with a painted surface; it will damage the paint.
- Never operate the engine without the proper amount and type of engine oil; doing so WILL result in severe engine damage.
- Timing belt maintenance is extremely important! Many models utilize an interference-type, non-freewheeling engine. If the timing belt breaks, the valves in the cylinder head may strike the pistons, causing potentially serious (also time-consuming and expensive) engine damage.
- Disconnecting the negative battery cable on some vehicles may interfere with the functions of the on-board computer system(s) and may require the computer to undergo a relearning process once the negative battery cable is reconnected.
- When servicing drum brakes, only disassemble and assemble one side at a time, leaving the remaining side intact for reference.
- Only an MVAC-trained, EPA-certified automotive technician should service the air conditioning system or its components.

ENGINE REPAIR

➡Disconnecting the negative battery cable on some vehicles may interfere with the operation of the on-board computer system. The computer may undergo a relearning process once the negative battery cable is reconnected.

Alternator

REMOVAL

Sedan and Wagon

1. Before servicing the vehicle, refer to the precautions in the beginning of this section.
2. Remove or disconnect the following:
   - Negative battery cable
   - Drive belt
   - Power steering hose bracket from the alternator bracket
   - Alternator mounting bolts
   - Alternator electrical connectors
   - Battery positive cable
   - Alternator from the vehicle

Coupe

1. Before servicing the vehicle, refer to the precautions in the beginning of this section.
2. Remove or disconnect the following:
   - Negative battery cable
   - Drive belt
   - Coolant tank reservoir
   - Alternator electrical connectors
   - Alternator mounting bolts
   - Battery positive cable
   - Alternator from the vehicle

90982G23

Location of the alternator upper and lower mounting bolts—sedan and wagon models

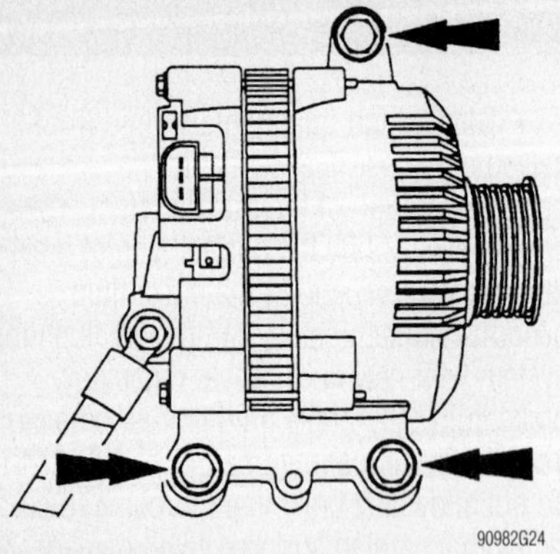

Location of the alternator upper and lower mounting bolts—coupe model

INSTALLATION

Sedan and Wagon

1. Install or connect the following:
   - Battery positive cable to the alternator. Tighten the nut to 7–9 ft. lbs. (9–12 Nm).
   - Alternator electrical connections
   - Alternator in position
   - Alternator mounting bolts. Tighten the upper mounting bolt to 29–40 ft. lbs. (40–55 Nm) and the lower mounting bolt to 15–22 ft. lbs. (20–30 Nm).
   - Power steering hose bracket to the alternator bracket
   - Drive belt
   - Negative battery cable

Coupe

1. Install or connect the following:
   - Battery positive cable to the alternator. Tighten the nut to 5–7 ft. lbs. (7–9 Nm).
   - Alternator in position
   - Alternator mounting bolts and tighten to 29–40 ft. lbs. (40–55 Nm)
   - Alternator electrical connections
   - Coolant tank reservoir
   - Drive belt
   - Negative battery cable

Ignition Timing

ADJUSTMENT

The Powertrain Control Module (PCM) controls ignition timing on these models. No adjustment is necessary or possible.

Engine Assembly

REMOVAL & INSTALLATION

1. Before servicing the vehicle, refer to the precautions in the beginning of this section.
2. Drain the cooling system.
3. Relieve the fuel system pressure.
4. Drain the engine oil.
5. Recover the A/C refrigerant, if equipped.
6. Remove or disconnect the following:
   - Battery and tray
   - Hood
   - Air cleaner outlet tube
   - Fuel pressure regulator vacuum hose
   - Intake manifold vacuum hoses
   - Exhaust Gas Recirculation (EGR) vacuum line
   - Positive Crankcase Ventilation (PCV) valve vacuum line
   - Vacuum tree
   - Throttle cable
   - Cruise control cable, if equipped
   - Shift cable
   - Constant Control Relay Module (CCRM)
   - EGR backpressure transducer
   - Transaxle range sensor connector, if equipped
   - Transaxle solenoid connector, if equipped
   - Turbine speed sensor connector, if equipped
   - Vehicle Speed (VSS) sensor
   - Heated Oxygen (HO_2S) sensor connectors
   - Ground strap
   - Engine cooling fan connector
   - Power steering hose brackets
   - Fuel line
   - Exhaust manifold heat shield
   - Starter bracket
   - Clutch cylinder fluid line, if equipped
   - Transaxle cooler lines, if equipped
   - A/C line bracket, if equipped
   - A/C condenser tube, if equipped
   - A/C accumulator tube, if equipped
   - Accessory drive belt and tensioner
   - Accessory drive belt idler pulley, if equipped
   - Power steering hoses
   - Front wheels
   - Left and right splash shields
   - Front subframe crossmember
   - Catalytic converter
   - Axle halfshafts
   - Starter motor
   - Oil pressure switch connector
   - A/C compressor, if equipped
   - Radiator
   - Heater hoses
   - Alternator harness connectors
7. Attach a support fixture to the engine.
8. Remove or disconnect the following:
   - Left front engine mount
   - Transaxle mount
   - Transaxle support crossmember
   - Upper transaxle mount
   - Right engine mount
9. Attach a hoist to the engine and remove the support fixture.
10. Raise the engine and transaxle from the vehicle.

To install:

11. Installation is the reverse of removal, but please note the following important steps.
    - Front isolator nuts and bolts: 50–68 ft. lbs. (67–93 Nm)
    - Front engine support isolator through-bolt: 50–68 ft. lbs. (67–93 Nm)
    - Front roll restrictor bolts: 48–65 ft. lbs. (64–89 Nm) and the nuts to 50–69 ft. lbs. (67–93 Nm)
    - Rear roll restrictor nuts: 28–38 ft. lbs. (38–51 Nm)
    - Torque converter nuts: 27 ft. lbs. (37 Nm)
12. Fill the crankcase to the correct level.
13. Fill the cooling system.
14. Recharge the A/C system, if equipped.
15. Start the engine and check for leaks.

Water Pump

REMOVAL & INSTALLATION

1. Before servicing the vehicle, refer to the precautions in the beginning of this section.
2. Drain the cooling system.
3. Remove or disconnect the following:
   - Negative battery cable
   - Splash shield
   - Drive belt
   - Air conditioning compressor
   - Water pump pulley
   - Water pump

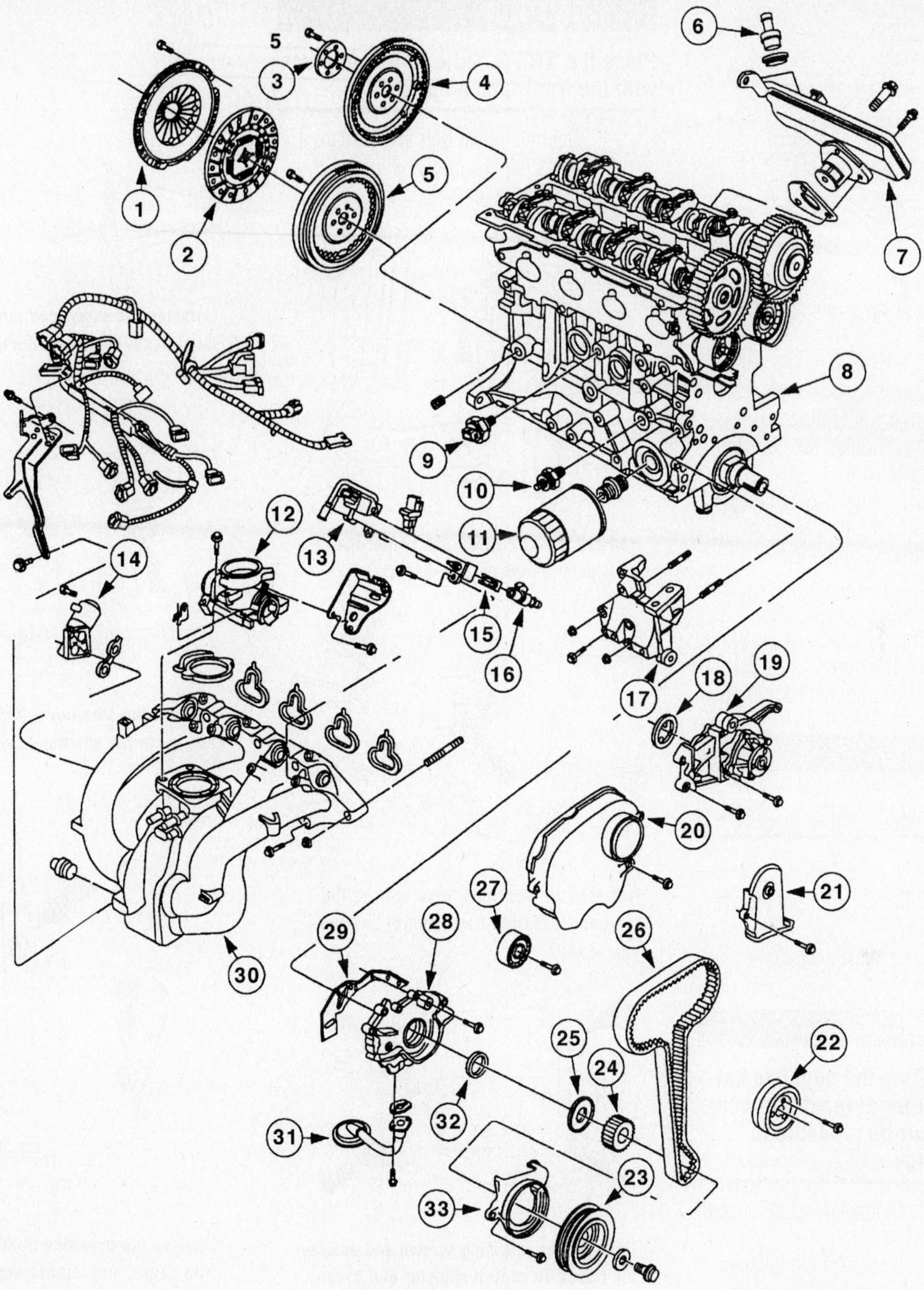

1. Clutch pressure plate
2. Clutch disc
3. Support
4. Flywheel (automatic trans.)
5. Flywheel (manual trans.)
6. PVC valve
7. Oil separator
8. Engine block
9. Knock sensor
10. Oil pressure sensor
11. Oil filter
12. Throttle body
13. Fuel injection supply manifold
14. Idle Air Control (IAC) valve
15. Retainer clip
16. Fuel injector
17. Bracket
18. Water pump housing gasket
19. Water pump
20. Upper timing belt cover
21. Center timing belt cover
22. Water pump pulley
23. Crankshaft pulley
24. Crankshaft sprocket
25. Timing belt guide
26. Timing belt
27. Timing belt idler pulley
28. Oil pump
29. Oil pump housing gasket
30. Intake manifold
31. Oil pump pick-up
32. Crankshaft front seal
33. Lower timing belt cover

9300MG10

Exploded view of peripheral engine component mounting, including water pump, oil pump and intake manifold—2.0L DOHC Zetec engine

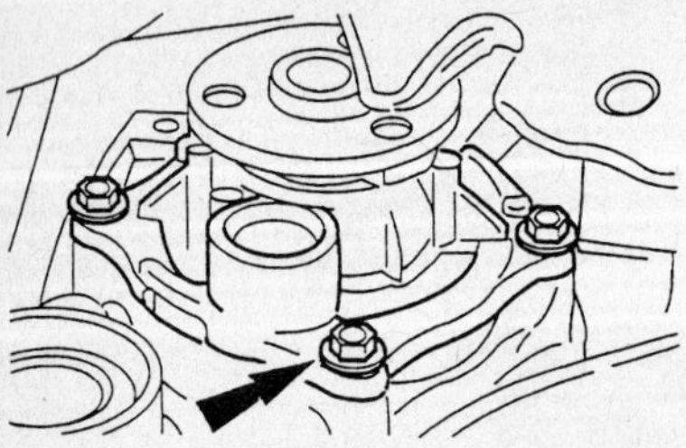

7922MG19

Remove the water pump mounting bolts—2.0L Zetec engine

To install:

4. Install or connect the following:
   - Water pump. Tighten the bolts to 17 ft. lbs. (24 Nm).
   - Water pump pulley. Tighten the bolts to 17 ft. lbs. (24 Nm).
   - Air conditioning compressor
   - Drive belt
   - Splash shield
   - Negative battery cable
5. Fill the cooling system.
6. Start the engine and check for leaks.

Heater Core

REMOVAL & INSTALLATION

1. Before servicing the vehicle, refer to the precautions in the beginning of this section.
2. Disconnect the negative battery cable.

**** CAUTION**

After disconnecting the negative battery cable, wait for at least 1 minute for the SRS or air bag module to deplete its energy.

3. Drain the cooling system into a clean container for reuse.
4. Remove or disconnect the following:
   - Air cleaner outlet tube, on coupe models only
   - Heater hoses from the heater core
5. Place the front wheels in the straight-ahead position. Lock the steering column.
6. Remove the steering wheel by removing or disconnecting the following:
   - SRS module-to-steering wheel bolts
   - SRS module (carefully), and disconnect the horn switch and the SRS electrical connectors

**** CAUTION**

Place the SRS module in a safe place with the front facing upward.

   - Steering wheel bolt and discard it
   - Press the steering wheel from the steering column

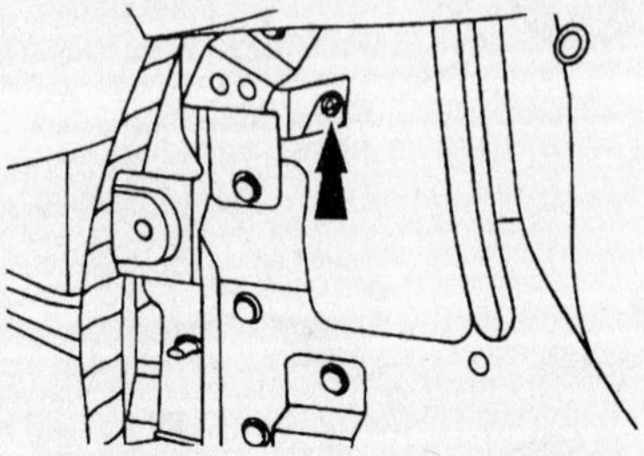

90980G42

Remove the screw from the center instrument panel finish panel—Escort

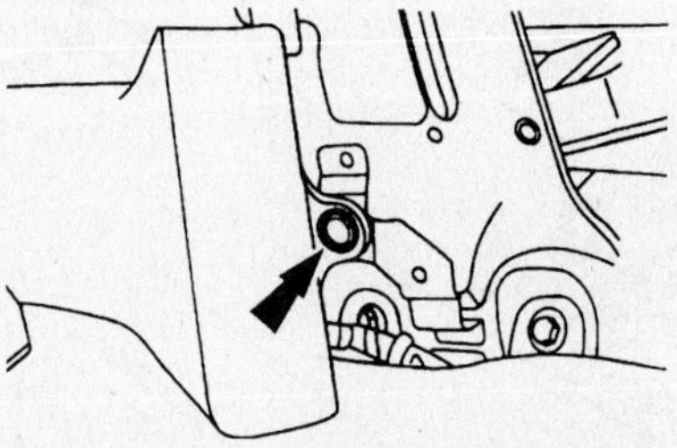

90980G44

Unfasten the pushpins, then remove the left-hand and right-hand control box covers—Escort

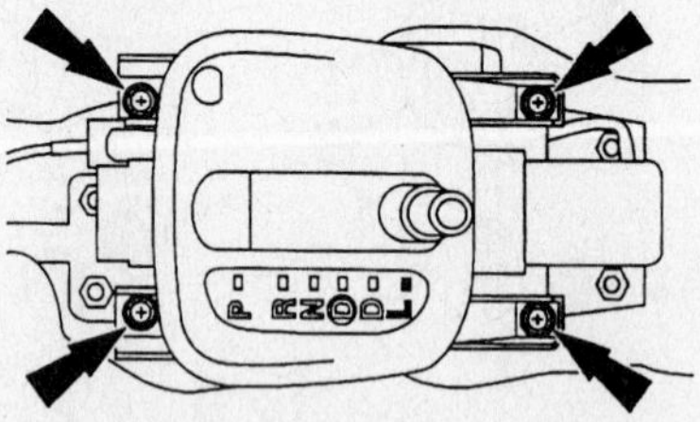

90980G45

Unfasten the retaining screws and position the transaxle control selector dial bezel sideways—Escort

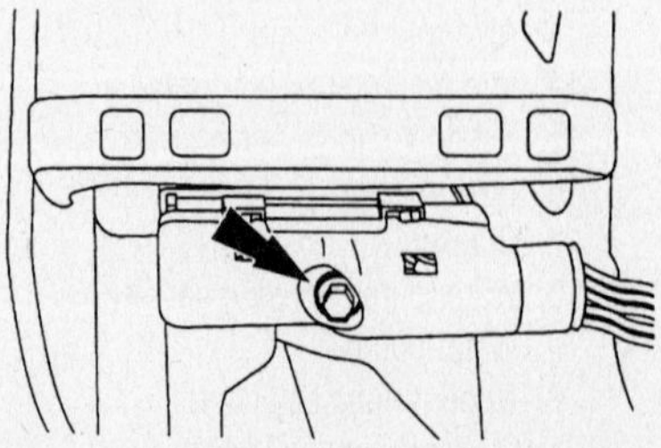

90980G43

Unfasten the PCM electrical connector bolt, unplug the connector and move it aside—Escort

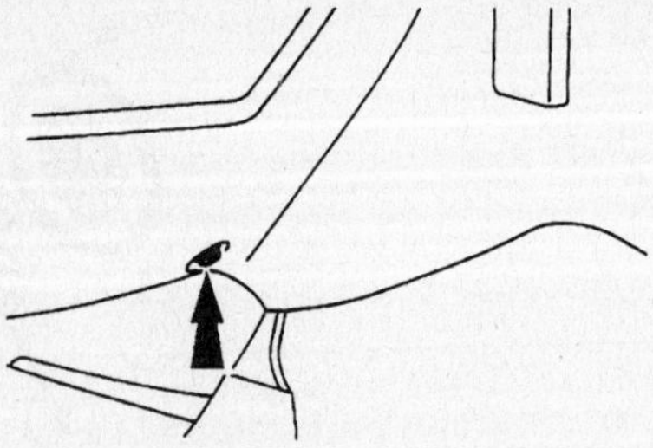

90980G46

Unfasten the screw and remove the instrument panel steering column cover—Escort

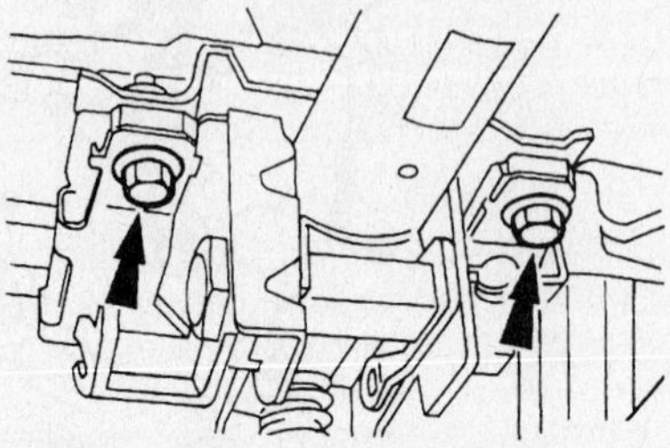

90980G47

Remove the steering column bracket bolts and lower the column—Escort

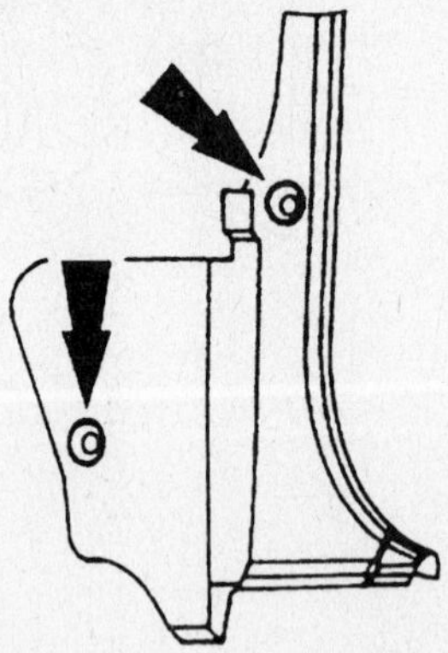

90980G48

Remove the pushpins from each cowl side trim panel, and remove the panels—Escort

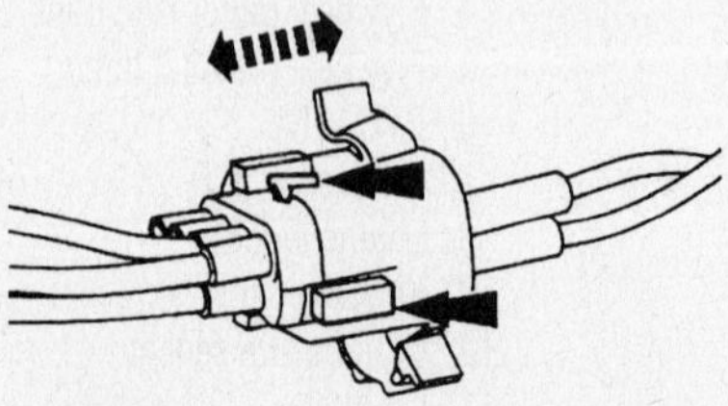

90980G49

Unplug the vacuum line harness connector—Escort

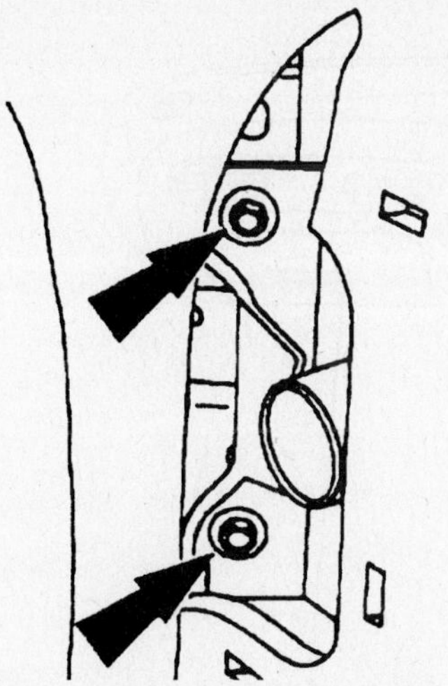

Unfasten both the left-hand and right-hand center instrument panel reinforcement bolts—Escort

Unfasten both the left-hand and right-hand lower instrument panel reinforcement bolts—Escort

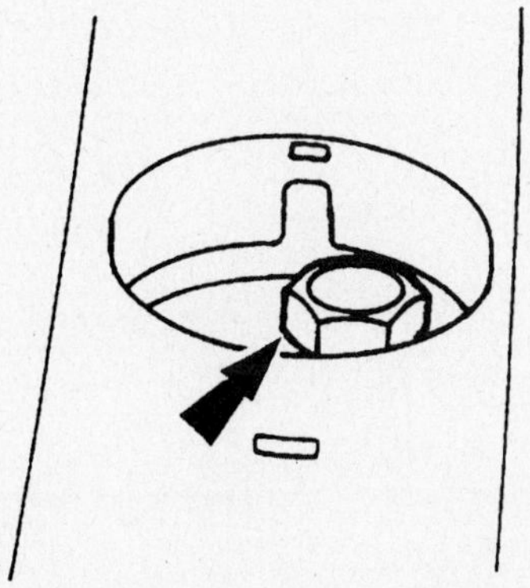

Remove the cover and unfasten the upper instrument panel reinforcement bolt—Escort

7. Remove the passenger's side SRS module by removing or disconnecting the following:

- Push inward on the 2 glove box door tabs and lower it
- SRS module's electrical connector
- SRS module-to-instrument panel bolts and the module

⁂ CAUTION

Place the SRS module in a safe place with the front facing upward.

8. Remove the instrument panel by removing or disconnecting the following:

- Floor console
- Screw located at the center instrument panel finish panel
- Pull the center instrument panel finish panel straight out from the instrument panel reinforcement (Coupe models only)
- Power point socket electrical connectors and remove the center instrument panel finish panel, if equipped
- Control box side cover pushpins and the covers located on both sides
- Radio antenna lead-in cable, located at the instrument panel reinforcement
- Radio antenna lead-in cable
- Transmission control selector dial bezel screws; then, place the transmission control selector dial bezel sideways
- Powertrain Control (PCM) electrical connector bolt and move the PCM aside
- 2 rear side PCM bracket nuts, the 2 bolts; then, the PCM and bracket as an assembly
- Rotate the temperature control switch to the COOL position and disconnect the heater control cable
- Hood latch control handle nut; then, position the hood latch control handle and cable aside
- Instrument panel steering column cover screw and release the cover
- Light switch rheostat resistor electrical connector and remove the instrument panel steering column cover (Coupe models only)
- Steering column shroud screws and the shrouds
- Steering column bracket bolts and lower the steering column
- Pull upward on the front door scuff plates
- Cowl side trim panel pushpins and the panels, located on both sides
- Interior fuse junction panel electrical connector
- Lower instrument panel reinforcement in-line electrical connector, located on the left side
- Vacuum line harness connector
- Lower instrument panel reinforcement in-line electrical connector, located on the right side
- Blower motor resistor electrical connector
- Instrument panels end panels, located on both sides
- Upper and lower instrument panel-to-chassis bolts, located on both sides
- Upper and lower instrument panel reinforcement-to-chassis bolts, located on both sides
- Cover and the upper instrument panel reinforcement bolt
- Slightly, pull the instrument panel rearward and disconnect the main electrical wiring connectors, located on the left side
- Instrument panel

9. Remove or disconnect the following:

- Antenna lead from the heater core housing
- Vacuum control motor vacuum connector
- Vacuum lines from the retainer at the evaporator housing
- Pushpins, the windshield defroster nozzle connector's screws and the connectors
- Air conditioning evaporator outlet duct clamp screw
- Lower heater core housing-to-chassis nut, the upper heater core housing-to-chassis nuts and the heater core housing
- Heater dash panel seal
- Heater core cover-to-housing screws and the cover
- Heater core from the housing

To install:

10. Install or connect the following:

- Heater core to the housing
- Heater core cover and the cover-to-housing screws
- Heater dash panel seal
- Lower heater core housing, the upper heater core housing-to-chassis nuts and the heater core housing-to-chassis nut
- Air conditioning evaporator outlet duct clamp screw
- Windshield defroster nozzle connector's screws, the connectors and the pushpins
- Vacuum lines to the retainer at the evaporator housing
- Vacuum control motor vacuum connector

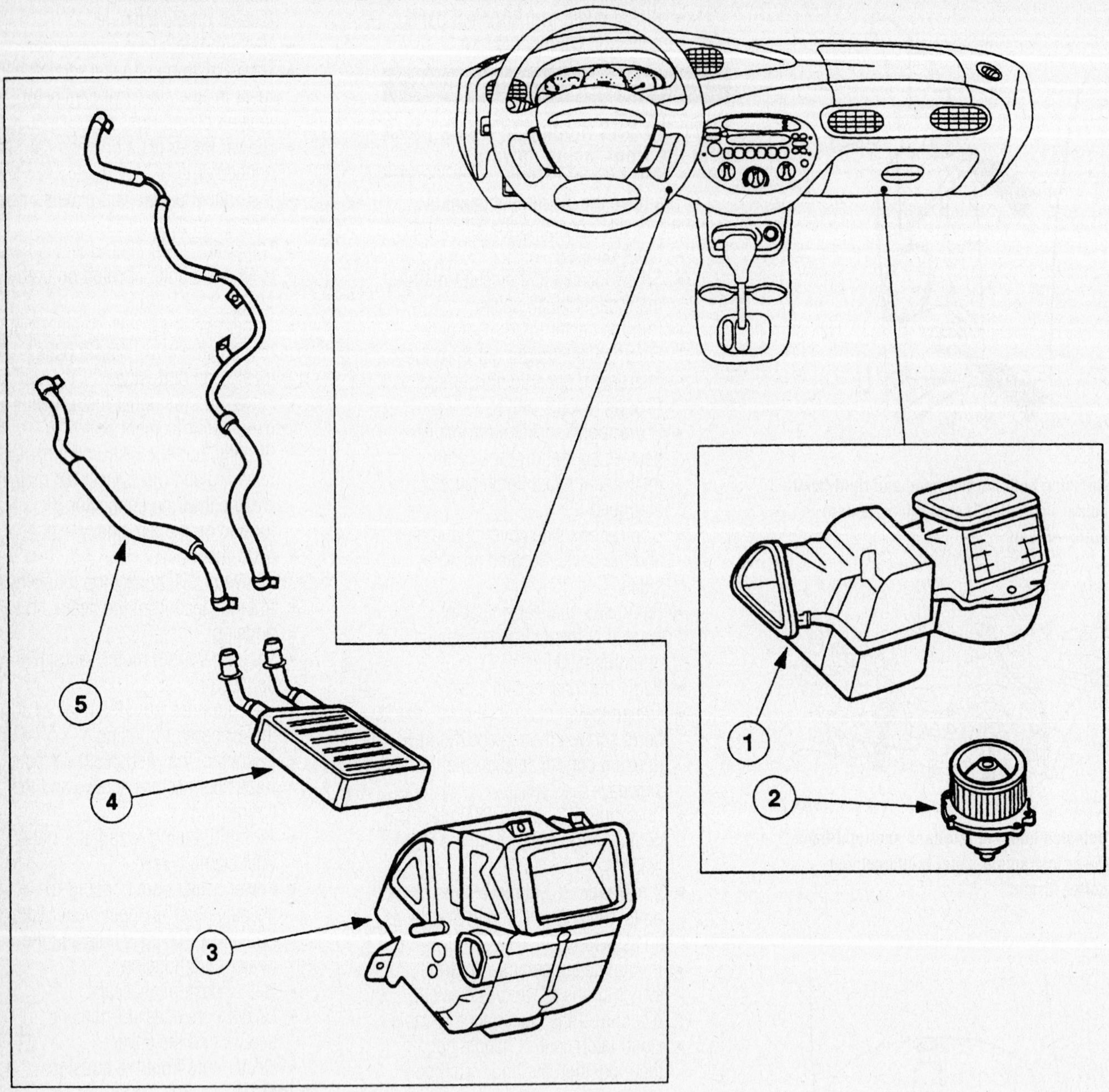

1 A/C Evaporator Housing
2 Blower Motor
3 Heater Core Housing
4 Heater Core
5 Heater Water Hoses

93111GB3

Exploded view of the heater core, heater housing and related components—Escort

- Antenna lead to the heater core housing

11. Install the instrument panel by installing or connecting the following:
    - Instrument panel
    - Main electrical wiring connectors (located on the left side), and install the instrument panel
    - Upper instrument panel reinforcement bolt and the cover
    - Upper and lower instrument panel reinforcement-to-chassis bolts (both sides)
    - Upper and lower instrument panel-to-chassis bolts (both sides)
    - Instrument panels end panels (both sides)
    - Blower motor resistor electrical connector
    - Lower instrument panel reinforcement in-line electrical connector, located on the right side
    - Vacuum line harness connector.
    - Lower instrument panel reinforcement in-line electrical connector, located on the left side
    - Interior fuse junction panel electrical connector
    - Cowl side trim panels and secure the panels with the pushpins, located on both sides
    - Front door scuff plates, located on both sides
    - Steering column and the steering column bracket bolts
    - Steering column shrouds and the shroud screws
    - Light switch rheostat resistor electrical connector and install the instrument panel steering column cover (coupe models only)
    - Instrument panel steering column cover and the cover screw
    - Hood latch control handle and cable and tighten the hood latch control handle nut
    - Heater control cable
    - PCM/bracket, the 2 PCM/bracket nuts and the 2 bolts
    - PCM electrical connector and the bolt
    - Transmission control selector dial bezel screws
    - Radio antenna lead-in cable
    - Radio antenna lead-in cable
    - Control box side covers and secure with the pushpins, located at both sides
    - Power point socket electrical connectors and install the center instrument panel finish panel, if equipped
    - Center instrument panel finish panel screw
    - Floor console
12. Install the passenger's side SRS module by installing or connecting the following:
    - SRS module and torque the module-to-instrument panel bolts to 62–97 inch lbs. (7–11 Nm)
    - SRS module's electrical connector
    - Glove box door
13. Install the steering wheel by installing or connecting the following:
    - Steering wheel to the steering column
    - New steering wheel bolt and torque it to 26–33 ft. lbs. (34–46 Nm)
    - SRS module-to-steering wheel bolts and torque the bolts to 89–123 inch lbs. (10–14 Nm)
    - SRS module and connect the electrical connector
14. Connect the heater hoses to the heater core.
15. On the Coupe, install the air cleaner outlet tube.
16. Refill the cooling system.
17. Connect the negative battery cable.
18. Operate the engine to normal operating temperatures; then, check the climate control operation and check for leaks.

Cylinder Head

REMOVAL & INSTALLATION

WARNING

To reduce the possibility of cylinder head warpage and/or distortion, do not remove the cylinder head while the engine is warm. Always allow the engine to cool entirely before disassembly.

1. Before servicing the vehicle, refer to the precautions in the beginning of this section.
2. Drain the cooling system.
3. Relieve the fuel system pressure.
4. Remove or disconnect the following:
    - Negative battery cable
    - Air cleaner assembly outlet tube
    - Timing belt
5. Tag and disconnect the vacuum hoses from the following components:
    - Positive Crankcase Ventilation Valve (PCV) valve
    - Throttle body
    - Fuel pressure sensor
    - Intake manifold
6. Remove or disconnect the following:
    - Speed control and accelerator cables from the control lever
    - Fuel charging electrical connectors at the main engine connector
    - Crankshaft Position (CKP) sensor and Heated Oxygen (HO_2S) sensor electrical connections
    - Fuel line
    - Power steering pump and bracket, then set them aside with the lines still attached
    - Alternator
    - Engine oil dipstick tube
    - Splash shield bolts and the shield
    - Air conditioning compressor electrical connection
    - Air conditioning compressor retainers and set the compressor aside with the lines still attached
    - Spark plug wires and the plugs
    - Valve cover
    - Upper radiator hose from the thermostat housing water hose connection
    - Heater hose from the thermostat housing
    - Camshafts
    - Ignition coil
    - Thermostat housing
    - Cylinder head bolts
    - Cylinder head and the gasket

To install:

7. Install a new head gasket on the cylinder block.

➡Always use new bolts when installing the cylinder head.

8. Lubricate the new cylinder bolts when engine oil.
9. Install the cylinder head and tighten the bolts in the sequence illustrated in the following steps:
    a. Step 1: 15 ft. lbs. (20 Nm)
    b. Step 2: 40 ft. lbs. (30 Nm)
    c. Step 3: Plus 105 degrees
10. Install or connect the following:
    - Thermostat housing
    - Ignition coil and the camshafts
    - Heater and upper radiator hoses
    - Valve cover
    - Spark plugs
    - Spark plug wires
    - Air conditioning compressor and tighten the retaining bolts
    - Air conditioning compressor electrical connection
    - Splash shield and tighten its retainers
    - Engine oil dipstick tube

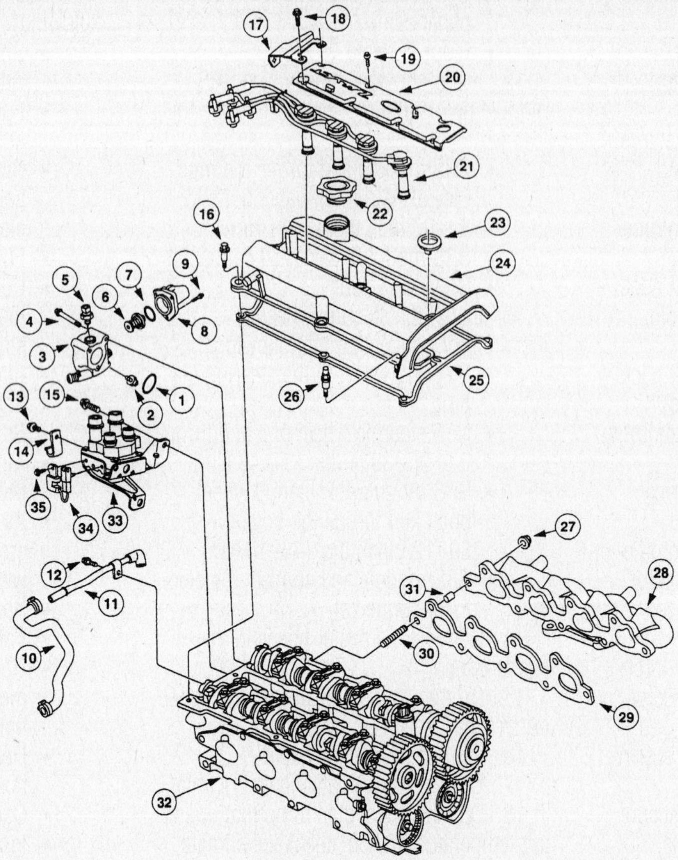

Item	Description
1	Water Thermostat Housing Oil Ring
2	Water Temperature Indicator Sender Unit
3	Water Thermostat Housing
4	Bolt (3 Req'd)
5	Engine Coolant Temperature Sensor
6	Water Thermostat
7	Water Thermostat Housing O-Ring
8	Water Outlet Connection
9	Bolt (3 Req'd)
10	Oil Separator Hose
11	Crankcase Ventilation Tube
12	Bolt
13	Bolt
14	Support Bracket
15	Bolt (3 Req'd)
16	Bolt (9 Req'd)

Item	Description
17	Bracket
18	Bolt (2 Req'd)
19	Bolt (6 Req'd)
20	Engine Appearance Cover
21	Ignition Wire (4 Req'd)
22	Oil Filler Cap
23	Grommet
24	Valve Cover
25	Valve Cover Gasket
26	Stud Bolt
27	Nut (9 Req'd)
28	Exhaust Manifold
29	Exhaust Manifold Gasket
30	Stud (9 Req'd)
31	Spacer
32	Cylinder Head
33	Ignition Coil Bracket
34	Radio Ignition Interference Capacitor
35	Bolt

9306MG77

Exploded view of the upper engine components—2.0L DOHC engine

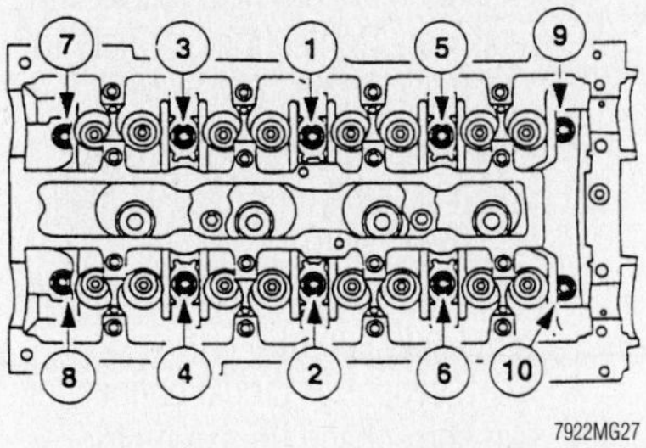

Cylinder head torque sequence—2.0L DOHC engine

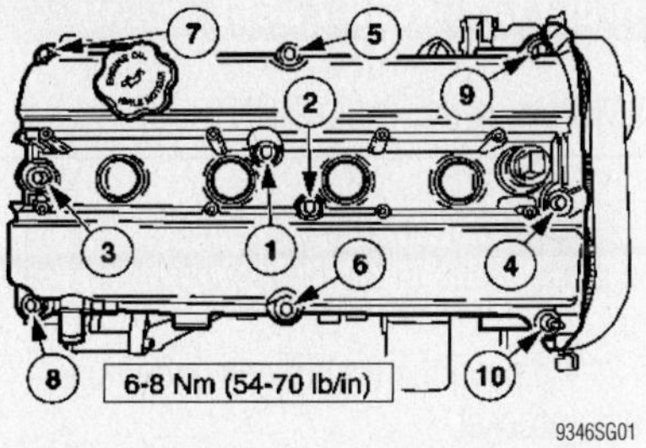

Valve cover torque sequence—DOHC Engine

- Alternator and the power steering reservoir and bracket
- Fuel line
- CKP sensor, HO_2sensor, and 2 main fuel charging electrical connections
- Accelerator cable and if equipped, the speed control cable to the control lever

11. Connect the vacuum hoses tagged and removed to the following components:
    - PCV valve
    - Throttle body
    - Fuel pressure sensor
    - Intake manifold
12. Install or connect the following:
    - Timing belt
    - Air cleaner outlet tube
    - Negative battery cable
13. Fill the cooling system.
14. Start the engine and check for leaks.

Rocker Arms

REMOVAL & INSTALLATION

This engine is not equipped with rocker arms. The camshafts act directly on the valves.

Intake Manifold

REMOVAL & INSTALLATION

1. Before servicing the vehicle, refer to the precautions in the beginning of this section.
2. Drain the cooling system.
3. Relieve the fuel system pressure.
4. Remove or disconnect the following:
    - Negative battery cable
    - Air cleaner outlet tube
    - Throttle Position (TP) sensor electrical connection
    - Bolt that secures the pipe located by the crankshaft pulley
    - Heater hoses from the core
    - Main engine control sensor wiring
    - Connectors from the mounting bracket
    - Vacuum hoses from the intake manifold by squeezing the tabs, twisting the hoses and pulling them away from the manifold
    - Crankcase ventilation hose from the valve cover
    - Drive belt
    - Alternator mounting bolts and move the alternator aside
    - Fuel lines
    - Intake manifold nuts and bolts in the sequence illustrated
    - Intake manifold and gasket

To install:

5. Clean all dirt and gasket residue from the intake manifold mating surfaces.

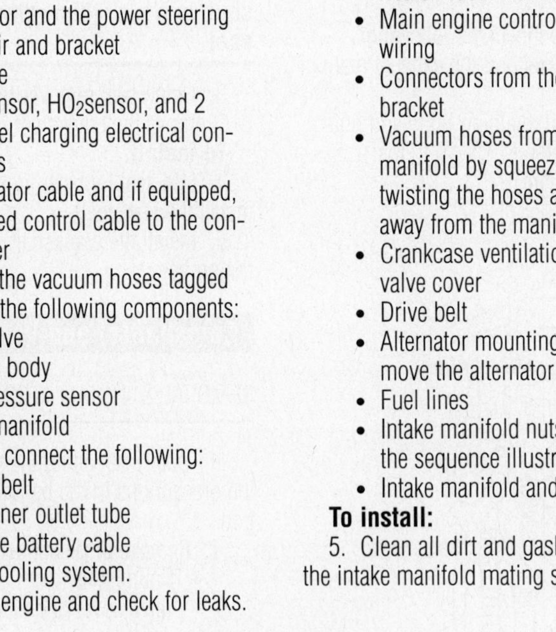

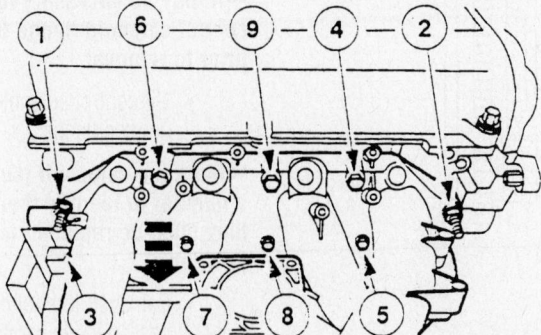

Remove the intake manifold bolts and nuts in this sequence—2.0L Zetec engine

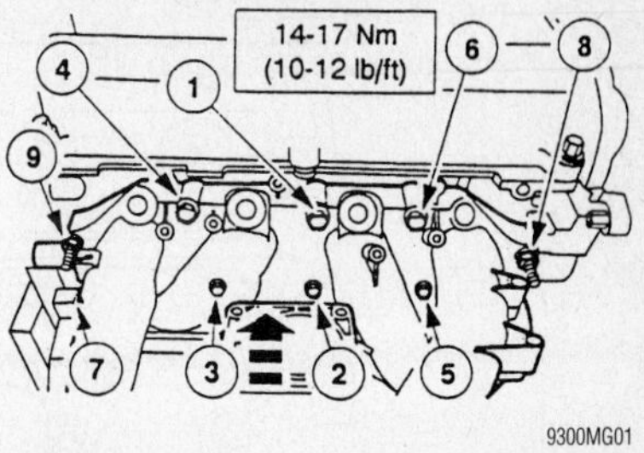

Intake manifold torque sequence—2.0L DOHC engine

6. Install or connect the following:
    - New gasket and place the manifold in position
    - Manifold bolts and nuts. Tighten in the sequence illustrated to 10–12 ft. lbs. (14–17 Nm).
    - Fuel lines
    - Alternator
    - Drive belt
    - Crankcase ventilation hose to the valve cover
    - Vacuum hoses to the manifold making sure the tabs are firmly engaged
    - Connectors in the mounting bracket
    - Main engine control sensor wiring
    - Heater hoses to the heater core
    - Bolt that secures the pipe located by the crankshaft pulley. Tighten the bolt to 71–97 inch lbs. (8–11 Nm).
    - TP sensor electrical connection
    - Air cleaner outlet tube
7. Fill the cooling system.
8. Connect the negative battery cable.

Exhaust Manifold

REMOVAL & INSTALLATION

1. Before servicing the vehicle, refer to the precautions in the beginning of this section.
2. Remove or disconnect the following:
    - Negative battery cable
    - Fan motor electrical connection
    - Fan shroud
    - Heated Oxygen (HO_2S) sensor electrical connection
    - Catalytic converter-to-exhaust manifold bolt and nut, then disconnect the converter from the manifold

⁂ WARNING

Do not break the threadlock seal on the engine oil dipstick tube. If the seal is broken, relock it to prevent any oil leaks from the block.

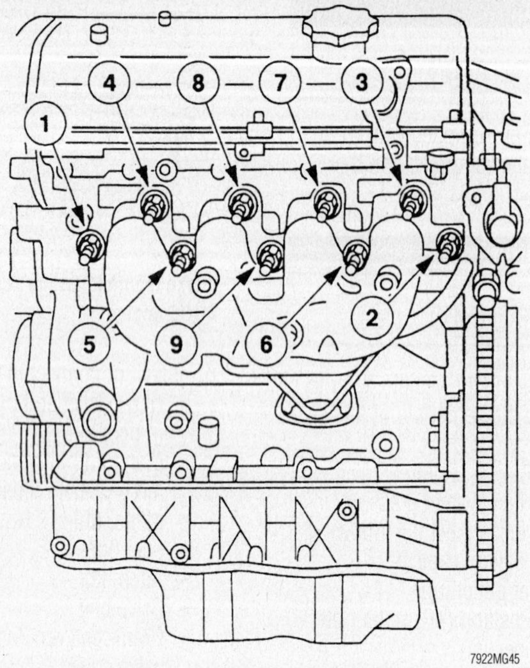

Remove the exhaust manifold nuts and bolts in the sequence illustrated—2.0L Zetec engine

- Engine oil dipstick tube bracket bolt
- Exhaust manifold heat shield bolts and nuts, then the shield
- 7 exhaust manifold bolts and 2 studs
- Manifold and gasket

To install:

3. Install or connect the following:
   - New gasket and the exhaust manifold
   - Manifold retaining bolts and nuts and tighten to 13–16 ft. lbs. (14–17 Nm)

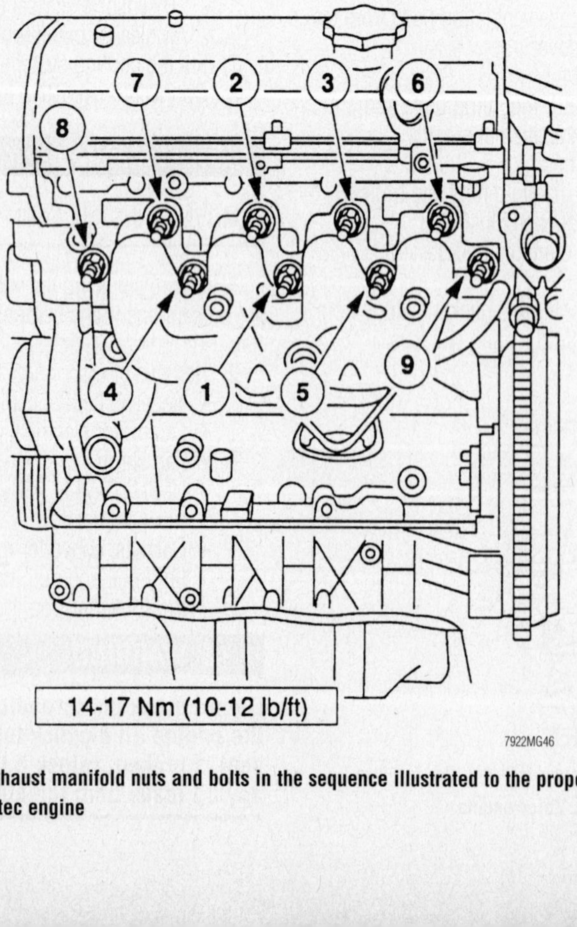

Tighten the exhaust manifold nuts and bolts in the sequence illustrated to the proper specification—2.0L Zetec engine

- Exhaust manifold heat shield and retainers. Tighten the retainers to 71–101 inch lbs. (8–11 Nm).
- Engine oil dipstick tube bracket bolt and tighten to 71–101 inch lbs. (8–11 Nm)
- Exhaust manifold-to-catalytic converter bolt and nut
- HO_2S sensor electrical connection
- Fan shroud and the fan motor wiring
- Negative battery cable

Front Crankshaft Seal

REMOVAL & INSTALLATION

1. Before servicing the vehicle, refer to the precautions in the beginning of this section.
2. Remove the timing belt and crankshaft sprocket.

**** WARNING**

Be careful not to damage the crankshaft surface when removing the seal.

3. Using seal remover tool T92C-6700-CH, remove the crankshaft front oil seal.

To install:

4. Use seal replacer tool T81P-6700-A, install the new seal.
5. Install the crankshaft sprocket and the timing belt.

Camshaft(s) and Valve Lifters

REMOVAL & INSTALLATION

1. Before servicing the vehicle, refer to the precautions in the beginning of this section.
2. Remove or disconnect the following:
   - Timing belt
   - Valve cover and camshaft sprockets

➡It may be necessary to rotate the oil control solenoid flange 90 degrees prior to removal.

   - Oil control solenoid flange bolts and the flange

➡Mark the camshaft journal caps with a number to identify their location, as they must be replaced in their original position.

   - Camshaft journal caps in several passes in the sequence illustrated, then remove the bolts and caps

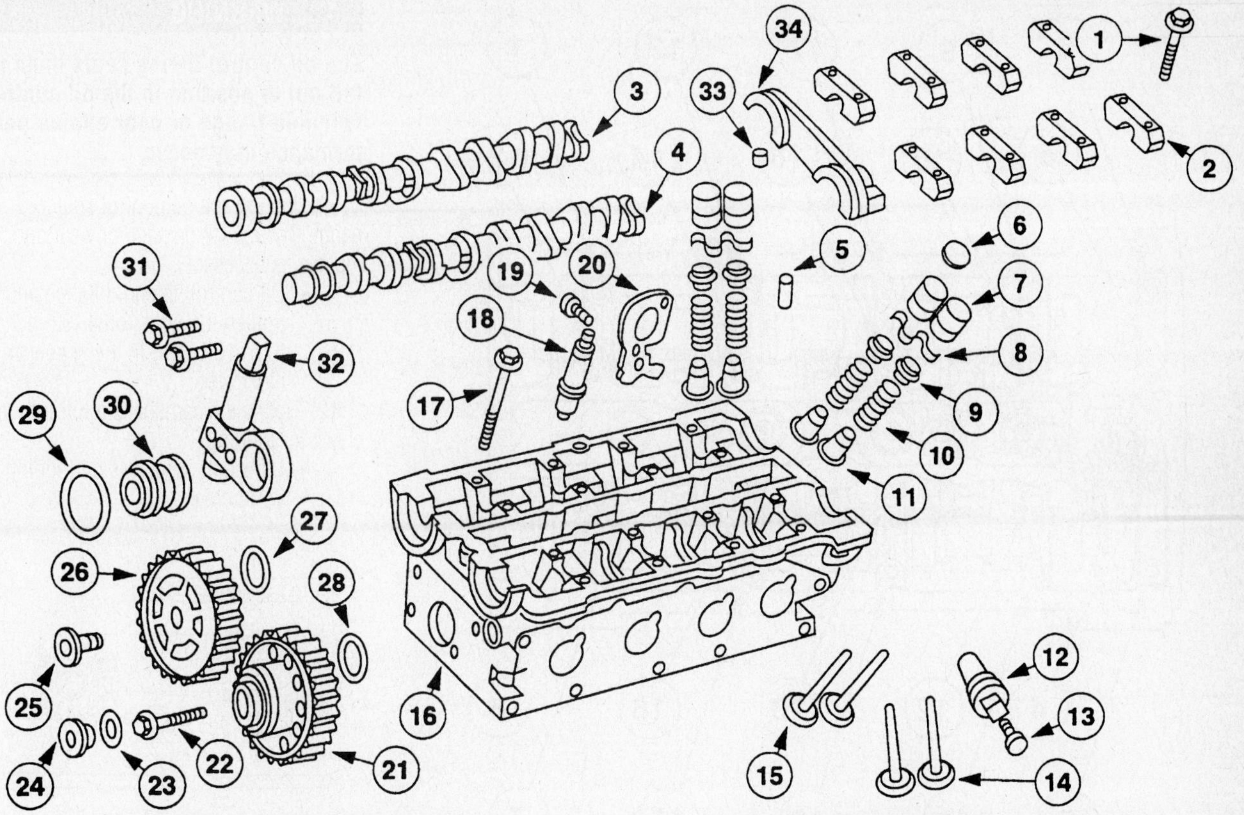

1	Bolt—Camshaft Bearing Cap	18	Spark Plug
2	Camshaft Bearing Cap	19	Engine Lifting Eye Bolt
3	Camshaft Intake	20	Engine Lifting Eye
4	Camshaft Exhaust	21	Camshaft Sprocket —Exhaust
5	Plug	22	Exhaust Camshaft Sprocket Bolt
6	Adjusting Shim—Valve Clearance	23	Blanking Plug O-Ring
7	Valve Tappets	24	Blanking Plug
8	Valve Spring Retainer Key	25	Bolt—Intake Camshaft
9	Valve Spring Retainers	26	Camshaft Sprocket Intake
10	Valve Spring (Color Coding: Exhaust - Blue, Intake - Red)	27	Camshaft Front Seal
11	Valve Stem Seal	28	O-Ring
12	Camshaft Position Sensor	29	Camshaft Front Seal
13	CMP Sensor Bolt	30	Oil Feed Ring
14	Intake Valves	31	Oil Feed Flange Bolts
15	Exhaust Valves	32	Oil Feed Flange
16	Cylinder Head	33	Guide Sleeve—Front Camshaft Bearing Cap
17	Cylinder Head Bolts	34	Fifth Camshaft Bearing Cap

9300MG06

Exploded view of the cylinder head, showing camshaft mounting—2.0L DOHC Zetec engine

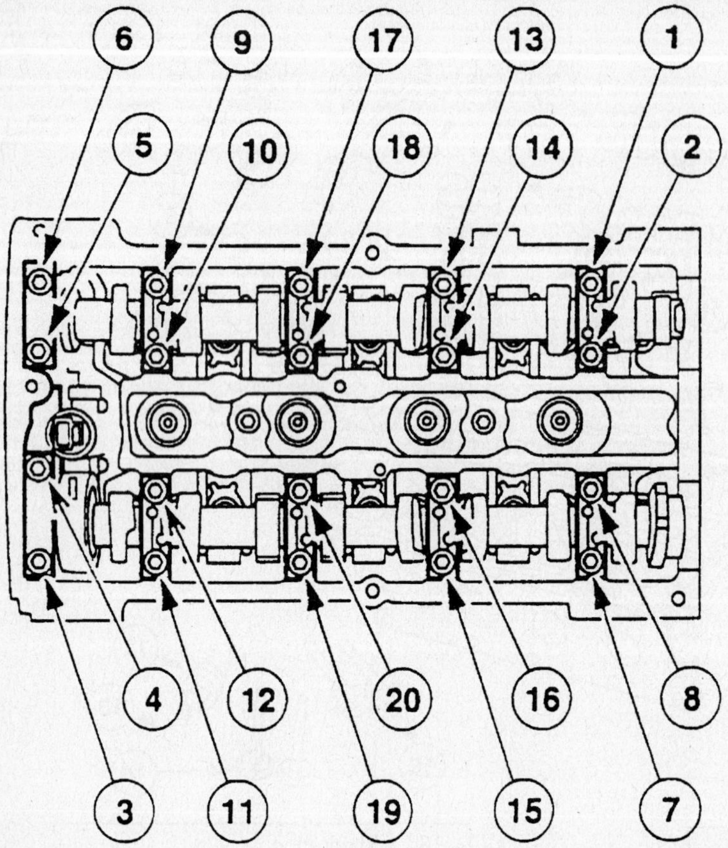

Loosen the camshaft journal caps in several passes in this sequence—2.0L Zetec engine

- Camshafts from the cylinder head
- Oil control sensor and bushing

3. Inspect the camshafts for damage and wear.

To install:

4. Be sure the valve clearance is correct.
5. Install the oil control solenoid bushing and flange on the exhaust camshaft.
6. Coat the surface of the front camshaft journal cap with gasket maker E2AZ-19562-B.
7. Place the camshafts in position and lubricate the bearing surfaces with engine assembly lubricant D9AZ-19579-D.
8. Install new camshaft front oil seals.
9. Apply a thin coat of silicone gasket and sealant F6AZ-19562-AA to the sealing surface of the of the front camshaft journal bearing cap.
10. Install or connect the following:
    - Caps and tighten the bolts in several 2-turn passes in the sequence illustrated to 10–12 ft. lbs. (13–17 Nm)

⁂ WARNING

The oil control O-ring seals must not fall out of position in the oil control solenoid flange or poor engine performance may occur.

11. Inspect the oil control solenoid flange O-rings for damage or wear and replace as necessary.
    - Oil control solenoid flange and tighten the bolts to 84–92 inch lbs. (9.5–10.5 Nm) in the sequence illustrated
12. Rotate the camshafts a full turn and check for binding.
    - Camshaft sprockets and timing belt
    - Valve cover

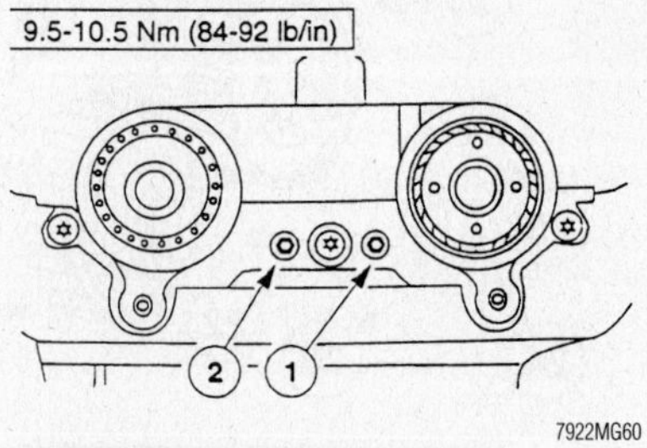

Tighten the oil control solenoid flange bolts in this sequence—2.0L Zetec engine

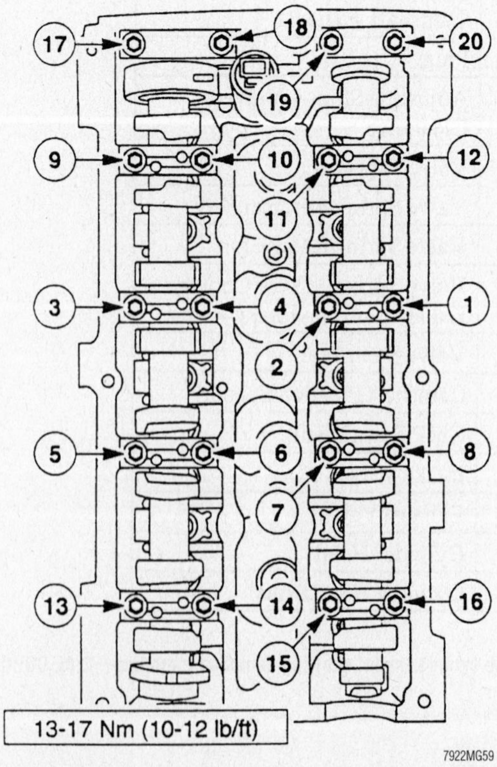

Tighten the camshaft journal caps in the sequence illustrated to the proper specification—2.0L Zetec engine

Valve Lash

ADJUSTMENT

1. Before servicing the vehicle, refer to the precautions in the beginning of this section.
2. Remove the valve cover.
3. Remove the timing belt.

➡Measure each valve's clearance at the base circle before removing the camshafts. The shims are not serviceable with the camshafts in place.

4. Measure the valve clearances, then remove the camshafts and shims.

The correct shims allow the following valve clearances:

- Intake valve clearance: 0.0043–0.0071 in. (0.11–0.18mm)
- Exhaust valve clearance: 0.0106–0.0134 in. (0.27–0.34mm)

➡A midrange clearance is the most desirable.

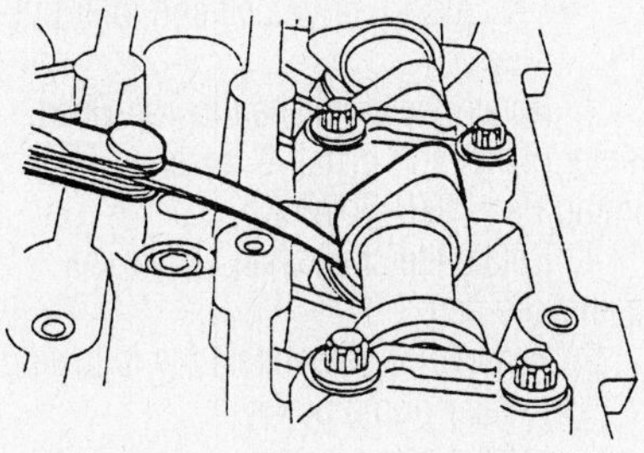

7922MG61

Measure each valve's clearance at the base circle before removing the camshafts—2.0L Zetec DOHC engine

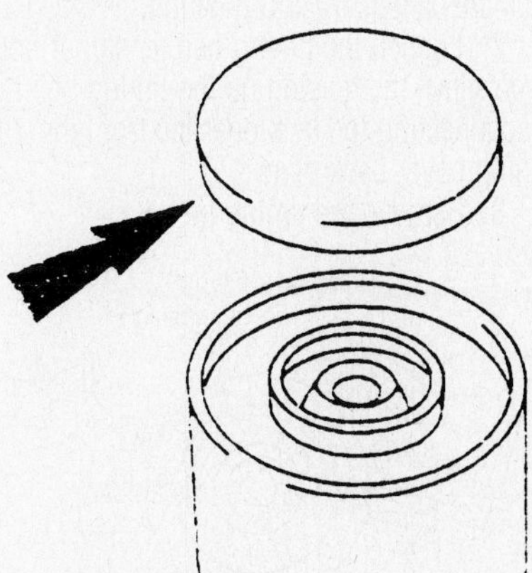

7922MG62

Example of a valve lifter shim (arrow)—2.0L Zetec DOHC engine

5. Select the shims using the following formula: shim thickness = measured clearance plus the base shim thickness minus the most desirable thickness.
6. Select the correct shims and mark their installation location.
7. Replace the shims and install the camshafts.
8. Check the new valve clearances.
9. Install the timing belt and covers.
10. Install the valve cover.
11. Start the vehicle and check for proper operation.

Starter Motor

REMOVAL & INSTALLATION

1. Before servicing the vehicle, refer to the precautions in the beginning of this section.
2. Remove or disconnect the following:
   - Negative battery cable
   - Air cleaner outlet tube
   - Lower starter motor bolt
   - Two starter motor upper bolts
3. Slide the starter motor from its mounting until you can gain access to the wiring.
   - Integral connector
   - Starter motor

To install:

4. Install or connect the following:
   - Integral connector and tighten the retaining nuts to 61 inch lbs. (7 Nm)
   - Starter motor to its mounting
   - Two starter motor upper bolts and tighten to 15–20 ft. lbs. (20–27 Nm)
   - Lower starter motor bolt and tighten to 15–20 ft. lbs. (20–27 Nm)
   - Air cleaner outlet tube
   - Negative battery cable

Oil Pan

REMOVAL & INSTALLATION

1. Before servicing the vehicle, refer to the precautions in the beginning of this section.
2. Drain the engine oil.
3. Remove or disconnect the following:
   - Catalytic converter
   - 17 oil pan bolts evenly
   - Oil pan

To install:

4. Apply a 0.1 in. (3.0mm) wide bead of silicone sealant F6AZ-19562-AA or its equivalent to the oil pan.
5. Install or connect the following:
   - Oil pan and the bolts. Tighten the oil pan bolts in the sequence illustrated to 15–22 ft. lbs. (20–30 Nm).

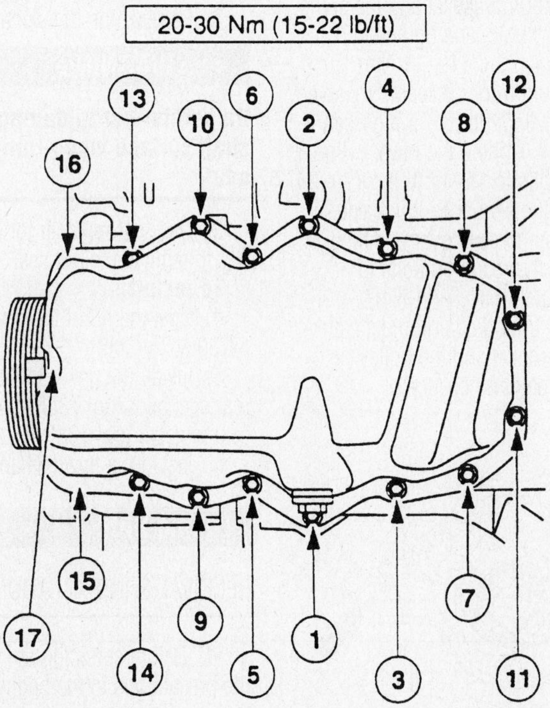

7922MG69

Tighten the oil pan bolts in the proper sequence to the correct torque specification—2.0L Zetec engine

- Oil pan drain plug

6. Fill the engine with the proper type and quantity of engine oil.
7. Start the vehicle and check for oil leaks.

Oil Pump

REMOVAL & INSTALLATION

1. Before servicing the vehicle, refer to the precautions in the beginning of this section.
2. Remove or disconnect the following:
   - Timing covers and belt
   - Crankshaft pulley, sprocket and the timing chain guide
   - Oil pan
   - Oil pump cover and screen bolts
   - Cover and screen
   - Lower cylinder block shims, if necessary
   - Lower cylinder block, the gasket and the crankshaft front oil seal, if necessary
   - Oil pump retaining bolts
   - Oil pump and gasket

To install:

➡The clearance between the lower cylinder block sealing surfaces on the oil pump and the cylinder block cannot exceed 0.012–0.031 in. (0.3–0.8mm).

3. Install or connect the following:
   - New oil pump gasket and the pump
   - Oil pump bolts and tighten to 88–97 inch lbs. (10–11 Nm)
   - Crankshaft front oil seal, if removed
   - Lower cylinder block and gasket, if removed. Tighten the lower cylinder block bolts to 15–17 ft. lbs. (20–24 Nm) in the sequence illustrated.
   - Oil pump screen cover and tube. Tighten the retaining bolts to 71–97 inch lbs. (8–11 Nm).
   - Oil pan, timing chain guide, crankshaft sprocket and the pulley
   - Timing belt and covers

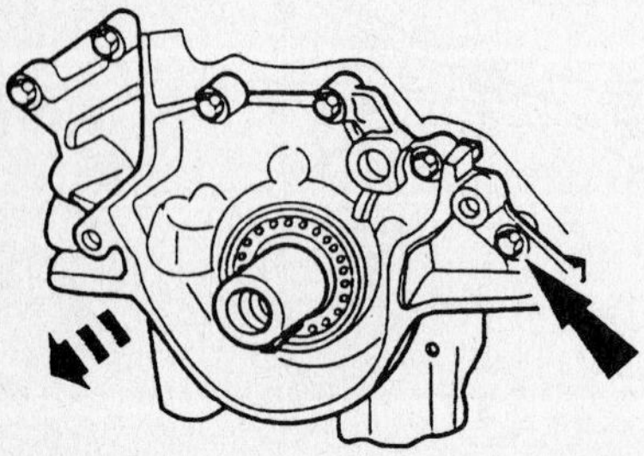

Remove the oil pump mounting bolts—2.0L Zetec engine

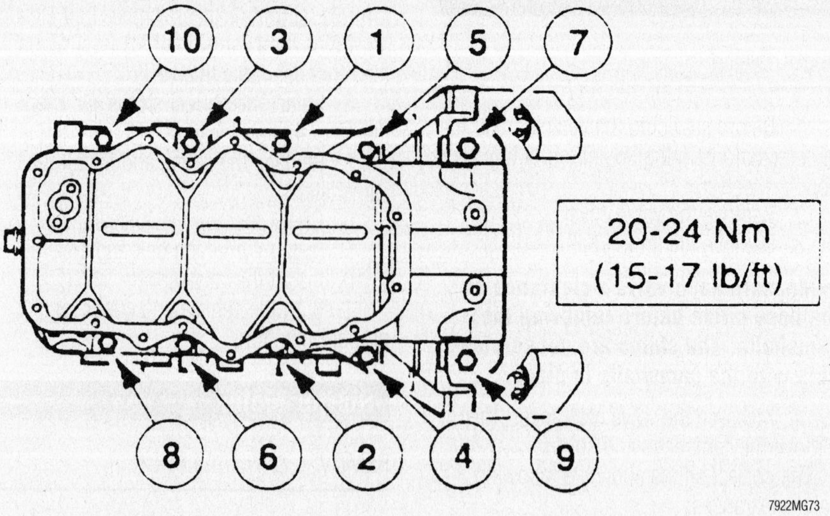

Tighten the lower cylinder block bolts in the sequence illustrated to the proper torque specification—2.0L Zetec engine

4. Fill the engine with the proper type and quantity of engine oil.
5. Start the vehicle and check for oil leaks.

Rear Main Seal

REMOVAL & INSTALLATION

1. Before servicing the vehicle, refer to the precautions in the beginning of this section.
2. Remove the transaxle and flywheel.

⁂ WARNING

Be careful not to damage the crankshaft surface when removing the seal.

3. Use seal replacer tool T92C-6700-Ch, to remove the old seal.

To install:

4. Coat the lip of the new seal with clean 5W30 engine oil.
5. Install the new seal using crankshaft rear seal pilot tool T88P-6701-B2 and Rear seal replacer tool T88P-6701-B1.
6. Install the flywheel and the transaxle.

Timing Belt

REMOVAL & INSTALLATION

1. Before servicing the vehicle, refer to the precautions in the beginning of this section.
2. Remove or disconnect the following:
   - Negative battery cable
   - Spark plugs
   - Catalytic converter
   - Right front wheel
   - Right inner splash shield
   - Accessory drive belt and idler pulley
3. Rotate the crankshaft to Top Dead Center (TDC) and install Crankshaft TDC Timing Peg T97P-6000-A.
4. Rotate the crankshaft clockwise against the peg.
5. Remove or disconnect the following:
   - Water pump pulley
   - Valve cover
   - Crankshaft pulley
   - Timing belt covers
6. Install Camshaft Alignment Timing Tool T94P-6256-CH into the slots of both camshafts at the rear of the cylinder head to lock the camshafts into position.
7. Loosen the timing belt tensioner bolt and relieve the tension on the timing belt by disconnecting the tensioner tab from the timing cover back plate.
8. Remove the timing belt.

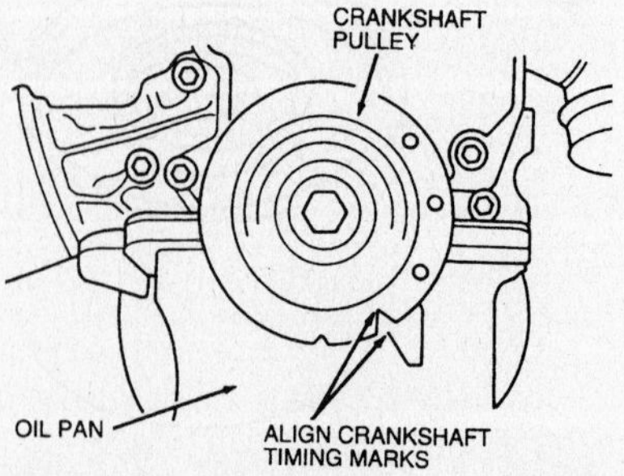

Crankshaft alignment position—Ford 2.0L (VIN 3) engines

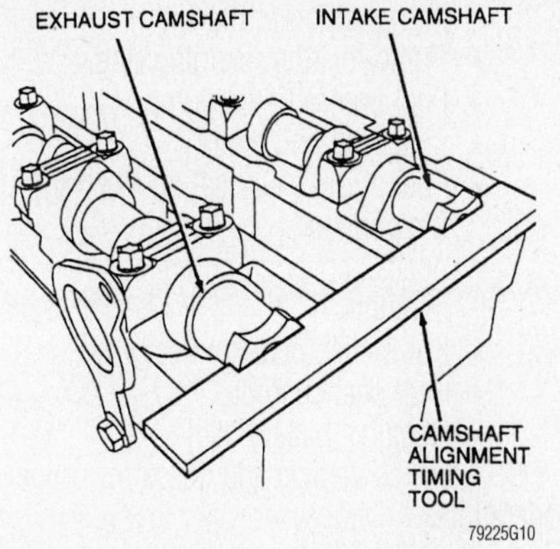

Placement of camshaft alignment timing tool T94P-6256-CH—Ford 2.0L (VIN 3) engines

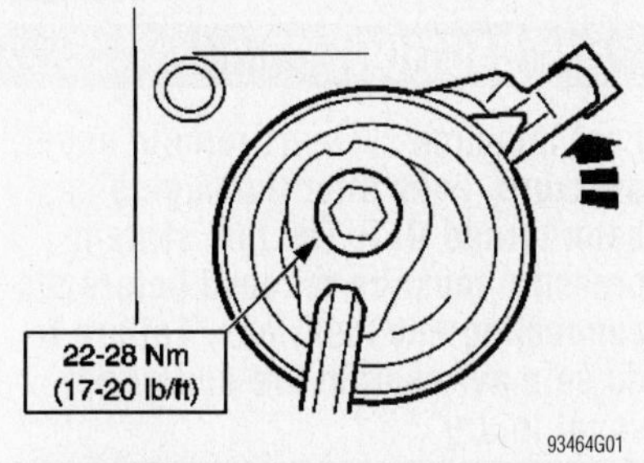

Timing belt tensioner index marks—Ford 2.0L (VIN 3) engines

To install:

➡Always replace the timing belt after loosening or removing it.

9. Install the timing belt.
10. Engage the timing belt tensioner tab into the timing cover back plate.
11. Adjust the timing belt tensioner until the index marks are aligned. Tighten the tensioner bolt to 17–20 ft. lbs. (22–28 Nm).
12. Remove the camshaft alignment tool and the crankshaft TDC peg.
13. Install or connect the following:
    - Timing belt covers
    - Crankshaft pulley. Tighten the bolt to 81–89 ft. lbs. (110–120 Nm).
    - Valve cover
    - Water pump pulley
    - Accessory drive belt and idler pulley
    - Right inner splash shield
    - Right front wheel
    - Catalytic converter
    - Spark plugs
    - Negative battery cable
14. Start the engine and check for proper operation.

Piston and Ring

POSITIONING

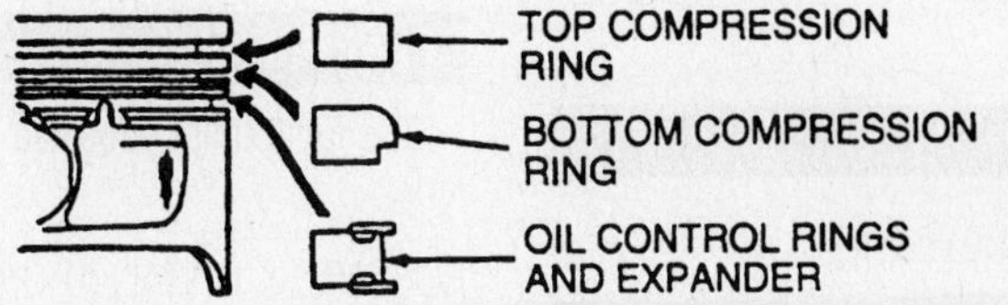

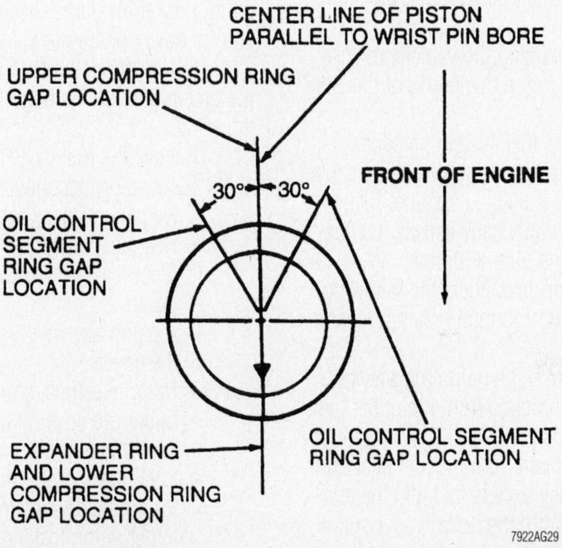

Ford 2.0L (VIN 3) engines—piston ring positioning, end-gap spacing and piston positioning. The small directional arrow must face the front of the engine.

FUEL SYSTEM

Fuel System Service Precautions

Safety is the most important factor when performing not only fuel system maintenance but any type of maintenance. Failure to conduct maintenance and repairs in a safe manner may result in serious personal injury or death. Maintenance and testing of the vehicle's fuel system components can be accomplished safely and effectively by adhering to the following rules and guidelines.

- To avoid the possibility of fire and personal injury, always disconnect the negative battery cable unless the repair or test procedure requires that battery voltage be applied.
- Always relieve the fuel system pressure prior to disconnecting any fuel system component (injector, fuel rail, pressure regulator, etc.), fitting or fuel line connection. Exercise extreme caution whenever relieving fuel system pressure, to avoid exposing skin, face and eyes to fuel spray. Please be advised that fuel under pressure may penetrate the skin or any part of the body that it contacts.
- Always place a shop towel or cloth around the fitting or connection prior to loosening to absorb any excess fuel due to spillage. Ensure that all fuel spillage (should it occur) is quickly removed from engine surfaces. Ensure that all fuel soaked cloths or towels are deposited into a suitable waste container.
- Always keep a dry chemical (Class B) fire extinguisher near the work area.
- Do not allow fuel spray or fuel vapors to come into contact with a spark or open flame.
- Always use a back-up wrench when loosening and tightening fuel line connection fittings. This will prevent unnecessary stress and torsion to fuel line piping.
- Always replace worn fuel fitting O-rings with new. Do not substitute fuel hose or equivalent, where fuel pipe is installed.

Fuel System Pressure

RELIEVING

1. Before servicing the vehicle, refer to the precautions in the beginning of this section.
2. Remove the Schrader valve cap at the end of the fuel rail and attach a fuel pressure gauge.
3. Open the manual relief valve on the fuel pressure gauge slowly to relive the fuel pressure

Fuel Filter

REMOVAL & INSTALLATION

The inline fuel filter is located in the engine compartment between the fuel tank and the fuel rail.

1. Before servicing the vehicle, refer to the precautions in the beginning of this section.
2. Relieve the fuel system pressure.
3. Disconnect the negative battery cable.
4. Position a suitable container below the fuel filter to collect any excess fuel that may leak from the filter and lines.
5. Remove or disconnect the following:
   - Fuel filter mounting clamp, loosen only
   - Filter from the mounting bracket
   - Retaining clips from the upper fuel filter hose
   - Upper hose from the fuel filter and drain any excess fuel into the container. Plug the hose.
   - Retaining clip from the fuel filter lower hose
   - Lower hose from the fuel filter and drain any excess fuel into the container. Plug the hose.
   - Fuel filter

To install:

6. Remove the plugs from the lower and upper hoses.
7. Install or connect the following:
   - Lower hose to the filter
   - Hose retaining clip
   - Upper hose to the filter
   - Hose retaining clip
   - Fuel filter and tighten the filter mounting clamp
   - negative battery cable
8. Start the engine and check for leaks.

Fuel Pump

The fuel pump is located inside the fuel tank.

REMOVAL & INSTALLATION

This procedure will require a new fuel pump gasket for pump installation, so be sure to have one before starting.

1. Before servicing the vehicle, refer to the precautions in the beginning of this section.
2. Relieve the fuel system pressure.
3. Remove or disconnect the following:
   - Negative battery cable
   - Rear seat cushion
   - Electrical connectors from the fuel pump
   - 4 pump access cover screws and the cover
   - Pump electrical connector located under the cover, if equipped
   - Fuel line clip(s)
   - Fuel line(s) from the pump
   - Pump locking retaining ring., using a fuel pump locking ring removal tool or a brass drift and a hammer
   - Fuel pump and the gasket from the tank

To install:

4. Install or connect the following:
   - New gasket and place the pump into position in the tank
   - Pump locking retaining ring
   - Fuel line(s) to the pump
   - Line retaining clip(s)
   - Fuel pump electrical connection(s), if equipped
   - Access cover and tighten the retainers
   - Pump electrical connector(s)
   - Rear seat cushion
   - Negative battery cable
5. Start the vehicle and check for proper operation.

Fuel Injector

REMOVAL & INSTALLATION

⁂ CAUTION

Fuel injection systems remain under pressure, even after the engine has been turned OFF. The fuel system pressure must be relieved before disconnecting any fuel lines. Failure to do so may result in fire and/or personal injury.

1. Before servicing the vehicle, refer to the precautions in the beginning of this section.
2. Relieve the fuel system pressure.
3. Remove or disconnect the following:
   - Negative battery cable
   - Fuel injection fuel rail
   - Injector retaining clips
4. Grasp the fuel injector body and pull up while gently rocking the fuel injector from side to side.
5. Once removed, inspect the fuel injector cap and body for signs of deterioration. Replace as required.
6. Remove the O-rings and discard. If an O-ring or end cap is missing, look in the intake manifold for the missing part.

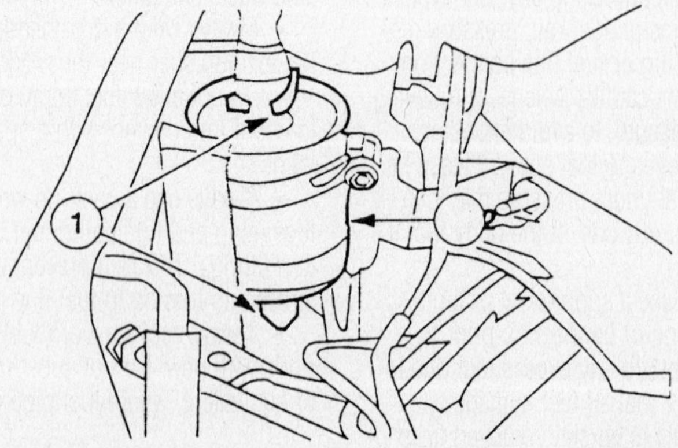

7922MG79

Detach the upper and lower hose clips before disconnecting hoses (1) from the fuel filter

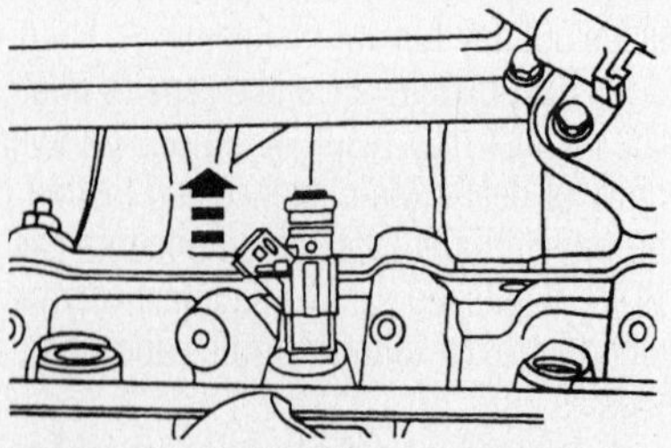

90985G07

Grasp the fuel injector's body and pull upward, while gently rocking the fuel injector from side to side

To install:

7. Install or connect the following:
   - New O-rings onto each injector and apply a small amount of clean engine oil to the O-rings
   - Injectors using a slight twisting downward motion
   - Injector retaining clips
   - Fuel injection fuel rail
   - Negative battery cable

8. Run the engine at idle for 2 minutes, then turn the engine **OFF** and check for fuel leaks and proper operation.

DRIVE TRAIN

Transaxle Assembly

REMOVAL & INSTALLATION

Manual

1. Before servicing the vehicle, refer to the precautions in the beginning of this section.
2. Remove or disconnect the following:
   - Both battery cables, negative cable first
   - Battery and tray
   - Engine air cleaner outlet tube
   - Constant Control Relay Module (CCRM) module and bracket
   - Slave cylinder line from the slave cylinder hose and plug the hose
   - Slave cylinder line retaining clip
   - Slave cylinder line from the bracket
   - Heated Oxygen (HO_2S) sensor electrical connection
   - Back-up light electrical connection
   - Electrical connector bracket bolt and the bracket
   - Vehicle Speed Sensor (VSS) electrical connection
   - Halfshafts
3. Install Engine Support Bar D88–6000A, and attach it to the engine lifting eyes with suitable chains or cables.
   - Insulator nuts from the left-hand engine support bracket
   - Front and rear upper transaxle-to-engine bolts
4. Drain the transaxle fluid and install the drain plug.
   - Air conditioning line from the retainer located on the engine support crossmember

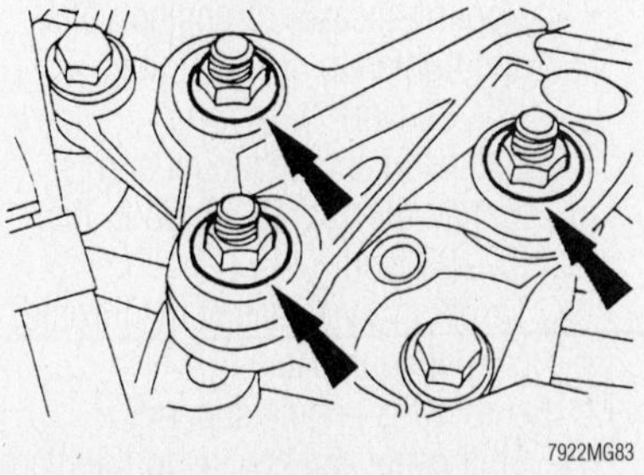

Location of the left-hand engine support bracket nuts—2.0L engines

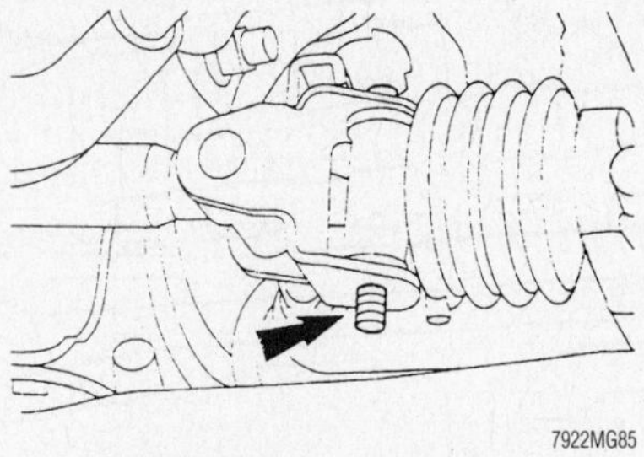

Loosen the transaxle gearshift bolt and remove the gearshift rod and clevis pin from the input shift shaft

   - Engine support crossmember bolts and nuts
   - Crossmember
   - Gearshift stabilizer bar-to-transaxle nut
   - Stabilizer bar and support from the transaxle
   - Transaxle gearshift rod nut
   - Transaxle gearshift bolt
   - Gearshift rod and clevis pin from the input shift shaft
   - Starter motor
   - Lower slave cylinder tube and the slave cylinder
   - Lower transaxle-to-engine bolts
5. Position and secure a suitable transaxle jack under the transaxle.
   - Catalytic converter
   - Middle transaxle-to-engine bolts
   - Transaxle from the vehicle

To install:

6. Apply a thin coating of suitable grease to the splines of the input shaft.
7. Place the transaxle onto a suitable transaxle jack. Be sure the transaxle is secure.
8. Install or connect the following:
   - Transaxle
   - Middle transaxle-to-engine bolts and tighten them to 23–38 ft. lbs. (38–51 Nm)
   - Lower engine-to-transaxle bolts and tighten them to 23–38 ft. lbs. (38–51 Nm)
   - clutch slave cylinder and tighten the nuts to 12–17 ft. lbs. (16–23 Nm)
   - Lower slave cylinder tube and tighten the fitting to 10–16 ft. lbs. (13–21 Nm)
   - Starter motor
   - Gearshift rod and clevis into position on the input shift shaft
   - Gearshift bolt and gearshift rod nut and tighten to 12–17 ft. lbs. (16–23 Nm)
   - Gearshift stabilizer bar and tighten the nut to 23–34 ft. lbs. (31–46 Nm)
   - Halfshafts and the catalytic converter
   - Engine support crossmember. Tighten the crossmember bolts and nuts to 47–65 ft. lbs. (64–89 Nm) and the insulator nuts to 28–37 ft. lbs. (38–51 Nm).
   - Front and rear upper transaxle-to-engine bolt and tighten it to 28–38 ft. lbs. (38–51 Nm)
   - Left-hand engine support insulator nuts and tighten them to 50–68 ft. lbs. (67–93 Nm)
9. Remove the engine support bar.
   - Electrical connector support bracket
   - All the electrical connectors
   - Slave cylinder line to the slave cylinder hose and attach the retaining clip. Tighten the fitting to 10–16 ft. lbs. (13–21 Nm).
10. Bleed the clutch hydraulic system.
11. Add the proper type and amount of fluid to the transaxle.
12. Install or connect the following:
    - CCRM and bracket assembly, if removed
    - CCRM retainers and attach the electrical connection
    - Engine air cleaner tube
    - Battery tray and the battery
    - Battery cables, negative cable last
13. Check for fluid leaks and proper clutch operation.
14. Road test the vehicle and check for proper transaxle operation.

Automatic

1. Before servicing the vehicle, refer to the precautions in the beginning of this section.
2. Remove or disconnect the following:
   - Battery cables, negative cable first
   - Battery and tray
   - Engine air cleaner assembly
   - Computer Control Relay Module

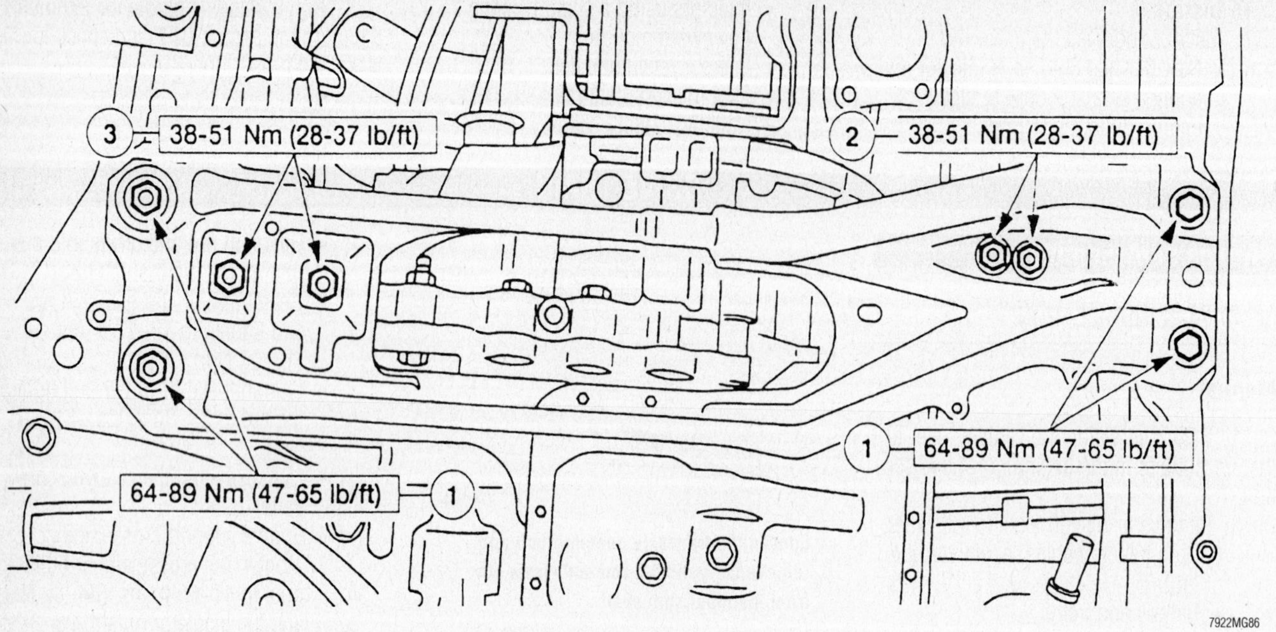

Install the engine support crossmember and tighten the retainers to specification

(CCRM) electrical connections, relay retainers, the relay and bracket from the engine
- Shift control cable and bracket retaining nut from the manual shift lever
- Shift cable and bracket clip, then set the bracket and cable assembly aside
- All electrical connections from the transaxle
- Starter motor
- Throttle valve actuating cable bolts and disconnect the cable from the throttle cam, if equipped

3. Install Engine Support D88L-6000-A, to the engine. The engine must be properly supported for transaxle removal.
4. Remove or disconnect the following:
   - Transaxle cooler lines at the transaxle
   - Left-hand engine mount
   - Upper transaxle housing bolts
   - Left-hand splash shields, if equipped
   - Transaxle plug and drain the fluid
   - Halfshafts. Install 2 transaxle plugs T88C-7025-AH, into the differential side gears

⁂ WARNING

Failure to install the transaxle plugs may cause the differential side gears to become improperly positioned. If the gears become misaligned, the differential will have to be removed from the transaxle to align them.

- Engine support crossmember and the catalytic converter
- Air conditioning line from the retainer located on the engine support crossmember
- Transaxle support crossmember bolts and nuts, the left-hand engine isolator nuts, then the crossmember
- Transaxle housing cover bolts and the cover
- 4 flywheel-to-torque converter nuts

5. Position a suitable transaxle jack under the transaxle. Secure the transaxle to the jack.
   - Remaining transaxle-to-engine bolts
6. Slowly lower the transaxle out of the vehicle.
7. Inspect all components including mounts and brackets.

To install:

➡Prior to installing the transaxle, lubricate the torque converter pilot hub with multi-purpose grease.

8. Align the torque converter studs to the flywheel.
9. Secure the transaxle on the transaxle jack.
10. Install or connect the following:
    - Transaxle. Tighten the bolts to 40–58 ft. lbs. (55–80 Nm).
    - Flywheel-to-converter nuts and tighten them to 26–36 ft. lbs. (35–49 Nm)
    - Transaxle housing cover and tighten the bolts to 40–58 ft. lbs. (55–80 Nm)
    - Engine support crossmember. Tighten the crossmember bolts and nuts to 47–65 ft. lbs. (64–89 Nm) and the insulator nuts to 28–37 ft. lbs. (38–51 Nm).
    - Front and rear upper transaxle-to-engine bolt and tighten it to 28–38 ft. lbs. (38–51 Nm)
11. Install or connect the following:
    - Halfshafts
    - Crossmember to the transaxle mounts and the chassis. Tighten the crossmember-to-transaxle mount nuts to 27–38 ft. lbs. (37–52 Nm). Tighten the crossmember-to-chassis nuts and bolts to 47–66 ft. lbs. (64–89 Nm).
    - Lower transaxle-to-engine oil pan bolts and tighten to 27–38 ft. lbs. (37–52 Nm)
    - Engine/transaxle splash shields
    - Starter motor
    - All the electrical connections
    - Upper transaxle-to-engine bolts and tighten to 40–58 ft. lbs. (55–80 Nm)
    - Left-hand engine mount and tighten the nuts to 50–68 ft. lbs. (67–93 Nm)
    - Throttle valve cable at the throttle cam, if equipped
12. Remove the engine support.
    - Shift cable and bracket to the manual shift lever, the shift cable and

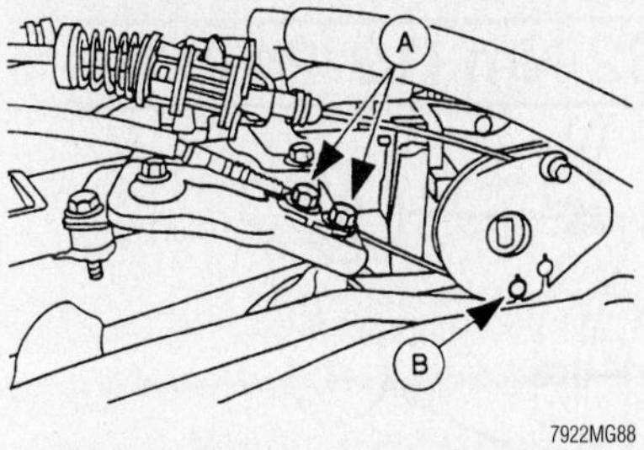

Connect the throttle valve cable at the throttle cam (B) and tighten the cable retaining bolts (A)

bracket clip. Tighten the nut to 12–16 ft. lbs. (16–22 Nm).

- Engine air cleaner assembly
- Battery tray and battery
- Both battery cables, negative cable last

13. Add the proper type and quantity of transaxle fluid.

14. Check the transaxle for leaks and for proper operation.

15. Check the MLP switch for proper adjustment, as follows:

a. Shift the transaxle into NEUTRAL, then align the marks on the transmission range switch and the manual control lever.

b. Install and finger-tighten the switch retaining bolts.

c. Attach an ohmmeter between terminals **A** and **B** on the switch as shown in the accompanying illustration.

d. Adjust the switch by rotating the switch housing on the manual control lever until there is no continuity between the terminals.

e. Hold the switch in place, then tighten its retaining bolts to 70–95 inch lbs. (8–11 Nm).

f. Remove the ohmmeter and attach the switch electrical connection.

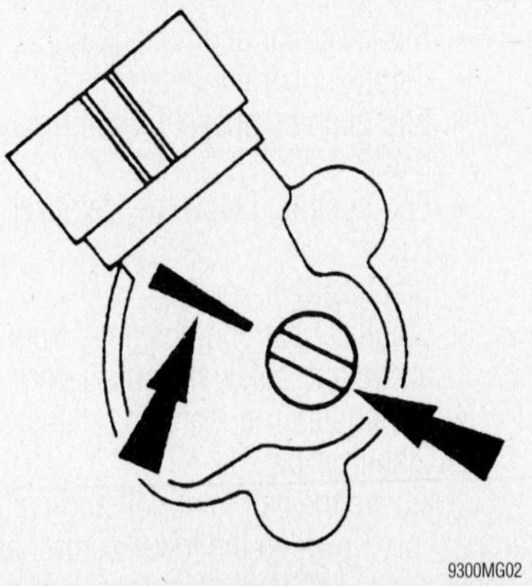

Align the marks on the MLP switch and the manual control lever as shown

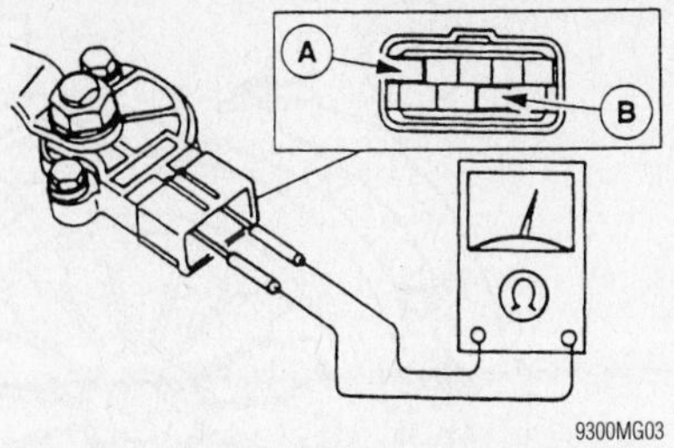

Attach an ohmmeter to terminals A and B on the transmission range switch

g. Place the manual control shift outer lever in position and tighten its retaining nut to 33–47 ft. lbs. (44–64 Nm).

16. Road test the vehicle and check for proper operation.

Clutch

ADJUSTMENT

Pedal Free-Play

1. Before servicing the vehicle, refer to the precautions in the beginning of this section.

2. Depress the clutch pedal until resistance can be felt, and measure the distance between the upper clutch height and where the resistance is felt. The free-play should be 0.0–0.40 in. (5.0–13.9mm).

3. If an adjustment is necessary, turn the locknut and equalizer bar-to-clutch release lever rod.

4. After adjustment, measure the disengagement height from the upper surface of the clutch pedal pad to the carpet. The distance should be 2.3 in. (59mm).

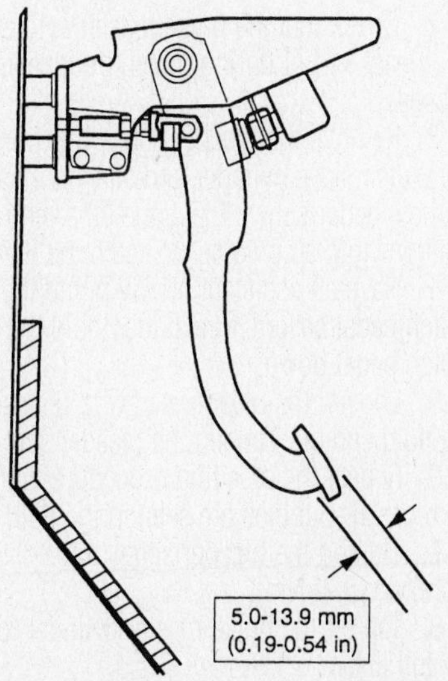

Clutch pedal free play measurement

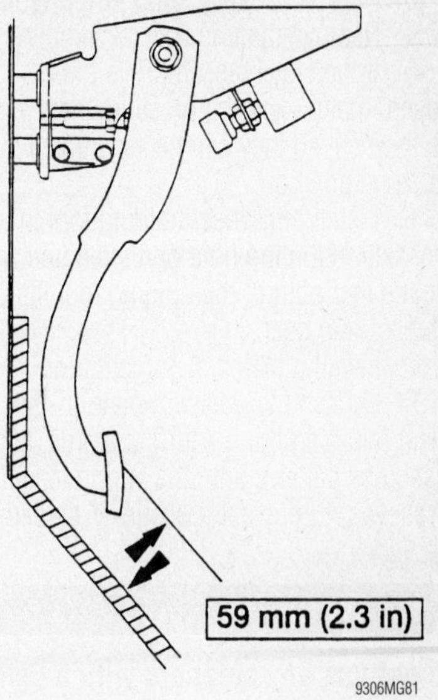

Clutch pedal free play adjustment

Pedal Height

1. Measure the distance from the upper surface of the pedal pad to the carpet. The measurement should be 8.35–8.54 in. (212–217mm).

2. If an adjustment is necessary, turn the locknut and the Clutch Pedal Position (CPP) switch until the pedal height is correct.

REMOVAL & INSTALLATION

1. Before servicing the vehicle, refer to the precautions in the beginning of this section.

2. Disconnect the negative battery cable.

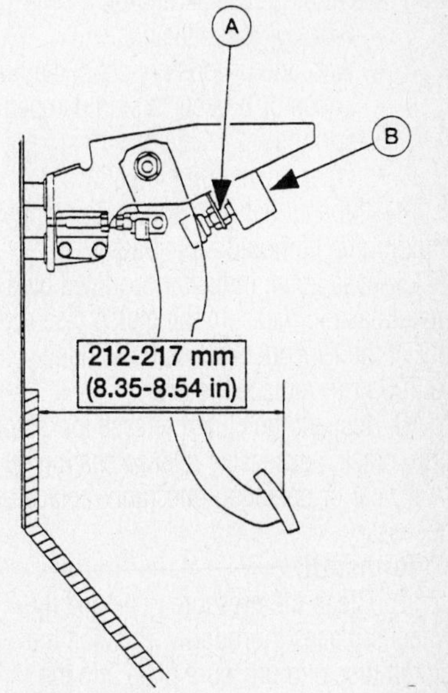

Clutch pedal height adjustment

3. Raise and safely support the vehicle.
4. Remove the transaxle assembly.
5. If the clutch assembly is to be reused, matchmark the pressure plate and the flywheel so they can be assembled in the same position.
6. Install a flywheel holding tool in a transaxle mounting hole on the engine and engage the tooth of the holding tool into the flywheel ring gear.
7. Install a clutch alignment tool.
8. Loosen the pressure plate-to-flywheel retaining bolts 1 turn at a time, in a crisscross pattern, until the spring tension is relieved, to prevent pressure plate damage.

⁂ WARNING

Do not use any cleaners with a petroleum base and do not immerse the clutch pressure plate in solvent.

9. Clean the clutch pressure plate with a suitable commercial alcohol base solvent.
10. Inspect the pressure plate surface for burns, scores, flatness or ridges. Reface or replace the pressure plate.
11. Inspect the pressure plate diaphragm fingers for wear. Replace the pressure plate if necessary.
12. Using a slide caliper, measure the depth of the rivet heads. If the rivet head is within 0.012 in. (0.3mm) from the clutch surface, replace the clutch.

➡Use emery cloth to remove minor imperfections from the clutch disc lining surface.

13. Inspect the clutch disc for the following:
   - Oil or grease saturation
   - Worn or loose facings
   - Warpage or loose rivets at the hub
   - Loose or broken torsion dampening springs
   - Wear or rust on the splines.
14. If the clutch disc shows any of these conditions, it should be replaced.
15. Use a dial indicator mounted on a metal base to measure the clutch disc run-out. If the run-out exceeds 0.0276 in. (0.700mm), replace the disc.
16. Inspect the clutch release for distortion, cracks, excessive release bearing surface wear or damaged tines and replace as necessary.

To install:

17. Clean the pressure plate and flywheel surfaces thoroughly. Position the clutch disc and pressure plate into the installed position and support them with a clutch aligning tool. If the clutch assembly is being reused, align the matchmarks that were made during the removal procedure.

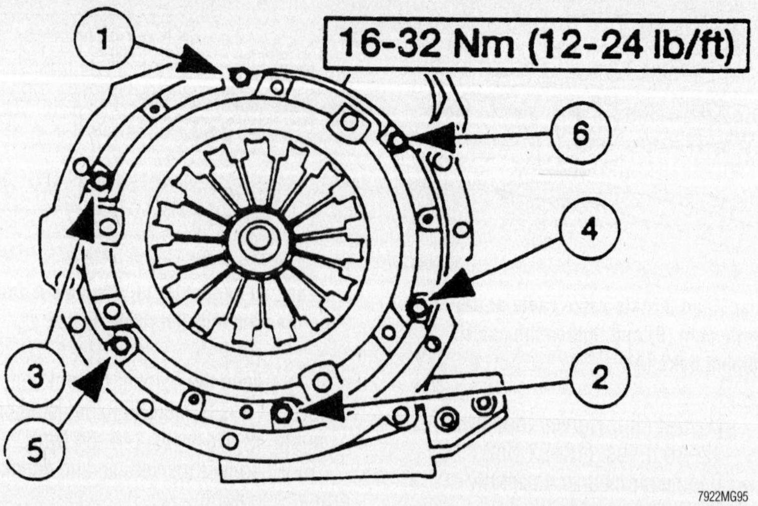

Tighten the pressure plate-to-flywheel retaining bolts in the sequence illustrated to specification

18. Install the pressure plate-to-flywheel retaining bolts. Tighten the bolts in the correct sequence to 12–24 ft. lbs. (16–32 Nm). Remove the alignment tool.
19. Install the transaxle assembly.
20. Lower the vehicle.
21. Bleed the hydraulic clutch system.
22. Adjust the clutch pedal free-play.
23. Connect the negative battery cable.
24. Road test the vehicle and check the clutch for proper operation.

Hydraulic Clutch System

BLEEDING

1. Before servicing the vehicle, refer to the precautions in the beginning of this section.
2. Check that the brake master cylinder is at least ¾ full during the entire bleeding process.
3. Remove the bleeder screw cap from the clutch slave cylinder and attach a hose to the bleeder screw. Place the other end of the hose into a container to catch the fluid.
4. Have an assistant slowly pump the clutch pedal several times, then hold the clutch pedal down.
5. Loosen the bleeder screw to release the fluid and air. Tighten the bleeder screw.
6. Repeat the bleeding procedure until no more air bubbles are seen in the fluid.
7. Tighten the bleeder screw to 52–78 inch lbs. (6–9 Nm).
8. Top off the brake master cylinder to the full line.
9. Check for proper clutch system operation.

Halfshafts

REMOVAL & INSTALLATION

1. Before servicing the vehicle, refer to the precautions in the beginning of this section.
2. Disconnect the negative battery cable.
3. With the vehicle sitting on the ground, carefully raise the staked portion of the halfshaft retaining nut using a suitable small chisel. Loosen the nut.
4. Remove or disconnect the following:
   - Wheel and tire assembly
   - Halfshaft retaining nut and discard
   - Cotter pin and nut from the tie rod end, then separate the tie rod end from the steering knuckle using a suitable removal tool. Discard the cotter pin.
5. Remove or disconnect the following components to remove the stabilizer bar link:
   - Stabilizer bar end nut and bolt
   - Stabilizer bar end retainer
   - End bushing above the stabilizer bar
   - End bushing below the stabilizer bar
   - Stabilizer bar spacer
   - Stabilizer bar end bushings from above and below the sub-frame
   - Lower stabilizer bar end retainer
   - Stabilizer bar
6. Remove the ball joint bolt and nut. Carefully pry down on the lower control arm to separate the ball joint stud from the steering knuckle.

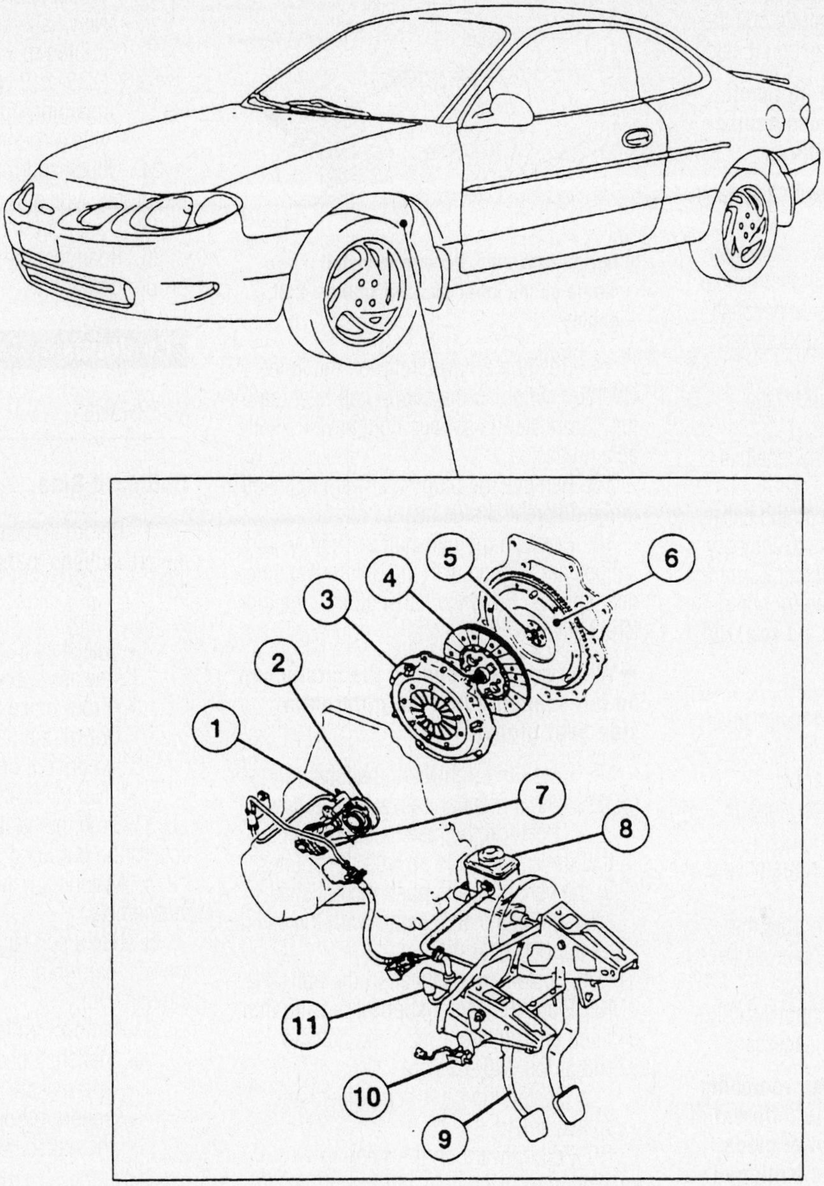

Item	Description
1	Clutch Slave Cylinder
2	Clutch Release Hub and Bearing
3	Clutch Pressure Plate
4	Clutch Disc
5	Pilot Bearing
6	Flywheel

Item	Description
7	Clutch Release Fork
8	Brake Master Cylinder Reservoir
9	Clutch Pedal
10	Clutch Pedal Position (CPP) Switch
11	Clutch Master Cylinder

9306MG78

Exploded view of the clutch system components

7. Pull outward on the steering knuckle/brake assembly. Carefully pull the halfshaft from the hub and position it aside.

➡Removal of the left side halfshaft requires removal of the crossmember to allow access with a prybar.

8. Support the transaxle with a suitable transaxle jack.
9. Unfasten the 4 transaxle crossmember retainers and remove the crossmember.
10. If removing the right side halfshaft, remove the right-hand shield and splash shield.
11. Position a drain pan under the transaxle.
12. Remove the left-hand halfshaft as follows:
   a. Insert a prybar between the halfshaft and the transaxle case. Gently pry outward to release the halfshaft from the differential side gear. Be careful not to damage the transaxle case, oil seal, CV-joint or CV-joint boot.
   b. Remove the halfshaft.
13. Remove the right-hand halfshaft as follows:
   a. On models with a manual transaxle, unfasten the center support bearing bolts.
   b. Lower the halfshaft and remove it from the transaxle.
   c. Separate the halfshaft from the center support bearing and remove the halfshaft from the vehicle.
   d. Inspect the center support bearing for damage and replace as necessary.

➡Install suitable plugs after removing the halfshafts to prevent the differential side gears from moving out of place. Should the gears become misaligned, the differential will have to be removed from the transaxle to align the gears.

To install:

➡Use new locknuts, split pins and circlips for assembly.

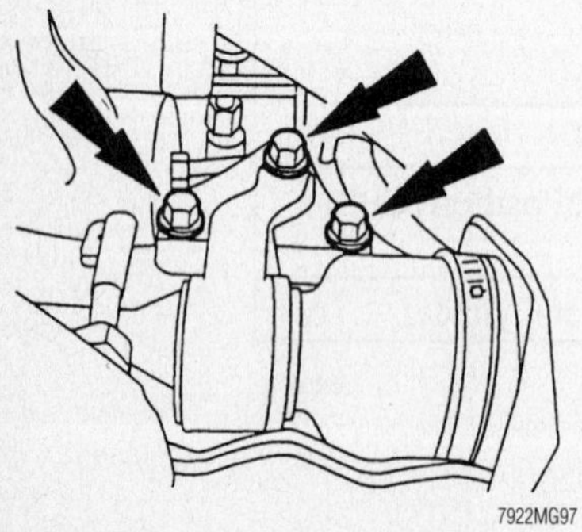

On models with a manual transaxle, unfasten the center support bearing bolts

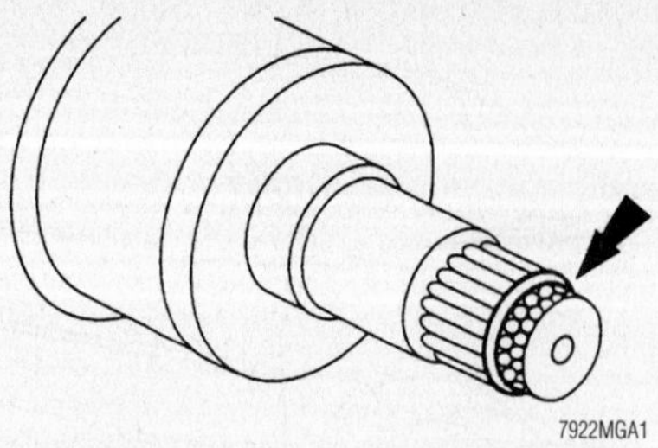

During assembly, be sure to install a new circlip on the inner CV-joint spline—2.0L engine

14. Position a new circlip on the inner CV-joint spline so the circlip gap is at the top. Lubricate the splines lightly with a suitable grease.
15. Remove the plugs that were installed in the differential side gears.
16. Position the halfshaft so the CV-joint splines are aligned with the differential side gear splines. Push the halfshaft into the differential.

➡When seated properly, the circlip can be felt snapping into the differential side gear groove.

17. Install the right-hand halfshaft on models with a manual transaxle as follows:
   a. Position the halfshaft and joint so that the splines line up with the splines in the halfshaft, and push the halfshaft, joint and halfshaft together with the center support bearing.
   b. Install the halfshaft in the transaxle.
18. Place the center support bearing into position and tighten it retaining bolts to 32–46 ft. lbs. (46–62 Nm).
   a. Install the right-hand shield and splash shield.
19. Pull outward on the steering knuckle/brake assembly and insert the halfshaft into the hub.
20. Pry downward on the lower control arm and position the lower ball joint stud in the steering knuckle.
21. Install the crossmember and the crossmember-to-frame bolts. Tighten the bolts to 69–93 ft. lbs. (94–126 Nm).
22. Remove the transaxle jack.
23. Install or connect the following:
   - Steering knuckle and ball joint nut and bolt. Tighten the nut and bolt to 32–43 ft. lbs. (43–59 Nm).
   - Stabilizer bar link in reverse order of removal. Tighten the bar end nut until the protruding bar end bolt length is 0.67–75 in. (17–19mm).
   - Tie rod end to the steering knuckle
   - Tie rod end nut and tighten to 32–41 ft. lbs. (43–56 Nm). Install a new cotter pin.
   - Wheel and tire assembly
   - New halfshaft retaining nut and tighten to 174–235 ft. lbs. (235–319 Nm). Stake the halfshaft retaining nut using a suitable chisel with a rounded cutting edge.
24. Check and refill the transaxle with the proper type and quantity of fluid.
25. Connect the negative battery cable.
26. Road test the vehicle and check for proper operation.

CV-Joints

OVERHAUL

Outboard Side

1. Before servicing the vehicle, refer to the precautions in the beginning of this section.
2. Remove or disconnect the following:
   - Halfshaft from the vehicle. Support the assembly in a vise with soft jaws.
   - Front brake anti-lock sensor indicator off the halfshaft using a punch to drive it off
   - Two halfshaft boot clamp and discard
3. Slide the boot back out of the way to access the outboard joint.
4. Matchmark the joint-to-halfshaft for reassembly.
5. Use a soft faced mallet to separate the outboard joint by gently tapping it off the halfshaft.
6. Remove or disconnect the following:
   - Halfshaft bearing retaining circlip and discard
   - snapring from the outboard side of the halfshaft

➡Do not remove the tape until after installation of the boot.

7. Wrap the outboard halfshaft splines with tape and then slide the boot from the shaft.
8. Clean and inspect the outboard bearings for damage, grit in the grease, pitting or cracks and replace as necessary.

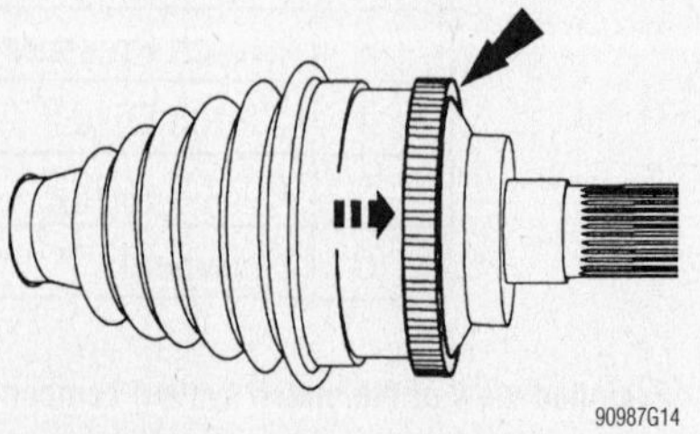

If equipped with ABS, use a punch to drive the front brake anti-lock sensor indicator off the halfshaft

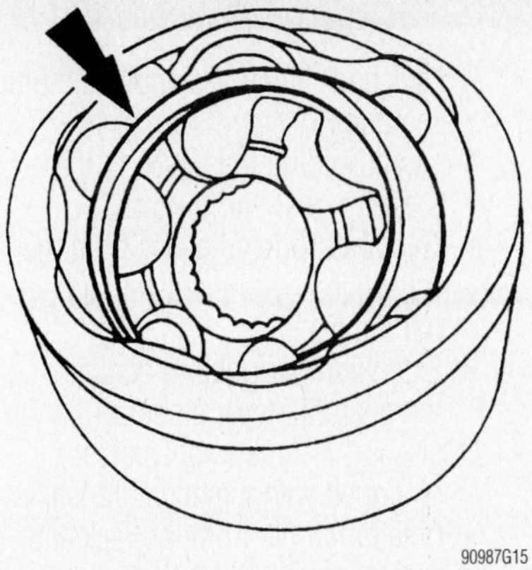

Clean and inspect the outboard bearings for damage

To install:

9. Lubricate the joint bearings with constant velocity grease E43Z-19590-A.
10. Install or connect the following:
   - Boot and remove the tape
   - Snapring on the outboard side of the halfshaft
   - New bearing retaining circlip
11. Use a soft faced mallet to install the outboard joint by gently tapping it onto the halfshaft.
12. Remove any excess grease from the mating surfaces and slide the boot forward onto the joint.
13. Insert a suitable cloth covered tool between the boot and the outer bearing race to allow trapped air to escape from the boot.
14. Use boot clamp pliers to install two new boot clamps.
15. If equipped with anti-lock brakes, use sensing ring replacer tool T94P-20202-B, to install the front brake anti-lock sensor indicator onto the halfshaft.
16. Install the halfshaft.

INBOARD SIDE

1. Before servicing the vehicle, refer to the precautions in the beginning of this section.

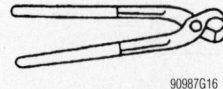

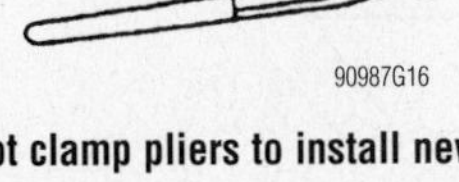

Use boot clamp pliers to install new boot clamps

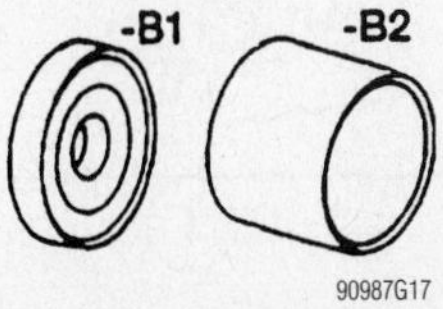

Use sensing ring replacer tool T94P-20202-B to install the front brake anti-lock sensor indicator onto the halfshaft

2. Remove or disconnect the following:
   - Halfshaft from the vehicle. Support the assembly in a vise with soft jaws.

➡The right-hand side halfshaft on models equipped with a manual transaxle do not have an inboard circlip.

   - Bearing retaining circlip from the inboard joint housing and discard
   - Joint boot clamps and discard
   - Joint boot from the joint housing
   - Joint housing from the tripod bearing and halfshaft
   - Tripod bearing snapring
3. Matchmark the tripod bearing-to-halfshaft for reassembly.
4. Remove the tripod bearing from the halfshaft.

➡Do not remove the tape until after installation of the boot.

5. Wrap the halfshaft splines with tape and then slide the boot from the shaft.
6. Clean and inspect the tripod bearing assembly and outboard joint housing for damage, grit in the grease, pitting or cracks and replace as necessary.

To install:

7. Lubricate the tripod bearing and inboard joint housing with constant velocity grease E43Z-19590-A.
8. Install or connect the following:
   - Boot and remove the tape
   - Tripod bearing onto the halfshaft
   - Tripod bearing snapring

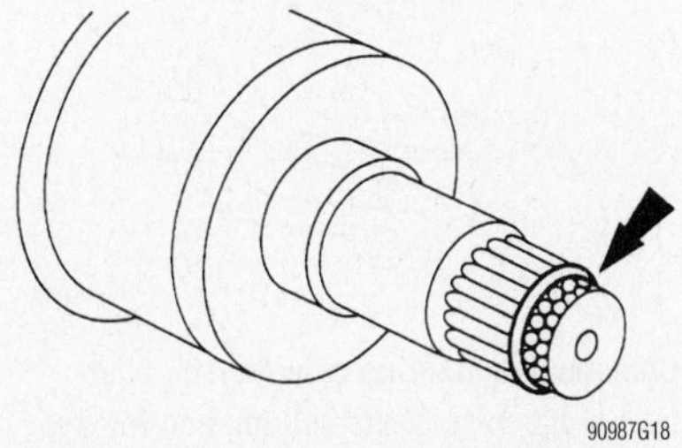

Remove and discard the bearing retaining circlip from the inboard joint housing

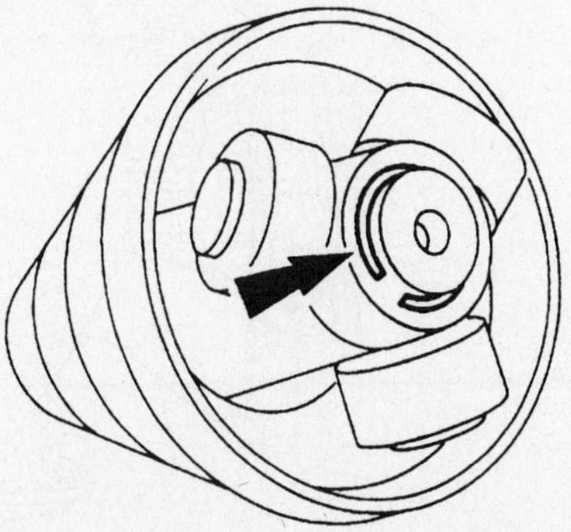

Location of the tripod bearing snapring

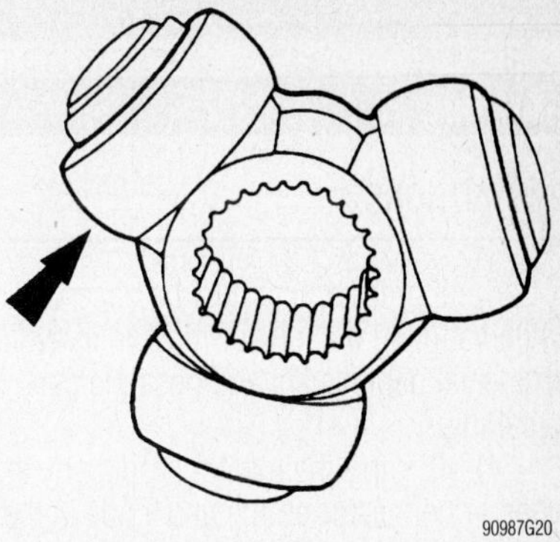

Clean and inspect the tripod bearing assembly . . .

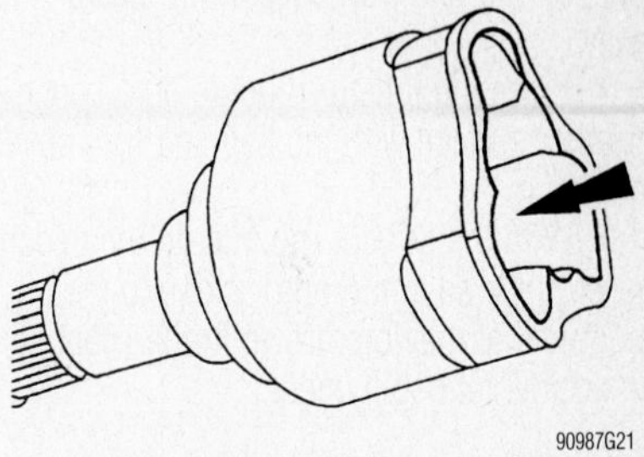

. . . and outboard joint housing for damage, grit in the grease, pitting or cracks, and replace as necessary

   - Inboard joint housing onto the tripod bearing
   - Boot onto the inboard joint housing
9. Insert a suitable cloth covered tool between the boot and the bearing to allow trapped air to escape from the boot.
10. Use boot clamp pliers to install two new boot clamps.

➡Measure the relaxed state of the halfshaft bearing circlip installed on the shaft. Refer to the accompanying illustration. The wrong size clip will not retain the halfshaft properly.

11. Install a new halfshaft bearing retainer circlip of the correct size onto the outboard halfshaft joint.
12. Install the halfshaft.

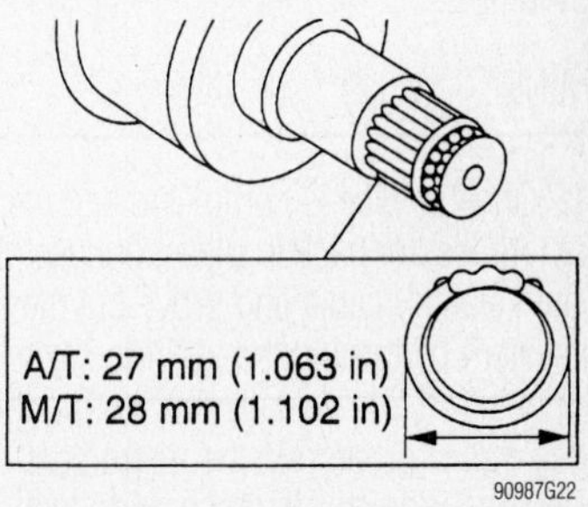

Measure the relaxed state of the halfshaft bearing circlip installed on the shaft, and make sure it is within specification

STEERING AND SUSPENSION

Air Bag

PRECAUTIONS

Several precautions must be observed when handling the inflator module to avoid accidental deployment and possible personal injury.

- Never carry the inflator module by the wires or connector on the underside of the module.
- When carrying a live inflator module, hold securely with both hands, and ensure that the bag and trim cover are pointed away.
- Place the inflator module on a bench or other surface with the bag and trim cover facing up.
- With the inflator module on the bench, never place anything on or close to the module that may be thrown in the event of an accidental deployment.

DISARMING

1. Before servicing the vehicle, refer to the precautions in the beginning of this section.
2. Disconnect both battery cables, negative cable first.
3. Wait at least 1 minute. This allows time for the back-up power supply to deplete its stored energy.
4. Remove the driver's side air bag module, then the passenger side if required.
5. Use caution when carrying live air bags. Always place the air bag with the cover up.
6. If the battery needs to be reconnected while one or both of the air bags are removed from the system, install Air Bag Simulator 105–00010, to the drivers side and/or passenger side air bag harness connectors as required. Before removing either air bag simulator, disconnect both battery cables and wait at least 1 minute before continuing.

ARMING

1. Once service is completed and the air bag modules are back in place, connect the negative battery cable and prove out the air bag system by turning the ignition key to the **RUN** position and visually monitoring the air bag indicator lamp in the instrument cluster. The indicator lamp should illuminate for approximately 6 seconds, then turn **OFF**. If the indicator lamp does not illuminate, stays **ON**, or flashes at any time, a fault has been detected by the air bag diagnostic monitor requiring immediate attention.

Rack and Pinion Steering Gear

REMOVAL & INSTALLATION

1. Before servicing the vehicle, refer to the precautions in the beginning of this section.
2. Turn the key to the **ACC** position.
3. Remove or disconnect the following:
   - Steering column tube boot nuts at the base of the column and the tube boots, from inside the passenger compartment
   - Steering column input shaft coupling-to-steering gear input shaft pinch bolt
   - Front wheel and tire assemblies
   - Separate the tie rod ends from the steering knuckles and discard the cotter pins
   - Right-hand lower splash shield
   - Crossmember
   - Pressure and return lines from the rack and pinion assembly and plug the lines
   - Strap holding the hoses to the steering gear and discard
   - Gearshift rod and clevis from the transaxle, if equipped with a manual transaxle
   - Extension bar nut and disconnect the gearshift lever stabilizer bar and support from the transaxle, if equipped with a manual transaxle
   - Retaining nuts from the rack and pinion mounting brackets
   - Pinion mounting brackets
   - Pushpin and position the right-hand boot aside
   - Rack and pinion assembly from the right-hand side of the vehicle

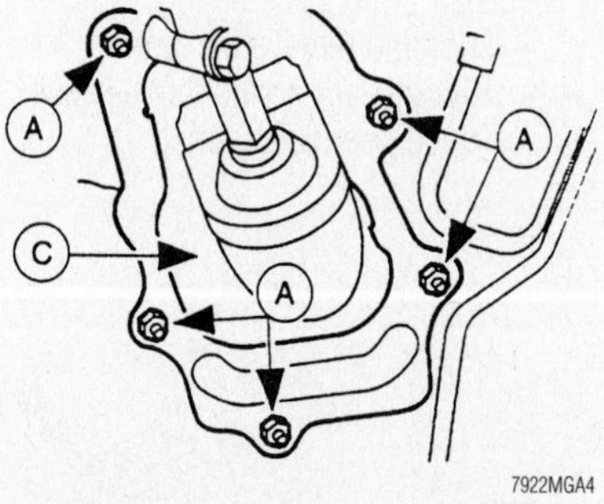

Unfasten the steering column tube boot nuts at the base of the column and remove the tube boots

To install:

4. Install or connect the following:
   - Rack and pinion assembly in its mounting location
5. Align the steering column input shaft coupling and the steering gear input shaft.
   - Rack and pinion mounting brackets. Tighten the retaining nuts to 28–38 ft. lbs. (37–57 Nm).
   - Gearshift stabilizer bar, support and the gearshift rod and clevis, if equipped with a manual transaxle

➡Install new Teflon seals on the power steering pressure hose fitting and be sure the threads are clean before connecting the hose.

- Plugs and connect the pressure and return lines to the rack and pinion assembly. Tighten the pressure hose to 21–25 ft. lbs. (28–33 Nm) and the return line fitting to 20–25 ft. lbs. (27.3–33.9 Nm).
- New strap to hold the power steering lines to the rack and pinion housing

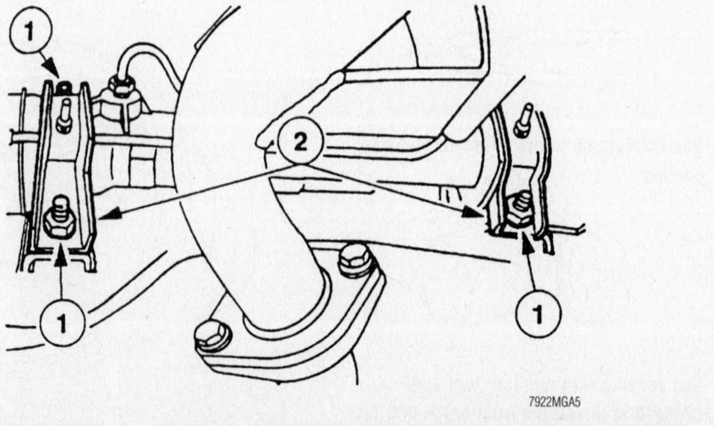

Unfasten the retaining nuts from the rack and pinion mounting brackets and remove the brackets

- Crossmember and tighten the bolts to 69–97 ft. lbs. (94–131 Nm)
- Right-hand boot shield into position
- Shield pushpin
- Right-hand splash shield
- Tie rod ends to the steering knuckles and the castellated nuts. Tighten to specification.
- New cotter pins
- Wheel and tire assemblies
- Steering column input shaft coupling-to-steering gear input shaft pinch bolt and tighten the bolt to 30–36 ft. lbs. (40–50 Nm)
- Steering column tube boots and the 5 retainers. Tighten the retainers to 18–52 inch lbs. (2–5.9 Nm).
- Negative battery cable

6. Fill and bleed the power steering system.
7. Check the alignment and adjust as required.
8. Start the engine and check for leaks.
9. Road test the vehicle and check for proper steering system operation.

Strut

REMOVAL & INSTALLATION

Front

1. Before servicing the vehicle, refer to the precautions in the beginning of this section.
2. Remove or disconnect the following:
   - Negative battery cable
   - Front wheel and tire assembly
   - Clip securing the brake hose to the strut (spring and shock) assembly.
   - Anti-lock brake harness cable and clip, if equipped with anti-lock brakes
   - 2 nuts and 2 bolts securing the strut assembly to the steering knuckle
   - 4 upper strut retaining nuts
   - Strut assembly from the vehicle

✲✲ CAUTION

Never remove the strut piston rod nut unless the coil spring is compressed. Always wear safety glasses when using a spring compressor.

3. Inspect all components and replace as needed.

To install:

4. Install or connect the following:
   - Strut assembly into the wheel housing. Be sure the direction indicator on the upper mounting bracket faces inboard.
   - Upper mounting bracket to the strut tower with the retaining nuts. Tighten the nuts to 22–30 ft. lbs. (29–40 Nm).
   - Strut assembly to the steering knuckle and the retaining bolts and nuts. Tighten to 69–93 ft. lbs. (93–127 Nm).
   - Brake hose on the strut assembly and secure it with the brake hose clip
   - Anti-lock harness cable and clip, if equipped with anti-lock brakes
   - Front wheel and tire assembly
   - Negative battery cable
5. Check the front wheel alignment.
6. Road test the vehicle and check for proper operation.

Rear

1. Before servicing the vehicle, refer to the precautions in the beginning of this section.

1	Front Shock Absorber Upper Mounting Bracket
2	Upper Spring Seat (Part of 18198)
3	Front Brake Hose
4	Disc Brake Caliper
5	Front Disc Brake Rotor
6	Front Wheel Knuckle
7	Tie Rod End
8	Front Stabilizer Bar

9	Front Suspension Lower Arm Mounting Bolt Bushing (Rear)
10	Front Wheel Spindle Tie Rod
11	Front Suspension Lower Arm
12	Front Suspension Lower Arm Mounting Bolt Bushing (Front)
13	Front Shock Absorber
14	Front Coil Spring

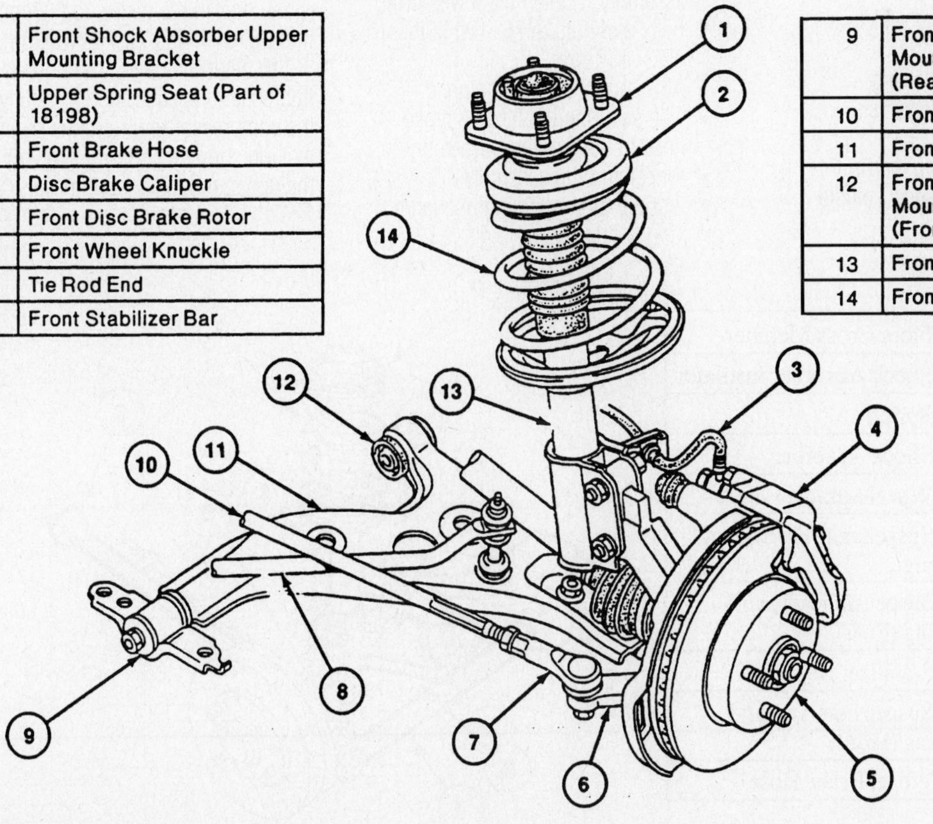

9300MG04

Identification of the front suspension components

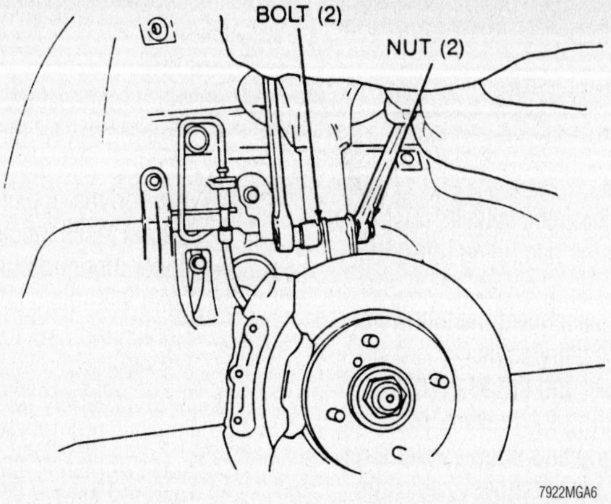

Remove the 2 nuts and 2 bolts securing the strut assembly to the steering knuckle

2. Remove or disconnect the following:
   - Negative battery cable
   - Package tray trim panel, on sedan and coupe models
   - Quarter trim pane, on wagon models
   - 2 upper strut retaining nuts
   - Wheel and tire assembly
   - Clip securing the brake hose to the rear strut assembly
   - ABS sensor bolt, if equipped
   - Nuts and bolts securing the rear strut assembly to the wheel spindle
   - strut assembly from the vehicle

To install:

3. Install or connect the following:
   - Strut assembly into the vehicle wheel housing
   - Nuts and bolts securing the strut assembly to the rear wheel spindle assembly. Tighten the lower strut bolts and nuts to 76–100 ft. lbs. (103–136 Nm).
   - ABS sensor bolt, if equipped
   - Clip securing the flexible brake hose to the rear strut assembly
   - Wheel and tire assembly
   - 2 upper strut retaining nuts and tighten to 34–46 ft. lbs. (47–62 Nm)
   - Trim panel
   - Negative battery cable
4. Check the rear wheel alignment.
5. Road test the vehicle and check for proper operation.

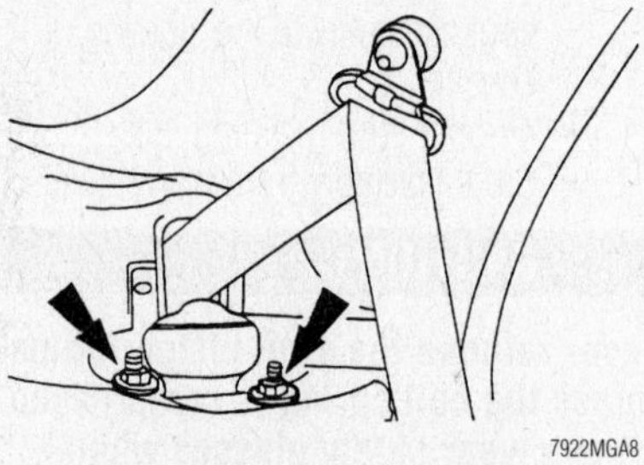

Location of the 2 rear strut upper retaining nuts (arrows) inside the vehicle

Coil Spring

REMOVAL & INSTALLATION

Front

1. Before servicing the vehicle, refer to the precautions in the beginning of this section.
2. Remove the strut assembly.
3. Place the strut assembly in a vise.
4. Install a suitable spring compressor tool onto the coil spring and compress the spring.
5. Unfasten the piston rod nut and remove the upper mounting bracket.
6. Remove the bound stopper, dust boot, rubber spring seat, spring, upper spring seat and thrust bearing.

To install:

7. Install the thrust bearing, upper spring seat, spring, rubber spring seat, dust boot and the bound stopper.
8. Install the piston rod nut and tighten it to 58–81 ft. lbs. (79–110 Nm).
9. After the piston rod nut has been tightened to specification, carefully remove the compressor tool from the spring while making sure the spring is properly seated in the upper and lower spring seats.

1	Rear Floor Cross Member
2	Rear Shock Absorber Insulator
3	Rear Spring
4	Rear Shock Absorber
5	Rear Wheel Spindle
6	Rear Suspension Tie Rod and Bushing
7	Rear Suspension Arm and Bushing (Rear)
8	Rear Stabilizer Bar
9	Rear Suspension Arm and Bushing (Front)
10	Rear Wheel Brake Hose

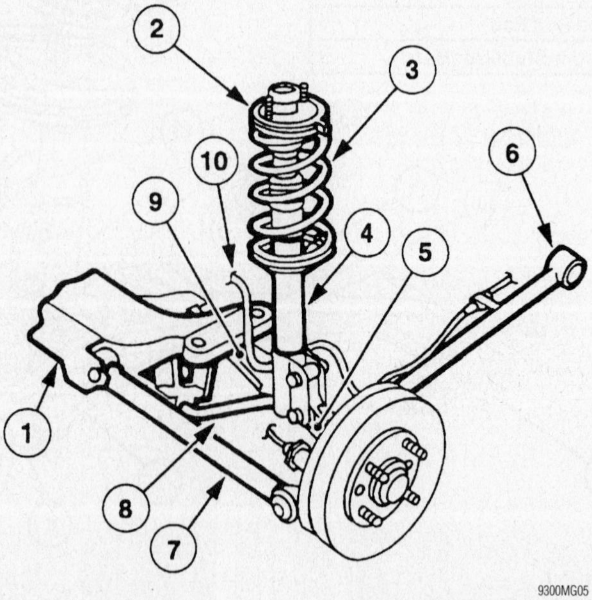

Identification of the rear suspension components

10. Install the strut assembly in the vehicle.

11. Have the vehicle alignment checked and if necessary, adjusted.

Rear

1. Remove the strut assembly.
2. Place the strut assembly in a vise.

⁂ CAUTION

Always take the necessary precautions when using a spring compressor

3. Install a suitable spring compressor tool onto the coil spring and compress the spring.
4. Remove or disconnect the following:
   - Rear strut assembly top mounting cover
   - Piston rod nut and the retainer
   - Strut insulator
   - Spring compressor
   - Rear strut dust boot and stopper seat
   - Spring and the rear strut insulator

To install:

5. Place the strut assembly in a vise.
6. Install or connect the following:
   - Strut insulator and spring
   - Stopper seat and dust boot
7. Use the spring compressor tool to compress the strut spring.
   - Strut insulator and the retainer
   - Piston rod nut and tighten it to 41–49 ft. lbs. (55–67 Nm)
   - Top mounting cover
8. Be sure the spring is properly aligned and carefully release the spring into the seats of the strut.
9. Remove the spring compressor from the coil spring.
10. Install the strut assembly.

Lower Ball Joint

REMOVAL & INSTALLATION

1. Before servicing the vehicle, refer to the precautions in the beginning of this section.
2. Remove or disconnect the following:
   - Wheel assembly
   - Nut and bolt attaching the lower ball joint to the steering knuckle
   - Ball joint from the steering knuckle
   - Lower control arm ball joint nuts and bolt
   - Ball joint

To install:

3. Install or connect the following:
   - Ball joint into position on the lower control arm
   - Ball joint-to-lower control arm retainers. Tighten the nuts and bolt to 69–86 ft. lbs. (93–117 Nm).
4. Apply Loctite® 290 thread locking compound to the ball joint nut and t hreads.
   - Ball joint to the steering knuckle
   - Ball joint nut and bolt. Tighten the nut and bolt to 32–43 ft. lbs. (43–59 Nm).
   - Wheel assembly
   - Check and adjust the wheel alignment as necessary

Control Arm

REMOVAL & INSTALLATION

Front

1. Before servicing the vehicle, refer to the precautions in the beginning of this section.
2. Remove or disconnect the following:
   - Front wheel and tire assembly
   - Stabilizer bar link nuts, retainers, bushings, bolts and sleeves
   - Pinch bolt and nut securing the ball joint to the steering knuckle
   - Lower ball joint from the steering knuckle by prying the lower control arm down with a prybar
   - Front lower control arm pivot bolt
   - Three lower control arm retaining bolts and the lower control arm
3. Inspect the lower control arm, lower control arm bushings and the lower ball joint. The ball joint and bushings can be replaced individually.

To install:

4. Install or connect the following:
   - Lower control arm into position at the bushings

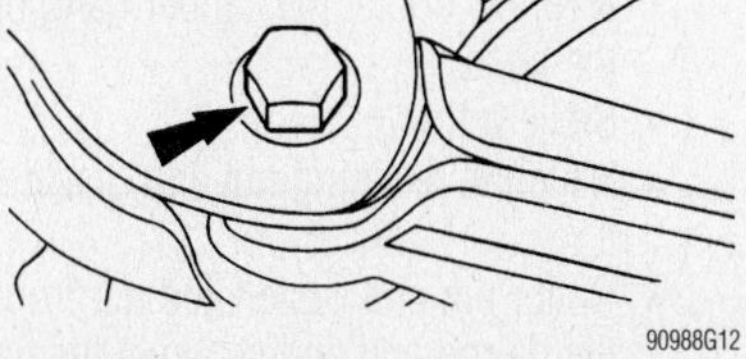

Unfasten of the lower control arm pivot bolt (arrow)

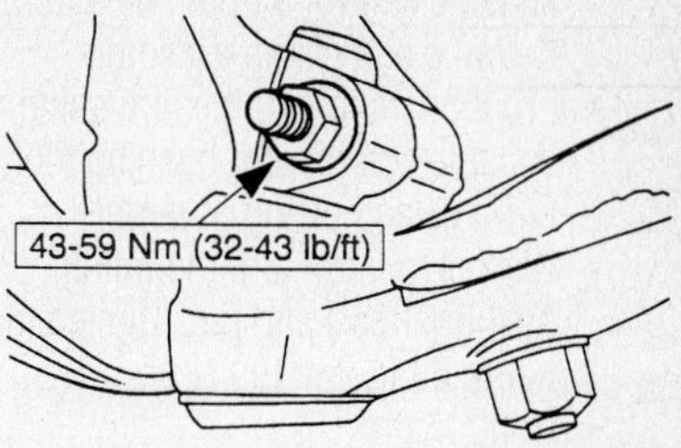

Engage the lower ball joint to the steering knuckle and tighten the pinch bolt and nut to specification

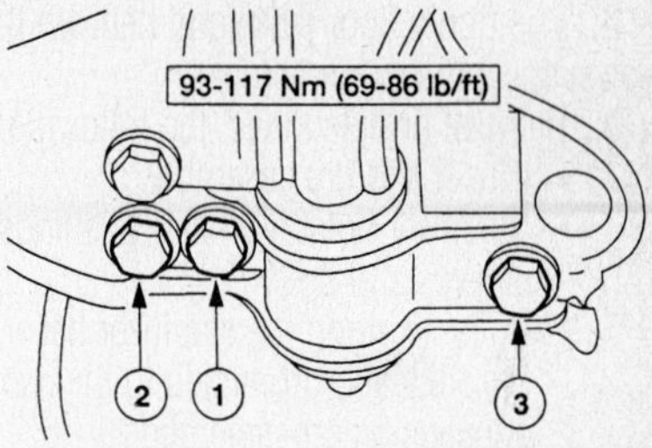

Tighten the lower control arm retaining bolts in the sequence shown to the proper specification

   - Lower ball joint to the steering knuckle and tighten the pinch bolt and nut to 32–43 ft. lbs. (43–59 Nm)
5. Use a transmission jack to raise the lower control arm so that it is parallel with the ground.
   - Control arm retaining bolts and tighten them to 69–86 ft. lbs. (93–117 Nm) in the order shown in the accompanying illustration
   - Control arm pivot bolt and tighten to 69–86 ft. lbs. (93–117 Nm)
   - Stabilizer bolts, washers, bushings, sleeves and nuts. Tighten the stabilizer nuts so 0.67–0.75 inches (17–19mm) of thread is exposed at the end of the bolt.
   - Wheel and tire assembly
6. Check the front wheel alignment.
7. Road test the vehicle and check for proper operation.

CONTROL ARM BUSHING REPLACEMENT

1. Before servicing the vehicle, refer to the precautions in the beginning of this section.
2. Remove or disconnect the following:
   - Control arm
   - Nut and control arm mounting bushing (rear) and washer
   - Front bushing from the arm using a

suitable control arm bushing tool, C-frame and clamp assembly
- Front bushing into the arm using a suitable control arm bushing tool, C-frame and clamp assembly
- Washer, control arm mounting bushing (rear) and nut. Tighten the nut to 69 ft. lbs. (93 Nm).
- Control arm

Rear

1. Before servicing the vehicle, refer to the precautions in the beginning of this section.
2. Position a floor jackstand beneath the rear suspension crossmember.
3. Remove or disconnect the following:
   - Wheel and tire assembly
   - Stabilizer nuts, washers, bushings, sleeves and bolts
   - Bolts securing the stabilizer bar brackets and grommets to the rear suspension crossmember
   - Stabilizer bar
   - Bolts securing the rear suspension crossmember to the vehicle frame
4. Lower the floor jackstand to allow the rear suspension crossmember to be lowered from the vehicle frame.
   - Control arm and bushing nut
   - Control arm and bushing bolt
   - Control arm

To install:

5. Install or connect the following:
   - Control arm into position

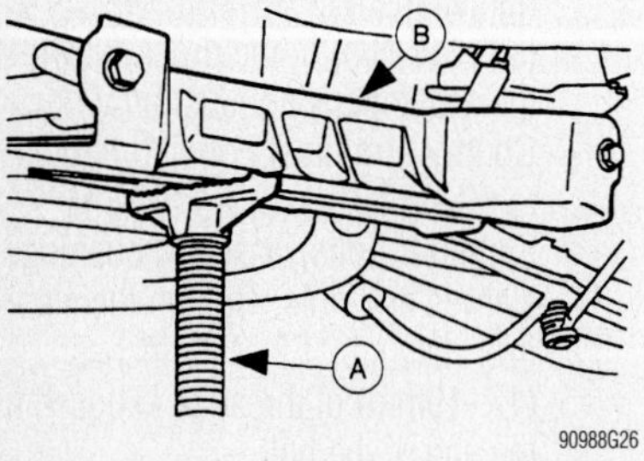

Support the rear suspension crossmember (B) with a floor jackstand (A)

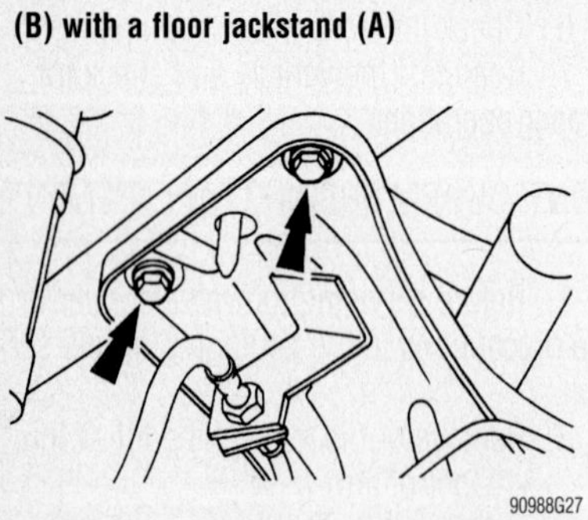

Unfasten the rear suspension crossmember bolts, then lower the floor jackstand and crossmember

   - Retaining bolt and tighten it to 50–70 ft. lbs. (68–95 Nm)
   - Control arm and bushing retaining nut, then tighten the nut to 64–86 ft. lbs. (87–116 Nm)
6. Raise the floor jack and place the crossmember into position.
   - Crossmember bolts and tighten them to 34–46 ft. lbs. (47–62 Nm)
   - Grommets onto the stabilizer bar and align the grommets to the positions painted on the bar
   - Stabilizer bar to the rear suspension crossmember and secure it in place with the straps and bolts. Tighten the bolts to 32–43 ft. lbs. (43–59 Nm).
   - Stabilizer bolts, washers, grommets, sleeves and nuts. Tighten the stabilizer nuts so 0.64–0.71 in. (16.2–18.2mm) of thread is exposed at the end of the bolt.
   - Wheel and tire assembly
7. Check the wheel alignment.

Wheel Bearings

ADJUSTMENT

The bearings on the front and rear wheels are a one piece cartridge design and cannot be adjusted. Wheel bearing play can be checked with a dial indicator. If wheel bearing play exceeds 0.002 in. (0.05mm) check the wheel hub retainer nut for proper torque. If the torque is correct, replacement of the wheel bearing is required.

REMOVAL & INSTALLATION

Front

1. Before servicing the vehicle, refer to the precautions in the beginning of this section.
2. With the vehicle sitting on the ground, carefully raise the staked portion of the halfshaft retaining nut using a suitable small chisel. Loosen the nut.
3. Remove or disconnect the following:
   - Wheel and tire assembly
   - Brake caliper and secure it out of the way with a piece of mechanic's wire. Do not let the caliper hang on the hose.
   - Brake rotor
   - Halfshaft retaining nut and discard it
   - Cotter pin and castellated nut from the tie rod end and separate the tie rod end from the steering knuckle using a suitable removal tool. Discard the cotter pin.
   - Separate the tie rod end from the wheel knuckle
   - ABS sensor bolt and the sensor, if equipped
   - Strut mounting nuts and the studs which attach the strut assembly to the steering knuckle
   - Strut from the steering knuckle
   - Lower ball joint pinch bolt. Carefully pry down on the lower control arm to separate the ball joint stud from the steering knuckle.
   - Wheel hub, knuckle and bearing assembly from the vehicle

To install:

4. Apply Loctite® 290 thread locking compound to the ball joint nut and threads.
5. Install or connect the following:
   - Wheel hub, knuckle and bearing assembly onto the ball joint and tighten the pinch bolt and nut to 32–43 ft. lbs. (43–59 Nm)
   - Tie rod end and tighten the nut to 25–33 ft. lbs. (34–46 Nm). Install a new cotter pin.
   - ABS sensor and tighten the bolt, if equipped
   - Knuckle to the strut assembly and tighten the strut mounting nuts to 69–93 ft. lbs. (93–127 Nm)
   - New halfshaft retaining nut and tighten to 174–235 ft. lbs. (235–319 Nm). Stake the retaining nut using a suitable chisel with the cutting edge rounded off.
   - Brake rotor and caliper
   - Wheel and tire assembly
6. Check the front wheel alignment.
7. Road test the vehicle and check for proper operation.

Rear

➡The wheel bearings are a cartridge design and are not serviceable. If bearing replacement is required, the bearings and hub must be replaced as an assembly.

1. Before servicing the vehicle, refer to the precautions in the beginning of this section.
2. Remove or disconnect the following:
   - Wheel and tire assembly
   - Hub grease cap
   - Brake drum or disc brake caliper and rotor, as necessary
   - Wheel hub retainer nut securing the hub to the spindle, unstake the nut prior to loosening

- Hub and bearing assembly. Discard the hub retainer nut.
- Disc brake shield retaining bolts and the shield, if equipped with disc brakes
- Backing plate, if equipped with drum brakes
- Strut-to-spindle retaining bolts and nuts
- Trailing arm bolt securing the trailing arm to the spindle
- Stabilizer bar link nuts, retainers, bushings, sleeves and bolts
- Control arm nut and bolt securing both control arms to the wheel spindle
- Spindle from the vehicle

3. Inspect all components. If the wheel spindle is damaged, replace it. Wheel bearings are sealed and must be replaced if damaged with the wheel hub.

To install:

4. Install or connect the following:
   - Wheel spindle in position to the strut
   - Spindle retaining bolts and nuts. Tighten to 69–93 ft. lbs. (93–127 Nm).
   - Trailing arm to the spindle and tighten retaining bolt to 69–93 ft. lbs. (93–127 Nm)
   - Both control arms in position to the spindle and tighten the retaining bolt and nut to 63–86 ft. lbs. (85–117 Nm)
   - Stabilizer bar link nuts, retainers, bushings, sleeves and bolts
   - Brake backing plate, if equipped with drum brakes
   - Brake shield and the disc brake shield retaining bolts, if equipped with disc brakes
   - Wheel hub and bearing assembly to the wheel spindle
   - New wheel hub retainer nut and tighten to 130–174 ft. lbs. (177–235 Nm)
5. Stake the wheel hub retainer nut using a cape or round end chisel. Do not use a sharp chisel to stake the hub nut.
   - Brake drum or the disc brake rotor and caliper, as equipped
   - Hub grease cap
   - Wheel and tire assembly
6. Pump the brake pedal several times to position the brake lining before attempting to move the vehicle.
7. Road test the vehicle and check for proper operation.

BRAKES

Brake Caliper

REMOVAL & INSTALLATION

Front

1. Before servicing the vehicle, refer to the precautions in the beginning of this section.
2. Remove or disconnect the following:
   - Wheels
   - Disc brake pads
   - Banjo bolt securing the brake hose to the brake caliper. Discard 2 copper sealing washers.
   - 2 brake caliper retaining bolts and the brake caliper

To install:

3. Install or connect the following:
   - Brake caliper and 2 brake caliper retaining bolts. Torque the bolts to 29–36 ft. lbs. (39–49 Nm).
   - Brake hose and banjo bolt to the brake caliper with 2 new copper sealing washers. Torque the banjo bolt to 16–22 ft. lbs. (22–29 Nm).
   - Disc brake pads
4. Bleed the brake system.
   - Wheels
5. Pump the brake pedal several times to position the brake pads to the brake rotor.
6. Road test the vehicle and check the brake system for proper operation.

Rear

1. Before servicing the vehicle, refer to the precautions in the beginning of this section.
2. Remove or disconnect the following:
   - Wheels
   - Disc brake pads
   - Parking brake cable bracket bolt and position the bracket aside
   - Parking brake cable from the operating lever
   - Banjo bolt securing the brake hose to the brake caliper. Discard 2 copper sealing washers.
   - Upper brake caliper retaining bolt
   - Brake caliper by sliding it off the mounting bracket

To install:

3. Install or connect the following:
   - Brake caliper on the mounting bracket and slide into position
   - Upper brake caliper retaining bolt. Torque the bolt to 33–43 ft. lbs. (45–59 Nm).
   - Brake hose to the brake caliper and secure with the banjo bolt using 2 new copper sealing washers. Torque the banjo bolt to 16–22 ft. lbs. (22–29 Nm).
   - Parking brake cable to the operating lever. Position the bracket and install the bracket bolt.
   - Disc brake pads
4. Bleed the brake system.
   - Wheels
5. Pump the brake pedal several times to position the brake pads to the brake rotor.
6. Road test the vehicle and check the brake system for proper operation.

Disc Brake Pads

REMOVAL & INSTALLATION

Front

1. Before servicing the vehicle, refer to the precautions in the beginning of this section.
2. Remove the master cylinder reservoir cap and check the fluid level in the reservoir. Remove brake fluid until the reservoir is ½ full. Discard the removed fluid.
3. Remove or disconnect the following:
   - Wheels
   - W-spring
   - 2 disc brake pad locating pins and the M-spring
   - Brake pads and shims from the brake caliper
4. Inspect the brake rotor. Resurface or replace if needed.

To install:

5. Use a suitable tool to push the piston into the brake caliper bore.
6. Apply a suitable grease between the shims and the disc brake pad guide plates.
7. Install or connect the following:
   - Brake pads and shims on the brake caliper
   - W-spring and 2 brake pad locating pins
   - M-spring
   - Wheels
8. Pump the brake pedal several times prior to moving the vehicle to position the brake pads.

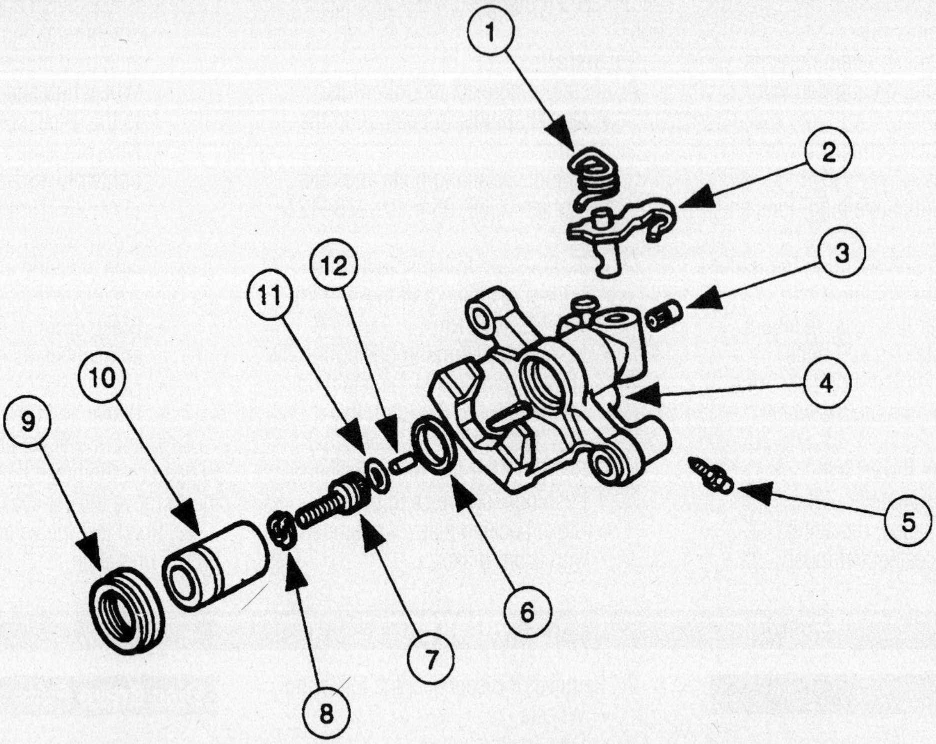

1	Rear Wheel Brake Caliper Spring	7	Rear Disc Brake Adjuster Spindle
2	Rear Brake Operating Lever	8	Snap Ring
3	Brake Adjuster Screw	9	Dust Seal
4	Rear Disc Brake Caliper	10	Piston
5	Wheel Cylinder Bleeder Screw	11	O-Ring
6	Piston Seal	12	Brake Connecting Link

93006G22

Rear disc brake assembly—Escort

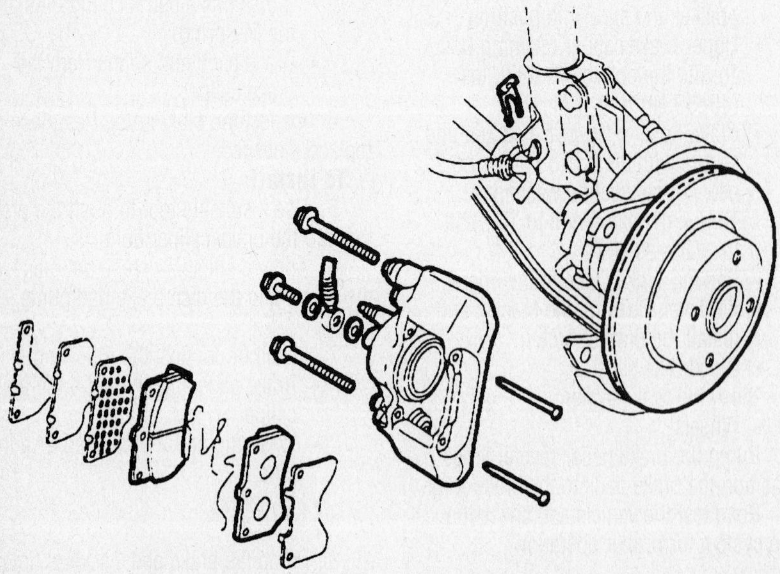

93006G29

Front disc brake pads—Escort

9. Check the fluid level in the master cylinder reservoir and fill as needed.
10. Road test the vehicle and check for proper brake system operation.

Rear

1. Before servicing the vehicle, refer to the precautions in the beginning of this section.
2. Remove the master cylinder reservoir cap and check the fluid level in the reservoir. Remove brake fluid until the reservoir is ½ full. Discard the removed fluid.
3. Remove or disconnect the following:
   - Wheels
4. If necessary, remove the screw plug and turn the adjustment gear counterclockwise with an Allen® wrench to pull the piston fully inward.
   - Lower brake caliper bolt
5. Using a small prybar, pivot the caliper on its mounting bracket to access

the brake pads. If the upper lock bolt requires lubrication or service, remove it and suspend the caliper with mechanics wire.

- Brake pads, shims, spring and guides

6. Inspect the brake rotor. Resurface or replace if needed.

To install:

7. Apply an appropriate brake pad grease between the shims and the brake pads.
8. Pivot the caliper on its mounting bracket and position the brake pads, shims, spring and guides to the brake rotor.
9. Lubricate and install the lower caliper bolt. Tighten the bolt to 33–43 ft. lbs. (45–59 Nm).
10. Turn the adjustment gear clockwise with an Allen wrench until the brake pads just touch the rotor, then loosen the gear ⅓ of a turn. Install the screw plug and torque to 12 ft. lbs. (16 Nm).
11. Install the wheels.
12. Pump the brake pedal several times prior to moving the vehicle to position the brake pads.
13. Check the brake fluid level in the master cylinder reservoir and fill if needed.
14. Road test the vehicle and check for proper brake system operation.

Brake Drums

REMOVAL & INSTALLATION

1. Before servicing the vehicle, refer to the precautions in the beginning of this section.
2. Remove or disconnect the following:
   - Wheels
   - 2 brake drum retaining screws
   - Brake drum from the hub
3. Inspect the drum and refinish or replace, as necessary. If refinishing, check the maximum inside diameter specification.

To install:

4. If needed, adjust the brake shoes to fit the brake drum.
5. Install or connect the following:
   - Brake drum on the hub
   - 2 brake drum retaining screws. Torque to 89–123 inch lbs. (10–14 Nm).
   - Wheels
6. Check the brake system for proper operation.

Brake Shoes

REMOVAL & INSTALLATION

1. Before servicing the vehicle, refer to the precautions in the beginning of this section.
2. Remove or disconnect the following:
   - Wheels
   - 2 brake drum retaining screws and the brake drum
   - 2 brake shoe return springs
   - Right-hand anti-rattle spring
   - 2 brake shoe hold-down springs

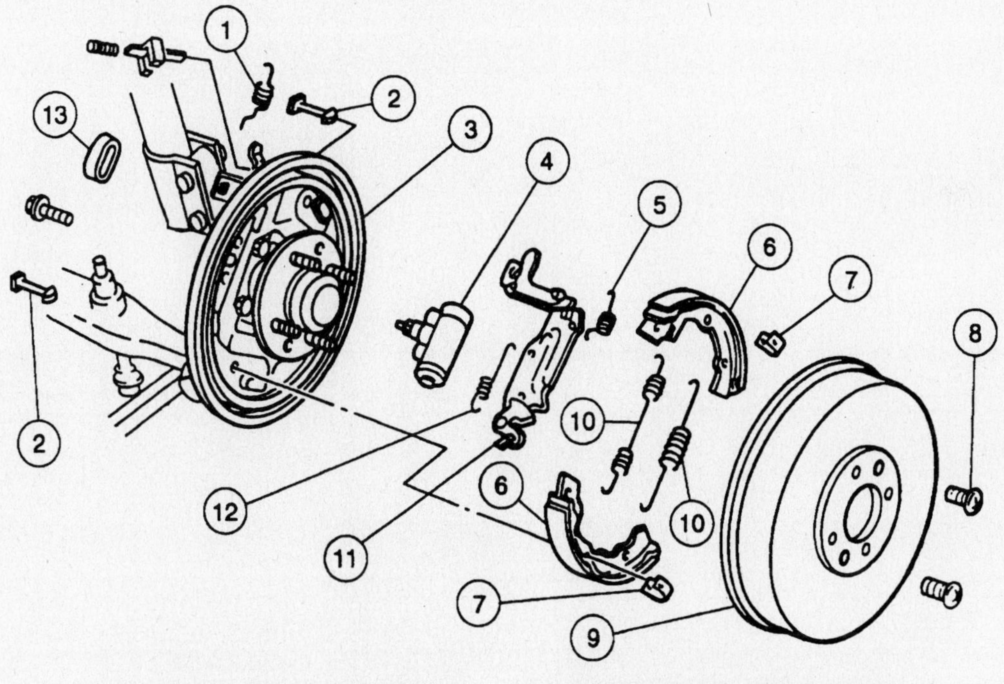

1	Parking Brake Link Spring		7	Brake Shoe Hold-Down Spring
2	Brake Shoe Hold-Down Spring Pin		8	Brake Drum Screw (2 Req'd)
			9	Brake Drum
3	Rear Brake Backing Plate		10	Brake Shoe Retracting Spring
4	Rear Wheel Cylinder		11	Parking Brake Lever
5	Right Hand Anti-Rattle Spring		12	Parking Brake Return Spring
6	Rear Brake Shoe and Lining		13	Brake Adjusting Hole Cover

93006G57

Brake shoes and related components—Escort

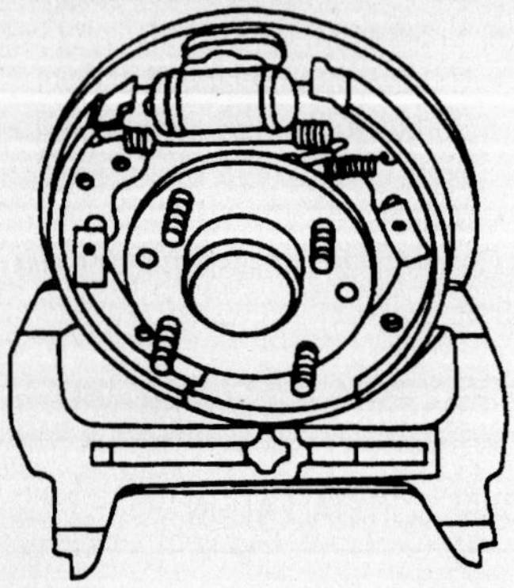

93006G58

Checking the adjustment of the brake shoes—Escort

- Leading and trailing shoes from the brake backing plate

To install:

3. Use a suitable high temperature grease to lightly lubricate the brake shoe contact points on the backing plate.
4. Remove or disconnect the following:
   - Trailing brake shoe on the backing plate and one of the brake shoe hold-down springs
   - Leading brake shoe on the backing plate and the other brake shoe hold-down spring
   - Right-hand anti-rattle spring
   - 2 brake shoe return springs
5. Using a brake adjusting gauge, measure the inside diameter of the brake drum.
6. Compare the brake drum measurement to the brake shoes.
7. Adjust the brake shoes by inserting a screwdriver into the knurled quadrant of the rear quad operating lever and adjust the shoes to the same measurement as the brake drum.
8. Install or connect the following:
   - Brake drum
   - 2 brake drum retaining screws
   - Wheels
9. Complete the brake shoe adjustment by sharply applying the brakes several times while driving the vehicle alternating between forward and reverse gears.
10. Check the brake system operation by making several stops while driving forward.

FORD AND LINCOLN

8

Excursion • Expedition • Navigator

SPECIFICATION CHARTS

ENGINE AND VEHICLE IDENTIFICATION

Engine						
Code ①	Liters (cc)	Cu. In.	Cyl.	Fuel Sys.	Type	Eng. Mfg.
S	6.8 (6802)	415	10	MFI	SOHC	Ford
6	4.6 (4588)	281	8	MFI	SOHC	Ford
F	7.3 (7292)	445	8	DI	OHV	Navistar
R	5.4 (5409)	330	8	EFI	DOHC	Ford
L	5.4 (5409)	330	8	EFI	SOHC	Ford
P	6.0 (5999)	365	8	DI ③	SOHC	Ford
W	4.6 (4588)	281	8	MFI	SOHC	Ford

Model Year	
Code ②	Year
1	2001
2	2002
3	2003
4	2004
5	2005

MFI: Multi-port Fuel Injection

DI: Direct Injection Turbo-Diesel

EFI: Electronic Fuel Injection

OHV: Overhead Valve

SOHC: Single Overhead Camshaft

DOHC: Dual Overhead Camshaft

① 8th digit of the Vehicle Identification Number (VIN)

② 10th digit of the Vehicle Identification Number (VIN)

③ Turbo

67197-NAVI-C01

GENERAL ENGINE SPECIFICATIONS

Year	Model	Engine Displacement Liters (VIN)	Net Horsepower @ rpm	Net Torque @ rpm (ft. lbs.)	Bore x Stroke (in.)	Com-pression Ratio	Oil Pressure @ rpm
2001	Excursion	5.4 (L)	235@4250	330@3000	3.55X4.17	9.0:1	40-70@1500
	Excursion	6.8 (S)	265@4250	410@2750	4.09X4.17	9.0:1	40-70@1500
	Excursion	7.3 (F)	210@3000	425@2000	4.11x4.18	17.5:1	40-70@3000
	Expedition	4.6 (6/W)	210@4400	290@3250	3.55x3.54	9.0:1	20-45@1500
	Expedition	5.4 (L)	235@4250	330@3000	3.55X4.17	9.0:1	40-70@1500
	Navigator	5.4 (L)	260@4250	330@3000	3.55X4.17	9.0:1	40-70@1500
2002	Excursion	5.4 (L)	235@4250	330@3000	3.55X4.17	9.0:1	40-70@1500
	Excursion	6.8 (S)	265@4250	410@2750	4.09X4.17	9.0:1	40-70@1500
	Excursion	7.3 (F)	210@3000	425@2000	4.11x4.18	17.5:1	40-70@3000
	Expedition	4.6 (6/W)	210@4400	290@3250	3.55x3.54	9.0:1	20-45@1500
	Expedition	5.4 (L)	235@4250	330@3000	3.55X4.17	9.0:1	40-70@1500
	Navigator	5.4 (L)	260@4250	330@3000	3.55X4.17	9.0:1	40-70@1500
2003	Excursion	5.4 (L)	235@4250	330@3000	3.55X4.17	9.0:1	40-70@1500
	Excursion	6.8 (S)	265@4250	410@2750	4.09X4.17	9.0:1	40-70@1500
	Excursion	7.3 (F)	210@3000	425@2000	4.11x4.18	17.5:1	40-70@3000
	Excursion	6.0 (P)	325@3300	560@2000	3.74x4.13	18.0:1	24@1200
	Expedition	4.6 (W)	210@4400	290@3250	3.55x3.54	9.3:1	40-60@2000
	Expedition	5.4 (L)	235@4250	330@3000	3.55X4.17	9.0:1	40-60@2000
	Expedition	5.4 (R)	235@4250	330@3000	3.55X4.23	9.5:1	40-60@2000
	Navigator	5.4 (L)	235@4250	330@3000	3.55X4.17	9.0:1	40-60@2000
	Navigator	5.4 (R)	260@4250	330@3000	3.55X4.17	9.0:1	40-60@2000
2004-05	Excursion	5.4 (L)	235@4250	330@3000	3.55X4.17	9.0:1	40-70@1500
	Excursion	6.8 (S)	265@4250	410@2750	4.09X4.17	9.0:1	40-70@1500
	Excursion	6.0 (P)	325@3300	560@2000	3.74x4.13	18.0:1	24@1200
	Expedition	4.6 (W)	210@4400	290@3250	3.55x3.54	9.3:1	40-60@2000
	Expedition	5.4 (L)	235@4250	330@3000	3.55X4.17	9.0:1	40-60@2000
	Expedition	5.4 (R)	235@4250	330@3000	3.55X4.23	9.5:1	40-60@2000
	Navigator	5.4 (L)	235@4250	330@3000	3.55X4.17	9.0:1	40-60@2000
	Navigator	5.4 (R)	260@4250	330@3000	3.55X4.17	9.0:1	40-60@2000

67197-NAVI-C02

GASOLINE ENGINE TUNE-UP SPECIFICATIONS

Year	Engine Displacement Liters (cc)	Engine ID/VIN	Spark Plug Gap (in.)	Ignition Timing (deg.) ① MT	Ignition Timing (deg.) ① AT	Fuel Pump (psi) ②	Idle Speed (rpm) MT	Idle Speed (rpm) AT	Valve Clearance In.	Valve Clearance Ex.
2001	4.6 (4588)	W	0.052-0.056	—	10B	30-45	—	①	HYD	HYD
	4.6 (4588)	6	0.052-0.056	—	10B	30-45	—	①	HYD	HYD
	5.4 (5409)	L	0.052-0.056	—	10B	28-45	—	①	HYD	HYD
	6.8 (6802)	5	0.052-0.055	—	10B	28-45	—	①	HYD	HYD
2002	4.6 (4588)	W	0.052-0.056	—	10B	30-45	—	①	HYD	HYD
	4.6 (4588)	6	0.052-0.056	—	10B	30-45	—	①	HYD	HYD
	5.4 (5409)	L	0.052-0.056	—	10B	28-45	—	①	HYD	HYD
	6.8 (6802)	5	0.052-0.055	—	10B	28-45	—	①	HYD	HYD
2003	4.6 (4588)	W	0.052-0.056	—	10B	30-45	—	①	HYD	HYD
	5.4 (5409)	R	0.052-0.056	—	10B	28-45	—	①	HYD	HYD
	5.4 (5409)	L	0.052-0.056	—	10B	28-45	—	①	HYD	HYD
2004-05	4.6 (4588)	W	0.052-0.056	—	10B	30-45	—	①	HYD	HYD
	5.4 (5409)	R	0.052-0.056	—	10B	28-45	—	①	HYD	HYD
	5.4 (5409)	L	0.052-0.056	—	10B	28-45	—	①	HYD	HYD

NOTE: The Vehicle Emission Control Information label often reflects specification changes changes made during production. The label figures must be used if they differ from those in this chart.

B: Before top dead center

HYD: Hydraulic

NA: Not Available

① Idle speed and timing are electronically controlled and cannot be adjusted

② With engine running

67197-NAVI-C03

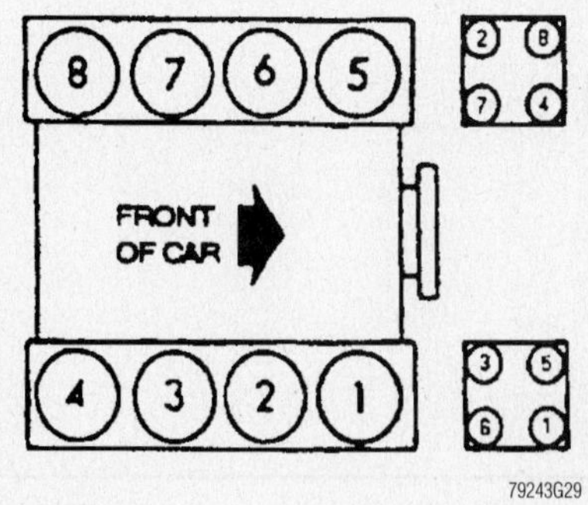

79243G29

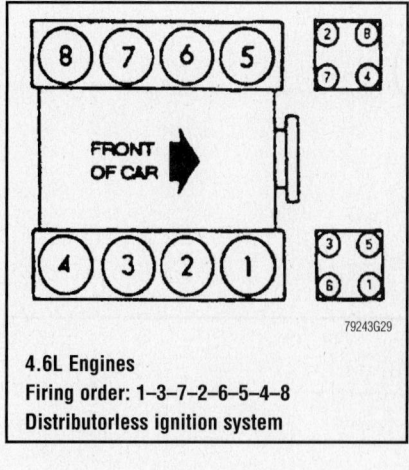

4.6L Engines
Firing order: 1–3–7–2–6–5–4–8
Distributorless ignition system

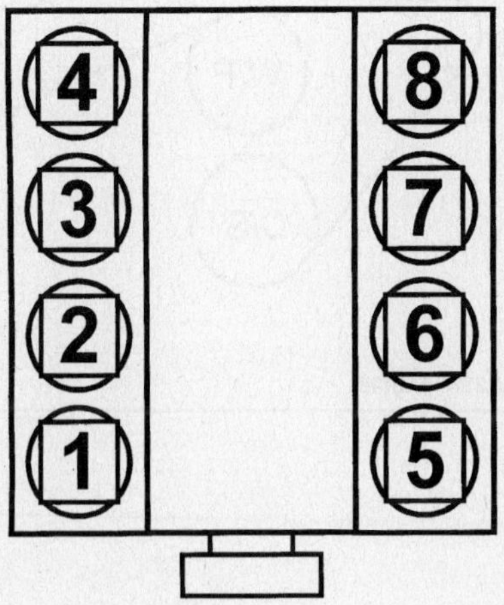

79243G58

5.4L Engines
Firing order: 1–3–7–2–6–5–4–8
Distributorless ignition system (one coil on each cylinder)

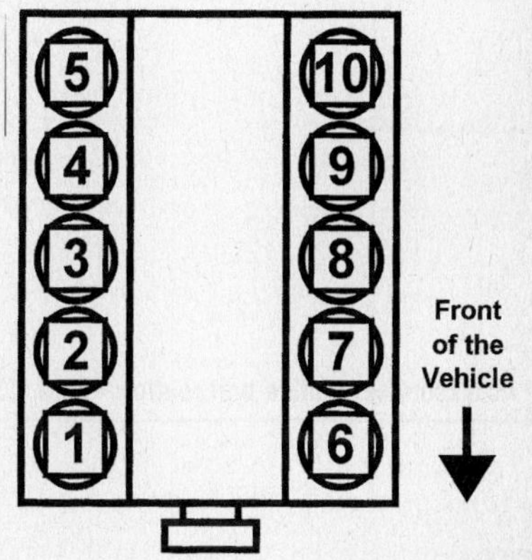

79243G57

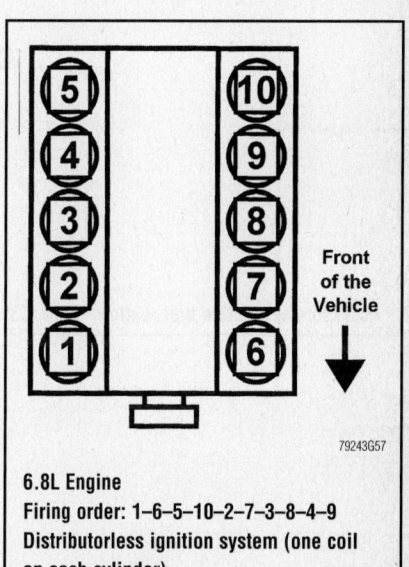

6.8L Engine
Firing order: 1–6–5–10–2–7–3–8–4–9
Distributorless ignition system (one coil on each cylinder)

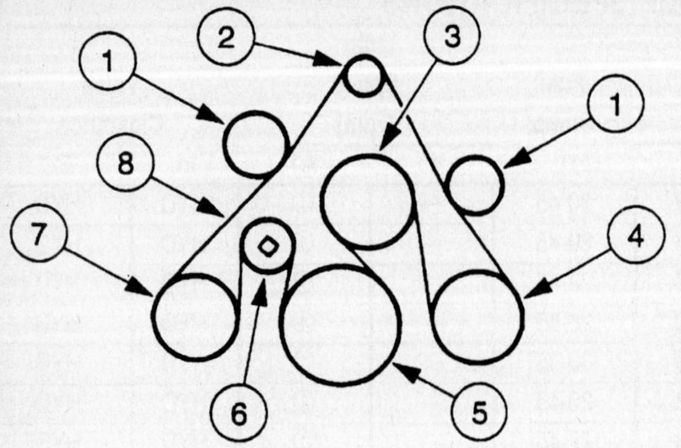

1. Idler
2. Alternator
3. Water Pump
4. Power Steering
5. Crankshaft
6. Drive Belt Tensioner
7. A/C Pulley
8. Drive Belt

79244G86

Accessory serpentine belt routing—Ford 4.6L, 5.4L, and 6.8L engines with A/C

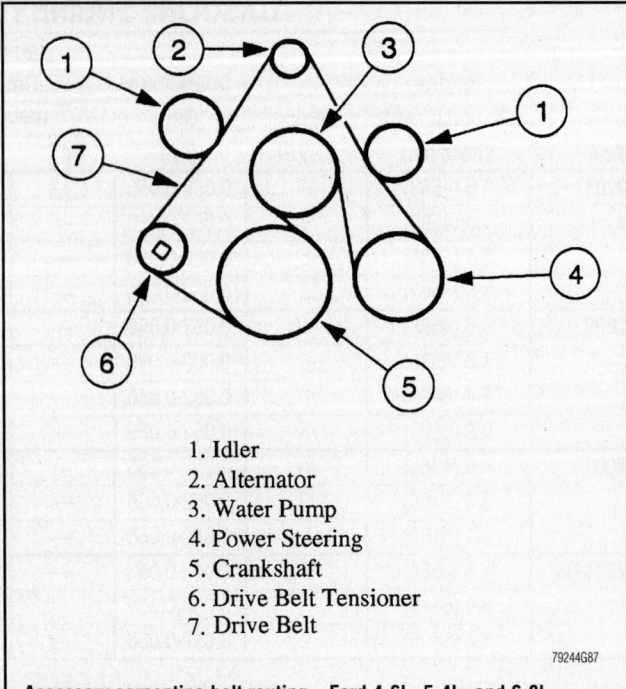

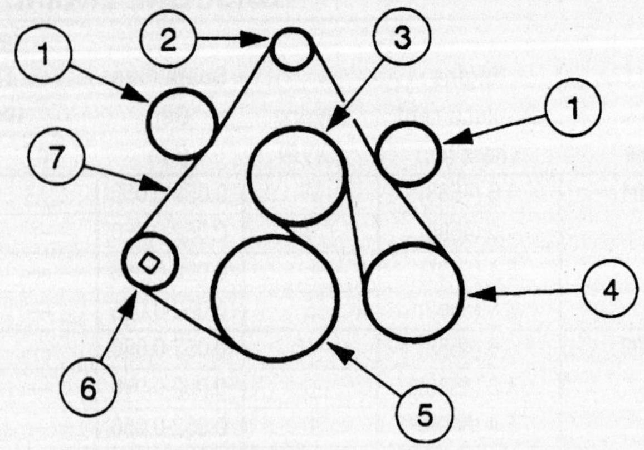

1. Idler
2. Alternator
3. Water Pump
4. Power Steering
5. Crankshaft
6. Drive Belt Tensioner
7. Drive Belt

79244G87

Accessory serpentine belt routing—Ford 4.6L, 5.4L, and 6.8L engines without A/C

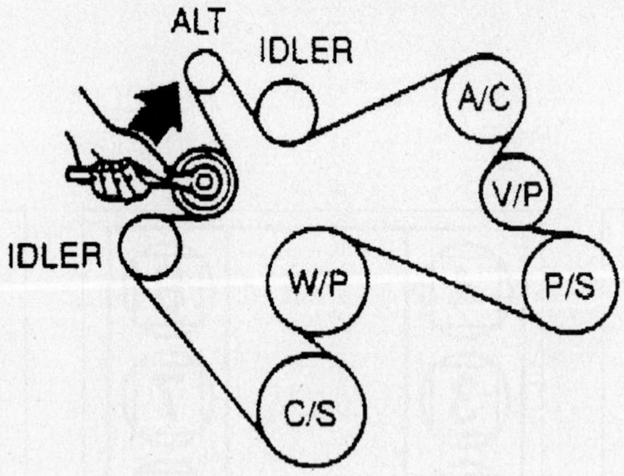

79244G84

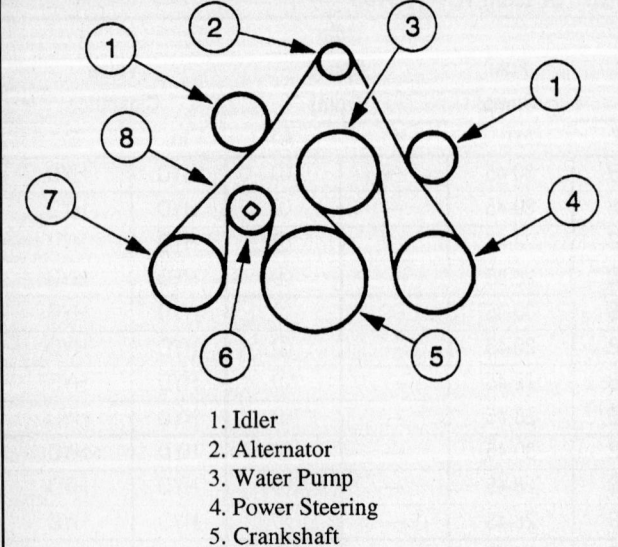

Accessory serpentine belt routing—Ford 7.3L turbo diesel engine

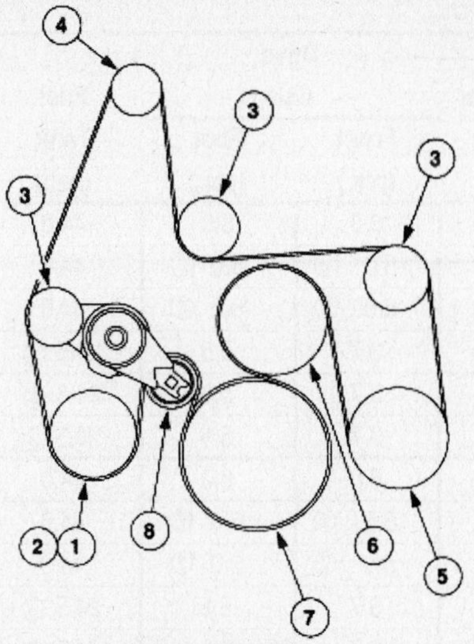

1 A/C clutch pulley
2 Idler pulley (without A/C)
3 Belt idler pulley
4 Generator pulley
5 Power steering pump pulley
6 Water pump pulley
7 Crankshaft pulley
8 Drive belt tensioner

67197-NAVI-G21

Accessory Drive—6.0L engine with single alternator

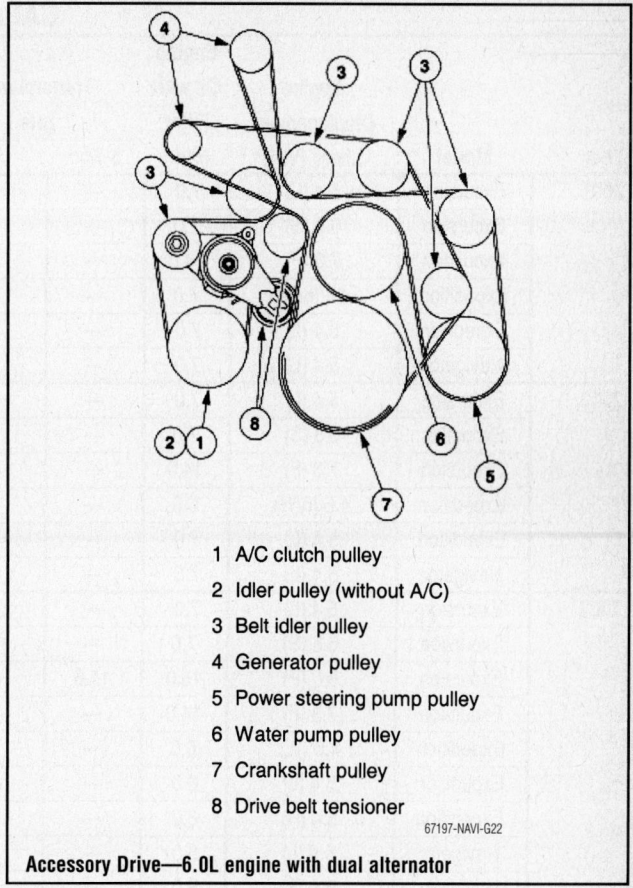

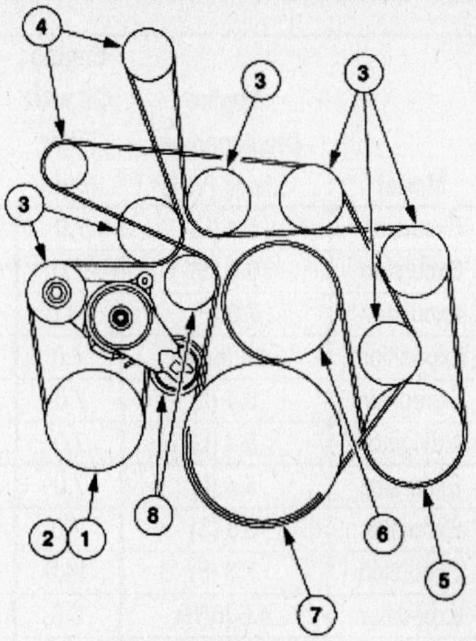

1 A/C clutch pulley
2 Idler pulley (without A/C)
3 Belt idler pulley
4 Generator pulley
5 Power steering pump pulley
6 Water pump pulley
7 Crankshaft pulley
8 Drive belt tensioner

67197-NAVI-G22

Accessory Drive—6.0L engine with dual alternator

DIESEL ENGINE TUNE-UP SPECIFICATIONS

Year	Engine Displacement cu. in. (VIN)	Valve Clearance Intake (in.)	Valve Clearance Exhaust (in.)	Intake Valve Opens (deg.)	Injection Pump Setting (deg.)	Injection Nozzle Pressure (psi) New	Injection Nozzle Pressure (psi) Used	Idle Speed (rpm)	Cranking Compression Pressure (psi)
2001	7.3 (F)	HYD	HYD	—	①	1875	1425	②	③
2002	7.3 (F)	HYD	HYD	—	①	1875	1425	②	③
2003	7.3 (F)	HYD	HYD	—	①	1875	1425	②	③
	6.0 (P)	HYD	HYD	—	①	NA	NA	②	NA
2004-05	7.3 (F)	HYD	HYD	—	①	1875	1425	②	③
	6.0 (P)	HYD	HYD	—	①	NA	NA	②	NA

NOTE: The Vehicle Emission Control Information label often reflects specification changes made during production.
The label figures must be used if they differ from those in this chart

HYD: Hydraulic

B: Before top dead center

NA: Not Available

① PCM controlled

② See underhood emission label

③ Compression pressure in the lowest cylinder must be at least 75% of the highest cylinder
Minimum pressure: 195 psi
Maximum pressure: 440 psi

67197-NAVI-C04

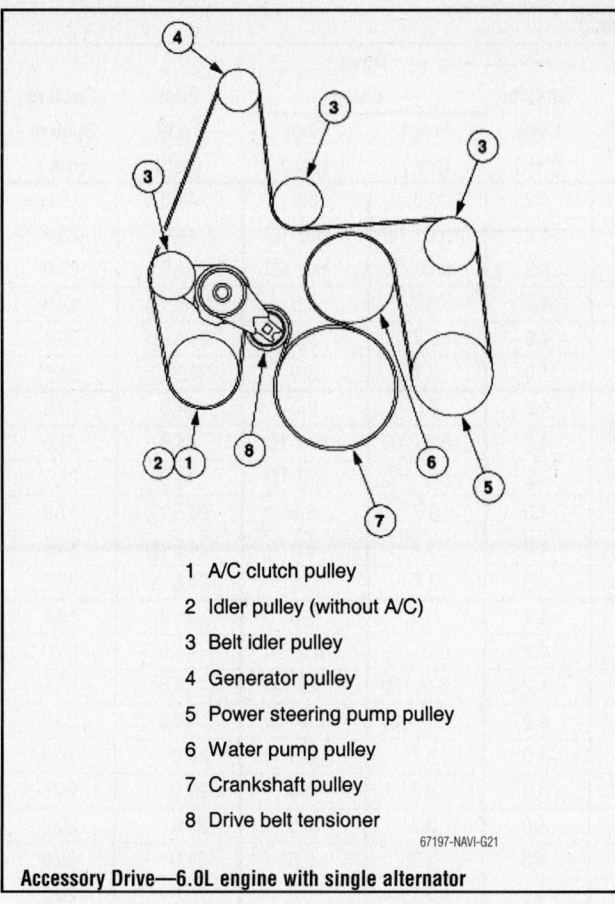

CAPACITIES

Year	Model	Engine Displacement Liters (VIN)	Engine Oil with Filter (qts.)	Transmission (pts.) 5-Spd	Transmission (pts.) Auto.	Transfer Case (pts.)	Drive Axle Front (pts.)	Drive Axle Rear (pts.)	Fuel Tank (gal.)	Cooling System (qts.)
2001	Excursion	5.4 (L)	7.0	—	①	4.2	6.0	6.0	44.0	19.8
	Excursion	6.8 (S)	7.0	—	①	4.2	6.0 ②	6.0 ②	44.0	23.0
	Excursion	7.3 (F)	14.0	—	①	4.2	6.0 ②	6.0 ②	44.0	23.0
	Expedition	4.6 (6/W)	6.0	—	①	4.0	3.7	5.5	24.5 ③	17.9
	Expedition	5.4 (L)	7.0	—	①	4.0	3.7	5.5	24.5 ③	20.8
	Navigator	5.4 (L)	7.0	—	①	4.0	3.7	5.5	24.5 ③	20.8
2002	Excursion	5.4 (L)	7.0	—	①	4.2	6.0	6.0	44.0	19.8
	Excursion	6.8 (S)	7.0	—	①	4.2	6.0 ②	6.0 ②	44.0	23.0
	Excursion	7.3 (F)	14.0	—	①	4.2	6.0 ②	6.0 ②	44.0	23.0
	Expedition	4.6 (6/W)	6.0	—	①	4.0	3.7	5.5	24.5 ③	17.9
	Expedition	5.4 (L)	7.0	—	①	4.0	3.7	5.5	24.5 ③	20.8
	Navigator	5.4 (L)	7.0	—	①	4.0	3.7	5.5	24.5 ③	20.8
2003	Excursion	5.4 (L)	7.0	—	①	4.2	6.0	6.0	44.0	19.8
	Excursion	6.8 (S)	7.0	—	①	4.2	6.0 ②	6.0 ②	44.0	23.0
	Excursion	6.0 (P)	15.0	11.6	④	4.2	6.0 ②	6.0 ②	44.0	27.5
	Excursion	7.3 (F)	14.0	—	①	4.2	6.0 ②	6.0 ②	44.0	23.0
	Expedition	4.6 (W)	6.0	—	①	4.0	3.7	5.5	28.0	17.9
	Expedition	5.4 (L)	6.0	—	①	4.0	3.7	5.5	28.0	20.8
	Expedition	5.4 (R)	6.0	—	①	4.0	3.7	5.5	28.0	20.8
	Navigator	5.4 (L)	6.0	—	①	4.0	3.7	5.5	28.0	20.8
	Navigator	5.4 (R)	6.0	—	①	4.0	3.7	5.5	28.0	20.8
2004-05	Excursion	5.4 (L)	7.0	—	①	4.2	6.0	6.0	44.0	19.8
	Excursion	6.8 (S)	7.0	—	①	4.2	6.0 ②	6.0 ②	44.0	23.0
	Excursion	6.0 (P)	15.0	11.6	④	4.2	6.0 ②	6.0 ②	44.0	27.5
	Excursion	7.3 (F)	14.0	—	①	4.2	6.0 ②	6.0 ②	44.0	23.0
	Expedition	4.6 (W)	6.0	—	①	4.0	3.7	5.5	28.0	17.9
	Expedition	5.4 (L)	6.0	—	①	4.0	3.7	5.5	28.0	20.8
	Expedition	5.4 (R)	6.0	—	①	4.0	3.7	5.5	28.0	20.8
	Navigator	5.4 (L)	6.0	—	①	4.0	3.7	5.5	28.0	20.8
	Navigator	5.4 (R)	6.0	—	①	4.0	3.7	5.5	28.0	20.8

NOTE: All capacities are approximate. Add fluid gradually and check to be sure a proper fluid level is obtained.

① With 4R70W: 28 pts.
With 4R100: 32.0 pts.
With E40D: 32.0 pts.

② Heavy duty: 7.5 pts.

③ Also available with a 30 gallon tank with 8 ft. box

④ 4R100: 34.2 pts.
TorqShift: 38.4 pts.

67197-NAVI-C05

VALVE SPECIFICATIONS

Year	Engine Displacement Liters (VIN)	Seat Angle (deg.)	Face Angle (deg.)	Spring Test Pressure (lbs. @ in.)	Spring Installed Height (in.)	Stem-to-Guide Clearance (in.) Intake	Stem-to-Guide Clearance (in.) Exhaust	Stem Diameter (in.) Intake	Stem Diameter (in.) Exhaust
2001	4.6 (6/W)	45	45.5	132@1.100	1.570	0.0008- 0.0027	0.0018- 0.0037	0.2750- 0.2746	0.2740- 0.2736
	5.4 (L)	45	45.5	150@1.10	1.570	0.0008- 0.0027	0.0018- 0.0037	0.275- 0.2746	0.274- 0.2736
	6.8 (5)	44.50- 45.25	45.25- 45.75	150@1.10	1.570	0.0008- 0.0027	0.0018- 0.0037	0.275- 0.2746	0.274- 0.2735
	7.3 (F)	①	①	200@1.38	②	0.0055	0.0055	0.3119- 0.3126	0.3119- 0.3126
2002	4.6 (6/W)	45	45.5	132@1.100	1.570	0.0008- 0.0027	0.0018- 0.0037	0.2750- 0.2746	0.2740- 0.2736
	5.4 (L)	45	45.5	150@1.10	1.570	0.0008- 0.0027	0.0018- 0.0037	0.275- 0.2746	0.274- 0.2736
	6.8 (5)	44.50- 45.25	45.25- 45.75	150@1.10	1.570	0.0008- 0.0027	0.0018- 0.0037	0.275- 0.2746	0.274- 0.2735
	7.3 (F)	①	①	200@1.38	②	0.0055	0.0055	0.3119- 0.3126	0.3119- 0.3126
2003	4.6 (W)	45.0- 44.50	45.75- 45.25	132@1.100	1.570	0.0008- 0.0027	0.0018- 0.0037	0.2754- 0.2746	0.2744- 0.2736
	5.4 (L)	45.0- 44.50	45.75- 45.25	150@1.10	1.570	0.0008- 0.0027	0.0018- 0.0037	0.2754- 0.2746	0.274- 0.2736
	5.4 (R)	45.0- 44.50	45.75- 45.25	150@1.10	1.676	0.0008- 0.0027	0.0018- 0.0037	0.2754- 0.2746	0.2744- 0.2736
	6.0 (P)	①	①	191@1.51	1.820	0.0055 max.	0.0055 max.	0.2720- 0.2735	0.2720- 0.2735
	7.3 (F)	①	①	200@1.38	②	0.0055	0.0055	0.3119- 0.3126	0.3119- 0.3126
2004-05	4.6 (W)	45.0- 44.50	45.75- 45.25	132@1.100	1.570	0.0008- 0.0027	0.0018- 0.0037	0.2754- 0.2746	0.2744- 0.2736
	5.4 (L)	45.0- 44.50	45.75- 45.25	150@1.10	1.570	0.0008- 0.0027	0.0018- 0.0037	0.2754- 0.2746	0.274- 0.2736
	5.4 (R)	45.0- 44.50	45.75- 45.25	150@1.10	1.676	0.0008- 0.0027	0.0018- 0.0037	0.2754- 0.2746	0.2744-
	6.0 (P)	①	①	191@1.51	1.820	0.0055 max.	0.0055 max.	0.2720- 0.2735	0.2720- 0.2735
	7.3 (F)	①	①	200@1.38	②	0.0055	0.0055	0.3119- 0.3126	0.3119- 0.3126

① Intake: 30 degrees
Exhaust: 37.5 degrees

② Intake: 1.767 in.
Exhaust: 1.833 in.

67197-NAVI-C06

CRANKSHAFT AND CONNECTING ROD SPECIFICATIONS

All measurements are given in inches.

Year	Engine Displacement Liters (VIN)	Crankshaft Main Brg. Journal Dia.	Crankshaft Main Brg. Oil Clearance	Crankshaft Shaft End-play	Crankshaft Thrust on No.	Connecting Rod Journal Diameter	Connecting Rod Oil Clearance	Connecting Rod Side Clearance
2001	4.6 (6/W)	2.6500-2.6570	0.0011-0.0026	0.0051-0.0120	5	2.0870-2.8670	0.0011-0.0026	0.0006-0.0177
	5.4 (L)	2.6568-2.6576	0.0009-0.0019	0.0015-0.0030	5	2.0859-2.0867	0.0010-0.0025	0.0006-0.0177
	6.8 (5)	2.6568-2.6576	0.0009-0.0019	0.0015-0.0030	5	2.0859-2.0867	0.0010-0.0025	0.0006-0.0177
	7.3 (F)	3.1228-3.1236	0.0018-0.0036	0.0025-0.0085	4	2.4980-2.4990	0.0015-0.0045	0.0120-0.0240
2002	4.6 (6/W)	2.6500-2.6570	0.0011-0.0026	0.0051-0.0120	5	2.0870-2.8670	0.0011-0.0026	0.0006-0.0177
	5.4 (L)	2.6568-2.6576	0.0009-0.0019	0.0015-0.0030	5	2.0859-2.0867	0.0010-0.0025	0.0006-0.0177
	6.8 (5)	2.6568-2.6576	0.0009-0.0019	0.0015-0.0030	5	2.0859-2.0867	0.0010-0.0025	0.0006-0.0177
	7.3 (F)	3.1228-3.1236	0.0018-0.0036	0.0025-0.0085	4	2.4980-2.4990	0.0015-0.0045	0.0120-0.0240
2003	4.6 (W)	2.4803	0.0011-0.0026	0.0051-0.0120	5	2.0870-2.8670	0.0011-0.0026	0.0006-0.0177
	5.4 (L)	2.6568-2.6576	0.0009-0.0019	0.0051-0.0120	5	2.0859-2.0867	0.0010-0.0025	0.0006-0.0177
	5.4 (R)	2.6568-2.6576	0.0009-0.0019	0.003-0.0148	5	2.0859-2.0867	0.0010-0.0025	0.0006-0.0177
	6.0 (P)	3.1500-3.1880	NA	0.0087	NA	2.7160-2.7170	NA	0.012-0.0240
	7.3 (F)	3.1228-3.1236	0.0018-0.0036	0.0025-0.0085	4	2.4980-2.4990	0.0015-0.0045	0.0120-0.0240
2004-05	4.6 (W)	2.4803	0.0011-0.0026	0.0051-0.0120	5	2.0870-2.8670	0.0011-0.0026	0.0006-0.0177
	5.4 (L)	2.6568-2.6576	0.0009-0.0019	0.0051-0.0120	5	2.0859-2.0867	0.0010-0.0025	0.0006-0.0177
	5.4 (R)	2.6568-2.6576	0.0009-0.0019	0.003-0.0148	5	2.0859-2.0867	0.0010-0.0025	0.0006-0.0177
	6.0 (P)	3.1500-3.1880	NA	0.0087	NA	2.7160-2.7170	NA	0.012-0.0240
	7.3 (F)	3.1228-3.1236	0.0018-0.0036	0.0025-0.0085	4	2.4980-2.4990	0.0015-0.0045	0.0120-0.0240

67197-NAVI-C07

PISTON AND RING SPECIFICATIONS

All measurements are given in inches.

Year	Engine Displacement Liters (VIN)	Piston Clearance	Ring Gap Top Compression	Ring Gap Bottom Compression	Ring Gap Oil Control	Ring Side Clearance Top Compression	Ring Side Clearance Bottom Compression	Ring Side Clearance Oil Control
2001	4.6 (6/W)	0.0005-0.0010	0.010-0.020	0.010-0.020	0.006-0.026	0.0016-0.0031	0.0012-0.0031	SNUG
	5.4 (L)	0.0000-0.0010	0.005-0.011	0.010-0.016	0.006-0.026	0.0012-0.0037	0.0012-0.0037	SNUG
	6.8 (5)	0.0000-0.0010	0.005-0.011	0.010-0.016	0.006-0.026	0.0012-0.0037	0.0012-0.0037	SNUG
	7.3 (F)	0.0044-0.0057	0.014-0.024	0.062-0.072	0.012-0.024	0.0013-0.0033	0.0013-0.0033	SNUG
2002	4.6 (6/W)	0.0005-0.0010	0.010-0.020	0.010-0.020	0.006-0.026	0.0016-0.0031	0.0012-0.0031	SNUG
	5.4 (L)	0.0000-0.0010	0.005-0.011	0.010-0.016	0.006-0.026	0.0012-0.0037	0.0012-0.0037	SNUG
	6.8 (5)	0.0000-0.0010	0.005-0.011	0.010-0.016	0.006-0.026	0.0012-0.0037	0.0012-0.0037	SNUG
	7.3 (F)	0.0044-0.0057	0.014-0.024	0.062-0.072	0.012-0.024	0.0013-0.0033	0.0013-0.0033	SNUG
2003	4.6 (W)	0.0005-0.0010	0.010-0.020	0.010-0.020	0.006-0.026	0.0016-0.0031	0.0012-0.0031	SNUG
	5.4 (L)	0.0000-0.0010	0.005-0.011	0.010-0.016	0.006-0.026	0.0012-0.0037	0.0012-0.0037	SNUG
	5.4 (R)	0.001-0.0020	0.005-0.011	0.0098-0.0157	0.0059-0.0256	0.0012-0.0020	0.0012-0.0031	SNUG
	6.0 (P)	0.0017-0.0036	0.011-0.0210	0.055-0.0650	0.009-0.0190	NA	NA	NA
	7.3 (F)	0.0044-0.0057	0.014-0.024	0.062-0.072	0.012-0.024	0.0013-0.0033	0.0013-0.0033	SNUG
2004-05	4.6 (W)	0.0005-0.0010	0.010-0.020	0.010-0.020	0.006-0.026	0.0016-0.0031	0.0012-0.0031	SNUG
	5.4 (L)	0.0000-0.0010	0.005-0.011	0.010-0.016	0.006-0.026	0.0012-0.0037	0.0012-0.0037	SNUG
	5.4 (R)	0.001-0.0020	0.005-0.011	0.0098-0.0157	0.0059-0.0256	0.0012-0.0020	0.0012-0.0031	SNUG
	6.0 (P)	0.0017-0.0036	0.011-0.0210	0.055-0.0650	0.009-0.0190	NA	NA	NA
	7.3 (F)	0.0044-0.0057	0.014-0.024	0.062-0.072	0.012-0.024	0.0013-0.0033	0.0013-0.0033	SNUG

NA: Not Available

67197-NAVI-C08

TORQUE SPECIFICATIONS

All readings in ft. lbs.

Year	Engine Displacement Liters (cc)	Cylinder Head Bolts	Main Bearing Bolts	Rod Bearing Bolts	Crankshaft Damper Bolts	Flywheel Bolts	Manifold Intake *	Manifold Exhaust	Spark Plugs	Oil Pan Drain Plug
2001	4.6 (6/W)	①	②	③	④	54-64	⑤	18-25	8	10
	5.4 (L)	①	⑥	③	④	54-64	⑤	18-25	8	10
	6.8 (5)	①	⑥	③	④	54-64	⑦	18-25	11	11
	7.3 (F)	⑧	⑨	⑩	212	89	24	45	—	11
2002	4.6 (6/W)	①	②	③	④	54-64	⑤	18-25	8	10
	5.4 (L)	①	⑥	③	④	54-64	⑤	18-25	8	10
	5.4 (R)	①	⑥	⑪	④	54-64	⑫	17-20	15	10
	6.8 (5)	①	⑥	③	④	54-64	⑦	18-25	11	11
	7.3 (F)	⑧	⑨	⑩	212	89	24	45	—	11
2003	4.6 (W)	①	②	③	④	54-64	⑤	18-25	8	10
	5.4 (L)	①	⑥	③	④	54-64	⑤	18-25	8	10
	5.4 (R)	①	⑥	⑪	④	54-64	⑫	17-20	15	10
	6.0 (P)	⑬	⑭	⑮	⑯	69	8	28 ⑰	—	18
	7.3 (F)	⑧	⑨	⑩	212	89	24	45	—	11
2004-05	4.6 (W)	①	⑱	⑲	④	54-64	⑤	18-25	8	10
	5.4 (L)	①	⑱	⑲	④	54-64	⑤	18-25	8	10
	5.4 (R)	①	⑬	⑲	④	54-64	⑫	17-20	15	10
	6.0 (P)	⑬	⑭	⑮	⑯	69	8	28 ⑰	—	18
	7.3 (F)	⑧	⑨	⑩	212	89	24	45	—	11

* NOTE: Applies to Lower Manifold only.

① Step 1: 30 ft. lbs.
Step 2: Plus 90 degrees
Step 3: Plus 90 degrees

② Vertical mounted bolts:
Step 1: 24 ft. lbs.
Step 2: Plus 90 degrees
Jack screws:
Step 1: 45 inch lbs.
Step 2: 98 inch lbs.

③ Step 1: 32 ft. lbs.
Step 2: Plus 105 degrees

④ Step 1: 66 ft. lbs.
Step 2: Loosen bolt
Step 3: 37 ft. lbs.
Step 4: Plus 90 degrees

⑤ Step 1: 18 inch lbs.
Step 2: 18 ft. lbs.

⑥ Vertical mounted bolts:
Step 1: 30 ft. lbs.
Step 2: Plus 90 degrees
Side Bolts:
Step 1: 22 ft. lbs.
Step 2: plus 90 degrees

⑦ Lower manofold:
Step 1: 18 inch lbs.
Step 2: 89 inch lbs.
Upper Manifold:
Step 1: 18 inch lbs.
Step 2: 18 ft. lbs.

⑧ Step 1: 65. Ft. lbs.
Step 2: 85 ft. lbs.
Step 3: 95 ft. lbs.

⑨ Step 1: 76 ft. lbs.
Step 2: 96 ft. lbs.

⑩ Step 1: 52 ft. lbs.
Step 2: 90 ft. lbs.

⑪ Step 1: 18 ft. lbs.
Step 2: 30 ft. lbs.
Step 3: Plus 90 degrees

⑫ Upper manifold:
Step 1: 18 inch lbs.
Step 2: 90 inch lbs.
Lower manifold:
Step 1: 89 inch lbs.
Step 2: plus an additional 90 degrees

⑬ See the procedure in the text

⑭ Vertical mounted bolts:
Step 1: 110 ft. lbs.
Step 2: 130 ft. lbs.
Step 3: 170 ft. lbs.
Side Bolts:
Step 1: 18 ft. lbs.

⑮ Step 1: 33 ft. lbs.
Step 2: 50 ft. lbs.

⑯ Step 1: 50 ft. lbs.
Step 2: Plus 90 degrees

⑰ Apply high temp nickel anti-seize to the threads

⑱ Cast Aluminum Block:
Vertical mounted bolts:
Step 1: 89 inch lbs.
Step 2: 18 ft. lbs.
Step 3: 30 ft. lbs.
Step 4: plus an additional 90 degrees
Side Bolts:
Step 1: 30 ft. lbs.
Step 2: plus 90 degrees
Cast Iron Block:
Step 1: 30 ft. lbs.
Step 2: Plus 90 degrees
Side Bolts:
Step 1: 44 ft. lbs.
Step 2: 89 inch lbs.

⑲ Step 1: 32 ft. lbs.
Step 2: Plus 90-120 degrees

67197-NAVI-C09

WHEEL ALIGNMENT

Year	Model		Caster Range (+/-Deg.)	Caster Preferred Setting (Deg.)	Camber Range (+/-Deg.)	Camber Preferred Setting (Deg.)	Toe-in (in.)
2001	Excursion	2WD	2.00	+4.00	1.00	+0.62	0.03+/-0.25
		4WD	2.00	+3.50	1.00	+0.25	0.03+/-0.25
	Expedition	2WD base	1.00	+6.10	0.70	-0.30	0.06+/-0.25
		2WD air	1.00	+6.10	0.70	-0.30	0.06+/-0.25
		4WD base	1.00	+5.10	0.70	-0.20	0.30+/-0.25
		4WD air	1.00	+5.00	0.40	-0.40	0.20+/-0.25
	Navigator	2WD	1.0	+6.10	0.75	-0.33	0.06+/-0.25
		4WD	1.0	+5.10	0.40	-0.40	0.20+/-0.25
2002	Excursion	2WD	2.00	+4.00	1.00	+0.62	0.03+/-0.25
		4WD	2.00	+3.50	1.00	+0.25	0.03+/-0.25
	Expedition	2WD base	1.00	+6.10	0.70	-0.30	0.06+/-0.25
		2WD air	1.00	+6.10	0.70	-0.30	0.06+/-0.25
		4WD base	1.00	+5.10	0.70	-0.20	0.30+/-0.25
		4WD air	1.00	+5.00	0.40	-0.40	0.20+/-0.25
	Navigator	2WD	1.0	+6.10	0.75	-0.33	0.06+/-0.25
		4WD	1.0	+5.10	0.40	-0.40	0.20+/-0.25
2003	Expedition	2WD base	1.00	+6.10	0.70	-0.30	0.06+/-0.25
		2WD air	1.00	+6.10	0.70	-0.30	0.06+/-0.25
		4WD base	1.00	+5.10	0.70	-0.20	0.30+/-0.25
		4WD air	1.00	+5.00	0.40	-0.40	0.20+/-0.25
	Navigator	2WD	1.0	+6.10	0.75	-0.33	0.06+/-0.25
		4WD	1.0	+5.10	0.40	-0.40	0.20+/-0.25
2004-05	Expedition	2WD base	1.00	+6.10	0.70	-0.30	0.06+/-0.25
		2WD air	1.00	+6.10	0.70	-0.30	0.06+/-0.25
		4WD base	1.00	+5.10	0.70	-0.20	0.30+/-0.25
		4WD air	1.00	+5.00	0.40	-0.40	0.20+/-0.25
	Navigator	2WD	1.0	+6.10	0.75	-0.33	0.06+/-0.25
		4WD	1.0	+5.10	0.40	-0.40	0.20+/-0.25

67197-NAVI-C10

TIRE, WHEEL AND BALL JOINT SPECIFICATIONS

Year	Model	OEM Tires Standard	OEM Tires Optional	Tire Pressures (psi) Front	Tire Pressures (psi) Rear	Wheel Size	Ball Joint Inspection	Lug Nut
2001	Expedition	P255/70R16	P265/70R17	35	35	7-JJ	0.030 in. ①	②
	Navigator	P245/75R16	P255/75R17	30	35	7.5	0.030 in. ①	②
2002	Expedition	P255/70R16	P265/70R17	35	35	7-JJ	0.030 in. ①	②
	Navigator	P245/75R16	P255/75R17	30	35	7.5	0.030 in. ①	②
2003	Expedition	P255/70R16	P265/70R17	35	35	7-JJ	0.030 in. ①	②
	Navigator	P245/75R16	P255/75R17	30	35	7.5	0.030 in. ①	②
2004-05	Expedition	P255/70R16	P265/70R17	35	35	7-JJ	0.030 in. ①	②
	Navigator	P245/75R16	P255/75R17	30	35	7.5	0.030 in. ①	②

OEM: Original Equipment Manufacturer

PSI: Pounds Per Square Inch

STD: Standard

OPT: Optional

① Both upper and lower

② 12mm nuts 100 ft. lbs.
14mm nuts 150 ft. lbs.

67197-NAVI-C11

BRAKE SPECIFICATIONS

All measurements in inches unless noted

Year	Model		Brake Disc Original Thickness	Brake Disc Minimum Thickness	Brake Disc Maximum Runout	Brake Drum Diameter Original Inside Diameter	Brake Drum Diameter Max. Wear Limit	Brake Drum Diameter Maximum Machine Diameter	Minimum Lining Thickness	Brake Caliper Bracket Bolts (ft. lbs.)	Brake Caliper Mounting Bolts (ft. lbs.)
2001	Excursion	F	1.023	0.964	0.0025	—	—	—	0.030	125-168	41
		R	0.700	0.657	0.0250	—	—	—	0.030	120	26
	Expedition	F	1.023	0.964	0.0025	—	—	—	0.030	125-168	41
		R	0.700	0.657	0.0250	—	—	—	0.030	120	26
	Navigator	F	1.023	0.964	0.0025	—	—	—	0.030	125-168	41
		R	0.700	0.657	0.0250	—	—	—	0.030	120	26
2001	Excursion	F	1.023	0.964	0.0025	—	—	—	0.030	125-168	41
		R	0.700	0.657	0.0250	—	—	—	0.030	120	26
	Expedition	F	1.023	0.964	0.0025	—	—	—	0.030	125-168	41
		R	0.700	0.657	0.0250	—	—	—	0.030	120	26
	Navigator	F	1.023	0.964	0.0025	—	—	—	0.030	125-168	41
		R	0.700	0.657	0.0250	—	—	—	0.030	120	26
2003	Expedition	F	1.023	0.106	0.0025	—	—	—	0.059	148	41
		R	0.700	0.075	0.0250	—	—	—	0.039	140	26
	Navigator	F	1.023	0.964	0.0025	—	—	—	0.059	148	41
		R	0.700	0.075	0.0250	—	—	—	0.039	140	26
2004-05	Expedition	F	1.023	0.106	0.0025	—	—	—	0.059	148	41
		R	0.700	0.075	0.0250	—	—	—	0.039	140	26
	Navigator	F	1.023	0.964	0.0025	—	—	—	0.059	148	41
		R	0.700	0.075	0.0250	—	—	—	0.039	140	26

NOTE: Due to changes made during production, refer to manufacturer's specifications if they differ from those in this chart

F: Front

R: Rear

67197-NAVI-C12

SCHEDULED MAINTENANCE INTERVALS
FORD—2001—03 EXPEDITION & LINCOLN—NAVIGATOR

		VEHICLE MILEAGE INTERVAL (x1000)																							
TO BE SERVICED	TYPE OF SERVICE	5	10	15	20	25	30	35	40	45	50	55	60	65	70	75	80	85	90	95	100	105	110	115	120
Accessory drive belt	S/I												✓												✓
Air cleaner filter ①	R						✓						✓						✓			✓			✓
Automatic transmission fluid	R						✓						✓						✓			✓			✓
Automatic transmission shift linkage	S/I & L	✓	✓	✓	✓	✓	✓	✓	✓	✓	✓	✓	✓	✓	✓	✓	✓	✓	✓	✓	✓	✓	✓	✓	✓
Brake caliper, slide rails	L			✓			✓			✓			✓			✓			✓			✓			✓
Brake system, hoses & lines	S/I			✓			✓			✓			✓			✓			✓			✓			✓
Clutch reservoir fluid level	S/I	✓	✓	✓	✓	✓	✓	✓	✓	✓	✓	✓	✓	✓	✓	✓	✓	✓	✓	✓	✓	✓	✓	✓	✓
Engine coolant ②	R										✓						✓						✓		
Engine cooling system hoses, clamps & coolant	S/I			✓			✓			✓			✓			✓			✓			✓			✓
Engine oil & filter	R	✓	✓	✓	✓	✓	✓	✓	✓	✓	✓	✓	✓	✓	✓	✓	✓	✓	✓	✓	✓	✓	✓	✓	✓
Exhaust system	S/I	✓	✓	✓	✓	✓	✓	✓	✓	✓	✓	✓	✓	✓	✓	✓	✓	✓	✓	✓	✓	✓	✓	✓	✓
Front wheel bearings	S/I & L						✓						✓						✓						✓
Front/rear axle driveshaft slip yoke	L						✓						✓						✓						✓
Front/rear axle fluid ③	R																				✓				
Fuel filter	R			✓			✓			✓			✓			✓			✓			✓			✓
Manual transmission fluid	R												✓												✓
Parking brake system	S/I						✓						✓						✓						✓
PCV valve	R												✓												✓
Rotate tires	S/I	✓	✓	✓	✓	✓	✓	✓	✓	✓	✓	✓	✓	✓	✓	✓	✓	✓	✓	✓	✓	✓	✓	✓	✓
Spark plugs	R																				✓				
Steering linkage, suspension, driveshaft U joints	S/I & L	✓	✓	✓	✓	✓	✓	✓	✓	✓	✓	✓	✓	✓	✓	✓	✓	✓	✓	✓	✓	✓	✓	✓	✓

R: Replace S/I: Service or Inspect

① Perform this at the mileage shown or every 30 months, whichever occurs first.

② Drain, flush and refill the cooling system initially at 50,000 miles or 48 months, whichever occurs first, then every 30,000 miles or 30 months thereafter.

③ The axle lubricant must be replaced every 100,000 miles or if the axle has been submerged under water. Otherwise the lube should not be checked or changed unless a repair is required.

FREQUENT OPERATION MAINTENANCE (SEVERE SERVICE)

If a vehicle is operated under any of the following conditions it is considered severe service:

- Towing a trailer or using a camper or car-top carrier.

- Repeated short trips of less than 5 miles in temperatures below freezing, or trips of less than 10 miles in any temperature.

- Extensive idling or low-speed driving for long distance as in heavy commercial use, such as delivery, taxi or police cars.

- Operating on rough, muddy or salt-covered roads.

- Operating on unpaved or dusty roads.

- Driving in extremely hot (over 90°) conditions.

Engine oil & filter: replace every 3000 miles.

Tires: rotate and inspect every 6000 miles.

Clutch reservoir fluid level: inspect every 6000 miles.

Automatic transmission shift linkage: lubricate every 6000 miles.

Steering linkage: suspension, U-joints: lubricate every 6000 miles.

Exhaust system: inspect for leaks of damage every 6000 miles.

Fuel filter: replace every 15,000 miles.

Automatic transmission fluid: change every 21,000 miles.

Crankcase emission air filter: replace every 60,000 miles.

PCV valve: replace every 60,000 miles.

Accessory drive belt: inspect every 60,000 miles.

Spark plugs: replace every 99,000 miles.

67197-NAVI-C13

SCHEDULED MAINTENANCE INTERVALS

FORD—EXCURSION (Except 6.0L Diesel Engine)

TO BE SERVICED	TYPE OF SERVICE	VEHICLE MILEAGE INTERVAL (x1000)																			
		5	10	15	20	25	30	35	40	45	50	55	60	65	70	75	80	85	90	95	100
Accessory drive belt	S/I																				✓
Air cleaner filter ①②	R						✓						✓						✓		
Automatic transmission fluid ③	R						✓						✓						✓		
Engine coolant ④⑤	R										✓						✓				
Engine cooling system hoses, clamps & coolant ⑥	S/I			✓			✓			✓			✓			✓			✓		
Engine oil & filter	R	✓	✓	✓	✓	✓	✓	✓	✓	✓	✓	✓	✓	✓	✓	✓	✓	✓	✓	✓	✓
Exhaust system	S/I			✓			✓			✓			✓			✓			✓		
Front wheel bearings	S/I & L																		✓		
Front/rear axle lubricant ⑦	R																				✓
Fuel filter ⑧	R						✓						✓						✓		
PCV valve	R	Every 120,000 miles																			
Rotate tires	S/I	✓	✓	✓	✓	✓	✓	✓	✓	✓	✓	✓	✓	✓	✓	✓	✓	✓	✓	✓	✓
Spark plugs	R																				✓
Steering linkage, suspension, driveshaft U joints	S/I & L	✓	✓	✓	✓	✓	✓	✓	✓	✓	✓	✓	✓	✓	✓	✓	✓	✓	✓	✓	✓

R: Replace S/I: Service or Inspect

① Perform this at the mileage shown or every 30 months, whichever occurs first.

② 7.3L DIT Diesel engine: the air filter should be replaced when the restriction gauge is in the red zone.

③ Except the E40D transmission.

④ Drain, flush and refill the cooling system initially at 50,000 miles or 48 months, whichever occurs first, then every 30,000 miles or 30 months thereafter.

⑤ 7.3L DIT Diesel engine: add 4 pints of FW-15 each time the coolant is replaced.

⑥ 7.3L DIT Diesel engine: add 8-10 oz. of FW-15 to the engine coolant every 15,000 miles.

⑦ The axle lubricant must be replaced every 100,000 miles of if the axle has been submerged under water. Otherwise the lube should not be checked or changed unless a repair is required.

⑧ 7.3L DIT Diesel engine: the fuel filter should be replaced when the restriction lamp is illuminated.

FREQUENT OPERATION MAINTENANCE (SEVERE SERVICE)

If a vehicle is operated under any of the following conditions it is considered severe service:

- Towing a trailer or using a camper or car-top carrier.

- Repeated short trips of less than 5 miles in temperatures below freezing, or trips of less than 10 miles in any temperature.

- Extensive idling or low-speed driving for long distances as in heavy commercial use, such as delivery, taxi or police cars.

- Operating on rough, muddy or salt-covered roads.

- Operating on unpaved or dusty roads.

Engine oil & filter: replace every 3000 miles.
Tires: rotate and inspect every 6000 miles.
Steering linkage, suspension, U-joints: lubricate every 6000 miles.
Exhaust system: inspect for leaks or damage every 12,000 miles.
Fuel filter: replace every 15,000 miles.
Automatic transmission fluid: change ever 21,000 miles.
Front wheel bearings (2WD): inspect and repack every 30,000 miles.
Rear axle lubricant (E-Super Duty only): replace every 30,000 miles.
Spark plugs (except 4.2L engine): replace every 60,000 miles.
PCV valve: replace every 60,000 miles.
Accessory drive belt: inspect every 60,000 miles.
Spark plugs: replace every 99,000 miles.

67197-NAVI-C14

SCHEDULED MAINTENANCE INTERVALS
FORD—2004 EXPEDITION & LINCOLN—NAVIGATOR

TO BE SERVICED	TYPE OF SERVICE	VEHICLE MILEAGE INTERVAL (x1000)												
		5	10	15	20	25	30	35	40	45	50	55	60	65
Engine oil & filter	R	✓	✓	✓	✓	✓	✓	✓	✓	✓	✓	✓	✓	✓
Tires	Rotate	✓	✓	✓	✓	✓	✓	✓	✓	✓	✓	✓	✓	✓
Wheels	I ①	✓	✓	✓	✓	✓	✓	✓	✓	✓	✓	✓	✓	✓
Auto trans. fluid	I			✓			✓			✓			✓	
Brake pads/shoes	I			✓			✓			✓			✓	
Coolant hoses	S/I			✓			✓			✓			✓	
Steering linkage	I			✓			✓			✓			✓	
Cabin air filter	R			✓			✓			✓			✓	
Ball joints (2wd)	L			✓			✓			✓			✓	
Exhaust system	I						✓						✓	
Engine air filter	R						✓						✓	
Fuel filter	R						✓						✓	
Auto trans fluid (4R100)	R						✓						✓	
Front wheel bearings (2wd)	R	at 150,000 miles, if not previously done so												
Spark plugs	R	every 100,000 miles												
PCV valve	R	every 120,000 miles												
Premium Gold coolant	R	every 5 years or 100,000 miles												
Auto trans fluid (exc. 4R100)	R	every 150,000 miles												
Differential fluid	R	every 150,000 miles												
Accessory drive belts	R	every 150,000 miles, if not previously done so												

R: Replace S: Service I: Inspect L: Lubricate

① Inspect for end play and noise

Special Operating Condition Requirements

When towing a trailer or using a camper or car-top carrier:

Change engine oil and install a new oil filter every 4,800 km (3,000 miles), 3 months or 200 hours of engine operation (whichever occurs first).

Change transfer case fluid every 96,000 km (60,000 miles).

Change manual transmission fluid as required.

Inspect and lubricate U-joints as required.

During extensive idling and/or low speed driving for long distances, as in heavy commercial use such as delivery, taxi, patrol car or livery:

Change engine oil and install a new oil filter every 4,800 km (3,000 miles), 3 months or 200 hours of engine operation (whichever occurs first).

Lube front lower control arm and steering linkage ball joints with zerk fittings (if equipped) every 4,800 km (3,000 miles) or 3 months.

Inspect brake system and check battery electrolyte level (Patrol cars) every 8,000 km (5,000 miles).

Install a new fuel filter every 24,000 km (15,000 miles).

Change automatic transmission fluid, lubricate 4x2 wheel bearings, install new grease seals and adjust bearings every 48,000 km (30,000 miles). If equipped, change the in-line service installed transmission fluid filter.

Install new spark plugs and change transfer case fluid every 96,000 km (60,000 miles).

Install a new cabin air filter as required.

When operating in dusty conditions such as unpaved or dusty roads:

Change engine oil and install a new oil filter every 4,800 km (3,000 miles) or 3 months.

Install a new fuel filter every 24,000 km (15,000 miles).

Change automatic transmission fluid every 48,000 km (30,000 miles). If equipped, change the in-line service installed transmission fluid filter.

Change transfer case fluid every 96,000 km (60,000 miles).

Install a new engine air filter as required.

Install a new cabin air filter as required.

When operating in off-road conditions:

Change automatic transmission fluid every 48,000 km (30,000 miles). If equipped, change the in-line service installed transmission fluid filter.

Change transfer case fluid every 96,000 km (60,000 miles).

Install a new cabin air filter as required.

Inspect and lubricate U-joints.

Inspect and lubricate steering linkage ball joints with zerk fittings.

67197-NAVI-C15

SCHEDULED MAINTENANCE INTERVALS

2004 Excursion With Gasoline Engines

TO BE SERVICED	TYPE OF SERVICE	VEHICLE MILEAGE INTERVAL (x1000)												
		5	10	15	20	25	30	35	40	45	50	55	60	65
Engine oil & filter	R	✓	✓	✓	✓	✓	✓	✓	✓	✓	✓	✓	✓	✓
Tires	Rotate	✓	✓	✓	✓	✓	✓	✓	✓	✓	✓	✓	✓	✓
Wheels	I ①	✓	✓	✓	✓	✓	✓	✓	✓	✓	✓	✓	✓	✓
Auto trans. fluid	I			✓			✓			✓			✓	
Brake pads, hoses, etc.	I			✓			✓			✓			✓	
Coolant hoses	I			✓			✓			✓			✓	
Steering linkage	I/L			✓			✓			✓			✓	
Cabin air filter	R			✓			✓			✓			✓	
Ball joints (2wd)	L			✓			✓			✓			✓	
Driveshaft	I			✓			✓			✓			✓	
4x4 front axle U-joints	L			✓			✓			✓			✓	
Exhaust system	I						✓						✓	
Engine air filter	R						✓						✓	
Fuel filter	R						✓						✓	
Auto trans fluid (4R100)	R						✓						✓	
4x2 front wheel bearings	Adj.												✓	
4x2 front wheel bearing grease seals	R												✓	
Accessory drive belts	I	Every 100,000 miles												
Front wheel bearings (2wd)	R	at 150,000 miles, if not previously done so												
4x4 front hub needle bearings	L	every 120,000 miles												
Spark plugs	R	every 100,000 miles												
PCV valve	R	every 120,000 miles												
Manual trans. Fluid	R	every 120,000 miles												
Premium Gold coolant	R	every 3 years or 100,000 miles and every 50,000 miles thereafter												
Coolant (exc. Premium Gold)	R	every 135,000 miles												
Auto trans fluid (exc. 4R100)	R	every 150,000 miles												
Differential fluid	R	at 150,000 miles then every 50,000 thereafter												
Transfer case fluid	R	every 150,000 miles												
Accessory drive belts	R	every 150,000 miles, if not previously done so												

R: Replace S: Service I: Inspect L: Lubricate Adj: adjust

① Inspect for end play and noise

Special Operating Condition Requirements

When towing a trailer or using a camper or car-top carrier:

Change engine oil and install a new oil filter every 4,800 km (3,000 miles), 3 months or 200 hours of engine operation (whichever occurs first).

Change transfer case fluid every 96,000 km (60,000 miles).

Change manual transmission fluid as required.

Inspect and lubricate U-joints as required.

During extensive idling and/or low speed driving for long distances, as in heavy commercial use such as delivery, taxi, patrol car or livery:

Change engine oil and install a new oil filter every 4,800 km (3,000 miles), 3 months or 200 hours of engine operation (whichever occurs first).

Lube front lower control arm and steering linkage ball joints with zerk fittings (if equipped) every 4,800 km (3,000 miles) or 3 months.

Inspect brake system and check battery electrolyte level (Patrol cars) every 8,000 km (5,000 miles).

Install a new fuel filter every 24,000 km (15,000 miles).

Change automatic transmission fluid, lubricate 4x2 wheel bearings, install new grease seals and adjust bearings every 48,000 km (30,000 miles). If equipped, change the in-line service installed transmission fluid filter.

Install new spark plugs and change transfer case fluid every 96,000 km (60,000 miles).

Install a new cabin air filter as required.

67197-NAVI-C16

SCHEDULED MAINTENANCE INTERVALS

2004 Excursion With Gasoline Engines

Footnotes Continued

When operating in dusty conditions such as unpaved or dusty roads:

Change engine oil and install a new oil filter every 4,800 km (3,000 miles) or 3 months.

Install a new fuel filter every 24,000 km (15,000 miles).

Change automatic transmission fluid every 48,000 km (30,000 miles). If equipped, change the in-line service installed transmission fluid filter.

Change transfer case fluid every 96,000 km (60,000 miles).

Install a new engine air filter as required.

Install a new cabin air filter as required.

When operating in off-road conditions:

Change automatic transmission fluid every 48,000 km (30,000 miles). If equipped, change the in-line service installed transmission fluid filter.

Change transfer case fluid every 96,000 km (60,000 miles).

Install a new cabin air filter as required.

Inspect and lubricate U-joints.

Inspect and lubricate steering linkage ball joints with zerk fittings.

67197-NAVI-C17

SCHEDULED MAINTENANCE INTERVALS

6.0L Deisel Engine

TO BE SERVICED	TYPE OF SERVICE	VEHICLE MILEAGE INTERVAL (x1000)												
		7.5	15	22.5	30	37.5	45	52.5	60	67.5	75	82.5	90	97.5
Engine oil & filter	R	✓	✓	✓	✓	✓	✓	✓	✓	✓	✓	✓	✓	✓
Tires	Rotate	✓	✓	✓	✓	✓	✓	✓	✓	✓	✓	✓	✓	✓
Air filter minder	I ①	✓	✓	✓	✓	✓	✓	✓	✓	✓	✓	✓	✓	✓
Wheels	I ②	✓	✓	✓	✓	✓	✓	✓	✓	✓	✓	✓	✓	✓
Brake pads, hoses, etc.	I		✓		✓		✓		✓		✓		✓	
Coolant hoses	I		✓		✓		✓		✓		✓		✓	
Steering linkage and suspension	I/L		✓		✓		✓		✓		✓		✓	
Cabin air filter	R		✓		✓		✓		✓		✓		✓	
Ball joints (2wd)	L		✓		✓		✓		✓		✓		✓	
Driveshaft	I/L		✓		✓		✓		✓		✓		✓	
4x4 front axle U-joints	L		✓		✓		✓		✓		✓		✓	
Exhaust system and heat shields	I		✓		✓		✓		✓		✓		✓	
Engine air filter	R				✓				✓				✓	
Fuel filters ③	R		✓		✓		✓		✓		✓		✓	
Auto trans fluid ④	R				✓				✓				✓	
4x2 front wheel bearings	L								✓					
4x2 front wheel bearing grease seals	R								✓					
Accessory drive belts	I													✓
4x4 front hub needle bearings	L								✓					
Manual trans. Fluid	R								✓					
Coolant	R										✓			✓
Rear differential fluid ⑤	R													✓
Transfer case fluid	R	every 150,000 miles												
4x2 front wheel bearings	R	every 150,000 miles												
Accessory drive belts	R	every 150,000 miles, if not previously done so												

R: Replace S: Service I: Inspect L: Lubricate Adj: adjust

① Reset after new filter is installed

② Inspect for end play and noise

③ Frame-mounted and engine

④ Including external and in-line filters

⑤ Dana axles using non-synthetic fluid only

Special Operating Condition Requirements

When towing a trailer or using a camper or car-top carrier:

Change engine oil and install a new oil filter every 4,800 km (3,000 miles), 3 months or 200 hours of engine operation (whichever occurs first).

Change transfer case fluid every 96,000 km (60,000 miles).

Change manual transmission fluid as required.

Inspect and lubricate U-joints as required.

67197-NAVI-C18

SCHEDULED MAINTENANCE INTERVALS

6.0L Deisel Engine

Footnotes Continued

During extensive idling and/or low speed driving for long distances, as in heavy commercial use such as delivery, taxi, patrol car or livery:

Change engine oil and install a new oil filter every 4,800 km (3,000 miles), 3 months or 200 hours of engine operation (whichever occurs first).

Lube front lower control arm and steering linkage ball joints with zerk fittings (if equipped) every 4,800 km (3,000 miles) or 3 months.

Inspect brake system and check battery electrolyte level (Patrol cars) every 8,000 km (5,000 miles).

Install a new fuel filter every 24,000 km (15,000 miles).

Change automatic transmission fluid, lubricate 4x2 wheel bearings, install new grease seals and adjust bearings every 48,000 km (30,000 miles). If equipped, change the in-line service installed transmission fluid filter.

Change transfer case fluid every 96,000 km (60,000 miles).

Install a new cabin air filter as required.

When operating in dusty conditions such as unpaved or dusty roads:

Change engine oil and install a new oil filter every 4,800 km (3,000 miles) or 3 months.

Install a new fuel filter every 24,000 km (15,000 miles).

Change automatic transmission fluid every 48,000 km (30,000 miles). If equipped, change the in-line service installed transmission fluid filter.

Change transfer case fluid every 96,000 km (60,000 miles).

Install a new engine air filter as required.

Install a new cabin air filter as required.

When operating in off-road conditions:

Change automatic transmission fluid every 48,000 km (30,000 miles). If equipped, change the in-line service installed transmission fluid filter.

Change transfer case fluid every 96,000 km (60,000 miles).

Install a new cabin air filter as required.

Inspect and lubricate U-joints.

Inspect and lubricate steering linkage ball joints with zerk fittings.

67197-NAVI-C19

PRECAUTIONS

Before servicing any vehicle, please be sure to read all of the following precautions, which deal with personal safety, prevention of component damage, and important points to take into consideration when servicing a motor vehicle:

- Never open, service or drain the radiator or cooling system when the engine is hot; serious burns can occur from the steam and hot coolant.
- Observe all applicable safety precautions when working around fuel. Whenever servicing the fuel system, always work in a well-ventilated area. Do not allow fuel spray or vapors to come in contact with a spark, open flame or excessive heat (a hot drop light, for example). Keep a dry chemical fire extinguisher near the work area. Always keep fuel in a container specifically designed for fuel storage; also, always properly seal fuel containers to avoid the possibility of fire or explosion. Refer to the additional fuel system precautions later in this section.
- Fuel injection systems often remain pressurized, even after the engine has been turned **OFF**. The fuel system pressure must be relieved before disconnecting any fuel lines. Failure to do so may result in fire and/or personal injury.
- Brake fluid often contains polyglycol ethers and polyglycols. Avoid contact with the eyes and wash your hands thoroughly after handling brake fluid. If you do get brake fluid in your eyes, flush your eyes with clean, running water for 15 minutes. If eye irritation persists, or if you have taken brake fluid internally, IMMEDIATELY seek medical assistance.
- The EPA warns that prolonged contact with used engine oil may cause a number of skin disorders, including cancer! You should make every effort to minimize your exposure to used engine oil. Protective gloves should be worn when changing oil. Wash your hands and any other exposed skin areas as soon as possible after exposure to used engine oil. Soap and water, or waterless hand cleaner should be used.
- All new vehicles are now equipped with an air bag system. The system must be disabled before performing service on or around system components, steering column, instrument panel components, wiring and sensors. Failure to follow safety and disabling procedures could result in accidental air bag deployment, possible personal injury and unnecessary system repairs.
- Always wear safety goggles when working with, or around, the air bag system. When carrying a non-deployed air bag, be sure the bag and trim cover are pointed away from your body. When placing a non-deployed air bag on a work surface, always face the bag and trim cover upward, away from the surface. This will reduce the motion of the module if it is accidentally deployed. Refer to the additional air bag system precautions later in this section.
- Clean, high quality brake fluid from a sealed container is essential to the safe and proper operation of the brake system. You should always buy the correct type of brake fluid for your vehicle. If the brake fluid becomes contaminated, completely flush the system with new fluid. Never reuse any brake fluid. Any brake fluid that is removed from the system should be discarded. Also, do not allow any brake fluid to come in contact with a painted surface; it will damage the paint.
- Never operate the engine without the proper amount and type of engine oil; doing so WILL result in severe engine damage.
- Timing belt maintenance is extremely important! Many models utilize an interference-type, non-freewheeling engine. If the timing belt breaks, the valves in the cylinder head may strike the pistons, causing potentially serious (also time-consuming and expensive) engine damage.
- Disconnecting the negative battery cable on some vehicles may interfere with the functions of the on-board computer system(s) and may require the computer to undergo a relearning process once the negative battery cable is reconnected.
- When servicing drum brakes, only disassemble and assemble one side at a time, leaving the remaining side intact for reference.
- Only an MVAC-trained, EPA-certified automotive technician should service the air conditioning system or its components.

GASOLINE ENGINE REPAIR

➡Disconnecting the negative battery cable on some vehicles may interfere with the functions of the on board computer systems and may require the computer to undergo a relearning process, once the negative battery cable is reconnected.

Distributor

These vehicles are equipped with a Distributorless Ignition System (DIS).

Alternator

REMOVAL

1. Before servicing the vehicle, refer to the precautions in the beginning of this section.
2. Remove or disconnect the following:
    - Negative battery cable
    - Air cleaner assembly, if necessary
    - Alternator bracket bolts, if necessary
    - Ignition wire pin-type retainer from the generator bracket, 4.6L engines
    - Drive belt from the alternator pulley
    - Electrical harness connectors at the alternator assembly
    - Positive battery cable, the nut and washer
    - 2 front alternator bolts
    - Rear alternator support bracket retaining bolts and the support bracket
    - Alternator from the vehicle

INSTALLATION

1. Before servicing the vehicle, refer to the precautions in the beginning of this section.
2. Install or connect the following:
    - Alternator in position
    - 2 front alternator retaining bolts, loosely
    - Alternator bracket and 3 alternator bracket bolts. Tighten to 84 inch lbs. (10 Nm).
    - Tighten 2 front alternator retaining bolts to 19 ft. lbs. (26 Nm)
    - 2 electrical harness connectors to the alternator assembly
    - Positive battery cable, the nut and washer. Tighten the nut to 72 inch lbs. (8 Nm).
    - Drive belt on the alternator pulley
    - Ignition wire pin-type retainer from the generator bracket, 4.6L engines
    - Alternator bracket bolts, if removed
    - Air cleaner assembly, if removed
    - Negative battery cable

3. Start the engine and check for proper charging system operation.

Ignition Timing

ADJUSTMENT

Distributorless Ignition System

Base timing for distributorless ignition engines is set at the factory at 10 degrees Before Top Dead Center (BTDC) and is not adjustable.

Engine Assembly

REMOVAL & INSTALLATION

4.6L, 5.4L SOHC and 6.8L Engines

2001 EXCURSION MODELS

1. Before servicing the vehicle, refer to the precautions in the beginning of this section.
2. Disconnect the negative battery cable.
3. Remove the hood.
4. Drain the cooling system.
5. Radiator grille supports
6. Remove the radiator.
7. Recover the A/C system refrigerant.
8. Remove the condenser
9. Disconnect the two 42 pin and 16 ping connectors.
10. Remove the intake manifold.
11. Remove the lower radiator hose from the oil water cooler outlet.
12. Remove the power steering pump and set aside.
13. Disconnect the A/C cycling switch.
14. Disconnect the A/C compressor electrical connector.
15. Disconnect the suction hose at the accumulator.
16. Disconnect the heater hoses at the heater core.
17. Disconnect the heated Oxygen Sensor (HO_2S) electrical connectors.
18. Remove the bolts and position the A/C compressor aside.
19. Remove the exhaust pipe.
20. Remove the transmission and clutch, if equipped with a manual transmission.
21. Drain the oil and remove the bypass filter.
22. Loosen the threaded shaft and position the oil cooler aside.
23. Remove the starter, if equipped with an automatic transmission.
24. Remove the transmission cooler hoses from the block bracket, if equipped with an automatic transmission.
25. Remove the stud bolt and position the ground strap aside.
26. Remove the nut and position the wiring bracket aside.
27. Remove the flywheel inspection cover, if equipped with an automatic transmission.
28. Remove the access plug, if equipped with an automatic transmission.
29. Remove the torque converter nuts, if equipped with an automatic transmission.
30. Support the transmission, using a block of wood and a floor jack, if equipped with an automatic transmission.
31. Remove the four lower transmission-to-engine bolts, if equipped with an automatic transmission.
32. Remove the nut retaining the ground strap, and on automatic transmission equipped vehicles remove the transmission oil filler tube.
33. Remove the two studs and the heater water tube.
34. Remove the three upper transmission to engine bolts, if equipped with an automatic transmission.
35. Install modular engine lifting bar 303-F047 tool.
36. Remove the engine support insulator nuts.
37. Support the transmission using a transmission jack, if equipped with an automatic transmission.
38. Remove the engine from the vehicle.

To install:

39. Install the engine in the vehicle.
40. Install the engine support insulator nuts and tighten to 66 ft. lbs. (90 Nm).
41. Remove modular engine lifting bar 303-F047 tool.
42. Install the transmission oil level tube, and the ground strap on all vehicles equipped with an automatic transmission.
43. Install the heater water tube and the two studs and tighten to 22 ft. lbs. (30 Nm).
44. Align the transmission to the engine and install the five lower transmission-to-engine bolts. Tighten to 40 ft. lbs. (54 Nm) , if equipped with an automatic transmission.
45. Install and tighten the four new nuts retaining the torque converter to 34 ft. lbs. (46 Nm) , if equipped with an automatic transmission.
46. Install the access plug, if equipped with an automatic transmission.
47. Install the flywheel inspection cover, if equipped with an automatic transmission.
48. Install the wiring bracket and tighten to 22 ft. lbs. (30 Nm).
49. Install the starter motor, if equipped with an automatic transmission.
50. Install the oil cooler to the oil filter adapter and install the oil bypass filter. Tighten to 44 ft. lbs. (60 Nm).
51. Install the clutch and transmission, if equipped with a manual transmission
52. Tighten the exhaust pipe nuts to 34 ft. lbs. (46 Nm).
53. Install the A/C compressor
54. Install the ground strap.
55. Install the transmission cooler line bracket, if equipped with an automatic transmission.
56. Install the remaining transmission-to-engine bolts, tighten to 40 ft. lbs. (54 Nm), if equipped with an automatic transmission.
57. Install the fuel line bracket.
58. Connect the transmission wiring harness
59. Connect the HO_2S electrical connectors.
60. Connect the heater hose.
61. Install the suction line to the accumulator.
62. Connect the A/C cycling switch.
63. Connect the A/C compressor clutch connector.
64. Connect the lower radiator hose to the oil cooler inlet.
65. Install the power steering pump.
66. Install the intake manifold.
67. Connect the 42-pin and 16-pin connectors.
68. Install the radiator and engine cooling fan
69. Install the condenser..
70. Install the engine air cleaner.
71. Fill the engine with clean engine oil.
72. Connect the negative battery cable.
73. Fill the engine cooling system.
74. Fill the power steering system..
75. Recharge the A/C system.
76. Install the hood.

2001 MODELS—EXCEPT 2001 EXCURSION

1. Before servicing the vehicle, refer to the precautions in the beginning of this section.
2. Disconnect the negative battery cable.
3. Remove the hood.
4. Drain the cooling system.
5. If equipped, turn the air suspension switch off.
6. Remove the radiator.
7. Recover the A/C system refrigerant.
8. Remove the condenser
9. Disconnect the climate control vacuum connector.
10. Disconnect the evaporative emission canister purge valve connector.

11. Remove the power steering pump reservoir and set aside.
12. Remove the intake manifold.
13. Disconnect the starter cables from the starter relay.
14. Disconnect the two 42 pin and single pin connectors, the 16 pin and single pin connectors.
15. Remove the junction block bracket.
16. Disconnect the heater hoses at the heater core.
17. Remove the drain plug and drain the engine oil and install the drain plug when finished
18. Disconnect the A/C compressor electrical connector.
19. Disconnect the Crankshaft Position (CKP) sensor electrical connector.
20. Remove the starter.
21. Remove the bolts and position the A/C compressor aside.
22. Remove the A/C muffler line and set aside.
23. Remove the transmission cooler hoses from the block mounted clip.
24. Remove the shift cable and bracket.
25. Remove the flexplate inspection cover.
26. Remove the cylinder block opening cover.
27. Remove the torque converter nuts.
28. Support the transmission, using a block of wood and a floor jack.
29. Remove the five lower transmission-to-engine bolts.
30. Remove the transmission filler tube-to-engine block bolt which is accessed through the ride side fender well.
31. Disconnect the Electronic Variable Orifice (EVO) connector.
32. Disconnect the oil pressure sender connector.
33. Disconnect the heated Oxygen Sensor (HO_2S) electrical connectors.
34. Remove the left side exhaust manifold to catalytic converter nuts and the right hand manifold studs.
35. Remove the left side exhaust manifold to catalytic converter nuts and position the Y pipe aside.
36. Remove the right side motor mount bolt.
37. Remove the left side mount through bolt.
38. Remove the oil bypass filter.
39. Remove the oil pressure sensor.
40. Remove the lower radiator hose from the oil filter adapter.
41. Remove the two upper transmission-to-engine bolts.

➡On some engines it will be necessary to shim between the modular engine lifting bar 303-F047 tool mounting brackets and the cylinder block.

42. Install modular engine lifting bar 303-F047 tool.
43. Remove the engine from the vehicle.

To install:

44. Install the engine in the vehicle
45. Remove modular engine lifting bar 303-F047 tool.
46. Align the transmission to the engine and install the five lower transmission-to-engine bolts. Tighten to 35 ft. lbs. (48 Nm).
47. Remove the floor jack and wood block supporting the transmission.
48. Install the torque converter nuts and tighten to 26 ft. lbs. (35 Nm).
49. Install the cylinder block opening cover.
50. Install the flexplate inspection cover and tighten the bolts to 25 ft. lbs. (34 Nm).
51. Install the shift cable and bracket.
52. Install the left side motor mount bolt and tighten to 68 ft. lbs. (92 Nm).
53. Install the right side motor mount nut and tighten to 68 ft. lbs. (92 Nm).
54. Install the starter.
55. Install the A/C compressor and tighten the bolts to 18 ft. lbs. (25 Nm).
56. Install the right side exhaust manifold studs and tighten the nuts.
57. Install the dual converter Y-pipe and tighten the four nuts to 30 ft. lbs. (40 Nm).
58. Install the EVO connector.
59. Install the lower radiator hose.
60. Install the oil bypass filter.
61. Install the oil pressure sensor.
62. Install the transmission cooler hoses to the block mounted clip.
63. Install the A/C muffler line.
64. Connect the A/C compressor electrical connector.
65. Connect the CKP sensor electrical connector.
66. Connect the HO_2S electrical connectors.
67. Install the transmission filler tube.
68. Install the water heater tube using a new O–ring.
69. Install the transmission wiring bracket.
70. Install the two upper transmission-to-engine bolts. Tighten to 35 ft. lbs. (48 Nm).
71. Install the transmission wiring harness connectors.
72. Install the intake manifold.
73. Install the climate control connector.
74. Install the condenser.
75. Install the radiator.
76. Install all remaining hoses, tubes and electrical connectors.
77. Install the power steering reservoir.
78. Install the engine air cleaner.
79. Fill the engine with clean engine oil.
80. Connect the negative battery cable.
81. Fill the engine cooling system.
82. Fill the power steering system..
83. Recharge the A/C system.
84. Install the hood.

2002–05 EXCURSION AND NAVIGATOR MODELS

1. Before servicing the vehicle, refer to the precautions in the beginning of this section.
2. Disconnect the negative battery cable.
3. If equipped, turn the air suspension switch off.
4. Remove the intake manifold.
5. Remove the hood.
6. Remove the engine air cleaner and reposition it to gain access to the Mass Air Flow (MAF) sensor electrical connector and disconnect the sensor connector.
7. Remove the radiator.
8. Recover the A/C system refrigerant.
9. Remove the condenser.
10. Disconnect the powertrain control module (PCM) electrical connectors.
11. Disconnect the engine wiring harness connector.
12. Remove the PCM mounting bolts and the PCM.
13. Remove the bolts and the PCM mounting bracket.
14. Remove the ground wire.
15. Disconnect the heater hoses at the heater core.
16. Disconnect the evaporator tubes from the manifold.
17. Remove the cable routing bracket from the heater outlet tube mounting studs.
18. On 5.4L engines disconnect the knock sensor connector.
19. Remove the wiring retainers from the heater outlet tube bracket.
20. Disconnect the hoses from the heater outlet tube.
21. Remove the heater outlet tube upper and lower studs on 4.6L engines.
22. Loosen the heater outlet tube upper and lower studs on 5.4L engines.
23. Remove the heater outlet tube and discard the O –ring.
24. Remove the wiring harness retainer from the front cover stud.
25. Remove the drive belt.
26. Remove the alternator.
27. Remove the power steering pump reservoir and set aside.
28. Remove the nuts and disconnect the transmission cooler tube and the power steering hose brackets.

29. Remove the power steering pump pulley using removal tool 211-016.
30. Disconnect the power steering pressure hose and drain the power steering fluid into a suitable container.
31. Remove the three bolts and position the power steering pump out of the way.
32. Disconnect the power steering pressure switch connector, if equipped.
33. Remove the drain plug and drain the engine oil and install the drain plug when finished.
34. Remove the oil filter.
35. Remove the lower radiator hose and degas bottle hose from the oil filter adapter.
36. Disconnect the Exhaust Gas Recirculation (EGR) valve tube lower fitting and remove the tube.
37. Disconnect the Crankshaft Position (CKP) sensor electrical connector.
38. Disconnect the A/C compressor electrical connector.
39. Remove the bolts and position the A/C compressor aside.
40. Remove the bolt and disconnect the wiring harness bracket.
41. Remove the starter.
42. Remove the transmission filler tube-to-engine block bolt which is accessed through the ride side fender well.
43. Disconnect the heated Oxygen Sensor (HO_2S) electrical connectors.
44. Remove the nuts and detach the dual converter Y-pipe.
45. Remove the shift cable and bracket.
46. Remove the flexplate inspection cover.
47. Remove the cylinder block opening cover.
48. Remove the torque converter nuts.
49. Support the transmission, using a block of wood and a floor jack.
50. Remove the five lower transmission-to-engine bolts.
51. Remove the right side motor mount.

➡If the left side engine mount through bolt nut is missing or damaged install a new nut using service part number W709375. If the left side engine mount through bolt nut cage is damaged or missing, remove the cage and install a new nut using service part number W520516-5301.

52. Remove the left side mount through bolt.

➡On 4x4 vehicles, disconnect the transfer case vent hose from the bracket and position aside to gain access to the upper transmission-to-engine bolts.

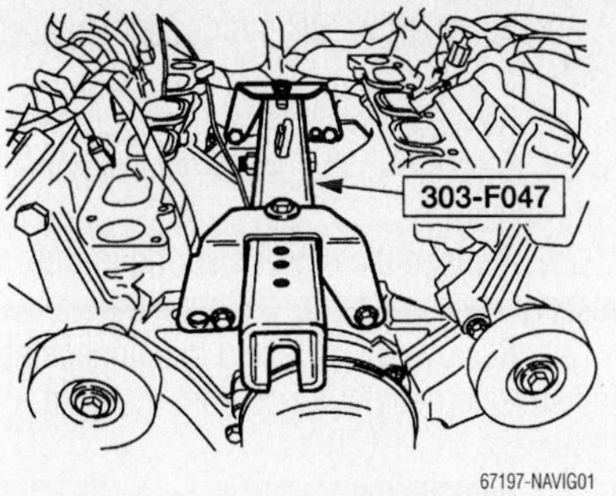

Install modular engine lifting bar 303-F047 tool

53. Remove the two upper transmission-to-engine bolts.

➡On some engines it will be necessary to shim between the modular engine lifting bar 303-F047 tool mounting brackets and the cylinder block.

54. Install modular engine lifting bar 303-F047 tool.
55. Remove the engine from the vehicle.

To install:

56. Install the engine in the vehicle
57. Remove modular engine lifting bar 303-F047 tool.
58. Remove the floor jack and wood block supporting the transmission.
59. Align the transmission to the engine and install the five lower transmission-to-engine bolts. Tighten to 35 ft. lbs. (48 Nm).

➡If the left side engine mount through bolt nut is missing or damaged install a new nut using service part number W709375. If the left side engine mount through bolt nut cage is damaged or missing, remove the cage and install a new nut using service part number W520516-5301.

60. Install the left side motor mount bolt and tighten to 148 ft. lbs. (200 Nm).
61. Install the right side motor mount nut and tighten to 129 ft. lbs. (175 Nm).
62. Install the four torque converter nuts and tighten to 26 ft. lbs. (35 Nm).
63. Install the cylinder block opening cover.
64. Install the flexplate inspection cover and tighten the bolts to 25 ft. lbs. (34 Nm).
65. Install the shift cable and bracket.
66. Install the dual converter Y-pipe and tighten the four nuts to 30 ft. lbs. (40 Nm).
67. Connect the HO_2S electrical connectors.
68. Install the transmission filler tube.
69. Install the starter.
70. Install the wiring harness bracket.
71. Install the A/C compressor and tighten the bolts to 18 ft. lbs. (25 Nm).
72. Connect the CKP sensor electrical connector.
73. Connect the A/C compressor electrical connector.
74. Connect the EGR valve tube and loosely install the lower fitting.
75. Install the lower radiator hose and degas bottle hose.
76. Install a new oil filter.
77. Connect the power steering pressure switch connector, if equipped.
78. Install the power steering pump and tighten the three bolts to 18 ft. lbs. (25 Nm).
79. Using the Teflon seal installer tool 211-D027, install a new O-ring seal on the power steering pressure hose fitting.
80. Connect the power steering pressure hose and tighten the fitting to 15 ft. lbs. (20 Nm).
81. Install the transmission cooler tube and the power steering hose brackets and tighten the nuts to 8 ft. lbs. (11 Nm) on 4.6L engines or 30 ft. lbs. (40 Nm) on 5.4L engines.
82. Install the power steering pump pulley using power steering pump pulley tool 211-185.
83. Install the power steering reservoir.
84. Install the alternator.
85. Install the drive belt.
86. Install the wiring harness retainer on the front cover stud.

➡On 4x4 vehicles, connect the transfer case vent hose to the bracket after installing the upper transmission-to-engine bolts.

87. Install the two upper transmission-to-engine bolts. Tighten to 35 ft. lbs. (48 Nm).
88. Install the heater outlet tube using a new O–ring.
89. Install the lower and upper heater outlet tube studs and tighten to 18 ft. lbs. (25 Nm) on 4.6L engines.

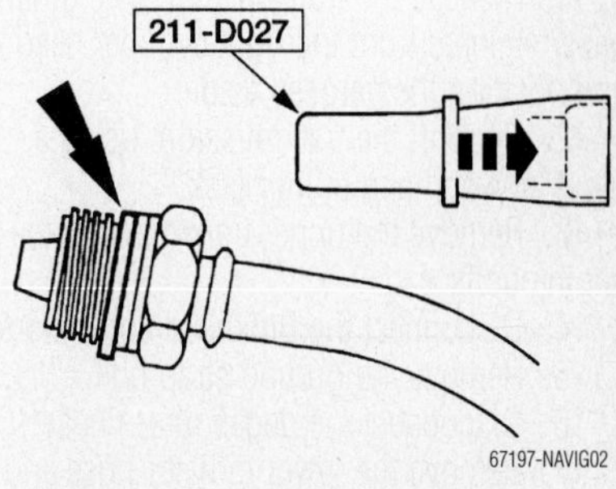

Use Teflon seal installer tool to install a new O-ring seal on the power steering pressure hose fitting—2004–05 models

90. Tighten the lower and upper heater outlet tube studs to 30 ft. lbs. (40 Nm) on 5.4L engines.
91. Install all remaining hoses, tubes and electrical connectors.
92. Install the PCM and attach the electrical connections.
93. Install the condenser.
94. Install the radiator.
95. Connect the MAF sensor electrical connections and install the engine air cleaner.
96. Install the intake manifold.
97. Fill the engine with clean engine oil.
98. Connect the negative battery cable.
99. Fill the engine cooling system.
100. Fill the power steering system..
101. Recharge the A/C system.
102. Install the hood.

5.4L DOHC Engine

2001–02 MODELS

⁂ CAUTION

Fuel injection systems remain under pressure, even after the engine has been turned OFF. The fuel system pressure must be relieved before disconnecting any fuel lines. Failure to do so may result in fire and/or personal injury.

1. Relieve the fuel system pressure.
2. Before servicing the vehicle, refer to the precautions in the beginning of this section.
3. Disconnect both battery cables, negative cable first.
4. Remove the hood.
5. Remove the engine air cleaner outlet tube
6. Remove the radiator.
7. Recover the A/C system refrigerant.
8. Remove the accessory drive belt.
9. Remove the alternator.
10. Remove the Exhaust Gas Recirculation (EGR) valve to exhaust manifold tube fittings and the tube.
11. Remove the transmission wiring harness retainers from the right cylinder head and position the harness aside.
12. Support the transmission, using a block of wood and a floor jack.
13. Remove the upper transmission-to-engine bolts.
14. Disconnect the Bulkhead connectors.
15. Remove the ground strap bolt.
16. Disconnect the degas return hose.
17. Remove the lower radiator hose and heater hoses.
18. Remove the starter-to-relay cable retainer nut.
19. Remove the A/C muffler line and set aside.
20. Remove the skid plate, if equipped.
21. Disconnect the power steering pressure sensor.
22. Remove the power steering pump reservoir and set aside.
23. Remove the nuts and disconnect the transmission cooler tubes from the front cover.
24. Remove the starter.
25. Disconnect the Crankshaft Position (CKP) sensor electrical connector.
26. Remove the bolt and the manifold and tube assembly.
27. Disconnect the A/C compressor electrical connector.
28. Remove the bolts and position the A/C compressor aside.
29. Disconnect the heated Oxygen Sensor (HO_2S) electrical connectors.
30. Disconnect the exhaust pipes from the manifolds.
31. Remove the left side mount through bolt.
32. Remove the right side motor mount bolt.
33. Remove the flexplate inspection cover.
34. Remove the torque converter nuts.
35. Remove the drain plug and drain the engine oil and install the drain plug when finished
36. Remove the oil filter.
37. Remove the five lower transmission-to-engine bolts.
38. Remove the transmission filler tube-to-engine block bolt which is accessed through the ride side fender well.
39. Install modular engine lifting bar 303-F047 tool.
40. Remove the engine from the vehicle.

To install:

41. Install the engine in the vehicle
42. Remove modular engine lifting bar 303-F047 tool.
43. Remove the floor jack and wood block supporting the transmission.
44. Install the transmission filler tube.
45. Align the transmission to the engine and install the six lower transmission-to-engine bolts. Tighten to 33 ft. lbs. (45 Nm).
46. Install a new oil filter.
47. Install the four torque converter nuts and tighten to 26 ft. lbs. (35 Nm).
48. Install the flexplate-to-torque converter nut access plug.
49. Install the inspection plate and tighten the bolts to 18 ft. lbs. (25 Nm)
50. Install the right side motor mount nut and tighten to 59 ft. lbs. (80 Nm).
51. Install the left side motor mount bolt and tighten to 59 ft. lbs. (80 Nm).
52. Install the starter.
53. Install the exhaust pipes to the manifolds. Tighten the nuts to 30 ft. lbs. (41 Nm).
54. Connect the HO_2S electrical connectors.
55. Install the A/C compressor and tighten the bolts to 22 ft. lbs. (30 Nm).
56. Connect the A/C clutch electrical connector.
57. Install the manifold tube assembly and bolt, tighten the bolt to 18 ft. lbs. (24 (Nm).
58. Connect the CKP sensor electrical connector.
59. Install the transmission cooler line brackets and tighten the nuts to 18 ft. lbs. (25 Nm).
60. Install the power steering pump and tighten the three bolts to 18 ft. lbs. (25 Nm).
61. Connect the power steering pressure sensor.
62. Install the skid plate and the bolts, if equipped.
63. Install the A/C muffler line.
64. Position the harness and install the nut.
65. Install the degas return hose.
66. Install the ground strap and bolt.
67. Connect the heater hose.
68. Install the lower radiator hose.
69. Install the bulkhead connectors.
70. Install the two upper transmission-to-engine bolts. Tighten to 33 ft. lbs. (45 Nm).
71. Install the wire harness retainers to the right cylinder head.
72. Connect the EGR valve tube nut to the and hand-tighten.
73. Install the alternator.
74. Install the drive belt.
75. Install the radiator.
76. Install the cooling fan assembly.
77. Install all remaining hoses.
78. Connect the transmission cooling fittings.
79. Install the front air deflector.
80. Install the intake manifold.
81. Tighten the EGR valve tube nut to 18 ft. lbs. (25 Nm).
82. Install the hood.
83. Fill the engine with clean engine oil.
84. Connect the negative battery cable.
85. Fill the engine cooling system.
86. Fill the power steering system..
87. Recharge the A/C system.

2003–05 MODELS

1. Before servicing the vehicle, refer to the precautions in the beginning of this section.
2. Disconnect the negative battery cable.

3. Remove the hood.
4. Remove the radiator.
5. Recover the A/C system refrigerant.
6. Remove the lower intake manifold.
7. Disconnect the powertrain control module (PCM) electrical connectors.
8. Remove the PCM mounting bolts and the PCM.
9. Remove the bolts and the PCM mounting bracket.
10. Disconnect the auxiliary heater hoses at the right side valve cover.
11. Disconnect the heater hoses at the heater core.
12. Disconnect the A/C hoses from the junction block.
13. Remove the safety clips and disconnect the spring lock couplers at the evaporator core.
14. Remove the drive belt.
15. Remove the alternator.
16. Remove the power steering pump pulley using power steering pump pulley tool 211-016.
17. Remove the nuts and disconnect the transmission cooler tube and the power steering hose brackets.
18. Disconnect the power steering pressure hose and drain the power steering fluid into a suitable container.
19. Remove the three bolts and position the power steering pump out of the way.
20. Disconnect the Exhaust Gas Recirculation (EGR) valve tube nut from the left side exhaust manifold and remove the tube.
21. Remove the engine noise shield.
22. Remove the positive crankcase ventilation (PCV) valve and hose assembly from the vehicle.
23. Remove the PVC tube.
24. Remove the bolt and disconnect the ground wire.
25. Disconnect the vacuum hose and remove the vacuum harness.
26. Disconnect the main engine harness electrical connector.
27. Remove the skid plate, if equipped.
28. Remove the transmission filler tube-to-engine block bolt which is accessed through the ride side fender well.
29. Disconnect the heated Oxygen Sensor (HO_2S) electrical connectors.
30. Remove the nuts and detach the dual converter Y-pipe.
31. Remove the starter.
32. Disconnect the A/C clutch electrical connector.
33. Remove the bolts and position the A/C compressor aside.
34. Disconnect the Crankshaft Position (CKP) sensor electrical connector.
35. Remove the oil filter.
36. Remove the drain plug and drain the engine oil and install the drain plug when finished.
37. Remove the lower radiator hose.
38. Remove the inspection plate.
39. Remove the flexplate-to-torque converter nut access plug.

➡Mark one stud and the flexplate for assembly reference.

40. Remove the four torque converter nuts.
41. Remove the six lower transmission-to-engine bolts.

➡The caged nut on the frame may break during bolt removal. If necessary, install a new nut.

42. Remove the bolt from the left side motor mount and the nut from the right side motor mount.
43. Support the transmission, using a block of wood and a floor jack.

➡On 4x4 vehicles, disconnect the transfer case vent hose from the bracket and position aside to gain access to the upper transmission-to-engine bolts.

44. Remove the two upper transmission-to-engine bolts.

➡The alternator bolts may be used to attach the modular engine lifting bar 303-F047 tool to the front of the engine. The two bolts used to attach the tool to the rear of the engine must be shorter than the transmission-to-engine mounting bolts.

45. Install modular engine lifting bar 303-F047 tool.
46. Using a floor crane, remove the engine from the vehicle.

To install:

47. Using a floor crane, install the engine in the vehicle. Make sure the marks on the flexplate and the torque converter stud made during removal are lined up.
48. Remove modular engine lifting bar 303-F047 tool.
49. Remove the floor jack and wood block supporting the transmission.
50. Install the right side motor mount nut and tighten to 66 ft. lbs. (90 Nm).
51. Install the left side motor mount bolt and tighten to 148 ft. lbs. (200 Nm).
52. Align the transmission to the engine and install the six lower transmission-to-engine bolts. Tighten to 33 ft. lbs. (45 Nm).
53. Install the four torque converter nuts and tighten to 26 ft. lbs. (35 Nm).
54. Install the flexplate-to-torque converter nut access plug.
55. Install the inspection plate and tighten the bolts to 25 ft. lbs. (34 Nm)
56. Install the lower radiator hose.
57. Install a new oil filter.
58. Connect the CKP sensor electrical connector.
59. Install the A/C compressor and tighten the bolts to 18 ft. lbs. (25 Nm).
60. Connect the A/C clutch electrical connector.
61. Install the starter.
62. Install the wire harness bracket and tighten the nut.
63. Install the dual converter Y-pipe and tighten the four nuts to 30 ft. lbs. (40 Nm).
64. Connect the HO_2S electrical connectors.
65. Install the transmission filler tube.
66. Install the skid plate and the bolts, if equipped.

➡On 4x4 vehicles, connect the transfer case vent hose to the bracket after installing the upper transmission-to-engine bolts.

67. Install the two upper transmission-to-engine bolts. Tighten to 33 ft. lbs. (45 Nm).
68. Connect the auxiliary heater hoses.
69. Connect the main engine harness electrical connector.
70. Connect the vacuum harness and the hose.
71. Connect the ground cable and tighten the bolt.
72. Install a new O-ring. Lubricate the O-ring with clean engine coolant. Install the PVC tube and tighten the bolt.
73. Attach the PVC hoses and install the clamps.
74. Install the engine noise shield.
75. Connect the EGR valve tube nut to the left side exhaust manifold and hand-tighten.
76. Install the power steering pump and tighten the three bolts to 18 ft. lbs. (25 Nm).
77. Using the Teflon seal installer tool 211-D027, install a new O-ring seal on the power steering pressure hose fitting.
78. Connect the power steering pressure hose and tighten the fitting to 15 ft. lbs. (20 Nm).
79. Install the transmission cooler tube and the power steering hose brackets and tighten the nuts to 18 ft. lbs. (25 Nm).
80. Install the power steering pump pulley using power steering pump pulley tool 211-185.
81. Install the alternator.
82. Install the drive belt.

83. Install all remaining hoses.
84. Install the PCM and attach the electrical connections.
85. Install the radiator and the radiator upper support brackets.
86. Install the degas hose to the radiator.
87. Install the front air deflector.
88. Install the splash shield.
89. Install the cooling fan assembly.
90. Install the lower intake manifold.
91. Tighten the EGR valve tube nut to 37 ft. lbs. (50 Nm).
92. Fill the engine with clean engine oil.
93. Connect the negative battery cable.
94. Fill the engine cooling system.
95. Fill the power steering system..
96. Recharge the A/C system.
97. Install the hood.

Water Pump

REMOVAL & INSTALLATION

1. Before servicing the vehicle, refer to the precautions in the beginning of this section.
2. Remove or disconnect the following:
   - Negative battery cable
   - Radiator, fan blade assembly and fan shroud
   - Accessory drive belt
   - Water pump pulley
   - Heater hose from the water pump
   - Water pump bolts and nuts. Note the locations of the bolts if different lengths.
   - Water pump stud bolt, the water pump and the water pump housing gasket. Discard the water pump housing gasket.

To install:

3. Before installing the water pump, be sure to completely clean the water pump mounting surfaces of all dirt, grime and old gasket material.

➡**All water pump housing bolts, nuts and studs are tightened to 15–22 ft. lbs. (20–30 Nm).**

4. Install or connect the following:
   - Water pump onto the engine with a new gasket. Install the water pump stud bolt temporarily finger-tight.
   - Water pump mounting nuts and bolts temporarily finger-tight, then tighten all water pump housing fasteners to 15–22 ft. lbs. (20–30 Nm).
   - Water outlet tube for the heater, if equipped
   - Water pump pulley and accessory drive belt
   - Fan shroud, fan blade assembly and the radiator
   - Coolant
   - Negative battery cable
5. Start the engine and check for any fluid leaks.
6. If necessary, bleed the cooling system.

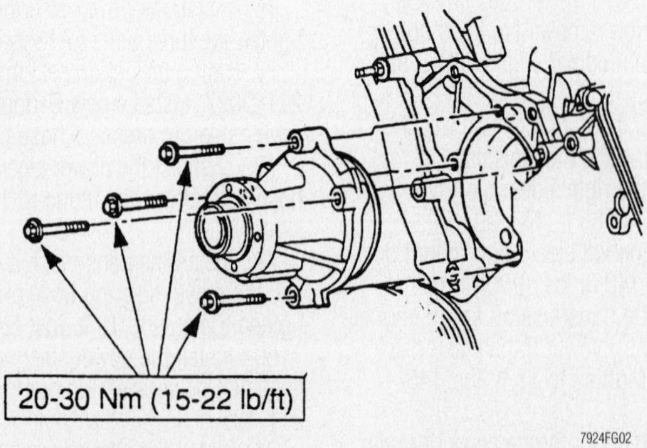

Exploded view of the water pump mounting—4.6L, 5.4L and 6.8L engines

Heater Core

REMOVAL & INSTALLATION

Ford Expedition

1. Before servicing the vehicle, refer to the precautions in the beginning of this section.
2. Disconnect the negative battery cable.

✲✲ CAUTION

After disconnecting the negative battery cable, wait 1 minute for the SRS module to deplete its energy.

3. Drain the cooling system into a clean container for reuse.
4. Remove the instrument panel by performing the following procedure:
   a. If equipped, remove the floor console assembly.
   b. Remove the lower steering column cover bolts and the cover.
   c. Remove both front door scuff plates.
   d. Remove both side cowl trim panels.
   e. Disconnect the electrical connector from the Brake Pedal Position (BPP) switch.
   f. Remove the radio ground and the GEM/CTM ground bolts.
   g. Disconnect the left side instrument panel main wiring harness connector.
   h. In the engine compartment, remove the bulkhead wiring harness connector bolts and disconnect the wiring connectors.
   i. In the driver's compartment, release the 6 locking tabs and remove the bulkhead electrical connector from the instrument panel.
   j. Disconnect the air bag diagnostic monitor electrical connector.
   k. Disconnect the inertia fuel shutoff switch electrical connector.
   l. Remove the right side ground bolts.
   m. Disconnect the right side instrument panel wiring harness connectors.
   n. Disconnect the electronic blend door actuator electrical connector.
   o. Disconnect the climate control head vacuum harness connector.
   p. Remove the steering column opening cover reinforcement nuts and the cover reinforcement.
   q. At the base of the steering column, disconnect the air bag sliding contact and the anti-theft sensor electrical connectors.
   r. At the steering column, disconnect the remaining electrical connectors.
   s. If equipped with a transmission range indicator, remove the bolt and disconnect the cable.
   t. Remove the steering column-to-instrument panel nuts and lower the steering column.
   u. Remove the right side front fender splash shield screws and move the shield away from the panel.
   v. Disconnect the antenna cable from the antenna base.
   w. Remove the instrument panel relay cover and disconnect the autolamp sensor electrical connector and/or the sunload sensor connector.
   x. Remove the glove box.

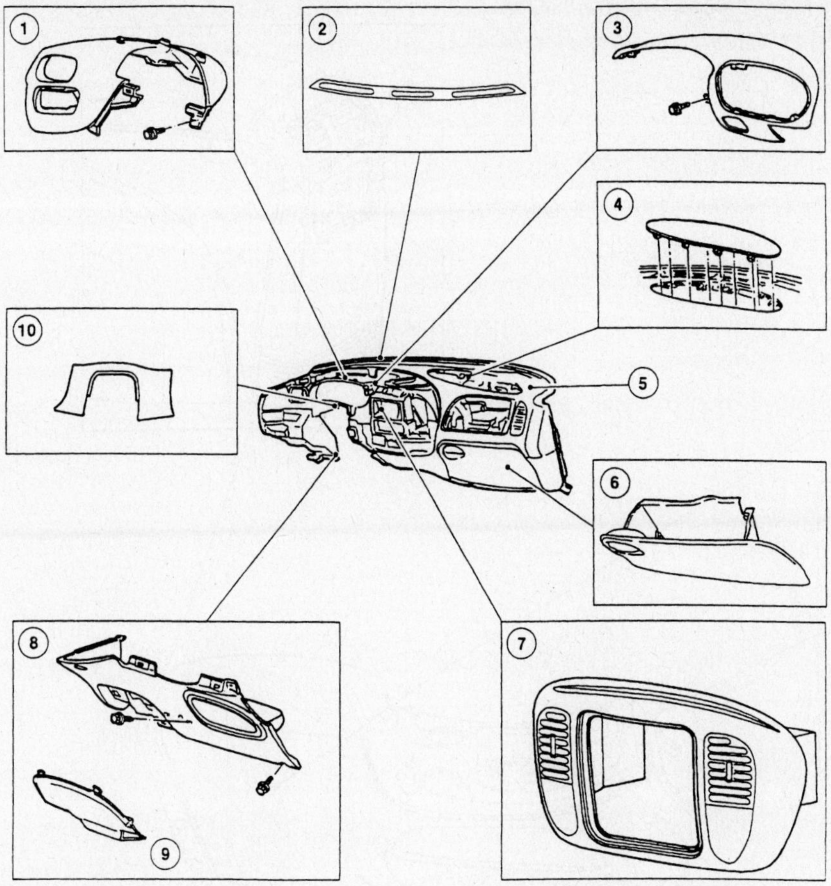

1. Instrument panel finish panel
2. Instrument panel defroster opening grille assembly
3. Instrument cluster panel
4. Instrument panel relay cover
5. Instrument panel
6. Glove compartment
7. Center instrument panel finish panel
8. Instrument panel steering column cover
9. Instrument panel fuse door
10. Steering column opening cover

93113GM2

Exploded view of the instrument panel components—Ford Expedition

y. At the passenger's air bag module, remove the screws, disconnect the electrical connector and remove the air bag module.

Place the air bag module in a safe place with the front facing upward.

a. Remove the right side assist handle screw covers, the screws and the handle.

b. At both doors, pull back the weatherstrip seals and remove the windshield garnish moldings.

c. Remove the instrument panel reinforcement bolt below the left side corner of the glove box.

d. Through the air bag module opening, remove the instrument panel bolts.

e. Remove the upper instrument panel cowl covers and bolts.

f. At the relay bracket, remove the instrument panel bolt.

g. At the lower left side of the cigar lighter, remove the instrument panel bolt.

h. At the both sides, remove the instrument panel-to-cowl side nuts.

i. At the steering column opening, remove the instrument panel bolts.

j. Remove the upper instrument panel floor brace bolt.

k. Using an assistant, remove the instrument panel.

5. If equipped with the 5.4L 4V engine, remove the junction block splash shield.

6. If equipped with the 5.4L 4V engine, remove the bolts and disconnect the cable ends from the starter relay.

7. If equipped with the 5.4L 4V engine, remove the junction block bracket.

8. Compress the holding tabs and disconnect the heater hoses from the heater core.

9. Remove the air conditioning plenum screw and the air conditioning plenum demister adapter.

10. Disconnect the vacuum line.

11. Remove the heater core bracket screws and the bracket.

12. Remove the 13 heater housing plenum camber cover screws and the heater housing plenum chamber cover.

13. Remove the blend door assembly from the heater housing.

14. Remove the heater core.

To install:

15. Install the heater core.

16. Install the blend door assembly to the heater housing.

17. Install the 13 heater housing plenum camber cover and the heater housing plenum chamber cover screws.

18. Install the heater core bracket and the bracket screws.

19. Connect the vacuum line.

20. Install the air conditioning plenum demister adapter and the air conditioning plenum screw.

21. Connect the heater hoses to the heater core.

22. Install the instrument panel by performing the following procedure:

a. Using an assistant, install the instrument panel.

b. Install the upper instrument panel floor brace bolt.

c. At the steering column opening, install the instrument panel bolts.

d. At the both sides, install the instrument panel-to-cowl side nuts.

e. At the lower left side of the cigar lighter, install the instrument panel bolt.

f. At the relay bracket, install the instrument panel bolt.

g. Install the upper instrument panel cowl bolts and covers.

h. Through the air bag module opening, install the instrument panel bolts.

i. Install the instrument panel reinforcement bolt below the left side corner of the glove box.

j. At both doors, install the windshield garnish moldings and the weatherstrip seals.

k. Install the right side assist handle, the screws and the handle screw covers.

l. At the passenger's air bag module, install the air bag module, connect the electrical connector and install the air bag module screws.

m. Install the glove box.

n. Connect the autolamp sensor electrical connector and/or the sunload sensor connector; then, install the instrument panel relay cover.

o. Connect the antenna cable to the antenna base.

p. Install the right side front fender splash shield and screws.

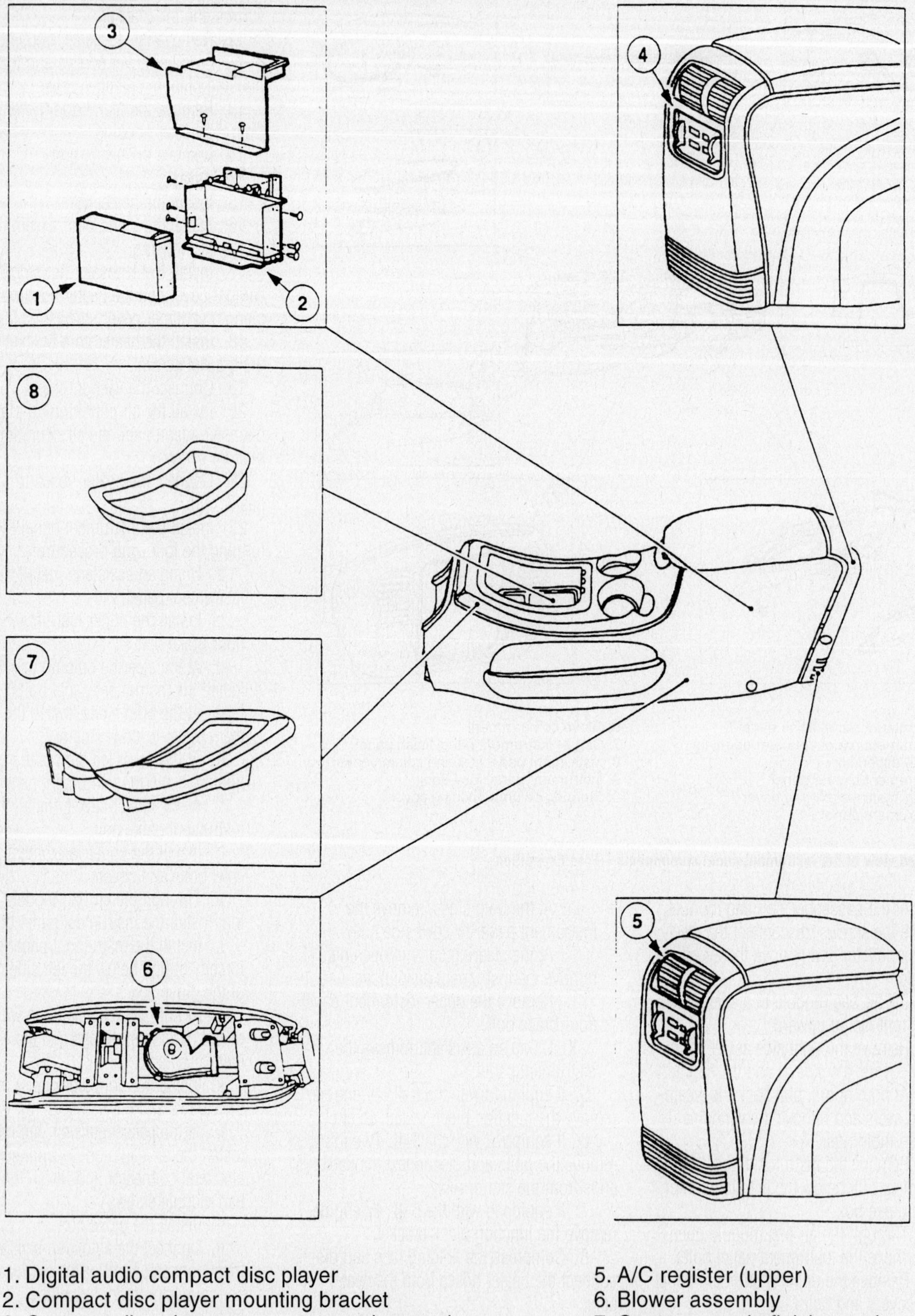

1. Digital audio compact disc player
2. Compact disc player mounting bracket
3. Compact disc player compartment trim panel
4. Radio and A/C integral control assembly
5. A/C register (upper)
6. Blower assembly
7. Center console finish panel
8. Console finish panel mat

93113GM3

Exploded view of the floor console components—Ford Expedition

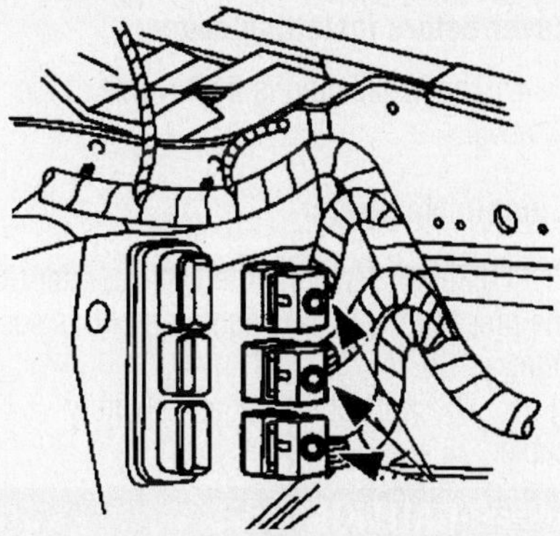

91190G40

Remove the bulkhead electrical connectors from inside the engine compartment—Ford Expedition

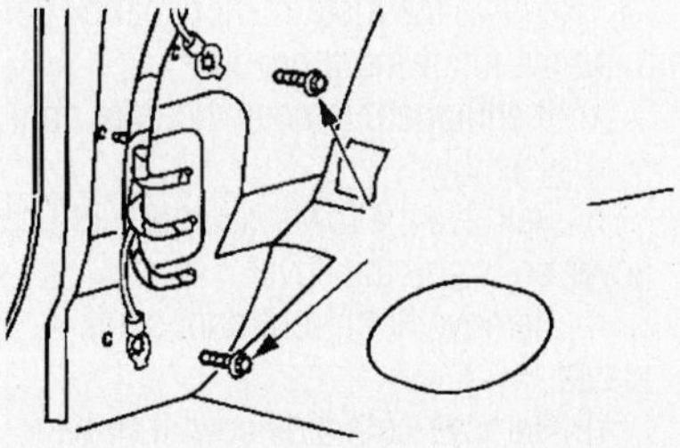

91190G41

Remove the audio unit ground and the GEM/CTM ground bolts—Ford Expedition

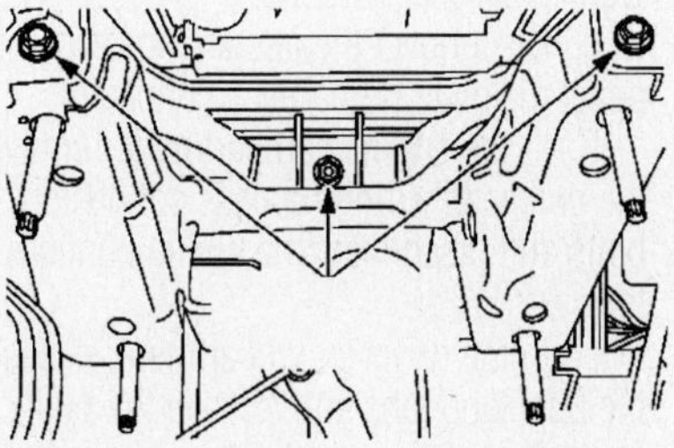

91190G42

Remove the instrument panel bolts through the steering column opening—Ford Expedition

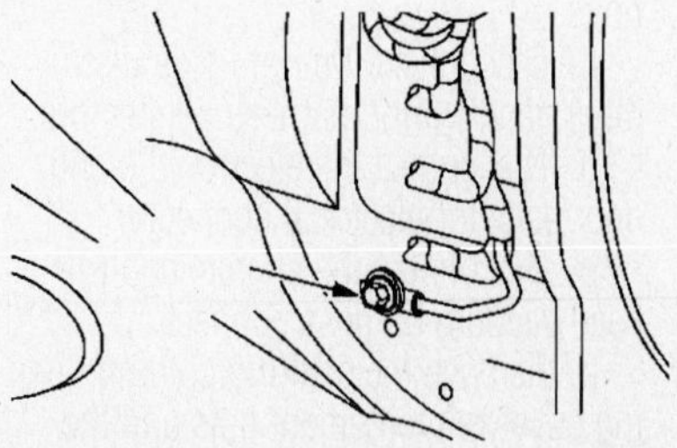

91190G43

Remove the passenger side ground bolt—Ford Expedition

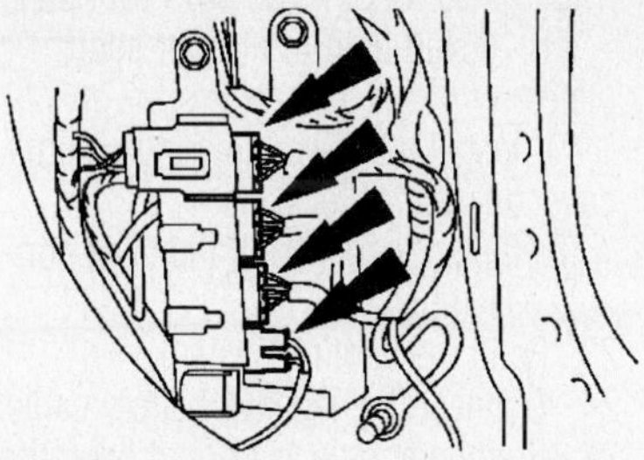

91190G44

Detach the passenger side instrument panel main harness connectors—Ford Expedition

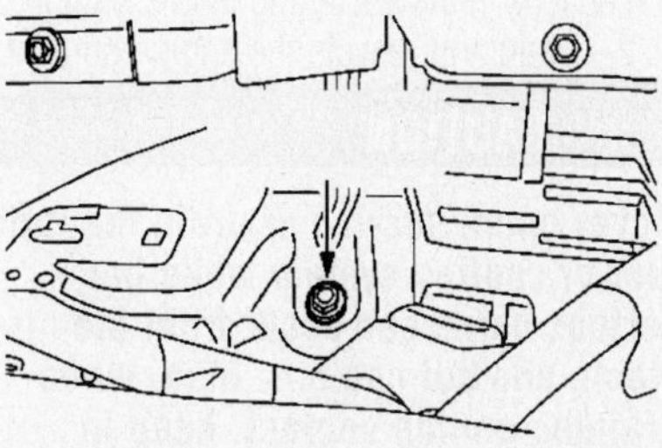

91190G45

Remove the instrument panel bolt on the relay bracket—Ford Expedition

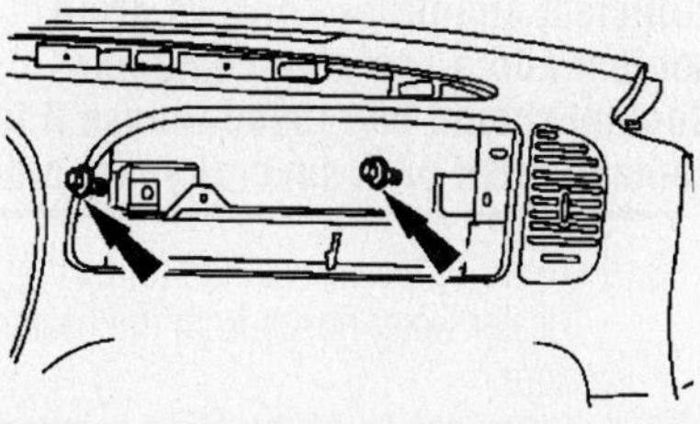

91190G46

Remove the instrument panel bolts through the passenger side air bag module opening—Ford Expedition

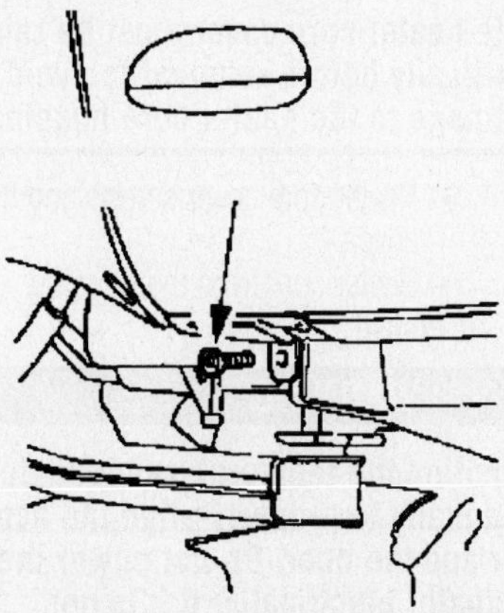

91190G47

Remove the instrument panel reinforcement bolt below the driver's side corner of the glove compartment—Ford Expedition

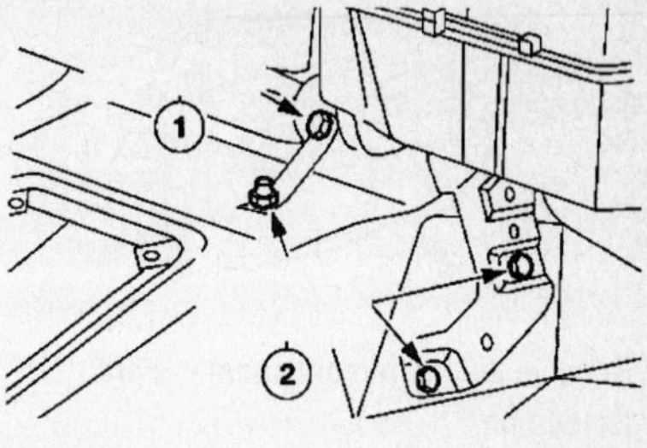

91190G48

Position the carpet aside and loosen the instrument panel floor brace—Ford Expedition

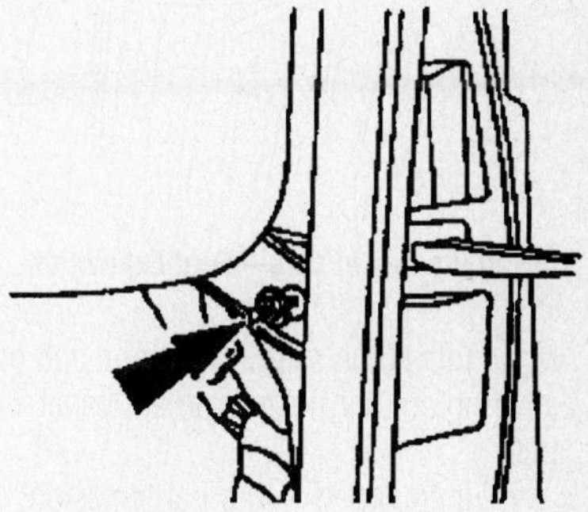

91190G49

Remove the passenger side instrument panel cowl side nut—Ford Expedition

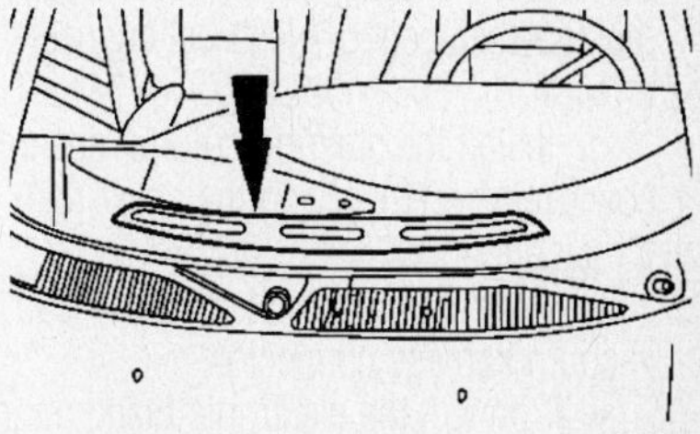

91190G50

On Navigator, remove the defroster grille assembly—Ford Expedition

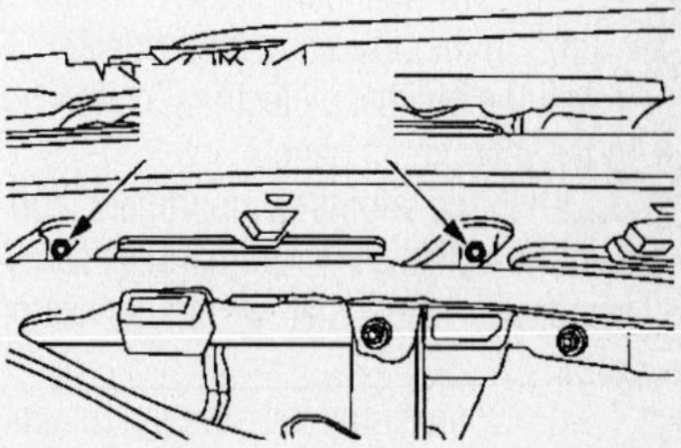

91190G51

Remove the cowl panel mounting bolts—Ford Expedition

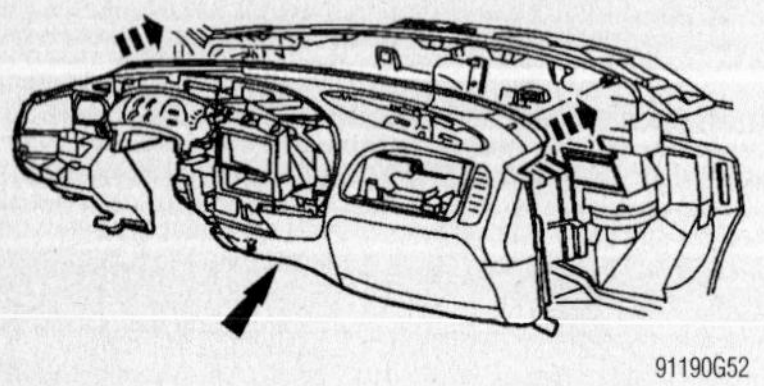

Remove the instrument panel —Ford Expedition

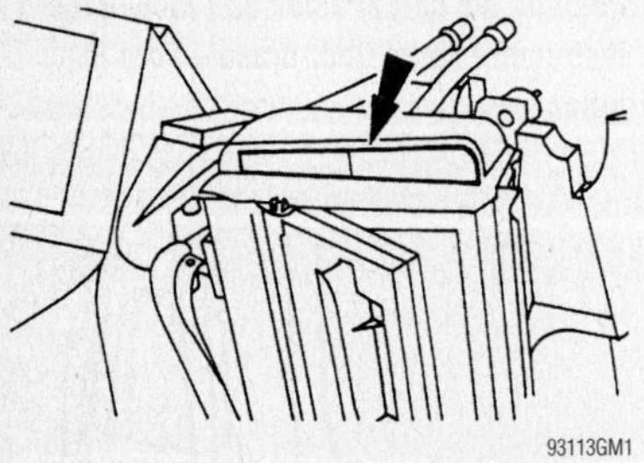

View of the heater core—Ford Expedition

q. Install the steering column and the steering column-to-instrument panel nuts.

r. If equipped with a transmission range indicator, connect the cable and install the bolt.

s. At the steering column, connect the remaining electrical connectors.

t. At the base of the steering column, connect the air bag sliding contact and the anti-theft sensor electrical connectors.

u. Install the steering column opening cover reinforcement and the cover reinforcement nuts.

v. Connect the climate control head vacuum harness connector.

w. Connect the electronic blend door actuator electrical connector.

x. Connect the right side instrument panel wiring harness connectors.

y. Install the right side ground bolts.

z. Connect the inertia fuel shutoff switch electrical connector.

aa. Connect the air bag diagnostic monitor electrical connector.

bb. In the driver's compartment, install the bulkhead electrical connector to the instrument panel.

cc. In the engine compartment, connect the bulkhead wiring harness connectors and the install wiring connector bolts.

dd. Connect the left side instrument panel main wiring harness connector.

ee. Install the radio ground and the GEM/CTM ground bolts.

ff. Connect the electrical connector to the Brake Pedal Position (BPP) switch.

gg. Install both side cowl trim panels.

hh. Install both front door scuff plates.

ii. Install the lower steering column cover and the cover bolts.

jj. If equipped, install the floor console assembly.

23. Refill the cooling system.

24. Connect the negative battery cable.

25. Run the engine to normal operating temperatures; then, check the climate control operation and check for leaks.

Ford Excursion

1. Before servicing the vehicle, refer to the precautions in the beginning of this section.

2. Drain and recycle the engine coolant.

⁂ CAUTION

Never open, service or drain the radiator or cooling system when hot; serious burns can occur from the steam and hot coolant. Also, when draining engine coolant, keep in mind that cats and dogs are attracted to ethylene glycol antifreeze and could drink any that is left in an uncovered container or in puddles on the ground. This will prove fatal in sufficient quantities. Always drain coolant into a sealable container. Coolant should be reused unless it is contaminated or is several years old.

3. Remove or disconnect the following:
   - Heater water hoses from the heater core.
   - Stops and lower the glove compartment door
   - Electronic blend door actuator and bracket assembly

⁂ WARNING

The heater core cover must be raised vertically before removal to avoid damage to the heater core housing.

   - Heater core cover screws and the cover
   - Heater core from the housing

To install:

⁂ WARNING

Position the temperature blend door manually to properly align the actuator and the door. Do not power the actuator electrically. If it is not engaged with the temperature blend door, damage to the actuator may occur.

➡Add gasket between housing and cover before installing cover.

4. The installation is the reverse of the removal.

Lincoln Navigator

1. Before servicing the vehicle, refer to the precautions in the beginning of this section.

2. Disconnect the negative battery cable.

⁂ CAUTION

After disconnecting the negative battery cable, wait 1 minute for the SRS module to deplete its energy.

3. Drain the cooling system into a clean container for reuse.

4. Remove the instrument panel by performing the following procedure:

a. If equipped, remove the floor console assembly.

b. Remove the lower steering column cover bolts and the cover.

c. Remove both front door scuff plates.

d. Remove both side cowl trim panels.

e. Disconnect the electrical connector from the Brake Pedal Position (BPP) switch.

f. Remove the radio ground and the GEM/CTM ground bolts.

g. Disconnect the left side instrument panel main wiring harness connector.

h. In the engine compartment, remove the bulkhead wiring harness connector bolts and disconnect the wiring connectors.

i. In the driver's compartment, release the 6 locking tabs and remove the bulkhead electrical connector from the instrument panel.

j. Disconnect the air bag diagnostic monitor electrical connector.

k. Disconnect the inertia fuel shutoff switch electrical connector.

l. Remove the right side ground bolts.

m. Disconnect the right side instrument panel wiring harness connectors.

n. Disconnect the electronic blend door actuator electrical connector.

o. Disconnect the climate control head vacuum harness connector.

p. Remove the steering column opening cover reinforcement nuts and the cover reinforcement.

q. At the base of the steering column, disconnect the air bag sliding contact

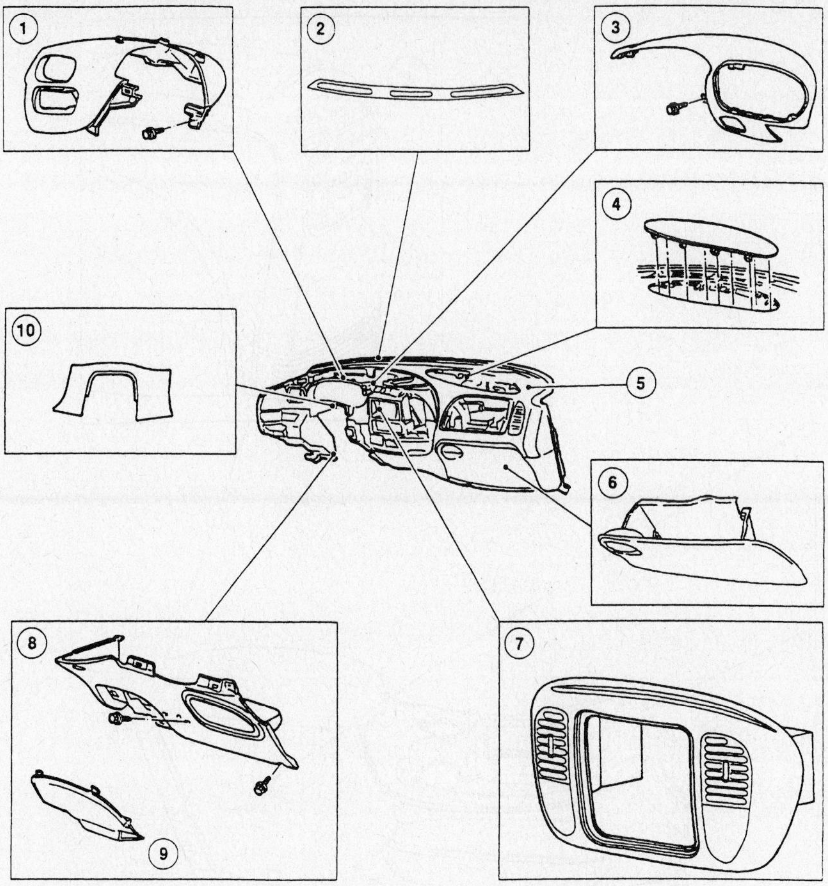

1. Instrument panel finish panel
2. Instrument panel defroster opening grille assembly
3. Instrument cluster panel
4. Instrument panel relay cover
5. Instrument panel
6. Glove compartment
7. Center instrument panel finish panel
8. Instrument panel steering column cover
9. Instrument panel fuse door
10. Steering column opening cover

93113GM2

Exploded view of the instrument panel components—Lincoln Navigator

and the anti-theft sensor electrical connectors.

r. At the steering column, disconnect the remaining electrical connectors.

s. If equipped with a transmission range indicator, remove the bolt and disconnect the cable.

t. Remove the steering column-to-instrument panel nuts and lower the steering column.

u. Remove the right side front fender splash shield screws and move the shield away from the panel.

v. Disconnect the antenna cable from the antenna base.

w. Remove the instrument panel relay cover and disconnect the autolamp sensor electrical connector and/or the sunload sensor connector.

x. Remove the glove box.

y. At the passenger's air bag module, remove the screws, disconnect the electrical connector and remove the air bag module.

Place the air bag module in a safe place with the front facing upward.

a. Remove the right side assist handle screw covers, the screws and the handle.

b. At both doors, pull back the weatherstrip seals and remove the windshield garnish moldings.

c. Remove the instrument panel reinforcement bolt below the left side corner of the glove box.

d. Through the air bag module opening, remove the instrument panel bolts.

e. Remove the instrument panel defroster grille assembly and the instrument panel cowl top bolts.

f. At the relay bracket, remove the instrument panel bolt.

g. At the lower left side of the cigar lighter, remove the instrument panel bolt.

h. At the both sides, remove the instrument panel-to-cowl side nuts.

i. At the steering column opening, remove the instrument panel bolts.

j. Remove the upper instrument panel floor brace bolt.

k. Using an assistant, remove the instrument panel.

5. If equipped with the 5.4L 4V engine, remove the junction block splash shield.
6. If equipped with the 5.4L 4V engine, remove the bolts and disconnect the cable ends from the starter relay.
7. If equipped with the 5.4L 4V engine, remove the junction block bracket.
8. Compress the holding tabs and disconnect the heater hoses from the heater core.
9. Remove the air conditioning plenum screw and the air conditioning plenum demister adapter.
10. Disconnect the vacuum line.
11. Remove the heater core bracket screws and the bracket.
12. Remove the 13 heater housing plenum camber cover screws and the heater housing plenum chamber cover.
13. Remove the blend door assembly from the heater housing.
14. Remove the heater core.

To install:

15. Install the heater core.
16. Install the blend door assembly to the heater housing.
17. Install the 13 heater housing plenum camber cover and the heater housing plenum chamber cover screws.
18. Install the heater core bracket and the bracket screws.
19. Connect the vacuum line.
20. Install the air conditioning plenum demister adapter and the air conditioning plenum screw.
21. Connect the heater hoses to the heater core.
22. Install the instrument panel by performing the following procedure:

a. Using an assistant, install the instrument panel.

b. Install the upper instrument panel floor brace bolt.

c. At the steering column opening, install the instrument panel bolts.

d. At the both sides, install the instrument panel-to-cowl side nuts.

e. At the lower left side of the cigar lighter, install the instrument panel bolt.

f. At the relay bracket, install the instrument panel bolt.

g. Install the instrument panel cowl top bolts and the instrument panel defroster grille assembly.

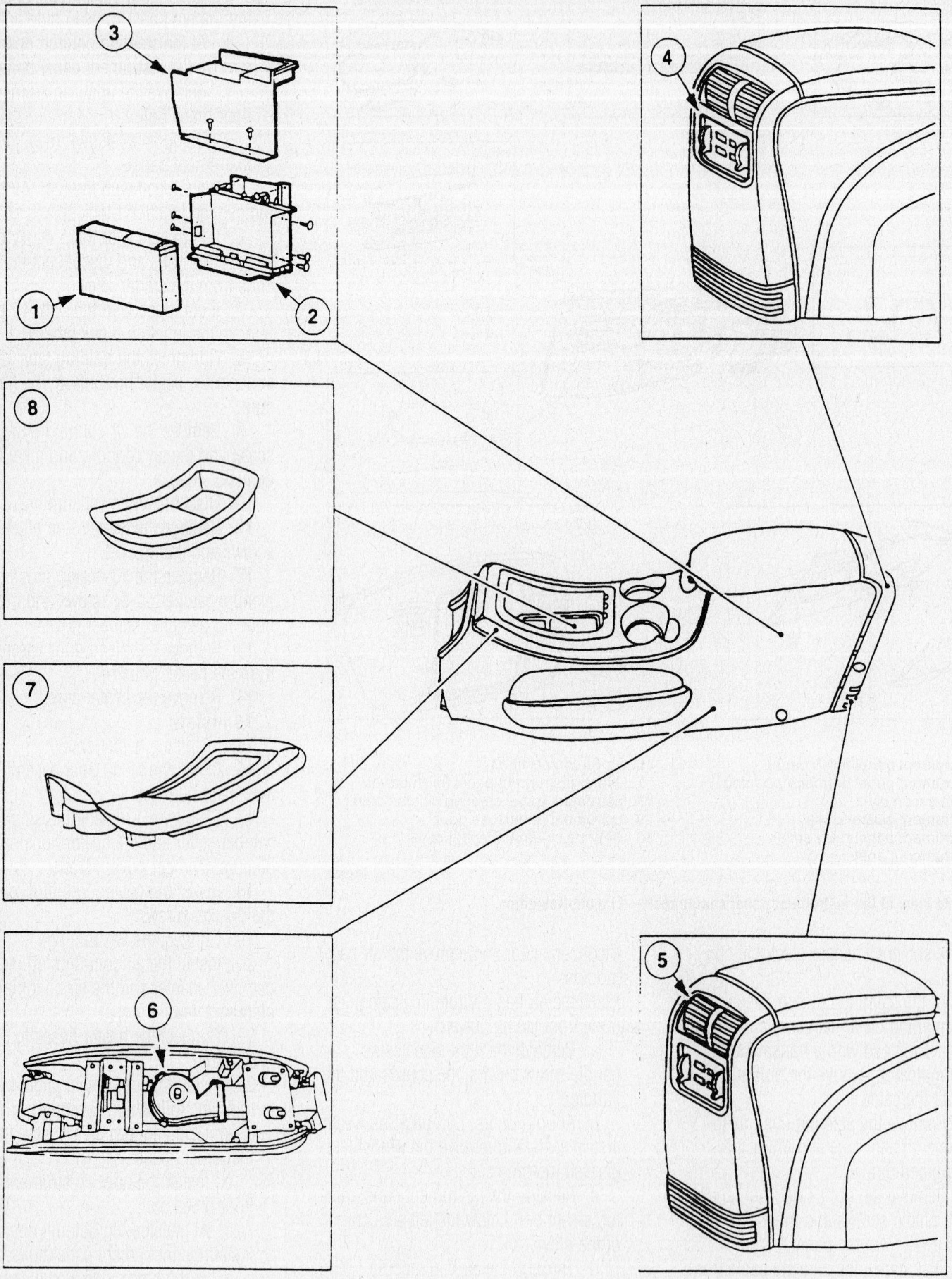

1. Digital audio compact disc player
2. Compact disc player mounting bracket
3. Compact disc player compartment trim panel
4. Radio and A/C integral control assembly
5. A/C register (upper)
6. Blower assembly
7. Center console finish panel
8. Console finish panel mat

93113GM3

Exploded view of the floor console components—Lincoln Navigator

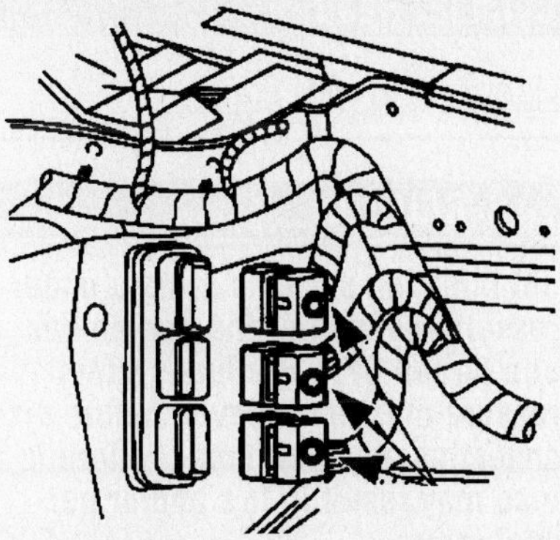

91190G40

Remove the bulkhead electrical connectors from inside the engine compartment—Lincoln Navigator

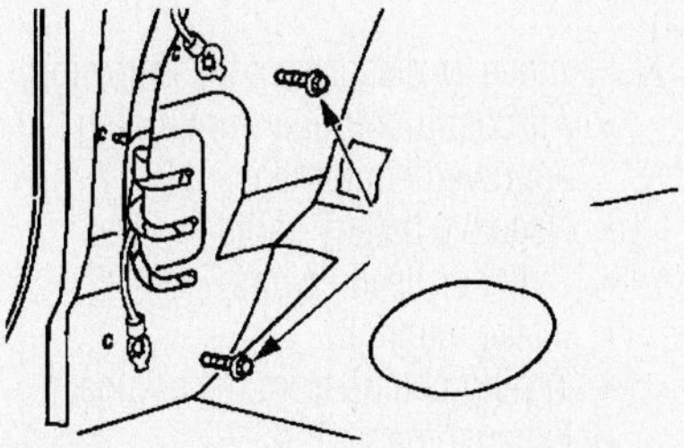

91190G41

Remove the audio unit ground and the GEM/CTM ground bolts—Lincoln Navigator

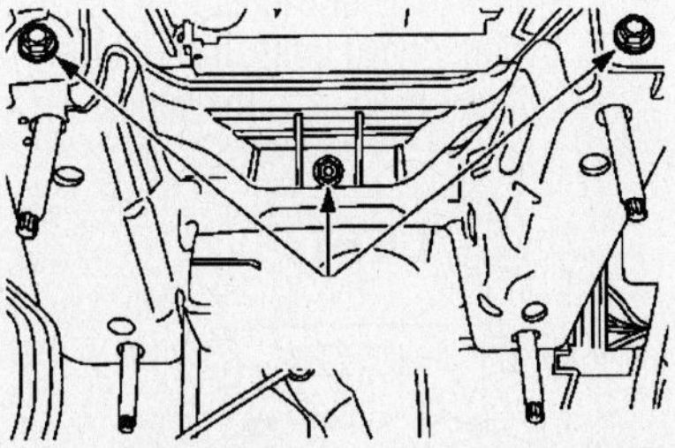

91190G42

Remove the instrument panel bolts through the steering column opening—Lincoln Navigator

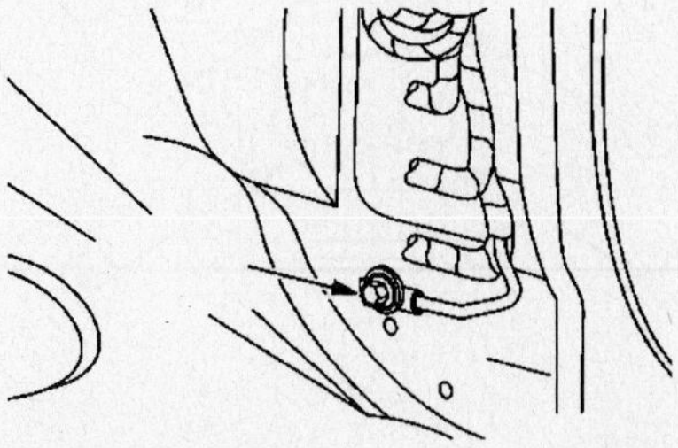

91190G43

Remove the passenger side ground bolt—Lincoln Navigator

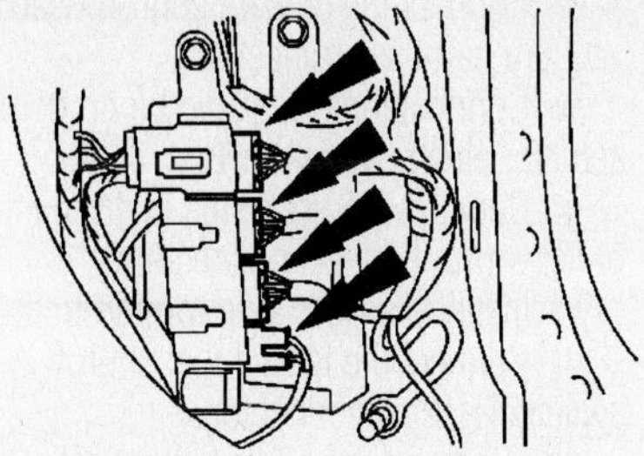

91190G44

Detach the passenger side instrument panel main harness connectors—Lincoln Navigator

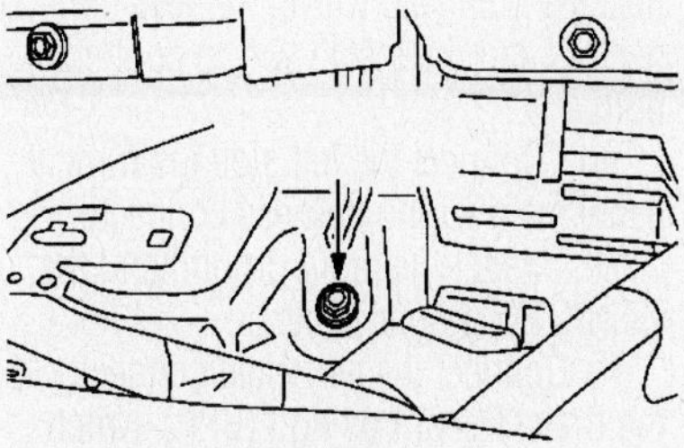

91190G45

Remove the instrument panel bolt on the relay bracket—Lincoln Navigator

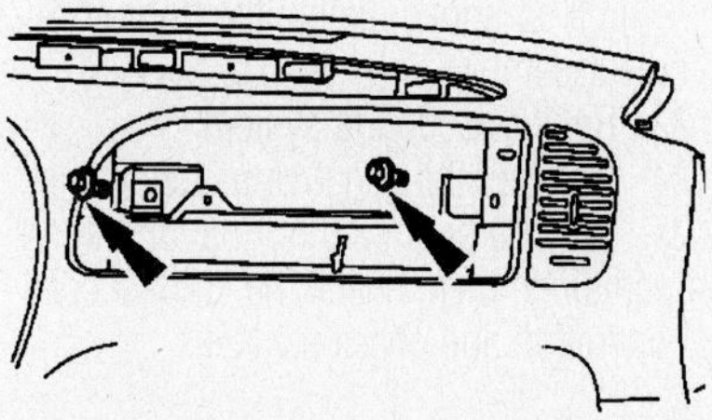

91190G46

Remove the instrument panel bolts through the passenger side air bag module opening—Lincoln Navigator

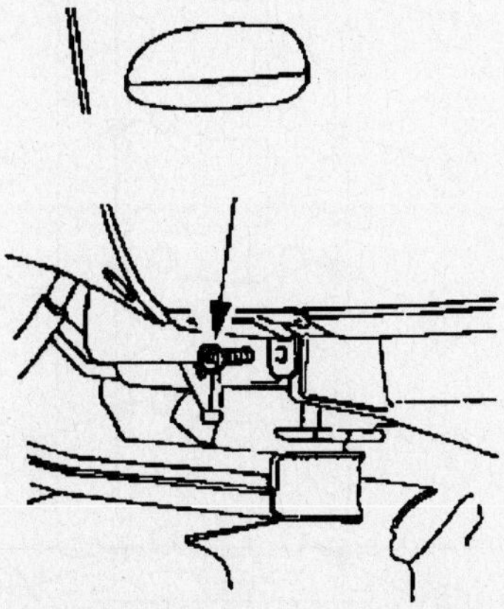

91190G47

Remove the instrument panel reinforcement bolt below the driver's side corner of the glove compartment—Lincoln Navigator

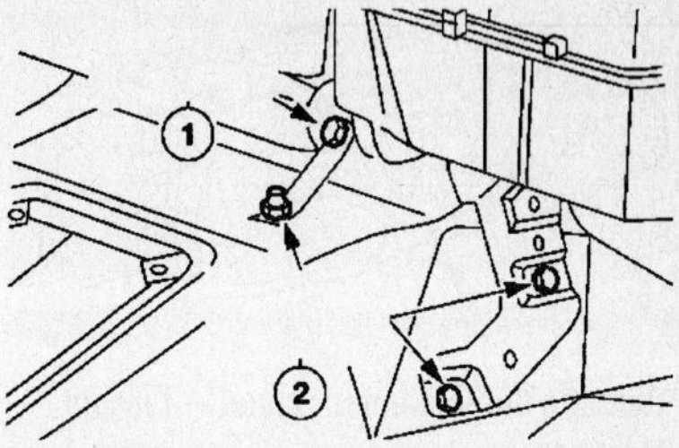

91190G48

Position the carpet aside and loosen the instrument panel floor brace—Lincoln Navigator

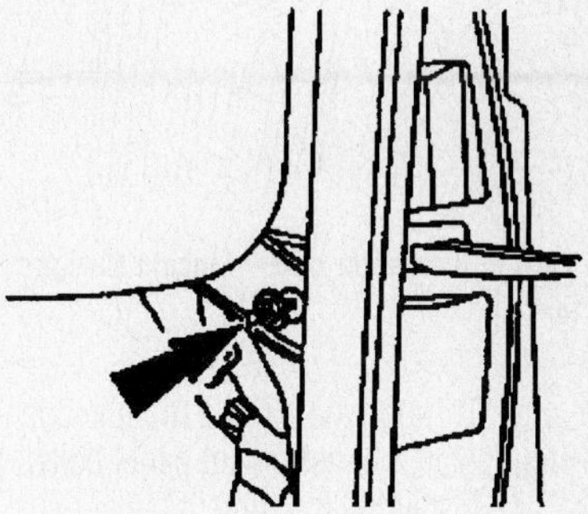

91190G49

Remove the passenger side instrument panel cowl side nut—Lincoln Navigator

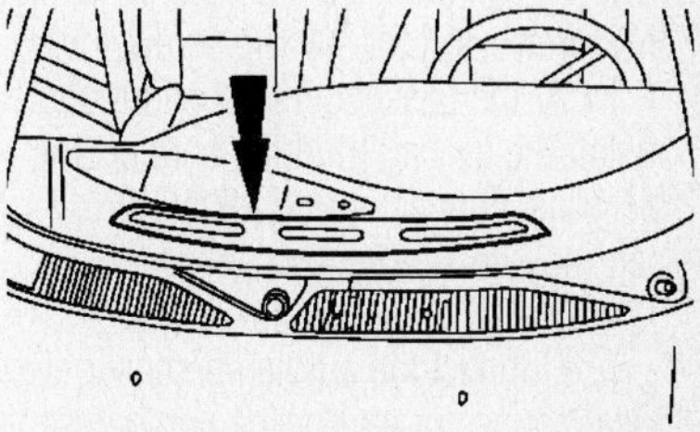

91190G50

On Navigator, remove the defroster grille assembly—Lincoln Navigator

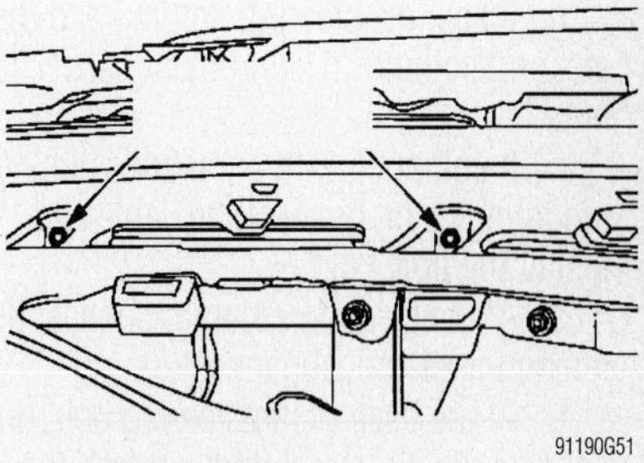

91190G51

Remove the cowl panel mounting bolts—Lincoln Navigator

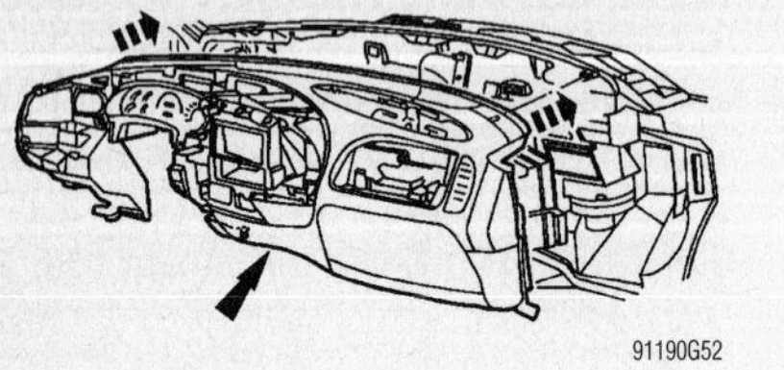

Remove the instrument panel —Lincoln Navigator

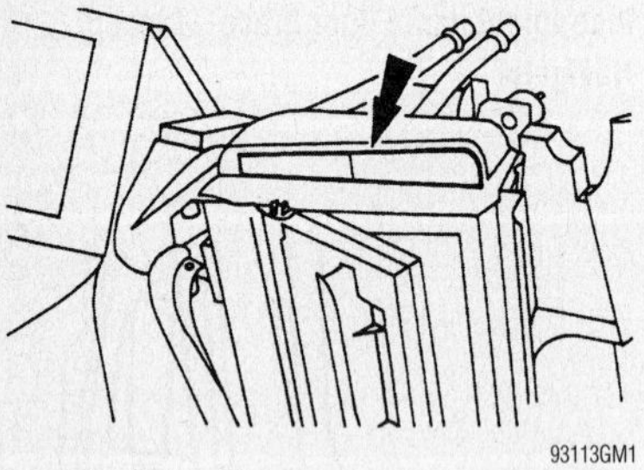

View of the heater core—Lincoln Navigator

h. Through the air bag module opening, install the instrument panel bolts.
i. Install the instrument panel reinforcement bolt below the left side corner of the glove box.
j. At both doors, install the windshield garnish moldings and the weatherstrip seals.
k. Install the right side assist handle, the screws and the handle screw covers.
l. At the passenger's air bag module, install the air bag module, connect the electrical connector and install the air bag module screws.
m. Install the glove box.
n. Connect the autolamp sensor electrical connector and/or the sunload sensor connector; then, install the instrument panel relay cover.
o. Connect the antenna cable to the antenna base.
p. Install the right side front fender splash shield and screws.
q. Install the steering column and the steering column-to-instrument panel nuts.
r. If equipped with a transmission range indicator, connect the cable and install the bolt.
s. At the steering column, connect the remaining electrical connectors.
t. At the base of the steering column, connect the air bag sliding contact and the anti-theft sensor electrical connectors.
u. Install the steering column opening cover reinforcement and the cover reinforcement nuts.
v. Connect the climate control head vacuum harness connector.
w. Connect the electronic blend door actuator electrical connector.
x. Connect the right side instrument panel wiring harness connectors.
y. Install the right side ground bolts.
z. Connect the inertia fuel shutoff switch electrical connector.
aa. Connect the air bag diagnostic monitor electrical connector.
bb. In the driver's compartment, install the bulkhead electrical connector to the instrument panel.
cc. In the engine compartment, connect the bulkhead wiring harness connectors and the install wiring connector bolts.
dd. Connect the left side instrument panel main wiring harness connector.
ee. Install the radio ground and the GEM/CTM ground bolts.
ff. Connect the electrical connector to the Brake Pedal Position (BPP) switch.
gg. Install both side cowl trim panels.
hh. Install both front door scuff plates.
ii. Install the lower steering column cover and the cover bolts.
jj. If equipped, install the floor console assembly.

23. Refill the cooling system.
24. Connect the negative battery cable.
25. Run the engine to normal operating temperatures; then, check the climate control operation and check for leaks.

Cylinder Head

REMOVAL & INSTALLATION

**** CAUTION**

Fuel injection systems remain under pressure, even after the engine has been turned OFF. The fuel system pressure must be relieved before disconnecting any fuel lines. Failure to do so may result in fire and/or personal injury.

➡To correctly tighten the cylinder head bolts, an angle torque wrench is needed.

1. Before servicing the vehicle, refer to the precautions in the beginning of this section.
2. Remove or disconnect the following:
   - Air conditioning system, using approved equipment
   - Negative battery cable
   - Cylinder head covers
   - Intake manifold
   - Timing chains from the engine
   - Exhaust manifolds
   - The 2 heater hose retaining bolts, then compress and slide the hose clamp back to remove the heater water hose.
   - Any remaining hoses, electrical connections or cables
   - Cylinder head bolts, then lift the cylinder head from the engine

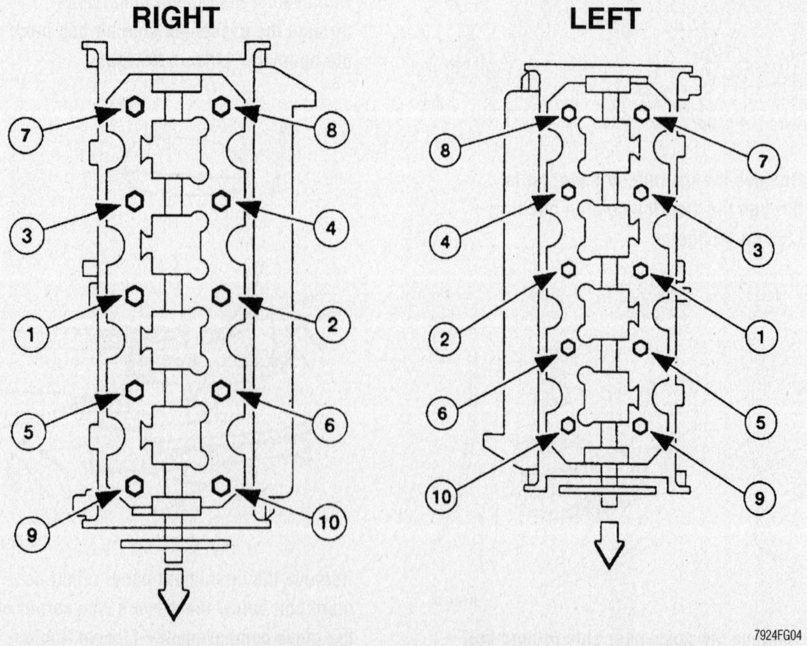

Tighten the cylinder head bolts using 3 steps in this sequence—4.6L and 5.4L engine

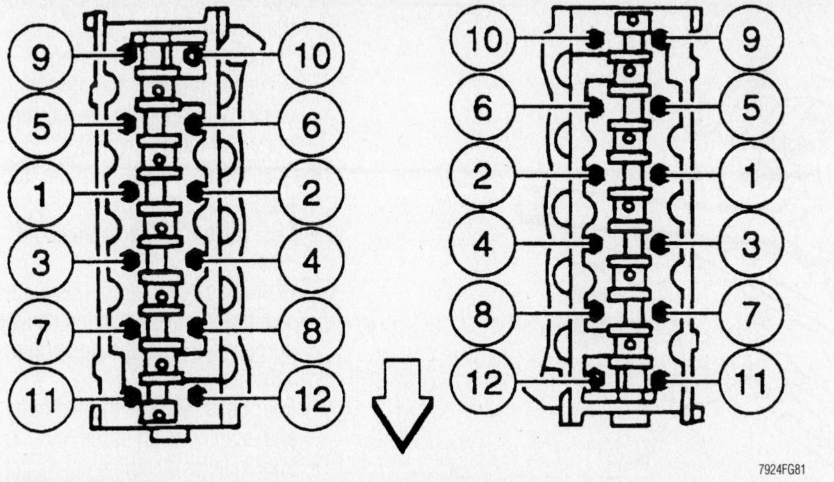

Tighten the cylinder head bolts using 3 steps in this sequence—6.8L engine

block. Discard the cylinder head gasket and clean the engine block surface.

To install:

⁂ WARNING

Cylinder head bolts must be replaced with new ones. They are torque-to-yield designed and cannot be reused.

3. Turn the crankshaft to position the keyway at the 12 o'clock position.
4. Clean and inspect the cylinder head for damage or warpage. Install the cylinder head gasket over the dowel pins. Then, install the cylinder head onto the engine block. Loosely install NEW cylinder head bolts.

➡Be sure to tighten the head bolts in 3 steps.

5. Tighten the cylinder head bolts in the correct sequence using 3 steps, as follows:
   a. Step 1—30 ft. lbs. (40 Nm).
   b. Step 2—tighten an additional 85–95 degrees.
   c. Step 3—tighten another 85–95 degrees.
6. Install or connect the following:
   - Heater hose
   - Exhaust manifolds
   - Timing chains
   - Intake manifold
   - Cylinder head covers
   - Electrical connections and cables
   - Refrigerant
   - Engine oil
   - Coolant
   - Negative battery cable
7. Start the engine and check for any fluid or vacuum leaks.

Rocker Arms

REMOVAL & INSTALLATION

4.6L and 5.4L Engines

1. Before servicing the vehicle, refer to the precautions in the beginning of this section.
2. Disconnect the negative battery cable.
3. Remove the camshaft covers.
4. Position the piston of the cylinder being serviced at the bottom of its travel.

➡Two different valve spring compressor tools are used for this procedure. Valve Spring Compressor (T91P-6565-A) is used on the exhaust camshaft and Valve Spring Compressor (T93P-6565-A) is used on the intake camshaft.

5. Compress the valve spring and remove the rocker arm.

To install:

6. Position the piston of the cylinder being serviced at the bottom of its travel.
7. Apply clean engine oil to the rocker arm, valve stem tip and tappet bore.

➡Valve tappet should have no more than 1/16 inch (1.5mm) of travel before installing the rocker arm.

8. Compress the valve spring using the correct tool and install the rocker arm.

6.8L Engine

1. Before servicing the vehicle, refer to the precautions in the beginning of this section.
2. Disconnect the negative battery cable.
3. Remove the camshaft covers.
4. Position the base circle of the camshaft lobe on the rocker arm to be serviced. Also, be sure the piston is not at the top of its travel near the valve.
5. Compress the valve spring and remove the rocker arm.

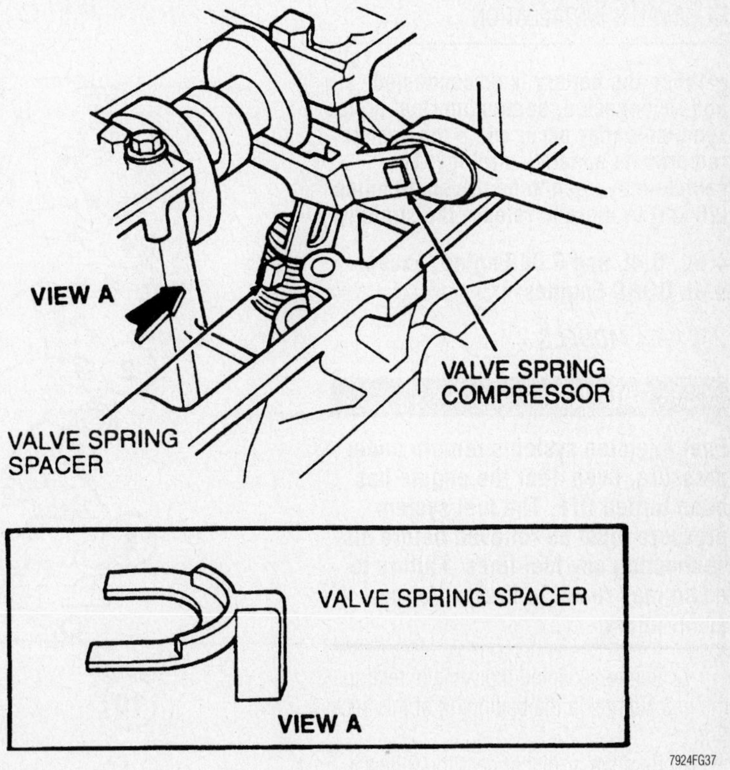

Using the proper tool, compress the valve spring and remove the rocker arm—4.6L and 5.4L engine

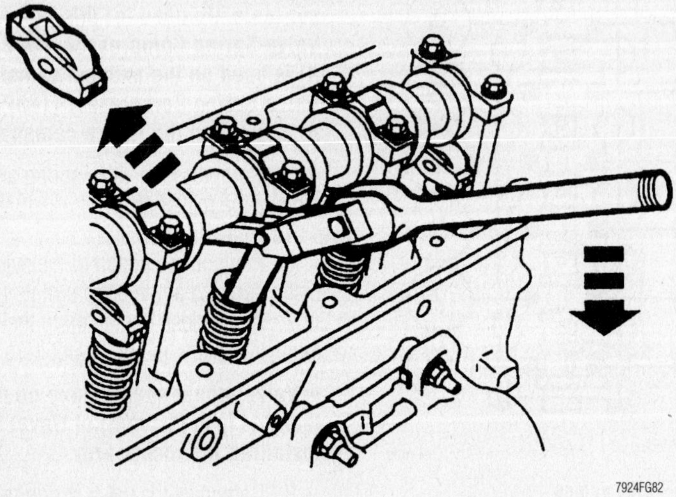

Compress the valve spring and remove the rocker arm—6.8L engine

To install:

6. Position the base circle of the camshaft lobe over the place where the rocker arm is to be installed.

7. Apply clean engine oil to the rocker arm, valve stem tip and tappet bore.

8. Compress the valve using the special tool and install the rocker arm.

9. Install the rocker arm covers and the remaining components.

Intake Manifold

REMOVAL & INSTALLATION

➡When the battery is disconnected and reconnected, some abnormal drive symptoms may occur while the vehicle relearns its adaptive strategy. The vehicle may need to be driven 10 miles (16 km) or more to relearn the strategy.

4.6L, 5.4L and 6.8L Engine, Except 5.4L DOHC Engines

2001–04 MODELS

✲✲ CAUTION

Fuel injection systems remain under pressure, even after the engine has been turned OFF. The fuel system pressure must be relieved before disconnecting any fuel lines. Failure to do so may result in fire and/or personal injury.

1. Before servicing the vehicle, refer to the precautions in the beginning of this section.

2. Remove or disconnect the following:

- Negative battery cable
- Engine cover, if equipped
- Fuel system pressure
- Coolant
- Upper radiator hose from the intake manifold
- Accelerator cable from the bracket and the throttle body cam
- Speed control actuator cable from the throttle body
- Throttle cable return spring
- All vacuum hoses, fuel lines and electrical wires from the throttle body and intake manifold
- Accelerator bracket from the throttle body
- EGR valve-to-exhaust manifold tube
- Brake booster vacuum hose bracket
- Two evaporative emission canister purge valve nuts and position the valve aside
- Four bolts, and the throttle body and adapter as an assembly
- Fuel injector electrical connectors
- Bolts and the eight ignition coils
- On the 6.8L engine: the radio interference capacitors from the left side of the intake manifold
- Alternator upper support bracket
- Heater hose from the intake manifold
- Water thermostat housing and thermostat. Discard the O-ring seal.
- Intake manifold bolts
- Intake manifold from the engine,

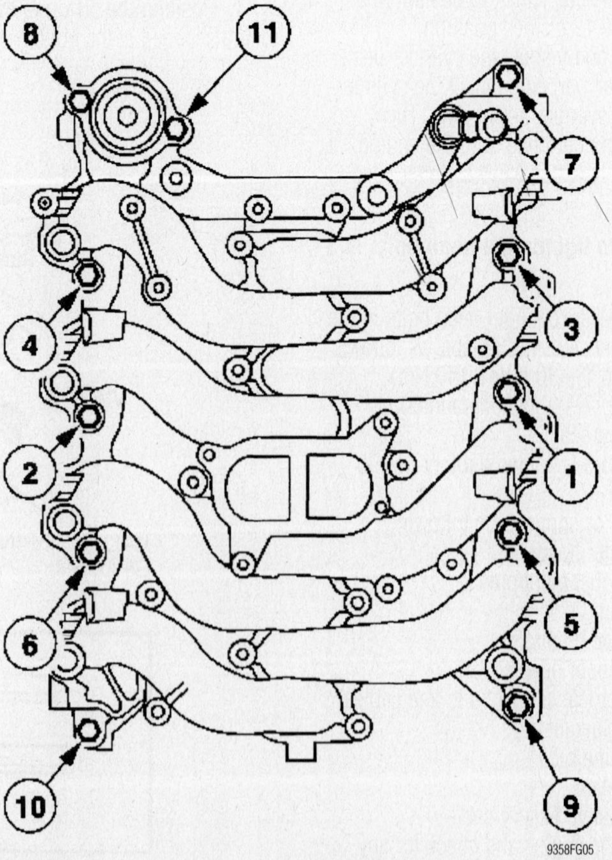

Tighten the upper intake manifold bolts in 2 steps using this sequence—2001–04 4.6L shown, 5.4L SOHC engine similar

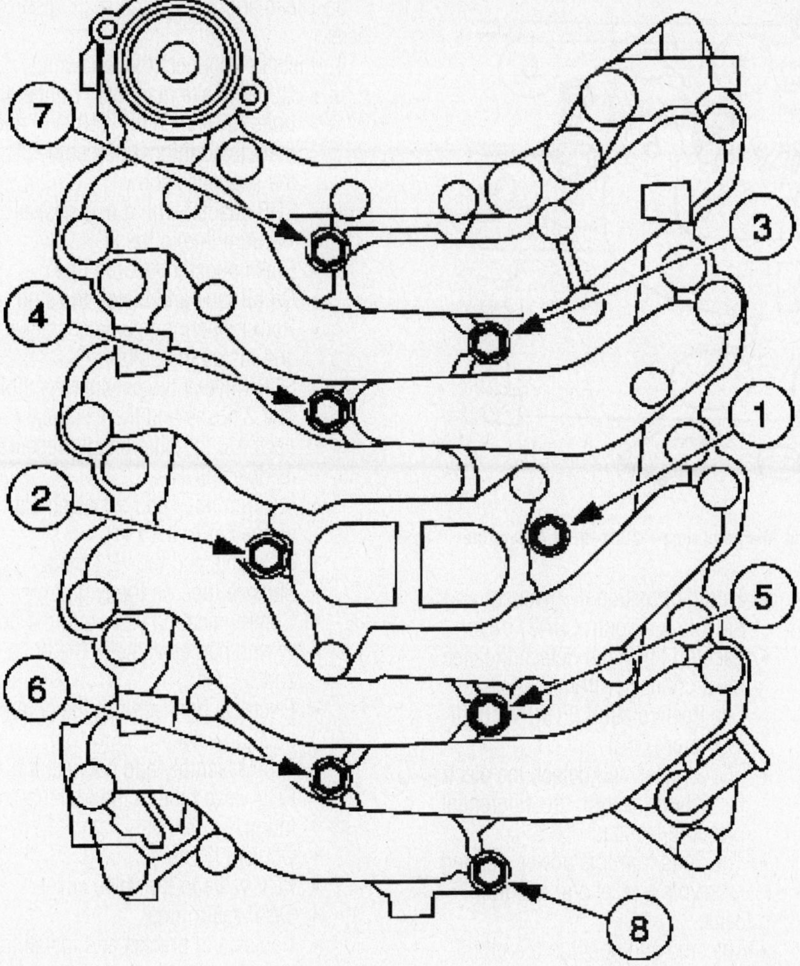

Tighten the lower intake manifold bolts in 2 steps using this sequence—2001–04 4.6L shown, 5.4L SOHC engine similar

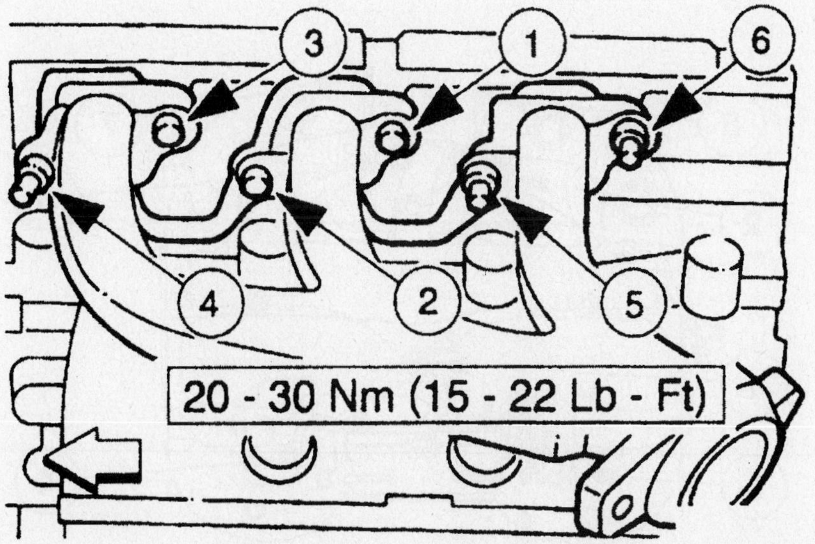

Tighten the lower intake manifold bolts in 2 steps using this sequence—2001–02 6.8L engine

then detach the Intake Manifold Tuning Valve (IMTV) electrical connector. Remove and discard the upper intake manifold gaskets.

- Upper-to-lower intake manifold bolts, then separate the upper intake manifold from the lower intake manifold. Discard the old gasket.

To install:

3. On the 6.8L engine, position the lower intake manifold gasket and the upper intake manifold onto the lower intake manifold, then loosely install the upper-to-lower intake manifold bolts.

➡Be sure to tighten the lower-to-upper manifold bolts in 2 steps.

4. Tighten the 8 lower-to-upper intake manifold bolts in 2 steps following the tightening sequence as follows:
   - 18 inch lbs. (2 Nm)
   - 89 inch. lbs. (25 Nm)
5. Position the 2 upper intake manifold gaskets on the cylinder heads. Set the upper intake manifold in place on the engine, then loosely install the intake manifold-to-cylinder head bolts.
6. Attach the IMTV electrical connector.

➡Check that the thermostat housing is in the correct position before the thermostat housing is installed.

7. Install the thermostat housing and start the 2 housing bolts.

➡Make certain to tighten the intake manifold in 2 steps.

8. Tighten the intake manifold bolts using the sequence shown, in 2 steps on all except 6.8L engines.
   - 18 inch lbs. (2 Nm)
   - 18 ft. lbs. (25 Nm)
9. Tighten the upper-to-lower intake manifold bolts using the sequence shown, in 2 steps on 6.8L engines.
   - 18 inch lbs. (2 Nm)
   - 18 ft. lbs. (25 Nm)
10. Install or connect the following:
    - Heater water hose
    - All electrical connections, fuel lines, vacuum tubes and coolant hoses to the intake manifold, fuel injectors and throttle body assembly.
    - Alternator support bracket
    - Ignition coils
    - New throttle body adapter gasket and the throttle body adapter assembly. Tighten the bolts to 89 inch lbs. (10 Nm).
    - All remaining hoses and electrical connections

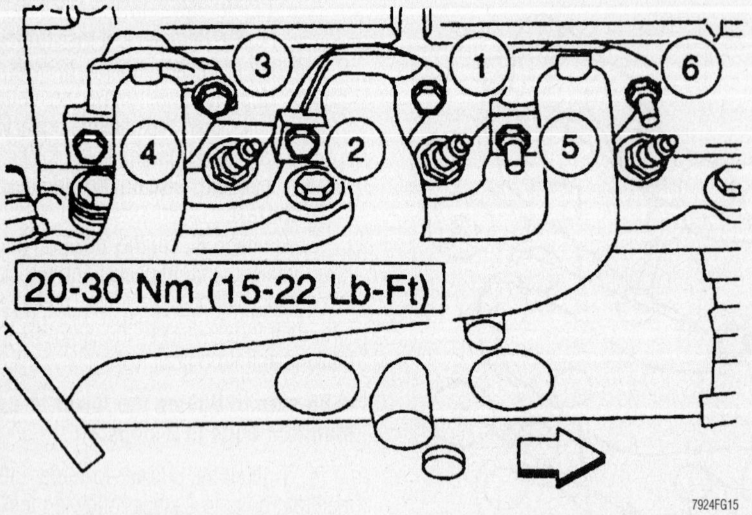

Tighten the upper intake manifold bolts in 2 steps using this sequence—2001–02 6.8L engine

- EGR valve-to-exhaust manifold tube. The tube fittings should be tightened to 26–33 ft. lbs. (35–45 Nm).
- Speed actuator cable, if equipped, and the accelerator cable to the throttle body.
- Coolant
- Negative battery cable. Start the engine and check for fuel, vacuum or coolant leaks.

5.4L DOHC Engines

UPPER INTAKE MANIFOLD

1. Before servicing the vehicle, refer to the precautions in the beginning of this section.
2. Remove or disconnect the following:
   - Negative battery cable
   - Air cleaner outlet tube
   - Accelerator cable, speed control cable, and the return spring
   - Bolts and position the cables and bracket out of the way
   - Evaporative emission return line
   - Positive Crankcase Ventilation (PCV) tube from the upper intake manifold
   - PCV valve from the valve cover
   - PCV valve tube from the water heated fitting and remove the tube assembly
   - Coolant lines and plug the lines.
   - Electrical connector from the communication valve
   - Bolts and the communication valve
   - Wiring harness shield bolts and 5 clips and remove the shield
   - Balance tube from the engine
   - Throttle Position (TP) sensor and the Idle Air Control (IAC) motor
   - Vacuum lines and detach the electrical connector from the Exhaust Gas Recirculation (EGR) vacuum regulator (EVR)
   - The 2 hoses and detach the electrical connector from the differential pressure feedback EGR
   - The bolts from the power steering reservoir bracket and position it aside
   - The stud and position the oil fill tube aside
   - Brake booster vacuum line
   - Vacuum line from the EGR valve
   - Bolts from the EGR adapter
   - Retaining bolts and remove the upper intake manifold

To install:

3. Clean and inspect the sealing surfaces.
4. Install or connect the following:
   - Upper intake manifold: Tighten the bolts to 89 inch lbs. (10 Nm) and then an additional 90 degrees in the sequence shown.
   - EGR adapter with a new gasket
   - Vacuum line to the EGR valve
   - Brake booster vacuum line
   - Oil fill tube and install the stud
   - Power steering reservoir bracket and install the bolts
   - EGR valve at the exhaust manifold
   - The 2 hoses and the electrical connector to the differential pressure feedback EGR
   - Vacuum lines and the electrical connector to the EVR
   - IAC valve and the TP sensor
   - Balance tube on the engine
   - Communication valve and the bolts
   - Wiring harness shield, the bolts and 5 clips
   - Electrical connector to the communication valve
   - Tube assembly and connect the PCV valve tube to the water-heated fitting
   - Coolant hoses
   - PCV valve in the valve cover
   - EVAP return line
   - Cables and bracket and install the bolts
   - Accelerator cable, speed control cable, and the return spring
   - Engine appearance cover and the 3 bolts
   - Engine air cleaner and outlet tube

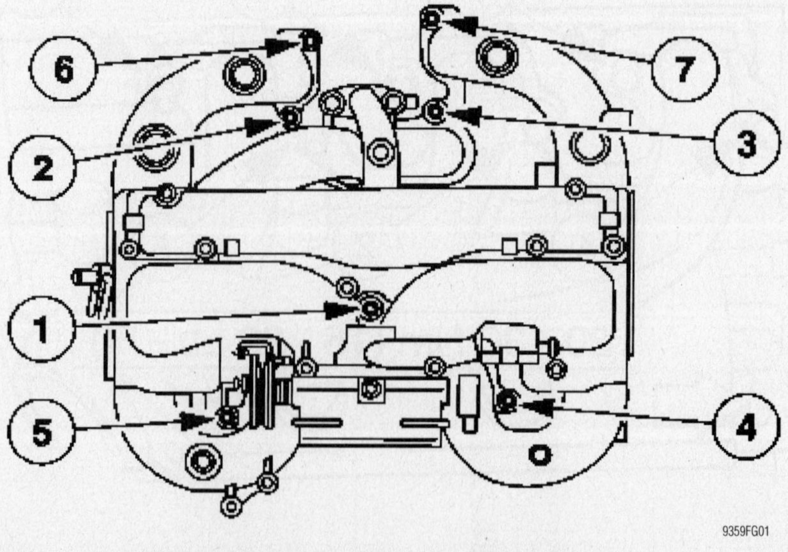

Tighten the upper intake manifold bolts as shown—5.4L DOHC engines

LOWER INTAKE MANIFOLD

✲✲ CAUTION

Fuel injection systems remain under pressure, even after the engine has been turned OFF. The fuel system pressure must be relieved before disconnecting any fuel lines. Failure to do so may result in fire and/or personal injury.

1. Before servicing the vehicle, refer to the precautions in the beginning of this section.
2. Remove or disconnect the following:
   - Negative battery cable
   - Coolant
   - Upper intake manifold
   - Fuel system pressure
   - Fuel lines
   - Engine water bypass hose
   - Electrical connector from the water temperature indicator sender
   - Upper radiator hose, the heater water inlet hose and the heated PCV water fitting inlet hose
   - The 4 bolts and the upper alternator support bracket
   - The 8 fuel injectors
   - Vacuum line from the fuel injector pressure regulator and position out of the way
   - Lower intake manifold
   - Radio ignition interference capacitors

To install:

3. Remove and inspect the gaskets, install new gaskets if necessary.
4. Clean the sealing surfaces.
5. Install or connect the following:
   - Radio ignition interference capacitors

➡Bolts should be hand-started, positions 7–12 first then 1–6.

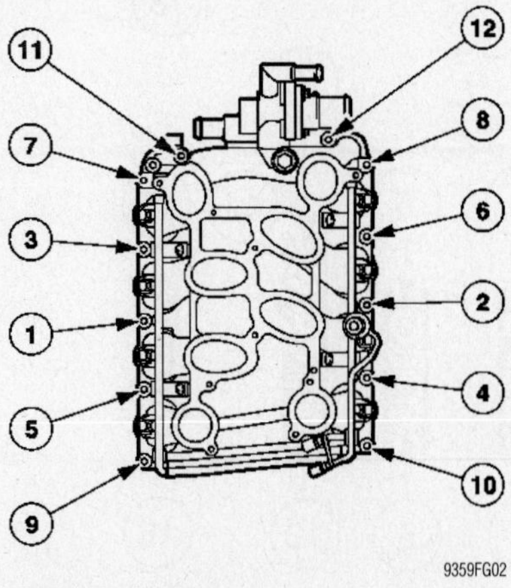

Tighten the lower intake manifold bolts as shown—5.4L DOHC engines

   - Lower intake manifold: Tighten the bolts to 89 inch lbs. (10 Nm) and then an additional 90 degrees in the sequence shown.
   - Water temperature indicator sensor
   - Vacuum harness and connect the vacuum line to the fuel injector pressure regulator
   - Fuel injectors
   - Upper generator support bracket and the 4 bolts
   - Heater water inlet hose, the upper radiator hose and the water heated fitting inlet hose
   - Engine water bypass return hose
   - Fuel lines
   - Upper intake manifold
   - Coolant

Exhaust Manifold

REMOVAL & INSTALLATION

1. Before servicing the vehicle, refer to the precautions in the beginning of this section.
2. Remove or disconnect the following:
   - Front fender splash shield
   - For the left-hand exhaust manifold: the Exhaust Gas Recirculation (EGR) valve-to-exhaust manifold tube and if equipped, the DPFE gas recirculation transducer hoses.
   - On the 4.6L and 5.4L engines: the catalytic converter-to-exhaust manifold bolts
   - On the 6.8L engine: the front exhaust pipe from the manifold
   - The exhaust manifold mounting nuts, then remove the exhaust manifold itself. Remove and discard the old gasket.
3. Clean and inspect the exhaust manifold for damage.

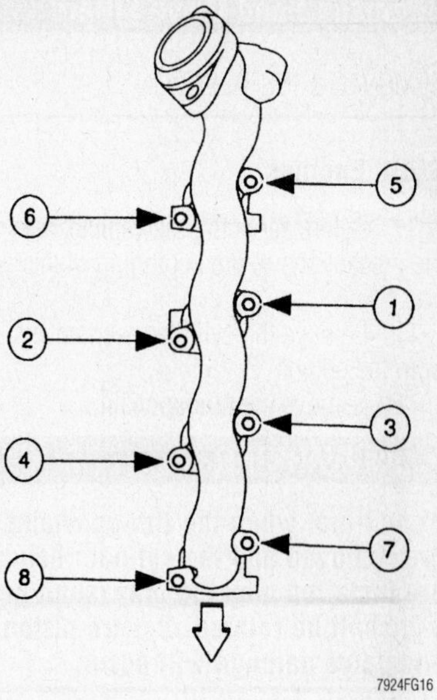

Tighten the exhaust manifold bolts in the sequence shown—4.6L engine shown, 5.4L engine similar

To install:

4. Position a new gasket and the exhaust manifold onto the engine block.
5. Install the mounting nuts and tighten following the sequence shown.
   - 18 ft. lbs. (25 Nm)
6. On the 6.8L engine, tighten the exhaust manifold-to-front pipe fasteners to 27–34 ft. lbs. (34–46 Nm).
7. On the 4.6L and 5.4L engines, attach the catalytic converter to the exhaust manifold, install the catalytic converter-to-exhaust manifold bolts and tighten to 25–34 ft. lbs. (34–46 Nm)
8. For the left-hand exhaust manifold, install the DPFE transducer hoses if equipped, and the EGR valve-to-exhaust manifold tube. Tighten the upper and lower fittings to 26–33 ft. lbs. (35–45 Nm).
9. Install the front fender splash shield.
10. Lower the vehicle to the ground.

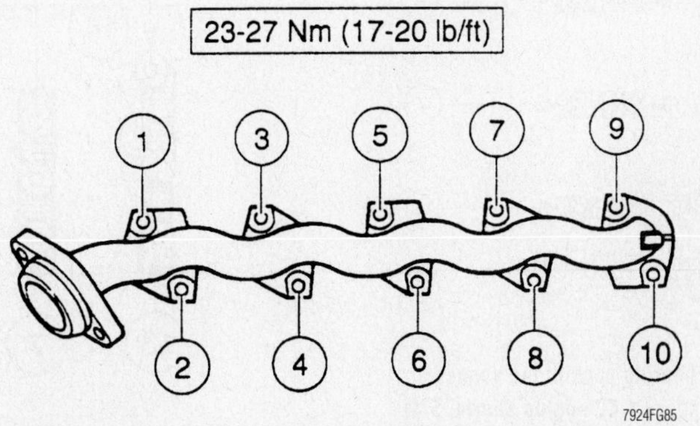

Tighten the exhaust manifold bolts in the sequence shown—right side of 6.8L engine shown

Camshaft and Valve Lifters

REMOVAL & INSTALLATION

SOHC Engines

1. Before servicing the vehicle, refer to the precautions in the beginning of this section.
2. Remove the cylinder head covers from the engine.
3. Remove the timing chain.

⁂ WARNING

At no time, when the timing chains are removed and the cylinder heads are installed may the crankshaft or camshaft be rotated. Severe piston and valve damage will occur.

4. On the 6.8L engine, remove the 6 bolts securing the balance shaft to the cylinder head and remove the shaft.
5. Remove the camshaft roller lifters.
6. On VIN W engines, remove the timing chain camshaft gear by removing the gear retaining bolt.

➡Keep the bearing caps in order so they can be installed in the same position.

7. Remove the camshaft bearing cap bolts, then lift the camshaft bearing caps off of the cylinder head.
8. Lift the camshaft from the cylinder head.

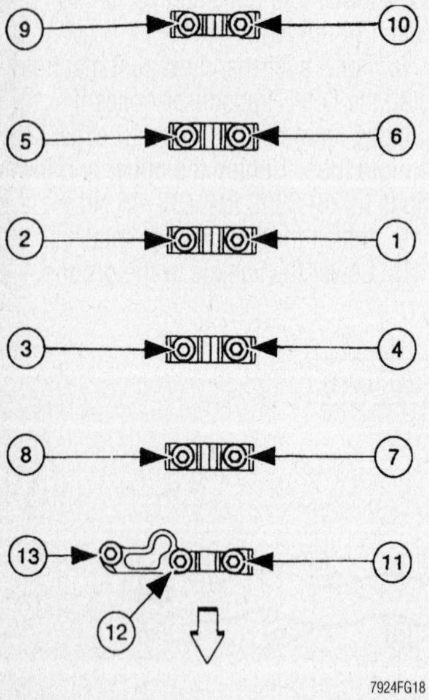

Tighten the bearing caps in the sequence shown—Windsor 4.6L engine shown, 5.4L engine similar

9. Remove the rocker arms and pull the lash adjusters out of their bores. Keep all the parts in order. They must be installed in their original positions.

To install:

10. Install the lash adjusters and rocker arms in their original positions.
11. Lubricate the camshaft journals and bearing caps with SAE 5W30 engine oil. On the 6.8L engine, lubricate the balance shaft journals and bearing caps with the same lubricant.
12. Lower the camshaft onto the camshaft bearing journals.
13. Install the camshaft bearing caps, then loosely install the bearing cap bolts.
14. Tighten the camshaft bearing cap mounting bolts, in the sequence shown for the particular engine, to 71–107 inch lbs. (8–12 Nm).
15. On the 6.8L engine, align the timing marks and position the balance shaft on the journals, then install the bearing caps. Tighten the bolts in sequence to 89 inch lbs. (10 Nm).
16. On engines equipped with bolt on sprockets, install the camshaft timing chain

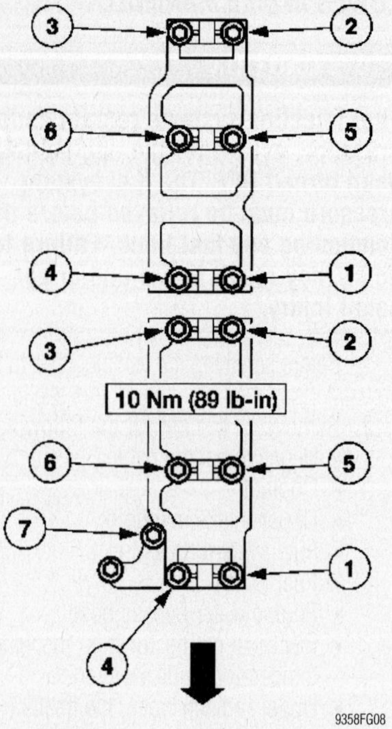

Tighten the bearing caps in the sequence shown—Romeo 4.6L engine shown, 5.4L engine similar

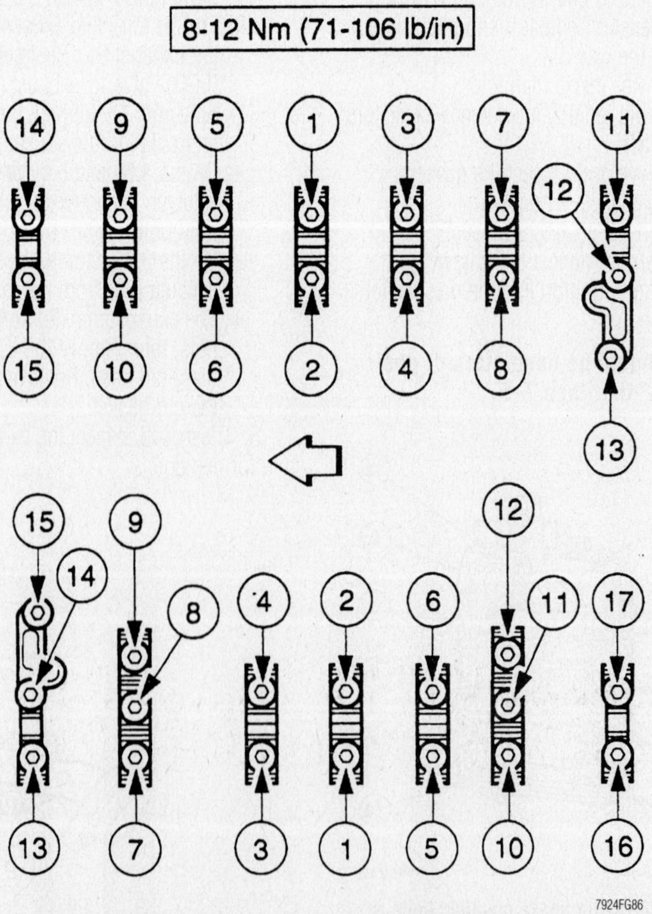

Camshaft bearing cap bolt tightening sequence—6.8L engine

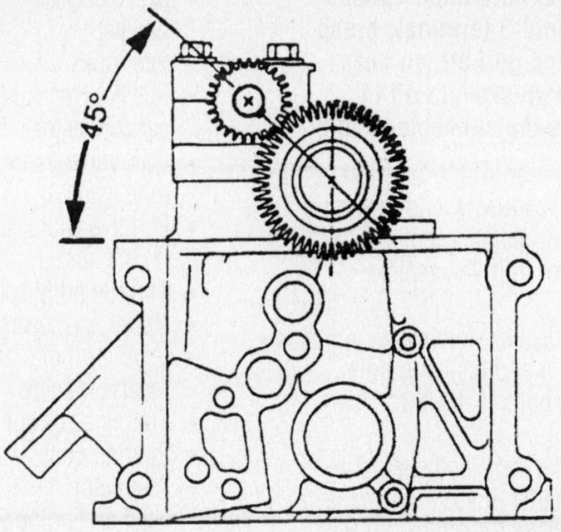

Be sure to align the balance shaft timing mark with the mark on the camshaft gear—6.8L engine

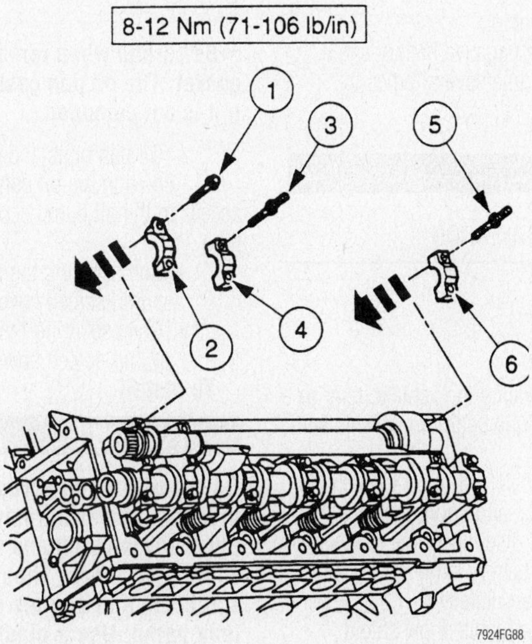

Tighten the balance shaft bearing cap bolts in the sequence shown—6.8L engine

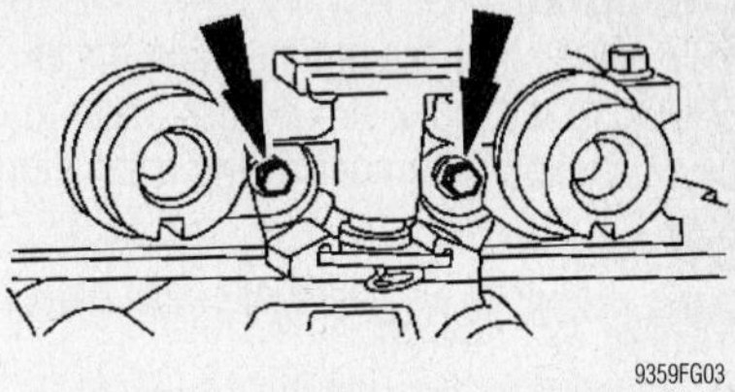

Remove the bolts shown before removing any other bearing cap bolts—DOHC models

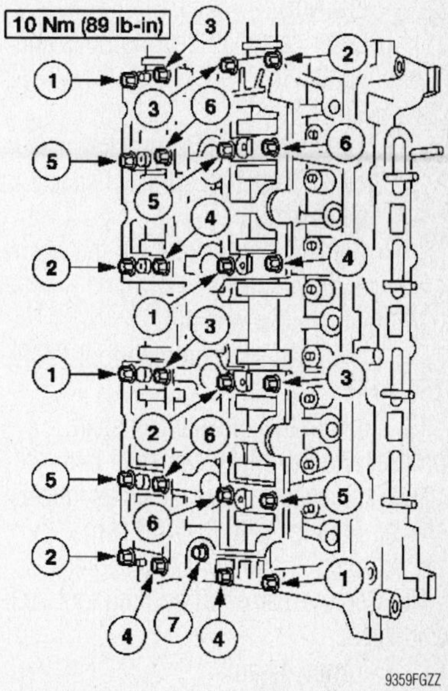

Tighten the camshaft cap bolts as shown to specification

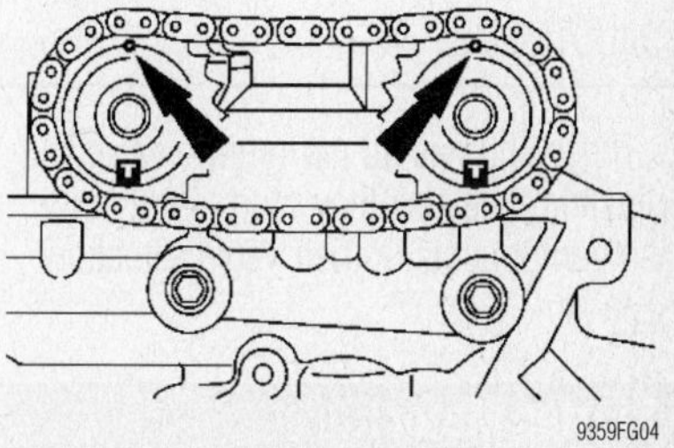

Make sure the timing marks are at 12 o'clock and index at 6 o'clock —DOHC models

gear by tightening the retaining bolts as follows:

a. M10 Step 1: 30 ft. lbs. (40 Nm).
b. M10 Step 2: Tighten 90 degrees.
c. M12 Step 1: 90 ft. lbs. (120 Nm).

17. Install the valve lifters.
18. Install the timing chain and sprockets, if applicable.
19. Install the cylinder head covers.

DOHC Engines

1. Before servicing the vehicle, refer to the precautions in the beginning of this section.
2. Remove or disconnect the following:
   - Roller followers
   - Left hand timing chain for the left hand side and both timing chains for the right hand side
3. Install camshaft holding tool T93P-6256-AHR.
   - Exhaust camshaft sprocket and the intake washer and the spacer
4. Remove camshaft holding tool T93P-6256-AHR.
5. Compress the timing chain tensioner and install a lock pin.
   - Timing chain, the sprocket, and intake camshaft sprocket spacer
   - Outer cam bearing bolts shown in the accompanying illustration

⁂ WARNING

The outer bolts on the outer cam bearing cap (exhaust) are longer and must be returned to the same location or engine damage may occur.

6. Make sure to identify the camshaft to cylinder head location. Caps are not interchangeable.
   - Bolts and the camshaft bearing cap assemblies
   - Camshafts

To install:

7. Make sure the timing marks are at 12 o'clock and index at 6 o'clock. Refer to the illustration for proper timing mark alignment.
8. Lubricate the camshafts. Use Super Premium SAE 5W-30.
9. Install or connect the following:
   - Camshafts
   - Camshaft bearing cap assemblies and bolts. Tighten the bolts in the sequence illustrated to 89 inch lbs. (10 Nm).
   - Outer camshaft bearing bolts and tighten to 71–106 inch lbs. (8–12 Nm)
   - Timing chain, the sprocket, and intake camshaft sprocket spacer as an assembly
   - Camshaft sprocket bolt, the washer and the spacer
10. Remove the lock pin.
11. Install camshaft holding tool T93P-6256-AHR.
12. Tighten the exhaust camshaft sprocket and the intake camshaft bolt washer and spacer in two steps as follows:
    a. Step 1: 30 ft. lbs. (40 Nm).
    b. Step 2: additional 90 degrees.
13. Remove camshaft holding tool T93P-6256-AHR.
    - Timing chains
    - Roller followers

Valve Lash

ADJUSTMENT

These engines do not require valve lash adjusting, because they utilize hydraulic lash components in their valve actuation systems.

Starter Motor

REMOVAL & INSTALLATION

1. Before servicing the vehicle, refer to the precautions in the beginning of this section.
2. Disconnect the negative battery cable.
3. Raise and safely support the vehicle.
4. Remove or disconnect the following:
   - Starter terminal cover
   - Terminal nut and separate the battery starter cable from the starter motor
   - Solenoid **S** terminal connector, if equipped with a starter mounted solenoid

➡**To disconnect the hard-shell connector from the solenoid S terminal, grasp the plastic shell and pull off; do not pull on the wire. Pull straight off to prevent damage to the connector and S terminal.**

5. Remove or disconnect the following:
   - Starter motor retaining bolts
   - Starter motor from the vehicle

To install:

6. Install or connect the following:
   - Starter motor and retaining bolts. Tighten the bolts to 15–20 ft. lbs. (20–27 Nm).
   - Battery starter cable and a terminal nut to the starter motor. Tighten the terminal nuts to 79 inch lbs. (9 Nm).
   - Solenoid **S** terminal connector, if equipped with a starter mounted solenoid
   - Starter solenoid safety cap, if equipped
7. Lower the vehicle.
8. Connect the negative battery cable.
9. Start the engine several times to check starter motor operation.

Oil Pan

REMOVAL & INSTALLATION

4X2 Models

SOHC ENGINES

1. Before servicing the vehicle, refer to the precautions in the beginning of this section.
2. Raise and safely support the vehicle.
3. Drain the oil and cooling system,
4. Remove or disconnect the following:
   - Negative battery cable
   - Radiator air deflector
   - Accelerator cable snow shield
   - Accelerator cable
   - Speed control actuator cable
   - Accelerator return spring
   - Accelerator cable bracket bolts and position the bracket and cables aside
   - Main vacuum harness
   - Throttle Position (TP) sensor and engine vacuum regulator connectors
   - Fuel pressure regulator vacuum line
   - Vapor management valve vacuum line
   - Differential pressure feedback Exhaust Gas Recirculation (EGR) connector
   - Brake booster vacuum line and bracket
   - Fuel lines
   - Idle Air Control (IAC) motor electrical connector
   - EGR valve-to-exhaust manifold tube aside
   - Positive Crankcase Ventilation (PCV) hose
   - Throttle body adapter
   - Upper and lower power steering reservoir bolts and position the reservoir aside
   - Fuel injector connections
   - Ignition coils
   - Alternator
5. Install an Engine Support Bracket on the engine using the alternator mounting bolts.
   - Right and left hand motor mount bolts
6. Raise the engine using the Engine Support.

➡**Be careful when removing the oil pan gasket. The oil pan gasket is reusable if it is not damaged.**

   - Oil pan bolts and pan
7. Position the oil pan aside to gain access to the oil pump screen cover and tube .
   - Bolt securing the rear of the oil pump screen cover and tube
   - Bolts securing the front of the oil pump screen cover and tube

To install:

✲✲ WARNING

Do not use metal scrapers, wire brushes, power abrasive discs or other abrasive means to clean the sealing surfaces. These tools cause scratches and gouges which make leak paths. Use a plastic scraping tool to remove all traces of old sealant.

8. Clean and inspect the mating surfaces.
9. Install or connect the following:
   - Oil pan gasket, the oil pan and the oil pump screen cover and tube together to the cylinder block
   - Bolts securing the front of the oil pump screen cover and tube and tighten to 71–107 inch lbs. (8–12 Nm)

➡**If the oil pan is not secured within four minutes, the sealant must be removed and the sealing area cleaned with Metal Surface Cleaner. Allow to**

dry until there is no sign of wetness, or four minutes, whichever is longer. Failure to follow this procedure may result in future oil leakage.

10. Apply the silicone Gasket and Sealant F7AZ-19554-EA at the rear oil seal retainer-to-cylinder block sealing surface.
11. Install the oil pan gasket and the oil pan and loosely install the 16 bolts. Tighten the bolts in three steps, in the sequence illustrated as follows:
    a. Step 1: 18 inch lbs. (2 Nm).
    b. Step 2: 15 ft. lbs. (20 Nm).
    c. Step 3: an additional 60 degrees.
12. Remove the Engine Support.
    - Right and left hand motor mount bolts and tighten to 50–68 ft. lbs. (68–92 Nm)
    - Alternator
    - Fan shroud
    - Throttle body adapter and the four bolts. Tighten the bolts to 89 inch lbs. (10 Nm).
    - EGR valve-to-exhaust manifold tube
    - All vacuum hoses, fuel lines and electrical connections
    - Accelerator cable bracket and the bolts
    - Accelerator cable, speed control actuator cable and the throttle return spring
    - Accelerator cable snow shield and the bolts
    - Upper radiator hose and position the clamp
    - Air cleaner assembly
    - Radiator air deflector
    - Negative battery cable
13. Refill and bleed the cooling system.
14. Refill the engine crankcase with clean engine oil.

DOHC ENGINES

1. Before servicing the vehicle, refer to the precautions in the beginning of this section.
2. Raise and safely support the vehicle.
3. Drain the oil.
4. Remove or disconnect the following:
    - Intake manifold
    - Fan shroud and the engine cooling fan
    - Alternator
5. Install an engine support device.
    - Two bolts from the engine mounts
6. Raise the engine 2.5 inches (63.5mm).

➡**Be careful when removing oil pan gasket. It is re-usable if not damaged.**

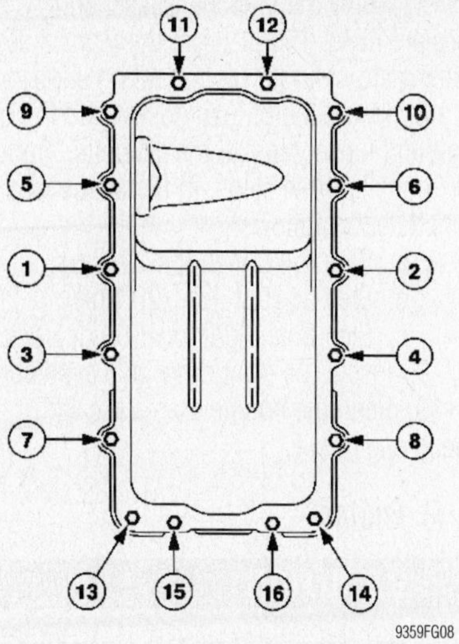

Tighten the oil pan bolts in the sequence shown

- Bolts, oil pan and gasket

7. Inspect the oil pan gasket for damage.

To install:

⁂ WARNING

Do not use metal scrapers, wire brushes, power abrasive discs or other abrasive means to clean the sealing surfaces. These tools cause scratches and gouges which make leak paths. Use a plastic scraping tool to remove all traces of old sealant.

8. Clean the oil pan mating surfaces.

➡**If the oil pan and the oil pan gasket are not secured within four minutes, the sealant must be removed and the sealing area cleaned with Metal Surface Cleaner F4AZ-19A536-RA. Allow to dry until there is no sign of wetness, or four minutes, whichever is longer. Failure to follow this procedure can cause future oil leakage.**

9. Apply the silicone Gasket and Sealant F7AZ-19554-EA at the rear oil seal retainer-to-cylinder block sealing surface.
10. Install the oil pan gasket and the oil pan and loosely install the 16 bolts. Tighten the bolts in three steps, in the sequence illustrated as follows:
    a. Step 1: 18 inch lbs. (2 Nm).
    b. Step 2: 15 ft. lbs. (20 Nm).
    c. Step 3: an additional 60 degrees.
11. Lower the vehicle, then the engine.
12. Remove the engine support
13. Install or connect the following:
    - Two bolts for the engine mounts and tighten to 59 ft. lbs. (80 Nm)
    - Alternator
    - Cooling fan
    - Lower intake manifold
14. Refill the engine crankcase with clean engine oil.

4X4 Models

SOHC ENGINES

1. Before servicing the vehicle, refer to the precautions in the beginning of this section.
2. Raise and safely support the vehicle.
3. Drain the oil.

⁂ CAUTION

The air suspension switch must be turned off prior to raising the vehicle. Failure to do so can result in unexpected inflation or deflation of the air springs, which could result in shifting of the vehicle during the repair operation.

4. Turn the air suspension switch off.
5. Support the front axle housing with a jackstand.
6. Remove or disconnect the following:
    - Negative battery cable
    - Front axle mount bolt
    - Front axle support bolts and the support

⁂ CAUTION

Use care when lowering the front axle housing, or the vacuum lines to the axle solenoid may become disconnected or damaged.

- Right and front axle housing mount bolts and lower the axle to allow clearance for the oil pan to be removed
- Oil pan and gasket

To install:

⁂ WARNING

Do not use metal scrapers, wire brushes, power abrasive discs or other abrasive means to clean the sealing surfaces. These tools cause scratches and gouges which make leak paths. Use a plastic scraping tool to remove all traces of old sealant.

7. Clean the oil pan mating surfaces.

➡**If the oil pan and the oil pan gasket are not secured within four minutes, the sealant must be removed and the sealing area cleaned with Metal Sur-**

face Cleaner F4AZ-19A536-RA. Allow to dry until there is no sign of wetness, or four minutes, whichever is longer. Failure to follow this procedure can cause future oil leakage.

8. Apply the silicone Gasket and Sealant F7AZ-19554-EA at the rear oil seal retainer-to-cylinder block sealing surface.
9. Install or connect the following:
   - New oil pan gasket and the oil pan and loosely install the bolts
10. Tighten the bolts in three steps in the sequence illustrated as follows:
    a. Step 1: 18 inch lbs. (2 Nm).
    b. Step 2: 15 ft. lbs. (20 Nm).
    c. Step 3: an additional 60 degrees.
11. Install or connect the following:
    - Front axle housing and loosely install the bolts
    - Axle support and tighten the bolts to 56 ft. lbs. (90 Nm)
12. Refill the engine crankcase with clean engine oil.
13. Turn on the air suspension switch.

DOHC ENGINES

1. Before servicing the vehicle, refer to the precautions in the beginning of this section.
2. Raise and safely support the vehicle.
3. Drain the oil.
4. Remove or disconnect the following:
   - Front drive axle assembly

➡Be careful when removing oil pan gasket. It is re-usable if not damaged.

   - Bolts, oil pan and gasket
5. Inspect the oil pan gasket for damage.

To install:

✲✲ WARNING

Do not use metal scrapers, wire brushes, power abrasive discs or other abrasive means to clean the sealing surfaces. These tools cause scratches and gouges which make leak paths. Use a plastic scraping tool to remove all traces of old sealant.

6. Clean the oil pan mating surfaces.

➡If the oil pan and the oil pan gasket are not secured within four minutes, the sealant must be removed and the sealing area cleaned with Metal Surface Cleaner F4AZ-19A536-RA. Allow to dry until there is no sign of wetness, or four minutes, whichever is longer. Failure to follow this procedure can cause future oil leakage.

7. Apply the silicone Gasket and Sealant F7AZ-19554-EA at the rear oil seal retainer-to-cylinder block sealing surface.
8. Install the oil pan gasket and the oil pan and loosely install the 16 bolts. Tighten the bolts in three steps, in the sequence illustrated as follows:
   a. Step 1: 18 inch lbs. (2 Nm).
   b. Step 2: 15 ft. lbs. (20 Nm).
   c. Step 3: an additional 60 degrees.
9. Install the front drive axle assembly.
10. Refill the engine crankcase with clean engine oil.

6.8L Engine

✲✲ CAUTION

Fuel injection systems remain under pressure, even after the engine has been turned OFF. The fuel system pressure must be relieved before disconnecting any fuel lines. Failure to do so may result in fire and/or personal injury.

1. Before servicing the vehicle, refer to the precautions in the beginning of this section.
2. Remove or disconnect the following:
   - Negative battery cable
   - Fuel system pressure
   - Coolant
   - Upper radiator hose from the intake manifold
   - Air cleaner outlet tube
   - Accelerator cable from the bracket and the throttle body cam
   - Speed control actuator cable from the throttle body
   - All vacuum hoses, fuel lines and electrical wires from the throttle body and intake manifold
   - Brake booster vacuum hose bracket
   - EGR valve-to-exhaust manifold tube and disconnect the vacuum line
   - Connector and vacuum line from the Engine Vacuum Regulator (EVR) solenoid
   - The 4 bolts and the throttle body adapter
   - Upper fan shroud mounting screws and position the shroud toward the engine.
   - The alternator and install the Modular Engine Support Bracket on the engine using the mounting holes.
   - Lower engine mount-to-frame nuts
   - Turbine Shaft Speed (TSS) and Output Shaft Speed (OSS) sensors from the transmission. Plug the openings.
3. Lower the vehicle to the floor and raise the engine using a hoist attached to the support bracket.
4. Install an engine support fixture with a J hook to keep the engine raised, then remove the hoist.
5. Remove or disconnect the following:
   - Engine oil and filter
   - Dual converter Y-pipe and the flywheel inspection cover
   - Driveshaft and the 2 transmission mounting nuts
   - Transmission. Be sure to support the transmission along the rails of the pan to avoid damage.
   - Oil pan mounting bolts and partially lower the pan

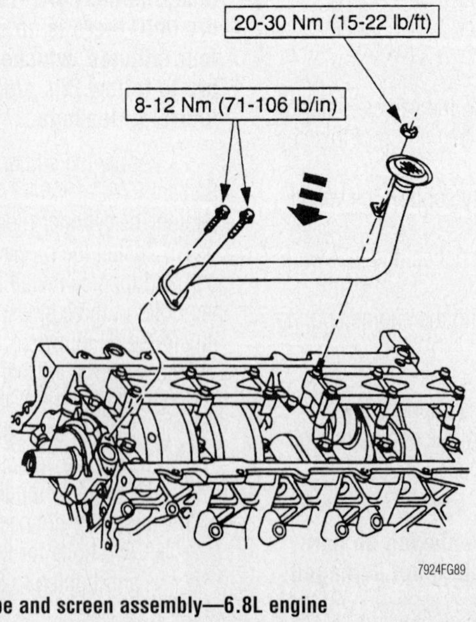

Oil pump pick-up tube and screen assembly—6.8L engine

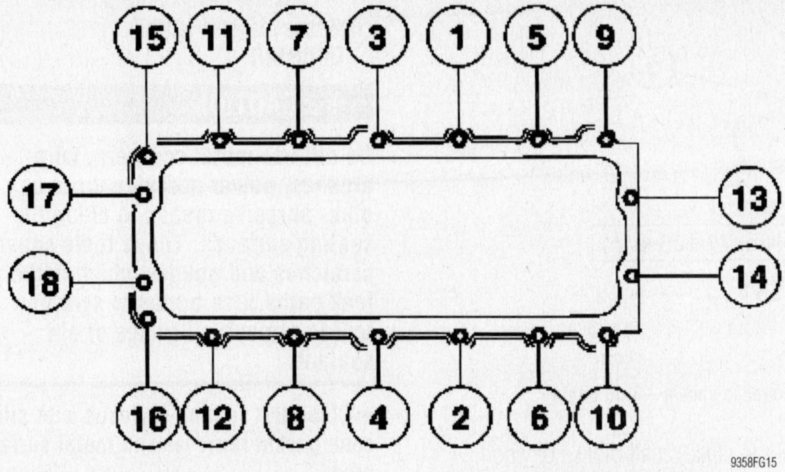

To prevent leaks, tighten the oil pan bolts in the order shown—2001-02 6.8L engine

- Oil pump pick-up tube and screen assembly and allow it to drop into the pan
- Oil pan towards the rear of the vehicle

To install:

WARNING

To prevent possible oil leaks, use only a plastic scraper to clean the oil pan mounting surface.

6. Clean the oil pan-to-engine mounting surface.
7. Place the oil pump pick-up tube and screen assembly in the oil pan, then position the pan and gasket near the engine.
8. Install the oil pump pick-up tube and screen assembly.
   a. Tighten the nut to 15–22 ft. lbs. (20–30 Nm).
   b. Tighten the 2 bolts to 71–106 inch lbs. (8–12 Nm).
9. Apply a bead of silicone sealant to the areas where the front cover and rear bearing cap meet the engine block.
10. Install the oil pan. Tighten the bolts in sequence using 3 steps as follows:
    - 18 inch lbs. (2 Nm)
    - 15 ft. lbs. (20 Nm)
    - Plus an additional 90°
11. Lower the transmission and install the 2 mounting nuts. Tighten the nuts to 60–80 ft. lbs. (81–108 Nm).
12. Install the driveshaft and the TSS and OSS sensors.
13. Install the flywheel cover and dual converter Y-pipe.
14. Install the oil bypass filter.
15. Lower the vehicle and remove the engine support fixture.
16. Install the engine mounting nuts. Tighten the nuts to 66 ft. lbs. (90 Nm).
17. Remove the modular engine support bracket and install the alternator.
18. Install the fan shroud.
19. Use a new gasket and install the throttle body adapter.
20. Tighten the bolts in 2 steps:
    - 71–88 inch lbs. (8–10 Nm)
    - 85–95 degrees
21. Connect the EVR solenoid harness and vacuum line.
22. Attach the vacuum line to the EGR valve.
23. Install the EGR valve-to-exhaust manifold tube. Tighten the fittings to 41 ft. lbs. (55 Nm).
24. Install the EGR transducer.
25. Install all remaining components in the reverse of the removal.

WARNING

Operating the engine without the proper amount and type of engine oil will result in severe engine damage.

26. Fill the engine with SAE 5W30 oil.
27. Fill and bleed the cooling system.

Oil Pump

REMOVAL & INSTALLATION

4.6L and 5.4L Engines

2001–04 MODELS

1. Before servicing the vehicle, refer to the precautions in the beginning of this section.
2. Disconnect the negative battery cable.
3. Remove or disconnect the following:
   - Crankshaft sprocket
   - Oil pan
   - Bolts and the oil pump

To install:

4. Clean and inspect the mating surfaces.
5. Install or connect the following:
   - Oil pump, and the bolts loosely. Tighten the bolts in the sequence illustrated to 89 inch lbs. (10 Nm).
   - Oil pan
   - Crankshaft sprocket

6.8L Engine

1. Before servicing the vehicle, refer to the precautions in the beginning of this section.
2. Disconnect the negative battery cable.
3. Remove the engine front cover and crankshaft sprocket.
4. Drain the engine oil into a suitable container.
5. Remove the oil pan.
6. Remove the 3 oil pump mounting bolts, then remove the oil pump from the engine.

To install:

7. Clean and inspect the mating surfaces.

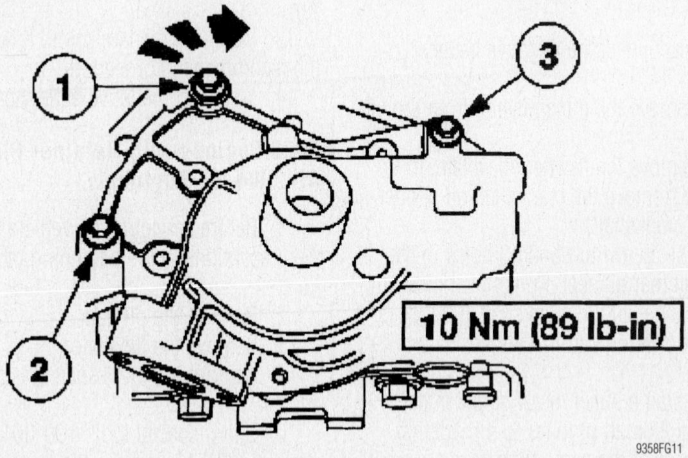

Tighten the oil pump mounting bolts in the sequence shown—4.6L and 5.4L engines

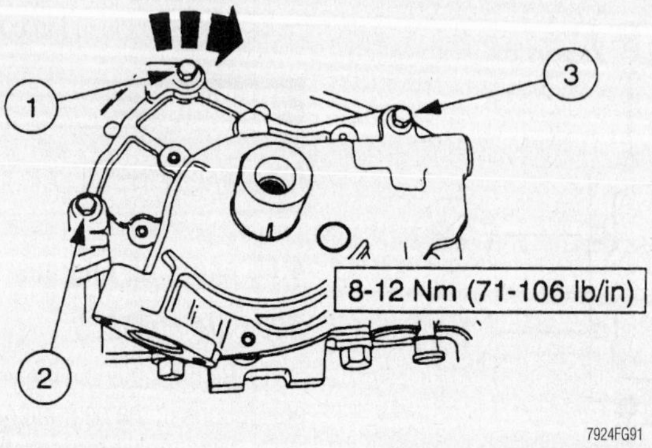

Be sure to tighten the oil pump mounting bolts in the sequence shown—6.8L engine

8. Install the oil pump and loosely install the oil pump mounting bolts. Tighten the bolts in the sequence shown to 71–106 inch lbs. (8–12 Nm).
9. Install the oil pan.
10. Install the crankshaft sprocket and timing chains.
11. Install the front cover.
12. Refill the engine oil with the recommended engine oil and amount.
13. Install the negative battery cable.

Rear Main Seal

REMOVAL & INSTALLATION

2001–02 Engines and 2003 Engines without Retainer Plate

If the crankshaft rear oil seal replacement is the only operation being performed, it can be done in the vehicle as detailed in the following procedure. If the oil seal is being replaced in conjunction with a rear main bearing replacement, the engine must be removed from the vehicle and installed on a work stand.

1. Before servicing the vehicle, refer to the precautions in the beginning of this section.
2. Disconnect the negative battery cable.
3. Remove the transmission from the vehicle.
4. Remove the flywheel/flexplate. If equipped, remove the crankshaft oil slinger from the crankshaft.
5. Use an awl to punch 2 holes in the crankshaft rear oil seal. Punch the holes on opposite sides of the crankshaft and just above the bearing cap-to-cylinder block split line.
6. Install a sheet metal screw in each hole. Use 2 small prybars to pry against both screws at the same time to remove the crankshaft rear oil seal. It may be necessary to place small blocks of wood against the cylinder block to provide a fulcrum point for the prybars. Use caution throughout this procedure to avoid scratching or otherwise damaging the crankshaft oil seal surface.
7. Clean the oil seal recess in the cylinder block and main bearing cap.

To install:

8. Clean, inspect and polish the rear oil seal rubbing surface on the crankshaft.
9. Coat the new oil seal and the crankshaft with a light film of engine oil.
10. Start the seal in the recess with the seal lip facing forward and install it with a seal driver. Keep the tool straight with the centerline of the crankshaft and install the seal until the tool contacts the cylinder block surface. Remove the tool and inspect the seal to be sure it was not damaged during installation.
11. If equipped, install the crankshaft oil slinger.
12. Position the flywheel on the crankshaft flange. Coat the threads of the flywheel attaching bolts with Loctite® and install the bolts. Tighten the bolts in sequence across from each other to 75–85 ft. lbs. (102–115 Nm).
13. Install the transmission, following the recommended procedure.
14. Install the negative battery cable.

2003 Engines with Retainer Plate and All 2004–05 Engines

1. Before servicing the vehicle, refer to the precautions in the beginning of this section.
2. Remove the oil pan.
3. Remove the flexplate.
4. Remove the crankshaft rear oil seal slinger.
5. Using special tools 100-001 with adapter 303-519, remove the crankshaft rear seal.
6. Remove the six bolts and the crankcase rear seal retainer.

To install:

CAUTION

Do not use metal scrapers, wire brushes, power abrasive discs or other abrasive means to clean the sealing surfaces. These tools cause scratches and gouges which make leak paths. Use a plastic scraping tool to remove all traces of old sealant.

➡Clean the sealing surfaces with silicone gasket remover and metal surface prep.

7. Follow the directions on the packaging. Failure to follow this procedure can cause future oil leakage.
8. Clean and inspect the mating surface.

CAUTION

Do not use metal scrapers, wire brushes, power abrasive discs or other abrasive means to clean the sealing surfaces. These tools cause scratches and gouges which make leak paths. Use a plastic scraping tool to remove all traces of old sealant.

➡If the rear crankshaft seal retaining plate is not secured within four minutes, the sealant must be removed and the sealing area cleaned. To clean the sealing area, follow the directions provided on the packaging of the silicone gasket remover and the metal surfacer prep. Failure to follow this procedure can cause future oil leakage.

➡The silicone must be applied on the groove along the retainer plate.

9. Apply a 4 mm bead of silicone gasket and sealant around the rear oil seal retainer sealing surface.
10. Install the rear seal retainer and loosely install the six bolts. Tighten the bolts in the sequence shown to 89 inch lbs. (10 Nm.)

➡Lubricate the inner lip of the rear crankshaft seal with clean engine oil.

11. Using the special tools, install the crankshaft rear seal.
12. Using the special tools, install the crankshaft rear oil slinger.
13. Install the oil pan.
14. Install the flexplate.

Timing Chain, Sprockets and Front Cover

REMOVAL & INSTALLATION

4.6L Engine

1. Before servicing the vehicle, refer to the precautions in the beginning of this section.
2. Remove or disconnect the following:
   - Negative battery cable
   - Radiator, fan blade and fan shroud assembly
   - Accessory drive belt
   - Water pump pulley
   - Electrical harness connectors from both ignition coils
   - Both ignition coils with their brackets attached
   - Left-hand and right-hand cylinder head covers
   - The 2 upper power steering pump retaining bolts
   - The 2 lower power steering pump retaining bolts and move the pump aside
   - Crankshaft Position (CKP) sensor electrical harness connector
   - CKP sensor retaining bolt and the sensor
   - Engine oil
   - The 4 oil pan-to-engine front cover retaining bolts
   - Crankshaft damper retaining bolt and washer from the crankshaft
   - Damper from the crankshaft
   - Camshaft Position (CMP) sensor retaining bolt and the sensor
   - Idler pulley bolt and the pulley
   - The 3 belt tensioner retaining bolts and the tensioner
   - The 8 engine front cover retaining bolts and the 7 nuts. Swing the top of the cover out off the dowel pins and remove the cover.
   - Sensor ring from the crankshaft
3. Use Camshaft Positioning Tool T91P-6256-A and Camshaft Positioning Adapters T92P-6256-A to position the camshaft.
4. Rotate the crankshaft until both camshaft keyways are 90 degrees from the cam cover surface. Be sure the copper links line up with the dots on the camshaft sprockets.

⁂ WARNING

At no time, when the timing chains are removed and the cylinder heads are installed may the crankshaft or

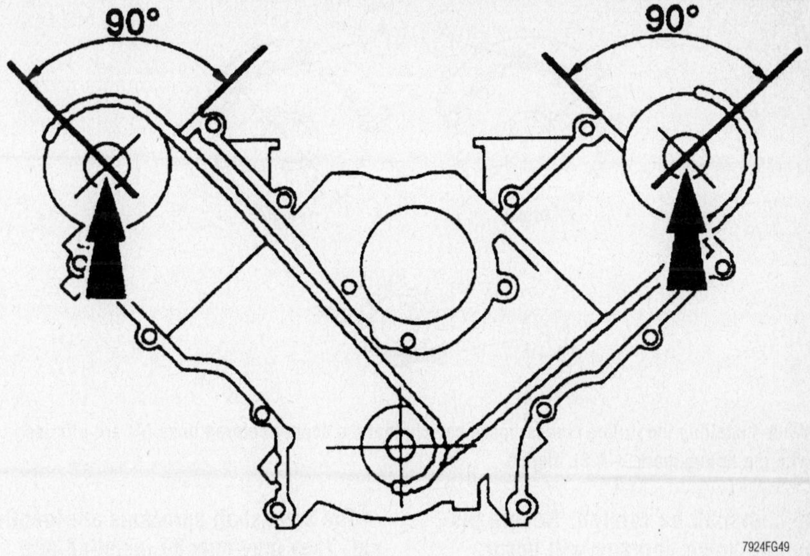

When removing the timing chains, rotate the crankshaft so that the camshaft keyways are positioned as shown—4.6L engines

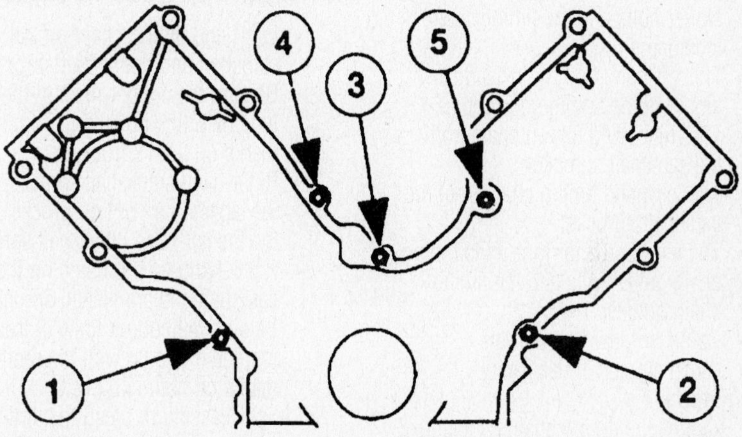

Tighten the first 5 front cover fasteners in the sequence shown—4.6L engine

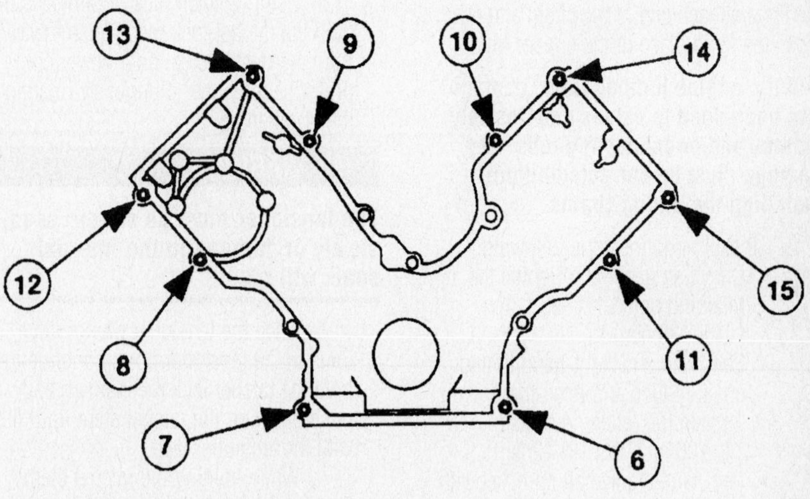

Continue tightening the remaining fasteners in the sequence shown here—4.6L engine

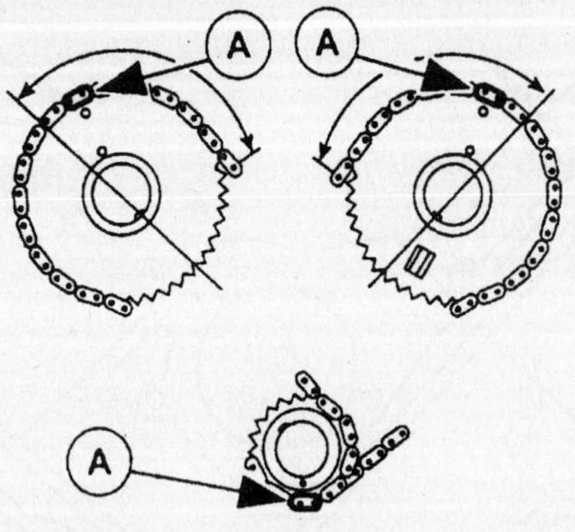

When installing the timing chains, make certain that the copper colored links (A) are aligned with the timing marks—4.6L engine

the camshaft be rotated. Severe piston and valve damage will occur.

5. Remove or disconnect the following:
- The 2 left-hand and right-hand tensioner bolts and the timing chain tensioners
- The left-hand and right-hand tensioner guides off the dowel pins
- The right-hand timing chain from the camshaft sprocket
- The left-hand timing chain from the camshaft sprocket
- The left-hand and right-hand timing chain guide bolts and the timing chain guides
- Camshaft gear bolt and the camshaft gear, if necessary

To install:

6. Examine the timing chains, looking for the copper links. If the copper links are not visible, lay the chain on a flat surface and pull the chain taught until the opposite sides of the chain contact one another. Mark the links at each end of the chain and use these marks in place of the copper links.

➡If the engine jumped time, damage has been done to valves and possibly pistons and/or connecting rods. Any damage must be corrected before installing the timing chains.

7. Install or connect the following:
- Camshaft gears and tighten the retaining bolt to 81–95 ft. lbs. (110–130 Nm)
- Left-hand and right-hand timing chain guides and retaining bolts. Tighten the retaining bolts to 71–106 inch lbs. (8–12 Nm).
- Left-hand crankshaft sprocket with the tapered part of the sprocket facing away from the engine block

➡The crankshaft sprockets are identical. They may only be installed one way, with the tapered part of the sprockets facing each other. Ensure that the keyway and timing marks on the crankshaft sprockets are aligned.

- Left-hand timing chain on the camshaft and crankshaft sprockets. Be sure the copper links of the timing chain line up with the timing marks on both sprockets.
- Right-hand crankshaft sprocket with the tapered part of the sprocket facing the left-hand crankshaft sprocket
- Right-hand timing chain on the camshaft and crankshaft sprockets. Be sure the copper links of the timing chain line up with the timing marks on both sprockets.

8. It is necessary to bleed the timing chain tensioners before installation. Proceed as follows:

a. Place the timing chain tensioner in a soft-jawed vise.

b. Using a small pick or similar tool, hold the ratchet lock mechanism away from the ratchet stem and slowly compress the tensioner plunger by rotating the vise handle.

⁂ WARNING

The tensioner must be compressed slowly or damage to the internal seals will result.

c. Once the tensioner plunger bottoms in the tensioner bore, continue to hold the ratchet lock mechanism and push down on the ratchet stem until flush with the tensioner face.

d. While holding the ratchet stem flush to the tensioner face, release the ratchet lock mechanism and install a paper clip or similar tool in the tensioner body to lock the tensioner in the collapsed position.

e. The paper clip must not be removed until the timing chain, tensioner, tensioner arm and timing chain guide are completely installed on the engine.

9. Install or connect the following:
- Left-hand and right-hand timing chain tensioner guides on the dowel pins
- Left-hand and right-hand timing chain tensioners in position and install the retaining bolts. Tighten the bolts to 15–22 ft. lbs. (20–30 Nm).

10. Remove the retaining pins from the timing chain tensioners.

11. Remove Camshaft Positioning Tool T91P-6256-A and Camshaft Positioning Adapters T92P-6256-A from the camshaft.

12. Install or connect the following:
- Crankshaft sensor ring on the crankshaft

13. Apply a bead of silicone sealer along the cylinder head-to-cylinder block and the oil pan-to-cylinder block sealing surfaces.
- Engine front cover carefully onto the dowel pins. Tighten the engine front cover bolts, in sequence, to 15–22 ft. lbs. (20–30 Nm).
- Idler pulley in position and the retaining bolt. Tighten the bolt to 15–22 ft. lbs. (20–30 Nm).
- Drive belt tensioner and the 3 retaining bolts. Tighten the bolts to 15–22 ft. lbs. (20–30 Nm).
- CMP sensor and the retaining bolt. Tighten the bolt to 106 inch lbs. (12 Nm).
- The 4 oil pan-to-engine front cover bolts and tighten, in sequence, in 2 steps: Tighten the bolts to 15 ft. lbs. (20 Nm). Rotate the bolts an additional 60 degrees.
- CKP sensor and the retaining bolt. Tighten the bolt to 106 inch lbs. (12 Nm).
- CKP sensor electrical harness connector
- Power steering pump and the 2 upper and 2 lower retaining bolts. Tighten the bolts to 15–20 ft. lbs. (20–30 Nm).
- Ignition coil and brackets on the engine front cover and the bracket bolts. Tighten the bolts to 15–22 ft. lbs. (20–30 Nm).
- Ignition coil and capacitor electrical harness connectors
- CMP electrical harness connector
- Water pump pulley and tighten the

bolts to 15–22 ft. lbs. (20–30 Nm).
- Radiator, fan blade and fan shroud assembly
- Accessory drive belt

14. Refill the engine oil with the recommended type of engine oil and the correct amount.
15. Connect the negative battery cable.
16. Start the engine and check for leaks.
17. Road test the vehicle and check for proper engine operation.

5.4L SOHC and 6.8L Engines

1. Before servicing the vehicle, refer to the precautions in the beginning of this section.
2. Remove or disconnect the following:
   - Negative battery cable
   - Radiator, fan blade and fan shroud assembly
   - Accessory drive belt
   - Water pump pulley
   - Electrical harness connectors from both ignition coils
   - Both ignition coils with their brackets attached
   - The 2 upper power steering pump retaining bolts
   - The 2 lower power steering pump retaining bolts and move the pump aside
   - Crankshaft Position (CKP) sensor electrical harness connector
   - Retaining bolt and the CKP sensor.
3. Drain the engine oil.
   - The 4 oil pan-to-engine front cover retaining bolts
   - The crankshaft damper retaining bolt and washer from the crankshaft
   - Crankshaft damper from the crankshaft
   - Camshaft Position (CMP) sensor retaining bolt and remove the CMP sensor
   - Idler pulley bolt and remove the pulley
   - The 3 belt tensioner bolts and remove the tensioner
   - The 8 engine front cover retaining bolts and the 7 nuts. Swing the top of the cover out off the dowel pins and remove the cover.
   - Sensor ring from the crankshaft
4. Use Crankshaft Holding Tool 303-448 (T93P-6303-A), to position the camshaft.
5. Rotate the crankshaft until both camshaft keyways are 90 degrees from the cam cover surface. Be sure the copper links line up with the dots on the camshaft sprockets.

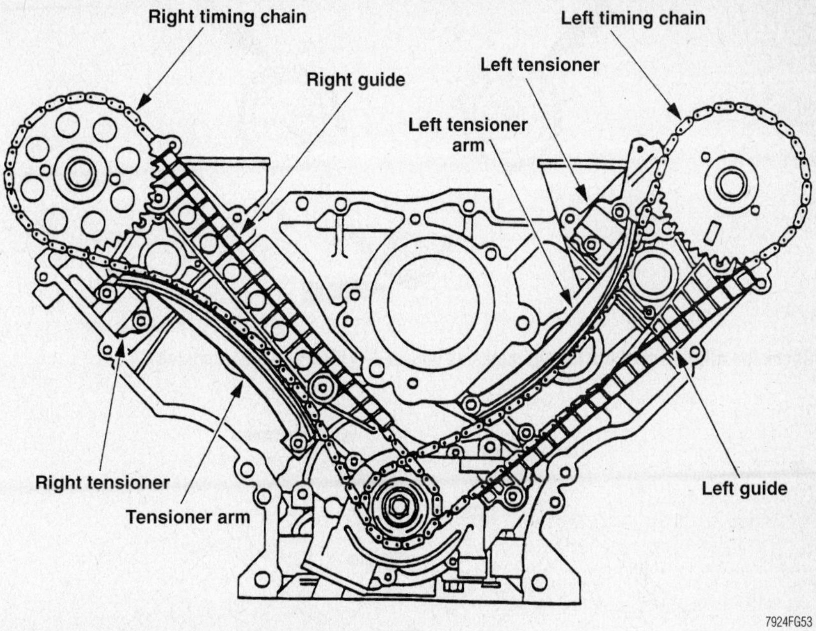

Timing chains and related components—5.4L and 6.8L engines

✲✲ WARNING

At no time, when the timing chains are removed and the cylinder heads are installed may the crankshaft or the camshaft be rotated. Severe piston and valve damage will occur.

6. Install Camshaft Holding Tool T96T-6256-B or equivalent, on the camshaft.
7. Remove or disconnect the following:
   - 2 left-hand and right-hand tensioner bolts and the timing chain tensioners
   - Left-hand and right-hand tensioner guides off the dowel pins
   - Right-hand timing chain from the camshaft sprocket
   - Left-hand timing chain from the camshaft sprocket
   - Left-hand and right-hand timing chain guide bolts, and the timing chain guides

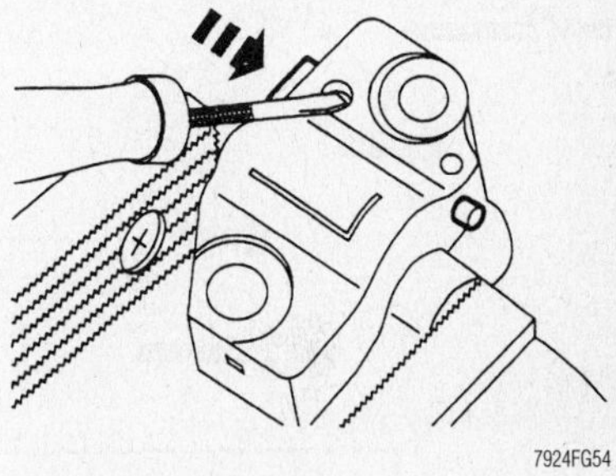

Compress the tensioner while holding the ratchet mechanism with a suitable tool—5.4L and 6.8L engines

To install:

8. Examine the timing chains, looking for the copper links. If the copper links are not visible, lay the chain on a flat surface

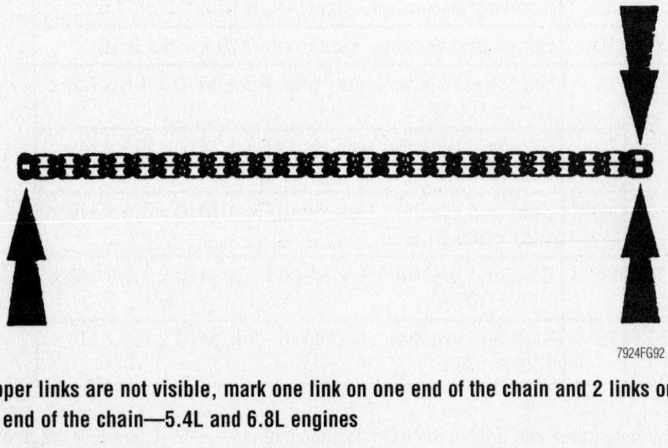

If the copper links are not visible, mark one link on one end of the chain and 2 links on the opposite end of the chain—5.4L and 6.8L engines

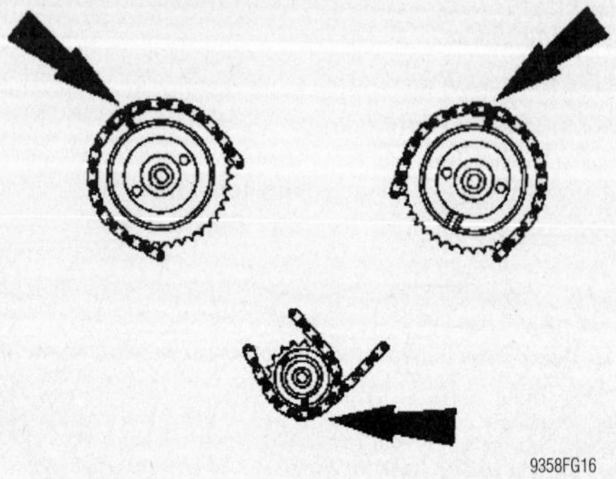

Check the alignment of the timing marks as shown——5.4L and 6.8L engines

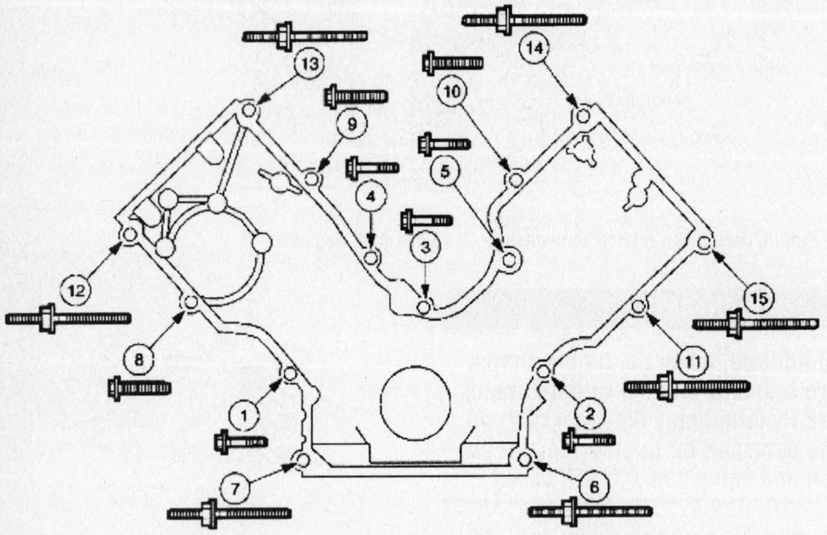

Item	Description
1	Bolt, Hex Flange Head Pilot, M8 x 1.25 x 53
2	Bolt, Hex Flange Head Pilot, M8 x 1.25 x 53
3	Bolt, Hex Flange Head Pilot, M8 x 1.25 x 53
4	Bolt, Hex Flange Head Pilot, M8 x 1.25 x 53
5	Bolt, Hex Flange Head Pilot, M8 x 1.25 x 53
6	Stud, Hex-Head Pilot, M10 x 1.5 x 1.5 x 103.1
7	Stud, Hex-Head Pilot, M10 x 1.5 x 1.5 x 103.1
8	Screw and Washer, Hex Pilot, M10 x 1.5 x 57.5
9	Screw and Washer, Hex Pilot, M10 x 1.5 x 57.5
10	Screw and Washer, Hex Pilot, M10 x 1.5 x 57.5
11	Stud and Washer, Hex-Head Pilot, M10 x 1.5 x M8 x 1.25 x 109.6
12	Stud and Washer, Hex-Head Pilot, M10 x 1.5 x M8 x 1.25 x 109.6
13	Stud and Washer, Hex-Head Pilot, M10 x 1.5 x M8 x 1.25 x 109.6
14	Stud and Washer, Hex-Head Pilot, M10 x 1.5 x M8 x 1.25 x 109.6
15	Stud and Washer, Hex-Head Pilot, M10 x 1.5 x M8 x 1.25 x 109.6

9358FG17

Install the timing cover and tighten the bolts to specification——5.4L and 6.8L engines

and pull the chain taught until the opposite sides of the chain contact one another. Mark the links at each end of the chain and use these marks in place of the copper links.

➡If the engine jumped time, damage has been done to valves and possibly pistons and/or connecting rods. Any damage must be corrected before installing the timing chains.

9. Install or connect the following:
   - Left-hand and right-hand timing chain guides and retaining bolts. Tighten the retaining bolts to 89 inch lbs. (10 Nm).
   - Left-hand crankshaft sprocket with the tapered part of the sprocket facing away from the engine block.

➡The crankshaft sprockets are identical. They may only be installed one way, with the tapered part of the sprockets facing each other. Ensure that the keyway and timing marks on the crankshaft sprockets are aligned.

   - Left-hand timing chain on the camshaft and crankshaft sprockets. Be sure the 1 copper link aligns with the mark on the crankshaft sprocket and the 2 copper links align with the mark on the camshaft sprocket.
   - Right-hand crankshaft sprocket with the tapered part of the sprocket facing the left-hand crankshaft sprocket
   - Right-hand timing chain on the camshaft and crankshaft sprockets. Be sure the copper links of the timing chain line up with the timing marks on both sprockets.

10. It is necessary to bleed the timing chain tensioners before installation. Proceed as follows:

    a. Place the timing chain tensioner in a soft-jawed vise.

    b. Using a small pick or similar tool, hold the ratchet lock mechanism away from the ratchet stem and slowly compress the tensioner plunger by rotating the vise handle.

⁂ WARNING

The tensioner must be compressed slowly or damage to the internal seals will result.

    c. Once the tensioner plunger bottoms in the tensioner bore, continue to hold the ratchet lock mechanism and push down on the ratchet stem until flush with the tensioner face.

d. While holding the ratchet stem flush to the tensioner face, release the ratchet lock mechanism and install a paper clip or similar tool in the tensioner body to lock the tensioner in the collapsed position.

e. The paper clip must not be removed until the timing chain, tensioner, tensioner arm and timing chain guide are completely installed on the engine.

11. Install the left-hand and right-hand timing chain tensioner guides on the dowel pins.

12. Place the left-hand and right-hand timing chain tensioners in position and install the retaining bolts. Tighten the bolts to 15–22 ft. lbs. (20–30 Nm).

13. Remove the retaining pins from the timing chain tensioners.

14. Remove Cam Holding Tool T96T-6256-B or equivalent, from the camshaft.

15. Install the crankshaft sensor ring on the crankshaft.

16. Apply silicone gasket along the cylinder head-to-cylinder block and engine oil pan-to-cylinder block sealing surfaces.

17. Install or connect the following:

- Engine front cover carefully onto the dowel pins
- Engine front cover bolts, in sequence, in the following manner: bolts 1 through 7: 15–22 ft. lbs. (20–30 Nm); bolts 6 through 15: 29–40 ft. lbs. (40–55 Nm)
- Idler pulley in position and install the retaining bolt. Tighten the bolt to 15–22 ft. lbs. (20–30 Nm).
- Drive belt tensioner in position and install 3 retaining bolts. Tighten the bolts to 15–22 ft. lbs. (20–30 Nm).
- CMP sensor in position and install the retaining bolt. Tighten the bolt to 106 inch lbs. (12 Nm).
- Damper on the crankshaft. Ensure the crankshaft key and keyway are aligned.
- 4 front oil pan-to-engine front cover retaining bolts and tighten in sequence in 2 steps: Tighten the bolts to 15 ft. lbs. (20 Nm); Rotate the bolts an additional 60 degrees.
- CKP sensor in position and install the retaining bolt. Tighten the bolt to 106 inch lbs. (12 Nm).
- CKP sensor electrical harness connector
- Power steering pump with 2 upper and 2 lower retaining bolts. Tighten the bolts to 15–20 ft. lbs. (20–30 Nm).
- Ignition coil and brackets on the engine front cover and install the bracket bolts. Tighten the bolts to 15–22 ft. lbs. (20–30 Nm).
- Ignition coil and capacitor electrical harness connectors
- CMP electrical harness connector
- Water pump pulley and tighten the bolts to 15–22 ft. lbs. (20–30 Nm)
- Radiator
- Negative battery cable

18. Refill the engine oil with the recommended type of engine oil and the correct amount.

19. Start the engine and check for leaks.

20. Road test the vehicle and check for proper engine operation.

5.4L DOHC Engines

1. Before servicing the vehicle, refer to the precautions in the beginning of this section.

2. Drain the oil.

3. Remove or disconnect the following:

- Negative battery cable
- Valve covers
- Cooling fan
- Water pump pulley
- Crankshaft pulley using Crankshaft Damper Remover 303-009 (T58P-6316-D)
- Nuts and the hose support, if equipped
- Power steering pump and position aside
- Air Conditioner (A/C) muffler bolt and position out of the way
- Crankshaft Position (CKP) sensor
- Crankshaft front oil seal using Front Cover Seal Remover 303-107 (T74P-6700-A)
- Camshaft Position (CMP) sensor
- Belt idler pulley
- Engine front cover bolts, studs, cover and the gaskets
- Sensor ring from the crankshaft

4. Use Crankshaft Holding Tool 303-448 (T93P-6303-A), to position the camshaft.

5. Rotate the crankshaft until both camshaft keyways are 90 degrees from the cam cover surface. Be sure the copper links line up with the dots on the camshaft sprockets.

⁂ WARNING

At no time, when the timing chains are removed and the cylinder heads are installed may the crankshaft or the camshaft be rotated. Severe piston and valve damage will occur.

6. Install Camshaft Holding Tool T96T-6256-B or equivalent, on the camshaft.

7. Remove or disconnect the following:

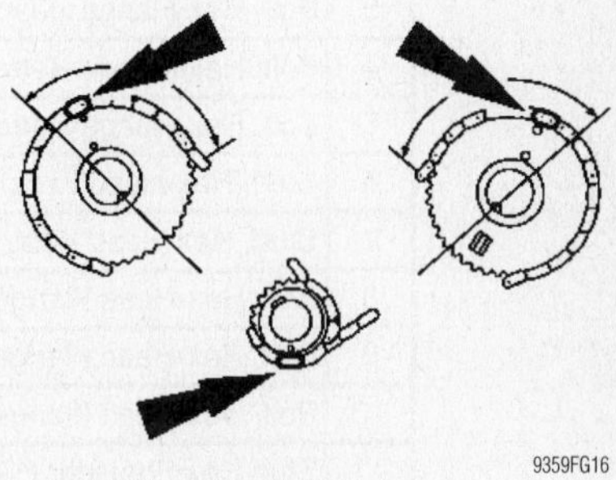

9359FG16

Check the alignment of the timing marks as shown——5.4L DOHC engines

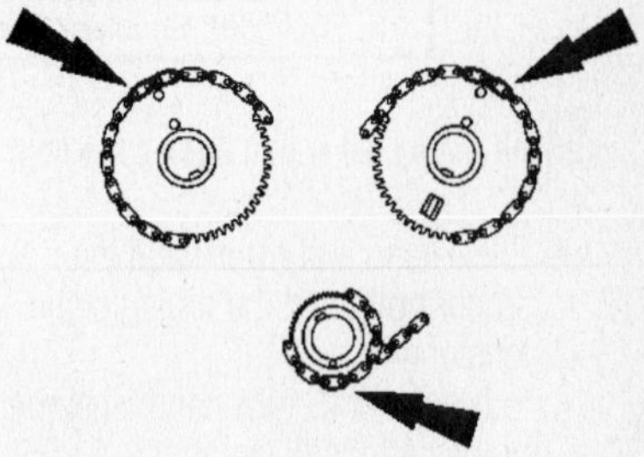

9359FG17

Verify the camshaft sprocket to copper link alignment—5.4L DOHC engines

7924FG92

If the copper links are not visible, mark one link on one end of the chain and 2 links on the opposite end of the chain—5.4L and 6.8L engines

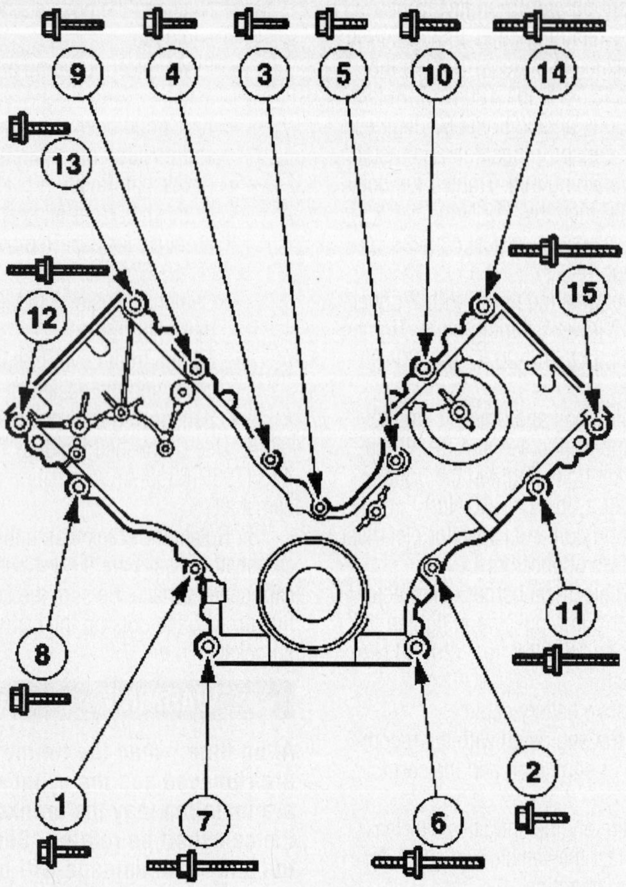

Item	Description
1	Bolt, Hex Flange Head Pilot, M8 x 1.25 x 50
2	Bolt, Hex Flange Head Pilot, M8 x 1.25 x 50
3	Bolt, Hex Flange Head Pilot, M8 x 1.25 x 50
4	Bolt, Hex Flange Head Pilot, M8 x 1.25 x 50
5	Bolt, Hex Flange Head Pilot, M8 x 1.25 x 50
6	Stud, Hex-Head Pilot, M10 x 1.5 x 30
7	Stud, Hex-Head Pilot, M10 x 1.5 x 30
8	Bolt, Hex-Head Flange, M8 x 1.25 x 50
9	Bolt, Hex-Head Flange, M8 x 1.25 x 50
10	Bolt, Hex-Head Flange, M8 x 1.25 x 50
11	Stud Hex Shoulder Pilot, M8 x 1.25 x 65 - M8 x 1.25 x
12	Stud Hex Shoulder Pilot, M8 x 1.25 x 65 - M8 x 1.25 x
13	Bolt Hex Shoulder Pilot, M8 x 1.25 x 65
14	Stud Hex Shoulder Pilot, M8 x 1.25 x 65 - M8 x 1.25 x
15	Bolt Hex Shoulder Pilot, M8 x 1.25 x 65

9359FG18

Install the timing cover and tighten the bolts to specification——5.4L DOHC engines

- 2 left-hand and right-hand tensioner bolts and the timing chain tensioners
- Left-hand and right-hand tensioner guides off the dowel pins
- Right-hand timing chain from the camshaft sprocket
- Left-hand timing chain from the camshaft sprocket
- Left-hand and right-hand timing chain guide bolts, and the timing chain guides

To install:

8. Examine the timing chains, looking for the copper links. If the copper links are not visible, lay the chain on a flat surface and pull the chain taught until the opposite sides of the chain contact one another. Mark the links at each end of the chain and use these marks in place of the copper links.

➡If the engine jumped time, damage has been done to valves and possibly pistons and/or connecting rods. Any damage must be corrected before installing the timing chains.

✲✲ WARNING

Do not compress the ratchet assembly. This will damage the ratchet assembly.

9. Compress the tensioner plunger, using the edge of a soft-jawed vise.
10. Use a small screwdriver or pick, push back and hold the ratchet mechanism.
11. While holding the ratchet mechanism, push the ratchet arm back into the tensioner housing.
12. Install a paper clip into the hole in the tensioner housing to hold the ratchet assembly and plunger in during installation.
13. Install the timing chain guides.
14. Remove tool T93P-6256-A.
15. Rotate the left hand camshaft sprocket until the timing mark is approximately at the 12 o' clock position. Rotate the right hand camshaft timing sprocket until the timing mark is approximately in the 11 o' clock position.
16. Install tool T93P-6256-A.

✲✲ WARNING

Unless otherwise instructed, at no time when the timing chains are removed and the cylinder heads are installed is the crankshaft or the camshaft to be rotated. Severe piston and valve damage will occur.

✲✲ WARNING

Rotate the crankshaft counterclockwise only. Do not rotate past position shown or severe piston or valve damage will occur.

17. Using Crankshaft Holding Tool 303-448 (T93P-6303-A), position the crankshaft, and then remove the tool.
18. Position the inner crankshaft sprocket with the long hub facing outward.

➡The outer hub will face inward.

19. Install the left hand timing chain

onto the crankshaft sprocket, aligning the one copper link on the timing chain with the slot on the crankshaft sprocket.

20. Verify the camshaft sprocket to copper link alignment.

21. Position the tensioner arms and tensioners, and install the bolts. Tighten the bolts to 18 ft. lbs. (25 Nm).

22. Remove the paper clips used to retain the ratchet assemblies.

23. Remove the camshaft holding tool.

24. Position the crankshaft sensor ring on the crankshaft.

25. Apply silicone gasket along the cylinder head-to-cylinder block and engine oil pan-to-cylinder block sealing surfaces.

26. Install or connect the following:

- Engine front cover carefully onto the dowel pins
- Engine front cover bolts, in sequence, in the following manner: bolts 1 through 5: 18 ft. lbs. (25 Nm); bolts 6 through 7: 37 ft. lbs. (50 Nm), bolts 1 through 15 to 18 ft. lbs. (25 Nm)
- Idler pulley in position and install the retaining bolt. Tighten the bolt to 18 ft. lbs. (25 Nm).
- CMP sensor in position and install the retaining bolt. Tighten the bolt to 106 inch lbs. (12 Nm).
- Damper on the crankshaft. Ensure the crankshaft key and keyway are aligned.
- 4 front oil pan-to-engine front cover retaining bolts and tighten in sequence in 3 steps: Tighten the bolts to 18 ft. lbs. (25 Nm), then to 15 ft. lbs. (20 Nm); then rotate the bolts an additional 60 degrees.
- Crankshaft front oil seal using Front Crankshaft Seal Installer 303-635 and Crankshaft Seal Installer/Aligner 303-335 (T88T-6701-A)
- CKP sensor in position and install the retaining bolt. Tighten the bolt to 106 inch lbs. (12 Nm).
- CKP sensor electrical harness connector
- A/C muffler and tighten the nut to 18 ft. lbs. (25 Nm).
- Power steering pump with the upper and lower retaining bolts. Tighten the bolts to 18 ft. lbs. (25 Nm).
- Crankshaft pulley, bolt and washer using Crankshaft Damper Replacer 303-102 (T74P-6316-B) and tighten the bolt in 4 steps as follows:

a. Step 1: 66 ft. lbs. (90 Nm).
b. Step 2: Loosen one full turn.
c. Step 3: 37 ft. lbs. (50 Nm).
d. Step 4: an additional 85-95 degrees.

27. Install or connect the following:

- Water pump pulley and tighten the bolts to 18 ft. lbs. (25 Nm)
- Cooling fan
- Valve covers
- Negative battery cable

28. Refill the engine oil with the recommended type of engine oil and the correct amount.

29. Start the engine and check for leaks.

30. Road test the vehicle and check for proper engine operation.

Piston and Ring Positioning

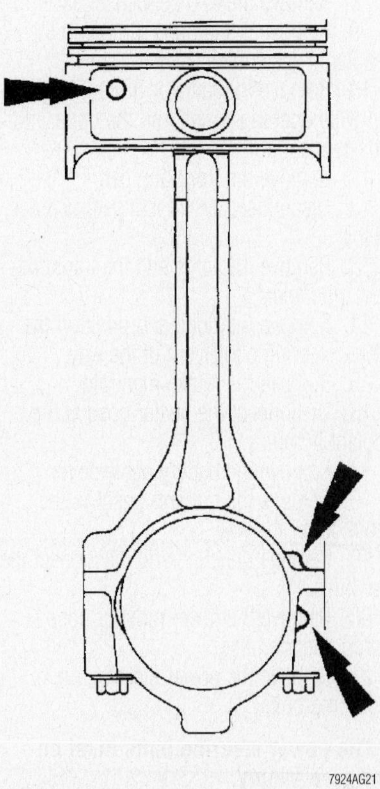

4.6L engines—piston and connecting rod front mark locations

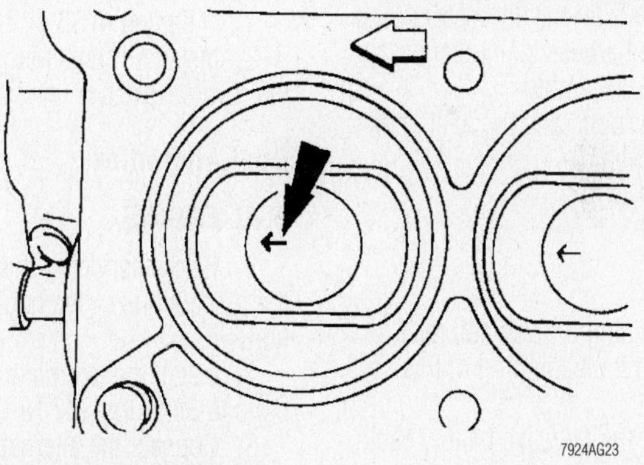

4.6L engines—piston-to-engine orientation

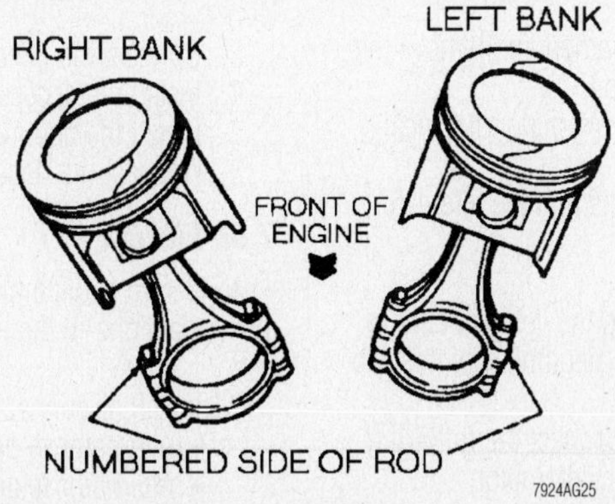

5.4L, 5.8L and 6.8L engines—piston and connecting rod assembly positioning

DIESEL ENGINE REPAIR

Alternator

REMOVAL

Single Alternator

6.0L ENGINE

1. Before servicing the vehicle, refer to the precautions in the beginning of this section.
2. Disconnect the negative battery cable.
3. Remove the cooling fan.
4. Remove the accessory drive belt.
5. Disconnect the alternator electrical connectors.
6. Remove the bolts and the alternator.

7.3L ENGINE

1. Before servicing the vehicle, refer to the precautions in the beginning of this section.
2. Remove or disconnect the following:
   - Negative battery cable
   - Air cleaner, if necessary
   - Drive belt from the alternator pulley
   - Electrical harness connectors at the alternator assembly
   - Alternator bolts and the alternator from the vehicle

Dual Alternator

6.0L ENGINE

1. Before servicing the vehicle, refer to the precautions in the beginning of this section.
2. Disconnect the negative battery cable.
3. Remove the cooling fan.
4. Remove the accessory drive belt.
5. Remove the nut and position the generator B+ cable aside.
6. Loosen the 2 clamps and remove the turbo-to-air cooler pipe.
7. Disconnect the alternator electrical connectors.
8. Remove the bolts and the alternator.

7.3L ENGINE

1. Before servicing the vehicle, refer to the precautions in the beginning of this section.
2. Remove or disconnect the following:
   - Single (upper) alternator
   - Electrical harness connectors at the alternator assembly
   - Alternator bolts and the alternator from the vehicle

INSTALLATION

Single Alternator

6.0L ENGINE

1. Before servicing the vehicle, refer to the precautions in the beginning of this section.
2. Install the alternator and tighten the bolts to 35 ft. lbs. (47 Nm)
3. Connect the alternator electrical connectors.
4. Install the accessory drive belt.
5. Install the cooling fan.
6. Connect the negative battery cable.

7.3L ENGINE

1. Before servicing the vehicle, refer to the precautions in the beginning of this section.
2. Install or connect the following:
   - Alternator in position
   - Alternator retaining bolts and tighten to 35 ft. lbs. (47 Nm)
3. Install all remaining components in the reverse order of removal.

Dual Alternator

6.0L ENGINE

1. Before servicing the vehicle, refer to the precautions in the beginning of this section.
2. Install the alternator and tighten the bolts to 35 ft. lbs. (47 Nm)
3. Connect the alternator electrical connectors.
4. Install the turbo-to-air cooler pipe and tighten the 2 clamps to 9 ft. lbs. (12 Nm).
5. Connect the generator B+ cable.
6. Install the accessory drive belt.
7. Install the cooling fan.
8. Connect the negative battery cable.

7.3L ENGINE

1. Before servicing the vehicle, refer to the precautions in the beginning of this section.
2. Install or connect the following:
   - Alternator in position
   - Alternator retaining bolts and tighten to 35 ft. lbs. (47 Nm)
   - Single (upper) alternator

Engine Assembly

REMOVAL & INSTALLATION

6.0L Engine

1. Before servicing the vehicle, refer to the precautions in the beginning of this section.
2. With the vehicle in NEUTRAL, position it on a hoist.
3. Disconnect the left hand and right hand battery ground cables.
4. On vehicles with manual transmission perform the following steps:

➡On vehicles equipped with manual transmissions, the transmission must be removed before the engine can be removed.

   a. Remove the transmission.
   b. Remove the clutch.
5. Remove the air cleaner assembly.
6. Remove the radiator.
7. Remove the charge air cooler.
8. Remove the A/C condenser assembly.
9. Remove the parking lamp and the headlamp assemblies.
10. Remove the radiator grille, the radiator grille opening panel, and the upper radiator core supports.
11. Remove the front bumper.
12. Disconnect the transmission cooler hoses.
13. Remove the bolts and the transmission oil cooler.
14. Remove the bolts and position the power steering cooler out of the way.
15. Remove the intake manifold.
16. Disconnect the heater hose at the coolant pump.
17. Remove the battery cable cover.
18. Remove the nut and position the cable out of the way.
19. Remove the clips and disconnect the fuel lines.
20. Remove the lower radiator hose clamp and the hose.
21. Remove the power steering upper mounting bolts.

➡The power steering bolts must be removed evenly.

22. Remove the bolts and position the power steering pump out of the way.

23. Remove the bolt and the left side ground cable.
24. Remove the nut and the battery cable bracket.
25. Remove the ground stud and the ground cable.
26. Disconnect the glow plug electrical connectors.
27. On vehicles with A/C perform the following steps:
a. Disconnect the A/C high pressure switch and the A/C clutch electrical connectors.
b. Disconnect the air conditioning manifold lines from the A/C compressor.
c. Remove the A/C compressor.
28. Disconnect the Crankshaft Position (CKP) sensor electrical connector.
29. On vehicles with automatic transmission, remove the automatic transmission fluid indicator and tube.
30. Remove the ground strap at the back of the right head.
31. Remove the solenoid cap and disconnect the starter wiring.
32. Disconnect the block heater electrical connector.
33. Position the block heater and starter wiring harness out of the way.
34. On vehicles with automatic transmission perform the following steps:
a. Remove the torque converter cover.
b. Remove the torque converter nuts.
35. Remove the bolts for the turbocharger adapter pipe.
36. Remove the motor mount nuts.
37. Loosen the nuts at the turbocharger adapter pipe flange.
38. On vehicles with automatic transmission, remove the nine bell housing bolts.
39. Remove the turbocharger adapter pipe.
40. Remove the fuel injection control module mounting bolts for access to the module electrical connectors.
41. Disconnect the electrical connectors and remove the fuel injection control module.

Install the front lifting brackets 303-D043-01—6.0L engine

42. Remove the nuts and the fuel injection control module bracket.
43. Remove the left rear valve cover stud.
44. Remove the transmission cooler line bracket.
45. Secure the turbocharger outlet pipe.
46. Remove the manufacturer's lifting eye.
47. Install the engine lifting eye 303-D043-02 on the right hand cylinder head.
48. Install the front lifting brackets 303-D043-01.
49. On vehicles with automatic transmission, position a suitable jack under the transmission.
50. Install the Heavy Duty Floor Crane and Diesel Engine Lifting Bracket on the engine.
51. Raise the engine high enough to clear the No. 1 crossmember and pull the engine forward and clear of the vehicle.

To install:

52. With the vehicle in NEUTRAL, position it on a hoist.
53. Raise the engine high enough to clear the No. 1 crossmember, then position the engine into the vehicle.
54. On vehicles with automatic transmission, align the torque converter studs with the holes in the engine flywheel, then lower the engine onto the engine mount towers.
55. On vehicles with manual transmission, lower the engine onto the engine mount towers.
56. Remove the Heavy Duty Floor Crane and the Diesel Engine Lifting Bracket.
57. on vehicles with automatic transmission perform the following steps:
a. Remove the transmission jack.
b. Install the transmission-to-engine mounting bolts and tighten to 35 ft. lbs. (47 Nm).
c. Install the torque converter-to-flywheel retaining nuts and tighten to 26 ft. lbs. (35 Nm).
d. Install the flywheel housing cover.

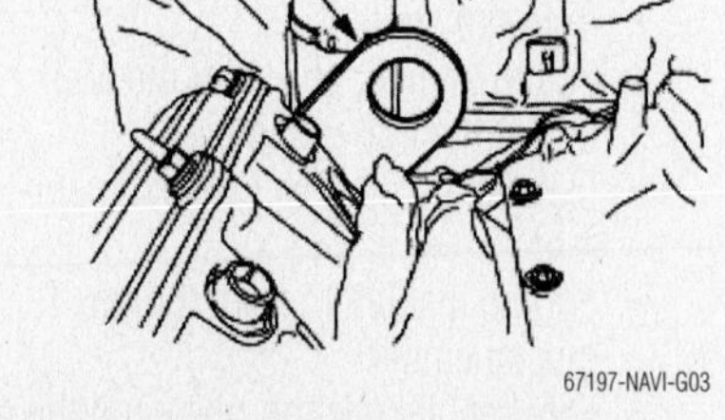

Install the engine lifting eye 303-D043-02 on the right hand cylinder head—6.0L engine

58. Install the left hand and right hand side engine mount retaining nuts and tighten to 76 ft. lbs. (103 Nm).
59. Remove the engine lifting eye from the right side cylinder head.
60. Install the manufacturer's lifting bracket.
61. Remove the two engine lift adapters.
62. Position the turbocharger outlet pipe.
63. Install the transmission cooler tube bracket and nut and tighten to 89 inch lbs. (10 Nm).
64. Install the left rear valve cover stud and tighten to 71 inch lbs. (8 Nm).
65. Install the fuel injection control module bracket and nuts and tighten to 71 inch lbs. (8 Nm).
66. Connect the fuel injection control module electrical connectors and position the module on the bracket.
67. Install the fuel injection control module bolts and nuts and tighten to 10 ft. lbs. (13 Nm).
68. Position the turbocharger adapter pipe.
69. Install the bolts for the turbocharger adapter pipe but do not tighten bolts at this time.
70. Position back the block heater and starter wiring.
71. Connect the block heater electrical connector.
72. Connect the starter wiring and install the solenoid cap.
73. Connect the ground strap on the right head and install the bolt and tighten to 89 inch lbs. (10 Nm).
74. On vehicles with automatic transmission, install the automatic transmission fluid indicator and tube and tighten to 71 inch lbs. (8 Nm).
75. Connect the CKP sensor electrical connector.
76. On vehicles with A/C perform the following steps:
a. Install the A/C compressor and tighten the bolts to 18 ft. lbs. (25 Nm).
b. Position the A/C compressor manifold and install the bolt and tighten to 15 ft. lbs. (21 Nm).
c. Connect the A/C high pressure switch and the clutch electrical connectors.
77. Connect the glow plug module electrical connectors.
78. Connect the ground cable and the ground stud and tighten to 35 ft. lbs. (47 Nm).
79. Install the battery cable bracket and the nut and tighten to 35 ft. lbs. (47 Nm).
80. Install the ground cable and bolt onto the left side of the engine block and tighten to 35 ft. lbs. (47 Nm).

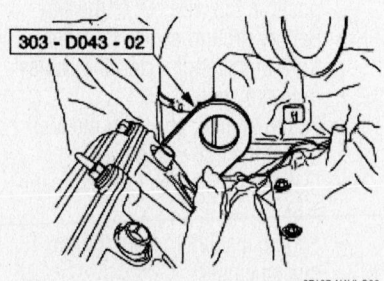

81. Position the power steering pump and install the lower bolts and tighten to 18 ft. lbs. (25 Nm).

➡**The lower bolts need to be installed evenly.**

82. Install the power steering pump upper bolts and tighten to 18 ft. lbs. (25 Nm).
83. Install the lower radiator hose and clamp.
84. Connect the fuel lines and install the clips.
85. Connect the battery crossover cable and tighten to 9 ft. lbs. (12 Nm).
86. Install the battery cable cover.
87. Connect the heater hose at the coolant pump.
88. Install the intake manifold.
89. Tighten the turbocharger adapter pipe at the flanges to 20 ft. lbs. (27 Nm).
90. Install the power steering cooler and bolts and tighten to 8 ft. lbs. (11 Nm).
91. Install the transmission oil cooler and tighten to 71 inch lbs. (8 Nm).
92. Connect the transmission cooler hoses.
93. Install the bumper.
94. Install the upper radiator core support, radiator grill opening panel and the radiator grill.
95. Install the headlamp and the parking lamp assemblies.
96. On vehicles with A/C, install the A/C condenser assembly.
97. Install the charge air cooler.
98. Install the radiator.
99. Install the air cleaner assembly.
100. On vehicles with manual transmission perform the following steps:
    a. Install the clutch assembly.
    b. Install the transmission.
101. Fill the cooling system.
102. Fill the motor with clean engine oil.
103. Connect the left and right hand battery cables.
104. Check and fill the automatic transmission, if equipped.

7.3L Engine

⁂ CAUTION

The fuel system remains under pressure, even after the engine has been turned OFF. The fuel system pressure must be relieved before disconnecting any fuel lines. Failure to do so may result in fire and/or personal injury.

1. Before servicing the vehicle, refer to the precautions in the beginning of this section.
2. Remove or disconnect the following:
    - Hood
    - Coolant
    - Engine oil
    - Negative battery cable
    - Air cleaner and intake duct assembly
    - Upper grille support bracket
    - Upper air conditioning condenser mounting bracket
    - Refrigerant, using approved equipment to remove the condenser
    - Radiator fan shroud halves
    - Fan and clutch assembly
    - Radiator hoses and the transmission cooler lines, if equipped
    - Condenser. Cap all openings immediately!
    - Radiator
    - Power steering pump and position it out of the way
    - Fuel supply line heater and alternator wires at the alternator
    - Oil pressure sending unit wire at the sending unit
    - Sender from the firewall and lay it on the engine
    - Accelerator cable and the speed control cable, if equipped, from the injection pump
    - Accelerator cable bracket with the cables attached, from the intake manifold and position it out of the way
    - Transmission kickdown rod from the injection pump, if equipped
    - Main wiring harness connector from the right side of the engine and the ground strap from the rear of the engine
    - Fuel system pressure
    - Fuel return hose from the left rear of the engine
    - The 2 upper transmission-to-engine attaching bolts
    - Heater hoses
    - Water temperature sender wire
    - Overheat light switch wire and position the wire out of the way
    - Battery ground cables from the front of the engine and the cables from the starter
    - Fuel inlet line, and plug the fuel line at the fuel pump
    - Exhaust pipe at the exhaust manifold
    - Engine insulators from the no. 1 crossmember
    - Flywheel inspection plate and the 4 converter-to-flywheel attaching nuts, if equipped with automatic transmission
3. Support the transmission on a jack.
4. Remove the 4 lower transmission attaching bolts.
5. Attach an engine lifting sling and remove the engine from the vehicle.

To install:

6. Lower the engine into vehicle.
7. Align the converter to the flexplate and the engine dowels to the transmission.
8. Install the engine mount bolts and tighten them to 80 ft. lbs. (109 Nm).
9. Remove the engine lifting sling.
10. Install the 4 lower transmission attaching bolts. Tighten the bolts to 65 ft. lbs. (88 Nm).
11. Remove transmission jack.
12. Raise and support the front end.
13. Install or connect the following:
    - 4 converter-to-flywheel attaching nuts, if equipped with an automatic transmission. Tighten the nuts to 34 ft. lbs. (47 Nm).
    - Flywheel inspection plate. Tighten the bolts to 60–90 inch lbs. (6.7–10 Nm).
    - Exhaust pipe at the exhaust manifold
    - Fuel inlet line
    - Battery ground cables to the front of the engine
    - Starter cables at the starter
    - Overheat light switch wire
    - Water temperature sender wire
    - Heater hoses
    - The 2 upper transmission-to-engine attaching bolts. Tighten the bolts to 65 ft. lbs. (88 Nm).
    - Fuel return hose at the left rear of the engine
    - Main wiring harness connector at the right side of the engine and the ground strap from the rear of the engine.
    - Transmission kickdown rod at the injection pump, if equipped
    - Accelerator cable and the speed control cable, if equipped, at the injection pump
    - Cable bracket with the cables attached, to the intake manifold
    - Oil pressure sending unit
    - Oil pressure sending unit wire at the sending unit
    - Fuel supply line heater and alternator wires at the alternator
    - Power steering pump
    - Radiator
    - Condenser
    - Radiator hoses and the transmission cooler lines, if equipped.
    - Fan and clutch assembly

- Radiator fan shroud halves
- Upper grille support bracket and upper air conditioning condenser mounting bracket
- Air cleaner and intake duct assembly

14. Refill the engine oil with the recommended type of engine oil and the correct amount.
15. Connect the negative battery cable.
16. If equipped with air conditioning, charge the system.
17. Fill the cooling system.
18. Install the hood.

Water Pump

REMOVAL & INSTALLATION

6.0L Engine

1. Before servicing the vehicle, refer to the precautions in the beginning of this section.
2. With the vehicle in NEUTRAL, position it on a hoist.
3. Remove the radiator and shroud assembly.
4. Disconnect the cooling fan electrical connector. Unclip and position the fan wiring aside.

➡Use a hole in the fan hub to prevent the fan from turning.

5. Using the special tool 303-591, loosen the cooling fan and clutch.
6. Remove the cooling fan assembly.
7. Remove the stator bolts and the stator.
8. Remove the accessory drive belt.
9. Remove the bolts and the water pump pulley.
10. Remove the bolts and the water pump.
11. Remove and discard the O-ring seal.

To install:

12. Clean and inspect the coolant pump mounting.

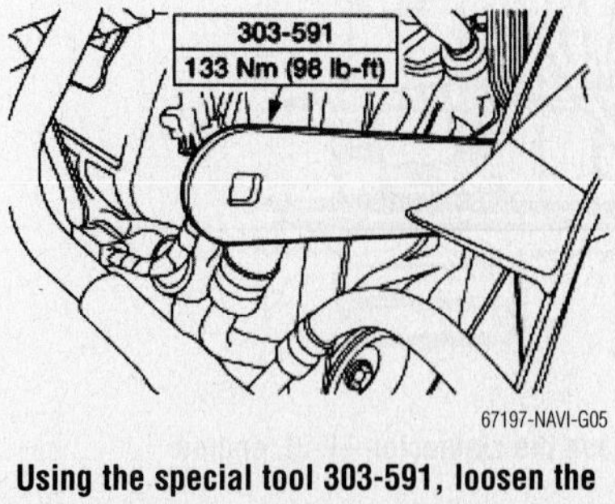

Using the special tool 303-591, loosen the cooling fan and clutch—6.0L engine

13. Install a new O-ring seal.
14. Install the water pump and tighten the bolts to 17 ft. lbs. (23 Nm).
15. Install the water pump pulley and tighten the bolts to 23 ft. lbs. (31 Nm).
16. Install the accessory drive belt.
17. Install the stator and tighten the bolts to 30 ft. lbs. (40 Nm).
18. Install the cooling fan assembly.

➡Use a hole in the fan hub to prevent the fan from turning.

19. Using the special tool 303-591, tighten the cooling fan and clutch to 98 ft. lbs. (133 Nm)
20. Connect the cooling fan electrical connector.
21. Install the radiator and shroud assembly.

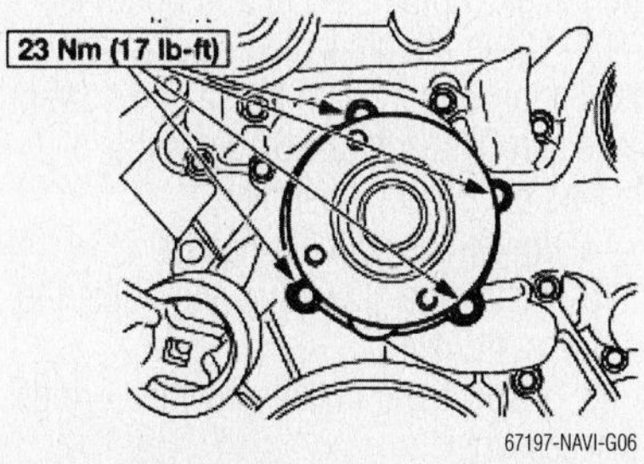

Location of the water pump retaining bolts—6.0L engine

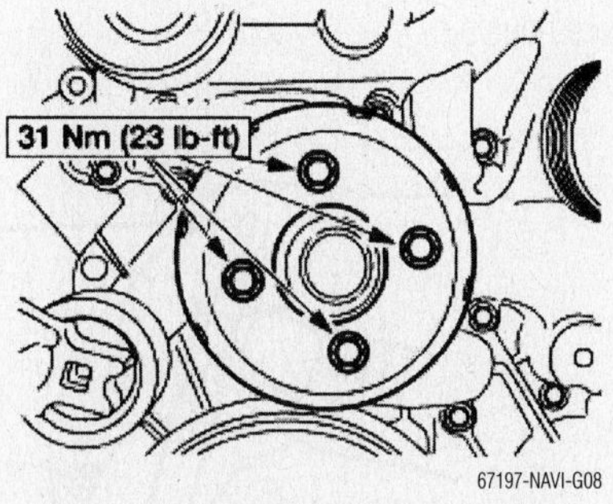

Location of the water pump pulley retaining bolts—6.0L engine

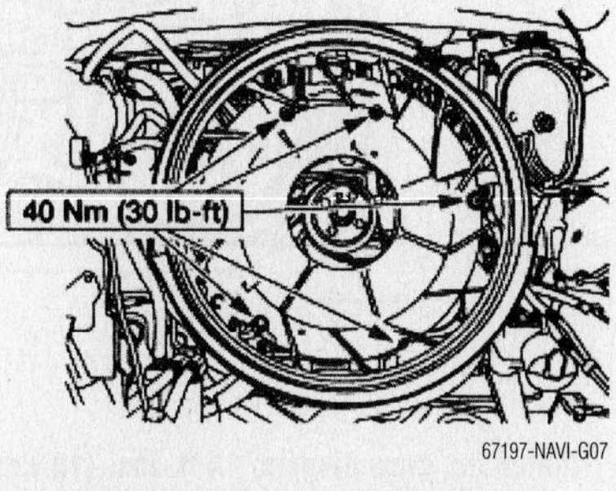

Location of the stator retaining bolts—6.0L engine

7.3L Engine

1. Before servicing the vehicle, refer to the precautions in the beginning of this section.
2. Disconnect both battery ground cables.
3. Drain the cooling system.
4. Remove or disconnect the following:
   - Radiator shroud, fan clutch and fan

➡The fan clutch bolts are right-hand thread. Remove them by turning counter-clockwise.

   - Water pump pulley bolts, loosen only at this time
   - Drive belt
   - Water pump pulley bolts and the pulley
   - Engine Coolant Temperature (ECT) sensor connector
   - Heater hose from the water pump
   - Bolts attaching the water pump to the front cover and lift off the pump

To install:

5. Thoroughly clean the mating surfaces of the pump and front cover.
6. Install or connect the following:
   - New gasket
   - Water pump over the dowel pins and into place on the front cover
   - Water pump attaching bolts. Tighten the bolts to 15 ft. lbs. (20 Nm).
   - Heater hose to the pump
   - ECT sensor connector
   - Water pump pulley and start the water pump pulley bolts
   - Drive belt
   - Tighten the pulley retaining bolts to 12–18 ft. lbs. (16–24 Nm)
   - Fan and fan shroud assembly
7. Fill and bleed the cooling system.
8. Connect the battery ground cables.
9. Start the engine and check for leaks.

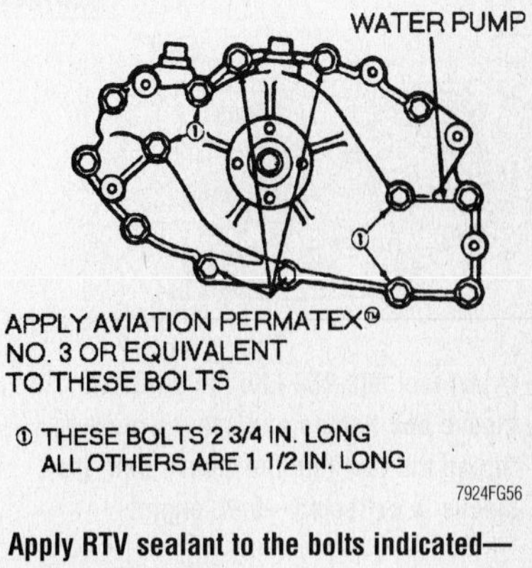

Apply RTV sealant to the bolts indicated—7.3L diesel engine

Glow Plugs

REMOVAL & INSTALLATION

6.0L Engine

1. Before servicing the vehicle, refer to the precautions in the beginning of this section.
2. If servicing the right hand glow plugs, remove the evaporator core housing.
3. Disconnect the glow plug electrical connector.
4. Remove the glow plug buss bar.
5. Remove the glow plug.

➡To install new glow plug sleeves, the cylinder heads must be removed from the vehicle.

6. If required, remove the cylinder heads from the engine.

➡The glow plug sleeve is made of stainless steel. Lubrication of the glow plug sleeve removal tap is necessary, or damage to the tap and excessive force will be needed to get the tap started.

7. Thread the glow plug sleeve removal tap 303-764 into the glow plug sleeve.
8. Install the glow plug sleeve removal tool 303-764 into the threaded sleeve and tighten. Use a wrench to thread the tool into the sleeve until the sleeve is extracted.
9. Clean the glow plug bore with a stiff wire brush. Use filtered compressed air to remove the debris from the glow plug recess.

To install:

10. Verify the glow plug bore is completely clean and dry.
11. Apply Threadlock 262 sealant to the glow plug sleeve in the two places illustrated.

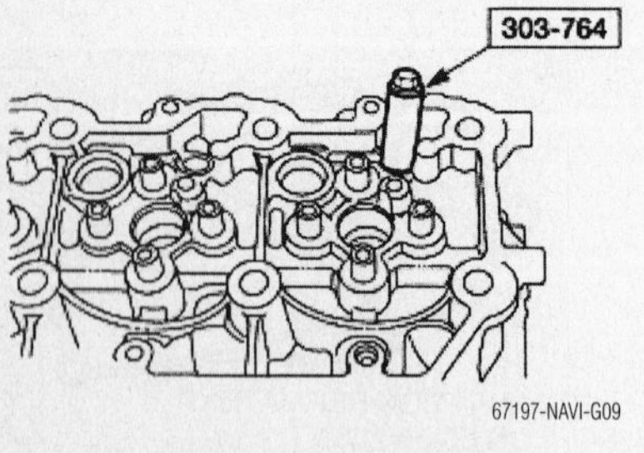

Insert tool 303-764 into the threaded sleeve and tighten and use a wrench to thread the tool into the sleeve until the sleeve is extracted—6.0L engine

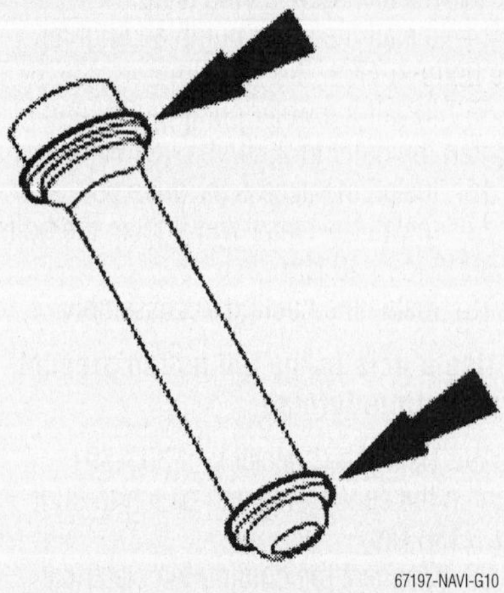

Apply Threadlock 262 sealant to the glow plug sleeve in the two places shown—6.0L engine

12. Install the glow plug sleeve 303-763 into the glow plug bore until it bottoms.
13. Clean the glow plug sleeve after installation with a nylon brush and solvent. Make sure Threadlock 262 sealant is cleaned out before it hardens.
14. Install the cylinder heads.
15. Install the glow plug and tighten to 14 ft. lbs. (19 Nm).
16. Inspect and install new the O-rings as necessary.
17. Install the glow plug buss bar.
18. Connect the glow plug electrical connector.
19. If removed, install the evaporator core housing.

7.3L Engine

✲✲ CAUTION

The red-striped wiring harness carries 115v direct current. Severe electrical shock may be received. DO NOT pierce.

1. Before servicing the vehicle, refer to the precautions in the beginning of this section.
2. Remove or disconnect the following:
   - Negative battery cable
   - Rocker arm cover
   - Glow plug electrical leads using a pair of pliers
   - Glow plugs by unscrewing them from the cylinder head with a 10mm socket and wrench
3. Inspect the tips of the plugs for any evidence of distortion or missing tip ends; replace them if necessary.

To install:

4. Install or connect the following:
   - Glow plug into the cylinder head. Tighten the glow plugs to 14 ft. lbs. (19 Nm).
   - Glow plug electrical connector. Be sure that the glow plug wiring is routed to avoid moving components in the engine bay.
   - Rocker arm cover

➡When the battery is disengaged and reconnected, some abnormal drive symptoms may occur while the vehicle

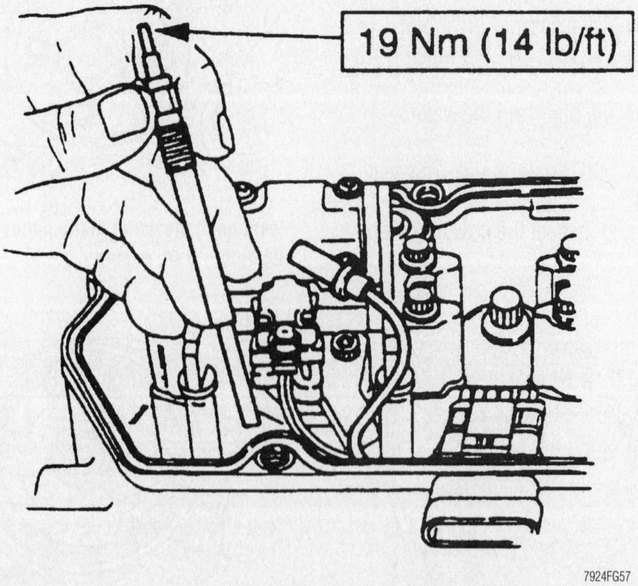

Tighten the glow plugs to 14 ft. lbs. (19 Nm) and attach the connector—7.3L engine

relearns its adaptive strategy. The vehicle may need to be driven 10 miles (16 km) or more to relearn this strategy.

- Negative battery cable

Cylinder Head

REMOVAL & INSTALLATION

6.0L Engine

RIGHT CYLINDER HEAD

1. Before servicing the vehicle, refer to the precautions in the beginning of this section.
2. With the vehicle in NEUTRAL, position it on a hoist.
3. Drain the engine oil.
4. Remove the intake manifold.
5. Remove the starter.
6. Remove the cylinder drain block.
7. Remove the evaporator core housing.
8. Disconnect the wiring retainers from the stud.

➡Mark the position of the valve cover bolts for valve cover installation.

9. Remove the bolts and the valve cover.
10. Using the special tool 303-755, disconnect the high-pressure oil rail supply line at the high-pressure oil rail.
11. Remove the bolts and the high-pressure oil rail.
12. Using the special tool 303-625, disconnect the high-pressure oil supply line.

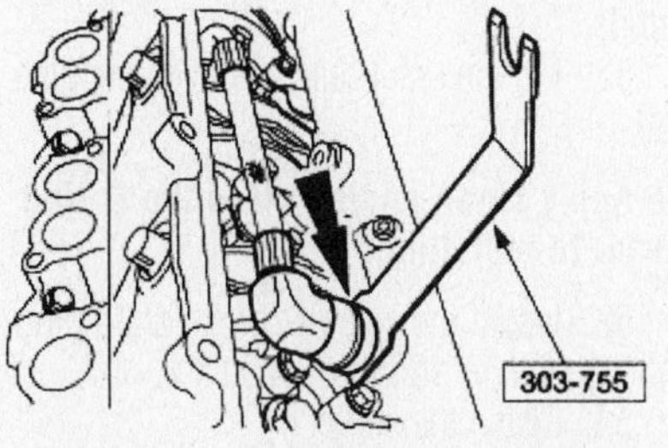

67197-NAVI-G11

Using the special tool 303-755, disconnect the high-pressure oil rail supply line at the high-pressure oil rail—6.0L engine

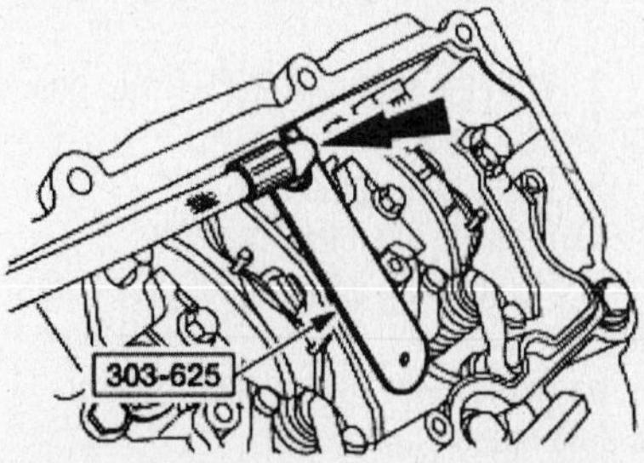

67197-NAVI-G12

Using the special tool 303-625, disconnect the high-pressure oil supply line—6.0L engine

✻✻ CAUTION

Do not attempt to put battery voltage to the fuel injector or damage to the fuel injector will occur.

13. Using a 19 mm socket, push the fuel injector electrical connector out of the rocker arm carrier.

✻✻ CAUTION

To prevent engine damage, do not use air tools to remove the fuel injectors. The clip that extracts the fuel injector can dislodge and fall into the oil drain hole.

➡There is no need to drain the fuel rail.

➡If engine oil is found in the engine coolant or engine coolant found in the combustion chambers, new injector sleeves may need to be installed.

14. Remove the bolt, fuel injector hold-down and fuel injector.
15. Remove the nuts and bolts from the turbocharger adapter pipe.
16. Remove the turbocharger adapter pipe.

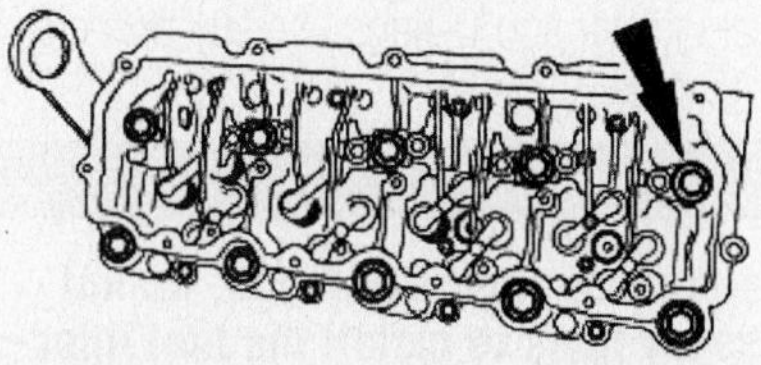

67197-NAVI-G13

Remove and discard the 10 inner cylinder head bolts—6.0L engine–right side shown, left side similar

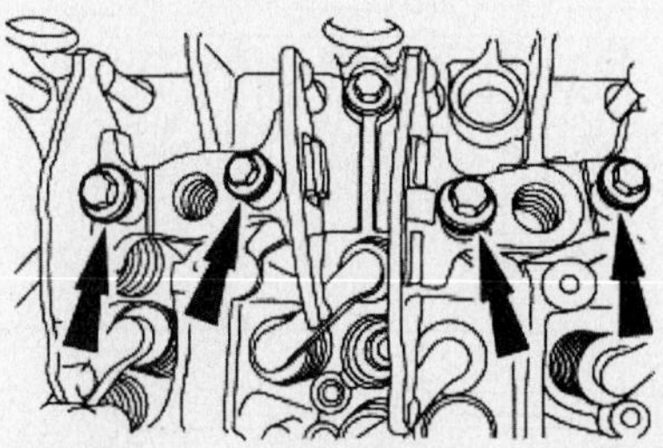

67197-NAVI-G14

Remove the eight bolts and the rocker arm assemblies—6.0L engine–right side shown, left side similar

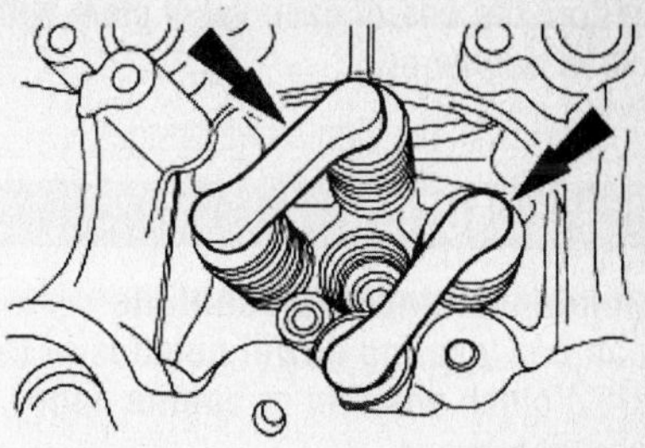

67197-NAVI-G15

Mark the eight valve bridges with a permanent marker and remove—6.0L engine–right side shown, left side similar

17. Remove the glow plug buss bar.
18. Remove the four glow plugs.
19. Remove and discard the crankcase to head tube assembly.
20. Remove and discard the 10 inner cylinder head bolts.
21. Remove the eight bolts and the rocker arm assemblies.
22. Mark the eight valve bridges with a permanent marker and remove.
23. Mark and remove the eight push rods.
24. Remove the outer cylinder head bolts.
25. Install the special tool 303-759 and the lifting crane. With the help of an assistant, remove the cylinder head from the vehicle.
26. Remove and discard the cylinder head gasket and dowels.
27. Clean and inspect the gasket sealing surfaces.

To install:

➡Use care to avoid scratching the blue compound on the cylinder head gasket. Install a new cylinder head gasket with the part number facing up and verify the top five holes and the head gasket push rod holes line up.

28. Install the dowels and the cylinder head gasket.
29. Using the special tool 303-759 and with the help of an assistant, install the cylinder head on the engine. Remove the special tools.
30. Install the outer cylinder head bolts finger-tight.

➡Higher mileage engines require push rods to be cleaned so the copper colored end of the push rod can be identified.

31. Apply clean engine oil to each end of the push rod. Insert them into their respective positions with the copper colored end up.

➡**Coat the end of each valve stem with clean engine oil.**

32. Install the eight valve bridges.

⁂ CAUTION

Rotate the crankshaft until the damper locating dowel notch is in the six o'clock position or engine damage can occur.

➡**Apply clean engine oil to the top center of each valve bridge.**

33. Install the rocker arm assemblies and eight bolts to 23 ft. lbs. (31 Nm).

⁂ CAUTION

Using too much engine oil on the threads of the cylinder head bolts can cause damage to the threads and poor sealing. Using anti-seize compounds, grease or any other lubricants other than engine oil on the cylinder head bolt threads can affect the true torque value of the bolts.

➡**Lightly lubricate the cylinder head bolt threads and flanges with clean engine oil.**

34. Install the 10 inner cylinder head retaining bolts finger-tight.

35. Tighten the head bolts in the following sequence:

a. Tighten bolts 1 through 10 to 65 ft. lbs. (88 Nm).

b. Tighten bolts 1, 3, 5, 7 and 9 to 85 ft. lbs. (115 Nm).

c. Tighten bolts in sequence 1 through 10, clockwise 90 degrees.

d. Tighten bolts in sequence 1 through 10, a second time, clockwise 90 degrees.

e. Tighten bolts in sequence 1 through 10, a third time, clockwise 90 degrees.

f. Tighten bolts 11 through 15 to 18 ft. lbs. (24 Nm).

g. Tighten bolts 11 through 15 to 23 ft. lbs. (31 Nm).

➡**Install a new lower O-ring.**

36. Install a new crankcase to head tube assembly and tighten to 33 ft. lbs. (45 Nm).

37. Install the four glow plugs and tighten to 14 ft. lbs. (19 Nm).

➡**Clean and inspect the glow plug buss bar O-rings and install new as necessary.**

➡**Apply clean engine oil to the O-rings before installing.**

38. Install the glow plug buss bar.

39. Position the turbocharger adapter pipe.

➡**Apply anti-seize lubricants to the bolt threads prior to installing the bolts.**

40. Install the nuts and bolts in the adapter pipe and tighten to 20 ft. lbs. (27 Nm).

41. Do not tighten until after the turbocharger is installed.

⁂ CAUTION

If the fuel injector oil inlet D-ring is damaged, a new fuel injector must be installed.

42. Install new O-rings and copper washer on the fuel injector. Lubricate the fuel injector and O-rings liberally with clean engine oil.

⁂ CAUTION

To prevent engine damage, do not use air tools to install the fuel injectors. The clip that extracts the injector can dislodge and fall into the oil drain hole.

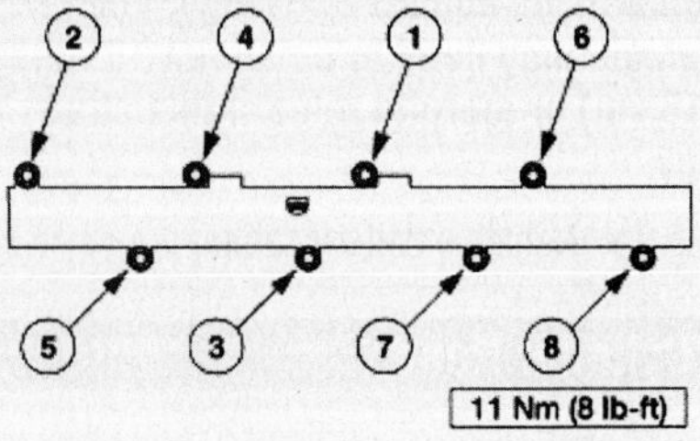

Tighten the high-pressure oil rail bolts in the sequence shown—6.0L engine–right side shown, left side similar

43. Install the fuel injector, fuel injector hold-down and bolt and tighten to 24 ft. lbs. (33 Nm).

44. Apply clean engine oil to the top fuel injector O-rings.

45. Install the high-pressure oil rail and bolts.

46. Install the high-pressure oil rail.

47. Install the bolts finger tight. Tighten the bolts in the sequence shown to 8 ft. lbs. (11 Nm).

48. Install the high-pressure oil line.

49. Install the valve cover and eleven bolts and tighten to 71 inch lbs. (8 Nm).

50. Install the evaporator core housing.

51. Connect the wiring retainers to the studs.

52. Fill the crankcase with clean engine oil.

➡**Apply clean engine oil to the O-ring prior to installing.**

53. Install the cylinder block drain plug and tighten to 15 ft. lbs. (20 Nm).

54. Install the starter.

55. Install a new oil filter.

56. Install the intake manifold.

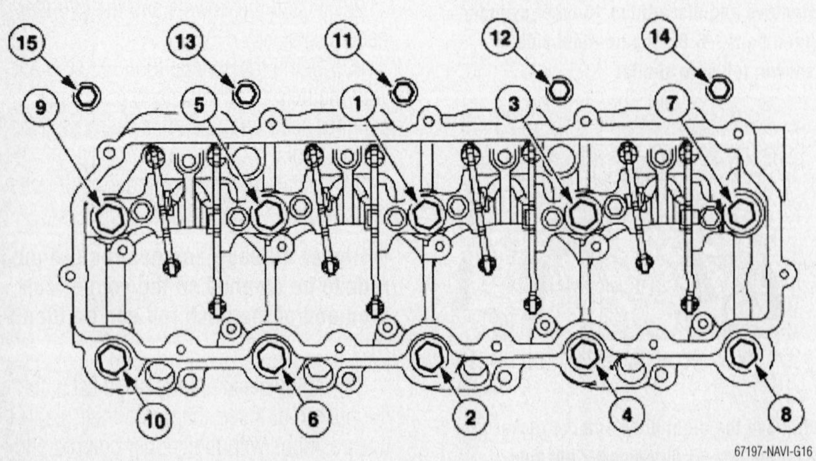

Tighten the head bolts in the sequence shown—6.0L engine–right side shown, left side similar

LEFT CYLINDER HEAD

1. Before servicing the vehicle, refer to the precautions in the beginning of this section.

2. With the vehicle in NEUTRAL, position it on a hoist.

3. Drain the engine oil.

4. Remove the intake manifold.

5. Remove the left cylinder block drain plug.

6. Remove the fuel injector control module mounting bracket.

7. Disconnect the glow plug connector and position aside.

8. Remove the nut and position the oil level indicator and tube aside.

➡Mark the position of the valve cover bolts for valve cover bolt installation.

9. Remove the eleven bolts and the valve cover. Remove the valve cover gasket.
10. Disconnect the high-pressure oil rail supply line at the high-pressure oil rail.
11. Remove the bolts and the high-pressure oil rail.
12. Disconnect and remove the high-pressure oil supply line.

CAUTION

Do not attempt to put battery voltage to the fuel injector or damage to the fuel injector will occur.

13. Using a 19 mm socket, push the fuel injector electrical connector out of the rocker arm carrier.

CAUTION

To prevent engine damage, do not use air tools to remove the fuel injectors. The clip that extracts the injector can dislodge and fall into the oil drain hole.

➡There is not need to drain the fuel rail.

➡If engine oil is found in the engine coolant or engine coolant is found in the combustion chambers, new injector sleeves may need to be installed.

14. Remove the bolt, fuel injector hold-down and fuel injector.
15. Remove the nuts and bolts from the turbocharger adapter pipe.
16. Remove the turbocharger adapter pipe.

➡Remove and discard the O-ring.

17. Remove the oil level indicator and tube.
18. Remove the bolts and turbocharger heat shield.
19. Remove the glow plug buss bar.
20. Remove the four glow plugs.
21. Disconnect the exhaust back pressure tube at the exhaust manifold.
22. Remove the retaining nuts. Remove the exhaust back pressure sensor and tube assembly.
23. Remove the retaining nut from the fuel line bracket and position the fuel lines aside.
24. Remove the bolt and the idler pulley.
25. Disconnect the heater hose from the front cover and position aside.
26. Remove the fan shroud mounting stud.

➡Remove and discard the sealing washers.

27. Remove the banjo fitting and the fuel line.
28. Remove and discard the crankcase-to-head tube assembly.
29. Remove and discard the inner cylinder head bolts.
30. Remove the eight bolts and the rocker arm assemblies.
31. Mark the eight valve bridges with a permanent marker and remove.
32. Mark and remove the eight push rods.
33. Remove the outer cylinder head bolts.
34. Install the special tool 303-579 and the lifting crane. With the help of an assistant, remove the cylinder from the vehicle.
35. Remove the cylinder head gasket and dowels.
36. Clean and inspect the gasket sealing surfaces.

To install:

➡Use care to avoid scratching the blue compound on the cylinder head gasket. Install the gasket with the part number facing upward.

37. Install the dowels and the cylinder head gasket.

➡Position the fuel lines in place before the cylinder head is installed.

38. Using the special tool 303-579 and with the help of an assistant, install the cylinder head on the engine. Remove the special tools.
39. Install the outer cylinder head bolts finger-tight.

➡Higher mileage engines require push rods to be cleaned so the cooper-colored end of the push rod can be identified.

40. Apply clean engine oil to each end of the push rod. Insert them into their respective positions with the copper-colored end up.

➡Coat the end of each valve stem with clean engine oil. Install the eight valve bridges.

CAUTION

Rotate the crankshaft until the damper locating dowel notch is in the six o'clock position or engine damage can occur. Install the rocker arm assemblies and eight bolts and tighten to 23 ft. lbs. (31 Nm).

➡Lightly lubricate the cylinder head bolt threads with clean engine oil.

41. Install the inner cylinder head retaining bolts finger-tight.
42. Tighten the head bolts in the following sequence:
   a. Tighten bolts 1 through 10 to 65 ft. lbs. (88 Nm).
   b. Tighten bolts 1, 3, 5, 7 and 9 to 85 ft. lbs. (115 Nm).
   c. Tighten bolts in sequence 1 through 10, clockwise 90 degrees.
   d. Tighten bolts in sequence 1 through 10, a second time, clockwise 90 degrees.
   e. Tighten bolts in sequence 1 through 10, a third time, clockwise 90 degrees.
   f. Tighten bolts 11 through 15 to 18 ft. lbs. (24 Nm).
   g. Tighten bolts 11 through 15 to 23 ft. lbs. (31 Nm).

➡Install a new lower O-ring.

43. Install a new crankcase to head tube assembly and tighten to 33 ft. lbs. (45 Nm)

➡Install new sealing washers.

44. Install the fuel line and the banjo fitting and tighten to 28 ft. lbs. (38 Nm).
45. Install the fan shroud mounting stud and tighten to 30 ft. lbs. (40 Nm).
46. Connect the heater hose.
47. Install the idler pulley and bolt and tighten to 35 ft. lbs. (47 Nm).
48. Position the exhaust backpressure sensor and tube assembly and install the retaining nuts and tighten to 23 ft. lbs. (31 Nm).
49. Tighten the exhaust backpressure tube fitting at the exhaust manifold to 22 ft. lbs. (30 Nm).
50. Position the fuel lines and install the fuel line bracket retaining bolt and tighten to 10 ft. lbs. (13 Nm).
51. Install the four glow plugs and tighten to 14 ft. lbs. (19 Nm).

➡Apply clean engine oil to the O-rings before installing.

52. Install the glow plug buss bar.

➡Install a new O-ring.

53. Install the oil level indicator and tube.
54. Install the turbocharger heat shield and bolts.
55. Position the turbocharger adapter pipe.

➡Do not torque until after the turbocharger is installed.

56. Install the nuts and bolts in the adapter pipe and tighten to 20 ft. lbs. (27 Nm).

⁂ CAUTION

If the fuel injector oil inlet D-ring is damaged, a new fuel injector must be installed.

57. Install new O-rings and copper washer on the fuel injector. Lubricate the fuel injector and O-rings liberally with clean engine oil.

⁂ CAUTION

To prevent engine damage, do not use air tools to install the fuel injectors. The clip that extracts the injector can dislodge and fall into the oil drain hole.

58. Install the fuel injector, fuel injector hold-down and bolt and tighten to 24 ft. lbs. (33 Nm).

⁂ CAUTION

To prevent engine damage, be sure the injector wiring is clear of all moving parts.

59. Install the fuel injector electrical connector into the rocker carrier.
60. Apply engine oil to the top fuel injector O-rings.
61. Install the high-pressure oil rail and bolts.
62. Install the high-pressure oil rail.
63. Install the bolts finger tight, then tighten the bolts in the sequence illustrated to 8 ft. lbs. (11 Nm).
64. Install the high-pressure oil line.
65. Clean and inspect the valve cover gasket. Install a new gasket if necessary.
66. Position the valve cover gasket. Install the valve cover and 11 bolts and tighten to 71 ft. lbs. (8 Nm).
67. Position back the oil level indicator and install the nut.
68. Connect the glow plug connector.
69. Install the fuel injection control module mounting bracket and tighten the retainers to 71 inch lbs. (8 Nm).
70. Fill the crankcase with clean engine oil.

➡Lightly lubricate the O-ring with clean engine oil before installing.

71. Install the cylinder block drain plug and tighten to 15 ft. lbs. (20 Nm).
72. Install a new oil filter.
73. Install the intake manifold.

7.3L Engine

RIGHT CYLINDER HEAD

⁂ CAUTION

The fuel system remains under pressure, even after the engine has been turned OFF. The fuel system pressure must be relieved before disconnecting any fuel lines. Failure to do so may result in fire and/or personal injury.

1. Before servicing the vehicle, refer to the precautions in the beginning of this section.
2. Drain the cooling system.
3. Remove or disconnect the following:
   - Negative battery cables
   - Radiator
   - Turbocharger assembly
   - Fuel lines from the rear of both cylinder heads and the fuel pump
   - Wiring from the alternator
   - Adjusting bolts and pivot bolts from the alternator and the vacuum pump and remove both units
   - Alternator and its bracket
   - Engine oil dipstick tube
   - Manifold Absolute Pressure (MAP) sensor and position it aside
   - Valve cover
   - Connectors from the injectors and glow plugs
   - Valve cover gasket
   - High pressure oil pump supply line to the right cylinder head
   - Exhaust back pressure line
   - Three glow plug relay bracket nuts and the ground wire
   - Heater hose from the cylinder head
   - Outer half of the heater distribution box
   - Four outboard fuel injector hold-down bolts, retaining screws and four oil deflectors

⁂ WARNING

Remove the oil drain plugs prior to removing the injectors or oil could enter the combustion chamber which could result in hydrostatic lock and severe engine damage.

   - Oil rail drain plugs
   - Fuel injectors using Injector Remover No. T94T–9000–AH1, or equivalent. Position the tool's fulcrum beneath the fuel injector hold-down plate and over the edge of the cylinder head. Install the remover screw in the threaded hole of the fuel injector plate. Tighten the screw to lift out the injector from its bore. Place the injector in

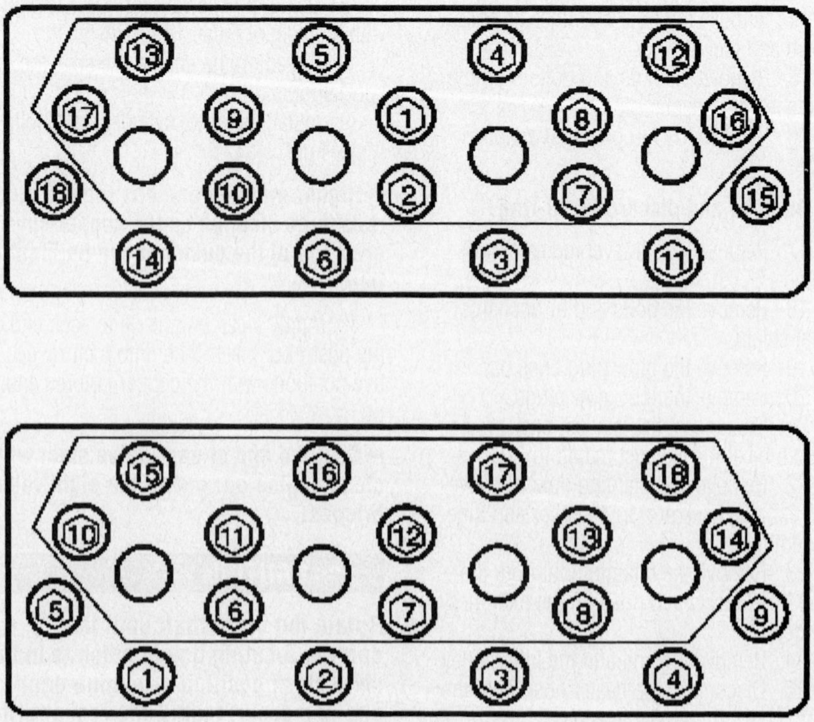

9355FG01

Cylinder head torque sequence—7.3L diesel engine

a suitable protective sleeve such as Rotunda Injector Protective Sleeve, No. 014–00933–2, and set the injector in a suitable holding rack.

4. Use a suitable vacuum tool, such as Rotunda Vacuum Pump, No. 021–00037, or equivalent to remove the oil and fuel left over in the injector bores.

5. Remove or disconnect the following:
- Rocker arms and pushrods, KEEP EVERYTHING IN ORDER
- Four glow plugs
- Right turbo exhaust inlet pipe
- Ground strap from the rear of the cylinder head
- Fuel return line at the front of the cylinder head
- Four inboard fuel injector shoulder bolts
- Cylinder head bolts and attach a Rotunda Cylinder Head Lifting Bracket, 014–00932–2, or equivalent

6. Carefully lift the cylinder head out of the engine compartment and remove the head gaskets.

To install:

➡To prepare a good seat for the fuel injector O-rings, use a suitable injector sleeve brush to clean any debris from the bore.

7. Carefully clean the cylinder block and head mating surfaces.

8. Install or connect the following:
- Cylinder head gasket on the engine block and carefully lower the cylinder head in place
- Cylinder head bolt and torque in 3 steps using the sequence shown in the illustration as follows:

a. Step 1: 65 ft. lbs. (88 Nm).
b. Step 2: 85 ft. lbs. (115 Nm).
c. Step 3: 95 ft. lbs. (129 Nm).

➡Lubricate the threads and the mating surfaces of the bolt heads and washers with engine oil.

- Fuel return line to the cylinder head
- Four inboard injector shoulder bolts. Tighten them to 9 ft. lbs. (12 Nm).

9. Install the fuel injectors using special tools as follows:

a. Lubricate the injectors with clean engine oil. Using new copper washers, carefully push the injectors square into the bore using hand pressure only to seat the O-rings.

b. Position the open end of Injector Replacer, No. T94T–9000–AH2, or equivalent between the fuel injector body and injector hold-down plate, while positioning the opposite end of the tool over the edge of the cylinder head.

c. Align the hole in the tool with the threaded hole in the cylinder head and install the bolt from the tool kit. Tighten the bolt to fully seat the injector, then remove the bolt and tool.

10. Install or connect the following:
- Four outboard fuel injector hold-down bolts, the four oil deflectors and retaining screws. Tighten them to 9 ft. lbs. (12 Nm).

11. Dip the pushrod ends in clean engine oil and install the pushrods with the copper colored ends toward the rocker arms, making sure the pushrods are fully seated in the tappet pushrod seats.
- Rocker arms and posts in their original positions. Apply Lubriplate® grease to the valve stem tips. Turn the engine over by hand until the timing mark is at the 11 o'clock position as viewed from the front.
- Rocker arm posts, bolts and tighten to 27 ft. lbs. (37 Nm)
- Valve covers
- Fuel rail drain plugs, tightening them to 8 ft. lbs. (11 Nm)
- Oil rail drain plugs, tightening them to 53 inch lbs. (6 Nm)
- Heater distribution box
- Heater hose to the cylinder head
- Glow plug relay bracket and ground wire
- Exhaust back pressure line
- Oil supply line to the cylinder head, tightening it to 19 ft. lbs. (26 Nm)
- Dipstick tube
- MAP sensor and screws
- Valve cover gasket
- Wiring to the fuel injectors and glow plugs
- Valve cover, tightening the bolts to 8 ft. lbs. (11 Nm)
- Injector wiring harness to the valve cover gasket
- Alternator, tightening the bracket bolts to 40–55 ft. lbs. (54–75 Nm)
- Alternator wiring
- Drive belt

12. The remainder of the installation is the reverse of the removal. Tighten the fuel pump-to-fuel line banjo bolt to 40 ft. lbs. (54 Nm).

13. Connect both negative battery cables.

14. Refill and bleed the cooling system.

15. Run the engine and check for fuel, coolant and exhaust leaks.

LEFT CYLINDER HEAD

⁂ CAUTION

The fuel system remains under pressure, even after the engine has been turned OFF. The fuel system pressure must be relieved before disconnecting any fuel lines. Failure to do so may result in fire and/or personal injury.

1. Before servicing the vehicle, refer to the precautions in the beginning of this section.

2. Drain the cooling system.

3. Using approved equipment, recover the refrigerant from the A/C system.

4. Remove or disconnect the following:
- Negative battery cables
- Radiator
- Turbocharger assembly
- Two crankcase breather screws and the breather
- Wiring from the air conditioning compressor
- Four left accessory bracket bolts
- Vacuum hose at the brake vacuum pump
- A/C lines from the compressor
- Power steering lines from the pump
- Left accessory bracket and accessories as an assembly
- Valve cover
- Fuel line assembly between the cylinder heads and fuel pump
- Fuel line nut from the intake manifold stud
- Fuel return line from the cylinder head
- High pressure oil pump supply line from the cylinder head

5. Raise the vehicle and support it safely on jackstands.
- Left turbo exhaust pipe from the manifold

6. Lower the vehicle.
- Oil rail drain plugs
- Four outboard fuel injector hold-down bolts, retaining screws and four oil deflectors

⁂ WARNING

Remove the oil drain plugs prior to removing the injectors or oil could enter the combustion chamber which could result in hydrostatic lock and severe engine damage.

- Oil rail drain plugs

7. Remove the fuel injectors using Injector Remover No. T94T–9000–AH1, or

equivalent. Position the tool's fulcrum beneath the fuel injector hold-down plate and over the edge of the cylinder head. Install the remover screw in the threaded hole of the fuel injector plate. Tighten the screw to lift out the injector from its bore. Place the injector in a suitable protective sleeve such as Rotunda Injector Protective Sleeve, No. 014–00933–2, and set the injector in a suitable holding rack.

8. Use a suitable vacuum tool, such as Rotunda Vacuum Pump, No. 021–00037, or equivalent to remove the oil and fuel left over in the injector bores.

9. Remove or disconnect the following:
- Rocker arms and pushrods, KEEP EVERYTHING IN ORDER
- Four glow plugs
- Left turbo exhaust inlet pipe
- Main engine harness connectors in the left fender well and position the harness aside
- Four inboard fuel injector shoulder bolts
- Cylinder head bolts and attach a Rotunda Cylinder Head Lifting Bracket, 014–00932–2, or equivalent

10. Carefully lift the cylinder head out of the engine compartment and remove the head gaskets.

To install:

➡To prepare a good seat for the fuel injector O-rings, use a suitable injector sleeve brush to clean any debris from the bore.

11. Carefully clean the cylinder block and head mating surfaces.

12. Install or connect the following:
- Cylinder head gasket on the engine block and carefully lower the cylinder head in place
- Cylinder head bolt and torque in 3 steps using the sequence shown in the illustration as follows:

a. Step 1: 65 ft. lbs. (88 Nm).
b. Step 2: 85 ft. lbs. (115 Nm).
c. Step 3: 95 ft. lbs. (129 Nm).

➡Lubricate the threads and the mating surfaces of the bolt heads and washers with engine oil.

13. Apply anti-seize paste and install the glow plugs, tightening them to 14 ft. lbs. (19 Nm).

14. Install or connect the following:
- Four outboard fuel injector hold-down bolts, the four oil deflectors and retaining screws. Tighten them to 9 ft. lbs. (12 Nm).

15. Dip the pushrod ends in clean engine oil and install the pushrods with the copper colored ends toward the rocker arms, making sure the pushrods are fully seated in the tappet pushrod seats.
- Rocker arms and posts in their original positions. Apply Lubriplate® grease to the valve stem tips. Turn the engine over by hand until the timing mark is at the 11 o'clock position as viewed from the front.
- Rocker arm posts, bolts and tighten to 27 ft. lbs. (37 Nm)
- Valve covers
- Four inboard injector shoulder bolts. Tighten them to 9 ft. lbs. (12 Nm).

16. Install the fuel injectors using special tools as follows:

a. Lubricate the injectors with clean engine oil. Using new copper washers, carefully push the injectors square into the bore using hand pressure only to seat the O-rings.

b. Position the open end of Injector Replacer, No. T94T–9000–AH2, or equivalent between the fuel injector body and injector hold-down plate, while positioning the opposite end of the tool over the edge of the cylinder head.

c. Align the hole in the tool with the threaded hole in the cylinder head and install the bolt from the tool kit. Tighten the bolt to fully seat the injector, then remove the bolt and tool.

17. Install or connect the following:
- Fuel rail drain plugs, tightening them to 8 ft. lbs. (11 Nm)
- Oil rail drain plugs, tightening them to 53 inch lbs. (6 Nm)
- Heater distribution box
- Valve cover gasket
- Wiring to the fuel injectors and glow plugs
- Valve cover, tightening the bolts to 8 ft. lbs. (11 Nm)
- Engine wiring harness

18. Raise and safely support the vehicle on jackstands.

19. Loosely install the turbo exhaust pipe to the manifold.

20. Lower the vehicle.

21. The remainder of the installation is the reverse of the removal. Tighten the fuel pump-to-fuel line banjo bolt to 40 ft. lbs. (54 Nm).

22. Connect both negative battery cables.

23. Refill and bleed the cooling system.

24. Run the engine and check for fuel, coolant and exhaust leaks.

Rocker Arms

REMOVAL & INSTALLATION

6.0L Engine

1. Before servicing the vehicle, refer to the precautions in the beginning of this section.

CAUTION

To prevent engine damage, always install new head bolts when installing the cylinder head, even if removing only one bolt. It will be necessary to install a new cylinder head gasket as well as all cylinder head bolts.

2. Remove the cylinder head.

➡Be careful when removing the rocker arm. Do not lose the rocker arm steel ball.

3. Push down on the rocker arm. Pull out on the bottom releasing the ball.

4. Remove the steel ball from the rocker arm.

5. Remove the rocker arm clip from the fulcrum plate.

6. Clean and inspect all parts for pitting or scuffing. Install new rocker arm and valve bridges, as needed.

To install:

7. Install the rocker arm clip on the fulcrum plate.

8. Place the steel ball in the rocker arm cup and lubricate with clean engine oil.

9. Push down on the rocker arm to compress the rocker arm clip. Insert the ball and release the pressure. Verify free movement of the rocker arm on the fulcrum.

10. Install the cylinder head.

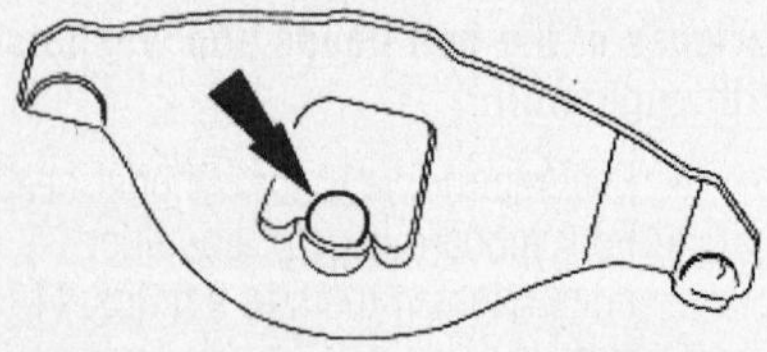

67197-NAVI-G18

Place the steel ball in the rocker arm cup and lubricate with clean engine oil —6.0L engine

7.3L Engine

1. Before servicing the vehicle, refer to the precautions in the beginning of this section.

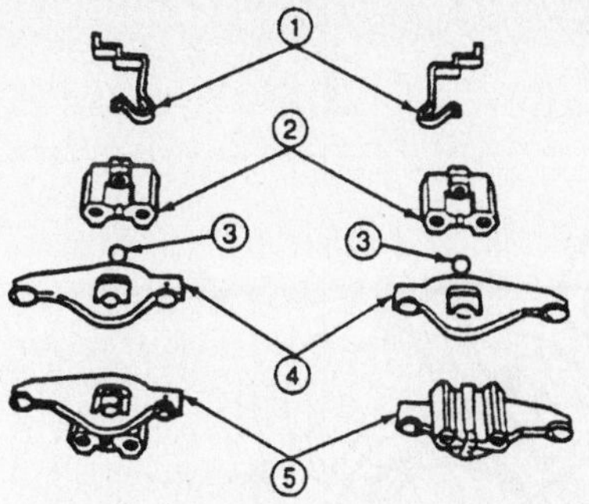

Item	Description
1	Snap Retaining Clip
2	Rocker Arm Pedestal
3	Rocker Arm Ball
4	Rocker Arm
5	Rocker Arm Assembly

7924FG58

Exploded view of the rocker arm assembly—7.3L diesel engine

2. Disconnect the ground cables from both batteries.
3. Remove the valve cover attaching screws and remove both valve covers.
4. Remove the valve rocker arm post mounting bolts. Remove the rocker arms and posts in order and mark them with tape so they can be installed in their original positions.
5. If the cylinder heads are to be removed, then the pushrods can now be removed. Make a holder for the pushrods out of a piece of wood or cardboard and remove the pushrods in order. It is very important that the pushrods be reinstalled in their original order. The pushrods can remain in position if no further disassembly is required.

To install:

6. If the pushrods were removed, install them in their original locations. Make sure they are fully seated in the tappet seats.

➡The copper colored end of the pushrod goes toward the rocker arm.

7. Apply a polyethylene grease to the valve stem tips. Install the rocker arms and posts in their original positions.
8. Turn the engine over by hand until the valve timing mark is at the 11 o'clock position, as viewed from the front of the engine. Install all of the rocker arm post attaching bolts and tighten to 20 ft. lbs. (27 Nm).
9. Install new valve cover gaskets and install the valve cover.
10. Install the battery cables, start the engine and check for leaks.

Turbocharger

REMOVAL & INSTALLATION

6.0L Engine

1. Before servicing the vehicle, refer to the precautions in the beginning of this section.

Remove the air cleaner assembly.

✲✲ CAUTION

Never remove the pressure relief cap while the engine is operating or when the cooling system is hot. Failure to follow these instructions can result in damage to the cooling system or engine or result in personal injury. To avoid having scalding hot coolant or steam blow out of the degas bottle when removing the pressure relief cap, wait until the engine has cooled, then wrap a thick cloth around the pressure relief cap and turn it slowly. Step back while the pressure is released from the cooling system. When certain all the pressure has been released, (still with a cloth) turn and remove the pressure relief cap. Failure to follow these instructions can result in personal injury.

✲✲ CAUTION

The coolant must be recovered in a suitable, clean container for reuse. If the coolant is contaminated, it must be recycled or disposed of correctly and the system filled with new coolant.

✲✲ CAUTION

Always fill the cooling system with the same type of coolant that was drained from the system. Do not mix coolant types. Do not add orange-colored Motorcraft Specialty Orange Engine Coolant VC-2, or equivalent meeting WSS-M97B44-D. Mixing coolants may degrade the coolant's corrosion protection. Do not add alcohol, methanol, brine, or any engine coolants mixed with alcohol or methanol antifreeze. These can cause engine damage from overheating or freezing.

2. Disconnect and plug the engine vent and radiator vent hoses.
3. Remove the bolts and position the degas bottle aside.
4. Loosen the clamp at the turbocharger.
5. Remove the bolts and remove the turbocharger intake tube.
6. Disconnect the charge air cooler inlet pipe.
7. Remove the push pins.
8. Disconnect the two wiring harness push pins and position out of the way.
9. Disconnect the turbocharger variable hydraulic control valve electrical connector.
10. Remove the bolts for the oil supply tube.
11. Remove and discard the gasket.
12. Remove the bolt and the wire retainer.
13. Using the special tool 303-755, remove the oil feed tube.
14. Remove the marmon clamp from the turbocharger outlet.
15. Remove the marmon clamp from the turbocharger inlet.
16. Remove the rear turbocharger mounting bolt.
17. Remove the front mounting bolts.
18. Position the turbocharger and remove the turbocharger drain tube.
19. Remove and discard the drain tube O-rings.
20. Remove the turbocharger.

To install:

21. Position the turbocharger on the turbocharger pedestal.

➡Install new O-rings and apply clean engine oil.

22. Position the turbocharger and install the turbocharger drain tube.
23. Install the turbocharger and the rear mounting bolt and tighten to 28 ft. lbs. (38 Nm).
24. Install the turbocharger front mounting bolts and tighten to 28 ft. lbs. (38 Nm).
25. Install the turbocharger inlet marmon clamp and tighten to 9 ft. lbs. (12 Nm).
26. Install the turbocharger exhaust marmon clamp and tighten to 89 inch lbs. (10 Nm).
27. Install the oil feed tube.
28. Position the wire retainer and install the bolt.
29. Pre-lubricate the oil inlet hole of the turbocharger assembly with clean engine oil and spin the compressor wheel several times to coat the bearing with oil.
30. Position the oil feed tube and install the bolts and tighten to 18 ft. lbs. (25 Nm).
31. Install a new gasket.

32. Connect the turbocharger variable vane hydraulic control valve electrical connector.
33. Position the wiring harness and connect the push pins.
34. Install the push pins.
35. Connect the charge air cooler inlet pipe and tighten to 9 ft. lbs. (12 Nm).
36. Install the turbocharger intake tube.

7.3L Engine

1. Before servicing the vehicle, refer to the precautions in the beginning of this section.
2. Remove or disconnect the following:
   - Negative battery cable
   - 2 air intake tube assembly bolts
   - Clamps at the turbocharger
   - Crankcase breather assembly
   - Engine air cleaner, air intake tube and hoses
   - Exhaust outlet clamp from the turbocharger
   - Engine charge exhaust pipe bolt from the transmission, if so equipped
   - Bolts and nuts from the catalytic converter-to-engine charge exhaust pipe, if so equipped
   - 2 bolts retaining the turbocharger exhaust inlet pipe to the left exhaust manifold, loosen only
   - Bolts retaining the left turbocharger exhaust inlet pipe to the turbocharger exhaust inlet adapter, on automatic transmissions
   - 2 bolts retaining the turbocharger exhaust inlet pipe to the right exhaust manifold, loosen only
   - Lower bolt retaining the right turbocharger exhaust inlet pipe to the turbocharger exhaust inlet adapter
   - Upper bolts retaining the right and left turbocharger exhaust inlet pipes to the turbocharger exhaust inlet adapter, on automatic transmissions
   - Right engine lift hook and bolt
   - Air inlet hose clamp at the turbocharger, hose and lay aside
   - Compressor manifold
   - 4 bolts retaining the turbocharger pedestal assembly to the cylinder block
   - Turbocharger assembly and all electrical connectors from it

➡If the turbocharger is not being removed for service, install the Fuel/Oil turbo Protector Cap Set T94T-9395-AH.

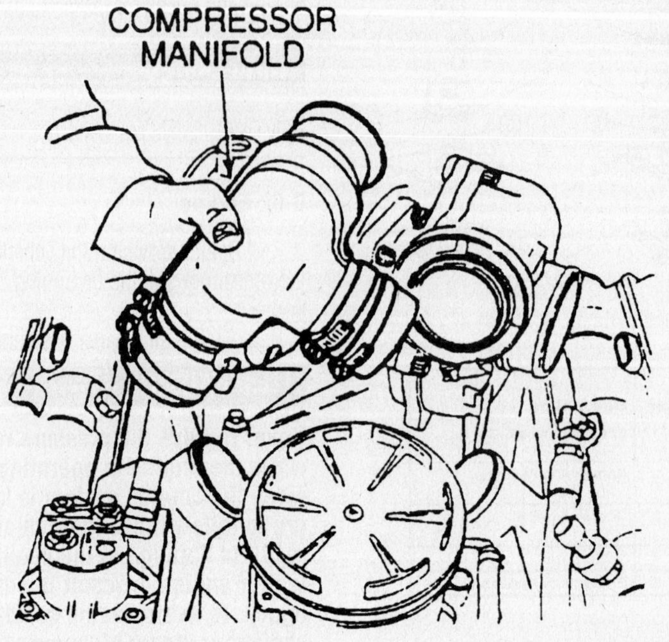

After loosening the clamps, the compressor manifold can be removed—7.3L diesel engine

   - Oil gallery O-rings

To install:

3. Install or connect the following:
   - Oil gallery O-rings
   - Turbocharger electrical connectors and the turbocharger assembly
   - 4 bolts retaining the turbocharger pedestal assembly to the engine block. Tighten the bolts to 18 ft. lbs. (25 Nm).
   - 4 bolts retaining the right and left turbocharger exhaust inlet pipes to the turbocharger exhaust inlet adapter.
   - Compressor manifold, intake manifold hoses and clamps. Be sure the compressor outlet seal is in position.
   - Right engine lift hook and bolt
   - 2 right and left lower bolts (retaining the turbocharger exhaust inlet pipes to the turbocharger exhaust inlet adapter) to 36 ft. lbs. (49 Nm)
   - 4 right and left bolts and nuts (retaining the turbocharger exhaust inlet pipes to the exhaust manifolds) to 36 ft. lbs. (49 Nm)
   - Catalytic converter to the engine charge exhaust pipe bolts and nuts
   - 2 right and left upper bolts (retaining the turbocharger exhaust inlet pipes to the turbocharger exhaust inlet adapter) to 36 ft. lbs. (49 Nm)
   - Exhaust outlet clamp to the turbocharger
   - Air intake tube and hose assembly
   - Negative battery cable

Intake Manifold

REMOVAL & INSTALLATION

6.0L Engine

1. Before servicing the vehicle, refer to the precautions in the beginning of this section.
2. Remove the auxiliary battery.
3. Remove the air cleaner assembly.
4. Remove the cooling fan-blade, clutch and shroud.
5. Remove the coolant reservoir.
6. Remove the upper radiator hose.
7. Remove the turbocharger-to-charge air cooler tube.
8. Remove the charge air cooler-to-engine tube.
9. on vehicles with dual alternators, perform the following steps:
   a. Remove the accessory drive belt.
   b. Remove the bolt and the accessory drive belt tensioner.
   c. Remove the accessory drive belt.
   d. Remove the bolts, bracket and accessory drive belt idler pulley.
   e. Remove the bolts and the accessory drive belt tensioner.
   f. Disconnect the wire retainer, generator electrical connector and B+ wire.

g. Remove the bolts and the generator with mounting bracket.

10. On vehicles with single alternator, perform the following steps:

a. Remove the accessory drive belt.

b. Disconnect the generator B+ wire and electrical connector.

c. Remove the three bolts and the alternator.

d. Remove the turbocharger and the turbocharger pedestal.

11. Remove the bolts and position the heater hose tube aside.

12. Remove and discard the O-ring.

13. Disconnect the Manifold Absolute Pressure (MAP) sensor hose.

14. Disconnect the engine coolant vent hose.

15. Remove the fuel injector control module.

16. Disconnect the wiring retainer, injector pressure regulator and Injector Control Pressure (ICP) sensor electrical connector.

17. Disconnect the Exhaust Gas Recirculation (EGR) valve electrical connector.

18. Disconnect the oil pressure sensor electrical connector.

19. Disconnect the oil temperature sensor electrical connector.

20. Disconnect the Throttle Position (TP) control module electrical connector.

21. Disconnect the TP sensor electrical connector.

22. Disconnect the pin-type retainer and water temperature sensor.

23. Disconnect the Intake Air Temperature (IAT) sensor electrical connector and wiring connector.

24. Disconnect the exhaust backpressure sensor and retaining clip.

25. Disconnect the eight fuel injectors. Remove the nut and the fuel injector wiring harness.

26. Disconnect the fuel line fittings.

➡It is necessary to remove the filter and drain the housing.

27. Remove the four bolts and the oil filter housing.

28. Remove the bolt and the oil filter return tube.

29. Remove the fuel line.

30. Remove the bolt, and the banjo fitting from the fuel line.

31. Discard the sealing washers and remove the fuel line.

32. Remove the nuts and the turbocharger heat shield.

➡Align the flat edge with the index feature located on the coolant supply port.

33. Pull the EGR cooler clamp forward, twist and then slide the EGR cooler hose rearward to remove.

34. Remove the EGR cooler V-clamp and gasket.

35. Remove the bolts and the intake manifold.

36. Remove the intake manifold gaskets.

37. Clean and inspect the gaskets. Install new gaskets if necessary.

38. Clean and inspect the sealing surfaces.

To install:

➡Locating tabs on the gaskets must be up and toward the center of the engine, or a leak will occur.

39. Install the intake manifold gaskets.

40. Install the intake manifold and bolts and tighten in the following sequence:

a. Loosely install the bolts 1-8.

b. Tighten bolts 9-16 to 8 ft. lbs. (11 Nm).

Tighten all bolts to 11 Nm (8 lb-ft) in the sequence shown.

41. Install the gasket and the EGR cooler V-clamp and tighten to 53 inch lbs. (8 Nm).

42. Slide the EGR cooler hose forward and rotate flat to lock.

43. Install the turbocharger heat shield and the nuts and tighten to 8 ft. lbs. (11 Nm).

44. Install the fuel line.

45. Install new sealing washers.

46. Install the fuel line.

47. Install the banjo fitting and the bolt. Tighten the upper bolt to 10 ft. lbs. (13 Nm) and the lower bolt to 28 ft. lbs. (38 Nm).

48. Install the oil filter return tube and bolt as follows:

a. On new oil filter return tubes, tighten to 53 inch lbs. (6 Nm).

b. On used oil filter return tubes, tighten to 27 inch lbs. (3 Nm).

49. Install the oil filter and the four bolts and tighten to 11 ft. lbs. (15 Nm).

50. Connect the fuel line fittings. Tighten the upper fitting to 19 ff. lbs. (26 Nm) and the lower fitting to 32 ft. lbs. (43 Nm).

51. Position the fuel injector harness and install the retaining nut and tighten to 8 ft. lbs. (11 Nm). Connect the eight fuel injectors electrical connectors.

52. Connect the retaining clip and exhaust backpressure sensor electrical connector.

53. Connect the wiring retainer and the IAT sensor electrical connector.

54. Connect the water temperature sensor and the pin-type retainer.

55. Connect the TP sensor electrical connector.

56. Connect the TP control module electrical connector.

57. Connect the oil temperature sensor electrical connector.

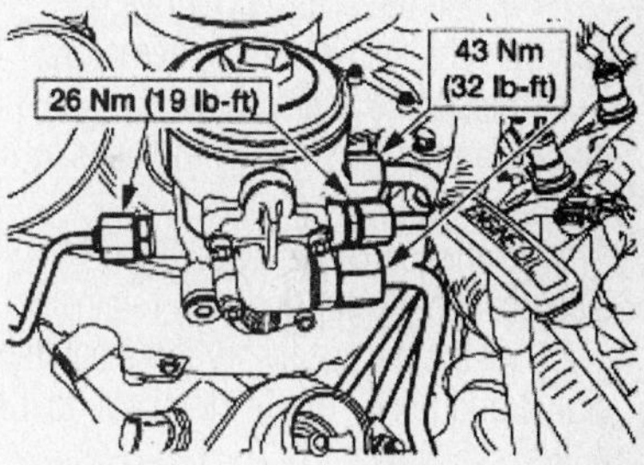

67197-NAVI-G20

Connect the fuel line fittings—6.0L engine

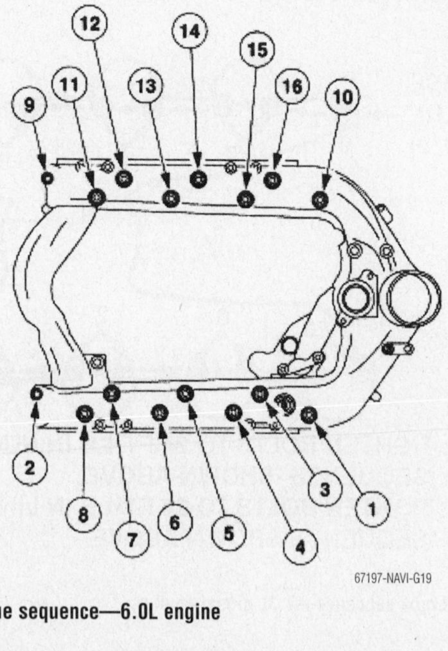

67197-NAVI-G19

Intake manifold torque sequence—6.0L engine

58. Connect the oil pressure sensor electrical connector.

59. Connect the EGR valve electrical connector.

60. Connect the wiring retainer, injector pressure regulator and ICP sensor electrical connectors.

61. Install the fuel injector control module.

62. Connect the engine coolant vent hose and clamp.

63. Connect the MAP sensor hose.

➡Install a new O-ring on the heater hose tube.

64. Install the heater hose tube and bolts and tighten to 10 ft. lbs. (13 Nm).

65. Install the turbocharger pedestal and turbocharger.

66. On models with single alternators perform the following:

a. Install the alternator and the three bolts. Tighten the bolts to 35 ft. lbs. (47 Nm).

b. Connect the alternator B+ wire land electrical connector and position the boot back.

c. Install the accessory drive belt.

67. On vehicles with dual alternators:

a. Install the alternator with mounting bracket and bolts. Tighten the bolts to 35 ft. lbs. (47 Nm).

b. Connect the B+ wire, alternator electrical connector and wire retainer.

c. Install the accessory drive belt tensioner and bolts. Tighten the bolts to 18 ft. lbs. (25 Nm).

d. Position the accessory drive belt idler pulley. Install the bracket and bolts and tighten to 18 ft. lbs. (25 Nm).

e. Install the accessory drive belt tensioner and bolt and tighten to 18 ft. lbs. (25 Nm).

f. Install the accessory drive belt.

68. Install the charge air cooler-to-engine tube and tighten the retainers to 9 ft. lbs. (12 Nm).

69. Install the turbocharger-to-charge air cooler tube and tighten the retainers to 9 ft. lbs. (12 Nm).

70. Install the upper radiator hose.

71. Install the coolant reservoir.

72. Install the auxiliary battery.

73. Install the air cleaner assembly.

74. Install the cooling fan blade, clutch and shroud.

7.3L Engine

CAUTION

The fuel system remains under pressure, even after the engine has been turned OFF. The fuel system pressure must be relieved before disconnecting any fuel lines. Failure to do so may result in fire and/or personal injury.

1. Before servicing the vehicle, refer to the precautions in the beginning of this section.

2. Remove or disconnect the following:

- Both battery ground cables
- Air cleaner and install clean rags into the air intake of the intake manifold. It is important that no dirt or foreign objects get into the intake.
- Fuel system pressure
- Injection pump
- Fuel return hose from No. 7 and No. 8 rear nozzles
- Return hose to the fuel tank
- Engine wiring harness from the engine

➡The engine harness ground cables must be removed from the back of the left cylinder head.

- Bolts attaching the intake manifold to the cylinder heads and the manifold
- Crankcase Depression Regulator (CDR) valve tube grommet from the valley pan
- Bolts attaching the valley pan strap to the front of the engine block and the strap
- Valley pan drain plug and the valley pan

To install:

3. Apply a 1/8 inch (3mm) bead of RTV sealer to each end of the cylinder block.

➡The RTV sealer should be applied immediately prior to the valley pan installation.

4. Install or connect the following:

- Valley pan drain plug, CDR valve tube and new grommet into the valley pan
- New O-ring and new back-up ring on the CDR valve
- Valley pan strap on the front of the valley pan
- Intake manifold and tighten the bolts to 24 ft. lbs. (33 Nm) using the tightening sequence
- Engine wiring harness and the engine ground wire located to the rear of the left cylinder head
- Injection pump
- No. 7 and No. 8 fuel return hoses and the fuel tank return hose
- Air cleaner
- Battery ground cables to both batteries

➡If necessary, purge the nozzle high pressure lines of air by loosening the connector 1/2 to 1 turn and cranking the engine until solid stream of fuel, devoid of any bubbles, flows from the connection.

5. Run the engine and check for oil and fuel leaks.

CAUTION

Keep eyes and hands away from the nozzle spray. Fuel spraying from the nozzle under high pressure can penetrate the skin.

6. Check and adjust the injection pump timing.

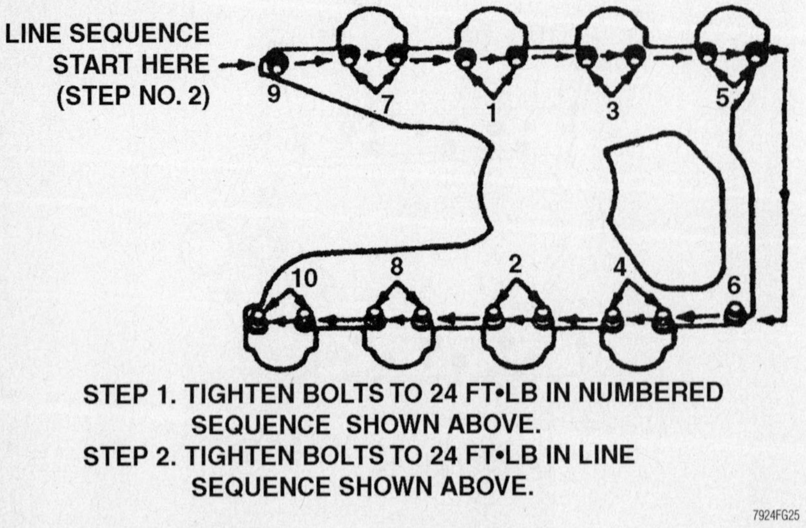
LINE SEQUENCE
START HERE
(STEP NO. 2)
9
7
1
3
5
10
8
2
4
6
STEP 1. TIGHTEN BOLTS TO 24 FT•LB IN NUMBERED SEQUENCE SHOWN ABOVE.
STEP 2. TIGHTEN BOLTS TO 24 FT•LB IN LINE SEQUENCE SHOWN ABOVE.
7924FG25

Intake manifold bolt torque sequence—7.3L diesel engine

Exhaust Manifold

REMOVAL & INSTALLATION

6.0L Engine

LEFT SIDE

1. Before servicing the vehicle, refer to the precautions in the beginning of this section.
2. If equipped, remove the exhaust manifold to Exhaust Gas Recirculation (EGR) valve tube.
3. On 4x4 vehicles, remove the front wheel opening moulding.

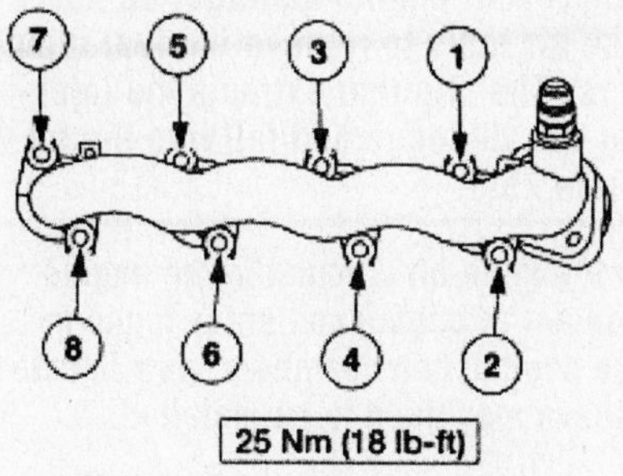

Tighten the left side exhaust manifold retainers in the sequence illustrated—6.0L engine

4. Remove the front fender splash shield
5. Remove the pipe-to-manifold nuts.
6. Remove the nuts and the exhaust manifold.
7. Remove the exhaust manifold gasket.
8. Clean and inspect the exhaust manifold.

To install:

9. Installation is the reverse of removal. Tighten the manifold retainers in the sequence illustrated to 18 ft. lbs. (25 Nm). If equipped, install the exhaust manifold to EGR valve tube and tighten the fasteners as shown in the illustration. Tighten the pipe-to-manifold bolts to 27–24 ft. lbs. (34–46 Nm).

RIGHT SIDE

1. Before servicing the vehicle, refer to the precautions in the beginning of this section.
2. Disconnect the battery ground cable.
3. Remove the pipe-to-manifold nuts.
4. Remove the right front wheel.
5. Remove the right front inner fender well.
6. Remove the nuts, the exhaust manifold and the gasket. Discard the gasket.
7. Clean and inspect the exhaust manifold.

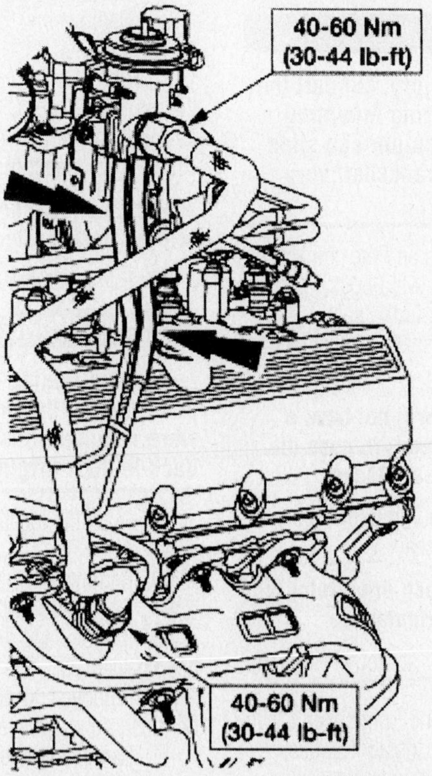

If equipped, install the exhaust manifold to EGR valve tube and tighten the fasteners as specified in this illustration—6.0L engine

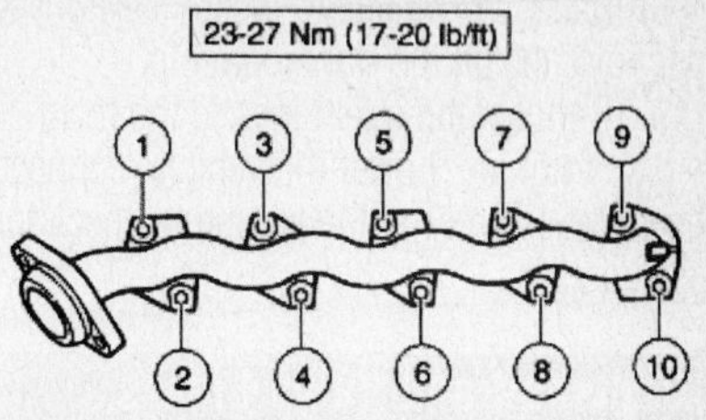

Tighten the right side exhaust manifold retainers in the sequence illustrated—6.0L engine

To install:

8. Installation is the reverse of removal. Tighten the manifold retainers in the sequence illustrated to 18 ft. lbs. (25 Nm). Tighten the pipe-to-manifold bolts to 27–24 ft. lbs. (34–46 Nm).

7.3L Engine

1. Before servicing the vehicle, refer to the precautions in the beginning of this section.
2. Disconnect the ground cables from both batteries.
3. Raise the vehicle and safely support it.
4. Disconnect the muffler inlet pipe from the exhaust manifolds.
5. If removing the right manifold, lower the vehicle. When removing the left manifold, raise the vehicle and remove the manifold from underneath. Bend the tabs on the manifold attaching bolts, then remove the bolts and manifold.

To install:

6. Before installing, clean all mounting surfaces on the cylinder heads and the manifold. Apply an anti-seize compound on the manifold both threads and install the left manifold, using a new gasket and new locking tabs.
7. Tighten the bolts to 45 ft. lbs. (61 Nm) and bend the tabs over the flats on the bolt heads to prevent the bolts from loosening.

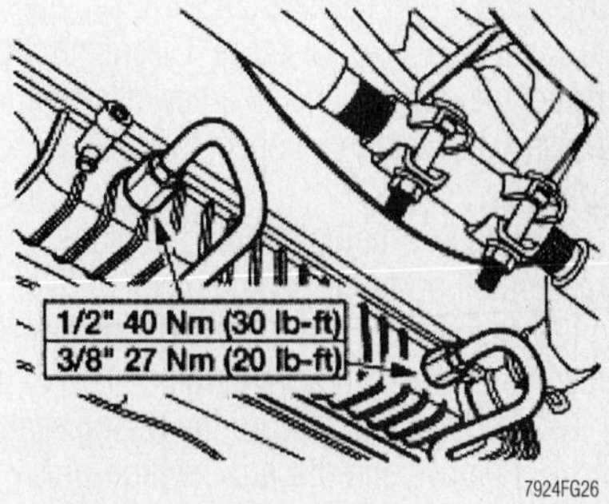

Tighten the manifold bolts in the proper sequence—7.3L diesel engine

8. Raise the vehicle to install the right manifold. Install the right manifold.

9. Connect the inlet pipes to the manifold and tighten. Lower the vehicle, connect the batteries and run the engine to check for exhaust leaks.

Camshaft and Valve Lifters

REMOVAL & INSTALLATION

6.0L Engine

1. Before servicing the vehicle, refer to the precautions in the beginning of this section.
2. Mount the engine on an engine stand.
3. Remove the serpentine belt idler.
4. Remove the serpentine belt tensioner.
5. Remove and discard the O-rings from the oil filter base.
6. Remove the Exhaust Gas Recirculation (EGR) cooler coolant supply port cover. Clean and inspect the gaskets. Install new gaskets if necessary. Clean and inspect the sealing surfaces.

✻✻ WARNING

In the event of a catastrophic engine failure, always install a new oil cooler cover assembly (with oil cooler). Foreign material cannot be removed from the oil cooler.

➡The oil cooler is replaced as an assembly.

7. Remove the oil cooler assembly. Clean and inspect the gaskets. Install a new gasket if necessary. Clean and inspect the sealing surfaces.
8. Remove the oil pump inlet strainer. Clean and inspect for tears and other damage.
9. Remove bolts and the turbocharger heat shield.
10. Remove the high-pressure oil pump cover. Use a thin gasket scraper to separate the cover from the crankcase. Clean and inspect the gaskets. Install a new gasket if necessary. Clean and inspect the sealing surfaces.
11. Remove the bolts from the high-pressure oil pump discharge pipe.
12. Disconnect and remove the high-pressure oil pump discharge pipe.
13. Remove and discard the D-ring seal.
14. Remove and discard the high-pressure pump O-ring seal.
15. Remove the bolts and the high-pressure oil pump.
16. Remove and discard the lower O-ring seal.
17. Remove the glow plug buss bar.
18. Remove the eight glow plugs.

➡Mark the location of the stud bolts.

19. Remove the valve covers. Clean and inspect the gaskets. Install a new gasket if necessary. Clean and inspect the sealing surfaces.
20. Remove the bolts and the coolant pump pulley.
21. Remove the coolant pump.
22. If equipped, remove the bolts and the dual generator pulley.
23. Prior to removing the crankshaft damper, check the crankshaft vibration damper runout.

➡Pry the crankshaft forward at the same point to eliminate possible error caused by crankshaft end play.

24. Rotate the crankshaft 90 degrees. Pry the crankshaft forward. Record the measurement. Repeat every 90 degrees. If the runout exceeds specification, install a new crankshaft vibration damper.

✻✻ WARNING

To prevent engine damage, you must always replace all four bolts when installing the vibration damper.

✻✻ CAUTION

To avoid personal injury, support the vibration damper during mounting bolt removal. The damper can slide off the nose of the crankshaft very easily.

25. Remove the bolts and the crankcase vibration damper. Discard the bolts.
26. Punch two holes in the seal. Remove the crankshaft seal with a slide-hammer.

➡Production engine will not have a wear sleeve. If equipped, remove the crankshaft damper wear sleeve.

27. Remove the oil pump body. Remove and discard the O-ring seal.

➡Mark the front of each drive rotor for correct reassembly orientation.

28. Remove the inner and outer oil pump drive rotors.
29. Remove the engine front cover. Clean and inspect the gaskets. Install new gaskets if necessary. Clean and inspect the sealing surfaces.
30. Using a quick-disconnect tool, disconnect the high-pressure oil rail supply line at the high-pressure oil rail.
31. Remove the bolts and the high pressure oil rail. Disconnect and remove the high-pressure oil supply line.

✻✻ WARNING

Do not attempt to put battery voltage to the fuel injector or damage to the fuel injector will occur.

32. Using a 19 mm socket, push the fuel injector electrical connector out of the rocker arm carrier.

✻✻ WARNING

To prevent engine damage, do not use air tools to remove the fuel injectors. The clip that extracts the injector can dislodge and fall into the oil drain hole.

➡If engine oil is found in the engine coolant or engine coolant is found in the combustion chambers, new injector sleeve may need to be installed.

33. Remove the bolt, the fuel injector hold down and the fuel injector.
34. Remove and discard the crankcase-to-head tube assembly.
35. Remove the inner head bolts from both cylinder heads.
36. Remove the 16 bolts and the rocker arm assemblies.
37. Remove the rocker arm carrier from the cylinder head. Clean and inspect the gaskets. Install new gaskets if necessary. Clean and inspect the sealing surfaces.

➡Mark the location of the valve bridges before removing.

38. Remove the 16 valve bridges.

✻✻ WARNING

To prevent engine damage, keep the push rods in the order in which they were removed. Install all push rods back in their original positions.

39. Mark the location and remove the 16 push rods.
40. Remove the 10 outer head bolts.
41. Remove the cylinder heads.
42. Remove and discard the cylinder head gasket.
43. Remove and discard the four cylinder head dowel sleeves.
44. Remove the bolts from the rear engine tube assembly.
45. Remove the bolt and the rear engine tube assembly.

⁂ WARNING

To prevent engine damage, keep the cam followers in the order in which they were removed. Install all cam followers back in their original positions.

46. Remove the bolts and the roller follower guides. Remove the hydraulic cam followers.
47. Install the special tool and measure the camshaft gear backlash. Install a new camshaft gear if backlash is not within specification.
48. Install the special tool and measure the camshaft end play. Install a new camshaft thrust plate if end play is not within specification.
49. Remove the bolt and the camshaft position (CMP) sensor.

⁂ WARNING

Do not knick or scratch the camshaft bearings with the camshaft lobes or engine damage will occur.

50. Remove the thrust plate mounting bolts and remove the camshaft and gear.

To install:

➡Check alignment of the oil holes after installing the bearings.

51. If removed, install the camshaft bearings.

⁂ WARNING

Do not nick or scratch the camshaft bearings with the camshaft lobes or engine damage can occur.

➡Apply clean engine oil to the camshaft prior to installing.

52. Using camshaft alignment tool 303-772, install the camshaft and gear assembly. Aligning it with the crankshaft. Install the thrust plate mounting bolts.
53. The remainder of installation is the reverse of removal.

7.3L Engine

➡Ford recommends removing the diesel engine from the vehicle for camshaft removal.

1. Before servicing the vehicle, refer to the precautions in the beginning of this section.
2. Remove or disconnect the following:
   - Intake manifold and valley pan, if equipped
   - Rocker covers
   - Either the rocker arm shafts or loosen the rockers on their pivots and remove the pushrods. The pushrods must be reinstalled in their original positions.
   - Valve lifters in sequence with a magnet. They must be replaced in their original positions.
   - Timing gear cover, timing gear and sprockets

➡A camshaft removal tool, Ford part no. T65L-6250-a and adapter 14-0314 is needed to remove the diesel camshaft.

   - Camshafts

To install:

3. liberally coat the camshaft with oil before installing it. Slide the camshaft into the engine very carefully so as not to scratch the bearing bores with the camshaft lobes. Install the camshaft thrust plate and tighten the attaching screws to 10–12 ft. lbs. (13–16 Nm). Measure the camshaft end-play. If the end-play is more than 0.009 inch (0.228mm), replace the thrust plate. Assemble the remaining components in the reverse order of removal.
4. Install or connect the following:
   - Timing gear and front cover
   - Valve lifters. They must be replaced in their original positions.
   - Pushrods, the rocker arms and the rocker arm covers
   - Intake manifold and valley pan, if equipped

Valve Lash

ADJUSTMENT

Valve lash on these engines is not adjustable.

Starter Motor

REMOVAL & INSTALLATION

6.0L Engine

1. Before servicing the vehicle, refer to the precautions in the beginning of this section.
2. Disconnect the battery ground cable.
3. Remove starter solenoid protective cap.
4. Disconnect the starter motor electrical connections.
5. Remove the bolts and the starter.

To install:

6. Installation is the reverse of the removal procedure. Tighten the starter bolts to 18 ft. lbs. (25 Nm)

7.3L Engine

1. Before servicing the vehicle, refer to the precautions in the beginning of this section.
2. Remove or disconnect the following:
   - Negative battery cable
3. Raise the front of the truck and install jackstands beneath the frame. Firmly apply the parking brake and place blocks in back of the rear wheels.
4. Remove or disconnect the following:
   - Wiring from the starter motor terminals
   - Starter motor retaining bolts, loosen
   - Starter retaining bolts while supporting the starter motor
   - Starter from the vehicle

To install:

5. The installation is the reverse of removal. Tighten the starter retaining bolts to 15–20 ft. lbs. (20–27 Nm)

Oil Pan

REMOVAL & INSTALLATION

6.0L Engine

1. Before servicing the vehicle, refer to the precautions in the beginning of this section.
2. With the vehicle in NEUTRAL, position it on a hoist.
3. Disconnect the negative battery cable.

⁂ WARNING

Never remove the pressure relief cap while the engine is operating or when the cooling system is hot. Failure to follow these instructions can result in damage to the cooling system or engine or result in personal injury. To avoid having scalding hot coolant or steam blow out of the degas bottle when removing the pressure relief cap, wait until the engine has cooled then wrap a thick cloth around the pressure relief cap and turn it slowly. Step back while the pressure is released from the cooling system. When certain all the pressure has been released, (still with a cloth) turn and remove the pressure relief cap. Failure to follow these instructions can result in personal injury.

⁂ CAUTION

The coolant must be removed in a suitable, clean container for reuse. If

the coolant is contaminated, it must be recycled or disposed of correctly and the system filled with new coolant.

➡Less than 80% of coolant capacity can be recovered with the engine in the vehicle. Dirty, rusty, or contaminated coolant requires replacement.

4. Place a suitable container below the radiator draincock. If equipped, disconnect the coolant return hose at the fluid cooler.
5. Remove the fill cap from the degas bottle.
6. Remove the oil pan drain plug and drain the engine oil.
7. Loosen the exhaust pipe retaining nuts.
8. Remove the motor mount retaining nuts.
9. Open the radiator draincock.
10. Disconnect the lower radiator hose.
11. Disconnect the transmission cooler tubes.
12. Remove the air cleaner assembly.
13. On vehicles with metal ducts, remove the charge air cooler duct.

➡Charge air cooler side shown, engine side similar.

14. On vehicles with blow molded ducts, loosen the clamps and remove the charge air cooler duct.
15. Remove the radiator support brackets.
16. Remove the 4 pin-type retainers and pull back the sight shield.
17. With the sight shield pulled back, remove the 3 pin-type retainers and the 2 wiring retainers and position the harness rearward out of the way.
18. Disconnect the upper radiator hose and the radiator overflow hose.
19. Remove the radiator and shroud as an assembly.
20. Disconnect the cooling fan electrical connector. Unclip and position the fan and wiring aside.

➡Use a hole in the fan hub to prevent the fan from turning.

21. Using the special tool 303-591, loosen the fan clutch. Remove the cooling fan.
22. Remove the bolts from the stator, remove the stator.
23. On vehicles with dual alternator perform the following:
    a. Remove the secondary accessory drive belt.
    b. Remove the bolt and the secondary accessory drive belt tensioner.
    c. Remove the primary accessory drive belt.
    d. Remove the bolts, bracket and secondary accessory drive belt idler pulley.
    e. Remove the bolts and the primary accessory drive belt tensioner.
    f. Disconnect the wire retainer, secondary generator electrical connector and B+ wire.
    g. Remove the bolts and the secondary generator with mounting bracket.
24. On vehicles with single alternator, remove the primary accessory drive belt.
25. Remove the bolts and position the alternator back.
26. Support the hood and disconnect the hood lift assemblies.
27. Install two lifting eyes 303-D030 on the right hand cylinder head.
28. Remove the retaining bolt for the fuel lines.
29. Disconnect the heater hose at the coolant pump.
30. Remove the fan shroud mounting stud.
31. Install one lifting eye 303-D030 on the left hand cylinder head.

⁂ CAUTION

Do not use the special tool to raise the engine. Damage to the special tool or vehicle may occur.

➡The ball studs may have to be removed.

➡This procedure requires a second bolt hook assembly. The tools are available through Rotunda tools with the following numbers: bolt hook 303-F070-6, handle 303-F070-8, bracket 303-F070-7 and washer 303-F070-12004.

32. Install the special tool.

➡The engine must be raised evenly.

33. Using a lifting crane, raise the engine until the turbo charger is about to touch the cowl. Secure the engine with the special tool.
34. Remove the transmission cooler line bracket.
35. Remove the bolts and position back the oil pan until the oil pick-up tube bolts are accessible.
36. Remove the bolts and let the oil pick-up tube go into the oil pan. Remove the oil pan.
37. Remove the press-in-place gasket and discard.
38. Clean and inspect the sealing surfaces.
39. Remove and discard the oil pick-up tube O-ring.

To install:

40. Install a new O-ring on the oil pick-up tube and position the oil pick-up tube in the oil pan.
41. Install a new press-in-place gasket into the upper oil pan.
42. Position the oil pan in the vehicle.
43. Install the oil pick-up tube and bolts. Tighten to 10 ft. lbs. (13 Nm).
44. Position the oil pan and install the bolts. Tighten to 10 ft. lbs. (13 Nm).
45. Install the transmission cooler tube bracket and nut. Tighten to 18 inch lbs. (10 Nm).
46. Using the special tool 303-F070, lower the engine.
47. Remove the lifting eyes 303-D030 and the special tool.
48. Install the fan shroud mounting stud. Tighten to 30 ft. lbs. (40 Nm).
49. Connect the heater hose at the coolant pump.
50. Install the retaining bolt for the fuel lines. Tighten to 10 ft. lbs. (10 Nm).
51. If removed, install the ball studs.
52. Connect the hood lifts and remove the hood support.
53. Position the primary alternator and install the bolts.
54. On a vehicle with a single alternator, install the primary accessory drive belt.
55. On vehicles with dual alternators, perform the following:
    a. Install the secondary alternator with mounting bracket and bolts. Tighten to 35 ft. lbs. (47 Nm).
    b. Connect the B+ wire, secondary generator electrical connector and wire retainer.
    c. Install the primary accessory drive belt tensioner and bolts. Tighten to 18 ft. lbs. (25 Nm).
    d. Position the secondary accessory drive belt idler pulley. Install the bracket and bolts. Tighten to 18 ft. lbs. (25 Nm).
    e. Install the primary accessory drive belt.
    f. Install the secondary accessory drive belt tensioner and bolt. Tighten to 18 ft. lbs. (25 Nm).
    g. Install the secondary accessory drive belt.
56. Install the stator and stator bolts. Tighten to 30 ft. lbs. (40 Nm).
57. Install the cooling fan clutch. Use the special tool 303-591 to tighten the cooling fan clutch. Tighten to 98 ft. lbs. (133 Nm).
58. Position and clip the cooling fan wiring. Connect the cooling fan electrical connector.
59. Install the radiator and shroud as an assembly.
60. Connect the upper radiator hose and the radiator overflow hose.

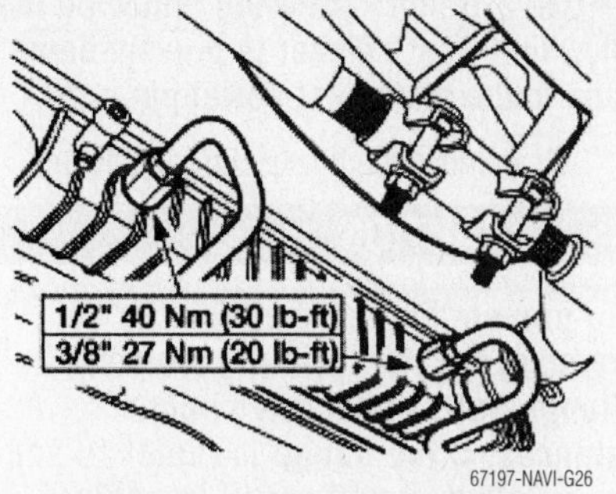

Connect the transmission cooler tubes. Tighten to the specifications illustrated—6.0L engine

61. Position the harness wiring, install the 2 wiring retainers and 3 pin-type retainers.
62. Position the sight shield and install the 4 pin-type retainers.
63. Install the radiator support brackets. Tighten to 22 ft. lbs. (30 Nm).
64. On vehicles with metal ducts, install the charge air cooler duct and tighten the clamps.
65. On vehicles with blow molded ducts, install the charge air cooler duct and tighten the clamps.
66. Install the air cleaner assembly.
67. Connect the transmission cooler tubes. Tighten to the specifications illustrated.
68. Close the radiator draincock. Connect the lower radiator hose.
69. Install the motor mount retaining nuts. Tighten to 76 ft. lbs. (103 Nm).
70. Tighten the exhaust pipe retaining nuts to 35 ft. lbs. (47 Nm).
71. Clean and inspect the oil pan drain plug and gasket, install new if necessary.
72. Install the oil pan drain plug and tighten to 18 ft. lbs. (25 Nm).
73. Connect the negative battery cables.
74. Fill the engine with clean engine oil.
75. Fill the coolant system.
76. Run the engine and check for leaks

7.3L Engine

1. Before servicing the vehicle, refer to the precautions in the beginning of this section.
2. Remove the engine.
3. Remove the oil pan bolts.
4. Lower the oil pan.

➡The oil pan is sealed to the crankcase with RTV silicone sealant in place of a gasket. It may be necessary to separate the pan from the crankcase with a utility knife. Also, the crankshaft may have to be turned to allow the pan to clear the crankshaft throws.

5. Clean the pan and crankcase mating surfaces thoroughly.

To install:

6. Apply a ⅛ in. (3mm) bead of RTV silicone sealant to the pan mating surfaces, and a ¼ in. (6mm) bead on the front and rear covers and in the corners; you have 15 minutes within which to install the pan!
7. Install the locating dowels into position.
8. Position the pan on the engine and install the pan bolts loosely.
9. Remove the dowels.
10. Tighten the pan bolts to:
    - ¼ in.-20 bolts: 84 inch lbs. (10 Nm)
    - 5⁄16 in.-18 bolts: 14 ft. lbs. (19 Nm)
    - ⅜ in.-16 bolts: 24 ft. lbs. (33 Nm)
11. Install the engine.

Oil Pump

REMOVAL & INSTALLATION

6.0L Engine

1. Before servicing the vehicle, refer to the precautions in the beginning of this section.
2. With the vehicle in NEUTRAL, position it on a hoist.
3. Disconnect the battery ground cable(s).
4. Remove the cooling fan
5. On vehicles with dual alternator, perform the following:
    a. Remove the dual generator accessory drive belt.
    b. Remove the bolts and the dual generator pulley.
6. Remove the accessory drive belt.
7. Check the crankshaft vibration damper runout.
8. Remove the paint from the face of the crankshaft vibration damper at four points 90 degrees apart.

Attach dial indicator 100-002 to the cylinder block. Position the tool on one of the unpainted surfaces.

Using a suitable tool, pry the crankshaft forward. Zero the dial indicator.

➡Pry the crankshaft forward only to eliminate possible error caused by crankshaft end play. Rotate the crankshaft 90 degrees. Pry the crankshaft forward. Record the measurement. Repeat at each unpainted surface.

9. If the runout exceeds specification, install a new crankshaft vibration damper.

⁂ CAUTION

To avoid personal injury, support the vibration damper during mounting bolt removal. The damper can slide off the nose of the crankshaft very easily.

10. Remove the bolts and the crankshaft vibration damper.
11. Discard the bolts.
12. Punch two holes in the seal.
13. Using the special tool 303-D060, remove the crankshaft seal.

➡Production engine will not have a wear sleeve.

14. If equipped, remove the crankshaft damper wear sleeve using tool 303-762.
15. Remove the bolts and the gerotor cover. Remove and discard the O-ring seal.

➡Mark the front of the inner and outer gerotor for correct reassembly.

16. Remove the inner and outer gerotors.
17. Inspect the oil pump components and replace as necessary.
18. Inspect the oil pump for excessive metal particles.
19. Inspect the oil pump for gouging, cracks or deep scratches.
20. Inspect the oil pump inner and outer gear rotors for damage or excessive wear.

To install:

21. Install the gears with marks pointing outward.

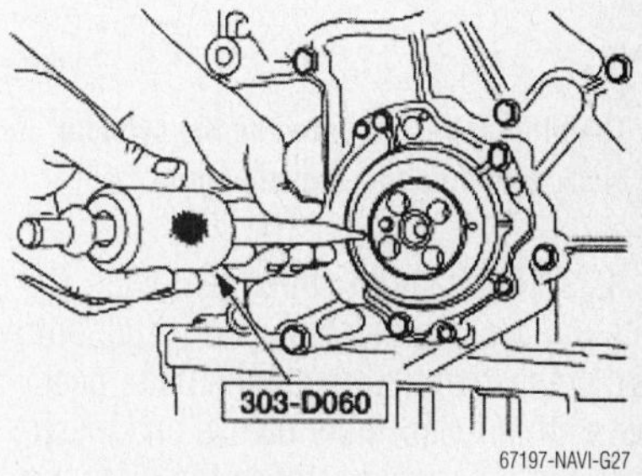

Using the special tool 303-D060, remove the crankshaft seal—6.0L engine

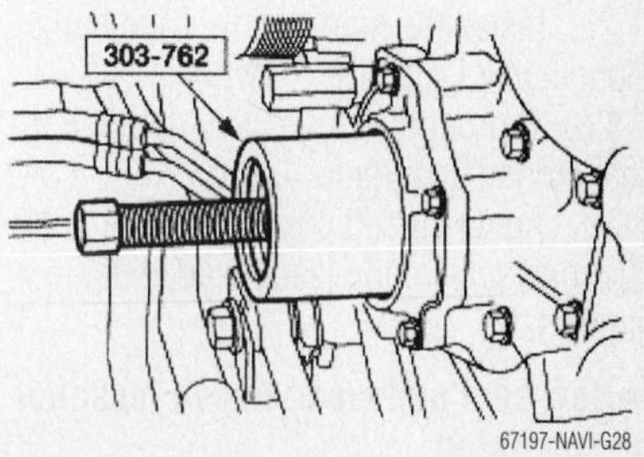

Remove the crankshaft damper wear sleeve using tool 303-762, if equipped—6.0L engine

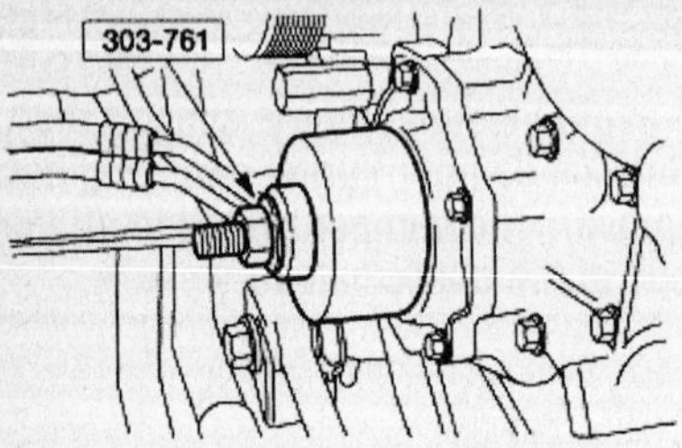

67197-NAVI-G29

Use tool 303-361 to install the oil seal and wear sleeve assembly—6.0L engine

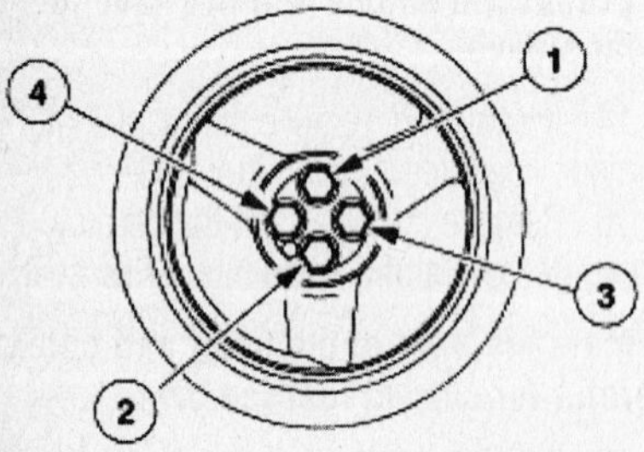

67197-NAVI-G30

Tighten the crankshaft vibration damper bolts in this sequence—6.0L engine

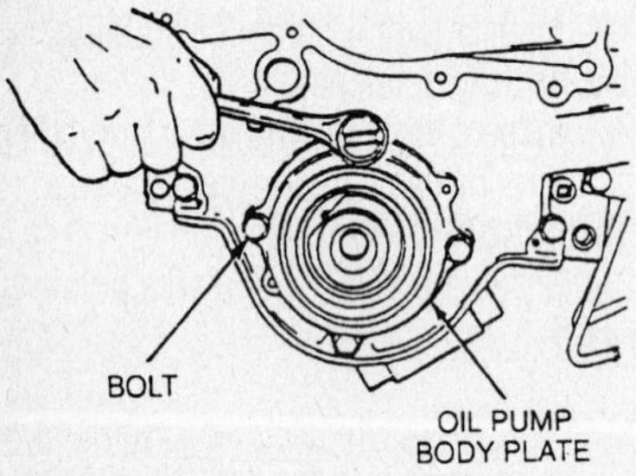

7924FG62

The oil pump is mounted on the cylinder block with 4 bolts—diesel engine

22. Lubricate the inner gear with lithium assembly grease and install onto the crankshaft. Lubricate the outer gear with lithium assembly grease and mesh with the inner gear rotor in the oil pump housing. Wipe off the excess assembly grease.
23. Install a new O-ring seal.
24. Install the gerotor cover and bolts. Tighten to 71 inch lbs. (8 Nm).
25. Thoroughly clean the crankshaft front seal mounting surface.
26. Apply Threadlock 262® to the outer circumference of the leading edge of the crankshaft.

➡New seal and wear sleeve must not be separated.

27. Using the special tool 303-361, install the oil seal and wear sleeve assembly.

✲✲ CAUTION

To prevent engine damage, you must always install four new bolts when installing the vibration damper.

➡Do not use anti-seize compounds, grease or any lubricants. Lubricants have an adverse effect on the torque results.

28. Install the crankshaft vibration damper and bolts.
29. Tighten the bolts in the sequence illustrated as follows:
    a. Tighten the bolts to 50 ft. lbs. (68 Nm).
    b. Tighten the bolts an additional 90 degrees.
30. Install the accessory drive belt.
31. On vehicles with dual alternator, perform the following:
    a. Install the dual alternator pulley and bolts. Tighten to 35 ft. lbs. (47 Nm).
    b. Install the dual generator accessory drive belt.
32. Install the cooling fan stator
33. Connect the battery ground cables.

7.3L Engine

1. Before servicing the vehicle, refer to the precautions in the beginning of this section.
2. Remove or disconnect the following:
    - Negative battery cables
    - Oil pan
    - Oil pick-up tube from the pump
    - Bolts and the oil pump

To install:

3. Assemble the pick-up tube and pump. Use a new gasket.
4. Install or connect the following:
    - Oil pump and tighten the bolts to 14 ft. lbs. (19 Nm)
    - Oil pick up tube
    - Oil pan
    - Negative battery cables

Rear Main Seal

REMOVAL & INSTALLATION

6.0L Engine

1. Before servicing the vehicle, refer to the precautions in the beginning of this section.
2. Remove the transmission.
3. Remove the bolts.
4. Remove the flexplate or flywheel.

➡Use extreme care when removing the flywheel front adapter to prevent damage to the alignment dowel pin.

5. Remove the flywheel front adapter.

✲✲ CAUTION

To prevent engine damage, do not remove the rear primary crankshaft flange bolts under any circumstances. If the flange is removed and reinstalled, it will result in engine vibration and premature transmission component wear.

6. Punch two holes in the rear main seal, across from each other.
7. Using the puller tool 100-001, remove the rear main seal.

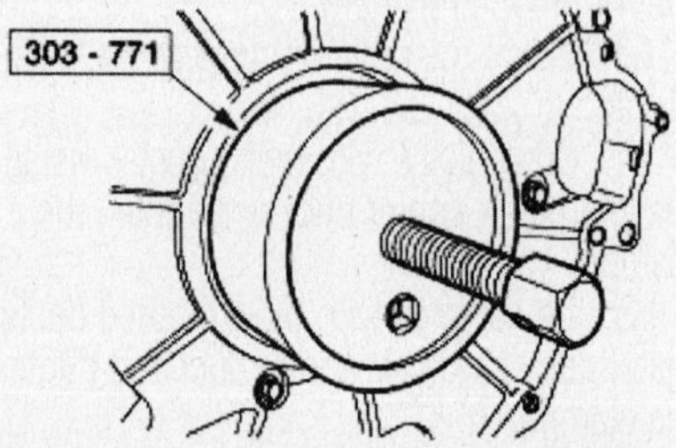

67197-NAVI-G31

If equipped with a crankshaft wear sleeve, use tool 303-771 to remove the crankshaft rear wear sleeve—6.0L engine

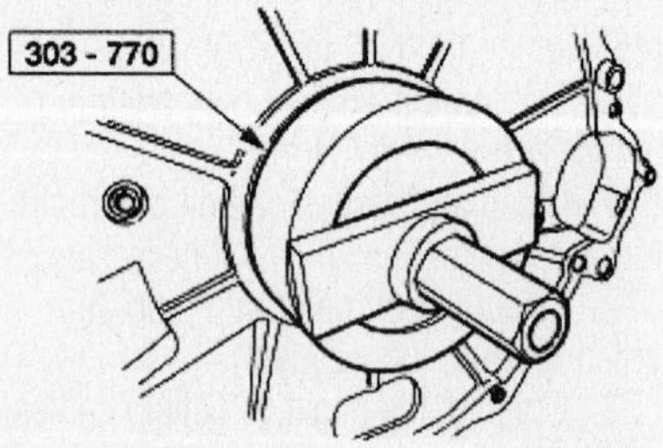

67197-NAVI-G32

Use tool 303-770 to install the crankshaft rear oil seal—6.0L engine

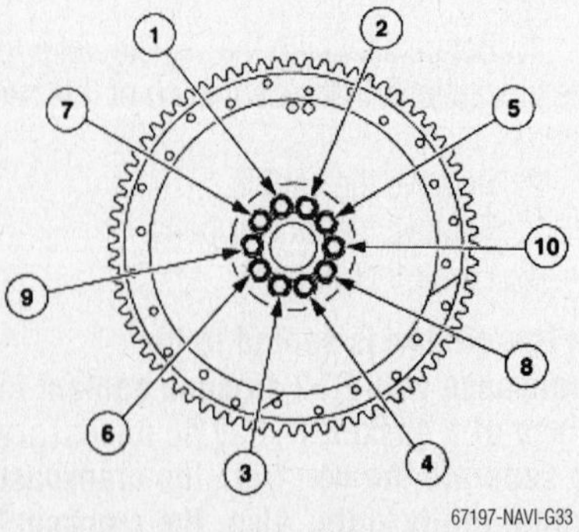

67197-NAVI-G33

Tighten the flexplate or flywheel bolts in this sequence—6.0L engine

➡Production engines will not have a wear sleeve.

8. If equipped with a crankshaft wear sleeve, use the tool 303-771 to remove the crankshaft rear wear sleeve.

9. Clean and inspect the crankshaft sealing surface.

To install:

➡The crankshaft rear oil seal and wear sleeve are installed as an assembly.

➡Lubricate the outer diameter of the rubber seal with a solution of dish soap and water (approximately 50/50 mix) prior to assembly. Do not use any other type of lubricant.

10. Apply a bead of Threadlock 262® around the circumference of the outer rear edge of the secondary crankshaft flange.

11. Using tool 303-770, install the crankshaft rear oil seal.

12. Install the flywheel front adapter.

13. Install the flexplate or flywheel.

14. Install the bolts. Snug all bolts to 44 inch lbs. (5 Nm), then tighten all bolts to 69 ft. lbs. (94 Nm) in the sequence illustrated.

15. Install the transmission.

7.3L Engine

1. Before servicing the vehicle, refer to the precautions in the beginning of this section.

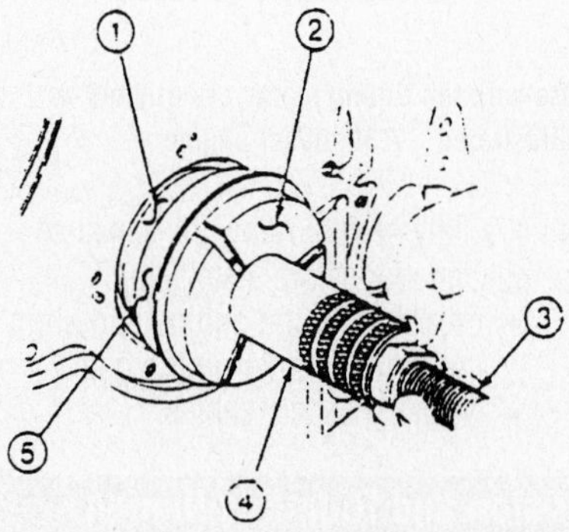

Item	Description
1	Crankshaft Wear Ring
2	Crankshaft Rear Wear Ring Remover
3	Forcing Screw
4	Remover Tube
5	Crankshaft Rear Wear Ring Remover Sleeve

7924FG61

Assemble the seal and wear ring removal tools, then remove the wear ring—7.3L diesel engine

2. Remove or disconnect the following:
   - Transmission
   - Flywheel
   - Crankshaft rear oil seal bolts and the seal

3. Clean the seal mating surfaces.

4. If installing the old seal, inspect it for damage.

5. Using crankshaft wear ring removal tool T94T-6701-AH1, forcing screw T84T-7025-B, remover tube T77J-7025-B and wear ring remover sleeve T94T-6701-AH2 (refer to the illustration), remove the wear ring.

To install:

6. Apply RTV silicone sealant to the seal retaining ring and the seal retaining bolts.

7. Using seal replacers T94T-6701-AH3 and T94T-AH4, driver sleeve T79T-6316-A4 (part of T79T-6316-A) and guide pins T94P-7000-P install the wear ring and oil seal.

8. Install or connect the following:
   - Seal retaining bolts
   - Flywheel
   - Transmission

Timing Gears, Front Cover and Seal

REMOVAL & INSTALLATION

6.0L Engine

1. Before servicing the vehicle, refer to the precautions in the beginning of this section.

2. Remove the intake manifold.

3. Disconnect the exhaust pressure tube at the exhaust.

4. Remove the nuts and the exhaust pressure bracket assembly.

➡Remove the thermostat housing only if a new front cover is being installed.

5. Remove the stud bolts and the thermostat housing.

6. Remove and discard the O-ring.

7. Disconnect the engine coolant fill hose.

8. Disconnect the lower radiator hose.

9. Remove the stator stand-off bolt.

10. Remove the four bolts and position aside the power steering pump.

11. Remove the bolts and the accessory drive belt tensioner.

12. Remove the bolts and the accessory drive idler pulleys.

➡Remove the coolant pump pulley only if a new front cover is being installed.

13. Remove the coolant pump pulley.

➡Remove the coolant pump only if a new front cover is being installed.

14. Remove the bolts and the coolant pump.

15. Remove and discard the O-ring.

16. Remove the nut and the battery cable bracket.

17. Remove the oil pump.

18. Check the crankshaft vibration damper runout.

19. Remove the paint from the face of the crankshaft vibration damper at four points 90 degrees apart.

20. Attach the special tool to the cylinder block. Position the special tool on one of the unpainted surfaces.

21. Using a suitable tool, pry the crankshaft forward. Zero the dial indicator.

➡Pry the crankshaft forward at the same point to eliminate possible error caused by crankshaft end play.

22. Rotate the crankshaft 90 degrees. Pry the crankshaft forward. Record the measurement. Repeat at each unpainted surface. If the runout exceeds specification, install a new crankshaft vibration damper.

**** WARNING**

To prevent engine damage, you must always install four new bolts when installing the vibration damper.

**** WARNING**

To avoid personal injury, support the vibration damper during mounting bolt removal. The damper can slide off the nose of the crankshaft very easily.

23. Remove the bolts and the crankcase vibration damper.

24. Discard the bolts.

25. Remove the bolts and the front cover.

**** WARNING**

Sealant is used where the crankcase and lower crankcase meet. Failure to cut the sealant could result in pulling the lower crankcase seal out while removing the front cover gasket.

26. Use a thin blade scraper to cut the sealant where the crankcase and the lower crankcase meet. Remove and discard the front cover gasket.

27. Clean and inspect the sealing surfaces.

28. Punch two holes in the seal.

29. Using the special tool, remove the crankshaft seal.

➡**Production engine will not have a wear sleeve.**

30. If equipped, remove the crankshaft damper wear sleeve.
31. Remove the thrust plate mounting bolts and remove the camshaft and gear.

To install:

32. Install the camshaft and gear assembly. Using the special tool, align the camshaft timing mark as shown. Install the thrust plate mounting bolts.
33. Thoroughly clean the crankshaft front seal mounting surface.
34. Apply Threadlock 262® to the outer circumference of the leading edge of the crankshaft.

➡**New seal and wear sleeve must not be separated.**

35. Using the special tool, install the oil seal and wear sleeve assembly.
36. If removed, install the front cover crankcase dowels into the cylinder block.

➡**Use guide studs to aid in installation. Studs must be fabricated locally.**

37. Install the guide studs.
38. Apply a bead of sealant at the seam where the crankcase and the lower crankcase meet.
39. Install a new engine front cover gasket.
40. Install the engine front cover and bolts.

⁂ WARNING

To prevent engine damage, you must always install four new bolts when installing the vibration damper.

➡**Do not use anti-seize compounds, grease or any lubricants. Lubricants have an adverse effect on the torque results.**

41. Install the crankshaft vibration damper and bolts.
42. Tighten the bolts in the sequence shown.
    a. Tighten the bolts to 68 Nm (50 ft. lbs.).
    b. Tighten the bolts an additional 90 degrees.
43. Install the oil pump.
44. Install the battery cable bracket and nut.

➡**Install a new O-ring on the coolant pump pulley.**

45. If removed, install the coolant pump and bolts.
46. If removed, install the coolant pump pulley and bolts.
47. Install the accessory drive idler pulleys and bolts.
48. Install the accessory drive belt tensioner and bolts.
49. Position back the power steering pump and install the bolts.
50. Install the stator stand-off bolt.
51. Connect the lower radiator coolant hose.
52. Connect the engine coolant hose.

➡**Install a new O-ring.**

53. If removed, install the thermostat housing and stud bolts.
54. Install the exhaust pressure bracket assembly and retaining nuts.
55. Connect the exhaust pressure tube fitting at the exhaust manifold.
56. Install the intake manifold.

7.3L Engine

➡**The crankshaft gear sprocket is not serviced separately from the crankshaft. Do not try to remove the sprocket or you will damage the crankshaft.**

1. Before servicing the vehicle, refer to the precautions in the beginning of this section.
2. Remove or disconnect the following:
    - Negative battery cables
    - Camshaft
    - Sprocket from the camshaft using a press
    - Thrust plate and sprocket key
3. Inspect the camshaft and related parts for wear and damage.

To install:

4. Clean the nose of the camshaft.
5. Install or connect the following:
    - Thrust plate
    - Key in the keyway on the camshaft
6. Heat the sprocket in an oven to 500°F (260°C).
7. Remove the sprocket from the oven, align the sprocket keyway with the camshaft key and install the sprocket on the camshaft until it is fully seated. Allow the camshaft assembly to cool before installation
    - Camshaft in the engine and align the timing marks on the gears
    - Negative battery cables

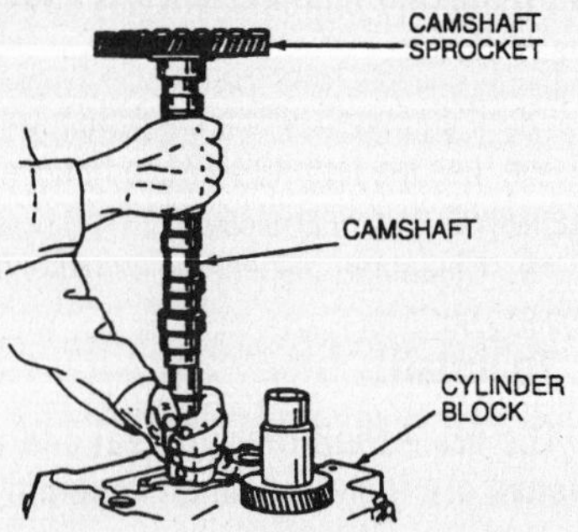

The camshaft gear is removed with the camshaft, then pressed off—7.3L diesel engine

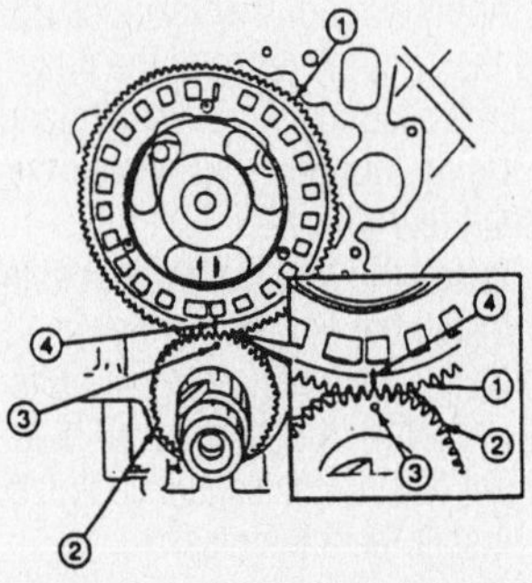

Item	Description
1	Camshaft Sprocket
2	Crankshaft Sprocket
3	Crankshaft Sprocket Timing Mark
4	Camshaft Sprocket Timing Mark

7924FG64

Be sure the timing marks are aligned as illustrated—7.3L diesel engine

Piston and Ring Positioning

1. Top compression ring is identified with one indentation mark and a 15 degree keystone profile.
2. The intermediate compression ring is identified with two indentation marks and a square profile.
3. Oil control ring.

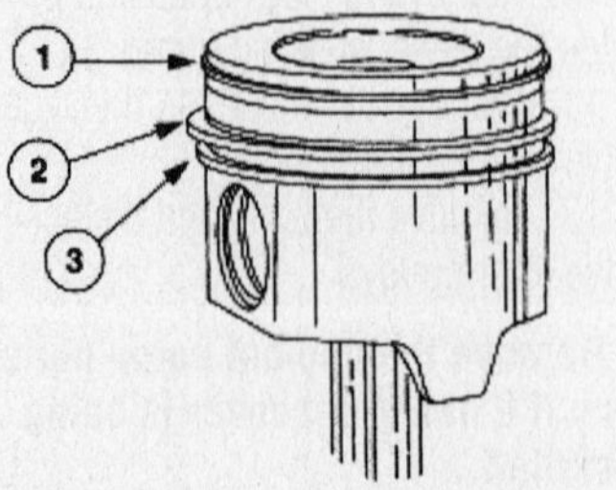

6.0L Diesel engines—piston ring locations

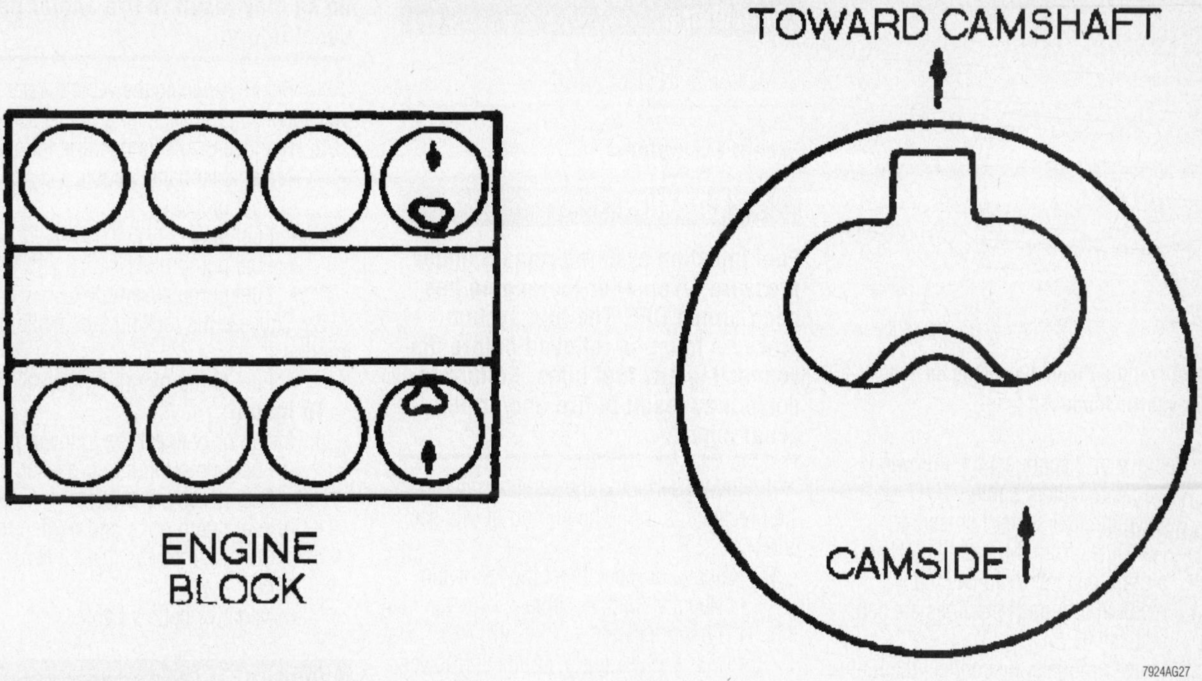

7.3L Diesel engines—piston-to-engine orientation

GASOLINE FUEL SYSTEM

Fuel System Service Precautions

Safety is the most important factor when performing not only fuel system maintenance, but any type of maintenance. Failure to conduct maintenance and repairs in a safe manner may result in serious personal injury or death. Maintenance and testing of the vehicle's fuel system components can be accomplished safely and effectively by adhering to the following rules and guidelines.

- To avoid the possibility of fire and personal injury, always disconnect the negative battery cable unless the repair or test procedure requires that battery voltage be applied.
- Always relieve the fuel system pressure prior to disconnecting any fuel system component (injector, fuel rail, pressure regulator, etc.), fitting or fuel line connection. Exercise extreme caution whenever relieving fuel system pressure, to avoid exposing skin, face and eyes to fuel spray. Please be advised that fuel under pressure may penetrate the skin or any part of the body that it contacts.
- Always place a shop towel or cloth around the fitting or connection prior to loosening to absorb any excess fuel due to spillage. Ensure that all fuel spillage (should it occur) is quickly removed from engine surfaces. Ensure that all fuel soaked cloths or towels are deposited into a suitable waste container.
- Always keep a dry chemical (Class B) fire extinguisher near the work area.
- Do not allow fuel spray or fuel vapors to come into contact with a spark or open flame.
- Always use a back-up wrench when loosening and tightening fuel line connection fittings. This will prevent unnecessary stress and torsion to fuel line piping. Always follow the proper torque specifications.
- Always replace worn fuel fitting O-rings with new. Do not substitute fuel hose or equivalent where fuel pipe is installed.

Fuel System Pressure

RELIEVING

➡A fuel pressure gauge is needed to correctly perform this procedure.

✱✱ CAUTION

Fuel injection systems remain under pressure, even after the engine has been turned OFF. The fuel system pressure must be relieved before disconnecting any fuel lines. Failure to do so may result in fire and/or personal injury.

1. Before servicing the vehicle, refer to the precautions in the beginning of this section.
2. Disconnect the negative battery cable and remove the fuel filler cap.
3. Remove the cap from the pressure relief valve on the fuel supply manifold. Install a fuel pressure gauge to the pressure relief valve.
4. Direct the gauge drain hose into a suitable container and depress the pressure relief button.
5. Remove the gauge and replace the cap on the pressure relief valve.

➡As an alternate method, disconnect the inertia switch and crank the engine for 15–20 seconds until the pressure is relieved.

Fuel Filter

REMOVAL & INSTALLATION

➡A fuel line disconnect tool is needed for this procedure.

✱✱ CAUTION

Fuel injection systems remain under pressure, even after the engine has been turned OFF. The fuel system pressure must be relieved before disconnecting any fuel lines. Failure to do so may result in fire and/or personal injury.

1. Before servicing the vehicle, refer to the precautions in the beginning of this section.

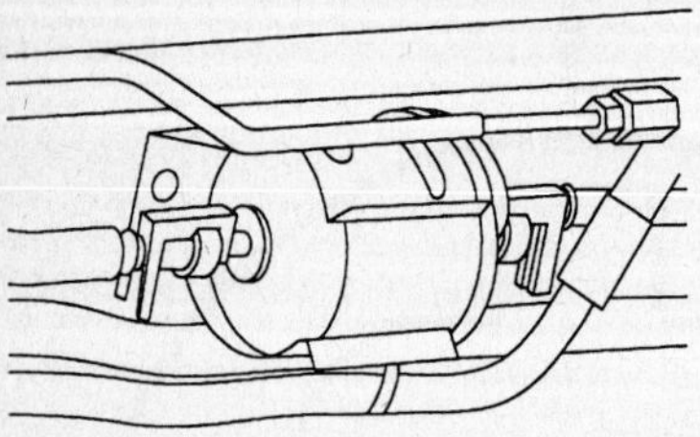

Typical fuel filter mounting along an under-vehicle frame rail

2. Remove or disconnect the following:
   - Negative battery cable
3. Relieve the fuel system pressure.
   - Fuel lines from the fuel filter. Have a drain pan handy to catch any residual fuel once the lines are separated.
4. On newer models, disconnect the fuel lines from the filter as follows:
   - Safety clip from the male hose
   - Install and push the fuel line disconnect tool into the female fitting.
5. Remove or disconnect the following:
   - Male and female fittings from the filter

➡Inspect the fuel lines for any damage after the fuel is finished draining.

   - Fuel filter from the bracket and the retainer, if equipped. Note the direction of the flow arrow so the replacement filter can be installed correctly.

To install:

6. Install or connect the following:
   - Fuel filter into the mounting bracket with the flow arrow pointing in the correct direction
   - Fuel lines to the fuel filter. Align and push the male tube into the female fitting until a click is heard. Pull on the fitting to ensure that it is fully engaged, then install the safety clip.
7. Lower the vehicle to the ground.

➡When the battery has been disconnected and reconnected, some abnormal drive symptoms may occur while the Powertrain Control Module (PCM) relearns its adaptive strategy. The vehicle may need to be driven 10 miles (16 km) or more to relearn the strategy.

8. Connect the negative battery cable.

Fuel Pump

REMOVAL & INSTALLATION

Except Excursion

⁂ CAUTION

Fuel injection systems remain under pressure, even after the engine has been turned OFF. The fuel system pressure must be relieved before disconnecting any fuel lines. Failure to do so may result in fire and/or personal injury.

1. Before servicing the vehicle, refer to the precautions in the beginning of this section.
2. Remove or disconnect the following:
   - Negative battery cable
   - Fuel pressure
   - Fuel tank skid plate bolts and lower the skid plate
   - Fuel
   - Fuel tank filler pipe hose from the tank
   - Fuel tank filler pipe vent hose from the tank
   - Fuel lines from the fuel pump
   - Front fuel tank connections
   - Rear Evaporative Emissions (EVAP) hose clamp and the hose
   - Electrical connector from the fuel pump
3. Support the fuel tank with a jack.
   - Fuel tank support strap bolts and the fuel tank straps
   - Fuel tank
   - Fuel pump lock ring
   - Fuel pump

To install:

4. Install or connect the following:
   - Fuel pump
   - Fuel pump lock ring and tighten to 66 ft. lbs. lbs. (89 Nm)
   - Fuel tank
5. Tighten the fuel tank strap bolts to 41 ft. lbs. (55 Nm).
6. Tighten the skid plate bolts to 41 ft. lbs. (55 Nm).
7. Connect the negative battery.

Excursion

⁂ CAUTION

Fuel injection systems remain under pressure, even after the engine has been turned OFF. The fuel system pressure must be relieved before disconnecting any fuel lines. Failure to do so may result in fire and/or personal injury.

1. Before servicing the vehicle, refer to the precautions in the beginning of this section.
2. Remove or disconnect the following:
   - Negative battery cable
   - Fuel pressure
   - Fuel tank
   - Fuel pump bolts
   - Fuel pump assembly
3. Squeeze the locking tabs while pushing down.
   - Fuel pump mounting gasket.

To install:

4. Install or connect the following:
   - Fuel pump mounting gasket.
   - Fuel pump assembly
   - Fuel pump bolts and tighten to 80–107 inch lbs. (9–12 Nm)
   - Fuel tank
   - Negative battery cable

Fuel Injector

REMOVAL & INSTALLATION

⁂ CAUTION

Fuel injection systems remain under pressure, even after the engine has been turned OFF. The fuel system pressure must be relieved before disconnecting any fuel lines. Failure to do so may result in fire and/or personal injury.

1. Before servicing the vehicle, refer to the precautions in the beginning of this section.
2. Remove or disconnect the following:
   - Negative battery cable
   - Engine cover, if equipped
   - Air cleaner assembly
3. Relieve the fuel pressure.
   - Upper intake manifold
   - Fuel pressure regulator valve vacuum hose
   - Fuel lines
   - Fuel injector electrical connectors
   - Fuel injector electrical harness from the fuel injector manifold
   - Fuel injection supply manifold
   - Fuel rail bolts and the rail

⁂ CAUTION

The fuel injectors are deposit-resistant. Do not clean. Use O-rings made of special fuel resistant material. Use of ordinary O-rings can cause the fuel system to leak.

- Fuel injector retaining clips and the fuel injectors. Inspect the fuel injector O-rings and if necessary.

To install:

➡Make sure the injector clips must engage the upper groove on the injectors or fuel leaks may occur.

4. Lubricate the new O-rings with clean engine oil to aid installation.
5. Install or connect the following:
   - Fuel injectors and the rail. Tighten the rail bolts to 89 inch lbs. (10 Nm).
   - Fuel injection supply manifold
   - Fuel injector electrical harness from the fuel injector manifold
   - Fuel injector electrical connectors
   - Fuel lines
   - Fuel pressure regulator valve vacuum hose
   - Upper intake manifold
   - Air cleaner assembly
   - Engine cover, if equipped
   - Negative battery cable

DIESEL FUEL SYSTEM

Fuel System Service Precautions

Safety is the most important factor when performing not only fuel system maintenance but any type of maintenance. Failure to conduct maintenance and repairs in a safe manner may result in serious personal injury or death. Maintenance and testing of the vehicle's fuel system components can be accomplished safely and effectively by adhering to the following rules and guidelines.

- To avoid the possibility of fire and personal injury, always disconnect the negative battery cable unless the repair or test procedure requires that battery voltage be applied.
- Always relieve the fuel system pressure prior to disconnecting any fuel system component (injector, fuel rail, pressure regulator, etc.), fitting or fuel line connection. Exercise extreme caution whenever relieving fuel system pressure, to avoid exposing skin, face and eyes to fuel spray. Please be advised that fuel under pressure may penetrate the skin or any part of the body that it contacts.
- Always place a shop towel or cloth around the fitting or connection prior to loosening to absorb any excess fuel due to spillage. Ensure that all fuel spillage (should it occur) is quickly removed from engine surfaces. Ensure that all fuel soaked cloths or towels are deposited into a suitable waste container.
- Always keep a dry chemical (Class B) fire extinguisher near the work area.
- Do not allow fuel spray or fuel vapors to come into contact with a spark or open flame.
- Always use a back-up wrench when loosening and tightening fuel line connection fittings. This will prevent unnecessary stress and torsion to fuel line piping. Always follow the proper torque specifications.
- Always replace worn fuel fitting O-rings with new. Do not substitute fuel hose or equivalent where fuel pipe is installed.

Fuel System Pressure

RELIEVING

6.0L Engine

1. Before servicing the vehicle, refer to the precautions in the beginning of this section.

✲✲ CAUTION

Before removing the fuel tank filler cap, turn the fuel tank filler cap 1/4 to

2. Remove the Schrader valve cap and install the fuel pressure gauge 310-012.
3. Open the manual valve slowly on the fuel pressure gauge 310-012 and relieve the fuel pressure. This will drain some fuel out of the system; place the fuel in a suitable container.

7.3L Engine

1. Before servicing the vehicle, refer to the precautions in the beginning of this section.

✲✲ CAUTION

Before removing the fuel tank filler cap, turn the fuel tank filler cap 1/4 to 3/4 turn counterclockwise and wait for the tank pressure to be relieved. Personal injury may result if the fuel tank filler cap is removed without the pressure fully relieved.

2. Remove the fuel tank filler cap to relieve any pressure in the fuel tank.
3. When servicing the fuel lines, loosen the fuel fitting to allow any residual fuel line pressure to be relieved.

Idle Speed

ADJUSTMENT

1. Before servicing the vehicle, refer to the precautions in the beginning of this section.
2. Place the transmission in **P**.
3. Bring the engine up to normal operating temperature.

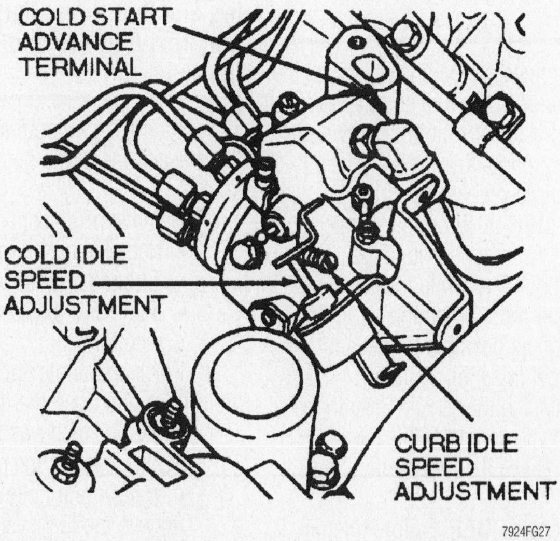

Raise or lower the curb idle speed by turning the curb idle speed adjusting screw—7.3L diesel engine

➡**Idle speed is measured with the transmission in D.**

4. Ensure that the curb idle adjusting screw is against the stop. If not, correct the vehicle linkage.

5. Check curb idle speed. Curb idle speed is specified on the Vehicle Emissions Control Information (VECI) decal on the underside of the vehicle's hood. Adjust the idle speed to specification using the idle speed adjusting screw.

6. Place the transmission in **P**. Rev the engine momentarily, then place the transmission in the specified gear and recheck the idle speed. Adjust again if necessary.

7. Remove the tachometer and close the hood.

Fuel Filter/Water Separator

DRAINING WATER

7.3L Engine

➡**Drain water from the water separator manual drain valve whenever the warning light comes ON or every 5000 miles (8000km). The "Water in Fuel" light will glow when approximately 3.5 oz. (103.5ml) of water accumulates in the separator.**

✲✲ CAUTION

The fuel system remains under pressure, even after the engine has been turned OFF. The fuel system pressure must be relieved before disconnecting any fuel lines. Failure to do so may result in fire and/or personal injury.

The diesel engines are equipped with a fuel/water separator in the fuel supply line. A "Water in Fuel" indicator light is provided on the instrument panel to alert the driver. The light should glow when the ignition switch is in the **start** position to indicate proper light and water sensor function. If the light glows continuously while the engine is running, the water must be drained from the separator as soon as possible to prevent damage to the fuel injection system.

1. Before servicing the vehicle, refer to the precautions in the beginning of this section.

2. Properly relieve the fuel system pressure.

3. Shut the engine **OFF**. Failure to shut the engine **OFF** before draining the separator will cause air to enter the system.

4. Unscrew the vent on the top center of the separator unit 2 ½ –3 turns.

5. Unscrew the drain screw on the bottom of the separator 1 ½ –2 turns and drain the water into an appropriate container.

6. After the water is completely drained, close the water drain finger-tight.

7. Tighten the vent until snug, then turn it an additional ¼ turn.

8. Start the engine and check the "Water in Fuel" indicator light; it should not be lit. If it is lit and continues to stay so, there is a problem somewhere else in the fuel system.

REMOVAL & INSTALLATION

6.0L Engine

1. Before servicing the vehicle, refer to the precautions in the beginning of this section.
2. Relieve the fuel system pressure.
3. Disconnect the fuel lines from the fuel filter.
4. Remove the fuel filter.
5. Loosen the fuel filter clamp screw.

To install:

6. Installation is the reverse of removal.

➡**Make sure that an audible click is heard when installing the fuel lines. Pull back on the fuel lines to confirm engagement.**

7.3L Engine

✲✲ CAUTION

The fuel system remains under pressure, even after the engine has been turned OFF. The fuel system pressure must be relieved before disconnecting any fuel lines. Failure to do so may result in fire and/or personal injury.

1. Before servicing the vehicle, refer to the precautions in the beginning of this section.
2. Remove or disconnect the following:
   - Negative battery cable
   - Turbocharger assembly
   - Baffle and the air inlet crossover manifold
3. Place a suitable container under the drain hose and open the filter drain.
   - The 2 capscrews securing the fuel filter base to the crankcase
   - Water drain hose from the filter
4. Relieve the fuel system pressure.
   - Fuel outlet hose, located between the fuel and filter housing, and the fuel return hose from the fuel pressure regulator valve
   - The 2 fuel supply hoses that connect the regulator block to the cylinder head fuel rails
   - Clamp at the fuel pump end of the hose (loosen only), which connects the fuel filter to the inlet of high pressure stage at the fuel pump.
   - Wiring harness from the right side of the filter housing
   - Electrical connections from the Water In Fuel (WIF) sensor and the fuel heater
   - Fuel filter
5. Use a prybar to remove the fuel filter cap and the filter element will come out with the cap.
6. Depress the element locking tabs and remove the element from the cap.

To install:

7. Clean the mating surfaces and install the filter element onto the cap, making sure the tabs engage.
8. Install or connect the following:
   - Filter gap and press down firmly, but gently, to engage it
   - Wiring connections to the fuel heater and WIF sensor
   - Wiring harness to the filter housing
   - Tighten the clamp at the fuel pump end of the hose, which connects

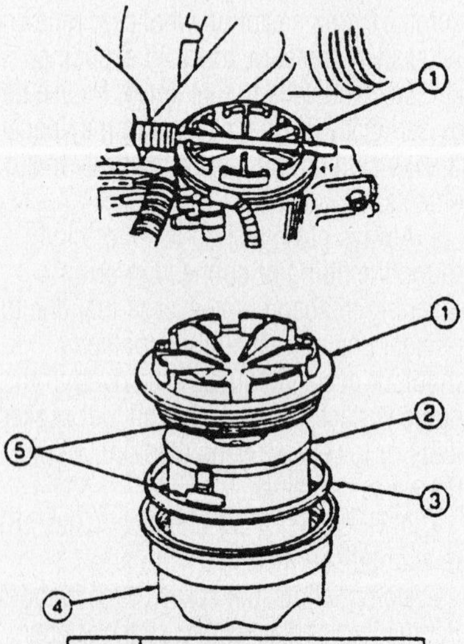

Item	Description
1	Fuel Filter Cap
2	Fuel Filter Element
3	Fuel Filter Bevel Cut Gasket
4	Fuel Filter Housing and Gland
5	Fuel Filter Element and Cap Locking Tabs

7924FG68

Use a prytool to remove the fuel filter cap to gain access to the filter element-7.3L diesel engine

the fuel filter to the inlet of high pressure stage at the fuel pump
- 2 fuel supply hoses that connect the regulator block to the cylinder head fuel rails
- Fuel outlet hose, located between the fuel and filter housing, and the fuel return hose from the fuel pressure regulator valve.
- Water drain hose to the filter
- The 2 capscrews securing the fuel filter base to the crankcase
- Air inlet crossover manifold and baffle
- Turbocharger assembly
- Negative battery cable

Fuel Conditioning Module

The Fuel Condition Module contains the fuel pump and water separator.

DRAINING

6.0L Diesel Engine

⁂ CAUTION

Smoking or open flame of any type must not be present when working near fuel or fuel vapor.

1. Disconnect both battery ground cables.
2. Raise and support the vehicle.
3. Open the fuel/water separator drain valve to release the fuel pressure.

REMOVAL & INSTALLATION

6.0L Diesel Engine

1. Before servicing the vehicle, refer to the precautions in the beginning of this section.

⁂ CAUTION

Smoking or open flame of any type must not be present when working near fuel or fuel vapor.

2. Disconnect both battery ground cables.
3. Raise and support the vehicle.
4. Open the fuel/water separator drain valve to release the fuel pressure.
5. Disconnect the electrical connectors.
6. Disconnect the fuel pump electrical connector.
7. Disconnect the fuel warmer electrical connector.
8. Disconnect the water-in-fuel electrical connector.
9. Disconnect the fuel hoses.
10. Remove the fuel hose retaining clips and discard. Disconnect the fuel hoses from the fuel pump.
11. Press in the retaining clips and release the fuel hoses.
12. Remove the mounting nuts and the fuel conditioning module.
13. To install, reverse the removal procedure.

High Pressure Oil Pump

REMOVAL & INSTALLATION

6.0L Engine

EARLY BUILD MODELS

1. Before servicing the vehicle, refer to the precautions in the beginning of this section.
2. Remove the intake manifold.
3. Remove the turbocharger heat shield.
4. Remove the bolts and the high-pressure oil pump cover.
5. Use a thin gasket scraper to separate the cover from the crankcase.
6. Remove and discard the press-in-place gasket.
7. Position the dial indicator with bracketry onto the oil pump drive and check the oil pump drive gear backlash. The reading should be 0.007–0.0124 inch (0.179–0.315mm).
8. Remove the bolts from the high-pressure oil pump discharge pipe.
9. Using the tool 303-755, disconnect and remove the high-pressure oil pump discharge pipe.
10. Remove and discard the D-shaped O-ring seal.
11. Remove and discard the high-pressure pump O-ring seal.
12. Remove the bolts and the high-pressure oil pump.
13. Remove and discard the lower O-ring seal.

To install:

14. Install a new lower O-ring seal.
15. Install the high-pressure oil pump and bolts. Tighten to 18 ft. lbs. (25 Nm).
16. Install the high-pressure pump O-ring seal.
17. Install the oil pump discharge pipe.
18. Install the bolts for the oil discharge pipe. Tighten to 71 inch lbs. (8 Nm).
19. Position the dial indicator with bracketry onto the oil pump drive and check the oil pump drive gear backlash. The reading should be 0.007–0.0124 inch (0.179–0.315mm).
20. Install a new D-ring seal on the high-pressure discharge pipe.
21. Install a new press-in-place gasket in the high-pressure pump cover.
22. Clean the cover mounting surface and apply sealer at the seams.
23. Install the high-pressure pump cover and bolts. Tighten to 8 ft. lbs. (11 Nm).
24. Install the turbocharger heat shield.
25. Install the intake manifold

LATE BUILD MODELS

1. Before servicing the vehicle, refer to the precautions in the beginning of this section.
2. Remove the intake manifold.
3. Remove the turbocharger heat shield.
4. Remove the bolts and the high-pressure oil pump cover.
5. Use a thin gasket scraper to separate the cover from the crankcase.
6. Remove and discard the press-in-place gasket.
7. Position the dial indicator with bracketry onto the oil pump drive and check the oil pump drive gear backlash. The reading should be 0.007–0.0124 inch (0.179–0.315mm).
8. Remove the bolts from the high-pressure oil pump discharge pipe. Position the high-pressure discharge pipe aside.
9. Remove and discard the high-pressure pump O-ring seal.
10. Remove the bolts and the high-pressure oil pump.
11. Remove and discard the lower O-ring seal.

To install:

12. Install a new lower O-ring seal.
13. Install the high-pressure oil pump and bolts. Tighten to 18 ft. lbs. (25 Nm).
14. Install the high-pressure pump O-ring seal.
15. Position back the high-pressure discharge tube and install the bolts. Tighten to 71 inch lbs. (8 Nm).
16. Position the dial indicator with bracketry onto the oil pump drive and check the oil pump drive gear backlash. The reading should be 0.007–0.0124 inch (0.179–0.315mm).
17. Install a new D-ring seal on the high-pressure discharge pipe.
18. Install a new press-in-place gasket in the high-pressure pump cover.
19. Clean the cover mounting surface and apply sealer at the seams.
20. Install the high-pressure pump cover and bolts. Tighten to 8 ft. lbs. (11 Nm).
21. Install the turbocharger heat shield.
22. Install the intake manifold

Injection Pump

REMOVAL & INSTALLATION

7.3L Engines

✻✻ CAUTION

The fuel system remains under pressure, even after the engine has been turned OFF. The fuel system pressure must be relieved before disconnecting any fuel lines. Failure to do so may result in fire and/or personal injury.

1. Before servicing the vehicle, refer to the precautions in the beginning of this section.
2. Remove or disconnect the following:
   - Negative battery cable
   - Turbocharger assembly
   - Fuel system pressure
   - Fuel line banjo bolt at the pump
   - Fuel line fittings at the rear of the cylinder heads
   - Fuel lines
   - The 3 hose clamps at the injection pump fittings
   - Water drain hose at the fuel filter
   - Filter and position it forward
   - Injection pump retaining bolts, then lift the pump out of the crankcase bore
   - Injection pump tappet from the crankcase bore

To install:

3. Rotate the engine so the injection pump eccentric is on the base circle.
4. Install or connect the following:
   - Injection pump tappet in the base of the injection pump
   - O-ring on the injection pump base
   - Injection pump and tighten the bolts to 19–27 ft. lbs. (26–37 Nm)
   - Fuel filter and connect the water drain hose
   - The 3 fuel hoses at the front of the injection pump
   - The fuel line clamps and the fuel filter retaining bolts
   - Fuel line assembly and new seal rings at the rear of the pump
   - Fuel line fittings at the rear of the cylinder heads
   - Fuel line banjo fitting at the pump. Tighten the bolt to 18 ft. lbs. (24 Nm).
   - Turbocharger assembly
   - Negative battery cable

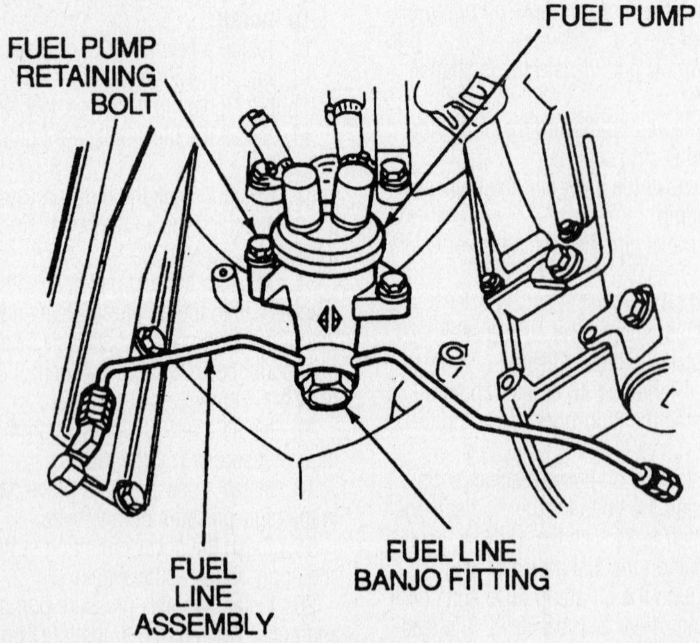

Injection pump assembly components—7.3L diesel engine

Fuel Injectors

REMOVAL & INSTALLATION

6.0L Engine

EARLY BUILD MODELS

1. Before servicing the vehicle, refer to the precautions in the beginning of this section.
2. If removing the right hand fuel injectors, remove the evaporator case.
3. Remove the valve cover.
4. Disconnect the fuel injector electrical connector.
5. Disconnect the high-pressure oil rail supply line at the high-pressure oil rail.
6. Remove the bolts and the high-pressure oil rail.
7. Disconnect and remove the high-pressure oil supply line.

✻✻ CAUTION

Do not attempt to apply battery voltage to the fuel injector or damage to the fuel injector will occur.

8. Using a 19 mm socket, push the fuel injector electrical connector out of the rocker arm carrier.

✻✻ CAUTION

To prevent engine damage, do not use air tools to remove/install the fuel injectors. The clip that extracts the injector can dislodge and fall into the oil drain hole.

➡There is no need to drain the fuel rail.

➡If engine coolant is found in the combustion chambers, It may be necessary to install a new injector sleeve.

9. Remove the bolt, fuel injector hold-down clamp and fuel injector.

To install:

✻✻ CAUTION

If the fuel injector oil inlet D-shaped O-ring is damaged, a new fuel injector must be installed.

10. Install new O-ring seals and copper washer on the fuel injector. Lubricate the fuel injector and O-ring seals liberally with clean engine oil.
11. Install the fuel injector, fuel injector hold-down clamp and bolt. Tighten to 24 ft. lbs. (33 Nm).
12. Install the fuel injector electrical connector into the rocker carrier.
13. Apply engine oil to the top fuel injector O-ring seals.
14. Install the high-pressure oil rail and bolts.
15. Install the high-pressure oil rail.
16. Install the bolts finger tight. Tighten the bolts in the sequence shown to 8 ft. lbs. (11 Nm).
17. Install the high-pressure oil line.

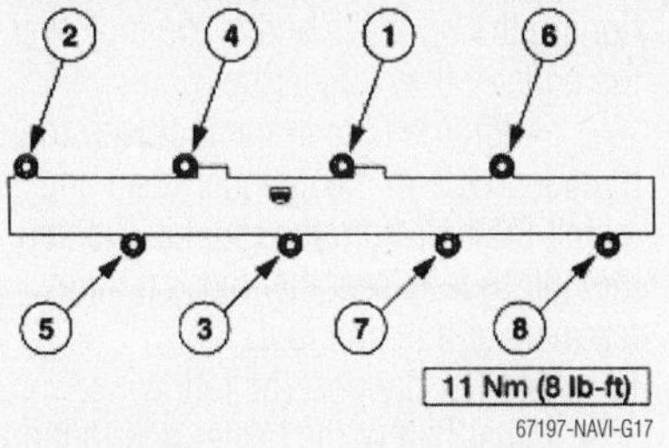

On early build models, tighten the high-pressure oil rail bolts in the sequence shown—6.0L engine–right side shown, left side similar

18. Connect the fuel injector electrical connector.
19. Install the valve covers.
20. If removed, install the evaporator case

LATE BUILD MODELS

1. Before servicing the vehicle, refer to the precautions in the beginning of this section.
2. If removing the right hand fuel injectors, remove the evaporator case.
3. Remove the valve cover.
4. Disconnect the fuel injector electrical connector.
5. Remove the bolts and the high-pressure oil rail.

⁂ CAUTION

Do not attempt to apply battery voltage to the fuel injector or damage to the fuel injector will occur.

6. Using a 19 mm socket, push the fuel injector electrical connector out of the rocker arm carrier.

⁂ CAUTION

To prevent engine damage, do not use air tools to remove/install the fuel injectors. The clip that extracts the injector can dislodge and fall into the oil drain hole.

➡There is no need to drain the fuel rail.

➡If engine coolant is found in the combustion chambers, it may be necessary to install a new injector sleeve.

7. Remove the bolt, fuel injector hold-down clamp and fuel injector.

To install:

⁂ CAUTION

If the fuel injector oil inlet D-shaped O-ring is damaged, a new fuel injector must be installed. Install new O-ring seals and copper washer on the fuel injector. Lubricate the fuel injector and O-ring seals liberally with clean engine oil.

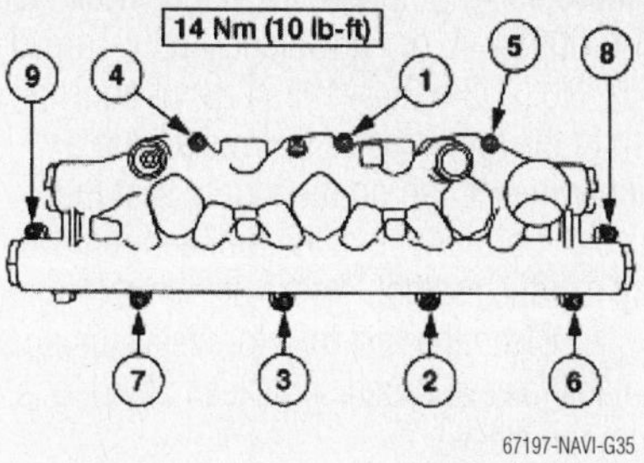

On late build models, tighten the high-pressure oil rail bolts in the sequence shown—6.0L engine–right side shown, left side similar

8. Install the fuel injector, fuel injector hold-down clamp and bolt. Tighten to 24 ft. lbs. (33 Nm).
9. Install the fuel injector electrical connector into the rocker carrier.
10. Apply engine oil to the top fuel injector O-ring seals.
11. Apply clean engine oil on the crankcase-to-head tube O-ring seal.

➡Apply clean engine oil on the tubes prior to installing the oil manifold.

12. Position the oil rail on the fuel injectors as follows:
    a. Place the oil rail on top of the carrier so that the four single ball tubes are engaging the injector lead angle.
    b. Insert three bolts, two on the ends of the straight side of the oil rail and one in the middle of the wavy side of the rail. Install the guide studs six to seven turns.
    c. Press the oil rail into the fuel injectors.
    d. Make sure that the oil rail mounting feet are flat against the mounting surface.
    e. Loosely install the six bolts.
13. Install the oil rail retaining bolts.
14. Remove the three guide bolts.
15. Install the bolts finger tight. Tighten the bolts in the sequence shown to 10 ft. lbs. (14 Nm).
16. Connect the fuel injector electrical connector.
17. Install the valve covers.
18. If removed, install the evaporator case.

7.3L Engines

⁂ CAUTION

Observe all applicable safety precautions when working around fuel. Whenever servicing the fuel system, always work in a well ventilated area. Do not allow fuel spray or vapors to come in contact with a spark or open flame. Keep a dry chemical fire extinguisher near the work area. Always keep fuel in a container specifically designed for fuel storage; also, always properly seal fuel containers to avoid the possibility of fire or explosion.

1. Before servicing the vehicle, refer to the precautions in the beginning of this section.

⁂ CAUTION

The red-striped wires on the DI Turbo carry 115 volts DC. A severe electrical shock may be given. Do not pierce the wires.

⁂ WARNING

Do not pierce the wires or damage to the harness could occur.

Special tools required:
- Slide Hammer, No. T50T–100–A, or equivalent
- Injector Remover, No. T94T–9000–AH1, or equivalent
- Injector Replacer, No. T94T–9000–AH2, or equivalent

2. Remove or disconnect the following:
    - Valve cover
    - Fuel injector electrical connector

⁂ WARNING

Remove the oil drain plugs prior to removing the injectors or oil could enter the combustion chamber which could result in hydrostatic lock and severe engine damage.

    - Oil rail drain plugs
    - Retaining screw and oil deflector. The shoulder bolt on the inboard side of the fuel injector does not require removal.
    - Outboard fuel injector retaining bolt
    - Heater distribution box screws, nuts and clip
    - Outer half of the case (to service No. 4 fuel injector only)

3. Remove the fuel injector using Injector Remover No. T94T–9000–AH1, or equivalent Position the tool's fulcrum beneath the fuel injector hold-down plate and over the edge of the cylinder head. Install the remover screw in the threaded hole of the fuel injector plate. Tighten the

screw to lift out the injector from its bore. Place the injector in a suitable protective sleeve such as Rotunda Injector Protective Sleeve, No. 014-00933-2, and set the injector in a suitable holding rack.

4. Remove the fuel injector sleeves, if required. Insert the Injector Sleeve Tap Plug, 014-00934-3 into the injector sleeve to prevent debris from entering the combustion chamber. Insert Injector Sleeve Tap Pilot into the fuel injector sleeve and tighten 1-1 ½ turns. Attach Slide Hammer T50T-100-A to the Injector Sleeve Tap 014-00934-1 and 014-00934-2 Injector Sleeve Tap Pilot and remove the fuel injector sleeve from the bore.

5. Use Rotunda Injector Sleeve Brush 104-00934-A, or equivalent to clean the injector bore of any sealant residue. Make sure to remove any debris.

To install:

6. If removed, install the fuel injector sleeves using Rotunda Sleeve Replacer, No. 014-00934-4, or equivalent. Apply Threadlock, No. 262-E2FZ-19554-B, or equivalent to the fuel injector sleeves. Using a rubber mallet, tap on the tool to seat the injector bore. Remove the tool and remove any residue sealant.

7. Clean the fuel injector sleeve using a suitable sleeve brush set. Clean any debris from the sleeve.

8. Clean the injector bore with a lint-free shop towel.

9. Install the fuel injectors using special tools as follows:

a. Lubricate the injectors with clean engine oil. Using new copper washers, carefully push the injectors square into the bore using hand pressure only to seat the O-rings.

b. Position the open end of Injector Replacer, No. T94T-9000-AH2, or equivalent between the fuel injector body and injector hold-down plate, while positioning the opposite end of the tool over the edge of the cylinder head.

c. Align the hole in the tool with the threaded hole in the cylinder head and install the bolt from the tool kit. Tighten the bolt to fully seat the injector, then remove the bolt and tool.

10. Install or connect the following:

- Outer half of the heater distribution box and retaining hardware (for No. 4 injector only)
- Oil deflector and bolt. Tighten the bolt to 108 inch lbs. (12 Nm).
- Fuel rail drain plug, tightening it to 96 inch lbs. (11 Nm)
- Oil rail drain plug, tightening it to 53 inch lbs. (6 Nm)
- Heater distribution box
- Fuel injector wiring harness
- Valve cover

DRIVE TRAIN

Transmission Assembly

REMOVAL & INSTALLATION

4R100 Transmission

EXCEPT EXCURSION

1. Before servicing the vehicle, refer to the precautions in the beginning of this section.

2. Place the transmission range selector lever in the NEUTRAL position.

➡All gasoline vehicles will have new adaptive shift strategies. Whenever the vehicle's battery has been disconnected for any type of service or repair the strategy parameters that are stored in the Keep Alive Memory (KAM) will be lost. The strategy will start to relearn once the battery is reconnected and the vehicle is driven. This is a temporary condition and will return to normal operating condition once the Powertrain Control Module (PCM) relearns all the parameters from the driving conditions. There is no set time frame for this process. If this concern is present during downshifts or converter clutch apply, it is not the fault of the shift strategy and will require diagnosis as outlined in the workshop manual. The customer needs to be notified that they may experience slightly different upshifts either soft or firm and that this is a temporary condition and will eventually return to normal operating condition.

3. Remove or disconnect the following:

- Negative battery cable
- Transmission fluid filler tube
- Transmission fluid cooler tubes at the cooler bypass valve. Catch the fluid in a suitable container.
- Transfer case assembly, if equipped

4. Mark the driveshaft yoke and axle flange so they may be installed in their original position.

- Rear driveshaft and install an appropriate plug in the transmission to prevent fluid leakage
- Shift cable from the transmission
- Shift cable from the manual lever
- Cable housing from the bracket and position aside
- Transmission Range (TR) sensor connector
- Shift cable at the transmission and position it aside
- Solenoid body connector
- Turbine Shaft Speed (TSS) sensor and the Output Shaft Speed (OSS) sensor
- Starter motor
- Cylinder block opening cover
- Torque converter-to-flexplate retaining nuts and discard

⁂ CAUTION

Be sure not to raise the back of the transmission too high. If it makes contact with the underbody, damage to the TSS sensor can occur.

5. Install a high lift transmission jack with the special jack adapter 014-0763 to the transmission.

- Exhaust and crossmember
- Six transmission-to-engine mounting bolts

6. Gently rock the transmission side-to-side to disengage it from the locator dowels.

7. Move the transmission and the transmission jack rearward to clear the engine flexplate.

- Transmission from the vehicle

To install:

8. Installation is the reverse of removal, please note the following specs:

9. Transmission-to-engine bolts to 45 ft. lbs. (61 Nm).

10. Torque converter-to-flexplate nuts to 26 ft. lbs. (35 Nm).

11. Flexplate inspection cover nuts to 25 ft. lbs. (34 Nm) on gasoline engines and 15 ft. lbs. (20 Nm) on diesel engines.

12. Fluid filler tube into the short fluid inlet tube. Tighten the nut to 15 ft. lbs. (22 Nm).

13. Refill all transmissions with the correct amount of Motorcraft MERCON® automatic transmission fluid.

14. Connect the negative battery cable.

EXCURSION

1. Before servicing the vehicle, refer to the precautions in the beginning of this section.

2. With the vehicle in NEUTRAL position it on a hoist.

➡When the battery has been disconnected and reconnected, some abnormal drive symptoms can occur while the vehicle relearns its adaptive strategy. The vehicle owner should to be notified that they may experience slightly different upshifts (either soft or firm) and that this is a temporary condition that will return to normal operating condition.

3. Disconnect the negative battery cable.
4. On 4x4 models, remove the transfer case assembly.
5. On 4x2 models, remove the rear driveshaft
6. Drain the transmission fluid.

CAUTION: Mixing the 4x2 and the 4x4 style transmission fluid filters and transmission pan assembly components can cause transmission damage.

7. Install the correct pan with gasket for this application. Alternately tighten the bolts.

⁂ CAUTION

Make sure securing straps or the transmission jack adapter do not touch the Cooler Bypass Valve (CBV). Do not use the CBV as a handle as damage to the CBV can cause a leak.

8. Install a suitable transmission jack and support the transmission.
9. On 4x4 models, remove the jack stand after installing a suitable transmission jack.
10. On 4x2 models, perform the following:
    a. Disconnect the wire loom from the crossmember.
    b. Remove the crossmember bolts.
    c. Remove the crossmember bolts.
    d. Remove the transmission mount nuts and the crossmember.
11. On 4x2 models, perform the following:
    a. Remove the right hand crossmember nuts.
    b. Remove the left hand crossmember nuts.
    c. Remove the nuts and the crossmember.
    d. Remove the transmission mount.
12. On models equipped with a transmission-mounted parking brake, perform the following:
    a. Disconnect the parking brake lever return spring from the parking brake lever.
    b. Apply penetrating oil to the adjusting clevis, jam nut and the threads on the front parking brake cable and conduit.
    c. Disconnect the front parking brake cable and conduit.
    d. Loosen the jam nut.
    e. Remove the clevis locking pin.
    f. Remove the clevis pin.
    g. Remove the adjusting clevis from the parking brake lever.
    h. Compress the retainer, and remove the front parking brake cable and conduit from the cable bracket.

➡If the vehicle is equipped with a Power Take-Off (PTO) unit, all or part of the PTO unit will need to be removed.

13. Disconnect the shift cable from the transmission.
14. Disconnect the shift cable from the manual lever.
15. Remove the shift cable bracket from the transmission and position aside.
16. Disconnect the digital Transmission Range (TR) sensor connector and the wire loom from the shift cable bracket.
17. Disconnect the solenoid body connector.
18. Disconnect the Turbine Shaft Speed (TSS) sensor and the Output Shaft Speed (OSS) sensor.
19. Remove the wiring harness from the transmission and position aside.
20. Remove the flexplate inspection cover.
21. Remove the bolts.
22. Remove the inspection plate.
23. Remove the starter motor.
24. Remove the cylinder block opening cover.
25. Remove and discard the torque converter-to-flexplate nuts.
26. Position a suitable drain pan and disconnect the transmission fluid cooler tubes from the cooler bypass valve.
27. Remove the seven transmission-to-engine mounting bolts and gently rock the transmission side-to-side to disengage it from the locator dowels.
28. Move the transmission and the transmission jack rearward to clear the engine flexplate.

⁂ WARNING

The torque converter is heavy and can result in injury if it falls out of the transmission. Secure the torque converter in the transmission. If the torque converter is dropped, a new one must be installed.

29. Use tool 307-346 to hold the torque converter in place.

⁂ CAUTION

Use care while removing the transmission to avoid obstructions.

30. Lower the transmission out of the vehicle.
31. On models equipped with a transmission-mounted parking brake, perform the following:
    a. Remove the transmission-mounted parking brake.
    b. Keep the parking brake vent in the upward position to prevent contamination of the brake shoes and linings.
    c. Remove the six bolts, parking brake assembly and the gasket from the extension housing.
    d. Discard the bolts and the gasket.
    e. Clean the mating surfaces.

To install:

⁂ CAUTION

Prior to the installation of the transmission, the fluid, cooler lines and the cooler bypass valve must be cleaned. A new transmission Oil-To-Air (OTA) cooler must be installed if the transmission was overhauled or exchanged due to a failure of the transmission. Transmission failure can occur if these procedures are not followed.

32. Inspect the wiring harness and the connectors for damage, terminal condition, corrosion and seal integrity. Repair or install new as required.
33. On models equipped with a transmission-mounted parking brake, perform the following:
    a. Install the parking brake, if removed.
    b. Position the parking brake assembly with a new gasket on the transmission extension housing.
    c. Install six new bolts and tighten to 41 ft. lbs. (55 Nm).

⁂ CAUTION

Prior to the installation of the transmission, the torque converter pilot hub must be lubricated or damage to the torque converter or the engine crankshaft can occur.

34. Lubricate the torque converter pilot hub with multi-purpose grease.
35. Raise the transmission into position.

CAUTION

Do not use the CBV as a handle. Damage to the CBV assembly can occur or damage to the case can result. Be careful not to raise the transmission up too far. The sensors can make contact with the underbody of the vehicle and cause damage to the sensors. Sensor failure or leakage can occur.

➡While raising the transmission up into the engine compartment, make sure to align the fluid filler tube with the stub tube on the transmission, using the dipstick as a guide.

36. Position the transmission.
37. Remove tool 307-346.

CAUTION

Do not allow the torque converter drive flats to disengage from the pump gear. Use care not to damage the flexplate and the converter pilot. The torque converter must rest squarely against the flexplate, indicating the converter pilot is not binding in the crankshaft.

38. While installing the transmission to the engine, align the torque converter studs with the mounting holes in the flexplate.
39. Install the transmission-to-engine bolts and tighten to 35 ft. lbs. (47 Nm).
40. Install the new torque converter-to-flexplate nuts and tighten to 26 ft. lbs. (35 Nm).
41. Install the cylinder block opening cover.
42. Install the flexplate inspection cover and bolts and tighten to 25 ft. lbs. (34 Nm).
43. Install the starter motor.
44. Connect the solenoid pack electrical connector.
45. Connect the digital TR sensor connector.
46. Connect the TSS and the OSS sensors.

➡If the vehicle is equipped with a PTO unit, all or part of the PTO unit will need to be installed.

47. Connect the shift cable.
48. Install the cable housing bracket.
49. Install the shift cable to the manual lever and tighten to 18 ft. lbs. (25 Nm).
50. Install the transmission fluid cooler tubes to the cooler bypass valve and tighten to 20 ft. lbs. (27 Nm).
51. On models equipped with a transmission-mounted parking brake perform the following:
    a. Position the front parking brake cable and conduit, and press the retainer into the cable bracket until it snaps into place.
    b. Set the adjusting clevis.
    c. Loosen the jam nut several turns.
    d. Position the parking brake lever in the applied position.
    e. Tighten or loosen the adjusting clevis until the adjusting clevis hole lines up with the parking brake lever hole, then loosen the adjusting clevis to 0.5 inch (13mm).
    f. Install the pins, and tighten the jam nut.
    g. Install the clevis pin through the adjusting clevis and the parking brake lever.
    h. Install the locking pin in the clevis pin. Tighten the jam nut to 17 ft. lbs. (23 Nm).
    i. Install the parking brake lever return spring.
52. On 4x4 models, perform the following:
    a. Support the extension housing with a jack stand and remove the transmission jack.
    b. Install the transfer case.
53. On 4x2 models, perform the following:
    a. Install the transmission mount and tighten to 70 ft. lbs. (95 Nm).
    b. Position the crossmember to the transmission mount and loosely install the nuts.
    c. Install the crossmember bolts and tighten to 60 ft. lbs. (81 Nm).
    d. Reconnect the wire harness to the frame.
54. On 4x2 models, remove the transmission jack and tighten the crossmember to the transmission mount nuts 69 ft. lbs. (94 Nm).
55. Install the rear driveshaft.
56. If the transmission was overhauled and the vehicle was equipped with an in-line fluid filter, install a new in-line fluid filter.
57. If the transmission was overhauled and the vehicle was not equipped with an in-line fluid filter, install a new in-line fluid filter kit.
58. If the transmission is being installed for a non-internal repair, do not install an in-line filter or filter kit.
59. If installing a new or a Ford-authorized remanufactured transmission, install the in-line transmission fluid filter that is supplied.
60. Prior to lowering the vehicle, install a new in-line transmission filter or a filter kit.
61. Connect the negative battery cable.
62. Refill all transmissions with the correct amount of automatic transmission fluid.

4R70W Transmission

1. Before servicing the vehicle, refer to the precautions in the beginning of this section.
2. Place the transmission range selector lever in the NEUTRAL position.

➡All gasoline vehicles will have new adaptive shift strategies. Whenever the vehicle's battery has been disconnected for any type of service or repair the strategy parameters that are stored in the Keep Alive Memory (KAM) will be lost. The strategy will start to relearn once the battery is reconnected and the vehicle is driven. This is a temporary condition and will return to normal operating condition once the Powertrain Control Module (PCM) relearns all the parameters from the driving conditions. There is no set time frame for this process. If this concern is present during downshifts or converter clutch apply, it is not the fault of the shift strategy and will require diagnosis as outlined in the workshop manual. The customer needs to be notified that they may experience slightly different upshifts either soft or firm and that this is a temporary condition and will eventually return to normal operating condition.

3. Remove or disconnect the following:
    - Negative battery cable
    - Transmission fluid filler tube
    - Transmission fluid cooler tubes at the cooler bypass valve. Catch the fluid in a suitable container.
    - Transfer case assembly, if equipped
4. Mark the driveshaft yoke and axle flange so they may be installed in their original position.
    - Starter motor
    - Torque converter-to-flexplate retaining nuts
    - Shift linkage
    - Transmission cooler lines
    - Transmission electrical connectors
    - Transmission Range (TR) sensor
    - Output Shaft Speed (OSS) sensor
    - Solenoid body assembly electrical connector
5. Install a high lift transmission jack with the special jack adapter 014-0763 to the transmission.

6. Remove or disconnect the following:
   - Three Way Converters (TWC) and crossmember
   - Fuel line bracket bolt
   - Transmission fill tube bolt from the back of the right hand cylinder head
   - Transmission fluid fill tube
   - 6 transmission to engine bolts

CAUTION

The torque converter is heavy and may result in injury if it falls out of the transmission. Secure the torque converter in the transmission.

7. Install a Torque Converter Holding Tool.
8. Lower the transmission from the vehicle.

To install:

9. Installation is the reverse of removal, please note the following specs:
10. Transmission-to-engine bolts to 41 ft. lbs. (55 Nm).
11. Fluid filler tube into the short fluid inlet tube. Tighten the nut to 16 ft. lbs. (23 Nm).
12. Refill all transmissions with the correct amount of Motorcraft MERCON® automatic transmission fluid.
13. Connect the negative battery cable.

Transfer Case Assembly

REMOVAL & INSTALLATION

Excursion

1. Before servicing the vehicle, refer to the precautions in the beginning of this section.
2. Shift the transfer case to the 2W HI position.
3. Remove the four bolts and the skid plate, if equipped.
4. Matchmark the driveshaft to maintain driveline balance.
5. Remove the rear driveshaft.
6. Matchmark the front driveshaft to the transfer case flange.
7. Support the front driveshaft with wire or a strap.
8. Remove and discard the four bolts and position the front driveshaft aside.
9. On models with a manual shift transfer case, perform the following:
   a. Remove the manual shift linkage, if equipped.
   b. Disconnect the switch electrical connector. Position the wire harness aside.
10. On models equipped with an electric shift transfer case, perform the following:
   a. Disconnect the gear motor encoder assembly electrical connector and the gear motor electrical connector.
   b. Position the wire harness aside.
   c. Disconnect the transfer case vent hose.
11. Drain the fluid into a suitable container. Install the plug when finished.
12. Position a suitable high-lift jack under the transfer case and secure it with safety straps.
13. Detach the wire harness from the crossmember.
14. Remove the crossmember-to-frame bolts.
15. Remove the nuts and the crossmember.
16. Remove the transmission mount.
17. Position a suitable jack stand under the extension housing.
18. Remove the transfer case-to-transmission bolts.
19. Separate the transfer case from the extension housing. Pull the transfer case rearward, then lower the transfer case from the vehicle.

CAUTION

Carefully clean the gasket surfaces as nicks and gouges can cause fluid leaks.

20. Remove the transfer case-to-transmission gasket. Clean the mating surfaces, using metal surface cleaner.

To install:

21. Install a new mounting gasket.
22. Secure the transfer case to the high-lift jack using a safety strap.
23. Raise the transfer case into position.
24. Install the bolts retaining the transfer case to the extension housing and tighten to 37 ft. lbs. (50 Nm).
25. Remove the jack stand from the extension housing.
26. Install the transmission mount and tighten to 70 ft. lbs. (95 Nm).
27. Position the crossmember and loosely install the two nuts.
28. Install the crossmember bolts and tighten to 60 ft. lbs. (81 Nm).
29. Attach the wire harness to the crossmember.
30. Remove the high-lift jack.
31. Tighten the transmission mount-to-crossmember nuts to 69 ft. lbs. (94 Nm).
32. On models with an electric shift transfer case, perform the following:
   a. Connect the vent hose.
   b. Connect the two gear motor encoder assembly electrical connectors.
33. On models with a manual shift transfer case, perform the following:
   a. Connect the 3-position mode switch harness connector.
   b. Connect the manual shift linkage.
34. Connect the front driveshaft to the transfer case, aligning the matchmarks made prior to removal and install the four new bolts. Tighten to 82 ft. lbs. (111 Nm).
35. Install the rear driveshaft.
36. Install the skid plate and the four bolts and tighten to 18 ft. lbs. (25 Nm).
   - Fill the transfer case.

Except Excursion

2001–02 4R70W AND 4R100 TRANSMISSIONS

1. Before servicing the vehicle, refer to the precautions in the beginning of this section.
2. Place the vehicle in NEUTRAL, raise and support the vehicle.
3. Drain the transfer case.
4. Remove the skid plate, if equipped.
5. Remove the rear driveshaft, matchmark the position for installation purposes before removal.
6. Remove the two nuts and the front driveshaft shield.
7. Remove the front driveshaft, matchmark the position for installation purposes before removal.
8. Matchmark the torsion bars to the torsion bar crossmember.
9. Measure and record the torsion bar adjuster bolt length.
10. Remove the bolts.
11. Using tools 204-203 and 204-185, remove the torsion bar adjustment nuts.
12. Tighten the tool 204-185 until the torsion bar adjuster lifts off the adjuster nut.
13. Remove the torsion bar adjusting brackets.

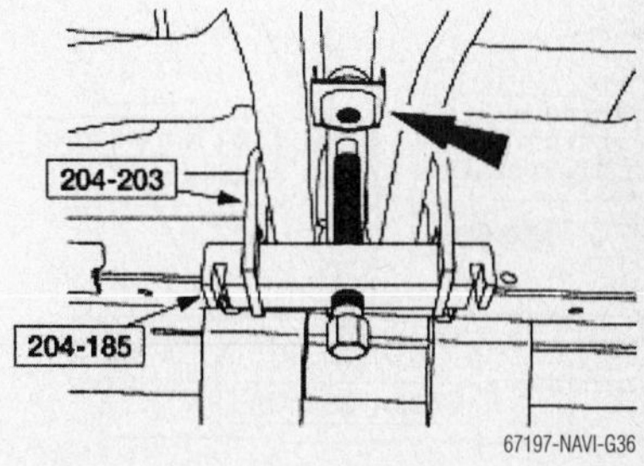

Use tools 204-203 and 204-185 to remove the torsion bar adjustment nuts—except Excursion

14. Move the torsion bars forward.
15. Remove the torsion bar adjusting brackets.
16. Pull the torsion bar rearward and separate it from the lower control arm.
17. Remove the right hand torsion bar only.
18. Remove the six bolts and the torsion bar crossmember.
19. On manual shift models, use tool 307-017, disconnect the shift rod.
20. On manual shift models, disconnect the 4WD indicator switch electrical connector, if equipped.
21. Disconnect the electric shift motor electrical connector, if equipped.
22. Disconnect the vent hose.
23. Using a suitable jack, support the transfer case. Secure the transfer case to the jack with a safety strap.
24. Remove the six transfer case-to-transmission bolts and separate the transfer case from the transmission.
25. Lower the transfer case from the vehicle.

CAUTION

Carefully clean the gasket surfaces as nicks and gouges can cause fluid leaks.

26. Remove the transfer case-to-transmission gasket.

To install:

27. Install a new transfer case-to-transmission gasket.
28. Place the transfer case in position.
29. Install the transfer case-to-transmission bolts and tighten to 35 ft. lbs. (47 Nm).
30. Remove the jack.
31. Connect the vent hose.
32. Connect the electric shift motor electrical connector, if equipped.
33. On manual shift models, Connect the 4WD indicator switch electrical connector, if equipped.

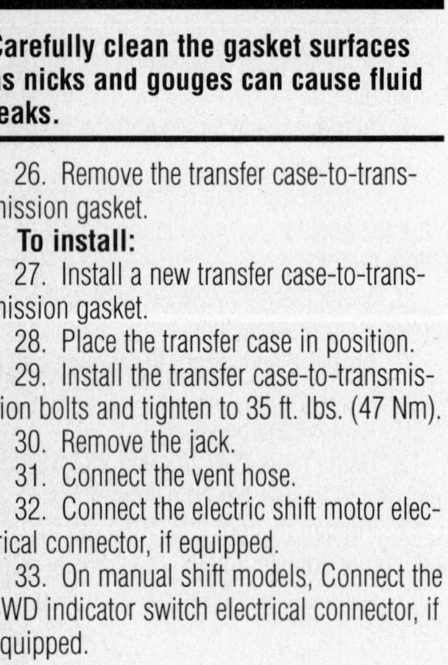

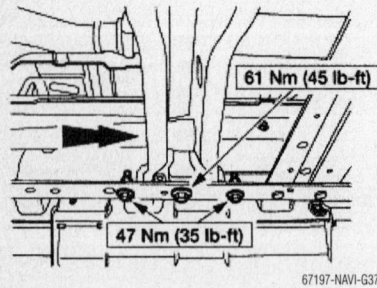

Tighten the torsion bar crossmember bolts as shown—except Excursion

34. On manual shift models, connect the shift rod.
35. Install the torsion bar crossmember. Tighten the bolts to the specifications shown in the accompanying illustration.
36. Install the right hand torsion bar.
37. Install the torsion bar adjusting brackets.
38. Install the torsion bar adjuster bolt and tighten until the length is the same as was recorded prior to removal..
39. Install the front driveshaft, align the matchmarks marked for installation purposes before removal.
40. Install the front driveshaft shield.
41. Install the rear driveshaft, align matchmarks marked for installation purposes before removal.
42. Fill the transfer case.

2003–05 4R70W TRANSMISSION

1. Before servicing the vehicle, refer to the precautions in the beginning of this section.
2. Place the vehicle in NEUTRAL, raise and support the vehicle.
3. Drain the transfer case.
4. Remove the two nuts and the front driveshaft shield.
5. Remove the front driveshaft, matchmark the position for installation purposes before removal.
6. Remove the rear driveshaft, matchmark the position for installation purposes before removal.
7. Disconnect the electrical connector.
8. Disconnect the vent hose.
9. Remove the bolts and the damper.
10. Using a suitable jack, support the transfer case. Secure the transfer case to the jack with a safety strap.
11. Remove the six transfer case-to-transmission bolts and separate the transfer case from the transmission.
12. Lower the transfer case from the vehicle.

CAUTION

Carefully clean the gasket surfaces as nicks and gouges can cause fluid leaks.

13. Remove the transfer case-to-transmission gasket.

To install:

14. Install a new transfer case-to-transmission gasket.
15. Place the transfer case in position.
16. Install the transfer case-to-transmission bolts and tighten to 35 ft. lbs. (47 Nm).
17. Remove the jack.
18. Connect the vent hose.
19. Connect the electrical connector.
20. Install the rear driveshaft, align matchmarks marked for installation purposes before removal.
21. Install the front driveshaft, align the matchmarks marked for installation purposes before removal.
22. Install the front driveshaft shield.
23. Fill the transfer case.

2003–05 4R100 TRANSMISSION

1. Before servicing the vehicle, refer to the precautions in the beginning of this section.
2. With the vehicle in NEUTRAL position the vehicle on a hoist.
3. Drain the transfer case fluid. Install the drain plug when finished draining.
4. Remove the transfer case-to-transmission brace.
5. Remove the bolts from the transmission.
6. Remove the nuts from the transfer case.
7. Remove the two nuts and the front driveshaft shield.
8. Remove the front driveshaft, matchmark the position for installation purposes before removal.
9. Remove the rear driveshaft, matchmark the position for installation purposes before removal.
10. Disconnect the electrical connector.
11. Disconnect the vent hose.

CAUTION

Never support the transmission from the bottom of the transmission pan.

12. Using a suitable jack stand, support the transmission. Secure the transfer case to the jack with a safety strap.
13. Using a suitable transmission jack, support the transfer case.
14. Remove the two exhaust heat shield-to-crossmember bolts.
15. Remove the crossmember-to-frame nuts and bolts.
16. Remove the transmission mount-to-crossmember nuts, then remove the crossmember.
17. Remove the bolts and the transmission mount.
18. Remove the bolt and the exhaust hanger bracket.
19. Remove the six transfer case-to-transmission extension housing bolts.
20. Separate the transfer case from the transmission extension housing, move the transfer case rearward off the output shaft, then lower the transfer case from the vehicle.

⁂ CAUTION

Carefully clean the gasket surfaces as nicks and gouges can cause fluid leaks.

21. Remove the transfer case-to-transmission gasket and clean the mating surfaces

To install:

22. Install a new transfer case-to-transmission gasket.
23. Place the transfer case in position.
24. Install the transfer case-to-transmission bolts and tighten to 30 ft. lbs. (40 Nm).
25. Install the exhaust hanger bracket and tighten to 30 ft. lbs. (40 Nm).
26. Install the transmission mount and tighten the bolts to 72 ft. lbs. (98 Nm).
27. Install the transmission mount-to-crossmember and tighten the nuts to 74 ft. lbs. (101 Nm).
28. Install the crossmember-to-frame nuts and bolts. Tighten to 66 ft. lbs. (90 Nm).
29. Install the two exhaust heat shield-to-crossmember bolts.
30. Remove the jack.
31. Connect the vent hose.
32. Connect the electrical connector.
33. Install the rear driveshaft, align matchmarks marked for installation purposes before removal.
34. Install the front driveshaft, align the matchmarks marked for installation purposes before removal.
35. Install the front driveshaft shield.
36. Install the transfer case-to-transmission brace. Tighten the retainers to 22 ft. lbs. (30 Nm).
37. Fill the transfer case.

Halfshaft

REMOVAL & INSTALLATION

1. Before servicing the vehicle, refer to the precautions in the beginning of this section.
2. Remove or disconnect the following:
   - Front wheels
   - Front hub cotter pin, retainer and nut
3. Using a floor hydraulic jack, support the lower suspension arm.
   - Upper ball joint cotter pin and castle nut
   - Knuckle from the front suspension upper arm
4. Lower the lower suspension arm and steering knuckle slightly to facilitate easier halfshaft removal.
5. Remove the 2 disc caliper mounting bolts, then lift the front disc caliper off of the front disc brake caliper anchor plate and position aside. Do not allow the caliper to hang by the brake hose; suspend it from the vehicle's frame with strong cord or wire.
6. Remove the 6 front halfshaft-to-differential bolts.

⁂ WARNING

Use care to avoid damaging the hub seal when removing the front halfshaft.

   - Inboard end of the halfshaft from the differential case or extension axle case. Separate the front halfshaft and joints from the hub, then remove the halfshaft and joints from the vehicle.

To install:

7. Slide the halfshaft outboard end into the hub, making sure that the splines engage.
8. Situate the inboard end of the halfshaft against the front differential flange and install the 6 halfshaft-to-differential bolts. Tighten the halfshaft bolts to 60 ft. lbs. (82 Nm).
9. Install the front disc brake caliper onto the rotor and anchor plate, then install and tighten the 2 caliper mounting bolts to 21–26 ft. lbs. (28–36 Nm).
10. Lift the lower suspension arm and steering knuckle up until the upper ball joint stud is inserted into the steering knuckle. Install the upper ball joint castle nut and tighten to 57–76 ft. lbs. (77–104 Nm). Install a new cotter pin.
11. Install the hub nut onto the halfshaft and tighten the hub nut to 221 ft. lbs. (300 Nm).
12. Install the hub nut retainer and a new cotter pin.
13. Install the front wheels and tighten the lug nuts in a star-shaped sequence to 83–112 ft. lbs. (113–153 Nm).
14. Lower the vehicle to the ground.

CV-Joint

OVERHAUL

➡Before continuing with this procedure, make sure to have available a new CV-joint boot kit, for each CV-joint being serviced. The outer CV-joint cannot be disassembled, only the boot can be replaced.

Inner CV-Joint And Boot

1. Before servicing the vehicle, refer to the precautions in the beginning of this section.
2. Remove the halfshaft assembly from the vehicle.
3. Clamp the halfshaft in a vise equipped with jaw caps to prevent damage to machined surfaces. Do not allow the vise jaws to contact the boot or its clamp.
4. Slide 2 inboard clamp protectors off the boot clamps.
5. Carefully remove 2 boot clamps, and slide the boot off the inner CV-joint and housing.
6. Remove the CV-joint retaining ring and remove the housing.
7. Mark the inner race and the ball cage for assembly.

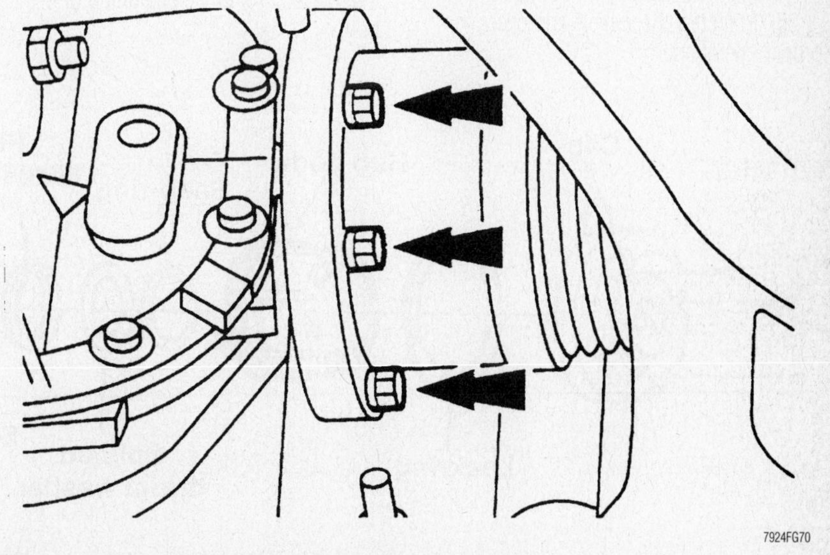

Halfshaft-to-differential mounting bolts (3 of the 6 bolts shown)

8. Remove 6 cage balls.
9. Remove the snapring.
10. Remove the inner race and ball cage.
11. Clean all parts in suitable parts cleaning solvent and inspect for wear.

To install:

12. Place the boot and the small boot clamp and protector on the shaft.
13. Place the ball cage on the shaft with the tapered end toward the outer CV-joint.

➡Line up the marks made at disassembly.

14. Position the inner race on the driveshaft in the position marked on disassembly.
15. Install the snapring.
16. Lubricate and position 6 balls with suitable CV-joint grease.
17. Place the boot protector and boot clamp on the CV-joint housing. Fill the housing with 8.29 ounces of suitable CV-joint grease.
18. Place the housing to the cage and bearings and install the retaining ring.
19. Remove any excess grease from the mating surface and position the boot and clamp.
20. Adjust the CV-joint to boot spacing to 16.43 in. (417.25mm).
21. After adjusting the CV-joint to boot spacing, insert a dull bladed screwdriver blade to relieve built up air pressure in the boot.
22. Use CV Boot Clamp Installer T95P-3514-A or equivalent, to install the boot clamps.
23. Place the clamp protectors over the boot clamps.
24. Install the halfshaft into the vehicle.
25. Road test the vehicle and check for proper operation.

Outer Boot

1. Remove the inner CV-joint and housing from the halfshaft.
2. Remove the inner boot from the halfshaft.
3. Remove the outer boot clamp protectors and carefully remove the boot clamps.
4. Remove the outer boot from the halfshaft and inspect the grease for contamination.
5. If the grease is contaminated, clean and inspect the joint for wear. Replace the joint and shaft if worn or damaged.

To install:

6. Place the new CV-joint boot on the shaft.
7. Using 5.82 ounces (165 grams) of suitable CV-joint grease, pack the outer CV-joint with grease, then spread the remaining grease inside the boot.
8. Clean the boot mounting surface and position the boot in the joint grooves.
9. Place the clamps in position and use CV Boot Clamp Installer T95P-3514-A or equivalent, to install the boot clamps.
10. Place the clamp protectors over the boot clamps.
11. Install the inner boot on the halfshaft.
12. Install the inner CV-joint and housing on the halfshaft.
13. Install the halfshaft in the vehicle.
14. Road test the vehicle and check for proper operation.

Locking Hubs

REMOVAL & INSTALLATION

1. Before servicing the vehicle, refer to the precautions in the beginning of this section.
2. Raise and safely support the vehicle.
3. Remove or disconnect the following:
   - Tire
   - 3 screws and separate the cap from the body
   - Lockring seated in the groove of the hub assembly
   - Body assembly from the brake rotor/hub
   - Snapring from the groove in the stub-shaft
   - 3 thrust washers from the stub-shaft
4. Pull the cam assembly to remove it.

To install:

5. Align the fixed cam retaining key on the cam assembly with the keyway on the spindle. Firmly push the cam assembly on the wheel retaining nut.
6. Install or connect the following:
   - Metal, plastic, then the splined washers on the stub-shaft
   - Snapring in the groove of the stub-shaft. It may be necessary to push the stub-shaft outward from the back of the knuckle assembly.

⁂ WARNING

Do not pack the hub assembly with grease. Too much grease will damage the hub assembly.

7. Rotate the moving cam assembly to the 1 o'clock position in relation to the fixed cam retaining key. Use any 1 of the 3 stops.
   - Body assembly onto the hub by lining up the 3 legs with the 3 pockets in the cam assembly. Be sure the assembly is in far enough to see the groove in the hub.
   - Large lockring in the groove on the hub. Ensure the lockring is seated completely.

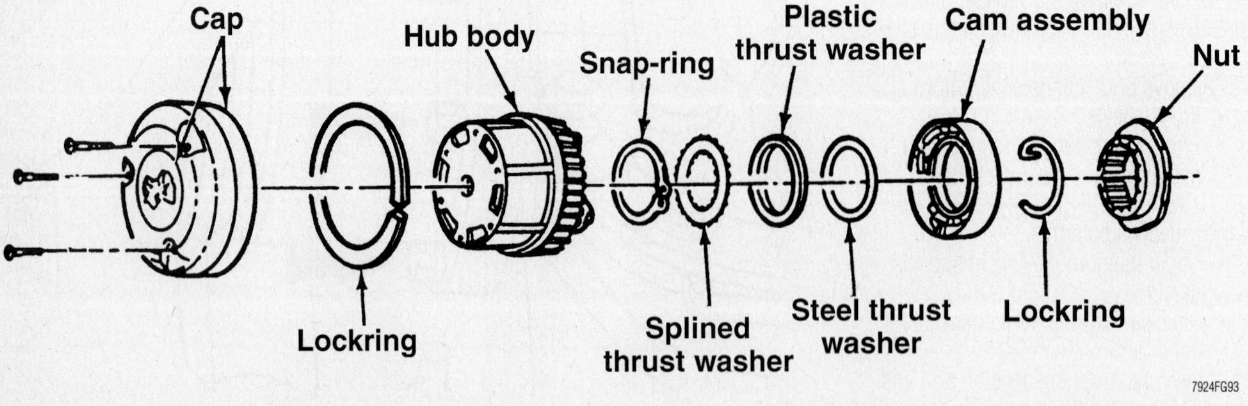

Exploded view of the typical automatic locking hub assembly

- Cap using the 3 screws. Tighten the screws to 35–53 inch lbs. (4–6 Nm).
- Tire and lower the vehicle to the floor

Axle Shaft, Bearing and Seal

REMOVAL & INSTALLATION

Dana 50 Independent Front Suspension

1. Before servicing the vehicle, refer to the precautions in the beginning of this section.
2. Raise and support the front end on jackstands.
3. Remove or disconnect the following:
   - Front wheels
   - Calipers
   - Hub/rotor assemblies
   - Nuts retaining the spindle to the steering knuckle. Tap the spindle with a plastic mallet to remove it from the knuckle.
   - Splash shield
4. On the left side, pull the shaft from the carrier, through the knuckle.
5. On the right side, remove and discard the keystone clamp from the shaft and joint assembly and the stub shaft. Slide the rubber boot onto the shaft and pull the shaft and joint assembly from the splines of the stub shaft.
6. Place the spindle in a soft-jawed vise clamped on the second step of the spindle.
7. Using a slide hammer and bearing puller, remove the needle bearing from the spindle.
8. Inspect all parts. If the spindle is excessively corroded or pitted it must be replaced. If the U-joints are excessively loose or don't move freely, they must be replaced. If any shaft is bent, it must be replaced.

To install:

9. Clean all dirt and grease from the spindle bearing bore. The bore must be free of nicks and burrs.
10. Insert a new spindle bearing in its bore with the printing facing outward. Drive it into place with drive T80T–4000–S for F-150 and F-250, or T80T–4000–R for the F-350, or their equivalents. Install a new bearing seal with the lip facing away from the bearing.
11. Pack the bearing and hub seal with grease. Install the hub seal with a driver.
12. Install or connect the following:
    - Thrust washer on the axle shaft
    - New slinger on the axle shaft
    - Rubber V-seal on the slinger. The seal lip should face the spindle.
    - Plastic spacer on the axle shaft. The chamfered side of the spacer should be inboard against the axle shaft.
13. Pack the thrust face of the seal in the spindle bore and the V-seal on the axle shaft with heavy duty, high temperature, waterproof wheel bearing grease.
14. On the right side, install the rubber boot and new keystone clamps on the stub shaft and slip yoke. The splines permit only one way of meshing so you'll have to properly align the missing spline in the slip yoke with the gapless male spline on the shaft. Slide the right shaft and joint assembly into the slip yoke, making sure that the splines are fully engaged. Slide the boot over the assembly and crimp the keystone clamp.
15. On the left side, slide the shaft and joint assembly through the knuckle and engage the splines in the carrier.
16. Install or connect the following:
    - Splash shield and spindle on the knuckle. Tighten the spindle nuts to 60 ft. lbs. (81 Nm).
    - Rotor on the spindle
    - Outer wheel bearing into the cup. Make sure that the grease seal lip totally encircles the spindle.
    - Wheel bearing, locknut, thrust bearing, snapring and locking hubs.
    - Caliper

Spindle and Front Axle Shaft

REMOVAL & INSTALLATION

Dana 60 Monobeam

1. Before servicing the vehicle, refer to the precautions in the beginning of this section.
2. Raise and support the front end on jackstands.
3. Remove or disconnect the following:
   - Caliper from the knuckle and wire it out of the way
   - Free-running hub
   - Front wheel bearings
   - Hub and rotor assembly
   - Spindle-to-knuckle bolts. Tap the spindle from the knuckle using a plastic mallet.
   - Splash shield and caliper support
4. Pull the axle shaft out through the knuckle.
5. Using a slide hammer and bearing cup puller, remove the needle bearing from the spindle.
6. Clean the spindle bore thoroughly and make sure that it is free of nicks and burrs. If the bore is excessively pitted or scored, the spindle must be replaced.

To install:

7. Insert a new spindle bearing in its bore with the printing facing outward. Drive it into place with driver T80T–4000–R, or its equivalent. Install a new bearing seal with the lip facing away from the bearing.
8. Pack the bearing with waterproof wheel bearing grease.
9. Pack the thrust face of the seal in the spindle bore and the V-seal on the axle shaft with waterproof wheel bearing grease.
10. Carefully guide the axle shaft through the knuckle and into the housing. Align the splines and fully seat the shaft.
11. Install or connect the following:
    - Bronze spacer on the shaft. The chamfered side of the spacer must be inboard.
    - Splash shield and caliper support
    - Spindle on the knuckle and install the bolts. Tighten the bolts to 50–60 ft. lbs. (68–81 Nm).
    - Hub/rotor assembly on the spindle
    - Wheel bearings
    - Free-running hub

Right Side Slip Yoke and Stub Shaft, Carrier, Carrier Oil Seal and Bearing

REMOVAL & INSTALLATION

Dana 50 Independent Front Axle

1. Before servicing the vehicle, refer to the precautions in the beginning of this section.

➡This procedure requires the use of special tools.

2. Raise and support the front end on jackstands.
3. Disconnect the front driveshaft from the carrier and wire it up out of the way.
4. Remove the left and right axle shafts and both spindles.
5. Support the carrier with a floor jack and unbolt the carrier from the support arm.
6. Place a drain pan under the carrier, separate the carrier from the support arm and drain the carrier.
7. Remove the carrier from the truck.
8. Place the carrier in holding fixture T57L–500–B with adapters T80T–4000–B.

9. Rotate the slip yoke and shaft assembly from the carrier.

10. Using a slide hammer/puller remove the caged needle bearing and oil seal as a unit. Discard the oil seal and bearing.

To install:

11. Clean the bearing bore thoroughly and make sure that it is free of nicks and burrs.

12. Insert a new bearing in its bore with the printing facing outward. Drive it into place with driver T83T–1244–A, or its equivalent. Install a new bearing seal with the lip facing away from the bearing. Coat the bearing and seal with waterproof wheel bearing grease.

13. Install the slip yoke and shaft assembly into the carrier so that the groove in the shaft is visible in the differential case.

14. Install the snapring in the groove in the shaft. It may be necessary to force the snapring into place with a small prybar. Don't strike the snapring!

15. Remove the carrier from the holding fixture.

16. Clean all traces of sealant from the carrier and support arm. Make sure the mating surfaces are clean. Apply a ¼ in. (6mm) wide bead of RTV sealant to the mating surface of the carrier. The bead must be continuous and should not pass through or outside of the holes. Install the carrier with 5 minutes of applying the sealer.

17. Position the carrier on the jack and raise it into position using guide pins to align it if you'd like. Install and hand-tighten the bolts. Tighten the bolts in a circular pattern to 30–40 ft. lbs. (41–54 Nm).

18. Install the support arm tab bolts and tighten them to 85–100 ft. lbs. (115–136 Nm).

19. Install all other parts in reverse order of removal.

Pinion Seal

REMOVAL & INSTALLATION

Independent Front Axle

1. Before servicing the vehicle, refer to the precautions in the beginning of this section.

➡A torque wrench capable of at least 225 ft. lbs. (305 Nm) is required for pinion seal installation.

2. Raise and safely support the vehicle with jackstands under the frame rails. Allow the axle to drop to rebound position for working clearance.

3. Mark the companion flanges and U-joints for correct reinstallation position.

4. Remove the driveshaft. Use a suitable tool to hold the companion flange. Remove the pinion nut and companion flange.

5. Use a slide hammer and hook or sheet metal screw to remove the oil seal.

To install:

6. Install a new pinion seal after lubricating the sealing surfaces. Use a suitable seal driver. Install the companion flange and pinion nut. Tighten the nut to 200–220 ft. lbs. (271–298 Nm).

Monobeam Front Axle

1. Before servicing the vehicle, refer to the precautions in the beginning of this section.

➡A torque wrench capable of at least 300 ft. lbs. (407 Nm) is required for pinion seal installation.

2. Raise and support the truck on jackstands.

3. Allow the axle to hang freely.

4. Matchmark and disconnect the driveshaft from the front axle.

5. Using a tool such as T75T–4851–B, or equivalent, hold the pinion flange while removing the pinion nut.

6. Using a puller, remove the pinion flange.

7. Use a puller to remove the seal, or punch the seal out using a pin punch.

To install:

8. Thoroughly clean the seal bore and make sure that it is not damaged in any way. Coat the sealing edge of the new seal with a small amount of 80W/90 oil and drive the seal into the housing using a seal driver.

9. Coat the inside of the pinion flange with clean 80W/90 oil and install the flange onto the pinion shaft.

10. Install the nut on the pinion shaft and tighten it to 250–300 ft. lbs. (339–407 Nm).

11. Connect the driveshaft.

STEERING AND SUSPENSION

Air Bag

⁂ CAUTION

Some vehicles are equipped with an air bag system. The system must be disabled before performing service on or around system components, steering column, instrument panel components, wiring and sensors. Failure to follow safety and disabling procedures could result in accidental air bag deployment, possible personal injury and unnecessary system repairs.

PRECAUTIONS

Several precautions must be observed when handling the inflator module to avoid accidental deployment and possible personal injury:

- Never carry the inflator module by the wires or connector on the underside of the module.
- When carrying a live inflator module, hold securely with both hands and ensure that the bag and trim cover are pointed away.
- Place the inflator module on a bench or other surface with the bag and trim cover facing up.
- With the inflator module on the bench, never place anything on or close to the module, which may be thrown in the event of an accidental deployment.

DISARMING

⁂ CAUTION

The Supplemental Inflatable Restraint (SIR) system must be disarmed before performing service around SIR system components or SIR system wiring. Failure to do so may cause accidental deployment of the air bag, resulting in unnecessary SIR system repairs and/or personal injury.

The positive battery cable must be disconnected for a minimum of 1 minute before beginning any air bag work to de-energize the back-up power supply. It is a good idea to disengage both the positive and negative battery cables to ensure that the Air Bag system is definitely discharged.

ARMING THE SYSTEM

⁂ WARNING

If the air bag simulators have been used, the air bag simulators must be removed and the air bags reconnected when the system is reactivated to avoid non-deployment in a collision resulting in possible personal injury.

1. Disconnect the positive battery cable.
2. Wait 1 minute, this is required for the back-up power supply in the air bag diagnostic monitor to deplete its stored energy.
3. Remove the air bag simulator from the air bag sliding contact connector at the top of the steering column. Reconnect the driver's side air bag module assembly. Position the driver's air bag module on the steering wheel and secure with the 2 bolts and washers. Tighten the bolt and washer assembly to 8–10 ft. lbs. (10–14 Nm).
4. Connect the positive battery cable.
5. Turn the ignition switch from the **OFF** to **RUN** and visually monitor the air bag warning indicator. The light will illuminate continuously for approximately 6 seconds and then turn off. If a fault occurs, the air bag indicator will either fail to light, remain lighted continuously or flash. The flashing may not occur until approximately 30 seconds after the ignition switch has been turned from **OFF** to **RUN**. This is the time needed for the air bag diagnostic monitor to complete testing the system. If the air bag indicator is inoperative, an air bag system fault exists, a tone will sound in a pattern of 5 sets of 5 beeps. If this occurs, the air bag indicator will need to be serviced before further diagnostics can be done.

Steering Gear

REMOVAL & INSTALLATION

Excursion

1. Before servicing the vehicle, refer to the precautions in the beginning of this section.

➡New O-ring seals must be installed any time the lines are disconnected from the steering gear.

2. Remove the air cleaner assembly.
3. Disengage the steering coupling shield from the line fitting and slide upward on the steering shaft.
4. Remove the pinch bolt. Turn the steering wheel as necessary to access the bolt.
5. Make sure the steering column is locked.
6. Disconnect the lines.
7. Discard the O-ring seals.
8. Remove the tie rod cotter pin and nut. Discard the cotter pin.
9. Using tool 211-003, disconnect the drag link.
10. Remove the bolts, washers and the steering gear.
11. Remove the pitman arm nut.
12. Using tool 211-003, remove the steering gear sector shaft arm.

To install:

13. Install the steering gear sector shaft arm.
14. Install the pitman arm nut. Tighten the nut to 199 ft. lbs. (270 Nm).
15. Install steering gear, bolts and washers. Tighten to 59 ft. lbs. (80 Nm).
16. Connect the drag link.
17. Install the tie rod nut and new cotter pin. Tighten to 66 ft. lbs. (90 Nm).
18. Install a new high pressure hose O-ring seal and a new return hose O-ring seal.
19. Connect the lines. Tighten to 26 ft. lbs. (35 Nm).
20. Make sure the steering column is locked.
21. Connect the pinch bolt. Tighten to 36 ft. lbs. (48 Nm).
22. Engage the steering coupling shield to the line fitting.
23. Install the air cleaner assembly.
24. Fill and leak check the system.

Expedition And Navigator

2001–02 MODELS

1. Before servicing the vehicle, refer to the precautions in the beginning of this section.

⁂ WARNING

The electrical power to the air suspension system must be shut off prior to hoisting, jacking or towing an air suspension vehicle. To do this, turn off the air suspension switch located in the right hand kick panel. Failure to do so can result in unexpected inflation or deflation of the air springs, which can result in shifting of the vehicle during these operations.

2. Remove the skid plate.
3. Remove the steering sector shaft arm drag link castellated nut and cotter pin.
4. Using tool 211-003, separate the steering sector shaft arm drag link.
5. Remove the dust shield.
6. Remove the pinch bolt.

⁂ CAUTION

Do not rotate the steering wheel when the steering column shaft is disconnected, or damage to the air bag sliding contact will result.

7. Slide the intermediate shaft off the steering gear input shaft.
8. Turn the ignition key to the locked position.
9. Disconnect the power steering lines at the steering gear.
10. Remove the bolts and the steering gear.

To install:

11. Install the steering gear. Tighten the bolts to 173–233 ft. lbs. (234–316 Nm).
12. Install a new high pressure hose O-ring seal and a new return hose O-ring seal.
13. Connect the power steering lines to the steering gear. Tighten to 17 ft. lbs. (23 Nm).
14. Slide the intermediate shaft Into the steering gear input shaft.

⁂ CAUTION

Do not rotate the steering wheel when the steering column shaft is disconnected, or damage to the air bag sliding contact will result.

15. Tighten the pinch bolt to 30–40 ft. lbs. (41–54 Nm).
16. Install the dust shield.
17. Attach the steering sector shaft arm drag link. Tighten to 173–233 ft. lbs. (234–316 Nm).
18. Install the steering sector shaft arm drag link castellated nut and new cotter pin. Tighten to 57–76 ft. lbs. (77–103 Nm).
19. Install the skid plate.
20. Fill and leak check the system

2003–05 4X2 MODELS

1. Before servicing the vehicle, refer to the precautions in the beginning of this section.
2. Place the front wheels in the straight-ahead position and the ignition switch in the OFF position.

⁂ WARNING

The electrical power to the air suspension system must be shut off prior to hoisting, jacking or towing an air suspension vehicle. To do this, turn off the air suspension switch located near the jack storage area in the rear of the passenger compartment. Failure to do so can result in unexpected inflation or deflation of the air springs, which can result in shifting of the vehicle during these operations.

3. If equipped, turn the air suspension switch to the OFF position.

4. With the vehicle in NEUTRAL, position on a hoist.

➡**The hex holding feature can be used to prevent turning of the stud while removing the nut.**

5. Remove the nuts and detach the tie-rods from the wheel knuckles.
6. Remove the bolts and the oil drip shield.
7. Detach the lines from the clip.

⁂ CAUTION

Do not allow the intermediate shaft to rotate while it is disconnected from the gear or damage to the clockspring can occur. If there is evidence that the intermediate shaft has rotated, the clockspring must be removed and recentered.

8. Remove the bolt and detach the intermediate shaft from the gear.
9. Remove the bolt and disconnect the power steering lines.
10. Discard the O-ring seals.
11. On Navigator models:
   a. Disconnect the electrical connector.
   b. Rotate the actuator until the connector housing is facing the steering gear input shaft.
12. Remove the nuts and bolts. Discard the nuts.

⁂ CAUTION

Make sure that the steering lines are clear from the removal path of the gear or damage to the lines can result.

13. Remove the nuts, bolts, brackets and the steering gear. Discard the nuts.

To install:

⁂ CAUTION

Make sure to tighten the bracket-to-steering gear bolts (M14) before tightening the bracket-to-crossmember bolts (M12) or damage to the steering gear can result.

14. Install the steering gear, bolts and new nuts. Tighten to 111 ft. lbs. (150 Nm).
15. Install the nuts and bolts. Tighten to 76 ft. lbs. (103 Nm).
16. On Navigator models:
   a. Rotate the actuator until the connector housing is facing the steering gear input shaft.
   b. Connect the electrical connector.
17. If equipped, turn the air suspension switch to the OFF position.
18. With the vehicle in NEUTRAL, position on a hoist.

➡**The hex holding feature can be used to prevent turning of the stud while removing the nut.**

19. Remove the nuts and detach the tie-rods from the wheel knuckles.
20. Remove the bolts and the oil drip shield.
21. Detach the lines from the clip.

⁂ CAUTION

Do not allow the intermediate shaft to rotate while it is disconnected from the gear or damage to the clockspring can occur. If there is evidence that the intermediate shaft has rotated, the clockspring must be removed and recentered.

22. Remove the bolt and detach the intermediate shaft from the gear.
23. Remove the bolt and disconnect the power steering lines.
24. Install a new high pressure hose O-ring seal and a new return hose O-ring seal.
25. Install the power steering lines. Tighten the bolt to 17 ft. lbs. (23 Nm).
26. Attach the intermediate shaft from the gear. Tighten the bolt to 22 ft. lbs. (30 Nm).

⁂ CAUTION

Do not allow the intermediate shaft to rotate while it is disconnected from the gear or damage to the clockspring can occur. If there is evidence that the intermediate shaft has rotated, the clockspring must be removed and recentered.

27. Attach the lines to the clip.
28. Install the oil drip shield.
29. Install the tie-rods to the wheel knuckles and tighten the nut to 111 ft. lbs. (150 Nm).
30. If equipped, turn the air suspension switch to the ON position.
31. Fill and leak check the system

2003–05 4X4 MODELS

1. Before servicing the vehicle, refer to the precautions in the beginning of this section.
2. Place the front wheels in the straight-ahead position and the ignition switch in the OFF position.

⁂ WARNING

The electrical power to the air suspension system must be shut off prior to hoisting, jacking or towing an air suspension vehicle. To do this, turn off the air suspension switch located near the jack storage area in the rear of the passenger compartment. Failure to do so can result in unexpected inflation or deflation of the air springs, which can result in shifting of the vehicle during these operations.

3. If equipped, turn the air suspension switch to the OFF position.
4. With the vehicle in NEUTRAL, position on a hoist.
5. Remove the cooling fan and shrouds.

➡**The hex holding feature can be used to prevent turning of the stud while removing the nut.**

6. Remove the nuts and detach the tie-rods from the wheel knuckles.
7. Remove the bolts and the oil drip shield.
8. Detach the lines from the clip.

⁂ CAUTION

Do not allow the intermediate shaft to rotate while it is disconnected from the gear or damage to the clockspring can occur. If there is evidence that the intermediate shaft has rotated, the clockspring must be removed and recentered.

9. Remove the bolt and detach the intermediate shaft from the gear.
10. Remove the bolt and disconnect the power steering lines.
11. Discard the O-ring seals.
12. On Navigator models:
   a. Disconnect the electrical connector.
   b. Rotate the actuator until the connector housing is facing the steering gear input shaft.
13. Remove the nuts and bolts. Discard the nuts.
14. Insert a block of wood between the lower arm and the frame.

⁂ CAUTION

Make sure that the steering lines are clear from the removal path of the gear or damage to the lines can result.

15. Remove the nuts, bolts, brackets and the steering gear. Discard the nuts.

To install:

⁂ CAUTION

Make sure to tighten the bracket-to-steering gear bolts (M14) before tightening the bracket-to-crossmember bolts (M12) or damage to the steering gear can result.

16. Install the steering gear, bolts and new nuts. Tighten to 111 ft. lbs. (150 Nm).
17. Install the nuts and bolts. Tighten to 76 ft. lbs. (103 Nm).
18. On Navigator models:
    a. Rotate the actuator until the connector housing is facing the steering gear input shaft.
    b. Connect the electrical connector.
19. If equipped, turn the air suspension switch to the OFF position.
20. With the vehicle in NEUTRAL, position on a hoist.

➡The hex holding feature can be used to prevent turning of the stud while removing the nut.

21. Remove the nuts and detach the tie-rods from the wheel knuckles.
22. Remove the bolts and the oil drip shield.
23. Detach the lines from the clip.

⁂ CAUTION

Do not allow the intermediate shaft to rotate while it is disconnected from the gear or damage to the clockspring can occur. If there is evidence that the intermediate shaft has rotated, the clockspring must be removed and recentered.

24. Remove the bolt and detach the intermediate shaft from the gear.
25. Remove the bolt and disconnect the power steering lines.
26. Install a new high pressure hose O-ring seal and a new return hose O-ring seal.
27. Install the power steering lines. Tighten the bolt to 17 ft. lbs. (23 Nm).
28. Attach the intermediate shaft from the gear. Tighten the bolt to 22 ft. lbs. (30 Nm).

⁂ CAUTION

Do not allow the intermediate shaft to rotate while it is disconnected from the gear or damage to the clockspring can occur. If there is evidence that the intermediate shaft has rotated, the clockspring must be removed and recentered.

29. Attach the lines to the clip.
30. Install the oil drip shield.
31. Install the tie-rods to the wheel knuckles and tighten the nut to 111 ft. lbs. (150 Nm).
32. If equipped, turn the air suspension switch to the ON position.
33. Fill and leak check the system

Shock Absorber

REMOVAL & INSTALLATION

Front

EXCEPT EXCURSION

1. Before servicing the vehicle, refer to the precautions in the beginning of this section.
2. If equipped with 2-wheel drive, perform the following:
    a. Hold the shock absorber stem and remove the nut from the top of the shock.
    b. Raise and safely support the vehicle.
    c. Remove the 2 lower retaining nuts and remove the shock absorber.
3. If equipped with 4-wheel drive, perform the following:
    a. Hold the shock absorber stem and remove the nut, washer and bushing from the top of the shock absorber stud.
    b. Raise and support the vehicle.
    c. Remove the lower retaining nut and bolt.
    d. Remove the shock absorber from the vehicle.

To install:

4. On 2-wheel drive models, perform the following:
    a. Install the washer and bushing to the top stem of the shock absorber.
    b. Place the shock absorber up through the coil spring.
    c. Install the 2 lower retaining nuts. Tighten the nuts to 22–29 ft. lbs. (30–40 Nm).
    d. Lower the vehicle.
    e. Install the bushing, washer and retaining nut to the top of the shock absorber stud. Tighten the nut to 34–46 ft. lbs. (47–63 Nm).
5. On 4-wheel drive models, perform the following:
    a. Install the washer and bushing to the top stem of the shock absorber.
    b. Place the shock absorber up through the coil spring.
    c. Install the lower retaining nut and bolt. Tighten to 57–76 ft. lbs. (77–104 Nm).
    d. Lower the vehicle.
    e. Install the shock absorber upper bushing, washer and retaining nut. Tighten the nut to 22–29 ft. lbs. (30–40 Nm).
6. Road test the vehicle and check for proper operation.

EXCURSION—4X2 MODELS

1. Before servicing the vehicle, refer to the precautions in the beginning of this section.
2. Remove the upper shock absorber retaining nut and upper shock absorber insulator.
3. Raise and support the vehicle.
4. Remove the lower shock absorber retaining nut and remove the shock absorber.

To install:

5. Installation is the reverse of removal. Using new fasteners, tighten the lower nut to 60 ft. lbs. (80 Nm) and the upper nut to 30 ft. lbs. (40 Nm).

EXCURSION—4X4 MODELS

1. Before servicing the vehicle, refer to the precautions in the beginning of this section.

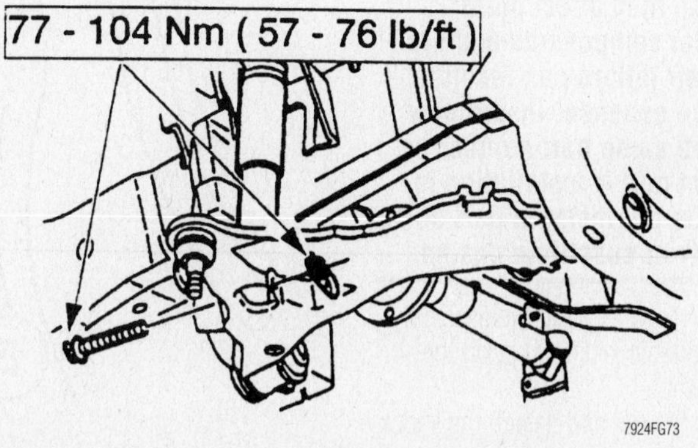

Lower shock absorber mounting—late models with torsion bar suspension

2. Remove the lower shock absorber retaining nut and remove the shock absorber.

3. Remove the upper shock absorber retaining nut and upper shock absorber insulator.

To install:

4. Installation is the reverse of removal. Using new fasteners, tighten the lower and upper nuts to 76 ft. lbs. (103 Nm).

Rear

1. Before servicing the vehicle, refer to the precautions in the beginning of this section.

2. Raise the vehicle and secure on support stands.

3. Remove the self-locking nut, steel washer, and rubber bushings at the upper end of the shock absorber.

4. Remove the bolt and nut at the lower end and remove the shock absorber. If needed, raise the rear axle assembly slightly with a jack.

To install:

5. When installing a new shock absorber, use new rubber bushings. Position the shock absorber on the mounting brackets with the stud end at the top. Install the upper bushing, steel washer and self-locking nut at the upper end, and the bolt and nut at the lower end.

6. Tighten the upper mounting studs to 63–84 ft. lbs. (85–115 Nm) and the lower mounting nuts to 63–84 ft. lbs. (85–115 Nm).

Coil Spring

REMOVAL & INSTALLATION

Front

EXCURSION

1. Before servicing the vehicle, refer to the precautions in the beginning of this section.

⁂ CAUTION

Suspension fasteners are critical parts because they affect performance of vital components and systems and their failure can result in major service expense. Install new parts with the same part number or an equivalent part if installation is necessary. Do not install a part of lesser quality or substitute design.

2. Remove the wheel and tire assembly.

3. Use a suitable jack to support the front axle assembly.

4. Remove the nut and detach the shock from the mounting stud.

5. Remove the upper spring retainer.

6. Lower the front axle until the spring is free of the upper spring seat.

7. Using an extension through the top of the spring, remove the lower spring retainer.

8. Remove the front spring.

To install:

9. Installation is the reverse of removal.. Tighten the lower spring retainer to 99 ft. lbs. (133 Nm), the upper retainer to 26 ft. lbs. (35 Nm) and the lower shock bolt to 60 ft. lbs. (80 Nm).

2001–2002 EXPEDITION AND NAVIGATOR

1. Before servicing the vehicle, refer to the precautions in the beginning of this section.

2. Remove or disconnect the following:
   - Wheels
   - Disc brake caliper and support aside with wire
   - Disc brake adapter
   - Rotor
   - Rotor splash shield
   - Shock absorber
   - Bracket supporting the brake hose
   - Sway bar link retaining nut and bushing from the lower control arm. Separate the sway bar link from the lower control arm.

3. Install a coil spring compressor and compress the coil spring enough to relieve the tension of the spring between the upper and lower control arms.
   - Cotter pin and castellated nut from the lower ball joint
   - Lower ball joint from the wheel spindle

4. Matchmark the lower control arm alignment cams for installation reference.
   - Lower control arm retaining nuts and bolts
   - Lower control arm and the compressed coil spring as an assembly

5. Loosen the coil spring compressor and remove the coil spring from the lower control arm.

6. Inspect the coil spring and replace as needed.

To install:

7. Install or connect the following:
   - Coil spring correctly in the saddle of the lower control arm
   - Coil spring compressor and compress the coil spring. The end of the coil spring **A**, must cover the hole designated **B** and be visible in the second hole designated as **C**, for proper installation.
   - Lower control arm to the frame
   - Retaining bolts, adjusting cams, and nuts

8. Align the matchmarks on the adjusting cams. The forward nut must be tightened first while the control arm is held at the curb position height. Tighten the nuts to 197–241 ft. lbs. (270–330 Nm).
   - Lower ball joint stud into the wheel spindle
   - Castellated nut and tighten to 83–113 ft. lbs. (113–153 Nm)
   - New cotter pin
   - Sway bar link to the lower control arm
   - Bushing and retaining nut. Tighten to 15–21 ft. lbs. (21–29 Nm).

9. Remove the coil spring compressor.

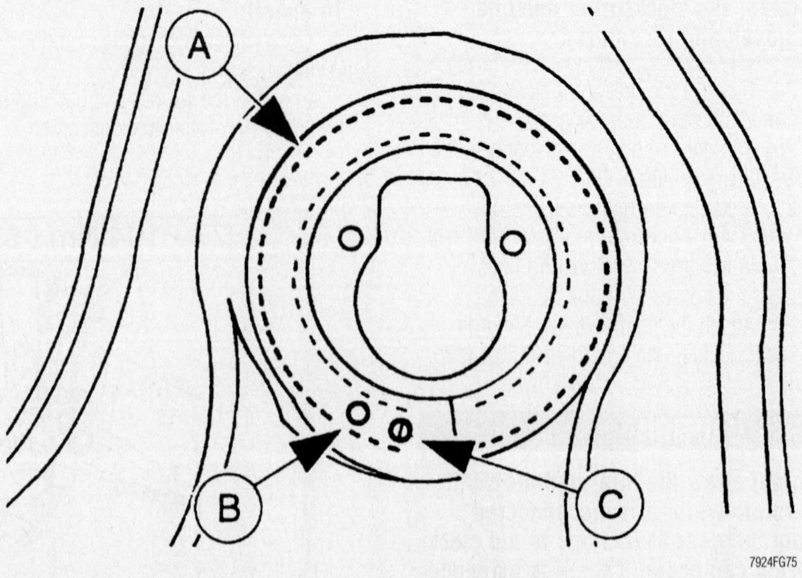

Be sure the coil spring is mounted correctly in the lower control arm

- Shock absorber. Tighten the lower bolts to 22 ft. lbs. (32 Nm) and the top nut to 45 ft. lbs. (61 Nm).
- Brake hose bracket
- Brake rotor splash shield
- 3 retaining bolts and tighten them to 90–107 inch lbs. (10–14 Nm)
- Disc brake rotor and caliper assemblies
- Wheel and tire assembly

10. Lower the vehicle.
11. Pump the brake pedal several times to position the brake pads prior to moving the vehicle.
12. Check the alignment and adjust if out of specification.
13. Road test the vehicle and check for proper operation.

2003–05 EXPEDITION AND NAVIGATOR

1. Before servicing the vehicle, refer to the precautions in the beginning of this section.
2. Remove the wheel and tire assembly.
3. Remove and discard the upper shock absorber nuts.
4. Remove the nut and detach the tie-rod from the wheel knuckle. Discard the nut.
5. Remove the nut, bolt and shock absorber/spring assembly. Discard the nut.
6. For reference during assembly, index the upper mount, spring and shock absorber.
7. Using a suitable spring compressor, compress the spring until the tension is released from the shock absorber.
8. While holding the shock rod, remove the nut and washer.
9. Remove the shock absorber. Discard the nut.
10. Remove the upper mount, dust shield and insulator.
11. Installation is the reverse of removal. Tighten the shock rod nut to 22 ft. lbs. (30 Nm).Tighten the shock absorber nut and bolt to 295 ft. lbs. (400 Nm). Tighten the tie rod nut to 111 ft. lbs. (150 Nm) and the new upper shock absorber nuts to 26 ft. lbs. (35 Nm).

Rear

2001–02 EXPEDITION AND NAVIGATOR

1. Before servicing the vehicle, refer to the precautions in the beginning of this section.
2. Raise and support the vehicle.
   - Wheel and tire assembly
   - Rear driveshaft
   - Rear Anti-lock Brake System (ABS) sensor electrical connector

➡Make sure the parking brake control is fully released.

3. Pull the front parking brake cable and conduit.
4. Insert a suitable retainer into the parking brake control.
5. Remove or disconnect the following:
   - Rear parking brake cable and conduit from the rear axle
   - Rear parking brake cable and conduit from the rear axle bracket assembly
   - Cable from the caliper
6. Repeat for the other side.
   - Disc brake caliper bolts
7. Repeat for the other side.
   - Axle vent tube
   - Stabilizer bar from the rear axle
   - Bolts from each stabilizer bar retainer
   - Retainers
8. Use a suitable jack to support the rear axle.
   - Lower shock absorber nut and bolt
9. Repeat for the other side.
   - Rear axle-to-trackbar assembly bolt
   - Rear suspension lower arm assembly-to-axle bolt
10. Repeat for the other side.
    - Rear suspension upper arm assembly-to-rear axle bolt
11. Repeat for the other side.
12. Lower the rear axle and remove the coil spring.

To install:

➡Do not tighten any nuts or bolts until the rear suspension is raised so that the rear suspension lower arms are parallel to the ground. Once in position, tighten all the nuts and bolts to specification.

13. Install or connect the following:
    - Rear coil spring
    - Rear suspension upper arm assemblies to the axle and the bolts. Tighten to 94–127 ft. lbs. (128–172 Nm).
    - Rear suspension lower arm assemblies to the axle and the bolts. Tighten to 94–127 ft. lbs. (128–172 Nm).
    - Trackbar assembly to the axle and the bolts. Tighten to 125–170 ft. lbs. (170–130 Nm).
    - Shock absorbers to the axle and the bolts. Tighten to 63–84 ft. lbs. (86–114 Nm).
    - Stabilizer bar to the axle and the stabilizer bar retainers and bolts. Tighten to 63–84 ft. lbs. (86–114 Nm).
    - Disc brake caliper and the bolts
    - Parking brake cables and conduit to the axle
    - Parking brake cable and conduit to the lever
    - Parking brake cable and conduit to the rear axle bracket assembly
    - Tire and wheel assembly
    - Lug nuts and lower the vehicle
    - Wheel cover.
14. Remove the pin from the parking brake control assembly.
    - Rear brake anti-lock sensor (ABS) electrical connector
    - Axle vent tube
    - Driveshaft
15. Lower the vehicle.

2003–05 EXPEDITION AND NAVIGATOR

1. Before servicing the vehicle, refer to the precautions in the beginning of this section.
2. Remove the wheel and tire assembly.
3. Remove and discard the upper shock absorber nuts.
4. Remove the nut, bolt and shock absorber/spring assembly. Discard the nut.
5. For reference during assembly, index the upper mount, spring and shock absorber.
6. Using a suitable spring compressor, compress the spring until the tension is released from the shock absorber.
7. While holding the shock rod, remove the nut and washer.
8. Remove the shock absorber. Discard the nut.
9. Remove the upper mount, dust shield and insulator.
10. Installation is the reverse of removal. Tighten the shock rod nut to 22 ft. lbs. (30 Nm).Tighten the shock absorber nut and bolt to 350 ft. lbs. (475 Nm). Tighten the new upper shock absorber nuts to 26 ft. lbs. (35 Nm).

Leaf Spring

REMOVAL & INSTALLATION

Excursion

1. Before servicing the vehicle, refer to the precautions in the beginning of this section.
2. Remove or disconnect the following:
3. Raise and support the vehicle.
   - Wheel and tire assembly
4. Support the rear axle with a suitable jack.

- U-bolt retaining nuts and the U-bolts.
- Rear spring upper plate.
- Nut and bolt from the rear spring front hanger bracket.
- Lower nut and bolt from the rear spring shackle bracket
- Rear spring assembly.
- Nut and bolt from the rear spring shackle assembly and the rear spring shackle

To install:

5. Installation is the reverse of removal, please use new fasteners and note the following torques:
   a. Nut and bolt for the rear spring shackle assembly to 185 ft. lbs. (285 Nm).
   b. Lower nut and bolt for the rear spring shackle bracket to 185 ft. lbs. (285 Nm).
   c. Nut and bolt for the rear spring front hanger bracket to 185 ft. lbs. (285 Nm).
   d. U-bolt retaining nuts and the U-bolts to 185 ft. lbs. (285 Nm).

Track Bar

REMOVAL & INSTALLATION

1. Before servicing the vehicle, refer to the precautions in the beginning of this section.

⁂ CAUTION

The electrical power to the air suspension system must be shut off prior to hoisting, jacking or towing an air suspension vehicle. This can be accomplished by turning off the air suspension switch located in the PASSENGER SIDE kick panel area. Failure to do so can result in unexpected inflation or deflation of the air springs, which can result in shifting of the vehicle during these operations.

2. Raise and support the vehicle.

⁂ WARNING

The air suspension height sensor has a plastic harness retainer to suspension that must be unclipped prior to removal.

3. Remove or disconnect the following:
   - Air suspension height sensor electrical connector
   - Metal retaining tabs and remove the air suspension height sensor from the ball studs
   - Passenger side trackbar bolt
   - Driver's side trackbar nut and bolt
   - Trackbar from the vehicle

To install:

4. The installation is the reverse of the removal. Tighten the upper and lower track bar bolts to 306 ft. lbs. (500 Nm) on 2001–02 models and 406 ft. lbs. (550 Nm) on all other models.

Torsion Bars

REMOVAL & INSTALLATION

1. Before servicing the vehicle, refer to the precautions in the beginning of this section.
2. Raise and safely support the vehicle.
3. Matchmark and measure the length of the torsion bar being removed.

➡Torsion bars are marked for left-hand or right-hand installation. If removing both torsion bars, make sure to reinstall in the correct position. If replacing a torsion bar, make sure that it is the correct torsion bar for the side being replaced.

4. Install Torsion Bar Adjuster T95T-5310-A and Adapter Plates T96T-5310-A, or equivalents.
5. Tighten the adjuster tool until it touches the torsion bar adjuster.
6. Remove the torsion bar adjuster bolt and nut.
7. Remove 6 torsion bar crossmember retaining bolts.
8. Remove the crossmember support.
9. Remove the torsion bar from the vehicle.

To install:

10. Place the torsion bar in position ensuring that it is the correct torsion bar for the side being installed.
11. Install the crossmember support and 6 retaining bolts. Tighten the retaining bolts to 40–50 ft. lbs. (53–72 Nm).
12. Install the torsion bar adjuster bolt and nut, then remove the torsion bar adjuster tool and adapters.
13. Tighten the adjusting bolt until the matchmarks are in alignment.
14. Lower the vehicle.
15. Measure the ride height and adjust if necessary.
16. Check the alignment and adjust if not within specification.
17. Road test the vehicle and check for proper operation.

ADJUSTMENT

1. Place the vehicle on a level surface.
2. Bounce the front end to normalize the ride height.
3. Measure the ride height on both sides of the vehicle from the frame to the ground.
4. Loosen the lock nut and add or delete turns until the vehicle is level and at the correct curb height.
5. Tighten the lock nut.
6. Road test the vehicle and recheck the measurements.

Ball Joint

REMOVAL & INSTALLATION

4X2 Models

EXCURSION

1. Before servicing the vehicle, refer to the precautions in the beginning of this section.
2. Remove the wheel and tire assembly.
3. Remove the disc brake caliper and the front disc brake hub and rotor.
4. Remove the front disc brake rotor shield.
5. Remove the ABS sensor retaining bolt, sensor harness retaining bolt and the sensor, if equipped and position out of the way.
6. Disconnect the tie rod end. Remove and discard the cotter pin.
7. Using a Pitman Arm Puller, remove the tie rod end.
8. Remove the pinch bolt and the camber adjuster.

⁂ CAUTION

To prevent damage to the ball joint seal and the ball joint socket, do not use a pickle fork-type remover to loosen the ball joints.

9. Remove and discard the cotter pin from the ball joint.
10. Loosen, but do not remove, the castellated nut.
11. Strike the lower end of the front axle to loosen the ball joint.
12. Remove the castellated nut and the front wheel spindle.
13. Position the front wheel spindle in a vise, and remove the snap ring from the lower ball joint.

⁂ CAUTION

To avoid damage to the components, do not use heat to aid ball joint removal.

14. Using the removal/installation tool 205-086 and suitable receiver cup, remove the lower ball joint from the front wheel spindle.
15. Using the removal/installation tool 205-086 and suitable receiver cup, remove the upper ball joint.

To install:

⁂ CAUTION

To avoid damage to components, do not use heat to aid installation.

16. Clean the wheel knuckle ball joint bores.

➡The lower ball joint must be installed first.

17. Using the removal/installation tool 205-086 with suitable receiver cups, install the lower ball joint.
18. Using the removal/installation tool 205-086 with suitable receiver cups, install the upper ball joint.
19. Install the snap ring in the groove at the bottom of the ball joint.
20. Install the front wheel spindle. Tighten the nut to 99 ft. lbs. (133 Nm). Install a new cotter pin.
21. Install the camber adjuster and pinch bolt. Tighten to 60 ft. lbs. (80 Nm).
22. Connect the tie rod end. Tighten the nut to 67 ft. lbs (90 Nm). Install a new cotter pin.
23. Install the ABS sensor, if equipped.
24. Install the front disc brake rotor shield.
25. Install the disc brake hub, rotor and caliper.
26. Install the wheel and tire assembly.

EXPEDITION AND NAVIGATOR

1. Before servicing the vehicle, refer to the precautions in the beginning of this section.
2. Remove the wheel and tire assembly.
3. Remove the disc brake caliper and the front disc brake hub and rotor.
4. Remove the front disc brake rotor shield.
5. Remove the ABS sensor retaining bolt, sensor harness retaining bolt and the sensor, if equipped and position out of the way.
6. Disconnect the tie rod end. Remove and discard the cotter pin.
7. Using tool 211-003, remove the tie rod end.

⁂ CAUTION

To prevent damage to the ball joint seal and the ball joint socket, do not use a pickle fork-type remover to loosen the ball joints.

8. Support the axle with a jack.
9. Remove and discard the cotter pin from the ball joint.
10. Using tool 211-003,separate the front wheel knuckle from the suspension lower arm. Then, loosely reinstall the castellated nut.
11. Remove the upper ball joint castellated nut. Discard the cotter pin and nut.

⁂ CAUTION

Do not damage the ball joint boot when installing the special tool.

12. Using the too 211-003, separate the front wheel knuckle from the front suspension upper arm.
13. Remove the hand-tightened lower ball joint castellated nut, then remove the front wheel knuckle.
14. Remove the snap ring from the ball joint. Discard the snap ring.
15. Using removal/installation tool 205-086 with suitable receiver cups, remove the ball joint.

To install:

⁂ CAUTION

To avoid damage to components, do not use heat to aid installation.

16. Clean the wheel knuckle ball joint bores.
17. Using the removal/installation tool 205-086 with suitable receiver cups, install the ball joint.
18. Install the snap ring in the groove at the bottom of the ball joint.
19. Install the front wheel spindle. Tighten the upper ball joint nut to 66 ft. lbs. (90 Nm) and the lower nut to 98 ft. lbs. (133 Nm). Install a new cotter pin.
20. Connect the tie rod end. Tighten the nut to 70 ft. lbs (95 Nm). Install a new cotter pin.
21. Install the ABS sensor, if equipped.
22. Install the disc brake hub, rotor and caliper.
23. Install the wheel and tire assembly.

4X4 Models

EXCURSION

1. Before servicing the vehicle, refer to the precautions in the beginning of this section.
2. Remove the wheel and tire assembly.
3. Remove the disc brake caliper and the front disc brake hub and rotor.
4. Remove the wheel hub and bearing.
5. Using a drift, drive the axle shaft main seal out of the wheel knuckle.
6. Remove the axle shaft and main seal.
7. Remove the tie-rod end castellated nut. Discard the cotter pin.
8. Using the tool 211-003, disconnect the tie-rod end from the wheel knuckle.
9. Remove the upper ball joint castellated nut and the insert.
10. Remove the wheel knuckle.
11. Position the front wheel spindle in a vise, and remove the snap ring from the lower ball joint.

⁂ CAUTION

To avoid damage to the components, do not use heat to aid ball joint removal.

12. Using the removal/installation tool 205-086 and suitable receiver cup, remove the lower ball joint from the front wheel spindle.
13. Using the removal/installation tool 205-086 and suitable receiver cup, remove the upper ball joint.

To install:

⁂ CAUTION

To avoid damage to components, do not use heat to aid installation.

14. Clean the wheel knuckle ball joint bores.

➡The lower ball joint must be installed first.

15. Using the removal/installation tool 205-086 with suitable receiver cups, install the lower ball joint.
16. Using the removal/installation tool 205-086 with suitable receiver cups, install the upper ball joint.
17. Install the snap ring in the groove at the bottom of the ball joint.
18. Position the wheel knuckle onto the axle housing.
19. Install the nut onto the lower ball joint. Do not tighten the nut at this time.
20. Install the insert and the castellated

nut onto the upper ball joint. Do not tighten the nut at this time.

21. Tighten the lower ball joint retaining nut. Pre-tighten the nut to 35 ft. lbs (47 Nm).

➡Do not loosen the castellated nut to install the cotter pin.

22. Install the cotter pin into the upper ball joint.
23. Tighten the upper ball joint castellated nut to 69 ft. lbs. (94 Nm). Install the cotter pin. If necessary, tighten the castellated nut until the cotter pin can be installed.
24. Tighten the lower ball joint nut to 150 ft. lbs. (204 Nm).
25. Install the tie-rod end onto the wheel knuckle.
26. Position the tie-rod end into the wheel knuckle. Tighten the castellated nut to 52 ft. lbs. (70 Nm). Install a new cotter pin.
27. Position a new main seal onto the axle shaft. Using the tools illustrated and a hammer, seat the main seal onto the axle shaft.
28. Position the axle shaft into the axle housing. Using the tools illustrated and a hammer, install the main seal into the wheel knuckle.

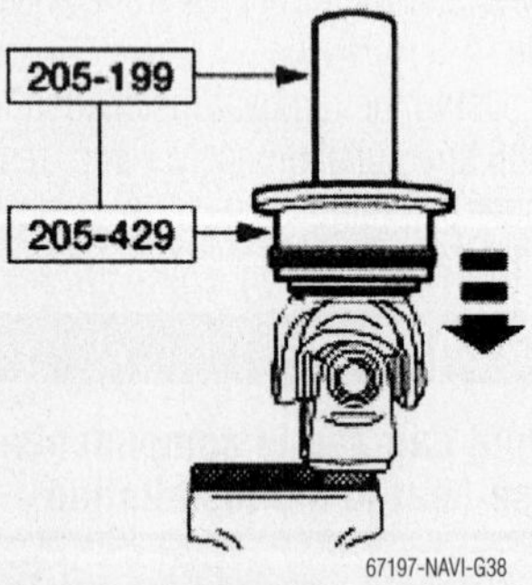

67197-NAVI-G38

Using the tools shown and a hammer, seat the main seal onto the axle shaft—4X4 Excursion

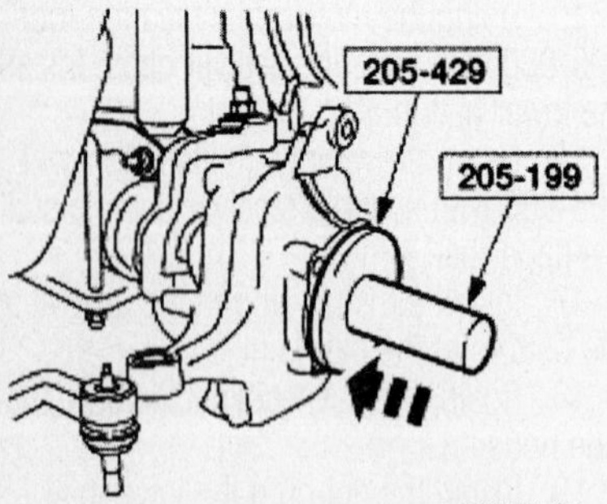

67197-NAVI-G39

Using the tools shown and a hammer, install the main seal into the wheel knuckle—4X4 Excursion

29. Install the wheel hub and bearing.
30. Install the front brake disc.
31. Install the wheel and tire assembly.

EXPEDITION AND NAVIGATOR

1. Before servicing the vehicle, refer to the precautions in the beginning of this section.
2. Remove the wheel and tire assembly.
3. Remove axle nut, cotter pin, retainer and nut.
4. Remove the brake caliper support bracket bolts and position the caliper support bracket and caliper aside.
5. Remove the brake disc.
6. Disconnect the ABS electrical connector.
7. Unclip the ABS harness routing clips.
8. Remove the tie-rod castellated nut and discard the cotter pin.

⁂ CAUTION

Do not use a hammer to separate the tie-rod from the wheel knuckle or damage to the wheel knuckle will result.

9. Using tool 211-003, separate the tie-rod end from the front wheel knuckle.
10. Using a suitable jack, support the front suspension lower arm.
11. Remove the lower ball joint castellated nut, discard the cotter pin and nut.
12. Using tool 211-003, separate the front wheel knuckle from the suspension lower arm. Then, loosely install the castellated nut.
13. Remove the upper ball joint castellated nut discard the cotter pin and nut.
14. Using tool 211-003, separate the front wheel knuckle from the front suspension upper arm.
15. Separate the halfshaft from the hub, then remove the hand-tightened lower ball joint castellated nut.
16. Remove the front wheel knuckle.
17. Position the halfshaft aside and support with wire.
18. Remove the snap ring from the ball joint and discard the snap ring.
19. Using a suitable ball joint remover tool and receiver cup, remove the ball joint.

To install:

⁂ CAUTION

To avoid damage to components, do not use heat to aid installation.

20. Clean the wheel knuckle ball joint bores.

➡The lower ball joint must be installed first.

21. Using a suitable ball joint installer tool and receiver cup, install the ball joint.
22. Install a new snap ring in the groove at the bottom of the ball joint. Make sure the snap ring is fully seated.

➡Always install new castellated nuts and cotter pins.

23. Install the remaining components in the reverse of removal. Please note the following torque specifications:
    a. Upper ball joint nut: 66 ft. lbs. (90 Nm).
    b. Lower ball joint nut: 98 ft. lbs. (133 Nm).
    c. Tie rod nut: 70 ft. lbs. (95 Nm).
    d. Axle shaft nut: 221 ft. lbs. (300 Nm).

Upper Control Arm

REMOVAL & INSTALLATION

Navigator And Expedition

2001–02 4X2 MODELS

1. Before servicing the vehicle, refer to the precautions in the beginning of this section.
2. Raise and safely support the vehicle.
3. Remove the wheel and tire assembly.
4. Support the lower control arm with a transmission jack or equivalent adjustable jack.
5. Matchmark the upper control arm adjusting cams.
6. If equipped with 4-wheel Anti-lock Brake System (ABS), remove the speed sensor harness bracket bolt and position the harness aside.
7. Remove the cotter pin and castellated nut from the upper ball joint. Discard the cotter pin.
8. Using Pitman Arm Puller T64P-3590-F or equivalent, separate the ball joint from the wheel spindle.
9. Remove 2 nuts and bolts securing the upper control arm to the body brackets.
10. Remove the upper control arm and ball joint assembly from the vehicle.

To install:

11. Align the matchmarks on the upper control arm adjusting cams.
12. Place the upper control arm in position to the body brackets and install 2 retaining bolts and nuts. Do not tighten at this time.

13. Install the ball joint stud to the wheel spindle and install the castellated nut. Tighten the nut to 56–77 ft. lbs. (76–104 Nm). Install a new cotter pin.

14. Raise the lower control arm to position the control arms at normal curb height and tighten the front upper control arm retaining nut first, to 84–112 ft. lbs. (113–153 Nm).

15. With the vehicle still at curb height, tighten the rear upper control arm retaining nut to 84–112 ft. lbs. (113–153 Nm).

16. Remove the transmission jack or equivalent from under the lower control arm.

17. If equipped with 4-wheel ABS brake system, position the speed sensor harness and install the bracket bolt. Tighten the speed sensor bracket bolt to 62–80 inch lbs. (7–9 Nm).

18. Install the wheel and tire assembly. Torque the lug nuts to 83–112 ft. lbs. (113–153 Nm).

19. Lower the vehicle.

20. Check the alignment and adjust if not within specification.

21. Road test the vehicle and check for proper operation.

ALL 2003–05 MODELS

1. Before servicing the vehicle, refer to the precautions in the beginning of this section.

2. Raise and safely support the vehicle.

3. Remove the wheel and tire assembly.

4. Remove the shock absorber and spring assembly.

5. On models with an air suspension, detach the height sensor from the upper arm.

➡Use the hex holding feature to prevent the stud from turning while removing the nut.

6. Remove and discard the front control arm nut. Discard the nut.

7. Remove the rear nut and bolt. Discard the nut.

8. Remove the upper arm.

To install:

➡Assemble the upper arm pivot nuts and bolts. Do not tighten until the installation procedure is complete and the weight of the vehicle is resting on the wheel and tire assemblies.

9. Installation is the reverse of removal. Tighten the control arm nuts/bolts to 111 ft. lbs. (150 Nm).

Lower Control Arm

REMOVAL & INSTALLATION

Expedition and Navigator

2001–02 4X2 MODELS

1. Before servicing the vehicle, refer to the precautions in the beginning of this section.

✻✻ WARNING

The electrical power to the air suspension system must be shut off prior to hoisting, jacking or towing an air suspension vehicle. To do this, turn off the air suspension switch located near the jack storage area in the rear of the passenger compartment. Failure to do so can result in unexpected inflation or deflation of the air springs, which can result in shifting of the vehicle during these operations.

✻✻ CAUTION

Do not use heat to loosen a seized wheel nut. Heat can damage the wheel and the wheel bearings.

2. Remove the tire and wheel assembly.

3. Remove the brake disc shield.

4. Remove the front shock absorber.

5. Remove the brake hose bracket screw and bracket from the front suspension lower arm.

6. Remove the front stabilizer bar link nut.

7. Using the coil spring compressor, compress the front coil spring.

8. Remove the lower ball joint castle nut. Discard the cotter pin.

9. Use a Pitman arm puller to separate the lower ball joint from the front wheel spindle.

➡The front suspension lower arm nut must be installed in its original position.

10. Index the front suspension lower arm nut.

11. Remove the front suspension lower arm and front coil spring.

12. Remove the two front suspension lower arm nuts.

13. Remove the two front suspension lower arm bolts.

14. Remove the front suspension lower arm and front coil spring.

To install:

15. Install the front coil spring into the front suspension lower arm.

- Coil spring compressor and compress the coil spring. The end of the coil spring **A**, must cover the hole designated **B** and be visible in the second hole designated as **C**, for proper installation.

➡Install the front suspension lower arm nuts in the original position.

➡The front suspension lower arm caster split adjuster is for production use only. Do not adjust.

➡The forward front suspension lower arm nut must be tightened first while the control arms are held at the curb position ride height.

16. Install the front suspension lower arm. Position the front suspension lower arm and front coil spring.

Install the bolts and nuts and tighten to 121–148 ft. lbs. (164–200 Nm).

17. Install the lower ball joint castle nut and tighten to 83–112 ft. lbs. (113–153 Nm). Install a new cotter pin.

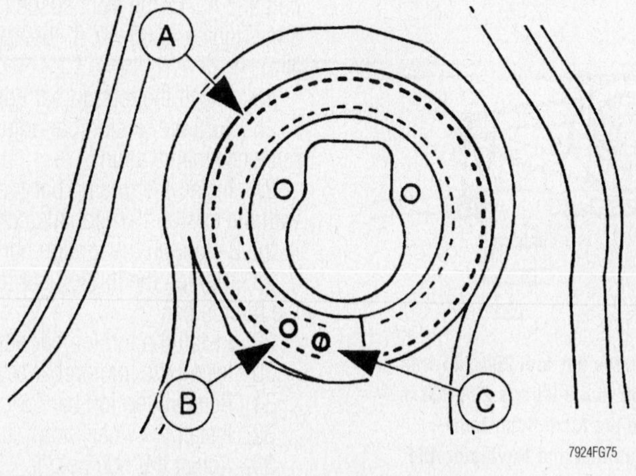

Be sure the coil spring is mounted correctly in the lower control arm

18. Install the front stabilizer bar link nut and tighten to 21 ft. lbs. (29 Nm).

19. Remove the Coil Spring Compressor.

20. Install the front shock absorber and the shock absorber lower nuts. Tighten to 29 ft. lbs. (40 Nm).

21. Install the upper shock absorber nut and washer. Tighten to 34–46 ft. lbs. (47–63 Nm).

22. Install the brake hose bracket screw.

23. Install the brake disc shield.

➡If equipped with air suspension, reactivate the system by turning on the air suspension switch.

24. Install the tire and wheel assembly.

25. Inspect and adjust the front end alignment

2001–02 4X4 MODELS

1. Before servicing the vehicle, refer to the precautions in the beginning of this section.

⁂ WARNING

The electrical power to the air suspension system must be shut off prior to hoisting, jacking or towing an air suspension vehicle. To do this, turn off the air suspension switch located near the jack storage area in the rear of the passenger compartment. Failure to do so can result in unexpected inflation or deflation of the air springs, which can result in shifting of the vehicle during these operations.

2. Remove the wheel hub.

3. Make an alignment mark on the torsion bar and the torsion bar crossmember support. Measure and record the length.

4. Relieve the torsion bar tension.

5. Remove the torsion bar adjuster bolt.

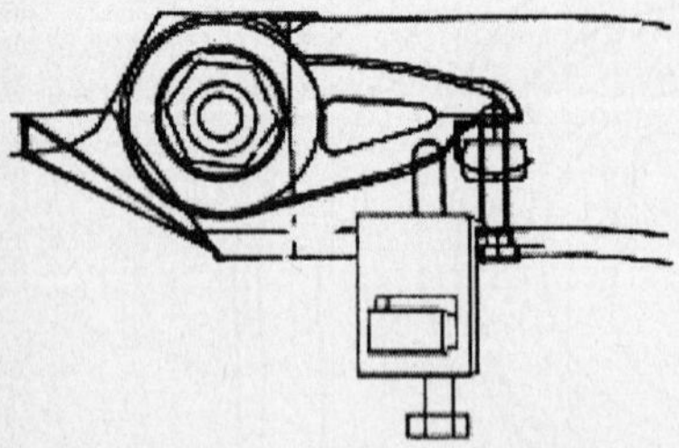

67197-NAVI-G40

Install the torsion bar tool 204-185 with the torsion bar tool adapters 204-203 in the B slots on the torsion bar tool—2001–02 Expedition and Navigator 4X4 models

6. Install the torsion bar tool 204-185 with the torsion bar tool adapters 204-203 in the B slots on the torsion bar tool.

7. Tighten the torsion bar tool until the torsion bar adjuster lifts off of the adjuster nut.

8. Remove the torsion bar adjuster nut and remove the torsion bar tool.

9. Remove the torsion bar adjuster.

10. Remove the stabilizer bar link nut from the front suspension lower arm.

11. Remove the front shock absorber lower bolt and nut from the front suspension lower arm.

12. Remove the lower ball joint castle nut. Discard the cotter pin.

13. Using a Pitman arm puller, separate the front suspension lower arm from the front wheel knuckle.

14. Separate the front suspension lower arm from the front wheel knuckle.

15. Remove the two front suspension lower arm nuts.

16. Remove the two front suspension lower arm bolts.

17. Remove the front suspension lower arm and torsion bar as an assembly.

18. Separate the front suspension lower arm from the torsion bar.

To install:

19. Attach the front suspension lower arm to the torsion bar.

20. Install the front suspension lower arm and torsion bar as an assembly.

21. Install the front suspension lower arm bolts and nuts. Tighten to 121 ft. lbs. (147 Nm).

22. Install the front suspension arm lower ball joint castle nut. Tighten the nut to 83–113 ft. lbs. (113–153 Nm). Install a new cotter pin.

23. Install the front stabilizer bar link to the front suspension lower arm. Tighten to 21 ft. lbs. (29 Nm).

24. Install the front shock absorber lower bolt and nut to the front suspension lower arm. Tighten to 65–87 ft. lbs. (88–118 Nm) Nm)

25. Install the torsion bar adjuster.

26. Turn the torsion bar adjuster until the reference marks align.

27. Install the torsion bar tool 204-185 with the torsion bar tool adapters 204-203 in the B slots on the torsion bar tool.

28. Tighten the torsion bar tool to load the torsion bar.

29. Install the torsion bar adjuster nut.

30. Install the torsion bar adjuster bolt.

31. Remove torsion bar tool.

32. Install the wheel hub.

33. Adjust the ride height.

ALL 2003–05 MODELS

1. Before servicing the vehicle, refer to the precautions in the beginning of this section.

2. If equipped, turn the air suspension switch to the OFF position.

3. Remove the wheel knuckle.

➡Use the hex holding feature to prevent the stud from turning while removing the nut.

4. Remove and discard the nuts/bolts using the illustrations below.

To install:

➡Assemble the lower arm pivot nuts and bolts. Do not tighten until the

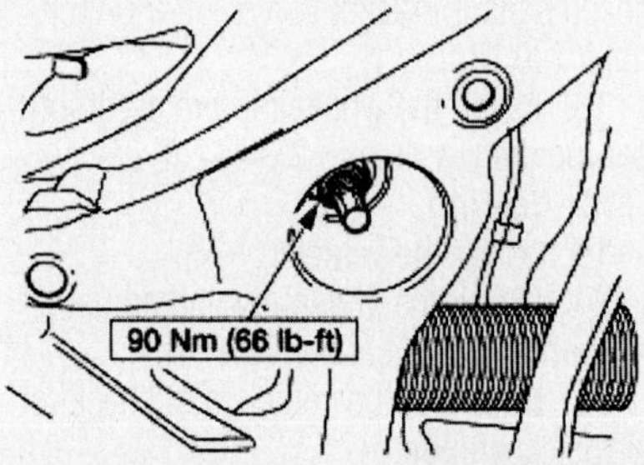

67197-NAVI-G41

Remove/discard this bolt first when removing the lower control arm . . .

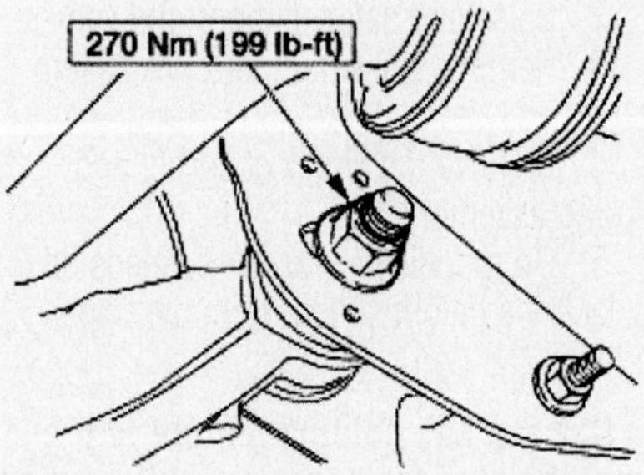

67197-NAVI-G42

. . . next remove/discard this bolt when removing the lower control arm

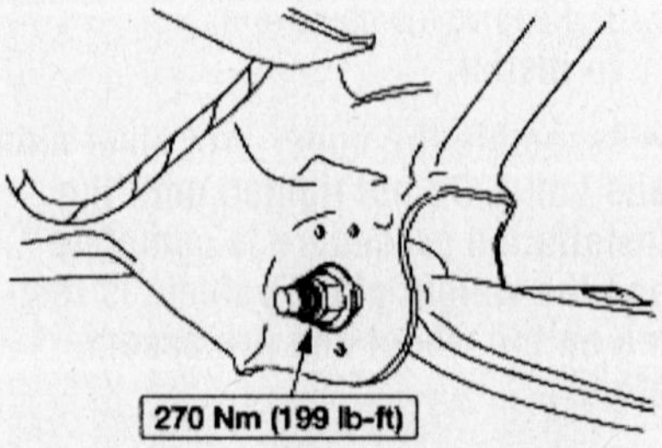

67197-NAVI-G43

. . . then remove/discard this bolt when removing the lower control arm . . .

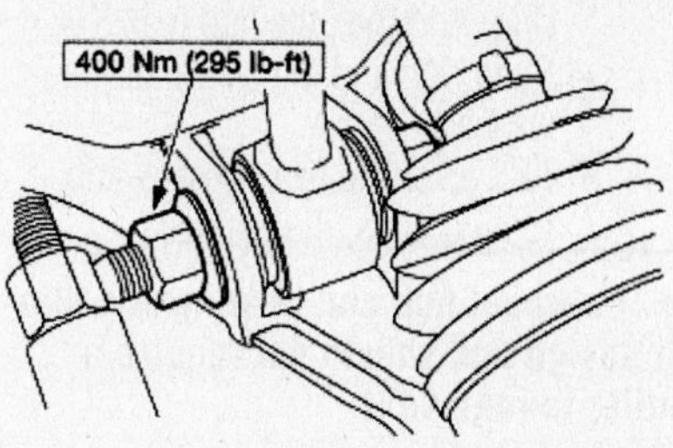

67197-NAVI-G44

. . . finally remove/discard this bolt when removing the lower control arm—All 2003–05 Expedition and Navigator models

installation procedure is complete and the weight of the vehicle is resting on the wheel and tire assemblies.

5. To install, reverse the removal procedure and tighten the retainers to the specifications shown in the illustrations.
6. Check and, if necessary, align the front end.

Front Wheel Bearings

ADJUSTMENT

⁂ WARNING

The electrical power to the air suspension system must be shut off prior to hoisting, jacking or towing an air suspension vehicle. Failure to shut the system off may lead to an unexpected inflation or deflation of the air springs, which may result in a shift of the vehicle.

➡On 4-wheel drive vehicles, the front wheel bearings are not adjustable.

1. Before servicing the vehicle, refer to the precautions in the beginning of this section.
2. Support the front end.
3. Remove the wheel cover, if equipped.
4. Remove the grease cap.

➡Check the wheel bearings for sufficient grease.

5. Remove the cotter pin and retaining washer. Back off the spindle nut. Discard the cotter pin.
6. Adjust the wheel bearings on Navigator and Expedition models as follows:
   a. Tighten the spindle nut to 30 ft. lbs. (40 Nm) while rotating the wheel and tire assembly to seat the wheel bearings.
   b. Back off the spindle nut 2 turns.
   c. Tighten the spindle nut to 17–24 ft. lbs. (23–34 Nm) while rotating the rotor clockwise.
   d. Loosen the nut 175 degrees.
   e. Tighten the nut to 17 inch lbs. (2 Nm).
7. Adjust the wheel bearings on Excursion models as follows:
   a. Tighten the spindle nut to 21 ft. lbs. (28 Nm) while rotating the wheel and tire assembly to seat the wheel bearings.
   b. Back off the spindle nut 120–180 degrees.
   c. Tighten the nut to 17 inch lbs. (2 Nm).
8. Install the retaining washer so the castellations are aligned with the cotter pin hole. Install a new cotter pin.
9. Check the wheel and tire assembly for proper rotation, then install the grease cap. If the wheel still does not rotate properly, inspect and clean or replace the wheel bearings and cups.
10. Install the wheel cover, if equipped.
11. Lower the vehicle.
12. Road test the vehicle and check for proper operation.

2 Wheel Drive

2001–02 EXPEDITION AND NAVIGATOR

⁂ WARNING

The electrical power to the air suspension system must be shut off prior to hoisting, jacking or towing an air suspension vehicle. Failure to shut the system off may lead to an unexpected inflation or deflation of the air springs, which may result in a shift of the vehicle.

The hub is part of the disc brake rotor and cannot be serviced separately. The inner and outer wheel bearing and races are serviced individually. Be sure to have a new hub grease seal when servicing the wheel bearings.

1. Before servicing the vehicle, refer to the precautions in the beginning of this section.
2. Remove or disconnect the following:
   - Wheels
   - Caliper
   - Brake pads and anti-rattle clips
   - Anchor plate
   - Hub grease cap, cotter pin, retainer washer and the spindle nut
   - Wheel bearing retainer washer and the outer wheel bearing
   - Brake hub and rotor assembly
   - Grease seal
   - Inner wheel bearing
3. Clean and inspect the wheel bearings and races for unusual wear or damage. Replace parts as necessary.
4. Inspect the hub and brake rotor assembly. If required, the hub and brake rotor assembly must be replaced as a unit.

To install:

5. If needed, pack the wheel bearing with a suitable high temperature wheel bearing grease before assembly.
6. Install or connect the following:
   - Inner wheel bearing in the hub and brake rotor assembly
   - New grease seal
   - Hub and rotor assembly on the wheel spindle
   - Outer wheel bearing
   - Retainer washer and the spindle nut
7. Adjust the wheel bearings (2WD only).
8. Install or connect the following:
   - Retaining washer, so the castellations are aligned with the cotter pin hole. Install a new cotter pin.
   - Anchor plate, and the 2 retaining bolts. Tighten the bolts to 125–168 ft. lbs. (170–230 Nm).
   - Anti-rattle clips, and the disc brake pads
   - Caliper
   - Wheels. Tighten the lug nuts to 83–112 ft. lbs. (113–153 Nm).
9. Check the wheel and tire assembly for proper rotation, then install the grease cap.
10. Lower the vehicle.
11. Road test the vehicle and check for proper operation.

2003–05 EXPEDITION AND NAVIGATOR

⁂ WARNING

The electrical power to the air suspension system must be shut off prior to hoisting, jacking or towing an air suspension vehicle. Failure to shut the system off may lead to an unexpected inflation or deflation of the air springs, which may result in a shift of the vehicle.

1. Before servicing the vehicle, refer to the precautions in the beginning of this section.
2. If equipped, turn the air suspension switch to the OFF position.
3. Disconnect the wheel speed sensor electrical connector.
4. Remove the wheel and tire assembly.

5. Remove the bolt and detach the brake line retainers.

CAUTION

Do not allow the caliper to hang from the brake hose or damage to the hose can result.

6. Remove the caliper, pads and anchor plate and set aside.
7. Remove the brake rotor.
8. Remove the dust cap.
9. Remove and discard the axle nut.
10. Remove the bolts and the wheel bearing and hub assembly.

To install:

CAUTION

If the original wheel bearing and hub is being reinstalled, make sure to install a new O-ring.

11. Install the remaining components in the reverse of removal. Please note the following torque specifications:
    a. Wheel bearing/hub assembly bolts: 148 ft. lbs. (200 Nm).
    b. Axle nut: 20 ft. lbs. (27 Nm).
    c. Brake anchor plate: 148 ft. lbs. (200 Nm).

EXCURSION

1. Before servicing the vehicle, refer to the precautions in the beginning of this section.
2. Remove the front wheel and tire assemblies.
3. Remove the front disc brake caliper and rotor, and position the caliper out of the way.
4. Remove the hub cap from the hub assembly.
5. Remove the cotter pin, adjusting nut and flat washer.

➡Inspect the condition of the spindle and nut threads to ensure a free turning nut when reassembling.

6. Remove the outer bearing cone and roller assembly, and pull the hub assembly from the spindle.
7. Using care not to damage the bearing cage, use a suitable slide hammer and bearing seal remover to remove the inner bearing cone and bearing seal.

To install:

CAUTION

Do not spin the bearing dry with compressed air.

➡Remove all traces of lubricant from the bearings, hub and axle spindle. Inspect bearings and bearing cups for pitting or unusual wear. If either bearings or bearing cups are worn or damaged, replace both bearings and bearing cups.

➡It is recommended that bearings and bearing cups be replaced in sets. If cups are worn or damaged, install the inner and outer bearing cups in the hub with an appropriate bearing cup driver tool. Check for proper seating of new bearing cups by trying to insert a 0.0015 inch (0.38-mm) feeler gauge between the bottom face of the cup and wheel hub seat. You should not be able to insert the feeler gauge.

8. Remove all burrs, nicks or scratches from the shoulder of the spindle and seal bore in the hub with emery cloth.
9. Pack the inside of the hub with lithium-base wheel bearing grease. Fill the hub until the grease is flush with the inside diameters of both bearing cups.
10. Pack the bearing cone and roller assemblies with wheel bearing grease. Use a bearing packer for this operation. If a packer is not available, work as much lubricant as possible between the rollers and cages.

CAUTION

Keep the hub centered on the spindle to prevent damage to the grease seal or spindle threads.

11. Place the inner bearing cone and roller assembly in the inner cup and install the wheel bearing hub seal, using a suitable seal replacer. Make sure seal is fully seated and lubricated.
12. Install the hub assembly.
13. Install the outer bearing cone and roller assembly and the flat washer on the spindle and install the adjusting nut. Install a new cotter pin.
14. Install the hub cap.
15. Install the front disc brake caliper and rotor.
16. Install the front wheel and tire assemblies.

4-Wheel Drive

EXCURSION

1. Before servicing the vehicle, refer to the precautions in the beginning of this section.
2. Remove or disconnect the following:
    - Front brake rotor
    - Hub lock by pulling out the retainer ring. And then the hub lock
    - Snap ring and the axle shaft thrust washers
    - ABS wheel sensor harness and routing clips, if equipped

➡The wheel hub and bearing is a slip fit design and should not require a puller to remove it.

- Four lock nuts
- Wheel hub and bearing
- Brake rotor shield
- Bolt and the Anti-lock Brake System (ABS) sensor, if equipped

3. Place the hub in a soft-jawed vise.
4. Install two nuts on the studs and use the inner nut to remove the studs.
5. Remove and discard the O-ring.

To install:

➡Any time the wheel hub is removed for any reason, a new O-ring seal must be installed. Failure to do so can cause a vacuum leak and loss of four wheel drive operations.

6. Install a new O-ring.
7. Place the hub in a soft-jawed vise.
8. Install two nuts on the studs and use the outer nut to install the studs.
9. Install or connect the following:
    - ABS sensor and the bolt. Tighten the bolt to 13 ft. lbs. (18 Nm).
    - Brake rotor shield
10. Apply a coat of High Temperature 4x4 Front Axle and Wheel Bearing Grease E8TZ-19590-A to the O-ring area of the wheel hub and bearing before installing the hub and bearing.
    - Wheel hub and bearing
    - Four lock nuts and tighten to 133 ft. lbs. (180 Nm)
    - ABS sensor harness and routing clips, if equipped

WARNING

The non-metallic thrust washer must be installed between the two metal thrust washers. Failure to do so will cause severe wear to the non-metallic thrust washer, allowing the axle shaft to travel further in and out during torque thrust causing damage to the wheel hub and bearing, the axle shaft end seal and the axle shaft.

- Three thrust washers onto the axle shaft
- Snap ring

➡Any time the hub lock is removed, a new O-ring seal must be installed.

Failure to do so can cause a vacuum leak and loss of four wheel drive functions.

- New O-ring seal
- Hub lock and the retainer ring
- Brake rotor

11. Perform a wheel-end vacuum leak test as follows:
    a. Install the vacuum pump and gauge line on the knuckle vacuum fitting and pump to 20 in-Hg.
    b. If the vacuum drop is less than 0.5 in-Hg in 30 seconds, the 4x4 hublock is working correctly.

2001–02 EXPEDITION AND NAVIGATOR

**** WARNING**

The electrical power to the air suspension system must be shut off prior to hoisting, jacking or towing an air suspension vehicle. Failure to shut the system off may lead to an unexpected inflation or deflation of the air springs, which may result in a shift of the vehicle.

The wheel bearings are of the cartridge design and are an integral part of the hub assembly. The bearings are permanently lubricated and require no maintenance or adjustments. If required, a new hub assembly must be installed.

1. Before servicing the vehicle, refer to the precautions in the beginning of this section.
2. Remove or disconnect the following:
    - Wheels
    - Caliper
    - Brake pads and anti-rattle clips
    - Anchor plate
    - Brake rotor
    - Rotor shield
    - Speed sensor retaining bolt and move the speed sensor and harness aside, if equipped with 4-wheel Anti-lock Brake System (ABS)
    - Hub nut cotter pin, retainer and the hub nut. Discard the cotter pin.
    - The 3 hub assembly retaining bolts from the inside of the steering knuckle
    - Hub assembly

➡If necessary, use a suitable puller to separate the hub assembly from the CV-joint. Use care not to over extend the CV-joint and boot when removing the hub assembly.

- Grease seal from the steering knuckle

To install:

3. Install or connect the following:
    - New grease seal using Bearing Cup Replacer 205-147 (T80T-4000-P), Knuckle Seal Replacer 205-361 (T96T-1175-A) and Threaded Drawbar 204-029 (T77F-1176-A)
    - Hub assembly to the steering knuckle and secure with the 3 retaining bolts. Tighten the bolts to 110–145 ft. lbs. (149–201 Nm).
    - CV-joint into the hub assembly
    - Speed sensor and secure with the retaining bolt, if equipped with 4-wheel ABS. Tighten the bolt to 60–84 inch lbs. (7–9 Nm).
    - Brake rotor shield and the 3 retaining bolts. Tighten the bolts to 80–107 inch lbs. (9–12 Nm).
    - Hub nut and tighten to 188–254 ft. lbs. (255–345 Nm)
    - Hub nut retainer and a new cotter pin
    - Brake rotor
    - Anchor plate in position and install the 2 retaining bolts. Tighten the bolts to 125–168 ft. lbs. (170–230 Nm).
    - Anti-rattle clips and the disc brake pads
    - Caliper
    - Wheels. Tighten the lug nuts to 83–112 ft. lbs. (113–153 Nm).
4. Lower the vehicle.
5. Pump the brake pedal several times to position the brake pads prior to moving the vehicle.
6. Road test the vehicle and check for proper operation.

2003–05 EXPEDITION AND NAVIGATOR

**** WARNING**

The electrical power to the air suspension system must be shut off prior to hoisting, jacking or towing an air suspension vehicle. Failure to shut the system off may lead to an unexpected inflation or deflation of the air springs, which may result in a shift of the vehicle.

1. Before servicing the vehicle, refer to the precautions in the beginning of this section.
2. If equipped, turn the air suspension switch to the OFF position.
3. Disconnect the wheel speed sensor electrical connector.
4. Remove the wheel and tire assembly..
5. Remove the bolt and detach the brake line retainers.

**** CAUTION**

Do not allow the caliper to hang from the brake hose or damage to the hose can result.

6. Remove the caliper, pads and anchor plate and set aside.
7. Remove the brake rotor.
8. Remove the dust cap.
9. Remove and discard the axle nut.
10. Remove the bolts and the wheel bearing and hub assembly.

To install:

**** CAUTION**

If the original wheel bearing and hub is being reinstalled, make sure to install a new O-ring.

11. Install the remaining components in the reverse of removal. Please note the following torque specifications:
    a. Wheel bearing/hub assembly bolts: 148 ft. lbs. (200 Nm).
    b. Axle nut: 20 ft. lbs. (27 Nm).
    c. Brake anchor plate: 148 ft. lbs. (200 Nm).

BRAKES

Brake Caliper

REMOVAL & INSTALLATION

Front

2001–02 MODELS

⁂ WARNING

The electrical power to the air suspension system must be shut off prior to hoisting, jacking or towing an air suspension vehicle. Failure to shut the system off may lead to an unexpected inflation or deflation of the air springs, which may result in a shift of the vehicle.

1. Before servicing the vehicle, refer to the precautions in the beginning of this section.
2. Remove the center cap.
3. Break the front wheel lug nuts loose.
4. Remove or disconnect the following:
   - Front wheels
   - Front brake hose bolt and the copper washers and plug the front brake hose
   - 2 front disc brake caliper slide pins, then lift the caliper off of the front caliper anchor plate

To install:

5. Install or connect the following:
   - Front disc brake caliper onto the caliper anchor plate. Install the 2 slide pins. Tighten the slider pins/bolts to 42 ft. lbs. (56 Nm) on Excursion models or 21–26 ft. lbs. (28–36 Nm) on all other models.
   - Front brake hose to the brake caliper, using new copper washers. Tighten the retaining bolt to 26 ft. lbs. (35 Nm) on Excursion models or 23–29 ft. lbs. (30–40 Nm) on all other models
6. Bleed the brake system.
7. Clean the wheel hub mounting surface.
8. Install the front wheels and snug the lug nuts to fully seat the wheel against the hub.
9. Lower the vehicle until some of the vehicle's weight rests on the front tires, then tighten the lug.
10. Lower the vehicle completely.
11. Make sure that the brakes are operating correctly.

2003–05 MODELS

⁂ WARNING

The electrical power to the air suspension system must be shut off prior to hoisting, jacking or towing an air suspension vehicle. Failure to shut the system off may lead to an unexpected inflation or deflation of the air springs, which may result in a shift of the vehicle.

1. Before servicing the vehicle, refer to the precautions in the beginning of this section.
2. Remove or disconnect the following:
   - Wheels
   - Brake pads
   - Front brake hose bolt and the copper washers and plug the front brake hose
   - 2 front disc brake caliper bolts, then lift the caliper off of the front caliper anchor plate

To install:

3. Install or connect the following:
   - 2 front disc brake caliper bolts, tighten the bolts to 41 ft. lbs. (55 Nm)
   - Front brake hose to the brake caliper, using new copper washers. Tighten the retaining bolt to 26 ft. lbs. (35 Nm).
   - Brake pads
   - Wheels

Rear

2001–02 MODELS–EXCEPT EXCURSION

⁂ WARNING

The electrical power to the air suspension system must be shut off prior to hoisting, jacking or towing an air suspension vehicle. Failure to shut the system off may lead to an unexpected inflation or deflation of the air springs, which may result in a shift of the vehicle.

1. Before servicing the vehicle, refer to the precautions in the beginning of this section.
2. Remove enough brake fluid from the brake master cylinder reservoir until it is ½ full.
3. Remove or disconnect the following:
   - Wheels
   - Banjo bolt connecting the brake hose to the disc brake caliper and plug the brake hose. Discard 2 copper sealing washers.
   - 2 brake caliper slide pins and lift the caliper off the anchor plate

To install:

4. Retract the disc brake caliper pistons fully in the piston bores using an old brake pad or block of wood and a C-clamp or equivalent.
5. Install or connect the following:
   - Disc brake caliper above the rotor and install it with a rotating motion. Make sure the inner and outer pads are properly positioned and the anti-rattle clips are correctly installed. The brake caliper bleed screw should be positioned on top of the caliper when assembled on the vehicle.
6. Lubricate the locating pins and the inside of the insulators with silicone grease.
   - Locating pins through the caliper insulators and hand-start the threads into the steering knuckle attaching holes. Tighten the locating pins to 20 ft. lbs. (27 Nm).
   - Brake hose to the disc brake caliper using 2 new copper sealing washers. Tighten the banjo bolt to 29 ft. lbs. (40 Nm).
7. Bleed the brake system, filling the brake master cylinder reservoir as required.
   - Wheels
8. Lower the vehicle.
9. Pump the brake pedal several times to position the brake pads prior to moving the vehicle.
10. Road-test the vehicle and check for proper brake system operation.

2001–02 MODELS–EXCURSION

⁂ WARNING

The electrical power to the air suspension system must be shut off prior to hoisting, jacking or towing an air suspension vehicle. Failure to shut the system off may lead to an unexpected inflation or deflation of the air springs, which may result in a shift of the vehicle.

1. Before servicing the vehicle, refer to the precautions in the beginning of this section.
2. Remove enough brake fluid from the brake master cylinder reservoir until it is ½ full.

3. Remove or disconnect the following:
- Wheels
- Stone shield
- 2 brake caliper slide pins and lift the caliper off the anchor plate
- Banjo bolt connecting the brake hose to the disc brake caliper and plug the brake hose. Discard 2 copper sealing washers.

To install:

4. Retract the disc brake caliper pistons fully in the piston bores using an old brake pad or block of wood and a C-clamp or equivalent.
5. Install or connect the following:
- Disc brake caliper above the rotor and install it with a rotating motion. Make sure the inner and outer pads are properly positioned and the anti-rattle clips are correctly installed. The brake caliper bleed screw should be positioned on top of the caliper when assembled on the vehicle.
- Brake hose to the disc brake caliper using 2 new copper sealing washers. Tighten the banjo bolt to 37 ft. lbs. (50 Nm).

6. Lubricate the locating pins and the inside of the insulators with silicone grease.
- Locating pins through the caliper insulators and hand-start the threads into the steering knuckle attaching holes. Tighten the locating pins to 27–41 ft. lbs. (36–55 Nm).

7. Bleed the brake system, filling the brake master cylinder reservoir as required.
- Wheels

8. Lower the vehicle.
9. Pump the brake pedal several times to position the brake pads prior to moving the vehicle.
10. Road-test the vehicle and check for proper brake system operation.

2003–05 MODELS

⁂ WARNING

The electrical power to the air suspension system must be shut off prior to hoisting, jacking or towing an air suspension vehicle. Failure to shut the system off may lead to an unexpected inflation or deflation of the air springs, which may result in a shift of the vehicle.

1. Before servicing the vehicle, refer to the precautions in the beginning of this section.
2. Remove or disconnect the following:
- Wheels
- Stone shield
- 2 rear disc brake caliper bolts, then lift the caliper off of the front caliper anchor plate
- Rear brake hose bolt and the copper washers and plug the front brake hose

To install:

3. Install or connect the following:
- 2 rear disc brake caliper bolts, tighten the bolts to 26 ft. lbs. (35 Nm)
- Rear brake hose to the brake caliper, using new copper washers. Tighten the retaining bolt to 37 ft. lbs. (50 Nm) on Excursion models or 26 ft. lbs. (35 Nm) on Expedition/Navigator models.
- Brake pads
- Wheels

Disc Brake Pads

REMOVAL & INSTALLATION

Front

2001–02 MODELS

1. Before servicing the vehicle, refer to the precautions in the beginning of this section.
2. Break the front wheel lug nuts loose, then raise and support the front of the vehicle safely on jackstands.
3. Remove the front wheels.
4. Remove the front disc brake calipers.
5. Note the position and orientation of the brake pads and anti-rattle clip. Remove the brake shoes and lings from the front disc brake caliper anchor plate, then remove the anti-rattle clips.

To install:

6. Thoroughly clean the caliper and spindle sliding areas.
7. Place a new anti-rattle clip on the lower end of the inboard shoe. Make sure the tabs on the clip are positioned correctly and the loop-type spring is away from the rotor.
8. Place the lower end of the inner brake pad in the spindle assembly pad abutment, against the anti-rattle clip and slide the upper end of the pad into position. Be sure the clip is still in position.
9. Check and make sure the caliper piston is fully bottomed in the cylinder bore. Use a large C-clamp, bearing on a piece of wood, to bottom the piston, if necessary.
10. Position the outer brake pad on the caliper and press the pad tabs into place with your fingers. If the pad cannot be pressed into place by hand, use a C-clamp. Be careful not to damage the lining with the clamp. Bend the tabs to prevent rattling.
11. Lightly lubricate the caliper sliding grooves with caliper pin grease.
12. Install the brake caliper onto the anchor plate.
13. Bleed the brake system.
14. Clean the wheel hub mounting surface.
15. Install the front wheels and snug the lug nuts to fully seat the wheel against the hub.
16. Lower the vehicle until some of the vehicle's weight rests on the front tires, then tighten the lug nuts.
17. Lower the vehicle completely.
18. Make sure that the brakes are operating correctly.

2003–05 MODELS

⁂ WARNING

The electrical power to the air suspension system must be shut off prior to hoisting, jacking or towing an air suspension vehicle. Failure to shut the system off may lead to an unexpected inflation or deflation of the air springs, which may result in a shift of the vehicle.

1. Before servicing the vehicle, refer to the precautions in the beginning of this section.
2. Using a suitable suction device, remove the brake fluid in the master cylinder reservoir until it is half filled.
3. Remove or disconnect the following:
- Wheels
- Anchor housing spring.

➡Ensure the anchor housing spring has one end with two tabs. If yes, the Left Hand (LH) side anchor housing spring must be installed with the two–tabbed end in the upper brake caliper cavity.

4. On the LH brake caliper, release the lower portion of the anchor housing spring as follows:
   a. Apply force at the center of the anchor housing spring and pull outward at the bottom of the anchor housing spring to remove it from the lower brake caliper cavity.
   b. Rotate the spring upward then remove it from the brake caliper.

➡On the Right Hand (RH) side, the anchor housing spring must be installed with the two–tabbed end in the lower brake caliper cavity.

5. For the RH brake caliper, release the upper portion of the anchor housing spring as follows:

a. Apply force at the center of the anchor housing spring and pull outward at the top of the anchor housing spring to remove it from the upper brake caliper cavity.

b. Rotate the spring downward then remove it from the brake caliper.

➡Never allow the brake caliper to hang from the brake hose.

6. Remove the brake caliper as follows:

a. Remove the zero-drag spring.

b. Remove and discard the brake caliper-to-anchor plate bolts, guide pins and boots.

7. Remove the brake pads from the brake caliper.

8. Compress the brake caliper pistons using a C–clamp or other suitable tool.

To install:

9. Clean the inner surfaces of the brake caliper where the brake pads attach.

10. Install new brake hardware as follows:

a. Install the guide pin bushings into the caliper bores.

b. Apply grease to the inside of the guide pin bushing. Do not apply grease to the guide pin threads.

c. Push the guide pins into the bushing.

11. Install or connect the following:

- Inboard brake pad into the brake caliper.
- Outboard brake pad into the brake caliper
- Brake caliper on the brake disc. Tighten the guide pins and install the dust caps.
- Zero-drag spring

➡If present, the 2-tabbed end of the anchor housing spring must be installed first.

12. Install the anchor housing spring as follows:

a. Insert tab of the anchor housing spring into the brake caliper cavity.

b. Twist tab into the brake caliper cavity (LH side-upper brake caliper cavity/RH side-lower brake caliper cavity).

c. Rotate the anchor housing spring and position the upper portion onto the anchor plate.

d. Position the other anchor housing spring portion onto the brake caliper anchor plate.

e. Push down and inward until the upper and lower ends of the anchor housing spring are latched and seated in the brake caliper cavities.

13. Verify that the anchor housing spring is correctly latched.

14. Bleed the brake system, filling the brake master cylinder reservoir as required.

15. Install the wheels.

16. Make sure that the brakes are operating correctly.

Rear

2001–02 MODELS

⁂ WARNING

The electrical power to the air suspension system must be shut off prior to hoisting, jacking or towing an air suspension vehicle. Failure to shut the system off may lead to an unexpected inflation or deflation of the air springs, which may result in a shift of the vehicle.

1. Before servicing the vehicle, refer to the precautions in the beginning of this section.

2. Remove enough brake fluid from the brake master cylinder reservoir until it is ½ full.

3. Remove or disconnect the following:

- Wheels
- Banjo bolt connecting the brake hose to the disc brake caliper and plug the brake hose. Discard 2 copper sealing washers.
- 2 brake caliper slide pins and lift the caliper off the anchor plate

To install:

4. Retract the disc brake caliper pistons fully in the piston bores using an old brake pad or block of wood and a C-clamp or equivalent.

5. Install or connect the following:

- Disc brake caliper above the rotor and install it with a rotating motion. Make sure the inner and outer pads are properly positioned and the anti-rattle clips are correctly installed. The brake caliper bleed screw should be positioned on top of the caliper when assembled on the vehicle.

6. Lubricate the locating pins and the inside of the insulators with silicone grease.

- Locating pins through the caliper insulators and hand-start the threads into the steering knuckle attaching holes. Tighten the locating pins to 20 ft. lbs. (27 Nm).
- Brake hose to the disc brake caliper using 2 new copper sealing washers. Tighten the banjo bolt to 29 ft. lbs. (40 Nm).

7. Bleed the brake system, filling the brake master cylinder reservoir as required.

- Wheels

8. Lower the vehicle.

9. Pump the brake pedal several times to position the brake pads prior to moving the vehicle.

10. Road-test the vehicle and check for proper brake system operation.

2003–05 MODELS

⁂ WARNING

The electrical power to the air suspension system must be shut off prior to hoisting, jacking or towing an air suspension vehicle. Failure to shut the system off may lead to an unexpected inflation or deflation of the air springs, which may result in a shift of the vehicle.

1. Before servicing the vehicle, refer to the precautions in the beginning of this section.

2. Remove enough brake fluid from the brake master cylinder reservoir until it is ½ full.

3. Remove or disconnect the following:

- Wheels
- Anchor housing spring by squeezing at the center of the spring until it unlatches from the brake caliper at both ends, then rotate the spring to remove it from the caliper housing
- Caps and brake caliper bolts
- Brake caliper without disconnecting the brake hose
- Brake pads

4. Thoroughly clean the areas of the caliper and caliper support assembly which contact each other during the sliding action of the caliper.

To install:

5. Compress the caliper piston using a C–clamp or other suitable tool.

6. Install or connect the following:

- Brake shoes on the disc brake caliper support bracket
- Anchor housing spring by placing the upper anchor housing spring end into the brake caliper cavity, then rotate the anchor housing spring and position the lower arm onto the anchor plate
- Upper arm onto the anchor plate, then press down and inward until it is correctly seated and latched into the brake caliper cavities
- Rear disc brake caliper onto the rear support bracket

7. Bleed the brake system, filling the brake master cylinder reservoir as required.

- Front wheels

8. Make sure that the brakes are operating correctly.

FORD

Focus

9

SPECIFICATION CHARTS

ENGINE AND VEHICLE IDENTIFICATION

Engine							Model Year	
Code ①	Liters (cc)	Cu. In.	Cyl.	Fuel Sys.	Type	Eng. Mfg.	Code ②	Year
3	2.0 (1988)	121	4	SFI	DOHC	Ford	1	2001
P	2.0 (1988)	121	4	SFI	SOHC	Ford	2	2002
5	2.0 (1988)	121	4	SFI	DOHC	Ford	3	2003
Z	2.3 (2261)	138	4	SFI	DOHC	Ford	4	2004
							5	2005

SFI: Sequential Fuel Injection

DOHC: Double Overhead Camshafts

SOHC: Single Overhead Camshaft

① 8th digit of the VIN

② 10th digit of the VIN

67197-FOCU-C01

GENERAL ENGINE SPECIFICATIONS

Year	Model	Engine Displacement Liters	Engine ID/VIN	Net Horsepower @ rpm	Net Torque @ rpm (ft. lbs.)	Bore x Stroke (in.)	Com-pression Ratio	Oil Pressure @ rpm
2001	Focus LX	2.0	P	110@5000	125@3750	3.34x3.46	9.2:1	35-65@2000
	Focus SE	2.0	P	110@5000	125@3750	3.34x3.46	9.2:1	35-65@2000
	Focus SE	2.0	3	130@5500	130@4000	3.34x3.46	10.0:1	35-65@2000
	Focus ZTS	2.0	3	130@5500	130@4000	3.34x3.46	10.0:1	35-65@2000
	Focus ZX3	2.0	3	130@5500	130@4000	3.34x3.46	10.0:1	35-65@2000
2002	Focus LX	2.0	P	110@5000	125@3750	3.34x3.46	9.2:1	35-65@2000
	Focus SE	2.0	P	110@5000	125@3750	3.34x3.46	9.2:1	35-65@2000
	Focus SE	2.0	3	130@5500	130@4000	3.34x3.46	10.0:1	35-65@2000
	Focus ZTS	2.0	3	130@5500	130@4000	3.34x3.46	10.0:1	35-65@2000
	Focus SVT	2.0	5	170@7000	198@5500	3.34x3.46	10.2:1	35-65@2000
	Focus ZX3	2.0	3	130@5500	130@4000	3.34x3.46	10.0:1	35-65@2000
2003	Focus LX	2.0	P	110@5000	125@3750	3.34x3.46	9.35:1	35-65@2000
	Focus SE	2.0	3	130@5500	135@4500	3.34x3.46	9.6:1	35-65@2000
	Focus ZTS	2.3	Z	144@5750	149@4200	3.44x3.70	9.9:1	35-65@2000
	Focus ZX3	2.0	3	130@5500	135@4500	3.34x3.46	9.6:1	35-65@2000
	Focus ZX5	2.0	3	130@5500	135@4500	3.34x3.46	9.6:1	35-65@2000
	Focus ZX3	2.3	Z	144@5750	149@4200	3.44x3.70	9.9:1	35-65@2000
	Focus ZX5	2.3	Z	144@5750	149@4200	3.44x3.70	9.9:1	35-65@2000
	Focus SVT	2.0	5	170@7000	198@5500	3.34x3.46	10.2:1	35-65@2000
	Focus ZTW	2.3	Z	144@5750	149@4200	3.44x3.70	9.9:1	35-65@2000
2004	Focus LX	2.0	P	110@5000	125@3750	3.34x3.46	9.35:1	35-65@2000
	Focus SE	2.0	3	130@5500	135@4500	3.34x3.46	9.6:1	35-65@2000
	Focus ZTS	2.3	Z	144@5750	149@4200	3.44x3.70	9.9:1	35-65@2000
	Focus ZX3	2.0	3	130@5500	135@4500	3.34x3.46	9.6:1	35-65@2000
	Focus ZX5	2.0	3	130@5500	135@4500	3.34x3.46	9.6:1	35-65@2000
	Focus ZX3	2.3	Z	144@5750	149@4200	3.44x3.70	9.9:1	35-65@2000
	Focus ZX5	2.3	Z	144@5750	149@4200	3.44x3.70	9.9:1	35-65@2000
	Focus SVT	2.0	5	170@7000	198@5500	3.34x3.46	10.2:1	35-65@2000
	Focus ZTW	2.3	Z	144@5750	149@4200	3.44x3.70	9.9:1	35-65@2000

SFI: Sequential Fuel Injection

67197-FOCU-C02

ENGINE TUNE-UP SPECIFICATIONS

Year	Engine Displacement Liters	Engine ID/VIN	Spark Plug Gap (in.)	Ignition Timing (deg.) MT	Ignition Timing (deg.) AT	Fuel Pump (psi)	Idle Speed (rpm) MT	Idle Speed (rpm) AT	Valve Clearance (in.) In.	Valve Clearance (in.) Ex.
2001	2.0	3	0.052-0.056	10B	10B	31-38 ①	②	②	0.004-0.007	0.010-0.013
	2.0	P	0.052-0.056	10B	10B	31-38 ①	②	②	HYD	HYD
2002	2.0	3	0.052-0.056	10B	10B	31-38 ①	②	②	0.004-0.007	0.010-0.013
	2.0	5	0.041-0.045	10B	10B	31-38 ①	②	②	0.008-0.011	0.013-0.015
	2.0	P	0.052-0.056	10B	10B	31-38 ①	②	②	HYD	HYD
2003	2.0	3	0.052-0.056	10B	10B	31-38 ①	②	②	0.004-0.007	0.010-0.013
	2.0	P	0.052-0.056	10B	10B	31-38 ①	②	②	HYD	HYD
	2.0	5	0.041-0.045	10B	10B	31-38 ①	②	②	0.008-0.011	0.013-0.015
	2.3	Z	0.041-0.045	10B	10B	31-38 ①	②	②	0.008-0.011	0.010-0.013
2004	2.0	3	0.052-0.056	10B	10B	31-38 ①	②	②	0.004-0.007	0.010-0.013
	2.0	P	0.052-0.056	10B	10B	31-38 ①	②	②	HYD	HYD
	2.0	5	0.041-0.045	10B	10B	31-38 ①	②	②	0.008-0.011	0.013-0.015
	2.3	Z	0.041-0.045	10B	10B	31-38 ①	②	②	0.008-0.011	0.010-0.013

NOTE: The Vehicle Emission Control Information label often reflects specification changes made during production. The label figures must be used if they differ from those in this chart.

B: Before Top Dead Center

HYD: Hydraulic

① Fuel pressure with engine running, pressure regulator vacuum hose connected

② Refer to Vehicle Emission Control Information label

67197-FOCU-C03

2.0L (VIN P) Engine
Firing order: 1–3–4–2
Distributorless ignition system

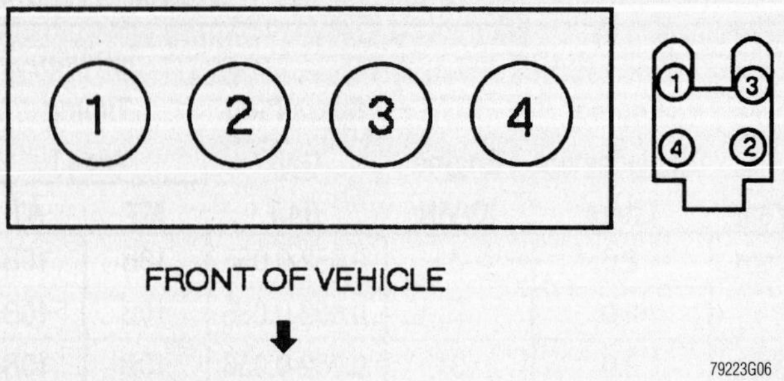

2.0L (VIN 3) Engine
Firing order: 1–3–4–2
Distributorless ignition system

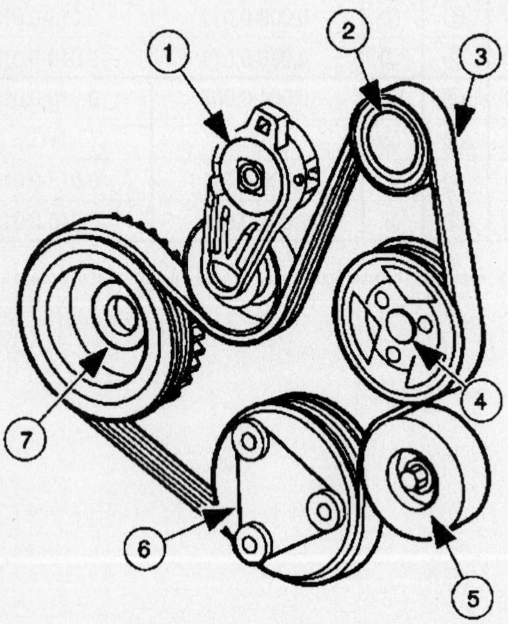

1	Belt tensioner
2	Generator pulley
3	Belt
4	Power steering pump pulley
5	Belt idler pulley
6	A/C clutch pulley
7	Crankshaft damper

93004G03

Accessory drive belt routing—2.0L (VIN P) engine with A/C

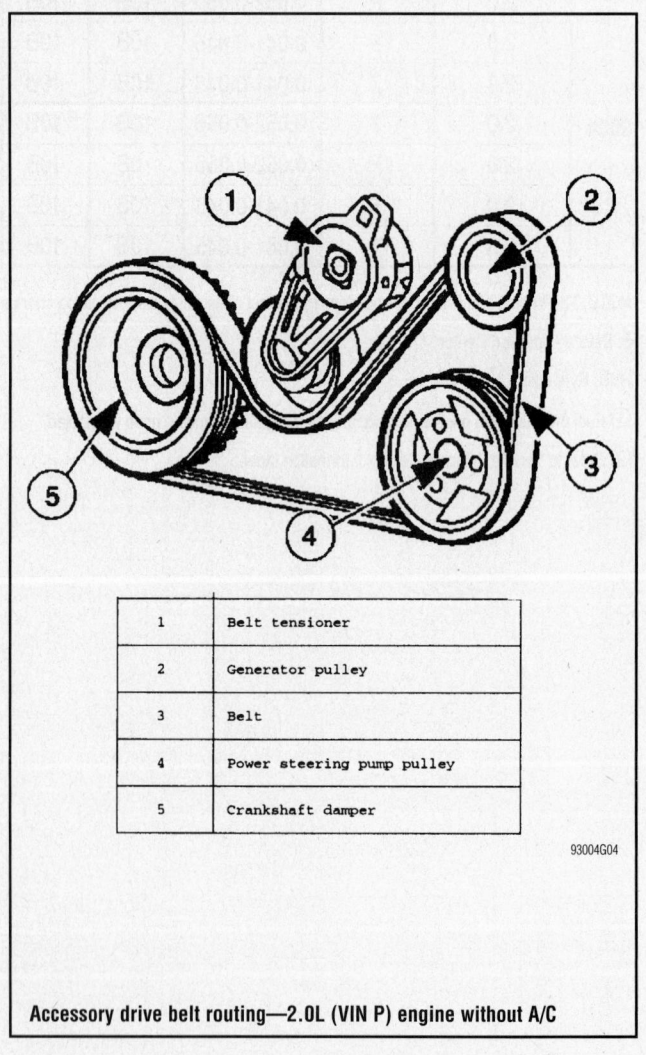

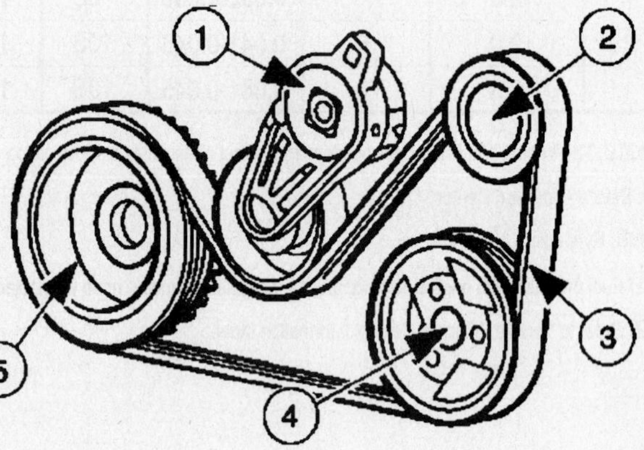

1	Belt tensioner
2	Generator pulley
3	Belt
4	Power steering pump pulley
5	Crankshaft damper

93004G04

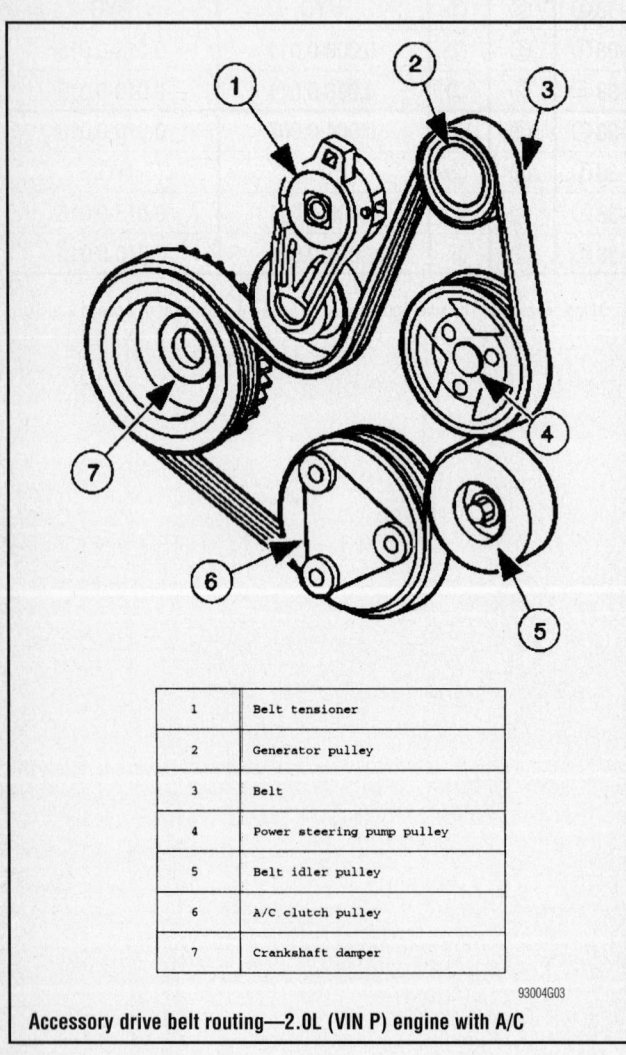

Accessory drive belt routing—2.0L (VIN P) engine without A/C

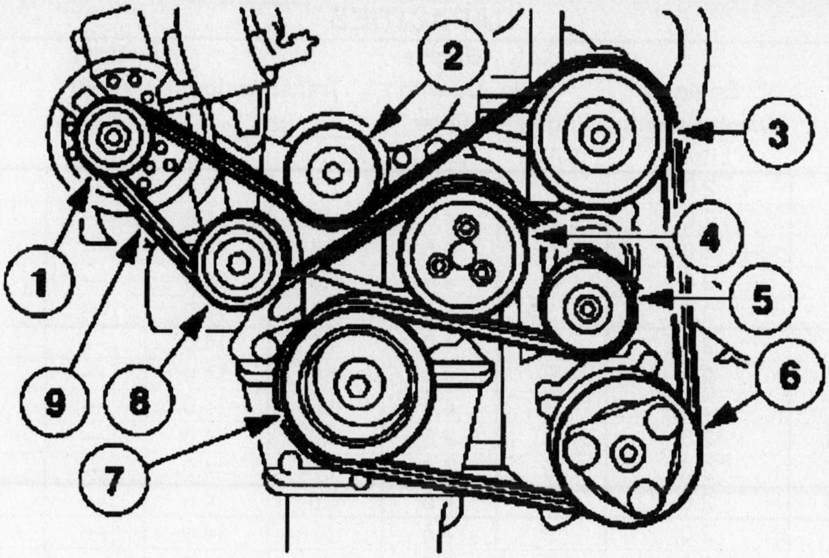

1	Generator pulley
2	Belt idler pulley
3	Power steering pump pulley
4	Water pump pulley
5	Belt tensioner
6	A/C clutch pulley
7	Crankshaft pulley damper
8	Belt idler pulley
9	Drive belt

Accessory drive belt routing—2.0L (VIN 3) engine

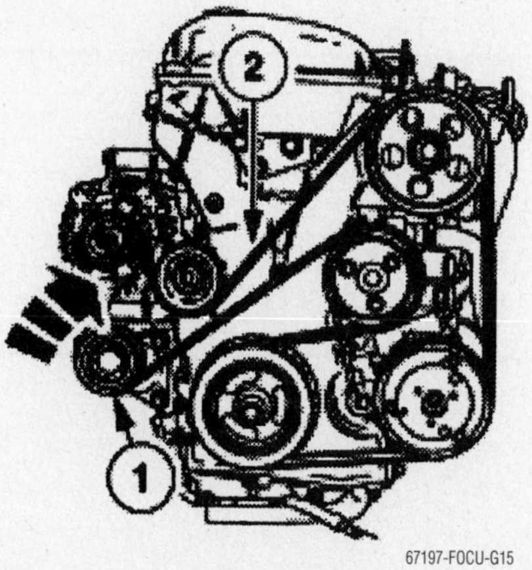

Accessory drive belt routing—2.3L (VIN Z) engine

CAPACITIES

Year	Model	Engine Displacement Liters	Engine ID/VIN	Engine Oil with Filter (qts.)	Transmission (pts.) Manual	Transmission (pts.) Auto.	Drive Front Axle (pts.)	Fuel Tank (gal.)	Cooling System (qts.)
2001	Focus LX	2.0	P	4.0	4.8	14.0	—	13.2	①
	Focus SE	2.0	P	4.0	4.8	14.0	—	13.2	①
	Focus SE	2.0	3	4.5	4.0	14.0	—	13.2	②
	Focus ZTS	2.0	3	4.5	4.0	14.0	—	13.2	②
	Focus ZX3	2.0	3	4.5	4.0	14.0	—	13.2	②
2002	Focus LX	2.0	P	4.0	4.8	14.0	—	13.2	①
	Focus SE	2.0	P	4.0	4.8	14.0	—	13.2	①
	Focus SE	2.0	3	4.5	4.0	14.0	—	13.2	②
	Focus ZTS	2.0	3	4.5	4.0	14.0	—	13.2	②
	Focus SVT	2.0	5	4.5	③	14.0	—	13.2	②
	Focus ZX3	2.0	3	4.5	4.0	14.0	—	13.2	②
2003	Focus LX	2.0	P	4.5	③	14.0	—	13.2	①
	Focus SE	2.0	3	4.5	③	14.0	—	13.2	②
	Focus ZTS	2.0	3	4.5	③	14.0	—	13.2	②
	Focus ZX3	2.0	3	4.5	③	14.0	—	13.2	②
	Focus ZX3	2.3	Z	4.5	③	14.0	—	14.0	②
	Focus ZX5	2.3	Z	4.5	③	14.0	—	14.0	②
	Focus SVT	2.0	5	4.5	③	14.0	—	13.2	②
	Focus ZTW	2.3	Z	4.5	③	14.0	—	14.0	②
2004	Focus LX	2.0	P	4.5	③	14.0	—	13.2	①
	Focus SE	2.0	3	4.5	③	14.0	—	13.2	②
	Focus ZTS	2.0	3	4.5	③	14.0	—	13.2	②
	Focus ZX3	2.0	3	4.5	③	14.0	—	13.2	②
	Focus ZX3	2.3	Z	4.5	③	14.0	—	14.0	②
	Focus ZX5	2.3	Z	4.5	③	14.0	—	14.0	②
	Focus SVT	2.0	5	4.5	③	14.0	—	13.2	②
	Focus ZTW	2.3	Z	4.5	③	14.0	—	14.0	②

Note: All capacities are approximates. Add fluid gradually and ensure a proper fluid level is obtained.

① Manual transaxle: 7.9
Automatic transaxle: 5.8

② Manual transaxle: 7.0
Automatic transaxle: 7.5

③ IB5: 5 pts.
MTX-75: 4 pts.

67197-FOCU-C04

CRANKSHAFT AND CONNECTING ROD SPECIFICATIONS

All measurements are given in inches.

Year	Engine Displacement Liters	Engine ID/VIN	Crankshaft Main Brg. Journal Dia.	Crankshaft Main Brg. Oil Clearance	Crankshaft Shaft End-play	Crankshaft Thrust on No.	Connecting Rod Journal Diameter	Connecting Rod Oil Clearance	Connecting Rod Side Clearance
2000	2.0	3	2.2827-2.2835	0.0008-0.0026	0.0040-0.0120	3	1.8461-1.8500	0.0006-0.0028	0.0040-0.0110
	2.0	P	2.2827-2.2835	0.0008-0.0026	0.0040-0.0120	3	1.7279-1.7287	0.0008-0.0026	0.0040-0.0110
2001	2.0	3	2.2827-2.2835	0.0008-0.0026	0.0040-0.0120	3	1.8461-1.8500	0.0006-0.0028	0.0040-0.0110
	2.0	P	2.2827-2.2835	0.0008-0.0026	0.0040-0.0120	3	1.7279-1.7287	0.0008-0.0026	0.0040-0.0110
2002	2.0	3	2.2827-2.2835	0.0008-0.0026	0.0040-0.0120	3	1.8461-1.8500	0.0006-0.0028	0.0040-0.0110
	2.0	5	2.2839-2.2849	0.0004-0.0023	0.0035-0.0102	3	1.8475-1.8488	0.0006-0.0027	0.0035-0.0126
	2.0	P	2.2827-2.2835	0.0008-0.0026	0.0040-0.0120	3	1.7279-1.7287	0.0008-0.0026	0.0040-0.0110
2003	2.0	3	2.2827-2.2835	0.0008-0.0026	0.0040-0.0120	3	1.8461-1.8500	0.0006-0.0028	0.0040-0.0110
	2.0	P	2.2827-2.2835	0.0008-0.0026	0.0040-0.0120	3	1.7279-1.7287	0.0008-0.0026	0.0040-0.0110
	2.0	5	2.2839-2.2849	0.0004-0.0023	0.0035-0.0102	3	1.8475-1.8488	0.0006-0.0027	0.0035-0.0126
	2.3	Z	2.0460-2.0470	0.0007-0.0013	0.0080-0.0160	3	2.0870-2.0880	0.0010-0.0020	0.0050-0.0140
2004	2.0	3	2.2827-2.2835	0.0008-0.0026	0.0040-0.0120	3	1.8461-1.8500	0.0006-0.0028	0.0040-0.0110
	2.0	P	2.2827-2.2835	0.0008-0.0026	0.0040-0.0120	3	1.7279-1.7287	0.0008-0.0026	0.0040-0.0110
	2.0	5	2.2839-2.2849	0.0004-0.0023	0.0035-0.0102	3	1.8475-1.8488	0.0006-0.0027	0.0035-0.0126
	2.3	Z	2.0460-2.0470	0.0007-0.0013	0.0080-0.0160	3	2.0870-2.0880	0.0010-0.0020	0.0050-0.0140

67197-FOCU-C05

VALVE SPECIFICATIONS

Year	Engine Displacement Liters	Engine ID/VIN	Seat Angle (deg.)	Face Angle (deg.)	Spring Test Pressure (lbs. @ in.)	Spring Installed Height (in.)	Stem-to-Guide Clearance (in.)		Stem Diameter (in.)	
							Intake	Exhaust	Intake	Exhaust
2001	2.0	3	45	45	NA	1.346	0.0007-0.0025	0.0014-0.0032	0.2373-0.2379	0.2366-0.2372
	2.0	P	45	45.6	200@1.09	1.420-1.540	0.0008-0.0027	0.0018-0.0037	0.3159-0.3167	0.3149-0.3156
2002	2.0	3	45	45	NA	1.346	0.0007-0.0025	0.0014-0.0032	0.2373-0.2379	0.2366-0.2372
	2.0	5	NA	NA	NA	NA	0.0007-0.0025	0.0007-0.0025	NA	NA
	2.0	P	45	45.6	200@1.09	1.420-1.540	0.0008-0.0027	0.0018-0.0037	0.3159-0.3167	0.3149-0.3156
2003	2.0	3	45	45	NA	1.346	0.0007-0.0025	0.0014-0.0032	0.2373-0.2379	0.2366-0.2372
	2.0	P	45	45.6	200@1.09	1.420-1.540	0.0008-0.0027	0.0018-0.0037	0.3159-0.3167	0.3149-0.3156
	2.0	5	NA	NA	NA	NA	0.0007-0.0025	0.0007-0.0025	NA	NA
	2.3	Z	45	45	38.667@1.492	1.492	0.0009	0.0011	0.2153-0.2159	0.2151-0.2157
2004	2.0	3	45	45	NA	1.346	0.0007-0.0025	0.0014-0.0032	0.2373-0.2379	0.2366-0.2372
	2.0	P	45	45.6	200@1.09	1.420-1.540	0.0008-0.0027	0.0018-0.0037	0.3159-0.3167	0.3149-0.3156
	2.0	5	NA	NA	NA	NA	0.0007-0.0025	0.0007-0.0025	NA	NA
	2.3	Z	45	45	38.667@1.492	1.492	0.0009	0.0011	0.2153-0.2159	0.2151-0.2157

NA: Not Available

67197-FOCU-C06

PISTON AND RING SPECIFICATIONS

All measurements are given in inches.

Year	Engine Displacement Liters	Engine ID/VIN	Piston Clearance	Ring Gap Top Compression	Ring Gap Bottom Compression	Ring Gap Oil Control	Ring Side Clearance Top Compression	Ring Side Clearance Bottom Compression	Ring Side Clearance Oil Control
2001	2.0	3	0.0010-0.0022	0.012-0.022	0.012-0.022	0.010-0.039	0.0015-0.0032	0.0015-0.0035	—
	2.0	P	0.0008-0.0027	0.010-0.030	0.010-0.030	0.016-0.066	0.0015-0.0032	0.0015-0.0035	—
2002	2.0	3	0.0010-0.0022	0.012-0.022	0.012-0.022	0.010-0.039	0.0015-0.0032	0.0015-0.0035	—
	2.0	5	0.0004-0.0011	0.011-0.019	0.011-0.019	0.015-0.055	NA	NA	NA
	2.0	P	0.0008-0.0027	0.010-0.030	0.010-0.030	0.016-0.066	0.0015-0.0032	0.0015-0.0035	—
2003	2.0	3	0.0010-0.0022	0.012-0.022	0.012-0.022	0.010-0.039	0.0015-0.0032	0.0015-0.0035	—
	2.0	P	0.0008-0.0027	0.010-0.030	0.010-0.030	0.016-0.066	0.0015-0.0032	0.0015-0.0035	—
	2.0	5	0.0004-0.0011	0.011-0.019	0.011-0.019	0.015-0.055	NA	NA	NA
	2.3	Z	0.0009-0.0017	0.006-0.012	0.012-0.018	0.007-0.027	0.0008-0.0013	0.0004-0.0011	0.0024-0.0055
2004	2.0	3	0.0010-0.0022	0.012-0.022	0.012-0.022	0.010-0.039	0.0015-0.0032	0.0015-0.0035	—
	2.0	P	0.0008-0.0027	0.010-0.030	0.010-0.030	0.016-0.066	0.0015-0.0032	0.0015-0.0035	—
	2.0	5	0.0004-0.0011	0.011-0.019	0.011-0.019	0.015-0.055	NA	NA	NA
	2.3	Z	0.0009-0.0017	0.006-0.012	0.012-0.018	0.007-0.027	0.0008-0.0013	0.0004-0.0011	0.0024-0.0055

NA: Not Available

67197-FOCU-C07

TORQUE SPECIFICATIONS

All readings in ft. lbs.

Year	Engine Displacement Liters	Engine ID/VIN	Cylinder Head Bolts	Main Bearing Bolts	Rod Bearing Bolts	Crankshaft Damper Bolts	Flywheel Bolts	Manifold Intake	Manifold Exhaust	Spark Plugs	Oil Pan Drain
2001	2.0	P	①	66-79	26-30	80-87	71-76	15-22	15-17	12-15	18
	2.0	3	②	55-65	③	80-87	71-76	11-12	10-12	10-12	18
2002	2.0	P	①	66-79	26-30	80-87	71-76	15-22	15-17	12-15	18
	2.0	5	②	④	⑤	85	82	⑥	⑦	11	18
	2.0	3	②	55-65	③	80-87	71-76	11-12	10-12	10-12	18
2003	2.0	P	①	66-79	26-30	80-87	71-76	15-22	15-17	12-15	18
	2.0	3	②	55-65	③	80-87	71-76	11-12	10-12	10-12	18
	2.0	5	②	④	⑤	85	82	⑥	⑦	11	18
	2.3	Z	⑧	NA	NA	⑨	⑩	13	⑪	11	21
2004	2.0	P	①	66-79	26-30	80-87	71-76	15-22	15-17	12-15	18
	2.0	3	②	55-65	③	80-87	71-76	11-12	10-12	10-12	18
	2.0	5	②	④	⑤	85	82	⑥	⑦	11	18
	2.3	Z	⑧	NA	NA	⑨	⑩	13	⑪	11	21

NA: Not Available

① Step 1: 37 ft. lbs.
Step 2: Loosen bolts 1/2 turn
Step 3: 37 ft. lbs.
Step 4: Plus 90 degrees
Step 5: Plus 90 degrees

② Step 1: 15 ft. lbs.
Step 2: 30 ft. lbs.
Step 3: Plus 90 degrees

③ Step 1: 22-25 ft. lbs.
Step 2: Plus 90 degrees

④ Step 1: 18 ft. lbs.
Step 2: Plus 60 degrees

⑤ Step 1: 26 ft. lbs.
Step 2: Plus 90 degrees

⑥ Studs: 44 inch lbs.
Bolts and nuts: 13 ft. lbs.

⑦ Studs: 44 inch lbs.
Nuts: 12 ft. lbs.
Bolts: 13 ft. lbs.

⑧ Step 1: 44 inch lbs.
Step 2: 11 ft. lbs.
Step 3: 33 ft. lbs.
Step 4: Plus 90 degrees
Step 5: Plus 90 degrees

⑨ Step 1: 74 ft. lbs.
Step 2: Plus 90 degrees

⑩ Step 1: 37 ft. lbs.
Step 2: 50 ft. lbs.
Step 3: 83 ft. lbs.

⑪ Studs: 13 ft. lbs.
Nuts: 40 ft. lbs.

67197-FOCU-C08

WHEEL ALIGNMENT

Year	Model		Caster Range (+/-Deg.)	Caster Preferred Setting (Deg.)	Camber Range (+/-Deg.)	Camber Preferred Setting (Deg.)	Toe-in (in.)
2001	Sedan	F	1.00	+2.93	1.00	-0.36	0 +/- 0.04
		R	—	—	0.30	-0.93	0.08 +/- 0.05
	Wagon	F	1.30	+2.46	1.25	-0.62	0 +/- 0.04
		R	—	—	0.75	-0.62	0.12 +/- 0.05
2002	Sedan	F	1.00	+2.93	1.00	-0.36	0 +/- 0.04
		R	—	—	0.30	-0.93	0.08 +/- 0.05
	Wagon	F	1.30	+2.46	1.25	-0.62	0 +/- 0.04
		R	—	—	0.75	-0.62	0.12 +/- 0.05
2003	Sedan	F	1.00	+2.93	1.00	-0.36	0 +/- 0.04
		R	—	—	0.30	-0.93	0.08 +/- 0.05
	Wagon	F	1.30	+2.46	1.25	-0.62	0 +/- 0.04
		R	—	—	0.75	-0.62	0.12 +/- 0.05
2004	Sedan	F	1.00	+2.93	1.00	-0.36	0 +/- 0.04
		R	—	—	0.30	-0.93	0.08 +/- 0.05
	Wagon	F	1.30	+2.46	1.25	-0.62	0 +/- 0.04
		R	—	—	0.75	-0.62	0.12 +/- 0.05

67197-FOCU-C09

TIRE, WHEEL AND BALL JOINT SPECIFICATIONS

Year	Model	OEM Tires Standard	OEM Tires Optional	Tire Pressures (psi) Front	Tire Pressures (psi) Rear	Wheel Size	Ball Joint Inspection	Lug Nut Torque (ft. lbs.)
2001	Focus LX	P185/65R14	—	32	32	5.5	①	94
	Focus ZX	P195/60/R15	P205/60/R16	32	32	5.5	①	94
2002	Focus LX	P185/65R14	—	32	32	5.5	①	94
	Focus SVT	P215/45ZR/17	—	32	32	5.5	①	94
	Focus ZX	P195/60/R15	P205/60/R16	32	32	5.5	①	94
2003	Focus LX	P185/65R14	—	32	32	5.5	①	94
	Focus SVT	P215/45ZR/17	—	32	32	5.5	①	94
	Focus ZX	P195/60/R15	P205/60/R16	32	32	5.5	①	94
2004	Focus LX	P185/65R14	—	32	32	5.5	①	94
	Focus SVT	P215/45ZR/17	—	32	32	5.5	①	94
	Focus ZX	P195/60/R15	P205/60/R16	32	32	5.5	①	94

OEM: Original Equipment Manufacturer

PSI: Pounds Per Square Inch

① Replace if any measurable movement is found

67197-FOCU-C10

BRAKE SPECIFICATIONS

All measurements in inches unless noted

			Brake Disc			Brake Drum Diameter			Minimum Lining Thickness		Brake Caliper Mounting Bolts
Year	Model		Original Thickness	Minimum Thickness	Maximum Runout	Original Inside Diameter	Max. Wear Limit	Maximum Machine Diameter	Front	Rear	(ft. lbs.)
2001	Focus	F	0.870	0.790	0.002	—	—	—	0.059	—	21
		R	0.390	0.320	0.002	7.99	8.03	—	—	①	26
2002	Focus Except SVT	F	0.870	0.790	0.002	—	—	—	0.059	—	21
		R	0.390	0.320	0.002	7.99	8.03	—	—	①	26
	Focus SVT	F	0.950	0.870	0.002	—	—	—	0.059	—	21
		R	0.390	0.350	0.002	—	—	—	—	0.059	26
2003	Focus Except SVT	F	0.870	0.790	0.002	—	—	—	0.059	—	21
		R	0.390	0.320	0.002	7.99	8.03	—	—	①	26
	Focus SVT	F	0.950	0.870	0.002	—	—	—	0.059	—	21
		R	0.390	0.350	0.002	—	—	—	—	0.059	26
2004	Focus Except SVT	F	0.870	0.790	0.002	—	—	—	0.059	—	21
		R	0.390	0.320	0.002	7.99	8.03	—	—	①	26
	Focus SVT	F	0.950	0.870	0.002	—	—	—	0.059	—	21
		R	0.390	0.350	0.002	—	—	—	—	0.059	26

① Drum brakes: 0.039 in.
Disc brakes: 0.059 in.

67197-FOCU-C11

SCHEDULED MAINTENANCE INTERVALS

Ford—Focus

TO BE SERVICED	TYPE OF SERVICE	VEHICLE MILEAGE INTERVAL (x1000)												
		5	10	15	20	25	30	35	40	45	50	55	60	65
Air cleaner filter	R						✓						✓	
Accessory drive belt	S/I												✓	
Brake system ①	S/I			✓			✓			✓			✓	
Clutch pedal operation	S/I						✓						✓	
Cooling fan operation	S/I		✓		✓		✓		✓		✓		✓	
Cooling system hoses and clamps	S/I			✓			✓			✓			✓	
CV-joint boots & axle seals	S/I						✓						✓	
Engine coolant	R	Ten years or 150,000 miles												
Engine oil & filter	R	✓	✓	✓	✓	✓	✓	✓	✓	✓	✓	✓	✓	✓
Exterior Lights	S/I	Check monthly												
PCV valve	S/I												✓	
Exhaust system & heat shields	S/I						✓						✓	
Parking brake system	S/I	Every 6 months												
Power steering fluid	S/I	Every 6 months												
Rotate tires	S/I	✓		✓		✓		✓		✓		✓		✓
Steering linkage	S/I						✓						✓	
Spark plugs	R	Change at 100,000 miles												
Suspension components	S/I						✓						✓	

R: Replace S/I: Inspect and service, if necessary L: Lubricate A: Adjust C: Clean

① Inspect the reservoir fluid level, rotor and or drum, brake lines, hoses, calipers and or wheel cylinders

FREQUENT OPERATION MAINTENANCE (SEVERE SERVICE)

If a vehicle is operated under any of the following conditions it is considered severe service:

- **Extremely dusty areas.**
- **50% or more of the vehicle operation is in 32°C (90°F) or higher temperatures, or constant operation in temperatures below 0°C (32°F).**
- **Prolonged idling (vehicle operation in stop and go traffic).**
- **Frequent short running periods (engine does not warm to normal operating temperatures).**
- **Police, taxi, delivery usage or trailer towing usage.**

Oil & oil filter change: change every 3000 miles.

Air filter element: change every 15,000 miles.

67197-FOCU-C12

PRECAUTIONS

Before servicing any vehicle, please be sure to read all of the following precautions, which deal with personal safety, prevention of component damage, and important points to take into consideration when servicing a motor vehicle:

- Never open, service or drain the radiator or cooling system when the engine is hot; serious burns can occur from the steam and hot coolant.
- Observe all applicable safety precautions when working around fuel. Whenever servicing the fuel system, always work in a well-ventilated area. Do not allow fuel spray or vapors to come in contact with a spark, open flame, or excessive heat (a hot drop light, for example). Keep a dry chemical fire extinguisher near the work area. Always keep fuel in a container specifically designed for fuel storage; also, always properly seal fuel containers to avoid the possibility of fire or explosion. Refer to the additional fuel system precautions later in this section.
- Fuel injection systems often remain pressurized, even after the engine has been turned **OFF**. The fuel system pressure must be relieved before disconnecting any fuel lines. Failure to do so may result in fire and/or personal injury.
- Brake fluid often contains polyglycol ethers and polyglycols. Avoid contact with the eyes and wash your hands thoroughly after handling brake fluid. If you do get brake fluid in your eyes, flush your eyes with clean, running water for 15 minutes. If eye irritation persists, or if you have taken brake fluid internally, IMMEDIATELY seek medical assistance.
- The EPA warns that prolonged contact with used engine oil may cause a number of skin disorders, including cancer. You should make every effort to minimize your exposure to used engine oil. Protective gloves should be worn when changing oil. Wash your hands and any other exposed skin areas as soon as possible after exposure to used engine oil. Soap and water, or waterless hand cleaner should be used.
- All new vehicles are now equipped with an air bag system, often referred to as a Supplemental Restraint System (SRS) or Supplemental Inflatable Restraint (SIR) system. The system must be disabled before performing service on or around system components, steering column, instrument panel components, wiring and sensors. Failure to follow safety and disabling procedures could result in accidental air bag deployment, possible personal injury and unnecessary system repairs.
- Always wear safety goggles when working with, or around, the air bag system. When carrying a non-deployed air bag, be sure the bag and trim cover are pointed away from your body. When placing a non-deployed air bag on a work surface, always face the bag and trim cover upward, away from the surface. This will reduce the motion of the module if it is accidentally deployed. Refer to the additional air bag system precautions later in this section.
- Clean, high quality brake fluid from a sealed container is essential to the safe and proper operation of the brake system. You should always buy the correct type of brake fluid for your vehicle. If the brake fluid becomes contaminated, completely flush the system with new fluid. Never reuse any brake fluid. Any brake fluid that is removed from the system should be discarded. Also, do not allow any brake fluid to come in contact with a painted surface; it will damage the paint.
- Never operate the engine without the proper amount and type of engine oil; doing so WILL result in severe engine damage.
- Timing belt maintenance is extremely important. Many models utilize an interference-type, non-freewheeling engine. If the timing belt breaks, the valves in the cylinder head may strike the pistons, causing potentially serious (also time-consuming and expensive) engine damage.
- Disconnecting the negative battery cable on some vehicles may interfere with the functions of the on-board computer system(s) and may require the computer to undergo a relearning process once the negative battery cable is reconnected.
- When servicing drum brakes, only disassemble and assemble one side at a time, leaving the remaining side intact for reference.

ENGINE REPAIR

➡Disconnecting the negative battery cable on some vehicles may interfere with the functions of the on board computer system. The computer may undergo a relearning process once the negative battery cable is reconnected.

Alternator

REMOVAL

2.0L (VIN P) Engine

1. Before servicing the vehicle, refer to the precautions in the beginning of this section.
2. Remove or disconnect the following:
   - Negative battery cable
   - Accessory drive belt
   - Power steering pipe brackets
   - Exhaust manifold heat shield
   - Alternator harness connectors
   - Alternator

2.0L (VIN 3 and 5) Engines

1. Before servicing the vehicle, refer to the precautions in the beginning of this section.
2. Remove or disconnect the following:
   - Negative battery cable
   - Accessory drive belt
   - Alternator harness connectors
   - Coolant expansion tank
   - Power steering reservoir
   - Engine wiring harness bracket
   - Ground cable
   - Evaporative Emissions (EVAP) canister purge valve
   - Alternator

2.3L Engine

1. Before servicing the vehicle, refer to the precautions in the beginning of this section.
2. Disconnect the negative battery cable.
3. Remove the generator heat shield top nut.
4. Raise and support the vehicle.
5. Remove the screws and the lower splash shield.
6. Release the accessory drive belt tension and remove the belt from the generator pulley.
7. Remove the generator heat shield lower nut and the heat shield.
8. Release the two air intake tube retainers and remove the air intake tube assembly.
9. Remove the generator B+ cable nut and disconnect the B+ cable.

10. Remove the two generator nuts and bolt.

11. Lower the vehicle.

12. Disconnect the generator electrical connector.

13. Remove the upper generator nut.

14. Raise and support the vehicle.

15. Remove the engine roll restrictor mount bolt.

16. Position the engine forward and remove the generator.

INSTALLATION

2.0L (VIN P) Engine

Install or connect the following:

- Alternator. Tighten the bolts to 35 ft. lbs. (45 Nm).
- Alternator harness connectors
- Exhaust manifold heat shield
- Power steering pipe brackets
- Accessory drive belt
- Negative battery cable

2.0L (VIN 3 and 5) Engines

Install or connect the following:

- Alternator. Tighten the bolts to 18 ft. lbs. (25 Nm).
- EVAP canister purge valve
- Ground cable
- Engine wiring harness bracket
- Power steering reservoir
- Coolant expansion tank
- Alternator harness connectors
- Accessory drive belt
- Negative battery cable

2.3L Engine

1. Installation is the reverse of the removal procedure. Use the following torque specifications:
    - Engine roll restrictor bolt: 35 ft. lbs. (48 Nm)
    - Generator nuts and bolt: 35 ft. lbs. (47 Nm)
    - Battery + terminal nut: 53 inch lbs. (6 Nm)
    - Generator heat shield nuts: 18 ft. lbs. (25 Nm)

Ignition Timing

ADJUSTMENT

The engines are equipped with a Distributorless Ignition System (DIS). No adjustment is necessary.

Engine Assembly

REMOVAL & INSTALLATION

Manual Transaxle

2.0L ENGINES EXCEPT SVT

1. Before servicing the vehicle, refer to the precautions in the beginning of this section.

2. Drain the cooling system.

3. Relieve the fuel system pressure.

4. Loosen the front strut center nuts 5 turns.

5. Remove or disconnect the following:
    - Battery, tray and cables
    - Mass Air Flow (MAF) sensor connector
    - Air cleaner housing and tubes
    - Accelerator cable
    - Cruise control cable, if equipped
    - Ignition coil connectors
    - Power steering pump pressure switch connector
    - Heated Oxygen (HO_2S) sensor connectors
    - Fuel injector harness connectors
    - Powertrain Control Module (PCM) connectors
    - Vehicle Speed (VSS) sensor connector
    - Reverse lamp switch connector
    - Clutch slave cylinder fluid pipe
    - Evaporative Emissions (EVAP) canister vacuum lines
    - Power brake booster vacuum line
    - Delta Pressure Feedback Electronic (DPFE) system sensor vacuum line
    - Exhaust Gas Recirculation (EGR) vacuum line
    - Intake manifold vacuum line
    - Fuel line
    - Coolant bypass hoses
    - Radiator hoses
    - Coolant expansion tank
    - Accessory drive belt
    - Power steering pump and reservoir
    - Power steering high pressure pipe
    - Radiator cooling fan
    - Catalytic converter
    - Exhaust flex pipe
    - Drive belt cover
    - Shift cable cover
    - Shift cables
    - A/C compressor, if equipped
    - Right engine mount
    - Lower ball joints
    - Axle halfshafts

6. Support the powertrain from below

7. Remove the front and rear engine mounts.

8. Raise the vehicle away from the powertrain.

To install:

9. Lower the vehicle over the powertrain.

10. Install or connect the following:
    - Front engine mount. Tighten the nuts to 59 ft. lbs. (80 Nm) and the bolts to 35 ft. lbs. (48 Nm).
    - Rear engine mount. Tighten the outer nuts to 35 ft. lbs. (48 Nm) and the center nut to 98 ft. lbs. (133 Nm).
    - Axle halfshafts
    - Lower ball joints. Tighten the pinch bolts to 37 ft. lbs. (50 Nm).
    - Right engine mount. Tighten the bolts to 35 ft. lbs. (48 Nm).
    - A/C compressor, if equipped
    - Shift cables
    - Shift cable cover
    - Drive belt cover
    - Exhaust flex pipe
    - Catalytic converter
    - Radiator cooling fan
    - Power steering high pressure pipe
    - Power steering pump and reservoir
    - Accessory drive belt
    - Coolant expansion tank
    - Radiator hoses
    - Coolant bypass hoses
    - Fuel line
    - Intake manifold vacuum line
    - EGR vacuum line
    - DPFE system sensor vacuum line
    - Power brake booster vacuum line
    - EVAP canister vacuum lines
    - Clutch slave cylinder fluid pipe
    - Reverse lamp switch connector
    - VSS sensor connector
    - PCM connectors
    - Fuel injector harness connectors
    - HO_2S sensor connectors
    - Power steering pump pressure switch connector
    - Ignition coil connectors
    - Cruise control cable, if equipped
    - Accelerator cable
    - Air cleaner housing and tubes
    - MAF sensor connector
    - Battery, tray and cables

11. Tighten the strut center nuts to 35 ft. lbs. (48 Nm).

12. Fill the cooling system.

13. Start the engine and check for leaks.

SVT

1. Before servicing the vehicle, refer to the precautions in the beginning of this section.

2. Release the fuel system pressure.

3. Release the coolant system pressure.

4. Remove the battery tray.
5. Remove the air cleaner.
6. Disconnect the ground cable from the inner fender.

➡The resonator is a push fit in the bracket.

7. Remove the air cleaner intake tube and resonator.
8. Raise and support the vehicle.
9. Drain the coolant. Allow the coolant to drain into a suitable container. Install the drain plug after draining.
10. Drain the engine oil. Allow the oil to drain into a suitable container. Install the drain plug after draining.
11. Partially lower the vehicle.
12. Loosen the strut and spring assembly top mount retaining nuts by three turns on both sides.
13. Remove the splash shield.
14. Disconnect the accelerator cable and the speed control cable (if equipped) from the throttle body.
15. Disconnect the intake manifold runner control (IMRC) actuator cable from the IMRC lever.
16. Remove the electronic ignition (EI) coil cover.
17. Disconnect the block heater harness (if equipped).
18. Disconnect the starter field wire electrical connector.
19. Disconnect the fuel injector wiring harness.
20. Disconnect the ground cable from the engine lifting eye.
21. Disconnect the generator electrical connector.
22. Disconnect the hydraulic line from the clutch slave cylinder.
23. Disconnect the wiring harness electrical connector.
24. Raise and support the vehicle.
25. Disconnect the vehicle speed sensor (VSS) electrical connector.
26. Disconnect the reversing lamp switch electrical connector.
27. Partially lower the vehicle.
28. Disconnect the vacuum hoses from the intake manifold.
29. Disconnect the brake booster pipe from the intake manifold.
30. Release the quick release coupling and disconnect the brake booster pipe.
31. Disconnect the fuel lines.
32. Disconnect the ground cable.
33. Disconnect the coolant hoses.
34. Disconnect the oil cooler hose.
35. Disconnect the gearshift cables from the transaxle.
36. Disconnect the gearshift cable and the selector cable from the selector levers.
37. Disconnect the retaining bracket from the transaxle.
38. Raise and support the vehicle.
39. Remove the accessory drive belt.
40. Disconnect the air conditioning compressor and secure it to the radiator crossmember.
41. Remove the power steering pump lower retaining bolts and disconnect the radiator lower coolant hose from the water pump housing.
42. Disconnect the power steering pressure (PSP) switch electrical connector.
43. Disconnect the crankshaft position (CKP) sensor connector.
44. Disconnect the coolant hose from the oil cooler.
45. Lower the vehicle.
46. Disconnect the coolant expansion tank and position it to one side.
47. Drain the power assisted steering (PAS) reservoir.
48. Disconnect the PAS reservoir and position it to one side.
49. Disconnect the power steering line bracket from the cylinder head.
50. Disconnect the power steering pump and bracket from the engine and position it to one side.
51. Raise and support the vehicle.
52. Remove the flexible exhaust pipe.
53. Remove the crossmember.

✲✲ CAUTION

Support the halfshaft. The inner joint must not be bent at more than 18 degrees. The outer joint must not be bent at more than 45 degrees.

✲✲ CAUTION

Do not damage the halfshaft oil seal.

➡Plug the transaxle to prevent oil loss or dirt ingress.

54. Remove the RH halfshaft and the intermediate shaft.
55. Allow the oil to drain into a suitable container.
56. Disconnect the LH halfshaft from the transaxle and secure it to one side. Allow the oil to drain into a suitable container. Discard the snap ring.
57. Position the workshop table with suitable wooden blocks under the engine and transmission assembly.
58. Carefully lower the vehicle until the engine and transmission assembly is positioned on the workshop table.
59. Secure the engine and transmission assembly to the assembly stand table with a retaining strap.
60. Remove the engine rear mount.
61. Remove the engine front mount.
62. Carefully raise the vehicle.
63. Pull the assembly stand forward.
64. Secure the engine and transmission assembly on an engine hoist.

✲✲ CAUTION

The catalytic converter must be supported while the support bracket is removed.

65. Remove the catalytic converter support bracket.
66. Remove the catalyst monitor sensor.
67. Disconnect the catalytic converter from the exhaust manifold. Discard the gasket and nuts.
68. Remove the heated oxygen (HO2S) sensor.
69. Remove the transaxle upper retaining bolts.
70. Disconnect the starter motor.
71. Remove the starter motor.
72. Disconnect the ground cable from the transaxle.
73. Remove the transaxle RH retaining bolts.
74. Remove the transaxle LH retaining bolts.
75. Separate the engine from the transaxle.

To install:

76. Install the transaxle. Tighten the bolts to 35 ft. lbs. (48 Nm).
77. Install the starter motor. Tighten the bolts to 26 ft. lbs. (35 Nm).
78. Attach the ground cable to the transaxle.
79. Connect the starter motor electrical connectors.

➡Install a new exhaust flange gasket and new nuts.

80. Attach the catalytic converter to the exhaust manifold. Tighten the nuts to 35 ft. lbs. (47 Nm).
81. Install the heated oxygen (HO2S) sensor.
82. Install the catalytic converter bracket. Tighten the bolts to 18 ft. lbs. (25 Nm).
83. Install the catalyst monitor sensor (CMS).
84. Raise and support the vehicle.
85. Position the engine and transmission assembly under the vehicle.
86. Carefully lower the vehicle.
87. Install the engine front mount. Do not tighten the bolts and nuts at this stage.

88. Install the engine rear mount. Do not tighten the bolts and nuts at this stage.
89. Remove the retaining strap from the engine and transmission assembly, and remove the assembly stand.
90. Raise and support the vehicle.

CAUTION

Support the halfshaft. The inner joint must not be bent more than 18 degrees. The outer joint must not be bent more than 45 degrees.

CAUTION

Do not damage the halfshaft oil seal.

➡**Install a new halfshaft center bearing cap and locknuts.**

91. Attach the RH halfshaft and the intermediate shaft to the transaxle.
92. Install the center bearing cap. Tighten the nuts to 18 ft. lbs. (25 Nm).
93. Attach the LH halfshaft to the transaxle.
94. Install the crossmember.
95. Install the flexible exhaust pipe.
96. Lower the vehicle.
97. Tighten the engine rear mount nuts and bolts. Tighten the bolts to 35 ft. lbs. (48 Nm), and the nuts to 59 ft. lbs. (80 Nm).
98. Tighten the engine front mount nuts. Tighten the small nuts to 35 ft. lbs. (48 Nm), and the large nut to 98 ft. lbs. (133 Nm).
99. Attach the power steering pump and the bracket to the engine. Tighten the bolts to 35 ft. lbs. (48 Nm).
100. Attach the power steering line bracket to the cylinder head. Tighten to 15 ft. lbs. (20 Nm).
101. Install the power assisted steering (PAS) reservoir.
102. Install the coolant expansion tank.
103. Raise and support the vehicle.
104. Connect the power steering pressure switch electrical connector.
105. Connect the crankshaft position sensor electrical connector.
106. Connect the radiator lower coolant hose to the coolant pump housing.
107. Connect the radiator lower coolant hose to the oil cooler.
108. Attach the air conditioning compressor. Tighten the bolts to 18 ft. lbs. (25 Nm).
109. Install the accessory drive belt.
110. Lower the vehicle.
111. Connect the oil cooler hose.
112. Attach the gearshift cables to the transaxle.
113. Attach the retaining bracket to the transaxle.
114. Attach the gearshift cable and the selector cable to the selector levers.
115. Connect the starter field wire electrical connector.
116. Attach the coolant hoses to the thermostat housing.
117. Attach the fuel pipes to the fuel rail.
118. Attach the brake servo vacuum hose to the intake manifold.
119. Attach the vacuum hoses to the throttle body.
120. Connect the reversing lamp switch electrical connector.
121. Clip on the wiring loom.
122. Connect the vehicle speed sensor (VSS) electrical connector.
123. Connect the wiring harness connectors.
124. Attach the clutch supply line to the clutch slave cylinder. Install the clip.
125. Connect the generator electrical connector.
126. Connect the fuel injector wiring harness.
127. Install the electronic ignition (EI) coil cover.
128. Attach the accelerator cable and the speed control cable (if equipped) to the throttle body.
129. Install the splash shield.
130. Connect the engine block heater (if equipped).
131. Connect the intake manifold runner control (IMRC).
132. Install the air cleaner intake tube and resonator.
133. Tighten the strut and spring assembly upper mounting retaining nuts to 18 ft. lbs. (25 Nm).
134. Attach the ground cable to the inner fender.
135. Install the air cleaner.
136. Install the battery tray.

➡**When the battery has been disconnected and reconnected, some abnormal drive symptoms may occur while the vehicle relearns its adaptive strategy. The vehicle may need to be driven 16 km (10 miles) or more to relearn the strategy.**

137. Install and connect the battery.
138. Fill the cooling system.
139. Fill the transaxle.
140. Fill the engine with engine oil.
141. Bleed the hydraulic clutch operating system.
142. Fill and bleed the PAS system.

2.3L ENGINE

1. Before servicing the vehicle, refer to the precautions in the beginning of this section.
2. With the vehicle in NEUTRAL, position it on a hoist.
3. Release the fuel system pressure.
4. Remove the battery tray.
5. Recover the A/C system.
6. Drain the cooling system.
7. Drain the engine oil. Install the drain plug.
8. Remove the accessory drive belt.
9. Remove the catalytic converter.
10. Remove the air intake resonator.
11. Disconnect the fuel tube.
12. Disconnect the vapor tube.
13. Disconnect the engine emissions hose and the exhaust gas recirculation (EGR) vacuum regulator solenoid electrical connector.
14. Disconnect the brake booster vacuum hose.
15. Disconnect the lower radiator hose, the heater hose and the overflow hose from the coolant bypass.
16. Disconnect the hose, remove the screw and the pin-type retainer.
17. Remove the snow shield.
18. Position aside the accelerator cable and speed control cable (if equipped).
19. Disconnect the accelerator and speed control cable (if equipped) from the throttle body.
20. Remove the three bolts and position aside the accelerator cable bracket aside.
21. Disconnect the fuel injection supply manifold wiring harness electrical connector.
22. Disconnect the EGR valve electrical connector.
23. Disconnect the gearshift cables from the transaxle.
24. Disconnect the shifter cable from the shift mast.
25. Disconnect the selector cable from the selector lever.
26. Disconnect the gearshift cables from the bracket.
27. Disconnect the shifter cable from the bracket, turning the abutment sleeves counterclockwise.
28. Disconnect the selector cable from the bracket, turning the abutment sleeves counterclockwise.
29. Remove the clutch slave cylinder supply line.
30. Remove the supply line and secure it to one side using cable tires.
31. Disconnect the reversing lamp switch electrical connector.

32. Disconnect the heater hose from the "T" fitting and position it aside.
33. Remove the coolant expansion tank.
34. Disconnect the power steering pressure (PSP) switch electrical connector.
35. Remove the power steering pump pulley.
36. Disconnect the PSP tube.
37. Disconnect the three main engine wiring harness electrical connectors.
38. Remove the wiring harness retainers from the valve cover.
39. Disconnect the cooling fan electrical connector.
40. Disconnect the upper radiator hose.
41. Remove the battery cable bracket nut.
42. Remove the generator heat shield top nut.
43. Remove the generator B+ cable nut and disconnect the B+ cable.
44. Remove the bolt and the ground cable.
45. Remove the generator cooling tube.
46. Remove the cooling fan.
47. Remove the two lower bellhousing bolts.
48. Remove the two oil pan-to-bellhousing bolts.
49. Disconnect the starter electrical connections.
50. Remove the nuts and disconnect the power steering pressure line brackets from the stud bolts.
51. Remove the three A/C compressor bolts.
52. Loosen the bolt and remove the A/C compressor.
53. Lower the A/C compressor to access the manifold bolt. Discard the O-ring seals.
54. Disconnect the field coil electrical connector.
55. Disconnect the LH brake hose from the support bracket.
56. Disconnect the LH caliper.

⁂ CAUTION

Suspend the caliper to prevent load from being placed on the brake hose.

57. Support the brake caliper.
58. Loosen the LH strut and spring assembly top mount nuts by four turns.
59. Disconnect the LH stabilizer bar at the strut.
60. Remove the LH tie-rod end nut.
61. Remove the RH tie-rod end nut.
62. Disconnect both of the tie-rods from the knuckles.
63. Disconnect both of the lower control arms from the knuckles.
64. Remove the mounting bracket nuts, then remove the bracket from the halfshaft intermediate bearing.
65. Remove the RH front drive halfshaft. Position the halfshaft aside and support with mechanic's wire.
66. Install a plug into the transaxle opening.
67. Remove the LH front drive halfshaft from the transaxle.
68. Install a plug into the transaxle opening.
69. Support the halfshaft using mechanic's wire.
70. Remove the lower engine mount.
71. Remove the three bolts and the starter.
72. Remove the starter isolator.
73. Disconnect the vehicle speed sensor (VSS) electrical connector.
74. Fasten the engine to the lift table.
75. Remove the RH motor mount nuts.
76. Remove the LH motor mount nut.
77. Lower the engine and transaxle assembly from the vehicle.
78. Separate the transaxle from the engine.

To install:

79. Install the transaxle. Make sure the transaxle is flush to the engine.

⁂ CAUTION

Failure to follow these steps will result in gear rattle at idle.

80. Install the two bolts at the dowels first. Tighten to 35 ft. lbs. (48 Nm).
81. Install the remaining transaxle RH flange bolts. Tighten to 35 ft. lbs. (48 Nm).
82. Connect the vehicle speed sensor (VSS) electrical connector.
83. Using the lift table, position the engine and transaxle assembly in the vehicle.
84. Install the RH motor mount nuts and tighten them to 59 ft. lbs. (80 Nm).
85. Install the LH motor mount nut and tighten it to 98 ft. lbs. (133 Nm).
86. Remove the lift table.
87. Install the lower motor mount. Tighten the bolts to 35 ft. lbs. (48 Nm).
88. Install the two lower engine-to-bellhousing bolts and tighten them to 35 ft. lbs. (48 Nm).
89. Install the remaining engine-to-bellhousing bolts and tighten them to 35 ft. lbs. (48 Nm).
90. Install the starter motor isolator.
91. Install the starter motor and tighten the bolts to 18 ft. lbs. (25 Nm).
92. Install the generator heat shield and the lower nut.

➡Install new O-ring seals.

93. Position the A/C compressor to the manifold and tube assembly and tighten the bolt to 15 ft. lbs. (21 Nm).
94. Install the A/C compressor and tighten the three bolts to 18 ft. lbs. (21 Nm).
95. Connect the field coil electrical connector.
96. Attach the power steering pressure tube brackets to the stud bolts and install the nuts.
97. Connect the starter motor electrical connections.
98. Install the cooling fan.
99. Position the battery cable and install the cable bracket nut.
100. Install the generator cooling tube.
101. Install the generator heat shield top nut.
102. Install the generator B+ cable and install the cable nut.
103. Install the upper radiator hose.
104. Position the ground cable and install the bolt.
105. Connect the cooling fan electrical connector.
106. Connect the wiring harness retainers to the valve cover studs.
107. Position the power steering pump (PSP) and install the bolts.
108. Connect the PSP tube.

➡Make sure the pulley is flush with the end of the power steering pump shaft.

109. Install the power steering pump pulley.
110. Connect the PSP switch electrical connector.
111. Connect the three main engine wiring harness connectors.
112. Install the coolant expansion tank.
113. Connect the heater hose to the "T" fitting.
114. Connect the exhaust gas recirculation (EGR) valve electrical connector.
115. Connect the fuel injection supply manifold wiring harness electrical connector.
116. Position the accelerator cable bracket.
117. Install the speed control cable (if equipped) and the throttle cable.
118. Attach the hose and install the screw and the pin-type retainer.
119. Install the snow shield.
120. Install the catalytic converter.
121. Connect the lower radiator hose, the heater hose and the overflow hose to the coolant bypass.
122. Connect the engine emissions hose

and EGR vacuum regulator solenoid electrical connector.

123. Install the vapor tube.

124. Connect the fuel tube.

125. Install the brake booster vacuum hose.

126. Fill the engine with clean engine oil.

➡Install a new snap ring.

127. Install the LH front drive halfshaft.

➡Install new nuts and a new center bearing cap.

128. Install the RH front halfshaft together with the intermediate shaft.

129. Install the mounting bracket for the front drive halfshaft intermediate bearing.

130. Install both of the lower control arms to the knuckles. Tighten the bolts to 37 ft. lbs. (50 Nm).

131. Install both of the tie-rod ends to the knuckles. Tighten the nuts to 35 ft. lbs. (47 Nm).

132. Connect the stabilizer bar at the strut.

133. Tighten the LH strut and spring assembly top mount nuts to 18 ft. lbs. (25 Nm).

134. Install the brake caliper. Tighten the bolts to 21 ft. lbs. (28 Nm).

135. Install the brake hose onto the support bracket.

136. Install the accessory drive belt.

137. Connect the reversing lamp switch electrical connector.

138. Connect the clutch slave cylinder supply line. Insert the clip.

➡The shift cable abutment sleeve is colored white.

➡The selector cable abutment sleeve is colored black.

139. Attach the gearshift cables to the bracket.

140. Attach the shifter cable to the bracket, turning the abutment sleeves counterclockwise to open. Position the cables into the metal holders.

141. Attach the selector cable to the bracket, turning the abutment sleeves counterclockwise to open. Position the cables into the metal holders.

142. Attach the shifter cable to the shift mast.

143. Attach the selector cable to the selector lever.

144. Check the shift cables at the metal holders. Pull outward on the cables. Cables should remain in the metal.

145. Adjust the shift cables.

146. Fill and bleed the clutch hydraulic system

147. Evacuate and recharge the A/C system.

148. Install the air intake resonator.

149. Install the battery tray.

150. Connect the battery ground cable.

151. Fill and bleed the cooling system.

152. Fill and bleed the power steering system.

Automatic Transaxle

2.0L SOHC ENGINE

1. Before servicing the vehicle, refer to the precautions in the beginning of this section.
2. Drain the cooling system.
3. Relieve the fuel system pressure.
4. Loosen the front strut center nuts 5 turns.
5. Remove or disconnect the following:
   - Battery, tray and cables
   - Mass Air Flow (MAF) sensor connector
   - Air cleaner housing and tubes
   - Accelerator cable
   - Cruise control cable, if equipped
   - Ignition coil connectors
   - Power steering pump pressure switch connector
   - Heated Oxygen (HO_2S) sensor connectors
   - Fuel injection wiring harness connector
   - Powertrain Control Module (PCM) connectors
   - Radiator cooling fan
   - Evaporative Emissions (EVAP) canister vacuum lines
   - Power brake booster vacuum line
   - Delta Pressure Feedback Electronic (DPFE) system sensor vacuum line
   - Exhaust Gas Recirculation (EGR) vacuum line
   - Intake manifold vacuum line
   - Fuel line
   - Coolant bypass hoses
   - Radiator hoses
   - Coolant expansion tank
   - Accessory drive belt
   - Power steering pump and reservoir
   - Power steering high pressure pipe
   - Alternator and bracket
   - Drive belt cover
   - A/C compressor, if equipped
   - Lower ball joints
   - Exhaust front pipe
   - Torque converter cover
   - Torque converter
   - Transaxle fluid cooler lines
   - Axle halfshafts
   - Intermediate shaft bracket
   - Starter motor
   - Crankshaft pulley
   - Right engine mount
   - Transaxle flange bolts. Support the transaxle
   - Engine front mount
   - Crankshaft Position (CKP) sensor
   - Engine

To install:

6. Install or connect the following:
   - Engine
   - CKP sensor
   - Engine front mount. Tighten the nuts to 59 ft. lbs. (80 Nm) and the bolts to 35 ft. lbs. (48 Nm).
   - Transaxle flange bolts. Tighten the large bolts to 35 ft. lbs. (48 Nm) and the small bolts to 16 ft. lbs. (22 Nm).
   - Right engine mount. Tighten the bolts to 35 ft. lbs. (48 Nm).
   - Crankshaft pulley. Tighten the bolt to 88 ft. lbs. (120 Nm).
   - Starter motor
   - Intermediate shaft bracket. Tighten the bolts to 35 ft. lbs. (48 Nm).
   - Axle halfshafts
   - Transaxle fluid cooler lines
   - Torque converter. Tighten the nuts to 27 ft. lbs. (37 Nm).
   - Torque converter cover
   - Exhaust front pipe
   - Lower ball joints. Tighten the pinch bolts to 37 ft. lbs. (50 Nm).
   - A/C compressor, if equipped
   - Drive belt cover
   - Alternator and bracket
   - Power steering high pressure pipe
   - Power steering pump and reservoir
   - Accessory drive belt
   - Coolant expansion tank
   - Radiator hoses
   - Coolant bypass hoses
   - Fuel line
   - Intake manifold vacuum line
   - EGR vacuum line
   - DPFE system sensor vacuum line
   - Power brake booster vacuum line
   - EVAP canister vacuum lines
   - Radiator cooling fan
   - PCM connectors
   - HO_2S sensor connectors
   - Fuel injection wiring harness connector
   - Power steering pump pressure switch connector
   - Ignition coil connectors
   - Cruise control cable, if equipped
   - Accelerator cable
   - Air cleaner housing and tubes
   - MAF sensor connector
   - Battery, tray and cables
7. Tighten the strut center nuts to 35 ft. lbs. (48 Nm).

8. Fill the cooling system.
9. Start the engine and check for leaks.

2.0L DOHC ENGINE

1. Before servicing the vehicle, refer to the precautions in the beginning of this section.
2. Drain the cooling system.
3. Relieve the fuel system pressure.
4. Loosen the front strut center nuts 5 turns.
5. Remove or disconnect the following:
   - Battery, tray and cables
   - Mass Air Flow (MAF) sensor connector
   - Positive Crankcase Ventilation (PCV) valve and hose
   - Air cleaner assembly and tubes
   - Accelerator cable
   - Cruise control cable, if equipped
   - Ignition coil connectors
   - Heated Oxygen (HO_2S) sensor connectors
   - Fuel injector harness connectors
   - Powertrain Control Module (PCM) connectors
   - Radiator cooling fan
   - Evaporative Emissions (EVAP) canister vacuum lines
   - Power brake booster vacuum line
   - Delta Pressure Feedback Electronic (DPFE) system sensor vacuum line
   - Exhaust Gas Recirculation (EGR) vacuum line
   - Intake manifold vacuum line
   - Fuel line
   - Coolant bypass hoses
   - Radiator hoses
   - Coolant expansion tank
   - Shift select cable and bracket
   - Transaxle fluid cooler lines
   - Drive belt cover
   - Power steering pressure switch connector
   - Accessory drive belt
   - A/C compressor, if equipped
   - Power steering pump
   - Power steering pressure pipe
   - Exhaust flex pipe
   - Right engine mount
   - Lower ball joints
   - Axle halfshafts and intermediate shaft
6. Support the powertrain from below
7. Remove the front and rear engine mounts.
8. Raise the vehicle away from the powertrain.

To install:

9. Lower the vehicle over the powertrain.
10. Install or connect the following:
    - Front engine mount. Tighten the nuts to 59 ft. lbs. (80 Nm) and the bolts to 35 ft. lbs. (48 Nm).
    - Rear engine mount. Tighten the outer nuts to 35 ft. lbs. (48 Nm) and the center nut to 98 ft. lbs. (133 Nm).
    - Axle halfshafts and intermediate shaft
    - Lower ball joints. Tighten the pinch bolts to 37 ft. lbs. (50 Nm).
    - Right engine mount. Tighten the bolts to 35 ft. lbs. (48 Nm).
    - Exhaust flex pipe
    - Power steering pressure pipe
    - Power steering pump
    - A/C compressor, if equipped
    - Accessory drive belt
    - Power steering pressure switch connector
    - Drive belt cover
    - Transaxle fluid cooler lines
    - Shift select cable and bracket
    - Coolant expansion tank
    - Radiator hoses
    - Coolant bypass hoses
    - Fuel line
    - Intake manifold vacuum line
    - EGR vacuum line
    - DPFE system sensor vacuum line
    - Power brake booster vacuum line
    - EVAP canister vacuum lines
    - Radiator cooling fan
    - PCM connectors
    - Fuel injector harness connectors
    - HO_2S sensor connectors
    - Ignition coil connectors
    - Cruise control cable, if equipped
    - Accelerator cable
    - Air cleaner assembly and tubes
    - PCV valve and hose
    - MAF sensor connector
    - Battery, tray and cables
11. Tighten the strut center nuts to 35 ft. lbs. (48 Nm).
12. Fill the cooling system.
13. Start the engine and check for leaks.

2.3L ENGINE

1. Before servicing the vehicle, refer to the precautions in the beginning of this section.
2. With the vehicle in NEUTRAL, position it on a hoist.
3. Release the fuel system pressure.
4. Remove the battery tray.
5. Recover the A/C system.
6. Drain the cooling system.
7. Drain the engine oil. Install the drain plug.
8. Remove the accessory drive belt.
9. Remove the catalytic converter.
10. Remove the air intake resonator.
11. Disconnect the fuel tube.
12. Disconnect the vapor tube.
13. Disconnect the engine emissions hose and the exhaust gas recirculation (EGR) vacuum regulator solenoid electrical connector.
14. Disconnect the brake booster vacuum hose.
15. Disconnect the lower radiator hose, the heater hose and the overflow hose from the coolant bypass.
16. Disconnect the hose, remove the screw and the pin-type retainer.
17. Remove the snow shield.
18. Position aside the accelerator cable and speed control cable (if equipped).
19. Disconnect the accelerator and speed control cable (if equipped) from the throttle body.
20. Remove the three bolts and position aside the accelerator cable bracket aside.
21. Disconnect the fuel injection supply manifold wiring harness electrical connector.
22. Disconnect the EGR valve electrical connector.
23. Disconnect the heater hose from the "T" fitting and position it aside.
24. Remove the coolant expansion tank.
25. Disconnect the power steering pressure (PSP) switch electrical connector.
26. Remove the power steering pump pulley.
27. Disconnect the PSP tube.
28. Disconnect the three main engine wiring harness electrical connectors.
29. Remove the wiring harness retainers from the valve cover.
30. Disconnect the cooling fan electrical connector.
31. Disconnect the upper radiator hose.
32. Remove the battery cable bracket nut.
33. Remove the generator heat shield top nut.
34. Remove the generator B+ cable nut and disconnect the B+ cable.
35. Remove the bolt and the ground cable.
36. Remove the generator cooling tube.
37. Disconnect the transmission shifter cable and position it aside.
38. Disconnect the transmission cooler hoses.
39. Remove the cooling fan.
40. Remove the two lower bellhousing bolts.
41. Remove the two oil pan-to-bellhousing bolts.
42. Disconnect the starter electrical connections.

43. Remove the nuts and disconnect the power steering pressure line brackets from the stud bolts.
44. Remove the three A/C compressor bolts.
45. Loosen the bolt and remove the A/C compressor.
46. Lower the A/C compressor to access the manifold bolt. Discard the O-ring seals.
47. Disconnect the field coil electrical connector.
48. Disconnect the LH brake hose from the support bracket.
49. Disconnect the LH caliper.

⁂ CAUTION

Suspend the caliper to prevent load from being placed on the brake hose.

50. Support the brake caliper.
51. Loosen the LH strut and spring assembly top mount nuts by four turns.
52. Disconnect the LH stabilizer bar at the strut.
53. Remove the LH tie-rod end nut.
54. Remove the RH tie-rod end nut.
55. Disconnect both of the tie-rods from the knuckles.
56. Disconnect both of the lower control arms from the knuckles.
57. Remove the mounting bracket nuts, then remove the bracket from the halfshaft intermediate bearing.
58. Remove the RH front drive halfshaft. Position the halfshaft aside and support with mechanic's wire.
59. Install a plug into the transaxle opening.
60. Remove the LH front drive halfshaft from the transaxle.
61. Install a plug into the transaxle opening.
62. Support the halfshaft using mechanic's wire.
63. Remove the lower engine mount.
64. Remove the three bolts and the starter.
65. Remove the starter isolator.

➡Mark one stud and the flexplate for assembly reference.

66. Remove the four torque converter nuts.
67. Fasten the engine to the lift table.
68. Remove the RH motor mount nuts.
69. Remove the LH motor mount nut.
70. Lower the engine and transaxle assembly from the vehicle.
71. Disconnect the output shaft speed (OSS) sensor electrical connector.
72. Disconnect the turbine shaft speed (TSS) sensor electrical connector.

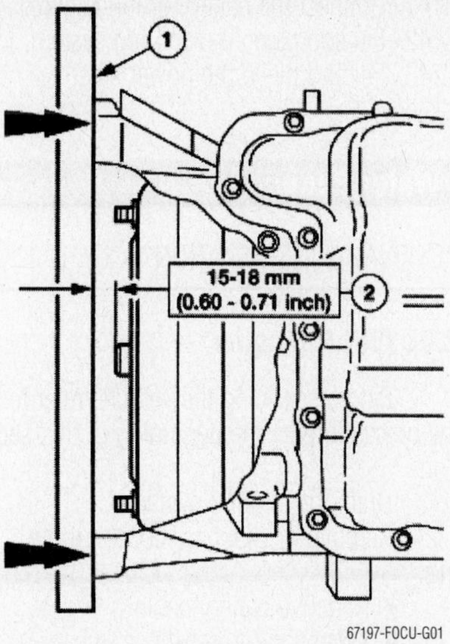

Use a straight edge (1) to check the torque converter installed depth (2)—2.3L engine

73. Disconnect the solenoid body and the transmission range (TR) sensor electrical connectors.
74. Separate the transaxle from the engine.

To install:

➡Lubricate the torque converter pilot hub with multi-purpose grease.

75. Check the installation depth of the torque converter. Lay a steel straightedge on the automatic transaxle flange. Check the installation depth between the transaxle flange and the torque converter centering spigot for the correct clearance. The dimension should be 0.60–0.71 inches (15–18 mm).
76. Position the engine and transaxle together and install five of the six top bolts.

➡Make sure to note the location of the different length bolts.

77. Install the LH flange bolts and tighten to 35 ft. lbs. (47 Nm).
78. Connect the output shaft speed (OSS) sensor electrical connector.
79. Connect the solenoid body and the transmission range (TR) sensor electrical connectors.
80. Connect the turbine shaft speed (TSS) sensor electrical connector.
81. Using the lift table, position the engine and transaxle assembly in the vehicle.
82. Install the RH motor mount nuts and tighten them to 59 ft. lbs. (80 Nm).
83. Install the LH motor mount nut and tighten it to 98 ft. lbs. (133 Nm).
84. Remove the lift table.
85. Install the lower motor mount. Tighten the bolts to 35 ft. lbs. (48 Nm).
86. Install the two lower engine-to-bell-housing bolts and tighten them to 35 ft. lbs. (48 Nm).
87. Install the remaining engine-to-bell-housing bolts and tighten them to 35 ft. lbs. (48 Nm).

➡If new parts are not being used, be sure to align the marks on the flexplate and the stud made during engine removal.

88. Install the four torque converter nuts and tighten them to 27 ft. lbs. (37 Nm).
89. Install the starter motor isolator.
90. Install the starter motor and tighten the bolts to 18 ft. lbs. (25 Nm).
91. Install the generator heat shield and the lower nut.

➡Install new O-ring seals.

92. Position the A/C compressor to the manifold and tube assembly and tighten the bolt to 15 ft. lbs. (21 Nm).
93. Install the A/C compressor and tighten the three bolts to 18 ft. lbs. (21 Nm).
94. Connect the field coil electrical connector.
95. Attach the power steering pressure tube brackets to the stud bolts and install the nuts.
96. Connect the starter motor electrical connections.
97. Install the cooling fan.
98. Connect the transmission cooler hoses.
99. Connect the transmission shifter cable.
100. Position the battery cable and install the cable bracket nut.
101. Install the generator cooling tube.
102. Install the generator heat shield top nut.
103. Install the generator B+ cable and install the cable nut.
104. Install the upper radiator hose.
105. Position the ground cable and install the bolt.
106. Connect the cooling fan electrical connector.
107. Connect the wiring harness retainers to the valve cover studs.
108. Position the power steering pump (PSP) and install the bolts.
109. Connect the PSP tube.

➡Make sure the pulley is flush with the end of the power steering pump shaft.

110. Install the power steering pump pulley.
111. Connect the PSP switch electrical connector.
112. Connect the three main engine wiring harness connectors.
113. Install the coolant expansion tank.
114. Connect the heater hose to the "T" fitting.
115. Connect the exhaust gas recirculation (EGR) valve electrical connector.
116. Connect the fuel injection supply manifold wiring harness electrical connector.
117. Position the accelerator cable bracket.
118. Install the speed control cable (if equipped) and the throttle cable.
119. Attach the hose and install the screw and the pin-type retainer.
120. Install the snow shield.
121. Install the catalytic converter.
122. Connect the lower radiator hose, the heater hose and the overflow hose to the coolant bypass.
123. Connect the engine emissions hose and EGR vacuum regulator solenoid electrical connector.
124. Install the vapor tube.
125. Connect the fuel tube.
126. Install the brake booster vacuum hose.
127. Fill the engine with clean engine oil.

➡Install a new snap ring.

128. Install the LH front drive halfshaft.

➡Install new nuts and a new center bearing cap.

129. Install the RH front halfshaft together with the intermediate shaft.
130. Install the mounting bracket for the front drive halfshaft intermediate bearing.
131. Install both of the lower control arms to the knuckles. Tighten the bolts to 37 ft. lbs. (50 Nm).
132. Install both of the tie-rod ends to the knuckles. Tighten the nuts to 35 ft. lbs. (47 Nm).
133. Connect the stabilizer bar at the strut.
134. Tighten the LH strut and spring assembly top mount nuts to 18 ft. lbs. (25 Nm).
135. Install the brake caliper. Tighten the bolts to 21 ft. lbs. (28 Nm).
136. Install the brake hose onto the support bracket.
137. Install the accessory drive belt.
138. Evacuate and recharge the A/C system.
139. Install the air intake resonator.
140. Install the battery tray.
141. Connect the battery ground cable.
142. Fill and bleed the cooling system.
143. Fill and bleed the power steering system.

Water Pump

REMOVAL & INSTALLATION

2.0L (VIN P) Engine

1. Before servicing the vehicle, refer to the precautions in the beginning of this section.
2. Drain the cooling system.
3. Remove or disconnect the following:
   - Negative battery cable
   - Accessory drive belt
   - Front cover
   - Timing belt
   - Timing belt tensioner
   - Water pump

To install:

4. Install or connect the following:
   - Water pump. Tighten the bolts to 18 ft. lbs. (25 Nm).
   - Timing belt tensioner
   - Timing belt
   - Accessory drive belt
   - Negative battery cable
5. Fill the cooling system.
6. Start the engine and check for leaks.

2.0L (VIN 3) Engines

1. Before servicing the vehicle, refer to the precautions in the beginning of this section.
2. Drain the cooling system.
3. Remove or disconnect the following:
   - Negative battery cable
   - Accessory drive belt
   - Water pump pulley
   - Front cover
   - Timing belt
   - Timing belt tensioner
   - Water pump

To install:

4. Install or connect the following:
   - Water pump. Tighten the bolts to 13 ft. lbs. (18 Nm).
   - Timing belt tensioner
   - Timing belt
   - Front cover
   - Water pump pulley
   - Accessory drive belt
   - Negative battery cable
5. Fill the cooling system.
6. Start the engine and check for leaks.

2.3L Engine

1. Before servicing the vehicle, refer to the precautions in the beginning of this section.
2. Drain the cooling system.
3. Remove the accessory drive belt.
4. Remove the coolant pump pulley bolts and the pulley.
5. Remove the coolant pump retaining bolts and the coolant pump.

To install:

➡Install a new coolant pump O-ring and lubricate with clean engine coolant.

6. Install the coolant pump and tighten the bolts to 89 inch lbs. (10 Nm).
7. Install the coolant pump pulley and tighten the bolts to 18 ft. lbs. (25 Nm).
8. Install the accessory drive belt.
9. Fill and bleed the cooling system.

Heater Core

REMOVAL & INSTALLATION

1. Before servicing the vehicle, refer to the precautions in the beginning of this section.
2. Drain the cooling system.
3. Recover the A/C refrigerant, if equipped.
4. Remove or disconnect the following:
   - Negative battery cable
   - Radio
   - Left and right lower panels
   - Data link electrical connector
   - Steering column upper shroud
   - Audio unit control switch, if equipped
   - Steering column lower shroud
   - Steering column
   - Instrument cluster
   - Glove compartment
   - Passenger air bag module
   - Climate control and audio trim panel
   - Ash tray
   - Heater core housing assembly panel
   - Center console
   - Instrument panel upper retaining screws
   - Instrument panel side trim panels
   - Instrument panel side bolts
   - Instrument panel bolts behind the instrument cluster, climate control and audio panel and passenger air bag module

- Defroster tubes
- Instrument panel
- Heater hoses
- A/C evaporator refrigerant lines
- Cross vehicle beam
- Rear footwell ventilation hoses
- Heater housing wiring harness
- Heater housing. Refer to the illustrations for housing fastener locations.

To install:

5. Install or connect the following:
   - Heater housing. Refer to the illustrations for housing fastener locations and torque specifications.
   - Heater housing wiring harness
   - Rear footwell ventilation hoses
   - Cross vehicle beam. Tighten the beam fasteners to 15 ft. lbs. (20 Nm) and the beam bracket bolts to 12 ft. lbs. (16 Nm).
   - A/C evaporator refrigerant lines
   - Heater hoses
   - Instrument panel
   - Defroster tubes
   - Instrument panel bolts behind the instrument cluster, climate control and audio panel and passenger air bag module
   - Instrument panel side bolts
   - Instrument panel side trim panels
   - Instrument panel upper retaining screws
   - Center console
   - Heater core housing assembly panel
   - Ash tray
   - Climate control and audio trim panel
   - Passenger air bag module
   - Glove compartment
   - Instrument cluster
   - Steering column. Tighten the nuts to 13 ft. lbs. (18 Nm) and the Torx® screw to 17 ft. lbs. (23 Nm)
   - Steering column lower shroud
   - Audio unit control switch, if equipped
   - Steering column upper shroud
   - Data link electrical connector
   - Left and right lower panels
   - Radio
   - Negative battery cable
6. Fill the cooling system.
7. Recharge the A/C system, if equipped.
8. Start the engine and check for leaks.

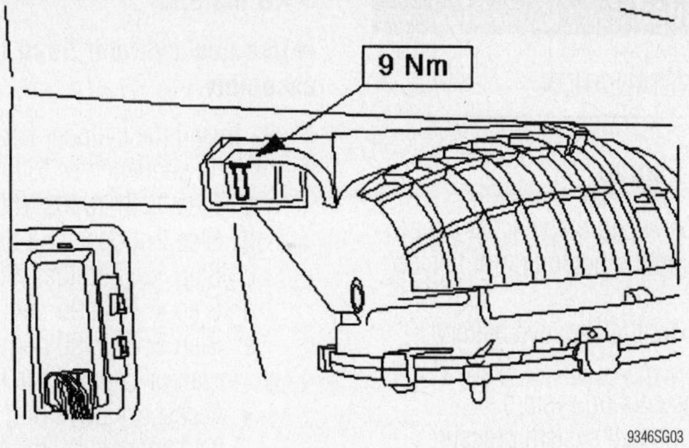

Heater housing fastener location

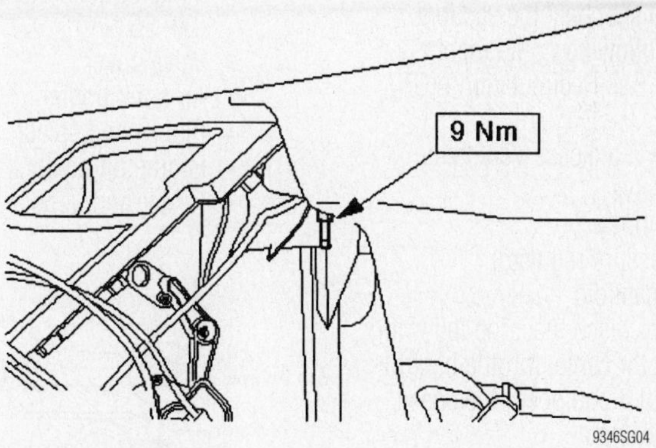

Heater housing fastener location

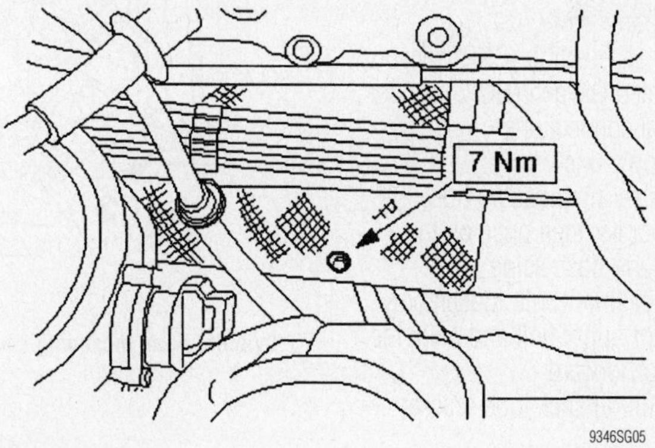

Heater housing fastener location

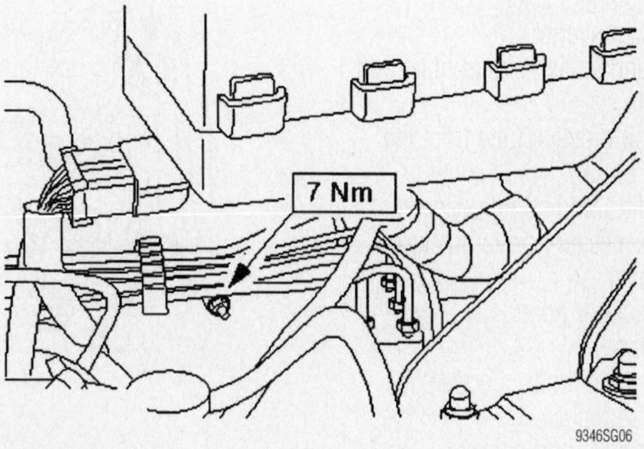

Heater housing fastener location

Cylinder Head

REMOVAL & INSTALLATION

2.0L (VIN P) Engine

1. Before servicing the vehicle, refer to the precautions in the beginning of this section.
2. Disconnect the negative battery cable.
3. Drain the cooling system.
4. Relieve the fuel system pressure.
5. Remove or disconnect the following:
   - Timing belt
   - Engine air cleaner intake tube
6. Tag and disconnect the vacuum hoses from the following components:
   - Exhaust Gas Recirculation (EGR) valve
   - Positive Crankcase Ventilation (PCV) valve
   - Throttle body
   - Fuel pressure regulator
   - Intake manifold
7. Remove or disconnect the following:
   - Accelerator cable, throttle control lever and if equipped, the speed control cable
   - Speed control cable bracket bolt, if equipped
   - 2 fuel charging wiring electrical connections
   - Crankshaft Position (CKP) sensor and Heated Oxygen (HO_2S) sensor electrical connections
   - Fuel supply line
   - Power steering pressure hose bracket bolts, then position the bracket and hose aside
   - Alternator lower bolt, loosen only
   - Alternator upper bolt and pivot the alternator forward
   - Engine oil dipstick tube bracket bolt
   - Exhaust Gas Recirculation (EGR) tube
   - Exhaust manifold heat shield
   - Catalytic converter
   - Right-hand splash shield bolts and the shield
   - Engine oil dipstick tube from the cylinder block
   - Air conditioning compressor
   - Engine accessory drive bracket
   - Valve cover
   - Upper radiator hose
   - Heater hose
   - Cylinder head bolts. Loosen the bolts in sequence.
   - Cylinder head and gasket

To install:

➡Use new cylinder head bolts for assembly.

8. Install the cylinder head and tighten the bolts in sequence as follows:
   a. Step 1: 37 ft. lbs. (50 Nm)
   b. Step 2: Loosen all bolts ½ turn
   c. Step 3: 37 ft. lbs. (50 Nm)
   d. Step 4: Plus 90 degrees
   e. Step 5: Plus 90 degrees
9. Install or connect the following:
   - Accessory drive bracket. Tighten the fasteners to 30–40 ft. lbs. (40–55 Nm)
   - Air conditioning line bracket and tighten the bolt to 15–18 ft. lbs. (20–25 Nm)
   - Valve cover
   - Air conditioning compressor
   - Right-hand splash shield
   - Engine oil dipstick tube
   - Heater and upper radiator hoses
   - Timing belt and alternator
   - Power steering pressure hose
   - CKP sensor and 2 main fuel charging electrical connections
   - Engine oil dipstick tube bolt
   - EGR valve tube
   - Fuel supply line
   - Catalytic converter
   - Heat shield
   - Speed control cable bracket, if equipped
   - Throttle control lever, accelerator cable and if equipped, the speed control cable
10. Connect the vacuum hoses to the following components:
   - EGR valve
   - PCV valve
   - Throttle body
   - Fuel pressure regulator
   - Intake manifold
   - Air cleaner outlet tube
   - Negative battery cable

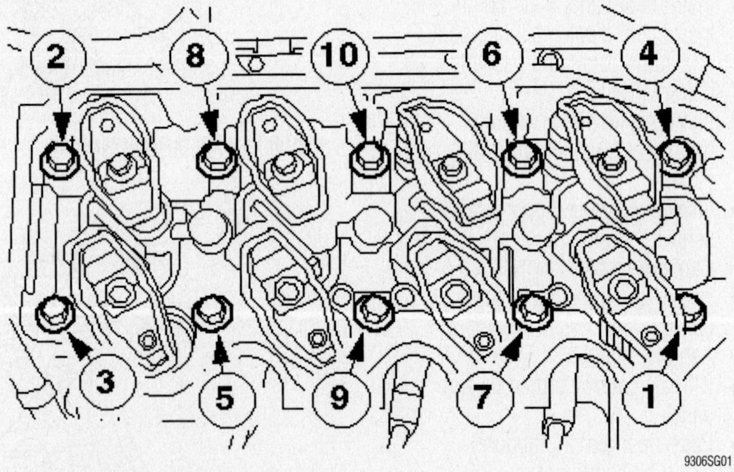

Cylinder head loosening sequence—2.0L (VIN P) engine

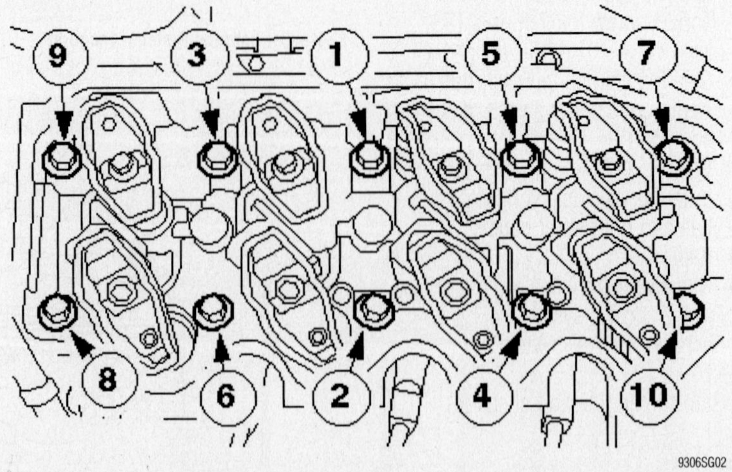

Cylinder head torque sequence—2.0L (VIN P) engine

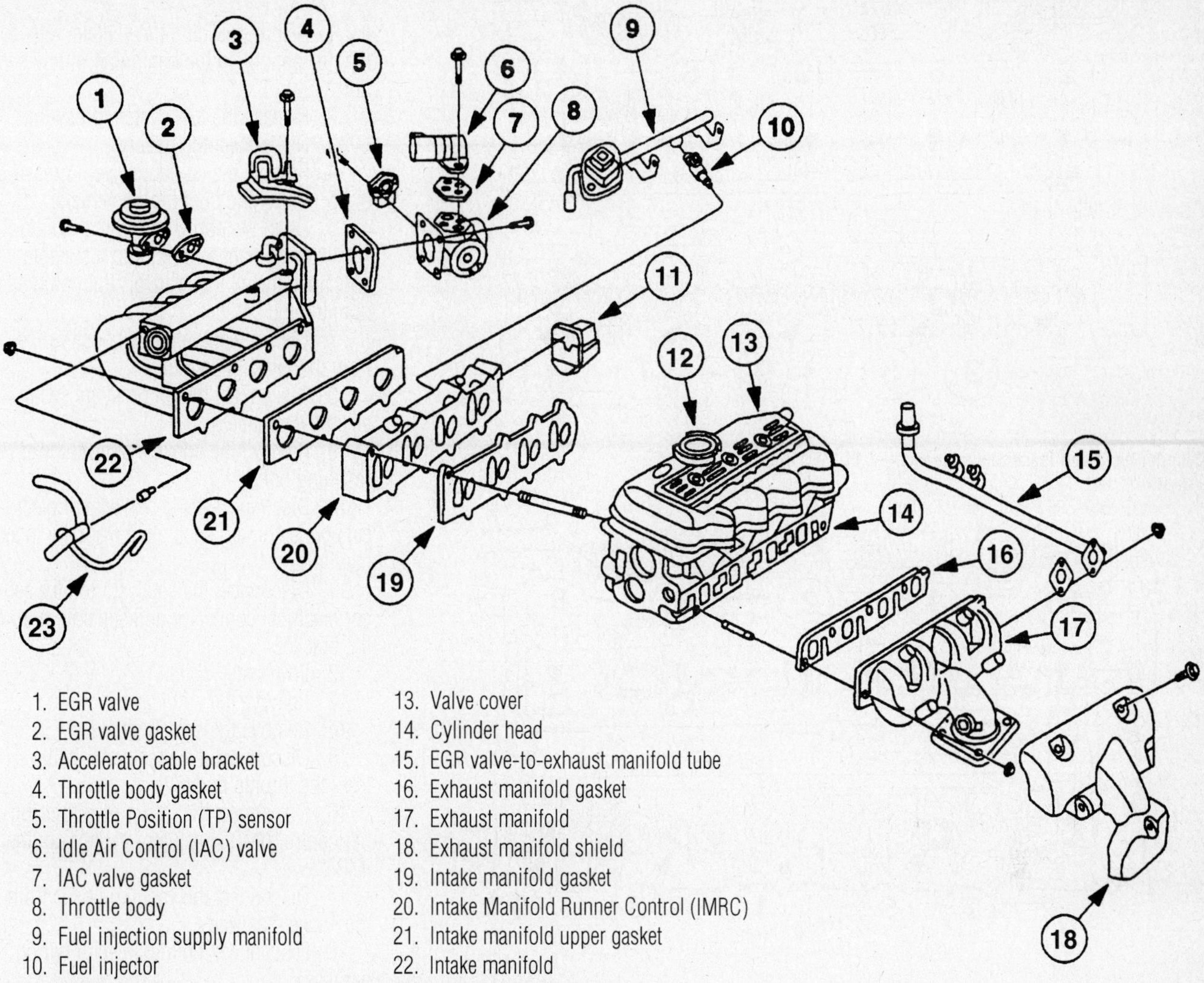

1. EGR valve
2. EGR valve gasket
3. Accelerator cable bracket
4. Throttle body gasket
5. Throttle Position (TP) sensor
6. Idle Air Control (IAC) valve
7. IAC valve gasket
8. Throttle body
9. Fuel injection supply manifold
10. Fuel injector
11. IMRC actuator
12. Oil filler cap
13. Valve cover
14. Cylinder head
15. EGR valve-to-exhaust manifold tube
16. Exhaust manifold gasket
17. Exhaust manifold
18. Exhaust manifold shield
19. Intake manifold gasket
20. Intake Manifold Runner Control (IMRC)
21. Intake manifold upper gasket
22. Intake manifold
23. Intake manifold tube

9300MG13

Exploded view of the cylinder head, exhaust manifold and intake manifold mounting—2.0L (VIN P) engine

11. Fill the cooling system.
12. Start the engine and check for leaks.

2.0L (VIN 3) Engine

1. Before servicing the vehicle, refer to the precautions in the beginning of this section.
2. Disconnect the negative battery cable.
3. Drain the cooling system.
4. Relieve the fuel system pressure.
5. Remove or disconnect the following:
   - Air cleaner assembly outlet tube
   - Timing belt
6. Tag and disconnect the vacuum hoses from the following components:
   - Positive Crankcase Ventilation (PCV) valve
   - Throttle body
   - Fuel pressure sensor
   - Intake manifold
7. Remove or disconnect the following:
   - Speed control and accelerator cables from the control lever
   - Fuel charging electrical connectors at the main engine connector
   - Crankshaft Position (CKP) sensor and Heated Oxygen (HO_2S) sensor electrical connections
   - Fuel line
   - Power steering pump and bracket
   - Alternator
   - Engine oil dipstick tube
   - Splash shield
   - Air conditioning compressor
   - Spark plug wires
   - Valve cover
   - Upper radiator hose
   - Heater hose
   - Camshafts
   - Ignition coil
   - Thermostat housing
   - Cylinder head bolts. Mark each bolt with a punch and loosen them in the sequence shown.
   - Cylinder head and the gasket

To install:

➡The cylinder head bolts may be reused twice. Replace bolts that have two punch marks.

8. Lubricate the cylinder bolts when engine oil.
9. Install the cylinder head and tighten the bolts in sequence as follows:
   a. Step 1: 15 ft. lbs. (20 Nm)
   b. Step 2: 30 ft. lbs. (40 Nm)
   c. Step 3: Plus 90 degrees
10. Install or connect the following:
    - Thermostat housing

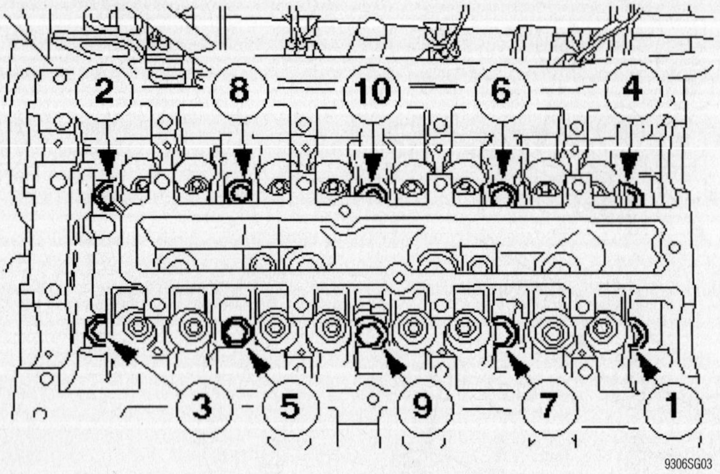

Cylinder head bolt loosening sequence—2.0L (VIN 3) engine

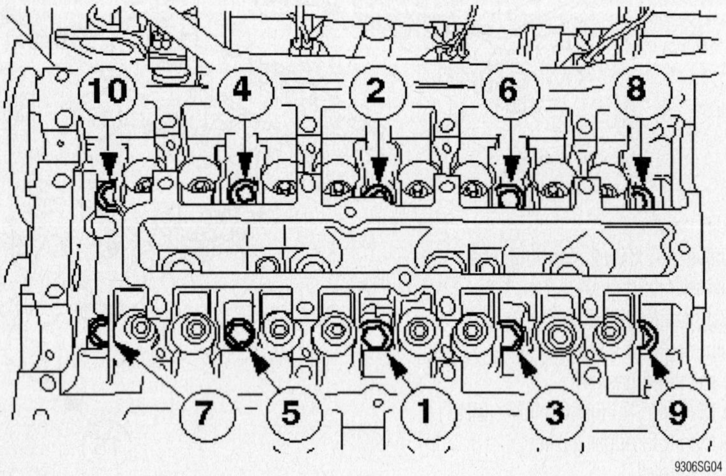

Cylinder head torque sequence—2.0L (VIN 3) engine

- Ignition coil and the camshafts
- Heater and upper radiator hoses
- Valve cover
- Spark plugs
- Spark plug wires
- Air conditioning compressor
- Splash shield
- Engine oil dipstick tube
- Alternator and the power steering reservoir and bracket
- Fuel line
- CKP sensor, HO_2sensor and 2 main fuel charging electrical connections
- Accelerator cable and if equipped, the speed control cable to the control lever

11. Connect the vacuum hoses to the following components:
    - PCV valve
    - Throttle body
    - Fuel pressure sensor
    - Intake manifold
12. Install or connect the following:
    - Timing belt
    - Air cleaner outlet tube
    - Negative battery cable
13. Fill the cooling system.
14. Start the engine and check for leaks.

2.0L (VIN 5) Engine

1. Before servicing the vehicle, refer to the precautions in the beginning of this section.
2. Release the fuel system pressure.
3. Disconnect the battery.
4. Release the coolant system pressure.
5. Raise and support the vehicle.
6. Drain the coolant from the radiator. Allow the coolant to drain into a suitable container. Install the radiator drain plug after draining the coolant.
7. Disconnect the brake booster pipe from the intake manifold.
8. Disconnect the oil pressure switch electrical connector.
9. Remove the intake manifold lower retaining bolt.
10. Disconnect the knock sensor electrical connector and separate it from the intake manifold.
11. Disconnect the catalyst monitor sensor electrical connector and separate it from the bracket.
12. Lower the vehicle.
13. Remove the air cleaner.
14. Remove the splash shield.
15. Disconnect the accelerator cable from the throttle body.
16. Disconnect the intake manifold runner control (IMRC) actuator cable from the IMRC lever.
17. Disconnect the vacuum hoses from the intake manifold.
18. Disconnect the fuel injector wiring harness.
19. Disconnect the camshaft position (CMP) sensor electrical connector.
20. Disconnect the fuel lines.
21. Disconnect the ground cable from the engine lifting eye.
22. Remove the exhaust manifold.

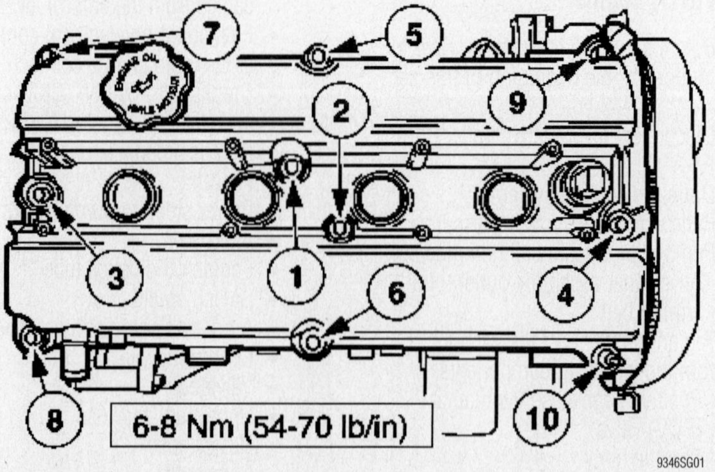

Valve cover torque sequence—2.0L (VIN 3) Engine

23. Disconnect the thermostat housing.
24. Disconnect the power steering pipe bracket from the cylinder head.
25. Disconnect the power steering pump bracket from the cylinder head and cylinder block.
26. Disconnect the generator.
27. Disconnect the wiring harness from the electronic ignition (EI) coil and from the engine coolant temperature (ECT) sensor.
28. Remove the upper bolt from the generator bracket.
29. Remove the camshafts.

➡Keep the valve tappets in order for installation.

30. Remove the valve tappets.

⁂ CAUTION

Cylinder head bolts can only be used twice, mark the bolts to indicate usage.

⁂ CAUTION

The cylinder head must be cooled to ambient temperature.

31. Remove the cylinder head bolts in the sequence shown.
32. Remove the cylinder head.
33. Using a spatula, remove gasket residue.
34. Thoroughly clean the threaded holes of the cylinder head bolts.
35. Check the cylinder head distortion.
36. Make up two locating studs as shown.
37. Install a new cylinder head gasket.
38. Insert the fabricated locating studs. Check that the locating studs are seated correctly.
39. Install the cylinder head.
40. Tighten the cylinder head bolts in sequence as follows:
    a. Step 1: 15 ft. lbs. (20 Nm)
    b. Step 2: 30 ft. lbs. (40 Nm)
    c. Step 3: Plus 90 degrees.

➡Coat the valve tappets with clean engine oil before installation.

41. Install the valve tappets in order.
42. Install the camshafts.
43. Check the valve clearance and, if necessary, adjust.
44. Attach the power steering pump bracket to the cylinder head and cylinder block.
45. Install the generator bracket bolt.
46. Connect the wiring harness to the electric ignition (EI) coil and to the engine coolant temperature (ECT) sensor.

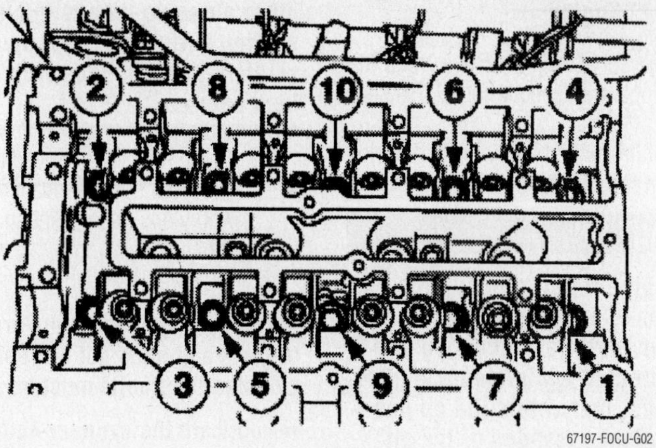

Cylinder head bolt removal sequence—2.0L (VIN 5) engine

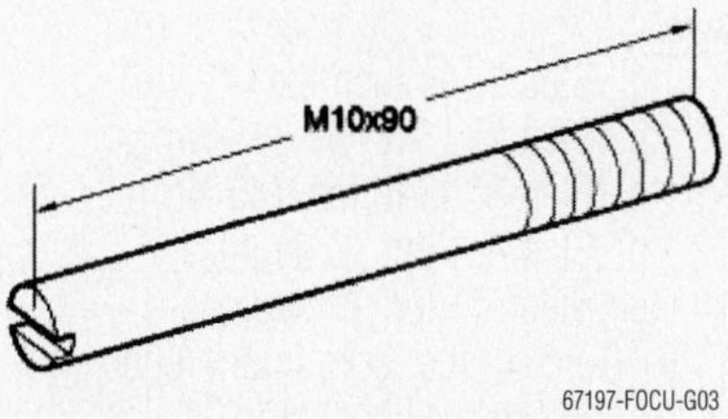

Locating stud—2.0L (VIN 5) engine

47. Attach the generator.
48. Attach the power steering pipe bracket to the cylinder head.
49. Install the intake manifold lower retaining bolt.
50. Connect the catalyst monitor sensor electrical connector and attach it to the bracket.
51. Connect the knock sensor electrical connector.
52. Install the exhaust manifold.

➡Install a new thermostat housing gasket.

53. Attach the thermostat housing. Tighten the bolts to 15 ft. lbs. (20 Nm).
54. Attach the fuel lines.
55. Attach the ground cable to the engine lifting bracket.
56. Connect the camshaft position (CMP) sensor connector.
57. Connect the fuel injector wiring harness.
58. Attach the accelerator cable and the speed control cable (if equipped) to the throttle body.
59. Install the splash shield.
60. Attach the intake manifold runner control (IMRC) actuator cable to the IMRC lever.
61. Attach the vacuum hoses to the intake manifold.
62. Install the air cleaner.
63. Raise and support the vehicle.
64. Connect the oil pressure switch electrical connector.
65. Attach the brake booster pipe to the intake manifold.
66. Drain and fill the engine with clean engine oil.
67. Connect the battery ground cable.
68. Fill and bleed the cooling system.

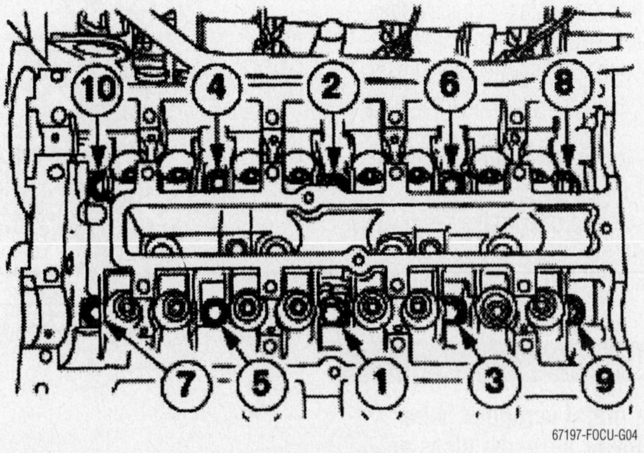

Cylinder head bolt torque sequence—2.0L (VIN 5) engine

2.3L (VIN Z) Engine

1. Before servicing the vehicle, refer to the precautions in the beginning of this section.
2. Disconnect the battery ground cable.
3. Drain the cooling system.

CAUTION

During engine repair procedures, cleanliness is extremely important. Any foreign material, including any material created while cleaning gasket surfaces, that enters the oil passages, coolant passages or the oil pan can cause engine failure.

CAUTION

The crankshaft, the crankshaft sprocket and the pulley are fitted together by friction, using diamond washers between the flange faces on each part. For that reason, the crankshaft sprocket is also unfastened if you loosen the pulley. Therefore the engine must be retimed each time the damper is removed. Otherwise severe damage can occur.

4. Remove the camshafts.
5. Remove the catalytic converter.
6. Remove the fuel injection supply manifold.
7. Remove the intake manifold.
8. Remove the fuel injection supply manifold spacers.
9. Disconnect the exhaust gas recirculation (EGR) vacuum regulator electrical connector.
10. Remove the hoses from the air control valve.
11. Disconnect the EGR valve electrical connector.
12. Remove the coolant hoses from the coolant bypass.
13. Remove the coolant bypass bolts and the coolant bypass.
14. Disconnect the EGR coolant hose.
15. Disconnect the radio interference capacitor.
16. Remove the bolts and the cylinder head. Discard the bolts.
17. Remove and discard the head gasket.

To install:

18. Inspect the cylinder head mating surfaces.

CAUTION

Do not use metal scrapers, wire brushes, power abrasive discs or other abrasive means to clean the sealing surfaces. These tools cause scratches and gouges which make leak paths.

19. Clean the cylinder head gasket surfaces with metal surface cleaner.
20. Apply silicone gasket and sealant to the locations shown.
21. Install a new head gasket.

➡The cylinder head bolts are torque-to-yield and must not be reused. New cylinder head bolts must be installed.

➡Lubricate the cylinder head bolts with clean engine oil.

22. Install the cylinder head and new bolts. Tighten the bolts in sequence as follows.
    a. Step 1: 44 inch lbs. (5 Nm)
    b. Step 2: 11 ft. lbs. (15 Nm)
    c. Step 3: 33 ft. lbs. (45 Nm)
    d. Step 4: Plus 90 degrees
    e. Step 5: Plus 90 degrees
23. Connect the radio interference capacitor electrical connector.
24. Install the EGR coolant hose.
25. Install the coolant bypass and tighten the bolts to 89 inch lbs. (10 Nm).

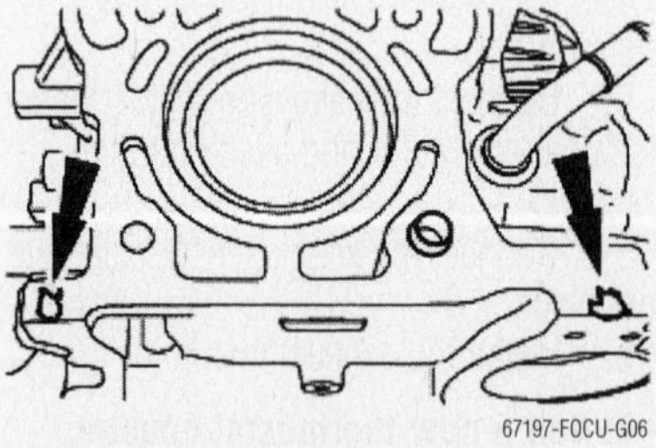

Apply silicone gasket sealant where indicated—2.3L (VIN Z) engine

26. Install the coolant hoses onto the coolant bypass.
27. Connect the EGR valve electrical connector.
28. Install the air control valve hoses.
29. Connect the EGR vacuum regulator electrical connector.
30. Install the fuel injector supply manifold spacers.
31. Install the intake manifold.
32. Install the fuel injector supply manifold.
33. Install the catalytic converter.
34. Install the camshafts.
35. Connect the battery ground cable.
36. Fill and bleed the cooling system.

Rocker Arms/Shafts

REMOVAL & INSTALLATION

2.0L (VIN P) Engine

1. Before servicing the vehicle, refer to the precautions in the beginning of this section.
2. Remove or disconnect the following:
    - Negative battery cable
    - Valve cover

➡Keep all valve train components in order for assembly.

- Rocker arm bolts
- Rocker arms and seats

To install:

3. Install or connect the following:
    - Seats and rocker arms in their original positions
    - Rocker arm bolts. Tighten to 17–22 ft. lbs. (23–30 Nm).
    - Valve cover
    - Negative battery cable

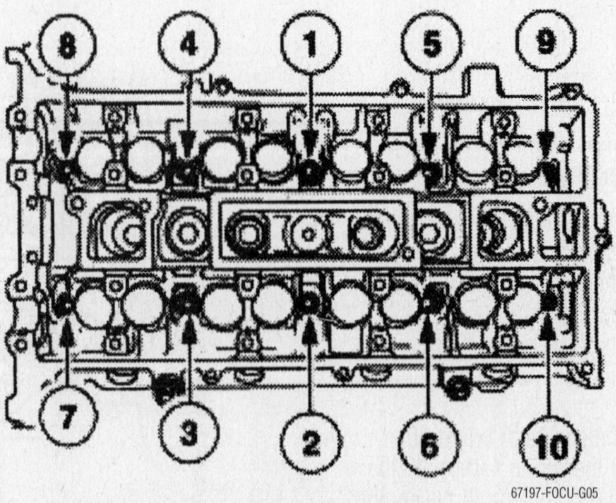

Cylinder head bolt torque sequence—2.3L (VIN Z) engine

2.0L (VIN 3 and 5) and 2.3L (VIN Z) Engines

These engines are not equipped with rocker arms. The camshafts act directly on the valves.

Intake Manifold

REMOVAL & INSTALLATION

2.0L (VIN P) Engine

1. Before servicing the vehicle, refer to the precautions in the beginning of this section.
2. Drain the cooling system.
3. Relieve the fuel system pressure.
4. Remove or disconnect the following:
   - Negative battery cable
   - Air cleaner outlet tube
5. Tag and disconnect the vacuum hoses from the following components:
   - Exhaust gas Recirculation (EGR) valve
   - Positive Crankcase Ventilation (PCV) valve
   - Throttle body
   - Fuel pressure regulator
   - Intake manifold
6. Remove or disconnect the following:
   - Accelerator cable, throttle control lever and if equipped, the speed control cable
   - Speed control cable bracket bolt, if equipped
   - Idle Air Control (IAC) valve and Throttle Position (TP) sensor electrical connections
   - Engine oil dipstick tube bracket bolt that attaches the tube to the intake manifold
   - EGR manifold tube located below the EGR valve
   - Engine oil dipstick tube from the block
   - Intake manifold lower nuts
   - Intake manifold upper nuts
   - Intake manifold

To install:

7. Clean and oil the intake manifold mounting studs.
8. Install or connect the following:
   - Intake manifold with a new gasket. Tighten the nuts to 15–22 ft. lbs. (20–30 Nm).
   - Engine oil dipstick tube
   - EGR manifold tube. Tighten it to 15–20 ft. lbs. (20–28 Nm).
   - IAC valve and TP sensor electrical connections
   - Speed control cable bracket bolt, if equipped. Tighten to 71–88 inch lbs. (8–10 Nm).
   - Throttle control lever, accelerator cable and if equipped, the speed control cable
9. Connect the vacuum hoses to the following components:
   - EGR valve
   - PCV valve
   - Throttle body
   - Fuel pressure regulator
   - Intake manifold
   - Air cleaner outlet tube
   - Negative battery cable
10. Fill the cooling system.
11. Start the engine and check for coolant leaks.

2.0L (VIN 3) Engine

1. Before servicing the vehicle, refer to the precautions in the beginning of this section.
2. Drain the cooling system.
3. Relieve the fuel system pressure.
4. Remove or disconnect the following:
   - Negative battery cable
   - Air cleaner outlet tube
   - Throttle Position (TP) sensor connector
   - Bolt that secures the pipe located by the crankshaft pulley
   - Heater hoses
   - Main engine control sensor wiring
   - Connectors from the mounting bracket
   - Intake manifold vacuum lines
   - Positive Crankcase Ventilation (PCV) hose
   - Drive belt
   - Alternator
   - Fuel line
   - Intake manifold

To install:

5. Install or connect the following:
   - Intake manifold. Tighten the fasteners to 13 ft. lbs. (18 Nm).
   - Fuel line
   - Alternator
   - Drive belt
   - PCV hose
   - Intake manifold vacuum lines
   - Connectors in the mounting bracket
   - Main engine control sensor wiring
   - Heater hoses
   - Bolt that secures the pipe located by the crankshaft pulley. Tighten the bolt to 71–97 inch lbs. (8–11 Nm).
   - TP sensor electrical connection
   - Air cleaner outlet tube
   - Negative battery cable
6. Fill the cooling system.
7. Start the engine and check for leaks.

2.0L (VIN 5) Engine

1. Before servicing the vehicle, refer to the precautions in the beginning of this section.
2. Remove the fuel injection supply manifold.
3. Raise and support the vehicle.
4. Remove the intake manifold lower retaining bolt.
5. Lower the vehicle.
6. Remove the engine lifting eye from the cylinder head.
7. Disconnect the idle air control (IAC) sensor electrical connector.
8. Disconnect the throttle position (TP) sensor electrical connector.
9. Remove the intake manifold runner control (IMRC) actuator cable from the IMRC lever.
10. Disconnect the IMRC actuator cable from the intake manifold assembly.
11. Disconnect the vacuum hoses from the throttle body.
12. Release the quick release coupling and pull out the brake booster pipe.
13. Separate the two sections of the intake manifold.
14. Disconnect the knock sensor (KS) electrical connector from the intake manifold inner section.
15. Remove the intake manifold inner section locknut and bolts. Discard the locknut.
16. Remove the intake manifold inner section retaining stud.
17. Remove the intake manifold sections.

To install:

➡Clean and inspect the intake manifold gasket. Install a new intake manifold gasket if necessary.

➡Install a new intake manifold locknut.

18. Install the intake manifold sections.
19. Install the intake manifold inner sec-

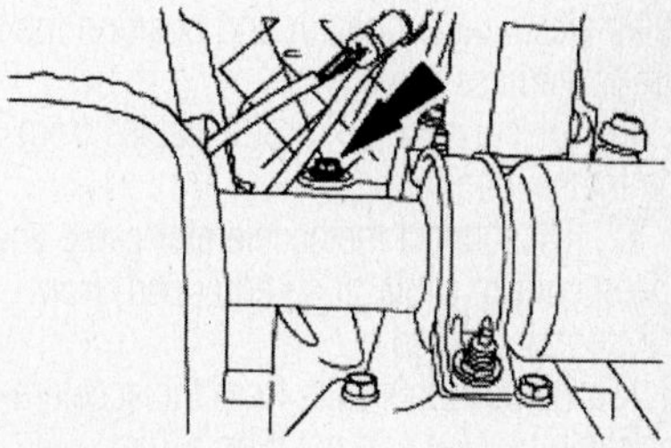

Intake manifold lower retaining bolt—2.0L (VIN 5) engine

tion retaining stud and tighten it to 88 inch lbs. (10 Nm).

20. Install the intake manifold inner section locknut and bolts. Tighten the fasteners to 13 ft. lbs. (18 Nm).

21. Connect the knock sensor (KS) electrical connector to the intake manifold inner section.

22. Connect the two sections of the intake manifold.

23. Install the brake booster pipe.

24. Connect the vacuum hoses from the throttle body.

25. Connect the IMRC actuator cable from the intake manifold assembly.

26. Install the intake manifold runner control (IMRC) actuator cable from the IMRC lever.

27. Connect the throttle position (TP) sensor electrical connector.

28. Connect the idle air control (IAC) sensor electrical connector.

29. Install the engine lifting eye from the cylinder head. Tighten the bolts to 26 ft. lbs. (35 Nm).

30. Raise the vehicle.

31. Install the intake manifold lower retaining bolt.

32. Lower the vehicle.

33. Install the fuel injection supply manifold.

2.3L (VIN Z) Engine

1. Before servicing the vehicle, refer to the precautions in the beginning of this section.

2. Remove the cooling fan motor and shroud.

3. Remove the lower intake manifold bolt.

4. Remove the accelerator control snow shield.

5. Remove the air cleaner outlet tube.

6. Loosen the retaining clamps.

7. Remove the emissions breather tube.

8. Remove the air cleaner outlet tube.

9. Disconnect the throttle position (TP) sensor electrical connector and position the wiring harness aside.

10. Disconnect the idle air control (IAC) valve electrical connector and position the wiring harness aside.

11. Disconnect the throttle cables from the intake manifold.

12. Disconnect the accelerator cable and speed control cable (if so equipped) from the throttle linkage.

13. Remove the bolts from the accelerator cable bracket and position aside.

14. Remove the evaporative emissions hoses.

15. Disconnect the temperature manifold absolute pressure (TMAP) sensor electrical connector.

16. Disconnect the electrical connectors and position the wiring harness aside.

17. Disconnect the outward intake manifold runner control (IMRC) actuator electrical connector.

18. Disconnect the vacuum hose from the fuel pressure regulator.

19. Disconnect the power brake booster vacuum tube.

20. Disconnect the quick release coupling and pull out the line.

21. Remove the two oil level indicator tube bolts.

➡There are two different size bolts used. Mark the location of the bolts to ensure installation in the correct location.

22. Remove the seven intake manifold bolts.

23. Raise the intake manifold enough to gain clearance to disconnect the knock sensor (KS) wiring harness.

24. Disconnect the positive crankcase ventilation (PCV) hose.

25. Remove the intake manifold.

To install:

26. Inspect and install new intake manifold gaskets, if necessary.

27. Install the PCV hose.

28. Connect the KS wiring harness.

➡Be sure to install the bolts in the previously marked locations.

29. Install the intake manifold and the seven mounting bolts. Tighten the bolts to 13 ft. lbs. (18 Nm).

30. Install the two oil level indicator tube bolts.

31. Reconnect the brake booster tube.

32. Install the fuel pressure regulator vacuum hose.

33. Connect the outward intake manifold runner control (IMRC) actuator electrical connector.

34. Connect the wiring harness and connectors.

35. Connect the TMAP sensor electrical connector.

36. Connect the evaporative emissions hoses.

37. Install the accelerator cable bracket.

38. Install the accelerator cable and speed control cable (if equipped).

39. Connect the IAC wiring harness and connector.

40. Connect the TP sensor electrical wiring harness and connector.

41. Install the air cleaner outlet tube.

42. Install the accelerator control snow shield.

43. Install the cooling fan motor and shroud.

44. Install the lower intake bolt.

Exhaust Manifold

REMOVAL & INSTALLATION

2.0L (VIN P) Engine

1. Before servicing the vehicle, refer to the precautions in the beginning of this section.
2. Remove or disconnect the following:
   - Negative battery cable
   - Drive belt
   - Heated Oxygen (HO_2S) sensor electrical connection
   - Exhaust manifold heat shield
   - Exhaust Gas Recirculation (EGR) tube
   - Power steering pressure hose bracket
   - Alternator
   - Catalytic converter
   - Exhaust manifold

To install:

3. Install or connect the following:
   - Exhaust manifold. Tighten the nuts to 15–18 ft. lbs. (20–24 Nm).
   - Catalytic converter. Tighten the nuts to 26–34 ft. lbs. (34–48 Nm).
   - Alternator
   - Power steering pressure hose bracket
   - EGR tube. Tighten the fasteners to 45–62 inch lbs. (5–7 Nm).
   - Exhaust manifold heat shield. Tighten the nuts to 45–62 inch lbs. (5–7 Nm).
   - HO_2S sensor electrical connection
   - Drive belt
   - Negative battery cable
4. Start the engine and check for leaks.

2.0L (VIN 3) Engine

1. Before servicing the vehicle, refer to the precautions in the beginning of this section.
2. Remove or disconnect the following:
   - Negative battery cable
   - Fan motor electrical connection
   - Fan shroud
   - Heated Oxygen (HO_2S) sensor electrical connection
   - Catalytic converter
   - Engine oil dipstick tube bracket bolt
   - Exhaust manifold heat shield
   - Exhaust manifold

To install:

3. Install or connect the following:
   - Exhaust manifold. Tighten the fasteners to 13 ft. lbs. (18 Nm).
   - Exhaust manifold heat shield. Tighten the retainers to 71–101 inch lbs. (8–11 Nm).
   - Engine oil dipstick tube bracket. Tighten the bolt to 71–101 inch lbs. (8–11 Nm).
   - Catalytic converter
   - HO_2sensor electrical connection
   - Fan shroud and the fan motor wiring
   - Negative battery cable
4. Start the engine and check for leaks.

2.0L (VIN 5) Engine

1. Before servicing the vehicle, refer to the precautions in the beginning of this section.
2. Remove the oil level indicator and tube.
3. Remove the radiator fan and shroud.
4. Remove the exhaust manifold heat shield lower retaining bolts.
5. Detach the exhaust manifold from the catalytic converter.
6. Lower the vehicle.
7. Remove the exhaust manifold heat shield.
8. Remove the exhaust manifold. Discard the gaskets.

To install:

➡**Install new exhaust manifold gaskets.**

9. Install the exhaust manifold. Tighten the fasteners to 13 ft. lbs. (18 Nm).
10. Install the heat shield and tighten the bolts to 88 inch lbs. (10 Nm).
11. Raise and support the vehicle.
12. Attach the catalytic converter and tighten the bolts to 35 ft. lbs. (47 Nm).
13. Install the radiator fan and shroud.
14. Install the oil level indicator and tube. Tighten the bolts to 15 ft. lbs. (20 Nm).

2.3L (VIN Z) Engine

➡**The exhaust manifold is integral to the catalytic converter.**

1. Before servicing the vehicle, refer to the precautions in the beginning of this section.
2. Position the vehicle on a hoist.
3. Relieve the fuel system pressure.
4. Remove the two catalytic converter-to-muffler assembly flange nuts. Discard the nuts.
5. Disconnect the catalyst monitor and the heated oxygen sensor (HO2S) electrical connectors.
6. Remove the two catalytic converter-to-engine bracket bolts.
7. Remove the support bracket bolts.
8. Remove the two lower heat shield screws.
9. Remove and discard the lower two catalytic converter nuts.
10. Lower the vehicle.
11. Disconnect the upper HO2S electrical connector.
12. Disconnect the wiring connector, then remove the bolt and the bracket.
13. Disconnect the fuel lines.
14. Remove the two bolts and the fuel canister assembly.
15. Remove the vapor management valve (VMV).
16. Disconnect the VMV electrical connector.
17. Remove the upper fuel line.
18. Remove the lower fuel line.
19. Remove the two VMV bolts and the VMV.
20. Remove the upper three catalytic converter nuts and studs. Discard the nuts.
21. Remove the PCV hose and the secondary air hose.
22. Disconnect the heat shield from the bulkhead, then remove the heat shield.
23. Position the power steering reservoir aside.
24. Remove the catalytic converter.

➡**If installing a new converter, remove the catalyst monitor sensors. Remove the HO2S and the catalyst monitor.**

➡**Make sure to apply anti-seize lubricant to the threads of the sensor before installation.**

➡**Clean the mating surfaces of the exhaust manifold and the catalytic converter mating surfaces.**

➡**Always install new fasteners and gaskets.**

➡**Do not tighten the fasteners until all components are assembled. Make sure to tighten all fasteners beginning at the front of the vehicle.**

25. To install, reverse the removal procedure.
26. Tighten the nuts to 41 ft. lbs. (55 Nm) in the sequence shown.

Front Crankshaft Seal

REMOVAL & INSTALLATION

All 2.0L Engines

1. Before servicing the vehicle, refer to the precautions in the beginning of this section.
2. Remove the timing belt and crankshaft sprocket.

**** WARNING**

Be careful not to damage the crankshaft surface when removing the seal.

3. Using seal remover tool 303-293, remove the crankshaft front oil seal.

To install:

4. Use seal replacer tools 303-684 and 303-395 to install the new seal.
5. Install the crankshaft sprocket and the timing belt.

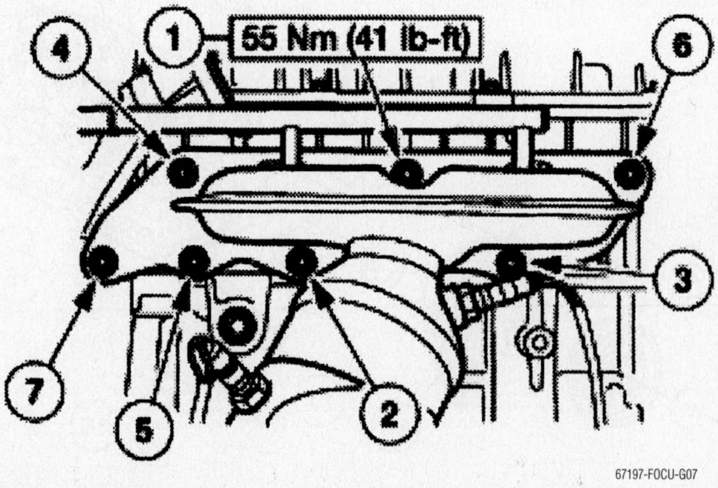

Exhaust manifold/catalytic converter torque sequence—2.3L (VIN Z) engine

Camshaft and Valve Lifters

REMOVAL & INSTALLATION

2.0L (VIN P) Engine

1. Before servicing the vehicle, refer to the precautions in the beginning of this section.
2. Disconnect the negative battery cable.
3. Remove the air cleaner and valve cover.
4. Remove the camshaft front seal as follows:
   a. Align the timing marks and remove the timing belt.
   b. Use cam sprocket holding/removing tool T74P-6256-B, and remove the camshaft sprocket.
   c. Unfasten the timing belt tensioner bolt and remove the tensioner.
   d. Remove the inner engine front cover.
   e. Use seal removal tool T92C-6700-CH, to remove the camshaft front seal.
5. Remove or disconnect the following:
   - Ignition coil and bracket
   - Rocker arms and valve tappets
   - Camshaft thrust plate bolts and the plate
   - Cup plug from the rear of the cylinder head and discard
   - Camshaft from the rear of the cylinder head

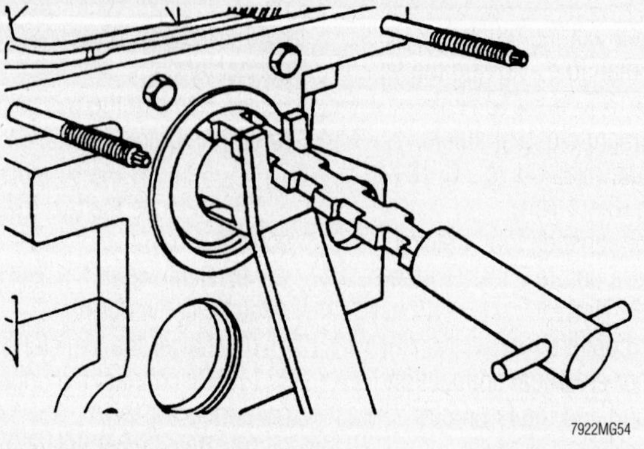

Removing the camshaft front seal—2.0L (VIN P) engine

To install:

➡Liberally coat the cam bore in the cylinder with clean 5W-30 motor oil.

6. Install or connect the following:
   - Camshaft through the rear of the cylinder head
   - Camshaft thrust plate and the bolts. Tighten the bolts to 71–115 inch lbs. (8–13 Nm).
   - New cup plug and the tappets
   - Rocker arms
   - Ignition coil bracket and coil
7. Install the camshaft front seal as follows:
   a. Apply a thin film of 5W-30 motor oil to the lip of the seal.
   b. Use seal replacer T81P-6292-A, to install the camshaft front seal.

1. Rocker arm
2. Rocker arm seats
3. Camshaft Position (CMP) sensor
4. Valve tappet
5. Valve tappet guide plate
6. Ignition coil
7. Radio interference capacitor
8. Ignition coil bracket
9. Cup plug
10. Camshaft
11. Water hose connection
12. Thermostat
13. Water hose connection gasket

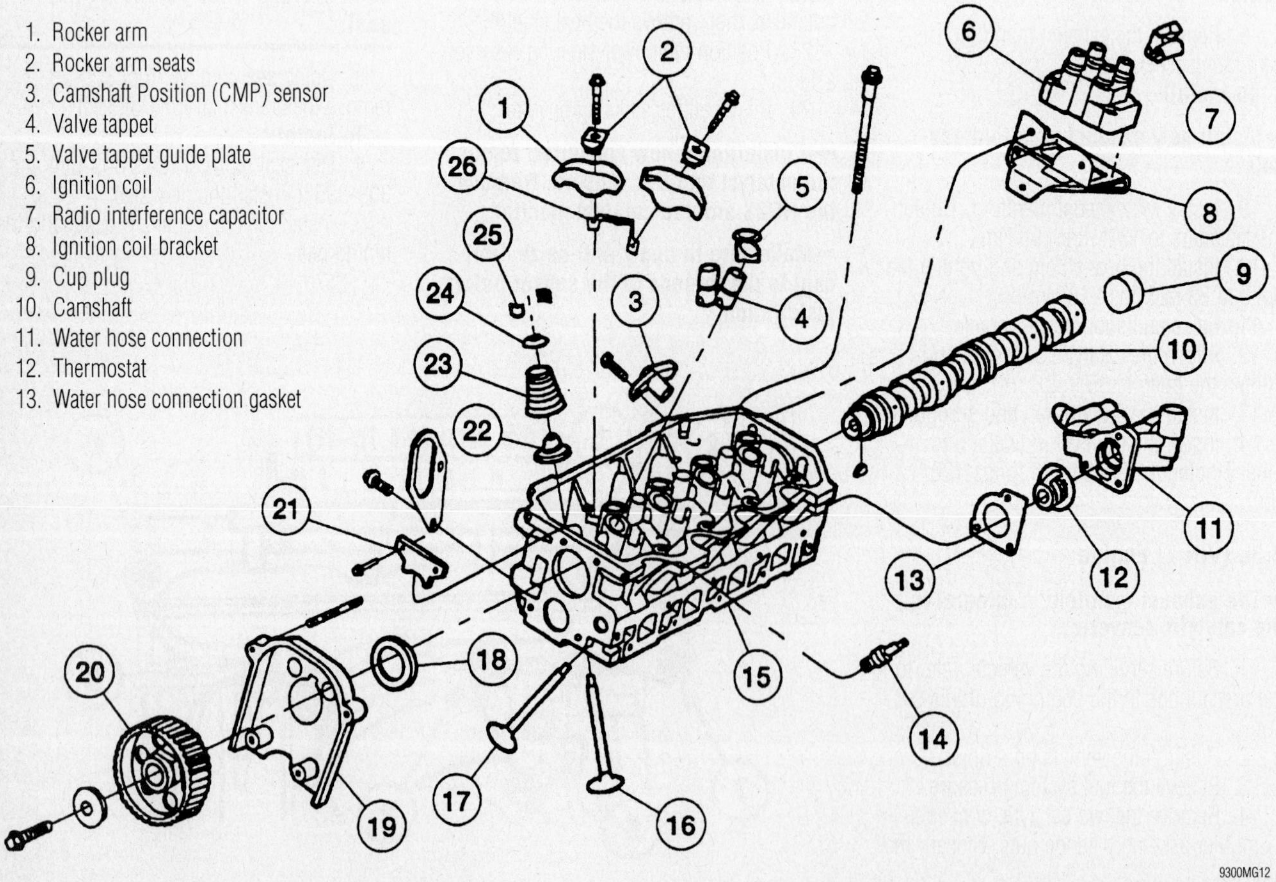

Exploded view of the cylinder head—2.0L (VIN P) engine

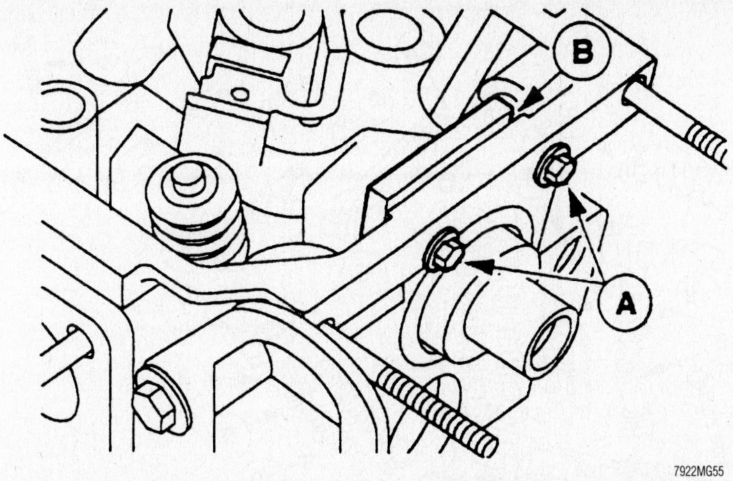

Remove the camshaft thrust plate retaining bolts (A) and the plate (B)—2.0L (VIN P) engine

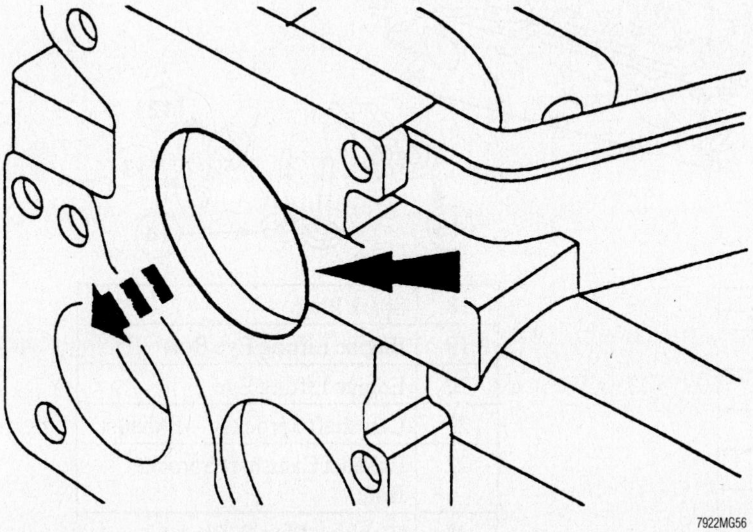

Remove and discard the cup plug located at the rear of the cylinder head—2.0L (VIN P) engine

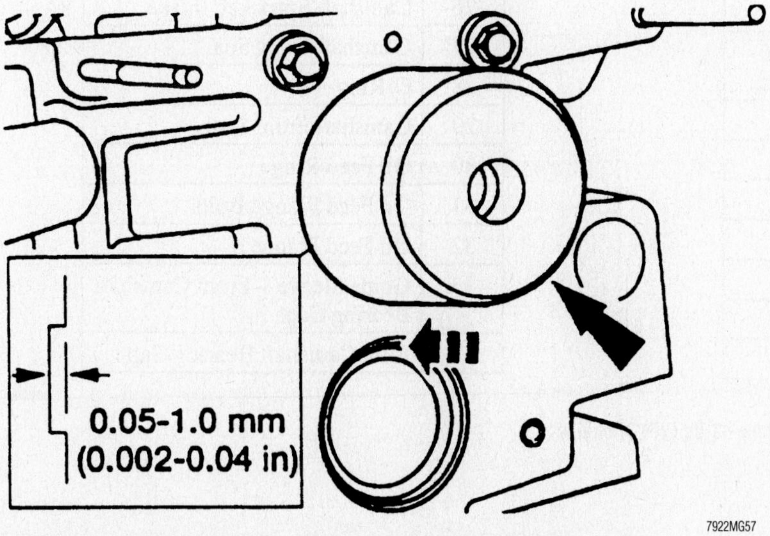

Use seal Replacer T81P-6292-A, to install the camshaft front seal 0.002–0.04 in. (0.05–1.0mm) below flush with the cylinder head front face—2.0L (VIN P) engine

➡The seal depth should be 0.002–0.04 in. (0.05–1.0mm) below flush with the cylinder head front face.

c. Install the inner engine front cover.

d. Install the timing belt tensioner. Tighten the tensioner bolt to 15–22 ft. lbs. (20–30 Nm).

e. Install the camshaft sprocket and the timing belt

8. Install the valve cover and the air cleaner assembly.

9. Connect the negative battery cable.

2.0L (VIN 3) Engine

1. Before servicing the vehicle, refer to the precautions in the beginning of this section.
2. Remove or disconnect the following:
   - Timing belt
   - Valve cover and camshaft sprockets
   - Oil control solenoid flange

➡Mark the camshaft journal caps with a number to identify their location.

- Camshaft journal caps. Loosen the bolts in sequence and in several passes.
- Camshafts
- Oil control sensor and bushing

To install:

3. Install the oil control solenoid bushing and flange on the exhaust camshaft.
4. Coat the surface of the front camshaft journal cap with gasket maker E2AZ-19562-B.
5. Place the camshafts in position and lubricate the bearing surfaces with engine assembly lubricant D9AZ-19579-D.
6. Install new camshaft front oil seals.
7. Apply a thin coat of silicone gasket and sealant F6AZ-19562-AA to the sealing surface of the of the front camshaft journal bearing cap.
8. Install the caps and tighten the bolts in several 2-turn passes in the sequence illustrated to 10–12 ft. lbs. (13–17 Nm).
9. Inspect the oil control solenoid flange O-rings for damage or wear and replace, as necessary.
10. Install the oil control solenoid flange and tighten the bolts to 84–92 inch lbs. (9.5–10.5 Nm) in the sequence.
11. Rotate the camshafts a full turn and check for binding.
12. Install or connect the following:
    - Camshaft sprockets and timing belt
    - Valve cover

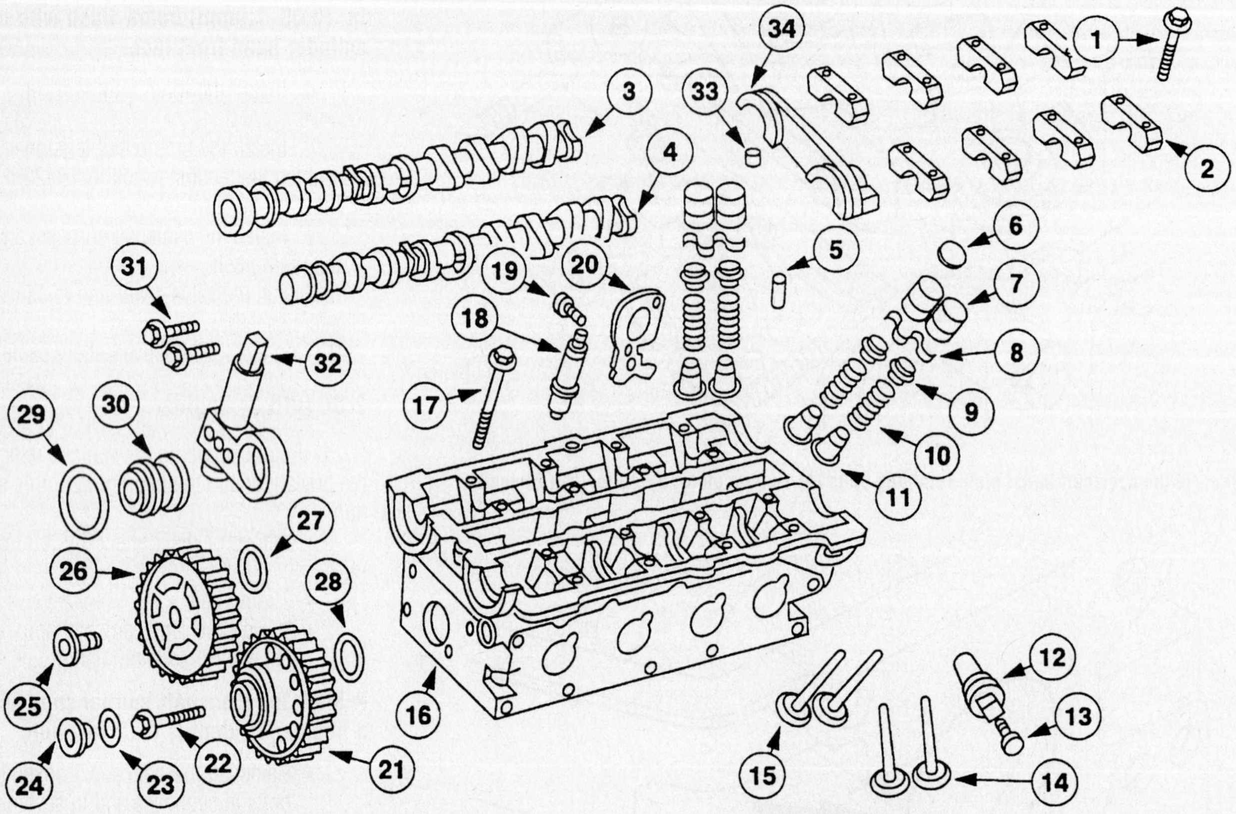

1	Bolt—Camshaft Bearing Cap	18	Spark Plug
2	Camshaft Bearing Cap	19	Engine Lifting Eye Bolt
3	Camshaft Intake	20	Engine Lifting Eye
4	Camshaft Exhaust	21	Camshaft Sprocket —Exhaust
5	Plug	22	Exhaust Camshaft Sprocket Bolt
6	Adjusting Shim—Valve Clearance	23	Blanking Plug O-Ring
7	Valve Tappets	24	Blanking Plug
8	Valve Spring Retainer Key	25	Bolt—Intake Camshaft
9	Valve Spring Retainers	26	Camshaft Sprocket Intake
10	Valve Spring (Color Coding: Exhaust = Blue, Intake = Red)	27	Camshaft Front Seal
11	Valve Stem Seal	28	O-Ring
12	Camshaft Position Sensor	29	Camshaft Front Seal
13	CMP Sensor Bolt	30	Oil Feed Ring
14	Intake Valves	31	Oil Feed Flange Bolts
15	Exhaust Valves	32	Oil Feed Flange
16	Cylinder Head	33	Guide Sleeve—Front Camshaft Bearing Cap
17	Cylinder Head Bolts	34	Fifth Camshaft Bearing Cap

9300MG06

Exploded view of the cylinder head and camshaft mounting—2.0L (VIN 3) engine

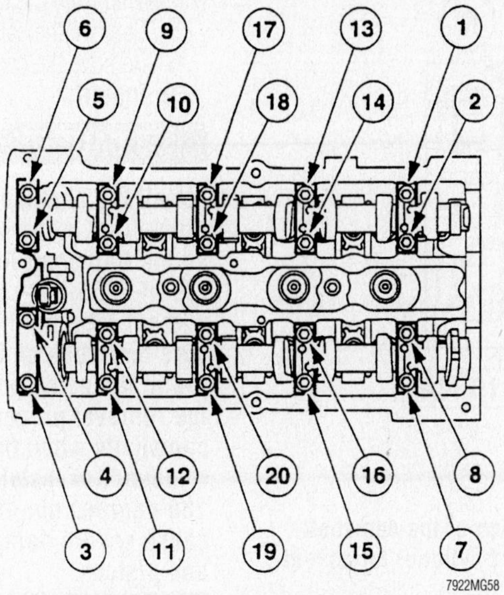

Camshaft journal cap loosening sequence—2.0L (VIN 3) engine

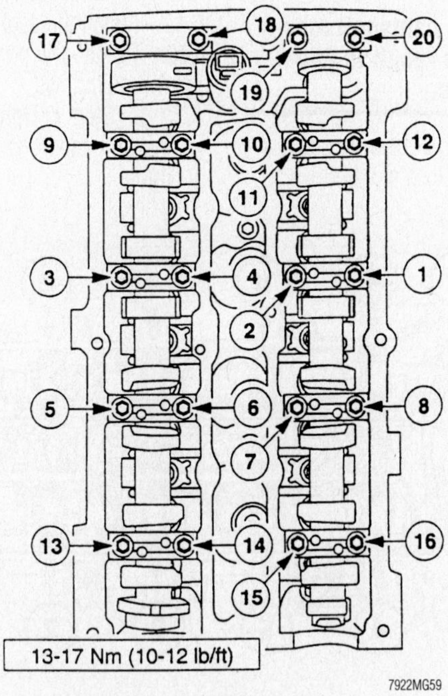

Camshaft journal cap torque sequence—2.0L (VIN 3) engine

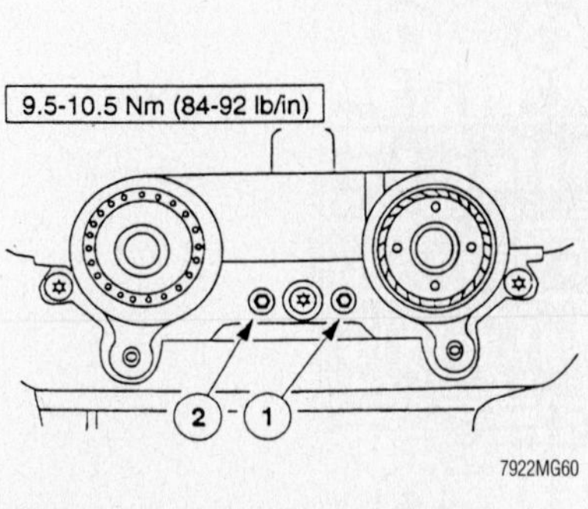

Tighten the oil control solenoid flange bolts in this sequence—2.0L (VIN 3) engine

2.0L (VIN 5) Engine

1. Before servicing the vehicle, refer to the precautions in the beginning of this section.
2. Remove the timing belt.

➡Hold the camshafts by the hexagon with an open ended wrench to stop them from rotating.

3. Remove the camshaft pulleys.
4. Remove the oil feed flange. Discard the camshaft oil seals and the oil feed flange oil seal.

⁂ CAUTION

Loosen the camshaft bearing caps in the sequence shown.

5. Working in several stages, evenly loosen each bolt two turns at a time in the indicated sequence and remove the bolts.
6. Remove the camshafts.
7. Remove the tappets.

➡Keep the valvetrain components in order for installation.

To install:

8. Install the tappets in their original locations.
9. Install the camshafts.
10. Working in several stages, install the camshaft bearing caps in the indicated sequence, tighten the bolts evenly one half turn at a time.
11. Tighten the bolts in the sequence shown to 11 ft. lbs. (15 Nm).
12. Install a new oil feed flange seal.
13. Install the oil feed flange and tighten the bolts to 11 ft. lbs. (15 Nm).
14. Using the Camshaft Oil Seal Installer 303-160, install the camshaft oil seals.
15. Lubricate the camshaft and oil seal lip with clean engine oil.
16. Install the camshaft pulleys.

➡Do not tighten the camshaft pulley retaining bolts at this stage.

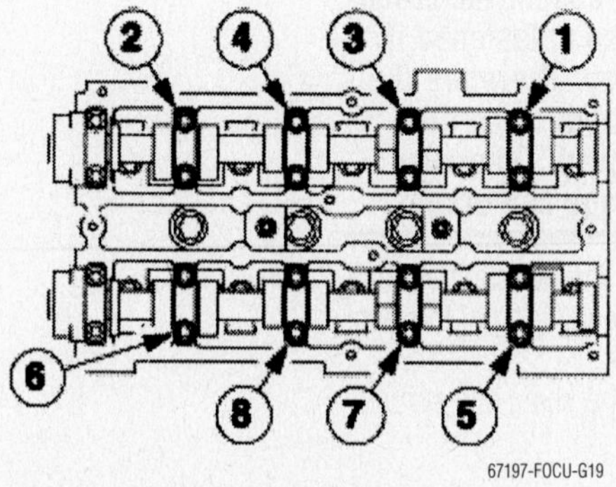

Camshaft bolt removal sequence—2.0L (VIN 5) engine

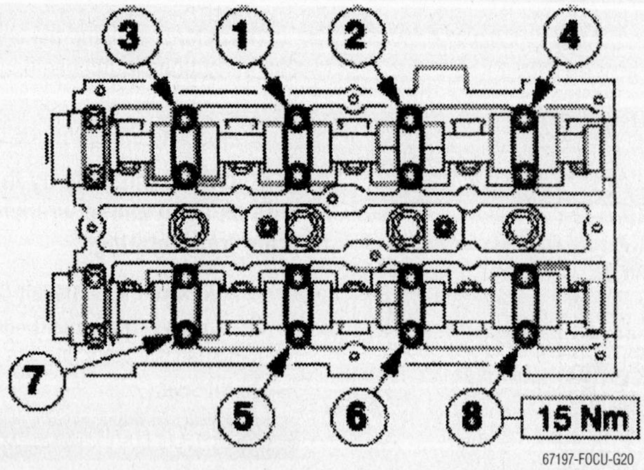

Camshaft torque sequence—2.0L (VIN 5) engine

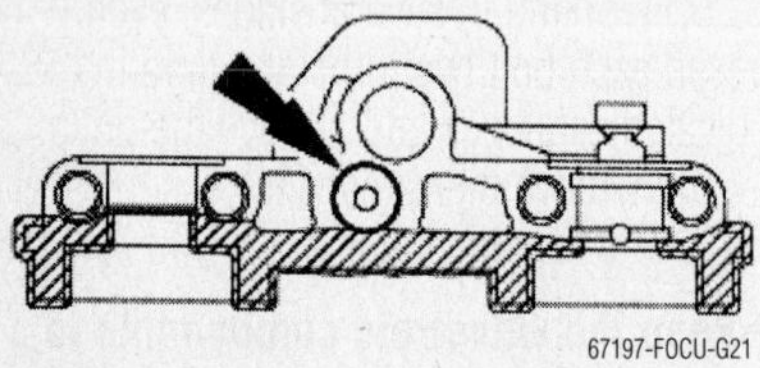

Apply sealant to the marked areas on the oil feed flange—2.0L (VIN 5) engine

17. Install the timing belt.

2.3L (VIN Z) Engine

⁂ CAUTION

During engine repair procedures, cleanliness is extremely important. Any foreign material, including any material created while cleaning gasket surfaces, that enters the oil passages, coolant passages or the oil pan can cause engine failure.

⁂ CAUTION

The crankshaft, the crankshaft sprocket and the pulley are fitted together by friction, using diamond washers between the flange faces on each part. For that reason, the crankshaft sprocket is also unfastened if you loosen the pulley. Therefore the engine must be retimed each time the damper is removed. Otherwise severe engine damage can occur.

1. Before servicing the vehicle, refer to the precautions in the beginning of this section.
2. Remove the timing chain and sprockets.

➡Mark the position of the camshaft lobes on the No. 1 cylinder for assembly reference.

⁂ CAUTION

Failure to follow the camshaft loosening procedure can result in damage to the camshafts.

3. Loosen the camshaft bearing cap bolts, in sequence, one turn at a time.
4. Repeat the first step until all tension is released from the camshaft bearing caps.
5. Remove the camshaft bearing caps.
6. Remove the camshafts.

To install:

⁂ CAUTION

Install the camshafts with the alignment slots in the camshafts lined up so the Camshaft Alignment Plate (T94P-6256-CH) can be installed without rotating the camshafts. Make sure the lobes on the No. 1 cylinder are in the same position as noted in the removal procedure. Rotating the camshafts when the timing chain is removed, or installing the camshafts 180 degrees out of position, can cause severe damage to the valves and pistons.

➡Lubricate the camshaft journals and bearing caps with clean engine oil.

7. Install the camshafts and bearing caps. Tighten the bolts in the sequence shown as follows:
   a. Step 1: Tighten the camshaft bearing bolt caps one turn at a time until tight.

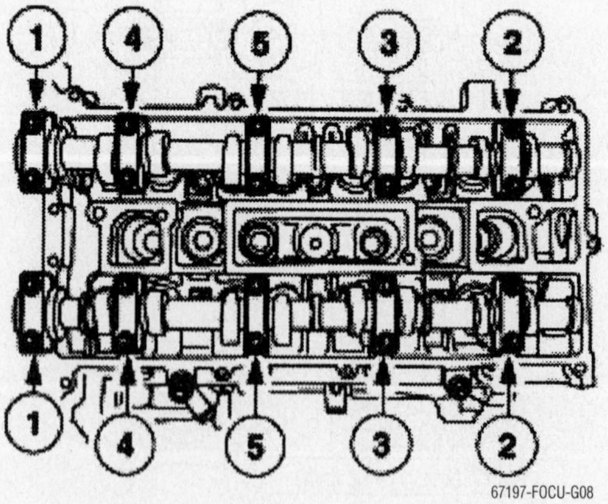

Camshaft bolt loosening sequence—2.3L (VIN Z) engine

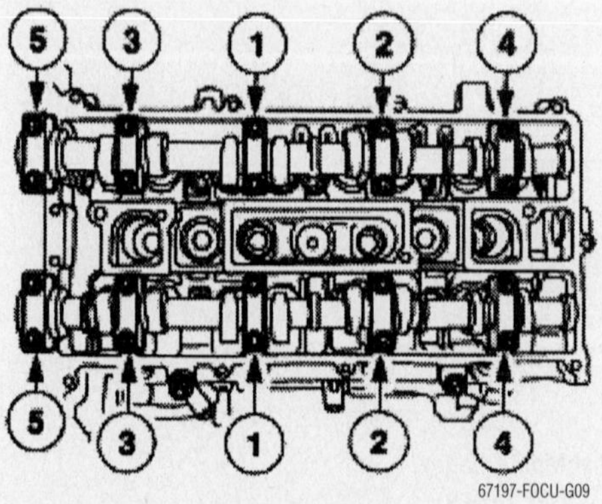

Camshaft bolt tightening sequence—2.3L (VIN Z) engine

b. Step 2: Tighten the bolts to 62 inch lbs. (7 Nm)
c. Step 3: Tighten the bolts to 12 ft. lbs. (16 Nm)
8. Install the camshaft drive gears and hand-tighten the bolts.
9. Install the timing chain and sprockets.

Valve Lash

ADJUSTMENT

2.0L (VIN P) Engine

The lash adjusters are hydraulic and are not adjustable.

2.0L (VIN 3) Engine

1. Before servicing the vehicle, refer to the precautions in the beginning of this section.
2. Remove the valve cover.
3. Rotate the crankshaft so that the valve is closed and measure the clearance between the lifter shim and the camshaft base circle. Intake valve clearance should be 0.004–0.007 inches. Exhaust valve clearance should be 0.010–0.013 inches.
4. Measure each valve and note the clearance.
5. If adjustment is necessary, remove the camshafts and replace the shims. To obtain the correct shim size, measure the current shim, add the measured clearance, and subtract 0.006 inches for intake valves or 0.012 inches for exhaust valves.
6. Install or connect the following:
   - Camshafts
   - Valve cover

2.0L (VIN 5) Engine

1. Before servicing the vehicle, refer to the precautions in the beginning of this section.
2. Remove the valve cover.
3. Rotate the crankshaft so that the valve is closed and measure the clearance between the lifter shim and the camshaft base circle. Intake valve clearance should be 0.008–0.0011 inches. Exhaust valve clearance should be 0.013–0.016 inches.
4. Measure each valve and note the clearance.
5. If adjustment is necessary, remove the camshafts and replace the shims. To obtain the correct shim size, measure the current shim, add the measured clearance, and subtract 0.009 inches for intake valves or 0.014 inches for exhaust valves.
6. Install or connect the following:
   - Camshafts
   - Valve cover

2.3L (VIN Z) Engine

1. Before servicing the vehicle, refer to the precautions in the beginning of this section.
2. Remove the valve cover.
3. Rotate the crankshaft so that the valve is closed and measure the clearance between the lifter shim and the camshaft base circle. Intake valve clearance should be 0.008–0.0011 inches. Exhaust valve clearance should be 0.010–0.013 inches.
4. Measure each valve and note the clearance.
5. If adjustment is necessary, remove the camshafts and replace the shims. To obtain the correct shim size, measure the current shim, add the measured clearance, and subtract 0.009 inches for intake valves or 0.011 inches for exhaust valves.
6. Install or connect the following:
   - Camshafts
   - Valve cover

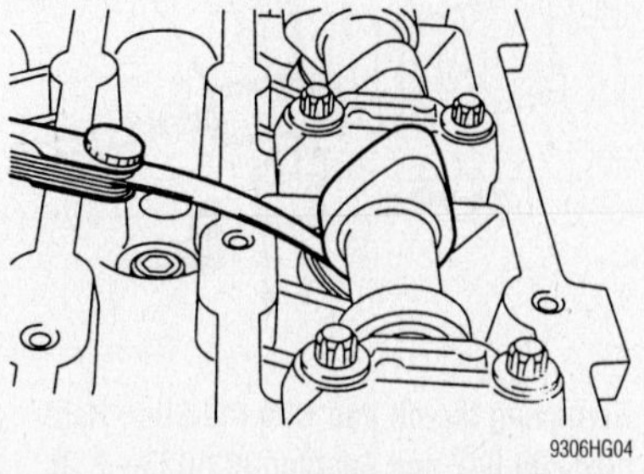

Checking valve clearance—2.0L engine

Starter Motor

REMOVAL & INSTALLATION

2.0L (VIN P) Engine

1. Before servicing the vehicle, refer to the precautions in the beginning of this section.
2. Remove or disconnect the following:
   - Negative battery cable
   - Air cleaner outlet tube
   - Two top starter motor bolts
   - **S** terminal wire and the **B** terminal nut and cable from the solenoid
   - Lower starter motor bolt
   - Starter

To install:

3. Install or connect the following:
   - Starter motor assembly and tighten the lower mounting bolt to 18–20 ft. lbs. (25–27 Nm)
   - **S** terminal wire, **B** terminal cable and nut. Tighten the nut to 80–120 inch lbs. (9–13.5 Nm).
   - Top starter motor bolts. Tighten the bolts to 18–20 ft. lbs. (25–27 Nm).
   - Air cleaner outlet tube
   - Negative battery cable

2.0L (VIN 3) Engine

1. Before servicing the vehicle, refer to the precautions in the beginning of this section.
2. Remove or disconnect the following:
   - Negative battery cable
   - Air cleaner
   - Starter electrical connectors
   - Starter

To install:

3. Install or connect the following:
   - Starter. Tighten the bolts to 26 ft. lbs. (35 Nm).
   - Starter electrical connectors
   - Air cleaner
   - Negative battery cable

2.0L (VIN 5) Engine

1. Before servicing the vehicle, refer to the precautions in the beginning of this section.
2. Remove the intake manifold.
3. Disconnect the starter motor electrical connectors.
4. Remove the starter motor.

To install:

5. Install the starter motor and tighten the bolts to 26 ft. lbs. (35 Nm).
6. Install the starter electrical connectors.
7. Install the intake manifold.

2.3L (VIN Z) Engine

1. Before servicing the vehicle, refer to the precautions in the beginning of this section.
2. Disconnect the battery ground cable.
3. With the vehicle in NEUTRAL, position it on a hoist.
4. Remove the terminal nuts and disconnect the wiring.

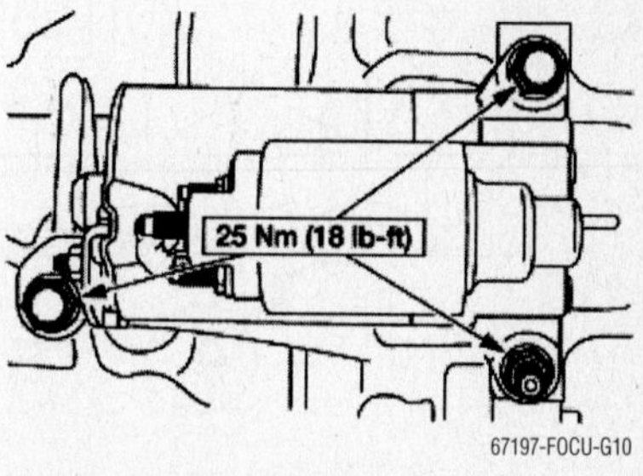

Starter motor—2.3L (VIN Z) engine

5. Remove the nuts and disconnect the power steering pressure line brackets from the stud bolts.
6. Remove the bolts and the starter motor.
7. To install, reverse the removal procedure. Tighten the starter bolts to 18 ft. lbs. (25 Nm).

Oil Pan

REMOVAL & INSTALLATION

2.0L (VIN P) Engine

1. Before servicing the vehicle, refer to the precautions in the beginning of this section.
2. Drain the engine oil.
3. Remove or disconnect the following:
   - Negative battery cable
   - Catalytic converter and brackets
   - Intake manifold bracket
   - Intermediate shaft bracket
   - Coolant pipe bolt
   - Oil pan

To install:

4. Apply silicone sealant to the oil pump and rear seal retainer joints.
5. Install or connect the following:
   - Oil pan. Tighten the bolts in sequence to 18 ft. lbs. (25 Nm).
   - Coolant pipe bolt
   - Intermediate shaft bracket
   - Intake manifold bracket
   - Catalytic converter and brackets
   - Negative battery cable
6. Fill the crankcase to the correct level.
7. Start the engine and check for leaks.

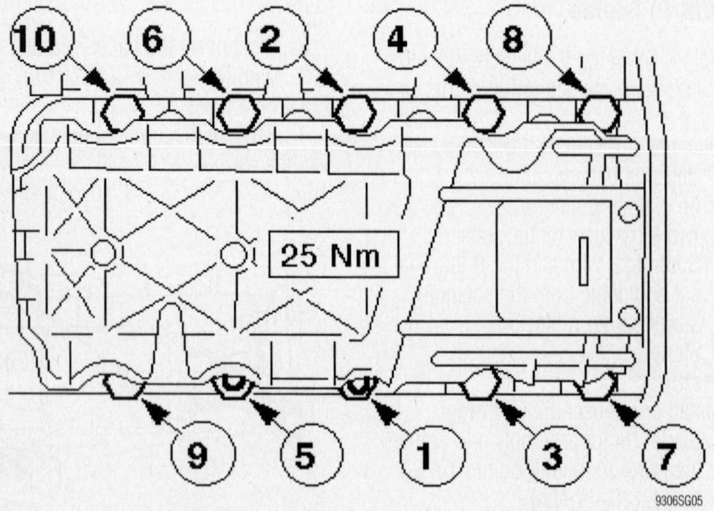

Oil pan torque sequence—2.0L (VIN P) engine

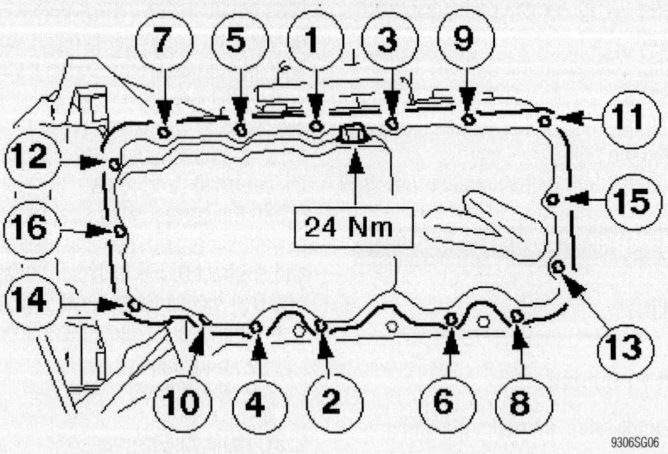

Oil pan torque sequence—2.0L (VIN 3) engine

2.0L (VIN 3) Engine

1. Before servicing the vehicle, refer to the precautions in the beginning of this section.
2. Drain the engine oil.
3. Remove or disconnect the following:
   - Negative battery cable
   - Catalytic converter
   - Oil pan

To install:

➡**Apply a bead of silicone gasket sealer to the oil pan flange.**

4. Install or connect the following:
   - Oil pan. Tighten the bolts in sequence and in several passes to 18 ft. lbs. (24 Nm).
   - Oil lever sensor connector
   - Catalytic converter
   - Negative battery cable
5. Fill the crankcase to the correct level.
6. Start the engine and check for leaks.

2.0L (VIN 5) Engine

1. Before servicing the vehicle, refer to the precautions in the beginning of this section.
2. Remove the catalytic converter.
3. Drain the engine oil.
4. Remove the oil pan retaining bolts.

⁂ CAUTION

To prevent damage to the sealing surface, only use the Slide Hammer 205-047 and Adapter 303-633 as shown to remove the oil pan.

5. Tighten the locknut against the oil pan and tap the special tool several time to release the oil pan from the lower crankcase.

To install:

➡**Do not damage the mating faces.**

6. Clean the mating faces. The mating faces must be free from oil and sealant residue.
7. Clean all traces of oil residue and oil sludge from the oil pan.

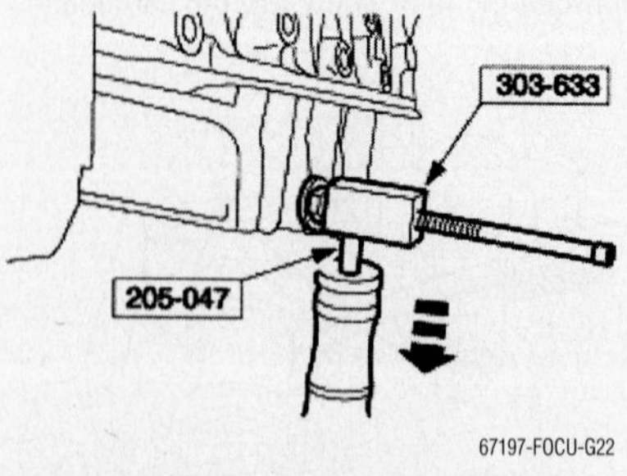

Removing the oil pan with the Slide Hammer 205-047 and Adapter 303-633—2.0L (VIN 5) engine

⁂ CAUTION

The lower crankcase may be damaged if sealant enters the blind holes.

8. Install ten M6 x 20 mm studs into the blind holes shown (engine shown removed for clarity).

➡Attach the oil pan within ten minutes of applying the sealant.

9. Apply a 3 mm diameter bead of sealant to the mating face of the oil pan.

➡The oil pan must not be removed once it comes into contact with the lower crankcase housing.

10. Attach the oil pan to the lower crankcase housing.
11. Remove the studs.
12. Install the oil pan retaining bolts and tighten them in sequence as follows:
    a. Step 1: 53 inch lbs. (6 Nm)
    b. Step 2: 88 inch lbs. (10 Nm)

➡Inspect the oil drain plug and gasket. Install a new oil drain plug and gasket if necessary.

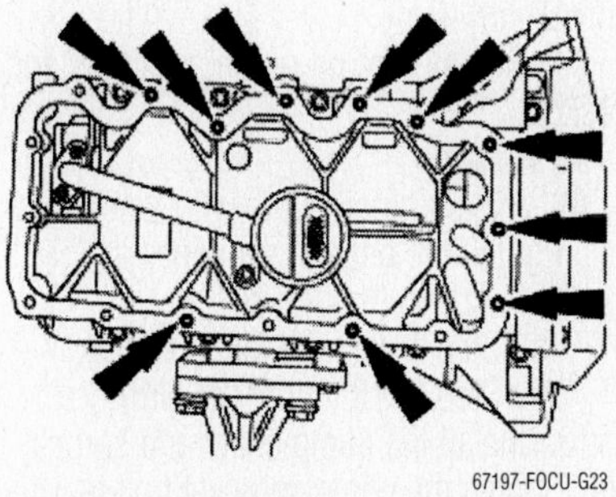

Install ten M6 x 20 mm studs into the blind holes indicated

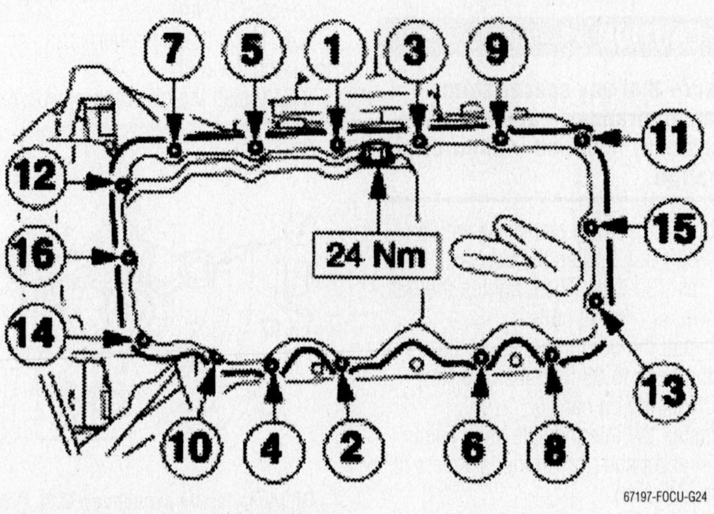

Oil pan torque sequence—2.0L (VIN 5) engine

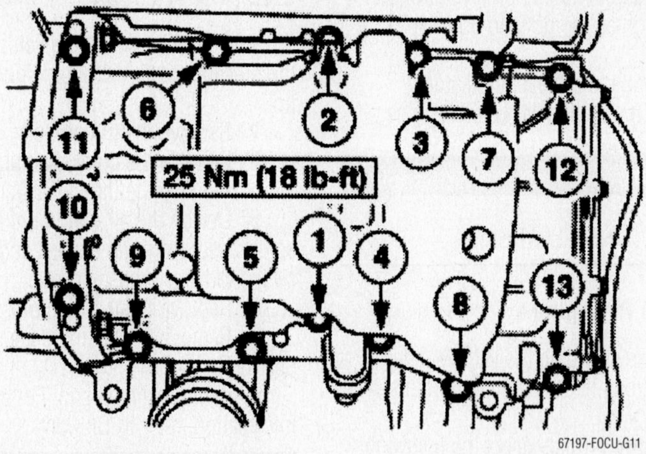

Oil pan bolt torque sequence—2.3L (VIN Z) engine

13. Install the oil pan drain plug and tighten it to 17 ft. lbs. (24 Nm).
14. Install the catalytic converter.
15. Fill the engine with clean oil.

2.3L (VIN Z) Engine

1. Before servicing the vehicle, refer to the precautions in the beginning of this section.
2. With the vehicle in NEUTRAL, position it on a hoist.
3. Drain the engine oil.
4. Remove the engine oil level indicator.
5. Remove the engine oil level indicator tube.
6. Remove the two nuts from the studs.
7. Remove the wiring harness from the studs and position aside.
8. Remove the nuts and studs.
9. Remove the oil pan-to-bellhousing bolts.
10. Remove the bolts and oil pan.

To install:

⁂ CAUTION

Do not use metal scrapers, wire brushes, power abrasive discs or other abrasive means to clean the sealing surfaces. These tools cause scratches and gouges, which make leak paths. Use a plastic scraping tool to remove traces of sealant.

11. Clean and inspect all mating surfaces.

➡If the oil pan is not secured within four minutes of sealant application the sealant must be removed and the sealing area cleaned with metal surface cleaner. Allow to dry until there is no sign of wetness, or four minutes, whichever is longer. Failure to follow this procedure can cause future oil leakage.

➡The oil pan must be installed and the bolts tightened within four minutes of applying the silicone gasket and sealant.

12. Apply a 2.5 mm bead of silicone gasket and sealant to the oil pan.
13. Install the oil pan.
14. Install the oil pan-to-bellhousing bolts and tighten them to 35 ft.. lbs. (48 Nm).
15. Install the nuts and the studs and tighten them to 89 inch lbs. (10 Nm).
16. Tighten the oil pan bolts in sequence to 18 ft. lbs. (25 Nm).
17. Install the wiring harness onto the studs.

➡Lubricate the O-ring with clean engine oil.

18. Install the engine oil level indicator tube and tighten the bolts to 89 inch lbs. (10 Nm).
19. Install the oil level indicator.
20. Fill the engine with clean engine oil.

Oil Pump

REMOVAL & INSTALLATION

2.0L (VIN P) Engine

1. Before servicing the vehicle, refer to the precautions in the beginning of this section.
2. Remove or disconnect the following:
    - Timing belt
    - Oil pan
    - Crankshaft Position (CKP) sensor
    - Oil pump pickup tube
    - Oil pump

To install:

3. Lubricate the crankshaft front oil seal lip with clean engine oil.
4. Install or connect the following:
    - Oil pump. Tighten the bolts to 10–12 ft. lbs. (13–16 Nm).
    - Oil pump pickup tube. Tighten the retaining bolts to 71–97 inch lbs. (8–11 Nm).
    - CKP sensor
    - Oil pan
    - Timing belt
5. Fill the crankcase to the correct level.
6. Start the engine and check for leaks.

2.0L (VIN 3) Engine

1. Before servicing the vehicle, refer to the precautions in the beginning of this section.
2. Drain the engine oil.
3. Remove or disconnect the following:
    - Negative battery cable
    - Timing belt
    - Oil pan
    - Oil pump pickup tube
    - Oil pump

To install:

4. Install or connect the following:
    - Oil pump. Tighten the bolts to 97 inch lbs. (11 Nm)
    - Oil pump pickup tube. Tighten the fasteners to 88 inch lbs. (10 Nm).
    - Oil pan
    - Timing belt
    - Negative battery cable
5. Fill the crankcase to the correct level.
6. Start the engine and check for leaks.

2.0L (VIN 5) Engine

1. Before servicing the vehicle, refer to the precautions in the beginning of this section.
2. Remove the timing belt.
3. Remove the crankshaft timing pulley.
4. Remove the thrust washer.
5. Detach the air conditioning (A/C) compressor from the bracket and secure it on the radiator crossmember.
6. Remove the A/C compressor bracket.
7. Remove the intermediate shaft center bearing bracket.
8. Remove the oil pan. For additional information, refer to Oil Pan .

✲✲ CAUTION

Note the location of any spacer shims between the transaxle and the lower crankcase to aid installation. The spacer shims must be installed in the same location during installation.

9. Remove the oil pick-up tube.
10. Remove the lower crankcase retaining bolts and the lower crankcase. Discard the gasket.
11. Remove the accessory drive belt tensioner.
12. Remove the oil pump. Discard the gasket.

To install:

➡Install a new oil pump gasket

13. Install the oil pump. Align the oil pump on both sides so that the mating face is between 0.3 mm—0.8 mm above the lower edge of the cylinder block.
14. Tighten the oil pump retaining bolts to 97 inch lbs. (11 Nm).
15. Install the accessory drive belt tensioner and tighten the bolt to 35 ft. lbs. (48 Nm).

✲✲ CAUTION

Make sure that any spacer shims between the transaxle and the lower crankcase are installed in their original location.

16. Install the lower crankcase with a new gasket. First, tighten the transaxle bolts to 22 ft. lbs. (30 Nm). Then, tighten the rest of the lower crankcase bolts.
17. Install the oil pump pick-up tube and tighten the bolts to 88 inch lbs. (10 Nm).
18. Install the oil pan.
19. Install the intermediate shaft center bearing bracket and tighten the fasteners to 18 ft. lbs. (25 Nm).
20. Install the air conditioning (A/C) compressor bracket and tighten the fasteners to 18 ft. lbs. (25 Nm).
21. Attach the A/C compressor to the bracket and tighten the bolts to 18 ft. lbs. (25 Nm).
22. Install the thrust washer and the crankshaft timing pulley.
23. Install the timing belt.

2.3L (VIN Z) Engine

1. Before servicing the vehicle, refer to the precautions in the beginning of this section.
2. Remove the engine from the vehicle and mount it on an engine stand.
3. Remove the oil pan.
4. Remove the oil pump pickup tube and screen.
5. Remove the front cover and the timing chain.
6. Release the tension on the tensioner spring.
7. Remove the tensioner and the shoulder bolt.
8. Remove the guide.

➡The oil pump chain sprocket must be held in place.

9. Remove the oil pump chain and sprockets.
10. Remove the oil pump assembly and gasket.

To install:

11. Install the oil pump with a new gasket. Tighten the bolts in sequence as follows:
    a. Step 1: 89 inch lbs. (10 Nm).
    b. Step 2: 17 ft. lbs. (23 Nm).
12. Install the pump chain and sprockets. Tighten the pump sprocket bolt to 18 ft. lbs. (25 Nm).
13. Install the chain guide, tensioner, and shoulder bolt. Tighten the bolts to 89 inch lbs. (10 Nm).
14. Hook the tensioner spring around the shoulder bolt.
15. Install the oil pump pickup tube and

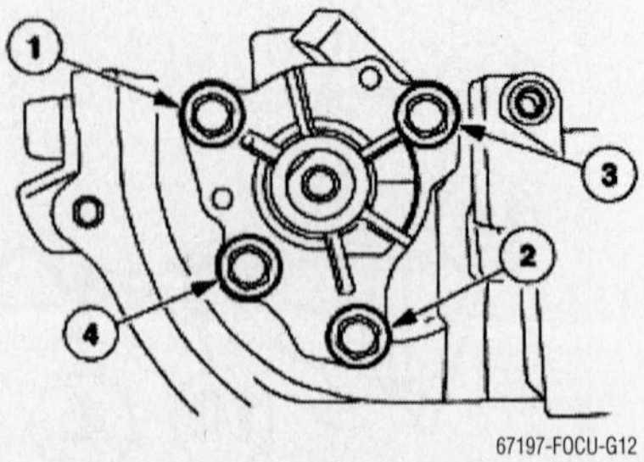

Oil pump torque sequence—2.3L (VIN Z) engine

screen with a new gasket. Tighten the bolts to 89 ft. lbs. (10 Nm).

16. Install the oil pan.
17. Install the timing chain and front cover.
18. Install the engine into the vehicle.

Rear Main Seal

REMOVAL & INSTALLATION

2.0L (VIN 3, 5 and P) Engines

1. Before servicing the vehicle, refer to the precautions in the beginning of this section.
2. Remove or disconnect the following:
    - Negative battery cable
    - Transaxle
    - Clutch, if equipped
    - Flexplate/flywheel
    - Oil seal

To install:

3. Install or connect the following:
    - Oil seal. Seat the seal flush with the rear of the crankshaft oil seal retainer.
    - Flexplate/flywheel. Tighten the bolts to 82 ft. lbs. (112 Nm).
    - Clutch, if equipped
    - Transaxle
    - Negative battery cable
4. Start the engine and check for leaks.

2.3L (VIN Z) Engine

1. Before servicing the vehicle, refer to the precautions in the beginning of this section.
2. Remove the transaxle.
3. Remove the flywheel or flexplate.

CAUTION

If the oil pan is not removed damage to the rear seal retainer joint can occur.

4. Remove the oil pan.
5. Remove the bolts and the crankshaft rear oil seal.

To install:

6. Install the crankshaft rear oil seal on the Crankshaft Rear Main Oil Seal Installer (T88P-6701-B1).
7. Install the Crankshaft Rear Main Oil Seal Installer and the crankshaft rear oil seal on the crankshaft.
8. Tighten the bolts in the sequence shown to 10 Nm (89 lb-in).
9. Remove the Crankshaft Rear Main Oil Seal Installer.
10. Install the oil pan.

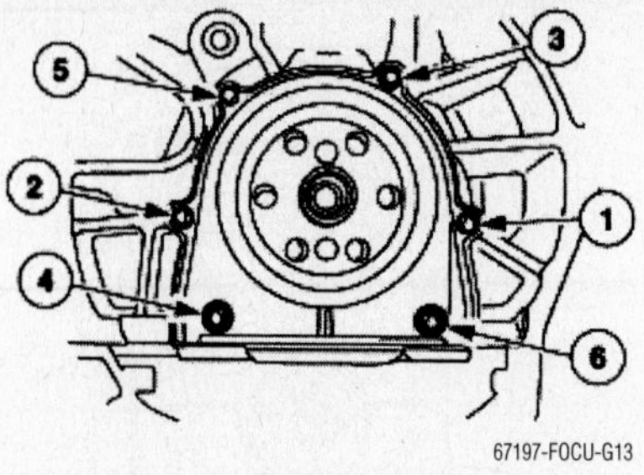

Seal retainer torque sequence—2.3L (VIN Z) engine

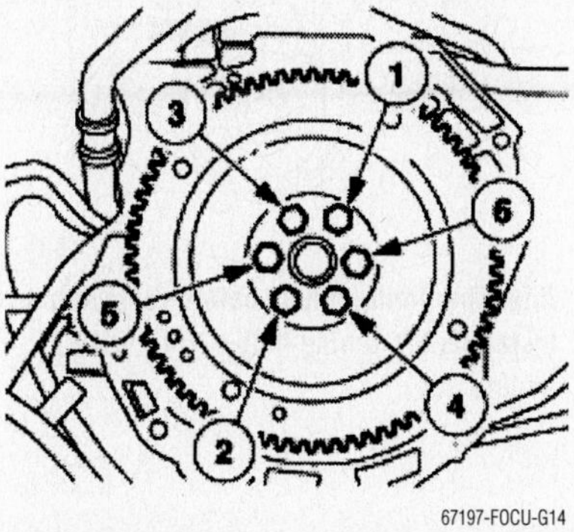

Flywheel/Flexplate torque sequence—2.3L (VIN Z) engine

11. Install the flywheel or flexplate. Tighten the bolts in the sequence shown as follows:.
    a. Step 1: Tighten to 37 ft. lbs. (50 Nm).
    b. Step 2: Tighten to 50 ft. lbs. (80 Nm).
    c. Step 3: Tighten to 83 ft. lbs. (112 Nm).
12. Install the transaxle.
13. Fill the engine with clean oil.

Timing Belt

REMOVAL & INSTALLATION

2.0L (VIN 3) Engine

1. Remove or disconnect the following:
    - Negative battery cable
    - Spark plugs
    - Catalytic converter
    - Right front wheel
    - Right inner splash shield
    - Accessory drive belt and idler pulley
2. Rotate the crankshaft to Top Dead Center (TDC) and install Crankshaft TDC Timing Peg T97P-6000-A.
3. Rotate the crankshaft clockwise against the peg.
4. Remove or disconnect the following:
    - Water pump pulley
    - Valve cover
    - Crankshaft pulley
    - Timing belt covers
5. Install Camshaft Alignment Timing Tool T94P-6256-CH into the slots of both camshafts at the rear of the cylinder head to lock the camshafts into position.
6. Loosen the timing belt tensioner bolt and relieve the tension on the timing belt by disconnecting the tensioner tab from the timing cover back plate.
7. Remove the timing belt.

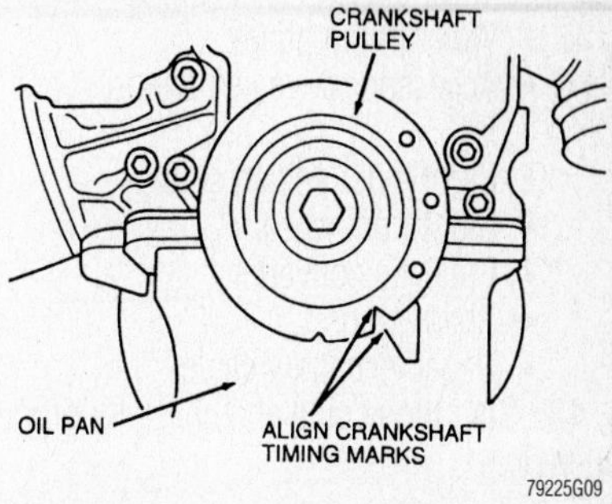

Crankshaft alignment position—Ford 2.0L (VIN 3) engines

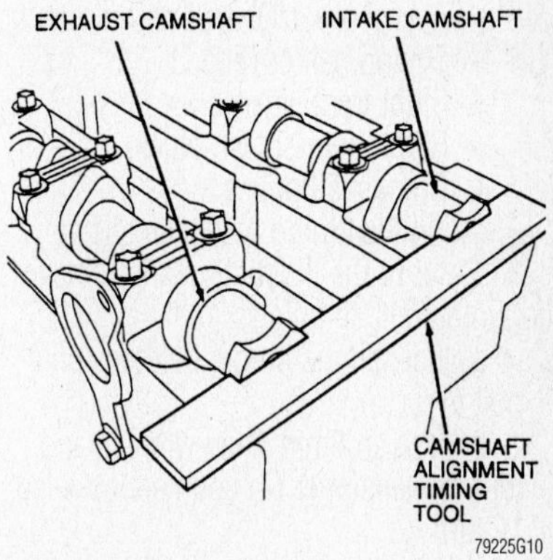

Placement of camshaft alignment timing tool T94P-6256-CH—Ford 2.0L (VIN 3) engines

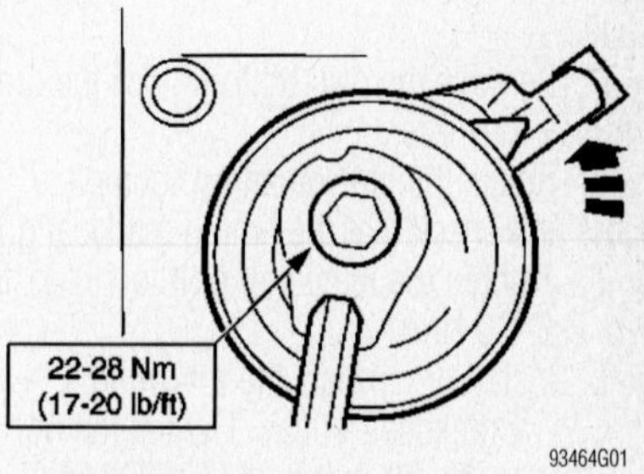

Timing belt tensioner index marks—Ford 2.0L (VIN 3) engines

To install:

➡Always replace the timing belt after loosening or removing it.

8. Install the timing belt.
9. Engage the timing belt tensioner tab into the timing cover back plate.
10. Adjust the timing belt tensioner until the index marks are aligned. Tighten the tensioner bolt to 17–20 ft. lbs. (22–28 Nm).
11. Remove the camshaft alignment tool and the crankshaft TDC peg.
12. Install or connect the following:
    - Timing belt covers
    - Crankshaft pulley. Tighten the bolt to 81–89 ft. lbs. (110–120 Nm).
    - Valve cover
    - Water pump pulley
    - Accessory drive belt and idler pulley
    - Right inner splash shield
    - Right front wheel
    - Catalytic converter
    - Spark plugs
    - Negative battery cable
13. Start the engine and check for proper operation.

2.0L (VIN P) Engine

1. Remove or disconnect the following:
    - Negative battery cable
    - Accessory drive belt and tensioner
    - Timing belt cover
    - Right front wheel
    - Right inner splash shield
    - Crankshaft pulley
2. Align the timing marks as shown.
3. Refer to the illustration and remove the timing belt as follows:
    a. Loosen the timing belt tensioner bolt (1).
    b. Use an 8mm Allen wrench, and turn the tensioner (2) counterclockwise ¼ turn.
    c. Insert a ⅛ inch drill bit in the hole (3) to lock the belt tensioner in place.
    d. Remove the timing belt (4).

To install:

4. Install the timing belt in a counterclockwise direction starting at the crankshaft.
5. Remove the drill bit to unlock the timing belt tensioner.
6. Rotate the crankshaft two complete turns and check that the timing marks align.
7. Tighten the tensioner bolt to 15–22 ft. lbs. (20–30 Nm).
8. Install or connect the following:
    - Crankshaft pulley. Tighten the bolt to 81–96 ft. lbs. (110–130 Nm).
    - Right inner splash shield

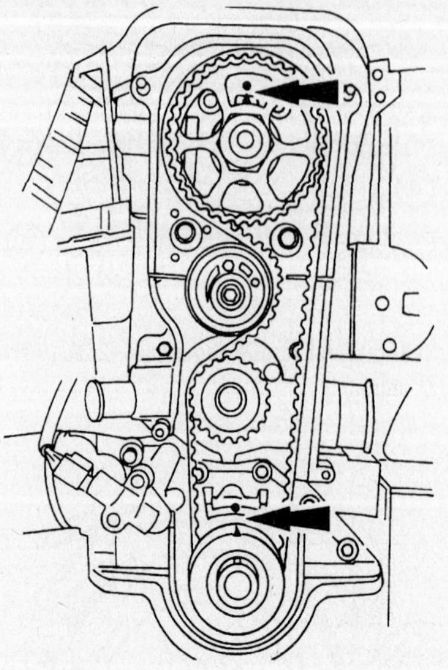

93005G01

Align the timing marks before removing or installing the timing belt—2.0L (VIN P) engine

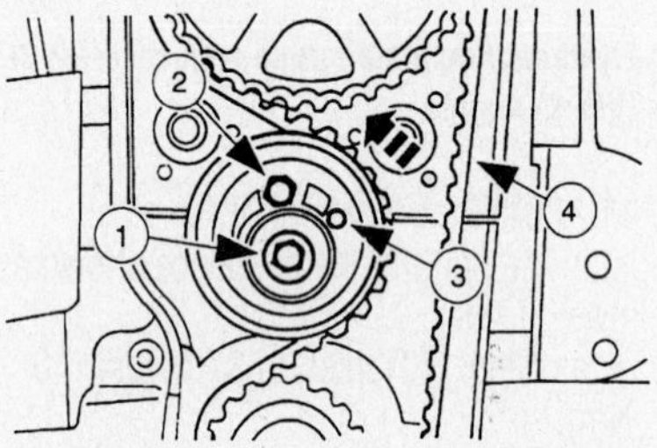

93005G02

Remove the timing belt by following these 4 numbered steps (refer to the text for an explanation)—2.0L (VIN P) engine

    - Right front wheel
    - Timing belt cover
    - Accessory drive belt and tensioner
    - Negative battery cable
9. Start the engine and check for proper operation.

2.0L (VIN 5) Engine

1. Before servicing the vehicle, refer to the precautions in the beginning of this section.
2. Disconnect the negative battery cable.
3. Raise and support the vehicle.
4. Detach the drive belt cover.
5. Loosen the coolant pump pulley retaining bolts.
6. Remove the accessory drive belt.
7. Remove the crankshaft pulley.
8. Remove the accessory drive belt idler pulley.
9. Remove the coolant pump pulley.
10. Detach the timing belt lower cover.
11. Lower the vehicle.
12. Detach the coolant expansion tank and position it to one side.
13. Detach the power steering fluid reservoir and position it to one side.
14. Position the trolley jack with the wooden block under the oil pan and raise the engine so that the engine front mount is free from load.

➡Mark the position of the engine front mount.

15. Remove the engine front mount.
16. Detach the timing belt upper cover. Leave the timing belt cover in its installed position.
17. Remove the engine front mount bracket.
18. Remove the timing belt upper cover.
19. Remove the electronic ignition (EI) coil cover.
20. Remove the engine upper cover.
21. Disconnect the variable cam timing (VCT) valve electrical connector.

➡Loosening sequence: from the outside to the inside, working diagonally.

22. Disconnect the spark plug connectors.
23. Detach the positive crankcase ventilation (PCV) hose.
24. Remove the valve cover.
25. Remove the spark plugs.
26. Turn the engine to TDC on cylinder number 1.
27. Loosen the timing belt tensioner bolt four turns.
28. Position the tensioner so that the locating tab is at approximately the 4 o'clock position.
29. Line up the hexagonal key slot in the tensioner adjusting washer with the pointer that is located behind the pulley.
30. Remove the timing belt.

➡Use an open ended wrench to prevent the camshaft from rotating.

31. Remove the intake camshaft pulley blanking plug.
32. Loosen the camshaft pulley retaining bolts.

To install:

➡The camshaft timing pulleys must be able to turn freely on the camshafts.

33. Turn the camshafts to the ignition position on cylinder No. 1 and insert the special tool Camshaft Alignment Plate 303-574 into the camshafts.

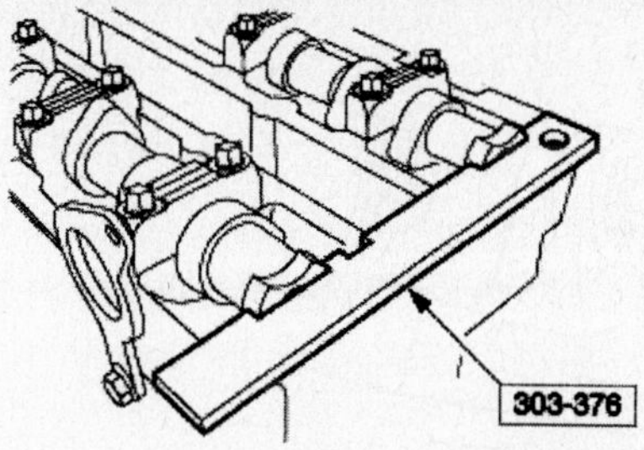

67197-FOCU-G25

Camshaft Alignment Tool 303-574 installation—2.0L (VIN 5) engine

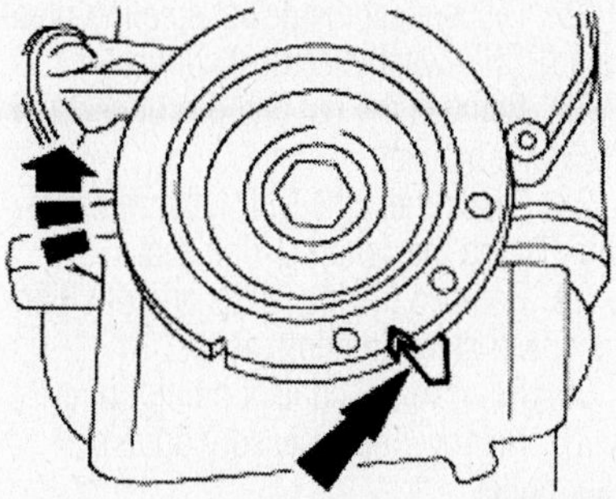

67197-FOCU-G26

Set the crankshaft to Top Dead Center (TDC)—2.0L (VIN 5) engine

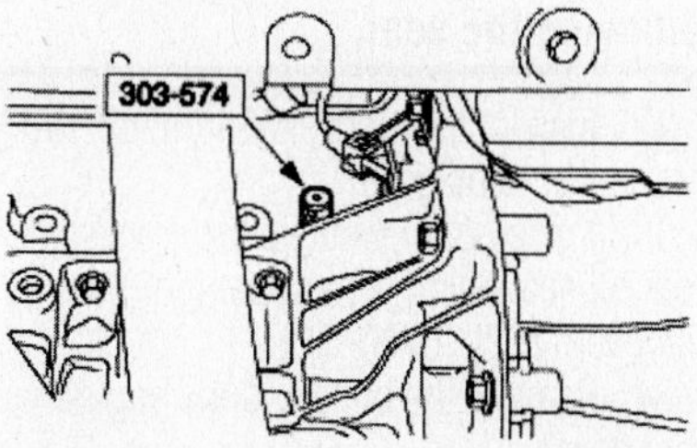

67197-FOCU-G27

Installation of the Crankshaft TDC Timing Peg—2.0L (VIN 5) engine

34. Rotate the crankshaft to TDC on cylinder No. 1.

35. Remove the blanking plug and install the Crankshaft TDC Timing Peg 303-574.

36. Starting from the crankshaft timing belt pulley and working counterclockwise install the timing belt, keeping it under tension.

CAUTION

Incorrect timing belt tension will result in incorrect valve timing.

37. Rotate the tensioner locating tab counterclockwise and insert the locating tab into the slot in the rear timing cover.

38. Position the hexagonal key slot in the tensioner adjusting washer to the 4 o'clock position.

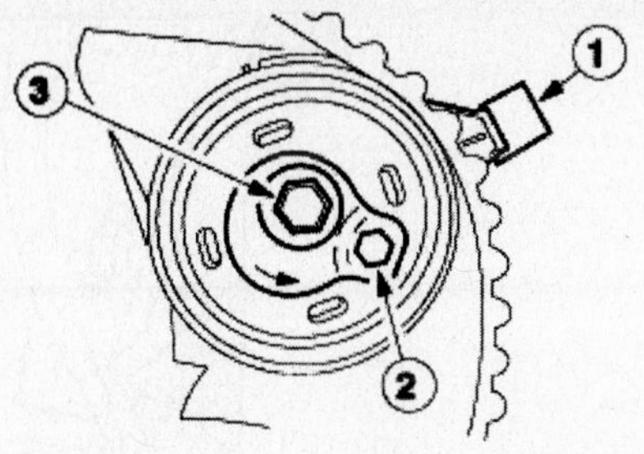

67197-FOCU-G28

Tensioner locating tab (1), hexagonal key slot (2) and attaching bolt (3)—2.0L (VIN 5) engine

39. Tighten the attaching bolt enough to seat the tensioner firmly against the rear timing cover, but still allow the tensioner adjusting washer to be rotated using a 6 mm hexagonal key.

CAUTION

Tension the timing belt working counterclockwise.

40. Using the hexagonal key, rotate the adjusting washer counterclockwise until the notch in the pointer is centered over the index line on the locating tab (the pointer will move clockwise during adjustment).

41. Tighten the bolt to 18 ft. lbs. (25 Nm), while holding the adjusting washer in position.

42. Use an open ended wrench to prevent the camshaft from rotating and tighten the camshaft pulley retaining bolts. Tighten the intake pulley bolt to 88 ft. lbs. (120 Nm) and the exhaust pulley bolt to 50 ft. lbs. (68 Nm).

43. Install the intake camshaft pulley blanking plug and tighten it to 27 ft. lbs. (37 Nm).

44. Remove the Crankshaft TDC Timing Peg.

45. Remove the Camshaft Alignment Plate from the camshafts.

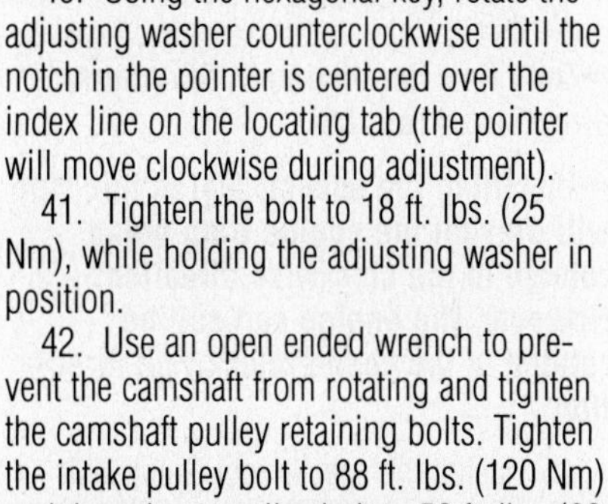

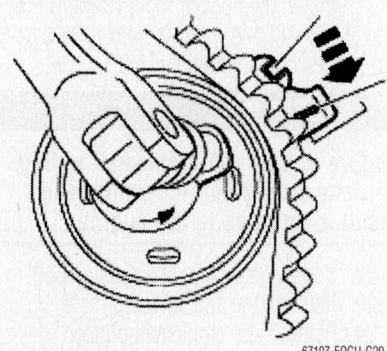

67197-FOCU-G29

Adjusting the timing belt tensioner—2.0L (VIN 5) engine

46. Turn the crankshaft two revolutions in the normal direction of rotation.

47. Check the valve timing by reinstalling the Camshaft Alignment Plate and the Crankshaft TDC Timing Peg. If necessary loosen the timing pulleys and correct the camshaft alignment.

48. Remove the Camshaft Alignment Plate and the Crankshaft TDC Timing Peg.

49. Install the blanking plug.

50. Install the valve cover and tighten the bolts as follows:

a. Step 1: 18 inch lbs. (2 Nm)
b. Step 2: 62 inch lbs. (7 Nm)

51. Attach the PCV hose to the valve cover.

52. Connect the VCT actuator electrical connector.

53. Coat the spark plug thread with anti-seize grease, screw in the spark plugs and push in the spark plug connector until it engages.

54. Connect the VCT valve electrical connector.

55. Install the engine upper cover.

56. Install the spark plug cover.

57. Position the upper timing belt cover and the center timing belt cover.

58. Install the engine front mount bracket. Tighten the bolts to 37 ft. lbs. (50 Nm).

59. Attach the timing belt upper cover. Tighten the bolts to 88 inch lbs. (10 Nm)

60. Install the engine front mount.

61. Attach the power steering pump fluid reservoir.

62. Attach the coolant expansion tank.

63. Remove the hydraulic jack and wooden block.

64. Raise and support the vehicle.

65. Attach the lower timing belt cover. Tighten the bolts to 62 inch lbs. (7 Nm).

66. Install the crankshaft pulley. Tighten the bolt to 85 ft. lbs. (115 Nm).

67. Attach the multi-groove belt idler pulley. Tighten the bolt to 29 ft. lbs. (40 Nm).

68. Attach the coolant pump pulley. Do not tighten the bolts at this stage.

69. Install the accessory drive belt.

70. Tighten the bolts on the coolant pump pulley to 17 ft. lbs. (24 Nm).

71. Attach the drive belt cover.

72. Lower the vehicle.

➡When the battery has been disconnected and reconnected, some abnormal drive symptoms may occur while the vehicle relearns its adaptive strategy. The vehicle may need to be driven 16 km (10 miles) or more to relearn the strategy.

73. Connect the battery negative cable.
74. Install the battery cover.
75. Check fluid levels and correct if necessary.
76. Check the routing of vacuum hoses and cables and secure them with cable ties.

Timing Chain, Sprockets, Front Cover and Seal

REMOVAL & INSTALLATION

2.3L (VIN Z) Engine

CAUTION

The crankshaft, the crankshaft sprocket and the pulley are fitted together by friction, using diamond washers between the flange faces on each part. For that reason, the crankshaft sprocket is also unfastened if you loosen the pulley. Therefore the engine must be retimed each time the damper is removed. Otherwise severe engine damage can occur.

1. Before servicing the vehicle, refer to the precautions in the beginning of this section.
2. With the vehicle in NEUTRAL, position it on a hoist.
3. Remove the fan shroud.
4. Remove the accessory drive belt.
5. Remove the valve cover.
6. Remove the degas bottle.
7. Remove the battery.
8. Remove the bolts and the battery tray.
9. Remove the nuts retaining the wiring harness.
10. Release the wiring harness retainers from the timing cover studs.
11. Disconnect the crankshaft position (CKP) sensor electrical connector.

CAUTION

Failure to position the No. 1 piston at top dead center (TDC) can result in damage to the engine. Turn the engine in the normal direction of rotation only.

12. Using the crankshaft pulley bolt, turn the crankshaft clockwise to position the No. 1 piston at TDC.

CAUTION

The special tool Camshaft Alignment Plate (T94P-6256-CH) is for camshaft alignment only. Using this tool to prevent engine rotation can result in engine damage.

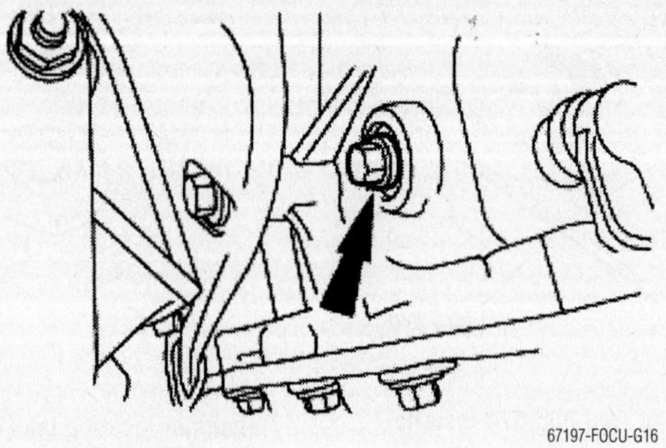

Plug bolt location—2.3L (VIN Z) engine

➡The camshaft timing slots are offset. If the special tool cannot be installed, rotate the crankshaft one complete revolution clockwise to correctly position the camshafts.

13. Install the special tool Camshaft Alignment Plate (T94P-6256-CH) in the slots on the rear of both camshafts.
14. Remove the plug bolt.

➡Only turn the engine in the normal direction of rotation.

➡Installing the special tool in this step will prevent the engine from being rotated in the clockwise direction. However, the engine can still be rotated in the counterclockwise direction.

15. Install the special tool Crankshaft Timing Peg (303-507) in place of the plug bolt.
16. Support the engine from above with a suitable support fixture.
17. Remove the LH motor mount center nut.
18. Remove the RH motor mount.
19. Loosen the lower engine mount nuts.
20. Lower the engine enough to allow for the clearance to install special tools.
21. Install the special tools Holding Fixture (T78P-4851-A) and Adapter (205-072-02) to the crankshaft pulley.

CAUTION

Failure to hold the crankshaft pulley in place during bolt loosening can result in damage to the engine.

22. Remove the crankshaft pulley bolt.
23. Remove the special tools.
24. Remove the crankshaft pulley.
25. Remove the accessory drive belt tensioner.
26. Disconnect the power steering pressure (PSP) switch electrical connector.
27. Remove the nut and disconnect the PSP tube bracket.
28. Disconnect the PSP tube. Remove and discard the Teflon O-ring seal.
29. Remove the four bolts and position the power steering pump aside.
30. Remove the coolant pump pulley.
31. Remove the accessory drive belt idler pulley.

CAUTION

Use care not to damage the engine front cover or the crankshaft when removing the seal.

32. Inspect the crankshaft front oil seal for wear or damage.
33. If necessary, remove the crankshaft front oil seal using the special tool Seal Remover (T92C-6700-CH).
34. Remove the bolts and the engine front cover.
35. Compress the timing chain tensioner, and insert a paper clip into the hole to retain the tensioner.
36. Remove the bolts and timing chain tensioner.
37. Remove the RH timing chain guide.
38. Remove the timing chain.
39. Remove the bolts and the LH timing chain guide.

CAUTION

Do not rely on the Camshaft Alignment Plate to prevent camshaft rotation. Damage to the tool or the camshaft can occur.

40. If necessary, remove the bolts and the camshaft sprockets. Use a wrench on the flats on the camshaft to prevent camshaft rotation.

To install:

41. Remove the Camshaft Alignment tool.

⁂ CAUTION

Do not rotate the camshafts. Damage to the valves and pistons can occur. If the camshaft sprockets were not removed, use the flats on the camshafts to prevent camshaft rotation and loosen the sprocket bolts.

42. If removed, install the camshaft sprockets and the bolts. Do not tighten the bolts at this time.
43. Install the LH timing chain guide and tighten the bolts to 89 inch lbs. (10 Nm).
44. Install the timing chain.
45. Install the RH timing chain guide.
46. Install the timing chain tensioner and tighten the bolts to 89 inch lbs. (10 Nm). Remove the paper clip to release the piston.
47. Install the Camshaft Alignment tool.

⁂ CAUTION

The special tool (T94P-6256-CH) is for camshaft alignment only. Using this tool to prevent engine rotation can result in engine damage.

48. Using a wrench on the flats on the camshafts to prevent camshaft rotation, tighten the bolts to 48 ft. lbs. (65 Nm).

⁂ CAUTION

Do not use metal scrapers, wire brushes, power abrasive disks or other abrasive means to clean sealing surfaces. These tools cause scratches and gouges which make leak paths.

49. Clean and inspect the mounting surfaces of the engine and the front cover.

➡The engine front cover must be installed and the bolts tightened within four minutes of applying the silicone gasket and sealant.

50. Apply a 2.5 mm bead of silicone gasket and sealant to the cylinder head and oil pan joint areas. Apply a 2.5 mm bead of silicone gasket and sealant to the front cover.
51. Install the engine front cover. Tighten the bolts in the sequence shown, to the following specifications:
    a. Tighten the 8 mm bolts to 89 inch lbs. (10 Nm).
    b. Tighten the 13 mm bolts to 35 ft. lbs. (48 Nm).

➡Lubricate the oil seal with clean engine oil.

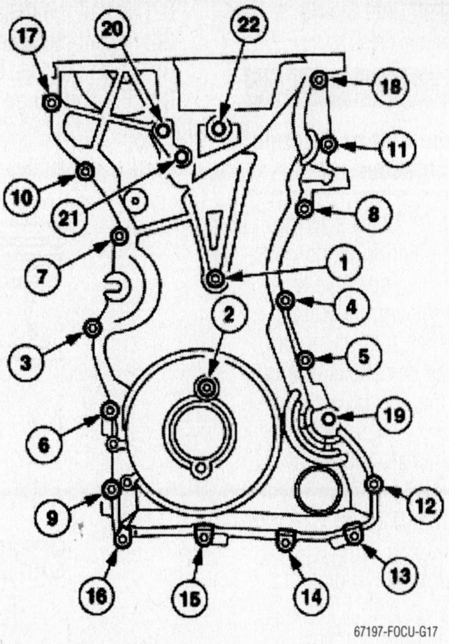

Front cover torque sequence—2.3L (VIN Z) engine

52. If a new seal is being installed, use the special tool Front Seal Installer (T74P-6150-A) to install the crankshaft front oil seal.
53. Position the power steering pump and tighten the four bolts to 18 ft. lbs. (25 Nm).
54. Using the special tool Teflon Seal Installer (D90P-3517-A), install a new O-ring on the PSP tube fitting.
55. Connect the power steering PSP tube fitting and tighten to 15 ft. lbs. (20 Nm).
56. Attach the PSP tube bracket and tighten the nut to 10 ft. lbs. (13 Nm).
57. Connect the PSP switch electrical connector.
58. Install the coolant pump pulley and tighten the bolts to 18 ft. lbs. (25 Nm).
59. Install the accessory drive belt idler pulley and tighten the bolt to 18 ft. lbs. (25 Nm).
60. Install the accessory drive belt tensioner and tighten the bolts to 18 ft. lbs. (25 Nm).

➡Do not reuse the crankshaft damper bolt.

➡Apply clean engine oil on seal area before installing.

61. Install the crankshaft pulley and hand-tighten the bolt.

⁂ CAUTION

Only hand-tighten the bolt or damage to the front cover can occur.

➡This step will correctly align the crankshaft pulley to the crankshaft.

62. Install a standard 6 mm (0.23 in) x 18 mm (0.7 in) bolt through the crankshaft pulley and thread it into the front cover.
63. Rotate the pulley as necessary to align the bolt holes.

⁂ CAUTION

Failure to hold the crankshaft pulley in place during bolt tightening can cause damage to the engine front cover.

64. Using the special tools to hold the crankshaft pulley in place, tighten the crankshaft pulley bolt as follows:
    a. Step 1: Tighten to 74 ft. lbs. (100 Nm).
    b. Step 2: Tighten an additional 90 degrees.
65. Remove the special tools.
66. Remove the 6 mm (0.23 in) x 18 mm (0.7 in) bolt.
67. Remove the Crankshaft Timing Peg.
68. Remove the Camshaft Alignment tool.

➡Only turn the engine in the normal direction of rotation.

69. Turn the engine two complete revolutions.
70. Install the Crankshaft Timing Peg.

➡Only turn the engine in the normal direction of rotation.

71. Turn the crankshaft until the No. 1 piston is at top dead center (TDC).

⁂ CAUTION

Only hand-tighten the bolt or damage to the front cover can occur.

72. Using the 6 mm (0.23 in) x 18 mm (0.7 in) bolt, check the position of the crankshaft pulley. If it is not possible to install the bolt, correct the engine timing.
73. Using the Camshaft Alignment tool, check the position of the camshafts. If it is not possible to install the special tool, correct the engine timing.
74. Remove the Camshaft Alignment tool.
75. Remove the 6 mm (0.23 in) x 18 mm (0.7 in) bolt.
76. Remove the Crankshaft Timing Peg.
77. Install the engine plug bolt and tighten it to 15 ft. lbs. (20 Nm).
78. Connect the CKP sensor electrical connector and the wiring harness pin-type retainers.
79. Raise the engine.
80. Install the RH motor mount and tighten the nuts to 59 ft. lbs. (80 Nm).
81. Install the LH motor mount nut and tighten it to 98 ft. lbs. (133 Nm).
82. Tighten the lower engine mount nuts to 35 ft. lbs. (48 Nm).
83. Remove the engine support fixture.
84. Install the battery tray and tighten the bolts to 108 inch lbs. (12 Nm).
85. Install the battery.
86. Install the valve cover.
87. Install the degas bottle.
88. Install the accessory drive belt.
89. Install the fan and shroud.
90. Fill and purge the power steering fluid.

Piston and Ring

POSITIONING

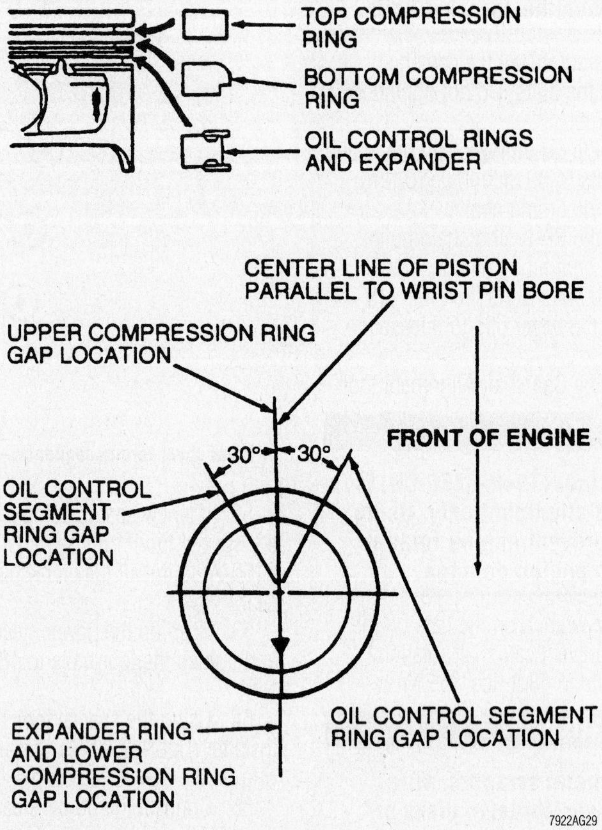

Ford 2.0L (VIN 3, 5 and P) and 2.3L (VIN Z) engines—piston ring positioning, end-gap spacing and piston positioning. The small directional arrow must face the front of the engine.

FUEL SYSTEM

Fuel System Service Precautions

Safety is the most important factor when performing not only fuel system maintenance but any type of maintenance. Failure to conduct maintenance and repairs in a safe manner may result in serious personal injury or death. Maintenance and testing of the vehicle's fuel system components can be accomplished safely and effectively by adhering to the following rules and guidelines.

- To avoid the possibility of fire and personal injury, always disconnect the negative battery cable unless the repair or test procedure requires that battery voltage be applied.
- Always relieve the fuel system pressure prior to disconnecting any fuel system component (injector, fuel rail, pressure regulator, etc.), fitting or fuel line connection. Exercise extreme caution whenever relieving fuel system pressure, to avoid exposing skin, face and eyes to fuel spray. Please be advised that fuel under pressure may penetrate the skin or any part of the body that it contacts.
- Always place a shop towel or cloth around the fitting or connection prior to loosening to absorb any excess fuel due to spillage. Ensure that all fuel spillage (should it occur) is quickly removed from engine surfaces. Ensure that all fuel soaked cloths or towels are deposited into a suitable waste container.
- Always keep a dry chemical (Class B) fire extinguisher near the work area.
- Do not allow fuel spray or fuel vapors to come into contact with a spark or open flame.
- Always use a backup wrench when loosening and tightening fuel line connection fittings. This will prevent unnecessary stress and torsion to fuel line piping.
- Always replace worn fuel fitting O-rings with new. Do not substitute fuel hose or equivalent, where fuel pipe is installed.

Before servicing the vehicle, make sure to refer to the precautions in the beginning of this section as well.

Fuel System Pressure

RELIEVING

1. Before servicing the vehicle, refer to the precautions in the beginning of this section.
2. Remove the fuel pump fuse from the battery junction box.

3. Start the engine and allow it to idle until it stalls.
4. Crank the engine for 5 seconds to ensure that the fuel pressure has been relieved.
5. Disconnect the negative battery cable.
6. When repairs are completed, connect the negative battery cable and replace the fuel pump fuse.

Fuel Filter

REMOVAL & INSTALLATION

1. Before servicing the vehicle, refer to the precautions in the beginning of this section.
2. Relieve the fuel system pressure.
3. Remove or disconnect the following:
    - Evaporative Emissions (EVAP) vapor line
    - Fuel filter lines
    - Fuel filter and bracket assembly
    - Fuel filter

To install:

4. Install or connect the following:
    - Fuel filter
    - Fuel filter and bracket assembly
    - Fuel filter lines
    - EVAP vapor line
5. Start the engine and check for leaks.

Fuel Pump

REMOVAL & INSTALLATION

1. Before servicing the vehicle, refer to the precautions in the beginning of this section.
2. Relieve the fuel system pressure.
3. Remove or disconnect the following:
    - Exhaust center muffler
    - Heat shield
    - Fuel filler and vent hoses
    - Fuel supply line
    - Evaporative Emissions (EVAP) vapor line
    - Rollover valve line
    - Fuel tank retainer strap. Support the fuel tank.
    - Fuel pump module harness connector
    - Fuel tank pressure sensor connector
    - Fuel tank
    - Fuel pump module

To install:

4. Install or connect the following:
    - Fuel pump module. Tighten the locknut to 59 ft. lbs. (80 Nm).
    - Fuel tank
    - Fuel tank pressure sensor connector
    - Fuel pump module harness connector
    - Fuel tank retainer strap. Tighten the bolt to 18 ft. lbs. (25 Nm).
    - Rollover valve line
    - EVAP vapor line
    - Fuel supply line
    - Fuel filler and vent hoses
    - Heat shield
    - Exhaust center muffler
5. Start the engine and check for leaks.

Fuel Injector

REMOVAL & INSTALLATION

2.0L (VIN 3, 5 and P) Engines

1. Before servicing the vehicle, refer to the precautions in the beginning of this section.
2. Relieve fuel system pressure.
3. Remove or disconnect the following:
    - Negative battery cable
    - Fuel line and retaining clip
    - Pressure regulator vacuum hose
    - Accelerator cable
    - Fuel injector electrical connectors
    - Fuel supply manifold with the injectors attached
    - Injectors from the supply manifold

To install:

4. Install or connect the following:
    - Fuel injectors using new O-ring seals
    - Fuel supply manifold with the injectors attached. Tighten the bolts to 88 inch lbs. (10 Nm).
    - Fuel injector electrical connectors
    - Accelerator cable
    - Pressure regulator vacuum hose
    - Fuel line and retaining clip
    - Negative battery cable

2.3L (VIN Z) Engine

WARNING

Do not smoke or carry lighted tobacco or open flame of any type when working on or near fuel-related components. Highly flammable vapors are always present and can ignite. Failure to follow these instructions can result in personal injury.

WARNING

This procedure involves fuel handling. Be prepared for fuel spillage at all times and always observe fuel handling precautions. Failure to follow these instructions can result in personal injury.

1. Before servicing the vehicle, refer to the precautions in the beginning of this section.
2. Disconnect the battery ground cable.
3. Release the fuel pressure.
4. Disconnect the fuel injector electrical connectors.
5. Disconnect the fuel pressure regulator electrical connector and the vacuum hose.
6. Disconnect the fuel injector harness retaining clips from the fuel injection supply manifold.
7. Disconnect the fuel tube.
8. Remove the bolts and the fuel injection supply manifold.

➡Remove and discard the fuel injector O-rings.

9. If necessary, remove the fuel injectors.
10. Install new O-rings and lubricate them with clean engine oil.
11. To install, reverse the removal procedure. Tighten the fuel supply manifold bolts to 18 ft. lbs. (25 Nm).

DRIVE TRAIN

Transaxle Assembly

REMOVAL & INSTALLATION

Manual

1. Before servicing the vehicle, refer to the precautions in the beginning of this section.
2. Attach a support fixture to the engine lifting eyes.
3. Loosen the front strut center nuts 5 turns.
4. Remove or disconnect the following:
    - Battery and tray
    - Mass Air Flow (MAF) sensor connector
    - Air cleaner assembly
    - Air intake pipe
    - Clutch slave cylinder fluid line
    - Shift cables
    - Exhaust flex pipe
    - Engine front mount
    - Lower ball joints
    - Axle halfshafts and intermediate shaft
    - Drive belt cover
    - Reverse light switch connector
    - Vehicle Speed (VSS) sensor connector
    - Rear engine mount and bracket
    - Starter motor
    - Transaxle flange bolts. Support the transaxle.
    - Transaxle

To install:

5. Install or connect the following:
    - Transaxle. Tighten the flange bolts to 35 ft. lbs. (48 Nm).
    - Starter motor
    - Rear engine mount bracket. Tighten the fastener to 59 ft. lbs. (80 Nm).
    - Rear engine mount. Tighten the outer nuts to 35 ft. lbs. (48 Nm) and the center nut to 98 ft. lbs. (133 Nm).
    - VSS sensor connector
    - Reverse light switch connector
    - Drive belt cover
    - Axle halfshafts and intermediate shaft
    - Lower ball joints. Tighten the pinch bolts to 37 ft. lbs. (50 Nm).
    - Right engine mount. Tighten the bolts to 35 ft. lbs. (48 Nm).
    - Exhaust flex pipe
    - Shift cables
    - Clutch slave cylinder fluid line
    - Air intake pipe
    - Air cleaner assembly
    - MAF sensor connector
    - Battery and tray
6. Tighten the strut center nuts to 35 ft. lbs. (48 Nm).

Automatic

1. Before servicing the vehicle, refer to the precautions in the beginning of this section.
2. Attach a support fixture to the engine lifting eyes.
3. Loosen the front strut center nuts 5 turns.
4. Remove or disconnect the following:
    - Battery and tray
    - Mass Air Flow (MAF) sensor connector
    - Positive Crankcase Ventilation (PCV) hose
    - Air intake pipe and resonator
    - Turbine Shaft Speed (TSS) sensor connector
    - Transmission range sensor connector
    - Transaxle harness connector
    - Starter motor
    - Shift cable and bracket
    - Front wheels
    - Exhaust front pipe
    - Right engine support
    - Outer tie rod ends
    - Lower ball joints
    - Axle halfshafts and intermediate shaft
    - Output Shaft Speed (OSS) sensor connector
    - Transaxle oil cooler lines
    - Torque converter
    - Transaxle oil dipstick tube
    - Rear engine mount and bracket

➡The transaxle flange bolts vary in length. Note their locations for installation.

- Transaxle flange bolts. Support the transaxle.
- Transaxle

To install:

5. Install or connect the following:
    - Transaxle. Tighten the flange bolts to 35 ft. lbs. (48 Nm).
    - Rear engine mount bracket. Tighten the fastener to 59 ft. lbs. (80 Nm).
    - Rear engine mount. Tighten the outer nuts to 35 ft. lbs. (48 Nm) and the center nut to 98 ft. lbs. (133 Nm).
    - Transaxle oil dipstick tube
    - Torque converter. Tighten the nuts to 27 ft. lbs. (37 Nm).
    - Transaxle oil cooler lines
    - OSS sensor connector
    - Axle halfshafts and intermediate shaft
    - Lower ball joints. Tighten the pinch bolts to 37 ft. lbs. (50 Nm).
    - Outer tie rod ends. Tighten the nuts to 35 ft. lbs. (48 Nm).
    - Right engine support. Tighten the bolts to 35 ft. lbs. (48 Nm).
    - Exhaust front pipe. Tighten the fasteners to 35 ft. lbs. (48 Nm).
    - Front wheels
    - Shift cable and bracket
    - Starter motor
    - Transaxle harness connector
    - Transmission range sensor connector
    - TSS sensor connector
    - Air intake pipe and resonator
    - PCV hose
    - MAF sensor connector
    - Battery and tray
6. Tighten the strut center nuts to 35 ft. lbs. (48 Nm).

Clutch

REMOVAL & INSTALLATION

1. Before servicing the vehicle, refer to the precautions in the beginning of this section.
2. Remove or disconnect the following:
    - Negative battery cable
    - Transaxle
    - Pressure plate. Loosen the bolts evenly in ½ turn steps.
    - Clutch disc

To install:

3. Install or connect the following:
    - Clutch disc and pressure plate. Tighten the pressure plate bolts evenly in ½ turns to 21 ft. lbs. (29 Nm).
    - Transaxle
    - Negative battery cable

Hydraulic Clutch System

BLEEDING

1. Before servicing the vehicle, refer to the precautions in the beginning of this section.

2. Remove or disconnect the following:
- Air cleaner assembly
- Mass Air Flow (MAF) sensor connector
- Air intake pipe

3. Remove fluid from the brake fluid reservoir until the level reaches the **MIN** mark.

4. Fill the reservoir of Special Tool 416-D002 with DOT 4 brake fluid.

5. Attach the tool to the slave cylinder bleed nipple.

6. Open the bleed nipple and pump 80 ml brake fluid into the clutch control system.

7. Tighten the bleed nipple to 88 inch lbs. (10 Nm) and remove the tool.

8. Have an assistant depress the clutch pedal 4–5 times and hold the pedal at full travel.

9. Close the bleeder before releasing the clutch pedal.

10. Repeat the procedure until no more air bubbles are seen.

11. Check the clutch for proper operation.

12. Check the brake fluid reservoir level and fill with DOT 4 brake fluid, as necessary.

13. Install or connect the following:
- Air intake pipe
- MAF sensor connector
- Air cleaner assembly

Halfshaft

REMOVAL & INSTALLATION

Left

➡The IB5 manual transmission is used with the 2.0L (VIN P) engine and the MTX75 manual transmission is used with the 2.0L (VIN 3) engine.

➡The hub nut may be reused 4 times. Mark the nut at removal and only use hub nuts with 3 or fewer marks for assembly.

1. Before servicing the vehicle, refer to the precautions in the beginning of this section.

2. Loosen the strut center nut 5 turns.

3. Remove or disconnect the following:
- Front wheel
- Hub retainer nut. Mark the nut.
- Lower ball joint

4. Press the stub shaft out of the wheel hub.

5. Separate the inner CV-joint from the transaxle as follows:
- If equipped with the IB5 manual transaxle, use Halfshaft remover 308-256
- If equipped with the MTX75 manual transaxle or with an automatic transaxle, use Halfshaft Remover 205-241 and a slide hammer

To install:

➡Replace the circlip for assembly.

6. Install the halfshaft inner joint so that the circlip is felt to seat. Draw the stub shaft into the wheel hub with Halfshaft Installer 205-379.

7. Install or connect the following:
- Lower ball joint. Tighten the pinch bolt to 37 ft. lbs. (50 Nm).
- Hub retainer nut. Tighten the nut to 232 ft. lbs. (316 Nm).
- Front wheel

8. Check the transaxle fluid and add, as necessary.

9. Tighten the strut center nuts to 35 ft. lbs. (48 Nm).

Right

➡The hub nut may be reused 4 times. Mark the nut at removal and only use hub nuts with 3 or fewer marks for assembly.

1. Before servicing the vehicle, refer to the precautions in the beginning of this section.

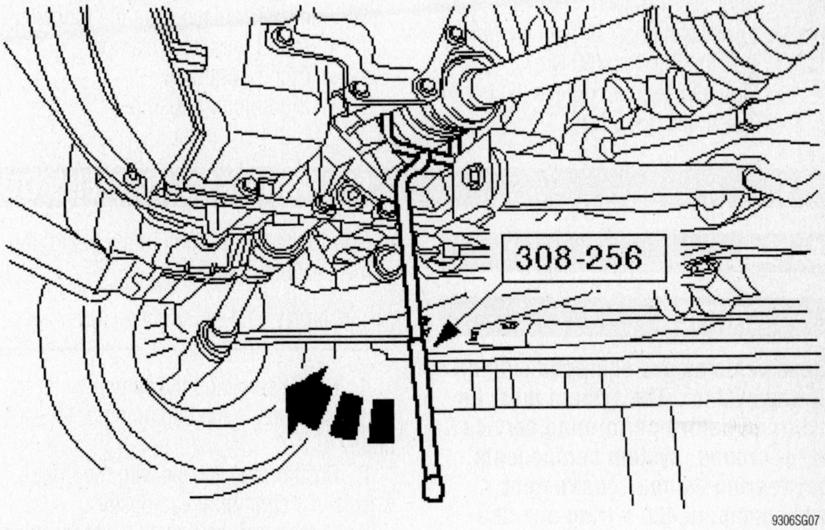

Left halfshaft removal—IB5 manual transaxle

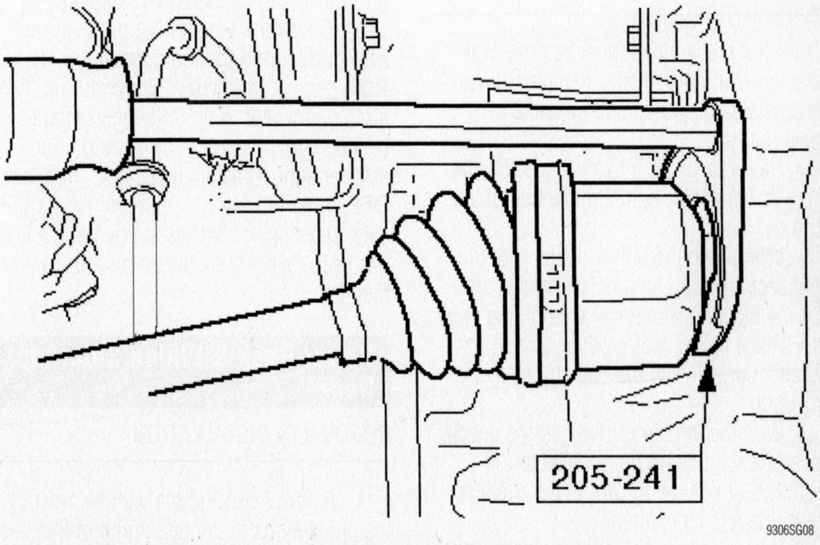

Left halfshaft removal—MTX75 manual transaxle and automatic transaxle

2. Loosen the strut center nut 5 turns.
3. Remove or disconnect the following:
    - Front wheel
    - Hub retainer nut. Mark the nut.
    - Lower ball joint
    - Intermediate shaft retaining clip
4. Press the stub shaft out of the wheel hub.
5. Remove the right axle halfshaft and the intermediate shaft as an assembly.

To install:

➡Use a new retaining clip for assembly.

6. Install or connect the following:
    - Axle halfshaft and intermediate shaft assembly. Tighten the nuts to 18 ft. lbs. (25 Nm).
    - Lower ball joint. Tighten the pinch bolt to 37 ft. lbs. (50 Nm).
    - Hub retainer nut. Tighten the nut to 232 ft. lbs. (316 Nm).
    - Front wheel
7. Check the transaxle fluid and add, as necessary.
8. Tighten the strut center nuts to 35 ft. lbs. (48 Nm).

CV-Joints

OVERHAUL

Inner Tripod Joint

1. Before servicing the vehicle, refer to the precautions in the beginning of this section.
2. Remove the axle halfshaft from the vehicle and place it in a vise.
3. Remove or disconnect the following:
    - Tripod joint boot clamps
    - Tripod joint boot
    - Tripod joint housing
    - Snapring
    - Tripod joint

To install:

➡Use new snaprings and boot clamps for assembly.

4. Install or connect the following:
    - Tripod joint
    - Snapring
    - Tripod joint housing. Fill the joint housing with grease.
    - Tripod joint boot
    - Tripod joint boot clamps
5. Install the halfshaft.

Outer CV-Joint

The outer CV-joint is serviced with the halfshaft as an assembly. The outer CV-joint boot can be serviced by removing the inner tripod joint.

STEERING AND SUSPENSION

Air Bag

⁂ CAUTION

These vehicles are equipped with an air bag system. The system must be disarmed before performing service on, or around, system components, the steering column, instrument panel components, wiring and sensors. Failure to follow the safety precautions and the disarming procedure could result in accidental air bag deployment, possible injury and unnecessary system repairs.

PRECAUTIONS

Several precautions must be observed when handling the inflator module to avoid accidental deployment and possible personal injury.

- Never carry the inflator module by the wires or connector on the underside of the module
- When carrying a live inflator module, hold securely with both hands, and ensure that the bag and trim cover are pointed away
- Place the inflator module on a bench or other surface with the bag and trim cover facing up
- With the inflator module on the bench, never place anything on or close to the module that may be thrown in the event of an accidental deployment

Before servicing the vehicle, also be sure to refer to the precautions in the beginning of this section as well

DISARMING

1. Before servicing the vehicle, refer to the precautions in the beginning of this section.
2. Position the vehicle with the front wheels in a straight-ahead position.
3. Disconnect the negative battery cable.
4. Disconnect the positive battery cable.
5. Wait at least 1 minute for the air bag backup power supply to drain before continuing.
6. Proceed with the repair.
7. Once repairs are completed, connect the battery cables, negative cable last.
8. Check the functioning of the air bag system by turning the ignition key to the **RUN** position and visually monitoring the air bag indicator lamp in the instrument cluster. The indicator lamp should illuminate for approximately 6 seconds, then turn **OFF**. If the indicator lamp does not illuminate, stays on, or flashes at any time, a fault has been detected by the air bag diagnostic monitor.

Power Rack & Pinion Steering Gear

REMOVAL & INSTALLATION

1. Before servicing the vehicle, refer to the precautions in the beginning of this section.
2. Center the steering wheel and turn the ignition to the **LOCK** position.
3. Remove or disconnect the following:
    - Negative battery cable
    - Lower instrument panel cover
    - Steering shaft pinch bolt
    - Front wheels
    - Outer tie rod ends
    - Stabilizer bar links
    - Power steering fluid cooler hose
    - Right engine mount
    - Steering gear heat shield
    - Power steering hose support clamp
    - Power steering hoses
    - 6 subframe bolts. Support the subframe.
    - Floor seal
    - Steering gear pinion extension
    - Steering gear

To install:

4. Install or connect the following:
    - Steering gear. Tighten the bolts to 59 ft. lbs. (80 Nm).
    - Steering gear pinion extension. Tighten the pinch bolt to 26 ft. lbs. (35 Nm).
    - Floor seal
5. Install Subframe Alignment Pins 502-002 and raise the subframe into position. Tighten the 4 rear bolts to 147 ft. lbs. (200 Nm) and the other bolts to 85 ft. lbs. (115 Nm).
6. Install or connect the following:
    - Power steering hoses
    - Power steering hose support clamp
    - Steering gear heat shield

- Right engine mount. Tighten the bolt to 37 ft. lbs. (50 Nm).
- Power steering fluid cooler hose
- Stabilizer bar links. Tighten the nuts to 37 ft. lbs. (50 Nm).
- Outer tie rod ends. Tighten the nuts to 35 ft. lbs. (48 Nm).
- Front wheels
- Steering shaft pinch bolt. Tighten the bolt to 21 ft. lbs. (28 Nm).
- Lower instrument panel cover
- Negative battery cable

7. Check the wheel alignment and adjust as necessary.

Strut

REMOVAL & INSTALLATION

1. Before servicing the vehicle, refer to the precautions in the beginning of this section.
2. Remove or disconnect the following:
   - Front wheel
   - Brake hose bracket
   - Stabilizer bar link
   - Wheel speed sensor, if equipped
   - Brake caliper and rotor
   - Outer tie rod end
   - Lower ball joint
   - Steering knuckle pinch bolt. Separate the knuckle from the strut.
   - Upper strut mount nuts
   - Strut assembly

To install:

3. Install or connect the following:
   - Strut assembly. Tighten the upper mount nuts to 18 ft. lbs. (25 Nm).
   - Steering knuckle. Tighten the pinch bolt to 66 ft. lbs. (90 Nm).
   - Lower ball joint. Tighten the pinch bolt to 37 ft. lbs. (50 Nm).
   - Outer tie rod end. Tighten the nut to 35 ft. lbs. (48 Nm).
   - Brake caliper and rotor
   - Wheel speed sensor, if equipped
   - Stabilizer bar link. Tighten the nut to 37 ft. lbs. (50 Nm).
   - Brake hose bracket
   - Front wheel
4. Check the wheel alignment and adjust as necessary.

Shock Absorber

REMOVAL & INSTALLATION

3 and Door Models

1. Before servicing the vehicle, refer to the precautions in the beginning of this section.
2. Remove or disconnect the following:
   - Luggage compartment interior trim panel
   - Upper shock absorber mounting nut
   - Lower shock absorber mounting bolt
   - Shock absorber

To install:

➡Tighten the shock absorber mounting fasteners with the suspension at curb height and the vehicle weight supported by the wheels.

3. Install or connect the following:
   - Shock absorber. Guide the rod into the locating hole.
   - Lower shock absorber mounting bolt. Tighten the bolt to 85 ft. lbs. (115 Nm).
   - Upper shock absorber mounting nut. Tighten the nut to 13 ft. lbs. (18 Nm).
   - Luggage compartment interior trim panel

Wagon

1. Before servicing the vehicle, refer to the precautions in the beginning of this section.
2. Remove the upper and lower mounting bolts and remove the shock absorber.

To install:

➡Tighten the shock absorber mounting fasteners with the suspension at curb height and the vehicle weight supported by the wheels.

3. Install the shock absorber and tighten the mounting bolts to 85 ft. lbs. (115 Nm).

Coil spring

REMOVAL & INSTALLATION

Front

1. Before servicing the vehicle, refer to the precautions at the beginning of this section.
2. Remove the strut from the vehicle.
3. Compress the coil spring using a suitable spring compressor until the spring comes away from the seat.
4. Remove the large center nut and slowly release the spring compressor.

To install:

5. Compress the spring and install it on the strut.
6. Install the upper strut mount. Tighten the nut to 35 ft. lbs. (48 Nm).

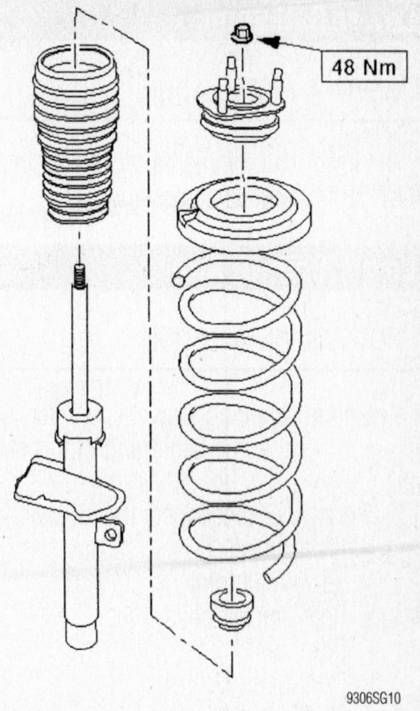

Front strut assembly exploded view

7. Install the strut assembly in the vehicle.

Rear

1. Before servicing the vehicle, refer to the precautions in the beginning of this section.
2. Raise and support the vehicle.
3. Install spring compressor 204-167 with Adapters 204-215.
4. Compress the coil spring and remove it.

To install:

5. Install the coil spring and remove the compressor.

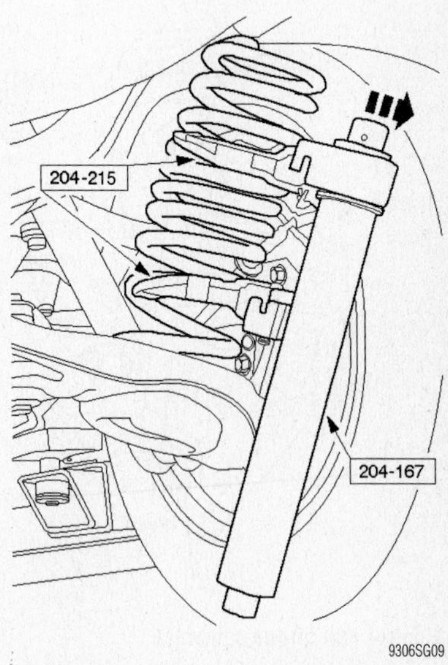

Coil spring compressor and adapters

Lower Ball Joint

REMOVAL & INSTALLATION

The lower ball joint is replaced with the lower control arm as an assembly.

Lower Control Arm

REMOVAL & INSTALLATION

1. Before servicing the vehicle, refer to the precautions in the beginning of this section.
2. Remove or disconnect the following:
   - Front wheel
   - Lower ball joint
   - Rear bracket bolts
   - Front bolt
   - Control arm

To install:

➡Use new nuts, bolts and ball bearing washers for assembly.

3. Install the control arm. Tighten the fasteners in sequence as follows:
   a. Step 1: Tighten nut No. 1 to 74 ft. lbs. (100 Nm) plus 60 degrees
   b. Step 2: Tighten nut No. 2 to 88 ft. lbs. (120 Nm)
   c. Step 3: Tighten bolt No. 3 to 88 ft. lbs. (120 Nm) plus 90 degrees
   d. Step 4: Check that bolt No. 3 is tightened to 125–169 ft. lbs. (170–230 Nm)
4. Install or connect the following:
   - Lower ball joint. Tighten the pinch bolt to 37 ft. lbs. (50 Nm).
   - Front wheel

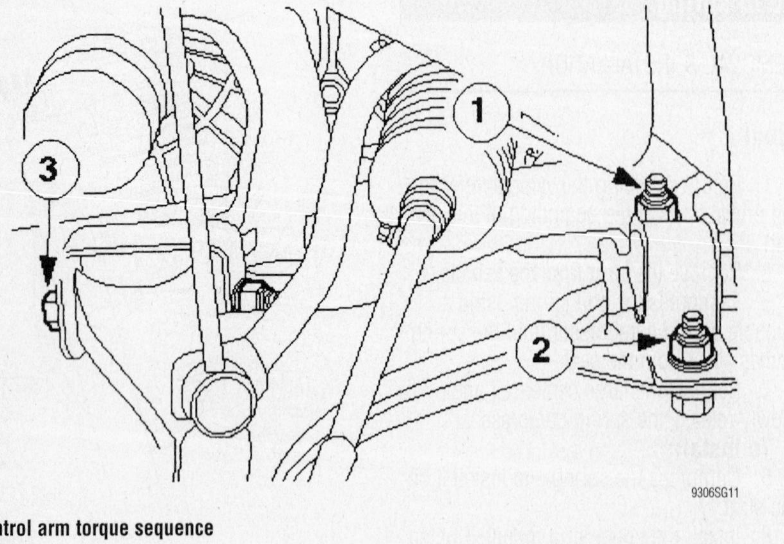

Control arm torque sequence

5. Check the wheel alignment and adjust as necessary.

CONTROL ARM BUSHING REPLACEMENT

The lower control arm bushings are replaced with the lower control arm as an assembly.

Wheel Bearings

ADJUSTMENT

The bearings on the front and rear wheels are a one piece cartridge design and cannot be adjusted. If wheel bearing play is excessive, check the wheel hub retainer nut for proper torque. If the torque is correct, replacement of the wheel bearing is required.

REMOVAL & REPLACEMENT

Front

➡The hub nut may be reused 4 times. Mark the nut at removal and only use hub nuts with 3 or fewer marks for assembly.

1. Before servicing the vehicle, refer to the precautions in the beginning of this section.
2. Loosen the strut center nut 5 turns.
3. Remove or disconnect the following:
   - Front wheel
   - Wheel speed sensor, if equipped
   - Hub retainer nut. Mark the nut.
   - Brake caliper and rotor
   - Outer tie rod end
   - Lower ball joint
   - Steering knuckle pinch bolt
4. Press the stub shaft out of the wheel hub.
5. Press the hub out of the wheel bearing.
6. Remove the snapring.
7. Press the bearing out of the hub.

To install:

8. Press the bearing into the hub.
9. Install the snapring.
10. Press the hub into the wheel bearing.
11. Draw the stub shaft into the wheel hub with Halfshaft installer 205-379.
12. Install or connect the following:
    - Steering knuckle. Tighten the pinch bolt to 66 ft. lbs. (90 Nm).
    - Lower ball joint. Tighten the pinch bolt to 37 ft. lbs. (50 Nm).
    - Outer tie rod end. Tighten the nut to 35 ft. lbs. (48 Nm).
    - Brake caliper and rotor
    - Hub retainer nut. Tighten the nut to 232 ft. lbs. (316 Nm).
    - Wheel speed sensor, if equipped
    - Front wheel
13. Tighten the strut center nuts to 35 ft. lbs. (48 Nm).
14. Check the wheel alignment and adjust as necessary.

Rear

1. Before servicing the vehicle, refer to the precautions in the beginning of this section.
2. Remove or disconnect the following:
   - Rear wheel
   - Dust cap
   - Hub nut
   - Brake drum
   - Wheel speed sensor tone ring, if equipped
   - Snapring
3. Remove the wheel bearing with a press.

To install:

➡Use a new wheel speed sensor tone ring for assembly.

4. Press the wheel bearing into the brake drum hub.
5. Install or connect the following:
   - Snapring
   - Wheel speed sensor tone ring, if equipped
   - Brake drum
   - Hub nut. Tighten the hub nut to 173 ft. lbs. (235 Nm).
   - Dust cap
   - Rear wheel

BRAKES

Brake Caliper

REMOVAL & INSTALLATION

Front

1. Before servicing the vehicle, refer to the precautions in the beginning of this section.
2. Remove or disconnect the following:
   - Front wheel
   - Brake hose from the support bracket
   - Outer brake pad retaining clip
   - Brake caliper from the steering knuckle
   - Brake pads
   - Brake hose from the caliper

To install:

3. Install or connect the following:
   - Brake hose to the caliper. Tighten the fitting to 11 ft. lbs. (15 Nm).
   - Brake pads
   - Brake caliper to the steering knuckle. Tighten the bolts to 21 ft. lbs. (28 Nm).
   - Outer brake pad retaining clip
   - Brake hose to the support bracket
   - Front wheel
4. Bleed the brakes and check for proper operation.

Rear

1. Before servicing the vehicle, refer to the precautions in the beginning of this section.
2. Remove the rear wheel and tire.
3. Detach the parking brake cable from the brake caliper.
4. Using a suitable brake hose clamp, clamp the brake hose.
5. Loosen the brake hose union.
6. Detach the brake caliper from the brake anchor plate.
7. Remove the brake caliper.
8. Disconnect the brake hose.

To install:

9. Loosely connect the brake hose to the brake caliper.

➡When the brake caliper piston is retracted into the piston housing, brake fluid will be displaced into the brake fluid reservoir.

10. Using the Rear Caliper Piston Adjuster 206-010 and Adapter 206-026, retract the brake caliper piston.

➡Make sure the brake caliper retaining bolt springs are correctly located.

11. Install the caliper and tighten the bolts to 26 ft. lbs. (35 Nm).
12. Tighten the brake hose union to 11 ft. lbs. (15 Nm).
13. Remove the brake hose clamp.
14. Attach the parking brake cable to the brake caliper.
15. Install the wheel and tire.
16. Bleed the brake system.

Disc Brake Pads

REMOVAL & INSTALLATION

Front

1. Before servicing the vehicle, refer to the precautions in the beginning of this section.
2. Remove or disconnect the following:
   - Front wheel
   - Brake hose from the support bracket
   - Outer brake pad retaining clip
   - Brake caliper
   - Inner and outer brake pads

To install:

3. Compress the caliper piston into the caliper bore.
4. Install or connect the following:
   - Inner and outer brake pads
   - Brake caliper. Tighten the bolts to 21 ft. lbs. (28 Nm).
   - Brake hose to the support bracket
   - Front wheel

Rear

1. Before servicing the vehicle, refer to the precautions in the beginning of this section.
2. Remove the rear wheel and tire.
3. Detach the parking brake cable from the brake caliper.
4. Loosen the brake hose union.
5. Detach the brake caliper from the brake anchor plate.
6. Remove the brake caliper.
7. Remove the brake pads.

To install:

➡When the brake caliper piston is retracted into the piston housing, brake fluid will be displaced into the brake fluid reservoir.

8. Using the Rear Caliper Piston Adjuster 206-010 and Adapter 206-026, retract the brake caliper piston.

➡Make sure the brake caliper retaining bolt springs are correctly located.

9. Install the brake pads.
10. Install the caliper and tighten the bolts to 26 ft. lbs. (35 Nm).
11. Attach the parking brake cable to the brake caliper.
12. Install the wheel and tire.
13. Bleed the brake system.

Brake Drums

REMOVAL & INSTALLATION

1. Before servicing the vehicle, refer to the precautions in the beginning of this section.
2. Remove or disconnect the following:
   - Rear wheel
   - Wheel speed sensor, if equipped
   - Brake drum, bearing and spindle assembly.

To install:

3. Install or connect the following:
   - Brake drum, bearing and spindle assembly. Tighten the bolts to 49 ft. lbs. (66 Nm).
   - Wheel speed sensor, if equipped
   - Rear wheel

Brake Shoes

REMOVAL & INSTALLATION

1. Before servicing the vehicle, refer to the precautions in the beginning of this section.
2. Remove or disconnect the following:
   - Rear wheel
   - Wheel speed sensor, if equipped
   - Brake drum, bearing and spindle assembly
   - Brake shoe hold-down springs and pins
   - Brake shoe assembly from the wheel cylinder and anchor block
   - Parking brake cable
   - Lower return spring
   - Upper return spring
   - Primary shoe from the strut and brake shoe adjuster
   - Parking brake return spring
   - Secondary shoe from the strut support

To install:

3. Install or connect the following:
   - Secondary shoe to the strut support
   - Parking brake return spring
   - Primary shoe to the strut and brake shoe adjuster. Rotate the adjuster fully clockwise.
   - Upper return spring
   - Lower return spring
   - Parking brake cable
   - Brake shoe assembly to the wheel cylinder and anchor block
   - Brake shoe hold-down springs and pins
   - Brake drum, bearing and spindle assembly. Tighten the bolts to 49 ft. lbs. (66 Nm).
   - Wheel speed sensor, if equipped
   - Rear wheel
4. Operate the brake pedal to adjust the rear brakes.

FORD AND MERCURY

10

Freestar • Monterey

SPECIFICATION CHARTS

VEHICLE AND ENGINE IDENTIFICATION CHART

Engine							Model Year	
Code	Liters	Cu. In.	Cyl.	Fuel Sys.	Engine Type	Eng. Mfg.	Code	Year
6	3.9	238	6	SEFI	OHV	Ford	4	2004
2	4.2	256	6	SEFI	OHV	Ford		

SEFI: Sequential Multi-port Fuel Injection

67197-FRST-C01

GENERAL ENGINE SPECIFICATIONS

Year	Engine Displacement Liters	Engine VIN	Net Horsepower @ rpm	Net Torque @ rpm (ft. lbs.)	Bore x Stroke (in.)	Com-pression Ratio	Oil Pressure @ rpm
2004	3.9	6	193@4500	240@3500	3.81x3.46	9.3:1	40-125@2500
	4.2	2	201@4250	263@3650	3.81x3.80	9.3:1	40-125@2500

67197-FRST-C02

GASOLINE ENGINE TUNE-UP SPECIFICATIONS

Year	Engine Displacement Liters	Engine VIN	Spark Plugs Gap (in.)	Ignition Timing (deg.)		Fuel Pump (psi)	Idle Speed (rpm)		Valve Clearance	
				MT	AT		MT	AT	In.	Ex.
2004	3.9	6	0.052-0.056	—	①	②	—	①	HYD	HYD
	4.2	2	0.052-0.056	—	①	②	—	①	HYD	HYD

NOTE: The Vehicle Emission Control Information label often reflects specification changes changes made during production.

The label figures must be used if they differ from those in this chart.

B: Before top dead center

HYD: Hydraulic

① Controlled by the Powertrain Control Module (PCM) and cannot be manually adjusted.

② Engine running: 58 psi

Key On, Engine Off (KOEO): 67 psi

67197-FRST-C03

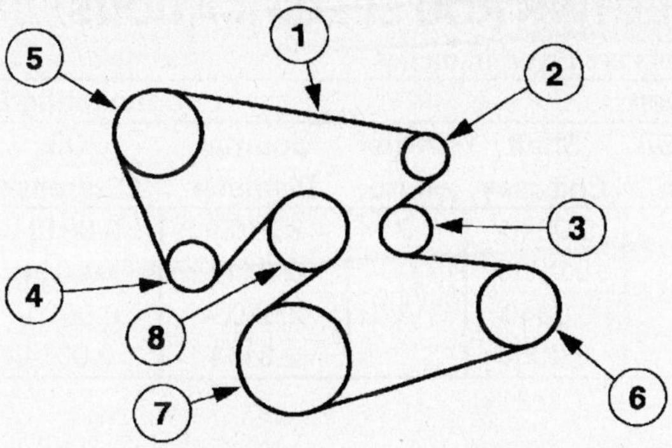

Item Description

Item	Description
1	Drive belt
2	Generator pulley
3	Idler pulley
4	Drive belt tensioner
5	Power steering pump pulley
6	A/C clutch pulley
7	Crankshaft vibration damper pulley
8	Coolant pump pulley

7197-FRST-CG01

3.9L and 4.2L engine accessory drive belt routing

CAPACITIES

Year	Model	Engine Displacement Liters	Engine ID/VIN	Engine Oil with Filter (qts.)	Transmission (pts.) 4-Spd	5-Spd	Auto.	Drive Axle Front (pts.)	Rear (pts.)	Fuel Tank (gal.)	Cooling System (qts.)
2004	Freestar	3.9	6	5.0	—	—	12.25	①	—	26	②
		4.2	2	5.0	—	—	12.25	①	—	26	②
	Monterey	3.9	6	5.0	—	—	12.25	①	—	26	②
		4.2	2	5.0	—	—	12.25	①	—	26	②

NOTE: All capacities are approximate. Add fluid gradually and check to be sure a proper fluid level is obtained.

① Included in transaxle capacity

② w/auxiliary heater: 16.0
wo/auxiliary heater: 15.0

67197-FRST-C04

PISTON AND RING SPECIFICATIONS

All measurements are given in inches.

Year	Engine Displ. Liters	Engine VIN	Piston Clearance	Ring Gap Top Comp.	Ring Gap Bottom Comp.	Ring Gap Oil Control	Ring Side Clearance Top Comp.	Ring Side Clearance Bottom Comp.	Ring Side Clearance Oil Control
2004	3.9	6	0.0007-0.0017	0.0067-0.0130	0.0118-0.0217	0.0059-0.0256	0.0012-0.0026	0.0012-0.0028	Snug
	4.2	2	0.0007-0.0017	0.0066-0.0129	0.0118-0.0217	0.0059-0.0255	0.0012-0.0031	0.0012-0.0028	Snug

67197-FRST-C05

CRANKSHAFT AND CONNECTING ROD SPECIFICATIONS

All measurements are given in inches.

Year	Engine Displ. Liters	Engine VIN	Crankshaft: Main Brg. Journal Dia.	Crankshaft: Main Brg. Oil Clearance	Crankshaft: Shaft End-play	Crankshaft: Thrust on No.	Connecting Rod: Journal Diameter	Connecting Rod: Oil Clearance	Connecting Rod: Side Clearance
2004	3.9	6	2.5188-2.5196	0.0010-0.0014	0.0040-0.0080	3	2.3103-2.3111	0.0010-0.0014	0.0047-0.0193
	4.2	2	2.5188-2.5196	0.0010-0.0014	0.0040-0.0080	3	2.3103-2.3111	0.0010-0.0014	0.0047-0.0193

67197-FRST-C06

VALVE SPECIFICATIONS

Year	Engine VIN	Engine Displacement Liters	Seat Angle (deg.)	Face Angle (deg.)	Spring Test Pressure (lbs. @ in.)	Spring Installed Height (in.)	Stem-to-Guide Clearance (in.) Intake	Stem-to-Guide Clearance (in.) Exhaust	Stem Diameter (in.) Intake	Stem Diameter (in.) Exhaust
2004	6	3.9	44.75	45.6	224@1.16	1.620	0.0008-0.0027	0.0018-0.0037	0.2746-0.2754	0.2735-0.2744
	2	4.2	44.75	45.6	224@1.16	1.620	0.0008-0.0027	0.0018-0.0037	0.2746-0.2754	0.2735-0.2744

67197-FRST-C07

TORQUE SPECIFICATIONS

All readings in ft. lbs.

Year	Engine VIN	Engine Displacement Liters	Cylinder Head Bolts	Main Bearing Bolts	Rod Bearing Bolts	Crankshaft Damper Bolts	Flywheel Bolts	Manifold Intake	Manifold Exhaust	Spark Plugs	Oil Pan Drain Plug
2004	6	3.9	①	②	③	118	59	④	18	12	19
	2	4.2	①	②	③	118	59	④	18	12	19

① Step 1: 14 ft. lbs.
Step 2: 29 ft. lbs.
Step 3: 36 ft. lbs.
Step 4: Loosen bolt no. 1 bolt 3 turns
Step 5: Tighten no. 1 bolt as listed in step 7
Step 6: Repeat steps 4 and 5 for each bolt in sequence
Step 7: Long bolts to 30 ft. lbs. plus 1/2 turn; short bolts to 18 ft. lbs. plus 1/2 turn

② Step 1: 37 ft. lbs.
Step 2: plus 120 degrees

③ Step 1: 18 ft. lbs.
Step 2: 33 ft. lbs.
Step 3: plus 105 degrees

④ Upper manifold :Step 1: 53 INCH lbs.
Step 2: 89 INCH lbs.
Lower manifold: Step 1: 44 INCH lbs.
Step 2: 89 INCH lbs.

67197-FRST-C08

WHEEL ALIGNMENT

Year	Model			Caster Range (+/-Deg.)	Caster Preferred Setting (Deg.)	Camber Range (+/-Deg.)	Camber Preferred Setting (Deg.)	Toe-in (Deg.)
2004	Freestar	Front	Left	0.75	+3.2	0.75	-0.40	-0.15+/-0.25
			Right	0.75	+3.2	0.75	-0.40	-0.15+/-0.25
		Rear		—	—	0.75	-0.30	-0.14+/-0.20
	Monterey	Front	Left	0.75	+3.2	0.75	-0.40	-0.15+/-0.25
			Right	0.75	+3.2	0.75	-0.40	-0.15+/-0.25
		Rear		—	—	0.75	-0.30	-0.14+/-0.20

67197-FRST-C09

BRAKE SPECIFICATIONS

All measurements in inches unless noted

Year	Model		Brake Disc Original Thickness	Brake Disc Minimum Thickness	Brake Disc Maximum Runout	Minimum Lining Thickness Front	Minimum Lining Thickness Rear	Brake Caliper Bracket Bolts (ft. lbs.)	Brake Caliper Mounting Bolts (ft. lbs.)
2004	Freestar	F	①	1.110	0.002	0.118	—	136	26
		R	①	0.720	0.002	—	0.118	75	24
	Monterey	F	①	1.110	0.002	0.118	—	136	26
		R	①	0.720	0.002	—	0.118	75	24

① Not available

67197-FRST-C10

TIRE AND WHEEL SPECIFICATIONS

Year	Model	OEM Tires Standard	OEM Tires Optional	Tire Pressures (psi) Front	Tire Pressures (psi) Rear	Wheel Size	Wheel Lug Nut Torque (Ft. Lbs.)
2004	Freestar	P225/60R16	P235/60R16	35	35	①	100
	Monterey	P225/60R16	P235/60R16	35	35	①	100

OEM: Original Equipment Manufacturer

PSI: Pounds Per Square Inch

STD: Standard

OPT: Optional

① Not available

67197-FRST-C11

SCHEDULED MAINTENANCE INTERVALS
FORD FREESTAR & MERCURY MONTEREY

		VEHICLE MILEAGE INTERVAL (x1000)																	
TO BE SERVICED	TYPE OF SERVICE	5	10	15	20	25	30	35	40	45	50	55	60	65	70	75	80	85	90
Engine oil & filter	R	✓	✓	✓	✓	✓	✓	✓	✓	✓	✓	✓	✓	✓	✓	✓	✓	✓	✓
Rotate tires	S/I	✓		✓			✓			✓			✓				✓		✓
Engine coolant strength hoses & clamps	S/I			✓			✓			✓			✓				✓		✓
Air cleaner filter	R						✓						✓						✓
Automatic transmission fluid & filter	R						✓						✓						✓
Engine coolant ①	R									✓							✓		
PCV valve	R												✓						✓
Spark plugs ②	R																		
Drive belts	S/I						✓						✓						✓
Exhaust system & heat shields	S/I						✓						✓						✓
Front & rear brakes	S/I			✓			✓			✓			✓				✓		✓
Fuel filter	R						✓						✓						✓

R: Replace S/I: Service or Inspect

① Engine coolant: change initially at 50,000 miles and every 30,000 miles thereafter.

② Spark plugs: replace every 100,000 miles.

FREQUENT OPERATION MAINTENANCE (SEVERE SERVICE)

If a vehicle is operated under any of the following conditions it is considered severe service:

- Extremely dusty areas.

- 50% or more of the vehicle operation is in 32°C (90°F) or higher temperatures, or constant operation in temperatures below 0°C (32°F).

- Prolonged idling (vehicle operation in stop and go traffic.

- Frequent short running periods (engine does not warm to normal operating temperatures).

- Police, taxi, delivery usage or trailer towing usage.

Engine oil & filter: replace every 3000 miles.

Rotate tires initially at 6000 miles and every 9000 miles thereafter.

Air cleaner filter: change every 15,000 miles.

Engine coolant strength, hoses & clamps: check every 15,000 miles.

Exhaust system: check every 15,000 miles.

Automatic transmission fluid & filter: change every 21,000 miles.

Special Operating Condition Requirements

When towing a trailer or using a camper or car-top carrier:

Change engine oil and install a new oil filter every 4,800 km (3,000 miles) or 3 months.

During extensive idling and/or low speed driving for long distances, as in heavy commercial use such as delivery, taxi, patrol car or livery:

Change engine oil and install a new oil filter, lube front lower control arm and steering linkage ball joints with Zerk fittings (if equipped) every 4,800 km (3,000 miles) or 3 months.

Inspect brake system and check battery electrolyte level (Patrol cars) every 8,000 km (5,000 miles).

Install a new fuel filter every 24,000 km (15,000 miles).

Change automatic transmission fluid, lubricate 4x2 wheel bearings, install new grease seals and adjust bearings every 48,000 km (30,000 miles).

Install new spark plugs and change transfer case fluid every 96,000 km (60,000 miles).

Install a new cabin air filter as required.

When operating in dusty conditions such as unpaved or dusty roads:

Change engine oil and install a new oil filter every 4,800 km (3,000 miles) or 3 months.

Install a new fuel filter every 24,000 km (15,000 miles).

Change automatic transmission fluid every 48,000 km (30,000 miles).

Install a new engine air filter as required.

Install a new cabin air filter as required.

When operating in off-road conditions:

Change automatic transmission fluid every 48,000 km (30,000 miles).

Install a new cabin air filter as required.

Inspect and lubricate U-joints.

Inspect and lubricate steering linkage ball joints with zerk fittings.

67197-FRST-C12

PRECAUTIONS

Before servicing any vehicle, please be sure to read all of the following precautions, which deal with personal safety, prevention of component damage, and important points to take into consideration when servicing a motor vehicle:

- Never open, service or drain the radiator or cooling system when the engine is hot; serious burns can occur from the steam and hot coolant.
- Observe all applicable safety precautions when working around fuel. Whenever servicing the fuel system, always work in a well-ventilated area. Do not allow fuel spray or vapors to come in contact with a spark, open flame, or excessive heat (a hot drop light, for example). Keep a dry chemical fire extinguisher near the work area. Always keep fuel in a container specifically designed for fuel storage; also, always properly seal fuel containers to avoid the possibility of fire or explosion. Refer to the additional fuel system precautions later in this section.
- Fuel injection systems often remain pressurized, even after the engine has been turned **OFF**. The fuel system pressure must be relieved before disconnecting any fuel lines. Failure to do so may result in fire and/or personal injury.
- Brake fluid often contains polyglycol ethers and polyglycols. Avoid contact with the eyes and wash your hands thoroughly after handling brake fluid. If you do get brake fluid in your eyes, flush your eyes with clean, running water for 15 minutes. If eye irritation persists, or if you have taken brake fluid internally, IMMEDIATELY seek medical assistance.
- The EPA warns that prolonged contact with used engine oil may cause a number of skin disorders, including cancer! You should make every effort to minimize your exposure to used engine oil. Protective gloves should be worn when changing oil. Wash your hands and any other exposed skin areas as soon as possible after exposure to used engine oil. Soap and water, or waterless hand cleaner should be used.
- All new vehicles are now equipped with an air bag system. The system must be disabled before performing service on or around system components, steering column, instrument panel components, wiring and sensors. Failure to follow safety and disabling procedures could result in accidental air bag deployment, possible personal injury and unnecessary system repairs.
- Always wear safety goggles when working with, or around, the air bag system. When carrying a non-deployed air bag, be sure the bag and trim cover are pointed away from your body. When placing a non-deployed air bag on a work surface, always face the bag and trim cover upward, away from the surface. This will reduce the motion of the module if it is accidentally deployed. Refer to the additional air bag system precautions later in this section.
- Clean, high quality brake fluid from a sealed container is essential to the safe and proper operation of the brake system. You should always buy the correct type of brake fluid for your vehicle. If the brake fluid becomes contaminated, completely flush the system with new fluid. Never reuse any brake fluid. Any brake fluid that is removed from the system should be discarded. Also, do not allow any brake fluid to come in contact with a painted surface; it will damage the paint.
- Never operate the engine without the proper amount and type of engine oil; doing so WILL result in severe engine damage.
- Timing belt maintenance is extremely important! Many models utilize an interference-type, non-freewheeling engine. If the timing belt breaks, the valves in the cylinder head may strike the pistons, causing potentially serious (also time-consuming and expensive) engine damage. Refer to the maintenance interval charts in the front of this manual for the recommended replacement interval for the timing belt, and to the timing belt section for belt replacement and inspection.
- Disconnecting the negative battery cable on some vehicles may interfere with the functions of the on-board computer system(s) and may require the computer to undergo a relearning process once the negative battery cable is reconnected.
- When servicing drum brakes, only disassemble and assemble one side at a time, leaving the remaining side intact for reference.
- Only an MVAC-trained, EPA-certified automotive technician should service the air conditioning system or its components.

ENGINE REPAIR

Alternator

REMOVAL & INSTALLATION

1. Disconnect the negative battery cable.
2. Disconnect the accessory drive belt.
3. Detach the alternator wiring connector.
4. Remove the positive cable nut.
5. Remove the alternator.

To install:

6. Position the alternator on the engine.
7. Install the alternator mounting bolts. Tighten the bolts to 18 ft. lbs. (25 Nm).
8. Tighten the positive cable nut to 71 INCH lbs. (8 Nm).
9. Connect the wiring connector.
10. Install and tension the accessory drive belt.
11. Connect the negative battery cable.

Engine Assembly

REMOVAL & INSTALLATION

1. Relieve the fuel system pressure.
2. Drain the engine coolant.
3. Recover the air conditioning refrigerant, into a refrigerant recovery station
4. Remove or disconnect the following:
   - Both battery cables
   - Wiper motor
   - Air cleaner assembly
   - Fuel supply manifold lines
   - Power steering pump
   - Starter

➡Do not allow the flex connector of the duel converter Y-pipe to hang unsupported or damage to the flex joint will result.

   - Dual converter Y-pipe and support it from the body
   - Accelerator and cruise control cables from throttle assembly
   - Accelerator cable bracket
   - Upper and lower radiator hoses from the engine
   - Heater water hoses and secure to body
   - Positive terminal from Power Distribution Box (PDB)
   - PDB 68-pin connector
   - Engine wiring retainers
   - Shift cable
   - Evaporative (EVAP) return tube

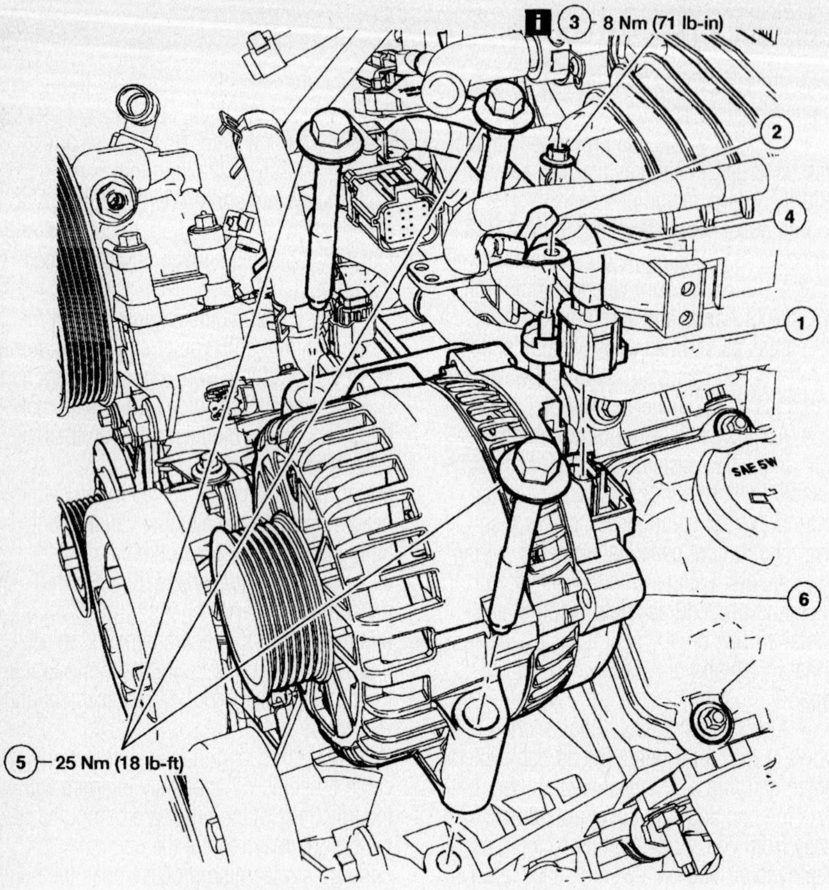

Item	Description	Item	Description
1	Generator electrical connector	4	Generator B+ cable
2	Generator B+ cable boot	5	Generator mounting bolts
3	Generator B+ cable nut	6	Generator

67197-FRST-CG03

Alternator mounting—3.9L engine, 4.2L similar

- EVAP canister connector and vacuum line
- Brake booster vacuum line
- Powertrain Control Module (PCM)
- Engine grounds (3)
- Radiator shield
- Passenger side engine roll restrictor
- A/C compressor and wire aside
- Power steering cooler line
- Transaxle cooler lines
- Steering column pinch bolt
- Oil pan-to-transaxle bolts
- Torque converter nuts (4)
- Brake calipers and wire aside
- ABS connectors
- Stabilizer bar links
- Lower strut bolts on both sides
- Position subframe support tool 014-00765 under subframe
- Subframe bolts and lower the engine, transaxle and subframe assembly
- Alternator brace
- A/C compressor and wire aside
- Driver side engine roll restrictor
- Label and remove remaining electrical connectors
- Transaxle fluid level tube
- Install suitable engine lifting device
- Bell housing bolts (4)
- Engine mount bolts
- Transaxle mount bolts
- Engine

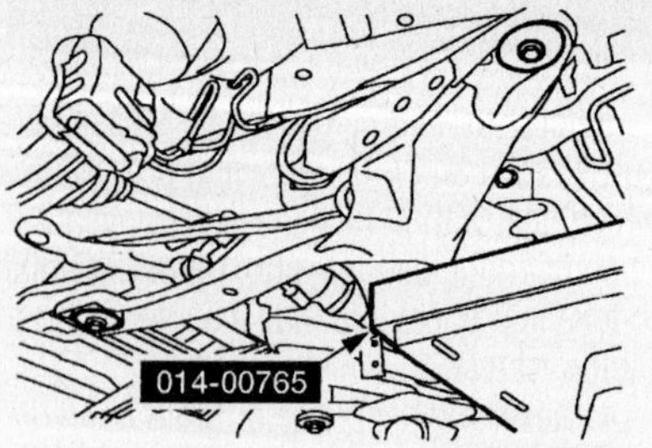

67197-FRST-CG06

Installing subframe support tool—Freestar and Monterey

To install:

5. Position the engine onto the transaxle and subframe assembly.
6. Install the transaxle mount bolts. Torque the bolts to 46 ft. lbs. (63 Nm).
7. Install the oil pan-to-transaxle bolt. Torque the bolt to 33 ft. lbs. (45 Nm).
8. Install the engine mount bolts. Torque the bolts to 66 ft. lbs. (90 Nm).
9. Install the bell housing bolts. Torque the bolts to 37 ft. lbs. (50 Nm).
10. Install or connect the following:
    - Transaxle fluid level tube
    - Under vehicle electrical connectors
    - Driver side roll restrictor bracket. Torque the bolts to 37 ft. lbs. (50 Nm).
    - A/C compressor. Torque the bolts to 18 ft. lbs. (25 Nm).

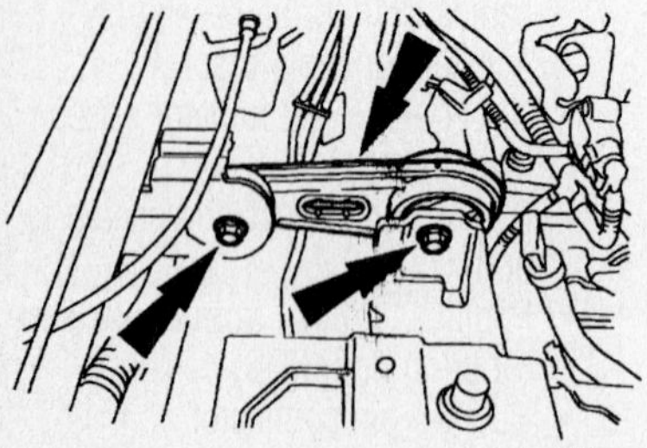

67197-FRST-CG04

Removing passenger side engine roll restrictor—Freestar and Monterey

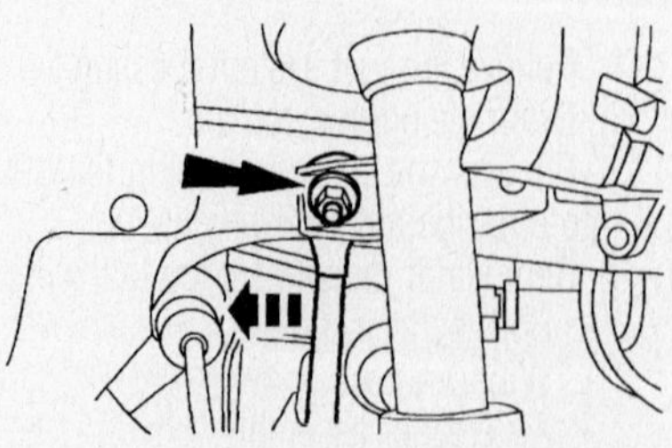

67197-FRST-CG05

Removing stabilizer bar links—Freestar and Monterey

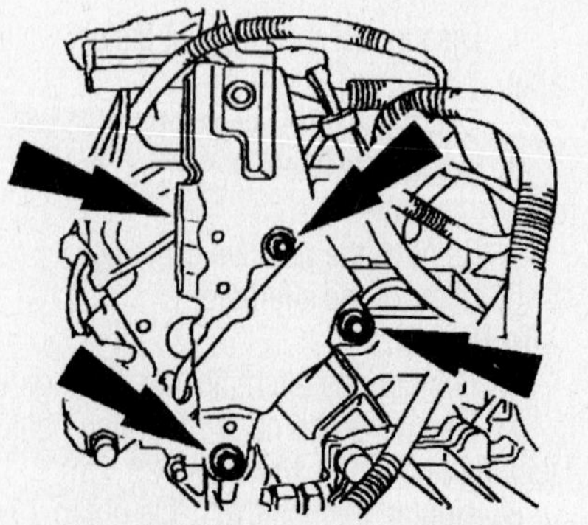

67197-FRST-CG07

Removing driver side engine roll restrictor—Freestar and Monterey

- Generator brace
- Raise the engine, transaxle and subframe assembly
- Subframe bolts. Torque the bolts to 66 ft. lbs. (90 Nm).
- Lower strut bolts. Torque the bolts to 91 ft. lbs. (124 Nm).
- Stabilizer bar link bolts. Torque the bolts to 70 ft. lbs. (95 Nm).
- ABS connectors
- Brake calipers. Torque the bolts to 26 ft. lbs. (35 Nm).
- Torque converter bolts. Torque the bolts to 27 ft. lbs. (36 Nm).
- Oil pan-to-transaxle bolts. Torque the bolts to 33 ft. lbs. (45 Nm).
- Steering column pinch bolt. Torque the bolts to 30 ft. lbs. (40 Nm).

11. The remainder of the installation is the reverse of removal.
12. Please note the following torque specifications:
    - Passenger side roll restrictor. Torque the bolts to 46 ft. lbs. (63 Nm).
13. Refill the engine, transaxle and cooling system with the correct amount of the appropriate fluids before starting the engine. Recharge the A/C system using approved recycling equipment.

Water Pump

REMOVAL & INSTALLATION

1. Remove or disconnect the following:
    - Negative battery cable
    - Coolant
    - Accessory drive belt
    - Lower radiator hose
    - Idler pulley
    - Generator bracket
    - Lower engine mount nuts

➡Install a suitable engine lifting device and raise the engine approximately 5 inches to gain access to the water pump pulley bolts.

- Water pump pulley
- Bypass hoses
- Water pump

To install:

⁂ WARNING

Be careful not to gouge the aluminum surfaces when scraping the old gasket material from the mating surfaces of the water pump and front cover.

2. Clean the gasket surfaces on the water pump and front cover.

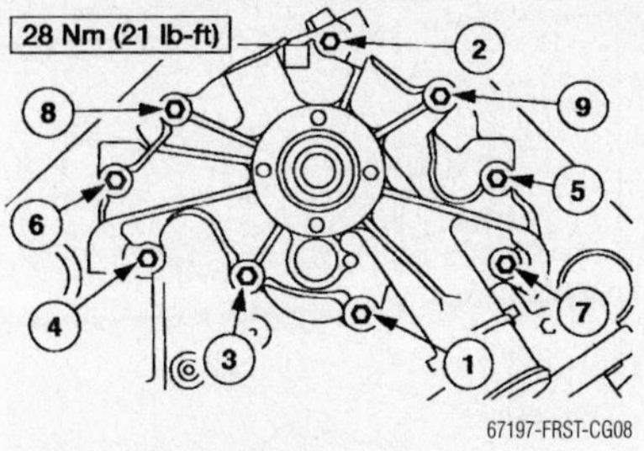

Water pump bolt tightening sequence—Freestar and Monterey

3. Position a new water pump housing gasket on the water pump sealing surface using gasket sealant to hold the gasket in place.
4. Install the water pump and tighten the bolts in sequence shown to 21 ft. lbs. (28 Nm).
5. The remainder of the installation is the reverse of removal.
6. Fill and bleed the cooling system.
7. Connect the negative battery cable.
8. Start the engine and check for leaks.

Heater Core

REMOVAL & INSTALLATION

Front System

1. Disconnect the negative battery cable.
2. Drain the cooling system into a clean container for reuse.
3. Disconnect the heater hoses from the heater core inlet and outlet tubes in the engine compartment.
4. Remove the steering column lower covers.
5. Remove the center storage compartment
6. Remove the upper instrument panel finish cover.
7. Remove the instrument panel bolts (9) at the top and driver side of the instrument panel.

➡Insert a long bolt or threaded rod in place of the left side instrument panel attaching bolt to support the instrument panel when removing the attaching bolts.

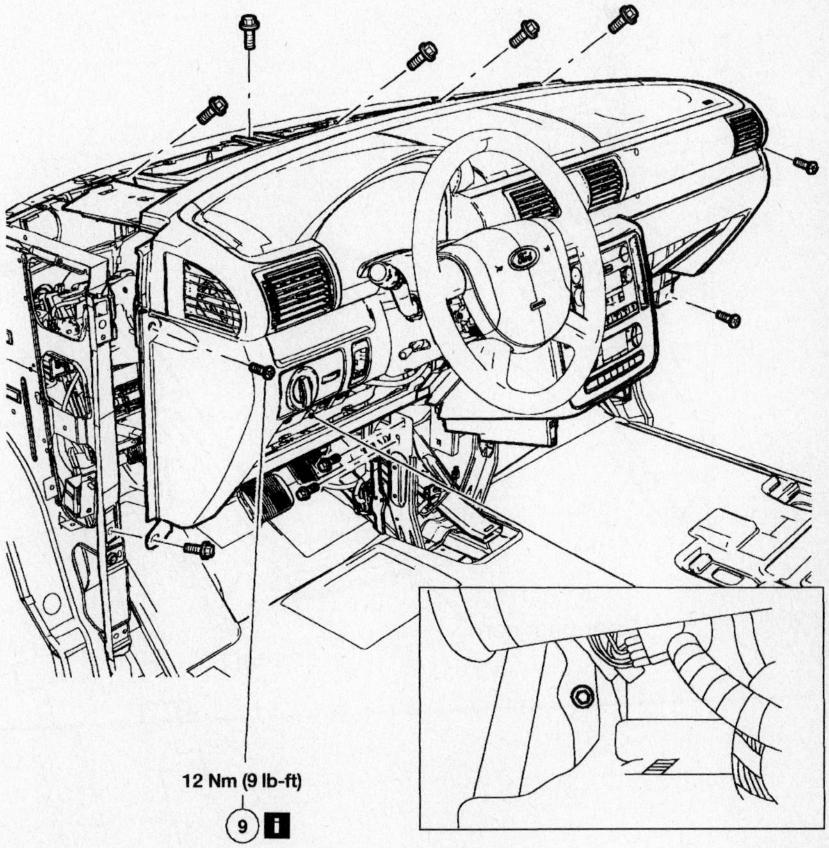

Removing instrument panel support bolts—Freestar and Monterey

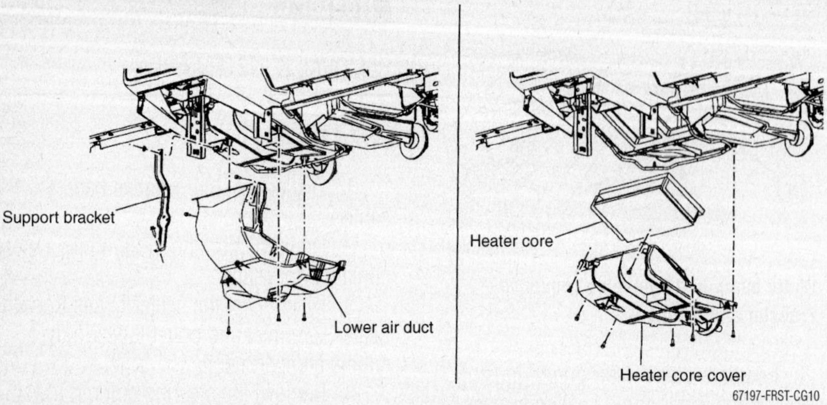

Removing center air duct and heater core—Freestar and Monterey

8. Loosen the passenger side instrument panel bolts approximately halfway.
9. Remove the center instrument panel support brace.
10. From inside the glove box, remove the attaching bolt.
11. Pull the instrument panel outward approximately 3 inches.
12. Remove the lower center air duct.
13. Disconnect the heater core wiring connector.
14. Remove the heater core cover and heater core.

To install:

15. Installation is the reverse of removal.
16. Please note the following torque specifications:
    - Upper instrument panel bolts. Torque the bolts to 44 INCH lbs. (5 Nm).
    - Side and lower instrument panel bolts. Torque the bolts to 9 ft. lbs. (12 Nm).
17. Refill the cooling system.
18. Connect the negative battery cable.
19. Run the engine to normal operating temperatures; then, check the climate control operation and check for leaks.

Rear Auxiliary System

1. Disconnect the negative battery cable.
2. Drain the cooling system into a clean container for reuse.
3. Remove the left side quarter trim panel.
4. Remove the temperature blend door actuator.
5. Remove the heater hoses.
6. Remove the heater core.

To install:

7. Installation is the reverse of the removal procedure.

Cylinder Head

REMOVAL & INSTALLATION

Left Side

1. Relieve the fuel system pressure.
2. Remove or disconnect the following:
   - Negative battery cable
   - Coolant
   - Dual converter Y-pipe.
   - Exhaust manifold

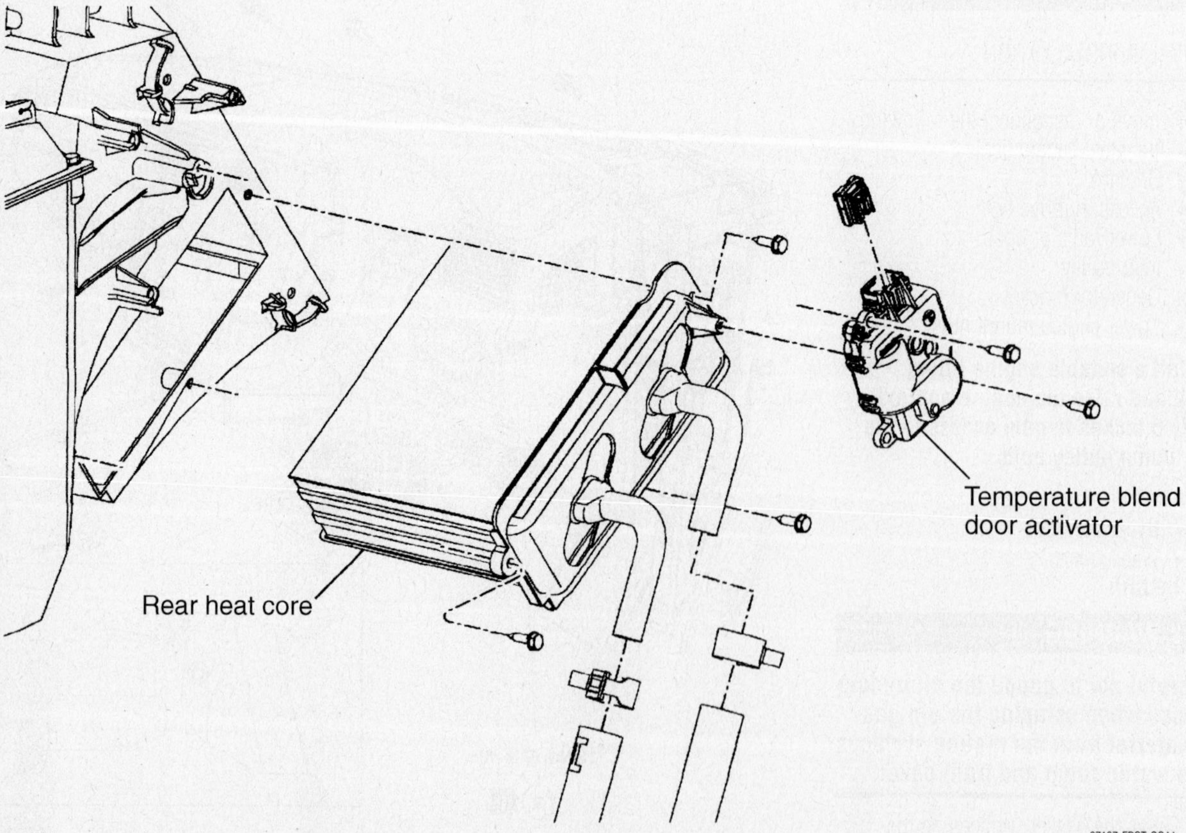

Removing rear auxiliary heater core—Freestar and Monterey

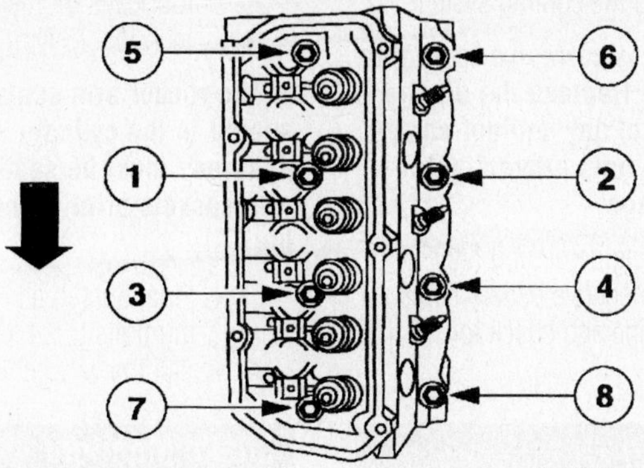

67197-FRST-CG12

Left side cylinder head bolt torque sequence—3.9L and 4.2L engines

- Upper and lower intake manifolds
- Alternator and mounting bracket
- A/C compressor and mounting bracket
- Valve covers

➡Pushrods must be installed in their original positions. Note pushrod location during removal.

- Pushrods
- Rotate tensioner clockwise and remove accessory drive belt
- Automatic belt tensioner assembly
- Lower exhaust manifold mounting studs
- Spark plugs
- Cylinder head bolts
- Cylinder head from the engine block and discard the gaskets

To install:

3. The cylinder head should be cleaned and inspected prior to installation.
4. Lightly oil all bolt and stud bolt threads before installation.
5. Clean all gasket mating surfaces thoroughly.
6. Install or connect the following:
   - New head gaskets on the cylinder block.

⁂ WARNING

Always use new cylinder head bolts when installing the cylinder head or damage to the engine may occur.

- Cylinder head on the cylinder block.

➡Long bolts go in the inside of the cylinder head and short bolts go on the outside.

- Tighten the cylinder head bolts in steps following the proper torque sequence. The first step is 14 ft. lbs. (20 Nm), the second step is 29 ft. lbs. (40 Nm), the third step is 36 ft. lbs. (50 Nm).
- Using the torque sequence, loosen the first long bolt 3 turns, then tighten the bolt to 30 ft. lbs. (40 Nm). Tighten the bolt an additional 180 degrees. Using the torque sequence, loosen the first short bolt 3 turns, then tighten the bolt to 18 ft. lbs. (25 Nm). Tighten the bolt an additional 180 degrees. Repeat these steps for each bolt in sequence using the proper torque for long and short bolts.
- Lower exhaust manifold mounting studs
- Pushrods. Dip each end in engine assembly lubricant.
- Rocker arms, seats and retaining bolts. Lubricate all rocker arm components with engine assembly lubricant. Tighten the bolts to 44 inch lbs. (5 Nm).

➡The rocker arm seats must be fully seated in the cylinder head and the pushrods must be seated in the rocker arm sockets prior to the final tightening.

- Final tighten all rocker arm retaining bolts to 24 ft. lbs. (32 Nm)
- Automatic belt tensioner assembly
- Accessory drive belt
- Valve covers
- A/C compressor and mounting bracket
- Alternator and mounting bracket
- Upper and lower intake manifolds
- Exhaust manifold
- Dual converter Y-pipe.

7. Fill and bleed the cooling system.

➡Engine coolant is corrosive to engine bearing material. Replace the engine oil after removal of any coolant-carrying component to help prevent potential bearing damage.

8. Change the engine oil and filter
9. Connect the negative battery cable.
10. Start the engine and check for leaks.

Right Side

1. Remove or disconnect the following:
   - Negative battery cable
   - Coolant
   - Dual converter Y-pipe.
   - Exhaust manifold
   - Upper and lower intake manifolds
   - Power steering pump and mounting bracket
   - Valve covers

➡Pushrods must be installed in their original positions. Note pushrod location during removal.

- Pushrods
- Rotate tensioner clockwise and remove accessory drive belt
- Automatic belt tensioner assembly
- Lower exhaust manifold mounting studs
- Spark plugs
- Cylinder head bolts
- Cylinder head from the engine block and discard the gaskets

To install:

2. The cylinder head should be cleaned and inspected prior to installation.

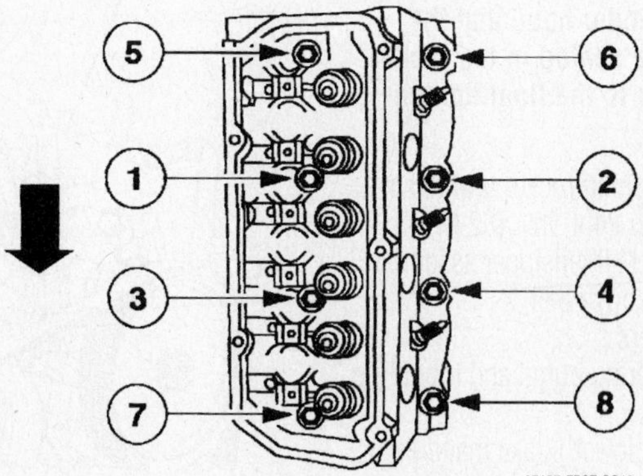

67197-FRST-CG12

Right side cylinder head bolt torque sequence—3.9L and 4.2L engines

3. Lightly oil all bolt and stud bolt threads before installation.
4. Clean all gasket mating surfaces thoroughly.
5. Install or connect the following:
   - New head gaskets on the cylinder block.

⁂ WARNING

Always use new cylinder head bolts when installing the cylinder head or damage to the engine may occur.

   - Cylinder head on the cylinder block.

➡Long bolts go in the inside of the cylinder head and short bolts go on the outside.

   - Tighten the cylinder head bolts in steps following the proper torque sequence. The first step is 14 ft. lbs. (20 Nm), the second step is 29 ft. lbs. (40 Nm), the third step is 36 ft. lbs. (50 Nm).
   - Using the torque sequence, loosen the first long bolt 3 turns, then tighten the bolt to 30 ft. lbs. (40 Nm). Tighten the bolt an additional 180 degrees. Using the torque sequence, loosen the first short bolt 3 turns, then tighten the bolt to 18 ft. lbs. (25 Nm). Tighten the bolt an additional 180 degrees. Repeat these steps for each bolt in sequence using the proper torque for long and short bolts.
   - Lower exhaust manifold mounting studs
   - Pushrods. Dip each end in engine assembly lubricant.
   - Rocker arms, seats and retaining bolts. Lubricate all rocker arm components with engine assembly lubricant. Tighten the bolts to 44 inch lbs. (5 Nm).

➡The rocker arm seats must be fully seated in the cylinder head and the pushrods must be seated in the rocker arm sockets prior to the final tightening.

   - Final tighten all rocker arm retaining bolts to 24 ft. lbs. (32 Nm)
   - Automatic belt tensioner assembly
   - Accessory drive belt
   - Valve covers
   - Power steering pump and mounting bracket
   - Upper and lower intake manifolds
   - Exhaust manifold
   - Dual converter Y-pipe.

6. Fill and bleed the cooling system.

➡Engine coolant is corrosive to engine bearing material. Replace the engine oil after removal of any coolant-carrying component to help prevent potential bearing damage.

7. Change the engine oil and filter
8. Connect the negative battery cable.
9. Start the engine and check for leaks.

Rocker Arms

REMOVAL & INSTALLATION

1. Remove the valve cover.
2. Remove the rocker arm retaining bolt.

➡Rocker the arms should be installed in their original location during assembly.

3. Remove the rocker arms. If more than 1 rocker arm is to be removed, identify each rocker arm location.

To install:

   - Rocker arms, seats and retaining bolts. Lubricate all rocker arm components with engine assembly lubricant. Tighten the bolts to 44 inch lbs. (5 Nm).

➡The rocker arm seats must be fully seated in the cylinder head and the pushrods must be seated in the rocker arm sockets prior to the final tightening.

   - Final tighten all rocker arm retaining bolts to 24 ft. lbs. (32 Nm)
4. Install the valve cover.

Intake Manifold

REMOVAL & INSTALLATION

Upper

1. Remove or disconnect the following:
   - Negative battery cable
   - Coolant
   - Fuel system pressure
   - Wiper motor and pivot assembly
   - Air cleaner outlet tube
   - Accelerator cable
   - Cruise control cable
   - Cable bracket
   - Evaporative (EVAP) return tube
   - Brake booster vacuum hose

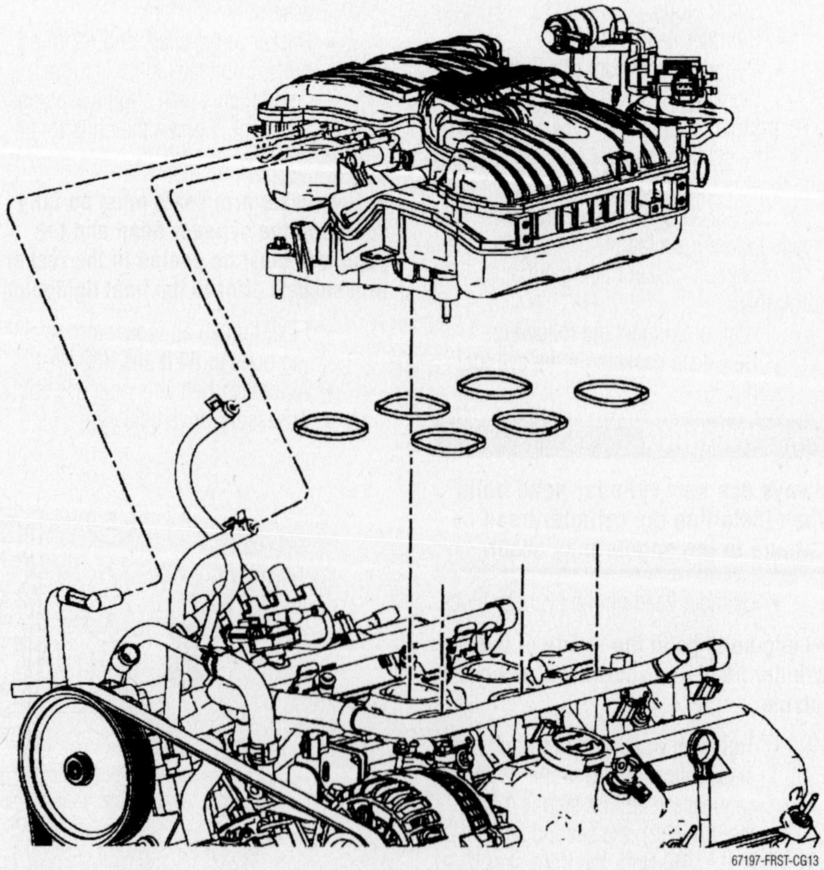

Upper intake manifold—3.9L and 4.2L engines

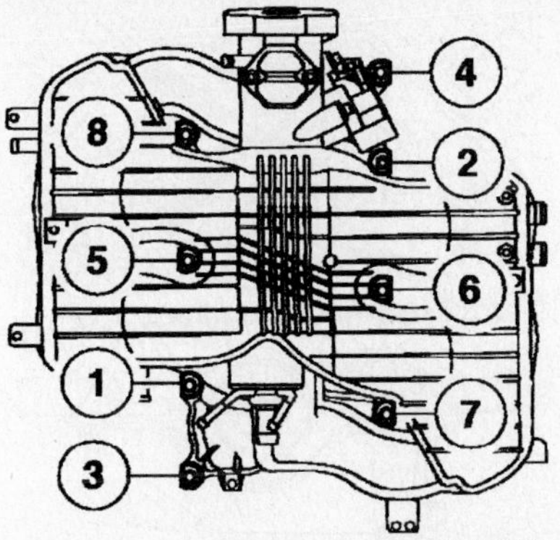

67197-FRST-CG14

Upper intake manifold tightening sequence—3.9L and 4.2L engines

- Exhaust Gas Recirculation (EGR) tubes and connector
- Remaining electrical connectors
- Spark plug wires
- Coolant hoses
- Intake manifold

To install:

2. Installation is the reverse of the removal procedure, using the following torque specifications.
   - Intake manifold bolts and tighten all bolts in steps in sequence to 53 inch lbs. (6 Nm), then to 89 inch lbs. (10 Nm).
3. Fill and bleed the engine cooling system.
4. Connect the negative battery cable.
5. Start the engine and check for leaks.

Lower

1. Remove or disconnect the following:
   - Coolant
   - Fuel system pressure
   - Upper intake manifold
   - Fuel supply manifold
   - Water bypass tube from the heater water outlet tube
   - Heater hoses from manifold
   - Upper radiator hose
   - Electrical wiring harnesses
   - Lower intake manifold retaining bolts. Note the location of the 6 long bolts and 8 short bolts.

➡The lower intake manifold is sealed at each corner with sealer. To break the seal it may be necessary to pry on the front of the intake manifold with a pry-bar. If it is necessary, use care to prevent damage to the machined surfaces.

- Lower intake manifold

To install:

2. Thoroughly clean all gasket mating surfaces.

➡When using silicone rubber sealer, assembly must occur within 5 minutes after sealer application. After this time, the sealer may start to set up and its sealing effectiveness may be reduced.

3. Install or connect the following:
   - Apply a 3mm bead of RTV silicone sealer at each corner where the cylinder head joins the engine block
   - Front and rear intake manifold seals
   - Lower intake manifold into position on the cylinder block, using new gaskets
   - Install intake bolts in their original locations. Tighten in 2 steps in sequence to 44 inch lbs. (5 Nm), then again to 89 inch lbs. (10 Nm).
   - Heater hoses
   - Water bypass hose
   - Fuel supply manifold
   - Upper intake manifold
   - Upper radiator hose
4. Fill and bleed the cooling system.
5. Start the engine and check for leaks.

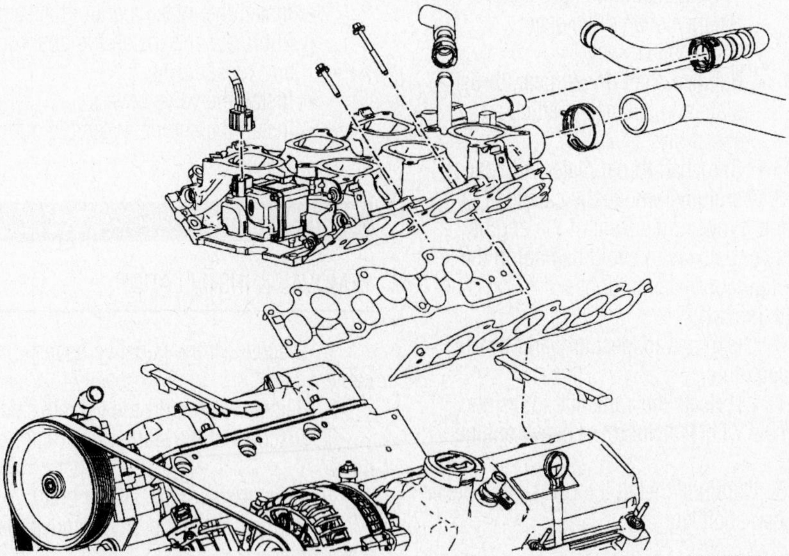

67197-FRST-CG15

Lower intake manifold and related components—3.9L and 4.2L engines

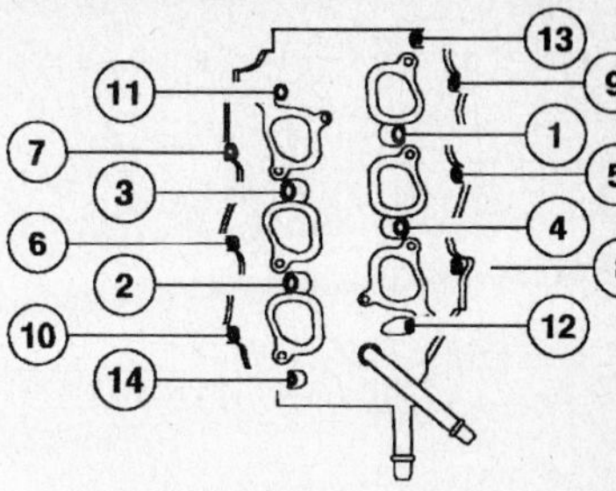

67197-FRST-CG16

Lower intake manifold bolt torque sequence—3.9L and 4.2L engines

Exhaust Manifold

REMOVAL & INSTALLATION

➡Spray the exhaust system fasteners with penetrating lubricant before removing them to help prevent broken studs and bolts. The use of a 6-point socket is highly recommended when removing exhaust system fasteners.

⁂ CAUTION

To prevent serious burns, allow the exhaust manifold to cool down before attempting to remove it.

Left Manifold

1. Disconnect the negative battery cable.
2. Remove the dipstick tube bolt.
3. Disconnect the EGR tube from the manifold.
4. Raise and support the vehicle safely on jackstands.
5. Disconnect the dual converter Y-pipe from the exhaust manifold.
6. Lower the vehicle.
7. Remove the exhaust manifold.

To install:

8. Clean all gasket mating surfaces thoroughly.
9. Install a new exhaust manifold gasket and the exhaust manifold on the cylinder head. Start 2 nuts to hold the manifold in position.
10. Install the remaining nuts. Tighten

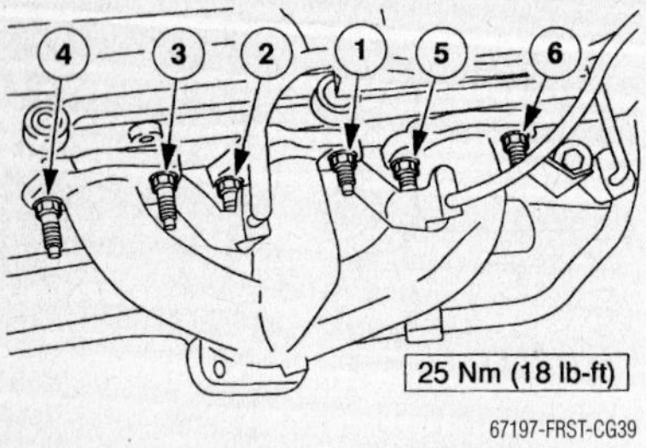

Left side exhaust manifold tightening sequence—3.9L and 4.2L engines

the nuts in the sequence shown to 18 ft. lbs. (25 Nm).

11. Raise and support the vehicle safely.
12. Connect the dual converter Y-pipe.
13. Lower the vehicle.
14. Install the EGR tube.
15. Install the disptick bolt.
16. Connect the negative battery cable.
17. Start the engine and check for exhaust leaks.

Right Manifold

1. Disconnect the negative battery cable.
2. Raise and support the vehicle safely on jackstands.
3. Disconnect the dual converter Y-pipe from the exhaust manifold.
4. Lower the vehicle.
5. Remove the heat shield.
6. Remove the exhaust manifold.

To install:

7. Clean all gasket mating surfaces thoroughly.
8. Install a new gasket and the exhaust manifold on the cylinder head. Start 2 nuts to hold the manifold in position.
9. Install the remaining nuts. Tighten the nuts in the sequence shown to 18 ft. lbs. (25 Nm).
10. Install the heat shield.

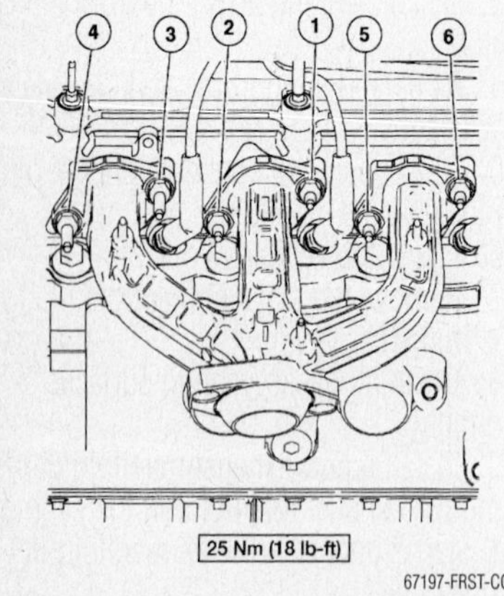

Right side exhaust manifold tightening sequence—3.9L and 4.2L engines

11. Raise and support the vehicle safely on jackstands.
12. Connect the dual converter Y-pipe.
13. Lower the vehicle.
14. Connect the negative battery cable.
15. Start the engine and check for exhaust leaks.

Camshaft and Valve Lifters

REMOVAL & INSTALLATION

1. Rotate the crankshaft until the No. 1 piston is at the TDC on its compression stroke and the timing marks are aligned.
2. Remove or disconnect the following:
   - Engine from the vehicle
   - Valve covers
   - Intake manifolds
   - Pushrods
   - Tappet guide plate and retainer
   - Tappets
   - Crankshaft pulley and damper
   - Engine front cover assembly
3. Check the camshaft end-play as follows:
   a. Push the camshaft toward the rear of the engine and install a dial indicator, so the indicator point is on the camshaft sprocket attaching screw.
   b. Zero the dial indicator. Position a small prybar between the camshaft sprocket or gear and block.
   c. Pull the camshaft forward and release it. Camshaft end-play should be 0.001–0.006 in. (0.025–0.15mm).
   d. If the camshaft end-play is not within specification, replace the thrust plate upon reassembly.
4. Remove or disconnect the following:
   - Timing chain and sprockets
   - Balance shaft drive gear
   - Camshaft keyway
   - Balance shaft driven gear, thrust plate and balance shaft as an assembly
   - Camshaft thrust plate and spacer
5. Carefully remove the camshaft by pulling it toward the front of the engine. Remove it slowly to avoid damaging the bearings, journals and lobes.

To install:

6. Clean and inspect all parts before installation.
7. Lubricate the camshaft lobes and journals with Molylube® or heavy engine oil.
8. Carefully install the camshaft, spacer and camshaft key.

➡If a new camshaft is being installed, recheck camshaft end-play.

Aligning balance shaft drive gear and driven gear—3.9L and 4.2L engines

9. Lubricate the engine thrust plate with engine assembly lubricant, then install the thrust plate. Tighten the retaining bolts to 71 inch lbs. (8 Nm).
10. Install the timing chain and sprockets.
11. Install the balance shaft assembly. Tighten the bolts to 71 inch lbs. (8 Nm).
12. Install the balance shaft drive gear, ensuring that the timing marks are correctly located as shown.

➡Check the camshaft sprocket bolt for blockage of the drilled oil passages prior to installation, and clean if necessary.

13. Install or connect the following:
    - Engine front cover
    - Crankshaft damper and pulley
    - Hydraulic tappets into their original bores
    - Align the valve tappet flats and install the tappet guide plate with the word **UP** facing you
    - Install the intake manifold assembly
    - Lubricate and install the pushrods and rocker arms
    - Install the valve covers
    - Install the engine assembly into the vehicle

Starter Motor

REMOVAL & INSTALLATION

1. Disconnect the negative battery cable.
2. Raise and support the vehicle safely.
3. Disconnect the starter electrical harness.
4. Remove the upper starter bolt.
5. Support the starter and remove the lower bolt.
6. Remove the starter from the vehicle.

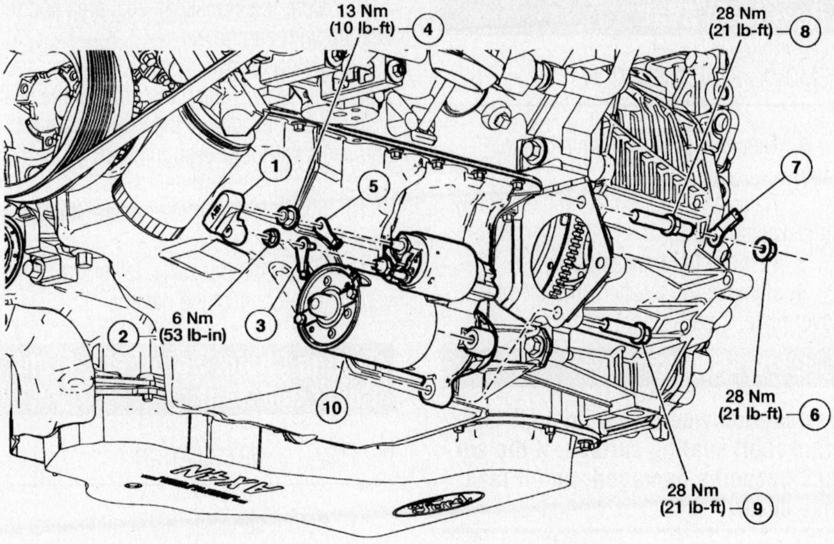

Item	Description	Item	Description
1	Starter motor solenoid terminal cover	6	Ground strap nut
2	Starter solenoid battery cable nut	7	Ground strap
3	Starter solenoid battery cable	8	Starter motor stud bolt
4	Starter solenoid wire nut	9	Starter motor bolt
5	Starter solenoid wire	10	Starter motor

67197-FRST-CG40

Starter motor mounting—3.9L and 4.2L engines

To install:

7. Position the starter in the vehicle.
8. Install the upper and lower bolts. Tighten to 21 ft. lbs. (28 Nm).
9. Connect the starter electrical harness.
10. Lower the vehicle.
11. Connect the negative battery cable.

Oil Pan

⁂ CAUTION

The EPA warns that prolonged contact with used engine oil may cause a number of skin disorders, including cancer! You should make every effort to minimize your exposure to used engine oil. Protective gloves should be worn when changing the oil. Wash your hands and any other exposed skin areas as soon as possible after exposure to used engine oil. Soap and water, or waterless hand cleaner, should be used.

REMOVAL & INSTALLATION

1. Disconnect the negative battery cable.
2. Raise and support the vehicle safely on jackstands.
3. Drain the engine oil.
4. Remove the oil filter.
5. Remove the dual converter Y-pipe assembly.
6. Remove the starter motor.
7. Remove the engine rear plate/converter housing cover.
8. Remove the retaining bolts and remove the oil pan.

To install:

9. Clean the gasket mating surfaces thoroughly.
10. Trial fit the oil pan to the cylinder block. Ensure that enough clearance has been provided to allow the oil pan to be installed without sealant being scraped off when pan is positioned under the engine.
11. Apply a bead of silicone sealer to the oil pan flange. Also apply a bead of sealer to the front cover/cylinder block joint and fill the grooves on both sides of the rear main seal cap.

➡When using silicone rubber sealer, assembly must occur within 5 minutes after sealer application. After this time, the sealer may start to harden and its sealing effectiveness may be reduced.

12. Install the oil pan and secure to the block with the attaching bolts. Tighten the bolts in sequence to 89 inch lbs. (10 Nm).
13. Install a new oil filter.
14. Install the engine rear plate/converter housing cover.
15. Install the starter motor.
16. Install the Y-pipe converter assembly.
17. Lower the vehicle.
18. Fill the engine with the proper type and amount of clean oil.
19. Connect the negative battery cable.
20. Start the engine and check for leaks.

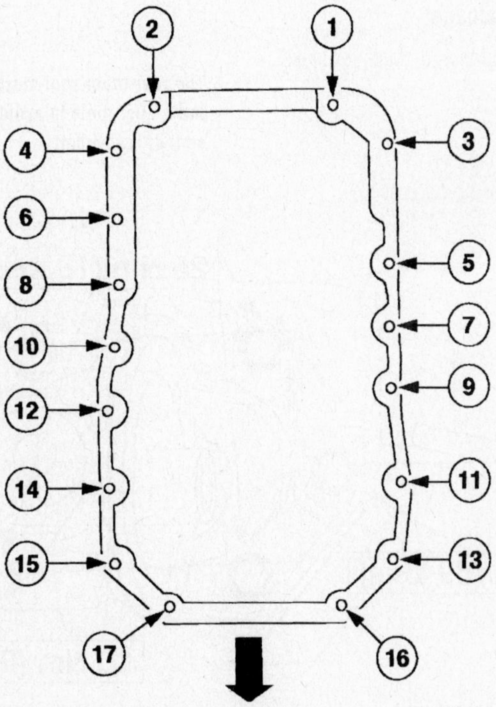

67197-FRST-CG18

Oil pan bolt torque sequence—3.9L and 4.2L engines

Oil Pump

REMOVAL & INSTALLATION

1. Disconnect the negative battery cable.
2. Drain the engine oil.
3. Remove the A/C compressor and wire aside.
4. Remove the A/C compressor mounting bracket
5. Remove the oil filter.
6. Remove the oil pump attaching bolts. Lift the oil pump from the engine.

To install:

7. Place the oil pump in the proper position with a new gasket and install the retaining bolt.
8. Tighten the oil pump retaining bolt to the proper torque as shown.
9. Install the oil filter.
10. Install the A/C compressor mounting bracket. Torque the bolts to 35 ft. lbs. 47 Nm).
11. Install the A/C compressor.
12. Fill the engine with clean oil.
13. Connect the negative battery cable.

➡Check for proper engine oil pressure immediately after starting the engine. If engine oil pressure is not within specification a few seconds after starting the engine, stop the engine and determine the reason for the low oil pressure condition. Running an engine with low oil pressure may result in serious engine damage.

14. Start the engine and check for leaks.

Rear Main Seal

REMOVAL & INSTALLATION

1. Disconnect the negative battery cable.
2. Raise and support the vehicle safely on jackstands.
3. Remove the transaxle.
4. Remove the flywheel and the rear cover plate, if necessary.

⁂ WARNING

Use caution when working near the crankshaft sealing surface. If the surface becomes damaged, an oil leak may occur.

5. Screw in the threaded end of a crankshaft rear seal replacer tool, then use the tool to remove the seal.

To install:

6. Inspect the crankshaft seal area for any damage that may cause the seal to leak. If damage is evident, service or replace the crankshaft as necessary.

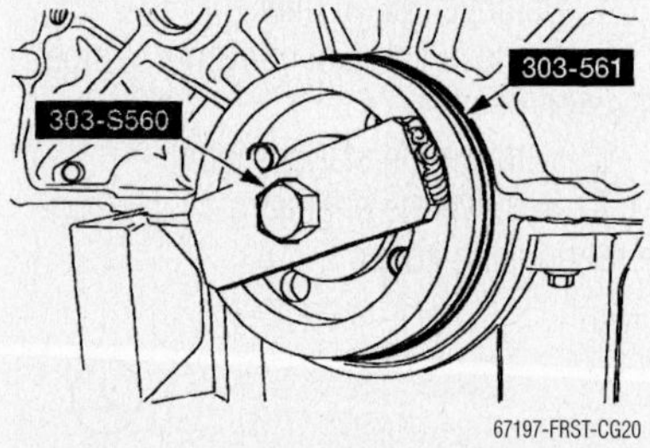

The rear main seal must be installed with the proper tools to avoid damaging the seal or crankshaft

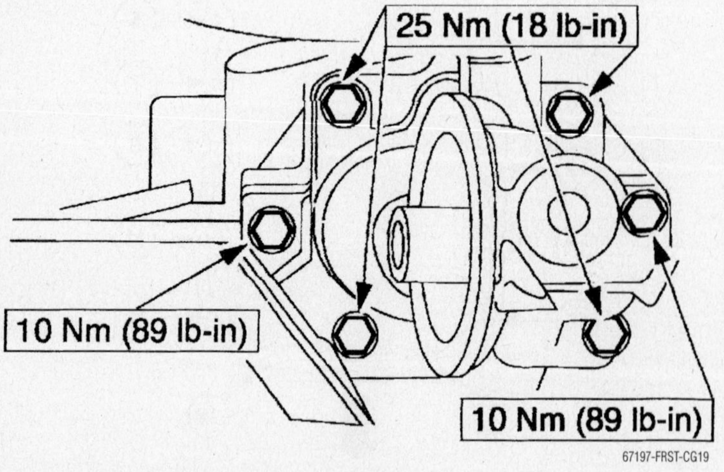

Oil pump bolt tightening torques—3.9L and 4.2L engines

7. Coat the crankshaft seal area and the seal lip with engine oil.
8. Using a crankshaft seal replacer tool, install the seal. Tighten the bolts of the seal installer tool evenly so the seal is straight and seats without misalignment.
9. Install the flywheel.
10. Install the rear cover plate, if necessary.
11. Install the transaxle, lower the vehicle and connect the battery.

Timing Chain, Sprockets, Front Cover and Seal

REMOVAL & INSTALLATION

1. Before servicing the vehicle, refer to the precautions in the beginning of this section.
2. Rotate the crankshaft until No. 1 cylinder is at TDC on compression stroke.
3. Remove or disconnect the following:
   - Negative battery cable
   - Coolant
   - Air cleaner assembly and air intake duct
   - Alternator
   - Coolant hoses
   - Water pump
   - Camshaft position sensor
   - Power steering pump and position aside
   - Camshaft synchronizer
   - Alternator mounting bracket
   - Accessory drive belt and tensioner
   - Oil filter
   - Right side ABS connector
   - Place safety stand under right side of subframe
   - Remove 2 subframe bolts and lower right side of subframe about 2 inches
   - Crankshaft pulley
   - Raise subframe and reinstall bolts
   - Crankshaft front seal
   - Front oil pan bolts
   - Front cover

➡Do not overlook the cover retaining bolt located behind the oil filter adapter. The front cover will break if pried on, and all retaining bolts are not removed.

   - Camshaft synchronizer drive gear
   - Camshaft gear
4. Ensure the crankshaft timing marks and keyways align as shown.
5. Compress and install a retaining pin to hold the timing chain tensioner.
6. Remove or disconnect the following:

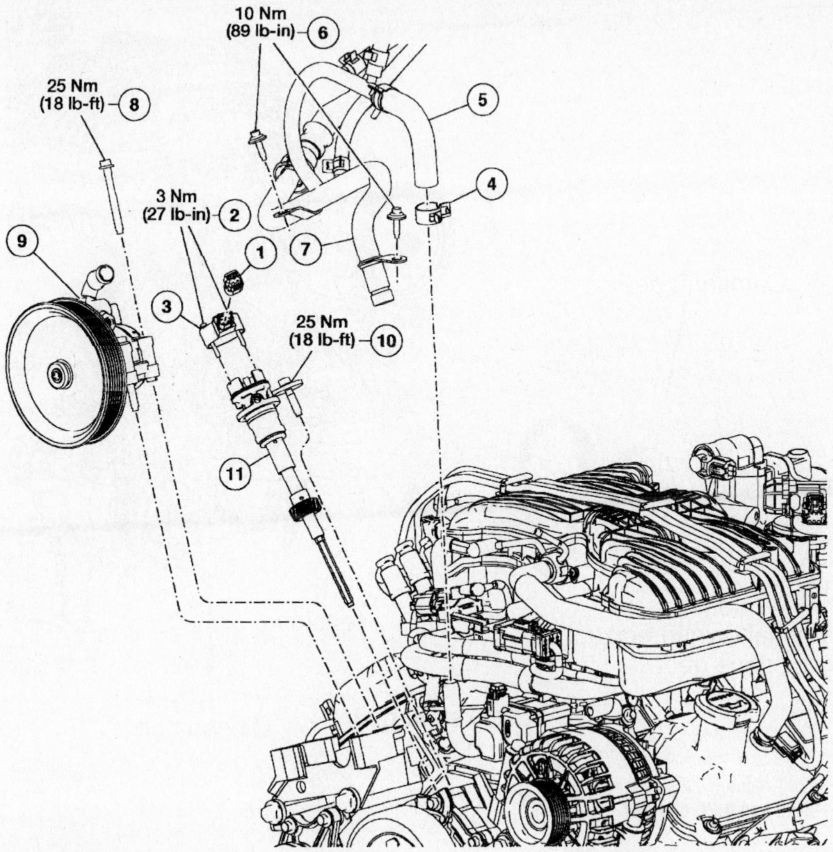

Description		Description		Description	
1	Camshaft position sensor (CMP) electrical connector	5	Coolant bypass hose	9	Power steering pump
2	CMP bolts	6	Coolant pump outlet tube bolts	10	Camshaft synchronizer-to-engine front cover bolt
3	CMP	7	Coolant pump outlet tube	11	Camshaft synchronizer
4	Clamp	8	Power steering pump bolts		

67197-FRST-CG21

Exploded view of camshaft synchronizer assembly—3.9L and 4.2L engines

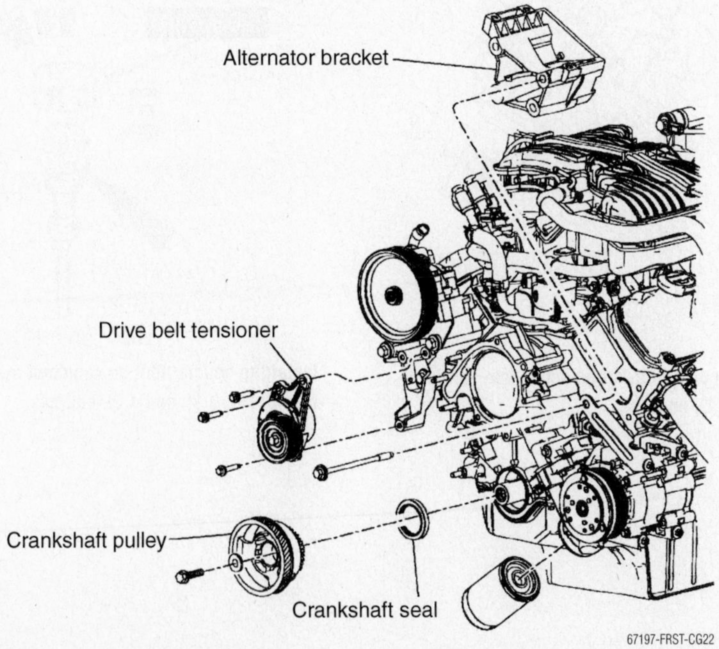

Exploded view of front crankshaft pulley—3.9L and 4.2L engines

- Camshaft gear, crankshaft gear and timing chain as an assembly
- Timing chain tensioner

To install:

7. Clean all the gasket mating surfaces.
8. Install the timing chain tensioner and tighten the bolt to 71 inch lbs. (11 Nm).
9. Retract the tensioner and insert a retaining pin.
10. Install or connect the following:
    - Camshaft sprocket, crankshaft sprocket and timing chain. Be sure the timing marks align.
    - Remove tensioner retaining pin
    - Camshaft synchronizer drive gear and tighten the bolt to 33 ft. lbs. (45 Nm)
    - Place silicone gasket sealant on front oil pan mating surface
    - Install water pump to front cover
    - Install front cover and torque bolts in sequence to 21 ft. lbs. (28 Nm)
    - Front oil pan bolts

➡Front cover bolt no. 12 is tightened to 89 inch lbs. (10 Nm)

- New crankshaft seal in the front cover and lubricate the seal lip with engine oil
- Remove right subframe bolts and lower subframe again about 2 inches
- Crankshaft pulley and tighten the bolt to 118 ft. lbs. (160 Nm)
- Raise subframe, install bolts and tighten to 66 ft. lbs. (90 Nm)
- Right side ABS connector
- Oil filter
- Accessory drive belt and tensioner
- Alternator mounting bracket
- Camshaft synchronizer

11. Coat the camshaft synchronizer gear with clean engine oil.
12. Install the special tool 303-630 onto the top of the camshaft synchronizer.

➡During installation the arrow on the special tool will rotate clockwise until the oil pump intermediate shaft engages the camshaft gear.

13. Install the camshaft synchronizer until the arrow on the special tool is at 54 degrees from the centerline of the engine as shown.
14. Install or connect the following:
    - Power steering pump
    - Camshaft position sensor
    - Coolant hoses
    - Alternator
    - Air cleaner assembly and air intake duct
15. Fill the crankcase with the proper

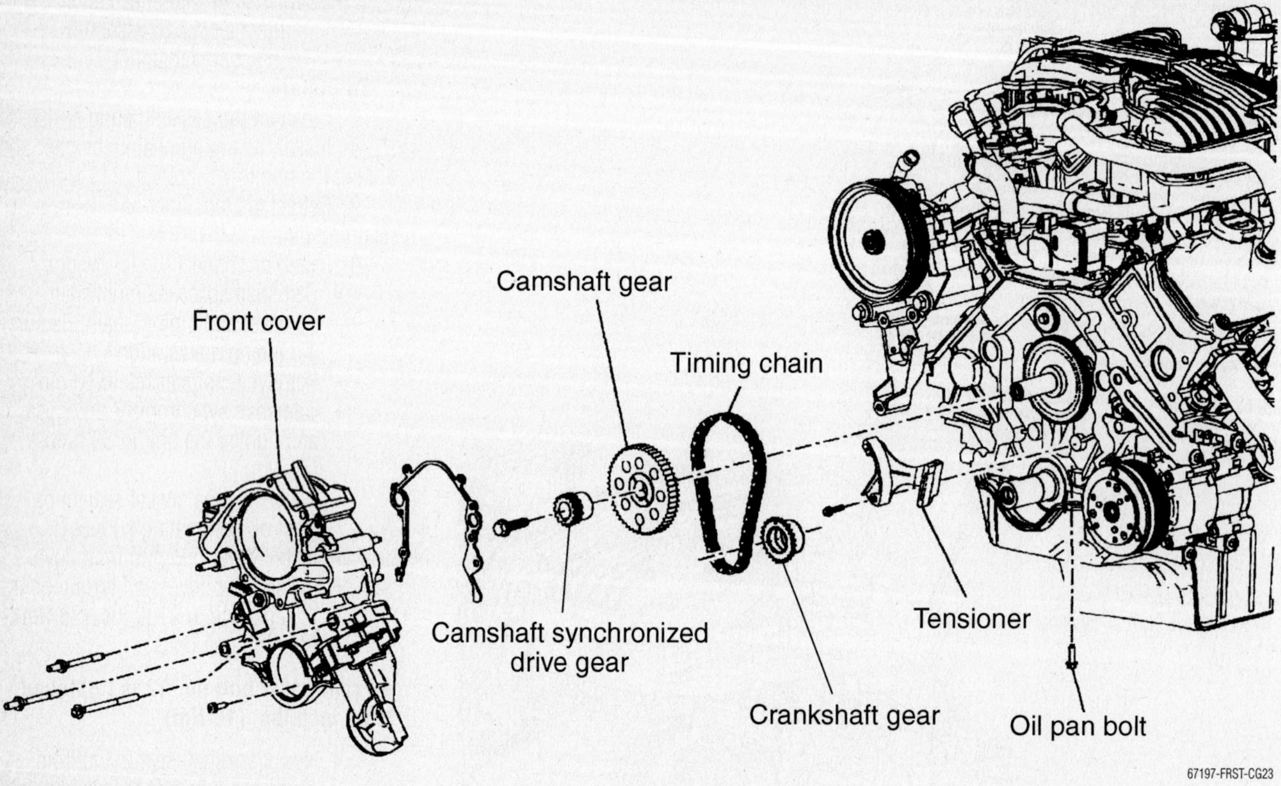

Exploded view of front cover and timing components—3.9L and 4.2L engines

type and quantity of engine oil. Fill and bleed the cooling system. Connect the negative battery cable.

16. Start the engine and check for leaks. Check the ignition timing and curb idle speed and adjust, as necessary.

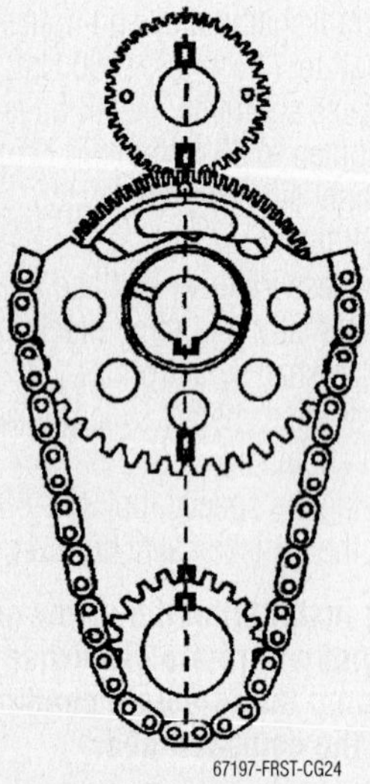

Aligning gear timing marks and keyway—3.9L and 4.2L engines

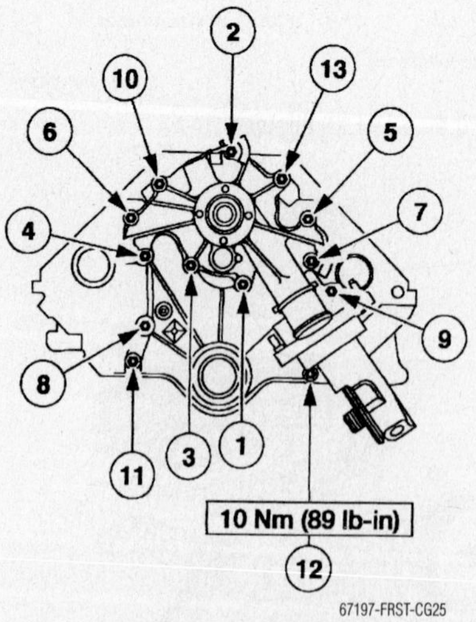

Front cover bolt torque sequence—3.9L and 4.2L engine

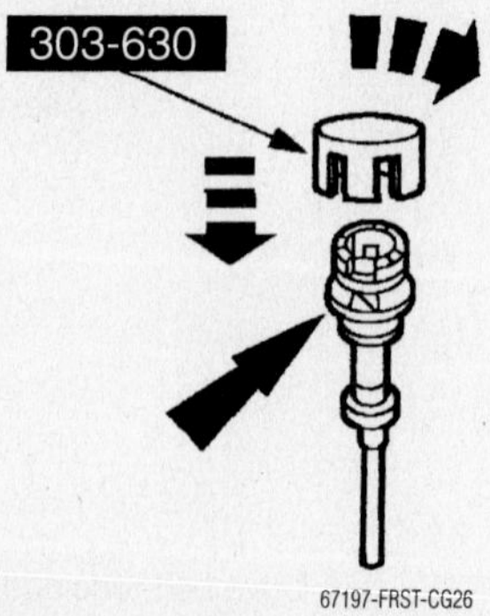

Installing special tool on camshaft synchronizer—3.9L and 4.2L engines

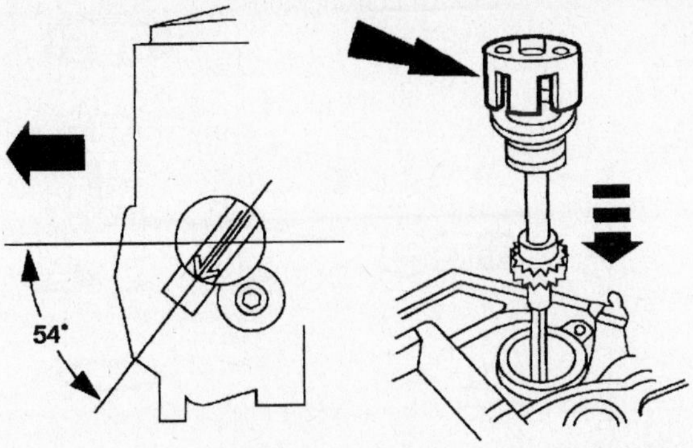

67197-FRST-CG27

Installing camshaft synchronizer—3.9L and 4.2L engines

Piston and Ring

POSITIONING

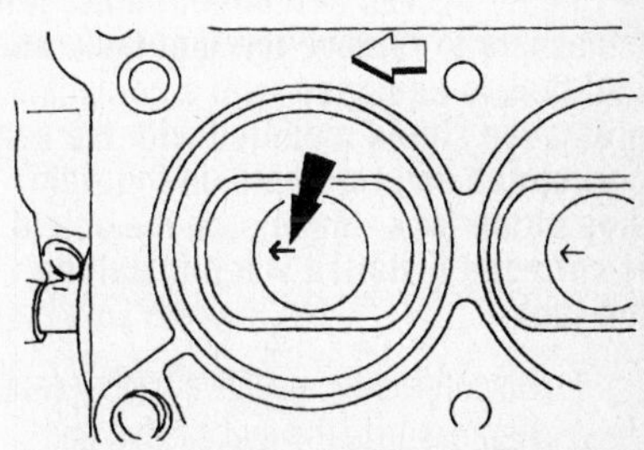

67197-FRST-CG28

Piston positioning in cylinder block—3.9L and 4.2L engines

FUEL SYSTEM

Fuel System Service Precautions

Safety is the most important factor when performing not only fuel system maintenance, but any type of maintenance. Failure to conduct maintenance and repairs in a safe manner may result in serious personal injury or death. Work on a vehicle's fuel system components can be accomplished safely and effectively by adhering to the following rules and guidelines.

- To avoid the possibility of fire and personal injury, always disconnect the negative battery cable unless the repair or test procedure requires that battery voltage by applied.
- Always relieve the fuel system pressure prior to disconnecting any fuel system component (injector, fuel rail, pressure regulator, etc.) fitting or fuel line connection. Exercise extreme caution whenever relieving fuel system pressure, to avoid exposing skin, face and eyes to fuel spray. Please be advised that fuel under pressure may penetrate the skin or any part of the body that it contacts.
- Always place a shop towel or rag around the fitting or connection prior to loosening to absorb any excess fuel due to spillage. Ensure that all fuel spillage is quickly remove from engine surfaces. Ensure that all fuel-soaked cloths or towels are deposited into a flame-proof waste container with a lid.
- Always keep a dry chemical (Class B) fire extinguisher near the work area.
- Do not allow fuel spray or fuel vapors to come into contact with a light bulb, spark or open flame.
- Always use a second wrench when loosening or tightening fuel line connections fittings. This will prevent unnecessary stress and torsion to fuel piping. Always follow the proper torque specifications.
- Always replace worn fuel fitting O-rings with new ones. Do not substitute fuel hose where rigid pipe is installed.

Fuel System Pressure

RELIEVING

Locate the fuel pump relay in the engine compartment fuse block at location K4/C-1051. Remove the fuel pump relay. Start the engine and allow it to idle until it stalls. Crank the engine for 5 seconds to ensure fuel supply manifold pressure is released. When vehicle service is complete, reinstall the fuel pump relay and turn the ignition on to pressurize the fuel system. Start the vehicle and check the system for leaks.

Fuel Filter

REMOVAL & INSTALLATION

1. Relieve the fuel system pressure.
2. Raise and support the vehicle safely on jackstands.
3. Place a rag under the fuel filter to catch any residual fuel that may leak out when the filter is removed.
4. Remove the quick-connect fittings at both ends of the fuel filter.
5. Install retainer clips in each fitting.
6. Note the flow arrow direction for installation reference.
7. Remove the fuel filter by pulling it from the bracket.

To install:

8. Install the fuel filter in its bracket, ensuring proper direction of flow as noted earlier.
9. Apply clean engine oil to end of fuel line and install the quick-connect fittings at both ends of the fuel filter. Ensure the fuel line clicks into place and is fully seated.
10. Start the engine and check the filter connections for leaks by running the tip of your finger around each connection.
11. Turn the engine off and lower the vehicle.

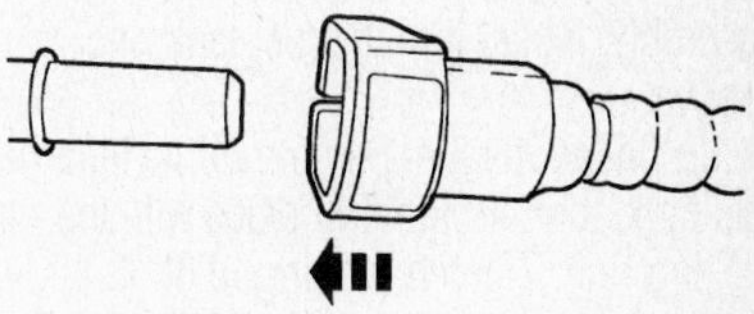

67197-FRST-CG41

Fuel line quick-connect fitting removal—3.9L and 4.2L engines

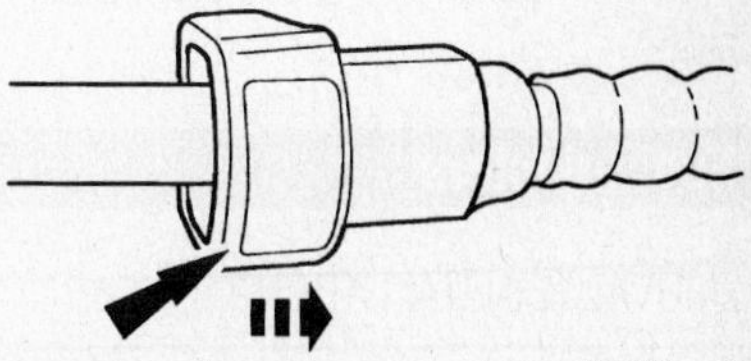

67197-FRST-CG42

Fuel line quick-connect fitting installation—3.9L and 4.2L engines

Fuel Pump

REMOVAL & INSTALLATION

➡To gain access to the fuel pump, it is necessary to remove the fuel tank. The fuel tank is equipped with a composite protective shield mounted with the support straps and retainers on the right side of the tank. Make sure the shield is correctly installed when installing the tank.

1. Disconnect the negative battery cable.
2. Depressurize the fuel system and drain the fuel into a suitable container.
3. Remove the fuel filler pipe.

⁂ CAUTION

Some fuel may remain in the filler pipe.

4. Support the fuel tank.
5. Disconnect the electrical connectors and fuel line fittings.
6. Disconnect the retaining straps and lower the tank.
7. Remove any dirt that has accumulated around the fuel pump module attaching flange to prevent it from entering the tank during service.
8. Turn the fuel pump module locking ring counterclockwise using a locking ring removal tool, and remove the locking ring.
9. Remove the fuel pump module.
10. Remove the seal gasket and discard it.

To install:

11. Put a light coating of grease on a new seal ring to hold it in place during assembly. Install it in the fuel tank ring groove.
12. Insert the fuel pump module into the fuel tank, then secure it in place with the locking ring. Tighten the ring to 60 ft. lbs. (81 Nm).
13. Install the tank in the vehicle.
14. Install a minimum of 10 gallons (38L) of fuel and check for leaks.
15. Check for fuel leaks at the fittings.
16. Start the engine and check for fuel leaks.

Fuel Injectors

REMOVAL & INSTALLATION

1. Disconnect the negative battery cable.
2. Relieve the fuel system pressure.
3. Disconnect the fuel charging wiring harness.

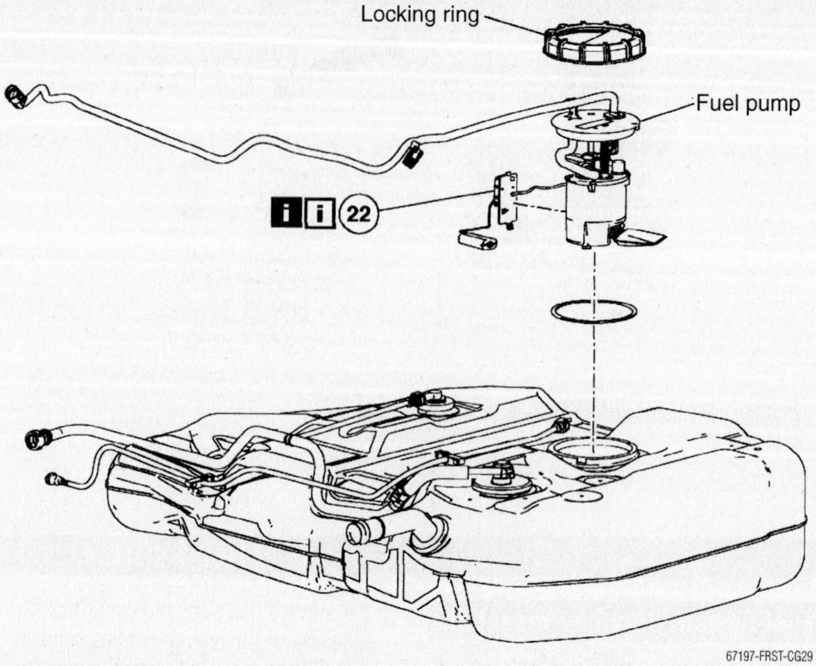

In-tank electric fuel pump and related components—Freestar and Monterey

4. Remove the fuel pressure vacuum connector and fuel supply line.
5. Remove the fuel injection supply manifold, with injectors attached.
6. Remove the retaining clips and separate the injectors from the supply manifold.

To install:

7. Inspect fuel injector O-rings for signs of deterioration and replace as required.

➡Never use silicone grease on fuel injectors.

8. Lubricate O-rings with clean engine oil and position fuel injectors on the supply manifold. Install the retaining clips.
9. Install the fuel injection supply manifold, making sure the injectors are seated properly. A slight rocking motion helps seat the injectors.
10. Install the fuel pressure vacuum connector and fuel supply line.
11. Install the fuel charging wiring connectors.
12. Start the engine and check for fuel leaks.

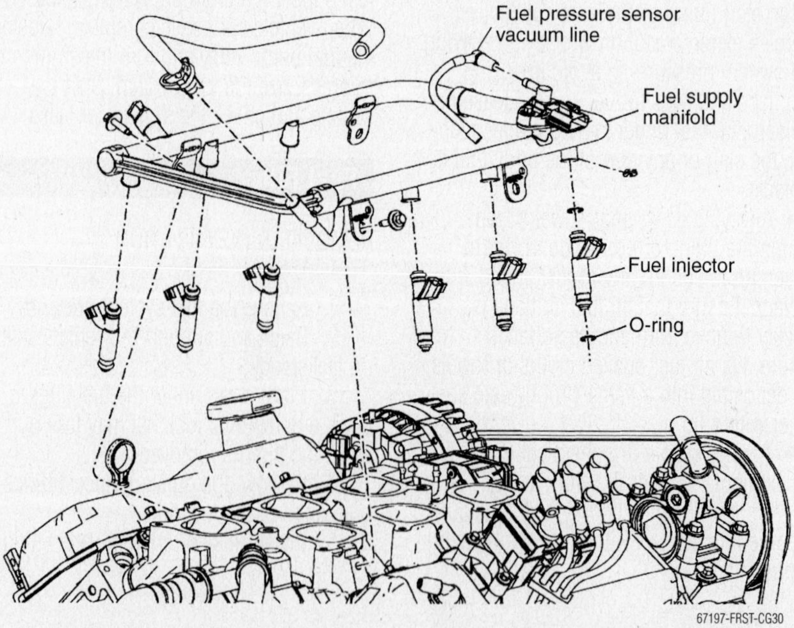

Fuel injectors and supply manifold—3.9L and 4.2L engines

DRIVE TRAIN

Transaxle Assembly

REMOVAL & INSTALLATION

1. Remove or disconnect the following:
   - Battery and battery tray
   - Air cleaner assembly
   - All transaxle electrical harnesses
   - Transaxle shift cable from the lever by unsnapping the shift cable end from the lever ball stud
   - Manual control lever
   - Hood
   - Wiper arm and pivot assembly
   - Passenger side air intake box
   - Air intake tube
   - Upper radiator shield
   - Hood latch and oil filler cap

➡Install engine lifting eyes to support the engine during transaxle removal.

2. Install an engine support kit and suitably support the engine.
3. Remove or disconnect the following:
   - Transaxle filer tube
   - Transaxle cooler lines
   - The 2 upper transaxle-to-engine bolts
   - Raise and support the vehicle
   - Anti-roll bracket under the vehicle
   - Rear transaxle support nut
   - Transaxle fluid
   - Catalytic converter assembly
   - Front wheels
4. Disconnect the left and right tie rod ends from the knuckles.
5. Disconnect the sway bar ends from the knuckles.

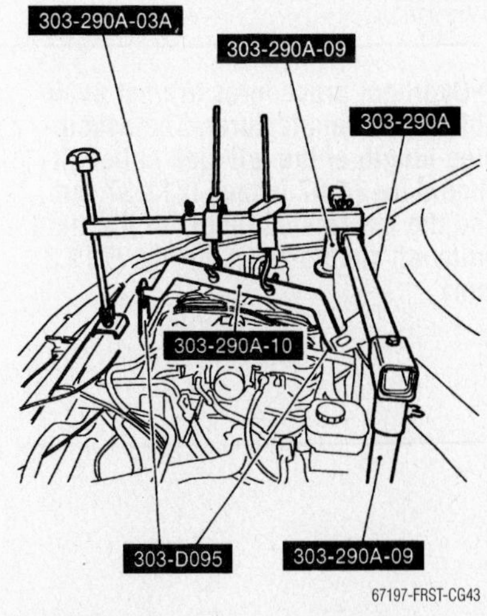

Installing engine support kit—3.9L and 4.2L engines

6. Remove the left and right knuckle pinch bolts.
7. Remove or disconnect the following:
   - Steering rack bolts
   - Power steering hose retainers
   - Front and rear engine mount nuts
   - Place a jack under the subframe to support it
   - 4 subframe mounting bolts
8. Partially lower the subframe and remove the sway bar mounting bolts. Remove the sway bar.
9. Lower the subframe from the vehicle.
10. Remove or disconnect the following:
    - Wire the steering rack in place
    - Starter
    - Halfshafts from the transaxle and support them out of the way
    - Torque converter nuts and discard them
    - Transaxle electrical connectors
11. Position a transaxle jack to support the transaxle and remove the transaxle mounting nuts.
12. Remove the oil pan-to-transaxle bolt.
13. Separate the transaxle from the engine block by carefully moving the transaxle rearward until enough clearance exists to remove the transaxle from the engine compartment.

To install:

➡If a different transaxle is being installed, transfer the heat shield to the new tranaxle.

14. Place the transaxle on a suitable jack and carefully raise it into position.
15. Install or connect the following:
    - Oil pan-to-transaxle bolt, and tighten the bolt to 46 ft. lbs. (62 Nm).
    - Engine bracket-to-transaxle bolts, and tighten the bolts to 46 ft. lbs. (62 Nm)
    - Transaxle-to-engine bolts, and tighten the bolts to 35 ft. lbs. (47 Nm)
    - Transaxle mounting bolts, and tighten the bolts to 46 ft. lbs. (62 Nm).
    - Transaxle electrical connectors
    - NEW torque converter nuts, and tighten them to 26 ft. lbs. (35 Nm)
    - Transaxle housing cover, and tighten the bolts to 89 inch lbs. (10 Nm)
    - Both halfshafts
    - Starter motor, and connect the electrical harness
    - Speedometer cable
    - Front subframe
    - Sway bar mounting bolts, and tighten the bolts to 46 ft. lbs. (62 Nm)
    - The 4 subframe bolts, and tighten them to 66 ft. lbs. (90 Nm)
    - Front and rear engine mount bolts, and tighten them to 66 ft. lbs. (90 Nm)
    - Power steering hose retainers
    - Steering rack bolts, and tighten them to 98 ft. lbs. 133 Nm).
    - Left and right knuckle pinch bolts, and tighten the bolts to 46 ft. lbs. (62 Nm).
    - Sway bar end mounting nuts, and tighten the bolts to 41 ft. lbs. (55 Nm).
    - Tie rod end nuts, and tighten the nuts to 41 ft. lbs. (55 Nm).
    - Catalytic converter
    - Rear transaxle mounting nut, and tighten to 66 ft. lbs. (90 Nm)
    - Anti-roll bracket
    - Upper transaxle-to-engine bolts, and tighten the bolts to 46 ft. lbs. (62 Nm)
    - Fluid cooler-to-transaxle lines
    - Front wheels and lower the vehicle
16. Remove the engine support kit.
17. Install or connect the following:
    - Hood latch
    - Radiator shield
    - Air intake tube and intake box
    - Wiper arm and pivot assembly
    - Hood
    - Transaxle shift cable to the manual lever ball stud
    - Transaxle electrical harnesses
    - Air cleaner assembly
    - Battery tray and battery
18. Fill the transaxle with proper amount of Mercon®V fluid.
19. Connect the positive, then the negative battery cable.

Halfshafts

REMOVAL & INSTALLATION

➡Do not begin this removal procedure unless a new wheel hub retainer nut, a new retainer circlip and a new lower ball joint-to-front wheel knuckle retaining bolt and nut are available. Once removed, these parts must not be reused during assembly. Their torque holding ability, or retention capability, is diminished during removal.

1. Remove or disconnect the following:
   - Front wheels

- ABS sensor
- Brake caliper and wire aside
- Axle hub nut and washer. Discard the nut.
- Lower control arm ball joint pinch bolt and nut and discard them

2. Separate the outboard CV-joint from the wheel hub using a front hub remover/replacer. Make sure the hub remover adapter is fully threaded onto the hub stud.

3. Separate the halfshaft from the hub, and pull the halfshaft out of the transaxle.

To install:

WARNING

Do not reuse the retainer circlip. A new circlip must be installed each time the inboard CV-joint stub shaft is installed into the transaxle differential.

4. Install a new retainer circlip on the inboard CV-joint stub shaft by starting one end in the groove and working the retainer circlip over the inboard shaft housing end and into the groove. This will avoid over-expanding the circlip.

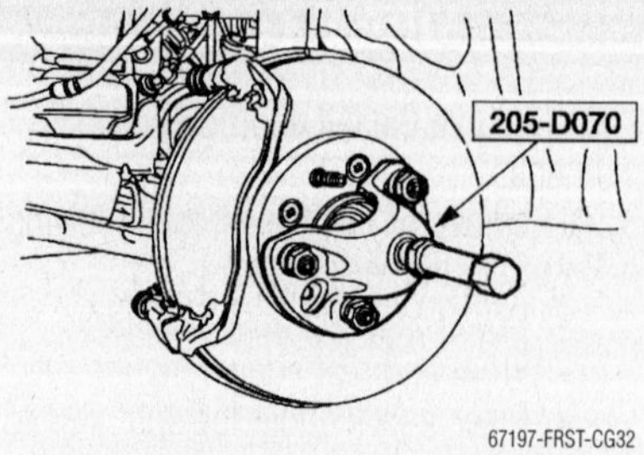

Remove the halfshaft from the transaxle using a puller

➡A non-metallic mallet may be used to aid in seating the retainer circlip into the differential side gear groove. If a mallet is necessary, tap only on the outboard CV-joint shaft.

5. Carefully align the splines of the inboard CV-joint stub shaft housing with the splines in the differential. Exerting some force, push the inboard CV-joint stub shaft housing into the differential until the retainer circlip is felt to seat in the differential side gear. Use care to prevent damage to the inboard CV-joint stub shaft and transaxle seal.

6. Carefully align the splines of the outboard CV-joint with the splines in the wheel hub and push the shaft into the wheel hub as far as possible.

7. Temporarily fasten the front disc brake rotor to the wheel hub with washers and 2 lug nuts. Insert a steel rod into the front disc brake rotor and rotate clockwise to contact the front wheel knuckle to prevent the front disc brake rotor from turning when the nut is tightened.

➡A new front axle wheel hub retaining nut must be installed.

8. Manually thread the front axle wheel hub retaining nut onto the outboard CV-joint stub shaft housing as far as possible.

➡A new bolt and nut must be used to connect the front suspension arm to the knuckle.

9. Connect the front suspension lower arm to the front wheel knuckle. Tighten the nut and bolt to 46 ft. lbs. (63 Nm).

10. Install the front brake anti-lock sensor.

➡Do not use power or impact tools to tighten the hub nut.

11. Tighten front axle wheel hub retaining nut to 111ft. lbs. (150 Nm).

12. Install the front wheels and lower the vehicle.

13. Fill the transaxle to the proper level with Mercon®V automatic transmission fluid.

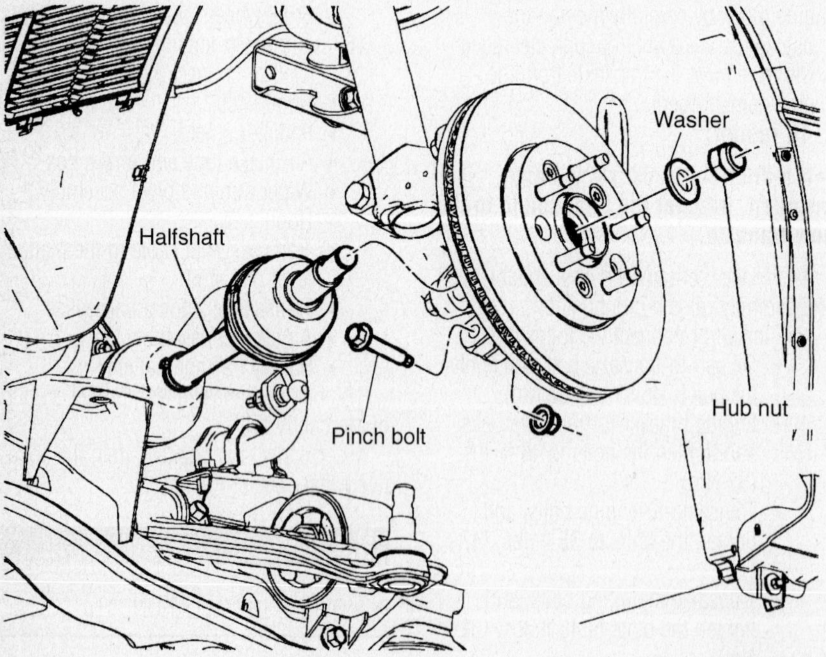

Exploded view of front halfshaft mounting

CV-Joint

OVERHAUL

➡Overhaul procedures are not available from manufacturer. The assembled length of the left side halfshaft should be 24.07 inches (611.37 mm), and the assembled length of the right halfshaft should be 28 inches (711.2 mm).

STEERING AND SUSPENSION

Air Bag (Supplemental Restraint) System

The Supplemental Restraint System (SRS) is designed to work in conjunction with the standard 3-point safety belts to reduce injury in a head-on collision.

✻✻ CAUTION

The SRS can actually cause physical injury or death if the safety belts are not used, or if the manufacturer's warnings are not followed. The manufacturer's warnings can be found in your owner's manual, or, in some cases, on your sun visor.

The SRS is comprised of the following components:

- Driver's side air bag module
- Passenger's side air bag module
- Right-hand and left-hand primary crash front air bag sensors
- Air bag diagnostic monitor computer
- Electrical wiring

The SRS primary crash front air bag sensors are hard-wired to the air bag modules and determine when the air bags are deployed. During a frontal collision, the sensors quickly inflate the 2 air bags to reduce injury by cushioning the driver and front passenger from striking the dashboard, windshield, steering wheel and any other hard surfaces. The air bag inflates so quickly (in a fraction of a second) that in most cases it is fully inflated before you actually start to move during a collision.

Since the SRS is a complicated and essentially important system, its components are constantly being tested by a diagnostic computer. The computer illuminates the air bag indicator light on the instrument cluster for approximately 6 seconds when the ignition switch is turned to the **RUN** position when the SRS is functioning properly. After being illuminated for the 6 seconds, the indicator light should then turn off.

If the air bag light does not illuminate at all, stays on continuously, or flashes at any time, a problem has been detected by the diagnostic computer.

✻✻ CAUTION

If at any time the air bag light indicates that the computer has noted a problem, immediately diagnose the problem. A faulty SRS can cause severe physical injury or death.

SERVICE PRECAUTIONS

Whenever working around, or on, the air bag supplemental restraint system, ALWAYS adhere to the following warnings and cautions.

- Always wear safety glasses when servicing an air bag vehicle and when handling an air bag module.
- Carry a live air bag module with the bag and trim cover facing away from your body, so that an accidental deployment of the air bag will have a small chance of personal injury.
- Place an air bag module on a table or other flat surface with the bag and trim cover pointing up.
- Wear gloves, a dust mask and safety glasses whenever handling a deployed air bag module. The air bag surface may contain traces of sodium hydroxide, a byproduct of the gas that inflates the air bag and which can cause skin irritation.
- Ensure to wash your hands with mild soap and water after handling a deployed air bag.
- All air bag modules with discolored or damaged cover trim must be replaced, not repainted.
- All component replacement and wiring service must be made with the negative and positive battery cables disconnected from the battery for a minimum of 1 minute prior to attempting service or replacement.
- NEVER probe the air bag electrical terminals. Doing so could result in air bag deployment, which can cause serious physical injury.
- If the vehicle is involved in a fender-bender that results in a damaged front bumper or grille, the air bag sensors should be inspected to ensure that they were not damaged.
- If at any time, the air bag light indicates that the computer has noted a problem, immediately diagnose the problem. A faulty SRS can cause severe physical injury or death.

DISARMING THE SYSTEM

1. Disconnect the negative battery cable from the battery.
2. Disconnect the positive battery cable from the battery.
3. Wait 1 minute. This time is required for the back-up power supply in the air bag diagnostic monitor to completely drain. The system is now disarmed.

ARMING THE SYSTEM

1. Connect the positive battery cable.
2. Connect the negative battery cable.
3. Stand outside the vehicle and carefully turn the ignition to the **RUN** position. Be sure that no part of your body is in front of the air bag module on the steering wheel, to prevent injury in case of an accidental air bag deployment.
4. Ensure the air bag indicator light turns off after approximately 6 seconds. If the light does not illuminate at all, does not turn off, or starts to flash, diagnose the problem. If the light does turn off after 6 seconds and does not flash, the SRS is working properly.

Rack and Pinion Steering Gear

REMOVAL & INSTALLATION

✻✻ WARNING

Do not allow the steering wheel to rotate when the intermediate shaft is disconnect, or damage to the clockspring can result. If it is suspected that the shaft has rotated, the clockspring must be removed and re-centered.

1. Remove or disconnect the following:
   - Front wheels
   - Tie rod end cotter pins and castle nuts
   - Tie rod ends from the knuckles
2. Position the dash opening weather seal for the steering column out of the way.
3. Remove or disconnect the following:
   - Pinch bolt retaining the steering column intermediate shaft coupling
   - Steering gear retaining nuts/bolts
4. Rotate the rack and pinion assembly to clear the bolts from the front crossmember, and pull toward the driver's side of the vehicle.
5. Place a drain pan under the vehicle and disconnect the power steering lines.
6. Remove the rack and pinion assembly through the driver's side of the vehicle.

To install:

7. Install new Teflon® O-rings on the power steering line fittings.
8. Place the rack and pinion retaining bolts in the gear housing.
9. Install or connect the following:
   - Rack and pinion assembly through the driver's side of the vehicle

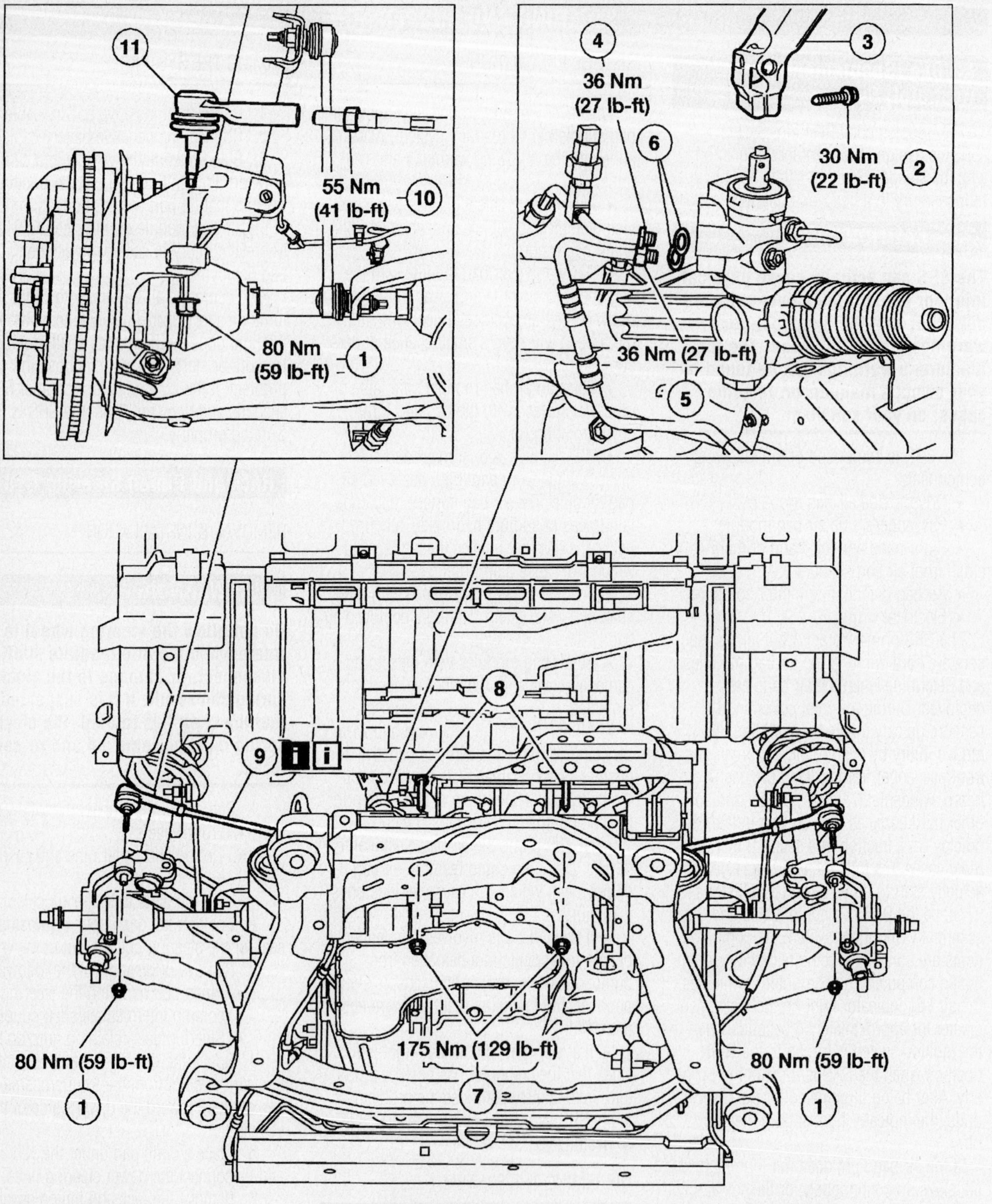

Item	***Description***	***Item***	***Description***	***Item***	***Description***
1	*Tie-rod end nuts*	*5*	*Power steering return line nut*	*9*	*Steering gear*
2	*Intermediate shaft-to-steering gear bolt*	*6*	*O-rings*	*10*	*Outer tie-rod end jam nuts*
3	*Intermediate shaft*	*7*	*Steering gear-to-crossmember nuts*	*11*	*Tie-rod end*
4	*Power steering pressure line nut*	*8*	*Steering gear-to-crossmember bolts*		

67197-FRST-CG33

Exploded view of the rack and pinion steering gear mounting on the front subframe of the vehicle—Freestar and Monterey

- Power steering lines on the rack and pinion assembly
- Rack and pinion assembly on the crossmember
- Tie rod ends to the knuckles. Tighten the castle nuts to 59 ft. lbs. (80 Nm) and install the cotter pins.
- Rack and pinion assembly retaining bolts, and tighten them to 129 ft. lbs. (175 Nm).
- Front wheels

10. Using a new pinch bolt, install the steering column intermediate shaft coupling on the rack input shaft. Tighten the pinch bolt to 22 ft. lbs. (30 Nm).
11. Position the steering column opening weather seal over the steering gear housing.
12. Lower the vehicle.
13. Fill the power steering oil reservoir.
14. Start the vehicle and check for leaks.
15. Check for proper wheel alignment and steering wheel position.

Strut

REMOVAL & INSTALLATION

➡Do not allow the axle shaft to move outward as damage to the CV joint may result.

1. Turn the ignition switch **OFF** and place the steering column in the unlocked position.
2. Remove or disconnect the following:
   - Front wheel
   - Stabilizer bar link nut
   - Support the steering knuckle and remove the pinch bolt
   - Upper strut mounting nuts
3. Push down on the steering knuckle until the strut is free of the steering knuckle and remove the strut.

To install:

4. Push down on the steering knuckle and insert the bottom of the strut into the steering knuckle.
5. Install the upper strut mounting nuts and tighten them to 26 ft. lbs. (35 Nm).
6. Install strut-to-wheel knuckle pinch bolt and tighten to 85 ft. lbs. (115 Nm).
7. Install or connect the following:
   - Stabilizer bar link nut and tighten to 66 ft. lbs. (90 Nm).
   - Rotor
   - Caliper
   - Wheels
8. Check the wheel alignment.

DISASSEMBLY

1. Position a suitable pass through socket wrench onto the strut shaft nut.
2. Place a 10 mm six point deep socket on the strut top retaining nut.
3. Remove the strut shaft while holding the retaining nut in place.
4. Remove the shock absorber.

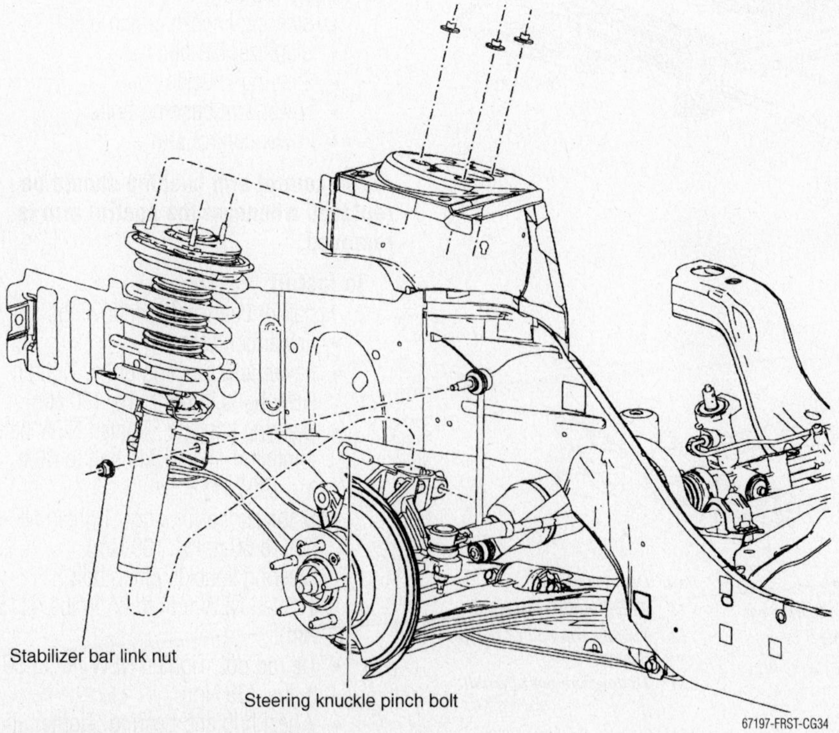

MacPherson strut mounting—Freestar and Monterey

5. Remove the upper shock mount and bearing assembly.
6. Remove the coil spring.
7. To reassemble, reverse the disassembly procedure. Tighten the strut top retaining nut to 37 ft. lbs. (50 Nm).

Shock Absorber

REMOVAL & INSTALLATION

1. Loosen the lug nuts on the rear wheels.
2. Raise and safely support the vehicle.
3. Remove the rear wheels.
4. Position a jack under the rear axle assembly and raise it slightly to put the suspension at normal ride height.
5. Remove the lower shock absorber bolt/nut and disconnect the shock from the rear axle.
6. Lower the rear axle slightly to help aid removal of the upper shock absorber bolt/nut.
7. Remove the shock absorber.

To install:

8. Attach the shock absorber to the upper mounting bracket and install a new retaining bolt/nut.
9. Slowly raise the rear axle assembly with a jack, and guide the lower shock absorber into the bracket on the rear axle assembly. Install a new retaining bolt/nut.
10. Raise the rear suspension to normal ride height and tighten the upper shock absorber retaining bolt to 76 ft. lbs. (103 Nm), and the lower bolt to 59 ft. lbs. (80 Nm).
11. Install the wheels.
12. Lower the vehicle.

Coil Springs

REMOVAL & INSTALLATION

1. Raise and safely support the vehicle.
2. Remove the rear wheels.
3. Remove the brake caliper and wire aside.
4. Remove the brake rotor.
5. Disconnect the ABS speed sensor.

➡The rear axle will need to be supported when the shock absorbers are removed.

6. Remove the wheel hub and bearing assembly.
7. Remove the shock absorber.
8. Slowly lower the rear axle assembly until the rear spring can be removed.
9. Remove the rear spring.

To install:

10. Position the rear spring insulator on the rear axle assembly and press the insulator downward into place. Verify rear spring insulator is properly seated into correct position.
11. Slowly raise the rear axle assembly with a jack, and guide the upper rear spring insulator onto the upper spring seat on the underbody.
12. Position the shock absorber on the lower rear axle assembly and tighten the upper shock absorber retaining bolt to 76 ft. lbs. (103 Nm), and the lower bolt to 59 ft. lbs. (80 Nm).
13. Install the wheel hub and bearing assembly and tighten the bolts to 85 ft. lbs. (115 Nm).
14. Connect the ABS speed sensor.
15. Install the brake rotor and caliper.
16. Install the wheels.
17. Lower the vehicle.

Stabilizer Bar

REMOVAL & INSTALLATION

1. Raise and safely support the vehicle.
2. Remove the wheels.
3. Remove the stabilizer bar link nuts.
4. Remove the stabilizer bar links.

➡The subframe will need to be supported when the mounting bolts are removed.

5. Remove the rear subframe mounting bolts and lower the rear of the subframe about 2-3 inches.
6. Remove the stabilizer bar brackets.
7. Remove the stabilizer bar.

To install:

8. Install or connect the following:
    - Stabilizer bar and brackets. Tighten the bracket bolts to 46 ft. lbs. (63 Nm).
    - NEW subframe bolts. Tighten the bracket bolts to 66 ft. lbs. (89 Nm).
    - Stabilizer bar links. Tighten the bracket bolts to 66 ft. lbs. (89 Nm).
    - ABS speed sensor
    - Brake rotor and caliper
    - Wheels
    - Lower the vehicle

Lower Ball Joint

REMOVAL & INSTALLATION

The lower ball joint and seal are an integral part of the lower control arm assembly, and can not be replaced separately. If the lower ball joint or seal is found to be defective, the lower control arm must be replaced as an assembly.

Lower Control Arm, Control Arm Bushing & Steering Knuckle

REMOVAL & INSTALLATION

➡Do not begin the removal procedure unless a new strut-to-lower arm nut, a new ball joint pinch bolt/nut and a new lower arm-to-front subframe bolt/nut are available.

1. Remove or disconnect the following:
    - Wheel
    - Brake caliper
    - Rotor
    - Axle shaft nut
    - Wheel hub and bearing
    - Tie rod nut
    - Steering knuckle pinch bolt
    - Stabilizer bar link nut
    - Steering knuckle
    - Lower arm bushing bolts
    - Lower control arm

➡The control arm bushing should be replaced whenever the control arm is removed.

To install:

2. Install or connect the following:
    - Lower control arm
    - Lower arm bushing bolts. Tighten NEW bolts to 66 ft. lbs. (90 Nm).
    - Steering knuckle. Tighten NEW ball joint stud-to-knuckle bolt to 66 ft. lbs. (90 Nm).
    - Stabilizer bar link nut. Tighten NEW bolt to 66 ft. lbs. (90 Nm).
    - Steering knuckle pinch bolt. Tighten NEW bolt to 85 ft. lbs. (115 Nm).
    - Tie rod nut. Tighten NEW nut to 59 ft. lbs. (80 Nm).
    - Wheel hub and bearing. Tighten the bolts to 85 ft. lbs. (115 Nm).

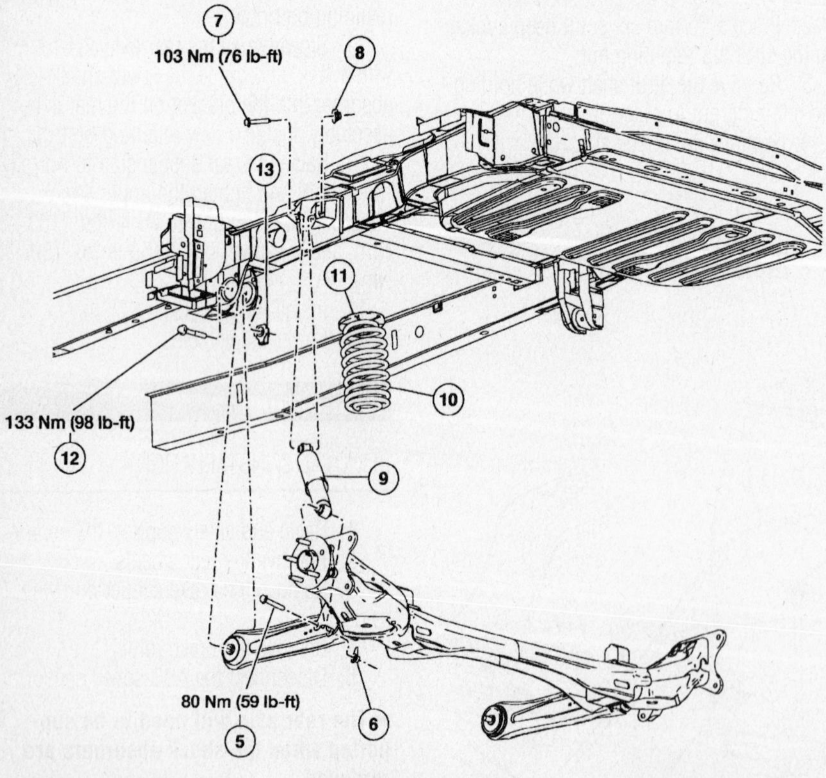

Item	Description	Item	Description	Item	Description
5	Shock absorber-to-axle bolt	8	Upper shock absorber nut	11	Spring insulator (2 req'd)
6	Shock absorber-to-axle flagnut	9	Shock absorber	12	Trailing arm bolt (2 req'd) Removal Note
7	Upper shock absorber bolt	10	Spring	13	Trailing arm nut (2 req'd)

67197-FRST-CG35

Exploded view of the rear coil spring and shock absorber mounting—Freestar and Monterey

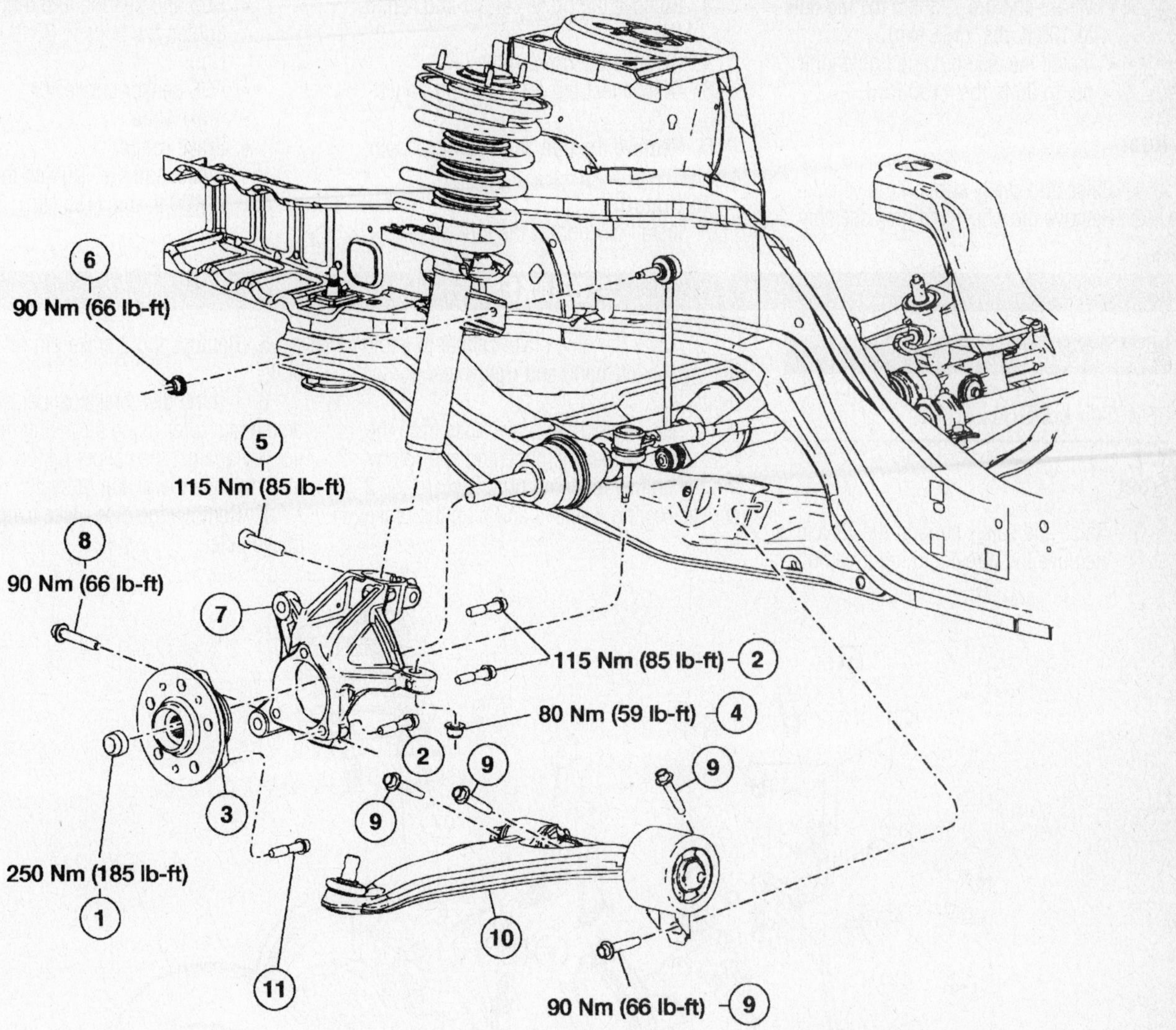

Item	Description	Item	Description	Item	Description
1	*Axle shaft nut*	*4*	*Tie rod end nut*	*8*	*Lower ball joint stud-to-wheel knuckle bolt and nut*
2	*Wheel hub and bearing assembly-to-knuckle bolt (3 req'd)*	*5*	*Knuckle pinch bolt*	*9*	*Lower arm bushing bolts*
3	*Wheel hub and bearing assembly*	*6*	*Stabilizer bar link nut*	*10*	*Lower arm (LH/RH)*
		7	*Knuckle (LH/RH)*	*11*	*Wheel stud*

67197-FRST-CG36

Exploded view of the front suspension—Freestar and Monterey

- Axle shaft nut. Tighten NEW bolt to 185 ft. lbs. (250 Nm).
- Rotor
- Brake caliper

3. Install the wheels.
4. Lower the vehicle.

Hub and Wheel Bearing

REMOVAL & INSTALLATION

The wheel bearing is integral with the wheel hub and can not be replaced separately. If the wheel bearing is found to be defective, the wheel hub must be replaced as an assembly.

Front

1. With the vehicle on the ground, remove and discard the axle shaft nut.
2. Raise and safely support the vehicle.
3. Remove the wheel and tire assembly.
4. Remove the brake caliper and wire it out of the way.
5. Remove the brake rotor.
6. Using a puller, remove the driveshaft from the hub and bearing housing.
7. Remove 3 hub attaching bolts, and remove the hub and bearing assembly.

To install:

8. Position the dust shield over the hub and fit the hub and bearing into the knuckle and seat it fully before installing the attaching bolts. Tighten the bolts to 85 ft. lbs. (115 Nm).
9. Install or connect the following:
   - Driveshaft
   - Brake rotor
   - Brake caliper

- Wheel and tire. Tighten the lug nuts to 100 ft. lbs. (135 Nm).
- Install the axle nut and tighten the nut to 96 ft. lbs. (130 Nm).

Rear

1. Raise and safely support the vehicle.
2. Remove the wheel and tire assembly.
3. Remove the brake caliper and wire it out of the way.
4. Remove the brake rotor.
5. Disconnect the ABS sensor connector.
6. Remove the hub and bearing assembly.

To install:

7. Install or connect the following:

- Hub and bearing assembly and tighten the bolts to 85 ft. lbs. (115 Nm).
- ABS sensor connector
- Brake rotor
- Brake caliper
- Wheel and tire. Tighten the lug nuts to 100 ft. lbs. (135 Nm).

BRAKES

Brake Caliper

REMOVAL & INSTALLATION

Front

1. Raise and safely support the vehicle.
2. Remove the wheel and tire assembly.
3. Mark the disc brake caliper to avoid mixing the left-hand and right-hand components.
4. Disconnect the brake hose from the disc brake caliper by loosening and removing the hollow retaining bolt. Discard the 2 copper sealing washers and plug the brake hose.
5. Remove the 2 brake pin retainer bolts.
6. Lift the disc brake caliper off of the disc brake rotor using a rotating motion. Do not pry against the caliper piston. Prying may damage the piston or seals.
7. Remove the disc brake caliper from the vehicle.

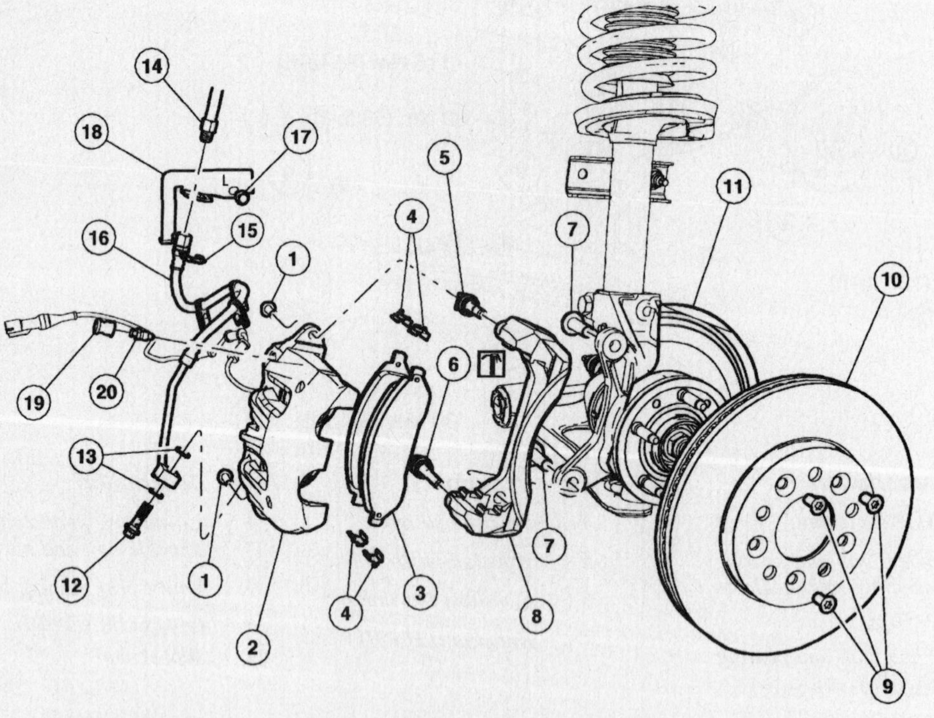

(1) 35 Nm (26 lb-ft)	(14) 17 Nm (13 lb-ft)
(7) 185 Nm (136 lb-ft)	(17) 17 Nm (13 lb-ft)
(12) 55 Nm (41 lb-ft)	(20) 10 Nm (7 lb-ft)

Item	Description	Item	Description	Item	Description
1	Brake caliper bolt (two required for each side)	7	Brake caliper anchor bracket bolts (2 required each side)	13	Copper washers
2	Brake caliper RH/LH	8	Brake caliper anchor bracket RH/LH	14	Brake line fitting
3	Brake pads	9	Brake disc screws (if equipped)	15	Retainer clip
4	Slippers (4 required each side)	10	Brake disc	16	Front brake hose RH/LH
5	Guide pin	11	Dust shield (RH/LH)	17	Brake hose bracket bolt
6	Locator pin	12	Flow bolt	18	Brake hose bracket
				19	Bleeder screw dust cover
				20	Bleeder screw

67197-FRST-CG37

Exploded view of the front disc brake caliper assembly—Freestar and Monterey

8. Remove the caliper anchor bracket, if necessary.

To install:

9. Install the brake caliper anchor, if removed. Tighten the anchor bolts to 136 ft. lbs. (185 Nm).

10. Retract the caliper piston fully into the caliper bore using a C-clamp and block of wood or equivalent.

11. Ensure that the disc brake pads are properly positioned and that the lining material is facing the rotor.

12. Place the disc brake caliper over the rotor and hand-start 2 brake pin retainer bolts. Starting with the bottom bolt first, tighten the brake pin retainer bolts-to-guide pin bolts to 26 ft. lbs. (35 Nm).

➡If both disc brake calipers were removed, make sure that they are mounted to the proper side. The brake bleeder on the caliper when properly installed should be on top of the caliper for proper bleeding of air.

13. Unplug and install the brake hose and hollow retaining bolt to the disc brake caliper using a new copper sealing washer on each side of the hose fitting.

14. Bleed the brake system and install the rubber bleeder screw caps when complete.

15. Install the wheel and tire assembly. Torque the lug nuts to 100 ft. lbs. (136 Nm).

16. Lower the vehicle.

17. Pump the brake pedal several times to position the brake pads before attempting to move the vehicle.

18. Check and fill the brake master cylinder as required.

19. Road-test the vehicle and check for proper brake operation.

Rear

1. Remove and discard ½ of the brake fluid from the brake master cylinder reservoir.

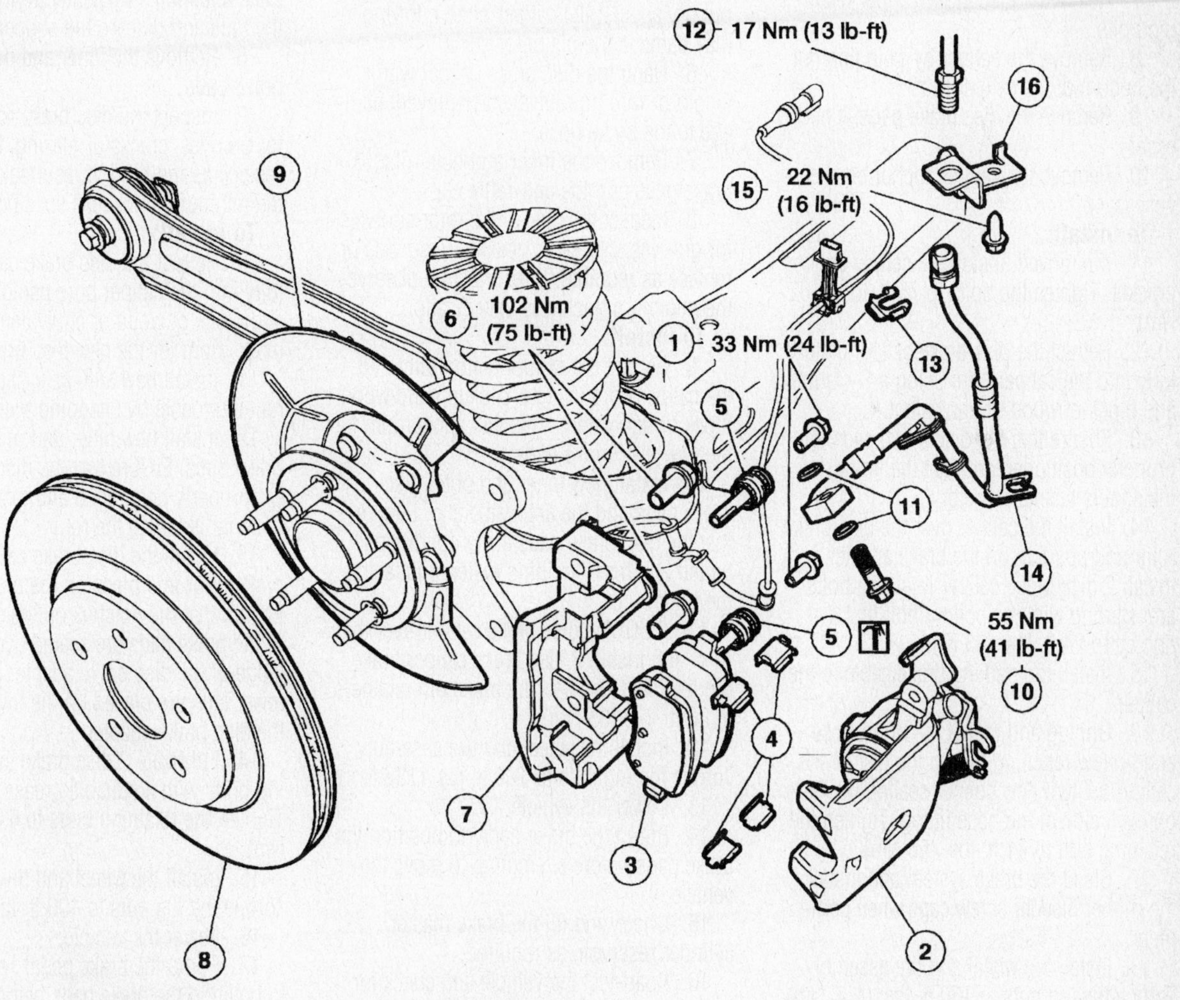

Item	Description	Item	Description	Item	Description
1	*Brake caliper bolts*	*6*	*Brake caliper anchor bracket bolt kit (2 bolts each side)*	*11*	*Copper washers*
2	*Brake caliper (RH/LH)*	*7*	*Brake caliper anchor bracket*	*12*	*Brake line fitting*
3	*Brake pads (kit)*	*8*	*Brake disc*	*13*	*Brake hose retainer*
4	*Slippers*	*9*	*Dust shield (RH/LH)*	*14*	*Brake hose (RH/LH)*
5	*Guide pin and boot*	*10*	*Flow bolt*	*15*	*Brake hose bracket bolt*
				16	*Brake hose bracket*

67197-FRST-CG38

Exploded view of the rear disc brake caliper assembly—Freestar and Monterey

2. Raise and safely support the vehicle.
3. Remove the wheel and tire assembly.
4. Disconnect the brake hose from the disc brake caliper by loosening and removing the hollow retaining bolt. Discard 2 copper sealing washers and plug the brake hose.
5. Using a C-clamp or equivalent, position the clamp frame on the inboard side of the disc brake caliper housing. Place the clamp screw on the outboard disc brake pad and tighten the clamp enough to press the caliper piston into the caliper housing releasing pressure on the disc brake pads.
6. Disconnect the parking brake cable from the caliper.
7. Remove 2 disc brake caliper retaining bolts.
8. Remove the caliper by swinging out the bottom of the caliper first.
9. Remove the disc brake pads, if necessary.
10. Remove the caliper anchor bracket, if necessary.

To install:

11. If removed, install the caliper anchor bracket. Tighten the bolts to 75 ft. lbs. (102 Nm).
12. Retract the disc brake caliper piston fully into the caliper bore using a C-clamp and block of wood or equivalent.
13. Ensure that the disc brake pads are properly positioned and that the lining material is facing the rotor.
14. Install the caliper over the disc brake rotor and position on the brake adapter. Install 2 disc brake caliper retaining bolts and starting with the bottom bolt first, tighten to 24 ft. lbs. (33 Nm).
15. Install the parking brake cable to the caliper.
16. Unplug and install the brake hose and hollow retaining bolt to the disc brake caliper using a new copper sealing washer on each side of the hose fitting. Tighten the retaining bolt to 41 ft. lbs. (55 Nm).
17. Bleed the brake system and install the rubber bleeder screw caps when complete.
18. Install the wheel and tire assembly. Torque the lug nuts to 100 ft. lbs. (136 Nm).
19. Lower the vehicle.
20. Pump the brake pedal several times to position the brake pads before attempting to move the vehicle.
21. Check and fill the brake master cylinder as required.
22. Road-test the vehicle and check for proper brake operation.

Disc Brake Pads

REMOVAL & INSTALLATION

Front

1. Remove ½ of the brake fluid from the brake master cylinder reservoir. Properly dispose of the brake fluid.
2. Raise and safely support the vehicle.
3. Remove the wheel and tire assembly.
4. Remove 2 disc brake caliper brake pin retainers. Do not remove the brake hose from the caliper.
5. Lift the disc brake caliper off of the disc brake rotor using a rotating motion. Do not pry against the caliper piston. Prying may damage the piston or seals.
6. Hang the disc brake caliper with a length of wire or equivalent to prevent damage to the brake hose.
7. Remove the inner and outer disc brake pads and the anti-rattle clip.
8. Inspect the disc brake rotor surfaces for grooves, cracks or glazing. Resurface or replace as required. If resurfacing, observe the minimum thickness specification.

To install:

9. Retract the caliper piston fully into the caliper bore using a C-clamp and wood block or equivalent. This will allow room for the new disc brake pads.
10. Install new inner and outer disc brake pads and the anti-rattle clip. Ensure that the disc brake pads are properly positioned and that the lining material is facing the rotor.
11. Place the disc brake caliper over the rotor and install 2 disc brake caliper brake pin retainers. Tighten the brake pin retainers to 26 ft. lbs. (35 Nm).
12. Install the wheel and tire assembly. Torque the lug nuts to 100 ft. lbs. (136 Nm).
13. Lower the vehicle.
14. Pump the brake pedal to position the brake pads before attempting to move the vehicle.
15. Check and fill the brake master cylinder reservoir, as required.
16. Road-test the vehicle and check for proper brake system operation.

Rear

1. Remove ½ of the brake fluid from the brake master cylinder reservoir. Properly dispose of the brake fluid.
2. Raise and safely support the vehicle.
3. Remove the wheel and tire assembly.
4. Using a C-clamp or equivalent, position the clamp frame on the inboard side of the disc brake caliper housing. Place the clamp screw on the outboard disc brake pad and tighten the clamp enough to press the caliper piston into the caliper housing releasing pressure on the disc brake pads.
5. Remove 2 disc brake caliper retaining bolts. Do not remove the disc brake caliper brake hose from the caliper.
6. Work the disc brake caliper off the brake rotor and disc brake adapter. Move the disc brake caliper aside and secure with wire or equivalent to prevent damage to the brake hose.
7. Remove the slippers from the anchor plate abutments by gently prying them off the rails and discard the slippers.
8. Remove the inner and outer disc brake pads.
9. Inspect the disc brake rotor surfaces for grooves, cracks or glazing. Resurface or replace as required. If resurfacing, observe the minimum thickness specification.

To install:

10. Retract the disc brake caliper piston fully into the caliper bore using a C-clamp and block of wood or equivalent. This will make room for the new disc brake pads.
11. Install new anti-wear slippers on the rail abutments by snapping them in place.
12. Install new inner and outer disc brake pads. Ensure that the disc brake pads are properly positioned and that the lining material is facing the rotor.
13. Install the disc brake caliper over the brake rotor and place on the brake adapter. Ensure that the notches on the upper ends of the brake pads are seated over the upper ledge of the disc brake adapter and the lower tabs are placed on the lower ledge of the disc brake adapter.
14. Lubricate 2 disc brake caliper retaining bolts with a suitable grease and install. Tighten the retaining bolts to 41ft. lbs. (55 Nm).
15. Install the wheel and tire assembly. Torque the lug nuts to 100 ft. lbs. (136 Nm).
16. Lower the vehicle.
17. Pump the brake pedal several times to position the brake pads before attempting to move the vehicle.
18. Check and fill the brake master cylinder reservoir, as required.
19. Road-test the vehicle and check for proper brake operation.

11

LINCOLN

LS

SPECIFICATION CHARTS

ENGINE AND VEHICLE IDENTIFICATION

			Engine			
Code ①	Liters (cc)	Cu. In.	Cyl.	Fuel Sys.	Type	Eng. Mfg.
S	3.0 (3049)	182	6	MFI	DOHC	Ford
A	3.9 (3947)	243	8	MFI	DOHC	Ford

MFI: Multi-Port Fuel Injection

DOHC: Double Overhead Camshaft

① 8th digit of the VIN

② 10th digit of the VIN

Model Year	
Code ②	Year
1	2001
2	2002
3	2003
4	2004
5	2005

67197-LILS -C01

GENERAL ENGINE SPECIFICATIONS

Year	Model	Engine Displacement Liters	Engine ID/VIN	Net Horsepower @ rpm	Net Torque @ rpm (ft. lbs.)	Bore x Stroke (in.)	Com-pression Ratio	Oil Pressure @ rpm
2001	LS6	3.0	S	210@6500	205@4750	3.50x3.13	10.5:1	20-45@1500
	LS8	3.9	A	252@6100	267@4300	NA	10.6:1	NA
2002	LS6	3.0	S	210@6500	205@4750	3.50x3.13	10.5:1	20-45@1500
	LS8	3.9	A	252@6100	267@4300	NA	10.6:1	NA
2003	LS6	3.0	S	210@6500	205@4750	3.50x3.13	10.5:1	20-45@1500
	LS8	3.9	A	280@6100	286@4300	NA	10.75:1	NA
2004	LS6	3.0	S	210@6500	205@4750	3.50x3.13	10.5:1	20-45@1500
	LS8	3.9	A	280@6100	286@4300	NA	10.75:1	NA

NA: Not Available

67197-LILS-C02

ENGINE TUNE-UP SPECIFICATIONS

Year	Engine Displacement Liters	Engine ID/VIN	Spark Plugs Gap (in.)	Ignition Timing (deg.) ①		Fuel Pump (psi)	Idle Speed (rpm) ①		Valve Clearance	
				MT	AT		MT	AT	In.	Ex.
2001	3.0	S	0.051-0.057	12-17B	12-17B	26-45	650-750	650-750	0.007-0.009	0.013-0.015
	3.9	A	0.039-0.043	—	10-20B	43	—	650-750	0.007-0.009	0.009-0.011
2002	3.0	S	0.051-0.057	12-17B	12-17B	26-45	650-750	650-750	0.007-0.009	0.013-0.015
	3.9	A	0.039-0.043	—	10-20B	43	—	650-750	0.007-0.009	0.009-0.011
2003	3.0	S	0.051-0.057	12-17B	12-17B	26-45	650-750	650-750	0.007-0.009	0.013-0.015
	3.9	A	0.039-0.043	—	10-20B	43	—	650-750	0.007-0.009	0.009-0.011
2004	3.0	S	0.051-0.057	12-17B	12-17B	26-45	650-750	650-750	0.007-0.009	0.013-0.015
	3.9	A	0.039-0.043	—	10-20B	43	—	650-750	0.007-0.009	0.009-0.011

The underhood specifications sticker often reflects tune-up specification changes in production. Sticker figures must be used if they disagree with those in this chart.

① Controlled by the engine computer

67197-LILS-C03

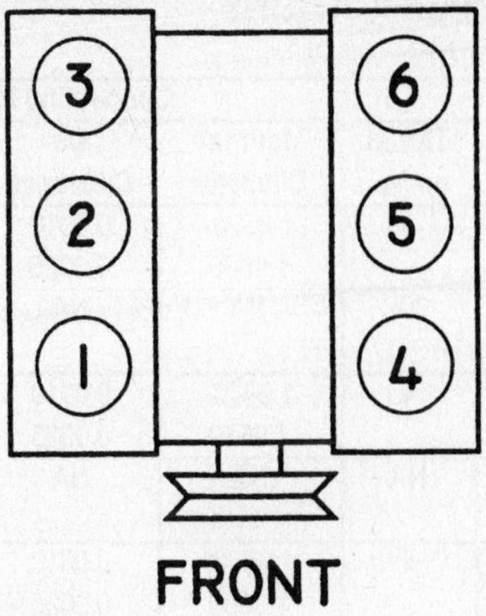

3.0L (VIN S) engine
Firing order:1–4–2–5–3–6
Distributorless ignition system (one coil on each cylinder)

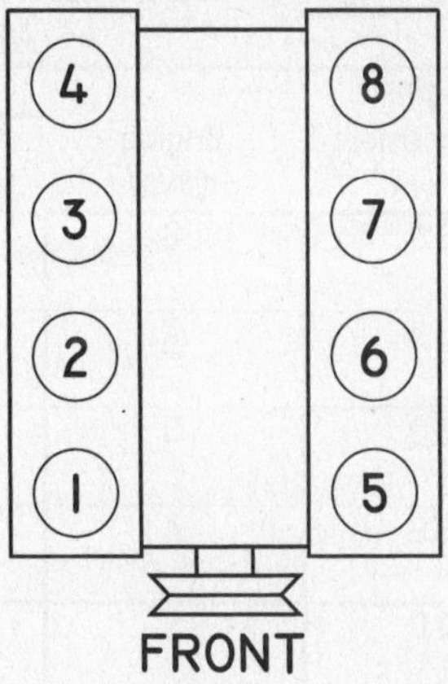

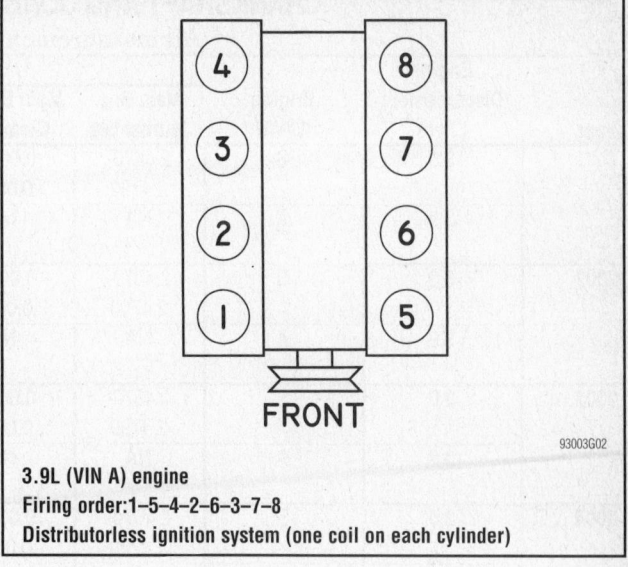

3.9L (VIN A) engine
Firing order:1–5–4–2–6–3–7–8
Distributorless ignition system (one coil on each cylinder)

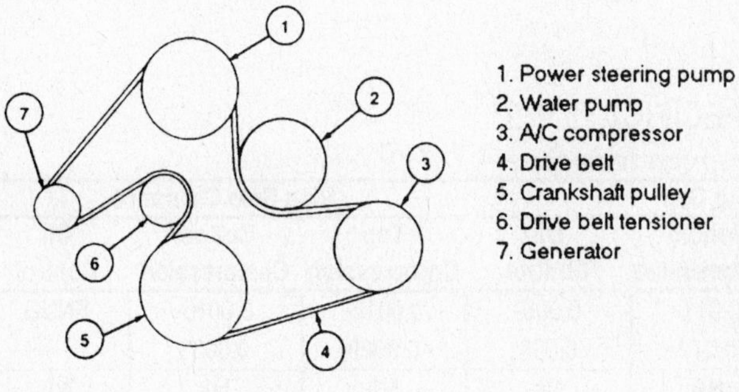

Serpentine accessory drive belt routing—3.0L (VIN S) engine

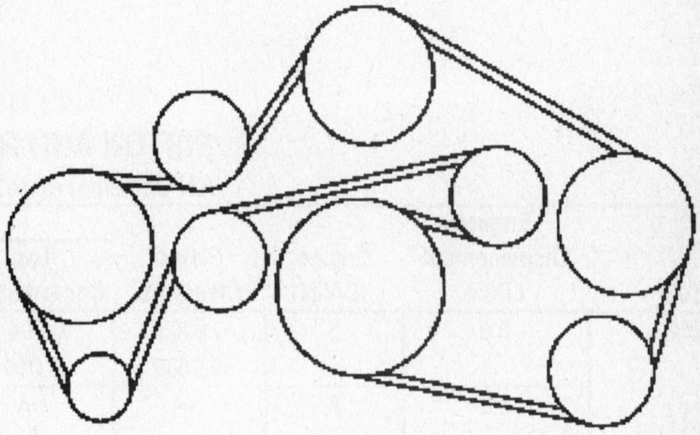

Serpentine accessory drive belt routing—3.9L (VIN A) engine

CAPACITIES

Year	Model	Engine Displacement Liters	Engine ID/VIN	Engine Oil with Filter (qts.)	Transmission (pts.) Manual	Transmission (pts.) Auto.	Drive Axle Rear (pts.)	Fuel Tank (gal.)	Cooling System (qts.)
2001	LS6	3.0	S	6.9	NA	23.8	3.0	18.0	10.6
	LS8	3.9	A	NA	—	23.8	3.0	18.0	11.3
2002	LS6	3.0	S	6.9	NA	23.8	3.0	18.0	10.6
	LS8	3.9	A	NA	—	23.8	3.0	18.0	11.3
2003	LS6	3.0	S	6.9	NA	23.8	3.0	18.0	10.6
	LS8	3.9	A	NA	—	23.8	3.0	18.0	11.3
2004	LS6	3.0	S	6.9	NA	23.8	3.0	18.0	10.6
	LS8	3.9	A	NA	—	23.8	3.0	18.0	11.3

N/A: Not Available

67197-LILS-C04

CRANKSHAFT AND CONNECTING ROD SPECIFICATIONS

All measurements are given in inches.

Year	Engine Displacement Liters	Engine ID/VIN	Crankshaft Main Brg. Journal Dia.	Crankshaft Main Brg. Oil Clearance	Crankshaft Shaft End-play	Crankshaft Thrust on No.	Connecting Rod Journal Diameter	Connecting Rod Oil Clearance	Connecting Rod Side Clearance
2001	3.0	S	2.4670-2.4790	0.0009-0.0018	0.0040-0.0090	4	1.9670-1.9680	0.0010-0.0025	0.0039-0.0118
	3.9	A	NA	NA	NA	NA	NA	NA	NA
2002	3.0	S	2.4670-2.4790	0.0009-0.0018	0.0040-0.0090	4	1.9670-1.9680	0.0010-0.0025	0.0039-0.0118
	3.9	A	NA	NA	NA	NA	NA	NA	NA
2003	3.0	S	2.4670-2.4790	0.0009-0.0018	0.0040-0.0090	4	1.9670-1.9680	0.0010-0.0025	0.0039-0.0118
	3.9	A	NA	NA	NA	NA	NA	NA	NA
2004	3.0	S	2.4670-2.4790	0.0009-0.0018	0.0040-0.0090	4	1.9670-1.9680	0.0010-0.0025	0.0039-0.0118
	3.9	A	NA	NA	NA	NA	NA	NA	NA

NA: Not Available

67197-LILS-C05

PISTON AND RING SPECIFICATIONS

All measurements are given in inches.

Year	Engine Displacement Liters	Engine ID/VIN	Piston Clearance	Ring Gap Top Compression	Ring Gap Bottom Compression	Ring Gap Oil Control	Ring Side Clearance Top Compression	Ring Side Clearance Bottom Compression	Ring Side Clearance Oil Control
2001	3.0	S	0.0005-0.0009	0.004-0.010	0.011-0.017	0.005-0.026	0.0015-0.0029	0.0015-0.0033	SNUG
	3.9	A	NA	NA	NA	NA	NA	NA	NA
2002	3.0	S	0.0005-0.0009	0.004-0.010	0.011-0.017	0.005-0.026	0.0015-0.0029	0.0015-0.0033	SNUG
	3.9	A	NA	NA	NA	NA	NA	NA	NA
2003	3.0	S	0.0005-0.0009	0.004-0.010	0.011-0.017	0.005-0.026	0.0015-0.0029	0.0015-0.0033	SNUG
	3.9	A	NA	NA	NA	NA	NA	NA	NA
2004	3.0	S	0.0005-0.0009	0.004-0.010	0.011-0.017	0.005-0.026	0.0015-0.0029	0.0015-0.0033	SNUG
	3.9	A	NA	NA	NA	NA	NA	NA	NA

NA: Not Available

67197-LILS-C06

VALVE SPECIFICATIONS

Year	Engine Displacement Liters	Engine ID/VIN	Seat Angle (deg.)	Face Angle (deg.)	Spring Test Pressure (lbs. @ in.)	Spring Free Length (in.)	Stem-to-Guide Clearance (in.)		Stem Diameter (in.)	
							Intake	Exhaust	Intake	Exhaust
2001	3.0	S	44.75	45.5	153@1.18	1.570	0.0007-0.0027	0.0017-0.0037	0.2350-0.2358	0.2343-0.2350
	3.9	A	NA	NA	NA	NA	NA	NA	NA	NA
2002	3.0	S	44.75	45.5	153@1.18	1.570	0.0007-0.0027	0.0017-0.0037	0.2350-0.2358	0.2343-0.2350
	3.9	A	NA	NA	NA	NA	NA	NA	NA	NA
2003	3.0	S	44.75	45.5	153@1.18	1.570	0.0007-0.0027	0.0017-0.0037	0.2350-0.2358	0.2343-0.2350
	3.9	A	NA	NA	NA	NA	NA	NA	NA	NA
2004	3.0	S	44.75	45.5	153@1.18	1.570	0.0007-0.0027	0.0017-0.0037	0.2350-0.2358	0.2343-0.2350
	3.9	A	NA	NA	NA	NA	NA	NA	NA	NA

NA: Not Available

67197-LILS-C07

TORQUE SPECIFICATIONS

All readings in ft. lbs.

Year	Engine Displacement Liters	Engine ID/VIN	Cylinder Head Bolts	Main Bearing Bolts	Rod Bearing Bolts	Crankshaft Damper Bolts	Flywheel Bolts	Manifold		Spark Plugs	Oil Pan Drain Plug
								Intake	Exhaust		
2001	3.0	S	①	②	③	④	54-64	⑤	13-16	7-15	17
	3.9	A	⑥	NA	NA	⑦	⑧	18	18	19	17
2002	3.0	S	①	②	③	④	54-64	⑤	13-16	7-15	17
	3.9	A	⑥	NA	NA	⑦	⑧	18	18	19	17
2003	3.0	S	①	②	③	④	54-64	⑤	13-16	7-15	17
	3.9	A	⑥	NA	NA	⑦	⑧	18	18	19	17
2004	3.0	S	①	②	③	④	54-64	⑤	13-16	7-15	17
	3.9	A	⑥	NA	NA	⑦	⑧	18	18	19	17

NA: Not Available

① Step 1: 28-31 ft. lbs.
Step 2: Plus 85-95 degrees
Step 3: Loosen one turn
Step 4: 28-31 ft. lbs.
Step 5: Plus 85-95 degrees
Step 6: Plus 85-95 degrees

② Step 1: Cap bolts 1-8 (outer) 17-20 ft. lbs.
Step 2: Cap bolts 9-16 (inner) 28-31 ft. lbs.
Step 3: Rotate bolts 1-16, 85-95 degrees
Step 4: Bolts 17-22; 15-22 ft. lbs.

③ Step 1: 30-33 ft. lbs.
Step 2: Plus 90-120 degrees

④ Step 1: 77-99 ft. lbs.
Step 2: Loosen 360 degrees
Step 3: Tighten to 35-39 ft. lbs.
Step 4: Plus 85-95 degrees

⑤ 71-106 inch lbs.

⑥ Step 1: Tighten M10 bolts to 15 ft. lbs.
Step 2: Tighten M10 bolts to 26 ft. lbs.
Step 3: Tighten M10 bolts to 33 ft. lbs.
Step 4: Tighten M10 bolts plus 90 degrees
Step 5: Tighten M10 bolts plus 90 degrees
Step 6: Tighten M8 bolts to 15 ft. lbs.
Step 7: Tighten M8 bolts plus 90 degrees

⑦ Step 1: 59 ft. lbs.
Step 2: Plus 80 degrees

⑧ Step 1: 11 ft. lbs.
Step 2: 81 ft. lbs.

67197-LILS-C08

WHEEL ALIGNMENT

Year	Model		Caster Range (+/-Deg.)	Caster Preferred Setting (Deg.)	Camber Range (+/-Deg.)	Camber Preferred Setting (Deg.)	Toe-in (in.)
2001	LS	F	0.70	0	0.70	0	0.08 +/- 0.13
		R	—	—	0.75	0	0.13 +/- 0.13
2002	LS	F	0.70	0	0.70	0	0.08 +/- 0.13
		R	—	—	0.75	0	0.13 +/- 0.13
2003	LS	F	0.70	0	0.70	0	0.08 +/- 0.13
		R	—	—	0.75	0	0.13 +/- 0.13
2004	LS	F	0.70	0	0.70	0	0.08 +/- 0.13
		R	—	—	0.75	0	0.13 +/- 0.13

67197-LILS-C09

TIRE, WHEEL AND BALL JOINT SPECIFICATIONS

Year	Model	OEM Tires Standard	OEM Tires Optional	Tire Pressures (psi) Front	Tire Pressures (psi) Rear	Wheel Size	Ball Joint Inspection	Lug Nut (ft. lbs.)
2001	LS6 and LS8	P215/60R16	P235/50R17	30	30	7-J/7.5J	U ① L ① ②	100
2002	LS6 and LS8	P215/60R16	P235/50R17	30	30	7-J/7.5J	U ① L ① ②	100
2003	LS6 and LS8	P215/60R16	P235/50R17	30	30	7-J/7.5J	U ① L ① ②	100
2004	LS6 and LS8	P215/60R16	P235/50R17	30	30	7-J/7.5J	U ① L ① ②	100

OEM: Original Equipment Manufacturer

PSI: Pounds Per Square Inch

STD: Standard

OPT: Optional

L: Lower

U: Upper

① Replace if any measurable movement is found.

② Do not lift car. Inspect the boss into which the grease fitting is threaded. Replace if the boss is flush or receded below the surface of the ball joint.

67197-LILS-C10

BRAKE SPECIFICATIONS

All measurements in inches unless noted

			Brake Disc			Minimum	Brake Caliper	
			Original	Minimum	Maximum	Lining	Bracket Bolts	Mounting Bolts
Year	Model		Thickness	Thickness	Runout	Thickness	(ft. lbs.)	(ft. lbs.)
2001	LS6	F	1.180	1.120	0.004	0.079	76	26
		R	0.810	0.740	0.004	0.039	76	25
	LS8	F	1.180	1.120	0.004	0.079	76	26
		R	0.810	0.740	0.004	0.039	76	25
2002	LS6	F	1.180	1.120	0.004	0.079	76	26
		R	0.810	0.740	0.004	0.039	76	25
	LS8	F	1.180	1.120	0.004	0.079	76	26
		R	0.810	0.740	0.004	0.039	76	25
2003	LS6	F	1.180	1.120	0.004	0.079	76	26
		R	0.810	0.740	0.004	0.039	76	25
	LS8	F	1.180	1.120	0.004	0.079	76	26
		R	0.810	0.740	0.004	0.039	76	25
2004	LS6	F	1.180	1.120	0.004	0.079	76	26
		R	0.810	0.740	0.004	0.039	76	25
	LS8	F	1.180	1.120	0.004	0.079	76	26
		R	0.810	0.740	0.004	0.039	76	25

67197-LILS-C11

SCHEDULED MAINTENANCE INTERVALS
Lincoln—LS

TO BE SERVICED	TYPE OF SERVICE	VEHICLE MILEAGE INTERVAL (x1000)												
		5	10	15	20	25	30	35	40	45	50	55	60	65
Air cleaner filter	R						✓						✓	
Accessory drive belt	S/I												✓	
Brake system ①	S/I			✓			✓			✓			✓	
Clutch pedal operation	S/I						✓						✓	
Cooling system hoses and clamps	S/I			✓			✓			✓			✓	
CV-joint boots & axle seals	S/I						✓						✓	
Engine coolant	R	Ten years or 150,000 miles												
Engine oil & filter	R	✓	✓	✓	✓	✓	✓	✓	✓	✓	✓	✓	✓	✓
Exterior Lights	S/I	Check monthly												
PCV valve	S/I												✓	
Exhaust system & heat shields	S/I						✓						✓	
Parking brake system	S/I	Every 6 months												
Power steering fluid	S/I	Every 6 months												
Rotate tires	S/I	✓		✓		✓		✓		✓		✓		✓
Steering linkage	S/I						✓						✓	
Spark plugs	R	Change at 100,000 miles												
Suspension components	S/I						✓						✓	

R: Replace S/I: Inspect and service, if necessary L: Lubricate A: Adjust C: Clean

① Inspect the reservoir fluid level, rotor and or drum, brake lines, hoses, calipers and or wheel cylinders

FREQUENT OPERATION MAINTENANCE (SEVERE SERVICE)

If a vehicle is operated under any of the following conditions it is considered severe service:

- Extremely dusty areas.

- 50% or more of the vehicle operation is in 32°C (90°F) or higher temperatures, or constant operation in temperatures below 0°C (32°F).

- Prolonged idling (vehicle operation in stop and go traffic).

- Frequent short running periods (engine does not warm to normal operating temperatures).

- Police, taxi, delivery usage or trailer towing usage.

Oil & oil filter change: change every 3000 miles.

Air filter element: change every 15,000 miles.

67197-LILS-C12

PRECAUTIONS

Before servicing any vehicle, please be sure to read all of the following precautions, which deal with personal safety, prevention of component damage, and important points to take into consideration when servicing a motor vehicle:

- Never open, service or drain the radiator or cooling system when the engine is hot; serious burns can occur from the steam and hot coolant.
- Observe all applicable safety precautions when working around fuel. Whenever servicing the fuel system, always work in a well-ventilated area. Do not allow fuel spray or vapors to come in contact with a spark, open flame, or excessive heat (a hot drop light, for example). Keep a dry chemical fire extinguisher near the work area. Always keep fuel in a container specifically designed for fuel storage; also, always properly seal fuel containers to avoid the possibility of fire or explosion. Refer to the additional fuel system precautions later in this section.
- Fuel injection systems often remain pressurized, even after the engine has been turned **OFF**. The fuel system pressure must be relieved before disconnecting any fuel lines. Failure to do so may result in fire and/or personal injury.
- Brake fluid often contains polyglycol ethers and polyglycols. Avoid contact with the eyes and wash your hands thoroughly after handling brake fluid. If you do get brake fluid in your eyes, flush your eyes with clean, running water for 15 minutes. If eye irritation persists, or if you have taken brake fluid internally, IMMEDIATELY seek medical assistance.
- The EPA warns that prolonged contact with used engine oil may cause a number of skin disorders, including cancer. You should make every effort to minimize your exposure to used engine oil. Protective gloves should be worn when changing oil. Wash your hands and any other exposed skin areas as soon as possible after exposure to used engine oil. Soap and water, or waterless hand cleaner should be used.
- All new vehicles are now equipped with an air bag system, often referred to as a Supplemental Restraint System (SRS) or Supplemental Inflatable Restraint (SIR) system. The system must be disabled before performing service on or around system components, steering column, instrument panel components, wiring and sensors. Failure to follow safety and disabling procedures could result in accidental air bag deployment, possible personal injury and unnecessary system repairs.
- Always wear safety goggles when working with, or around, the air bag system. When carrying a non-deployed air bag, be sure the bag and trim cover are pointed away from your body. When placing a non-deployed air bag on a work surface, always face the bag and trim cover upward, away from the surface. This will reduce the motion of the module if it is accidentally deployed. Refer to the additional air bag system precautions later in this section.
- Clean, high quality brake fluid from a sealed container is essential to the safe and proper operation of the brake system. You should always buy the correct type of brake fluid for your vehicle. If the brake fluid becomes contaminated, completely flush the system with new fluid. Never reuse any brake fluid. Any brake fluid that is removed from the system should be discarded. Also, do not allow any brake fluid to come in contact with a painted surface; it will damage the paint.
- Never operate the engine without the proper amount and type of engine oil; doing so WILL result in severe engine damage.
- Timing belt maintenance is extremely important. Many models utilize an interference-type, non-freewheeling engine. If the timing belt breaks, the valves in the cylinder head may strike the pistons, causing potentially serious (also time-consuming and expensive) engine damage.
- Disconnecting the negative battery cable on some vehicles may interfere with the functions of the on-board computer system(s) and may require the computer to undergo a relearning process once the negative battery cable is reconnected.
- When servicing drum brakes, only disassemble and assemble one side at a time, leaving the remaining side intact for reference.

ENGINE REPAIR

➡Disconnecting the negative battery cable on some vehicles may interfere with the functions of the on board computer system. The computer may undergo a relearning process once the negative battery cable is reconnected.

Alternator

REMOVAL

3.0L Engine

1. Before servicing the vehicle, refer to the precautions in the beginning of this section.
2. Remove or disconnect the following:
   - Negative battery cable
   - Accessory drive belt
   - Lower splash shield
   - Alternator mounting bolts
   - Alternator harness connectors
   - Alternator

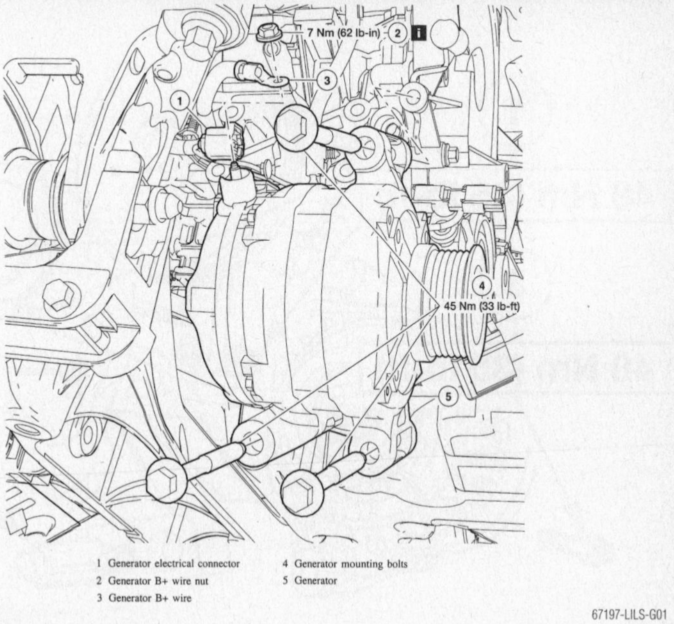

Alternator mounting—3.0L engine

3.9L Engine

1. Before servicing the vehicle, refer to the precautions in the beginning of this section.
2. Remove or disconnect the following:
   - Negative battery cable
   - Air intake tube
   - Accessory drive belt
   - Lower splash shield
   - Alternator mounting bolts
   - Alternator harness connectors
   - Alternator

INSTALLATION

3.0L Engine

Install or connect the following:
- Alternator
- Alternator harness connectors. Tighten the battery cable terminal nut to 70 inch lbs. (8 Nm).
- Alternator mounting bolts. Tighten the bolts to 35 ft. lbs. (48 Nm).
- Lower splash shield
- Accessory drive belt
- Negative battery cable

3.9L Engine

1. Install the alternator harness connectors, then install the alternator. Tighten the bolts in sequence as follows:
   a. Step 1: Tighten bolt No. 1 to 35 ft. lbs. (48 Nm)
   b. Step 2: Tighten bolt No. 2 to 15 ft. lbs. (20 Nm) plus 90 degrees
   c. Step 3: Tighten bolt No. 3 to 35 ft. lbs. (48 Nm)
2. Install or connect the following:
   - Lower splash shield
   - Accessory drive belt
   - Air intake tube
   - Negative battery cable

Ignition Timing

ADJUSTMENT

This vehicle is equipped with a Distributorless Ignition System (DIS). The ignition timing is not adjustable. It is controlled by the PCM.

Engine Assembly

REMOVAL & INSTALLATION

3.0L Engine

1. Before servicing the vehicle, refer to the precautions in the beginning of this section.
2. Drain the cooling system.
3. Relieve the fuel system pressure.
4. Recover the A/C refrigerant.
5. Drain the engine oil.
6. Remove or disconnect the following:
   - Negative battery cable
   - Air cleaner housing and outlet tube
   - Engine appearance cover
   - Upper radiator shield
   - Upper radiator support brackets
   - A/C pressure switch connector
   - Power steering reservoir
   - Fuel line
   - Vapor Management Valve (VMV) cover
   - VMV vacuum hose
   - Cowl leaf screens
   - Chassis vacuum lines
   - Cross vehicle support bar
   - Fresh air intake housing
   - Intake manifold rear main vacuum hose
   - Accelerator cable and cruise control cable
   - Cable bracket
   - Ground strap
   - Main engine wiring harness connector
   - Main transmission wiring harness connector
   - 2 fuel charging harness connectors
   - A/C line mounting bracket
   - Hydraulic cooling fan reservoir
   - Left, right and center splash shields
   - A/C compressor manifold and tube assembly
   - Coolant hoses
   - Exhaust front pipe and heat shields
   - Driveshaft
   - Shift cable and bracket
   - Front wheels
   - Front wheel speed sensor connectors
   - Front brake calipers
   - Lower stabilizer bar links
   - Upper ball joints
   - Lower strut mount bolts
   - Starter motor wiring harness connectors
   - Power Steering Pressure (PSP) switch connector
   - Steering shaft pinch bolt
   - Torque converter, if equipped
7. Support the engine, transmission, front and center crossmembers and the cooling system with a powertrain lift and transmission support bracket.
8. Support the rear of the vehicle with safety stands.
9. Remove or disconnect the following:
   - Transmission crossmember bolts
   - Subframe bolts
   - Crossmember bolts
   - Powertrain assembly
10. Attach a hoist to the engine.
11. Remove or disconnect the following:
    - Wire harness retainers
    - Starter motor
    - Heated Oxygen (HO_2S) sensor bracket
    - Motor mount nuts
    - Accessory drive belt
    - Power steering pump
    - Hydraulic cooling fan pump
    - Upper radiator hose
    - Transmission oil cooler lines, if equipped

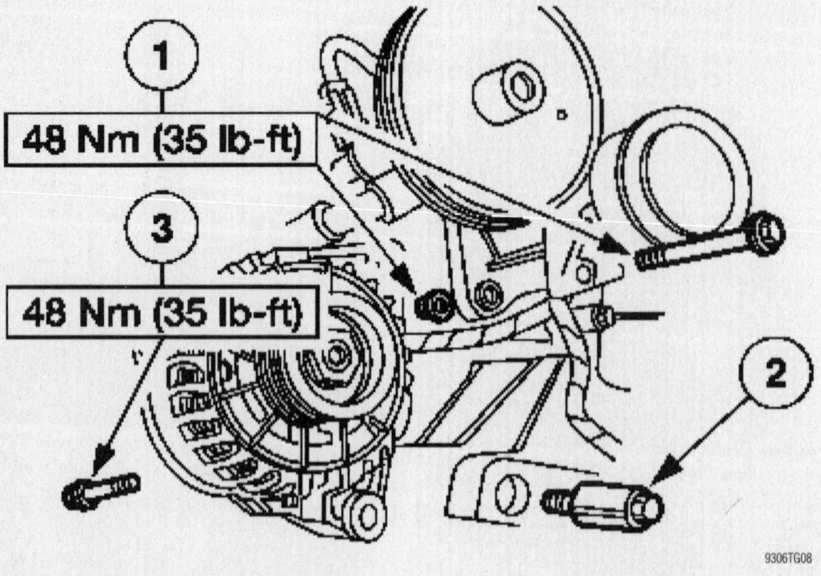

Alternator torque sequence—3.9L engine

12. Lift the engine and transmission out of the subframe.

13. Remove or disconnect the following:
- Oil cooler hoses
- Transmission flange bolts
- Transmission from the engine

To install:

14. Install or connect the following:
- Transmission to the engine. Tighten the flange bolts to 35 ft. lbs. (48 Nm).
- Powertrain on the subframe. Tighten the motor mount nuts to 46 ft. lbs. (63 Nm).
- Oil cooler hoses
- Transmission oil cooler lines, if equipped
- Upper radiator hose
- Hydraulic cooling fan pump. Tighten the bolts to 18 ft. lbs. (25 Nm).
- Power steering pump. Tighten the bolts to 18 ft. lbs. (25 Nm).
- Accessory drive belt
- HO_2S sensor bracket. Tighten the nut to 89 inch lbs. (10 Nm).
- Starter motor. Tighten the bolts to 18 ft. lbs. (25 Nm).
- Wire harness retainers
- Powertrain assembly. Tighten the crossmember bolts to 76 ft. lbs. (103 Nm).
- Torque converter, if equipped.
- Steering shaft pinch bolt. Tighten the bolts to 18 ft. lbs. (25 Nm).
- PSP switch connector
- Starter motor wiring harness connectors
- Lower strut mount bolts. Tighten the bolts to 129 ft. lbs. (175 Nm).
- Upper ball joints. Tighten the nuts to 66 ft. lbs. (90 Nm).
- Lower stabilizer bar links. Tighten the nuts to 41 ft. lbs. (55 Nm).
- Front brake calipers
- Front wheel speed sensor connectors
- Front wheels
- Shift cable and bracket
- Driveshaft
- Exhaust front pipe and heat shields
- Coolant hoses
- A/C compressor manifold and tube assembly. Tighten the bolt to 15 ft. lbs. (21 Nm).
- Left, right and center splash shields
- Hydraulic cooling fan reservoir
- A/C line mounting bracket
- 2 fuel charging harness connectors
- Main transmission wiring harness connector. Tighten the bolt to 89 inch lbs. (10 Nm).
- Main engine wiring harness connector. Tighten the bolt to 89 inch lbs. (10 Nm).
- Ground strap. Tighten the bolt to 89 inch lbs. (10 Nm).
- Cable bracket. Tighten the bolts to 89 inch lbs. (10 Nm).
- Accelerator cable and cruise control cable
- Intake manifold rear main vacuum hose
- Fresh air intake housing
- Cross vehicle support bar. Tighten the bolts to 15 ft. lbs. (20 Nm).
- Chassis vacuum lines
- Cowl leaf screens
- VMV vacuum hose
- VMV cover
- Fuel line
- Power steering reservoir
- A/C pressure switch connector
- Upper radiator support brackets. Tighten the bolts to 89 inch lbs. (10 Nm).
- Upper radiator shield
- Engine appearance cover
- Air cleaner housing and outlet tube
- Negative battery cable

15. Fill the crankcase to the correct level.

16. Fill the cooling system.

17. Recharge the A/C system.

18. Start the engine and check for leaks.

3.9L Engine

1. Before servicing the vehicle, refer to the precautions in the beginning of this section.

2. Drain the cooling system.

3. Relieve the fuel system pressure.

4. Recover the A/C refrigerant.

5. Drain the engine oil.

6. Remove or disconnect the following:
- Negative battery cable
- Air cleaner inlet tube
- Upper radiator shield
- Upper radiator support brackets
- A/C pressure switch connector
- Power steering return line clip
- Power steering reservoir
- Vapor Management Valve (VMV) cover
- VMV vacuum hose and canister purge hose
- Main vacuum supply hose
- Cowl vent screens
- Cross vehicle support bar
- Degas bottle hose
- Accelerator cable
- Cruise control cable
- Ground strap
- Fresh air filter and housing
- Powertrain harness connectors at right strut tower
- Fresh air filter panel
- Main engine wiring harness connector
- Main transmission wiring harness connector
- Heater hoses at the water control valve. Note the locations for assembly.
- Hydraulic cooling fan reservoir
- Water control valve harness connector
- Front wheels
- Inner splash shields
- Wheel speed sensor connectors and harness clips
- Brake calipers
- Lower stabilizer bar links
- Upper ball joints
- Lower strut mount bolts
- Left, right and center splash shields
- A/C suction and discharge lines
- Shift cable and bracket
- Power steering line frame rail clip
- Rack and pinion harness connectors
- Steering shaft bolt and coupling
- Starter motor harness connectors and ground cable
- Alternator harness connectors
- Lower transmission flange bolts
- Torque converter
- Inner air deflector
- Engine block heater, if equipped

7. Support the engine, transmission, front and center crossmembers and the cooling system with a powertrain lift and transmission support bracket.

8. Support the rear of the vehicle with safety stands.

9. Remove or disconnect the following:
- Transmission crossmember bolts
- Front crossmember bolts
- Center crossmember bolts
- Powertrain assembly
- A/C compressor manifold and tube assembly
- Power steering pump return hose
- Hydraulic cooling fan return hose
- Lower radiator hose
- Upper radiator hoses
- Knock Sensor (KS) connector
- Heater hose
- Transmission cooler lines and bracket
- Power steering pressure line and bracket
- Hydraulic cooling fan pressure line and bracket

10. Attach a hoist to the engine.

11. Remove the motor mount nuts and lift the powertrain out of the subframe.
12. Remove or disconnect the following:
    - Wiring harness retainers
    - Upper transmission flange bolts
    - Transmission from the engine

To install:

13. Install or connect the following:
    - Transmission to the engine. Tighten the upper flange bolts to 35 ft. lbs. (48 Nm).
    - Wiring harness retainers. Tighten the nuts to 89 inch lbs. (10 Nm).
    - Powertrain to the subframe. Tighten the mount nuts to 30 ft. lbs. (40 Nm).
    - Hydraulic cooling fan pressure line and bracket
    - Power steering pressure line and bracket
    - Transmission cooler lines and bracket
    - Heater hose
    - Knock sensor connector
    - Upper radiator hoses
    - Lower radiator hose
    - Hydraulic cooling fan return hose
    - Power steering pump return hose
    - A/C compressor manifold and tube assembly. Tighten the bolt to 15 ft. lbs. (21 Nm).
    - Powertrain assembly. Tighten the front and center crossmember bolts to 76 ft. lbs. (103 Nm) and the transmission crossmember bolts to 30 ft. lbs. (40 Nm).
    - Engine block heater, if equipped
    - Inner air deflector
    - Torque converter. Tighten the nuts to 28 ft. lbs. (38 Nm).
    - Lower transmission flange bolts. Tighten the bolts to 35 ft. lbs. (47 Nm).
    - Alternator harness connectors
    - Starter motor harness connectors and ground cable
    - Steering shaft bolt and coupling. Tighten the coupling pinch bolt to 26 ft. lbs. (35 Nm) and the shaft bolt to 22 ft. lbs. (30 Nm).
    - Rack and pinion harness connectors
    - Power steering line frame rail clip
    - Shift cable and bracket
    - A/C suction and discharge lines
    - Left, right and center splash shields
    - Lower strut mount bolts. Tighten the bolts to 129 ft. lbs. (175 Nm).
    - Upper ball joints. Tighten the nuts to 66 ft. lbs. (90 Nm).
    - Lower stabilizer bar links. Tighten the nuts to 41 ft. lbs. (55 Nm).
    - Brake calipers
    - Wheel speed sensor connectors and harness clips
    - Inner splash shields
    - Front wheels
    - Water control valve harness connector
    - Hydraulic cooling fan reservoir
    - Heater hoses at the water control valve
    - Main transmission wiring harness connector
    - Main engine wiring harness connector
    - Fresh air filter panel
    - Powertrain harness connectors at right strut tower
    - Fresh air filter and housing
    - Ground strap
    - Cruise control cable
    - Accelerator cable
    - Degas bottle hose
    - Cross vehicle support bar. Tighten the bolts to 15 ft. lbs. (20 Nm).
    - Cowl vent screens
    - Main vacuum supply hose
    - VMV vacuum hose and canister purge hose
    - VMV cover
    - Power steering reservoir
    - Power steering return line clip
    - A/C pressure switch connector
    - Upper radiator support brackets
    - Upper radiator shield
    - Air cleaner inlet tube
    - Negative battery cable
14. Fill the crankcase to the correct level.
15. Fill the cooling system.
16. Recharge the A/C system.
17. Start the engine and check for leaks.

Water Pump

REMOVAL & INSTALLATION

3.0L Engine

1. Before servicing the vehicle, refer to the precautions in the beginning of this section.
2. Drain the cooling system.
3. Remove or disconnect the following:
    - Negative battery cable
    - Air cleaner outlet tube
    - Engine vent hose
    - Upper radiator hose
    - Heater supply hose
    - Water pump hose
    - Lower radiator hose
    - Water crossover assembly
    - Water inlet hose
    - Accessory drive belt and idler pulley
    - Bracket assembly
    - Water pump

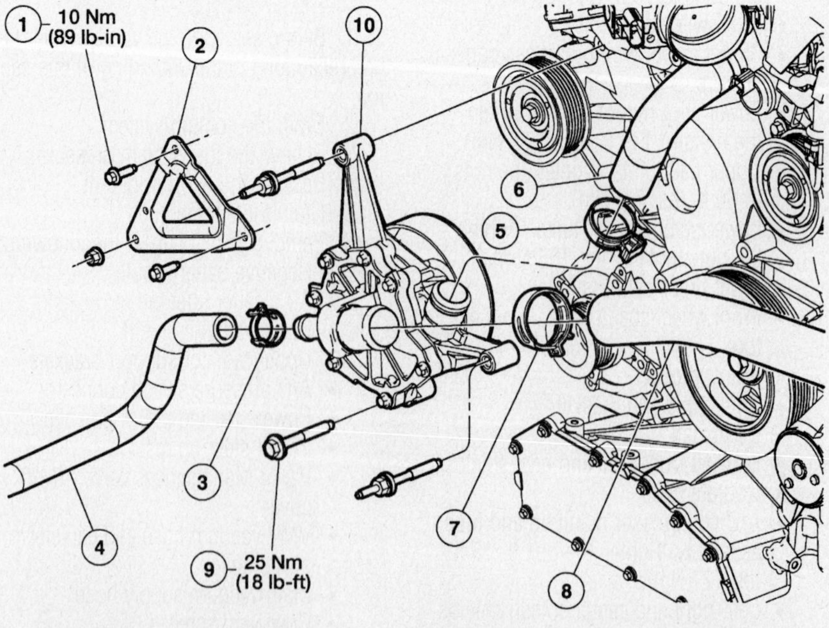

1 Intake manifold support bracket bolts
2 Intake manifold support bracket
3 Coolant pump-to-radiator hose clamp
4 Coolant pump-to-radiator hose
5 Coolant pump-to-coolant inlet pipe hose clamp
6 Coolant pump-to-coolant inlet pipe hose
7 Coolant pump-to-thermostat housing hose clamp
8 Coolant pump-to-thermostat housing hose

67197-LILS-G02

Water pump mounting—3.0L engine

To install:

4. Install or connect the following:
   - Water pump. Tighten the bolts to 18 ft. lbs. (25 Nm).
   - Bracket assembly. Tighten the fasteners to 89 inch lbs. (10 Nm).
   - Accessory drive belt and idler pulley
   - Water inlet hose
   - Water crossover assembly
   - Lower radiator hose
   - Water pump hose
   - Heater supply hose
   - Upper radiator hose
   - Engine vent hose
   - Air cleaner outlet tube
   - Negative battery cable
5. Fill the cooling system.
6. Start the engine and check for leaks.

3.9L Engine

1. Before servicing the vehicle, refer to the precautions in the beginning of this section.
2. Drain the cooling system.
3. Remove or disconnect the following:
   - Accessory drive belt
   - Water pump pulley
   - Water pump

To install:

4. Install or connect the following:
   - Water pump. Tighten the bolts to 71 inch lbs. (8 Nm) plus 90 degrees.
   - Water pump pulley. Tighten the bolts to 89 inch lbs. (10 Nm) plus 45 degrees.
   - Accessory drive belt
5. Fill the cooling system.
6. Start the engine and check for leaks.

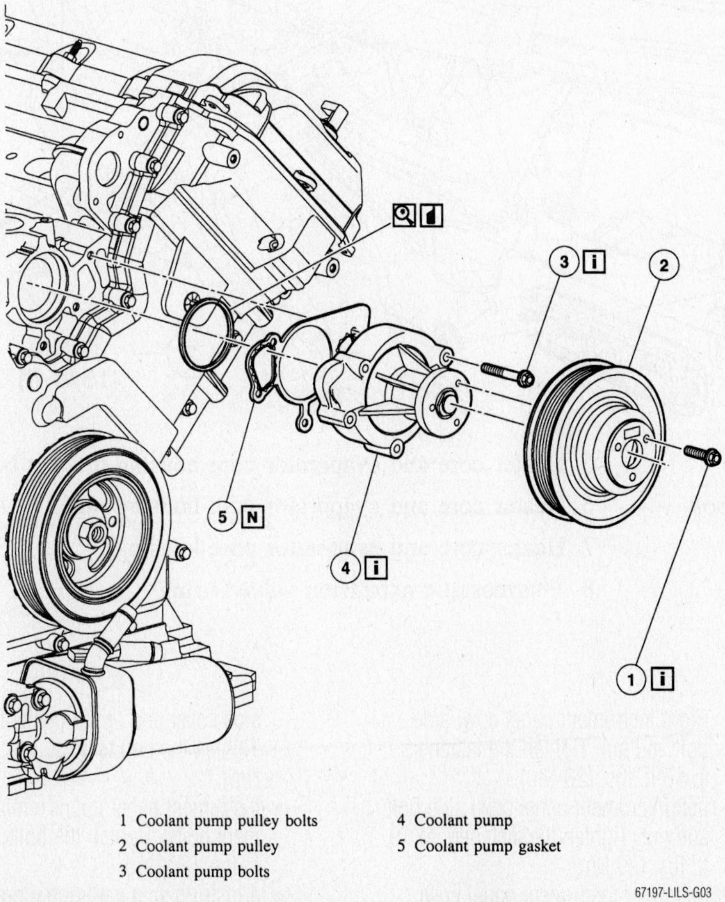

Exploded view of the water pump—3.9L engine

Heater Core

REMOVAL & INSTALLATION

1. Before servicing the vehicle, refer to the precautions in the beginning of this section.
2. Drain the cooling system.
3. Recover the A/C refrigerant.
4. Remove or disconnect the following:
   - Negative battery cable
   - Heater hose assembly
   - Cabin air filter plenum
   - Thermostatic expansion valve manifold and tube assembly
   - Driver's side air bag module
   - Floor console and A/C duct
   - Shift lever assembly, if equipped with automatic transmission
   - Left and right instrument panel insulators
   - Left and right door sill scuff plates
   - Left and right door weatherstrips
   - Left and right A pillar lower trim panels
   - Left and right windshield side garnish moldings
   - Instrument panel defroster opening grille assembly
   - Instrument panel cowl top screws
   - Instrument panel upper reinforcement bolts
   - Left and right instrument panel side finish panels
   - Hood release handle and cable
   - Upper right bulkhead electrical connector
   - Right instrument panel electrical connectors
   - Passenger side tunnel electrical connector
   - Steering column intermediate shaft
   - Left junction box electrical connectors
   - Left instrument panel electrical connectors
   - Ignition shift interlock connector, if equipped
   - 4 instrument panel tunnel brace bolts
   - Left outer instrument panel cowl side cover and reinforcement bolt
   - Left instrument panel cowl side bolt and nut
   - Right instrument panel cowl side bolt and nut
   - Instrument panel
   - Evaporator housing electrical connector
   - Cowl top attachment bolt
   - Evaporator housing attachment bolt
   - 3 evaporator nuts and washers in the engine compartment
   - Rear seat floor ducts
   - Evaporator core housing
   - Evaporator core housing air inlet
   - Bypass door harness connector
   - Heater core

To install:

5. Install or connect the following:
   - Heater core
   - Bypass door harness connector
   - Evaporator core housing air inlet
   - Evaporator core housing
   - Rear seat floor ducts
   - 3 evaporator nuts and washers in the engine compartment

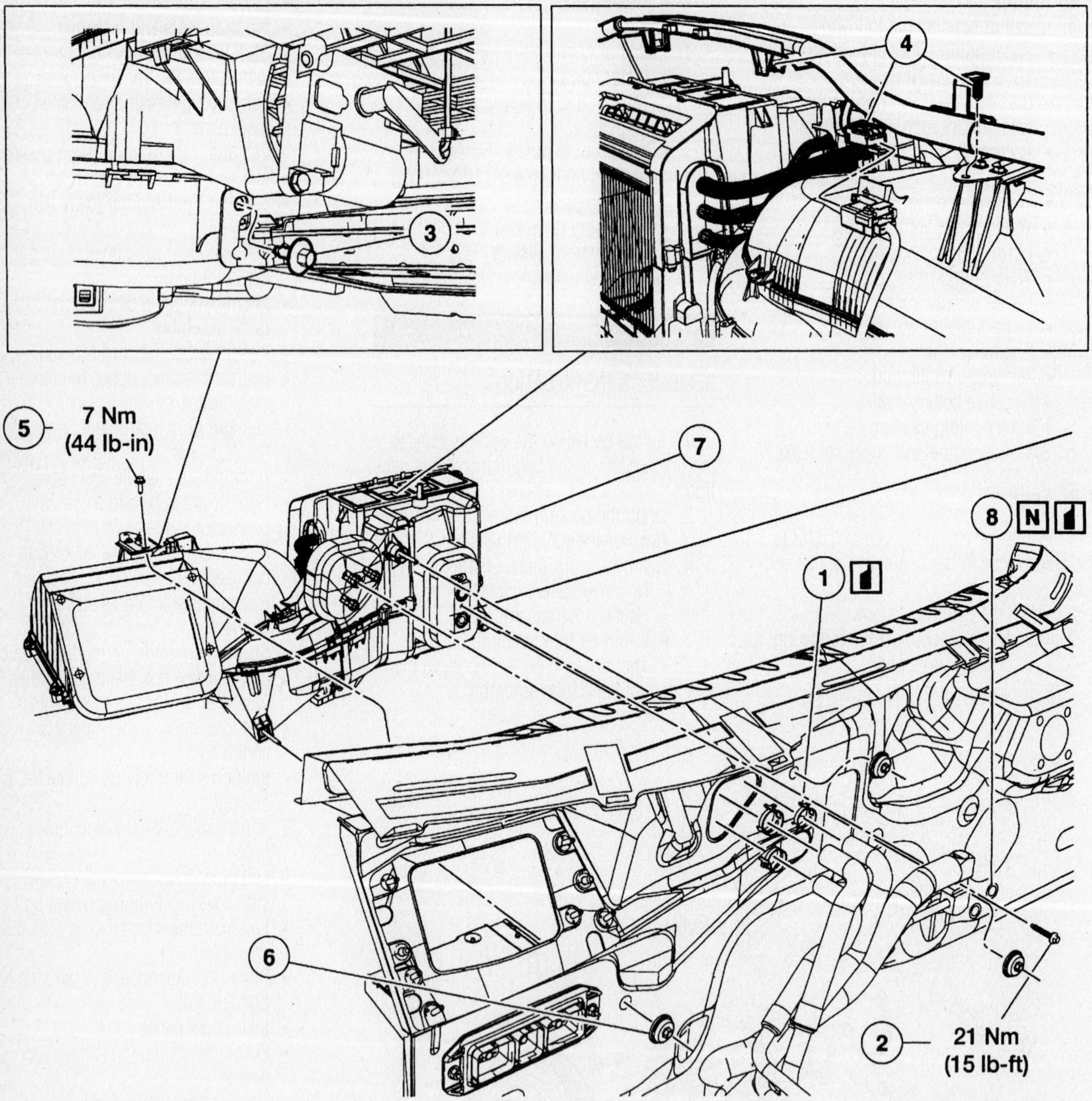

1 Heater hose clamp
2 Thermostatic expansion valve manifold bolt
3 PCM bracket bolt
4 Wire harness
5 Heater core and evaporator core housing bracket bolt
6 Heater core and evaporator core housing nut
7 Heater core and evaporator core housing
8 Thermostatic expansion valve O-ring

67197-LILS-G04

Heater core and related components (1 of 2)

- Evaporator housing attachment bolt
- Cowl top attachment bolt
- Evaporator housing electrical connector
- Instrument panel
- Instrument panel cowl top screws. Tighten to 27 inch lbs. (3 Nm).
- Right instrument panel cowl side bolt and nut. Tighten the fasteners to 15 ft. lbs. (20 Nm).
- Left instrument panel cowl side bolt and nut. Tighten the fasteners to 15 ft. lbs. (20 Nm).
- Left outer instrument panel cowl side cover and reinforcement bolt. Tighten the bolt to 15 ft. lbs. (20 Nm).
- Instrument panel upper reinforcement bolts. Tighten the bolts to 15 ft. lbs. (20 Nm).
- 4 instrument panel tunnel brace

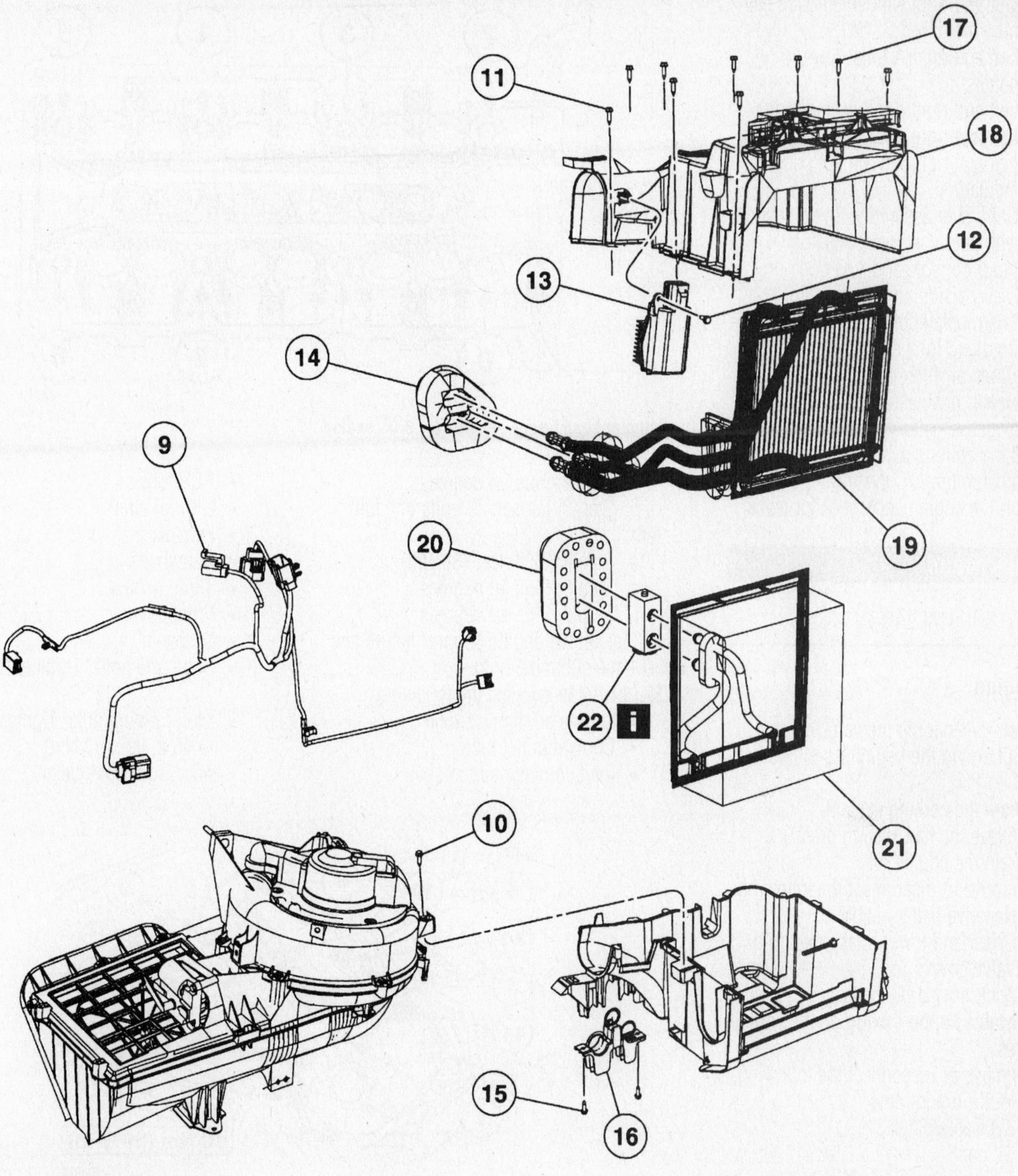

9 Wire harness
10 Heater core and evaporator core housing screw
11 Heater core and evaporator core housing screw
12 Blower motor speed control screw
13 Blower motor speed control
14 Heater core tube seal
15 Heater core tube bracket screw

67197-LILS-G05

Heater core and related components (2 of 2)

bolts. Tighten the bolts to 15 ft. lbs. (20 Nm).

- Ignition shift interlock connector, if equipped
- Left instrument panel electrical connectors
- Left junction box electrical connectors
- Steering column intermediate shaft. Tighten the pinch bolt to 26 ft. lbs. (35 Nm).
- Passenger side tunnel electrical connector
- Right instrument panel electrical connectors
- Upper right bulkhead electrical connector
- Hood release handle and cable
- Left and right instrument panel side finish panels
- Instrument panel defroster opening grille assembly

- Left and right windshield side garnish moldings
- Left and right A pillar lower trim panels
- Left and right door weatherstrips
- Left and right door sill scuff plates
- Left and right instrument panel insulators
- Shift lever assembly, if equipped with automatic transmission
- Floor console and A/C duct
- Driver's side air bag module
- Thermostatic expansion valve manifold and tube assembly
- Cabin air filter plenum
- Heater hose assembly
- Negative battery cable

6. Fill the cooling system.
7. Recharge the A/C system.
8. Start the engine and check for leaks.

Cylinder Head

REMOVAL & INSTALLATION

3.0L Engine

1. Before servicing the vehicle, refer to the precautions in the beginning of this section.
2. Drain the cooling system.
3. Relieve the fuel system pressure.
4. Drain the engine oil.
5. Remove or disconnect the following:
   - Negative battery cable
   - Upper and lower intake manifolds
   - Valve covers
   - Accessory drive belts
6. Install a support fixture to the engine lifting eyes.
7. Remove or disconnect the following:
   - Motor mount nuts
   - Subframe bolts
   - Oil pan
   - Front cover
   - Timing chains
   - Camshafts
   - Exhaust manifolds
   - Ground strap
   - Positive Crankcase Ventilation (PCV) tube
   - Ignition noise suppressor
   - Coolant outlet tube
   - Engine oil dipstick tube
8. Install the subframe bolts and remove the engine support fixture.
9. Remove the cylinder heads.

To install:

10. Install the cylinder heads. Tighten the bolts in sequence as follows:
    a. Step 1: 22 ft. lbs. (30 Nm)

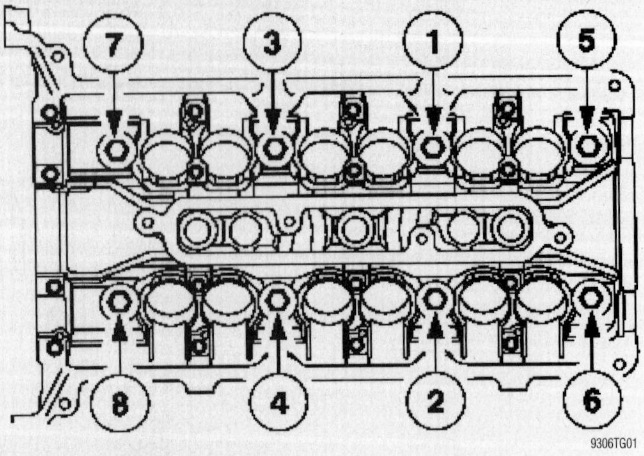

Cylinder head torque sequence—3.0L engine

    b. Step 2: Plus 90 degrees
    c. Step 3: Loosen all bolts one full turn
    d. Step 4: 22 ft. lbs. (30 Nm)
    e. Step 5: Plus 90 degrees
    f. Step 6: Plus 90 degrees

11. Install the engine support fixture and remove the subframe bolts.
12. Install or connect the following:
    - Engine oil dipstick tube
    - Coolant outlet tube
    - Ignition noise suppressor
    - PCV tube
    - Ground strap
    - Exhaust manifolds
    - Camshafts
    - Timing chains
    - Front cover
    - Oil pan
    - Subframe bolts. Tighten the bolts to 76 ft. lbs. (103 Nm).
    - Motor mount nuts. Tighten the nuts to 46 ft. lbs. (63 Nm).
    - Accessory drive belts

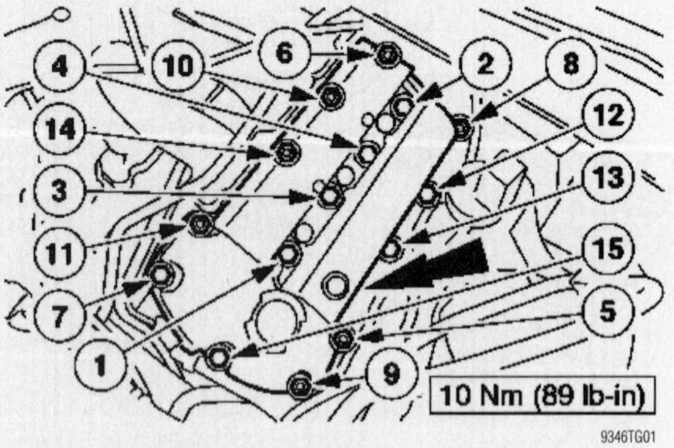

Left valve cover torque sequence—3.0L engine

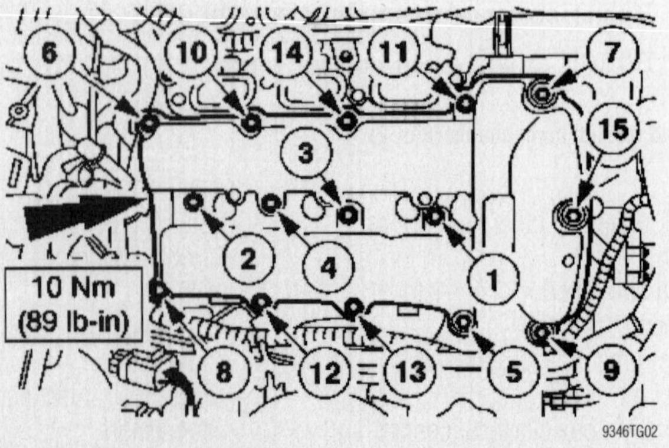

Right valve cover torque sequence—3.0L engine

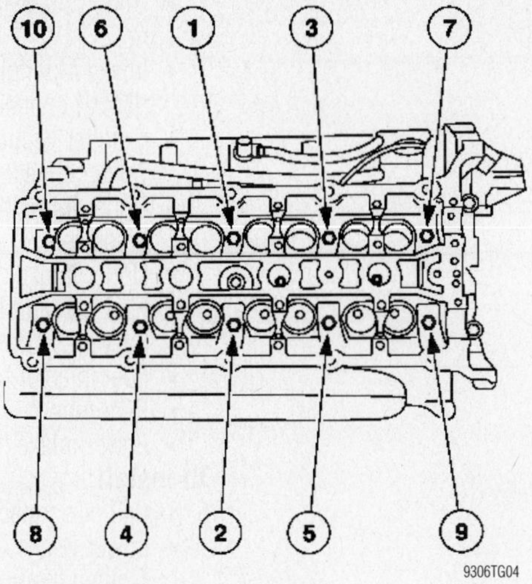

Right cylinder head torque sequence—3.9L engine

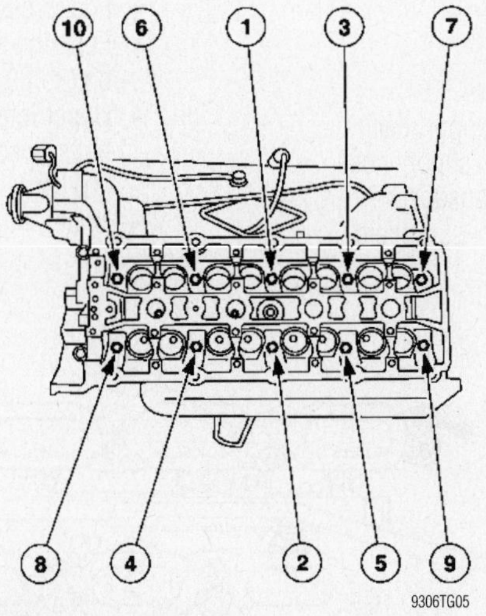

Left cylinder head torque sequence—3.9L engine

- Valve covers
- Upper and lower intake manifolds
- Negative battery cable

13. Fill the crankcase to the correct level.
14. Fill the cooling system.
15. Start the engine and check for leaks.

3.9L Engine

1. Before servicing the vehicle, refer to the precautions in the beginning of this section.
2. Drain the cooling system.
3. Relieve the fuel system pressure.
4. Remove or disconnect the following:
   - Negative battery cable
   - Engine appearance cover
   - Intake manifold
   - Valve covers
   - Accessory drive belts
   - Front cover
   - Timing chains
   - Camshafts
   - Water outlet pipe
   - Cylinder head temperature sensor connector
   - Exhaust front pipes
   - Exhaust Gas Recirculation (EGR) tube
   - Bolts and stud bolts at the rear of the cylinder heads
   - Cylinder heads. Loosen the bolts in reverse of the tightening sequence.

To install:

5. Install the cylinder heads. Tighten the bolts in sequence as follows:
   a. Step 1: M10 bolts to 15 ft. lbs. (20 Nm)
   b. Step 2: M10 bolts to 26 ft. lbs. (35 Nm)
   c. Step 3: M10 bolts to 33 ft. lbs. (45 Nm)
   d. Step 4: M10 bolts plus 90 degrees
   e. Step 5: M10 bolts plus 90 degrees
   f. Step 6: M8 bolts to 15 ft. lbs. (20 Nm)
   g. Step 7: M8 bolts plus 90 degrees
6. Install or connect the following:
   - Bolts and stud bolts at the rear of the cylinder heads. Tighten the bolts to 37 ft. lbs. (50 Nm).
   - EGR tube
   - Exhaust front pipes
   - Cylinder head temperature sensor connector
   - Water outlet pipe
   - Camshafts
   - Timing chains
   - Front cover
   - Accessory drive belts
   - Valve covers
   - Intake manifold

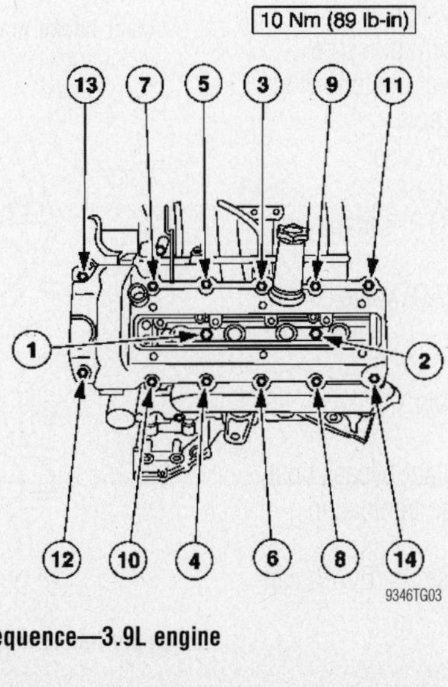

Left valve cover torque sequence—3.9L engine

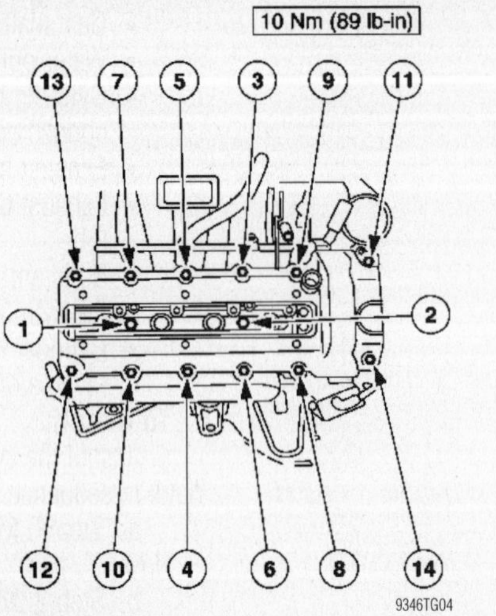

Right valve cover torque sequence—3.9L engine

- Engine appearance cover
- Negative battery cable

7. Fill the cooling system.
8. Start the engine and check for leaks.

Rocker Arms/Shafts

REMOVAL & INSTALLATION

The vehicles covered in this section are not equipped with rocker arms/shafts. The camshaft directly actuates the valves.

Intake Manifold

REMOVAL & INSTALLATION

3.0L Engine

1. Before servicing the vehicle, refer to the precautions in the beginning of this section.
2. Drain the cooling system.
3. Relieve the fuel system pressure.
4. Remove or disconnect the following:
   - Negative battery cable
   - Engine appearance cover
   - Air cleaner outlet tube
   - Throttle Position (TP) sensor connector
   - Idle Air Control (IAC) valve connector
   - Accelerator cable
   - Cruise control cable
   - Cable bracket
   - Throttle body coolant bypass hose
   - Positive Crankcase Ventilation (PCV) hose
   - Evaporative Emissions (EVAP) canister purge hose
   - Exhaust Gas Recirculation (EGR) vacuum line
   - EGR tube
   - Cowl vent screen
   - Cross vehicle support bar
   - Chassis vacuum supply hose
   - EGR pressure transducer
   - Fuel pressure sensor shield
   - Upper intake manifold vacuum hose
   - Intake Manifold Tuning Valve (IMTV) connector
   - Exhaust Vacuum Regulator (EVR) valve connector and vacuum line
   - Upper intake manifold support brackets
   - Upper intake manifold
   - Fuel line and bracket
   - Fuel pressure sensor vacuum line
   - Fuel injector harness connectors
   - PCV tube
   - Lower intake manifold assembly

To install:

5. Install or connect the following:
   - Lower intake manifold assembly. Tighten the bolts in sequence to 89 inch lbs. (10 Nm).
   - PCV tube
   - Fuel injector harness connectors
   - Fuel pressure sensor vacuum line
   - Fuel line and bracket. Tighten the bolt to 89 inch lbs. (10 Nm).
   - Upper intake manifold. Tighten the bolts in sequence to 89 inch lbs. (10 Nm).
   - Upper intake manifold support

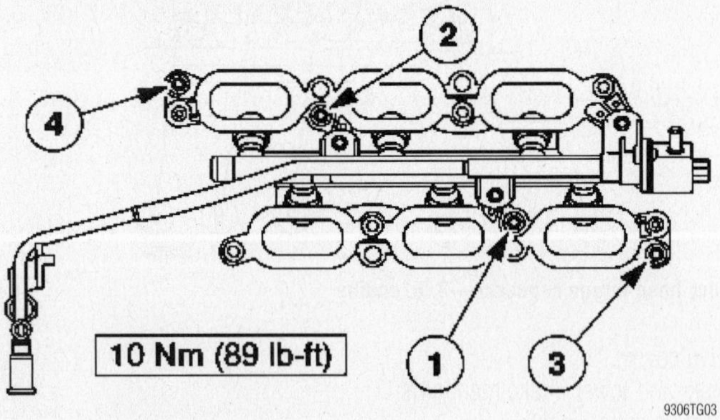

Lower intake manifold torque sequence—3.0L engine

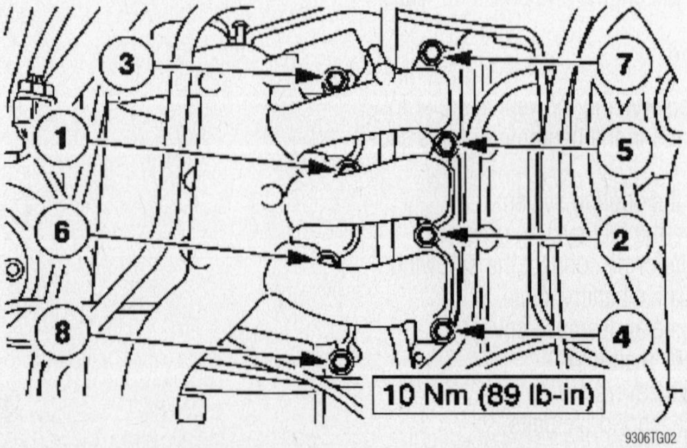

Upper intake manifold torque sequence—3.0L engine

brackets. Tighten the bolts to 89 inch lbs. (10 Nm).
- EVR valve connector and vacuum line
- IMTV connector
- Upper intake manifold vacuum hose
- Fuel pressure sensor shield
- EGR pressure transducer
- Chassis vacuum supply hose
- Cross vehicle support bar. Tighten the bolts to 15 ft. lbs. (20 Nm).
- Cowl vent screen
- EGR tube
- EGR vacuum line
- EVAP canister purge hose
- PCV hose
- Throttle body coolant bypass hose
- Cable bracket
- Cruise control cable
- Accelerator cable
- IAC valve connector
- TP sensor connector
- Air cleaner outlet tube
- Engine appearance cover
- Negative battery cable

6. Fill the cooling system.
7. Start the engine and check for leaks.

3.9L Engine

1. Before servicing the vehicle, refer to the precautions in the beginning of this section.
2. Drain the cooling system.
3. Relieve the fuel system pressure.
4. Remove or disconnect the following:
   - Negative battery cable
   - Air cleaner outlet tube
   - Cowl vent screen
   - Cross vehicle support bar
   - Accelerator cable
   - Cruise control cable
   - Intake manifold vacuum hoses
   - Exhaust Gas Recirculation (EGR) valve
   - Camshaft Position (CMP) sensor connector
   - Evaporative Emissions (EVAP) canister purge valve line
   - Fuel pressure sensor connector and vacuum line
   - Fuel line
   - Knock Sensor (KS) connector
   - Cylinder head temperature sensor connector
   - Sensor connector bracket
   - Wiring harness and hose bracket
   - Left bank fuel injector connectors
   - Idle Air Control (IAC) valve connector
   - Throttle Position (TP) sensor connector
   - Positive Crankcase Ventilation (PCV) tube
   - Throttle body coolant hoses
   - Right bank fuel injector connectors
   - Delta Pressure Feedback Electronic (DPFE) system sensor
   - Intake manifold. Loosen the bolts in reverse of the tightening sequence.

To install:

5. Install or connect the following:
   - Intake manifold. Tighten the bolts in sequence to 18 ft. lbs. (25 Nm).
   - DPFE system sensor
   - Right bank fuel injector connectors
   - Throttle body coolant hoses
   - PCV tube
   - TP sensor connector
   - IAC valve connector
   - Left bank fuel injector connectors
   - Wiring harness and hose bracket
   - Sensor connector bracket
   - Cylinder head temperature sensor connector
   - KS sensor connector
   - Fuel line
   - Fuel pressure sensor connector and vacuum line
   - EVAP canister purge valve line
   - CMP sensor connector
   - EGR valve
   - Intake manifold vacuum hoses
   - Cruise control cable
   - Accelerator cable
   - Cross vehicle support bar. Tighten the bolts to 15 ft. lbs. (20 Nm).
   - Cowl vent screen
   - Air cleaner outlet tube
   - Negative battery cable

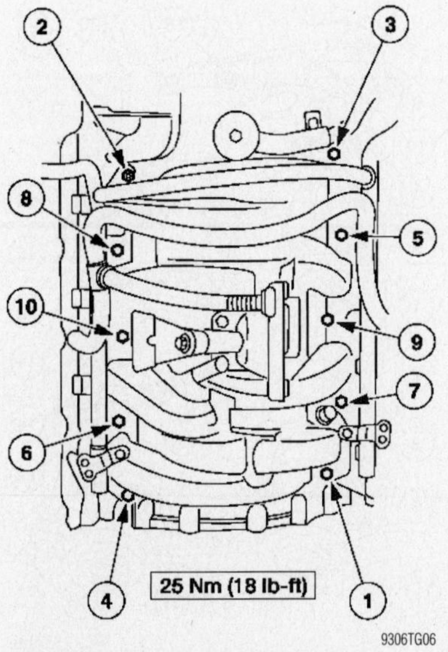

Intake manifold torque sequence—3.9L engine

6. Fill the cooling system.
7. Start the engine and check for leaks.

Exhaust Manifolds

REMOVAL & INSTALLATION

3.0L Engine

1. Before servicing the vehicle, refer to the precautions in the beginning of this section.
2. Remove or disconnect the following:
   - Negative battery cable
   - Heat shields
   - Lower splash shields
   - Exhaust front pipes
   - Secondary air tubes
   - Exhaust Gas Recirculation (EGR) tube
   - Exhaust manifolds

To install:

3. Install the exhaust manifolds. Tighten the fasteners in sequence as follows:
   a. Step 1: 15 ft. lbs. (20 Nm)
   b. Step 2: 15 ft. lbs. (20 Nm)
4. Install or connect the following:
   - Exhaust Gas Recirculation (EGR) tube
   - Secondary air tubes
   - Exhaust front pipes
   - Lower splash shields
   - Heat shields
   - Negative battery cable
5. Start the engine and check for leaks.

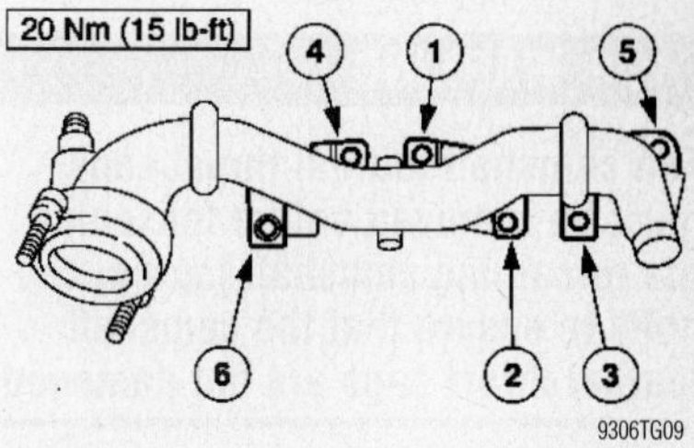

Right exhaust manifold torque sequence—3.0L engine

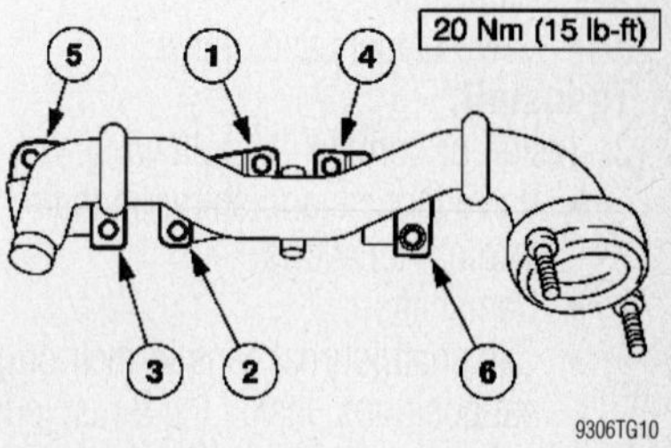

Left exhaust manifold torque sequence—3.0L engine

3.9L Engine

1. Before servicing the vehicle, refer to the precautions in the beginning of this section.
2. Remove or disconnect the following:
    - Negative battery cable
    - Power steering pump reservoir
    - Oil dipstick tube
    - Exhaust front pipes
    - Exhaust Gas Recirculation (EGR) tube
    - Exhaust manifolds

To install:

3. Install or connect the following:
    - Exhaust manifolds. Tighten the bolts to 18 ft. lbs. (25 Nm).
    - EGR tube
    - Exhaust front pipes
    - Oil dipstick tube
    - Power steering pump reservoir
    - Negative battery cable
4. Start the engine and check for leaks.

Camshaft and Valve Lifters

REMOVAL & INSTALLATION

3.0L Engine

1. Before servicing the vehicle, refer to the precautions in the beginning of this section.
2. Remove or disconnect the following:
    - Negative battery cable
    - Valve covers
    - Front cover
    - Timing chains

⁂ WARNING

The camshaft journal thrust caps must be removed before loosening the remaining camshaft journal cap bolts to ensure that the camshaft journal thrust caps are not damaged.

➡Keep all valvetrain components in order for assembly.

- Camshaft journal thrust caps
- Remaining camshaft journal caps
- Camshafts
- Valve tappets and shims

To install:

3. Install or connect the following:
    - Valve tappets and shims in their original locations
    - Camshafts
    - Camshaft journal caps in their original positions. Install the thrust journal caps last. Tighten the bolts in sequence to 89 inch lbs. (10 Nm).

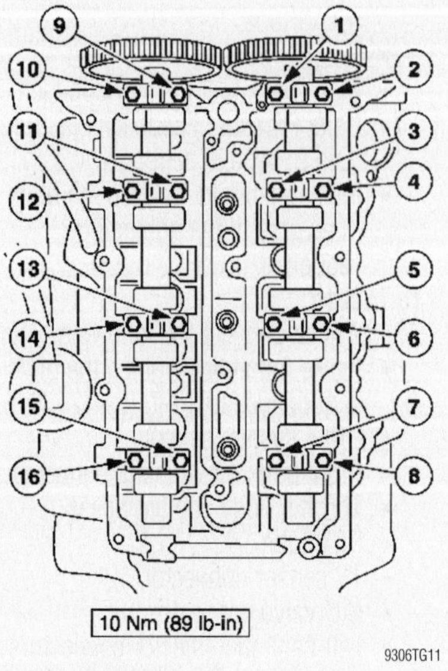

Camshaft journal cap torque sequence—3.0L engine

- Timing chains
- Front cover
- Valve covers
- Negative battery cable

4. Start the engine and check for leaks.

3.9L Engine

1. Before servicing the vehicle, refer to the precautions in the beginning of this section.
2. Remove or disconnect the following:
    - Negative battery cable
    - Valve covers
    - Front cover
    - Timing chains

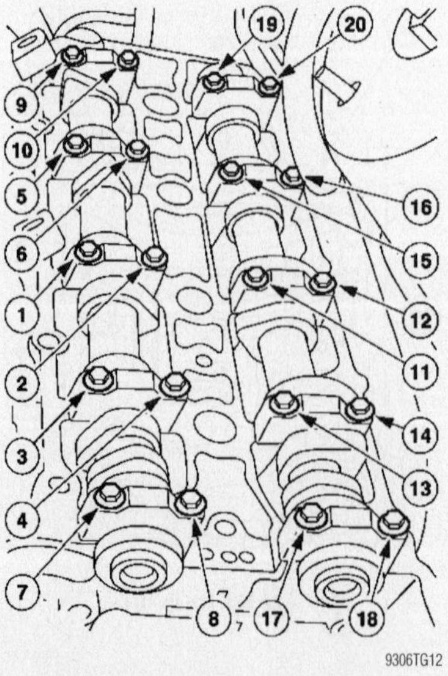

Camshaft journal bearing cap torque sequence—3.9L engine

➡Keep all valvetrain components in order for assembly.

- Camshaft journal bearing caps
- Camshafts
- Valve tappets and shims

To install:

3. Install or connect the following:
    - Valve tappets and shims
    - Camshafts
4. Install the camshaft journal bearing caps in their original positions. Tighten the bolts in sequence as follows:
    a. Step 1: Finger tight
    b. Step 2: 53 inch lbs. (6 Nm)
    c. Step 3: Plus 90 degrees
5. Install or connect the following:
    - Timing chains
    - Front cover
    - Valve covers
    - Negative battery cable
6. Start the engine and check for leaks.

Valve Lash

ADJUSTMENT

3.0L Engine

1. Before servicing the vehicle, refer to the precautions in the beginning of this section.
2. Remove or disconnect the following:
    - Negative battery cable
    - Engine appearance covers
    - Ignition coils
    - Valve covers
3. Measure the valve clearance while the camshaft lobe is pointed away from the valve shim. Rotate the crankshaft as necessary for each valve to be measured.

➡Keep all valvetrain components in order for assembly.

4. Remove the camshaft thrust cap and the rear camshaft bearing journal cap from the camshaft that requires shim adjustment.
5. Install Service Tool set 303-659 in place of the bearing journal caps, with the taller tool in place of the rear bearing cap.
6. Remove the center camshaft bearing journal caps.
7. Remove the valve shims with compressed air and replace as required to achieve the correct adjustment.
8. Valve clearance should be 0.007–0.009 in. (0.18–0.23mm) for intake valves or 0.013–0.015 in. (0.33–0.38mm) for exhaust valves.
9. Install the center bearing journal caps.
10. Remove the service tools.

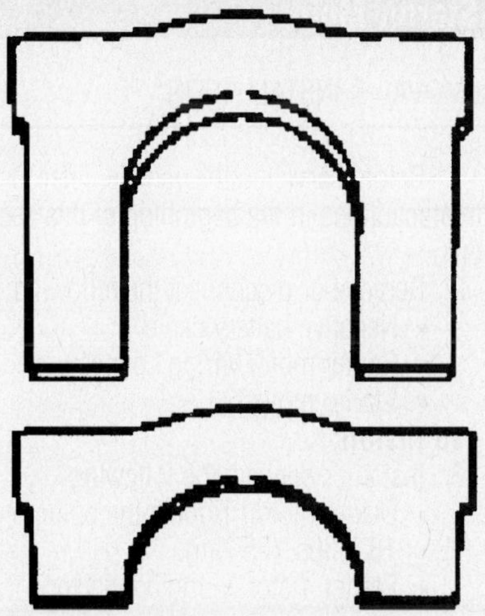

9306TG13

Camshaft lift tools 303-659 needed to adjust the valve lash on 3.0L engines

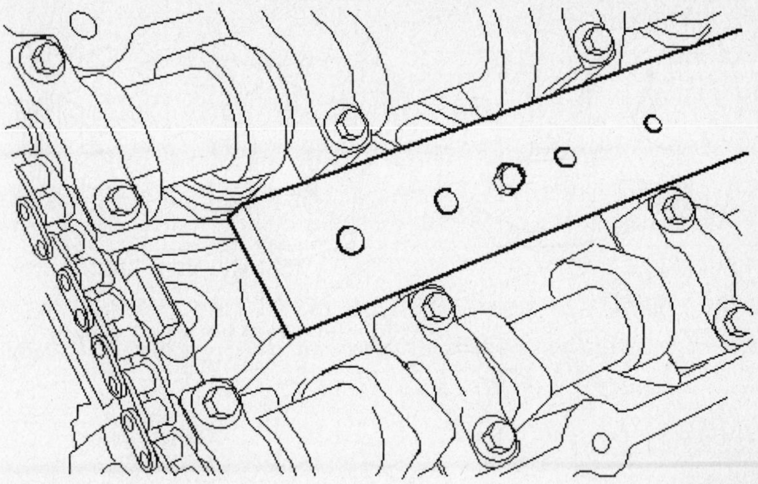

9301JG15

Valve adjustment tool base plate—3.9L engine

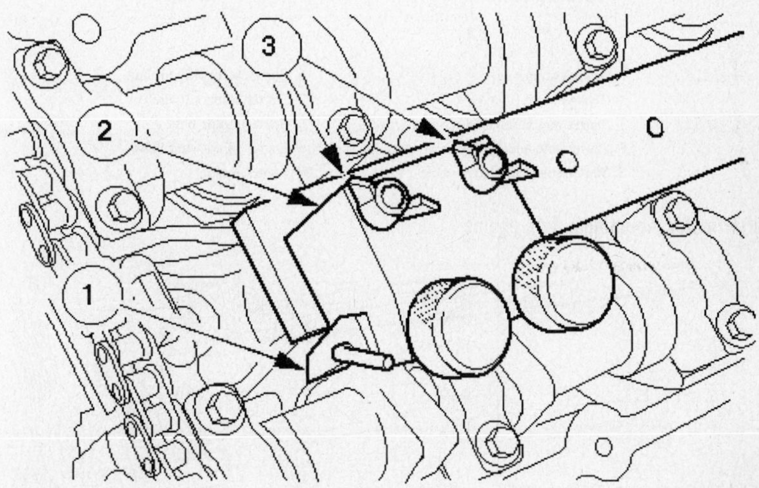

9301JG16

Valve adjustment tool attachment—3.9L engine

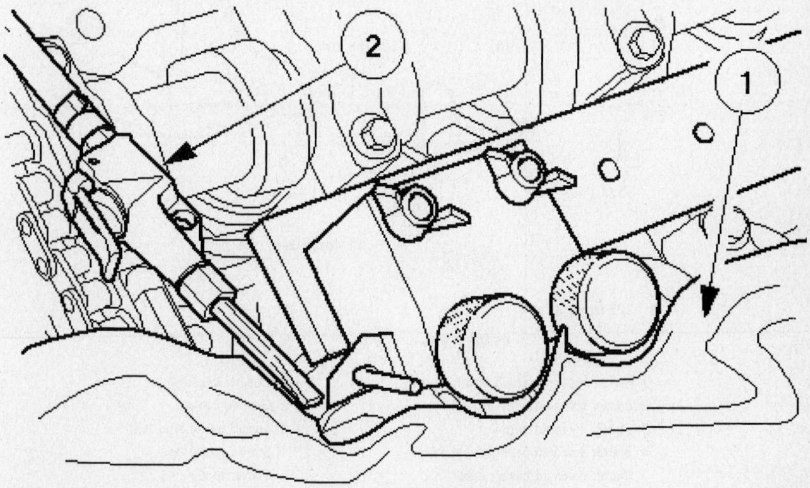

9301JG17

Remove the shims with compressed air—3.9L engine

11. Install or connect the following:
    - Rear bearing journal cap and the thrust journal cap. Tighten the bearing journal caps in sequence to 89 inch lbs. (10 Nm).
    - Valve covers
    - Ignition coils
    - Engine appearance covers
    - Negative battery cable

3.9L Engine

1. Before servicing the vehicle, refer to the precautions in the beginning of this section.
2. Remove or disconnect the following:
    - Negative battery cable
    - Engine appearance covers
    - Ignition coils
    - Valve covers
3. Measure the valve clearance while the camshaft lobe is pointed away from the valve shim. Rotate the crankshaft as necessary for each valve to be measured.
4. Valve clearance should be 0.007–0.009 in. (0.18–0.23mm) for intake valves or 0.009–0.011 in. (0.23–0.28mm) for exhaust valves.
5. If adjustment is necessary, compress the valves with the special tools and remove the shim with compressed air. Repeat for each valve to be adjusted.
6. Install or connect the following:
    - Valve covers
    - Ignition coils
    - Engine appearance covers
    - Negative battery cable

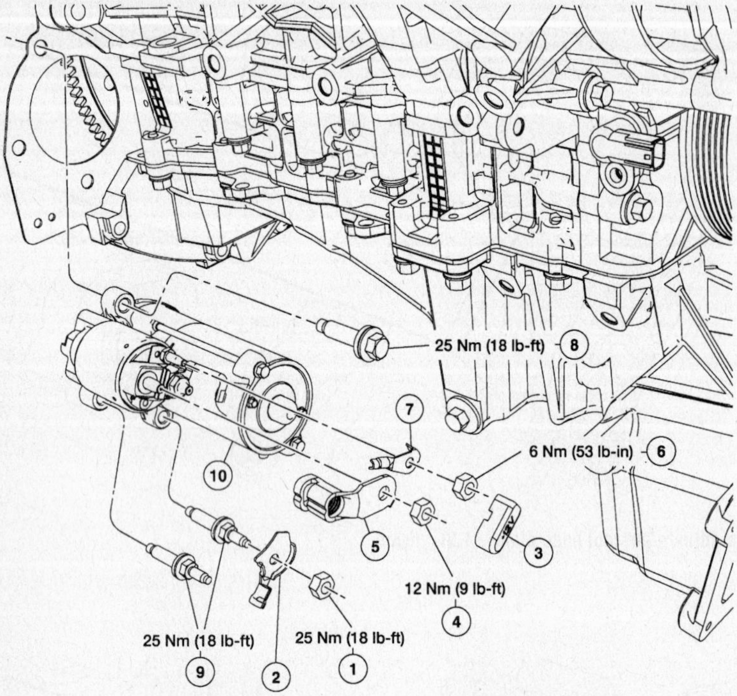

1 Ground strap nut	6 Starter solenoid wire nut
2 Ground strap	7 Starter solenoid wire
3 Starter motor solenoid terminal cover	8 Starter motor bolt
4 Starter solenoid battery cable nut	9 Starter motor stud bolts
5 Starter solenoid battery cable	10 Starter motor

67197-LILS-G07

Starter motor mounting—3.0L engine

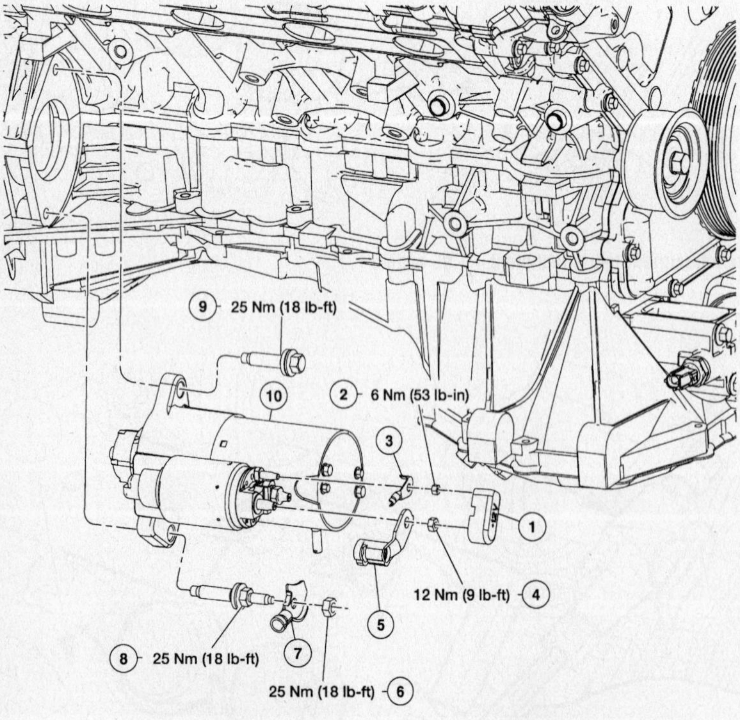

1 Starter motor solenoid terminal cover	6 Ground cable nut
2 Starter solenoid wire nut	7 Ground cable
3 Starter solenoid wire	8 Starter motor stud bolt
4 Starter solenoid battery cable nut	9 Starter motor bolt
5 Starter solenoid battery cable	10 Starter motor

67197-LILS-G08

Starter motor mounting—3.9L engine

Starter Motor

REMOVAL & INSTALLATION

1. Before servicing the vehicle, refer to the precautions in the beginning of this section.
2. Remove or disconnect the following:
   - Negative battery cable
   - Starter motor wiring connectors
   - Starter motor

To install:

3. Install or connect the following:
   - Starter motor. Tighten the bolts to 18 ft. lbs. (25 Nm).
   - Starter motor wiring connectors
   - Negative battery cable

Oil Pan

REMOVAL & INSTALLATION

3.0L Engine

1. Before servicing the vehicle, refer to the precautions in the beginning of this section.
2. Drain the engine oil.
3. Install a support fixture to the engine lifting eyes.
4. Remove or disconnect the following:
   - Negative battery cable
   - Air cleaner outlet tube
   - Accessory drive belt
   - Alternator
   - Center and right splash shields
   - A/C compressor
   - Electronic Thermactor Air (ETA) bracket, if equipped
   - Rack and pinion steering gear
   - Lower control arm through bolts
   - Transmission cooler line bracket
   - Engine mount nuts
   - Left and right subframe bolts. Pry the subframe down for clearance.
   - Wiring harness bracket
   - Oil pan

To install:

5. Install or connect the following:
   - Oil pan. Tighten the pan bolts in sequence to 18 ft. lbs. (25 Nm) and the transmission bolts to 35 ft. lbs. (47 Nm).
   - Wiring harness bracket
   - Left and right subframe bolts. Tighten the bolts to 76 ft. lbs. (103 Nm).
   - Engine mount nuts. Tighten the nuts to 46 ft. lbs. (63 Nm).
   - Transmission cooler line bracket

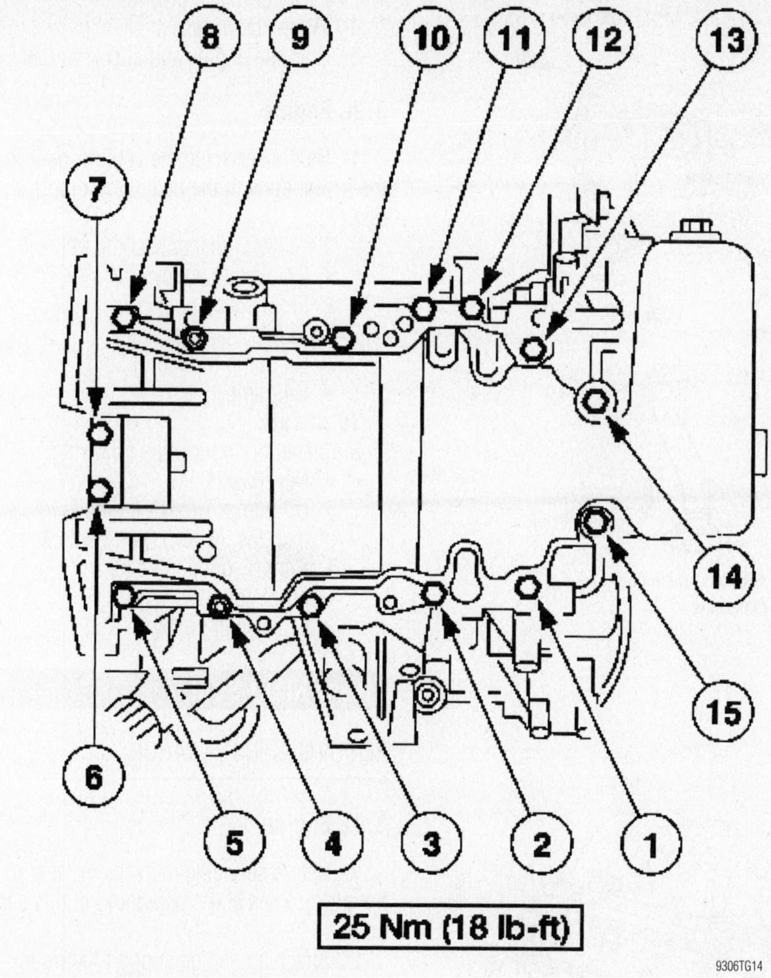

Oil pan torque sequence—3.0L engine

- Lower control arm through bolts. Tighten the bolts to 129 ft. lbs. (175 Nm).
- Rack and pinion steering gear. Tighten the nuts to 76 ft. lbs. (103 Nm).
- ETA bracket, if equipped
- A/C compressor
- Center and right splash shields
- Alternator
- Accessory drive belt
- Air cleaner outlet tube
- Negative battery cable

6. Fill the crankcase to the correct level.
7. Start the engine and check for leaks.

3.9L Engine

1. Before servicing the vehicle, refer to the precautions in the beginning of this section.
2. Drain the engine oil.
3. Remove the oil pan.

To install:

4. Install the oil pan. Tighten the bolts in sequence as follows:
   a. Step 1: 44 inch lbs. (5 Nm)
   b. Step 2: 108 inch lbs. (12 Nm)
5. Fill the crankcase to the correct level.
6. Start the engine and check for leaks.

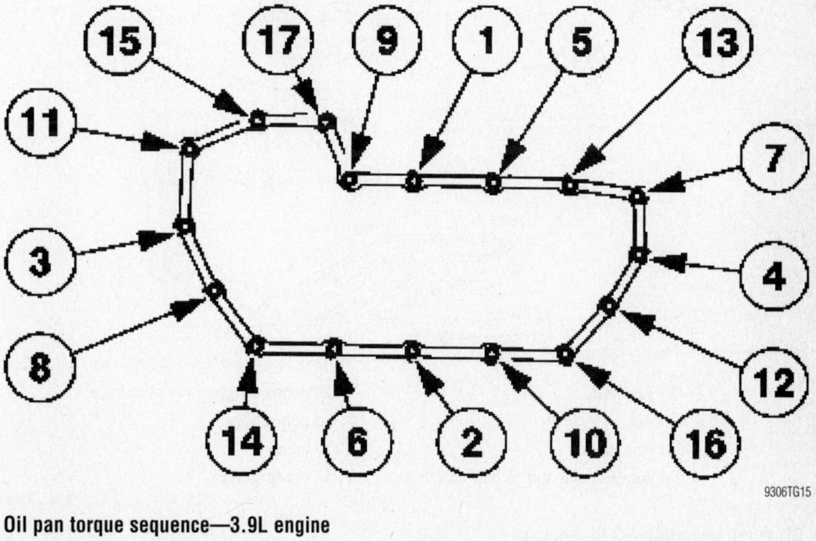

Oil pan torque sequence—3.9L engine

Oil Pump

REMOVAL & INSTALLATION

3.0L Engine

1. Before servicing the vehicle, refer to the precautions at the beginning of this section.
2. Remove or disconnect the following:
   - Upper intake manifold
   - Valve covers
   - Accessory drive belt
   - Power steering pump
   - Alternator
   - Water pump
   - A/C compressor and bracket
   - Crankshaft pulley
   - Oil pan
   - Oil pump screen and tube
   - Front cover
   - Timing chains
   - Crankshaft timing gears
   - Oil pump

To install:

3. Install or connect the following:
   - Oil pump. Tighten the bolts to 89 inch lbs. (10 Nm).
   - Crankshaft timing gears
   - Timing chains
   - Front cover
   - Oil pump screen and tube. Tighten the bolts to 89 inch lbs. (10 Nm).
   - Oil pan
   - Crankshaft pulley
   - A/C compressor and bracket
   - Water pump
   - Alternator
   - Power steering pump
   - Accessory drive belt
   - Valve covers

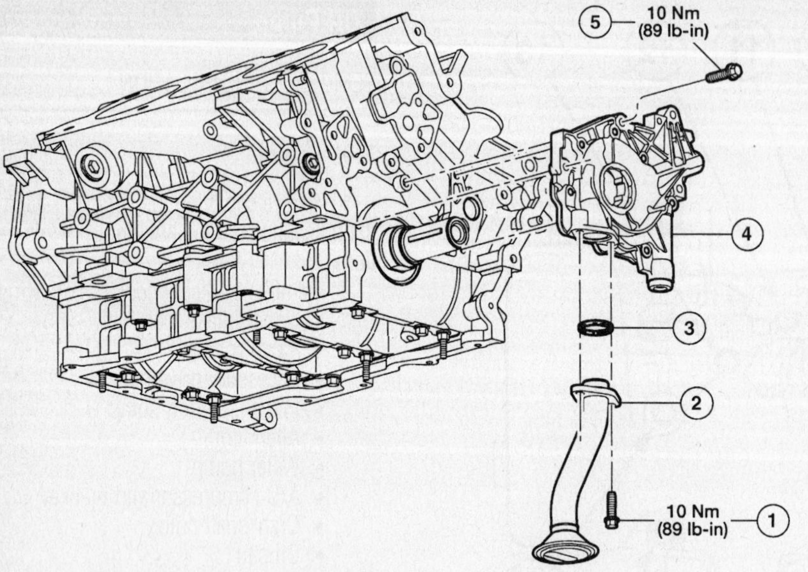

1 Screen and pick-up tube mounting bolts
2 Screen and pick-up tube
3 Screen and pick-up tube O-ring seal
4 Oil pump mounting bolts
5 Oil pump

67197-LILS-G09

Exploded view of the oil pump assembly—3.0L engine

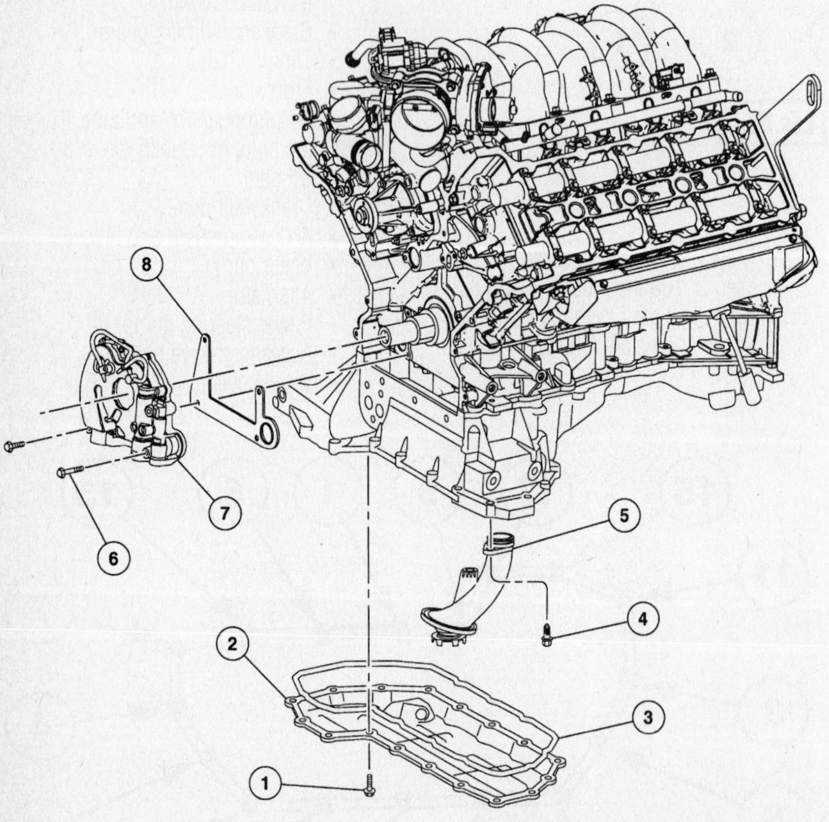

1 Oil pan bolts
2 Oil pan
3 Oil pan gasket
4 Oil pump screen and pickup tube bolts
5 Oil pump screen and pickup tube
6 Oil pump bolts
7 Oil pump
8 Oil pump gasket

67197-LILS-G10

Oil pump mounting—3.9L engine

- Upper intake manifold

4. Fill the crankcase.
5. Start the engine and check for leaks.

3.9L Engine

1. Before servicing the vehicle, refer to the precautions in the beginning of this section.
2. Remove or disconnect the following:
   - Crankshaft pulley
   - Front cover
   - Primary timing chains
   - Oil pump mounting bolts
   - Oil pump

To install:

3. Install or connect the following:
   - New gasket
   - Oil pump. Tighten the bolts to 53 inch lbs. (6 Nm) plus 90 degrees.
   - Primary timing chains
   - Front cover
   - Crankshaft pulley

Rear Main Seal

REMOVAL & INSTALLATION

3.0L Engine

1. Before servicing the vehicle, refer to the precautions at the beginning of this section.
2. Attach an engine support fixture to the engine lifting eyes.
3. Remove or disconnect the following:
   - Negative battery cable
   - Transmission
   - Clutch, if equipped
   - Flywheel
   - Rear crankshaft seal

To install:

4. Install or connect the following:
   - Rear main seal flush with the cylinder block surface
   - Flywheel. Tighten the bolts to 59 ft. lbs. (80 Nm).
   - Clutch, if equipped
   - Transmission
   - Negative battery cable
5. Start the engine and check for leaks.

3.9L Engine

1. Before servicing the vehicle, refer to the precautions at the beginning of this section.
2. Attach an engine support fixture to the engine lifting eyes.
3. Remove or disconnect the following:
   - Negative battery cable
   - Transmission

- Flywheel
- Rear crankshaft seal

To install:

4. Install or connect the following:
   - Rear main seal flush with the cylinder block surface
   - Flywheel
5. Tighten the flywheel bolts in a crossing pattern as follows:
   a. Step 1: 11 ft. lbs. (15 Nm)
   b. Step 2: 81 ft. lbs. (110 Nm)
6. Install or connect the following:
   - Transmission
   - Negative battery cable
7. Start the engine and check for leaks.

Timing Chain, Sprockets, Front Cover and Seal

REMOVAL & INSTALLATION

3.0L Engine

1. Before servicing the vehicle, refer to the precautions at the beginning of this section.
2. Remove or disconnect the following:
   - Upper intake manifold
   - Valve covers
   - Accessory drive belt
   - Power steering pump
   - Alternator
   - Water pump
   - A/C compressor and bracket
   - Crankshaft pulley
   - Oil pan
   - Oil pump screen and tube
   - Crankshaft Position (CKP) sensor connector
   - Camshaft Position (CMP) sensor connector
   - Front cover
   - CKP sensor pulse ring
3. Rotate the crankshaft so that the keyway is at the 11 o'clock position to locate the crankshaft at Top Dead Center (TDC) for No. 1 cylinder.
4. Verify that the alignment arrows on the camshafts are aligned. If not, rotate the crankshaft 1 complete revolution and recheck.
5. Rotate the crankshaft so that the keyway is at the 3 o'clock position. This positions the right cylinder head camshafts to the neutral position.
6. Remove or disconnect the following:
   - Right timing chain tensioner
   - Right timing chain tensioner arm
   - Right timing chain and crankshaft sprocket

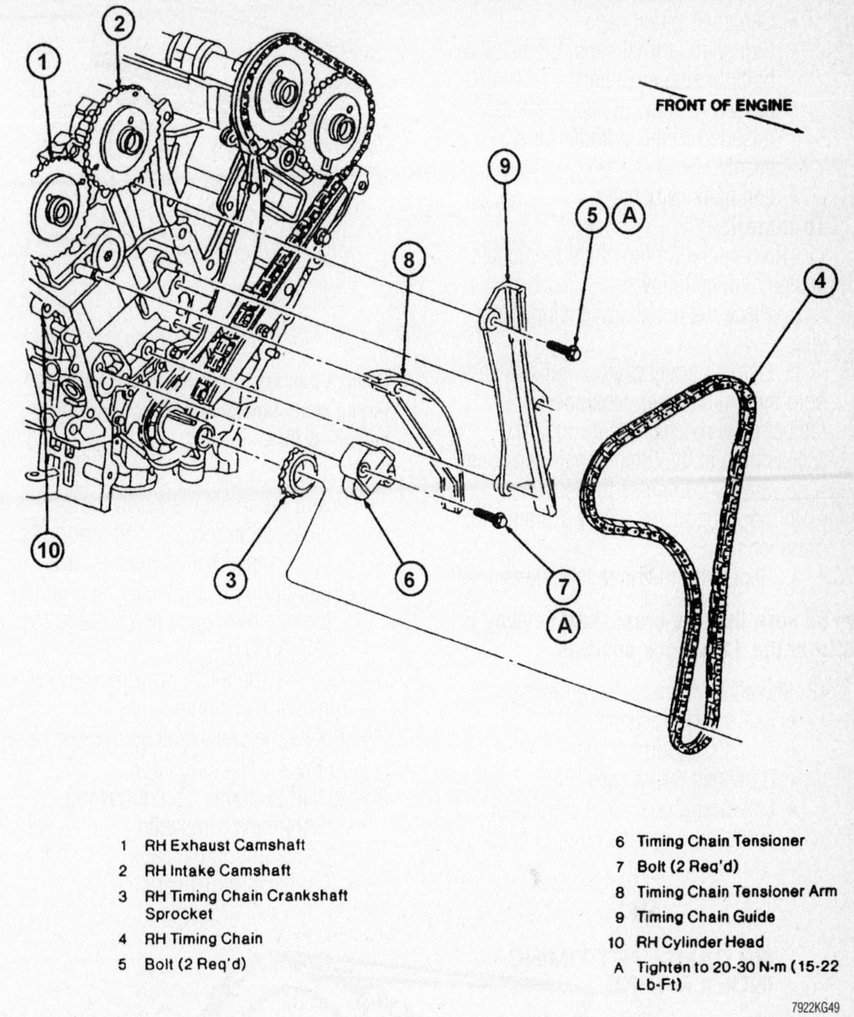

Exploded view of the right cylinder head timing chain and related components—3.0L engine—left side similar

WARNING

The camshaft thrust caps must be removed before loosening the remaining camshaft journal cap bolts to ensure that the thrust caps are not damaged.

➡The camshaft journal caps and cylinder heads are numbered to ensure that they are assembled in their original positions.

- Camshaft thrust caps
- Camshaft journal caps. Loosen the bolts in sequence and in several passes to allow the camshaft to be raised from the cylinder head evenly.
- Right bank camshafts

7. Rotate the crankshaft 2 revolutions and locate the crankshaft keyway at the 11 o'clock position. This will position the left cylinder head camshafts to their neutral position.
8. Verify that the alignment arrows on the camshafts are aligned.
9. Remove or disconnect the following:
   - Left cylinder head timing chain tensioner
   - Left timing chain tensioner arm
   - Left timing chain and crankshaft sprocket

WARNING

The camshaft thrust caps must be removed before loosening the remaining camshaft journal cap bolts to ensure that the thrust caps are not damaged.

➡The camshaft journal caps and cylinder heads are numbered to ensure that they are assembled in their original positions.

10. Remove or disconnect the following:
   - Camshaft thrust caps
   - Camshaft journal caps. Loosen the bolts in sequence and in several passes to allow the camshaft to be raised from the cylinder head evenly.
   - Left bank camshafts

To install:

11. Prepare the timing chain tensioners for installation as follows:
   a. Place the left chain tensioner in a vise.
   b. Using a small prytool, release and hold the timing chain tensioner ratchet/pawl mechanism through the access hole in the timing chain tensioner.
   c. Slowly compress the tensioner.
   d. Lock the piston with a 1.5mm wire or paperclip.
   e. Repeat for the right chain tensioner.

➡Be sure that the crankshaft keyway is still at the 11 o'clock position.

12. Install or connect the following:
   - Left bank camshafts
   - Camshaft journal caps
   - Camshaft thrust caps
   - Left timing chain and crankshaft sprocket. Align the colored links with the index marks on the camshaft and crankshaft sprockets.
   - Left timing chain tensioner arm
   - Left timing chain tensioner. Tighten the retaining bolts to 15–22 ft. lbs. (20–30 Nm).
13. Remove the retaining wire from the left timing chain tensioner.
14. Rotate the crankshaft so that the keyway is in the 3 o'clock position.
15. Install or connect the following:
   - Right bank camshafts
   - Camshaft journal caps
   - Camshaft thrust caps
   - Right timing chain and crankshaft sprocket. Align the colored links with the index marks on the camshaft and crankshaft sprockets.
   - Right timing chain tensioner arm
   - Right timing chain tensioner. Tighten the retaining bolts to 15–22 ft. lbs. (20–30 Nm).
16. Remove the retaining wire from the right timing chain tensioner.
17. Install the CKP sensor pulse ring.
18. Replace the crankshaft seal in the front cover with a new one. Apply clean engine oil to the seal lip.

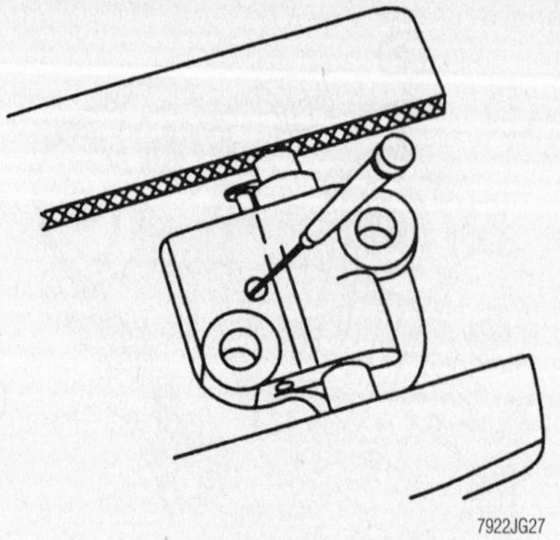

Using a thin prytool, release and hold the timing chain tensioner ratchet/pawl mechanism—3.0L engine

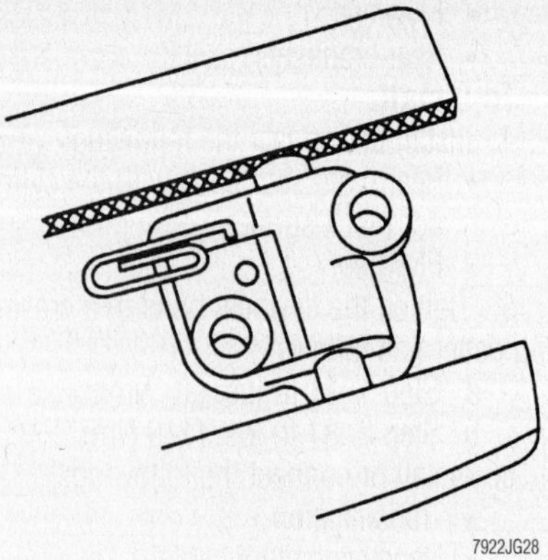

Retain the piston with a 1.5mm wire or paperclip—3.0L engine

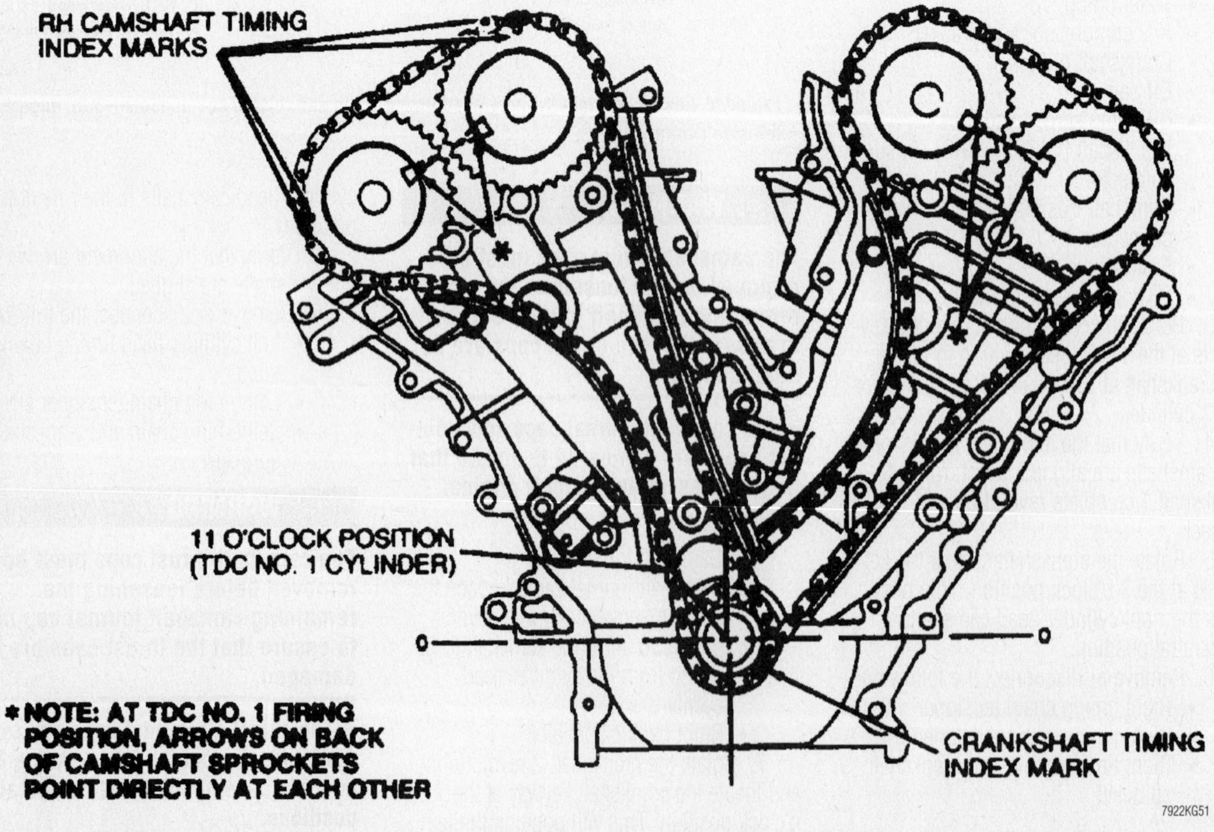

Timing mark alignment—3.0L engine

19. Apply silicone sealer to the 6 critical areas shown in View **A**, to the cylinder block.

20. Place new front cover gaskets onto the dowel pins on the cylinder block and heads.

21. Place the front cover into position.

22. Install the 6 front cover retaining bolts and stud bolts where the silicone sealer was applied.

23. Tighten the bolts and stud bolts until the front cover contacts the cylinder block and heads and, then turn the bolts and stud bolts an additional ¼ turn.

24. Install the remaining front cover retaining bolts and stud bolts.

25. Tighten all of the front cover retaining bolts and stud bolts in sequence to 15–22 ft. lbs. (20–30 Nm).

26. Install or connect the following:

- CMP sensor connector
- CKP sensor connector
- Oil pump screen and tube
- Oil pan

27. Install the crankshaft damper and tighten the bolt as follows:

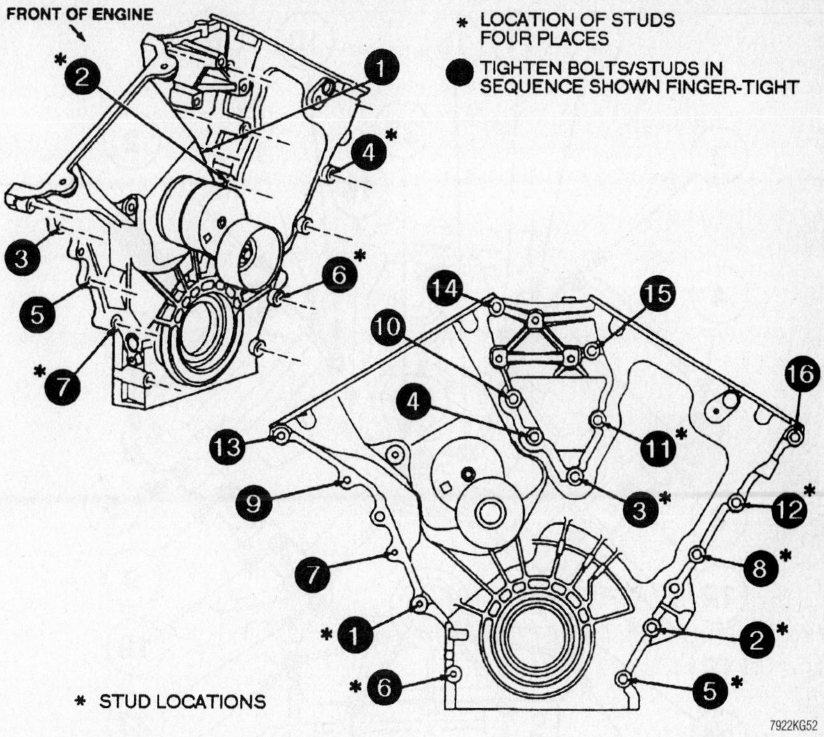

Front cover torque sequence—3.0L engine

a. Step 1: 78–99 ft. lbs. (105–135 Nm).

b. Step 2: Loosen one full turn.

c. Step 3: 35–39 ft. lbs. (47–53 Nm).

d. Step 4: Plus an 85–95 degrees.

28. Install or connect the following:

- A/C compressor and bracket
- Water pump
- Alternator
- Power steering pump
- Accessory drive belt
- Valve covers
- Upper intake manifold

29. Start the engine and check for leaks.

3.9L Engine

1. Before servicing the vehicle, refer to the precautions in the beginning of this section.

2. Drain the cooling system.

3. Remove or disconnect the following:

- Negative battery cable
- Engine appearance cover and brackets
- Valve covers
- Engine cooling fan assembly
- Accessory drive belt
- Water pump pulley
- Alternator
- Lower radiator hose and pipe
- Heater hose
- Idler pulleys
- Crankshaft pulley

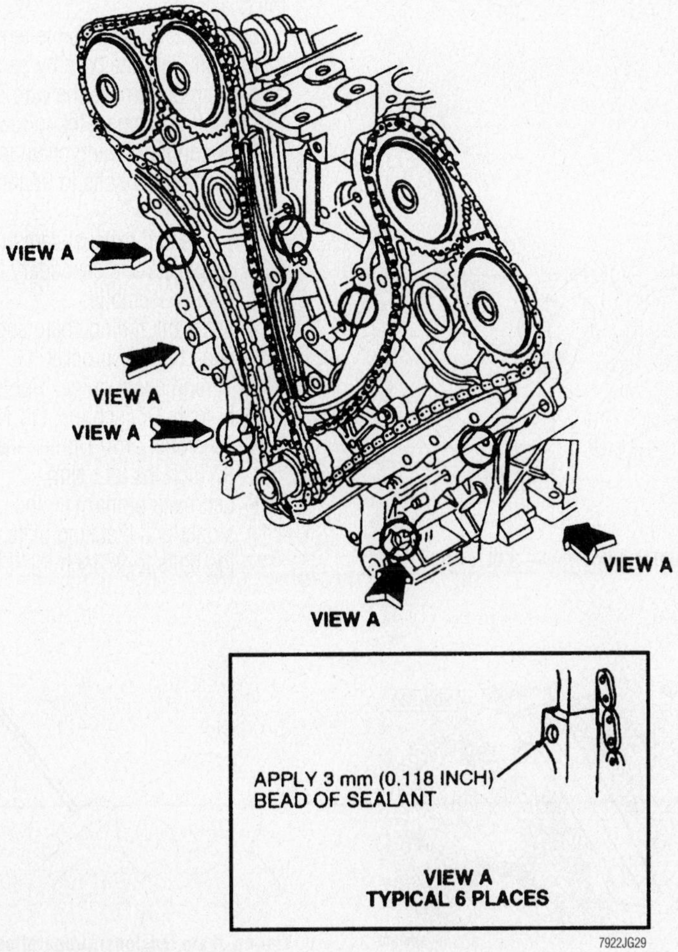

Apply sealant to the places indicated—3.0L engine

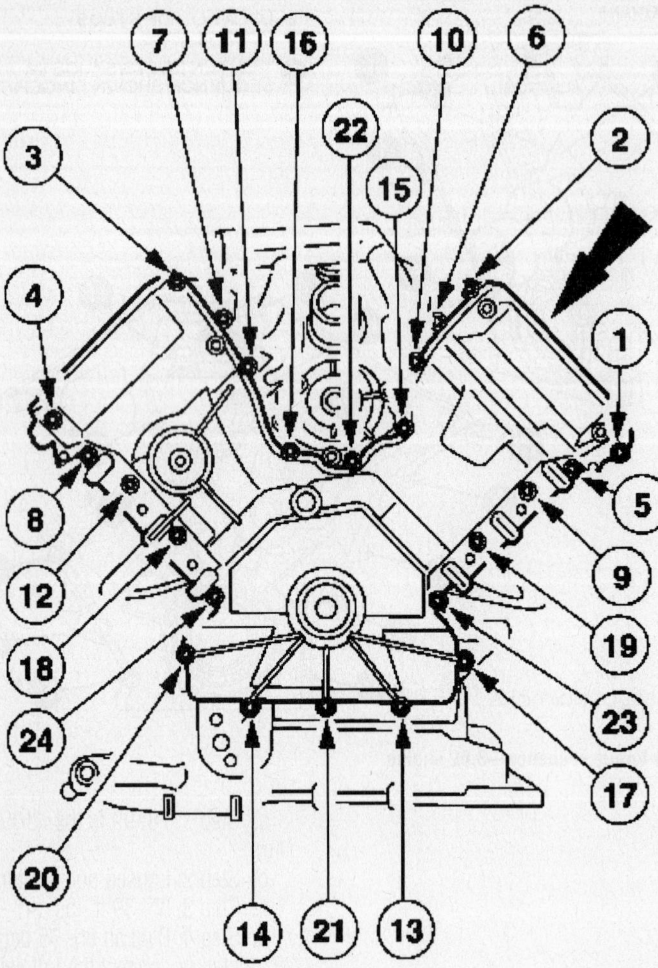

Front cover loosening sequence—3.9L engine

- Front crankshaft seal
- Power steering reservoir hose
- Power steering pump and bracket
- Hydraulic cooling fan reservoir hose and bracket
- Hydraulic cooling fan pump and bracket
- Front cover wiring harness clips
- Front cover. Loosen the bolts in sequence.
- Torque converter access panel
- Crankshaft Position (CKP) sensor

4. Rotate the crankshaft to 45 degrees After Top Dead Center (ATDC). The crankshaft keyway will be in the 6 o'clock position. Check that the camshaft lobes are facing upwards. If not, rotate the crankshaft 1 full turn.

5. Install Crankshaft Holding Tool 303-645 in place of the CKP sensor.

6. Install Camshaft Locking Tool 303-530 to the right bank camshafts.

7. Loosen the exhaust and intake camshaft sprocket bolts and slide the sprockets forward on the bolts.

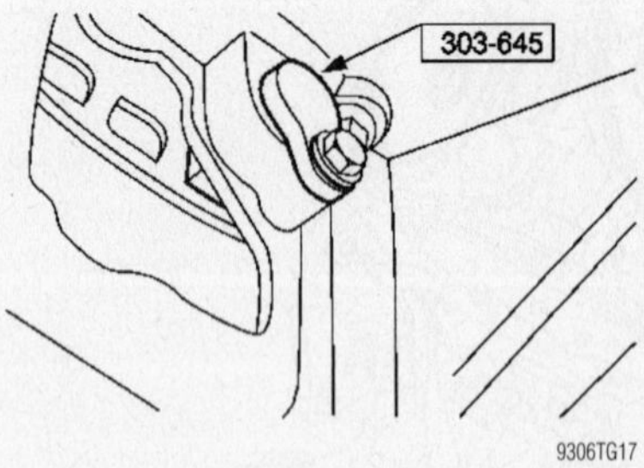

Crankshaft Holding Tool—3.9L engine

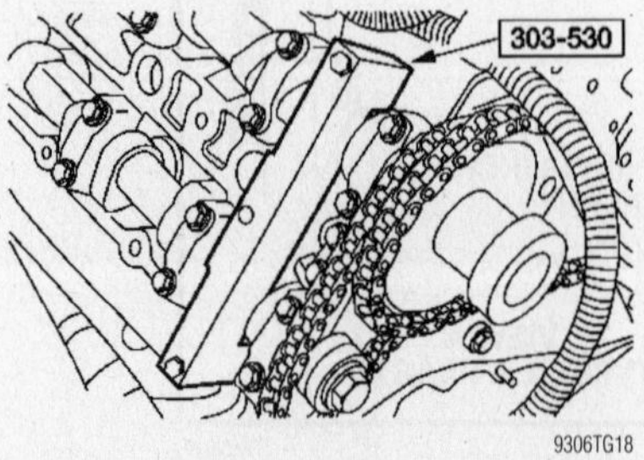

Camshaft Locking Tool—3.9L engine

8. Remove or disconnect the following:
   - Right bank primary timing chain tensioner and blanking plate
   - Tensioner arm
   - Timing chain guide
   - Right bank timing chain and crankshaft timing sprocket

9. Remove the locking tool from the right bank camshafts and install it on the left bank camshafts.

10. Remove or disconnect the following:
    - Left bank primary timing chain tensioner and blanking plate
    - Tensioner arm
    - Timing chain guide
    - Left bank timing chain and crankshaft timing sprocket
    - Intake and exhaust camshaft sprocket bolts
    - Intake and exhaust camshaft sprockets and secondary timing chain assemblies
    - Secondary timing chain tensioners

To install:

11. Prepare the primary and secondary timing chain tensioners for installation as follows:

    a. Step 1: Use a thin wire to unseat the check valve

    b. Step 2: Compress the tensioner piston fully into the bore by hand

    c. Step 3: Remove the wire

12. Install or connect the following:
    - Secondary timing chain tensioners. Tighten the bolts to 97 inch lbs. (11 Nm).
    - Intake and exhaust camshaft sprockets and secondary timing chain assemblies
    - Left bank timing chain and crankshaft timing sprocket
    - Timing chain guide. Tighten the bolts to 97 inch lbs. (11 Nm).
    - Tensioner arm. Tighten the bolts to 97 inch lbs. (11 Nm).
    - Left bank primary timing chain tensioner and blanking plate. Tighten the bolts to 97 inch lbs. (11 Nm).

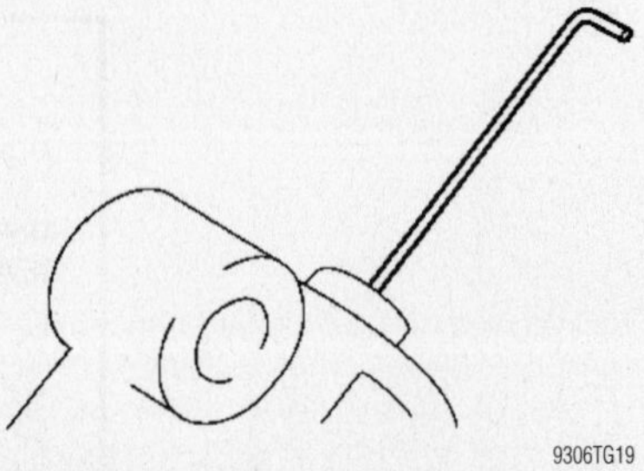

Timing chain tensioner preparation—3.9L engine

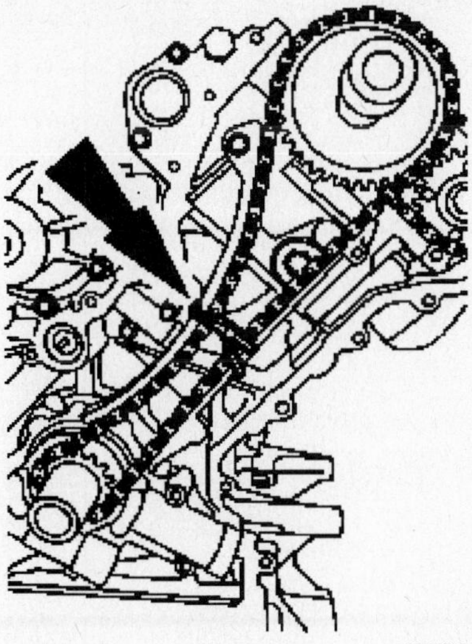

Use a tie strap to remove slack from the primary timing chain—3.9L engine

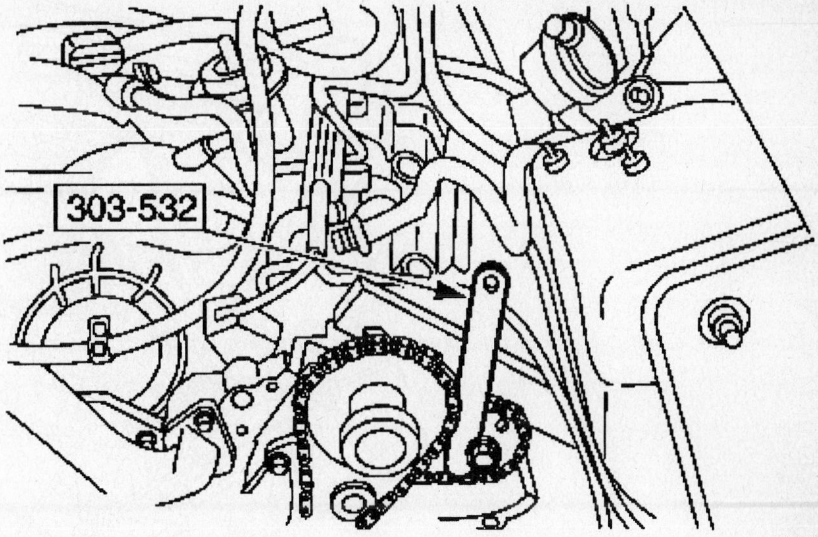

Timing Chain Tensioning Tool—3.9L engine

13. Install a tie strap to take up the slack in the timing chain.

14. Use Timing Chain Tensioning Tool 303-532 to apply tension to the left bank exhaust camshaft.

➡Tighten the exhaust camshaft bolt first.

15. Tighten the left bank camshaft sprocket bolts as follows:
   a. Step 1: 15 ft. lbs. (20 Nm)
   b. Step 2: Plus 90 degrees

16. Remove the locking tool from the left bank camshafts and install it on the right bank camshafts.

17. Install or connect the following:
   - Right bank timing chain and crankshaft timing sprocket. Ensure that the crankshaft sprocket is installed as shown.
   - Timing chain guide. Tighten the bolts to 97 inch lbs. (11 Nm).
   - Tensioner arm. Tighten the bolts to 97 inch lbs. (11 Nm).
   - Right bank primary timing chain tensioner and blanking plate. Tighten the bolts to 97 inch lbs. (11 Nm).

18. Install a tie strap to take up the slack in the timing chain.

19. Use Timing Chain Tensioning Tool 303-532 to apply tension to the right bank exhaust camshaft.

➡Tighten the exhaust camshaft bolt first.

20. Tighten the right bank camshaft sprocket bolts as follows:
   a. Step 1: 15 ft. lbs. (20 Nm)

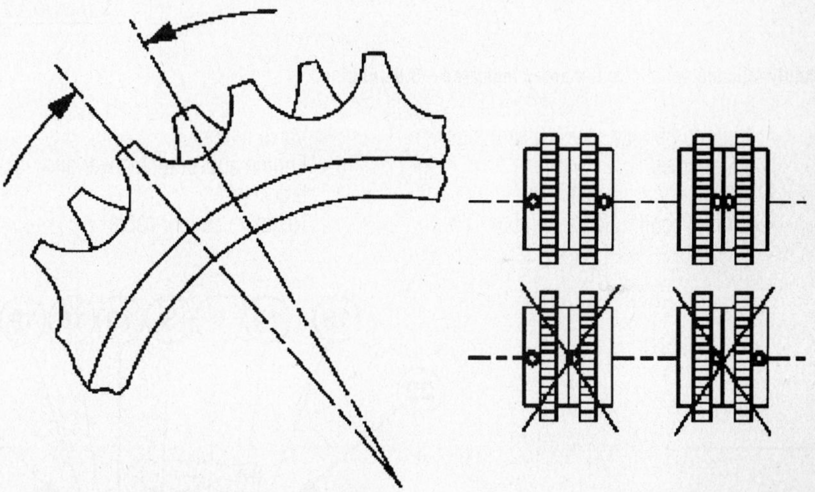

Crankshaft sprocket alignment—3.9L engine

   b. Step 2: Plus 90 degrees

21. Remove the camshaft locking tool.

22. Apply silicone sealant to the areas indicated and install the front cover.

23. Tighten the front cover bolts in sequence as follows:
   a. Step 1: 44 inch lbs. (5 Nm)
   b. Step 2: 89 inch lbs. (10 Nm)

24. Remove the crankshaft locking tool.

25. Install or connect the following:
   - CKP sensor
   - Torque converter access panel
   - Front cover wiring harness clips
   - Hydraulic cooling fan pump and bracket. Tighten the bolts to 18 ft. lbs. (25 Nm).
   - Hydraulic cooling fan reservoir hose and bracket
   - Power steering pump and bracket. Tighten the bolts to 18 ft. lbs. (25 Nm).
   - Power steering reservoir hose
   - Front crankshaft seal

26. Install the crankshaft pulley and tighten the bolt as follows:
   a. Step 1: 59 ft. lbs. (80 Nm)
   b. Step 2: Loosen the bolt 2 complete turns
   c. Step 3: 37 ft. lbs. (50 Nm)
   d. Step 4: Plus 90 degrees

27. Install or connect the following:
   - Idler pulleys. Tighten the bolts to 18 ft. lbs. (25 Nm).
   - Heater hose
   - Lower radiator hose and pipe
   - Alternator
   - Water pump pulley. Tighten the

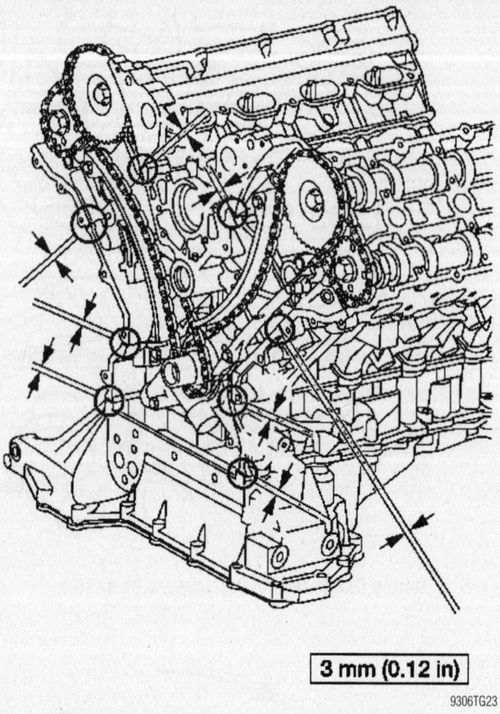

Apply silicone sealant to the areas indicated—3.9L engine

bolts to 89 inch lbs. (10 Nm) plus 45 degrees.

- Accessory drive belt
- Engine cooling fan assembly
- Valve covers
- Engine appearance cover and brackets
- Negative battery cable

28. Fill the cooling system.
29. Start the engine and check for leaks.

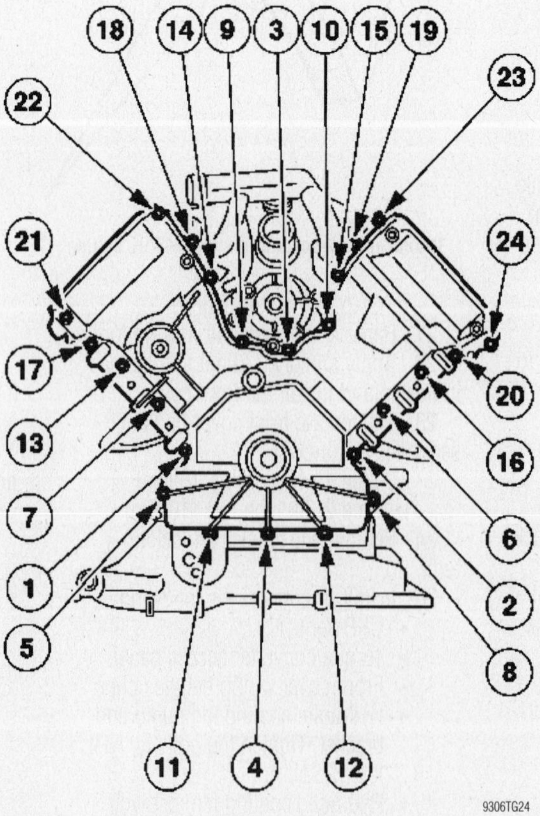

Front cover torque sequence—3.9L engine

Piston and Ring

POSITIONING

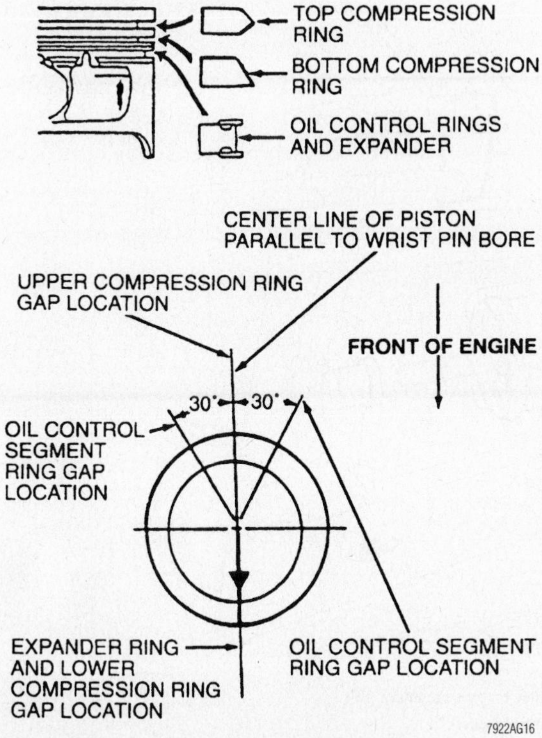

Piston ring end-gap spacing—3.0L (VIN S) engine

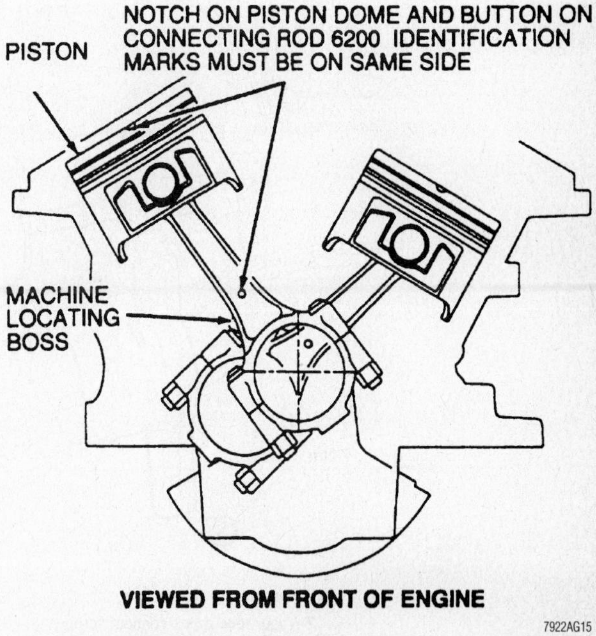

Piston ring end-gap spacing—3.0L (VIN S) engine

FUEL SYSTEM

Fuel System Service Precautions

Safety is the most important factor when performing not only fuel system maintenance but any type of maintenance. Failure to conduct maintenance and repairs in a safe manner may result in serious personal injury or death. Maintenance and testing of the vehicle's fuel system components can be accomplished safely and effectively by adhering to the following rules and guidelines.

- To avoid the possibility of fire and personal injury, always disconnect the negative battery cable unless the repair or test procedure requires that battery voltage be applied.
- Always relieve the fuel system pressure prior to disconnecting any fuel system component (injector, fuel rail, pressure regulator, etc.), fitting or fuel line connection. Exercise extreme caution whenever relieving fuel system pressure, to avoid exposing skin, face and eyes to fuel spray. Please be advised that fuel under pressure may penetrate the skin or any part of the body that it contacts.
- Always place a shop towel or cloth around the fitting or connection prior to loosening to absorb any excess fuel due to spillage. Ensure that all fuel spillage (should it occur) is quickly removed from engine surfaces. Ensure that all fuel soaked cloths or towels are deposited into a suitable waste container.
- Always keep a dry chemical (Class B) fire extinguisher near the work area.
- Do not allow fuel spray or fuel vapors to come into contact with a spark or open flame.
- Always use a back-up wrench when loosening and tightening fuel line connection fittings. This will prevent unnecessary stress and torsion to fuel line piping.
- Always replace worn fuel fitting O-rings with new. Do not substitute fuel hose or equivalent where fuel pipe is installed.

Fuel System Pressure

RELIEVING

1. Before servicing the vehicle, refer to the precautions in the beginning of this section.
2. Disconnect the negative battery cable.
3. Connect the fuel injection pressure test equipment JD 209 to the valve on the fuel supply manifold.
4. Insert the drain/bleed tube into the fuel container.
5. Follow the manufacturer's instructions and depressurize the fuel system.

Fuel Filter

REMOVAL & INSTALLATION

1. Before servicing the vehicle, refer to the precautions in the beginning of this section.
2. Relieve the fuel system pressure.
3. Remove or disconnect the following:
   - Negative battery cable
   - Fuel filter bracket cover
   - Fuel lines
   - Fuel filter

To install:

4. Install or connect the following:
   - Fuel filter into the bracket making sure the flow direction is correct. Tighten the clamp to 15–25 inch lbs. (2–3 Nm).

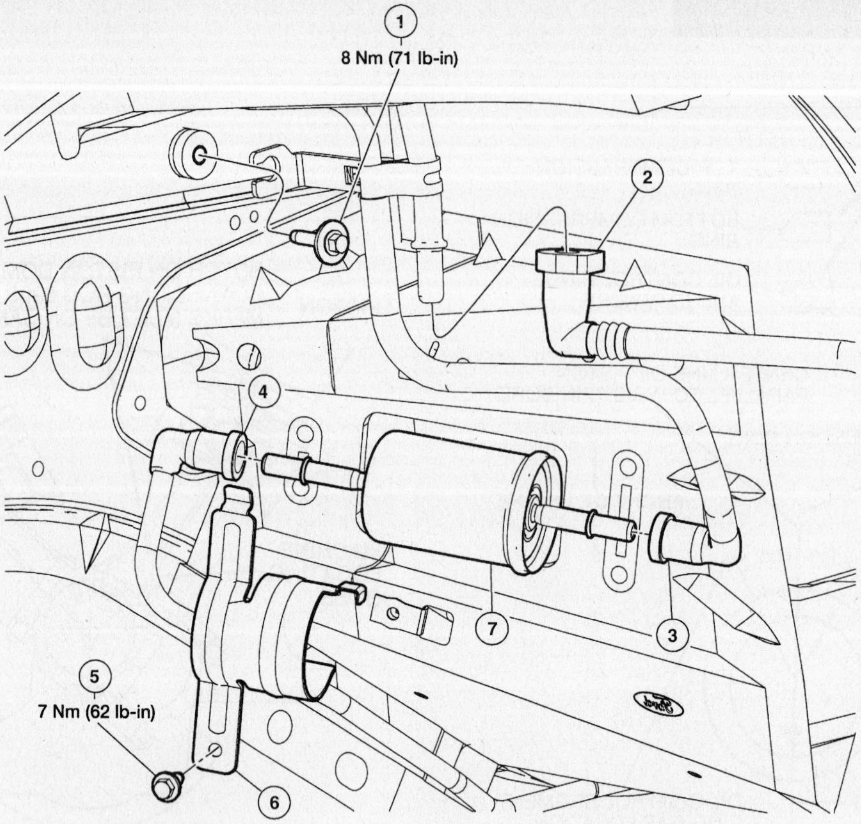

1 Fuel hose mounting bolt
2 Vapor tube quick-connect fitting
3 Fuel tube quick-connect fitting
4 Fuel tube quick-connect fitting
5 Fuel filter bracket mounting bolt
6 Fuel filter bracket
7 Fuel filter

67197-LILS-G11

Fuel filter location and mounting

- Fuel lines. Tighten the fittings to 22 ft. lbs. (30 Nm).

5. Start the engine and check for leaks.

Fuel Pump

REMOVAL & INSTALLATION

1. Before servicing the vehicle, refer to the precautions in the beginning of this section.
2. Relieve the fuel system pressure.
3. Drain the fuel tank.
4. Remove or disconnect the following:
   - Negative battery cable
   - Trunk liner
   - Trunk seal retainer
   - Rear lamp assembly interior trim finisher
   - Left and right side liners
   - Fuel feed and return lines
   - Fuel filler and vent hoses
   - Fuel tank wiring connectors
   - Fuel filler cap
   - Fuel tank retaining straps
   - Fuel tank
   - Fuel pump module

To install:

5. Install or connect the following:
   - Fuel pump module
   - Fuel tank
   - Fuel tank retaining straps
   - Fuel filler cap
   - Fuel tank wiring connectors
   - Fuel filler and vent hoses
   - Fuel feed and return lines
   - Left and right side liners
   - Rear lamp assembly interior trim finisher
   - Trunk seal retainer
   - Trunk liner
   - Negative battery cable
6. Fill the fuel tank with at least 10 gallons (38L) of fuel.
7. Start the engine and check for leaks.

Fuel Injector

REMOVAL & INSTALLATION

1. Before servicing the vehicle, refer to the precautions in the beginning of this section.
2. Relieve fuel system pressure.
3. Remove or disconnect the following:
   - Negative battery cable
   - Engine appearance covers
   - Fuel lines
   - Fuel pressure regulator
   - Fuel cross over elbow
   - Fuel injector connectors
   - Fuel injector clamping plates
   - Fuel injectors

To install:

4. Install or connect the following:
   - Fuel injectors. Use new O-ring seals.
   - Fuel injector clamping plates
   - Fuel injector connectors
   - Fuel cross over elbow
   - Fuel pressure regulator
   - Fuel lines
   - Engine appearance covers
   - Negative battery cable
5. Start the engine and check for leaks.

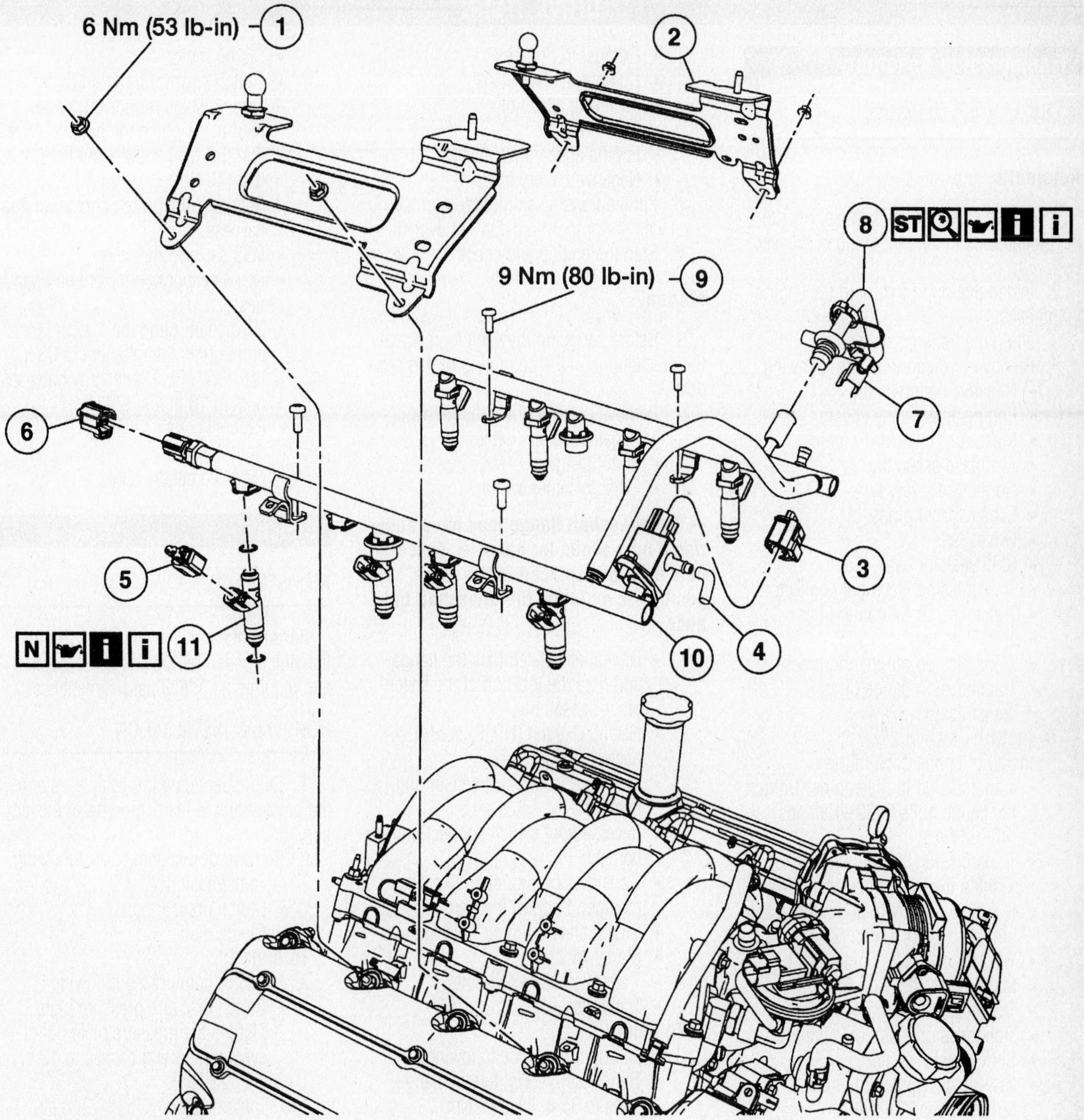

1 Engine appearance cover mounting bracket nuts
2 Engine appearance cover mounting brackets
3 Fuel pressure sensor electrical connector
4 Fuel pressure sensor vacuum connector
5 Fuel injector electrical connector
6 Fuel temperature sensor electrical connector
7 Fuel tube retaining clip
8 Fuel tube spring lock coupling
9 Fuel injection supply manifold bolts
10 Fuel injection supply manifold and fuel injectors
11 Fuel injectors

67197-LILS-G13

Exploded view of the fuel injector mounting—3.9L engine shown

DRIVE TRAIN

Transmission Assembly

REMOVAL & INSTALLATION

Automatic

1. Before servicing the vehicle, refer to the precautions in the beginning of this section.
2. Install a support fixture to the engine lifting eyes.
3. Drain the transmission fluid.
4. Remove or disconnect the following:
   - Negative battery cable
   - Engine appearance covers
   - Mass Air Flow (MAF) meter
   - Air intake assembly
   - Coolant recovery tank
   - Exhaust front pipes
   - Driveshaft
   - Shift selector cable
   - Transmission electrical connectors
   - Transmission oil cooler lines
   - Torque converter
   - Transmission mount and bracket
   - Transmission flange bolts
   - Transmission

To install:

5. Install or connect the following:
   - Transmission to the engine. Tighten the flange bolts to 32–42 ft. lbs. (43–57 Nm).
   - Transmission mount and bracket. Tighten the mount bolts to 22–30 ft. lbs. (30–40 Nm) and the bracket bolts to 16–21 ft. lbs. (22–28 Nm).
   - Torque converter. Tighten the bolts to 32–42 ft. lbs. (43–57 Nm).
   - Transmission oil cooler lines
   - Transmission electrical connectors
   - Shift selector cable
   - Driveshaft. Tighten the bolts to 55–65 ft. lbs. (75–88 Nm).
   - Exhaust front pipes
   - Coolant recovery tank
   - Air intake assembly
   - MAF meter
   - Engine appearance covers
   - Negative battery cable
6. Fill the transmission to the correct level with the proper fluid. Do not over-fill.
7. Start the engine and check for leaks.

Manual

1. Before servicing the vehicle, refer to the precautions in the beginning of this section.
2. Remove or disconnect the following:
   - Negative battery cable
   - Shift linkage
   - Exhaust center section

➡The driveshaft flange may have additional nuts added for balance. These nuts must be replaced in the same position to maintain the driveshaft balance.

   - Driveshaft. Matchmark the flange and note the location of the fasteners for assembly.
   - Heated Oxygen (HO_2S) sensor connectors and harness
   - Vehicle Speed (VSS) sensor connector
   - Reverse light switch connector and harness
   - Transmission mount and crossmember. Support the transmission.
   - Clutch slave cylinder fluid line
   - Starter motor
   - Transmission flange bolts
   - Transmission

To install:

3. Install or connect the following:
   - Transmission. Tighten the flange bolts to 35 ft. lbs. (47 Nm).
   - Starter motor
   - Clutch slave cylinder fluid line
   - Transmission mount and crossmember. Tighten the bolts to 41 ft. lbs. (55 Nm) and the nut to 30 ft. lbs. (40 Nm).
   - Reverse light switch connector and harness
   - VSS sensor connector
   - HO_2S sensor connectors and harness
   - Driveshaft. Align the matchmarks and tighten the retainers to 63 ft. lbs. (85 Nm). Tighten any balance nuts to 18 ft. lbs. (24 Nm).
   - Exhaust center section
   - Shift linkage
   - Negative battery cable

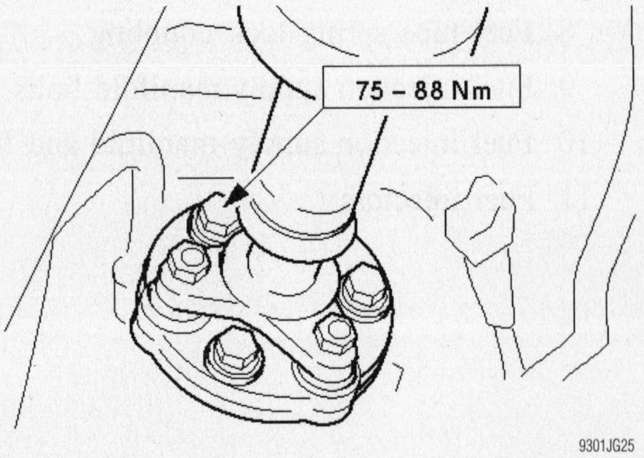

Driveshaft flange bolts

Clutch

ADJUSTMENTS

Because the clutch system is hydraulic, the clutch pedal free-play is self-adjusting and requires no additional maintenance.

REMOVAL & INSTALLATION

1. Before servicing the vehicle, refer to the precautions in the beginning of this section.
2. Remove or disconnect the following:
   - Transmission
   - Clutch pressure plate
   - Clutch disc

To install:

3. Install or connect the following:
   - Clutch disc and pressure plate. Tighten the pressure plate bolts evenly in several passes to 17 ft. lbs. (23 Nm).
   - Transmission
4. Bleed the hydraulic clutch system, if required.
5. Check the clutch system for proper operation.

Hydraulic Clutch System

BLEEDING

1. Before servicing the vehicle, refer to the precautions in the beginning of this section.
2. Remove the inspection cover.
3. Connect a hose to the bleeder valve fitting on the clutch slave cylinder. Sub-

merge the other end of the hose into a container of clean brake fluid.

4. Open the bleeder valve and have an assistant depress the clutch pedal.

5. Close the bleeder before releasing the clutch pedal.

6. Repeat the procedure until no more air bubbles are present.

7. Install the inspection cover to the bell housing.

8. Top off the clutch master cylinder fluid reservoir and install the diaphragm and cap.

Halfshaft

REMOVAL & INSTALLATION

1. Before servicing the vehicle, refer to the precautions in the beginning of this section.

2. Remove or disconnect the following:
   - Negative battery cable
   - Rear wheel
   - Brake caliper
   - Wheel speed sensor
   - Hub carrier pivot bolt
   - Hub retaining nut

3. Press the stub shaft out of the hub and pry the inner joint out of the differential.

To install:

4. Install or connect the following:
   - Halfshaft inner joint to the differential
   - Halfshaft in the wheel hub by applying Loctite® 270 thread locking compound to the splines
   - Hub carrier pivot bolt by aligning the bolt head matchmarks. Tighten it to 66–81 ft. lbs. (90–110 Nm).
   - Hub retaining nut. Tighten the new nut to 221 ft. lbs. (300 Nm).
   - Wheel speed sensor
   - Brake caliper
   - Rear wheel
   - Negative battery cable

5. Check the wheel alignment and adjust as necessary.

➡If the hub is removed for any reason, a new bearing assembly must be installed. Never attempt to re-use a bearing.

CV-Joint

OVERHAUL

The CV-joints are serviced with the axle halfshaft as an assembly.

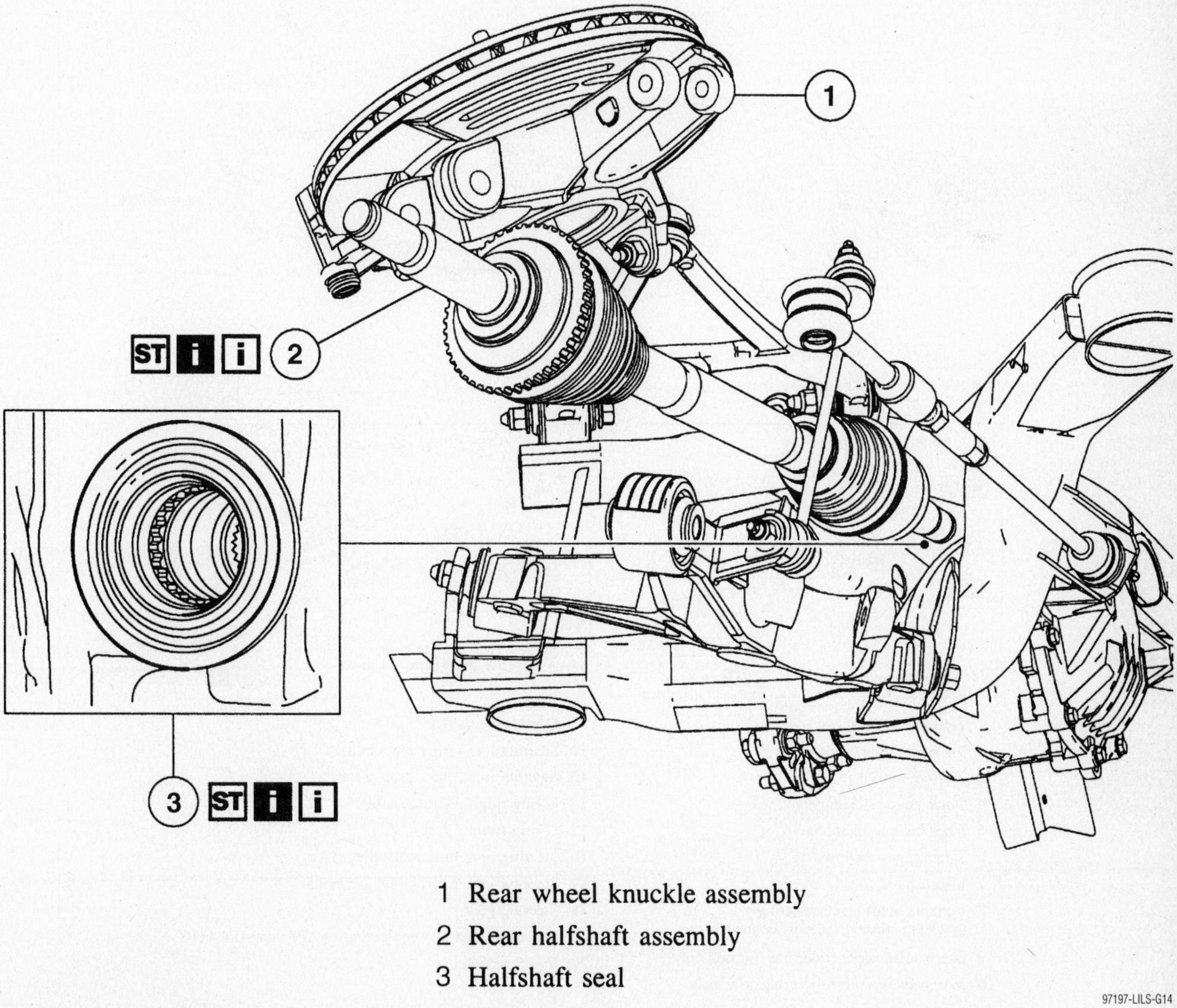

Rear halfshaft mounting

STEERING AND SUSPENSION

Air Bag

**** CAUTION**

Some vehicles are equipped with an air bag system. The system must be disarmed before performing service on, or around, system components, the steering column, instrument panel components, wiring and sensors. Failure to follow the safety precautions and the disarming procedure could result in accidental air bag deployment, possible injury and unnecessary system repairs.

PRECAUTIONS

Several precautions must be observed when handling the inflator module to avoid accidental deployment and possible personal injury.

- Never carry the inflator module by the wires or connector on the underside of the module.
- When carrying a live inflator module, hold securely with both hands, and ensure that the bag and trim cover are pointed away.
- Place the inflator module on a bench or other surface with the bag and trim cover facing up.
- With the inflator module on the bench, never place anything on or close to the module that may be thrown in the event of an accidental deployment.

DISARMING

Proper SRS disarming can be obtained by disconnecting and isolating the negative battery cable. Allow the air bag system capacitor at least 2 minutes to discharge before removing any air bag system components.

Power Rack and Pinion Steering Gear

REMOVAL & INSTALLATION

1. Before servicing the vehicle, refer to the precautions in the beginning of this section.

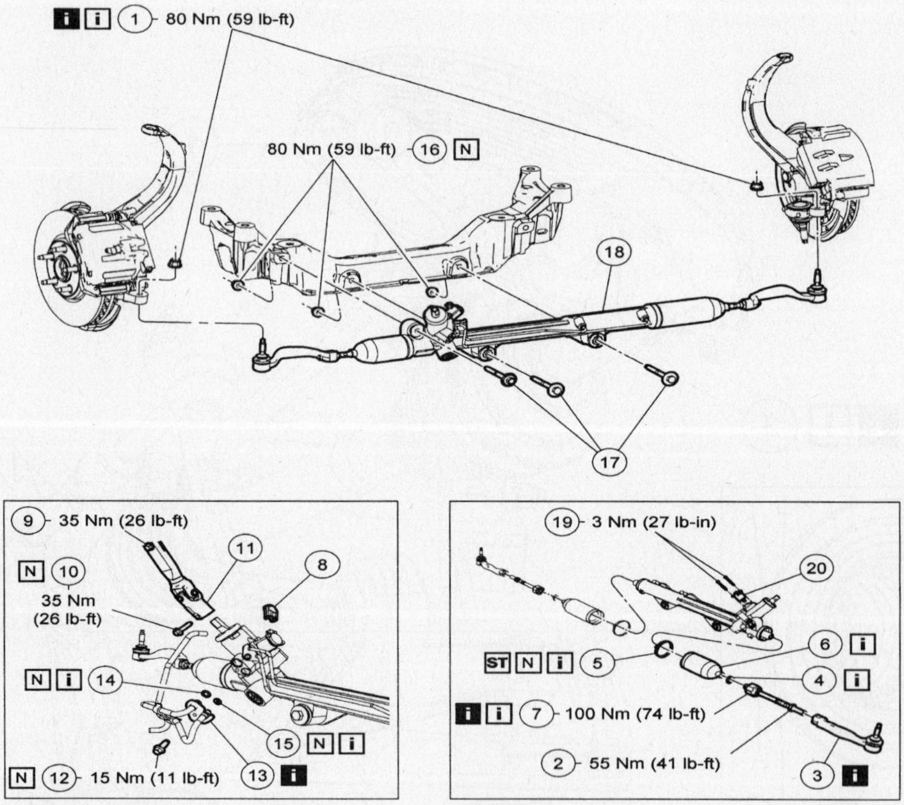

1 Tie-rod end nuts
2 Tie-rod end jam nuts
3 Tie-rod end
4 Outer bellows clamp
5 Inner bellows clamp
6 Steering gear bellows
7 Inner tie-rod
8 Variable assist power steering (VAPS) actuator electrical connector
9 Intermediate shaft slider bolt (loosen only)
10 Intermediate shaft-to-steering gear bolt
11 Intermediate shaft
12 Steering line clamp plate bolt
13 Steering lines and clamp plate
14 O-ring (high pressure hose)
15 O-ring (return hose)
16 Steering gear-to-crossmember nuts
17 Steering gear-to-crossmember bolts
18 Steering gear
19 Variable assist power steering (VAPS) actuator bolts
20 VAPS actuator

67197-LILS-G15

Steering rack and pinion gear mounting

2. Lock the steering wheel in the straight-ahead position.
3. Remove or disconnect the following:
   - Negative battery cable
   - Front wheels
   - Steering column intermediate shaft
   - Outer tie rod ends
   - Power steering lines
   - Steering rack and pinion gear

To install:

4. Install or connect the following:
   - Steering rack and pinion gear. Tighten the bolts to 76 ft. lbs. (103 Nm).
   - Power steering lines
   - Outer tie rod ends. Tighten the nuts to 52–63 ft. lbs. (71–85 Nm).
   - Steering column intermediate shaft
   - Front wheels
   - Negative battery cable
5. Fill the power steering fluid reservoir.
6. Start the engine and check for leaks.
7. Check the wheel alignment and adjust, as necessary.

Strut

REMOVAL & INSTALLATION

Front

1. Before servicing the vehicle, refer to the precautions in the beginning of this section.
2. Remove or disconnect the following:
   - Front wheel
   - Stabilizer bar link
   - Lower strut mounting bolt
   - Upper strut mount cover and fasteners
   - Strut and spring assembly

To install:

➡Use new fasteners for assembly.

3. Install or connect the following:
   - Strut and spring assembly. Tighten the upper mount nuts to 21 ft. lbs. (29 Nm).
   - Upper strut mount cover
   - Lower strut mounting bolt. Tighten the bolts to 129 ft. lbs. (175 Nm).
   - Stabilizer bar link. Tighten the nut to 41 ft. lbs. (55 Nm).
   - Front wheel

Rear

1. Before servicing the vehicle, refer to the precautions in the beginning of this section.
2. Remove or disconnect the following:
   - Trunk trim covers

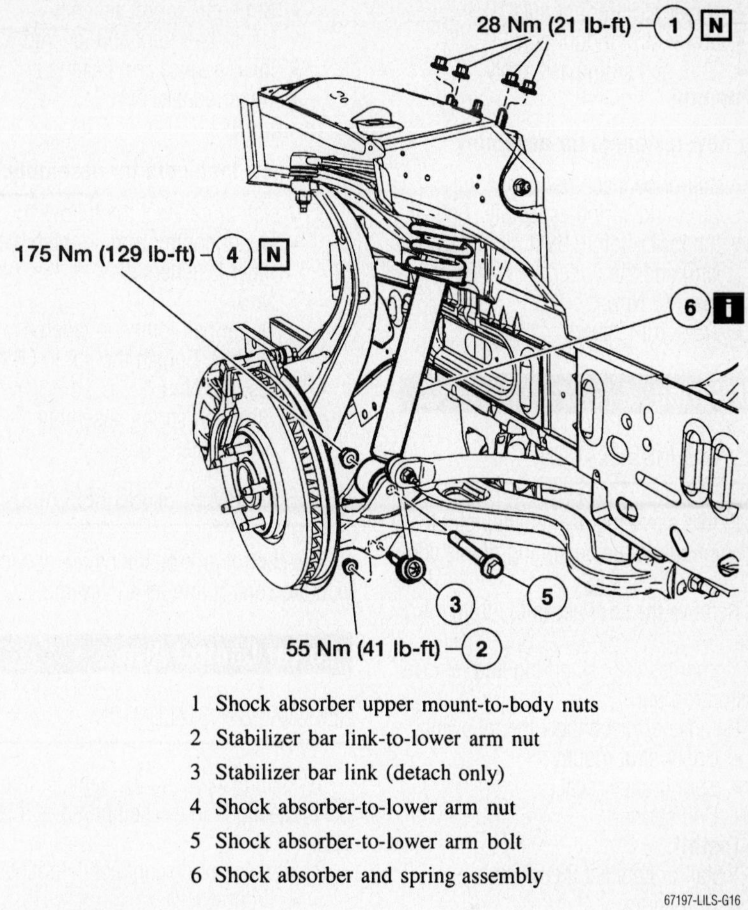

Front strut mounting

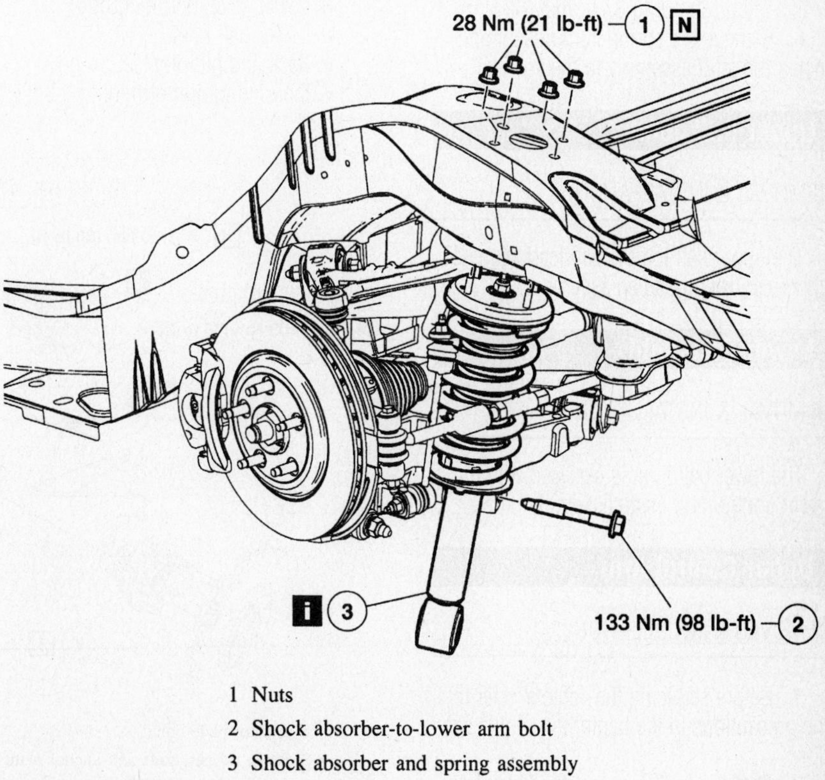

Rear strut and spring assembly mounting

- Upper strut mount nuts
- Lower strut mount bolt
- Strut and spring assembly

To install:

➡Use new fasteners for assembly.

3. Install or connect the following:
    - Strut and spring assembly. Tighten the lower bolt to 98 ft. lbs. (133 Nm) and the upper nuts to 21 ft. lbs. (28 Nm).
    - Trunk trim covers

Coil Spring

REMOVAL & INSTALLATION

1. Before servicing the vehicle, refer to the precautions in the beginning of this section.
2. Remove the strut assembly from the vehicle.
3. Compress the coil spring and remove the piston rod nut.
4. Remove or disconnect the following:
    - Upper strut mount
    - Spring upper seat
    - Coil spring

To install:

5. Install or connect the following:
    - Coil spring
    - Spring upper seat
    - Upper strut mount. Tighten the piston rod nut to 37 ft. lbs. (50 Nm).
6. Remove the spring compressor and install the strut assembly to the vehicle.

Upper Ball Joint

REMOVAL & INSTALLATION

The upper ball joint is serviced with the upper control arm as an assembly.

Lower Ball Joint

REMOVAL & INSTALLATION

The lower ball joint is serviced with the lower control arm as an assembly.

Upper Control Arm

REMOVAL & INSTALLATION

1. Before servicing the vehicle, refer to the precautions in the beginning of this section.
2. Remove or disconnect the following:
    - Front wheel
    - Strut and spring assembly
    - Upper ball joint and tapered washer
    - Inner control arm fasteners
    - Upper control arm

To install:

➡Use new fasteners for assembly.

3. Install or connect the following:
    - Upper control arm. Tighten the inner fasteners to 35 ft. lbs. (48 Nm).
    - Upper ball joint and tapered washer. Tighten the nut to 66 ft. lbs. (90 Nm).
    - Strut and spring assembly
    - Front wheel

CONTROL ARM BUSHING REPLACEMENT

The control arm bushings are serviced with the control arm as an assembly.

Lower Control Arm

REMOVAL & INSTALLATION

1. Before servicing the vehicle, refer to the precautions in the beginning of this section.
2. Remove or disconnect the following:
    - Front wheel
    - Splash shield
    - Stabilizer bar link
    - Lower strut mounting bolt
    - Lower ball joint
    - Rack and pinion steering gear
    - Inner control arm mounting bolts
    - Lower control arm

To install:

➡Use new fasteners for assembly.

3. Install or connect the following:
    - Lower control arm. Tighten the inner mounting bolts to 129 ft. lbs. (175 Nm).
    - Rack and pinion steering gear
    - Lower ball joint. Tighten the nut to 111 ft. lbs. (150 Nm).
    - Lower strut mounting bolt. Tighten the bolt to 129 ft. lbs. (175 Nm).
    - Stabilizer bar link. Tighten the nut to 41 ft. lbs. (55 Nm).
    - Splash shield
    - Front wheel
4. Check the wheel alignment and adjust as necessary.

CONTROL ARM BUSHING REPLACEMENT

The control arm bushings are serviced with the control arm as an assembly.

Wheel Bearings

ADJUSTMENT

The wheel bearings are not adjustable.

REMOVAL & REPLACEMENT

Front

1. Before servicing the vehicle, refer to the precautions in the beginning of this section.

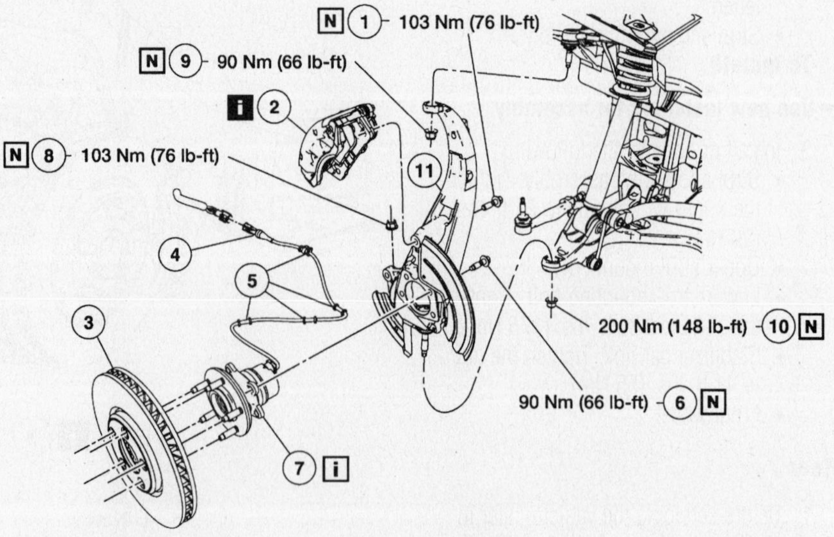

1 Anchor plate bolt
2 Brake caliper, pads and anchor plate
3 Brake disc
4 Speed sensor electrical connector (disconnect)

67197-LILS-G20

Front hub and bearing exploded view

2. Remove or disconnect the following:
   - Front wheel
   - Brake caliper and rotor
   - Wheel speed sensor connector
   - Hub and bearing assembly

➡The hub and bearing assembly is not pressed into the knuckle. Do not use a slide hammer or press to remove the hub and bearing assembly. Damage to the hub and bearing assembly may result.

To install:

➡Do not remove the wheel speed sensor from the hub and bearing assembly unless it is being replaced. If installing a new hub and bearing assembly, a new wheel speed sensor must be installed.

➡Use new fasteners for assembly.

3. Install or connect the following:
   - Hub and bearing assembly. Tighten the bolts to 66 ft. lbs. (90 Nm).
   - Wheel speed sensor connector
   - Brake caliper and rotor
   - Front wheel

Rear

1. Before servicing the vehicle, refer to the precautions in the beginning of this section.
2. Remove or disconnect the following:
   - Rear wheel
   - Wheel speed sensor
   - Brake caliper and rotor
   - Hub, bearing and knuckle assembly
   - Disc brake dust shield
3. Press the hub from the knuckle and bearing assembly.
4. Remove the snapring and press the bearing assembly out of the knuckle.

To install:

5. Press the bearing assembly into the knuckle.
6. Install the snapring and press the hub into the knuckle and bearing assembly.
7. Install or connect the following:
   - Disc brake dust shield. Use aluminum rivets.
   - Hub, bearing and knuckle assembly
   - Brake caliper and rotor
   - Wheel speed sensor
   - Rear wheel
8. Check the wheel alignment and adjust as necessary.

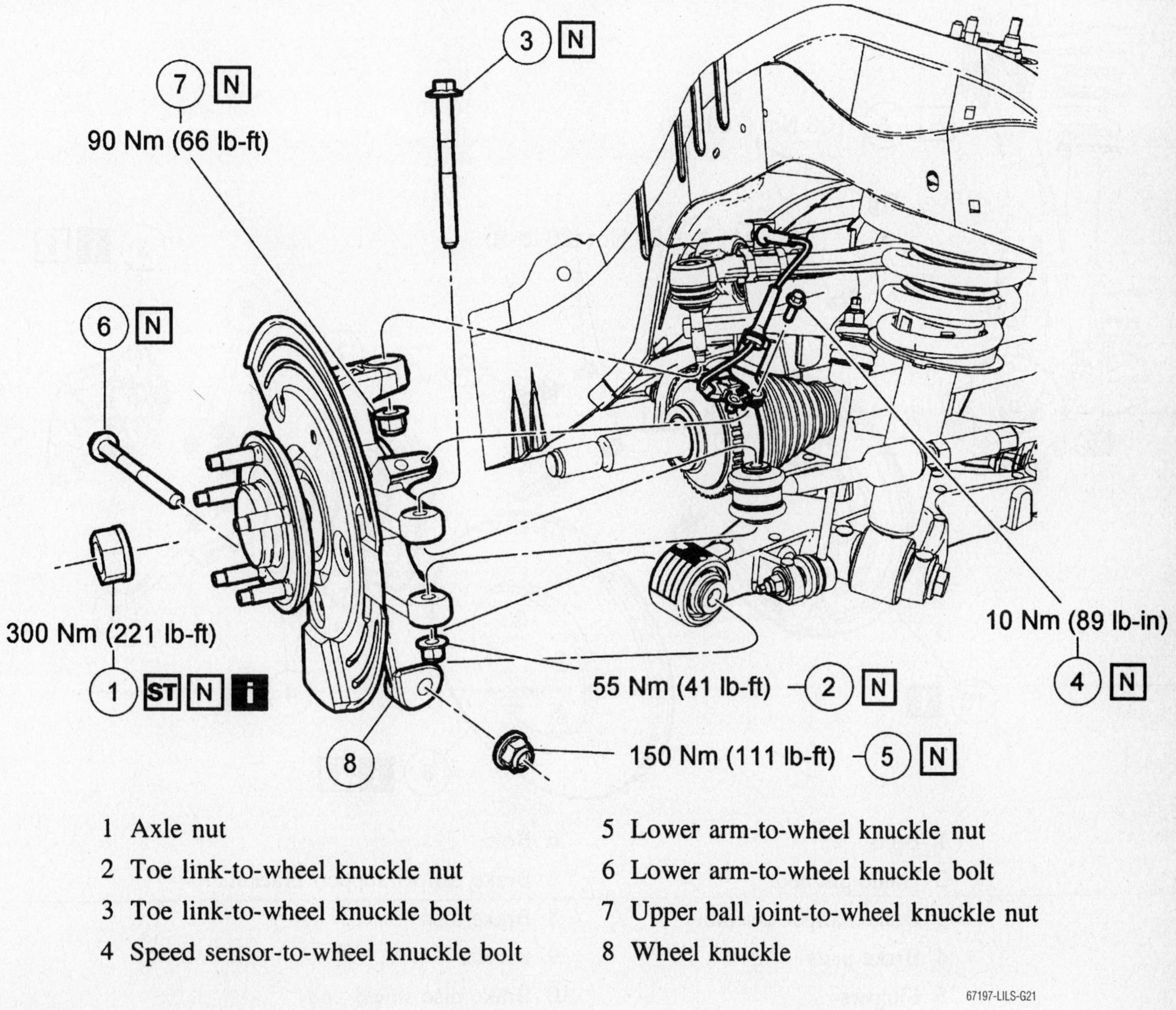

Exploded view of the rear wheel bearing assembly

BRAKES

Brake Caliper

REMOVAL & INSTALLATION

Front

1. Before servicing the vehicle, refer to the precautions in the beginning of this section.
2. Remove or disconnect the following:
   - Front wheel
   - Brake fluid hose
   - Caliper mounting bolts
   - Brake caliper

To install:

3. Install or connect the following:
   - Brake caliper. Tighten the mounting bolts to 26 ft. lbs. (35 Nm).
   - Brake fluid hose. Use new copper washers and tighten the bolt to 35 ft. lbs. (47 Nm).
   - Front wheel
4. Bleed the brake system.
5. Before attempting to move the vehicle, pump the brake pedal to seat the pads against the rotors. Make sure the vehicle has a firm brake pedal. Check the level of the brake fluid and add DOT 3 or 4 brake fluid if necessary.

Rear

1. Before servicing the vehicle, refer to the precautions in the beginning of this section.
2. Remove or disconnect the following:
   - Rear wheel
   - Parking brake cable
   - Caliper mounting bolts
   - Brake fluid hose
   - Brake caliper

To install:

3. Install or connect the following:
   - Brake fluid hose. Use new copper washers and tighten the bolt to 36 ft. lbs. (48 Nm).
   - Brake caliper. Tighten the mounting bolts to 25 ft. lbs. (33 Nm).
   - Parking brake cable
   - Rear wheel
4. Bleed the brake system.
5. Before attempting to move the vehicle, pump the brake pedal to seat the pads against the rotors. Make sure the vehicle has a firm brake pedal. Check the

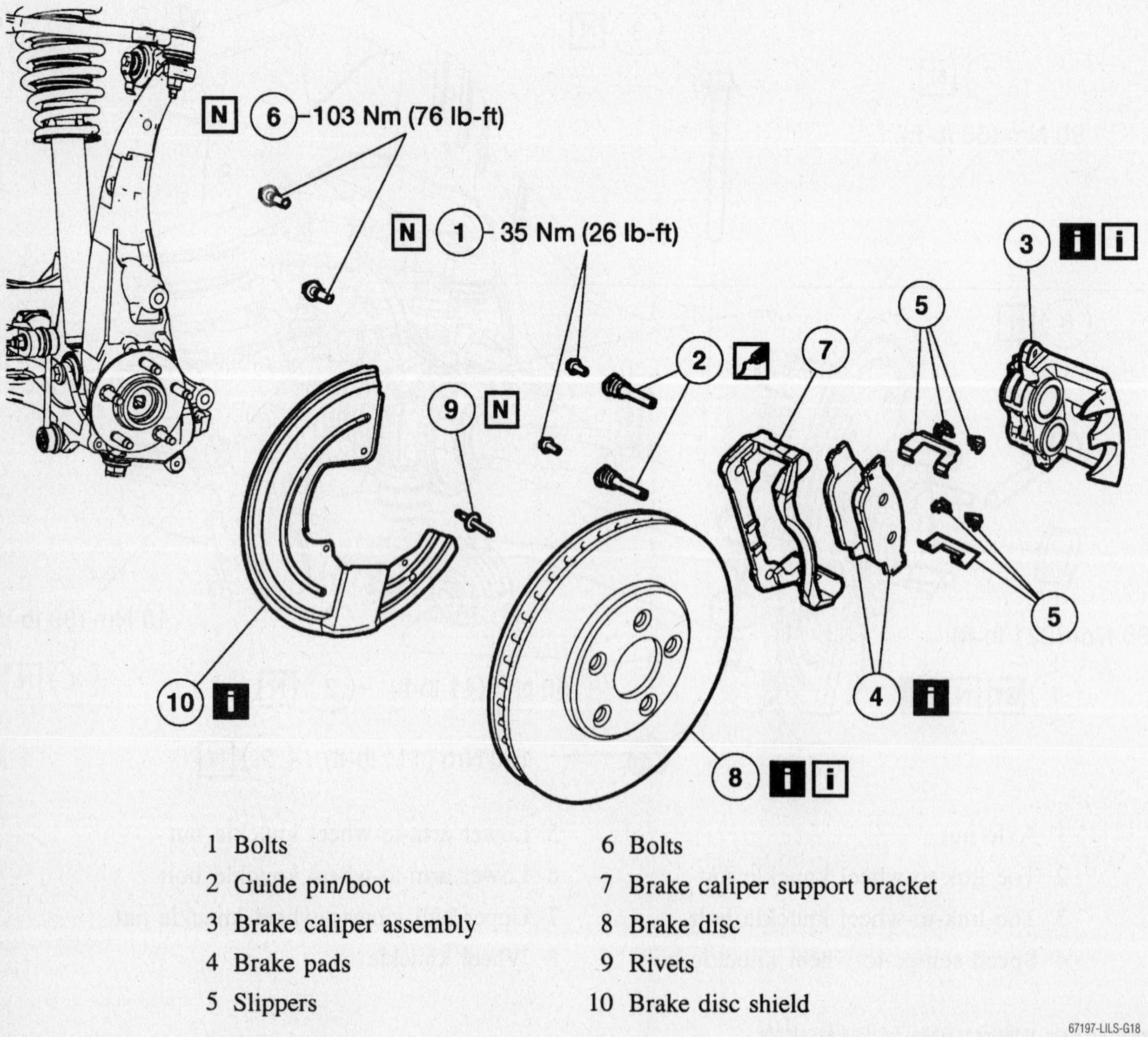

Exploded view of the front disc brake components

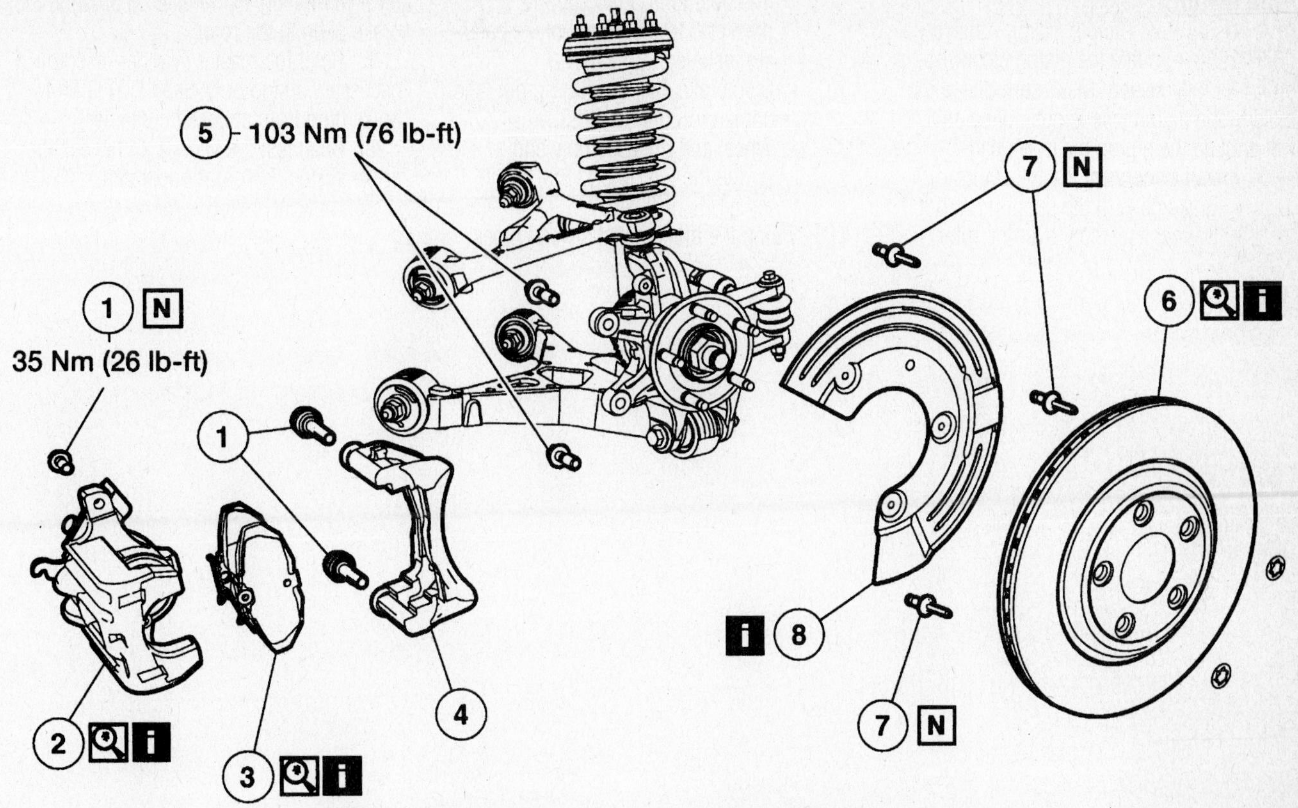

1 Guide pin assembly	5 Brake caliper support bracket
2 Brake caliper	6 Brake disc
3 Brake pads (part of kit)	7 Rivets
4 Bolts	8 Brake disc shield

67197-LILS-G19

Exploded view of the rear disc brake components

level of the brake fluid and add DOT 3 or 4 brake fluid if necessary.

Disc Brake Pads

REMOVAL & INSTALLATION

Front

1. Before servicing the vehicle, refer to the precautions in the beginning of this section.
2. Remove or disconnect the following:
   - Master cylinder reservoir cap and check the fluid level in the reservoir
   - Brake fluid until the reservoir is ½ full. Discard the removed fluid.
   - Wheel and tire assembly
   - Disc brake caliper locating pins. Lift the caliper assembly from the anchor plate and rotor.
3. Suspend the caliper inside the fender housing with wire. Do not allow the caliper to hang from the brake hose.
   - Inner and outer brake pads. Inspect the rotor braking surfaces for scoring and machine as necessary.

To install:

4. Use a C-clamp and an old brake pad or block of wood to seat the caliper piston in its bore.
5. Remove any rust buildup from the inside of the caliper in the brake pad contact area.
6. Install or connect the following:
   - Inner pad in the caliper piston
   - Outer pad onto the anchor plate. Make sure the clips are properly seated.
   - Disc brake caliper onto the anchor plate
   - Caliper locating pins and torque to 26 ft. lbs. (35 Nm)
   - Wheel and tire assembly and torque lugs nuts to 85–104 ft. lb. (115–142 Nm)
7. Pump the brake pedal several times prior to moving the vehicle to position the brake pads to the rotor.
8. Refill the master cylinder reservoir as necessary, using only clean DOT 3 or 4 brake fluid from a closed container.

Rear

1. Before servicing the vehicle, refer to the precautions in the beginning of this section.
2. Remove or disconnect the following:
   - Master cylinder reservoir cap and check the fluid level in the reservoir
   - Brake fluid until the reservoir is ½ full. Discard the removed fluid.
   - Wheel and tire assembly
   - Disc brake caliper locating pins at the support bracket
   - Caliper
   - Disc brake pads
3. Inspect the rotor braking surfaces for scoring and machine as necessary.

To install:

4. Using Rear Caliper Piston Adjuster T87P-2588-A, rotate the piston clockwise until it is fully seated. Make sure one of the slots in the piston face is positioned so it will engage the nib on the brake pad.
5. Install or connect the following:
   - Brake pads in the support bracket
   - Caliper assembly over the rotor into position on the support bracket. Make sure the brake pads are installed correctly.
   - Disc brake caliper locating pin, and tighten to 25 ft. lbs. (33 Nm)
   - Wheel and tire assembly and torque the lug nuts to 85–104 ft. lbs. (115–142 Nm)
6. Pump the brake pedal several times prior to moving the vehicle, to position the brake pads to the rotor.
7. Refill the master cylinder reservoir if necessary, using only clean DOT 3 or 4 brake fluid from a closed container.
8. Road test the vehicle and check the brake system for proper operation.

FORD

Mustang

12

SPECIFICATION CHARTS

ENGINE AND VEHICLE IDENTIFICATION

Engine							Model Year	
Code ①	Liters (cc)	Cu. In.	Cyl.	Fuel Sys.	Engine Type	Eng. Mfg.	Code ②	Year
4	3.8 (3802)	232	6	SFI	OHV	Ford	1	2001
R	4.6 (4593)	281	8	SFI	DOHC	Ford	2	2002
V	4.6 (4593)	281	8	SFI	DOHC	Ford	3	2003
X	4.6 (4593)	281	8	SFI	SOHC	Ford	4	2004
Y	4.6 (4593)	281	8	SFI	DOHC	Ford	5	2005

OHV: Overhead Valve

SOHC: Single Overhead Camshaft

DOHC: Double Overhead Camshaft

SFI: Sequential Fuel Injection

① 8th position of VIN

② 10th position of VIN

67197-MUST-C01

GENERAL ENGINE SPECIFICATIONS

All measurements are given in inches.

Year	Model	Engine Displacement Liters	Engine Series (ID/VIN)	Net Horsepower @ rpm	Net Torque @ rpm (ft. lbs.)	Bore x Stroke (in.)	Com-pression Ratio	Oil Pressure @ rpm
2001	Mustang	3.8	4	190@5250	220@3000	3.81x3.39	9.0:1	40-60@2500
		4.6	V	320@6000	317@4750	3.55x3.54	9.5:1	20-45@1500
		4.6	X	260@5000	302@4000	3.55x3.54	9.0:1	20-45@1500
2002	Mustang	3.8	4	190@5250	220@3000	3.81x3.39	9.0:1	40-60@2500
		4.6	V	320@6000	317@4750	3.55x3.54	9.5:1	20-45@1500
		4.6	X	260@5000	302@4000	3.55x3.54	9.0:1	20-45@1500
2003	Mustang	3.8	4	190@5250	220@3000	3.81x3.39	9.0:1	40-60@2500
		4.6	V	320@6000	317@4750	3.55x3.54	9.5:1	20-45@1500
		4.6	X	260@5000	302@4000	3.55x3.54	9.0:1	20-45@1500
2004	Mustang	3.8	4	193@5500	225@2800	3.81x3.39	9.3:1	40-60@2500
		4.6	R	305@5800	320@4200	3.55x3.54	10.1:1	20-45@1500
		4.6	X	260@5250	302@4000	3.55x3.54	9.37:1	20-45@1500
		4.6	Y	390@6000	390@3500	3.55x3.54	8.5:1	20-45@1500

67197-MUST-C02

ENGINE TUNE-UP SPECIFICATIONS

Year	Engine Displacement Liters	Engine ID/VIN	Spark Plug Gap (in.)	Ignition Timing (deg.) MT	Ignition Timing (deg.) AT	Fuel Pump (psi) ①	Idle Speed (rpm) MT	Idle Speed (rpm) AT	Valve Clearance Intake	Valve Clearance Exhaust
2001	3.8	4	0.054	②	②	28-54	②	②	HYD	HYD
	4.6	V	0.054	10B	—	35-45	②	—	HYD	HYD
	4.6	X	0.054	10B	10B	35-45	②	②	HYD	HYD
2002	3.8	4	0.054	②	②	28-54	②	②	HYD	HYD
	4.6	V	0.054	10B	—	35-45	②	—	HYD	HYD
	4.6	X	0.054	10B	10B	35-45	②	②	HYD	HYD
2003	3.8	4	0.054	②	②	28-54	②	②	HYD	HYD
	4.6	V	0.054	10B	—	35-45	②	—	HYD	HYD
	4.6	X	0.054	10B	10B	35-45	②	②	HYD	HYD
2004	3.8	4	0.052-0.056	B	B	27-37	②	②	HYD	HYD
	4.6	R	0.052-0.056	10B	10B	27-37	②	②	HYD	HYD
	4.6	X	0.052-0.056	10B	10B	27-37	②	②	HYD	HYD
	4.6	Y	0.052-0.056-	10B	10B	27-37	②	B	HYD	HYD

NOTE: The Vehicle Emission Control Information label often reflects specification changes made during production.
The label figures must be used if they differ from those in this chart.

B: Before Top Dead Center

HYD: Hydraulic

① Fuel pressure with engine running, pressure regulator vacuum hose connected

② Refer to Vehicle Emission Control Information label

67197-MUST-C03

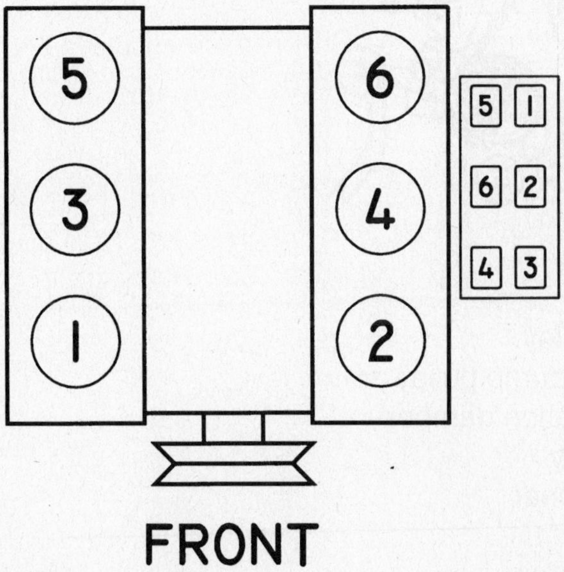

93003G01

3.8L Engine
Firing order: 1–4–2–5–3–6
Distributorless ignition system

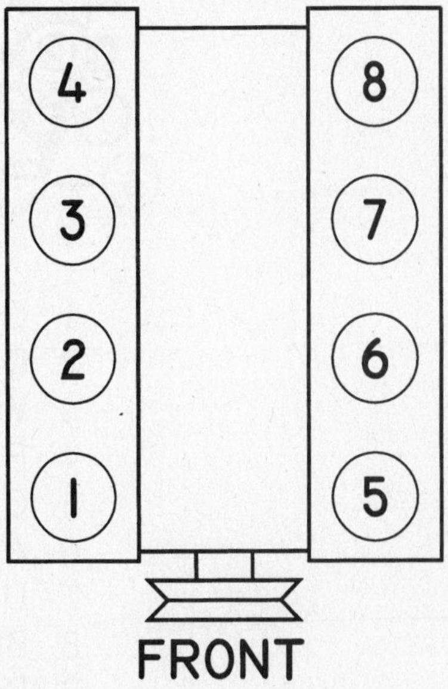

93003G02

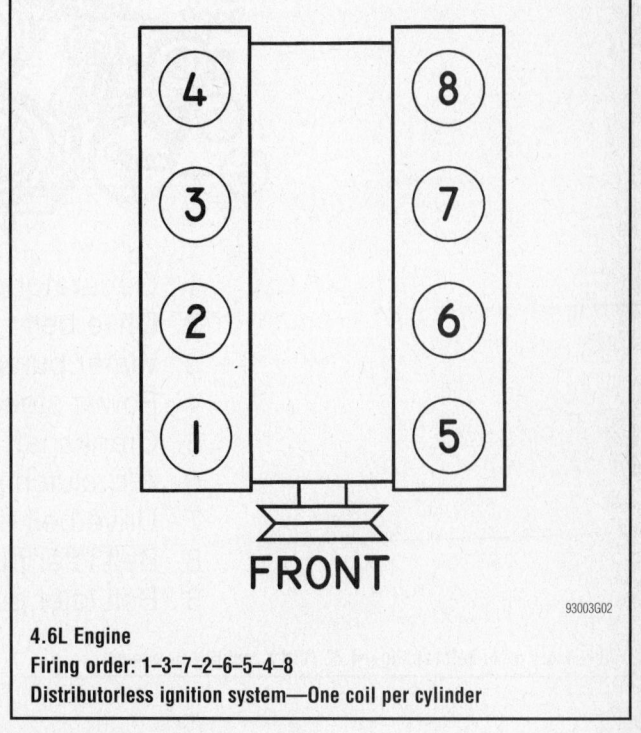

4.6L Engine
Firing order: 1–3–7–2–6–5–4–8
Distributorless ignition system—One coil per cylinder

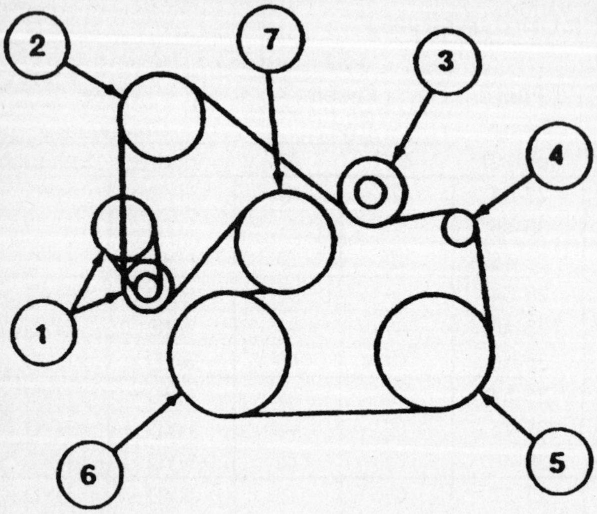

1 Tensioner
2 A/C compressor or idler pulley
3 Idler pulley
4 Alternator
5 Power steering pump
6 Crankshaft
7 Water pump

67197-MUST-G01

Accessory drive belt routing—3.8L (VIN 4) engine (2001–03 model)

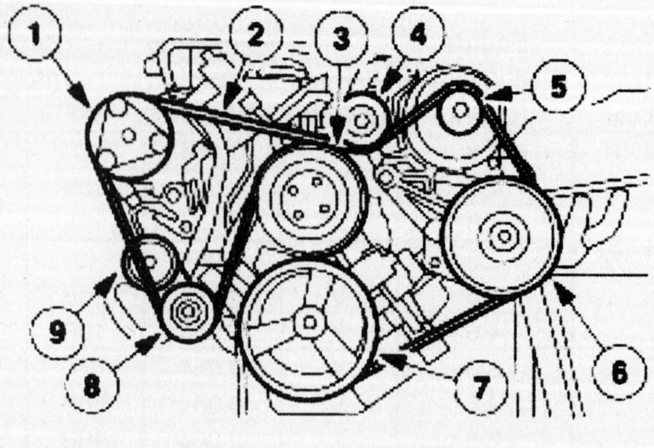

1 A/C clutch
2 Drive belt
3 Water pump
4 Idler pulley
5 Alternator
6 Power steering pump
7 Vibration damper
8 Tensioner
9 Tensioner

67197-MUST-G02

Accessory drive belt routing—3.8L (VIN 4) engine (2004 model)

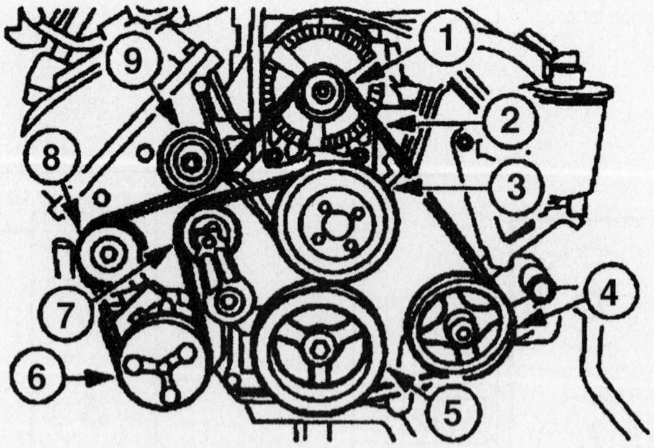

1 Generator pulley
2 Drive belt
3 Water pump pulley
4 Power steering pump pulley
5 Crankshaft vibration damper
6 A/C clutch pulley
7 Drive belt tensioner
8 Belt idler pulley
8 Belt idler pulley

67197-MUST-G03

Accessory drive belt routing—4.6L (VIN X and VIN R) engine

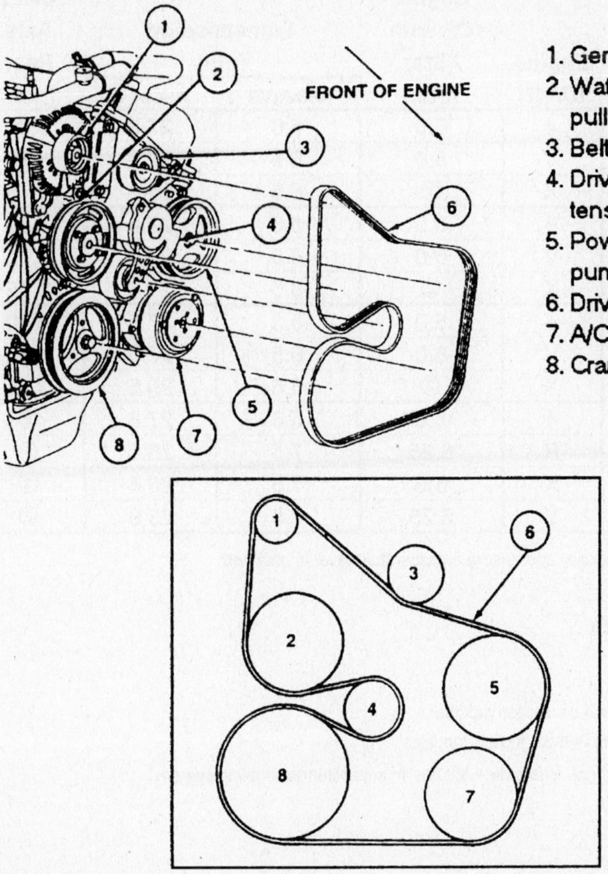

Accessory drive belt routing—4.6L (VIN V) engine

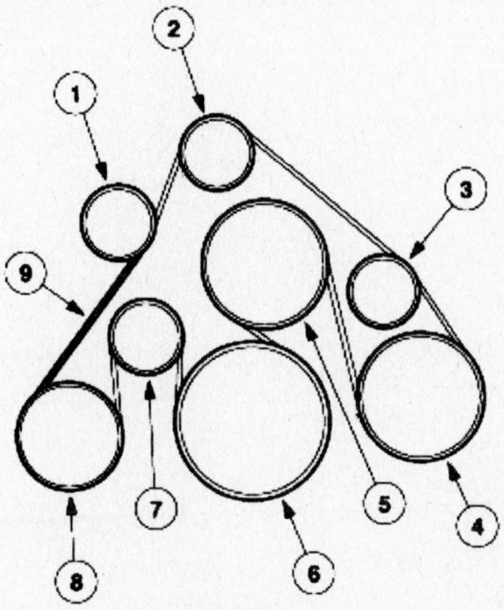

Accessory drive belt routing—4.6L (VIN Y) engine

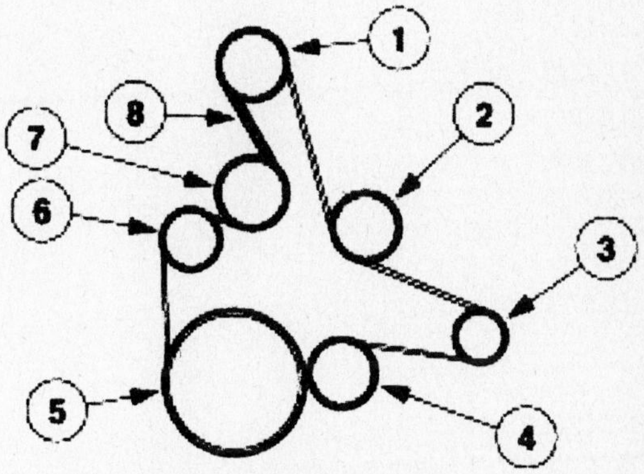

Supercharger accessory drive belt routing—4.6L (VIN Y) engine

CAPACITIES

Year	Model	Engine Displacement Liters	Engine ID/VIN	Engine Oil with Filter (qts.)	Transmission (pts) Manual	Transmission (pts) Auto. ①	Drive Axle Rear (pts.)	Fuel Tank (gal.)	Cooling System (qts.)
2001	Mustang	3.8	4	5.0	5.6	27.8	3.50	15.7	11.8
		4.6	V	6.0	6.5	27.8	3.75	15.7	14.1
		4.6	X	②	6.5	25.6	3.75	15.7	14.1
2002	Mustang	3.8	4	5.0	5.6	27.8	3.50	15.7	11.8
		4.6	V	6.0	6.5	27.8	3.75	15.7	14.1
		4.6	X	②	6.5	25.6	3.75	15.7	14.1
2003	Mustang	3.8	4	5.0	5.6	27.8	3.50	15.7	11.8
		4.6	V	6.0	6.5	27.8	3.75	15.7	14.1
		4.6	X	②	6.5	25.6	3.75	15.7	14.1
2004	Mustang	3.8	4	5.0	5.6	27.8	③	15.7	11.8
		4.6	R	6.25	7.5	25.6	③	15.7	14.1
		4.6	X	6.5	7.5	25.6	③	15.7	14.1
		4.6	Y	6.25	7.5	25.6	③	15.7	④

NOTE: All capacities are approximate. Add fluid gradually and ensure a proper fluid level is obtained.

① Includes torque converter

② If equipped with an automatic transmission: 6.7 qts.
If equipped with a manual transmission: 6.4 qts.

③ If equipped with 7.5 inch ring gear: 3.25-3.50 pts.
If equipped with 8.8 inch ring gear: 3.5-3.75 pts. + 4 oz. friction modifier.
If equipped with 8.8 inch ring gear IRS: 2.6-2.9 pts. + 4 oz. friction modifier.

④ 4.6L Supercharged engine coolant capacity is 4.0 gal. in engine + 4.0 qts. In supercharger cooling system.

67197-MUST-C04

VALVE SPECIFICATIONS

Year	Engine Displacement Liters	Engine ID/VIN	Seat Angle (deg.)	Face Angle (deg.)	Spring Test Pressure (lbs. @ in.)	Spring Installed Height (in.)	Stem-to-Guide Clearance (in.) Intake	Stem-to-Guide Clearance (in.) Exhaust	Stem Diameter (in.) Intake	Stem Diameter (in.) Exhaust
2001	3.8	4	44.7	45.7	224@1.16	1.620	0.0450-0.0900	0.0015-0.0033	0.2738-0.2751	0.2728-0.2741
	4.6	V	45	45.5	160@1.03	1.660	0.0008-0.0027	0.0018-0.0037	0.2746-0.275	0.2736-0.2744
	4.6	X	45	45.5	161@1.03	1.570	0.0008-0.0027	0.0018-0.0037	0.2746-0.2754	0.2736-0.2744
2002	3.8	4	44.7	45.7	224@1.16	1.620	0.0450-0.0900	0.0015-0.0033	0.2738-0.2751	0.2728-0.2741
	4.6	V	45	45.5	160@1.03	1.660	0.0008-0.0027	0.0018-0.0037	0.2746-0.275	0.2736-0.2744
	4.6	X	45	45.5	161@1.03	1.570	0.0008-0.0027	0.0018-0.0037	0.2746-0.2754	0.2736-0.2744
2003	3.8	4	44.7	45.7	224@1.16	1.620	0.0450-0.0900	0.0015-0.0033	0.2738-0.2751	0.2728-0.2741
	4.6	V	45	45.5	160@1.03	1.660	0.0008-0.0027	0.0018-0.0037	0.2746-0.275	0.2736-0.2744
	4.6	X	45	45.5	161@1.03	1.570	0.0008-0.0027	0.0018-0.0037	0.2746-0.2754	0.2736-0.2744
2004	3.8	4	44.7	45.7	224@1.16	1.620	0.0450-0.0900	0.0015-0.0033	0.2738-0.2751	0.2728-0.2741
	4.6	R	45	45.5	160@1.03	1.660	0.0008-0.0027	0.0018-0.0037	0.2746-0.275	0.2736-0.2744
	4.6	X	45	45.5	162@1.13	1.67-1.690	0.0008-0.0027	0.0018-0.0037	0.2746-0.275	0.2736-0.2744
	4.6	Y	45	45.5	160@1.03	1.423	0.0008-0.0027	0.0018-0.0037	0.2746-0.2754	0.2736-0.2744

67197-MUST-C05

PISTON AND RING SPECIFICATIONS

All measurements are given in inches.

	Engine			Ring Gap			Ring Side Clearance		
	Displacement	Engine	Piston	Top	Bottom	Oil	Top	Bottom	Oil
Year	Liters	ID/VIN	Clearance	Compression	Compression	Control	Compression	Compression	Control
2001	3.8	4	0.0007-0.0017	0.011-0.012	0.009-0.020	0.015-0.058	0.0016-0.0034	0.0016-0.0034	SNUG
	4.6	V	0.0000-0.0010	0.010-0.020	0.009-0.019	0.006-0.026	0.0003-0.0009	0.0012-0.0031	SNUG
	4.6	X	0.0005-0.0010	0.009-0.019	0.009-0.019	0.006-0.026	0.0016-0.0035	0.0018-0.0031	SNUG
2002	3.8	4	0.0007-0.0017	0.011-0.012	0.009-0.020	0.015-0.058	0.0016-0.0034	0.0016-0.0034	SNUG
	4.6	V	0.0000-0.0010	0.010-0.020	0.009-0.019	0.006-0.026	0.0003-0.0009	0.0012-0.0031	SNUG
	4.6	X	0.0005-0.0010	0.009-0.019	0.009-0.019	0.006-0.026	0.0016-0.0035	0.0018-0.0031	SNUG
2003	3.8	4	0.0007-0.0017	0.011-0.012	0.009-0.020	0.015-0.058	0.0016-0.0034	0.0016-0.0034	SNUG
	4.6	V	0.0000-0.0010	0.010-0.020	0.009-0.019	0.006-0.026	0.0003-0.0009	0.0012-0.0031	SNUG
	4.6	X	0.0005-0.0010	0.009-0.019	0.009-0.019	0.006-0.026	0.0016-0.0035	0.0018-0.0031	SNUG
2004	3.8	4	0.0007-0.0017	0.007-0.013	0.012-0.021	0.001-0.025	0.0012-0.0032	0.0012-0.0032	SNUG
	4.6	R	0.0000-0.0010	0.006-0.002	0.002-0.0022	0.006-0.026	0.0003-0.0009	0.0012-0.0031	SNUG
	4.6	X	0.0000-0.0010	0.005-0.011	0.0018-0.0022	0.006-0.026	0.002-0.0040	0.0012-0.0031	SNUG
	4.6	Y	0.0000-0.0010	0.0020	0.0200	0.0260	0.0014-0.0031	0.0012-0.003	0.002-0.008

67197-MUST-C06

CRANKSHAFT AND CONNECTING ROD SPECIFICATIONS

All measurements are given in inches.

Year	Engine Displacement Liters	Engine ID/VIN	Crankshaft: Main Brg. Journal Dia.	Crankshaft: Main Brg. Oil Clearance	Crankshaft: Shaft End-play	Crankshaft: Thrust on No.	Connecting Rod: Journal Diameter	Connecting Rod: Oil Clearance	Connecting Rod: Side Clearance
2001	3.8	4	①	0.0010-0.0014	0.0040-0.0080	3	2.3103-2.3111	0.0010-0.0014	0.0047-0.0144
	4.6	V	2.6567-2.6577	0.0001-0.0018	0.0051-0.0119	5	2.0859-2.0867	0.0011-0.0027	0.0006-0.0177
	4.6	X	2.6569-2.6576	0.0011-0.0026	0.0051-0.0119	5	2.0861-2.0867	0.0011-0.0027	0.0006-0.0177
2002	3.8	4	①	0.0010-0.0014	0.0040-0.0080	3	2.3103-2.3111	0.0010-0.0014	0.0047-0.0144
	4.6	V	2.6567-2.6577	0.0001-0.0018	0.0051-0.0119	5	2.0859-2.0867	0.0011-0.0027	0.0006-0.0177
	4.6	X	2.6569-2.6576	0.0011-0.0026	0.0051-0.0119	5	2.0861-2.0867	0.0011-0.0027	0.0006-0.0177
2003	3.8	4	①	0.0010-0.0014	0.0040-0.0080	3	2.3103-2.3111	0.0010-0.0014	0.0047-0.0144
	4.6	V	2.6567-2.6577	0.0001-0.0018	0.0051-0.0119	5	2.0859-2.0867	0.0011-0.0027	0.0006-0.0177
	4.6	X	2.6569-2.6576	0.0011-0.0026	0.0051-0.0119	5	2.0861-2.0867	0.0011-0.0027	0.0006-0.0177
2004	3.8	4	①	0.0010-0.0014	0.0040-0.0080	3	2.3103-2.3111	0.0010-0.0014	0.0047-0.0144
	4.6	R	2.6567-2.6577	0.0001-0.0018	0.0051-0.0119	5	2.0859-2.0867	0.0011-0.0027	0.0006-0.0177
	4.6	X	2.6567-2.6577	0.0001-0.0018	0.0051-0.0119	5	2.0859-2.0867	0.0011-0.0027	0.0006-0.0177
	4.6	Y	2.6569-2.6576	0.0011-0.0026	0.0051-0.0119	5	2.0861-2.0867	0.0011-0.0027	0.0006-0.0177

① Journals 1, 2, 3: 2.5194-2.5186 in.
Journal 4: 2.5100-2.5092 in.

67197-MUST-C07

TORQUE SPECIFICATIONS

All readings in ft. lbs.

Year	Engine Displacement Liters	Engine ID/VIN	Cylinder Head Bolts	Main Bearing Bolts	Rod Bearing Bolts	Crankshaft Damper Bolts	Flywheel Bolts	Manifold Intake	Manifold Exhaust	Spark Plugs	Oil Pan Drain Plug
2001	3.8	4	①	②	③	118	54-64	④	15-22	7-15	22-30
	4.6	V	⑤	⑥	⑦	114-121	54-64	⑧	13-16	7-15	22-30
	4.6	X	⑨	⑩	⑪	114-121	54-64	15-22	15-22	7-15	22-30
2002	3.8	4	①	②	③	118	54-64	④	15-22	7-15	22-30
	4.6	V	⑤	⑥	⑦	114-121	54-64	⑧	13-16	7-15	22-30
	4.6	X	⑨	⑩	⑪	114-121	54-64	15-22	15-22	7-15	22-30
2003	3.8	4	①	②	③	118	54-64	④	15-22	7-15	22-30
	4.6	V	⑤	⑥	⑦	114-121	54-64	⑧	13-16	7-15	22-30
	4.6	X	⑨	⑩	⑪	114-121	54-64	15-22	15-22	7-15	22-30
2004	3.8	4	①	②	③	118	54-64	④	15-22	7-15	22-30
	4.6	V	⑤	⑥	⑦	114-121	54-64	⑧	13-16	7-15	22-30
	4.6	V	⑤	⑥	⑦	114-121	54-64	⑧	13-16	7-15	22-30
	4.6	X	⑨	⑩	⑪	114-121	54-64	15-22	15-22	7-15	22-30

① Do not reuse cylinder head bolts.
Step 1: 15 ft. lbs.
Step 2: 29 ft. lbs.
Step 3: 37 ft. lbs.
Step 4: Loosen bolts one at a time and retorque as follows:
Long bolts: 11-18 ft. lbs.
Short bolts: 7-15 ft. lbs.
Step 5: Tighten 85-95 degrees

② Step 1: 37 ft. lbs.
Step 2: Tighten 115-125 degrees

③ Step 1: 18 ft. lbs.
Step 2: 33 ft. lbs.
Step 3: Tighten 90-120 degrees

④ Upper intake manifold bolts:
Step 1: 8 ft. lbs.
Step 2: 15 ft. lbs.
Step 3: 24 ft. lbs
Lower intake manifold bolts:
Step 1: 13 ft. lbs.
Step 2: 16 ft. lbs.

⑤ Step 1: 30 ft. lbs.
Step 2: Tighten 90 degrees
Step 3: Loosen all bolts one (1) turn
Step 4: 30 ft. lbs.
Step 5: Tighten 90 degrees
Step 6: Tighten an additional 90 degrees

⑥ Step 1: Main bearing cap bolts: 6-9 ft. lbs.
Step 2: Main bearing cap bolts, outer: 16-21 ft. lbs.
Step 3: Main bearing cap bolts, inner: 27-32 ft. lbs.
Step 4: Tighten main bearing cap bolts 85-95 degrees
Step 5: Main cap adjusting screws: 4 ft. lbs. then 7.5 ft. lbs.
Step 6: Main cap side bolts: 7 ft. lbs. then 14-17 ft. lbs.

⑦ Step 1: 5ft. lbs.
Step 2: 10 ft. lbs.
Step 3: 18-25 ft. lbs.
Step 4: Tighten 85-95 degrees

⑧ Step 1: Four inside short bolts: 9-11 ft. lbs.
Step 2: All other bolts: 13-16 ft. lbs.
Step 3: Tighten 85-95 degrees

⑨ Do not reuse cylinder head bolts.
Step 1: 27-32 ft. lbs.
Step 2: Tighten each bolt 85-95 degrees
Step 3: Repeat Step 2

⑩ Do not reuse main cap bolts.
Step 1: Main bearing cap bolts: 22-25 ft. lbs.
Step 2: Tighten each bolt 85-95 degrees
Step 3: Main bearing cap adjust screws: 4 ft. lbs. then 6-8 ft. lbs.
Step 4: Main bearing cap side bolts: 7 ft. lbs. then 14-17 ft. lbs.

⑪ Do not reuse rod bolts.
Step 1: 12 ft. lbs.
Step 2: Tighten 85-95 degrees

67197-MUST-C08

WHEEL ALIGNMENT

Year	Model		Caster Range (+/-Deg.)	Caster Preferred Setting (Deg.)	Camber Range (+/-Deg.)	Camber Preferred Setting (Deg.)	Toe-in (in.)
2001	Mustang	F	0.75	0	0.75	0	0.13 +/- 0.13
		R	—	—	0.50	0	0.13 +/- 0.07
2002	Mustang	F	0.75	0	0.75	0	0.13 +/- 0.13
		R	—	—	0.50	0	0.13 +/- 0.07
2003	Mustang	F	0.75	0	0.75	0	0.13 +/- 0.13
		R	—	—	0.50	0	0.13 +/- 0.07
2004	Mustang	F	0.75	0	0.75	0	0.13 +/- 0.13
		R	—	—	0.50	0	0.13 +/- 0.07
	Mustang Cobra	F	0.75	0	0.75	0	0.13 +/- 0.13
		R	—	—	0.50	0	0.13 +/- 0.07

67197-MUST-C09

TIRE, WHEEL AND BALL JOINT SPECIFICATIONS

Year	Model	OEM Tires Standard	OEM Tires Optional	Tire Pressures (psi) Front	Tire Pressures (psi) Rear	Wheel Size	Ball Joint Inspection	Wheel Lug Torque (ft. lbs.)
2001	Mustang, base	P205/65HR15	None	30	30	7-JJ	①	95
	Mustang GT	P225/55HR16	P245/45ZR17	30	30	8-JJ	①	95
2002	Mustang, base	P205/65HR15	None	30	30	7-JJ	①	95
	Mustang GT	P225/55HR16	P245/45ZR17	30	30	8-JJ	①	95
2003	Mustang, base	P205/65HR15	None	30	30	7-JJ	①	95
	Mustang GT	P225/55HR16	P245/45ZR17	30	30	8-JJ	①	95
2004	Mustang, base	P225/55R16	None	30	30	-	②	95
	Mustang GT	245/45ZR17	None	30	30	-	②	95
	Mustang Cobra	245/45ZR17	None	30	30	-	②	95

OEM: Original Equipment Manufacturer

PSI: Pounds Per Square Inch

① Replace if any measurable movement is found

② Replace if movement exceeds the maximum measurement of 1/32 inches (0.8mm)

67197-MUST-C10

BRAKE SPECIFICATIONS

All measurements in inches unless noted

			Brake Disc			Brake Drum Diameter				Brake Caliper
Year	Model		Original Thickness	Minimum Thickness	Maximum Run-out	Original Inside Diameter	Max. Wear Limit	Maximum Machine Diameter	Minimum Lining Thickness	Mounting Bolts (ft. lbs.)
2001	Mustang	F	1.030	0.970	0.001	—	—	—	0.125	64
		R	0.550	0.500	0.002	—	—	—	0.123	26
	Mustang	F	1.100	1.040	0.001	—	—	—	0.125	64
	Cobra	R	0.710	0.660	0.002	—	—	—	0.123	26
2002	Mustang	F	1.030	0.970	0.001	—	—	—	0.125	64
		R	0.550	0.500	0.002	—	—	—	0.123	26
	Mustang	F	1.100	1.040	0.001	—	—	—	0.125	64
	Cobra	R	0.710	0.660	0.002	—	—	—	0.123	26
2003	Mustang	F	1.030	0.970	0.001	—	—	—	0.125	64
		R	0.550	0.500	0.002	—	—	—	0.123	26
	Mustang	F	1.100	1.040	0.001	—	—	—	0.125	64
	Cobra	R	0.710	0.660	0.002	—	—	—	0.123	26
2004	Mustang	F	1.020	0.970	0.002	—	—	—	0.118	23
		R	0.590	0.500	0.004	—	—	—	0.118	25
	Mustang	F	1.100	1.030	0.002	—	—	—	0.118	23
	Cobra	R	0.590	0.500	0.004	—	—	—	0.118	25

NOTE: Follow specifications stamped on rotor or drum if figures differ from those in this chart.

NA: Not Available

F: Front

R: Rear

67197-MUST-C11

SCHEDULED MAINTENANCE INTERVALS

Ford Mustang

TO BE SERVICED	TYPE OF SERVICE	VEHICLE MILEAGE INTERVAL (x1000)												
		5	10	15	20	25	30	35	40	45	50	55	60	65
Engine oil & filter	R	✓	✓	✓	✓	✓	✓	✓	✓	✓	✓	✓	✓	✓
Adjust clutch pedal by lifting pedal	S/I	✓	✓	✓	✓	✓	✓	✓	✓	✓	✓	✓	✓	✓
Rotate tires	S/I	✓		✓		✓		✓		✓		✓		✓
Cooling system, hoses, clamps & coolant strength	S/I			✓			✓			✓			✓	
Air cleaner element	R						✓						✓	
Automatic transmission fluid & filter	R						✓						✓	
Engine coolant ①	R						✓						✓	
Spark plugs	R													
Accessory drive belt(s)	S/I						✓						✓	
Brake lines, hoses & connections	S/I						✓						✓	
Exhaust heat shields	S/I						✓						✓	
Front & rear brakes	S/I						✓						✓	
PCV valve	R												✓	
Rear axle lubricant ②	R													

R: Replace S/I: Service or Inspect

① Engine coolant: change engine coolant at 48,000 to 50,000 miles and thereafter every 30,000 miles.

② Replace every 100,000 miles.

FREQUENT OPERATION MAINTENANCE (SEVERE SERVICE)

If a vehicle is operated under any of the following conditions it is considered severe service:

- Extremely dusty areas.

- 50% or more of the vehicle operation is in 32°C (90°F) or higher temperatures, or constant operation in temperatures below 0°C (32°F).

- Prolonged idling (vehicle operation in stop and go traffic).

- Frequent short running periods (engine does not warm to normal operating temperatures).

- Police, taxi, delivery usage or trailer towing usage.

Oil & filter change: change every 3000 miles.

Rotate tires at 6000 miles & every 9000 miles thereafter.

Automatic transmission fluid & filter: change every 21,000 miles.

67197-MUST-C12

PRECAUTIONS

Before servicing any vehicle, please be sure to read all of the following precautions, which deal with personal safety, prevention of component damage, and important points to take into consideration when servicing a motor vehicle:

- Never open, service or drain the radiator or cooling system when the engine is hot; serious burns can occur from the steam and hot coolant.
- Observe all applicable safety precautions when working around fuel. Whenever servicing the fuel system, always work in a well-ventilated area. Do not allow fuel spray or vapors to come in contact with a spark, open flame, or excessive heat (a hot drop light, for example). Keep a dry chemical fire extinguisher near the work area. Always keep fuel in a container specifically designed for fuel storage; also, always properly seal fuel containers to avoid the possibility of fire or explosion. Refer to the additional fuel system precautions later in this section.
- Fuel injection systems often remain pressurized, even after the engine has been turned **OFF**. The fuel system pressure must be relieved before disconnecting any fuel lines. Failure to do so may result in fire and/or personal injury.
- Brake fluid often contains polyglycol ethers and polyglycols. Avoid contact with the eyes and wash your hands thoroughly after handling brake fluid. If you do get brake fluid in your eyes, flush your eyes with clean, running water for 15 minutes. If eye irritation persists, or if you have taken brake fluid internally, IMMEDIATELY seek medical assistance.
- The EPA warns that prolonged contact with used engine oil may cause a number of skin disorders, including cancer. You should make every effort to minimize your exposure to used engine oil. Protective gloves should be worn when changing oil. Wash your hands and any other exposed skin areas as soon as possible after exposure to used engine oil. Soap and water, or waterless hand cleaner should be used.
- All new vehicles are now equipped with an air bag system, often referred to as a Supplemental Restraint System (SRS) or Supplemental Inflatable Restraint (SIR) system. The system must be disabled before performing service on or around system components, steering column, instrument panel components, wiring and sensors. Failure to follow safety and disabling procedures could result in accidental air bag deployment, possible personal injury and unnecessary system repairs.
- Always wear safety goggles when working with, or around, the air bag system. When carrying a non-deployed air bag, be sure the bag and trim cover are pointed away from your body. When placing a non-deployed air bag on a work surface, always face the bag and trim cover upward, away from the surface. This will reduce the motion of the module if it is accidentally deployed. Refer to the additional air bag system precautions later in this section.
- Clean, high quality brake fluid from a sealed container is essential to the safe and proper operation of the brake system. You should always buy the correct type of brake fluid for your vehicle. If the brake fluid becomes contaminated, completely flush the system with new fluid. Never reuse any brake fluid. Any brake fluid that is removed from the system should be discarded. Also, do not allow any brake fluid to come in contact with a painted surface; it will damage the paint.
- Never operate the engine without the proper amount and type of engine oil; doing so WILL result in severe engine damage.
- Timing belt maintenance is extremely important. Many models utilize an interference-type, non-freewheeling engine. If the timing belt breaks, the valves in the cylinder head may strike the pistons, causing potentially serious (also time-consuming and expensive) engine damage.
- Disconnecting the negative battery cable on some vehicles may interfere with the functions of the on-board computer system(s) and may require the computer to undergo a relearning process once the negative battery cable is reconnected.
- When servicing drum brakes, only disassemble and assemble one side at a time, leaving the remaining side intact for reference.
- Only an MVAC-trained, EPA-certified automotive technician should service the air conditioning system or its components.

ENGINE REPAIR

➡Disconnecting the negative battery cable on some vehicles may interfere with the functions of the on board computer system. The computer may undergo a relearning process once the negative battery cable is reconnected.

Distributor

The 3.8L and 4.6L engines use Distributorless Ignition Systems (DIS).

Alternator

REMOVAL

3.8L Engine

1. Before servicing the vehicle, refer to the precautions at the beginning of this section.
2. Disconnect the negative battery cable.
3. Remove the accessory drive belt.
4. Disconnect the alternator electrical connectors.
5. Remove the alternator mounting bolts.
6. Remove the alternator from the vehicle.

4.6L Engines (Non-Supercharged)

1. Before servicing the vehicle, refer to the precautions at the beginning of this section.
2. Disconnect the negative battery cable.
3. Remove the air intake scoop bracket, if equipped.
4. Remove the accessory drive belt.
5. Remove the upper alternator bracket.
6. Disconnect the alternator electrical connectors.
7. Remove the alternator bolts.
8. Remove the alternator from the vehicle.

4.6L Engines (Supercharged)

1. Before servicing the vehicle, refer to the precautions at the beginning of this section.
2. Disconnect the negative battery cable.
3. Remove the accessory drive belt.
4. Raise the vehicle.
5. Disconnect the alternator electrical connectors.
6. Remove the alternator bolts.
7. Remove the alternator from the vehicle.

INSTALLATION

3.8L Engine

1. Position the alternator onto engine. Tighten the lower bolt to 35 ft. lbs. (47 Nm) and the upper bolt to 18 ft. lbs. (25 Nm).
2. Connect the alternator electrical connectors.
3. Install the accessory drive belt.
4. Connect the negative battery cable.

4.6L Engines (Non-Supercharged)

1. Position the alternator onto engine. Tighten the mounting bolts to 18 ft. lbs. (25 Nm).
2. Connect the alternator electrical connectors.
3. Install the upper alternator bracket. Tighten the bolts to 89 inch lbs. (10 Nm).
4. Install the accessory drive belt.
5. Install the air intake scoop bracket, if equipped.
6. Connect the negative battery cable.

4.6L Engines (Supercharged)

1. Position the alternator onto engine. Tighten the lower bolt to 18 ft. lbs. (25 Nm) and the upper bolt to 37 ft. lbs. (50 Nm).
2. Connect the alternator electrical connectors.
3. Lower the vehicle.
4. Install the accessory drive belt.
5. Connect the negative battery cable.

Ignition Timing

ADJUSTMENT

The ignition timing is controlled by the Powertrain Control Module (PCM). No adjustment is necessary.

Engine Assembly

REMOVAL & INSTALLATION

3.8L Engine

1. Before servicing the vehicle, refer to the precautions at the beginning of this section.
2. Discharge the air conditioning system.
3. Drain the engine cooling system.
4. Drain the engine oil and remove the oil filter.
5. Disconnect the negative battery cable.
6. Relieve the fuel system pressure.
7. Remove the hood.
8. Remove cooling fan assembly.
9. Remove radiator and hoses.
10. Disconnect heater hoses.
11. Disconnect Intake Air Temperature (IAT) sensor connector.
12. Remove air intake duct.
13. Remove the accessory drive belt.
14. Disconnect alternator wiring connectors.
15. Remove coolant recovery tank.
16. Remove power steering pump and bracket.
17. Disconnect power steering pressure switch connector.
18. Remove A/C compressor and bracket.
19. Disconnect cruise control cable, if equipped.
20. Disconnect accelerator cable and remove bracket.
21. Disconnect fuel supply line.
22. Disconnect fuel return line, if equipped.
23. Disconnect fuel charging wiring harness connectors.
24. Disconnect main vacuum source hose.
25. Disconnect evaporative emissions hose.
26. Disconnect Heated Oxygen (HO_2S) sensor connectors.
27. Remove dual converter Y-pipe.
28. Remove starter.
29. Remove starter wiring harness retainers.
30. Remove transmission flange bolts.
31. Disconnect torque converter, automatic transmission models only.
32. Disconnect transmission oil cooler lines, automatic transmission models only.
33. Remove left and right engine support insulators.
34. Attach an engine hoist and remove the engine from the vehicle.

To install:

35. Lower the engine into the vehicle.
36. Install left and right engine support insulators. Tighten through bolts to 35–50 ft. lbs. (47–68 Nm).
37. Install Transmission flange bolts. Tighten bolts to 25–33 ft. lbs. (34–46 Nm).
38. Connect torque converter, automatic transmission models only. Tighten the nuts to 20–33 ft. lbs. (27–46 Nm).
39. Connect transmission oil cooler lines, automatic transmission models only.
40. Install starter wiring harness retainers.
41. Install starter.
42. Install dual converter Y-pipe.
43. Connect HO_2S sensor connectors.

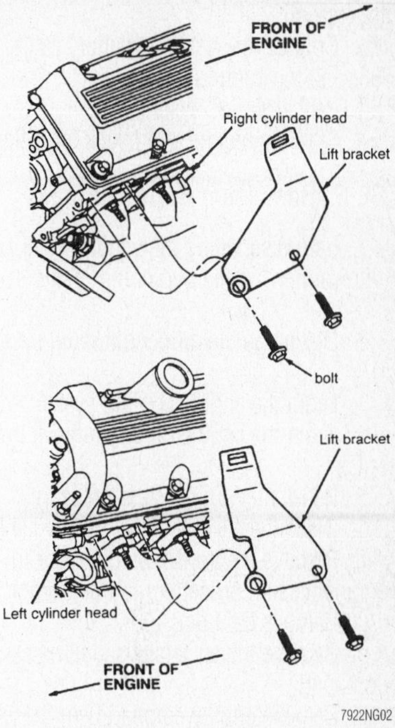

Install the engine lifting brackets on both sides of the engine—4.6L SOHC engine shown, other engines similar

44. Connect evaporative emissions hose.
45. Connect main vacuum source hose.
46. Connect fuel charging wiring harness connectors.
47. Connect fuel supply line.
48. Connect fuel return line, if equipped.
49. Connect accelerator cable and install bracket.
50. Connect cruise control cable, if equipped.
51. Install A/C compressor and bracket.
52. Connect power steering pressure switch connector.
53. Install power steering pump and bracket.
54. Install coolant recovery tank.
55. Connect alternator wiring connectors.
56. Install the accessory drive belt.
57. Install air intake duct.
58. Connect IAT sensor connector.
59. Connect heater hoses.
60. Install radiator and hoses.
61. Install cooling fan assembly.
62. Install hood.

⁂ WARNING

Be sure to check engine to see that all electrical connectors, hoses and cables are properly connected and secure.

63. Install new oil filter and fill engine with fresh oil.
64. Connect the negative battery cable.
65. Fill the cooling system.
66. Recharge A/C system.
67. Start the engine and check for leaks.

4.6L Engines (Non-Supercharged)

1. Before servicing the vehicle, refer to the precautions at the beginning of this section.
2. Discharge the air conditioning system.
3. Drain the engine cooling system.
4. Drain the engine oil and remove the oil filter.
5. Relieve the fuel system pressure.
6. Disconnect the negative battery cable.
7. Remove the transmission. Refer to the transmission procedure in this section.
8. Remove the hood.
9. Remove the air intake scoop bracket, if equipped.
10. Remove engine compartment brace, if necessary.
11. Disconnect Intake Air Temperature (IAT) sensor connector.
12. Disconnect Mass Airflow (MAF) sensor connector.
13. Remove air cleaner and intake duct assembly.
14. Remove coolant recovery tank assembly.
15. Remove cooling fan assembly.
16. Disconnect oil filter adapter hose, if necessary.
17. Disconnect oil pressure sensor connector.
18. Remove radiator hoses.
19. Disconnect heater hoses.
20. Disconnect cruise control cable, if equipped.
21. Disconnect accelerator cable.
22. Remove the accessory drive belt.
23. Disconnect alternator wiring connectors.
24. Disconnect fuel supply line.
25. Disconnect fuel return line, if equipped.
26. Disconnect fuel charging and engine control wiring harness connectors.
27. Separate wiring harness from cowl panel.
28. Disconnect evaporative emissions hose.
29. Disconnect vacuum hoses and label, if necessary.
30. Disconnect A/C compressor suction and discharge lines.
31. Disconnect secondary air injection switching valve, on some DOHC engines.
32. Remove power steering reservoir.
33. Disconnect power steering fluid cooler hoses.
34. Disconnect power steering pressure and return hoses.
35. Disconnect Heated Oxygen (HO_2S) sensor connectors. Place wiring harness out of way.
36. Remove dual converter H-pipe.
37. Remove starter motor.
38. Position transmission cooler lines out of way, automatic transmission only.
39. Disconnect engine ground cable.
40. Remove body ground strap bolt.
41. Remove left and right engine support insulators on 2001–03 models. Remove left and right engine mount nuts on 2004 models.
42. Install left and right engine lifting brackets.
43. Attach an engine hoist and remove engine from the vehicle.

To install:

44. Lower the engine into the vehicle.
45. Install left and right engine support insulators on 2001–03 models and tighten through bolts to 15–22 ft. lbs. (20–30 Nm). Install left and right engine mount retaining nuts on 2004 models and tighten retaining nuts 111 ft. lbs. (150 Nm).
46. Remove engine lifting brackets.
47. Position body ground strap and tighten bolt.
48. Connect engine ground cable.
49. Install starter motor.
50. Install dual converter H-pipe.
51. Connect Heated Oxygen (HO_2S) sensor connectors.
52. Connect power steering pressure and return hoses.
53. Connect power steering fluid cooler hoses.
54. Install power steering reservoir.
55. Connect secondary air injection switching valve, if equipped.
56. Connect A/C compressor suction and discharge lines.
57. Connect vacuum hoses.
58. Connect evaporative emissions hose.
59. Position and secure wiring harness onto cowl panel.
60. Connect fuel charging and engine control wiring harness connectors.
61. Connect fuel return line, if equipped.
62. Connect fuel supply line.
63. Connect alternator wiring connectors.
64. Install the accessory drive belt.
65. Connect accelerator cable.
66. Connect cruise control cable, if equipped.
67. Connect heater hoses.
68. Install radiator hoses.
69. Connect oil pressure sensor connector.
70. Connect oil filter adapter hose, if necessary.
71. Install cooling fan assembly.
72. Install coolant recovery tank assembly.
73. Install air cleaner and intake duct assembly.
74. Connect MAF sensor connector.
75. Connect IAT sensor connector.
76. Install engine compartment brace, if necessary.
77. Install the air intake scoop bracket, if equipped.
78. Install hood.
79. Install transmission.

**** WARNING**

Be sure to check engine to see that all electrical connectors, hoses and cables are properly connected and secure.

80. Install a new oil filter and fill engine with fresh oil.
81. Fill the cooling system.
82. Connect the negative battery cable.
83. Recharge the air conditioning.
84. Start the engine and check for leaks.

4.6L Engine (Supercharged)

1. Before servicing the vehicle, refer to the precautions at the beginning of this section.
2. Remove battery and battery tray, making sure to disconnect negative battery cable first.
3. Relieve the fuel system pressure.
4. Drain the engine cooling system.
5. Drain the engine oil and remove the oil filter.
6. Drain the supercharger cooling system.
7. Discharge the air conditioning system.
8. Remove hood and hood prop.
9. Disconnect Intake Air Temperature (IAT) sensor connector.
10. Disconnect breather hose.
11. Disconnect and label vacuum hoses.
12. Remove air cleaner assembly.
13. Disconnect Mass Airflow (MAF) connector.
14. Disconnect Throttle Position (TP) sensor and Idle Air Control (IAC) valve connectors.
15. Disconnect A/C pressure switch connector.
16. Disconnect fuel supply line.

17. Disconnect accelerator cable.
18. Disconnect cruise control cable, if equipped.
19. Remove accelerator/cruise control cable bracket and place out of the way of engine.
20. Remove throttle body and spacer assembly.
21. Disconnect 16-pin and 42-pin wiring connectors.
22. Separate wiring harness from cowl panel.
23. Remove heater hoses.
24. Remove radiator and supercharger cooling system hoses.
25. Remove A/C compressor suction and discharge lines.
26. Disconnect dual function pressure switch wiring connector.
27. Remove coolant recovery tank and support bracket.
28. Remove power steering reservoir and tube bracket.
29. Remove wiring harness support brackets.
30. Remove supercharger coolant recovery tank.
31. Remove electric cooling fan and shroud assemblies.
32. Disconnect power steering tube and bracket bolt, then remove hose from grommet.
33. Remove auxiliary crank pulley.
34. Remove clutch.
35. Disconnect Heated Oxygen (HO_2S) sensor connectors.
36. Disconnect transmission wiring connector, then remove harness with bracket.
37. Disconnect exhaust pipe at right and left exhaust manifolds. Move exhaust pipe out of way.
38. Disconnect oil pressure sender wiring connector.
39. Disconnect alternator wiring connectors.
40. Disconnect engine ground cable.
41. Disconnect oil filter adapter hose.
42. Remove engine mount retaining nuts.
43. Disconnect clutch cable and position out of way.
44. Remove starter motor.
45. Remove left and right ignition coil covers.
46. Install left and right engine lifting brackets.
47. Attach an engine hoist and remove the engine from the vehicle.

To install:

48. Lower the engine into the vehicle.
49. Install left and right engine mount retaining nuts. Tighten retaining nuts 111 ft. lbs. (150 Nm).
50. Remove engine lifting brackets.
51. Install ignition coil covers.
52. Install starter motor.
53. Connect clutch cable and tighten retaining screws.
54. Connect oil filter adapter hose.
55. Connect engine ground cable.
56. Connect alternator wiring connectors.
57. Connect oil pressure sender wiring connector.
58. Connect exhaust pipe to right and left exhaust manifolds and tighten manifold flange nuts.
59. Install transmission wiring harness bracket, then connect to transmission wiring connector.
60. Connect Heated Oxygen (HO_2S) sensor connectors.
61. Install clutch.
62. Install auxiliary crank pulley.
63. Secure power steering hose in grommet. Position power steering tube and bracket. Tighten power steering tube bracket bolt.
64. Install electric cooling fan and shroud assemblies.
65. Install supercharger coolant recovery tank.
66. Install wiring harness support brackets.
67. Install power steering reservoir and tube bracket.
68. Install coolant recovery tank and support bracket.
69. Connect dual function pressure switch wiring connector.
70. Install A/C compressor suction and discharge lines.
71. Install radiator and supercharger cooling system hoses.
72. Install heater hoses.
73. position and secure wiring harness to cowl panel.
74. Connect 16-pin and 42-pin wiring connectors.
75. Install throttle body and spacer assembly.
76. Install accelerator/cruise control cable bracket.
77. Connect cruise control cable, if equipped.
78. Connect accelerator cable.
79. Connect fuel supply line.
80. Connect A/C pressure switch connector.
81. Connect TP sensor and IAC valve connectors.
82. Connect MAF connector.
83. Install air cleaner assembly.
84. Connect vacuum hoses.
85. Connect breather hose.
86. Connect IAT sensor connector.
87. Install hood and hood prop.

WARNING

Be sure to check engine to see that all electrical connectors, hoses and cables are properly connected and secure.

88. Install a new oil filter and fill engine with fresh oil.
89. Fill the supercharger cooling system.
90. Fill the cooling system.
91. Recharge air conditioning system.
92. Install battery tray and install battery. Connect positive battery cable first, negative battery cable second.
93. Start the engine and check for leaks.

Water Pump

REMOVAL & INSTALLATION

3.8L Engine

1. Before servicing the vehicle, refer to the precautions at the beginning of this section.
2. Drain the engine cooling system.
3. Remove electric cooling fan assembly.
4. Loosen water pump pulley bolts.
5. Remove the accessory drive belt.
6. Remove power steering pump pulley.
7. Remove water pump pulley.
8. Remove ignition coil and bracket, if necessary.
9. Remove power steering pump bracket.
10. Disconnect lower radiator hose from water pump.
11. Disconnect heater water outlet tube.
12. Remove water pump. Clean and inspect gasket mating surfaces.

To install:

➡The threads of the No. 1 water pump retaining bolt must be coated with a Teflon® sealant prior to installation.

13. Install water pump. Use a new gasket and tighten the bolts and studs to 15–22 ft. lbs. (20–30 Nm). Tighten the nuts to 71–106 inch lbs. (8–12 Nm).
14. Connect lower radiator hose.
15. Connect heater water outlet tube. Use a new O-ring seal and tighten the bolts to 71–106 inch lbs. (8–12 Nm).
16. Install power steering pump bracket.
17. Install ignition coil and bracket, if necessary.

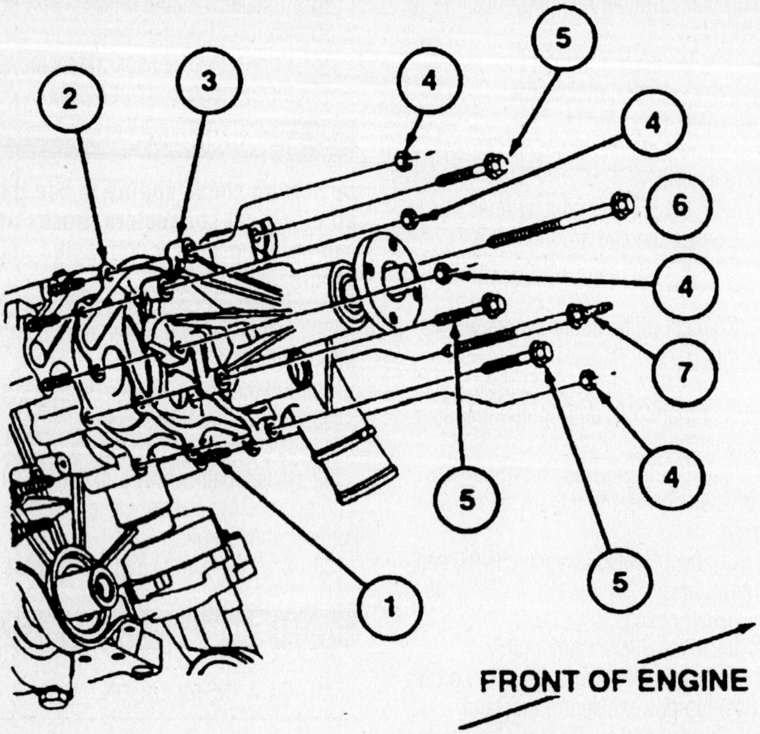

1. Mounting stud
2. Water pump housing gasket
3. Water pump
4. Mounting nuts
5. Short mounting bolts
6. Long mounting bolt
7. Mounting stud bolt

7922NG03

Exploded view of the water pump mounting on the 3.8L engine—during removal, note the original positions of the different length bolts for reassembly

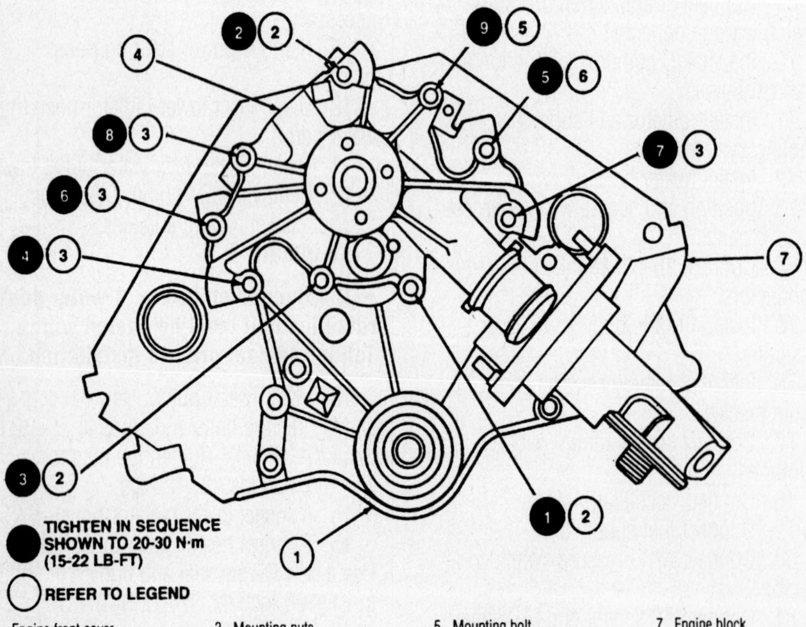

1. Engine front cover
2. Mounting bolts
3. Mounting nuts
4. Water pump
5. Mounting bolt
6. Mounting stud bolt
7. Engine block

7922NG04

Water pump torque sequence—3.8L engine

18. Install water pump pulley. Tighten the bolts to 15–21 ft. lbs. (21–29 Nm).
19. Install power steering pump pulley.
20. Install water pump pulley. Tighten the bolts to 15–21 ft. lbs. (21–29 Nm).
21. Install the accessory drive belt.
22. Install electric cooling fan assembly.
23. Fill the cooling system.
24. Start the engine and check for coolant leaks.

4.6L Engines (Non-Supercharged)

1. Before servicing the vehicle, refer to the precautions at the beginning of this section.
2. Remove air intake scoop, if equipped.
3. Drain the engine cooling system.
4. Remove electric cooling fan assembly.
5. Loosen water pump pulley bolts.
6. Remove the accessory drive belt.
7. Remove water pump pulley.
8. Remove water pump.

To install:

9. Install water pump. Use a new O-ring seal and tighten the bolts in a crossing pattern to 15–22 ft. lbs. (20–30 Nm).
10. Install water pump pulley. Tighten the bolts to 15–21 ft. lbs. (21–29 Nm).
11. Install the accessory drive belt.
12. Install electric cooling fan assembly.
13. Install air intake scoop, if equipped.
14. Fill the cooling system.
15. Start the engine and check for coolant leaks.

4.6L Engines (Supercharged)

1. Before servicing the vehicle, refer to the precautions at the beginning of this section.
2. Drain the engine cooling system.
3. Drain the supercharger cooling system.
4. Remove nut and move wiring harness aside.
5. Remove stud and move coolant hose assembly aside.
6. Lower the vehicle.
7. Disconnect radiator hoses.
8. Remove top radiator shield.
9. Remove coolant recovery tank.
10. Remove supercharger drive belt cover.
11. Remove supercharger drive belt.
12. Remove supercharger cooling system hose.
13. Remove fasteners and move power steering reservoir aside.
14. Remove nut and move A/C muffler bracket aside.

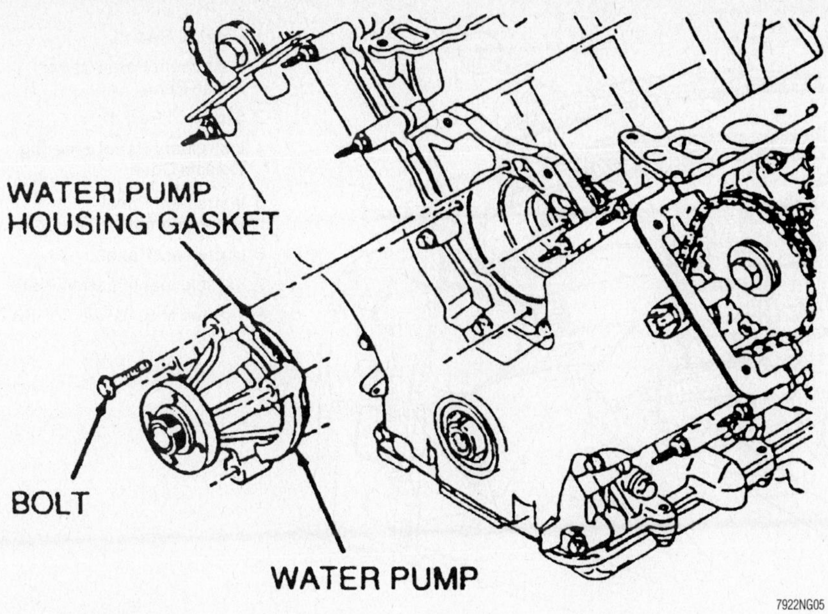

Exploded view of the water pump mounting—4.6L engines

15. Remove belt idler bracket support.
16. Loosen water pump pulley bolts.
17. Remove the accessory drive belt.
18. Remove water pump pulley.
19. Remove water pump.

To install:

20. Install water pump. Use a new O-ring seal and tighten the bolts in a crossing pattern to 18 ft. lbs. (25 Nm).
21. Install water pump pulley. Tighten the bolts to 18 ft. lbs. (25 Nm).
22. Install the accessory drive belt.
23. Install belt idler bracket support.
24. Place A/C muffler bracket in position and tighten mounting nut.
25. Place power steering reservoir in position and install fasteners.
26. Install supercharger cooling system hose.
27. Install supercharger drive belt.
28. Install supercharger drive belt cover.
29. Install coolant recovery tank.
30. Install top radiator shield.
31. Connect radiator hoses.
32. Place coolant hose assembly in position and tighten stud.
33. Place wiring harness in position and tighten nut..
34. Fill the cooling system.
35. Start the engine and check for coolant leaks.

Heater Core

REMOVAL & INSTALLATION

1. Before servicing the vehicle, refer to the precautions at the beginning of this section.
2. Drain the cooling system.
3. Recover the A/C refrigerant, if equipped.
4. Disconnect the negative battery cable.
5. Remove radio.
6. Remove CD player, if equipped.
7. Remove center console.
8. Remove left front wheel.
9. Remove left front inner fender.
10. Disconnect left bulkhead harness connector.
11. Remove left and right air bag modules.
12. Disconnect antenna cable connector.
13. Remove upper right instrument panel support bolt.
14. Disconnect climate control vacuum harness connector.
15. Remove left and right scuff plates.
16. Remove left and right A-pillar lower trim panels.
17. Remove left and right windshield side garnish moldings.
18. Remove pin-type retainers on convertible models only.
19. Position left and right door weatherstrips out of way.
20. Disconnect right main wire harness connectors.
21. Disconnect climate control wire harness connector.
22. Remove instrument panel steering column cover.
23. Remove instrument panel reinforcement.
24. Remove pinch bolt, then separate steering column shaft from intermediate shaft.
25. Disconnect left main wiring harness connectors.
26. Disconnect Generic Electronic Module (GEM) connectors.
27. Disconnect Electronic Crash Sensor (ECS) module connector.
28. Disconnect shift interlock cable, if equipped.
29. Disconnect shifter assembly harness connector, if equipped.
30. Disconnect temperature control cable.
31. Remove 4 center instrument panel support bolts.
32. Remove left instrument panel support bolt and nut.
33. Remove right instrument panel support bolt.
34. Remove instrument panel upper finish panel.
35. Remove upper instrument panel support bolts.
36. Remove instrument panel.
37. Disconnect heater hoses.
38. Disconnect vacuum supply hose.
39. Remove A/C accumulator/drier.
40. Disconnect A/C liquid line.
41. Remove evaporator housing.
42. Remove foam weather seal.
43. Remove heater core cover.
44. Remove heater core.

To install:

45. Install heater core.
46. Install heater core cover.
47. Install a new foam weather seal.
48. Install evaporator housing. Tighten the fasteners to 71 inch lbs. (8 Nm).
49. Connect A/C liquid line.
50. Install A/C accumulator/drier.
51. Connect vacuum supply hose.
52. Connect heater hoses.
53. Install instrument panel.
54. Install upper instrument panel support bolts. Tighten the bolts to 80 inch lbs. (9 Nm).
55. Install instrument panel upper finish panel.
56. Install right instrument panel support bolt. Tighten the bolt to 80 inch lbs. (9 Nm).
57. Install left instrument panel support bolt and nut. Tighten the bolt to 80 inch lbs. (9 Nm) and the nut to 35 ft. lbs. (48 Nm).
58. Install 4 center instrument panel support bolts. Tighten the bolts to 80 inch lbs. (9 Nm).
59. Connect temperature control cable.
60. Connect shifter assembly harness connector, if equipped.
61. Connect shift interlock cable, if equipped.
62. Connect ECS module connector.

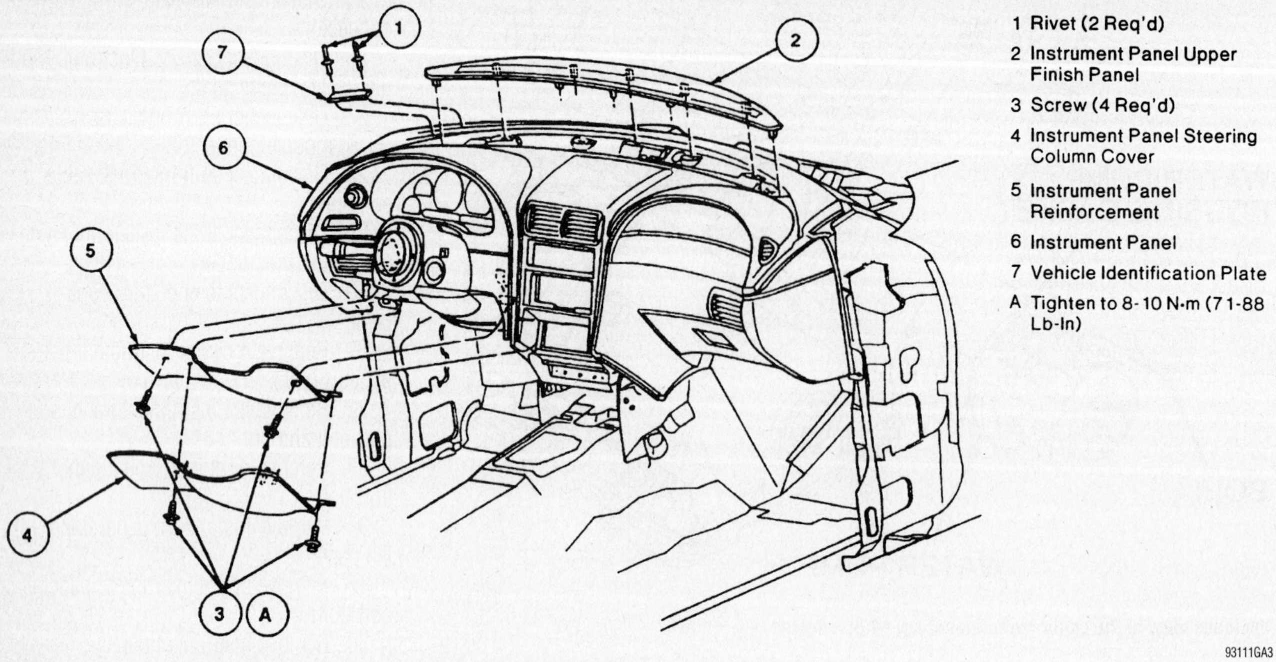

Exploded view of the instrument panel, cover and reinfircement—Mustang

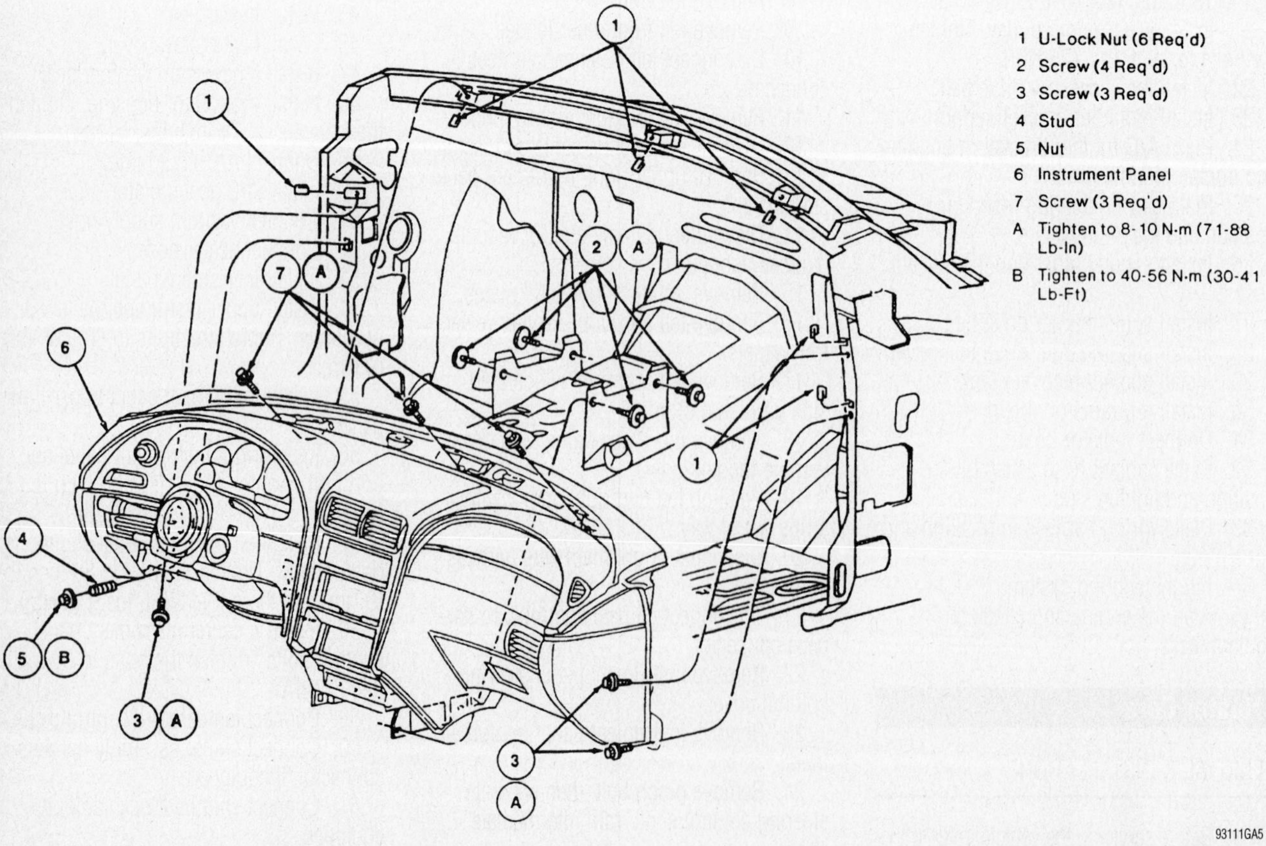

Exploded view of the instrument panel and related components—Mustang

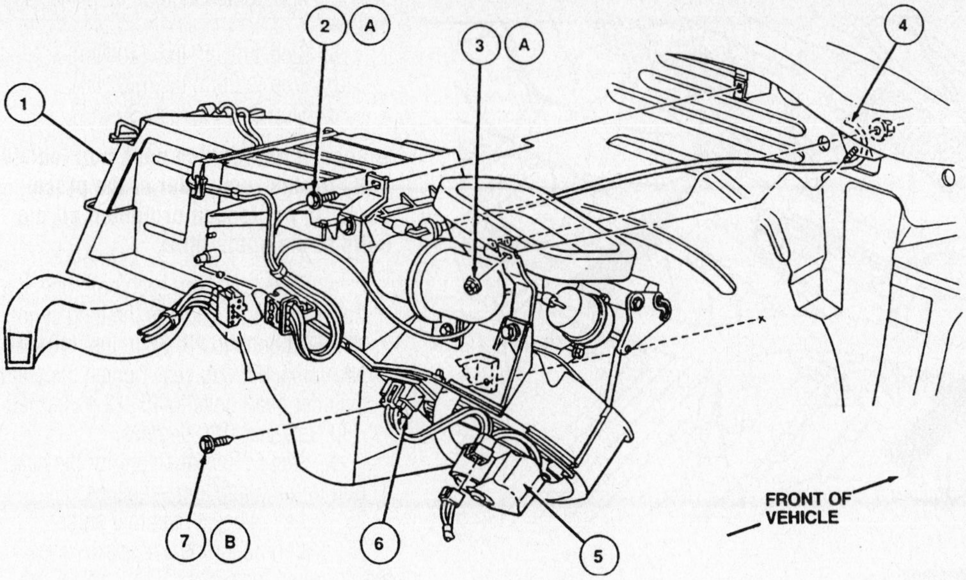

Exploded view of the heater/air conditioning housing assembly—Mustang

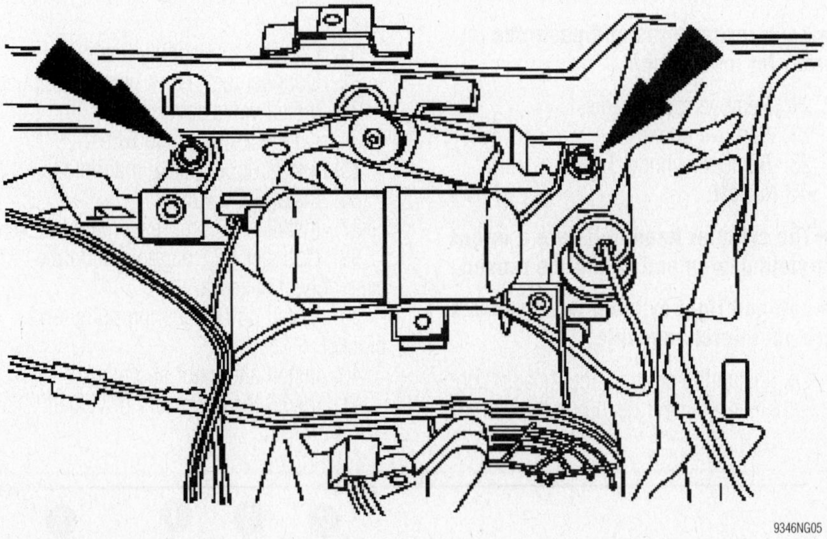

Evaporator case interior fasteners—Mustang

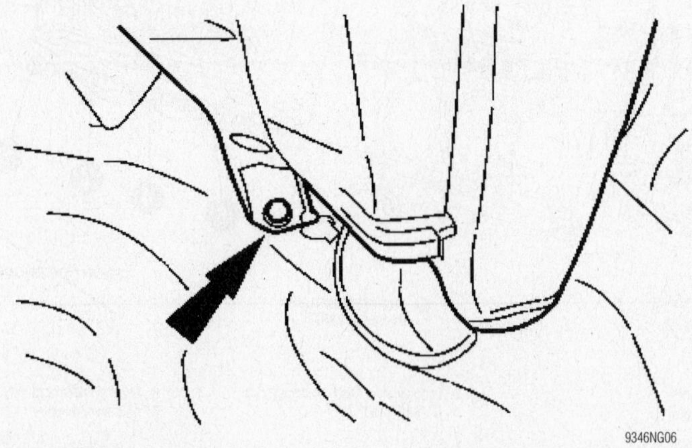

Evaporator case screw—Mustang

63. Connect GEM connectors.
64. Connect left main wiring harness connectors.
65. Connect steering column shaft to intermediate shaft and install pinch bolt. Tighten the pinch bolt to 35 ft. lbs. (47 Nm).
66. Install instrument panel reinforcement. Tighten the screws to 80 inch lbs. (9 Nm).
67. Install instrument panel steering column cover. Tighten the screws to 80 inch lbs. (9 Nm).
68. Connect climate control wire harness connector.
69. Connect right main wire harness connectors.
70. Place left and right door weatherstrips back in correct position.
71. Install pin-type retainers on convertible models only.
72. Install left and right windshield side garnish moldings.
73. Install left and right A-pillar lower trim panels.
74. Install left and right scuff plates.
75. Connect climate control vacuum harness connector.
76. Connect antenna cable connector.
77. Install upper right instrument panel support bolt. Tighten the bolt to 80 inch lbs. (9 Nm).
78. Install left and right air bag modules.
79. Connect left bulkhead harness connector.
80. Install left front inner fender.
81. Install left front wheel.
82. Install center console.
83. Install CD player, if equipped.

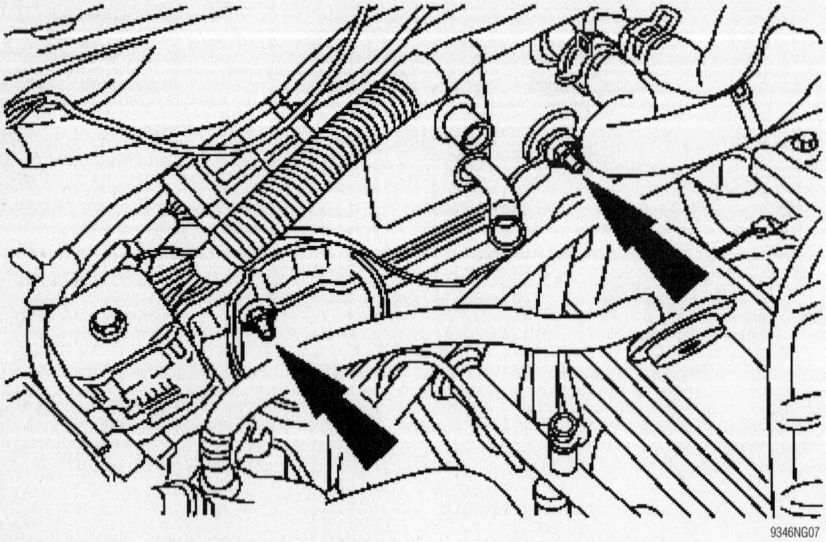

Evaporator case under-hood nuts—Mustang

84. Install radio.
85. Connect the negative battery cable.
86. Fill the cooling system.
87. Recharge the A/C system.
88. Start the engine and check for leaks.

Cylinder Head

REMOVAL & INSTALLATION

3.8L Engine

1. Before servicing the vehicle, refer to the precautions at the beginning of this section.
2. Drain the cooling system.
3. Disconnect the negative battery cable.
4. Remove air intake tube.
5. Remove the accessory drive belt.
6. Remove alternator and bracket.
7. Remove power steering pump and bracket.
8. Disconnect A/C manifold and tube assembly, if necessary.
9. Remove A/C compressor and bracket.
10. Remove Positive Crankcase Ventilation (PCV) valve.
11. Remove upper intake manifold.
12. Remove spark plug wires.
13. Remove valve covers.
14. Disconnect fuel supply line.
15. Disconnect fuel return line, if equipped.
16. Disconnect fuel injector wiring connectors.
17. Remove fuel injection supply manifold.
18. Remove lower intake manifold.
19. Remove exhaust manifolds.
20. Remove exhaust manifold studs.

➡Keep rocker arms and pushrods in order for installation.

21. Remove rocker arms.
22. Remove pushrods.
23. Remove cylinder heads.

To install:

➡The cylinder head bolts are a torque-to-yield design and cannot be reused.

➡Left and right cylinder head gaskets are not interchangeable.

24. Install the new cylinder heads. Lubricate the new cylinder head bolt threads with clean oil and tighten them in sequence as follows:
    a. Step 1: 15 ft. lbs. (20 Nm).
    b. Step 2: 29 ft. lbs. (40 Nm).
    c. Step 3: 36 ft. lbs. (50 Nm).

➡Loosen and tighten each bolt individually for the remainder of the procedure. Do not loosen or tighten all the bolts at the same time.

    d. Step 4: Loosen each bolt, one at a time, 2–3 turns. Tighten the long cylinder head bolts to 29–37 ft. lbs. (40–50 Nm) plus 180 degrees. Tighten the short cylinder head bolts to 15–22 ft. lbs. (20–30 Nm) plus 180 degrees.
    e. Step 5: Repeat step 4 for the next bolt in the tightening sequence.
25. Install exhaust manifold studs.
26. Install pushrods and rocker arms in their original locations.
27. Install exhaust manifolds.
28. Install lower intake manifold.
29. Install fuel injection supply manifold.
30. Connect fuel injector wiring connectors.
31. Connect fuel supply line.
32. Connect fuel return line, if equipped.
33. Install valve covers.
34. Connect spark plug wires.
35. Install upper intake manifold.
36. Install PCV valve.
37. Install A/C compressor and bracket.
38. Connect A/C manifold and tube assembly, if necessary.
39. Install power steering pump and bracket.
40. Install alternator and bracket.
41. Install the accessory drive belt.

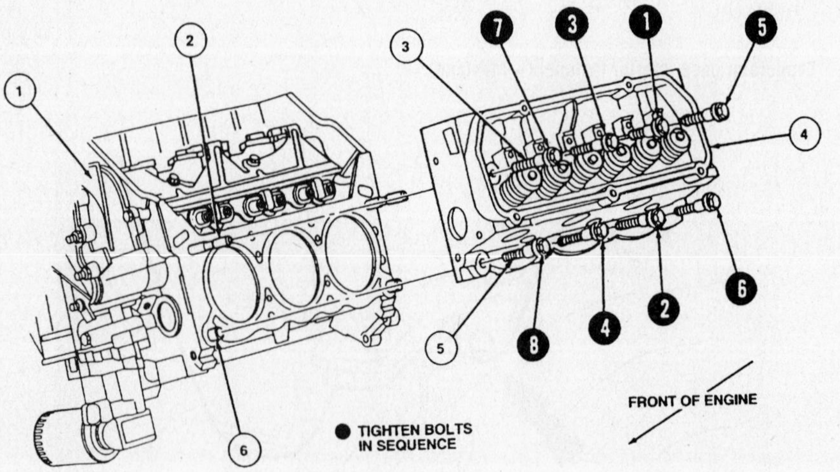

1. Engine block
2. Locating pin
3. Long cylinder head mounting bolts
4. Cylinder head
5. Short cylinder head mounting bolts
6. Alignment dowel

7922NG07

Cylinder head torque sequence—3.8L engine

42. Install air intake tube.
43. Connect the negative battery cable.
44. Fill the cooling system.
45. Start the engine and check for leaks.
46. Recharge A/C system.

4.6L SOHC Engine

1. Before servicing the vehicle, refer to the precautions at the beginning of this section.
2. Drain the cooling system.
3. Relieve the fuel system pressure.
4. Disconnect the negative battery cable.
5. Remove electric cooling fan.
6. Disconnect fuel supply line.
7. Disconnect fuel return line, if equipped.
8. Remove air intake tube.
9. Remove windshield wiper governor, if necessary.
10. Remove the accessory drive belt.
11. Remove spark plug wires, if equipped.
12. Remove ignition coils.
13. Disconnect Camshaft Position (CMP) sensor connector.
14. Remove power steering reservoir and bracket as an assembly.
15. Remove alternator and bracket.
16. Remove water pump pulley.
17. Disconnect Mass Air Flow (MAF) sensor connector.
18. Disconnect fuel injector wiring connectors.
19. Disconnect Crankshaft Position (CKP) sensor connector.
20. Disconnect A/C compressor clutch connector.
21. Disconnect cruise control cable.
22. Disconnect accelerator cable and remove bracket.
23. Disconnect canister purge valve connector, if necessary.
24. Drain coolant from engine block.
25. Remove power steering pump.
26. Remove left and right exhaust manifolds.
27. Remove oil pan.
28. Remove crankshaft pulley.
29. Disconnect oil pressure sensor connector, if necessary.
30. Disconnect Exhaust Gas Recirculation (EGR) tube.
31. Remove PCV valve and hose.
32. Disconnect Heated Oxygen (HO_2S) sensor connectors.
33. Remove dual converter H-pipe.
34. Remove starter motor wiring harness retainer, if necessary.
35. Remove valve covers.
36. Remove rocker arms.
37. Remove throttle body and adapter as an assembly.
38. Remove heater hoses.
39. Remove upper radiator hose and adapter.
40. Remove intake manifold.
41. Remove front cover.
42. Remove crankshaft sensor ring.
43. Remove timing chains.
44. Remove heater water tube.
45. Remove hydraulic lash adjusters.
46. Remove cylinder heads.

To install:

➡The cylinder head bolts are a torque-to-yield design and cannot be reused.

47. Rotate the crankshaft so that the keyway is at the 9 o'clock position.
48. Rotate each camshaft to a stable position where the valves do not extend below the cylinder head face.
49. Install the new cylinder heads. Lubricate the new cylinder head bolt threads with clean oil and tighten them in sequence as follows:
    a. Step 1: 28–31 ft. lbs. (37–43 Nm).
    b. Step 2: Plus 85–95 degrees.
    c. Step 3: Loosen all bolts one full turn.
    d. Step 4: 28–31 ft. lbs. (37–43 Nm).
    e. Step 5: Plus 85–95 degrees.
    f. Step 6: Plus 85–95 degrees.
50. Install hydraulic lash adjusters.
51. Install heater water tube.
52. Install timing chains.
53. Install crankshaft sensor ring.
54. Install front cover.
55. Install intake manifold.
56. Install upper radiator hose and adapter.
57. Install heater hoses.
58. Install throttle body and adapter as an assembly.
59. Install rocker arms.
60. Install valve covers.
61. Install starter motor wiring harness retainer.
62. Install dual converter H-pipe.
63. Connect HO_2S sensor connectors.
64. Install PCV valve and hose.
65. Connect EGR tube.
66. Connect oil pressure sensor connector.
67. Install crankshaft pulley.
68. Install oil pan.
69. Install power steering pump.
70. Connect canister purge valve connector.
71. Connect accelerator cable and install bracket.
72. Connect cruise control cable.
73. Connect A/C compressor clutch connector.

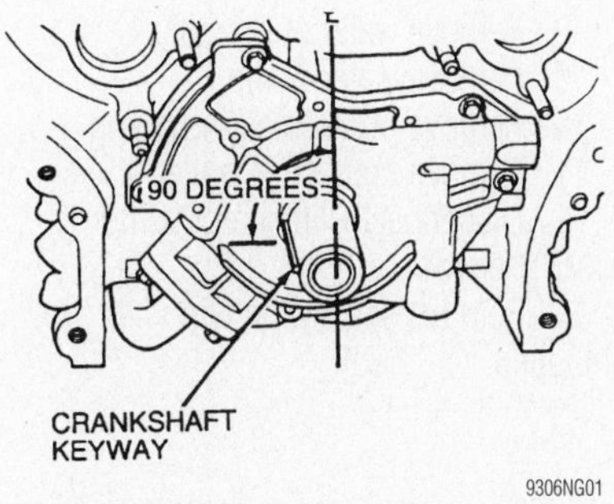

Place the crankshaft keyway at 9 o'clock for cylinder head installation—4.6L SOHC engine

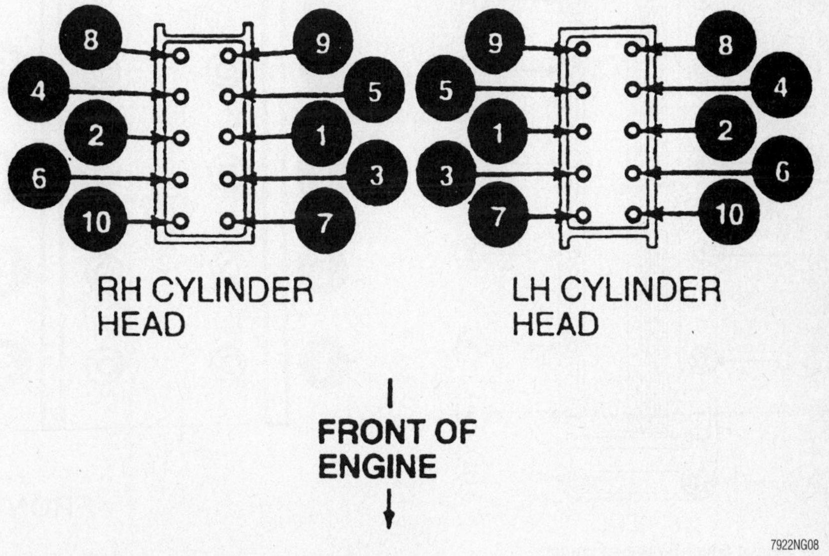

Cylinder head torque sequence—4.6L SOHC engine

74. Connect CKP sensor connector.
75. Connect fuel injector wiring connectors.
76. Connect MAF sensor connector.
77. Install water pump pulley.
78. Install alternator and bracket.
79. Install left and right exhaust manifolds.
80. Install power steering reservoir and bracket.
81. Connect CMP sensor connector.
82. Install ignition coils.
83. Remove spark plug wires, if equipped.
84. Install the accessory drive belt.
85. Install windshield wiper governor, if necessary.
86. Install air intake tube.
87. Connect fuel supply line.
88. Connect fuel return line, if equipped.
89. Install electric cooling fan.
90. Connect the negative battery cable.
91. Fill the cooling system.
92. Start the engine and check for leaks.

4.6L DOHC Engine (Non-Supercharged)

1. Before servicing the vehicle, refer to the precautions at the beginning of this section.
2. Remove the engine from the vehicle and mount it on a suitable workstand.
3. Remove accessory drive belt and tensioner.
4. Remove idler pulley.
5. Disconnect Engine Coolant Temperature (ECT) sensor connector.
6. Remove water bypass tube.
7. Remove alternator and support bracket.
8. Remove water pump pulley.
9. Remove water pump.
10. Remove power steering pump.
11. Remove crankshaft pulley.
12. Disconnect Camshaft Position (CMP) sensor.
13. Remove spark plug wires, if equipped.
14. Remove ignition coils.
15. Disconnect fuel pressure sensor connector and vacuum line.
16. Remove spark plugs.
17. Remove valve covers.
18. Remove rocker arms.
19. Remove Exhaust Gas Recirculation (EGR) vacuum regulator valve.
20. Remove EGR tube.
21. Disconnect fuel injector wiring connectors.
22. Remove intake manifold.
23. Remove left and right exhaust manifolds.
24. Remove front cover.
25. Remove Crankshaft Position (CKP) sensor pulse wheel.
26. Remove timing chains.
27. Remove hydraulic lash adjusters.
28. Remove cylinder heads.

To install:

➡**The cylinder head bolts are a torque-to-yield design and cannot be reused.**

29. Use new gaskets and install the cylinder heads.
30. Tighten the cylinder head bolts in sequence as follows:
    a. Step 1: 28–31 ft. lbs. (37–43 Nm).
    b. Step 2: Tighten the bolts 85–95 degrees.
    c. Step 3: Loosen all bolts 1 full turn.
    d. Step 4: Tighten all bolts to 28–31 ft. lbs. (37–43 Nm).

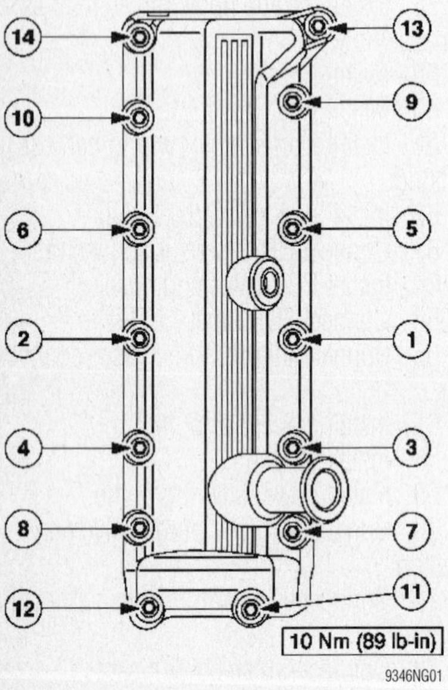

Right valve cover torque sequence—4.6L SOHC Engine

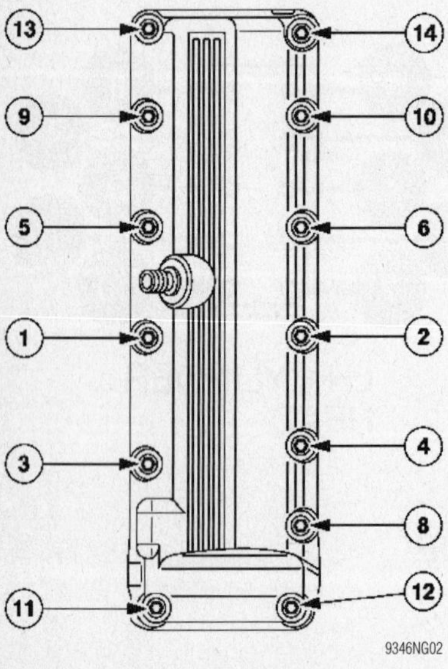

Left valve cover torque sequence—4.6 SOHC Engine

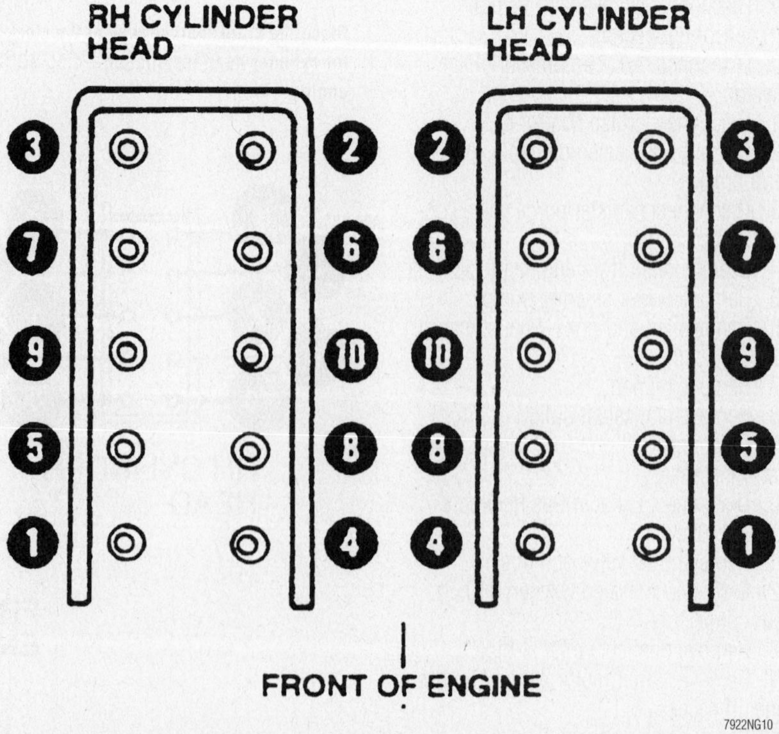

Cylinder head loosening sequence—4.6L DOHC engine

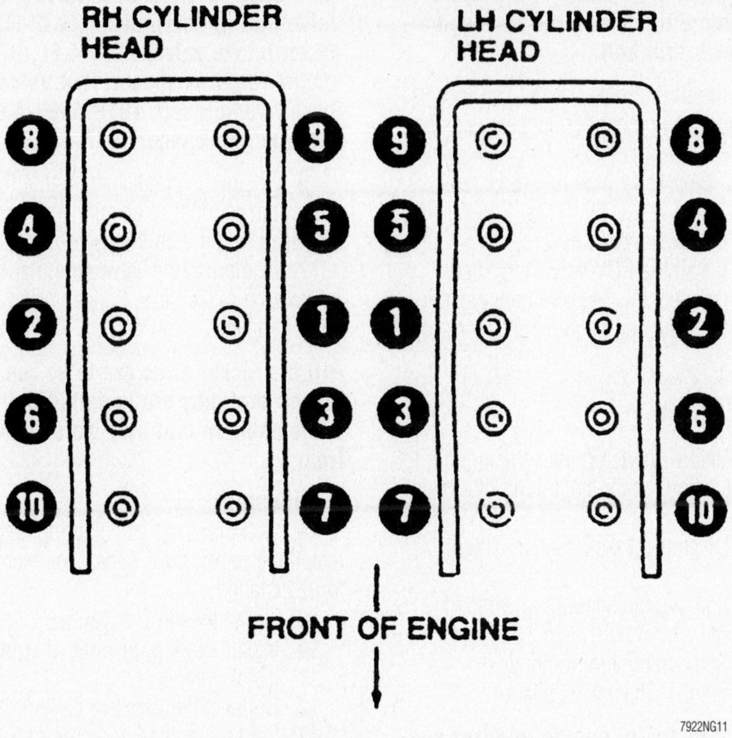

Cylinder head torque sequence—4.6L DOHC engine

e. Step 5: Tighten the bolts 85–95 degrees.

f. Step 6: Tighten the bolts 85–95 degrees.

31. Install hydraulic lash adjusters.
32. Install timing chains.
33. Install CKP sensor pulse wheel.
34. Install front cover.
35. Install left and right exhaust manifolds.
36. Install intake manifold.
37. Connect fuel injector wiring connectors.

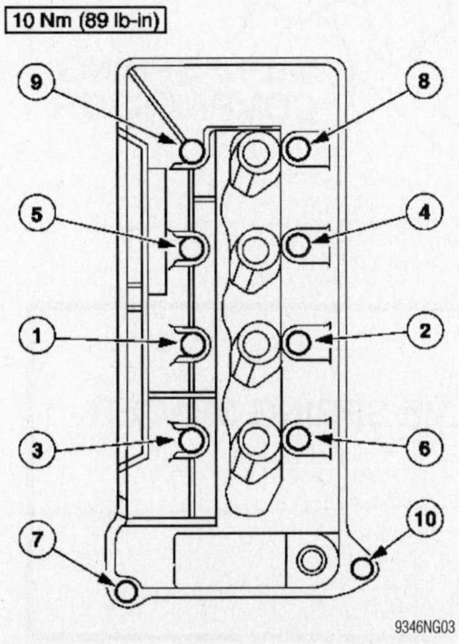

Right valve cover torque sequence—4.6L DOHC Engine

38. Install EGR tube.
39. Install EGR vacuum regulator valve.
40. Install rocker arms.
41. Install valve covers.
42. Install spark plugs.
43. Connect fuel pressure sensor connector and vacuum line.
44. Install ignition coils.
45. Install spark plug wires, if equipped.
46. Connect CMP sensor.
47. Install crankshaft pulley.
48. Install water pump.

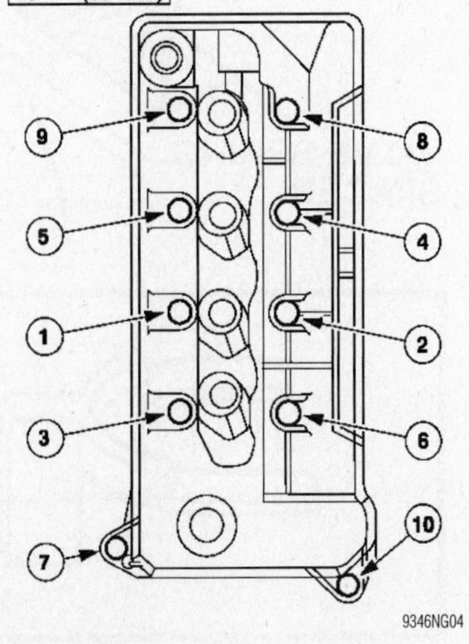

Left valve cover torque sequence—4.6L DOHC Engine

49. Install water pump pulley.
50. Install power steering pump.
51. Install alternator and support bracket.
52. Install water bypass tube.
53. Connect ECT sensor connector.
54. Install idler pulley.
55. Install accessory drive belt and tensioner.

4.6L DOHC Engine (Supercharged)

1. Before servicing the vehicle, refer to the precautions at the beginning of this section.
2. Remove the engine from the vehicle and mount it on a suitable workstand.
3. Disconnect Engine Coolant Temperature (ECT) sensor connector.
4. Remove water bypass tube.
5. Disconnect Camshaft Position (CMP) sensor.
6. Disconnect fuel pressure sensor connector and vacuum line.
7. Remove ignition coils.
8. Remove spark plugs.
9. Remove valve covers.
10. Remove supercharger belt idler support bracket assembly.
11. Remove the accessory drive belt.
12. Remove water pump pulley.
13. Remove water pump.
14. Remove crankshaft pulley.
15. Remove alternator and support bracket.
16. Remove power steering pump.
17. Remove idler pulley.
18. Remove rocker arms.
19. Remove Exhaust Gas Recirculation (EGR) valve.
20. Remove EGR tube.
21. Disconnect fuel injector wiring connectors.
22. Remove intake manifold, supercharger, and fuel supply manifold as an assembly.
23. Remove left and right exhaust manifolds.
24. Remove front cover.
25. Remove Crankshaft Position (CKP) sensor pulse wheel.
26. Remove timing chains.
27. Remove hydraulic lash adjusters.
28. Remove cylinder heads.

To install:

➡The cylinder head bolts are a torque-to-yield design and cannot be reused.

29. Use new gaskets and install the cylinder heads.
30. Tighten the cylinder head bolts in sequence as follows:

a. Step 1: 28–31 ft. lbs. (37–43 Nm).

b. Step 2: Tighten the bolts 85–95 degrees.
c. Step 3: Loosen all bolts 1 full turn.
d. Step 4: Tighten all bolts to 28–31 ft. lbs. (37–43 Nm).
e. Step 5: Tighten the bolts 85–95 degrees.
f. Step 6: Tighten the bolts 85–95 degrees.

31. Install hydraulic lash adjusters.
32. Install timing chains.
33. Install CKP sensor pulse wheel.
34. Install front cover.
35. Install left and right exhaust manifolds.
36. Install intake manifold, supercharger, and fuel supply manifold as an assembly.
37. Connect fuel injector wiring connectors.
38. Install EGR valve.
39. Install EGR tube.
40. Install rocker arms.
41. Install idler pulley.
42. Install power steering pump.
43. Install alternator and support bracket.
44. Install crankshaft pulley.
45. Install water pump.
46. Install water pump pulley.
47. Install the accessory drive belt.
48. Install supercharger belt idler support bracket assembly.
49. Install valve covers.
50. Install spark plugs.
51. Install ignition coils.
52. Connect CMP sensor.
53. Connect fuel pressure sensor connector and vacuum line.
54. Install water bypass tube.

Rocker Arms

REMOVAL & INSTALLATION

3.8L Engine

1. Before servicing the vehicle, refer to the precautions at the beginning of this section.
2. Disconnect the negative battery cable.
3. Remove the Positive Crankcase Ventilation (PCV) valve and hose.
4. Remove valve covers.
5. Remove the rocker arms.

➡Keep rocker arms in order for installation.

To install:

➡The rocker arm bolts are tightened with the valves closed. Rotate the crankshaft as necessary to position the lifter on the base circle of the camshaft lobe before tightening the corresponding rocker arm bolt.

6. Install the rocker arms. The rocker arm bolts are tightened in two steps as follows:
   a. Step 1: 44 inch lbs. (5 Nm).
   b. Step 2: 23–29 ft. lbs. (30–40 Nm).
7. Install the valve covers.
8. Install the PCV valve and hose.
9. Connect the negative battery cable.
10. Start the engine and check for proper operation.

4.6L Engines

1. Before servicing the vehicle, refer to the precautions at the beginning of this section.
2. Disconnect the negative battery cable.
3. Remove the spark plug wires, if equipped.
4. Remove the ignition coils.
5. Remove the valve covers.

➡The 4.6L DOHC engine requires special tool Valve Spring Compressor T91P-6565-A for exhaust valves, and Valve Spring Compressor T93P-6565-AR for intake valves. The 4.6L SOHC engine requires special tool Valve Spring Compressor T91P-6565-A and Valve Spring Spacer T91P-6565-AH.

6. Rotate the crankshaft so that the piston on the cylinder to be serviced is at bottom dead center with the valves closed.
7. Compress the valve spring and remove the rocker arm. Repeat for each arm to be removed.

➡If the rocker arms are to be reused, ensure that they are installed in the same position that they were removed from.

To install:

8. Compress the valve spring and install the rocker arm. Repeat for each arm to be installed.
9. Install the valve covers.
10. Install spark plug wires, if equipped.
11. Install ignition coils.
12. Connect the negative battery cable.
13. Start the engine and check for proper operation.

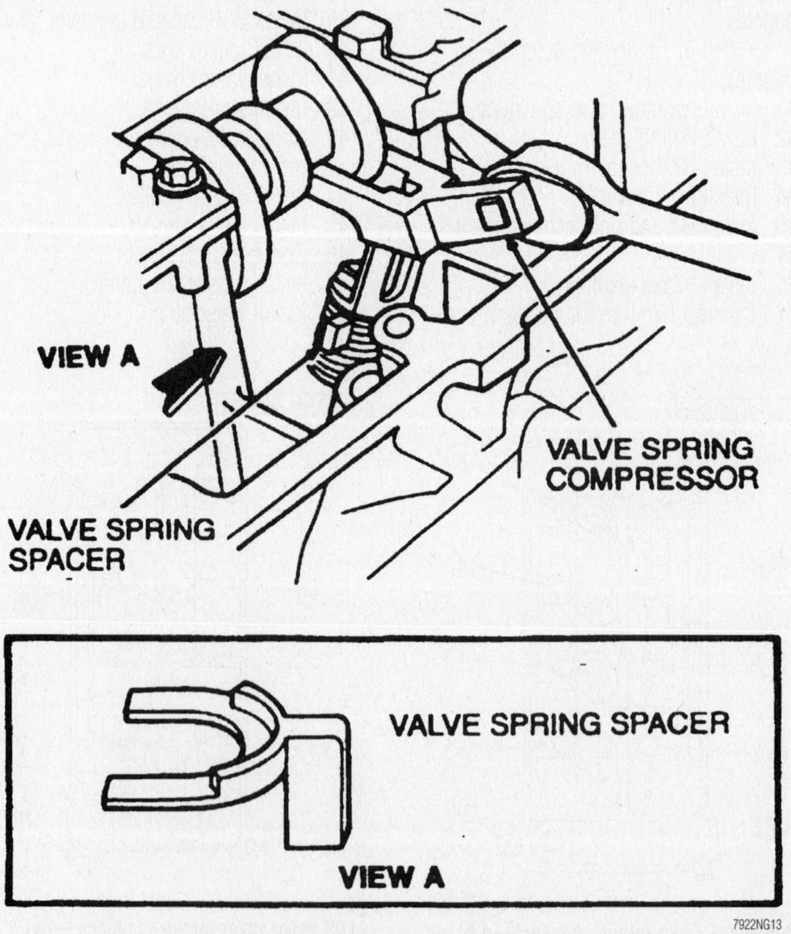

Valve spring compression tool and spacer—4.6L SOHC engine

Supercharger

REMOVAL & INSTALLATION

1. Before servicing the vehicle, refer to the precautions at the beginning of this section.
2. Disconnect the negative battery cable.
3. Remove supercharger drive belt.
4. Remove PCV hoses.
5. Remove coolant supply and return manifold.
6. Remove coolant supply and return tubes and seal.
7. Remove supercharger and Charge Air Cooler (CAC) assembly.

To install:

8. Install supercharger and CAC assembly onto the engine. Use new gaskets, if necessary, and tighten the bolts in sequence as follows:
   a. Step 1: 18 inch lbs. (2 Nm).
   b. Step 2: 18 ft. lbs. (25 Nm).
9. Install coolant supply and return tubes and seals.
10. Install coolant supply and return manifold. Tighten the manifold fasteners as follows:
    a. Step 1: 89 inch lbs. (10 Nm).
    b. Step 2: Tighten an additional 90 degrees.
11. Install PCV hoses.
12. Install supercharger drive belt.
13. Connect the negative battery cable.

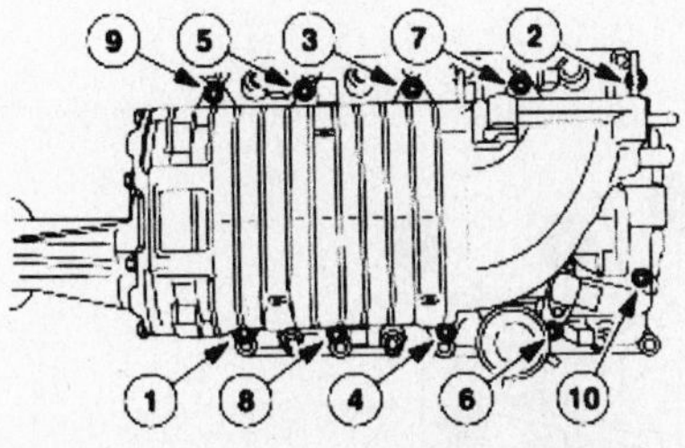

Supercharger assembly-to-lower intake manifold torque sequence—Cobra

Intake Manifold

REMOVAL & INSTALLATION

3.8L Engine

1. Before servicing the vehicle, refer to the precautions at the beginning of this section.
2. Drain the cooling system.
3. Relieve the fuel system pressure.
4. Disconnect the negative battery cable.
5. Disconnect Intake Air Temperature (IAT) sensor.
6. Remove air intake tube.
7. Disconnect cruise control cable.
8. Disconnect accelerator cable and remove bracket.
9. Disconnect differential pressure feedback Exhaust Gas Recirculation (EGR) and vacuum regulator solenoid electrical connectors, if equipped.
10. Remove spark plug wires and ignition coil assembly.
11. Disconnect and label all area vacuum lines.
12. Remove Positive Crankcase Ventilation (PCV) valve and hose.
13. Disconnect Throttle Position (TP) sensor connector.
14. Disconnect Idle Air Control (IAC) valve connector.
15. Remove EGR valve.
16. Remove engine control sensor wiring harness retainer bracket and position harness out of the way.
17. Remove upper intake manifold.

➡**Be sure to record the locations of the long bolts and short bolts in the upper and lower intake manifolds.**

18. Disconnect fuel supply line.
19. Disconnect fuel return line, if equipped.
20. Disconnect Engine Coolant Temperature (ECT) sensor connector.
21. Remove fuel supply manifold and injectors.
22. Remove upper radiator hose.
23. Remove heater hose.
24. Remove lower intake manifold.

To install:

25. Install the lower intake manifold. Use new gaskets and tighten the bolts in sequence as follows:

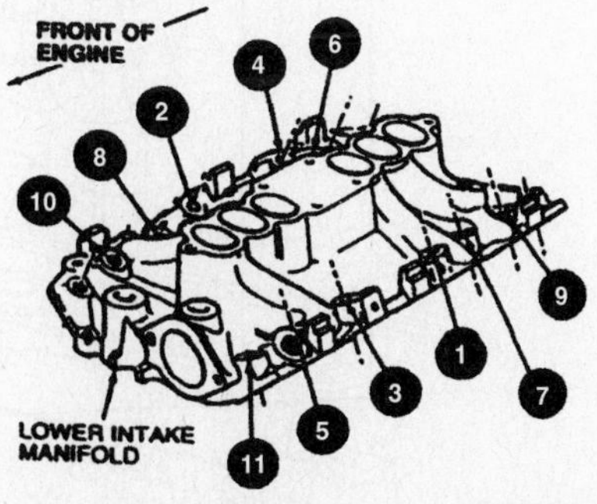

Lower intake manifold torque sequence—3.8L engine (2001–03 models)

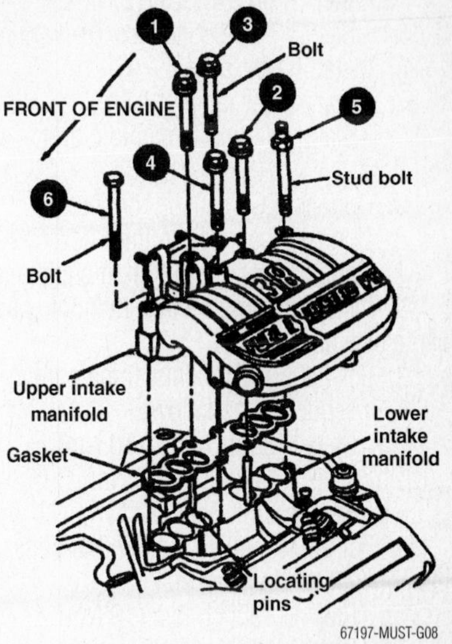

Upper intake manifold torque sequence—3.8L engine (2001–03 models)

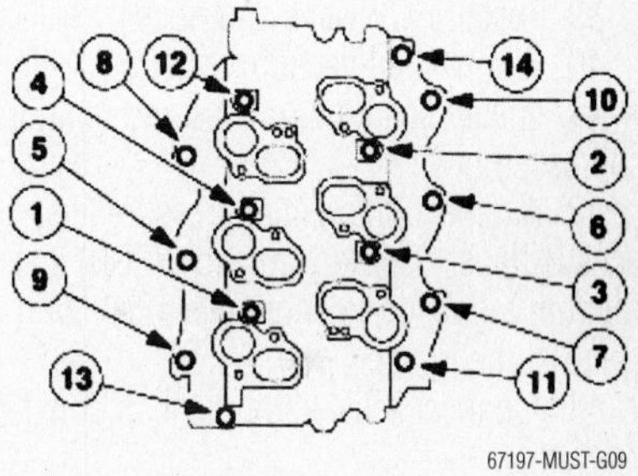

Lower intake manifold torque sequence—3.8L engine (2004 models)

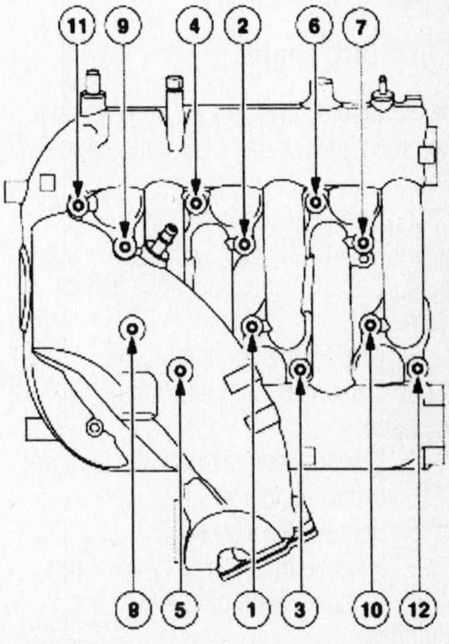

Upper intake manifold torque sequence—3.8L engine (2004 models)

a. Step 1: 45 inch lbs. (5 Nm).
b. Step 2: 71–106 inch lbs. (8–12 Nm).
26. Install heater hose.
27. Install upper radiator hose.
28. Install fuel supply manifold and injectors.
29. Connect ECT sensor connector.
30. Connect fuel supply line.
31. Connect fuel return line, if equipped.
32. Install the upper intake manifold and use a new gasket.
33. For 2001–03 models, tighten the bolts, in sequence, as follows:
a. Step 1: 89 inch lbs. (10 Nm).
b. Step 2: 15 ft. lbs. (20 Nm).
c. Step 3: 24 ft. lbs. (32 Nm).
34. For 2004 models, tighten the bolts, in sequence, as follows:
a. Step 1: 89 inch lbs. (10 Nm).
b. Step 2: additional 90 degrees.
35. Install engine control sensor wiring harness retainer bracket.
36. Install EGR valve.
37. Connect IAC valve connector.
38. Connect TP sensor connector.
39. Install PCV valve and hose.
40. Connect all vacuum lines.
41. Install spark plug wires and ignition coil assembly.
42. Connect differential pressure feedback Exhaust Gas Recirculation (EGR) and vacuum regulator solenoid electrical connectors, if equipped.
43. Connect accelerator cable and install bracket.
44. Connect cruise control cable.
45. Install Air intake tube.
46. Connect IAT sensor connector.
47. Connect the negative battery cable.
48. Fill the cooling system.
49. Start the engine and check for leaks.

4.6L SOHC Engine

1. Before servicing the vehicle, refer to the precautions at the beginning of this section.
2. Drain the cooling system.
3. Relieve the fuel system pressure.
4. Disconnect the negative battery cable.
5. Disconnect fuel supply line.
6. Disconnect fuel return line, if equipped.
7. Disconnect Intake Air Temperature (IAT) sensor connector.
8. Remove air intake tube.
9. Remove the accessory drive belt.
10. Remove spark plug wires, if equipped.
11. Disconnect Camshaft Position (CMP) sensor connector.
12. Remove alternator and bracket.
13. Disconnect fuel injector wiring connectors.
14. Disconnect oil pressure sensor connector.
15. Remove Exhaust Gas Recirculation (EGR) tube.
16. Disconnect cruise control cable.
17. Disconnect accelerator cable and remove bracket.
18. Disconnect throttle body vacuum adapter hose.
19. Remove heater hose.
20. Remove upper radiator hose adapter.
21. Remove intake manifold by removing the mounting bolts in the same sequence as the installation.

To install:

22. Install intake manifold. Use new gaskets and tighten the bolts, in sequence, to 15–22 ft. lbs. (20–30 Nm).
23. Install upper radiator hose adapter. Use a new O-ring and tighten the bolts to 15–22 ft. lbs. (20–30 Nm).
24. Install heater hose.
25. Connect throttle body vacuum adapter hose.
26. Connect accelerator cable and install bracket.
27. Connect cruise control cable.
28. Install EGR tube. Tighten the nut to 26–33 ft. lbs. (35–45 Nm).
29. Connect oil pressure sensor connector.
30. Connect fuel injector wiring connectors.
31. Install alternator and bracket.
32. Connect CMP sensor connector.
33. Install spark plug wires, if equipped.
34. Install the accessory drive belt.
35. Install air intake tube.
36. Connect IAT sensor connector.
37. Connect fuel supply line.
38. Connect fuel return line, if equipped.
39. Connect the negative battery cable.
40. Fill the engine cooling system.
41. Start the engine and check for leaks.

4.6L DOHC Engine (Non-Supercharged)

1. Before servicing the vehicle, refer to the precautions at the beginning of this section.
2. Drain the cooling system.
3. Relieve the fuel system pressure.
4. Disconnect the negative battery cable.
5. Remove the air intake scoop bracket, if equipped.
6. Remove engine compartment brace, if equipped.
7. Remove air intake tube.
8. Disconnect cruise control cable.

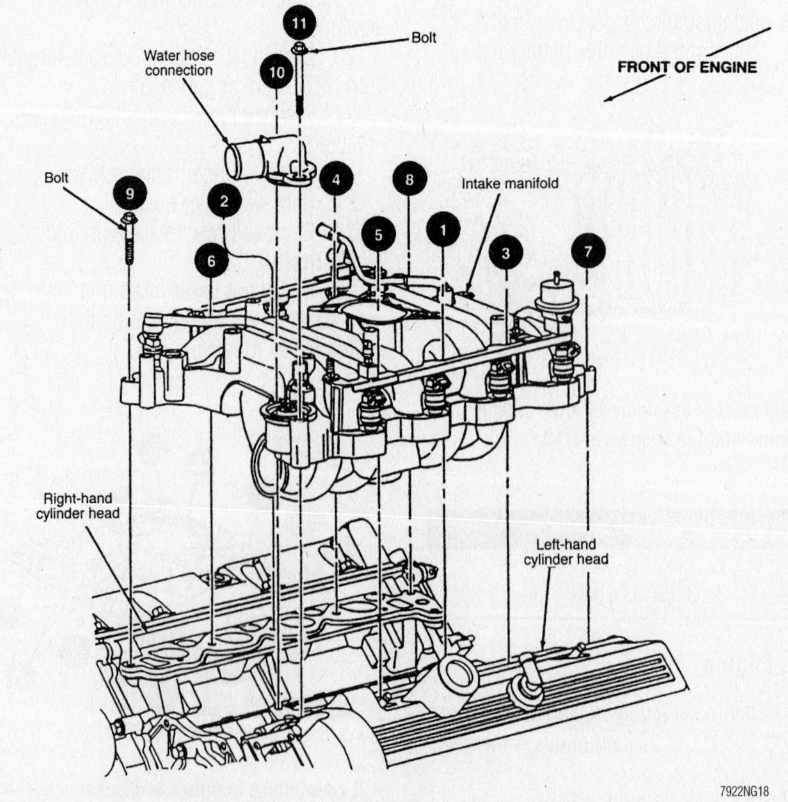

Intake manifold torque sequence—4.6L SOHC engine

9. Disconnect accelerator cable and bracket.
10. Remove Positive Crankcase Ventilation (PCV) valve and hose.
11. Remove Exhaust Gas Recirculation (EGR) valve.
12. Remove manifold bolts, in sequence, and remove upper intake manifold.
13. Disconnect fuel supply line.
14. Disconnect fuel return line, if equipped.
15. Disconnect, and label, all intake manifold area vacuum lines.
16. Remove spark plug wires, if equipped.
17. Remove alternator and bracket.
18. Remove upper radiator hose.
19. Disconnect Throttle Position (TP) sensor connector.
20. Disconnect Engine Coolant Temperature (ECT) sensor connector.
21. Disconnect Idle Air Control (IAC) valve connector.
22. Disconnect coolant temperature gauge sender connector.
23. Remove water bypass tube.
24. Disconnect fuel injector wiring connectors.
25. Disconnect fuel pressure sensor vacuum and wiring connectors.
26. Remove manifold bolts, in sequence, and remove lower intake manifold.

To install:

27. Install lower intake manifold with new gaskets. Tighten the bolts, in sequence, to 89 inch lbs. (10 Nm).
28. Connect fuel pressure sensor vacuum and wiring connectors.

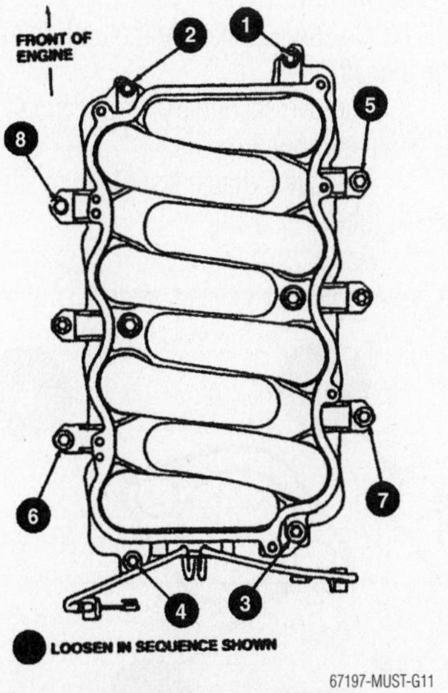

Lower intake manifold loosening sequence—4.6L DOHC engine (Non-Supercharged)

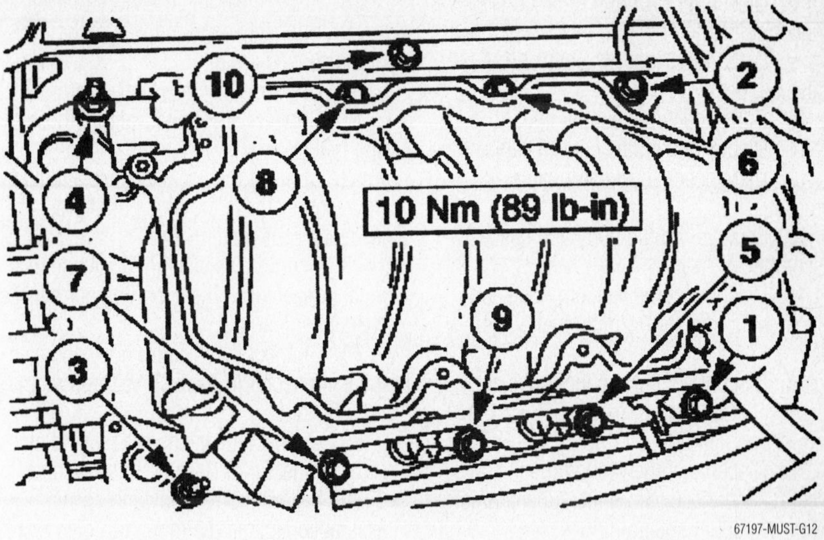

Lower intake manifold torque sequence—4.6L DOHC engine (Non-Supercharged)

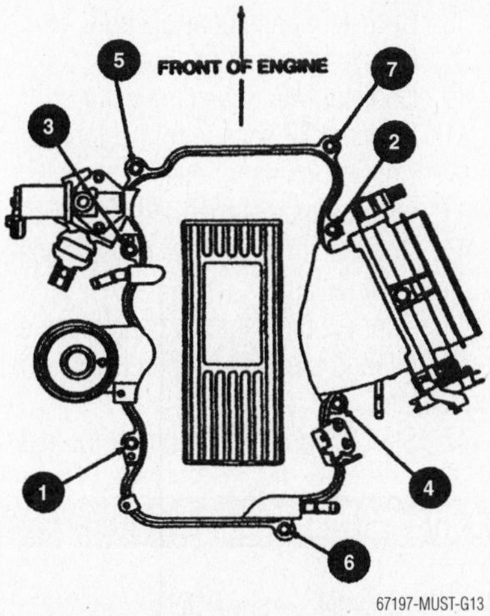

Upper intake manifold torque sequence—4.6L DOHC engines (2001–03 models)

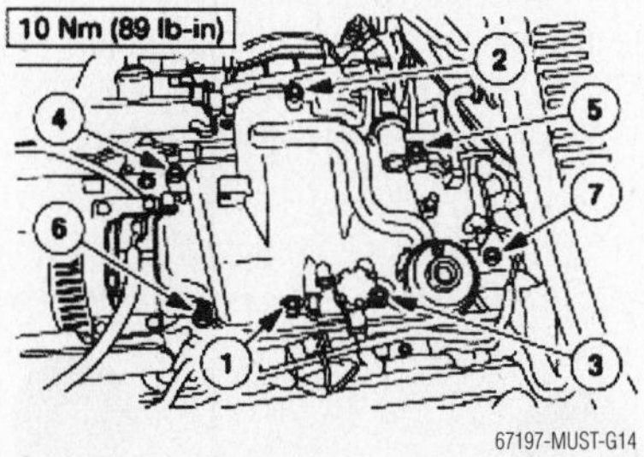

Upper intake manifold torque sequence—4.6L DOHC engines (2004 Non-Supercharged)

29. Connect fuel injector wiring connectors.
30. Install alternator and bracket.
31. Install water bypass tube.
32. Connect coolant temperature gauge sender connector.
33. Connect IAC valve connector.
34. Connect ECT sensor connector.
35. Connect TP sensor connector.
36. Install upper radiator hose.
37. Install spark plug wires, if equipped.
38. Install the accessory drive belt.
39. Connect fuel supply line.
40. Connect fuel return line, if equipped.
41. Install upper intake manifold and new gasket. Tighten the bolts, in sequence, to 71–106 inch lbs. (8–12 Nm).
42. Install EGR valve.
43. Install PCV valve and hose.
44. Connect accelerator cable and install bracket.
45. Connect cruise control cable.
46. Connect all intake manifold area vacuum lines.
47. Install air intake tube.
48. Install engine compartment brace, if equipped.
49. Install the air intake scoop bracket, if equipped.
50. Connect the negative battery cable.
51. Fill the engine cooling system.
52. Start the engine and check for leaks.

4.6L DOHC Engine (Supercharged)

1. Before servicing the vehicle, refer to the precautions at the beginning of this section.
2. Drain the supercharger cooling system.
3. Relieve the fuel system pressure.
4. Disconnect the negative battery cable.
5. Disconnect supercharger coolant hose to the overflow tank.
6. Remove supercharger belt.
7. Disconnect coolant hose.

8. Disconnect Throttle Position (TP) sensor connector.
9. Disconnect Idle Air Control (IAC) valve connector.
10. Disconnect fuel supply line.
11. Disconnect cruise control cable.
12. Disconnect accelerator cable and bracket.
13. Remove throttle body and spacer assembly.
14. Disconnect the vacuum hose and electrical connector from the fuel pulse damper.
15. Disconnect the vacuum hose and electrical connector from the EGR vacuum regulator solenoid.
16. Disconnect the vacuum hose and electrical connector from the supercharger bypass vacuum solenoid.
17. Disconnect the vacuum hose and electrical connector from the differential pressure feedback EGR system.
18. Disconnect EGR valve vacuum hose and tube.
19. Disconnect Positive Crankcase Ventilation (PCV) valve hose.
20. Disconnect vacuum hoses at the back of the supercharger and position them out of the way.
21. Remove vacuum accessory bracket.
22. Disconnect Barometric (BARO) pressure sensor electrical connector.
23. Separate fuel charge wiring harness from the fuel injection supply manifold and position out of the way.
24. Remove manifold bolts, in sequence, and remove lower intake manifold, supercharger, and fuel supply manifold as an assembly.

To install:

25. Install lower intake manifold with new gaskets. Tighten the bolts, in sequence, to 89 inch lbs. (10 Nm).
26. Secure fuel charge wiring harness onto the fuel injection supply manifold.
27. Connect BARO sensor electrical connector.
28. Install vacuum accessory bracket.
29. Connect vacuum hoses at the back of the supercharger.
30. Connect PCV valve hose.
31. Connect EGR valve vacuum hose and tube.
32. Connect the vacuum hose and electrical connector to the differential pressure feedback EGR system.
33. Connect the vacuum hose and electrical connector to the supercharger bypass vacuum solenoid.
34. Connect the vacuum hose and electrical connector to the EGR vacuum regulator solenoid.
35. Connect the vacuum hose and electrical connector to the fuel pulse damper.
36. Install throttle body and spacer assembly. Tighten the mounting nuts to 18 ft. lbs. (25 Nm).
37. Connect accelerator cable and install bracket.
38. Connect cruise control cable.
39. Connect fuel supply line.
40. Connect IAC valve connector.
41. Connect TP sensor connector.
42. Connect coolant hose.
43. Install supercharger belt.
44. Connect supercharger coolant hose to the overflow tank.
45. Connect the negative battery cable.
46. Fill and bleed the supercharger cooling system.
47. Start the engine and check for leaks.

Exhaust Manifold

REMOVAL & INSTALLATION

3.8L Engine

1. Before servicing the vehicle, refer to the precautions at the beginning of this section.
2. Disconnect the negative battery cable.
3. Disconnect, and label, spark plug wires.
4. Remove secondary air injection diverter valve, if equipped.
5. Remove secondary air injection tubes, if equipped.
6. Remove Exhaust Gas Recirculation (EGR) tube.
7. Remove oil dipstick tube.
8. Remove Heated Oxygen (HO_2S) sensors.
9. Disconnect dual converter Y-pipe.
10. Remove exhaust manifolds. Note the locations of the studs and bolts for reassembly.

To install:

11. Install exhaust manifolds. Tighten the bolts in sequence to 22–26 ft. lbs. (30–36 Nm).
12. Connect dual converter Y-pipe.
13. Install HO_2S sensors. Tighten the sensors to 28–33 ft. lbs. (37–45 Nm).
14. Install oil dipstick tube.
15. Install EGR tube.
16. Install secondary air injection tubes, if equipped.
17. Install secondary air injection diverter valve, if equipped.
18. Connect spark plug wires.
19. Connect the negative battery cable.

4.6L Engines

1. Before servicing the vehicle, refer to the precautions at the beginning of this section.
2. Disconnect the negative battery cable.
3. Remove engine compartment brace, if equipped.
4. Disconnect dual converter H-pipe.
5. Remove starter.
6. Disconnect steering column intermediate shaft.
7. Remove secondary air injection tubes, if equipped.
8. Remove Exhaust Gas Recirculation (EGR) tube.

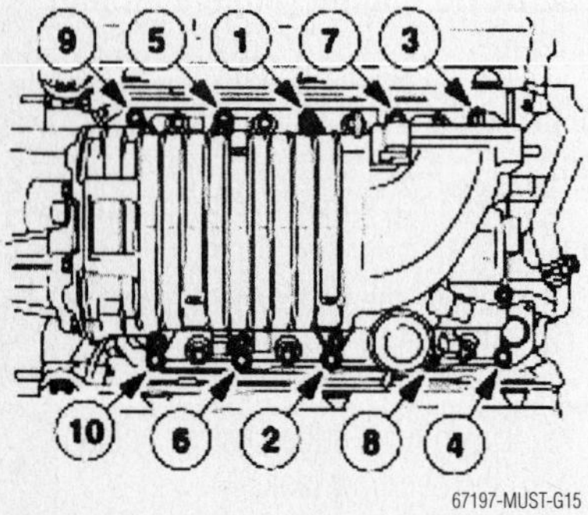

Lower intake manifold torque sequence—4.6L DOHC engine (Supercharged)

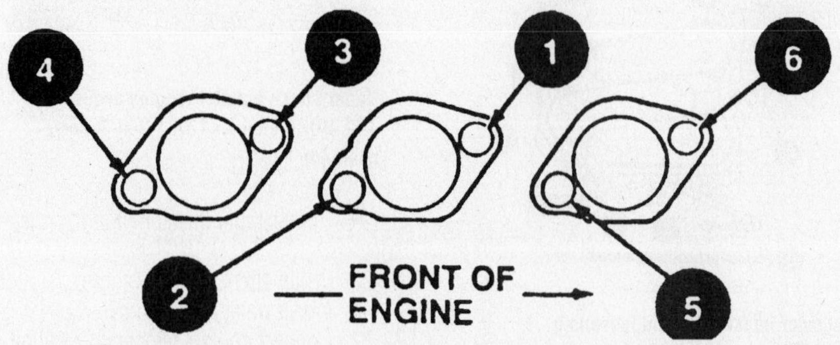

Exhaust manifold torque sequence—3.8L engine

9. Remove oil dipstick tube.
10. Remove oil filter.
11. Disconnect left and right motor mounts.
12. Raise the engine about 1.5 inches for clearance and remove the exhaust manifolds.

To install:

13. Install exhaust manifolds. Use new gaskets and tighten the nuts in sequence to 15 ft. lbs. (20 Nm).
14. Connect left and right motor mounts.
15. Install oil filter.
16. Install EGR tube.
17. Install oil dipstick tube.
18. Install secondary air injection tubes, if equipped.
19. Connect steering column intermediate shaft.
20. Install starter.
21. Connect dual converter H-pipe.
22. Install engine compartment brace, if equipped.
23. Connect the negative battery cable.

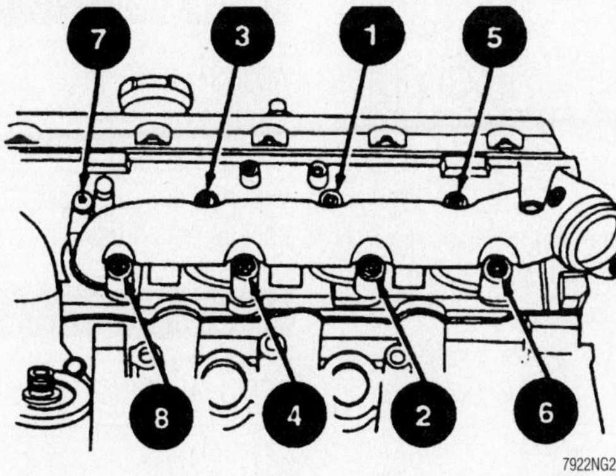

Exhaust manifold torque sequence—4.6L SOHC engine

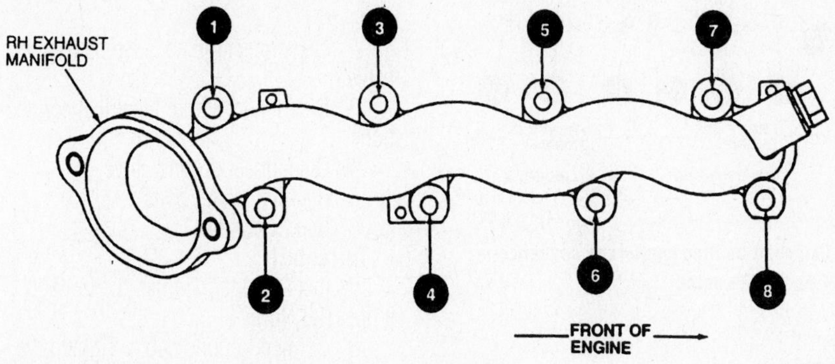

Left exhaust manifold torque sequence—4.6L DOHC engine

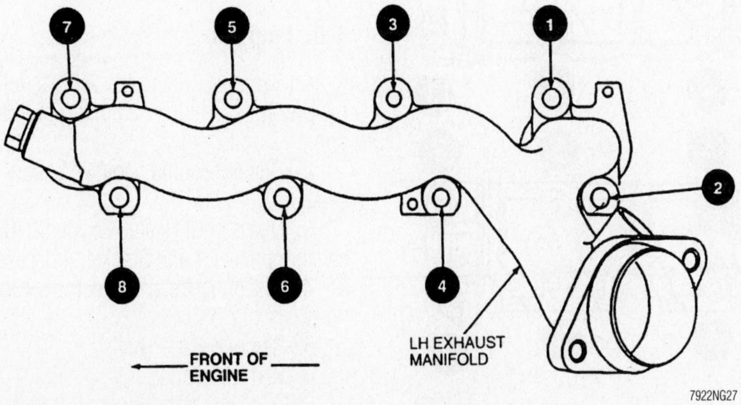

Right exhaust manifold torque sequence—4.6L DOHC engine

Camshaft and Valve Lifters

REMOVAL & INSTALLATION

3.8L Engine

VALVE LIFTERS

➡Keep all valvetrain parts in order for installation.

1. Before servicing the vehicle, refer to the precautions at the beginning of this section.
2. Drain the cooling system.
3. Disconnect the negative battery cable.
4. Remove Positive Crankcase Ventilation (PCV) valve and hose.
5. Remove valve covers.
6. Remove rocker arms and pushrods.
7. Remove upper and lower intake manifolds.
8. Remove lifter guide plates and retainers.
9. Remove valve lifters.

To install:

10. Install valve lifters in their original locations.
11. Install lifter guide plates and retainers. Tighten the bolts to 88–124 inch lbs. (10–14 Nm).
12. Install upper and lower intake manifolds.
13. Install rocker arms and pushrods in their original locations.
14. Install valve covers.
15. Install PCV valve and hose.
16. Connect the negative battery cable.
17. Fill the cooling system.
18. Start the engine and check for leaks.

CAMSHAFT

1. Before servicing the vehicle, refer to the precautions at the beginning of this section.
2. Drain the cooling system.
3. Recover the A/C refrigerant.
4. Remove radiator and cooling fan.
5. Remove A/C condenser.
6. Remove upper and lower intake manifolds.
7. Remove valve lifters.
8. Remove front cover.
9. Remove timing chain and gears.
10. Remove camshaft thrust plate and spacer.
11. Remove camshaft.

To install:

12. Install camshaft.
13. Install camshaft thrust plate. Tighten the bolts to 71–124 inch lbs. (8–14 Nm).

14. Install timing chain and gears.
15. Install front cover.
16. Install valve lifters.
17. Install upper and lower intake manifolds.
18. Install A/C condenser.
19. Install radiator and cooling fan.
20. Fill the cooling system.
21. Recharge the A/C system.
22. Run the engine and check for leaks.

4.6L Engines

VALVE LIFTERS

➡Keep all valvetrain parts in order for installation.

1. Before servicing the vehicle, refer to the precautions at the beginning of this section.
2. Remove valve covers.
3. Remove rocker arms.
4. Remove hydraulic lifters.

To install:

5. Inspect each lifter. If the plunger travel exceeds 0.059 inches (1.5 mm), replace the lifter.
6. Install hydraulic lifters in their original positions.
7. Install rocker arms.
8. Install valve covers.

CAMSHAFTS

1. Before servicing the vehicle, refer to the precautions at the beginning of this section.
2. Remove valve covers.
3. Remove rocker arms.
4. Remove oil pan (SOHC engine).
5. Remove front cover.
6. Remove timing chains and sprockets.
7. Remove camshaft bearing caps.
8. Remove camshafts.

To install:

➡On 4.6L DOHC engines, the outboard exhaust camshaft bearing cap bolts are of a different length than the other bearing cap bolts.

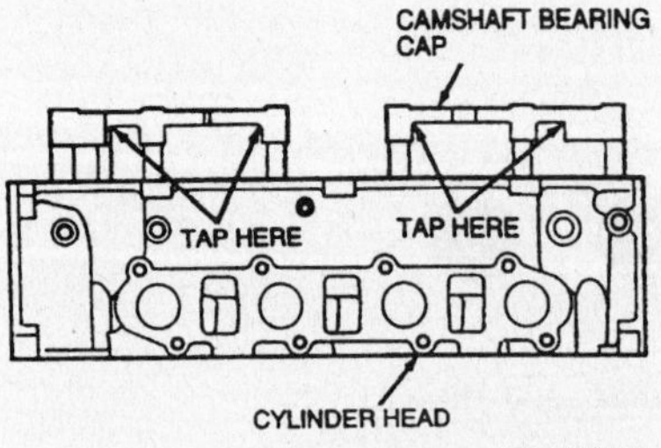

To remove the camshaft bearing caps, tap the caps with a rubber or leather mallet where shown—4.6L engines

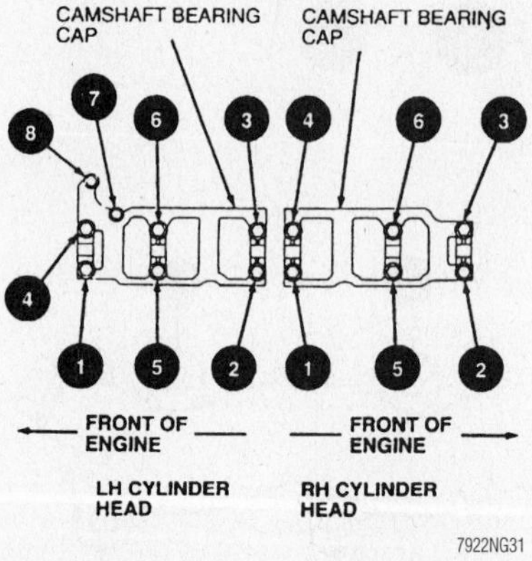

Camshaft bearing cap torque sequence—4.6L SOHC engine

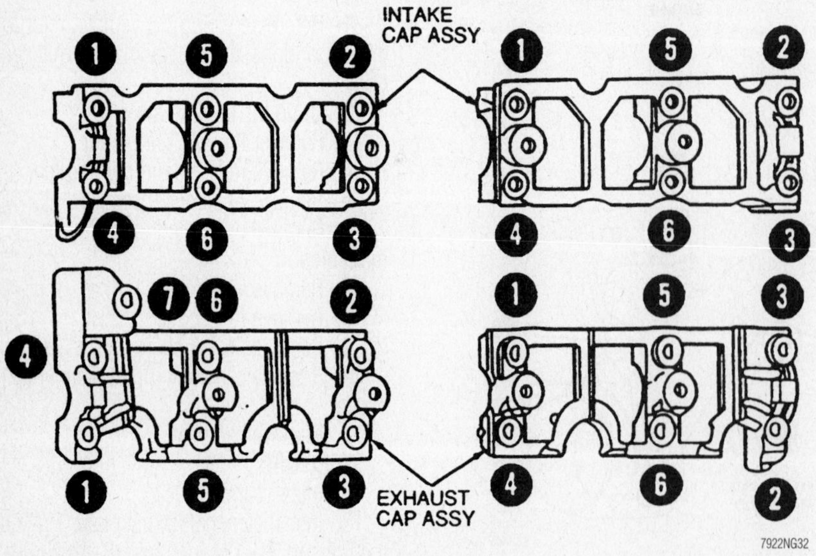

Camshaft bearing cap torque sequence—4.6L DOHC engine

9. Install the camshafts. Tighten the bearing cap bolts in sequence as follows:
   a. Step 1: 71–106 inch lbs. (8–12 Nm).
   b. Step 2: Loosen all bolts 2 turns.
   c. Step 3: 71–106 inch lbs. (8–12 Nm).
10. Install timing chains and sprockets.
11. Install front cover.
12. Install oil pan (SOHC engine).
13. Install rocker arms.
14. Install valve covers.
15. Start the engine and check for leaks.

Valve Lash

ADJUSTMENT

The 3.8L and 4.6L engines are equipped with hydraulic lash adjusters. Valve clearance is not adjustable.

Starter Motor

REMOVAL & INSTALLATION

3.8L Engine

1. Before servicing the vehicle, refer to the precautions at the beginning of this section.
2. Disconnect the negative battery cable.
3. Disconnect starter electrical connections.
4. Disconnect ground cable.
5. Remove starter bolts.
6. Remove starter.

To install:

7. Install starter. Tighten the bolts to 17 ft. lbs. (23 Nm).
8. Connect ground cable. Tighten the bolt to 17 ft. lbs. (23 Nm).
9. Connect starter electrical connections.
10. Connect the negative battery cable.

4.6L Engines

1. Before servicing the vehicle, refer to the precautions at the beginning of this section.
2. Disconnect the negative battery cable.
3. Disconnect Heated Oxygen (HO_2S) sensor connector and bracket, if necessary.
4. Disconnect starter electrical connections.
5. Remove starter bolts.
6. Remove starter.

To install:

7. Install starter. Tighten the bolts to 17 ft. lbs. (23 Nm).
8. Connect starter electrical connections.
9. Connect HO_2S sensor connector and bracket, if necessary.
10. Connect the negative battery cable.

Oil Pan

REMOVAL & INSTALLATION

3.8L Engine

1. Before servicing the vehicle, refer to the precautions at the beginning of this section.
2. Drain the engine oil.
3. Disconnect the negative battery cable.
4. Remove air cleaner outlet pipe.
5. Remove radiator shroud cover.
6. Install engine lifting brackets and attach an engine support fixture.
7. On convertible models, remove 16 splash shield pin-type retainers and remove the cross brace support, if equipped.
8. On coupe models, remove the front subframe brace, if equipped.
9. Remove oil filter.
10. Disconnect left and right motor mounts.
11. Remove starter motor.
12. Disconnect steering column intermediate shaft.
13. Remove subframe mounting bolts, and lower subframe for clearance.
14. Remove oil pan.

To install:

15. Apply a bead of silicone sealant to the oil pan mounting flange. Install the oil pan and tighten the bolts as follows:
    a. Step 1: 44 inch lbs. (5 Nm).
    b. Step 2: 88 inch lbs. (10 Nm).
    c. Step 3: Tighten the bell housing bolts to 33 ft. lbs. (45 Nm).

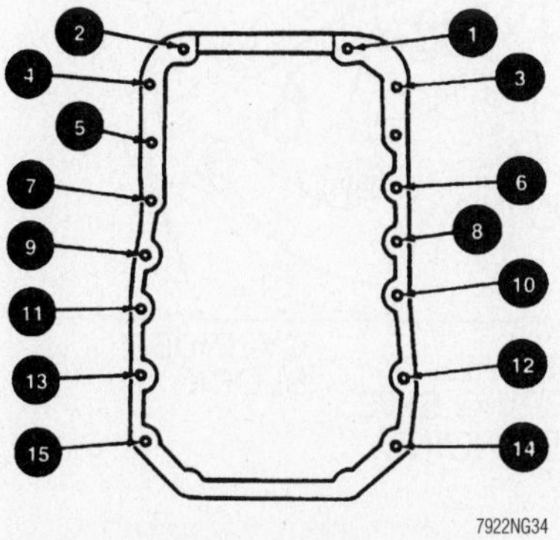

Oil pan torque sequence—3.8L engine

16. Raise the subframe and tighten the upper bolts to 85 ft. lbs. (115 Nm). Tighten the lower bolts to 68 ft. lbs. (90 Nm).
17. Connect steering column intermediate shaft.
18. Install starter motor.
19. Connect left and right motor mounts.
20. Install oil filter.
21. On coupe models, install the front subframe brace, if equipped and tighten the upper bolts to 30 ft. lbs. (41 Nm).
22. On convertible models, install the cross brace support, and install 16 splash shield pin-type retainers, if equipped.
23. Install radiator shroud cover.
24. Install air cleaner outlet pipe.
25. Connect the negative battery cable.
26. Fill the engine with oil.
27. Run the engine and check for leaks.

4.6L SOHC Engine

1. Before servicing the vehicle, refer to the precautions at the beginning of this section.
2. Drain the engine oil.
3. Disconnect the negative battery cable.
4. Remove air cleaner outlet pipe.
5. Remove radiator shroud cover.
6. Remove engine compartment brace, if equipped.
7. Install engine lifting brackets and attach an engine support fixture.
8. Remove front subframe crossmember brace.
9. Compress the front coil springs.
10. Place a safety stand under the subframe.
11. Remove left and right motor mount nuts.
12. Remove transmission housing cover.
13. Slightly lower the front subframe.
14. Remove the oil pan.

To install:

15. Install the oil pan. Use a new gasket and tighten the bolts in sequence as follows:
    a. Step 1: 18 inch lbs. (2 Nm).
    b. Step 2: 15 ft. lbs. (20 Nm).
    c. Step 3: Plus 60 degrees.
16. Raise the subframe into position and tighten the bolts to 85 ft. lbs. (115 Nm).
17. Install transmission housing cover.
18. Install left and right motor mount nuts. Tighten the nuts to 95–126 ft. lbs. (128–172 Nm).
19. Install front subframe crossmember brace. Tighten the bolts to 30–40 ft. lbs. (40–55 Nm).
20. Install engine compartment brace, if necessary.

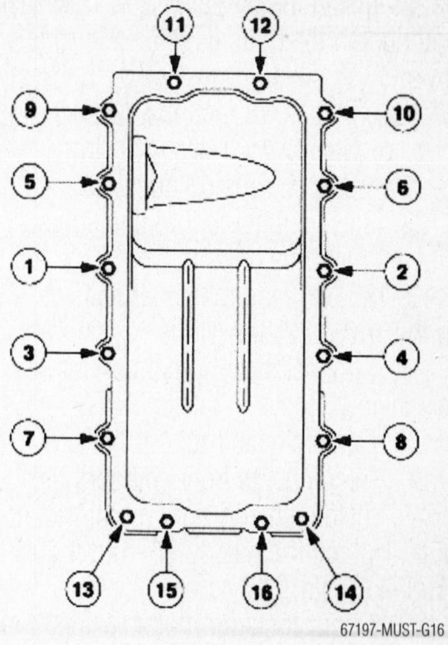

Oil pan torque sequence—4.6L engines

21. Install radiator shroud cover.
22. Install air cleaner outlet pipe.
23. Connect the negative battery cable.
24. Fill the engine with oil.
25. Start the engine and check for leaks.

4.6L DOHC Engine

1. Before servicing the vehicle, refer to the precautions at the beginning of this section.
2. Disconnect the negative battery cable.
3. Remove air intake scoop, if equipped.
4. Remove air cleaner outlet pipe.
5. Remove radiator shroud cover.
6. Drain the engine oil.
7. Remove oil filter.
8. Install engine lifting brackets and attach an engine support fixture.
9. Remove transmission.
10. Remove A/C manifold and tube assembly (accumulator to compressor).
11. Remove A/C line.
12. Remove left and right motor mount nuts.
13. Disconnect wiring from starter motor and position harness out of the way, if necessary.
14. Compress front coil springs.
15. Disconnect stabilizer bar links.
16. Disconnect steering column intermediate shaft.
17. Place a safety stand under the subframe, and loosen subframe bolts.
18. Lower subframe for clearance.
19. Remove subframe brace, if equipped.
20. Remove oil pan.

To install:

21. Apply silicone sealant to the joints where the front cover and the rear seal retainer meet the cylinder block.

22. Install the oil pan. Use a new gasket and tighten the bolts in sequence as follows:
   a. Step 1: 18 inch lbs. (2 Nm).
   b. Step 2: 15 ft. lbs. (20 Nm).
   c. Step 3: Plus 60 degrees.
23. Raise subframe into position. Tighten the large bolts to 83–113 ft. lbs. (113–153 Nm) and the small bolts to 66–97 ft. lbs. (90–132 Nm).
24. Connect steering column intermediate shaft.
25. Connect stabilizer bar links.
26. Decompress front coil springs.
27. Install left and right motor mount nuts. Tighten the nuts to 95–126 ft. lbs. (128–172 Nm).
28. Connect starter motor wiring, if necessary.
29. Install A/C line.
30. Install A/C manifold and tube assembly (accumulator to compressor).
31. Remove engine lifting brackets and attach an engine support fixture.
32. Install radiator shroud cover.
33. Install air cleaner outlet pipe.
34. Install air intake scoop, if equipped.
35. Install oil filter.
36. Install transmission.
37. Connect the negative battery cable.
38. Fill the engine with oil.
39. Start the engine and check for oil leaks.

Oil Pump

REMOVAL & INSTALLATION

3.8L Engine

1. Before servicing the vehicle, refer to the precautions at the beginning of this section.
2. Remove the oil filter.
3. Remove oil pump body and gears.

To install:

4. Install oil pump. Use a new O-ring seal. Tighten the large bolts to 18 ft. lbs. (25 Nm) and the small bolts to 89 inch lbs. (10 Nm).
5. Install oil filter.
6. Check for leaks and proper operation.

4.6L Engines

1. Before servicing the vehicle, refer to the precautions at the beginning of this section.
2. Remove valve covers.
3. Remove oil pan.
4. Remove front cover.
5. Remove timing chains and sprockets.

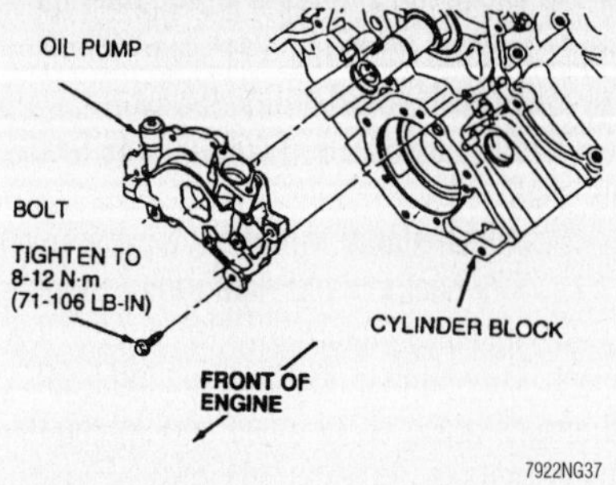

Exploded view of the oil pump mounting—4.6L engines

6. Remove oil pump.

To install:

7. Install oil pump. Tighten the bolts to 71–106 inch lbs. (8–12 Nm).
8. Install timing chains and sprockets.
9. Install front cover.
10. Install oil pan.
11. Install valve covers.
12. Check for leaks and proper operation.

Rear Main Seal

REMOVAL & INSTALLATION

3.8L Engine

1. Before servicing the vehicle, refer to the precautions at the beginning of this section.
2. Disconnect the negative battery cable.
3. Remove transmission.
4. Remove clutch pressure plate and disc, if equipped.
5. Remove flywheel.
6. Remove engine rear plate, if equipped.
7. Remove rear main seal. Use Crankshaft Rear Seal Remover T95P-6701-EH.

To install:

8. Install rear main seal. Use Rear Crankshaft Seal Installer T82L-6701-A.
9. Install engine rear plate, if equipped.
10. Install flywheel. Tighten the bolts to 54–64 ft. lbs. (73–87 Nm).
11. Install clutch pressure plate and disc, if equipped.
12. Install transmission.
13. Connect the negative battery cable.

4.6L Engines

1. Before servicing the vehicle, refer to the precautions at the beginning of this section.
2. Disconnect the negative battery cable.
3. Remove transmission.
4. Remove clutch pressure plate and disc, if equipped.
5. Remove flywheel.
6. Remove oil slinger. Use Rear Crankshaft Oil Slinger Remover T95P-6701-AH.
7. Remove rear oil seal retainer.
8. Remove rear oil seal.

To install:

9. Install the rear oil seal retainer. Apply silicone sealant as shown. Tighten the bolts in sequence to 71–106 inch lbs. (8–12 Nm).

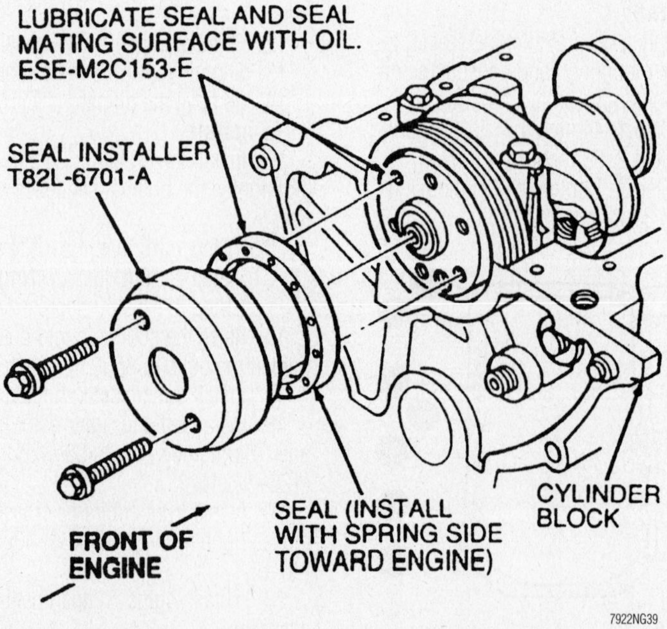

Installing the rear main seal—3.8L engine

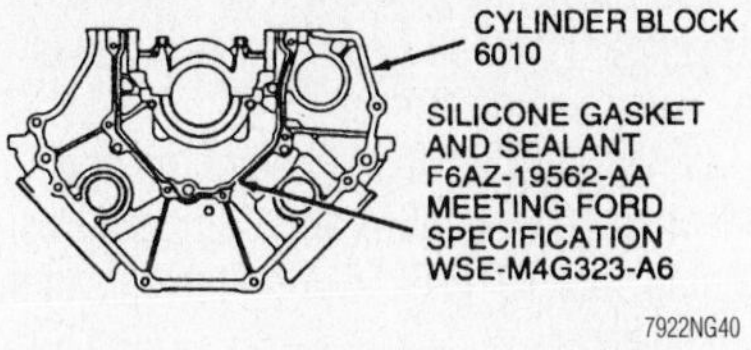

Apply silicone sealant to the engine block when installing the oil seal retainer—4.6L engines

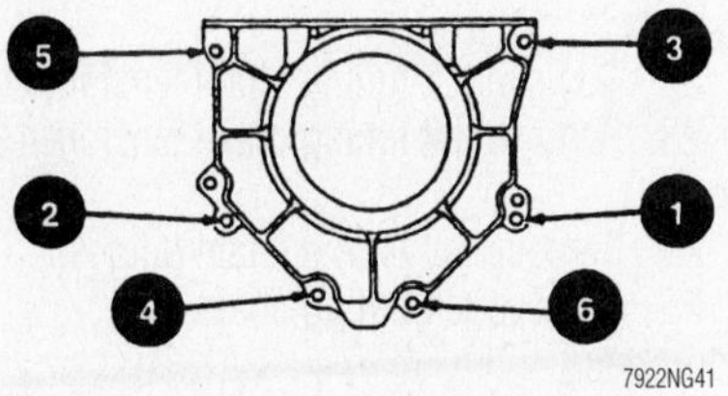

Oil seal retainer torque sequence—4.6L engines

10. Install rear oil seal. Use Rear Crankshaft Seal Replacer T95P-6701-BH.
11. Install oil slinger. Use Rear Crankshaft Oil Slinger Replacer T95P-6701-CH.
12. Install flywheel. Tighten the bolts in a crossing pattern to 54–64 ft. lbs. (73–87 Nm).
13. Install clutch pressure plate and disc, if equipped.
14. Install transmission.
15. Connect the negative battery cable.
16. Run the engine and check for leaks.

Timing Chain, Sprockets, Front Cover and Seal

REMOVAL & INSTALLATION

3.8L Engine

1. Before servicing the vehicle, refer to the precautions at the beginning of this section.
2. Drain the cooling system and the engine oil.
3. Rotate the crankshaft so that the No. 1 cylinder is at Top Dead Center (TDC) of the compression stroke.
4. Disconnect the negative battery cable.
5. Disconnect Intake Air Temperature (IAT) sensor connector.
6. Remove air cleaner and air intake tube.
7. Remove cooling fan and shroud.
8. Remove the accessory drive belt.
9. Remove power steering pump and bracket.
10. Remove A/C compressor front bracket, if equipped.
11. Remove oil filter.
12. Remove radiator hoses.
13. Remove heater water outlet tube.
14. Remove crankshaft pulley and damper.
15. Remove front crankshaft seal.

➡Note the position of the Camshaft Position (CMP) sensor connector. The installation procedure requires that the connector be located in the same position.

16. Remove Camshaft Position (CMP) sensor housing.
17. Remove Crankshaft Position (CKP) sensor.
18. Remove oil pan.
19. Remove engine front cover.
20. Remove distributor drive gear.
21. Remove timing chain and sprockets.

To install:

22. Compress the timing chain vibration damper and install a retaining pin.
23. Install timing chain and sprockets with the timing marks aligned as shown.
24. Install distributor drive gear. Tighten the bolt to 30–36 ft. lbs. (40–50 Nm) and remove the vibration damper retaining pin.
25. Install front cover. Tighten the bolts in sequence to 15–22 ft. lbs. (20–30 Nm).
26. Install oil pan.
27. Install CKP sensor.
28. Install front crankshaft seal.
29. Install crankshaft pulley and damper. Tighten the bolt to 103–132 ft. lbs. (140–180 Nm).
30. Install heater water outlet tube.
31. Install radiator hoses.
32. Install oil filter.
33. Install A/C compressor front bracket, if equipped. Tighten the bolts to 30–45 ft. lbs. (41–61 Nm).
34. Install power steering pump and bracket. Tighten the bolts to 30–45 ft. lbs. (41–61 Nm).
35. Install Synchro Positioning Tool T96T-12200-A to the CMP sensor housing and turn it clockwise until the tool boss engages the notch in the housing assembly.
36. Install the CMP sensor housing so that the CMP sensor connector is in the position noted earlier. Tighten the hold down bolt to 15–22 ft. lbs. (20–30 Nm).
37. Install the accessory drive belt.
38. Install cooling fan and shroud.
39. Install air cleaner and air intake tube.
40. Connect IAT sensor connector.

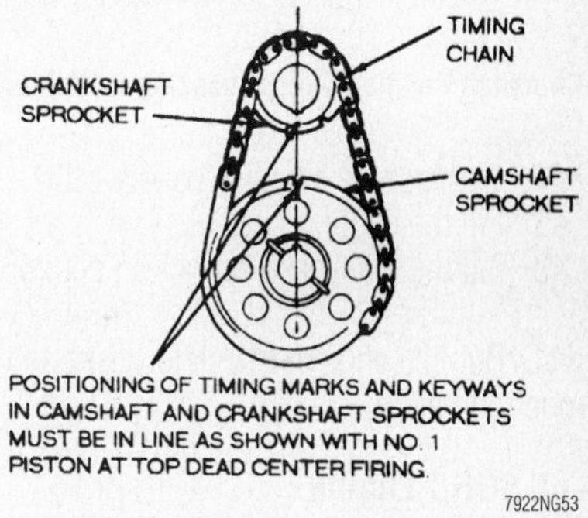

Camshaft timing marks—3.8L engine

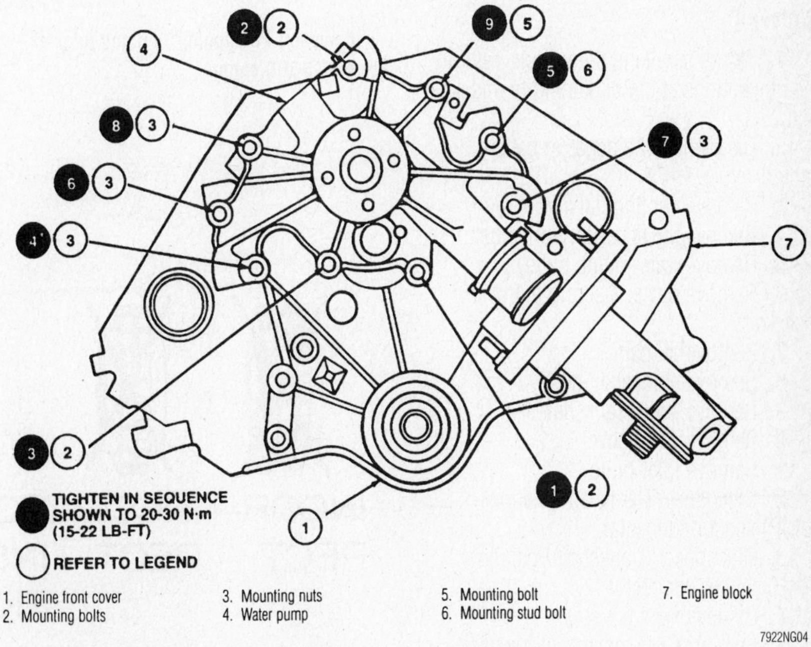

Front cover torque sequence—3.8L engine

Camshaft Position Sensor

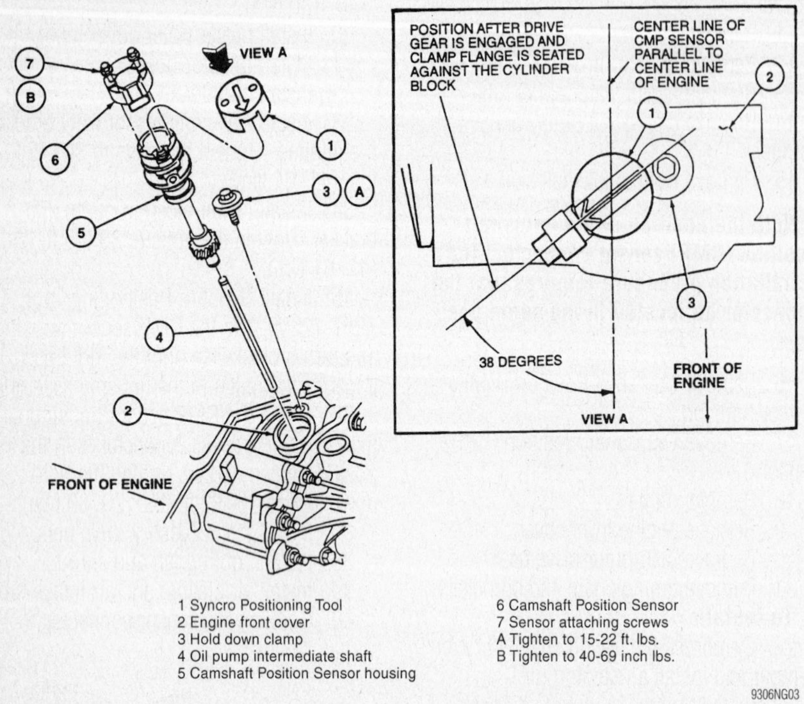

Camshaft Position Sensor housing installation—3.8L engine

41. Connect the negative battery cable.
42. Fill the cooling system.
43. Fill the crankcase with clean engine oil.
44. Run the engine. Check for leaks and proper operation.

4.6L SOHC Engine

➡This is not a free wheeling engine. Do not rotate the crankshaft or camshafts with the timing chains removed.

1. Before servicing the vehicle, refer to the precautions at the beginning of this section.
2. Disconnect the negative battery cable.
3. Remove cooling fan and shroud.
4. Remove the accessory drive belt.
5. Remove water pump pulley.
6. Remove power steering pump and reservoir.
7. Remove oil pan.
8. Remove crankshaft pulley.
9. Remove front crankshaft seal.
10. Remove valve covers.
11. Remove idler pulley.
12. Disconnect Camshaft Position (CMP) sensor connector.
13. Disconnect Crankshaft Position (CKP) sensor connector.
14. Remove front cover.
15. Remove CKP sensor pulse wheel.

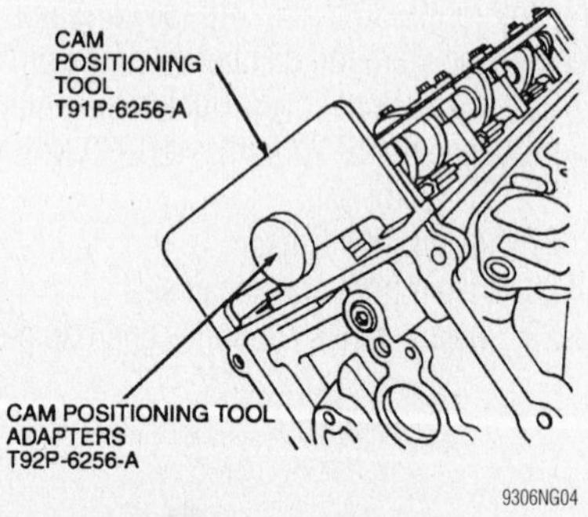

Camshaft Positioning Tool and Adapter—4.6L SOHC engine

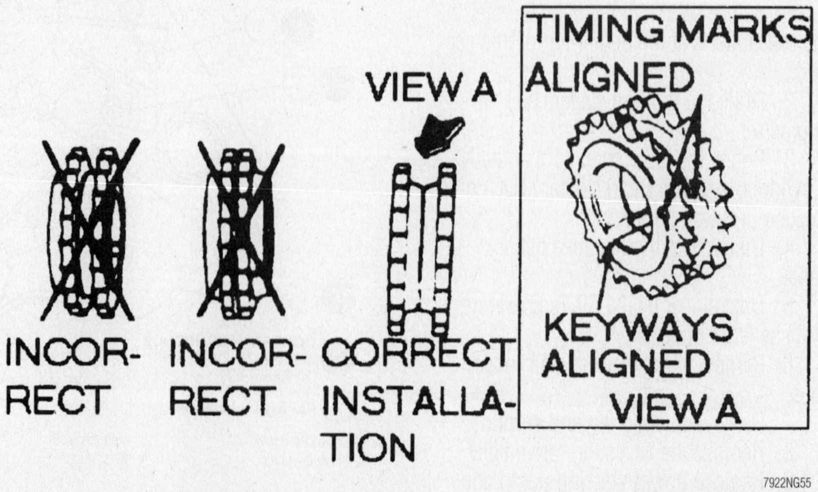

Crankshaft sprocket positioning—4.6L SOHC engine

16. Rotate the crankshaft so that the No. 1 cylinder is at Top Dead Center (TDC) of the compression stroke.
17. Install Camshaft Positioning Tool Adapters T92P-6256-A and Camshaft Positioning Tool T91P-6256-A on the flats of the camshafts.
18. Remove right timing chain tensioner.
19. Remove right timing chain tensioner arm.
20. Remove right timing chain guides.
21. Remove right timing chain and sprockets.
22. Remove left timing chain tensioner.
23. Remove left timing chain tensioner arm.
24. Remove left timing chain guides.
25. Remove left timing chain and sprockets.

To install:

26. Install timing chain guides. Tighten the bolts to 71–106 inch lbs. (8–12 Nm).
27. Install camshaft sprockets. Do not tighten the bolts at this time.
28. Install crankshaft sprockets.
29. Install left timing chain with the copper links aligned with the timing marks on the crankshaft and camshaft sprockets.
30. Install right timing chain with the copper links aligned with the timing marks on the crankshaft and camshaft sprockets.
31. Compress the timing chain tensioners as shown. Install locking pins.
32. Install tensioner arms and tensioners. Tighten the tensioner mounting bolts to 15–22 ft. lbs. (20–30 Nm).
33. Install crankshaft Holding Tool T93P-6303-A.
34. Use a C-clamp on the tensioner arm and chain guide to remove slack from the timing chain.

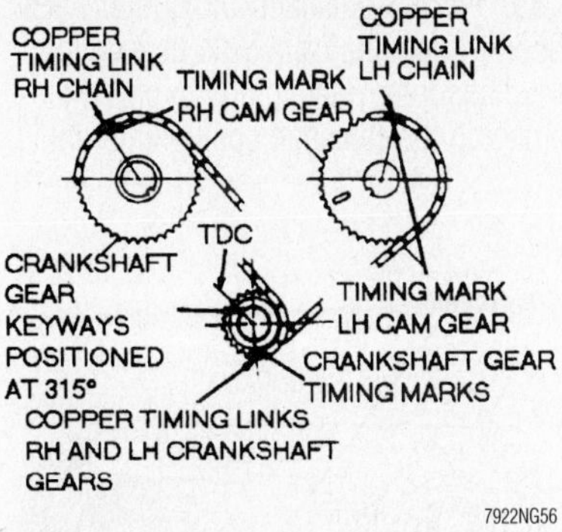

Timing chain and sprocket alignment—4.6L SOHC engine

Remove slack from the timing chains with a C-clamp—4.6L SOHC engine

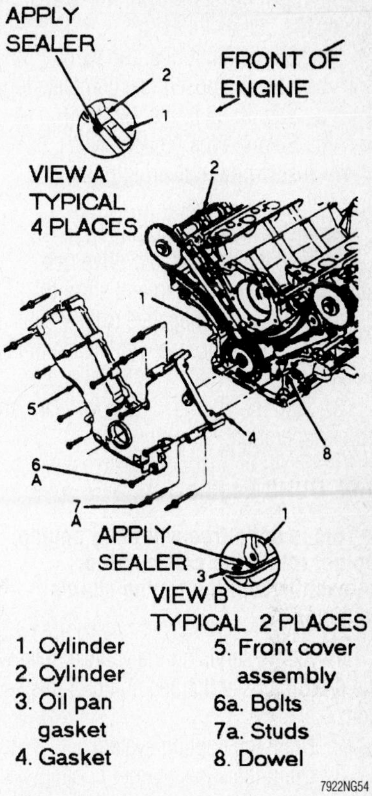

Apply sealer when installing the front cover—4.6L SOHC engine

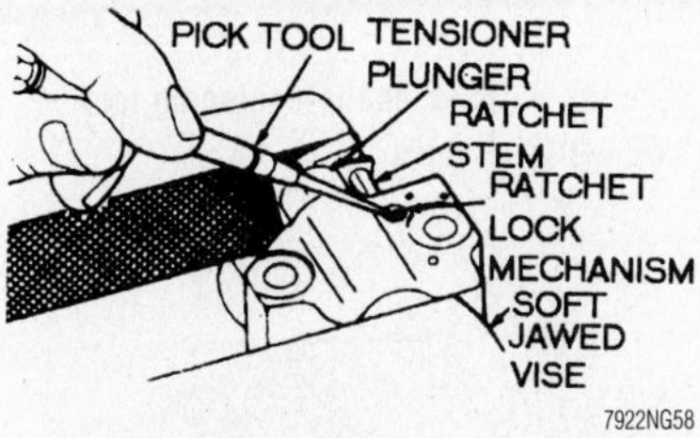

Timing chain tensioner bleeding procedure—4.6L SOHC engine

35. Remove the tensioner locking pins.
36. Tighten the camshaft sprocket bolts to 81–95 ft. lbs. (110–130 Nm).
37. Remove the C-clamps, Camshaft Positioning Tools and Adapters, and the Crankshaft Holding Tool.
38. Install CKP sensor pulse wheel.
39. Install front cover. Tighten the bolts in sequence to 15–22 ft. lbs. (20–30 Nm).
40. Connect CKP sensor connector.
41. Connect CMP sensor connector.
42. Install idler pulley.
43. Install valve covers.
44. Install front crankshaft seal.
45. Install oil pan.

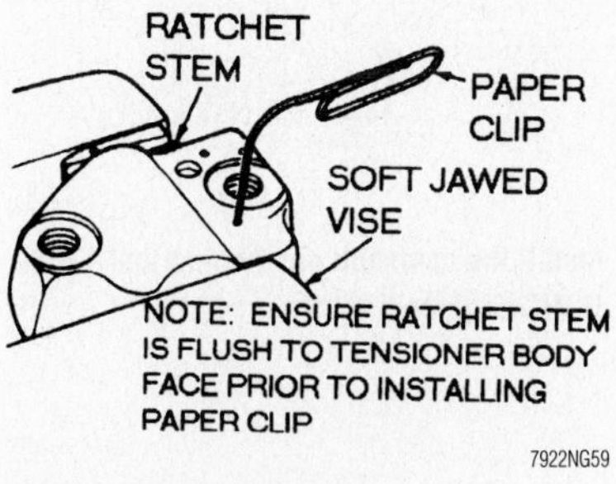

Timing chain tensioner locking procedure—4.6L SOHC engine

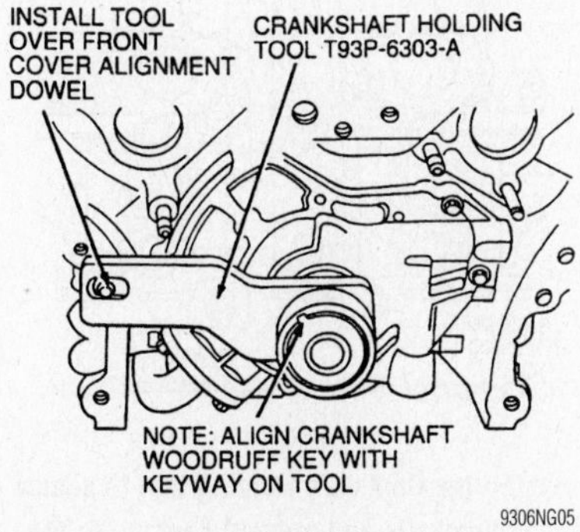

Crankshaft Holding Tool—4.6L SOHC engine

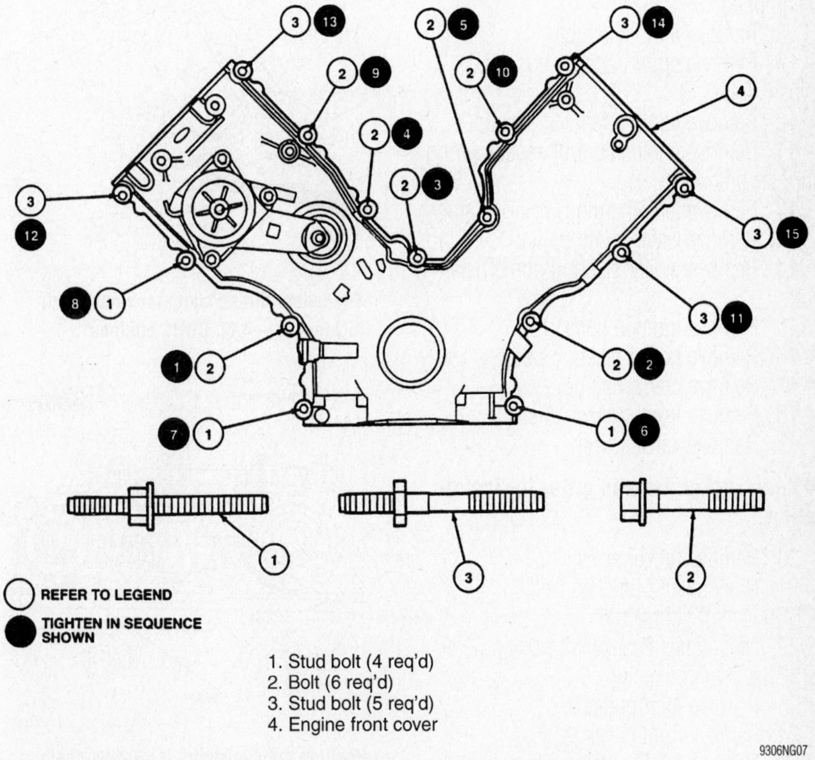

Front cover torque sequence and bolt identification—4.6L SOHC engine

46. Install the crankshaft pulley and tighten the bolt as follows:
    a. Step 1: 66 ft. lbs. (90 Nm).
    b. Step 2: Loosen one complete turn.
    c. Step 3: 36 ft. lbs. (50 Nm).
    d. Step 4: Plus 90 degrees.
47. Install power steering pump and reservoir.
48. Install water pump pulley.
49. Install the accessory drive belt.
50. Install cooling fan and shroud.
51. Connect the negative battery cable.
52. Fill the crankcase with clean engine oil.
53. Start the engine. Check for leaks and proper operation.

4.6L DOHC Engine

➡This is not a free wheeling engine. Do not rotate the crankshaft or camshafts with the timing chains removed.

1. Before servicing the vehicle, refer to the precautions at the beginning of this section.
2. Drain the cooling system.
3. Drain the supercharger cooling system, if equipped.
4. Remove supercharger drive belt, if equipped.
5. Remove the accessory drive belt.
6. Disconnect the negative battery cable.
7. Remove engine compartment brace, if equipped.
8. Remove air intake tube.
9. Remove upper radiator hose and bypass tube.
10. Remove cooling fan and shroud.
11. Remove engine control sensor wiring support bracket.
12. Remove fuel charging wiring retainer.
13. Remove water pump pulley.
14. Remove power steering pump reservoir.
15. Remove Ignition coils.
16. Remove power steering pump.
17. Remove crankshaft pulley.
18. Remove front crankshaft seal.
19. Remove valve covers.

➡Keep rocker arms in order for installation.

20. Remove rocker arms.
21. Disconnect Camshaft Position (CMP) sensor connector.
22. Disconnect Crankshaft Position (CKP) sensor connector.
23. Remove idler pulley.
24. Remove front cover.
25. Remove CMP sensor pulse wheel.
26. Rotate the crankshaft so that the No. 1 cylinder is at Top Dead Center (TDC) of the compression stroke.
27. Remove right primary timing chain tensioner.
28. Remove right primary timing chain tensioner arm and chain guide.
29. Remove right primary timing chain and sprockets.
30. Remove left primary timing chain tensioner.
31. Remove left primary timing chain tensioner arm and chain guide.
32. Remove left primary timing chain and sprockets.
33. Compress the secondary chain tensioners and install locking pins.
34. Remove the left and right secondary timing chains and sprockets.

To install:

35. Position the camshafts with the keys facing downward.
36. Install the secondary timing chains and sprockets. Do not tighten the sprocket bolts at this time.
37. Tension the secondary timing chains with the Secondary Timing Chain Tensioning tool T93P-6256-BH.
38. Install Camshaft Positioning tool T93P-6256-A in the rear D-slots of the camshaft.
39. Install Camshaft Holding tool T93P-6256-AH onto the camshafts to keep the camshafts from rotating and to prevent damaging the camshaft positioning tool.

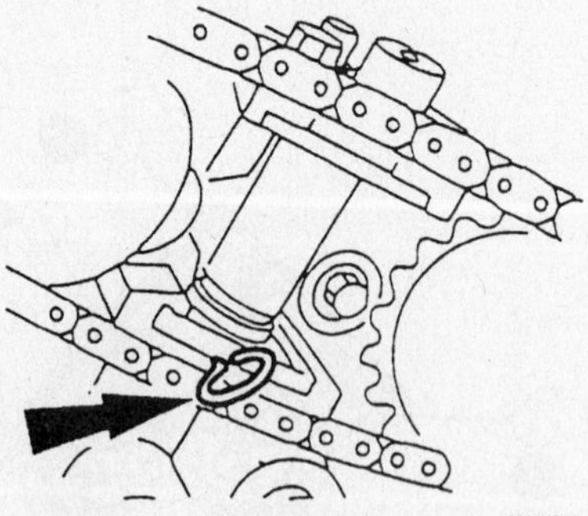

Secondary timing chain tensioner and locking pin—4.6L DOHC engines

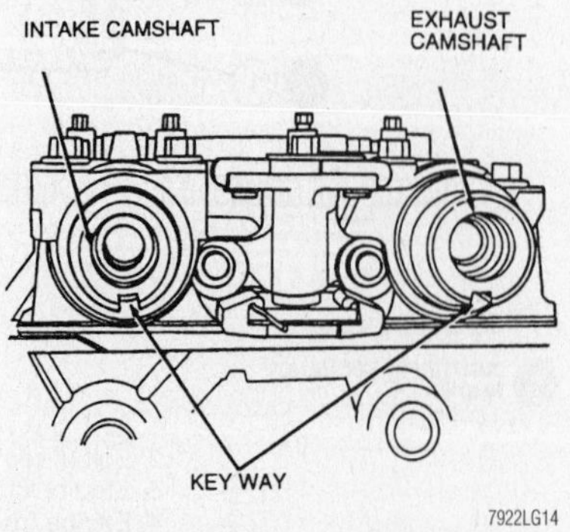

Position the camshafts for timing chain installation—4.6L DOHC engines

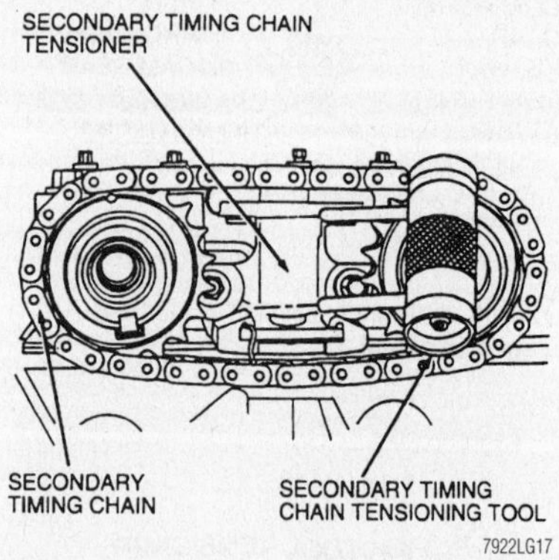

Secondary timing chain tensioning tool—4.6L DOHC engines

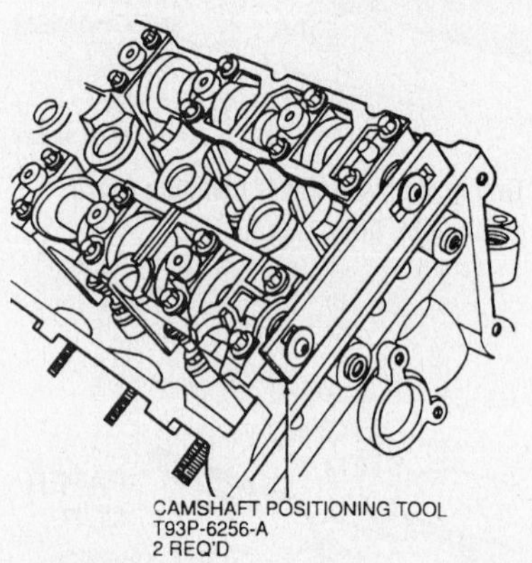

Install the camshaft positioning tool—4.6L DOHC engines

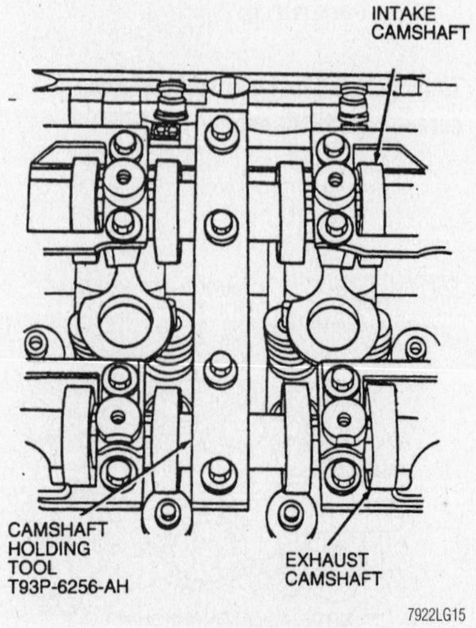

Install the Camshaft Holding tool to secure the camshafts and prevent damage to the camshaft positioning tool—4.6L DOHC engines

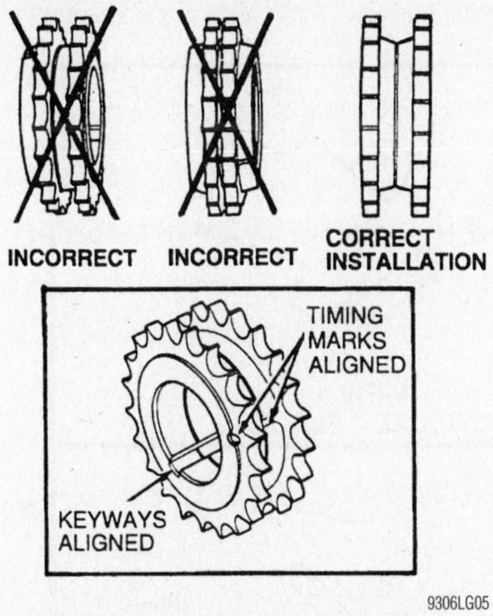

Crankshaft timing sprocket installation—4.6L DOHC engines

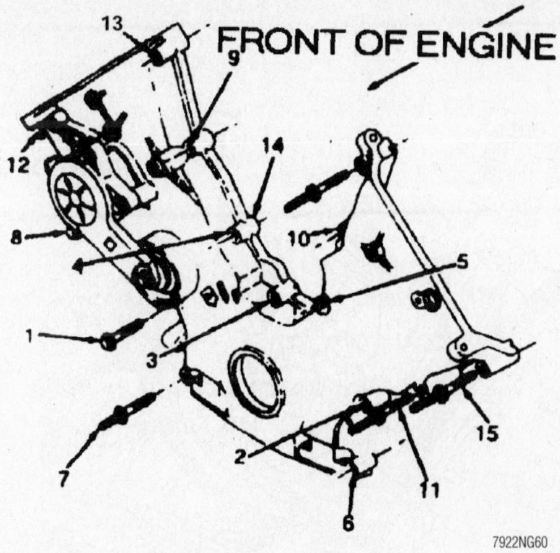

Front cover torque sequence—4.6L DOHC engines

40. Tighten the secondary timing chain sprocket bolts to 81–95 ft. lbs. (110–130 Nm).
41. Remove the secondary timing chain tensioner locking pins and remove the tensioner tool.
42. Compress the primary timing chain tensioners and install locking pins.
43. Install the crankshaft timing sprockets by aligning the timing marks.
44. Align the copper colored primary timing chain links with the sprocket timing marks as shown.
45. Install left primary timing chain and sprockets.
46. Install left primary timing chain tensioner arm and chain guide.
47. Install left primary timing chain tensioner.
48. Install right primary timing chain and sprockets.
49. Install right primary timing chain tensioner arm and chain guide.
50. Install right primary timing chain tensioner.
51. Tighten the camshaft sprocket bolts to 81–95 ft. lbs. (110–130 Nm).
52. Remove the primary timing chain tensioner locking pins.
53. Remove the Camshaft Holding and Camshaft Positioning tools.
54. Install the CMP pulse wheel
55. Install the front cover. Tighten the bolts in sequence as follows:
   a. Step 1: 14 ft. lbs. (20 Nm).
   b. Step 2: Plus 60 degrees.
56. Install idler pulley.
57. Install rocker arms in their original positions.
58. Install valve covers.
59. Install front crankshaft seal.
60. Connect CKP sensor connector.
61. Connect CMP sensor connector.

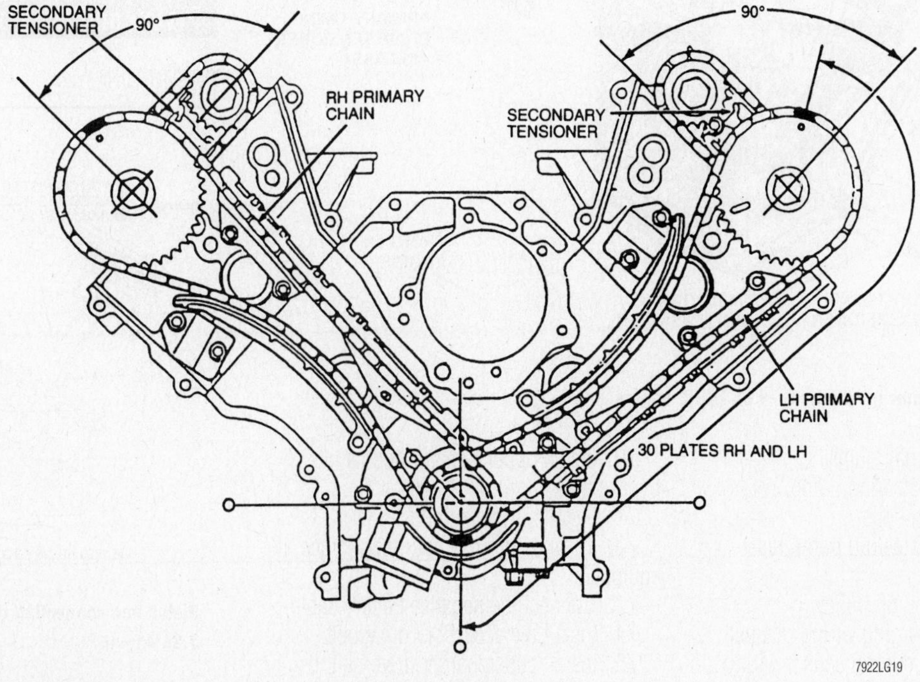

Correct alignment of the primary timing chains and sprockets—4.6L DOHC engines

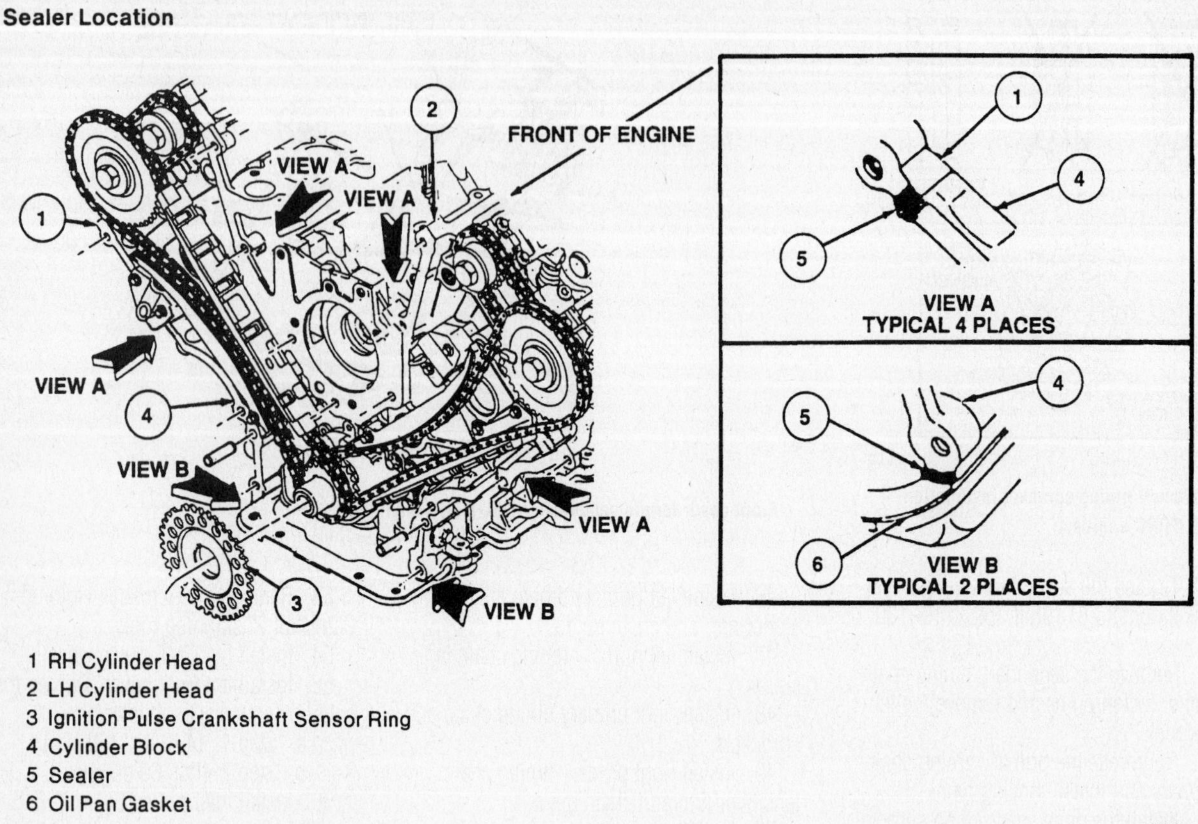

Apply sealer to these locations—4.6L DOHC engines

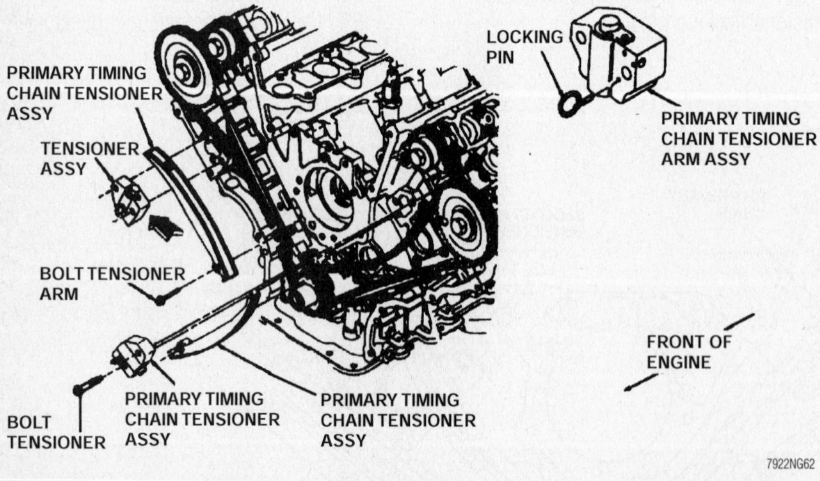

Timing chain tensioner installation—4.6L DOHC engine

62. Install Crankshaft pulley.
63. Install power steering pump.
64. Install ignition coils.
65. Install power steering pump reservoir.
66. Install water pump pulley.
67. Install fuel charging wiring retainer.
68. Install engine control sensor wiring support bracket.
69. Install cooling fan and shroud.
70. Install upper radiator hose and bypass tube.
71. Install air intake tube.
72. Install engine compartment brace, if equipped.
73. Connect the negative battery cable.
74. Install the accessory drive belt.
75. Install supercharger drive belt, if equipped.
76. Fill the cooling system.
77. Fill the supercharger cooling system, if equipped.
78. Start the engine. Check for leaks and proper operation.

Piston and Ring

POSITIONING

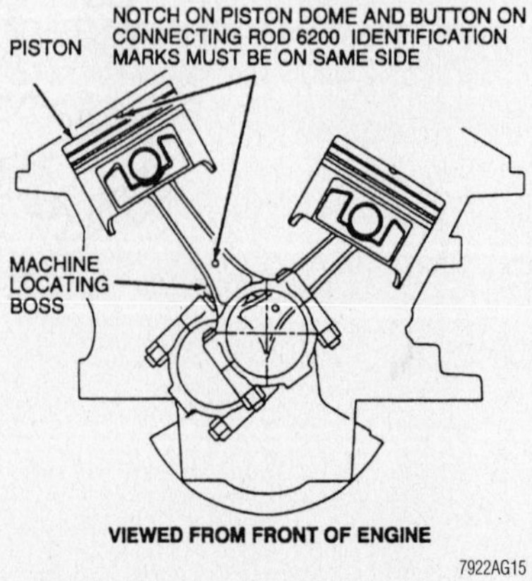

Piston and connecting rod positioning—3.8L engine

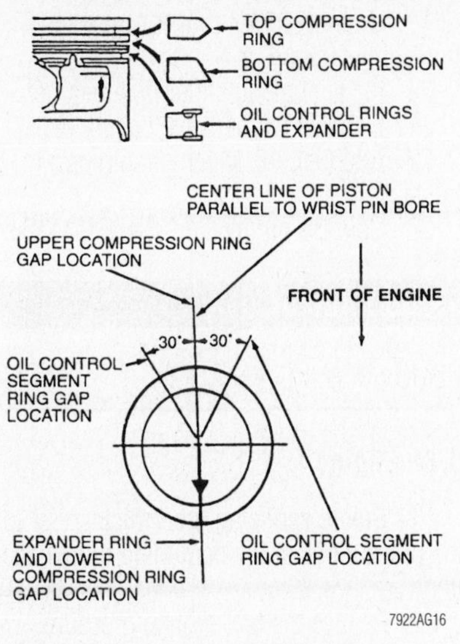

Piston ring end-gap spacing—3.8L engine

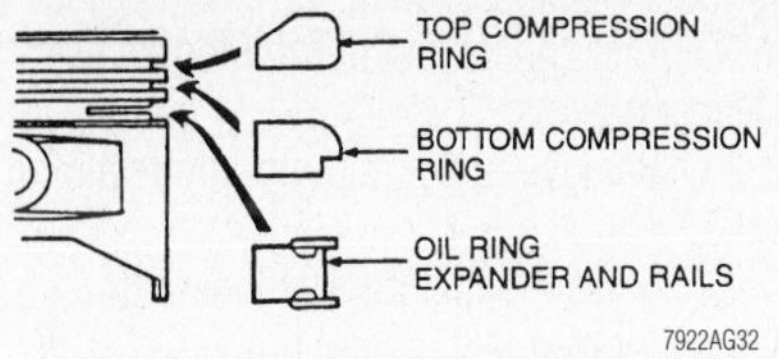

Piston ring positioning—4.6L SOHC engine

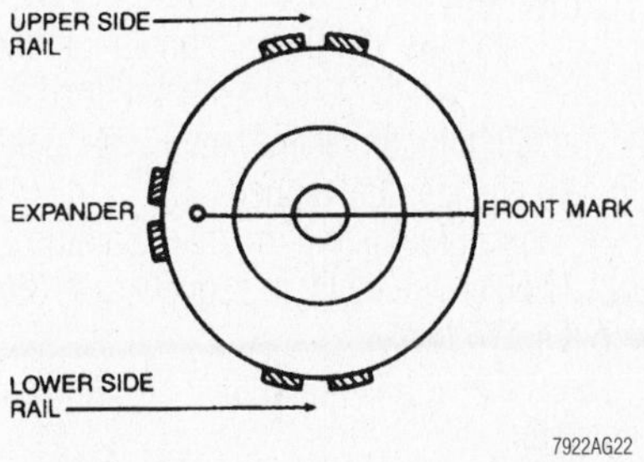

Piston ring end-gap spacing—4.6L SOHC engine

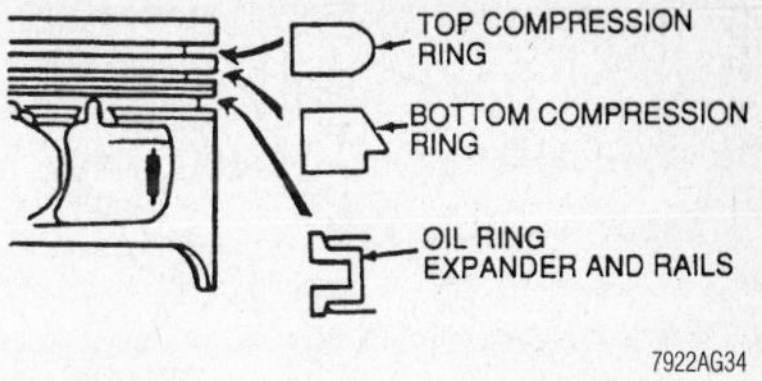

Piston ring positioning—4.6L DOHC engine

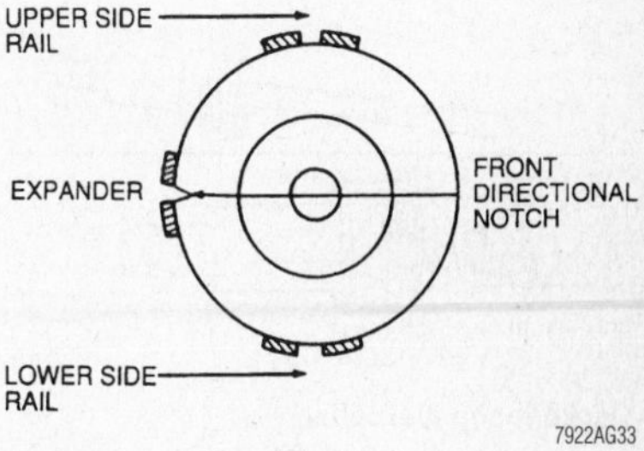

Piston ring end-gap spacing—4.6L DOHC engine

FUEL SYSTEM

Fuel System Service Precautions

Safety is the most important factor when performing not only fuel system maintenance, but any type of maintenance. Failure to conduct maintenance and repairs in a safe manner may result in serious personal injury or death. Work on a vehicle's fuel system components can be accomplished safely and effectively by adhering to the following rules and guidelines.

- To avoid the possibility of fire and personal injury, always disconnect the negative battery cable unless the repair or test procedure requires that battery voltage be applied.
- Always relieve the fuel system pressure prior to disconnecting any fuel system component (injector, fuel rail, pressure regulator, etc.) fitting or fuel line connection. Exercise extreme caution whenever relieving fuel system pressure to avoid exposing skin, face and eyes to fuel spray. Please be advised that fuel under pressure may penetrate the skin or any part of the body that it contacts.
- Always place a shop towel or cloth around the fitting or connection prior to loosening to absorb any excess fuel due to spillage. Ensure that all fuel spillage is quickly removed from engine surfaces. Ensure that all fuel-soaked cloths or towels are deposited into a flame-proof waste container with a lid.
- Always keep a dry chemical (Class B) fire extinguisher near the work area.
- Do not allow fuel spray or fuel vapors to come into contact with a spark or open flame.
- Always use a second wrench when loosening or tightening fuel line connection fittings. This will prevent unnecessary stress and torsion on fuel piping. Always follow the proper torque specifications.
- Always replace worn fuel fitting O-rings with new ones. Do not substitute fuel hose where rigid pipe is installed.

Fuel System Pressure

RELIEVING

Most Sequential Fuel Injection (SFI) engines , except 2004 4.6L SOHC engines, are equipped with a pressure relief valve located on the fuel supply manifold. Remove the fuel tank cap and attach fuel pressure gauge T80L-9974-B, or equivalent, to the valve on the fuel rail. Place the tube from the tool into a small container and open the valve on the tool to release the fuel pressure. Be sure to drain the fuel into a suitable container and to avoid gasoline spillage. If a pressure gauge is not available, disconnect the vacuum hose from the fuel pressure regulator and attach a hand-held vacuum pump. Apply about 25 in. Hg (84 kPa) of vacuum to the regulator to vent the fuel system pressure into the fuel tank through the fuel return hose.

For 2004 4.6L SOHC engines, remove the fuel pump fuse; located in the engine compartment fuse box. Start the engine and allow it to run at idle until it stalls. After the engine stalls, crank the engine for approximately five seconds to ensure the pressure in the fuel injection supply manifold has been released. After the fuel pressure has been fully released, turn the ignition switch to the "OFF" position.

➡This procedure will remove the fuel pressure from the lines, but not the fuel. Take precautions to avoid the risk of fire and use clean rags to soak up any spilled fuel when the lines are disconnected.

Fuel Filter

REMOVAL & INSTALLATION

➡Always replace fuel line fitting plastic clips.

1. Before servicing the vehicle, refer to the precautions at the beginning of this section.
2. Relieve the fuel system pressure.
3. Disconnect the negative battery cable.
4. Disconnect fuel line fittings.

➡The filter may be secured in place with hairpin clips, duckbill clips, or push connect fittings.

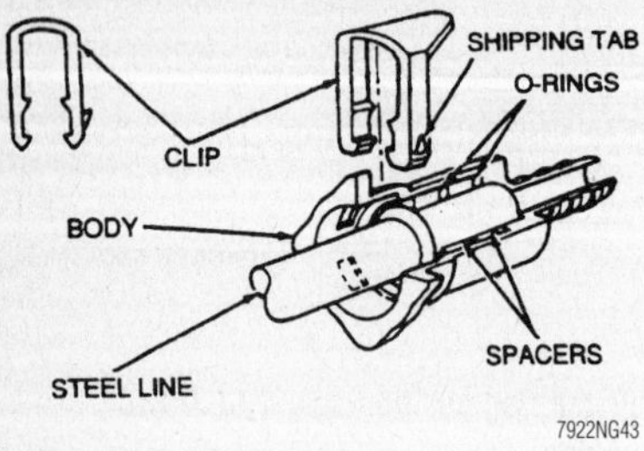

Hairpin clip fuel fitting

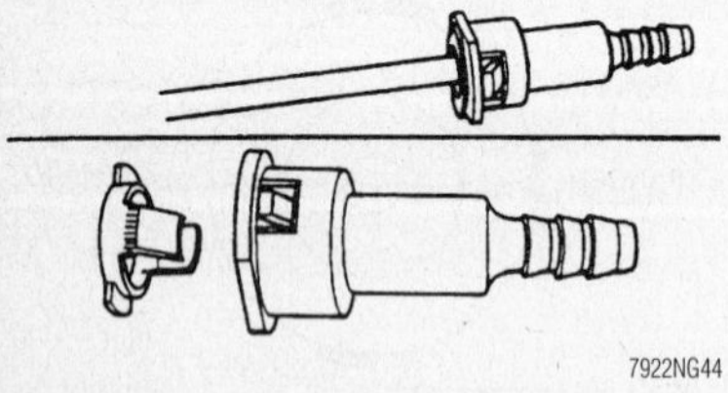

Duckbill clip fuel fitting

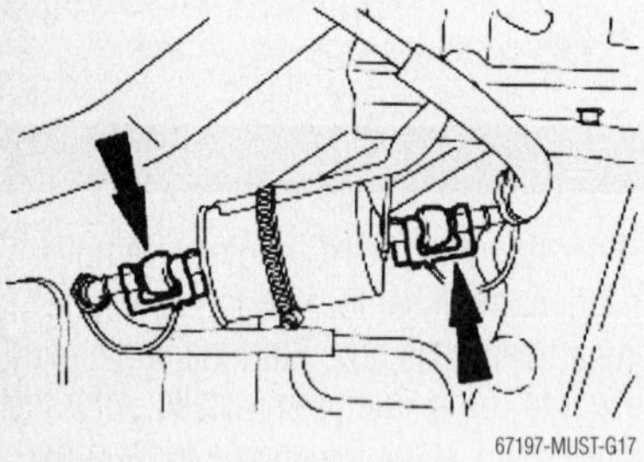

Push connect fuel fitting

5. Disengage the filter from the fuel filter bracket clamp.
6. Remove fuel filter.

To install:

7. Install fuel filter. Note the flow direction arrow.
8. Secure the fuel filter in the bracket clamp.
9. Install fuel line fittings.
10. Connect the negative battery cable.
11. Start the engine and check for leaks.

Fuel Pump

REMOVAL & INSTALLATION

1. Before servicing the vehicle, refer to the precautions at the beginning of this section.
2. Relieve the fuel system pressure.
3. Disconnect the negative battery cable.
4. Drain the fuel tank.
5. Remove fuel tank filler pipe retainer.
6. Disconnect EVAP canister tube and hose, if necessary.
7. Disconnect fuel tank vent hose.
8. Remove fuel tank support straps.
9. Disconnect fuel lines.
10. Disconnect fuel pump module electrical connector.
11. Remove fuel pump module retaining bolts.
12. Raise the fuel pump module to access locking tabs, if equipped. Squeeze the locking tabs together, if necessary, and remove the fuel pump module from the fuel tank.

To install:

13. Install the fuel pump module into the retainer. Use a new O-ring seal and push the module into the retainer until both locking tabs engage, if equipped.
14. Install fuel pump module retaining bolts. Tighten the bolts, in sequence, to 89 inch lbs. (10 Nm).
15. Connect fuel pump module electrical connector.
16. Connect fuel lines.
17. Install fuel tank support straps. Tighten the bolts to 22–29 ft. lbs. (30–40 Nm).

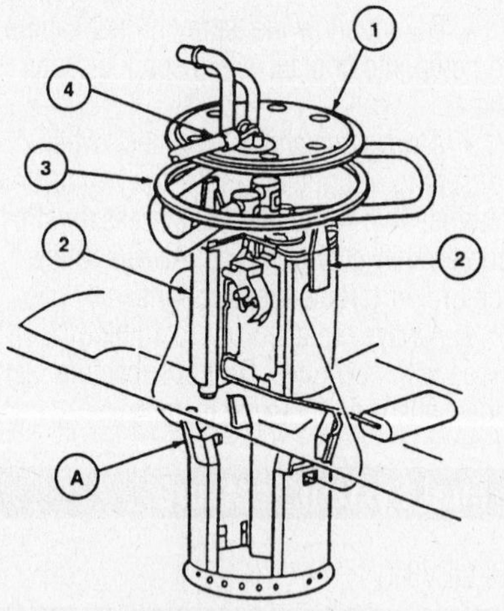

1 Fuel Pump Module
2 Locking Tab (Part of 9H307)
3 O-Ring Seal
4 Connector (Part of 14405 Wiring Assy)
A Fuel Pump Must Be Snapped Into Retainer (2 Places)

9306NG08

Fuel pump module assembly

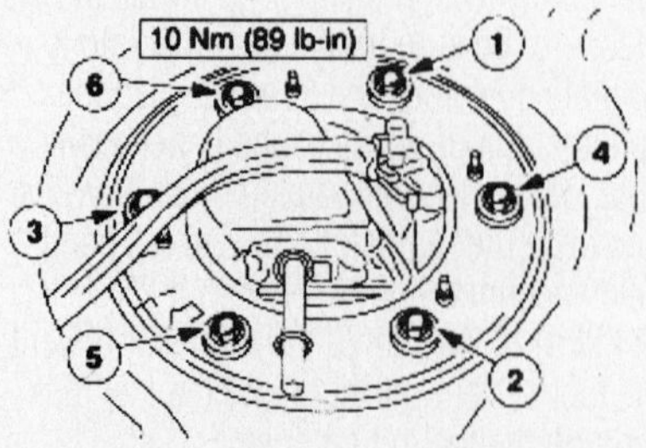

Fuel pump module torque sequence

18. Connect fuel tank vent hose.
19. Connect EVAP canister tube and hose, if necessary.
20. Install fuel tank filler pipe retainer.
21. Connect the negative battery cable.
22. Add fuel (10 gallons minimum) to the tank.
23. Start the engine and check for leaks.

Fuel Injector

REMOVAL & INSTALLATION

3.8L Engine

1. Before servicing the vehicle, refer to the precautions at the beginning of this section.
2. Relieve the fuel system pressure.
3. Disconnect the negative battery cable.
4. Remove upper intake manifold.
5. Disconnect fuel supply line.
6. Disconnect fuel return line, if equipped.
7. Disconnect fuel injector connectors.
8. Disconnect fuel pressure sensor vacuum hose and electrical connector.
9. Remove fuel supply manifold with injectors attached.
10. Remove fuel injectors from the supply manifold.

To install:

11. Install new O-rings.
12. Install fuel injectors to the supply manifold.
13. Install fuel supply manifold with injectors. Tighten the bolts to 89 inch lbs. (10 Nm).
14. Connect fuel pressure sensor vacuum hose and electrical connector.
15. Connect fuel injector connectors.
16. Connect fuel supply line.
17. Connect fuel return line, if equipped.
18. Install upper intake manifold.
19. Connect the negative battery cable.
20. Start the engine and check for leaks.

4.6L SOHC Engine

1. Before servicing the vehicle, refer to the precautions at the beginning of this section.
2. Relieve the fuel system pressure.
3. Disconnect the negative battery cable.
4. Remove air intake tube.
5. Disconnect fuel supply line.
6. Disconnect fuel return line, if equipped.
7. Disconnect accelerator cable.

8. Disconnect cruise control cable, if equipped.
9. Remove throttle body.
10. Disconnect Idle Air Control (IAC) valve connector and hose.
11. Disconnect Positive Crankcase Ventilation (PCV) hose.
12. Disconnect main chassis vacuum supply hose.
13. Disconnect fuel injector connectors.
14. Disconnect fuel pressure sensor vacuum hose and electrical connector.
15. Disconnect ground wire, if necessary.
16. Disconnect Exhaust Vacuum Regulator (EVR) solenoid vacuum lines.
17. Remove Exhaust Gas Recirculation (EGR) pressure transducer bracket.
18. Disconnect EGR tube and vacuum line.
19. Remove fuel supply manifold with injectors attached.
20. Remove fuel injectors from the supply manifold.

To install:

21. Install new O-rings.
22. Install fuel injectors to the supply manifold.
23. Install fuel supply manifold with injectors attached. Tighten the bolts to 89 inch lbs. (10 Nm).
24. Install EGR tube and vacuum line.
25. Install EGR pressure transducer bracket.
26. Connect EVR solenoid vacuum lines.
27. Connect ground wire, if necessary.
28. Connect fuel pressure sensor vacuum hose and electrical connector.
29. Connect fuel injector connectors.
30. Connect main chassis vacuum supply hose.
31. Connect PCV hose.
32. Connect IAC valve connector and hose.
33. Install throttle body.
34. Connect cruise control cable, if equipped.
35. Connect accelerator cable.
36. Connect fuel supply line.
37. Connect fuel return line, if equipped.
38. Install air intake tube.
39. Connect the negative battery cable.
40. Start the engine and check for leaks.

4.6L DOHC Engine

1. Before servicing the vehicle, refer to the precautions at the beginning of this section.
2. Relieve the fuel system pressure.
3. Disconnect the negative battery cable.
4. Remove upper intake manifold.
5. Disconnect fuel supply line.
6. Disconnect fuel return line, if equipped.
7. Disconnect fuel injector connectors.
8. Disconnect fuel pressure sensor vacuum hose and electrical connector.
9. Remove fuel supply manifold with injectors attached.
10. Remove fuel injectors from the supply manifold.

To install:

11. Install new O-rings.
12. Install fuel injectors to the supply manifold.
13. Install fuel supply manifold with injectors attached. Tighten the bolts to 89 inch lbs. (10 Nm).
14. Connect fuel pressure sensor vacuum hose and electrical connector.
15. Connect fuel injector connectors.
16. Connect fuel supply line.
17. Connect fuel return line, if equipped.
18. Install upper intake manifold.
19. Connect the negative battery cable.
20. Start the engine and check for leaks.

DRIVE TRAIN

Transmission

REMOVAL & INSTALLATION

Manual

➡The clutch pedal must be held in the uppermost position during clutch release cable removal and installation. Failure to properly position and support the clutch pedal can result in damage to the self-adjusting mechanism.

1. Before servicing the vehicle, refer to the precautions at the beginning of this section.
2. Lift the clutch pedal and secure in place with safety wire.
3. Disconnect the negative battery cable.
4. Remove shift lever and boot.
5. Remove driveshaft.
6. Remove clutch release lever cover, if equipped.
7. Disconnect clutch cable.
8. Disconnect Heated Oxygen (HO_2S) sensor connectors.
9. Remove dual converter "Y" or "H" pipe.
10. Disconnect backup lamp switch connector.
11. Disconnect Vehicle Speed (VSS) sensor connector.
12. Remove flywheel cover.
13. Remove starter motor.
14. Support the transmission with a jack and remove the rear transmission support crossmember.
15. Lower the transmission jack and remove the transmission flange bolts.
16. Slide the transmission input shaft out of the clutch and lower the transmission away from the vehicle.

To install:

17. Install transmission. Tighten the flange bolts to 26 ft. lbs. (35 Nm) on the T56 transmission, and 55 ft. lbs. (75 Nm) on the other transmissions.
18. Install transmission crossmember. Tighten the transmission mount bolts to 43 ft. lbs. (58 Nm) and the crossmember bolts to 30 ft. lbs. (41 Nm).
19. Install starter motor. Tighten the bolts to 17 ft. lbs. (23 Nm).
20. Install flywheel cover. Tighten the bolts to 20 ft. lbs. (27 Nm).
21. Connect VSS sensor connector.
22. Connect backup lamp switch connector.
23. Install dual converter "Y" or "H" pipe.
24. Connect HO_2S sensor connectors.
25. Connect clutch cable.
26. Install driveshaft.
27. Install shift lever and boot.
28. Connect the negative battery cable.
29. Remove the safety wire from the clutch pedal and check for proper operation.

Automatic

1. Before servicing the vehicle, refer to the precautions at the beginning of this section.
2. Drain the transmission fluid.
3. Remove the torque converter access cover and drain the torque converter.
4. Disconnect the negative battery cable.
5. Disconnect Heated Oxygen (HO_2S) sensor connectors.
6. Remove dual converter "Y" or "H" pipe.
7. Remove torque converter nuts.
8. Remove driveshaft.
9. Disconnect shift cable.

10. Disconnect transmission wiring connectors.
11. Disconnect and plug transmission fluid cooler lines.
12. Remove starter motor.
13. Support the transmission with a jack and remove the rear transmission support crossmember.
14. Lower the transmission jack and remove the transmission flange bolts.
15. Lower the transmission away from the vehicle.

To install:

16. Install transmission. Tighten the flange bolts to 35 ft. lbs. (47 Nm).
17. Transmission crossmember. Tighten the transmission mount bolts to 59 ft. lbs. (80 Nm) and the crossmember bolts to 41 ft. lbs. (55 Nm).
18. Install starter motor. Tighten the bolts to 18 ft. lbs. (25 Nm).
19. Connect transmission fluid cooler lines
20. Connect transmission wiring connectors.
21. Connect shift cable.
22. Install driveshaft.
23. Install new torque converter nuts. Tighten the nuts to 20–33 ft. lbs. (27–46 Nm).
24. Install torque converter access cover. Tighten the bolts to 25 ft. lbs. (34 Nm).
25. Install dual converter "Y" or "H" pipe.
26. Connect HO_2S sensor connectors.
27. Connect the negative battery cable.
28. Fill the transmission with fluid.
29. Start the engine. Check for leaks and proper operation.

Clutch

ADJUSTMENTS

The clutch is equipped with a self-adjusting mechanism. Pull the clutch pedal up to activate the adjuster.

REMOVAL & INSTALLATION

➡The clutch pedal must be held in the uppermost position during clutch release cable removal and installation. Failure to properly position and support the clutch pedal can result in damage to the self-adjusting mechanism.

1. Before servicing the vehicle, refer to the precautions at the beginning of this section.
2. Disconnect the negative battery cable.
3. Lift the clutch pedal and secure in place with safety wire.
4. Remove transmission.
5. Remove starter, if necessary.
6. Remove pressure plate bolts, loosen them evenly in several passes to avoid distortion of the pressure plate.
7. Remove pressure plate and clutch disk.

To install:

8. Install clutch disk and pressure plate. Tighten the pressure plate bolts evenly in several passes to 35 ft. lbs. (47 Nm) for 3.8L engines or to 33 ft. lbs. (45 Nm) plus an additional 60 degrees for 4.6L engines.
9. Install starter, if necessary.
10. Install transmission.

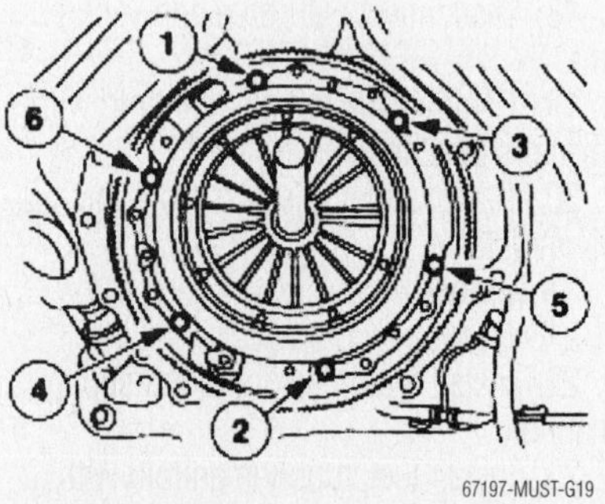

Clutch pressure plate bolts torque sequence

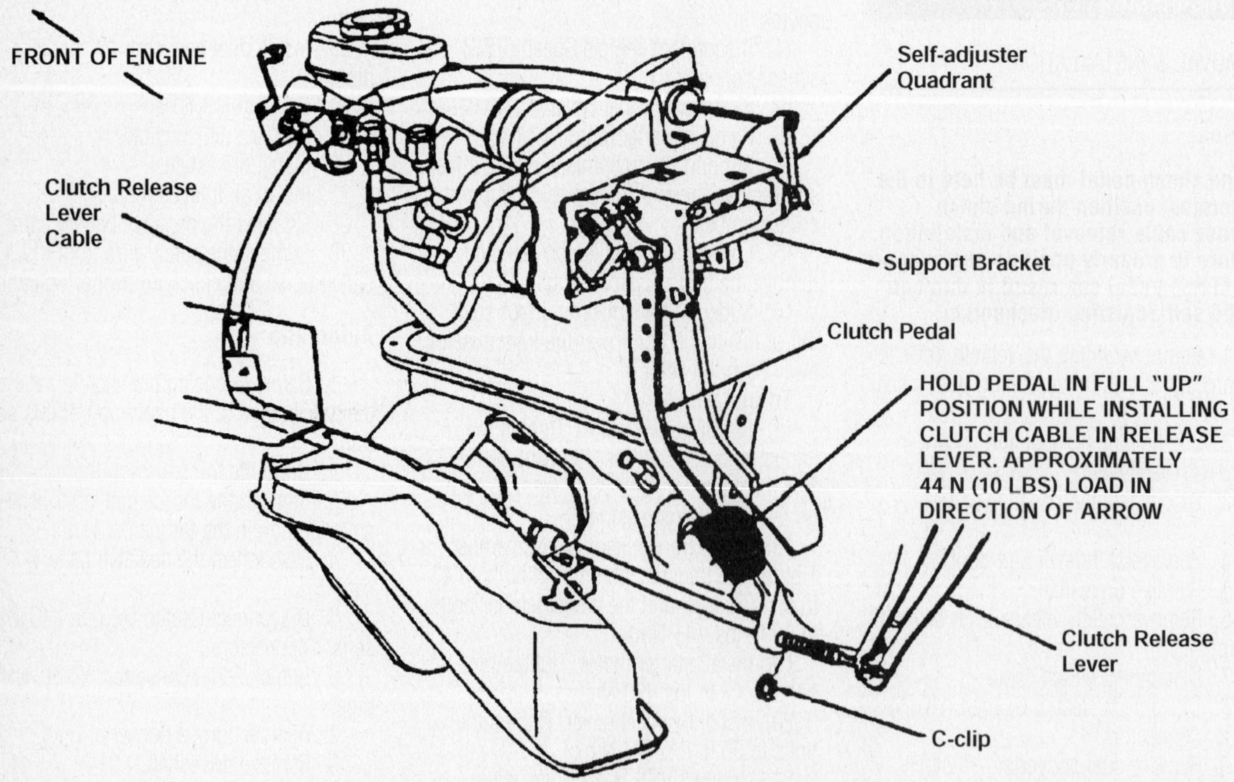

Clutch self-adjusting system component identification—4.6L engine—3.8L engine is similar

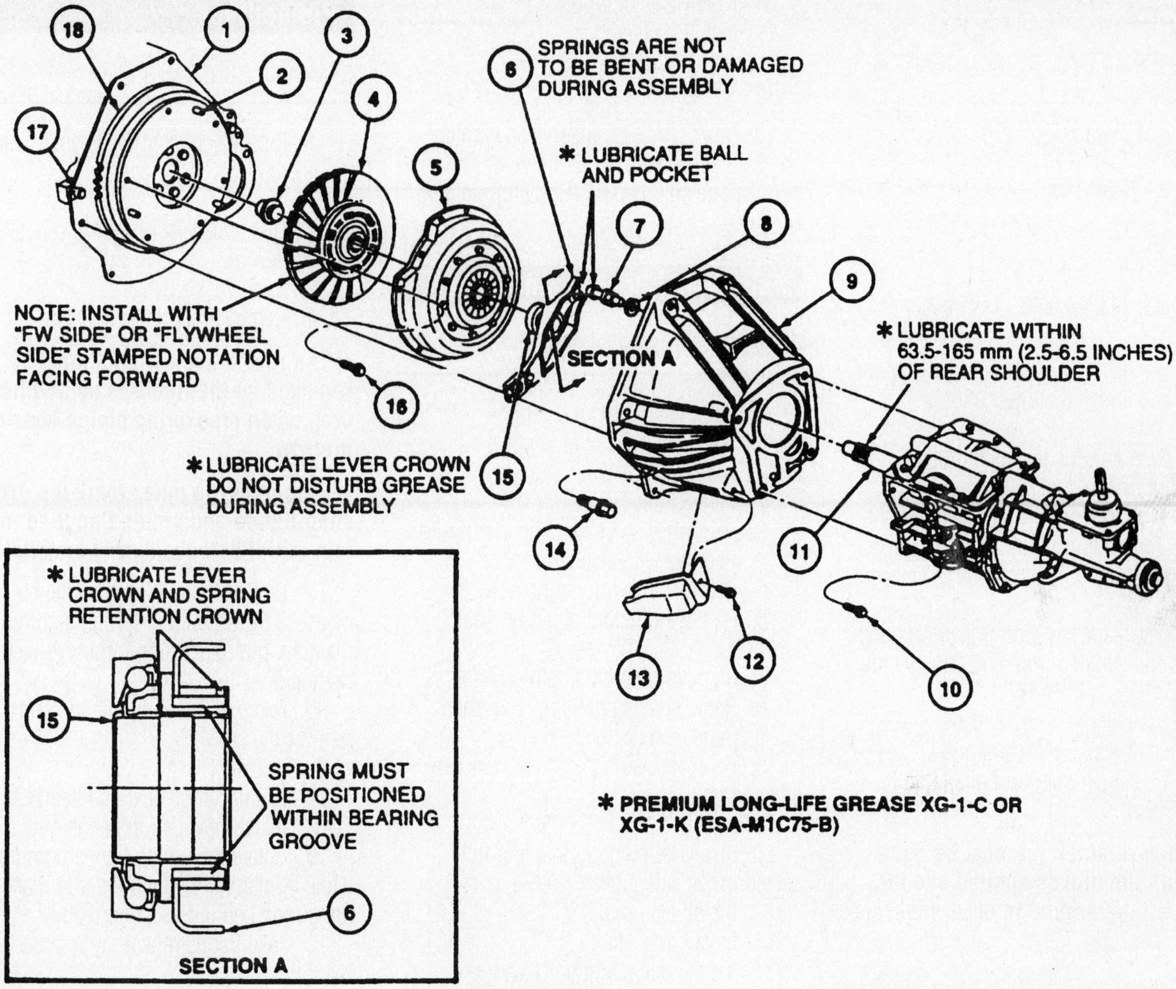
SPRINGS ARE NOT TO BE BENT OR DAMAGED DURING ASSEMBLY
* LUBRICATE BALL AND POCKET
NOTE: INSTALL WITH "FW SIDE" OR "FLYWHEEL SIDE" STAMPED NOTATION FACING FORWARD
SECTION A
* LUBRICATE WITHIN 63.5-165 mm (2.5-6.5 INCHES) OF REAR SHOULDER
* LUBRICATE LEVER CROWN DO NOT DISTURB GREASE DURING ASSEMBLY
* LUBRICATE LEVER CROWN AND SPRING RETENTION CROWN
SPRING MUST BE POSITIONED WITHIN BEARING GROOVE
SECTION A
* PREMIUM LONG-LIFE GREASE XG-1-C OR XG-1-K (ESA-M1C75-B)

1. Rear face of engine block
2. Flywheel-to-clutch cover alignment dowel
3. Pilot bearing
4. Clutch disc
5. Clutch pleasure plate
6. Clutch release lever
7. Clutch release lever stud
8. Lockwasher
9. Flywheel housing
10. Bolt
11. Input shaft
12. Bolt
13. Clutch release lever dust shield
14. Bolt
15. Clutch release bearing
16. Bolt
17. Flywheel housing-to-engine block dowel

7922NG46

Exploded view of the clutch system components

11. Connect the negative battery cable.
12. Remove the clutch pedal safety wire and check for proper operation.

Halfshaft

REMOVAL & INSTALLATION

SVT Cobra

1. Before servicing the vehicle, refer to the precautions at the beginning of this section.
2. Before raising the vehicle, matchmark the rear shock absorber to indicate curb height.
3. Remove the rear wheel.
4. Remove hub retainer nut.
5. Remove disc brake caliper and rotor.
6. Remove wheel speed sensor.
7. Disconnect the tie rod link.
8. Place a support under the lower control arm.
9. Remove lower shock absorber mounting bolt.
10. Disconnect the upper control arm from the knuckle.
11. Press the outboard CV-joint from the hub. Lower the knuckle until the stub shaft clears the hub.
12. Use Halfshaft Removal Tool 205-475 to remove the halfshaft from the differential.

To install:

➡Use new nuts, bolts, circlips, and split pins for assembly.

13. Install Differential Seal Protector 205-461.

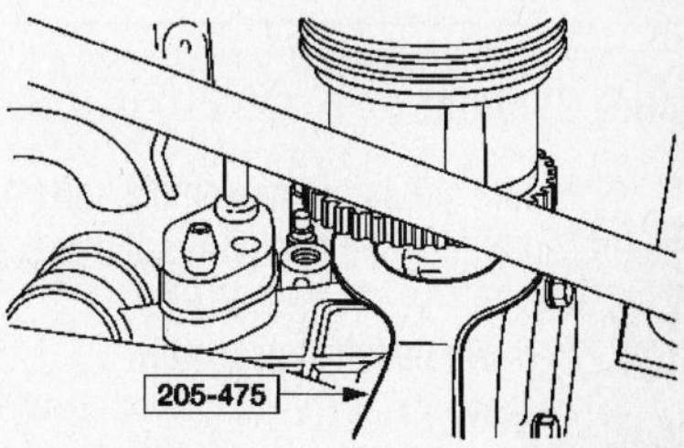
205-475

9306NG09

Halfshaft Removal Tool—SVT Cobra

14. Install the axle inner stub shaft into the differential until the splines are past the seal.
15. Remove the Differential Seal Protector.
16. Install the inner stub shaft into the differential until the circlip seats.
17. Install outer stub shaft into the knuckle.

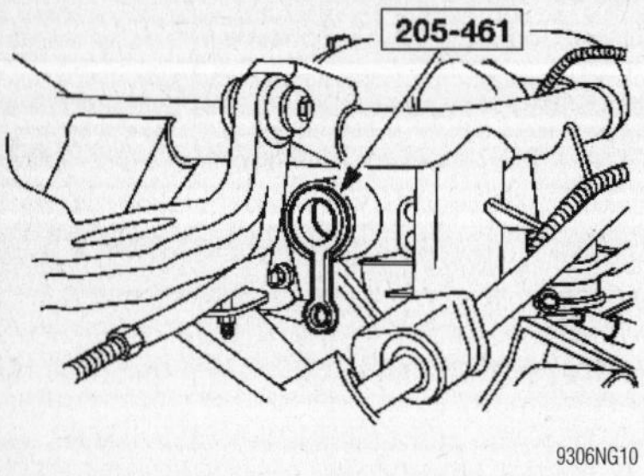

Differential Seal Protector—SVT Cobra

18. Connect the upper control arm to the knuckle and install the lower shock absorber bolt.
19. Lower control arm to align the matchmark on the shock absorber.
20. Install lower shock absorber mounting bolt. Torque it to 98 ft. lbs. (133 Nm).
21. Install upper control arm bolt. Torque it to 66 ft. lbs. (90 Nm).
22. Remove the lower control arm support.
23. Connect the tie rod link. Tighten the nut to 35 ft. lbs. (47 Nm).
24. Install wheel speed sensor.
25. Install disc brake rotor and caliper. Tighten the caliper support bolts to 76 ft. lbs. (103 Nm).

➡The hub retainer nut must be tightened with the brakes applied and the wheels off the ground to ensure correct bearing seating.

26. Install hub nut. Tighten it to 240 ft. lbs. (325 Nm).
27. Install wheel.
28. Check the rear wheel alignment and adjust as necessary.

CV-Joints

REMOVAL & REPLACEMENT

Inner CV-Joint

1. Before servicing the vehicle, refer to the precautions at the beginning of this section.
2. Remove halfshaft assembly.
3. Remove inner CV-joint boot clamps.
4. Remove CV-joint boot, slide it away from the joint.
5. Remove snapring.
6. Remove the inner CV-joint from the halfshaft.

To install:

7. CV-joint, fill it with fresh grease and slide it onto the halfshaft.
8. Install new snapring.
9. Install new CV-joint clamps.
10. Install CV-joint boot.
11. Install halfshaft.

Outer CV-Joint

The outer CV-joint is serviced with the halfshaft as an assembly. The outer CV-joint boot can be serviced by removing the inner CV-joint.

Axle Shaft, Bearing and Seal

REMOVAL & INSTALLATION

1. Before servicing the vehicle, refer to the precautions at the beginning of this section.
2. Remove rear wheel.
3. Remove disc brake caliper and rotor.
4. Remove wheel speed sensor.
5. Remove axle housing cover.
6. Remove differential pinion shaft lockbolt.
7. Remove differential pinion shaft.
8. Remove axle retaining U-washer.
9. Remove axle shaft.
10. Remove bearing and seal, using a slide hammer

To install:

11. Install bearing, so that it is fully seated in the axle tube.
12. Install axle seal.
13. Install axle shaft.
14. Install axle retaining U-washer.
15. Install differential pinion shaft. Tighten the lockbolt to 15–30 ft. lbs. (20–41 Nm).
16. Install axle housing cover. Tighten the cover bolts to 18–28 ft. lbs. (24–38 Nm).
17. Install wheel speed sensor.
18. Install disc brake rotor and caliper. Tighten the caliper mounting bolts to 65–87 ft. lbs. (87–119 Nm).
19. Install rear wheel.
20. Fill the differential with gear lubricant. Tighten the filler plug to 15–30 ft. lbs. (20–41 Nm).

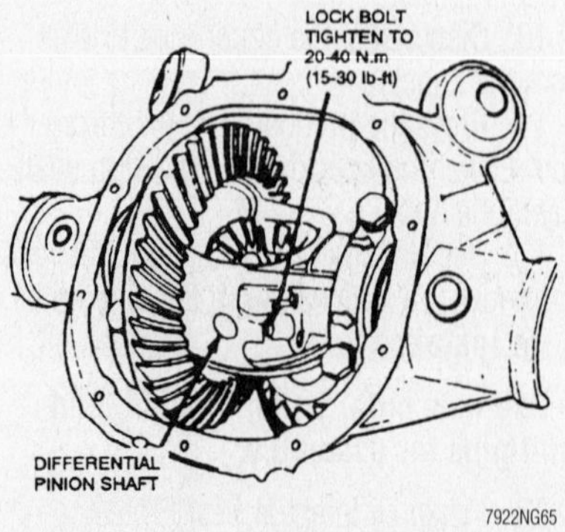

Differential pinion shaft and lockbolt

Pinion Seal

REMOVAL & INSTALLATION

1. Before servicing the vehicle, refer to the precautions at the beginning of this section.
2. Disconnect driveshaft and position it out of the way.
3. Remove rear wheels.
4. Remove rear brake calipers.

➡The rear brake calipers must be removed so that there is no additional drag when measuring pinion bearing preload.

➡Remember to index mark the driveshaft flange and pinion flange to maintain initial balance during installation.

5. Use an inch lb. torque wrench and measure the amount of torque required to maintain pinion rotation through several revolutions.
6. Remove the pinion flange and remove the seal.

To install:

7. Install pinion seal and flange.
8. Install new pinion flange nut.
9. Rotate the pinion flange occasionally while tightening the flange nut to make sure the pinion bearings seat correctly.
10. Take frequent bearing preload torque readings.
11. If the preload recorded prior to disassembly is **lower** than the specification for used bearings, then tighten the pinion flange nut to specification. If the preload recorder prior to disassembly is **higher** than the specification for used bearings, then tighten the pinion flange nut to the original reading as recorded.
12. The pinion bearing preload specifications are as follows:
    a. Used bearings: 8–14 inch lbs. (0.9–1.5 Nm).
    b. New bearings: 16–29 inch lbs. (1.8–3.3 Nm).

✲✲ CAUTION

Never loosen the pinion nut to reduce bearing preload. If it is necessary to reduce bearing preload, install a new collapsible spacer and pinion nut.

13. Connect the driveshaft and install new bolts. Tighten the bolts to 83 ft. lbs. (112 Nm).
14. Install brake calipers.
15. Install rear wheels.
16. Fill the differential with gear lubricant and check for leaks.

STEERING AND SUSPENSION

Air Bag

⁂ CAUTION

Some vehicles are equipped with an air bag system. The system must be disarmed before performing service on, or around, system components, the steering column, instrument panel components, wiring and sensors. Failure to follow the safety precautions and the disarming procedure could result in accidental air bag deployment, possible injury and unnecessary system repairs.

PRECAUTIONS

Several precautions must be observed when handling the inflator module to avoid accidental deployment and possible personal injury.

- Never carry the inflator module by the wires or connector on the underside of the module.
- When carrying a live inflator module, hold securely with both hands, and ensure that the bag and trim cover are pointed away.
- Place the inflator module on a bench or other surface with the bag and trim cover facing up.
- With the inflator module on the bench, never place anything on or close to the module that may be thrown in the event of an accidental deployment.

DISARMING

1. Before servicing the vehicle, refer to the precautions at the beginning of this section.
2. Disconnect both battery cables from the battery, negative cable first.
3. Wait 1 minute before proceeding with the service procedure. This is the time required for the back-up power supply in the air bag diagnostic monitor to deplete its stored energy.
4. After service is completed, reconnect the battery cables, negative cable last.
5. Turn the ignition switch to the **RUN** position. The air bag indicator should light continuously for approximately 6 seconds, then turn **OFF**. If the indicator fails to light, flashes or remains lit continuously, there is a fault in the air bag system.

Power Rack and Pinion Steering Gear

REMOVAL & INSTALLATION

1. Before servicing the vehicle, refer to the precautions at the beginning of this section.
2. Disconnect the negative battery cable.
3. Remove front wheels.
4. Disconnect steering column intermediate shaft coupling.
5. Disconnect outer tie rod ends.
6. Remove steering gear retaining bolts.
7. Disconnect power steering pressure and return lines.
8. Remove steering gear.

To install:

9. Attach the pressure and return lines to the steering gear before attaching the steering gear to the subframe. Use new plastic seals and tighten the line fittings to 20–25 ft. lbs. (27–34 Nm).

➡The power steering fluid lines are designed to swivel when properly tightened. Do not overtighten the fittings.

10. Install steering gear. Tighten the bolts to 31–39 ft. lbs. (41–54 Nm).
11. Connect outer tie rod ends. Tighten the nuts to 36–46 ft. lbs. (48–63 Nm).
12. Connect steering column intermediate shaft coupling. Tighten the pinch bolt to 21–29 ft. lbs. (28–40 Nm).
13. Install front wheels.
14. Connect the negative battery cable.
15. Fill the power steering system with the proper type and quantity of fluid.
16. Check the front end alignment and adjust as necessary.

Strut

REMOVAL & INSTALLATION

Front

1. Before servicing the vehicle, refer to the precautions at the beginning of this section.
2. Support the front of the vehicle on jackstands placed under the control arms.
3. Remove front wheel.
4. Remove disc brake caliper.
5. Remove wheel speed sensor and bracket.
6. Remove upper strut retaining fasteners.
7. Remove wheel spindle attachment bolts.
8. Compress the strut assembly and remove it from the vehicle.

To install:

9. If replacing the strut, transfer the upper mounting bracket.
10. Install strut assembly. Tighten the

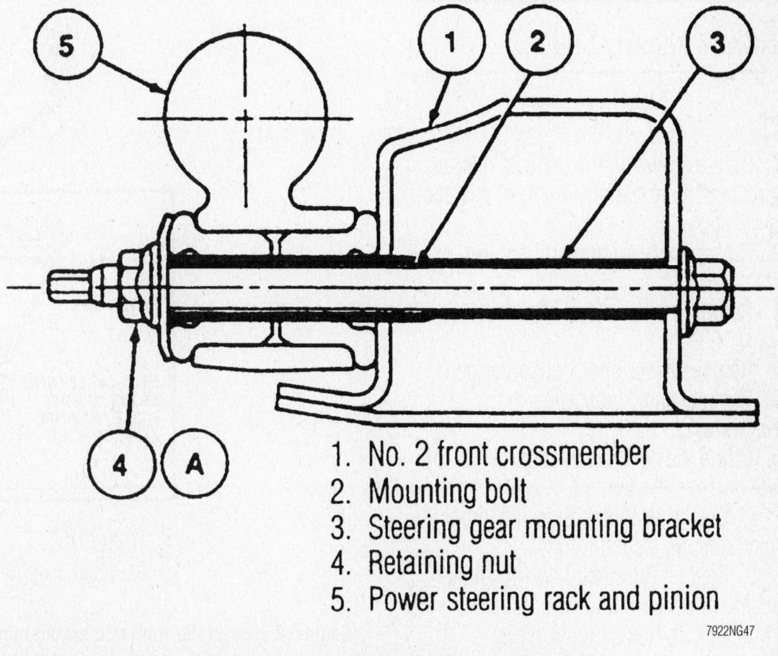

Power rack and pinion steering gear mounting on the No. 2 crossmember

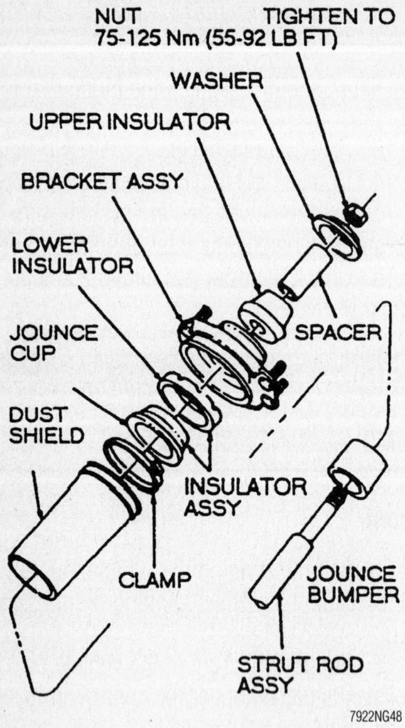

Exploded view of the front strut upper mounting

spindle bolts to 141–191 ft. lbs. (190–259 Nm).

11. Install upper strut retaining fasteners. Tighten to 25–34 ft. lbs. (34–46 Nm).
12. Install wheel speed sensor and bracket.
13. Install disc brake caliper.
14. Install front wheel.
15. Remove the jackstands and check the alignment. Adjust as necessary.

Shock Absorber

REMOVAL & INSTALLATION

Rear

1. Before servicing the vehicle, refer to the precautions at the beginning of this section.
2. Pull aside rear compartment trim panels.
3. Remove upper shock absorber retaining nut.
4. Remove lower shock absorber bolt.
5. Remove shock absorber.

To install:

6. Prime the new shock absorber as follows:
   a. Step 1: With the shock absorber right side up, extend it fully.
   b. Step 2: Turn the shock absorber upside down and fully compress it.
   c. Step 3: Repeat for 3 cycles.

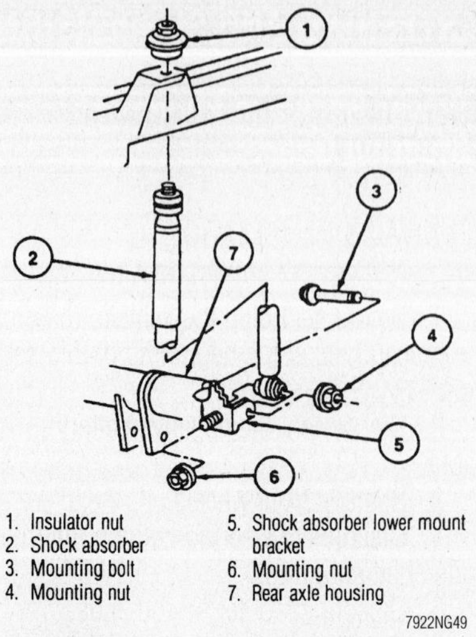

1. Insulator nut
2. Shock absorber
3. Mounting bolt
4. Mounting nut
5. Shock absorber lower mount bracket
6. Mounting nut
7. Rear axle housing

Exploded view of the rear shock absorber mounting

7. Install the shock absorber and tighten the lower bolt as follows:
   a. SVT Cobra: 98 ft. lbs. (133 Nm).
   b. All others: 57–75 ft. lbs. (76–103 Nm).
8. Install upper shock absorber nut. Tighten it to 25–33 ft. lbs. (34–46 Nm).
9. Place rear compartment trim panels back into position.

Coil Spring

REMOVAL & INSTALLATION

Front

1. Before servicing the vehicle, refer to the precautions at the beginning of this section.
2. Install an internal spring compressor to the spring to be serviced. Tighten the compressor to relieve spring pressure on the lower control arm.
3. Remove front wheel.
4. Disconnect outer tie rod end.
5. Disconnect ball joint stud from front wheel spindle.
6. Disconnect stabilizer bar link.
7. Loosen lower control arm mounting bolts.
8. Remove coil spring.

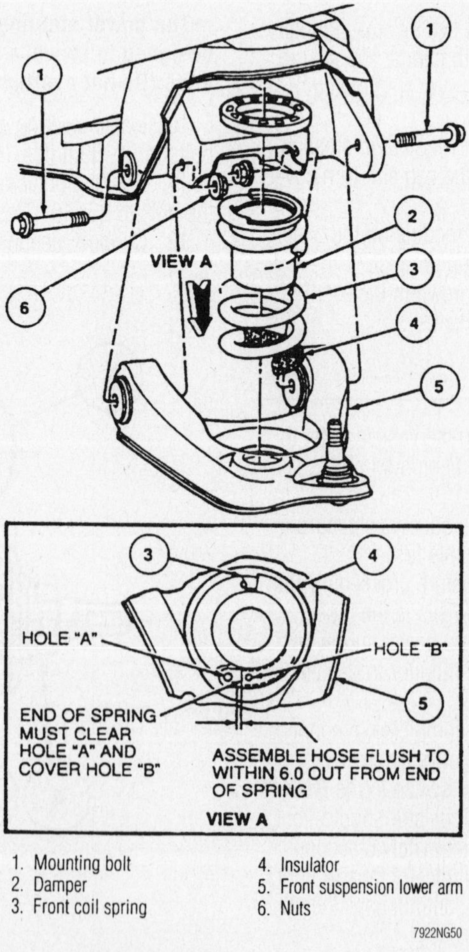

1. Mounting bolt
2. Damper
3. Front coil spring
4. Insulator
5. Front suspension lower arm
6. Nuts

Exploded view of the front coil spring mounting

To install:

9. If replacing the coil spring, transfer the spring compressor to the new spring.
10. Install coil spring.
11. Connect ball joint stud to front wheel spindle. Use a jack to raise the control arm to a normal position and tighten the bolts to 109–149 ft. lbs. (148–202 Nm).
12. Connect stabilizer bar link. Tighten the nut to 11–16 ft. lbs. (16–22 Nm).
13. Tighten lower control arm mounting bolts to 148 ft. lbs. (200 Nm).
14. Connect outer tie rod end. Tighten the nut to 36–46 ft. lbs. (48–63 Nm).
15. Install front wheel.
16. Position the coil spring and remove the spring compressor.

Rear

EXCEPT SVT COBRA

1. Before servicing the vehicle, refer to the precautions at the beginning of this section.
2. Raise and support the vehicle safely under the frame. Support the body at the rear body crossmember.
3. Disconnect wheel speed sensor wire from parking brake cable retaining bracket, on side of lower control arm.
4. Remove and discard parking brake cable bracket bolt. Separate bracket from lower control arm.
5. If equipped, remove the stabilizer bar.
6. Support the axle with a jack.
7. Place another jack under the lower arm axle pivot bolt. Remove and discard the bolt and nut. Lower the jack slowly until the coil spring load is relieved.
8. Remove the coil spring and insulator from the vehicle.

To install:

9. Place the upper spring insulator on top of the spring. Place the lower spring insulator on the lower arm.
10. Position the coil spring on the lower arm spring seat with the pigtail on the lower arm at the rear of the vehicle and pointing toward the left side of the vehicle.
11. Slowly raise the jack until the arm is in position. Insert a new rear pivot bolt and nut.
12. Raise the axle to curb height. Tighten the pivot bolt to 71–97 ft. lbs. (97–132 Nm).
13. Position parking brake cable bracket onto lower control arm. Install a new bracket bolt.
14. Connect wheel speed sensor wire to retaining bracket on side of lower control arm.
15. If equipped, install the stabilizer bar.
16. Remove the crossmember supports and lower the vehicle.

SVT COBRA

1. Before servicing the vehicle, refer to the precautions at the beginning of this section.
2. Support the rear subframe with a jack.
3. Remove rear wheels.
4. Remove both mufflers.
5. Remove driveshaft.
6. Disconnect parking brake cables and remove brackets.
7. Disconnect brake fluid lines.
8. Remove wheel speed sensors.
9. Disconnect tie rod links.
10. Disconnect shock absorbers at lower control arms.
11. Disconnect lower control arms.
12. Loosen the front subframe bolts and remove the rear subframe bolts.
13. Lower the rear subframe and remove the coil springs.

To install:

➡Use new nuts, bolts and split pins for assembly.

14. Install the coil springs and raise the rear subframe. Install new subframe bolts and tighten them to 76 ft. lbs. (103 Nm).
15. Connect lower control arms.
16. Connect shock absorbers. Tighten the bolts to 98 ft. lbs. (133 Nm).
17. Connect tie rod links.
18. Install wheel speed sensors.
19. Connect brake fluid lines.
20. Connect parking brake cables and install brackets.
21. Install driveshaft.
22. Install both mufflers.
23. Install rear wheels.
24. Check the wheel alignment and adjust as necessary.

Lower Ball Joint

REMOVAL & INSTALLATION

1. Before servicing the vehicle, refer to the precautions at the beginning of this section.
2. Remove front wheel.
3. Remove disc brake caliper and rotor.
4. Remove wheel speed sensor.
5. Disconnect outer tie rod end.
6. Support the lower control arm with a jackstand and remove the spindle.
7. Press the lower ball joint out of the control arm.

To install:

8. Press the lower ball joint into the control arm so that the joint is fully seated in the control arm.
9. Install spindle. Tighten the ball joint nut to 109–149 ft. lbs. (148–202 Nm). Tighten the strut bolts to 141–191 ft. lbs. (190–259 Nm).
10. Connect outer tie rod end. Tighten the nut to 36–46 ft. lbs. (48–63 Nm).
11. Install disc brake rotor and caliper. Tighten the caliper mounting bolts to 96 ft. lbs. (130 Nm).
12. Install wheel speed sensor.
13. Install front wheel.

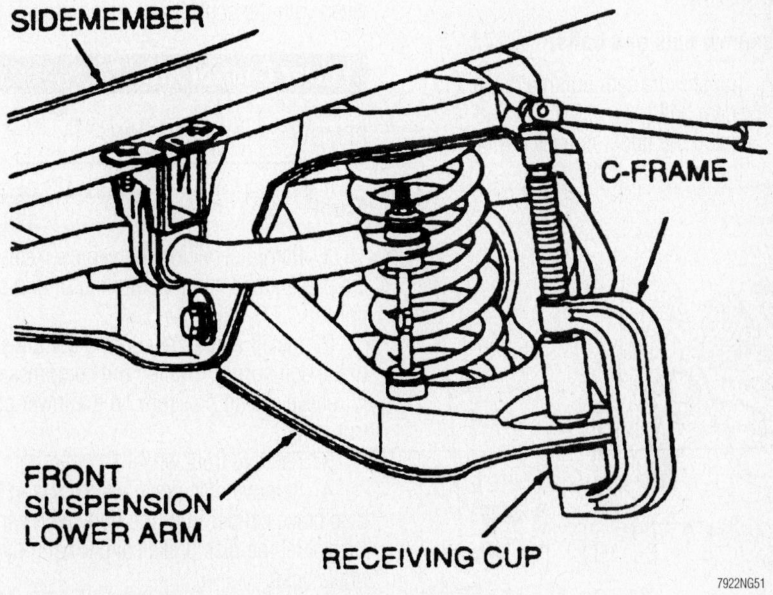

Use the C-clamp, cup and adapters to press the ball joint out of the lower control arm

Upper Control Arm

REMOVAL & INSTALLATION

Rear

EXCEPT SVT COBRA

1. Before servicing the vehicle, refer to the precautions at the beginning of this section.
2. Before raising the vehicle, matchmark the shock absorbers to indicate curb height.
3. Support the axle housing with a jack.
4. Remove the upper control arms.

To install:

➡Use new nuts and bolts.

5. Install the upper control arms.
6. Align the matchmarks on the shock absorbers and tighten the control arm bolts as follows:
   a. Step 1: Tighten the mounting bracket bolt to 76 ft. lbs. (103 Nm).
   b. Step 2: Tighten the axle housing bolt to 66 ft. lbs. (90 Nm).

SVT COBRA

1. Before servicing the vehicle, refer to the precautions at the beginning of this section.
2. Before raising the vehicle, matchmark the shock absorbers to indicate curb height.
3. Remove the coil springs.
4. Remove the subframe front bolts and lower the subframe from the vehicle.
5. Matchmark the upper control arm cam bolt to the knuckle.
6. Remove the upper control arm.

To install:

➡Use new nuts and bolts.

7. Transfer the cam bolt matchmark to the new cam bolt.
8. Install the upper control arm. Do not tighten the fasteners at this time.

Cam bolt matchmark—SVT Cobra

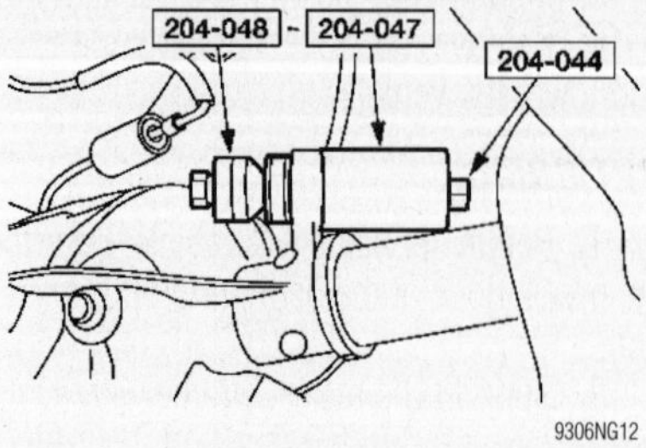

Axle bushing removal

9. Install the coil springs.
10. Raise the suspension to align the shock absorber matchmarks and tighten the upper control arm inner bolts to 66 ft. lbs. (90 Nm).
11. Align the cam bolt matchmark and tighten the nut to 66 ft. lbs. (90 Nm).
12. Check the wheel alignment and adjust as necessary.

CONTROL ARM BUSHING REPLACEMENT

Rear

EXCEPT SVT COBRA

The inboard control arm bushing is serviced with the control arm as an assembly.

1. Before servicing the vehicle, refer to the precautions at the beginning of this section.
2. Remove the upper control arm.
3. Press the bushing out of the axle housing.
4. Press a new bushing into the axle housing and install the upper control arm.

SVT COBRA

The upper control arm bushing are serviced with the control arm as an assembly.

Lower Control Arm

REMOVAL & INSTALLATION

Front

1. Before servicing the vehicle, refer to the precautions at the beginning of this section.
2. Install an internal spring compressor to the coil spring. Tighten the compressor to relieve spring pressure on the lower control arm.
3. Remove front wheel.
4. Remove disc brake caliper. Hang the disc brake caliper from the body with wire. Do not let the disc brake caliper hang by the brake hose.
5. Disconnect outer tie rod end.
6. Disconnect stabilizer bar link.
7. Remove power steering gear.
8. Remove lower control arm mounting bolts.
9. Remove coil spring.
10. Disconnect lower ball joint.
11. Remove lower control arm.

To install:

12. Connect the lower ball joint to the spindle and install the nut hand tight.
13. Install coil spring.
14. Install lower control arm mounting bolts. Use a jack to raise the control arm to curb height and tighten the bolts to 141–191 ft. lbs. (191–259 Nm). Tighten the ball joint nut to 109–149 ft. lbs. (148–202 Nm).
15. Install power steering gear.
16. Connect stabilizer bar link. Tighten the nut to 11–16 ft. lbs. (16–22 Nm).
17. Connect outer tie rod end. Tighten the nut to 36–46 ft. lbs. (48–63 Nm).
18. Install front disc brake caliper.
19. Install front wheel.
20. Position the coil spring and remove the spring compressor.
21. Check the front end alignment.

Rear

EXCEPT SVT COBRA

1. Before servicing the vehicle, refer to the precautions at the beginning of this section.
2. Before raising the vehicle, matchmark the shock absorbers to indicate curb height.
3. Raise and support the vehicle safely under the frame. Support the body at the rear body crossmember.
4. If necessary, remove the muffler assembly.
5. If equipped, remove the stabilizer bar.
6. Support the axle with a jack.
7. Place another jack under the lower arm axle pivot bolt. Remove and discard the bolt and nut. Lower the jack slowly until the coil spring load is relieved.
8. Remove coil spring.
9. Remove lower control arm.

To install:

➡Use new nuts and bolts.

10. Install lower control arm to the axle housing.
11. Install coil spring.
12. Raise the control arm and install the pivot bolt.
13. Raise the axle to align the matchmarks on the shock absorbers. Tighten the lower control arm bolts to 71–97 ft. lbs.

(97–132 Nm) for 2001–03 models, and 111 ft. lbs. (150 Nm) for 2004 models.

14. install the stabilizer bar, if equipped.

15. Install the muffler assembly, if necessary.

SVT COBRA

1. Before servicing the vehicle, refer to the precautions at the beginning of this section.
2. Before raising the vehicle, matchmark the shock absorbers to indicate curb height.
3. Remove coil springs. Refer to the coil spring procedure in this section.
4. Remove subframe front bolts and lower it from the vehicle.
5. Remove lower control arm.

To install:

➡Use new nuts and bolts.

6. Install lower control arm. Do not tighten the fasteners at this time.
7. Install rear subframe. Tighten the rear subframe bolts to 76 ft. lbs. (103 Nm).
8. Install coil springs.
9. Raise the suspension to align the shock absorber matchmarks.
10. Tighten the lower control arm inboard bolts to 184 ft. lbs. (250 Nm) and the knuckle bolt to 85 ft. lbs. (115 Nm).
11. Check the wheel alignment and adjust as necessary.

CONTROL ARM BUSHING REPLACEMENT

Front and Rear

ALL MODELS

The lower control arm bushings are serviced with the lower control arm as an assembly.

Wheel Bearings

ADJUSTMENT

The front wheel bearings are an integral part of the hub assembly. They require no periodic maintenance or adjustment. If the bearings are found to be defective, they must be replaced along with the hub assembly.

The rear wheel bearings are not adjustable.

REMOVAL & INSTALLATION

Front

1. Before servicing the vehicle, refer to the precautions at the beginning of this section.
2. Remove front wheel.
3. Remove disc brake caliper and rotor.
4. Remove grease cap.
5. Remove spindle retainer nut.
6. Remove hub and bearing assembly.

To install:

7. Install hub and bearing assembly. Tighten the spindle retainer nut to 221–295 ft. lbs. (300–400).
8. Install grease cap.
9. Install disc brake caliper and rotor.
10. Install front wheel

Rear

SVT COBRA

1. Before servicing the vehicle, refer to the precautions at the beginning of this section.
2. Before raising the vehicle, matchmark the shock absorbers to indicate curb height.
3. Remove rear wheel.
4. Disconnect parking brake cable.
5. Remove brake caliper and disc.
6. Remove hub retainer.
7. Disconnect shock absorber from the lower suspension arm.
8. Disconnect tie rod link from knuckle.
9. Matchmark the upper control arm cam bolt to the knuckle.
10. Remove knuckle from the vehicle.
11. Remove dust shield and press the hub out of the bearing.
12. Remove snapring and press the bearing out of the knuckle.

To install:

➡Use new nuts, bolts, snaprings, and split pins for assembly.

13. Install bearing so that it is fully seated in the knuckle bore.
14. Install snapring.
15. Support the bearing inner race and press the hub into the bearing.
16. Install the dust shield and tighten the bolts to 89 inch lbs. (10 Nm).
17. Transfer the cam bolt matchmark to a new cam bolt.
18. Install knuckle. Do not tighten the fasteners at this time.
19. Connect tie rod link. Tighten the nut to 35 ft. lbs. (47 Nm).
20. Connect shock absorber. Tighten the bolt to 98 ft. lbs. (133 Nm).
21. Install hub retainer. Do not tighten the nut at this time.
22. Raise the suspension to align the shock absorber matchmarks and tighten the lower control arm bolt to 85 ft. lbs. (115 Nm).
23. Align the cam bolt matchmarks and tighten the nut to 66 ft. lbs. (90 Nm).
24. Install the brake disc and caliper. Install the parking brake cable.

➡The hub retainer nut must be tightened with the brakes applied and the wheels off the ground to ensure correct bearing seating.

25. Tighten the hub retainer to 240 ft. lbs. (325 Nm) and install the wheel.
26. Check the wheel alignment and adjust as necessary.

BRAKES

Brake Caliper

REMOVAL & INSTALLATION

Front

1. Before servicing the vehicle, refer to the precautions at the beginning of this section.
2. Remove the wheel and tire assembly.
3. Mark the disc brake caliper to ensure that it will be installed to the same side if both calipers are being removed.
4. Remove the banjo bolt that connects the brake hose fitting to the disc brake caliper.
5. Disconnect the brake hose and plug. Discard 2 copper sealing washers.
6. Remove lower brake caliper locating pin, if equipped.
7. Remove brake caliper mounting bolts, if equipped.
8. Rotate the brake caliper approximately 90° away from the anchor assembly.
9. If equipped with locating pins, slide the brake caliper away from anchor assembly until the brake caliper disengages from the upper locating pin.
10. Remove the disc brake caliper.

To install:

11. If required, compress the caliper piston(s) using a C-clamp and an old disc brake pad or block of wood.
12. If brake caliper is equipped with locating pins, engage the upper caliper locating pin into the insulator.
13. Make sure the anti-rattle is properly installed on caliper.
14. Make sure the brake pads are properly installed.
15. Position the brake caliper assembly into place on the anchor plate.

➡If equipped, new locating pins are recommended due to the thread locking compound on the threads.

16. Install lower disc brake caliper locating pin, or upper and lower caliper bolts, and torque to 23 ft. lbs. (31 Nm) for dual piston calipers. Install lower disc brake caliper locating pin and torque the bolts to 65 ft. lbs. (88 Nm) for single piston calipers.
17. If equipped, install disc brake caliper locating pin and clip by pressing the caliper down to compress the bias springs and the locating pin into position and secure with the washer and clip.
18. Connect the brake hose to the disc brake caliper using 2 new copper sealing washers, 1 on each side of the fitting. Install the banjo bolt and torque to 30 ft. lbs. (40 Nm).
19. Bleed the brake system and check for leaks. Replace the bleeder screw rubber cap.
20. Install wheel and tire assembly. Torque the lug nuts in a star pattern to 85–105 ft. lbs. (115–142 Nm).
21. Pump the brake pedal several times to position the brake pads before the vehicle is moved.
22. Road test the vehicle and check for proper brake system operation.

Rear

1. Before servicing the vehicle, refer to the precautions at the beginning of this section.
2. Remove wheel and tire assembly.
3. Remove banjo bolt securing the brake hose to the disc brake caliper.
4. Remove brake hose from the caliper and plug the hose. Discard 2 copper sealing washers.
5. Remove retaining clip from the parking brake rear cable at the disc brake caliper. Release the tension from the parking brake cable and disengage the cable end from the parking brake lever on the brake caliper.
6. Hold the disc brake caliper locating pin hex-heads with an open-end wrench and remove 2 brake pin retainer bolts.
7. Rotate the caliper assembly away from the disc support bracket and remove the from the vehicle.
8. Remove caliper locating pins and boots from support bracket.

To install:

9. If required, use Caliper Piston Adjuster T87P-2588-A to rotate the caliper piston into the caliper bore. Align the tool tips with the slots in the piston and rotate the tool clockwise until the piston is seated.

➡Apply a suitable silicone dielectric compound to the inside of the locating pin boots and the caliper locating pins.

10. Install brake caliper locating pins and boots in the disc support bracket.
11. Install disc brake pads and anti-rattle clips.
12. Apply a drop of suitable thread locking compound to the threads of the brake pin retainer bolts and install the brake pin retainers. Torque the brake pin retainer bolts to 23–26 ft. lbs. (31–35 Nm) while holding the brake caliper locating pins with an open-end wrench.
13. Connect the end of the parking brake cable to the parking brake lever on the disc brake caliper.
14. Connect the brake hose to the disc brake caliper using 2 new copper sealing washers on the fitting.
15. Install the banjo bolt through both washers and the brake line fitting and torque to 30 ft. lbs. (40 Nm).
16. Properly bleed the brake system and check for leaks.
17. Install wheel and tire assembly. Torque the lug nuts in a star pattern to 85–105 ft. lbs. (115–142 Nm).
18. Pump the brake pedal several times to position the brake pads before moving the vehicle.
19. Road test the vehicle and check the brake system for proper operation.

Disc Brake Pads

REMOVAL & INSTALLATION

Front

WITH SINGLE PISTON CALIPER

1. Before servicing the vehicle, refer to the precautions at the beginning of this section.
2. Remove ½ of the brake fluid from the brake master cylinder reservoir. Properly dispose of the used brake fluid.
3. Remove wheel and tire assembly.
4. Remove the front brake caliper assembly. Hang the disc brake caliper from the body with wire. Do not let the disc brake caliper hang by the brake hose.
5. Remove the outer and inner disc brake pads and the anti-rattle clip from the disc brake caliper
6. Clean any residue from the caliper anchor plate and disc brake pad contact areas.
7. Inspect the disc brake rotor for scoring and wear. Replace or machine, as necessary.
8. Inspect the piston boot and the caliper pin boots for damage. Replace as necessary.

To install:

9. Use a large C-clamp and a wood block to push the caliper piston back into its bore.
10. Remove the protective paper from the insulators on the back of the disc brake pads, if equipped.
11. Install inner and outer disc brake

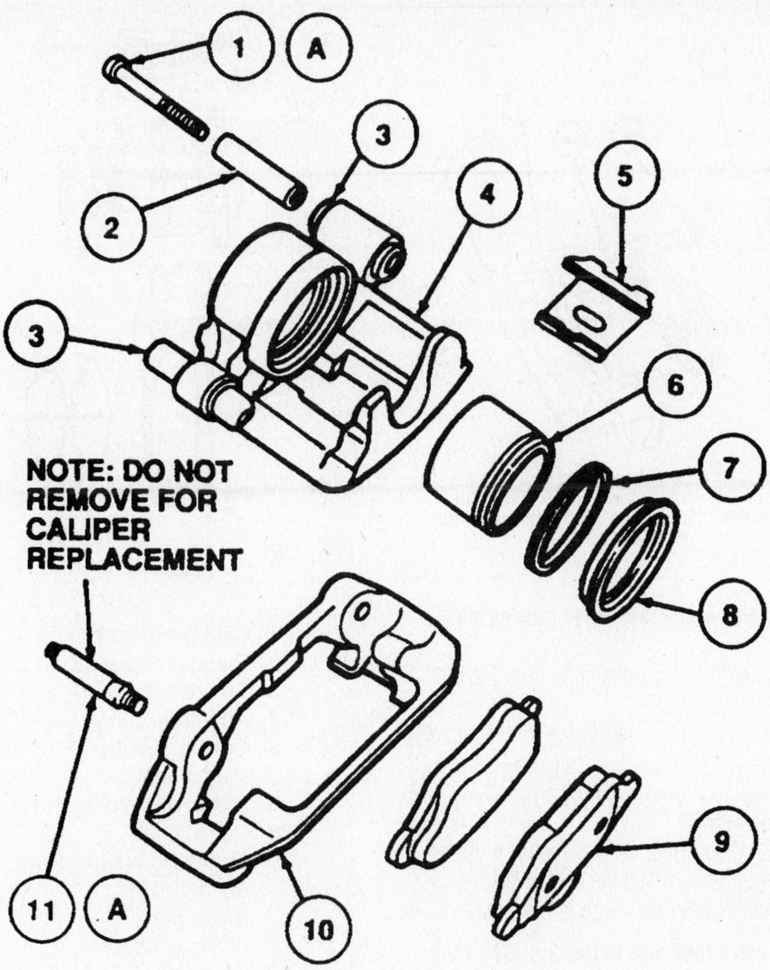

1	Upper Caliper Brake Bolt
2	Front Caliper Sleeve
3	Insulator
4	Caliper Housing (Part of 2B119)
5	Disc Brake Pad Anti-Rattle Clip
6	Caliper Piston
7	Piston Seal
8	Dust Boot
9	Brake Shoe and Lining
10	Front Disc Brake Caliper Anchor Plate
11	Disc Brake Caliper Locating Pin
A	Tighten to 88 N·m (65 Lb-Ft)

93006G30

Disc brake pads and related components—Mustang with single piston caliper

pads and the anti-rattle clip in the disc brake caliper.

12. Install the disc brake caliper and pads.

13. Install wheel and tire assembly and torque the lug nuts in a star pattern to 85–105 ft. lbs. (115–142 Nm).

14. Repeat the procedure for the opposite disc brake caliper assembly.

15. Pump the brake pedal prior to moving the vehicle to seat the brake pads.

16. Fill the brake master cylinder reservoir with clean DOT 3 brake fluid from a closed container.

17. If the disc brake calipers were replaced or repaired, be sure to bleed the system.

18. Road test the vehicle and check the brake system for proper operation.

WITH DUAL PISTON CALIPER

1. Before servicing the vehicle, refer to the precautions at the beginning of this section.

2. Remove and discard ½ of the brake fluid from the brake master cylinder reservoir. Properly dispose of the used brake fluid.

3. Remove the wheel and tire assembly.

4. Remove disc brake caliper assembly from anchor plate. Hang the disc brake caliper from the body with wire. Do not let the disc brake caliper hang by the brake hose.

5. Remove the outer and inner disc brake pads from the disc brake caliper.

6. Clean any residue from the caliper anchor plate and disc brake pad contact areas.

7. Inspect the disc brake rotor for scoring and wear. Replace or machine, as necessary.

8. Inspect the piston boots and the caliper pin boots for damage. Replace as necessary.

To install:

9. Use a large C-clamp and a wood block to push the caliper pistons back into their bores.

10. Remove the protective paper from the insulators on the back of the disc brake pads, if equipped.

11. Install the inner and outer disc brake pads in the disc brake caliper. Make sure the pads are seated properly.

12. Install the disc brake caliper and pads.

13. Install the wheel and tire assembly. Torque the lug nuts in a star pattern to 85–105 ft. lbs. (115–142 Nm).

14. Repeat the procedure for the opposite disc brake caliper assembly.

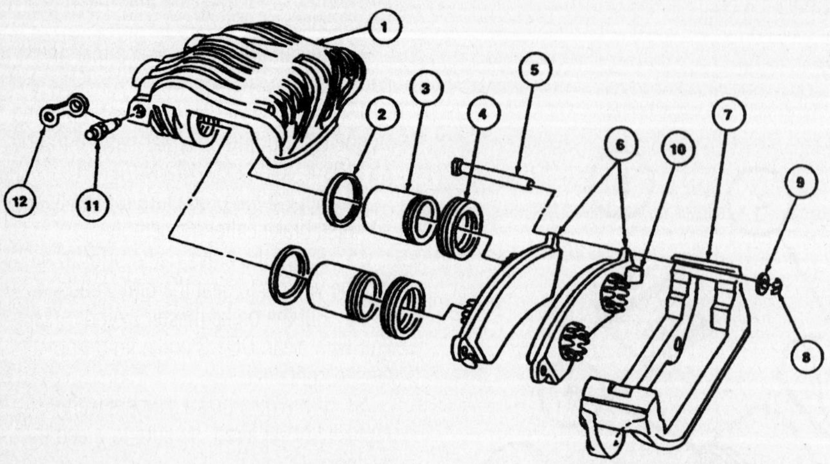

1 Caliper housing
2 Piston seal
3 Caliper piston
4 Dust boot
5 Front brake locating pin
6 Brake shoe and lining
7 Front disc brake caliper anchor plate
8 Clip
9 Washer
10 Anti-rattle clip and insulator
11 Wheel cylinder bleeder screw
12 Cover

67197-MUST-G20

Disc brake pads and related components—Mustang with dual piston caliper and locating pins

15. Pump the brake pedal prior to moving the vehicle to position the brake pads.
16. Fill the brake master cylinder reservoir with clean DOT 3 brake fluid from a closed container.
17. If the disc brake calipers were replaced or repaired, be sure to bleed the system.
18. Road test the vehicle and check the brake system for proper operation.

Rear

1. Before servicing the vehicle, refer to the precautions at the beginning of this section.
2. Remove ½ of the brake fluid from the brake master cylinder reservoir. Properly dispose of the used brake fluid.
3. Remove wheel and tire assembly.
4. Remove screw retaining the brake hose bracket to the shock absorber bracket, if necessary.
5. Remove the rear disc brake caliper. Hang the caliper from the body with wire. Do not let the disc brake caliper hang by the brake hose.
6. Remove inner and outer disc brake pads and the anti-rattle clips from the disc support bracket.
7. Inspect the disc brake rotor for scoring and wear. Replace or machine, as necessary.

To install:

8. Using Rear Caliper Piston Adjuster T87P-2588-A, rotate the caliper piston clockwise until the piston is fully seated.

➡Ensure that one of the 2 slots in the rear brake piston face is positioned so it will engage the nib on the back of the inner brake pad, if equipped.

9. Install disc brake pad anti-rattle clips.
10. Install inner and outer disc brake pads on the rear disc support bracket.
11. Install rear disc brake caliper and pads. Make sure the disc brake pads and anti-rattle clips are correctly installed.
12. Install wheel and tire assembly. Torque the lug nuts in a star pattern to 85–105 ft. lbs. (115–142 Nm).
13. Repeat the procedure for the opposite disc brake caliper assembly.
14. Pump the brake pedal prior to moving the vehicle to position the brake pads.

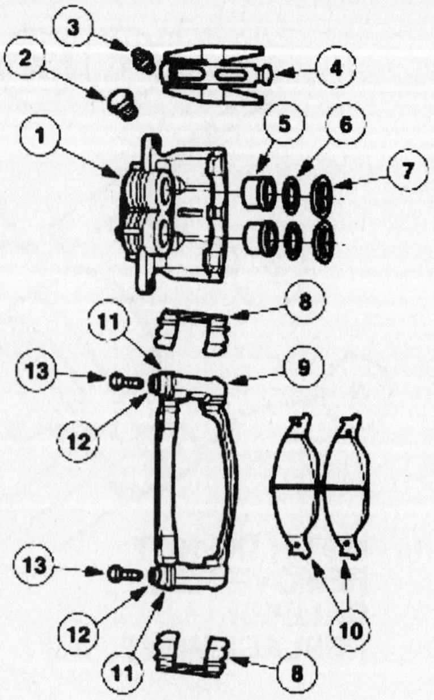

1 Disc brake caliper
2 Bleeder screw cap
3 Bleeder screw
4 Anti-rattle spring
5 Caliper piston
6 Piston seal
7 Piston dust boot
8 Pad slipper
9 Front disc brake caliper anchor plate
10 Brake pads
11 Guide pin boot
12 Guide pin
13 Caliper bolt

67197-MUST-G21

Disc brake pads and related components—Mustang with dual piston caliper and guide pins

15. Fill the brake master cylinder reservoir with clean DOT 3 brake fluid from a closed container.
16. If the disc brake calipers were replaced or repaired, be sure to bleed the system.
17. Road test the vehicle and check the brake system for proper operation.

FORD

Ranger

13

SPECIFICATION CHARTS

ENGINE AND VEHICLE IDENTIFICATION

Engine							Model Year	
Code ①	Liters (cc)	Cu. In.	Cyl.	Fuel Sys.	Type	Eng. Mfg.	Code ②	Year
D	2.3 (2261)	138	4	MFI	DOHC	Ford	1	2001
C	2.5 (2500)	152	4	MFI	SOHC	Ford	2	2002
U	3.0 (2999)	183	6	MFI	OHV	Ford	3	2003
E	4.0 (4000)	244	6	MFI	SOHC	Ford	4	2004
							5	2005

MFI: Multi-port Fuel Injection

OHV: Overhead Valve

DOHC: Dual Overhead Camshafts

SOHC: Single Overhead Camshaft

① 8th digit of the Vehicle Identification Number (VIN)

② 10th digit of the Vehicle Identification Number (VIN)

67197-RANG-C01

GENERAL ENGINE SPECIFICATIONS

Year	Model	Engine Displacement Liters	Engine (VIN)	Net Horsepower @ rpm	Net Torque @ rpm (ft. lbs.)	Bore x Stroke (in.)	Compression Ratio	Oil Pressure @ rpm
2001	Ranger	2.3	D	143@5200	154@3750	3.44x3.70	NA	29-39@3000
	Ranger	2.5	C	147@5000	147@5000	3.50x3.14	9.3:1	40-60@2500
	Ranger	3.0	U	147@5000	162@3250	3.50x3.14	9.3:1	40-60@2500
	Ranger	4.0	E	160@4000	225@2500	3.81x3.39	9.0:1	40-60@2000
2002	Ranger	2.3	D	143@5200	154@3750	3.44x3.70	NA	29-39@3000
	Ranger	3.0	U	147@5000	147@5000	3.50x3.14	9.3:1	40-60@2500
	Ranger	4.0	E	160@4000	225@2500	3.81x3.39	9.0:1	40-60@2000
2003	Ranger	2.3	D	143@5200	154@3750	3.44x3.70	NA	29-39@3000
	Ranger	3.0	U	147@5000	147@5000	3.50x3.14	9.3:1	40-60@2500
	Ranger	4.0	E	160@4000	225@2500	3.81x3.39	9.0:1	40-60@2000
2004	Ranger	2.3	D	143@5200	154@3750	3.44x3.70	NA	29-39@3000
	Ranger	3.0	U	147@5000	147@5000	3.50x3.14	9.3:1	40-60@2500
	Ranger	4.0	E	160@4000	225@2500	3.81x3.39	9.0:1	40-60@2000

MFI: Multi-port Fuel Injection

67197-RANG-C02

GASOLINE ENGINE TUNE-UP SPECIFICATIONS

Year	Engine Displacement Liters	Engine VIN	Spark Plug Gap (in.)	Ignition Timing (deg.) ① MT	Ignition Timing (deg.) ① AT	Fuel Pump (psi)	Idle Speed (rpm) MT	Idle Speed (rpm) AT	Valve Clearance In.	Valve Clearance Ex.
2001	2.3	D	0.041-0.045	10B	10B	64-72	①	①	HYD	HYD
	2.5	C	0.044	10B	10B	64-72	①	①	HYD	HYD
	3.0	U	0.044	10B	10B	64-72	①	①	HYD	HYD
	4.0	E	0.052-0.056	10B	10B	64-72	①	①	HYD	HYD
2002	2.3	D	0.041-0.045	10B	10B	64-72	①	①	HYD	HYD
	3.0	U	0.042-0.046	10B	10B	64-72	①	①	HYD	HYD
	4.0	E	0.062-0.068	10B	10B	64-72	①	①	HYD	HYD
2003	2.3	D	0.041-0.045	10B	10B	64-72	①	①	HYD	HYD
	3.0	U	0.042-0.046	10B	10B	64-72	①	①	HYD	HYD
	4.0	E	0.062-0.068	10B	10B	64-72	①	①	HYD	HYD
2004	2.3	D	0.041-0.045	10B	10B	64-72	①	①	HYD	HYD
	3.0	U	0.042-0.046	10B	10B	64-72	①	①	HYD	HYD
	4.0	E	0.062-0.068	10B	10B	64-72	①	①	HYD	HYD

NOTE: The Vehicle Emission Control Information label often reflects specification changes changes made during production. The label figures must be used if they differ from those in this chart.

B: Before top dead center

HYD: Hydraulic

NA: Not Available

① Electronically controlled and cannot be adjusted

67197-RANG-C03

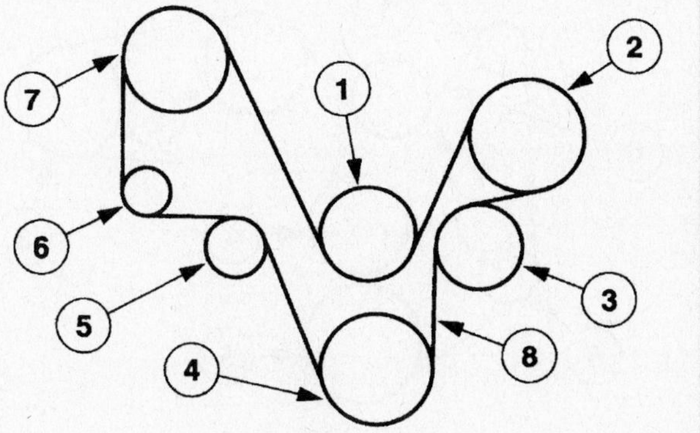

1 Fan pulley
2 Power steering pump pulley
3 Water pump pulley
4 Crankshaft pulley
5 Belt tensioner pulley
6 Generator pulley
7 A/C clutch pulley
8 Drive belt

67197RANGG97

Accessory drive belt routing—2.3L with A/C

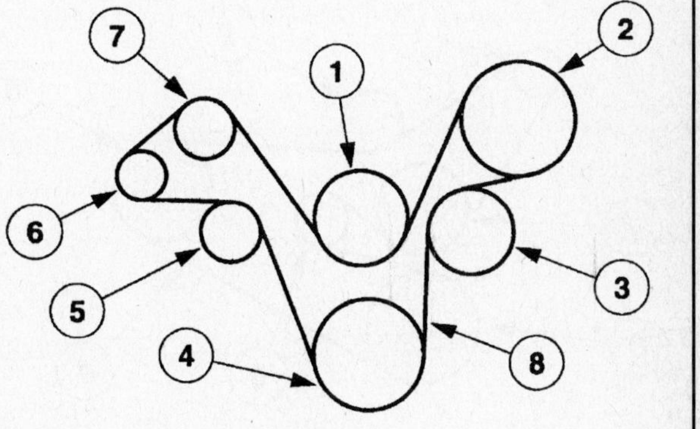

1 Fan pulley
2 Power steering pump pulley
3 Water pump pulley
4 Crankshaft pulley
5 Belt tensioner pulley
6 Generator pulley
7 Belt idler pulley
8 Drive belt

67197RANGG98

Accessory drive belt routing—2.3L without A/C

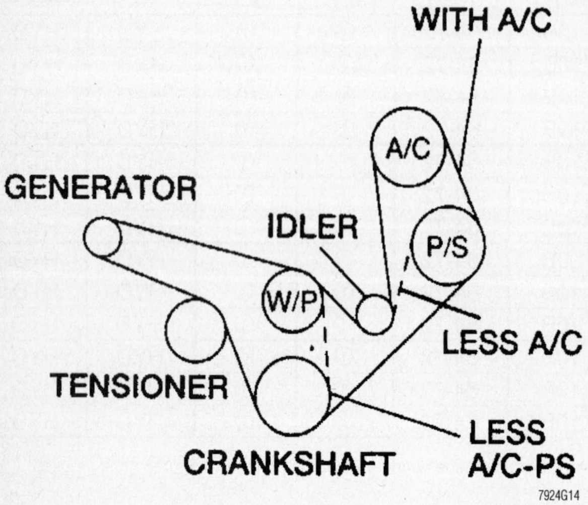

Accessory drive belt routing—2.5L

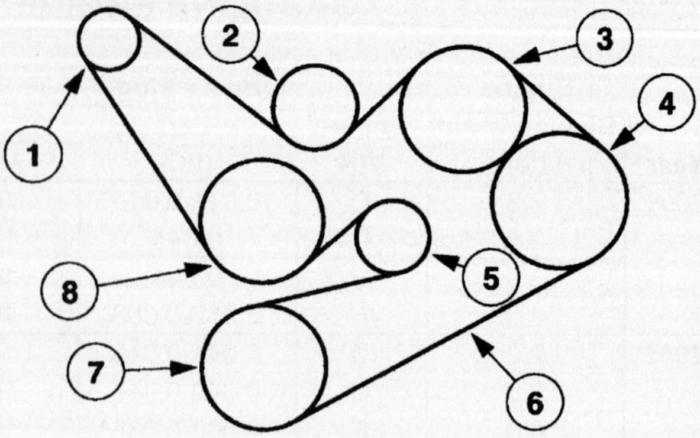

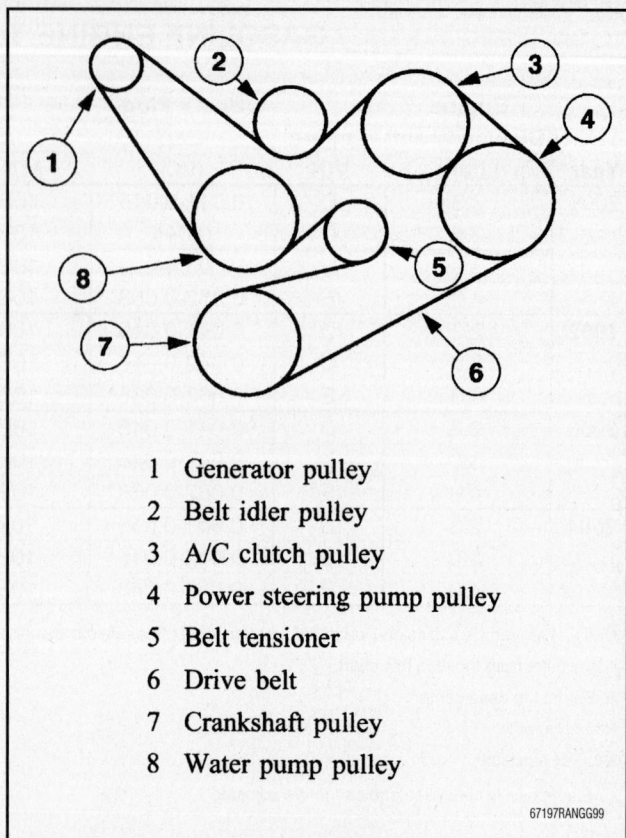

1 Generator pulley
2 Belt idler pulley
3 A/C clutch pulley
4 Power steering pump pulley
5 Belt tensioner
6 Drive belt
7 Crankshaft pulley
8 Water pump pulley

67197RANGG99

Accessory drive belt routing—3.0L with A/C

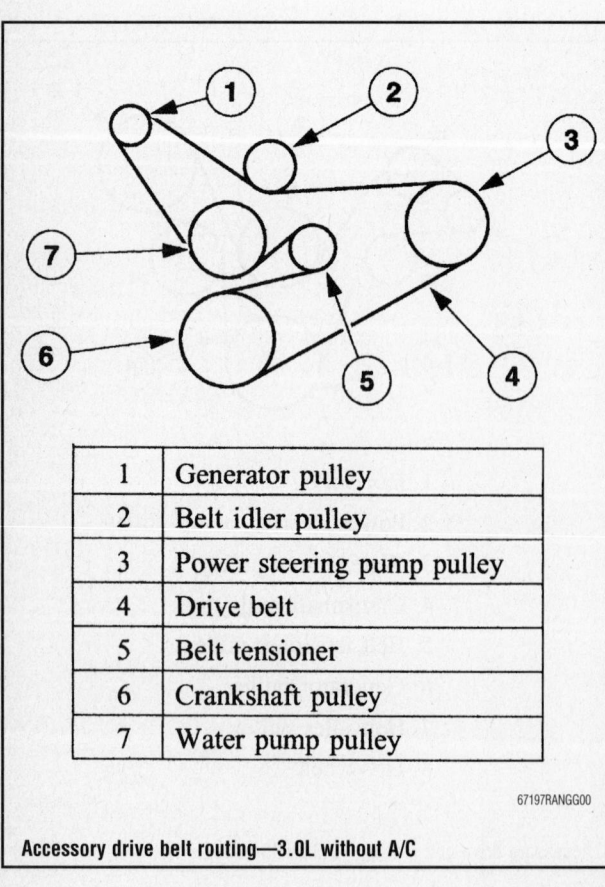

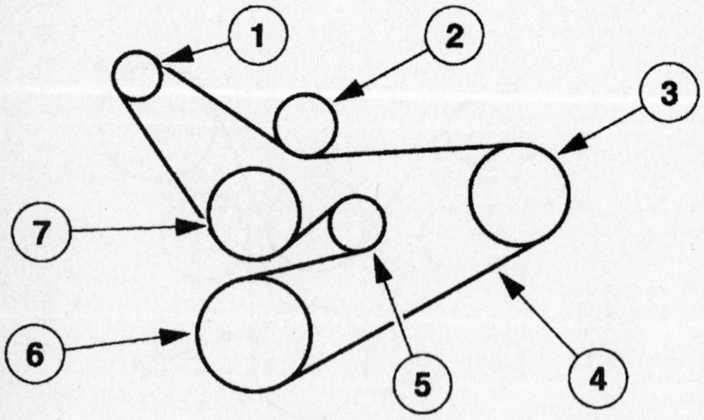

1	Generator pulley
2	Belt idler pulley
3	Power steering pump pulley
4	Drive belt
5	Belt tensioner
6	Crankshaft pulley
7	Water pump pulley

67197RANGG00

Accessory drive belt routing—3.0L without A/C

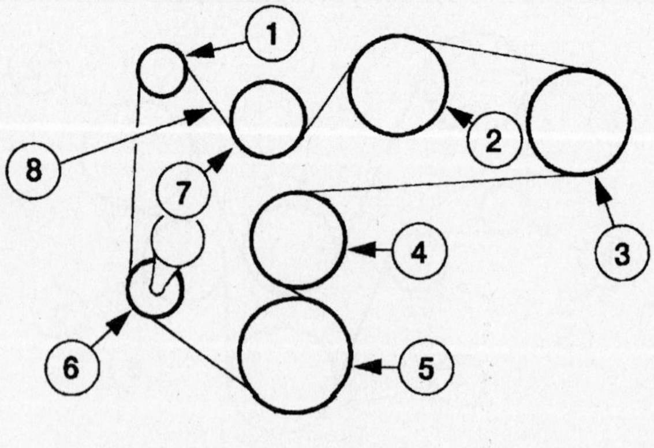

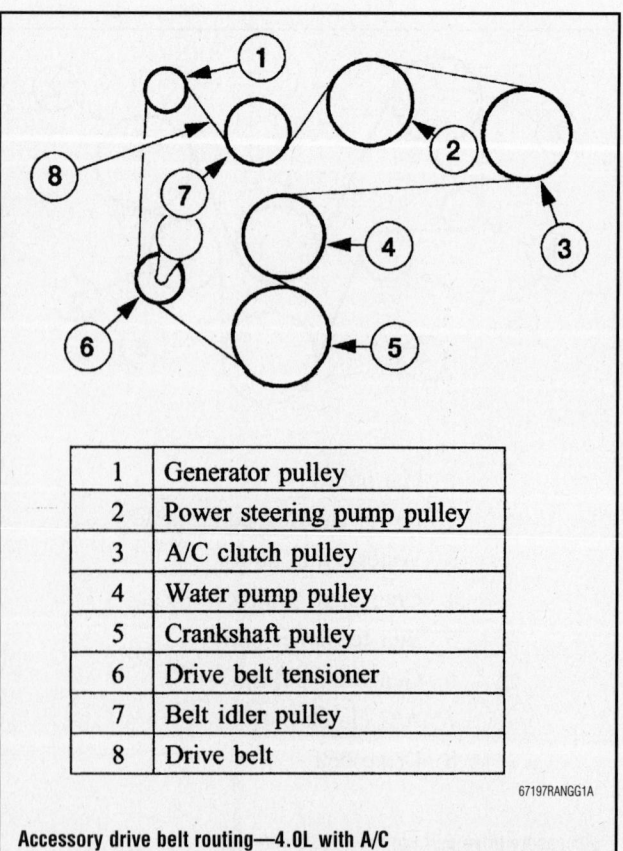

1	Generator pulley
2	Power steering pump pulley
3	A/C clutch pulley
4	Water pump pulley
5	Crankshaft pulley
6	Drive belt tensioner
7	Belt idler pulley
8	Drive belt

67197RANGG1A

Accessory drive belt routing—4.0L with A/C

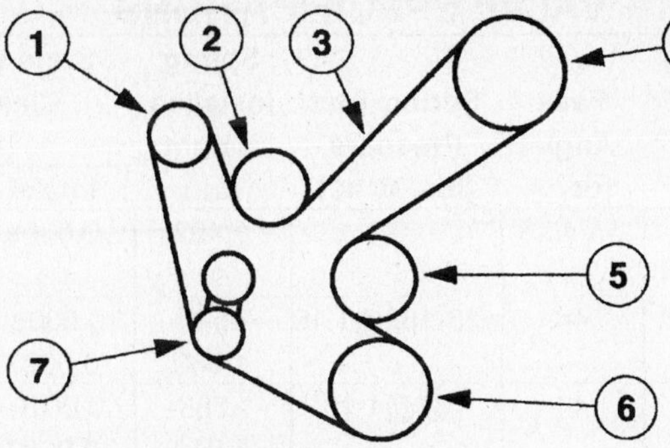

1	Generator pulley
2	Belt idler pulley
3	Drive belt
4	Power steering pump pulley
5	Water pump pulley
6	Crankshaft pulley
7	Belt tensioner

67197RANGG2A

Accessory drive belt routing—4.0L without A/C

CAPACITIES

Year	Model	Engine Displacement Liters	Engine VIN	Engine Oil with Filter (qts.)	Transmission (pts.) 5-Spd	Transmission (pts.) Auto.	Transfer Case (pts.)	Drive Axle Front (pts.)	Drive Axle Rear (pts.)*	Fuel Tank (gal.)	Cooling System (qts.)
2001	Ranger	2.3	D	4.0	3.0	19.8	—	—	5.0	①	②
	Ranger	2.5	C	5.0	3.0	19.8	—	—	5.0	①	②
	Ranger	3.0	U	4.5	3.0	③	2.5	3.3	5.0	①	④
	Ranger	4.0	E	5.0	3.0	③	2.5	3.25	5.0	①	⑤
2002	Ranger	2.3	D	4.0	3.0	19.8	—	—	5.0	①	②
	Ranger	3.0	U	4.5	3.0	③	2.5	2.7	5.0	①	④
	Ranger	4.0	E	5.0	3.0	③	2.5	2.7	5.0	①	⑤
2003	Ranger	2.3	D	4.0	3.0	19.8	—	—	5.0	①	②
	Ranger	3.0	U	4.5	3.0	③	2.5	2.7	5.0	①	④
	Ranger	4.0	E	5.0	3.0	③	2.5	2.7	5.0	①	⑤
2004	Ranger	2.3	D	4.0	3.0	19.8	—	—	5.0	①	②
	Ranger	3.0	U	4.5	3.0	③	2.5	3.25	5.0	①	④
	Ranger	4.0	E	5.0	3.0	③	2.5	3.25	5.0	①	⑤

NOTE: All capacities are approximate. Add fluid gradually and check to be sure a proper fluid level is obtained.

* For limited slip axles, add 4 oz. of friction modifier, exc. for 1-ton models

① Std: 16.5
Long Wheelbase: 20.0
Super Cab: 19.5

② w/MT: 11.2
w/AT: 10.9

③ 2wd: 20.0
4wd: 20.6

④ w/MT: 15.2
w/AT: 14.8

⑤ w/MT: 13.5
w/AT: 13.2

67197-RANG-C04

VALVE SPECIFICATIONS

Year	Engine Displacement Liters	Engine VIN	Seat Angle (deg.)	Face Angle (deg.)	Spring Test Pressure (lbs. @ in.)	Spring Installed Height (in.)	Stem-to-Guide Clearance (in.) Intake	Stem-to-Guide Clearance (in.) Exhaust	Stem Diameter (in.) Intake	Stem Diameter (in.) Exhaust
2001	2.3	D	44.5-45	45-45.5	①	1.492	0.0009	0.0011	0.2153-0.2159	0.2151-0.2157
	2.5	C	44.75	44	118-132@1.16	1.540-1.580	0.0008-0.0027	0.0018-0.0037	0.2746-0.2754	0.2736-0.2744
	3.0	U	45	44	185@1.16	1.580-1.610	0.0010-0.0027	0.0015-0.0032	0.3126-0.3134	0.3121-0.3129
	4	E	45	45	202-225@1.413-1.445	1.569-1.601	0.0010-0.0020	0.0010-0.0020	0.2740-0.2748	0.2730-0.2740
2002	2.3	D	45	45	①	1.492	0.0009	0.0011	0.2153-0.2159	0.2151-0.2157
	3.0	U	45	44	185@1.16	1.580-1.610	0.0010-0.0027	0.0015-0.0032	0.3126-0.3134	0.3121-0.3129
	4.0	E	45	45	202-225@1.413-1.445	1.569-1.601	0.0010-0.0020	0.0010-0.0030	0.2742-0.2748	0.2736-0.2742
2003	2.3	D	45	45	①	1.492	0.0009	0.0011	0.2153-0.2159	0.2151-0.2157
	3.0	U	45	44	185@1.16	1.580-1.610	0.0010-0.0027	0.0015-0.0032	0.3126-0.3134	0.3121-0.3129
	4.0	E	45	45	202-225@1.413-1.445	1.569-1.601	0.0010-0.0020	0.0010-0.0030	0.2742-0.2748	0.2736-0.2742
2004	2.3	D	45	45	①	1.492	0.0009	0.0011	0.2153-0.2159	0.2151-0.2157
	3.0	U	45	44	185@1.16	1.580-1.610	0.0010-0.0027	0.0015-0.0032	0.3126-0.3134	0.3121-0.3129
	4.0	E	45	45	202-225@1.413-1.445	1.569-1.601	0.0010-0.0020	0.0010-0.0030	0.2742-0.2748	0.2736-0.2742

① Intake: 97.0@1.201
Exhaust: 93.3@1.201

67197-RANG-C05

PISTON AND RING SPECIFICATIONS

All measurements are given in inches.

Year	Engine Displacement Liters	Engine VIN	Piston Clearance	Ring Gap Top Compression	Ring Gap Bottom Compression	Ring Gap Oil Control	Ring Side Clearance Top Compression	Ring Side Clearance Bottom Compression	Ring Side Clearance Oil Control
2001	2.3	D	0.0009-0.0017	0.006-0.0012	0.012-0.0180	0.007-0.0270	0.0008-0.0013	0.0004-0.0011	0.0025-0.0054
	2.5	C	0.0010-0.0020	0.008-0.014	0.013-0.019	0.010-0.030	0.0014-0.0030	0.0014-0.0030	SNUG
	3.0	U	0.0012-0.0023	0.010-0.020	0.010-0.020	0.010-0.049	0.0602-0.0612	0.0602-0.0612	SNUG
	4.0	E	0.0008-0.0019	0.015-0.023	0.015-0.023	0.015-0.055	0.0010-0.0030	0.0010-0.0030	SNUG
2002	2.3	D	0.0009-0.0017	0.006-0.0012	0.012-0.0180	0.007-0.0270	0.0008-0.0013	0.0004-0.0011	0.0025-0.0054
	3.0	U	0.0012-0.0023	0.010-0.020	0.010-0.020	0.010-0.049	0.0602-0.0612	0.0602-0.0612	SNUG
	4.0	E	0.0008-0.0019	0.015-0.023	0.015-0.023	0.015-0.055	0.0010-0.0030	0.0010-0.0030	SNUG
2003	2.3	D	0.0009-0.0017	0.006-0.0012	0.012-0.0180	0.007-0.0270	0.0008-0.0013	0.0004-0.0011	0.0025-0.0054
	3.0	U	0.0012-0.0023	0.010-0.020	0.010-0.020	0.010-0.049	0.0602-0.0612	0.0602-0.0612	SNUG
	4.0	E	0.0008-0.0019	0.015-0.023	0.015-0.023	0.015-0.055	0.0010-0.0030	0.0010-0.0030	SNUG
2004	2.3	D	0.0009-0.0017	0.006-0.0012	0.012-0.0180	0.007-0.0270	0.0008-0.0013	0.0004-0.0011	0.0025-0.0054
	3.0	U	0.0012-0.0023	0.010-0.020	0.010-0.020	0.010-0.049	0.0602-0.0612	0.0602-0.0612	SNUG
	4.0	E	0.0008-0.0019	0.015-0.023	0.015-0.023	0.015-0.055	0.0010-0.0030	0.0010-0.0030	SNUG

67197-RANG-C06

CRANKSHAFT AND CONNECTING ROD SPECIFICATIONS

All measurements are given in inches.

	Engine		Crankshaft				Connecting Rod		
Year	Displacement Liters	Engine VIN	Main Brg. Journal Dia.	Main Brg. Oil Clearance	Shaft End-play	Thrust on No.	Journal Diameter	Oil Clearance	Side Clearance
2001	2.3	D	2.0465-2.2059	0.0007-0.0013	0.0080-0.0160	3	1.9606-1.9685	0.0010-0.0020	0.0767-0.1200
	2.5	C	2.2051-2.2059	0.0008-0.0015	0.0040-0.0080	3	2.0464-2.0472	0.0008-0.0015	0.0035-0.0115
	3.0	U	2.5190-2.5198	0.0010-0.0014	0.0040-0.0080	3	2.1253-2.1261	0.0010-0.0014	0.0060-0.0140
	4.0	E	2.2430-2.2440	0.0008-0.0015	0.0020-0.0125	3	2.7252-2.7260	0.0003-0.0024	0.0036-0.0106
2002	2.3	D	2.0465-2.2059	0.0007-0.0013	0.0080-0.0160	3	1.9606-1.9685	0.0010-0.0020	0.0767-0.1200
	3.0	U	2.5190-2.2059	0.0010-0.0015	0.0040-0.0080	3	2.1253-2.0472	0.0010-0.0015	0.0060-0.0115
	4.0	E	2.2430-2.2440	0.0008-0.0015	0.0020-0.0125	3	2.7252-2.7260	0.0003-0.0024	0.0036-0.0106
2003	2.3	D	2.0465-2.2059	0.0007-0.0013	0.0080-0.0160	3	1.9606-1.9685	0.0010-0.0020	0.0767-0.1200
	3.0	U	2.5190-2.2059	0.0010-0.0015	0.0040-0.0080	3	2.1253-2.0472	0.0010-0.0015	0.0060-0.0115
	4.0	E	2.2430-2.2440	0.0008-0.0015	0.0020-0.0125	3	2.7252-2.7260	0.0003-0.0024	0.0036-0.0106
2004	2.3	D	2.0465-2.2059	0.0007-0.0013	0.0080-0.0160	3	1.9606-1.9685	0.0010-0.0020	0.0767-0.1200
	3.0	U	2.5190-2.2059	0.0010-0.0015	0.0040-0.0080	3	2.1253-2.0472	0.0010-0.0015	0.0060-0.0115
	4.0	E	2.2430-2.2440	0.0008-0.0015	0.0020-0.0125	3	2.7252-2.7260	0.0003-0.0024	0.0036-0.0106

67197-RANG-C07

TORQUE SPECIFICATIONS

All readings in ft. lbs.

	Engine Displacement Liters	Engine VIN	Cylinder Head Bolts	Main Bearing Bolts	Rod Bearing Bolts	Crankshaft Damper Bolts	Flywheel Bolts	Manifold Intake *	Manifold Exhaust	Spark Plugs	Oil Pan Drain Plug
2001	2.3	D	①	NA	NA	②	③	13	40	9	21
	2.5	C	④	⑤	⑥	103-133	54-64	19-28	14-21	5-10	NA
	3.0	U	⑦	60	26	107	54-64	24	⑧	8-10	NA
	4.0	E	⑨	72	⑩	⑪	75-85	7	16	15	18
2002	2.3	D	①	NA	NA	⑫	⑬	13	40	9	21
	3.0	U	⑦	60	26	107	54-64	24	⑧	8-10	NA
	4.0	E	⑭	72	⑩	⑪	75-85	7	16	13	18
2003	2.3	D	①	NA	NA	⑫	⑬	13	40	9	21
	3.0	U	⑦	60	26	107	54-64	24	⑧	8-10	NA
	4.0	E	⑭	72	⑩	⑪	75-85	7	16	13	18
2004	2.3	D	①	NA	NA	⑫	⑬	13	40	9	21
	3.0	U	⑦	60	26	107	59	21	⑧	11	NA
	4.0	E	⑭	72	⑩	⑪	⑮	7	16	13	18

NA: Information not available

* NOTE: Applies to Lower Manifold only.

① Step 1: 44 inch lbs.
Step 2: 11 ft. lbs.
Step 3: 33 ft. lbs.
Step 4: +90 degrees
Step 5: +90 degrees

② Step 1: 30 ft. lbs.
Step 2: +90 degrees

③ Step 1: 51-59 ft. lbs.
Step 2: 50 ft. lbs.
Step 3: 83 ft. lbs.

④ Step 1: 52 ft. lbs.
Step 2: recheck at 52 ft. lbs.
Step 3: +90 degrees

⑤ Step 1: 51-59 ft. lbs.
Step 2: 76-84 ft. lbs.

⑥ Step 1: 26-30 ft. lbs.
Step 2: 31-36 ft. lbs.

⑦ Step 1: 59 ft. lbs.
Step 2: Back off 1 full turn
Step 3: 34-40 ft. lbs.
Step 4: 63-73 ft. lbs.

⑧ Step 1: 89 inch lbs.
Step 2: 17 ft. lbs.

⑨ Step 1: 24 ft. lbs.
Step 2: plus 90 degrees

⑩ Step 1: 15 ft. lbs.
Step 2: +90 degrees

⑪ Step 1: 37 ft. lbs.
Step 2: plus 90 degrees

⑫ Step 1: 74 ft. lbs.
Step 2: +90 degrees

⑬ Step 1: 37 ft. lbs.
Step 2: 50 ft. lbs.
Step 3: 83 ft. lbs.

⑭ 8mm bolts: 24 ft. lbs.
12mm bolts: 24 ft. lbs. +80 degrees, +80 degrees more

⑮ Step 1: 10 ft.lbs.
Step 2: 52 ft. lbs.

67197-RANG-C08

BRAKE SPECIFICATIONS

All measurements in inches unless noted

		Brake Disc			Brake Drum Diameter				Brake Caliper	
Year	Model	Original Thickness	Minimum Thickness	Maximum Runout	Original Inside Diameter	Max. Wear Limit	Maximum Machine Diameter	Minimum Lining Thickness	Bracket Bolts (ft. lbs.)	Mounting Bolts (ft. lbs.)
2001	Ranger ①	0.850	0.810	0.0030	9.00	9.09	9.06	0.030	72-97	21-26
	②	0.850	0.810	0.0030	10.00	10.09	10.06	0.030	72-97	21-26
	③	0.850	0.810	0.0030	10.00	10.09	10.06	0.030	72-97	21-26
2002	Ranger ①	0.850	0.810	0.0030	9.00	9.09	9.06	0.030	85	24
	②	0.850	0.810	0.0030	10.00	10.09	10.06	0.030	85	24
	③	0.850	0.810	0.0030	10.00	10.09	10.06	0.030	85	24
2003	Ranger ①	0.850	0.810	0.0030	9.00	9.09	9.06	0.030	85	24
	②	0.850	0.810	0.0030	10.00	10.09	10.06	0.030	85	24
	③	0.850	0.810	0.0030	10.00	10.09	10.06	0.030	85	24
2004	Ranger	NA	④	NA	NA	NA	⑤	⑥	85	24

NOTE: Due to changes made during production, refer to manufacturer's specifications if they differ from those in this chart

NA: Not Available

F: Front

R: Rear

① With 9 inch brakes

② 4x2 with 10 inch brakes

③ 4x4 with 10 inch brakes

④ Molded into the disc

⑤ Molded into the drum

⑥ Front: 0.04 in.
Rear: 0.03 in.

67197-RANG-C09

TIRE, WHEEL AND BALL JOINT SPECIFICATIONS

		OEM Tires		Tire Pressures (psi)		Wheel	Ball Joint	Lug
Year	Model	Standard	Optional	Front	Rear	Size	Inspection	Nut
2001	2wd, XL	P205/75R14SL	P225/70R14SL	35	35	6-JJ	0.030 in. ①	100
	2wd XLT	P205/75R14SL	P225/70R14SL	35	35	6-JJ	0.030 in. ①	100
	Splash, 2wd	P235/60R15SL	P235/75R15SL	35	35	6-JJ	0.030 in. ①	100
	4wd, XL	P215/75R15SL	P235/75R15SL	35	35	6-JJ	0.030 in. ①	100
	4wd, XLT	P215/75R15SL	P235/75R15SL P235/70R16SL P265/75R15SL	35	35	7-JJ	0.030 in. ①	100
	Splash 4wd	P235/75R15SL	P235/75R15SL	35	35	7-JJ	0.030 in. ①	100
2002	2wd, exc. Tremor	P235/75R15	none	②	②	7-JJ	0.030 in. ①	100
	Tremor	P235/70R16	none	②	②	7-JJ	0.030 in. ①	100
	4wd exc. FX4	P245/75R16	none	②	②	7-JJ	0.030 in. ①	100
	FX4	31x10.5/15	none	②	②	8-JJ	0.030 in. ①	100
2003	2wd, exc. Tremor	P235/75R15	none	②	②	7-JJ	0.030 in. ①	100
	Tremor	P235/70R16	none	②	②	7-JJ	0.030 in. ①	100
	4wd exc. FX4	P245/75R16	none	②	②	7-JJ	0.030 in. ①	100
	FX4	31x10.5/15	none	②	②	8-JJ	0.030 in. ①	100
2004	2wd XL	P225/70R15	none	②	②	6-JJ	0.030 in. ①	100
	2wd XLT	P225/70R15	none	②	②	7-JJ	0.030 in. ①	100
	Edge 2wd	P235/75R15	none	②	②	7-JJ	0.030 in. ①	100
	Edge 4wd	P245/75R16	none	②	②	7-JJ	0.030 in. ①	100
	4wd XLT FX4	P245/75R16	none	②	②	7-JJ	0.030 in. ①	100
	Tremor 2wd	P235/75R16	none	②	②	7-JJ	0.030 in. ①	100
	Tremor 4wd	P245/75R16	none	②	②	7-JJ	0.030 in. ①	100

OEM: Original Equipment Manufacturer

PSI: Pounds Per Square Inch

STD: Standard

OPT: Optional

① Both upper and lower

② See placard on door post

67197-RANG-C10

WHEEL ALIGNMENT

Year	Model		Caster Range (+/-Deg.)	Caster Preferred Setting (Deg.)	Camber Range (+/-Deg.)	Camber Preferred Setting (Deg.)	Toe-in (in.)
2001	Exc. Splash	2wd	1.0	④	0.70	-0.50	0.06+/-0.25
		4wd	1.0	②	0.70	-0.50	0.12+/-0.25
	Splash		1.0	②	0.70	-0.50	0.12+/-0.25
2002	Ranger	2wd	1.0	④	0.70	-0.50	0.06+/-0.25
		4wd	1.0	②	0.70	-0.50	0.12+/-0.25
		Rear	—	—	0.75	0	0+/-0.30
2003	Ranger	2wd	1.0	④	0.70	-0.50	0.06+/-0.25
		4wd	1.0	②	0.70	-0.50	0.12+/-0.25
		Rear	—	—	0.75	0	0+/-0.30
2004	Ranger	2wd	1.0	⑤	0.70	⑥	0.06+/-0.25
		4wd	1.0	⑦	0.70	-0.50	0.12+/-0.25
		Rear	—	—	0.75	0	0+/-0.30

① Left: +4.1
Right: +4.5

② Left: +3.9
Right: +4.4

③ Left: +4.6
Right: +5.0

④ Left: +4.0
Right: +4.4

⑤ Left: +3.0
Right: +3.4

⑥ Left: -0.5
Right: -0.7

⑦ Left: +2.9
Right: +3.6

67197-RANG-C11

SCHEDULED MAINTENANCE INTERVALS
2001-02 FORD RANGER

TO BE SERVICED	TYPE OF SERVICE	VEHICLE MILEAGE INTERVAL (x1000)												
		5	10	15	20	25	30	35	40	45	50	55	60	65
Engine oil & filter	R	✓	✓	✓	✓	✓	✓	✓	✓	✓	✓	✓	✓	✓
Tires	Rotate	✓	✓	✓	✓	✓	✓	✓	✓	✓	✓	✓	✓	✓
Auto trans. fluid	I			✓			✓			✓			✓	
Brake pads/shoes	I			✓			✓			✓			✓	
Coolant hoses	S/I			✓			✓			✓			✓	
Steering linkage	I			✓			✓			✓			✓	
Cabin air filter	R			✓			✓			✓			✓	
Ball joints (2wd)	L			✓			✓			✓			✓	
Exhaust system	I						✓						✓	
Engine air filter	R						✓						✓	
Fuel filter ①	R						✓						✓	
Auto trans fluid (4-speed)	R						✓						✓	
Green coolant ②	R									✓				
Wheel bearings (2wd)	L												✓	
Manual trans. fluid	R												✓	
Spark plugs	R	every 100,000 miles												
PCV valve	R	every 100,000 miles												
Orange coolant	R	every 150,000 miles												
Auto trans fluid (5-speed)	R	every 150,000 miles												
Differential fluid	R	every 150,000 miles												
Accessory drive belts	R	every 150,000 miles												
Transfer case fluid	R	every 150,000 miles												

R: Replace S: Service I: Inspect L: Lubricate

① Recommended, but not required in Calif.

② Change every 30,000 miles or 36 months thereafter

FREQUENT OPERATION MAINTENANCE (SEVERE SERVICE)

If a vehicle is operated under any of the following conditions it is considered severe service:

- Towing a trailer or using a camper or car-top carrier.

- Repeated short trips of less than 5 miles in temperatures below freezing, or trips of less than 10 miles in any temperature.

- Extensive idling or low-speed driving for long distance as in heavy commercial use, such as delivery, taxi or police cars.

- Operating on rough, muddy or salt-covered roads.

- Operating on unpaved or dusty roads.

- Driving in extremely hot (over 90°) conditions.

Engine oil & filter: replace every 3000 miles.
Air cleaner filter: service or inspect every 6000 miles.
Exhaust system: check every 6000 miles.
Automatic transmission fluid & filter: change every 30,000 miles.
Transfer case fluid: change every 60,000 miles
Fule filter: change every 15,000 miles
Spark plugs: change every 60,000 miles
2wd front wheel bearings: lubricate every 30,000 miles

67197-RANG-C12

SCHEDULED MAINTENANCE INTERVALS
2003 FORD RANGER

TO BE SERVICED	TYPE OF SERVICE	VEHICLE MILEAGE INTERVAL (x1000)												
		5	10	15	20	25	30	35	40	45	50	55	60	65
Engine oil & filter	R	✓	✓	✓	✓	✓	✓	✓	✓	✓	✓	✓	✓	✓
Tires	Rotate	✓	✓	✓	✓	✓	✓	✓	✓	✓	✓	✓	✓	✓
Auto trans. fluid	I			✓			✓			✓			✓	
Brake pads/shoes	I			✓			✓			✓			✓	
Coolant hoses	S/I			✓			✓			✓			✓	
Steering linkage	I			✓			✓			✓			✓	
Suspension and driveshaft	I			✓			✓			✓			✓	
Cabin air filter	R			✓			✓			✓			✓	
Ball joints (2wd)	I			✓			✓			✓			✓	
Exhaust system	I						✓						✓	
Engine air filter	R						✓						✓	
Fuel filter ①	R						✓						✓	
Auto trans fluid ②	R						✓						✓	
Green coolant	R									✓				
Accessory drive belts	I	every 100,000 miles												
Spark plugs	R	every 100,000 miles												
PCV valve	R	every 100,000 miles												
Yellow coolant ③	R	first 100,000 miles or 5 years												
Auto trans fluid and filter ④	R	every 150,000 miles												
Rear axle fluid	R	every 150,000 miles												
Accessory drive belts ⑤	R	every 150,000 miles												

R: Replace S: Service I: Inspect L: Lubricate

① Recommended, but not required in Calif.

② If equipped with AX4S, 4F50N, 4R100, 4F27E

③ Every 3 years or 50,000 miles thereafter

④ All except AX4S, 4F50N, 4R100, 4F27E

⑤ If not previously replaced

Special Operating Condition Requirements

When towing a trailer or using a camper or car-top carrier:
Change engine oil and install a new oil filter every 4,800 km (3,000 miles) or 3 months.
Change transfer case fluid every 96,000 km (60,000 miles).
Change manual transmission fluid as required.
Inspect and lubricate U-joints as required.

During extensive idling and/or low speed driving for long distances, as in heavy commercial use such as delivery, taxi, patrol car or livery:
Change engine oil and install a new oil filter, lube front lower control arm and steering linkage ball joints with zerk fittings (if equipped) every 4,800 km (3,000 miles) or 3 months.
Inspect brake system and check battery electrolyte level (Patrol cars) every 8,000 km (5,000 miles).
Install a new fuel filter every 24,000 km (15,000 miles).
Change automatic transmission fluid, lubricate 4x2 wheel bearings, install new grease seals and adjust bearings every 48,000 km (30,000 miles).
Install new spark plugs and change transfer case fluid every 96,000 km (60,000 miles).
Install a new cabin air filter as required.

When operating in dusty conditions such as unpaved or dusty roads:
Change engine oil and install a new oil filter every 4,800 km (3,000 miles) or 3 months.
Install a new fuel filter every 24,000 km (15,000 miles).
Change automatic transmission fluid every 48,000 km (30,000 miles).
Change transfer case fluid every 96,000 km (60,000 miles).
Install a new engine air filter as required.
Install a new cabin air filter as required.

When operating in off-road conditions:
Change automatic transmission fluid every 48,000 km (30,000 miles).
Change transfer case fluid every 96,000 km (60,000 miles).
Install a new cabin air filter as required.
Inspect and lubricate U-joints.
Inspect and lubricate steering linkage ball joints with zerk fittings.

67197-RANG-C13

SCHEDULED MAINTENANCE INTERVALS

2004 FORD RANGER

TO BE SERVICED	TYPE OF SERVICE	VEHICLE MILEAGE INTERVAL (x1000)												
		5	10	15	20	25	30	35	40	45	50	55	60	65
Engine oil & filter	R	✓	✓	✓	✓	✓	✓	✓	✓	✓	✓	✓	✓	✓
Tires	Rotate	✓	✓	✓	✓	✓	✓	✓	✓	✓	✓	✓	✓	✓
Auto trans. fluid	I			✓			✓			✓			✓	
Brake pads/shoes	I			✓			✓			✓			✓	
Coolant hoses	S/I			✓			✓			✓			✓	
Steering linkage	I			✓			✓			✓			✓	
Suspension and driveshaft	I			✓			✓			✓			✓	
Cabin air filter	R			✓			✓			✓			✓	
Ball joints (2wd)	I/L			✓			✓			✓			✓	
Exhaust system	I						✓						✓	
Engine air filter	R						✓						✓	
Fuel filter	R						✓						✓	
Auto trans fluid ①	R						✓						✓	
Wheel bearings (2wd)	R	at 150,000 miles, if not previously replaced, including seals												
Coolant, exc. Premium Gold	R	every 105,000 miles												
Premium Gold coolant ③	R	at 5 years or 100,000 miles												
Spark plugs	R	every 100,000 miles												
PCV valve	R	every 100,000 miles												
Auto trans fluid ④	R	every 150,000 miles												
Fuel injectors	Clean	every 100,000 miles												
Rear axle fluid	R	every 150,000 miles												
Accessory drive belts	I	every 100,000 miles												
Accessory drive belts	R	every 150,000 miles, if not previously replaced												

R: Replace S: Service I: Inspect L: Lubricate

① Vehicles equipped with AX4S, 4F50N, 4R100, 4F27E

② Change every 30,000 miles or 36 months thereafter

③ After the initial change, every 3 years or 50,000 miles

④ Except vehicles equipped with AX4S, 4F50N, 4R100, 4F27E

Special Operating Condition Requirements

When towing a trailer or using a camper or car-top carrier:

Change engine oil and install a new oil filter every 4,800 km (3,000 miles), 3 months or 200 hours of engine operation (whichever occurs first).

Change transfer case fluid every 96,000 km (60,000 miles).

Change manual transmission fluid as required.

Inspect and lubricate U-joints as required.

During extensive idling and/or low speed driving for long distances, as in heavy commercial use such as delivery, taxi, patrol car or livery:

Change engine oil and install a new oil filter every 4,800 km (3,000 miles), 3 months or 200 hours of engine operation (whichever occurs first).

Lube front lower control arm and steering linkage ball joints with zerk fittings (if equipped) every 4,800 km (3,000 miles) or 3 months.

Inspect brake system and check battery electrolyte level (Patrol cars) every 8,000 km (5,000 miles).

Install a new fuel filter every 24,000 km (15,000 miles).

Change automatic transmission fluid, lubricate 4x2 wheel bearings, install new grease seals and adjust bearings every 48,000 km (30,000 miles). If equipped, change the in-line service installed transmission fluid filter.

Install new spark plugs and change transfer case fluid every 96,000 km (60,000 miles).

Install a new cabin air filter as required.

67197-RANG-C14

SCHEDULED MAINTENANCE INTERVALS
FORD RANGER
Footnotes Continued

When operating in dusty conditions such as unpaved or dusty roads:

Change engine oil and install a new oil filter every 4,800 km (3,000 miles) or 3 months.

Install a new fuel filter every 24,000 km (15,000 miles).

Change automatic transmission fluid every 48,000 km (30,000 miles). If equipped, change the in-line service installed transmission fluid filter.

Change transfer case fluid every 96,000 km (60,000 miles).

Install a new engine air filter as required.

Install a new cabin air filter as required.

When operating in off-road conditions:

Change automatic transmission fluid every 48,000 km (30,000 miles). If equipped, change the in-line service installed transmission fluid filter.

Change transfer case fluid every 96,000 km (60,000 miles).

Install a new cabin air filter as required.

Inspect and lubricate U-joints.

Inspect and lubricate steering linkage ball joints with zerk fittings.

67197-RANG-C15

PRECAUTIONS

Before servicing any vehicle, please be sure to read all of the following precautions, which deal with personal safety, prevention of component damage, and important points to take into consideration when servicing a motor vehicle:

- Never open, service or drain the radiator or cooling system when the engine is hot; serious burns can occur from the steam and hot coolant.
- Observe all applicable safety precautions when working around fuel. Whenever servicing the fuel system, always work in a well-ventilated area. Do not allow fuel spray or vapors to come in contact with a spark, open flame, or excessive heat (a hot drop light, for example). Keep a dry chemical fire extinguisher near the work area. Always keep fuel in a container specifically designed for fuel storage; also, always properly seal fuel containers to avoid the possibility of fire or explosion. Refer to the additional fuel system precautions later in this section.
- Fuel injection systems often remain pressurized, even after the engine has been turned **OFF**. The fuel system pressure must be relieved before disconnecting any fuel lines. Failure to do so may result in fire and/or personal injury.
- Brake fluid often contains polyglycol ethers and polyglycols. Avoid contact with the eyes and wash your hands thoroughly after handling brake fluid. If you do get brake fluid in your eyes, flush your eyes with clean, running water for 15 minutes. If eye irritation persists, or if you have taken brake fluid internally, IMMEDIATELY seek medical assistance.
- The EPA warns that prolonged contact with used engine oil may cause a number of skin disorders, including cancer! You should make every effort to minimize your exposure to used engine oil. Protective gloves should be worn when changing oil. Wash your hands and any other exposed skin areas as soon as possible after exposure to used engine oil. Soap and water, or waterless hand cleaner should be used.
- All new vehicles are now equipped with an air bag system, often referred to as a Supplemental Restraint System (SRS) or Supplemental Inflatable Restraint (SIR) system. The system must be disabled before performing service on or around system components, steering column, instrument panel components, wiring and sensors. Failure to follow safety and disabling procedures could result in accidental air bag deployment, possible personal injury and unnecessary system repairs.
- Always wear safety goggles when working with, or around, the air bag system. When carrying a non-deployed air bag, be sure the bag and trim cover are pointed away from your body. When placing a non-deployed air bag on a work surface, always face the bag and trim cover upward, away from the surface. This will reduce the motion of the module if it is accidentally deployed. Refer to the additional air bag system precautions later in this section.
- Clean, high quality brake fluid from a sealed container is essential to the safe and proper operation of the brake system. You should always buy the correct type of brake fluid for your vehicle. If the brake fluid becomes contaminated, completely flush the system with new fluid. Never reuse any brake fluid. Any brake fluid that is removed from the system should be discarded. Also, do not allow any brake fluid to come in contact with a painted surface; it will damage the paint.
- Never operate the engine without the proper amount and type of engine oil; doing so WILL result in severe engine damage.
- Timing belt maintenance is extremely important! Many models utilize an interference-type, non-freewheeling engine. If the timing belt breaks, the valves in the cylinder head may strike the pistons, causing potentially serious (also time-consuming and expensive) engine damage. Refer to the maintenance interval charts in the front of this manual for the recommended replacement interval for the timing belt, and to the timing belt section for belt replacement and inspection.
- Disconnecting the negative battery cable on some vehicles may interfere with the functions of the on-board computer system(s) and may require the computer to undergo a relearning process once the negative battery cable is reconnected.
- When servicing drum brakes, only disassemble and assemble one side at a time, leaving the remaining side intact for reference.

ENGINE REPAIR

Alternator

REMOVAL

2.3L Engines

1. Before servicing the vehicle, refer to the precautions in the beginning of this section.
2. Remove or disconnect the following:
   - Battery ground cable
   - Air cleaner outlet tube
   - Accessory drive belt
   - Mounting bolts and alternator
   - Nut and the electrical connectors.
3. To install, reverse the removal procedure. Torque all mounting bolts to 18 ft. lbs. (25Nm).

2.5L Engines

1. Before servicing the vehicle, refer to the precautions in the beginning of this section.
2. Remove or disconnect the following:
   - Negative battery cable
   - Drive belt
   - Electrical connections to the alternator
   - Alternator

To install:

3. Install or connect the following:
   - Alternator. Torque the bolts to 40 ft. lbs. (55 Nm).

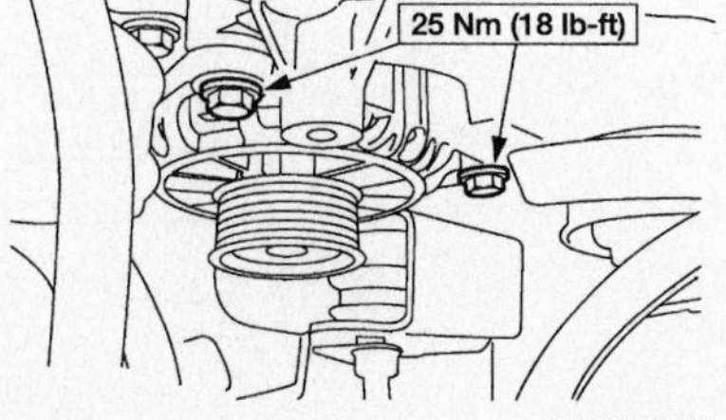

Alternator mounting bolts—2.3L engine

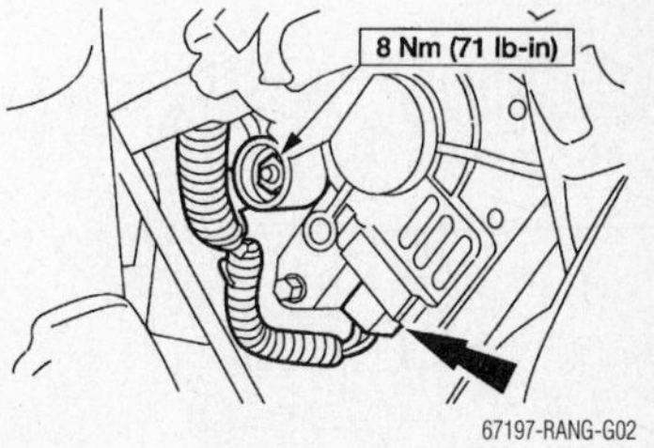

Alternator wiring connections—2.3L engine

- Electrical connectors to the alternator
- Drive belt
- Negative battery cable

3.0L Engines

1. Before servicing the vehicle, refer to the precautions in the beginning of this section.
2. Remove or disconnect the following:
   - Negative battery cable
   - Air cleaner outlet tube
   - Drive belt
   - Electrical connectors from the alternator
   - Wiring harness to alternator push pin
   - Alternator

To install:

3. Install or connect the following:
   - Alternator. Torque the bolts to 40 ft. lbs. (55 Nm) for 2001–03; 35 ft. lbs for (48 Nm) fir 2004.
   - Pushpin for the alternator wiring harness
   - Electrical connectors to the alternator
   - Drive belt
   - Air cleaner outlet tube
   - Negative battery cable

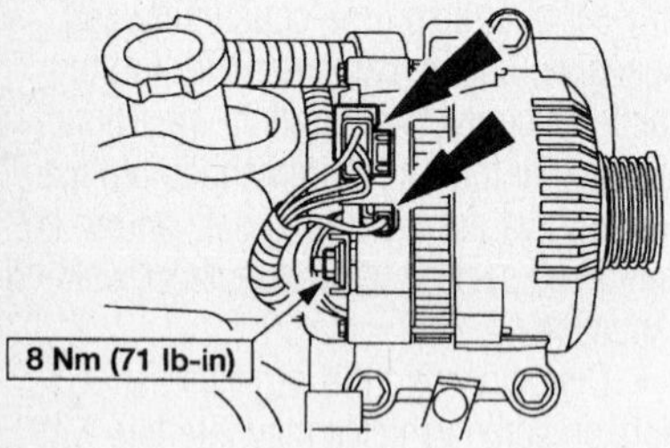

67197-RANG-G03

Alternator wiring connections—3.0L engine

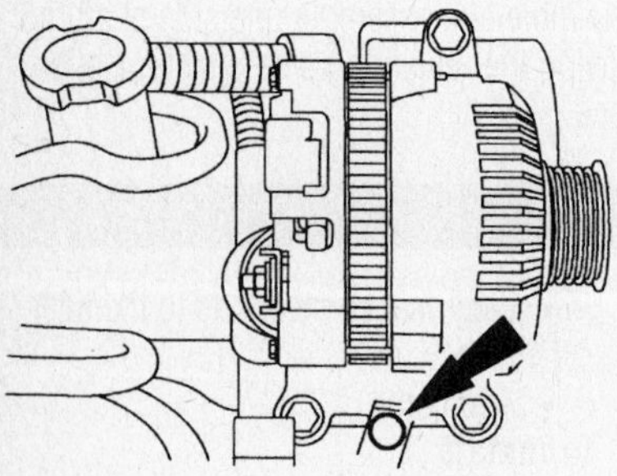

67197-RANG-G04

Wiring harness pin-type retainer—3.0L engine

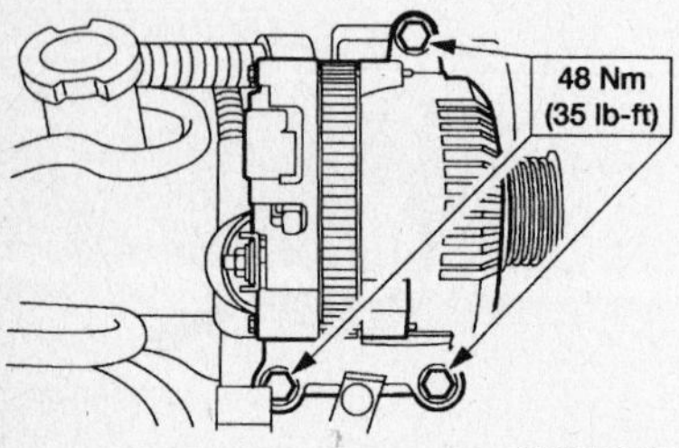

67197-RANG-G05

Alternator mounting bolts—3.0L engine

4.0L OHC Engine

1. Before servicing the vehicle, refer to the precautions in the beginning of this section.
2. Remove or disconnect the following:
   - Negative battery cable
   - Air cleaner outlet tube
   - Accessory drive belt
   - Electrical connectors
   - Wiring harness-to-generator pin-type retainer
   - Stud bolt, the bolts and the alternator
3. To install, reverse the removal procedure. Torque all mounting bolts to 35 ft. lbs. (47Nm).

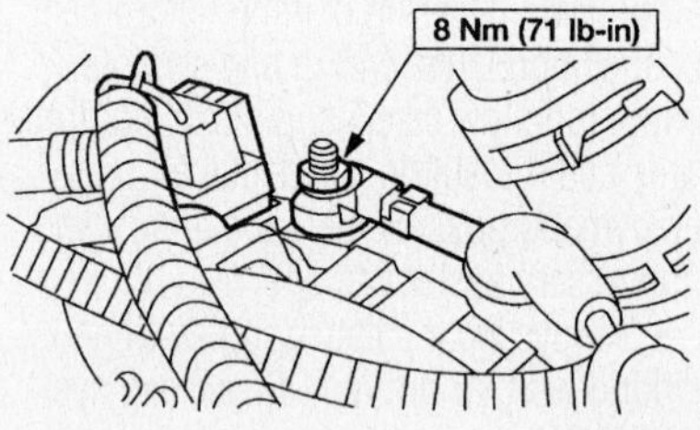

67197-RANG-G06

Alternator wiring connections—4.0L SOHC engine

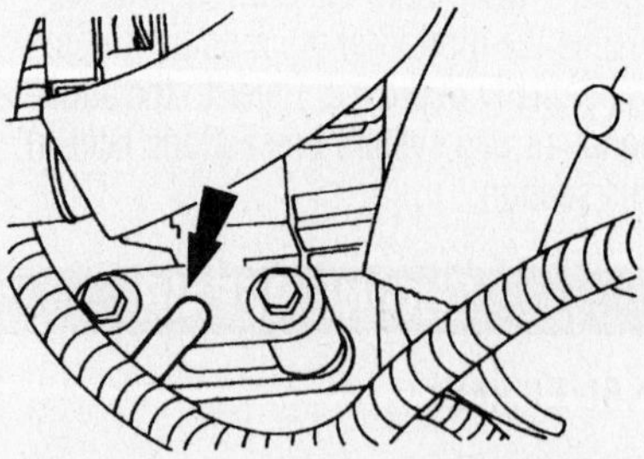

67197-RANG-G07

Alternator wiring pin-type retainer—4.0L SOHC engine

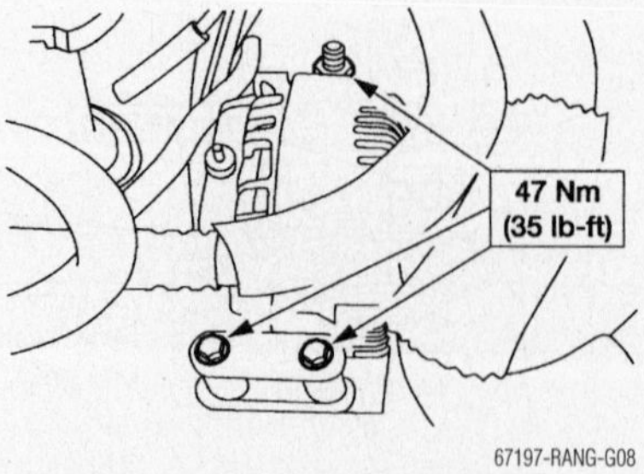

67197-RANG-G08

Alternator mounting bolts—4.0L SOHC engine

Ignition Timing

ADJUSTMENT

The ignition timing is preset and is not adjustable.

Engine Assembly

REMOVAL & INSTALLATION

2.3L Engines

1. Before servicing the vehicle, refer to the precautions in the beginning of this section.
2. Relieve the fuel system pressure.
3. Drain the cooling system.
4. Drain the engine oil.
5. Properly discharge the air conditioning system.
6. Remove or disconnect the following:
   - Hood
   - Accelerator control snow shield
   - Air cleaner tube
   - Upper radiator hose
   - Lower radiator hose
   - Fan and shroud
   - PCM electrical connector. Remove the retaining nut on the harness clamp. Position the harness on the engine.
   - Ground stud for the PCM
   - Heater hoses
   - All vacuum hoses
   - Coolant reservoir hoses
   - Air conditioning compressor clutch
   - MAF electrical connector
   - Air conditioning compressor manifold, plug the lines and the compressor ports
   - Accelerator and speed control cables
   - Power steering return hose
   - PSP switch electrical connector
   - High pressure power steering hose
   - Fuel supply hose
   - 42-pin electrical connector
   - VMV vacuum regulator solenoid supply hose
   - Evaporative purge hose
   - Brake booster vacuum hose and the engine ground strap

- Positive battery cable
- Solenoid control wire at the starter
- Starter wiring harness clamp bolt and position it out of the way.
- RH splash shield
- Alternator electrical connections
- Block heater electrical connector
- Front heated oxygen sensor electrical connector at the bell housing
- Oil pressure sensor electrical connector
- Engine wiring pushpins and position the engine wiring harnesses out of the way.
- Oil filter
- With automatic transmission, the bolt retaining the transmission cooling tubes to the engine. Remove the bracket.
- Transmission dust shield
- Starter motor
- Heated oxygen sensor electrical connector at the rear of the transmission
- Transmission wiring harness
- Vehicle speed sensor, transmission range sensor, backup light switch and the transmission electrical connectors. Disconnect the pushpins and position the harness forward to the engine.
- Oil filter adapter

➡Leave two side bolts in until the engine is ready to be removed.

- Nine of the transmission-to-engine bolts
- With automatic transmission, the transmission fluid indicator and tube assembly
- Starter dust shield

➡Mark one stud and the flexplate for assembly reference.

- With automatic transmission, the four torque converter nuts

7. Support the transmission with a floor jack.

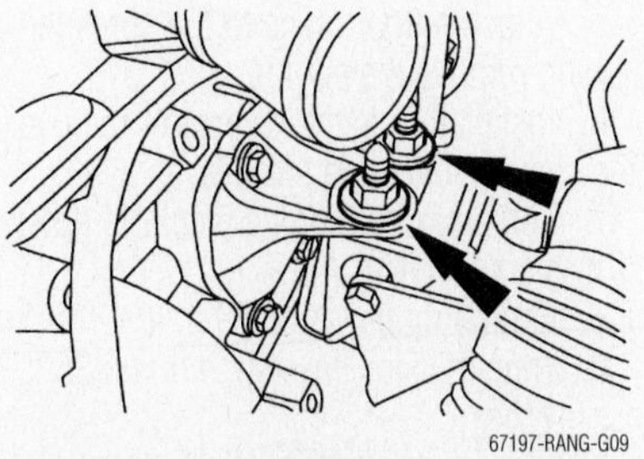

Engine support insulator nuts, left side shown, right side similar—2.3L engine

8. Support the engine with a floor crane using a spreader bar.
9. Remove the two side transmission-to-engine bolts.
10. Remove the four engine support insulator.
11. Remove the engine from the vehicle.
12. Installation is the reverse of removal. Observe the following torques:
    - Torque converter bolts: 26 ft. lbs. (35Nm)
    - Nine transmission-to-engine bolts 35 ft. lbs. (48Nm)
    - Oil filter adapter: 18 ft. lbs. (25Nm)
    - Starter: 30 ft. lbs. (40Nm)
    - Engine support nuts: 75 ft. lbs. (102Nm)

2.5L Engines

1. Before servicing the vehicle, refer to the precautions in the beginning of this section.
2. Relieve the fuel system pressure.
3. Drain the cooling system.
4. Drain the engine oil.
5. Properly discharge the air conditioning system.
6. Remove or disconnect the following:
    - Hood
    - Air cleaner outlet tube
    - Accelerator control splash shield
    - Upper radiator hose
    - Lower radiator hose
    - The two bolts and position the fan shroud on the fan
    - Water pump pulley, fan and shroud
    - Radiator overflow tube
    - Transmission cooler lines
    - Mass air flow sensor
    - Heater hoses
    - Power connection from the alternator
    - Vacuum connection at the vacuum reservoir
    - Throttle body heater hose
    - Air conditioning cycling switch
    - Connector from the PCM
    - Ground wire stud from the powertrain control module
    - Power steering cut-out switch
    - Peanut fitting from the air conditioning condenser core
    - Air conditioning high pressure cut-out switch
    - Air conditioning manifold hose
    - Accelerator cable and speed control cable
    - Brake booster vacuum hose and vacuum tube at the upper intake manifold assembly
    - EVR vacuum supply hose and the vacuum reservoir vacuum line
    - Fuel line
    - Air conditioning compressor
    - Power steering pressure and return hoses
    - Nut retaining the wiring harness
    - Block heater
    - Engine ground cable
    - With automatic transmission, the two transmission harness connectors
    - With manual transmission, the transmission harness and the heated oxygen sensor
    - Starter motor
    - Nut retaining the starter harness and transmission cooler brackets
    - Three way catalytic converter

➡The torque converter nuts are accessed through the starter motor hole.

- With automatic transmission, remove the four torque converter nuts and six bolts

7. Support the transmission with a floor jack.
8. Remove the differential pressure feedback EGR sensor.
9. Support the engine with a floor crane.
10. Remove the two upper transmission-to-engine bolts.
11. On vehicles equipped with automatic transmission, disconnect then separate the heated exhaust gas oxygen sensor connector from the bracket located on the bell housing. Remove the four nuts.

Remove the engine from the vehicle.

12. Installation is the reverse of removal. Observe the following torques:
    - Engine mount nuts: 85 ft. lbs. (115Nm)
    - Transmission-to-engine bolts: 39 ft. lbs. (51 Nm)

3.0L Engines

1. Before servicing the vehicle, refer to the precautions in the beginning of this section.
2. Relieve the fuel system pressure.
3. Drain the cooling system.
4. Drain the engine oil.
5. Properly discharge the air conditioning system.
6. Remove or disconnect the following:
    - Hood
    - Air cleaner outlet tube
    - Upper and the lower radiator hoses

✲✲ WARNING

The fan clutch has left-hand threads.

- The fan clutch and blade as an assembly
- Drive belt
- Fan shroud
- Radiator
- Air conditioning manifold and tube. Remove the nut and position the line aside.
- Air conditioning compressor wiring
- Air conditioning compressor and the air conditioning compressor mounting bracket
- Heater hoses
- Ground cable
- Fuel lines
- Snow shield
- Accelerator cable and the speed control actuator cable
- All vacuum lines
- 42-pin connector
- Powertrain control module connector
- Nut from the powertrain control module harness
- Stud bolt and the powertrain control module ground strap
- Alternator wiring and position aside
- Both heated oxygen sensors
- Transmission harness connectors
- MAF sensor
- LH heated oxygen sensor
- Dual converter Y pipe
- Starter motor and the starter grounding stud bolt
- Torque converter nuts
- 8 transmission-to-engine bolts

7. Install the lifting eyes.
8. Remove the four nuts.
9. Support the transmission.
10. Remove the engine from the vehicle.
11. Installation is the reverse of removal. Observe the following torques:

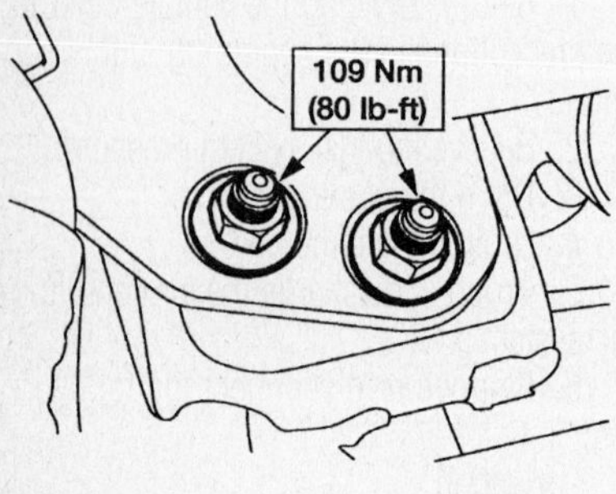

Engine support nuts, one side shown—3.0L engine

- Engine mount nuts: 80 ft. lbs. (109Nm)
- Transmission-to-engine bolts: 33 ft. lbs. (45Nm)
- Torque converter nuts: 2001–03 26 ft. lbs. (35Nm); 2004 35 ft. lbs. (47 Nm).

4.0L OHC Engines

1. Before servicing the vehicle, refer to the precautions in the beginning of this section.

✲✲ CAUTION

If the fuel supply manifold is used as a leverage device, damage may occur to the supply manifold. Care must be taken when working around the fuel supply manifold.

2. Remove or disconnect the following:

- Accelerator cable from engine
- Speed control cable from engine
- Radiator, the fan blade, and the fan shroud
- Accessory bracket bolts and position bracket aside
- Alternator wiring
- Wiring harness retainer and position generator wiring away from engine
- Engine electrical connector
- PCM connector
- PCM ground wire
- Engine ground wire
- Brake booster vacuum hose
- Air conditioning high pressure switch electrical connector
- Bolt and position the air conditioning lines aside

➡Heater hose will be removed with engine.

- Heater hoses
- Fuel line
- Starter motor
- Engine oil
- Oil drain plug
- Transmission portion of wiring harness
- RH and LH heated oxygen sensor connectors
- Transmission control connector
- Output shaft speed sensor connector
- Digital transmission range sensor connector
- Catalyst monitor sensor electrical connector

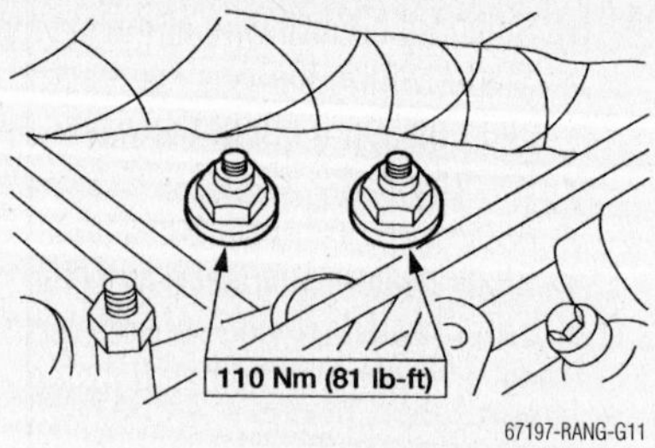

Left side engine insulator nuts—4.0L SOHC engine

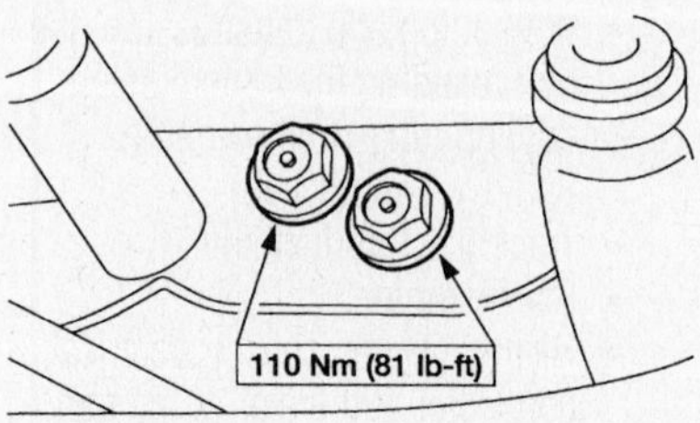

Right side engine insulator nuts—4.0L SOHC engine

- Transmission/transfer case portion of the wiring harness from any routing clips or pushpins. Route transmission/transfer case portion of the wiring harness to top of engine.
- Bolt, and position the transmission cooling line bracket aside
- Air conditioning line bracket nut and position it aside
- Power steering return hose
- Power steering pressure hose
- Vapor management valve hose connector
- Eight bolts and the LH and the RH engine support insulator nuts

➡The lifting eyes should be installed on the exhaust manifold studs for number three and number four cylinders.

3. Install the lifting eyes.
4. Install the spreader bar to the lifting eyes.
5. Attach a floor crane to the spreader bar and remove the engine.
6. Installation is the reverse of removal. Observe the following torques:
7. Left and right engine insulator nuts: 81 ft. lbs. (110 Nm)
8. Engine mount nuts: 59 ft. lbs. (80Nm)
9. Transmission-to-engine bolts: 35 ft. lbs. (47Nm)
10. Torque converter nuts: 35 ft. lbs. (47Nm)

Water Pump

REMOVAL & INSTALLATION

2.3L Engine

1. Before servicing the vehicle, refer to the precautions in the beginning of this section.
2. Drain the cooling system.
3. Remove the drive belt.
4. Remove the water pump pulley.
5. Remove the water pump.

➡Lubricate the water pump O-ring, with MERPOL®, or equivalent..

6. To install, reverse the removal procedure. Torque the water pump mount bolts to 89 inch lbs. (10Nm). Torque the pulley bolts to 18 ft. lbs. (25Nm).

2.5L Engines

1. Before servicing the vehicle, refer to the precautions in the beginning of this section.
2. Drain the cooling system.
3. Remove or disconnect the following:
   - Negative battery cable
   - Drive belt
   - Fan clutch and shroud
   - Water pump pulley
   - Heater hose from the water pump inlet tube
   - Lower radiator hose
   - Water pump inlet tube
   - Water pump and discard the gasket

To install:

4. Clean the mating surface with the water pump connects to the engine.
5. Install or connect the following:
   - Water pump with a new O-ring. Torque the bolts to 15 ft. lbs. (20 Nm).
   - Inlet tube. Torque the bolts to 89 inch lbs. (10 Nm).
   - Lower radiator hose
   - Water pump pulley
   - Heater hose from the water pump inlet tube

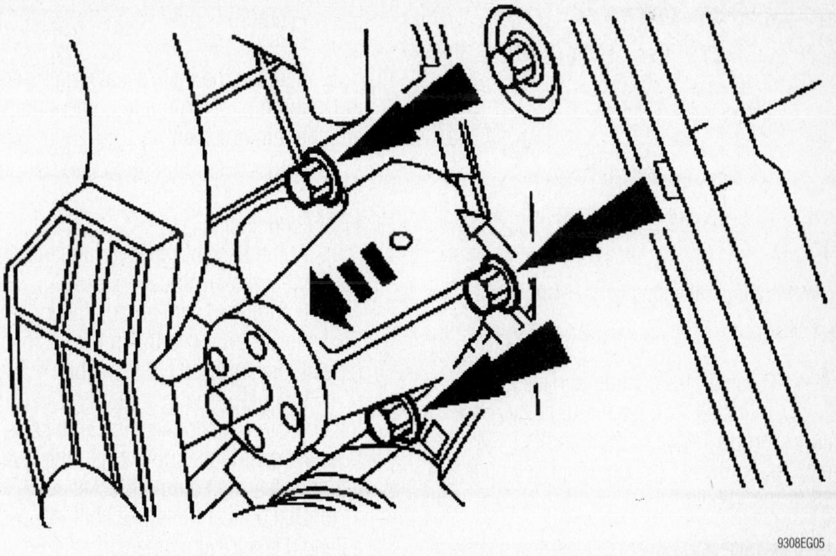

Exploded view of the water pump 2.5L engines

   - Fan clutch and shroud
   - Drive belt
   - Negative battery cable
6. Fill the cooling system.
7. Start the vehicle and check for leaks, repair if necessary.

3.0L Engines

1. Before servicing the vehicle, refer to the precautions in the beginning of this section.
2. Drain the cooling system.
3. Remove or disconnect the following:
   - Negative battery cable
   - Air cleaner outlet tube
   - Fan and radiator shroud
   - Water bypass tube
   - Drive belt
   - Heater hose
   - Water pump pulley
   - Lower radiator hose
   - Air conditioning compressor and bracket assembly and move them aside
   - Water pump

To install:

4. Clean the mating surfaces where the water pump attaches to the engine.
5. Install or connect the following:
   - Water pump. Torque the bolts to 89 inch lbs. (10 Nm).
   - Air conditioning compressor mounting bracket. Torque the bolts to 44 ft. lbs. (61 Nm).
   - Water pump pulley. Torque the bolts to 20 ft. lbs. (28 Nm).
   - Drive belt
   - Heater hose
   - Lower radiator hose
   - Fan and shroud
   - Air cleaner outlet tube
   - Negative battery cable
6. Fill the cooling system.
7. Start the vehicle and check for leaks, repair if necessary.

4.0L OHC Engine

1. Before servicing the vehicle, refer to the precautions in the beginning of this section.
2. Drain the cooling system.
3. Remove or disconnect the following:
   - Fan shroud
   - Accessory drive belt
   - Idler pulley
   - Water bypass hose
   - Heater hose
   - Lower radiator hose
   - Water pump pulley
   - Water pump

⁂ WARNING

Use care when scraping the water pump-to-engine block mating surfaces. Gouges in the aluminum could form leak paths.

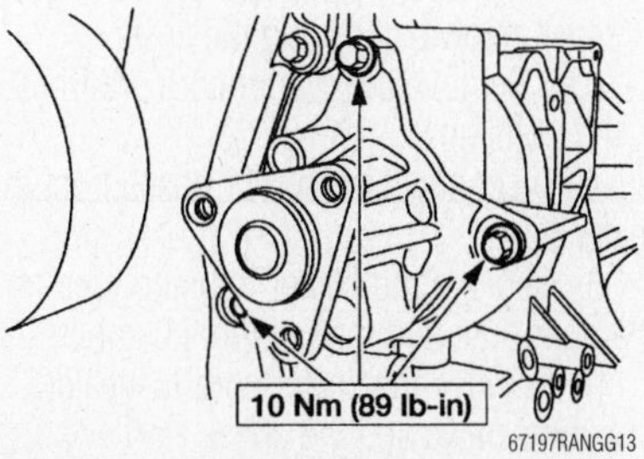

Water pump mounting bolts—2.3L engine

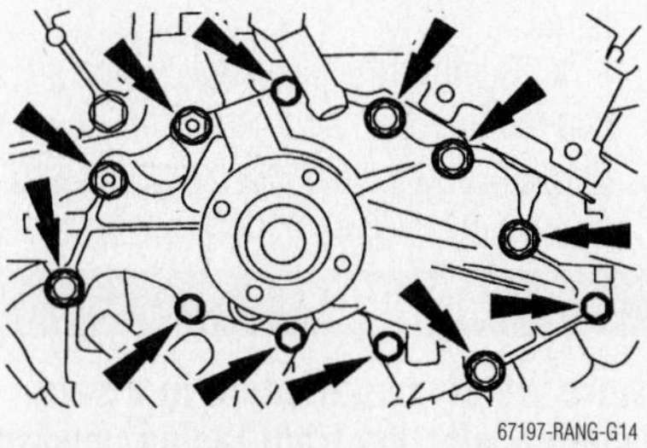

Water pump mounting bolts—3.0L engine

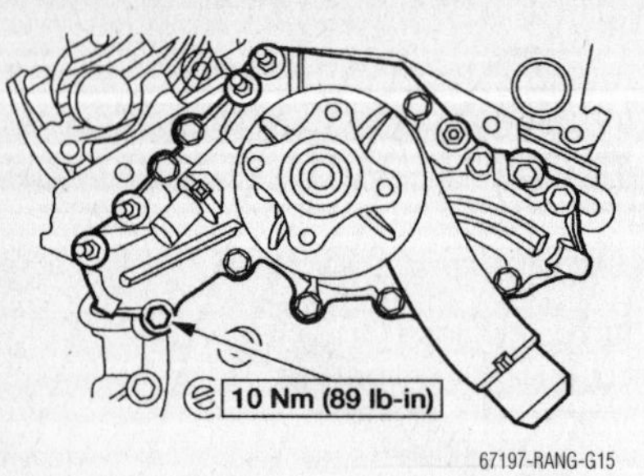

Water pump mounting bolts—4.0L SOHC engine

4. Clean all the sealing surfaces.
5. To install, reverse the removal procedure. Torque the water pump bolts to 89 inch lbs. (10Nm). Torque the pulley bolts to 18 ft. lbs. (25Nm).

Heater Core

REMOVAL & INSTALLATION

2001–03

1. Before servicing the vehicle, refer to the precautions in the beginning of this section.
2. Disconnect the negative battery cable.

**** CAUTION**

After disconnecting the negative battery cable, wait for 1 minute for the SRS module to deplete its energy.

3. Drain the cooling system into a clean container for reuse.
4. Remove the steering column by performing the following procedure:
   a. Position the front wheels in the straight-ahead direction.
   b. At the both sides of the steering wheel, remove the cover plugs, the steering wheel-to-air bag module screws, disconnect the air bag electrical connector and carefully remove the air bag module.

**** CAUTION**

Safely store the air bag module with the front side facing upward.

   c. Remove the steering wheel-to-steering column nut.
   d. Using a steering wheel puller, press the steering wheel from the steering column.
   e. Remove the parking brake release handle screws and move the release handle aside.
   f. Remove the hood release screws and move the hood release aside.
   g. Remove the 2 instrument panel-to-steering column cover screws and the cover.
   h. Remove the instrument panel steering column opening reinforcement bolts and the reinforcement.
   i. Remove the ignition switch bolt and disconnect the ignition switch electrical connector.
   j. At the base of the steering column, disconnect the electrical connectors.
   k. If equipped with an automatic transmission, remove the transmission range indicator bolt and the cable.
   l. If equipped with an automatic transmission, disconnect the shift cable from the steering column shift tube lever and the steering column bracket.
   m. Disconnect the brake shift interlock solenoid electrical connector.
   n. Remove the air bag sliding contact.
   o. Remove the upper intermediate steering shaft-to-column shaft bolt and discard the bolt.
   p. Remove the lower steering column-to-instrument panel nuts and the steering column.
5. Remove the instrument panel by performing the following procedure:
   a. Remove the parking brake release handle screws and move the release handle aside.
   b. Disconnect the Brake Pedal Position (BPP) switch electrical connector.
   c. Remove both front door scuff plates.
   d. Remove the push pins and remove both cowl side trim panels.
   e. At the right side cowl panel, disconnect the electrical connectors and ground wires.
   f. Remove both sides windshield garnish moldings.
   g. Remove the instrument panel fuse door.
   h. Disconnect the power distribution box from its bracket and move it aside.
   i. In the engine compartment, loosen the bulkhead wiring harness bolts and disconnect the electrical connectors.
   j. Pull the bulkhead electrical connector handle and disconnect the wiring harness.
   k. Remove the passenger's side air bag module-to-instrument panel screws, disconnect the electrical connector and remove the air bag module.

**** WARNING**

Store the air bag module in a safe location with the front facing upward.

   l. Disconnect the blend door actuator's electrical connector.
   m. Disconnect the climate control vacuum harness connector.
   n. Disconnect the radio's antenna connector.
   o. Remove the glove compartment.
   p. Remove the instrument panel defroster grille.
   q. Remove the upper instrument panel bolts.
   r. Under the steering column, remove the instrument panel brace bolt.
   s. Remove both the right and left instrument panel-to-cowl bolts.
   t. Pull the instrument panel away from the dash.
   u. Loosen the instrument panel-to-body harness bolt and disconnect the harness.
   v. Using an assistant, remove the instrument panel.
6. Remove the evaporator core by performing the following procedure:
   a. Discharge and recover the air conditioning system refrigerant.
   b. Remove the refrigerant lines from the evaporator core. Discard the O-rings.
   c. If equipped, remove the air conditioning vacuum reservoir tank/bracket screws and reposition the tank.
   d. If equipped, disconnect the speed control servo connector; then, remove the bolt and reposition the speed control servo.
   e. If equipped with a 3.0L or 4.0L engine, remove the support bracket.
   f. Disengage the windshield washer hose retainer and move it aside.
   g. Disconnect the vacuum hose and the retainer; then, move the hose aside.
   h. Remove the passenger's compartment nut.
   i. At the back of the engine, remove the hose support bolts.
   j. Remove the evaporator housing-to-chassis nuts.
   k. Remove the air conditioning accumulator bracket screws.
   l. Remove the evaporator housing cover screws, clips and the cover.
   m. Remove the evaporator core from the housing.
7. Disconnect the heater hoses from the heater core.
8. Remove the heater housing plenum chamber nuts and the plenum chamber.
9. Remove the heater core-to-heater housing screws and the cover.
10. Remove the heater core.

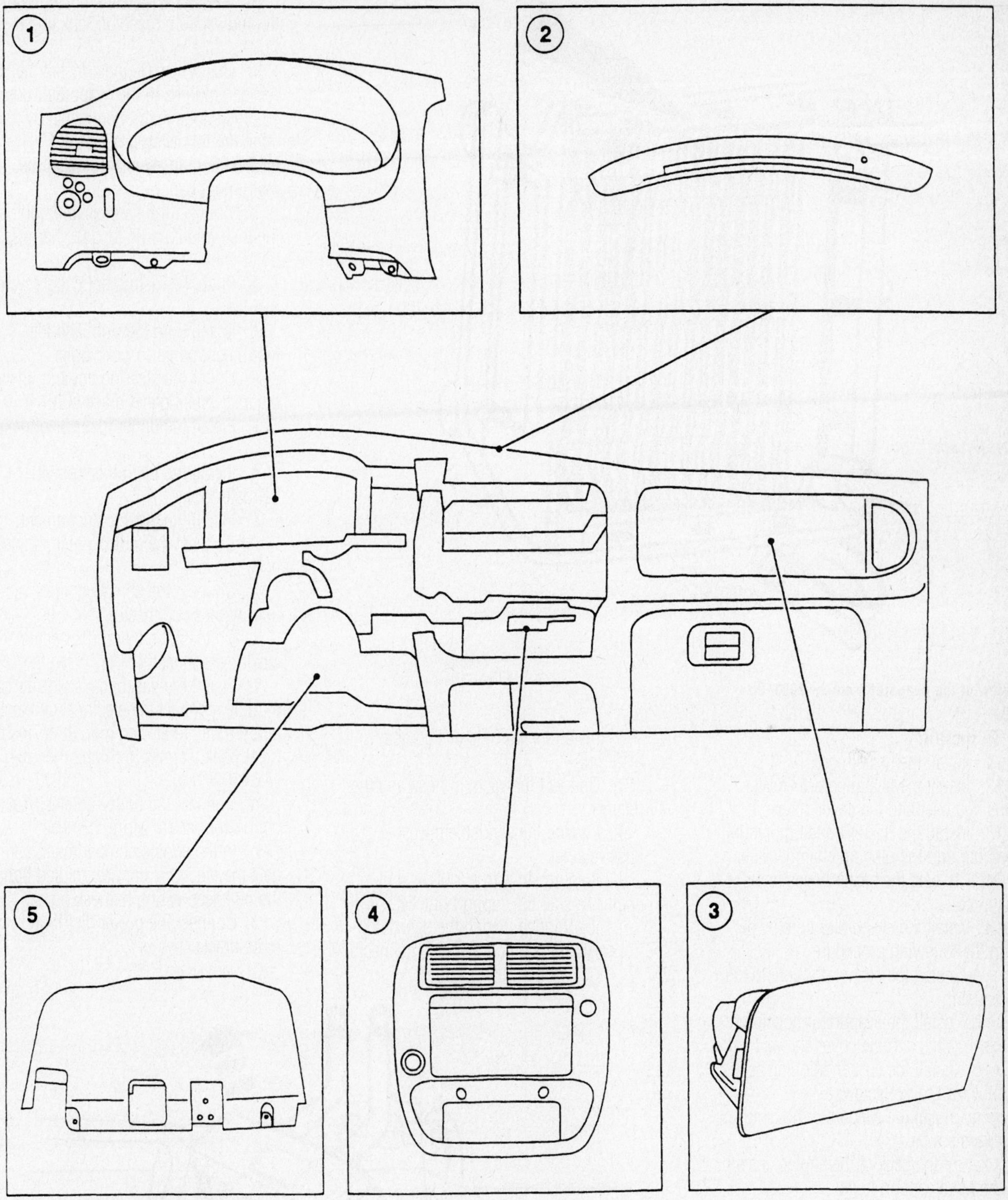

1 Instrument Panel Finish Panel
2 Instrument Panel Defroster Opening Grille Assembly
3 Passenger Side Air Bag Module
4 Instrument Panel Center Finish Panel
5 Instrument Panel Steering Column Cover

93113GL3

View of the instrument panel components—2001–03

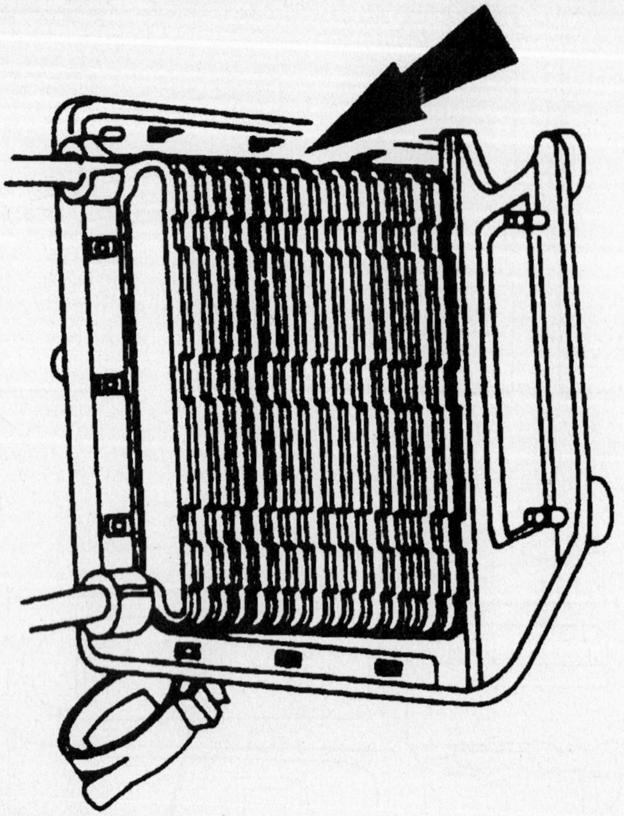

93113GL4

View of the evaporator core—2001–03

To install:

11. Install the heater core.
12. Install the heater core-to-heater housing cover and the cover screws.
13. Install the heater housing plenum chamber and the plenum chamber nuts.
14. Connect the heater hoses to the heater core.
15. Install the evaporator core by perform the following procedure:
    a. Install the evaporator core to the housing.
    b. Install the evaporator housing cover, clips and the cover screws.
    c. Install the air conditioning accumulator bracket screws.
    d. Install the evaporator housing-to-chassis nuts.
    e. At the back of the engine, install the hose support bolts.
    f. Install the passenger's compartment nut.
    g. Connect the vacuum hose and the retainer.
    h. Engage the windshield washer hose retainer.
    i. If equipped with a 3.0L or 4.0L engine, install the support bracket.
    j. If equipped, install the speed control servo bolt and connect the connector.
    k. If equipped, install the air conditioning vacuum reservoir tank and the bracket screws.
    l. Using new O-rings, install the refrigerant lines to the evaporator core.
16. Install the instrument panel by performing the following procedure:
    a. Using an assistant, install the instrument panel.
    b. Connect the harness and tighten the instrument panel-to-body harness bolt.
    c. Push the instrument panel toward the dash.
    d. Install both the right and left instrument panel-to-cowl bolts.
    e. Under the steering column, install the instrument panel brace bolt.
    f. Install the upper instrument panel bolts.
    g. Install the instrument panel defroster grille.
    h. Install the glove compartment.
    i. Connect the radio's antenna connector.
    j. Connect the climate control vacuum harness connector.
    k. Connect the blend door actuator's electrical connector.
    l. Install the passenger's side air bag module, connect the electrical connector and torque the air bag module-to-instrument panel screws to 67–92 inch lbs. (7.6–10.4 Nm).
    m. Connect the bulkhead electrical connector handle wiring harness.
    n. In the engine compartment, connect the electrical connectors and tighten the bulkhead wiring harness bolts.
    o. Connect the power distribution box to its bracket.

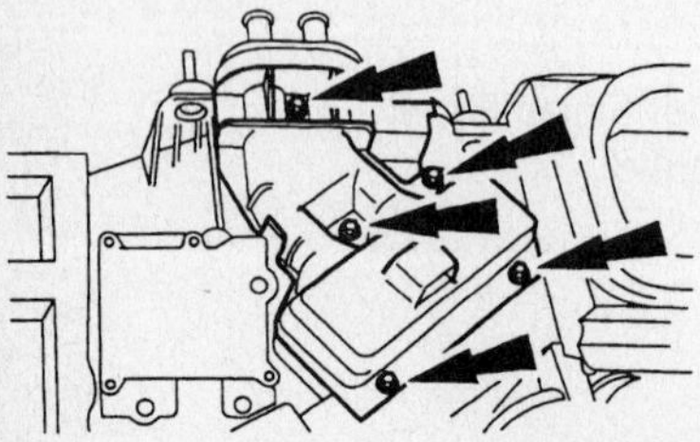

93113GL6

View of the heater core cover—2001–03

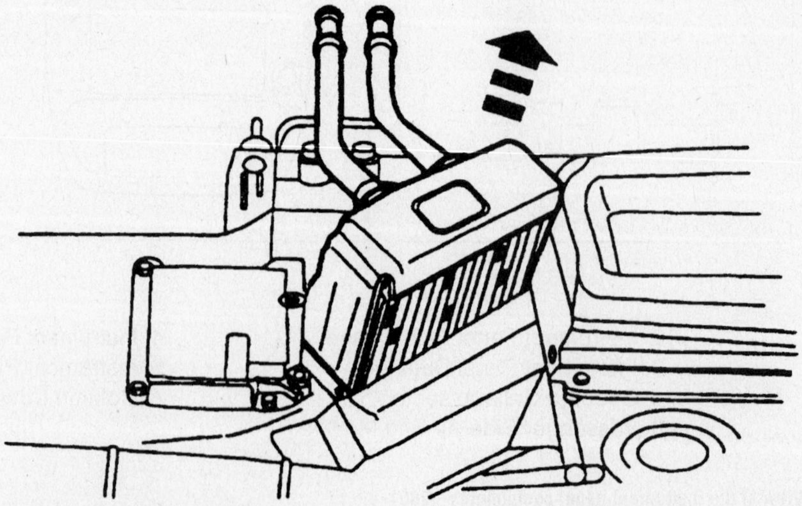

93113GL5

View of the heater core—2001–03

p. Install the instrument panel fuse door.

q. Install both sides windshield garnish moldings.

r. At the right side cowl panel, connect the electrical connectors and ground wires.

s. Install both cowl side trim panels and the push pins.

t. Install both front door scuff plates.

u. Connect the brake pedal position (BPP) switch electrical connector.

v. Install the parking brake release handle and the release handle aside screws.

17. Install the steering column by performing the following procedure:

18. Install the steering column by performing the following procedure:

a. Install the lower steering column and the steering column-to-instrument panel nuts; then, torque the nuts to 10–13 ft. lbs. (13–17 Nm).

b. Using a new bolt, install the upper intermediate steering shaft-to-column shaft bolt and torque to 19–25 ft. lbs. (26–34 Nm).

c. Install the air bag sliding contact.

d. Connect the brake shift interlock solenoid electrical connector.

e. If equipped with an automatic transmission, connect the shift cable from the steering column shift tube lever and the steering column bracket.

f. If equipped with an automatic transmission, install the transmission range indicator cable and bolt.

g. At the base of the steering column, connect the electrical connectors.

h. Connect the ignition switch electrical connector and install the ignition switch bolt.

i. Install the instrument panel steering column opening reinforcement and the reinforcement bolts.

j. Install the instrument panel-to-steering column cover and the 2 cover screws.

k. Install the hood release and the hood release screws.

l. Install the parking brake release handle and the release handle screws.

m. Install the steering wheel to the steering column.

n. Install the steering wheel-to-steering column nut and torque the nut to 25–34 ft. lbs. (34–46 Nm).

o. At the both sides of the steering wheel, install the air bag module, connect the air bag electrical connector, install the steering wheel-to-air bag module screws and the cover plugs.

19. Refill the cooling system.

20. Connect the negative battery cable.

21. Evacuate and charge the air conditioning system.

22. Run the engine to normal operating temperatures; then, check the climate control operation and check for leaks.

2004

1. Before servicing the vehicle, refer to the precautions in the beginning of this section.
2. Recover the refrigerant.
3. Remove the suction accumulator.
4. On trucks with the 2.3L engine:
   a. Remove the A/C compressor.
   b. Remove the engine oil indicator and tube.
5. On vehicles with the 3.0L or 4.0L engine Position the coolant reservoir and windshield washer reservoir aside.
6. Detach the speed control servo.
7. Disconnect the blower motor and blower motor resistor electrical connectors.
8. Disconnect the heater hoses. Using suitable tools, clamp-off the heater
9. Detach the pin-type retainer and position aside the windshield washer hose.
10. Disconnect and detach the heater control valve vacuum hose.
11. Disconnect the vacuum supply hose near the evaporator core housing.
12. Disconnect the condenser to evaporator line spring lock coupling from the passenger compartment evaporator core inlet.
13. Disconnect the vacuum hose connector and remove the nut.
14. Remove the nuts and the evaporator core housing.
15. Remove the driver side and passenger side air bag modules.
16. Remove the front seats.
17. Remove the screws and position the parking brake release handle aside.
18. Remove the left and right lower cowl kick panels.
19. Position the parking brake assembly aside.
20. Remove the screws and position the hood release handle aside.
21. Remove the instrument panel steering column cover.
22. Remove the instrument panel steering column opening cover reinforcement.
23. Disconnect the brake pedal position (BPP) switch electrical connector from the steering column shaft.
24. If equipped, disconnect the clutch pedal position (CPP) switch electrical connector.
25. If equipped, disconnect the shift cable from the steering column.
26. Disconnect the upper intermediate steering shaft.
27. Remove the left and right side garnish moldings.
    a. Remove the bolt covers and remove the bolts.
    b. Remove the assist handle.
    c. Remove the windshield side garnish molding.
28. Remove the door moldings.
29. On regular cab vehicles
    a. Remove the screws.
    b. Remove the scuff plate.
30. Disconnect the electrical connectors and the ground wire from the RH side lower kick panel.

➡To avoid damaging the bulkhead electrical connectors, be sure the release tab is fully depressed before pulling release lever into the disconnect position.

31. Disconnect the LH side bulkhead electrical connector. Press the release tab and pull the release lever.
32. Remove the audio unit. Insert the removal tool. Remove and support the audio unit.
33. Disconnect the audio unit electrical connector and antenna cable.
34. Lower the glove compartment. Press the release tabs inward while lowering the compartment.
35. Through the glove compartment opening, disconnect the blend door actuator electrical connector.
36. Through the glove compartment opening, disconnect the climate control vacuum harness connector.
37. Raise and secure the glove compartment. Press the release tabs inward while raising the glove compartment.
38. Remove the instrument panel defroster opening grille.
39. Remove the instrument panel cowl top bolts.
40. If equipped, remove the floor console.
41. If not equipped with the high-series floor console, remove the cup holders.
42. Release the clips and remove the restraints control module (RCM) cover.
43. If not equipped with high-series floor console, remove the consolette mat.
44. Remove the screws and remove the restraints control module (RCM) cover.
45. Remove the consolette base.
46. Remove the gearshift lever.
47. Remove the screws and remove the manual transmission consolette.

48. Disconnect the RCM electrical connector.
49. Pull the floor carpeting back.
50. On 4x2 vehicles disconnect the instrument panel main harness.
51. On 4x4 vehicles disconnect the instrument panel main harness. From underneath the vehicle, release the instrument panel main harness at the transfer case.
52. Remove the instrument panel side finish panel.
53. Disconnect the door harness electrical connector.
54. Remove the RH side instrument panel bolt. If necessary, transfer the components to the new instrument panel.
55. Remove the LH instrument panel cowl side bolts.
56. Remove the instrument panel.
57. Remove the powertrain control module (PCM).
58. Remove the PCM heat sink.
59. Remove the four nuts from the engine side of the dash panel. Position the plenum chamber on the vehicle floor.
60. Remove the heater core cover.
61. Remove the heater core.
62. To install, reverse the removal procedure. During installation, be sure to install a new oval foam seal around the heater core inlet and outlet tubes.
63. Lubricate the refrigerant system with the correct amount of clean PAG oil or equivalent. Install new O-ring seals lubricated in clean mineral oil. Lubricate the coolant hoses with plain water only, if needed.
64. Evacuate, leak test, and charge the refrigerant system.

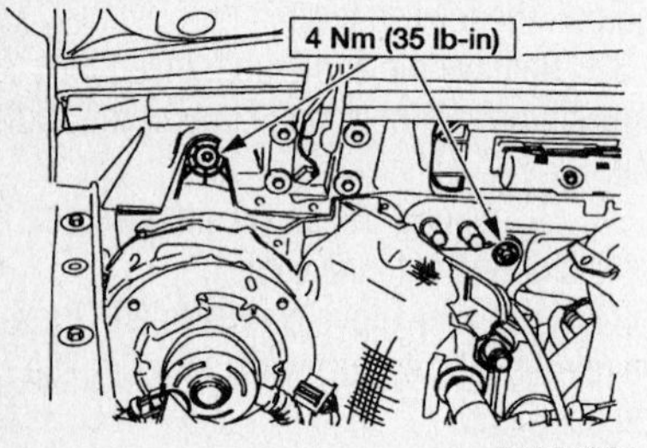

2004 evaporator core housing bolts

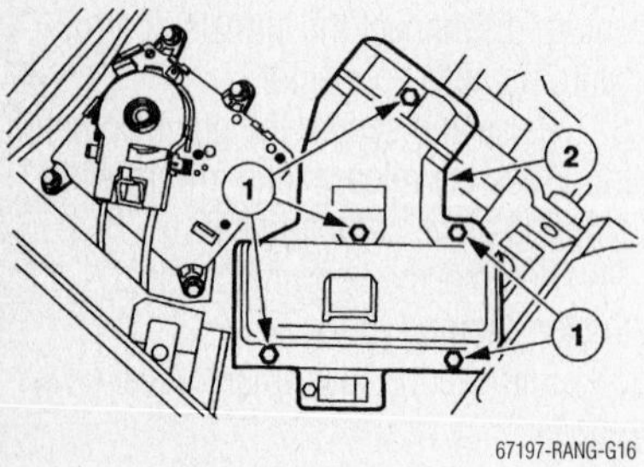

2004 heater core cover bolts

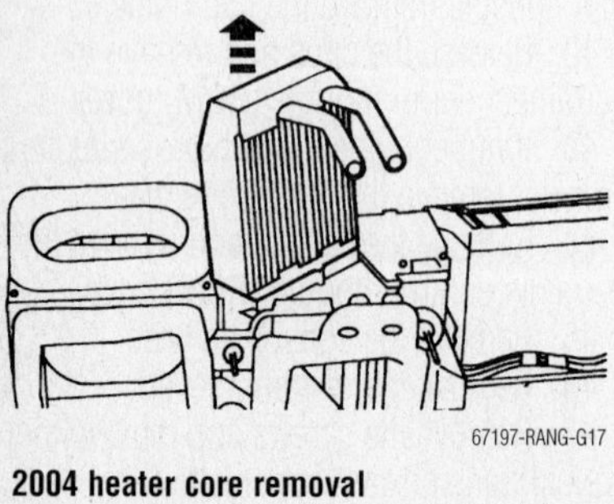

2004 heater core removal

Cylinder Head

REMOVAL & INSTALLATION

2.3L Engines

1. Before servicing the vehicle, refer to the precautions in the beginning of this section.
2. Relieve the fuel system pressure.
3. Drain the cooling system.
4. Properly discharge the air conditioning system.
5. Remove or disconnect the following:
   - Negative battery cable
   - Drive belt.
   - Engine oil level indicator assembly.
   - Engine oil level indicator.
   - Engine oil level indicator tube.
   - Water outlet tube.
   - Water outlet tube.
   - Air conditioning compressor.

➡The generator will be removed with the accessory bracket.

   - Accessory bracket.
   - Right motor mount.
   - Coolant hose from the thermostat.
   - Coolant hose from the EGR valve.
   - Coolant tube assembly.
   - Exhaust manifold and gasket.
   - Block heater (if so equipped).
   - Water outlet.
   - EGR valve.
   - Power steering pump and reservoir as an assembly.
   - Idle air control (IAC) valve.
   - Throttle position (TP) sensor.
   - Manifold absolute pressure (MAP) sensor.
   - Swirl control valve monitor electrical connector.
   - CKP sensor and the wiring harness pin-type retainers.
   - Knock sensor (KS).
   - Electric thermostat.
   - Swirl control valve.
   - CMP sensor electrical connector and disconnect the PCV hose from the intake manifold.
   - Engine wiring harness pin-type retainers from the intake manifold.
   - Engine wiring harness connector bracket. Position the engine wiring harness aside.
   - EGR tube.
   - Fuel supply line clip from the front of the intake manifold. Disconnect the vacuum hose from the intake manifold.
   - Intake manifold assembly.
   - Fuel injector electrical connectors. Detach the wiring harness pin-type retainers.
   - Ignition coil and the cylinder head temperature (CHT) sensor electrical connectors.
   - Engine wiring harness anchors from the valve cover studs. Remove the engine wiring harness.
   - Ignition coil.
   - Bypass hose.
   - Thermostat housing.
   - Knock sensor and the engine vent cover.
   - Left motor mount.
   - Fuel injector supply manifold with the injectors and the ground strap.
   - Water pump pulley.
   - Water pump.
   - CMP sensor.
   - CHT sensor.
   - Spark plugs.
   - Valve cover.
   - CKP sensor.
   - Crankshaft vibration damper

➡There is one front cover bolt behind the cooling fan drive pulley. To remove this bolt, align one of the cooling fan drive pulley access holes with the bolt head to access the bolt.

   - Front cover.
   - Timing chain tensioner.
   - Timing chain guides.
   - Timing chain assembly.

➡Use a wrench on the flats between cylinders No. 1 and No. 2 to hold the camshaft in place.

   - Camshaft drive sprockets.
   - Oil pump chain tensioner and guide.

➡The oil pump chain sprocket must be held in place.

   - Oil pump chain and sprockets.

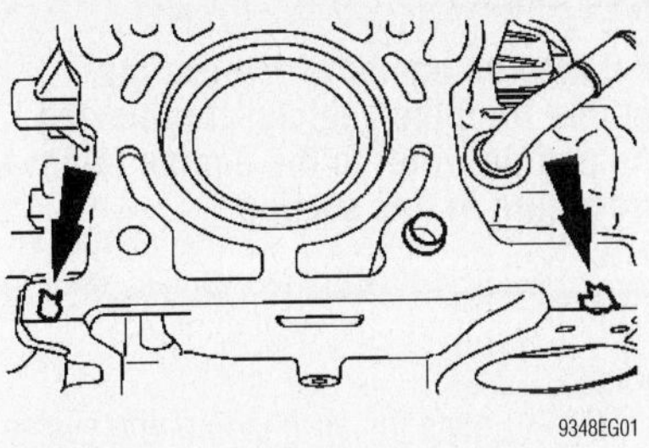

RTV sealer application—2.3L cylinder head

➡**Note the position of the lobes on the No. 1 cylinder before removing the camshafts for assembly reference.**

6. Loosen the camshaft bearing cap bolts in sequence, one turn at a time. Repeat the first step until all tension is released from the camshaft bearing caps. Remove the camshaft bearing caps.

7. Remove or disconnect the following:
   - Camshafts.
   - Cylinder head bolts and the cylinder head.
   - Cylinder head gasket.

8. Installation is the reverse of removal. Apply RTV sealer to the places shown. The head must be installed within 4 minutes of application. Observe the following torques:

   a. Cylinder head:
   - Step 1: Tighten the bolts to 5 Nm (44 inch lbs.)
   - Step 2: Tighten the bolts to 15 Nm (11 ft. lbs.)
   - Step 3: tighten the bolts to 45 Nm (33 ft. lbs.)
   - Step 4: Tighten the bolts an additional 90 degrees (1/4 turn)
   - Step 5: Tighten the bolts an additional 90 degrees (1/4 turn)

   b. Camshafts:

➡**Install the camshafts with the alignment notches in the camshaft lined up so the camshaft alignment plate can be installed without rotating the camshafts. Make sure the lobes on the No. 1 cylinder are in the same position as noted in the disassembly procedure. Rotating the camshafts, or installing the camshafts 180 degrees out of position can cause severe damage to the valves and pistons. Lubricate the camshaft journals and bearing caps with clean engine oil. Install the camshafts and bearing caps. Tighten the bolts in the sequence shown in three stages.**

   - Step 1: Tighten the camshaft bearing caps one turn at a time until tight.
   - Step 2: Tighten the bolts to 7 Nm (62 inch lbs.)
   - Step 3: Tighten the bolts to 16 Nm (12 ft. lbs.)

   c. Crankshaft vibration damper:

➡**Do not reuse the crankshaft pulley bolt. Tighten the bolt in two stages.**

   - Step 1: Tighten the bolt to 40 Nm (30 ft. lbs.)
   - Step 2: Tighten the bolt and additional 90 degrees (1/4 turn).

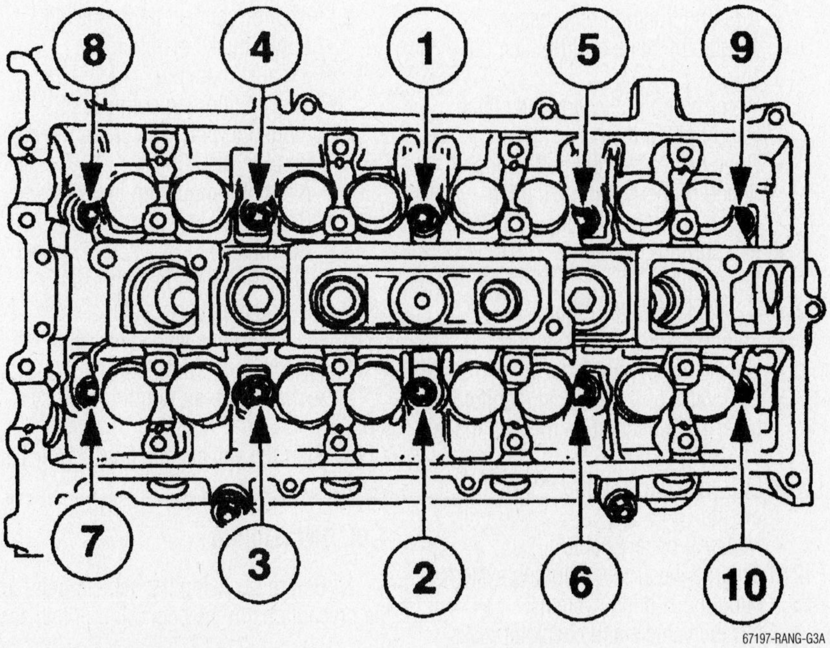

Head bolt torque sequence—2.3L

2.5L Engines

1. Before servicing the vehicle, refer to the precautions in the beginning of this section.
2. Relieve the fuel system pressure.
3. Drain the cooling system.
4. Properly discharge the air conditioning system.
5. Remove or disconnect the following:
   - Negative battery cable
   - Loosen the water pump pulley bolts
   - Drive belt
   - Water pump pulley
   - Fan and clutch assembly
   - Intake manifolds
   - Ignition wires from the spark plugs
   - Spark plugs
   - Oil level indicator tube
   - Exhaust Gas Recirculation (EGR) valve to the exhaust manifold tube
   - Valve cover
   - Engine control wiring from the air conditioning compressor
   - Air conditioning compressor mounting bracket with the power steering pump attached and move them aside
   - Engine control sensor wiring from the alternator
   - Lower radiator hose
   - Water pump inlet tube
   - Upper radiator hose
   - Alternator
   - Alternator mounting bracket
   - Ignition wire and bracket
   - Outer timing belt cover
   - Timing belt
   - Exhaust manifold
   - Cylinder head and discard the bolts and the gasket

To install:

6. Clean the mating surface where the cylinder head attaches to the engine.
7. Install a new gasket and the cylinder head.
8. Torque the new cylinder head bolts in stages as follows:
   a. Step 1: 51 ft. lbs. (70 Nm).
   b. Step 2: An additional 51 ft. lbs. (70 Nm).
   c. Step 3: Plus and additional 90 degrees.
9. Install or connect the following:
   - Exhaust manifold. Torque the bolts to 15 ft. lbs. (20 Nm) plus an additional 30 ft. lbs. (40 Nm).
   - Timing belt tensioner and timing belt

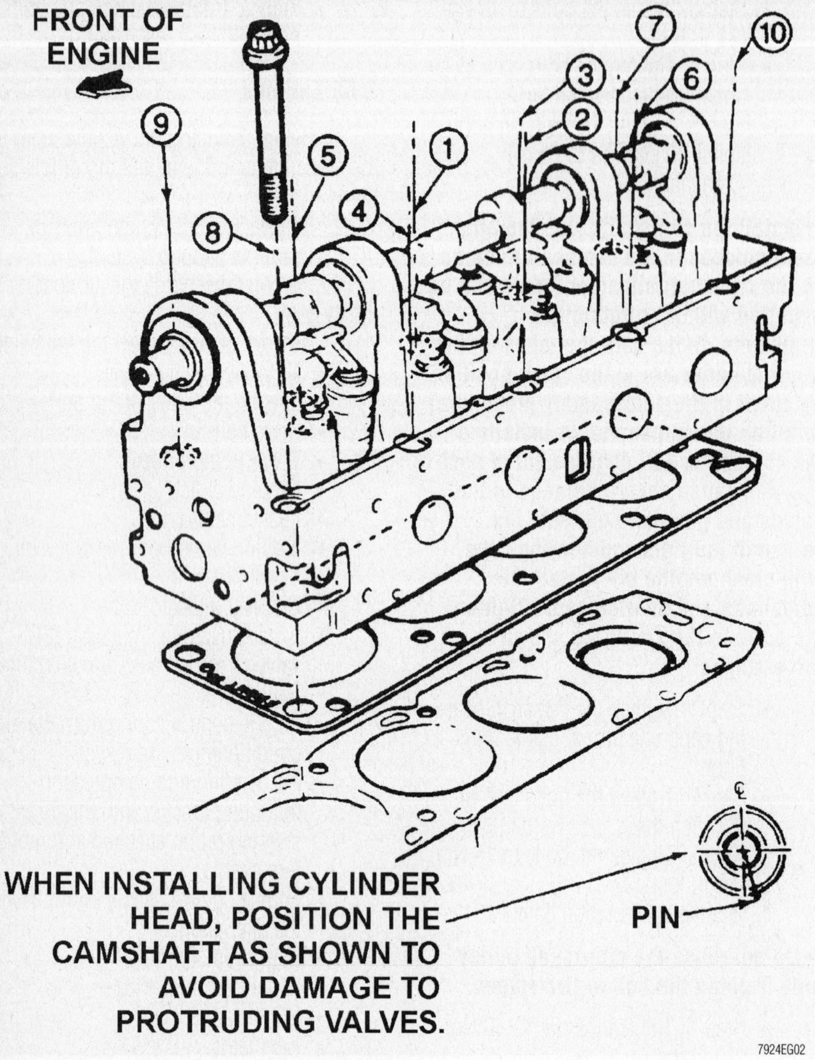
FRONT OF ENGINE
WHEN INSTALLING CYLINDER HEAD, POSITION THE CAMSHAFT AS SHOWN TO AVOID DAMAGE TO PROTRUDING VALVES.
PIN
7924EG02

Cylinder head bolt torque sequence—2.5L engines

- Timing belt cover
- Ignition wires and coil

10. Install the alternator bracket. Torque the bolts in 4 stages as follows:
    a. Step 1: Hand tighten bolt No. 1.
    b. Step 2: Torque bolt No. 2 to 40 ft. lbs. (55 Nm).
    c. Step 3: Torque bolt No. 3 to 40 ft. lbs. (55 Nm).
    d. Step 4: Torque bolt No. 1 to 40 ft. lbs. (55 Nm).
11. Install or connect the following:
    - Water pump inlet tube with a new O-ring to the water pump. Torque the bolts to 89 inch lbs. (10 Nm).
    - Lower radiator hose
    - Alternator
    - Upper radiator hose and heater hose
    - Air conditioning compressor mounting bracket with the power steering pump attached. Torque the bolts to 40 ft. lbs. (55 Nm).
    - Air conditioning compressor. Torque the bolts to 20 ft. lbs. (28 Nm).
    - Water pump pulley and fan clutch. Hand tighten the bolts
    - Drive belt. When the belt is positioned properly, torque the fan clutch bolts to 16 ft. lbs. (23 Nm).
    - Fan shroud
    - Sparks plugs
    - Oil level indicator tube
    - Engine control sensor wiring
    - Upper intake manifold
    - EGR valve to the exhaust manifold tube. Torque the bolts to 34 ft. lbs. (47 Nm).
    - EGR transducer to the rear of the engine
    - Negative battery cable
12. Recharge the air conditioning system.
13. Filling the cooling system.
14. Start the vehicle and check for leaks, repair if necessary.

3.0L Engine

➡It may be easier to remove the engine from the vehicle. If removing the engine, refer to the engine removal procedure in this section.

1. Before servicing the vehicle, refer to the precautions in the beginning of this section.
2. Evacuate the air conditioning system.
3. Drain the cooling system.
4. Drain the engine oil.
5. Remove or disconnect the following:
    - Negative battery cable
    - Lower intake manifold
    - Air conditioning compressor
    - Alternator
    - Power steering pump
    - Alternator mounting bracket
    - Air conditioning compressor mounting bracket
    - Exhaust manifolds
    - Cylinder head and discard the bolts and gasket

To install:

➡The "V" in the cylinder head gasket must face the front of the engine.

6. Clean the mating surfaces where the head attaches to the engine.
7. Install a new cylinder head gasket and the cylinder head to the engine.
8. Torque the new cylinder head bolts in stages as follows:
    a. Step 1: 59 ft. lbs. (80 Nm).
    b. Step 2:Loosen the bolts one full turn.
    c. Step 3:40 ft. lbs. (55 Nm).
    d. Step 4: 63 ft. lbs. (85 Nm).
9. Install or connect the following:
    - Lower intake manifold
    - Exhaust manifold
    - Air conditioning compressor mounting bracket. Torque the bolts to 44 ft. lbs. (66 Nm).
    - Alternator mounting bracket
    - Power steering pump
    - Alternator
    - Air conditioning compressor
    - Negative battery cable
10. Fill the engine with clean oil
11. Fill the cooling system.
12. recharge the air conditioning system
13. Start the vehicle and check for leaks, repair if necessary.

4.0L OHC Engine

1. Before servicing the vehicle, refer to the precautions in the beginning of this section.

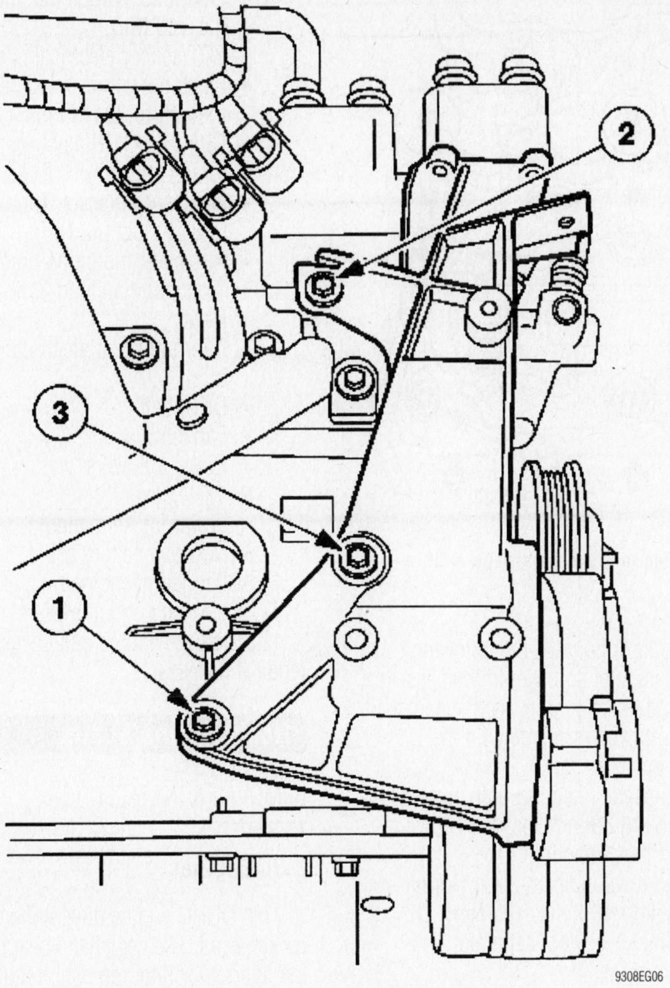

Alternator bracket bolt tightening sequence 2.5L engines

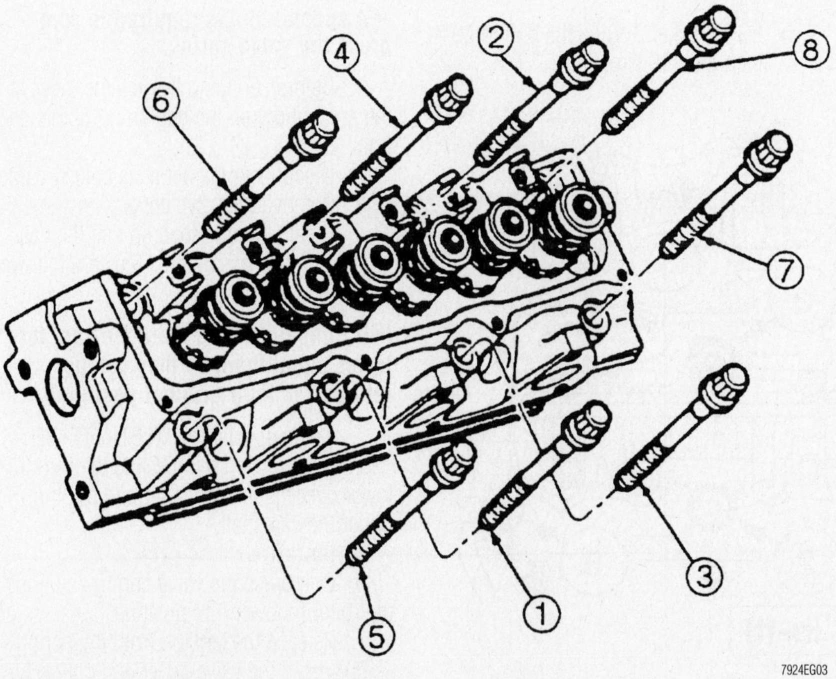

Cylinder head bolt torque sequence 3.0L engine

➡If only one cylinder head is to be removed, only follow the procedures that apply. The following tools, or their equivalents are absolutely necessary to properly perform this procedure:

- Cam Chain Tensioner tool T97T-6K254-A
- Cam Gear Removal tool T97T-6256-F
- Cam Gear Torque adapter T97T-6256-G
- Camshaft Gear Positioning/Holding tool T97T-6256-B
- Camshaft Gear Positioning/Holding tool adapter T97T-6256-A
- Camshaft holding tool T97T-6256-C
- Crankshaft holding tool T97T-6303-A
- Camshaft holding tool adapter T97T-6256-D

2. Before servicing the vehicle, refer to the precautions in the beginning of this section.
3. Properly relieve the fuel system pressure.
4. Drain the cooling system.
5. Remove or disconnect the following:
   - Negative battery cable
   - Lower intake manifold
   - Fan blade and shroud
   - Valve cover
   - Roller followers, if equipped
   - Drive belt
   - Upper radiator hose and tube
   - Alternator electrical connectors
   - Alternator mounting bracket
   - Engine accessory bracket and move it aside
   - Camshaft Position (CMP) electrical connector
   - Crankshaft Position (CKP) sensor electrical connector
   - Engine Coolant Temperature (ECT) sensor electrical connector
   - Coil pack electrical connector
   - Exhaust Gas Recirculation (EGR) valve electrical connector
   - EGR valve bracket and move it aside
   - Heater hoses
   - Fuel injector electrical connectors
   - Water bypass hose
   - Thermostat housing
   - Spark plug wires
   - Fuel injection supply manifold
   - Fuel injectors
   - Crankcase vent separator spring
   - Oil dipstick housing
   - Exhaust manifold
   - Hydraulic chain tensioner

The correct cylinder head bolt loosening sequence must be used to prevent warpage–4.0L SOHC engine

- Cassette retaining bolt
- Camshaft sprocket
- Cylinder head and discard the gasket

To install:

6. Thoroughly clean all gasket mating surfaces. Remove all traces of old gasket material, oil, grease or dirt.
7. Insure that the rubber band is holding the right-hand chain to the cassette.
8. Install a new head gasket and the cylinder head.
9. Torque the new cylinder head bolts in sequence as follows:
   a. Step 1: 26 ft. lbs. (34 Nm).
   b. Step 2: Plus 90 degrees.
   c. Step 3: Plus an additional 90 degrees.
10. Install or connect the following:
   - Camshaft sprocket in the cassette and make certain that the camshaft sprocket turns freely on the camshaft
   - Cassette retaining bolt. Torque the bolt to 89 inch lbs. (10 Nm).
   - Exhaust manifold
   - Oil level indicator tube. Torque the bolt to 18 ft. lbs. (25 Nm).
   - Crankcase vent separator and spring
   - Thermostat housing. Torque the bolts to 8 ft. lbs. (11 Nm).
   - Water bypass hose
   - Heater hoses
   - EGR bracket. Torque the bolt to 89 inch lbs. (10 Nm).

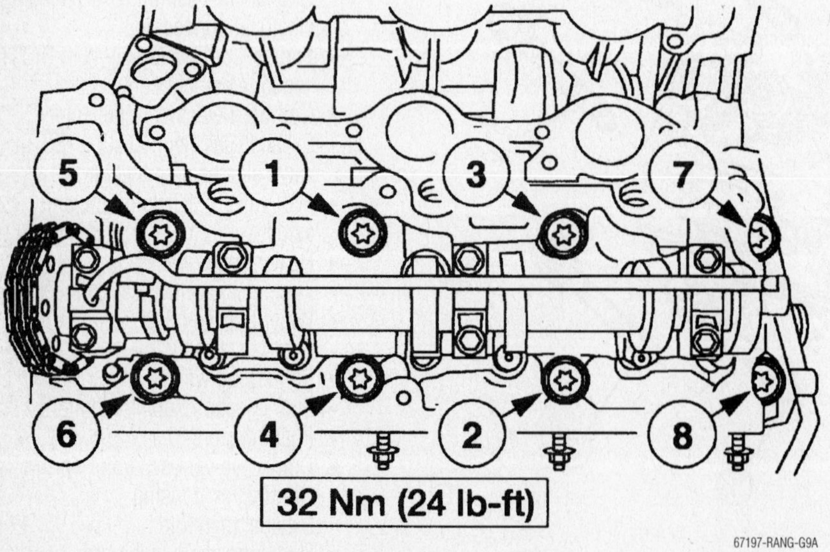

Cylinder head bolt torque sequence 4.0L SOHC engine

   - EGR tube. Torque the nut to 30 ft. lbs. (40 Nm).
   - ECT sensor electrical connector
   - Electrical harness retainer. Torque the bolt to 89 inch lbs. (10 Nm).
   - CKP and CMP electrical connectors
   - Accessory bracket. Torque the bolts to 31 ft. lbs. (42 Nm).
   - Alternator mounting bracket. Torque the bolts to 31 ft. lbs. (42 Nm).
   - Alternator and electrical connectors
   - Drive belt
   - Fan shroud
   - Roller followers
   - Valve cover
   - Lower intake manifold
   - Negative battery cable
11. Change the engine oil and filter.
12. Refill the cooling system.
13. Start the engine and check for leaks, repair if necessary.

Rocker Arms/Shafts

REMOVAL & INSTALLATION

2.3L Engines

This DOHC engine does not employ rocker arms. The camshafts bear directly on the lifters. For lifter removal, remove the camshafts.

2.5L Engines

➡A special tool is required to compress the valve spring.

1. Before servicing the vehicle, refer to the precautions in the beginning of this section.
2. Disconnect the negative battery cable.
3. Remove the valve cover.
4. Rotate the camshaft so that the base circle of the cam is against the cam follower you intend to remove.

➡If removing more than one cam follower, label them so they can be returned to their original position.

5. Using special tool T88T-6565-BH depress the valve spring. Slide the cam follower over the lash adjuster and out from under the camshaft.

To install:

6. Compress the valve spring and slide the roller follower into position.
7. Release the tension from the spring.
8. Install the valve cover and connect the negative battery cable.

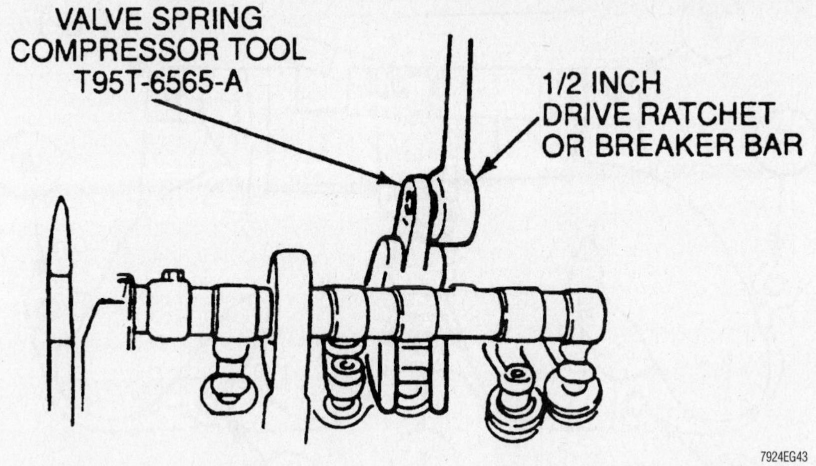

To remove the cam follower (rocker arm), use the special tool to depress the valve spring, then remove the cam follower—2.5L engines

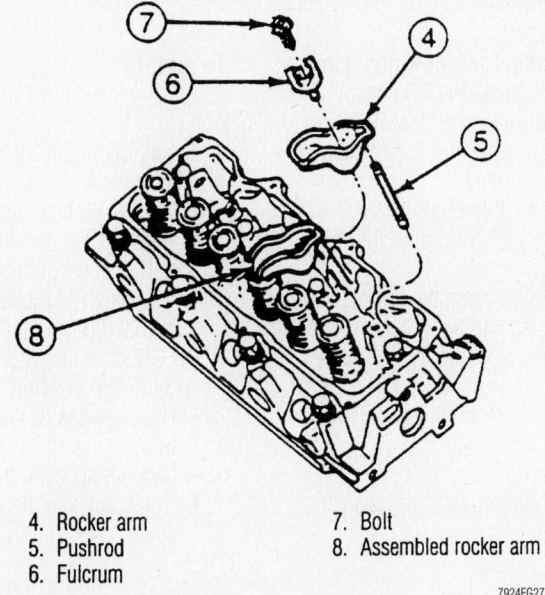

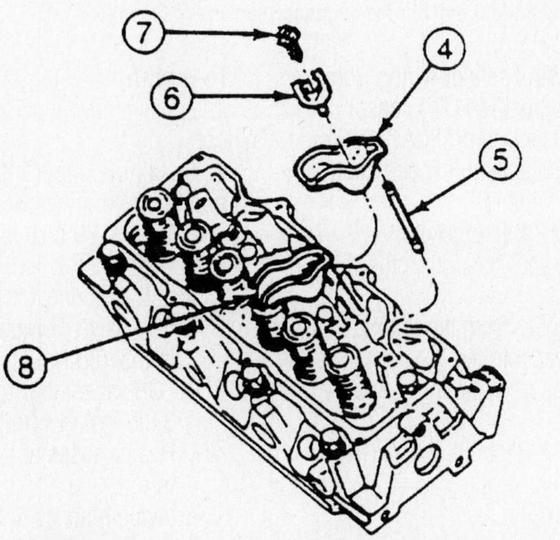

4. Rocker arm
5. Pushrod
6. Fulcrum
7. Bolt
8. Assembled rocker arm

7924EG27

Exploded view of the rocker arm assembly—3.0L engine

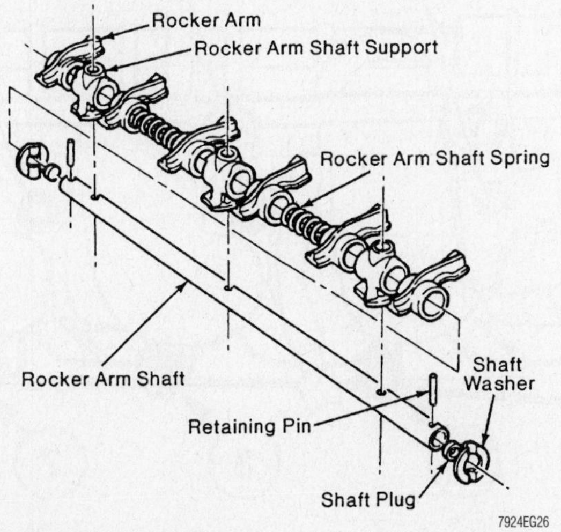

Rocker arm and shaft assembly—4.0L SOHC engine

3.0L Engines

1. Before servicing the vehicle, refer to the precautions in the beginning of this section.
2. Remove or disconnect the following:
   - Negative battery cable
   - Rocker arm covers
   - Retaining bolt at each rocker arm
3. The rocker arm and pushrod may then be removed from the engine. Keep all rocker arms and pushrods in order so they may be installed in their original locations.

To install:

4. Lubricate the rocker arm assemblies with SAE 50W engine oil.
5. Ensure that the fulcrums are properly seated into the cylinder head. Torque the rocker arm fulcrum bolts to 19 ft. lbs. (26 Nm).
6. Install the rocker arm covers and connect the negative battery cable.

4.0L SOHC Engines

➡A special tool is required to compress the valve spring.

1. Before servicing the vehicle, refer to the precautions in the beginning of this section.
2. Disconnect the negative battery cable.
3. Remove the valve cover.
4. Rotate the camshaft so that the base circle of the cam is against the cam follower you intend to remove.

➡If removing more than one cam follower, label them so they can be returned to their original position.

5. Using special tool T97T-6565-A depress the valve spring. Slide the cam follower over the lash adjuster and out from under the camshaft.

To install:

6. Compress the valve spring and slide the roller follower into position.
7. Release the tension from the spring.
8. Install the valve cover and connect the negative battery cable.

Intake Manifold

REMOVAL & INSTALLATION

2.3L Engine

1. Before servicing the vehicle, refer to the precautions in the beginning of this section.
2. Relieve the fuel system pressure.
3. Drain the cooling system.

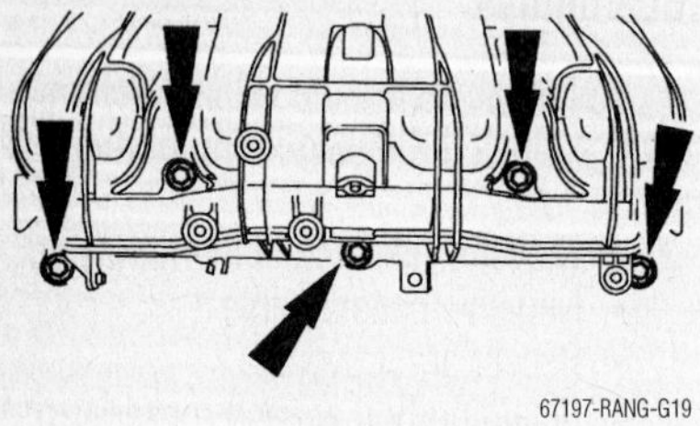

2.3L engine intake manifold bolts

4. Properly discharge the air conditioning system.
5. Remove or disconnect the following:
   - Negative battery cable
   - Water outlet tube.
   - Water outlet tube.

➡The alternator will be removed with the accessory bracket.

   - Accessory bracket.
   - Coolant hose from the thermostat.
   - Coolant hose from the EGR valve.
   - Coolant tube assembly.
   - Block heater (if so equipped).
   - Water outlet.
   - EGR valve.
   - Idle air control (IAC) valve.
   - Throttle position (TP) sensor.
   - Manifold absolute pressure (MAP) sensor.
   - Swirl control valve monitor electrical connector.
   - Electric thermostat.
   - Swirl control valve.
   - CMP sensor electrical connector and disconnect the PCV hose from the intake manifold.
   - Engine wiring harness pin-type retainers from the intake manifold.
   - Engine wiring harness connector bracket. Position the engine wiring harness aside.
   - EGR tube.
   - Fuel supply line clip from the front of the intake manifold. Disconnect the vacuum hose from the intake manifold.
   - Intake manifold assembly.

6. Installation is the reverse of removal. Torque the bolts to 13 ft. lbs. (18Nm). There is no special torque sequence.

2.5L Engine

1. Before servicing the vehicle, refer to the precautions in the beginning of this section.
2. Remove or disconnect the following:
   - Negative battery cable
   - Intake air temperature (IAT) sensor
   - Air cleaner outlet tube
   - Accelerator control splash shield

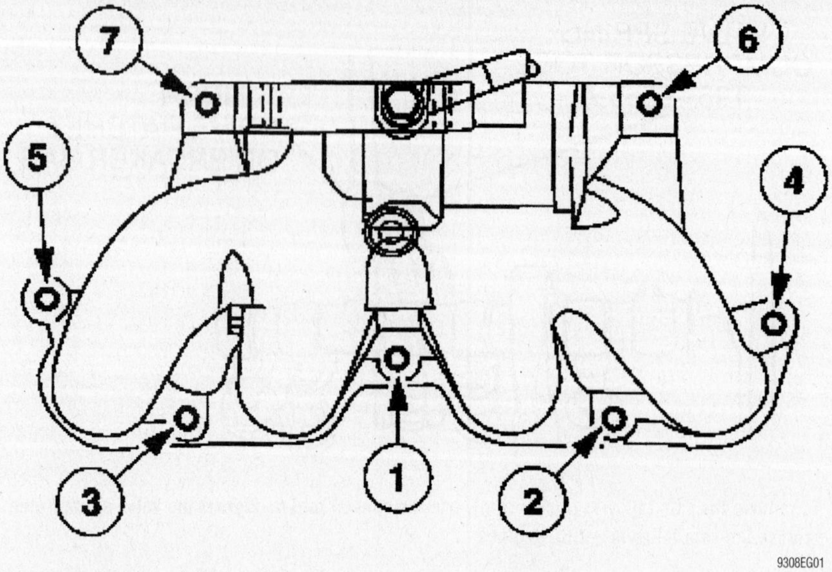

Tighten the lower manifold bolts in the sequence shown—2.5L engine

   - Engine control sensor wiring from the throttle position (TP) sensor and the idle air control (IAC) valve
   - Accelerator cable and speed control cable, if equipped
   - Accelerator cable bracket
   - Crankcase vent hose from the valve cover
   - Vacuum hoses from the intake manifold vacuum tee
   - Heater hose from the intake manifold
   - Exhaust Gas Recirculation (EGR) tube
   - EGR valve
   - Upper intake manifold and discard the gasket

To install:

3. Install a new upper intake manifold gasket.
4. Install the upper intake manifold. Torque the bolts in sequence as follows:
   a. Step 1: 89 inch lbs. (10 Nm).
   b. Step 2: 28 ft. lbs. (38 Nm).
5. Install or connect the following:
   - EGR valve. Torque the bolts to 22 ft. lbs. (30 Nm).
   - EGR valve tube. Torque the bolts to 34 ft. lbs. (47 Nm).
   - Heater hoses to the intake manifold
   - Vacuum hoses to the tee
   - Crankcase vent hose to the valve cover

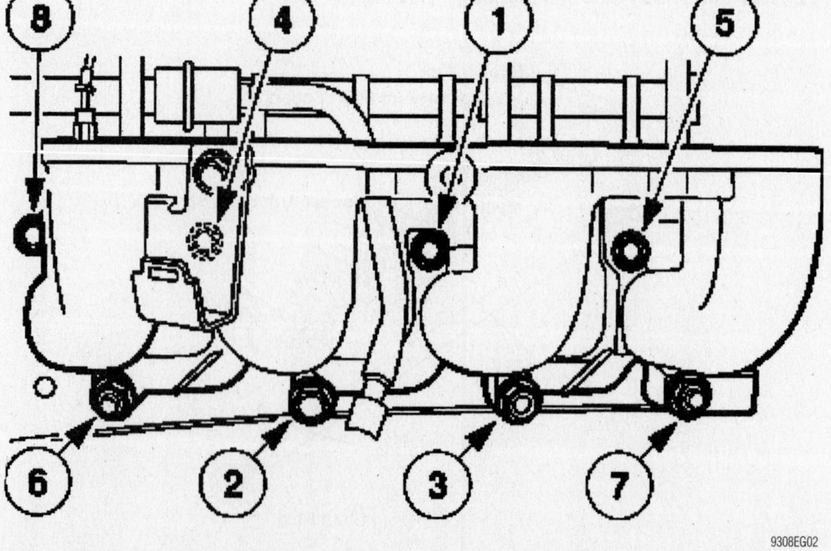

Tighten the upper manifold bolts in the sequence shown—2.5L engine

- Accelerator cable bracket. Torque the bolts to 20 ft. lbs. (28 Nm).
- Accelerator cable and speed control cable, if equipped
- IAC valve and TP sensor electrical connectors
- Air cleaner outlet tube
- IAT sensor
- Negative battery cable

6. Start the engine and check for leaks, repair if necessary.

3.0L Engine

1. Before servicing the vehicle, refer to the precautions in the beginning of this section.
2. Remove or disconnect the following:
   - Negative battery cable
   - Intake Air Temperature (IAT) sensor
   - Air cleaner outlet tube
   - Accelerator control splash shield
   - Accelerator cable and speed control cable, if equipped
   - Engine control sensor wiring from the Throttle Position (TP) sensor and the Idle Air Control (IAC) valve and Exhaust Gas Recirculation (EGR) transducer
   - EGR tube from the valve
   - EGR vacuum lines
   - 42 pin connector bracket
   - Throttle body and gasket
   - Ignition coil and move it aside
   - Evaporative Emissions (EVAP) hose
   - Upper intake manifold bolts and discard them
   - Crankcase vent hose
   - Upper intake manifold and discard the gasket

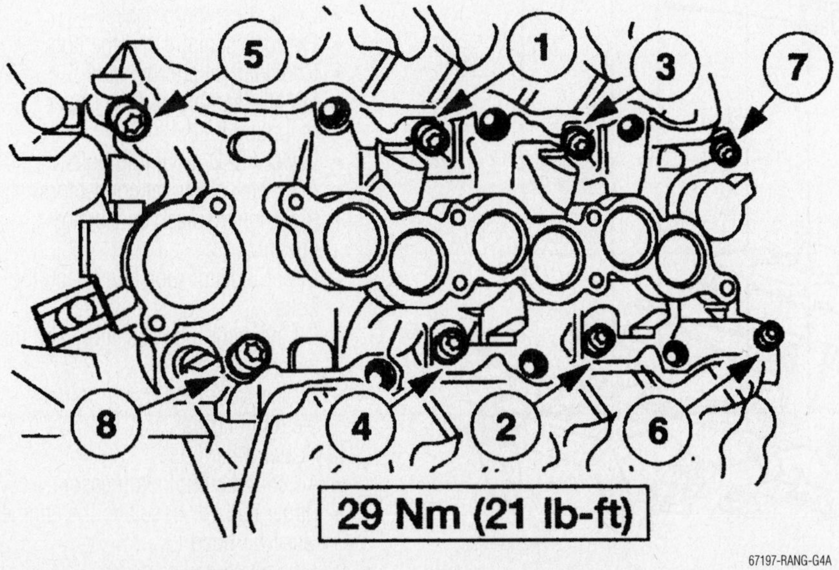

Tighten the lower manifold bolts in the sequence shown—3.0L engine

To install:

3. Clean all mating surfaces.
4. Install or connect the following:
   - New upper intake manifold
   - Intake manifold
   - Crankcase vent hose
5. Torque the upper intake manifold bolts in sequence as follows:
   a. Step 1: 15 ft. lbs. (20 Nm).
   b. Step 2: 18 ft. lbs. (25 Nm).
6. Install or connect the following:
   - EVAP hose
   - EGR tube
   - EGR transducer. Torque the bolts to 89 inch lbs. (10 Nm).
   - Ignition coil. Torque the bolts to 15 ft. lbs. (20 Nm).
   - New gasket and throttle body. Torque the bolts to 22 ft. lbs. (30 Nm).
   - 42 pin connector
   - EGR vacuum lines
   - EGR transducer, IAC valve and TP sensor electrical connectors
   - Accelerator cable and speed control, if equipped
   - Accelerator cable splash shield
   - Air cleaner outlet tube
   - IAT sensor
   - Negative battery cable
7. Start the vehicle and check for leaks, repair if necessary.

4.0L OHC Engine

2001–02

1. Before servicing the vehicle, refer to the precautions in the beginning of this section.
2. Remove or disconnect the following:
   - Negative battery cable
   - Air cleaner-to-intake tube
   - Accelerator splash shield
   - Accelerator and, if equipped with cruise control, speed control cables from the throttle control cam
   - Accelerator cable retaining bracket from the upper intake manifold
   - Label and disengage all vacuum and electrical connections on the intake manifold.
   - Upper intake manifold attaching bolts
   - Lift up on the manifold and remove both fuel Vapor Management Valve (VMV) hoses
   - Upper intake manifold and discard the gasket

To install:

➡Ford does not specify a sequence, but it is recommended that you start tightening in the middle and work your way out to the ends. Repeat the tightening sequence several times until the specified torque is reached.

3. Position the upper manifold on the lower manifold.
4. Install or connect the following:
   - Attach both VMV hoses to the manifold
   - Upper manifold attaching bolts. Torque the bolts to 62 inch lbs. (7 Nm).
   - Attach any vacuum and electrical connections that were removed
   - Accelerator cable bracket to the intake and the cable (or cables if equipped with cruise control) to the throttle cam
   - Accelerator splash shield
   - Air cleaner-to-intake supply tube
   - Negative battery cable
5. Start the vehicle and check for leaks, repair if necessary.

2003–04

1. Disconnect the battery ground cable.
2. Remove the bolts and the shield.
3. Remove the air cleaner outlet pipe.
4. Disconnect the idle air control (IAC) valve, throttle position (TP) sensor electrical connectors and the TP sensor wiring pin-type retainer.
5. Disconnect the MAF sensor wiring pin-type retainer.
6. Detach the accelerator and speed control cables from the throttle body.
7. Detach the accelerator and speed control cables from the bracket, and position the cables aside.

8. Disconnect the exhaust gas recirculation (EGR) valve vacuum hose and tube fitting.

9. Disconnect the EGR vacuum regulator solenoid valve electrical connector and vacuum hose.

10. Disconnect the hose.

11. Loosen the clamp and disconnect the brake booster vacuum hose.

CAUTION

It is important to twist the spark plug wire boots while pulling upward to avoid possible damage to the spark plug wire.

➡Mark the spark plug wire locations before removing them.

12. Disconnect the RH spark plug wires from the coil. Remove the spark plug wire routing clip pin-type retainer and position the wires aside.

13. Remove the wiring harness bracket retainer, then position the wiring harness aside.

14. Remove the accelerator cable routing clip pin-type retainer and the wiring harness pin-type retainer.

15. Remove the bolts.

16. Remove the bolts and position the coil and bracket aside.

17. Disconnect the vacuum hoses.

18. Remove the nut.

19. Disconnect the powertrain control module (PCM) electrical connector.

20. Remove the retainer and position the ground wires aside.

21. Detach the electrical connector retainer.

22. Remove the intake manifold bolts and lift up the intake manifold.

23. Remove the heated positive crankcase ventilation (PCV) hose retainers and remove the heated PCV fitting.

24. Remove the intake manifold.

25. To install, reverse the removal procedure. Torque the fasteners to 89 inch lbs. (10 Nm).

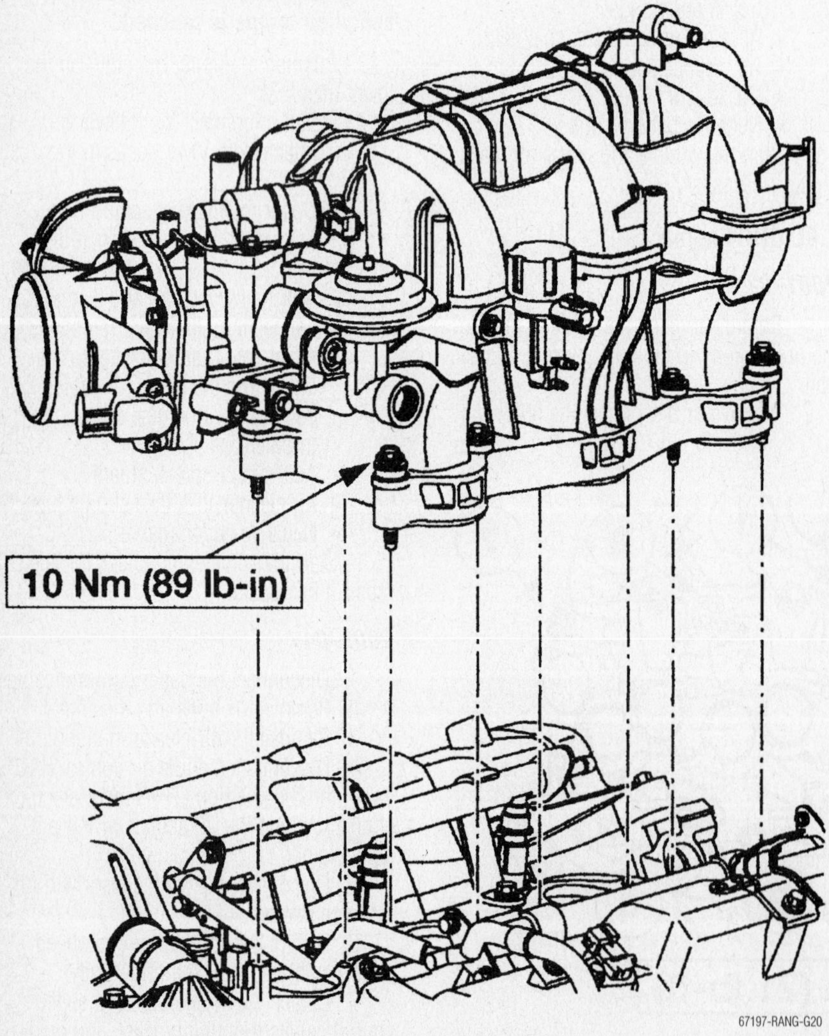

Intake manifold installation–2003–04 4.0L SOHC engine

Exhaust Manifold

REMOVAL & INSTALLATION

2.3L Engine

1. Before servicing the vehicle, refer to the precautions in the beginning of this section.
2. Remove or disconnect the following:
   - Negative battery cable
   - Exhaust flange nuts
   - Drive belt
   - Coolant
   - Upper radiator hose and the engine reservoir hose
   - Air conditioning compressor
   - Heater hose
   - Oil indicator and the upper bolt for the tube assembly
   - Lower bolt and remove the oil indicator tube assembly
   - Front radiator tube
3. Remove the pushpins and position the right inner fender splash shield out of the way.
4. Remove or disconnect the following:
   - Alternator electrical connectors
   - Lower front end accessory drive (FEAD) mounting bolts
   - Upper mounting bolt and the FEAD assembly
   - Two nuts and position the coolant tube out of the way
   - Exhaust manifold
   - Exhaust manifold gasket

To install:

5. Install or connect the following:
   - Exhaust manifold gasket
   - Exhaust manifold and the nuts
   - Coolant tube and the nuts
   - FEAD assembly and the upper mounting bolts
   - Lower FEAD mounting bolts
   - Alternator electrical connectors
   - Right inner splash shield and pushpins
   - Upper radiator tube and install the bolts
   - Oil indicator tube assembly and the lower bolt
   - Oil indicator tube upper bolt and the oil indicator
   - Heater water hose
   - Air conditioning compressor
   - Upper radiator hose and the engine reservoir hose
6. Fill the cooling system.

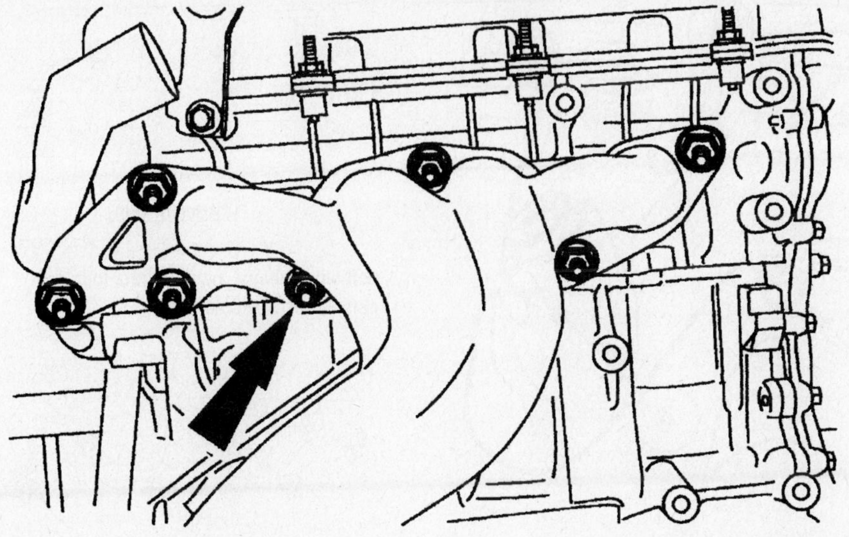

Exhaust manifold retaining nuts—2.3L engine

7. Install the serpentine drive belt.
8. Install the exhaust flange nuts.
9. Connect the battery ground cable.

2.5L Engines

1. Before servicing the vehicle, refer to the precautions in the beginning of this section.
2. Remove or disconnect the following:
   - Negative battery cable
   - Intake Air Temperature (IAT) sensor
   - Air cleaner outlet tube
   - Differential Pressure Feedback (DPFE) sensor and move it aside
   - Exhaust Gas Recirculation (EGR) transducer lines at the tube
   - Loosen and remove the EGR valve-to-exhaust manifold tube
   - Catalytic converter from the exhaust manifold
   - Rear engine lifting eye nuts
   - Exhaust manifold and discard the gasket

To install:

3. Clean the mating surfaces on the exhaust manifold and the cylinder head.
4. Install a new gasket and the exhaust manifold. Torque the bolts in sequence as follows:
   a. 16 ft. lbs. (23 Nm).
   b. 59 ft. lbs. (80 Nm).
5. Install or connect the following:
   - Rear engine lifting eye. Torque the bolts to 15 ft. lbs. (20 Nm).
   - Catalytic converter to the exhaust manifold
   - EGR valve to the exhaust manifold tube
   - EGR transducer lines
   - DPFE sensor
   - Air cleaner outlet tube
   - IAT sensor
   - Negative battery cable
6. Start the vehicle and check for leaks, repair if necessary.

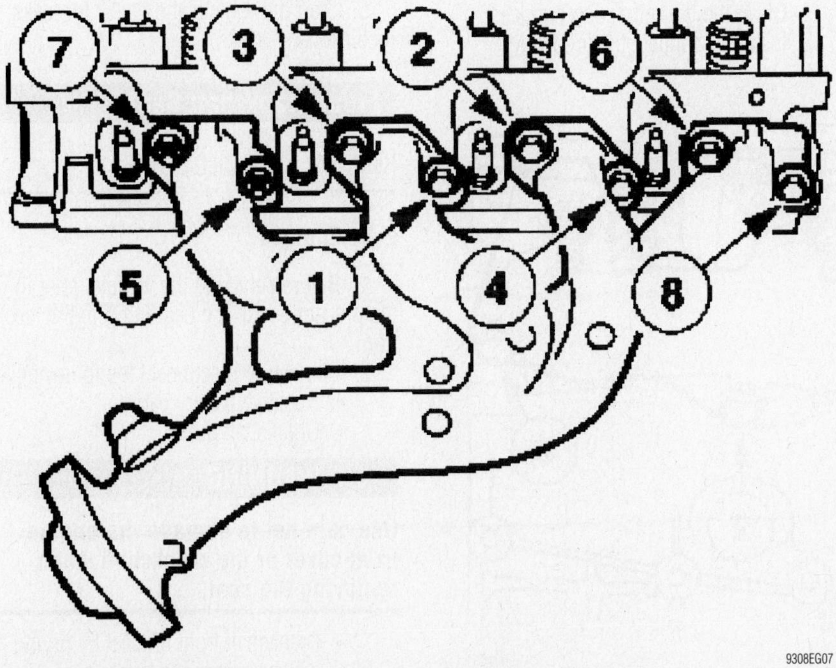

Tighten the exhaust manifold bolts in 2 stages—2.5L engine

3.0L Engine

LEFT SIDE

1. Before servicing the vehicle, refer to the precautions in the beginning of this section.
2. Install or connect the following:
   - Negative battery cable
   - Exhaust flange nuts
   - Exhaust Gas Recirculation (EGR) valve from the exhaust manifold tube
   - Oil lever indicator and bracket
   - Exhaust manifold and discard the gasket

To install:

3. Clean the mating surfaces for the exhaust manifold and cylinder head.
4. Install a new gasket and the exhaust manifold. Torque the bolts in sequence to:
   a. 89 inch lbs. (10 Nm).
   b. 15 ft. lbs. (20 Nm).
5. Install or connect the following:
   - Oil lever indicator tube and bracket. Torque the bolt to 12 ft. lbs. (16 Nm).
   - EGR valve to the exhaust manifold tube. Torque the fastener to 26 ft. lbs. (35 Nm).
   - Exhaust flange. Torque the nuts to 25 ft. lbs. (34 Nm).
   - Negative battery cable
6. Start the vehicle and check for leaks, repair if necessary.

RIGHT SIDE

1. Before servicing the vehicle, refer to the precautions in the beginning of this section.
2. Remove or disconnect the following:
   - Negative battery cable
   - Exhaust manifold flange
   - Ignition coil support bracket

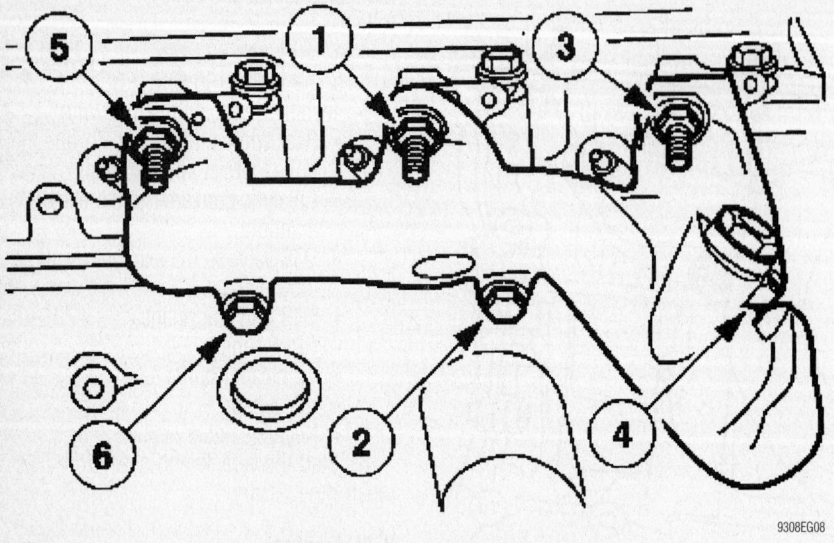

Tighten the exhaust manifold bolts in sequence–3.0L left side

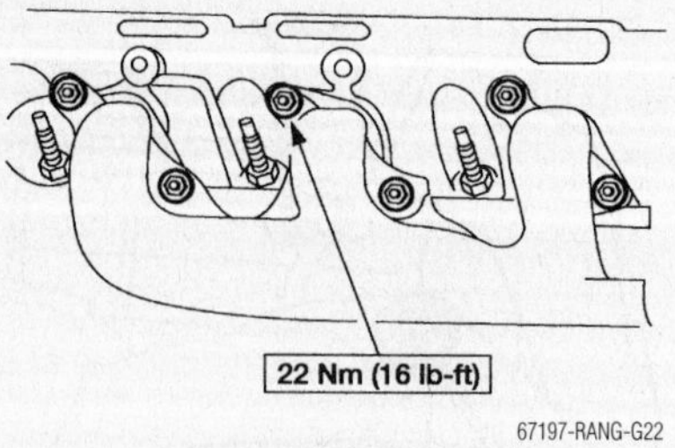

Left side exhaust manifold retaining nuts—4.0L SOHC engine

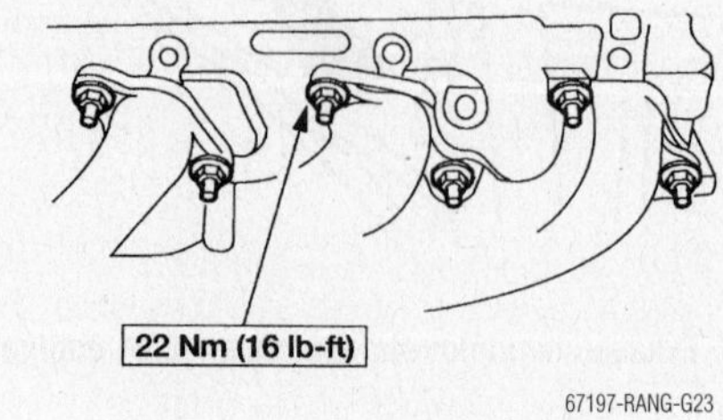

Right side exhaust manifold retaining nuts—4.0L SOHC engine

- Exhaust manifold and discard the gasket

To install:

3. Clean the mating surfaces for the exhaust manifold and cylinder head
4. Install a new gasket and the exhaust manifold. Torque the bolts is sequence to:
   a. 89 inch lbs. (10 Nm).
   b. 18 ft. lbs. (25 Nm).
5. Install or connect the following:
   - Ignition coil support bracket. Torque the bolts to 15 ft. lbs. (20 Nm).
   - Exhaust flange nuts. Torque the nuts to 33 ft. lbs. (46 Nm).
   - Negative battery cable
6. Start the vehicle and check for leaks, repair if necessary.

4.0L OHC Engine

1. Before servicing the vehicle, refer to the precautions in the beginning of this section.
2. Remove or disconnect the following:
   - Negative battery cable
   - Exhaust inlet pipe-to-manifold attaching bolts
   - Differential Pressure Feedback EGR (DPFE) transducer hoses, left side manifold only
   - Exhaust Gas Recirculation (EGR) tube from the manifold and valve, left side manifold only
   - Exhaust manifold and discard the gasket

To install:

3. Clean the gasket mating surfaces.
4. Install or connect the following:
   - New gasket and the exhaust manifold. Torque the bolts to 16 ft. lbs. (22 Nm).
   - EGR tube to the manifold. Torque the fastener to 30 ft. lbs. (40 Nm) left side manifold only
   - DPFE transducer hoses, left side manifold only
   - Exhaust inlet pipe-to-manifold attaching bolts. Torque the bolts to 30 ft. lbs. (40 Nm).
   - Negative battery cable
5. Start the vehicle and check for leaks, repair if necessary.

Front Crankshaft Seal

REMOVAL & INSTALLATION

2.3L Engines

1. Before servicing the vehicle, refer to the precautions in the beginning of this section.
2. Remove or disconnect the following:
   - Negative battery cable
   - Crankshaft pulley

WARNING

Use care not to damage the engine front cover or the crankshaft when removing the seal.

   - Crankshaft front oil seal by prying the seal out of the front cover

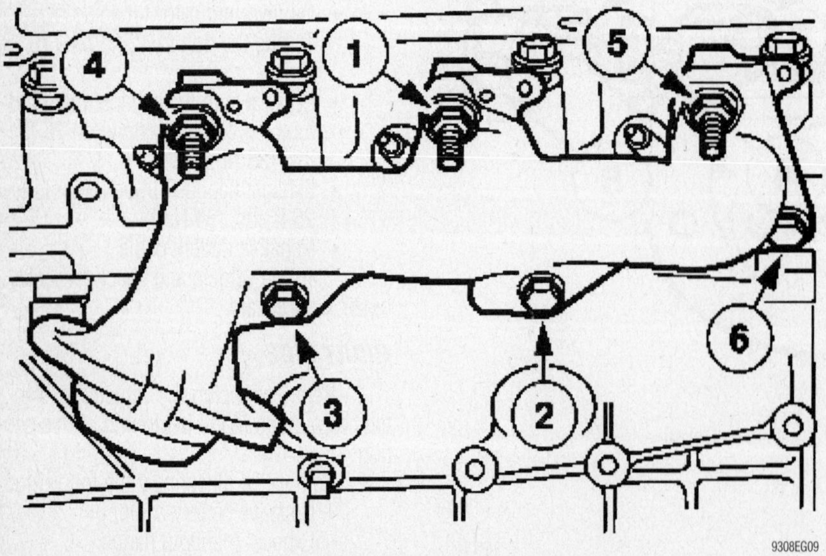

Tighten the right side exhaust manifold bolts in the proper sequence–3.0L

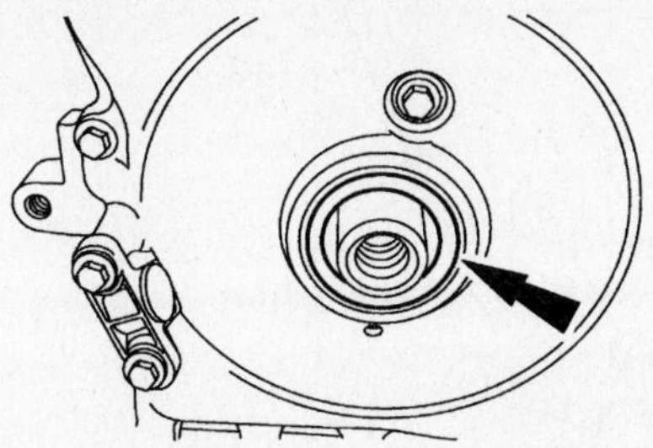

67197-RANG-G24

Front crankshaft seal—2.3L engine

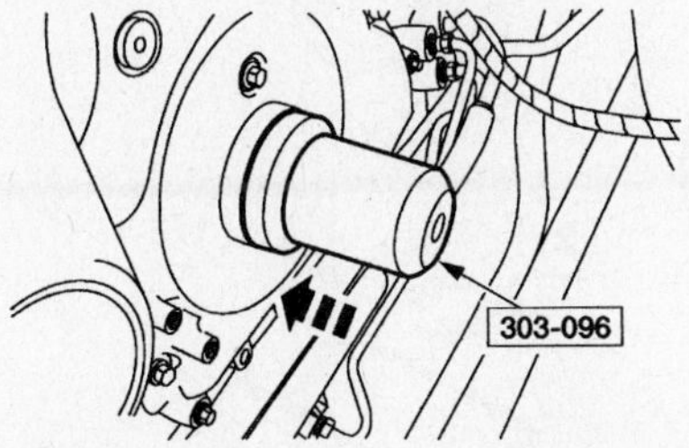

67197-RANG-G25

Front crankshaft seal installation—2.3L engine

To install:

3. Using the special tool, install the crankshaft front oil seal.
4. Install the crankshaft pulley.
5. Tighten the crankshaft damper in two stages:
    - Step 1: Tighten to 40 Nm (30 ft. lbs.)
    - Step 2: Tighten an additional 90 degrees

2.5L Engines

1. Before servicing the vehicle, refer to the precautions in the beginning of this section.
2. Remove or disconnect the following:
    - Negative battery cable
    - Timing belt cover
    - Drive belt
3. Align the crankshaft and camshaft timing marks and remove the timing belt.
    - Crankshaft pulley center bolt and slide the pulley off of the crankshaft
    - Crankshaft key

⁂ WARNING

Do not damage the crankshaft sealing surface while removing the oil seal.

- Crankshaft Seal Remover tool T74P-6700-B on the crankshaft and into the oil seal.
- Oil seal and clean the seal journal

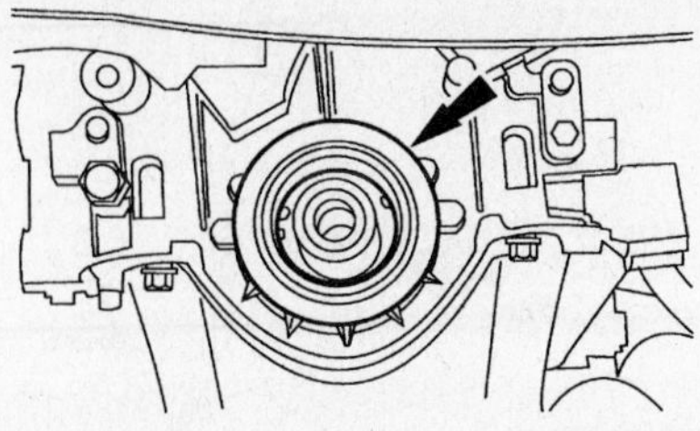

67197-RANG-G26

Front crankshaft seal—3.0L engine

To install:

4. Apply clean engine oil to the rubber lip of the new seal to aid installation.
5. Using Cam Bearing Adapter Tube T72C-6250, or equivalent, and crankshaft center bolt, carefully install the new oil seal until flush with the engine.
6. Install or connect the following:
    - Key and crankshaft pulley, washer and bolt. Torque the bolt to 92–121 ft. lbs. (125–165 Nm).
    - Timing belt
    - Timing belt and cover
    - Drive belt
    - Negative battery cable
7. Start the vehicle and check for leaks, repair if necessary.

3.0L Engine

1. Before servicing the vehicle, refer to the precautions in the beginning of this section.
2. Remove the crankshaft vibration damper.
3. Remove the crankshaft front seal.

To install:

4. Install the crankshaft front seal.
    a. Lubricate the seal lip with clean engine oil.

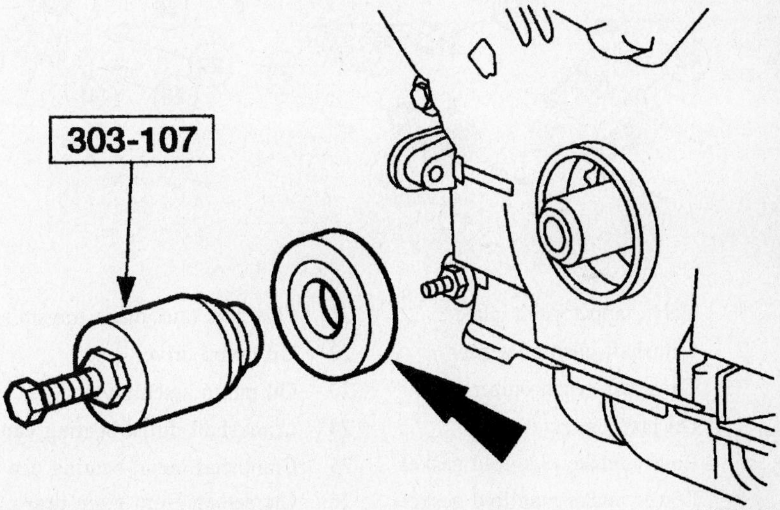

67197-RANG-G28

Front crankshaft seal removal—4.0L SOHC engine

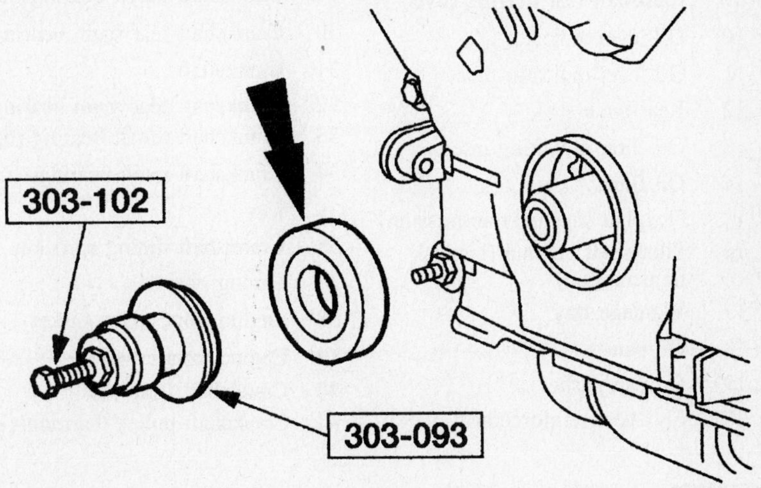

67197-RANG-G29

Front crankshaft seal installation—4.0L SOHC engine

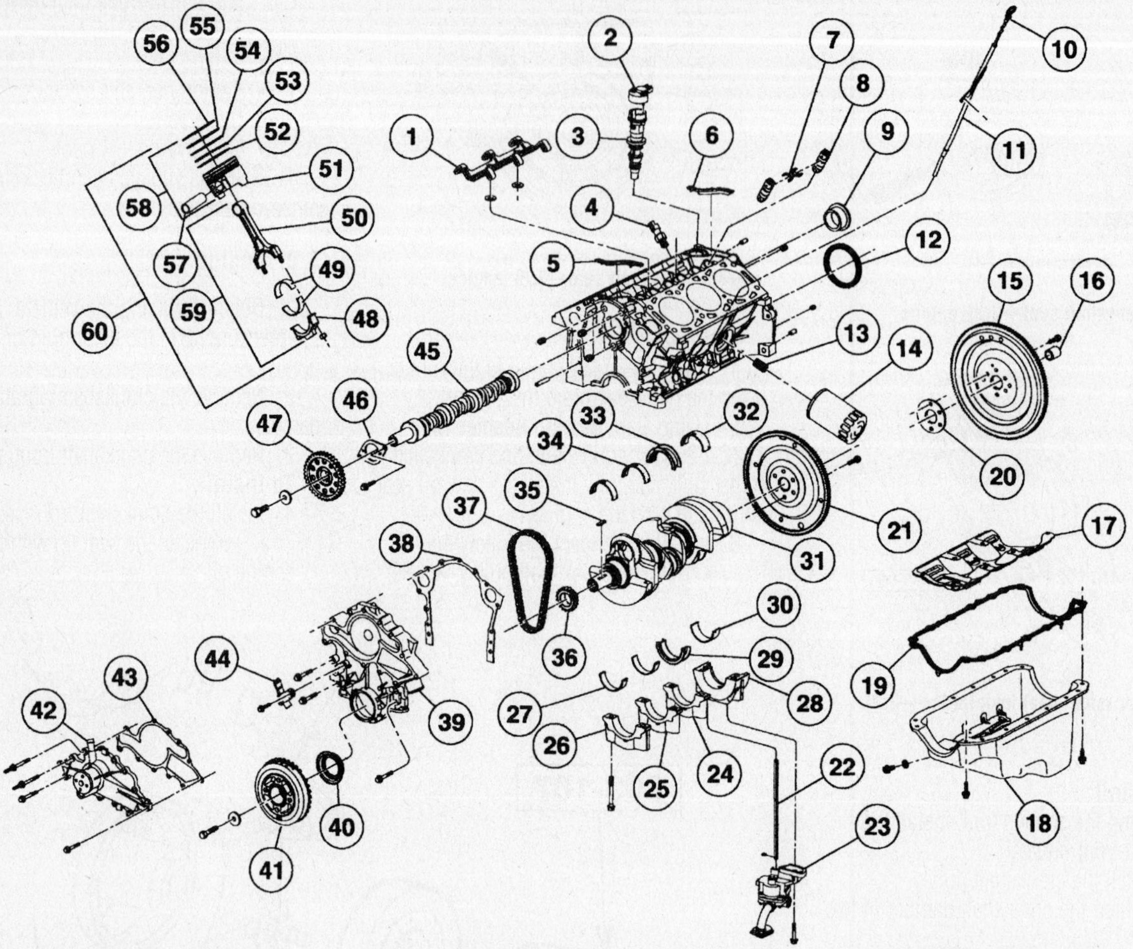

1 Valve tappet guide plate
2 Camshaft position sensor
3 Camshaft synchronizer
4 Oil pressure sender
5 Lower intake manifold gasket
6 Lower intake manifold gasket
7 Rocker arm center cover
8 Valve tappet
9 Camshaft rear bearing cover
10 Oil level indicator
11 Oil level indicator tube
12 Rear main seal
13 Oil filter mounting insert
14 Oil filter
15 Flexplate (manual transmission)
16 Pilot shaft bearing (manual transmission)
17 Windage tray
18 Oil pan
19 Oil pan gasket
20 Flywheel reinforcement plate
21 Flywheel (automatic transmission)
22 Oil pump drive
23 Oil pump assembly
24 Crankshaft thrust bearing cap
25 Crankshaft main bearing cap
26 Crankshaft front main bearing caps
27 Crankshaft main bearing
28 Crankshaft rear main bearing cap
29 Crankshaft thrust bearing (lower)
30 Crankshaft rear main bearing (lower)
31 Crankshaft
32 Crankshaft rear main bearing (upper)
33 Crankshaft thrust bearing (upper)
34 Crankshaft main bearings (upper)
35 Key
36 Crankshaft timing sprocket
37 Timing chain
38 Engine front cover gasket
39 Engine front cover
40 Crankshaft front seal
41 Crankshaft pulley (harmonic balancer)
42 Coolant pump
43 Coolant pump gasket
44 Crankshaft position (CKP) sensor
45 Camshaft
46 Camshaft thrust plate
47 Camshaft sprocket
48 Connecting rod bearing cap
49 Connecting rod bearings
50 Piston connecting rod
51 Piston
52 Oil ring (lower)
53 Oil ring baffle
54 Oil ring (upper)
55 Piston compression ring (lower)
56 Piston ring (upper)
57 Piston pin
58 Piston assembly
59 Connecting rod assembly
60 Piston and connecting rod assembly

67197-RANG-G30

Lower engine components—3.0L engine

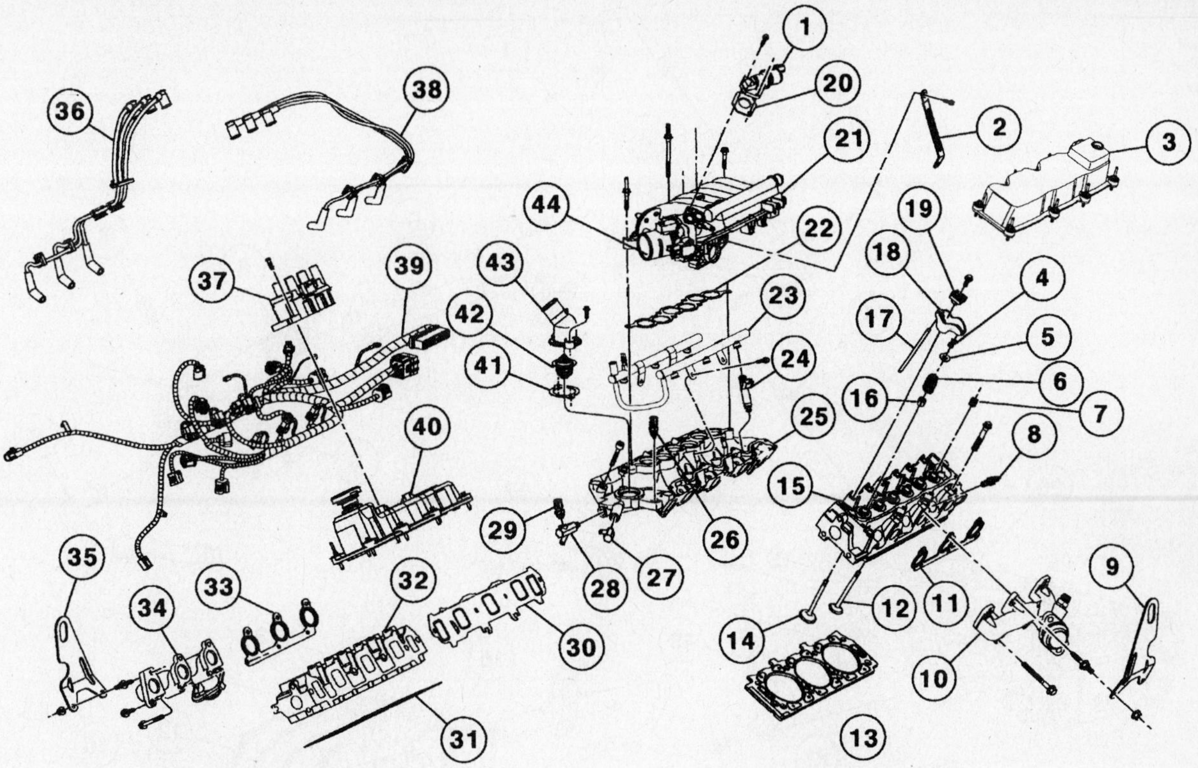

1 Idle air control valve (IAC)
2 Intake manifold support bracket
3 Valve cover — LH
4 Valve spring retainer key
5 Valve spring retainer
6 Valve spring
7 Valve stem seal
8 Spark plug
9 Engine lifting bracket
10 Exhaust manifold — LH
11 Exhaust manifold gasket
12 Exhaust valve
13 Cylinder head gasket
14 Intake valve
15 Cylinder head
16 Valve stem seal
17 Pushrod
18 Rocker arm
19 Rocker arm pivot
20 Idle air control valve gasket
21 Intake manifold — upper
22 Upper intake manifold gasket
23 Fuel injection supply manifold
24 Fuel injector
25 Intake manifold — lower
26 Engine coolant temperature (ECT) sensor
27 Heater connection elbow
28 Engine coolant hose connection
29 Engine coolant temperature (ECT) sensor
30 Lower intake manifold gasket
31 Cylinder head gasket — RH
32 Cylinder head — RH
33 Exhaust manifold gasket — RH
34 Exhaust manifold — RH
35 Engine lifting bracket
36 Ignition wire set — RH
37 Ignition coil
38 Ignition wire set — LH
39 Engine wiring harness assembly
40 Valve cover — RH
41 Thermostat housing gasket
42 Thermostat
43 Thermostat housing
44 Throttle body

67197-RANG-G31

Major upper engine components—3.0L engine

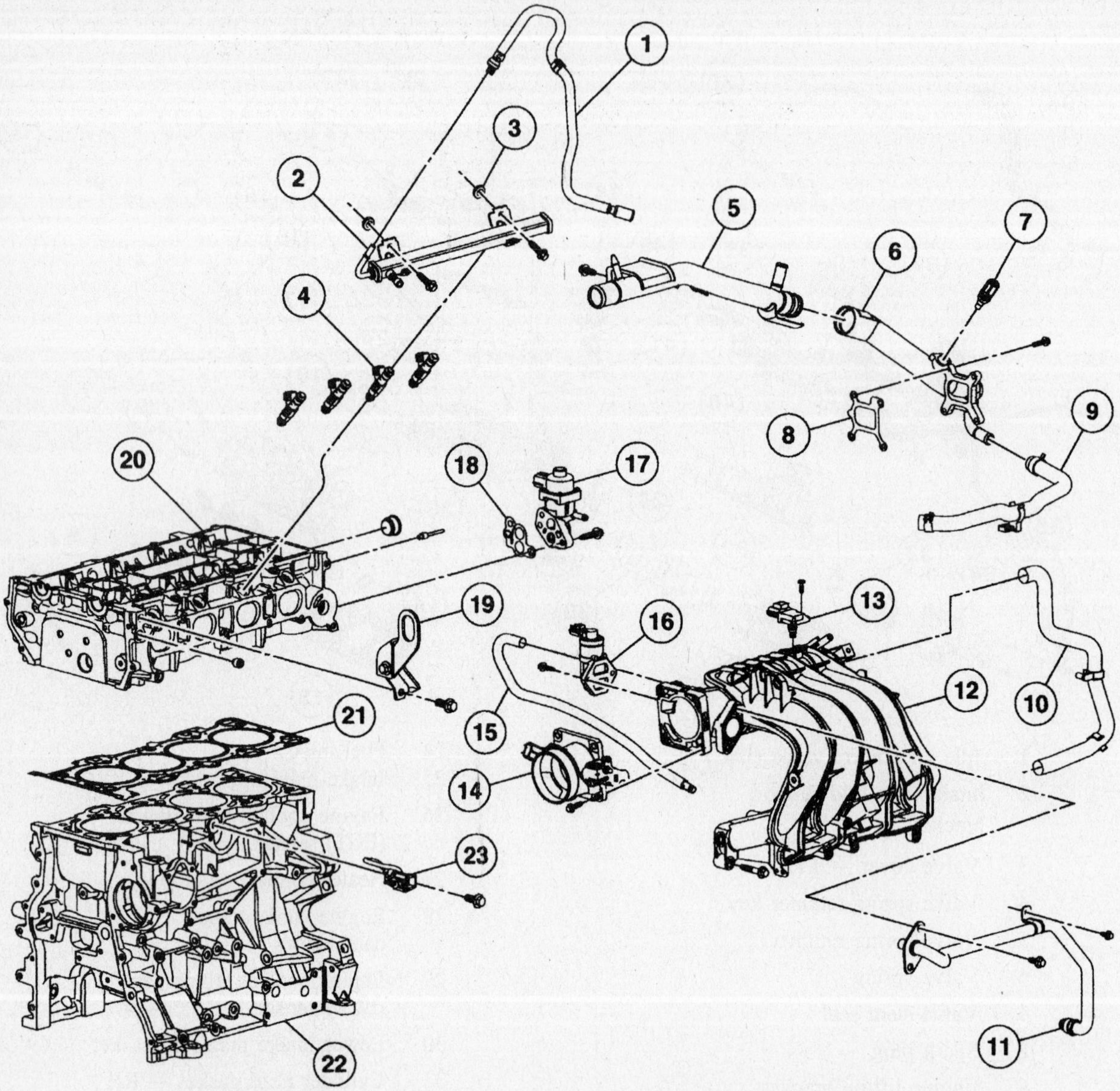

1	High pressure fuel line (supply)	11	EGR valve-to-exhaust manifold tube
2	Fuel injection supply manifold	12	Intake manifold
3	Spacer (2 req'd)	13	Absolute pressure sensor
4	Fuel injector (4 req'd)	14	Throttle body assembly
5	Water outlet adapter assembly (front)	15	Evaporative emission hose
6	Water outlet adapter assembly (rear)	16	Idle air control valve assembly
7	Water temperature indicator sender unit	17	EGR valve
8	Water outlet connector gasket	18	EGR valve mounting gasket
9	Water bypass hose and clamp assembly	19	Engine lifting eye
10	Crankcase ventilation tube	20	Cylinder head
		21	Cylinder head gasket
		22	Cylinder block
		23	Knock sensor

67197-RANG-G32

Intake manifold components—2.3L engine

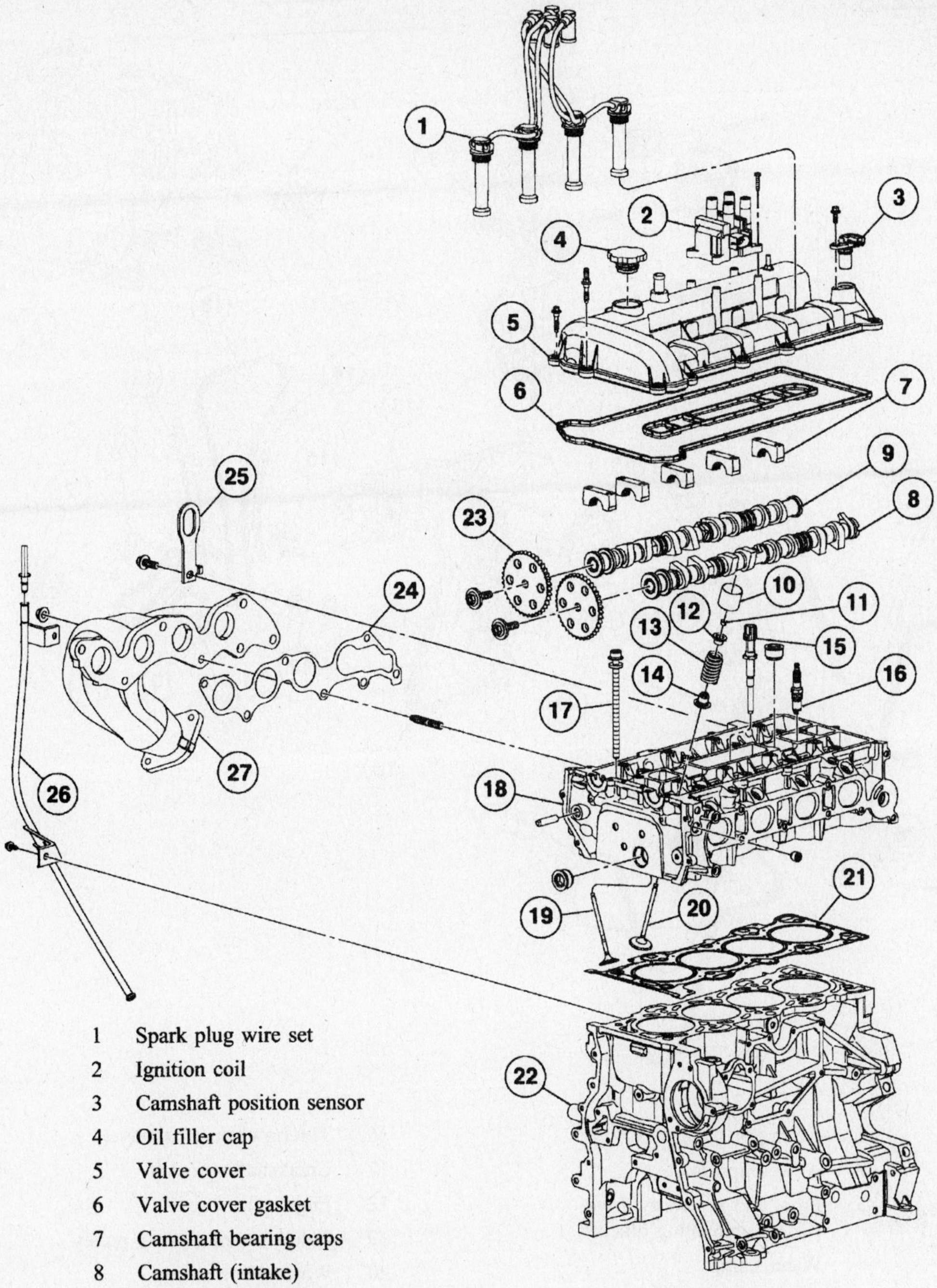

1 Spark plug wire set
2 Ignition coil
3 Camshaft position sensor
4 Oil filler cap
5 Valve cover
6 Valve cover gasket
7 Camshaft bearing caps
8 Camshaft (intake)
9 Camshaft (exhaust)
10 Valve tappet (16 req'd)
11 Valve spring retainer key (16 req'd)
12 Valve spring retainer (16 req'd)
13 Valve spring (16 req'd)
14 Valve stem seal (16 req'd)
15 Cylinder head temperature (CHT) sensor
16 Spark plug (4 req'd)
17 Cylinder head bolt (10 req'd)
18 Cylinder head
19 Exhaust valve (8 req'd)
20 Intake valve (8 req'd)
21 Head gasket
22 Cylinder block
23 Camshaft sprocket (2 req'd)
24 Exhaust manifold gasket
25 Engine lifting eye
26 Oil level indicator tube assembly
27 Exhaust manifold

67197-RANG-G33

Cylinder head and related components—2.3L engine

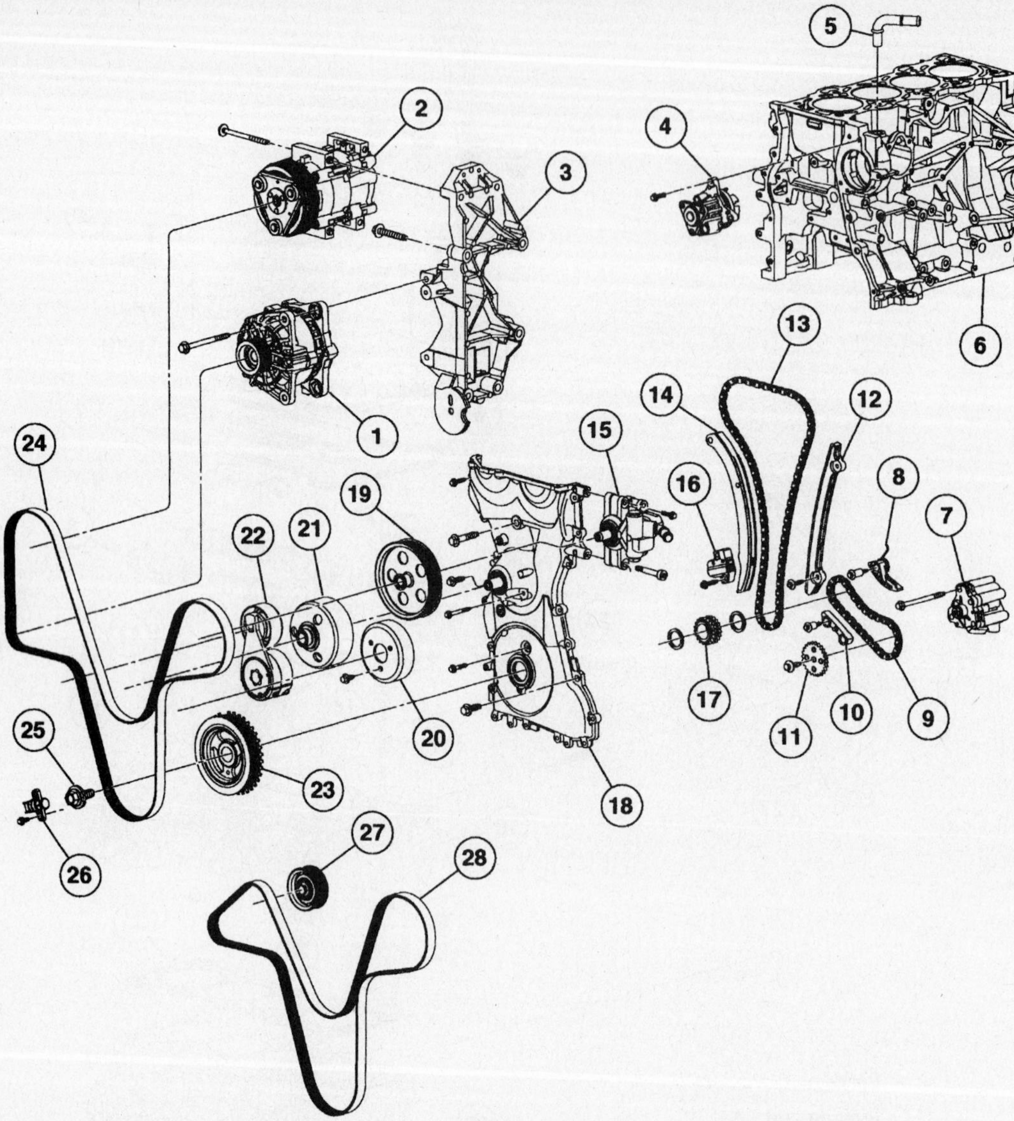

1	Generator
2	A/C compressor
3	Front end accessory drive (FEAD) mounting bracket
4	Water pump
5	Water bypass tube
6	Cylinder block
7	Oil pump
8	Oil pump chain tensioner
9	Oil pump chain
10	Oil pump chain guide
11	Oil pump drive gear
12	Timing chain guide
13	Timing chain
14	Timing chain tensioner arm
15	Power steering pump
16	Timing chain tensioner
17	Crankshaft sprocket
18	Engine front cover
19	Power steering pump pulley
20	Water pump pulley
21	Fan drive pulley
22	Drive belt tensioner
23	Crankshaft damper
24	Accessory drive belt (with A/C)
25	Crankshaft pulley bolt
26	Crankshaft position sensor
27	Drive belt pulley idler (without A/C)
28	Accessory drive belt (without A/C)

67197-RANG-G34

Engine block front components—2.3L engine

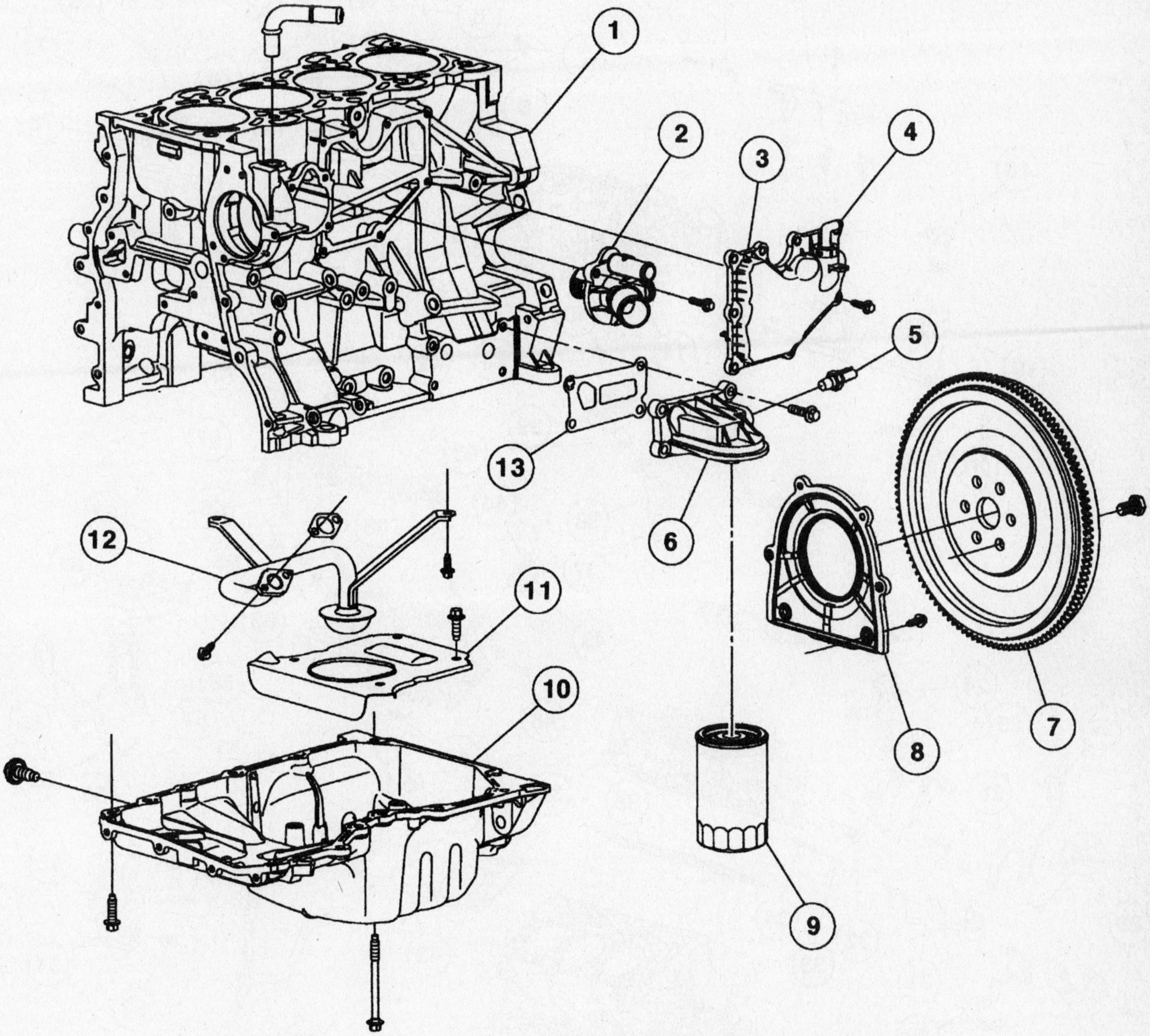

1	Cylinder block	8	Crankshaft rear oil seal and retainer
2	Water thermostat assembly	9	Oil filter
3	Crankcase vent oil separator	10	Oil pan
4	PCV valve	11	Oil pan baffle
5	Oil pressure sensor	12	Oil pump screen and pickup tube
6	Oil filter adapter	13	Oil filter adapter gasket
7	Flywheel		

67197-RANG-G35

Lower engine block components—2.3L engine

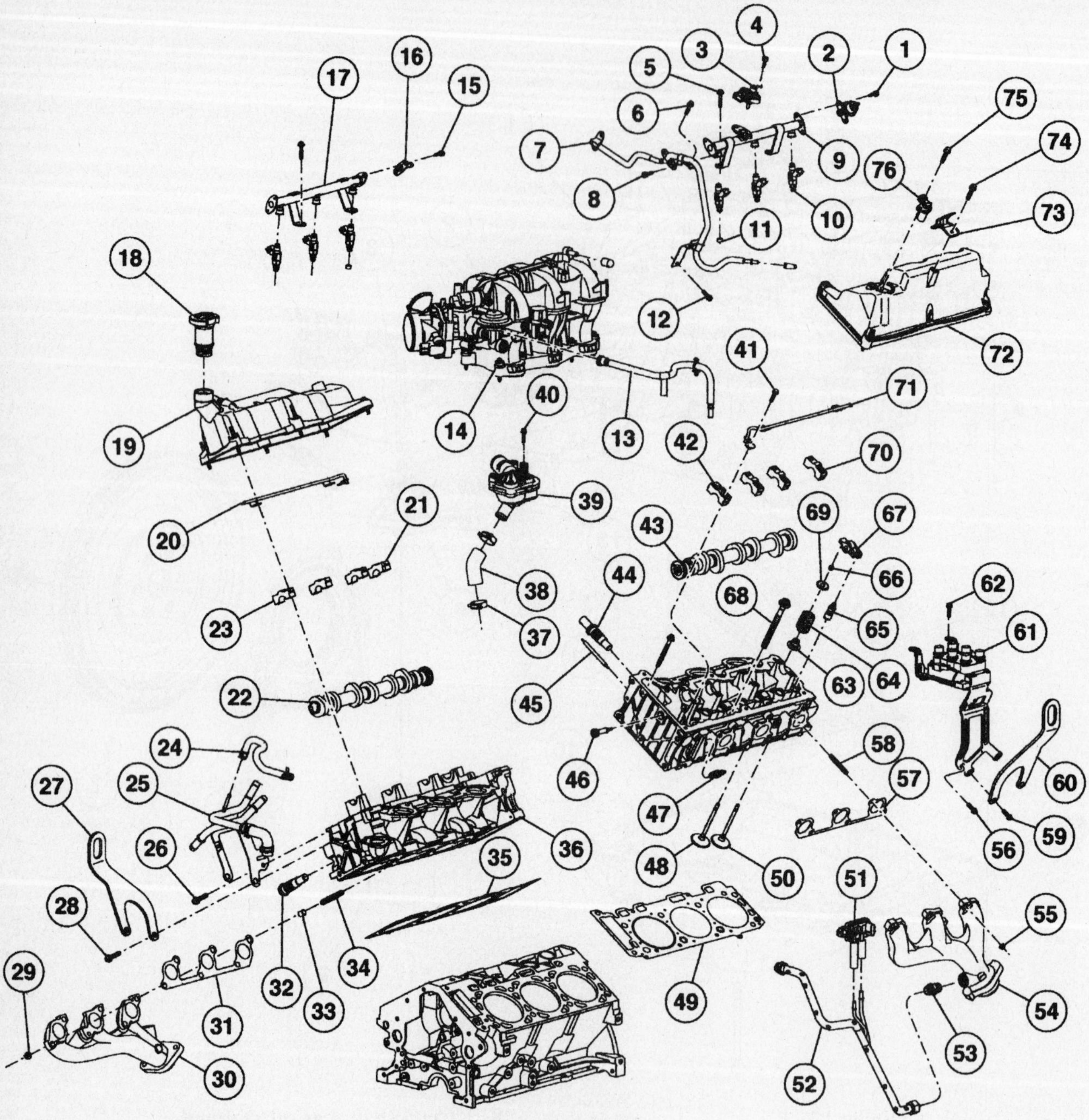

1 Bolt (2 req'd)
2 Fuel injection pulse damper
3 Fuel pressure/temperature sensor
4 Bolt (2 req'd)
5 Bolt (4 req'd)
6 Bolt
7 Fuel supply tube
8 Bolt (4 req'd)
9 Fuel injection supply manifold LH
10 Fuel injector (6 req'd)
11 Fuel injector adapter (6 req'd)
12 Bolt
13 Fuel vapor tube
14 Intake manifold

67197-RANG-G36A

Upper engine components—4.0L SOHC engine

15 Bolt (2 req'd)
16 Cover
17 Fuel injection supply manifold RH
18 Oil filler pipe
19 Valve cover RH
20 Oil supply tube RH
21 Camshaft bearing cap (3 req'd)
22 Camshaft RH
23 Camshaft bearing cap
24 Heated positive crankcase ventilation (PCV) coolant hose
25 Heater coolant inlet tube assembly
26 Bolt
27 Lifting eye RH
28 Bolt (2 req'd)
29 Nut (6 req'd)
30 Exhaust manifold RH
31 Exhaust manifold gasket RH
32 Timing chain tensioner RH
33 Spacer
34 Stud (6 req'd)
35 Cylinder head gasket RH
36 Cylinder head RH
37 Clamp (2 req'd)
38 Coolant hose
39 Thermostat housing
40 Bolt (3 req'd)
41 Bolt (16 req'd)
42 Camshaft bearing cap
43 Camshaft LH
44 Timing chain tensioner LH
45 Oil volume reduction plug
46 Bolt
47 Spark plug (6 req'd)
48 Exhaust valve (6 req'd)
49 Cylinder head gasket LH
50 Intake valve (6 req'd)
51 Differential pressure feedback exhaust gas recirculation (EGR) system
52 EGR tube
53 Adapter
54 Exhaust manifold LH
55 Nut (6 req'd)
56 Bolt
57 Exhaust manifold gasket LH
58 Stud (6 req'd)
59 Bolt (2 req'd)
60 Lifting eye LH
61 Ignition coil assembly
62 Bolt (2 req'd)
63 Valve stem seal (12 req'd)
64 Valve spring (12 req'd)
65 Lash adjuster (12 req'd)
66 Valve spring retainer key (24 req'd)
67 Roller follower (12 req'd)
68 Cylinder head bolts (16 req'd)
69 Valve spring retainer seat (12 req'd)
70 Camshaft bearing cap (3 req'd)
71 Oil supply tube LH
72 Valve cover LH
73 Bracket
74 Bolt
75 Bolt
76 Camshaft position (CMP) sensor

67197-RANG-G36B

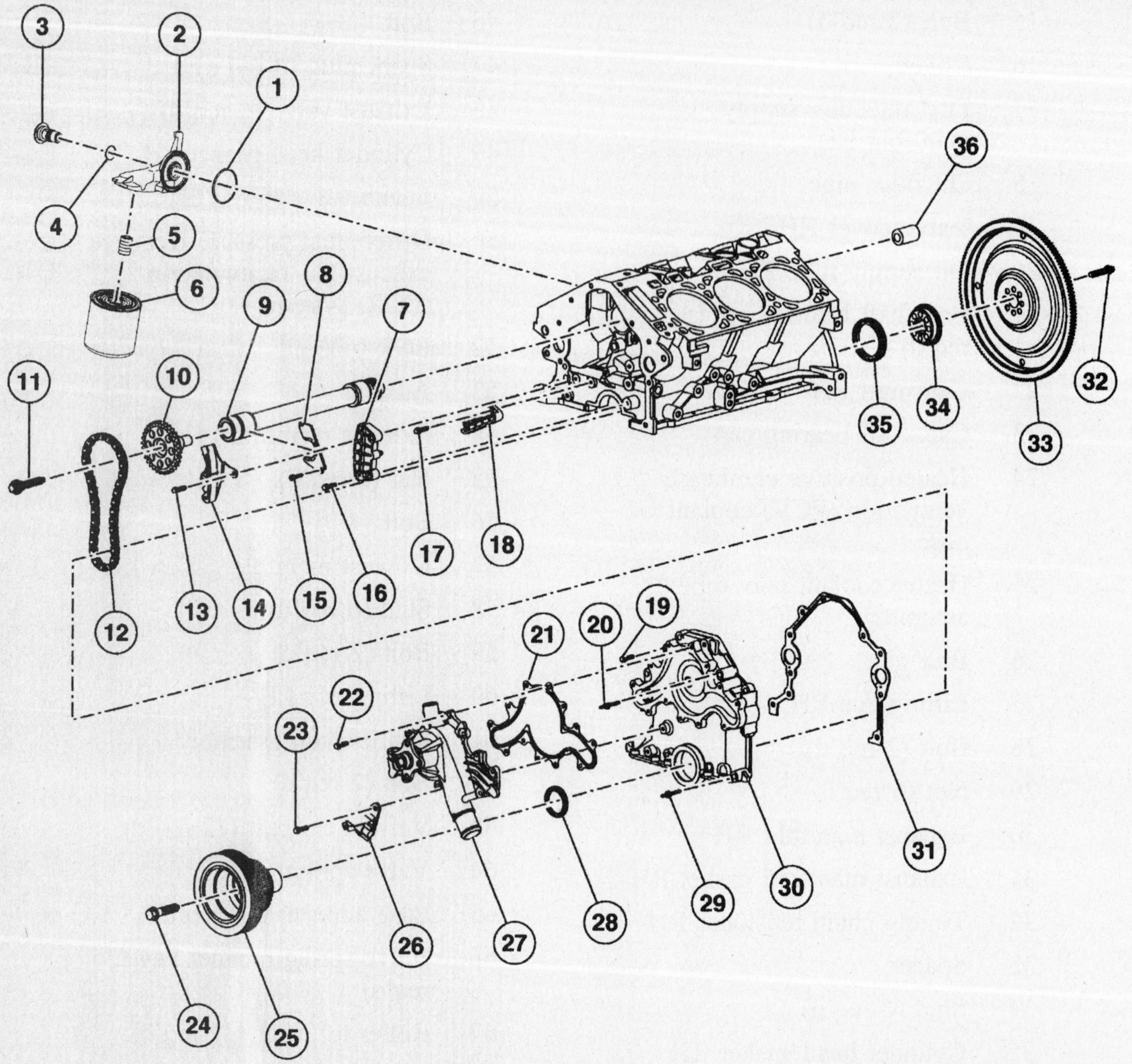

1	Gasket	14	Jackshaft chain tensioner	26	Crankshaft position (CKP) sensor
2	Oil filter adapter	15	Bolt (2 req'd)	27	Coolant pump
3	Bolt	16	Bolt (2 req'd)	28	Crankshaft front seal
4	O-ring	17	Bolt (2 req'd)	29	Stud (4 req'd)
5	Adapter	18	Balance shaft chain guide (4x4 only)	30	Front cover
6	Oil filter	19	Bolt (5 req'd)	31	Gasket
7	Jackshaft chain guide	20	Stud	32	Bolt (8 req'd)
8	Jackshaft thrust plate	21	Gasket	33	Flexplate
9	Jackshaft	22	Bolt (12 req'd)	34	Spacer
10	Jackshaft sprocket	23	Bolt (2 req'd)	35	Crankshaft rear seal
11	Bolt	24	Bolt	36	Spacer
12	Jackshaft chain	25	Crankshaft pulley		
13	Bolt (2 req'd)				

67197-RANG-G37

Front and rear engine components—4.0L SOHC engine

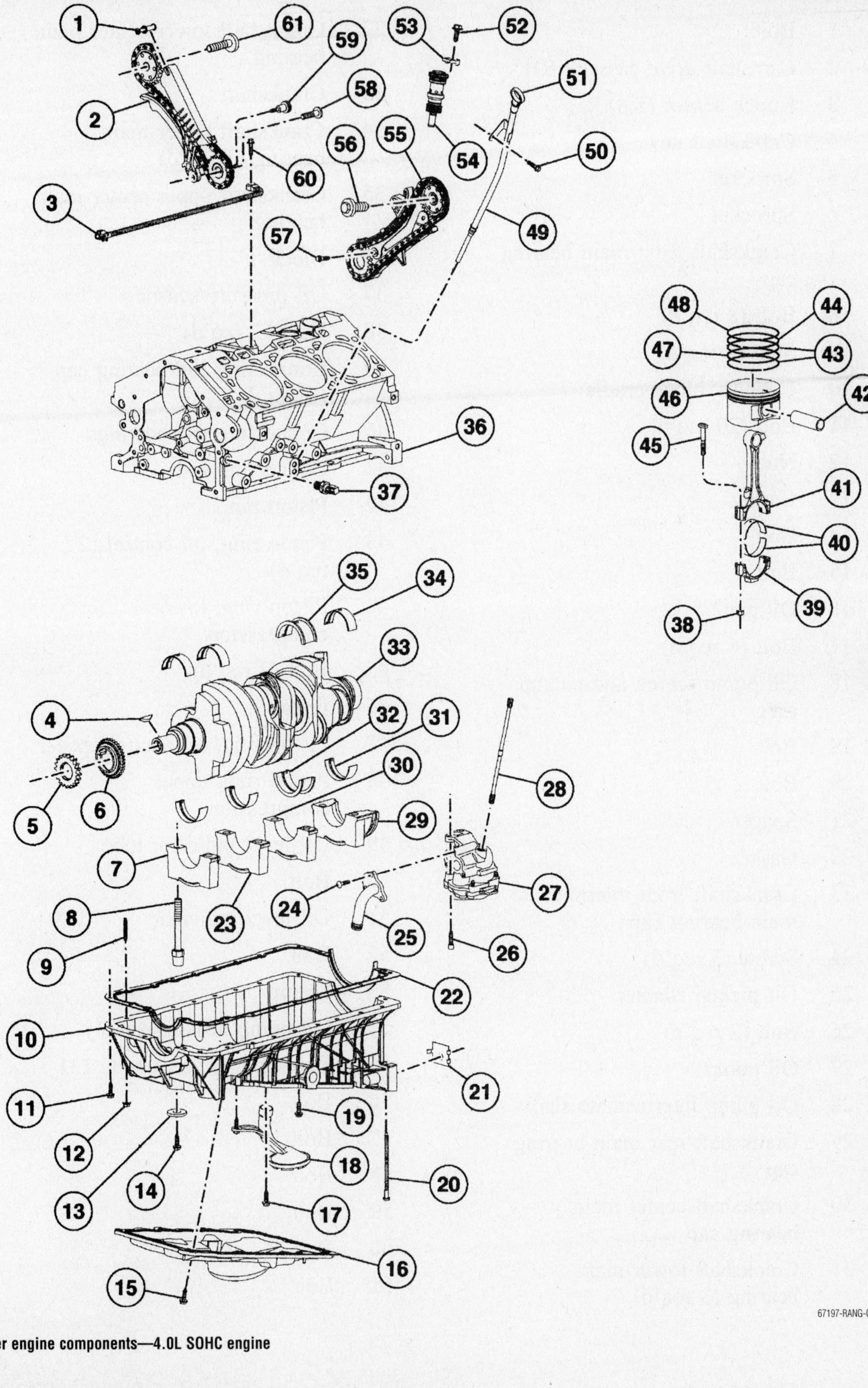

Lower engine components—4.0L SOHC engine

1 Bolt
2 Camshaft drive cassette RH
3 Knock sensor (KS)
4 Crankshaft key
5 Sprocket
6 Sprocket
7 Crankshaft front main bearing cap
8 Bolt (8 req'd)
9 Stud (2 req'd)
10 Cylinder block cradle
11 Bolt (20 req'd)
12 Nut
13 Gasket
14 Bolt
15 Bolt
16 Oil pan
17 Bolt (6 req'd)
18 Oil pump screen and pickup tube
19 Bolt
20 Bolt
21 Spacer
22 Gasket
23 Crankshaft front intermediate main bearing cap
24 Screw (3 req'd)
25 Oil pickup adapter
26 Bolt (2 req'd)
27 Oil pump
28 Oil pump intermediate shaft
29 Crankshaft rear main bearing cap
30 Crankshaft center main bearing cap
31 Crankshaft lower main bearing (3 req'd)
32 Crankshaft lower center main bearing
33 Crankshaft
34 Crankshaft upper main bearing (3 req'd)
35 Crankshaft upper center main bearing
36 Block
37 Oil pressure sensor
38 Nut (12 req'd)
39 Connecting rod bearing cap (6 req'd)
40 Connecting rod bearings
41 Connecting rod
42 Piston pin
43 Piston ring, oil control (2 req'd)
44 Piston ring, lower compression
45 Bolt (2 req'd)
46 Piston
47 Piston ring, oil control spacer
48 Piston ring, upper compression
49 Oil level indicator tube
50 Bolt
51 Oil level indicator
52 Bolt
53 Clamp
54 Oil pump drive assembly
55 Camshaft drive cassette LH
56 Bolt
57 Bolt
58 Bolt
59 Bolt
60 Bolt
61 Bolt

67197-RANG-G38B

b. Using a seal installer, install the crankshaft front seal.
5. Install the crankshaft damper.

4.0L OHC Engine

1. Before servicing the vehicle, refer to the precautions in the beginning of this section.
2. Remove or disconnect the following:
   - Negative battery cable
   - Crankshaft pulley
3. Using a seal remover, remove the crankshaft front oil seal.

To install:

4. Lubricate the seal lip with clean engine oil.
5. Using a seal driver, install the crankshaft front oil seal.
6. Install the crankshaft pulley.

Camshaft and Valve Lifters

➡Although Ford suggests that this component is removable while the engine is installed in the vehicle, depending on the particular options with which your truck is equipped, working clearance may be extremely tight and this procedure may be much easier to perform with the engine removed. Before commencing, read through this procedure and make certain enough clearance, or working room, exists with the engine in the vehicle; if there is not enough space, the engine should be removed.

REMOVAL & INSTALLATION

2.3L Engine

1. Before servicing the vehicle, refer to the precautions in the beginning of this section.
2. Relieve the fuel system pressure.
3. Drain the cooling system.
4. Properly discharge the air conditioning system.
5. Remove or disconnect the following:
   - Negative battery cable
   - Drive belt.
   - Engine oil level indicator assembly.
   - Engine oil level indicator.
   - Engine oil level indicator tube.
   - Water outlet tube.
   - Water outlet tube.
   - Air conditioning compressor.

➡The generator will be removed with the accessory bracket.

   - Accessory bracket.
   - Right motor mount.
   - Coolant hose from the thermostat.

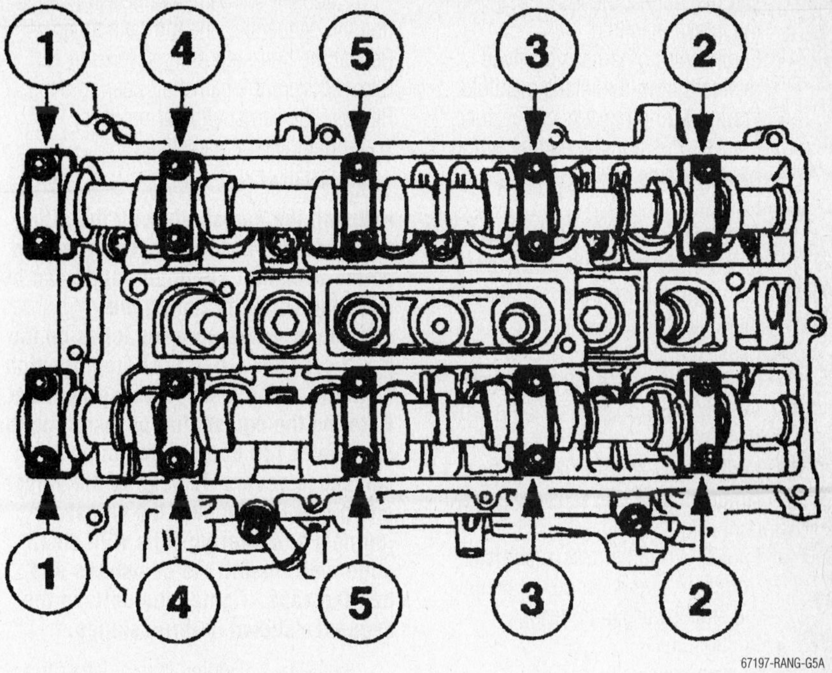

Camshaft cap loosening sequence—2.3L

   - Coolant hose from the EGR valve.
   - Coolant tube assembly.
   - Exhaust manifold and gasket.
   - Block heater (if so equipped).
   - Water outlet.
   - EGR valve.
   - Power steering pump and reservoir as an assembly.
   - Idle air control (IAC) valve.
   - Throttle position (TP) sensor.
   - Manifold absolute pressure (MAP) sensor.
   - Swirl control valve monitor electrical connector.
   - CKP sensor and the wiring harness pin-type retainers.
   - Knock sensor (KS).
   - Electric thermostat.
   - Swirl control valve.
   - CMP sensor electrical connector

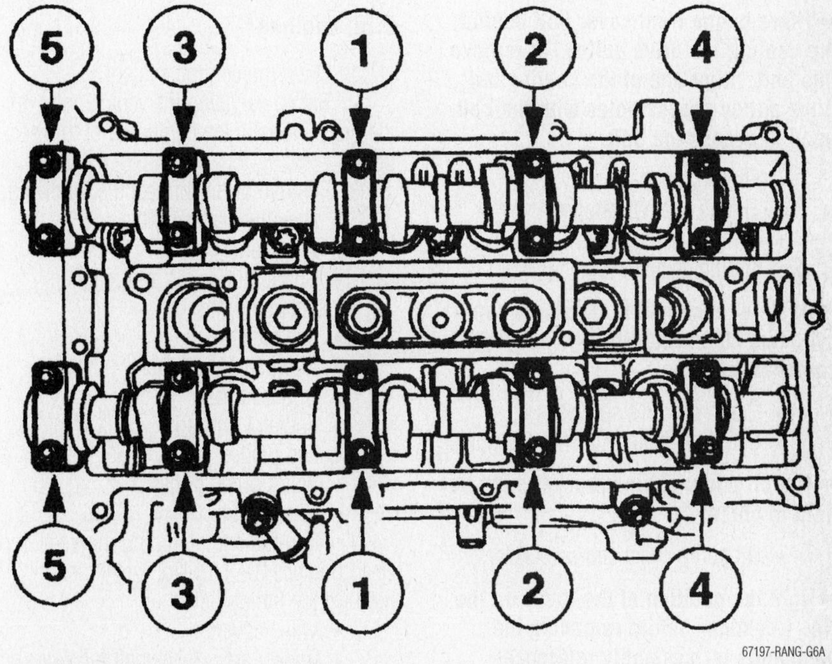

Camshaft cap torque sequence—2.3L

and disconnect the PCV hose from the intake manifold.
- Engine wiring harness pin-type retainers from the intake manifold.
- Engine wiring harness connector bracket. Position the engine wiring harness aside.
- EGR tube.
- Fuel supply line clip from the front of the intake manifold. Disconnect the vacuum hose from the intake manifold.
- Intake manifold assembly.
- Fuel injector electrical connectors. Detach the wiring harness pin-type retainers.
- Ignition coil and the cylinder head temperature (CHT) sensor electrical connectors.
- Engine wiring harness anchors from the valve cover studs. Remove the engine wiring harness.
- Ignition coil.
- Bypass hose.
- Thermostat housing.
- Knock sensor and the engine vent cover.
- Left motor mount.
- Fuel injector supply manifold with the injectors and the ground strap.
- Water pump pulley.
- Water pump.
- CMP sensor.
- CHT sensor.
- Spark plugs.
- Valve cover.
- CKP sensor.
- Crankshaft vibration damper

➡There is one front cover bolt behind the cooling fan drive pulley. To remove this bolt, align one of the cooling fan drive pulley access holes with the bolt head to access the bolt.

- Front cover.
- Timing chain tensioner.
- Timing chain guides.
- Timing chain assembly.

➡Use a wrench on the flats between cylinders No. 1 and No. 2 to hold the camshaft in place.

- Camshaft drive sprockets.
- Oil pump chain tensioner and guide.

➡The oil pump chain sprocket must be held in place.

- Oil pump chain and sprockets.

➡Note the position of the lobes on the No. 1 cylinder before removing the camshafts for assembly reference.

6. Loosen the camshaft bearing cap bolts in sequence, one turn at a time. Repeat the first step until all tension is released from the camshaft bearing caps. Remove the camshaft bearing caps.
7. Remove the camshafts.
8. Installation is the reverse of removal.

➡Install the camshafts with the alignment notches in the camshaft lined up so the camshaft alignment plate can be installed without rotating the camshafts. Make sure the lobes on the No. 1 cylinder are in the same position as noted in the disassembly procedure. Rotating the camshafts, or installing the camshafts 180 degrees out of position can cause severe damage to the valves and pistons. Lubricate the camshaft journals and bearing caps with clean engine oil. Install the camshafts and bearing caps. Tighten the bolts in the sequence shown in three stages.

- Step 1: Tighten the camshaft bearing caps one turn at a time until tight.
- Step 2: Tighten the bolts to 7 Nm (62 inch lbs.)
- Step 3: Tighten the bolts to 16 Nm (12 ft. lbs.)

a. Crankshaft vibration damper:

➡Do not reuse the crankshaft pulley bolt. Tighten the bolt in two stages.

- Step 1: Tighten the bolt to 40 Nm (30 ft. lbs.)
- Step 2: Tighten the bolt and additional 90 degrees (1/4 turn).

2.5L Engines

1. Drain the cooling system.
2. Before servicing the vehicle, refer to the precautions in the beginning of this section.
3. Remove or disconnect the following:
   - Negative battery cable
   - Air cleaner
   - Spark plug wires and retainers
   - Vacuum lines
   - Drive belts
   - Alternator and bracket
   - Upper radiator hose
   - Radiator shroud
   - Fan blades
   - Water pump pulley
   - Fan shroud
4. Align the engine timing marks at Top Dead Center (TDC) for No. 1 cylinder. Remove the timing belt.
   - Valve covers
   - Rocker arms (camshaft followers)
   - Camshaft drive gear and belt guide using a suitable puller. Remove the front oil seal with Front Seal Replacer T74P-6150-A
   - Camshaft retainer located on the rear mounting stand
   - Front motor mount bolts
   - Lower radiator hose from the radiator
   - Automatic transmission cooler lines, if equipped
5. Position a piece of wood on a floor jack and raise the engine carefully as far as it will go. Place blocks of wood between the engine mounts and crossmember pedestals.
6. Remove camshaft by carefully withdrawing it toward the front of the engine. Caution should be used to prevent damage to cam bearings, lobes and journals.
7. Check the camshaft journals and lobes for wear. Inspect the cam bearings, if worn (unless the proper bearing installing tool is on hand), the cylinder head must be removed for new bearings to be installed by a machine shop.

To install:

8. Install or connect the following:
   - Camshaft and lower the engine to its original position
   - Transmission cooler lines, if equipped
   - Lower radiator hose
   - Front motor mount. Torque the bolts to 65 ft. lbs. (88 Nm).
   - Camshaft retainer on the rear mounting stand
   - Camshaft drive gear and belt guide
   - Valve covers
   - Timing belt. Make certain that the timing marks are properly aligned
   - Fan shroud
   - Water pump pulley
   - Fan blades
   - Upper radiator hose and radiator shroud
   - Alternator and bracket
   - Drive belts
   - Vacuum lines
   - Spark plugs wires and retainers
   - Air cleaner
   - Negative battery cable
9. Fill the cooling system.
10. Start the engine and check for leaks, repair if necessary.

3.0L Engine

1. Before servicing the vehicle, refer to the precautions in the beginning of this section.
2. Properly relieve the fuel system pressure.

3. Drain the cooling system.
4. Drain the engine oil.
5. Evacuate the air conditioning system.
6. Remove or disconnect the following:
   - Negative battery cable
   - Air cleaner hoses
   - Fan, spacer and shroud
   - Radiator

7. Rotate the crankshaft so that No. 1 piston is at Top Dead Center (TDC) on the compression stroke.
   - Air conditioning condenser
   - Fuel lines from the fuel supply manifold
   - Vacuum hoses
   - Electrical wiring
   - Engine front cover
   - Water pump
   - Alternator
   - Power steering pump. Do not disconnect the hoses
   - Air conditioning compressor. Do not disconnect the hoses
   - Throttle body
   - Fuel injection wire harness

8. Turn the engine by hand to TDC of the power stroke on No. 1 cylinder.
   - Spark plug wires from the plugs
   - Distributor cap with the spark plug wires as an assembly, if equipped

9. Matchmark the rotor, distributor body and engine. Disconnect the distributor wiring harness and remove the distributor, if equipped.
   - Rocker arm covers
   - Intake manifold
   - Loosen the rocker arm bolts enough to pivot the rocker arms out of the way and remove the pushrods. Identify them for installation
   - Lifters and identify them for installation
   - Crankshaft pulley/damper
   - Starter
   - Oil pan
   - Camshaft gear attaching bolt and washer, then slide the gear off the camshaft
   - Camshaft thrust plate

10. Carefully slide the camshaft out of the engine block, using caution to avoid any damage to the camshaft bearings.

To install:

11. Oil the camshaft journals and cam lobes with heavy SJ engine oil (50W). Install the spacer ring with the chamfered side toward the camshaft, then insert the camshaft key.
12. Install or connect the following:
   - Camshaft using caution to avoid any damage to the camshaft bearings
   - Thrust plate. Torque the screws to 84 inch lbs. (10 Nm).

13. Rotate the camshaft and crankshaft as necessary to align the timing marks. Install the camshaft gear and chain. Torque the bolt to 46 ft. lbs. (62 Nm).
14. Coat the tappets with 50W engine oil and place them in their original locations.
15. Apply 50W engine oil to both ends of the pushrods. Install the pushrods in their original locations.
16. Pivot the rocker arms into position. Torque the fulcrum bolts to 96 inch lbs. (11 Nm).
17. Rotate the engine until both timing marks are at the top of their sprockets and aligned. Torque the following fulcrum bolts to 18 ft. lbs. (24 Nm):
   a. No.1 intake.
   b. No.2 exhaust.
   c. No.4 intake.
   d. No.5 exhaust.
18. Rotate the engine until the camshaft timing mark is at the bottom of the sprocket and the crankshaft timing mark is at the top of the sprocket, and both are aligned. Torque the following fulcrum bolts to 18 ft. lbs. (24 Nm):
   a. No.1 exhaust.
   b. No.2 intake.
   c. No.3 intake and exhaust.
   d. No.4 exhaust.
   e. No.5 intake.
   f. No.6 intake and exhaust.
19. Torque all the bolts to 24 ft. lbs. (33 Nm).
20. Turn the engine by hand to 0 degrees Before Top Dead center (BTDC) of the power stroke on No. 1 cylinder.
21. Install or connect the following:
   - Engine front cover and water pump assembly
   - Oil pan
   - Crankshaft damper/pulley and tighten the retaining bolt to 107 ft. lbs. (145 Nm).
   - Intake manifold
   - Starter
   - Crankshaft pulley and damper
   - Rocker arm covers
   - Rotor and distributor cap, if equipped
   - Spark plug wires
   - Fuel lines to the fuel supply manifold
   - Fuel injection wire harness
   - Throttle body
   - Air conditioning compressor
   - Power steering pump
   - Alternator
   - Water pump
   - Engine front cover
   - All electrical connectors and vacuum lines
   - Air conditioning condenser
   - Radiator
   - Fan, spacer and shroud
   - Air cleaner hoses
   - Negative battery cable

22. Recharge the air conditioning system.
23. Refill the cooling system.
24. Replace the oil filter and refill the engine with the specified amount of engine oil.
25. Start the engine and check the ignition timing and idle speed. Adjust if necessary. Run the engine at fast idle and check for coolant, fuel, vacuum or oil leaks.

4.0L OHC Engine

1. Before servicing the vehicle, refer to the precautions in the beginning of this section.
2. Remove or disconnect the following:
   - Negative battery cable for safety
   - Valve cover
   - Hydraulic camshaft tensioner

➡**The right-hand camshaft sprocket bolt uses left-hand threads.**

3. For the right-hand camshaft use the Cam Gear Torque Adapter tool T97T-6256-F, to remove the camshaft sprocket bolt.
4. For the left-hand camshaft, remove the sprocket bolt.

➡**When removing the followers, label them so that they may be returned to their original positions.**

5. Using the Valve Spring Compressor tool ST1330-A, remove the camshaft roller followers.
6. Remove or disconnect the following:
   - Camshaft bearing cap bolts and the oil rail
   - Camshaft

To install:

7. Lubricate all of the moving parts with SAE 50W engine oil.
8. Install camshaft onto the cylinder head.
9. Position the oil rail and install the bearing caps and bolts. Torque the bolts in 2 steps:
   a. Step 1—53.5 inch lbs. (6 Nm).
   b. Step 2—11–12.5 ft. lbs. (15–17 Nm).
10. Install or connect the following:
   - Camshaft followers
   - Camshaft sprocket bolt and hand tighten the bolt

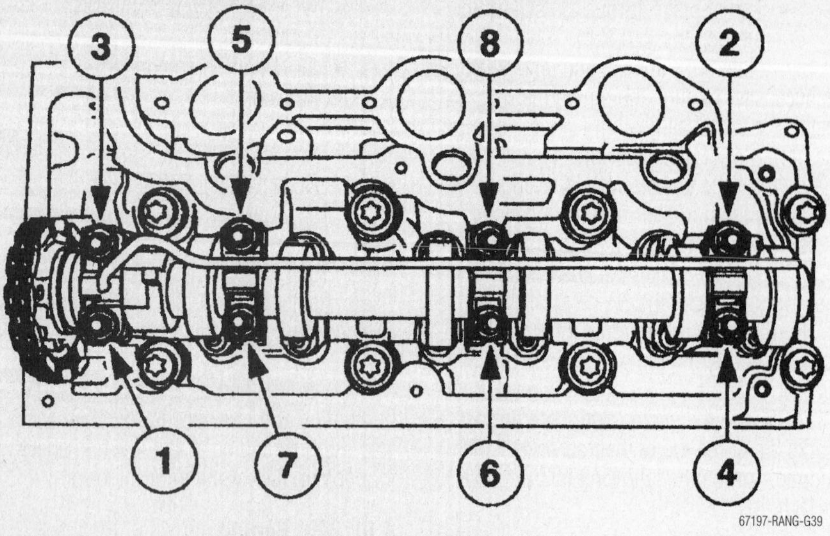

Camshaft bolt removal sequence—4.0L SOHC engine

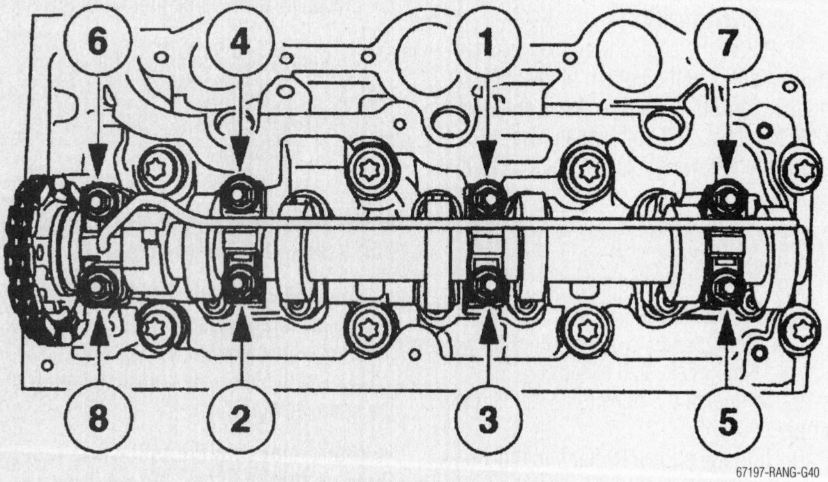

Camshaft bolt tightening sequence—4.0L SOHC engine

- Camshaft Chain Tensioner T97T-6K254-A in the hole that the hydraulic chain tensioner was in

11. Turn the crankshaft one revolution clockwise until No. 1 piston is Top Dead Center (TDC).
12. Install or connect the following:
    - Crankshaft Holding tool T97T-6303-A on the crankshaft to keep it from turning
    - Position the timing slot on the rear of the camshaft to fit Camshaft Holding tool T97T-6256-C and install the holding tool on the rear of the head
    - Camshaft Gear Holding tool T97T-6256-B and Camshaft Gear Holding tool T97T-6256-A on the front of the cylinder head to securely hold the camshaft gear
    - Tighten the camshaft sprocket bolt to 63 ft. lbs. (85 Nm).
13. Remove the Camshaft Chain Tensioner tool and install the hydraulic chain tensioner, tighten the tensioner to 35–39 ft. lbs. (47–53 Nm).
14. Remove the special tools from the engine.
15. Install or connect the following:
    - Valve cover
    - Negative battery cable
16. Start the engine check for leaks and repair if necessary.

Oil Pan

REMOVAL & INSTALLATION

2.3L Engine

1. Before servicing the vehicle, refer to the precautions in the beginning of this section.
2. Drain the engine oil.
3. Remove or disconnect the following:
    - Engine from the vehicle
    - Engine oil level indicator assembly
    - Engine oil pan bolts and oil pan

To install:

4. Clean and inspect all mating surfaces.

➡The oil pan must be installed and the bolts tightened with four minutes of applying the silicone gasket and sealant.

5. Apply a 2.5 mm bead of silicone gasket and sealant to the oil pan. Install the oil pan. Tighten the oil pan in the sequence shown.
6. Lubricate the O-ring with clean engine oil and install the engine oil level indicator assembly.
7. Install the engine into the vehicle.

2.5L Engines

1. Before servicing the vehicle, refer to the precautions in the beginning of this section.
2. Drain the engine oil.
3. Remove or disconnect the following:
    - Negative battery cable
    - Engine from the vehicle and place it on a suitable engine stand
    - Oil pan and discard the gasket

To install:

4. Clean the mating surface on the oil pan.
5. Install or connect the following:
    - Oil pan gasket
    - Apply a bead of silicone sealant to the oil pan
    - Oil pan. Torque the bolts in sequence to 141 inch lbs. (16 Nm).
    - Engine
    - Negative battery cable
6. Fill the engine with clean oil.
7. Start the vehicle and check for leaks, repair if necessary.

3.0L Engine

2WD

1. Before servicing the vehicle, refer to the precautions in the beginning of this section.
2. Drain the engine oil.
3. Remove or disconnect the following:
    - Negative battery cable
    - Oil level dipstick tube
    - Fan shroud. Leave the fan shroud over the fan assembly
    - Motor mount nuts from the frame

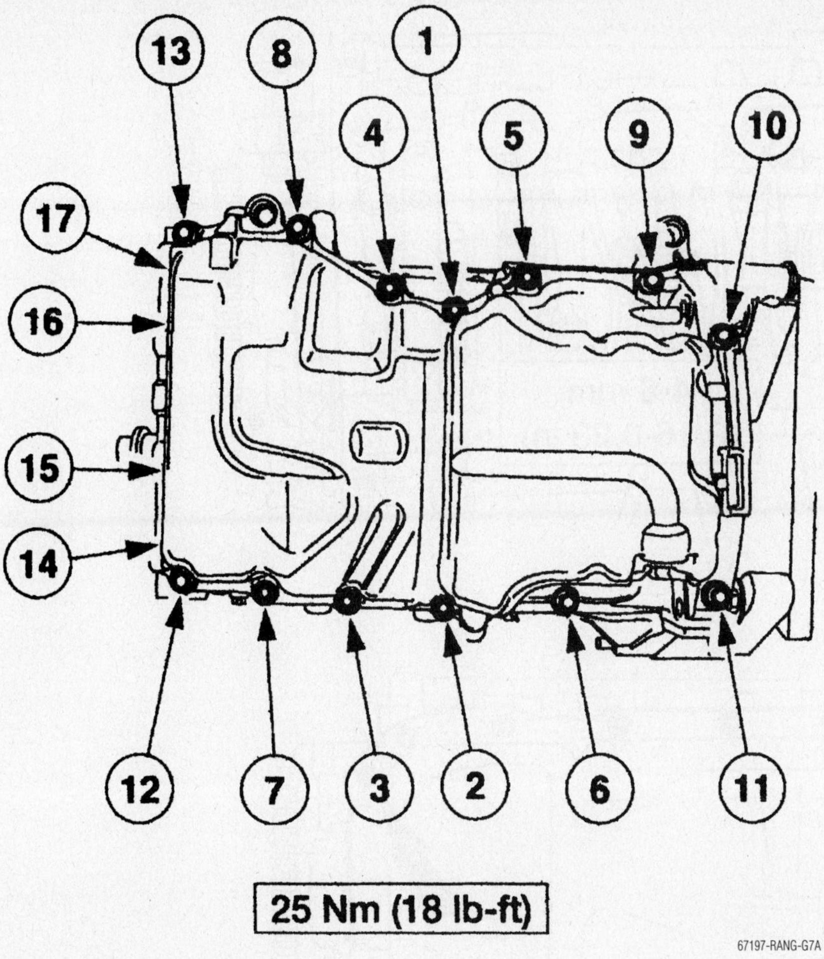

Oil pan torque sequence—2.3L

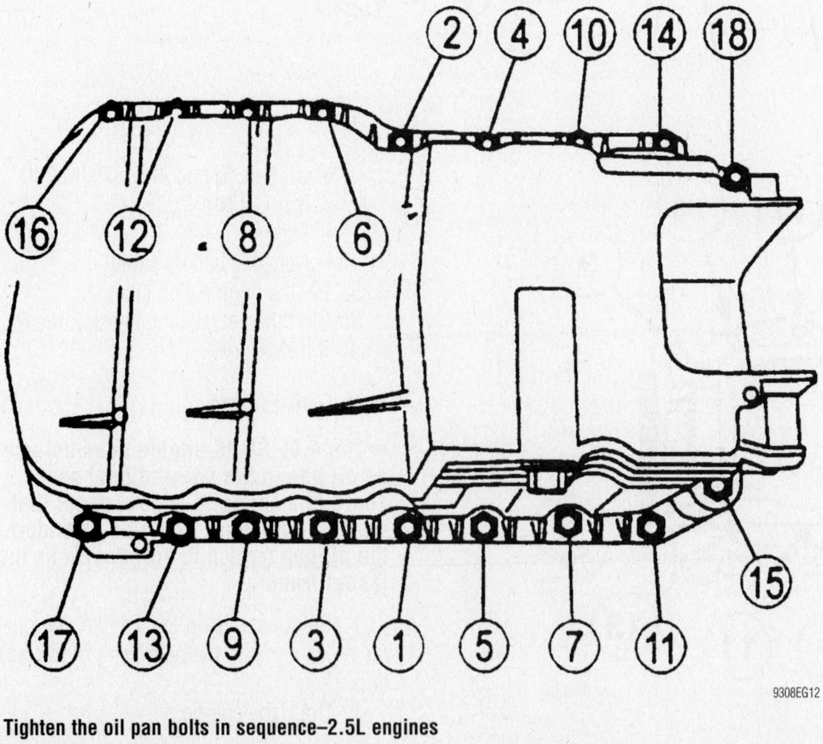

Tighten the oil pan bolts in sequence–2.5L engines

⁂ WARNING

On models equipped with distributor ignition, failure to remove the distributor will damage or break it when the engine is lifted.

- Starter
- Transmission inspection cover
- Right hand axle I-beam. The brake caliper must be removed and secured out of the way.
- Oil pan attaching bolts, using a suitable lifting device, raise the engine about 2 in. (5cm)
- Oil pan and discard the gasket

➡The oil pan fits tightly between the transmission spacer plate and oil pump pick-up tube. Use care when removing the oil pan from the engine.

4. Clean all gasket surfaces on the engine and oil pan. Remove all traces of old gasket and/or sealer.

To install:

5. Apply a 1/8 (4mm) bead of RTV sealer to the junctions of the rear main bearing cap and block, and the front cover and block. The sealer sets in 15 minutes, so work quickly!
6. Apply adhesive to the gasket surfaces and install the oil pan gasket.
7. Install or connect the following:
   - Oil pan on the engine block. Torque the bolts EVENLY to 9 ft. lbs. (12 Nm) working from the center to the end position on the oil pan.
   - Right hand axle I-beam
   - Brake caliper
   - Transmission inspection cover
   - Starter
   - Fan shroud
   - Motor mount retaining nuts
   - Oil level dipstick tube
   - Negative battery cable
8. Fill the engine with clean oil.
9. Start the vehicle and check for leaks, repair if necessary.

4WD

1. Before servicing the vehicle, refer to the precautions in the beginning of this section.
2. Drain the engine oil.
3. Remove or disconnect the following:
   - Negative battery cable
   - Engine from the vehicle and place it on a suitable engine stand
   - Oil pan and discard the gasket

To install:

4. Install or connect the following:
   - New oil pan gasket and secure the gasket with trim adhesive

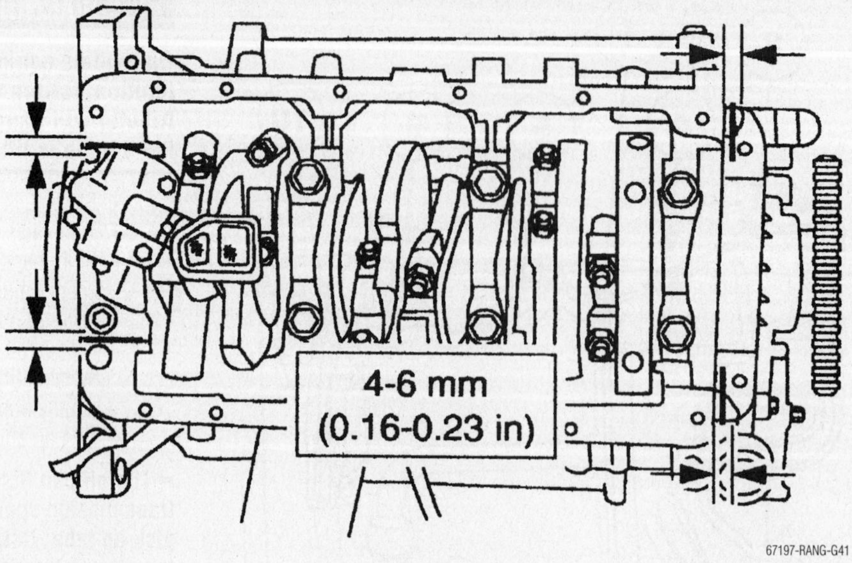

Oil pan sealer application—3.0L engine

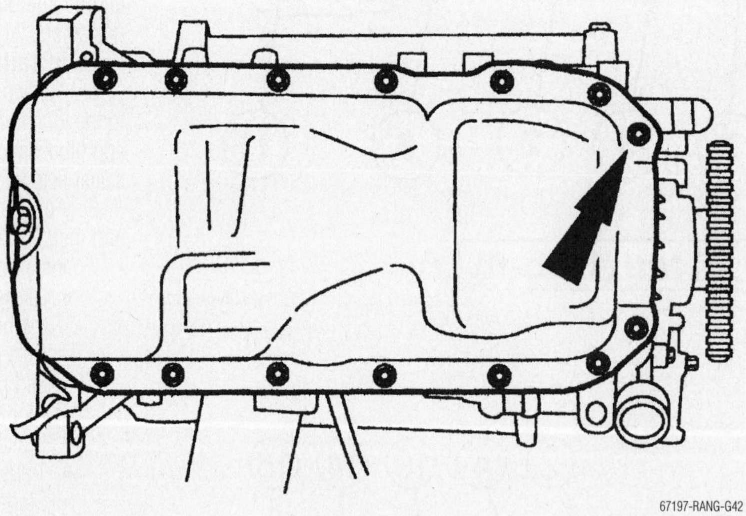

Oil pan bolts—3.0L engine with 2-wheel drive

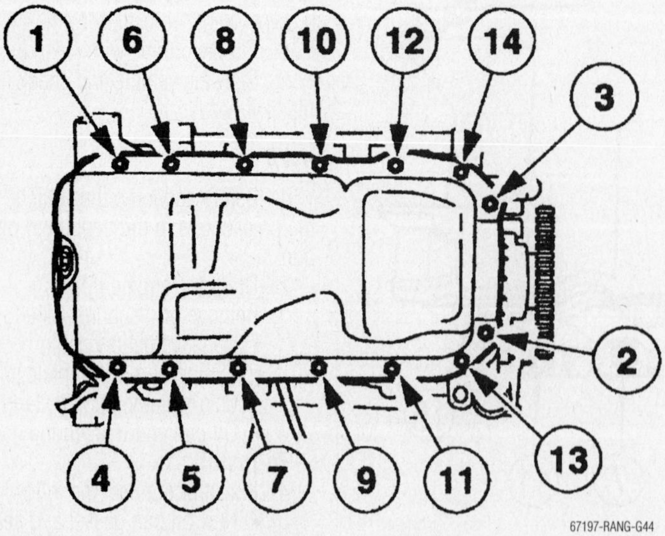

Oil pan bolts—3.0L engine with 4-wheel drive

- Oil pan. Torque the bolts to 9 ft. lbs. (12 Nm).
- Engine
- Negative battery cable

5. Fill the engine with clean oil.
6. Start the vehicle and check for leaks, repair if necessary.

4.0L OHC Engine

➡The 4.0L SOHC engine does not use an oil pan in the conventional sense. There is a separate access panel that unbolts from what would be considered the oil pan (which is now known as the ladder frame).

1. Before servicing the vehicle, refer to the precautions in the beginning of this section.
2. Drain the engine oil.

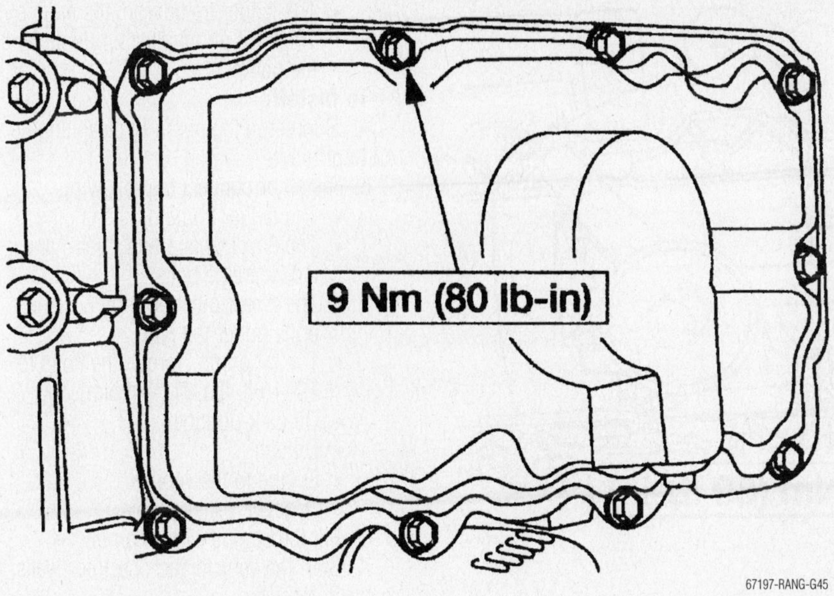

Oil pan installation—4.0L SOHC engine

3. Remove or disconnect the following:
   - Negative battery cable
   - Oil pan and discard the gasket

To install:

4. Install or connect the following:
   - New gasket and oil pan. Torque the bolts to 80 inch lbs. (9 Nm).
   - Negative battery cable
5. Fill the engine with clean oil.
6. Start the vehicle and check for leaks, repair if necessary.

Oil Pump

REMOVAL & INSTALLATION

2.3L Engine

➨The oil pump is located on the front of the engine and is turned by the timing belt.

1. Before servicing the vehicle, refer to the precautions in the beginning of this section.
2. Remove or disconnect the following:
   - Negative battery cable
   - Timing chain
   - Oil pump chain and sprockets
   - Oil pan
   - Oil pump pickup tube and gasket
   - Oil pump assembly and gasket

To install:

3. Turn the crankshaft clockwise to position the No. 1 piston.
4. Remove the plug bolt.
5. Install the Engine Timing Peg 303-507.

➨Clean the gasket surface with metal surface cleaner.

6. Install a new gasket and the oil pump assembly. Tighten the bolts in the sequence shown in two stages.
   - Step 1: Tighten the bolts to 10 Nm (80 inch lbs.)
   - Step 2: Tight the bolts to 23 Nm (17 ft. lbs.)
7. Install a new oil pump pickup tube gasket and the pickup tube. Tighten the bolts in the sequence shown

2.5L Engines

➨The oil pump is located on the front of the engine and is turned by the timing belt.

1. Before servicing the vehicle, refer to the precautions in the beginning of this section.
2. Remove or disconnect the following:
   - Negative battery cable
   - Timing belt
   - Camshaft Position (CMP) sensor electrical connector
   - Oil pump sprocket
   - CMP sensor

➨Use a prybar or drift through one of the holes in the pump sprocket to keep it from turning while loosening the bolt.

   - 4 bolts retaining the oil pump to the engine block
   - Oil pump from the front of the engine and discard the gasket
3. Inspect the oil pump and O-rings and replace as necessary.
4. Clean all gasket mating surfaces thoroughly.

To install:

5. Prime the oil pump and with 8 ounces (236 ml) of new engine oil and lubricate the O-rings.
6. Install or connect the following:
   - New gasket on the oil pump
   - Oil pump. Torque the bolts to 89 inch lbs. (10 Nm).
   - CMP sensor. Torque the bolts to 61 inch lbs. (7 Nm).
   - Oil pump sprocket bolt. Torque the bolt to 40 ft. lbs. (55 Nm).
   - CMP sensor electrical connector
   - Timing belt
   - Negative battery cable
7. Fill the engine with clean oil.
8. Start the vehicle and check for leaks, repair if necessary.

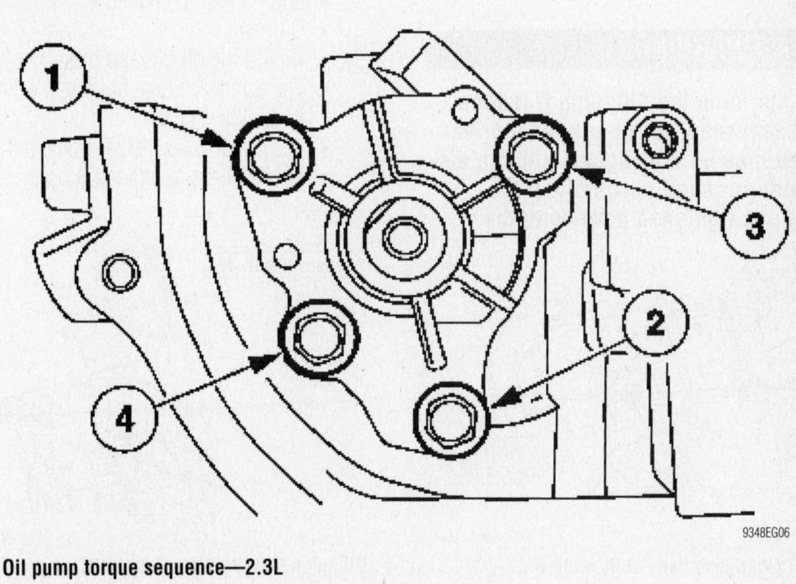

Oil pump torque sequence—2.3L

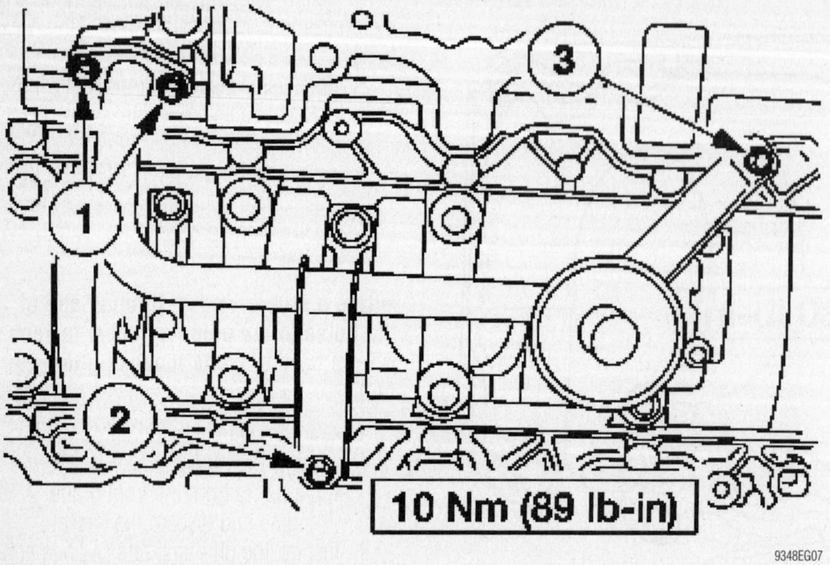

Oil pump pickup tube torque sequence—2.3L

3.0 L Engines

1. Before servicing the vehicle, refer to the precautions in the beginning of this section.
2. Drain the engine oil.
3. Remove or disconnect the following:
   - Negative battery cable
   - Oil pan
   - Oil pick-up and tube assembly from the pump
   - Oil pump retainer bolts and the oil pump

To install:

4. Prime the oil pump with clean engine oil by filling either the inlet or outlet port. Rotate the pump shaft to distribute the oil within the pump body.
5. Install the oil pump and tighten the mounting bolts to 30–40 ft. lbs. (41–54 Nm).

⁂ WARNING

Do not force the oil pump if it does not seat readily. The oil pump driveshaft may be misaligned with the distributor or shaft assembly. If the pump is tightened down with the driveshaft misaligned, damage to the pump could occur. To align, rotate the intermediate driveshaft into a new position.

6. Install or connect the following:
   - Oil pick-up and tube assembly
   - Oil pan
7. Fill the engine with clean oil.
8. Start the vehicle and check for leaks, repair if necessary.

4.0L SOHC Engines

➡The oil pump cannot be removed with the engine in the vehicle.

1. Before servicing the vehicle, refer to the precautions in the beginning of this section.
2. Drain the engine oil.
3. Remove or disconnect the following:
   - Engine from the vehicle
   - Oil pan
   - Unbolt the oil pick-up tube
   - The 8 ladder frame bolts that were under the oil pan
   - The 2 rear outer ladder frame bolts
   - The 7 left-hand and the 8 right-hand ladder frame bolts
   - The ladder frame from the engine
   - The 2 oil pump attaching bolts and the pump.

To install:

4. Submerge the pump in clean engine oil to prime it.
5. Install or connect the following:
   - The ladder frame on the engine
   - The 8 right-hand and 7 left-hand ladder frame bolts
   - The 2 rear outer and the 8 frame bolts under the pan
   - The oil pump. Torque the bolts to 13–15 ft. lbs. (17–21 Nm).
   - Oil pick-up tube
   - Oil pan
   - Engine to the vehicle
   - Negative battery cable
6. Fill the engine with clean oil.
7. Start the vehicle and check for leaks, repair if necessary.

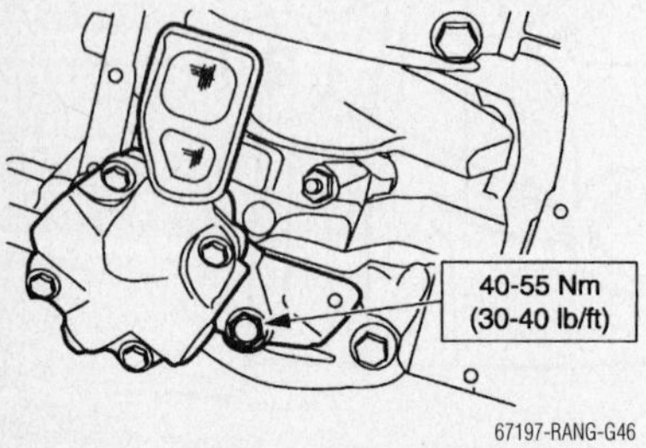

Oil pump installed—3.0L engine

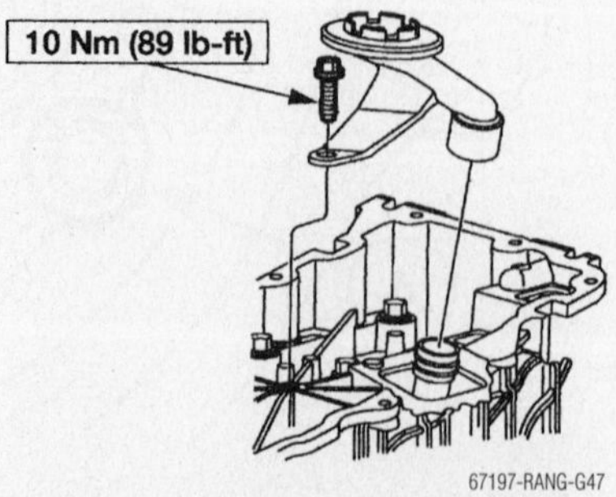

Oil pump installation—4.0L SOHC engine

Rear Main Seal

REMOVAL & INSTALLATION

2.3L Engine

1. Before servicing the vehicle, refer to the precautions in the beginning of this section.
2. Remove or disconnect the following:
   - Flywheel or flexplate
   - Bolts and the crankshaft rear oil seal

To install

3. Install or connect the following:
   - Rear oil seal on the Crankshaft Rear Main Oil Seal Installer
   - Crankshaft Rear Main Oil Seal Installer and the crankshaft rear oil seal on the crankshaft
4. Tighten the bolts in the sequence shown to 10 Nm (89 inch lbs.)
5. Remove the Crankshaft Rear Main Oil Seal Installer.
6. Install the flywheel or flexplate.

2.5L Engine

1. Before servicing the vehicle, refer to the precautions in the beginning of this section.
2. Remove the flywheel or flexplate

➡Clean the crankshaft rear oil seal and cylinder block prior to removing the rear oil seal.

3. Screw in the Jet Plug Remover.
4. Remove the seal.

To install

➡Apply 5W-30 motor oil to seal and seal edge.

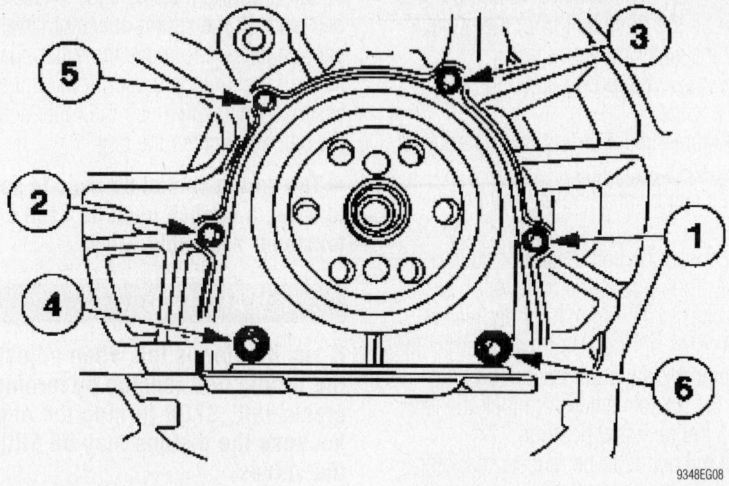

Rear main seal torque sequence—2.3L

5. Install crankshaft rear oil seal on Rear Main Seal Replacer.
6. Install the Rear Main Seal Replacer and the crankshaft rear oil seal on the crankshaft.
7. Alternate bolt tightening to crankshaft rear oil seal.
8. Install the flywheel or flexplate.

3.0L Engines

1. Before servicing the vehicle, refer to the precautions in the beginning of this section.
2. Remove the flexplate or flywheel.

⁂ WARNING

Use care to avoid scratching or damaging the oil seal surface or leakage may occur.

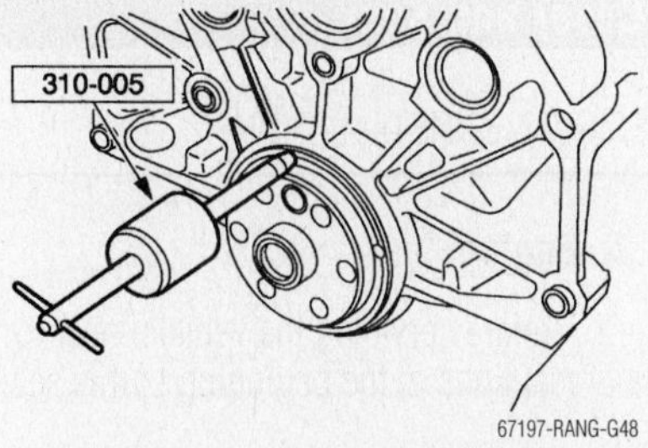

Rear main seal removal—3.0L engine

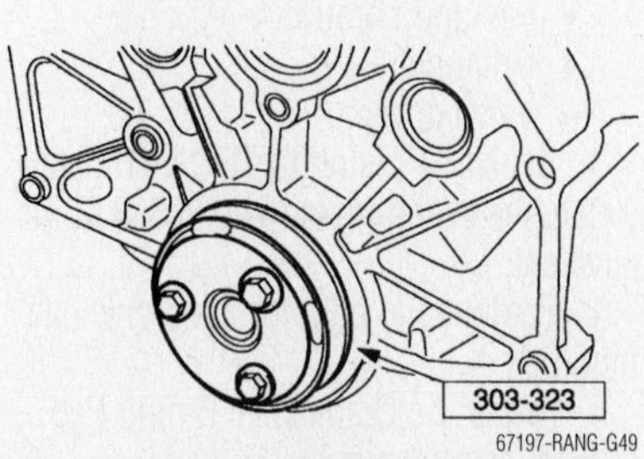

Rear main seal installation—3.0L engine

3. Using a sharp awl, punch one hole into the crankshaft rear oil seal metal surface between the seal lip and the cylinder block.
4. Screw the threaded end of the special tool into the oil seal. Use the special tool to remove the crankshaft rear oil seal.

To install:

5. Lubricate the outer lips and the inner seal on the crankshaft rear oil seal with clean engine oil.
6. Using the special tool, install the crankshaft rear oil seal. Alternate bolt tightening to correctly seat the crankshaft rear oil seal.
7. Install the flexplate or flywheel.

4.0L SOHC Engine

1. Before servicing the vehicle, refer to the precautions in the beginning of this section.
2. Remove the flexplate or flywheel.

⁂ WARNING

Avoid scratching or damaging the oil crankshaft seal running surface during removal of the crankshaft rear oil seal.

3. Using the special tool, remove the crankshaft rear oil seal.

To install:

➡Be sure the crankshaft rear sealing surface is clean and free of any rust or corrosion. To clean the crankshaft rear sealing surface, use extra-fine emery cloth or extra-fine 0000 steel wool with metal surface cleaner.

4. Lubricate the crankshaft rear oil seal with clean engine oil and install on the special tool.

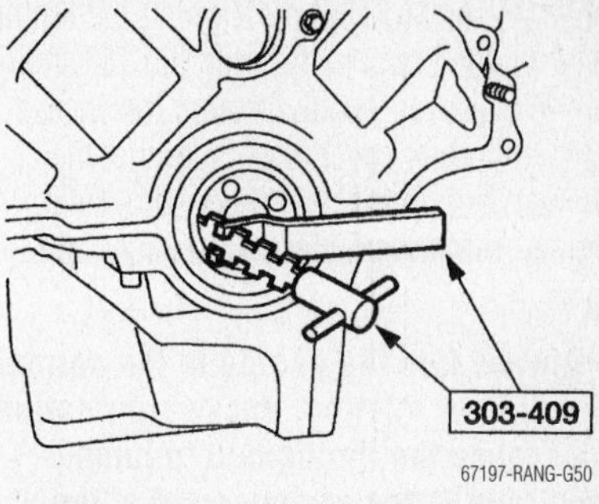

Rear main seal removal—4.0L SOHC engine

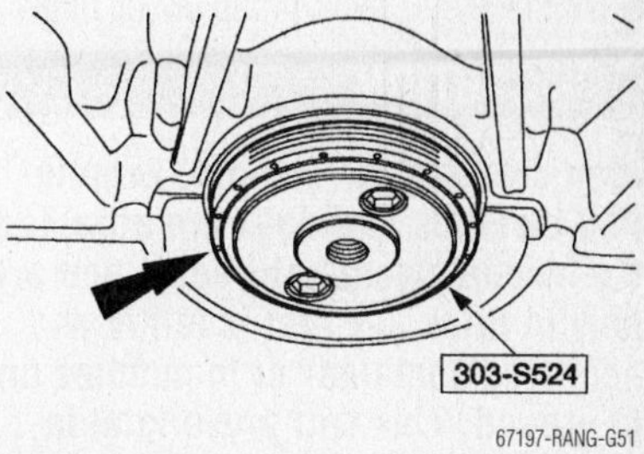

Front part of installation tool—4.0L SOHC engine

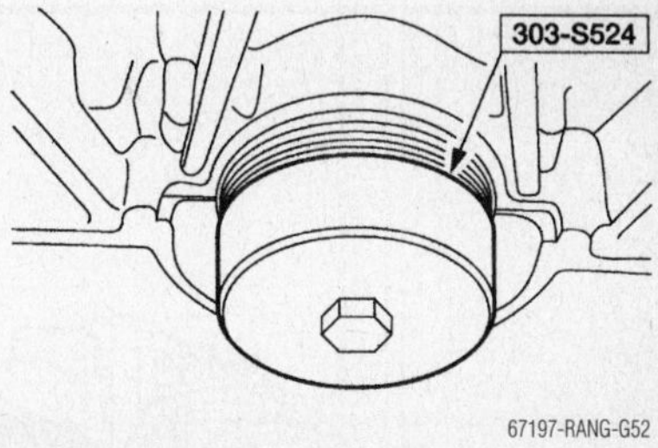

Rear main seal installation—4.0L SOHC engine

5. Using the special tool, install the crankshaft rear oil seal.
6. Install the flexplate or flywheel.

Timing Belt and Cover

REMOVAL & INSTALLATION

2.5L Engine

1. Before servicing the vehicle, refer to the precautions in the beginning of this section.
2. Rotate the engine so that No. 1 cylinder is at Top Dead Center (TDC) on the compression stroke. Check that the timing marks are aligned on the camshaft and crankshaft pulleys. An access plug is provided in the cam belt cover so that the camshaft timing can be checked without

removal of the cover or any other parts. Set the crankshaft to TDC by aligning the timing mark on the crank pulley with the TDC mark on the belt cover. Look through the access hole in the belt cover to be sure that the timing mark on the cam drive sprocket is aligned with the pointer on the inner belt cover.

➡**Always turn the engine in the normal direction of rotation. Backward rotation may cause the timing belt to jump time, due to the arrangement of the belt tensioner.**

3. Drain cooling system. Remove the upper radiator hose as necessary. Remove the fan blade and water pump pulley bolts.

✲✲ CAUTION

When draining the coolant, keep in mind that cats and dogs are attracted by ethylene glycol antifreeze, and are likely to drink any that is left in an uncovered container or in puddles on the ground. This will prove fatal in sufficient quantity. Always drain the coolant into a sealable container. Coolant should be reused unless it is contaminated or several years old.

4. Loosen the alternator retaining bolts and remove the drive belt from the pulleys. Remove the water pump pulley.

5. Remove the power steering pump and set it aside.

6. Remove the 4 timing belt outer cover retaining bolts and remove the cover. Remove the crankshaft pulley and belt guide.

7. Loosen the belt tensioner pulley assembly, then position a camshaft belt adjuster tool T74P-6254-A, or equivalent, on the tension spring roll pin and retract the belt tensioner away from the timing belt. Tighten the adjustment bolt to lock the tensioner in the retracted position.

8. If the belt is to be reused, mark the direction of rotation on the belt for installation reference.

9. Remove the timing belt.

To install:

10. Install the new belt over the crankshaft sprocket and then counterclockwise over the auxiliary and camshaft sprockets, making sure the lugs on the belt properly engage the sprocket teeth on the pulleys. Be careful not to rotate the pulleys when installing the belt.

11. Release the timing belt tensioner pulley, allowing the tensioner to take up the belt slack. If the spring does not have enough tension to move the roller against the belt (belt hangs loose), it might be necessary to manually push the roller against the belt and tighten the bolt.

➡**The spring cannot be used to set belt tension; a wrench must be used on the tensioner assembly.**

✲✲ WARNING

If any binding is felt when adjusting the timing belt tension by turning the crankshaft, STOP turning the engine, because the pistons may be hitting the valves.

12. Rotate the crankshaft 2 complete turns by hand (in the normal direction of rotation) to remove slack from the belt. Tighten the tensioner adjustment to 26–33 ft. lbs. (35–45 Nm) and pivot bolts to 30–40 ft. lbs. (40–55 Nm). Be sure the belt is seated properly on the pulleys and that the timing marks are still in alignment when No. 1 cylinder is again at TDC/compression.

13. Install the crankshaft pulley and belt guide.

14. Install the timing belt cover.

15. Install the water pump pulley and fan blades. Install the upper radiator hose if necessary. Refill the cooling system.

16. Install the accessory drive belts.

17. Start the engine and check the ignition timing. Adjust the timing, if necessary.

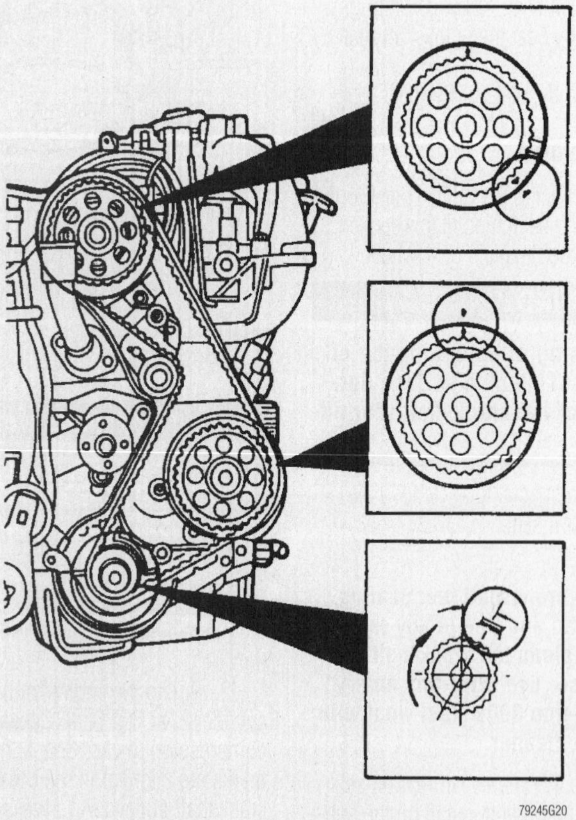

Camshaft, auxiliary shaft and crankshaft timing belt sprocket alignment mark locations—2.5L engines

Timing Chain, Sprockets, Front Cover and Seal

REMOVAL & INSTALLATION

2.3L Engine

1. Before servicing the vehicle, refer to the precautions in the beginning of this section.
2. Remove or disconnect the following:
   - Negative battery cable
   - Fan and shroud
   - Drive belt
   - Valve cover
3. Set No. 1 piston to TDC and install the Camshaft Alignment Plate 303-376, or equivalent.
4. Remove the plug for the crankshaft timing peg.
5. Install the Crankshaft Timing Peg 303-507, or equivalent.

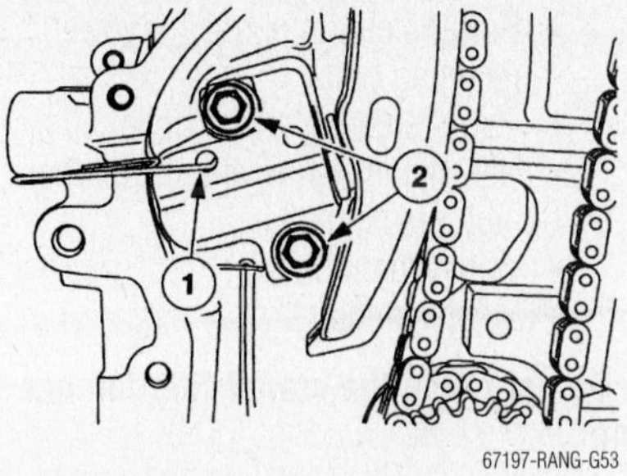

Timing chain tensioner removal; 1-paper clip, 2-bolts—2.3L engine

6. Install an M6 bolt into the crankshaft pulley to verify the engine timing.
7. Remove or disconnect the following:
   - Camshaft pulley
   - Crankshaft position sensor
   - Crankshaft position sensor
   - Belt tensioner
   - Water pump pulley
   - Power steering high pressure hose. Remove the nylon O-ring.
   - Power steering return hose
   - Power steering pump

➡This step is needed only if a new front cover is being installed.

8. Using a three-jaw puller, remove the fan drive pulley.

➡There is one bolt behind the cooling fan drive pulley. This bolt can be accessed by lining up one of the holes in the pulley with the bolt.

9. Remove the bolts and the engine front cover.
10. Compress the timing chain tensioner and remove the tensioner.

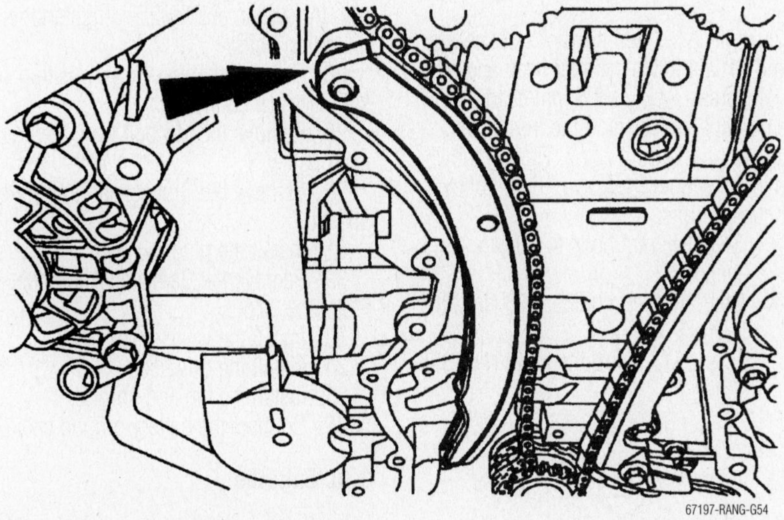

Right side timing chain guide—2.3L engine

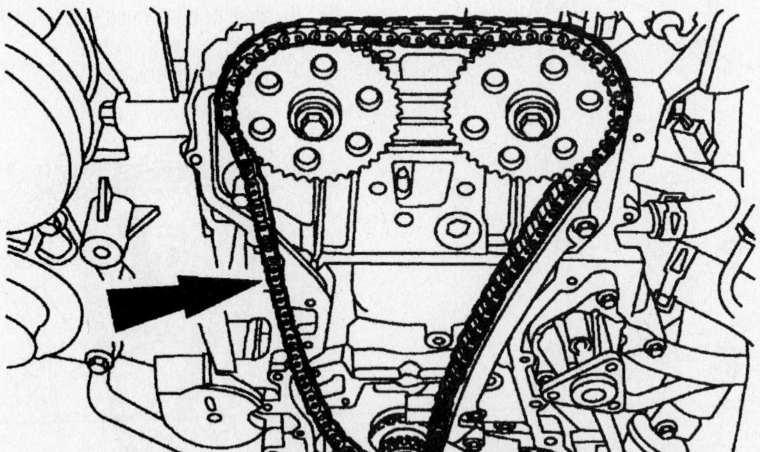

Timing chain removal—2.3L engine

11. Remove the right-hand timing chain guide.
12. Remove the timing chain.
13. Remove the bolts and the left-hand timing chain guide.

⁂ WARNING

Do not rely on the Camshaft Alignment Plate to prevent camshaft rotation. Damage to the tool or the camshaft can occur.

14. If necessary, remove the bolts and the camshaft sprockets. Use the flats on the camshaft to prevent camshaft rotation.

To install:

15. Remove the special tool.

⁂ WARNING

Do not rotate the camshafts. Damage to the valves and pistons can occur.

If the camshaft sprockets were not removed, use the flats on the camshafts to prevent camshaft rotation and loosen the sprocket bolts.

16. If removed, install the camshaft sprockets and the bolts. Do not tighten the bolts at this time.
17. Install or connect the following:
   - Left-hand timing chain guide and bolts
   - Timing chain
   - Right-hand timing chain guide
   - Timing chain tensioner and release the piston
   - Timing chain tensioner and the bolts
18. Remove the drill rod to release the piston.
19. Install the special tool.

⁂ WARNING

Do not rely on the Camshaft Alignment Plate to prevent camshaft rotation. Damage to the tool or the camshafts can result. Using the flats on the camshafts to prevent camshaft rotation, tighten the bolts.

➡This step is needed only if a new front cover is being installed.

20. Install the fan drive pulley using a nut and bolt with flat washers.
21. Clean and inspect the mounting surfaces of the engine and the front cover.

➡The engine front cover must be installed and the bolts tightened within four minutes of applying the silicone gasket and sealant.

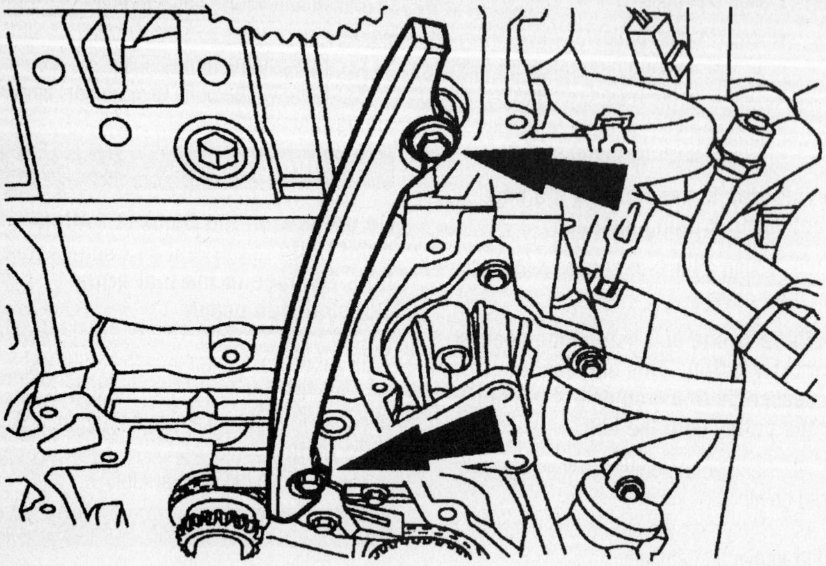

67197-RANG-G56

Left timing chain guide—2.3L engine

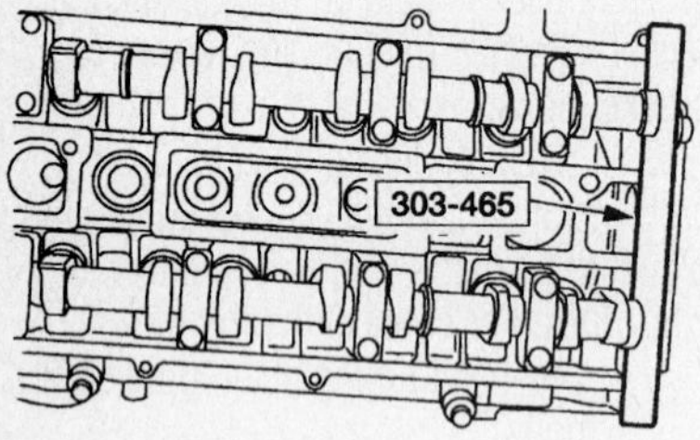

67197-RANG-G57

Camshaft alignment plate installed—2.3L engine

22. Apply a 2.5 mm bead of silicone gasket and sealant to the cylinder head and oil pan joint areas. Apply a 2.5 mm bead of silicone gasket and sealant to the front cover.

23. Install the front cover. Tighten the bolts in the sequence shown, to the following specifications:
- Step 1: 8 mm bolts to 10 Nm (89 inch lbs.)
- Step 2: 10 mm bolts to 25 Nm (18 ft. lbs.)
- Step 3: 13 mm bolts to 48 Nm (35 ft. lbs.)

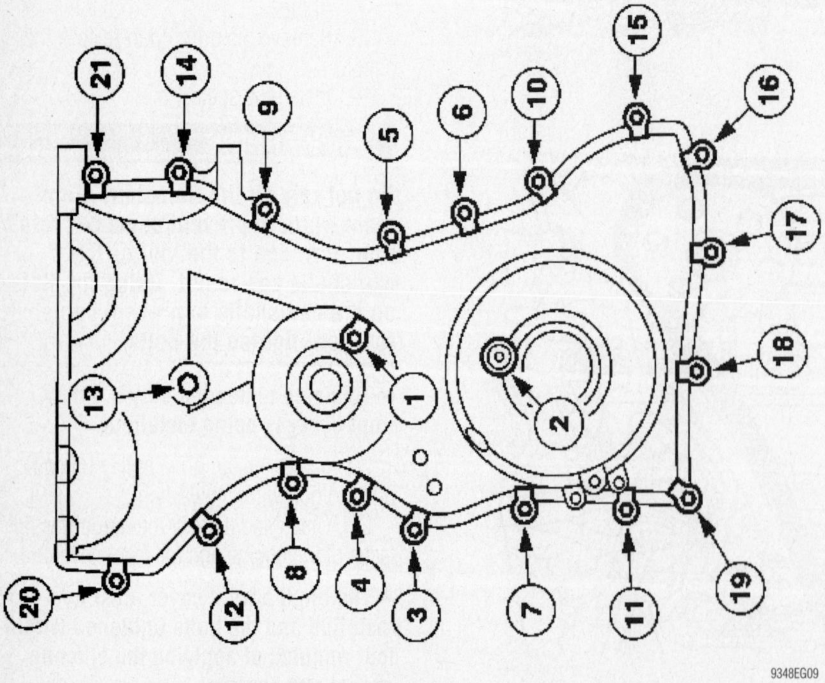

9348EG09

Front cover torque sequence—2.3L

24. Install or connect the following:
- Power steering pump and lower retaining bolt
- Power steering return hose
- New nylon O-ring and install the high pressure line.
- Water pump pulley
- Belt tensioner

➡**Do not reuse the crankshaft damper bolt.**

- Crankshaft pulley and hand-tighten the bolt

25. Install an M6 bolt in the crankshaft pulley. Tighten the crankshaft retaining bolt in two stages.
- Step 1: 40 Nm (30 ft. lbs.)
- Step 2: Rotate the bolt an additional 90 degrees.

26. Install the crankshaft position sensor, do not tighten the bolts at this time.

27. Adjust the crankshaft position sensor with the Alignment Tool, and tighten the mounting bolts.

28. Connect the crankshaft position sensor electrical connector.

29. Remove the M6 bolt from the crankshaft pulley.

30. Remove the Crankshaft Timing Peg 303-507.

31. Install the plug.

32. Remove the Camshaft Alignment Plate 303-376.

33. Install the valve cover.

34. Install the drive belt.

35. Install the fan and shroud.

36. Connect the battery ground cable.

3.0L Engines

1. Before servicing the vehicle, refer to the precautions in the beginning of this section.

2. Remove or disconnect the following:
- Negative battery cable
- Engine front cover

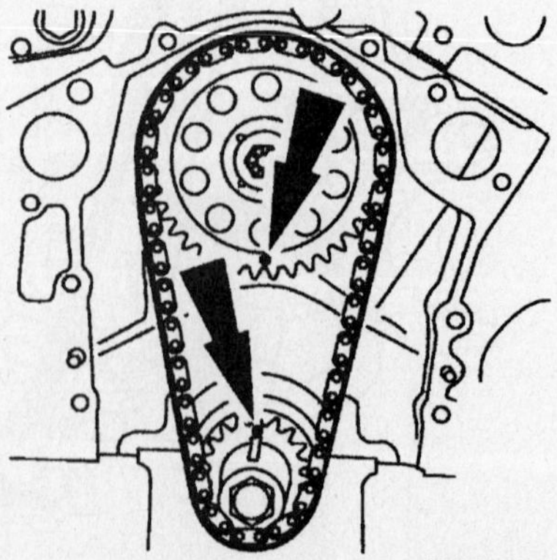

67197-RANG-G58

Timing mark alignment—3.0L engine

- Rotate the crankshaft and align the timing marks
- Timing chain tensioner, 4.0L engine only
- Sprocket bolt
- Timing chain, camshaft sprocket and crankshaft sprocket as an assembly

To install:

3. Install or connect the following:
   - Timing chain, camshaft and crankshaft sprockets as an assembly
4. Align the timing marks.
5. Install or connect the following:
   - Timing chain tensioner
   - Sprocket bolt. Torque the bolt to 51 ft. lbs. (70 Nm).
   - Engine front cover
   - Negative battery cable

4.0L OHC Engine

TIMING DRIVE COMPONENTS

1. Before servicing the vehicle, refer to the precautions in the beginning of this section.
2. With the vehicle in neutral, position it on a hoist.
3. Remove the intake manifold.
4. Remove the fuel supply manifold.
5. Remove the accessory drive belt.
6. Remove the thermostat housing.
7. Remove the roller followers.

➡You must retime the LH and RH camshafts when either camshaft is disturbed. Turn the crankshaft clockwise to position the number one cylinder at top dead center (TDC).

➡The special tool must be installed on the damper and should contact the engine block to position the engine at TDC.

8. Install the special tool.

➡The right-hand camshaft sprocket bolt is a left-hand threaded bolt.

➡If necessary, use camshaft gear torque adapter to loosen the camshaft sprocket bolt.

9. Using the special tool, loosen the RH camshaft sprocket bolt.

➡The camshaft timing slots are off-center.

10. Position the camshaft timing slots below the centerline of the camshaft to correctly fit the special tools. Install the special tools on the front of the RH cylinder head.
11. Remove the RH lower splash shield.
12. Remove the RH camshaft tensioner.

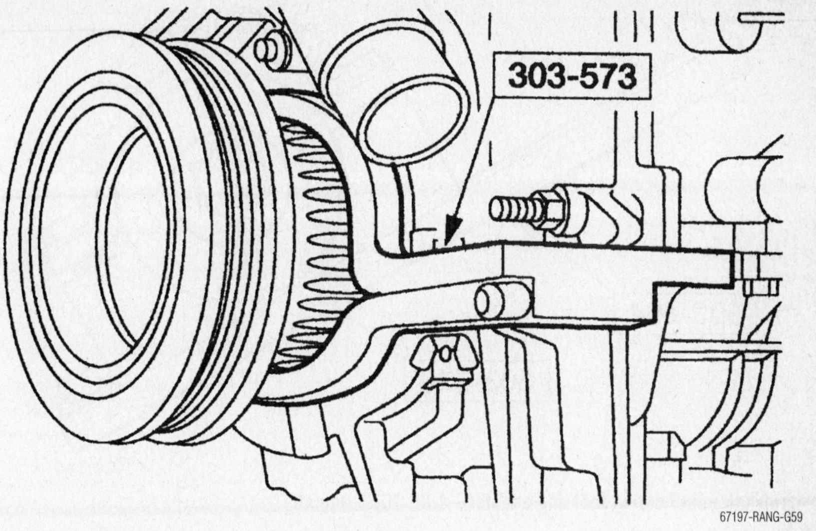

TDC positioning tool installed—4.0L SOHC engine

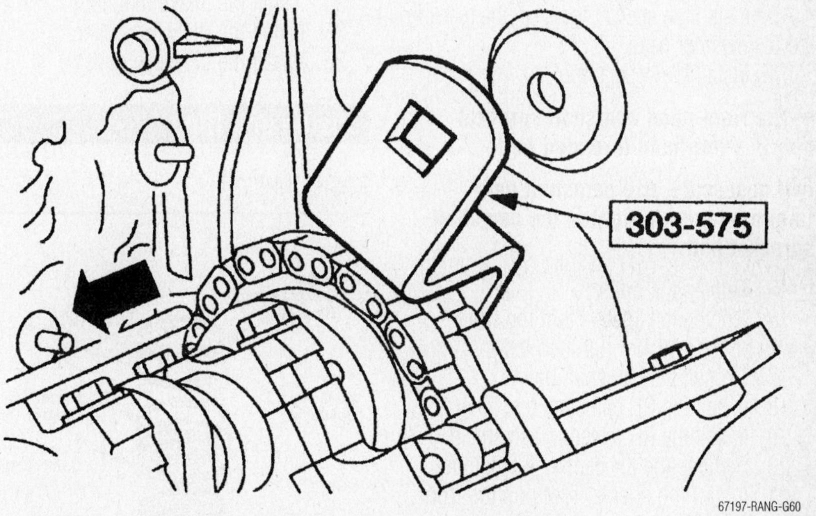

Loosening the right side camshaft sprocket bolt—4.0L SOHC engine

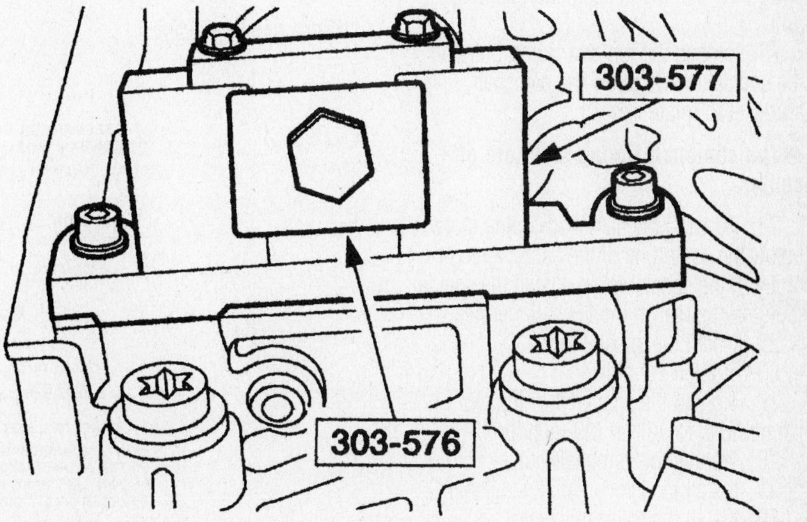

Camshaft holding tool installed—4.0L SOHC engine

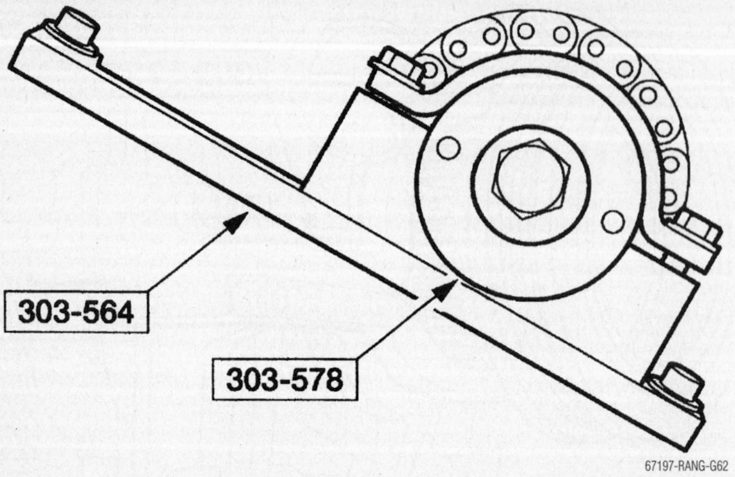

Camshaft gear holding tool and adapter—4.0L SOHC engine

➡Leave the top two special tool clamp bolts loose.

13. Install the special tools on the rear of the RH cylinder head.
14. Install the special tool.

➡The right-hand camshaft sprocket bolt is a left-hand threaded bolt.

➡If necessary, use camshaft gear torque adapter to tighten the camshaft sprocket bolt.

15. Tighten the bolts.
16. Tighten the special tool top two clamp bolts to 10 Nm (89 inch lbs.).
17. Tighten the camshaft bolt.
18. Install the RH camshaft tensioner.
19. Install the RH lower splash shield.
20. Remove the LH camshaft tensioner.
21. Install the special tools on the front of the LH cylinder head and tighten the top two clamp bolts to 10 Nm (89 inch lbs.).
22. Loosen the LH camshaft sprocket bolt.
23. Loosen the top two clamp bolts on the special tool to allow the camshaft sprocket to rotate freely.

➡The camshaft timing slots are off-center.

24. Position the camshaft timing slots below the centerline of the camshaft to correctly fit the special tools. Install the special tools on the rear of the LH cylinder head.
25. Install the special tool.
26. Tighten the bolts.
27. Tighten the special tool top two clamp bolts to 10 Nm (89 inch lbs.).
28. Tighten the camshaft bolt.
29. Install the LH camshaft tensioner.
30. Install the roller followers.
31. Install the thermostat housing.
32. Install the accessory drive belt.
33. Install the fuel supply manifold.
34. Install the intake manifold.
35. Install the RH valve cover.
36. Install the LH valve cover.

Piston and Ring

POSITIONING

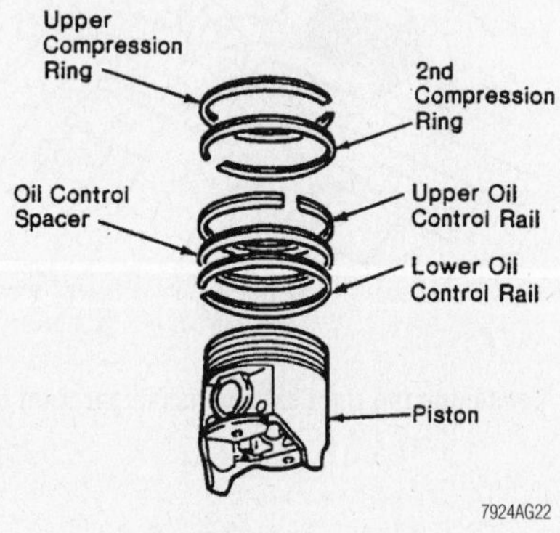

Piston ring positioning

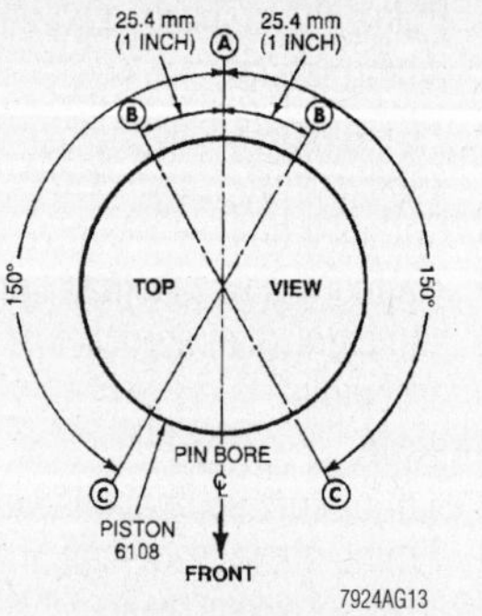

Piston ring end gap spacing

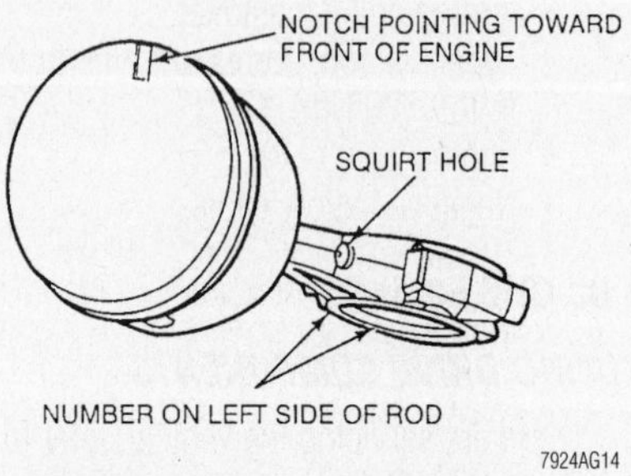

Piston and connecting rod positioning on 2.5L engines

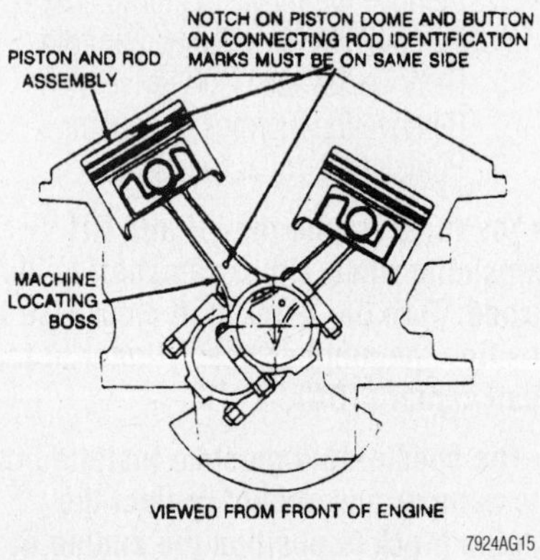

Piston and connecting rod positioning on 3.0L engines

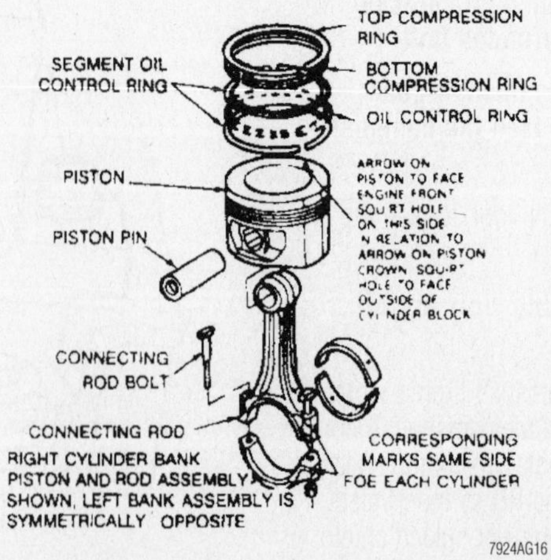

Piston and connecting rod positioning on 4.0L engines

FUEL SYSTEM

Fuel System Service Precautions

Safety is the most important factor when performing not only fuel system maintenance, but any type of maintenance. Failure to conduct maintenance and repairs in a safe manner may result in serious personal injury or death. Work on a vehicle's fuel system components can be accomplished safely and effectively by adhering to the following rules and guidelines.

- To avoid the possibility of fire and personal injury, always disconnect the negative battery cable unless the repair or test procedure requires that battery voltage be applied.
- Always relieve the fuel system pressure prior to disconnecting any fuel system component (injector, fuel rail, pressure regulator, etc.) fitting or fuel line connection. Exercise extreme caution whenever relieving fuel system pressure, to avoid exposing your skin, face and eyes to fuel spray. Please be advised that fuel under pressure may penetrate the skin or any part of the body that it contacts.
- Always place a shop towel or cloth around the fitting or connection prior to loosening to absorb any excess fuel due to spillage. Ensure that all fuel spillage is quickly remove from engine surfaces. Ensure that all fuel-soaked cloths or towels are deposited into a flame-proof waste container with a lid.
- Always keep a dry chemical (Class B) fire extinguisher near the work area.
- Do not allow fuel spray or fuel vapors to come into contact with a light bulb, spark or open flame.
- Always use a second wrench when loosening or tightening fuel line connection fittings. This will prevent unnecessary stress and torsion to fuel piping. Always follow the proper torque specifications.
- Always replace worn fuel fitting O-rings with new ones. Do not substitute fuel hose where rigid pipe is installed.

Relieving Fuel System Pressure

All Sequential Fuel Injection (SFI) fuel injected engines are equipped with a pressure relief valve located on the fuel supply manifold. Remove the fuel tank cap and attach fuel pressure gauge T80L-9974-B, or equivalent, to the valve to release the fuel pressure. Be sure to drain the fuel into a suitable container and to avoid gasoline spillage. If a pressure gauge is not available, disconnect the vacuum hose from the fuel pressure regulator and attach a hand-held vacuum pump. Apply about 25 in. Hg (84 kPa) of vacuum to the regulator to vent the fuel system pressure into the fuel tank through the fuel return hose. Note that this procedure will remove the fuel pressure from the lines, but not the fuel. Take precautions to avoid the risk of fire and use clean rags to soak up any spilled fuel when the lines are disconnected.

An alternate method of relieving the fuel system pressure involves disconnecting the inertia switch.

Fuel Filter

REMOVAL & INSTALLATION

1. Before servicing the vehicle, refer to the precautions in the beginning of this section.
2. Properly relieve the fuel system pressure.
3. Remove or disconnect the following:
   - Negative battery cable
   - Push connect and R-clip fittings from the fuel filter
   - Fuel filter

To install:

4. Install or connect the following:
   - Fuel filter. Torque the nut to 17 ft. lbs. (23 Nm).
   - R-clip and push connect fittings
   - Negative battery cable
5. Start the vehicle, check for leaks and repair if necessary.

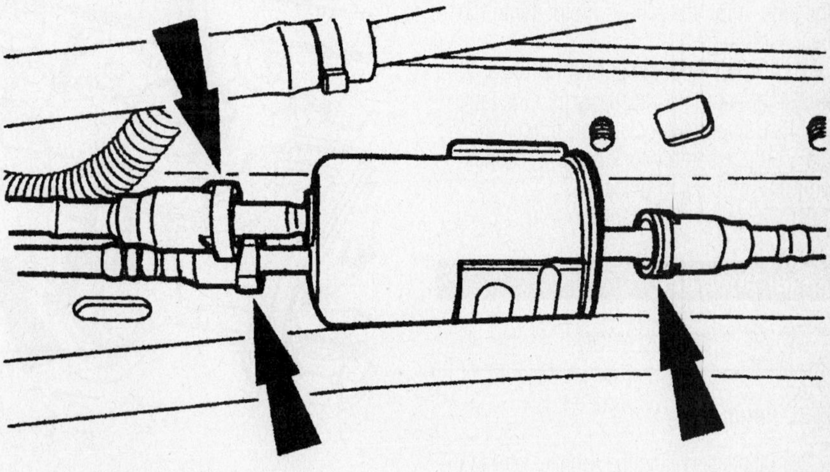

Fuel filter connections

Fuel Pump

REMOVAL & INSTALLATION

2001–03

1. Before servicing the vehicle, refer to the precautions in the beginning of this section.
2. Properly relieve the fuel system pressure.
3. Remove or disconnect the following:
   - Negative battery cable
   - Fuel tank
   - Fuel tank pump locking retainer ring
   - Fuel pump mounting gasket and discard the gasket
   - Fuel pump

To install:

4. Install or connect the following:
   - Fuel pump and a new mounting gasket
   - Fuel tank pump locking retainer ring. Torque the ring to 66 ft. lbs. (90 Nm).
   - Fuel tank

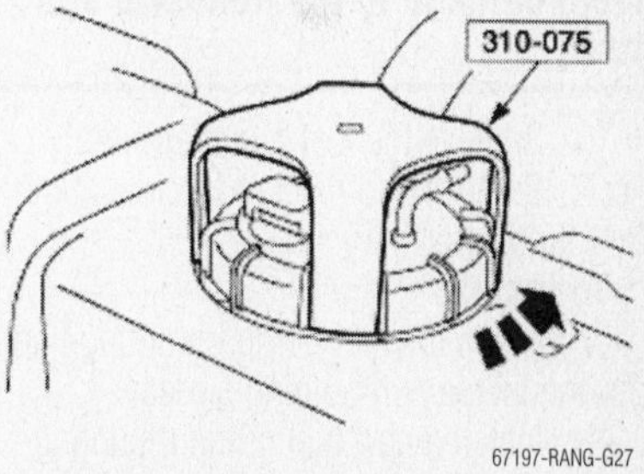

Lock ring wrench—2001–03 models

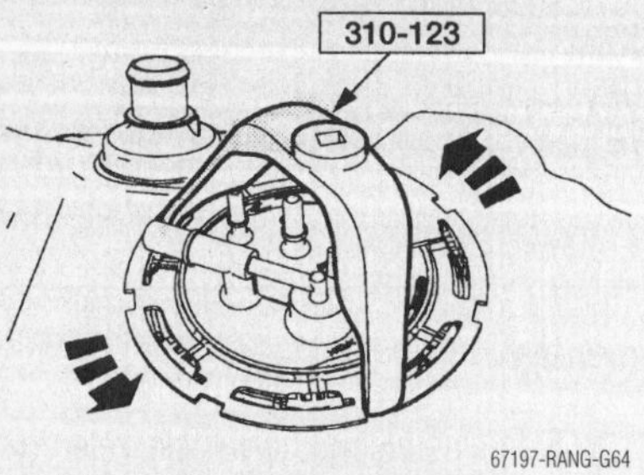

Lock ring tool—2004 models

- Negative battery cable

5. Start the vehicle, check for leaks and repair if necessary.

2004

1. Before servicing the vehicle, refer to the precautions in the beginning of this section.
2. Remove the fuel tank.
3. Clean the area around the fuel pump mounting flange.
4. Using the special tool, remove the fuel tank pump assembly locking retainer ring.

⁂ WARNING

The fuel pump assembly must be removed and handled carefully to avoid damage to the float arm and filter.

5. Remove the fuel pump assembly.
6. Remove and discard the fuel pump mounting gasket.

To install:

7. Clean the fuel pump mounting flange and the fuel tank mounting surface.
8. Install a new fuel pump mounting gasket.
9. Install the fuel pump and sender assembly with the float toward the rear of the tank. Align the arrows molded into the tank and flange.
10. Install the locking ring while compressing the pump assembly into the tank.
11. Using the special tool, tighten the fuel pump assembly locking ring retainer ring until it locks in place.
12. Install the fuel tank.

Fuel Injectors

REMOVAL & INSTALLATION

2.3L Engine

1. Before servicing the vehicle, refer to the precautions in the beginning of this section.

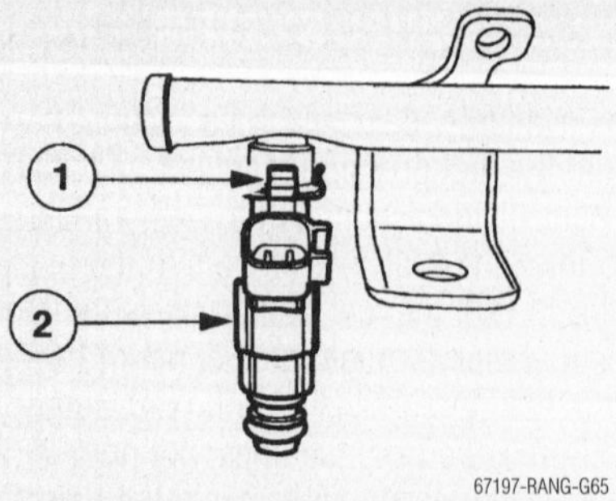

Fuel injector-to-fuel rail installation. 1-retaining clip; 2-injector—2.3L engine

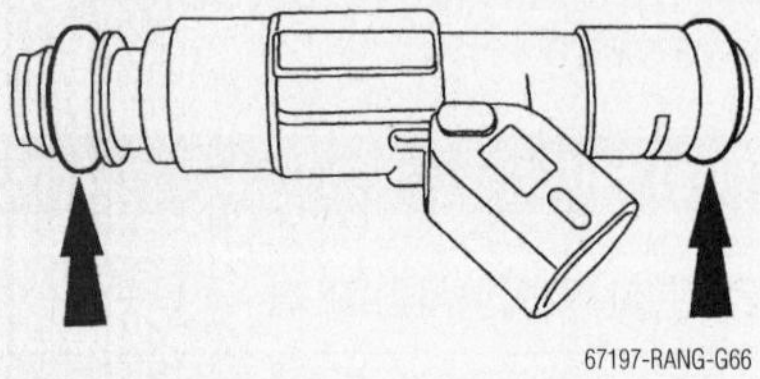

Fuel injector O-rings—2.3L engine

2. Properly relieve the fuel system pressure.
3. Remove or disconnect the following:
   - Negative battery cable
   - Upper intake manifold
   - Fuel injector connectors
   - Fuel injector harness from the fuel injector supply manifold
   - Fuel line spring lock
   - Fuel line
   - Fuel injection supply manifold
   - Fuel injector retaining clip
   - Fuel injector

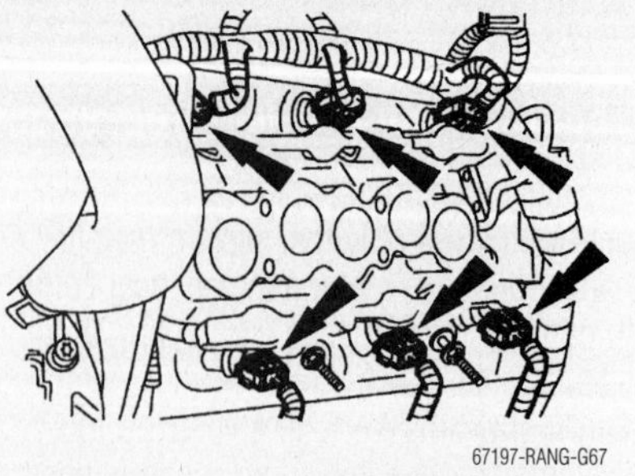

Fuel injector wiring connectors—3.0L engine

⁂ WARNING

Use O-ring seals that are made of special fuel-resistant material. Use of ordinary O-ring seals can cause the fuel system to leak. Do not reuse the O-ring seals.

4. Installation is the reverse of removal. Install new O-rings. Lubricate the O-rings with clean engine oil. Torque the supply manifold bolts to 18 ft. lbs. (25 Nm).

2.5L Engines

1. Before servicing the vehicle, refer to the precautions in the beginning of this section.
2. Properly relieve the fuel system pressure.
3. Remove or disconnect the following:
   - Negative battery cable
   - Fuel injection supply manifold
   - Fuel injectors by gently twisting them

Fuel rail bolts—3.0L engine

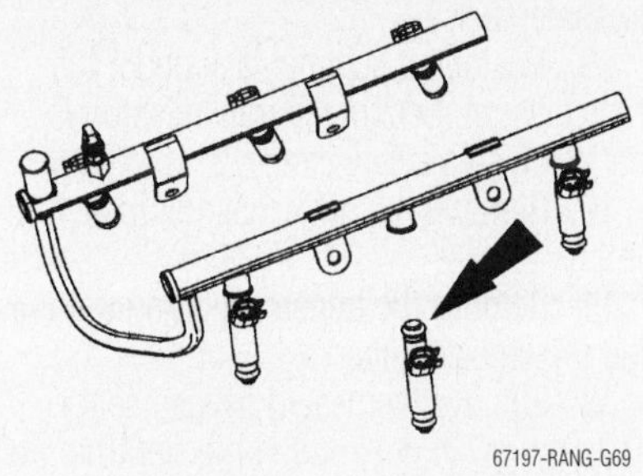

Fuel rail and injectors—3.0L engine

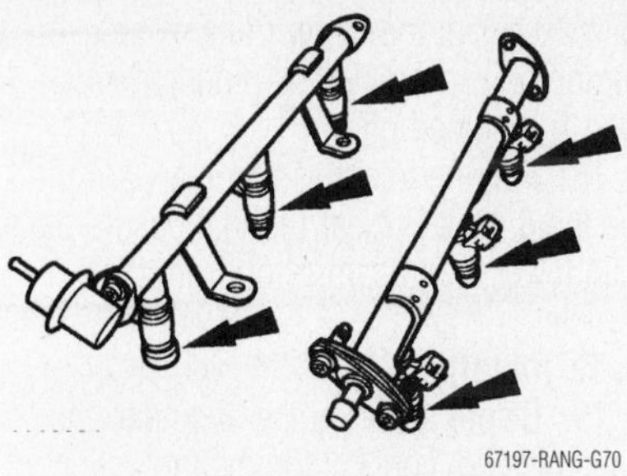

Fuel rail and injectors—4.0L engine

- Inspect the O-rings and replace as needed

To install:

4. Install or connect the following:
    - Fuel injectors
    - Fuel injector supply manifold
    - Negative battery cable
5. Start the vehicle, check for leaks and repair if necessary.

3.0L and 4.0L Engines

1. Before servicing the vehicle, refer to the precautions in the beginning of this section.
2. Properly relieve the fuel system pressure.

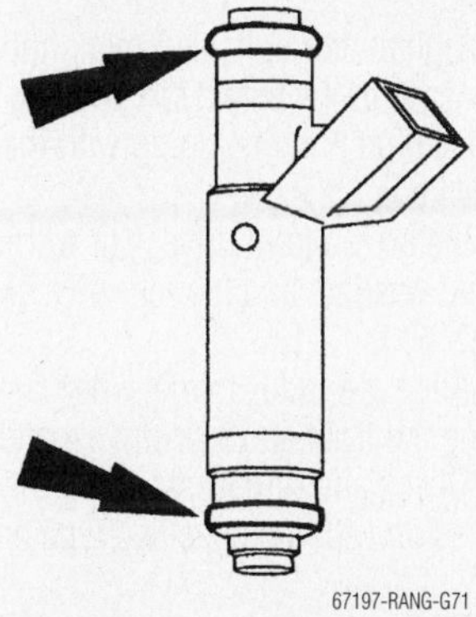

Fuel injector O-rings—4.0L shown; 3.0L similar

3. Remove or disconnect the following:
    - Negative battery cable
    - Upper intake manifold
    - Engine control sensor wiring from the fuel injectors
    - Fuel lines
    - Fuel injection supply manifold and injectors as an assembly
    - Vacuum line
    - Fuel injectors from the supply manifold
    - Inspect the O-rings and replace them as needed

To install:

4. Install or connect the following:
    - Fuel injectors
    - Vacuum line
    - Fuel injection supply manifold. Torque the bolts to 89 inch lbs. (10 Nm).
    - Fuel line
    - Engine control sensor wiring to the fuel injectors
    - Upper intake manifold
    - Negative battery cable
5. Start the vehicle, check for leaks and repair if necessary.

DRIVE TRAIN

Transmission Assembly

REMOVAL & INSTALLATION

Manual Transmission

2001–03

1. Before servicing the vehicle, refer to the precautions in the beginning of this section.
2. Remove or disconnect the following:
    - Negative battery cable
    - Upper gearshift lever and the outer gearshift lever boot and console assembly as an assembly
3. If transmission disassembly is necessary, remove the drain plug, and drain the transmission fluid. Install the drain plug after draining all of the fluid.
4. Remove or disconnect the following:
    - Electrical connector from the reverse lamp switch
    - Electrical connector from the vehicle speed sensor (VSS)
    - Heated oxygen sensor (HO_2S) electrical connector from the bracket
    - Wiring harness from the bracket
    - Electrical connectors from the heated oxygen sensors (HO_2S)
    - Starter motor

➡The driveshaft centering socket yoke fits tightly on the rear axle pinion flange pilot. Never hammer on the driveshaft or any of its components to disconnect the yoke from the flange. Pry only in the area shown, with a suitable tool, to disconnect the yoke from the flange.

➡If equipped, always disconnect the front driveshaft from the transfer case first. Otherwise, the weight of the driveshaft can cause the boot to tear.

- Rear driveshaft, and the front driveshaft, if so equipped

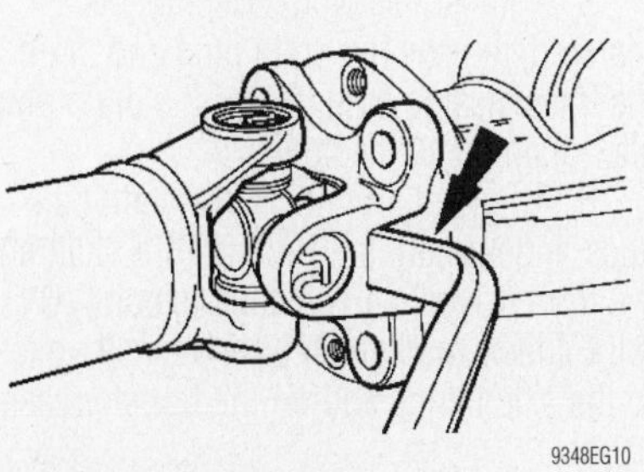

Pry here for driveshaft removal

- Bolts retaining the exhaust inlet crossover pipe to the exhaust manifold
- Bolts retaining the catalytic converter to the muffler. Discard the exhaust converter outlet gasket.
- Exhaust hanger from the insulator. Position the exhaust assembly aside.
- On 4-wheel drive vehicles, the transfer case
- Clutch hydraulic line from the clutch slave cylinder

⁂ WARNING

Secure the transmission to the jack with a suitable safety strap. Failure to follow these instructions may result in personal injury.

5. Using a suitable transmission jack, support the transmission. Secure the transmission to the jack with a suitable safety strap.
6. Loosen, but do not remove the nuts retaining the transmission insulator to the crossmember.
7. Remove the six bolts retaining the crossmember to the frame.
8. Remove the nuts and the crossmember.

➡Lower the transmission enough to gain access to the upper bolts retaining the transmission to the engine.

9. Remove the nine bolts retaining the transmission to the engine.
10. Remove the transmission from the vehicle.
11. Installation is the reverse of removal. Observe the following torques:
    - Transmission-to-engine bolts: 44 ft. lbs. (60Nm)
    - Crossmember-to-frame: 46 ft. lbs. (63Nm)
    - Transmission insulator-to-crossmember: 72 ft. lbs. (98Nm)

2004

2-WHEEL DRIVE

1. Before servicing the vehicle, refer to the precautions in the beginning of this section.
2. Remove the upper gearshift lever, the outer gearshift lever boot and the console as an assembly.
3. With the vehicle in NEUTRAL, position it on a hoist.
4. If transmission disassembly is required, remove the drain plug and drain the transmission fluid. Install the drain plug after draining all the fluid.
5. To maintain initial driveshaft balance, index-mark the driveshaft yoke to the axle flange, so they can be installed in their original positions. Remove the rear driveshaft.
6. Disconnect the wire harness from the crossmember.
7. Place a suitable jack under the transmission. Secure the transmission to the jack with a safety strap.
8. Remove the six crossmember bolts.
9. Remove the transmission mount nuts and the crossmember.
10. Remove the heated oxygen senor (HO_2S) bracket nut and the bracket from the extension housing.
11. Remove the transmission mount bolts and the transmission mount.
12. For 3.0L and 4.0L engines, remove the catalytic converter Y-pipe.
13. Disconnect the vehicle speed sensor (VSS) electrical connector and the reverse lamp switch electrical connector. Then unclip the wiring harness from the transmission.
14. Remove the starter motor. Using mechanics wire, position the starter aside.
15. Using the special tool, disconnect the clutch hydraulic line.
16. Using a suitable jack, support the engine.
17. Lower the transmission enough to gain access to the upper transmission-to-engine bolts. Remove the nine transmission-to-engine bolts.
18. Pull the transmission rearward until the input shaft is clear of the pressure plate, then lower the transmission from the vehicle.

To install:

19. Before securing the engine to the transmission, connect the hydraulic line to the clutch slave cylinder.
20. Install the transfer case with a new gasket.
21. Tighten the bolts that retain the transfer case to the extension housing in a clockwise direction beginning with the upper LH bolt.
22. Install the front driveshaft with new bolts and washers and the rear driveshaft with new bolts.
23. Align the index marks when installing the front and rear driveshafts.
24. Check and, if necessary, fill the transmission with the specified type and quantity of fluid.
25. To install, reverse the removal procedure. Observe the following torques:
    - Drain plug: 36 ft. lbs. (48 Nm)
    - Crossmember bolts: 72 ft. lbs. (98Nm)
    - Transmission mount nuts: 72 ft. lbs. (98 Nm)
    - Oxygen sensor bracket nut: 29 ft. lbs. (39 Nm)
    - transmission mount bolts: 73 ft. lbs. (99 Nm)
    - Transmission-to-engine bolts: 44 ft. lbs. (60 Nm)

4-WHEEL DRIVE

1. Before servicing the vehicle, refer to the precautions in the beginning of this section.
2. Remove the upper gearshift lever, the outer gearshift lever boot and the console as an assembly.
3. With the vehicle in NEUTRAL, position it on a hoist.
4. Remove the skid plate, if equipped.
5. If transmission disassembly is required, remove the drain plug and drain the transmission fluid. Install the drain plug after draining all the fluid.
6. To maintain initial driveshaft balance, index-mark the front output shaft and the front driveshaft constant velocity (CV) joint. Index-mark the rear driveshaft yokes to the axle flange and on the transfer case flange.
7. Remove the transfer case.
8. Disconnect the wire harness from the crossmember.
9. Position a suitable jack under the transmission. Secure the transmission to the jack with a safety strap.
10. Remove the six crossmember bolts.
11. Remove the transmission mount nuts and the crossmember
12. Remove the heated oxygen senor (HO_2S) bracket nut and the bracket from the extension housing.
13. Remove the catalytic converter Y-pipe.
14. Disconnect the vehicle speed sensor (VSS) electrical connector and the reverse lamp switch electrical connector. Then unclip the wiring harness from the transmission.
15. Remove the starter motor. Using mechanics wire, position the starter aside.
16. Using the special tool, disconnect the clutch hydraulic line.
17. Using a suitable jack, support the engine.
18. Lower the transmission enough to gain access to the upper transmission-to-engine bolts. Remove the nine transmission-to-engine bolts.
19. Pull the transmission rearward until the input shaft is clear of the pressure plate, then lower the transmission from the vehicle.

To install:

20. Before securing the engine to the transmission, connect the hydraulic line to the clutch slave cylinder.
21. Install the transfer case with a new gasket.
22. Tighten the bolts that retain the transfer case to the extension housing in a clockwise direction beginning with the upper LH bolt.
23. Install the front driveshaft with new bolts and washers and the rear driveshaft with new bolts.
24. Align the index marks when installing the front and rear driveshafts.
25. Check and, if necessary, fill the transmission with the specified type and quantity of fluid.
26. To install, reverse the removal procedure. Observe the following torques:
    - Skid plate: 18 ft. lbs. (24 Nm)
    - Drain plug: 36 ft. lbs. (48 Nm)
    - Crossmember bolts: 46 ft. lbs. (63 Nm)
    - Transmission mount nuts: 72 ft. lbs. (98 Nm)
    - Oxygen sensor bracket: 29 ft. lbs. (39 Nm)
    - Transmission-to-engine bolts: 44 ft. lbs. (60Nm)

Automatic Transmission

2001–03

1. Before servicing the vehicle, refer to the precautions in the beginning of this section.
2. Place the selector lever in NEUTRAL position.
3. Remove or disconnect the following:
   - Negative battery cable
   - Fluid level indicator
   - The two bolts retaining the fan shroud to the radiator.

➡**If transmission disassembly is required, drain the transmission fluid.**

- With 4wd, the transfer case

➡**Mark the driveshaft yoke and axle flange, so they may be installed in their original alignment.**

- Rear driveshaft
- Starter motor
- Torque converter access cover

➡**Mark the torque converter and the flexplate for correct alignment at reinstallation.**

- The four converter nuts
- Shift cable
- Transmission wiring harness
- Three way catalytic converter
- Left HO_2S sensor
- Front exhaust crossover pipe
- Transmission cooler lines

4. Position a High-Lift Jack under the transmission. Raise and support the transmission.
5. Remove or disconnect the following:
   - Crossmember.
   - Transmission mount
   - Transmission upper fill tube

➡**Lower the High-Lift Transmission Jack to gain access to screws.**

- On 4x4 models, the vent tube assembly

⁂ WARNING

Install the Torque Converter Holding Tool before lowering the transmission from the vehicle. Secure the transmission to the transmission jack with a safety chain. Failure to follow these instructions can result in personal injury.

6. Lower the transmission.
7. Installation is the reverse of removal. Observe the following torques:
   - Transmission-to-engine bolts: 41 ft. lbs. (55Nm)
   - Exhaust bracket bolts: 81 ft. lbs. (110Nm)
   - Crossmember-to-frame: 87 ft. lbs. (118Nm)
   - Transmission mount-to-crossmember: 81 ft. lbs. (110Nm)
   - Converter-to-flexplate: 30 ft. lbs. (40Nm)
   - Rear driveshaft-to-flange bolts: 95 ft. lbs. (129Nm)

2004

1. Before servicing the vehicle, refer to the precautions in the beginning of this section.

➡**If the transmission is to be removed for a period of time, support the engine with a safety stand and a wood block.**

2. With the vehicle in NEUTRAL, position it on a hoist.

➡**When the battery has been disconnected and reconnected, some abnormal drive symptoms can occur while the vehicle relearns its adaptive strategy.**

3. Disconnect the battery ground cable
4. On 4.0L SOHC vehicles, remove the fluid level indicator tube bolt and remove the tube and indicator.
5. If transmission disassembly is required, drain the transmission fluid.
6. On 4x4 vehicles, remove the transfer case.
7. To maintain initial driveshaft balance, mark the driveshaft yoke and axle flange so they can be installed in their original alignment.
8. Remove the rear driveshaft.
   a. Remove the four bolts.
   b. Remove the driveshaft.
9. Remove the starter motor.
10. With a 2.3L engine:
    a. When removing the torque converter nuts, the crankshaft must be rotated only in the clockwise direction, otherwise engine damage can occur. The crankshaft, crankshaft sprocket and the pulley are fitted together by friction between the flange faces on each part. For that reason, the crankshaft sprocket can also be moved when the crankshaft pulley is turned in the counterclockwise direction.
    b. It may be necessary to gain access to the flexplate nuts through the wheel well.
    c. Mark the torque converter and the flexplate for correct alignment at reinstallation.
    d. Remove and discard the four torque converter nuts. Rotate the flexplate to access to all the nuts.
11. With 3.0L and 4.0L SOHC engines:
    a. Mark the torque converter and the flexplate for correct alignment at reinstallation. Remove the four nuts. Rotate the flexplate to access to all the nuts.
    b. Disconnect the shift cable.
    c. Disconnect the transmission wiring harness from the case.
12. Disconnect the transmission wiring harness. Remove the three way catalytic converter.
13. With a 2.3L engine, remove the rear engine cover plate.

➡**Care should be taken not to bend or damage the cooler lines.**

14. Hold the case fitting and remove the transmission cooler lines.
15. Remove the nuts.
16. Position a transmission jack under the transmission. Raise and support the transmission.
17. Remove the crossmember.
18. Remove the transmission mount.
19. With a 2.3L engine, remove the rear vibration damper.
20. 2.3L and 3.0L engines, remove the transmission upper fill tube.
21. Lower the jack to gain access to screws. Remove the transmission-to-engine bolts.
22. With a 2.3L engine, remove the lower screws.
23. Remove the HO_2S connector bracket from the transmission.
24. On 4x4 vehicles, remove the vent tube assembly.

⁂ CAUTION

The torque converter is heavy and may result in injury if it falls out of the transmission. Secure the torque converter in the transmission. Failure to follow these instructions may result in personal injury. Install a converter locking tool before lowering the transmission from the vehicle.

⁂ CAUTION

Secure the transmission to the transmission jack with a safety chain. Failure to follow these instructions may result in personal injury.

25. Lower the transmission.
26. If the transmission is being overhauled or if installing a new or remanufac-

tured transmission, carry out transmission fluid cooler backflushing and cleaning.

To install:

27. On 4x4 vehicles, install the vent tube assembly.

⁂ CAUTION

Secure the transmission to the transmission jack with a safety chain. Failure to follow these instructions may result in personal injury.

28. Raise and position the transmission.
29. Remove the holding tool.
30. With the 4.0L SOHC engine:
    a. Align the flexplate to the converter marks made at removal.
    b. Install the transmission-to-engine screws. Torque to 35 ft. lbs. (48 Nm).
31. With the 2.3L and 3.0L engines
    a. Align the flexplate to the converter marks made at removal.

➡Align the flexplate to the converter marks made at removal.

    b. Install the transmission-to-engine screws. Torque to 35 ft. lbs. (48 Nm).
    c. Install the upper fluid filler tube and bracket screw.
32. With the 2.3L engine:
    a. Install the lower transmission-to-engine screws. Torque to 35 ft. lbs. (48 Nm).
    b. Install the HO_2S connector bracket.
    c. Install the rear engine cover plate.
33. On 4x4 vehicles, install the transfer case.
34. Install the exhaust bracket. Torque the bolts to 73 ft. lbs. (99 Nm).
35. Install the crossmember. Tighten the bolts to 74 ft. lbs. (101 Nm).
36. Install the transmission mount into the crossmember and torque the nuts to 73 ft. lbs. (99 Nm).
37. With 3.0L engines, install the rear vibration damper. Torque the nuts to 22 ft. lbs. (30 Nm).
38. With 3.0L and 4.0L SOHC engines:

➡Prior to installing the cooler lines to the case, inspect the O-rings. If damaged new O-rings will need to be installed.

    a. Hold the case fitting and install the transmission cooler lines. Torque to 19 ft. lbs. (28 Nm).
    b. Install four new torque converter nuts. Rotate the crankshaft as needed to gain access to all the nuts. Torque to 26 ft. lbs. (35 Nm).
39. With the 2.3L vehicles: Install four new torque converter nuts. Rotate the crankshaft as needed to gain access to all the nuts.®Torque to 26 ft. lbs. (35 Nm).

➡When installing the torque converter nuts, the crankshaft must be rotated only in the clockwise direction, otherwise engine damage can occur. The crankshaft, the crankshaft sprocket and the pulley are fitted together by friction between the flange faces on each part. For that reason, the crankshaft sprocket can also be moved when the crankshaft pulley is turned in the counterclockwise direction.

40. Install the starter motor.
41. Install the catalytic converter assembly.
42. Position the transmission wiring harness in place.
43. Connect the transmission wiring harness.
44. Install the shift cable.
45. Align the driveshaft yoke and the axle shaft marks made at removal to maintain driveline balance. Install the rear driveshaft. Install the driveshaft bolts. Torque to 83 ft. lbs. (112 Nm).
46. Use the following guidelines for installing the in-line transmission fluid filter:
    a. If the transmission was overhauled and the vehicle was equipped with an in-line fluid filter, install a new in-line fluid filter.
    b. If the transmission was overhauled and the vehicle was not equipped with an in-line fluid filter, install a new in-line fluid filter kit.
    c. If the transmission is being installed for a non-internal repair, do not install an in-line filter or filter kit.
    d. If installing a new or re-manufactured transmission, install the in-line transmission fluid filter that is supplied.
    e. Prior to lowering the vehicle, install a new in-line transmission filter or a filter kit.
47. With the 4.0L SOHC engine, install the transmission fill tube and indicator as an assembly.

➡When the battery has been disconnected and reconnected, some abnormal drive symptoms can occur while the vehicle relearns its adaptive strategy.

48. Connect the battery ground cable.
49. Fill the transmission with clean automatic transmission fluid to the specified level.
50. Check the transmission for correct operation.
51. Verify that the shift cable is correctly adjusted.

Clutch

REMOVAL & INSTALLATION

1. Before servicing the vehicle, refer to the precautions in the beginning of this section.
2. Remove or disconnect the following:
    - Negative battery cable
    - Transmission

➡If the clutch disc and pressure plate are to be reinstalled, bolts must be removed evenly or permanent damage to the diaphragm spring will occur resulting in complete clutch release.

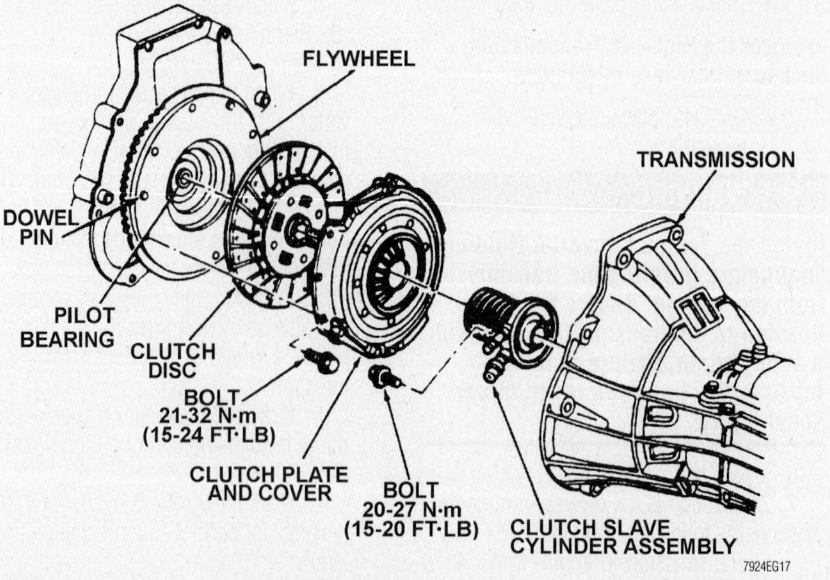

Clutch disc, pressure plate and bearing assembly

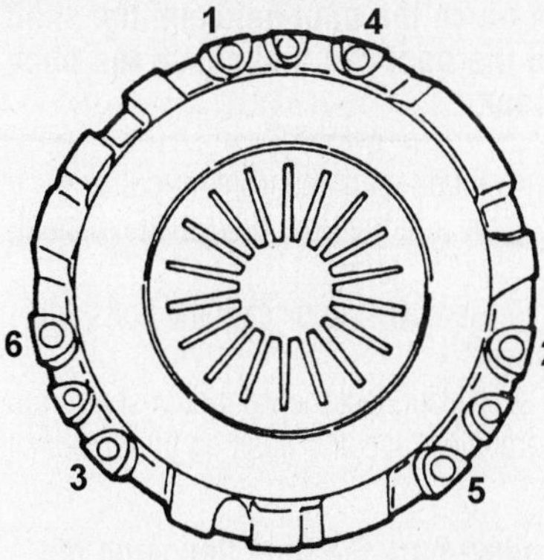

Tighten the bolts gradually in the correct sequence to avoid warping the pressure plate

- Bolts, clutch pressure plate and the clutch disc.

➡If the parts are to be reused, index-mark the clutch pressure plate to the flywheel.

To installation:

3. Lubricate the transmission input shaft pilot bearing with front axle grease.
4. Using a suitable press, press downward on the pressure plate fingers until the adjusting ring moves freely.
5. Rotate the adjusting ring counterclockwise to compress the tension springs. Hold the adjusting ring in this position.
6. Release the pressure on the fingers. The adjusting ring will stay in the reset position.
7. Position the clutch disc on the flywheel.

➡If reusing the clutch pressure plate and flywheel, align the marks made during removal.

8. Align the clutch disc and the clutch pressure plate. Install the bolts and tighten in a star pattern sequence to 24 ft. lbs. (35Nm) for 2001–02; 20 ft. lbs. (27 Nm) for 2003–04 models.
   - Install the transmission.

ADJUSTMENT

Because the clutch is hydraulically driven, there is no adjustment required.

In the event the clutch pedal develops a squeak or uneven feel when depressing, spray the pedal bushing assembly with penetrating oil and work the pedal back-and-forth.

Hydraulic Clutch System

BLEEDING

The following procedure is recommended for bleeding the clutch hydraulic system installed on the vehicle. It is recommended that the original clutch tube, with quick-connect fitting be replaced when servicing the hydraulic system, because air can be trapped in the quick-connect fitting and prevent complete bleeding of the system. The replacement tube does not include a quick-connect fitting.

1. Before servicing the vehicle, refer to the precautions in the beginning of this section.
2. Clean the dirt and grease from the dust cap.
3. Remove the cap and diaphragm and fill the reservoir to the top with approved brake fluid C6AZ-19542-AA or BA, (ESA-M6C25-A).

➡To keep brake fluid from entering the clutch housing, route a suitable rubber tube of appropriate inside diameter from the bleed screw to a container.

4. Loosen the bleed screw, located in the slave cylinder body, next to the inlet connection. Fluid will now begin to move from the master cylinder down the tube to the slave cylinder.

➡The reservoir must be kept full at all times during the bleeding operation, to ensure no additional air enters the system.

5. Observe the bleed screw outlet. When the slave cylinder is full, a steady stream of fluid will flow from the outlet port. Tighten the bleed screw.
6. Depress the clutch pedal to the floor and hold for 1–2 seconds. Release the pedal as rapidly as possible. The pedal must be released completely. Pause for 1–2 seconds. Repeat 10 times.
7. Check the fluid level in the reservoir. The fluid should be level with the step when the diaphragm is removed.
8. Hold the pedal to the floor, slightly open the bleed screw to allow any additional air to escape. Close the bleed screw, then release the pedal.
9. Check the fluid in the reservoir. The hydraulic system should now be fully bled, and should actuate the clutch.
10. Check the vehicle by starting, pushing the clutch pedal to the floor and selecting reverse gear. There should be no grating of gears. If there is, and the hydraulic system still contains air; repeat the bleeding procedure.

Transfer Case Assembly

REMOVAL & INSTALLATION

2001–03

1. Before servicing the vehicle, refer to the precautions in the beginning of this section.
2. Place the transmission in neutral.
3. Remove or disconnect the following:
   - Skid plate
   - Damper
   - Transfer case harness connector and position it aside
4. If transfer case disassembly is necessary, remove the drain plug and drain the fluid.

➡Index-mark the front output shaft assembly and the front driveshaft constant velocity (CV) joint. Always disconnect the front driveshaft from the transfer case first. Otherwise, the weight of the driveshaft can pinch the boot between the shaft and the boot can and cause the boot to tear.

- Front driveshaft from the transfer case and position the driveshaft aside. Remove and discard the bolts and washers.

➡Index-mark the front flange on the rear driveshaft and the flange on the transfer case.

- Rear driveshaft

➡Secure the transfer case to the jack with safety straps.

5. Position a high lift jack under the transfer case.
6. Remove or disconnect the following:
   - Five bolts retaining the transfer case to the extension housing
   - Transfer case rearward and off of the transmission output shaft
7. Remove and discard the front extension housing gasket and clean the mating surfaces.

To install:

8. Installation is the reverse of removal. Take note of the following:
   - Install the transfer case with a new gasket.
   - Tighten the bolts that retain the transfer case to the extension housing in a clockwise direction beginning with the upper LH bolt.

- Install the front and the rear driveshafts with new bolts. If new bolts are not available, coat the threads of the original bolts with Threadlock and Sealer E0AZ-19554-AA or equivalent meeting Ford specification WSK-M2G351-A5.
- When installing the front driveshaft, always connect it to the axle first and then connect it to the transfer case.
- Align the index marks when installing the front and rear driveshafts.

9. Observe the following torques:
   - Nut retaining the flange to the rear output shaft: 262 ft. lbs. (355Nm)
   - Bolt retaining the rear driveshaft to the flange: 82 ft. lbs. (111Nm)
   - Bolt retaining the motor assembly and connector to the transfer case cover: 89 inch lbs. (10Nm)
   - Bolt retaining the skid plate to the frame: 18 ft. lbs. (24Nm)
   - Bolt retaining the damper to the transfer case: 30 ft. lbs. (40Nm)
   - Bolt retaining the driveshaft CV joint to the front output shaft assembly: 22 ft. lbs. (30Nm)
   - Bolt retaining the transfer case to the extension housing: 40 ft. lbs. (54Nm)
   - Bolt retaining the front adapter to the transfer case: 30 ft. lbs. (40Nm)
   - Bolt retaining the transfer case to the transfer case cover: 27 ft. lbs. (36Nm)
   - Drain plug: 18 ft. lbs. (24Nm)
   - Fill plug: 18 ft. lbs. (24 Nm)

2004

1. Before servicing the vehicle, refer to the precautions in the beginning of this section.
2. With the vehicle in NEUTRAL, raise and support the vehicle.
3. Remove the skid plate.
4. Remove the damper, if so equipped.
5. Disconnect the transfer case harness connector and position it aside.
6. If transfer case disassembly is necessary, remove the drain plug and drain the fluid. Install the drain plug when all of the fluid has drained.

➡Index-mark the front output shaft assembly and the front driveshaft constant velocity (CV) joint.

✲✲ WARNING

Always disconnect the front driveshaft from the transfer case first. Otherwise, the weight of the driveshaft can pinch the boot between the shaft and the boot can and cause the boot to tear.

7. Index-mark the front output shaft assembly and the front driveshaft constant velocity (CV) joint.
8. Remove and discard the bolts and washers.
9. Disconnect the front driveshaft from the transfer case and position the driveshaft aside.

➡Index-mark the front flange on the rear driveshaft and the flange on the transfer case.

10. Remove the rear driveshaft.

✲✲ CAUTION

Secure the transfer case to the jack with safety straps.

11. Position a high lift jack under the transfer case.
12. Remove the five bolts retaining the transfer case to the extension housing.
13. Slide the transfer case rearward and off of the transmission output shaft.
14. Remove and discard the front extension housing gasket, and clean the mating surfaces.

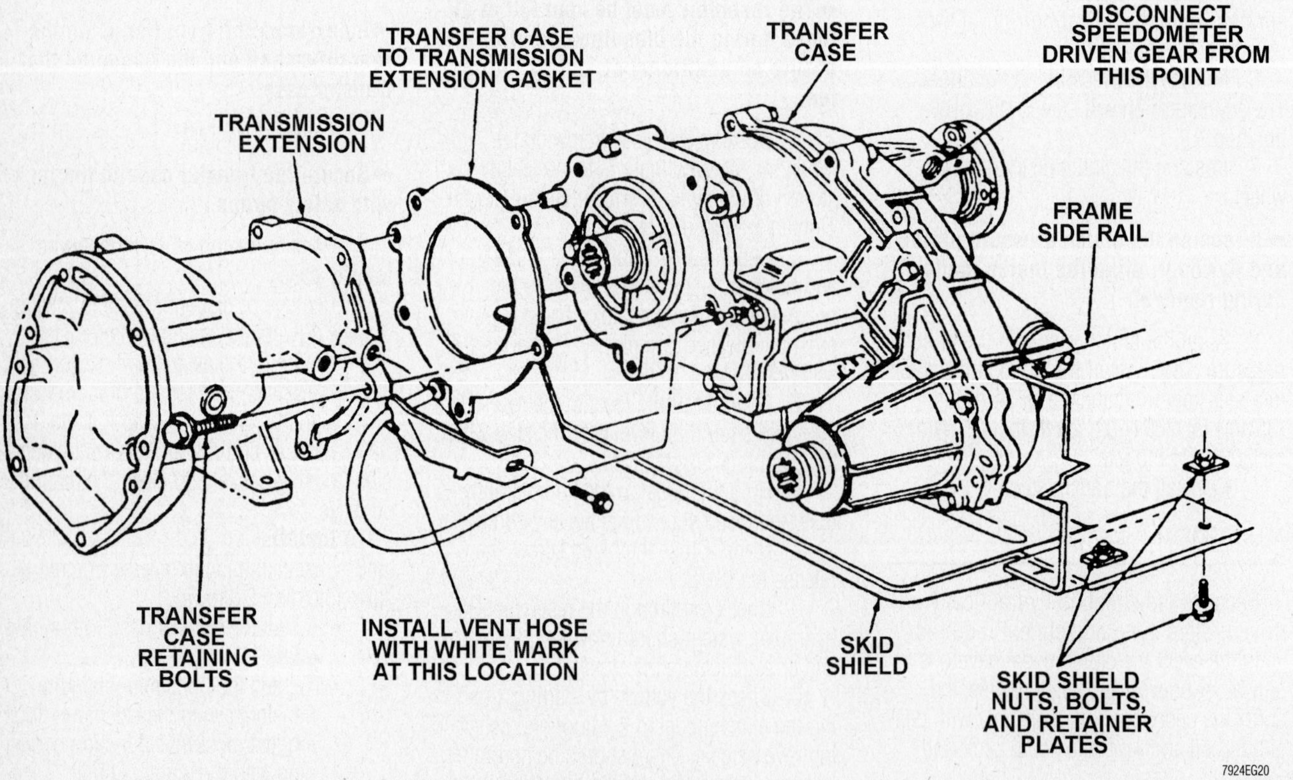

Exploded view of the 13-54 electronic shift transfer case-to-transmission mounting

To install:

15. Install the transfer case with a new gasket.
16. Tighten the bolts that retain the transfer case to the extension housing in a clockwise direction beginning with the upper LH bolt.
17. Install the front driveshaft with new bolts and washers and the rear driveshaft with new bolts. If new bolts are not available, coat the threads of the original bolts with Threadlock and Sealer E0AZ-19554-AA, or equivalent.

➡When installing the front driveshaft, always connect it to the axle first and then connect it to the transfer case.

➡Align the index marks when installing the front and rear driveshafts.

18. The remainder of installation is the reverse of the removal procedure.
19. Check and, if necessary, fill the transfer case with the specified type and quantity of fluid.

Halfshaft

REMOVAL & INSTALLATION

1. Before servicing the vehicle, refer to the precautions in the beginning of this section.
2. With the vehicle in NEUTRAL, raise and support the vehicle.
3. Remove the front wheel and tire assembly.

➡Do not reuse the torque prevailing design hub nut and washer assembly.

4. Remove and discard the hub nut and washer assembly.

⁂ WARNING

Do not allow the disc brake caliper to hang suspended from the brake hose. Provide a suitable support.

5. Remove the front disc brake caliper, anchor plate, and pads as an assembly, and position the assembly aside.
6. Remove the brake disc.

⁂ WARNING

Do not use a hammer to separate the outboard front wheel halfshaft joint from the wheel hub. Damage to the outboard CV joint stub shaft threads and internal CV joint components may result.

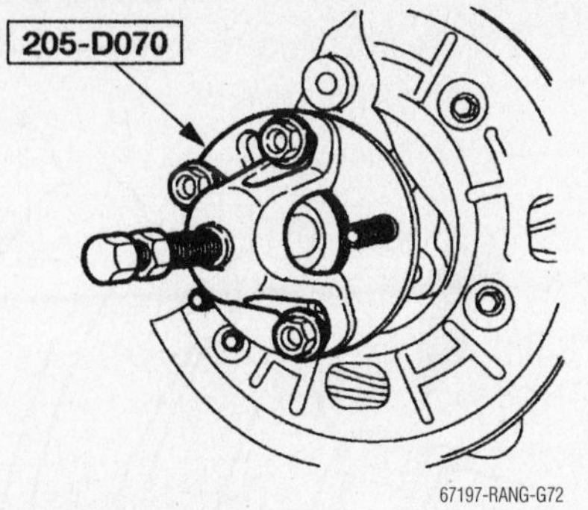

Halfshaft removal tool

7. Using the special tool, separate the outboard front wheel halfshaft joint from the wheel hub. Remove the special tool.
8. Support the front suspension lower arm.
9. Remove the nut and bolt retaining the upper ball joint to the front wheel knuckle.
10. Rotate the front wheel knuckle.
11. Compress the outboard front wheel halfshaft joint.
12. Remove the outboard front wheel halfshaft joint from the wheel hub.
13. Using the special tools, 205-241 and 100-001, or equivalent, separate the inboard front wheel halfshaft joint from the front axle housing.
14. Remove the halfshaft assembly from the vehicle with both hands. Do not damage the axle seal.

To install:

⁂ WARNING

Install the halfshaft with a new hub nut and washer assembly. Do not use power or impact tools to tighten the hub nut and washer assembly.

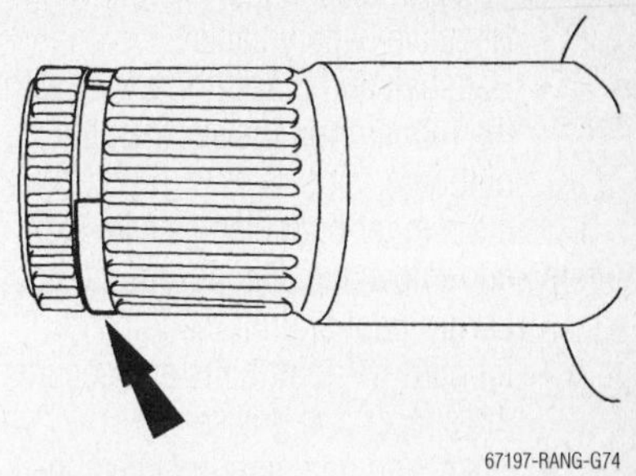

Circlip installed

➡Install a new retainer circlip in the groove in the LH inboard CV joint housing stub shaft before installing the halfshaft in the vehicle. To prevent the new retainer circlip from over-expanding when installing it, start one end in the groove and work the circlip over the shaft and into the groove.

15. To install, reverse the removal procedure. Observe the following torques:
   - Hub nut: 162 ft. lbs. (220 Nm)
   - Upper ball joint-to-knuckle nut: 41 ft. lbs. (55 Nm)

CV-Joints

OVERHAUL

2001–03

1. Before servicing the vehicle, refer to the precautions in the beginning of this section.

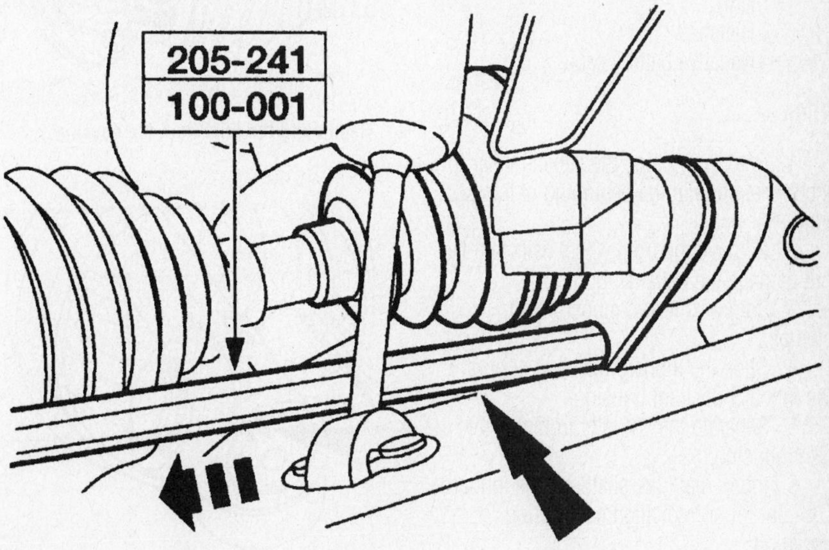

Separating the halfshaft from the axle housing

2. Remove or disconnect the following:
   - Negative battery cable
   - Halfshaft and place it in a vice with the inboard joint lower than the outboard joint

3. Cut the inner boot clamps with side cutters and remove the clamp from the boot.
   - Larger boot end off the joint
   - Inboard CV-joint bolts and separate the spacer and grease cap
   - Snap-ring retaining the interconnecting shaft end to the CV-joint cage
   - CV-joint and discard the washer

➡**The outboard CV-joint is non-serviceable other than to replace the boot.**

To install:

4. Install or connect the following:
   - Slide the boot over the shaft
5. Fill the CV-joint area with grease.
   - Assemble the outer boot to the outboard CV-joint and interconnecting shaft. Make certain that the boot is seated in the grooves on the outer race and on the shaft
   - New clamps to the boot
   - New inner boot to the shaft
   - New washer to the end of the shaft
   - Assemble the inboard CV-joint to the interconnecting shaft spline until it rests on the washer
   - Snap-ring
6. Fill the CV-joint area with grease.
   - Boot into position and make certain that it is seated in the grooves on the boot adapter and the shaft
   - New clamps and tighten the clamps with crimping pliers
   - Spacer to the CV-joint end pilot. Torque the bolts to 25 ft. lbs. (34 Nm).
   - Halfshaft
   - Negative battery cable

2004

1. Before servicing the vehicle, refer to the precautions in the beginning of this section.
2. Remove the front wheel halfshaft. Do not damage the halfshaft boot.
3. Remove the two inboard boot clamps.
4. Slide the inboard halfshaft boot off the inboard CV joint housing.
5. Separate the CV joint from the CV joint housing.
6. Index-mark the shaft and the inboard CV joint for correct alignment during assembly.
7. Remove the snap ring.

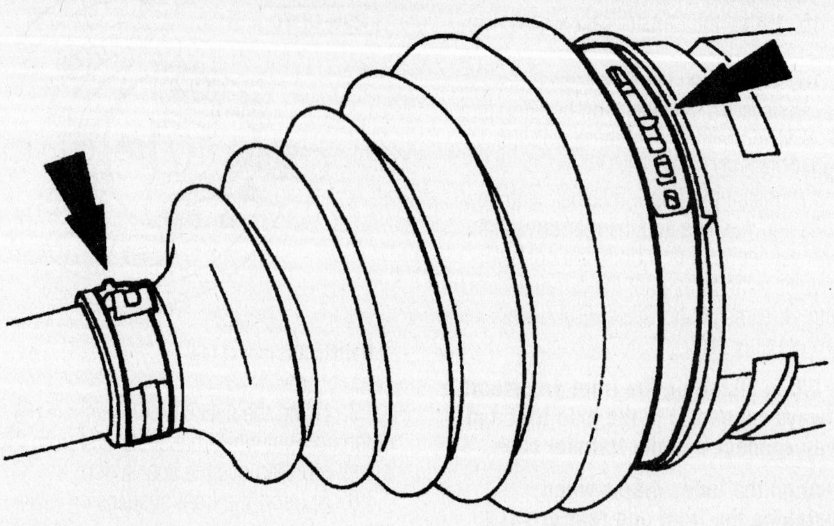
67197-RANG-G75

Inboard boot clamps

8. Remove the CV joint.
9. Remove the inboard halfshaft boot from the shaft assembly.
10. Remove the front wheel excluder seal, if necessary. Discard the seal. Tap uniformly around the seal to separate it from the joint.
11. Remove the two outboard boot clamps.
12. Remove the outboard halfshaft boot.
13. If the grease is contaminated, clean and inspect the joint for wear. Install a new outboard CV joint and shaft assembly if worn/damaged.
14. Inspect the assembly for contaminated grease.

To assemble

15. Pack the outboard CV joint with grease. Use Ford High Temp Constant Velocity Joint Grease E43Z-19590-A or equivalent meeting Ford specification ESP-M1C207-A. Spread any remaining grease from the service kit evenly inside the outboard halfshaft boot.
16. Clean the halfshaft boot mounting surfaces of excess grease before positioning the halfshaft boot into place.
17. Position the outboard halfshaft boot.
18. Position the boot clamps on the outboard halfshaft boot.
19. Tighten the through-bolt until the installer is in the closed position.
20. Install the outboard CV joint boot clamps. There are special tools made for this procedure.
21. Position the boot clamp on the halfshaft.
22. Position the inboard halfshaft boot.
23. Align the index marks on the halfshaft and the CV joint.

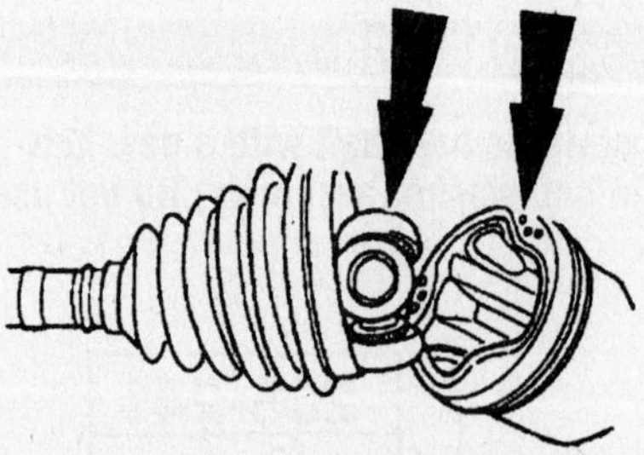
67197-RANG-G76

Separating the joint from the housing

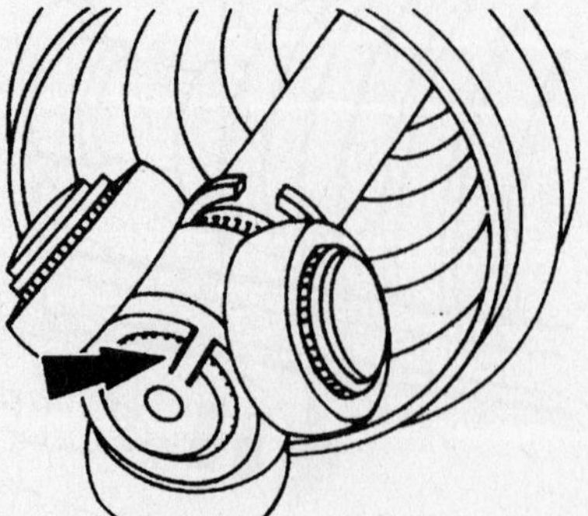
67197-RANG-G77

Make an alignment mark for reassembly

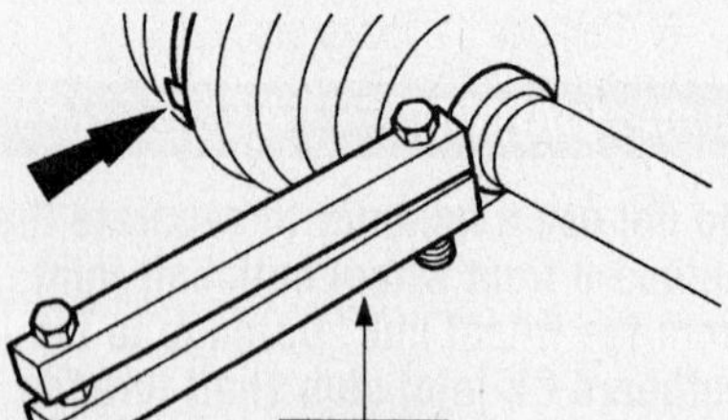

67197-RANG-G78

Boot clamp crimping tool

24. Install the CV joint on the halfshaft.
25. Install the snap ring.
26. Lubricate the three CV joint needle bearings. Use Ford High Temp Constant Velocity Joint Grease E43Z-19590-A or equivalent meeting Ford specification ESP-M1C207-A.
27. Fill the inboard CV joint housing with 235 grams (8.3 oz.) of grease. Use Ford High Temp Constant Velocity Joint Grease E43Z-19590-A or equivalent meeting Ford specification ESP-M1C207-A.
28. Position the CV joint housing onto the CV joint.
29. Remove any excess grease from the inboard halfshaft boot mating surface before positioning it into place.
30. Position the inboard halfshaft boot into place.
31. Position the boot clamp.
32. Insert a dulled screwdriver blade to relieve built-up air pressure in the halfshaft boot.
33. Using the special tool, install the inboard boot clamps.
34. Using the special tool, install the new front wheel excluder seal, if removed. Seat the metal ring at the seal's inner diameter flat against the CV joint housing.
35. Install the front wheel halfshaft.

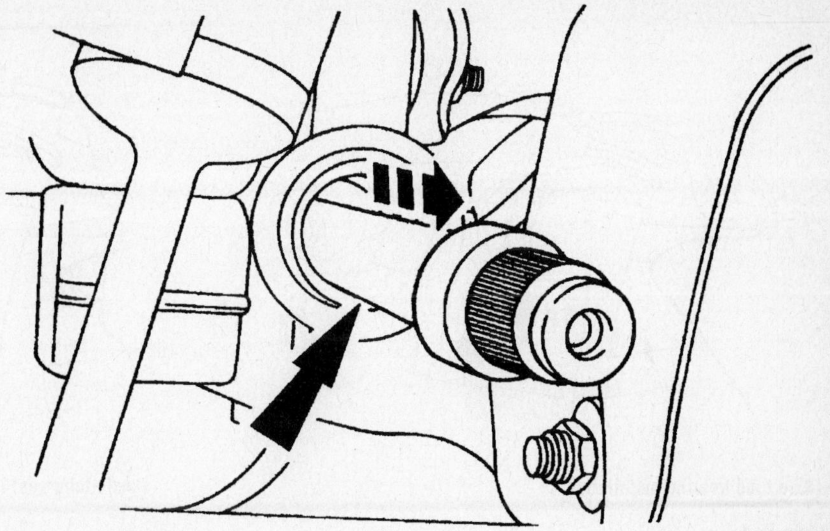

Right side axle shaft removal

Front Axle Tube Bearing

REMOVAL & INSTALLATION

1. Before servicing the vehicle, refer to the precautions in the beginning of this section.
2. Remove or disconnect the following:
   - Right-hand halfshaft
   - Right-hand axle shaft
   - Axle seal, with a slide hammer
   - Axle tube bearing, with a slide hammer
3. Clean the bearing and seal surfaces of any foreign debris.

To install:

4. Use an axle bearing replacer and the handle to replace the RH axle tube bearing.
5. Check the bearing depth as shown.
6. Use an axle seal replacer and the handle to replace the axle tube seal.

➡Care should be taken not to damage the axle seal surface.

7. Install the axle shaft.
8. Refill the front drive axle to proper level using SAE 80W90.
9. Install the RH halfshaft.

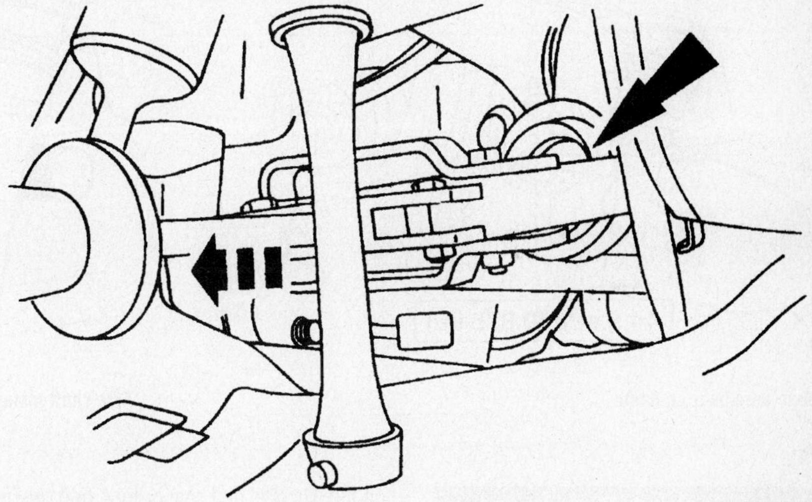

Axle seal removal

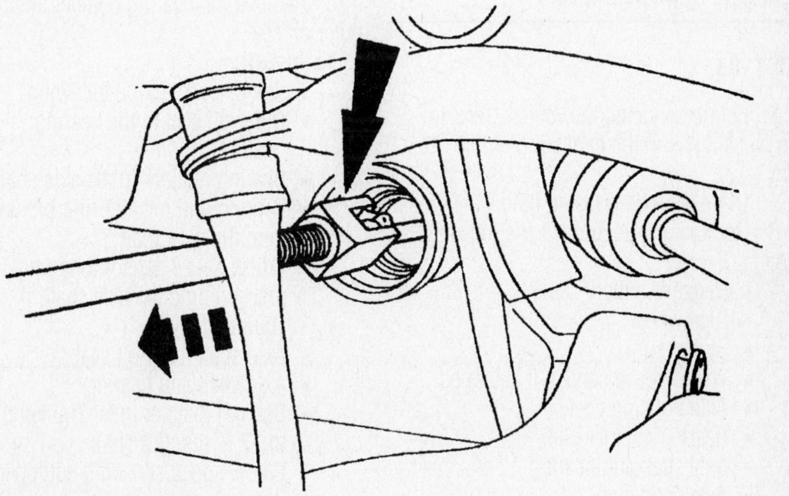

Axle tube bearing removal

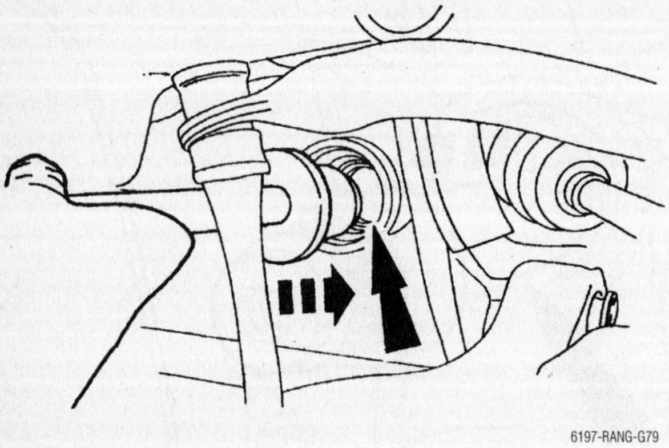

Axle tube bearing installation

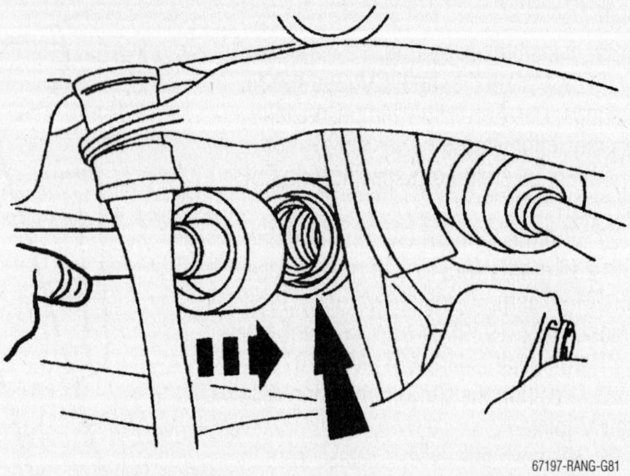

Axle tube seal installation

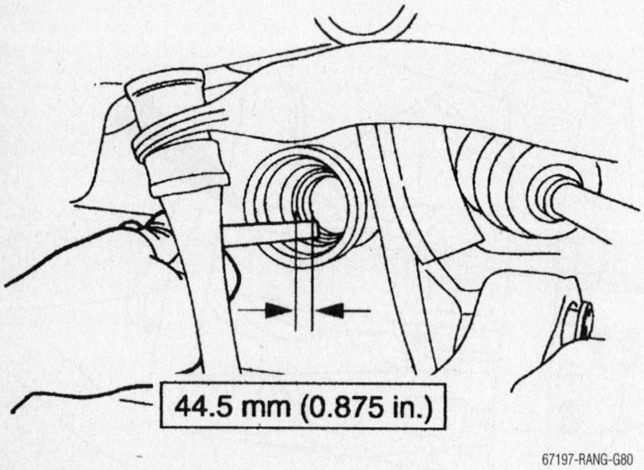

Axle tube bearing depth

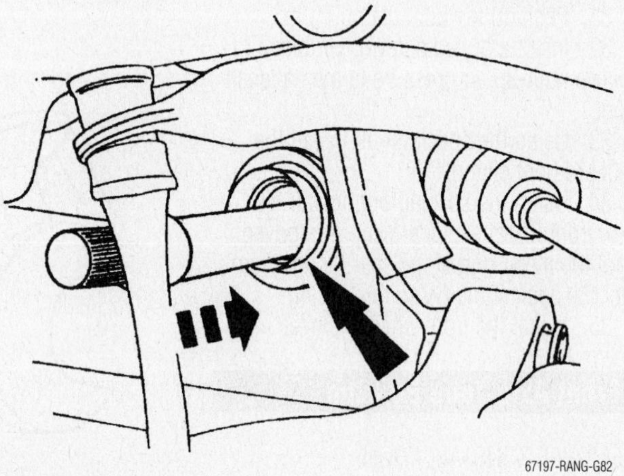

Axle shaft installation

Rear Axle Shaft, Bearing and Seal

REMOVAL & INSTALLATION

2001–03

1. Before servicing the vehicle, refer to the precautions in the beginning of this section.
2. Drain the axle housing fluid.
3. Remove or disconnect the following:
   - Negative battery cable
   - Rear wheel
   - Brake drum
   - Wheel speed sensor, if equipped
   - Axle housing cover
   - Bearing retainer nuts
   - Axle shaft and bearing
   - Axle shaft inner oil seal
4. If equipped with ABS, grind a flat spot on the wheel speed sensor tone ring, then split the ring with a chisel.
5. Press the wheel bearing off the axle shaft.
6. Remove the bearing retainer and the outer oil seal.

To install:

7. Install or connect the following:
   - Outer oil seal to the bearing retainer
   - Bearing retainer to the axle shaft
   - Bearing and retainer ring pressed onto the axle shaft
   - Wheel speed sensor tone ring pressed onto the axle shaft, if equipped
   - Axle shaft inner oil seal
   - Axle shaft and bearing
   - Bearing retainer nuts. Tighten them to 17 ft. lbs. (23 Nm).
   - Wheel speed sensor, if equipped
   - Brake drum
   - Rear wheel
   - Negative battery cable
8. Fill the rear differential to the correct level.

2004

FORD 7½ INCH RING GEAR

1. Before servicing the vehicle, refer to the precautions in the beginning of this section.
2. Raise and support the vehicle.
3. Remove the wheel and tire assembly.
4. Remove the 10 differential housing cover bolts and drain the lubricant from the rear axle housing.
5. Remove the differential housing cover.
6. Remove the rear brake drums.
7. Remove and discard the differential pinion shaft lock bolt.
8. Remove the differential pinion shaft.

➡Do not damage the rubber O-rings in the axle shaft grooves.

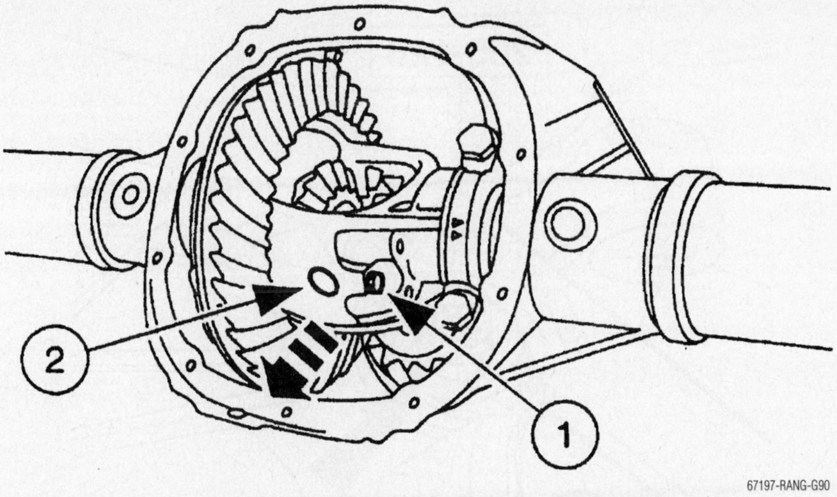

Differential pinion shaft removal. 1-lock bolt; 2-pinion shaft

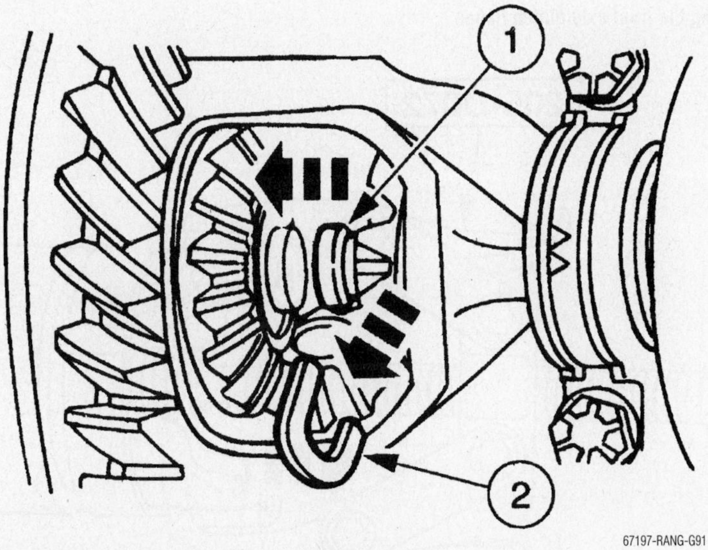

Removing the U-washers. 1-axle shaft; 2-U-washer

9. Push the axle shafts inboard.
10. Remove the U-washers.

➡Do not damage the wheel bearing oil seal.

11. Remove the two axle shafts.

➡If only a new seal needs to be installed, use care to avoid damaging the seal bore.

12. Using a suitable seal remover, remove the axle shaft oil seal. Discard the oil seal.
13. Inspect the rear wheel bearing and axle shaft for wear or damage.
14. If necessary, using a slidehammer, remove the rear wheel bearing.

To install:

15. Lubricate the new rear wheel bearing with lubricant.
16. Using a driver, install the rear wheel bearing.
17. Lubricate the lip of the new wheel bearing oil seal with grease.
18. Using a driver, install the wheel bearing oil seal.

➡Make sure the machined surfaces on both the rear axle housing and the differential housing cover are clean and free of oil before installing the new silicone sealant. The inside of the rear axle must be covered when cleaning the machined surface to prevent contamination.

19. Clean the gasket mating surface of the rear axle and the differential housing cover.
20. Lubricate the lip of the wheel bearing oil seal with grease.

➡Do not damage the wheel bearing oil seal.

21. Install the axle shafts.

➡Do not damage the rubber O-rings in the axle shaft grooves.

22. Position the two U-washers on the button end of the axle shafts.
23. Pull the axle shafts outward.

➡If a new pinion shaft lock bolt is unavailable, coat the threads with threadlock and sealer prior to installation.

24. Install the differential pinion shaft.
    a. Align the hole in the differential pinion shaft with the case lock bolt hole.
    b. Install a new differential pinion shaft lock bolt. Torque to 20 ft. lbs. (33 Nm).
25. Install the rear brake drums.
26. Apply a new continuous bead of sealant of the specified thickness to the differential housing cover.

➡The differential housing cover must be installed within 15 minutes of application of the silicone, or new sealant must be applied. If possible, allow one hour before filling with lubricant to make sure the silicone sealant has correctly cured.

27. Install the different housing cover.
28. Install the 10 differential housing cover bolts. Torque to 33 ft. lbs. (45Nm).
29. Fill the rear axle housing with 2.4 liters (5 pints) of lubricant.
30. Install the wheels and tires.
31. Lower the vehicle.

FORD 8.8 INCH RING GEAR

1. Before servicing the vehicle, refer to the precautions in the beginning of this section.
2. Raise and support the vehicle.
3. Remove the rear wheel and tire assembly.
4. Remove the brake drum.
5. Remove the differential housing cover and drain the lubricant.
6. Remove and discard the pinion shaft bolt.
7. Remove the differential pinion shaft.

➡Do not damage the rubber O-ring in the U-washer groove.

8. Push the axle shaft inboard.
9. Remove the U-washer.

10. Do not damage the wheel bearing oil seal.
11. Remove the axle shaft.

➡**If only a new seal needs to be installed, use care to avoid damaging the seal bore. If the wheel bearing oil seal is leaking, the differential housing vent may be plugged with foreign material.**

12. Using a suitable seal remover, remove the axle shaft oil seal. Discard the oil seal.
13. Inspect the rear wheel bearing and axle shaft for wear or damage.
14. Using the special tools, remove the rear wheel bearing.

To install:

15. Lubricate the new rear wheel bearing with rear axle lubricant.
16. Using the special tools, install the rear wheel bearing.
17. Lubricate the lip of the new wheel bearing oil seal with grease.
18. Using the special tools, install the wheel bearing oil seal.
19. Install the axle shaft.
20. Lubricate the lip of the wheel bearing oil seal with grease.

➡**Do not damage the wheel bearing oil seal.**

21. Install the axle shaft.
22. Do not damage the rubber O-ring in the U-washer groove.
23. Position the U-washer on the button end of the axle shaft.
24. Pull the axle shaft outward.

➡**If a new bolt is unavailable, coat the bolt threads with threadlock prior to installation.**

25. Align the bolt hole in the differential pinion shaft with the bolt hole in the case.
26. Install the new bolt. Torque to 22 ft. lbs. (33 Nm).
27. Install the brake drum.
28. Install the differential housing cover and fill the differential housing with the specified lubricant.
29. Install the rear wheel and tire assembly.
30. Lower the vehicle.

Front Pinion Seal

REMOVAL & INSTALLATION

➡**This operation disturbs the differential pinion bearing preload. Carefully reset the preload during assembly.**

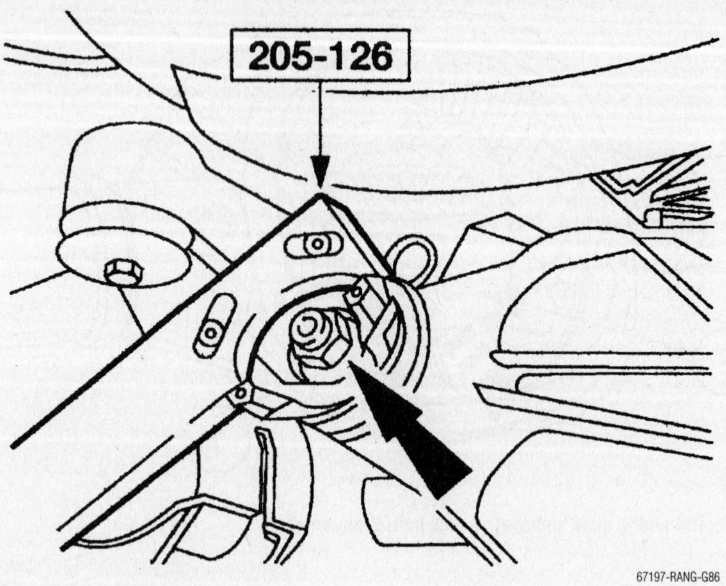

Holding the front axle pinion flange

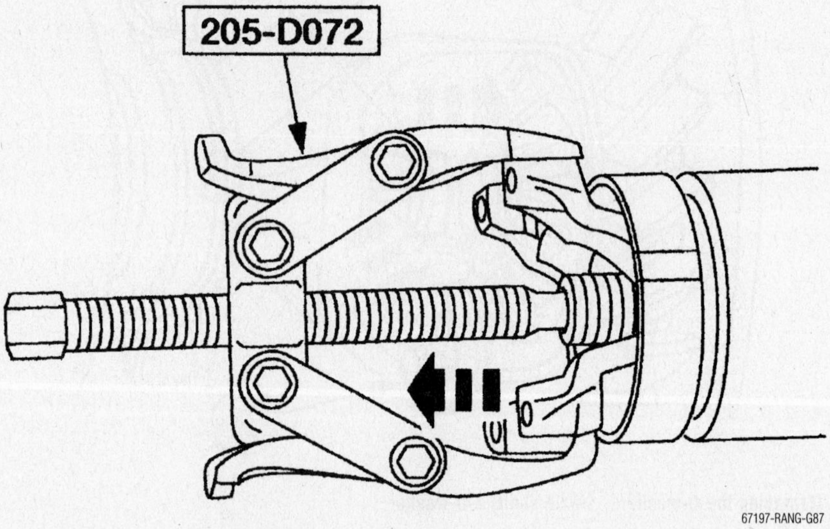

Removing the front axle pinion flange

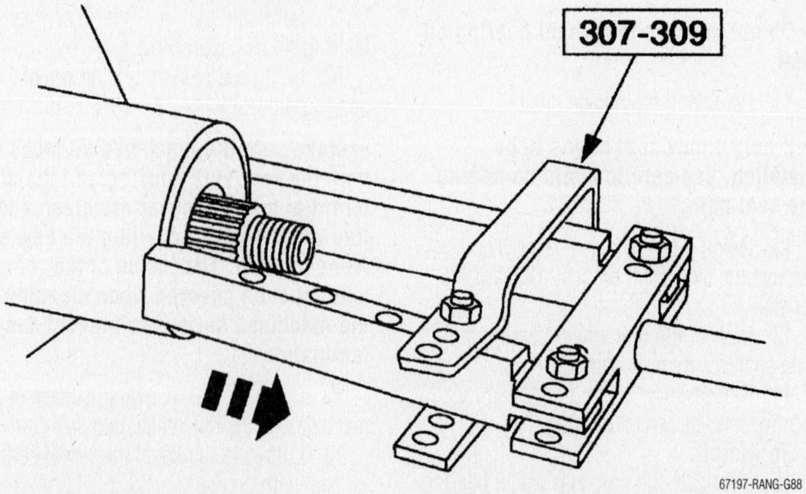

Removing the front axle pinion seal

⁂ CAUTION

The electrical power to the air suspension system must be shut off prior to hoisting, jacking or towing an air suspension vehicle. This can be accomplished by turning off the air suspension switch located in the rear jack storage area. Failure to do so can result in unexpected inflation or deflation of the air springs, which can result in shifting of the vehicle during these operations.

1. Before servicing the vehicle, refer to the precautions in the beginning of this section.
2. Index-mark the front driveshaft and pinion flange.
3. Remove or disconnect the following:
   - Front driveshaft from the pinion flange, and position it aside

➡Do not allow the driveshaft to hang unsupported.

4. Using a Nm (inch-pound) torque wrench, measure the torque required to maintain pinion rotation. Record the measurement.
5. Index-mark the pinion flange and the pinion stem.
6. Hold the pinion flange while removing the nut.
7. Place a drain pan under the differential housing.
8. Using a puller, remove the pinion flange.
9. Inspect the pinion flange for burrs and damage. Inspect the end of the pinion flange that contacts the bearing cone, the nut counterbore, and the seal surface for nicks. Discard the pinion flange as necessary.
10. Using a seal remover and impact slide hammer, remove the pinion seal.
11. Remove the front axle drive pinion shaft oil slinger and the differential pinion bearing.
12. Remove and discard the collapsible spacer.

To install:

13. Verify that the splines on the pinion stem are free of burrs. If burrs are evident, remove them with a fine crocus cloth. Work in a rotating motion to wipe the pinion clean.
14. Clean the pinion seal bore.
15. Install a new collapsible spacer.
16. Install the original differential pinion bearing and the front axle drive pinion shaft oil slinger.
17. Lubricate the pinion seal. Use Motorcraft SAE 80W90 Thermally Stable 4x4 Axle Lubricant meeting Ford specification WSP-M2C197-A.
18. Install the pinion seal.
19. Lubricate the pinion flange splines. Use Motorcraft SAE 80W90 Thermally Stable 4x4 Axle Lubricant meeting Ford specification WSP-M2C197-A.

➡Never use a metal hammer on the pinion flange or install the flange with power tools. If necessary, use a plastic hammer to tap on a tight fitting flange.

- Align the index marks and install the pinion flange.
- Install the new nut hand-tight.

➡Do not loosen the nut to reduce preload. Install a new collapsible spacer and nut if preload reduction is necessary.

20. Use the special tool to hold the pinion flange while tightening the nut to set the preload.
21. Tighten the nut, rotating the pinion occasionally to ensure the differential pinion bearings are seating correctly. Take frequent differential pinion bearing preload readings by rotating the pinion with a Nm (inch-pound) torque wrench. The final reading must be 0.56 Nm (5 inch lbs.) more than the initial reading taken during removal.
22. Align the index marks and position the front driveshaft.
23. Install the universal joint spider retainers and bolts.
24. Check the fluid level and, if necessary, fill the axle to specification. Use Motorcraft SAE 80W90 Thermally Stable 4x4 Axle Lubricant meeting Ford specification WSP-M2C197-A.
25. Lower the vehicle.
26. If so equipped, reactivate the air suspension.

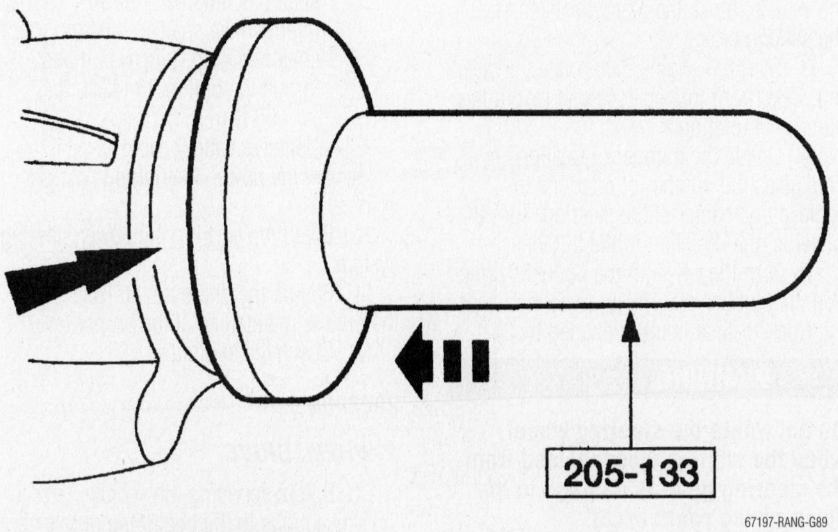

Installing the front axle pinion seal

Rear Pinion Seal

REMOVAL & INSTALLATION

1. Before servicing the vehicle, refer to the precautions in the beginning of this section.
2. Drain the axle housing fluid.
3. Remove or disconnect the following:
   - Negative battery cable
   - Rear wheels
   - Driveshaft
   - Brake calipers and pads or brake drum

➡The brake calipers and pads or brake drum must be removed so that there is no additional drag when measuring pinion bearing preload.

4. Use an inch lb. torque wrench and measure and record the amount of torque required to maintain pinion rotation through several revolutions.
5. Remove or disconnect the following:
   - Pinion flange
   - Pinion seal
   - Pinion bearing
   - Collapsible spacer

To install:

➡Use a new collapsible spacer and flange nut for assembly.

6. Install or connect the following:
   - Collapsible spacer
   - Pinion bearing
   - Pinion seal
   - Pinion flange

7. Rotate the pinion flange occasionally while tightening the flange nut to make sure the pinion bearings seat correctly.

8. Take frequent bearing preload torque readings. Tighten the flange nut to achieve the preload torque readings originally recorded.

CAUTION

Never loosen the pinion nut to reduce bearing preload. If it is necessary to reduce bearing preload, install a new collapsible spacer and pinion nut.

9. Install or connect the following:
   - Driveshaft
   - Brake calipers and pads or brake drum
   - Wheels
   - Negative battery cable

10. Fill the differential with gear lubricant and check for leaks.

STEERING AND SUSPENSION

Air Bag

PRECAUTIONS

- Always wear safety glasses when servicing an air bag vehicle, and when handling an air bag.
- Never attempt to service the steering wheel or steering column on an air bag-equipped vehicle without first properly disarming the air bag system. The air bag system should be properly disarmed whenever ANY service procedure in this manual indicates that you should do so.
- When carrying a live air bag module, always make sure the bag and trim cover are pointed away from your body. In the unlikely event of an accidental deployment, the bag will then deploy with minimal chance of injury.
- When placing a live air bag on a bench or other surface, always face the bag and trim cover up, away from the surface. This will reduce the motion of the air bag if it is accidentally deployed.
- If you should come in contact with a deployed air bag, be advised that the air bag surface may contain deposits of sodium hydroxide, which is a product of the gas combustion and is irritating to the skin. Always wear gloves and safety glasses when handling a deployed air bag, and wash your hands with mild soap and water afterwards.

DISARMING THE SYSTEM

1. Before servicing the vehicle, refer to the precautions in the beginning of this section.
2. Disconnect the negative battery cable from the battery.
3. Disconnect the positive battery cable from the battery.
4. Wait 1 minute. This time is required for the back-up power supply in the air bag diagnostic monitor to completely drain. The system is now disarmed.

ARMING THE SYSTEM

1. Before servicing the vehicle, refer to the precautions in the beginning of this section.
2. Connect the positive battery cable.
3. Connect the negative battery cable.
4. Stand outside the vehicle and carefully turn the ignition to the **RUN** position. Be sure that no part of your body is in front of the air bag module on the steering wheel, to prevent injury in case of an accidental air bag deployment.
5. Ensure the air bag indicator light turns off after approximately 6 seconds. If the light does not illuminate at all, does not turn off, or starts to flash, test the system.

Power Rack and Pinion Steering Gear

REMOVAL & INSTALLATION

2001–02

WARNING

If equipped, always turn off the Automatic Ride Control (ARC) service switch before lifting the vehicle off of the ground. Failure to do so could damage the ARC system components.

1. Before servicing the vehicle, refer to the precautions in the beginning of this section.
2. Raise and safely support the front of the vehicle, block the rear wheels and apply the parking brake.
3. Start the engine then rotate the steering wheel from lock-to-lock and record the number of rotations.
4. Divide the number of rotations by 2. This gives the number of rotations to achieve true center of the steering. Turn the wheel in one direction to the full lock.
5. Turn the wheel in the opposite direction the number of turns equal to true steering (lock-to-lock number divided by 2).

WARNING

Do not rotate the steering wheel when the shaft is disconnected from the steering gear as damage to the clock spring could occur.

6. Drain the power steering fluid reservoir.
7. Remove or disconnect the following:
   - Negative battery cable
   - Bolt retaining the lower steering column shaft to the steering gear input shaft
   - Stabilizer bar
   - Quick-connect fittings for the power steering pressure and return hoses at the steering gear housing
   - Nuts securing the power steering cooler and remove the cooler
   - Outer tie rod ends
   - Nuts, bolts and washer assemblies retaining the steering gear housing to the front crossmember
   - Steering gear from the vehicle

To install:

8. Install or connect the following:
   - Position the steering gear to the front crossmember and install the nuts, bolts and washer assemblies. Torque to 94–127 ft. lbs. (128–172 Nm).
   - Power steering cooler retaining bolts
   - Power steering lines to the steering gear housing and torque the fittings to 20–25 ft. lbs. (27–34 Nm).
   - Outer tie rod ends and ensure that the steering shaft or gear input shaft has not been rotated
   - Intermediate shaft-to-steering input shaft retaining (pinch) bolt and torque the bolt to 30–42 ft. lbs. (41–56 Nm)
   - Negative battery cable
9. Fill the power steering pump reservoir.
10. Bleed the air from the power steering system.
11. Ensure that there are no leaks and the fluid is maintained at the proper level.
12. Check the alignment.

2003–04

2-WHEEL DRIVE

1. Before servicing the vehicle, refer to the precautions in the beginning of this section.

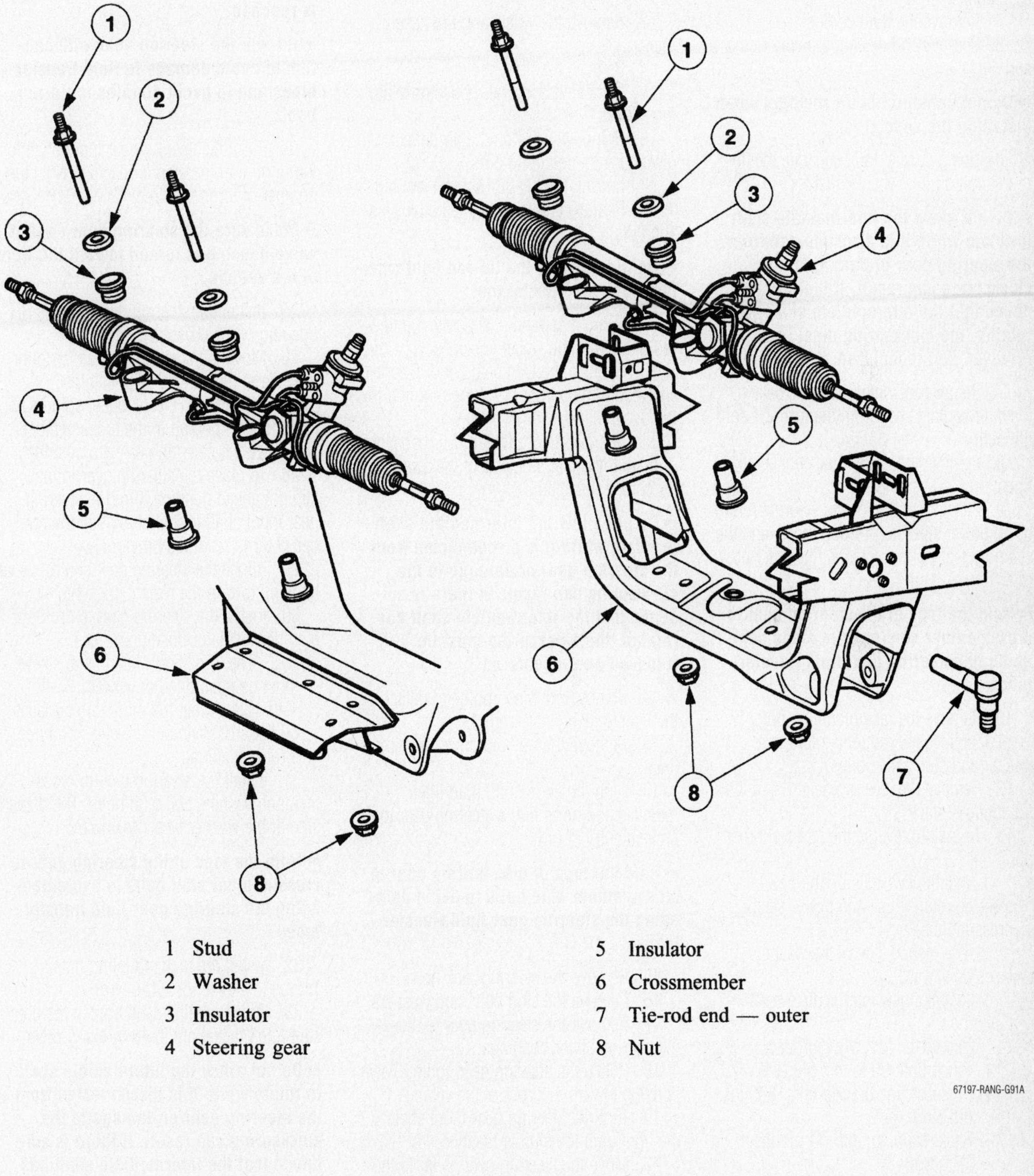

Steering gear mounting—2-wheel drive and 4-wheel drive, 2003–04 models

2. Turn the wheel to the straight-ahead position and turn the ignition switch to the OFF position.
3. Remove the front wheel and tire assemblies.
4. Remove the fluid cooler.
5. Remove and discard the cotter pins and nuts.

➡**Do not damage the tie-rod boot when installing the special tool.**

6. Using special tool, separate the tie-rod ends from the wheel knuckles.

➡**Do not allow the intermediate shaft to rotate while it is disconnected from the steering gear or damage to the clockspring can result. If there is evidence that the intermediate shaft has rotated, the clockspring must be removed and recentered.**

7. Remove the pinch bolt and detach the intermediate shaft from the gear. Discard the bolt.
8. Remove the nut and disconnect the lines.
9. Plug or cap the power steering return hose, power steering pressure hose, and the steering gear ports to prevent the entry of dirt.

➡**Hold the tops of the steering gear to crossmember stud bolts to avoid damaging the steering gear fluid transfer tubes.**

10. Remove the rack mounting nuts.
11. Remove the mounting stud, nut, washer and stop assemblies.
12. Remove the steering gear. Clean the mounting surfaces.
13. To install, reverse the removal procedure.
    a. Install new seals on the power steering return hose and power steering pressure hose.
    b. The dished side of the washer faces downward.
    c. Install a new intermediate shaft pinch bolt.
14. Observe the following torques:
    - Pinch bolt: 35 ft. lbs. (48 Nm)
    - Pressure line bracket nut: 18 ft. lbs. (25 Nm)
    - Rack retaining nuts: 111 ft. lbs. (150 Nm)
15. Fill and leak check the power steering system.
16. Check and, if necessary, adjust the wheel alignment.

4-WHEEL DRIVE

1. Before servicing the vehicle, refer to the precautions in the beginning of this section.
2. Turn the wheel to the straight-ahead position and turn the ignition switch to the OFF position.
3. Remove the wheel and tire assemblies.
4. Remove the fluid cooler.
5. Remove the four air deflector retaining screws.
6. Pull downward on the air deflector to disengage the retaining pins.
7. Loosen the LH tie-rod end jam nut.
8. Remove and discard the cotter pins and nuts.

➡**Do not damage the tie-rod boot when installing the special tool.**

9. Using a separator, separate the tie-rod ends from the wheel knuckles
10. Remove the LH tie-rod end. Count and record the number of turns required to remove the tie-rod end.
11. Remove the front stabilizer bar. Note or mark the driver side end of the sway bar for correct installation.

➡**Do not allow the intermediate shaft to rotate while it is disconnected from the steering gear or damage to the clockspring can result. If there is evidence that the intermediate shaft has rotated, the clockspring must be removed and recentered.**

12. Remove the pinch bolt and detach the intermediate shaft from the gear.
13. Remove the nut and disconnect the lines.
14. Plug the ends of all fluid lines removed and ports in the steering gear to prevent entry of dirt.

➡**Hold the tops of the steering gear to crossmember stud bolts to avoid damaging the steering gear fluid transfer tubes.**

15. Remove the steering rack nuts.
16. Remove the stud bolts and washers.
17. Remove the steering gear to crossmember insulator bushings.
18. Rotate the steering gear control valve housing toward the front of the vehicle.
19. Turn the steering gear input shaft to the right until the stop is reached.
20. Move the steering gear as far to the RH side of the vehicle as possible.
21. Move the LH front wheel spindle tie-rod forward to clear the frame crossmember.
22. Remove the steering gear from the vehicle.

To install:

23. Using special tool 211-027, install new seals on the power steering return hose and power steering pressure hose.

➡**Make sure the steering gear input shaft is turned to the left until the stop is reached.**

➡**Handle the steering gear with caution to avoid damage to fluid transfer tubes and to avoid dimples in tie-rod boots.**

24. Turn the steering gear input shaft to the right until the stop is reached. Note the number of turns required.

➡**Make sure the steering gear control valve housing is turned toward the front of the vehicle.**

25. Install the steering gear into the RH opening of the crossmember.
26. Move the steering gear as far to the RH side of the vehicle as possible.
27. Move the LH front wheel spindle tie-rod into the opening in the crossmember and move the steering gear into position.
28. To place the steering gear in the straight ahead position, turn the steering gear input shaft to the left by half the number of turns recorded previously.
29. Rotate the steering gear control valve housing toward the rear of the vehicle.
30. Install the steering gear to crossmember insulator bushings.
    a. The large end of the metal sleeve must be positioned downward.
    b. Check that the mounting surfaces on the crossmember are clean and free of foreign material.
31. Install the steering gear to crossmember washers and stud bolts. The dished side of the washer faces downward.

➡**Hold the tops of the steering gear to crossmember stud bolts to avoid damaging the steering gear fluid transfer tubes.**

32. Install the rack mounting nuts. Torque to 111 ft. lbs. (150 Nm).
33. Install the lines and tighten the pressure line bracket nut to 18 ft. lbs. (25 Nm).

➡**Do not allow the intermediate shaft to rotate while it is disconnected from the steering gear or damage to the clockspring can result. If there is evidence that the intermediate shaft has rotated, the clockspring must be removed and recentered.**

34. Connect the intermediate shaft to the steering gear input shaft. Install a new lower steering column pinch bolt. Torque to 35 ft. lbs. (48 Nm).

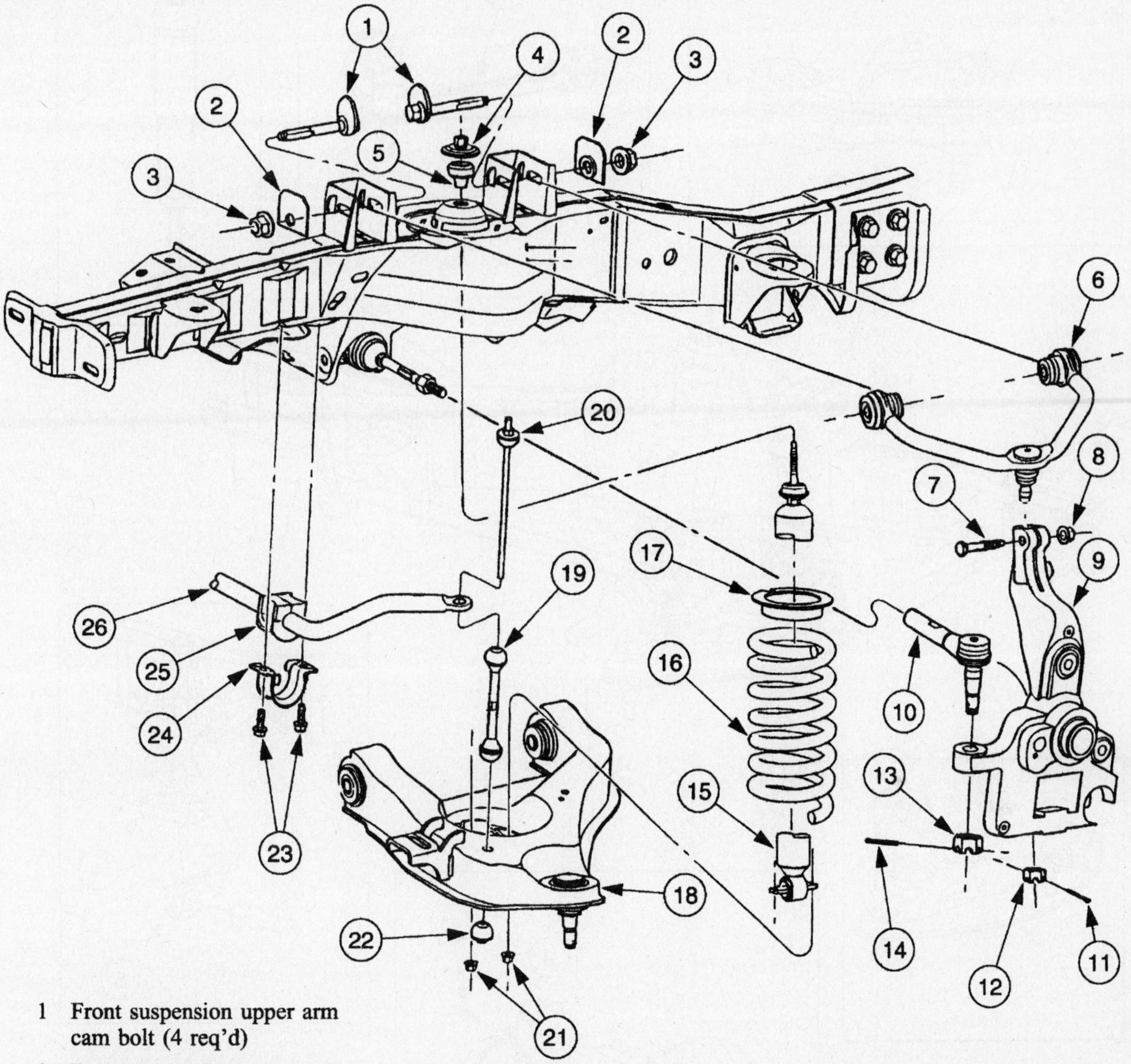

1 Front suspension upper arm cam bolt (4 req'd)
2 Front suspension upper arm cam assy (2 req'd)
3 Front suspension upper arm cam assy nut (2 req'd)
4 Front shock absorber upper nut/washer assy (2 req'd)
5 Front shock absorber upper bushing (2 req'd)
6 Front suspension upper arm
6 Front suspension upper arm
7 Front wheel spindle pinch bolt (2 req'd)
8 Front wheel spindle pinch bolt nut
9 Front wheel spindle
10 Tie-rod end
11 Cotter pin
12 Lower ball joint castellated nut (2 req'd)
13 Tie-rod end castellated nut (2 req'd)
14 Cotter pin
15 Front shock absorber
16 Front coil spring
17 Front spring insulator
18 Front suspension lower arm
18 Front suspension lower arm
19 Front stabilizer bar link
20 Front stabilizer bar stud and bushing assy (2 req'd)
21 Front shock absorber lower nut (4 req'd)
22 Front stabilizer bar nut and washer assy (2 req'd)
23 Front stabilizer bar mounting bolts (4 req'd)
24 Stabilizer bar bracket (2 req'd)
25 Front stabilizer bar bushing assy (2 req'd)
26 Front stabilizer bar

67197-RANG-G92

2-wheel drive front suspension—2004 models shown

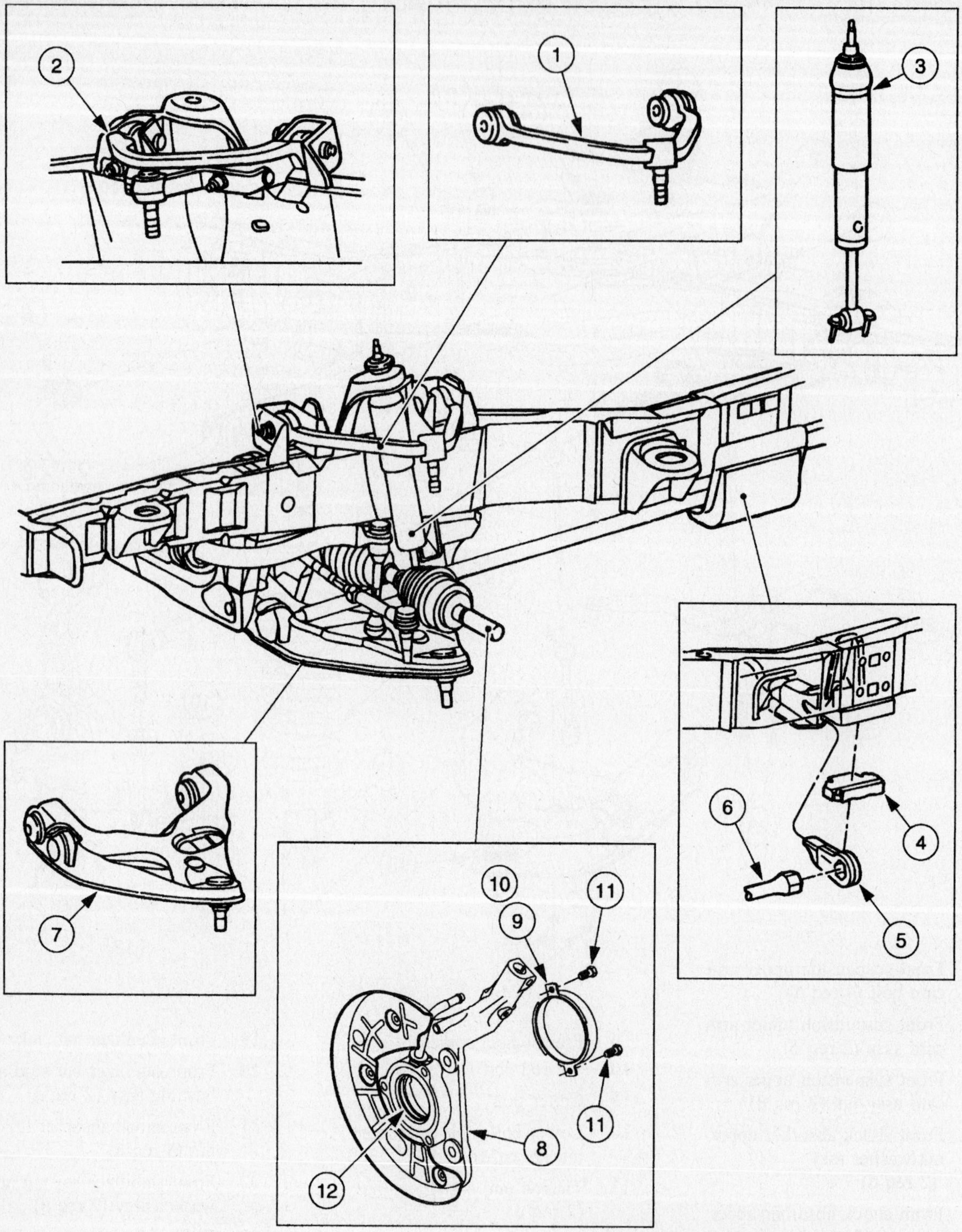

1 Upper arm, bushing and joint assembly (LH)
2 Upper arm, bushing and joint assembly (RH)
3 Shock assembly
4 Torsion bar adjuster plate
5 Torsion bar adjuster
6 Torsion bar (LH)
7 Lower arm, bushing and joint assembly (LH)
8 Knuckle assembly
9 Protection shield (RH)
10 Protection shield (LH)
11 Screw (self-tapping)
12 Oil seal

67197-RANG-G93

4-wheel drive front suspension—2004 models shown

35. Install the power steering fluid cooler.

36. Install the front stabilizer bar. Orient the front stabilizer bar as noted during removal.

37. Install the LH tie-rod end on the front wheel spindle tie-rod. Rotate the tie-rod end the number of turns recorded during removal.

38. Position the tie-rod ends on the steering knuckles. Install the castellated nuts and new cotter pins. Check that the brake dust shields are not bent and are not in contact with the outer tie-rod boot seals. Torque to 52 ft. lbs. (70 Nm).

39. Tighten the tie-rod end jam nut. Torque to 59 ft. lbs. (80 Nm).

40. Position the air deflector, and install the retaining screws.

41. Install the front wheel and tire assemblies.

42. Fill and leak check the system.

43. Check and, if necessary, adjust the wheel alignment.

Shock Absorber

REMOVAL & INSTALLATION

Front

➡Low pressure gas shocks are charged with nitrogen gas. Do not attempt to open, puncture or apply heat to them. Prior to installing a new shock absorber, hold it upright and extend it fully. Invert it and fully compress and extend it at least 3 times. This will bleed trapped air.

1. Before servicing the vehicle, refer to the precautions in the beginning of this section.
2. Remove or disconnect the following:
   - Negative battery cable
   - Upper shock-to-frame attaching nut, washer and insulator assembly
   - Lower shock-to-control arm attaching nuts
   - Slightly compress the shock absorber by hand and remove it from the vehicle

To install:

3. Install or connect the following:
   - Position the lower washer and insulator on the shock absorber rod and position the shock absorber to the upper frame bracket mount
   - Position the upper insulator and washer on the shock absorber rod and install the attaching nut loosely.
   - Position the lower shock absorber mounting studs into the control arm and install the attaching nuts loosely.
   - Torque the lower shock attaching nuts to 15–21 ft. lbs. (21–29 Nm), and the upper shock attaching bolts to 30–40 ft. lbs. (40–55 Nm).
   - Negative battery cable

Rear

➡Low pressure gas shocks are charged with nitrogen gas. Do not attempt to open, puncture or apply heat to them. Prior to installing a new shock absorber, hold it upright and extend it fully. Invert it and fully compress and extend it at least 3 times. This will bleed trapped air.

1. Before servicing the vehicle, refer to the precautions in the beginning of this section.
2. Remove or disconnect the following:
   - Upper shock-to-frame attaching nut
   - Lower shock nut
   - Slightly compress the shock absorber by hand and remove it from the vehicle

To install:

3. Install or connect the following:
   - Shock absorber upper end and nut
   - Shock absorber lower end and nut
   - Torque the upper and lower shock attaching nuts to 53 ft. lbs. (72Nm)

Coil Spring

REMOVAL & INSTALLATION

1. Before servicing the vehicle, refer to the precautions in the beginning of this section.
2. Remove or disconnect the following:
   - Wheel and tire assembly
   - Shock absorber
   - Front stabilizer bar link nut
3. Use a coil spring compressor to compress the coil spring.
4. Remove the cotter pin and castellated nut.
5. Separate the lower ball joint from the front wheel spindle.
6. Position the front wheel spindle out of the way and remove the coil spring.

To install:

➡The end of the coil spring must cover the first hole and should not be visible in the second hole.

7. Install the coil spring in the lower arm.

⁂ WARNING

Always install the cotter pin into the lower ball joint castellated nut from outboard to inboard. Failure to do so will result in damage to the wheel and tire assembly.

8. Install the lower ball joint.
9. Install the front stabilizer bar link nut.
10. Remove the Coil Spring Compressor.
11. Install the front shock absorber and the two lower nuts.
12. Install the upper shock absorber bushing and nut/washer assembly.
13. Install the wheel and tire assembly.

Leaf Springs

REMOVAL & INSTALLATION

1. Before servicing the vehicle, refer to the precautions in the beginning of this section.
2. Remove or disconnect the following:
   - Negative battery cable
   - Rear wheels
   - U-bolts from the rear spring plate
   - Hardware from the spring to bracket at the front of the rear spring
   - Upper and lower shackle bolts at the rear of the spring
   - Spring and shackle from the bracket

To install:

3. Install or connect the following:
   - Spring and shackle to the bracket
   - Upper and lower shackle bolts at the rear of the spring. Torque the nuts to 87 ft. lbs. (118 Nm).
   - U-bolts to the spring plate. Torque the nuts 83 ft. lbs. (113 Nm).
   - Rear wheels
   - Negative battery cable

Torsion Bar

REMOVAL & INSTALLATION

⁂ CAUTION

The electrical power to the air suspension system must be shut off prior to hoisting, jacking or towing an air suspension vehicle. This can be accomplished by turning off the air suspension switch located in the rear jack storage area. Failure to do so can result in unexpected inflation or deflation of the air springs or shocks, which can result in shifting of the vehicle during these operations.

1. Before servicing the vehicle, refer to the precautions in the beginning of this section.
2. Remove or disconnect the following:
3. Remove the torsion bar cover plate

➡**Before relieving the torsion bar tension, measure and record the measurement of the torsion bar adjustment bolt. This measurement will be used as the preset depth for the new torsion bar adjustment bolt during installation.**

4. Relieve the torsion bar tension.
   a. Position the Torsion Bar Tool and adapters.
   b. Tighten the Torsion Bar Tool until the torsion bar adjuster lifts off the adjustment bolt.

**** CAUTION**

The torsion bar adjustment bolt is coated with dry adhesive; and must be replaced if it is backed off or removed. Failure to do so can cause the adjustment bolt to loosen during operation and cause a loss of vehicle alignment.

   c. Remove the torsion bar adjustment bolt and nut.
   d. Loosen the Torsion Bar Tool until the tension is removed from the torsion bar.
5. Mark the torsion bar and the adjuster for proper installation.
6. Remove the torsion bar insulator.
7. Grasp the torsion bar, and pull it free from the front suspension lower arm.

To install:

8. Position the torsion bar and the torsion bar adjuster.
9. Align the marks on the torsion bar and the torsion bar adjuster, then install the torsion bar adjuster.
10. Position the torsion bar insulator.
11. Install the Torsion Bar Tool and the adapters.
12. Tighten the Torsion Bar Tool until the new adjustment bolt and nut can be installed.
13. Turn the adjustment bolt until the preliminary adjustment measurement (recorded length of the old adjustment bolt) is reached.
14. Install the torsion bar cover plate. Torque the bolts to 46 ft. lbs. (63Nm).
15. If equipped with air suspension, reactivate the system by turning on the air suspension switch.
16. Lower the vehicle.
17. Adjust the ride height.
18. Check the alignment.

Upper Ball Joint

REMOVAL & INSTALLATION

The ball joints are integral with the control arm. If the ball joint is defective, the entire control arm must be replaced.

Lower Ball Joint

REMOVAL & INSTALLATION

The ball joints are integral with the control arm. If the ball joint is defective, the entire control arm must be replaced.

Upper Control Arm

REMOVAL & INSTALLATION

Coil Spring Suspension

1. Before servicing the vehicle, refer to the precautions in the beginning of this section.
2. Remove or disconnect the following:
   - Wheel and tire assembly
   - Brake disc shield
3. Use a jack to support the front suspension lower arm.
4. Mark the position of the front suspension upper arm adjustment cams.
5. Remove the upper ball joint retaining nut and pinch bolt.
6. Separate the ball joint from the front wheel spindle.
7. Remove the front suspension upper arm.
8. Installation is the reverse of removal. Align the marks made during removal on the front suspension upper arm adjustment cam. The forward front suspension upper arm nut must be tightened first while the arm is held at the curb position ride height.

Observe the following torques:

- Control arm attaching nuts: 98 ft. lbs. (133Nm)
- Pinch bolt: 46 ft. lbs. (63Nm)

Torsion Bar Suspension

2-WHEEL DRIVE

**** WARNING**

The electrical power to the air suspension system must be shut off prior to hoisting, jacking or towing an air suspension vehicle. This can be accomplished by turning off the air suspension switch located in the rear jack storage area. Failure to do so can result in unexpected inflation or deflation of the air springs or shocks, which can result in shifting of the vehicle during these operations.

1. Before servicing the vehicle, refer to the precautions in the beginning of this section.
2. Raise the vehicle on a hoist.
3. Remove the wheel and tire assembly
4. Use a suitable jack stand to support the front suspension lower arm. ®

➡**To avoid possible damage to the front wheel spindle, secure the spindle to keep it from tilting before removing the pinch bolt and nut.**

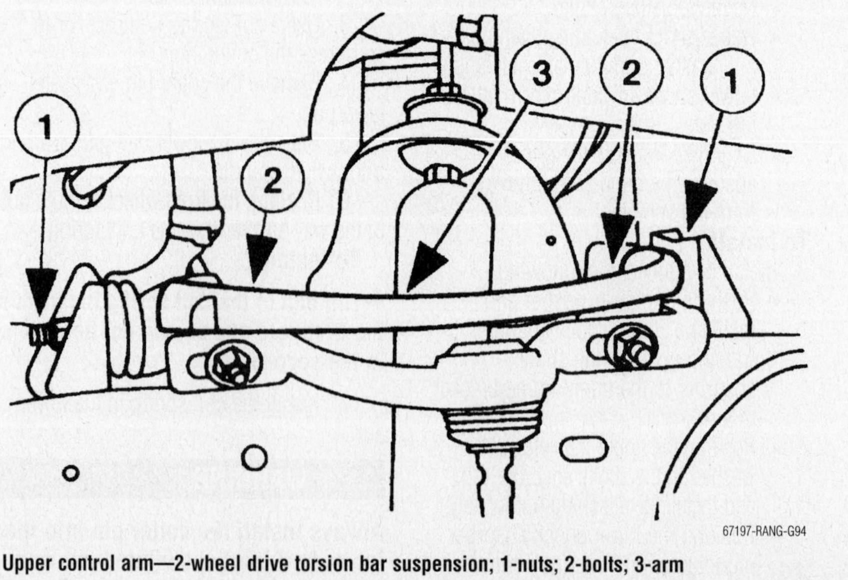

Upper control arm—2-wheel drive torsion bar suspension; 1-nuts; 2-bolts; 3-arm

5. Remove the pinch bolt and nut from the front wheel spindle.
6. Remove the front suspension upper arm:
   a. Remove the two nuts.
   b. Remove the two bolts.
   c. Remove the front suspension upper arm.

To install:

7. Position the front suspension upper arm.
8. Install the two bolts and two nuts. Torque to 83–112 ft. lbs. (113–153 Nm).
9. Position the front wheel spindle.
10. Install the pinch bolt and nut. Torque to 35–46 ft. lbs. (47–63 Nm).
11. Remove the jack stand from under the front suspension lower arm.
12. Install the tire and wheel assembly

➡If equipped with air suspension, reactivate the system by turning on the air suspension switch.

13. Lower the vehicle.
14. Inspect the front end ride height
15. Inspect and adjust the front end alignment

4-WHEEL DRIVE

1. Before servicing the vehicle, refer to the precautions in the beginning of this section.
2. Raise the vehicle on a hoist.
3. Remove the wheel and tire assembly.
4. Use a suitable jack stand to support the front suspension lower arm.
5. Remove the pinch bolt.

✲✲ WARNING

Before separating the front suspension upper arm from the front wheel knuckle, secure the front wheel knuckle to prevent it from tilting outward. Failure to do so can cause damage to the front axle shaft.

6. Separate the front suspension upper arm from the front wheel knuckle.
7. Remove the front suspension upper arm:
   a. Remove the two nuts and alignment plates.
   b. Remove the two bolts and the cams.
   c. Remove the front suspension upper arm.

To Install:

➡When installing the front suspension upper arm, replace the alignment plates with new alignment cams.

8. Install the front suspension upper arm.
   a. Position the front suspension arm bushing joint.
   b. Install the two bolts, four cams and two nuts. Torque to 83–112 ft. lbs. (113–153 Nm).
9. Install the pinch bolt and nut.
   a. Position the upper arm into the front wheel knuckle.
   b. Install the pinch bolt and nut. Torque to 41 ft. lbs. (55 Nm).
10. Remove the jack stand from under the front suspension lower arm.
11. Install the wheel and tire assembly.
12. Lower the vehicle.
13. Check the wheel alignment.

UPPER CONTROL ARM BUSHING REPLACEMENT

The control arm bushings are not serviceable. If they require service, the upper or lower arm must be replaced.

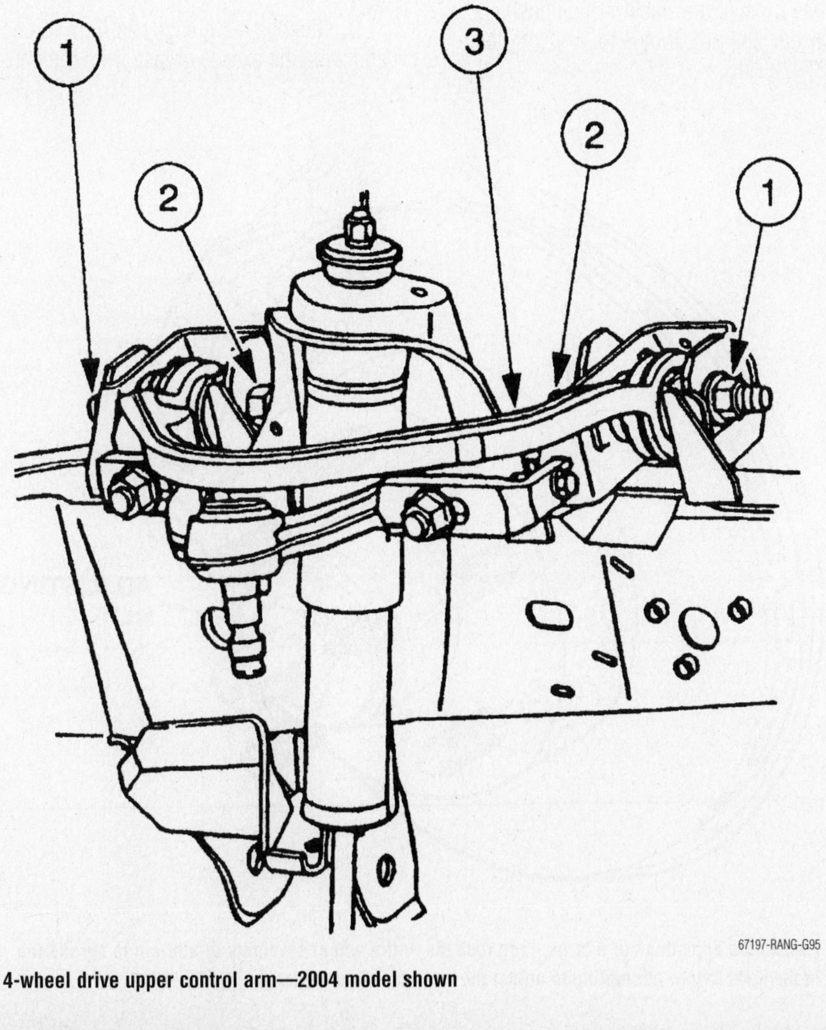

4-wheel drive upper control arm—2004 model shown

Lower Control Arm

REMOVAL AND & INSTALLATION

Coil Spring Suspension

1. Before servicing the vehicle, refer to the precautions in the beginning of this section.
2. Remove or disconnect the following:
   - Negative battery cable
   - Front wheel
   - Brake rotor shield
   - Shock absorber
   - Stabilizer bar link hardware
3. Using a spring compressor tool, compress the coil spring.
   - Lower ball joint from the spindle
   - Lower control arm bolts
   - Lower control arm and coil spring

To install:

4. Install or connect the following:
   - Coil spring to the lower control arm

➡The end of the coil spring must cover the first hole and should not be visible in the second hole.

   - Lower arm and front coil spring
   - The two front suspension lower

arm bolts and nuts. Do not tighten the nuts at this time.

➡**On the RH front suspension lower arm, install the rear bolt adjustment cam, and nut in the center of the frame slot.**

✱✱ CAUTION

Always install the cotter pin into the lower ball joint castellated nut from outboard to inboard, with the fingers bent together at a right angle. Failure to do so will result in damage to the wheel and tire assembly.

- Lower ball joint. Torque the nut to 113 ft. lbs. (153Nm).

5. Remove the Coil Spring Compressor.
6. Install or connect the following:
   - Front stabilizer bar link nut. Torque the nut to 21 ft. lbs. (29Nm).
   - Shock absorber and the two lower nuts
   - Upper shock absorber bushing and nut/washer assembly
7. Support the lower control arm with a jackstand. Torque the bolts to 129 ft. lbs. (175 Nm).
   - Brake disc shield
   - Wheel and tire assembly
8. Inspect and adjust the front end alignment.

Torsion Bar Suspension

1. Before servicing the vehicle, refer to the precautions in the beginning of this section.
2. Raise the vehicle on a hoist.
3. Remove the wheel and tire assembly.
4. Remove the stabilizer link nut, washer and bushing.
5. Remove the front shock absorber-to-front suspension lower arm nuts.
6. Remove the torsion bar
7. Remove the lower ball joint castellated nut.

➡**Do not use a hammer to separate the ball joint from the wheel knuckle or damage to the wheel knuckle will result. Do not damage the ball joint boot while installing the special tool.**

8. Using the special tool, separate the front suspension lower arm from the front wheel knuckle/spindle.
9. Remove the front suspension lower arm bolts and nuts.
10. Remove the front suspension lower arm.

To install:

➡**Tighten the front suspension lower arm pivot bolts and nuts until snug. Do not tighten to specification until the installation procedure is complete.**

11. Position the front suspension lower arm to the front suspension crossmember.
12. Install the pivot bolts and nuts and tighten until snug.

✱✱ WARNING

Install the cotter pin into the lower ball joint from outboard to inboard with the fingers bent together at a right angle. Failure to do so will cause damage to the wheel and tire assembly.

13. Position the lower ball joint into the front wheel knuckle/spindle.
14. Install the new castellated nut. Torque to 83–112 ft. lbs. (113–153 Nm).
15. Install a new cotter pin.
16. Install the front shock absorber-to-front suspension lower arm nuts. Torque to 15–21 ft. lbs. (21–29 Nm).
17. Install the stabilizer link bushing, washer, and nut. Torque to 15–21 ft. lbs. (21–29 Nm).

➡**Whenever the torsion bar or torsion bar adjuster is removed, the vehicle ride height must be checked.**

18. Install the torsion bar.
19. Install the tire and wheel assembly.
20. Lower the vehicle.
21. Tighten the front suspension lower arm nuts. Torque to 111–148 ft. lbs. (150–200 Nm).
22. Inspect and adjust the front end alignment.

LOWER CONTROL ARM BUSHING REPLACEMENT

The control arm bushings are not serviceable. If they require service, the upper or lower arm must be replaced.

Wheel Bearings

ADJUSTMENT

2-Wheel Drive Vehicles

1. Before servicing the vehicle, refer to the precautions in the beginning of this section.
2. Remove the grease cap from the hub and wipe the excess grease from the end of

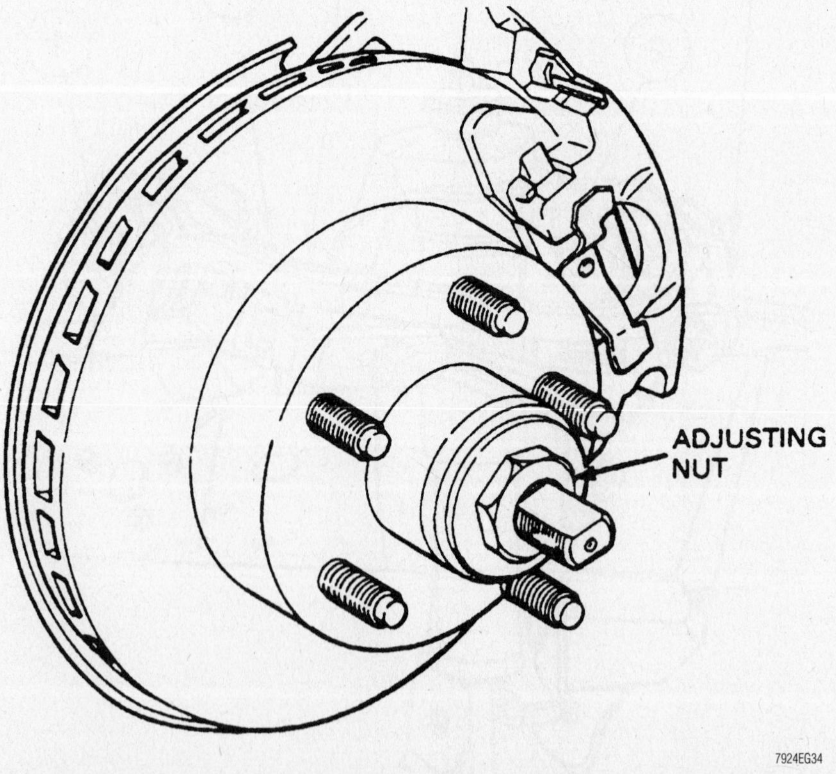

Loosen the adjusting nut 3 turns, then rock the entire wheel assembly in-and-out to spread the brake pads before attempting to adjust the bearing—2wd vehicles

the spindle. Remove the cotter pin and retainer. Discard the cotter pin.

3. Loosen the adjusting nut 3 turns.

✲✲ WARNING

Obtain running clearance between the disc brake rotor surface and shoe linings by rocking the entire wheel assembly in and out several times in order to push the caliper and brake pads away from the rotor. An alternate method to obtain proper running clearance is to tap lightly on the caliper housing. Be sure not to tap on any other area that may damage the disc brake rotor or the brake lining surfaces. Do not pry on the phenolic caliper piston. The running clearance must be maintained throughout the adjustment procedure. If proper clearance cannot be maintained, the caliper must be removed from its mounting.

4. While rotating the wheel assembly, tighten the adjusting nut to 17–25 ft. lbs. (23–34 Nm) in order to seat the bearings. Loosen the adjusting nut a half turn. Retighten the adjusting nut 18–20 inch lbs. (2.0–2.2 Nm).

5. Place the retainer on the adjusting nut. The castellations on the retainer must be in alignment with the cotter pin holes in the spindle. Once this is accomplished install a new cotter pin and bend the ends to insure its being locked in place.

6. Check for proper wheel rotation. If correct, install the grease cap.

7. Lower the vehicle and tighten the lug nuts to 100 ft. lbs., (136 Nm) if the wheel was removed. Before driving the vehicle, pump the brake pedal several times to restore normal brake pedal travel.

✲✲ CAUTION

If the wheel was removed, retighten the wheel lug nuts to specification after about 500 miles (804km) of driving. Failure to do this could result in the wheel coming off while the vehicle is in motion causing loss of vehicle control or collision.

4-Wheel Drive

1. Before servicing the vehicle, refer to the precautions in the beginning of this section.

2. Remove or disconnect the following:
   - Wheel assembly
   - Retainer washers from the lug nut studs and remove the automatic locking hub assembly from the spindle
   - Snapring and spacer from the end of the spindle shaft
   - Pull the locking cam assembly and the 2 plastic spacers off of the wheel bearing adjusting nut

3. Use a magnet and remove the locking key from under the adjusting nut. If required, rotate the adjusting nut slightly to relieve pressure against the locking key.

✲✲ WARNING

To prevent damage to the adjusting nut and spindle threads on vehicles equipped with automatic hubs, look into the spindle keyway under the adjusting nut and remove the separate locking key before removing the adjusting nut.

4. Loosen the wheel bearing locknut using a 2⅜ inch (60.3mm) hex socket, such as Hex Locknut Wrench T70T-4252-B.

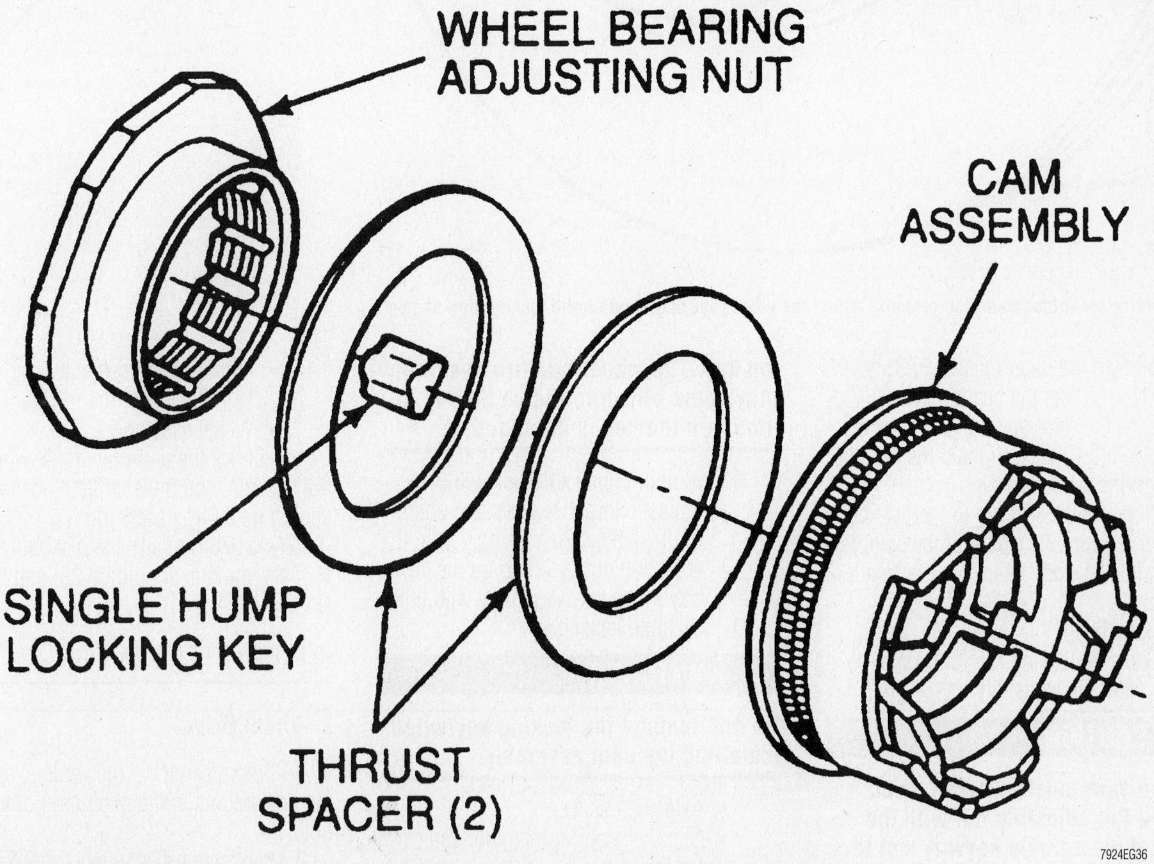

Exploded view of the wheel bearing adjusting nut and related components—automatic locking hub shown

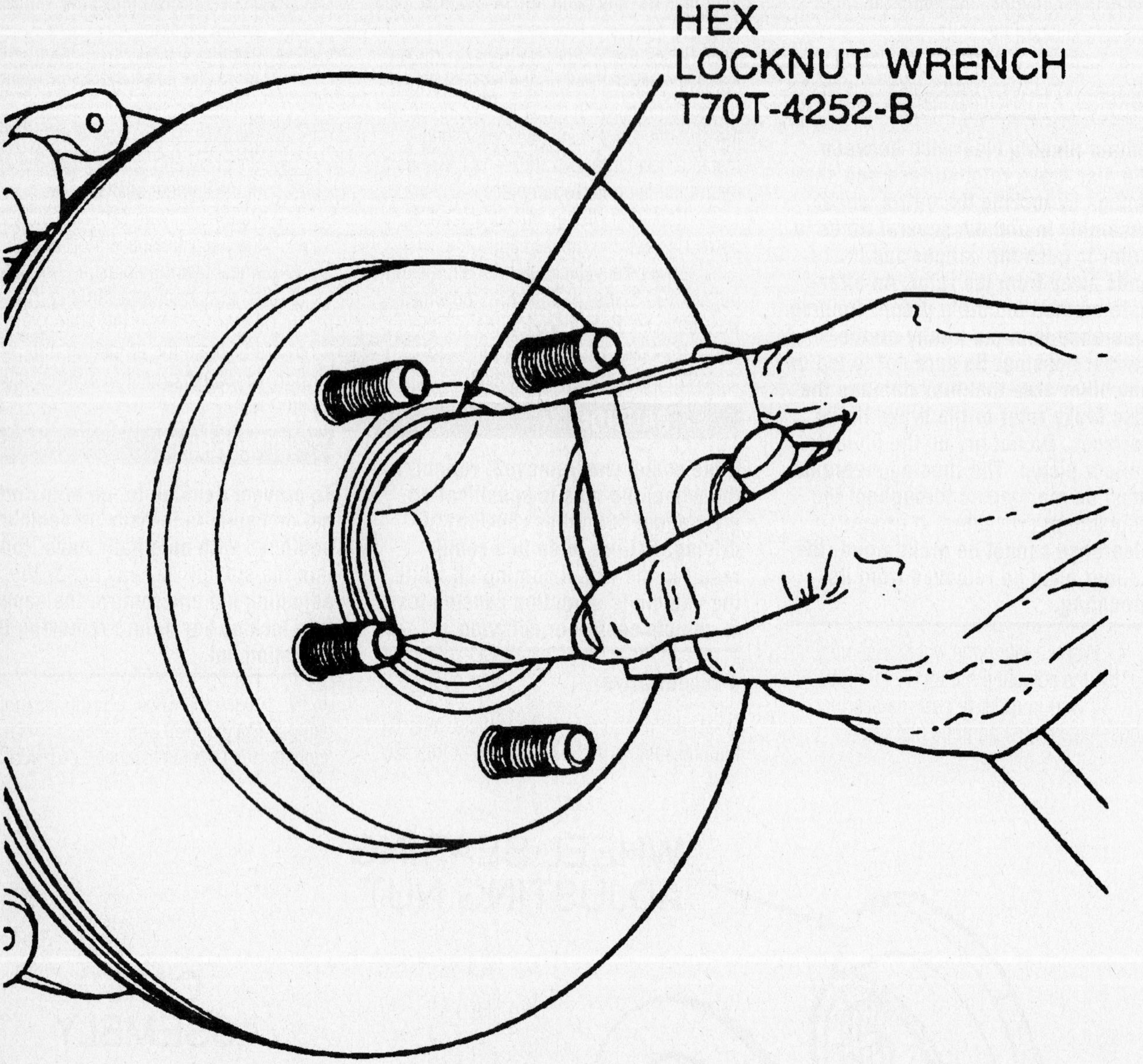

An oversize socket is needed to properly adjust the wheel bearing—automatic locking hub shown

5. Tighten the inner locknut to 35 ft. lbs. (47 Nm) to seat the bearings.

6. Spin the rotor and back off the inner locknut ¼ turn (90°). Retighten the locknut to 16 inch lbs. (1.8 Nm).

7. Align the closest lug in the bearing adjusting nut with the center of the spindle keyway slot. Advance the nut to the next if required.

To install:

8. Separate locking key in the spindle keyway under the adjusting nut.

⁂ CAUTION

Extreme care must be taken when aligning the adjusting nut with the center of the spindle keyway slot to prevent damage to the separate locking key. The wheel and tire assembly may come off while the vehicle is in motion if the key is damaged.

9. Install or connect the following:

- 2 plastic thrust spacers and push or press the cam assembly onto the adjusting nut by lining up the keyway in the cam assembly with the separate locking key

⁂ WARNING

Do not damage the locking key when installing the cam assembly.

- Axle shaft spacer
- Clip the snapring onto the end of the spindle
- Manual hub assembly over the spindle. Install the retainer washers
- Wheel assembly

10. Check the end-play of the wheel and tire assembly on the spindle. End-play should be 0.001–0.003 in. (0.025–0.076mm) and the maximum torque to rotate the hub should be 25 inch lbs. (2.8 Nm).

REMOVAL & INSTALLATION

2-Wheel Drive

1. Before servicing the vehicle, refer to the precautions in the beginning of this section.

2. Remove or disconnect the following:

- Disc brake caliper anchor plate

- Hub grease cap
- Cotter pin
- Nut retainer
- Spindle nut
- Wheel outer bearing retainer washer
- Outer front wheel bearing
- Brake disc and hub
- Hub grease seal
- Inner wheel bearing

To install:

3. Thoroughly clean and inspect the front wheel bearings and the brake disc and hub.
4. Lubricate the front wheel bearings.
5. Install the inner front wheel bearing.
6. Install a new wheel hub grease seal.
7. Position the brake disc and hub.
8. Assemble all parts and adjust the bearings.

4-Wheel Drive

1. Before servicing the vehicle, refer to the precautions in the beginning of this section.
2. Remove or disconnect the following:
   - Negative battery cable
   - Wheel assembly
   - Retainer washers from the lug nut studs and remove the automatic locking hub assembly from the spindle
   - Snapring and spacer from the end of the spindle shaft
   - Pull the locking cam assembly and the 2 plastic spacers off of the wheel bearing adjusting nut
3. Use a magnet and remove the locking key from under the adjusting nut. If required, rotate the adjusting nut slightly to relieve pressure against the locking key

⁂ WARNING

To prevent damage to the adjusting nut and spindle threads on vehicles equipped with automatic hubs, look into the spindle keyway under the adjusting nut and remove the separate locking key before removing the adjusting nut.

   - Wheel bearing locknut using a 2⅜ inch (60.3mm) hex socket, such as Hex Locknut Wrench T70T-4252-B
   - Outer bearing cone and roller assembly from the hub
   - Hub and rotor from the spindle
   - Grease seal, using seal removal tool 1175-AC and discard
   - Inner bearing cone and roller assembly from the hub
4. Clean the inner and outer bearing assemblies in solvent. Inspect the bearings and the cones for wear and damage. Replace defective parts, as required.
5. If the cups are worn or damaged, remove them with front hub remover tool T81P-1104-C and tool T77F-1102-A.
6. Wipe the old grease from the spindle. Check the spindle for excessive wear or damage. Replace defective parts, as required.

To install:

7. If the inner and outer cups were removed, use bearing driver handle tool T80-4000-W and replace the cups. Be sure to seat the cups properly in the hub.
8. Use a bearing packer tool and properly repack the wheel bearings with the proper grade and type of grease. If a bearing packer is not available, work as much of the grease as possible between the rollers and cages. Also, grease the cone surfaces.
9. Install or connect the following:
   - Inner bearing cone and roller assembly in the inner cup. A light film of grease should be included between the lips of the new grease seal.
   - Grease seal by driving in place with Hub Seal Replacer tool T83T-1175-B and Driver Handle T80T-4000-W
   - Hub and rotor assembly onto the spindle. Keep the hub centered on the spindle to prevent damage to the spindle and the retainer
   - Outer bearing cone and roller assembly
   - Rotor onto the spindle
   - Outer wheel bearing in the rotor
   - Adjusting nut. Torque the nut to 35 ft. lbs. (47 Nm) to seat the bearings. Adjust the bearing as needed.
   - Thrust spacers and press the cam assembly on the locknut by aligning the key in the fixed cam with the keyway of the front spindle
   - Axle shaft spacer
   - Snapring on the end of the shaft
   - Locking hub assembly over the front spindle
   - Align the 3 hub legs to the cam pockets and install the retainer washers
   - Wheel assembly
   - Negative battery cable

BRAKES

Brake Caliper

REMOVAL & INSTALLATION

1. Before servicing the vehicle, refer to the precautions in the beginning of this section.
2. Loosen the wheel lug nuts.
3. Raise and safely support the front of the vehicle. Remove the wheel.
4. Place an 8 in. (203mm) C-clamp on the caliper and tighten the clamp to bottom the caliper pistons in their bores. Remove the clamp.
5. Remove the two caliper slide pin bolts and lift the caliper from the anchor plate.

➡Use care to retain as much of the original caliper slide pin grease as possible.

6. Position the caliper on a frame member or suspend it with some wire. Do not allow the caliper to hang by the brake hose.
7. Disconnect and plug the brake hose at the caliper. Remove the caliper from the rotor.

To install:

8. Position the caliper over the brake pads and align the slide pin mounting holes.
9. Install the slide pin bolts and tighten them to 21–26 ft. lbs. (30–36 Nm).
10. Install the caliper brake hose using new washers. Tighten the bolt to 29 ft. lbs. (40
11. Install the wheel and snug the lug nuts.
12. Lower the vehicle and tighten the lug nuts to 100 ft. lbs. (135 Nm).

➡The first couple of times you apply the brakes, the pedal may go to the floor. Continue to pump the brake pedal until it feels firm.

13. Start the engine and apply the brakes several times to readjust the caliper pistons. Ensure that the pedal feels firm before operating the vehicle.

Disc Brake Pads

REMOVAL & INSTALLATION

1. Before servicing the vehicle, refer to the precautions in the beginning of this section.
2. Raise and safely support the front of the vehicle. Remove the wheel.
3. Place an 8 in. (203mm) C-clamp on the caliper and tighten the clamp to bottom the caliper pistons in their bores. Remove the clamp.
4. Remove the two caliper slide pin bolts and lift the caliper from the anchor plate.

➡Use care to retain as much of the original caliper slide pin grease as possible.

5. Position the caliper on a frame member or suspend it with some wire. Do not allow the caliper to hang by the brake hose.
6. Remove the brake pads and, if necessary, the anti-rattle clips from the anchor plate.
7. Remove the shims, if any, from the brake pads for re-use.

To install:

8. If removed, install the anti-rattle clips.
9. Install the brake pads to the anchor plate.
10. Position the caliper over the brake pads and align the slide pin mounting holes.
11. Install the slide pin bolts and tighten them to 21–26 ft. lbs. (30–36 Nm).
12. Install the wheel and snug the lug nuts.
13. Lower the vehicle and tighten the lug nuts to 100 ft. lbs. (135 Nm).

➡The first couple of times you apply the brakes, the pedal may go to the floor. Continue to pump the brake pedal until it feels firm.

14. Start the engine and apply the brakes several times to readjust the caliper pistons. Ensure that the pedal feels firm before operating the vehicle.

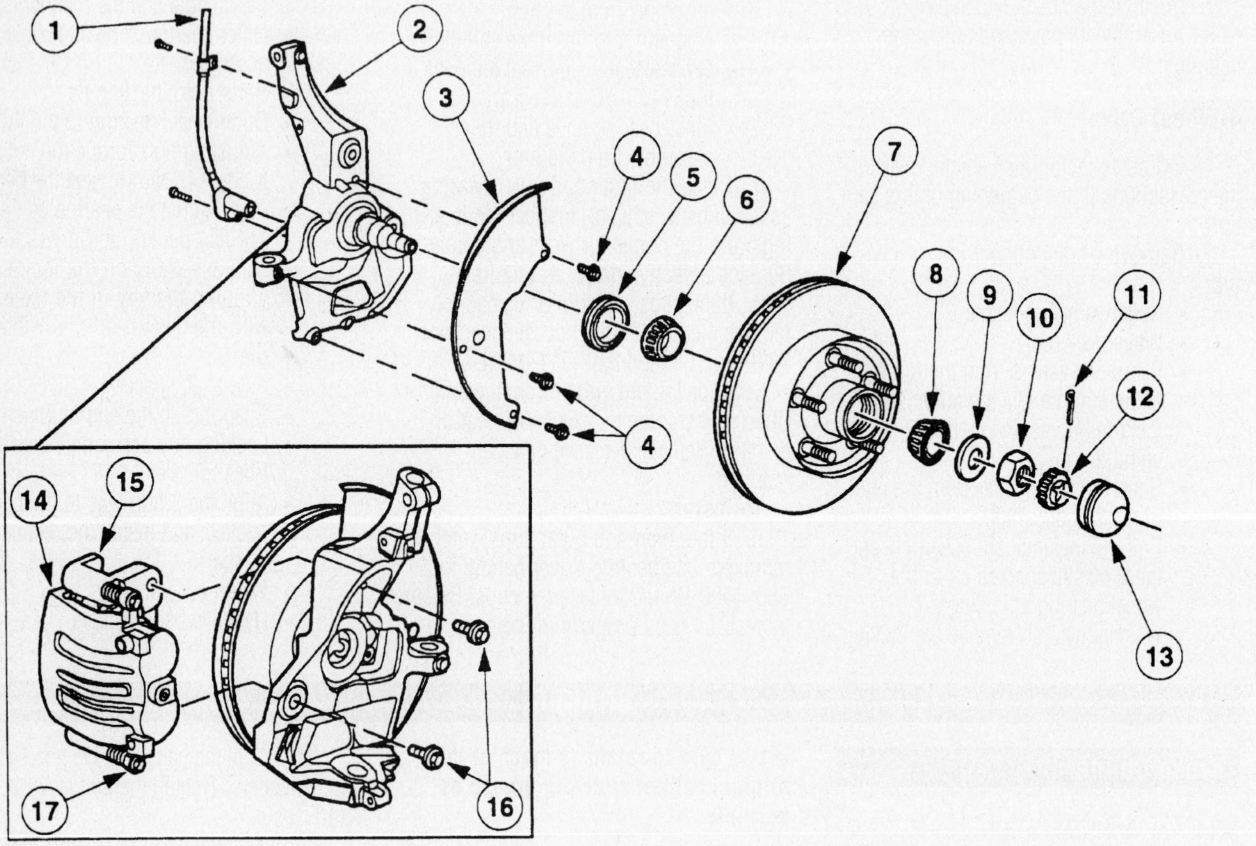

1 Front Brake Anti-Lock Sensor
2 Front Wheel Spindle
3 Front Disc Brake Rotor Shield
4 Rotor Shield Bolt
5 Grease Seal
6 Front Wheel Bearing
7 Front Disc Brake Hub and Rotor
8 Front Wheel Bearing
9 Front Wheel Outer Bearing Retainer Washer
10 Hub Spindle Nut
11 Cotter Pin
12 Nut Retainer
13 Hub Grease Cap
14 Disc Brake Caliper
15 Front Disc Brake Caliper Anchor Plate
16 Caliper Anchor Plate Bolts
17 Disc Brake Caliper Bolt

93026G22

Exploded view of the 2WD front disc brake assembly

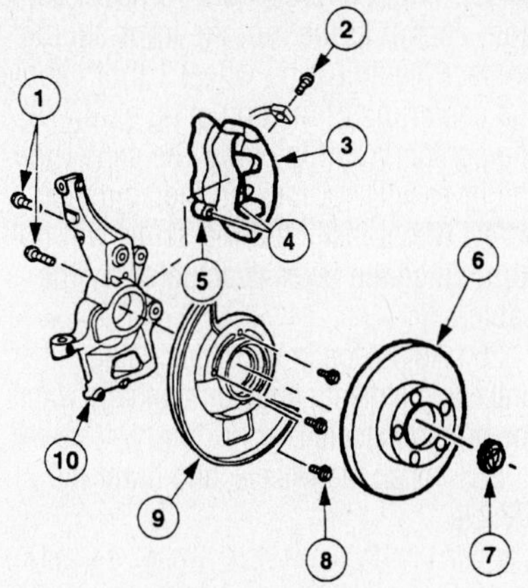

1 Front Disc Brake Caliper Anchor Plate Bolt (2 Req'd)
2 Front Brake Hose Bolt
3 Disc Brake Caliper
4 Pads
5 Front Disc Brake Caliper Anchor Plate
6 Front Disc Brake Rotor
7 Front Axle Wheel Hub Retainer
8 Front Disc Brake Rotor Shield Bolt (3 Req'd)
9 Front Disc Brake Rotor Shield
10 Front Wheel Knuckle

93026G23

Exploded view of the 4WD front disc brake assembly

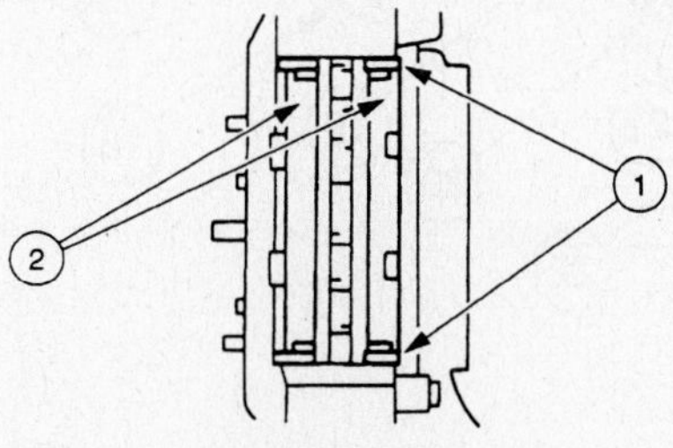

1 stainless slippers
2 pads

93026G24

Position of the front disc brake components

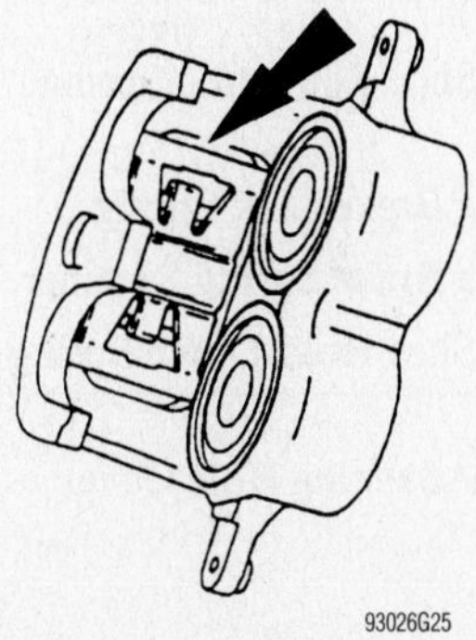

93026G25

View of the front disc brake anti-rattle spring

Brake Drums

REMOVAL & INSTALLATION

1. Before servicing the vehicle, refer to the precautions in the beginning of this section.
2. Raise and safely support the vehicle. Remove the wheel and tire assembly.
3. Remove the retaining nuts, if equipped, and remove the brake drum.
4. Inspect the brake drum surface for wear, scoring and runout. Machine or replace, as necessary.

To install:

5. Install the brake drum and secure in place with the retainer nuts, if equipped.
6. Adjust the rear brakes.
7. Install the wheel. Lower the vehicle.

Brake Shoes

REMOVAL & INSTALLATION

1. Before servicing the vehicle, refer to the precautions in the beginning of this section.
2. Raise and safely support the vehicle. Remove the wheel and tire assembly and the brake drum.
3. Pull backward on the adjusting lever cable to disengage the adjusting lever from the adjusting screw. Move the outboard side of the adjusting screw upward and back off the pivot nut as far as it will go.
4. Pull the adjusting lever, cable and automatic adjuster spring down and toward the rear to unhook the pivot hook from the large hole in the secondary shoe web. Do not pry the pivot hook from the hole.
5. Remove the automatic adjuster spring and adjusting lever.
6. Remove the secondary shoe-to-anchor spring using a suitable brake spring removal/installation tool. Using the tool, remove the primary shoe-to-anchor spring and unhook the cable anchor. Remove the anchor pin plate, if equipped.
7. Remove the cable guide from the secondary shoe.
8. Remove the shoe hold-down springs, shoes, adjusting screw, pivot nut and socket. Note the color and position of each hold-down spring so they can be reassembled in the same position.
9. Remove the parking brake link and spring. Disconnect the parking brake cable from the parking brake lever.
10. Remove the secondary brake shoe. On 9 in. (22.8cm) rear brakes, remove the parking brake lever from the shoe. On 10 in. (25.4cm) rear brakes, remove the retainer clip and spring washer and remove the parking brake lever.

To install:

11. Clean the backing plate ledge pads and sand lightly. Apply a light coating of high temperature lithium grease to the points where the brake shoes touch the backing plate. Lubricate the adjusting cable eye and the anchor pin area.
12. Install the parking brake lever on the secondary shoe. On 10 in. (25.4cm) brakes, secure with the spring washer and retaining clip.
13. Position the brake shoes on the backing plate and install the hold-down spring pins, springs and cups. Install the parking brake link, spring and washer. Connect the parking brake cable to the parking brake lever.
14. Install the anchor pin plate, if equipped, and place the cable anchor over the anchor pin with the crimped side toward the backing plate.
15. Install the primary shoe-to-anchor spring using the brake spring removal/installation tool.
16. Install the cable guide on the secondary shoe with the flanged hole fitted into the hole in the secondary shoe. Thread the cable around the cable guide groove.

➡Make sure the cable is positioned in the groove and not between the guide and shoe web.

17. Install the secondary shoe-to-anchor (long) spring.

➡Make sure the cable end is not cocked or binding on the anchor pin when installed. All parts should be flat on the anchor pin.

18. Apply high temperature lithium grease to the threads and the socket end of the adjusting screw. Turn the adjusting screw into the adjusting pivot nut to the end of the threads and then loosen, ½ turn.
19. Place the adjusting socket on the screw and install the assembly between the shoe ends with the adjusting screw nearest the secondary shoe.

➡Be sure to install the adjusting screw on the same side of the vehicle from which it came. To prevent incorrect installation, the socket end of each adjusting screw is stamped with R or L, to indicate installation on the right or left side of the vehicle. The adjusting pivot nuts have lines machined around the body of the nut, 2 lines indicating the right side nut and 1 line indicating the left side nut.

20. Hook the cable hook into the hole in the adjusting lever from the outboard plate side. The adjusting levers are also stamped with an **R** or **L** to indicate right or left side installation.

21. Place the hooked end of the adjuster spring in the large hole in the primary shoe web and connect the loop end of the spring to the adjuster lever hole.

22. Pull the adjuster lever, cable and automatic adjuster spring down toward the rear to engage the pivot hook in the large hole in the secondary shoe web.

23. After installation, check the action of the adjuster by pulling the section of the cable between the cable guide and the adjusting lever toward the secondary shoe web far enough to lift the lever past a tooth on the adjusting screw wheel. The lever should snap into position behind the next tooth and releasing the cable should cause the adjuster spring to return the lever to its original position. This return action will turn the adjusting screw 1 tooth.

24. If pulling the cable does not produce the action described previously, or if lever action is sluggish instead of positive and sharp, check the position of the lever on the adjusting screw toothed wheel. With the brake in a vertical position, anchor at the top, the lever should contact the adjusting wheel 1 tooth above the centerline of the adjusting screw. If the contact point is below the centerline, the lever will not lock on the adjusting screw wheel teeth and the screw will not turn, since the lever is actuated by the cable.

25. Adjust the brake shoes using either a brake adjustment gauge or manually with the drums installed.

26. Install the wheels, and lower the vehicle.

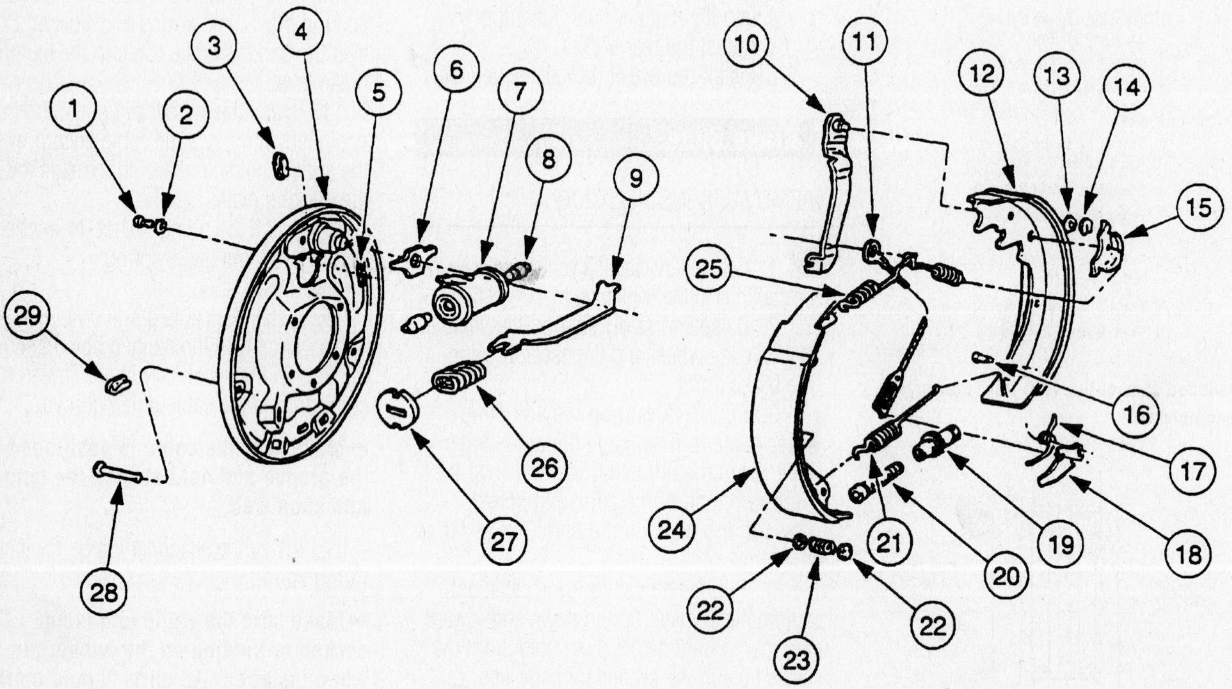

- 1 Wheel Cylinder-to-Backing Plate Bolt (2 Req'd)
- 2 Washer
- 3 Inspection Hole Cover
- 4 Brake Backing Plate
- 5 Lining Inspection Hole
- 6 Anchor Pin Guide Plate
- 7 Rear Wheel Cylinder
- 8 Wheel Cylinder Brake Shoe Link
- 9 Parking Brake Strut
- 10 Parking Brake Lever
- 11 Brake Shoe Adjusting Lever Cable
- 12 Rear Brake Shoe and Lining, Secondary
- 13 Washer
- 14 Parking Brake Lever Pin Retainer
- 15 Cable Guide
- 16 Adjusting Lever Pin
- 17 Adjusting Lever Return Spring
- 18 Brake Shoe Adjusting Lever
- 19 Brake Shoe Adjusting Screw Nut
- 20 Brake Adjuster Screw
- 21 Brake Shoe Adjusting Screw Spring
- 22 Brake Shoe Hold-Down Spring Cup
- 23 Brake Shoe Hold-Down Spring
- 24 Rear Brake Shoe and Lining, Primary
- 25 Brake Shoe Retracting Spring, Short
- 26 Parking Brake Link Spring
- 27 Parking Brake Spring Retainer
- 28 Brake Shoe Hold-Down Spring Pin
- 29 Brake Adjusting Hole Cover

93026G21

Exploded view of the rear brake shoes and components

FORD AND MERCURY

14

Sable • Taurus

SPECIFICATION CHARTS

ENGINE AND VEHICLE IDENTIFICATION

Engine						
Code ①	Liters (cc)	Cu. In.	Cyl.	Fuel Sys.	Engine Type	Eng. Mfg.
S	3.0 (3049)	182	6	SFI	DOHC	Ford
U	3.0 (2982)	181	6	SFI	OHV	Ford

Model Year	
Code ②	Year
1	2001
2	2002
3	2003
4	2004
5	2005

OHV: Overhead Valves

DOHC: Double Overhead Camshafts

SFI: Sequential Fuel Injection

① 8th digit of the Vehicle Identification Number (VIN)

② 10th digit of the Vehicle Identification Number (VIN)

7197-TAUR-C01

GENERAL ENGINE SPECIFICATIONS

Year	Model	Engine Displacement Liters	Engine ID/VIN	Net Horsepower @ rpm	Net Torque @ rpm (ft. lbs.)	Bore x Stroke (in.)	Com-pression Ratio	Oil Pressure @ rpm
2001	Sable	3.0	U	155@4900	185@3950	3.50x3.15	9.3:1	40-60@2500
		3.0	S	200@5750	200@4500	3.50x3.13	10.0:1	20-45@1500
	Taurus	3.0	U	155@4900	185@3950	3.50x3.15	9.3:1	40-60@2500
		3.0	S	200@5750	200@4500	3.50x3.13	10.0:1	20-45@1500
2002	Sable	3.0	U	155@4900	185@3950	3.50x3.15	9.3:1	40-60@2500
		3.0	S	200@5750	200@4500	3.50x3.13	10.0:1	20-45@1500
	Taurus	3.0	U	155@4900	185@3950	3.50x3.15	9.3:1	40-60@2500
		3.0	S	200@5750	200@4500	3.50x3.13	10.0:1	20-45@1500
2003	Sable	3.0	U	155@4900	185@3950	3.50x3.15	9.3:1	40-60@2500
		3.0	S	200@5750	200@4500	3.50x3.13	10.0:1	20-45@1500
	Taurus	3.0	U	155@4900	185@3950	3.50x3.15	9.3:1	40-60@2500
		3.0	S	200@5750	200@4500	3.50x3.13	10.0:1	20-45@1500
2004	Sable	3.0	U	155@4900	185@3950	3.50x3.15	9.3:1	40-60@2500
		3.0	S	200@5750	200@4500	3.50x3.13	10.0:1	20-45@1500
	Taurus	3.0	U	155@4900	185@3950	3.50x3.15	9.3:1	40-60@2500
		3.0	S	200@5750	200@4500	3.50x3.13	10.0:1	20-45@1500

SFI: Sequential Fuel Injection

7197-TAUR-C02

ENGINE TUNE-UP SPECIFICATIONS

Year	Engine Displacement Liters	Engine ID/VIN	Spark Plug Gap (in.)	Ignition Timing (deg.)	Fuel Pump (psi) ①	Idle Speed (rpm)	Valve Clearance	
							Intake	Exhaust
2001	3.0	U	0.042-0.046	10B	26-45	②	HYD	HYD
	3.0	S	0.052-0.056	10B	26-45	②	HYD	HYD
2002	3.0	U	0.042-0.046	10B	26-45	②	HYD	HYD
	3.0	S	0.052-0.056	10B	26-45	②	HYD	HYD
2003	3.0	U	0.042-0.046	10B	26-45	②	HYD	HYD
	3.0	S	0.052-0.056	10B	26-45	②	HYD	HYD
2004	3.0	U	0.042-0.046	10B	26-45	②	HYD	HYD
	3.0	S	0.052-0.056	10B	26-45	②	HYD	HYD

NOTE: The Vehicle Emission Control Information label often reflects specification changes made during production. The label figures must be used if they differ from those in this chart.

B: Before Top Dead Center

HYD: Hydraulic

① Fuel pressure with engine running, pressure regulator vacuum hose connected

② Refer to Vehicle Emission Control Information label

7197-TAUR-C03

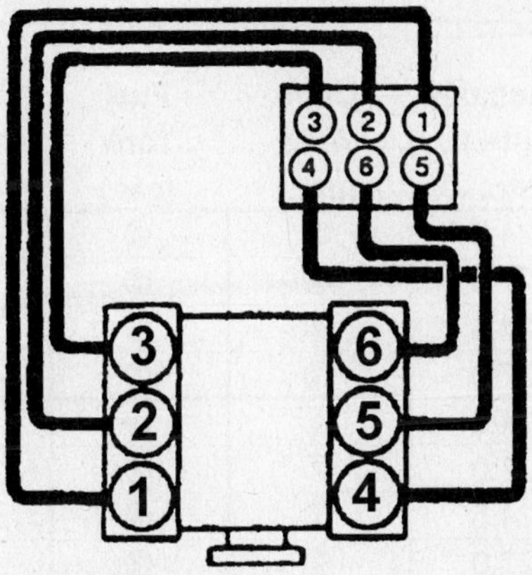

3.0L (VIN U) engine
Firing order: 1–4–2–5–3–6
Distributorless ignition system

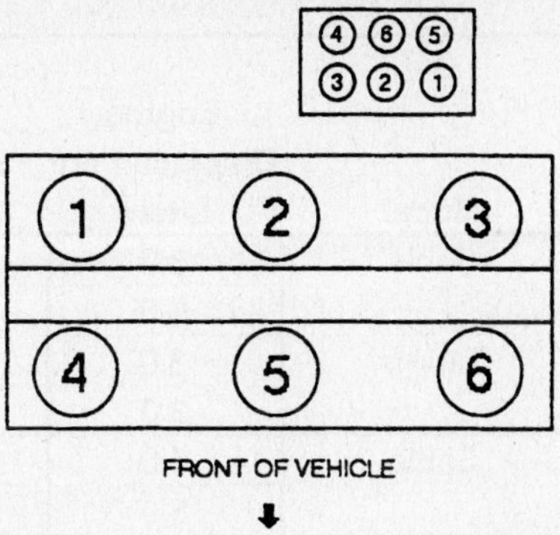

3.0L (VIN S) engine
Firing order: 1–4–2–5–3–6
Distributorless ignition system

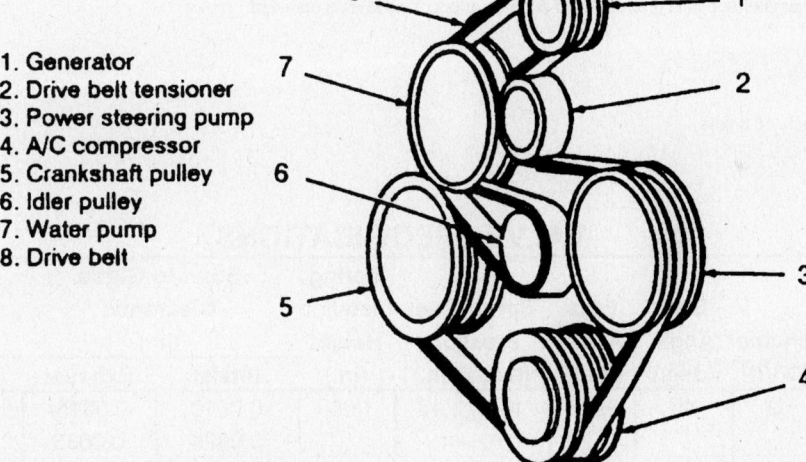

1. Generator
2. Drive belt tensioner
3. Power steering pump
4. A/C compressor
5. Crankshaft pulley
6. Idler pulley
7. Water pump
8. Drive belt

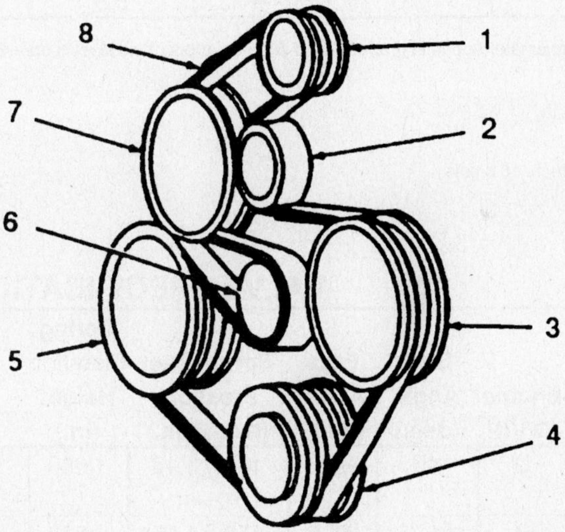

Serpentine accessory drive belt routing—3.0L (VIN U) engine

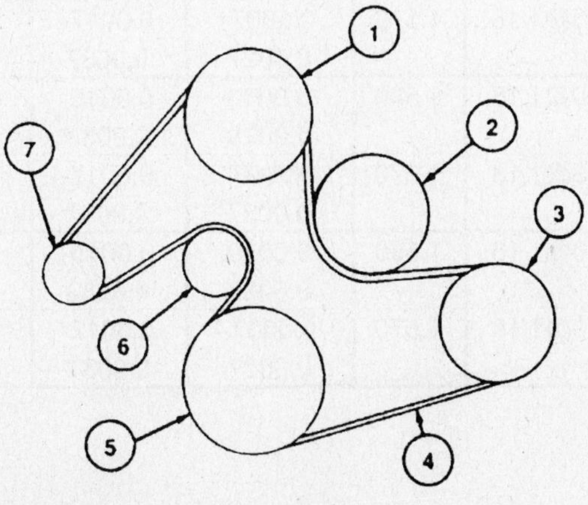

1. Power steering pump
2. Water pump
3. A/C compressor
4. Drive belt
5. Crankshaft pulley
6. Drive belt tensioner
7. Generator

Serpentine accessory drive belt routing—3.0L (VIN S) engine

CAPACITIES

Year	Model	Engine Displacement Liters	Engine ID/VIN	Engine Oil with Filter (qts.)	Transaxle (pts.) Auto. ①	Drive Axle (pts.)	Fuel Tank (gal.)	Cooling System (qts.)
2001	Sable	3.0	U	4.5	24.5	②	③	11.6
		3.0	S	5.8	27.0	②	③	10.5
	Taurus	3.0	U	4.5	24.5	②	③	11.6
		3.0	S	5.8	27.0	②	③	10.5
2002	Sable	3.0	U	4.5	24.5	②	③	11.6
		3.0	S	5.8	27.0	②	③	10.5
	Taurus	3.0	U	4.5	24.5	②	③	11.6
		3.0	S	5.8	27.0	②	③	10.5
2003	Sable	3.0	U	4.5	24.5	②	③	11.6
		3.0	S	5.8	27.0	②	③	10.5
	Taurus	3.0	U	4.5	24.5	②	③	11.6
		3.0	S	5.8	27.0	②	③	10.5
2004	Sable	3.0	U	4.5	24.5	②	③	11.6
		3.0	S	5.8	27.0	②	③	10.5
	Taurus	3.0	U	4.5	24.5	②	③	11.6
		3.0	S	5.8	27.0	②	③	10.5

NOTE: All capacities are approximate. Add fluid gradually and ensure a proper fluid level is obtained.

① Includes torque converter

② Included in transaxle capacity

③ Standard tank: 16.0 gals.
Optional extended range tank: 18.6 gals.

7197-TAUR-C04

VALVE SPECIFICATIONS

Year	Engine Displacement Liters	Engine ID/VIN	Seat Angle (deg.)	Face Angle (deg.)	Spring Test Pressure (lbs. @ in.)	Spring Installed Height (in.)	Stem-to-Guide Clearance (in.) Intake	Stem-to-Guide Clearance (in.) Exhaust	Stem Diameter (in.) Intake	Stem Diameter (in.) Exhaust
2001	3.0	U	45	44	180@1.16	1.580	0.0010 0.0028	0.0015-0.0033	0.3126-0.3134	0.3121-0.3129
	3.0	S	44.75	45.5	153@1.18	1.570	0.0007-0.0027	0.0017-0.0037	0.2350-0.2358	0.2343-0.2350
2002	3.0	U	45	44	180@1.16	1.580	0.0010 0.0028	0.0015-0.0033	0.3126-0.3134	0.3121-0.3129
	3.0	S	44.75	45.5	153@1.18	1.570	0.0007-0.0027	0.0017-0.0037	0.2350-0.2358	0.2343-0.2350
2003	3.0	U	45	44	180@1.16	1.580	0.0010 0.0028	0.0015-0.0033	0.3126-0.3134	0.3121-0.3129
	3.0	S	44.75	45.5	153@1.18	1.570	0.0007-0.0027	0.0017-0.0037	0.2350-0.2358	0.2343-0.2350
2004	3.0	U	45	44	180@1.16	1.580	0.0010 0.0028	0.0015-0.0033	0.3126-0.3134	0.3121-0.3129
	3.0	S	44.75	45.5	153@1.18	1.570	0.0007-0.0027	0.0017-0.0037	0.2350-0.2358	0.2343-0.2350

7197-TAUR-C05

PISTON AND RING SPECIFICATIONS

All measurements are given in inches.

Year	Engine Displacement Liters	Engine ID/VIN	Piston Clearance	Ring Gap Top Compression	Ring Gap Bottom Compression	Ring Gap Oil Control	Ring Side Clearance Top Compression	Ring Side Clearance Bottom Compression	Ring Side Clearance Oil Control
2001	3.0	U	0.0014-0.0022	0.010-0.020	0.010-0.020	0.010-0.049	0.0012 0.0031	0.0012 0.0031	SNUG
	3.0	S	0.0005-0.0009	0.004-0.010	0.011-0.017	0.006-0.026	0.0015-0.0029	0.0015-0.0033	SNUG
2002	3.0	U	0.0014-0.0022	0.010-0.020	0.010-0.020	0.010-0.049	0.0012 0.0031	0.0012 0.0031	SNUG
	3.0	S	0.0005-0.0009	0.004-0.010	0.011-0.017	0.006-0.026	0.0015-0.0029	0.0015-0.0033	SNUG
2003	3.0	U	0.0014-0.0022	0.010-0.020	0.010-0.020	0.010-0.049	0.0012 0.0031	0.0012 0.0031	SNUG
	3.0	S	0.0005-0.0009	0.004-0.010	0.011-0.017	0.006-0.026	0.0015-0.0029	0.0015-0.0033	SNUG
2004	3.0	U	0.0014-0.0022	0.010-0.020	0.010-0.020	0.010-0.049	0.0012 0.0031	0.0012 0.0031	SNUG
	3.0	S	0.0005-0.0009	0.004-0.010	0.011-0.017	0.006-0.026	0.0015-0.0029	0.0015-0.0033	SNUG

7197-TAUR-C06

CRANKSHAFT AND CONNECTING ROD SPECIFICATIONS

All measurements are given in inches.

Year	Engine Displacement Liters	Engine ID/VIN	Crankshaft Main Brg. Journal Dia.	Crankshaft Main Brg. Oil Clearance	Crankshaft Shaft End-play	Crankshaft Thrust on No.	Connecting Rod Journal Diameter	Connecting Rod Oil Clearance	Connecting Rod Side Clearance
2001	3.0	U	2.5190-2.5198	0.0009 0.0027	0.0040 0.0080	3	2.1253-2.1261	0.0009 0.0027	0.0060 0.0140
	3.0	S	2.4670-2.4790	0.0009-0.0018	0.0040 0.0090	4	1.9670-1.9680	0.0010 0.0025	0.0039-0.0118
2002	3.0	U	2.5190-2.5198	0.0009 0.0027	0.0040 0.0080	3	2.1253-2.1261	0.0009 0.0027	0.0060 0.0140
	3.0	S	2.4670-2.4790	0.0009-0.0018	0.0040 0.0090	4	1.9670-1.9680	0.0010 0.0025	0.0039-0.0118
2003	3.0	U	2.5190-2.5198	0.0009 0.0027	0.0040 0.0080	3	2.1253-2.1261	0.0009 0.0027	0.0060 0.0140
	3.0	S	2.4670-2.4790	0.0009-0.0018	0.0040 0.0090	4	1.9670-1.9680	0.0010 0.0025	0.0039-0.0118
2004	3.0	U	2.5190-2.5198	0.0009 0.0027	0.0040 0.0080	3	2.1253-2.1261	0.0009 0.0027	0.0060 0.0140
	3.0	S	2.4670-2.4790	0.0009-0.0018	0.0040 0.0090	4	1.9670-1.9680	0.0010 0.0025	0.0039-0.0118

7197-TAUR-C07

TORQUE SPECIFICATIONS

All readings in ft. lbs.

Year	Engine Displacement Liters	Engine ID/VIN	Cylinder Head Bolts	Main Bearing Bolts	Rod Bearing Bolts	Crankshaft Damper Bolts	Flywheel Bolts	Manifold Intake	Manifold Exhaust	Spark Plugs	Oil Pan Drain Plug
2001	3.0	U	①	56-62	23-28	93-121	54-64	②	15-18	7-15	10
	3.0	S	③	④	⑤	⑥	54-64	⑦	13-16	7-15	20
2002	3.0	U	①	56-62	23-28	93-121	54-64	②	15-18	7-15	10
	3.0	S	③	④	⑤	⑥	54-64	⑦	13-16	7-15	20
2003	3.0	U	①	56-62	23-28	93-121	54-64	②	15-18	7-15	10
	3.0	S	③	④	⑤	⑥	54-64	⑦	13-16	7-15	20
2004	3.0	U	①	56-62	23-28	93-121	54-64	②	15-18	7-15	10
	3.0	S	③	④	⑤	⑥	54-64	⑦	13-16	7-15	20

① Step 1: 35-39 ft. lbs.
Step 2: Loosen one turn
Step 3: 20-24 ft. lbs.
Step 4: Rotate 85-95 degrees
Step 5: Repeat Step 4

② Upper intake: 15-22 ft. lbs.
Lower Intake:
Step 1: 15-22 ft. lbs.
Step 2: 20-23 ft. lbs.

③ Step 1: 28-31 ft. lbs.
Step 2: Rotate 85-95 degrees
Step 3: Loosen one turn
Step 4: 28-31 ft. lbs.
Step 5: Rotate 85-95 degrees
Step 6: Repeat Step 5

④ Step 1: Cap bolts 1-8 (outer) 17-20 ft. lbs.
Step 2: Cap bolts 9-16 (inner) 28-31 ft. lbs.
Step 3: Rotate bolts 1-16, 85-95 degrees
Step 4: Bolts 17-22; 15-22 ft. lbs.

⑤ Step 1: 30-33 ft. lbs.
Step 2: Rotate 90-120 degrees

⑥ Step 1: 77-99 ft. lbs.
Step 2: Loosen 360 degrees
Step 3: Tighten to 35-39 ft. lbs.
Step 4: Rotate 85-95 degrees

⑦ 71-106 inch lbs.

⑧ Step 1: 20-23 ft. lbs.
Step 2: Rotate 85-95 degrees

7197-TAUR-C08

WHEEL ALIGNMENT

Year	Model		Caster Range (+/-Deg.)	Caster Preferred Setting (Deg.)	Camber Range (+/-Deg.)	Camber Preferred Setting (Deg.)	Toe-in (in.)
2001	Taurus Sedan	F	0.70	0	0.70	0	0.10 +/- 0.13
		R	—	—	0.70	0	0.18 +/- 0.13
	Taurus Wagon	F	0.70	0	0.70	0	0.10 +/- 0.13
		R	—	—	1.20	0	0.18 +/- 0.13
	Sable	F	0.70	0	0.70	0	0.20 +/- 0.25
		R	—	—	0.70	0	0.36 +/- 0.25
	Sable Wagon	F	0.70	0	0.70	0	0.20 +/- 0.25
		R	—	—	0.12	0	0.36 +/- 0.25
2002	Taurus Sedan	F	0.70	0	0.70	0	0.10 +/- 0.13
		R	—	—	0.70	0	0.18 +/- 0.13
	Taurus Wagon	F	0.70	0	0.70	0	0.10 +/- 0.13
		R	—	—	1.20	0	0.18 +/- 0.13
	Sable	F	0.70	0	0.70	0	0.20 +/- 0.25
		R	—	—	0.70	0	0.36 +/- 0.25
	Sable Wagon	F	0.70	0	0.70	0	0.20 +/- 0.25
		R	—	—	0.12	0	0.36 +/- 0.25
2003	Taurus Sedan	F	0.70	0	0.70	0	0.10 +/- 0.13
		R	—	—	0.70	0	0.18 +/- 0.13
	Taurus Wagon	F	0.70	0	0.70	0	0.10 +/- 0.13
		R	—	—	1.20	0	0.18 +/- 0.13
	Sable	F	0.70	0	0.70	0	0.20 +/- 0.25
		R	—	—	0.70	0	0.36 +/- 0.25
	Sable Wagon	F	0.70	0	0.70	0	0.20 +/- 0.25
		R	—	—	0.12	0	0.36 +/- 0.25
2004	Taurus Sedan	F	0.70	0	0.70	0	0.10 +/- 0.13
		R	—	—	0.70	0	0.18 +/- 0.13
	Taurus Wagon	F	0.70	0	0.70	0	0.10 +/- 0.13
		R	—	—	1.20	0	0.18 +/- 0.13
	Sable	F	0.70	0	0.70	0	0.20 +/- 0.25
		R	—	—	0.70	0	0.36 +/- 0.25
	Sable Wagon	F	0.70	0	0.70	0	0.20 +/- 0.25
		R	—	—	0.12	0	0.36 +/- 0.25

7197-TAUR-C09

TIRE, WHEEL AND BALL JOINT SPECIFICATIONS

Year	Model	OEM Tires Standard	OEM Tires Optional	Tire Pressure (psi) Front	Tire Pressure (psi) Rear	Wheel Size	Ball Joint Inspection	Lug Nut Torque (ft lbs)
2001	Sable GS, LS Sedan	P205/65R16	None	33	33	7J	0.030 in. ①	95
	Sable LS Premium Sedan	P215/60R16	None	33	33	7J	0.030 in. ①	95
	Sable Wagon	P215/60R16	None	31	31	7J	0.030 in. ①	95
	Taurus	P215/60R16	None	33	33	6.5-JJ	0.030 in. ①	95
2002	Sable GS, LS Sedan	P205/65R16	None	33	33	7J	0.030 in. ①	95
	Sable LS Premium Sedan	P215/60R16	None	33	33	7J	0.030 in. ①	95
	Sable Wagon	P215/60R16	None	31	31	7J	0.030 in. ①	95
	Taurus	P215/60R16	None	33	33	6.5-JJ	0.030 in. ①	95
2003	Sable GS, LS Sedan	P205/65R16	None	33	33	7J	0.030 in. ①	95
	Sable LS Premium Sedan	P215/60R16	None	33	33	7J	0.030 in. ①	95
	Sable Wagon	P215/60R16	None	31	31	7J	0.030 in. ①	95
	Taurus	P215/60R16	None	33	33	6.5-JJ	0.030 in. ①	95
2004	Sable GS, LS Sedan	P205/65R16	None	33	33	7J	0.030 in. ①	95
	Sable LS Premium Sedan	P215/60R16	None	33	33	7J	0.030 in. ①	95
	Sable Wagon	P215/60R16	None	31	31	7J	0.030 in. ①	95
	Taurus	P215/60R16	None	33	33	6.5-JJ	0.030 in. ①	95

OEM: Original Equipment Manufacturer

PSI: Pounds Per Square Inch

① Maximum radial tolerance in inches

7197-TAUR-C10

BRAKE SPECIFICATIONS
Ford Taurus, Mercury Sable

All measurements in inches unless noted

Year	Model		Brake Disc Original Thickness	Brake Disc Minimum Thickness	Brake Disc Maximum Run-out	Brake Drum Original Inside Diameter	Brake Drum Max. Wear Limit	Brake Drum Maximum Machine Diameter	Minimum Lining Thickness	Brake Caliper Bracket Bolts (ft. lbs.)	Brake Caliper Mounting Bolts (ft. lbs.)
2001	Sable	F	1.020	0.974	0.002	—	—	—	0.039	65-85	23-28
		R	0.550	0.500	0.004	8.85	0.05	8.92	0.039	65-87	23-25
	Taurus	F	1.020	0.974	0.002	—	—	—	0.039	65-85	23-28
		R	0.940	0.500	0.002	8.85	0.05	8.92	0.039	65-87	23-25
2002	Sable	F	1.020	0.974	0.002	—	—	—	0.039	65-85	23-28
		R	0.550	0.500	0.004	8.85	0.05	8.92	0.039	65-87	23-25
	Taurus	F	1.020	0.974	0.002	—	—	—	0.039	65-85	23-28
		R	0.940	0.500	0.002	8.85	0.05	8.92	0.039	65-87	23-25
2003	Sable	F	1.020	0.974	0.002	—	—	—	0.039	65-85	23-28
		R	0.550	0.500	0.004	8.85	0.05	8.92	0.039	65-87	23-25
	Taurus	F	1.020	0.974	0.002	—	—	—	0.039	65-85	23-28
		R	0.940	0.500	0.002	8.85	0.05	8.92	0.039	65-87	23-25
2004	Sable	F	1.020	0.974	0.002	—	—	—	0.039	65-85	23-28
		R	0.550	0.500	0.004	8.85	0.05	8.92	0.039	65-87	23-25
	Taurus	F	1.020	0.974	0.002	—	—	—	0.039	65-85	23-28
		R	0.940	0.500	0.002	8.85	0.05	8.92	0.039	65-87	23-25

NOTE: Follow specifications stamped on rotor or drum if figures differ from those in this chart.

F: Front

R: Rear

7197-TAUR-C11

SCHEDULED MAINTENANCE INTERVALS

Ford—Taurus & Mercury—Sable

TO BE SERVICED	OF SERVIC	VEHICLE MILEAGE INTERVAL (X1000)																			
		5	10	15	20	25	30	35	40	45	50	55	60	65	70	75	80	85	90	95	100
Engine oil & filter	R	✓	✓	✓	✓	✓	✓	✓	✓	✓	✓	✓	✓	✓	✓	✓	✓	✓	✓	✓	✓
Rotate tires	S/I	✓		✓		✓		✓		✓		✓		✓		✓		✓		✓	
Engine coolant protection, hoses & clamps	S/I			✓			✓			✓			✓			✓			✓		
Passenger compartment air filter	R				✓				✓				✓				✓				✓
Air cleaner filter	R						✓						✓						✓		
Automatic transaxle fluid & filter	R						✓						✓						✓		
Brake lines & connections	S/I						✓						✓						✓		
Exhaust heat shields	S/I						✓						✓						✓		
Front and rear disc brake pads & rotors	S/I						✓						✓						✓		
Accessory drive belt(s)	S/I												✓								
Engine coolant ①	R										✓						✓				
Spark plugs (exc. 3.0L FF) ②	R																				✓
Spark plugs (3.0L FF)	R						✓						✓								
PCV valve (except 3.0L 4-valve)	R												✓								
PCV valve (3.0L 4-valve)	R																				✓

①Engine coolant: change initially at 50,000 miles & thereafter every 30,000 miles
②Platinum tip spark plugs: change every 100,000 miles
R: Replace S/I: Service and Inspect

FREQUENT OPERATION MAINTENANCE (SEVERE SERVICE)

If a vehicle is operated under any of the following conditions it is considered severe service:

- Extremely dusty areas.
- 50% or more of the vehicle operation is in 32°C (90°F) or higher temperatures, or constant operation in temperatures below 0°C (32°F).
- Prolonged idling (vehicle operation in stop and go traffic)..
- Frequent short running periods (engine does not warm to normal operating temperatures).
- Police, taxi, delivery usage or trailer towing usage.

Oil & oil filter: change every 3000 miles
Rotate tires at 6000 miles & every 9000 miles thereafter
Air cleaner element service or inspect every 15,000 miles
Automatic transaxle fluid & filter: change every 21,000 miles

7197-TAUR-C12

PRECAUTIONS

Before servicing any vehicle, please be sure to read all of the following precautions, which deal with personal safety, prevention of component damage, and important points to take into consideration when servicing a motor vehicle:

- Never open, service or drain the radiator or cooling system when the engine is hot; serious burns can occur from the steam and hot coolant.
- Observe all applicable safety precautions when working around fuel. Whenever servicing the fuel system, always work in a well-ventilated area. Do not allow fuel spray or vapors to come in contact with a spark, open flame, or excessive heat (a hot drop light, for example). Keep a dry chemical fire extinguisher near the work area. Always keep fuel in a container specifically designed for fuel storage; also, always properly seal fuel containers to avoid the possibility of fire or explosion. Refer to the additional fuel system precautions later in this section.
- Fuel injection systems often remain pressurized, even after the engine has been turned **OFF**. The fuel system pressure must be relieved before disconnecting any fuel lines. Failure to do so may result in fire and/or personal injury.
- Brake fluid often contains polyglycol ethers and polyglycols. Avoid contact with the eyes and wash your hands thoroughly after handling brake fluid. If you do get brake fluid in your eyes, flush your eyes with clean, running water for 15 minutes. If eye irritation persists, or if you have taken brake fluid internally, IMMEDIATELY seek medical assistance.
- The EPA warns that prolonged contact with used engine oil may cause a number of skin disorders, including cancer. You should make every effort to minimize your exposure to used engine oil. Protective gloves should be worn when changing oil. Wash your hands and any other exposed skin areas as soon as possible after exposure to used engine oil. Soap and water, or waterless hand cleaner should be used.
- All new vehicles are now equipped with an air bag system, often referred to as a Supplemental Restraint System (SRS) or Supplemental Inflatable Restraint (SIR) system. The system must be disabled before performing service on or around system components, steering column, instrument panel components, wiring and sensors. Failure to follow safety and disabling procedures could result in accidental air bag deployment, possible personal injury and unnecessary system repairs.
- Always wear safety goggles when working with, or around, the air bag system. When carrying a non-deployed air bag, be sure the bag and trim cover are pointed away from your body. When placing a non-deployed air bag on a work surface, always face the bag and trim cover upward, away from the surface. This will reduce the motion of the module if it is accidentally deployed. Refer to the additional air bag system precautions later in this section.
- Clean, high quality brake fluid from a sealed container is essential to the safe and proper operation of the brake system. You should always buy the correct type of brake fluid for your vehicle. If the brake fluid becomes contaminated, completely flush the system with new fluid. Never reuse any brake fluid. Any brake fluid that is removed from the system should be discarded. Also, do not allow any brake fluid to come in contact with a painted surface; it will damage the paint.
- Never operate the engine without the proper amount and type of engine oil; doing so will result in severe engine damage.
- Timing belt maintenance is extremely important. Many models utilize an interference-type, non-freewheeling engine. If the timing belt breaks, the valves in the cylinder head may strike the pistons, causing potentially serious (also time-consuming and expensive) engine damage
- Disconnecting the negative battery cable on some vehicles may interfere with the functions of the on-board computer system(s) and may require the computer to undergo a relearning process once the negative battery cable is reconnected.
- When servicing drum brakes, only disassemble and assemble one side at a time, leaving the remaining side intact for reference.

ENGINE REPAIR

➡Disconnecting the negative battery cable on some vehicles may interfere with the functions of the on board computer system. The computer may undergo a relearning process once the negative battery cable is reconnected.

Alternator

REMOVAL

2001–02 3.0L (VIN U) Engine

1. Before servicing the vehicle, refer to the precautions at the beginning of this section.
2. Remove or disconnect the following:
   - Negative battery cable
   - Accessory drive belt
   - Alternator wiring connectors
   - Alternator brace
   - Alternator

2003–04 3.0L (VIN U) Engine

1. Before servicing the vehicle, refer to the precautions in the beginning of this section.
2. Disconnect the battery ground cable.
3. Release the accessory drive belt tension and remove the belt from the alternator pulley.
4. Remove the alternator B+ nut and the B+ cable.

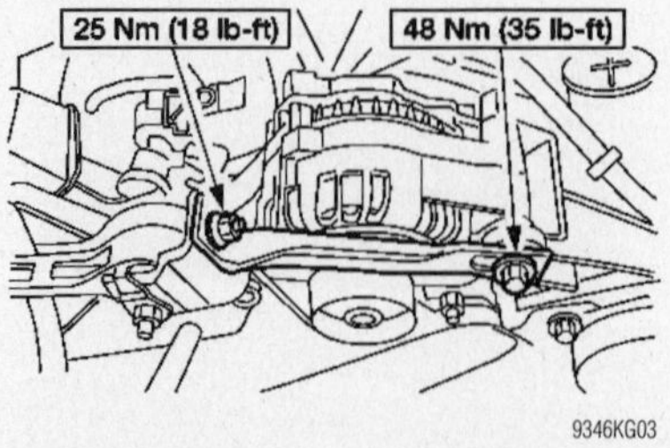

Alternator bracket–3.0L (VIN U) engine

5. Position the B+ cable cover aside.
6. Position the wiring harness aside.
7. Remove the bolts and the alternator.
8. Disconnect the voltage regulator electrical connector.

3.0L (VIN S) Engine

1. Before servicing the vehicle, refer to the precautions in the beginning of this section.
2. Remove or disconnect the following:
   - Negative battery cable
   - Right front wheel
   - Right fender splash shield
   - Accessory drive belt

➡The crankshaft pulley retainer has left-hand threads.

   - Crankshaft pulley
   - Alternator harness connectors
   - Alternator

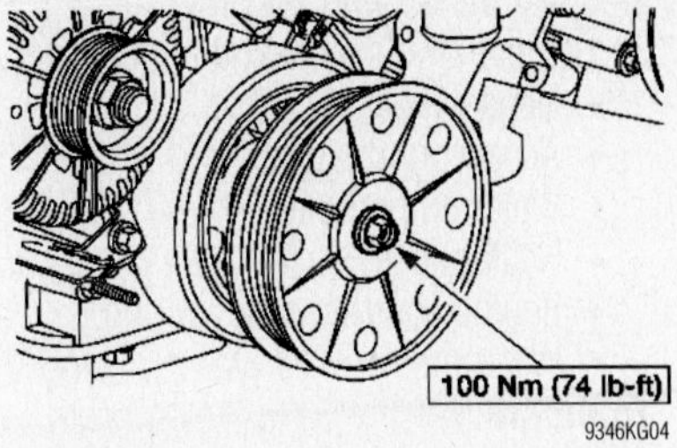

Remove the crankshaft pulley to provide clearance—3.0L (VIN S) engine

INSTALLATION

2001–02 3.0L (VIN U) Engine

➡Do not tighten the alternator pivot and bracket bolts until the accessory drive belt is installed.

1. Install or connect the following:
   - Alternator
   - Alternator brace. Tighten the bolts to 76–97 inch lbs. (8–11 Nm), and the nut to 15–22 ft. lbs. (20–30 Nm).
   - Alternator wiring connectors
   - Accessory drive belt
2. Tighten the alternator pivot bolt to 30–40 ft. lbs. (40–55 Nm), and the bracket bolt to 15–22 ft. lbs. (20–30 Nm).
3. Install the negative battery cable. Start the engine and check for proper operation.

2003–04 3.0L (VIN U) Engine

1. Connect the voltage regulator electrical connector.
2. Install the alternator and tighten the bolts to 35 ft. lbs. (48 Nm).
3. Install the B+ cable and tighten the nut to 71 inch lbs. (8 Nm).
4. Install the accessory drive belt.

3.0L (VIN S) Engine

1. Install or connect the following:
   - Alternator and tighten the fasteners to 18 ft. lbs. (25 Nm)
   - Alternator harness connectors

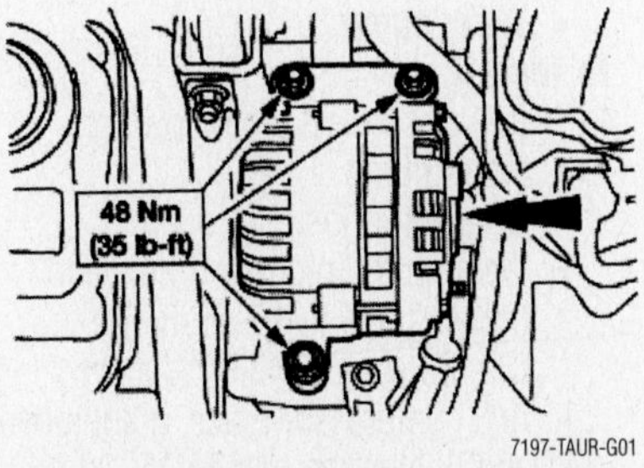

Alternator and mounting bolts—2003-04 3.0L (VIN U) Engine

➡The crankshaft pulley retainer has left-hand threads.

   - Crankshaft pulley and tighten to 74 ft. lbs. (100 Nm)
   - Accessory drive belt
   - Right fender splash shield
   - Right front wheel
   - Negative battery cable
2. Start the engine and check for proper operation.

Ignition Timing

ADJUSTMENT

The base ignition timing is set at 10 degrees Before Top Dead Center (BTDC) and is not adjustable.

Engine Assembly

REMOVAL & INSTALLATION

1. Before servicing the vehicle, refer to the precautions at the beginning of this section.
2. Disconnect the negative battery cable.
3. Drain the engine oil.
4. Drain the cooling system.
5. Recover the A/C refrigerant.
6. On the 3.0L (VIN S) engine, remove or disconnect the following:
   - Windshield wipers
   - Cowl extension
   - Emission vacuum control connector at the right side of the dash panel
   - Hood
   - Steering column pinch bolt
   - Mass Airflow (MAF) sensor connector
   - Intake Air Temperature (IAT) sensor
   - Air cleaner assembly
   - Idle Air Control (IAC) valve connector
   - Throttle Position (TP) sensor connector
   - Fuel lines
   - Intake manifold vacuum hoses
   - Ground straps
   - Powertrain Control Module (PCM) connector
   - Engine control sensor wiring harness connectors from the bracket located at the top of the transaxle
   - Evaporative emission canister purge valve connector
   - Crankcase vent hose
   - Accelerator cable
   - Cruise control actuator
   - Shift cable and lever
   - Secondary air injection pump relay connector
   - Main emission vacuum control connector near the fan shroud
   - Radiator hoses
   - Heater hoses
   - Transaxle oil cooler lines
   - Power steering return hose
   - Alternator wiring harness connectors
   - A/C compressor lines
   - A/C pressure cutoff switch, if equipped
   - Front wheels
   - Stabilizer bar links
   - Lower ball joints
   - Outer tie rod ends
   - Halfshafts
   - Radiator splash shield
   - Heated Oxygen (HO_2S) sensors
   - Dual converter Y-pipe
   - Power steering pressure switch connector
   - Power steering pressure hose
7. Support the powertrain from below and remove the subframe bolts.
8. Raise the vehicle away from the powertrain.
9. Attach an engine hoist to the powertrain.
10. Remove the left, right, and rear powertrain support insulators and lift the powertrain away from the subframe.

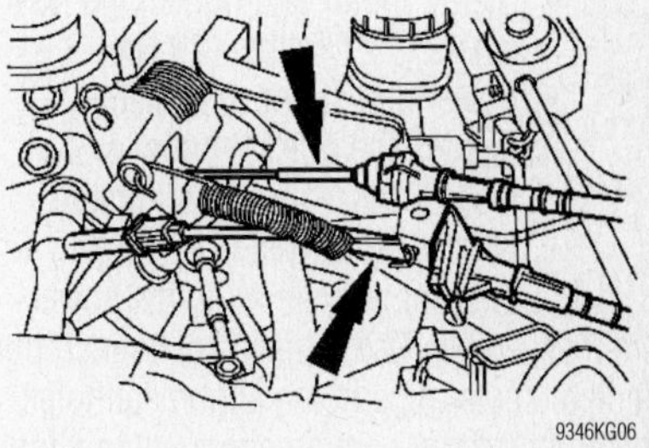

Throttle and cruise control cables—3.0L (VIN U) engine shown

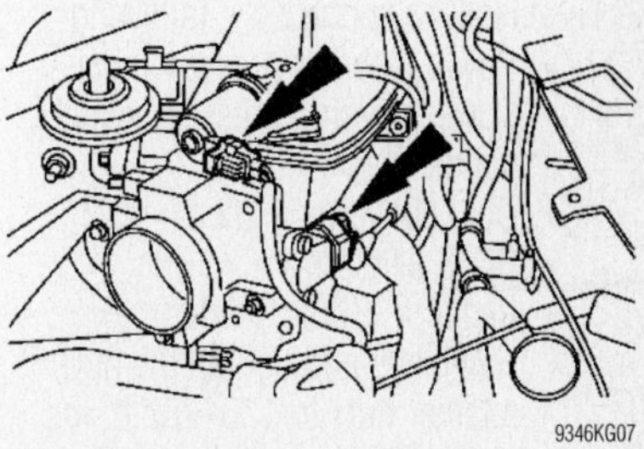

Throttle Position (TP) sensor and Intake Air Control (IAC) valve connectors—3.0L (VIN U) engine shown

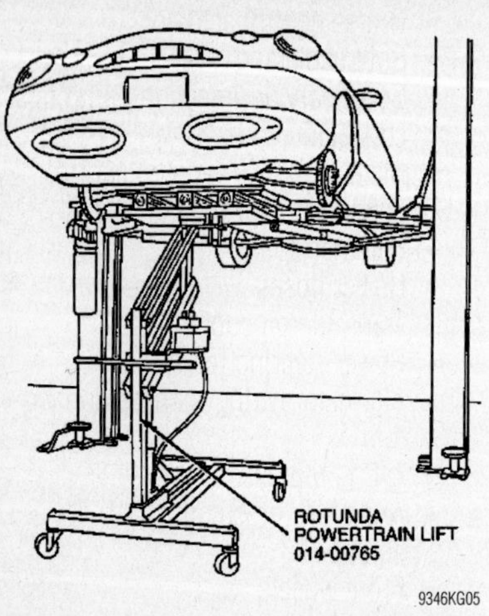

Support the powertrain when removing the subframe bolts

11. Support the transaxle from below.
12. Remove or disconnect the following:
    - Starter motor
    - Torque converter
    - Transaxle flange bolts
13. Separate the engine from the transaxle.

To install:

14. Install the engine to the transaxle. For 3.0L (VIN S) engines, tighten the flange bolts to 25–33 ft. lbs. (33–46 Nm). For all other engines, tighten the flange bolts to 30–44 ft. lbs. (40–60 Nm). Tighten the torque converter nuts to 20–34 ft. lbs. (27–46 Nm).
15. Install or connect the following:
    - Starter motor and tighten the fasteners to 21 ft. lbs. (29 Nm)
    - Powertrain support insulators. Tighten the bracket bolts to 65 ft. lbs. (88 Nm) and the subframe bolts to 90 ft. lbs. (122 Nm).
16. Lower the vehicle on to the powertrain assembly. Use 2 pieces of ¾ inch outside diameter pipe in the alignment holes behind the front subframe mounts to align the subframe to the body. Tighten the subframe mounting bolts to 57–76 ft. lbs. (77–103 Nm).
17. Install or connect the following:
    - Power steering pressure hose
    - Power steering pressure switch connector
    - Dual converter Y-pipe
    - HO_2S sensors
    - Radiator splash shield
    - Halfshafts and tighten the hub retainer nuts to 170–202 ft. lbs. (230–275 Nm)
    - Outer tie rod ends and tighten the nuts to 35–46 ft. lbs. (47–63 Nm)
    - Lower ball joints and tighten the nuts to 50–68 ft. lbs. (68–92 Nm)
    - Stabilizer bar links and tighten the nuts to 30–40 ft. lbs. (40–55 Nm)
    - Front wheels
    - A/C pressure cutoff switch, if equipped
    - A/C compressor lines
    - Alternator wiring harness connectors
    - Power steering return hose
    - Transaxle oil cooler lines
    - Heater hoses
    - Radiator hoses
    - Main emission vacuum control connector near the fan shroud
    - Secondary air injection pump relay connector
    - Shift cable and lever
    - Cruise control actuator
    - Accelerator cable
    - Crankcase vent hose
    - Evaporative emission canister purge valve connector
    - Engine control sensor wiring harness connectors from the bracket located at the top of the transaxle
    - PCM connector
    - Ground straps
    - Intake manifold vacuum hoses
    - Fuel lines
    - IAC valve connector
    - TP sensor connector
    - Air cleaner assembly
    - IAT sensor
    - MAF sensor connector
    - Steering column pinch bolt
    - Hood
18. On the 3.0L (VIN S) engine, install or connect the following:
    - Windshield wipers
    - Cowl extension
    - Emission vacuum control connector at the right side of the dash panel
19. Fill the cooling system, fill the crankcase and recharge the A/C system.
20. Connect the negative battery cable.
21. Start the engine. Check for leaks and proper operation.

➡Whenever the vehicle's subframe is removed or lowered, the wheel alignment should be checked.

Water Pump

REMOVAL & INSTALLATION

3.0L (VIN U) Engine

1. Before servicing the vehicle, refer to the precautions at the beginning of this section.
2. Drain the cooling system.
3. Remove or disconnect the following:
    - Negative battery cable
    - Accessory drive belt and tensioner
    - Water pump pulley
    - Heater hose
    - Engine control sensor wiring
    - Water pump

To install:

➡The water pump bolts are of different lengths and must be installed in the correct locations.

4. Install the water pump. Tighten bolts 1, 2, 3, 4, 5, 6, 7, 8, 9 and 15 to 15–22 ft. lbs. (20–30 Nm). Tighten bolts 10, 11, 12, 13 and 14 to 72–106 inch lbs. (8–12 Nm). Refer to the accompanying illustration for the bolt locations.
5. Install or connect the following:
    - Engine control sensor wiring
    - Heater hose
    - Water pump pulley and tighten the bolts to 15–22 ft. lbs. (20–30 Nm)
    - Accessory drive belt and tensioner
    - Negative battery cable
6. Fill the cooling system and check for leaks.

3.0L (VIN S) Engine

1. Before servicing the vehicle, refer to the precautions at the beginning of this section.
2. Drain the cooling system.
3. Remove or disconnect the following:
    - Splash shield
    - Lower heater hose
    - Radiator lower tube bolt
    - Radiator upper front tube bolt
    - Air cleaner assembly
    - Battery and tray
    - Water pump belt
    - Upper radiator hose
    - Engine vent hose
    - Transaxle 10 pin connector
    - Radiator bypass hose assembly
    - Thermostat housing
    - A/C compressor brace, if equipped
    - Water pump mounting nuts
    - Connector hose
    - Water pump

To install:

4. Install or connect the following:
    - Water pump
    - Connector hose
    - Water pump mounting nuts and tighten them to 15–22 ft. lbs. (20–30 Nm)
    - A/C compressor brace, if equipped, and tighten the nuts to 15–22 ft. lbs. (20–30 Nm)

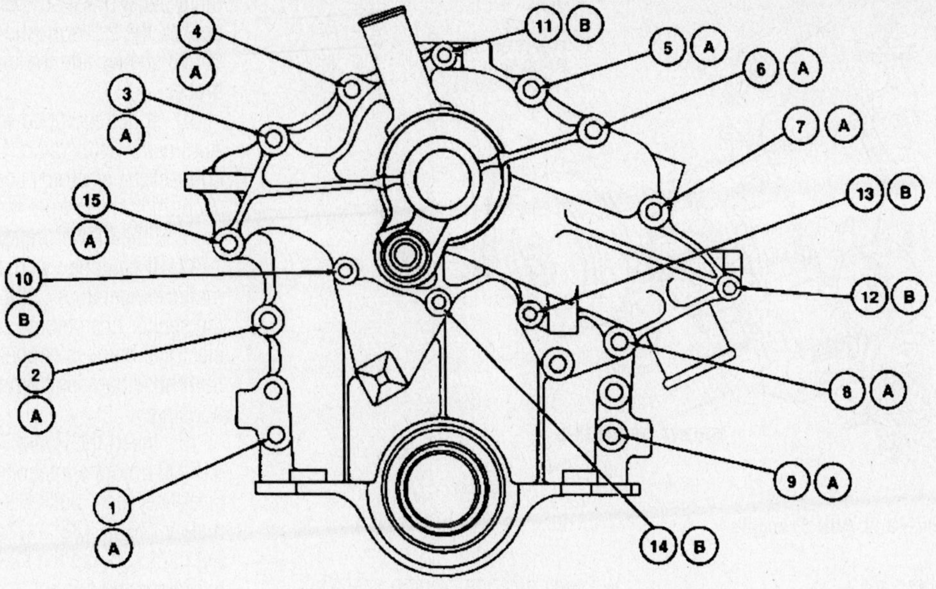

	Fasteners		Torque Specifications	
Fastener And Hole No.	Size	Fastener Application	Nm	Lb-Ft
1A	M8 x 1.25 x 43.5	F/C TO BLOCK	20-30	15-22
2A	M8 x 1.25 x 73	W/P & F/C TO BLOCK	20-30	15-22
3A	M8 x 1.25 x 104.3	W/P & F/C TO BLOCK	20-30	15-22
4A	M8 x 1.25 x 71.3	F/C TO BLOCK	20-30	15-22
5A	M8 x 1.25 x 71.3	W/P & F/C TO BLOCK	20-30	15-22
6A	M8 x 1.25 x 71.3	W/P & F/C TO BLOCK	20-30	15-22
7A	M8 x 1.25 x 104.3	W/P & F/C TO BLOCK	20-30	15-22
8A	M8 x 1.25 x 104.3	W/P & F/C TO BLOCK	20-30	15-22
9A	M8 x 1.25 x 52	F/C TO BLOCK	20-30	15-22
10B	M6 x 1 x 28.5	W/P TO F/C	8-12	71-106 (lb-in)
11B	M6 x 1 x 28.5	W/P TO F/C	8-12	71-106 (lb-in)
12B	M6 x 1 x 28.5	W/P TO F/C	8-12	71-106 (lb-in)
13B	M6 x 1 x 28.5	W/P TO F/C	8-12	71-106 (lb-in)
14B	M6 x 1 x 28.5	W/P TO F/C	8-12	71-106 (lb-in)
15A	M8 x 1.25 x 71.3	W/P & F/C TO BLOCK	20-30	15-22

W/P—Water Pump
F/C—Engine Front Cover

9300KG01

Exploded view of the water pump and timing front cover bolt locations—3.0L (VIN U) engine

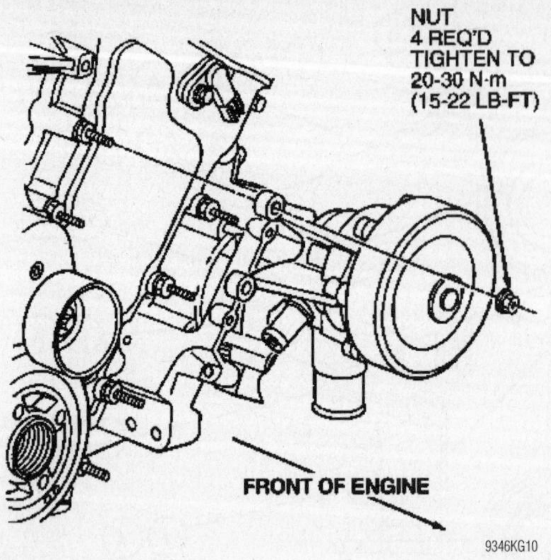

Water pump mounting—3.0L (VIN S) engine

- Thermostat housing
- Radiator bypass hose assembly
- Transaxle 10 pin connector
- Engine vent hose
- Upper radiator hose
- Water pump belt
- Battery and tray
- Air cleaner assembly
- Radiator upper front tube bolt
- Radiator lower tube bolt
- Lower heater hose
- Splash shield

5. Fill the cooling system.
6. Start the engine and check for leaks.

Heater Core

REMOVAL & INSTALLATION

1. Disconnect the negative battery cable.

※※ CAUTION

After disconnecting the negative battery cable, wait for at least 1 minute for the SRS or air bag module to deplete its energy.

2. Place the front wheels in the straight-ahead position.
3. Lock the steering column.
4. Remove or disconnect the following:
   - SRS bolt covers from both sides of the steering wheel
   - SRS module-to-steering wheel bolts
   - SRS module and disconnect the electrical connector
   - Steering wheel bolt and discard it
5. Press the steering wheel from the steering column.
   - Lower steering column shaft bolt
   - 2 lower instrument panel cover-to-instrument panel screws and unsnap the lower cover from the instrument panel
6. Turn ignition switch to the RUN position
7. Insert a ⅛ in. (3mm) wire or pin punch in the lower steering column shroud hole, under the ignition switch; then, press on the pin while pulling out on the ignition switch lock cylinder and remove it from the steering column lock cylinder housing.
8. Remove or disconnect the following:
   - 3 steering column shroud screws and the shrouds
   - Shift control selector lever boot
   - Gearshift lever pin from the manual control lever and remove the lever
   - Wiring connector at the bottom of the steering column and remove the wiring from the column, (if equipped with an overdrive lockout switch on the manual control lever)
   - Electrical connectors
   - Multi-function switch screws and move it aside
   - Shift indicator cable from the shifter tube
   - Shifter indicator-to-column adjustment cable screw
   - Interlock cable and actuator (if equipped with a column shift)
   - 4 steering column-to-instrument panel bracket nuts and steering column
   - Push pins and the lower instrument cover from the instrument panel reinforcement (at the passenger's side)
   - Console finish panel
9. Under the steering column, 2 instrument panel brace screws, the courtesy lamp socket, the 2 Diagnostic Link Connector (DLC) screws and the instrument panel brace.
10. If not equipped with an Electronic Automatic Temperature Control (EATC), disconnect the electrical connectors and the vacuum harness from the evaporator housing and the blower motor.
11. If equipped with an Electronic Automatic Temperature Control (EATC), remove the sensor hose/elbow and disconnect the electrical harness connectors and the vacuum hose harness from the evaporator housing.
12. Insert the Radio Removing tools 415-001into the integrated control panel faceplate; then, push the tools in approximately 1½ in. (38mm) to release the retaining clips. Spread the tools slightly and pull the integrated control panel from the instrument panel.
13. Remove or disconnect the following:
   - Electrical connectors and the automatic temperature control sensor hose and elbow, if equipped
   - 6 console center finish panel screws and the panel, (if equipped with a floor shift)
   - 4 instrument panel finish panel screws and panel, located at the right side of the steering column, (if equipped with a column shift)
   - Instrument cluster finish panel-to-instrument panel clips and pull the finish panel straight out
   - 4 instrument cluster-to-instrument panel screws, the electrical connectors and the instrument cluster
   - Upper finish panel (located at the top of the instrument panel)
   - Light sensor amplifier connector, (if equipped with an autolamp)
   - Instrument panel finish end panels on both sides of the instrument panel
   - Front scuff plates and pull the door weatherstrip away from the instrument panel (on both sides of the instrument panel)
   - 3 instrument panel-to-upper cowl top panel screws
   - 2 screws at each end of the instrument panel
   - Instrument panel electrical connectors and the parking brake switch
   - Instrument panel
14. Drain the cooling system into a clean container for reuse
15. Remove or disconnect the following:
   - Heater hoses from the heater core and plug the openings

- 4 air conditioning electronic blend door actuator-to-heater/air conditioning housing assembly screws and the actuator
- Metal cover, disengage the spring from the heater core cover and from the lever
- Gently, depress the locking ramp and remove the lever from the secondary air temperature control door end

WARNING

Do not attempt to bend any part of the lever, for it is brittle and will break.

16. Rotate the primary air conditioning air temperature control door shaft downward, swing the metal link counterclockwise and remove it from the pin.
17. Remove the 3 heater core cover-to-heater/air conditioning housing assembly screws, the cover and the seal.
18. Press on the heater core tubes and remove the heater core from the heater/air conditioning housing assembly.

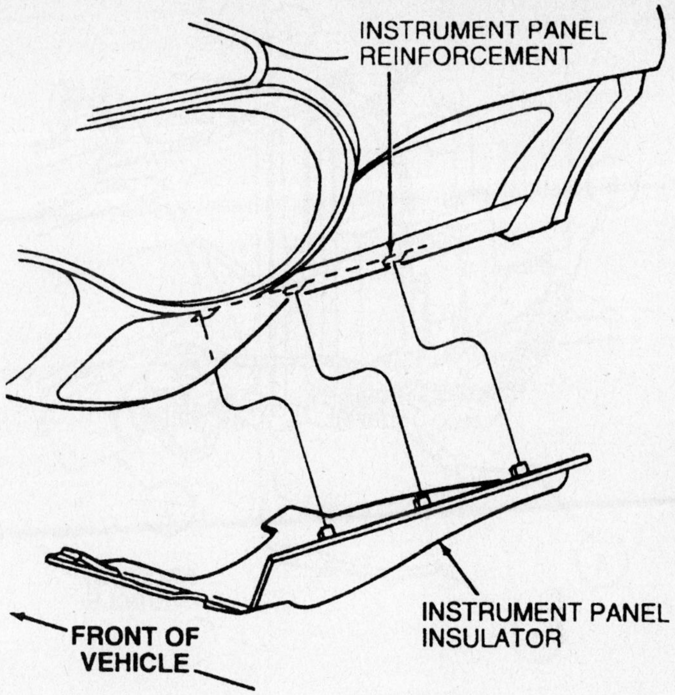

Remove the instrument panel insulator from the instrument panel reinforcement—Taurus and Sable

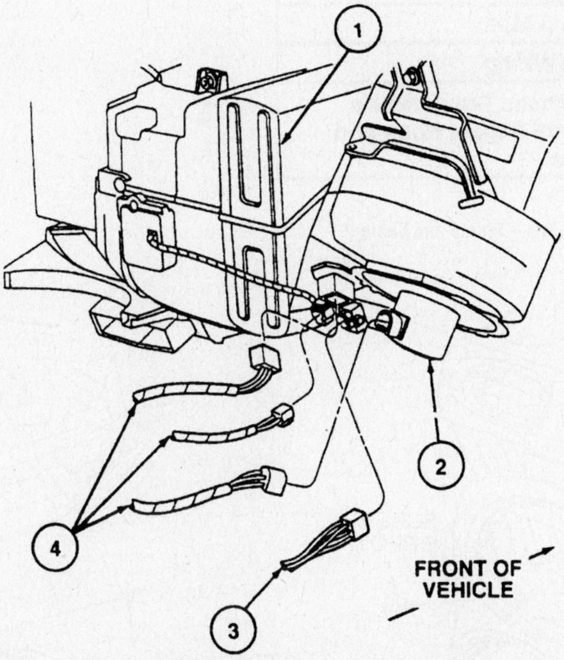

Item	Description
1	A/C Evaporator Housing
2	Blower Motor
3	Vacuum Hose Harness
4	Main Wiring

91170G06

Manual air conditioning/heater equipped vehicles disconnection points—Taurus and Sable

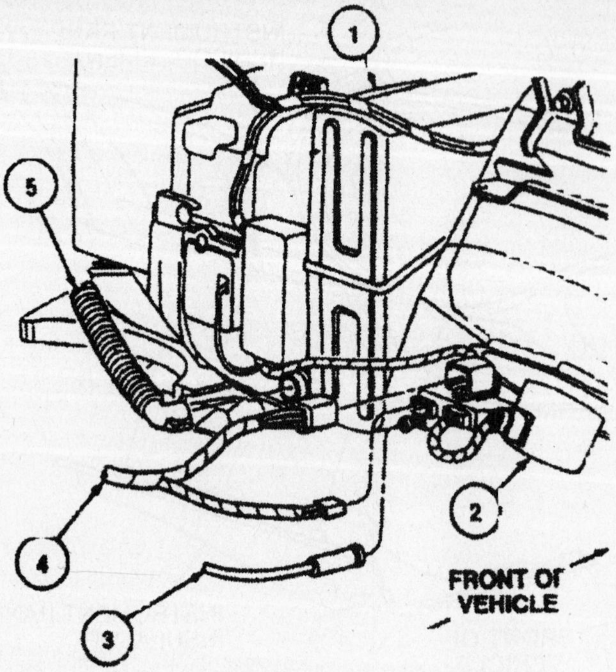

Item	Description
1	A/C Evaporator Housing
2	Blower Motor
3 & 4	Main Wiring
5	Automatic Temperature Control Sensor Hose and Elbow

91170G07

EATC equipped vehicles disconnection points—Taurus and Sable

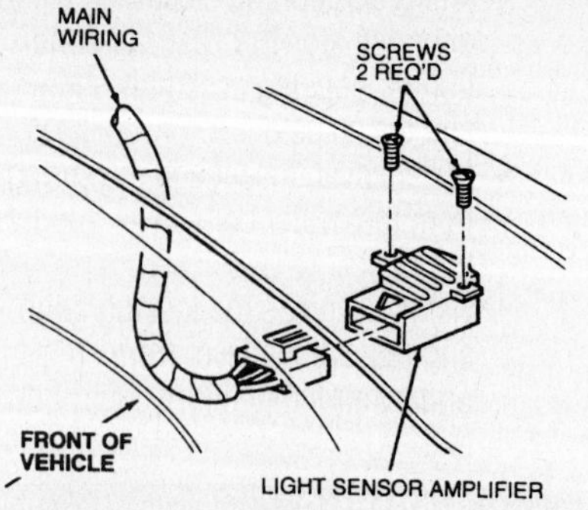

91170G10

If equipped with autolamps, detach the light sensor amplifier—Taurus and Sable

To install:

19. Install or connect the following:
    - Heater core to the heater/air conditioning housing assembly
    - Heater core cover, the seal and the 3 heater core cover-to-heater/air conditioning housing assembly screws
    - Lever to the secondary air conditioning air temperature control door end
    - Spring to the lever, engage the spring to the heater core cover and install the metal cover
    - Air conditioning electronic blend door actuator and the 4 actuator-to-heater/air conditioning housing assembly screws
    - Heater hoses to the heater core
    - Instrument panel with the help of an assistant

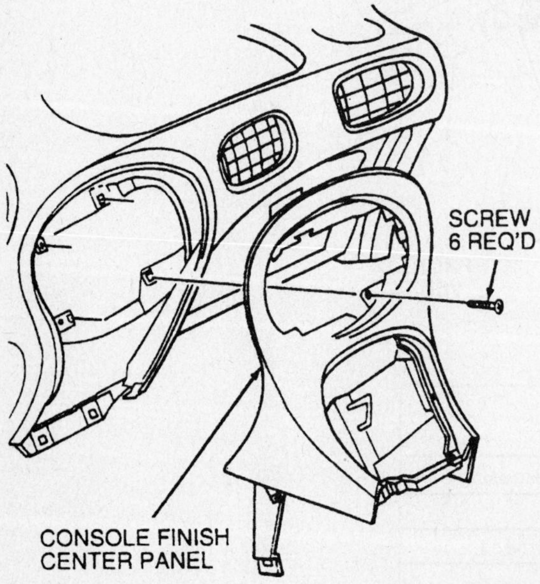

91170G08

On floor shift vehicles, remove the console center finish panel—Taurus and Sable

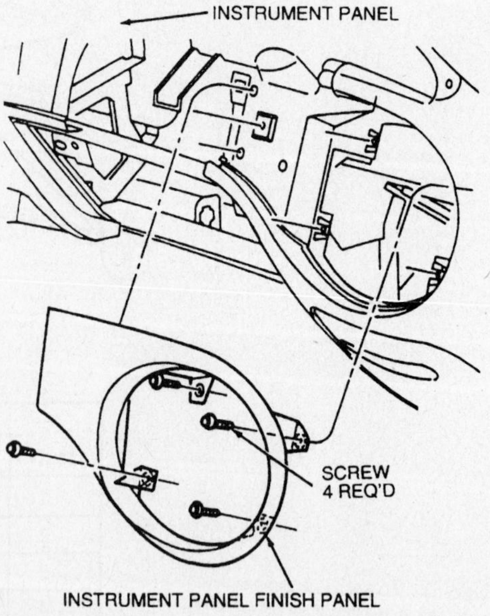

91170G09

On floor shift vehicles, remove the finish panel from around the integrated control panel—Taurus and Sable

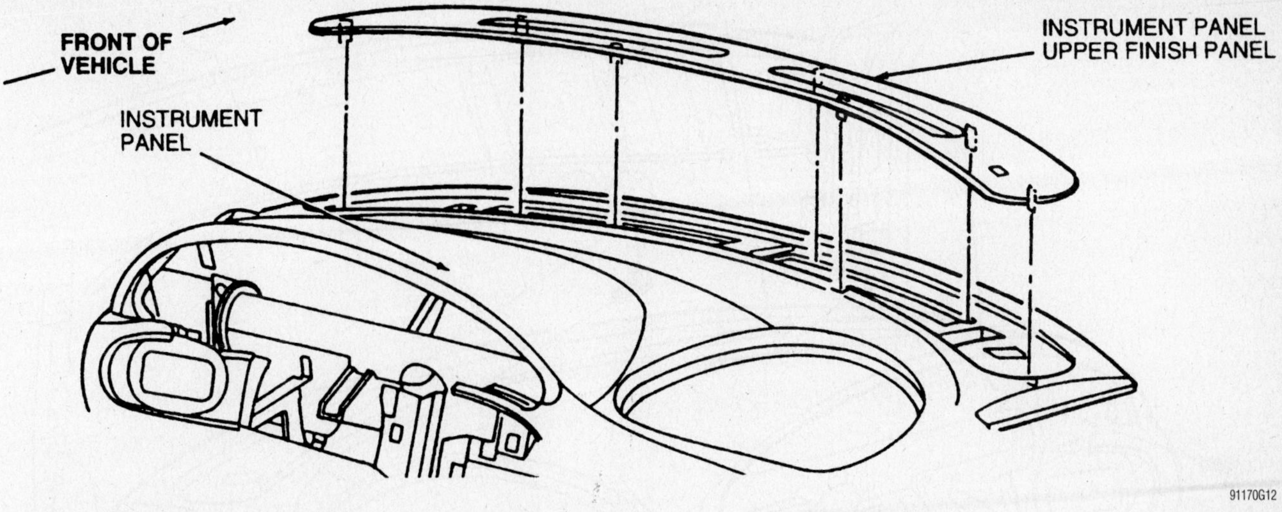

Unsnap the upper finish panel from the instrument panel—Taurus and Sable

- Instrument panel electrical connectors and the parking brake switch
- 2 screws at each end of the instrument panel with the help of an assistant
- 3 instrument panel-to-upper cowl top panel screws
- Front scuff plates and the door weatherstrip (on both sides of the instrument panel)
- Instrument panel finish end panels (on both sides of the instrument panel)
- Electrical connector to the light sensor amplifier (if equipped with an autolamp)
- Upper finish panel (at the top of the instrument panel)
- Instrument cluster, electrical connectors and the 4 instrument cluster-to-instrument panel screws
- Instrument cluster finish panel and

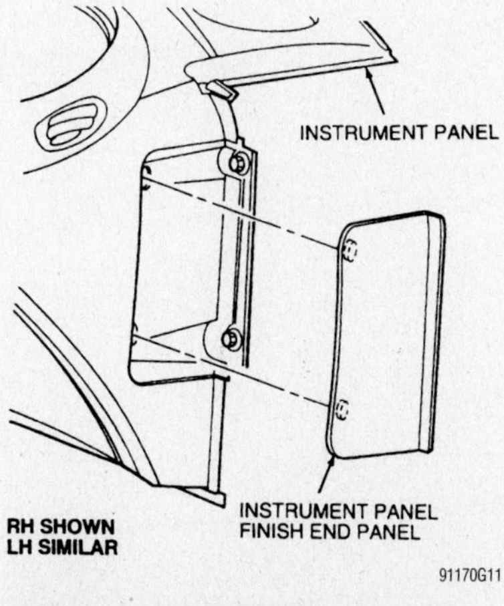

Remove the instrument panel finish end panel—Taurus and Sable

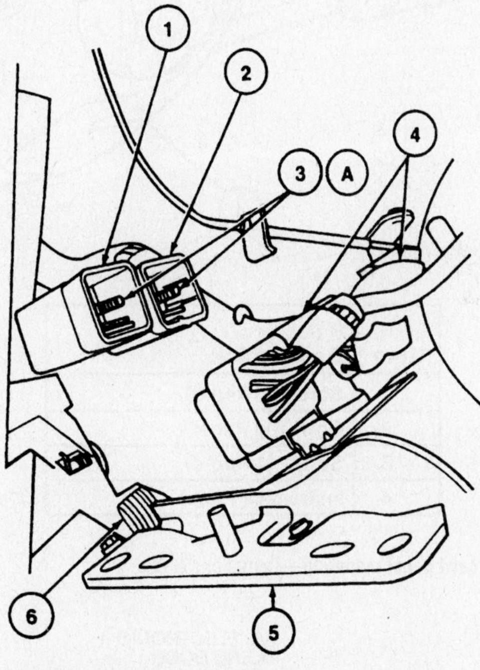

Item	Description
1	Wiring Assy Electrical Connector
2	Wiring Assy Electrical Connector
3	Bolt (Part of 14290 and 14A005)
4	Wiring Assy Electrical Connector
5	Instrument Panel
6	Parking Brake Control
A	Tighten to 4-6 N·m (36-53 Lb-In)

91170G13

Detach the electrical connectors—Taurus and Sable

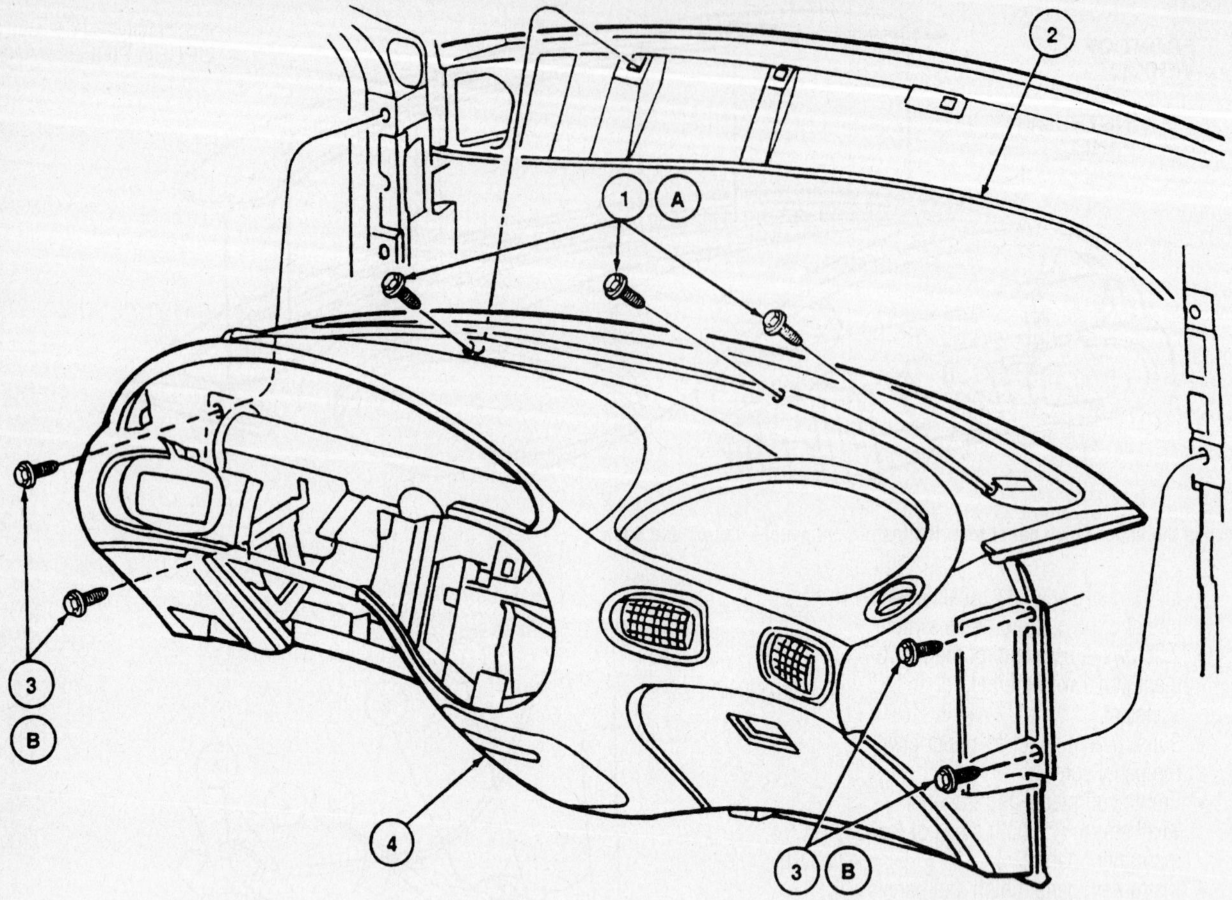

Item	Description
1	Screw (3 Req'd)
2	Cowl Top Panel
3	Screw (4 Req'd)
4	Instrument Panel

Item	Description
A	Tighten to 7-9 N·m (62-79 Lb-In)
B	Tighten to 19-25 N·m (15-18 Lb-Ft)

91170G14

Instrument panel-to-cowl panel mounting—Taurus and Sable

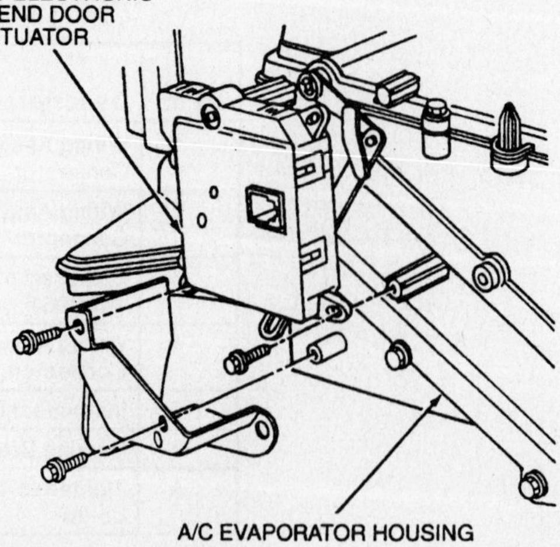

91176G05

Remove the 4 blend door actuator retaining screws—Taurus and Sable

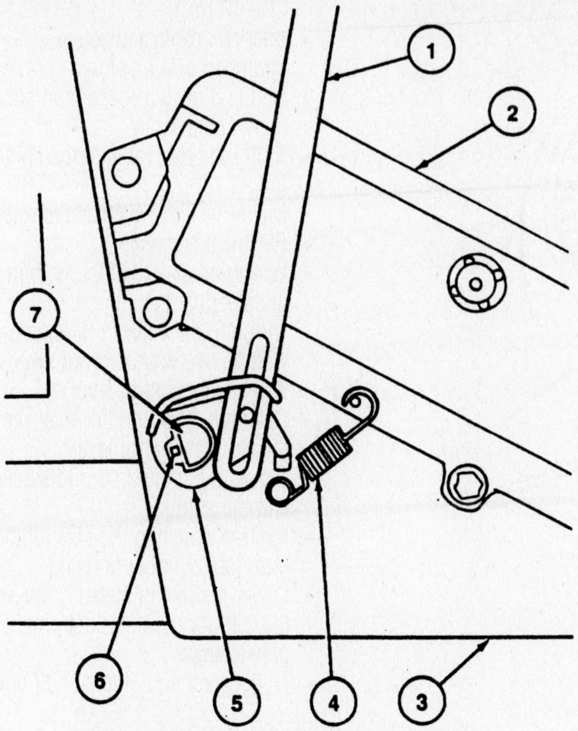

Item	Description
1	Metal Link (Part of 19B555)
2	Heater Core Cover
3	A/C Evaporator Housing
4	Spring (Part of 19B555)
5	Lever (Part of 19B555)
6	Locking Ramp (Part of Secondary A/C Air Temperature Control Door Shaft)
7	Secondary A/C Air Temperature Control Door Shaft (Part of 19B555)

91176G06

Secondary air temperature door connections—Taurus and Sable

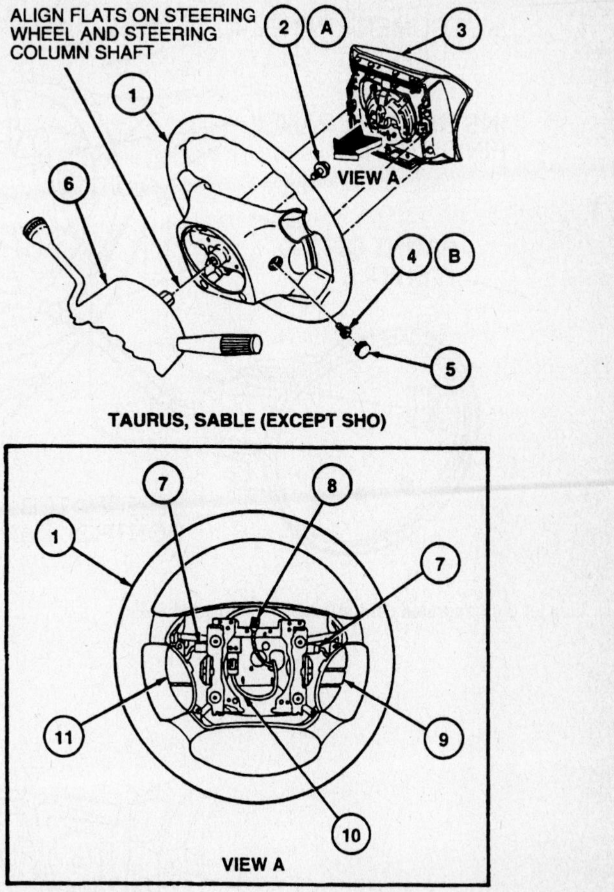

1 Steering Wheel
2 Bolt
3 Driver Side Air Bag Module
4 Screw (2 Req'd)
5 Steering Wheel Spoke Cover
6 Steering Column Tube
7 Electrical Connector
8 Air Bag Electrical Connector
9 Speed Control Actuator Switch (Right Hand)
10 Speed Control/Horn Wire Connector
11 Speed Control Actuator Switch (Left Hand)
A Tighten to 34-46 N·m (26-33 Lb-Ft)
B Tighten to 10-14 N·m (89-123 Lb-In)

93111G90

Exploded view of the SRS module and the steering wheel assembly—Taurus and Sable

engage the finish panel-to-instrument panel clips

- Instrument panel finish panel and 4 panel screws, located at the right side of the steering column (if equipped with a column shift)
- Console center finish panel and the 6 panel screws (if equipped with a floor shift)
- Electrical connectors and the automatic temperature control sensor hose and elbow, if equipped
- Integrated control panel faceplate

WARNING

Do not use excessive force when installing the radio, the retaining clips can become damaged.

- Sensor hose/elbow, the electrical harness connectors and the vacuum hose harness to the evaporator housing (if equipped with an Electronic Automatic Temperature Control (EATC)
- Connect the electrical connectors and the vacuum harness to the evaporator housing and the blower motor (if not equipped with an EATC)
- Instrument panel brace and the 2 Diagnostic Link Connector (DLC) screws
- Courtesy lamp socket and install the 2 instrument panel brace screws
- Console finish panel (if equipped with a floor shift)
- Lower instrument cover to the instrument panel reinforcement (located on the passenger's side)
- Steering column and torque the 4 steering column-to-instrument

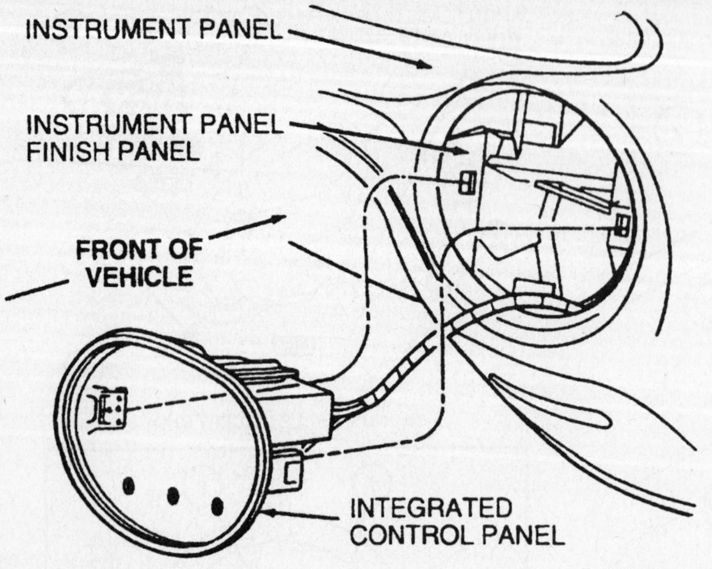

View of the integrated control panel—Taurus and Sable

panel bracket nuts to 10–13 ft. lbs. (13–17 Nm)

- Interlock cable and actuator (if equipped with a column shift)
- Shifter indicator-to-column adjustment cable screw
- Shift indicator cable to the shifter tube
- Multi-function switch and the screws
- Electrical connectors
- Wiring connector at the bottom of the steering column and install the wiring to the column (if equipped with an overdrive lockout switch on the manual control lever)
- Gearshift lever and the lever pin to the manual control lever
- 3 steering column shroud screws and the shrouds
- Ignition switch to the steering column lock cylinder housing
- Lower instrument panel cover and the 2 lower cover-to-instrument panel screws
- Lower steering column shaft bolt

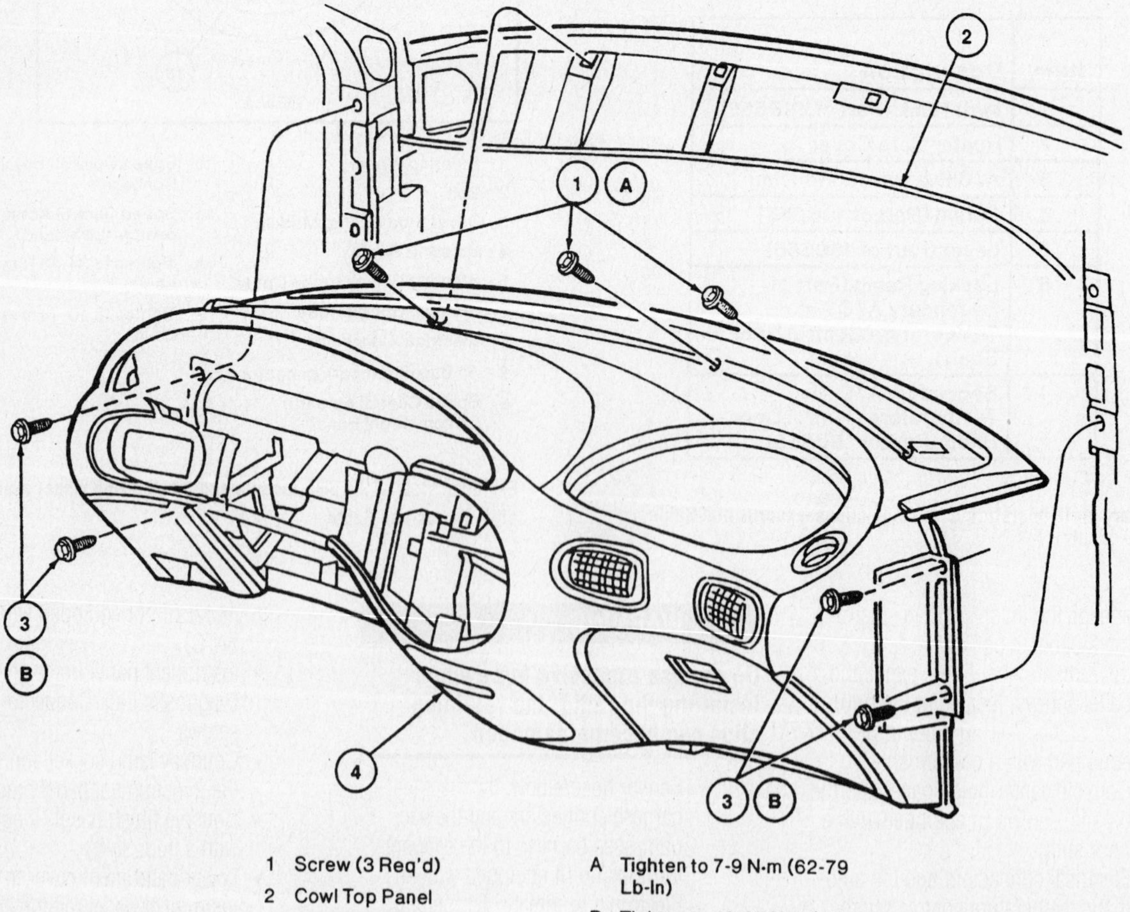

1 Screw (3 Req'd)
2 Cowl Top Panel
3 Screw (4 Req'd)
4 Instrument Panel

A Tighten to 7-9 N·m (62-79 Lb-In)
B Tighten to 19-25 N·m (15-18 Lb-Ft)

93111G92

Exploded view of the instrument panel—Taurus and Sable

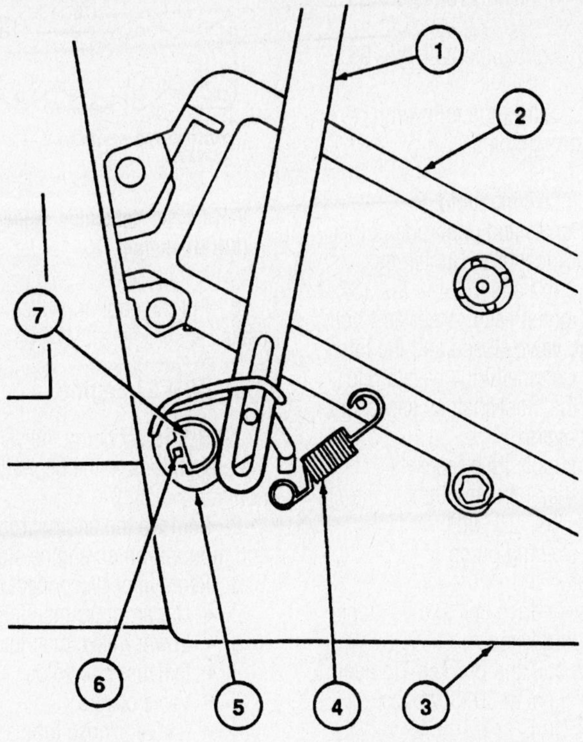

1 Metal Link
2 Heater Core Cover
3 A/C Evaporator Housing
4 Spring
5 Lever
6 Locking Ramp (Part of Secondary A/C Air Temperature Control Door Shaft)
7 Secondary A/C Air Temperature Control Door Shaft

93111G93

View of the temperature control mechanism—Taurus and Sable

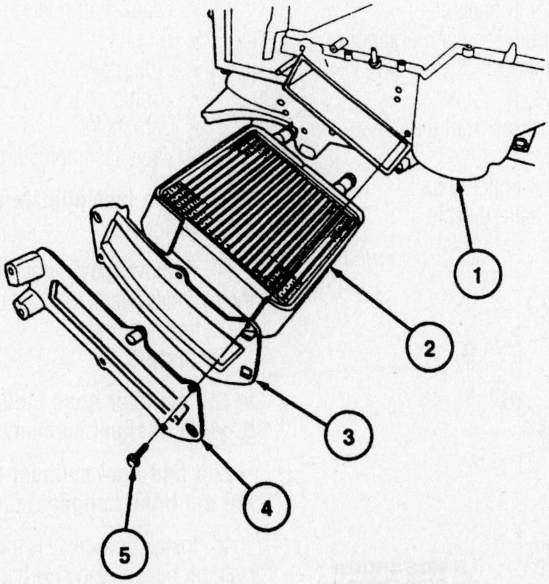

1 A/C Evaporator Housing
2 Heater Core
3 Heater Core Cover Seal
4 Heater Core Cover
5 Screw

93111G94

Exploded view of the heater core—Taurus and Sable

and torque it to 17–20 ft. lbs. (22–26 Nm)
- Steering wheel to the steering column
- New steering wheel bolt and torque it to 26–33 ft. lbs. (34–46 Nm)
- SRS module and connect the electrical connector
- SRS module-to-steering wheel bolts and torque the bolts to 89–123 inch lbs. (10–14 Nm)
- SRS bolt covers
- Negative battery cable

20. Fill the cooling system.
21. Operate the engine to normal operating temperatures; then, check the climate control operation and check for leaks.

Cylinder Head

REMOVAL & INSTALLATION

3.0L (VIN U) Engine

1. Before servicing the vehicle, refer to the precautions at the beginning of this section.
2. Drain the cooling system.
3. Set the No. 1 cylinder to Top Dead Center (TDC) of the compression stroke.
4. Remove or disconnect the following:
   - Negative battery cable
   - Air cleaner outlet hose
   - Vacuum hoses from the intake manifold
   - Exhaust Gas Recirculation (EGR) valve
   - EGR transducer
   - EGR vacuum regulator solenoid
   - Intake Air Temperature (IAT) sensor connector
   - Throttle Position (TP) sensor connector
   - Intake Air Control (IAC) valve connector
   - Camshaft Position (CMP) sensor and housing
   - Engine Coolant Temperature (ECT) sensor connector
   - Accelerator cable
   - Cruise control cable
   - Fuel lines
   - Upper intake manifold
   - Fuel injector wiring harness
   - Coolant hoses
   - Ignition coil and bracket
   - Spark plug wires
   - Accessory drive belt and tensioner
   - Alternator
   - Power steering pump
   - Oil dipstick tube
   - Heater supply tube brackets
   - Valve covers
   - Rocker arms and pushrods. Keep

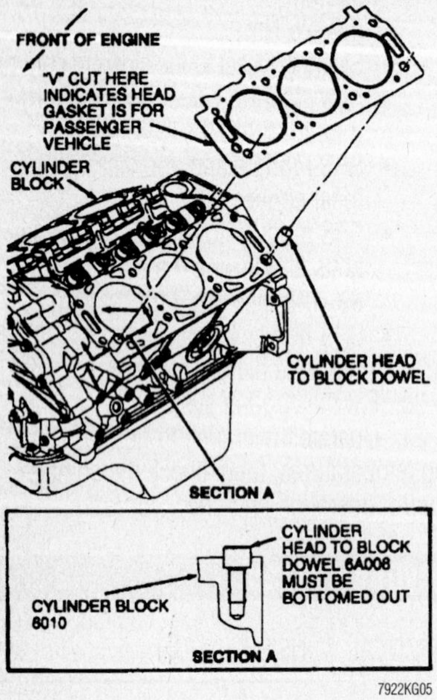

Install the cylinder head gasket with the "UP" designation on gasket facing the cylinder head—3.0L (VIN U) engines

the rocker arms and pushrods in order for installation.

- Lower intake manifold
- Spark plugs
- Exhaust manifolds
- Cylinder heads

To install:

➡The cylinder head bolts are a torque-to-yield design and cannot be reused.

5. Install the cylinder heads with new gaskets.
6. Tighten the bolts in sequence as follows:
   a. Step 1: 36–39 ft. lbs. (47–53 Nm).
   b. Step 2: Loosen the bolts 1 complete turn.
   c. Step 3: Tighten the bolts to 20–24 ft. lbs. (27–33 Nm).
   d. Step 4: Tighten the bolts 85–95 degrees.
   e. Step 5: Tighten the bolts 85–95 degrees.
7. Install or connect the following:
   - Exhaust manifolds
   - Spark plugs
   - Lower intake manifold
   - Rocker arms and pushrods in their original locations. Tighten the rocker arm bolts to 24 ft. lbs. (32 Nm). Tighten each rocker arm bolt with the valve closed and the lifter on the camshaft lobe base circle. Rotate the crankshaft as necessary.
   - Valve covers
   - Heater supply tube brackets
   - Oil dipstick tube and tighten the nut to 13 ft. lbs. (18 Nm)
   - Power steering pump
   - Alternator
   - Accessory drive belt and tensioner
   - Spark plug wires
   - Ignition coil and bracket. Tighten the fasteners to 30–39 ft. lbs. (40–55 Nm).
   - Coolant hoses
   - Fuel injector wiring harness
   - Upper intake manifold
   - Fuel lines
   - Cruise control cable
   - Accelerator cable
   - ECT sensor connector
   - CMP sensor and housing
   - IAC valve connector
   - TP sensor connector
   - IAT sensor connector
   - EGR vacuum regulator solenoid
   - EGR transducer
   - EGR valve
   - Vacuum hoses from the intake manifold
   - Air cleaner outlet hose
   - Negative battery cable

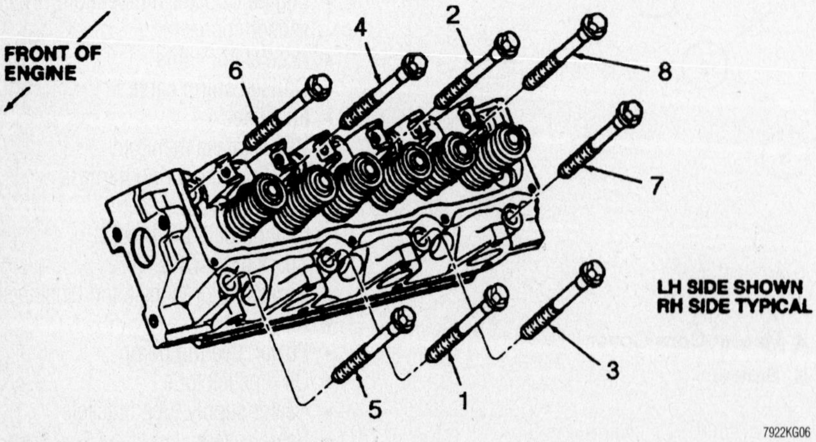

Cylinder head torque sequence—3.0L (VIN U) engine

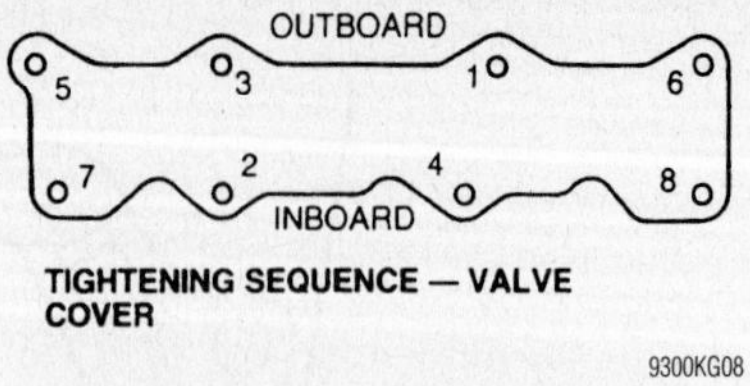

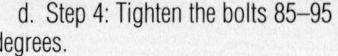

Valve cover tightening sequence—3.0L (VIN U) engine

8. Fill the cooling system and check for leaks.

3.0L (VIN S) Engine

1. Before servicing the vehicle, refer to the precautions at the beginning of this section.
2. Remove the engine from the vehicle and mount it on an engine stand.
3. Remove or disconnect the following:
   - Upper intake manifold
   - Lower intake manifold
   - Exhaust manifolds
   - Valve covers
   - Radiator hose tube and bracket
   - Accessory drive belt
   - Crankcase ventilation tube
   - Heater hose bypass tube
   - Power steering pump
   - A/C compressor and bracket
   - Water pump

➡The crankshaft accessory drive pulley shaft has left-hand threads. Rotate the pulley shaft clockwise to remove.

   - Crankshaft pulley
   - Oil pan
   - Front cover
   - Timing chains
   - Camshafts
   - Valve lash adjusters

➡Keep the lash adjusters in order for installation.

4. Remove the cylinder head bolts from the cylinder heads in sequence and remove the cylinder heads.

To install:

➡The cylinder head bolts are a torque-to-yield design and cannot be reused.

➡Left and right cylinder head gaskets are not interchangeable.

5. Install new gaskets and install the cylinder heads. Use new bolts and tighten as follows:
   a. Step 1: Tighten the bolts in sequence to 28–31 ft. lbs. (37–43 Nm).
   b. Step 2: Plus 85–95 degrees.
   c. Step 3: Loosen the bolts in reverse sequence 1 full turn.

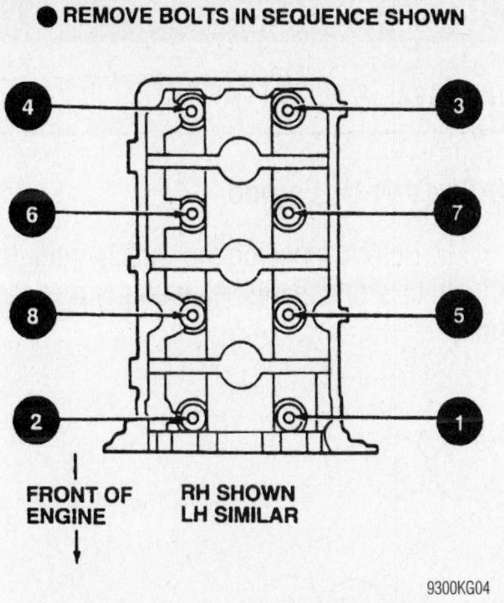

Cylinder head bolt removal sequence–right side head shown, left side similar—3.0L (VIN S) engine

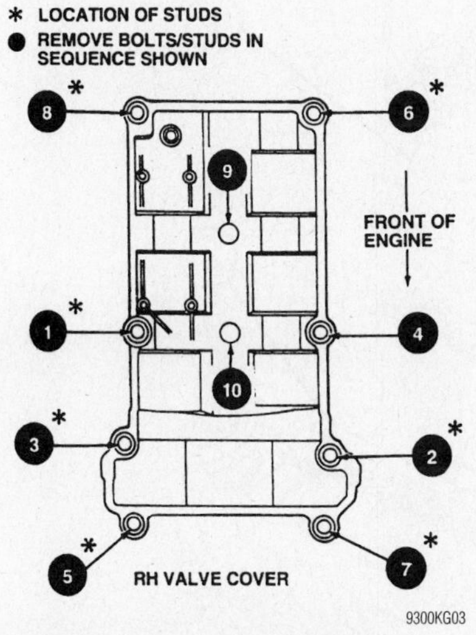

Right side valve cover bolt removal sequence—3.0L (VIN S) engine

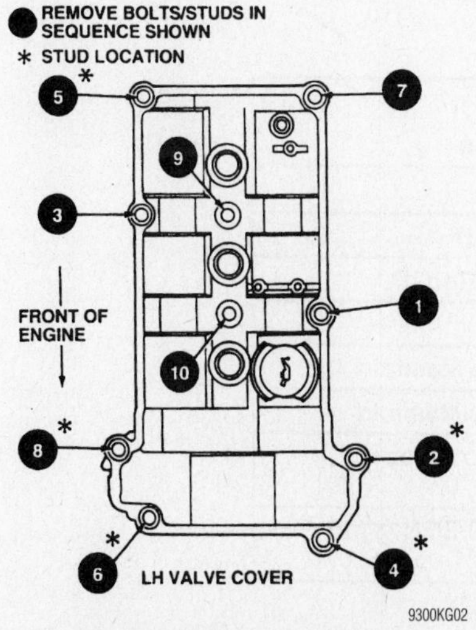

Left side valve cover bolt removal sequence—3.0L (VIN S) engine

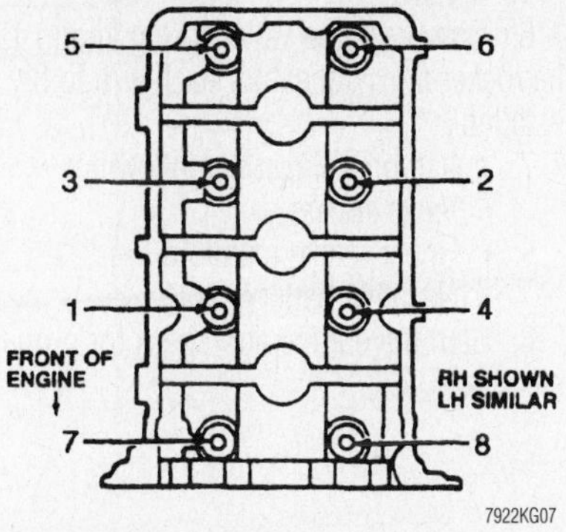

Cylinder head torque sequence—3.0L (VIN S) engine

d. Step 4: Tighten the bolts in sequence to 28–31 ft. lbs. (37–43 Nm).

e. Step 5: Tighten the bolts in sequence 85–95 degrees.

f. Step 6: Tighten the bolts in sequence 85–95 degrees.

6. Install or connect the following:
   - Valve lash adjusters in their original locations
   - Camshafts
   - Timing chains
   - Front cover
   - Oil pan
   - Crankshaft pulley
   - Water pump
   - A/C compressor and bracket
   - Power steering pump
   - Heater hose bypass tube
   - Crankcase ventilation tube
   - Accessory drive belt
   - Radiator hose tube and bracket
   - Valve covers
   - Exhaust manifolds
   - Lower intake manifold

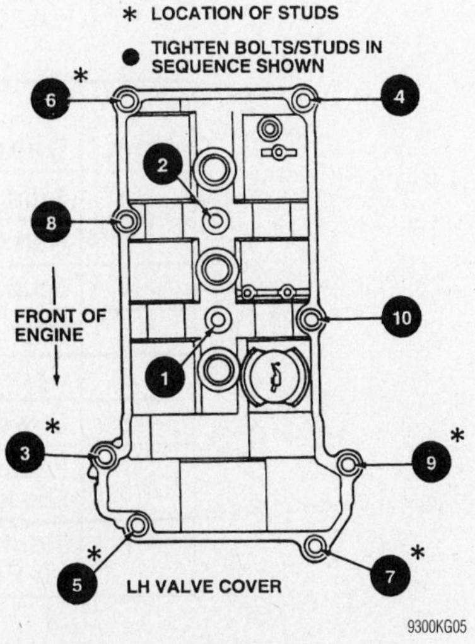

Left side valve cover bolt tightening sequence—3.0L (VIN S) engine

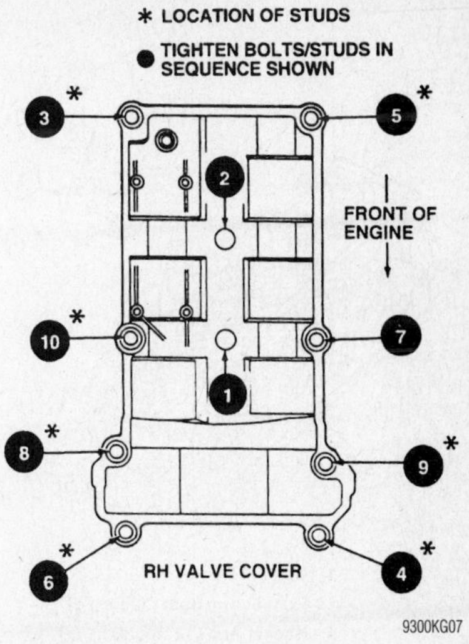

Right side valve cover bolt tightening sequence—3.0L (VIN S) engine

   - Upper intake manifold
   - Engine assembly into the vehicle.

7. Run the engine and check for leaks.

Rocker Arms/Shafts

REMOVAL & INSTALLATION

3.0L (VIN U) Engine

1. Before servicing the vehicle, refer to the precautions at the beginning of this section.
2. Remove or disconnect the following:
   - Negative battery cable
   - Upper intake manifold
   - Valve covers
   - Rocker arms

➡**Keep the rocker arms in order for installation.**

To install:

➡**The rocker arm bolts are tightened with the valves closed. Rotate the crankshaft as necessary to position the lifter on the base circle of the camshaft lobe before tightening the corresponding rocker arm bolt.**

3. Install the rocker arms. The rocker arm bolts are tightened in two steps as follows:
   a. Step 1: 96 inch lbs. (11 Nm).
   b. Step 2: 24 ft. lbs. (32 Nm).
4. Install or connect the following:
   - Valve covers
   - Upper intake manifold
   - Negative battery cable
5. Start the engine and check for proper operation.

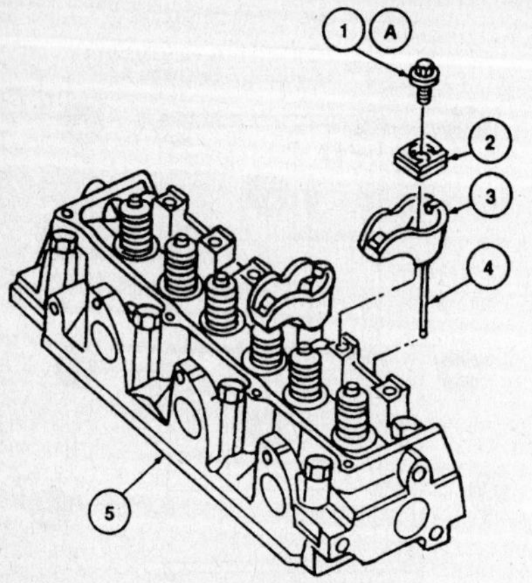

1 Bolt (12 Req'd)
2 Rocker Arm Seat (12 Req'd)
3 Rocker Arm (12 Req'd)
4 Push Rod (12 Req'd)
5 Cylinder Head (2 Req'd)
6 2.15-4.69 mm (0.085-0.185 inch)
A Tighten in Two Steps:
7-15 N·m (5-11 Lb-Ft)
26-38 N·m (19-28 Lb-Ft)

7922KG22

Exploded view of the rocker arms and related components—3.0L (VIN U) engines

3.0L (VIN S) Engine

1. Before servicing the vehicle, refer to the precautions at the beginning of this section.
2. Remove or disconnect the following:
   - Negative battery cable
   - Upper intake manifold
   - Valve covers
3. Rotate the crankshaft so that the camshaft lobe on the valve to be serviced is pointing directly away from the valve.
4. Install special tool Valve Spring Compressor 303-473.
5. Compress the valve spring and remove the rocker arm. Repeat for each arm to be removed.

➡If the rocker arms are to be reused, ensure that they are installed in the same position that they were removed from.

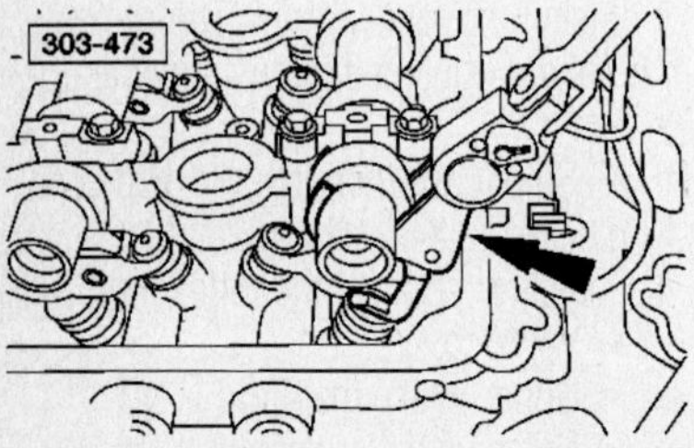

9346KG01

Valve spring compressor 303-473—3.0L (VIN S) engine

To install:

6. Compress the valve spring and install the rocker arm. Repeat for each arm to be installed.
7. Install or connect the following:
   - Valve covers
   - Upper intake manifold
   - Negative battery cable
8. Start the engine and check for proper operation.

Intake Manifold

REMOVAL & INSTALLATION

3.0L (VIN U) Engine

1. Before servicing the vehicle, refer to the precautions at the beginning of this section.

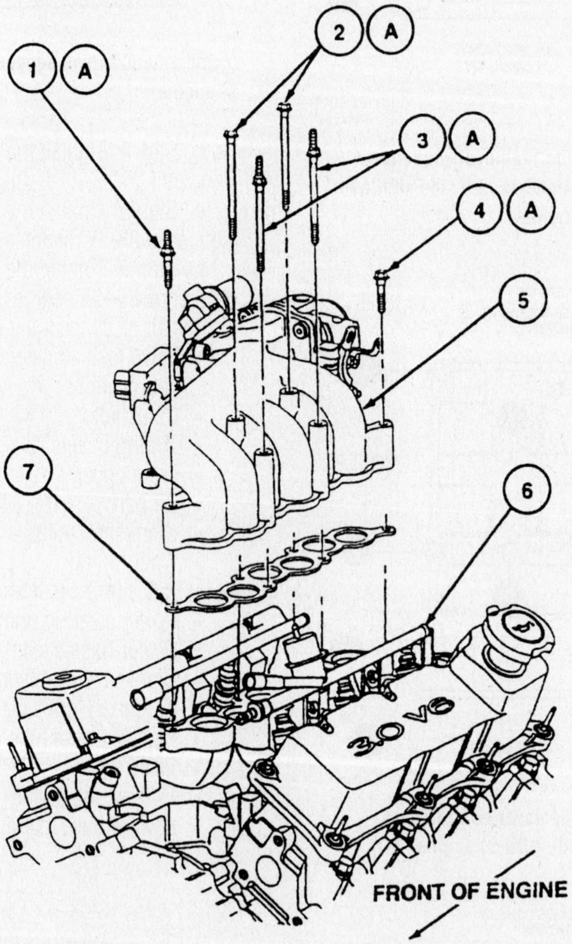

Item	Description
1	Stud Bolt
2	Bolt (2 Req'd)
3	Stud Bolt (2 Req'd)
4	Bolt
5	Upper Intake Manifold
6	Lower Intake Manifold
7	Intake Manifold Upper Gasket
A	Tighten to 20-30 N·m (15-22 Lb-Ft)

9300KG09

Exploded view of the upper intake manifold mounting—3.0L (VIN U) engine

2. Drain the cooling system.
3. Remove or disconnect the following:
   - Negative battery cable
   - Air cleaner and outlet tube
   - Cruise control cable
   - Accelerator cable and bracket
   - Fuel lines
   - Vacuum lines
   - Exhaust Gas Recirculation (EGR) valve
   - EGR transducer
   - EGR vacuum solenoid
   - Throttle Position (TP) sensor connector
   - Intake Air Control (IAC) valve connector
   - Camshaft Position (CMP) sensor connector
   - Engine Coolant Temperature (ECT) sensor connector
   - Coolant temperature indicator sender connector
   - Coolant hoses
   - Upper alternator bracket
   - Engine control sensor wiring harness and bracket
   - Upper intake manifold

➡Note the position of the CMP sensor electrical connector. Installation requires that the connector be located in the same position.

➡Before removing the CMP sensor, position the engine at Top Dead Center (TDC) of the No. 1 cylinder compression stroke.

4. Remove or disconnect the following:
   - CMP sensor and housing
   - Spark plug wires
   - Ignition coil
   - Valve covers
   - No. 3 cylinder intake valve rocker arm and pushrod
   - Lower intake manifold

To install:

5. Install the lower intake manifold. Use new gaskets and tighten the bolts in sequence as follows:
   a. Step 1: 15–22 ft. lbs. (20–30 Nm).
   b. Step 2: 19–24 ft. lbs. (26–32 Nm).

➡A special Synchro Positioning tool T95T-12200-A must be used when installing the CMP sensor housing.

6. Attach the Synchro Position tool T95T-12200-A as follows:
   a. Engage the CMP sensor housing vane into the radial slot of the tool.
   b. Rotate the tool on the CMP sensor housing until the tool boss engages the notch in the CMP sensor housing.
   c. Install the CMP sensor housing so the drive gear engagement occurs when the arrow on the locator tool is pointed about 75 degrees counterclockwise from the rear face of the cylinder block. This step will locate the CMP sensor electrical connector in the same position as was noted on removal.
   d. Install the hold-down clamp and tighten the bolt to 14–22 ft. lbs. (19–30 Nm).
7. Remove the Synchro Position tool.
8. Install or connect the following:
   - CMP sensor
   - No. 3 cylinder intake valve rocker arm and pushrod
   - Valve covers
   - Ignition coil and tighten the bolts to 30–40 ft. lbs. (40–55 Nm)
   - Spark plug wires
   - Upper intake manifold and tighten the bolts to 15–22 ft. lbs. (20–30 Nm)
   - Engine control sensor wiring harness and bracket
   - Upper alternator bracket
   - Coolant hoses
   - Coolant temperature indicator sender connector
   - ECT sensor connector
   - CMP sensor connector
   - IAC valve connector
   - TP sensor connector
   - EGR vacuum solenoid
   - EGR transducer
   - EGR valve
   - Vacuum lines
   - Fuel lines
   - Accelerator cable and bracket
   - Cruise control cable
   - Air cleaner and outlet tube
   - Negative battery cable
9. Fill the cooling system.
10. Start the engine and check for leaks and proper operation.

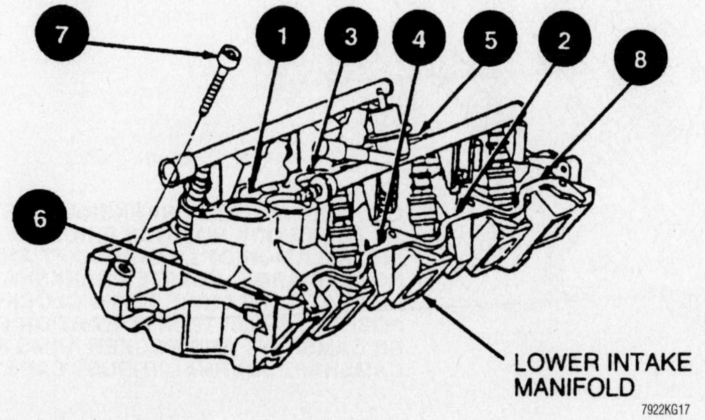

Intake manifold torque sequence—3.0L (VIN U) engine

3.0L (VIN S) Engine

1. Before servicing the vehicle, refer to the precautions at the beginning of this section.
2. Disconnect or remove:
   - Negative battery cable
   - Windshield wipers and cowl top inner panels
   - Cruise control cable
   - Accelerator cable and bracket
   - Vacuum hoses
   - Intake Air Control (IAC) valve supply hose
   - IAC connector
   - Throttle Position (TP) sensor connector
   - Exhaust Gas Recirculation (EGR) valve
   - Secondary air injection valve bracket
   - Upper intake manifold
   - Fuel lines
   - Fuel injector wiring harness
   - Intake Manifold Runner Control (IMRC) cable
   - Spark plug wires
   - Lower intake manifold

To install:

3. Install the lower intake manifold. Use new gaskets and tighten the bolts in sequence to 71–106 inch lbs. (8–12 Nm).
4. Install or connect the following:
   - Spark plug wires
   - IMRC cable
   - Fuel injector wiring harness
   - Fuel lines
5. Install the upper intake manifold. Use new gaskets and tighten the bolts in sequence to 71–106 inch lbs. (8–12 Nm).
6. Install or connect the following:
   - Secondary air injection valve bracket

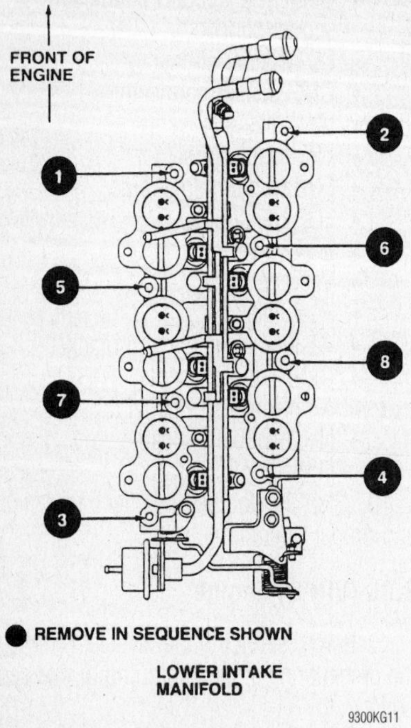
FRONT OF ENGINE
1
2
5
6
7
8
3
4
REMOVE IN SEQUENCE SHOWN
LOWER INTAKE MANIFOLD
9300KG11

Lower intake manifold bolt removal sequence—3.0L (VIN S) engine

- EGR valve
- TP sensor connector
- IAC connector
- IAC valve supply hose
- Vacuum hoses
- Accelerator cable and bracket

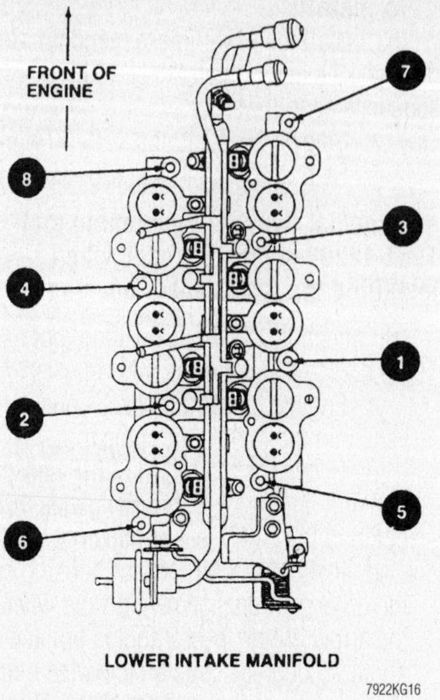
FRONT OF ENGINE
7
8
3
4
1
2
5
6
LOWER INTAKE MANIFOLD
7922KG16

Lower intake manifold torque sequence—3.0L (VIN S) engine

- Cruise control cable
- Windshield wipers and cowl top inner panels
- Negative battery cable

7. Run the engine and check for leaks and proper engine operation.

Exhaust Manifold

REMOVAL & INSTALLATION

3.0L (VIN U) Engine

RIGHT SIDE

1. Before servicing the vehicle, refer to the precautions at the beginning of this section.
2. Remove or disconnect the following:
   - Negative battery cable
   - Cowl vent screen and extension
   - Heated Oxygen (HO_2S) sensor connector
   - Exhaust Gas Recirculation (EGR) tube
   - Exhaust manifold heat shield

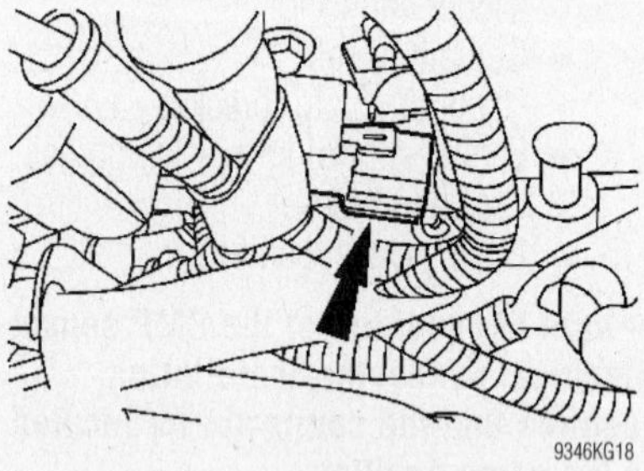
9346KG18

Right side Heated Oxygen (HO_2S) sensor connector—3.0L (VIN U) engine

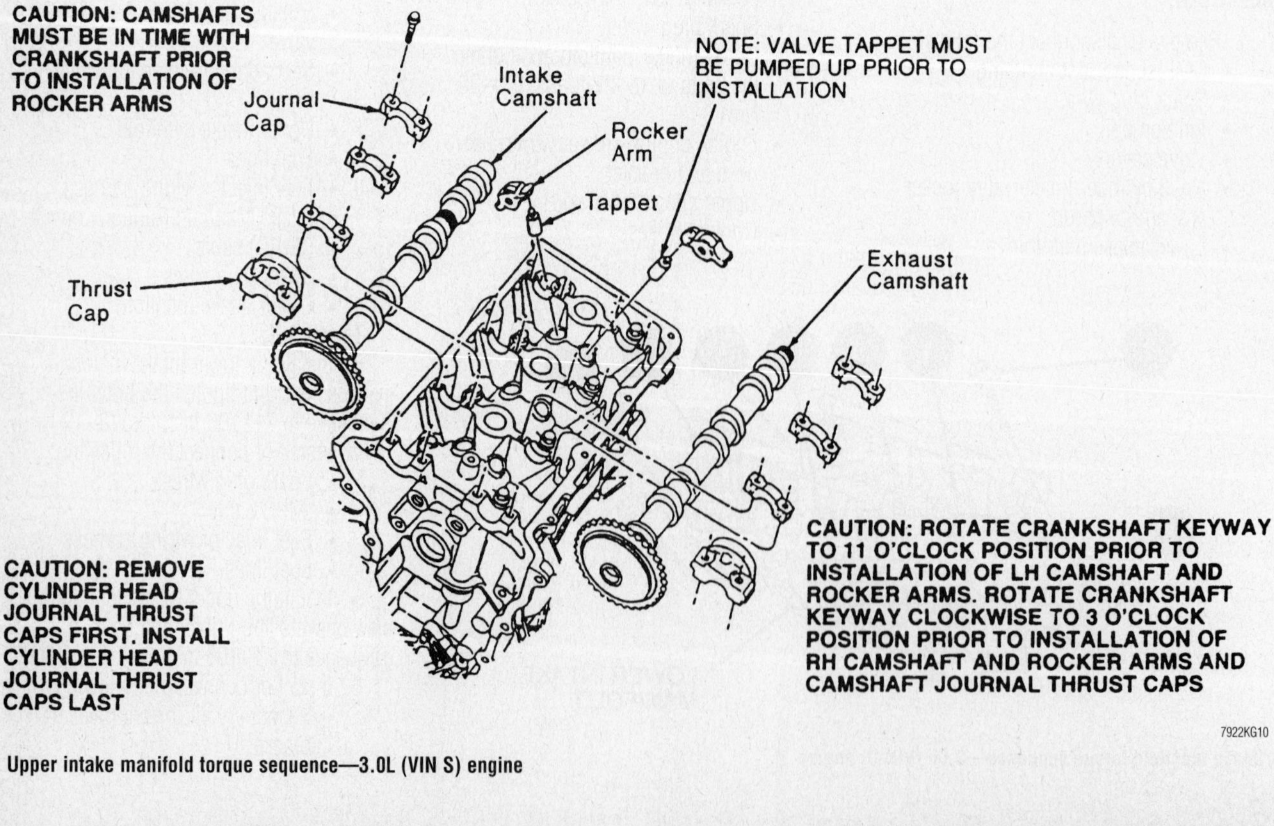
CAUTION: CAMSHAFTS MUST BE IN TIME WITH CRANKSHAFT PRIOR TO INSTALLATION OF ROCKER ARMS
Journal Cap
Intake Camshaft
NOTE: VALVE TAPPET MUST BE PUMPED UP PRIOR TO INSTALLATION
Rocker Arm
Tappet
Thrust Cap
Exhaust Camshaft
CAUTION: REMOVE CYLINDER HEAD JOURNAL THRUST CAPS FIRST. INSTALL CYLINDER HEAD JOURNAL THRUST CAPS LAST
CAUTION: ROTATE CRANKSHAFT KEYWAY TO 11 O'CLOCK POSITION PRIOR TO INSTALLATION OF LH CAMSHAFT AND ROCKER ARMS. ROTATE CRANKSHAFT KEYWAY CLOCKWISE TO 3 O'CLOCK POSITION PRIOR TO INSTALLATION OF RH CAMSHAFT AND ROCKER ARMS AND CAMSHAFT JOURNAL THRUST CAPS
7922KG10

Upper intake manifold torque sequence—3.0L (VIN S) engine

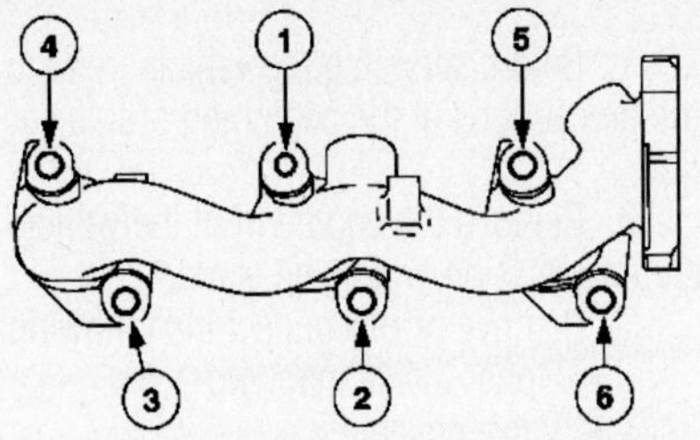

9346KG19

Right exhaust manifold torque sequence—3.0L (VIN U) engine

- Catalytic converter
- Exhaust manifold

To install:

3. Install the exhaust manifold with a new gasket. Tighten the bolts in sequence as follows:
   a. Step 1: 89 inch lbs. (10 Nm)
   b. Step 2: 16 ft. lbs. (22 Nm)
4. Install or connect the following:
   - Catalytic converter and tighten the bolts to 30 ft. lbs. (40 Nm)
   - Exhaust manifold heat shield and tighten the bolts to 89 inch lbs. (10 Nm)
   - EGR tube
   - HO_2S sensor connector
   - Cowl vent screen and extension
   - Negative battery cable
5. Start the engine and check for leaks.

LEFT SIDE

1. Before servicing the vehicle, refer to the precautions in the beginning of this section.
2. Remove or disconnect the following:
   - Negative battery cable
   - Heated Oxygen (HO_2S) sensor connector
   - Oil dipstick tube
   - Power steering pressure line
   - Secondary air injection tube, if equipped
   - Dual converter Y-pipe
   - Exhaust manifold

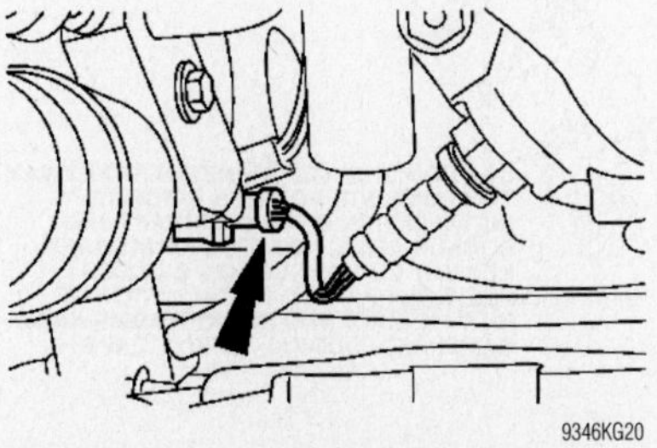

9346KG20

Left side Heated Oxygen (HO_2S) sensor connector—3.0L (VIN U) engine

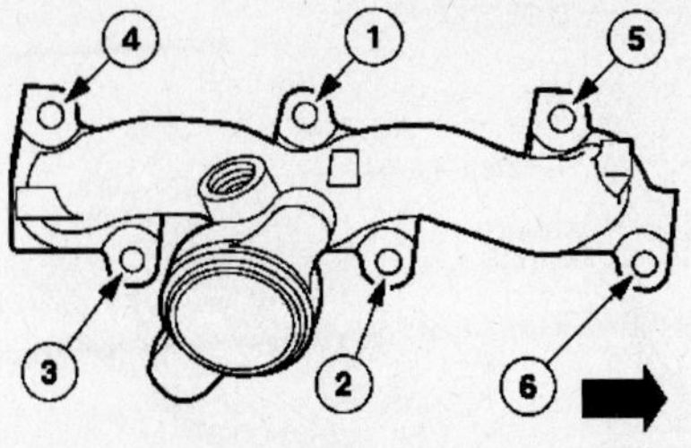

9346KG21

Left exhaust manifold torque sequence—3.0L (VIN U) engine

To install:

3. Install the exhaust manifold with a new gasket. Tighten the bolts in sequence as follows:
   a. Step 1: 89 inch lbs. (10 Nm)
   b. Step 2: 16 ft. lbs. (22 Nm)
4. Install or connect the following:
   - Dual converter Y-pipe and tighten the bolts to 30 ft. lbs. (40 Nm)
   - Secondary air injection tube, if equipped
   - Power steering pressure line
   - Oil dipstick tube
   - HO_2S sensor connector
   - Negative battery cable
5. Start the engine and check for leaks.

3.0L (VIN S) Engine

1. Before servicing the vehicle, refer to the precautions at the beginning of this section.
2. Remove or disconnect the following:
   - Negative battery cable
   - Upper intake manifold
   - Right exhaust manifold heat shield
   - Ignition coil
   - Exhaust Gas Recirculation (EGR) tube
   - Heated Oxygen (HO_2S) sensors
   - Secondary air injection tube
   - Dual converter Y-pipe
   - Exhaust manifolds

To install:

3. Install or connect the following:
   - Exhaust manifolds. Use new gaskets and tighten the bolts in

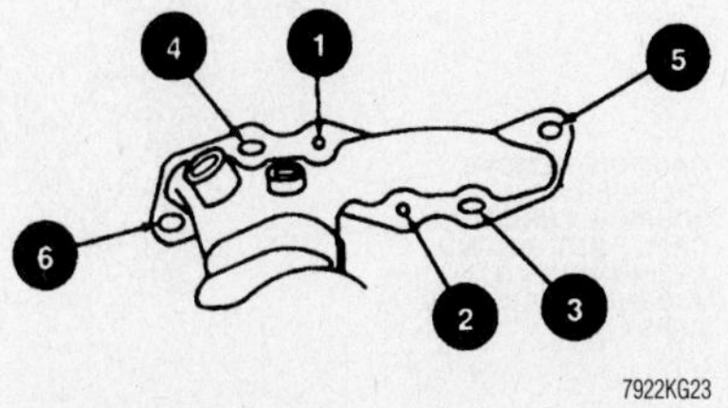

7922KG23

Right side exhaust manifold mounting bolt tightening sequence—3.0L (VIN S) engine

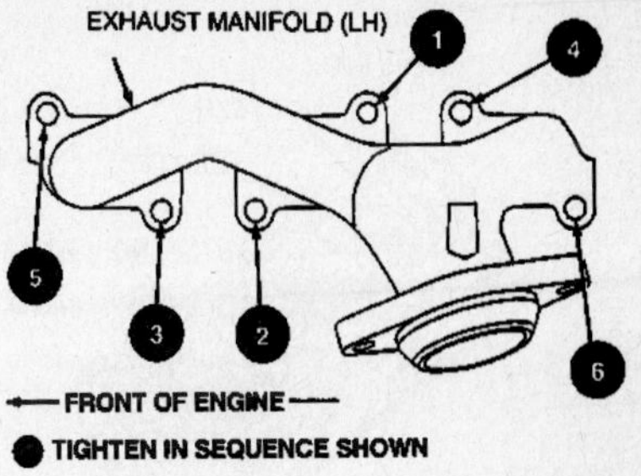

9300KG13

Left side exhaust manifold mounting bolt tightening sequence—3.0L (VIN S) engine

sequence to 13–16 ft. lbs. (18–22 Nm).
   - Dual converter Y-pipe
   - Secondary air injection tube
   - HO_2S sensors
   - EGR tube
   - Ignition coil
   - Right exhaust manifold heat shield
   - Upper intake manifold
   - Negative battery cable
4. Run the engine and check for exhaust leaks and proper operation.

Camshaft and Valve Lifters

REMOVAL & INSTALLATION

3.0L (VIN U) Engine

➡If the rocker arms, pushrods or lifters are to be reused, they must be installed in the same positions they were removed from.

1. Before servicing the vehicle, refer to the precautions at the beginning of this section.
2. Remove the engine from the vehicle and mount it on an engine stand.
3. Remove or disconnect the following:
   - Valve covers
   - Rocker arms and pushrods
   - Accessory drive belt and tensioner
   - Alternator and brackets
   - Intake manifold
   - Valve lifter guide plate
   - Valve lifters
   - Crankshaft damper
   - Oil pan
   - Front cover
   - Timing chain and gears
   - Camshaft thrust plate
   - Camshaft

To install:

➡If replacing the camshaft, the valve lifters must also be replaced.

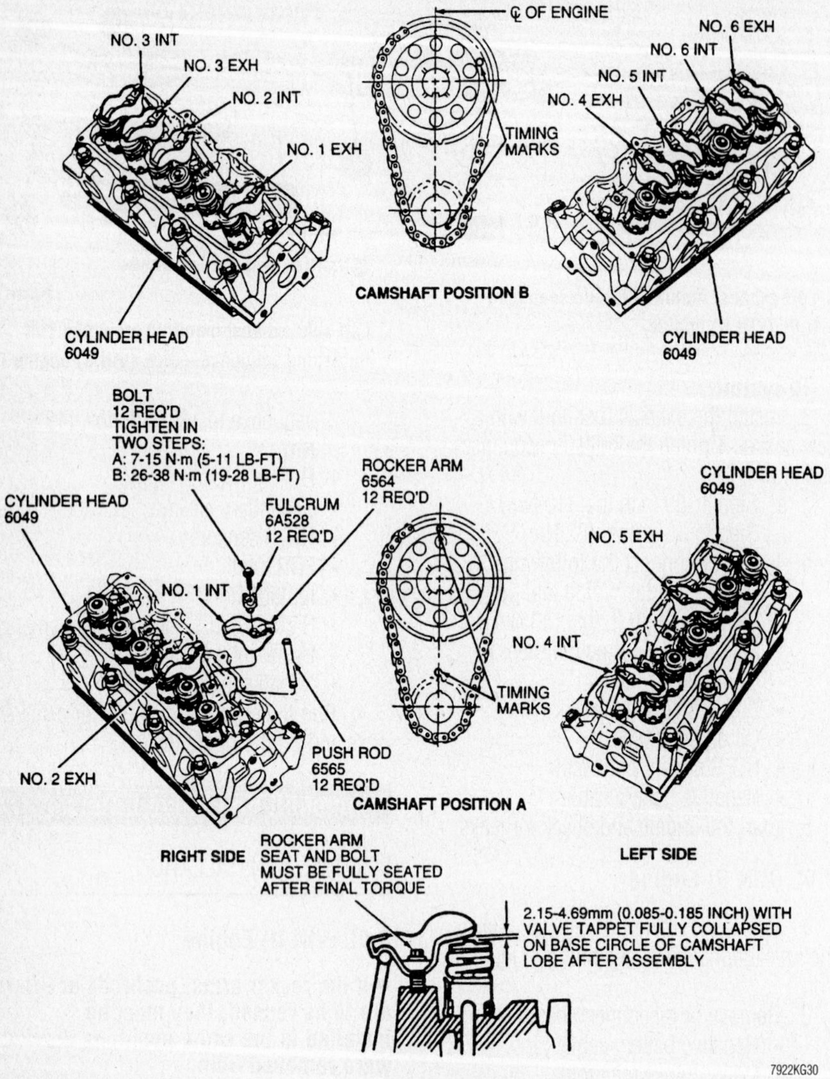

Tighten the rocker arm nuts according to the position of the camshaft—3.0L (VIN U) engine

4. Install or connect the following:
   - Camshaft
   - Camshaft thrust plate and tighten the bolts to 84 inch lbs. (10 Nm)
   - Timing chain and gears. Align the timing marks on the sprockets.
   - Front cover
   - Oil pan
   - Crankshaft damper
   - Valve lifters
   - Valve lifter guide plate
   - Intake manifold
   - Alternator and brackets
   - Accessory drive belt and tensioner
   - Rocker arms and pushrods
   - Valve covers
5. Install the engine assembly into the vehicle.
6. Start the engine and check for leaks and proper engine operation.

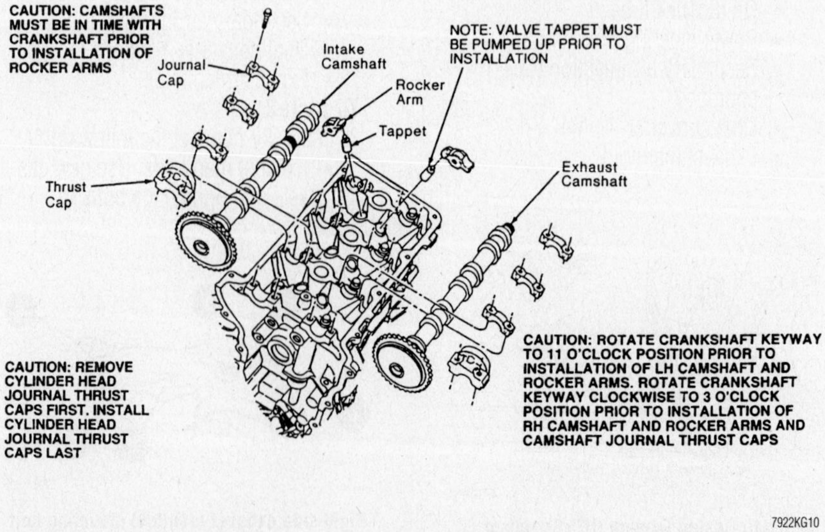

Exploded view of camshaft mounting—3.0L (VIN S) engine

3.0L (VIN S) Engine

1. Before servicing the vehicle, refer to the precautions at the beginning of this section.
2. Remove the engine from the vehicle and mount it on an engine stand.
3. Remove or disconnect the following:
   - Upper intake manifold
   - Valve covers

➡If the rocker arms are to be reused, they must be kept in order so that they can be installed in their original positions.

- Rocker arms
- Accessory drive belt
- Power steering pump
- Water pump
- A/C compressor and bracket

➡The crankshaft accessory drive pulley shaft has left-hand threads. Rotate the pulley shaft clockwise to remove.

- Crankshaft pulley
- Oil pan
- Front cover
- Timing chains

➡The camshaft journal thrust caps must be removed first, before loosening the remaining camshaft journal cap bolts, to ensure that the camshaft journal thrust caps are not damaged.

4. Loosen the camshaft journal cap bolts in sequence and in several passes to allow the camshaft to raise off the cylinder head evenly.
5. Remove the journal caps and the camshafts.

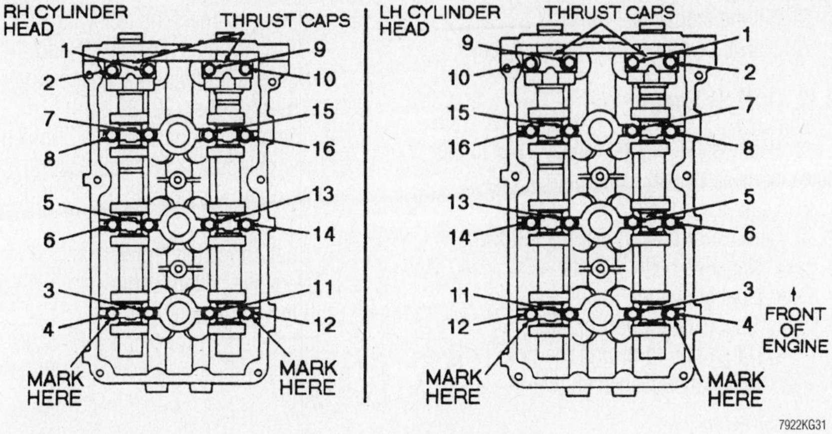

Camshaft journal bolt removal sequence—3.0L (VIN S) engine

➡If the hydraulic lifters are to be reused, they must be kept in order so that they can be installed in their original positions.

6. Remove the hydraulic lifters.

To install:

7. Rotate the crankshaft so the keyway is at the 11 o'clock position for installation of the camshafts.
8. Install or connect the following:
   - Hydraulic lifters in their original positions
   - Camshafts with the timing marks (marked RFF) on the back of the camshaft sprockets aligned

➡The camshaft journal caps and cylinder heads are numbered to ensure that they are assembled in their original positions.

➡Do not install the camshaft journal thrust caps until the rocker arms and timing chains have been installed and the camshaft journal caps are tightened into position.

   - Camshaft journal caps in their original positions
   - Timing chains

9. Tighten the camshaft journal cap bolts, in sequence, to 71–106 inch lbs. (8–12 Nm).
10. Install the thrust caps and tighten the bolts to 71–106 inch lbs. (8–12 Nm).
11. Install or connect the following:
   - Front cover
   - Oil pan
   - Crankshaft pulley
   - A/C compressor and bracket
   - Water pump
   - Power steering pump
   - Accessory drive belt
   - Rocker arms in their original positions
   - Valve covers
   - Upper intake manifold
   - Engine assembly into the vehicle.
12. Run the engine and check for leaks and proper operation.

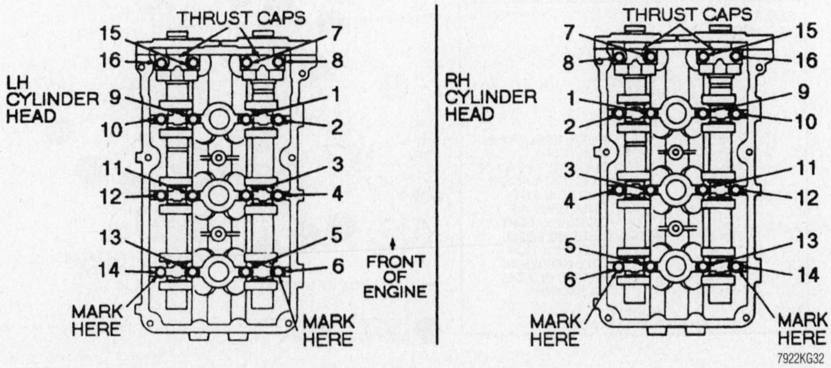

Camshaft journal bolt tightening sequence—3.0L (VIN S) engine

Valve Lash

ADJUSTMENT

3.0L (VIN U and S) Engines

The lash adjusters (valve tappets), are hydraulic and are not adjustable.

Starter Motor

REMOVAL & INSTALLATION

1. Before servicing the vehicle, refer to the precautions at the beginning of this section.
2. Remove or disconnect the following:

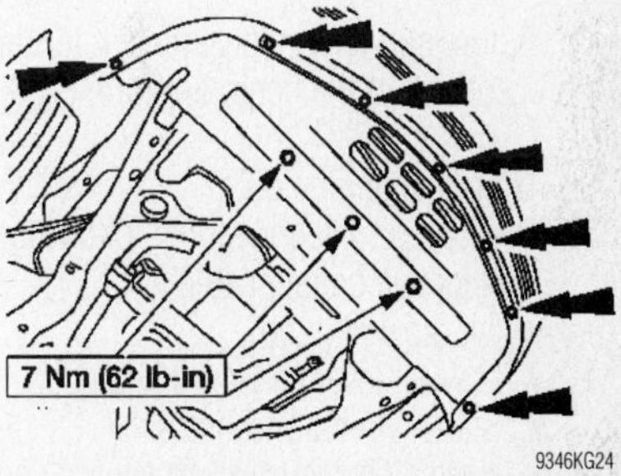

Remove the front splash shield for access

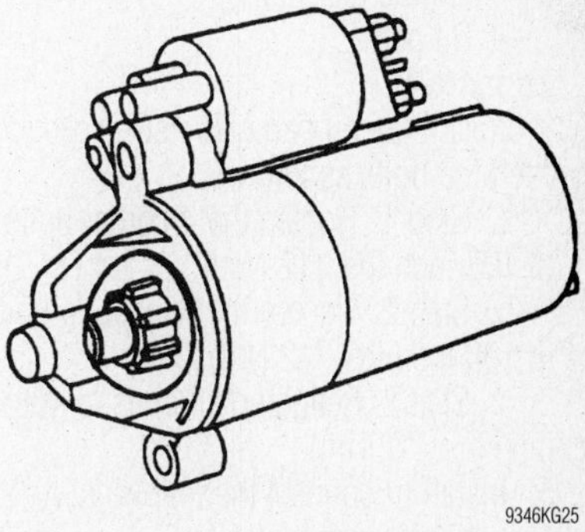

Starter motor—3.0L (VIN U) engine

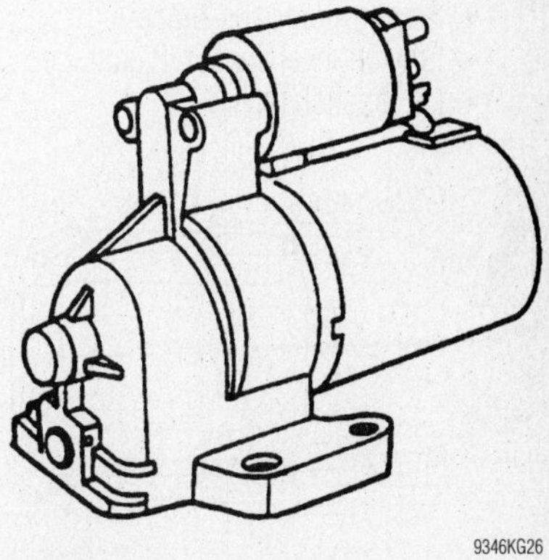

Starter motor—3.0L (VIN S) engine

- Negative battery cable
- Splash shield
- Starter electrical connectors
- Starter

To install:

3. Install or connect the following:
   - Starter and tighten the bolts to 16–21 ft. lbs. (21–29 Nm)
   - Starter electrical connectors and tighten the battery cable nut to 80–123 inch lbs. (9–14 Nm)
   - Splash shield
   - Negative battery cable

Oil Pan

REMOVAL & INSTALLATION

3.0L (VIN U) Engine

1. Before servicing the vehicle, refer to the precautions at the beginning of this section.
2. Drain the engine oil.
3. Remove or disconnect the following:
   - Negative battery cable
   - Oil dipstick tube
   - Low oil level sensor, if equipped
   - Starter motor and brace
   - Heated Oxygen (HO_2S) sensor connectors
   - Dual converter Y-pipe
   - Engine rear plate
   - Oil pan

To install:

4. Install the oil pan with new gaskets. Tighten the bolts as follows:
   a. Step 1: Tighten the 4 corner bolts to 108 inch lbs. (12 Nm).
   b. Step 2: Tighten the remaining bolts to 108 inch lbs. (12 Nm).
   c. Step 3: Retighten all bolts to 108 inch lbs. (12 Nm).
5. Install or connect the following:
   - Engine rear plate
   - Dual converter Y-pipe
   - HO_2S sensor connectors
   - Starter motor and brace
   - Low oil level sensor, if equipped
   - Oil dipstick tube

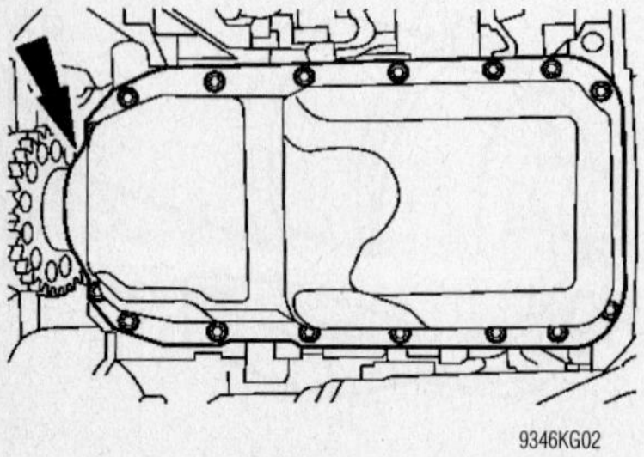
9346KG02

Oil pan—3.0L (VIN U) engine

- Negative battery cable

6. Fill the crankcase.
7. Start the engine and check for leaks.

3.0L (VIN S) Engine

1. Before servicing the vehicle, refer to the precautions at the beginning of this section.
2. Drain the engine oil.
3. Remove or disconnect the following:
   - Negative battery cable
   - Dual converter Y-pipe
   - Transaxle support bracket
   - Oil pan by following the bolt removal sequence shown

To install:

4. Install or connect the following:
   - Oil pan. Use a new gasket and tighten the bolts in sequence to 15–22 ft. lbs. (20–30 Nm).
   - Transaxle support bracket. Tighten the nuts to 71–106 inch lbs. (8–12 Nm) and the bolts to 15–22 ft. lbs. (20–30 Nm).
   - Dual converter Y-pipe
   - Negative battery cable
5. Fill the crankcase.
6. Start the engine and check for leaks and proper operation.

Oil Pan Removal Sequence

Item	Description
1	Upper Cylinder Block
2	Lower Cylinder Block (Part of 6010)
3	Oil Pan
4	Stud Bolt (5 Req'd)
5	Bolt (10 Req'd)
6	Oil Pan Gasket
7	Engine Front Cover

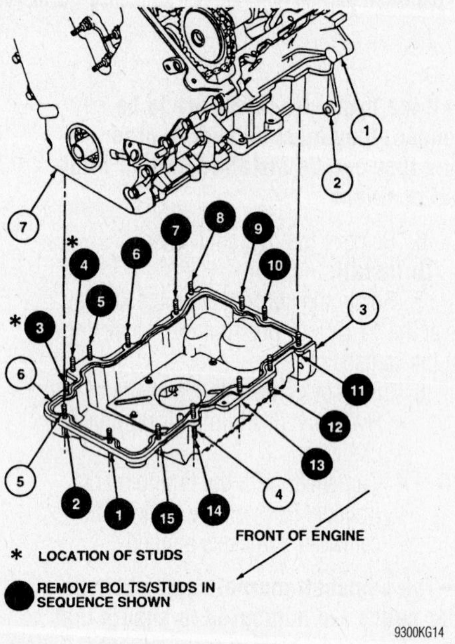
FRONT OF ENGINE
* LOCATION OF STUDS
REMOVE BOLTS/STUDS IN SEQUENCE SHOWN
9300KG14

Oil pan bolt removal sequence—3.0L (VIN S) engine

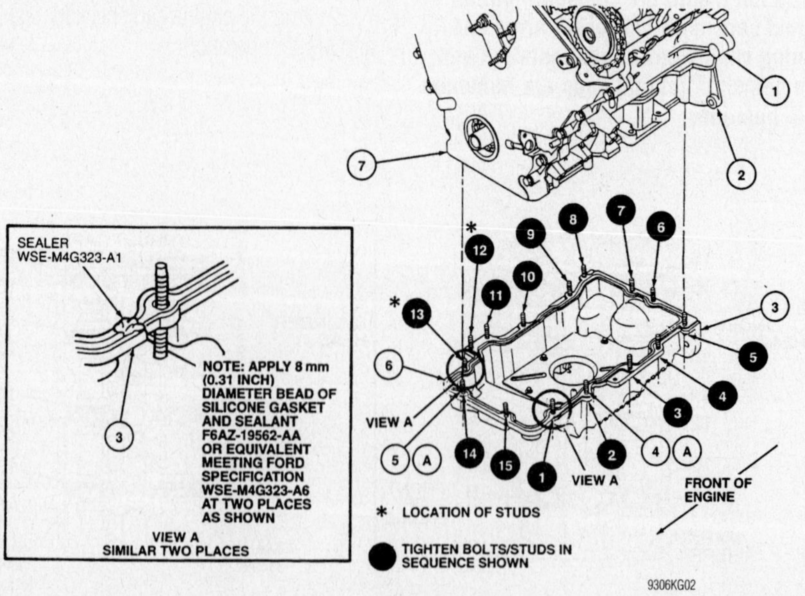
SEALER
WSE-M4G323-A1
NOTE: APPLY 8 mm (0.31 INCH) DIAMETER BEAD OF SILICONE GASKET AND SEALANT F6AZ-19562-AA OR EQUIVALENT MEETING FORD SPECIFICATION WSE-M4G323-A6 AT TWO PLACES AS SHOWN
VIEW A
SIMILAR TWO PLACES
VIEW A
VIEW A
FRONT OF ENGINE
* LOCATION OF STUDS
TIGHTEN BOLTS/STUDS IN SEQUENCE SHOWN
9306KG02

Oil pan bolt tightening sequence—3.0L (VIN S) engine

Oil Pump

REMOVAL & INSTALLATION

3.0L (VIN U) Engine

1. Before servicing the vehicle, refer to the precautions at the beginning of this section.
2. Drain the engine oil.
3. Remove or disconnect the following:
   - Negative battery cable
   - Oil dipstick tube
   - Low oil level sensor, if equipped
   - Starter motor and brace
   - Heated Oxygen Sensor (HO_2S) connectors
   - Dual converter Y-pipe
   - Engine rear plate
   - Oil pan
   - Oil pump
4. Separate the intermediate shaft from the oil pump.

To install:

5. Insert the intermediate shaft into the oil pump until the snapring seats.
6. Install or connect the following:
   - Oil pump and tighten the bolt to 30–40 ft. lbs. (40–55 Nm)
   - Oil pan
   - Engine rear plate
   - Dual converter Y-pipe
   - HO_2S sensor connectors
   - Starter motor and brace
   - Low oil level sensor, if equipped
   - Oil dipstick tube
   - Negative battery cable
7. Fill the crankcase.
8. Start the engine and check for leaks and proper operation.

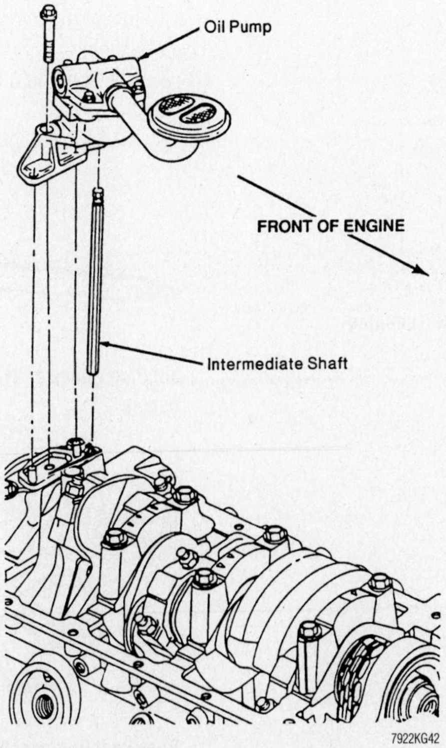

Exploded view of the oil pump mounting—3.0L (VIN U) engine

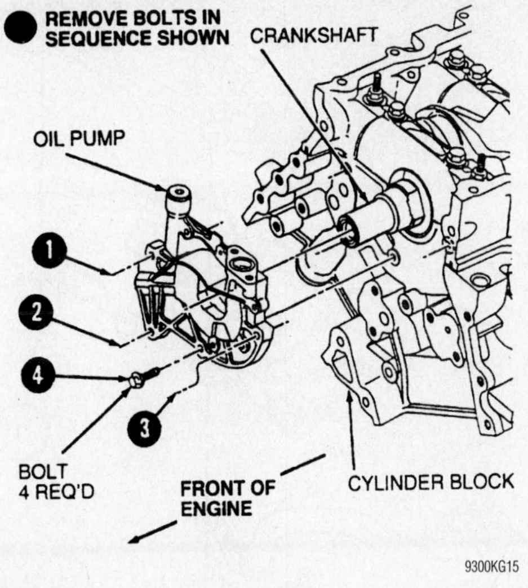

Oil pump bolt removal sequence—3.0L (VIN S) engine

3.0L (VIN S) Engine

1. Before servicing the vehicle, refer to the precautions at the beginning of this section.
2. Remove the engine from the vehicle and mount it on an engine stand.
3. Remove or disconnect the following:
   - Upper intake manifold
   - Valve covers
   - Accessory drive belt
   - Power steering pump
   - Alternator
   - Water pump
   - A/C compressor and bracket

➡The crankshaft accessory drive pulley shaft has left-hand threads. Rotate the pulley shaft clockwise to remove.

   - Crankshaft pulley
   - Oil pan
   - Oil pump screen and tube
   - Front cover
   - Timing chains
   - Crankshaft timing gears
   - Oil pump

To install:

4. Install or connect the following:
   - Oil pump and tighten the bolts in sequence to 71–106 inch lbs. (8–12 Nm)
   - Crankshaft timing gears
   - Timing chains
   - Front cover
   - Oil pump screen and tube. Tighten the bolts to 71–106 inch lbs. (8–12 Nm), and the nut to 15–22 ft. lbs. (20–30 Nm).
   - Oil pan
   - Crankshaft pulley

➡The air conditioning mounting bracket bolts are torque-to-yield bolts and must be replaced when removed.

   - A/C compressor and bracket
   - Water pump

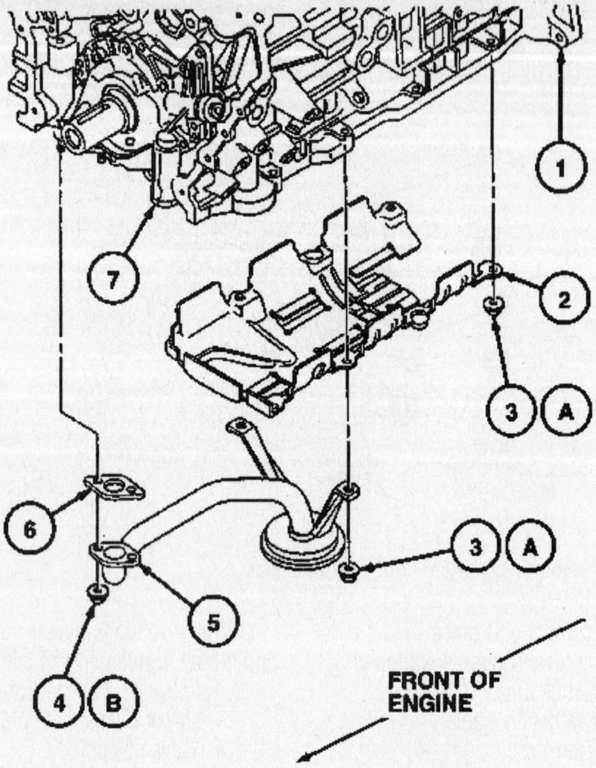

Item	Description
1	Cylinder Block
2	Oil Pan Baffle
3	Nut (5 Req'd)
4	Nut (2 Req'd)
5	Oil Pump Screen Cover and Tube
6	Oil Pump Inlet Tube Gasket
7	Oil Pump
A	Tighten to 14-25 Nm (11-18 Lb-Ft)
B	Tighten to 8-14 Nm (71-123 Lb-In)

9300KG16

Exploded view of the oil pump pickup tube and screen mounting—3.0L (VIN S) engine

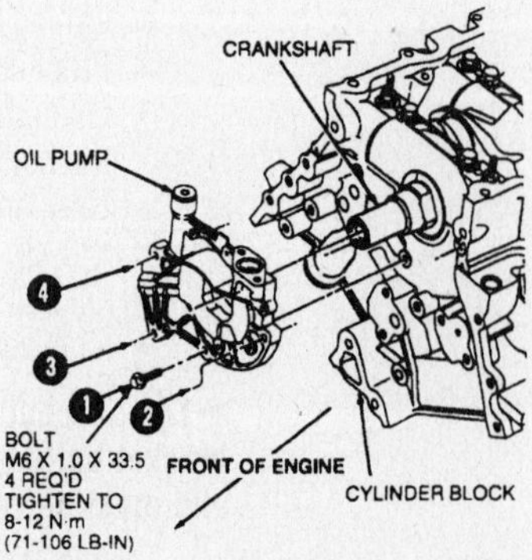

7922KG43

Oil pump bolt tightening sequence—3.0L (VIN S) engine

- Alternator
- Power steering pump
- Accessory drive belt
- Valve covers
- Upper intake manifold
- Engine in the vehicle.

5. Fill the crankcase.
6. Start the engine and check for leaks and proper operation.

Rear Main Seal

REMOVAL & INSTALLATION

3.0L (VIN U) Engine

1. Before servicing the vehicle, refer to the precautions at the beginning of this section.
2. Remove the negative battery cable, transaxle and flexplate.
3. Use a sharp awl and punch a hole into the rear main seal metal surface between the seal lip and the cylinder block.
4. Screw the threaded end of Jet Plug Remover 310-005 into the seal. Use the Jet Plug Remover to remove the rear main seal.

To install:

5. Position the rear oil seal on Rear Seal Replacer 303-323. Position the tool and seal on the rear of the engine. Tighten the bolts alternately to seat the rear main oil seal.
6. Install the flexplate and tighten the bolts to 54–64 ft. lbs. (73–87 Nm).
7. Install the transaxle and the negative battery cable.
8. Check all fluid levels and fill as needed.
9. Start the engine and check for leaks.

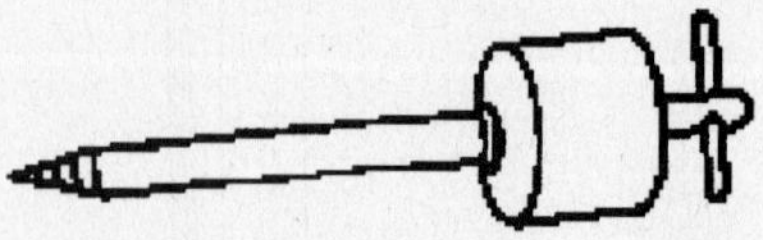

9346KG27

Jet Plug Remover 310-005—3.0L (VIN U) engine

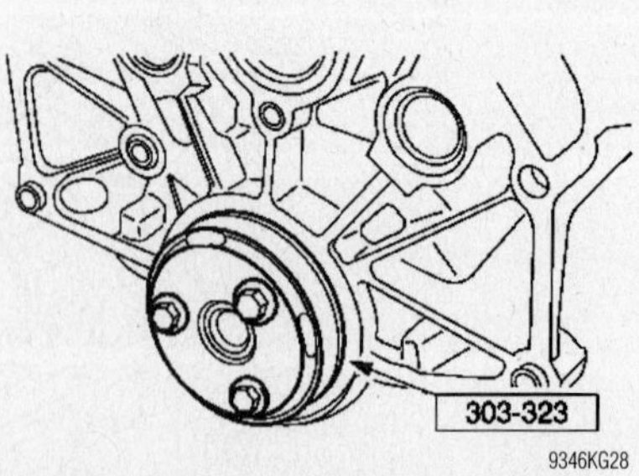

9346KG28

Rear Seal Replacer 303-323—3.0L (VIN U) engine

3.0L (VIN S) Engine

1. Before servicing the vehicle, refer to the precautions at the beginning of this section.
2. Attach an engine support fixture to the engine lifting eyes.
3. Remove or disconnect the following:
   - Negative battery cable
   - Transaxle
   - Flexplate
4. Use Seal Remover T95P-6701-EH and a slide hammer to remove the rear crankshaft seal.

To install:

5. Install or connect the following:
   - Rear main seal flush with the cylinder block surface. Use Rear Main Seal Replacer T82L-6701-A and Adapter Bolts T91P-6701-A.
   - Flexplate and tighten the bolts to 54–64 ft. lbs. (73–87 Nm)
   - Transaxle
   - Negative battery cable
6. Run the engine and check for leaks.

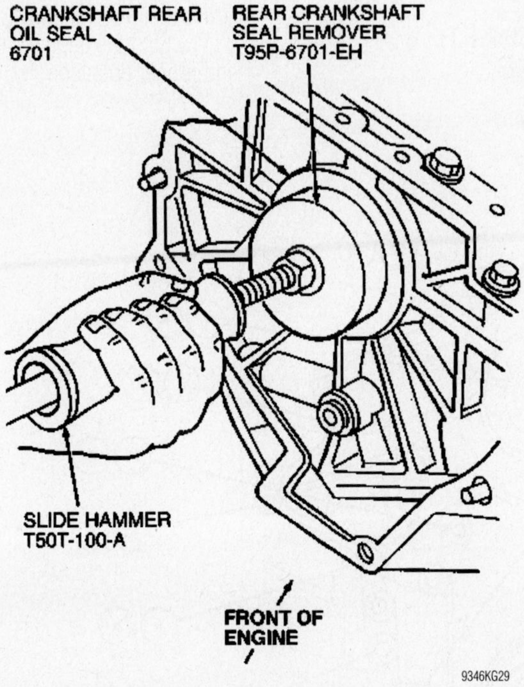

Rear Crankshaft Seal Remover T95P-6701-EH—3.0L (VIN S) engine

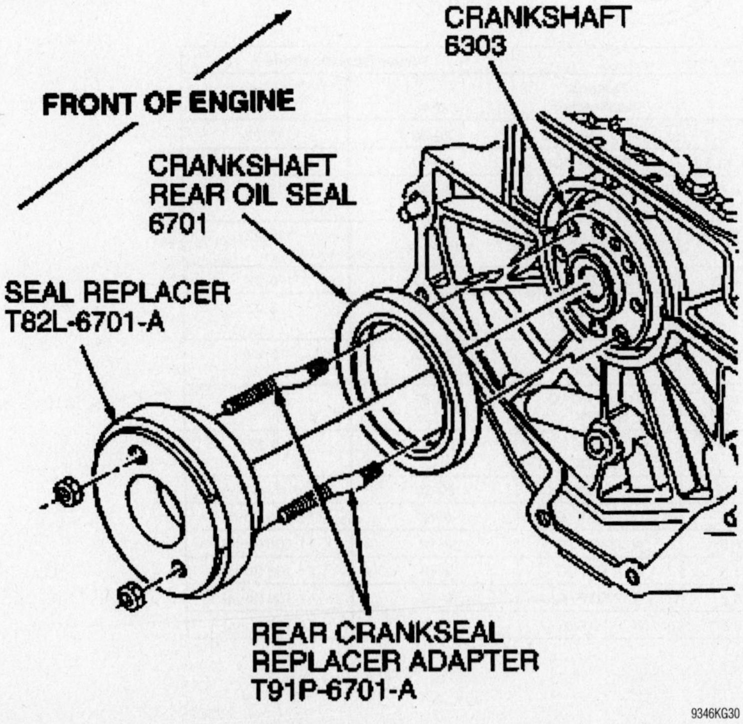

Rear Seal Replacer T82L-6701-A and Adapters T91P-6701-A—3.0L (VIN S) engine

Timing Chain, Sprockets and Front Cover and Seal

REMOVAL & INSTALLATION

3.0L (VIN U) Engine

1. Before servicing the vehicle, refer to the precautions at the beginning of this section.
2. Drain the engine oil and the cooling system.
3. Remove or disconnect the following:
   - Negative battery cable
   - Accessory drive belt
   - Accessory drive belt tensioner and idler pulley
   - Water pump pulley
   - Crankshaft pulley
   - Crankshaft Position (CKP) sensor
   - Coolant hoses
   - Dual converter Y-pipe
   - Oil pan
   - Front crankshaft seal
   - Front cover
4. Rotate the crankshaft until the No. 1 piston is at Top Dead Center (TDC) and the timing sprocket marks are aligned.
5. Remove the timing chain and sprockets.

To install:

➡The camshaft bolt has a drilled oil passage in it for timing chain lubrication. Prior to installation, clean the passage and be sure it is clear. Never replace the camshaft bolt with a standard bolt.

6. Install the timing chain and sprockets with the timing marks aligned. Tighten the camshaft sprocket bolt to 46 ft. lbs. (63 Nm).
7. Use a new gasket and install the front cover. Apply sealant to bolts 1, 2 and 3. Tighten bolts 1–10 to 19 ft. lbs. (25 Nm) and bolts 11–15 to 84 inch lbs. (10 Nm).
8. Install or connect the following:
   - Front crankshaft seal
   - Oil pan
   - Dual converter Y-pipe
   - Coolant hoses
   - CKP sensor

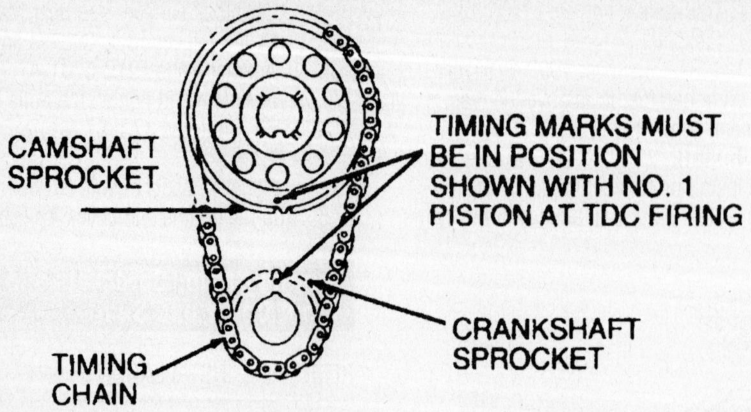

Timing chain alignment marks—3.0L (VIN U) engine

- Crankshaft pulley and tighten the damper bolt to 107 ft. lbs. (145 Nm)
- Water pump pulley and tighten the bolts to 16 ft. lbs. (21 Nm)
- Accessory drive belt tensioner and idler pulley
- Accessory drive belt
- Negative battery cable

9. Fill the cooling system.
10. Fill the crankcase.
11. Start the engine and check for leaks and proper operation.

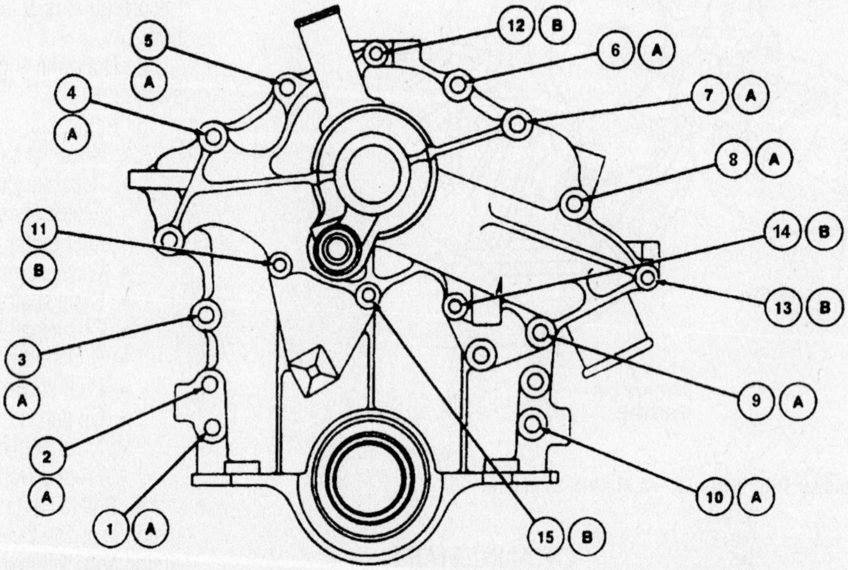

Fastener And Hole No.	Fasteners		Torque Specifications	
	Size	Fastener Application	N·m	LB-FT
1A	M8 x 1.25 x 43.5	F/C TO BLOCK	20-30	15-22
2A	M8 x 1.25 x 43.5	F/C TO BLOCK	20-30	15-22
3A	M8 x 1.25 x 73	W/P & F/C TO BLOCK	20-30	15-22
4A	M8 x 1.25 x 104.3	W/P & F/C TO BLOCK	20-30	15-22
5A	M8 x 1.25 x 73	F/C TO BLOCK	20-30	15-22
6A	M8 x 1.25 x 73	W/P & F/C TO BLOCK	20-30	15-22
7A	M8 x 1.25 x 73	W/P & F/C TO BLOCK	20-30	15-22
8A	M8 x 1.25 x 104.3	W/P & F/C TO BLOCK	20-30	15-22
9A	M8 x 1.25 x 104.3	W/P & F/C TO BLOCK	20-30	15-22
10A	M8 x 1.25 x 52	F/C TO BLOCK	20-30	15-22
11B	M6 x 1 x 28.5	W/P TO F/C	8-12	71-106 (lb-in)
12B	M6 x 1 x 28.5	W/P TO F/C	8-12	71-106 (lb-in)
13B	M6 x 1 x 28.5	W/P TO F/C	8-12	71-106 (lb-in)
14B	M6 x 1 x 28.5	W/P TO F/C	8-12	71-106 (lb-in)
15B	M6 x 1 x 28.5	W/P TO F/C	8-12	71-106 (lb-in)

W/P—Water Pump
F/C—Engine Front Cover

7922KG48

Timing chain front cover bolt location and identification—3.0L (VIN U) engine

3.0L (VIN S) Engine

1. Before servicing the vehicle, refer to the precautions at the beginning of this section.
2. Remove the engine from the vehicle and mount it on an engine stand.
3. Remove or disconnect the following:
   - Upper intake manifold
   - Valve covers

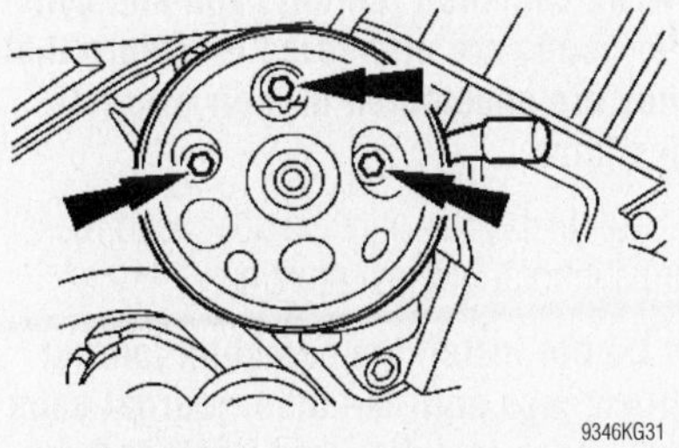

9346KG31

Access the power steering pump bolts through the holes in the pulley—3.0L (VIN S) engine

   - Accessory drive belt
   - Power steering pump
   - Alternator
   - Water pump
   - A/C compressor and bracket

➡The crankshaft accessory drive pulley shaft has left-hand threads. Rotate the pulley shaft clockwise to remove.

   - Crankshaft pulley
   - Crankshaft damper
   - Oil pan
   - Oil pump screen and tube
   - Crankshaft Position (CKP) sensor connector
   - Camshaft Position (CMP) sensor connector
   - Front cover
   - CKP sensor pulse ring. Note the keyway alignment for installation.

➡The Crankshaft Position Sensor (CKP) pulse ring is used on several different engines. Note the keyway alignment for installation.

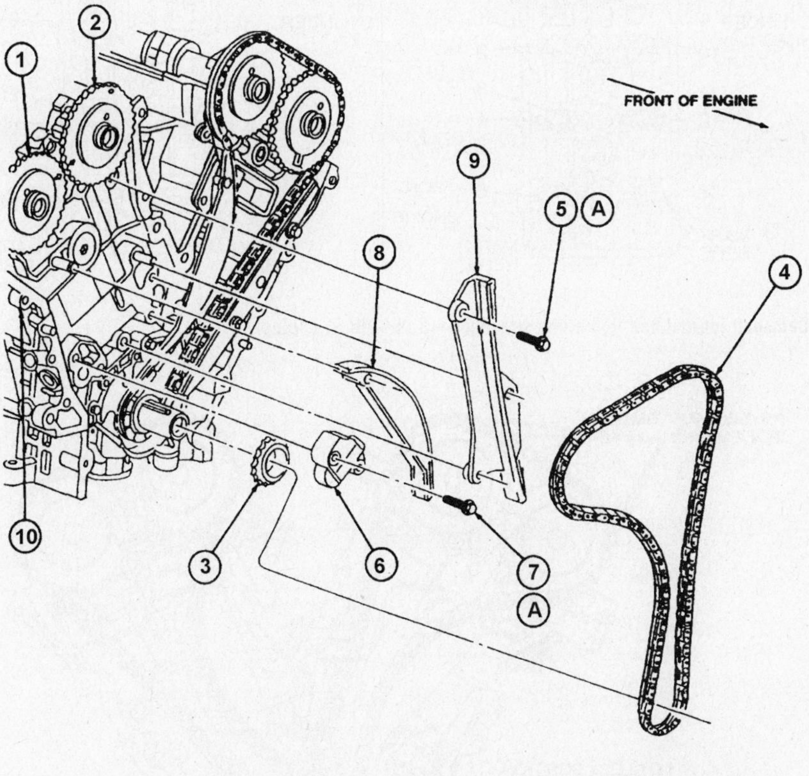

1 RH Exhaust Camshaft
2 RH Intake Camshaft
3 RH Timing Chain Crankshaft Sprocket
4 RH Timing Chain
5 Bolt (2 Req'd)
6 Timing Chain Tensioner
7 Bolt (2 Req'd)
8 Timing Chain Tensioner Arm
9 Timing Chain Guide
10 RH Cylinder Head
A Tighten to 20-30 N-m (15-22 Lb-Ft)

7922KG49

Exploded view of the right cylinder head timing chain and related components—3.0L (VIN S) engine—left side similar

4. Rotate the crankshaft so that the keyway is at the 11 o'clock position to locate the crankshaft at Top Dead Center (TDC) for No. 1 cylinder.
5. Verify that the alignment arrows on the camshafts are aligned. If not, rotate the crankshaft 1 complete revolution and recheck.
6. Rotate the crankshaft so that the keyway is at the 3 o'clock position. This positions the right cylinder head camshafts to the neutral position.
7. Remove or disconnect the following:
   - Right timing chain tensioner

➡The camshaft thrust caps must be removed before loosening the remaining camshaft journal cap bolts to ensure that the thrust caps are not damaged.

   - Camshaft thrust caps
   - Camshaft journal caps. Loosen the bolts in sequence and in several passes to allow the camshaft to be raised from the cylinder head evenly.
   - Rocker arms. Keep the rocker arms in order for installation.
   - Right timing chain tensioner arm
   - Right timing chain and crankshaft sprocket
8. Rotate the crankshaft 2 revolutions and locate the crankshaft keyway at the 11 o'clock position. This will position the left cylinder head camshafts to their neutral position.
9. Verify that the alignment arrows on the camshafts are aligned.
10. Remove the left cylinder head timing chain tensioner retaining bolts and the timing chain tensioner.

➡The camshaft thrust caps must be removed before loosening the remaining camshaft journal cap bolts to ensure that the thrust caps are not damaged.

11. Remove or disconnect the following:
   - Camshaft thrust caps
   - Camshaft journal caps. Loosen the bolts in sequence and in several passes to allow the camshaft to be raised from the cylinder head evenly.
   - Rocker arms. Keep the rocker arms in order for installation.
   - Left timing chain tensioner arm
   - Left timing chain and crankshaft sprocket

To install:

12. Prepare the timing chain tensioners for installation as follows:

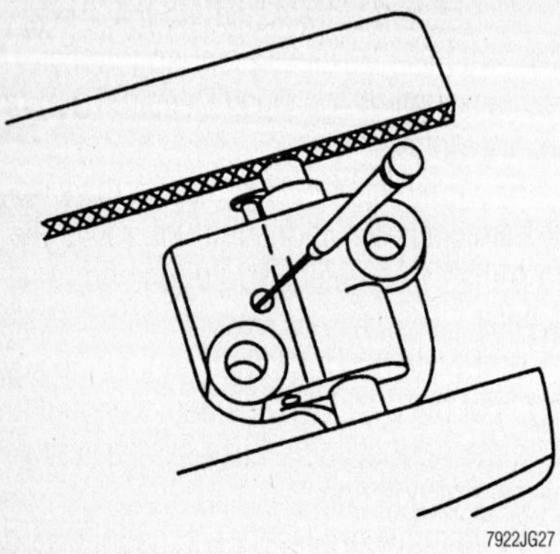

Using a thin prytool, release and hold the timing chain tensioner ratchet/pawl mechanism—3.0L (VIN S) engine

a. Place the left chain tensioner in a vise.

b. Using a small prytool, release and hold the timing chain tensioner ratchet/pawl mechanism through the access hole in the timing chain tensioner.

c. Slowly compress the tensioner.

d. Lock the piston with a 1.5mm wire or paperclip.

e. Repeat for the right chain tensioner.

➡Be sure that the crankshaft keyway is still at the 11 o'clock position.

13. Install or connect the following:

- Left timing chain and crankshaft sprocket. Align the colored links with the index marks on the camshaft and crankshaft sprockets.
- Left timing chain tensioner arm
- Left timing chain tensioner and tighten the retaining bolts to 15–22 ft. lbs. (20–30 Nm)
- Right timing chain and crankshaft sprocket. Align the colored links with the index marks on the camshaft and crankshaft sprockets.

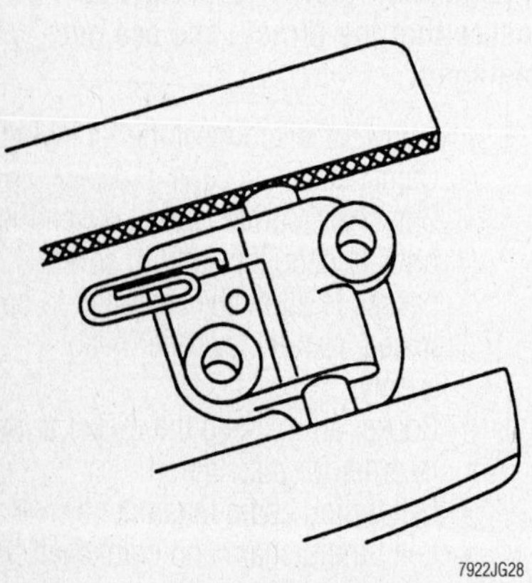

Retain the piston with a 1.5mm wire or paperclip—3.0L (VIN S) engine

- Right timing chain tensioner arm
- Right timing chain tensioner and tighten the retaining bolts to 15–22 ft. lbs. (20–30 Nm)

➡The crankshaft keyway must be in the 11 o'clock position to install the left cylinder head rocker arms.

➡The camshaft journal caps and cylinder heads are numbered to ensure that they are assembled in their original positions.

14. Install the left cylinder head rocker arms in their original positions.

➡Do not install the camshaft journal thrust caps until the other journal caps have been installed and tightened.

15. Tighten the left camshaft journal caps in the order shown and in several passes to 71–106 inch lbs. (8–12 Nm).

16. Install the left camshaft journal thrust caps and tighten the bolts to 71–106 inch lbs. (8–12 Nm).

17. Remove the retaining wire from the left timing chain tensioner.

➡The crankshaft keyway must be in the 3 o'clock position to install the right cylinder head rocker arms.

18. Rotate the crankshaft so that the keyway is in the 3 o'clock position.

➡The camshaft journal caps and cylinder heads are numbered to ensure that they are assembled in their original positions.

19. Install the right cylinder head rocker arms in their original positions.

➡Do not install the camshaft journal thrust caps until the other journal caps have been installed and tightened.

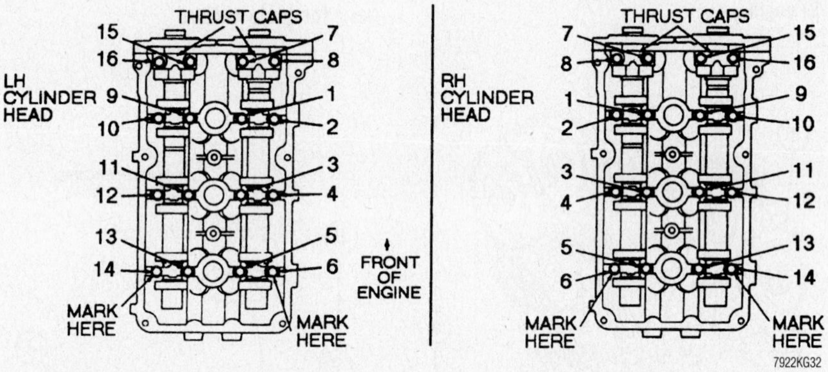

Camshaft journal bolt tightening sequence—3.0L (VIN S) engine

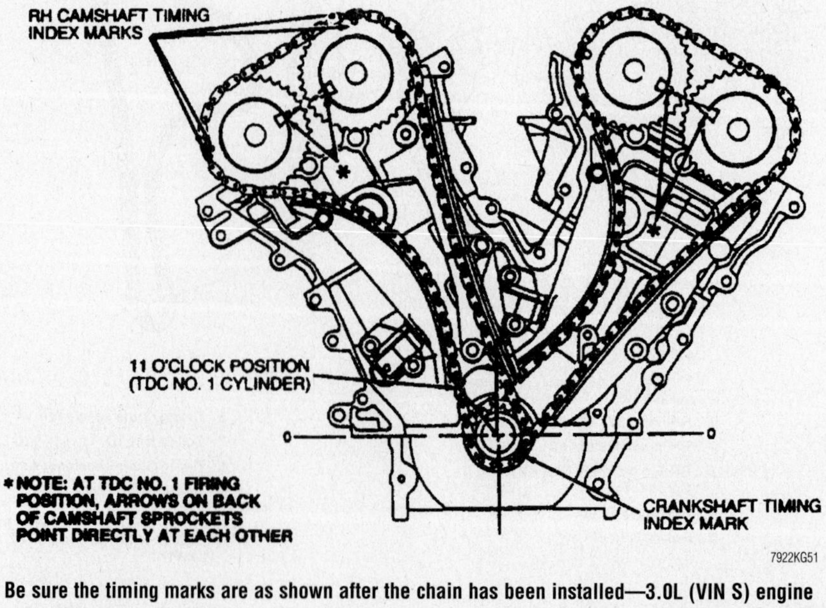

Be sure the timing marks are as shown after the chain has been installed—3.0L (VIN S) engine

20. Tighten the right camshaft journal caps in the order shown and in several passes to 71–106 inch lbs. (8–12 Nm).

21. Install the right camshaft journal thrust caps and tighten the bolts to 71–106 inch lbs. (8–12 Nm).

22. Remove the retaining wire from the right timing chain tensioner.

➡The Crankshaft Position Sensor (CKP) pulse ring is used on several different engines. The keyway position for the 3.0L (VIN S) engine will be marked 30, 30RFF or with an ORANGE stripe.

23. Install the CKP sensor pulse ring. Align the keyway with the proper slot.

24. Replace the crankshaft seal in the front cover with a new one. Apply clean engine oil to the seal lip.

25. Apply silicone sealer to the 6 critical areas shown in View **A**, to the cylinder block to prevent oil seepage.

26. Place new front cover gaskets onto the dowel pins on the cylinder block and heads.

27. Place the front cover into position.

28. Install the 6 front cover retaining bolts and stud bolts where the silicone sealer was applied.

29. Tighten the bolts and stud bolts until the front cover contacts the cylinder block and heads, then turn the bolts and stud bolts an additional ¼ turn.

30. Install the remaining front cover retaining bolts and stud bolts.

31. Tighten all of the front cover retaining bolts and stud bolts in sequence to 15–22 ft. lbs. (20–30 Nm).

➡The air conditioning mounting bracket bolts are torque-to-yield bolts and must be replaced.

32. Install or connect the following:
- CMP sensor connector
- CKP sensor connector
- Oil pump screen and tube
- Oil pan

33. Install the crankshaft damper and tighten the bolt as follows:

a. Step 1: Tighten the bolt to 78–99 ft. lbs. (105–135 Nm).

b. Step 2: Loosen the bolt one full turn.

c. Step 3: Tighten the bolt to 35–39 ft. lbs. (47–53 Nm).

d. Step 4: Tighten the bolt 85–95 degrees.

34. Install or connect the following:
- Crankshaft pulley and tighten coun-

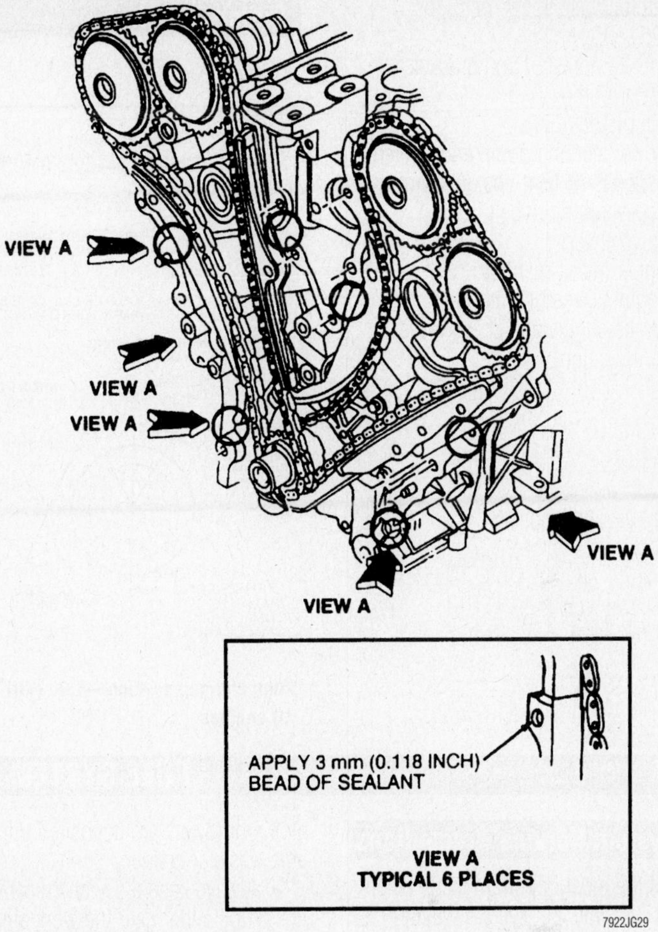

To prevent oil leakage, apply sealant to the places indicated—3.0L (VIN S) engine

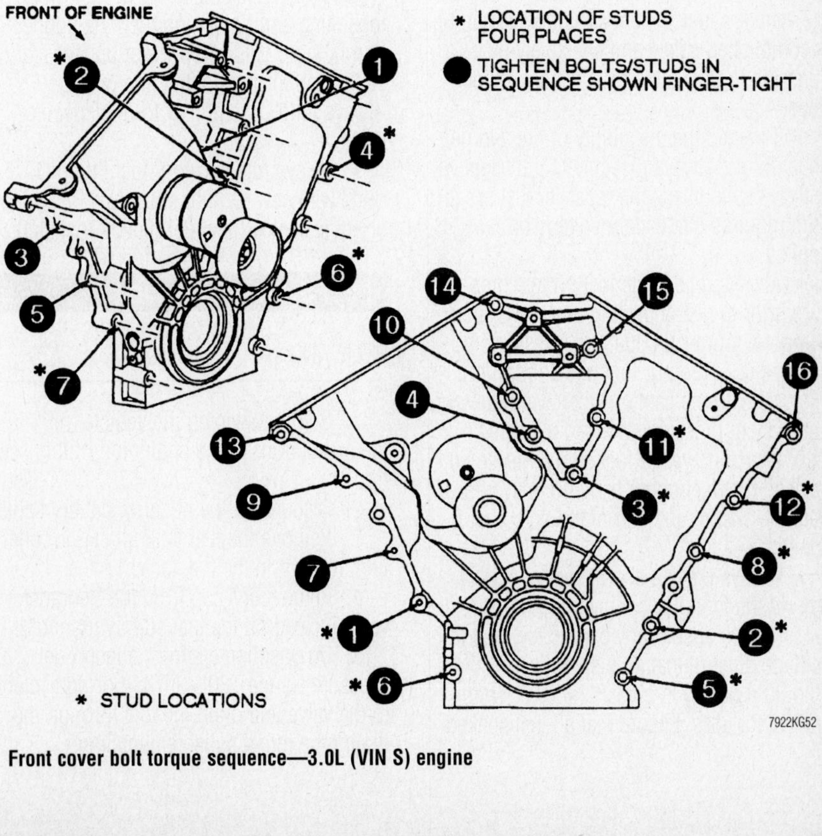

Front cover bolt torque sequence—3.0L (VIN S) engine

terclockwise to 70–77 ft. lbs. (95–105 Nm)

- A/C compressor and bracket
- Water pump
- Alternator
- Power steering pump and tighten the bolts to 18 ft. lbs. (25 Nm)
- Accessory drive belt
- Valve covers
- Upper intake manifold
- Engine assembly into the vehicle

35. Run the engine and check for leaks and proper operation.

Piston and Ring

POSITIONING

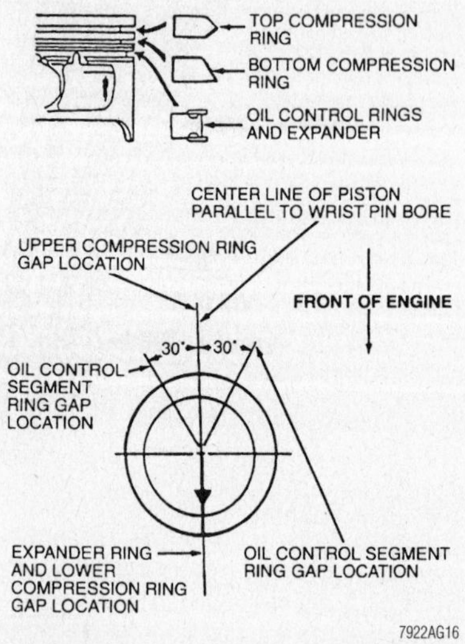

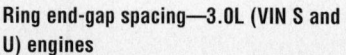

Ring end-gap spacing—3.0L (VIN S and U) engines

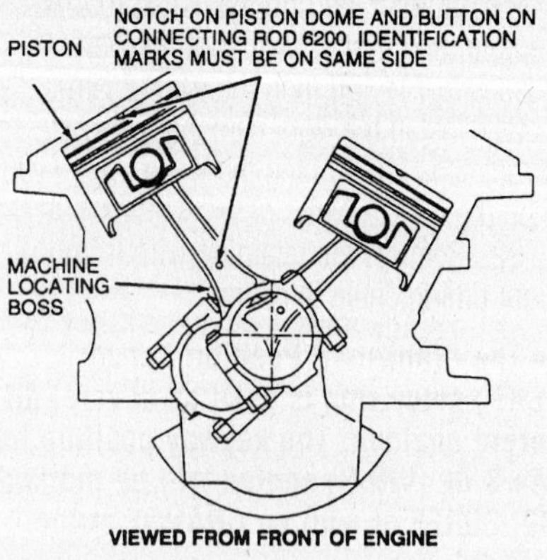

Piston and connecting rod positioning—3.0L (VIN S and U) engines

FUEL SYSTEM

Fuel System Service Precautions

Safety is the most important factor when performing not only fuel system maintenance but any type of maintenance. Failure to conduct maintenance and repairs in a safe manner may result in serious personal injury or death. Maintenance and testing of the vehicle's fuel system components can be accomplished safely and effectively by adhering to the following rules and guidelines.

- To avoid the possibility of fire and personal injury, always disconnect the negative battery cable unless the repair or test procedure requires that battery voltage be applied.
- Always relieve the fuel system pressure prior to disconnecting any fuel system component (injector, fuel rail, pressure regulator, etc.), fitting or fuel line connection. Exercise extreme caution whenever relieving fuel system pressure, to avoid exposing skin, face and eyes to fuel spray. Please be advised that fuel under pressure may penetrate the skin or any part of the body that it contacts.
- Always place a shop towel or cloth around the fitting or connection prior to loosening to absorb any excess fuel due to spillage. Ensure that all fuel spillage (should it occur) is quickly removed from engine surfaces. Ensure that all fuel soaked cloths or towels are deposited into a suitable waste container.
- Always keep a dry chemical (Class B) fire extinguisher near the work area.
- Do not allow fuel spray or fuel vapors to come into contact with a spark or open flame.
- Always use a back-up wrench when loosening and tightening fuel line connection fittings. This will prevent unnecessary stress and torsion to fuel line piping. Always follow the proper torque specifications.
- Always replace worn fuel fitting O-rings with new. Do not substitute fuel hose or equivalent, where fuel pipe is installed.

Fuel System Pressure

RELIEVING

1. Before servicing the vehicle, refer to the precautions at the beginning of this section.
2. Disconnect the negative battery cable.
3. Remove the fuel tank fill cap to relieve the pressure in the fuel tank.
4. Remove the cap from the Schrader valve located on the fuel supply manifold.
5. On gasoline engines, attach Fuel Pressure Gauge T80L-9974-A or equivalent, to the valve and drain the fuel through the drain tube into a suitable container.
6. On flex-fuel engines, connect Fuel Pressure Gauge T80L-9974-A or equivalent and Fuel Pressure Test Kit 134-R0035 or equivalent, to the Schrader valve. Drain the fuel through the drain tube into a suitable container.
7. After the fuel system pressure is relieved, remove the fuel pressure gauge and install the cap on the Schrader valve.
8. Install the fuel tank fill cap.
9. Connect the negative battery cable after system repairs are completed.

Fuel Filter

REMOVAL & INSTALLATION

1. Before servicing the vehicle, refer to the precautions at the beginning of this section.
2. Disconnect the negative battery cable.
3. Relieve the fuel system pressure.
4. Disconnect the fuel lines.
5. Loosen the filter retaining clamp and remove the fuel filter.

To install:

6. Install the fuel filter with the flow arrow facing the proper direction and tighten the filter retaining clamp.
7. Push the fuel lines on to the filter fittings until an audible click is heard.
8. Connect the negative battery cable.

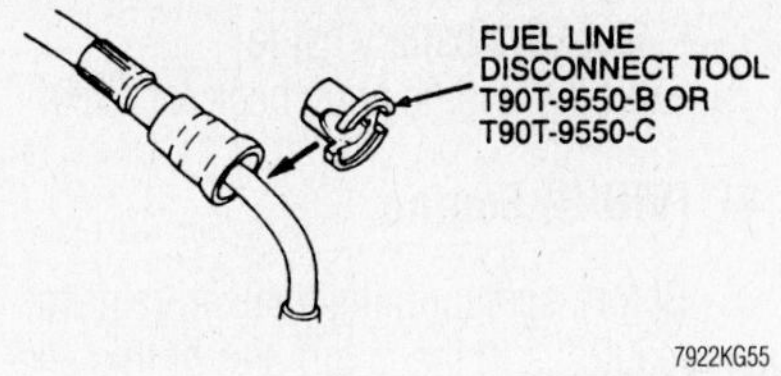

Push connect fitting and removal tool

9. Start the engine and check for fuel leaks and proper operation.

Fuel Pump

REMOVAL & INSTALLATION

1. Before servicing the vehicle, refer to the precautions at the beginning of this section.
2. Relieve the fuel system pressure.
3. Drain the fuel tank.
4. Remove or disconnect the following:
   - Negative battery cable
   - Fuel lines
   - Fuel pump module electrical connector
   - Fuel tank
   - Fuel pump module

To install:

5. Install the fuel pump module carefully to ensure the filter and hoses and float rod are not damaged. Use a new O-ring seal.
6. Align the fuel pump module and the fuel tank retainer and push the fuel pump module into the fuel tank retainer. When the fuel pump module is properly engaged, a definite click will be heard engaging 2 locking tabs on the outside of the fuel pump.
7. Install or connect the following:
   - Fuel pump module locking ring
   - Fuel tank
   - Fuel pump module electrical connector
   - Fuel lines
   - Negative battery cable
8. Add a minimum of 10 gallons of clean fuel to the tank.
9. Start the engine and check for leaks.

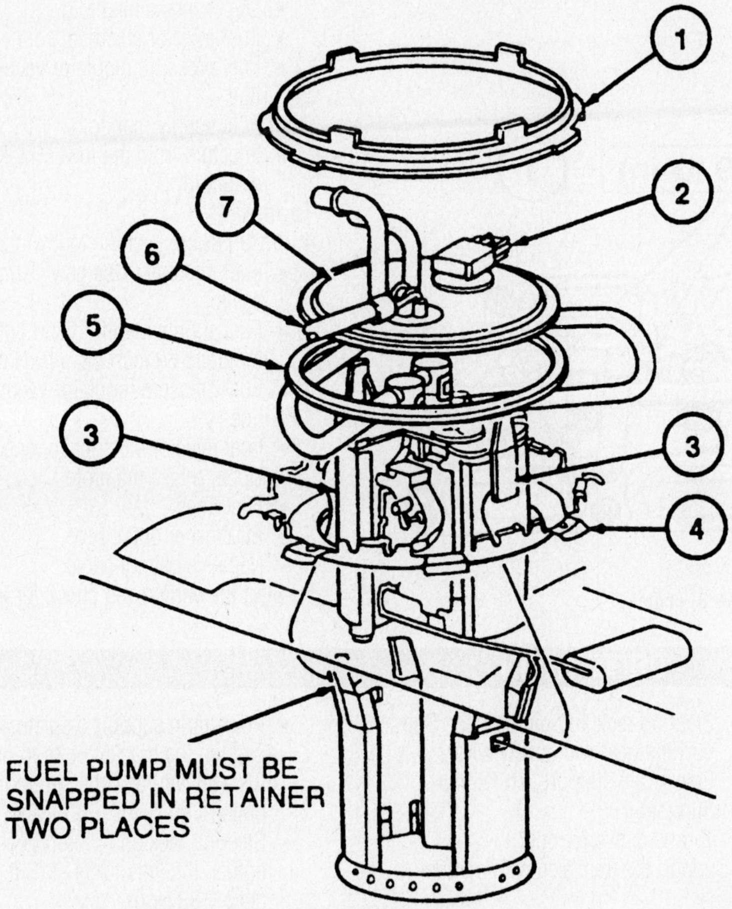

1 Fuel Pump Locking Retainer Ring
2 Fuel Tank Pressure / Vacuum Transducer
3 Locking Tab
4 Fuel Tank
5 O-Ring Seal
6 Connector
7 Fuel Pump Module

7922KG58

Exploded view of the fuel pump module mounting

Fuel Injector

REMOVAL AND INSTALLATION

3.0L (VIN U) Engine

1. Before servicing the vehicle, refer to the precautions in the beginning of this section.
2. Relieve fuel system pressure.
3. Remove or disconnect the following:
   - Negative battery cable
   - Air cleaner outlet tube
   - Fuel lines
   - Upper intake manifold
   - Fuel injector electrical connectors
   - Fuel pressure regulator vacuum line
   - Fuel supply manifold with the injectors attached
   - Injectors from the supply manifold

To install:

4. Install or connect the following:
   - Fuel injectors. Use new O-ring seals.
   - Fuel supply manifold with the injectors attached and tighten the bolts to 89 inch lbs. (10 Nm)
   - Fuel pressure regulator vacuum line
   - Fuel injector electrical connectors
   - Upper intake manifold
   - Fuel lines

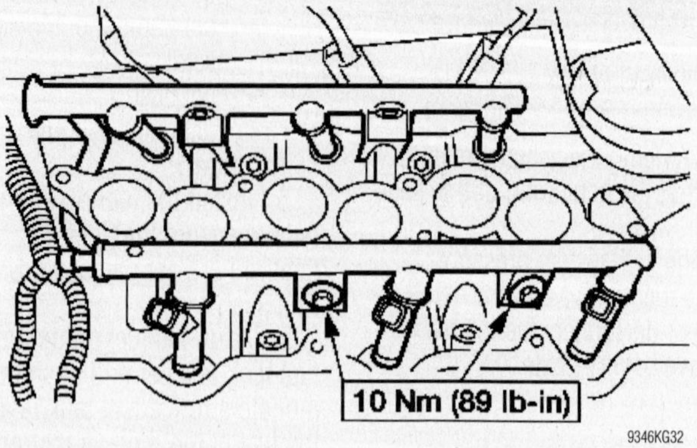

Fuel supply manifold—3.0L (VIN U) engine

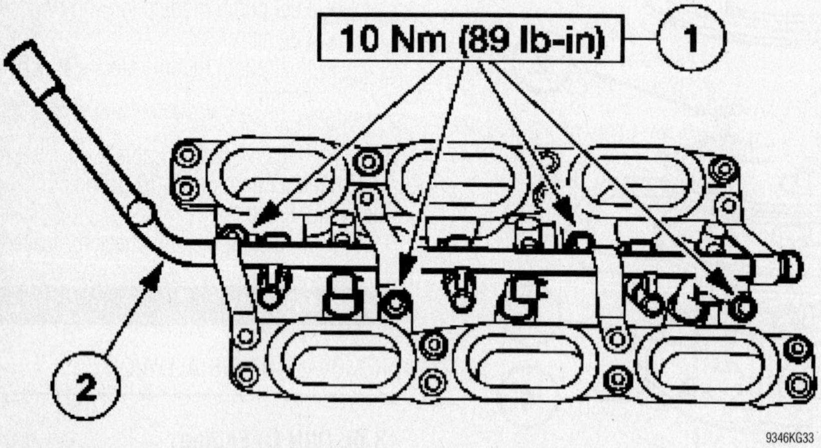

Retaining bolts (1) and Fuel supply Manifold (2)—3.0L (VIN S) engine

- Air cleaner outlet tube
- Negative battery cable

5. Start the engine and check for leaks.

3.0L (VIN S) Engine

1. Before servicing the vehicle, refer to the precautions in the beginning of this section.
2. Relieve fuel system pressure.
3. Remove or disconnect the following:
   - Negative battery cable
   - Air cleaner outlet tube
   - Fuel lines
   - Upper intake manifold
   - Fuel injector electrical connectors
   - Fuel pressure regulator vacuum line
   - Fuel supply manifold
   - Injectors from the lower intake manifold

To install:

4. Install or connect the following:
   - Fuel injectors. Use new O-ring seals.
   - Fuel supply manifold and tighten the bolts 89 inch lbs. (10 Nm)
   - Fuel pressure regulator vacuum line
   - Fuel injector electrical connectors
   - Upper intake manifold
   - Fuel lines
   - Air cleaner outlet tube
   - Negative battery cable
5. Start the engine and check for leaks.

DRIVE TRAIN

Transaxle

REMOVAL & INSTALLATION

1. Before servicing the vehicle, refer to the precautions at the beginning of this section.
2. Attach a powertrain support to the engine lifting eyes.
3. Remove or disconnect the following:
   - Battery and tray
   - Air cleaner assembly
   - Transaxle electrical connectors
   - Shift cable
   - Transaxle cooler lines
   - Front wheels
   - Stabilizer bar links
   - Lower ball joints
   - Axle halfshafts
   - Heated Oxygen Sensor (HO_2S) connectors
   - Dual converter Y-pipe
   - Starter
   - Steering rack and pinion gear. Support the gear with safety wire.
   - Powertrain support insulators
   - Subframe
   - Torque converter nuts
4. Support the transaxle with a transmission jack.
5. Remove the transaxle flange bolts and remove the transaxle.

To install:

6. Install the transaxle. For 3.0L (VIN S) engines, tighten the flange bolts to 25–33 ft. lbs. (33–46 Nm). For all other engines, tighten the flange bolts to 30–44 ft. lbs. (40–60 Nm).
7. Tighten the torque converter nuts to 20–34 ft. lbs. (27–46 Nm).
8. Install the subframe. Use 2 pieces of ¾ inch outside diameter pipe in the alignment holes behind the front subframe mounts to align the subframe to the body. Tighten the subframe mounting bolts to 57–76 ft. lbs. (77–103 Nm).
9. Install or connect the following:
   - Powertrain support insulators. Tighten the bracket bolts to 65 ft. lbs. (88 Nm) and the subframe bolts to 90 ft. lbs. (122 Nm).
   - Steering rack and pinion gear and tighten the nuts to 84–113 ft. lbs. (113–133 Nm)
   - Starter and tighten the fasteners to 21 ft. lbs. (29 Nm)
   - Dual converter Y-pipe
   - HO_2S sensor connectors
   - Axle halfshafts and tighten the hub retainer nuts to 170–202 ft. lbs. (230–275 Nm)
   - Lower ball joints and tighten the nuts to 51–67 ft. lbs. (68–92 Nm)
   - Stabilizer bar links and tighten the nuts to 35–46 ft. lbs. (47–63 Nm)
   - Front wheels
   - Transaxle cooler lines
   - Shift cable
   - Transaxle electrical connectors
   - Air cleaner assembly
   - Battery and tray

10. Start the engine. Check for leaks and proper operation.

➡**Whenever the vehicle subframe is removed or lowered, the wheel alignment should be checked.**

Halfshaft

REMOVAL & INSTALLATION

1. Before servicing the vehicle, refer to the precautions at the beginning of this section.
2. Remove or disconnect the following:
   - Front wheels
   - Lower ball joints
   - Outer tie rod ends
   - Stabilizer bar links
   - Height sensors, if equipped with air suspension
   - Wheel speed sensors
   - Steering knuckle from the halfshaft
   - Halfshafts

To install:

➡**Use new nuts, bolts and circlips.**

3. Install or connect the following:
   - Halfshafts
   - Steering knuckle to the halfshaft
   - Hub retainer nut and tighten to 170–202 ft. lbs. (230–275 Nm)
   - Wheel speed sensors
   - Height sensors, if equipped with air suspension
   - Stabilizer bar links and tighten the nuts to 57–75 ft. lbs. (77–103 Nm)
   - Outer tie rod ends and tighten the nuts to 35–46 ft. lbs. (47–63 Nm)
   - Lower ball joints and tighten the fasteners to 50–68 ft. lbs. (68–92 Nm)
   - Front wheels

CV-Joint

REMOVAL AND REPLACEMENT

Inner Tripod Joint

The inner CV-joint is serviced with the axle shaft as an assembly. The inner CV-joint boot can be serviced by removing the outer CV-joint.

Outer CV-Joint

1. Before servicing the vehicle, refer to the precautions in the beginning of this section.

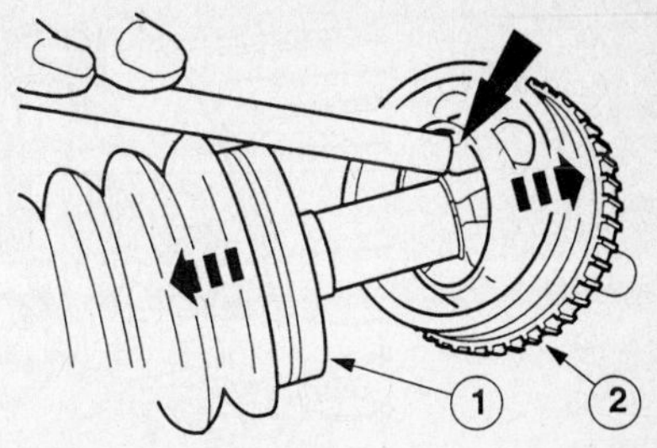

1. CV-Joint boot
2. CV-Joint

9306HG07

Removing the outer CV-joint

2. Place the halfshaft in a vise.
3. Remove the CV-joint boot clamps and slide the boot away from the joint.
4. Drive the CV-joint off the halfshaft with a brass drift and a hammer.

To install:

5. Replace the snapring.
6. Fill the CV-joint with fresh grease and slide the joint on to the halfshaft.
7. Use new clamps and install the CV-joint boot.

STEERING AND SUSPENSION

Air Bag

⁂ CAUTION

Some vehicles are equipped with an air bag system. The system must be disarmed before performing service on, or around, system components, the steering column, instrument panel components, wiring and sensors. Failure to follow the safety precautions and the disarming procedure could result in accidental air bag deployment, possible injury and unnecessary system repairs.

PRECAUTIONS

Several precautions must be observed when handling the inflator module to avoid accidental deployment and possible personal injury.

- Never carry the inflator module by the wires or connector on the underside of the module.
- When carrying a live inflator module, hold securely with both hands, and ensure that the bag and trim cover are pointed away.
- Place the inflator module on a bench or other surface with the bag and trim cover facing up.
- With the inflator module on the bench, never place anything on or close to the module which may be thrown in the event of an accidental deployment.

DISARMING

1. Before servicing the vehicle, refer to the precautions in the beginning of this section.
2. Position the vehicle with the front wheels in a straight-ahead position.
3. Disconnect the negative battery cable.
4. Disconnect the positive battery cable.
5. Wait at least 1 minute for the air bag back-up power supply to drain before continuing.
6. Proceed with the repair.

ARMING

1. After service is completed, connect the battery cables, negative cable last.
2. Check the functioning of the air bag system by turning the ignition key to the **RUN** position and visually monitoring the air bag indicator lamp in the instrument cluster. The indicator lamp should illuminate for approximately 6 seconds, then turn **OFF**. If the indicator lamp does not illuminate, stays ON, or flashes at any time, a fault has been detected by the air bag diagnostic monitor.

Power Rack and Pinion Steering Gear

REMOVAL & INSTALLATION

1. Before servicing the vehicle, refer to the precautions at the beginning of this section.
2. Remove or disconnect the following:
   - Negative battery cable
   - Intermediate steering shaft
   - Front wheels
   - Dual converter Y-pipe
   - Outer tie rod ends
   - Left stabilizer link
   - Heat shield
   - Power steering hose bracket
   - Auxiliary actuator connector, if equipped
   - Power steering pressure switch connector

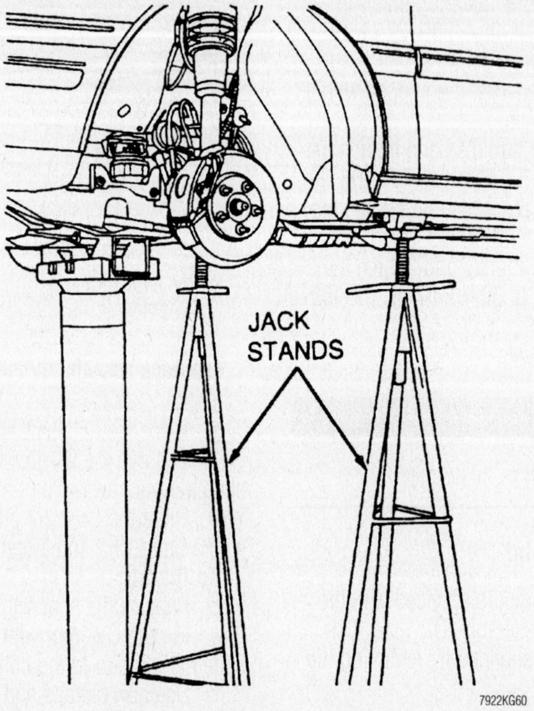

Support the rear of the subframe with 2 tall jackstands while removing the steering gear

- Rear subframe bolts and lower the rear of the subframe about 4 inches
- Steering gear mounting nuts, then move the steering gear to the left to access the power steering hoses
- Power steering hoses
- Steering gear through the left wheel opening

To install:

3. Install or connect the following:
   - Steering gear. Use new seals on the hydraulic fittings. Tighten the mounting nuts to 85–100 ft. lbs. (115–135 Nm).
   - Rear subframe bolts and tighten them to 57–76 ft. lbs. (77–103 Nm)
   - Power steering pressure switch connector
   - Auxiliary actuator connector, if equipped
   - Power steering hose bracket
   - Heat shield
   - Left stabilizer link
   - Outer tie rod ends and tighten the nuts to 35 ft. lbs. (48 Nm)
   - Dual converter Y-pipe
   - Front wheels
   - Intermediate steering shaft
   - Negative battery cable

4. Fill and bleed the power steering system. Check the system for leaks and proper operation.

5. Check the alignment and adjust as necessary.

Strut

REMOVAL & INSTALLATION

Front

1. Before servicing the vehicle, refer to the precautions at the beginning of this section.
2. Remove or disconnect the following:
   - Negative battery cable
   - Front wheels
   - Height sensor and wiring, if equipped with air suspension
   - Brake hose bracket
   - Disc brake caliper and rotor
   - Wheel speed sensor and wiring
   - Outer tie rod end
   - Stabilizer link
   - Steering knuckle pinch bolt
   - Strut assembly

To install:

3. Install or connect the following:
   - Strut assembly. Tighten the upper nuts to 22–29 ft. lbs. (30–40 Nm), and the knuckle pinch bolt to 73–97 ft. lbs. (98–132 Nm).
   - Stabilizer link and tighten the nut to 55–75 ft. lbs. (75–101 Nm)
   - Outer tie rod end and tighten the nut to 35 ft. lbs. (48 Nm)

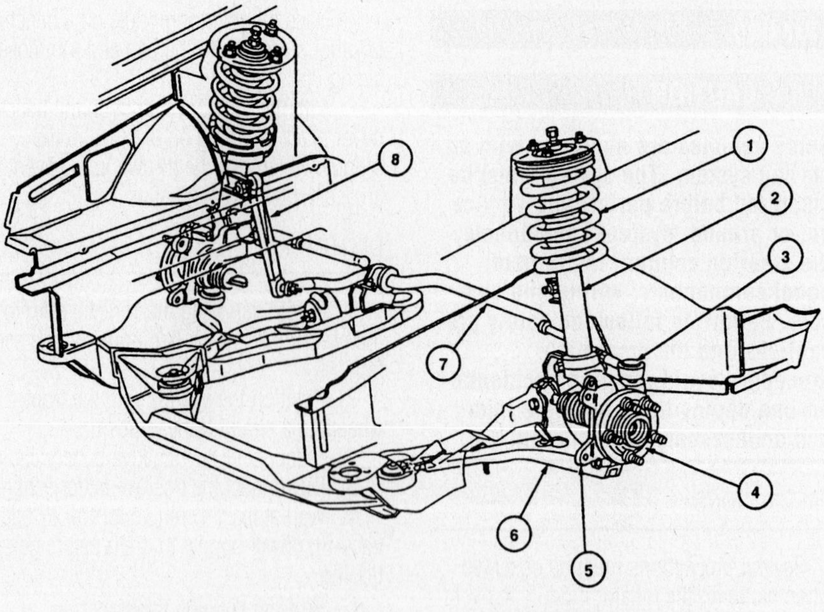

Item	Description
1	Front Coil Spring
2	Front Shock Absorber
3	Tie Rod End
4	Wheel Hub

Item	Description
5	Front Wheel Knuckle
6	Front Suspension Lower Arm
7	Front Stabilizer Bar
8	Stabilizer Bar Link

7922KG61

Front suspension component identification

- Wheel speed sensor and wiring
- Disc brake caliper and rotor and tighten the caliper anchor bracket bolts to 65–87 ft. lbs. (88–118 Nm)
- Brake hose bracket
- Height sensor and wiring, if equipped with air suspension
- Front wheels
- Negative battery cable

4. Road test the vehicle and check for proper operation.

Rear

1. Before servicing the vehicle, refer to the precautions at the beginning of this section.
2. Remove or disconnect the following:
   - Rear package tray trim panel
   - Rear wheels
   - Brake load sensor
   - Brake hose
   - Stabilizer bar bracket and link
   - Tension strut
   - Spindle pinch bolt
   - Strut assembly

To install:

➡Use new mounting nuts and bolts.

3. Install or connect the following:
   - Strut assembly. Use a new pinch

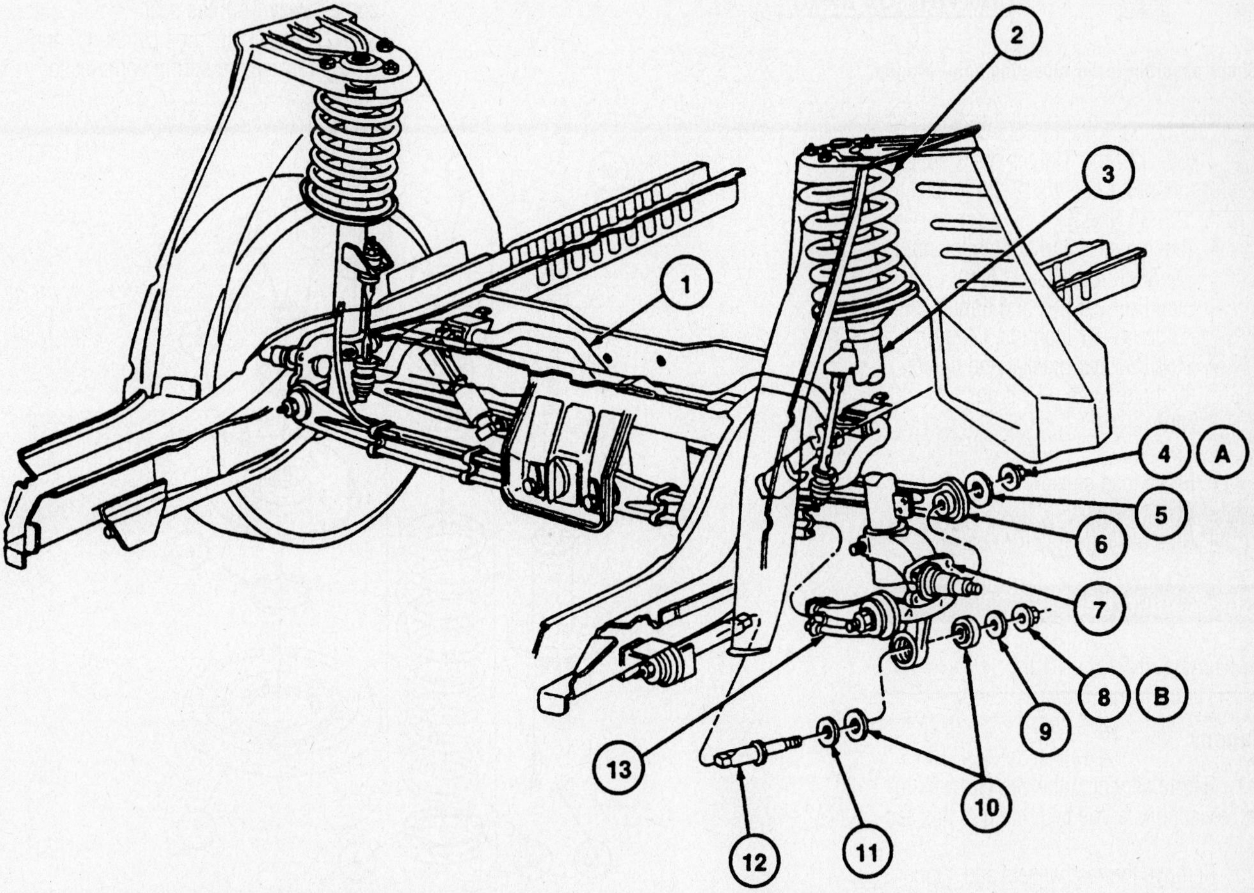

Item	Description
1	Rear Stabilizer Bar
2	Rear Spring
3	Shock Absorber
4	Nut
5	Washer
6	Lower Suspension Arm (Rear)
7	Rear Wheel Spindle
8	Nut (4 Req'd)
9	Washer (2 Req'd)

Item	Description
10	Rear Suspension Tie Rod Bushing (4 Req'd)
11	Washer (2 Req'd)
12	Rear Suspension Tension Strut and Bushing (2 Req'd)
13	Rear Suspension Lower Arm (Front)
A	Tighten to 68-92 N·m (50-67 Lb-Ft)
B	Tighten to 46.7-63.3 N·m (35-46 Lb-Ft)

7922KG62

Rear suspension component identification—sedan models

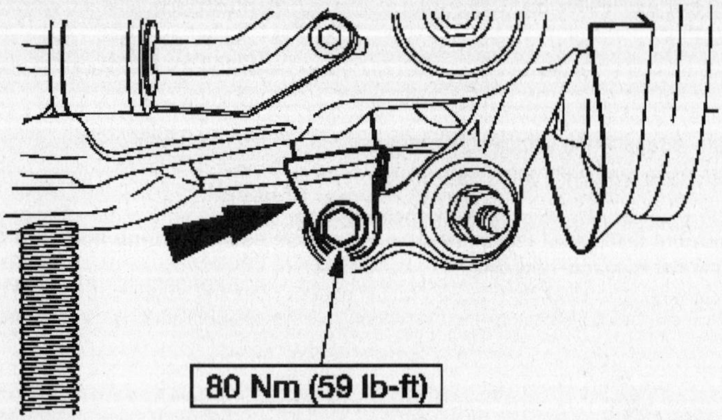

Shock absorber lower mounting bolt—Wagon

bolt and tighten to 50–67 ft. lbs. (68–92 Nm). Tighten the 3 upper mounting nuts to 19–25 ft. lbs. (25–34 Nm).
- Tension strut and tighten the nut to 35–46 ft. lbs. (68–92 Nm)
- Stabilizer bar link and tighten the nut to 60–81 inch lbs. (7–9 Nm)
- Stabilizer bar bracket and tighten the bolts to 25 –33 ft. lbs. (34–46 Nm)
- Brake hose
- Brake load sensor
- Rear wheels
- Rear package tray trim panel

Shock Absorber

REMOVAL & INSTALLATION

Wagons

1. Before servicing the vehicle, refer to the precautions at the beginning of this section.
2. Remove the rear wheels and support the rear control arms on jackstands.
3. Remove or disconnect the following:
   - Rear compartment access panels
   - Upper shock mounting nuts and insulators
   - Lower shock mounting bolts
   - Shock absorbers

To install:

➡Use new mounting fasteners and insulators.

4. Install or connect the following:
   - Shock absorbers. Tighten the lower bolt to 50–68 ft. lbs. (68–92 Nm), and the upper nuts to19–25 ft. lbs. (26–34 Nm).
   - Rear compartment access panels
   - Rear wheels

Coil Spring

REMOVAL & INSTALLATION

Struts

1. Before servicing the vehicle, refer to the precautions at the beginning of this section.
2. Remove the strut from the vehicle.
3. Compress the coil spring using a suitable spring compressor until the spring comes away from the seat.
4. Remove the large center nut and slowly release the spring compressor.

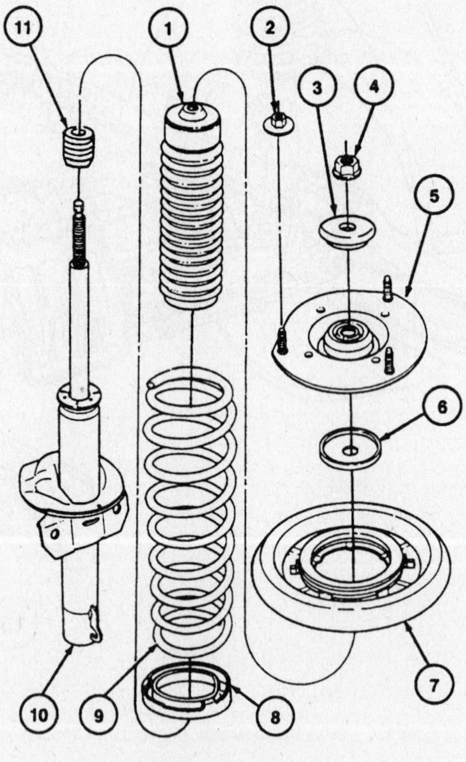

Item	Description
1	Dust Boot (Part of 18124)
2	Nut (3 Req'd)
3	Washer
4	Nut
5	Front Shock Absorber Mounting Bracket
6	Washer
7	Front Suspension Bearing and Seal
8	Front Spring Insulator (Part of 18124)
9	Front Coil Spring
10	Front Shock Absorber
11	Jounce Bumper (Part of 18124)

7922KG63

Exploded view of the front strut and coil spring assembly

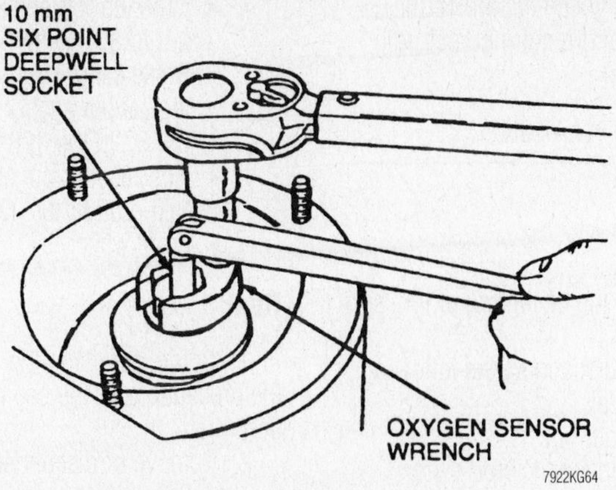

Hold the strut rod while loosening or tightening the nut

To install:

5. Compress the spring and install it on the strut.
6. Install the lower washer and mounting bracket.
7. Install the upper washer and a new nut. Tighten the nut to 39–53 ft. lbs. (53–72 Nm).
8. Install the strut assembly in the vehicle.

Wagons with rear shock absorbers

1. Before servicing the vehicle, refer to the precautions at the beginning of this section.
2. Remove or disconnect the following:
   - Rear shock absorber
   - Stabilizer bar link and bracket
   - Brake hose bracket
   - Upper ball joint
3. Install spring keepers on the coil springs.
4. Slowly lower the lower control arm until the tension is relaxed on the coil spring. Remove the coil spring and the upper and lower spring insulators.

To install:

➡Use new mounting nuts and bolts.

5. Install or connect the following:
   - Coil spring with upper and lower spring insulators
   - Upper ball joint and tighten the nut to 50–68 ft. lbs. (68–92 Nm)
   - Brake hose bracket
   - Stabilizer bar link and bracket
   - Rear shock absorber

Lower Ball Joints

REMOVAL & INSTALLATION

The lower ball joint is an integral part of the steering knuckle. If the lower ball joint is found to be defective, the entire steering knuckle must be replaced.

Upper Control Arm

REMOVAL AND INSTALLATION

Rear

WAGON ONLY

1. Before servicing the vehicle, refer to the precautions at the beginning of this section.
2. Support the lower control arm on a jackstand.
3. Remove or disconnect the following:
   - Rear wheel
   - Brake hose bracket
   - Upper ball joint
   - Upper control arm

To install:

➡Use new mounting nuts and bolts.

4. Install or connect the following:
   - Upper control arm. Tighten the ball joint nut to 50–67 ft. lbs. (68–92 Nm), then tighten the control arm mounting bolts to 73–97 ft. lbs. (98–132 Nm).
   - Brake hose bracket
   - Rear wheel
5. Check the wheel alignment and adjust as necessary.

CONTROL ARM BUSHING REPLACEMENT

The control arm bushings are serviced with the control arm as an assembly.

Lower Control Arm

REMOVAL AND INSTALLATION

Front

1. Before servicing the vehicle, refer to the precautions at the beginning of this section.
2. Remove or disconnect the following:
   - Front wheel
   - Wheel speed sensor wiring harness
   - Lower ball joint
   - Lower control arm

To install:

3. Install or connect the following:
   - Lower control arm. Tighten the front bolt to 57–75 ft. lbs. (77–103 Nm), and the rear bolt to 72–97 ft. lbs. (98–132 Nm).
   - Lower ball joint. Use a new nut and tighten to 50–67 ft. lbs. (68–92 Nm).
   - Wheel speed sensor wiring harness
   - Front wheel
4. Check the wheel alignment and adjust as necessary.

Rear

SEDAN

1. Before servicing the vehicle, refer to the precautions at the beginning of this section.
2. Remove or disconnect the following:
   - Rear wheel
   - Parking brake cable
   - Proportioning valve
   - Lower control arm

To install:

➡Use new bolts and nuts.

➡The rear suspension lower control arms are marked BOTTOM on the lower edge. The flange edge of the right side rear suspension arm and bushing stamping must face the front of the vehicle. The other three must face the rear of the vehicle.

➡The rear suspension arms have two adjustment cams that fit inside the bushings at the arm-to-body attachment. Each adjustment cam is installed from the front on the rear suspension arm and bushing.

3. Install or connect the following:
   - Lower control arm and tighten the bolts to 50–67 ft. lbs. (68–92 Nm)
   - Proportioning valve
   - Parking brake cable
   - Rear wheel

4. Check the wheel alignment and adjust as necessary.

WAGON

1. Before servicing the vehicle, refer to the precautions at the beginning of this section.
2. Remove or disconnect the following:
   - Rear wheel
   - Coil spring
   - Wheel spindle
   - Lower control arm

To install:

➡Use new nuts and bolts.

➡Tighten the control arm fasteners with the vehicle weight resting on the wheels.

3. Install or connect the following:
   - Lower control arm
   - Wheel spindle
   - Coil spring
   - Rear wheel
4. Tighten the control arm-to-body bolt to 40–52 ft. lbs. (54–71 Nm), and the arm-to-spindle bolt to 50–67 ft. lbs. (68–92 Nm).
5. Check the wheel alignment and adjust as necessary.

CONTROL ARM BUSHING REPLACEMENT

The control arm bushings are serviced with the control arm as an assembly.

Wheel Bearings

ADJUSTMENT

There is no adjustment for the front or rear wheel bearings due to the nature of their design. These bearings are permanently lubricated and require no periodic maintenance.

REMOVAL & INSTALLATION

Front

1. Before servicing the vehicle, refer to the precautions at the beginning of this section.
2. Remove or disconnect the following:
   - Front wheel
   - Hub retainer nut
   - Brake caliper and rotor
   - Outer tie rod end
   - Stabilizer bar link
   - Wheel speed sensor
   - Lower ball joint
   - Steering knuckle
   - Hub and bearing assembly

To install:

➡Use new nuts, bolts, and split pins.

➡The knuckle must be clean enough to allow the wheel hub to be completely seated by hand. Do not press or draw the wheel hub into place.

3. Install or connect the following:
   - Hub and bearing assembly and tighten the bolts to 61–78 ft. lbs. (83–107 Nm)
   - Steering knuckle and tighten the pinch bolt to 72–97 ft. lbs. (98–132 Nm)
   - Lower ball joint and tighten the nut to 50–67 ft. lbs. (68–92 Nm)
   - Wheel speed sensor
   - Stabilizer bar link and tighten the nut to 57–75 ft. lbs. (77–103 Nm)
   - Outer tie rod end and tighten the nut to 35–46 ft. lbs. (47–63 Nm)
   - Brake caliper and rotor and tighten the caliper anchor bracket bolts to 65–87 ft. lbs. (88–118 Nm)
   - Hub retainer nut and tighten to 170–202 ft. lbs. (230–275 Nm)
   - Front wheel

Rear

1. Before servicing the vehicle, refer to the precautions at the beginning of this section.
2. Remove or disconnect the following:
   - Rear wheel
   - Brake hose bracket
   - Brake caliper and rotor, if equipped with rear disc brakes
   - Brake drum, if equipped with rear drum brakes
   - Hub and bearing assembly grease cap and retaining nut
   - Hub and bearing assembly

To install:

➡Use new retaining nuts and grease caps

3. Install or connect the following:
   - Hub and bearing assembly and tighten the retaining nut to 188–254 ft. lbs. (255–345 Nm)
   - Grease cap
   - Brake caliper and rotor, if equipped with rear disc brakes
   - Brake drum, if equipped with rear drum brakes
   - Brake hose bracket
   - Rear wheel

BRAKES

Brake Caliper

REMOVAL & INSTALLATION

Front

1. Before servicing the vehicle, refer to the precautions in the beginning of this section.
2. Remove brake fluid from the brake master cylinder reservoir until the reservoir is ½ full.
3. Remove the wheel and tire assembly.
4. Mark the disc brake caliper to ensure that it is reinstalled in the correct location.
5. Remove the hollow bolt connecting the brake hose to the disc brake caliper and plug the brake hose. Discard the 2 copper sealing washers.
6. Remove the caliper locating pins and lift the caliper off the rotor using a rotating motion.

To install:

7. Retract the disc brake caliper piston fully in the piston bore, using an old brake pad or block of wood and a C-clamp.

➡Make sure the clip-on insulators are attached to the brake pads.

8. Install the disc brake pads to the caliper. Make sure the brake pad insulators are correctly attached to the brake pad plate.
9. Position the disc brake caliper and pad assembly above the rotor and install it with a rotating motion. Make sure the inner and outer pads are properly positioned and the outer anti-rattle spring is properly positioned.
10. Lubricate the locating pins and the inside of the insulators with silicone grease. Torque the locating pins to 25 ft. lbs. (34 Nm).
11. Remove the plug and install the brake hose to the disc brake caliper. Use 2 new copper washers and torque the hollow bolt to 30–40 ft. lbs. (41–54 Nm).
12. Bleed the brake system, filling the master cylinder as required.
13. Install the wheel and tire assembly;

torque the nuts to 85–104 ft. lbs. (115–142 Nm).

14. Pump the brake pedal several times to position the brake pads prior to moving the vehicle.

15. Road test the vehicle and check for proper brake system operation.

Rear

1. Before servicing the vehicle, refer to the precautions in the beginning of this section.
2. Remove brake fluid from the brake master cylinder reservoir until the reservoir is ½ full.
3. Remove the wheel and tire assembly.
4. Remove the retaining bolt and disconnect the brake hose from the caliper assembly. Discard the copper sealing washers.
5. Remove the retaining clip from the parking brake at the caliper. Disengage the parking brake cable end from the lever arm.
6. Lift the rear disc brake caliper away from the rear disc support bracket.
7. Remove the disc brake caliper locating pins and boots from the rear disc support bracket.

To install:

8. Using rear caliper piston adjuster tool T87P-2588-A, rotate the rear disc brake piston and adjuster clockwise until fully seated.

➡Make sure one of the 2 slots in the rear disc brake piston and adjuster face is positioned so it will engage the nib on the disc brake pad.

9. Apply silicone dielectric compound to the inside of the slider pin boots and the slider pins.
10. Position the slider pins and boots in the support bracket. Position the caliper assembly on the support bracket. Make sure the brake pads are installed correctly.
11. Remove the residue from the pin retainer threads and apply 1 drop of threadlock and sealer. Install the pin retainers and torque to 23–26 ft. lbs. (31–35 Nm).
12. Attach the cable end to the parking brake lever. Install the cable retaining clip on the caliper assembly.
13. Using new washers, connect the brake flex hose to the caliper. Torque the retaining bolt to 40 ft. lbs. (54 Nm).
14. Bleed the brake system, filling the master cylinder as required.
15. Install the wheel and tire assembly; torque the nuts to 85–104 ft. lbs. (115–142 Nm).
16. Pump the brake pedal several times to position the brake pads prior to moving the vehicle.
17. Road test the vehicle and check for proper brake system operation.

Disc Brake Pads

REMOVAL & INSTALLATION

Front

1. Before servicing the vehicle, refer to the precautions in the beginning of this section.
2. Remove the master cylinder reservoir cap and check the fluid level in the reservoir. Remove brake fluid until the reservoir is ½ full. Discard the removed fluid.
3. On Continentals, turn the air suspension switch, located in the left side of the luggage compartment, to the **OFF** position.
4. Remove the wheel and tire assembly.
5. Remove the disc brake caliper locating pins. Lift the caliper assembly from the anchor plate and rotor using a rotating motion.
6. Suspend the caliper inside the fender housing with wire. Do not allow the caliper to hang from the brake hose.
7. Remove the inner and outer brake pads. Inspect the rotor braking surfaces for scoring and machine as necessary.

To install:

8. Use a C-clamp and an old brake pad or block of wood to seat the caliper piston in its bore.
9. Remove any rust buildup from the inside of the caliper in the brake pad contact area.
10. Install the inner pad in the caliper piston.
11. Install the outer pad onto the anchor plate. Make sure the clips are properly seated.

➡Make sure the insulators are installed on the brake pads.

12. Install the disc brake caliper onto the anchor plate.
13. Install caliper locating pins and torque to 23–28 ft. lbs. (31–38 Nm).
14. Install wheel and tire assembly and torque lugs nuts to 85–104 ft. lb. (115-142 Nm)
15. Pump the brake pedal several times prior to moving the vehicle to position the brake pads to the rotor.
16. Refill the master cylinder reservoir as necessary, using only clean DOT 3 brake fluid from a closed container.
17. Road test the vehicle and check the brake system for proper operation.

Rear

1. Before servicing the vehicle, refer to the precautions in the beginning of this section.
2. Remove the master cylinder reservoir cap and check the fluid level in the reservoir. Remove brake fluid until the reservoir is ½ full. Discard the removed fluid.
3. Remove the wheel and tire assembly.

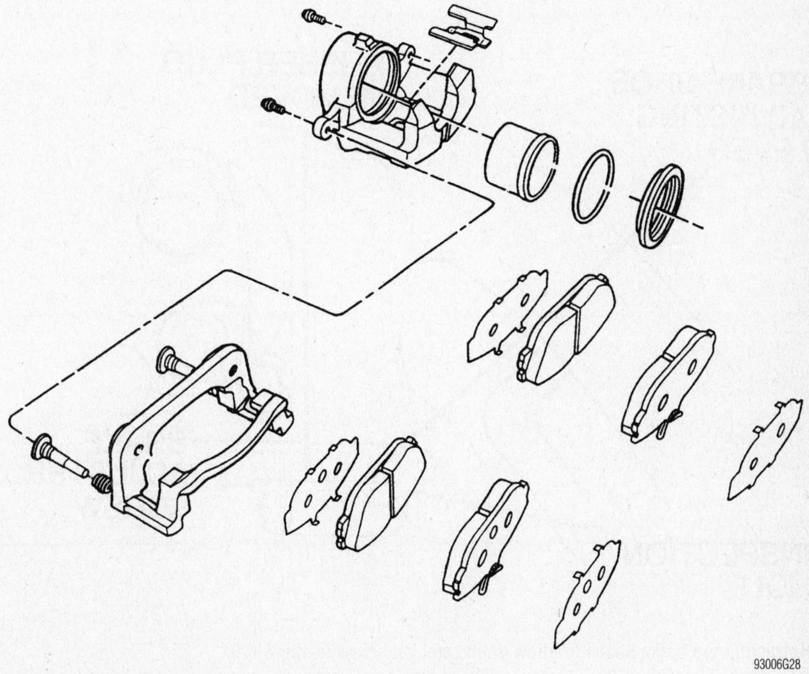

Front disc brake caliper, pads and related components

4. Remove the screw retaining the brake hose bracket to the frame side rail.

5. Remove the retaining clip from the parking brake cable at the disc brake caliper. Remove the cable end from the parking brake lever.

6. Remove the upper disc brake caliper locating pin at the support bracket. Rotate the caliper away from the rotor.

7. Remove the disc brake pads.

8. Inspect the rotor braking surfaces for scoring and machine as necessary.

To install:

9. Using Rear Caliper Piston Adjuster T87P-2588-A, rotate the piston clockwise until it is fully seated. Make sure one of the slots in the piston face is positioned so it will engage the nib on the brake pad.

10. Install the brake pads in the support bracket. Rotate the caliper assembly over the rotor into position on the support bracket. Make sure the brake pads are installed correctly.

11. Remove the residue from the rear brake pin retainer bolt threads and apply 1 drop of a suitable threadlock sealer. Install and torque the disc brake caliper locating pin to 23–26 ft. lbs. (31–35 Nm).

12. Attach the cable end to the parking brake lever. Install the cable retaining clip on the caliper assembly. Position the brake flex hose and bracket assembly to the side rail, and install the retaining screw. Torque to 11 ft. lbs. (16 Nm).

13. Install the wheel and tire assembly and torque lug nuts to 85–104 ft. lbs. (115–142 Nm).

14. Pump the brake pedal several times prior to moving the vehicle, to position the brake pads to the rotor.

15. Refill the master cylinder reservoir if necessary, using only clean DOT 3 brake fluid from a closed container.

16. Road test the vehicle and check the brake system for proper operation.

Brake Drums

REMOVAL & INSTALLATION

1. Before servicing the vehicle, refer to the precautions in the beginning of this section.

2. Remove the wheel and tire assembly.

3. Remove the brake drum.

➡If the brake drum cannot be removed easily, remove the brake tube-to-axle retention bracket and pry rubber plug from rear brake backing plate inspection hole. This will allow sufficient room for insertion of a screwdriver and brake tools to disengage brake shoe adjusting lever and back off the brake adjuster screw.

4. Inspect the drum for scoring and/or other wear. Machine or replace, as necessary.

To install:

5. Measure the brake drum inside diameter using D81L-1103-A brake adjustment gauge.

6. Using the brake adjustment gauge, adjust the brake shoes to the same dimensions as the brake drum.

7. Position the brake drum over the brake shoes on the axle hub.

8. Install the wheel and tire assembly. Torque the lug nuts to 85–104 ft. lbs. (115–141 Nm).

9. Pump the brake pedal several times to position the brake shoes and complete the adjustment.

10. Road test the vehicle and check for proper brake system operation.

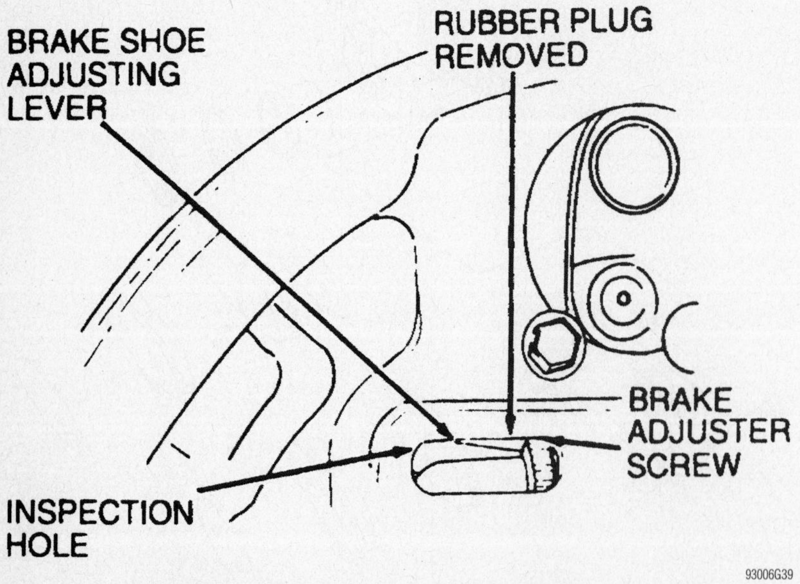

Retracting the brake shoes to allow drum removal—Taurus and Sable

Brake Shoes

REMOVAL & INSTALLATION

1. Before servicing the vehicle, refer to the precautions in the beginning of this section.

2. Remove the wheel and tire assembly.

3. Remove the brake drum.

4. Remove the parking brake cable from the parking brake lever.

5. Remove the 2 brake shoe hold-down springs and pins.

6. Lift the brake shoes, springs and adjuster assembly off the backing plate and wheel cylinder assembly. When removing the assembly, be careful not to bend the adjusting lever.

7. Remove the retracting springs from the lower brake attachments and upper shoe-to-adjusting lever attachment points.

8. Remove the horseshoe retaining clip and spring washer and slide the lever off the parking brake lever pin on the trailing shoe. Discard the horseshoe clip.

To install:

9. Apply a light coating of disc brake caliper slide grease at the points where the brake shoes contact the backing plate.

10. Apply a thin coat of lubricant to the adjuster screw threads and socket end of the adjusting screw. Install the stainless steel washer over the socket end of the adjusting screw and install the socket. Turn the adjusting screw into the adjusting pivot nut to the limit of the threads and then back off ½ turn.

11. Assemble the parking brake lever to the trailing shoe by installing the spring washer and a new horseshoe retaining clip. Crimp the clip until it retains the lever to the shoe securely.

12. Position the trailing shoe on the backing plate and attach the rear parking brake cable.

13. Position the leading shoe on the backing plate and attach the lower brake shoe adjusting spring to the brake shoes.

14. Install the adjuster assembly in the

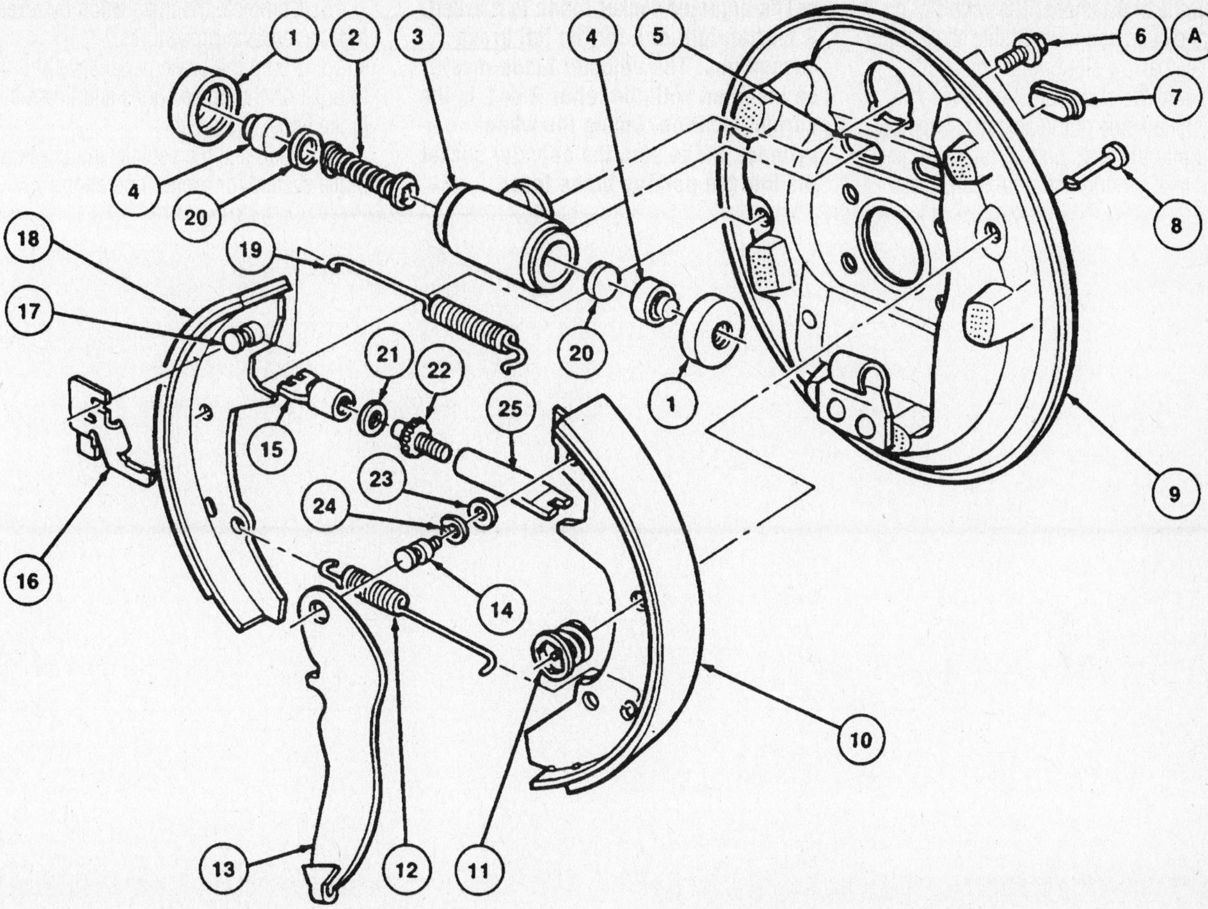

1	Boot	14	Parking Brake Lever Pin (Inner)
2	Spring Expander	15	Brake Shoe Adjusting Screw Socket
3	Rear Wheel Cylinder	16	Brake Shoe Adjusting Lever
4	Piston and Insert	17	Parking Brake Lever Pin
5	Shoe Adjustment Access Hole	18	Leading Shoe and Lining
6	Wheel Cylinder Retaining Bolt (2 Req'd)	19	Brake Shoe Adjusting Screw Spring
7	Brake Adjusting Hole Cover	20	Cup
8	Brake Shoe Hold-Down Spring Pin	21	Washer
9	Rear Brake Backing Plate	22	Brake Adjuster Screw
10	Trailing Shoe and Lining	23	Washer
11	Brake Shoe Hold-Down Spring	24	Parking Brake Lever Pin Retainer
12	Brake Shoe Retracting Spring	25	Adjusting Pivot Nut
13	Parking Brake Lever	A	Tighten to 12-18 N·m (107-159 Lb-In)

93006G56

Brake shoes and related components—Taurus and Sable

slots on the brake shoes. The wide slot on the dual slotted end must fit into the leading shoe. The narrow slot on the dual slotted end fits into the shoe adjusting lever. The single slotted side of the adjuster assembly must fit into the slots on the trailing shoe and the rear parking brake cable bracket.

➡**The adjuster socket blade is marked R for the right or L for the left brake assemblies. The adjuster blade must be installed with the letter R or L in the upright position, facing the wheel cylinder. Make sure the adjuster socket fits into the parking brake lever.**

15. Complete the installation by reversing the removal procedures.
16. Pump the brake pedal several times to position the brake shoes and finish the brake shoe adjustment.
17. Road test the vehicle and check the brake system for proper operation.

FORD

Thunderbird

15

SPECIFICATION CHARTS

ENGINE AND VEHICLE IDENTIFICATION

Engine						
Code ①	Liters (cc)	Cu. In.	Cyl.	Fuel Sys.	Type	Eng. Mfg.
A	3.9 (3947)	243	8	MFI	DOHC	Ford

Model Year	
Code ②	Year
2	2002
3	2003
4	2004
5	2005

MFI: Multi-Port Fuel Injection

DOHC: Double Overhead Camshaft

① 8th digit of the VIN

② 10th digit of the VIN

67197-THUN-C01

GENERAL ENGINE SPECIFICATIONS

Year	Model	Engine Displacement Liters	Engine ID/VIN	Net Horsepower @ rpm	Net Torque @ rpm (ft. lbs.)	Bore x Stroke (in.)	Com-pression Ratio	Oil Pressure @ rpm
2002	Thunderbird	3.9	A	252@6100	267@4300	3.38x3.34	10.6:1	NA
2003	Thunderbird	3.9	A	280@6000	286@4000	3.38x3.34	10.6:1	NA
2004	Thunderbird	3.9	A	280@6000	286@4000	3.38x3.34	10.6:1	61-73@400

NA: Not Available

67197-THUN-C02

ENGINE TUNE-UP SPECIFICATIONS

Year	Engine Displacement Liters	Engine ID/VIN	Spark Plugs Gap (in.)	Ignition Timing (deg.) ①		Fuel Pump (psi)	Idle Speed (rpm) ①		Valve Clearance	
				MT	AT		MT	AT	In.	Ex.
2002	3.9	A	0.039-0.043	—	10-20B	43	—	650-750	0.007-0.009	0.009-0.011
2003	3.9	A	0.039-0.043	—	10-20B	43	—	650-750	0.007-0.009	0.009-0.011
2004	3.9	A	0.039-0.043	—	10-20B	43	—	650-750	0.007-0.009	0.013-0.015

The underhood specifications sticker often reflects tune-up specification changes in production. Sticker figures must be used if they disagree with those in this chart

① Controlled by the engine computer

67197-THUN-C03

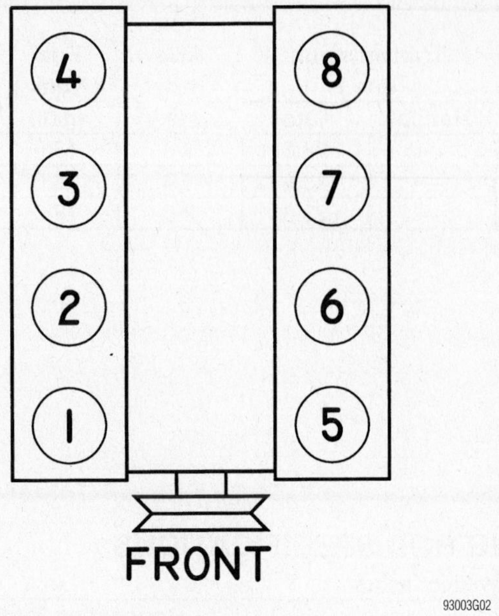

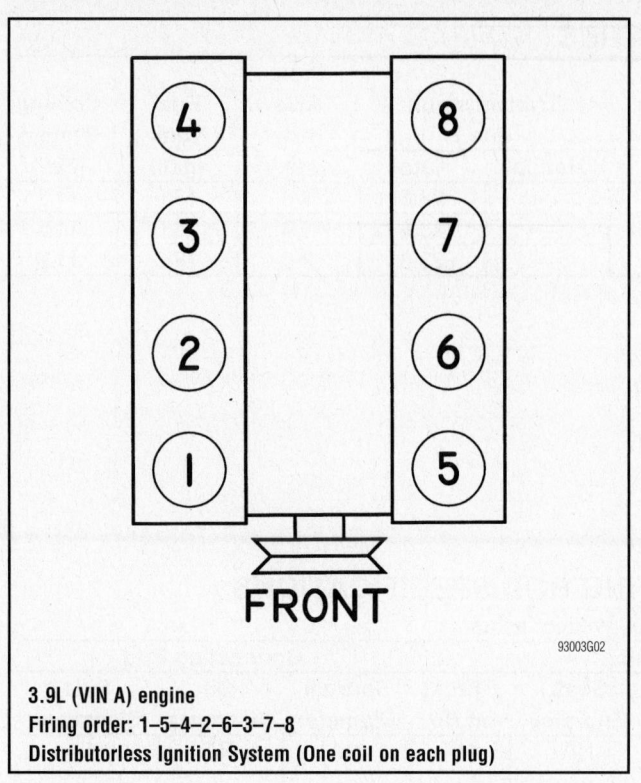

3.9L (VIN A) engine
Firing order: 1–5–4–2–6–3–7–8
Distributorless Ignition System (One coil on each plug)

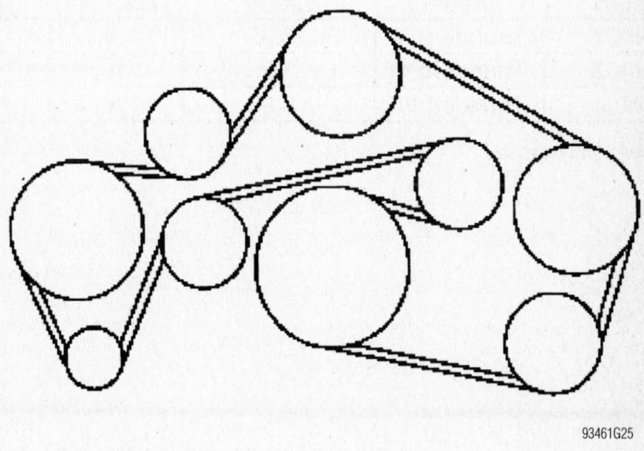

Serpentine accessory drive belt routing—2002–2003 3.9L (VIN A) engine

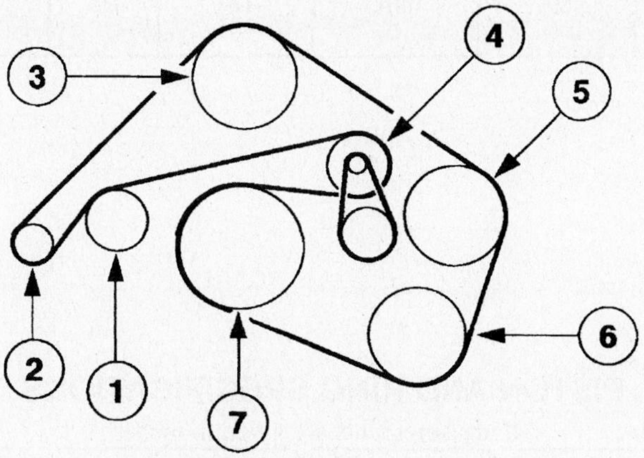

1 Belt idler pulley—unflanged
2 Generator pulley
3 Water pump pulley
4 Drive belt tensioner
5 Power steering pump pulley
6 A/C clutch pulley
7 Crankshaft vibration damper

67197-THUN-G00

Serpentine accessory drive belt routing—2004 3.9L (VIN A) engine

CAPACITIES

Year	Model	Engine Displacement Liters	Engine ID/VIN	Engine Oil with Filter (qts.)	Transmission (pts.) Manual	Transmission (pts.) Auto.	Drive Axle Rear (pts.)	Fuel Tank (gal.)	Cooling System (qts.)
2002	Thunderbird	3.9	A	NA	—	23.8	3.0	18.0	11.3
2003	Thunderbird	3.9	A	NA	—	23.8	3.0	18.0	11.3
2004	Thunderbird	3.9	A	6.9	—	23.8	2.6	18.0	11.9

N/A: Not Available

67197-THUN-C04

CRANKSHAFT AND CONNECTING ROD SPECIFICATIONS

All measurements are given in inches.

Year	Engine Displacement Liters	Engine ID/VIN	Crankshaft Main Brg. Journal Dia.	Crankshaft Main Brg. Oil Clearance	Crankshaft Shaft End-play	Crankshaft Thrust on No.	Connecting Rod Journal Diameter	Connecting Rod Oil Clearance	Connecting Rod Side Clearance
2002	3.9	A	NA	NA	NA	NA	NA	NA	NA
2003	3.9	A	NA	NA	NA	NA	NA	NA	NA
2004	3.9	A	NA	NA	NA	NA	NA	NA	NA

NA: Not Available

67197-THUN-C05

PISTON AND RING SPECIFICATIONS

All measurements are given in inches.

Year	Engine Displacement Liters	Engine ID/VIN	Piston Clearance	Ring Gap Top Compression	Ring Gap Bottom Compression	Ring Gap Oil Control	Ring Side Clearance Top Compression	Ring Side Clearance Bottom Compression	Ring Side Clearance Oil Control
2002	3.9	A	NA	NA	NA	NA	NA	NA	NA
2003	3.9	A	NA	NA	NA	NA	NA	NA	NA
2004	3.9	A	NA	NA	NA	NA	NA	NA	NA

NA: Not Available

67197-THUN-C06

VALVE SPECIFICATIONS

Year	Engine Displacement Liters	Engine ID/VIN	Seat Angle (deg.)	Face Angle (deg.)	Spring Test Pressure (lbs. @ in.)	Spring Free Length (in.)	Stem-to-Guide Clearance (in.) Intake	Stem-to-Guide Clearance (in.) Exhaust	Stem Diameter (in.) Intake	Stem Diameter (in.) Exhaust
2002	3.9	A	NA	NA	NA	NA	NA	NA	NA	NA
2003	3.9	A	NA	NA	NA	NA	NA	NA	NA	NA
2004	3.9	A	14	45	NA	NA	0.001	0.001	0.197	0.196

NA: Not Available

67197-THUN-C07

TORQUE SPECIFICATIONS

All readings in ft. lbs.

Year	Engine Displacement Liters	Engine ID/VIN	Cylinder Head Bolts	Main Bearing Bolts	Rod Bearing Bolts	Crankshaft Damper Bolts	Flywheel Bolts	Manifold Intake	Manifold Exhaust	Spark Plugs	Oil Pan Drain Plug
2002	3.9	A	①	NA	NA	②	③	18	18	19	NA
2003	3.9	A	①	NA	NA	②	③	18	18	19	NA
2004	3.9	A	①	NA	NA	②	③	18	18	19	NA

NA: Not Available

① Step 1: Tighten M10 bolts to 15 ft. lbs.
Step 2: Tighten M10 bolts to 26 ft. lbs.
Step 3: Tighten M10 bolts to 33 ft. lbs.
Step 4: Tighten M10 bolts plus 90 degrees
Step 5: Tighten M10 bolts plus 90 degrees
Step 6: Tighten M8 bolts to 15 ft. lbs.
Step 7: Tighten M8 bolts plus 90 degrees

② Step 1: 59 ft. lbs.
Step 2: Plus 80 degrees

③ Step 1: 11 ft. lbs.
Step 2: 81 ft. lbs.

67197-THUN-C08

WHEEL ALIGNMENT

Year	Model		Caster Range (+/-Deg.)	Caster Preferred Setting (Deg.)	Camber Range (+/-Deg.)	Camber Preferred Setting (Deg.)	Toe-in (in.)
2002	Thunderbird	F	0.70	0	0.70	0	0.08 +/- 0.13
		R	—	—	0.75	0	0.13 +/- 0.13
2003	Thunderbird	F	0.70	0	0.70	0	0.08 +/- 0.13
		R	—	—	0.75	0	0.13 +/- 0.13
2004	Thunderbird	F	0.70	0	0.70	0	0.08 +/- 0.13
		R	—	—	0.75	0	0.13 +/- 0.13

67197-THUN-C09

TIRE, WHEEL AND BALL JOINT SPECIFICATIONS

		OEM Tires		Tire Pressures (psi)		Wheel	Ball Joint	Wheel Lug
Year	Model	Standard	Optional	Front	Rear	Size	Inspection	Nut Torque
2002	Thunderbird	P235/50VR17	—	30	30	7-J/7.5J	U ① L① ②	③
2003	Thunderbird	P235/50VR17	—	30	30	7-J/7.5J	U ① L① ②	③
2004	Thunderbird	P235/50VR17	—	30	30	7-J/7.5J	U ① L① ②	③

OEM: Original Equipment Manufacturer

PSI: Pounds Per Square Inch

STD: Standard

OPT: Optional

L: Lower

U: Upper

① Replace if any measurable movement is found.

② Do not lift car. Inspect the boss into which the grease fitting is threaded. Replace if the boss is flush or receded below the surface of the ball joint.

③ 100 ft. lbs.

67197-THUN-C10

BRAKE SPECIFICATIONS

All measurements in inches unless noted

			Brake Disc				Brake Caliper	
			Original	Minimum	Maximum	Minimum Lining	Bracket Bolts	Mounting Bolts
Year	Model		Thickness	Thickness	Runout	Thickness	(ft. lbs.)	(ft. lbs.)
2002	Thunderbird	F	1.180	1.120	0.004	0.079	76	26
		R	0.810	0.740	0.004	0.039	76	25
2003	Thunderbird	F	1.180	1.120	0.004	0.079	76	26
		R	0.810	0.740	0.004	0.039	76	25
2004	Thunderbird	F	1.180	1.120	0.004	0.079	76	26
		R	0.810	0.740	0.004	0.039	76	25

67197-THUN-C11

SCHEDULED MAINTENANCE INTERVALS

Ford—Thunderbird

TO BE SERVICED	TYPE OF SERVICE	VEHICLE MILEAGE INTERVAL (x1000)												
		5	10	15	20	25	30	35	40	45	50	55	60	65
Air cleaner filter	R						✓						✓	
Accessory drive belt	S/I												✓	
Brake system ①	S/I			✓			✓			✓			✓	
Cooling system hoses and clamps	S/I			✓			✓			✓			✓	
CV-joint boots & axle seals	S/I						✓						✓	
Engine coolant	R	Eight years or 150,000 miles												
Engine oil & filter	R	✓	✓	✓	✓	✓	✓	✓	✓	✓	✓	✓	✓	✓
Exterior Lights	S/I	Check monthly												
PCV valve	S/I												✓	
Exhaust system & heat shields	S/I						✓						✓	
Parking brake system	S/I	Every 6 months												
Power steering fluid	S/I	Every 6 months												
Rotate tires	S/I	✓		✓		✓		✓		✓		✓		✓
Steering linkage	S/I						✓						✓	
Spark plugs	R	Change at 100,000 miles												
Suspension components	S/I						✓						✓	

R: Replace S/I: Inspect and service, if necessary L: Lubricate A: Adjust C: Clean

① Inspect the reservoir fluid level, rotor and or drum, brake lines, hoses, calipers and or wheel cylinders

FREQUENT OPERATION MAINTENANCE (SEVERE SERVICE)

If a vehicle is operated under any of the following conditions it is considered severe service:

- Extremely dusty areas.

- 50% or more of the vehicle operation is in 32°C (90°F) or higher temperatures, or constant operation in temperatures below 0°C (32°F).

- Prolonged idling (vehicle operation in stop and go traffic).

- Frequent short running periods (engine does not warm to normal operating temperatures).

- Police, taxi, delivery usage or trailer towing usage.

Oil & oil filter change: change every 3000 miles.

Air filter element: change every 15,000 miles.

Special Operating Condition Requirements

When towing a trailer or using a camper or car-top carrier:

Change engine oil and install a new oil filter every 4,800 km (3,000 miles) or 3 months.

During extensive idling and/or low speed driving for long distances, as in heavy commercial use such as delivery, taxi, patrol car or livery:

Change engine oil and install a new oil filter, lube front lower control arm and steering linkage ball joints with Zerk fittings (if equipped) every 4,800 km (3,000 miles) or 3 months.

Inspect brake system and check battery electrolyte level (Patrol cars) every 8,000 km (5,000 miles).

Install a new fuel filter every 24,000 km (15,000 miles).

Change automatic transmission fluid, lubricate 4x2 wheel bearings,

Install new spark plugs and change transfer case fluid every 96,000 km (60,000 miles).

Install a new cabin air filter as required.

When operating in dusty conditions such as unpaved or dusty roads:

Change engine oil and install a new oil filter every 4,800 km (3,000 miles) or 3 months.

Install a new fuel filter every 24,000 km (15,000 miles).

Change automatic transmission fluid every 48,000 km (30,000 miles).

Install a new engine air filter as required.

Install a new cabin air filter as required.

When operating in off-road conditions:

Change automatic transmission fluid every 48,000 km (30,000 miles).

Install a new cabin air filter as required.

Inspect and lubricate U-joints.

Inspect and lubricate steering linkage ball joints with zerk fittings.

67197-THUN-C12

PRECAUTIONS

Before servicing any vehicle, please be sure to read all of the following precautions, which deal with personal safety, prevention of component damage, and important points to take into consideration when servicing a motor vehicle:

- Never open, service or drain the radiator or cooling system when the engine is hot; serious burns can occur from the steam and hot coolant.
- Observe all applicable safety precautions when working around fuel. Whenever servicing the fuel system, always work in a well-ventilated area. Do not allow fuel spray or vapors to come in contact with a spark, open flame, or excessive heat (a hot drop light, for example). Keep a dry chemical fire extinguisher near the work area. Always keep fuel in a container specifically designed for fuel storage; also, always properly seal fuel containers to avoid the possibility of fire or explosion. Refer to the additional fuel system precautions later in this section.
- Fuel injection systems often remain pressurized, even after the engine has been turned **OFF**. The fuel system pressure must be relieved before disconnecting any fuel lines. Failure to do so may result in fire and/or personal injury.
- Brake fluid often contains polyglycol ethers and polyglycols. Avoid contact with the eyes and wash your hands thoroughly after handling brake fluid. If you do get brake fluid in your eyes, flush your eyes with clean, running water for 15 minutes. If eye irritation persists, or if you have taken brake fluid internally, IMMEDIATELY seek medical assistance.
- The EPA warns that prolonged contact with used engine oil may cause a number of skin disorders, including cancer. You should make every effort to minimize your exposure to used engine oil. Protective gloves should be worn when changing oil. Wash your hands and any other exposed skin areas as soon as possible after exposure to used engine oil. Soap and water, or waterless hand cleaner should be used.
- All new vehicles are now equipped with an air bag system, often referred to as a Supplemental Restraint System (SRS) or Supplemental Inflatable Restraint (SIR) system. The system must be disabled before performing service on or around system components, steering column, instrument panel components, wiring and sensors. Failure to follow safety and disabling procedures could result in accidental air bag deployment, possible personal injury and unnecessary system repairs.
- Always wear safety goggles when working with, or around, the air bag system. When carrying a non-deployed air bag, be sure the bag and trim cover are pointed away from your body. When placing a non-deployed air bag on a work surface, always face the bag and trim cover upward, away from the surface. This will reduce the motion of the module if it is accidentally deployed. Refer to the additional air bag system precautions later in this section.
- Clean, high quality brake fluid from a sealed container is essential to the safe and proper operation of the brake system. You should always buy the correct type of brake fluid for your vehicle. If the brake fluid becomes contaminated, completely flush the system with new fluid. Never reuse any brake fluid. Any brake fluid that is removed from the system should be discarded. Also, do not allow any brake fluid to come in contact with a painted surface; it will damage the paint.
- Never operate the engine without the proper amount and type of engine oil; doing so WILL result in severe engine damage.
- Timing belt maintenance is extremely important. Many models utilize an interference-type, non-freewheeling engine. If the timing belt breaks, the valves in the cylinder head may strike the pistons, causing potentially serious (also time-consuming and expensive) engine damage. Refer to the maintenance interval charts in the front of this manual for the recommended replacement interval for the timing belt.
- Disconnecting the negative battery cable on some vehicles may interfere with the functions of the on-board computer system(s) and may require the computer to undergo a relearning process once the negative battery cable is reconnected.
- When servicing drum brakes, only disassemble and assemble one side at a time, leaving the remaining side intact for reference.

ENGINE REPAIR

➡Disconnecting the negative battery cable on some vehicles may interfere with the functions of the on board computer system. The computer may undergo a relearning process once the negative battery cable is reconnected.

Alternator

REMOVAL

1. Before servicing the vehicle, refer to the precautions in the beginning of this section.
2. Remove or disconnect the following:
   - Negative battery cable
   - Air intake tube
   - Accessory drive belt
   - Lower splash shield
   - Alternator mounting bolts
   - Alternator harness connectors
   - Alternator

INSTALLATION

1. Install the alternator harness connectors, then install the alternator. Tighten the bolts as follows:
   a. Step 1: Tighten bolt No. 1 to 15 ft. lbs. (21 Nm)
   b. Step 2: Tighten nut No. 3 to 33 ft. lbs. (45 Nm)
2. Install or connect the following:
   - Lower splash shield
   - Accessory drive belt
   - Air intake tube
   - Negative battery cable

Ignition Timing

ADJUSTMENT

This vehicle is equipped with a Distributorless Ignition System (DIS). The ignition timing is not adjustable. It is controlled by the PCM.

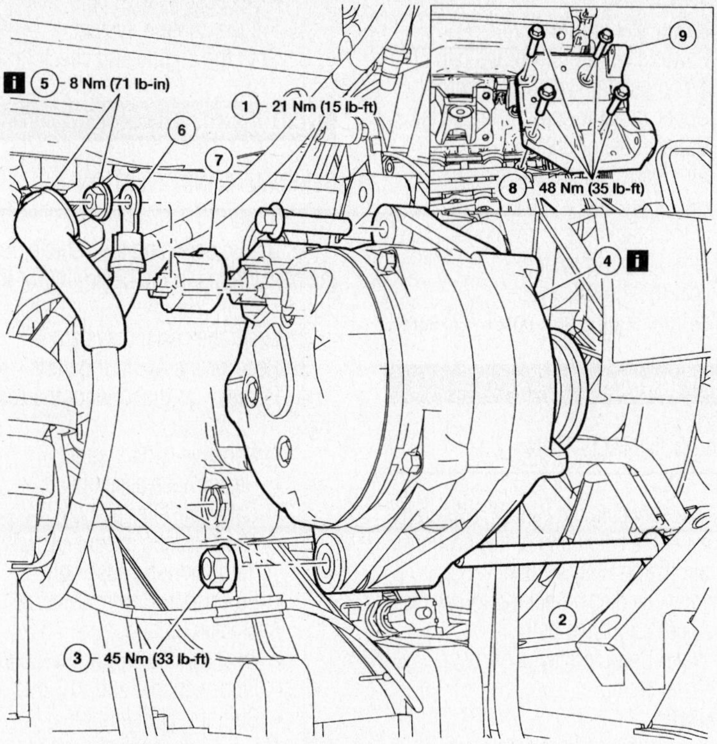

1 Generator upper bolt
2 Generator lower bolt
3 Generator lower nut
4 Generator
5 Generator B+ wire nut
6 Generator B+ wire
7 Connector
8 Generator bracket bolts
9 Generator bracket

67197-THUN-G01

Alternator mounting—3.9L engine

Engine Assembly

REMOVAL & INSTALLATION

1. Before servicing the vehicle, refer to the precautions in the beginning of this section.
2. Drain the cooling system.
3. Relieve the fuel system pressure.
4. Recover the A/C refrigerant.
5. Drain the engine oil.
6. Remove or disconnect the following:
   - Negative battery cable
   - Air cleaner inlet tube
   - Upper radiator shield
   - Upper radiator support brackets
   - A/C pressure switch connector
   - Power steering return line clip
   - Power steering reservoir
   - Vapor Management Valve (VMV) cover
   - VMV vacuum hose and canister purge hose
   - Main vacuum supply hose
   - Cowl vent screens
   - Cross vehicle support bar
   - Degas bottle hose
   - Accelerator cable
   - Cruise control cable
   - Ground strap
   - Fresh air filter and housing
   - Powertrain harness connectors at right strut tower
   - Fresh air filter panel
   - Main engine wiring harness connector
   - Main transmission wiring harness connector
   - Heater hoses at the water control valve. Note the locations for assembly.
   - Hydraulic cooling fan reservoir
   - Water control valve harness connector
   - Driveshaft
   - Front wheels
   - Inner splash shields
   - Wheel speed sensor connectors and harness clips
   - Brake calipers
   - Lower stabilizer bar links
   - Upper ball joints
   - Lower strut mount bolts
   - Left, right and center splash shields
   - A/C suction and discharge lines
   - Shift cable and bracket
   - Power steering line frame rail clip
   - Rack and pinion harness connectors
   - Steering shaft bolt and coupling
   - Starter motor harness connectors and ground cable
   - Alternator harness connectors
   - Lower transmission flange bolts
   - Torque converter
   - Inner air deflector
   - Engine block heater, if equipped
7. Support the engine, transmission, front and center crossmembers and the cooling system with a powertrain lift and transmission support bracket.
8. Support the rear of the vehicle with safety stands.
9. Remove or disconnect the following:
   - Transmission crossmember bolts
   - Front crossmember bolts
   - Center crossmember bolts
   - Powertrain assembly
   - A/C compressor manifold and tube assembly
   - Power steering pump return hose
   - Hydraulic cooling fan return hose
   - Lower radiator hose
   - Upper radiator hoses
   - Knock Sensor (KS) connector
   - Heater hose
   - Transmission cooler lines and bracket
   - Power steering pressure line and bracket
   - Hydraulic cooling fan pressure line and bracket
10. Attach a hoist to the engine.
11. Remove the motor mount nuts and lift the powertrain out of the subframe.
12. Remove or disconnect the following:
    - Wiring harness retainers
    - Upper transmission flange bolts
    - Transmission from the engine

To install:

13. Install or connect the following:
    - Transmission to the engine. Tighten the upper flange bolts to 35 ft. lbs. (48 Nm).
    - Wiring harness retainers. Tighten the nuts to 89 inch lbs. (10 Nm).
    - Powertrain to the subframe. Tighten the mount nuts to 30 ft. lbs. (40 Nm).
    - Hydraulic cooling fan pressure line and bracket
    - Power steering pressure line and bracket
    - Transmission cooler lines and bracket
    - Heater hose
    - Knock sensor connector
    - Upper radiator hoses
    - Lower radiator hose
    - Hydraulic cooling fan return hose
    - Power steering pump return hose

- A/C compressor manifold and tube assembly. Tighten the bolt to 15 ft. lbs. (21 Nm).
- Powertrain assembly. Tighten the front and center crossmember bolts to 76 ft. lbs. (103 Nm) and the transmission crossmember bolts to 30 ft. lbs. (40 Nm).
- Engine block heater, if equipped
- Inner air deflector
- Torque converter. Tighten the nuts to 28 ft. lbs. (38 Nm).
- Lower transmission flange bolts. Tighten the bolts to 35 ft. lbs. (47 Nm).
- Alternator harness connectors
- Starter motor harness connectors and ground cable
- Steering shaft bolt and coupling. Tighten the coupling pinch bolt to 26 ft. lbs. (35 Nm) and the shaft bolt to 22 ft. lbs. (30 Nm).
- Rack and pinion harness connectors
- Power steering line frame rail clip
- Shift cable and bracket
- A/C suction and discharge lines
- Left, right and center splash shields
- Lower strut mount bolts. Tighten the bolts to 129 ft. lbs. (175 Nm).
- Upper ball joints. Tighten the nuts to 66 ft. lbs. (90 Nm).
- Lower stabilizer bar links. Tighten the nuts to 41 ft. lbs. (55 Nm).
- Brake calipers
- Wheel speed sensor connectors and harness clips
- Driveshaft
- Inner splash shields
- Front wheels
- Water control valve harness connector
- Hydraulic cooling fan reservoir
- Heater hoses at the water control valve
- Main transmission wiring harness connector
- Main engine wiring harness connector
- Fresh air filter panel
- Powertrain harness connectors at right strut tower
- Fresh air filter and housing
- Ground strap
- Cruise control cable
- Accelerator cable
- Degas bottle hose
- Cross vehicle support bar. Tighten the bolts to 15 ft. lbs. (20 Nm).
- Cowl vent screens
- Main vacuum supply hose
- VMV vacuum hose and canister purge hose
- VMV cover
- Power steering reservoir
- Power steering return line clip
- A/C pressure switch connector
- Upper radiator support brackets
- Upper radiator shield
- Air cleaner inlet tube
- Negative battery cable

14. Fill the crankcase to the correct level.
15. Fill the cooling system.
16. Recharge the A/C system.
17. Start the engine and check for leaks.

Water Pump

REMOVAL & INSTALLATION

1. Before servicing the vehicle, refer to the precautions in the beginning of this section.
2. Drain the cooling system.
3. Remove or disconnect the following:
   - Accessory drive belt
   - Water pump pulley
   - Water pump

To install:

4. Install or connect the following:
   - Water pump. Tighten the bolts to 71 inch lbs. (8 Nm) plus 90 degrees.
   - Water pump pulley. Tighten the bolts to 89 inch lbs. (10 Nm) plus 45 degrees.
   - Accessory drive belt
5. Fill the cooling system.
6. Start the engine and check for leaks.

Heater Core

REMOVAL & INSTALLATION

1. Before servicing the vehicle, refer to the precautions in the beginning of this section.
2. Drain the cooling system.
3. Recover the A/C refrigerant.
4. Remove or disconnect the following:
   - Negative battery cable
   - Heater hose assembly
   - Cabin air filter plenum
   - Thermostatic expansion valve manifold and tube assembly
   - Powertrain Control Module (PCM) bracket bolt
   - Driver's side air bag module
   - Floor console and A/C duct
   - Shift lever assembly
   - Glove box assembly
   - Left and right instrument panel insulators
   - Left and right door sill scuff plates
   - Left and right door weatherstrips
   - Left and right A pillar lower trim panels

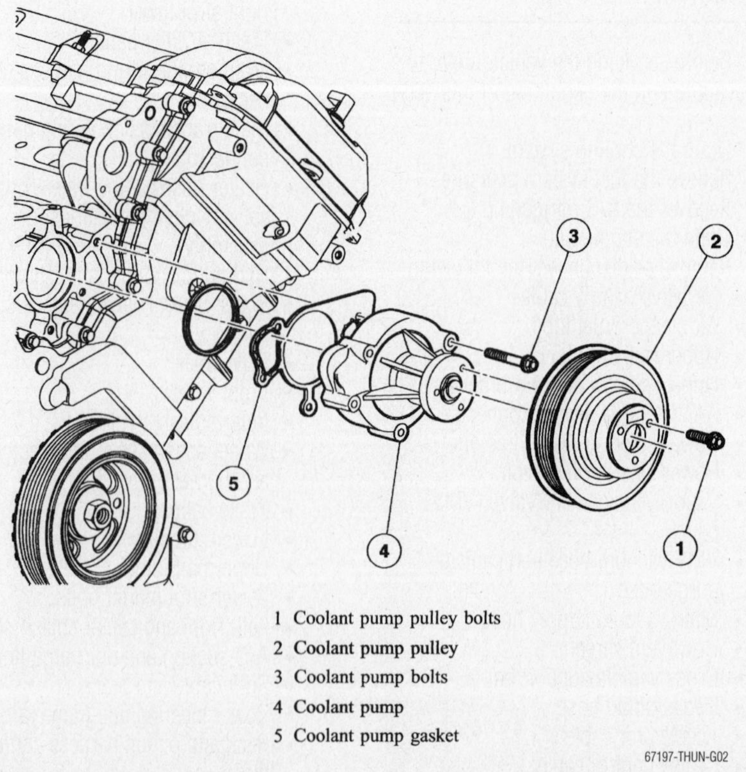

Water pump mounting—3.9L engine

- Left and right windshield side garnish moldings
- Instrument panel defroster opening grille assembly
- Instrument panel cowl top screws
- Instrument panel upper reinforcement bolts
- Left and right instrument panel side finish panels
- Hood release handle and cable
- Upper right bulkhead electrical connector
- Right instrument panel electrical connectors
- Passenger side tunnel electrical connector
- Steering column intermediate shaft
- Left junction box electrical connectors
- Left instrument panel electrical connectors
- Ignition shift interlock connector, if equipped
- 4 instrument panel tunnel brace bolts
- Left outer instrument panel cowl side cover and reinforcement bolt
- Left instrument panel cowl side bolt and nut
- Right instrument panel cowl side bolt and nut
- Instrument panel
- Evaporator housing electrical connector
- Cowl top attachment bolt
- Evaporator housing attachment bolt
- 3 evaporator nuts and washers in the engine compartment
- Rear seat floor ducts
- Evaporator core housing
- Evaporator core housing air inlet
- Bypass door harness connector
- Heater core

To install:

5. Install or connect the following:
   - Heater core
   - Bypass door harness connector
   - Evaporator core housing air inlet
   - Evaporator core housing
   - Rear seat floor ducts
   - 3 evaporator nuts and washers in the engine compartment
   - Evaporator housing attachment bolt
   - Cowl top attachment bolt
   - Evaporator housing electrical connector
   - Instrument panel
   - Instrument panel cowl top screws. Tighten to 27 inch lbs. (3 Nm).
   - Right instrument panel cowl side bolt and nut. Tighten the fasteners to 15 ft. lbs. (20 Nm).

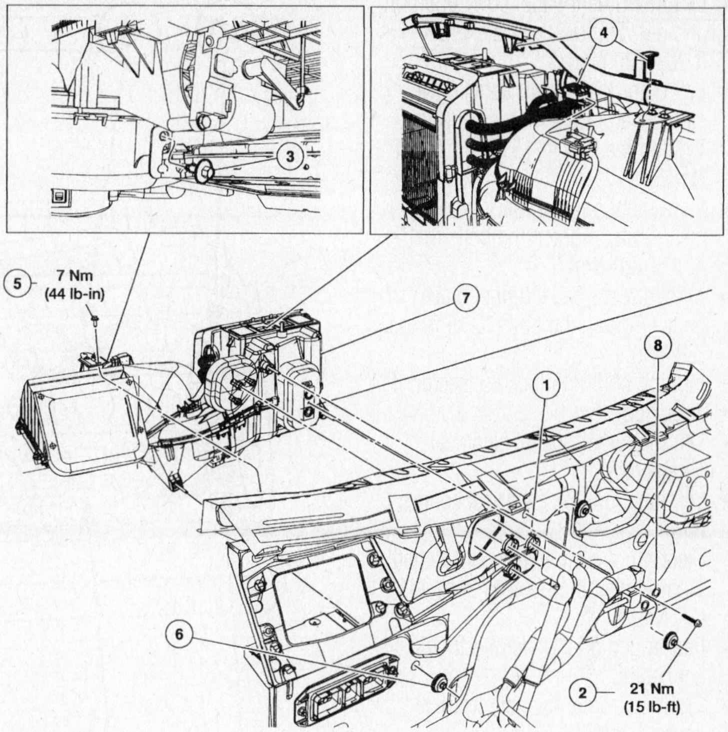

1 Heater hose clamp
2 Thermostatic expansion valve manifold bolt
3 PCM bracket bolt
4 Wire harness
5 Heater core and evaporator core housing bracket bolt
6 Heater core and evaporator core housing nut
7 Heater core and evaporator core housing
8 Thermostatic expansion valve O-ring

67197-THUN-G03

Heater core and evaporator core housing mounting—Thunderbird

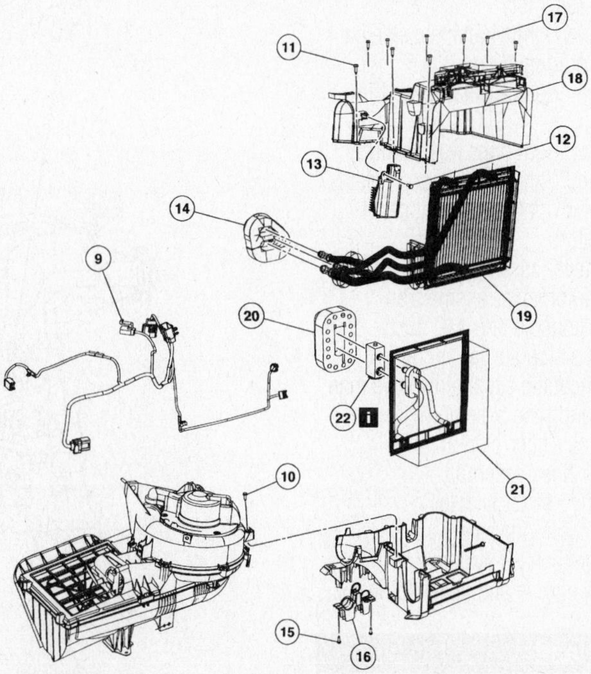

9 Wire harness
10 Heater core and evaporator core housing screw
11 Heater core and evaporator core housing screw
12 Blower motor speed control screw
13 Blower motor speed control
14 Heater core tube seal
15 Heater core tube bracket screw
16 Heater core tube bracket
17 Heater core and evaporator core housing screw
18 Heater core and evaporator core housing cover
19 Heater core
20 Thermostatic expansion valve seal
21 Evaporator core
22 Thermostatic expansion valve

67197-THUN-G04

Exploded view of heater core—Thunderbird

- Left instrument panel cowl side bolt and nut. Tighten the fasteners to 15 ft. lbs. (20 Nm).
- Left outer instrument panel cowl side cover and reinforcement bolt. Tighten the bolt to 15 ft. lbs. (20 Nm).
- Instrument panel upper reinforcement bolts. Tighten the bolts to 15 ft. lbs. (20 Nm).
- 4 instrument panel tunnel brace bolts. Tighten the bolts to 15 ft. lbs. (20 Nm).
- Ignition shift interlock connector, if equipped
- Left instrument panel electrical connectors
- Left junction box electrical connectors
- Steering column intermediate shaft. Tighten the pinch bolt to 26 ft. lbs. (35 Nm).
- Passenger side tunnel electrical connector
- Right instrument panel electrical connectors
- Upper right bulkhead electrical connector
- Hood release handle and cable
- Left and right instrument panel side finish panels
- Instrument panel defroster opening grille assembly
- Left and right windshield side garnish moldings
- Left and right A pillar lower trim panels
- Left and right door weatherstrips
- Left and right door sill scuff plates
- Left and right instrument panel insulators
- Shift lever assembly, if equipped with automatic transmission
- Floor console and A/C duct
- Driver's side air bag module
- Thermostatic expansion valve manifold and tube assembly
- Cabin air filter plenum
- Heater hose assembly
- Negative battery cable

6. Fill the cooling system.
7. Recharge the A/C system.
8. Start the engine and check for leaks.

Cylinder Head

REMOVAL & INSTALLATION

1. Before servicing the vehicle, refer to the precautions in the beginning of this section.
2. Drain the cooling system.

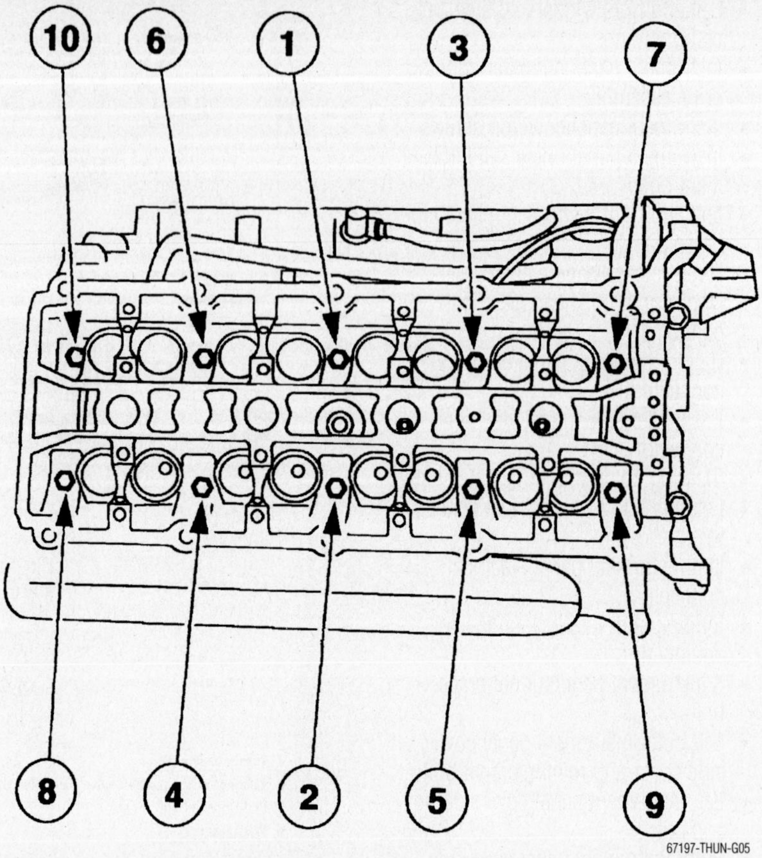

Right cylinder head torque sequence—3.9L engine

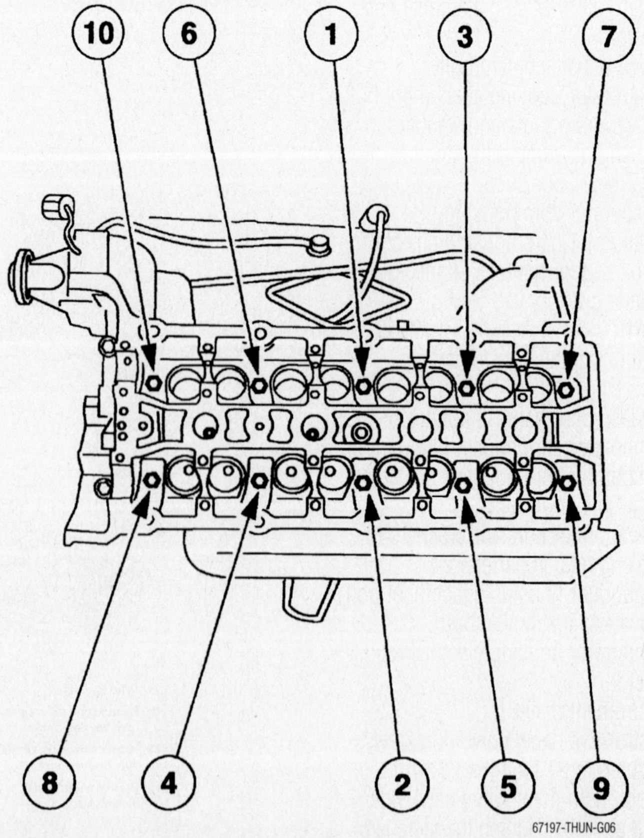

Left cylinder head torque sequence—3.9L engine

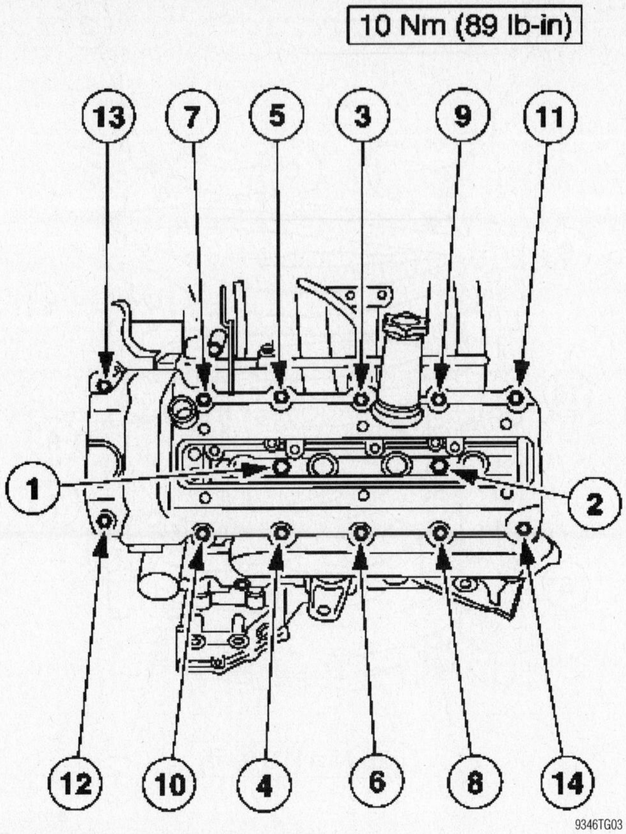

Left valve cover torque sequence—3.9L Engine

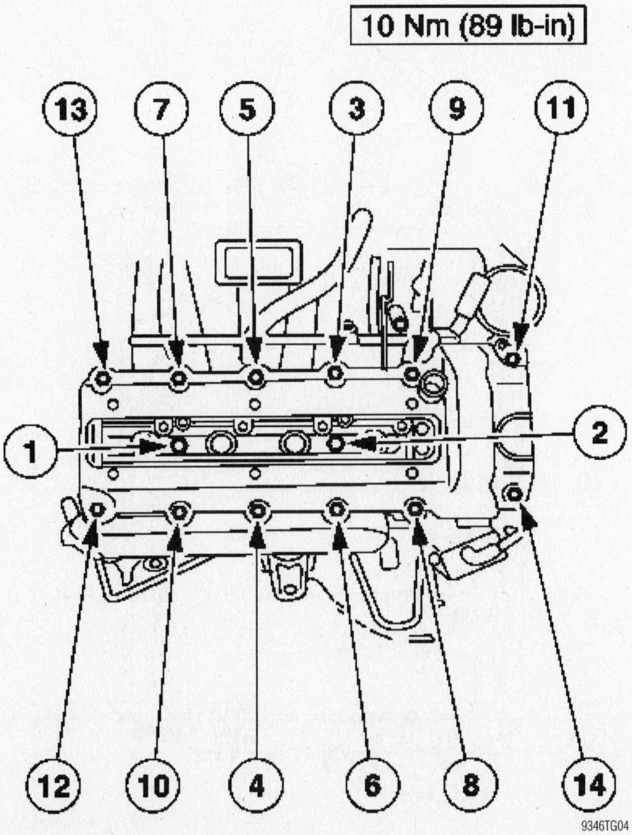

Right valve cover torque sequence—3.9L Engine

3. Relieve the fuel system pressure.
4. Remove or disconnect the following:
   - Negative battery cable
   - Engine appearance cover
   - Intake manifold
   - Valve covers
   - Accessory drive belts
   - Front cover
   - Timing chains
   - Camshafts
   - Water outlet pipe
   - Cylinder head temperature sensor connector
   - Exhaust front pipes
   - Exhaust Gas Recirculation (EGR) tube
   - Bolts and stud bolts at the rear of the cylinder heads
   - Cylinder heads. Loosen the bolts in reverse of the tightening sequence.

To install:

5. Install the cylinder heads. Tighten the NEW bolts in sequence as follows:
   a. Step 1: M10 bolts to 15 ft. lbs. (20 Nm)
   b. Step 2: M10 bolts to 26 ft. lbs. (35 Nm)
   c. Step 3: M10 bolts to 33 ft. lbs. (45 Nm)
   d. Step 4: M10 bolts plus 90 degrees
   e. Step 5: M10 bolts plus 90 degrees
   f. Step 6: M8 bolts to 15 ft. lbs. (20 Nm)
   g. Step 7: M8 bolts plus 90 degrees
6. Install or connect the following:
   - Bolts and stud bolts at the rear of the cylinder heads. Tighten the bolts to 37 ft. lbs. (50 Nm).
   - EGR tube
   - Exhaust front pipes
   - Cylinder head temperature sensor connector
   - Water outlet pipe
   - Camshafts
   - Timing chains
   - Front cover
   - Accessory drive belts
   - Valve covers
   - Intake manifold
   - Engine appearance cover
   - Negative battery cable
7. Fill the cooling system.
8. Start the engine and check for leaks.

Rocker Arms/Shafts

REMOVAL & INSTALLATION

The vehicles covered in this section are not equipped with rocker arms/shafts. The camshaft directly actuates the valves.

Intake Manifold

REMOVAL & INSTALLATION

1. Before servicing the vehicle, refer to the precautions in the beginning of this section.
2. Drain the cooling system.
3. Relieve the fuel system pressure.
4. Remove or disconnect the following:
   - Negative battery cable
   - Air cleaner outlet tube
   - Cowl vent screen
   - Cross vehicle support bar
   - Fuel line
   - Exhaust Gas Recirculation (EGR) valve
   - Throttle Position (TP) sensor connector
   - Fuel pressure sensor connector and vacuum line
   - Throttle plate motor connector
   - Vapor management valve connector
   - Fuel rail temperature sensor connector
   - Accelerator cable
   - Cruise control cable
   - Intake manifold vacuum hoses
   - Idle Air Control (IAC) valve connector
   - Positive Crankcase Ventilation (PCV) tube
   - Knock Sensor (KS) connector
   - Camshaft Position (CMP) sensor connector
   - Evaporative Emissions (EVAP) canister purge valve line
   - Appearance covers and support brackets
   - Fuel rail
   - Fuel injector connectors
   - Throttle body coolant hoses
   - Delta Pressure Feedback Electronic (DPFE) system sensor
   - Intake manifold. Loosen the bolts in reverse of the tightening sequence.

To install:

5. Install or connect the following:
   - Intake manifold. Tighten the bolts in sequence to 18 ft. lbs. (25 Nm) on 2002–2003 models, or 15 ft. lbs. (20 Nm) on 2004 models.
   - Delta Pressure Feedback Electronic (DPFE) system sensor
   - Throttle body coolant hoses
   - Fuel injector connectors
   - Fuel rail
   - Appearance covers and support brackets

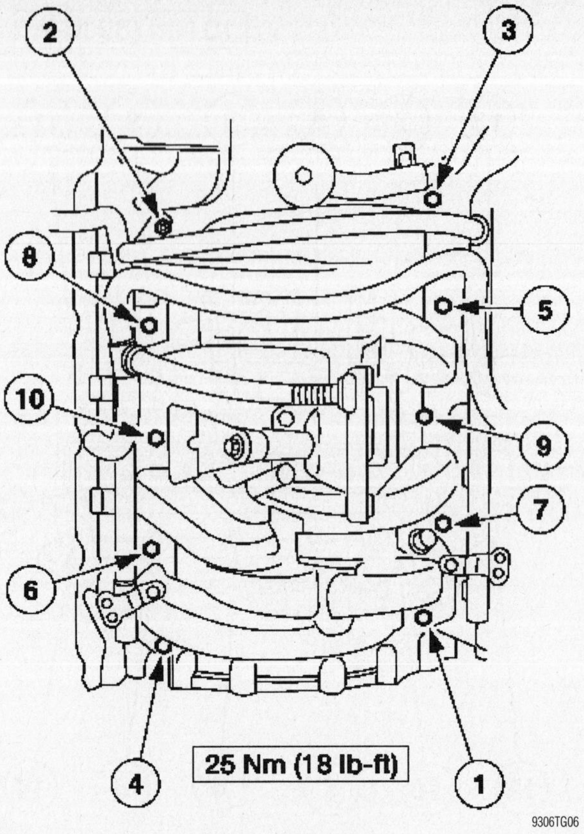

Intake manifold torque sequence—2002–2003 3.9L engine

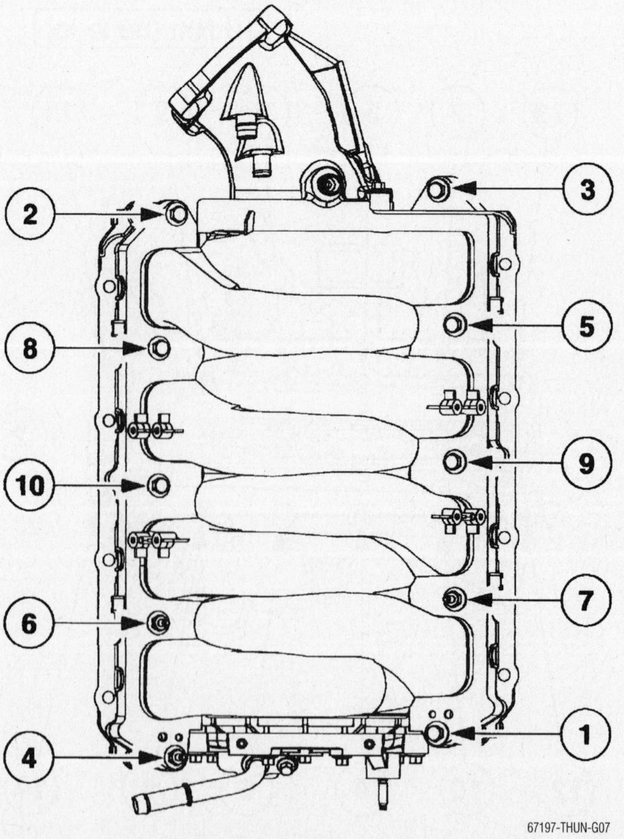

Intake manifold torque sequence—2004 3.9L engine

- Evaporative Emissions (EVAP) canister purge valve line
- Camshaft Position (CMP) sensor connector
- Knock Sensor (KS) connector
- Positive Crankcase Ventilation (PCV) tube
- Idle Air Control (IAC) valve connector
- Intake manifold vacuum hoses
- Cruise control cable
- Accelerator cable
- Fuel rail temperature sensor connector
- Vapor management valve connector
- Throttle plate motor connector
- Fuel pressure sensor connector and vacuum line
- Throttle Position (TP) sensor connector
- Exhaust Gas Recirculation (EGR) valve
- Fuel line
- Cross vehicle support bar. Tighten the bolts to 15 ft. lbs. (20 Nm).
- Cowl vent screen
- Air cleaner outlet tube
- Negative battery cable

6. Fill the cooling system.
7. Start the engine and check for leaks.

Exhaust Manifolds

REMOVAL & INSTALLATION

1. Before servicing the vehicle, refer to the precautions in the beginning of this section.
2. Remove or disconnect the following:
   - Negative battery cable
   - Power steering pump reservoir
   - Oil dipstick tube
   - Exhaust front pipes
   - Exhaust Gas Recirculation (EGR) tube
   - Heat shield
   - Exhaust manifolds

To install:

3. Install or connect the following:
   - Exhaust manifolds. Tighten the bolts in sequence to 18 ft. lbs. (25 Nm).
   - Heat shield
   - EGR tube
   - Exhaust front pipes
   - Oil dipstick tube
   - Power steering pump reservoir
   - Negative battery cable
4. Start the engine and check for leaks.

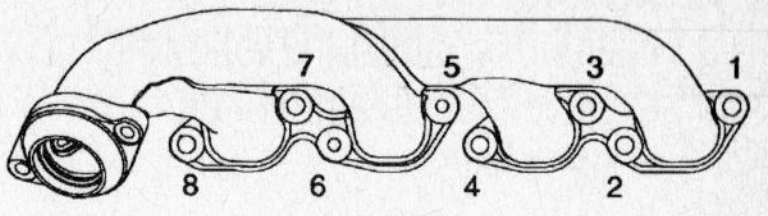

Exhaust manifold torque sequence—3.9L engine

Camshaft and Valve Lifters

REMOVAL & INSTALLATION

1. Before servicing the vehicle, refer to the precautions in the beginning of this section.
2. Remove or disconnect the following:
   - Negative battery cable
   - Valve covers
   - Front cover
   - Timing chains

➡**Keep all valve train components in order for assembly.**

   - Camshaft journal bearing caps
   - Camshafts
   - Valve tappets and shims

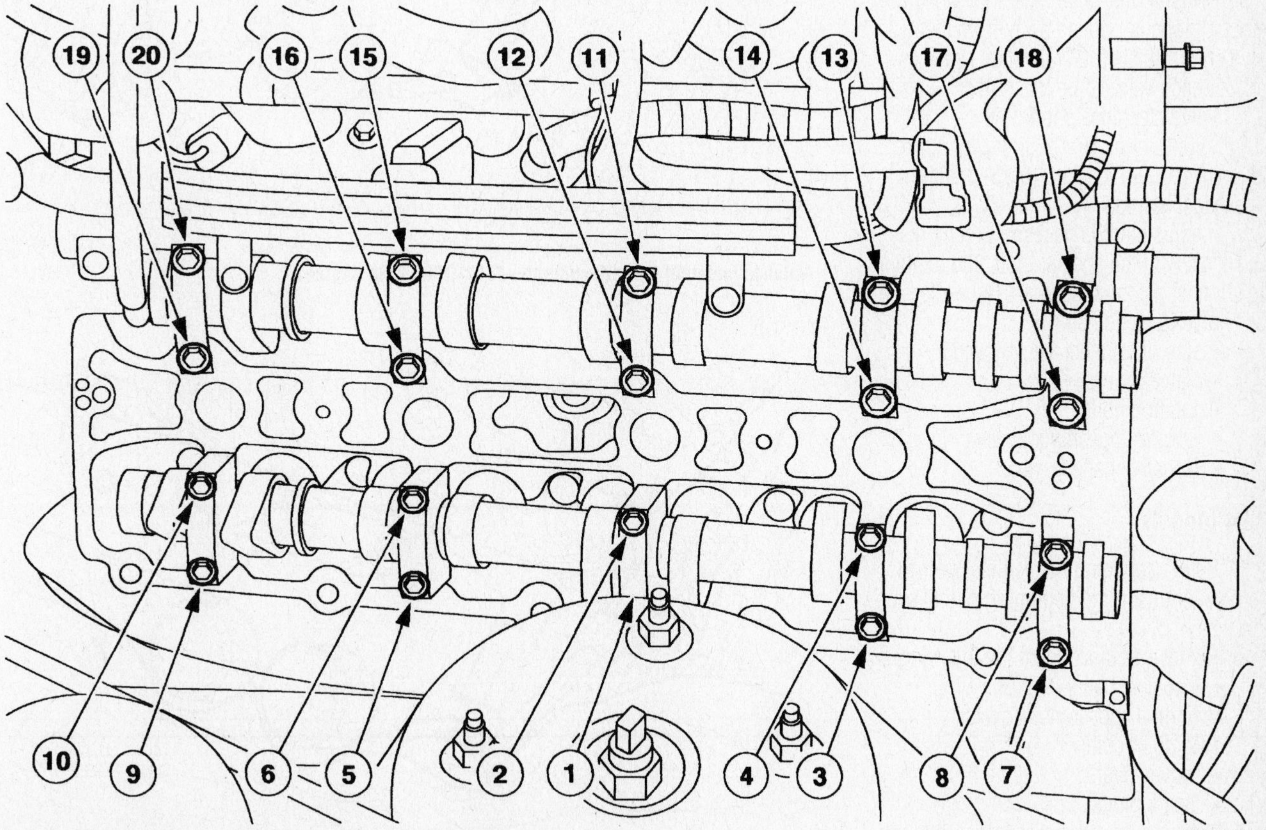

Camshaft journal bearing cap torque sequence—3.9L engine

To install:

3. Install or connect the following:
   - Valve tappets and shims
   - Camshafts
4. Install the camshaft journal bearing caps in their original positions. Tighten the bolts in sequence as follows:
   a. Step 1: Finger tight
   b. Step 2: 53 inch lbs. (6 Nm)
   c. Step 3: Plus 90 degrees
5. Install or connect the following:
   - Timing chains
   - Front cover
   - Valve covers
   - Negative battery cable
6. Start the engine and check for leaks.

Valve Lash

ADJUSTMENT

2002–2003 Models

1. Before servicing the vehicle, refer to the precautions in the beginning of this section.
2. Remove or disconnect the following:
   - Negative battery cable
   - Engine appearance covers
   - Ignition coils
   - Valve covers
3. Measure the valve clearance while the camshaft lobe is pointed away from the valve shim. Rotate the crankshaft as necessary for each valve to be measured.
4. Valve clearance should be 0.007–0.009 in. (0.18–0.23mm) for intake valves or 0.009–0.011 in. (0.23–0.28mm) for exhaust valves.
5. If adjustment is necessary, compress the valves with the special tools and remove the shim with compressed air. Repeat for each valve to be adjusted.
6. Install or connect the following:
   - Valve covers
   - Ignition coils
   - Engine appearance covers
   - Negative battery cable

2004 Models

1. Before servicing the vehicle, refer to the precautions in the beginning of this section.
2. Remove or disconnect the following:
   - Negative battery cable
   - Engine appearance covers
   - Ignition coils
   - Valve covers
   - Spark plugs
3. Measure the valve clearance while the camshaft lobe is pointed away from the

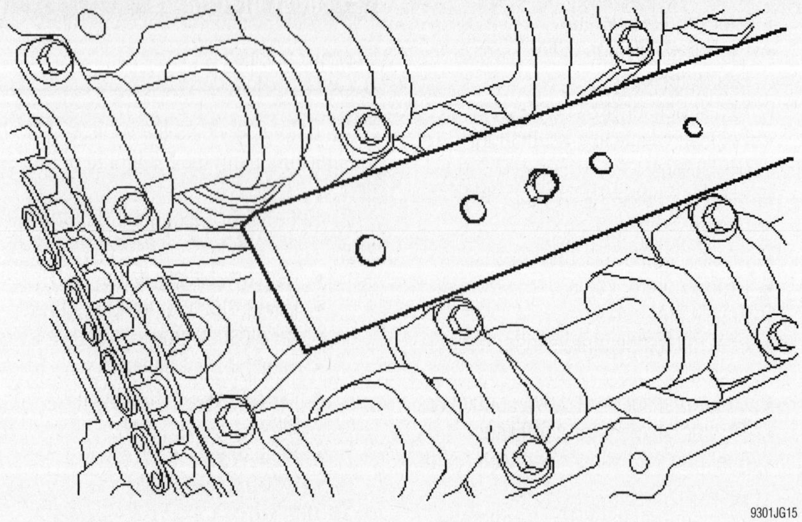

Valve adjustment tool base plate—2002–2003 3.9L engine

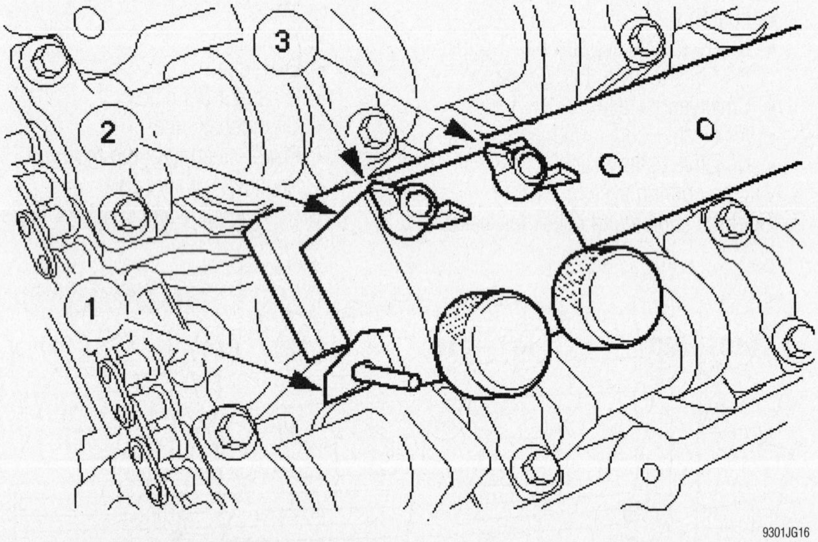

Valve adjustment tool attachment—2002–2003 3.9L engine

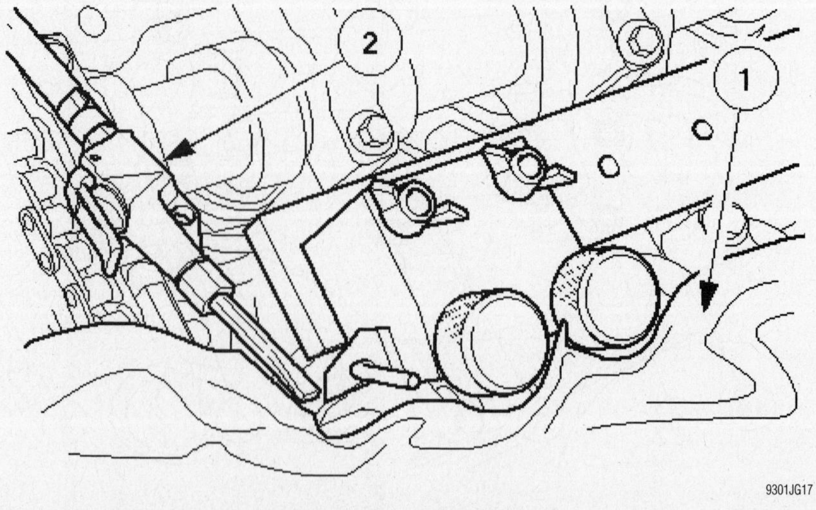

Remove the shims with compressed air—2002–2003 3.9L engine

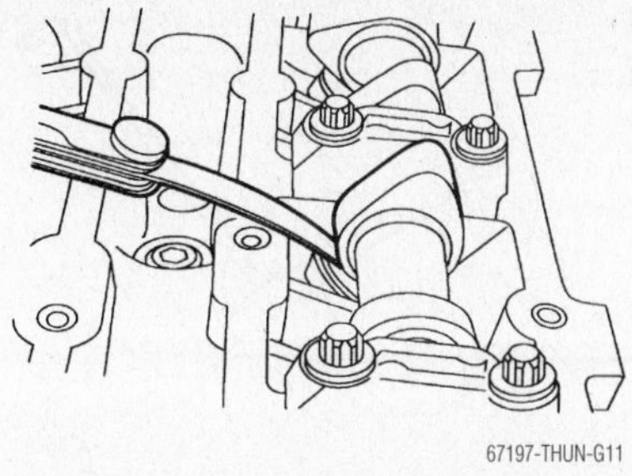

Measuring valve clearance—2004 3.9L engine

valve shim. Rotate the crankshaft as necessary for each valve to be measured.

4. Valve clearance should be 0.007–0.009 in. (0.18–0.23mm) for intake valves or 0.013–0.015 in. (0.33–0.38mm) for exhaust valves.

5. If adjustment is necessary, remove the timing chains and camshafts. Remove the shim and replace the shim with the correct shim to achieve the desired clearance. Repeat for each valve to be adjusted.

6. Install or connect the following:
- Camshafts
- Timing chains
- Spark plugs
- Valve covers
- Ignition coils
- Engine appearance covers
- Negative battery cable

Starter Motor

REMOVAL & INSTALLATION

1. Before servicing the vehicle, refer to the precautions in the beginning of this section.

2. Remove or disconnect the following:
- Negative battery cable
- Starter motor wiring connectors
- Starter motor

To install:

3. Install or connect the following:
- Starter motor. Tighten the bolts to 18 ft. lbs. (25 Nm).
- Starter motor wiring connectors
- Negative battery cable

Oil Pan

REMOVAL & INSTALLATION

1. Before servicing the vehicle, refer to the precautions in the beginning of this section.

2. Drain the engine oil.

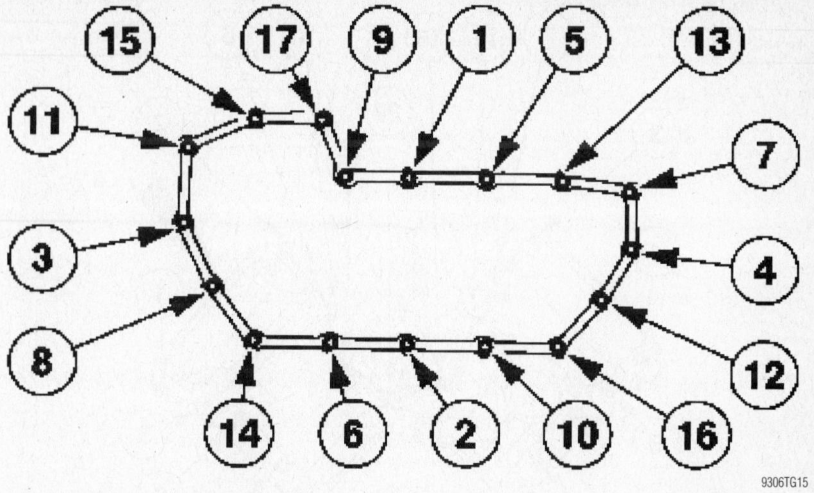

Oil pan torque sequence—3.9L engine

3. Remove the engine crossmember

4. Remove the oil pan.

To install:

5. Install the oil pan. Tighten the bolts in sequence as follows:
   a. Step 1: 44 inch lbs. (5 Nm)
   b. Step 2: 108 inch lbs. (12 Nm)

6. Install the engine crossmember and tighten the bolts to 36 ft. lbs. (49 Nm).

7. Fill the crankcase to the correct level.

8. Start the engine and check for leaks.

Oil Pump

REMOVAL INSTALLATION

1. Before servicing the vehicle, refer to the precautions in the beginning of this section.

2. Remove or disconnect the following:
- Crankshaft pulley
- Front cover
- Primary timing chains
- Oil pump mounting bolts
- Oil pump

To install:

3. Install or connect the following:
- New gasket
- Oil pump. Tighten the bolts to 53 inch lbs. (6 Nm) plus 90 degrees.
- Primary timing chains
- Front cover
- Crankshaft pulley

Rear Main Seal

REMOVAL & INSTALLATION

1. Before servicing the vehicle, refer to the precautions at the beginning of this section.

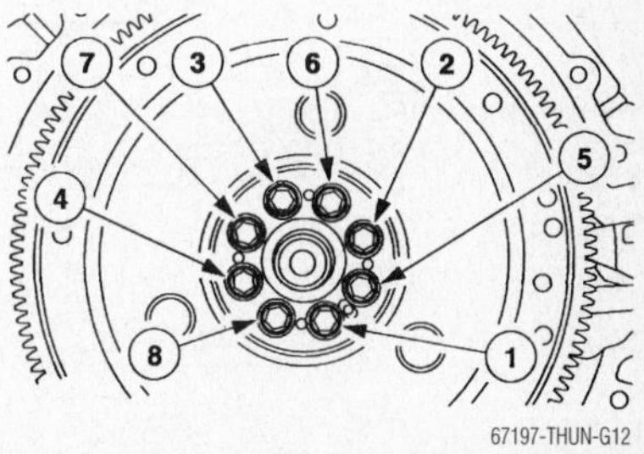

Flywheel torque sequence—3.9L engine

2. Attach an engine support fixture to the engine lifting eyes.

3. Remove or disconnect the following:
- Negative battery cable
- Transmission
- Flywheel
- Rear crankshaft seal

To install:

4. Install or connect the following:
- Rear main seal flush with the cylinder block surface
- Flywheel

5. Tighten the flywheel bolts in sequence as follows:
   a. Step 1: 11 ft. lbs. (15 Nm)
   b. Step 2: 81 ft. lbs. (110 Nm)

6. Install or connect the following:
- Transmission
- Negative battery cable

7. Start the engine and check for leaks.

Timing Chain, Sprockets, Front Cover and Seal

REMOVAL & INSTALLATION

1. Before servicing the vehicle, refer to the precautions in the beginning of this section.

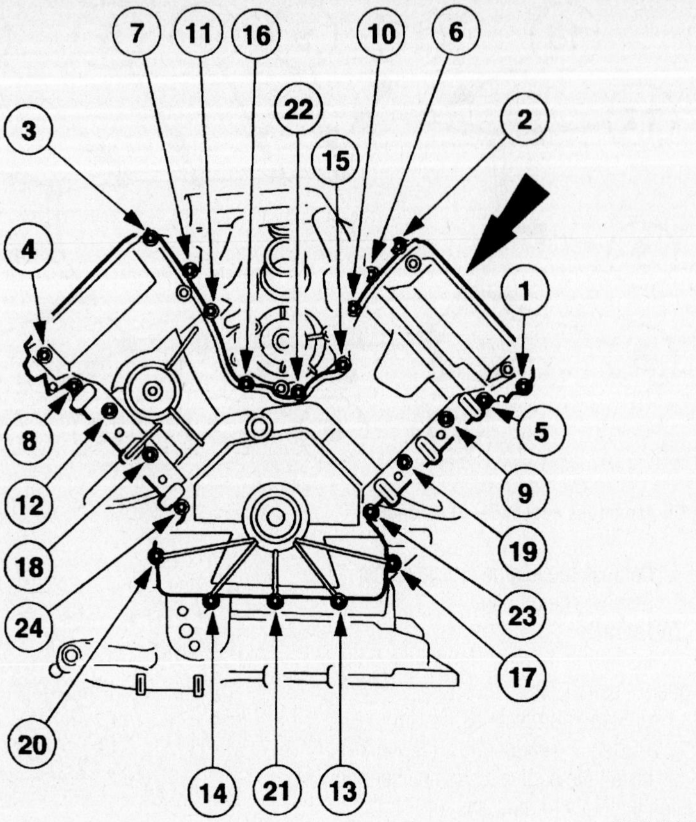

Front cover bolt removal sequence—3.9L engine

2. Drain the cooling system.
3. Drain the engine oil.
4. Remove or disconnect the following:
   - Negative battery cable
   - Engine appearance cover and brackets
   - Valve covers
   - Engine cooling fan assembly
   - Accessory drive belt
   - Lower splash shield
   - Oil filter
   - Oil cooler
   - Crankshaft pulley and discard the bolt
   - Front crankshaft seal
   - A/C compressor
   - Lower radiator hose and pipe
   - Water pump
   - Heater hose
   - Power steering reservoir hose
   - Power Steering Pressure (PSP) switch connector
   - Power steering pump and bracket
   - Alternator and bracket
   - Front cover wiring harness clips
   - Front cover. Loosen the bolts in sequence.

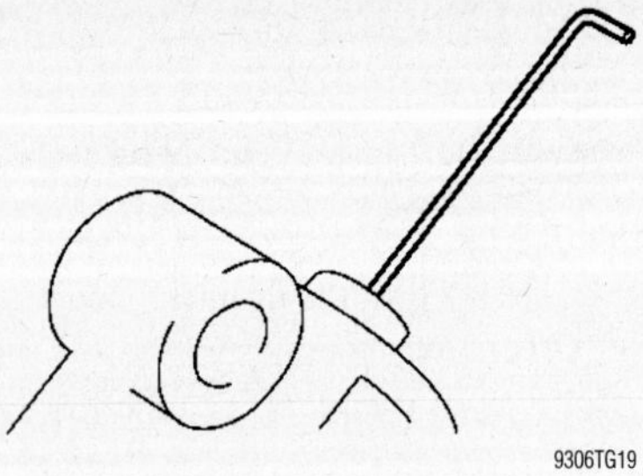

Timing chain tensioner preparation—3.9L engine

   - Left and right side Variable Camshaft Timing (VCT) housings
   - Crankshaft Position (CKP) sensor
   - Torque converter access panel

5. Rotate the crankshaft to 45 degrees After Top Dead Center (ATDC). The crankshaft keyway will be in the 6 o'clock position. Check that the camshaft lobes are facing upward. If not, rotate the crankshaft 1 full turn.
6. Install Crankshaft Holding Tool 303-645 in place of the CKP sensor.
7. Install Camshaft Locking Tool 303-530 to the right bank camshafts.
8. Remove the right side exhaust camshaft sprocket and VCT bolts and discard.
9. Insert a drill rod into the right side timing chain tensioner to lock the piston.
10. Remove the timing chain tensioner and tensioner arm.
11. Remove the timing chain guide, right bank timing chain and camshaft timing sprockets.
12. Remove the locking tool from the right bank camshafts and install it on the left bank camshafts.

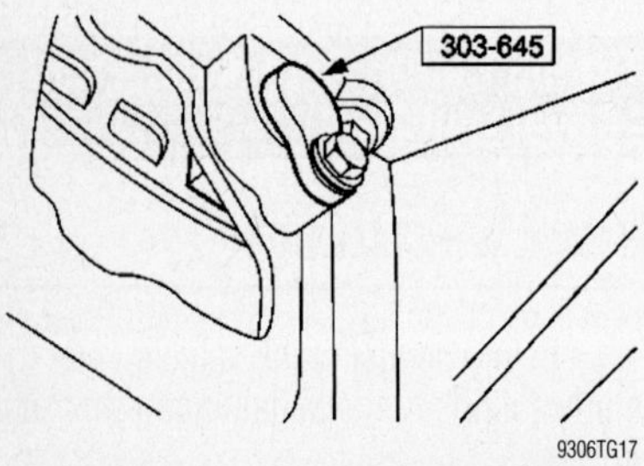

Crankshaft Holding Tool—3.9L engine

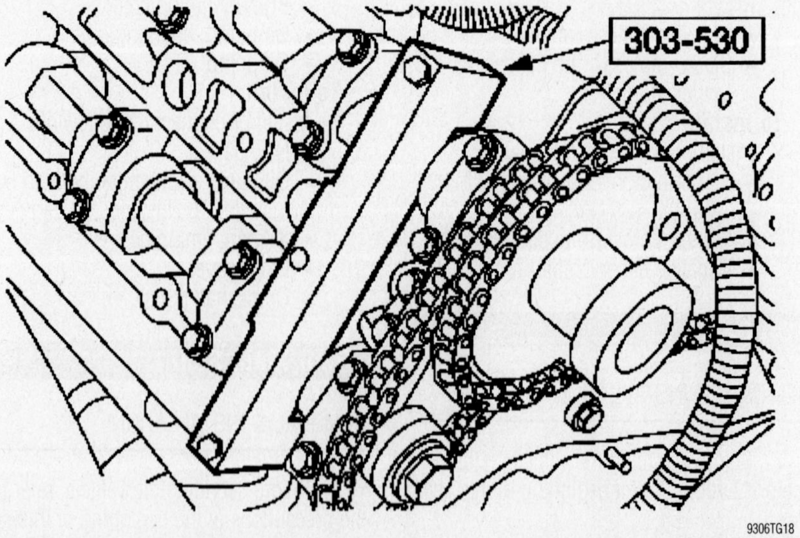

Camshaft Locking Tool—3.9L engine

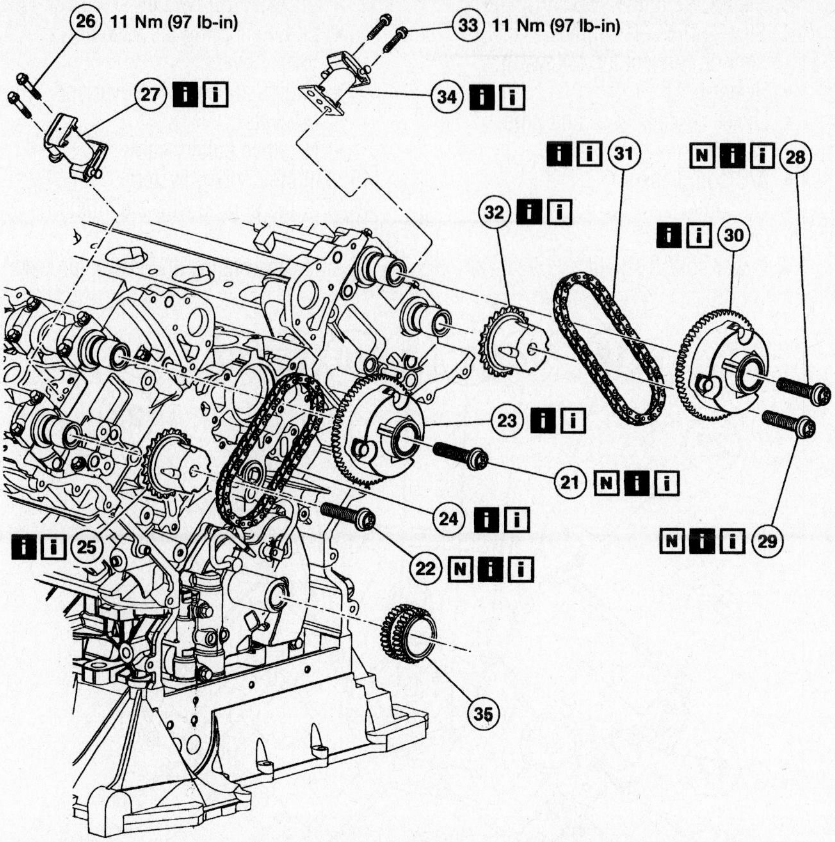

21 RH VCT unit bolt
22 RH exhaust camshaft sprocket bolt
23 RH VCT unit
24 RH secondary timing chain
25 RH exhaust camshaft sprocket
26 RH secondary timing chain tensioner bolts
27 RH secondary timing chain tensioner
28 LH VCT unit bolt
29 LH exhaust camshaft sprocket bolt
30 LH VCT unit
31 LH secondary timing chain
32 LH exhaust camshaft sprocket
33 LH secondary timing chain tensioner bolts
34 LH secondary timing chain tensioner
35 Crankshaft sprocket

Exploded view of timing chain sprocket assembly—3.9L engine

13. Remove the left side exhaust camshaft sprocket and VCT bolts and discard.
14. Insert a drill rod into the left side timing chain tensioner to lock the piston.
15. Remove the timing chain tensioner and tensioner arm.
16. Remove the timing chain guide, left bank timing chain and camshaft timing sprockets.

To install:

17. Install the left bank secondary timing chain and camshaft timing sprockets as an assembly.
18. Hand tighten the NEW sprockets bolts.
19. Remove the locking pin from the tensioner.
20. Install the right bank secondary timing chain and camshaft timing sprockets as an assembly.
21. Hand tighten the NEW sprockets bolts.
22. Remove the locking pin from the tensioner.
23. To reset the timing chain tensioners, compress the tensioners and install a drill rod to lock the pistons.

CAUTION

Ensure the camshaft holding tool is installed on the left camshaft.

24. Place the primary timing chain on the left camshaft and crankshaft gears.
25. Install the left side timing chain guide. Torque the bolts to 97 inch lbs. (11 Nm).
26. Install the tensioner arm. Torque the bolts to 97 inch lbs. (11 Nm).
27. Install the timing chain tensioner and torque the bolts to 97 inch lbs. (11 Nm). Remove the drill rod.
28. Install special tool 303-532 onto the exhaust camshaft sprocket and apply 89 inch lbs. (10 Nm) of force in a counter-clockwise direction, then tighten the exhaust camshaft sprocket bolt in two stages: Step 1 to 26 ft. lbs. (35 Nm), step 2 and additional 90°.
29. Tighten the intake camshaft sprocket bolt in two stages: Step 1 to 26 ft. lbs. (35 Nm), step 2 an additional 90°.
30. Remove the locking tool from the left camshaft.
31. Install the locking tool on the right camshaft.
32. Place the primary timing chain on the right camshaft and crankshaft gears.
33. Install the right side timing chain guide. Torque the bolts to 97 inch lbs. (11 Nm).
34. Install the tensioner arm. Torque the bolts to 97 inch lbs. (11 Nm).
35. Install the timing chain tensioner and torque the bolts to 97 inch lbs. (11 Nm). Remove the drill rod.
36. Install special tool 303-532 onto the exhaust camshaft sprocket and apply 89 inch lbs. (10 Nm) of force in a counter-clockwise direction, then tighten the exhaust camshaft sprocket bolt in two stages: Step 1 to 26 ft. lbs. (35 Nm), step 2 and additional 90°.
37. Tighten the intake camshaft sprocket bolt in two stages: Step 1 to 26 ft. lbs. (35 Nm), step 2 an additional 90°.
38. Remove the camshaft locking tool.
39. Install or connect the following:
   - CKP sensor

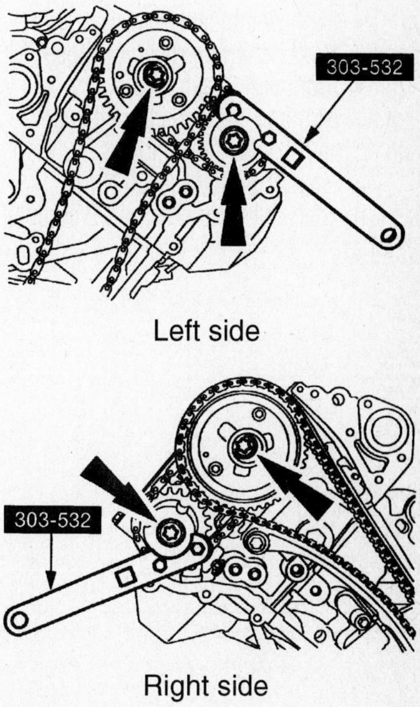

Installing special tool to exhaust camshafts—3.9L engine

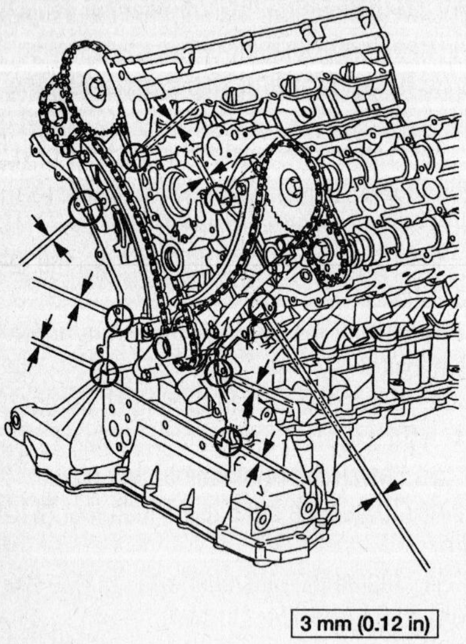

Apply silicone sealant to the areas indicated—3.9L engine

- Torque converter access panel
- VCT housings. Tighten the bolts to 15 ft. lbs. (20 Nm), and the nut to 89 inch lbs. (10 Nm).

40. Apply silicone sealant to the areas indicated and install the front cover.

41. Tighten the front cover bolts in sequence as follows:
   a. Step 1: 44 inch lbs. (5 Nm)
   b. Step 2: 89 inch lbs. (10 Nm)

42. Install or connect the following:
- Front cover wiring harness clips
- Alternator and bracket
- Power steering pump and bracket
- PSP switch connector
- Power steering reservoir hose
- Front crankshaft seal

43. Install the crankshaft pulley and tighten the NEW bolt as follows:
   a. Step 1: 59 ft. lbs. (80 Nm)
   b. Step 2: Loosen the bolt 2 complete turns
   c. Step 3: 37 ft. lbs. (50 Nm)
   d. Step 4: Plus 90 degrees

44. Install or connect the following:
- Heater hose
- Lower radiator hose and pipe
- Water pump
- A/C compressor
- Oil cooler
- Oil filter
- Lower splash shield
- Accessory drive belt
- Engine cooling fan assembly
- Valve covers
- Engine appearance cover and brackets
- Negative battery cable

45. Fill the cooling system.

46. Fill the engine with clean engine oil.

47. Start the engine and check for leaks.

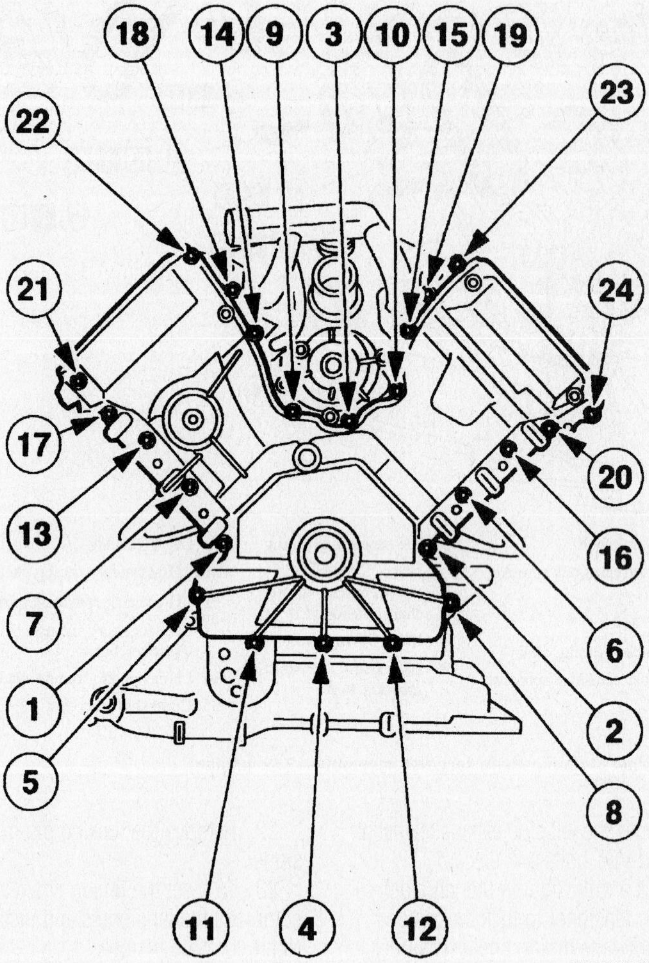

Front cover torque sequence—3.9L engine

FUEL SYSTEM

Fuel System Service Precautions

Safety is the most important factor when performing not only fuel system maintenance but also any type of maintenance. Failure to conduct maintenance and repairs in a safe manner may result in serious personal injury or death. Maintenance and testing of the vehicle's fuel system components can be accomplished safely and effectively by adhering to the following rules and guidelines.

- To avoid the possibility of fire and personal injury, always disconnect the negative battery cable unless the repair or test procedure requires that battery voltage be applied.
- Always relieve the fuel system pressure prior to disconnecting any fuel system component (injector, fuel rail, pressure regulator, etc.), fitting or fuel line connection. Exercise extreme caution whenever relieving fuel system pressure, to avoid exposing skin, face and eyes to fuel spray. Please be advised that fuel under pressure may penetrate the skin or any part of the body that it contacts.
- Always place a shop towel or cloth around the fitting or connection prior to loosening to absorb any excess fuel due to spillage. Ensure that all fuel spillage (should it occur) is quickly removed from engine surfaces. Ensure that all fuel soaked cloths or towels are deposited into a suitable waste container.
- Always keep a dry chemical (Class B) fire extinguisher near the work area.
- Do not allow fuel spray or fuel vapors to come into contact with a spark or open flame.
- Always use a back-up wrench when loosening and tightening fuel line connection fittings. This will prevent unnecessary stress and torsion to fuel line piping.
- Always replace worn fuel fitting O-rings with new. Do not substitute fuel hose or equivalent where fuel pipe is installed.

Fuel System Pressure

RELIEVING

1. Before servicing the vehicle, refer to the precautions in the beginning of this section.
2. Disconnect the negative battery cable.
3. Connect the fuel injection pressure test equipment JD 209 to the valve on the fuel supply manifold.
4. Insert the drain/bleed tube into the fuel container.
5. Follow the manufacturer's instructions and depressurize the fuel system.

Fuel Filter

REMOVAL & INSTALLATION

1. Before servicing the vehicle, refer to the precautions in the beginning of this section.
2. Relieve the fuel system pressure.
3. Remove or disconnect the following:
   - Negative battery cable
   - Left front wheel rear splash shield
   - Fuel lines
   - Vapor line fittings
   - Fuel filter bracket cover
   - Fuel filter

To install:

4. Install or connect the following:
   - Fuel filter into the bracket making sure the flow direction is correct. Tighten the clamp to 15–25 inch lbs. (2–3 Nm).
   - Vapor line fittings
   - Fuel lines. Tighten the fittings to 22 ft. lbs. (30 Nm).
   - Splash shield
5. Start the engine and check for leaks.

Fuel Pump

REMOVAL & INSTALLATION

2002–2003 Models

1. Before servicing the vehicle, refer to the precautions in the beginning of this section.
2. Relieve the fuel system pressure.
3. Drain the fuel tank.
4. Remove or disconnect the following:
   - Negative battery cable
   - Trunk liner
   - Trunk seal retainer
   - Rear lamp assembly interior trim finisher

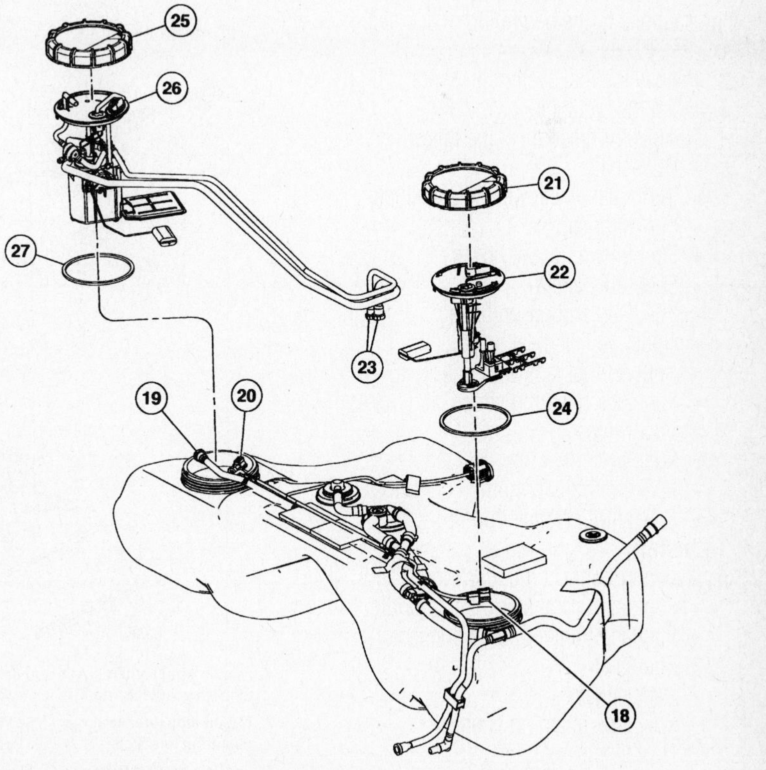

18 Fuel transfer pump electrical connector
19 Fuel delivery tube quick connect fitting
20 Fuel delivery module electrical connector
21 Fuel transfer pump lock ring
22 Fuel transfer pump
23 Fuel crossover hose quick connect fittings
24 Fuel transfer pump gasket
25 Fuel delivery module lock ring
26 Fuel delivery module
27 Fuel delivery module gasket

67197-THUN-G18

Fuel and transfer pump mounting—2004 Thunderbird

- Left and right side liners
- Fuel feed and return lines
- Fuel filler and vent hoses
- Fuel tank wiring connectors
- Fuel filler cap
- Fuel tank retaining straps
- Fuel tank
- Fuel pump module

To install:

5. Install or connect the following:
   - Fuel pump module
   - Fuel tank
   - Fuel tank retaining straps
   - Fuel filler cap
   - Fuel tank wiring connectors
   - Fuel filler and vent hoses
   - Fuel feed and return lines
   - Left and right side liners
   - Rear lamp assembly interior trim finisher
   - Trunk seal retainer
   - Trunk liner
   - Negative battery cable
6. Fill the fuel tank with at least 10 gallons (38L) of fuel.
7. Start the engine and check for leaks.

2004 Models

1. Before servicing the vehicle, refer to the precautions in the beginning of this section.
2. Relieve the fuel system pressure.
3. Drain the fuel tank.
4. Remove or disconnect the following:
   - Driveshaft
   - Right rear wheel well splash shield
   - Fuel filler cap
   - Fuel filler and vent hoses
   - Fuel tube shield
   - Fuel feed and return lines
   - Under vehicle heat shield
   - Support fuel tank with a jack
   - Fuel tank retaining straps
   - Vapor lines
   - Fuel tank
   - Fuel and transfer pump connectors
   - Fuel and/or transfer pump

To install:

5. Install or connect the following:
   - Fuel and/or transfer pump module
   - Fuel and transfer pump connectors
   - Fuel tank
   - Vapor lines
   - Fuel tank retaining straps
   - Heat shield
   - Fuel feed and return lines
   - Fuel tube shield
   - Fuel filler and vent hoses
   - Fuel filler cap
   - Right rear wheel well splash shield
   - Driveshaft
   - Negative battery cable
6. Fill the fuel tank with at least 10 gallons (38L) of fuel.
7. Start the engine and check for leaks.

Fuel Rail And Injectors

REMOVAL & INSTALLATION

1. Before servicing the vehicle, refer to the precautions in the beginning of this section.
2. Relieve fuel system pressure.
3. Remove or disconnect the following:
   - Negative battery cable
   - Engine appearance covers and mounting brackets
   - Fuel lines
   - Fuel pressure sensor connectors
   - Fuel rail
   - Fuel injector connectors
   - Fuel injectors

To install:

4. Install or connect the following:
   - Fuel injectors. Use new O-ring seals.
   - Fuel injector connectors
   - Fuel rail. Tighten the mounting bolts to 80 inch lbs. (9 Nm).
   - Fuel pressure sensor connectors
   - Fuel lines
   - Engine appearance covers and mounting brackets
   - Negative battery cable
5. Start the engine and check for leaks.

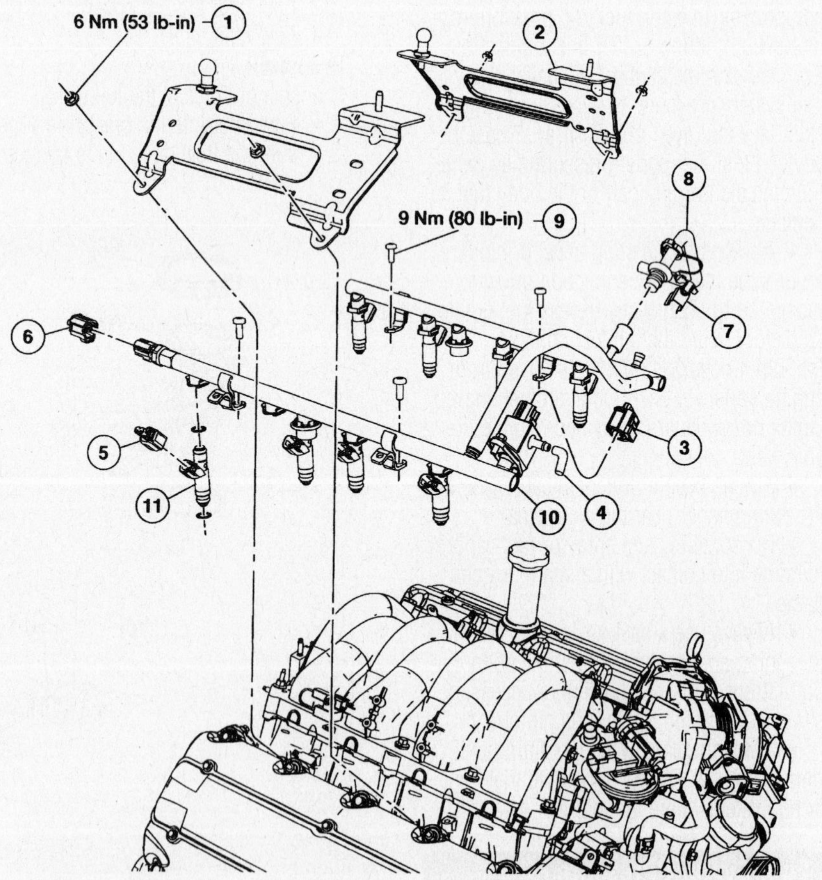

1 Engine appearance cover mounting bracket nuts
2 Engine appearance cover mounting brackets
3 Fuel pressure sensor electrical connector
4 Fuel pressure sensor vacuum connector
5 Fuel injector electrical connector
6 Fuel temperature sensor electrical connector
7 Fuel tube retaining clip
8 Fuel tube spring lock coupling
9 Fuel injection supply manifold bolts
10 Fuel injection supply manifold and fuel injectors
11 Fuel injectors

67197-THUN-G19

Fuel rail and injector assembly—3.9L engine

DRIVE TRAIN

Transmission Assembly

REMOVAL & INSTALLATION

1. Before servicing the vehicle, refer to the precautions in the beginning of this section.
2. Install a support fixture to the engine lifting eyes.
3. Drain the transmission fluid.
4. Remove or disconnect the following:
   - Negative battery cable
   - Exhaust heat shields
   - Exhaust front pipes
   - Driveshaft
5. Support the transmission with a suitable jack.
6. Remove or disconnect the following:
   - Shift selector cable
   - Transmission electrical connectors
   - Transmission oil cooler lines

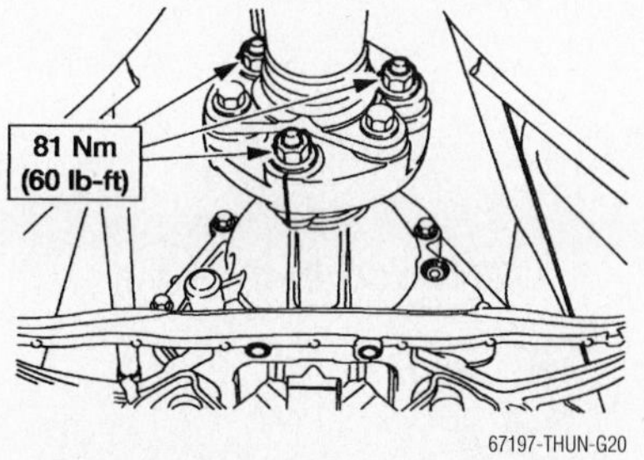

Driveshaft flange bolts

   - Torque converter
   - Transmission mount
   - Transmission-to-engine bolts
   - Transmission

To install:

7. Install or connect the following:
   - Transmission to the engine. Tighten the bolts to 35 ft. lbs. (48 Nm).
   - Transmission mount. Tighten the mount screws to 41 ft. lbs. (55 Nm) and the center screw to 30 ft. lbs. (40 Nm).
   - Torque converter. Tighten the bolts to 35 ft. lbs. (48 Nm).
   - Transmission oil cooler lines
   - Transmission electrical connectors
   - Shift selector cable
   - Driveshaft. Tighten the bolts to 60 ft. lbs. (81 Nm).
   - Exhaust front pipes
   - Exhaust heat shields
   - Negative battery cable
8. Fill the transmission to the correct level with the proper fluid. Do not over-fill.
9. Start the engine and check for leaks.

Halfshaft

REMOVAL & INSTALLATION

1. Before servicing the vehicle, refer to the precautions in the beginning of this section.
2. Remove or disconnect the following:
   - Negative battery cable
   - Rear wheel
   - Brake caliper
   - Wheel speed sensor
   - Axle hub retaining nut
3. Press the stub shaft out of the hub and pry the inner joint out of the differential.

To install:

4. Install or connect the following:
   - Halfshaft inner joint to the differential
   - Halfshaft in the wheel hub by applying Loctite® 270 thread locking compound to the splines
   - Axle hub retaining nut. Tighten the new nut to 221 ft. lbs. (300 Nm).
   - Wheel speed sensor
   - Brake caliper
   - Rear wheel
   - Negative battery cable
5. Check the wheel alignment and adjust as necessary.

➡**If the hub is removed for any reason, a new bearing assembly must be installed. Never attempt to re-use a bearing.**

CV-Joint

OVERHAUL

The CV-joints are serviced with the axle halfshaft as an assembly.

STEERING AND SUSPENSION

Air Bag

**** CAUTION**

Some vehicles are equipped with an air bag system. The system must be disarmed before performing service on, or around, system components, the steering column, instrument panel components, wiring and sensors. Failure to follow the safety precautions and the disarming procedure could result in accidental air bag deployment, possible injury and unnecessary system repairs.

PRECAUTIONS

Several precautions must be observed when handling the inflator module to avoid accidental deployment and possible personal injury.

- Never carry the inflator module by the wires or connector on the underside of the module.
- When carrying a live inflator module, hold securely with both hands, and ensure that the bag and trim cover are pointed away.
- Place the inflator module on a bench or other surface with the bag and trim cover facing up.
- With the inflator module on the bench, never place anything on or close to the module that may be thrown in the event of an accidental deployment.

DISARMING

Proper SRS disarming can be obtained by disconnecting and isolating the negative battery cable. Allow the air bag system capacitor at least 2 minutes to discharge before removing any air bag system components.

Power Rack and Pinion Steering Gear

REMOVAL & INSTALLATION

**** WARNING**

Do not allow the steering wheel to rotate when the intermediate shaft is disconnected, or damage to the clockspring can result. If it is suspected that the shaft has rotated, the clockspring must be removed and re-centered.

1. Before servicing the vehicle, refer to the precautions in the beginning of this section.
2. Lock the steering wheel in the straight-ahead position.
3. Remove or disconnect the following:
   - Negative battery cable
   - Front wheels

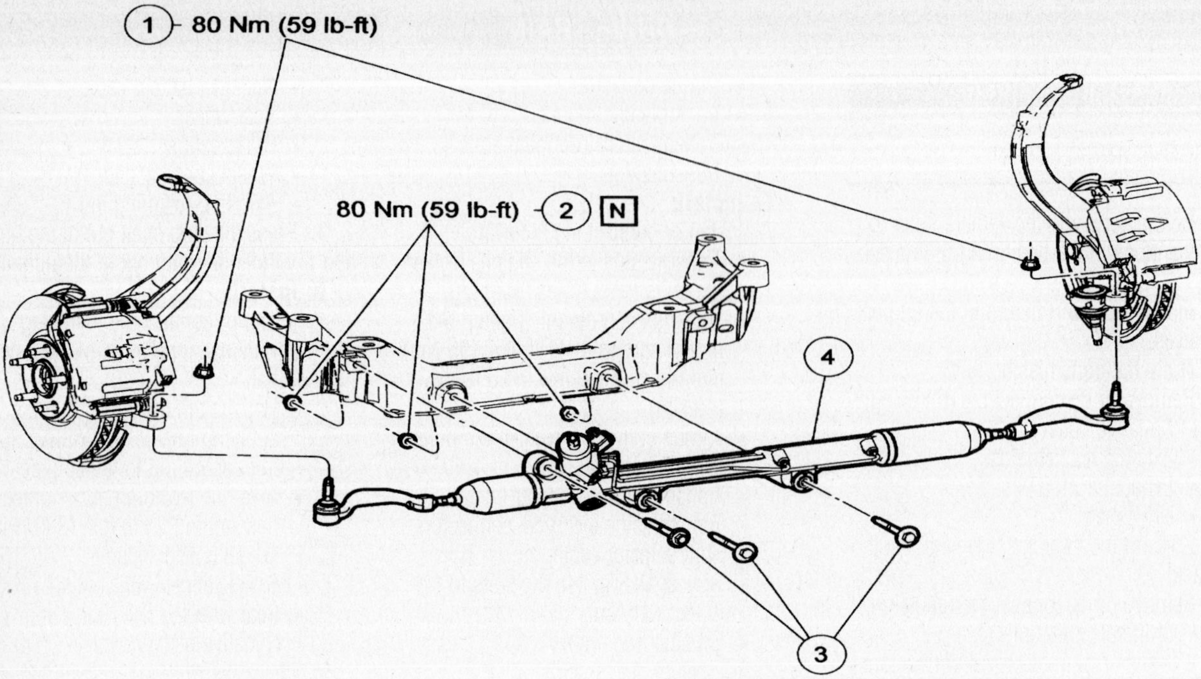

1 Tie-rod end nuts
2 Steering gear-to-crossmember nuts
3 Steering gear-to-crossmember bolts
4 Steering gear

67197-THUN-G21

Steering gear mounting

- Outer tie rod ends
- Variable Assist Power Steering (VASP) connector
- Steering column intermediate shaft
- Power steering lines
- Steering rack and pinion gear

To install:

4. Install or connect the following:
   - Steering rack and pinion gear. Tighten the steering gear-to-nuts to bolts to 59 ft. lbs. (80 Nm).
   - Power steering lines
   - Outer tie rod ends. Tighten the nuts to 59 ft. lbs. (80 Nm).
   - Steering column intermediate shaft
   - VASP connector
   - Front wheels
   - Negative battery cable
5. Fill the power steering fluid reservoir.
6. Start the engine and check for leaks.
7. Check the wheel alignment and adjust, as necessary.

Strut

REMOVAL & INSTALLATION

Front

1. Before servicing the vehicle, refer to the precautions in the beginning of this section.

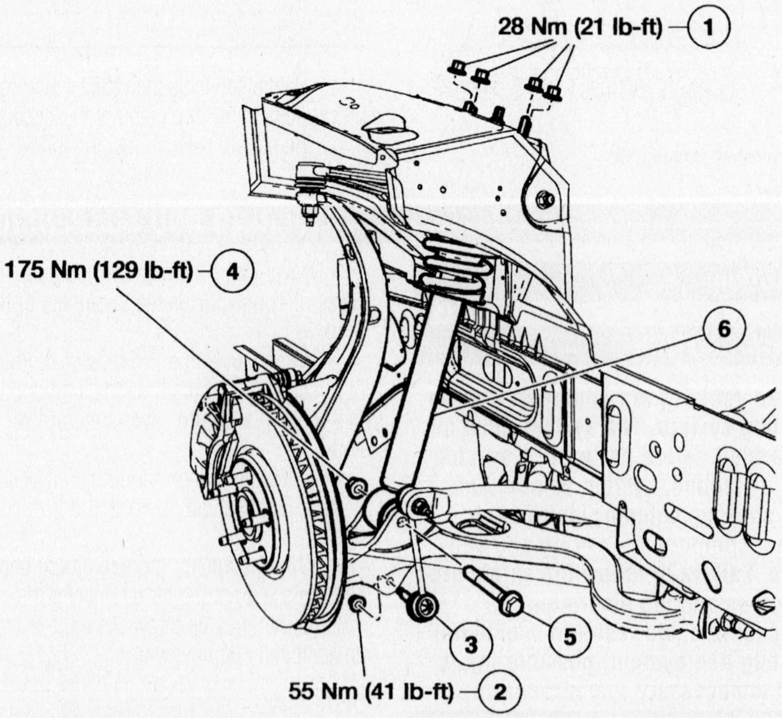

1 Shock absorber upper mount-to-body nuts
2 Stabilizer bar link-to-lower arm nut
3 Stabilizer bar link (detach only)
4 Shock absorber-to-lower arm nut
5 Shock absorber-to-lower arm bolt
6 Shock absorber and spring assembly

67197-THUN-G22

Front strut mounting

2. Remove or disconnect the following:
   - Front wheel
   - Stabilizer bar link
   - Lower strut mounting bolt
   - Upper strut mount cover and fasteners
   - Strut and spring assembly

To install:

➡Use new fasteners for assembly.

3. Install or connect the following:
   - Strut and spring assembly. Tighten the upper mount nuts to 21 ft. lbs. (28 Nm).
   - Upper strut mount cover
   - Lower strut mounting bolt. Tighten the bolts to 129 ft. lbs. (175 Nm).
   - Stabilizer bar link. Tighten the nut to 41 ft. lbs. (55 Nm).
   - Front wheel

Rear

1. Before servicing the vehicle, refer to the precautions in the beginning of this section.
2. Remove or disconnect the following:
   - Trunk trim covers
   - Upper strut mount nuts
   - Lower strut mount bolt
   - Strut and spring assembly

To install:

➡Use new fasteners for assembly.

3. Install or connect the following:
   - Strut and spring assembly. Tighten the lower bolt to 98 ft. lbs. (133 Nm) and the upper nuts to 21 ft. lbs. (28 Nm).
   - Trunk trim covers

Coil Spring

REMOVAL & INSTALLATION

1. Before servicing the vehicle, refer to the precautions in the beginning of this section.
2. Remove the strut assembly from the vehicle.
3. Compress the coil spring and remove the piston rod nut.
4. Remove or disconnect the following:
   - Upper strut mount
   - Spring upper seat
   - Coil spring

To install:

5. Install or connect the following:
   - Coil spring
   - Spring upper seat
   - Upper strut mount. Tighten the piston rod nut to 37 ft. lbs. (50 Nm).
6. Remove the spring compressor and install the strut assembly to the vehicle.

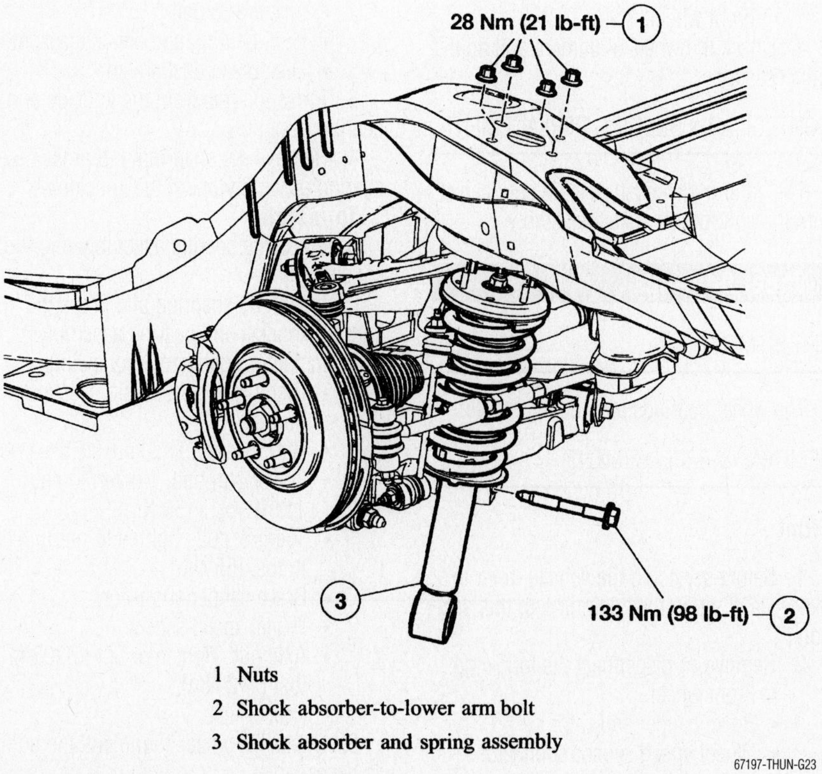

Rear strut mounting

Upper Ball Joint

REMOVAL & INSTALLATION

The upper ball joint is serviced with the upper control arm as an assembly.

Lower Ball Joint

REMOVAL & INSTALLATION

The lower ball joint is serviced with the lower control arm as an assembly.

Upper Control Arm

REMOVAL & INSTALLATION

1. Before servicing the vehicle, refer to the precautions in the beginning of this section.
2. Remove or disconnect the following:
   - Front wheel
   - Strut and spring assembly
   - Upper ball joint and tapered washer
   - Inner control arm fasteners

➡To remove the upper arm on the left side, position the power steering reservoir aside. Disconnect the master cylinder electrical connector, remove master cylinder attaching nuts and position the master cylinder aside. This will allow access to the upper arm-to-body nut.

   - Upper control arm

To install:

➡Use new fasteners for assembly.

3. Install or connect the following:
   - Upper control arm. Tighten the inner fasteners to 35 ft. lbs. (48 Nm), with the weight of the vehicle resting on the suspension.
   - Upper ball joint and tapered washer. Tighten the nut to 66 ft. lbs. (90 Nm).
   - Master cylinder. Tighten the nut to 18 ft. lbs. (25 Nm).
   - Power steering reservoir. Tighten the nut to 35 ft. lbs. (48 Nm).
   - Strut and spring assembly
   - Front wheel

CONTROL ARM BUSHING REPLACEMENT

The control arm bushings are serviced with the control arm as an assembly.

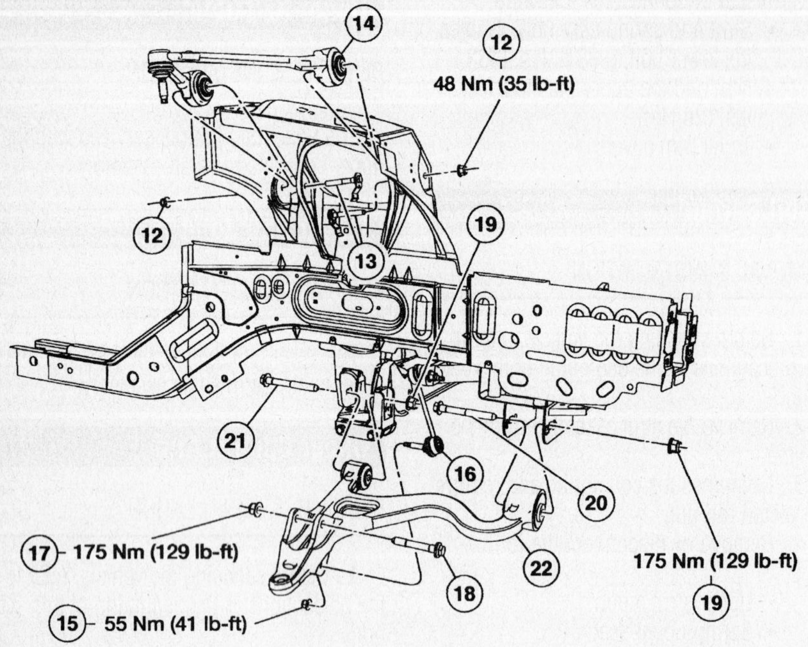

12 Upper arm-to-body nuts
13 Upper arm-to-body bolts
14 Upper arm
15 Stabilizer bar link-to-lower arm nut
16 Stabilizer bar link (detach only)
17 Shock absorber-to-lower arm nut
18 Shock absorber-to-lower arm bolt
19 Lower arm-to-frame nuts
20 Lower arm-to-frame bolt
21 Lower arm-to-frame bolt
22 Lower arm RH/LH

67197-THUN-G24

Upper and lower control arms

Lower Control Arm

REMOVAL & INSTALLATION

1. Before servicing the vehicle, refer to the precautions in the beginning of this section.
2. Remove or disconnect the following:
   - Front wheel
   - Splash shield
   - Stabilizer bar link
   - Lower strut mounting bolt
   - Lower ball joint
   - Rack and pinion steering gear
   - Inner control arm mounting bolts
   - Lower control arm

To install:

➡Use new fasteners for assembly.

3. Install or connect the following:
   - Lower control arm. Tighten the inner mounting bolts to 129 ft. lbs. (175 Nm) with the weight of the vehicle resting on the suspension.
   - Rack and pinion steering gear
   - Lower ball joint. Tighten the nut to 148 ft. lbs. (200 Nm).
   - Lower strut mounting bolt. Tighten the bolt to 129 ft. lbs. (175 Nm).
   - Stabilizer bar link. Tighten the nut to 41 ft. lbs. (55 Nm).
   - Splash shield
   - Front wheel
4. Check the wheel alignment and adjust as necessary.

CONTROL ARM BUSHING REPLACEMENT

The control arm bushings are serviced with the control arm as an assembly.

Wheel Bearings

ADJUSTMENT

The wheel bearings are not adjustable.

REMOVAL & REPLACEMENT

Front

1. Before servicing the vehicle, refer to the precautions in the beginning of this section.
2. Remove or disconnect the following:
   - Front wheel
   - Brake caliper and rotor
   - Wheel speed sensor connector
   - Hub and bearing assembly

➡The hub and bearing assembly is not pressed into the knuckle. Do not use a slide hammer or press to remove the hub and bearing assembly. Damage to the hub and bearing assembly may result.

To install:

➡Do not remove the wheel speed sensor from the hub and bearing assembly unless it is being replaced. If installing a new hub and bearing assembly, a new wheel speed sensor must be installed.

➡Use new fasteners for assembly.

3. Install or connect the following:
   - Hub and bearing assembly. Tighten the bolts to 66 ft. lbs. (90 Nm).
   - Wheel speed sensor connector
   - Brake caliper and rotor
   - Front wheel

Rear

1. Before servicing the vehicle, refer to the precautions in the beginning of this section.
2. Remove or disconnect the following:
   - Rear wheel
   - Axle nut
   - Wheel speed sensor
   - Brake caliper and rotor
   - Toe link bolt
   - Lower arm bolt
   - Hub, bearing and knuckle assembly
   - Disc brake dust shield
3. Press the hub from the knuckle and bearing assembly.
4. Remove the snapring and press the bearing assembly out of the knuckle.

To install:

5. Press the bearing assembly into the knuckle.
6. Install the snapring and press the hub into the knuckle and bearing assembly.
7. Install or connect the following:
   - Disc brake dust shield. Use aluminum rivets.
   - Hub, bearing and knuckle assembly
   - Lower arm bolt. Tighten to nut to 111 ft. lbs. (150 Nm).
   - Toe link bolt. Tighten to nut to 41 ft. lbs. (55 Nm).
   - Brake caliper and rotor
   - Wheel speed sensor
   - Axle nut. Tighten to nut to 302 ft. lbs. (410 Nm).
   - Rear wheel
8. Check the wheel alignment and adjust as necessary.

BRAKES

Brake Caliper

REMOVAL & INSTALLATION

Front

1. Before servicing the vehicle, refer to the precautions in the beginning of this section.
2. Remove or disconnect the following:
   - Front wheel
   - Brake fluid hose
   - Caliper mounting bolts
   - Brake caliper

To install:

3. Install or connect the following:
   - Brake caliper. Tighten the mounting bolts to 26 ft. lbs. (35 Nm).
   - Brake fluid hose. Use new copper washers and tighten the bolt to 35 ft. lbs. (47 Nm).
   - Front wheel
4. Bleed the brake system.
5. Before attempting to move the vehicle, pump the brake pedal to seat the pads against the rotors. Make sure the vehicle has a firm brake pedal. Check the level of the brake fluid and add DOT 3 or 4 brake fluid if necessary.

Rear

1. Before servicing the vehicle, refer to the precautions in the beginning of this section.
2. Remove or disconnect the following:
   - Rear wheel
   - Parking brake cable
   - Caliper mounting bolts
   - Brake fluid hose
   - Brake caliper

To install:

3. Install or connect the following:
   - Brake fluid hose. Use new copper washers and tighten the bolt to 36 ft. lbs. (48 Nm).
   - Brake caliper. Tighten the mounting bolts to 25 ft. lbs. (33 Nm).
   - Parking brake cable
   - Rear wheel
4. Bleed the brake system.
5. Before attempting to move the vehicle, pump the brake pedal to seat the pads against the rotors. Make sure the vehicle has a firm brake pedal. Check the level of the brake fluid and add DOT 3 or 4 brake fluid if necessary.

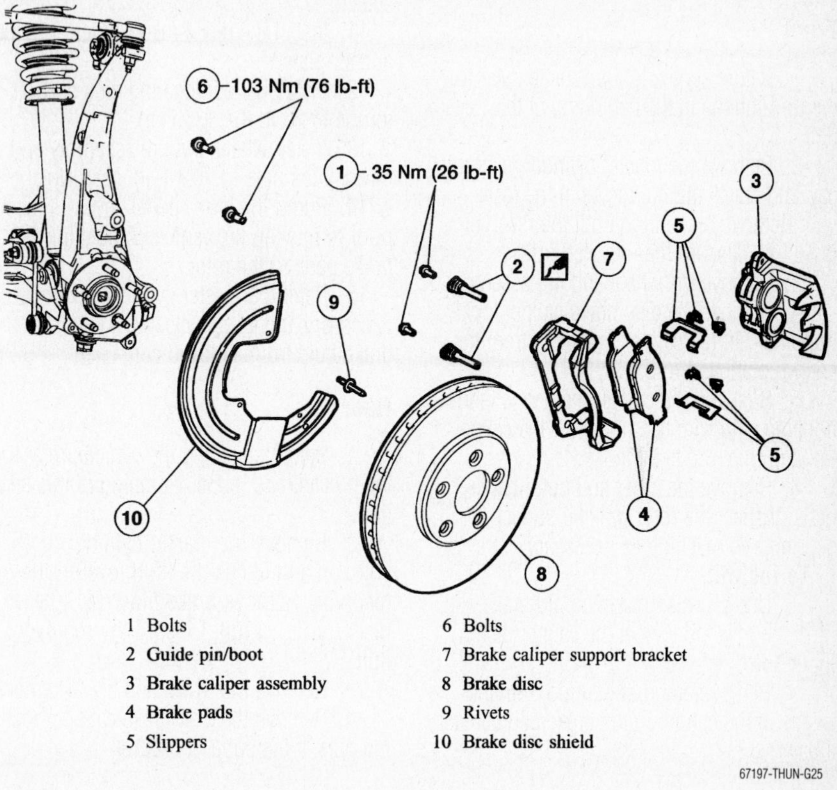

Exploded view of front brake assembly

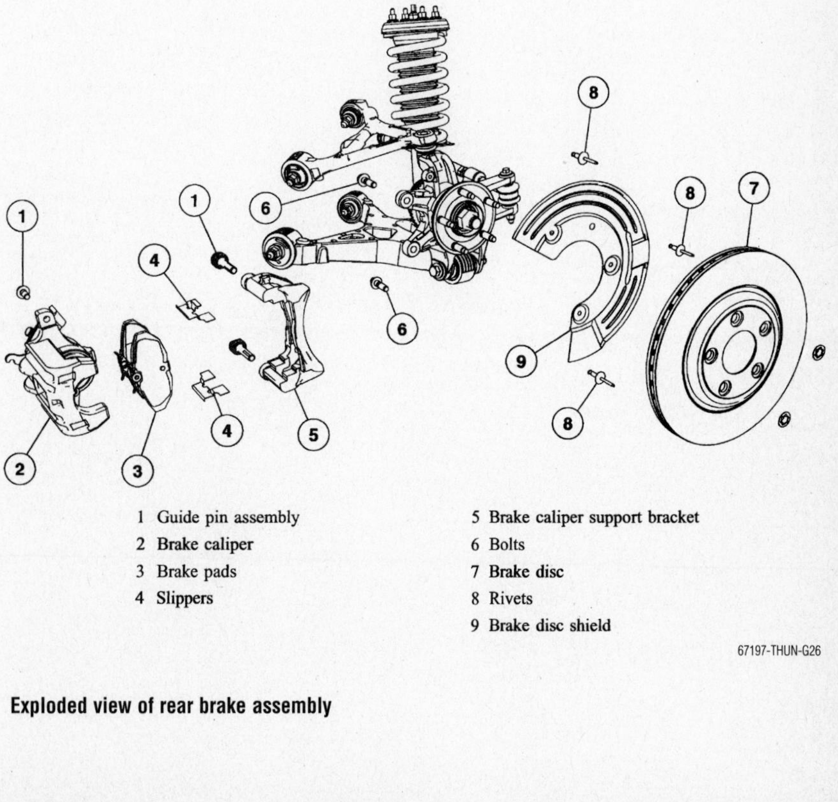

Exploded view of rear brake assembly

Disc Brake Pads

REMOVAL & INSTALLATION

Front

1. Before servicing the vehicle, refer to the precautions in the beginning of this section.
2. Remove the master cylinder reservoir cap and check the fluid level in the reservoir. Remove brake fluid until the reservoir is ½ full. Discard the removed fluid.
3. Remove the wheel and tire assembly.
4. Remove the disc brake caliper locating pins. Lift the caliper assembly from the anchor plate and rotor.
5. Suspend the caliper inside the fender housing with wire. Do not allow the caliper to hang from the brake hose.
6. Remove the inner and outer brake pads. Inspect the rotor braking surfaces for scoring and machine as necessary.

To install:

7. Use a C-clamp and an old brake pad or block of wood to seat the caliper piston in its bore.
8. Remove any rust buildup from the inside of the caliper in the brake pad contact area.
9. Install the inner pad in the caliper piston.
10. Install the outer pad onto the anchor plate. Make sure the clips are properly seated.
11. Install the disc brake caliper onto the anchor plate.
12. Install caliper locating pins and torque to 26 ft. lbs. (35 Nm).
13. Install wheel and tire assembly and torque lugs nuts to 100 ft. lb. (135 Nm).
14. Pump the brake pedal several times prior to moving the vehicle to position the brake pads to the rotor.
15. Refill the master cylinder reservoir as necessary, using only clean DOT 3 or 4 brake fluid from a closed container.

Rear

1. Before servicing the vehicle, refer to the precautions in the beginning of this section.
2. Remove the master cylinder reservoir cap and check the fluid level in the reservoir. Remove brake fluid until the reservoir is ½ full. Discard the removed fluid.
3. Remove the wheel and tire assembly.
4. Remove the disc brake caliper locating pins at the support bracket.
5. Remove the caliper.
6. Remove the disc brake pads.
7. Inspect the rotor braking surfaces for scoring and machine as necessary.

To install:

8. Using Rear Caliper Piston Adjuster T87P-2588-A, rotate the piston clockwise until it is fully seated. Make sure one of the slots in the piston face is positioned so it will engage the nib on the brake pad.
9. Install the brake pads in the support bracket. Place the caliper assembly over the rotor into position on the support bracket. Make sure the brake pads are installed correctly.
10. Install and torque the disc brake caliper locating pin to 25 ft. lbs. (33 Nm).
11. Install the wheel and tire assembly and torque the lug nuts to 85–104 ft. lbs. (115–142 Nm).
12. Pump the brake pedal several times prior to moving the vehicle, to position the brake pads to the rotor.
13. Refill the master cylinder reservoir if necessary, using only clean DOT 3 or 4 brake fluid from a closed container.
14. Road test the vehicle and check the brake system for proper operation.

MERCURY

16

Villager

SPECIFICATION CHARTS

VEHICLE AND ENGINE IDENTIFICATION CHART

Engine Code							Model Year	
Code	Liters (cc)	Cu. In.	Cyl.	Fuel Sys.	Engine Type	Eng. Mfg.	Code	Year
T	3.3 (3275)	200	6	SEFI	SOHC	Nissan	1	2001
							2	2002

MFI: Multi-port Fuel Injection

SEFI: Sequential Multi-port Fuel Injection

67197-VILL-C01

GENERAL ENGINE SPECIFICATIONS

Year	Engine Displacement Liters (VIN)	Net Horsepower @ rpm	Net Torque @ rpm (ft. lbs.)	Bore x Stroke (in.)	Com-pression Ratio	Oil Pressure @ rpm
2001	3.3 (T)	195@4500	190@3800	3.60x3.27	8.9:1	60-65@2000
2002	3.3 (T)	195@4500	190@3800	3.60x3.27	8.9:1	60-65@2000

MFI: Multiport fuel injection

SEFI: Sequential Multi-port Fuel Injection

67197-VILL-C02

ENGINE TUNE-UP SPECIFICATIONS

Year	Engine Displacement Liters (VIN)	Spark Plug Gap (in.)	Ignition Timing (deg.)		Fuel Pump (psi) ①	Idle Speed (rpm)		Valve Clearance	
			MT	AT		MT	AT ②	In.	Ex.
2001	3.3 (T)	0.043	—	13-17B	34	—	650-750	HYD	HYD
2002	3.3 (T)	0.043	—	13-17B	34	—	650-750	HYD	HYD

NOTE: The Vehicle Emission Control Information label must be used if they differ from those in this chart.

B: Before top dead center

HYD: Hydraulic

① System pressure at idle with vacuum hose connected should increase to 43 psi when disconnected

② Transmission in Neutral

67197-VILL-C03

3.3L engines
Firing order: 1–2–3–4–5–6
Distributor rotation: Counterclockwise

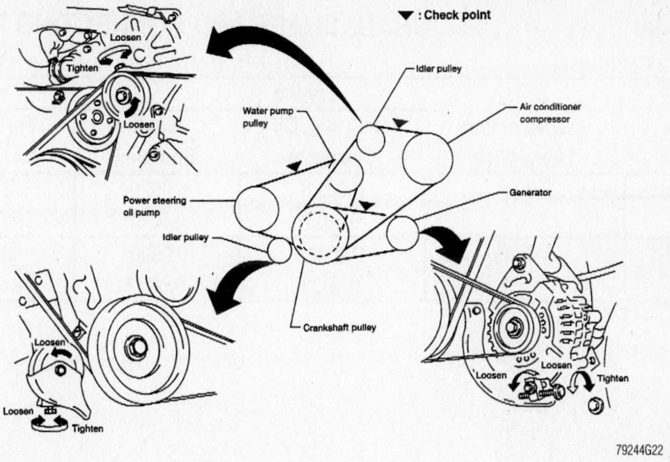

Accessory V-belt routing—3.3L engines

CAPACITIES

Year	Model	Engine Displacement Liters (VIN)	Engine Oil with Filter (qts.)	Transmission (pts.) 4-Spd	5-Spd	Auto.	Drive Axle Front (pts.)	Rear (pts.)	Fuel Tank (gal.)	Cooling System (qts.) ②
2001	Villager	3.3 (T)	4.0	—	—	20.0	①	—	20	11.25
2002	Villager	3.3 (T)	4.0	—	—	20.0	①	—	20	11.25

NOTE: All capacities are approximate. Add fluid gradually and check to be sure a proper fluid level is obtained.

① Included in transaxle capacity

② Includes reservoir tank.

67197-VILL-C04

VALVE SPECIFICATIONS

Year	Engine Displacement Liters (VIN)	Seat Angle (deg.)	Face Angle (deg.)	Spring Test Pressure (lbs. @ in.)	Spring Installed Height (in.)	Stem-to-Guide Clearance (in.) Intake	Exhaust	Stem Diameter (in.) Intake	Exhaust
2001	3.3 (T)	45	45	①	②	0.0008-0.0021	0.0012-0.0019	0.2742-0.2748	0.3136-0.3138
2002	3.3 (T)	45	45	①	②	0.0008-0.0021	0.0012-0.0019	0.2742-0.2748	0.3136-0.3138

① Outer spring: 118@1.81
Inner spring: 57.3@0.984

② Spring height measured unloaded
Minimum length. outer spring: 2.016
Minimum length. inner spring: 1.736

67197-VILL-C05

CRANKSHAFT AND CONNECTING ROD SPECIFICATIONS

All measurements are given in inches.

		Engine			Connecting Rod		
Year	Displacement Liters (VIN)	Main Brg. Oil Clearance	Shaft End-play	Thrust on No.	Journal Diameter	Oil Clearance	Side Clearance
2001	3.3 (T)	0.0011-0.0022	0.0020-0.0067	3	1.9667-1.9675	0.0006-0.0021	0.0079-0.00138
2002	3.3 (T)	0.0011-0.0022	0.0020-0.0067	3	1.9667-1.9675	0.0006-0.0021	0.0079-0.00138

67197-VILL-C06

PISTON AND RING SPECIFICATIONS

All measurements are given in inches.

			Ring Gap			Ring Side Clearance		
Year	Engine Displacement Liters (VIN)	Piston Clearance	Top Compression	Bottom Compression	Oil Control	Top Compression	Bottom Compression	Oil Control
2001	3.3 (T)	①	0.0083-0.0122	0.0197-0.0236	0.0079-0.0236	0.0016-0.0031	0.0012-0.0028	0.0006-0.0073
2002	3.3 (T)	①	0.0083-0.0122	0.0197-0.0236	0.0079-0.0236	0.0016-0.0031	0.0012-0.0028	0.0006-0.0073

① Journals 1, 2 and 6: 0.0010 - 0.0018 in.
Journals 3 and 4: 0.0006 - 0.0010 in.
Journal 5: 0.0012 - 0.0016 in.

67197-VILL-C07

TORQUE SPECIFICATIONS

All readings in ft. lbs.

Year	Engine Displacement Liters (VIN)	Cylinder Head Bolts	Main Bearing Bolts	Rod Bearing Bolts	Crankshaft Damper Bolts	Flywheel Bolts	Manifold Intake	Manifold Exhaust	Spark Plugs	Oil Pan Drain Plug
2001	3.3 (T)	①	②	②	141-156	61-69	①	13-16	14-22	24
2002	3.3 (T)	①	②	②	141-156	61-69	①	13-16	14-22	24

① Intake manifold and cylinder heads are installed at the same time.
Step 1: Cylinder head bolts to 22 ft. lbs.
Step 2: Cylinder head bolts to 43 ft. lbs.
Step 3: Loosen all bolts completely
Step 4: Cylinder head bolts to 7 ft. lbs.
Step 5: Intake manifold bolts to 2.9 (4Nm) ft. lbs.
Step 6: Intake manifold bolts to 13 ft. lbs.
Step 7: Intake manifold bolts to 14 ft. lbs.
Step 8: Loosen all maifold bolts completely
Step 9: Cylinder head bolts to 26 inch lbs.
Step 10: Cylinder head bolts to 47 ft. lbs. Or, an additional 65 degrees
Step 11: Cylinder head sub-bolts to 104 inch lbs.
Step 12: Intake manifold bolts to 36 inch lbs.
Step 13: Intake manifold bolts to 78 inch lbs.
Step 14: Intake manifold bolts to 84 inch lbs.

② Step 1: 34-37 ft. lbs.
Step 2: 67-74 ft. lbs.

67197-VILL-C08

TIRE, WHEEL AND BALL JOINT SPECIFICATIONS

Year	Model	OEM Tires Standard	OEM Tires Optional	Tire Pressures (psi) Front	Tire Pressures (psi) Rear	Wheel Size	Ball Joint Inspection	Lug Nut
2001	Villager	P215/70R15	P225/60R16	30	30	Std: 5.5-JJ Opt: 6.5-JJ	①	80
2002	Villager	P215/70R15	P225/60R16	30	30	Std: 5.5-JJ Opt: 6.5-JJ	①	80

OEM: Original Equipment Manufacturer

PSI: Pounds Per Square Inch

① Replace if any measurable movement is found.

67197-VILL-C09

WHEEL ALIGNMENT

Year	Model		Caster Range (Deg.)	Caster Preferred Setting (Deg.)	Camber Range (Deg.)	Camber Preferred Setting (Deg.)	Toe-in (in.)
2001	Villager	F	0.75	2.75	0.75	-0.25	0.04 +/- 0.04
		R	—	—	0.75	-1.00	0.04 +/- 0.16
2002	Villager	F	0.75	2.75	0.75	-0.25	0.04 +/- 0.04
		R	—	—	0.75	-1.00	0.04 +/- 0.16

67197-VILL-C10

BRAKE SPECIFICATIONS

All measurements in inches unless noted

Year	Model		Brake Disc Original Thickness	Brake Disc Minimum Thickness	Brake Disc Maximum Runout	Brake Drum Diameter Original Inside Diameter	Brake Drum Diameter Max, Wear Limit	Brake Drum Diameter Maximum Machine Diameter	Minimum Lining Thickness	Brake Caliper Bracket-to-Hub Bolt (ft. lbs.)	Brake Caliper Mounting Pin or Bolt (ft. lbs.)
2001	Villager	F	1.002	0.940	0.0028	—	—	—	0.079	—	18-25
		R	—	—	—	9.84	9.90	9.86	0.079	—	—
2002	Villager	F	1.002	0.940	0.0028	—	—	—	0.079	—	18-25
		R	—	—	—	9.84	9.90	9.86	0.079	—	—

NOTE: Due to changes made during production, refer to the manufacturer's specifications if they differ from those in this chart

F: Front

R: Rear

67197-VILL-C11

SCHEDULED MAINTENANCE INTERVALS
MERCURY—VILLAGER

		VEHICLE MILEAGE INTERVAL (x1000)												
TO BE SERVICED	TYPE OF SERVICE	5	10	15	20	25	30	35	40	45	50	55	60	65
Engine oil & filter	R	✓	✓	✓	✓	✓	✓	✓	✓	✓	✓	✓	✓	✓
Rotate tires	S/I	✓		✓		✓		✓		✓		✓		✓
Engine coolant strength hoses & clamps	S/I			✓			✓			✓			✓	
Air cleaner filter	R						✓						✓	
Automatic transmission fluid & filter	R						✓						✓	
Engine coolant ①	R						✓						✓	
PCV valve	R												✓	
Spark plugs ②	R													
Drive belts ③	S/I			✓			✓			✓			✓	
Timing belt ④	S/I													
Exhaust system & heat shields	S/I			✓			✓			✓			✓	
Drive shaft boots	S/I			✓			✓			✓		✓		
Front & rear brake components	S/I	✓	✓	✓	✓	✓	✓	✓	✓	✓	✓	✓	✓	✓

R: Replace S/I: Service or Inspect

① Engine coolant: change every 30,000 miles or 36 months.

② Replace every 105,000 miles

③ Inspect every 15, 000 miles or 12 months and replace every 60,000 miles or 48 months.

④ Replace every 105,000 miles

FREQUENT OPERATION MAINTENANCE (SEVERE SERVICE)

If a vehicle is operated under any of the following conditions it is considered severe service:

- Extremely dusty areas.

- 50% or more of the vehicle operation is in 32°C (90°F) or higher temperatures, or constant operation in temperatures below 0°C (32°F).

- Prolonged idling (vehicle operation in stop and go traffic.

- Frequent short running periods (engine does not warm to normal operating temperatures).

- Police, taxi, delivery usage or trailer towing usage.

Engine oil & filter: replace every 3000 miles.

Rotate tires initially at 6000 miles and every 9000 miles thereafter.

Air cleaner filter: change every 15,000 miles.

Engine coolant strength, hoses & clamps: check every 15,000 miles.

Exhaust system: check every 15,000 miles.

Automatic transmission fluid & filter: change every 21,000 miles.

67197-VILL-C12

PRECAUTIONS

Before servicing any vehicle, please be sure to read all of the following precautions, which deal with personal safety, prevention of component damage, and important points to take into consideration when servicing a motor vehicle:

- Never open, service or drain the radiator or cooling system when the engine is hot; serious burns can occur from the steam and hot coolant.
- Observe all applicable safety precautions when working around fuel. Whenever servicing the fuel system, always work in a well-ventilated area. Do not allow fuel spray or vapors to come in contact with a spark, open flame or excessive heat (a hot drop light, for example). Keep a dry chemical fire extinguisher near the work area. Always keep fuel in a container specifically designed for fuel storage; also, always properly seal fuel containers to avoid the possibility of fire or explosion. Refer to the additional fuel system precautions later in this section.
- Fuel injection systems often remain pressurized, even after the engine has been turned **OFF**. The fuel system pressure must be relieved before disconnecting any fuel lines. Failure to do so may result in fire and/or personal injury.
- Brake fluid often contains polyglycol ethers and polyglycols. Avoid contact with the eyes and wash your hands thoroughly after handling brake fluid. If you do get brake fluid in your eyes, flush your eyes with clean, running water for 15 minutes. If eye irritation persists, or if you have taken brake fluid internally, IMMEDIATELY seek medical assistance.
- The EPA warns that prolonged contact with used engine oil may cause a number of skin disorders, including cancer! You should make every effort to minimize your exposure to used engine oil. Protective gloves should be worn when changing oil. Wash your hands and any other exposed skin areas as soon as possible after exposure to used engine oil. Soap and water, or waterless hand cleaner should be used.
- All new vehicles are now equipped with an air bag system. The system must be disabled before performing service on or around system components, steering column, instrument panel components, wiring and sensors. Failure to follow safety and disabling procedures could result in accidental air bag deployment, possible personal injury and unnecessary system repairs.
- Always wear safety goggles when working with, or around, the air bag system. When carrying a non-deployed air bag, be sure the bag and trim cover are pointed away from your body. When placing a non-deployed air bag on a work surface, always face the bag and trim cover upward, away from the surface. This will reduce the motion of the module if it is accidentally deployed. Refer to the additional air bag system precautions later in this section.
- Clean, high quality brake fluid from a sealed container is essential to the safe and proper operation of the brake system. You should always buy the correct type of brake fluid for your vehicle. If the brake fluid becomes contaminated, completely flush the system with new fluid. Never reuse any brake fluid. Any brake fluid that is removed from the system should be discarded. Also, do not allow any brake fluid to come in contact with a painted surface; it will damage the paint.
- Never operate the engine without the proper amount and type of engine oil; doing so WILL result in severe engine damage.
- Timing belt maintenance is extremely important! Many models utilize an interference-type, non-freewheeling engine. If the timing belt breaks, the valves in the cylinder head may strike the pistons, causing potentially serious (also time-consuming and expensive) engine damage.
- Disconnecting the negative battery cable on some vehicles may interfere with the functions of the on-board computer system(s) and may require the computer to undergo a relearning process once the negative battery cable is reconnected.
- When servicing drum brakes, only disassemble and assemble one side at a time, leaving the remaining side intact for reference.
- Only an MVAC-trained, EPA-certified automotive technician should service the air conditioning system or its components.

ENGINE REPAIR

Distributor

REMOVAL

1. Before servicing the vehicle, refer to the precautions in the beginning of this section.
2. Remove or disconnect the following:
   - Negative battery cable
   - Distributor cap
   - Distributor wiring harness connector
3. Matchmark the rotor to the distributor housing and the distributor housing to the cylinder head.
4. Remove the distributor hold-down bolt and the distributor.

INSTALLATION

Timing Not Disturbed

1. Install or connect the following:
   - Distributor and align the matchmarks made during removal. Tighten the hold-down bolt to 10–12 ft. lbs. (14–17 Nm).
   - Distributor wiring harness connector
   - Distributor cap
   - Negative battery cable
2. Check the ignition timing and adjust, as necessary.

Timing Disturbed

1. Set the engine to Top Dead Center (TDC) of the compression stroke for the No. 1 cylinder.
2. Align the index mark on the distributor shaft with the protrusion on the distributor housing.
3. Install the distributor and check that the distributor rotor is aligned.

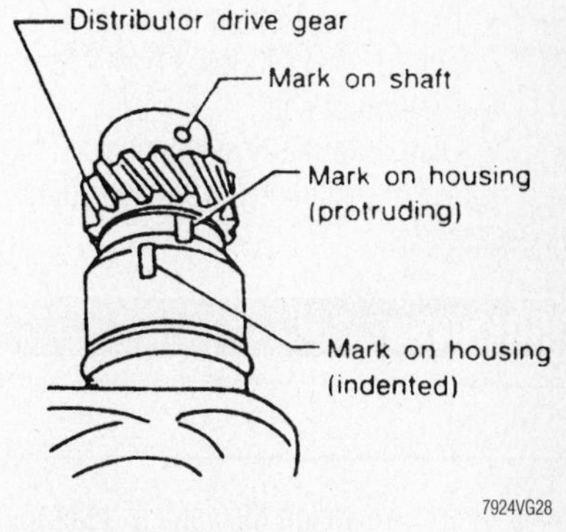

Distributor shaft alignment

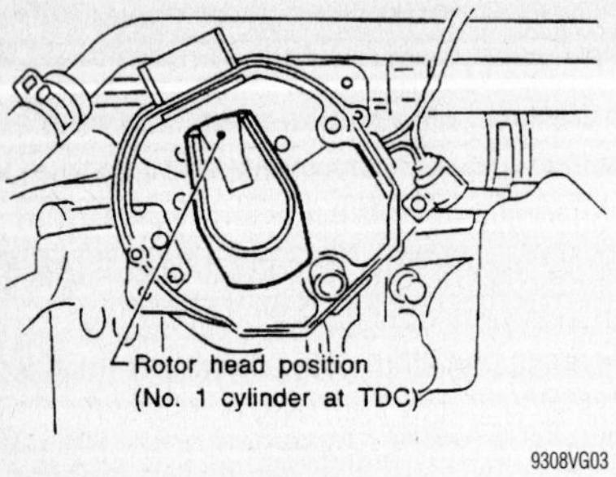

Distributor rotor alignment

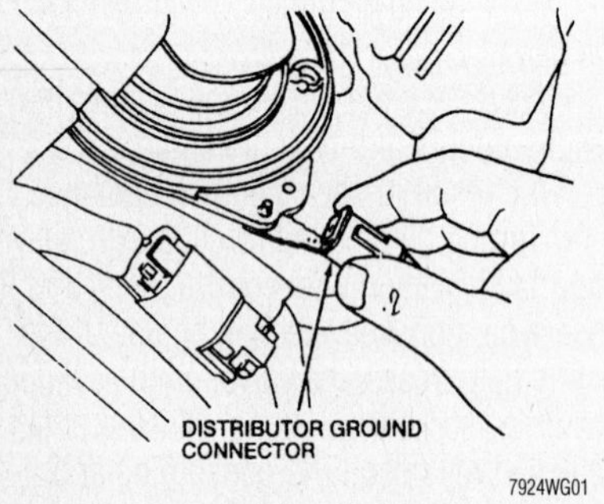

Disengage the distributor ground connector when removing the distributor

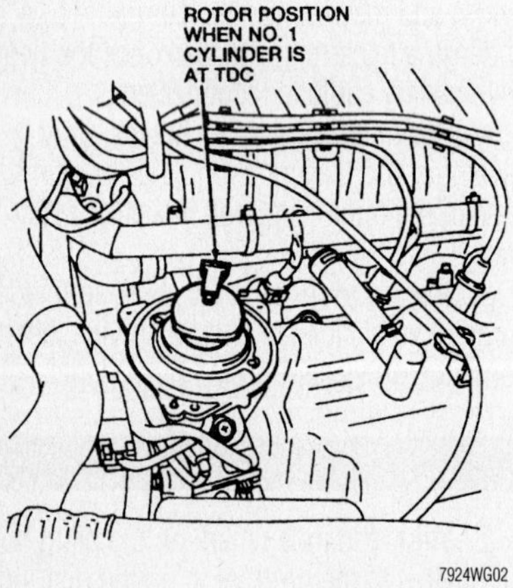

Note the position of the rotor when the No. 1 piston is at TDC on the compression stroke

4. Install or connect the following:
   - Distributor. Tighten the hold-down bolt to 10–12 ft. lbs. (14–17 Nm).
   - Distributor cap
   - Distributor harness connector

5. Check the ignition timing and adjust, as necessary.

Alternator

REMOVAL

1. Before servicing the vehicle, refer to the precautions in the beginning of this section.

2. Remove or disconnect the following:
   - Negative battery cable
   - Idler adjusting bolt, loosen
   - A/C belt
   - Engine undercover
   - Alternator electrical connectors and bracket
   - Alternator mounting bolts
   - Alternator belt
   - Alternator

INSTALLATION

1. Install the components in the reverse order of removal. Tighten the fasteners to the following specifications:
   a. Alternator mounting bolts to 16–22 ft. lbs. (22–29 Nm).
   b. Alternator bracket bolt to 12–15 ft. lbs. (16–20 Nm).

Ignition Timing

ADJUSTMENT

1. Before servicing the vehicle, refer to the precautions in the beginning of this section.
2. Check for trouble codes and make necessary repairs if needed.
3. Apply the parking brake and be sure that the vehicle is in PARK.
4. Start and run the engine until it reaches normal operating temperature.
5. Run the engine at about 2000 rpm for 2 minutes under no-load.
6. Turn off all electrical loads.
7. Disconnect the Throttle Position (TP) sensor electrical connector.
8. Be sure the engine speed is 700–800 rpm.
9. Rev the engine 2 or 3 times to 2,000–3,000 rpm and return the engine to idle speed.
10. Connect a timing light to the distributor end and check the ignition timing. Be sure that the timing pointer is pointing to the 15° BTDC mark on the crankshaft pulley.

➡Each notch on the crankshaft pulley represents 5°.

11. If the timing is not within the specification, loosen the distributor mounting bolt and adjust the distributor until the timing is at the proper specification.
12. Tighten the distributor mounting bolt to 10–12 ft. lbs., (14–17 Nm).
13. Stop the engine and connect the TP sensor.

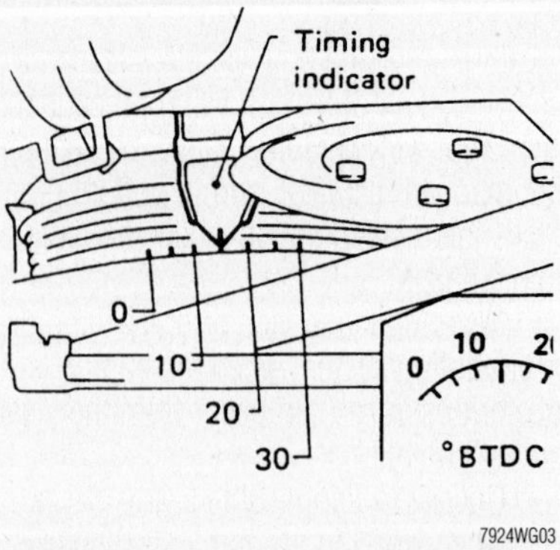

Adjust the timing so the pointer on the engine indicates 15° before top dead center (3 notches from TDC) on the crankshaft pulley.

Engine Assembly

REMOVAL & INSTALLATION

1. Before servicing the vehicle, refer to the precautions in the beginning of this section.
2. Properly relieve the fuel system pressure.
3. Drain the coolant and crankcase.
4. Remove or disconnect the following:
   - Negative battery cable
   - Front wheels
   - All vacuum hoses, fuel lines, wires, harnesses and connectors that would interfere with engine removal
   - Exhaust tube
   - Ball joints
   - Drive shafts
5. Recover the refrigerant from the A/C system
   - A/C compressor manifold
   - Power steering pump
6. Support the engine using a suitable lift.
   - Left hand engine mount bolts
   - Right hand engine mount
   - Rear A/C refrigerant line bracket, if equipped
   - Crossmember
7. Lower the engine and transaxle assembly and remove it from the vehicle.

To install:

8. Installation is the reverse of removal. Refer to the accompanying engine mounting illustration for all necessary torque specifications.

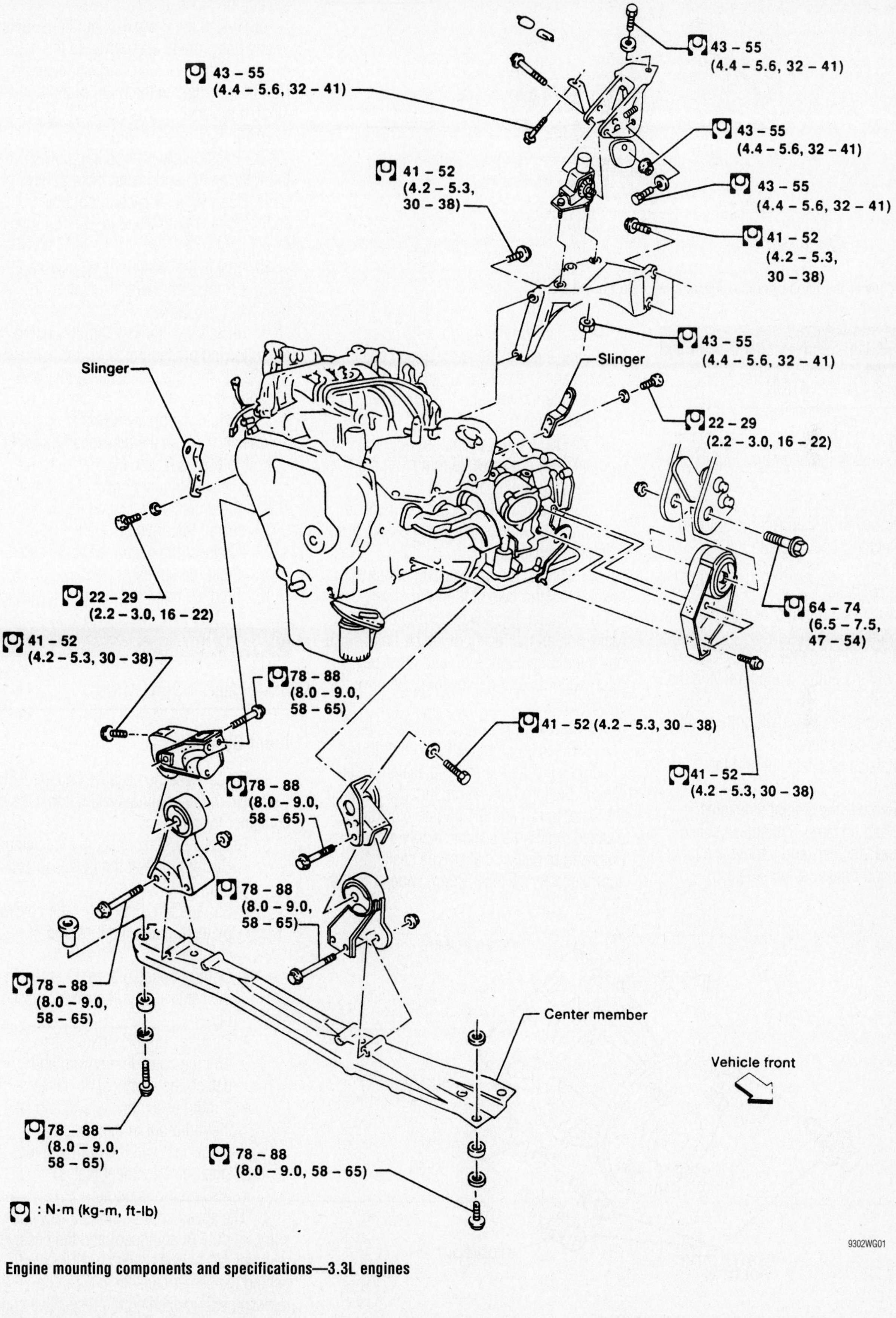

Engine mounting components and specifications—3.3L engines

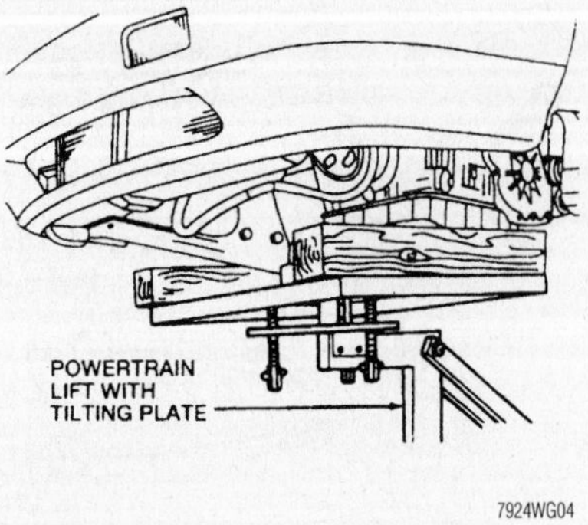

Carefully lower the engine/transaxle assembly from the vehicle.

Water Pump

REMOVAL & INSTALLATION

1. Before servicing the vehicle, refer to the precautions in the beginning of this section.
2. Drain the coolant.
3. Remove or disconnect the following:
   - Negative battery cable
   - Radiator hoses and fan shroud
   - Drive belts
   - Water pump pulley using strap wrench 303–D055–(D85L–6000–A) to hold the pulley while removing the bolts
4. Remove the crankshaft pulley using the following procedure:
   a. Raise and safely support the vehicle.
   b. Remove the 5 right side inner engine and transmission splash shield bolts and 2 screws and remove the inner engine and transmission shield.
   c. Remove the 4 right side outer engine and transmission splash shield bolts and 2 screws and remove the right side outer engine and transmission splash shields.
   d. Use a strap wrench to hold the crankshaft pulley while removing the crankshaft pulley bolt.
   e. Use a crankshaft damper remover to draw the crankshaft pulley off the front of the crankshaft.
5. Remove the 5 lower engine front cover bolts and take of the front cover.
6. Remove the 6 water pump bolts. Make note of the locations of the bolts since one should be a stud/bolt and must be returned to its original location. Remove the water pump.

To install:

7. Clean all parts well. The bolt threads should be cleaned of any old sealer or corrosion. Be sure the mating surfaces between the water pump and the engine block are cleaned of any old sealant. Apply a continuous bead of gasket maker type sealer approximately ⅛ inch (3mm) wide onto the water pump and position the water pump on the engine block.
8. Install the 6 water pump bolts. Refer to any notes made at removal so the bolts can be returned to their original locations. Do not over-tighten the water pump bolts. Tighten the water pump bolts evenly to 12–15 ft. lbs. (16–21 Nm).
9. Position the water pump pulley on the water pump and install the 4 pulley bolts. Use a strap wrench to hold the pulley as the bolts are tightened to 12–15 ft. lbs. (16–21 Nm).
10. Install the front engine cover and the 5 lower front cover bolts. Tighten to 27–44 inch lbs. (3–5 Nm).
11. Install the crankshaft pulley using the following procedure:
    a. Install the crankshaft pulley and pulley bolt.
    b. Hold the pulley with a strap wrench. Tighten the crankshaft pulley bolt to 90–98 ft. lbs. (123–132 Nm).
    c. Install the inner and outer engine and transmission splash shields.
12. Install the drive belts.
13. Connect the negative battery cable.
14. Refill the cooling system.
15. Start the engine and check for leaks.

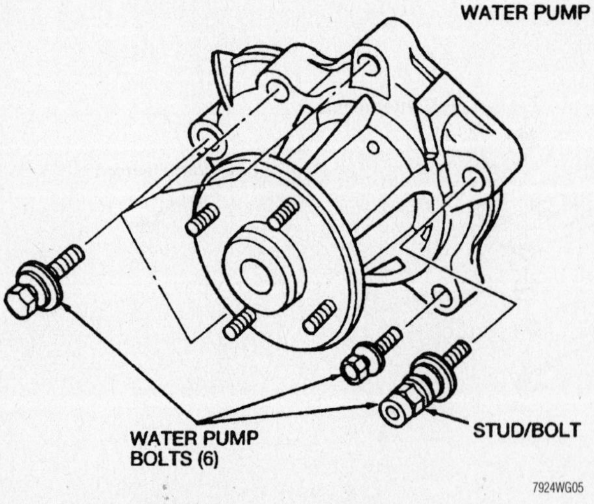

Water pump mounting. Note the location of the stud/bolt

Heater Core

REMOVAL & INSTALLATION

Front System

1. Disconnect the negative battery cable.
2. Drain the cooling system into a clean container for reuse.
3. Remove or disconnect the following:
   - Heater hoses at the bulkhead and plug
   - Storage bin, then both side covers by the bin and the footlamp, if equipped
   - Control console bezel (1 screw in the center), then the ashtray assembly
   - Climate control console screws, pull the console rearward and detach the electrical connectors
   - 4 radio assembly screws and take the radio out of the vehicle
   - Floor duct and the right and left knee reinforcement plates
   - ABS control module.
4. The speed control module, keyless entry module (if equipped) and the passive restraint (air bag) module are all located behind the center console and can be removed after detaching the respective con-

nectors and removing the retaining nuts or screws.

⁂ WARNING

The control modules are very sensitive to static electricity and can be damaged if exposed to static or stray electrical impulses.

5. Remove or disconnect the following:
   - Center air duct
   - 2 ground wire bolts, the U-bracket and the 2 console brackets
   - Glove box and lamp
   - Accelerator pedal and pedal stop
   - Floor air duct
   - Temperature blend sir door actuator and mode door actuator by unfastening the attaching bracket bolts and detaching the electrical connections
   - Ceter distribution duct
   - 4 evaporator/blower assembly screws, the 4 heater assembly screws and the heater assembly
   - Hater pipe plate from the assembly
   - Hater core retainer, disengage the shut-off valve control rod
   - Hater core from the assembly

To install:

6. Install or connect the following:
   - Hater core to the case, the retainer and pipe plate
   - Hater assembly in the vehicle and attach the 4 retaining screws
   - Cnter distribution duct, the blend air and mode door actuators
   - Foor air duct
   - Acelerator stop and pedal
   - Gove box and lamp, then the center console and U-brackets
   - Cnter air duct, the passive restraint, the keyless entry, the speed control and the ABS modules, as removed
   - Remaining center console components
   - Heater hoses to the heater core
7. Refill the cooling system.
8. Connect the negative battery cable.
9. Run the engine to normal operating temperatures; then, check the climate control operation and check for leaks.

Rear Auxiliary System

➡The rear heater/air conditioning assembly must be removed as a complete unit in order to remove the heater core and/or evaporator core.

1. Disconnect the negative battery cable.

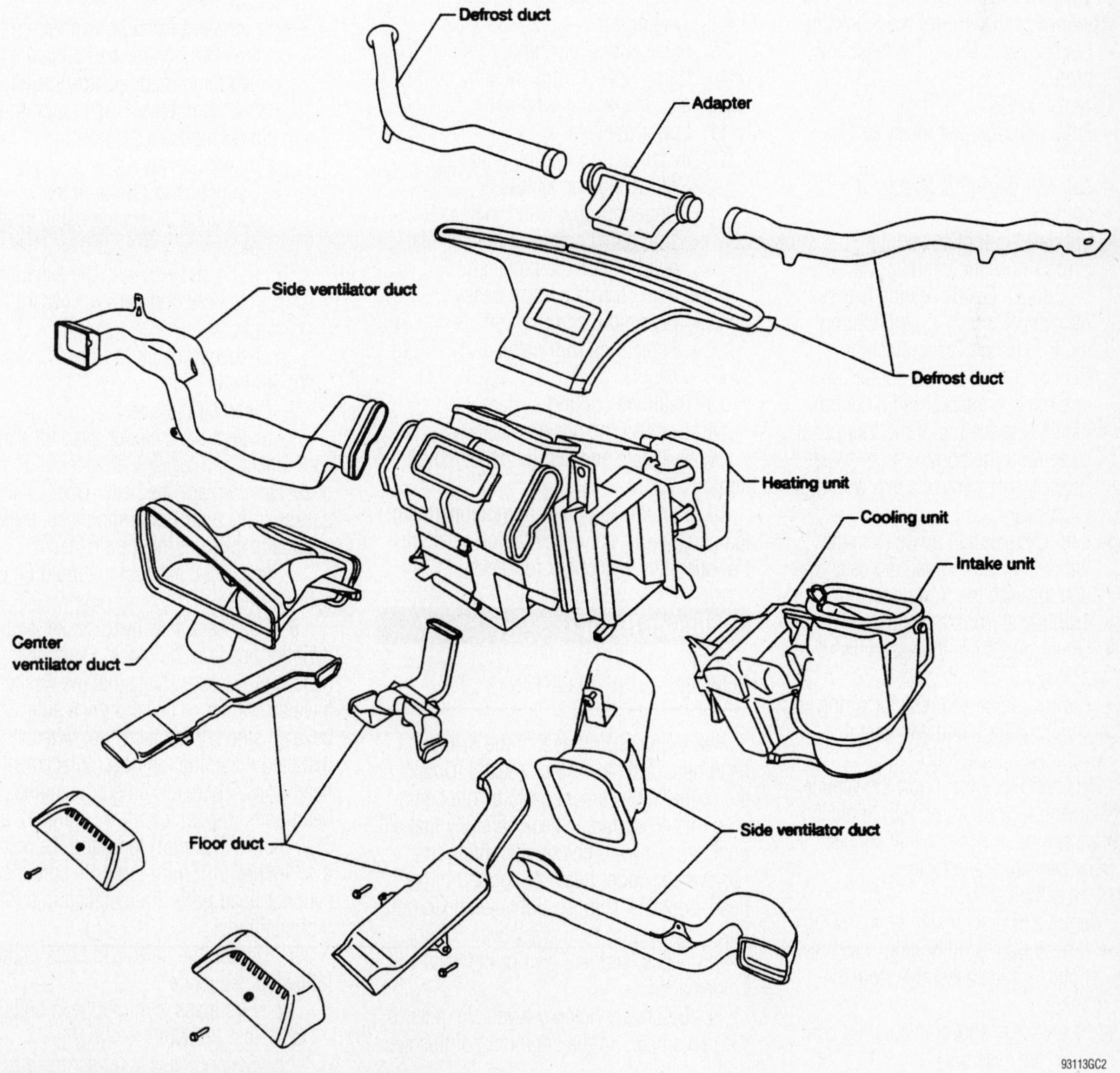

Exploded view the front heater/air conditioning assembly

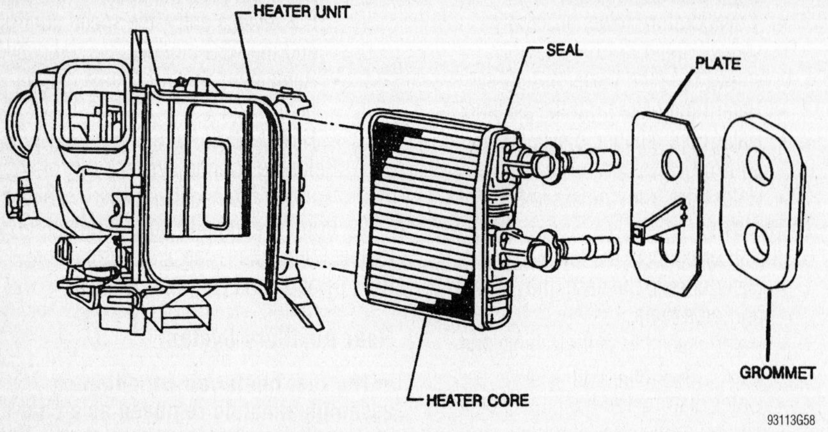

View the front heater core and heater housing assembly

2. Drain the cooling system into a clean container for reuse.
3. Discharge and recover the air conditioning system refrigerant.
4. Remove or disconnect the following:
   - Heater hoses at the bulkhead and plug
   - Center seats
   - 2 left half seat belt lower anchor bolts
   - Left rear cargo net retainers, if equipped
   - Lift gate scuff plate and the 3 screws from the left rear quarter trim panel. Gently pry the rear seat remote control (if equipped) from the trim panel. Disconnect the remote control wiring connector and remove the rear radio control panel. Pull the top of the trim panel away from the body.
   - Rear climate control panel wiring, if equipped
   - Left front lap belt guide from the left quarter trim panel and pass the belt through the trim panel
   - Trim panel from the vehicle
   - Upper duct from the assembly (6 screws)
   - Blower motor and resistor wiring
   - Temperature blend and vent door actuator connectors
5. Raise and safely support the vehicle. Use the spring lock coupling tool to disconnect and plug the refrigerant line connections from beneath the vehicle.
6. Lower the vehicle.
7. Remove or disconnect the following:
   - 4 heater/air conditioning assembly bolts and the assembly from the vehicle
   - Heater core and/or evaporator core from the assembly

To install:

8. Install or connect the following:
   - Heater core and/or evaporator core into the assembly and the 4 retaining bolts
9. Raise and safely support the vehicle.
10. Using new O-rings, reconnect the refrigerant lines to the evaporator.
11. Lower the vehicle.
12. Install or connect the following:
    - All wiring connectors
    - Upper air duct with the 6 screws
    - Trim panel and pass the lap seat belt through the panel slot
    - Rear climate control panel
    - Rear radio and rear remote control
    - Remaining trim panel and components
13. Refill the cooling system.
14. Connect the negative battery cable.
15. Evacuate and charge the air conditioning system.
16. Run the engine to normal operating temperatures; then, check the climate control operation and check for leaks.

Cylinder Head

REMOVAL & INSTALLATION

The factory specifies that the cylinder head bolts ARE NOT to be reused. Obtain the proper replacement parts before beginning this procedure. Check carefully that all bolts are removed before attempting to remove a cylinder head. A tab, part of the head, contains 1 lightly tightened head bolt that is external to the valve cover. Do not overlook this "hidden" bolt or the head will be damaged.

1. Before servicing the vehicle, refer to the precautions in the beginning of this section.
2. Properly relieve the fuel system pressure.
3. Drain the coolant.
4. Remove or disconnect the following:
   - Negative battery cable
   - Air intake tube
   - Timing belt
   - Upper intake manifold (plenum)
   - Fuel feed and return hoses from the fuel rail
   - Fuel injector's electrical connections
   - Fuel rail and injectors as an assembly
   - Intake manifold (lower)
   - Camshaft sprockets
   - Rear timing belt cover
   - Distributor
   - Harness clamp from the right hand rocker cover
   - Exhaust tube from the left hand manifold
   - Left hand exhaust manifold from the right hand exhaust manifold
   - Left hand manifold-to-bracket bolt
   - A/C compressor, alternator and their brackets
   - Rocker covers
   - Cylinder head bolts in the sequence illustrated using several passes
   - Cylinder head with the exhaust manifold and gasket. Discard the gasket.
   - Exhaust manifold from the head

To install:

5. Clean all parts well.
6. Inspect the cylinder head for damage, cracks and leakage of water and oil. If necessary, replace the head. Check the head gasket surface for burrs and nicks. If the head is cracked, it must be replaced.
7. Install the exhaust manifold on the cylinder head.
8. Position a new head gasket and the cylinder head on the block. Examine the head bolt washers. Note that the washers have a chamfer or bevel on one side. The beveled side should face "up" when installed. Examine the new replacement head bolts. There are different lengths. The head bolts in positions 4, 7, 9 and 12 are 5.00 inches (127mm) long and the rest are 4.17 inches (106mm) long. Be sure the new cylinder head bolts are installed in the correct positions.
9. Tighten the new head bolts in the following sequence:
   a. First pass: cylinder head bolts to 22 ft. lbs. (29 Nm).
   b. Second pass: cylinder head bolts to 43 ft. lbs. (59 Nm).

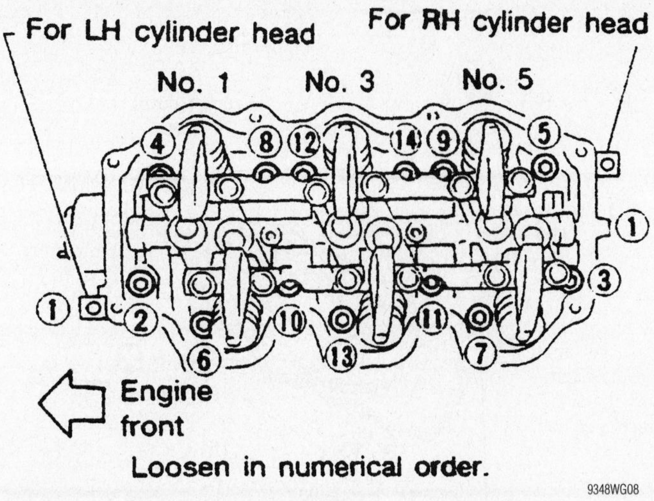

Remove the cylinder head bolts in the sequence shown—3.3L engines

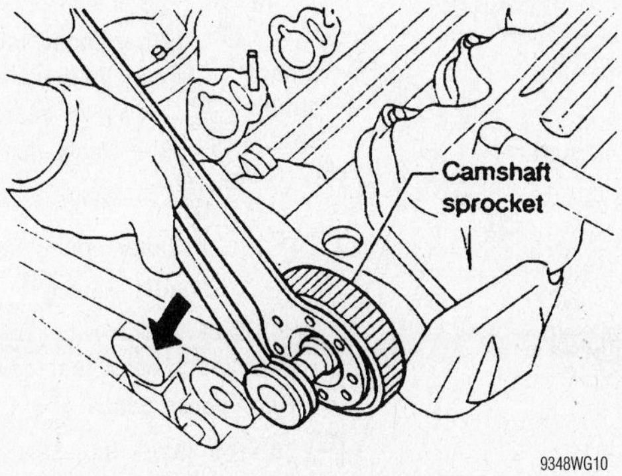

Hold the camshaft sprocket while removing the sprocket retaining bolt—3.3L engines

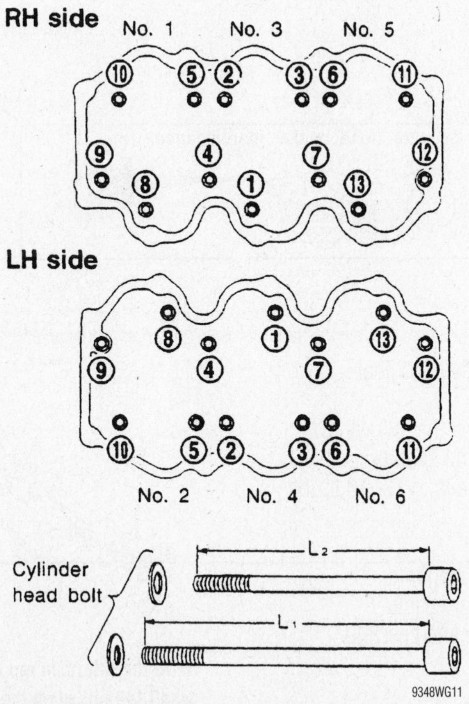

Tighten the cylinder head bolts is sequence as shown—3.3L engines

c. Third pass: Loosen all of the cylinder head bolts completely.

d. Fourth pass: cylinder head bolts to 7 ft. lbs. (10 Nm).

e. Fifth pass: intake manifold bolts and nuts to 2.9 ft. lbs. (4 Nm).

f. Sixth pass: intake manifold bolts and nuts to 13 ft. lbs. (18 Nm).

g. Seventh pass: intake manifold bolts and nuts to 12–14 ft. lbs. (16–20 Nm).

h. Eight pass: Loosen all of the intake manifold bolts and nuts completely.

i. Ninth pass: cylinder head bolts to 22 ft. lbs. (29 Nm).

j. Tenth pass: cylinder head bolts to 40–47 ft. lbs. (54–64 Nm).

k. Eleventh pass: cylinder head sub-bolts to 6.7–8.7 ft. lbs. (9–12 Nm).

l. Twelfth pass: intake manifold bolts and nuts to 2.9 ft. lbs. (4 Nm).

m. Thirteenth pass: intake manifold bolts and nuts to 6.5 ft. lbs. (9 Nm).

n. Fourteenth pass: intake manifold bolts and nuts to 6–7 ft. lbs. (8–10 Nm).

- Rocker covers
- A/C compressor, alternator brackets
- A/C compressor and alternator
- Left hand manifold-to-bracket bolt
- Left hand exhaust manifold to the right hand exhaust manifold
- Exhaust tube to the left hand manifold
- Harness clamp to the right hand rocker cover
- Distributor
- Rear timing belt cover
- Camshaft sprockets
- Intake manifold (lower)
- Fuel rail and injectors as an assembly
- Fuel injector's electrical connections
- Fuel feed and return hoses to the fuel rail
- Upper intake manifold (plenum)
- Timing belt
- Air intake tube
- Negative battery cable

10. Fill the cooling system. An oil and filter change is recommended.

11. Start the vehicle and check for leaks. Check the ignition timing and adjust as required.

Rocker Arms/Shafts

REMOVAL & INSTALLATION

1. Before servicing the vehicle, refer to the precautions in the beginning of this section.

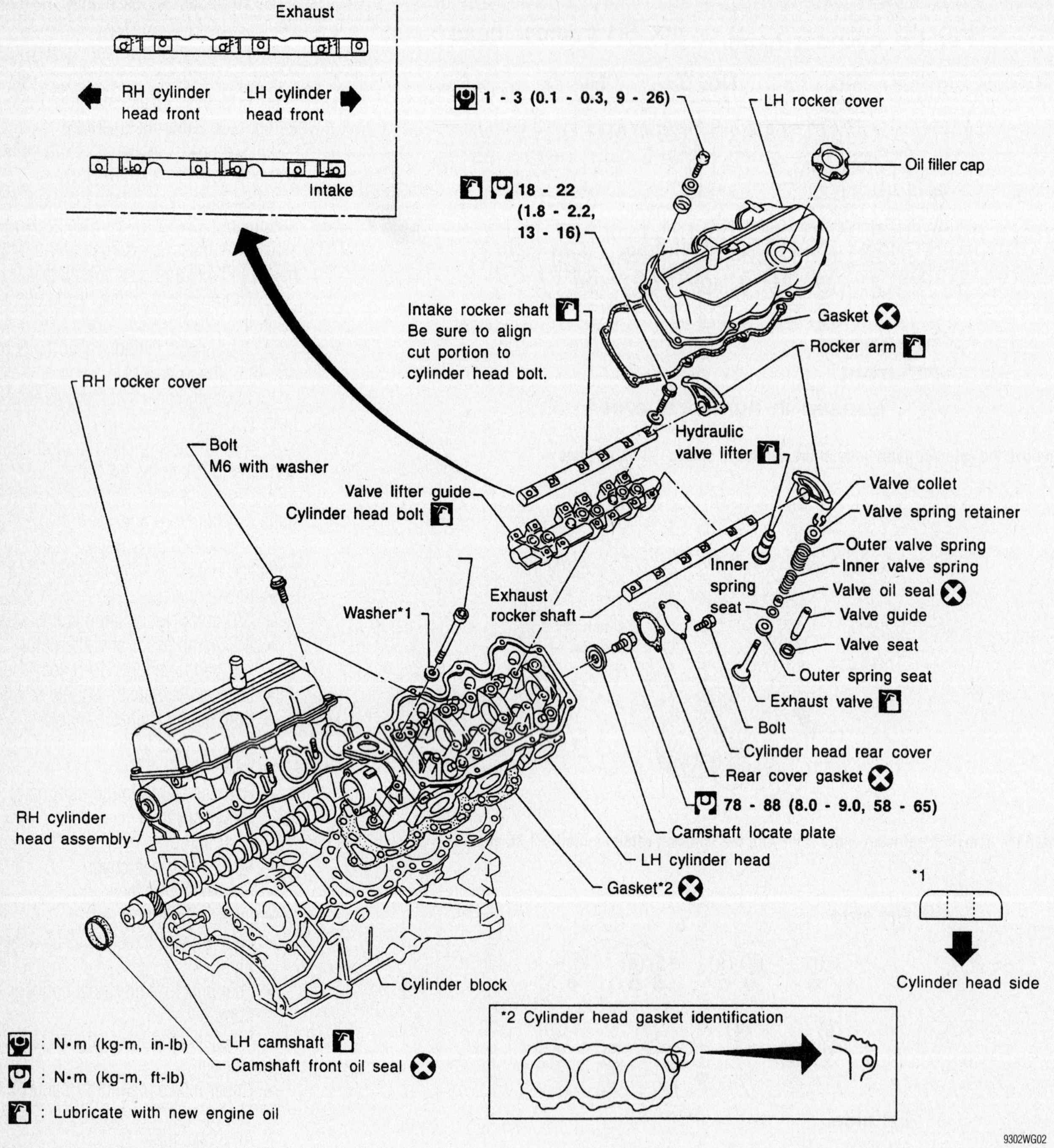

Rocker arm and shaft components

2. Remove or disconnect the following:
   - Negative battery cable
   - Upper intake manifold
   - Valve covers
   - Rocker arm and shaft assemblies
   - Rocker arms from the shafts

➡Keep all valvetrain components in order for assembly.

To install:

3. Lubricate all contact points with clean engine oil and assemble the rocker arms to the shafts in their original positions.
4. Install or connect the following:
   - Rocker arm and shaft assemblies. Tighten the bolts to 13–16 ft. lbs. (18–22 Nm).
   - Valve covers
   - Upper intake manifold
   - Negative battery cable
5. Start the engine and check for leaks.

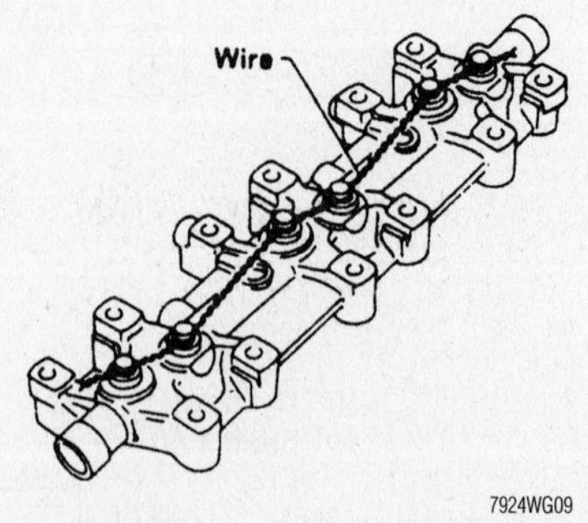

Wire the lifters on top of the guide so they won't fall out when the guide is removed from the head

Intake Manifold

REMOVAL & INSTALLATION

1. Before servicing the vehicle, refer to the precautions in the beginning of this section.
2. Drain the cooling system.
3. Relieve the fuel system pressure.
4. Remove or disconnect the following:
   - Negative battery cable
   - Air intake duct
   - Idle Air Control (IAC) valve connectors
   - Throttle Position (TP) sensor and switch connectors
   - Exhaust Gas Recirculation (EGR) solenoid valve connector
   - Evaporative Emissions (EVAP) canister vacuum and purge hoses
   - Water, heater and Positive Crankcase Ventilation (PCV) valve hoses
   - Vacuum hoses from the EVAP canister, brake cylinder, pressure regulator and EGR tube
   - Spark plug wires
   - Distributor cap
   - 3 left bank injector connectors
   - Thermal transmitter
   - Ground harness
   - Breather pipe
   - Upper manifold
   - Fuel feed and return lines from the fuel rail
   - Right injector harness connectors
   - Fuel rail and injectors
   - Coolant temperature switch harness connector
   - Water hose from the thermostat
   - Lower manifold bolts in the sequence illustrated.
   - Manifold gasket and discard

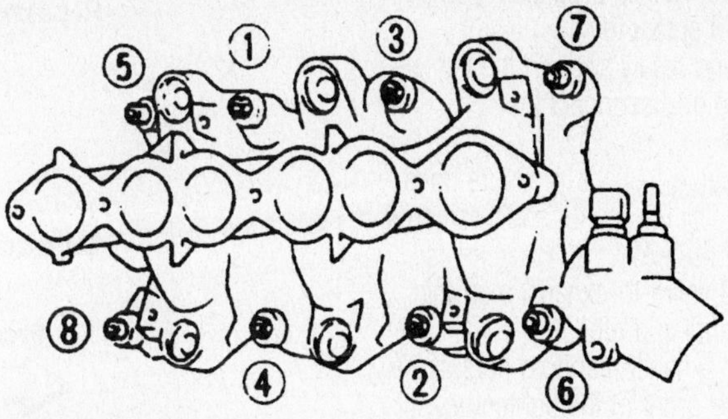

Intake manifold tightening sequence—3.3L engine

To install:

5. Install the lower intake manifold with a new gasket.
   a. Step 1: 35 inch lbs. (4 Nm)
   b. Step 2: 78 inch lbs. (9 Nm)
   c. Step 3: 70–84 inch lbs. (8–10 Nm)
6. Install or connect the following:
   - ECT sensor connector
   - Fuel supply manifold
   - Right bank injector connectors
   - Fuel lines
   - Upper intake manifold
   - Breather pipe
   - Upper intake manifold ground cable
   - Thermal transmitter
   - Left bank injector connectors
   - Distributor
   - Spark plug wires
   - EGR tube
   - Fuel pressure regulator vacuum hose
   - Brake booster vacuum hose
   - EVAP canister vacuum and purge hoses
   - PCV valve and hose
   - Heater hoses
   - Radiator hoses
   - EGR temperature sensor connector
   - EGR solenoid valve connector
   - Ignition coil and power transistor connectors
   - TP sensor and switch connectors
   - IAC valve connector
   - Cruise control cable
   - Accelerator cable
   - Air intake duct
   - Negative battery cable
7. Fill the cooling system.
8. Start the engine and check for leaks.

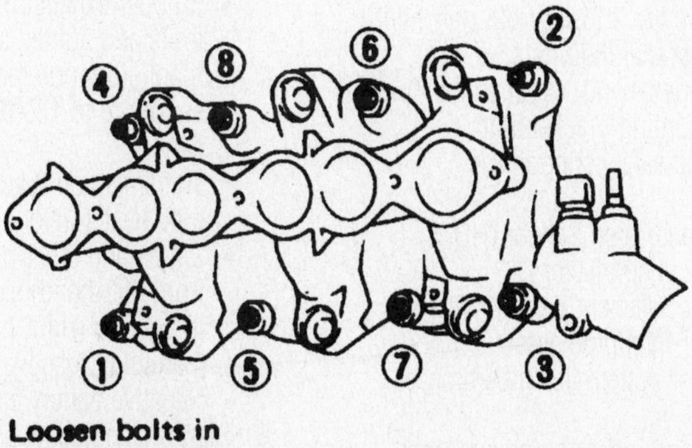

Intake manifold loosening sequence—3.3L engine

Exhaust Manifold

REMOVAL & INSTALLATION

Rear (Right-Hand) Exhaust Manifold

1. Before servicing the vehicle, refer to the precautions in the beginning of this section.
2. Remove or disconnect the following:
   - Negative battery cable
   - Radiator overflow hose from the radiator
   - Radiator coolant-recovery reservoir off of the bracket
   - Reservoir
   - Air cleaner intake tube and the engine air intake resonator
   - 6 rear (right-hand) exhaust manifold crossover tube heat-shield bolts and the heat shields
   - 2 nuts and the 1 bolt securing the rear (right-hand) exhaust manifold tube to the front (left-hand) exhaust manifold. Discard the gasket.
   - Transmission fluid level indicator tube heat shield
3. Disengage the following electrical connectors:
   - Idle switch
   - Throttle Position (TP) sensor

- Exhaust Gas Recirculation (EGR) control solenoid

4. Raise and safely support the vehicle.
5. Remove or disconnect the following:
   - EGR valve-to-back-pressure transducer valve tube nut and position it out of the way
   - 2 EGR valve-to-exhaust manifold tube nuts and tube
   - 6 rear exhaust manifold nuts in the reverse order of the tightening sequence
6. Safely lower the vehicle, remove the exhaust manifold and discard the exhaust manifold gasket.

To install:

7. Raise and safely support the vehicle.
8. Be sure that both the exhaust manifold and the cylinder head mating surfaces are clean of any old gasket material.
9. Install or connect the following:
   - Rear (right-hand) exhaust manifold gasket onto the exhaust manifold mounting studs
10. Lower the vehicle safely.
    - Rear (right-hand) exhaust manifold onto the studs
    - 6 rear (right-hand) exhaust manifold nuts. Tighten the nuts in sequence to 13–16 ft. lbs. (18–22 Nm).
    - EGR valve-to-exhaust manifold tube and tube nuts
    - EGR valve-to-back-pressure transducer valve tube nut
11. Lower the vehicle carefully.
12. Reconnect the following electrical connectors:
    - EGR solenoid
    - TP sensor
    - Idle switch
    - Transmission fluid level indicator tube heat shield
    - New gasket between the front (left-hand) exhaust manifold and the rear exhaust manifold crossover tube
    - 2 nuts and the 1 bolt securing the rear (right-hand) exhaust manifold crossover tube to the front (left-hand) exhaust manifold. Tighten the rear exhaust manifold crossover tube-to-front (left-hand) exhaust manifold nuts and bolt.
    - Rear (right-hand) exhaust manifold crossover tube heat shield with the 6 mounting bolts
    - Rear (right-hand) exhaust manifold crossover tube bolts
    - Air cleaner intake tube and the engine air intake resonator
    - Radiator coolant recovery reservoir
    - Radiator overflow hose to the radiator
    - Negative battery cable
13. Start the engine and check for leaks and proper operation.

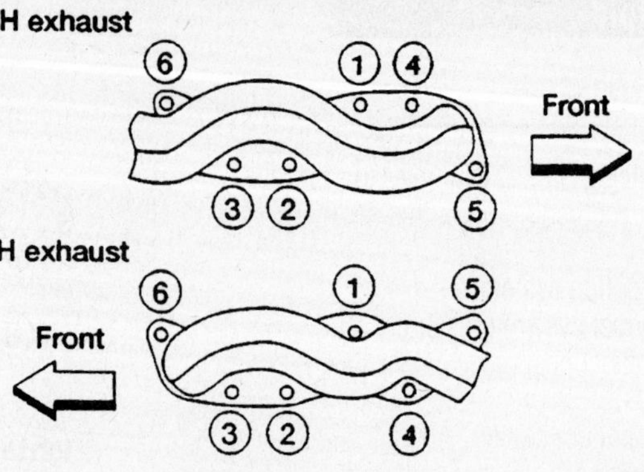

To avoid warping the exhaust manifolds, use this sequence when loosening the bolts—3.3L engine

Front (Left-Hand) Exhaust Manifold

1. Before servicing the vehicle, refer to the precautions in the beginning of this section.
2. Remove or disconnect the following:
   - Negative battery cable and wait at least 90 seconds before performing any work. This allows time for the SRS or air bag system to deplete its back up energy supply.
   - 2 nuts and the 1 bolt securing the front (left-hand) exhaust manifold to the rear (right-hand) exhaust manifold crossover tube. Discard the gasket.
3. Remove the transmission fluid level indicator tube heat shield.
   - 6 front (left-hand) exhaust manifold nuts in 2 steps in the reverse order of the tightening sequence. Do not remove the 3 lower front (left-hand) exhaust manifold nuts.
   - Front (left-hand) exhaust manifold-to-mounting bracket bolt
4. Raise and safely support the vehicle.
   - Heated Oxygen Sensor (HO_2S) electrical connector
   - 3 front (left-hand) exhaust manifold-to-inlet pipe nuts
   - Exhaust system flex tube bracket bolt
   - Left-hand inner engine and transmission splash shield bolts and screws
   - Left-hand inner engine and transmission splash shield
   - 3 lower exhaust manifold nuts
   - Front (left-hand) exhaust manifold and discard the exhaust manifold gasket

To install:

5. Be sure that both the exhaust manifold and the cylinder head mating surfaces are clean of any old gasket material.
6. Install or connect the following:
   - New front exhaust manifold gasket in place
   - Front (left-hand) exhaust manifold
   - 3 lower exhaust manifold mounting nuts. Do not tighten the nuts at this time.
   - Left-hand inner engine and transmission splash shield with their mounting bolts and screws
   - Exhaust system flex tube bracket bolt
   - 3 exhaust manifold-to-exhaust inlet pipe nuts
   - HO_2S electrical connector
7. Lower the vehicle.
   - Front (left-hand) exhaust manifold-to-mounting bracket bolt
   - 3 upper exhaust manifold mounting bolts and tighten all 6 exhaust manifold mounting bolts in sequence to 13–16 ft. lbs. (18–22 Nm)
   - Transmission fluid level indicator tube heat shield
   - 2 nuts and the 1 bolt securing the front (left-hand) exhaust manifold to the rear (right-hand) exhaust manifold crossover tube
   - Negative battery cable
8. Start the engine, check for leaks and road test for proper operation.

Starter

REMOVAL & INSTALLATION

1. Before servicing the vehicle, refer to the precautions in the beginning of this section.
2. Remove or disconnect the following:
   - Battery negative cable
   - Air cleaner
   - Nut attaching the positive cable to the starter
   - Positive cable from the starter
   - S-terminal connector
   - 2 starter bolts and the starter

To install:

3. Installation is the reverse of removal.
4. Tighten the starter bolts to 17–19 ft. lbs. (23–26 Nm) and the nut that attaches the positive battery cable to the starter to 87–104 inch lbs. (10–12 Nm).

Front Crankshaft Seal

REMOVAL & INSTALLATION

1. Before servicing the vehicle, refer to the precautions in the beginning of this section.
2. Remove or disconnect the following:
   - Negative battery cable
   - Drive belts
   - Radiator hoses
   - Crankshaft pulley
   - Front cover
   - Timing belt
   - Crankshaft timing sprocket
   - Crankshaft seal using a suitable prytool

To install:

3. Install or connect the following:
   - Crankshaft seal using a driver and a hammer until its flush with the housing
   - Crankshaft timing sprocket
   - Timing belt
   - Front cover and tighten the bolts to 26–43 inch lbs. (3–5 Nm)
   - Crankshaft pulley and tighten the bolt to 141–156 ft. lbs. (191–211 Nm)
   - Radiator hoses
   - Drive belts
   - Negative battery cable
4. Fill the cooling system, start the vehicle and check for leaks.

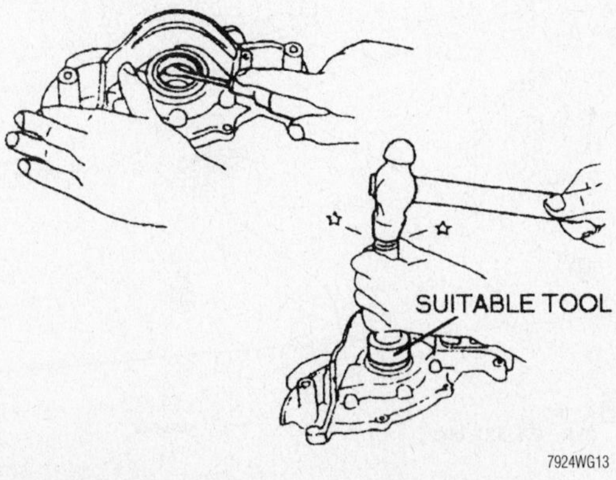

Removal and installation of the front crankshaft seal—3.3L engines

Camshaft And Valve Lifters

REMOVAL & INSTALLATION

1. Before servicing the vehicle, refer to the precautions in the beginning of this section.
2. Drain the cooling system.
3. Remove or disconnect the following:
   - Negative battery cable
   - Upper intake manifold
   - Valve covers

➡Keep all valvetrain components in order for assembly.

   - Rocker arm and shaft assemblies
   - Valve lifter guide and valve lifters. Attach a wire to the top of the lifters so that they will not drop from the lifter guide.
   - Radiator
   - Accessory drive belts
   - Front cover
   - Timing belt
   - Camshaft sprockets
   - Camshaft seals
   - Rear timing cover
   - Distributor
   - Cylinder head rear covers
   - Camshaft locating plates
   - Camshafts

To install:

4. Install or connect the following:
   - Camshafts
   - Camshaft locating plates. Tighten the bolts to 58–65 ft. lbs. (78–88 Nm).
   - Cylinder head rear covers
   - Distributor
   - Rear timing cover
   - Camshaft seals
   - Camshaft sprockets. Tighten the bolts to 58–65 ft. lbs. (78–88 Nm).
   - Timing belt
   - Front cover
   - Accessory drive belts
   - Radiator
   - Valve lifter guide and valve lifters
   - Rocker arm and shaft assemblies. Tighten the bolts to 13–16 ft. lbs. (18–22 Nm).
   - Valve covers
   - Upper intake manifold
   - Negative battery cable
5. Fill the cooling system.
6. Start the engine and check for leaks.

Valve Lash

ADJUSTMENT

The engines covered in this section use hydraulic valve lifters that automatically adjust the valve lash. No periodic adjustment is needed.

Oil Pan

REMOVAL & INSTALLATION

1. Before servicing the vehicle, refer to the precautions in the beginning of this section.
2. Drain the engine oil.
3. Remove or disconnect the following:
   - Negative battery cable
   - Front engine mount (support) insulator through-bolt
   - Rear engine mount (support) through-bolt
   - 2 rear refrigerant/heater pipe hold down bracket bolts
   - 4 crossmember (also called a transverse member) bolts, and remove the crossmember.
   - Exhaust inlet pipe
   - 4 rear transaxle-to-engine brace bolts and the 5 front transaxle-to-engine brace bolts
   - Front transaxle-to-engine brace
   - Low oil level sensor electrical connector
   - 18 oil pan bolts in the reverse order of the tightening sequence, working from the outside, towards the center bolts.
   - Oil pan and discard the seals

To install:

4. Clean all parts well. Be sure that all old sealing material is removed from the oil pan and engine mating surfaces.
5. Position new oil pan seals. Apply Loctite® Ultra Gray 599 Silicone Sealer, or equivalent, to the ends of the oil pan seals.
6. Apply a bead of Loctite® Ultra Gray 599 Silicone Sealer or equivalent to the oil pan gasket rail inboard of the bolt holes.
7. Install or connect the following:
   - Oil pan on the engine block. Tighten the 18 oil pan bolts in sequence, working from the inside, towards the outer bolts. Do not over-tighten. Tighten to 62–70 inch lbs. (7–8 Nm).
   - Low oil level sensor electrical connector.
   - Front and rear transaxle braces. Tighten all bolts to 22–30 ft. lbs. (30–40 Nm).
   - Exhaust inlet pipe
   - Crossmember and tighten the bolts to 58–65 ft. lbs. (78–88 Nm).
   - Both engine support through-bolts and tighten to 58–65 ft. lbs. (78–88 Nm).
8. Remove the support jack from under the crankshaft pulley.
9. Lower the vehicle.
10. Fill the engine with the specified engine oil to the required level.
11. Connect the negative battery cable. Start the engine and check for leaks.

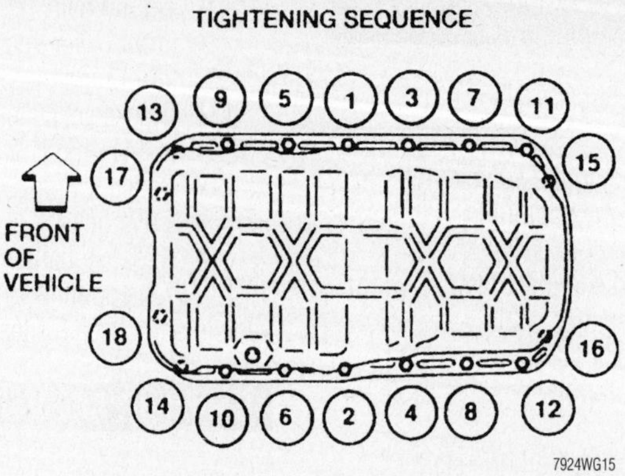

Tighten the 18 oil pan bolts in sequence, working from the inside, towards the outer bolts

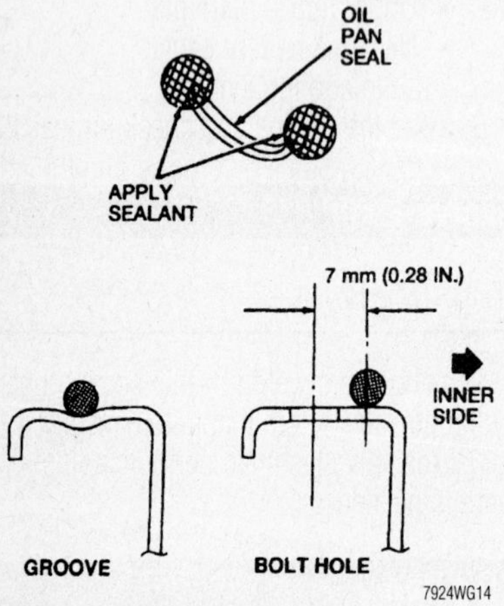

Apply RTV silicone sealer to the seal ends and to the oil pan gasket rail

Oil Pump

REMOVAL & INSTALLATION

1. Before servicing the vehicle, refer to the precautions in the beginning of this section.
2. Drain the engine oil.
3. Drain the cooling system.
4. Remove or disconnect the following:
   - Negative battery cable
   - Oil pan
5. After removing the oil pan, reinstall the crossmember and the mount bolts.

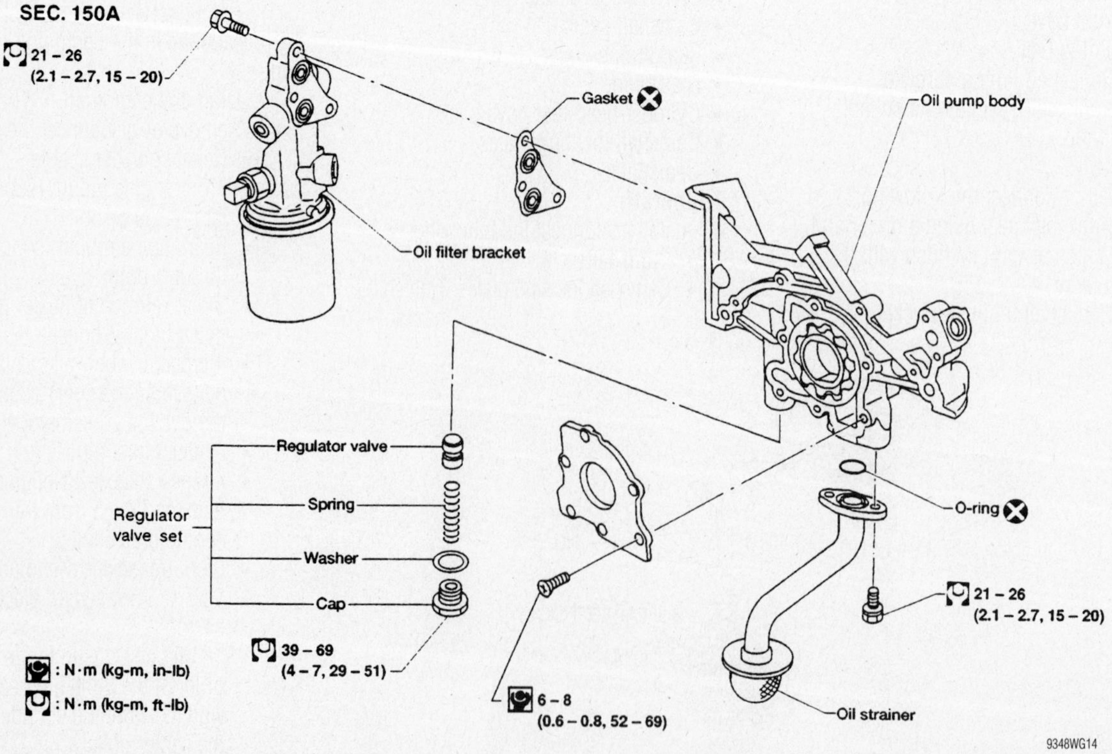

Exploded view of the oil pump assembly—3.3L engine

- Timing belt
- Crankshaft sprocket and timing belt plate
- Oil pump

To install:

6. Install or connect the following:
   - Oil pump-to-body bolts to 52–69 inch lbs. (6–8 Nm)
   - Timing belt plate and tighten the bolts to 52–69 inch lbs. (6–8 Nm)
   - Crankshaft sprocket
   - Timing belt
   - Oil pan
   - Negative battery cable
7. Fill the cooling system.
8. Fill the crankcase to the correct evel.
9. Start the engine and check for leaks.

Rear Main Seal

REMOVAL & INSTALLATION

3.3L Engine

1. Before servicing the vehicle, refer to the precautions in the beginning of this section.
2. Disconnect the negative battery cable.
3. Remove the transaxle from the vehicle.
4. Remove the flexplate from the crankshaft.
5. Remove the rear oil seal retainer.

⁂ WARNING

Do not scratch the seal bore of the oil seal retainer when removing the oil seal.

6. Remove the oil seal from the seal retainer.

To install:

7. Apply clean engine oil to the lip and outer surface of the new seal to aid during installation.
8. Install the seal in the retainer using a suitable seal driver.

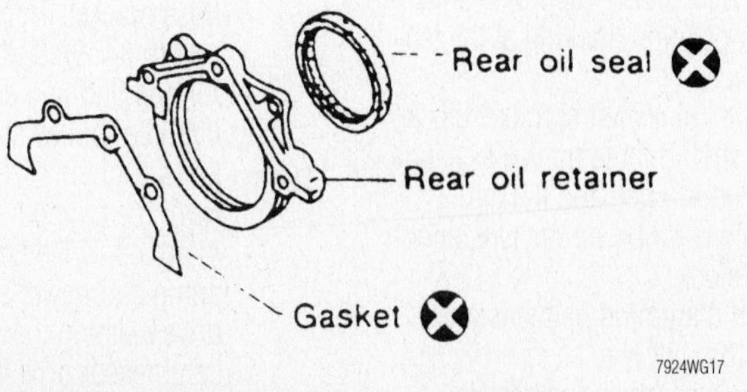

Exploded view of the oil seal, retainer and gasket—3.3L engine

9. Using a new gasket install the retainer on the engine. Tighten the bolts to 52–61 inch lbs. (6–7 Nm).
10. Install the flexplate. Tighten the bolts to 61–69 ft. lbs. (83–93 Nm).
11. Install the transaxle and remaining components.

Timing Belt

REMOVAL & INSTALLATION

3.3L (VIN T) engines

On this vehicle, right side refers to the "rear" components (near the firewall) and left side refers to the "front" components (near the radiator).

1. If the timing belt is to be removed, it is good practice to turn the crankshaft until the engine is at Top Dead Center (TDC) of the No. 1 cylinder, compression stroke (firing position), before beginning work. This should align all timing marks and serve as a reference for all work that follows. After verifying that the engine is at TDC for the No. 1 cylinder, do not crank the engine or allow the crankshaft or camshaft sprockets to be turned otherwise engine timing will be lost.
2. Drain the cooling system.
3. Remove or disconnect the following:
   - Negative battery cable
   - Alternator drive belt, water pump and power steering pump belt and the air conditioning compressor belt, if equipped
   - 3 air conditioning compressor drive belt idler pulley bolts and the idler pulley, if equipped with air conditioning
   - Upper radiator hose bracket bolt
   - Upper hose with the bracket from the vehicle
   - Water bypass hose from between the thermostat housing and the lower water hose connection
   - Main wiring harness from the upper engine front cover
   - 8 upper engine front cover bolts and the upper cover
   - Right side front wheel and tire assembly
   - 4 right side engine and transmission splash shield bolts and 2 screws, and right side outer engine and transaxle splash shield
4. Use a strap wrench to hold the water pump pulley. Remove the 4 pulley bolts, and the water pump pulley.
5. Use a strap wrench to hold the crankshaft pulley. Remove the center pulley bolt, and the crankshaft pulley using a harmonic balancer (damper) puller to draw the pulley from the front of the crankshaft.
   - 5 lower engine front cover bolts, then remove the lower engine front cover
6. Be sure that the timing marks between the crankshaft sprocket and the oil pump housing align.
7. If the timing belt is to be reused, mark an arrow on the belt indicating the direction of rotation. The directional arrow is necessary to ensure that the timing belt, if it to be reused, can reinstalled in the same direction.
8. Loosen the timing belt tensioner nut and slip the timing belt off of the sprockets.
9. If necessary, the camshaft sprockets can be removed. A special spanner tool is designed to hold the sprocket to keep it from turning while the center bolt is being loosened. Use care if using substitutes.

➡The sprockets are not interchangeable.

10. If necessary, the crankshaft sprocket can be removed. The outer timing belt guide (looks like a large washer) and the crankshaft sprocket simply pull off the front of the crankshaft.

➡Be careful, there are 2 crankshaft keys. Use care not to loose them.

To install:

11. Clean all parts well. If removed, inspect the crankshaft sprocket for warping or abnormal wear. Check the sprocket teeth for wear, deformation, chipping or other damage. Replace as necessary. Clean the sprocket mounting surface to ease installation. Install the key. Slip the sprocket onto the crankshaft. Tap it in place with a suitably-sized socket.
12. If removed, inspect the camshaft sprockets for damage and wear. Replace as required. The sprockets should be marked **L3** to designate the front, or left side camshaft and **R3** to designate the rear, or right side camshaft. Use care to install the

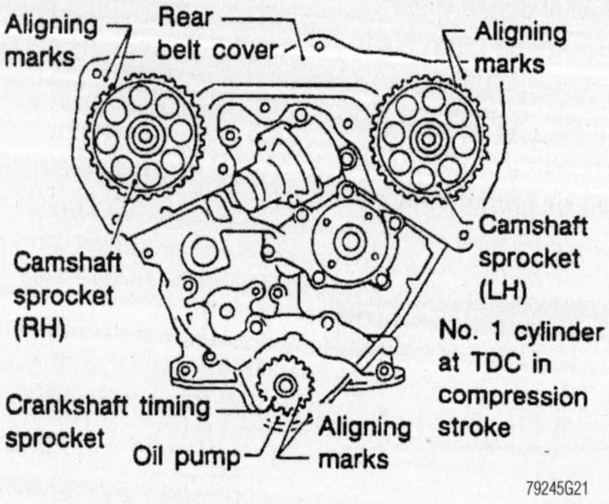

Use a shop rag to clean the alignment marks for the timing belt— 3.3L (VIN T) engines

sprockets properly. A special spanner tool is designed to hold the sprocket to keep it from turning while the center bolt is being tightened. Use care if using a substitute. Tighten the camshaft sprocket center bolts to 61 ft. lbs. (83 Nm). Verify that the timing marks on the camshaft sprockets and the timing marks on the rear cover (called the seal plate) are aligned.

13. Use an Allen wrench to turn the timing belt tensioner clockwise until the belt tensioner spring is fully extended. Temporarily tighten the tensioner nut to 32–43 ft. lbs. (43–58 Nm).

14. If a new timing belt is to be installed, look for a printed arrow on the belt. Be sure the arrow is pointing away from the engine. If the original timing belt is to be reused, be sure that the directional arrow that was marked at disassembly is facing the correct direction.

15. A new Original Equipment Manufacture (OEM) timing belt should have 3 white timing marks on it that indicate the correct timing positions of the camshafts and the crankshaft. These marks are to help ensure that the engine is properly timed. When the engine is properly timed, each white timing mark on the timing belt will be aligned with the corresponding camshaft and crankshaft timing mark on the sprocket. Because the white timing marks are not evenly spaced, the technician needs to use care in installing the belt. There should be 40 timing belt teeth between the timing marks on the front and rear camshaft sprockets and 43 teeth between the timing mark on the front camshaft sprocket and the timing mark on the crankshaft sprocket.

16. Verify that the camshaft timing marks are aligned with the timing marks on the rear cover (seal plate) and that the crankshaft sprocket timing mark is aligned with the timing mark on the oil pump housing.

17. Install the timing belt starting at the crankshaft sprocket and moving around the camshaft sprockets following a counterclockwise path. Do not allow any slack in the timing belt between the sprockets. After all of the timing marks are aligned with the timing belt installed, slip the timing belt onto the belt tensioner.

18. While holding the timing belt tensioner with an Allen wrench, loosen the tensioner nut. Allow the tensioner to put pressure on the timing belt. Use an Allen wrench to turn the timing belt tensioner 70–80 degrees clockwise and tighten the timing belt tensioner nut to 32–43 ft. lbs. (43–58 Nm).

⁂ WARNING

If any binding is felt when adjusting the timing belt tension by turning the crankshaft, STOP turning the engine, because the pistons may be hitting the valves.

19. Rotate the crankshaft clockwise twice and align the No. 1 piston to TDC on the compression stroke (firing position).

20. Apply 22 lbs. (10kg) of force on the timing belt between the rear camshaft sprocket and the timing belt tensioner. An assistant may be needed. While holding the timing belt tensioner steady with an Allen wrench, loosen the timing belt tensioner nut. Remove the Allen wrench and adjust the timing belt tensioner using the following procedure:

a. Install a 0.0138 in. (0.35mm) thick and 0.500 in. (12.7mm) wide feeler gauge where the timing belt just starts to go around the tensioner (approximately the 4 o'clock position, looking at the tensioner).

b. Turn the crankshaft sprocket clockwise, which should force the feeler gauge between the timing belt and the tensioner, up to a position on the tensioner of about 1 o'clock.

c. Tighten the timing belt tensioner nut to 61 ft. lbs. (83 Nm.

d. Turn the crankshaft clockwise to rotate the feeler gauge out from between the timing belt tensioner and the timing belt.

21. Rotate the crankshaft clockwise twice, and once again align the No. 1 piston to TDC on the compression stroke (firing position).

22. Apply 22 lbs. (10kg) of force on the timing belt between the front and rear camshaft sprockets. Measure the amount of belt deflection. Belt deflection should be between 0.51–0.59 in. (13–15mm). If belt deflection is out of specification, repeat Steps 29 through 33. If the timing belt deflection cannot be adjusted into specification, the timing belt will have to be replaced.

23. Install or connect the following:

- Lower engine front cover and the 5 lower cover bolts. Do not over tighten. Tighten to 27–44 inch lbs. (3–5 Nm).
- Outer timing belt guide next to the crankshaft sprocket with the dished side facing away from the cylinder block. Install the crankshaft pulley. Use a strap wrench to keep the crankshaft pulley from turning and tighten the center bolt to 148 ft. lbs. (201 Nm).
- Water pump pulley on the pump. Install the 4 bolts. Use a strap wrench to keep the water pump pulley from turning and tighten the 4 water pump pulley bolts to 89 inch lbs. (10 Nm).
- Right side outer engine and transaxle splash shield, and secure with the 4 bolts and 2 screws
- Right side front wheel. Tighten the lug nuts to 72–87 ft. lbs. (98–118 Nm).
- Upper engine timing belt front cover, and tighten the 8 bolts to 27–44 inch lbs. (3–5 Nm)
- Main wiring harness on the upper engine front cover
- Water bypass hose between the thermostat housing and water connection
- Upper radiator hose between the radiator and the water hose connection. Secure the hoses with clamps. Install the upper radiator hose bracket. Tighten the bracket bolt to 34–58 ft. lbs. (46–65 Nm).
- Air conditioning compressor drive belt idler pulley and install the 3 bolts. Tighten to 15 ft. lbs. (21 Nm), if equipped
- Alternator drive belt, the water pump and power steering pump drive belt and the air conditioning compressor drive belt, if equipped.
- Negative battery cable

24. Fill the cooling system.

25. Start the engine and allow it to warm to operating temperature. Check and adjust the ignition timing. Road test to verify correct engine operation.

Piston and Ring Positioning

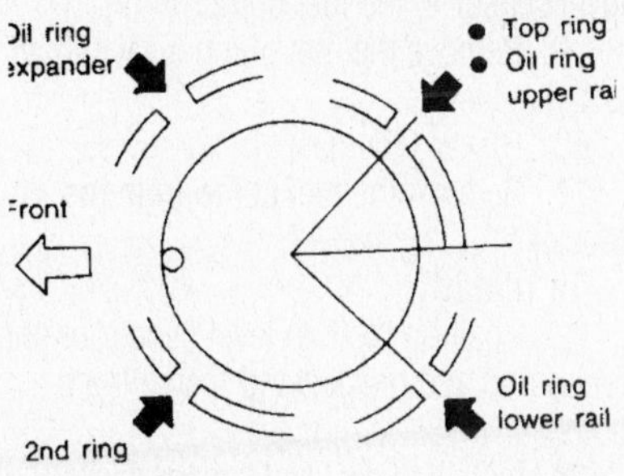

3.3L engines piston ring end-gap spacing

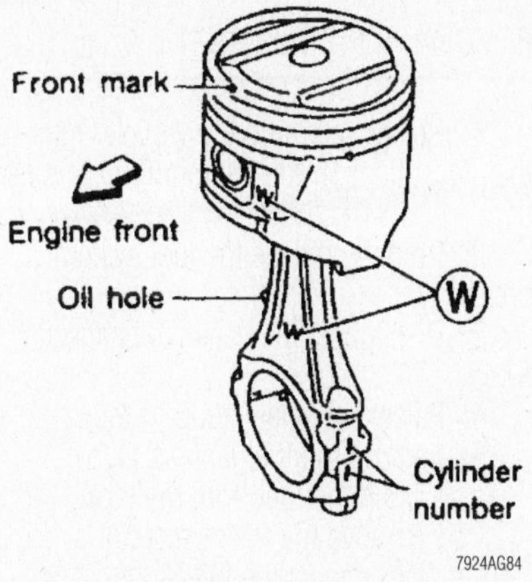

3.3L engines piston and connecting rod assembly positioning

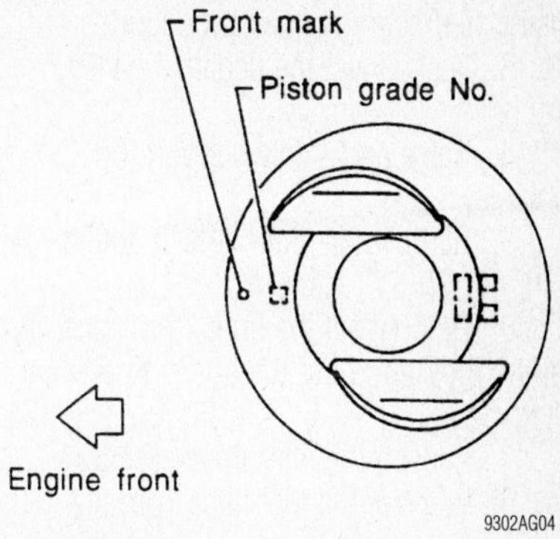

3.3L engine piston positioning

FUEL SYSTEM

Fuel System Service Precautions

Safety is the most important factor when performing not only fuel system maintenance but any type of maintenance. Failure to conduct maintenance and repairs in a safe manner may result in serious personal injury or death. Maintenance and testing of the vehicle's fuel system components can be accomplished safely and effectively by adhering to the following rules and guidelines.

- To avoid the possibility of fire and personal injury, always disconnect the negative battery cable unless the repair or test procedure requires that battery voltage be applied.
- Always relieve the fuel system pressure prior to disconnecting any fuel system component (injector, fuel rail, pressure regulator, etc.), fitting or fuel line connection. Exercise extreme caution whenever relieving fuel system pressure, to avoid exposing skin, face and eyes to fuel spray. Please be advised that fuel under pressure may penetrate the skin or any part of the body that it contacts.
- Always place a shop towel or cloth around the fitting or connection prior to loosening to absorb any excess fuel due to spillage. Ensure that all fuel spillage (should it occur) is quickly removed from engine surfaces. Ensure that all fuel soaked cloths or towels are deposited into a suitable waste container.
- Always keep a dry chemical (Class B) fire extinguisher near the work area.
- Do not allow fuel spray or fuel vapors to come into contact with a spark or open flame.
- Always use a back-up wrench when loosening and tightening fuel line connection fittings. This will prevent unnecessary stress and torsion to fuel line piping. Always follow the proper torque specifications.
- Always replace worn fuel fitting O-rings with new. Do not substitute fuel hose or equivalent, where fuel pipe is installed.

Fuel System Pressure

RELIEVING

1. Before servicing the vehicle, refer to the precautions in the beginning of this section.
2. Remove the left side engine compartment relay panel cover.
3. Locate and remove the fuel pump relay from the relay panel.
4. Start the engine.
5. Allow the engine to run until it stalls from fuel starvation. After the engine stalls, crank the engine over 2 more times to ensure all pressure has been released.
6. Turn the ignition switch to the **OFF** position and install the fuel pump relay.
7. Most service work that follows fuel pressure relief also requires that the negative battery cable (ground) be disconnected before service work begins. This also prevents accidental fuel pump energizing that could pressurize the system.

Fuel Filter

REMOVAL & INSTALLATION

In-Line—Except California

1. Before servicing the vehicle, refer to the precautions in the beginning of this section.
2. Relieve the fuel system pressure using the recommended procedure.
3. Disconnect the negative battery cable.
4. Raise and safely support the vehicle.
5. Remove the fuel hose clamps.
6. Disconnect and plug the hoses to prevent leakage.
7. Remove the fuel filter from the bracket.

To install:

8. Install the fuel filter into the bracket with the arrow facing up, in the direction of the fuel travel to the engine.
9. Reconnect the fuel hoses.
10. Install and tighten the hose clamps. Verify that the clamps are properly tightened. System operating pressure is approximately 36 psi (248 kPa) and fuel will leak is connections are not properly made.
11. Lower the vehicle.
12. Reconnect the negative battery cable.
13. Check for leaks.

In-Line—California

1. Before servicing the vehicle, refer to the precautions in the beginning of this section.

2. Relieve the fuel system pressure using the recommended procedure.
3. Disconnect the negative battery cable.
4. Raise and safely support the vehicle.
5. Remove the filter splash shield olts.
6. Disconnect the lines from each end of the filter and plug the hoses to prevent leakage.
7. Loosen the filter bracket nuts.
8. Remove the fuel filter from the bracket.

To install:

9. Install the fuel filter into the bracket with the arrow facing forward. Tighten the bracket bolts to 44 inch lbs. (5 Nm).
10. Reconnect the fuel hoses.
11. Lower the vehicle.
12. Reconnect the negative battery cable.
13. Check for leaks.

Fuel Pump

REMOVAL & INSTALLATION

1. Before servicing the vehicle, refer to the precautions in the beginning of this section.
2. Properly relieve the fuel system pressure.
3. Disconnect the negative battery cable.
4. Raise and safely support the vehicle.
5. Remove the fuel tank as follows:
   a. Drain the fuel from the tank.
   b. Remove the filler protector.
   c. Disconnect the filler tube.
   d. Detach any electrical connectors related to the fuel pump and fuel level sending unit.
   e. Detach the fuel line quick connectors.
   f. Safely support the fuel tank.
   g. Remove the tank mounting straps, then lower the tank out of the vehicle.
6. Remove the 6 fuel pump bolts.
7. Lift the fuel pump out of the fuel tank. Use care. The fuel level sensor and fuel pump and bracket must be tipped to remove it from the fuel tank. Do not lift the fuel sensor and pump assembly straight out of the fuel tank or damage to the level sensor may occur.
8. Remove the 2 bolts attaching the level sensor to the fuel pump.
9. Remove the fuel pump level sensor and the gasket.
10. Discard the gasket.
11. Remove the fuel pump from the bracket.

To install:

12. Position the fuel level sensor on the fuel pump and bracket and install the 2 bolts.
13. Install a new level sensor gasket. Carefully install the level sensor and pump assembly.
14. Install the 6 fuel pump bolts. Do not over-tighten the bolts. Tighten the bolts to just 17–23 inch lbs. (2–3 Nm).

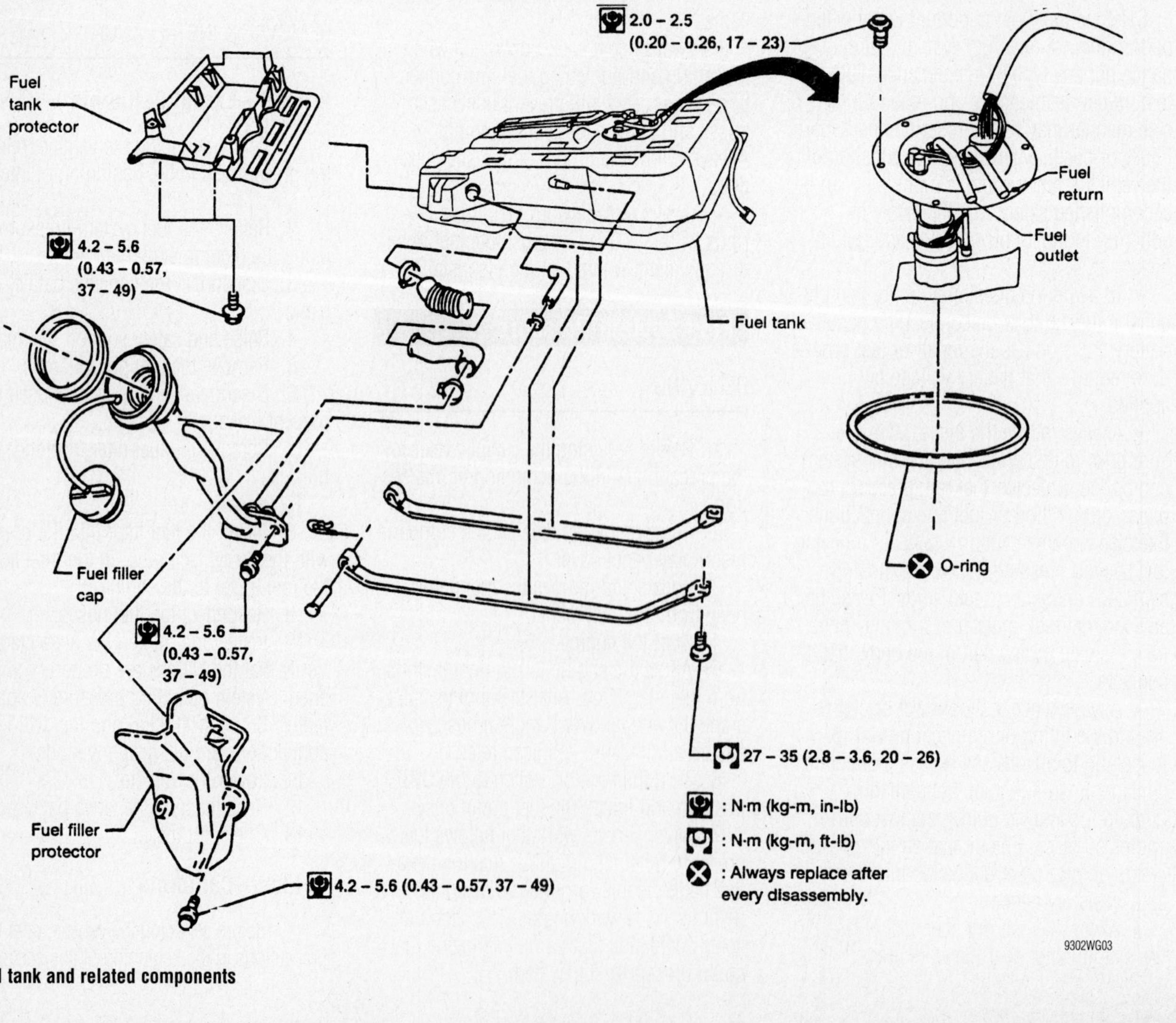

Fuel tank and related components

15. Install the fuel tank in the reverse order of removal, be sure to tighten the tank mounting straps to 20–26 ft. lbs. (27–35 Nm).
16. Lower the vehicle. Refill the fuel tank as required.
17. Connect the negative battery cable. Verify that the fuel pump relay has been properly installed. Start the engine and check for proper operation.

Fuel Injector

REMOVAL & INSTALLATION

1. Before servicing the vehicle, refer to the precautions in the beginning of this section.
2. Disconnect the negative battery cable.
3. If removing a rear injector, remove the upper intake manifold.
4. Disengage the injector electrical connection.

➡When removing the fuel injectors, use a screwdriver head socket to remove the injector cap screws.

5. Remove the injector cap screws and the cap.
6. Pull the injector from the fuel rail.
7. remove and discard the injector O-rings.

To install:

➡Use new insulators and O-ring seals for assembly.

8. Install or connect the following:
   - Fuel injectors with the rail and tighten the fasteners to 8–11 ft. lbs. (11–15 Nm)
   - Fuel injector caps. Tighten the screws to 26–33 inch lbs. (3–4 Nm).
   - Injector electrical connections.
   - Intake manifold, if removed.
   - Negative battery cable.
9. Start the vehicle and check for leaks.

DRIVE TRAIN

Automatic Transaxle Assembly

REMOVAL & INSTALLATION

1. Before servicing the vehicle, refer to the precautions in the beginning of this section.
2. Remove or disconnect the following:
   - Negative battery cable
   - Battery and tray
   - Resonator
   - Terminal cord assembly harness connector
   - Vacuum lines
   - Starter motor
   - Transaxle fluid from the unit
   - Halfshafts
   - Transaxle cooler hose and control cable
   - Front exhaust manifold
   - Crankshaft Position (CKP) sensor
   - Engine gusset and torque converter undercover
   - Bolts from the drive plate from the torque converter. Rotate the crankshaft to access all the bolts.
3. Support the transaxle with a suitable jack.
   - Front mount
   - Rear mount
   - Bolts attaching the transaxle to the engine
4. Carefully separate the transaxle assembly from the engine assembly. Lower the assembly from the vehicle.

To install:

5. Be sure that the transaxle is secured firmly to the transaxle jack.
6. Carefully raise the transaxle into the vehicle and align the transaxle to the engine assembly, making sure that the alignment dowels are positioned properly.
7. Install or connect the remaining components in the reverse order of removal. Refer to the accompanying transaxle torque specification illustration for bolt locations and their specifications.
8. Connect the negative battery cable.
9. Fill the transaxle with the correct amount and type of fluid.
10. Start the engine.
11. Check for leaks and proper operation.

Halfshaft

REMOVAL & INSTALLATION

1. Before servicing the vehicle, refer to the precautions in the beginning of this section.
2. Raise and safely support the vehicle.
3. Remove or disconnect the following:
   - Wheel
   - Fender splash shield
   - Cotter pin, nut retainer, and the hub retainer washers from the front hub assembly
   - Lower ball joint from the knuckle
   - Sway bar from the lower control arm at the sway bar link nut

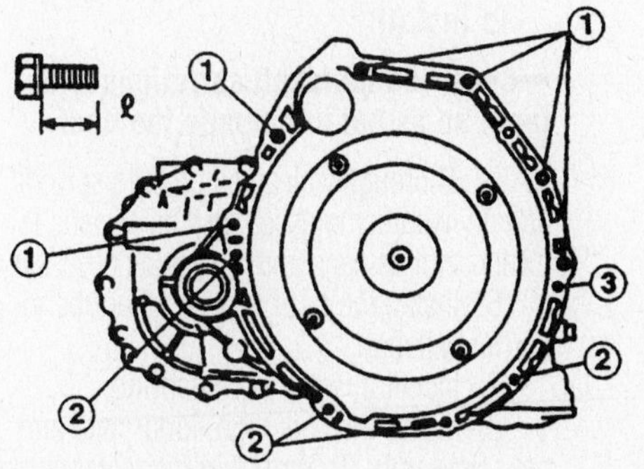

Bolt No.	Tightening torque N·m (kg-m, ft-lb)	ℓ mm (in)
1	39 - 49 (4.0 - 5.0, 29 - 36)	60 (2.36)
2	30 - 40 (3.1 - 4.1, 22 - 30)	25 (0.98)
3*	30 - 40 (3.1 - 4.1, 22 - 30)	25 (0.98)

*: TORX bolt

9348WG17

Transaxle torque specification and locations—3.3L engine

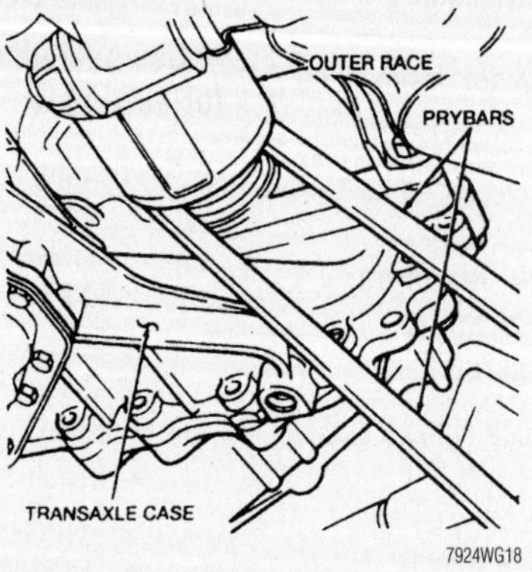

Removing the left side halfshaft by gently prying with 2 prybars to unseat the circlip

- Halfshaft and CV-joint from the wheel hub

4. Position a drain pan under the transaxle since some fluid may run out when the inner joint is disengaged from the transaxle.

5. A prybar is used to separate the inner CV-joint from the transaxle. Use great care that the prybar does not damage the transaxle case, differential oil seal, outer race or boot. If removing the left side halfshaft, position prybars on both sides of the outer race, between the outer race and the transaxle case. Gently pry outward to unseat the circlip.

6. When removing the right side halfshaft, it is not be necessary to remove the halfshaft bearing retainer bracket from the cylinder block. Remove the 3 bearing retainer bolts and pull the right side halfshaft CV-joint with the bearing retainer from the differential side gear.

7. Support the halfshafts and remove them from the vehicle. Use care not to damage the boots. Place the halfshafts on a flat, protected work area.

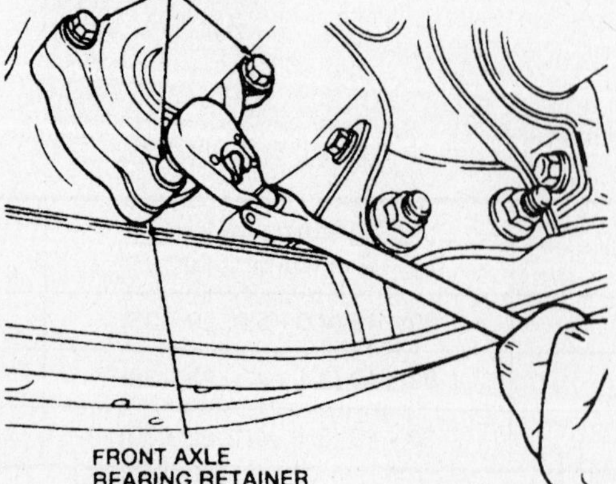

Right side halfshaft bearing retainer bracket

To install:

⁂ CAUTION

Do not reuse the circlip used on the left side halfshaft.

8. To prevent over-expanding the circlip, install the circlip carefully, starting one end in the shaft groove, then working the circlip over the CV-joint splined end. Always use a new circlip. No circlip is used on the right side halfshaft.

9. Inspect the CV-joint boots. If service is required, replace the CV-joint boots.

10. Inspect the differential oil seals. If damaged, the factory recommends using a hook-type puller and slide hammer arrangement to remove the seals. A seal driver is used to install the replacement differential oil seals.

11. If installing the left side halfshaft and CV-joint assembly, position the CV-joint so the splines are aligned with the differential side gear splines, then push the halfshaft joint into the differential case. As the circlip locks into the differential side gear groove, a click will be felt.

12. If installing the right side halfshaft and CV-joint assembly, simply push the CV-joint into the differential side gear. Position the bearing retainer onto the bearing retainer bracket that should still be on the cylinder block. Install the 3 bolts and tighten to 8–14 ft. lbs. (13–19 Nm).

13. Install or connect the following:
    - Halfshaft
    - Lower ball joint and tighten the lower ball joint stud nut to 52–63 ft. lbs. (71–86 Nm). Secure the nut with a new cotter pin.
    - Sway bar link to the lower control arm and tighten the link nut to 12–16 ft. lbs. (16–22 Nm).
    - Wheel outer bearing retainer, washer and axle nut. Tighten the hub nut to 174–231 ft. lbs. (235–314 Nm). Install the nut retainer and secure with a new cotter pin.
    - Splash shield
    - Wheel. Tighten the lug nuts to 72–87 ft. lbs. (98–118 Nm).

14. Lower the vehicle.

15. Check the transaxle fluid level.

16. Road test the vehicle to verify correct operation and no noise or vibration.

CV-Joint

OVERHAUL

Inner

1. Remove the boot bands.
2. Matchmark the slide joint housing and inner race, prior to separating the joint assembly.
3. Pry off the snapring and remove the ball cage, inner race and balls as a unit.
4. Remove the snapring and withdraw the boot.

To install:

➡Cover the halfshaft serrations with tape, so as not to damage the boot.

5. Thoroughly clean all parts in solvent and dry with compressed air. Check parts for evidence of damage and replace as necessary.
6. Install the boot and new boot band on the halfshaft.
7. Install a new inner snapring.
8. Install the ball cage, inner race and balls as a unit. Confirm that the matchmarks are aligned.
9. Install a new outer snapring.
10. Pack the CV-joint with 5.0–6.0 ounces (165–175 g) of grease.

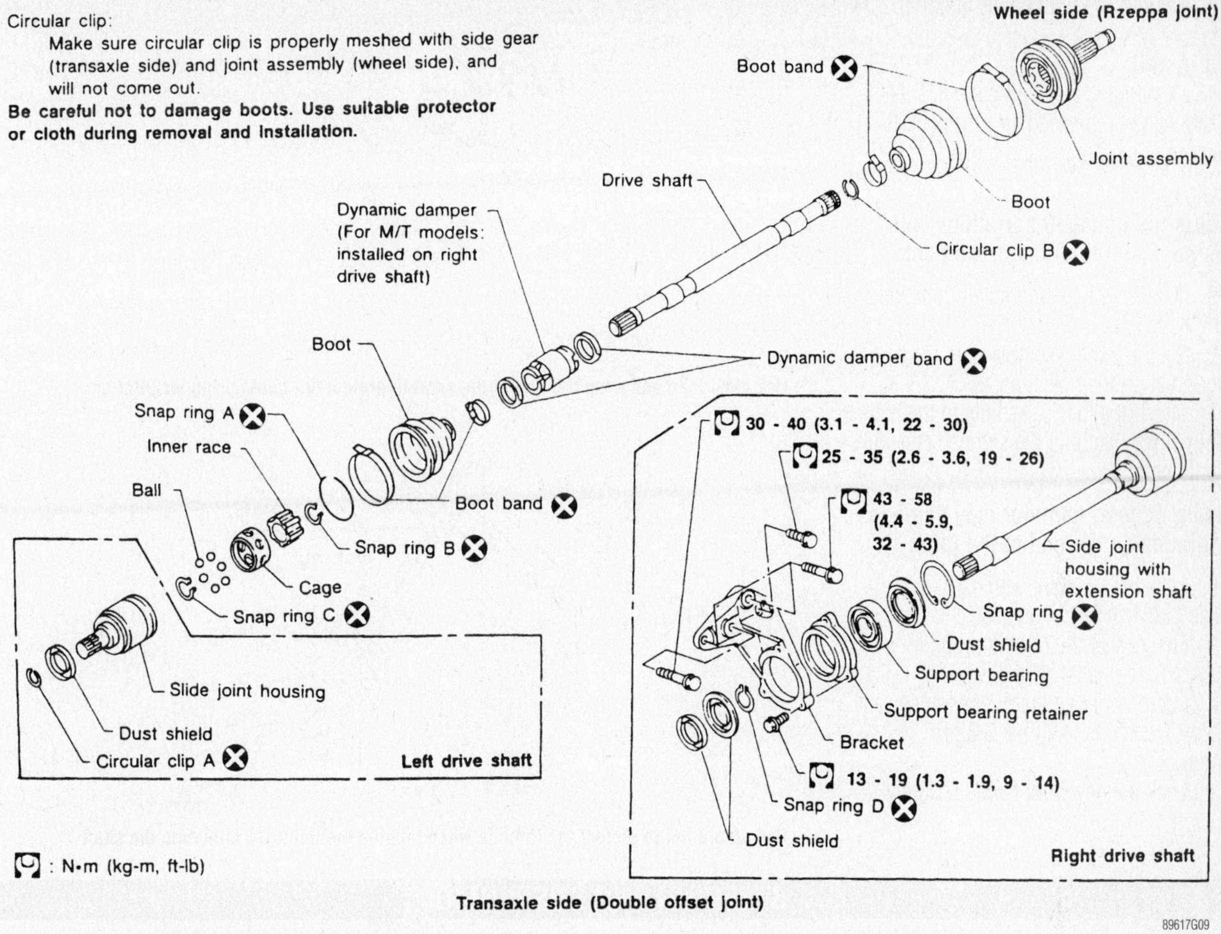

Exploded view of the halfshafts and related components

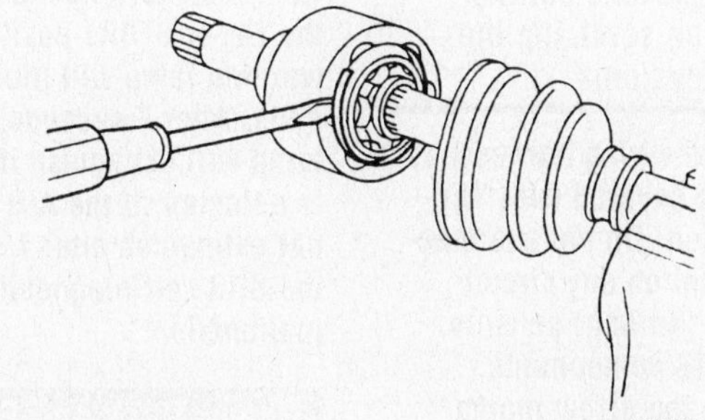

The inner CV-joint uses a large C-clip to retain the ball and cage assembly in the outer housing

11. Ensure that the boot is properly installed on the halfshaft groove.
12. Set the boot so that it does not swell or deform when its length is 3.86 in. (98mm).
13. Lock the new boot bands securely.

Outer

The joint on the wheel side cannot be disassembled.

1. Prior to separating the joint assembly, matchmark the halfshaft and joint assembly.
2. Separate the joint using a slide hammer.

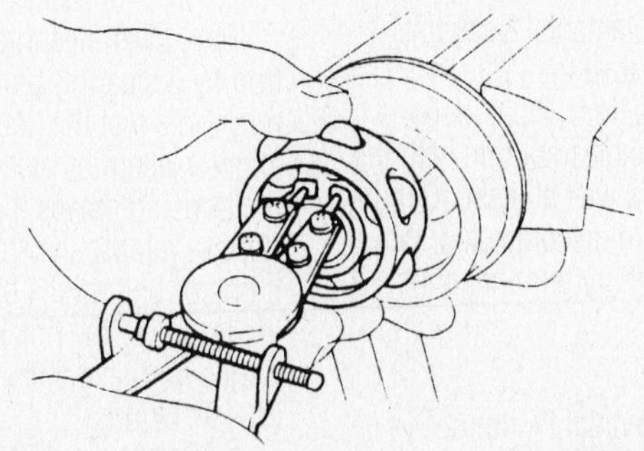

After the outer housing is removed, the ball and cage assembly can slide from the shaft by removing the C-clip

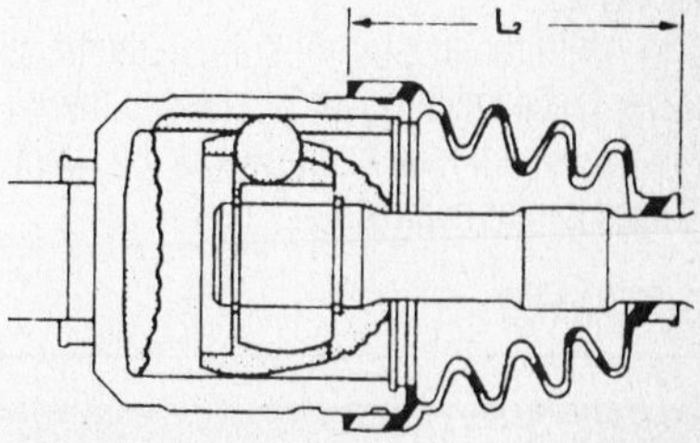

Make sure to properly position the boot before tightening the boot clamps

3. Remove the boot bands and the boot.

To install:

4. Thoroughly clean all parts in solvent and dry with compressed air. Check parts for evidence of damage and replace as necessary.

➡Cover the halfshaft serrations with tape, so as not to damage the boot.

5. Install the boot and small boot band on the halfshaft.
6. Set the joint assembly onto the halfshaft and align the matchmarks.
7. Attach the joint assembly to the halfshaft by lightly tapping the serrated end with a plastic hammer.

➡Using a metal hammer may damage the threads on the end of the joint.

8. Pack the CV-joint with 4.76–5.11 ounces (135–145 g) of grease.
9. Ensure that the boot is properly installed on the halfshaft groove.
10. Set the boot so that it does not swell or deform when its length is 3.82 in. (97mm).
11. Lock the new boot bands securely.

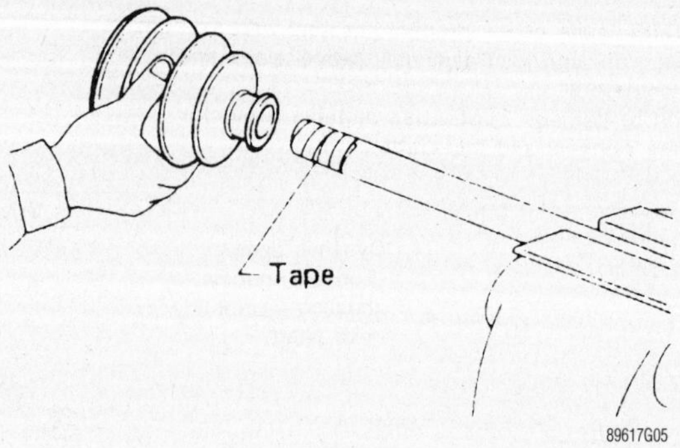

Use vinyl tape and wrap the end of the shaft to protect the boot during installation

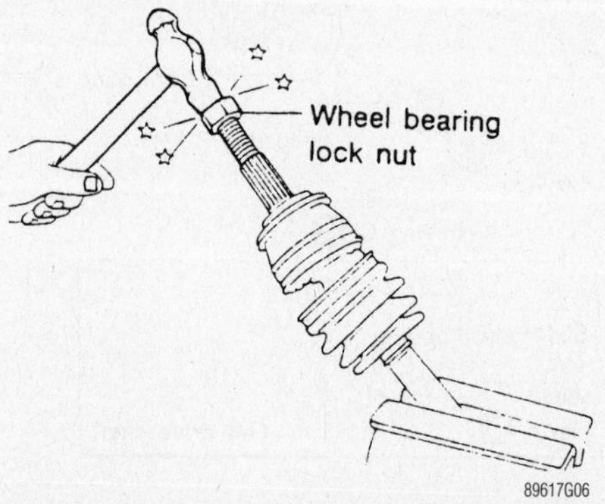

Use an old nut to protect the threads when tapping the outer CV-joint onto the shaft

STEERING AND SUSPENSION

Air Bag

PRECAUTIONS

Several precautions must be observed when handling the inflator module to avoid accidental deployment and possible personal injury.

- Never carry the inflator module by the wires or connector on the underside of the module.
- When carrying a live inflator module, hold securely with both hands, and ensure that the bag and trim cover are pointed away.
- Place the inflator module on a bench or other surface with the bag and trim cover facing up.
- With the inflator module on the bench, never place anything on or close to the module which may be thrown in the event of an accidental deployment.

DISARMING

CAUTION

To avoid rendering the Supplemental Restraint System (SRS) inoperative, which could lead to personal injury or death in the event of a severe frontal collision, extreme caution must be taken when servicing the electrical related systems.

➡All SRS electrical wiring harnesses and connectors are covered with YELLOW outer insulation. Do not use electrical test equipment on any circuit related to the SRS (air bag) sensors. When installing SRS components, always install with the arrow marks facing the front of the vehicle.

Disarming

To disarm the Supplemental Restraint System (SRS) system turn the ignition switch to the **OFF** position. Then, disconnect the both battery cables starting with the negative cable first and wait at least 10 minutes after the cables are disconnected. Be sure to insulate the battery terminal ends.

Arming

To arm the Supplemental Restraint System (SRS) system turn the ignition switch to **OFF** position. Connect the both battery cables starting with the positive cable first.

➡The SRS or air bag system is equipped with a self-diagnostic operation. After turning the ignition key to the ON or START position, the AIR BAG warning lamp will illuminate for 7 seconds. After 7 seconds, the AIR BAG lamp will extinguish if no malfunction is detected. If the AIR BAG lamp does not extinguish after 7 seconds, check the SRS self-diagnostic system for a malfunction.

Power Rack and Pinion

REMOVAL & INSTALLATION

The power steering gear is held in position by 2 steering gear brackets and insulators. Note that the housing may move slightly when the steering wheel is turned. If the housing moves more than 0.080 inch (2mm), replace the steering gear insulators. If one or both of the brackets move, check the torque of the bracket bolts. The correct torque for these bolts is 54–72 ft. lbs. (73–97 Nm).

1. Before servicing the vehicle, refer to the precautions in the beginning of this section.

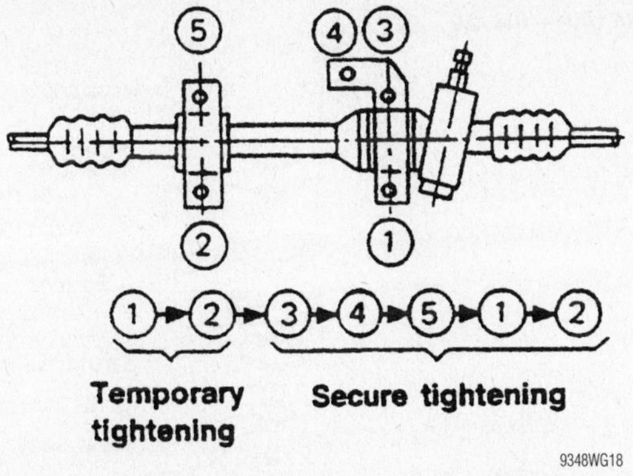

Tighten the power steering rack mounting bolts in the sequence shown

2. Place a drain pan under the steering rack.

3. Remove or disconnect the following:
   - Brake master cylinder remote reservoir bracket screws. Position the reservoir out of the way and secure with wire.
   - Junction block/high pressure line from the steering rack. Position the junction block and line out of the way.
   - Both front wheels
   - Front sway bar
   - Tie rod ends from the steering knuckles
   - Lower steering column shaft clamp bolt
   - Power steering fluid return hose and position out of the way.
   - The steering rack clamp bracket bolts

4. Lower the steering rack from the vehicle.

To install:

5. Carefully slide the steering gear rack and pinion assembly in place from the left side of the vehicle. Position the input shaft so it is just below the lower steering column shaft clamp.

6. Raise the steering gear until the plastic aligning tab on the input shaft enters the clamp bolt gap on the lower column shaft. Do not install the clamp bolt yet.

7. Examine the steering gear brackets. They should be marked UP with arrows pointing to one end of the bracket. Be sure the brackets are installed correctly. Tighten the steering gear bracket bolts to 54–72 ft. lbs. (73–97Nm) in sequence, working counterclockwise from the number 1 bolt (upper right side).

8. Install or connect the following:
   - Fluid return line to the steering gear
   - Steering column shaft clamp bolt. Tighten the bolt to 17–22 ft lbs. (24–29 Nm). Install the dust cover.
   - Tie rod ends
   - Stabilizer bar
   - Wheel. Tighten the lug nuts to 72–87 ft. lbs. (98–118 Nm).
   - Junction block. Tighten the high-pressure line to 11–18 ft. lbs. (15–25 Nm).
   - Brake master cylinder reservoir

9. Check for leaks and proper operation.

MacPherson Strut

REMOVAL & INSTALLATION

1. Before servicing the vehicle, refer to the precautions in the beginning of this section.

2. Disconnect the negative battery cable.

3. Matchmark the front strut upper mounting bracket and the chassis strut tower.

4. Raise and safely support the vehicle.

5. Remove the front wheel.

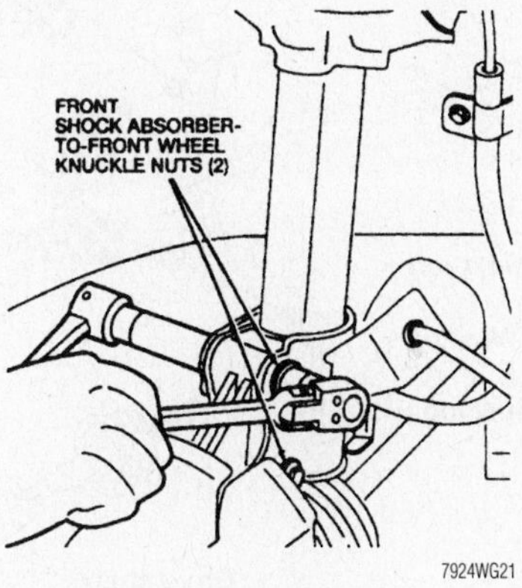

The strut is attached to the knuckle with 2 large bolts

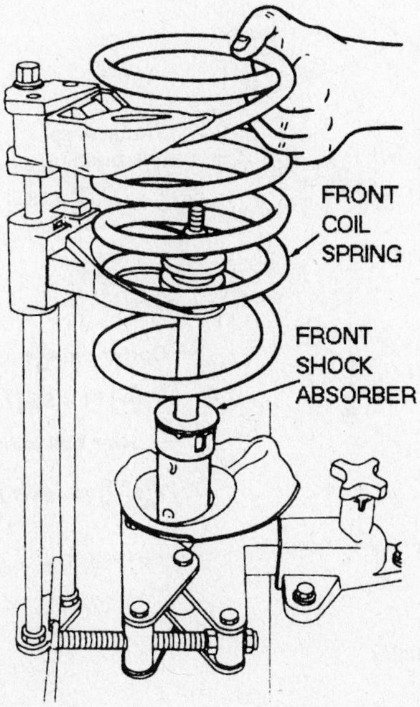

Compress the coil spring in an approved spring compressor

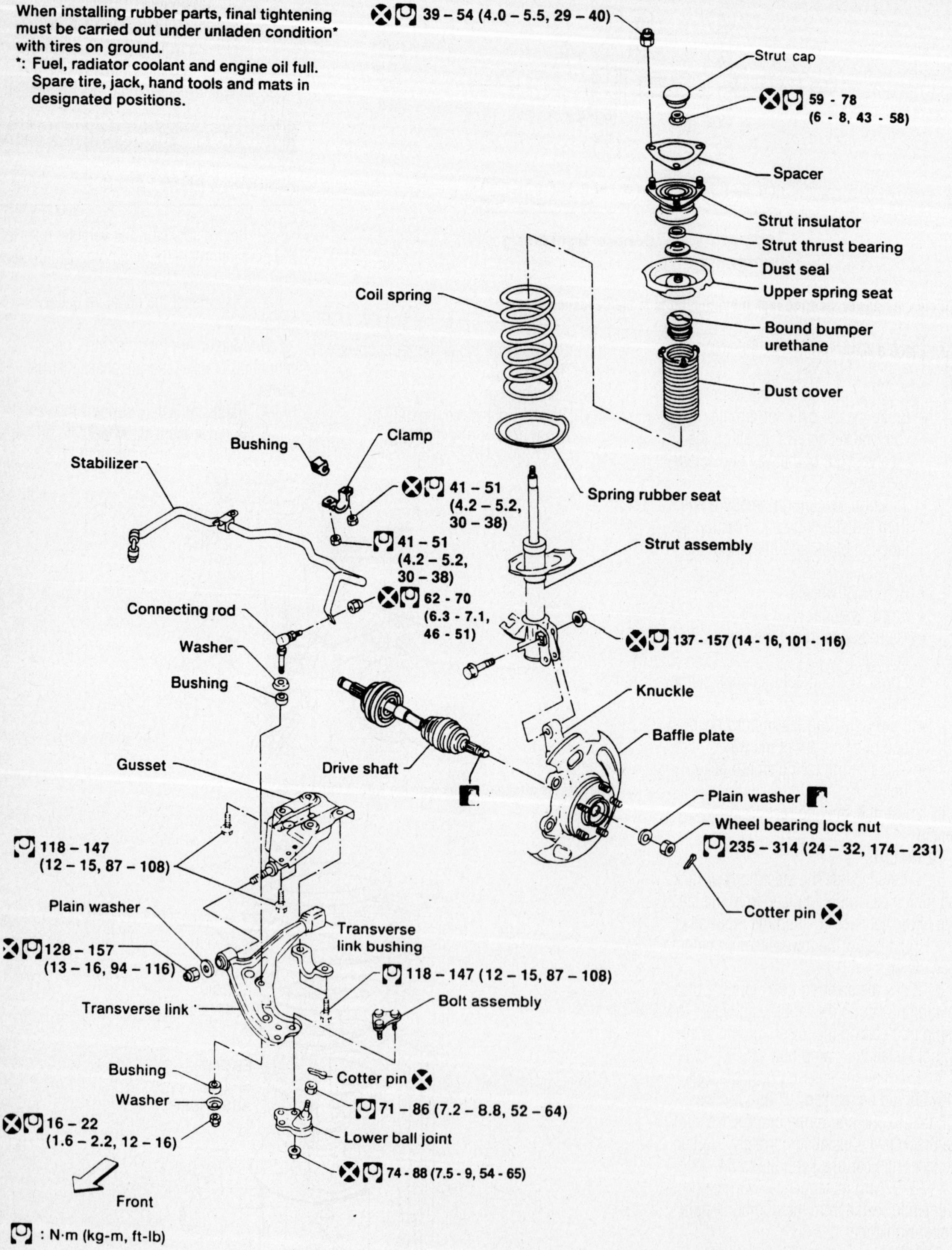

When installing rubber parts, final tightening must be carried out under unladen condition* with tires on ground.
*: Fuel, radiator coolant and engine oil full. Spare tire, jack, hand tools and mats in designated positions.
39 – 54 (4.0 – 5.5, 29 – 40)
Strut cap
59 - 78 (6 - 8, 43 - 58)
Spacer
Strut insulator
Strut thrust bearing
Dust seal
Upper spring seat
Coil spring
Bound bumper urethane
Dust cover
Bushing
Clamp
Stabilizer
Spring rubber seat
41 – 51 (4.2 – 5.2, 30 – 38)
41 – 51 (4.2 – 5.2, 30 – 38)
Strut assembly
Connecting rod
62 - 70 (6.3 - 7.1, 46 - 51)
Washer
137 - 157 (14 - 16, 101 - 116)
Bushing
Knuckle
Baffle plate
Gusset
Drive shaft
Plain washer
Wheel bearing lock nut
235 – 314 (24 – 32, 174 – 231)
118 – 147 (12 – 15, 87 – 108)
Plain washer
Cotter pin
Transverse link bushing
128 – 157 (13 – 16, 94 – 116)
118 – 147 (12 – 15, 87 – 108)
Transverse link
Bolt assembly
Bushing
Cotter pin
Washer
71 – 86 (7.2 – 8.8, 52 – 64)
16 – 22 (1.6 – 2.2, 12 – 16)
Lower ball joint
74 - 88 (7.5 - 9, 54 - 65)
Front
: N·m (kg-m, ft-lb)
9302WG04

Coil spring and strut assembly

6. If equipped, remove the 2 front brake anti-lock sensor cable bracket bolts and position the anti-lock sensor cable out of the way.
7. Detach the brake tube from the strut.
8. Support the control arm.
9. Matchmark the knuckle to the strut so it can installed in the same position. This is important for the camber angle of the front wheel.
10. Remove the strut-to-steering knuckle bolts.
11. Support the strut and remove the 3 upper strut-to-chassis nuts. Remove the strut from the vehicle.

✱✱ WARNING

Never loosen the strut center nut until the spring is compressed or serious injury or vehicle damage may occur.

12. Place the strut and coil spring assembly in a suitable vise and remove the strut nut cover.
13. Slightly loosen, but **do not** remove the front strut nut.

If desired, use the following steps to remove the coil spring from the strut.

14. Using an approved coil spring compressor, compress the coil spring.
15. Remove the strut assembly top nut.
16. Remove the following components from the strut assembly:
    - Upper mounting bracket
    - Strut bearing
    - The bearing seat.
    - Upper coil spring seat and dust boot
    - Coil spring
17. Slowly release the tension of the coil spring compressor and remove the coil spring from the compressor tool.
18. Remove the coil spring insulator and slide the jounce bumper off of the strut assembly.

To install:

19. Slide the jounce bumper onto the strut assembly and install the coil spring insulator.
20. Carefully compress the coil spring with an approved coil spring compressor.
21. Reinstall the following components to the strut assembly:
    - Coil spring

➡Install the coil spring to the strut assembly with the end of the spring in the lower coil spring seat indentation.

    - Upper coil spring seat and dust boot
    - Bearing seat and the bearing
    - Upper mounting bracket
22. Install and tighten the strut assembly nut and tighten the nut to 43–58 ft. lbs. (59–78 Nm).
23. Install the strut assembly onto the vehicle and tighten the following:
    - Strut-to-body nuts: 29–40 ft. lbs. (39–54 Nm)
    - Strut-to-knuckle bolts: 101–116 ft. lbs. (137–157 Nm)
24. Reattach the brake tube to the strut assembly.
25. Install and tighten the 2 front brake anti-lock sensor cable bracket bolts.
26. Reinstall the tire and wheel assembly.
27. Connect the negative battery cable and the adjustable strut electrical connectors, if equipped.
28. Check and/or adjust the wheel alignment.

Shock Absorber

REMOVAL & INSTALLATION

1. Before servicing the vehicle, refer to the precautions in the beginning of this section.
2. Raise and safely support the vehicle.
3. Support the rear axle and slightly lower the vehicle enough to lessen tension on the shock absorber.
4. Remove the lower shock absorber retaining nut and washer.
5. Disconnect the lower end of the shock absorber from the mounting stud.
6. Remove the shock absorber upper end retaining nut and washer.
7. Remove the shock absorber from the vehicle.

To install:

8. Install the shock absorber onto the upper and lower mounting studs of the vehicle.
9. Install the washers and retaining nuts. Tighten the upper and lower retaining nuts to 22–30 ft. lbs. (30–41 Nm).
10. Lower the vehicle.

Lower Ball Joints

REMOVAL & INSTALLATION

To check if ball joint replacement is required, raise and safely support the vehicle clear of the floor and try to rock the wheel up and down. If any play is felt, have an assistant rock the wheel while observing the front suspension lower arm ball joint at the bottom of the steering knuckle. If any movement is seen, the ball joint should be replaced. If not, any wheel play indicates wheel bearing wear.

1. Before servicing the vehicle, refer to the precautions in the beginning of this section.
2. Raise and safely support the vehicle.
3. Remove the tire and wheel.
4. Remove and discard the ball joint cotter pin. Loosen the ball joint attaching nut from the steering knuckle. Because of tight clearance, the nut likely cannot be removed until the ball joint stud is loosened and lowered slightly.
5. Strike the front knuckle with a hammer while pulling down on the lower control arm. There should now be enough clearance to allow removal of the ball joint stud nut. Separate the ball joint from the steering knuckle.
6. Remove the 3 bolts attaching the ball joint to the control arm.
7. Remove the ball joint from the control arm.

To install:

8. Install the ball joint to the control arm and install the attaching bolts.
9. Tighten the bolts to 54–65 ft. lbs. (74–88 Nm).
10. Install the ball joint into the steering

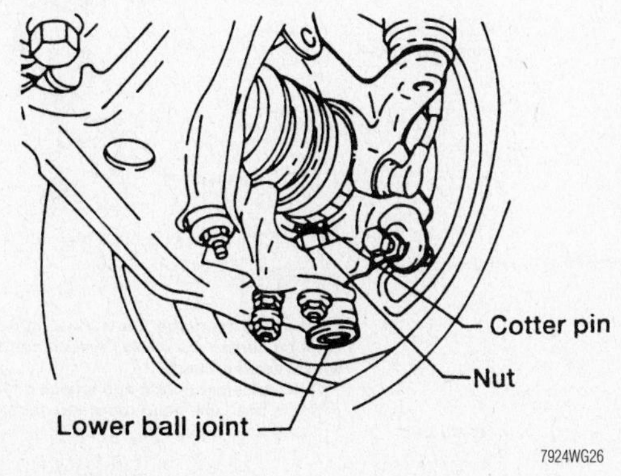

Loosen the nut on the lower ball joint stud

knuckle, just enough to get the nut started on the stud. Then, push the ball joint stud fully in place. Tighten the nut to 52–63 ft. lbs. (71–86 Nm). Secure the nut with a new cotter pin.

11. Install the tire and wheel.
12. Lower the vehicle.
13. A front end alignment check is recommended.

Lower Control Arm

REMOVAL & INSTALLATION

1. Before servicing the vehicle, refer to the precautions in the beginning of this section.
2. Remove the wheel.
3. Disconnect the ball joint.
4. Disconnect the stabilizer bar from the control arm.
5. Remove the 2 rear arm bolts and the mounting bracket.
6. Remove the lower arm nut.
7. Pull the rear of the arm down and gently pry the arm forward and off the gusset.
8. Installation is the reverse of removal. Observe the following torques:
   - Stabilizer bar-to-lower arm: 12–16 ft. lbs. (16–22 Nm)
   - Lower arm rear bolts: 87–108 ft. lbs. (118–147 Nm)
   - Lower arm nuts: 94–115 ft. lbs. (128–156 Nm)
   - Ball stud nut: 56–80 ft. lbs. (76–109 Nm)

BUSHING REPLACEMENT

The bushings are press-fit types. Support the arm in a press, using the proper adapters. Ford tool numbers are: T93P-5493-A, T75L-1165-B and -DA.

Wheel Bearings

ADJUSTMENT

The wheel bearings are not adjustable. If the bearings become loose or make noise, they must be replaced using the following procedure.

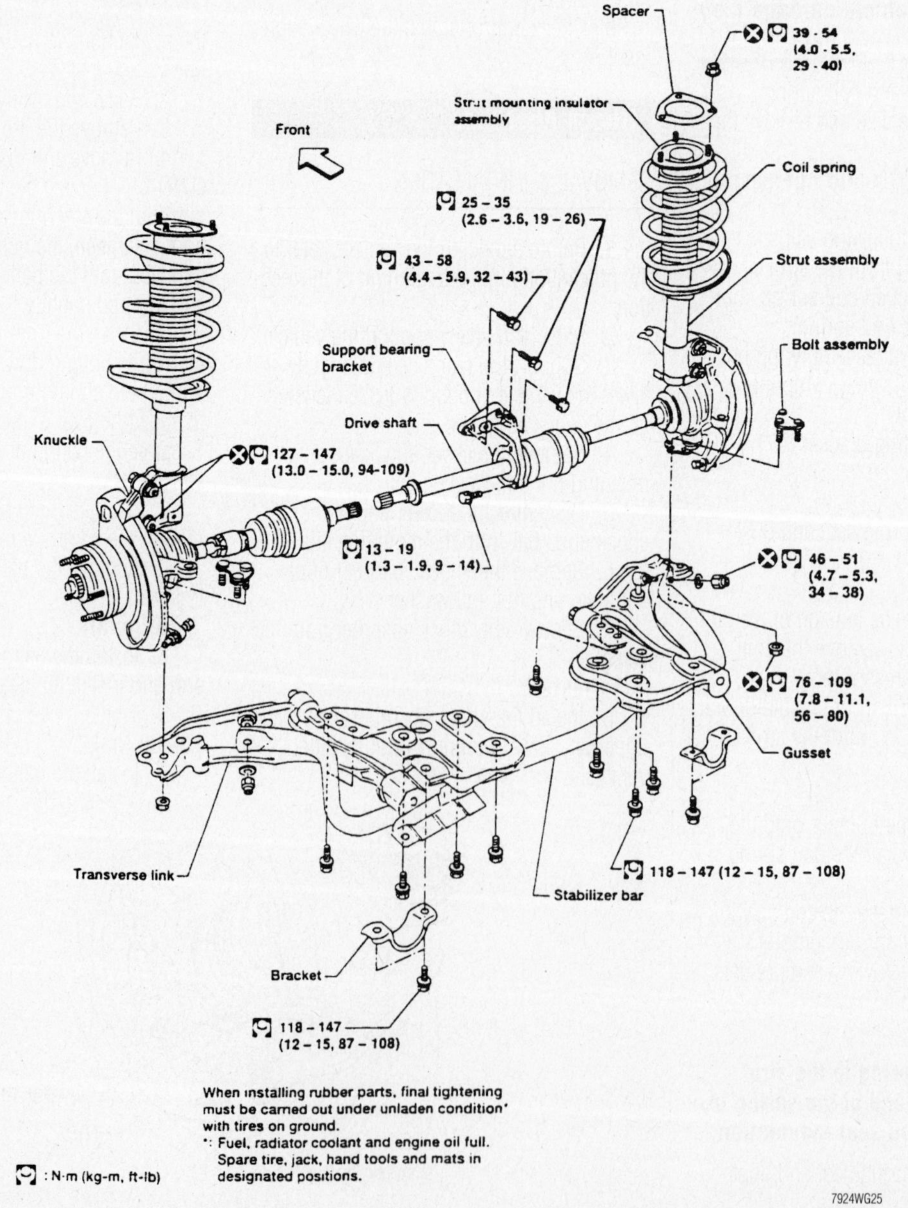

Exploded view of the front suspension and drive axles

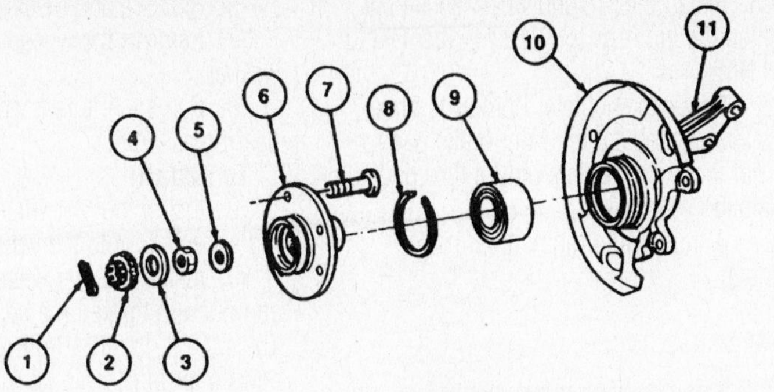

1. Cotter pin
2. Nut retainer
3. Insulator
4. Front axle wheel hub retainer
5. Front wheel outer bearing retainer washer
6. Wheel hub
7. Wheel hub bolt
8. Snap ring
9. Front wheel bearing
10. Front disc brake rotor shield
11. Front wheel knuckle

7924WG23

Exploded view of the knuckle, hub and bearing

REMOVAL & INSTALLATION

Front

1. Before servicing the vehicle, refer to the precautions in the beginning of this section.
2. Raise and safely support the vehicle.
3. Remove the wheel and tire.
4. Remove the brake caliper assembly. DO NOT disconnect the brake hose. Hang the caliper on a piece of wire from a near by support such as the strut.
5. Remove the brake rotor.
6. Remove and discard the cotter pin from the end of the outboard CV-joint stub shaft. Remove the hub nut retainer, washer and the hub nut. There should be another washer under the hub nut that acts as a front wheel bearing outer bearing retainer.
7. Disengage the lower ball joint stud from the steering knuckle using the following procedure.
   a. Remove and discard the cotter pin from the front lower ball joint.
   b. Loosen the lower ball joint nut until it contacts the front halfshaft joint.
   c. Strike the front knuckle with a hammer while pulling down on the lower control arm until the ball joint stud separates from the knuckle.
   d. Remove the ball joint nut.
   e. Disengage the lower ball joint stud from the steering knuckle.
8. Disengage the outer tie rod end stud from the steering knuckle using the following procedure.
   a. Remove and discard the cotter pin from the outer tie rod end stud.
   b. Remove the outer tie rod end retaining nut.
   c. Use a tie rod end puller to carefully press the tie rod end from the steering knuckle.
9. Remove the front ABS sensor bolt.
10. Remove the 2 front strut-to-front knuckle nuts and remove the 2 bolts. Disengage the strut from the steering knuckle.
11. Use a 2-jaw puller to separate the front halfshaft outboard CV-joint stub shaft from the knuckle/bearing assembly.
12. Remove the front wheel hub, knuckle and wheel bearing assembly from the vehicle.
13. If the knuckle is being replaced with a service part, change over the steering stop bolt and jam nut from the old knuckle to the replacement part.
14. To remove the front wheel bearing, jig up a puller to bear against the front wheel bearing inner race and pull the race from the hub/knuckle assembly.
15. Use a shop press to press out damaged wheel studs and also to press out the outer bearing race.
16. Use a shop press to press out the inner bearing race.

To install:

17. If the front wheel bearings were removed, assemble the ABS sensing ring, if removed and the disc brake dust shield under the steering knuckle. Use a shop press to push in new front wheel bearing inner and outer races. Support the knuckle and press the front wheel bearing into the knuckle and install the snapring retainer. Support the bearing assemblies and press the hub onto the knuckle and wheel bearing assembly.
18. Install the hub, knuckle and bearings as an assembly. Position the assembly on the halfshaft outer CV-joint stub axle end. Guide the knuckle into the front strut and install the 2 knuckle-to-strut bolts and nuts. Tighten the nuts to 83–91 ft. lbs. (113–123 Nm).
19. Install the ABS sensor bolt. Do not over-tighten. Tighten to just 16–21 inch lbs. (1.8–2.4 Nm).
20. Install the outer tie rod end to the steering knuckle. Tighten the nut to 22–29 ft. lbs. (29–39 Nm). If the cotter pin holes

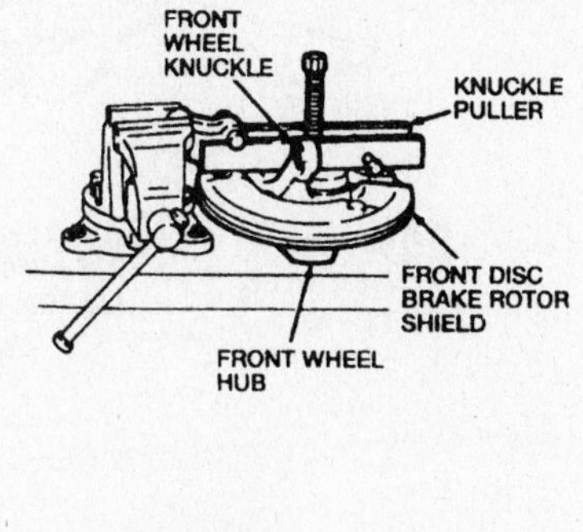

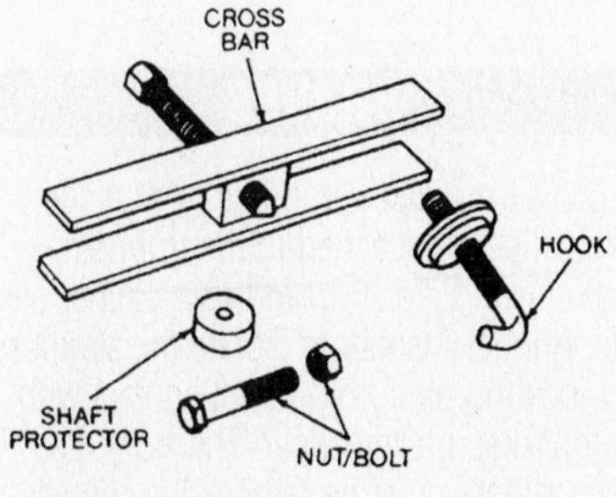

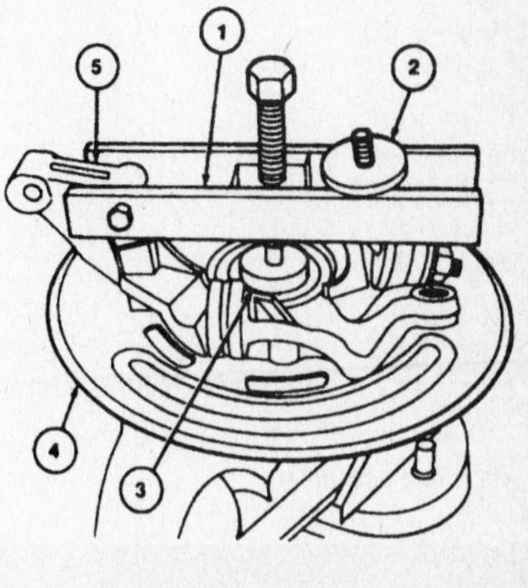

1. Knuckle puller
2. Knuckle puller adapter
3. Step plate adapter
4. Front disc brake rotor shield
5. Front wheel knuckle

7924WG27

Example of a puller set up to bear against the front wheel bearing inner race

do not align, tighten the nut slightly until they do. Never loosen the nut to align the holes. Secure the nut with a new cotter pin.

21. Start the lower ball joint stud to the steering knuckle and partially install the nut, then push the ball joint stud fully in place. Tighten the ball joint stud nut to 52–63 ft. lbs. (71–86 Nm). Secure the nut with a new cotter pin.

22. Install the front wheel outer bearing retaining washer and the hub retainer nut. Tighten to 174–231 ft. lbs. (235–314 Nm). Install the nut retainer, insulator and a new cotter pin.

23. Install the front brake rotor and install the disc brake caliper.

24. If removed, install the steering stop bolt.

25. Install the tire and wheel assembly. Tighten the lug nuts to 72–87 ft. lbs. (98 to 118 Nm).

26. Lower the vehicle. Pump the brake pedal slowly to seat the front brake pads. Do not move the vehicle until a firm pedal is obtained.

27. A front end alignment is recommended.

Rear

1. Before servicing the vehicle, refer to the precautions in the beginning of this section.
2. Raise and safely support the vehicle.
3. Remove the rear wheel(s).
4. Remove the brake drum.
5. Remove the grease cap for the hub.
6. Remove and discard the cotter pin.
7. Remove the wheel bearing nut and washer.
8. Remove the rear wheel hub and bearing assembly.

To install:

9. Install the rear wheel hub and bearing assembly onto the vehicle.
10. Install the rear wheel bearing washer and nut and tighten the bearing nut to 159–210 ft. lbs. (216–284 Nm). Install a new cotter pin.
11. Install the wheel hub grease cap. Install the brake drum.
12. Install the rear wheel(s) and lug nuts. Tighten the lug nuts, in a star sequence, to 72–87 ft. lbs. (98–118 Nm).
13. Lower the vehicle.

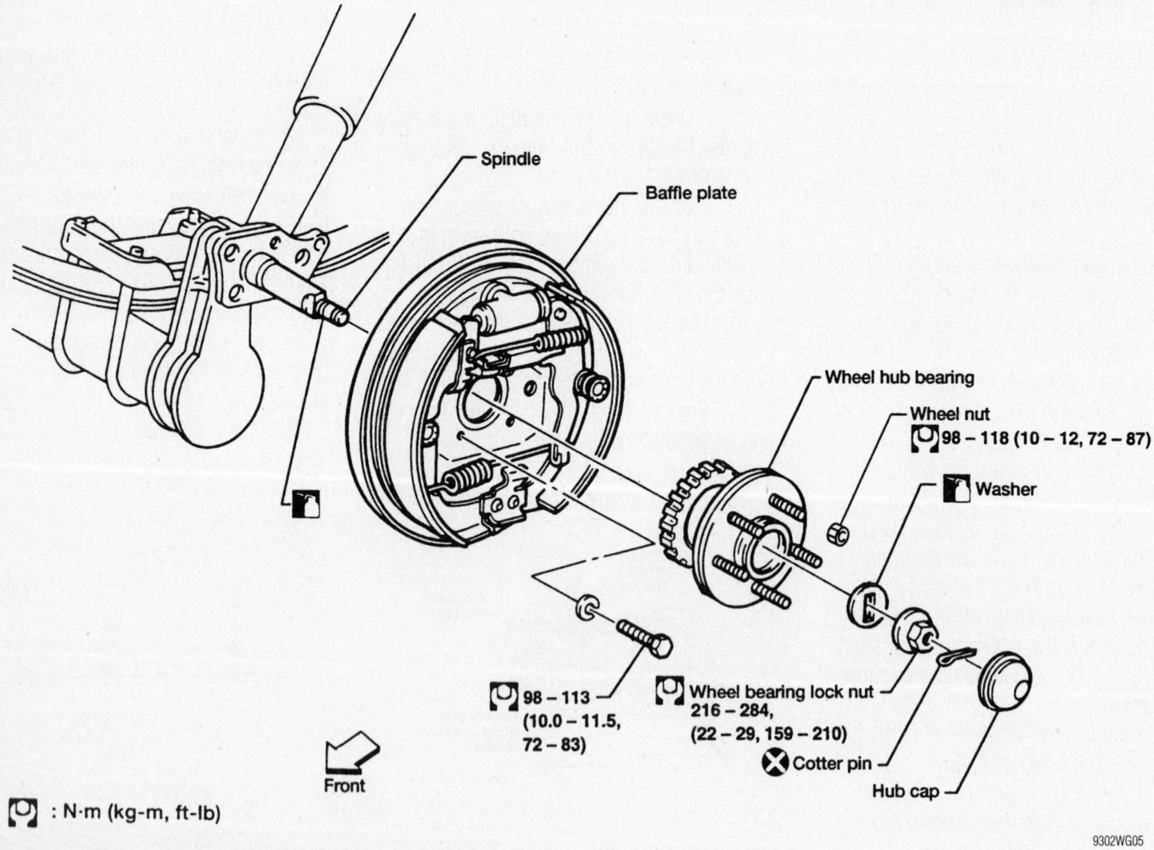

Rear hub assembly

BRAKES

Brake Caliper

REMOVAL & INSTALLATION

The front disc brake caliper slides on 2 stainless steel locating pins. The front disc brakes use a conventional pin slider-type front disc brake caliper with a 10.875 inch (27.6cm) front disc rotor. The front disc brake caliper is attached to the front suspension with 2 Torx® head brake caliper bolts. Rubber insulators isolate the stainless steel locating pins from direct contact with the front disc brake caliper. The front disc brake calipers must be removed to replace the front brake pads.

1. Before servicing the vehicle, refer to the precautions in the beginning of this section.
2. Remove or disconnect the following:
   - Wheel and tire

➡If the brake caliper is being removed for brake pad replacement only, DO NOT disconnect the brake hose.

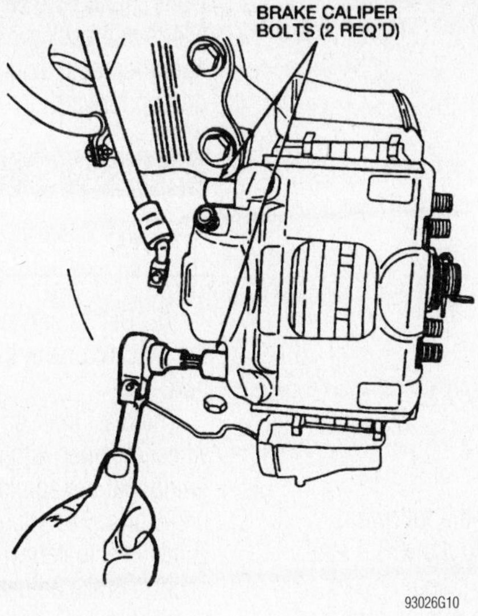

Caliper pin bolt removal

- 2 caliper pin bolts. Most applications will require a Torx® T-40 bit to remove the 2 brake caliper bolts.

3. If the brake caliper is being removed just for brake service, with the brake hose still attached to the caliper, use a length of wire to support the caliper from the front shock absorber. Do not let the caliper hang by the brake hose. If the caliper is being completely removed from the vehicle for overhaul, use care not to drip brake fluid on the paint.

➡If both calipers are being completely removed from the vehicle at the same time, mark them Left and Right so the calipers can be reinstalled to their original locations. The reason for this is that the bleeder screws must be positioned on the top of the front disc brake caliper when installed on the vehicle.

To install:

4. Clean all parts well. Use a C-clamp and a used brake pad to push the caliper piston fully in the piston bore. Inspect the caliper pins and clean any dirt and debris.
5. Install the caliper onto the rotor. Make sure the inboard and outboard brake pads are properly positioned.
6. Lubricate the stainless steel locating pins with a Silicone Dielectric Compound such as Ford DZAZ-19A331-A or equivalent silicone grease. Install the 2 caliper pin bolts and torque to 18–25 ft. lbs. (24–34 Nm).
7. If disconnected, install the brake hose using a new replacement copper washer, install the banjo bolt and torque to 12–14 ft. lbs. (17–20 Nm).
8. If the brake hose had been disconnected, bleed the brake system.
9. Install the wheel and tire.
10. Torque the lug nuts to 72–87 ft. lbs. (98–118 Nm).
11. Check the master cylinder reservoir and add fresh DOT 3 brake fluid as required.
12. Lower the vehicle. Pump the brake pedal slowly until a firm brake pedal is obtained, indicating that the brake pads are properly seated, before attempting to move the vehicle. Road-test and check for proper brake operation.

Disc Brake Pads

REMOVAL & INSTALLATION

Front

1. Before servicing the vehicle, refer to the precautions in the beginning of this section.
2. Remove or disconnect the following:
   - Wheels
   - Bottom guide pin from the caliper and swing the caliper cylinder body upward; support the caliper with a wire
   - Brake pad retainers and the pads

To install:

3. Compress the piston of the disc brake caliper.
4. Install or connect the following:
   - Brake pads and caliper assembly. Torque the guide pin to 23–30 ft. lbs. (31–41 Nm).
   - Wheels
5. Apply the brakes a few times to seat the pads. Check the master cylinder and add fluid if necessary. Bleed the brakes, if necessary.

Rear

1. Before servicing the vehicle, refer to the precautions in the beginning of this section.

➡Do not press the piston into the bore as performed on the front disc brakes. Due to the parking brake mechanism, the caliper piston must be turned into the bore using a special tool.

2. Remove or disconnect the following:
   - Rear wheels
3. Release the parking brake.
   - Parking brake cable bracket bolt
   - Pin bolts and lift off the caliper body
   - Pad springs and then remove the pads and shims

To install:

4. Clean the piston end of the caliper body and the area around the pin holes. Be careful not to get oil on the rotor.
5. Using the proper tool, carefully turn the piston clockwise back into the caliper

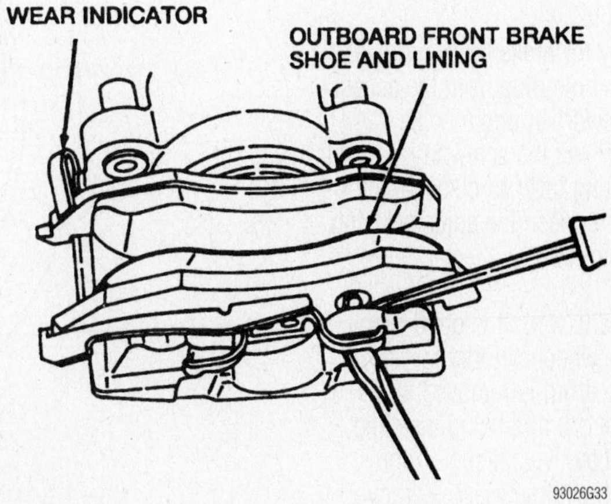

Replacing the disc brake pads

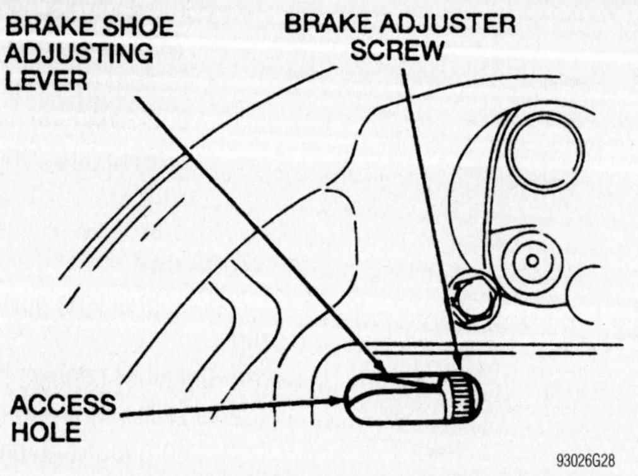

Brake shoe adjustment may need to be loosened to remove the brake drum

body. Take care not to damage the piston boot.

6. Coat the pad contact area on the mounting support with a silicone based grease.
7. Install or connect the following:
   - Pads, shims, and the pad springs. Always use new shims.
   - Caliper body in the mounting support and tighten the pin bolts to 28–38 ft. lbs. (38–52 Nm)
   - Wheels
8. Bleed the system if necessary.

Brake Drums

REMOVAL & INSTALLATION

1. Before servicing the vehicle, refer to the precautions in the beginning of this section.

The rear drum brakes used on these vehicles are conventional expanding shoe-type with the brake shoe lining applied to the inside of the rotating drum. An incremental brake adjuster screw is designed to actuate whenever sufficient wear occurs.

2. Remove or disconnect the following:
   - Wheel and tire
   - Brake drum by pulling it from the wheel studs
3. If necessary for brake drum removal, pry off the access hole plug from the access hole. Insert a screwdriver and a brake adjustment tool. Press the screwdriver against the adjusting lever to disengage it from the adjuster. Loosen the adjuster using the brake adjusting tool.

To install:

4. Clean all parts well. It is good practice to inspect the wheel cylinder for leaks anytime the brake drum is removed. If a new replacement brake drum is being installed, inspect it for a protective coating on the machined inside braking surface. Remove any coating with suitable solvent.
5. Install or connect the following:
   - Brake drum onto the wheel studs
6. In most all cases, manual brake adjustment IS NOT recommended. Adjustment is performed by driving the vehicle and applying the brakes.
   - Tire and wheel and torque the fasteners to 72–87 ft. lbs. (98–118 Nm)
7. Adjust the rear brake shoes by sharply applying the brakes several times while driving the vehicle alternately forwards and backwards. Check the brake operation by making several stops while driving forward.

Brake Shoes

REMOVAL & INSTALLATION

1. Before servicing the vehicle, refer to the precautions in the beginning of this section.

The rear drum brakes use an internal rear wheel cylinder with expanding shoes and lining that are applied against a rotating brake drum. An incremental brake adjuster screw is actuated whenever sufficient wear occurs. Brake adjustment takes place in forward or reverse braking but not with parking brake application.

2. Remove or disconnect the following:
   - Wheels
   - Brake drum
   - Parking brake rear cable and conduit from the parking brake lever
   - 2 brake shoe hold-down springs

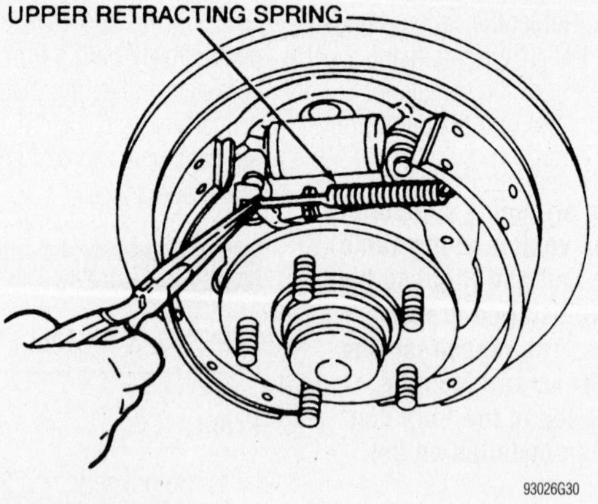

Remove the upper retracting spring

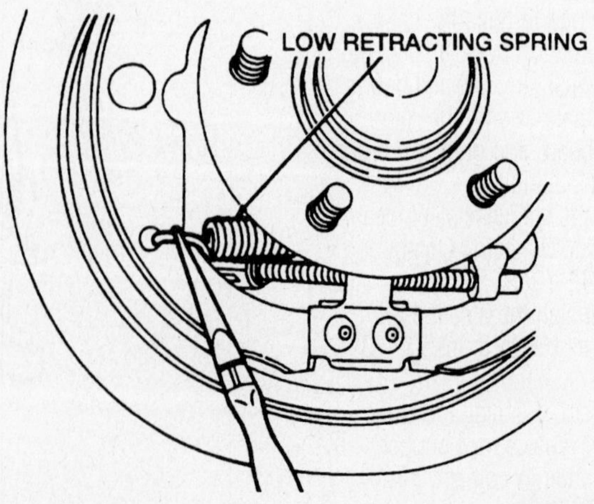

Remove the lower retracting spring

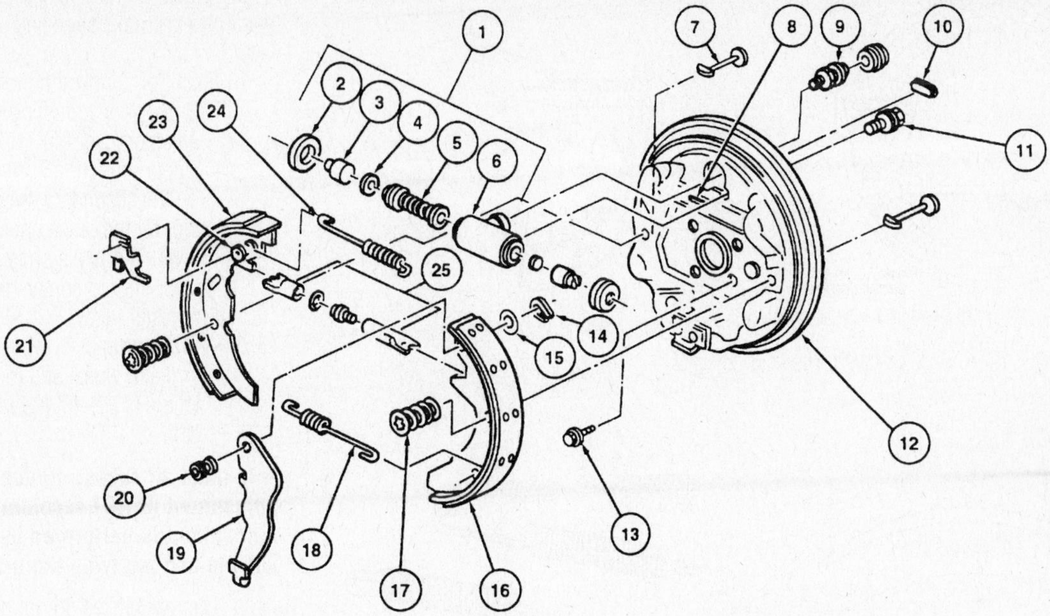

1 Rear Wheel Cylinder	13 Rear Brake Backing Plate Bolts (4 Req'd)
2 Dust Boot (2 Req'd)	14 Parking Brake Lever Clip
3 Wheel Cylinder Piston (2 Req'd)	15 Spring Washer
4 Cup (2 Req'd)	16 Secondary Brake Shoe and Lining
5 Wheel Cylinder Piston Cup Spring	17 Brake Shoe Hold-Down Spring
6 Wheel Cylinder Housing	18 Lower Retracting Spring
7 Brake Shoe Hold-Down Pin (2 Req'd)	19 Parking Brake Lever
8 Access Hole	20 Parking Brake Lever Pin
9 Rear Brake Bleeder Screw	21 Brake Shoe Adjusting Lever
10 Access Hole Plug	22 Adjuster Lever Pin
11 Rear Wheel Cylinder Bolt (2 Req'd)	23 Primary Brake Shoe and Lining
12 Rear Brake Backing Plate	24 Upper Retracting Spring
	29 Brake Adjuster Screw

93026G29

Rear drum brake assembly and related components

and the 2 brake shoe hold-down pins

- Upper retracting spring
- Lower retracting spring
- Brake adjuster screw
- Rear brake shoes and linings from the brake backing plate
- Parking brake lever clip and washer
- Parking brake lever from the secondary brake shoe and lining

To install:

3. Clean all parts well.

4. Inspect the wheel cylinder for signs of leaking. Service as required.

5. Inspect the retracting springs for heat damage, bends or damage to the coils or shank or loss of tension. A good retracting spring will make a full thud when dropped on a concrete floor. A heat-damaged retracting spring that has lost tension will make a distinctive ringing sound when dropped on a concrete floor.

6. Check the brake backing plate for signs of scoring. The shoe contact points must be smooth and have a light coating of lithium grease. Verify that the brake lining thickness is between 0.059–0.232 in. (1.5–5.9mm). Failure to replace worn rear brake shoes will result in a scored drum.

7. Inspect the brake drum for scratches, scoring, bell mouth and out-of-round conditions. Remove minor scores on a brake drum with sandpaper. Do not refinish brake drums to remove scoring marks. A brake drum surface that is highly polished can cause the brakes to lock up. Remove polished surfaces with sandpaper or refinish the brake drum. Refinish a brake drum that is out-of-round enough to cause vehicle vibration or noise when braking. Remove only enough surface metal to true-up the brake drum. Brake drum maximum inside diameter is shown on each drum. If the maximum inside diameter shown on the brake drum is exceeded through wear or refinishing, replace the brake drum. After a brake drum is refinished, wipe the refinished surface with a cloth soaked in clean denatured alcohol. If one brake drum is refinished, the brake drum on the opposite side of the vehicle should also be refinished to the same diameter. The standard inner brake drum diameter is 9.840 inches (250.0mm). Replace the brake drum if worn beyond 9.900 inches (251.5mm).

8. Install or connect the following:

- Parking brake lever to the secondary brake shoe and lining with a new parking brake lever clip
- Secondary (rear) shoe on the back-

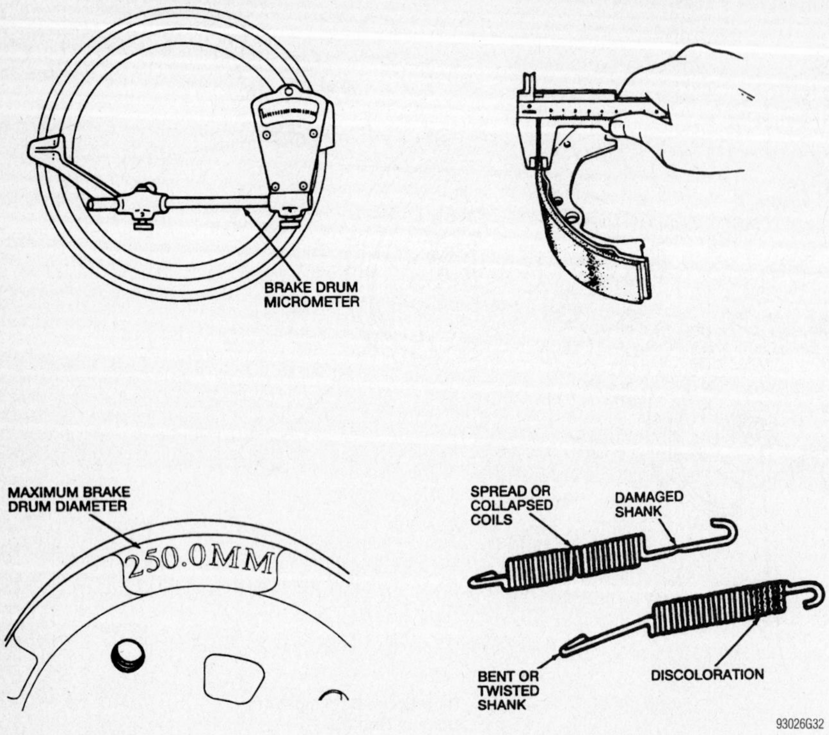

These size checks should be made and the retracting springs' condition checked. Replace questionable parts

ing plate and the brake shoe hold-down spring and pin

- Primary (front) shoe on the backing plate and the brake shoe hold-down spring and pin
- Parking brake rear cable and conduit to the parking brake lever
- Lower retracting spring to the rear brake shoes

9. Apply a light coat of high-quality grease to the threaded areas of the adjuster nut and adjuster socket. Turn the adjuster nut all the way down on the brake adjuster screw, then loosen the adjuster ½ turn. Install the adjuster screw in the slots on the rear brake shoes. The wider slot on the socket must fit in the slot on the primary (front) brake shoe. The slot on the adjuster nut end must fit into the slots in the secondary (rear) brake shoe and parking brake lever.

10. Install or connect the following:
   - Brake shoe adjusting lever on the adjuster lever pin
   - Upper retracting spring in the slot on the secondary shoe and in the slot on the brake shoe adjusting lever. The brake shoe adjusting lever should contact the brake adjuster screw.
   - Brake drum
   - Tire and wheel and torque the fasteners to 72–87 ft. lbs. (98–118 Nm).

➡In most all cases, manual brake adjustment IS NOT recommended. Adjustment is performed by driving the vehicle and applying the brakes.

11. The rear brakes do not require adjustment when being serviced to obtain a firm brake pedal feel. To achieve a firm brake pedal after servicing the rear brakes, sharply apply the brake pedal several times while driving the vehicle alternately forwards and backwards. Check the brake operation by making several stops while driving forward. The self-adjusting mechanism will sufficiently adjust the rear brake shoes without any manual tightening at the brake shoe adjuster. If the rear brake shoes are manually adjusted, the additional action of the brake shoe adjuster can cause the brakes to become over-tightened and result in binding or overheated rear brakes.

FORD

Windstar

17

SPECIFICATIONS AND MAINTENANCE CHARTS

VEHICLE AND ENGINE IDENTIFICATION CHART

Engine Code						
Code	Liters (cc)	Cu. In.	Cyl.	Fuel Sys.	Engine Type	Eng. Mfg.
4	3.8 (3802)	231	6	SEFI	OHV	Ford

SEFI: Sequential Multi-port Fuel Injection

Model Year	
Code	Year
1	2001
2	2002
3	2003

67197-WIND-C01

GENERAL ENGINE SPECIFICATIONS

Year	Engine Displacement Liters	Engine VIN	Net Horsepower @ rpm	Net Torque @ rpm (ft. lbs.)	Bore x Stroke (in.)	Com-pression Ratio	Oil Pressure @ rpm
2001	3.8	4	200@5000	225@3000	3.81x3.39	9.3:1	40-60@2500
2002	3.8	4	200@5000	225@3000	3.81x3.39	9.3:1	40-60@2500
2003	3.8	4	200@5000	225@3000	3.81x3.39	9.4:1	40-125@2500①

① Hot

67197-WIND-C02

GASOLINE ENGINE TUNE-UP SPECIFICATIONS

Year	Engine Displacement Liters	Engine VIN	Spark Plugs Gap (in.)	Ignition Timing (deg.)		Fuel Pump (psi)	Idle Speed (rpm)		Valve Clearance	
				MT	AT		MT	AT	In.	Ex.
2001	3.8	4	0.052-0.056	—	①	②	—	①	HYD	HYD
2002	3.8	4	0.052-0.056	—	①	②	—	①	HYD	HYD
2003	3.8	4	0.052-0.056	—	①	②	—	①	HYD	HYD

NOTE: The Vehicle Emission Control Information label often reflects specification changes changes made during production.

The label figures must be used if they differ from those in this chart.

B: Before top dead center

HYD: Hydraulic

① Controlled by the Powertrain Control Module (PCM) and cannot be manually adjusted.

② Engine running: 28-45 psi

Key On, Engine Off (KOEO): 35-45 psi

67197-WIND-C03

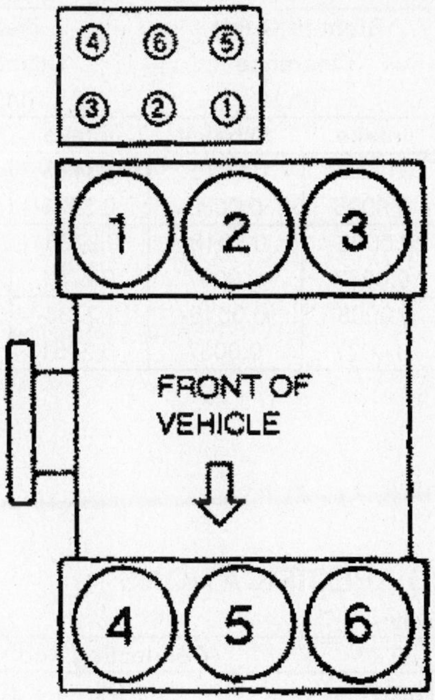

3.8L Engine
Firing order: 1–4–2–5–3–6
Distributorless ignition system

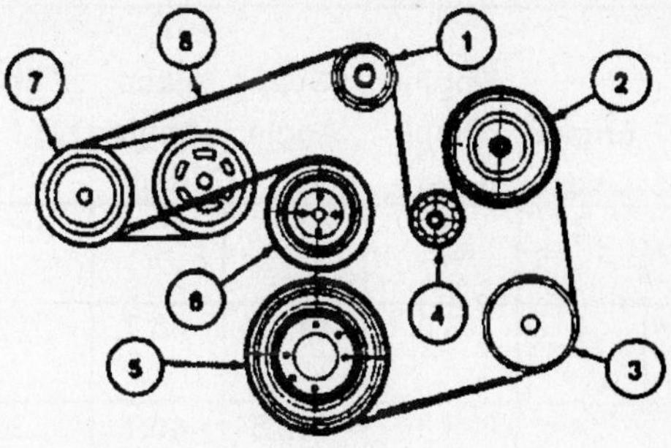

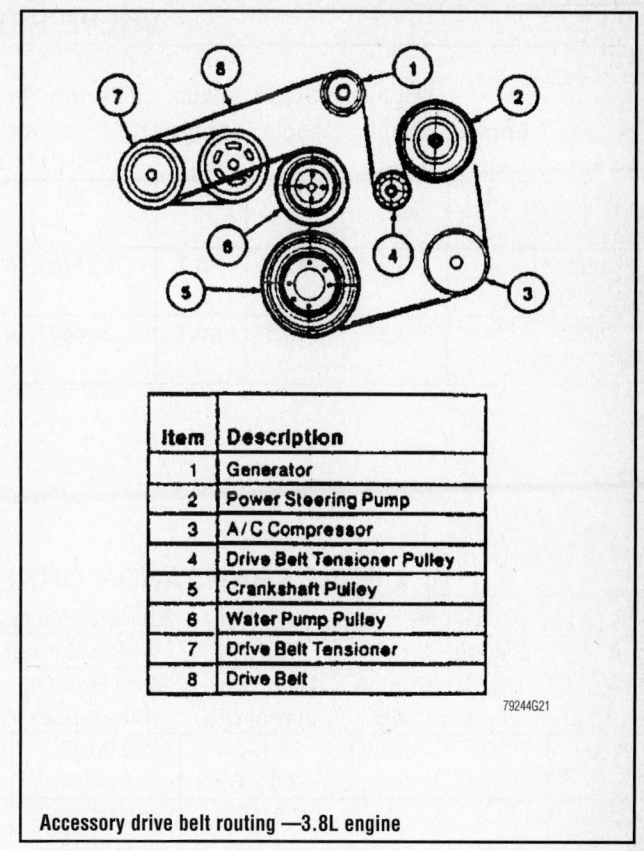

Item	Description
1	Generator
2	Power Steering Pump
3	A/C Compressor
4	Drive Belt Tensioner Pulley
5	Crankshaft Pulley
6	Water Pump Pulley
7	Drive Belt Tensioner
8	Drive Belt

Accessory drive belt routing —3.8L engine

CAPACITIES

Year	Model	Engine Displacement Liters	Engine ID/VIN	Engine Oil with Filter (qts.)	Transmission (pts.) 4-Spd	5-Spd	Auto.	Drive Axle Front (pts.)	Rear (pts.)	Fuel Tank (gal.)	Cooling System (qts.)
2001	Windstar	3.8	4	5.0	—	—	24.5	①	—	26	16.0
2002	Windstar	3.8	4	5.0	—	—	24.5	①	—	26	16.0
2003	Windstar	3.8	4	5.0	—	—	24.5	①	—	26	16.0

NOTE: All capacities are approximate. Add fluid gradually and check to be sure a proper fluid level is obtained.

① Included in transaxle capacity

67197-WIND-C04

VALVE SPECIFICATIONS

Year	Engine VIN	Engine Displ. Liters	Seat Angle (deg.)	Face Angle (deg.)	Spring Test Pressure (lbs. @ in.)	Spring Installed Height (in.)	Stem-to-Guide Clearance (in.)		Stem Diameter (in.)	
							Intake	Exhaust	Intake	Exhaust
2001	4	3.8	44.75	45.8	198-220@1.18	1.970	0.0010-0.0028	0.0015-0.0033	0.3415-0.3423	0.3410-0.3418
2002	4	3.8	44.75	45.7	224@1.16	1.620	0.0008-0.0027	0.0018-0.0037	0.2738-0.2751	0.2728-0.2741
2003	4	3.8	44.75	45.7	224@1.16	1.620	0.0008-0.0027	0.0018-0.0037	0.2738-0.2751	0.2728-0.2741

67197-WIND-C05

CRANKSHAFT AND CONNECTING ROD SPECIFICATIONS

All measurements are given in inches.

Year	Engine Displ. Liters	Engine VIN	Crankshaft				Connecting Rod		
			Main Brg. Journal Dia.	Main Brg. Oil Clearance	Shaft End-play	Thrust on No.	Journal Diameter	Oil Clearance	Side Clearance
2001	3.8	4	2.5190-2.5198	0.0010-0.0014	0.0040-0.0080	3	2.3103-2.3111	0.0010-0.0015	0.0043-0.0192
2002	3.8	4	2.5190-2.5198	0.0010-0.0014	0.0040-0.0080	3	2.3103-2.3111	0.0010-0.0015	0.0043-0.0192
2003	3.8	4	2.5190-2.5198	0.0010-0.0014	0.0040-0.0080	3	2.3103-2.3111	0.0010-0.0015	0.0047-0.0192

67197-WIND-C06

PISTON AND RING SPECIFICATIONS

All measurements are given in inches.

Year	Engine Displ. Liters	Engine VIN	Piston Clearance	Ring Gap			Ring Side Clearance		
				Top Comp.	Bottom Comp.	Oil Control	Top Comp.	Bottom Comp.	Oil Control
2001	3.8	4	0.0007-0.0017	0.0098-0.0161	0.0150-0.0252	0.0059-0.0650	0.0012-0.0031	0.0012-0.0031	Snug
2002	3.8	4	0.0007-0.0017	0.0098-0.0161	0.0150-0.0252	0.0059-0.0650	0.0012-0.0031	0.0012-0.0031	Snug
2003	3.8	4	0.0007-0.0017	0.0067-0.0130	0.0118-0.0217	0.0059-0.0256	0.0012-0.0031	0.0012-0.0031	Snug

67197-WIND-C07

TORQUE SPECIFICATIONS

All readings in ft. lbs.

Year	Engine VIN	Engine Displ. Liters	Cylinder Head Bolts	Main Bearing Bolts	Rod Bearing Bolts	Crankshaft Damper Bolts	Flywheel Bolts	Manifold Intake	Manifold Exhaust	Spark Plugs	Oil Pan Drain Plug
2001	4	3.8	①	④	②	104-132	54-64	③	15-22	7-15	19
2002	4	3.8	①	④	②	118	59	③	18	11	19
2003	4	3.8	①	④	⑤	118	59	③	18	11	19

① See text for torque procedure

② Step 1: 30-34 ft. lbs.
Step 2: Tighten an additional 90-120 degrees

③ Upper: 89 inch lbs.
Lower: Step 1, 44 inch lbs.; step 2, 89 inch lbs.

④ Step 1: 37 ft. lbs.
Step 2: plus 115-125 degrees

⑤ Step 1: 18 ft. lbs.
Step 2: 33 ft. lbs.
Step 3: an additional 105 degrees

67197-WIND-C08

WHEEL ALIGNMENT

Year	Model			Caster Range (+/-Deg.)	Caster Preferred Setting (Deg.)	Camber Range (+/-Deg.)	Camber Preferred Setting (Deg.)	Toe-in (in.)
2001	Windstar	Front	Left	0.75	+3.3	0.50	-0.44	-0.15+/-0.25
			Right	0.75	+3.8	0.50	-0.44	-0.15+/-0.25
		Rear		—	—	0.50	-0.10	0.06+/-0.34
2002	Windstar	Front	Left	0.75	+3.3	0.50	-0.44	-0.15+/-0.25
			Right	0.75	+3.8	0.50	-0.44	-0.15+/-0.25
		Rear		—	—	0.50	-0.10	0.06+/-0.34
2003	Windstar	Front	Left	0.75	+3.3	0.50	-0.44	-0.15+/-0.25
			Right	0.75	+3.8	0.50	-0.44	-0.15+/-0.25
		Rear		—	—	0.50	-0.10	0.06+/-0.34

67197-WIND-C09

TIRE, WHEEL AND BALL JOINT SPECIFICATIONS

Year	Model	OEM Tires Standard	OEM Tires Optional	Tire Pressures (psi) Front	Tire Pressures (psi) Rear	Wheel Size	Ball Joint Inspection	Lug Nut Torque (ft. lbs.)
2001	Windstar	P215/70R15	P205/70R15	①	①	NA	②	100
2002	Windstar	P215/70R15	P205/70R15	①	①	NA	②	100
2003	Windstar	P215/70R15	P205/70R15	①	①	NA	②	100

NA: Information not available

OEM: Original Equipment Manufacturer

PSI: Pounds Per Square Inch

① See placard on vehicle

② Replace if any perceptible movement is noticed

67197-WIND-C10

BRAKE SPECIFICATIONS

All measurements in inches unless noted

Year	Model	Brake Disc Original Thickness	Brake Disc Minimum Thickness	Brake Disc Maximum Runout	Brake Drum Diameter Original Inside Diameter	Brake Drum Diameter Max. Wear Limit	Brake Drum Diameter Maximum Machine Diameter	Minimum Lining Thickness	Brake Caliper Bracket-to-Hub Bolt (ft. lbs.)	Brake Caliper Mounting Pin or Bolt (ft. lbs.)
2001	All	1.020	0.974	0.003	9.84	9.90	9.90	①	85	25
2002	All	1.063	1.010	0.003	9.84	9.90	9.90	①	85	25
2003	All	1.063	1.010	0.003	9.84	9.90	9.90	①	85	25

NOTE: Due to changes made during production, refer to the manufacturer's specifications if they differ from those in this chart

① With drum brakes: 0.059

With disc brakes: 0.125

67197-WIND-C11

SCHEDULED MAINTENANCE INTERVALS

2001-02 FORD—WINDSTAR

TO BE SERVICED	TYPE OF SERVICE	VEHICLE MILEAGE INTERVAL (x1000)												
		5	10	15	20	25	30	35	40	45	50	55	60	65
Engine oil & filter	R	✓	✓	✓	✓	✓	✓	✓	✓	✓	✓	✓	✓	✓
Rotate tires	S/I	✓		✓		✓		✓		✓		✓		✓
Engine coolant strength	S/I			✓			✓			✓			✓	
Air cleaner filter	R						✓						✓	
Automatic transmission fluid & filter	R						✓						✓	
Engine coolant ①	R						✓						✓	
PCV valve	R												✓	
Spark plugs ②	R						✓						✓	
Drive belts	S/I						✓						✓	
Exhaust system & heat shields	S/I						✓						✓	
Front & rear brakes	S/I						✓						✓	

R: Replace S/I: Service or Inspect

① Engine coolant: change initially at 50,000 miles and every 30,000 miles thereafter.

② Spark plugs: replace every 100,000 miles.

FREQUENT OPERATION MAINTENANCE (SEVERE SERVICE)

If a vehicle is operated under any of the following conditions it is considered severe service:

- Extremely dusty areas.

- 50% or more of the vehicle operation is in 32°C (90°F) or higher temperatures, or constant operation in temperatures below 0°C (32°F).

- Prolonged idling (vehicle operation in stop and go traffic.

- Frequent short running periods (engine does not warm to normal operating temperatures).

- Police, taxi, delivery usage or trailer towing usage.

Engine oil & filter: replace every 3000 miles.

Rotate tires initially at 6000 miles and every 9000 miles thereafter.

Air cleaner filter: change every 15,000 miles.

Engine coolant strength, hoses & clamps: check every 15,000 miles.

Exhaust system: check every 15,000 miles.

Automatic transmission fluid & filter: change every 21,000 miles.

67197-WIND-C12

SCHEDULED MAINTENANCE INTERVALS
2003 FORD—WINDSTAR

TO BE SERVICED	TYPE OF SERVICE	VEHICLE MILEAGE INTERVAL (x1000)												
		5	10	15	20	25	30	35	40	45	50	55	60	65
Engine oil & filter	R	✓	✓	✓	✓	✓	✓	✓	✓	✓	✓	✓	✓	✓
Rotate tires	S/I	✓		✓		✓		✓		✓		✓		✓
Automatic transmission fluid	I			✓			✓			✓			✓	
Engine cooling system & hoses	I			✓			✓			✓			✓	
Steering linkage	I			✓			✓			✓			✓	
Ball joints	I			✓			✓			✓			✓	
Cabin filter	R			✓			✓			✓			✓	
Engine coolant (yellow)	R	every 100,000 miles												
Engine coolant (green)	R									✓				
Fuel filter	R						✓						✓	
Air cleaner filter	R						✓						✓	
Automatic transmission fluid & filter	R						✓						✓	
PCV valve	R	every 100,000 miles												
Spark plugs	R	every 100,000 miles												
Accessory drive belts	S/I	every 100,000 miles												
Exhaust system & heat shields	S/I						✓						✓	
Front & rear brakes	S/I			✓			✓			✓			✓	

R: Replace S/I: Service or Inspect

② Spark plugs: replace every 100,000 miles.

Special Operating Condition Requirements

When towing a trailer or using a camper or car-top carrier:

Change engine oil and install a new oil filter every 4,800 km (3,000 miles), 3 months or 200 hours of engine operation (whichever occurs first).

Change transfer case fluid every 96,000 km (60,000 miles).

Change manual transmission fluid as required.

Inspect and lubricate U-joints as required.

During extensive idling and/or low speed driving for long distances, as in heavy commercial use such as delivery, taxi, patrol car or livery:

Change engine oil and install a new oil filter every 4,800 km (3,000 miles), 3 months or 200 hours of engine operation (whichever occurs first).

Lube front lower control arm and steering linkage ball joints with zerk fittings (if equipped) every 4,800 km (3,000 miles) or 3 months.

Inspect brake system and check battery electrolyte level (Patrol cars) every 8,000 km (5,000 miles).

Install a new fuel filter every 24,000 km (15,000 miles).

Change automatic transmission fluid, lubricate 4x2 wheel bearings, install new grease seals and adjust bearings every 48,000 km (30,000 miles). If equipped, change the in-line service installed transmission fluid filter.

Install new spark plugs and change transfer case fluid every 96,000 km (60,000 miles).

Install a new cabin air filter as required.

When operating in dusty conditions such as unpaved or dusty roads:

Change engine oil and install a new oil filter every 4,800 km (3,000 miles) or 3 months.

Install a new fuel filter every 24,000 km (15,000 miles).

Change automatic transmission fluid every 48,000 km (30,000 miles). If equipped, change the in-line service installed transmission fluid filter.

Change transfer case fluid every 96,000 km (60,000 miles).

Install a new engine air filter as required.

Install a new cabin air filter as required.

When operating in off-road conditions:

Change automatic transmission fluid every 48,000 km (30,000 miles). If equipped, change the in-line service installed transmission fluid filter.

Change transfer case fluid every 96,000 km (60,000 miles).

Install a new cabin air filter as required.

Inspect and lubricate U-joints.

Inspect and lubricate steering linkage ball joints with zerk fittings.

67197-WIND-C13

PRECAUTIONS

Before servicing any vehicle, please be sure to read all of the following precautions, which deal with personal safety, prevention of component damage, and important points to take into consideration when servicing a motor vehicle:

- Never open, service or drain the radiator or cooling system when the engine is hot; serious burns can occur from the steam and hot coolant.
- Observe all applicable safety precautions when working around fuel. Whenever servicing the fuel system, always work in a well-ventilated area. Do not allow fuel spray or vapors to come in contact with a spark, open flame, or excessive heat (a hot drop light, for example). Keep a dry chemical fire extinguisher near the work area. Always keep fuel in a container specifically designed for fuel storage; also, always properly seal fuel containers to avoid the possibility of fire or explosion. Refer to the additional fuel system precautions later in this section.
- Fuel injection systems often remain pressurized, even after the engine has been turned **OFF**. The fuel system pressure must be relieved before disconnecting any fuel lines. Failure to do so may result in fire and/or personal injury.
- Brake fluid often contains polyglycol ethers and polyglycols. Avoid contact with the eyes and wash your hands thoroughly after handling brake fluid. If you do get brake fluid in your eyes, flush your eyes with clean, running water for 15 minutes. If eye irritation persists, or if you have taken brake fluid internally, IMMEDIATELY seek medical assistance.
- The EPA warns that prolonged contact with used engine oil may cause a number of skin disorders, including cancer! You should make every effort to minimize your exposure to used engine oil. Protective gloves should be worn when changing oil. Wash your hands and any other exposed skin areas as soon as possible after exposure to used engine oil. Soap and water, or waterless hand cleaner should be used.
- All new vehicles are now equipped with an air bag system. The system must be disabled before performing service on or around system components, steering column, instrument panel components, wiring and sensors. Failure to follow safety and disabling procedures could result in accidental air bag deployment, possible personal injury and unnecessary system repairs.
- Always wear safety goggles when working with, or around, the air bag system. When carrying a non-deployed air bag, be sure the bag and trim cover are pointed away from your body. When placing a non-deployed air bag on a work surface, always face the bag and trim cover upward, away from the surface. This will reduce the motion of the module if it is accidentally deployed. Refer to the additional air bag system precautions later in this section.
- Clean, high quality brake fluid from a sealed container is essential to the safe and proper operation of the brake system. You should always buy the correct type of brake fluid for your vehicle. If the brake fluid becomes contaminated, completely flush the system with new fluid. Never reuse any brake fluid. Any brake fluid that is removed from the system should be discarded. Also, do not allow any brake fluid to come in contact with a painted surface; it will damage the paint.
- Never operate the engine without the proper amount and type of engine oil; doing so WILL result in severe engine damage.
- Timing belt maintenance is extremely important! Many models utilize an interference-type, non-freewheeling engine. If the timing belt breaks, the valves in the cylinder head may strike the pistons, causing potentially serious (also time-consuming and expensive) engine damage. Refer to the maintenance interval charts in the front of this manual for the recommended replacement interval for the timing belt, and to the timing belt section for belt replacement and inspection.
- Disconnecting the negative battery cable on some vehicles may interfere with the functions of the on-board computer system(s) and may require the computer to undergo a relearning process once the negative battery cable is reconnected.
- When servicing drum brakes, only disassemble and assemble one side at a time, leaving the remaining side intact for reference.
- Only an MVAC-trained, EPA-certified automotive technician should service the air conditioning system or its components.

ENGINE REPAIR

Alternator

REMOVAL & INSTALLATION

1. Disconnect the negative battery cable.
2. Disconnect the alternator wiring harness.
3. Detach the alternator drive belt.
4. Remove the three alternator attaching bolts.
5. Remove the alternator.

To install:

6. Position the alternator on the engine.
7. Install the three alternator attaching bolts. Tighten the bolts 18 ft. lbs. (25 Nm).
8. Install and tension the alternator drive belt.

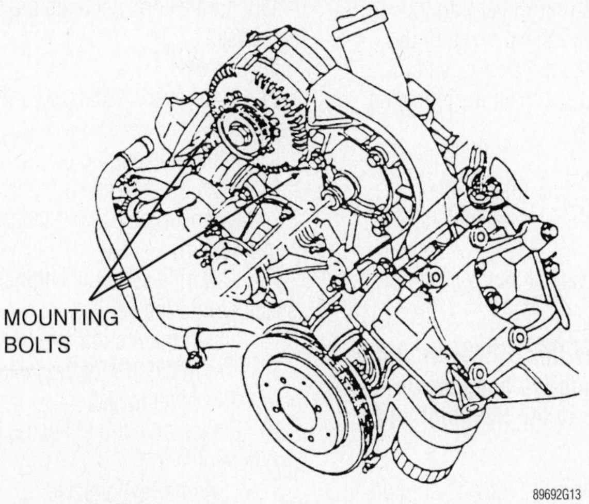

Alternator mounting—3.8L engine

9. Connect the alternator wiring harness. Tighten the output terminal nut to 80–97 inch lbs. (9–11 Nm).

10. Connect the negative battery cable.

Engine Assembly

REMOVAL & INSTALLATION

1. Remove or disconnect the following:
   - Negative battery cable
   - Coolant
   - Refrigerant, into a refrigerant recovery station
   - Cowl top vent panel
   - Wiring from the alternator
   - Air cleaner assembly
   - Upper and lower radiator hoses from the engine
   - Heater water hoses and secure to body
   - Air conditioning discharge and suction hoses, then secure them to the engine. Cap the open lines to prevent moisture from entering the system.
   - Accelerator cable and speed control cable from the throttle body lever
   - Accelerator cable bracket from the throttle body
   - Fuel supply and return lines from the fuel injection supply manifold
   - All engine wiring harnesses from the engine and secure to the body
   - All vacuum hoses from the engine
   - Gear shift cable from the transaxle

➡Damage to the steering column air bag wiring can result if the steering wheel is allowed to rotate freely. The wire is wound like a watch spring and can be over-tightened and break if the steering wheel is rotated too far in either direction.

2. Lock the steering wheel with the wheels in the straight-ahead position by turning the ignition to the **OFF** position.

3. Remove or disconnect the following:
   - Wheels
   - Engine oil
   - Transaxle cooler lines at the transaxle. Secure the lines to the radiator.
   - Heated oxygen sensor wiring harness

➡Do not allow the flex connector of the duel converter Y-pipe to hang unsupported or damage to the flex joint will result.

   - Dual converter Y-pipe and support it from the body

➡The routing of the battery ground cable to the cylinder block is critical. It should go between the transaxle and the bracket. Take note during disassembly.

   - Starter motor wires and secure out of the way
   - Starter motor
   - Engine rear plate and torque converter-to-flywheel nuts
   - Power steering cooler lines
   - Upper bolt from the sway bar links
   - Dust boot from the steering rack pinion support
   - Steering coupling pinch bolt from the steering column intermediate shaft at the steering gear
   - Intermediate shaft from the steering gear
   - Front stabilizer bar links
   - Front suspension lower arms from the knuckles at the ball joint
   - Tie rod ends from the knuckle
   - Front axle wheel hub retainers from the halfshaft ends
   - Halfshafts from the front wheel knuckle

4. Support the front subframe, engine and transaxle assembly.

5. Remove or disconnect the following:
   - The 4 retaining bolts, and lower the engine transaxle and front subframe from the vehicle
   - Power steering pressure hose from the power steering pump

6. Attach an engine hoist and lift the engine slightly.

7. Remove the engine support insulators.

8. Lift the engine and transaxle assembly from the front subframe.

9. Lower the engine and transaxle.

10. Support the transaxle on a level stationary surface and separate the engine from the transaxle.

To install:

11. If removed, install the transaxle on the engine.

12. Install the engine on the subframe.

13. Install the engine support insulators.

14. Connect the power steering hose to the pump.

15. Carefully raise the engine/transaxle assembly into the vehicle.

16. Install the 4 subframe bolts.

17. The remainder of the installation is the reverse of removal.

18. Please note the following torque specifications:
   - Engine-to-transaxle: 37 ft. lbs. (50 Nm)
   - Torque converter nuts: 20–34 ft. lbs. (27–46 Nm)
   - Subframe-to-body bolts: 57–76 ft. lbs. (77–103 Nm)
   - Steering coupling pinch bolt: 25–34 ft. lbs. (34–46 Nm)
   - Dual converter Y-pipe-to-exhaust manifold: 25–34 ft. lbs. (34–46 Nm)
   - Flex pipe retaining bolts: 25–34 ft. lbs. (34–46 Nm)
   - Accelerator cable bracket retaining bolts: 71–106 inch lbs. (8–12 Nm)
   - Front engine support insulator-to-subframe 50–68 ft. lbs. (68–92 Nm)
   - Transmission insulator-to-subframe: 65–87 ft. lbs. (88–119 Nm)
   - Rear engine and transaxle support insulator-to-subframe: 56–75 ft. lbs. (76–103 Nm)

19. Refill the engine, transaxle and cooling system with the correct amount of the appropriate fluids before starting the engine.

Water Pump

REMOVAL & INSTALLATION

1. Remove or disconnect the following:
   - Negative battery cable
   - Coolant
   - Drive belts
   - Lower radiator hose
   - Lower nut on both front engine supports
   - Alternator
   - Power steering pressure line from the pump
   - Power steering reservoir filler cap
   - Water bypass hose and oil cooler hose from the heater water outlet tube
   - Heater water outlet tube from the water pump
   - Air conditioning bracket brace

2. Raise the engine approximately 2 inches (51mm) to provide necessary clearance for water pump removal.

3. Remove the water pump pulley.

4. Remove the drive belt tensioner from the power steering pump brace.

5. Remove the power steering pump brace and place the pump and brace aside in the engine compartment.

6. Remove the water pump.

To install:

7. Clean all gasket mating surfaces thoroughly.

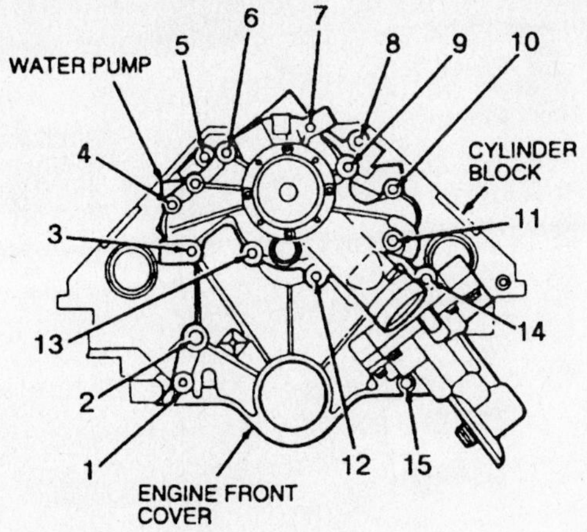

1. M8 x 1.25 x 98
2. M8 x 1.25 x 98
3. M8 x 1.25 x 131
4. M8 x 1.25 x 131
5. M8 x 1.25 x 25
6. M8 x 1.25 x 35
7. M8 x 1.25 x 35
8. M8 x 1.25 x 25
9. M8 x 1.25 x 61.5
10. M8 x 1.25 x 141
11. M8 x 1 x 131
12. M8 x 1 x 35
13. M8 x 1 x 35
14. M8 x 1 x 105
15. M8 x 1 x 20

7924GG02

Because of their varying lengths, be sure to install the water pump bolts in the correct bolt holes—3.8L engine

⁂ WARNING

Be careful not to gouge the aluminum surfaces when scraping the old gasket material from the mating surfaces of the water pump and front cover.

8. Cover the threads of the No. 1 engine front cover stud with Teflon® tape.
9. Install or connect the following:
   - New water pump housing gasket on the water pump sealing surface using gasket sealant to hold the gasket in place
   - Water pump and tighten the bolts to 18 ft. lbs. (25 Nm)
   - Drive belt tensioner
   - Water pump pulley
   - Air conditioning bracket brace
   - Heater water outlet tube
   - Water bypass hose and oil cooler hose
   - Power steering reservoir filler cap
   - Power steering pressure line
   - Alternator
   - Lower nut on both front engine supports
   - Lower radiator hose
   - Drive belts
10. Fill and bleed the cooling system.
11. Connect the negative battery cable.

Heater Core

REMOVAL & INSTALLATION

Front System

1. Disconnect the negative battery cable.
2. Drain the cooling system into a clean container for reuse.

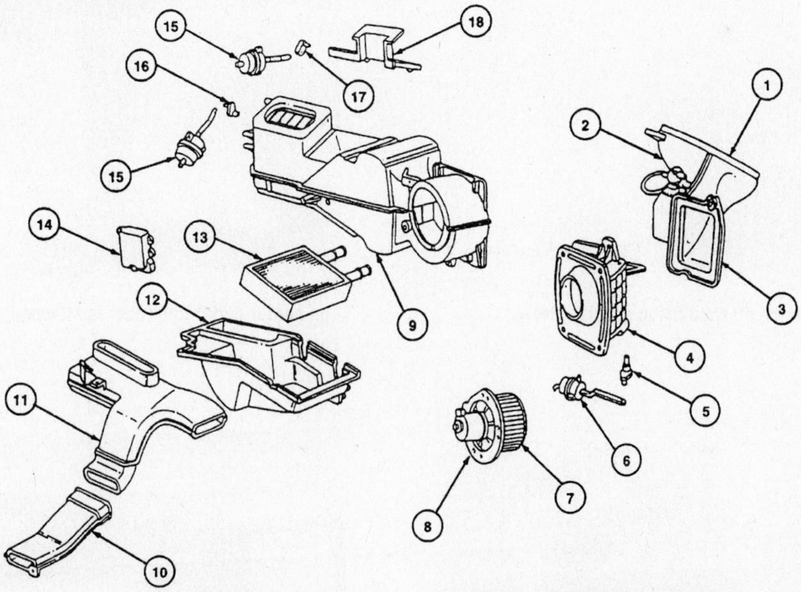

Item	Description
1	Outside Air Seal
2	A/C Air Inlet Duct
3	A/C Air Inlet Door Inner Seal
4	A/C Recirculating Air Duct
5	A/C Damper Door Shaft
6	Vacuum Control Motor
7	Blower Motor Wheel
8	Blower Motor
9	A/C Evaporator Housing
10	Rear Seat Airflow Duct

Item	Description
11	Heater Outlet Floor Duct
12	Heater Housing Core Plate
13	Heater Core
14	A/C Electronic Door Actuator Motor
15	Vacuum Control Motor
16	Shaft — Heater Air Damper Door
17	Windshield Defroster Door Shaft
18	Windshield Defroster Duct Connector

89696G06

Front heater housing assembly

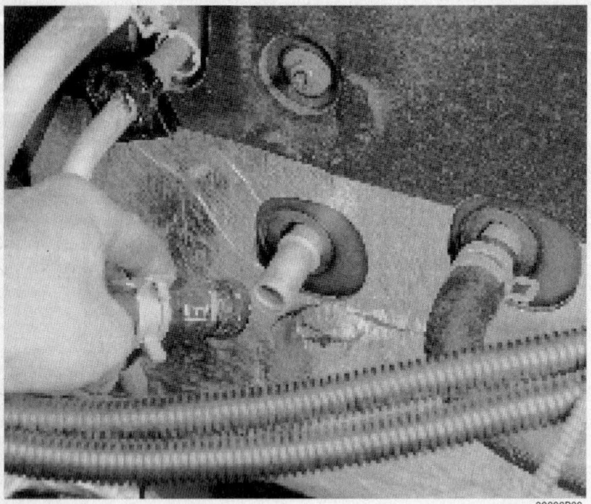

89696P09

Be careful when removing the hoses from the heater core, as the tubes are easily damaged

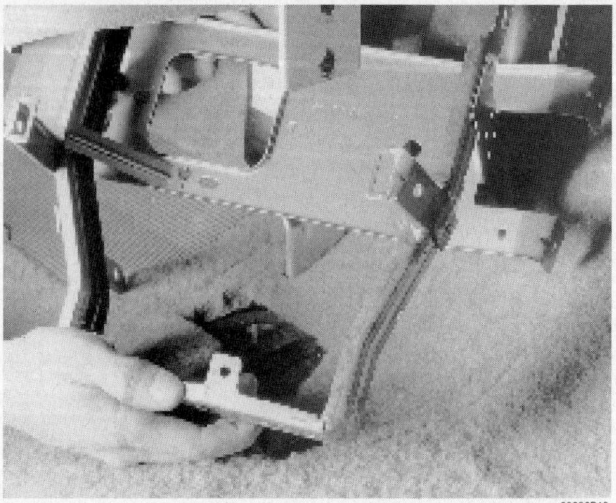

89696P12

The instrument panel support bracket must be removed to lower the heater core

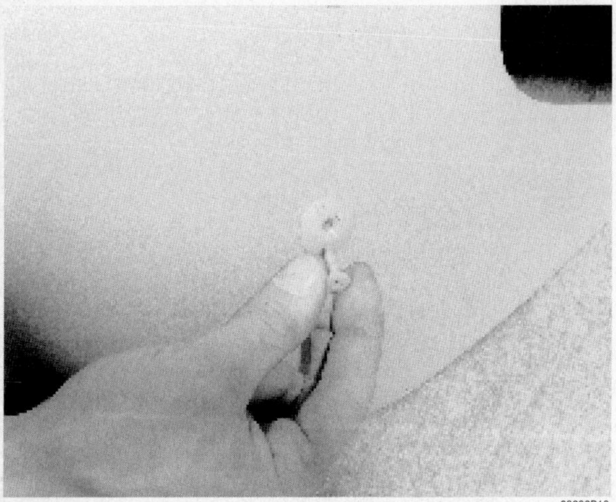

89696P10

The center instrument panel is attached using special plastic screws

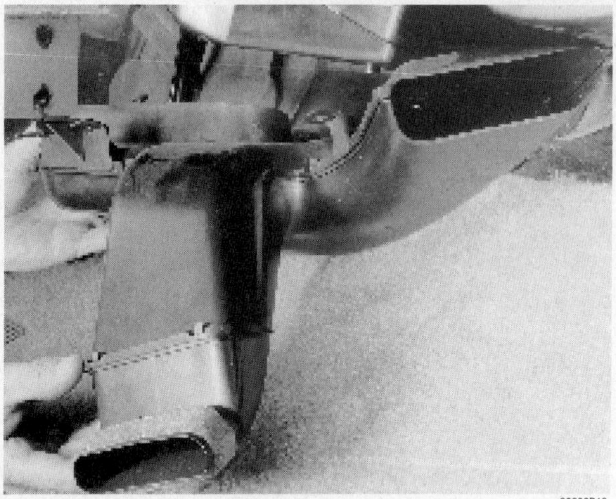

89696P13

The heater outlet floor duct is attached to the bottom of the housing core

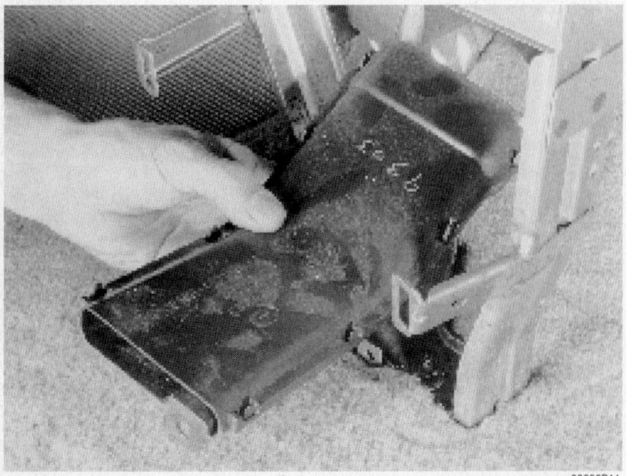

89696P11

The rear seat airflow duct carries air to the rear of the passenger compartment

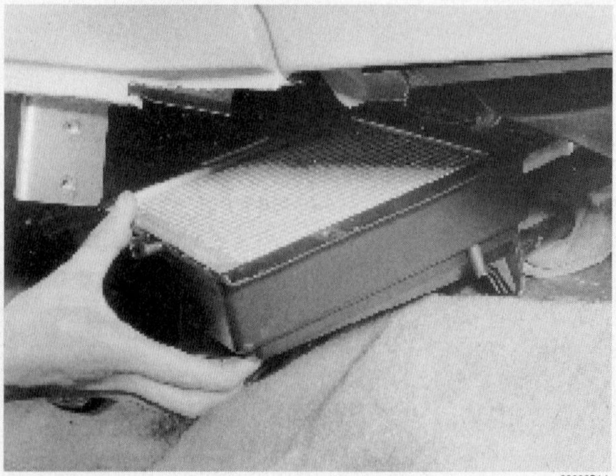

89696P14

The heater core is located in a box at the bottom of the housing core

89696P15

A foam liner is used to cushion the heater core

3. Remove the cowl top vent panel for clearance.

4. Disconnect the heater hoses from the heater core inlet and outlet tubes in the engine compartment.

5. Remove the cassette box/center instrument support trim.

6. Pull out and remove the ashtray cup holder by depressing the service lever.

7. Remove the floor/rear seat lower air duct.

8. Disconnect the keyless entry wiring harness, if equipped.

9. Remove the center instrument panel support brackets.

10. Disconnect the climate control vacuum harness connector.

11. Remove the heater floor/rear seat upper air duct.

12. Remove the heater core cover retaining screws and remove the heater core cover.

13. Remove the heater core.

14. Remove the seal from the heater core tubes and discard.

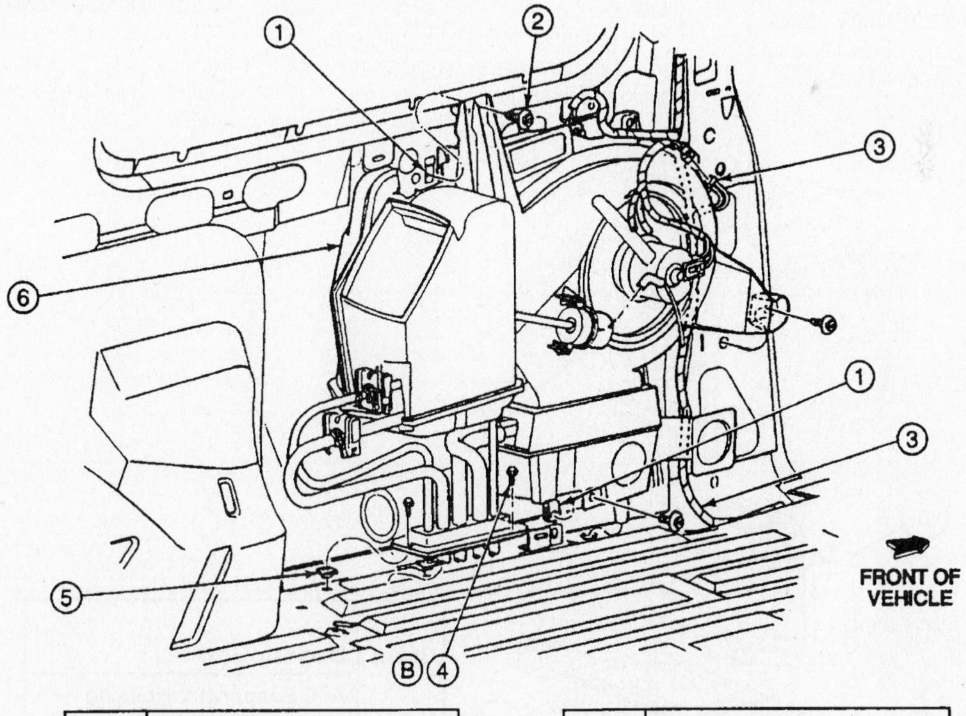

Item	Description
1	U-Nut
2	Screw and Washer Assembly
3	Body Main Wiring
4	Screw and Washer Assembly

Item	Description
5	Nut Insert
6	Heater and A/C Assembly

89696G04

Rear heater and air conditioning assembly

To install:

15. Install the seal to the heater core tubes and discard.
16. Install the heater core.
17. Install the heater core cover and the heater core cover retaining screws.
18. Install the heater floor/rear seat upper air duct.
19. Connect the climate control vacuum harness connector.
20. Install the center instrument panel support brackets.
21. Connect the keyless entry wiring harness, if equipped.
22. Install the floor/rear seat lower air duct.
23. Install the ashtray cup holder by depressing the service lever.
24. Install the cassette box/center instrument support trim.
25. Connect the heater hoses to the heater core inlet and outlet tubes in the engine compartment.
26. Install the cowl top vent panel for clearance.
27. Refill the cooling system.
28. Connect the negative battery cable.
29. Run the engine to normal operating temperatures; then, check the climate control operation and check for leaks.

Rear Auxiliary System

1. Disconnect the negative battery cable.
2. Drain the cooling system into a clean container for reuse.
3. Discharge and recover the rear auxiliary air conditioning refrigerant.
4. Raise and support the vehicle safely.
5. Place a drain pan under the heater water hose connections for the rear heater assembly.
6. Drain and recycle the engine coolant from the auxiliary heater core and hoses.
7. Disconnect the 2 air conditioning spring lock couplings using a quick disconnect coupling tool.
8. Lower the vehicle.
9. Remove all rear passenger's seating.
10. Remove the 3 push pin retainers from the lower edge of the auxiliary heater and air conditioning service cover.
11. Lift the service cover outward and upward to remove.
12. Label and disconnect the electrical harnesses.
13. Label and disconnect the vacuum lines.
14. Remove the body side trim panels.
15. Disconnect and plug all air conditioning refrigerant lines.

➡It is extremely important that the air conditioning lines be plugged to prevent the entry of dirt or moisture.

16. Remove the lower air conditioning recirculation air duct and heater extension air duct.
17. Remove the auxiliary heater and air conditioning assembly.
18. Remove the housing cover screws and the cover.
19. Remove the heater core and the heater core case seal.

To install:

20. Install the heater core and the heater core case seal.
21. Install the housing cover and screws.
22. Position the auxiliary heater and air conditioning assembly in the vehicle. Guide the heater water hoses through the floor grommet. Tighten the retaining screws to 16–23 inch lbs. (2–3 Nm).
23. Connect all air conditioning refrigerant lines.
24. Install the lower air conditioning recirculation air duct and heater extension air duct. Tighten the retaining screws to 16–23 inch lbs. (2–3 Nm).
25. Install the body side trim panels.
26. Connect the vacuum lines.
27. Connect the electrical harnesses.

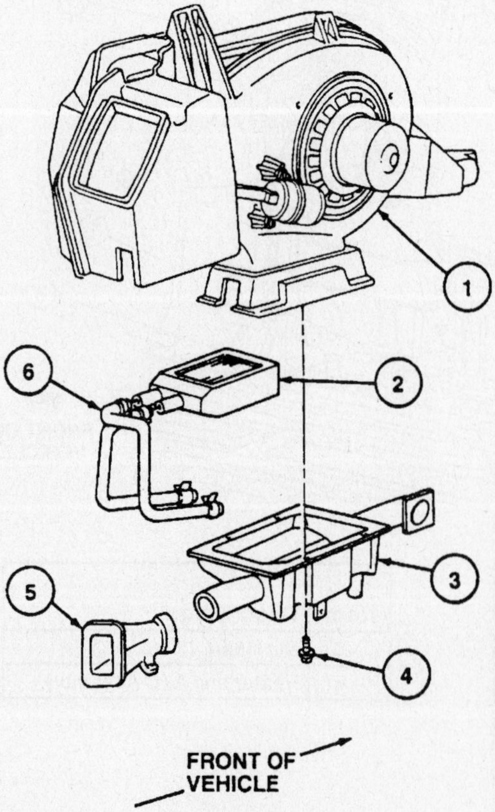

Item	Description
1	A / C Evaporator Housing Assy
2	Heater Core
3	Heater Housing Core Plate
4	Screw
5	Rear Seat Airflow Duct
6	Heater Water Hose

89696G05

Rear heater housing assembly

28. Install the auxiliary heater and air conditioning service cover and secure with 3 push pin retainers.
29. Install all rear passenger's seating.
30. Raise and support the vehicle safely.
31. Connect the heater water hoses to the heater core tubes.
32. Lower the vehicle.
33. Refill the cooling system.
34. Connect the negative battery cable.
35. Evacuate and charge the rear auxiliary air conditioning refrigerant.
36. Run the engine to normal operating temperatures; then, check the climate control operation and check for leaks.

Cylinder Head

REMOVAL & INSTALLATION

1. Remove or disconnect the following:
   - Negative battery cable
   - Coolant
   - Air cleaner assembly
   - Cowl top vent panel
   - Accessory drive belt
2. If the front cylinder head is being removed, perform the following:
   a. Remove the oil filler cap.
   b. Remove the air conditioning compressor mounting bracket and set the air conditioning compressor aside with the refrigerant lines still connected.
   c. Remove the power steering pump and bracket. Leave the power steering hoses connected and place the pump aside in the engine compartment.
   d. Remove the alternator and mounting bracket.
3. If the rear cylinder head is being removed, perform the following:
   a. Remove the accessory drive belt tensioner.
   b. Remove the PCV valve.
   c. Remove the power steering line bracket.
   d. Remove the tensioner bracket.
   e. Remove the coil pack assembly.
4. Remove or disconnect the following:
   - Upper intake manifold
   - Valve cover
   - Fuel system pressure
   - Fuel charging assembly
   - Lower intake manifold
   - Exhaust manifolds

➡Pushrods must be installed in their original positions. Note pushrod location during removal.

- Pushrods
- Cylinder head bolts
- Cylinder head from the engine block and discard the gaskets

To install:

5. The cylinder head should be cleaned and inspected prior to installation.
6. Lightly oil all bolt and stud bolt threads before installation.
7. Clean all gasket mating surfaces thoroughly.
8. Position new head gaskets on the cylinder block, noting the **UP** position mark on the gasket face, using the dowels in the engine block for alignment. If the dowels are damaged, they must be replaced.

⁂ WARNING

Always use new cylinder head bolts when installing cylinder head or damage to the engine may occur.

9. Position the cylinder head on the cylinder block.
10. Lubricate the cylinder head bolts with engine oil and install. Tighten the cylinder head bolts following the proper torque sequence as follows:
    - Step 1: 15 ft. lbs. (20 Nm)
    - Step 2: 29 ft. lbs. (40 Nm)
    - Step 3: 37 ft. lbs. (50 Nm)

➡Do not loosen all of the cylinder head bolts at once. Only work on 1 bolt at a time or damage to the engine may occur.

In sequence, loosen each cylinder head bolt 2–3 turns and retighten in 2 steps.

- Long bolts, step 1: 29–37 ft. lbs. (40–50 Nm); step 2: tighten an additional 175–185 degrees.
- Short bolts, step 1: 15–22 ft. lbs. (20–30 Nm); second 2: tighten an additional 175–185 degrees.

11. Dip each pushrod end in engine assembly lubricant. Install the pushrods in their original positions.
12. Lubricate all rocker arm components with engine assembly lubricant.
13. For the rocker arm being installed, rotate the engine clockwise until the valve tappet rests on the heel (base circle) of the camshaft lobe.
14. Install the rocker arms, seats and bolts and tighten to 44 inch lbs. (5 Nm).
15. Perform the previous 2 steps for each rocker arm.
16. After all rocker arms have been installed, final tighten all bolts to 22–29 ft. lbs. (30–40 Nm).
17. Install or connect the following:
    - Exhaust manifolds
    - Lower intake manifold
    - Fuel injection charging assembly
    - Valve covers with new gaskets. Tighten the bolts to 71–97 inch lbs. (8–11 Nm).
    - Upper intake manifold
    - Spark plugs and ignition wires
18. If the front cylinder head is being installed, perform the following:
    a. Install the alternator and mounting bracket.
    b. Install the power steering pump and bracket.
    c. Install the air conditioning compressor bracket.
    d. Install the oil filler cap.

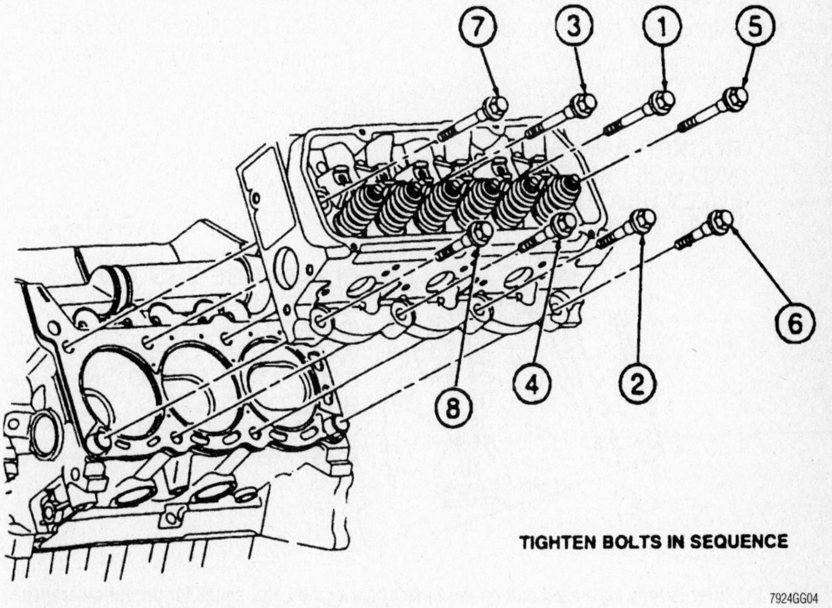

Cylinder head bolt torque sequence—3.8L engine

19. If the rear cylinder head is being installed, perform the following:
    a. Install the coil pack assembly.
    b. Install the tensioner bracket.
    c. Install the power steering line bracket.
    d. Install the PCV valve.
    e. Install the accessory drive belt tensioner.
    f. Rotate the tensioner clockwise and install the accessory drive belt.
20. Install the cowl top vent panel.
21. Install the air cleaner assembly.
22. Fill and bleed the cooling system.
23. Connect the negative battery cable.

Rocker Arms

REMOVAL & INSTALLATION

1. Remove the valve cover.
2. Remove the rocker arm retaining bolt.

➡Rocker the arms should be installed in their original location during assembly.

3. Remove the rocker arms. If more than 1 rocker arm is to be removed, identify each rocker arm location.

To install:

4. Lubricate the pushrods and rocker arms with engine assembly lubricant. Lubricate the retaining bolts with engine oil.

➡Prior to final tightening, the rocker arm seats must be fully seated into the cylinder head. The pushrods must be fully seated in the rocker arm and valve tappet sockets.

5. Install the rocker arms into position with the pushrods and snug the retaining bolt.
6. Rotate the crankshaft until the lifter for the rocker arm being installed, is on the base circle (heel) of the cam lobe.
7. Tighten the rocker arm retaining bolt to 44 inch lbs. (5 Nm)
8. Finally, tighten the bolt with the camshaft in any position to 20–28 ft. lbs. (26–38 Nm).
9. Install the valve cover.

Intake Manifold

REMOVAL & INSTALLATION

Upper Manifold

1. Remove or disconnect the following:
   - Air cleaner outlet tube
   - Accelerator cable and speed control actuator cable at the throttle body
   - Accelerator cable bracket and position it aside
   - Vacuum lines
   - Necessary electrical harnesses
   - Crankcase ventilation tube from the PCV valve
   - Throttle body
   - Idle air control valve
   - Intake manifold retaining bolts, noting their positions

➡Keep the intake manifold bolts in order, so they can be installed in their original positions.

   - Upper intake manifold

To install:

2. Inspect the intake gasket to ensure the seals are completely installed in the manifold groove and the seals show no signs of damage.
3. Install or connect the following:
   - Intake manifold and tighten the bolts to 71–106 inch lbs. (8–12 Nm) in the sequence shown
   - Idle air control valve
   - Throttle body
   - Crankcase ventilation tube to the PCV valve
   - Electrical harnesses
   - Vacuum lines
   - Accelerator cable bracket and tighten the bolts to 71–106 inch lbs. (8–12 Nm)
   - Accelerator cable and speed control actuator cable at the throttle body
   - Air cleaner outlet tube

Lower Manifold

1. Remove or disconnect the following:
   - Coolant
   - Upper intake manifold
   - Water bypass hose from the heater water outlet tube
   - Bypass hose from the lower intake manifold
   - Fuel system pressure
   - Electrical wiring harnesses
   - Fuel injectors and fuel charging assembly
   - Vacuum motor and bracket assemblies
   - Valve assembly and linkage from the IMRC lever and bushing by using a prytool
   - Tube retaining bolts
   - Old bushing from the lever
   - EGR valve and adapter
   - Lower intake manifold retaining bolts

➡The lower intake manifold is sealed at each corner with sealer. To break the seal it may be necessary to pry on the front of the intake manifold with a prybar. If it is necessary, use care to prevent damage to the machined surfaces.

   - Lower intake manifold

To install:

2. Thoroughly clean all gasket mating surfaces.

➡When using silicone rubber sealer, assembly must occur within 15 minutes after sealer application. After this time, the sealer may start to set up and its sealing effectiveness may be reduced.

3. Install or connect the following:
   - New bushings into the IMRC levers
   - Apply a 3mm bead of RTV silicone sealer at each corner where the cylinder head joins the engine block
   - Front and rear intake manifold seals

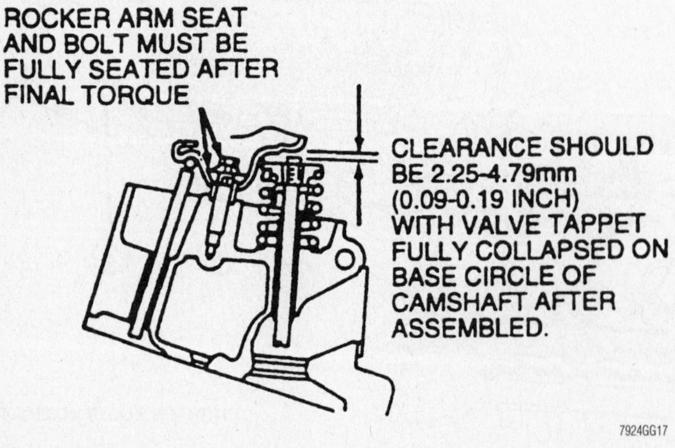
ROCKER ARM SEAT AND BOLT MUST BE FULLY SEATED AFTER FINAL TORQUE
CLEARANCE SHOULD BE 2.25-4.79mm (0.09-0.19 INCH) WITH VALVE TAPPET FULLY COLLAPSED ON BASE CIRCLE OF CAMSHAFT AFTER ASSEMBLED.
7924GG17

When the lifter is fully collapsed and on the base circle of the cam, check for proper clearance between the tip of the valve and the rocker arm—3.8L engine

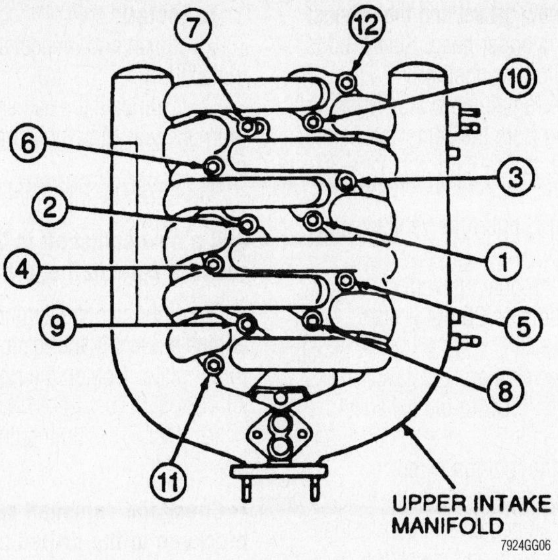

Upper intake manifold bolt torque sequence—3.8L engine

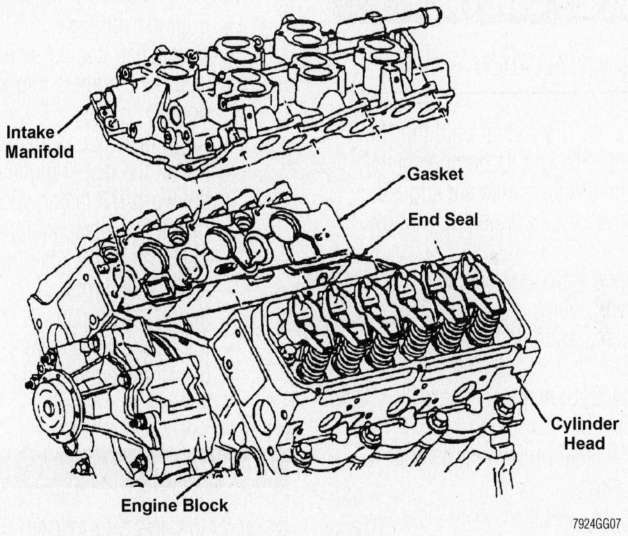

Lower intake manifold and related components—3.8L engine

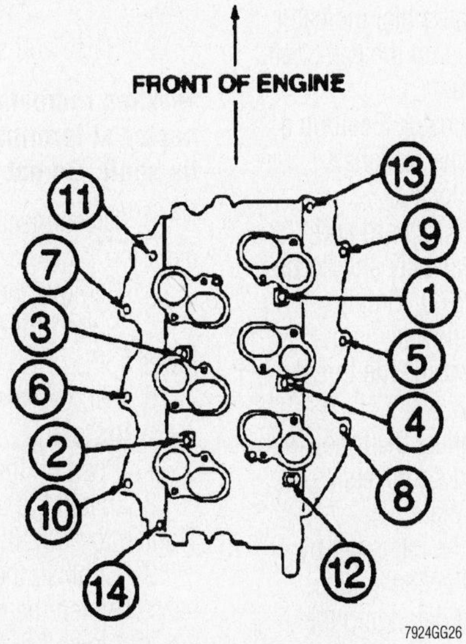

Lower intake manifold bolt torque sequence—3.8L engine

- Lower intake manifold into position on the cylinder block, using new gaskets
- Apply pipe sealant to the intake bolts and install in their original locations. Tighten in sequence to 44 inch lbs.; then to 89 inch lbs. (8–12 Nm).
- EGR valve and adapter
- IMRC vacuum motors, and tighten the retaining bolts to 71–106 inch lbs. (8–12 Nm)
- Fuel injectors and charging assembly. Tighten retaining bolts to 71–97 inch lbs. (8–11 Nm).
- Water bypass tube to the lower intake manifold. Tighten the retaining bolts to 71–97 inch lbs. (8–11 Nm).
- Water bypass tube hose to the outlet tube and tighten the hose clamp securely
- Electrical wiring harnesses
- Vacuum lines to the IMRC motors
- Upper radiator hose to water hose connection and tighten the hose clamp securely
- Upper intake manifold

4. Fill and bleed the cooling system.
5. Start the engine and check for leaks.

Exhaust Manifold

REMOVAL & INSTALLATION

➡Spray the exhaust system fasteners with penetrating lubricant before removing them to help prevent broken studs and bolts. The use of a 6-point socket is highly recommended when removing exhaust system fasteners.

✱✱ CAUTION

To prevent serious burns, allow the exhaust manifold to cool down before attempting to remove it.

Rear Manifold

1. Disconnect the negative battery cable.
2. Remove the cowl vent panel.
3. Remove the engine air cleaner and air cleaner outlet tube.
4. Disconnect the ignition wires from the rear cylinder head and ignition coil.
5. Remove the spark plugs from the rear cylinder head.
6. Raise and support the vehicle safely on jackstands.
7. Disconnect the dual converter Y-pipe from the exhaust manifold.

8. Lower the vehicle.
9. Remove the exhaust manifold.

To install:

10. Clean all gasket mating surfaces thoroughly.

➡A slight warpage in the exhaust manifold may cause a misalignment between the bolt holes in the cylinder head and exhaust manifold. Elongate the holes in the exhaust manifold as necessary to correct the misalignment. Do not elongate the pilot hole.

11. Install a new exhaust manifold gasket and the exhaust manifold on the cylinder head. Start 2 bolts to hold the manifold in position.
12. Install the remaining bolts. Tighten the bolts beginning from the center port and working outward to 15–22 ft. lbs. (20–30 Nm).
13. Raise and support the vehicle safely.
14. Connect the dual converter Y-pipe and tighten the bolts to 25–34 ft. lbs. (34–47 Nm).
15. Lower the vehicle.
16. Install the spark plugs in the rear cylinder head.
17. Connect the ignition wires.
18. Install the engine air cleaner and air cleaner outlet tube.
19. Install the cowl vent panel.
20. Connect the negative battery cable.
21. Start the engine and check for exhaust leaks.

Front Manifold

1. Disconnect the negative battery cable.
2. Remove the oil level indicator tube.
3. Label and disconnect the ignition wires from the front cylinder head.
4. Disconnect the EGR-to-exhaust manifold tube.
5. Raise and support the vehicle safely on jackstands.
6. Disconnect the dual converter Y-pipe from the exhaust manifold.
7. Lower the vehicle.
8. Remove the exhaust manifold.

To install:

9. Clean all gasket mating surfaces thoroughly.

➡A slight warpage in the exhaust manifold may cause a misalignment between the bolt holes in the cylinder head and exhaust manifold. Elongate the holes in the exhaust manifold as necessary to correct the misalignment. Do not elongate the pilot hole.

10. Install a new gasket and the exhaust manifold on the cylinder head. Start 2 bolts to hold the manifold in position.
11. Install the remaining bolts. Tighten the bolts, starting from the center port and working outward, to 15–22 ft. lbs. (20–30 Nm).
12. Raise and support the vehicle safely on jackstands.
13. Connect the dual converter Y-pipe and tighten the bolts to 25–34 ft. lbs. (34–47 Nm).
14. Lower the vehicle.
15. Connect the EGR to the exhaust manifold tube.
16. Connect the ignition wires.
17. Install the oil level indicator tube.
18. Connect the negative battery cable.
19. Start the engine and check for exhaust leaks.

Camshaft and Valve Lifters

REMOVAL & INSTALLATION

1. Rotate the crankshaft until the No. 1 piston is at the TDC on its compression stroke and the timing marks are aligned.
2. Remove or disconnect the following:
   - Engine from the vehicle
   - Valve covers
   - Intake manifolds
   - Pushrod
   - Tappet guide plate
   - Tappets
   - Crankshaft pulley and damper
   - Oil pan
   - Engine front cover assembly
3. Check the camshaft end-play as follows:
   a. Push the camshaft toward the rear of the engine and install a dial indicator, so the indicator point is on the camshaft sprocket attaching screw.
   b. Zero the dial indicator. Position a small prybar between the camshaft sprocket or gear and block.
   c. Pull the camshaft forward and release it. Camshaft end-play should be 0.001–0.006 in. (0.025–0.15mm).
   d. If the camshaft end-play is not within specification, replace the thrust plate upon reassembly.
4. Remove or disconnect the following:
   - Timing chain and sprockets
   - Camshaft thrust plate
5. Carefully remove the camshaft by pulling it toward the front of the engine. Remove it slowly to avoid damaging the bearings, journals and lobes.

To install:

6. Clean and inspect all parts before installation.
7. Lubricate the camshaft lobes and journals with Molylube® or heavy engine oil.
8. Carefully install the camshaft.

➡If a new camshaft is being installed, recheck camshaft end-play.

9. Lubricate the engine thrust plate with engine assembly lubricant, then install the thrust plate. Tighten the retaining bolts to 71–124 inch lbs. (8–14 Nm).
10. Install the timing chain and sprockets.

➡Check the camshaft sprocket bolt for blockage of the drilled oil passages prior to installation, and clean if necessary.

11. Install or connect the following:
   - Engine front cover
   - Crankshaft damper and pulley
   - Hydraulic tappets into their original bores
   - Align the valve tappet flats and install the tappet guide plate with the word **UP** facing you
   - Install the intake manifold assembly
   - Lubricate and install the pushrods and rocker arms
   - Install the oil pan
   - Install the valve covers
   - Install the engine assembly into the vehicle

Starter Motor

REMOVAL & INSTALLATION

1. Disconnect the negative battery cable.
2. Raise and support the vehicle safely.

➡When removing the hard shell connector at terminal "S", grasp the plastic shell. Do not pull on the wire.

3. Disconnect the starter electrical harness.
4. Remove the upper starter bolt.
5. Support the starter and remove the lower bolt.
6. Remove the starter from the vehicle.

To install:

7. Position the starter in the vehicle.
8. Install the upper and lower bolts. Tighten to 20 ft. lbs. (28 Nm).
9. Connect the starter electrical harness. Tighten the starter cable nut to 80–124 inch lbs. (9–14 Nm).

➡**When installing the hard shell connector, be careful to push it straight on and make sure it locks in position with a notable click or detent.**

10. Lower the vehicle.
11. Connect the negative battery cable.

Oil Pan

⁂ CAUTION

The EPA warns that prolonged contact with used engine oil may cause a number of skin disorders, including cancer! You should make every effort to minimize your exposure to used engine oil. Protective gloves should be worn when changing the oil. Wash your hands and any other exposed skin areas as soon as possible after exposure to used engine oil. Soap and water, or waterless hand cleaner, should be used.

REMOVAL & INSTALLATION

1. Disconnect the negative battery cable.
2. Raise and support the vehicle safely on jackstands.
3. Drain the engine oil.
4. Remove the oil filter.
5. Remove the dual converter Y-pipe assembly.
6. Remove the starter motor.
7. Remove the engine rear plate/converter housing cover.
8. Remove the retaining bolts and remove the oil pan.

To install:

9. Clean the gasket mating surfaces thoroughly.
10. Trial fit the oil pan to the cylinder block. Ensure that enough clearance has been provided to allow the oil pan to be installed without sealant being scraped off when pan is positioned under the engine.
11. Apply a bead of silicone sealer to the oil pan flange. Also apply a bead of sealer to the front cover/cylinder block joint and fill the grooves on both sides of the rear main seal cap.

➡**When using silicone rubber sealer, assembly must occur within 15 minutes after sealer application. After this time, the sealer may start to harden and its sealing effectiveness may be reduced.**

12. Install the oil pan and secure to the block with the attaching screws. Tighten the screws to 80–106 inch lbs. (9–12 Nm). Torque the oil pan-to-transaxle bolts to 33 ft. lbs. (45 Nm).
13. Install a new oil filter.
14. Install the engine rear plate/converter housing cover.
15. Install the starter motor.
16. Install the Y-pipe converter assembly.
17. Lower the vehicle.
18. Fill the engine with the proper type and amount of clean oil.
19. Connect the negative battery cable.
20. Start the engine and check for leaks.

Oil Pump

REMOVAL & INSTALLATION

➡**The oil pump, oil pressure relief valve and drive intermediate shaft are contained in the front cover assembly.**

1. Disconnect the negative battery cable.
2. If necessary for access, remove the oil filter.
3. Remove the oil pump and filter body-to-engine front cover retaining bolts, then remove the oil pump and filter body from the engine front cover.
4. Inspect the oil pump body seal, oil pump and filter body, and engine front cover for distortion. Replace damaged components as necessary.

To install:

5. Position the oil pump and filter body on the engine front cover, then install the retaining bolts.
6. Tighten the 4 large engine front cover retaining bolts to 17–23 ft. lbs. (23–31 Nm), then tighten the remaining retaining bolts to 71–97 inch lbs. (8–11 Nm).
7. If removed, install the oil filter.
8. Connect the negative battery cable.

➡**Check for proper engine oil pressure immediately after starting the engine. If engine oil pressure is not within specification a few seconds after starting the engine, stop the engine and determine the reason for the low oil pressure condition. Running an engine with low oil pressure may result in serious engine damage.**

9. Start the engine and check for leaks.

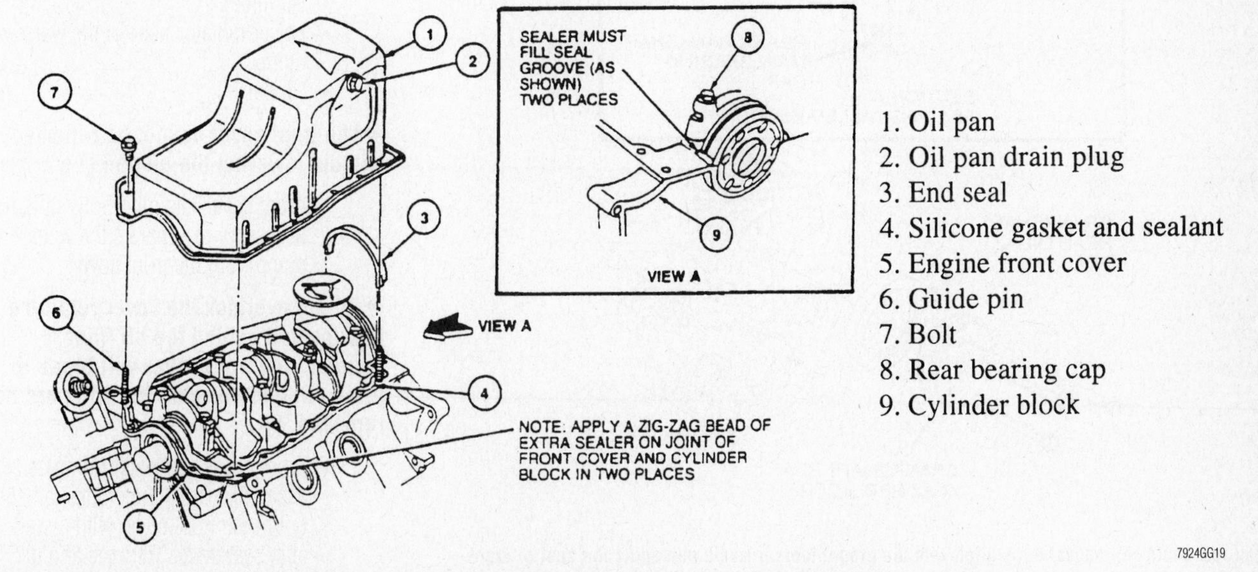

Exploded view of the oil pan and related components. Apply silicone gasket sealant in the places shown—3.8L engine

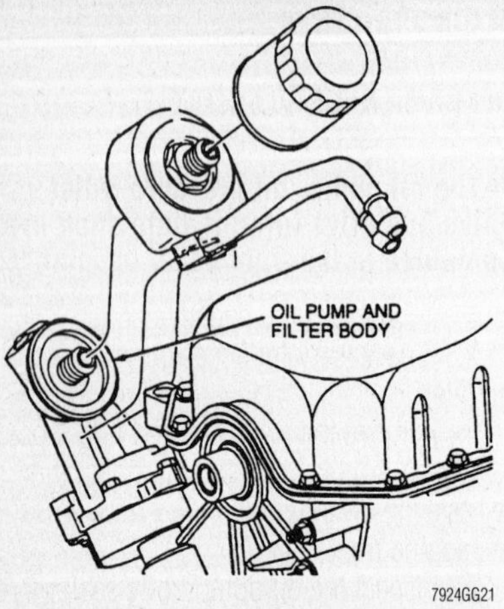

The oil pump assembly is mounted to the side of the 3.8L engine

Rear Main Seal

REMOVAL & INSTALLATION

1. Disconnect the negative battery cable.
2. Raise and support the vehicle safely on jackstands.
3. Remove the transaxle.
4. Remove the flywheel and the rear cover plate, if necessary.
5. Using a sharp awl, punch 1 hole into the crankshaft rear oil seal metal surface between the seal lip and the cylinder block.

✲✲ WARNING

Use caution when working near the crankshaft sealing surface. If the surface becomes damaged, an oil leak may occur.

6. Screw in the threaded end of a crankshaft rear seal replacer tool, then use the tool to remove the seal.

To install:

7. Inspect the crankshaft seal area for any damage that may cause the seal to leak. If damage is evident, service or replace the crankshaft as necessary.
8. Coat the crankshaft seal area and the seal lip with engine oil.
9. Using a crankshaft seal replacer tool, install the seal. Tighten the bolts of the seal installer tool evenly so the seal is straight and seats without misalignment.
10. Install the flywheel.
11. Install the rear cover plate, if necessary.
12. Install the transaxle, lower the vehicle and connect the battery.

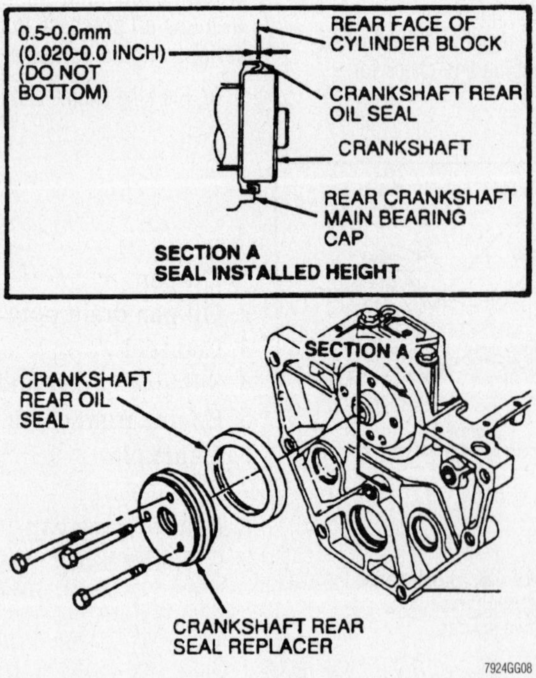

The rear main seal must be installed with the proper tools to avoid damaging the seal or crankshaft

Timing Chain, Sprockets, Front Cover and Seal

REMOVAL & INSTALLATION

1. Before servicing the vehicle, refer to the precautions in the beginning of this section.
2. Remove or disconnect the following:
   - Negative battery cable
   - Coolant
   - Air cleaner assembly and air intake duct
   - Fan/clutch assembly and shroud
   - Accessory drive belt idlers, drive belts and the water pump pulley
   - Power steering pump bracket retaining bolts. Leaving the hoses connected, place the pump/bracket assembly aside in a position to prevent fluid from leaking out.
   - Compressor front support bracket but leave the compressor in place
   - Coolant bypass hose and heater hose at the water pump
   - Upper radiator hose at the thermostat housing
   - Coil wire from the distributor cap
   - Cap with the secondary wires attached
   - Distributor hold-down clamp and lift the distributor out of the front cover
   - Crankshaft damper and pulley

➡If the crankshaft pulley and vibration damper have to be separated, mark the damper and pulley so they may be reassembled in the same relative position. This is important as the damper and pulley are initially balanced as a unit. If the crankshaft damper is being replaced, check if the original damper has balance pins installed. If so, new balance pins must be installed on the new damper in the same position as the original damper. The crankshaft pulley, new or original, must also be installed in the same relative position as originally installed.

   - Oil filter
   - Lower radiator hose at the water pump
   - Oil pan

➡The front cover cannot be removed without lowering the oil pan.

   - Front cover retaining bolts. It is not necessary to separate the water pump from the front cover.

➡Do not overlook the cover retaining bolt located behind the oil filter adapter. The front cover will break if pried on, and all retaining bolts are not removed.

   - Front cover and water pump as an assembly. Drive the crankshaft seal out of the front cover with a suitable seal driver. Remove and discard the cover gasket.

➡The front cover contains the oil pump, water pump and crankshaft seal. If a new front cover is to be installed, remove the water pump and oil pump from the old front cover and install them on the new cover along with a new crankshaft seal.

- Camshaft bolt and washer from the end of the camshaft
- Distributor drive gear, camshaft sprocket, crankshaft sprocket and timing chain

➡If the crankshaft sprocket is difficult to remove, pry the sprocket off the shaft using a pair of large prybars positioned on both sides of the sprocket.

To install:

3. Clean all gasket mating surfaces. If reusing the front cover, replace the front cover oil seal.
4. If removed, install the timing chain vibration damper. Tighten the mounting bolts to 71–123 inch lbs. (8–14 Nm).
5. Rotate the crankshaft to position the No. 1 piston at TDC and the crankshaft keyway at the 12 o' clock position.
6. Lubricate the timing chain with engine oil.
7. Install or connect the following:
   - Camshaft sprocket, crankshaft sprocket and timing chain. Be sure the timing marks align.
   - Distributor drive gear. Install the bolt and washer assembly on the end of the camshaft and tighten to 30–37 ft. lbs. (40–50 Nm).
   - New crankshaft seal in the front cover and lubricate the seal lip with engine oil
   - New gasket on the cylinder block and install the front cover using dowels for proper alignment. Install the front cover retaining bolts and tighten to 15–22 ft. lbs. (20–30 Nm).
   - Oil pan
   - Lower radiator hose
   - Oil filter
8. Coat the crankshaft damper sealing surface with clean engine oil. Apply a small amount of silicone sealer to the crankshaft keyway.
9. Position the crankshaft pulley key in the crankshaft keyway and install the damper, using a suitable installation tool.
10. Install or connect the following:
    - Damper washer and retaining bolt and tighten to 103–132 ft. lbs. (140–180 Nm).
    - Crankshaft pulley and tighten the retaining bolts to 20–28 ft. lbs. (26–38 Nm).
    - Coolant bypass hose
    - Distributor with the rotor pointing at the No. 1 distributor cap tower
    - Distributor cap and coil wire
    - Upper radiator hose at the thermostat housing
    - Heater hose
    - Compressor and mounting brackets. Tighten the retaining bolts to 30–45 ft. lbs. (41–61 Nm).
    - Power steering pump and mounting bracket. Tighten the retaining bolts to 30–45 ft. lbs. (41–61 Nm).
    - Water pump pulley. Position the accessory drive belts over the pulleys.
    - Fan/clutch assembly and fan shroud. Cross-tighten the fan/clutch assembly retaining bolts to 12–18 ft. lbs. (16–24 Nm).
11. Fill the crankcase with the proper type and quantity of engine oil. Fill and bleed the cooling system. Connect the negative battery cable.
12. Start the engine and check for leaks. Check the ignition timing and curb idle speed and adjust, as necessary.

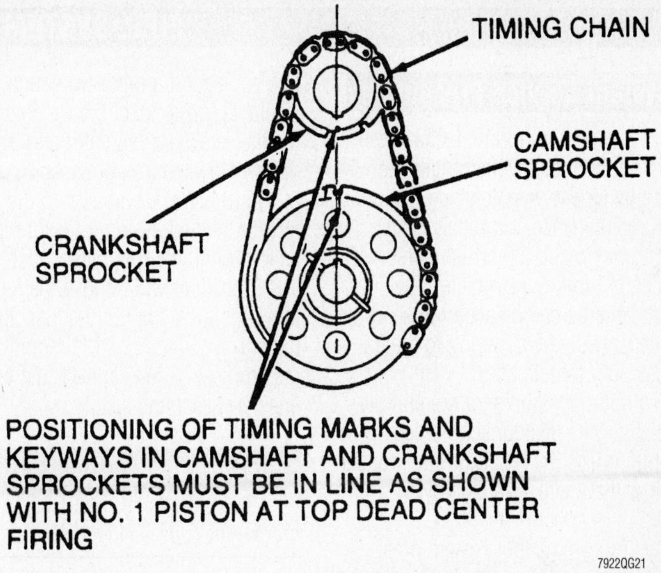
TIMING CHAIN
CAMSHAFT SPROCKET
CRANKSHAFT SPROCKET
POSITIONING OF TIMING MARKS AND KEYWAYS IN CAMSHAFT AND CRANKSHAFT SPROCKETS MUST BE IN LINE AS SHOWN WITH NO. 1 PISTON AT TOP DEAD CENTER FIRING
7922QG21

The timing marks should be facing each other, when the timing chain is installed correctly—3.8L engines

Piston and Ring

POSITIONING

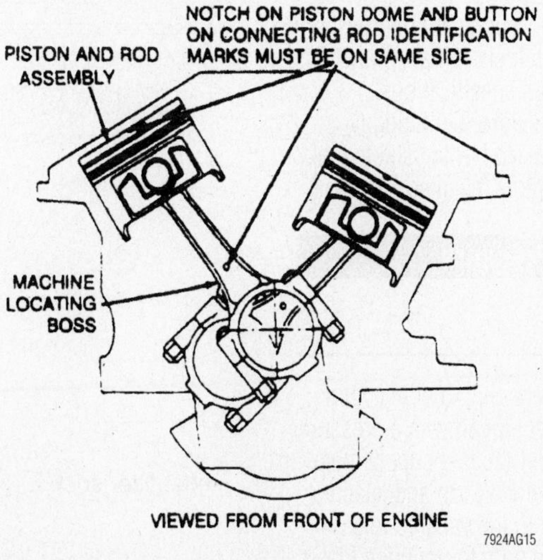
NOTCH ON PISTON DOME AND BUTTON ON CONNECTING ROD IDENTIFICATION MARKS MUST BE ON SAME SIDE
PISTON AND ROD ASSEMBLY
MACHINE LOCATING BOSS
VIEWED FROM FRONT OF ENGINE
7924AG15

3.8L engines—piston and connecting rod assembly positioning

FUEL SYSTEM

Fuel System Service Precautions

Safety is the most important factor when performing not only fuel system maintenance, but any type of maintenance. Failure to conduct maintenance and repairs in a safe manner may result in serious personal injury or death. Work on a vehicle's fuel system components can be accomplished safely and effectively by adhering to the following rules and guidelines.

- To avoid the possibility of fire and personal injury, always disconnect the negative battery cable unless the repair or test procedure requires that battery voltage by applied.
- Always relieve the fuel system pressure prior to disconnecting any fuel system component (injector, fuel rail, pressure regulator, etc.) fitting or fuel line connection. Exercise extreme caution whenever relieving fuel system pressure, to avoid exposing skin, face and eyes to fuel spray. Please be advised that fuel under pressure may penetrate the skin or any part of the body that it contacts.
- Always place a shop towel or rag around the fitting or connection prior to loosening to absorb any excess fuel due to spillage. Ensure that all fuel spillage is quickly removed from engine surfaces. Ensure that all fuel-soaked cloths or towels are deposited into a flame-proof waste container with a lid.
- Always keep a dry chemical (Class B) fire extinguisher near the work area.
- Do not allow fuel spray or fuel vapors to come into contact with a light bulb, spark or open flame.
- Always use a second wrench when loosening or tightening fuel line connections fittings. This will prevent unnecessary stress and torsion to fuel piping. Always follow the proper torque specifications.
- Always replace worn fuel fitting O-rings with new ones. Do not substitute fuel hose where rigid pipe is installed.

Fuel System Pressure

RELIEVING

All Sequential Electronic Fuel Injection (SEFI) engines are equipped with a pressure relief valve located on the fuel supply manifold. Remove the fuel tank cap and attach a fuel pressure gauge to the valve to release the fuel pressure. Be sure to drain the fuel into a suitable container and to avoid gasoline spillage. If a pressure gauge is not available, disconnect the vacuum hose from the fuel pressure regulator and attach a hand-held vacuum pump. Apply about 25 in. Hg (84 kPa) of vacuum to the regulator to vent the fuel system pressure into the fuel tank through the fuel return hose. Note that this procedure will remove the fuel pressure from the lines, but not the fuel. Take precautions to avoid the risk of fire and use clean rags to soak up any spilled fuel when the lines are disconnected.

Fuel Filter

REMOVAL & INSTALLATION

Although the manufacturer does not specify a replacement interval for fuel filters, we at Chilton feel the fuel filter should be replaced every 30,000 miles (48,000 km) under normal conditions or 15,000 miles (24,000 km) under severe conditions. Those intervals are industry standards.

1. Relieve the fuel system pressure.
2. Raise and support the vehicle safely on jackstands.
3. Place a rag under the fuel filter to catch any residual fuel that may leak out when the filter is removed.
4. Remove the push-connect fittings at both ends of the fuel filter.
5. Install retainer clips in each fitting.
6. Note the flow arrow direction for installation reference.
7. Remove the fuel filter by pulling it from the bracket.

To install:

8. Install the fuel filter in its bracket, ensuring proper direction of flow as noted earlier.
9. Install push-connect fittings at both ends of the fuel filter.
10. Start the engine and check the filter connections for leaks by running the tip of your finger around each connection.
11. Turn the engine off and lower the vehicle.

Fuel Pump

REMOVAL & INSTALLATION

➡To gain access to the fuel pump, it is necessary to remove the fuel tank.

1. Depressurize the fuel system and remove the fuel tank from the vehicle.
2. Remove any dirt that has accumulated around the fuel pump module attaching flange to prevent it from entering the tank during service.
3. Turn the fuel pump module locking ring counterclockwise using a locking ring removal tool or a brass drift, and remove the locking ring.
4. Remove the fuel pump module.
5. Remove the seal gasket and discard it.

To install:

6. Put a light coating of grease on a new seal ring to hold it in place during

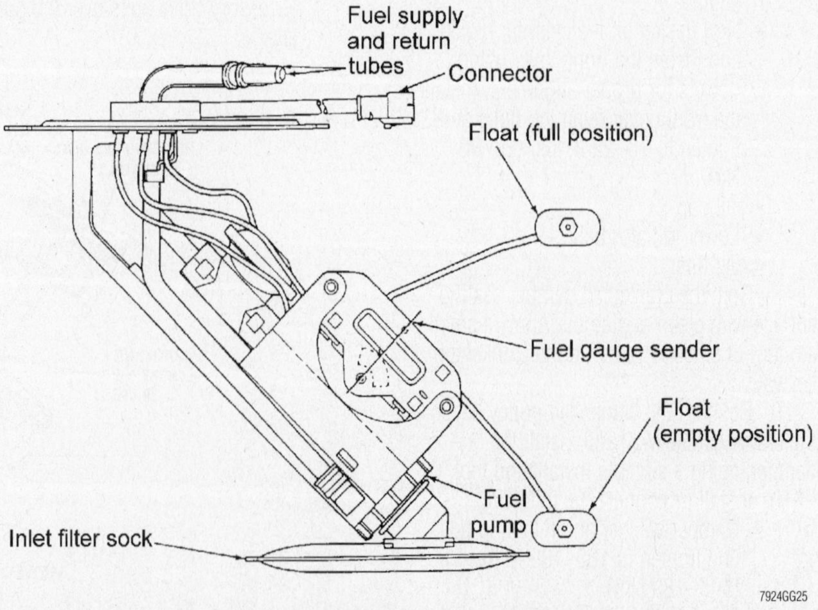

In-tank electric fuel pump and related components

assembly. Install it in the fuel tank ring groove.

7. Insert the fuel pump module into the fuel tank, then secure it in place with the locking ring. Tighten the ring until secure.

8. Install the tank in the vehicle.

9. Install a minimum of 10 gallons (38L) of fuel and check for leaks.

10. Install a pressure gauge on the throttle body valve and turn the ignition **ON** for 3 seconds. Turn the key **OFF**, then repeat the key cycle 5 to 10 times until the pressure gauge shows at least 30 psi. (207 kPa).

11. Check for fuel leaks at the fittings.

12. Remove the pressure gauge.

13. Start the engine and check for fuel leaks.

Fuel Injectors

REMOVAL & INSTALLATION

1. Remove the upper intake manifold.
2. Remove the fuel injection supply manifold.
3. Carefully remove the fuel charging wiring harness connectors from the fuel injectors.
4. Pull fuel injector body up while gently rocking fuel injector from side to side.

To install:

5. Inspect fuel injector O-rings for signs of deterioration and replace as required.

➡Never use silicone grease on fuel injectors.

6. Lubricate O-rings with clean engine oil and install fuel injectors using a slight twisting motion.
7. Install fuel injection supply manifold.
8. Install the fuel charging wiring connectors.
9. Install the upper intake manifold.

DRIVE TRAIN

Transaxle Assembly

REMOVAL & INSTALLATION

1. Remove or disconnect the following:
   - Battery and battery tray
   - Hood and cowl vent
   - Air cleaner assembly
   - All transaxle electrical harnesses
   - Transaxle shift cable from the lever by unsnapping the shift cable end from the lever ball stud
   - Transaxle fluid cooler lines

➡Leave the 2 lower engine-to-transaxle bolts in place to hold the transaxle secure against the engine block until a suitable jack can be placed under the transaxle to support it during removal.

   - Upper transaxle-to-engine bolts
   - Engine electrical harness bracket
   - Battery cable bracket

➡Install engine lifting eyes to support the engine during transaxle removal.

2. Install an engine support kit and suitably support the engine.
3. Raise and support the vehicle.
4. Remove or disconnect the following:
   - Transaxle fluid
   - Front wheels
   - Halfshafts
   - Bolts retaining the rear engine support to the transaxle
   - Front subframe
   - Speedometer cable from the vehicle speed sensor
   - Starter
   - Transaxle housing cover
   - The 4 flexplate-to-converter nuts
5. Support the transaxle with a suitable jack and remove the remaining transaxle-to-engine bolts.
6. Remove the engine bracket-to-transaxle bolts.
7. Separate the transaxle from the engine block by carefully moving the transaxle rearward until enough clearance exists to remove the transaxle from the engine compartment.
8. Slowly lower the transaxle from the engine compartment.

To install:

9. Place the transaxle on a suitable jack and position it in place.
10. Install or connect the following:
    - Engine bracket-to-transaxle bolts, and tighten the bolts to 39–53 ft. lbs. (53–72 Nm)
    - Lower transaxle-to-engine bolts, and tighten the bolts to 39–53 ft. lbs. (53–72 Nm)
    - Flexplate-to-torque converter bolts, and tighten them to 20–34 ft. lbs. (27–46 Nm)
    - Transaxle housing cover, and tighten the bolts to 80–106 inch lbs. (9–12 Nm)
    - Starter motor, and connect the electrical harness
    - Speedometer cable
    - Front subframe
    - The 4 bolts retaining the rear engine support, and tighten them to 39–53 ft. lbs. (53–72 Nm)
    - Both halfshafts
    - Front wheels and lower the vehicle
11. Remove the engine support kit.
12. Install or connect the following:
    - Transaxle electrical harnesses
    - Upper transaxle-to-engine bolts and tighten to 39–53 ft. lbs. (53–72 Nm)
    - Fluid cooler-to-transaxle lines
    - Transaxle shift cable to the manual lever ball stud
    - Air cleaner assembly
    - Cowl vent and hood
    - Battery tray and battery
13. Fill the transaxle with proper amount of Mercon® fluid.
14. Connect the positive, then the negative battery cable.

Halfshafts

REMOVAL & INSTALLATION

➡Do not begin this removal procedure unless a new wheel hub retainer nut, a new retainer circlip and a new lower ball joint-to-front wheel knuckle retaining bolt and nut are available. Once removed, these parts must not be reused during assembly. Their torque holding ability, or retention capability, is diminished during removal.

1. Remove or disconnect the following:
   - Front wheels
   - Axle hub nut. Discard the nut.
   - Ball joint-to-front wheel knuckle retaining nut. Drive the bolt out of the front wheel knuckle using a punch and hammer.
   - Front brake anti-lock sensor and position it out of the way
   - Ball joint from the knuckle

⁂ WARNING

Use care to prevent damage to the CV-joint boot.

   - Stabilizer bar link at the front stabilizer bar

➡Make sure the CV-joint puller does not contact the transaxle shaft speed sensor. Damage to the sensor will result.

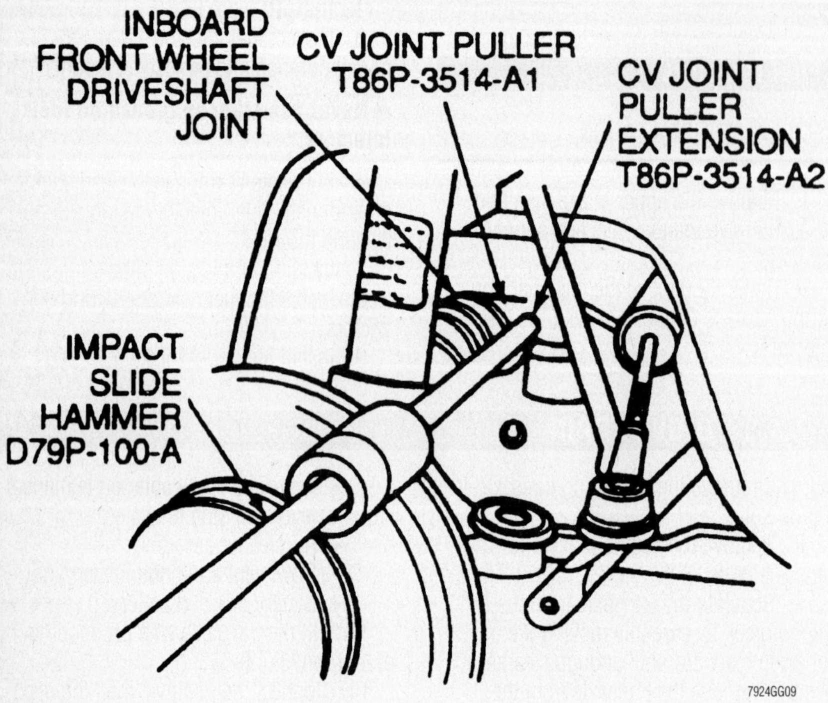

Remove the halfshaft from the transaxle using a CV-joint puller, extension and impact slide hammer

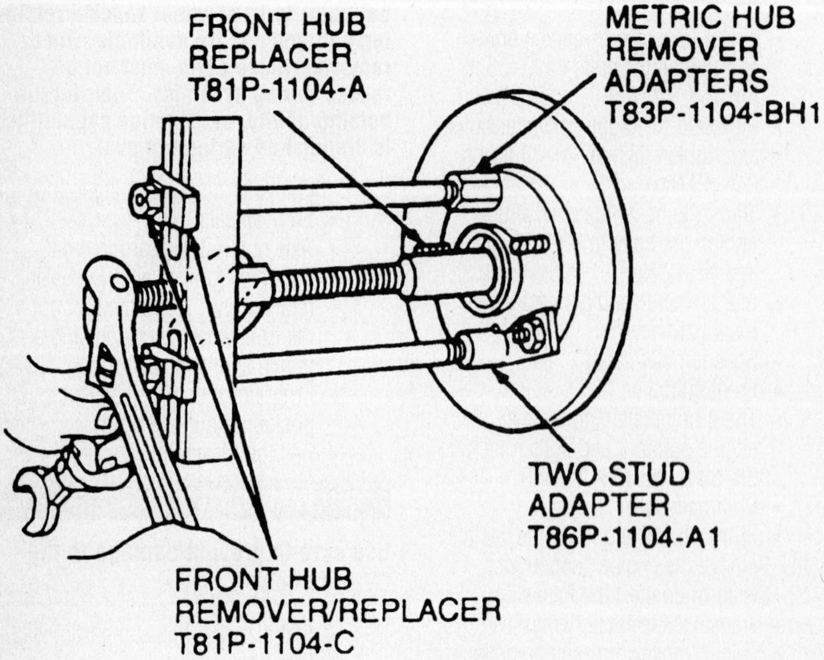

The front hub adapter must be used to remove the hub without damage

2. Install a CV-joint puller between the inboard CV-joint and the transaxle case.
3. Install a CV-joint extension into the puller and hand-tighten.
4. Using a slide hammer, remove the driveshaft from the transaxle.

WARNING

Do not allow the halfshaft to hang unsupported. Damage to the CV-joint may result. Do not wrap wire around the joint boot. Damage to the boot may result.

5. Support the end of the halfshaft assembly by suspending it from the chassis using a length of wire.

WARNING

Never use a hammer to separate the outboard CV-joint from the wheel hub. Damage to the outboard CV joint threads and internal components may result.

6. Separate the outboard CV-joint from the wheel hub using a front hub remover/replacer. Make sure the hub remover adapter is fully threaded onto the hub stud.

WARNING

Do not move vehicle without the outboard CV-joint properly installed as damage to the bearing may occur.

7. Remove the halfshaft assembly from the vehicle.

To install:

WARNING

Do not reuse the retainer circlip. A new circlip must be installed each time the inboard CV-joint stub shaft is installed into the transaxle differential.

8. Install a new retainer circlip on the inboard CV-joint stub shaft by starting one end in the groove and working the retainer circlip over the inboard shaft housing end and into the groove. The will avoid over-expanding the circlip.

➡A non-metallic mallet may be used to aid in seating the retainer circlip into the differential side gear groove. If a mallet is necessary, tap only on the outboard CV-joint shaft.

9. Carefully align the splines of the inboard CV-joint stub shaft housing with

the splines in the differential. Exerting some force, push the inboard CV-joint stub shaft housing into the differential until the retainer circlip is felt to seat in the differential side gear. Use care to prevent damage to the inboard CV-joint stub shaft and transaxle seal.

10. Carefully align the splines of the outboard CV-joint with the splines in the wheel hub and push the shaft into the wheel hub as far as possible.

11. Temporarily fasten the front disc brake rotor to the wheel hub with washers and 2 lug nuts. Insert a steel rod into the front disc brake rotor and rotate clockwise to contact the front wheel knuckle to prevent the front disc brake rotor from turning when the nut is tightened.

➡A new front axle wheel hub retaining nut must be installed.

12. Manually thread the front axle wheel hub retaining nut onto the outboard CV-joint stub shaft housing as far as possible.

➡A new bolt and nut must be used to connect the front suspension arm to the knuckle.

13. Connect the front suspension lower arm to the front wheel knuckle. Tighten the nut and bolt to 40–55 ft. lbs. (54–75 Nm).

14. Install the front brake anti-lock sensor.

15. Connect the front stabilizer bark link and tighten to 35–45 ft. lbs. (47–65 Nm).

➡Do not use power or impact tools to tighten the hub nut.

16. Tighten front axle wheel hub retaining nut to 157–212 ft. lbs. (213–288 Nm).

17. Install the front wheels and lower the vehicle.

18. Fill the transaxle to the proper level with Mercon® automatic transmission fluid.

CV-Joint

OVERHAUL

Inboard Joint

1. Remove the clamps.

2. Separate the front wheel driveshaft joint boot from the inboard CV joint housing.

3. Check the CV joint grease for contamination by rubbing it between two fingers. Any gritty feeling indicates contamination.

➡Other than the front wheel driveshaft joint boot, the interconnecting shaft and inboard CV joint housing are not repairable. Install a new assembly if worn/damaged.

4. To install a new inboard CV joint housing assembly front wheel driveshaft joint boot, remove the front wheel driveshaft joint and boot.

5. Slide the front wheel driveshaft joint boot off the interconnecting shaft.

6. Install the front wheel driveshaft joint boot.

7. Position the boot into the small boot groove.

8. Using the special tool, install the small clamp.

9. Position the special tool on the clamp ear, and tighten the tool through bolt until the tool closes completely.

10. Install the front wheel driveshaft joint boot and the joint.

11. Fill the inboard CV joint housing with 475 grams (16.75 ounces) of grease. Spread the remaining grease evenly inside the front wheel driveshaft joint boot. Use constant velocity joint grease, only.

➡Remove all excess grease from the CV joint external surface and the front wheel driveshaft joint boot sealing surface.

12. Seat the front wheel driveshaft joint boot in the inboard CV joint housing boot groove.

13. Set the halfshaft assembled length to specification. Halfshaft assembled length, left side: 612.6 mm (24.5 in); right side: 746.4 mm (29.85 in). Measure the entire assembly length. Push in or pull out on the inner joint as necessary to adjust the halfshaft assembled length to specification.

14. Hold the inner joint to prevent the assembled length from changing, and insert a small flat blade screwdriver between the boot and the joint to equalize the pressure.

➡Make sure the front wheel driveshaft joint boot seats in the groove.

15. Install the clamps as tight as possible by hand.

16. Using the special tool, tighten the clamp.

17. Position the special tool on the clamp ear, and tighten the tool through bolt until the tool closes completely.

18. Move the CV joints through their full range of travel at various angles. The joints must flex, extend and compress smoothly.

Outboard Joint

⁂ CAUTION

Do not allow the vise jaws to contact the front wheel driveshaft joint boot or the clamp. Use a vise equipped with soft jaws or wood blocks.

1. Clamp the halfshaft assembly in a vise.

2. Remove the clamps.

3. Separate the front wheel driveshaft joint boot from the front wheel driveshaft joint.

4. Check the CV joint grease for contamination by rubbing it between two fingers. Any gritty feeling indicates contamination.

5. If the grease is contaminated/additional disassembly is necessary, proceed as follows.

⁂ CAUTION

Do not allow the front wheel driveshaft joint to fall.

6. Using a brass drift and a hammer, separate the front wheel driveshaft joint from the interconnecting shaft.

7. Give a sharp tap to the inner bearing race to dislodge the circlip.

⁂ CAUTION

Do not reuse the circlip.

➡Install a new stop ring, located just below the circlip, only if the stop ring is damaged/worn, or additional halfshaft disassembly is necessary.

8. Remove and discard the circlip. Remove the stop ring if damaged/worn, or additional halfshaft disassembly is necessary.

9. If necessary, remove the boot from the shaft.

⁂ CAUTION

Do not remove the front brake anti-lock sensor indicator unless it is necessary to install a new one.

10. Position the special tool on a press bed, and place the front wheel driveshaft joint on the special tool.

11. Press the damaged front brake anti-lock sensor indicator off of the front wheel driveshaft joint and discard it.

CAUTION

Do not damage the front brake anti-lock sensor indicator. Tooth damage will affect brake performance.

12. If removed, position the special tool on a press bed, and place the new front brake anti-lock sensor indicator on the special tool.
13. Position the front wheel driveshaft joint in the special tool.
14. Place a steel plate across the front wheel driveshaft joint back face.

➡Installation is complete when the front wheel driveshaft joint bottoms out in the special tool.

15. Press the front brake anti-lock sensor indicator on the front wheel driveshaft joint.

➡Make sure the front wheel driveshaft joint boot seats in the shaft boot groove.

16. If removed, install the clamp and the front wheel driveshaft joint boot.
17. Install the clamps as tight as possible by hand.
18. Using the special tool, tighten the clamp.
19. Position the special tool on the clamp ear, and tighten the tool through bolt until the tool closes completely.

➡Make sure the stop ring seats in the groove.

20. If removed, install the stop ring.

CAUTION

Do not reuse the circlip.

CAUTION

Do not over-expand or twist the circlip during installation.

21. Install a new circlip.
22. Start one end in the groove and work the circlip over the shaft and into the groove. This will avoid over-expanding the circlip.
23. Fill the front wheel driveshaft joint with 180 grams (6.3 ounces) of grease. Spread the remaining grease evenly inside the front wheel driveshaft joint boot. Use constant velocity joint grease only!

➡The front wheel driveshaft joint has seated when the circlip locks in the groove cut in the inner race.

24. Using a non-metallic hammer, tap the front wheel driveshaft joint onto the interconnecting shaft. Make sure the front wheel driveshaft joint has locked on the interconnecting shaft by attempting to pull the joint off the shaft.

➡Remove all excess grease from the front wheel driveshaft joint external surface and the front wheel driveshaft joint boot mating surface.

25. Seat the front wheel driveshaft joint boot in the front wheel driveshaft joint boot groove.
26. Install the clamp as tight as possible by hand.
27. Using the special tool, tighten the clamp.
28. Position the special tool on the clamp ear, and tighten the tool through bolt until the tool closes completely.

STEERING AND SUSPENSION

Air Bag (Supplemental Restraint) System

The Supplemental Restraint System (SRS) is designed to work in conjunction with the standard 3-point safety belts to reduce injury in a head-on collision.

CAUTION

The SRS can actually cause physical injury or death if the safety belts are not used, or if the manufacturer's warnings are not followed. The manufacturer's warnings can be found in your owner's manual, or, in some cases, on your sun visor.

The SRS is comprised of the following components:

- Driver's side air bag module
- Passenger's side air bag module
- Right-hand and left-hand primary crash front air bag sensors
- Air bag diagnostic monitor computer
- Electrical wiring

The SRS primary crash front air bag sensors are hard-wired to the air bag modules and determine when the air bags are deployed. During a frontal collision, the sensors quickly inflate the 2 air bags to reduce injury by cushioning the driver and front passenger from striking the dashboard, windshield, steering wheel and any other hard surfaces. The air bag inflates so quickly (in a fraction of a second) that in most cases it is fully inflated before you actually start to move during a collision.

Since the SRS is a complicated and essentially important system, its components are constantly being tested by a diagnostic computer. The computer illuminates the air bag indicator light on the instrument cluster for approximately 6 seconds when the ignition switch is turned to the **RUN** position when the SRS is functioning properly. After being illuminated for the 6 seconds, the indicator light should then turn off.

If the air bag light does not illuminate at all, stays on continuously, or flashes at any time, a problem has been detected by the diagnostic computer.

CAUTION

If at any time the air bag light indicates that the computer has noted a problem, immediately diagnose the problem. A faulty SRS can cause severe physical injury or death.

SERVICE PRECAUTIONS

Whenever working around, or on, the air bag supplemental restraint system, ALWAYS adhere to the following warnings and cautions.

- Always wear safety glasses when servicing an air bag vehicle and when handling an air bag module.
- Carry a live air bag module with the bag and trim cover facing away from your body, so that an accidental deployment of the air bag will have a small chance of personal injury.
- Place an air bag module on a table or other flat surface with the bag and trim cover pointing up.
- Wear gloves, a dust mask and safety glasses whenever handling a deployed air bag module. The air bag surface may contain traces of sodium hydroxide, a byproduct of the gas that inflates the air bag and which can cause skin irritation.
- Ensure to wash your hands with mild soap and water after handling a deployed air bag.
- All air bag modules with discolored or damaged cover trim must be replaced, not repainted.
- All component replacement and wiring service must be made with the negative and positive battery cables disconnected from the battery for a minimum of 1 minute prior to attempting service or replacement.
- NEVER probe the air bag electrical terminals. Doing so could result in air bag deployment, which can cause serious physical injury.

• If the vehicle is involved in a fender-bender that results in a damaged front bumper or grille, the air bag sensors should be inspected to ensure that they were not damaged.

• If at any time, the air bag light indicates that the computer has noted a problem, immediately diagnose the problem. A faulty SRS can cause severe physical injury or death.

DISARMING THE SYSTEM

1. Disconnect the negative battery cable from the battery.
2. Disconnect the positive battery cable from the battery.
3. Wait 1 minute. This time is required for the back-up power supply in the air bag diagnostic monitor to completely drain. The system is now disarmed.

ARMING THE SYSTEM

1. Connect the positive battery cable.
2. Connect the negative battery cable.
3. Stand outside the vehicle and carefully turn the ignition to the **RUN** position. Be sure that no part of your body is in front of the air bag module on the steering wheel, to prevent injury in case of an accidental air bag deployment.
4. Ensure the air bag indicator light turns off after approximately 6 seconds. If the light does not illuminate at all, does not turn off, or starts to flash, diagnose the problem. If the light does turn off after 6 seconds and does not flash, the SRS is working properly.

Rack and Pinion Steering Gear

REMOVAL & INSTALLATION

1. Remove or disconnect the following:
   - Front wheels
   - Tie rod end cotter pins and castle nuts
   - Tie rod ends from the knuckles
   - Front stabilizer bar
2. Position the dash opening weather seal for the steering column out of the way.
3. Remove or disconnect the following:
   - Pinch bolt retaining the steering column intermediate shaft coupling
   - Steering gear retaining nuts/bolts
   - Rear subframe bolts

➡**Use wire to support exhaust components unless you are removing them completely.**

4. Support the exhaust system flex tube and remove the flex tube-to-dual converter Y-pipe attachment.

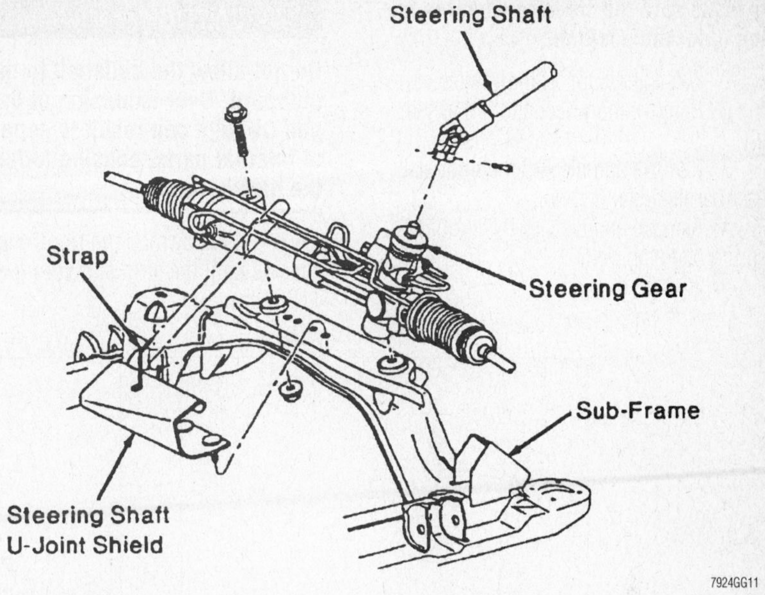

Exploded view of the rack and pinion steering gear mounting on the front subframe of the vehicle

5. Lower the vehicle slightly until the rear subframe separates from the body approximately 4 inches (10cm).
6. Remove the heat shield band and fold the heat shield down.
7. Rotate the rack and pinion assembly to clear the bolts from the front subframe, and pull toward the driver's side of the vehicle.
8. Place a drain pan under the vehicle and disconnect the power steering lines.
9. Remove the rack and pinion assembly through the driver's side of the vehicle.

To install:

10. Install new Teflon® O-rings on the power steering line fittings.
11. Place the rack and pinion retaining bolts in the gear housing.
12. Install or connect the following:
    - Rack and pinion assembly through the driver's side of the vehicle
    - Power steering lines on the rack and pinion assembly
    - Rack and pinion assembly on the subframe
    - Strap on the heat shield
    - Tie rod ends to the knuckles. Tighten the castle nuts and install the cotter pins.
    - Stabilizer bar
    - Rack and pinion assembly retaining bolts, and tighten them to 85–99 ft. lbs. (115–135 Nm).
13. Raise the vehicle until the subframe contacts the body.
14. Install or connect the following:
    - Rear subframe retaining bolts, and tighten them to 83–112 ft. lbs. (113–153 Nm).
    - Exhaust system flex tube-to-dual converter Y-pipe
    - Front wheels
15. Using a new pinch bolt, install the steering column intermediate shaft coupling on the rack input shaft. Tighten the pinch bolt to 25–33 ft. lbs. (34–46 Nm).
16. Position the steering column opening weather seal over the steering gear housing.
17. Lower the vehicle.
18. Fill the power steering oil reservoir.
19. Start the vehicle and check for leaks.
20. Check for proper wheel alignment and steering wheel position.

Struts

REMOVAL & INSTALLATION

✲✲ CAUTION

Suspension fasteners are critical parts because they affect performance of vital components and systems and their failure can result in major service expense. They must be exchanged with the same part number or an equivalent part if an exchange is necessary. Do not use an alternative part of lesser quality or substitute design. Torque values must be used as specified during reassembly to make sure these parts are correctly retained.

➡Make sure the steering wheel is in the unlocked position.

1. Raise and support the vehicle.
2. Remove the wheel and tire assembly.
3. Remove and discard the front stabilizer bar link retaining nut.
4. Remove and discard the strut and spring retaining bolt.

✲✲ CAUTION

Do not allow the halfshaft to move outboard. Over-extension of the tripod CV joint can result in separation of internal parts, causing failure of the halfshaft.

5. Push down on the front wheel knuckle until the strut and spring assembly is free of the front wheel knuckle. Support the front wheel knuckle to prevent the front axle shaft from moving outboard.
6. Partially lower the vehicle.
7. Remove and discard the three strut and spring mounting bracket retaining nuts.
8. Remove the strut and spring assembly.

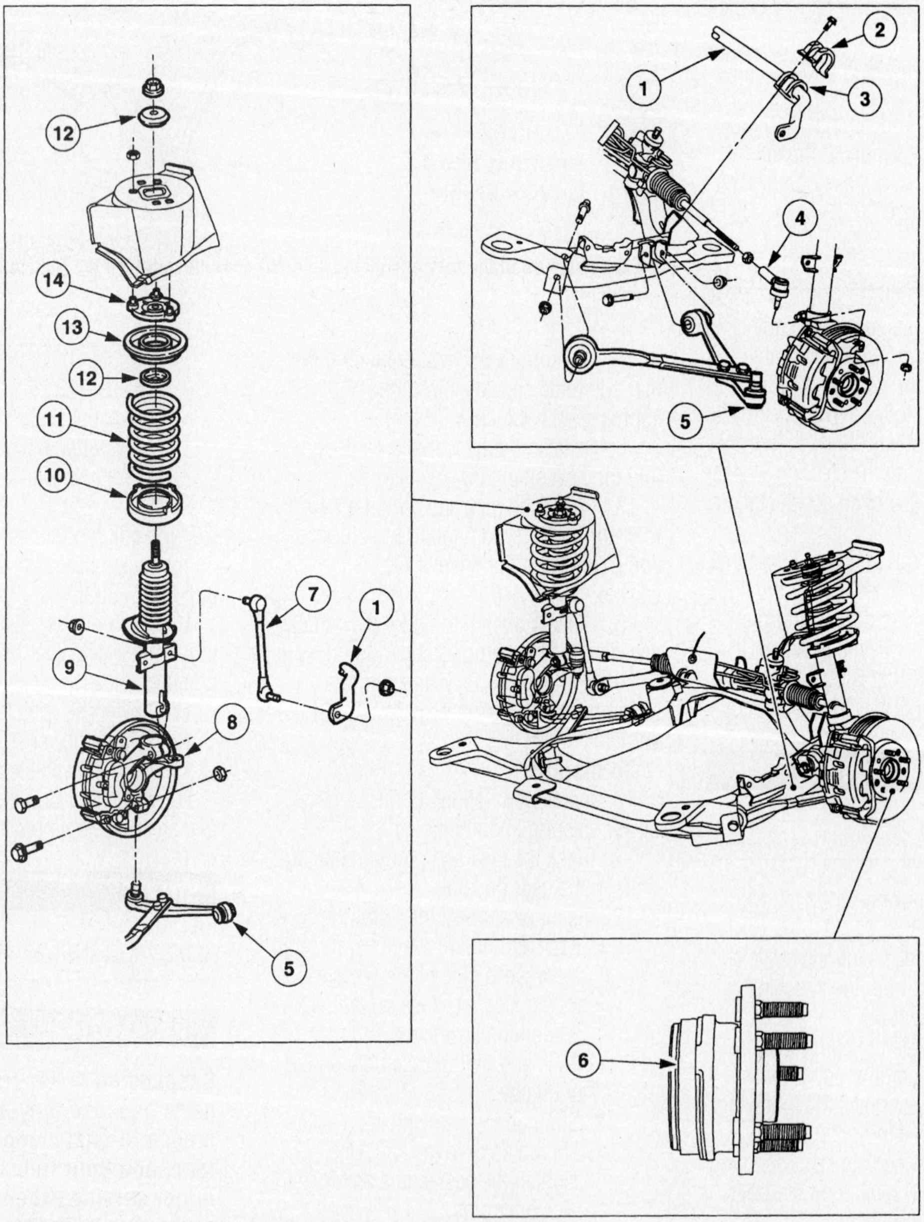

1 Front stabilizer bar
2 Front stabilizer bar bracket
3 Front stabilizer bar insulator
4 Tie-rod end
5 Front suspension lower arm
5 Front suspension lower arm
6 Front wheel hub (includes bearing)
7 Front stabilizer bar link
8 Front wheel knuckle
9 Front strut assy
10 Front spring insulator
11 Front coil spring
12 Front strut assy washer
13 Front suspension bearing and seal
14 Front strut assy mounting bracket

67197-WIND-G02

Front suspension components

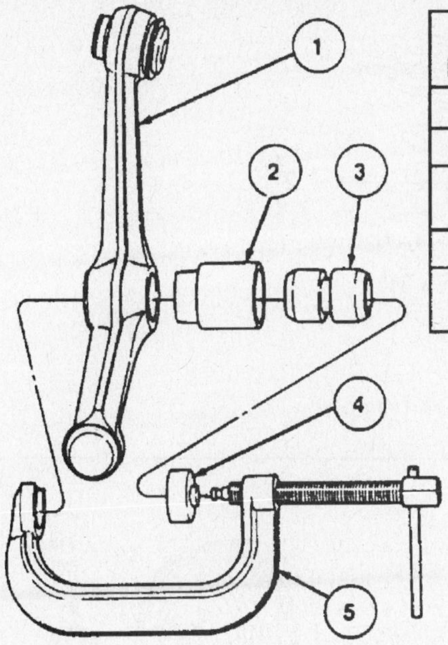

Item	Description
1	Front Suspension Lower Arm
2	Sleeve
3	Front Suspension Lower Arm Rear Strut Bushing
4	Bushing Driver
5	C-Frame and Clamp Assembly

Installing the lower arm bushing using a C-clamp and the proper adapters

➡Use only vegetable oil. Any mineral or petroleum based oil or brake fluid will deteriorate the rubber.

4. Using a bushing driver, press the bushing into the lower arm using a C-clamp as a press.
5. Install the lower arm on the vehicle.

Hub and Wheel Bearing

ADJUSTMENT

The wheel bearings on the Ford Windstar are not adjustable. If the bearings become loose or make noise they must be replaced as an assembly.

REMOVAL & INSTALLATION

Front

CAUTION

Suspension fasteners are critical parts because they affect performance of vital components and systems and their failure can result in major service expense. They must be exchanged with the same part number or an equivalent part if an exchange is necessary. Do not use an alternative part of lesser quality or substitute design. Torque values must be used as specified during reassembly to make sure these parts are correctly retained.

➡Make sure the steering wheel is in the unlocked position.

1. If equipped, remove the front wheel center cap.

CAUTION

A new wheel hub retainer must be installed, if it is loosened or removed. Failure to install a new retainer after removal can result in loss of the front wheel or damage to the wheel hub and bearing assembly.

CAUTION

The front axle wheel hub retainer must be tightened to specification immediately during installation. If the retainer is not tightened immediately, the nylon lock will set incorrectly, leading to incorrect torque readings and bearing failure. Any front wheel hub retainer that is not immediately tightened to specification or is loosened must be removed and a new retainer installed.

➡The Windstar uses a wheel hub and bearing assembly and is not repaired separately.

2. Remove and discard the wheel hub retainer.
3. Raise and support the vehicle.
4. Remove the wheel and tire assembly.
5. Remove the front brake disc.
6. Remove the ABS sensor retaining bolt, ABS sensor harness retaining bolt and the ABS sensor. Position out of the way.
7. Remove and discard the tie-rod end cotter pin and the castellated nut.
8. Using a puller, remove the tie-rod end.
9. Remove and discard the front stabilizer bar link nut and separate the front stabilizer bar link.
10. Remove and discard the ball joint pinch bolt and nut.
11. Using a pry bar between the front subframe and the front suspension lower arm, push down until the lower ball joint is free of the front wheel knuckle.

CAUTION

Do not allow the halfshaft to move outboard. Over-extension of the tripod CV joint could result in separation of internal parts, causing failure of the halfshaft.

12. Using a hub puller, remove the front driveshaft and joint from the wheel hub.

➡The wheel hub is a slip-fit design and should not require a puller to remove.

13. Remove and discard the three wheel hub retaining bolts and remove the wheel hub.

➡Apply a small patch of Loctite®242 or equivalent meeting Ford specification WSK-M2G351-A5 to the last five front wheel driveshaft joint threads prior to installing the wheel hub retainer nut.

14. Using new fasteners, follow the removal procedure in reverse order. Observe the following torques:
    - Hub retaining bolts: 85 ft. lbs. (115 Nm)
    - Ball joint pinch bolt: 46 ft. lbs. (63 Nm)
    - Stabilizer bar link nut: 66 ft. lbs. (90 Nm)
    - Tie rod end nut: 41 ft. lbs. (55 Nm)
    - Wheel hub retainer nut: 185 ft. lbs. (250 Nm)

Rear

1. Raise the vehicle on a hoist.
2. Remove the wheel and tire assembly.
3. Remove the brake drum.
4. Remove the hub cap grease seal.

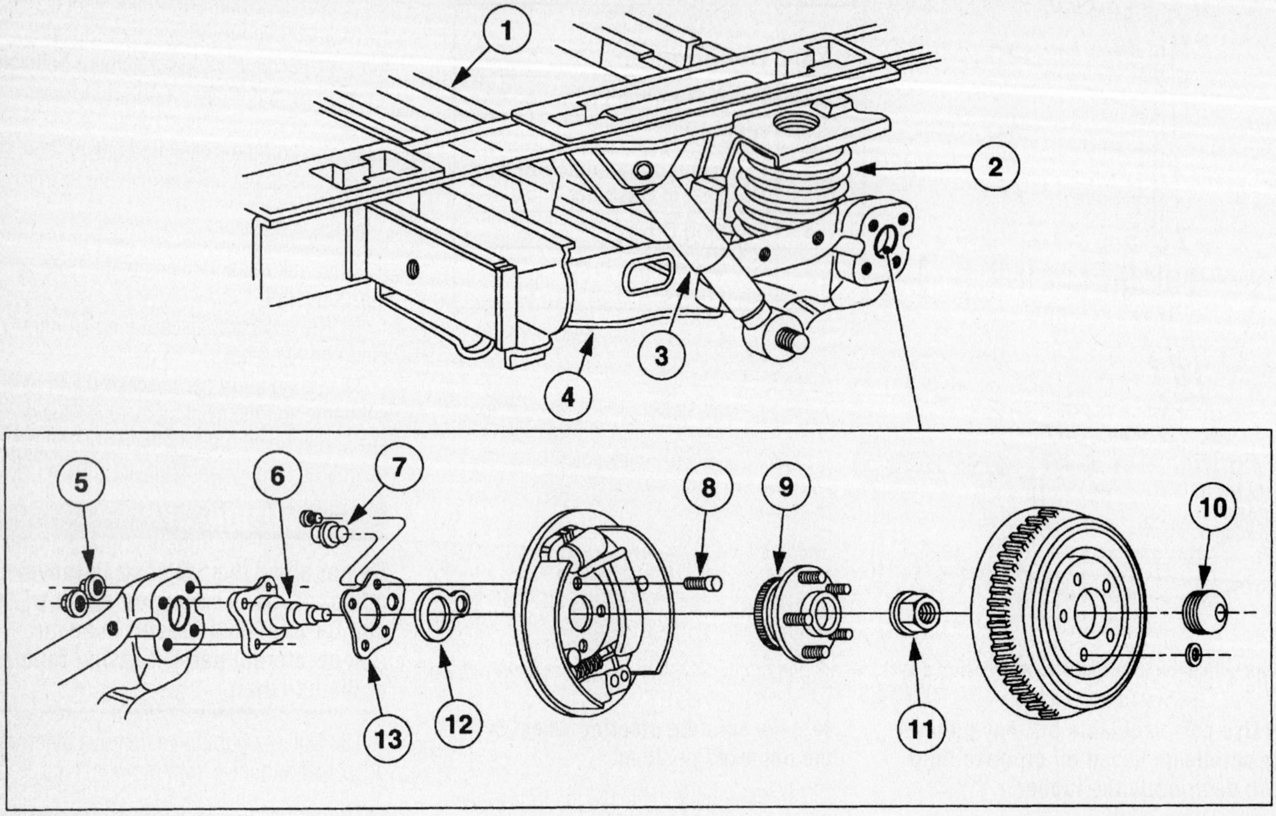

1 Rear axle assy
2 Rear spring
3 Shock absorber
4 Trailing arm
5 Nut, rear wheel spindle-to-rear axle
6 Rear wheel spindle
7 Rear brake anti-lock sensor
8 Bolt, rear wheel spindle-to-rear axle
9 Hub and bearing assy
10 Hub grease cap
11 Retainer and washer assy
12 Rear wheel gasket
13 Anti-lock brake sensor bracket
14 Wheel studs

67197-WIND-G01

Rear hub assembly

⁂ WARNING

Discard the rear wheel hub retainer and washer assembly. It is a torque prevailing design and cannot be reused.

5. Remove the retainer and washer assembly.

6. Remove the wheel hub from the rear wheel spindle.

7. Thoroughly clean any grease from the surrounding surfaces. If a new hub assembly is being installed, remove the protective coating using carburetor degreaser.

8. To install, reverse the removal procedure. Use a new retainer and washer assembly. Torque the nut to 221 ft. lbs. (300 Nm).

BRAKES

Brake Caliper

REMOVAL & INSTALLATION

1. Raise and support the vehicle.
2. Remove the wheel and tire assembly.
3. Remove the disc brake caliper bolts.

**** CAUTION**

Brake fluid contains polyglycol ethers and polyglycols. Avoid contact with eyes. Wash hands thoroughly after handling. If brake fluid contacts eyes, flush eyes with running water for 15 minutes, get medical attention if irritation persists. If taken internally, drink water and induce vomiting. Get medical attention immediately.

**** CAUTION**

Brake fluid is harmful to painted and plastic surfaces. If brake fluid is spilled onto a painted or plastic surface, wash it with water immediately.

4. Remove the front brake hose bolt and discard the copper washers.
5. Remove the brake caliper.
6. Inspect the disc brake caliper. If leaks or damaged boots are found, disassembly is required.

To install:

7. Install the brake caliper.
8. Install the front brake hose bolt. Use new copper washers; connect the front brake hose. Torque the bolt to 35 ft. lbs. (48 Nm).

**** CAUTION**

Make sure guide pin boots are correctly seated or damage to guide pins can occur.

➡Tighten the lower caliper bolt first.

9. Install the disc brake caliper.
10. Install the disc brake caliper bolts. Torque to 26 ft. lbs. (35 Nm).
11. Bleed the caliper.
12. Install the wheel and tire assembly.
13. Fill the brake master cylinder reservoir with clean High Performance DOT 3 Motor Vehicle Brake Fluid C6AZ-19542-AB or equivalent DOT 3 fluid meeting Ford specification ESA-M6C25-A. Install brake master cylinder filler cap.
14. Inspect the brake system operation.

Disc Brake Pads

REMOVAL & INSTALLATION

1. Remove the brake master cylinder filler cap. Check the brake fluid level in the brake master cylinder reservoir. Remove fluid until the brake master cylinder reservoir is half full.
2. Raise and support the vehicle.
3. Remove the wheel and tire assembly.

**** WARNING**

Install a new pad if worn to or past the specified thickness above the metal backing plate or rivets. Install new pads in complete axle sets.

4. Inspect the pads for wear and contamination.

**** WARNING**

Do not pry in caliper sight hole to retract pistons as this can damage the pistons and boots.

**** WARNING**

When removing the disc brake caliper, never allow it to hang from the brake hose. Provide a suitable support.

5. Remove two front disc brake caliper bolts.
6. Lift the disc brake caliper off the front disc brake caliper anchor plate.
7. Inspect the disc brake caliper. If leaks or damaged boots are found, disassembly is required.
8. Remove the front disc brake caliper anchor plate stainless steel slippers and discard.
9. Inspect the front disc brake anchor plate assembly. Check the guide pin boots for damage. Check the guide pins for binding and damage. Replace worn or damaged pins. Lube pins with Silicone Brake Caliper Grease and Dielectric Compound D7AZ-19A331-A or equivalent meeting Ford specification ESE-M1C171-A.

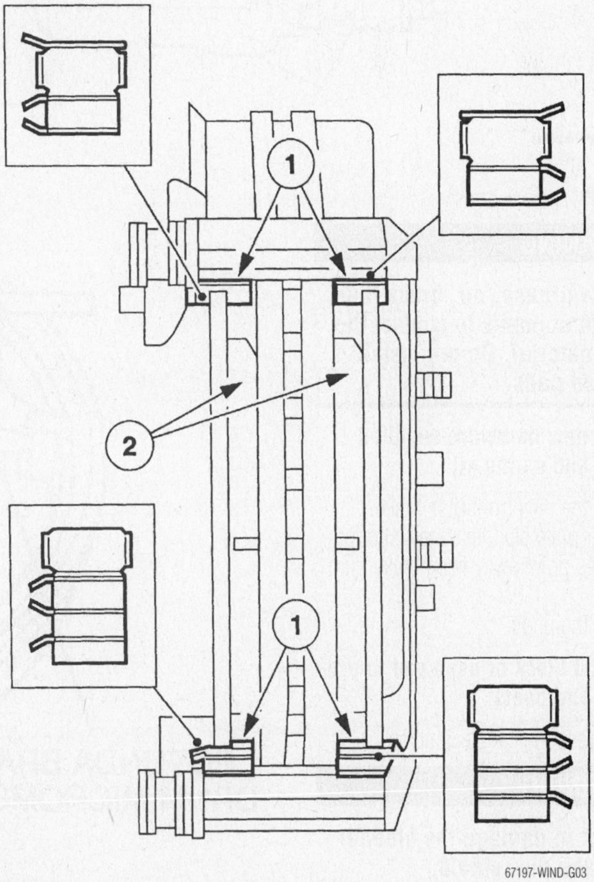

Pads and slippers

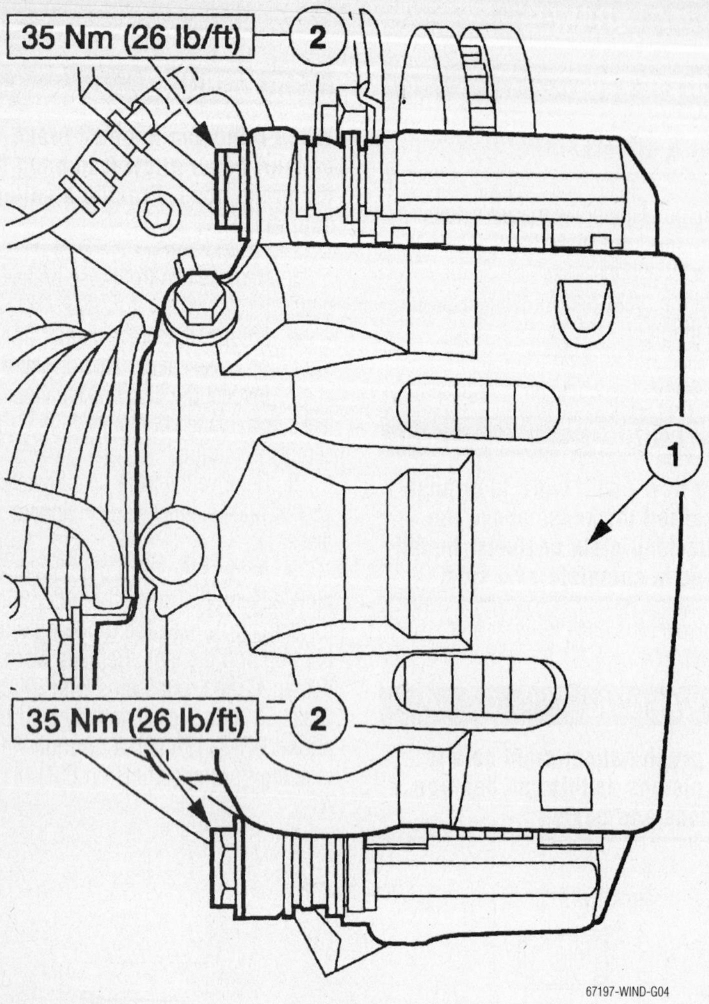

Caliper installation

To install:

✻✻ WARNING

Do not allow grease, oil, brake fluid or other contaminants to contact the pad lining material. Do not install contaminated pads.

➡Install all new hardware supplied with pad kit and spring kit.

10. Install the new front disc brake caliper anchor plate stainless steel slippers. Make sure taps point away from brake disc as illustrated.
11. Install the pads.

➡Use a wood block or used pad to protect pistons and boots.

12. Compress the caliper pistons.

✻✻ WARNING

Use care not to damage the bleeder screw or brake disc shield.

13. Install the disc brake caliper.
14. Install the disc brake caliper bolts; tighten the top bolt first.
15. Install the wheel and tire assembly.
16. Lower the vehicle.
17. Fill the brake master cylinder reservoir with clean High Performance DOT 3 Brake Fluid C6AZ-19542-AB or equivalent DOT 3 fluid meeting Ford specification ESA-M6C25-A. Install brake master cylinder filler cap.
18. Inspect brake operation.

Brake Drums

REMOVAL & INSTALLATION

1. Raise and safely support the vehicle.
2. Remove the wheel and tire assembly.
3. Remove the retainers holding the drum to the hub, if installed and discard.
4. Grasp the drum and remove.
5. If the drum will not slide off with light force, the brake shoes will need to be backed off. Remove the rubber plug on the backing plate and insert a screwdriver and a brake adjusting tool into the slot. Hold the adjuster lever away from the adjuster wheel with the screwdriver and back of the adjuster wheel with the brake adjusting tool.
6. Remove the brake drum. Inspect the drum for wear and/or damage. Machine or

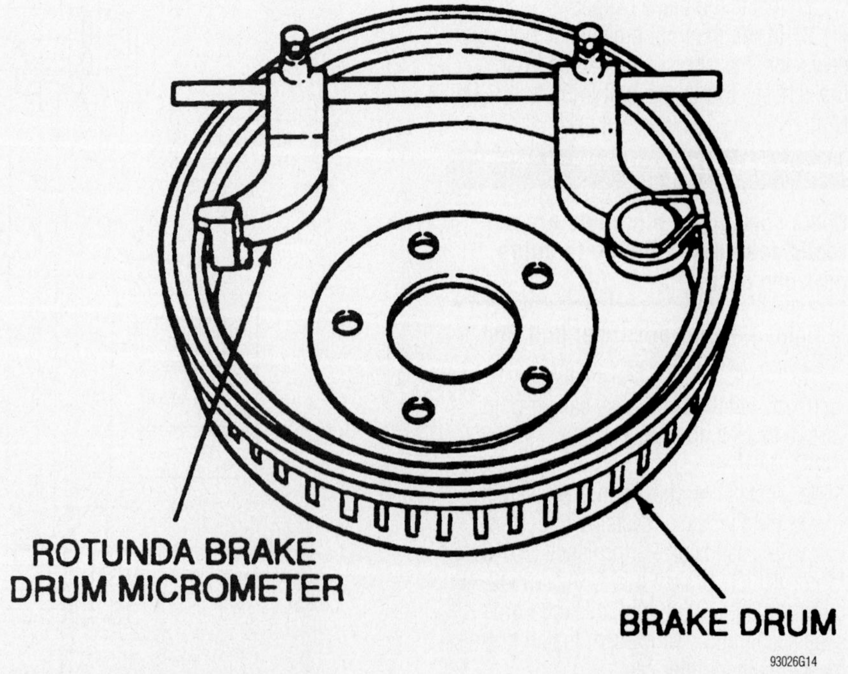

Measuring the brake shoes and drum

replace as necessary. If machining, observe the maximum diameter specification.

To install:

7. If a new brake drum is being installed, remove the protective coating from the inner brake surface.

8. Use a suitable brake adjustment gauge to measure the inside diameter of the brake drum.

9. Adjust the brake shoes to match the inside diameter of the brake drum.

10. Slide the brake drum onto the hub. Make sure that the brake shoes are not tight to the brake drum.

11. Install the rubber plug in the access hole. Retainers do not need to be reused to hold the drum.

12. Install the wheel and tire assembly. Torque the lug nuts to 85–105 ft. lbs. (115–142 Nm).

13. Lower the vehicle.

14. Check and fill the brake master cylinder as required.

15. Road-test the vehicle and check for proper brake operation.

Brake Shoes

REMOVAL & INSTALLATION

1. Remove the brake drum.

2. Inspect the rear brake assembly for the following:

- The rear wheel cylinder for leakage.

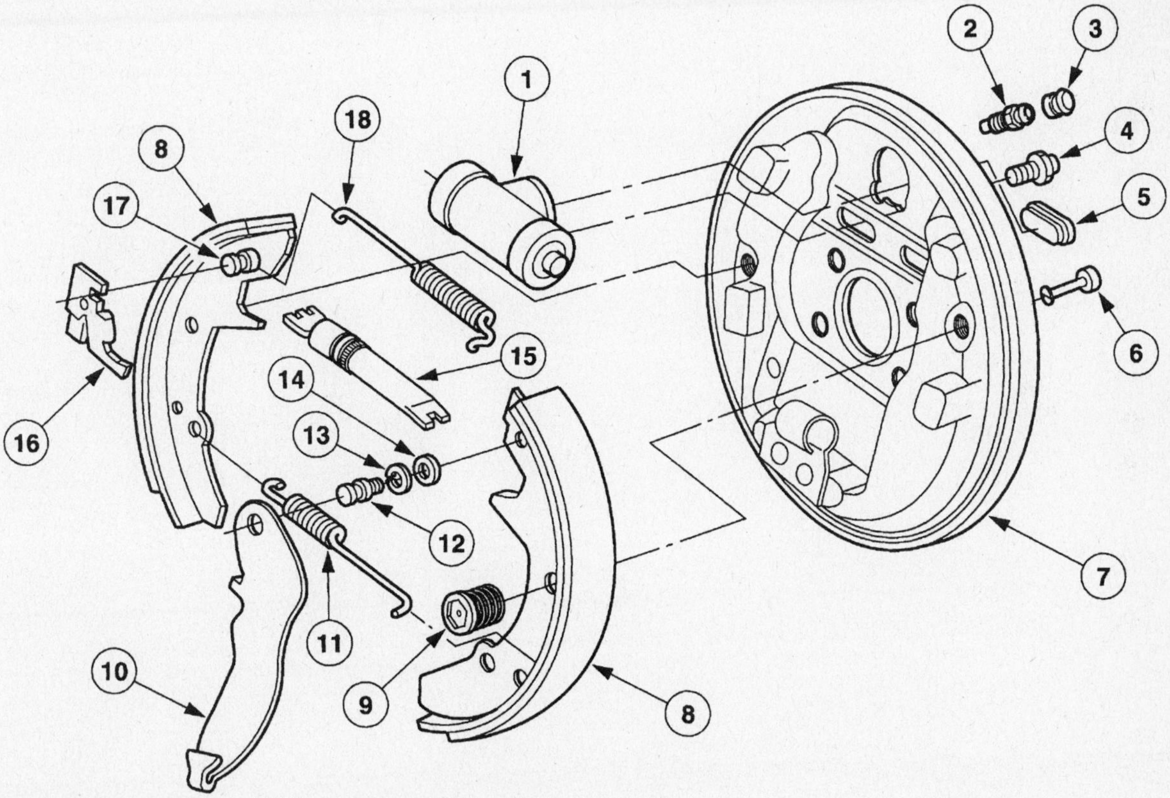

1 Rear wheel cylinder
2 Bleed screw
3 Bleeder dust cap
4 Wheel cylinder retaining screw (2 req'd)
5 Brake adjusting hole cover
6 Brake shoe hold-down spring pin
7 Brake backing plate (LH)
7 Brake backing plate (RH)
8 Shoe and lining
9 Brake shoe hold-down spring
10 Parking brake lever (LH)
10 Parking brake lever (RH)
11 Brake shoe retracting spring
12 Parking brake lever pin (inner)
13 Parking brake lever pin retainer
14 Washer
15 Brake shoe adjusting screw (LH)
15 Brake shoe adjusting screw (RH)
16 Brake adjusting lever (LH)
16 Brake adjusting lever (RH)
17 Parking brake lever pin (outer)
18 Brake adjusting screw spring

67197-WIND-G05

Rear drum brake assembly

- The rear brake shoes and linings for contamination, and install new as necessary.
- The springs for heat discoloration and install new as necessary.
- The adjusting lever contact with the brake adjuster screw.

3. Inspect the rear brake shoes and linings for minimum thickness, above the backing plate or rivets, and install new as necessary.
4. Remove the brake shoe adjusting screw spring.
5. Remove the brake adjuster lever.
6. Remove the brake adjuster screw.
7. Remove the brake shoe hold-down spring.
8. Remove the brake shoe and the retracting spring.
9. Remove the rear brake shoe hold-down spring.
10. Remove the rear brake shoes and linings.
11. Remove and discard the parking brake lever clip.
12. Remove the rear brake shoes and linings from the parking brake lever.
13. To install, reverse the removal procedure.

GLOSSARY

ABS: Anti-lock braking system. An electro-mechanical braking system which is designed to minimize or prevent wheel lock-up during braking.

ABSOLUTE PRESSURE: Atmospheric (barometric) pressure plus the pressure gauge reading.

ACCELERATOR PUMP: A small pump located in the carburetor that feeds fuel into the air/fuel mixture during acceleration.

ACCUMULATOR: A device that controls shift quality by cushioning the shock of hydraulic oil pressure being applied to a clutch or band.

ACTUATING MECHANISM: The mechanical output devices of a hydraulic system, for example, clutch pistons and band servos.

ACTUATOR: The output component of a hydraulic or electronic system.

ADVANCE: Setting the ignition timing so that spark occurs earlier before the piston reaches top dead center (TDC).

ADAPTIVE MEMORY (ADAPTIVE STRATEGY): The learning ability of the TCM or PCM to redefine its decision-making process to provide optimum shift quality.

AFTER TOP DEAD CENTER (ATDC): The point after the piston reaches the top of its travel on the compression stroke.

AIR BAG: Device on the inside of the car designed to inflate on impact of crash, protecting the occupants of the car.

AIR CHARGE TEMPERATURE (ACT) SENSOR: The temperature of the airflow into the engine is measured by an ACT sensor, usually located in the lower intake manifold or air cleaner.

AIR CLEANER: An assembly consisting of a housing, filter and any connecting ductwork. The filter element is made up of a porous paper, sometimes with a wire mesh screening, and is designed to prevent airborne particles from entering the engine through the carburetor or throttle body.

AIR INJECTION: One method of reducing harmful exhaust emissions by injecting air into each of the exhaust ports of an engine. The fresh air entering the hot exhaust manifold causes any remaining fuel to be burned before it can exit the tailpipe.

AIR PUMP: An emission control device that supplies fresh air to the exhaust manifold to aid in more completely burning exhaust gases.

AIR/FUEL RATIO: The ratio of air-to-gasoline by weight in the fuel mixture drawn into the engine.

ALDL (assembly line diagnostic link): Electrical connector for scanning ECM/PCM/TCM input and output devices.

ALIGNMENT RACK: A special drive-on vehicle lift apparatus/measuring device used to adjust a vehicle's toe, caster and camber angles.

ALL WHEEL DRIVE: Term used to describe a full time four wheel drive system or any other vehicle drive system that continuously delivers power to all four wheels. This system is found primarily on station wagon vehicles and SUVs not utilized for significant off road use.

ALTERNATING CURRENT (AC): Electric current that flows first in one direction, then in the opposite direction, continually reversing flow.

ALTERNATOR: A device which produces AC (alternating current) which is converted to DC (direct current) to charge the car battery.

AMMETER: An instrument, calibrated in amperes, used to measure the flow of an electrical current in a circuit. Ammeters are always connected in series with the circuit being tested.

AMPERAGE: The total amount of current (amperes) flowing in a circuit.

AMPLIFIER: A device used in an electrical circuit to increase the voltage of an output signal.

AMP/HR. RATING (BATTERY): Measurement of the ability of a battery to deliver a stated amount of current for a stated period of time. The higher the amp/hr. rating, the better the battery.

AMPERE: The rate of flow of electrical current present when one volt of electrical pressure is applied against one ohm of electrical resistance.

ANALOG COMPUTER: Any microprocessor that uses similar (analogous) electrical signals to make its calculations.

ANODIZED: A special coating applied to the surface of aluminum valves for extended service life.

ANTIFREEZE: A substance (ethylene or propylene glycol) added to the coolant to prevent freezing in cold weather.

ANTI-FOAM AGENTS: Minimize fluid foaming from the whipping action encountered in the converter and planetary action.

ANTI-WEAR AGENTS: Zinc agents that control wear on the gears, bushings, and thrust washers.

ANTI-LOCK BRAKING SYSTEM: A supplementary system to the base hydraulic system that prevents sustained lock-up of the wheels during braking as well as automatically controlling wheel slip.

ANTI-ROLL BAR: See stabilizer bar.

ARC: A flow of electricity through the air between two electrodes or contact points that produces a spark.

ARMATURE: A laminated, soft iron core wrapped by a wire that converts electrical energy to mechanical energy as in a motor or relay. When rotated in a magnetic field, it changes mechanical energy into electrical energy as in a generator.

ATDC: After Top Dead Center.

ATF: Automatic transmission fluid.

ATMOSPHERIC PRESSURE: The pressure on the Earth's surface caused by the weight of the air in the atmosphere. At sea level, this pressure is 14.7 psi at 32°F (101 kPa at 0°C).

ATOMIZATION: The breaking down of a liquid into a fine mist that can be suspended in air.

AUXILIARY ADD-ON COOLER: A supplemental transmission fluid cooling device that is installed in series with the heat exchanger (cooler), located inside the radiator, to provide additional support to cool the hot fluid leaving the torque converter.

AUXILIARY PRESSURE: An added fluid pressure that is introduced into a regulator or balanced valve system to control valve movement. The auxiliary pressure itself can be either a fixed or a variable value. (See balanced valve; regulator valve.)

AWD: All wheel drive.

AXIAL FORCE: A side or end thrust force acting in or along the same plane as the power flow.

AXIAL PLAY: Movement parallel to a shaft or bearing bore.

AXLE CAPACITY: The maximum load-carrying capacity of the axle itself, as specified by the manufacturer. This is usually a higher number than the GAWR.

AXLE RATIO: This is a number (3.07:1, 4.56:1, for example) expressing the ratio between driveshaft revolutions and wheel revolutions. A low numerical ratio allows the engine to work easier because it doesn't have to turn as fast. A high numerical ratio means that the engine has to turn more rpm's to move the wheels through the same number of turns.

BACKFIRE: The sudden combustion of gases in the intake or exhaust system that results in a loud explosion.

BACKLASH: The clearance or play between two parts, such as meshed gears.

BACKPRESSURE: Restrictions in the exhaust system that slow the exit of exhaust gases from the combustion chamber.

BAKELITE®: A heat resistant, plastic insulator material commonly used in printed circuit boards and transistorized components.

BALANCED VALVE: A valve that is positioned by opposing auxiliary hydraulic pressures and/or spring force. Examples include mainline regulator, throttle, and governor valves. (See regulator valve.)

BAND: A flexible ring of steel with an inner lining of friction material. When tightened around the outside of a drum, a planetary member is held stationary to the transmission/transaxle case.

BALL BEARING: A bearing made up of hardened inner and outer races between which hardened steel balls roll.

BALL JOINT: A ball and matching socket connecting suspension components (steering knuckle to lower control arms). It permits rotating movement in any direction between the components that are joined.

BARO (BAROMETRIC PRESSURE SENSOR): Measures the change in the intake manifold pressure caused by changes in altitude.

BAROMETRIC MANIFOLD ABSOLUTE PRESSURE (BMAP) SENSOR: Operates similarly to a conventional MAP sensor; reads intake manifold pressure and is also responsible for determining altitude and barometric pressure prior to engine operation.

BAROMETRIC PRESSURE: (See atmospheric pressure.)

BALLAST RESISTOR: A resistor in the primary ignition circuit that lowers voltage after the engine is started to reduce wear on ignition components.

BATTERY: A direct current electrical storage unit, consisting of the basic active materials of lead and sulfuric acid, which converts chemical energy into electrical energy. Used to provide current for the operation of the starter as well as other equipment, such as the radio, lighting, etc.

BEAD: The portion of a tire that holds it on the rim.

BEARING: A friction reducing, supportive device usually located between a stationary part and a moving part.

BEFORE TOP DEAD CENTER (BTDC): The point just before the piston reaches the top of its travel on the compression stroke.

BELTED TIRE: Tire construction similar to bias-ply tires, but using two or more layers of reinforced belts between body plies and the tread.

BEZEL: Piece of metal surrounding radio, headlights, gauges or similar components; sometimes used to hold the glass face of a gauge in the dash.

BIAS-PLY TIRE: Tire construction, using body ply reinforcing cords which run at alternating angles to the center line of the tread.

BI-METAL TEMPERATURE SENSOR: Any sensor or switch made of two dissimilar types of metal that bend when heated or cooled due to the different expansion rates of the alloys. These types of sensors usually function as an on/off switch.

BLOCK: See Engine Block.

BLOW-BY: Combustion gases, composed of water vapor and unburned fuel, that leak past the piston rings into the crankcase during normal engine operation. These gases are removed by the PCV system to prevent the buildup of harmful acids in the crankcase.

BOOK TIME: See Labor Time.

BOOK VALUE: The average value of a car, widely used to determine trade-in and resale value.

BOOST VALVE: Used at the base of the regulator valve to increase mainline pressure.

BORE: Diameter of a cylinder.

BRAKE CALIPER: The housing that fits over the brake disc. The caliper holds the brake pads, which are pressed against the discs by the caliper pistons when the brake pedal is depressed.

BRAKE HORSEPOWER (BHP): The actual horsepower available at the engine flywheel as measured by a dynamometer.

BRAKE FADE: Loss of braking power, usually caused by excessive heat after repeated brake applications.

BRAKE HORSEPOWER: Usable horsepower of an engine measured at the crankshaft.

BRAKE PAD: A brake shoe and lining assembly used with disc brakes.

BRAKE PROPORTIONING VALVE: A valve on the master cylinder which restricts hydraulic brake pressure to the wheels to a specified amount, preventing wheel lock-up.

BREAKAWAY: Often used by Chrysler to identify first-gear operation in D and 2 ranges. In these ranges, first-gear operation depends on a one-way roller clutch that holds on acceleration and releases (breaks away) on deceleration, resulting in a freewheeling coast-down condition.

BRAKE SHOE: The backing for the brake lining. The term is, however, usually applied to the assembly of the brake backing and lining.

BREAKER POINTS: A set of points inside the distributor, operated by a cam, which make and break the ignition circuit.

BRINNELLING: A wear pattern identified by a series of indentations at regular intervals. This condition is caused by a lack of lube, overload situations, and/or vibrations.

BTDC: Before Top Dead Center.

BUMP: Sudden and forceful apply of a clutch or band.

BUSHING: A liner, usually removable, for a bearing; an anti-friction liner used in place of a bearing.

CALIFORNIA ENGINE: An engine certified by the EPA for use in California only; conforms to more stringent emission regulations than Federal engine.

CALIPER: A hydraulically activated device in a disc brake system, which is mounted straddling the brake rotor (disc). The caliper contains at least one piston and two brake pads. Hydraulic pressure on the piston(s) forces the pads against the rotor.

CAPACITY: The quantity of electricity that can be delivered from a unit, as from a battery in ampere-hours, or output, as from a generator.

CAMBER: One of the factors of wheel alignment. Viewed from the front of the car, it is the inward or outward tilt of the wheel. The top of the tire will lean outward (positive camber) or inward (negative camber).

CAMSHAFT: A shaft in the engine on which are the lobes (cams) which operate the valves. The camshaft is driven by the crankshaft, via a belt, chain or gears, at one half the crankshaft speed.

CAPACITOR: A device which stores an electrical charge.

CARBON MONOXIDE (CO): A colorless, odorless gas given off as a normal byproduct of combustion. It is poisonous and extremely dangerous in confined areas, building up slowly to toxic levels without warning if adequate ventilation is not available.

CARBURETOR: A device, usually mounted on the intake manifold of an engine, which mixes the air and fuel in the proper proportion to allow even combustion.

CASTER: The forward or rearward tilt of an imaginary line drawn through the upper ball joint and the center of the wheel. Viewed from the sides, positive caster (forward tilt) lends directional stability, while negative caster (rearward tilt) produces instability.

CATALYTIC CONVERTER: A device installed in the exhaust system, like a muffler, that converts harmful byproducts of combustion into carbon dioxide and water vapor by means of a heat-producing chemical reaction.

CENTRIFUGAL ADVANCE: A mechanical method of advancing the spark timing by using flyweights in the distributor that react to centrifugal force generated by the distributor shaft rotation.

CENTRIFUGAL FORCE: The outward pull of a revolving object, away from the center of revolution. Centrifugal force increases with the speed of rotation.

CETANE RATING: A measure of the ignition value of diesel fuel. The higher the cetane rating, the better the fuel. Diesel fuel cetane rating is roughly comparable to gasoline octane rating.

CHECK VALVE: Any one-way valve installed to permit the flow of air, fuel or vacuum in one direction only.

CHOKE: The valve/plate that restricts the amount of air entering an engine on the induction stroke, thereby enriching the air/fuel ratio.

CHUGGLE: Bucking or jerking condition that may be engine related and may be most noticeable when converter clutch is engaged; similar to the feel of towing a trailer.

CIRCLIP: A split steel snapring that fits into a groove to hold various parts in place.

CIRCUIT BREAKER: A switch which protects an electrical circuit from overload by opening the circuit when the current flow exceeds a pre-determined level. Some circuit breakers must be reset manually, while most reset automatically.

CIRCUIT: Any unbroken path through which an electrical current can flow. Also used to describe fuel flow in some instances.

CIRCUIT, BYPASS: Another circuit in parallel with the major circuit through which power is diverted.

CIRCUIT, CLOSED: An electrical circuit in which there is no interruption of current flow.

CIRCUIT, GROUND: The non-insulated portion of a complete circuit used as a common potential point. In automotive circuits, the ground is composed of metal parts, such as the engine, body sheet metal, and frame and is usually a negative potential.

CIRCUIT, HOT: That portion of a circuit not at ground potential. The hot circuit is usually insulated and is connected to the positive side of the battery.

CIRCUIT, OPEN: A break or lack of contact in an electrical circuit, either intentional (switch) or unintentional (bad connection or broken wire).

CIRCUIT, PARALLEL: A circuit having two or more paths for current flow with common positive and negative tie points. The same voltage is applied to each load device or parallel branch.

CIRCUIT, SERIES: An electrical system in which separate parts are connected end to end, using one wire, to form a single path for current to flow.

CIRCUIT, SHORT: A circuit that is accidentally completed in an electrical path for which it was not intended.

CLAMPING (ISOLATION) DIODES: Diodes positioned in a circuit to prevent self-induction from damaging electronic components.

CLEARCOAT: A transparent layer which, when sprayed over a vehicle's paint job, adds gloss and depth as well as an additional protective coating to the finish.

CLUTCH: Part of the power train used to connect/disconnect power to the rear wheels.

CLUTCH, FLUID: The same as a fluid coupling. A fluid clutch or coupling performs the same function as a friction clutch by utilizing fluid friction and inertia as opposed to solid friction used by a friction clutch. (See fluid coupling.)

CLUTCH, FRICTION: A coupling device that provides a means of smooth and positive engagement and disengagement of engine torque to the vehicle powertrain. Transmission of power through the clutch is accomplished by bringing one or more rotating drive members into contact with complementing driven members.

COAST: Vehicle deceleration caused by engine braking conditions.

COEFFICIENT OF FRICTION: The amount of surface tension between two contacting surfaces; identified by a scientifically calculated number.

COIL: Part of the ignition system that boosts the relatively low voltage supplied by the car's electrical system to the high voltage required to fire the spark plugs.

COMBINATION MANIFOLD: An assembly which includes both the intake and exhaust manifolds in one casting.

COMBINATION VALVE: A device used in some fuel systems that routes fuel vapors to a charcoal storage canister instead of venting them into the atmosphere. The valve relieves fuel tank pressure and allows fresh air into the tank as the fuel level drops to prevent a vapor lock situation.

COMBUSTION CHAMBER: The part of the engine in the cylinder head where combustion takes place.

COMPOUND GEAR: A gear consisting of two or more simple gears with a common shaft.

COMPOUND PLANETARY: A gearset that has more than the three elements found in a simple gearset and is constructed by combining members of two planetary gearsets to create additional gear ratio possibilities.

COMPRESSION CHECK: A test involving removing each spark plug and inserting a gauge. When the engine is cranked, the gauge will record a pressure reading in the individual cylinder. General operating condition can be determined from a compression check.

COMPRESSION RATIO: The ratio of the volume between the piston and cylinder head when the piston is at the bottom of its stroke (bottom dead center) and when the piston is at the top of its stroke (top dead center).

COMPUTER: An electronic control module that correlates input data according to prearranged engineered instructions; used for the management of an actuator system or systems.

CONDENSER: An electrical device which acts to store an electrical charge, preventing voltage surges.

2. A radiator-like device in the air conditioning system in which refrigerant gas condenses into a liquid, giving off heat.

CONDUCTOR: Any material through which an electrical current can be transmitted easily.

CONNECTING ROD: The connecting link between the crankshaft and piston.

CONSTANT VELOCITY JOINT: Type of universal joint in a halfshaft assembly in which the output shaft turns at a constant angular velocity without variation, provided that the speed of the input shaft is constant.

CONTINUITY: Continuous or complete circuit. Can be checked with an ohmmeter.

CONTROL ARM: The upper or lower suspension components which are mounted on the frame and support the ball joints and steering knuckles.

CONVENTIONAL IGNITION: Ignition system which uses breaker points.

CONVERTER: (See torque converter.)

CONVERTER LOCKUP: The switching from hydrodynamic to direct mechanical drive, usually through the application of a friction element called the converter clutch.

COOLANT: Mixture of water and anti-freeze circulated through the engine to carry off heat produced by the engine.

CORROSION INHIBITOR: An inhibitor in ATF that prevents corrosion of bushings, thrust washers, and oil cooler brazed joints.

COUNTERSHAFT: An intermediate shaft which is rotated by a mainshaft and transmits, in turn, that rotation to a working part.

COUPLING PHASE: Occurs when the torque converter is operating at its greatest hydraulic efficiency. The speed differential between the impeller and the turbine is at its minimum. At this point, the stator freewheels, and there is no torque multiplication.

CRANKCASE: The lower part of an engine in which the crankshaft and related parts operate.

CRANKSHAFT: Engine component (connected to pistons by connecting rods) which converts the reciprocating (up and down) motion of pistons to rotary motion used to turn the driveshaft.

CURB WEIGHT: The weight of a vehicle without passengers or payload, but including all fluids (oil, gas, coolant, etc.) and other equipment specified as standard.

CURRENT: The flow (or rate) of electrons moving through a circuit. Current is measured in amperes (amp).

CURRENT FLOW CONVENTIONAL: Current flows through a circuit from the positive terminal of the source to the negative terminal (plus to minus).

CURRENT FLOW, ELECTRON: Current or electrons flow from the negative terminal of the source, through the circuit, to the positive terminal (minus to plus).

CV-JOINT: Constant velocity joint.

CYCLIC VIBRATIONS: The off-center movement of a rotating object that is affected by its initial balance, speed of rotation, and working angles.

CYLINDER BLOCK: See engine block.

CYLINDER HEAD: The detachable portion of the engine, usually fastened to the top of the cylinder block and containing all or most of the combustion chambers. On overhead valve engines, it contains the valves and their operating parts. On overhead cam engines, it contains the camshaft as well.

CYLINDER: In an engine, the round hole in the engine block in which the piston(s) ride.

DATA LINK CONNECTOR (DLC): Current acronym/term applied to the federally mandated, diagnostic junction connector that is used to monitor ECM/PC/TCM inputs, processing strategies, and outputs including diagnostic trouble codes (DTCs).

DEAD CENTER: The extreme top or bottom of the piston stroke.

DECELERATION BUMP: When referring to a torque converter clutch in the applied position, a sudden release of the accelerator pedal causes a forceful reversal of power through the drivetrain (engine braking), just prior to the apply plate actually being released.

DELAYED (LATE OR EXTENDED): Condition where shift is expected but does not occur for a period of time, for example, where clutch or band engagement does not occur as quickly as expected during part throttle or wide open throttle apply of accelerator or when manually downshifting to a lower range.

DETENT: A spring-loaded plunger, pin, ball, or pawl used as a holding device on a ratchet wheel or shaft. In automatic transmissions, a detent mechanism is used for locking the manual valve in place.

DETENT DOWNSHIFT: (See kickdown.)

DETERGENT: An additive in engine oil to improve its operating characteristics.

DETONATION: An unwanted explosion of the air/fuel mixture in the combustion chamber caused by excess heat and compression, advanced timing, or an overly lean mixture. Also referred to as "ping".

DEXRON®: A brand of automatic transmission fluid.

DIAGNOSTIC TROUBLE CODES (DTCs): A digital display from the control module memory that identifies the input, processor, or output device circuit that is related to the powertrain emission/driveability malfunction detected. Diagnostic trouble codes can be read by the MIL to flash any codes or by using a handheld scanner.

DIAPHRAGM: A thin, flexible wall separating two cavities, such as in a vacuum advance unit.

DIESELING: The engine continues to run after the car is shut off; caused by fuel continuing to be burned in the combustion chamber.

DIFFERENTIAL: A geared assembly which allows the transmission of motion between drive axles, giving one axle the ability to rotate faster than the other, as in cornering.

DIFFERENTIAL AREAS: When opposing faces of a spool valve are acted upon by the same pressure but their areas differ in size, the face with the larger area produces the differential force and valve movement. (See spool valve.)

DIFFERENTIAL FORCE: (See differential areas)

DIGITAL READOUT: A display of numbers or a combination of numbers and letters.

DIGITAL VOLT OHMMETER: An electronic diagnostic tool used to measure voltage, ohms and amps as well as several other functions, with the readings displayed on a digital screen in tenths, hundredths and thousandths.

DIODE: An electrical device that will allow current to flow in one direction only.

DIRECT CURRENT (DC): Electrical current that flows in one direction only.

DIRECT DRIVE: The gear ratio is 1:1, with no change occurring in the torque and speed input/output relationship.

DISC BRAKE: A hydraulic braking assembly consisting of a brake disc, or rotor, mounted on an axle shaft, and a caliper assembly containing, usually two brake pads which are activated by hydraulic pressure. The pads are forced against the sides of the disc, creating friction which slows the vehicle.

DISPERSANTS: Suspend dirt and prevent sludge buildup in a liquid, such as engine oil.

DOUBLE BUMP (DOUBLE FEEL): Two sudden and forceful applies of a clutch or band.

DISPLACEMENT: The total volume of air that is displaced by all pistons as the engine turns through one complete revolution.

DISTRIBUTOR: A mechanically driven device on an engine which is responsible for electrically firing the spark plug at a pre-determined point of the piston stroke.

DOHC: Double overhead camshaft.

DOUBLE OVERHEAD CAMSHAFT: The engine utilizes two camshafts mounted in one cylinder head. One camshaft operates the exhaust valves, while the other operates the intake valves.

DOWEL PIN: A pin, inserted in mating holes in two different parts allowing those parts to maintain a fixed relationship.

DRIVELINE: The drive connection between the transmission and the drive wheels.

DRIVE TRAIN: The components that transmit the flow of power from the engine to the wheels. The components include the clutch, transmission, driveshafts (or axle shafts in front wheel drive), U-joints and differential.

DRUM BRAKE: A braking system which consists of two brake shoes and one or two wheel cylinders, mounted on a fixed backing plate, and a brake drum, mounted on an axle, which revolves around the assembly.

DRY CHARGED BATTERY: Battery to which electrolyte is added when the battery is placed in service.

DVOM: Digital volt ohmmeter

DWELL: The rate, measured in degrees of shaft rotation, at which an electrical circuit cycles on and off.

DYNAMIC: An application in which there is rotating or reciprocating motion between the parts.

EARLY: Condition where shift occurs before vehicle has reached proper speed, which tends to labor engine after upshift.

EBCM: See Electronic Control Unit (ECU).

ECM: See Electronic Control Unit (ECU).

ECU: Electronic control unit.

ELECTRODE: Conductor (positive or negative) of electric current.

ELECTROLYSIS: A surface etching or bonding of current conducting transmission/transaxle components that may occur when grounding straps are missing or in poor condition.

ELECTROLYTE: A solution of water and sulfuric acid used to activate the battery. Electrolyte is extremely corrosive.

ELECTROMAGNET: A coil that produces a magnetic field when current flows through its windings.

ELECTROMAGNETIC INDUCTION: A method to create (generate) current flow through the use of magnetism.

ELECTROMAGNETISM: The effects surrounding the relationship between electricity and magnetism.

ELECTROMOTIVE FORCE (EMF): The force or pressure (voltage) that causes current movement in an electrical circuit.

ELECTRONIC CONTROL UNIT: A digital computer that controls engine (and sometimes transmission, brake or other vehicle system) functions based on data received from various sensors. Examples used by some manufacturers include Electronic Brake Control Module (EBCM), Engine Control Module (ECM), Powertrain Control Module (PCM) or Vehicle Control Module (VCM).

ELECTRONIC IGNITION: A system in which the timing and firing of the spark plugs is controlled by an electronic control unit, usually called a module. These systems have no points or condenser.

ELECTRONIC PRESSURE CONTROL (EPC) SOLENOID: A specially designed solenoid containing a spool valve and spring assembly to control fluid mainline pressure. A variable current flow, controlled by the ECM/PCM, varies the internal force of the solenoid on the spool valve and resulting mainline pressure. (See variable force solenoid.)

ELECTRONICS: Miniaturized electrical circuits utilizing semiconductors, solid-state devices, and printed circuits. Electronic circuits utilize small amounts of power.

ELECTRONIFICATION: The application of electronic circuitry to a mechanical device. Regarding automatic transmissions, electrification is incorporated into converter clutch lockup, shift scheduling, and line pressure control systems.

ELECTROSTATIC DISCHARGE (ESD): An unwanted, high-voltage electrical current released by an individual who has taken on a static charge of electricity. Electronic components can be easily damaged by ESD.

ELEMENT: A device within a hydrodynamic drive unit designed with a set of blades to direct fluid flow.

ENAMEL: Type of paint that dries to a smooth, glossy finish.

END BUMP (END FEEL OR SLIP BUMP): Firmer feel at end of shift when compared with feel at start of shift.

END-PLAY: The clearance/gap between two components that allows for expansion of the parts as they warm up, to prevent binding and to allow space for lubrication.

ENERGY: The ability or capacity to do work.

ENGINE: The primary motor or power apparatus of a vehicle, which converts liquid or gas fuel into mechanical energy.

ENGINE BLOCK: The basic engine casting containing the cylinders, the crankshaft main bearings, as well as machined surfaces for the mounting of other components such as the cylinder head, oil pan, transmission, etc.

ENGINE BRAKING: Use of engine to slow vehicle by manually downshifting during zero-throttle coast down.

ENGINE CONTROL MODULE (ECM): Manages the engine and incorporates output control over the torque converter clutch solenoid. (Note: Current designation for the ECM in late model vehicles is PCM.)

ENGINE COOLANT TEMPERATURE (ECT) SENSOR: Prevents converter clutch engagement with a cold engine; also used for shift timing and shift quality.

EP LUBRICANT: EP (extreme pressure) lubricants are specially formulated for use with gears involving heavy loads (transmissions, differentials, etc.).

ETHYL: A substance added to gasoline to improve its resistance to knock, by slowing down the rate of combustion.

ETHYLENE GLYCOL: The base substance of antifreeze.

EXHAUST MANIFOLD: A set of cast passages or pipes which conduct exhaust gases from the engine.

FAIL-SAFE (BACKUP) CONTROL: A substitute value used by the PCM/TCM to replace a faulty signal from an input sensor. The temporary value allows the vehicle to continue to be operated.

FAST IDLE: The speed of the engine when the choke is on. Fast idle speeds engine warm-up.

FEDERAL ENGINE: An engine certified by the EPA for use in any of the 49 states (except California).

FEEDBACK: A circuit malfunction whereby current can find another path to feed load devices.

FEELER GAUGE: A blade, usually metal, of precisely predetermined thickness, used to measure the clearance between two parts.

FILAMENT: The part of a bulb that glows; the filament creates high resistance to current flow and actually glows from the resulting heat.

FINAL DRIVE: An essential part of the axle drive assembly where final gear reduction takes place in the powertrain. In RWD applications and north-south FWD applications, it must also change the power flow direction to the axle shaft by ninety degrees. (Also see axle ratio).

FIRING ORDER: The order in which combustion occurs in the cylinders of an engine. Also the order in which spark is distributed to the plugs by the distributor.

FIRM: A noticeable quick apply of a clutch or band that is considered normal with medium to heavy throttle shift; should not be confused with harsh or rough.

FLAME FRONT: The term used to describe certain aspects of the fuel explosion in the cylinders. The flame front should move in a controlled pattern across the cylinder, rather than simply exploding immediately.

FLARE (SLIPPING): A quick increase in engine rpm accompanied by momentary loss of torque; generally occurs during shift.

FLAT ENGINE: Engine design in which the pistons are horizontally opposed. Porsche, Subaru and some old VW are common examples of flat engines.

FLAT RATE: A dealership term referring to the amount of money paid to a technician for a repair or diagnostic service based on that particular service versus dealership's labor time (NOT based on the actual time the technician spent on the job).

FLAT SPOT: A point during acceleration when the engine seems to lose power for an instant.

FLOODING: The presence of too much fuel in the intake manifold and combustion chamber which prevents the air/fuel mixture from firing, thereby causing a no-start situation.

FLUID: A fluid can be either liquid or gas. In hydraulics, a liquid is used for transmitting force or motion.

FLUID COUPLING: The simplest form of hydrodynamic drive, the fluid coupling consists of two look-alike members with straight radial varies referred to as the impeller (pump) and the turbine. Input torque is always equal to the output torque.

FLUID DRIVE: Either a fluid coupling or a fluid torque converter. (See hydrodynamic drive units.)

FLUID TORQUE CONVERTER: A hydrodynamic drive that has the ability to act both as a torque multiplier and fluid coupling. (See hydrodynamic drive units; torque converter.)

FLUID VISCOSITY: The resistance of a liquid to flow. A cold fluid (oil) has greater viscosity and flows more slowly than a hot fluid (oil).

FLYWHEEL: A heavy disc of metal attached to the rear of the crankshaft. It smoothes the firing impulses of the engine and keeps the crankshaft turning during periods when no firing takes place. The starter also engages the flywheel to start the engine.

FOOT POUND (ft. lbs., lbs. ft. or sometimes, ft. lb.): The amount of energy or work needed to raise an item weighing one pound, a distance of one foot.

FREEZE PLUG: A plug in the engine block which will be pushed out if the coolant freezes. Sometimes called expansion plugs, they protect the block from cracking should the coolant freeze.

FRICTION: The resistance that occurs between contacting surfaces. This relationship is expressed by a ratio called the coefficient of friction (CL).

FRICTION, COEFFICIENT OF: The amount of surface tension between two contacting surfaces; expressed by a scientifically calculated number.

FRONT END ALIGNMENT: A service to set caster, camber and toe-in to the correct specifications. This will ensure that the car steers and handles properly and that the tires wear properly.

FRICTION MODIFIER: Changes the coefficient of friction of the fluid between the mating steel and composition clutch/band surfaces during the engagement process and allows for a certain amount of intentional slipping for a good "shift-feel".

FRONTAL AREA: The total frontal area of a vehicle exposed to air flow.

FUEL FILTER: A component of the fuel system containing a porous paper element used to prevent any impurities from entering the engine through the fuel system. It usually takes the form of a canister-like housing, mounted in-line with the fuel hose, located anywhere on a vehicle between the fuel tank and engine.

FUEL INJECTION: A system replacing the carburetor that sprays fuel into the cylinder through nozzles. The amount of fuel can be more precisely controlled with fuel injection.

FULL FLOATING AXLE: An axle in which the axle housing extends through the wheel giving bearing support on the outside of the housing. The front axle of a four-wheel drive vehicle is usually a full floating axle, as are the rear axles of many larger (1 ton and over) pick-ups and vans.

FULL-TIME FOUR-WHEEL DRIVE: A four-wheel drive system that continuously delivers power to all four wheels. A differential between the front and rear driveshafts permits variations in axle speeds to control gear wind-up without damage.

FULL THROTTLE DETENT DOWNSHIFT: A quick apply of accelerator pedal to its full travel, forcing a downshift.

FUSE: A protective device in a circuit which prevents circuit overload by breaking the circuit when a specific amperage is present. The device is constructed around a strip or wire of a lower amperage rating than the circuit it is designed to protect. When an amperage higher than that stamped on the fuse is present in the circuit, the strip or wire melts, opening the circuit.

FUSIBLE LINK: A piece of wire in a wiring harness that performs the same job as a fuse. If overloaded, the fusible link will melt and interrupt the circuit.

FWD: Front wheel drive.

GAWR: (Gross axle weight rating) the total maximum weight an axle is designed to carry.

GCW: (Gross combined weight) total combined weight of a tow vehicle and trailer.

GARAGE SHIFT: initial engagement feel of transmission, neutral to reverse or neutral to a forward drive.

GARAGE SHIFT FEEL: A quick check of the engagement quality and responsiveness of reverse and forward gears. This test is done with the vehicle stationary.

GEAR: A toothed mechanical device that acts as a rotating lever to transmit power or turning effort from one shaft to another. (See gear ratio.)

GEAR RATIO: A ratio expressing the number of turns a smaller gear will make to turn a larger gear through one revolution. The ratio is found by dividing the number of teeth on the smaller gear into the number of teeth on the larger gear.

GEARBOX: Transmission

GEAR REDUCTION: Torque is multiplied and speed decreased by the factor of the gear ratio. For example, a 3:1 gear ratio changes an input torque of 180 ft. lbs. and an input speed of 2700 rpm to 540 Ft. lbs. and 900 rpm, respectively. (No account is taken of frictional losses, which are always present.)

GEARTRAIN: A succession of intermeshing gears that form an assembly and provide for one or more torque changes as the power input is transmitted to the power output.

GEL COAT: A thin coat of plastic resin covering fiberglass body panels.

GENERATOR: A device which produces direct current (DC) necessary to charge the battery.

GOVERNOR: A device that senses vehicle speed and generates a hydraulic oil pressure. As vehicle speed increases, governor oil pressure rises.

GROUND CIRCUIT: (See circuit, ground.)

GROUND SIDE SWITCHING: The electrical/electronic circuit control switch is located after the circuit load.

GVWR: (Gross vehicle weight rating) total maximum weight a vehicle is designed to carry including the weight of the vehicle, passengers, equipment, gas, oil, etc.

HALOGEN: A special type of lamp known for its quality of brilliant white light. Originally used for fog lights and driving lights.

HARD CODES: DTCs that are present at the time of testing; also called continuous or current codes.

HARSH(ROUGH): An apply of a clutch or band that is more noticeable than a firm one; considered undesirable at any throttle position.

HEADER TANK: An expansion tank for the radiator coolant. It can be located remotely or built into the radiator.

HEAT RANGE: A term used to describe the ability of a spark plug to carry away heat. Plugs with longer nosed insulators take longer to carry heat off effectively.

HEAT RISER: A flapper in the exhaust manifold that is closed when the engine is cold, causing hot exhaust gases to heat the intake manifold providing better cold engine operation. A thermostatic spring opens the flapper when the engine warms up.

HEAVY THROTTLE: Approximately three-fourths of accelerator pedal travel.

HEMI: A name given an engine using hemispherical combustion chambers.

HERTZ (HZ): The international unit of frequency equal to one cycle per second (10,000 Hertz equals 10,000 cycles per second).

HIGH-IMPEDANCE DVOM (DIGITAL VOLT-OHMMETER): This styled device provides a built-in resistance value and is capable of limiting circuit current flow to safe milliamp levels.

HIGH RESISTANCE: Often refers to a circuit where there is an excessive amount of opposition to normal current flow.

HORSEPOWER: A measurement of the amount of work; one horsepower is the amount of work necessary to lift 33,000 lbs. one foot in one minute. Brake horsepower (bhp) is the horsepower delivered by an engine on a dynamometer. Net horsepower is the power remaining (measured at the flywheel of the engine) that can be used to turn the wheels after power is consumed through friction and running the engine accessories (water pump, alternator, air pump, fan etc.)

HOT CIRCUIT: (See circuit, hot; hot lead.)

HOT LEAD: A wire or conductor in the power side of the circuit. (See circuit, hot.)

HOT SIDE SWITCHING: The electrical/electronic circuit control switch is located before the circuit load.

HUB: The center part of a wheel or gear.

HUNTING (BUSYNESS): Repeating quick series of up-shifts and downshifts that causes noticeable change in engine rpm, for example, as in a 4-3-4 shift pattern.

HYDRAULICS: The use of liquid under pressure to transfer force of motion.

HYDROCARBON (HC): Any chemical compound made up of hydrogen and carbon. A major pollutant formed by the engine as a by-product of combustion.

HYDRODYNAMIC DRIVE UNITS: Devices that transmit power solely by the action of a kinetic fluid flow in a closed recirculating path. An impeller energizes the fluid and discharges the high-speed jet stream into the turbine for power output.

HYDROMETER: An instrument used to measure the specific gravity of a solution.

HYDROPLANING: A phenomenon of driving when water builds up under the tire tread, causing it to lose contact with the road. Slowing down will usually restore normal tire contact with the road.

HYPOID GEARSET: The drive pinion gear may be placed below or above the centerline of the driven gear; often used as a final drive gearset.

IDLE MIXTURE: The mixture of air and fuel (usually about 14:1) being fed to the cylinders. The idle mixture screw(s) are sometimes adjusted as part of a tune-up.

IDLER ARM: Component of the steering linkage which is a geometric duplicate of the steering gear arm. It supports the right side of the center steering link.

IMPELLER: Often called a pump, the impeller is the power input (drive) member of a hydrodynamic drive. As part of the torque converter cover, it acts as a centrifugal pump and puts the fluid in motion.

INCH POUND (inch lbs.; sometimes in. lb. or in. lbs.): One twelfth of a foot pound.

INDUCTANCE: The force that produces voltage when a conductor is passed through a magnetic field.

INDUCTION: A means of transferring electrical energy in the form of a magnetic field. Principle used in the ignition coil to increase voltage.

INITIAL FEEL: A distinct firmer feel at start of shift when compared with feel at finish of shift.

INJECTOR: A device which receives metered fuel under relatively low pressure and is activated to inject the fuel into the engine under relatively high pressure at a predetermined time.

INPUT: In an automatic transmission, the source of power from the engine is absorbed by the torque converter, which provides the power input into the transmission. The turbine drives the input(turbine)shaft.

INPUT SHAFT: The shaft to which torque is applied, usually carrying the driving gear or gears.

INTAKE MANIFOLD: A casting of passages or pipes used to conduct air or a fuel/air mixture to the cylinders.

INTERNAL GEAR: The ring-like outer gear of a planetary gearset with the gear teeth cut on the inside of the ring to provide a mesh with the planet pinions.

ISOLATION (CLAMPING) DIODES: Diodes positioned in a circuit to prevent self-induction from damaging electronic components.

IX ROTARY GEAR PUMP: Contains two rotating members, one shaped with internal gear teeth and the other with external gear teeth. As the gears separate, the fluid fills the gaps between gear teeth, is pulled across a crescent-shaped divider, and then is forced to flow through the outlet as the gears mesh.

IX ROTARY LOBE PUMP: Sometimes referred to as a gerotor type pump. Two rotating members, one shaped with internal lobes and the other with external lobes, separate and then mesh to cause fluid to flow.

JOURNAL: The bearing surface within which a shaft operates.

JUMPER CABLES: Two heavy duty wires with large alligator clips used to provide power from a charged battery to a discharged battery mounted in a vehicle.

JUMPSTART: Utilizing the sufficiently charged battery of one vehicle to start the engine of another vehicle with a discharged battery by the use of jumper cables.

KEY: A small block usually fitted in a notch between a shaft and a hub to prevent slippage of the two parts.

KICKDOWN: Detent downshift system; either linkage, cable, or electrically controlled.

KILO: A prefix used in the metric system to indicate one thousand.

KNOCK: Noise which results from the spontaneous ignition of a portion of the air-fuel mixture in the engine cylinder caused by overly advanced ignition timing or use of incorrectly low octane fuel for that engine.

KNOCK SENSOR: An input device that responds to spark knock, caused by over advanced ignition timing.

LABOR TIME: A specific amount of time required to perform a certain repair or diagnostic service as defined by a vehicle or after-market manufacturer .

LACQUER: A quick-drying automotive paint.

LATE: Shift that occurs when engine is at higher than normal rpm for given amount of throttle.

LIGHT-EMITTING DIODE (LED): A semiconductor diode that emits light as electrical current flows through it; used in some electronic display devices to emit a red or other color light.

LIGHT THROTTLE: Approximately one-fourth of accelerator pedal travel.

LIMITED SLIP: A type of differential which transfers driving force to the wheel with the best traction.

LIMP-IN MODE: Electrical shutdown of the transmission/ transaxle output solenoids, allowing only forward and reverse gears that are hydraulically energized by the manual valve. This permits the vehicle to be driven to a service facility for repair.

LIP SEAL: Molded synthetic rubber seal designed with an outer sealing edge (lip) that points into the fluid containing area to be sealed. This type of seal is used where rotational and axial forces are present.

LITHIUM-BASE GREASE: Chassis and wheel bearing grease using lithium as a base. Not compatible with sodium-base grease.

LOAD DEVICE: A circuit's resistance that converts the electrical energy into light, sound, heat, or mechanical movement.

LOAD RANGE: Indicates the number of plies at which a tire is rated. Load range B equals four-ply rating; C equals six-ply rating; and, D equals an eight-ply rating.

LOAD TORQUE: The amount of output torque needed from the transmission/transaxle to overcome the vehicle load.

LOCKING HUBS: Accessories used on part-time four-wheel drive systems that allow the front wheels to be disengaged from the drive train when four-wheel drive is not being used. When four-wheel drive is desired, the hubs are engaged, locking the wheels to the drive train.

LOCKUP CONVERTER: A torque converter that operates hydraulically and mechanically. When an internal apply plate (lockup plate) clamps to the torque converter cover, hydraulic slippage is eliminated.

LOCK RING: See Circlip or Snapring

MAGNET: Any body with the property of attracting iron or steel.

MAGNETIC FIELD: The area surrounding the poles of a magnet that is affected by its attraction or repulsion forces.

MAIN LINE PRESSURE: Often called control pressure or line pressure, it refers to the pressure of the oil leaving the pump and is controlled by the pressure regulator valve.

MALFUNCTION INDICATOR LAMP (MIL): Previously known as a check engine light, the dash-mounted MIL illuminates and signals the driver that an emission or driveability problem with the powertrain has been detected by the ECM/PCM. When this occurs, at least one diagnostic trouble code (DTC) has been stored into the control module memory.

MANIFOLD ABSOLUTE PRESSURE (MAP) SENSOR: Reads the amount of air pressure (vacuum) in the engine's intake manifold system; its signal is used to analyze engine load conditions.

MANIFOLD VACUUM: Low pressure in an engine intake manifold formed just below the throttle plates. Manifold vacuum is highest at idle and drops under acceleration.

MANIFOLD: A casting of passages or set of pipes which connect the cylinders to an inlet or outlet source.

MANUAL LEVER POSITION SWITCH (MLPS): A mechanical switching unit that is typically mounted externally to the transmission/transaxle to inform the PCM/ECM which gear range the driver has selected.

MANUAL VALVE: Located inside the transmission/transaxle, it is directly connected to the driver's shift lever. The position of the manual valve determines which hydraulic circuits will be charged with oil pressure and the operating mode of the transmission.

MANUAL VALVE LEVER POSITION SENSOR (MVLPS): The input from this device tells the TCM what gear range was selected.

MASS AIR FLOW (MAF) SENSOR: Measures the airflow into the engine.

MASTER CYLINDER: The primary fluid pressurizing device in a hydraulic system. In automotive use, it is found in brake and hydraulic clutch systems and is pedal activated, either directly or, in a power brake system, through the power booster.

MacPherson STRUT: A suspension component combining a shock absorber and spring in one unit.

MEDIUM THROTTLE: Approximately one-half of accelerator pedal travel.

MEGA: A metric prefix indicating one million.

MEMBER: An independent component of a hydrodynamic unit such as an impeller, a stator, or a turbine. It may have one or more elements.

MERCON: A fluid developed by Ford Motor Company in 1988. It contains a friction modifier and closely resembles operating characteristics of Dexron.

METAL SEALING RINGS: Made from cast iron or aluminum, their primary application is with dynamic components involving pressure sealing circuits of rotating members. These rings are designed with either butt or hook lock end joints.

METER (ANALOG): A linear-style meter representing data as lengths; a needle-style instrument interfacing with logical numerical increments. This style of electrical meter uses relatively low impedance internal resistance and cannot be used for testing electronic circuitry.

METER (DIGITAL): Uses numbers as a direct readout to show values. Most meters of this style use high impedance internal resistance and must be used for testing low current electronic circuitry.

MICRO: A metric prefix indicating one-millionth (0.000001).

MILLI: A metric prefix indicating one-thousandth (0.001).

MINIMUM THROTTLE: The least amount of throttle opening required for upshift; normally close to zero throttle.

MISFIRE: Condition occurring when the fuel mixture in a cylinder fails to ignite, causing the engine to run roughly.

MODULE: Electronic control unit, amplifier or igniter of solid state or integrated design which controls the current flow in the ignition primary circuit based on input from the pick-up coil. When the module opens the primary circuit, high secondary voltage is induced in the coil.

MODULATED: In an electronic-hydraulic converter clutch system (or shift valve system), the term modulated refers to the pulsing of a solenoid, at a variable rate. This action controls the buildup of oil pressure in the hydraulic circuit to allow a controlled amount of clutch slippage.

MODULATED CONVERTER CLUTCH CONTROL (MCCC): A pulse width duty cycle valve that controls the converter lockup apply pressure and maximizes smoother transitions between lock and unlock conditions.

MODULATOR PRESSURE (THROTTLE PRESSURE): A hydraulic signal oil pressure relating to the amount of engine load, based on either the amount of throttle plate opening or engine vacuum.

MODULATOR VALVE: A regulator valve that is controlled by engine vacuum, providing a hydraulic pressure that varies in relation to engine torque. The hydraulic torque signal functions to delay the shift pattern and provide a line pressure boost. (See throttle valve.)

MOTOR: An electromagnetic device used to convert electrical energy into mechanical energy.

MULTIPLE-DISC CLUTCH: A grouping of steel and friction lined plates that, when compressed together by hydraulic pressure acting upon a piston, lock or unlock a planetary member.

MULTI-WEIGHT: Type of oil that provides adequate lubrication at both high and low temperatures.

needed to move one amp through a resistance of one ohm.

MUSHY: Same as soft; slow and drawn out clutch apply with very little shift feel.

MUTUAL INDUCTION: The generation of current from one wire circuit to another by movement of the magnetic field surrounding a current-carrying circuit as its ampere flow increases or decreases.

NEEDLE BEARING: A bearing which consists of a number (usually a large number) of long, thin rollers.

NITROGEN OXIDE (NOx): One of the three basic pollutants found in the exhaust emission of an internal combustion engine. The amount of NOx usually varies in an inverse proportion to the amount of HC and CO.

NONPOSITIVE SEALING: A sealing method that allows some minor leakage, which normally assists in lubrication.

02 SENSOR: Located in the engine's exhaust system, it is an input device to the ECM/PCM for managing the fuel delivery and ignition system. A scanner can be used to observe the fluctuating voltage readings produced by an 02 sensor as the oxygen content of the exhaust is analyzed.

O-RING SEAL: Molded synthetic rubber seal designed with a circular cross-section. This type of seal is used primarily in static applications.

OBD II (ON-BOARD DIAGNOSTICS, SECOND GENERATION): Refers to the federal law mandating tighter control of 1996 and newer vehicle emissions, active monitoring of related devices, and standardization of terminology, data link connectors, and other technician concerns.

OCTANE RATING: A number, indicating the quality of gasoline based on its ability to resist knock. The higher the number, the better the quality. Higher compression engines require higher octane gas.

OEM: Original Equipment Manufactured. OEM equipment is that furnished standard by the manufacturer.

OFFSET: The distance between the vertical center of the wheel and the mounting surface at the lugs. Offset is positive if the center is outside the lug circle; negative offset puts the center line inside the lug circle.

OHM'S LAW: A law of electricity that states the relationship between voltage, current, and resistance. Volts = amperes x ohms

OHM: The unit used to measure the resistance of conductor-to-electrical flow. One ohm is the amount of resistance that limits current flow to one ampere in a circuit with one volt of pressure.

OHMMETER: An instrument used for measuring the resistance, in ohms, in an electrical circuit.

ONE-WAY CLUTCH: A mechanical clutch of roller or sprag design that resists torque or transmits power in one direction only. It is used to either hold or drive a planetary member.

ONE-WAY ROLLER CLUTCH: A mechanical device that transmits or holds torque in one direction only.

OPEN CIRCUIT: A break or lack of contact in an electrical circuit, either intentional (switch) or unintentional (bad connection or broken wire).

ORIFICE: Located in hydraulic oil circuits, it acts as a restriction. It slows down fluid flow to either create back pressure or delay pressure buildup downstream.

OSCILLOSCOPE: A piece of test equipment that shows electric impulses as a pattern on a screen. Engine performance can be analyzed by interpreting these patterns.

OUTPUT SHAFT: The shaft which transmits torque from a device, such as a transmission.

OUTPUT SPEED SENSOR (OSS): Identifies transmission/transaxle output shaft speed for shift timing and may be used to calculate TCC slip; often functions as the VSS (vehicle speed sensor).

OVERDRIVE: (1.) A device attached to or incorporated in a transmission/transaxle that allows the engine to turn less than one full revolution for every complete revolution of the wheels. The net effect is to reduce engine rpm, thereby using less fuel. A typical overdrive gear ratio would be .87:1, instead of the normal 1:1 in high gear. (2.) A gear assembly which produces more shaft revolutions than that transmitted to it.

OVERDRIVE PLANETARY GEARSET: A single planetary gearset designed to provide a direct drive and overdrive ratio. When coupled to a three-speed transmission/transaxle configuration, a four-speed/overdrive unit is present.

OVERHEAD CAMSHAFT (OHC): An engine configuration in which the camshaft is mounted on top of the cylinder head and operates the valve either directly or by means of rocker arms.

OVERHEAD VALVE (OHV): An engine configuration in which all of the valves are located in the cylinder head and the camshaft is located in the cylinder block. The camshaft operates the valves via lifters and pushrods.

OVERRUNCLUTCH: Another name for a one-way mechanical clutch. Applies to both roller and sprag designs.

OVERSTEER: The tendency of some vehicles, when steering into a turn, to over-respond or steer more than required, which could result in excessive slip of the rear wheels. Opposite of under-steer.

OXIDATION STABILIZERS: Absorb and dissipate heat. Automatic transmission fluid has high resistance to varnish and sludge buildup that occurs from excessive heat that is generated primarily in the torque converter. Local temperatures as high as 6000F (3150C) can occur at the clutch plates during engagement, and this heat must be absorbed and dissipated. If the fluid cannot withstand the heat, it burns or oxidizes, resulting in an almost immediate destruction of friction materials, clogged filter screen and hydraulic passages, and sticky valves.

OXIDES OF NITROGEN: See nitrogen oxide (NOx).

OXYGEN SENSOR: Used with a feedback system to sense the presence of oxygen in the exhaust gas and signal the computer which can use the voltage signal to determine engine operating efficiency and adjust the air/fuel ratio.

PARALLEL CIRCUIT: (See circuit, parallel.)

PARTS WASHER: A basin or tub, usually with a built-in pump mechanism and hose used for circulating chemical solvent for the purpose of cleaning greasy, oily and dirty components.

PART-TIME FOUR WHEEL DRIVE: A system that is normally in the two wheel drive mode and only runs in four-wheel drive when the system is manually engaged because more traction is desired. Two or four wheel drive is normally selected by a lever to engage the front axle, but if locking hubs are used, these must also be manually engaged in the Lock position. Otherwise, the front axle will not drive the front wheels.

PASSIVE RESTRAINT: Safety systems such as air bags or automatic seat belts which operate with no action required on the part of the driver or passenger. Mandated by Federal regulations on all vehicles sold in the U.S. after 1990.

PAYLOAD: The weight the vehicle is capable of carrying in addition to its own weight. Payload includes weight of the driver, passengers and cargo, but not coolant, fuel, lubricant, spare tire, etc.

PCM: Powertrain control module.

PCV VALVE: A valve usually located in the rocker cover that vents crankcase vapors back into the engine to be reburned.

PERCOLATION: A condition in which the fuel actually "boils," due to excessive heat. Percolation prevents proper atomization of the fuel causing rough running.

PICK-UP COIL: The coil in which voltage is induced in an electronic ignition.

PING: A metallic rattling sound produced by the engine during acceleration. It is usually due to incorrect ignition timing or a poor grade of gasoline.

PINION: The smaller of two gears. The rear axle pinion drives the ring gear which transmits motion to the axle shafts.

PINION GEAR: The smallest gear in a drive gear assembly.

PISTON: A disc or cup that fits in a cylinder bore and is free to move. In hydraulics, it provides the means of converting hydraulic pressure into a usable force. Examples of piston applications are found in servo, clutch, and accumulator units.

PISTON RING: An open-ended ring which fits into a groove on the outer diameter of the piston. Its chief function is to form a seal between the piston and cylinder wall. Most automotive pistons have three rings: two for compression sealing; one for oil sealing.

PITMAN ARM: A lever which transmits steering force from the steering gear to the steering linkage.

PLANET CARRIER: A basic member of a planetary gear assembly that carries the pinion gears.

PLANET PINIONS: Gears housed in a planet carrier that are in constant mesh with the sun gear and internal gear. Because they have their own independent rotating centers, the pinions are capable of rotating around the sun gear or the inside of the internal gear.

PLANETARY GEAR RATIO: The reduction or overdrive ratio developed by a planetary gearset.

PLANETARY GEARSET: In its simplest form, it is made up of a basic assembly group containing a sun gear, internal gear, and planet carrier. The gears are always in constant mesh and offer a wide range of gear ratio possibilities.

PLANETARY GEARSET (COMPOUND): Two planetary gearsets combined together.

PLANETARY GEARSET (SIMPLE): An assembly of gears in constant mesh consisting of a sun gear, several pinion gears mounted in a carrier, and a ring gear. It provides gear ratio and direction changes, in addition to a direct drive and a neutral.

PLY RATING: A. rating given a tire which indicates strength (but not necessarily actual plies). A two-ply/four-ply rating has only two plies, but the strength of a four-ply tire.

POLARITY: Indication (positive or negative) of the two poles of a battery.

PORT: An opening for fluid intake or exhaust.

POSITIVE SEALING: A sealing method that completely prevents leakage.

POTENTIAL: Electrical force measured in volts; sometimes used interchangeably with voltage.

POWER: The ability to do work per unit of time, as expressed in horsepower; one horsepower equals 33,000 ft. lbs. of work per minute, or 550 ft. lbs. of work per second.

POWER FLOW: The systematic flow or transmission of power through the gears, from the input shaft to the output shaft.

POWER-TO-WEIGHT RATIO: Ratio of horsepower to weight of car.

POWERTRAIN: See Drivetrain.

POWERTRAIN CONTROL MODULE (PCM): Current designation for the engine control module (ECM). In many cases, late model vehicle control units manage the engine as well as the transmission. In other settings, the PCM controls the engine and is interfaced with a TCM to control transmission functions.

Ppm: Parts per million; unit used to measure exhaust emissions.

PREIGNITION: Early ignition of fuel in the cylinder, sometimes due to glowing carbon deposits in the combustion chamber. Preignition can be damaging since combustion takes place prematurely.

PRELOAD: A predetermined load placed on a bearing during assembly or by adjustment.

PRESS FIT: The mating of two parts under pressure, due to the inner diameter of one being smaller than the outer diameter of the other, or vice versa; an interference fit.

PRESSURE: The amount of force exerted upon a surface area.

PRESSURE CONTROL SOLENOID (PCS): An output device that provides a boost oil pressure to the mainline regulator valve to control line pressure. Its operation is determined by the amount of current sent from the PCM.

PRESSURE GAUGE: An instrument used for measuring the fluid pressure in a hydraulic circuit.

PRESSURE REGULATOR VALVE: In automatic transmissions, its purpose is to regulate the pressure of the pump output and supply the basic fluid pressure necessary to operate the transmission. The regulated fluid pressure may be referred to as mainline pressure, line pressure, or control pressure.

PRESSURE SWITCH ASSEMBLY (PSA): Mounted inside the transmission, it is a grouping of oil pressure switches that inputs to the PCM when certain hydraulic passages are charged with oil pressure.

PRESSURE PLATE: A spring-loaded plate (part of the clutch) that transmits power to the driven (friction) plate when the clutch is engaged.

PRIMARY CIRCUIT: The low voltage side of the ignition system which consists of the ignition switch, ballast resistor or resistance wire, bypass, coil, electronic control unit and pick-up coil as well as the connecting wires and harnesses.

PROFILE: Term used for tire measurement (tire series), which is the ratio of tire height to tread width.

PROM (PROGRAMMABLE READ-ONLY MEMORY): The heart of the computer that compares input data and makes the engineered program or strategy decisions about when to trigger the appropriate output based on stored computer instructions.

PULSE GENERATOR: A two-wire pickup sensor used to produce a fluctuating electrical signal. This changing signal is read by the controller to determine the speed of the object and can be used to measure transmission/transaxle input speed, output speed, and vehicle speed.

PSI: Pounds per square inch; a measurement of pressure.

PULSE WIDTH DUTY CYCLE SOLENOID (PULSE WIDTH MODULATED SOLENOID): A computer-controlled solenoid that turns on and off at a variable rate producing a modulated oil pressure; often referred to as a pulse width modulated (PWM) solenoid. Employed in many electronic automatic transmissions and transaxles, these solenoids are used to manage shift control and converter clutch hydraulic circuits.

PUSHROD: A steel rod between the hydraulic valve lifter and the valve rocker arm in overhead valve (OHV) engines.

PUMP: A mechanical device designed to create fluid flow and pressure buildup in a hydraulic system.

QUARTER PANEL: General term used to refer to a rear fender. Quarter panel is the area from the rear door opening to the tail light area and from rear wheel well to the base of the trunk and roof-line.

RACE: The surface on the inner or outer ring of a bearing on which the balls, needles or rollers move.

RACK AND PINION: A type of automotive steering system using a pinion gear attached to the end of the steering shaft. The pinion meshes with a long rack attached to the steering linkage.

RADIAL TIRE: Tire design which uses body cords running at right angles to the center line of the tire. Two or more belts are used to give tread strength. Radials can be identified by their characteristic sidewall bulge.

RADIATOR: Part of the cooling system for a water-cooled engine, mounted in the front of the vehicle and connected to the engine with rubber hoses. Through the radiator, excess combustion heat is dissipated into the atmosphere through forced convection using a water and glycol based mixture that circulates through, and cools, the engine.

RANGE REFERENCE AND CLUTCH/BAND APPLY CHART: A guide that shows the application of clutches and bands for each gear, within the selector range positions. These charts are extremely useful for understanding how the unit operates and for diagnosing malfunctions.

RAVIGNEAUX GEARSET: A compound planetary gearset that features matched dual planetary pinions (sets of two) mounted in a single planet carrier. Two sun gears and one ring mesh with the carrier pinions.

REACTION MEMBER: The stationary planetary member, in a planetary gearset, that is grounded to the transmission/transaxle case through the use of friction and wedging devices known as bands, disc clutches, and one-way clutches.

REACTION PRESSURE: The fluid pressure that moves a spool valve against an opposing force or forces; the area on which the opposing force acts. The opposing force can be a spring or a combination of spring force and auxiliary hydraulic force.

REACTOR, TORQUE CONVERTER: The reaction member of a fluid torque converter, more commonly called a stator. (See stator.)

REAR MAIN OIL SEAL: A synthetic or rope-type seal that prevents oil from leaking out of the engine past the rear main crankshaft bearing.

RECIRCULATING BALL: Type of steering system in which recirculating steel balls occupy the area between the nut and worm wheel, causing a reduction in friction.

RECTIFIER: A device (used primarily in alternators) that permits electrical current to flow in one direction only.

REDUCTION: (See gear reduction.)

REGULATOR VALVE: A valve that changes the pressure of the oil in a hydraulic circuit as the oil passes through the valve by bleeding off (or exhausting) some of the volume of oil supplied to the valve.

REFRIGERANT 12 (R-12) or 134 (R-134): The generic name of the refrigerant used in automotive air conditioning systems.

REGULATOR: A device which maintains the amperage and/or voltage levels of a circuit at predetermined values.

RELAY: A switch which automatically opens and/or closes a circuit.

RELAY VALVE: A valve that directs flow and pressure. Relay valves simply connect or disconnect interrelated passages without restricting the fluid flow or changing the pressure.

RELIEF VALVE: A spring-loaded, pressure-operated valve that limits oil pressure buildup in a hydraulic circuit to a predetermined maximum value.

RELUCTOR: A wheel that rotates inside the distributor and triggers the release of voltage in an electronic ignition.

RESERVOIR: The storage area for fluid in a hydraulic system; often called a sump.

RESIN: A liquid plastic used in body work.

RESIDUAL MAGNETISM: The magnetic strength stored in a material after a magnetizing field has been removed.

RESISTANCE: The opposition to the flow of current through a circuit or electrical device, and is measured in ohms. Resistance is equal to the voltage divided by the amperage.

RESISTOR SPARK PLUG: A spark plug using a resistor to shorten the spark duration. This suppresses radio interference and lengthens plug life.

RESISTOR: A device, usually made of wire, which offers a preset amount of resistance in an electrical circuit.

RESULTANT FORCE: The single effective directional thrust of the fluid force on the turbine produced by the vortex and rotary forces acting in different planes.

RETARD: Set the ignition timing so that spark occurs later (fewer degrees before TDC).

RHEOSTAT: A device for regulating a current by means of a variable resistance.

RING GEAR: The name given to a ring-shaped gear attached to a differential case, or affixed to a flywheel or as part of a planetary gear set.

ROADLOAD: grade.

ROCKER ARM: A lever which rotates around a shaft pushing down (opening) the valve with an end when the other end is pushed up by the pushrod. Spring pressure will later close the valve.

ROCKER PANEL: The body panel below the doors between the wheel opening.

ROLLER BEARING: A bearing made up of hardened inner and outer races between which hardened steel rollers move.

ROLLER CLUTCH: A type of one-way clutch design using rollers and springs mounted within an inner and outer cam race assembly.

ROTARY FLOW: The path of the fluid trapped between the blades of the members as they revolve with the rotation of the torque converter cover (rotational inertia).

ROTOR: (1.) The disc-shaped part of a disc brake assembly, upon which the brake pads bear; also called, brake disc. (2.) The device mounted atop the distributor shaft, which passes current to the distributor cap tower contacts.

ROTARY ENGINE: See Wankel engine.

RPM: Revolutions per minute (usually indicates engine speed).

RTV: A gasket making compound that cures as it is exposed to the atmosphere. It is used between surfaces that are not perfectly machined to one another, leaving a slight gap that the RTV fills and in which it hardens. The letters RTV represent room temperature vulcanizing.

RUN-ON: Condition when the engine continues to run, even when the key is turned off. See dieseling.

SEALED BEAM: A automotive headlight. The lens, reflector and filament from a single unit.

SEATBELT INTERLOCK: A system whereby the car cannot be started unless the seatbelt is buckled.

SECONDARY CIRCUIT: The high voltage side of the ignition system, usually above 20,000 volts. The secondary includes the ignition coil, coil wire, distributor cap and rotor, spark plug wires and spark plugs.

SELF-INDUCTION: The generation of voltage in a current-carrying wire by changing the amount of current flowing within that wire.

SEMI-CONDUCTOR: A material (silicon or germanium) that is neither a good conductor nor an insulator; used in diodes and transistors.

SEMI-FLOATING AXLE: In this design, a wheel is attached to the axle shaft, which takes both drive and cornering loads. Almost all solid axle passenger cars and light trucks use this design.

SENDING UNIT: A mechanical, electrical, hydraulic or electromagnetic device which transmits information to a gauge.

SENSOR: Any device designed to measure engine operating conditions or ambient pressures and temperatures. Usually electronic in nature and designed to send a voltage signal to an on-board computer, some sensors may operate as a simple on/off switch or they may provide a variable voltage signal (like a potentiometer) as conditions or measured parameters change.

SERIES CIRCUIT: (See circuit, series.)

SERPENTINE BELT: An accessory drive belt, with small multiple v-ribs, routed around most or all of the engine-powered accessories such as the alternator and power steering pump. Usually both the front and the back side of the belt comes into contact with various pulleys.

SERVO: In an automatic transmission, it is a piston in a cylinder assembly that converts hydraulic pressure into mechanical force and movement; used for the application of the bands and clutches.

SHIFT BUSYNESS: When referring to a torque converter clutch, it is the frequent apply and release of the clutch plate due to uncommon driving conditions.

SHIFT VALVE: Classified as a relay valve, it triggers the automatic shift in response to a governor and a throttle signal by directing fluid to the appropriate band and clutch apply combination to cause the shift to occur.

SHIM: Spacers of precise, predetermined thickness used between parts to establish a proper working relationship.

SHIMMY: Vibration (sometimes violent) in the front end caused by misaligned front end, out of balance tires or worn suspension components.

SHORT CIRCUIT: An electrical malfunction where current takes the path of least resistance to ground (usually through damaged insulation). Current flow is excessive from low resistance resulting in a blown fuse.

SHUDDER: Repeated jerking or stick-slip sensation, similar to chuggle but more severe and rapid in nature, that may be most noticeable during certain ranges of vehicle speed; also used to define condition after converter clutch engagement.

SIMPSON GEARSET: A compound planetary gear train that integrates two simple planetary gearsets referred to as the front planetary and the rear planetary.

SINGLE OVERHEAD CAMSHAFT: See overhead camshaft.

SKIDPLATE: A metal plate attached to the underside of the body to protect the fuel tank, transfer case or other vulnerable parts from damage.

SLAVE CYLINDER: In automotive use, a device in the hydraulic clutch system which is activated by hydraulic force, disengaging the clutch.

SLIPPING: Noticeable increase in engine rpm without vehicle speed increase; usually occurs during or after initial clutch or band engagement.

SLUDGE: Thick, black deposits in engine formed from dirt, oil, water, etc. It is usually formed in engines when oil changes are neglected.

SNAP RING: A circular retaining clip used inside or outside a shaft or part to secure a shaft, such as a floating wrist pin.

SOFT: Slow, almost unnoticeable clutch apply with very little shift feel.

SOFTCODES: DTCs that have been set into the PCM memory but are not present at the time of testing; often referred to as history or intermittent codes.

SOHC: Single overhead camshaft.

SOLENOID: An electrically operated, magnetic switching device.

SPALLING: A wear pattern identified by metal chips flaking off the hardened surface. This condition is caused by foreign particles, overloading situations, and/or normal wear.

SPARK PLUG: A device screwed into the combustion chamber of a spark ignition engine. The basic construction is a conductive core inside of a ceramic insulator, mounted in an outer conductive base. An electrical charge from the spark plug wire travels along the conductive core and jumps a preset air gap to a grounding point or points at the end of the conductive base. The resultant spark ignites the fuel/air mixture in the combustion chamber.

SPECIFIC GRAVITY (BATTERY): The relative weight of liquid (battery electrolyte) as compared to the weight of an equal volume of water.

SPLINES: Ridges machined or cast onto the outer diameter of a shaft or inner diameter of a bore to enable parts to mate without rotation.

SPLIT TORQUE DRIVE: In a torque converter, it refers to parallel paths of torque transmission, one of which is mechanical and the other hydraulic.

SPONGY PEDAL: A soft or spongy feeling when the brake pedal is depressed. It is usually due to air in the brake lines.

SPOOLVALVE: A precision-machined, cylindrically shaped valve made up of lands and grooves. Depending on its position in the valve bore, various interconnecting hydraulic circuit passages are either opened or closed.

SPRAG CLUTCH: A type of one-way clutch design using cams or contoured-shaped sprags between inner and outer races. (See one-way clutch.)

SPRUNG WEIGHT: The weight of a car supported by the springs.

SQUARE-CUT SEAL: Molded synthetic rubber seal designed with a square- or rectangular-shaped cross-section. This type of seal is used for both dynamic and static applications.

SRS: Supplemental restraint system

STABILIZER (SWAY) BAR: A bar linking both sides of the suspension. It resists sway on turns by taking some of added load from one wheel and putting it on the other.

STAGE: The number of turbine sets separated by a stator. A turbine set may be made up of one or more turbine members. A three-element converter is classified as a single stage.

STALL: In fluid drive transmission/transaxle applications, stall refers to engine rpm with the transmission/transaxle engaged and the vehicle stationary; throttle valve can be in any position between closed and wide open.

STALL SPEED: In fluid drive transmission/transaxle applications, stall speed refers to the maximum engine rpm with the transmission/transaxle engaged and vehicle stationary, when the throttle valve is wide open. (See stall; stall test.)

STALL TEST: A procedure recommended by many manufacturers to help determine the integrity of an engine, the torque converter stator, and certain clutch and band combinations. With the shift lever in each of the forward and reverse positions and with the brakes firmly applied, the accelerator pedal is momentarily pressed to the wide open throttle (WOT) position. The engine rpm reading at full throttle can provide clues for diagnosing the condition of the items listed above.

STALL TORQUE: The maximum design or engineered torque ratio of a fluid torque converter, produced under stall speed conditions. (See stall speed.)

STARTER: A high-torque electric motor used for the purpose of starting the engine, typically through a high ratio geared drive connected to the flywheel ring gear.

STATIC: A sealing application in which the parts being sealed do not move in relation to each other.

STATOR (REACTOR): The reaction member of a fluid torque converter that changes the direction of the fluid as it leaves the turbine to enter the impeller vanes. During the torque multiplication phase, this action assists the impeller's rotary force and results in an increase in torque.

STEERING GEOMETRY: Combination of various angles of suspension components (caster, camber, toe-in); roughly equivalent to front end alignment.

STRAIGHT WEIGHT: Term designating motor oil as suitable for use within a narrow range of temperatures. Outside the narrow temperature range its flow characteristics will not adequately lubricate.

STROKE: The distance the piston travels from bottom dead center to top dead center.

SUBSTITUTION: Replacing one part suspected of a defect with a like part of known quality.

SUMP: The storage vessel or reservoir that provides a ready source of fluid to the pump. In an automatic transmission, the sump is the oil pan. All fluid eventually returns to the sump for recycling into the hydraulic system.

SUN GEAR: In a planetary gearset, it is the center gear that meshes with a cluster of planet pinions.

SUPERCHARGER: An air pump driven mechanically by the engine through belts, chains, shafts or gears from the crankshaft. Two general types of supercharger are the positive displacement and centrifugal type, which pump air in direct relationship to the speed of the engine.

SUPPLEMENTAL RESTRAINT SYSTEM: See air bag.

SURGE: Repeating engine-related feeling of acceleration and deceleration that is less intense than chuggle.

SWITCH: A device used to open, close, or redirect the current in an electrical circuit.

SYNCHROMESH: A manual transmission/transaxle that is equipped with devices (synchronizers) that match the gear speeds so that the transmission/transaxle can be downshifted without clashing gears.

SYNTHETIC OIL: Non-petroleum based oil.

TACHOMETER: A device used to measure the rotary speed of an engine, shaft, gear, etc., usually in rotations per minute.

TDC: Top dead center. The exact top of the piston's stroke.

TEFLON SEALING RINGS: Teflon is a soft, durable, plastic-like material that is resistant to heat and provides excellent sealing. These rings are designed with either scarf-cut joints or as one-piece rings. Teflon sealing rings have replaced many metal ring applications.

TERMINAL: A device attached to the end of a wire or cable to make an electrical connection.

TEST LIGHT, CIRCUIT-POWERED: Uses available circuit voltage to test circuit continuity.

TEST LIGHT, SELF-POWERED: Uses its own battery source to test circuit continuity.

THERMISTOR: A special resistor used to measure fluid temperature; it decreases its resistance with increases in temperature.

THERMOSTAT: A valve, located in the cooling system of an engine, which is closed when cold and opens gradually in response to engine heating, controlling the temperature of the coolant and rate of coolant flow.

THERMOSTATIC ELEMENT: A heat-sensitive, spring-type device that controls a drain port from the upper sump area to the lower sump. When the transaxle fluid reaches operating temperature, the port is closed and the upper sump fills, thus reducing the fluid level in the lower sump.

THROTTLE POSITION (TP) SENSOR: Reads the degree of throttle opening; its signal is used to analyze engine load conditions. The ECM/PCM decides to apply the TCC, or to disengage it for coast or load conditions that need a converter torque boost.

THROTTLE PRESSURE/MODULATOR PRESSURE: A hydraulic signal oil pressure relating to the amount of engine load, based on either the amount of throttle plate opening or engine vacuum.

THROTTLE VALVE: A regulating or balanced valve that is controlled mechanically by throttle linkage or engine vacuum. It sends a hydraulic signal to the shift valve body to control shift timing and shift quality. (See balanced valve; modulator valve.)

THROW-OUT BEARING: As the clutch pedal is depressed, the throwout bearing moves against the spring fingers of the pressure plate, forcing the pressure plate to disengage from the driven disc.

TIE ROD: A rod connecting the steering arms. Tie rods have threaded ends that are used to adjust toe-in.

TIE-UP: Condition where two opposing clutches are attempting to apply at same time, causing engine to labor with noticeable loss of engine rpm.

TIMING BELT: A square-toothed, reinforced rubber belt that is driven by the crankshaft and operates the camshaft.

TIMING CHAIN: A roller chain that is driven by the crankshaft and operates the camshaft.

TIRE ROTATION: Moving the tires from one position to another to make the tires wear evenly.

TOE-IN (OUT): A term comparing the extreme front and rear of the front tires. Closer together at the front is toe-in; farther apart at the front is toe-out.

TOP DEAD CENTER (TDC): The point at which the piston reaches the top of its travel on the compression stroke.

TORQUE: Measurement of turning or twisting force, expressed as foot-pounds or inch-pounds.

TORQUE CONVERTER: A turbine used to transmit power from a driving member to a driven member via hydraulic action, providing changes in drive ratio and torque. In automotive use, it links the driveplate at the rear of the engine to the automatic transmission.

TORQUE CONVERTER CLUTCH: The apply plate (lockup plate) assembly used for mechanical power flow through the converter.

TORQUE PHASE: Sometimes referred to as slip phase or stall phase, torque multiplication occurs when the turbine is turning at a slower speed than the impeller, and the stator is reactionary (stationary). This sequence generates a boost in output torque.

TORQUE RATING (STALL TORQUE): The maximum torque multiplication that occurs during stall conditions, with the engine at wide open throttle (WOT) and zero turbine speed.

TORQUE RATIO: An expression of the gear ratio factor on torque effect. A 3:1 gear ratio or 3:1 torque ratio increases the torque input by the ratio factor of 3. Input torque (100 ft. lbs.) x 3 = output torque (300 ft. lbs.)

TRACTION: The amount of usable tractive effort before the drive wheels slip on the road contact surface.

TORSION BAR SUSPENSION: Long rods of spring steel which take the place of springs. One end of the bar is anchored and the other arm (attached to the suspension) is free to twist. The bars' resistance to twisting causes springing action.

TRACK: Distance between the centers of the tires where they contact the ground.

TRACTION CONTROL: A control system that prevents the spinning of a vehicle's drive wheels when excess power is applied.

TRACTIVE EFFORT: The amount of force available to the drive wheels, to move the vehicle.

TRANSAXLE: A single housing containing the transmission and differential. Transaxles are usually found on front engine/front wheel drive or rear engine/rear wheel drive cars.

TRANSDUCER: A device that changes energy from one form to another. For example, a transducer in a microphone changes sound energy to electrical energy. In automotive air-conditioning controls used in automatic temperature systems, a transducer changes an electrical signal to a vacuum signal, which operates mechanical doors.

TRANSMISSION: A powertrain component designed to modify torque and speed developed by the engine; also provides direct drive, reverse, and neutral.

TRANSMISSION CONTROL MODULE (TCM): Manages transmission functions. These vary according to the manufacturer's product design but may include converter clutch operation, electronic shift scheduling, and mainline pressure.

TRANSMISSION FLUID TEMPERATURE (TFT) SENSOR: Originally called a transmission oil temperature (TOT) sensor, this input device to the ECM/PCM senses the fluid temperature and provides a resistance value. It operates on the thermistor principle.

TRANSMISSION INPUT SPEED (TIS) SENSOR: Measures turbine shaft (input shaft) rpm's and compares to engine rpm's to determine torque converter slip. When compared to the transmission output speed sensor or VSS, gear ratio and clutch engagement timing can be determined.

TRANSMISSION OIL TEMPERATURE (TOT) SENSOR: (See transmission fluid temperature (TFT) sensor.)

TRANSMISSION RANGE SELECTOR (TRS) SWITCH: Tells the module which gear shift position the driver has chosen.

TRANSFER CASE: A gearbox driven from the transmission that delivers power to both front and rear driveshafts in a four-wheel drive system. Transfer cases usually have a high and low range set of gears, used depending on how much pulling power is needed.

TRANSISTOR: A semi-conductor component which can be actuated by a small voltage to perform an electrical switching function.

TREAD WEAR INDICATOR: Bars molded into the tire at right angles to the tread that appear as horizontal bars when $\frac{1}{16}$ in. of tread remains.

TREAD WEAR PATTERN: The pattern of wear on tires which can be "read" to diagnose problems in the front suspension.

TUNE-UP: A regular maintenance function, usually associated with the replacement and adjustment of parts and components in the electrical and fuel systems of a vehicle for the purpose of attaining optimum performance.

TURBINE: The output (driven) member of a fluid coupling or fluid torque converter. It is splined to the input (turbine) shaft of the transmission.

TURBOCHARGER: An exhaust driven pump which compresses intake air and forces it into the combustion chambers at higher than atmospheric pressures. The increased air pressure allows more fuel to be burned and results in increased horsepower being produced.

TURBULENCE: The interference of molecules of a fluid (or vapor) with each other in a fluid flow.

TYPE F: Transmission fluid developed and used by Ford Motor Company up to 1982. This fluid type provides a high coefficient of friction.

TYPE 7176: The preferred choice of transmission fluid for Chrysler automatic transmissions and transaxles. Developed in 1986, it closely resembles Dexron and Mercon. Type 7176 is the recommended service fill fluid for all Chrysler products utilizing a lockup torque converter dating back to 1978.

U-JOINT (UNIVERSAL JOINT): A flexible coupling in the drive train that allows the driveshafts or axle shafts to operate at different angles and still transmit rotary power.

UNDERSTEER: The tendency of a car to continue straight ahead while negotiating a turn.

UNIT BODY: Design in which the car body acts as the frame.

UNLEADED FUEL: Fuel which contains no lead (a common gasoline additive). The presence of lead in fuel will destroy the functioning elements of a catalytic converter, making it useless.

UNSPRUNG WEIGHT: The weight of car components not supported by the springs (wheels, tires, brakes, rear axle, control arms, etc.).

UPSHIFT: A shift that results in a decrease in torque ratio and an increase in speed.

VACUUM: A negative pressure; any pressure less than atmospheric pressure.

VACUUM ADVANCE: A device which advances the ignition timing in response to increased engine vacuum.

VACUUM GAUGE: An instrument used for measuring the existing vacuum in a vacuum circuit or chamber. The unit of measure is inches (of mercury in a barometer).

VACUUM MODULATOR: Generates a hydraulic oil pressure in response to the amount of engine vacuum.

VALVES: Devices that can open or close fluid passages in a hydraulic system and are used for directing fluid flow and controlling pressure.

VALVE BODY ASSEMBLY: The main hydraulic control assembly of the transmission/transaxle that contains numerous valves, check balls, and other components to control the distribution of pressurized oil throughout the transmission.

VALVE CLEARANCE: The measured gap between the end of the valve stem and the rocker arm, cam lobe or follower that activates the valve.

VALVE GUIDES: The guide through which the stem of the valve passes. The guide is designed to keep the valve in proper alignment.

VALVE LASH (clearance): The operating clearance in the valve train.

VALVE TRAIN: The system that operates intake and exhaust valves, consisting of camshaft, valves and springs, lifters, pushrods and rocker arms.

VAPOR LOCK: Boiling of the fuel in the fuel lines due to excess heat. This will interfere with the flow of fuel in the lines and can completely stop the flow. Vapor lock normally only occurs in hot weather.

VARIABLE DISPLACEMENT (VARIABLE CAPACITY) VANE PUMP: Slipper-type vanes, mounted in a revolving rotor and contained within the bore of a movable slide, capture and then force fluid to flow. Movement of the slide to various positions changes the size of the vane chambers and the amount of fluid flow. **Note:** GM refers to this pump design as variable displacement, and Ford terms it variable capacity.

VARIABLE FORCE SOLENOID (VFS): Commonly referred to as the electronic pressure control (EPC) solenoid, it replaces the cable/linkage style of TV system control and is integrated with a spool valve and spring assembly to control pressure. A variable computer-controlled current flow varies the internal force of the solenoid on the spool valve and resulting control pressure.

VARIABLE ORIFICE THERMAL VALVE: Temperature-sensitive hydraulic oil control device that adjusts the size of a circuit path opening. By altering the size of the opening, the oil flow rate is adapted for cold to hot oil viscosity changes.

VARNISH: Term applied to the residue formed when gasoline gets old and stale.

VCM: See Electronic Control Unit (ECU).

VEHICLE SPEED SENSOR (VSS): Provides an electrical signal to the computer module, measuring vehicle speed, and affects the torque converter clutch engagement and release.

VESPEL SEALING RINGS: Hard plastic material that produces excellent sealing in dynamic settings. These rings are found in late versions of the 4T60 and in all 4T60-E and 4T80-E transaxles.

VISCOSITY: The ability of a fluid to flow. The lower the viscosity rating, the easier the fluid will flow. 10 weight motor oil will flow much easier than 40 weight motor oil.

VISCOSITY INDEX IMPROVERS: Keeps the viscosity nearly constant with changes in temperature. This is especially important at low temperatures, when the oil needs to be thin to aid in shifting and for cold-weather starting. Yet it must not be so thin that at high temperatures it will cause excessive hydraulic leakage so that pumps are unable to maintain the proper pressures.

VISCOUS CLUTCH: A specially designed torque converter clutch apply plate that, through the use of a silicon fluid, clamps smoothly and absorbs torsional vibrations.

VOLT: Unit used to measure the force or pressure of electricity. It is defined as the pressure

VOLTAGE: The electrical pressure that causes current to flow. Voltage is measured in volts (V).

VOLTAGE, APPLIED: The actual voltage read at a given point in a circuit. It equals the available voltage of the power supply minus the losses in the circuit up to that point.

VOLTAGE DROP: The voltage lost or used in a circuit by normal loads such as a motor or lamp or by abnormal loads such as a poor (high-resistance) lead or terminal connection.

VOLTAGE REGULATOR: A device that controls the current output of the alternator or generator.

VOLTMETER: An instrument used for measuring electrical force in units called volts. Voltmeters are always connected parallel with the circuit being tested.

VORTEX FLOW: The crosswise or circulatory flow of oil between the blades of the members caused by the centrifugal pumping action of the impeller.

WANKEL ENGINE: An engine which uses no pistons. In place of pistons, triangular-shaped rotors revolve in specially shaped housings.

WATER PUMP: A belt driven component of the cooling system that mounts on the engine, circulating the coolant under pressure.

WATT: The unit for measuring electrical power. One watt is the product of one ampere and one volt (watts equals amps times volts). Wattage is the horsepower of electricity (746 watts equal one horsepower).

WHEEL ALIGNMENT: Inclusive term to describe the front end geometry (caster, camber, toe-in/out).

WHEEL CYLINDER: Found in the automotive drum brake assembly, it is a device, actuated by hydraulic pressure, which, through internal pistons, pushes the brake shoes outward against the drums.

WHEEL WEIGHT: Small weights attached to the wheel to balance the wheel and tire assembly. Out-of-balance tires quickly wear out and also give erratic handling when installed on the front.

WHEELBASE: Distance between the center of front wheels and the center of rear wheels.

WIDE OPEN THROTTLE (WOT): Full travel of accelerator pedal.

WORK: The force exerted to move a mass or object. Work involves motion; if a force is exerted and no motion takes place, no work is done. Work per unit of time is called power. Work = force x distance = ft. lbs. 33,000 ft. lbs. in one minute = 1 horsepower

ZERO-THROTTLE COAST DOWN: A full release of accelerator pedal while vehicle is in motion and in drive range.

Commonly Used Abbreviations

2

2WD Two Wheel Drive

4

4WD Four Wheel Drive

A

A/C Air Conditioning
ABDC After Bottom Dead Center
ABS Anti-lock Brakes
AC Alternating Current
ACL Air cleaner
ACT Air Charge Temperature
AIR Secondary Air Injection
ALCL Assembly Line Communications Link
ALDL Assembly Line Diagnostic Link
AT Automatic Transaxle/Transmission
ATDC After Top Dead Center
ATF Automatic Transmission Fluid
ATS Air Temperature Sensor
AWD All Wheel Drive

B

BAP Barometric Absolute Pressure
BARO Barometric Pressure
BBDC Before Bottom Dead Center
BCM Body Control Module
BDC Bottom Dead Center
BPT Backpressure Transducer
BTDC Before Top Dead Center
BVSV Bimetallic Vacuum Switching Valve

C

CAC Charge Air Cooler
CARB California Air Resources Board
CAT Catalytic Converter
CCC Computer Command Control
CCCC Computer Controlled Catalytic Converter
CCCI Computer Controlled Coil Ignition
CCD Computer Controlled Dwell
CDI Capacitor Discharge Ignition
CEC Computerized Engine Control
CFI Continuous Fuel Injection
CIS Continuous Injection System
CIS-E Continuous Injection System - Electronic
CKP Crankshaft Position
CL Closed Loop
CMP Camshaft Position
CPP Clutch Pedal Position
CTOX Continuous Trap Oxidizer System
CTP Closed Throttle Position
CVC Constant Vacuum Control
CYL Cylinder

D

DBC Dual Bed Catalyst
DC Direct Current
DFI Direct Fuel Injection
DIS Distributorless Ignition System
DLC Data Link Connector
DMM Digital Multimeter
DOHC Double Overhead Camshaft
DRB Diagnostic Readout Box
DTC Diagnostic Trouble Code
DTM Diagnostic Test Mode
DVOM Digital Volt/Ohmmeter

E

EBCM Electronic Brake Control Module
ECM Engine Control Module
ECT Engine Coolant Temperature
ECU Engine Control Unit or Electronic Control Unit
EDIS Electronic Distributorless Ignition System
EEC Electronic Engine Control
EEPROM Electrically Erasable Programmable Read Only Memory
EFE Early Fuel Evaporation
EGR Exhaust Gas Recirculation
EGRT Exhaust Gas Recirculation Temperature
EGRVC EGR Valve Control
EPROM Erasable Programmable Read Only Memory
EVAP Evaporative Emissions
EVP EGR Valve Position

F

FBC Feedback Carburetor
FEEPROM Flash Electrically Erasable Programmable Read Only Memory
FF Flexible Fuel
FI Fuel Injection
FT Fuel Trim
FWD Front Wheel Drive

G

GND Ground

H

HAC High Altitude Compensation
HEGO Heated Exhaust Gas Oxygen sensor
HEI High Energy Ignition
HO2 Sensor Heated Oxygen Sensor

I

IAC Idle Air Control
IAT Intake Air Temperature
ICM Ignition Control Module
IFI Indirect Fuel Injection
IFS Inertia Fuel Shutoff
ISC Idle Speed Control
IVSV Idle Vacuum Switching Valve

Commonly Used Abbreviations

K

KOEO Key On, Engine Off
KOER Key ON, Engine Running
KS Knock Sensor

M

MAF Mass Air Flow
MAP Manifold Absolute Pressure
MAT Manifold Air Temperature
MC Mixture Control
MDP Manifold Differential Pressure
MFI Multiport Fuel Injection
MIL Malfunction Indicator Lamp or Maintenance
MST Manifold Surface Temperature
MVZ Manifold Vacuum Zone

N

NVRAM Nonvolatile Random Access Memory

O

O2 Sensor Oxygen Sensor
OBD On-Board Diagnostic
OC Oxidation Catalyst
OHC Overhead Camshaft
OL Open Loop

P

P/S Power Steering
PAIR Pulsed Secondary Air Injection
PCM Powertrain Control Module
PCS Purge Control Solenoid
PCV Positive Crankcase Ventilation
PIP Profile Ignition Pick-up
PNP Park/Neutral Position
PROM Programmable Read Only Memory
PSP Power Steering Pressure
PTO Power Take-Off
PTOX Periodic Trap Oxidizer System

R

RABS Rear Anti-lock Brake System
RAM Random Access Memory
ROM Read Only Memory
RPM Revolutions Per Minute
RWAL Rear Wheel Anti-lock Brakes
RWD Rear Wheel Drive

S

SBC Single Bed Converter
SBEC Single Board Engine Controller
SC Supercharger
SCB Supercharger Bypass
SFI Sequential Multiport Fuel Injection
SIR Supplemental Inflatible Restraint
SOHC Single Overhead Camshaft
SPL Smoke Puff Limiter
SPOUT Spark Output
SRI Service Reminder Indicator
SRS Supplemental Restraint System
SRT System Readiness Test
SSI Solid State Ignition
ST Scan Tool
STO Self-Test Output

T

TAC Thermostatic Air Clearner
TBI Throttle Body Fuel Injection
TC Turbocharger
TCC Torque Converter Clutch
TCM Transmission Control Module
TDC Top Dead Center
TFI Thick Film Ignition
TP Throttle Position
TR Sensor Transaxle/Transmission Range Sensor
TVV Thermal Vacuum Valve
TWC Three-way Catalytic Converter

V

VAF Volume Air Flow, or Vane Air Flow
VAPS Variable Assist Power Steering
VRV Vacuum Regulator Valve
VSS Vehicle Speed Sensor
VSV Vacuum Switching Valve

W

WOT Wide Open Throttle
WU-TWC Warm Up Three-way Catalytic Converter

ENGLISH TO METRIC CONVERSION: TORQUE

To convert foot-pounds (ft. lbs.) to Newton-meters (Nm), multiply the number of ft. lbs. by 1.36
To convert Newton-meters (Nm) to foot-pounds (ft. lbs.), multiply the number of Nm by 0.7376

ft. lbs.	Nm	ft. lbs.	Nm	ft. lbs.	Nm	ft. lbs.	Nm
0.1	0.1	34	46.2	76	103.4	118	160.5
0.2	0.3	35	47.6	77	104.7	119	161.8
0.3	0.4	36	49.0	78	106.1	120	163.2
0.4	0.5	37	50.3	79	107.4	121	164.6
0.5	0.7	38	51.7	80	108.8	122	165.9
0.6	0.8	39	53.0	81	110.2	123	167.3
0.7	1.0	40	54.4	82	111.5	124	168.6
0.8	1.1	41	55.8	83	112.9	125	170.0
0.9	1.2	42	57.1	84	114.2	126	171.4
1	1.4	43	58.5	85	115.6	127	172.7
2	2.7	44	59.8	86	117.0	128	174.1
3	4.1	45	61.2	87	118.3	129	175.4
4	5.4	46	62.6	88	119.7	130	176.8
5	6.8	47	63.9	89	121.0	131	178.2
6	8.2	48	65.3	90	122.4	132	179.5
7	9.5	49	66.6	91	123.8	133	180.9
8	10.9	50	68.0	92	125.1	134	182.2
9	12.2	51	69.4	93	126.5	135	183.6
10	13.6	52	70.7	94	127.8	136	185.0
11	15.0	53	72.1	95	129.2	137	186.3
12	16.3	54	73.4	96	130.6	138	187.7
13	17.7	55	74.8	97	131.9	139	189.0
14	19.0	56	76.2	98	133.3	140	190.4
15	20.4	57	77.5	99	134.6	141	191.8
16	21.8	58	78.9	100	136.0	142	193.1
17	23.1	59	80.2	101	137.4	143	194.5
18	24.5	60	81.6	102	138.7	144	195.8
19	25.8	61	83.0	103	140.1	145	197.2
20	27.2	62	84.3	104	141.4	146	198.6
21	28.6	63	85.7	105	142.8	147	199.9
22	29.9	64	87.0	106	144.2	148	201.3
23	31.3	65	88.4	107	145.5	149	202.6
24	32.6	66	89.8	108	146.9	150	204.0
25	34.0	67	91.1	109	148.2	151	205.4
26	35.4	68	92.5	110	149.6	152	206.7
27	36.7	69	93.8	111	151.0	153	208.1
28	38.1	70	95.2	112	152.3	154	209.4
29	39.4	71	96.6	113	153.7	155	210.8
30	40.8	72	97.9	114	155.0	156	212.2
31	42.2	73	99.3	115	156.4	157	213.5
32	43.5	74	100.6	116	157.8	158	214.9
33	44.9	75	102.0	117	159.1	159	216.2

METRIC TO ENGLISH CONVERSION: TORQUE

To convert foot-pounds (ft. lbs.) to Newton-meters (Nm), multiply the number of ft. lbs. by 1.36

To convert Newton-meters (Nm) to foot-pounds (ft. lbs.), multiply the number of Nm by 0.7376

Nm	ft. lbs.	Nm	ft. lbs.	Nm	ft. lbs.	Nm	ft. lbs.	Nm	ft. lbs.
0.1	0.1	34	25.0	76	55.9	118	86.8	160	117.6
0.2	0.1	35	25.7	77	56.6	119	87.5	161	118.4
0.3	0.2	36	26.5	78	57.4	120	88.2	162	119.1
0.4	0.3	37	27.2	79	58.1	121	89.0	163	119.9
0.5	0.4	38	27.9	80	58.8	122	89.7	164	120.6
0.6	0.4	39	28.7	81	59.6	123	90.4	165	121.3
0.7	0.5	40	29.4	82	60.3	124	91.2	166	122.1
0.8	0.6	41	30.1	83	61.0	125	91.9	167	122.8
0.9	0.7	42	30.9	84	61.8	126	92.6	168	123.5
1	0.7	43	31.6	85	62.5	127	93.4	169	124.3
2	1.5	44	32.4	86	63.2	128	94.1	170	125.0
3	2.2	45	33.1	87	64.0	129	94.9	171	125.7
4	2.9	46	33.8	88	64.7	130	95.6	172	126.5
5	3.7	47	34.6	89	65.4	131	96.3	173	127.2
6	4.4	48	35.3	90	66.2	132	97.1	174	127.9
7	5.1	49	36.0	91	66.9	133	97.8	175	128.7
8	5.9	50	36.8	92	67.6	134	98.5	176	129.4
9	6.6	51	37.5	93	68.4	135	99.3	177	130.1
10	7.4	52	38.2	94	69.1	136	100.0	178	130.9
11	8.1	53	39.0	95	69.9	137	100.7	179	131.6
12	8.8	54	39.7	96	70.6	138	101.5	180	132.4
13	9.6	55	40.4	97	71.3	139	102.2	181	133.1
14	10.3	56	41.2	98	72.1	140	102.9	182	133.8
15	11.0	57	41.9	99	72.8	141	103.7	183	134.6
16	11.8	58	42.6	100	73.5	142	104.4	184	135.3
17	12.5	59	43.4	101	74.3	143	105.1	185	136.0
18	13.2	60	44.1	102	75.0	144	105.9	186	136.8
19	14.0	61	44.9	103	75.7	145	106.6	187	137.5
20	14.7	62	45.6	104	76.5	146	107.4	188	138.2
21	15.4	63	46.3	105	77.2	147	108.1	189	139.0
22	16.2	64	47.1	106	77.9	148	108.8	190	139.7
23	16.9	65	47.8	107	78.7	149	109.6	191	140.4
24	17.6	66	48.5	108	79.4	150	110.3	192	141.2
25	18.4	67	49.3	109	80.1	151	111.0	193	141.9
26	19.1	68	50.0	110	80.9	152	111.8	194	142.6
27	19.9	69	50.7	111	81.6	153	112.5	195	143.4
28	20.6	70	51.5	112	82.4	154	113.2	196	144.1
29	21.3	71	52.2	113	83.1	155	114.0	197	144.9
30	22.1	72	52.9	114	83.8	156	114.7	198	145.6
31	22.8	73	53.7	115	84.6	157	115.4	199	146.3
32	23.5	74	54.4	116	85.3	158	116.2	200	147.1
33	24.3	75	55.1	117	86.0	159	116.9	201	147.8

ENGLISH/METRIC CONVERSION: TEMPERATURE

To convert Fahrenheit (F°) to Celsius (C°), take F° temperature and subtract 32, multiply the result by 5 and divide the result by 9

To convert Celsius (C°) to Fahrenheit (F°), take C° temperature and multiply it by 9, divide the result by 5 and add 32

F°	C°	F°	C°	C°	F°	C°	F°
-40	-40.0	150	65.6	-38	-36.4	46	114.8
-35	-37.2	155	68.3	-36	-32.8	48	118.4
-30	-34.4	160	71.1	-34	-29.2	50	122
-25	-31.7	165	73.9	-32	-25.6	52	125.6
-20	-28.9	170	76.7	-30	-22	54	129.2
-15	-26.1	175	79.4	-28	-18.4	56	132.8
-10	-23.3	180	82.2	-26	-14.8	58	136.4
-5	-20.6	185	85.0	-24	-11.2	60	140
0	-17.8	190	87.8	-22	-7.6	62	143.6
1	-17.2	195	90.6	-20	-4	64	147.2
2	-16.7	200	93.3	-18	-0.4	66	150.8
3	-16.1	205	96.1	-16	3.2	68	154.4
4	-15.6	210	98.9	-14	6.8	70	158
5	-15.0	212	100.0	-12	10.4	72	161.6
10	-12.2	215	101.7	-10	14	74	165.2
15	-9.4	220	104.4	-8	17.6	76	168.8
20	-6.7	225	107.2	-6	21.2	78	172.4
25	-3.9	230	110.0	-4	24.8	80	176
30	-1.1	235	112.8	-2	28.4	82	179.6
35	1.7	240	115.6	0	32	84	183.2
40	4.4	245	118.3	2	35.6	86	186.8
45	7.2	250	121.1	4	39.2	88	190.4
50	10.0	255	123.9	6	42.8	90	194
55	12.8	260	126.7	8	46.4	92	197.6
60	15.6	265	129.4	10	50	94	201.2
65	18.3	270	132.2	12	53.6	96	204.8
70	21.1	275	135.0	14	57.2	98	208.4
75	23.9	280	137.8	16	60.8	100	212
80	26.7	285	140.6	18	64.4	102	215.6
85	29.4	290	143.3	20	68	104	219.2
90	32.2	295	146.1	22	71.6	106	222.8
95	35.0	300	148.9	24	75.2	108	226.4
100	37.8	305	151.7	26	78.8	110	230
105	40.6	310	154.4	28	82.4	112	233.6
110	43.3	315	157.2	30	86	114	237.2
115	46.1	320	160.0	32	89.6	116	240.8
120	48.9	325	162.8	34	93.2	118	244.4
125	51.7	330	165.6	36	96.8	120	248
130	54.4	335	168.3	38	100.4	122	251.6
135	57.2	340	171.1	40	104	124	255.2
140	60.0	345	173.9	42	107.6	126	258.8
145	62.8	350	176.7	44	111.2	128	262.4

LENGTH CONVERSION

To convert inches (in.) to millimeters (mm), multiply the number of inches by 25.4
To convert millimeters (mm) to inches (in.), multiply the number of millimeters by 0.04

Inches	Millimeters	Inches	Millimeters	Inches	Millimeters	Inches	Millimeters
0.0001	0.00254	0.005	0.1270	0.09	2.286	4	101.6
0.0002	0.00508	0.006	0.1524	0.1	2.54	5	127.0
0.0003	0.00762	0.007	0.1778	0.2	5.08	6	152.4
0.0004	0.01016	0.008	0.2032	0.3	7.62	7	177.8
0.0005	0.01270	0.009	0.2286	0.4	10.16	8	203.2
0.0006	0.01524	0.01	0.254	0.5	12.70	9	228.6
0.0007	0.01778	0.02	0.508	0.6	15.24	10	254.0
0.0008	0.02032	0.03	0.762	0.7	17.78	11	279.4
0.0009	0.02286	0.04	1.016	0.8	20.32	12	304.8
0.001	0.0254	0.05	1.270	0.9	22.86	13	330.2
0.002	0.0508	0.06	1.524	1	25.4	14	355.6
0.003	0.0762	0.07	1.778	2	50.8	15	381.0
0.004	0.1016	0.08	2.032	3	76.2	16	406.4

ENGLISH/METRIC CONVERSION: LENGTH

To convert inches (in.) to millimeters (mm), multiply the number of inches by 25.4
To convert millimeters (mm) to inches (in.), multiply the number of millimeters by 0.04

Inches		Millimeters	Inches		Millimeters	Inches		Millimeters
Fraction	Decimal	Decimal	Fraction	Decimal	Decimal	Fraction	Decimal	Decimal
1/64	0.016	0.397	11/32	0.344	8.731	11/16	0.688	17.463
1/32	0.031	0.794	23/64	0.359	9.128	45/64	0.703	17.859
3/64	0.047	1.191	3/8	0.375	9.525	23/32	0.719	18.256
1/16	0.063	1.588	25/64	0.391	9.922	47/64	0.734	18.653
5/64	0.078	1.984	13/32	0.406	10.319	3/4	0.750	19.050
3/32	0.094	2.381	27/64	0.422	10.716	49/64	0.766	19.447
7/64	0.109	2.778	7/16	0.438	11.113	25/32	0.781	19.844
1/8	0.125	3.175	29/64	0.453	11.509	51/64	0.797	20.241
9/64	0.141	3.572	15/32	0.469	11.906	13/16	0.813	20.638
5/32	0.156	3.969	31/64	0.484	12.303	53/64	0.828	21.034
11/64	0.172	4.366	1/2	0.500	12.700	27/32	0.844	21.431
3/16	0.188	4.763	33/64	0.516	13.097	55/64	0.859	21.828
13/64	0.203	5.159	17/32	0.531	13.494	7/8	0.875	22.225
7/32	0.219	5.556	35/64	0.547	13.891	57/64	0.891	22.622
15/64	0.234	5.953	9/16	0.563	14.288	29/32	0.906	23.019
1/4	0.250	6.350	37/64	0.578	14.684	59/64	0.922	23.416
17/64	0.266	6.747	19/32	0.594	15.081	15/16	0.938	23.813
9/32	0.281	7.144	39/64	0.609	15.478	61/64	0.953	24.209
19/64	0.297	7.541	5/8	0.625	15.875	31/32	0.969	24.606
5/16	0.313	7.938	41/64	0.641	16.272	63/64	0.984	25.003
21/64	0.328	8.334	21/32	0.656	16.669	1/1	1.000	25.400
			43/64	0.672	17.066			

THOMSON
DELMAR LEARNING

Manual ISBN 1-4018-7412-6/Part No. 27412

With the *Chilton® 2005 Labor Guide*, professional technicians gain access to labor times for vehicle brands and models that conform to current Automotive Aftermarket Industry Association standards. Thousands of labor times for 1981 through 2005 domestic and imported vehicles reflect technicians' use of aftermarket tools and training. Updates based on technical hotline input, Original Equipment Manufacturer (OEM) warranty times, and technical editor evaluation include more diagnostic labor times than ever before. Labor operations have been rewritten to conform to the most recent industry standards. Prior model coverage has been re-evaluated by experts to ensure accuracy. Chilton labor times are accepted by insurance and extended warranty companies.

Labor Guide Manual Benefits:

- 2,500 pages of Chilton labor times
- each OEM is arranged alphabetically by section for easy reference
- improved indexing means easier access to today's repair industry standards

Hardcover manual is 8 7/8" x 11", ©2005

Labor Guide CD-ROM Benefits:

- easy-to-use software to create and print professional-quality estimates and invoices
- three user-defined levels of labor rates correspond to different types of job scenarios, for "real-world" application
- functions as a database of aftermarket labor times for monitoring warranty and insurance claims
- software keeps track of customers and prior estimates for time-saving recall
- customizable application allows service writers to add labor operations and times, and parts companies to add labor times to existing parts ordering systems

CD-ROM ISBN 1-4018-7818-0/Part No. 27818

Previous Year Editions

Chilton 2004 Labor Guide Manual, **ISBN 1-4018-4356-5/Part No. 24356**

Chilton 2004 Labor Guide CD-ROM, **ISBN 1-4018-4357-3/Part No. 24357**

CHILTON MECHANICAL SERVICE MANUALS - ANNUAL EDITIONS

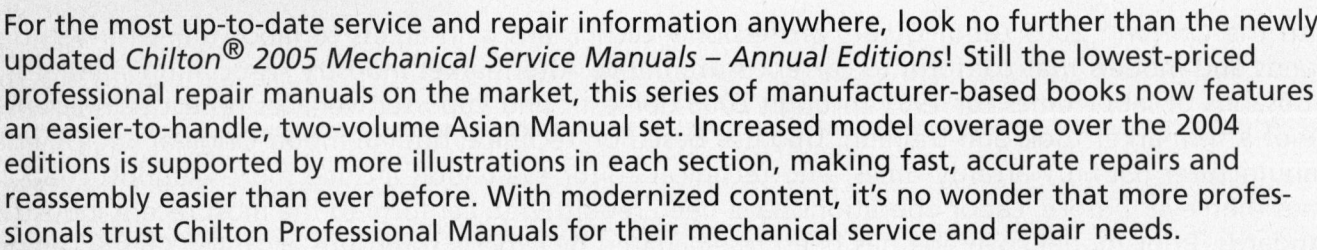

For the most up-to-date service and repair information anywhere, look no further than the newly updated *Chilton® 2005 Mechanical Service Manuals – Annual Editions*! Still the lowest-priced professional repair manuals on the market, this series of manufacturer-based books now features an easier-to-handle, two-volume Asian Manual set. Increased model coverage over the 2004 editions is supported by more illustrations in each section, making fast, accurate repairs and reassembly easier than ever before. With modernized content, it's no wonder that more professionals trust Chilton Professional Manuals for their mechanical service and repair needs.

Mechanical Service Manual Benefits:

- all books are grouped by manufacturer to make accessing information simple
- step-by-step procedures from drive train to chassis and related components help yield fast accurate results
- comprehensive, technically-detailed content is organized by model and system, and is supported by exploded-view illustrations, diagrams, and specification charts for added clarity
- most mechanical systems are included, such as engines, suspensions, steering components, and more
- special tools are described and clearly illustrated so that performing repairs is as easy and quick as possible

Chilton 2005 Ford Mechanical Service Manual
ISBN 1-4018-6719-7/Part No. 26719
Chilton 2005 General Motors Mechanical Service Manual
ISBN 1-4018-7146-1/Part No. 27146
Chilton 2005 Chrysler Mechanical Service Manual
ISBN 1-4018-6718-9/Part No. 26718
Chilton 2005 Asian Mechanical Service Manual (Complete Set of 2 manuals)
ISBN 1-4018-7180-1/Part No.
Chilton 2005 Asian Mechanical Service Manual, Acura - Mazda
ISBN 1-4018-6716-2/Part No. 26716
Chilton 2005 Asian Mechanical Service Manual, Mitsubishi - Toyota
ISBN 1-4018-6717-0/Part No. 26717
Chilton 2005 European Mechanical Service Manual
ISBN 1-4018-6720-0/Part No. 26720

Manuals are 8 1/2" x 11", ©2005

THOMSON
DELMAR LEARNING

The *Chilton® Perennial Editions* contain repair and maintenance information for popular mechanical systems that may not be available elsewhere. They offer a wide range of repair information on cars, trucks, vans, and SUVs dating back to the early 1960s, and as current as 2002. Information for 1993 and later model years includes scheduled maintenance interval charts.

Benefits:

- covers the most common vehicle models found in the repair aftermarket today
- gain quick understanding of systems using exploded-view illustrations, diagrams, and charts
- simplify tough jobs with easy-to-follow removal and installation instructions for heater core and other components
- obtain complete coverage of repair procedures from drive train to chassis and associated components

Auto Repair Manual, 1998-2002, 1,426 pages
ISBN 0-8019-9362-8/Part No. 9362
Auto Repair Manual, 1993-1997, 2,064 pages
ISBN 0-8019-7919-6/Part No. 7919
Auto Repair Manual, 1988-1992, 1,284 pages
ISBN 0-8019-7906-4/Part No. 7906
Auto Repair Manual, 1980-1987, 1,344 pages
ISBN 0-8019-7670-7/Part No. 7670

Import Car Repair Manual, 1998-2002, 1,792 pps
ISBN 0-8019-9363-6/Part No. 9363
Import Car Repair Manual, 1993-1997, 2,080 pps
ISBN 0-8019-7920-X/Part No. 7920
Import Car Repair Manual, 1988-1992, 1,632 pages
ISBN 0-8019-7907-2/Part No. 7907
Import Car Repair Manual, 1980-1987, 1,488 pages
ISBN 0-8019-7672-3/Part No. 7672

Truck & Van Repair Manual, 1998-2002, 1,408 pages
ISBN 0-8019-9364-4/Part No. 9364
Truck & Van Repair Manual, 1993-1997, 2,096 pages
ISBN 0-8019-7921-8/Part No. 7921
Truck & Van Repair Manual, 1991-1995, 1,664 pages
ISBN 0-8019-7911-0/Part No. 7911
Truck & Van Repair Manual, 1986-1990, 1,536 pages
ISBN 0-8019-7902-1/Part No. 7902
Truck & Van Repair Manual, 1979-1986, 1,440 pages
ISBN 0-8019-7655-3/Part No. 7655

SUV Repair Manual, 1998-2002, 1,292 pages
ISBN 0-8019-9365-2/Part No. 9365

Hardcover manuals are 8 1/2" x 11".

Chilton Collector's Editions - *Reference Manuals for Vintage Vehicles*
Auto Repair Manual, 1964-1971, ISBN 0-8019-5974-8/Part No. 5974,
Truck & Van Repair Manual, 1961-1971, ISBN 0-8019-6198-X/Part No. 6198
Truck & Van Repair Manual, 1971-1978, ISBN 0-8019-7012-1/Part No. 7012

ASE Test Preparation Series

Thomson Delmar Learning

ISBN 1-4018-5182-7
Part No. 25182
(Complete Set: A1-A8, L1, P2 X1, C1)

Thomson Delmar Learning has developed comprehensive ASE Test Preparation Manuals to help automotive technicians increase their success on these certification programs. The material covers the topics one might find during the test process. The booklets include many review questions and answers, as well as detailed descriptions of the repairs involved. Designed to look like the actual test, participants will feel more comfortable with practice, which will translate into greater success in taking the actual tests. The design of the Delmar Learning product also includes helpful test taking hints and student preparation ideas designed to enhance success.

BENEFITS
- The history of the ASE
- Test-taking strategies
- Tasks lists and overview
- Sample test questions
- ASE-style exams
- Explanations to the answers (right and wrong)
- Glossary of terms

(A1) Automotive Engine Repair, 2E

1-4018-2040-9
Part No. 22040
General Engine Diagnosis, Cylinder Head and Valve Train Diagnosis and Repair, Engine Block Diagnosis and Repair, Lubrication and Cooling Systems Diagnosis and Repair, and Fuel, Electrical, Ignition and Exhaust Systems Inspection and Service.

(A2) Automotive Transmissions and Transaxles, 2E

1-4018-2041-7
Part No. 22041
General Transmission/ Transaxle Diagnosis (Mechanical/Hydraulic Systems and Electronic Systems), Transmission/Transaxle Maintenance and Adjustment, In-Vehicle Transmission/Transaxle Repair, Off-Vehicle Transmission/Transaxle Repair.

(A3) Automotive Manual Drive Trains and Axles, 2E

1-4018-2042-5
Part No. 22042
Clutch Diagnosis and Repair, Transmission Diagnosis and Repair, Transaxle Diagnosis and Repair, Drive Shaft/Half Shaft and Universal Joint/Constant Velocity (CV) Joint Diagnosis and Repair (Front and Rear Wheel Drive), Rear Axle Diagnosis and Repair, Four Wheel Drive/All Wheel Drive Component Diagnosis and Repair.

(A4) Automotive Suspension and Steering, 2E

1-4018-2043-3
Part No. 22043
Steering Systems Diagnosis and Repair (Steering Columns and Manual Steering Gears, Power Assisted Steering Units, Steering Linkage), Suspension Systems Diagnosis and Repair (Front Suspensions, Rear Suspensions, Miscellaneous Services), Wheel Alignment Diagnosis, Adjustment and Repair, and Wheel and Tire Diagnosis and Repair.

(A5) Automotive Brakes, 2E

1-4018-2044-1
Part No. 22044
Hydraulic System Diagnosis and Repair, Drum Brake Diagnosis and Repair, Disc Brake Diagnosis and Repair, Power Assist Units Diagnosis and Repair, Miscellaneous Systems Diagnosis and Repair, Antilock Brake Systems (ABS) Diagnosis and Repair.

(A6) Automotive Electrical-Electronic Systems, 2E

1-4018-2045-X
Part No. 22045
General Electrical/Electronic Systems Diagnosis, Battery Diagnosis and Service, Starting Systems Diagnosis and Repair, Charging Systems Diagnosis and Repair, Lighting Systems Diagnosis and Repair, Gauges, Warning Devices and Driver Information Systems Diagnosis and Repair, Horn and Wiper/Washer Diagnosis and Repair.

(A7) Automotive Heating and Air Conditioning, 2E

1-4018-2046-8
Part No. 22046
The manual for A7 includes the following topics: A/C System Diagnosis and Repair, Refrigeration System Component Diagnosis and Repair, Heating and Engine Cooling Systems Diagnosis and Repair, Operating Systems and Related Controls Diagnosis and Repair, Refrigerant Recovery, Recycling, Handling and Retrofit.

(A8) Automotive Engine Performance, 2E

1-4018-2047-6
Part No. 22047
The manual for A8 includes the following topics: General Engine Diagnosis, Ignition System Diagnosis and Repair, Fuel, Air Induction, and Exhaust Systems Diagnosis and Repair, Emissions Control Systems Diagnosis and Repair (Including OBDII), Computerized Engine controls Diagnosis and Repair (Including OBDII), Engine Electrical Systems diagnosis and Repair.

(L1) Automotive Advance Engine Performance, 2E

1-4018-2049-2
Part No. 22049
The manual for L1 includes the following topics: General Powertrain Diagnosis, Computerized Powertrain Controls Diagnosis (Including OBDII), Ignition System Diagnosis, Fuel Systems and Air Induction Systems Diagnosis, Emission Control Systems Diagnosis, I/M Failure Diagnosis.

(P2) Automobile Parts Specialist, 2E

1-4018-2048-4
Part No. 22048
The manual for P2 includes the following topics: General Operations, Customer Relations and Sales Skills, Vehicle Systems Knowledge, Vehicle Identification, Cataloging Skills, Inventory Management, Merchandising.

(X1) Exhaust Systems

1-4018-2050-6
Part No. 22050
Exhaust Systems includes the following topics: Exhaust Systems Inspection and Repair, Emissions Systems Diagnosis, Exhaust System Fabrication, Exhaust System Installation, Exhaust System Repair Regulations.

(C1) Service Consultant

See next page for details

ASE Test Preparation Series in Español!

Thomson Delmar Learning

ISBN 1-4018-1530-8
(Complete Set: A1-A8, L1, P2, X1)

Now available in Español – the first of its kind for Spanish-speaking technicians! This comprehensive package of ASE test preparation booklets are intended for any Spanish-speaking automotive technician who is preparing to take an ASE examination. The series includes questions that relate to each competency required for certification by ASE. In addition to a multitude of questions, the reason why each answer is right or wrong is explained, along with task lists and overview, test-taking strategies, and more.

(A1) Reparación de Motores, 2A Edición
1-4018-1014-4/Part No. 21014

(A2) Transmision Automática/ Eje de Transmision Automática, 2A Edición
1-4018-1015-2/Part No. 21015

(A3) Tren de y Mando Ejes Manuales, 2A Edición
1-4018-1016-0/Part No. 21016

(A4) Suspensión y Dirección, 2A Edición
1-4018-1017-9/Part No. 21017

(A5) Frenos, 2A Edición
1-4018-1018-7/Part No. 21018

(A6) Sistemas Eléctricos/ Electrónicos, 2A Edición
1-4018-1019-5/Part No. 21019

(A7) Calefacción y Aire Acondicionado, 2A Edición
1-4018-1020-9/Part No. 21020

(A8) Funcionamiento de Motores, 2A Edición
1-4018-1021-7/Part No. 21021

(L1) Especialista en el Funciommiato Avansado de Motores, 2A Edición
1-4018-1022-5/Part No. 21022

(P2) Especialista en Partes de Automovil, 2A Edición
1-4018-1023-3/Part No. 21023

(X1) Sistemas de Escape, 2A Edición
1-4018-1024-1/Part No. 21024

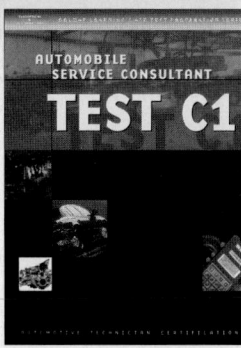

Automotive ASE Preparation Video Series

Thomson Delmar Learning

ISBN 0-7668-3168-X *(Complete Set of 12 Tapes)*
ISBN 0-7668-8042-7 *(Complete Set of 3 CD-ROMs)*

Thomson Delmar Learning's Automotive ASE Test Prep Videos present test takers with a review of the A1-A8, L1, and P2 tests prior to taking the exam. Each tape summarizes key topics and key task areas through live action and animation. Actual technicians, authentic automotive shops, and late-model vehicles are featured for an up-to-date look and feel. Safety is emphasized throughout each tape. An overview tape introduces test takers to the ASE testing style.

BENEFITS OF THE VIDEO SERIES

- lively, easy to follow videos emphasize safety throughout
- covers major task areas and topics for each of the ASE exams
- accompanying Instructor's Guide helps users comprehend and retain information presented

Complete Set of 12 Tapes (with Instructor's Guide), ©2001

Tape 1: Overview of ASE, **0-7668-2484-5**
Tape 2: A1 Engine Repair, **0-7668-2485-3**
Tape 3: A2 Automatic Transmission, **0-7668-2498-5**
Tape 4: A3 Manual Transmission, **0-7668-2499-3**
Tape 5: A4 Steering and Suspension, **0-7668-2500-0**
Tape 6: A5 Automotive Brakes, **0-7668-2501-9**
Tape 7: A6 Electricity/Electronics, **0-7668-2493-4**
Tape 8: A7 Air Conditioning, **0-7668-2486-1**
Tape 9: A8 Engine Performance, **0-7668-2494-2**
Tape 10: P2 Parts Specialist, **0-7668-2487-X**
Tape 11: L1 Advanced Engine Performance (Part 1), **0-7668-2491-8**
Tape 12: L1 Advanced Engine Performance (Part 2), **0-7668-2492-6**

BUNDLES

Bundle 1: Specialty Topics (Set of 4 Tapes) includes Overview of ASE, A1 Engine Repair, A7 Air Conditioning, and P2 Parts Specialist, **0-7668-2483-7**
Bundle 2: Engine Performance/Electronics (Set of 4 Tapes) includes L1 Part 1, L1 Part 2, A6 Electricity/ Electronics, and A8 Engine Performance, **0-7668-2490-X**
Bundle 3: Undercar (Set of 4 Tapes) includes A2 Automatic Transmissions, A3 Manual Transmissions, A4 Steering and Suspension, and A5 Automotive Brakes, **0-7668-2497-7**

CD-ROM COURSEWARE

Based on the ASE Test Prep Series, the CD-ROMs offer the following in addition to the video content:

- Gradebook
- Pre-test/Post-test
- Ability to modify
- Video Glossary
- Variety of question types
- Remediation
- Video File Server compatible

CD-ROM 1: Specialty Topics CD-ROM includes Overview of ASE, A1 Engine Repair, A7 Air Conditioning, and P2 Parts Specialist, **0-7668-2489-6**
CD-ROM 2: Engine Performance/Electronics CD-ROM includes L1 Part 1, L1 Part 2,
A6 Electricity/ Electronics, and A8 Engine Performance, **0-7668-2496-9**
CD-ROM 3: Undercar CD-ROM includes A2 Automatic Transmissions, A3 Manual Transmissions, A4 Steering and Suspension, and A5 Automotive Brakes, **0-7668-2503-5**

The ASE "Passing Lane" Package

Thomson Delmar Learning

ISBN 0-7668-4338-6

(Complete Set: A1-A8, L1, P2)

The most comprehensive test preparation for Automotive Tests A1-A8, L1, and P2. Combining the most thorough ASE Test Preparation books with the latest in ASE videos, this package provides a program of self-study for the automotive ASE Tests.

EACH BOOK IN THE SERIES BENEFITS:

- test-taking strategies
- tasks lists and overview
- sample test questions
- ASE-style exams
- explanations to the answers
- glossary of terms

EACH VIDEO IN THE SERIES BENEFITS:

- lively, easy to follow videos emphasize safety throughout
- covers major task areas and topics for each of the ASE exams
- accompanying Activity Sheets help comprehend and retain information

(A1) Automotive Engine Repair Book/Video, 0-7668-4181-2
(A2) Automotive Transmissions and Transaxles Book/Video, 0-7668-4182-0
(A3) Automotive Manual Drive Trains and Axles Book/Video, 0-7668-4183-9
(A4) Automotive Suspension and Steering Book/Video, 0-7668-4184-7
(A5) Automotive Brakes Book/Video, 0-7668-4185-5
(A6) Automotive Electrical-Electronics Systems Book/Video, 0-7668-4186-3
(A7) Automotive Heating and Air Conditioning Book/Video, 0-7668-4187-1
(A8) Automotive Engine Performance Book/Video, 0-7668-4188-X
(L1) Automotive Advanced Engine Performance Book/Video, 0-7668-4189-8
(P2) Automobile Parts Specialist Book/Video, 0-7668-4190-1

Automotive Technician Certification Test Preparation Manual, 2E

Don Knowles

ISBN 0-7668-1948-5/ Part No. 11948

The second edition of Certified ASE Master Technician Don Knowles' popular ASE test preparation book adds coverage of the L1 Advanced Engine Performance test to its coverage of automotive tests A1 through A8. All nine tests covered in this book reflect year 2000 task lists, including the updated composite vehicle in the L1 test. This revised edition contains at least one practice question for every ASE task in the tests. Also included is the updated and expanded coverage of electronic automatic transmissions, electronically controlled automatic transmissions, electronically controlled 4 wheel drive and steering, ABS systems, wiring diagrams, and repairing electronic components.

BENEFITS

- a new section has been added on computer-controlled automatic transmissions and transaxles including those used in OBD II vehicles
- new information has been included on electronically-controlled 4WD systems and ABS systems
- the chapter on Electrical/Electronic Systems has been expanded to include information on reading wiring diagrams and inspecting, testing, and repairing electronic components
- a complete chapter has been added to prepare technicians for the Advanced Engine Performance (L1) test

CONTENTS

Engine Repair Automatic Transmission/ Transaxle. Manual Drive Train and Axles. Suspension and Steering. Brakes. Electrical/Electronic Systems. Heating, Ventilation, and Air Conditioning Systems. Engine Performance. Advanced Engine Performance.

788 pp, 8½" x 11", softcover, ©2001

ATC AUTOMOTIVE TECHNICIAN CERTIFICATION CHALLENGE.COM
ONLINE ASE CERTIFICATION TEST PREPARATION

ATC Plus CHALLENGE.COM
ONLINE ASE CERTIFICATION TEST PREPARATION

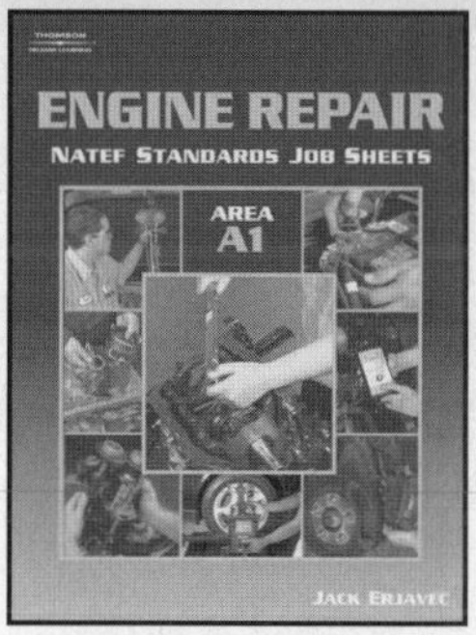

Prepare to Pass the ASE Exam Online

ATCChallenge.com

Thomson Delmar Learning

Updated to the Latest ASE Task Lists!

Thomson Delmar Learning's online ASE test preparation web site has been carefully reviewed and researched by ASE master technicians to include fully updated content on tests A1-A8, L1, P2, X1, and X1. The site offers two different study options so users can choose their study method each time they sign on.

- practice questions provide helpful hints, insight into right and wrong answers, and links back to further reading for each task area
- sample tests prepare users for test day by using ASE-style questions - reflecting the type of questions and task areas on the actual exam - making this the most up-to-date and realistic ASE test preparation study aid available
- one year secure access through any web-enabled computer
- a complete task list including an overview of each task to further enhance study
- automotive dictionary with more than 5,000 terms and Spanish translation
- technical support provided by Thomson Delmar Learning

ATCChallenge.com Plus

Thomson Delmar Learning

ATCChallenge.com Plus is the ideal way to gain the expertise required to pass the *A5* (brakes) and *A7* (heating and air conditioning) ASE exams. Thoroughly reviewed and researched by master automotive technicians, this total online courseware solution may be used effectively in professional training and education courses as well as by individuals preparing for selected ASE exams.

While this version includes all the features that *ATCChallenge.com* contains, it also brings with it a variety of new tools for use by students and educators. The biggest addition from the original version is the immediate remediation to the ASE-style questions that brings the user to a file or a video clip that explains the answer to each question.

- combines the *Today's Technician Series*, Erjavec's *Automotive Technology*, and Delmar Learning's *Automotive Video Series* to bring technicians the most comprehensive ASE coverage available in one place
- individuals can track their scores by ASE task, by test, or by question; create personalized testbanks; and generate their own progress reports; making *ATCChallenge.com Plus* an ideal self-study guide and test preparation tool
- a single training director or administrator can easily track average and/or raw scores at the shop-level, by region, or system-wide to gain a measure of learning achievement and program effectiveness
- a complete task list, with an overview of each task to enhances the learning experience, ensures that users are 100% prepared to pass each ASE test

Call Your Thomson Delmar Learning Sales Rep for Part Numbers & Pricing

Visit **ATCChallenge.com** to see the latest modules and a free demo!

ATC Challenge 3.0 CD-ROM

Thomson Delmar Learning

ISBN 0-7668-2982-0

These exciting interactive CD-ROMs have been designed to prepare technicians for successful completion of the Automotive ASE task areas (A1-A8, L1, P2, and F1). This multimedia software assesses strengths and weaknesses by identifying topics needing further study while allowing users to review ASE task areas at their own pace. Explanations, hints, notes, and a glossary aid the user in comprehension, critical thinking and retention. These CD-ROMs offer hundreds of ASE-style questions, a test taking strategy section and LAN compatibility. Not only is *ATC Challenge 3.0* the ultimate in test preparation, but it is also an excellent learning tool!

CD-ROM, ©2001

Site License Available for Multiple Unit Purchases or Multiple Workstations for ATC Challenge 3.0:
User 1: Full Price (List or Net)
Users 2-5:
$80/workstation + Full Price
Users 6-10:
$70/workstation + Full Price
Users 11-20:
$60/workstation + Full Price
Users 21+:
$50/workstation + Full Price

ATC Challenge for P2

Thomson Delmar Learning

ISBN 0-7668-1827-6

This interactive CD-ROM contains material that will help prepare technicians for the Automotive Parts Specialist (P2) certification exam.
CD-ROM, ©2000

NATEF Standards Job Sheets

Thomson Delmar Learning

ISBN 0-7668-6375-1

(Complete Set: A1-A8)

Each of our eight *NATEF (National Automotive Technicians Education Foundation) Standards Job Sheets* workbooks has been thoughtfully designed to assist users in gaining valuable job preparedness skills and mastering specific technical competencies required for success as a professional automotive technician. The entire series is based on current NATEF standards.

Central to each manual are well-designed and easy-to-read job sheets, each of which contains specific, performance-based objectives, lists of required tools and materials, safety precautions, plus step-by-step procedures to lead users to completion of shop activities.

KEY FEATURES

- easy to use in any automotive education or training program in which NATEF coverage is desired
- completed Job Sheets may be kept as records, providing tangible evidence that instructors are addressing all NATEF tasks while paving the way for program certification

JOB SHEETS AVAILABLE FOR:

(A1) Automotive Engine Repair,
(A2) Automatic Transmissions and Transaxles, 0-7668-6368-9
(A3) Manual Drive Trains and Axles, 0-7668-6369-7
(A4) Automotive Suspension and Steering, 0-7668-6370-0
(A5) Automotive Brakes, 0-7668-6371-9
(A6) Automotive Electrical and 0-7668-6367-0
Electronic Systems, 0-7668-6372-7
(A7) Automotive Heating and Air Conditioning, 0-7668-6373-5
(A8) Automotive Engine Performance, 0-7668-6374-3

All share the following information: 8½" x 11", softcover, ©2002

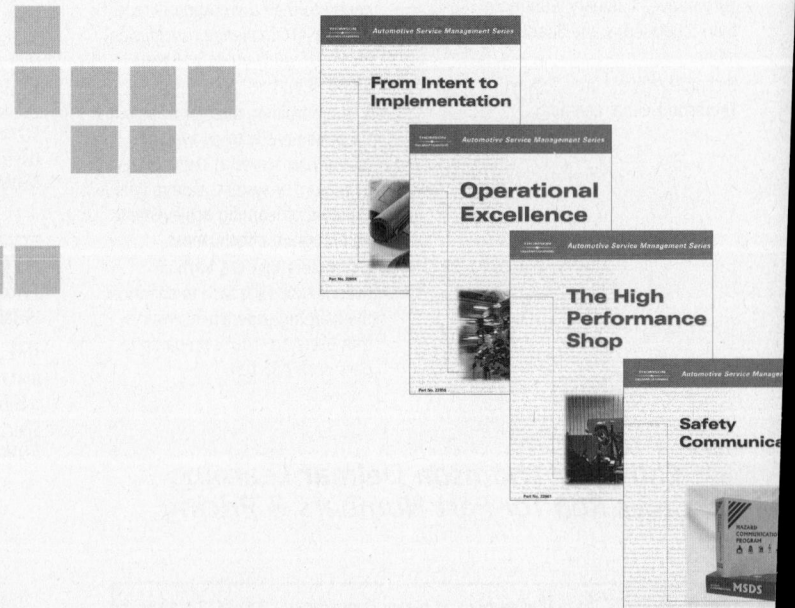

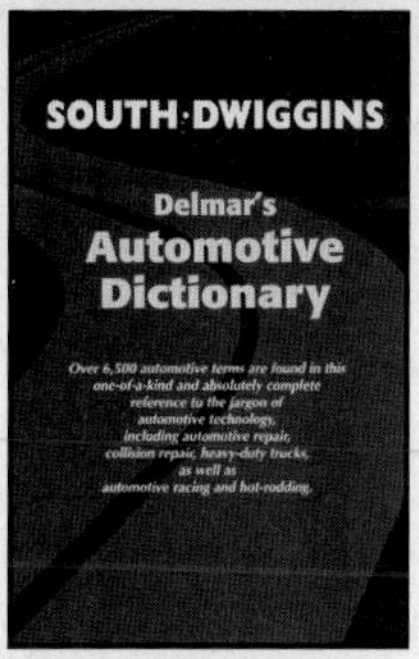

Delmar's Automotive Dictionary

David W. South & Boyce Dwiggins

ISBN 0-8273-7405-4

This handy, ready-reference dictionary provides the automotive engineer, technician, mechanic, student, enthusiast or layperson with a single source for the most up-to-date definitions available of technical, professional and informal terminology used in today's automotive world. It is descriptive and covers the wide scope of terms pertinent to the automotive field. With multiple definitions and aids, and proper pronunciation of terms, this dictionary is a must for all!

BENEFITS

- over 3000 terms comprehensively covering more than 100 subject areas
- enhanced by a list of acronyms and abbreviations
- up-to-date definitions of today's automotive terminology
- aids for proper pronunciation
- each term has multiple definitions

281 pp, 6" x 9", softcover, ©1997

Practical Problems in Mathematics for Automotive Technicians, 5E

George Morre, Todd Sformo & Larry Sformo

ISBN 0-8273-7944-7

By showing how to apply math solutions to everyday problems, this all-in-one math reference transforms the "remove it and replace it" mechanic into a complete automotive technician. The book builds from math basics to cover more complex topics--not to mention such workplace issues as invoices and scale reading of test meters. Each easy-to-read chapter features step-by-step instructions, diagrams, charts and examples to make the problem-solving process a snap.

256 pp, 7⅞" x 9¼", softcover, ©1998
Instructor's Manual **0-8273-7945-5**

Math for the Automotive Trade, 3E

John C. Peterson & William deKryger

ISBN 0-8273-6712-0

Math for Automotive Trades, 3E provides excellent examples and problems that reflect technological requirements of workers in automotive technology. The text has three parts: review of basic mathematics skills, math applications to specific automotive situations, and an examination of measurement aspects beginning with angle and linear measurements and ending with an extensive look at measurement tools used in the automotive trade.

345 pp, 8½" x 11", softcover, ©1995
Instructor's Manual **0-8273-6713-9**